W9-AWL-825

Cover (clockwise from top left):
Streptococci: Tantalum and tungsten surface replica of critical-point-dried group A streptococci, revealing the distribution of the surface molecules and cell division in parallel with the growing chain. Magnification, ×96,000. Courtesy V. A. Fischetti.
Spore-forming bacteria: Morphology of *Clostridium botulinum* type A cells viewed by phase-contrast microscopy. The spore-bearing cells of other serotypes of *C. botulinum* and *C. tetani* also have characterized morphologies typically with swelling of the rod-shaped vegetative cell. See chapter 55.
Staphylococci: Localization of *atl* gene products on the cell surface of *S. aureus* during the division cycle as determined by scanning electron microscopy. See chapter 36.
Listeria: The surface of a macrophage infected with *Listeria monocytogenes*. The single listeria bacillus is being propelled by a polymerizing actin tail. Reproduced from the Journal of Cell Biology (**109:**1597–1608, 1989), with permission of the publisher.

Copyright © 2000   ASM Press
American Society for Microbiology
1325 Massachusetts Ave., N.W.
Washington, DC 20005-4171

Library of Congress Cataloging-in-Publication Data
Gram-positive pathogens / edited by Vincent A. Fischetti ... [et al.]/
   p. cm.
   Includes bibliographical references and index.
   ISBN 1-55581-166-3
   1. Gram-positive bacterial infections. 2. Gram-positive bacteria.
I. Fischetti, Vincent A.
   [DNLM: 1. Gram-Positive Bacteria. 2. Bacterial Vaccines. 3. Drug
Resistance, Microbial. 4. Gram-Positive Bacterial Infections. QW
142 G745 2000]
QR201.G76G73 2000
616'.014—dc21
DNLM/DLC
for Library of Congress   99-35911
                                                                    CIP

Gram-Positive Pathogens

EDITED BY

Vincent A. Fischetti
Rockefeller University, New York, NY 10021

Richard P. Novick
*Department of Molecular Pathogenesis, Skirball Institute for Biomedical Medicine,
New York University Medical School, New York, NY 10016*

Joseph J. Ferretti
University of Oklahoma, Oklahoma City, OK 73104

Daniel A. Portnoy
*Department of Molecular and Cell Biology, University of California, Berkeley,
Berkeley, CA 94720-3202*

Julian I. Rood
Department of Microbiology, Monash University, Clayton, Victoria 3168, Australia

ASM PRESS WASHINGTON, D.C.

Gram-Positive Pathogens

CONTENTS

CONTRIBUTORS

STAFFAN ARVIDSON
Microbiology and Tumorbiology Center, Karolinska
Institutet, S-171 77 Stockholm, Sweden

SHARON BALTER
Centers for Disease Control and Prevention, Atlanta, GA
30333

JOANNIS BASSIAS
Mikrobielle Genetik, Universität Tübingen, D-72076
Tübingen, Germany

ARNOLD S. BAYER
Division of Infectious Diseases, Department of Medicine,
UCLA School of Medicine, Harbor-UCLA Medical Center,
St. John's Cardiovascular Research Center, 1000 West
Carson St., Torrance, CA 90509

BLAINE L. BEAMAN
Department of Medical Microbiology and Immunology,
University of California School of Medicine, Davis, CA
95616

DEBRA E. BESSEN
Department of Epidemiology and Public Health, Yale
University School of Medicine, New Haven, CT 06520

TERRY J. BEVERIDGE
Canadian Bacterial Disease Network, Department of
Microbiology, College of Biological Science, University of
Guelph, Guelph, Ontario, Canada N1G 2W1

ALAN L. BISNO
Department of Medicine, University of Miami School of
Medicine, Miami, FL 33101, and Medical Service, Miami
Veterans Affairs Medical Center, 1201 NW 16th St., Miami,
FL 33125

REGINE BÖCKMANN
Theodor-Boveri-Institut (Biozentrum) für Biowissenschaften
der Universität Würzburg, Lehrstuhl für Mikrobiologie,
D97074 Würzburg, Germany

GREGORY A. BOHACH
Department of Microbiology, Molecular Biology, and
Biochemistry, University of Idaho, Moscow, ID 83844-3052

ALEXANDRE BOLOTIN
Génétique Microbienne, Institut National de la Recherche
Agronomique, Department of Microbiology, Domaine de
Vilvert, 78352 Jouy en Josas, France

D. E. BRILES
Department of Microbiology and Department of Pediatrics,
University of Alabama at Birmingham, 658 BBRB, 845 19th
St. S., Birmingham, AL 35294

REINHOLD BRÜCKNER
Mikrobielle Genetik, Universität Tübingen, D-72076
Tübingen, Germany

MICHAEL CAPARON
Department of Molecular Microbiology, Box 8230,
Washington University School of Medicine, 660 S. Euclid
Ave., St. Louis, MO 63110-1093

GURSHARAN S. CHHATWAL
Department of Microbial Pathogenesis and Vaccine
Research, Technical University/GBF, National Centre for
Biotechnology, D-38106 Braunschweig, Germany

P. PATRICK CLEARY
Department of Microbiology, University of Minnesota
Medical School, Minneapolis, MN 55455

L. VINCENT COLLINS
Department of Rheumatology, University of Göteborg,
Göteborg, Sweden

PASCALE COSSART
Unité des Interactions Bactéries-Cellules, Institut Pasteur, 28
rue du Dr. Roux, 75724 Paris Cedex 15, France

PATRICE COURVALIN
Unité des Agents Antibactériens, Institut Pasteur, 25, rue du
Dr. Roux, 75724 Paris Cedex 15, France

DAVID CUE
Department of Microbiology, University of Minnesota
Medical School, Minneapolis, MN 55455

MADELEINE W. CUNNINGHAM
Department of Microbiology and Immunology, University of
Oklahoma Health Sciences Center, Oklahoma City, OK
73190

PRISCILLA E. DOMBEK
Department of Microbiology, University of Minnesota
Medical School, Minneapolis, MN 55455

PATRICK DUWAT
Génétique Appliquée - URLGA, Institut National de la
Recherche Agronomique, Department of Microbiology,
Domaine de Vilvert, 78352 Jouy en Josas, France

JEROME ETIENNE
EA1655, Faculté de Médecine Laennec, Rue Guillaume
Paradin, 69372 Lyon Cedex 08, France

ALYCE FINELLI
Infectious Diseases Section, Yale University School of
Medicine, 803 LCI, 333 Cedar St., New Haven, CT 06520

NEVILLE FIRTH
School of Biological Sciences, University of Sydney, Sydney,
NSW 2006, Australia

WERNER FISCHER
Institut für Biochemie, Facultät der Medizin, Universität
Erlangen-Nürnberg, Nürnberg, Germany

VINCENT A. FISCHETTI
Laboratory of Bacterial Pathogenesis and Immunology, The
Rockefeller University, 1230 York Ave., New York, NY
10021

TIMOTHY J. FOSTER
Microbiology Department, Moyne Institute of Preventive
Medicine, Trinity College, Dublin 2, Ireland

HENRY S. FRAIMOW
Division of Infectious Diseases, Graduate Hospital,
Philadelphia, PA 19146

NANCY E. FREITAG
Department of Immunology and Microbiology, Wayne State
University School of Medicine, Detroit, MI 48201

ROBERT P. GAYNES
NNIS Surveillance Activity, Hospital Infections Program,
Centers for Disease Control and Prevention, Mailstop E-55,
1600 Clifton Rd., Atlanta, GA 30333

MICHAEL S. GILMORE
Department of Microbiology and Immunology, University of
Oklahoma Health Sciences Center, Oklahoma City, OK
73104

WERNER GOEBEL
Theodor-Boveri-Institut (Biozentrum) für Biowissenschaften
der Universität Würzburg, Lehrstuhl für Mikrobiologie,
D97074 Würzburg, Germany

KHOOSHEH GOSINK
Department of Infectious Diseases, St. Jude Children's
Research Hospital, Memphis, TN 38139

ALEXANDRA GRUSS
Génétique Appliquée - URLGA, Institut National de la
Recherche Agronomique, Department of Microbiology,
Domaine de Vilvert, 78352 Jouy en Josas, France

AMIT GUPTA
Department of Microbiology and Immunology, University of
Illinois, Chicago, IL 60612-7344

KARIN HAMMER
Technical University of Denmark, Building 301, DK-2800
Lyngby, Denmark

LYNN E. HANCOCK
Department of Microbiology and Immunology, University of
Oklahoma Health Sciences Center, Oklahoma City, OK
73104

CHRISTINE HEILMANN
Institute of Medical Microbiology, University of Muenster,
Domagkstrasse 10, D-48149 Muenster, Germany

SUSAN K. HOLLINGSHEAD
Department of Microbiology, University of Alabama at
Birmingham, Birmingham, AL 35294-2170

MAGNUS HÖÖK
Center for Extracellular Matrix Biology, Institute of
Biosciences and Technology, Texas A&M University, 2121
W. Holcombe Blvd., Houston, TX 77030

DAVID C. HOOPER
Division of Infectious Diseases, Massachusetts General
Hospital, Harvard Medical School, Boston, MA 02114-2696

JOHN J. IANDOLO
Department of Microbiology and Immunology, University of
Oklahoma Health Sciences Center, Oklahoma City, OK
73190

HOWARD F. JENKINSON
Department of Oral and Dental Science, University of
Bristol Dental School, Bristol BS1 2LY, United Kingdom

ERIC A. JOHNSON
Department of Food Microbiology and Toxicology, Food
Research Institute, University of Wisconsin, Madison, WI
53706

DENNIS L. KASPER
Channing Laboratory, Department of Medicine, Brigham &
Women's Hospital, Harvard Medical School, Boston, MA
02115

DOUGLAS S. KERNODLE
Division of Infectious Diseases, Department of Medicine,
Vanderbilt University School of Medicine, Nashville, TN
37232-2605, and Infectious Diseases Section, Department of
Veterans Affairs Medical Center, Nashville, TN 37212-2637

THERESA M. KOEHLER
Department of Microbiology and Molecular Genetics,
University of Texas-Houston Medical School, Houston, TX
77030

JÜRGEN KREFT
Theodor-Boveri-Institut (Biozentrum) für Biowissenschaften
der Universität Würzburg, Lehrstuhl für Mikrobiologie,
D97074 Würzburg, Germany

HOWARD K. KURAMITSU
Department of Oral Biology and Department of
Microbiology, State University of New York, Buffalo, NY
14214

CHIA Y. LEE
Department of Microbiology, Molecular Genetics and
Immunology, University of Kansas Medical Center, Kansas
City, KS 66160

JEAN C. LEE
Channing Laboratory, Department of Medicine, Brigham
and Women's Hospital and Harvard Medical School, Boston,
MA 02115

STUART B. LEVY
Center for Adaptation Genetics and Drug Resistance and
Department of Molecular Biology and Microbiology, Tufts
University School of Medicine, 136 Harrison Ave., Boston,
MA 02111

GERARD LINA
EA1655, Faculté de Médecine Laennec, Rue Guillaume
Paradin, 69372 Lyon Cedex 08, France

JOHN F. LOVE
Department of Biochemistry, Boston University School of
Medicine, Boston, MA 02118

FRANKLIN D. LOWY
Department of Medicine, Montefiore Medical Center, and
Departments of Medicine, Microbiology and Immunology,
Albert Einstein College of Medicine, Bronx, NY 10467

DAVID M. LYERLY
TechLab, Inc., 1861 Pratt Drive, Corporate Research Center,
Blacksburg, VA 24060-6364

DENA LYRAS
Department of Microbiology, Monash University, Wellington
Road, Clayton, Victoria 3168, Australia

LAWRENCE C. MADOFF
Channing Laboratory, Department of Medicine, Brigham &
Women's Hospital, Harvard Medical School, Boston, MA
02115

HORST MALKE
Institute for Molecular Biology, Friedrich Schiller University
Jena, D-07745 Jena, Germany

BRUCE A. McCLANE
Department of Molecular Genetics and Biochemistry,
University of Pittsburgh School of Medicine, Pittsburgh, PA
15661

JOHN K. McCORMICK
Department of Microbiology, University of Minnesota
Medical School, Minneapolis, MN 55455

LAURA M. McMURRY
Center for Adaptation Genetics and Drug Resistance and
Department of Molecular Biology and Microbiology, Tufts
University School of Medicine, 136 Harrison Ave., Boston,
MA 02111

W. MICHAEL McSHAN
Department of Microbiology and Immunology, BMSB 1053,
University of Oklahoma Health Sciences Center, 940 S. L.
Young Blvd., Oklahoma City, OK 73190

J. SCOTT MONCRIEF
TechLab, Inc., 1861 Pratt Drive, Corporate Research Center,
Blacksburg, VA 24060-6364, and Fralin Biotech Center,
Virginia Polytechnic Institute and State University,
Blacksburg, VA 24061

JUDY K. MORONA
Molecular Microbiology Unit, Women's and Children's
Hospital, North Adelaide, S.A., 5006, Australia

JOHN R. MURPHY
Evans Department of Clinical Research and Department of
Medicine, Boston University School of Medicine, Boston
Medical Center, Boston, MA 02118

IAIN A. MURRAY
Krebs Institute for Biomolecular Research, Department of
Molecular Biology and Biotechnology, University of
Sheffield, Western Bank, Sheffield S10 2TN, United
Kingdom

M. H. NAHM
Department of Pediatrics and Division of Immunology,
Allergy, and Rheumatology, University of Rochester, 601
Elmwood Ave., Rochester, NY 14642-8777

VICTOR NIZET
Department of Pediatrics, University of California, San
Diego, La Jolla, CA 92093

RICHARD P. NOVICK
Skirball Institute of Biomolecular Medicine, New York
University Medical Center, 540 First Ave., New York, NY
10016

HAROLD R. OSTER
Department of Medicine, University of Miami School of
Medicine, Miami, FL 33101, and Infectious Diseases, Miami
Veterans Affairs Medical Center, 1201 NW 16th St., Miami,
FL 33125

ERIC G. PAMER
Infectious Diseases Section, Yale University School of
Medicine, 803 LCI, 333 Cedar St., New Haven, CT 06520

VIJAYKUMAR PANCHOLI
Laboratory of Bacterial Pathogenesis and Immunology, The
Rockefeller University, 1230 York Ave., New York, NY
10021

LAWRENCE C. PAOLETTI
Channing Laboratory, Department of Medicine, Brigham &
Women's Hospital, Harvard Medical School, Boston, MA
02115

JAMES C. PATON
Molecular Microbiology Unit, Women's and Children's
Hospital, 72 King William Road, North Adelaide, S.A.,
5006, Australia

GEORG PETERS
Institute of Medical Microbiology, University of Muenster,
Domagkstrasse 10, D-48149 Muenster, Germany

DANIEL A. PORTNOY
Department of Molecular and Cell Biology and The School
of Public Health, University of California, Berkeley,
Berkeley, CA 94720

KLAUS T. PREISSNER
Department of Biochemistry, Faculty of Medicine, University
of Giessen, Giessen, Germany

RICHARD A. PROCTOR
Department of Medical Microbiology and Immunology, 407
SMI, University of Wisconsin Medical School, Madison, WI
53706

STEVEN J. PROJAN
Infectious Diseases, Wyeth-Ayerst Research, 401 N.
Middletown Road, Pearl River, NY 10965

JULIAN I. ROOD
Department of Microbiology, Monash University, Wellington
Road, Clayton, Victoria 3168, Australia

CRAIG E. RUBENS
Department of Pediatrics, Children's Hospital and Medical Center, University of Washington, Seattle, WA 98105

R. R. B. RUSSELL
Department of Oral Biology, Dental School, University of Newcastle, Newcastle upon Tyne NE2 4BW, United Kingdom

WALTER F. SCHLECH III
Division of Infectious Diseases, Department of Medicine, Dalhousie University, Halifax, Nova Scotia, Canada

PATRICK M. SCHLIEVERT
Department of Microbiology, University of Minnesota Medical School, Minneapolis, MN 55455

ANNE SCHUCHAT
Centers for Disease Control and Prevention, Atlanta, GA 30333

KAREN J. SHAW
Schering-Plough Research Institute, 2015 Galloping Hill Rd. 4700, Kenilworth, NJ 07033-0539

SIMON SILVER
Department of Microbiology and Immunology, University of Illinois, Chicago, IL 60612-7344

RONALD A. SKURRAY
School of Biological Sciences, University of Sydney, Sydney, NSW 2006, Australia

DENNIS L. STEVENS
Infectious Disease Section, Veterans Affairs Medical Center, Boise, ID 83702, and Department of Medicine, University of Washington School of Medicine, Seattle, WA 98195

E. SWIATLO
Division of Infectious Diseases, University of Mississippi Medical Center, and Research and Education (151), VA Medical Center, 1500 E. Woodrow Wilson Drive, Jackson, MS 39216-5199

SUSANNE R. TALAY
Department of Microbial Pathogenesis and Vaccine Research, Technical University/GBF, National Centre for Biotechnology, 38106 Braunschweig, Germany

ANDREJ TARKOWSKI
Department of Rheumatology and Department of Clinical Immunology, University of Göteborg, Guldhedsgatan 10, S-413 46 Göteborg, Sweden

FRED C. TENOVER
Nosocomial Pathogens Laboratory Branch, Hospital Infections Program, Centers for Disease Control and Prevention, Mailstop G-08, 1600 Clifton Rd., Atlanta, GA 30333

ALEXANDER TOMASZ
The Rockefeller University, 1230 York Ave., New York, NY 10021

ELAINE TUOMANEN
Department of Infectious Diseases, St. Jude Children's Research Hospital, Memphis, TN 38139

FRANÇOIS VANDENESCH
EA 1655, Faculté de Médecine Laennec, Rue Guillaume Paradin, 69372 Lyon Cedex 08, France

KEITH E. WEAVER
Division of Basic Biomedical Sciences, School of Medicine, University of South Dakota, Vermillion, SD 57069

BERNARD WEISBLUM
Department of Pharmacology, University of Wisconsin Medical School, 1300 University Ave., Madison, WI 53706

JEFFREY N. WEISER
Department of Pediatrics and Department of Microbiology, Children's Hospital of Philadelphia, and University of Pennsylvania School of Medicine, Philadelphia, PA 19104-6076

JERROLD WEISS
Inflammation Program, Departments of Internal Medicine and Microbiology, The University of Iowa College of Medicine, 200 Hawkins Dr., Iowa City, IA 52240

MICHAEL R. WESSELS
Channing Laboratory and Division of Infectious Diseases, Brigham and Women's Hospital, Harvard Medical School, Boston, MA 02115

CYNTHIA G. WHITNEY
Centers for Disease Control and Prevention, Atlanta, GA 30333

TRACY D. WILKINS
TechLab, Inc., 1861 Pratt Drive, Corporate Research Center, Blacksburg, VA 24060-6364, and Fralin Biotech Center, Virginia Polytechnic Institute and State University, Blacksburg, VA 24061

WOLFGANG WITTE
Robert Koch Institute, Wernigerode Branch, Burgstrasse 37, G-38855 Wernigerode, Germany

GERARD D. WRIGHT
Department of Biochemistry, McMaster University, Hamilton, Ontario L8N 3Z5, Canada

MICHAEL YEAMAN
Division of Infectious Diseases, Department of Medicine, UCLA School of Medicine, Harbor-UCLA Medical Center, St. John's Cardiovascular Research Center, 1000 West Carson St., Torrance, CA 90509

MARIA K. YEUNG
Department of Pediatric Dentistry/Microbiology, University of Texas Health Science Center at San Antonio, 7703 Floyd Curl Dr., San Antonio, TX 78284

JANET YOTHER
Department of Microbiology, University of Alabama at Birmingham, Birmingham, AL 35294

PREFACE

Gram-positive bacteria are structurally distinct from their gram-negative relatives. These differences, reflected in the lack of an outer membrane and related secretory systems and the presence of a thick peptidoglycan layer, have enabled these organisms to develop novel approaches to pathogenesis by acquiring (among others) a unique family of surface proteins, toxins, and enzymes. This volume is a compendium of research directed to the gram-positive bacterial pathogen at all levels, and we think that to date no other single publication accomplishes this objective. This book deals with the mechanisms of gram-positive bacterial pathogenicity, including the current knowledge on gram-positive structure and mechanisms of antibiotic resistance. It emphasizes streptococci, staphylococci, listeria, and spore-forming pathogens, with chapters written by many of the leading researchers in these areas. The chapters systematically dissect these organisms biologically, genetically, and immunologically in an attempt to understand the strategies used by these bacteria to cause human disease. It is hoped that the insights gained from understanding these strategies will lead to rational design of therapeutics and vaccines to control infections caused by gram-positive bacteria.

VINCENT A. FISCHETTI
RICHARD P. NOVICK
JOSEPH J. FERRETTI
DANIEL A. PORTNOY
JULIAN I. ROOD

THE GRAM-POSITIVE CELL WALL

SECTION EDITOR: Vincent A. Fischetti

I

THE GRAM-POSITIVE CELL WALL structure differs in many respects from its gram-negative counterpart. While much has been reported regarding the complexity of the gram-negative envelope, until recently similar information has not been available for the gram-positive wall. The following chapters are designed to give the reader a good impression of the gram-positive bacterial cell wall structure and the increasing complexity of its surface protein coat, offering the most comprehensive understanding to date of the gram-positive envelope.

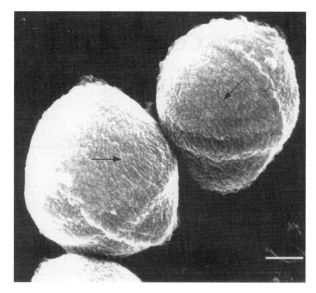

FIGURE 1 SEM micrograph of *S. epidermidis* showing the concentric circular structures of the cell wall surface (arrows) that suggest the circular arrangement of polymers in the cell wall. Bar = 250 nm. Previously published in *J. Electron Microsc.* **27**:147–148 (1978) and reproduced with permission.

Freeze-Etching

Freeze-etching is an extension of the shadow-casting procedure. A drawback of shadow-casting is that no internal detail can be discerned, only the surface. Freeze-etching gets around this problem by employing freeze-fracturing on the sample, which is first snap-frozen in a cryogen held at −196°C in liquid nitrogen. (Until recently the cryogen was Freon 22, but environmental concerns make propane a better option.) The freezing is so rapid that the bacteria are embedded in vitreous ice, ensuring that all molecular motion is immediately stopped. This hard, frozen specimen is now fractured under vacuum by giving it a blow with the fracturing knife. The fracture occurs through those molecular regions that are loosely bonded together. Once the fracture has taken place and the fracture surface is exposed, the specimen is etched under vacuum for ~60 s; under vacuum the surface ice of the specimen sublimes, thereby exposing more of the specimen and giving it more topography. At this time, the exposed surface is shadow-cast (usually with platinum) at an angle. Next, directly above the sample, carbon vapor is also cast on the specimen face so that it is completely covered by a thin layer of carbon. This results in a platinum-shadowed carbon replica of the fractured material that can be floated off the sample once the vacuum is broken. To help accomplish this, the sample is treated with harsh acids and oxidants that hydrolyze the organic material and break the adhesion between replica and sample. The replica mimics the original fracture face with good fidelity. The great advantages of freeze-etching over shadow-casting are as follows. (i) Since the specimen is fractured, inside views of bacteria can be seen. (ii) The snap-freezing ensures good preservation so that the natural state of the structure is preserved. (iii) Weakly bonded regions within the structure are identified and seen. (iv) Since the replica is so thin, normal inelastic electron scattering (which detracts from image quality) is reduced, and higher image resolution is achieved.

FIGURE 2 TEM micrograph of a freeze-fractured and freeze-etched *Bacillus licheniformis* showing a closing septum (small arrows) between two daughter cells. Cross-fractures of the cell wall show little infrastructure within the wall's matrix. The large arrow with circle denotes shadow direction. Bar = 100 nm. Reproduced by permission of the author.

For gram-positive bacteria, the fracture usually runs through the hydrophobic domain of the plasma (cytoplasmic) membrane and reveals intramembranous proteins (4, 9). Some fractures run over and through the surface of the cell wall and usually show a relatively featureless matrix. Occasionally, more exciting and informative fracture faces are seen (e.g., the inner surface of the wall, which normally lies adjacent to the outer leaflet of the plasma membrane). Fractures of growing (Fig. 2) and completed (Fig. 3) septa are especially important since they reveal how the new wall polymers in gram-positive walls are laid down.

FIGURE 3 TEM freeze-fracture image of a cell wall septum from a *Staphylococcus aureus* wall preparation that was previously treated with trichloroacetic acid to remove all associated teichoic acid and proteins. Because this wall now contains only peptidoglycan, the concentric circular arrangement (small arrows) must be the arrangement of peptidoglycan. The large arrow with circle denotes shadow direction. Bar = 100 nm. Previously published in *J. Bacteriol.* **150**:844–850 (1982) and reproduced with permission.

Negative Staining

The technique of negative staining is not nearly as complicated as freeze-etching. Specimens are laid on TEM grids and floated on a drop of a heavy metal salt for 10 to 15 s. Usually 2% (wt/vol) solutions of uranyl acetate, ammonium molybdate, or potassium tungstenate are used. The stain is removed by touching the grid to filter paper; as the stain drains off, it collects around the specimen and negatively stains the sample. The specimen shows up white against a dark background.

Negative stains of intact gram-negative bacteria are more informative than those of gram-positive bacteria because gram-negative cell walls are not as strong as gram-positive walls. Gram-negative cells are therefore more flattened (and their substance more spread out) by surface tension as they dry down on the TEM grid; inelastic electron scattering is reduced, and more detail is seen. Because gram-positive cells are not as readily flattened, they make poorer specimens for negative staining. Most information has come from negative stains of cell wall fragments derived by the physical breakage of intact bacteria (e.g., by French pressure cell extrusion or sonication). Polar caps of gram-positive rods are readily identified and, interestingly, the concentric arrangement of polymers can be identified in images of septa (Fig. 4) which are similar to those seen in freeze-etching (Fig. 3). Of all TEM techniques, negative staining can provide the highest resolution of gram-positive cell walls.

Conventional Embeddings

Conventional embedding is a tried-and-true technique that was first used in the 1950s (12). Because it eventually provides thin sections of the embedded bacteria, the procedure reveals how the various layers of the cell envelope are juxtaposed on top of one another (Fig. 5). This method revealed that gram-positive walls consist of an amorphous

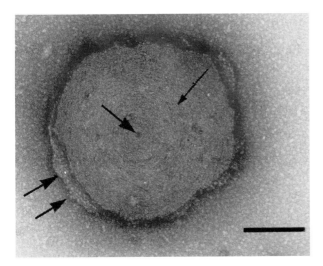

FIGURE 4 TEM negative stain of a septum from a mechanically disrupted culture of *Bacillus anthracis*. The small arrow (upper right) points to the concentric circular arrangement of the peptidoglycan fibers; the large arrow points to the central hole left in this almost completed septum; and the double arrows point to a section of septal wall from the other daughter cell. Bar = 250 nm.

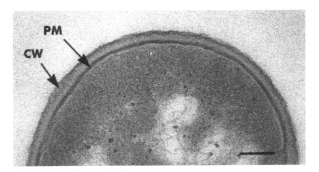

FIGURE 5 TEM micrograph of a thin section of a conventionally embedded *S. epidermis* showing an amorphously textured cell wall (CW) above the plasma membrane (PM). Bar = 100 nm.

amalgamation of peptidoglycan and ancillary secondary polymers such as teichoic and teichuronic acids (4).

Before thin sectioning can begin, the cells must be infiltrated with unpolymerized plastics and cured (polymerized) into a hard solid block that will resist tearing as the cells are sectioned by an extremely sharp glass or diamond knife. Virtually all suitable plastics are water insoluble, and the bacteria need to be dehydrated before embedding in plastic. How does one dehydrate with harsh organic solvents without denaturing protein and nucleic acids and without extracting the phospholipids of membranes? Electron microscopists attempt to overcome these severe problems by using chemical cross-linking agents to fix the bacteria before beginning dehydration and plastic embedding procedures. Often 2% (vol/vol) glutaraldehyde in a suitable buffer is used to fix cellular protein, followed by 2% (wt/vol) osmium tetroxide in buffer to fix lipids. The two fixatives cannot be used together because glutaraldehyde is a strong reducing agent and osmium tetroxide is an oxidant. Because all cells consist of elements of relatively low atomic number (i.e., C, H, O, N, S, and P), which do not scatter electrons effectively to produce a TEM image, heavy metal stains are also used before embedding. Osmium tetroxide is not only a fixative; it is also a good stain (owing to its osmium atom) that bonds to unsaturated carbon atoms of lipids. After osmium tetroxide treatment, the bacteria are also stained with 2% (wt/vol) uranyl acetate since the massive uranium atom imparts more contrast to the cellular material. After this point, the cells are systematically dehydrated by immersion in a graded series of an organic solvent (e.g., ethanol, acetone, propylene oxide) and then infiltrated with unpolymerized plastic (e.g., epoxy, acrylate, polyester). Next the plastic block is cured by increasing its temperature or subjecting it to UV radiation in the presence of a cross-linking agent. The blocks are trimmed and thin-sectioned in an ultramicrotome, and the sections are stained with uranyl acetate and lead citrate (to further increase contrast) and viewed by TEM.

Freeze-Substitution

For at least a decade, electron microscopists have realized that conventional embeddings do not provide a highly accurate picture of thin-sectioned bacteria. No matter how hard they tried to preserve cells with chemical fixatives, substantial quantities of lipid, protein, and nucleic acids were extracted during the procedure, and many structural components were compacted. Accordingly, difficult, tedious, and expensive ultra-low-temperature techniques were devised to preserve delicate structures. Methods to view frozen thin foils of bacterial components and frozen thin sections of intact cells were developed, but because they were of unstained material, microscopy was difficult. Cryo-specimen holders and cryo-chambers had to be incorporated into microscopes, and difficult-to-use ancillary equipment had to be built. Expense and dedicated expertise confined these cryo-techniques to only a few electron microscopy facilities worldwide, and this type of cryo-TEM had little impact on the imaging of gram-positive cell walls.

It was left to the cryo-technique of freeze-substitution to reformulate our perception of gram-positive cell walls (7, 8, 16, 29, 36). This technique combines the best traits of cryo-TEM and conventional embedding (17, 18, 26–28). A thin layer of cells is snap-frozen in liquid propane at −196°C and immediately immersed in a substitution medium consisting of 2% (wt/vol) osmium tetroxide and 2% (wt/vol) uranyl acetate dissolved in anhydrous acetone. A molecular sieve to absorb sublimed water is also added, and the entire suspension is held at −80°C for 2 days. When the cells are snap-frozen, they are encased in noncrystalline, vitreous ice in a matter of microseconds; all molecular motion in the bacteria is immediately stopped (including the action of degrading enzymes), and delicate, hydrated structure is preserved. During substitution, the cells are fixed, stained, and dehydrated (all at the same time) at such a low temperature (−80°C) that fine structure is maintained. Once substitution has been completed, the bacteria can be returned to ambient temperature and infiltrated with the same plastics used in conventional embedding. Thin sections of cell walls subjected to freeze-substitution are no longer featureless (Fig. 6). They have a dark-stained region immediately above the plasma membrane and a highly fibrous surface (7, 8, 16, 19, 36).

GENERAL CHEMISTRY OF THE GRAM-POSITIVE CELL WALL

Peptidoglycan

Most gram-positive cell walls possess an extensive meshwork of peptidoglycan that is typically 15 to 30 nm thick. Since one layer of peptidoglycan is ~1 nm thick (4), this implies that bacteria such as *Bacillus subtilis*, which has a wall 25 nm thick, consist of 25 layers of peptidoglycan. A certain proportion of all peptidoglycan polymers are cross-linked to adjacent strands so that a huge macromolecular murein sacculus completely surrounds the cell. For *B. subtilis*, ~50% of the peptidoglycan fibers are cross-linked owing to direct bonding between the *meso*-diaminopimelic acid (at position C_3 along the peptide stem) and the terminal D-alanine (at position C_4) of adjacent peptide stems emanating from the *N*-acetyl muramyl residues; this is A1γ chemotype peptidoglycan (32). *Mycobacterium tuberculosis* also has A1γ peptidoglycan (32). There is a great diversity of gram-positive peptidoglycan chemotypes (i.e., more than 100 chemotypes are known; 32) that rely on subtly different amino acid substituents in the peptide stems and on different linkage units between the stems. For example, *S. aureus* has a pentaglycine linkage unit between an L-lysine (C_3) (which replaces the diaminopimelic acid of A1γ chemotype) and the terminal D-alanine (C_4); this is the A1α chemotype. This chemo-

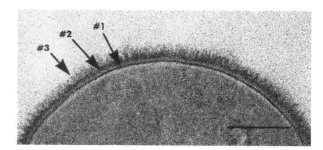

FIGURE 6 TEM micrograph of a thin section of freeze-substituted *Bacillus subtilis* showing the tripartite fine structure in the cell wall that demonstrates cell wall turnover. Region #1 is densely stained because the newly laid down peptidoglycan is condensed and reactive to the staining reagents. Region #2 is more lightly stained because this is the stress-bearing peptidoglycan that is highly stretched and therefore not as dense or condensed as region #1. Region #3 is undergoing hydrolysis by peptidoglycan hydrolases (autolysins) and is more fibrous. Because covalent bonds are being clipped, there are many reactive groups with which the staining reagents can interact, and this region is darkly stained. Bar = 100 nm.

type can subtly alter when *S. aureus* is stressed by methicillin or high NaCl content (15, 39). (The chemotype classification system provides a wealth of chemical information about each type of peptidoglycan; the Roman capital letters [i.e., A or B] indicate the cross-linking class; the Arabic numbers [i.e., 1 to 5] indicate the type of interpeptide bridge or lack of it; and the Greek letters [i.e., α to δ] indicate the amino acid present in position C_3 of the stem peptide.)

Secondary Polymers

Attached to the peptidoglycan, especially to the *N*-acetyl muramyl residues, are a variety of secondary polymers. In the simplest case, as exemplified by *B. subtilis* and *B. licheniformis*, these are teichoic (a highly phosphorylated polymer) and teichuronic (phosphate is substituted by uronic acid) acids; the ratio of these two secondary polymers is modulated in *B. subtilis* by available environmental phosphate, whereas in *B. licheniformis*, a mixture of both polymers is constant. Some gram-positive bacteria, such as some *Clostridium* and *Sporosarcina* species, do not possess secondary polymers, whereas others have a complex variety of them beyond the two polymeric acids that were previously mentioned. For example, *M. tuberculosis* possesses mycolic acids, arabinogalactans, and glycolipids that are linked to the peptidoglycan (14) as are the M protein of *Streptococcus pyogenes* (see chapter 2, this volume) and the nonteichoic acid Lancefield antigens of *Streptococcus* and *Enterococcus*. To compound the complexity, certain walls (e.g., those of *Streptococcus pneumoniae*) contain both teichoic acids (linked to peptidoglycan muramyl moieties) and lipoteichoic acids (linked to the plasma membrane by the "lipo" substituent) (41).

ELECTRON MICROSCOPY

It is clear that the chemistry of gram-positive walls can be complicated and that many types of secondary polymers can be associated with the peptidoglycan. Yet, peptidoglycan is common to all and is the primary fabric that controls most aspects of cell shape and, as it is laid down, of growth. No matter the peptidoglycan chemotype, TEM has not found major differences in the structural formats of most gram-positive walls. Typically, these cell walls are homogeneous with little differentiation of polymers within the wall's matrix (Fig. 5 and 6). I believe that the peptidoglycan strands (and associated secondary polymers) are arranged in a circumferential manner at right angles to the long cell axis (4, 37, 38). Yet, there is little hint of this arrangement unless actively growing regions of the wall (e.g., septa) are investigated (2, 30). In this case, small concentric ridges emanating out from the center of the septum are seen by freeze-etching and negatively stained preparations (Fig. 3 and 4). These ridges are presumed to be the arrangement of the new wall polymers as they are fabricated and inserted into the growing septum (1, 2). Once the septum is completed and the two daughter cells have separated from one another, this region becomes a new hemispherical cap (or pole) in rod-shaped bacteria (Fig. 7) or the newly formed one-half of the cell wall of an *Enterococcus* sp. (Fig. 8). In the latter, a septal scar denotes the old from the new cell wall. Intuitively, the concentric arrangement of polymers seen in septa is the typical arrangement in cell walls of coccoid cells and substantiates the circumferential arrangement believed to exist in rod-shaped bacteria.

CELL WALL TURNOVER

Gram-positive walls undergo a process of turnover for cell growth to occur. This must be carefully accomplished since there are substantial hydrostatic turgor pressures within gram-positive bacteria. In fact, these cells would lyse because of this pressure if it were not for the strong fabrication of their walls. Therefore, the "make before break" concept of wall turnover has been invoked to explain cellular growth (21). New polymers are exported, compacted, and linked both together and to pre-existing wall polymers at the inner face of the cell wall by penicillin-binding proteins (e.g., see references 25 and 30). These new polymers are packed tightly together to a rela-

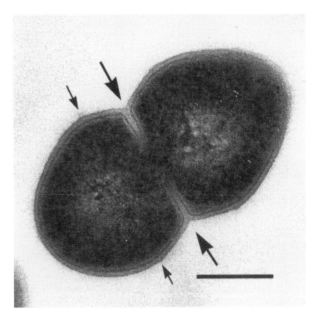

FIGURE 8 Thin section of a conventionally embedded *Enterococcus hirae* cell showing a growing septum (large arrows) and a septal scar (small arrows). All new cell wall growth occurs between the septal scars of both daughter cells as the growth of the septum progresses. Bar = 500 nm.

tively high density and are not under stress from cellular turgor pressure. This is because the older peptidoglycan (which is directly overlying these new polymers) is the stress-bearing material. Being stressed, the older peptidoglycan is stretched almost to its breaking limit and is (therefore) a low-density fabric. Directly above the stress-bearing region is the oldest peptidoglycan, which is being solubilized by its constituent autolysins.

New material is constantly being put in at the inner wall face, and old material is constantly being removed from the outer wall face, but more material is put in than is removed. This wall turnover, from inner to outer faces, combined with the expansive forces exerted by turgor pressure, ensures wall expansion and therefore cell growth. A dynamic equilibrium between polymer renewal at the inner face and polymer hydrolysis at the outer face is highly controlled to ensure cellular growth and division. If the equilibrium is disturbed, the consequences can be severe; e.g., the antibiotics penicillin and cephalosporin bind to penicillin-binding proteins and inhibit the renewal of wall polymers at the inner face. Since the autolysins continue to hydrolyze surface peptidoglycan (e.g., see reference 34), the wall grows thinner until turgor pressure physically lyses the cell.

Freeze-substitution has captured the process of wall turnover in gram-positive bacteria (19). The wall is discriminated into three separate regions (Fig. 6). (i) The inner face is a darkly staining region because of the high density of its polymers and the abundance of reactive groups (which bind the stain). (ii) The middle region is more lightly stained and has low density because it is stress-bearing. (iii) The outer face is highly fibrous because autolysins are breaking up and solubilizing the polymeric matrix in this region. This same general format can be seen in a wide number of such cell walls (Fig. 9). It is important

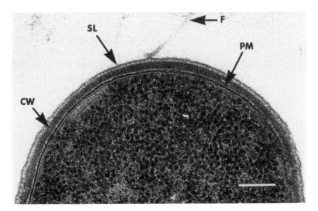

FIGURE 7 TEM micrograph of a thin section of a cell pole of a conventionally embedded *Bacillus thuringiensis* showing its S-layer (SL) above the cell wall (CW) and plasma membrane (PM). A flagellum (F) emanating from the surface can also be seen. Bar = 100 nm.

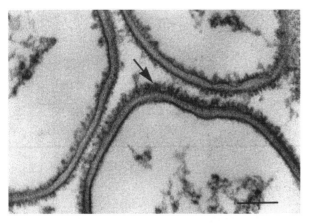

FIGURE 9 Thin section of isolated cell walls from *S. aureus* that were processed by the freeze-substitution technique after the culture was mechanically disrupted. As with the freeze-substituted *B. subtilis* cells shown in Fig. 6, these cell walls also have a fibrous surface (arrow) and tripartite structure. Bar = 100 nm. Previously published in *J. Bacteriol.* **169**:2482–2487 (1987) and reproduced by permission.

to recognize that, because secondary polymers (e.g., teichoic acid) are attached to the peptidoglycan, they too will be removed (as muramyl fragments) by autolytic action. For example, if the isolated walls seen in Fig. 9 are treated with trichloroacetic acid to remove teichoic acids, the surface fibrils are absent (36).

MYCOBACTERIAL WALLS—AN EXAMPLE OF A MORE UNUSUAL WALL TYPE

Because mycobacterial walls possess a number of unusual and unique substances, they are difficult to preserve. These walls contain 30 to 60% (by weight) lipid (14). Substances such as glycolipids, glycopeptidolipids, glycophospholipids, lipo-oligosaccharides, and even mycolic acids resist cross-linking by chemical fixatives and are prone to extraction by the organic solvents used during the dehydration procedure of conventional embeddings. Yet, with care, conventional embeddings have shown a remarkable degree of cell wall complexity since a triple-layered structure is seen (31). The outermost wall structure consists of an irregular electron-dense outer layer, followed by an electron-translucent region (sometimes referred to as the electron-transparent zone), which sits on top of a densely stained peptidoglycan layer. The electron-translucent region, which some researchers consider to be a lipid capsule, has been especially difficult to preserve (14, 26).

There have not been many attempts to elucidate the mycobacterial wall by freeze-substitution. Because of the high lipid content in the walls, these bacteria are highly hydrophobic, which affects the freezing rate; good vitrification of the cells by freezing is difficult (26–28). Persistence and high freezing rates have finally produced good freeze-substitution images (26). Although the tripartite wall of conventional embeddings is still seen, the wall is approximately two-thirds the conventional thickness (Fig. 10). Since many of the lipids are extracted from mycobacterial walls during conventional embeddings, strong hydrophobic interactions (which compress the cell wall) are reduced and the wall expands during a conventional em-

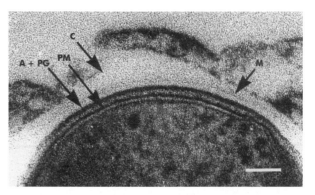

FIGURE 10 Thin section of a freeze-substituted *Mycobacterium kansasii* that shows the complexity of this bacterium's cell envelope. C, capsule or electron-transparent region; M, mycolates; A + PG, arabinogalactan plus peptidoglycan; PM, plasma membrane. Bar = 50 nm. More details of this image can be found in references 14 and 28. Reproduced with permission of the author.

bedding (28). Figure 10 shows a freeze-substituted wall of *Mycobacterium kansasii* and the possible arrangement of its various constituents (see references 14 and 28 for more details).

S-LAYERED WALLS

It is not unusual for gram-positive walls to have S-layers. These are paracrystalline surface arrays that self-assemble onto the surfaces of bacteria (and archaea) into oblique (p1, p2), square (p4), or hexagonally (p3, p6) packed lattices composed of proteins or glycoproteins with a molecular mass of 50 to 120 kDa (6, 24, 33a). They are frequently found on *Bacillus*, *Clostridium*, *Lactobacillus*, and *Sporosarcina* spp. and are seen as an extra surface layer on top of the cell wall (Fig. 7). Although the S-layers of gram-positive pathogens have not been extensively studied, especially by freeze-substitution (20), it is clear that the S-layer of *B. anthracis* (along with the bacterium's two toxins and poly-γ-D-glutamic acid capsule) contributes to this microorganism's virulence (23). It has also been demonstrated that *Clostridium difficile* possesses an S-layer (35). Freeze-etching has been especially useful in the detection of gram-positive S-layers because a major surface fracture plane follows the outer face of S-layers, and the etching clearly reveals the lattice type (Fig. 11).

A GRAM-POSITIVE PERIPLASM

Unlike gram-negative cell envelopes, gram-positive envelopes do not have a clearly defined periplasmic space (7, 8). Yet, many of the functional attributes of periplasmic components (e.g., proteases, DNases, autolysins, newly formed wall components, alkaline phosphatase) can be closely associated with the fabric of their walls. In fact, those bacilli that possess S-layers also have a pool of S-layer protein intermingled with the wall polymers that, presumably, can be called upon for the repair or new assembly of their S-layer (10). Accordingly, even without a dedicated physical space for a periplasm, gram-positive bacteria possess periplasmic components interdigitated among the polymers that make up their cell walls. By definition, then,

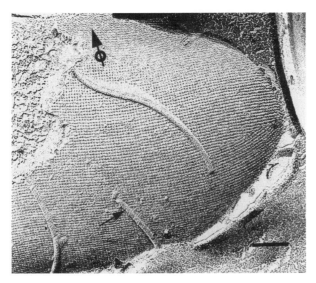

FIGURE 11 Freeze-etched surface of an *Aneurinibacillus thermoaerophilus* 10155/C$^+$ showing the S-layer with its square lattice (p4; $a = b = 10$ nm), which is made up of a glycoprotein. See reference 20 for more details. Flagella can be seen on top of the S-layer (small arrow). The large arrow with circle denotes shadow direction. Bar = 100 nm.

these bacteria have a periplasm (residing within the cell wall), but unlike gram-negative bacteria, they have no clearly defined periplasmic space (7).

I thank Kazuobu Amako and Paul Messner for supplying their excellent images (K.A. supplied Fig. 1, 3, and 9, and P. M. supplied Fig. 11). Research was funded by a National Centres of Excellence grant through the Canadian Bacterial Disease Network and a grant from the Natural Sciences and Engineering Research Council of Canada (NSERC). The NSERC Guelph Regional STEM Facility is partially funded by an NSERC-Major Facilities Access grant.

REFERENCES

1. **Amako, K., and A. Umeda.** 1977. Scanning electron microscopy of *Staphylococcus aureus*. *J. Ultrastruct. Res.* **58:**34–40.

2. **Amako, K., A. Umeda, and K. Murata.** 1982. Arrangement of peptidoglycan in the cell wall of *Staphylococcus* spp. *J. Bacteriol.* **150:**844–850.

3. **Arthur, M., and P. Courvalin.** 1993. Genetics and mechanisms of glycopeptide resistance in enterococci. *Antimicrob. Agents Chemother.* **37:**1563–1571.

4. **Beveridge, T. J.** 1981. Ultrastructure, chemistry, and function of the bacterial wall. *Int. Rev. Cytol.* **12:**229–317.

5. **Beveridge, T. J.** 1988. Wall ultrastructure: how little we know, p. 3–20. *In* P. Actor, L. Daneo-Moore, M. Higgins, M. R. J. Salton, and G. D. Shockman (ed.), *Antibiotic Inhibition of Bacterial Cell Surface Assembly and Function.* American Society for Microbiology, Washington, D.C.

6. **Beveridge, T. J.** 1994. Bacterial S-layers. *Curr. Opin. Struct. Biol.* **4:**204–212.

7. **Beveridge, T. J.** 1995. The periplasmic space and the periplasm in gram-positive and gram-negative bacteria. *ASM News* **61:**125–130.

8. **Beveridge, T. J., and L. L. Graham.** 1991. Surface layers of bacteria. *Microbiol. Rev.* **55:**684–705.

9. **Beveridge, T. J., T. J. Popkin, and R. M. Cole.** 1993. Electron microscopy, p. 42–71. *In* P. Gerhardt (ed.), *Methods for General and Molecular Bacteriology.* American Society for Microbiology, Washington, D.C.

10. **Breitwieser, A., K. Gruber, and U. B. Sleytr.** 1992. Evidence for an S-layer protein pool in the peptidoglycan of *Bacillus stearothermophilus*. *J. Bacteriol.* **174:**8008–8015.

11. **Brudney, K., and J. Dobkin.** 1991. Resurgent tuberculosis in New York City. Human immunodeficiency virus, homelessness and the decline of tuberculosis control programs. *Am. Rev. Respir. Dis.* **144:**745–749.

12. **Chapman, G. B., and J. Hillier.** 1953. Electron microscopy of ultrathin sections of bacteria. I. Cellular division in *Bacillus cereus*. *J. Bacteriol.* **66:**362–373.

13. **Culliton, B. J.** 1992. Drug resistant TB may bring epidemic. *Nature* **356:**473.

14. **Daffé, M., and P. Draper.** 1998. The envelope layers of mycobacteria with reference to their pathogenicity. *Adv. Microb. Physiol.* **39:**131–203.

15. **deJonge, B. L. M., Y.-S. Chang, D. Gage, and A. Tomasz.** 1992. Peptidoglycan composition of a highly methicillin-resistant *Staphylococcus aureus*. *J. Biol. Chem.* **267:**11248–11254.

16. **Graham, L. L.** 1991. Freeze-substitution studies of bacteria. *Electron Microsc. Rev.* **5:**77–103.

17. **Graham, L. L., and T. J. Beveridge.** 1990. Evaluation of freeze-substitution and conventional embedding protocols for routine electron microscopic processing of eubacteria. *J. Bacteriol.* **172:**2141–2149.

18. **Graham, L. L., and T. J. Beveridge.** 1990. Effect of chemical fixatives on accurate preservation of *Escherichia coli* and *Bacillus subtilis* structure in cells prepared by freeze-substitution. *J. Bacteriol.* **172:**2150–2159.

19. **Graham, L. L., and T. J. Beveridge.** 1994. Structural differentiation of the *Bacillus subtilis* cell wall. *J. Bacteriol.* **176:**1413–1421.

20. **Kadurugamuwa, J. L., A. Mayer, P. Messner, M. Sára, U. B. Sleytr, and T. J. Beveridge.** 1998. S-layered *Aneurinibacillus* and *Bacillus* spp. are susceptible to the lytic action of *Pseudomonas aeruginosa* membrane vesicles. *J. Bacteriol.* **180:**2306–2311.

21. **Koch, A. L., and R. J. Doyle.** 1985. Inside-to-outside growth and turnover of the cell wall of gram-positive rods. *J. Theor. Biol.* **117:**137–157.

22. **Koval, S. F., and T. J. Beveridge.** Electron microscopy. *In* J. Lederberg (ed.), *Encyclopedia of Microbiology*, 2nd ed., in press. Academic Press, Inc., New York, N.Y.

23. **Mesnage, S., E. Tosi-Couture, M. Mock, P. Gounon, and A. Fouet.** 1997. Molecular characterization of the *Bacillus anthracis* main S-layer component; evidence that it is the major cell-associated antigen. *Mol. Microbiol.* **23:**1147–1155.

24. **Messner, P., and U. B. Sleytr.** 1992. Crystalline bacterial cell-surface layers. *Adv. Microb. Physiol.* **33:**213–275.

25. **Murray, T., D. L. Popham, and P. Setlow.** 1997. Identification and characterization of *pbpA* encoding *Bacillus subtilis* penicillin-binding protein 2A. *J. Bacteriol.* **179:**3021–3029.

26. **Paul, T. R., and T. J. Beveridge.** 1992. Reevaluation of envelope profiles and cytoplasmic ultrastructure of mycobacteria processed by conventional embedding and freeze-substitution protocols. *J. Bacteriol.* **174:**6508–6517.

27. **Paul, T. R., and T. J. Beveridge.** 1993. Ultrastructure of mycobacterial surfaces by freeze-substitution. *Zentrabl. Bakteriol.* **279:**450–457.

28. **Paul, T. R., and T. J. Beveridge.** 1994. Preservation of surface lipids and ultrastructure of *Mycobacterium kansasii* using freeze-substitution. *Infect. Immun.* **62:**1542–1550.

29. **Paul, T. R., L. L. Graham, and T. J. Beveridge.** 1993. Freeze-substitution and conventional electron microscopy of medically-important bacteria. *Rev. Med. Microbiol.* **4:**65–72.

30. **Paul, T. R., A. Venter, L. C. Blaszczak, T. R. Parr, H. Labishinski, and T. J. Beveridge.** 1995. Localization of penicillin-binding proteins to the splitting system of *Staphylococcus aureus* septa using a mercury-penicillin V derivative. *J. Bacteriol.* **177:**3631–3640.

31. **Rastogi, N., C. Fréhel, and H. L. David.** 1986. Triple-layered structure of mycobacterial cell wall: evidence for the existence of a polysaccharide-rich outer layer in 18 mycobacterial species. *Curr. Microbiol.* **13:**237–242.

32. **Schleifer, K. H., and O. Kandler.** 1972. Peptidoglycan types of bacterial cell walls and their taxonomic implications. *Bacteriol. Rev.* **36:**407–477.

33. **Sieradzki, K., and A. Tomasz.** 1997. Inhibition of cell wall turnover and autolysis by vancomycin in a highly vancomycin-resistant mutant of *Staphylococcus aureus. J. Bacteriol.* **179:**2557–2566.

33a. **Sleytr, U. B., and T. J. Beveridge.** 1999. Bacterial S-layers. *Trends Microbiol.* **7:**253–260.

34. **Sugai, M., S. Yamada, S. Nakashima, H. Komatsuzawa, A. Matsumoto, T. Oshida, and H. Suginaka.** 1997. Localized perforation of the cell wall by a major autolysin: *atl* gene products and the onset of penicillin-induced lysis of *Staphylococcus aureus. J. Bacteriol.* **179:**2958–2962.

35. **Takeoka, A., K. Takumi, T. Koga, and T. Kawata.** 1991. Purification and characterization of S-layer proteins from *Clostridium difficile* GAI 0714. *J. Gen. Microbiol.* **137:**261–267.

36. **Umeda, A., Y. Ueki, and K. Amako.** 1987. Structure of the *Staphylococcus aureus* cell wall determined by the freeze-substitution method. *J. Bacteriol.* **169:**2482–2487.

37. **Verwer, R. W. H., and N. Nanninga.** 1976. Electron microscopy of isolated cell walls of *Bacillus subtilis* var. *niger. Arch. Microbiol.* **109:**195–197.

38. **Verwer, R. W. H., N. Nanninga, W. Keck, and U. Schwarz.** 1978. Arrangement of glycan chains in the sacculus of *Escherichia coli. J. Bacteriol.* **136:**723–729.

39. **Vijaranakul, U., M. J. Nadakavukaren, B. deJonge, B. J. Wilkinson, and R. K. Jayaswal.** 1995. Increased cell size and shortened peptidoglycan interpeptide bridge of NaCl-stressed *Staphylococcus aureus* and their reversal by glycine betaine. *J. Bacteriol.* **177:**5116–5121.

40. **Woodford, N., A. P. Johnson, D. Morrison, and D. C. E. Speller.** 1995. Current perspectives on glycopeptide resistance. *Clin. Microbiol. Rev.* **8:**585–615.

41. **Yother, J., K. Leopold, J. White, and W. Fischer.** 1998. Generation and properties of *Streptococcus pneumoniae* mutant which does not require choline or analogs for growth. *J. Bacteriol.* **180:**2093–2101.

Surface Proteins on Gram-Positive Bacteria

VINCENT A. FISCHETTI

2

As arms, legs, hair, and fur are used in higher species for their survival in the environment, surface appendages are used by bacteria for similar purposes. Surface molecules in bacteria range from complex structures like flagella that propel the organism in aqueous environments, to less sophisticated polysaccharides and proteins. All of these molecules serve to benefit the organism for survival in a hostile environment, such as the waters of a rushing stream, the blood of an infected animal, the surface of an object, or the surface of a mucosal epithelium. While it was previously believed that bacteria were simple single-cell organisms with little complexity, it is now becoming apparent that they are highly evolved, advanced particles that possess a wide array of surface molecules that serve to manipulate the organism in its environment. For human pathogens, surface molecules have been finely tuned to allow the organism to attach to specific cells, invade those cells, and persist in infected tissues.

In an effort to emphasize the complexity of bacterial surface molecules and their use in the everyday life of the bacterium, this chapter will focus on those surface proteins found on gram-positive bacteria. For an extensive review of the subject, see reference 54a.

GRAM-POSITIVE CELL WALL

From electron microscopic analysis of the gram-positive cell envelope (see chapter 1, this volume) and a large number of elegant chemical and immunological analyses, a picture of the gram-positive cell wall has emerged (Fig. 1). The structure differs significantly from the gram-negative cell wall in two ways: (i) the presence of a thicker and more cross-linked peptidoglycan and (ii) the lack of an outer membrane. Because of these differences, surface molecules in gram-positive organisms vary from those in gram-negative organisms, in which specialized systems are required to transport and anchor molecules through the outer membrane (86). In general, surface proteins in gram-positive bacteria can be separated into three categories: (i) those that anchor at their C-terminal ends (through an LPXTG motif), (ii) those that bind by way of charge or hydrophobic interactions, and (iii) those that bind via their N-terminal region (lipoproteins) (Fig. 1).

C-TERMINAL-ANCHORED PROTEINS

Decades ago, relatively harsh treatments of extraction were used to remove surface proteins from gram-positive bacteria, resulting in the isolation and characterization of molecular fragments rather than complete molecules. It was not until the early 1980s that the genes for surface proteins on gram-positive bacteria were first published (protein A from *Staphylococcus aureus* [27, 89] and M protein from *Streptococcus pyogenes* [33]). While at first glance these molecules had little in common, it was soon realized that the C-terminal end of the protein A sequence contained a sequence error. Upon correction (89), it became apparent that this region of protein A exhibited very close sequence homologies with the M protein molecule. Since that time >65 proteins from gram-positive bacteria that anchor through their C-terminal region have been reported (Table 1). As these sequences were published in the ensuing years, it became obvious that there was a common theme within the structure of these C-terminal-anchored molecules.

M Protein Structure

Of the C-terminal-anchored proteins, the M protein may be considered an archetypical molecule, a characteristic of which is the presence of regions with sequence repeats and a conserved anchor. As seen in Fig. 2, the M protein is composed of four repeat blocks, each differing in size and sequence (33). The A-repeats are each composed of 14 amino acids in which the central blocks are identical and the end blocks slightly diverge from the central consensus repeats. The B-repeats, composed of 25 amino acids each, are arranged in the same way. The C-repeats, composed of 2.5 blocks of 42 amino acids, are not as similar to each other as are the A- and B-repeats. There are also four short D-repeats that show some homology. These repeat segments make up the central helical rod region of the M molecule because of the high helical potential ascribed to

Gram-Positive Pathogens, ed. by V. A. Fischetti et al.

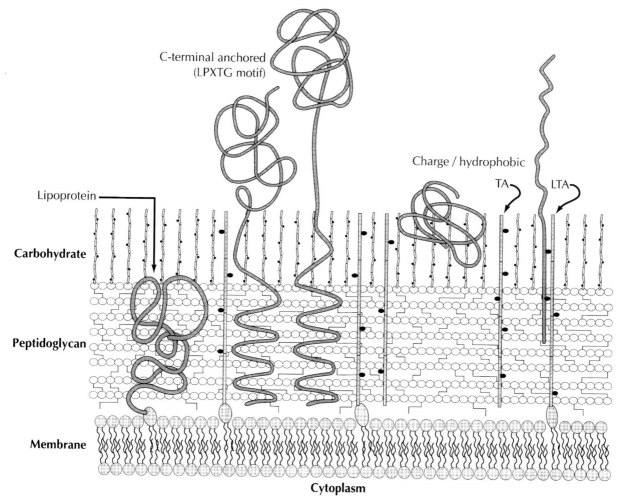

FIGURE 1 Major surface structures of the cell wall of gram-positive bacteria. Linked to the surface of the peptidoglycan, many gram-positive organisms have polysaccharide structures that in some cases are used for their immunological classification. Surface proteins are linked by three mechanisms. (i) Lipoproteins have a lipid linked through a cysteine at the N terminus. (ii) C-terminal-anchored proteins are attached and stabilized in the peptidoglycan through a C-terminal complex containing an LPXTG motif. (Most surface proteins are anchored in this way.) (iii) Certain surface proteins are attached through hydrophobic and/or charge interactions to the cell surface. (Some proteins are bound ionically to the lipoteichoic acid.) The teichoic acids (TA) are a common feature of the gram-positive cell wall. TA is usually composed of a repeating carbohydrate-phosphate polymer linked through a phosphodiester linkage to the peptidoglycan. Lipoteichoic acid (LTA) is composed of a similar polymer linked to the cytoplasmic membrane through a fatty acid (see chapter 19, this volume).

the amino acids found within this segment, as determined by conformational analysis (21, 66).

A close examination of the sequence within the M molecule revealed a repeating seven-residue periodicity of nonpolar amino acids, a basic characteristic of α-helical coiled-coil proteins like mammalian tropomyosin. In general, α-helical coiled-coil proteins are constructed from a reiterating seven-residue amino acid pattern (a-b-c-d-e-f-g)$_n$ in which residues in positions a and d are hydrophobic and form the core residues in the coiled-coil, and the intervening residues are primarily helix-promoting. The arrangement of amino acids of the M6 protein within the seven-residue pattern is shown in Fig. 3. The seven-residue periodicity from amino acids 12 through 362 strongly sug-

gests that this region is in a coiled-coil conformation and forms the helical central rod region of the molecule. Discontinuities in the heptad pattern (seen especially in the B-repeat region), which have been found in the sequence of other M proteins and other coiled-coil molecules, probably account for the flexibility of the M molecules observed in electron micrographs (66). Based on these irregularities in the heptad pattern, the central rod region is divided into three subregions that correlate with the A-, B-, and C-repeat blocks (17).

Electron and Fluorescent Microscopy

Although early electron micrographs of the surface of gram-positive bacteria revealed certain structures, it was

TABLE 1 Sequenced surface proteins[a] from gram-positive bacteria

Name/gene	Function/name	Organism	LPXTG[b]	Reference	GenBank accession no.
C-terminal-linked surface proteins					
1. emm6	M protein	Streptococcus pyogenes	LPSTG	33	M11338
2. emm5	M protein	S. pyogenes	LPSTG	53	M20374
3. emm12	M protein	S. pyogenes	LPSTG	69	U02342
4. emm24	M protein	S. pyogenes	LPSTG	54	M19031
5. emm49	M protein	S. pyogenes	LPSTG	28	M23689
6. emm57	M protein	S. pyogenes	LPSTG	50	X60959
7. emm2	M protein	S. pyogenes	LPSTG	3	X61276
8. emm3	M protein	S. pyogenes	LPSTG	42	Z21845
9. ARP2	IgA-binding protein	S. pyogenes	LPSTG	3	X61276
10. ARP4	IgA-binding protein	S. pyogenes	LPSTG	22	X15198
11. Mrp4	IgG/fibrinogen binding	S. pyogenes	LPSTG	60	M87831
12. FcRA	Fc-binding protein	S. pyogenes	LPSTG	29	M22532
13. Prot H	Human IgG Fc binding	S. pyogenes	LPSTG	26	M29398
14. SCP	C5a peptidase	S. pyogenes	LPTTN	7	J05229
15. T6	Protease-resistant protein	S. pyogenes	LPSTG	74	M32978
16. sof22	Serum opacity factor	S. pyogenes	LPASG	67	U02290
17. Sfb	Fibronectin binding	S. pyogenes	LPATG	84	X67947
18. ZAG	Binds γ_2 M, Alb, IgG	Streptococcus zooepidemicus	LPTTG	38	U02290
19. PrtF	Fibronectin binding	S. pyogenes	LPATG	76	L10919
20. PAM	Plasmin binding	S. pyogenes	LPSTG	1	Z22219
21. Prot L	Light-chain binding	Peptostreptococcus magnus	LPKAG	41	M86697
22. PAB	Human serum albumin binding	P. magnus	LPEAG	13	X77864
23. bac	IgA binding protein	Group B streptococcus	LPYTG	30	X58470
24. bca	Alpha C antigen	Group B streptococcus	LPATG	52	M97256
25. fnbA	Fibronectin binding	Streptococcus dysgalactiae	LPQTG	49	Z22150
26. fnbB	Fibronectin binding	Streptococcus dysgalactiae	LPAAG	49	Z22151
27. Fnz	Fibronectin binding	Streptococcus equi	LPQTS	55	Y17116
28. SeM	M-like	S. equi	LPSTG	87	U73162
29. SzPSe	M-like	S. equi	LPQTS	87	U73163
30. Prot G	IgG-binding protein	Group G streptococcus	LPTTG	59	X06173
31. EmmG1	M protein	Group G streptococcus	LPSTG	11	M95774
32. DG12	Albumin-binding protein	Group G streptococcus	LPSTG	78	M95520
33. GfbA	Fibronectin binding	Group G streptococcus	LPATG	44	U31115
34. MRP	Surface protein[a]	Streptococcus suis	LPNTG	79	X64450
35. PAc	Surface protein	Streptococcus mutans	LPNTG	58	X14490
36. spaP	Surface protein	S. mutans	LPNTG	43	X17390
37. spaA	Surface protein	Streptococcus sobrinus	LPATG	88	D90354
38. wapA	Wall-associated protein A	S. mutans	LPSTG	16	M19347
39. fruA	Fructosidase	S. mutans	LPDTG	5	L03358
40. Sec10	Surface protein	Enterococcus faecalis	LPQTG	40	M64978
41. Asc10	Surface protein	E. faecalis	LPKTG	40	M64978
42. asa1	Aggregation substance	E. faecalis	LPQTG	24	X17214
43. Prot A	IgG-binding protein	Staphylococcus aureus	LPETG	27	X00342
44. FnBPb	Fibronectin binding	S. aureus	LPETG	77	J04151
45. FnBPa	Fibronectin binding	S. aureus	LPETG	39	X62992

(Continued on next page)

TABLE 1 Sequenced surface proteins[a] from gram-positive bacteria (*Continued*)

Name/gene	Function/name	Organism	LPXTG[b]	Reference	GenBank accession no.
46. *cna*	Collagen-binding protein	*S. aureus*	**LPKTG**	65	M81736
47. *clfA*	Clumping factor	*S. aureus*	**LPDTG**	51	Z18852
48. EDIN	Epidermal cell inhibitor	*S. aureus*	**LPRGT**	36	NA[c]
49. *prtM*	Cell wall protease	*Lactobacillus paracasei*	**LPKTA**	32	M83946
50. *wg2*	Cell wall protease	*Streptococcus cremoris*	**LPKTG**	45	M24767
51. *inlA*	Internalization protein	*Listeria monocytogenes*	**LPTTG**	23	M67471
52. Fimbriae	Type 1 fimbriae	*Actinomyces viscosus*	**LPLTG**	95	M32067
53. Fimbriae	Type 2 fimbriae	*Actinomyces naeslundii*	**LPLTG**	94	M21976
54. FimA	Type 2 fimbriae	*A. naeslundii*	**LPLTG**	96	AF019629
55. Hyal1	Hyaluronidase	*Streptococcus pneumoniae*	**LPQTG**	2	L20670
56. *nanA*	Neuraminidase	*S. pneumoniae*	**LPETG**	6	X72967
57. *glnA*	Glutamine synthetase	Group B streptococcus	**LPATL**	81	U61271
58. Protein F2	Fibronectin binding	*S. pyogenes*	**LPATG**	37	U31980
59. Fbe	Fibrinogen binding	*Staphylococcus epidermidis*	**LPDTG**	56	Y17116
60. InIC2	Internalin-like	*L. monocytogenes*	**LPTAG**	14	U77368
61. InID	Internalin-like	*L. monocytogenes*	**LPTAG**	14	U77368
62. InIE	Internalin-like	*L. monocytogenes*	**LPITG**	14	U77368
63. InIF	Internalin-like	*L. monocytogenes*	**LPKTG**	14	U77367
64. SfBP1	Fibronectin binding	*S. pyogenes*	**LPXTG**	70	AF071083
65. *dex*	Dextranase	*S. sobrinus*	**LPKTG**	91	M96978
66. *pad1*	Pheromone	*E. faecalis*	**LPHTG**	92	X62658
67. FAI	Fibrinogen/albumin/IgG	Group C streptococcus	**LPSTG**	83	NA
68. SfbII	Fibronectin binding	*S. pyogenes*	**LPASG**	47	X83303

Surface proteins attached by other mechanisms

1. *ftf*	Fructosyl transferase	*Streptococcus salivarius*		68	
2. *pspA*	Surface protein	*S. pneumoniae*		97	
3. *aas*	Adhesin/autolytic	*Streptococcus saprophiticus*		31	
4. FbpA	Fibrinogen binding	*S. aureus*		9	
5. FBP54	Fibronectin/fibrinogen binding	*S. pyogenes*		12	
6. Fib	Fibronectin binding	*S. aureus*		4	
7. SEN	Surface enolase	*S. pyogenes*		64	
8. SDH	Surface dehydrogenase	*S. pyogenes*		63	

[a] Surface proteins: proteins that have been identified to have a C-terminal-anchor motif but the function is unknown.

[b] L, 68/68; P, 68/68; (X); T, 59/68; G, 61/68.

[c] NA, not available.

not known whether they were proteinaceous or polysaccharides. It was not until Swanson et al. (82) published the first electron micrographs of the M protein on the surface of group A streptococci that it was realized that the molecule was an elongated structure. Although other proteins were also reported to be on the streptococcal surface, they were not apparent in these electron micrographs. Thus, direct visualization of surface molecules is limited to molecules with certain physicochemical characteristics. In the case of the M protein, it is the α-helical coiled-coil structure that allowed its visualization by electron microscopy (17).

In experiments designed to answer questions about the synthesis and placement of M protein on the cell wall,

Swanson et al. (82) found that, after trypsinization to remove existing M protein on living streptococci, newly synthesized M protein was first seen by electron microscopy on the cell in the position of the newly forming septum. In similar experiments designed after the classical experiments of Cole and Hahn (10) using fluorescein-labeled anti-M antibodies, trypsinized streptococci placed in fresh medium for 10 min revealed M protein first at the newly forming septum (Fig. 4). On organisms examined after 40 min or more of incubation, M protein was not observed in the position of the old wall, suggesting that the M molecule is produced only where the new cell wall is synthesized. This confirmed the observation of Swanson et al. (82). Whether this is also true of the many other C-terminal-

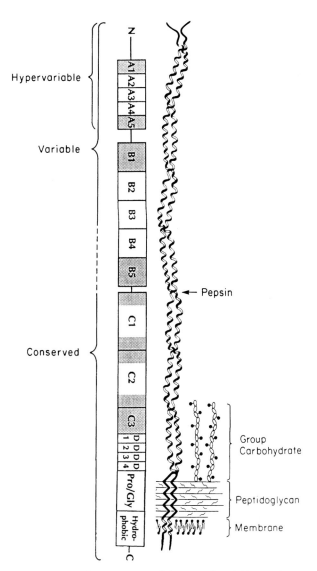

FIGURE 2 Characteristics of the complete M6 protein sequence. Blocks A, B, C, and D designate the location of the sequence repeat blocks. Numbers above the block indicate the number of amino acids per block. Block C3 is half the size of blocks C1 and C2. Shadowed blocks indicate those in which the sequence diverges from the central consensus sequence. Pro/Gly denotes the proline- and glycine-rich region likely located in the peptidoglycan. Membrane anchor is a 19-hydrophobic-amino-acid region adjacent to a 6-amino-acid charged tail. Pepsin identifies the position of the pepsin-sensitive site after amino acid 228. The C-terminal end is located within the cell wall and membrane.

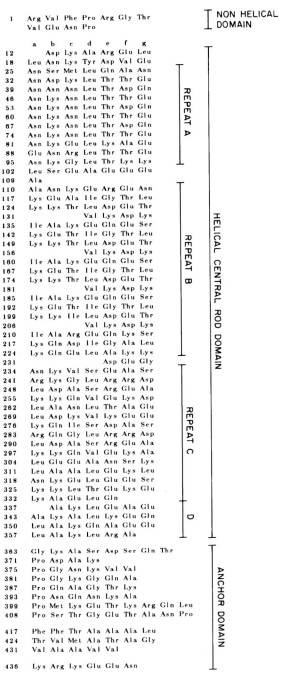

	a	b	c	d	e	f	g	
1	Arg	Val	Phe	Pro	Arg	Gly	Thr	NON HELICAL DOMAIN
	Val	Glu	Asn	Pro				
12		Asp	Lys	Ala	Arg	Glu	Leu	
18	Leu	Asn	Lys	Tyr	Asp	Val	Glu	
25	Asn	Ser	Met	Leu	Gln	Ala	Asn	
32	Asn	Asp	Lys	Leu	Thr	Thr	Glu	
39	Asn	Asn	Asn	Leu	Thr	Asp	Gln	
46	Asn	Lys	Asn	Leu	Thr	Thr	Glu	REPEAT A
53	Asn	Lys	Asn	Leu	Thr	Asp	Gln	
60	Asn	Lys	Asn	Leu	Thr	Thr	Glu	
67	Asn	Lys	Asn	Leu	Thr	Asp	Gln	
74	Asn	Lys	Asn	Leu	Thr	Thr	Glu	
81	Asn	Lys	Glu	Leu	Lys	Ala	Glu	
88	Glu	Asn	Arg	Leu	Thr	Thr	Glu	
95	Asn	Lys	Gly	Leu	Thr	Lys	Lys	
102	Leu	Ser	Glu	Ala	Glu	Glu	Glu	
109	Ala							
110	Ala	Asn	Lys	Glu	Arg	Glu	Asn	
117	Lys	Glu	Ala	Ile	Gly	Thr	Leu	
124	Lys	Lys	Thr	Leu	Asp	Glu	Thr	
131				Val	Lys	Asp	Lys	
135	Ile	Ala	Lys	Glu	Gln	Glu	Ser	
142	Lys	Glu	Thr	Ile	Gly	Thr	Leu	
149	Lys	Lys	Thr	Leu	Asp	Glu	Thr	
156				Val	Lys	Asp	Lys	
160	Ile	Ala	Lys	Glu	Gln	Glu	Ser	REPEAT B
167	Lys	Glu	Thr	Ile	Gly	Thr	Leu	
174	Lys	Lys	Thr	Leu	Asp	Glu	Thr	
181				Val	Lys	Asp	Lys	
185	Ile	Ala	Lys	Glu	Gln	Glu	Ser	
192	Lys	Glu	Thr	Ile	Gly	Thr	Leu	
199	Lys	Lys	Ile	Leu	Asp	Glu	Thr	
206				Val	Lys	Asp	Lys	
210	Ile	Ala	Arg	Glu	Gln	Lys	Ser	
217	Lys	Gln	Asp	Ile	Gly	Ala	Leu	
224	Lys	Gln	Glu	Leu	Ala	Lys	Lys	
231					Asp	Glu	Gly	
234	Asn	Lys	Val	Ser	Glu	Ala	Ser	
241	Arg	Lys	Gly	Leu	Arg	Arg	Asp	
248	Leu	Asp	Ala	Ser	Arg	Glu	Ala	
255	Lys	Lys	Gln	Val	Glu	Lys	Asp	
262	Leu	Ala	Asn	Leu	Thr	Ala	Glu	
269	Leu	Asp	Lys	Val	Lys	Glu	Glu	
276	Lys	Gln	Ile	Ser	Asp	Ala	Ser	
283	Arg	Gln	Gly	Leu	Arg	Arg	Asp	REPEAT C
290	Leu	Asp	Ala	Ser	Arg	Glu	Ala	
297	Lys	Lys	Gln	Val	Glu	Lys	Ala	
304	Leu	Glu	Glu	Ala	Asn	Ser	Lys	
311	Leu	Ala	Ala	Leu	Glu	Lys	Leu	
318	Asn	Lys	Glu	Leu	Glu	Glu	Ser	
325	Lys	Lys	Leu	Thr	Glu	Lys	Glu	
332	Lys	Ala	Glu	Leu	Gln			
337		Ala	Lys	Leu	Glu	Ala	Glu	
343	Ala	Lys	Ala	Leu	Lys	Glu	Gln	D
350	Leu	Ala	Lys	Gln	Ala	Glu	Glu	
357	Leu	Ala	Lys	Leu	Arg	Ala		
363	Gly	Lys	Ala	Ser	Glu	Asp	Ser	Gln Thr
371	Pro	Asp	Ala	Lys				
375	Pro	Gly	Asn	Lys	Val	Val		
381	Pro	Gly	Lys	Val	Glu	Ala		ANCHOR DOMAIN
387	Pro	Gln	Ala	Gly	Thr	Lys		
393	Pro	Asn	Gln	Asn	Lys	Ala		
399	Pro	Met	Lys	Glu	Thr	Lys	Arg	Gln Leu
408	Pro	Ser	Thr	Gly	Glu	Thr	Ala	Asn Pro
417	Phe	Phe	Thr	Ala	Ala	Ala	Leu	
424	Thr	Val	Met	Ala	Thr	Ala	Gly	
431	Val	Ala	Ala	Val	Val			
436	Lys	Arg	Lys	Glu	Glu	Asn		

FIGURE 3 Complete M6 protein sequence arranged to highlight the seven-residue periodicity found in the helical central rod region. Region assignments are based on sequence and conformational analyses. Arrangement of the sequence is based on the position of amino acids in a seven-residue periodicity designated by letters a through g beginning at residue 12 and continuing through residue 362, with interruptions at residues 109, 131, 156, 181, 206, 231, and 337. Alignment from residue 363 to 416 is used essentially to highlight the regularity of the position of prolines in the sequence. No periodicity is found from residue 417 to the end. Three major regions are indicated (nonhelical, helical, and anchor). The pepsin-sensitive site is between Ala-228 and Lys-229.

anchored surface proteins has not yet been determined; however, it is likely to be the case.

C-Terminal Anchor Region

An examination of the C-terminal end of those surface molecules that anchor at that end revealed, without exception, that all have a similar arrangement of amino acids. Up to seven charged amino acids are found at the C terminus and are composed of a mixture of both negative and positive charged residues. Immediately N-terminal to this short charged region is a segment of 15 to 22 predomi-

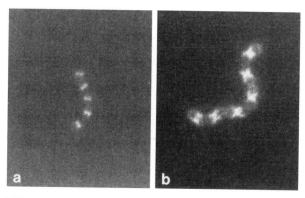

FIGURE 4 Fluorescein-labeled anti-M6 antibody analysis of the appearance of M6 protein on streptococcal cell walls. M6 streptococci were treated with trypsin to remove surface M protein. Cells, reincubated at 37°C, were removed at intervals, fixed, and stained with fluorescein-labeled anti-M6 antibody. (a) After 10 min of incubation the M protein is located within a thin band at the position of the newly forming septum. Magnification, ×5,000. (b) Location of the M protein after 40 min of incubation. Note that no fluorescein label is seen in the position of the old wall. Magnification, ×6,000.

nantly hydrophobic amino acids sufficient to span the cytoplasmic membrane of the bacterium, placing the charged end in the cytoplasm (Fig. 2). In all these proteins, the sequences found in the hydrophobic and charged regions are not necessarily the same, but the chemical characteristics of the amino acids used to compose them are conserved. Beginning about eight amino acids N-terminal from the hydrophobic domain is a heptapeptide with the consensus sequence LPXTG, which is extraordinarily conserved among all C-terminal-anchored proteins examined so far (20). While several amino acid substitutions are seen in position 3 of the heptapeptide (predominantly A, Q, E, T, N, D, K, and L), positions 1, 2, 4, and 5 are nearly completely conserved (100% in the L and P positions and >92% in the T and G positions). This conservation is also maintained at the DNA level. The preservation of this heptapeptide and the high homology within the hydrophobic and charged regions suggest that the method of anchoring these molecules within the bacterial cell is also highly conserved. Evidence suggests that this motif serves as an enzyme recognition sequence. The identification of the cleaving enzyme is currently under intensive investigation.

Thus, C-terminal-anchored surface proteins in gram-positive bacteria (which could number >25 different molecules in a single organism) are synthesized and exported at the septum, where new cell wall is also being produced and translocated to the surface (10, 17). Therefore, to coordinate the export and anchoring of all these proteins, the C-terminal hydrophobic domain and charged tail function as a temporary stop, precisely positioning the LPXTG motif at the outer surface of the cytoplasmic membrane (Fig. 5). This motif is then cleaved, ultimately resulting in the attachment of the surface-exposed segment of the protein to a cellular substrate (75). This idea is supported by studies indicating that the C-terminal hydrophobic domain and charged tail are missing from the streptococcal M protein extracted from the streptococcal cell wall (61, 75). Studies also suggest that the substrate in the peptidoglycan to which the protein is linked is the cross bridge (e.g., glycine

in *S. aureus*, alanine in *S. pyogenes*) of the pentapeptide (73). However, if this is the case, then it would be anticipated that not all the molecules translocated to the surface would be attached by this mechanism; this process would compete for the normal cross bridges in the assembly of the peptidoglycan, compromising the integrity of the wall. An alternative scenario is that the region of the protein located within the peptidoglycan (N-terminal to the LPXTG motif) would be stabilized within the cell wall by the cross-linked peptidoglycan. Support for this hypothesis is presented in the next section. C-terminal-anchored proteins are relatively stable, since in most cases free protein is not found in the growth supernatant. It should be emphasized, however, that in some circumstances, molecules are also partially released into the growth medium, as illustrated by the recovery of opacity factor from the supernatant of strains of *S. pyogenes*. The reason for the release of some molecules and the stability of others is as yet undetermined, but it is not unreasonable to assume that the differences may be associated with the type of amino acid found in the X position in the LPXTG motif.

Wall-Associated Region

Immediately N-terminal from the LPXTG motif is the wall-associated region, which spans about 50 to as many as 125 amino acid residues and is found in nearly all the proteins analyzed. Although this region does not exhibit a high degree of sequence identity among the known surface proteins, it is characterized by a high percentage of proline/glycine and threonine/serine residues. For some proteins (like the M protein) the concentration of proline/glycine is significantly higher than that of threonine/serine, while in others this relationship is either reversed or nearly equal. Because of its proximity to the hydrophobic domain (which is located in the cell membrane), this region would be positioned within the peptidoglycan layer of the cell wall (61) (Fig. 1). The reason for the presence of these particular amino acids at this location is unknown. One hypothesis suggests that the prolines and glycines, with their ability to initiate bends and turns within proteins, allow the peptidoglycan to become more easily cross-linked around these folds, thus stabilizing the molecule within the cell wall (61). The function of the threonines and serines within this region is not immediately apparent. While these amino acids are commonly used as O-linked glycosylation sites in eukaryotic proteins, such substitutions have not as yet been established in these bacteria.

Surface-Exposed Region

As the C-terminal region is characteristically conserved among the C-terminal-anchored surface proteins, the surface exposed region is characteristically unique. Despite their differences, these molecules appear to fall into three groups: (i) those with several repeating sequence blocks, (ii) those with repeat blocks located close to and within the cell wall region, and (iii) those without any repeat sequences (Fig. 6). Streptococcal M protein and protein G may be considered molecules representative of those containing several sequence repeats. However, the most commonly found structures are those with repeats located close to the cell wall and an extended nonrepetitive region N-terminal to this segment.

Conformational analysis of 12 representative surface molecules by the algorithm of Garnier et al. (25) revealed that, in general, those proteins containing repeat sequences

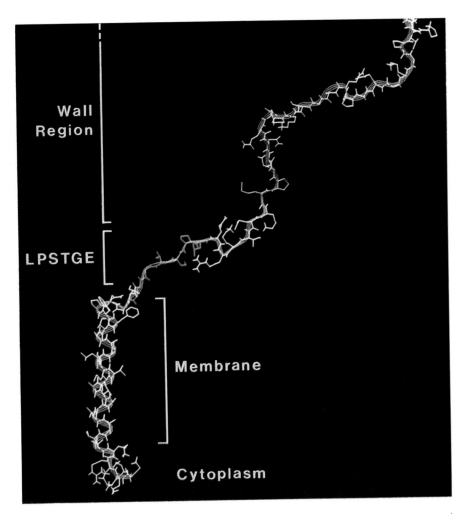

FIGURE 5 Computer-generated model of the C-terminal end of the M protein sequence (residues 371 to 441). A comparable region is found in all C-terminal-anchored surface proteins from gram-positive bacteria (see Table 1). The predicted location of this segment of the molecule is shown in the cytoplasm, membrane, and peptidoglycan. The space between the membrane and peptidoglycan (wall region) may be considered the "periplasm" of the gram-positive bacterium. The figure was generated on a Steller computer using the Quanta 2.1A program for energy minimization.

were predominantly helical within the region containing repeat segments (Fig. 7). Conversely, regions and molecules without repeat blocks were predominantly composed of amino acids exhibiting high β-sheet, β-turn, and/or random coil potential. Thus, within most of these molecules, the presence of repeat segments usually predicts the location of a helical domain. Possibly one of the pressures for the maintenance of repeat blocks is to preserve the helix potential within specific regions of these molecules, the presence of which may determine an extended protein structure, as has been shown for the M protein (66). An exception to this is found in the Sec10 protein, a predominantly helical molecule with limited repeat segments.

Seven-Residue Periodicity

As shown above, most of the surface molecules containing repeat sequences were found to be α-helical in those regions composed of repeats, while molecules without repeat blocks exhibited a high degree of β-sheet, β-turn, and random coil. However, little information was available to

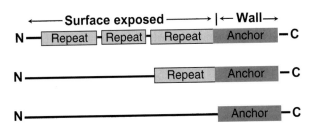

FIGURE 6 Characteristics of C-terminal-anchored surface proteins. These proteins fall into three basic categories: those with multiple repeats, single repeats, or no repeats. Proteins with a single repeat located close to the cell wall are the most common. Anchor is the region of the molecule located within the cell wall carbohydrate and peptidoglycan.

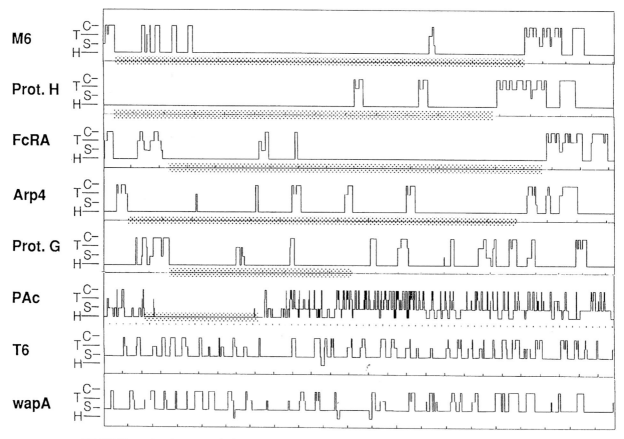

FIGURE 7 Conformational characteristics of the surface molecules from gram-positive bacteria. The sequences were analyzed by the Garnier-Robson algorithm supplied with the EuGene protein analysis package. The location within these molecules of regions exhibiting random coil (C), β-turn (T), β-sheet (S), and α-helix (H) are designated. Sequences were derived from the references listed in Table 1. Shaded areas are those containing a seven-residue periodicity based on the Matcher program. Proteins T6 and wapA exhibit no extended helical regions or seven-residue periodicity.

indicate that, except for M6 protein, any of these surface proteins are also in a coiled-coil conformation. To attempt to answer this question, an algorithm (Matcher) was developed (19) and used to determine if a seven-residue periodicity also exists in those molecules containing α-helical regions. When this algorithm was applied to representative sequences, extended regions of seven-residue periodicity were found in Arp4, FcRA, protein H, PAc, and spaP proteins (Fig. 7). This strongly suggested that these molecules may attain a coiled-coil conformation within these α-helical segments. It is apparent that only those proteins that contain sequence repeats and exhibit high α-helical potential show the presence of a seven-residue pattern of hydrophobic amino acids. On the other hand, even though the protein G molecule exhibits extensive helix potential, only a limited portion was found to contain the heptad pattern. Interestingly, however, the PAc protein from *Streptococcus mutans* has only one small segment between residues 120 and 520 that displays repeat blocks, and this segment exhibits strong helix potential. This is precisely the region that was also found to contain a regular seven-residue heptad pattern. Taken together, these data strongly suggest that the coiled-coil conformation may also be a common characteristic of surface proteins found on gram-positive organisms.

Size Variation

To fully appreciate the diversity among certain surface proteins on gram-positive organisms, it is important to understand a property of the M protein molecule, namely, size variation. Using a broadly cross-reactive monoclonal antibody as a probe, the size of the M protein, extracted by solubilizing the cell wall from a number of streptococcal strains, was examined by Western blot. M protein derived from 20 different serotypes exhibited variations in size ranging from 41 to 80 kDa in molecular mass, depending on the strain (17). Similar size variation was observed when streptococcal strains of the same serotype (M6) isolated over a period of 40 years were examined in a similar way. This variation in size may be explained by the observation that the M sequence contains extensive repeats at both the protein and DNA levels. Long, reiterated DNA sequences are likely to serve as substrates for recombinational events or for replicative slippage, generating deletions and duplications within the M protein gene that lead to the production of M proteins of different sizes.

Sequence analysis of the M6 gene isolates from local streptococcal outbreaks revealed that they are clonally related and not the result of separate acquisitions of unrelated M6 strains during the course of the infection (18).

The observation that size variants occur among these clinical isolates suggests that the size mutant constituted the major organism in the streptococcal population at the time of isolation and that a given size mutant has a selective advantage under clinical conditions. Perhaps serological pressure as a result of a local immune response forces the appearance of the size variant.

Although sequence repeats have been reported in several surface proteins on gram-positive bacteria, size variation has not been reported or tested for all these proteins. In addition to M protein, size variation has been observed in PspA from *Streptococcus pneumoniae* (90), HagA from *Porphyromonas gingivalis* (46), protein A from *S. aureus* (8), and Sfbp1 from *S. pyogenes* (70).

Multifunctional Proteins

Of the >65 proteins identified on the gram-positive surface, the great majority contain domains that bind molecules found in body secretions (Table 1), including immunoglobulins (IgA and IgG), albumin, fibronectin, and fibrinogen. Other proteins have been identified because of the presence of enzymatic domains. Based on current data, it is clear that the majority of these proteins are multifunctional, in which the identified function is limited to only one segment of the molecule. In most cases, the function of the other region(s) has not been identified. For example, as described above, it appears that many of the surface proteins contain a repeat region located close to the cell wall and a nonrepetitive segment N-terminal to it (Fig. 6). In most cases the binding domain (e.g., fibronectin, immunoglobulin) has been localized to the repeat segment while the N-terminal region exhibits no known function. In the case of streptococcal opacity factor, both domains have been defined (67); the repeats bind fibronectin while the N-terminal domain has the catalytic function that cleaves ApoA1 of high-density lipoprotein, resulting in the opacity of serum. In the case of highly repetitive proteins such as M protein and protein G, each repeat domain has a specific function. In M protein, the A-repeats bind albumin, the B-repeats bind fibrinogen (35, 72), and the C-repeats bind factor H of complement (34) and keratinocytes (57). In protein G, one repeat domain binds IgG while the second binds albumin. The multifunctional characteristic of surface proteins is not limited to molecules anchored via their C-terminal region. Proteins bound through hydrophobic/charge interactions and N-terminally anchored proteins also exhibit this characteristic (see below).

Thus, in general, surface proteins on most gram-positive organisms are multifunctional molecules with at least two and in some cases three or more independent functions. Given the fact that at least 25 or more of these surface proteins may ultimately be found on the cell surface, the potential complexity associated with the bacterial surface becomes enormous.

PROTEINS ANCHORED BY CHARGE AND/OR HYDROPHOBIC INTERACTIONS

While the great majority of proteins identified on the surface of gram-positive bacteria anchor through their C termini, a few have been identified to bind through less-defined charge and/or hydrophobic interactions (Table 1).

Surface Glycolytic Enzymes

Through an in-depth analysis of the individual proteins identified in a cell wall extract of group A streptococci, it was discovered that some proteins were composed of enzymes normally found in the glycolytic pathway located in the cytosol (62). In all, five enzymes have been identified (triosphosphate isomerase, glyceraldehyde-3-phosphate dehydrogenase [GAPDH], phosphoglycerate kinase, phosphoglycerate mutase, and α-enolase), which form a short contiguous segment of the glycolytic pathway. Interestingly, they are a complex of enzymes involved in the production of ATP (64a).

Support for the specific translocation of these enzymes to the cell surface and not a nonspecific release through cell lysis comes from the inability to identify cytoplasmic markers or other glycolytic enzymes in the growth medium or on the cell surface.

These enzymes have been identified on the surface of nearly all streptococcal groups, and certain enzymes have been identified on the surface of fungi such as *Candida albicans* and parasites such as trypanosomes and schistosomes (Table 2). Given the right substrate, streptococci have the capacity to produce ATP on the cell's surface, which further increases the complexity and the potential of such a bacterial surface.

Two of the best-characterized surface glycolytic enzymes on the streptococcus are GAPDH and α-enolase (62, 64). Like other surface proteins on gram-positive bacteria, the GAPDH is a multifunctional protein with binding affinities for fibronectin and lysozyme as well as for cytoskeletal proteins like actin and myosin. GAPDH has also been shown to have ADP-ribosylating activity in addition to its GAPDH activity. α-Enolase is also multifunctional in its normal enzymatic activity and the ability to specifically bind plasmin(ogen) (64). It is also the major plasmin-binding protein for streptococci. Evidence has shown that there is one gene for each of these enzymes, which are produced in the cytosol after which a proportion (roughly 30 to 40%) is translocated to the surface. Since these molecules are not synthesized with an N-terminal signal sequence, it is unclear how they are transported to the cell surface. Also, because they do not contain an apparent anchor motif, it is not known how they are attached to the cell surface. Since they can be removed with chaotropic agents, it is reasonable to assume that they are bound through charge and/or hydrophobic interactions.

TABLE 2 Glycolytic enzymes found on the surface of microorganisms

Enzyme[a]	Bacteria	Fungus	Parasite
GAPDH	Streptococci S. pneumoniae S. aureus Mycobacterium avium	Candida albicans	Schistosome Trypanosome
PGK	Streptococci	C. albicans	Trypanosome
TPI	Streptococci		Trypanosome
PGM	Streptococci		
α-Enolase	Streptococci	C. albicans	

[a]GAPDH, glyceraldehyde-3-phosphate dehydrogenase; PGK, phosphoglycerate kinase; TPI, triosphosphate isomerase; PGM, phosphoglycerate mutase.

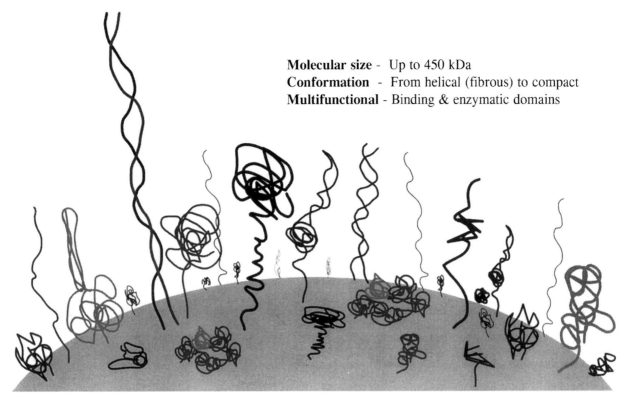

Molecular size - Up to 450 kDa
Conformation - From helical (fibrous) to compact
Multifunctional - Binding & enzymatic domains

FIGURE 8 Appearance of the surface of gram-positive bacteria exhibiting a wide array of protein molecules. Each molecule depicted may have thousands of identical copies densely packed on the surface.

Other Proteins

PspA, a surface protein and a virulence determinant for *S. pneumoniae* (93), has many of the structural characteristics found in M protein, such as sequence repeats and α-helical coiled-coil conformation, but it does not contain a C-terminal-anchor motif. It has been shown that after surface translocation, PspA anchoring requires charge interactions between the membrane-associated choline-containing lipoteichoic acid and the C-terminal repeat region of the PspA molecule (Fig. 1).

N-TERMINAL-ANCHORED LIPOPROTEINS

While lipoproteins (proteins containing a lipid covalently linked to an N-terminal cysteine) have been described in gram-negative bacteria, they have only recently been identified in gram-positive bacteria (for review, see reference 80). While most are predicted to be lipoproteins by genetic analysis based on the consensus sequence for lipoproteins, they usually comprise molecules that are not surface exposed. For example, these proteins are involved in transport systems such as the *mal* and *ami* operons of *S. pneumoniae* and the Msm system of *S. mutans*. A few of these lipoproteins have been identified by their ability to act as adhesins, allowing these organisms to adhere to a variety of substrates. SsaB, a lipoprotein from *Streptococcus sanguis* that shares sequence homology with other streptococcal adhesins, has been implicated in the interaction with a salivary receptor and the coaggregation with *Actinomyces naeslundii*. Many of these so-called adhesins have also been found to participate in various transport systems

for the streptococcus. Lipoproteins have also been implicated in the spore cell cycle of *Bacillus subtilis* (15), conjugation in *Enterococcus faecalis* (85), and enzymatic functions (48, 71).

Thus, it is expected that these proteins will be located predominantly on the surface of the cell membrane, with the majority of the molecule located within the peptidoglycan (Fig. 1). In some cases it is possible that certain domains will be surface exposed, allowing them to function as adhesins or even surface enzymes.

CONCLUSION

Based on current information, the surface of the gram-positive bacterial cell wall is highly complex (Fig. 8) and could even be considered to be a distinct organelle, since it is composed of proteins with specific binding functions and enzymatic activity combined with the ability to generate energy. Considering the fact that there could be more than 25 different proteins on the cell surface, each with the potential of up to three functional domains, more than 75 independent activities could potentially be present on the cell surface, much more than had ever been previously anticipated.

Since extensive cytoplasmic domains are not present within the surface proteins thus far identified in gram-positive bacteria, it is unlikely that the binding of these molecules to specific ligands in the bacterial cell surface induces a cytoplasmic signal to activate a gene product. It is more likely that binding initiates a conformational signal on the bacterial cell surface to perform a specific function.

For example, streptococci usually infect the pharynx, particularly the tonsils, through contact with contaminated saliva. Upon entering the oral cavity the organism first encounters the mucus which coats the mucosal epithelium. Soluble components found in the mucus, such as IgA, IgG, albumin, fibronectin, etc., are able to interact with their respective binding proteins on the streptococcal surface. This binding may set up a set of conformational events on the surface of the bacterial particle to drive the organism through the mucus to the epithelial surface. Perhaps the energy required for this is derived from those surface glycolytic enzymes necessary to generate ATP, or an enzyme is activated to solubilize the mucin. This must all occur quickly or the organism will be swept into the gut and eliminated. Thus, the molecules necessary to initiate infection are poised and ready on the bacterial surface. Therefore, before arriving on the mucosal epithelium the organism may be considered an inert particle with a wide array of surface proteins capable of driving the organism to the mucosal surface. When stable contact is made, the organism has time to send signals into the cell to initiate infection. Attempting to sort out when and how these proteins function during the infection process will be the challenge for future studies. With this understanding, however, new strategies could be developed in controlling diseases by this class of bacteria.

Work was supported in part by U.S. Public Health Service Grant AI11822 and a grant from SIGA Pharmaceuticals.

REFERENCES

1. **Berge, A., and U. Sjobring.** 1993. PAM, a novel plasminogen-binding protein from *Streptococcus pyogenes*. *J. Biol. Chem.* **268:**25417–25424.

2. **Berry, A. M., R. A. Lock, S. M. Thomas, D. P. Rajan, D. Hansman, and J. C. Paton.** 1994. Cloning and nucleotide sequence of the *Streptococcus pneumoniae* hyaluronidase gene and purification of the enzyme from recombinant *Escherichia coli*. *Infect. Immun.* **62:**1101–1108.

3. **Bessen, D. E., and V. A. Fischetti.** 1992. Nucleotide sequences of two adjacent M and M-like protein genes of group A streptococci: different RNA transcript levels and identification of a unique IgA-binding protein. *Infect. Immun.* **60:**124–135.

4. **Boden, M. K., and J.-I. Flock.** 1994. Cloning and characterization of a gene for a 19kDa fibrinogen-binding protein from *Staphylococcus aureus*. *Mol. Microbiol.* **12:**599–606.

5. **Burne, R. A., and J. E. C. Penders.** 1993. Characterization of the *Streptococcus mutans* GS-5 *fruA* gene encoding Exo-beta-D-fructosidase. *Infect. Immun.* **60:**4621–4632.

6. **Camara, M., G. J. Boulnois, P. W. Andrew, and T. J. Mitchell.** 1994. A neuraminidase from *Streptococcus pneumoniae* has the features of a surface protein. *Infect. Immun.* **62:**3688–3695.

7. **Chen, C. C., and P. P. Cleary.** 1990. Complete nucleotide sequence of the streptococcal C5a peptidase gene of *Streptococcus pyogenes*. *J. Biol. Chem.* **265:**3161–3167.

8. **Cheung, A. L., and V. A. Fischetti.** 1988. Variation in the expression of cell wall proteins of *Staphylococcus aureus* grown on solid and liquid media. *Infect. Immun.* **56:**1061–1065.

9. **Cheung, A. I., S. J. Projan, R. E. Edelstein, and V. A. Fischetti.** 1995. Cloning, expression, and nucleotide sequence of a *Staphylococcus aureus* (*fbpA*) encoding a fibrinogen-binding protein. *Infect. Immun.* **63:**1914–1920.

10. **Cole, R. M., and J. J. Hahn.** 1962. Cell wall replication in *Streptococcus pyogenes*. *Science* **135:**722–724.

11. **Collins, C. M., A. Kimura, and A. L. Bisno.** 1992. Group G streptococcal M protein exhibits structural features analogous to those of class I M protein of group A streptococci. *Infect. Immun.* **60:**3689–3696.

12. **Courtney, H. S., Y. Li, J. B. Dale, and D. L. Hasty.** 1994. Cloning, sequencing, and expression of a fibronectin/fibrinogen-binding protein from group A streptococci. *Infect. Immun.* **62:**3937–3946.

13. **de Chateau, M., and L. Bjorck.** 1994. Protein PAB, a mosaic albumin-binding bacterial protein representing the first contemporary example of module shuffling. *J. Biol. Chem.* **269:**12147–12151.

14. **Dramsi, S., P. Dehoux, M. Lebrun, P. L. Goossens, and P. Cossart.** 1998. Identification of four new members of the internalin multigene family of *Listeria monocytogenes* EGD. *Infect. Immun.* **65:**1615–1625.

15. **Errington, J., L. Appleby, R. A. Daniel, H. Goodfellow, S. R. Partridge, and M. D. Yudkin.** 1992. Structure and function of the *spoIIIJ* gene of *Bacillus subtilis*: a vegetatively expressed gene that is essential for epsilon^G activity at an intermediate stage of sporulation. *J. Gen. Microbiol.* **138:**2609–2618.

16. **Ferretti, J. J., R. R. B. Russell, and M. L. Dao.** 1989. Sequence analysis of the wall-associated protein precursor of *Streptococcus mutans* antigen A. *Mol. Microbiol.* **3:**469–478.

17. **Fischetti, V. A.** 1989. Streptococcal M protein: molecular design and biological behavior. *Clin. Microbiol. Rev.* **2:**285–314.

18. **Fischetti, V. A., M. Jarymowycz, K. F. Jones, and J. R. Scott.** 1986. Streptococcal M protein size mutants occur at high frequency within a single strain. *J. Exp. Med.* **164:**971–980.

19. **Fischetti, V. A., G. M. Landau, P. H. Sellers, and J. P. Schmidt.** 1993. Identifying periodic occurrences of a template with applications to protein structure. *Inform. Process. Lett.* **45:**11–18.

20. **Fischetti, V. A., V. Pancholi, and O. Schneewind.** 1990. Conservation of a hexapeptide sequence in the anchor region of surface proteins of Gram-positive cocci. *Mol. Microbiol.* **4:**1603–1605.

21. **Fischetti, V. A., D. A. D. Parry, B. L. Trus, S. K. Hollingshead, J. R. Scott, and B. N. Manjula.** 1988. Conformational characteristics of the complete sequence of group A streptococcal M6 protein. *Proteins* **3:**60–69.

22. **Frithz, E., L.-O. Heden, and G. Lindahl.** 1989. Extensive sequence homology between IgA receptor and M protein in *Streptococcus pyogenes*. *Mol. Microbiol.* **3:**1111–1119.

23. **Gaillard, J. L., P. Berche, C. Frehel, E. Gouin, and P. Cossart.** 1991. Entry of *L. monocytogenes* into cells is mediated by a repeat protein analogous to surface antigens from gram-positive, extracellular pathogens. *Cell* **65:**1127–1141.

24. **Galli, D., F. Lottspeich, and R. Wirth.** 1990. Sequence analysis of *Enterococcus faecalis* aggregation substance encoded by the sex pheromone plasmid pAD1. *Mol. Microbiol.* **4:**895–904.

25. **Garnier, J., D. J. Osguthorpe, and B. Robson.** 1978. Analysis of the accuracy and implications of simple methods for predicting the secondary structure of globular proteins. *J. Mol. Biol.* **120:**97–120.

26. **Gomi, H., T. Hozumi, S. Hattori, C. Tagawa, F. Kishimoto, and L. Bjorck.** 1990. The gene sequence and some properties of protein H. *J. Immunol.* **144:**4046–4052.

27. **Guss, B., M. Uhlen, B. Nilsson, M. Lindberg, J. Sjoquist, and J. Sjodahl.** 1984. Region X, the cell-wall-attachment part of staphylococcal protein A. *Eur. J. Biochem.* **138:**413–420.

28. **Haanes, E. J., and P. P. Cleary.** 1989. Identification of a divergent M protein gene and an M protein related gene family in serotype 49 *Streptococcus pyogenes. J. Bacteriol.* **171:**6397–6408.

29. **Heath, D. G., and P. P. Cleary.** 1989. Fc-receptor and M protein genes of group A streptococci are products of gene duplication. *Proc. Natl. Acad. Sci. USA* **86:**4741–4745.

30. **Heden, L.-O., E. Frithz, and G. Lindahl.** 1991. Molecular characterization of an IgA receptor from Group B streptococci: sequence of the gene, identification of a proline-rich region with unique structure and isolation of N-terminal fragments with IgA-binding capacity. *Eur. J. Immunol.* **21:**1481–1490.

31. **Hell, W., H.-G.W. Meyer, and S. G. Gatermann.** 1998. Cloning of aas, a gene encoding a staphylococcus saprophyticus surface protein with adhesive and autolytic properties. *Mol. Microbiol.* **29:**871–881.

32. **Holck, A., and H. Naes.** 1992. Cloning, sequencing and expression of the gene encoding the cell-envelope-associated proteinase from *Lactobacillus paracasei* subsp. *paracasei* NCDO 151. *J. Gen. Microbiol.* **138:**1353–1364.

33. **Hollingshead, S. K., V. A. Fischetti, and J. R. Scott.** 1986. Complete nucleotide sequence of type 6 M protein of the group A streptococcus: repetitive structure and membrane anchor. *J. Biol. Chem.* **261:**1677–1686.

34. **Horstmann, R. D., H. J. Sievertsen, J. Knobloch, and V. A. Fischetti.** 1988. Antiphagocytic activity of streptococcal M protein: selective binding of complement control protein Factor H. *Proc. Natl. Acad. Sci. USA* **85:**1657–1661.

35. **Horstmann, R. D., H. J. Sievertsen, M. Leippe, and V. A. Fischetti.** 1992. Role of fibrinogen in complement inhibition by streptococcal M protein. *Infect. Immun.* **60:**5036–5041.

36. **Inoue, S., M. Sugai, Y. Murooka, S.-Y. Paik, Y.-M. Hong, H. Ohgai, and H. Suginaka.** 1991. Molecular cloning and sequencing of the epidermal cell differentiation inhibitor gene from *Staphylococcus aureus. Biochem. Biophys. Res. Commun.* **174:**459–464.

37. **Jaffe, J., S. Natanson-Yaron, M. G. Caparon, and E. Hanski.** 1996. Protein F2, a novel fibronectin-binding protein from *Streptococcus pyogenes*, possesses two binding domains. *Mol. Microbiol.* **21:**373–384.

38. **Johnsson, H., H. Lindmark, and B. Guss.** 1995. A protein G-related cell surface protein in *Streptococcus zooepidemicus. Infect. Immun.* **63:**2968–2975.

39. **Jonsson, K., C. Signas, H.-P. Muller, and M. Lindberg.** 1991. Two different genes encode fibronectin binding proteins in *Staphylococcus aureus. Eur. J. Biochem.* **202:**1041–1048.

40. **Kao, S.-M., S. B. Olmsted, A. S. Viksnins, J. C. Gallo, and G. M. Dunny.** 1991. Molecular and genetic analysis of a region of plasmid pCF10 containing positive control genes and structural genes encoding surface proteins involved in pheromone-inducible conjugation in *Enterococcus faecalis. J. Bacteriol.* **173:**7650–7664.

41. **Kastern, W., U. Sjobring, and L. Bjorck.** 1992. Structure of peptostreptococcal protein L and identification of a repeated immunoglobulin light chain binding domain. *J. Biol. Chem.* **267:**12820–12825.

42. **Katsukawa, C.** 1994. Cloning and nucleotide sequence of type 3 M protein gene (emm3) consisting of an N-terminal variable portion and C-terminal conserved C repeat regions: relation to other genes of *Streptococcus pyogenes. J. Jpn. Assoc. Infect. Dis.* **68:**698–705.

43. **Kelly, C., P. Evans, L. Bergmeier, S. F. Lee, F. A. Progulske, A. C. Harris, A. Aitken, A. S. Bleiweis, and T. Lerner.** 1990. Sequence analysis of a cloned streptococcal surface antigen I/II. *FEBS Lett.* **258:**127–132.

44. **Kline, J. B., Xu, S., Bisno, A. L., and Collins, C. M.** 1996. Identification of a fibronectin-binding protein (GfbA) in pathogenic group G streptococci. *Infect. Immun.* **64:**2122–2129.

45. **Kok, J., K. J. Leenhouts, A. J. Haandrikman, A. M. Ledeboer, and G. Venema.** 1988. Nucleotide sequence of the cell wall proteinase gene of *Streptococcus cremoris* Wg2. *Appl. Environ. Microbiol.* **54:**231–238.

46. **Kozarov, E., J. Whitlock, H. Dong, E. Carrasco, and A. Progulske-Fox.** 1998. The number of direct repeats in hagA is variable among *Porphyromonas gingivalis* strains. *Infect. Immun.* **66:**4721–4725.

47. **Kreikemeyer, B., S. R. Talay, and G. S. Chhatwal.** 1995. Characterization of a novel fibronectin-binding surface protein in group A streptococci. *Mol. Microbiol.* **17:**137–145.

48. **Lansing, M., S. Lellig, A. Mausolf, I. Martini, F. Crescenzi, M. O'Regan, and P. Prehm.** 1993. Hyaluronate synthases: cloning and sequencing of the gene from *Streptococcus* sp. *Biochem. J.* **289:**179–184.

49. **Lindgren, P.-E., M. J. McGavin, C. Signas, B. Guss, S. Gurusiddappa, M. Hook, and M. Lindberg.** 1993. Two different genes coding for fibronectin binding proteins from *Streptococcus dysgalactiae*. The complete nucleotide sequences and characterization of the binding domain. *Eur. J. Biochem.* **214:**819–827.

50. **Manjula, B. N., K. M. Khandke, T. Fairwell, W. A. Relf, and K. S. Sripakash.** 1991. Heptad motifs within the distal subdomain of the coiled-coil rod region of M protein from rheumatic fever and nephritis associated serotypes of group A streptococci are distinct from each other: nucleotide sequence of the M57 gene and relation of the deduced amino acid sequence of other M proteins. *J. Protein Chem.* **10:**369–383.

51. **McDevitt, D., P. Francois, P. Vaudaux, and T. J. Foster.** 1994. Molecular charcterization of the clumping factor (fibrinogen receptor) of *Staphylococcus aureus. Mol. Microbiol.* **11:**237–248.

52. **Michel, J. L., L. C. Madoff, K. Olson, D. E. Kling, D. L. Kasper, and F. M. Ausubel.** 1992. Large, identical, tandem repeating units in the C protein alpha antigen gene, bca, of group B streptococci. *Proc. Natl. Acad. Sci. USA* **89:**10060–10064.

53. **Miller, L., L. Gray, E. H. Beachey, and M. A. Kehoe.** 1988. Antigenic variation among group A streptococcal M proteins: Nucleotide sequence of the serotype 5 M pro-

tein gene and its relationship with genes encoding types 6 and 24 M proteins. *J. Biol. Chem.* **263:**5668–5673.

54. **Mouw, A. R., E. H. Beachey, and V. Burdett.** 1988. Molecular evolution of streptococcal M protein: cloning and nucleotide sequence of type 24 M protein gene and relation to other genes of *Streptococcus pyogenes. J. Bacteriol.* **170:**676–684.

54a. **Navarre, W. W., and O. Schneewind.** 1999. Surface proteins of gram-positive bacteria and mechanisms of their targeting to the cell wall envelope. *Microbiol. Mol. Biol. Rev.* **63:**174–229.

55. **Nilsson, A., L. Frykberg, J.-I. Flock, L. Pei, M. Lindberg, and B. Guss.** 1998. A fibrinogen-binding protein of *Staphylococcus epidermis. Infect. Immun.* **66:**2666–2673.

56. **Nilsson, M., L. Frykberg, J. I. Flock, L. Pei, M. Lindberg, and B. Guss.** 1998. A fibrinogen-binding protein of *Staphylococcus epidermidis. Infect. Immun.* **66:**2666–2673.

57. **Okada, N., M. K. Liszewski, J. P. Atkinson, and M. Caparon.** 1995. Membrane cofactor protein (CD46) is a keratinocyte receptor for the M protein of group A streptococcus. *Proc. Natl. Acad. Sci. USA* **92:**2489–2493.

58. **Okahashi, N., C. Sasakawa, S. Yoshikawa, S. Hamada, and T. Koga.** 1989. Molecular characterization of a surface protein antigen gene from serotype c *Streptococcus mutans* implicated in dental caries. *Mol. Microbiol.* **3:**673–678.

59. **Olsson, A., M. Eliasson, B. Guss, B. Nilsson, U. Hellman, M. Lindberg, and M. Uhlen.** 1987. Structure and evolution of the repetitive gene encoding streptococcal protein G. *Eur. J. Biochem.* **168:**319–324.

60. **O'Toole, P., L. Stenberg, M. Rissler, and G. Lindahl.** 1992. Two major classes in the M protein family in group A streptococci. *Proc. Natl. Acad. Sci. USA* **89:**8661–8665.

61. **Pancholi, V., and V. A. Fischetti.** 1988. Isolation and characterization of the cell associated region of group A streptococcal M6 protein. *J. Bacteriol.* **170:**2618–2624.

62. **Pancholi, V., and V. A. Fischetti.** 1992. A major surface protein on group A streptococci is a glyceraldehyde-3-phosphate dehydrogenase with multiple binding activity. *J. Exp. Med.* **176:**415–426.

63. **Pancholi, V., and V. A. Fischetti.** 1993. Glyceraldehyde-3-phosphate dehydrogenase on the surface of group A streptococci is also an ADP ribosylating enzyme. *Proc. Natl. Acad. Sci. USA* **90:**8154–8158.

64. **Pancholi, V., and V. A. Fischetti.** 1998. α-Enolase, a novel strong plasmin(ogen) binding protein of the surface of pathogenic streptococci. *J. Biol. Chem.* **273:**14503–14515.

64a. **Pancholi, V., and V. A. Fischetti.** Unpublished results.

65. **Patti, J. M., H. Jonsson, B. Guss, L. M. Switalski, K. Wiberg, M. Lindberg, and M. Hook.** 1992. Molecular characterizations and expression of a gene encoding a *Staphylococcus aureus* collagen adhesin. *J. Biol. Chem.* **267:**4766–4772.

66. **Phillips, G. N., P. F. Flicker, C. Cohen, B. N. Manjula, and V. A. Fischetti.** 1981. Streptococcal M protein: alpha-helical coiled-coil structure and arrangement on the cell surface. *Proc. Natl. Acad. Sci. USA* **78:**4689–4693.

67. **Rakonjac, J. V., J. C. Robbins, and V. A. Fischetti.** 1995. DNA sequence of the serum opacity factor of group A streptococci: identification of a fibronectin-binding repeat domain. *Infect. Immun.* **63:**622–631.

68. **Rathsam, C., P. M. Giffard, and N. A. Jacques.** 1993. The cell-bound fructosyltransferase of *Streptococcus salivarius*: the carboxyl terminus specifies attachment in a *Streptococcus gordonii* model system. *J. Bacteriol.* **175:**4520–4527.

69. **Robbins, J. C., J. G. Spanier, S. J. Jones, W. J. Simpson, and P. P. Cleary.** 1987. *Streptococcus pyogenes* type 12 M protein regulation by upstream sequences. *J. Bacteriol.* **169:**5633–5640.

70. **Rocha, C., and V. A. Fischetti.** 1999. Isolation and characterization of a fibronectin binding protein on the surface of group A streptococci. *Infect. Immun.* **67:**2720–2728.

71. **Rothe, B., P. Roggentin, R. Frank, H. Blocker, and R. Schauer.** 1989. Cloning, sequencing and expression of sialidase gene from *Clostridium sordellii. J. Gen. Microbiol.* **135:**3087–3096.

72. **Ryc, M., E. H. Beachy, and E. Whitnack.** 1989. Ultrastructural localization of the fibrinogen-binding domain of streptococcal M protein. *Infect. Immun.* **57:**2397–2404.

73. **Schneewind, O., A. Fowler, and K. F. Faull.** 1995. Structure of the cell wall anchor of surface proteins in *Staphylococcus aureus. Science* **268:**103–106.

74. **Schneewind, O., K. F. Jones, and V. A. Fischetti.** 1990. Sequence and structural characterization of the trypsin-resistant T6 surface protein of group A streptococci. *J. Bacteriol.* **172:**3310–3317.

75. **Schneewind, O., P. Model, and V. A. Fischetti.** 1992. Sorting of protein A to the staphylococcal cell wall. *Cell* **70:**267–281.

76. **Sela, S., A. Aviv, A. Tovi, I. Burstein, M. G. Caparon, and E. Hanski.** 1993. Protein F: an adhesin of *Streptococcus pyogenes* binds fibronectin via two distinct domains. *Mol. Microbiol.* **10:**1049–1055.

77. **Signas, C., G. Raucci, K. Jonsson, P. Lindgren, G. M. Anantharamaiah, M. Hook, and M. Lindberg.** 1989. Nucleotide sequence of the gene for fibronectin-binding protein from *Staphylococcus aureus*: use of this peptide sequence in the synthesis of biologically active peptides. *Proc. Natl. Acad. Sci. USA* **86:**699–703.

78. **Sjobring, U.** 1992. Isolation and molecular characterization of a novel albumin binding protein from group G streptococci. *Infect. Immun.* **60:**3601–3608.

79. **Smith, H. E., U. Vecht, A. L. J. Gielkens, and M. A. Smits.** 1992. Cloning and nucleotide sequence of the gene encoding the 136-kilodalton surface protein (muramidase-released protein) of *Streptococcus suis* type 2. *Infect. Immun.* **60:**2361–2367.

80. **Sutcliffe, I. C., and R. R. B. Russell.** 1995. Lipoproteins of gram-positive bacteria. *J. Bacteriol.* **177:**1123–1128.

81. **Suvorov, A. N., A. E. Flores, and P. Ferrieri.** 1997. Cloning of the glutamine synthetase gene from group B streptococci. *Infect. Immun.* **65:**191–196.

82. **Swanson, J., K. C. Hsu, and E. C. Gotschlich.** 1969. Electron microscopic studies on streptococci. I. M antigen. *J. Exp. Med.* **130:**1063–1091.

83. **Talay, S. R., M. P. Grammel, and G. S. Chhatwal.** 1996. Structure of a group C streptoccal protein that binds to fibrinogen, albumin and immunoglobulin G via overlapping modules. *Biochem. J.* **315:**577–582.

84. **Talay, S. R., P. Valentin-Weigand, P. G. Jerlstrom, K. N. Timmis, and G. S. Chhatwal.** 1992. Fibronectin-binding protein of *Streptococcus pyogenes*: sequence of

the binding domain involved in adherence of streptococci to epithelial cells. *Infect. Immun.* **60:**3837–3844.

85. **Tanimoto, K., F. Y. An, and D. B. Clewell.** 1993. Characterization of the *traC* determinant of the *Enterococcus faecalis* hemolysin-bacteriocin plasmid pAD1: binding of sex pheromone. *J. Bacteriol.* **175:**5260–5264.

86. **Thanassi, D. G., E. T. Saulino, M. J. Lombardo, R. Roth, J. Heuser, and S. J. Hultgren.** 1998. The PapC usher forms an oligomeric channel: implications for pilus biogenesis across the outer membrane. *Proc. Natl. Acad. Sci. USA* **95:**3146–3151.

87. **Timoney, J. F., S. C. Artiushin, and J. S. Boschwitz.** 1997. Comparison of the sequences and functions of *Streptococcus equi* M-like proteins SeM and SzPSe. *Infect. Immun.* **65:**3600–3605.

88. **Tokuda, M., N. Okahashi, I. Takahashi, M. Nakai, S. Nagaoka, M. Kawagoe, and T. Koga.** 1991. Complete nucleotide sequence of the gene for a surface protein antigen of *Streptococcus sobrinus. Infect. Immun.* **59:**3309–3312.

89. **Uhlen, M., B. Guss, B. Nilsson, S. Gatenbeck, L. Philipson, and M. Lindberg.** 1984. Complete sequence of the staphylococcal gene encoding protein A. *J. Biol. Chem.* **259:**1695–1702. (Erratum, **259:**13628.)

90. **Waltman, W. D., L. S. McDaniel, B. M. Gray, and D. E. Briles.** 1990. Variation in the molecular weight of PspA (pneumococcal surface protein A) among *Streptococcus pneumoniae. Microb. Pathog.* **8:**61–69.

91. **Wanda, S.-Y., and R. Curtiss III.** 1994. Purification and characterization of *Streptococcus sobrinus* dextranase produced in recombinant *Escherichia coli* and sequence analysis of the dextranase gene. *J. Bacteriol.* **176:**3839–3850.

92. **Weidlich, G., R. Wirth, and D. Galli.** 1992. Sex pheromone plasmid pADI encoded surface exclusion protein of *Enterococcus faecalis. Mol. Gen. Genet.* **233:**161–168.

93. **Wu, H. Y., M. H. Nahm, Y. Guo, M. W. Russel, and D. E. Briles.** 1997. Intranasal immunization of mice with PspA (pneumococcal surface protein A) can prevent intranasal carriage, pulmonary infection, and sepsis with *Streptococcus pneumoniae. J. Infect. Dis.* **175:**839–846.

94. **Yeung, M. K., and J. O. Cisar.** 1988. Cloning and nucleotide sequence of a gene for *Actinomyces naeslundii* 45 type 2 fimbriae. *J. Bacteriol.* **170:**3803–3809.

95. **Yeung, M. K., and J. O. Cisar.** 1990. Sequence homology between the subunits of two immunologically and functionally distinct types of fimbriae of *Actinomyces* spp. *J. Bacteriol.* **172:**2462–2468.

96. **Yeung, M. K., J. A. Donkersloot, J. O. Cisar, and P. A. Ragsdale.** 1998. Identification of a gene involved in assembly of *Actinomyces naeslundii* T14V type 2 fimbriae. *Infect. Immun.* **66:**1482–1491.

97. **Yother, J., and D. E. Briles.** 1992. Structural properties and evolutionary relationships of PspA, a surface protein of *Streptococcus pneumoniae*, as revealed by sequence analysis. *J. Bacteriol.* **174:**601–609.

THE STREPTOCOCCUS

SECTION EDITORS: Vincent A. Fischetti and Joseph J. Ferretti

A S A SPECIES, THE STREPTOCOCCI ARE A DIVERSE GROUP, ranging from commensal organisms which occupy various niches of the body to pathogens having the capacity to infect a wide range of tissue sites. They are also diverse in their resistance to current antibiotic therapy, from the multiresistant enterococci to the barely resistant *S. pyogenes*. Thus, it is not surprising, when looking back over the past century of research on pathogenic organisms, that the streptococcus appears as one of the best studied of the gram-positive pathogens.

Whereas published results steadily increased during the first three-quarters of this period, the introduction of molecular techniques in the last quarter of the century has inspired a nearly logarithmic increase in new information. Despite this wealth of knowledge, numerous questions still remain unanswered. The first part of the century relied on the biology of the organism to answer questions of pathogenesis; however, in the latter part, molecular techniques overshadowed biological studies in a mad scramble to understand the genetic complexity of these bacteria. Now, with the availability of whole streptococcal genomes, we have returned to the biology to ask pertinent questions regarding pathogenicity. For ultimately, in the next millennium, it will be the biology coupled with elegant genetic manipulations that will result in our understanding of the complexities inherent in the development of streptococcal-related diseases.

The chapters to follow are the best of a mixture of genetics, biology, and immunology, designed to both ask and answer questions pertaining to the pathogenicity of the streptococci. The information gained from these studies and the ideas they stimulate for future experiments will allow us to develop new tools to prevent infection by not only these organisms but perhaps gram-positive bacteria in general.

A. GROUP A STREPTOCOCCI

Intracellular Invasion by *Streptococcus pyogenes*: Invasins, Host Receptors, and Relevance to Human Disease

DAVID CUE, PRISCILLA E. DOMBECK, AND P. PATRICK CLEARY

3

The major reservoir for *Streptococcus pyogenes* infections is the human oral-nasal mucosa. Other animals are almost never the source of this pathogen. Although the most common infection of consequence in temperate climates is pharyngitis, the past decade has witnessed a dramatic increase in systemic disease in many regions of the world. Historically, *S. pyogenes* has been associated with sepsis and fulminate systemic infection, but the mechanism by which these streptococci cross mucosal or epidermal barriers is not understood. The discovery that *S. pyogenes* (19) and *Streptococcus agalactiae* (28) can be internalized by mammalian epithelial cells at high frequencies offers potential explanations for changes in epidemiology and the ability of these species to breach such barriers. In this chapter, the invasins and pathways used by *S. pyogenes* to reach the intracellular state are reviewed, and the relationship between intracellular invasion and human disease is discussed.

S. pyogenes has evolved a variety of both surface-bound and extracellular factors that alter the inflammatory response and impair phagocytic clearance of the bacteria. The more than 100 serotypes use both similar and different strategies at the biochemical level to colonize their host and avoid protective defenses. These are reviewed elsewhere (11) and in other chapters of this volume. Intracellular invasion is dependent on at least two classes of surface proteins, the M proteins (1, 9, 12) and fibronectin (Fn)-binding proteins (18, 21, 27). The M proteins serve many functions in the pathogenesis of *S. pyogenes*, including resistance to phagocytosis, adherence, and intracellular invasion (11). The Fn-binding proteins are also adhesins and invasins, but direct involvement in virulence has not been demonstrated. The structure and function of these proteins in the context of intracellular invasion are described below.

THE IMPACT OF INTRACELLULAR INVASION ON *S. PYOGENES* INFECTION

Many geographic and temporal clusters of sepsis and toxic shock, caused by a few serotypes of *S. pyogenes*, have been reported (5, 31). An M1 subclone, designated M1inv+, was associated with many such clusters (2, 22). Surpris-

ingly, devastating soft tissue infections were often reported in patients who had not experienced previous trauma or wounds that might initiate systemic spread of the organism (31). Cleary and coworkers (3, 19) considered the possibility that intracellular invasion of mucosal surfaces provided a window for streptococci to reach deeper tissue, even the bloodstream. Indeed, they discovered that the M1inv+ subclone invaded A549 epithelial cells at significantly higher frequency than did other subclones of M1 streptococci. The source of M1 strains, i.e., uncomplicated disease or more invasive disease (blood isolated), did not correlate with efficiency of ingestion by epithelial cells. Only the M1inv+ genotype was associated with high-frequency invasion. The primary genotypic difference between the highly invasive M1inv+ strain and less invasive strains is 70 kb of DNA sequence provided by two distinct prophages (3). One prophage, T14, encodes the SpeA erythrogenic toxin. The presence of these prophages raised the possibility that one might encode a factor that promotes high-frequency internalization by epithelial cells. Attempts to transmit high-frequency invasion to a poorly ingested laboratory strain failed (3); therefore, it is not clear whether the phage have anything to do with intracellular invasion or whether the genetic background of the laboratory strain used in these experiments was inappropriate.

Schrager et al. (30) questioned the relationship between intracellular invasion and systemic disease. They reported that an M18 strain that is highly encapsulated with hyaluronic acid was inefficiently ingested by immortalized keratinocytes, whereas a nonencapsulated mutant of that strain invaded keratinocytes at 2 to 4 times the frequency of the parent culture. In a mouse model for systemic infection, where streptococci are injected under the skin, the encapsulated parent culture was more virulent than the unencapsulated mutant. Therefore, these investigators discounted a role for intracellular invasion in systemic infections. Cleary et al. (4) reached a different conclusion with regard to the impact of capsule on invasion of epithelial cells. They reported that spontaneous, unencapsulated, M protein-deficient variants of an M1inv+ strain were poorly

Gram-Positive Pathogens, ed. by V. A. Fischetti et al.
© 2000 American Society for Microbiology, Washington, D.C.

internalized by A549 epithelial cells. These variants are not expected to invade epithelial cells efficiently, however, since high-frequency invasion of the parent culture is dependent on the M1 protein (9). Although capsules of M1inv+ streptococci are small relative to M18 cultures, it is likely that elimination of hyaluronic acid biosynthesis by mutation would also enhance ingestion of M1inv+ streptococci by epithelial cells. Capsule may retard binding of fibronectin or other mammalian agonists that are required for efficient ingestion of streptococci. Capsule expression is highly variable from strain to strain and may have little impact on persistence of streptococci in tissue, because production of hyaluronidases by streptococcal cultures is likely to disrupt this barrier to host macromolecules. In conclusion, capsules do not enhance invasion of epithelial cells by S. pyogenes, nor do they appear to be an absolute impediment to invasion.

ARE TONSILS A RESERVOIR FOR S. PYOGENES?

Although the relationship between high-frequency intracellular invasion and systemic streptococcal disease is still uncertain, the intracellular state may significantly increase the capacity of this bacterium to disseminate in human populations. When strains from carriers and from patients with uncomplicated pharyngitis and sepsis were compared, it was observed that those from carriers were internalized by HEp2 cells at the highest frequency (20). A highly variable M1inv+ culture could be enriched for more invasive streptococci by in vitro serial passage through human epithelial cells (4). Cycling of streptococci between the interior and exterior of the mucosal epithelium may select for variants that are more efficiently internalized. From 30 to 40% of children continue to shed streptococci after treatment with penicillin (25). In vitro, intracellular S. pyogenes can resist at least 100 μg/ml of penicillin (7); therefore, penicillin treatment may further select for strains that can be efficiently internalized by epithelial cells. High-frequency intracellular invasion may increase the rate of antibiotic therapy failure and therefore increase the size of the human reservoir that can disseminate the organism to others in the population. As the reservoir enlarges, the probability of serious, systemic infection will also increase. Thus, strains or serotypes that are less able to acquire an antibiotic-free niche may be less often associated with severe disease. A thorough analysis of a cluster of toxic shock in southern Minnesota showed that nearly 40% of the school children in nearby communities were carriers of a serotype M3 clone that was responsible for systemic disease in adults (5). These investigators suggested that school children served as the reservoir for S. pyogenes that produced toxic shock in adults with underlying physical disabilities.

The best and most direct evidence that intracellular bacteria are an important source for dissemination of streptococci and the cause of recurrent tonsillitis is based on microscopic studies of surgically excised tonsils (25). This study found that 13 of 14 tonsils from children plagued by recurrent tonsillitis harbored streptococci within epithelial cells in the tonsillar crypts. S. pyogenes was also observed in macrophagelike cells at high frequency in these specimens. Tonsils from control subjects who had tonsils removed for other reasons did not contain streptococci. Intracellular streptococci were also observed in cultured tonsils infected with S. pyogenes (25). The capacity of

M1inv+ streptococci to efficiently invade primary keratinized tonsillar epithelial cells was confirmed in vitro (7). Ingestion of streptococci by these cells is dependent on M protein and agonists such as fibronectin or laminin.

S. PYOGENES INVASINS AND HOST RECEPTORS

Invasion of cultured mammalian cells by S. pyogenes is analogous in many respects to invasion by Yersinia pseudotuberculosis. This organism can invade eukaryotic cells by at least three different mechanisms, the most efficient of which is mediated by the bacterial InvA protein (10, 17). InvA is an outer membrane protein capable of high-affinity binding to multiple β1 integrins. Integrins are a mammalian family of heterodimeric, transmembrane glycoproteins that bind to extracellular matrix (ECM) proteins (e.g., fibronectin [Fn] and laminin [Ln]) and blood-clotting proteins (e.g., fibrinogen) (16). Engagement of integrins by InvA results in activation of host signal transduction pathways and rearrangements of host cell cytoskeletal components. Bacterial uptake occurs by a zipperlike mechanism wherein cytoskeletal rearrangements push or "zipper" the host cell cytoplasmic membrane around the bacterial cell (10).

InvA is an example of a subclass of adhesin molecules generally referred to as invasins. Typically, invasins are proteins expressed on the surfaces of bacterial cells that recognize, directly or indirectly, specific host cell receptors (10). Factors that can contribute to the ability of an adhesin molecule to function as an invasin include the affinity of the invasin for its receptor, receptor density, and the biological function of the host receptor (10, 17). In general, invasins are capable of inducing reorganization of the host cell cytoskeleton by interaction with specific host cell receptors. Integrins are often exploited for microbial entry into mammalian cells. This may be due to their ubiquitous expression and their ability to affect cytoskeletal arrangement (16, 17).

While studies of intracellular invasion by streptococci are less advanced than those of Y. pseudotuberculosis, recent results indicate that S. pyogenes has evolved multiple mechanisms for invasion of a wide variety of mammalian cells. Several streptococcal invasins have been identified, including two related Fn-binding proteins and four different M proteins. Additionally, multiple integrins have been implicated in bacterial internalization.

HIGH-AFFINITY Fn-BINDING PROTEINS CAN FUNCTION AS INVASINS

S. pyogenes expresses an array of cell surface proteins that facilitate bacterial colonization of a variety of human tissues (6). Recent studies have demonstrated that several of these adhesins can also function as invasins (Table 1). The proteins SfbI and F1 (PrtFI) are two closely related streptococcal adhesins that bind Fn with high affinity. Approximately 70% of S. pyogenes isolates carry the gene encoding SfbI/PrtFI, although many do not express the gene under laboratory conditions (20, 21, 23). Prebinding of soluble Fn by PrtFI-bearing streptococci can promote bacterial adherence to cultured cells and dermal tissue. This led Okada et al. (24) to propose that Fn promotes bacterial adherence by serving as an adapter or molecular bridge, connecting bacteria to host tissues. Experimental evidence from several

TABLE 1 *S. pyogenes* cell surface proteins implicated in mediation of intracellular invasion

Protein	Cell line[a]	Reference
SfbI	HEp2	21
F1	HEp2	18
M1	He1A, A549, HEp2, HTE[b]	7–9
M3	HEp2, HaCaT	1
M6	HEp2, HaCaT, L cells	1, 18
M18	L cells	1

[a]Cultured cell lines for which streptococcal invasion is influenced by the indicated bacterial protein.

[b]Primary human tonsillar epithelial cells.

laboratories is consistent with SfbI/PrtFI's promoting intracellular invasion via a similar mechanism.

Molinari et al. (21) demonstrated that SfbI can mediate invasion of HEp2 cells. Streptococcal invasion can be ablated by antiserum raised against SfbI or by preincubation of host cells with recombinant SfbI. Latex beads coated with recombinant SfbI readily adhere to and are efficiently ingested by epithelial cells, demonstrating that the interaction of SfbI with host cells is sufficient for internalization. Fn binding by SfbI is apparently required for invasion, because beads coated with a recombinant SfbI peptide lacking the Fn-binding domains do not efficiently adhere to HEp2 cells.

Jadoun et al. (18) reported that exogenous Fn can stimulate invasion of HEp2 cells by streptococci that express PrtFI. Ozeri et al. (27) reported that invasion of HeLa cells by PrtFI[+] bacteria is dependent on addition of serum or purified Fn. In both studies, antibodies against Fn inhibited invasion. Also, rPrtFI peptides, containing at least one Fn-binding domain, were found to inhibit PrtFI-mediated invasion. PrtFI binds to a 70-kDa N-terminal fragment of Fn. This region of Fn differs from the region bound by epithelial cells. The 70-kDa fragment can inhibit Fn-mediated invasion by competing with intact Fn for PrtFI binding. Only intact Fn molecules are capable of supporting bacterial invasion (27).

Antibody directed against the integrin $\beta 1$ subunit specifically blocks Fn-PrtF1-mediated invasion of HeLa cells (27). This result suggests that one or more $\beta 1$-containing integrins are involved in bacterial internalization. In contrast, invasion of GD25 (embryonic mouse stem) cells appears to be mediated by integrin $\alpha v \beta 3$ (the major Fn-binding integrin of this cell line) since invasion is inhibited by a peptide that specifically blocks Fn binding by this integrin (27). Collectively, these results are consistent with the proposal that Fn functions as a molecular bridge between SfbI/PrtFI and host cell Fn receptors.

M PROTEINS AS INVASINS

While the studies cited above demonstrated that SfbI/PrtFI can function as invasins, reports from other laboratories indicate that these proteins are but two of several invasins expressed by different isolates of *S. pyogenes* (Table 1). For example, serotype M1 strains typically lack the gene encoding SfbI/PrtFI but can invade human epithelial cells with high efficiency (3, 23).

Invasion of cultured cells by the M1inv+ strain *S. pyogenes* 90-226 is dependent on bacterial expression of M1

protein. Inactivation of the gene encoding this protein, *emm1*, decreases invasion of HeLa, A549, and HEp2 cells by 50-fold (9). M1 expression is also required for invasion of primary cultures of human tonsillar epithelial cells but not for invasion of a mouse macrophage cell line, J774A.1 (7). Also, latex beads coated with M1 are readily ingested by HeLa cells (9) (Fig. 1B). Collectively, these results indicate that invasion of epithelial cells by strain 90-226 is mediated in large part by M1 protein.

Expression of M1 is not sufficient for invasion, though, because bacterial internalization is dependent on exogenous serum, serum Fn, or Ln (8). M1 appears to be the major Fn-binding protein expressed by this strain, as inactivation of *emm1* reduces binding by 88%. Also, purified M1 protein can bind Fn in vitro (8). Fn binding is not a general property of M proteins. Only two types, M1 and M3, have been demonstrated to bind Fn (8, 29). Interestingly, M3 protein was recently found to promote serum-dependent invasion of HEp2 cells (1). Some *S. pyogenes* isolates express an M-like protein, protein H (PrtH), that also has Fn-binding activity (13). PrtH has not been demonstrated to promote intracellular invasion, and the gene encoding PrtH is not carried by the M1inv+ clone described above.

Several researchers have reported very low (0.1 to 1%) invasion by serotype M1 isolates. These results are, in part, attributable to the concentration or quality of the invasion agonists used in the studies. For example, Jadoun et al. (18) reported inefficient invasion of HEp2 cells by the M1[+] strain AP1. However, their experiments were performed in the absence of serum. We have recently found that AP1 can be efficiently ingested by A549 cells if Fn is present. Greco et al. (15) reported that commercially acquired Fn did not promote invasion by serotype M1 strains, but Fn acquired from the same commercial supplier also failed to promote invasion by strain 90-226 (8).

The invasion efficiencies of M1 strains are not dictated solely by the presence of invasion agonists, however. LaPenta et al. (19) and Cleary et al. (3) reported that M1 isolates exhibit widely varying invasion efficiencies, even when experiments are performed in the presence of serum. These variations are likely due, at least in part, to varying levels of M1 protein synthesis, as levels of *emm1* transcription can vary considerably between isolates and even between different cultures of the same isolate (4). While this possibility has yet to be thoroughly investigated, it is consistent with the finding of Ozeri et al. (27) that the number of integrin-binding sites on bacteria can greatly affect invasion efficiency.

In the presence of Fn, internalization of M1 streptococci appears to depend on the interaction of M1, Fn, and the epithelial cell Fn receptor integrin $\alpha 5 \beta 1$. A monoclonal antibody that specifically blocks Fn binding by this integrin can ablate invasion of A549, HeLa, and human tonsillar epithelial cells (7, 8). Low-molecular-weight nonpeptidyl $\alpha 5 \beta 1$ antagonists are also effective at inhibiting invasion of A549 and human tonsillar epithelial cells (7). The inhibitory effects of $\alpha 5 \beta 1$ antagonists are not due to generalized effects on either bacterial or host cells, because the inhibitors do not abrogate Ln-mediated invasion. Rather, the inhibitory effects of $\alpha 5 \beta 1$ antagonists are observed only when bacteria are exposed to either serum or purified Fn. Thus, as in the case of PrtFI[+] strains, Fn appears to function as a bridging molecule in promoting invasion by M1[+] bacteria.

M3, M6, and M18 proteins have recently been implicated in mediating streptococcal invasion. Jadoun et al.

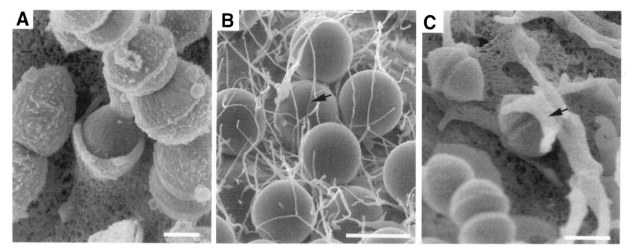

FIGURE 1 (A) Scanning electron micrograph of a chain of streptococci partially internalized by an A549 cell. Bar = 0.3 μm. (B) Uptake of latex beads coated with M1 protein by HeLa cells. Bar = 3 μm. (C) Streptococci in contact with HeLa cell microvilli. Bar = 0.5 μm. Arrows show M1 protein-coated beads (B) and a streptococcus interacting with a microvillus (C).

(18) studied adherence and invasion by an *S. pyogenes* strain that expresses PrtFI and M6 protein. PrtFI is the major adhesin/invasin for this strain, but disruption of the gene encoding M6, *emm6.1*, decreases both adherence and invasion. Disruption of *emm6.1* in a PrtFI mutant resulted in further decreases in both adherence and invasion. Fluckiger et al. (12) also demonstrated a role for M6 protein in invasion by showing that rabbit anti-M6 immunoglobulin G can block bacterial invasion of pharyngeal epithelial cells.

Expression of M18 was determined to promote invasion of mouse fibroblast L cells but not HaCaT or HEp2 cells. M3 expression permits serum-dependent invasion of both HEp2 and HaCaT cells but does not facilitate invasion of L cells. It thus appears that M proteins of different serotypes often recognize different receptors on the surfaces of mammalian cells (1). These studies indicate that, while several types of M proteins can facilitate invasion by *S. pyogenes*, there is no single mechanism underlying M protein-mediated invasion.

BACTERIAL-HOST CELL INTERACTIONS LEADING TO ADHERENCE AND INVASION

Several studies have demonstrated that adherence of streptococci to host cells is not necessarily sufficient for internalization to occur. For example, M18 promotes bacterial adherence to HEp2 and HaCaT cells, but adherent bacteria are inefficiently internalized (1). The M1$^-$ derivative of strain 90–226 readily adheres to cultured cells, suggesting that adherence by the wild-type strain is partially attributable to one or more unidentified adhesins. The latter cannot promote bacterial invasion of epithelial cells, because less than 1% of adherent M1$^-$ bacteria are internalized. The aforementioned adhesin may well facilitate invasion of other cell types, however. The fact that the unidentified adhesin cannot, on its own, facilitate invasion does not preclude the possibility that it contributes to M1-mediated invasion. Adherence via the adhesin may stabilize bacterial-host cell interactions, thereby promoting the interaction of M1, integrins, and integrin ligands. In this

context, efficient invasion would require the expression of two bacterial receptors, the unidentified adhesin and M1 (Fig. 2). A similar type of two-receptor mechanism seems to account for Fn-dependent invasion by *Neisseria gonorrhoeae* (33). On the other hand, M1-coated beads are internalized by HeLa cells, indicating that there is no absolute requirement for other bacterial adhesins (9). Testing of the two-receptor model awaits identification of the adhesin.

The abilities of M1$^+$ and M1$^-$ bacteria to adhere to host cells are comparable in the absence of invasion agonists. Invasion by the two strains is also comparable under this condition. The addition of an agonist (e.g., serum, Fn, or Ln) stimulates adherence of the M1$^+$ bacterium by approximately twofold. These agonists do not appreciably affect either adherence or invasion by the M1$^-$ strain. As a whole, these results suggest that M1 functions as a host factor-dependent adhesin that specifically targets bacterial interaction with integrins. Experimental results are consistent with Fn's promoting bacterial interaction with integrin $\alpha5\beta1$, whereas bacteria are targeted to a different $\beta1$ integrin when Ln is present (8).

ZIPPER PHAGOCYTOSIS OF STREPTOCOCCI

Cytoskeletal events accompanying uptake of a variety of pathogenic bacteria have been studied. Ruffling of the membrane precedes the entry of *Shigella* and *Salmonella* spp. In contrast, internalization by a zipper mechanism is specific and is mediated by interactions between surface invasins, ligands, and host cell receptors. These interactions lead to pseudopod formation, which requires the extension of actin filaments beneath the host cell membrane (10).

Scanning electron microscopy revealed that adherent, M1$^+$ bacteria are often in close contact with host cell microvilli, suggesting that this association may be an initial step in internalization (Fig. 1C). Microvilli frequently extend across bacterial surfaces and appear to entrap streptococci. Some microvilli seem to change into pseudopodialike structures, although these structures may have arisen independently (Fig. 1C). In contrast to M1$^+$ strep-

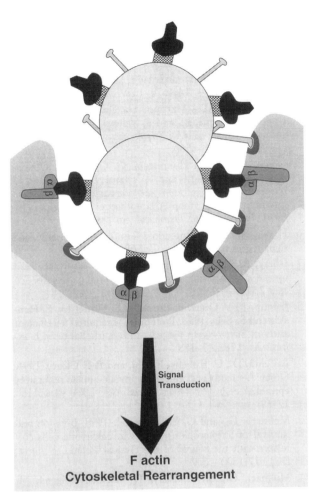

Signal Transduction

F actin
Cytoskeletal Rearrangement

FIGURE 2 Two-receptor model for high-frequency epithelial cell invasion. Distinct adhesins and invasins can mediate adherence of streptococci to epithelial cells. Serotype M1 streptococci make use of an unidentified adhesin (depicted by the rodlike structure) to adhere efficiently to A549 lung epithelial cells. Adherence alone is not sufficient to induce endocytosis, however. Invasion is triggered when an ECM protein (black mushroomlike structure), either Fn or Ln, makes contact between a streptococcal ECM-binding protein (stippled) and an integrin receptor ($\alpha\beta$) on target cells. This contact triggers signals that initiate actin polymerization and internalization of the bacteria. The ECM-binding protein can be either M1 or SfbI/PrtFI. While Ln-mediated invasion is dependent on expression of M1 protein, a direct interaction between M1 and Ln has not been demonstrated.

tococci, M1$^-$ bacteria are rarely observed to interact with microvilli, and no pseudopodialike structures were seen (9).

De novo actin polymerization is essential for uptake of streptococci, because invasion is blocked by cytochalasin D, an inhibitor of actin polymerization (15, 19). Confocal immunofluorescent microscopy has confirmed that vacuoles containing internalized streptococci are surrounded by polymerized actin. This actin is occasionally associated with membrane-spanning chains and seems to disappear soon after entry. These results suggest that streptococci, like *Listeria* and *Yersinia* spp., are internalized by a zipperlike mechanism (10). *Listeria* spp. are able to escape from vacuoles, and subsequent actin tail formation allows spread to

neighboring cells (10). Immunofluorescent microscopy did not reveal actin tails associated with streptococci at 5 or 7 h postinfection (9).

FATE OF INTRACELLULAR STREPTOCOCCI

Transmission electron microscopy has shown streptococci enclosed in vacuoles within host cells (14, 19). The ultimate fate of internalized bacteria remains controversial, however. Results from several studies have indicated that there is no observable growth of intracellular streptococci, and infected cell lines are eventually cleared of bacteria (14, 25, 30, 32). Consistent with this finding are transmission electron micrographs taken 8 h postinfection, showing degraded streptococci inside phagolysosomes (14). Österlund and Engstrand (26), however, reported that *S. pyogenes* can remain viable within HEp2 cells for up to 7 days, and we have found that M1$^+$ bacteria can be routinely recovered from infected cells after several days' exposure to antibiotics (7). Thus, streptococci seem able to persist within epithelial cells for extended periods. Also, microscopic evidence suggests that intracellular streptococci may have a limited ability to escape phagolysosomal killing and replicate. A three-dimensional image constructed from confocal microscopy sections showed a streptococcal chain extending through a HeLa cell, an image consistent with intracellular replication (9).

Salmonella spp. remain enclosed within vacuoles and are able to replicate intracellularly after a lag time. Certain *Salmonella*-containing vacuoles do not fuse to lysosomes and instead follow a secretory pathway (10). These vacuoles are associated with the lysosomal membrane glycoprotein LAMP2 but not with several other lysosomal markers. Vacuoles containing streptococci are associated with a similar lysosomal protein, LAMP1 (9). Two hours after infection, vacuoles associated with LAMP1 are more prevalent than those associated with polymerized actin. Vacuoles associated with both proteins were not observed, indicating that the association with LAMP1 probably occurs at some time after entry. Although other lysosomal markers were not examined, a fraction of streptococci-containing vacuoles may also bypass the endocytic route and traffic to the plasma membrane. Regrowth of bacteria after removal of antibiotics from infected monolayers indicates that some streptococci-containing vacuoles fail to mature and fuse with lysosomes (25, 26). The mechanism by which these streptococci are transported to the cell surface has not been studied.

SUMMARY

S. pyogenes has evolved multiple mechanisms for invasion of epithelial cells. Strains of this species differ substantially in invasion frequency and in dependence on invasion agonists. High-frequency internalization is mediated by ECM proteins, such as Fn and Ln, that form bridges between a surface invasin and the appropriate integrin receptor on the target mammalian cell. Because adherence and invasion can be independent events, mediated by distinct surface proteins and distinct integrins, a two-receptor process is proposed for high-efficiency internalization. The spectrum of target host cells, invasion efficiency, and requirement for serum agonists is determined by the ECM-binding proteins displayed on the bacterial surface. No single invasin or surface protein accounts for high-efficiency invasion of epi-

Capsular Polysaccharide of Group A *Streptococcus*

MICHAEL R. WESSELS

4

In her early studies of group A streptococci (GAS), Rebecca Lancefield noted an association between virulence and a distinctive appearance of the bacterial colonies on solid media. Isolates that were highly virulent for mice and that grew well in fresh human blood typically formed large colonies with a translucent, liquid appearance (mucoid) or an irregular, collapsed appearance (matte). By contrast, avirulent isolates that grew poorly in human blood formed compact, opaque colonies (glossy). Strains that grew as mucoid or matte colonies usually produced large amounts of M protein, which was given the designation "M" by Lancefield because of this association with colony morphology (31, 56). Later work by Armine Wilson demonstrated that the mucoid or matte appearance of such strains was in fact due to elaboration of capsular polysaccharide, not M protein (63). Wilson showed that the mucoid or matte colony type was converted to a nonmucoid or glossy colony by growth on medium containing hyaluronidase, which digested the hyaluronic acid capsule, whereas growth on medium containing trypsin, which digested M protein, had no such effect. Furthermore, while many strains that produced abundant capsular polysaccharide also were rich in M protein, expression of the two surface products was not always linked; certain mucoid or highly encapsulated strains produced little or no M protein, and certain strains rich in M protein produced little or no capsule (46, 56).

Both M protein and capsule have since been shown to contribute independently to GAS virulence. The association of mucoid strains of GAS with both invasive infection and acute rheumatic fever suggested a role for the capsular polysaccharide in virulence. Anecdotal observations that linked the capsule to pathogenesis of streptococcal disease were supported by a study that characterized the colony morphology of more than 1,000 GAS clinical isolates received by a streptococcal reference lab in the 1980s. In that study, only 3% of GAS isolates from patients with pharyngitis had a mucoid colony morphology, whereas 21% of isolates from patients with invasive disease syndromes such as bacteremia, necrotizing fasciitis, or streptococcal toxic shock syndrome and 42% of isolates associated with acute rheumatic fever were mucoid (27). The overrepresentation of mucoid strains among isolates associated with invasive infection and rheumatic fever suggested that the capsule contributed to enhanced virulence. The occurrence of outbreaks of acute rheumatic fever associated with mucoid strains of GAS at several locations in the United States in the 1980s supported the same inference (35, 57, 62). These clinical and epidemiologic observations implicating the capsule in pathogenesis have been corroborated by extensive experimental studies that have demonstrated a definite role of the hyaluronic acid capsule as a virulence factor in GAS infection.

BIOCHEMISTRY AND GENETICS OF CAPSULE PRODUCTION

Hyaluronic Acid Biosynthesis

The GAS capsular polysaccharide is composed of hyaluronic acid, a high-M_r ($>10^9$) linear polymer made up of $\beta(1 \rightarrow 4)$-linked disaccharide repeat units of D-glucuronic acid$(1 \rightarrow 3)$-β-D-N-acetylglucosamine. The polysaccharide is synthesized from the nucleotide sugar precursors UDP-glucuronic acid and UDP-N-acetylglucosamine by a membrane-associated enzyme or enzyme complex. High-M_r hyaluronic acid can be produced by incubation of cell-free membrane extracts of GAS with the two substrate UDP-sugars in the presence of divalent cations (36, 53, 54). However, enzymatically active hyaluronic acid synthase from GAS has not yet been purified.

A genetic locus required for hyaluronic acid production in GAS was located independently by three laboratories using transposon mutagenesis to produce acapsular mutants from a mucoid GAS strain. Transposon insertions that produced the acapsular phenotype were mapped to a locus that was highly conserved among GAS strains (10, 15, 60). Characterization of this region of the GAS chromosome revealed a cluster of three genes, *hasABC*, whose products are involved in hyaluronic acid biosynthesis. The first gene, *hasA*, encodes hyaluronic acid synthase, a protein with a predicted M_r of 47,900 that includes several predicted membrane-spanning domains, consistent with evidence

Gram-Positive Pathogens, ed. by V. A. Fischetti et al.
© 2000 American Society for Microbiology, Washington, D.C.

that the enzyme is membrane-associated (11, 17). The *hasA* gene product shares significant similarity with the NodC protein of *Rhizobium meliloti*, which is required for synthesis of a *N*-acetylglucosamine-containing polysaccharide, with chitin synthases, and with hyaluronan synthases from other microbial and higher animal species (12, 58). The second gene, *hasB*, encodes UDP-glucose dehydrogenase, a protein with a predicted M_r of 45,500 that catalyzes the oxidation of UDP-glucose to UDP-glucuronic acid (16). The third gene in the cluster, *hasC*, encodes a predicted 33.7-kDa protein identified as UDP-glucose pyrophosphorylase (7). The function of this enzyme is to catalyze the condensation of UTP with glucose-1-phosphate to form UDP-glucose. Thus, the reaction catalyzed by the *hasC* product yields a substrate for UDP-glucose dehydrogenase encoded by *hasB*, whose reaction product is, in turn, a substrate for hyaluronate synthase encoded by *hasA* (Fig. 1). While the enzyme protein encoded by *hasC* is enzymatically active, it is not required for hyaluronic acid synthesis by GAS. Selective inactivation of *hasC* resulted in no decrement in hyaluronic acid synthesis by a highly encapsulated strain of GAS; the implication is that another source of UDP-glucose is available within the cell (2). Furthermore, expression of recombinant *hasA* and *hasB* (without *hasC*) resulted in synthesis of hyaluronic acid in *Escherichia coli* and *Enterococcus faecalis* (10, 11). Together, these observations have led to the conclusion that *hasA* and *B* are the only two genes uniquely required for bacterial synthesis of hyaluronic acid. The three genes *hasABC* are transcribed as a single message of approximately 4.1 kb from a promoter immediately upstream of *hasA* (8). The identification of a potential *rho*-independent terminator at the 3′ terminus of *hasC* suggests that no additional genes are included in the operon.

Examination of GAS chromosomal DNA flanking the *has* gene cluster has revealed no additional genes involved in capsule synthesis or surface expression (Fig. 2). Down-stream of *hasC* is an open reading frame with sequence similarity to the *recF* gene in other bacteria (13). Inactivation of the *recF* homolog in GAS results in increased susceptibility to UV irradiation, supporting the identification of the GAS gene product as a DNA repair enzyme. Further downstream from *recF* is an open reading frame of unknown function that resembles a membrane protein and *guaB*, which encodes inosine monophosphate dehydrogenase, an enzyme in the purine biosynthesis pathway (4). Upstream of *hasA* are three open reading frames of unknown function; introduction of a large deletion encompassing most of the first two open reading frames resulted in no decrement in capsule production, evidence that neither of these genes is involved in capsule expression (2). Further upstream are a homolog of phosphatidylglycerophosphate synthase (*pgsA*) and a homolog of the ATP-binding component of ABC transporters (*stpA*) (2). While ABC transport systems have been associated with capsule gene clusters in several bacterial species, it is unlikely that the product of *stpA* plays such a role because (i) it is separated from the *has* genes by four unrelated open reading frames and (ii) the construction of a viable mutant in *stpA* has not been possible despite the successful inactivation of polysaccharide transporters in several other species. Therefore, the *stpA* product probably serves some other function that is essential for bacterial viability.

The simplicity of the capsule gene cluster in GAS contrasts with the size and complexity of capsule synthesis loci in other encapsulated gram-positive bacteria, which generally consist of at least 12 genes (30, 39, 41, 47, 48). One exception is the type 3 pneumococcus whose capsule production is strikingly similar to that in GAS with respect to the structure of the polysaccharide and the capsule synthesis genes; in type 3 pneumococcus, capsule synthesis involves three genes that are homologs of *hasABC* (14, 23). It has been suggested that, in contrast to the complex cellular machinery required for capsular polysaccharide polymerization and export in most bacteria, a single membrane protein functions as both the polymerase and exporter of the polysaccharide in both GAS and type 3 pneumococcus (2).

Regulation of Hyaluronic Acid Biosynthesis

The *has* operon is highly conserved among GAS strains. A study of nine GAS isolates that came from diverse settings and whose dates of isolation spanned 75 years found only six mutations within a 300-nucleotide central region of the *hasA* gene (12); all were single-nucleotide substitutions, and none resulted in a change in the amino acid sequence of the protein. Despite the high degree of conservation of the hyaluronate synthase gene, individual GAS strains vary widely in the amount of capsule they produce. Since the *has* genes are conserved, differences in capsule expression among strains are likely to reflect differences in the regulation of *has* gene transcription. Experimental support for this hypothesis comes from the observation that *has* gene transcripts are detected during exponential growth of well-encapsulated strains of GAS, but not in strains that do not produce detectable hyaluronic acid (8). Analysis of the *has* operon promoter using a reporter system in which the promoter was fused to a promoterless chloramphenicol acetyl transferase (CAT) gene showed that full activity of the *has* promoter required no more than 12 nucleotides of chromosomal DNA upstream of the promoter's −35 site (1). This observation is consistent with the finding that many strains of GAS (10 of 23) contain the insertion el-

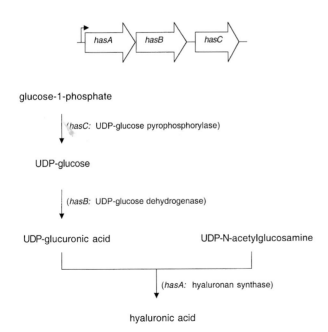

FIGURE 1 Schematic diagram of the *has* operon and the function of each gene product in the biosynthetic pathway for synthesis of hyaluronic acid in GAS.

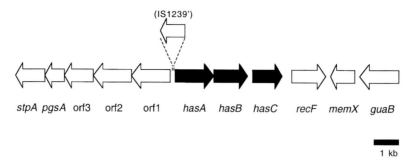

FIGURE 2 Map of the region of the GAS chromosome that includes the *has* operon encoding enzymes required for hyaluronic acid synthesis. The other genes shown appear not to be involved in capsular polysaccharide synthesis or surface expression. In some strains, insertion sequence IS*1239'* is present approximately 50 nucleotides upstream of the *has* operon promoter.

ement IS*1239'* approximately 50 nucleotides upstream of the −35 site without any apparent effect on capsule production (2).

One factor that influences the amount of capsule produced by a particular strain is the structure of the *has* operon promoter itself. Experiments using CAT reporter gene fusions revealed threefold higher activity of the *has* promoter from a highly encapsulated M type 18 strain than that from a poorly encapsulated M type 3 strain (1). Construction of hybrid promoters by site-directed mutagenesis demonstrated that nucleotide substitutions in two regions accounted for the differences in promoter activity between the two serotypes: three nucleotides in the −35, −10 spacer region and four nucleotides in the +2 to +8 positions relative to the start site of *hasA* transcription. The characteristic fine structure of the *has* operon promoter in type 18 strains may account for the observation that M18 strains are typically mucoid (i.e., highly encapsulated) since the precise sequence of the *has* promoter appears to be serotype specific.

Capsule production varies not only among strains but also in an individual strain under different circumstances. For example, during growth in liquid batch culture, GAS produce large capsules during exponential phase and then shed their capsules into the medium during stationary phase. Studies using a *has* promoter-CAT fusion integrated into the GAS chromosome demonstrated a rapid increase in promoter activity to a maximal level during the exponential phase of growth, with a sharp fall as the organisms entered stationary phase. The production of cell-associated capsular polysaccharide followed an identical but slightly delayed pattern, consistent with the expected lag for changes in the expression of enzyme proteins that are reflected in the rate of polysaccharide synthesis. These results indicate that the changing levels of encapsulation observed during different phases of growth are due to transcriptional regulation of *has* operon expression. Further studies on the effect of growth rate on capsule expression in a chemostat continuous culture system found that levels of CAT activity and capsule production were at least three times higher in cells maintained at a constant rapid growth rate (doubling time of 1.4 h) than in cells maintained at a slow growth rate (doubling time of 11 h) (1a). Up-regulation of capsule production is likely to occur during infection when the organism moves from the pharyngeal mucosa or external skin into the blood or deep tissues, where both temperature and nutrient availability are optimal for bacterial growth. Thus, the regulation of capsule expression in re-

lation to growth rate may be a useful adaptation to survival in the host, since it is also under these circumstances that the capsule is required to defend the organism against complement-mediated phagocytic killing.

The dynamic changes in GAS capsule expression with changes in growth rate and the observation that some strains of GAS appear more highly encapsulated after animal passage have suggested that capsule gene transcription is regulated by cellular mechanisms in addition to those attributable to the intrinsic structure of the *has* operon promoter. Mga, the *trans*-acting regulatory protein that influences expression of several GAS proteins, appears not to regulate capsule expression, at least in most strains of GAS. Inactivation of Mga was associated with reduced *has* gene transcription in an M1 strain (6); however, no effect on capsule production and/or *has* transcription was observed upon Mga inactivation in strains of M6, M18, or M49 (34, 43, 44). RofA, a regulatory protein involved in control of protein F expression, also is not known to affect capsule production (21).

Transposon mutagenesis of a poorly encapsulated M3 strain of GAS resulted in the identification of a novel regulatory locus whose inactivation dramatically increased capsule production (34). The locus consists of two genes, *csrR* and *csrS*, whose predicted products resemble the regulator and sensor proteins, respectively, of bacterial two-component regulatory systems (Fig. 3). Targeted inactivation of *csrR* by allelic exchange mutagenesis resulted

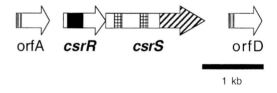

■ Predicted receiver domain
▦ Predicted membrane-spanning regions
▨ Predicted transmitter domain

FIGURE 3 Diagram of the *csrR*/*csrS* chromosomal locus encoding a putative two-component regulatory system that regulates hyaluronic acid synthesis. Sequences corresponding to regions of the predicted proteins with properties characteristic of the response regulator (CsrR) or sensor (CsrS) components are indicated.

in a sixfold increase in capsule production. The increase in hyaluronic acid synthesis was accompanied by a parallel increase in *has* gene transcription—evidence that CsrR regulates transcription of the capsule synthesis genes. Mutation of *csrR* appeared not to affect expression of M protein or hemolytic activity. Exactly how the *csrR/csrS* system controls capsule expression during infection has not been determined; however, the likely importance of this regulatory system in virulence is supported by the markedly greater resistance to in vitro phagocytosis displayed by the (highly encapsulated) *csrR* mutant than by the wild-type parent strain and the 500-fold increase in the mutant's virulence for mice after intraperitoneal challenge.

THE CAPSULE AND HOST IMMUNE DEFENSES

Immunogenicity of Hyaluronic Acid

In 1937, Kendall et al. (29) reported the purification of capsular polysaccharide from mucoid strains of GAS. Their analysis revealed that an apparently identical capsular material was produced by different M types of GAS. The GAS polysaccharide contained equal amounts of *N*-acetylglucosamine and glucuronic acid as exclusive component monosaccharides; this finding suggested that it might be structurally identical to a polysaccharide isolated from bovine vitreous humor and human umbilical cord. Immunization of rabbits and horses with mucoid GAS cells failed to elicit antibodies detectable by precipitin reactions with the polysaccharide. Later studies have confirmed that the "serologically inactive" polysaccharide produced by GAS is hyaluronic acid, a high-M_r polysaccharide with a repeating unit structure identical to that of hyaluronic acid isolated from animal sources. Presumably because it is recognized as a "self" antigen, hyaluronic acid is poorly immunogenic in several animal species. Fillit and colleagues (19, 20) evoked antibodies to hyaluronic acid by immunization of rabbits with encapsulated streptococci of group A or C and by immunization of mice with hyaluronidase-treated hyaluronic acid linked to liposomes. Such antibodies, however, have not been shown to have any opsonic activity against GAS in vitro or to confer protection against GAS infection in vivo.

Capsule and Resistance to Phagocytosis

GAS, like other gram-positive pathogens, is resistant to direct bacteriolysis by complement and antibodies. Therefore, clearance of the organisms from the blood or deep tissues depends largely on uptake and killing by phagocytes. This concept was supported by the observations of Todd, Lancefield, and others that virulence of GAS isolates in both clinical settings and experimental animal infections was closely paralleled by the capacity of the strains to resist killing by human blood leukocytes. GAS strains that produced large amounts of M protein multiplied in fresh human blood, while those that lacked M protein did not (32, 55). In addition, however, early studies demonstrated a relationship between surface expression of the hyaluronic acid capsule and resistance to phagocytosis. Mucoid and highly encapsulated strains tended to be resistant, and hyaluronidase treatment increased their susceptibility to phagocytosis in vitro, a result that suggested a protective effect of the capsule (22, 45, 52). More recent work with acapsular mutants has provided definitive evidence that the

capsule is a major determinant of resistance to phagocytic killing. In several assays measuring complement-dependent phagocytic killing by human blood leukocytes, acapsular mutants of type 18 and type 24 GAS were significantly more susceptible than the corresponding encapsulated wild-type strains (9, 40, 60).

One mechanism through which capsular polysaccharides may confer resistance to phagocytosis is the inhibition of complement activation. The type III capsular polysaccharide of group B *Streptococcus*, for example, prevents deposition of C3b on the bacterial surface through inhibition of alternative complement pathway activation by sialic acid residues present in the polysaccharide (18, 37). In GAS, however, the presence of the capsular polysaccharide appears to have no effect on the amount of C3b deposited on the bacterial surface. Similar amounts of C3 were detected on acapsular mutants and on encapsulated wild-type strains after incubation of the organisms in serum (9). The increased resistance of well-encapsulated GAS strains to complement-dependent phagocytic killing, therefore, must reflect the capacity of the capsule to block access of neutrophils to opsonic complement components bound to bacterial surface rather than inhibition of C3b deposition (Fig. 4).

Resistance of GAS to phagocytosis has been assessed most often with the bactericidal test of Lancefield, in which a small inoculum of the test strain is rotated for 3 h in heparinized fresh human blood (32). Strains that exhibit a significant increase in CFU during rotation in blood are considered resistant. An effect of the capsule is evident in this assay—acapsular mutants show less net growth than encapsulated wild-type strains—but the effect of M protein is more striking (40, 60). M protein-deficient strains not only fail to grow but typically are reduced in number after a 3-h incubation. By contrast, when resistance to phagocytosis is assessed with a different assay, the effect of capsule is more dramatic than that of M protein. In this second assay system, a larger bacterial inoculum is rotated for 1 h with human peripheral blood leukocytes in the presence of 10% normal human serum as a complement source (5, 61). In the latter assay, acapsular mutant strains and poorly encapsulated wild-type strains are killed (~90% reduction in CFU) despite the presence of M protein, while well-encapsulated strains do not change or increase in number. A mutant that is deficient in M protein but that produces wild-type amounts of capsule is killed less efficiently than a capsule-deficient mutant in the same background (40). In a study of isogenic mutants deficient in capsule, M pro-

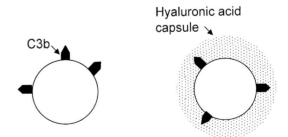

FIGURE 4 The hyaluronic acid capsule and resistance to complement-mediated phagocytosis. The capsule does not prevent deposition of C3b on the bacterial cell wall, but rather interferes with the interaction of bound C3b with phagocyte receptors.

tein, or both in the background of a mucoid type 18 GAS strain, mouse virulence correlated better with results in the second assay system than with the Lancefield whole-blood assay (40). It appears, therefore, that the role of capsule in resistance to phagocytosis as a correlate of virulence may be better assessed in the opsonophagocytic assay in 10% serum than in the whole-blood assay of Lancefield.

CAPSULE AS A VIRULENCE FACTOR

Experimental Infection Models

In the 1940s, investigators attempted to determine experimentally whether the hyaluronic acid capsule contributed to the virulence of GAS. Hirst (25) found that decapsulating GAS by treatment with leech extract containing hyaluronidase had no effect on virulence of the bacteria in mice. However, Kass and Seastone (28) found reduced virulence of decapsulated GAS in mice, perhaps because they administered hyaluronidase to the infected mice at frequent intervals to prevent regeneration of the capsule after challenge. During the 1990s, more direct proof that the capsule contributes to virulence has come from several studies using acapsular mutant strains of GAS derived by transposon mutagenesis or by targeted inactivation of the gene(s) required for hyaluronic acid synthesis (Table 1). In the first such study, an acapsular mutant was derived by Tn916 mutagenesis from a mucoid M type 18 strain of GAS (61). The mutant strain, TX4, did not produce detectable hyaluronic acid but made wild-type amounts of type 18 M protein. Loss of capsule expression in the mutant strain was associated with a >100-fold increase in 50% lethal dose LD_{50} in mice after intraperitoneal challenge. Further characterization of strain TX4 revealed a deletion of chromosomal DNA adjacent to the site of Tn916 insertion of approximately 9.5 kb, including the has operon required for hyaluronic acid synthesis (see above). To exclude the possibility that the reduced virulence of the mutant might be due in part to loss of an unrecognized phenotype encoded by the deleted DNA rather than entirely to the capsule synthesis genes, a second Tn916 mutant was characterized. The second mutant, TX72, harbored a single copy of Tn916 within hasA, the gene encoding hyaluronate synthase, and no associated chromosomal deletion (60). Virulence testing of TX72 in parallel with TX4 demonstrated that the two strains were similarly attenuated in the mouse model of lethal GAS infection. This result firmly established the role of the capsule in an experimental infection model. Both the simple insertion mutation in TX72 and the insertion-deletion mutation in TX4 were transduced independently to a highly encapsulated M type 24 strain (Vaughn), and both resulted in loss of capsule ex-

pression in the recipient strain. The type 24 acapsular mutants, 24–72 and 24-4, also exhibited reduced virulence in mice after intraperitoneal challenge, with respective LD_{50} values approximately 900 and 600 times higher than that for the type 24 wild-type strain (59). Virulence of both acapsular mutants (TX-4 and 24-4) was similarly reduced from that of their respective parent strains (M18 and M24) in a chicken embryo model of lethal GAS infection—results indicating that the role of the capsule in virulence was not limited to a single animal species (49).

Because the capsule is required for GAS resistance to complement-mediated phagocytic killing (see above), it is not surprising that capsule expression enhances virulence in systemic infection models in which the animals succumb to overwhelming bacteremia. In addition, however, the hyaluronic acid capsule contributes in a pivotal fashion to the capacity of the organisms to produce invasive soft tissue infection. A mouse model of invasive GAS soft tissue infection has been developed in which the bacterial inoculum is delivered by superficial injection just below the skin surface, in a small volume, and with minimal trauma to the tissues (3, 51). Mice challenged with the highly encapsulated M24 strain Vaughn developed dermal necrosis with underlying purulent inflammation and secondary bacteremia, while mice challenged with the acapsular mutant strain 24-4 developed no lesion at all or minor superficial inflammation and no bacteremia (51). Similar results were observed in the same model with a moderately encapsulated M3 strain of GAS (originally isolated from a patient with necrotizing fasciitis) that was compared with two isogenic acapsular mutants derived by allelic exchange mutagenesis of the hasA (hyaluronate synthase) gene. In the latter experiments, the majority of animals challenged with the encapsulated M3 parent strain developed spreading, necrotic soft tissue infection and died, whereas the acapsular mutants produced only mild local infection and no deaths (3). The dramatic impact of the capsule on virulence in this model is correlated with equally striking histopathologic findings. In animals challenged with an acapsular mutant, a small focus of neutrophilic inflammation and necrosis is confined by a well-formed abscess. By contrast, after challenge with the encapsulated wild-type strain, acute inflammatory cells extend throughout the subcutaneous tissue and are associated with thrombosis of blood vessels, with infarction and necrosis of the overlying dermis—a histopathologic picture very similar to that in patients with GAS necrotizing fasciitis. Examination of the histologic sections by immunofluorescence microscopy with a GAS-specific antibody showed that the acapsular GAS cells were confined to the abscess cavity, while the encapsulated wild-type organisms were widely dispersed throughout the subcutaneous tissues.

Capsule and GAS Entry into Epithelial Cells

Several laboratories have documented the entry of GAS into human epithelial cells, a process that has been proposed to represent a virulence mechanism—either a step in bacterial invasion into deep tissues or a means of escape of the organism into an intracellular sanctuary shielded from antibiotics and host immune defenses (24, 33, 42, 51). However, while there is irrefutable evidence that the capsule plays a key role in invasive infection in vivo, there is equally clear evidence that the capsule prevents GAS entry into epithelial cells. In studies of GAS entry into cultured human keratinocytes from pharyngeal epithelium or external skin, poorly encapsulated strains entered cells far more

TABLE 1 Animal models in which the hyaluronic acid capsule has been shown to enhance virulence of group A streptococci

Type of infection	Animal species	Reference
Systemic infection	Mouse	60, 61
	Chicken (embryo)	49
Pharyngeal colonization	Mouse	26, 59
Pneumonia	Mouse	26
Soft tissue infection	Mouse	3, 51

efficiently than well-encapsulated strains, regardless of M type or clinical source of the isolates (51). Acapsular mutants entered more than 1,000 times more efficiently than corresponding encapsulated parent strains. The apparently opposite effects of the capsule on invasive infection in vivo compared with those on epithelial cell "invasion" in vitro suggest that bacterial entry into epithelial cells is not a step in productive infection, but rather a host response that limits the spread of poorly encapsulated strains. By contrast, virulent encapsulated strains resist ingestion by eukaryotic cells, both epithelial cells and professional phagocytes, and invade deep tissues by an extracellular route. This hypothesis is supported by the finding that clinical isolates from patients with invasive GAS infection "invade" epithelial cells at much lower rates than throat or skin isolates (38). While it is difficult to completely discount the possibility that epithelial cells serve as a sanctuary site for intracellular GAS during some phase of colonization or latent infection, persistence of intracellular GAS for prolonged periods seems unlikely, since the viability of GAS within epithelial cells falls rapidly after bacterial entry (24, 51).

Capsule and Mucosal Colonization

Studies by at least two groups have demonstrated a critical role for the hyaluronic acid capsule in pharyngeal colonization in mice. BALB/c mice inoculated intranasally with the stable (transposon insertion plus chromosomal deletion) acapsular mutant strain 24-4 rapidly cleared the organism from the pharynx, and no mice died (59). In contrast, throat cultures of mice challenged with the revertible (simple transposon insertion) acapsular mutant 24-72 yielded encapsulated revertants, and the rate of persistent throat colonization and lethal systemic infection was similar to that in animals challenged with the encapsulated type 24 parent strain. A study of an acapsular mutant derived from the mouse-virulent M50 strain, B514, yielded similar results: intranasal challenge with the wild-type strain resulted in chronic throat colonization in C57BL/10SnJ mice. After intranasal challenge with an acapsular mutant derived by plasmid insertion mutagenesis of the *hasA* gene, encapsulated revertants accounted for all GAS recovered from throat cultures 4 days after challenge (26). Intratracheal inoculation of C3HeB/FeJ mice with the wild-type strain produced pneumonia and secondary systemic infection in approximately 50% of the animals within 72 h, while only 2 animals challenged with the acapsular mutant became ill, 5 and 6 days after challenge, respectively. In both cases only encapsulated revertants were recovered from cultures of the throat, lungs, and blood (26). The findings in both studies of impaired colonization by the acapsular mutant strains and the rapid emergence of encapsulated revertants together provide strong evidence that encapsulation confers a powerful survival advantage for GAS in the upper airway.

Capsule and GAS Attachment to Epithelial Cells

The influence of the hyaluronic acid capsule on the capacity of GAS to colonize the pharyngeal epithelium may be explained, at least in part, by the protective effect of the capsule in preventing opsonophagocytic killing, although the importance of resistance to opsonophagocytosis at a mucosal site is less well defined than in blood or deep tissues. In addition, the capsule is an important factor in the process of GAS attachment to epithelial cells, both by modulating interactions of other bacterial surface molecules with the epithelium and by direct interaction of GAS capsular hyaluronic acid with eukaryotic cells. The capsule may influence adherence by interfering with the interaction of any of a variety of GAS surface molecules that have been implicated as adhesins: M and M-like proteins, fibronectin-binding proteins, lipoteichoic acid, and glyceraldehyde-3-phosphate dehydrogenase. In general, acapsular mutant strains of GAS attach in higher numbers to skin or throat keratinocytes than do the corresponding encapsulated wild-type strains. For type 24 GAS, the capsule prevents M protein-mediated adherence to keratinocytes, since inactivation of the *emm24* gene reduced adherence in the background of an acapsular mutant, but not in the background of the same strain expressing wild-type amounts of capsule (50). It is likely that the capsule exerts a similar modulating influence on bacterial adherence mediated by other GAS surface molecules.

Not only does the capsule influence the binding interactions of other GAS surface structures, but there is evidence that the capsule itself can serve as a ligand for attachment of GAS to CD44 on epithelial cells. CD44 is a hyaluronic acid-binding glycoprotein found on a variety of human cells, including epithelial and hematopoietic cells. The interaction of CD44 with hyaluronic acid appears to have important functions in cell migration, organization of epithelia, lymphocyte homing, and tumor metastasis. Because CD44 is expressed on the surface of tonsillar epithelial cells, it may serve as a receptor for binding of hyaluronic acid on the surface of GAS. This hypothesis was supported by the finding that a monoclonal antibody to CD44 reduced attachment of wild-type strains of GAS to human pharyngeal keratinocytes by ~90%, but had no effect on attachment of acapsular mutant strains (50). Further evidence that CD44 acts as a receptor for GAS came from the observation that transfection of K562 cells with cDNA encoding either of two epithelial isoforms of CD44 conferred on the cells the capacity to bind GAS. Cells expressing CD44 bound a variety of GAS wild-type strains, including several in which the level of capsule expression was below the limit of detection in a dye-binding assay; however, inactivation of the hyaluronate synthase gene in such a poorly encapsulated GAS strain completely abrogated binding (50). These results indicate that capsule production, albeit at low levels, may be far more prevalent among clinical isolates of GAS than has been generally appreciated. Furthermore, strains that appear to be unencapsulated by conventional criteria may produce sufficient hyaluronic acid to mediate adherence to CD44 on eukaryotic cells. It remains to be determined whether ligation of CD44 on human cells (epithelial or other cell types) results in signal transduction and induction of cellular events important in GAS disease pathogenesis or in immune responses to infection.

CONCLUSIONS

Careful observations by microbiologists and clinicians in the 1920s and 1930s pointed to a link between the hyaluronic acid capsule and GAS disease pathogenesis. More recent studies have defined the genetic locus that directs hyaluronic acid biosynthesis and have characterized the molecular mechanisms through which the capsule enhances GAS virulence. It is now appreciated that the *has* capsule synthesis operon is both highly conserved and widely distributed among GAS strains, attesting to the adaptive role served by the hyaluronic acid capsule in the coevolution of GAS with the human host. As a poorly immunogenic "self" antigen, the capsular polysaccharide appears to have persisted in an invariant form in GAS, presumably because it has not been subject to the selective pressure of an effective, polysaccharide-specific, humoral immune response in the infected host that has led to emergence of multiple capsular types in other pathogenic bacteria. Studies of acapsular mutant strains have demonstrated that the capsule protects GAS from complement-mediated phagocytic killing and is essential for full virulence in a variety of experimental infection models. The GAS capsule also influences attachment of the bacteria to human epithelial cells, both by modulating interaction of M protein and other potential adhesins and by itself serving as a ligand for attachment of GAS to CD44 on epithelial cells. The latter mechanism may be of particular importance in GAS colonization of the pharynx, as it appears to participate in adherence of diverse GAS strains, including those that produce only small amounts of capsule. Finally, the successful adaptation of GAS to survival in the human host involves regulation of capsule expression that is dependent on the fine structure of the *has* operon promoter, on the *csrR*/*csrS* two-component regulatory system, and, likely, on additional mechanisms yet to be uncovered.

REFERENCES

1. **Albertí, S., C. D. Ashbaugh, and M. R. Wessels.** 1998. Structure of the *has* operon promoter and regulation of hyaluronic acid capsule expression in group A *Streptococcus*. *Mol. Microbiol.* **28:**343–353.

1a. **Albertí, S., et al.** Unpublished data.

2. **Ashbaugh, C. D., S. Albertí, and M. R. Wessels.** 1998. Molecular analysis of the capsule gene region of group A *Streptococcus*: the *hasAB* genes are sufficient for capsule expression. *J. Bacteriol.* **180:**4955–4959.

3. **Ashbaugh, C. D., H. B. Warren, V. J. Carey, and M. R. Wessels.** 1998. Molecular analysis of the role of the group A streptococcal cysteine protease, hyaluronic acid capsule, and M protein in a murine model of human invasive soft-tissue infection. *J. Clin. Invest.* **102:**550–560.

4. **Ashbaugh, C. D., and M. R. Wessels.** 1995. Identification, cloning, and expression of the group A streptococcal *guaB* gene encoding inosine monophosphate dehydrogenase. *Gene* **165:**57–60.

5. **Baltimore, R. S., D. L. Kasper, C. J. Baker, and D. K. Goroff.** 1977. Antigenic specificity of opsonophagocytic antibodies in rabbit anti-sera to group B streptococci. *J. Immunol.* **118:**673–678.

6. **Cleary, P. P., L. McLandsborough, L. Ikeda, D. Cue, J. Krawczak, and H. Lam.** 1998. High-frequency intracellular infection and erythrogenic toxin A expression undergo phase variation in M1 group A streptococci. *Mol. Microbiol.* **28:**157–167.

7. **Crater, D. L., B. A. Dougherty, and I. van de Rijn.** 1995. Molecular characterization of hasC from an operon required for hyaluronic acid synthesis in group A streptococci. *J. Biol. Chem.* **270:**28676–28680.

8. **Crater, D. L., and I. van de Rijn.** 1995. Hyaluronic acid synthesis operon (*has*) expression in group A streptococci. *J. Biol. Chem.* **270:**18452–18458.

9. **Dale, J. B., R. G. Washburn, M. B. Marques, and M. R. Wessels.** 1996. Hyaluronate capsule and surface M protein in resistance to opsonization of group A streptococci. *Infect. Immun.* **64:**1495–1501.

10. **DeAngelis, P. L., J. Papaconstantinou, and P. H. Weigel.** 1993. Isolation of a *Streptococcus pyogenes* gene locus that directs hyaluronan biosynthesis in acapsular mutants and in heterologous bacteria. *J. Biol. Chem.* **268:**14568–14571.

11. **DeAngelis, P. L., J. Papaconstantinou, and P. H. Weigel.** 1993. Molecular cloning, identification, and sequence of the hyaluronan synthase gene from group A *Streptococcus pyogenes*. *J. Biol. Chem.* **268:**19181–19184.

12. **DeAngelis, P. L., N. Yang, and P. H. Weigel.** 1994. The *Streptococcus pyogenes* hyaluronan synthase: sequence comparison and conservation among various group A strains. *Biochem. Biophys. Res. Commun.* **199:**1–10.

13. **DeAngelis, P. L., N. Yang, and P. H. Weigel.** 1995. Molecular cloning of the gene encoding RecF, a DNA repair enzyme, from *Streptococcus pyogenes*. *Gene* **156:**89–91.

14. **Dillard, J. P., M. W. Vandersea, and J. Yother.** 1995. Characterization of the cassette containing genes for type 3 capsular polysaccharide biosynthesis in *Streptococcus pneumoniae*. *J. Exp. Med.* **181:**973–983.

15. **Dougherty, B. A., and I. van de Rijn.** 1992. Molecular characterization of a locus required for hyaluronic acid capsule production in group A streptococci. *J. Exp. Med.* **175:**1291–1299.

16. **Dougherty, B. A., and I. van de Rijn.** 1993. Molecular characterization of *hasB* from an operon required for hyaluronic acid synthesis in group A streptococci. *J. Biol. Chem.* **10:**7118–7124.

17. **Dougherty, B. A., and I. van de Rijn.** 1994. Molecular characterization of *hasA* from an operon required for hyaluronic acid synthesis in group A streptococci. *J. Biol. Chem.* **269:**169–175.

18. **Edwards, M. S., D. L. Kasper, H. J. Jennings, C. J. Baker, and A. Nicholson-Weller.** 1982. Capsular sialic acid prevents activation of the alternative complement pathway by type III, group B streptococci. *J. Immunol.* **128:**1278–1283.

19. **Fillit, H. M., M. Blake, C. MacDonald, and M. McCarty.** 1988. Immunogenicity of liposome-bound hyaluronate in mice. At least two different antigenic sites on hyaluronate are identified by mouse monoclonal antibodies. *J. Exp. Med.* **168:**971–982.

20. **Fillit, H. M., M. McCarty, and M. Blake.** 1986. Induction of antibodies to hyaluronic acid by immunization of rabbits with encapsulated streptococci. *J. Exp. Med.* **164:**762–776.

21. **Fogg, G. C., C. M. Gibson, and M. G. Caparon.** 1994. The identification of *rofA*, a positive-acting regulatory component of *prtF* expression: use of an m gamma delta-based shuttle mutagenesis strategy in *Streptococcus pyogenes*. *Mol. Microbiol.* **11:**671–684.

22. **Foley, S. M. J., and W. B. Wood.** 1959. Studies on the pathogenicity of group A streptococci. II. The antiphagocytic effects of the M protein and the capsular gel. *J. Exp. Med.* **110:**617–628.

23. **Garcia, E., and R. Lopez.** 1997. Molecular biology of the capsular genes of *Streptococcus pneumoniae*. *FEMS Microbiol. Lett.* **149:**1–10.

24. Greco, R., L. De Martino, G. Donnarumma, M. P. Conte, L. Seganti, and P. Valenti. 1995. Invasion of cultured human cells by *Streptococcus pyogenes*. *Res. Microbiol.* **146**:551–560.

25. Hirst, G. K. 1941. The effect of a polysaccharide splitting enzyme on streptococcal infection. *J. Exp. Med.* **73**: 493–506.

26. Husmann, L. K., D.-L. Yung, S. K. Hollingshead, and J. R. Scott. 1997. Role of putative virulence factors of *Streptococcus pyogenes* in mouse models of long-term throat colonization and pneumonia. *Infect. Immun.* **65**:1422–1430.

27. Johnson, D. R., D. L. Stevens, and E. L. Kaplan. 1992. Epidemiologic analysis of group A streptococcal serotypes associated with severe systemic infections, rheumatic fever, or uncomplicated pharyngitis. *J. Infect. Dis.* **166**:374–382.

28. Kass, E. H., and C. V. Seastone. 1944. The role of the mucoid polysaccharide (hyaluronic acid) in the virulence of group A hemolytic streptococci. *J. Exp. Med.* **79**: 319–330.

29. Kendall, F., M. Heidelberger, and M. Dawson. 1937. A serologically inactive polysaccharide elaborated by mucoid strains of group A hemolytic streptococcus. *J. Biol. Chem.* **118**:61–69.

30. Kolkman, M. A. B., W. Wakarchuk, P. J. M. Nuijten, and B. A. M. van der Zeijst. 1997. Capsular polysaccharide synthesis in *Streptococcus pneumoniae* serotype 14: molecular analysis of the complete *cps* locus and identification of genes encoding glycosyltransferases required for the biosynthesis of the tetrasaccharide subunit. *Mol. Microbiol.* **26**:197–208.

31. Lancefield, R. C. 1940. Type specific antigens, M and T, of matt and glossy variants of group A hemolytic streptococci. *J. Exp. Med.* **71**:521–537.

32. Lancefield, R. C. 1957. Differentiation of group A streptococci with a common R antigen into three serological types with special reference to the bactericidal test. *J. Exp. Med.* **107**:525–544.

33. LaPenta, D., C. Rubens, E. Chi, and P. P. Cleary. 1994. Group A streptococci efficiently invade human respiratory epithelial cells. *Proc. Natl. Acad. Sci. USA* **91**:12115–12119.

34. Levin, J. C., and M. R. Wessels. 1998. Identification of *csrR/csrS*, a genetic locus that regulates hyaluronic acid capsule synthesis in group A *Streptococcus*. *Mol. Microbiol.* **30**:209–219.

35. Marcon, M. J., M. M. Hribar, D. M. Hosier, D. A. Powell, M. T. Brady, A. C. Hamoudi, and E. L. Kaplan. 1988. Occurrence of mucoid M-18 *Streptococcus pyogenes* in a central Ohio pediatric population. *J. Clin. Microbiol.* **26**:1539–1542.

36. Markovitz, A., J. A. Cifonelli, and A. Dorfman. 1959. The biosynthesis of hyaluronic acid by group A Streptococcus. VI. Biosynthesis from uridine nucleotides in cell-free extracts. *J. Biol. Chem.* **234**:2343–2350.

37. Marques, M. B., D. L. Kasper, M. K. Pangburn, and M. R. Wessels. 1992. Prevention of C3 deposition is a virulence mechanism of type III group B *Streptococcus* capsular polysaccharide. *Infect. Immun.* **60**:3986–3993.

38. Molinari, G., and G. S. Chhatwal. 1998. Invasion and survival of *Streptococcus pyogenes* in eukaryotic cells correlates with the source of the clinical isolates. *J. Infect. Dis.* **177**:1600–1607.

39. Morona, J. K., R. Morona, and J. C. Paton. 1997. Characterization of the locus encoding the *Streptococcus pneumoniae* type 19F capsular polysaccharide biosynthetic pathway. *Mol. Microbiol.* **23**:751–763.

40. Moses, A. E., M. R. Wessels, K. Zalcman, S. Alberti, S. Natanson-Yaron, T. Menes, and E. Hanski. 1997. Relative contributions of hyaluronic acid capsule and M protein to virulence in a mucoid strain of group A *Streptococcus*. *Infect. Immun.* **65**:64–71.

41. Munoz, R., M. Mollerach, R. Lopez, and E. Garcia. 1997. Molecular organization of the genes required for the synthesis of type 1 capsular polysaccharide of *Streptococcus pneumoniae*: formation of binary encapsulated pneumococci and identification of cryptic dTDP-rhamnose biosynthesis genes. *Mol. Microbiol.* **25**:79–92.

42. Österlund, A., R. Popa, T. Nikkila, A. Scheynius, and L. Engstrand. 1997. Intracellular reservoir of *Streptococcus pyogenes* in vivo: a possible explanation for recurrent pharyngotonsillitis. *Laryngoscope* **107**:640–647.

43. Perez-Casal, J., M. G. Caparon, and J. R. Scott. 1991. Mry, a *trans*-acting positive regulator of the M protein gene of *Streptococcus pyogenes* with similarity to the receptor proteins of two-component regulatory systems. *J. Bacteriol.* **173**:2617–2624.

44. Podbielski, A., M. Woischnik, B. Pohl, and K. H. Schmidt. 1996. What is the size of the group A streptococcal *vir* regulon? The Mga regulator affects expression of secreted and surface virulence factors. *Med. Microbiol. Immunol.* **185**:171–181.

45. Rothbard, S. 1948. Protective effect of hyaluronidase and type-specific anti-M serum on experimental group A *Streptococcus* infections in mice. *J. Exp. Med.* **88**:325–342.

46. Rothbard, S., and R. F. Watson. 1948. Variation occurring in group A streptococci during human infection. Progressive loss of M substance correlated with increasing susceptibility to bacteriostasis. *J. Exp. Med.* **87**:521–533.

47. Rubens, C. E., R. F. Haft, and M. R. Wessels. 1995. Characterization of the capsular polysaccharide genes of group B streptococci. *Dev. Biol. Stand.* **85**:237–244.

48. Sau, S., N. Bhasin, W. E. R, J. C. Lee, T. J. Foster, and C. Y. Lee. 1997. The *Staphylococcus aureus* allelic genetic loci for serotype 5 and 8 capsule expression contain the type-specific genes flanked by common genes. *Microbiology* **143**:2395–2405.

49. Schmidt, K.-H., E. Gunther, and H. S. Courtney. 1996. Expression of both M protein and hyaluronic acid capsule by group A streptococcal strains results in a high virulence for chicken embryos. *Med. Microbiol. Immunol.* **184**: 169–173.

50. Schrager, H. M., S. Albertí, C. Cywes, G. J. Dougherty, and M. R. Wessels. 1998. Hyaluronic acid capsule modulates M protein-mediated adherence and acts as a ligand for attachment of group A *Streptococcus* to CD44 on human keratinocytes. *J. Clin. Invest.* **101**:1708–1716.

51. Schrager, H. M., J. G. Rheinwald, and M. R. Wessels. 1996. Hyaluronic acid capsule and the role of streptococcal entry into keratinocytes in invasive skin infection. *J. Clin. Invest.* **98**:1954–1958.

52. Stollerman, G. H., M. Rytel, and O. Ortiz. 1963. Accessory plasma factors involved in the bactericidal test for type-specific antibody to group A streptococci. II. Human

plasma cofactor(s) enhancing opsonization of encapsulated organisms. *J. Exp. Med.* **117:**1–17.

53. **Stoolmiller, A. C., and A. Dorfman.** 1969. The biosynthesis of hyaluronic acid by *Streptococcus. J. Biol. Chem.* **244:**236–246.

54. **Sugahara, K., N. B. Schwartz, and A. Dorfman.** 1979. Biosynthesis of hyaluronic acid by *Streptococcus. J. Biol. Chem.* **254:**6252–6261.

55. **Todd, E. W.** 1927. The influence of sera obtained from cases of streptococcal septicaemia on the virulence of the homologous cocci. *Br. J. Exp. Pathol.* **8:**361–368.

56. **Todd, E. W., and R. C. Lancefield.** 1928. Variants of hemolytic streptococci; their relation to type specific substance, virulence, and toxin. *J. Exp. Med.* **48:**751.

57. **Veasy, L. G., S. E. Wiedmeier, G. S. Orsmond, H. D. Ruttenberg, M. M. Boucek, S. J. Roth, V. F. Tait, J. A. Thompson, J. A. Daly, E. L. Kaplan, and H. R. Hill.** 1987. Resurgence of acute rheumatic fever in the intermountain area of the United States. *N. Engl. J. Med.* **316:**1298–1315.

58. **Weigel, P. H., V. C. Hascall, and M. Tammi.** 1997. Hyaluronan synthases. *J. Biol. Chem.* **272:**13997–14000.

59. **Wessels, M. R., and M. S. Bronze.** 1994. Critical role of the group A streptococcal capsule in pharyngeal colonization and infection in mice. *Proc. Natl. Acad. Sci. USA* **91:**12238–12242.

60. **Wessels, M. R., J. B. Goldberg, A. E. Moses, and T. J. DiCesare.** 1994. Effects on virulence of mutations in a locus essential for hyaluronic acid capsule expression in group A streptococci. *Infect. Immun.* **62:**433–441.

61. **Wessels, M. R., A. E. Moses, J. B. Goldberg, and T. J. DiCesare.** 1991. Hyaluronic acid capsule is a virulence factor for mucoid group A streptococci. *Proc. Natl. Acad. Sci. USA* **88:**8317–8321.

62. **Westlake, R., T. Graham, and K. Edwards.** 1990. An outbreak of acute rheumatic fever in Tennessee. *Pediatr. Infect. Dis. J.* **9:**97–100.

63. **Wilson, A. T.** 1959. The relative importance of the capsule and the M antigen in determining colony form of group A streptococci. *J. Exp. Med.* **109:**257–270.

Toxins and Superantigens of Group A Streptococci

JOHN K. McCORMICK AND PATRICK M. SCHLIEVERT

5

Despite extensive research on group A streptococci, a clear understanding of streptococcal pathogenesis is not complete, and the molecular basis for the diversity and severity of streptococcal disease is poorly understood. Since early in this century, the incidence of serious streptococcal disease has greatly decreased, likely owing to the widespread use of antibiotics (35). However, since the mid-1980s there has been a resurgence in frequency and severity of streptococcal toxic shock syndrome (TSS) (10, 59) and other invasive diseases caused by group A streptococci (8, 41, 52) as well as rheumatic fever (23). Today, group A streptococci continue to generate significant morbidity and mortality through a variety of human diseases, including pharyngitis, tonsillitis, and skin infections and more serious diseases such as acute rheumatic and scarlet fevers, bacteremia, streptococcal TSS, and necrotizing fasciitis/myositis. A recent study in Ontario, Canada, determined the annual incidence of invasive infection due to group A streptococci to be 1.5 cases per 100,000 persons (11).

Group A streptococci produce a variety of virulence factors that include potent exotoxins, many of which belong to the large family of pyrogenic toxin superantigens (PTSAgs) that are primarily produced by group A streptococci and *Staphylococcus aureus* (4). Included within the PTSAgs group are the staphylococcal enterotoxins (SEs), which commonly cause food poisoning and cases of nonmenstrual staphylococcal TSS (51), and toxic shock syndrome toxin-1 (TSST-1), the cause of menstrual (and many cases of nonmenstrual) staphylococcal TSS (55). PTSAgs produced by group A streptococci, also known as scarlet fever toxins, erythrogenic toxins, or lymphocyte mitogens, were named streptococcal pyrogenic exotoxins (SPEs) owing to their distinctive fever-producing ability (61). The characteristic rash seen in scarlet fever and streptococcal TSS is likely due to an amplified hypersensitivity reaction resulting from superantigenicity. Rash is not seen in individuals or laboratory animals that have not had previous exposure to group A streptococci. Since the appearance of a rash was the defining feature of erythrogenic toxins, Watson (61) proposed the name streptococcal pyrogenic exotoxins because he initially thought the SPEs

were a separate family. Besides their ability to cause scarlet fever, SPEs are highly associated with streptococcal TSS (8, 40, 59). Owing to their structural similarities, it is not surprising that PTSAgs produced by both *S. aureus* and group A streptococci can cause TSS. However, it was not until the late 1980s that a description of well-defined cases led to the recognition of a severely toxic streptococcal disease called streptococcal TSS (10, 59). Also, it is noteworthy that two physicians injected themselves with 3 to 5 μg of SPEs and induced all of the required clinical features of streptococcal TSS, providing direct evidence that these toxins can cause this illness in humans. There are four well-characterized and serologically distinct SPEs, designated types A, B, C, and F, as well as streptococcal superantigen (SSA), streptococcal mitogenic exotoxin Z (SMEZ) and SMEZ-2, and SPEs G, H, and J that are all related.

This chapter focuses on the true exotoxins of group A streptococci, including the SPEs and the hemolysins streptolysin O and streptolysin S, with regard to their structure, function, and genetics, as well as their roles in human disease.

STRUCTURE AND FUNCTION OF THE SPEs

SPEs have low molecular weights (generally below 30,000), are heat stable, and are relatively resistant to proteolytic degradation (Table 1). Each toxin has been purified, and the corresponding gene has been cloned (15, 17, 19, 22, 47, 62). They are ribosomally synthesized with a classical signal peptide that is cleaved after secretion to release the mature toxin. With the exception of SPE B, these toxins do not undergo major modifications after release from the cell. The SPE B toxin, which has been shown to be a protease, is made as a precursor of 43,000 Da and then undergoes a series of proteolytic cleavages to result in the final 28,000-Da protein (17). SPEs are most easily purified from culture medium by ethanol precipitation (4 volumes), resolubilization in pH 4.0 acetate-buffered saline, dialysis against water, and, finally, preparative thin-layer isoelectric focusing (4). SPE A can be produced from streptococci,

Gram-Positive Pathogens, ed. by V. A. Fischetti et al.
© 2000 American Society for Microbiology, Washington, D.C.

TABLE 1 Biological properties of streptococcal pyrogenic exotoxin superantigens

Biological property	Streptococcal superantigens				
	SPE A	SPE B	SPE C	SPE F	SSA
Size (Da)	25,787[a]	40,314[b] (27,580)	24,354	25,363	26,892
pI	4.5–5.5	8–9	6.7	7	7.6
Pyrogenicity	+	+	+	ND[c]	+
Enhancement of lethal endotoxin shock	+	+	+	ND	+
Enhancement of cardiotoxicity	+	ND	+	ND	ND
Superantigenicity	+	+[d]	+	+	+
Lethality	+	ND	+	ND	ND
Immune cell lethality in the presence of endotoxin	+	ND	ND	ND	ND

[a]SPE A occurs as two molecular forms with pIs of 4.5 and 5.5.
[b]SPE B occurs as three major molecular forms with pIs of 8.0, 8.5, and 9.0.
[c]ND, not done.
[d]Activity is debated in the literature.

Escherichia coli, Bacillus subtilis, or *S. aureus* clones. The remaining toxins are purified from streptococci or *E. coli* clones.

The PTSAgs can be classified into subfamilies based on primary amino acid sequence comparisons. SPE A, SSA, and SEs B and C constitute one subfamily (Fig. 1). Not only do these proteins share significant amino acid homologies, but they may also share one or more antibody epitopes (4). SEs A, D, and E form another subfamily (4), but TSST-1, SPE B, and SPE F share little sequence relatedness to the other toxins while SPE C is minimally related to SPE A (Fig. 1) (15).

PTSAgs (including SPEs) share many biological properties (Table 1), including pyrogenicity, enhancement of endotoxin shock, superantigenicity, and lethality when administered in subcutaneously implanted miniosmotic pumps. SPEs were initially defined by their capacity to induce fever in rabbits (61); the time of maximum temperature (4 h) differed from that seen with endotoxin (3 h). All PTSAgs are pyrogenic, with SPEs being among the most potent pyrogens known. Injection of 10 μg/kg of SPE A or SPE C into a rabbit results in a peak fever of 2 to 3°C at 4 h after injection (27, 28). The minimum pyrogenic dose of these toxins in rabbits is 0.15 μg/kg. The ability of the toxins to induce fever appears to depend on both stimulation of the hypothalamic fever response control center, possibly through release of endogenous pyrogens from central nervous system astrocytes (4), and induction and systemic release of potent host pyrogens that act centrally on the hypothalamus.

SPEs also enhance host susceptibility to lethal endotoxin shock by up to one millionfold (28). The enhancement of lethal shock may result in part from a poisoning of liver Kupffer cells, thus blocking the normal clearance of other toxic substances such as endotoxin (54). Studies have shown that SPEs cause a transient inhibition of RNA synthesis in isolated liver cells (54), although it is unclear whether this is a primary or secondary action of the toxins. Alternatively, the amplification of susceptibility to endotoxin may result from synergistic release of tumor necrosis factor (TNF) from macrophages and T cells as a result of exposure to both SPEs and endotoxin. The dramatic enhancement of endotoxin toxicity may be the most severe symptom of TSS-like illness in which the rabbits die of a capillary leak syndrome induced by cytokines (52). Ani-

mals that do not succumb to the enhanced endotoxin shock that are challenged with SPEs and then with subtoxic amounts of streptolysin O or endotoxin develop highly significant myocardial necrosis after 4 to 5 days (56). This effect of amplification of myocardial necrosis is not shared with other PTSAgs from *S. aureus.*

The SPEs (and other PTSAgs) are also potent T-lymphocyte mitogens that have recently been referred to as superantigens owing to the mechanism of stimulation. Kappler and Marrack and their colleagues (63) first used the term superantigen to describe the massive T-cell-receptor (TCR) Vβ-restricted primary response to certain bacterial toxins. Superantigens use this common but very efficient mechanism of T-cell stimulation. Leung et al. (33) noted four hallmarks that distinguish the superantigens from conventional peptide antigens. First, superantigens elicit a strong primary response that is never seen with conventional peptide antigens. Second, the conventional peptide antigen binds in the groove of the major histocompatibility complex (MHC) molecule while the superantigen binds to the outer surface of both the MHC molecule and the Vβ chain of the TCR (63). Third, although all known superantigens require MHC class II proteins for T-cell presentation, the T-cell response is not class II MHC-restricted. MHC restriction is seen for all conventional peptide antigens. Fourth, superantigens associate with class II MHC molecules in an unprocessed form. Conventional peptide antigens are processed before being presented in the MHC peptide-binding groove.

The SPEs are potent stimulators of lymphocytes from a variety of species, including mice, rabbits, and humans (4), and are active at concentrations of toxin 100,000 times lower than that of the T-cell mitogen concanavalin A. However, unlike concanavalin A, SPEs do not stimulate all T cells to proliferate. Stimulation is specific for the Vβ TCR region on T cells. Studies suggest that the T-cell alteration by the toxins contributes significantly to the development of TSS-like illnesses through massive release of cytokines (63) and consequent capillary leak. These cytokines include interleukin-1 and TNF-α from peripheral macrophages (13, 45) and interferon-γ and TNF-β (lymphotoxin) from T cells. These properties have also been proposed to contribute to enhancement of delayed-hypersensitivity skin reactions to provoke positive Dick tests (scarlet fever rash) in the antibody-negative host (53)

```
SPE A      --QQDPDPSQLHRSS-LVKNLQNIYFLYEGDPVTHENVKSVDQLLSHDLI 47
SSA        SSQPDPTPEQLNKSSQFTGVMGNLRCLYDNHFVEGTNVRSTGQLLQHDLI 50
SEB        ESQPDPKPDELHKSSKFTGLMENMKVLYDDNHVSAINVKSIDQFLYFDLI 50
SEC1       ESQPDPTPDELHKASKFTGLMENMKVLYDDHYVSATKVKSVDKFLAHDLI 50
             |  | | |   |       |    ||    |    | |      |||
Consensus  esQpDPtPdeLhksSkftglmeNmkvLYddh-VsatnVkSvdqfL-hDLI
                                                |                |
SPE C              DSKKDISNVKSDLLYAYTITPYDYKDCR-VNFSTTHTLN 38


SPE A      YNVSG---PNYDKLKTELKNQEMATLFKDKNVDIYGVEYYHLCYLCENAE 94
SSA        FPIKDLKLKNYDSVKTEFNSKDLAAKYKNKDVDIFGSNYYYNCYYSE--- 97
SEB        YSIKDTKLGNYDNVRVEFKNKDLADKYKDKYVDVFGANYYQCYFSKKTN 100
SEC1       YNISDKKLKNYDKVKTELLNEGLAKKYKDEVVDVYGSNYYVNCYFSSK-D 99
                   |||     |       |      |   ||  |   ||   ||
Consensus  ynikd-klkNYDkvktEfknkdlA-kyKdk-VDifGsnYYynCYfsek--
                  |        |            |      ||   |     |
SPE C      IDTQKYRGKDY-YISSEMSYEASQKFKRDDHVDVFGL-FYI---LNSHTG 83


SPE A      ---------RSACIYGGVTNHEGNHLEIPK--KIVVKVSIDGIQSLSFDI 133
SSA        --GNSCKNAKKTCMYGGVTEHHRNQIE-GKFPNITVKVYEDNENILSFDI 144
SEB        DINSHQTDKRKTCMYGGVTEHNGNQLDK--YRSITVRVFEDGKNLLSFDV 148
SEC1       NVGKVTGGK--TCMYGGITKHEGNHFDNGNLQNVLIRVYENKRNTISFEV 147
                     |   || | |   |                  |        ||
Consensus  --g-----krktCmYGGvTeHegNhld-gk--nitvkVyedg-n-lSFdi
                      || |   |
SPE C      E-----------YIYGGITPAQNNKVNHKLLGNLFIS--GESQQNLNNKI 120


SPE A      ETNKKMVTAQELDYKVRKYLTDNKQLYTNGPSKYETGYIKFIPKNKESFW 183
SSA        TTNKKQVTVQELDCKTRKILVSRKNLYEFNNSPYETGYIKFIESSGDSFW 194
SEB        QTNKKKVTAQELDYLTRHYLVKNKKLYEFNNSPYETGYIKFIENE-NSFW 197
SEC1       QTDKKSVTAQELDIKARNFLINKKNLYEFNSSPYETGYIKFIENNGNTFW 197
             | || || ||||    |  |    | ||     | ||||||||    ||
Consensus  qTnKK-VTaQELDyktRkyLv-nKnLYefnnSpYETGYIKFIenngnsFW
             |  || ||||      |          |    | |  |
SPE C      ILEKDIVTFQEIDFKIRKYLMDNYKIYD-ATSPYVSGRIEIGTKDGKHEQ 169


SPE A      FDFFPEPE--FTQSKYLMIYKDNETLDSNT-SQIEVYLTTK 221
SSA        YDMMPAPGAIFDQSKYLMLYNDNKTVSSSA-IAIEVHLTKK 234
SEB        YDMMPAPGDKFDQSKYLMMYNDNKMVDSKD-VKIEVYLTTKKK 239
SEC1       YDMMPAPGDKFDQSKYLMMYNDNKTVDSKS-VKIEVHLTTKNG 239
             |   | |   | |||||| | ||     |     ||| || |
Consensus  yDmmPaPgdkFdQSKYLMmYnDNktvdSk--vkIEVhLTtK--
             |        |    | |                      |
SPE C      IDLFDSPNEG-TRSDIFAKYKDNRIINMKNFSHFDIYLEK 208
```

FIGURE 1 Amino acid alignment of various mature pyrogenic toxin superantigens. Identical residues are shown in bold and indicated (|). Gaps (−) were introduced to maximize homology. A consensus sequence for SPE A, SSA, SEB, and SEC is shown with conserved residues shown in lowercase and identical residues shown in uppercase and in bold. SPE C, which is minimally related to these PTSAgs, is shown below.

and suppression of immunoglobulin M synthesis such that most individuals who develop TSS remain susceptible even after recovering from the illness (4, 12).

SPEs as well as other PTSAgs are lethal when administered to rabbits in subcutaneously implanted miniosmotic pumps. These pumps are designed to release a constant amount of toxin over a 7-day period. The 100% lethal dose (LD$_{100}$) of SPE A, SEB, or SEC by this route is approximately 200 μg/1-2 kg of rabbit. In contrast, the LD$_{100}$ of SPE C, TSST-1, or SEA is 100 μg/1 to 2 kg of rabbit. The basis for this difference among toxins is unclear. Animals challenged with SPEs or other PTSAgs by the miniosmotic pump route succumb in three major groups. One group succumbs after 1 to 2 days' exposure. A second group dies between days 5 to 8, and a small group of animals succumbs on day 14 to 15. The explanation for this timing is unclear but may relate to timing of release of the various cytokines. SPEs and other PTSAgs are much more lethal when given in miniosmotic pumps than when they are administered in a bolus intravenous injection; we have given up to 1.5 mg of SPE A or TSST-1 per animal as a bolus without inducing lethality. Finally, administration of multiple PTSAgs types (e.g., SPEs A, B, and C in combination) is much more toxic to rabbits than is giving the same dose of any one toxin.

The presence of different SPE and SSA genes appears to be a variable trait in *Streptococcus pyogenes*, except for SPE B and SPE F, which both appear to be present in all strains (17, 18). Interestingly, SPE B and SPE F have been shown to have specific enzymatic activities (discussed below). Although the gene for SPE B is present in virtually all *S. pyogenes* strains tested, the protein is detectable in only 59% of strains, a phenomenon that is currently unexplained but is likely due to some type of regulatory process. Both SPE A and SPE C are encoded by bacteriophage (15, 22).

STREPTOCOCCAL PYROGENIC TOXIN TYPE A (SPE A)

In 1927, Frobisher and Brown (14) showed that a filterable agent from scarlet fever isolates could convert nonscarlatinal streptococci to toxigenic strains. Other investigators later confirmed these results (66). The gene encoding the exotoxin (*speA*) was later shown to be encoded on the 36-kb bacteriophage T12 (21, 22), and in 1986 the structural gene was sequenced (62). McShan et al. (36) has shown that the bacteriophage encoding SPE A integrates into the serine tRNA gene in the streptococcal genome. SPE A is made in group A streptococci in highly variable amounts, ranging from approximately 1 ng/ml to 3 μg/ml. The reason for this variability is unknown. When expressed in *S. aureus*, SPE A is partially under the control of the global regulator termed the accessory gene regulator (*agr*). SPE A is stable when made in *S. aureus*, and thus the toxin is resistant to staphylococcal proteases. Because of the *agr* control, similarity in sequence to SEs B and C, stability in *S. aureus* (unlike SPEs B and C), and presence on bacteriophage, it has been suggested that SPE A was at one time a staphylococcal toxin that was transferred to group A streptococci by bacteriophage (27).

SPE A is likely a significant virulence factor in that pure SPE A could induce symptoms of streptococcal TSS without necrotizing fasciitis/myositis in rabbits (48) and in humans owing to laboratory accidents. Pure SPE A

administered to rabbits also conferred significant protection from TSS with necrotizing fasciitis/myositis after challenge with viable M1 and M3 streptococci (52), the two most common type M serotypes associated with streptococcal TSS. In that study it was proposed that production of SPE A in nonimmune animals delayed the influx of polymorphonuclear leukocytes (owing to release of cytokines) necessary to contain the infection. The necrotizing fasciitis/myositis likely resulted from production of other virulence factors such as SPE B (cysteine protease) and hemolysins by the invading organisms, since SPE A alone could not cause this effect. The lack of inflammation seen in nonimmune rabbits is consistent with a lack of inflammation seen in both staphylococcal TSS cases and many patients with streptococcal TSS. In contrast to the nonimmune animals, rabbits immune to SPE A developed a significant polymorphonuclear leukocyte inflammatory response that contained the invading organisms, and the animals remained healthy. In mice challenged with an invasive M1 strain, mortality was correlated with peak serum levels of SPE A (58).

In addition to the above-mentioned properties, SPE A has the ability to form a complex with endotoxin that is lethal to T cells (31). This may explain the disappearance of immune system cells in general in patients who succumb to TSS.

SPE A is produced by many *S. pyogenes* strains isolated from patients with streptococcal TSS (4, 6, 9, 18, 40). Four naturally occurring alleles of *speA* exist (42), and three of these (*speA1*, *speA2*, and *speA3*) differ by a single amino acid, with the *speA4* allele differing by 26 amino acids. M1 and M3 streptococcal strains expressing *speA2* and *speA3* caused the majority of streptococcal TSS in the 1980s (42).

SPE A is clearly related to PTSAgs of staphylococcal origin and also to SPE C and SSA (Fig. 1). Crystallography of various staphylococcal enterotoxins has revealed a common three-dimensional structure, and computer modeling has proposed a similar folding for SPE A (48) (Fig. 2). This toxin is kidney shaped and can be divided structurally into two domains designated A and B. Major structural features of SPE A include a 26-residue amino-terminal segment on top of domain A, of which only the last portion is highly organized. The remainder of the amino part of the molecule (up to residue 106) folds into domain B, composed mainly of a β-barrel structure with a short α-helix segment at the bottom and an apparently unorganized 12-residue "cysteine loop" segment. In the SE structure, a disulfide bridge holds this unorganized structure in a loop. A similar sulfhydryl linkage has been proposed for SPE A (29, 48), although SPE A has three cysteines that potentially could form two alternative cross-links.

Three studies have examined regions involved in the biological activities of SPE A by mutagenesis. Hartwig and Fleischer (16) reported five amino acid substitutions that decreased binding to the class II MHC molecule. Kline and Collins (29) generated 22 mutant forms of SPE A1 using a PCR-based strategy, including two mutations that changed SPE A1 to SPE A2 or SPE A3. Most mutations were based on residues of SEB known to be important for biological activity. The major conclusions reached by these authors were that the class II MHC-binding domains of SPE A1 and SEB are similar but not identical, that a disulfide bond exists between cysteines 87 and 98 and this bond is required for activity, and that the SPE A3 allele is more mitogenic and has a stronger affinity for the class II MHC molecule than does SPE A1.

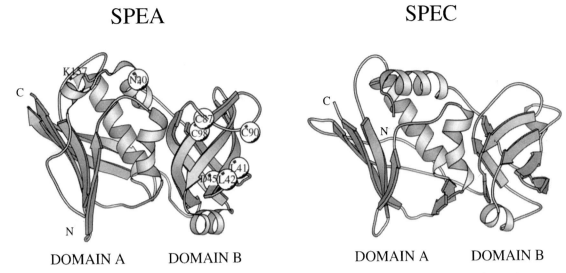

FIGURE 2 Ribbon diagram of the modeled three-dimensional structure of SPE A and SPE C. Domains A and B are indicated, and organized structures such as β-strands and α-helices are represented. Alpha carbons of the mutated residues are represented as spheres for SPE A.

As defined by the three-dimensional structure of SPE A, several directed single, double, triple, and penta mutants were made in the SPE A molecule (positions shown in Fig. 2) (48). Mutant SPE A molecules were analyzed in vivo for their lethal activity and in vitro for their superantigenic ability (Table 2). This work indicated that superantigenic activity and the ability to induce lethality and enhancement of host susceptibility to endotoxin were separable. Results also confirmed that Cys87 and Cys98 were required for mitogenicity and suggested that a disulfide bond is likely to be present between these two residues. Two double, one triple, and one penta mutant were able to induce antibodies in rabbits capable of neutralizing the biological activity of SPE A (mitogenicity, endotoxin susceptibility enhance-

ment, and lethality when administered in subcutaneous miniosmotic pumps). These mutants may be useful as vaccine candidates (48).

STREPTOCOCCAL PYROGENIC TOXIN TYPE B (SPE B, OR CYSTEINE PROTEASE)

speB was first cloned in 1988, and protein expressed from *E. coli* was shown to retain its mitogenic activity (5). Analysis of the amino acid sequence revealed that SPE B was nearly identical to streptococcal proteinase precursor (17), and subsequent studies revealed that SPE B had protease activity (25).

TABLE 2 Properties of SPE A and mutants[a]

Protein	Lethality in miniosmotic pumps	Superantigenicity	Pyrogenicity	Enhancement of	
				Endotoxin shock	Cardiotoxicity
SPE A	+	+	+	+	+
K16N	+	+	+	+	+
N20D	−	−	−	−	+/−
D45N	−	−	−	−	−
C87S	ND[b]	−	−	+	+
C90S	+	+	+	+	+
C98S	−	−	−	−	−
K138E	+	+	+	+	+
K157E	−	+	+	−	+/−
S195A	+	+	+	+	+
N20D/C98S	−	−	−	−	−
N20D/K157E	−	−	−	−	−
N20D/D45N/C98C	−	−	−	−	−
N20D/L41A/L42A/D45N/C98S	−	−	−	−	−

[a] Data taken in part from Roggiani et al. (48).
[b] ND, not done.

Evidence has shown that SPE B is an important virulence factor. For example, Lukomski et al. (34) reported that a SPE B-deficient M3 strain was less virulent in mice after intraperitoneal challenge than was the wild-type M3 strain. Also, SPE B-deficient mutants were shown to have enhanced in vitro internalization into human epithelial and endothelial cells (7). In 1979, Schlievert et al. (53) showed that all nephritogenic streptococci made SPE B. Alternatively, in a recent study using isogenic gene replacement strains and a mouse model of invasive soft tissue infection, speB mutants had no apparent effect on the ability of group A streptococci to cause local tissue injury and invasive infection (1).

Evidence also exists that SPE B is a superantigen (5), although this activity is still debated (reviewed in reference 39). Tomai et al. (60) have shown that cloned SPE B caused significant skewing (overstimulation) of T cells bearing human Vβ2, consistent with the toxin's having superantigen activity.

STREPTOCOCCAL PYROGENIC TOXIN TYPE C (SPE C)

Compared with SPE A and SPE B, considerably less is known about SPE C. This is likely due to several factors, including the poor immunogenicity of SPE C. Also, SPE C is produced in smaller amounts and is less stable, and difficulties are encountered during purification owing to its strong association with hyaluronic acid. Mature SPE C is approximately 24,000 Da and is encoded on a lysogenic phage (15), similar to SPE A. Although this toxin belongs to the PTSAgs family and has similar biological properties, the toxin shares only 25 to 30% primary amino acid similarity to other PTSAgs (Fig. 1). SPE C is unusual in that, with regular inoculations, it takes up to 2 years for laboratory animals to develop immune responses to this toxin.

SPE C is also epidemiologically associated with the development of severe invasive streptococcal diseases such as streptococcal TSS (30). In addition, group A streptococci associated with guttate psoriasis consistently produce SPE C, either alone or together with SPEs A and B (32). Finally, all M18 strains associated with rheumatic fever, including strains involved in the recent resurgence of the illness, produce SPE C, although a causative role of SPE C in rheumatic fever remains to be determined (53). The crystal structure of SPE C has recently been solved (Fig. 2), and it resembles the structures of other PTSAgs (49). From the structural data, SPE C is thought to bind to MHC class II molecules through a zinc-binding site and is believed to form a dimer. Zinc was hypothesized to stabilize the SPE C dimer but was not essential for dimer formation.

STREPTOCOCCAL PYROGENIC TOXIN TYPE F (SPE F, OR MITOGENIC FACTOR)

SPE F was originally named mitogenic factor (64) and was later named SPE F because of its ability to induce cytokine production (44). SPE F was cloned, and no significant homology was found with any of the other SPE molecules (19). Among streptococci, speF appears to be present only in group A streptococci (65) and may be present in all group A streptococci (44). Purified SPE F induced cytokine production from peripheral blood mononuclear cells, including interferon-γ, and TNF-β, but not interleukin-2 or

TNF-α. Also, SPE F required antigen-presenting cells for T-cell proliferation, and T cells bearing Vβ 2, 4, 8, 15, and 19 were preferentially activated (44).

Recently, SPE F was shown to have nuclease activity, and a protein containing an amino acid change from histidine to arginine at position 122 was inactive for DNase activity (20). It is not clear at this time if this mutation has any effect on superantigenicity. All group A streptococci produce DNases; however, aside from SPE F, these enzymes have not been well characterized.

STREPTOCOCCAL SUPERANTIGEN (SSA)

SSA was originally characterized from a serotype M3 strain as a result of a search for novel superantigens (37). This toxin has high similarity in sequence to SPE A and the staphylococcal enterotoxins (Fig. 1) (37). The toxin is produced from M3 isolates of streptococci associated with streptococcal TSS and other M types of streptococci, but not from M1 isolates (47). Because SSA is a superantigen and has a high similarity to other PTSAgs, this toxin likely has a similar three-dimensional structure. Three alleles of ssa have been found. ssa-1 and ssa-3 differ by a single base pair, but the two codons encode the same amino acid (47). Both ssa-1 and ssa-3 encode SSA-1. ssa-2 encodes SSA-2 with an amino acid change from arginine to serine at position 28. Differences in toxicity between SSA-1 and SSA-2 have not yet been studied.

STREPTOCOCCAL CYTOLYTIC TOXINS

Group A streptococci produce two known hemolysin exotoxins, streptolysin O (SLO) and streptolysin S (SLS). Although both are exotoxins, neither is a suspected superantigen. Of these two hemolysins, only SLO is well characterized (26). SLO is a thiol-activated cytolysin that is secreted from nearly all group A streptococci. At least two active forms of SLO exist, with respective molecular weights between 50,000 and 70,000 and pIs between 6.0–6.5 and 7.0–7.8. SLO is active in a reduced state but is rapidly inactivated in the presence of oxygen. SLO binds to cholesterol present in eukaryotic membranes, where it forms multimeric transmembrane pores (3, 57). The ability to form these pores may result in lysis of susceptible cells.

In a recent study, SLO was shown to work synergistically, along with an adherence mechanism, to cause membrane damage to keratinocyte skin cells, resulting in release of proinflammatory cytokines (50). The adherence mechanism may promote an increased local concentration of SLO at the keratinocyte membrane or may allow for delivery of SLO at a specific area of the membrane.

SLS is also produced by essentially all strains of S. pyogenes and is the agent responsible for the beta-hemolytic zones used routinely in the identification of group A streptococci. Despite this important diagnostic feature, the molecular identity of SLS is still not known. This hemolysin is an oxygen-stable and nonimmunogenic toxin. A recent study using transposon mutagenesis created a mutant deficient in SLS production that remained normal with respect to other exoprotein expression (2). These mutants were markedly less virulent in a mouse model of subcutaneous infection. The location of the transposon was in the putative promoter region of a yet-uncharacterized open reading frame, which may represent the SLS gene (2).

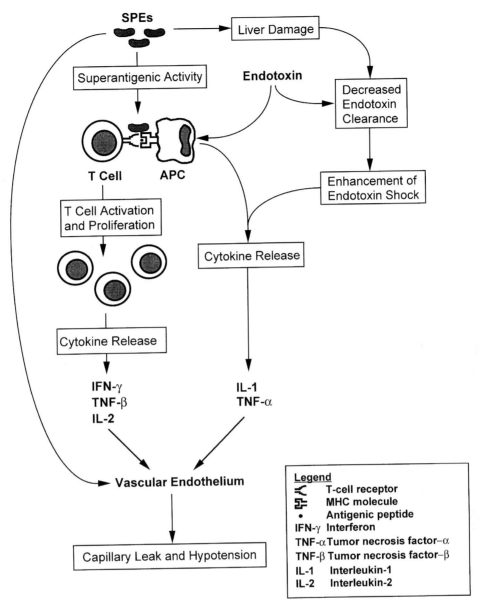

FIGURE 3 Proposed mechanism of streptococcal TSS development by SPEs.

CONCLUSIONS

The increased incidence of severe invasive group A streptococcal diseases and streptococcal TSS remains unexplained; however, SPEs are clearly implicated in streptococcal TSS. A model for the action of SPEs in streptococcal TSS would involve a very complex toxic process in which death would be the final endpoint (Fig. 3). Cytokine mediators, released in massive amounts as a result of both superantigenicity and the enhancement of susceptibility to endotoxin, and the direct toxicity of PTSAgs are likely to contribute to capillary leak, which is ultimately responsible for hypotension, shock, and death. When considering the design of effective therapeutic strategies to combat streptococcal TSS, it is important to realize that no single streptococcal superantigen is responsible for all cases of streptococcal TSS. It has been reported that uncharacter-

ized SPEs exist (43), and they have also been detected in certain group B, C, F, and G streptococcal strains that may mimic the role of the group A streptococcal SPEs. Other group A streptococcal superantigens have been described, including SMEZ (24) and SMEZ-2, which was cloned during an attempt to clone SMEZ, as well as SPE-G, SPE-H, and SPE-J, which were identified from the *S. pyogenes* M1 database (46).

Most research on SPEs has focused on their role in streptococcal TSS, although these superantigens may also be involved in other important biological processes, such as colonization or circumvention of local or systemic immune responses. A recent study has suggested that SPE A may contribute to more chronic streptococcal diseases by induction of Th2-derived immunoregulatory and hematopoietic cytokines (38). Furthermore, although rabbits are the accepted animal model for TSS, the amount of SPE required to induce death appears to be 100 to 200 μg, while in humans 2 μg may be lethal.

Taken together, many unanswered questions remain to be clarified to fully understand the immunobiology of SPEs and their roles in human streptococcal diseases.

This work was supported by USPHS research grant HL36611 from the National Heart, Lung, and Blood Institute. J. K. M. was supported by USPHS grant HL51987. We gratefully acknowledge Gregory Vath and Douglas Ohlendorf, Department of Biochemistry, University of Minnesota, for SPE A and C structures.

REFERENCES

1. **Ashbaugh, C. D., H. B. Warren, V. J. Carey, and M. R. Wessels.** 1998. Molecular analysis of the role of the group A streptococcal cysteine protease, hyaluronic acid capsule, and M protein in a murine model of human invasive soft-tissue infection. *J. Clin. Invest.* **102:**550–560.

2. **Betschel, S. D., S. M. Borgia, N. L. Barg, D. E. Low, and J. C. S. De Azavedo.** 1998. Reduced virulence of group A streptococcal TN916 mutants that do not produce streptolysin S. *Infect. Immun.* **66:**1671–1679.

3. **Bhakdi, S., J. Tranum-Jensen, and A. Sziegoleit.** 1985. Mechanism of membrane damage by streptolysin-O. *Infect. Immun.* **47:**52–60.

4. **Bohach, G. A., D. J. Fast, R. D. Nelson, and P. M. Schlievert.** 1990. Staphylococcal and streptococcal pyrogenic toxins involved in toxic shock syndrome and related illnesses. *Crit. Rev. Microbiol.* **17:**251–272.

5. **Bohach, G. A., A. R. Hauser, and P. M. Schlievert.** 1988. Cloning of the gene, *speB*, for streptococcal pyrogenic exotoxin type B in *Escherichia coli. Infect. Immun.* **56:**1665–1667.

6. **Brieman, R. F., J. P. Davis, R. R. Facklam, B. M. Gray, C. W. Hoge, E. L. Kaplan, E. A. Mortimer, P. M. Schlievert, B. Schwartz, D. L. Stevens, and J. K. Todd.** 1993. Defining severe invasive streptococcal infections: rationale and consensus definition. *JAMA* **269:**390–391.

7. **Burns, E. H., Jr., S. Lukomski, J. Rurangirwa, A. Podbielski, and J. M. Musser.** 1998. Genetic inactivation of the extracellular cysteine protease enhances *in vitro* internalization of group A streptococci by human epithelial and endothelial cells. *Microb. Pathog.* **24:**333–339.

8. **Cleary, P. P., E. L. Kaplan, J. P. Handley, A. Wlazlo, M. H. Kim, A. R. Hauser, and P. M. Schlievert.** 1992. Clonal basis for resurgence of serious *Streptococcus pyogenes* disease in the 1980s. *Lancet* **339:**518–521.

9. **Cockerill, F. R., R. L. Thompson, J. M. Musser, P. M. Schlievert, J. Talbot, K. E. Holley, W. S. Harmsen, D. M. Ilstrup, P. C. Kohner, M. H. Kim, B. Frankfort, J. M. Manahan, J. M. Steckelberg, F. Roberson, and W. R. Wilson.** 1998. Molecular, serological, and clinical features of 16 consecutive cases of invasive streptococcal disease. Southeastern Minnesota Streptococcal Working Group. *Clin. Infect. Dis.* **26:**1448–1458.

10. **Cone, L. A., D. R. Woodard, P. M. Schlievert, and G. S. Tomory.** 1987. Clinical and bacteriologic observations of a toxic shock-like syndrome due to *Streptococcus pyogenes. N. Engl. J. Med.* **317:**146–149.

11. **Davies, H. D., A. McGeer, B. Schwartz, K. Green, D. Cann, A. E. Simor, D. E. Low, and The Ontario Group A Streptococcal Study Group.** 1996. Invasive group A streptococcal infections in Ontario, Canada. *N. Engl. J. Med.* **35:**547–554.

12. **Davis, J. P., M. T. Osterholm, C. M. Helms, J. M. Vergeront, L. A. Wintermeyer, J. C. Forfang, L. A. Judy, J. Rondeau, and W. L. Schell.** 1982. Tri-state toxic-shock syndrome study. II. Clinical and laboratory findings. *J. Infect. Dis.* **145:**441–448.

13. **Fast, D. J., P. M. Schlievert, and R. D. Nelson.** 1989. Toxic shock syndrome-associated staphylococcal and streptococcal pyrogenic toxins are potent inducers of tumor necrosis factor production. *Infect. Immun.* **57:**291–294.

14. **Frobisher, M., and J. H. Brown.** 1927. Transmissible toxigenicity of streptococci. *Bull. Johns Hopkins Hosp.* **41:**167–173.

15. **Goshorn, S. C., G. A. Bohach, and P. M. Schlievert.** 1988. Cloning and characterization of the gene, *speC*, for pyrogenic exotoxin type C from *Streptococcus pyogenes. Mol. Gen. Genet.* **212:**66–70.

16. **Hartwig U. F., and B. Fleischer.** 1993. Mutations affecting MHC class II binding of the superantigen streptococcal erythrogenic toxin A. *Int. Immunol.* **5:**869–875.

17. **Hauser, A. R., and P. M. Schlievert.** 1990. Nucleotide sequence of the streptococcal pyrogenic exotoxin type B gene and toxin relationship to streptococcal proteinase precursor. *J. Bacteriol.* **172:**4536–4542.

18. **Hauser, A. R., D. L. Stevens, E. L. Kaplan, and P. M. Schlievert.** 1991. Molecular analysis of pyrogenic exotoxins from *Streptococcus pyogenes* isolates associated with toxic shock-like syndrome. *J. Clin. Microbiol.* **29:**1562–1567.

19. **Iwasaki, M., H. Igarashi, Y. Hinuma, and T. Yutsudo.** 1993. Cloning, characterization and overexpression of a *Streptococcus pyogenes* gene encoding a new type of mitogenic factor. *FEBS Lett.* **331:**187–192.

20. **Iwasaki, M., H. Igarashi, and T. Yutsudo.** 1997. Mitogenic factor secreted by *Streptococcus pyogenes* is a heat-stable nuclease requiring His^{122} for activity. *Microbiology* **143:**2449–2455.

21. **Johnson L. P., and P. M. Schlievert.** 1983. A physical map of the group A streptococcal pyrogenic exotoxin bacteriophage T12 genome. *Mol. Gen. Genet.* **189:**251–255.

22. **Johnson, L. P., and P. M. Schlievert.** 1984. Group A streptococcal phage T12 carries the structural gene for pyrogenic exotoxin type A. *Mol. Gen. Genet.* **194:**52–56.

23. **Johnson, L. P., D. L. Stevens, and E. L. Kaplan.** 1992. Epidemiologic analysis of group A streptococcal serotypes associated with severe systemic infections, rheumatic fever, or uncomplicated pharyngitis. *J. Infect. Dis.* **166:**374–382.

24. **Kamezawa, Y., T. Nakahara, S. Nakano, Y. Abe, J. Nozaki-Renard, and T. Isono.** 1997. Streptococcal mitogenic exotoxin Z, a novel acidic superantigenic toxin produced by a T1 strain of *Streptococcus pyogenes. Infect. Immun.* **65:**3828–3833.

25. **Kapur, V., M. W. Majesky, L.-L. Li, R. A. Black, and J. M. Musser.** 1993. Cleavage of interleukin 1b (IL-1b) precursor to produce active IL-1b by a conserved extracellular cysteine protease from *Streptococcus pyogenes. Proc. Natl. Acad. Sci. USA* **90:**7676–7680.

26. **Kehoe M. A., L. Miller, J. A. Walker, and G. J. Boulnois.** 1987. Nucleotide sequence of the streptolysin O (SLO) gene: structural homologies between SLO and

other membrane-damaging, thiol-activated toxins. *Infect. Immun.* **55:**3228–3232.

27. **Kim, M. H., and P. M. Schlievert.** 1997. Molecular genetics, structure, and immunobiology of streptococcal pyrogenic exotoxins A and C, p 257–279. *In* D. Y. M. Leung, B. T. Huber, and P. M. Schlievert (ed.), *Superantigens. Molecular Biology, Immunology, and Relevance to Human Disease.* Marcel Dekker, Inc., New York, N.Y.

28. **Kim, Y. B., and D. W. Watson.** 1970. A purified group A streptococcal pyrogenic exotoxin: physicochemical and biological properties including the enhancement of susceptibility to endotoxin lethal shock. *J. Exp. Med.* **131:**611–628.

29. **Kline, J. B., and C. M. Collins.** 1996. Analysis of the superantigenic activity of mutant and allelic forms of streptococcal pyrogenic exotoxin A. *Infect. Immun.* **64:**861–869.

30. **Leggiadro, R. J., M. C. Bugnitz, B. A. Peck, G. S. Luedtke, M. H. Kim, E. L. Kaplan, and P. M. Schlievert.** 1993. Group A streptococcal bacteremia in a mid-south children's hospital. *South. Med. J.* **86:**615–618.

31. **Leonard, B. A. B., and P. M. Schlievert.** 1992. Immune cell lethality induced by streptococcal pyrogenic exotoxin A and endotoxin. *Infect. Immun.* **60:**3747–3755.

32. **Leung D. Y., J. B. Travers, R. Giorno, D. A. Norris, R. Skinner, J. Aelion, L. V. Kazemi, M. H. Kim, A. E. Trumble, M. Kotb, and P. M. Schlievert.** 1995. Evidence for a streptococcal superantigen-driven process in acute guttate psoriasis. *J. Clin. Invest.* **96:**2106–2112.

33. **Leung, D. Y. M., B. T. Huber, and P. M. Schlievert.** 1997. Historical perspectives of superantigens and their biological activities, p. 1–13. *In* D. Y. M. Leung, B. T. Huber, and P. M. Schlievert (ed.), *Superantigens. Molecular Biology, Immunology, and Relevance to Human Disease.* Marcel Dekker, Inc., New York, N.Y.

34. **Lukomski, S., E. H. Burns, Jr., P. R. Wyde, A. Podbielski, J. Rurangirwa, D. K. Moore-Poveda, and J. M. Musser.** 1998. Genetic inactivation of an extracellular cysteine protease (SpeB) expressed by *Streptococcus pyogenes* decreases resistance to phagocytosis and dissemination to organs. *Infect. Immun.* **66:**771–776.

35. **Massell, B. F., C. G. Chute, A. M. Walker, and G. S. Kurland.** 1988. Penicillin and the marked decrease in morbidity and mortality from rheumatic fever in the United States. *N. Engl. J. Med.* **318:**280–286.

36. **McShan, W. M., Y. F. Tang, and J. J. Ferretti.** 1997. Bacteriophage T12 of *Streptococcus pyogenes* integrates into the gene encoding a serine tRNA. *Mol. Microbiol.* **23:**719–728.

37. **Mollick, J. A., G. G. Miller, J. M. Musser, R. G. Cook, D. Grossman, and R. R. Rich.** 1993. A novel superantigen isolated from pathogenic strains of *Streptococcus pyogenes* with amino terminal homology to staphylococcal enterotoxins B and C. *J. Clin. Invest.* **92:**710–719.

38. **Müller-Alouf, H., D. Gerlach, P. Desreumaux, C. Leportier, J. E. Alouf, and M. Capron.** 1997. Streptococcal pyrogenic exotoxin A (SPE A) superantigen induced production of hematopoietic cytokines, IL-12 and IL-13 by human peripheral blood mononuclear cells. *Microb. Pathog.* **23:**265–272.

39. **Musser, J. M.** 1997. Streptococcal superantigen, mitogen factor, and pyrogenic exotoxin B expressed by *Streptococcus pyogenes.* Structure and function, p 281–310. *In*

D. Y. M. Leung, B. T. Huber, and P. M. Schlievert (ed.), *Superantigens. Molecular Biology, Immunology, and Relevance to Human Disease.* Marcel Dekker, Inc., New York, N.Y.

40. **Musser, J. M., A. R. Hauser, M. Kim, P. M. Schlievert, K. Nelson, and R. K. Selander.** 1991. *Streptococcus pyogenes* causing toxic shock-like syndrome and other invasive diseases: clonal diversity and pyrogenic exotoxin expression. *Proc. Natl. Acad. Sci. USA* **88:**2668–2672.

41. **Musser, J. M., V. Kapur, J. Szeto, X. Pan, D. S. Swanson, and D. R. Martin.** 1995. Genetic diversity and relationships among *Streptococcus pyogenes* strains expressing serotype M1 protein: recent intercontinental spread of a subclone causing episodes of invasive disease. *Infect. Immun.* **63:**994–1003.

42. **Nelson, K., P. M. Schlievert, R. K. Selander, and J. M. Musser.** 1991. Characterization and clonal distribution of four alleles of the *speA* gene encoding pyrogenic exotoxin A (scarlet fever toxin) in *Streptococcus pyogenes. J. Exp. Med.* **174:**1271–1274.

43. **Newton, D., A. Norrby-Teglund, A. McGeer, D. E. Low, P. M. Schlievert, and M. Kotb.** 1997. Novel superantigens from streptococcal toxic shock syndrome *Streptococcus pyogenes* isolates. *Adv. Exp. Med. Biol.* **418:**525–529.

44. **Norrby-Teglund, A., D. Newton, M. Kotb, S. E. Holm, and M. Norgren.** 1994. Superantigenic properties of the group A streptococcal exotoxin SpeF (MF). *Infect. Immun.* **62:**5227–5233.

45. **Parsonnet, J., R. K. Hickman, D. D. Eardley, and G. B. Pier.** 1985. Induction of human interleukin-1 by toxic-shock-syndrome toxin-1. *J. Infect. Dis.* **151:**514–522.

46. **Proft, T., S. L. Moffatt, C. J. Berkahn, and J. D. Fraser.** 1999. Identification and characterization of novel superantigens from *Streptococcus pyogenes. J. Exp. Med.* **189:**89–101.

47. **Reda, K. B., V. Kapur, J. A. Mollick, J. G. Lamphear, J. M. Musser, and R. R. Rich.** 1994. Molecular characterization and phylogenetic distribution of the streptococcal superantigen gene (*ssa*) from *Streptococcus pyogenes. Infect. Immun.* **62:**1867–1874.

48. **Roggiani, M., J. A. Stoehr, B. A. B. Leonard, and P. M. Schlievert.** 1997. Analysis of toxicity of streptococcal pyrogenic exotoxin A mutants. *Infect. Immun.* **65:**2868–2875.

49. **Roussel, A., B. F. Anderson, H. M. Baker, J. D. Fraser, and E. N. Baker.** 1997. Crystal structure of the streptococcal superantigen SPE-C: dimerization and zinc binding suggests a novel mode of interaction with MHC class II molecules. *Nat. Struct. Biol.* **4:**635–643.

50. **Ruiz, N., B. Wang, A. Pentland, and M. Caparon.** 1998. Streptolysin O and adherence synergistically modulate proinflammatory responses of keratinocytes to group A streptococci. *Mol. Microbiol.* **27:**337–346.

51. **Schlievert, P. M.** 1986. Staphylococcal enterotoxin B and toxic-shock syndrome toxin-1 are significantly associated with non-menstrual TSS. *Lancet* **i:**1149–1150.

52. **Schlievert, P. M., A. P. Assimacopoulos, and P. P. Cleary.** 1996. Severe invasive group A streptococcal disease: clinical description and mechanisms of pathogenesis. *J. Lab. Clin. Med.* **127:**13–22.

53. **Schlievert, P. M., K. M. Bettin, and D. W. Watson.** 1979. Production of pyrogenic exotoxins by groups of

streptococci: association with group A. *J. Infect. Dis.* **140:** 676–681.

54. **Schlievert, P. M., K. M. Bettin, and D. W. Watson.** 1980. Inhibition of ribonucleic acid synthesis by group A streptococcal pyrogenic exotoxin. *Infect. Immun.* **27:** 542–548.

55. **Schlievert, P. M., K. N. Shands, B. B. Dan, G. P. Schmid, and R. D. Nishimura.** 1981. Identification and characterization of an exotoxin from *Staphylococcus aureus* associated with toxic-shock syndrome. *J. Infect. Dis.* **143:**509–516.

56. **Schwab, J. H., D. W. Watson, and W. J. Cromartie.** 1955. Further studies of group A streptococcal factors with lethal and cardiotoxic properties. *J. Infect. Dis.* **96:**14–18.

57. **Sekiya, K., R. Satoh, H. Danbara, and Y. Futaesaku.** 1993. A ring-shaped structure with a crown formed by streptolysin O on the erythrocyte membrane. *J. Bacteriol.* **175:**5953–5961.

58. **Sriskandan, S., D. Moyes, L. K. Buttery, T. Krausz, T. J. Evans, J. Polak, and J. Cohen.** 1996. Streptococcal pyrogenic exotoxin A (SPE A) release, distribution, and role in a murine model of fasciitis and multi-organ failure due to *Streptococcus pyogenes*. *J. Infect. Dis.* **173:**1399–1407.

59. **Stevens, D. L., M. H. Tanner, J. Winship, R. Swarts, K. M. Ries, P. M. Schlievert, and E. Kaplan.** 1989. Severe group A streptococcal infections associated with a toxic shock-like syndrome and scarlet fever toxin A. *N. Engl. J. Med.* **321:**1–7.

60. **Tomai, M. A., P. M. Schlievert, and M. Kotb.** 1992. Distinct T-cell receptor V beta gene usage by human T lymphocytes stimulated with the streptococcal pyrogenic exotoxins and pep M5 protein. *Infect. Immun.* **60:**701–705.

61. **Watson, D. W.** 1960. Host-parasite factors in group A streptococcal infections: pyrogenic and other effects on immunologic distinct exotoxins related to scarlet fever toxins. *J. Exp. Med.* **111:**255–283.

62. **Weeks, C. R., and J. J. Ferretti.** 1986. Nucleotide sequence of the type A streptococcal exotoxin (erythrogenic toxin) gene from *Streptococcus pyogenes* bacteriophage T12. *Infect. Immun.* **52:**144–150.

63. **White, J., A. Herman, A. M. Pullen, R. Kubo, J. W. Kappler, and P. Marrack.** 1989. The $V\beta$-specific superantigen staphylococcal enterotoxin B: stimulation of mature T-cells and clonal deletion in neonatal mice. *Cell* **56:** 27–35.

64. **Yutsudo, T., H. Murai, J. Gonzalez, T. Takao, Y. Shimonishi, Y. Takeda, H. Igarashi, and Y. Hinuma.** 1992. A new type of mitogenic factor produced by *Streptococcus pyogenes*. *FEBS Lett.* **308:**30–34.

65. **Yutsudo, T., K. Okumura, M. Iwasake, A Hara, S. Kamitani, W. Minamide, H. Igarashi, and Y. Hinuma.** 1994. The gene encoding a new mitogenic factor in a *Streptococcus pyogenes* strain is distributed only in group A streptococci. *Infect. Immun.* **62:**4000–4004.

66. **Zabriskie, J. B.** 1964. The role of temperate bacteriophage in the production of erythrogenic toxin by group A streptococci. *J. Exp. Med.* **119:**761–780.

Genetics of Group A Streptococci

MICHAEL CAPARON

6

Streptococcus pyogenes (the group A streptococcus) is remarkable in terms of the large number of very different diseases it can cause in humans. These range from superficial and self-limiting diseases of the pharynx (e.g., pharyngitis, commonly known as strep throat) and skin (impetigo) to infections that involve increasingly deeper layers of tissue and are associated with increasing degrees of destruction of tissue (e.g., erysipelas, cellulitis, necrotizing fasciitis, and myositis). The organism has the ability to spread rapidly through tissue and to penetrate into the vasculature to cause lethal sepsis. Other diseases result from the production of toxins that spread through tissue or systemically from a site of local infection (scarlet fever and toxic shock syndrome). Still other diseases are the result of an immunopathological response on the part of the host that is triggered by a streptococcal infection. These diseases include rheumatic fever, acute glomerulonephritis, certain types of psoriasis, and potentially even some forms of obsessive-compulsive disorder.

S. *pyogenes* is even more remarkable in terms of the very large number of factors that have been identified as potential virulence determinants for these various diseases. These include surface proteins (M proteins, fibronectin-binding proteins, surface dehydrogenases, C5a peptidase), the hyaluronic acid capsule, secreted degradative enzymes (several distinct DNases, a cysteine protease, NADase, hyaluronidase), and many different secreted toxins (streptolysin S, streptolysin O, the pyrogenic exotoxins, streptococcal superantigen, streptokinase). This represents only a partial list of the potential virulence factors this bacterium can produce, and many of these are considered in more detail in other chapters of this volume.

Until recently, the function that any of these potential virulence factors contributed to the pathogenesis of any specific streptococcal disease was only poorly understood. A major reason for this deficiency was the lack of sophisticated genetic systems that could be applied to the analysis of a specific virulence factor according to modern molecular criteria. A succinct statement of these criteria is Falkow's "molecular Koch's postulates" (13). These postulates state that (i) the phenotype under investigation should be associated with pathogenic members of a species; (ii) the gene(s) associated with the virulence trait should be identified and isolated by molecular methods; (iii) specific inactivation of an identified gene(s) should lead to a measurable loss in pathogenicity; and (iv) reintroduction of the unmodified wild-type gene should lead to a restoration in pathogenicity. An exciting development is that, over the past few years, the work of many different research groups has contributed to the development of sophisticated group A streptococcal genetic systems that now allow the many different diseases that this organism can cause to be studied at this level of molecular resolution. The goal of this chapter is to present an overview of these methods and their applications.

GENETIC EXCHANGE

Transduction

A key element of any genetic system involves some system of genetic exchange between different bacterial hosts that allows the construction of an altered genome in the target host, which can then be subjected to an analysis of its virulence phenotypes. Unlike several other species of streptococci, the group A streptococci are not naturally competent for the uptake of exogenous DNA. Conjugative DNA transfer does occur in group A streptococci; however, it is restricted to the transfer of conjugative plasmids and conjugative transposons (see below), and there is no evidence for mobilization of chromosomal markers. There is also no evidence that any important virulence traits are encoded by these types of mobile elements, although they are undoubtedly important in the transmission of resistance to various antibiotics.

Even though there is no evidence to support genetic exchange by transformation or by conjugation, analysis of several polymorphic loci, most notably the genes that encode the M proteins, has provided considerable evidence for horizontal transfer of genetic material among natural populations of S. *pyogenes*. In this regard, considerable attention has focused on the contribution of phage to genetic

Gram-Positive Pathogens, ed. by V. A. Fischetti et al.
© 2000 American Society for Microbiology, Washington, D.C.

transfer. *S. pyogenes* strains are rich in phage, and several of these have been demonstrated to encode virulence factors, including the pyrogenic exotoxins SPE-A and SPE-C (40). Thus, it is not surprising that transduction was the first mechanism of genetic exchange that was exploited in the manipulation of the *S. pyogenes* genome. The most highly developed of these phage are derivatives of phage A25. This lytic phage is classified as a Bradley group B phage that recognizes peptidoglycan of groups A, C, and G streptococci as its cellular receptor. It has a 35-kb double-stranded genome with circular permutation and terminal repetition (31), and it is proficient for transducing markers in vitro. (For a detailed description of a method for transduction, see reference 7.) The most useful derivative of A25 for use in transduction is that developed by Malke (28), which has two distinct temperature-sensitive lesions ($A25_{ts1-2}$) and becomes defective for growth at 37°C. This feature is useful for transduction, since it allows the production of a transducing lysate of the donor host at the permissive temperature (30°C) but prevents the killing of transduced hosts when infection takes place at the nonpermissive temperature.

A limitation to the use of transduction for the construction of mutated chromosomes for virulence studies is that the method is restricted to the exchange of preexisting markers between different *S. pyogenes* hosts. Thus, it cannot be applied to the construction of novel mutations or used when a suitable selectable marker is not linked to the preexisting mutation of choice. However, transduction has been most useful in linkage analysis of transposon Tn916-generated mutations and in analysis of Tn916-generated mutations when the mutant chromosome contains multiple copies of the transposable element. In both cases, transduction is used to cross the transposon-containing locus back into a wild-type background. Since Tn916 does not transpose at high frequency during transduction, the resulting transductants arise by homologous recombination, with the result that the transposon serves as a selectable marker to cross the mutated locus back into a wild-type background (7). The phenotypes of the resulting transductants can then be analyzed and compared with that of the original mutant. In the case of a mutant with multiple insertions, it is possible to generate transductants with just one of the mutated loci so that the contribution of each individual transposon insertion to the generation of the mutant phenotype can be analyzed.

Transformation

One of the most significant advances in genetic technology for *S. pyogenes* has been the development of methods for transformation. This has allowed the introduction of heterologous DNA into an *S. pyogenes* host and has opened the door for a large number of techniques for mutagenesis and allelic exchange that are described in greater detail below. The breakthrough technology that began the era of transformation for *S. pyogenes* was the introduction by several companies of reasonably priced instruments for the introduction of DNA by electroporation. As is true for most methods of transformation, success with electroporation-based transformation depends on careful attention to growth conditions. Electroporation of streptococci generally requires that the cell walls be weakened. Most successful methods have been adapted from the technique originally developed by Dunny and colleagues (10a) for electroporation of the enterococci. This method uses cells from the early exponential stages of growth in medium supplemented with glycine. The addition of glycine is thought to contribute to a decreased level of cross-linking in the cell wall, and the exact stage of growth and concentration of glycine is determined empirically. (For a detailed description of the method, see reference 7.) Alternative conditions have been described by Simon and Ferretti (45), and a modified medium developed by Husmann et al. (21) has proved useful for transformation of strains that have difficulty growing in the more widely used media.

PLASMID TECHNOLOGY

The development of transformation systems allowed the modification of *S. pyogenes* hosts through the introduction of plasmids proficient for episomal replication in *S. pyogenes*. The most widely used strategy involves shuttle-type plasmids. The plasmids are first manipulated in *Escherichia coli* using standard technologies, purified from *E. coli*, and then used to transform an *S. pyogenes* host for analysis of the properties of the resulting strain. The main applications for this strategy have been in complementation studies in which the wild-type allele of the gene of interest is introduced in *trans* into an *S. pyogenes* host that contains a defined mutation in the target gene and for expression of an *S. pyogenes* gene in a heterologous host. In the latter case, heterologous expression is utilized in functional studies to validate the contribution of a given gene to a given phenotype. The heterologous host can be an *S. pyogenes* strain that naturally lacks the phenotype under investigation or can be another streptococcal or enterococcal species. In one of the first examples of this strategy, Scott et al. (44) introduced a copy of the gene that encodes the M6.1 protein (*emm6.1*) on a mobilizable plasmid vector, used this to transform a naturally competent streptococcal species, and then mobilized the plasmid by conjugation into an *emm*-deficient *S. pyogenes* host. Analysis of the resulting strain demonstrated that the introduction of *emm6.1* converted the host strain from being sensitive to killing in a bactericidal assay to being resistant to killing in the assay. It should also be noted that, using plasmid-based gene transfer techniques and the mutagenesis techniques described below, the contribution of the M6.1 protein to the phenotype of resistance to killing by phagocytic cells was the first formal application of the molecular Koch's postulates in *S. pyogenes* (36). Additional applications using transforming vectors for expression in heterologous hosts have included the demonstration that protein F is sufficient to confer a fibronectin-binding phenotype to non-fibronectin-binding *S. pyogenes* (20) and enterococcal hosts and that the *has* operon is sufficient to confer the ability to produce hyaluronate to both acapsular *S. pyogenes* and enterococcal hosts (9).

The strategies outlined above make extensive use of *E. coli* molecular biology to manipulate the plasmid before its introduction and analysis in *S. pyogenes*. Therefore, the most useful vectors have been those with the ability to replicate in both *E. coli* and *S. pyogenes*. The plasmid vectors most commonly used have been based on one of two replicons. The first of these vectors were based on the pWV01 replicon and its relatives (24). This plasmid was isolated from a lactococcal species and is the prototype member of a family of "promiscuous replicons" that have the remarkable property that they can replicate in both gram-negative (*E. coli*) and gram-positive (*S. pyogenes*) hosts. This is a unique property that is not shared by other gram-positive- or gram-negative-derived plasmids. In fact, many plasmids of gram-positive origin have not been reported to replicate reliably in *S. pyogenes*. These include vectors based on staphylococcal replicons, including pC194, pE194, and pUB110, that have been useful for genetic analysis in many other gram-positive species. The second class of commonly used vectors in *S. pyogenes* was based on the pAMβ1 replicon originally isolated from *Enterococcus faecalis* (8). Unlike pWV01, the pAMβ1 replicon cannot replicate in *E. coli*, and this requires that the vector also contain an origin of replication that is proficient in an *E. coli* host. The plasmid pAT28 developed by Trieu-Cuot et al. (47) is typical of the pAMβ1-derived vectors and contains the high-copy pUC replicon derived from ColE1 for replication in *E. coli*. It has several other useful features, including the multiple cloning site and the *lacZα* reporter gene of pUC18, which allows screening for insertion of cloned DNA by α-complementation in *E. coli*. Several other significant differences exist between the pWV01- and pAMβ1-derived vectors. Most notable among these is the fact that pWV01 replicates by using a rolling-circle-type mechanism more characteristic of the single-stranded bacteriophages and likely exists in a single-stranded conformation for extended periods of time in replicating cells. In contrast, pAMβ1 replicates via a double-stranded θ-type mechanism. Also, as a general rule, pWV01-based vectors transform *S. pyogenes* at a much higher efficiency than do pAMβ1-derived vectors. Because of their different modes of replication, this difference may involve ways in which the restriction system of *S. pyogenes* recognizes the two different plasmids. However, it should be noted that very little is currently understood about restriction in *S. pyogenes*, and that, in general, it has not posed a significant barrier to the introduction of DNA purified from any number of *E. coli* K12 hosts. Despite its lower transformation efficiency, the pAMβ1-derived vectors have proved very useful in obtaining expression of cloned genes that were only poorly expressed by pWV01-based vectors (15).

An alternative to using replicating plasmids to express genes has recently been developed (29a). This vector system contains an *E. coli* origin of replicon to facilitate manipulation in *E. coli* and contains the integrase gene (*int*) and attachment site (*attP*) from the T12 temperate bacteriophage of *S. pyogenes*. Following its introduction into *S. pyogenes*, *int* catalyzes the site-specific recombination of the entire circular molecule into the chromosomal phage T12 attachment site (*attB*), which is located in a serine tRNA gene (29a). Because the vector lacks the phage excisionase, integration is highly stable. This vector should be very useful under circumstances when it is desirable to study the gene in question at low copy number.

A fourth plasmid has become very useful in genetic manipulation of *S. pyogenes*. This is a derivative of a pWV01-type replicon that is temperature sensitive for replication in both *E. coli* and *S. pyogenes*. Developed by Maguin and coworkers (27), the plasmid is now known as pG+host4 and contains four amino acid substitutions in the RepA protein, which is responsible for nicking one DNA strand at the plus origin to initiate replication. Other derivatives of this plasmid include pJRS233 (37) and pG+host5 (4), which respectively add low-copy (pSC101) and high-copy (ColE1) *E. coli* origins of replication so that they are not temperature sensitive for replication in *E. coli*. In addition, pJRS233 includes the multiple cloning site of pBluescriptSK+. The modification to the standard electro-

poration conditions described by Perez-Casal et al. (37) greatly increases the transformation efficiency of these plasmids in *S. pyogenes*. The main application of the pG+host-derived plasmids has been in the construction of in-frame deletion alleles of chromosomally encoded genes and in the delivery of transposable elements for mutagenesis (see below).

DIRECTED MUTAGENESIS

The techniques developed for directed mutagenesis (so-called reverse genetics) in *S. pyogenes* have proved invaluable for the analysis of the contributions that specific genes make to pathogenesis. In these techniques, a defined mutation is constructed to inactivate a specific gene to construct an isogenic mutant for analysis. These techniques have become particularly useful with the availability of almost the complete genome sequence of *S. pyogenes* strain ATCC 700294 (1).

Allelic Replacement Using Linear DNA

The allelic replacement method of mutagenesis is straightforward and first involves cloning the gene of interest into any suitable plasmid vector in *E. coli*. With the available genome information, this has become a simple task of analyzing the sequence and designing suitable primers for PCR amplification, followed by insertion of amplification product into an *E. coli* cloning vector. As a general rule, *S. pyogenes* DNA is most stable when cloned on low-copy vectors in *E. coli*, and several different vectors have been used with great success. The only restriction is that the vector must not contain a gene that encodes a β-lactamase. Because this DNA will eventually be introduced into *S. pyogenes*, because no *S. pyogenes* isolate containing a β-lactamase has yet been described in nature, and because β-lactam antibiotics are the drug of choice for treatment of *S. pyogenes* infections, it is unethical to introduce a β-lactamase into *S. pyogenes*. Also, as a general rule, the larger the fragment amplified, the greater the frequency with which the allelic exchange will occur, although this method has been successful with fragments in the 1-kb range. Once the gene of interest has been cloned, its sequence is examined for a suitable unique restriction site. A selectable marker is then introduced into this site to construct the mutant allele. The markers that have been most successful have been those derived from gram-positive organisms that can also be used for selection in *E. coli*. They include *ermAM* (resistance to erythromycin) (GenBank accession no. M20334), *aphA3* (resistance to kanamycin) (GenBank accession no. V01547), *tetM* (resistance to tetracycline) (GenBank accession no. X92947), and *aad9* (resistance to spectinomycin) (GenBank accession no. M69221). A modified omega mutagenesis cassette constructed by introduction of the *aphA3* kanamycin-resistance determinant into the original omega interposon (ΩKm-2) has been widely used in this mutagenesis technique (35). The advantages to the use of this element are that it has a cassettelike structure that facilitates its manipulation and strong transcription and translation termination signals so that it generates a strong polar mutation. The plasmid that contains this mutant allele is then converted to a linear molecule by digestion with a restriction endonuclease that cuts only in the vector sequences. The linear molecule is then used to transform the target *S. pyogenes* host with selection for the inserted resistance marker. Because the introduced molecule is linear, preservation of circular chromosomal structure requires that all resistant transformants arise by two homologous recombination events flanking each side of the inserted resistance marker. The end result is the exchange of the inactivated allele for the wild-type allele (Fig. 1). While this type of recombination does not occur at high frequency, the method is usually successful because the frequency of nonhomologous recombination is typically extremely low. Insertion into the expected locus is confirmed by using Southern blot and PCR-based analyses, and the resulting mutant can be subjected to functional studies. However, in interpreting the resulting functional data, it should be kept in mind that this method of mutagenesis generates strong polar mutations. Thus, expression of distal genes is also likely to be affected should the gene of interest be located in a polycistronic transcriptional unit.

A variation of this technique can be used when a convenient restriction site is not located within the target gene or when it is desirable to subject a large cloned region to rapid, high-resolution mutagenesis. This is desirable in the latter case because genes for complex phenotypes and pathways are often clustered together on the chromosome. This method is a technique for shuttle mutagenesis that uses a version of the E. coli transposon mini-γδ (mγδ-200) that has been modified to contain a kanamycin-resistance gene that can be selected for in both E. coli and S. pyogenes. In this technique, the ability of mγδ to easily and efficiently

generate a large number of random transposon insertions into a segment of DNA cloned on a plasmid is used to obtain a series of mutations along the entire length of cloned streptococcal DNA. Mutated plasmids containing insertions at desired locations are then crossed into the streptococcal chromosome as described above. This method has been particularly useful for the identification of regulatory genes linked to a target gene of interest (16). For a detailed description of this technique, consult Hanski et al. (19).

Directed Insertional Inactivation

Perhaps the most commonly used technique for directed mutagenesis utilizes mutations that are constructed as the result of a single homologous recombination event. For this technique, instead of cloning a large segment of chromosome that includes the target gene, an internal segment of the target gene is cloned; this segment should not include either the 5′ or 3′ ends of the target gene. A selectable marker for S. pyogenes is also included on the plasmid, but in this case it is located adjacent to the cloned segment rather than introduced to interrupt the cloned segment. The resulting plasmid is purified from E. coli and used to transform S. pyogenes with selection for resistance to the introduced marker. Because the commonly used E. coli replicons do not replicate in S. pyogenes, resistant transformants most frequently arise as a result of a single homologous recombination event between the internal seg-

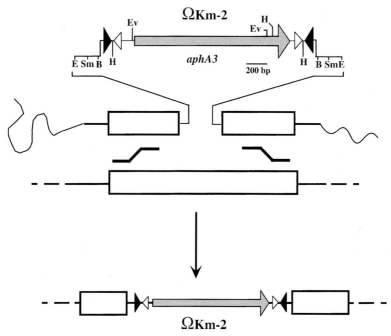

FIGURE 1 Strategy for allelic replacement mutagenesis using an omega interposon. The ωKm-2 interposon contains a kanamycin-resistance gene that can be selected for in both gram-positive and gram-negative organisms. Additional features of ΩKm-2 include a cassettelike structure with several convenient restriction sites and strong transcription (open triangles) and translation (closed triangles) termination signals such that insertion of the element into a gene cloned in E. coli (top half of the figure) is strongly polar. The E. coli plasmid vector is converted to a linear molecule by digestion outside the cloned streptococcal gene (vector DNA is represented by the nonstraight lines) and introduced into S. pyogenes with selection for resistance to kanamycin. Recombination between homologous sequences (indicated by the lines between introduced DNA and chromosome) results in the exchange of the ΩKm-2-inactivated allele for the chromosomal allele (shown below the arrow). Abbreviations: E, EcoRI; Sm, SmaI; B, BamHI; H, HindIII; Ev, EcoRV.

ment of the gene introduced on the plasmid and the identical sequence on the streptococcal chromosome. The resulting chromosomal structure is generated by integration of the entire plasmid and consists of a partial duplication of the target gene, which now flanks the integrated plasmid sequences (Fig. 2A). As a consequence of using only an internal fragment of the target gene, one of the duplicated copies will lack its 3′ end and the other will lack its 5′ end; thus, both partial copies should be inactive. The end result is the directed insertional inactivation of the target gene. A number of modified *E. coli* plasmids have been developed to simplify this mutagenesis technique (33, 39, 46).

The principal advantage of this method is that single recombination events occur at a higher frequency than do double recombination events. In addition, this approach requires the cloning of a much smaller segment of the chromosome and can work with fragments as small as 0.5 kb. However, there are several disadvantages of the method. These include the fact that the mutations are not necessarily stable, because a second recombination event between the duplicated gene segments can result in excision of the plasmid vector and regeneration of a wild-type structure. Also, the mutations generated by this technique are strongly polar. However, it is possible to modify the technique to address the issue of polarity directly. In this case, an additional control strain is constructed: instead of cloning a segment internal to the target gene, a segment is cloned that has its 5′ end anchored within the gene but its 3′ end anchored at a location distal to the 3′ end of the coding region of the gene. Integration of this construction into the target gene in a wild-type strain also results in a partial duplication. However, because the cloned segment overlaps the 3′ end of the gene, the first copy of the gene will be regenerated and the integrated vector will be located adjacent to this intact copy such that it will still be polar on expression of any distal genes (Fig. 2B). If the phenotype under analysis is the result of a polar effect and the loss of expression of a distal gene, then this control strain should also demonstrate the mutant phenotype, even though it has an intact and functional copy of the target gene. On the other hand, an unaltered wild-type phenotype in the control strain indicates that the mutant phenotype is solely the result of insertional inactivation of the target gene.

An additional modification of this method can be used to map the promoter and *cis*-acting control regions of a target operon. This is based on the technique developed by Piggot and colleagues (37a) for analysis of promoters in *Bacillus subtilis* and is similar in concept to the method described above for placing a polar insertion downstream of a target gene. However, in this technique, the segment of DNA cloned into the integrational vector is anchored within the coding region at its 3′ end but is anchored upstream of the start of the gene at its 5′ end. Integration of this construct into the target locus will again generate a duplication. However, since the 3′ end of the cloned segment ends within the gene, the first copy will be truncated and will be inactive. This is followed by the integrated vector and then by the second intact copy of the target gene, which is now preceded exactly by the 5′ flanking region cloned into the vector. If this 5′ flanking region includes the promoter and other *cis*-acting control regions, the target gene will be expressed. If this segment lacks these elements, the target gene will not be expressed (Fig. 2C). By using a nested set of insertions containing different lengths of the upstream control region, it is possible to map the sequences required for expression and regulation of the target gene with a high degree of precision. This method has been used to examine regulation of *mga*, the regulator of the genes that encode the M proteins and the C5a peptidase (33).

In-Frame Deletion

An in-frame deletion mutation has the advantage that it is both stable and nonpolar. It has the disadvantage that it is somewhat more time-consuming to construct. In-frame deletions have now been constructed in the genes for many different *S. pyogenes* virulence factors, including the C5a peptidase (22), the M protein, the cysteine proteinase, the *has* operon (2), and streptolysin O (42). Most in-frame deletion mutations have been constructed by the method pioneered in the Cleary lab (22) that utilizes derivatives of the temperature-sensitive plasmid pG+host4. In this method, the gene of interest is cloned into the vector in *E. coli*. Molecular techniques are then used to delete a large central region of the gene while preserving its reading frame. The resulting construct is introduced into an *S. pyogenes* host at a temperature that is permissive for replication of the plasmid. The culture is then shifted to the nonpermissive temperature while maintaining selection for the resistance determinant of the plasmid. This selects for chromosomes into which the nonreplicating plasmid has inserted by homologous recombination between the in-frame deletion allele and the resident wild-type allele. At this stage the chromosome will contain both the wild-type and deletion alleles. At some frequency, a second homologous recombination will occur that will result in the excision of the integrated plasmid. Depending on the recombination junctions, either the wild-type or the deletion allele will remain in the chromosome (Fig. 3). Shifting the culture back to the permissive temperature enriches for chromosomes from which the plasmid has excised, since replication of the integrated plasmid creates a second origin of replication for the chromosome that is usually deleterious to growth. The culture is once again shifted to the nonpermissive temperature, but this time without selection for the plasmid. This enriches for segregants that have lost the excised and now nonreplicating plasmid. These segregants are plated as single colonies, which are then examined by PCR to identify chromosomes that contain the in-frame deletion allele. For a detailed description of this technique, see Ji et al. (22).

TRANSPOSON MUTAGENESIS

The techniques described above have been extremely useful for the analysis of virulence in *S. pyogenes*. However, they require a detailed preexisting knowledge of the target gene. While the available genomic information is very helpful for analysis of the chromosomal structure of known genes and for identification of homologs of known virulence genes from other organisms, the directed mutagenesis approach is most successfully applied to testing hypotheses that address the functions of previously identified genes. It is not well suited for the identification of novel genes that contribute to virulence. This is most successfully done using a traditional genetic approach, in which a virulence phenotype is established and random mutations are generated to identify genes that influence this phenotype. In *S. pyogenes*, the most successful method of random muta-

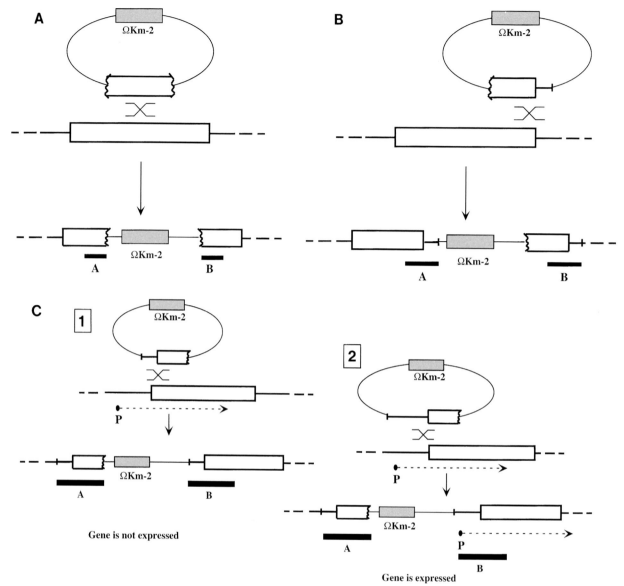

FIGURE 2 Directed insertional mutagenesis of targeted genes. (A) A DNA segment internal to the targeted gene (shown by the box enclosed by wavy lines) is cloned onto an *E. coli* plasmid that cannot replicate in *S. pyogenes*. The plasmid is introduced into an *S. pyogenes* host as a circular molecule (top of figure) with selection for a resistance marker on the plasmid (ΩKm-2). A single homologous recombination event between the chromosome and the circular molecule (shown by the X) results in the integration of the plasmid into the chromosome and a partial duplication of the gene (the solid bars labeled A and B represent the duplicated segment) in which neither of the two copies is complete. (B) Generation of a polar insertion 3′ to the target gene. If the segment of DNA cloned on the integrational plasmid contains sequences that include the 3′ terminus of the target gene, the resulting structure will also contain a partial duplication of the gene (the solid bars labeled A and B represent the duplicated segment), but in this case the 5′ copy will be intact and will now be flanked at its 3′ end by a polar element. This strategy can be used to test if the insertion generated in the target gene (see above) is polar on distal genes. (C) Mapping the *cis*-acting control regions of the target gene. In this strategy, the technique is modified by including different regions of DNA extending 5′ to the target gene. The plasmid is integrated into the target locus as described above, and the end product is also a partial duplication of the target gene (the solid bars labeled A and B represent the duplicated segment); however, it is the distal copy that is intact. If this region does not include the *cis*-acting control regions (represented by the broken arrow and the closed circle labeled P), the distal intact copy will not be expressed (scenario labeled 1). In contrast, if the cloned segment includes the *cis*-acting control regions, then the distal copy will be expressed (scenario labeled 2).

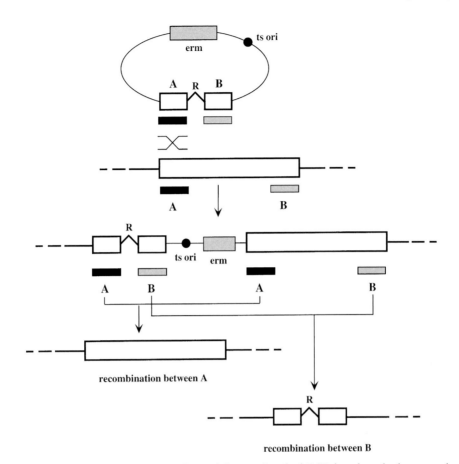

FIGURE 3 Construction of an in-frame deletion. Standard PCR-based methods are used to generate a deletion of the internal region of a copy of the target gene that has been cloned on an *E. coli*-streptococcal shuttle vector that is temperature sensitive for replication (ts ori). The deletion is constructed to maintain the reading frame of the gene (the bent line connecting the 5′ region labeled A and the 3′ region labeled B). After its introduction into *S. pyogenes*, growth at a temperature nonpermissive for replication of the plasmid with selection for the antibiotic-resistant determinant of the plasmid (erm) selects for chromosomes in which the plasmid has integrated by homologous recombination (X). The two regions of homology flanking the deletion are represented by the solid and gray bars labeled A and B. Recombination between the A regions is shown, and the product is shown below the first arrow. A second homologous recombination event can occur between the 5′ homologous regions (labeled A) or the 3′ homologous regions (labeled B), which results in excision of the plasmid and either restoration of the wild-type structure or replacement by the deletion allele (these products are illustrated below the second set of arrows). Growth at a temperature permissive for replication of the plasmid enriches for chromosomes from which the plasmid has been excised. Presence of the wild-type or deletion allele in any one isolate is easily determined by assay for the unique restriction site engineered into the deletion allele (indicated by the R above the bent line).

genesis has used the insertion of a transposable element at multiple loci in the chromosome. This section considers the various types of transposons that have been successfully used to identify novel genes in *S. pyogenes*.

Tn*916*

The first transposable element used to identify novel virulence genes in *S. pyogenes* was Tn*916* (Fig. 4). This element is the prototype of the family of transposons discovered by Clewell's group that are now known as the conjugative transposons (16a). Conjugative transposons have the remarkable property that they are self-transmissible and transpose from a locus in a donor chromosome to a different locus in a recipient chromosome,

when these chromosomes are located in different cells. The conjugal transfer event requires cell-cell contact and can occur between very distantly related species and even between gram-positive and gram-negative hosts. The biology of these unique elements has been extensively investigated and is the subject of several excellent reviews (for example, reference 43), and so will only be briefly mentioned here. The transposon moves by an excision-insertion-type mechanism in which the first step is the excision of the element from the donor locus by a λ-like pathway that involves a recombination event between sequences adjacent to the ends of the element. However, in contrast to movement of λ, in which the substrate sequences must be identical, the substrate sequences used by Tn*916* are almost always non-

Tn*916*

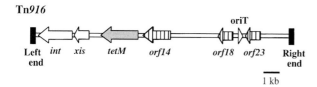

Tn*917*-LTV3

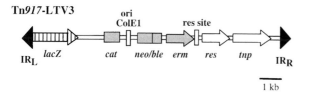

TnSpc (Tn*4001*.Spc)

FIGURE 4 Transposon mutagenesis of *S. pyogenes*. The several transposons that have been used for mutagenesis of *S. pyogenes* are shown. Antibiotic-resistance genes are represented by the gray bars; transposon ends and/or terminal inverted repeats are shown in black; genes that are essential for transposition are shown by white bars, and genes that are nonessential for transposition are shown by striped bars. Tn*916* (GenBank accession no. U09422) is the prototype conjugative transposon and contains at least 24 different open reading frames. Of these, one is required for resistance to tetracycline, two are essential for transposition, and the rest are likely involved in conjugal transfer. Tn*917*-LTV3 is a highly engineered derivative of the Tn*3*-like transposon Tn*917*. This element transposes via a replicative mechanism and has been modified to include a promoterless *lacZ* reporter gene to generate random transcriptional fusions and an *E. coli* ColE1 plasmid origin of replication to facilitate the cloning and analysis of inactivated loci. TnSpc is a derivative of the Tn*5*-like transposon Tn*4001*. This element transposes via a "cut-and-paste" mechanism and consists of the left and right inverted repeats and transposase of IS*256* and a spectinomycin-resistance gene.

homologous. The circular intermediate is nicked at a discrete site, and a single strand is directionally transferred into the donor cell. Transfer requires direct cell-cell contact and may involve some degree of zygote formation. This may explain that, while Tn*916* does not appear to mobilize chromosomal markers, transfer of unlinked chromosomal markers can occur.

The ability of the recombination reaction to occur between nonhomologous sequences allows the element to utilize a large number of different sites for insertion. However, there are preferred target sites for insertion. These preferred sites do not necessarily share the same specific sequence but likely share a similar conformation and are the sites of bent DNA. Interestingly, of the insertions in *S. pyogenes* that have identified virulence genes, a large percentage are not in the coding region of the genes but have been located 5' to the coding region in the transcriptional control region of the genes. From a practical standpoint, mutagenesis with Tn*916* requires a donor strain and an antibiotic-resistant derivative of the *S. pyogenes* host of interest. The donors can be distantly related to *S. pyogenes* and are most often either *E. faecalis* or *B. subtilis*, and it is

usually a simple procedure to select a spontaneous high-level streptomycin-resistant mutant of the *S. pyogenes* strain under analysis. (For a detailed method for Tn*916* mutagenesis of *S. pyogenes*, consult Caparon and Scott [7].)

The advantage of using Tn*916* is that it can be applied to probably any isolate of *S. pyogenes*, including those that may be difficult to transform. The disadvantages are that Tn*916* is a large element for a transposon (over 18 kb in size), does transpose into preferred sites, and frequently generates sites with multiple insertions. This latter issue can often be resolved through analysis of the resulting mutants by transduction (see above). In addition to mutagenesis, Tn*916* has been engineered as a vector for the introduction of foreign DNA into *S. pyogenes*, including promoter-reporter gene fusions for the analysis of gene expression. The transposon has served as a useful mutagenesis agent and has been used to identify novel genes involved in regulation, e.g., *mga* (6), expression of streptolysin S (3, 25, 32), and production of the capsule (10, 49), among others.

Tn*917*

Tn*917* was first described by Clewell's group in *E. faecalis* (46a) and has been highly engineered by Youngman and colleagues (4a) and applied with great success to the mutagenesis of *Listeria monocytogenes* and *B. subtilis*. This transposable element is a member of the large Tn*3* family, which also includes the δγ element of *E. coli* discussed above. The members of this family contain fairly large terminal repeats, but they are much smaller than the conjugative transposons. They produce stable insertions, generate a 5-bp target-site duplication upon insertion, and transpose at high frequency and with a high degree of randomness. Their movement involves a replicative pathway, which requires that donor and recipient molecules undergo replication, and proceeds through the cointegration of donor and host molecules, which contain two directly repeated copies of the transposon at the fusion junctions. Resolution of the cointegrate occurs via site-specific recombination catalyzed by a transposon-encoded enzyme called resolvase that recognizes a specific site in each copy of the transposon (the *res* site). The end result is that a single copy of the transposon is located at a random site in the recipient molecule.

The requirement for replication means that the transposon must be delivered into the target *S. pyogenes* host on a replicating plasmid. A technical problem is that the plasmid must have a counterselectable marker so that chromosomes that have the transposon inserted can be distinguished from hosts in which the transposon remains on the plasmid. This problem was solved with the advent of temperature-sensitive plasmid vectors for *S. pyogenes*. Eichenbaum and Scott (12) developed a mutagenesis vector (pJRS290) that is based on pG+host4 and contains a highly engineered derivative of Tn*917* developed by Youngman and coworkers (Tn*917*-LTV3) (4a) (Fig. 4). This derivative of Tn*917* has several useful features, including an *E. coli* plasmid origin of replication that facilitates the direct cloning of chromosomal DNA adjacent to the inserted transposon. Using this vector, Tn*917*-LTV3 was found to transpose efficiently and with a high degree of randomness in *S. pyogenes*.

Tn*4001*

The transposable element Tn*4001* was originally isolated as an agent of transmissible gentamicin resistance in *Staph-*

ylococcus aureus. It is a composite-type transposon and a member of the Tn*5* family. These transposable elements share a common structure in which directly repeated copies of an insertion sequence flank an antibiotic-resistance gene. Each insertion sequence itself is bounded by a short inverted repeat encoded by the transposase gene, which is the only transposon-encoding gene required for movement. The insertion sequences are usually capable of transposition independent of the entire element. For Tn*4001*, two copies of IS*256* flank a central gentamicin-resistance gene. The pathway of transposition of this class of elements involves a cut-and-paste mechanism catalyzed by transposase, which recognizes as its substrates the short inverted repeats of the insertion sequences. Studies have shown that Tn*4001* has a broad host range, including staphylococci, oral streptococci, and mycoplasmas, and that it chooses its targets for insertion with a high degree of randomness. Since the cut-and-paste pathway does not require replication of the donor molecule, the element can easily be delivered into the host of choice via a nonreplicating suicide vector. However, a number of problems have limited its use. The most significant problem is that transposition of the entire element occurs at a somewhat low frequency relative to the frequency of transposition of the individual insertion sequences. As a result, a chromosome with a Tn*4001* element will frequently also have multiple copies of IS*256* inserted at other loci. To address this problem, derivatives of Tn*4001* have been constructed that essentially contain only the inverted repeats of IS*256* oriented to flank the transposase gene and a selectable marker (26). The resulting elements are relatively small in size (~2 kb) and transpose at a high frequency characteristic of an individual copy of IS*256*. The organization of the element also prevents independent transposition of the insertion sequence and ensures that the resulting population of insertions is homogeneous, although a significant percentage of the resulting mutants may have insertions at two different loci. A spectinomycin-resistant version of this transposon (called TnSpc for simplicity) (Fig. 4) was used to identify three novel genes required for expression of the cysteine protease of *S. pyogenes,* including a transcriptional regulator (*ropB*) and a chaperone (*ropA*). Furthermore, since the terminal inverted repeats are quite short (26 bp for IS*256*), it is likely that specialized derivatives of the transposon can be developed adjacent to an end that will allow the translational fusion of the inserted target to the reporter.

Analysis of Transposon Mutants

Regardless of the type of transposable element used, a number of additional tests should be performed to ensure that the mutant phenotype is the direct result of insertion of the transposon. This involves determining the sequence of the locus into which the transposon has inserted. Since the transposons commonly used in *S. pyogenes* contain antibiotic-resistance markers that can also be used for selection in *E. coli,* it is relatively simple to obtain a clone of the insertion locus. This is done by construction in *E. coli* of a plasmid library of the chromosome of the mutant with direct selection for the transposon-encoded marker. It is also possible to obtain sequences flanking the inserted transposon by using one of a number of inverse and vectorette PCR techniques (for an example, see Lyon et al. [26]). The availability of significant amounts of genomic sequence means that it is usually only necessary to generate a small amount of sequence data that can then be compared with the genome sequence database to obtain large amounts of sequence information. With these data in hand, the next step is to construct a mutant in the locus in the wild-type parental strain using one of the strategies for directed mutagenesis outlined above. If the original insertion is responsible for the mutant phenotype, then it is expected that all isolates obtained via directed mutagenesis will also be mutants. It should also be kept in mind when interpreting results that transposon-generated mutations are polar. It is also possible to conduct a linkage analysis of Tn*916* insertions without first cloning the locus by transducing the insertion locus into a wild-type strain (see above). In the case of TnSpc, it is possible to use PCR techniques to delete transposase after the insertionally inactivated locus has been cloned into *E. coli* (26). This prevents the element from further transposition and allows the mutant locus to be crossed back into a wild-type background using the strategy of allelic replacement via linear DNA described above.

ANALYSIS OF GENE EXPRESSION

Gene regulation phenomena play a key role in pathogenesis. The interaction between the host and the microbe is dynamic and often is a progression through a number of discrete steps. Each of these steps is characterized by the expression of specific bacterial genes required for survival and multiplication and the host's response to the action of the products of these microbial genes. As a consequence, the microorganism is continually challenged to adapt to new and changing environmental conditions. Pathogens have taken advantage of the dynamic nature of this interaction and have evolved to recognize changes in specific environmental conditions as markers that define a particular host compartment or stage of infection. The pathogen makes use of this information to modulate expression of virulence genes required for survival in the host compartment. Thus, an understanding of the in vitro conditions that regulate expression of a specific virulence gene provides insight into how the gene contributes to virulence in vivo (for review see reference 30). Regulation of virulence genes often involves control at the level of transcription. For *S. pyogenes,* transcriptional regulation has often been analyzed through the use of reporter genes.

The basic strategy for using reporter genes to analyze gene expression involves fusing the promoter for the virulence gene of interest to a reporter gene that encodes a gene whose product can easily be quantified. This feature is of particular utility when the assays for quantification of the product of the streptococcal gene under analysis are time-consuming, expensive, or cumbersome. The reporter gene lacks its own promoter, so its product becomes an accurate relative indicator of the steady-state level of transcription initiation from the target promoter. It should be kept in mind that because the message is a chimera between the initiation signals of the target gene and the reporter gene, it will not be subject to the same posttranscriptional and posttranslational controls. Furthermore, the half-life of the reporter gene's translation product will also not reflect that of the native polypeptide. Thus, reporter genes are most appropriately used to quantify the strength of initiation of transcription of the target promoter.

The reporter genes that have been successfully used in *S. pyogenes* include those encoding chloramphenicol acetyltransferase (Cat), β-galactosidase (Lac), β-glucuronidase (Gus), and alkaline phosphatase (Pho). The use of luciferase has recently been described (39a). These gene products are stable, highly active enzymes that are highly specific for their substrates. Their utility is enhanced by the availability of synthetic substrates with isotopic, colorimetric, fluorogenic, or light-emitting properties that allow very sensitive detection. Also, the reporter genes encoding these enzymes are derived from gram-positive bacteria or have been extensively modified to optimize their expression in gram-positive hosts and differ extensively from their counterparts used for analysis in *E. coli*. The most successful applications of Cat have employed *cat86*, originally isolated from *Bacillus pumilis* that had been modified by the deletion of an attenuator sequence and the substitution of an ATG start codon for the native gene's TTG start codon. The advantages of Cat are that *cat86* is very stably maintained in *S. pyogenes* and that the assays for its enzymatic acetylation of chloramphenicol are very sensitive (5, 29). Its disadvantages are that the most sensitive assays involve radioactivity, are expensive, are somewhat labor-intensive relative to those for other reporter genes, and require the preparation of cell-free extracts. Both Lac and Gus utilize *E. coli* genes that have been modified by the introduction of ribosome binding sites recognized by gram-positive bacteria (14, 23). The advantages of using these as reporter genes are that assays for their enzymatic activities are rapid and easy to perform; many different samples can be analyzed simultaneously; a wide variety of different substrates are available; and it is possible to analyze both permeabilized cells and cell extracts. Their disadvantages are that they are not as sensitive or as stably maintained in *S. pyogenes* as is *cat86*.

More recently, a chimeric reporter protein based on fusion of two naturally secreted proteins was developed. This reporter contains as its enzymatic partner the alkaline phosphatase of *E. faecalis* (PhoZ) (41) (Fig. 5). This enzyme is secreted from *E. faecalis* and is a lipoprotein that becomes tethered to the outer leaflet of the cell membrane. The enzymatic C-terminal domain, not including its lipoprotein secretion signal domain, is fused to the N-terminal domain of protein F of *S. pyogenes*. Protein F is a cell wall-associated fibronectin-binding adhesin whose N-terminal domain contains the sequence that targets the protein for secretion; its central domain contains the regions responsible for binding to fibronectin; and its C-terminal domain is responsible for targeting the protein to a sorting pathway that covalently couples a sequence near its C terminus to the cell wall (34). Fusion of the N-terminal secretion domain of protein F to the C-terminal enzymatic domain of PhoZ results in a very stable, highly enzymatically active chimeric protein that—because it lacks the postsecretion attachment domains of either protein—is freely secreted from the streptococcal cell into the surrounding medium (18). This feature makes quantitative analysis of the secreted chimeric protein very simple, requiring neither permeabilization nor preparation of cytoplasmic extracts. Because of the widespread popularity of alkaline phosphatase enzymes in a large number of histological and biochemical techniques, there is an excellent selection of sensitive substrates with a variety of useful characteristics, including those with fluorescent and light-emitting properties.

Numerous strategies have been used to introduce reporter fusion constructs into *S. pyogenes* for analysis. Plasmid vectors have the advantages that they are easy to manipulate and can be used to generate and analyze a large number of permutations and variations of promoter structure to probe the *cis*- and *trans*-acting control regions. A plasmid-based reporter will be present in multiple copies, and this amplification will result in increased sensitivity in detecting the activity of the promoter under analysis. This latter feature can also make plasmid-based systems sensitive to multiple-copy-number-derived artifacts, such as titration of *trans*-acting regulatory components, when these components are present in limiting quantities. Furthermore, any sequence-directed local effects of chromosomal structure on expression of a given promoter will not likely be replicated in a plasmid environment. Integration of reporter constructs into the chromosome can alleviate many of these latter types of potential artifacts but with a trade-off in ease of use and sensitivity. Integration has most often been performed by introducing the reporter construct on a nonreplicating *E. coli*-based plasmid to target integration into the target gene's native locus by homologous recombination promoted by the cloned promoter segment. The use of a reporter gene can be combined with the recombinational strategy described above for the analysis of the *cis*-acting regions of a promoter. Other reporter gene delivery vectors have included Tn916-based shuttle transposons and the incorporation of a promoterless *lacZ* near the end of Tn917 for the generation of libraries with fusions in genes generated at random. Reporter genes have been successfully applied to the analysis of expression of many different *S. pyogenes* genes, including *mga* (17, 33), *rofA* (15), various *emm* genes (5, 29, 38), *prtF* (15, 48), and the *has* operon (1a).

Techniques for Heterologous Expression

Structure-function studies on the role of putative virulence factors require techniques for heterologous expression, both to control the timing and relative levels of expression of a given gene and to study the structure-function relationships of specific domains of a given virulence-associated protein. The former case requires the availability of a heterologous promoter whose activity can be tightly, easily, and quantitatively controlled by some external factor. The latter case requires a method for the expression of various domains of the polypeptide under analysis in the context of an unrelated polypeptide. This allows an independent analysis of the functionality of that domain separate from other regions of the original polypeptide. Methods to accomplish both these strategies of heterologous expression have been developed for analysis of virulence in *S. pyogenes*.

An approach for using a regulated heterologous promoter to direct expression of the gene under analysis has been developed that is based on the *nisA* promoter of *Lactococcus lactis*. In lactococci, this promoter controls the expression of a large multigene operon involved in the synthesis of the lantibiotic nisin. An interesting feature of the control of the *nisA* promoter is that its activity is tightly regulated and is stimulated in response to nisin. The signal transduction pathway is well characterized and includes the histidine protein kinase NisK and cognate response regulator. The method for using this regulated promoter involves the introduction of *nisK* and *nisR* into an *S. pyogenes* strain under the control of the constitutively active *nisR* promoter (11). The gene of interest is introduced under control of the *nisA* promoter, which now makes expression of the gene of interest sensitive to the concentration of exogenous nisin that is added to the cul-

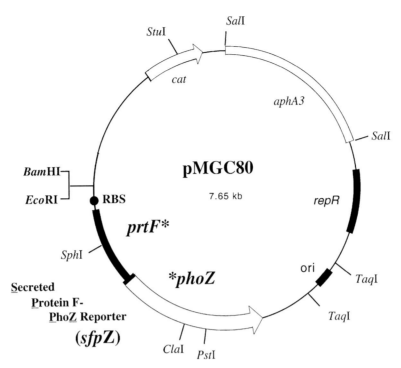

FIGURE 5 A chimeric secreted alkaline phosphatase reporter gene. The plasmid pMGC80 is a pWV01-based *E. coli*-streptococcal shuttle vector whose genes encode resistance to chloramphenicol (*cat*) and kanamycin (*aphA3*). The plasmid contains a promoterless reporter gene formed by the fusion of the N-terminal region of the cell wall-associated protein F (*prtF**) to the enzymatic domain of the enterococcal alkaline phosphatase (*phoZ**). Since the chimeric protein (secreted protein F-PhoZ reporter, or SfpZ) lacks the C-terminal cell wall attachment domain of protein F and the N-terminal lipoprotein tethering domain of PhoZ, it is freely secreted from the cell. The SfpZ chimera retains the enzymatic activity of PhoZ and is easily quantified in culture supernatants. Restriction sites for *Bam*HI and *Eco*RI can be used to place the promoter of interest in an orientation to direct transcription of *spfZ*, which is then translated using the ribosome binding site of *prtF* (RBS).

ture. Tightly regulated expression of a *gusA* reporter gene and induction of up to 60-fold over background using a broad range of nisin concentrations have been reported for *S. pyogenes* (11).

The ability to express a domain of a target protein within the context of a heterologous protein is highly desirable for the structure-function analysis of adhesins. A technique for this strategy that uses the framework of the M6.1 protein has been developed. It is based on the work of Pozzi and coworkers (39b), who demonstrated that the central domain of the M protein could be replaced with a heterologous sequence without affecting expression and surface presentation of the resulting chimera. Essentially, the N-terminal secretion and C-terminal attachment domains are used to display the introduced heterologous domain on the streptococcal cell surface. A successful application of this strategy has been the analysis of the two distinct fibronectin-binding domains of protein F (34). For a more detailed explanation of this technique, see Hanski et al. (19).

CONCLUDING REMARKS

Rapid progress has been made in recent years in the development of sophisticated techniques for genetic analysis in *S. pyogenes*. Much of this effort has been directed at the

development of methods for the mutagenesis of known genes. Much progress has also been made in the development of strategies for the identification of novel genes. It is likely that the widespread application of these techniques to the virulence properties of *S. pyogenes* will enrich our understanding of streptococcal pathogenesis with insight at the molecular level and will help to establish and clarify the contributions of specific genes. Additional use and development of methods for analysis of gene expression and heterologous expression will allow analyses of virulence factors at much higher levels of resolution than was previously possible.

REFERENCES

1. **Advanced Center for Genome Technology, University of Oklahoma.** *Streptococcus pyogenes* Genome Sequencing Project. http://www.genome.ou.edu/strep.html

1a. **Alberti, S., C. D. Ashbaugh, and M. R. Wessels.** 1998. Structure of the *has* operon promoter and regulation of hyaluronic acid capsule expression in group A streptococcus. *Mol. Microbiol.* **28:**343–353.

2. **Ashbaugh, C. D., H. B. Warren, V. J. Carey, and M. R. Wessels.** 1998. Molecular analysis of the role of the group A streptococcal cysteine protease, hyaluronic acid capsule

and M protein in a murine model of human invasive soft-tissue infection. *J. Clin. Invest.* **102:**550–560.

3. Betschel, S. D., S. M. Borgia, N. L. Barg, D. E. Low, and J. C. S. De Azavedo. 1998. Reduced virulence of group A streptococcal Tn916 mutants that do not produce streptolysin S. *Infect. Immun.* **66:**1671–1679.

4. Biswas, I., A. Gruss, S. D. Ehrlich, and E. Maguin. 1993. High-efficiency gene inactivation and replacement system for gram-positive bacteria. *J. Bacteriol.* **175:**3628–3635.

4a. Camilli, A., D. A. Portnoy, and P. Youngman. 1990. Insertional mutagenesis of *Listeria monocytogenes* with a novel Tn917 derivative that allows direct cloning of DNA flanking transposon insertion. *J. Bacteriol.* **172:**3738–3744.

5. Caparon, M. G., R. T. Geist, J. Perez-Casal, and J. R. Scott. 1992. Environmental regulation of virulence in group A streptococci: transcription of the gene encoding M protein is stimulated by carbon dioxide. *J. Bacteriol.* **174:**5693–5701.

6. Caparon, M. G., and J. R. Scott. 1987. Identification of a gene that regulates expression of M protein, the major virulence determinant of group A streptococci. *Proc. Natl. Acad. Sci. USA* **84:**8677–8681.

7. Caparon, M. G., and J. R. Scott. 1991. Genetic manipulation of the pathogenic streptococci. *Methods Enzymol.* **204:**556–586.

8. Clewell, D. B., Y. Yagi, G. M. Dunny, and S. K. Schultz. 1974. Characterization of three plasmid deoxyribonucleic acid molecules in a strain of *Streptococcus faecalis*: identification of a plasmid determining erythromycin resistance. *J. Bacteriol.* **117:**283–289.

9. DeAngelis, P. L., J. Papaconstantinou, and P. H. Weigel. 1993. Isolation of a *Streptococcus pyogenes* gene locus that directs hyaluronan biosynthesis in acapsular mutants and in heterologous bacteria. *J. Biol. Chem.* **268:**14568–14571.

10. Dougherty, B. A., and I. van de Rijn. 1992. Molecular characterization of a locus required for hyaluronic acid capsule production in group A streptococci. *J. Exp. Med.* **175:**1291–1299.

10a. Dunny, G. M., L. N. Lee, and D. J. LeBlanc. 1991. Improved electroporation and cloning vector system for gram-positive bacteria. *Appl. Environ. Microbiol.* **57:**1194–1201.

11. Eichenbaum, Z., M. J. Federle, D. Marra, W. M. de Vos, O. P. Kuipers, M. Kleerebezem, and J. R. Scott. 1998. Use of the lactococcal *nisA* promoter to regulate gene expression in gram-positive bacteria: comparison of induction level and promoter strength. *Appl. Environ. Microbiol.* **64:**2763–2769.

12. Eichenbaum, Z., and J. R. Scott. 1997. Use of Tn917 to generate insertion mutatins in the group A streptococcus. *Gene* **186:**213–217.

13. Falkow, S. 1988. Molecular Kock's postulates applied to microbial pathogenicity. *Rev. Infect. Dis.* **10:**S274–S276.

14. Ferrari, E., D. J. Henner, M. Perego, and J. A. Hoch. 1988. Transcription of *Bacillus subtilis* subtilisin and expression of subtilisin in sporulation mutants. *J. Bacteriol.* **170:**289–295.

15. Fogg, G. C., and M. G. Caparon. 1997. Constitutive expression of fibronectin binding in *Streptococcus pyogenes* as

a result of anaerobic activation of *rofA*. *J. Bacteriol.* **179:**6172–6180.

16. Fogg, G. C., C. M. Gibson, and M. G. Caparon. 1993. Identification of *rofA*, a positive-acting regulatory component of *prtF* expression: use of a mγδ-based shuttle mutagenesis strategy in *Streptococcus pyogenes*. *Mol. Microbiol.* **11:**671–684.

16a. Gawron-Burke, C., and D. B. Clewell. 1982. A transposon in *Streptococcus faecalis* with fertility properties. *Nature* **300:**281–284.

17. Geist, R. T., N. Okada, and M. G. Caparon. 1993. Analysis of *Streptococcus pyogenes* promoters by using novel Tn916-based shuttle vectors for the construction of transcriptional fusions to chloramphenicol acetyltransferase. *J. Bacteriol.* **175:**7561–7570.

18. Granok, A., and M. Caparon. Unpublished data.

19. Hanski, E., G. Fogg, A. Tovi, N. Okada, I. Burstein, and M. Caparon. 1994. Molecular analysis of *Streptococcus pyogenes* adhesion. *Methods Enzymol.* **253:**269–305.

20. Hanski, E., P. A. Horwitz, and M. G. Caparon. 1992. Expression of protein F, the fibronectin-binding protein of *Streptococcus pyogenes* JRS4, in heterologous streptococcal and enterococcal strains promotes their adherence to respiratory epithelial cells. *Infect. Immun.* **60:**5119–5125.

21. Husmann, L. K., D. L. Yung, S. K. Hollingshead, and J. R. Scott. 1997. Role of putative virulence factors of *Streptococcus pyogenes* in mouse models of long-term throat colonization and pneumonia. *Infect. Immun.* **65:**1422–1430.

22. Ji, Y., L. McLandsborough, A. Kondagunta, and P. P. Cleary. 1996. C5a peptidase alters clearance and trafficking of group A streptococci by infected mice. *Infect. Immun.* **64:**503–510.

23. Karow, M. L., and P. J. Piggot. Construction of *gusA* transcriptional fusion vectors for *Bacillus subtilis* and their utilization for studies of spore formation. *Gene* **163:**69–74.

24. Leenhouts, K. J., B. Tolner, S. Bron, J. Kok, G. Venema, and J. F. Seegers. 1991. Nucleotide sequence and characterization of the broad-host-range lactococcal plasmid pWVO1. *Plasmid* **26:**55–66.

25. Liu, S., S. Sela, G. Cohen, J. Jadoun, A. Cheung, and I. Ofek. 1997. Insertional inactivation of streptolysin S expression is associated with altered riboflavin metabolism in *Streptococcus pyogenes*. *Microb. Pathog.* **22:**227–234.

26. Lyon, W. R., C. M. Gibson, and M. G. Caparon. 1998. A role for trigger factor and an rgg-like regulator in the transcription, secretion and processing of the cysteine proteinase of *Streptococcus pyogenes*. *EMBO J.* **17:**6263–6275.

27. Maguin, E., P. Duwat, T. Hege, D. Ehrlich, and A. Gruss. 1992. New thermosensitive plasmid for gram-positive bacteria. *J. Bacteriol.* **174:**5633–5638.

28. Malke, H. 1969. Transduction of *Streptococcus pyogenes* K 56 by temperature-sensitive mutants of the transducing phage A 25. *Z. Naturforsch. B* **24:**1556–1561.

29. McIver, K. S., A. S. Heath, and J. R. Scott. 1995. Regulation of virulence by environmental signals in group A streptococci: influence of osmolarity, temperature, gas exchange, and iron limitation on *emm* transcription. *Infect. Immun.* **63:**4540–4542.

29a. McShan, W. M., R. E. McLaughlin, A. Norstrand, and J. J. Ferretti. Vectors containing streptococcal bacteriophage integrases for site-specific gene insertion. *Methods Cell Sci.*, in press.

30. **Mekalanos, J. J.** 1992. Environmental signals controlling expression of virulence determinants in bacteria. *J. Bacteriol.* **174**:1–7.

31. **Moynet, D. J., A. E. Colon-Whitt, G. B. Calandra, and R. M. Cole.** 1985. Structure of eight streptococcal bacteriophages. *Virology* **142**:263–269.

32. **Nida, K., and P. P. Cleary.** 1983. Insertional inactivation of streptolysin S expression in *Streptococcus pyogenes*. *J. Bacteriol.* **155**:1156–1161.

33. **Okada, N., R. T. Geist, and M. G. Caparon.** 1993. Positive transcriptional control of *mry* regulates virulence in the group A streptococcus. *Mol. Microbiol.* **7**:893–903.

34. **Ozeri, V., A. Tovi, I. Burstein, S. Natanson-Yaron, M. G. Caparon, K. M. Yamada, S. K. Akiyama, I. Vlodavsky, and E. Hanski.** 1996. A two-domain mechanism for group A streptococcal adherence through protein F to the extracellular matrix. *EMBO J.* **15**:898–998.

35. **Perez-Casal, J., M. G. Caparon, and J. R. Scott.** 1991. Mry, a trans-acting positive regulator of the M protein gene of *Streptococcus pyogenes* with similarity to the receptor proteins of two-component regulatory systems. *J. Bacteriol.* **173**:2617–2624.

36. **Perez-Casal, J., M. G. Caparon, and J. R. Scott.** 1992. Introduction of the *emm6* gene into an *emm*-deletion strain of *Streptococcus pyogenes* restores its ability to resist phagocytosis. *Res. Microbiol.* **143**:549–558.

37. **Perez-Casal, J., E. Maguin, and J. R. Scott.** 1993. An M protein with a single C repeat prevents phagocytosis of *Streptococcus pyogenes*: use of a temperature-sensitive shuttle vector to deliver homologous sequences to the chromosome of *S. pyogenes*. *Mol. Microbiol.* **8**:809–819.

37a.**Piggot, P. J., C. A. M. Curtis, and H. DeLencastre.** 1984. Use of integrational plasmid vectors to demonstrate the poly-cistronic nature of a transcriptional unit (*spoIIA*) required for sporulation of *Bacillus subtilis*. *J. Gen. Microbiol.* **120**:2123–2136.

38. **Podbielski, A., J. A. Peterson, and P. P. Cleary.** 1992. Surface protein-CAT reporter fusions demonstrate differential gene expression in the *vir* regulon of *Streptococcus pyogenes*. *Mol. Microbiol.* **6**:2253–2265.

39. **Podbielski, A., B. Spellerberg, M. Woischnik, B. Pohl, and R. Lutticken.** 1996. Novel series of plasmid vectors for gene inactivation and expression analysis in group A streptococci (GAS). *Gene* **177**:137–147.

39a.**Podbielski, A., M. Woischnik, B. A. B. Leonard, and K.-H. Schmidt.** 1999. Characterization of *ura*, a global negative regulator gene in group A streptococci. *Mol. Microbiol.* **31**:1051–1064.

39b.**Pozzi, G., M. Contorni, M. R. Oggioni, R. Manganell, M. Tommasino, F. Cavalieri, and V. A. Fischetti.** 1992. Delivery and expression of a heterologous antigen on the surface of streptococci. *Infect. Immun.* **60**:1902–1907.

40. **Rajo, J. V., and P. M. Schlievert.** 1998. Mechanisms of pathogenesis of staphylococcal and streptococcal superantigens. *Curr. Top. Microbiol. Immunol.* **225**:81–97.

41. **Rothschild, C. B., R. P. Ross, and A. Claiborne.** 1991. Molecular analysis of the gene encoding alkaline phosphatase in *Streptococcus faecalis* 10C1, p. 45–48. *In* G. M. Dunny, P. P. Cleary, and L. L. McKay (ed.), *Genetics and Molecular Biology of Streptococci, Lactococci and Enterococci*. American Society for Microbiology, Washington, D.C.

42. **Ruiz, N., B. Wang, A. Pentland, and M. Caparon.** 1998. Streptolysin O and adherence synergistically modulate proinflammatory responses of keratinocytes to group A streptococci. *Mol. Microbiol.* **27**:337–346.

43. **Scott, J. R., F. Bringel, D. Marra, G. Van Alstine, and C. K. Rudy.** 1994. Conjugative transposition of Tn*916*: preferred targets and evidence for conjugative transfer of a single strand and for a double-stranded circular intermediate. *Mol. Microbiol.* **11**:1099–1108.

44. **Scott, J. R., P. C. Guenther, L. M. Malone, and V. A. Fischetti.** 1986. Conversion of an M⁻ group A streptococcus to M⁺ by transfer of a plasmid containing an M6 gene. *J. Exp. Med.* **164**:1641–1651.

45. **Simon, D., and J. J. Ferretti.** 1991. Electrotransformation of *Streptococcus pyogenes* with plasmid and linear DNA. *FEMS Microbiol. Lett.* **82**:219–224.

46. **Tao, L., D. J. LeBlanc, and J. J. Ferretti.** 1992. Novel streptococcal-integration shuttle vectors for gene cloning and inactivation. *Gene* **120**:105–110.

46a.**Tomich, P. K., F. Y. An, and D. B. Clewell.** 1980. Properties of erythromycin-inducible transposon Tn*917* in *Streptococcus faecalis*. *J. Bacteriol.* **141**:1366–1374.

47. **Trieu-Cuot, P., C. Carlier, C. Poyart-Salmeron, and P. Courvalin.** 1990. A pair of mobilizable shuttle vectors conferring resistance to spectinomycin for molecular cloning in *Escherichia coli* and in gram-positive bacteria. *Nucleic Acids Res.* **18**:4296.

48. **VanHeyningen, T., G. Fogg, D. Yates, E. Hanski, and M. Caparon.** 1993. Adherence and fibronectin-binding are environmentally regulated in the group A streptococcus. *Mol. Microbiol.* **9**:1213–1222.

49. **Wessels, M. R., A. E. Moses, J. B. Goldberg, and T. J. DiCesare.** 1991. Hyaluronic acid capsule is a virulence factor for mucoid group A streptococci. *Proc. Natl. Acad. Sci. USA* **88**:8317–8321.

Cross-Reactive Antigens of Group A Streptococci

MADELEINE W. CUNNINGHAM

7

Cross-reactive antigens are molecules on the group A streptococcus that mimic host molecules and during infection induce an immune response against host tissues. Molecular mimicry is the term used to describe immunological cross-reactivity between host and bacterial antigens. Immunological cross-reactions between streptococcal and host molecules have been identified by antibodies or T cells that react with streptococcal components and tissue antigens. The advent of monoclonal antibodies and T-cell clones/hybridomas has greatly facilitated the identification of host and streptococcal antigens responsible for immunological cross-reactions associated with immunization, infection, and autoimmune sequelae. The identification of cross-reactive antigens in group A streptococci is important in our understanding of the pathogenesis of autoimmune sequelae, such as rheumatic fever and glomerulonephritis, that may occur following group A streptococcal infection.

Molecular mimicry between host and bacterial antigens was first defined as identical amino acid sequences shared between different molecules present in tissues and the bacterium (48, 49, 112). The investigation of molecular mimicry through the use of monoclonal antibodies has identified other types of molecular mimicry. The second type of mimicry involves antibody recognition of similar structures such as alpha-helical coiled-coil molecules like streptococcal M protein and host proteins myosin, keratin, tropomyosin, vimentin, and laminin, which share regions containing 40% identity or less and whose cross-reactive sites are not completely identical (1, 3, 26–28, 30, 34–36, 39, 41, 46, 75–77). A third type of molecular mimicry is revealed in immunological cross-reactions between molecules as diverse as DNA and proteins (36, 46, 99) or carbohydrates and peptides (113–116). Studies of the cross-reactive antigens of the group A streptococcus have contributed greatly to our knowledge about molecular mimicry and the antibody molecules involved. The antibody molecules are described as polyreactive or cross-reactive to indicate recognition of multiple antigens. As described, the basis of the immunological cross-reactions may be identical or homologous amino acid sequences shared between two different proteins or may be epitopes shared between two entirely different chemical structures.

HISTORICAL PERSPECTIVE

Cross-reactive antigens and antibodies were first associated with acute rheumatic fever and group A streptococci when it was discovered that rheumatic fever sera or anti-group A streptococcal antisera reacted with human heart or skeletal muscle tissues (128–130). Antibodies against group A streptococci in rheumatic fever serum were shown to be absorbed from the sera with human heart extracts, and, conversely, the antiheart antibodies were shown to be absorbed with group A streptococci or streptococcal membranes (70). Rabbit antisera produced against group A streptococcal cell walls reacted with human heart tissue, and antibodies produced against human heart tissue reacted with group A streptococcal antigens (65, 67, 69). In rheumatic fever, heart-reactive antibodies appeared to persist in patients with rheumatic recurrences, and there appeared to be a relationship between high titers of antiheart antibodies and recurrence of rheumatic fever (128, 130). Heart-reactive antibodies were reported to decline within 5 years of the initial rheumatic fever attack.

To put the early work in perspective, Kaplan and Suchy (69) implicated the streptococcal cell wall containing the M protein as the cross-reactive antigen recognized by the antiheart antibodies, while studies by Zabriskie and Freimer (128, 129) implicated the streptococcal membrane as the site of the cross-reactive antigen. Beachey and Stollerman (8) and Widdowson and colleagues (124–126) reported an M-associated non-type-specific antigen and an M-associated protein, respectively, which were thought to be associated with immunological cross-reactions in acute rheumatic fever. In 1977, van de Rijn et al. (121) demonstrated that highly purified peptides from group A streptococcal membranes reacted with the antiheart antibodies in rheumatic fever serum. Taken together, this evidence suggested that cross-reactive antigens in group A streptococci were located in both cell wall and membrane.

Evidence also supported the hypothesis that the group A polysaccharide was a cross-reactive antigen. Goldstein and colleagues (52) showed that glycoproteins in heart valves contained the N-acetylglucosamine determinant, which they proposed to be responsible for immune cross-

Gram-Positive Pathogens, ed. by V. A. Fischetti et al.
© 2000 American Society for Microbiology, Washington, D.C.

reactivity of group A streptococci with heart tissues. Dudding and Ayoub (45) demonstrated persistence of anti-group A carbohydrate antibody in patients with rheumatic valvular disease. In Russia, Lyampert and colleagues (82–84) also published studies suggesting that the group A streptococcal polysaccharide antigen induced responses against host tissues. McCarty (91) suggested that the terminal O-linked N-acetylglucosamine might cross-react with antibodies against the group A carbohydrate and host tissues. Additional evidence suggested that the hyaluronic acid capsule of the group A streptococcus also might induce responses against joint tissues (90, 104). Administration of a peptidoglycan-polysaccharide complex prepared from group A streptococci to rats induced carditis and arthritis (20, 21, 108–111). The studies in the 1960s and 1970s left little doubt that group A streptococci induced autoantibodies against heart and other host tissues and that streptococcal components could induce inflammatory lesions resembling arthritis and carditis in animal tissues.

CROSS-REACTIVE MONOCLONAL ANTIBODIES RECOGNIZE MYOSIN AND OTHER ALPHA-HELICAL COILED-COIL MOLECULES IN TISSUES

Monoclonal antibodies (MAbs) cross-reactive with group A streptococci and human heart tissues were produced from mice immunized with streptococcal cell wall and membrane components (29, 36, 46, 77) and from patients with rheumatic carditis (1, 113). Figure 1 illustrates the cross-reactivity of an antistreptococcal MAb with myocardium. In 1985, Krisher and Cunningham (77) identified myosin as a cross-reactive host tissue antigen that provided the link between streptococci and heart. In 1987, it was demonstrated that cardiac myosin could induce myocarditis in genetically susceptible mice (96). Myosin is an alpha-helical coiled-coil molecule that was shown to be an important tissue target of the cross-reactive MAbs (29). Other host tissue antigens recognized by antistreptococcal

mouse and human MAbs included the alpha-helical coiled-coil molecules tropomyosin (1, 46), keratin (1, 113, 117), vimentin (1, 36), laminin (3, 4), DNA (36, 46), and N-acetyl-β-D-glucosamine (1, 113, 115, 116), the immunodominant epitope of the group A polysaccharide. The tissue targets identified by the cross-reactive MAbs may be important in the manifestations of sequelae of group A streptococcal rheumatic fever: arthritis, carditis, chorea, and erythema marginatum (38, 64). A summary of the cross-reactive human and mouse MAb specificities has been published in previous reviews (22, 24).

The cross-reactive antibodies were divided into three major subsets based on their cross-reactivity with (i) myosin and other alpha-helical molecules, (ii) DNA, or (iii) N-acetylglucosamine. Figure 2 illustrates the subsets of cross-reactive antistreptococcal/antimyosin MAbs. All three subsets were identified among MAbs from mice immunized with group A streptococcal components, but in humans the predominant subset reacted with the N-acetylglucosamine epitope and myosin and related molecules. This result is not surprising, since patients with rheumatic fever do not develop antinuclear antibodies during the course of their disease. Elevated levels of polyreactive antimyosin antibodies found in acute rheumatic fever (ARF) sera (34) and in animals immunized with streptococcal membranes or walls (65, 69, 70, 129) most likely account for the reactivity of these sera with myocardium and other tissues. Similar types of cross-reactive mouse and human antibodies have been investigated by Lange (80) and by Wu et al. (127), respectively.

Although the role of the cross-reactive antibodies in disease is not clear, cytotoxic mouse and human MAbs have identified the extracellular matrix protein laminin as a potential tissue target present in basement membrane surrounding myocardium and valve surface endothelium (3, 4, 50). Cross-reactive antibodies may become trapped in extracellular matrix, which may act like a sieve to capture antibody and lead to inflammation in host tissues.

Polyspecific or cross-reactive autoantibodies have emerged as a theme in autoimmunity and molecular mimicry (2, 17, 36, 78). The V-D-J region genes of the human and mouse cross-reactive MAbs have been sequenced (1, 3, 92), but there is no consensus sequence to explain the molecular basis of polyspecificity and cross-reactivity. It is worth noting that the three groups of reactivities in mice did not have specific antibody V gene families or VH and VL gene combinations that correlate with a specific reac-

FIGURE 1 Reaction of mouse antistreptococcal MAb with human tissue section of myocardium in indirect immunofluorescence assay. Mouse IgM (20 μg/ml) was unreactive (not shown). MAbs were tested at 20 μg/ml. From reference 46 with permission from *Journal of Immunology*.

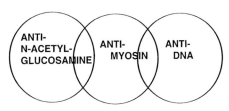

FIGURE 2 Subsets of cross-reactive antistreptococcal and antimyosin antibodies. Human and mouse antistreptococcal/antimyosin MAbs have been divided into three subsets based on their reactivity with myosin and N-acetylglucosamine, the dominant group A carbohydrate epitope; with DNA and the cell nucleus, a property found among mouse MAbs; and with myosin and a family of alpha-helical coiled-coil molecules. From reference 23 with permission from Indiana University School of Dentistry Press.

tivity. The V genes of cross-reactive human MAbs from rheumatic fever were encoded by a heterogeneous group of VH3 family genes (VH-3, VH-8, VH-23, VH-30) and a VH4-59 gene segment (1). Wu et al. (127) have also reported a similar group of V gene sequences for their human antistreptococcal/antimyosin antibodies produced from Epstein-Barr virus-transformed B cell lines. Many sequences were found to be either in germ line configuration or highly homologous with a previously sequenced germ line V gene. It has been proposed that a germ line antibody may be polyreactive because of conformational rearrangement and configurational change, permitting binding of diverse molecules (123).

MONOCLONAL ANTIBODIES IDENTIFY STREPTOCOCCAL CROSS-REACTIVE ANTIGENS

Antimyosin MAbs identified cross-reactive antigens in streptococcal membranes (31, 36) and walls (36). The streptococcal M protein was shown to react with the heart- or myosin-cross-reactive MAbs (31, 32, 46). In addition, a 60-kDa protein present in the cell membrane (5, 7) was cross-reactive with myosin in the heart, and a 67-kDa protein that was immunologically similar to class II major histocompatibility complex (MHC) molecules was also identified (73). The data suggested that cross-reactive antigens were present in both the wall and membrane of the group A streptococcus. The data support previous evidence from both Kaplan and colleagues (65–67, 69, 70) and Zabriskie and colleagues (128–130). Although these data were once thought to be conflicting, it is clear that the cross-reactive antibodies recognize more than one antigen in the streptococcal cell. The data previously reported by Goldstein et al. (52) are also supported by the evidence that a subset of cross-reactive MAbs recognized N-acetyl-β-D-glucosamine, the immunodominant epitope of the group A polysaccharide. N-acetylglucosamine is a major epitope of some of the cross-reactive mouse MAbs and virtually all of the human cross-reactive MAbs investigated (1, 113, 115, 116). The studies link together the cross-reactivity of the group A carbohydrate epitope GlcNAc, human cardiac myosin, and streptococcal M protein. Antigenic redundancy due to cross-reactivity may be an important factor in triggering disease in a susceptible host. The cross-reactive antigens are now seen as separate entities recognized by MAbs that recognize more than one antigen molecule. Thus, the previous studies, which before seemed conflicting, were all correct. The MAbs have allowed dissection of the group A streptococcal cross-reactive antigens, which would have not been possible using polyclonal sera. The following sections provide more evidence about the identification and analysis of the cross-reactive antigens of the group A streptococcus. The cross-reactive antigens are M proteins, N-acetylglucosamine/group A polysaccharide, 60-kDa wall-membrane antigen, and 67-kDa antigen.

M Proteins and Rheumatic Fever

Investigation of the M proteins over the past 15 years has provided important information about the sequence and primary structure of the molecule. The hypothesis that M proteins and myosin have immunological similarities was supported by the structural studies of Manjula and colleagues (86–88), which demonstrated the 7-amino-acid-residue periodicity common among group A streptococcal

M proteins and shared with proteins such as tropomyosin, myosin, desmin, vimentin, and keratin. Figure 3 illustrates the 7-residue periodicity and the homology characteristic of alpha-helical coiled-coil proteins such as tropomyosin and M6 protein. Studies by Dale, Beachey, Bronze, and colleagues using polyclonal sera and affinity-purified antibodies demonstrated immunological cross-reactivity between M proteins and myosin (14–16, 39, 40, 41). Studies using cross-reactive MAbs also identified immunological cross-reactivity between streptococcal M proteins, both PepM and recombinant molecules, and cardiac and skeletal myosins (27, 32, 36, 46).

Since streptococcal M proteins have been investigated for potential epitopes that were recognized by the cross-reactive MAbs, affinity-purified antimyosin antibody from ARF sera was reacted with peptides of streptococcal M5 protein. Affinity-purified antimyosin antibodies from ARF reacted with an M5 amino acid sequence (residues 184 to 188) near the pepsin cleavage site in M5 and M6 proteins (32). The epitope, located in the B-repeat region of M5 and M6 proteins, appeared to be a B-cell epitope for antibody-mediated cross-reactivity with myosin (28, 32). An M5 peptide, containing the B-repeat region epitope Gln-Lys-Ser-Lys-Gln (QKSKQ), was shown to inhibit antimyosin antibodies in ARF (32). Furthermore, an M5 peptide that contained the QKSKQ sequence induced antibodies in BALB/c mice against cardiac or skeletal myosins and the LMM fragment of myosin as well. This evidence further supported the previous findings indicating that the QKSKQ sequence is important in the antibody-mediated cross-reactions with myosin in ARF (32). M5 residues 164 to 197 were demonstrated to induce antibodies against sarcolemmal membrane of heart tissue (105), and M5 residues 84 to 116 induced heart-reactive antibody and reacted with antimyosin antibody purified from ARF sera (32, 105). Studies by Kraus and colleagues (76) demonstrated a vimentin cross-reactive epitope present in the M12 protein, and M protein peptides containing brain cross-reactive epitopes were localized to the M5 protein sequence 134 to 184 and the M19 sequence 1 to 24 (14, 105). A summary of the currently known myosin cross-reactive B-cell epitopes in M5 protein is shown in Fig. 4, which identifies the A-, B-, and C-repeat regions of M protein.

Immunization of BALB/c mice with each of 23 peptides of M5 protein has revealed that M5 peptides NT3–NT7, B2B3B, and C1A–C3 induce anti-human cardiac myosin antibodies as shown in Fig. 5. Amino acid sequences of the overlapping synthetic peptides spanning the M5 protein molecule have been reported elsewhere (28). Titers of the anti-M5 peptide sera were 10 times greater against cardiac myosin than against skeletal myosin, tropomyosin, vimentin, and laminin (28). Certain M5 peptides appeared to induce a cardiac myosin-directed response in BALB/c mice. Furthermore, mice developed mild myocarditis when immunized with M5 peptides NT4, NT5, NT6, B1A, and B3B (28). Peptides from the C-repeat region of M5 protein did not produce myocardial lesions when administered to BALB/c mice.

The C-repeat region of class I M proteins contains the class I epitope, which is identified by reactivity with anti-M protein MAb 10B6 (9, 10, 62, 63). Class I M protein serotypes were streptococcal strains associated with pharyngitis and rheumatic fever. Responses against the class I epitope were stronger in patients with rheumatic fever than in those with uncomplicated disease (11, 101). MAb 10B6, which recognized the C-repeat epitope, also reacted with skeletal and cardiac myosins and the HMM subfragment of

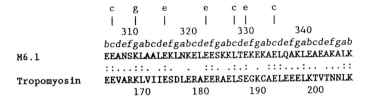

```
                c   g   e       e   c e   c
                |   |   |       |   | |   |
               310     320     330     340
            bcdefgabcdefgabcdefgabcdefgabcdefgab
M6.1        EEANSKLAALEKLNKELEESKKLTEKEKAELQAKLEAEAKALK
            ::...::. .:.  . ::.  . :::...::...:::.
Tropomyosin EEVARKLVIIESDLERAEERAELSEGKCAELEEELKTVTNNLK
               170      180     190      200
```

FIGURE 3 Sequence alignment of streptococcal M6 protein and human cardiac tropomyosin in a region exhibiting significant homology. Lowercase letters a to g directly above the sequence designate the position of these amino acids within the 7-residue periodicity in both segments. Lowercase letters at the top of the figure designate identities at external locations in the heptad repeat. Double dots indicate identities, and single dots indicate conservative substitutions. Within this segment of the streptococcal M6 molecule, 31% homology is observed with tropomyosin. Since both molecules are alpha-helical coiled-coil proteins, they contain the 7-residue repeat pattern in which positions a and d are usually hydrophobic. Similar homologies are seen between M proteins and myosin heavy chains and any of the three laminin chains. From reference 46 with permission from *Journal of Immunology*.

myosin (101). The C-repeat class I epitope in M5 contains the amino acid sequence KGLRRDLDASREAK, which shares homology with RRDL, a conserved amino acid sequence found in cardiac and skeletal myosins. The myosin sequence RRDL is located in the HMM subfragment of cardiac and skeletal myosin heavy chain (101).

It was reported that reactivity of immunoglobulin G (IgG) from ARF sera was greater to the class I peptide sequence than was IgG from uncomplicated pharyngitis (35). This was confirmed in another study by Mori and colleagues (95). However, the lowered immune response to the class I epitope in uncomplicated pharyngitis was most likely because the patients with uncomplicated pharyngitis had been treated with penicillin (61). Antibodies against the class I epitope were affinity purified from ARF sera and shown to react with myosin (101). Whether the myosin-reactive antibodies against the class I epitope and C-repeat region play a role in the disease pathogenesis is not known; however, the C-repeat peptides do not produce heart lesions in animal models of carditis (28).

Studies by Huber and Cunningham (56) demonstrated that streptococcal M5 peptide NT4 containing the amino acid sequence GLKTENEGLKTENEGLKTE could produce myocarditis in MRL/++ mice, an autoimmune-prone strain. The studies demonstrated that the myocarditis was mediated by CD4+ T cells and class II

MHC molecules. Antibodies against the IA^k MHC molecule or antibody against CD4+ T cells abrogated the myocarditis (56). Amino acid sequence homology between the NT4 peptide of M5 and human cardiac myosin demonstrates 80% identity. Figure 6 illustrates the homology between the M5 peptide NT4 sequence and the cardiac myosin sequence. The mimicking sequence in NT4 is repeated four times in the M5 protein and is present in cardiac myosin only once. Repeated regions of the M protein that mimic cardiac myosin may be important in breaking tolerance in the susceptible host and producing autoimmune disease. The data support the hypothesis that epitopes in streptococcal M protein that mimic cardiac myosin may be important in breaking tolerance to this potent autoantigen. Table 1 summarizes the M protein amino acid sequences observed to induce inflammatory myocardial lesions in mice.

Streptococcal M proteins mimic not only epitopes in cytoskeletal proteins but also epitopes in other strong bacterial and viral antigens, namely, heat shock protein (Hsp) 65 (100) and coxsackieviral capsid proteins (27, 55, 56, 59). These immunological cross-reactions may be vital to the survival of the host and suggest that antibody molecules may recognize and neutralize more than a single infectious agent. Such antibodies may be an important first line of defense and would be highly advantageous for the host. Multiple pieces of evidence from experiments using synthetic peptides have shown that sites in the M protein mimic a site(s) in the VP1 capsid protein of coxsackievirus (25, 27). Further evidence demonstrated that some of the antistreptococcal/antimyosin MAbs actually neutralized enteroviruses and were cytotoxic for heart cells (27). This was extremely interesting, because coxsackieviruses cause autoimmune myocarditis in susceptible hosts (51, 57, 58). The Hsp 65 antigen has been shown to play a role in the development of arthritis and diabetes (18, 19). Cross-reactive epitopes shared between streptococcal M proteins and Hsp 65 play a role in arthritis sequelae. Antibodies against Hsp 60 have been implicated in cytotoxicity against endothelium (106). Shared epitopes among pathogens may be important in molecular mimicry, may break tolerance to cryptic host molecules, and may influence the development of autoimmune diseases.

M Proteins and Cross-Reactive T Cells

M proteins have also been shown to stimulate human T-lymphocyte responses (42, 54, 97, 98, 102, 103, 118–120),

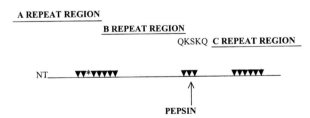

FIGURE 4 Antimyosin antibody cross-reactive sites in M5 protein A-, B-, and C-repeat regions of the molecule. Arrows point to sites determined to induce human cardiac myosin cross-reactive antibody (28). The asterisk marks the location of peptide NT4 containing several repeats of an epitope in cardiac myosins that causes myocarditis in MRL/++ and BALB/c mouse strains (28, 56). Site QKSKQ is the epitope determined to react with antimyosin antibody in ARF sera (32). From reference 1a with permission from Oxford University Press.

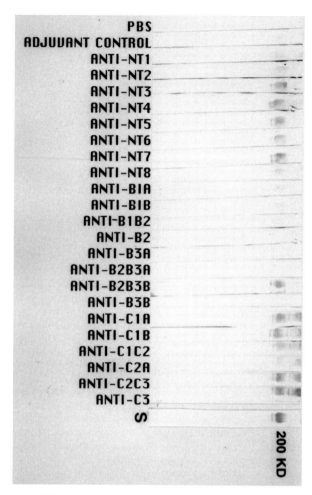

TABLE 1 M5 protein sequences that produce myocarditis in mice[a]

Peptide	Amino acid sequence	Residues
A-repeat region		
NT4	GLKTENEGLKTENEGLKTE	40–58
NT5	KKEHEAENDKLKQQRDTL	59–76
NT6	QRDTLSTQKETLEREVQN	72–89
B-repeat region		
B1A	TRQELANKQQESKENEKAL	111–129
B3B	GALKQELAKKDEANKISD	202–219

[a]NT4 produced myocarditis in BALB/c (28) and MRL/++ (56) mice; the other peptides produced myocarditis in BALB/c mice. From *Effects of Microbes on the Immune System* with permission from Lippincott-Williams & Wilkins.

FIGURE 5 Reactivity of the antistreptococcal M5 peptide sera in the Western immunoblot of human cardiac myosin. The 200-kDa protein band of the purified human cardiac myosin, shown in the stained (S) portion of the Western blot, reacted most strongly with antipeptide sera from mice immunized with M5 peptides NT3–NT7, B2B3B, and C1A, C1B, C2C3, and C3. Sera were tested at a 1:1000 dilution. A control antimyosin MAb, CCM-52 (a gift from William Clark, Cardiovascular Research Institute, Michael Reese Hospital and Medical Center, Chicago, Ill.), reacted with the 200-kDa band present in our purified preparation of human cardiac myosin heavy chain. Purification of the human cardiac myosin heavy chain to homogeneity has been previously described by Dell et al. (44). The Western blot confirms the data seen in the enzyme-linked immunosorbent assay with human cardiac myosin. From reference 28.

including cytotoxic T lymphocytes from patients with ARF (42, 60). PepM protein (pepsin-extracted M protein fragment) has been reported to be a superantigen that stimulated Vβ2-, Vβ4-, and Vβ8-bearing T-cell subsets (118–120, 122). Other reports show that neither recombinant nor native forms of M proteins are superantigenic (43, 47, 107). A superantigenic site reported for the M5 protein has been localized to the amino acid sequence located in M5 residues 157 to 197 (KEQENKETIGTLKKILDE-TVKDKLAKE QKSKQNIGALKQEL) in the B3-repeat region (122). The site has a significant amount of sequence homology with other superantigens. The role of M protein or other superantigens in ARF may be to activate large numbers of T cells, including some that are cross-reactive and may lead to ARF.

Studies of T-cell epitopes in rheumatic fever and in animal models focus on the M5 protein molecule because serotype M5 has often been associated with ARF outbreaks (12). T- and B-cell epitopes of the M5 protein were defined in previous studies by Robinson and colleagues (102, 103), by Pruksakorn et al. (97, 98), and in my laboratory (28).

Guilherme et al. (54) recently isolated cross-reactive T cells from mitral valves, papillary muscle, and left atrium of rheumatic fever patients previously infected with M5 group A streptococci. T cell lines were responsive to several peptides of the streptococcal serotype M5 protein and proteins from heart tissue extracts. Sequences from the A- and B-repeat regions of M5 protein that stimulated the valvular T cells are shown in Table 2. In BALB/c mice, myosin-cross-reactive T-cell epitopes from the A-, B-, and C-repeat regions of the M5 protein were identified using synthetic peptides (28). Six dominant myosin cross-reactive T-cell epitopes were located in the M5 molecule (28). Table 2 summarizes (i) the dominant myosin-cross-reactive T-cell epitopes of M5 protein in BALB/c mice (28), (ii) the M5 sequences recognized by T-cell clones from rheumatic heart valves (54), and (iii) M5 peptides reported by Pruksakorn et al. (97, 98) to stimulate human

```
Streptococcal M Protein M5 Peptide NT4        GLKTENEG
                                              : :::
                    cardiac myosin            KLQTENGE
```

FIGURE 6 Sequence homology between human cardiac myosin and peptide NT4, which causes myocarditis (28, 56). The homologous sequence repeats four times in the streptococcal M5 protein and in NT4 and once in cardiac myosin. Repeated sequences in M proteins that mimic cardiac myosin may be important in inducing inflammatory heart disease. Adapted from reference 56 with permission from *Journal of Immunology*.

TABLE 2 Summary of myosin or heart-cross-reactive T-cell epitopes of streptococcal M5 protein[a]

Peptide	Sequence[b]	Origin of T cell clone or response[a]
1–25	TVTRGTISDPQRAKEALDKYELENH	ARF/valve (54)
81–96	DKLKQQRDTLSTQKETLEREVQN	ARF/valve (54)
163–177	ETIGTLKKILDETVK	ARF/valve (54)
337–356	LRRDLDASREAKKQVEKAL	Normal/PBL (97)
347–366	AKKQVEKALEEANSKLAALE	Mice (98)/normal PBL (97) ARF/PBL (97)
397–416	LKEQLAKQAEELAKLRAGKA	ARF/PBL (97)
NT4 40–58	GLKTENEGLKTENEGLKTE	BALB/c/lymph node[d]
NT5 59–76	KKEHEAENDKLKQQRDTL	
B1B2 137–154	VKDKIAKEQENKETIGTL	
B2 150–167	TIGTLKKILDETVKDKIA	
C2A 254–271	EASRKGLRRDLDASREAK	
C3 293–308	KGLRRDLDASREAKKQ	

[a]From reference 27.

[b]The amino terminal TVTRGTIS sequence (peptide 1–25) was taken from the M5 amino acid sequence published by Manjula et al. (85) and deviates from the M5 sequence published by Miller et al. (94) at positions 1 and 8. Sequences for peptides 81–96 and 163–177 were taken from the PepM5 sequence as reported by Manjula et al. (85). These two sequences are given as 67–89 and 174–188, respectively, in the sequence reported by Miller et al. (94). All other sequences shown are from the M5 gene sequence reported by Miller et al. (94). Underlined sequences are those shared among the epitopes/sequences compared in Tables 1 and 2.

[c]PBL, peripheral blood lymphocytes; ARF, acute rheumatic fever.

[d]BALB/c mice immunized with purified human cardiac myosin and the recovered lymph node lymphocytes were stimulated with each of the peptides in tritiated thymidine uptake assays.

T cells from normal control subjects and rheumatic fever patients that were responsive to myosin peptides. An important correlation seen in Table 2 is that M5 peptides NT5 (KKEHEAENDKLKQQRDTL) and B1B2/B2 (VKDKIAKEQENKETIGTL and TIGTLKKILDETVKDKIA), which were dominant cross-reactive T-cell epitopes in the BALB/c mouse, contain sequences similar to those recognized by T cells from rheumatic valves (54). Peptides NT4 and NT5 produced inflammatory infiltrates in the myocardium of animals immunized with those peptides (28). The collective evidence on cross-reactive T cells suggests that amino acid sequences in M5 protein that share homology with cardiac myosin may break tolerance and promote T-cell-mediated inflammatory heart disease in animals and humans (28).

M Proteins and Poststreptococcal Acute Glomerulonephritis

Sera from patients with poststreptococcal acute glomerulonephritis were shown by Kefalides and colleagues (72) to contain antibodies against laminin, collagen, and other macromolecules found in the glomerular basement membrane. The epitope recognized in collagen was identified and shown to be located in the 7-S domain of type IV collagen (71). Streptococcal antigens, immunoglobulins, and complement were detected in the kidney glomeruli in acute glomerulonephritis (93). Studies by Lange, Markowitz, and colleagues suggested that the glomerular basement membrane antigens share epitopes with streptococcal M12 protein (89), and others have shown immunologic cross-reactivity between the group A streptococcus and glomeruli (13, 53, 75, 79). Kraus and Beachey (74) discovered a renal autoimmune epitope (Ile-Arg-Leu-Arg)

in M protein. Evidence suggested that certain M protein serotypes were associated with nephritis and that molecular mimicry or immunological cross-reactivity between glomeruli and M protein could be one mechanism by which antiglomerular antibodies were produced during infection. In animal models of nephritis induced by nephritogenic streptococci (M type 12), antiglomerular antibodies eluted from kidney glomeruli reacted with the type 12 streptococcal M protein (81). Furthermore, an antiglomerular MAb reacted with the M12 protein (53). These studies support immune-mediated mechanisms in poststreptococcal acute glomerulonephritis. Streptococcal and renal antigens may be an important source of mimicking antigen present in the kidney and may play a role in binding Ig and complement and producing nephritis.

Novel Cross-Reactive Protein Antigens in Group A Streptococci

Antimyosin antibodies from ARF were used to identify group A streptococcal cross-reactive antigens other than M proteins. The cross-reactive antigens were detected in a *Streptococcus pyogenes* strain that had spontaneously lost its *emm* gene. The affinity-purified antimyosin antibodies were used to screen gene libraries as well as the antigen preparations from this M protein-deficient streptococcus. A 60-kDa streptococcal actinlike protein (6), a 60-kDa wall-membrane protein (5, 7), and a 67-kDa protein from a cloned gene product (73) were identified. All of the antigens were unique and interesting. The 60-kDa actinlike molecule behaved biochemically as an actin and produced actinlike filaments observed by electron microscopy (6). The 67-kDa cloned gene product recognized by antimyosin antibody from ARF sera was novel in that it shared enough sequence homology with MHC molecules that it reacted with sera against mouse MHC class II molecules and exhibited a hydrophilicity plot almost identical to that of MHC class II molecules (73). The sequence homology between the 67-kDa molecule and MHC class II molecules is shown in Fig. 7. The gene for the 67-kDa protein was present only in pathogenic streptococci groups A, C, and G (73). The gene for the 67-kDa protein was not found in other streptococcal groups, *Escherichia coli*, or staphylococci when they were screened for hybridization with the gene probe. The role of the 67-kDa protein as a potential virulence factor in the pathogenesis of streptococcal infections or sequelae is not yet known.

N-Acetyl-β-D-Glucosamine: Dominant Epitope of Group A Polysaccharide

The structure of the group A polysaccharide is a polymer of rhamnose with N-acetyl-D-glucosamine linked to the rhamnose backbone. The N-acetylglucosamine is the immunodominant epitope of the group A carbohydrate that distinguishes S. *pyogenes* from other streptococcal species. The N-acetylglucosamine and the group A polysaccharide have been identified in cross-reactions between streptococci and heart tissues (52, 83, 84). Recently, Shikhman, Adderson, and colleagues reported that a subset of cross-reactive antimyosin/antistreptococcal MAbs from immunized mice and virtually all human MAbs from patients with rheumatic carditis reacted strongly with N-acetylglucosamine epitope of the group A streptococcal carbohydrate (1, 113, 115, 116). Synthetic peptides from keratin (113, 116), coxsackievirus (115), or human cardiac myosin (1) reacted with MAbs against N-acetylglucosamine and myosin as previously described. In addition, a synthetic cytokeratin peptide SFGSGFGGGY

mimicked N-acetylglucosamine in its reaction with antibodies to N-acetylglucosamine and lectins. Mimicry between the group A carbohydrate epitope and peptide molecules is a form of mimicry between chemically diverse molecules. Based on results with peptides altered at a single amino acid residue, it was deduced that aromatic and hydrophobic interactions were important in cross-reactive antibody binding to the peptide. The data clearly link myosin and N-acetylglucosamine with antiheart cross-reactivity.

Dudding and Ayoub (45) described the elevation and persistence of anti-group A polysaccharide antibody in rheumatic valvulitis. Studies of the human antistreptococcal/antimyosin MAbs that reacted strongly with N-acetylglucosamine have revealed that one of the MAbs was cytotoxic for human endothelium and reacted with valvular endothelium in tissue sections of human valve (50). The cytotoxic MAb recognized the extracellular matrix protein laminin that is part of the basement membrane underlying the valvular endothelium. These data suggest that human antibody cross-reactivity with valve tissues may be established through antimyosin/anti-N-acetylglucosamine/antilaminin reactivity. Antibodies that react with valve endothelium may lead to inflammation at the valve surface and promote T-cell infiltration of the valve in rheumatic heart disease.

SUMMARY AND CONCLUSIONS

Cross-reactive antigens, cross-reactive antibodies, and cross-reactive T cells against group A streptococci are important in the pathogenesis of autoimmune sequelae following infection. Three subsets of cross-reactive antibodies have been defined. MAbs identified host tissue antigens as alpha-helical coiled-coil molecules such as

```
                 Similarity: 48%              Identity: 19%

    DQ     1  DFVYQFKGMCYFT...NGTERVRLVTRYIYNREEYARFDSDVGVYRAVTP  47
              : :. |.::.. .    :|.|:.:|.    .||  .:|:: ::||.:..
  67KD    50  ERIHLFEELPLAGGSLDGIEKPHLGFVTRGGREMENHFECMWDMYRSIPS  99

    DQ    48  LGRPDA....EYWNSQKEVLERTRAELDTVCRHNYEVELRTTLQRRVEPT  93
              |: |:|   |:: :|:  :....|    :. : : .||.:. ..
  67KD   100  LEIPGASYLDEFYWLDKDDPNSSNCRLIHKRGNRVDDDGQYTLGKQSKEL 149

    DQ    94  VTISPSRTEALNHHNLLVCSVTDFYPAQIKV...................124
              |: .   .|.|..:.:   .||:...:.:
  67KD   150  VHLIMKTEESLGDQTIEEFFSEDFFKSNFWIYWATMFAFEKWQFCCRNAG 199

    DQ   125  ..RWFRNDQEETAGVVSTPLIRNGDW.......................148
              .:| .|: :: ::..| .: : .::
  67KD   200  YAMRFIHHIDGLPDFTSLKFNKYNQYDSMVKPIIAYLESHDVDIQFDTKV 249

    DQ   149  ...............TFQILV.....MLEMTPQ...........RGDVYT 167
                             |:::|    :|:||:     .... .|.
  67KD   250  TDIQVEQTAGKKVAKTIHMTVSGEAKAIELTPDDLVFVTNGSITESSTYG 299

    DQ   168  CHVEHPSLESPITVEWR......AQSE............SAQSKMLSGIG 199
              :| | :. ..::...|.   |||  .|:|.::|:. :
  67KD   300  SHHEVAKPTKALGGSWNLWENLAAQSDDFGHPKVFYQDLPAESWFVSATA 349

    DQ   200  GFVLGLIFLGLGLIIHHRSQKGLL 223
              .: . |  :: :.|: :.|.
  67KD   350  TIKHPAIEPYIERLTHRDLHDGKV 373
```

FIGURE 7 Amino acid sequence homology between the group A streptococcal 67-kDa antigen, identified with affinity-purified antimyosin antibody from ARF sera, and the human MHC antigen DQ. As shown, there is 19% identity and 48% homology between the class II DQ molecule and the 67-kDa antigen. The hydrophilicity plot is nearly identical for the two molecules. From reference 73.

myosin, tropomyosin, keratin, vimentin, and laminin. Laminin, an extracellular matrix molecule present in basement membrane, may trap cross-reactive antibodies at the cell surface and lead to tissue injury. Cross-reactive group A streptococcal antigens include M proteins, the group A polysaccharide, and other novel wall-membrane antigens that mimic host antigens such as MHC class II molecules. Cross-reactive antipolysaccharide antibodies recognize peptide sequences in alpha-helical proteins as well as the N-acetylglucosamine molecule. Among the antipolysaccharide antibodies, a few reacted with valvular endothelium and were cytotoxic for endothelial cells. Amino acid sequences of streptococcal M5 protein that have been shown to be pathogenic in animals have also been reported to stimulate T cells from rheumatic hearts. The identification of human cardiac myosin-cross-reactive B- and T-cell epitopes of M5 protein has been a step forward in understanding the cross-reactive epitopes that produce disease in animals and humans. Although the identity of cross-reactive antigens is much clearer now than in the past, there are still many lessons to be learned from the cross-reactive antigens about how mimicry of host tissues leads to autoimmune sequelae following streptococcal disease.

This work was supported by grants HL35280 and HL56267 from the National Heart, Lung and Blood Institute.

REFERENCES

1. **Adderson, E. E., A. R. Shikhman, K. E. Ward, and M. W. Cunningham.** 1998. Molecular analysis of polyreactive monoclonal antibodies from rheumatic carditis: human anti-N-acetylglucosamine/anti-myosin antibody V region genes. *J. Immunol.* **161:**2020–2031.

1a. **Agoub, E. M., M. Keth, and M. W. Cunningham.** Rheumatic fever pathogenesis. *In* D. Stevens and E. Kaplan (ed.), *Streptococcal Infections: Clinical Aspects, Microbiology, and Molecular Pathogenesis.* Oxford University Press, New York, N.Y., in press.

2. **Andrezejewski, C., Jr., J. Rauch, B. D. Stollar, and R. S. Schwartz.** 1980. Antigen binding diversity and idiotypic cross-reactions among hybridoma autoantibodies to DNA. *J. Immunol.* **126:**226–231.

3. **Antone, S. M., E. E. Adderson, N. M. J. Mertens, and M. W. Cunningham.** 1997. Molecular analysis of V gene sequences encoding cytotoxic anti-streptococcal/anti-myosin monoclonal antibody 36.2.2 that recognizes the heart cell surface protein laminin. *J. Immunol.* **159:**5422–5430.

4. **Antone, S. M., and M. W. Cunningham.** 1992. Cytotoxicity linked to a streptococcal monoclonal antibody which recognizes laminin, p. 189–191. *In* G. Orefici (ed.), *New Perspectives on Streptococci and Streptococcal Infections. Proceedings of the XI Lancefield International Symposium.* (*Zbl. Bakt. Suppl.* **22.**) Gustav Fischer Verlag, New York, N.Y.

5. **Barnett, L. A., and M. W. Cunningham.** 1990. A new heart-cross-reactive antigen in *Streptococcus pyogenes* is not M protein. *J. Infect. Dis.* **162:**875–882.

6. **Barnett, L. A., and M. W. Cunningham.** 1992. Evidence for actinlike proteins in an M protein-negative strain of *Streptococcus pyogenes. Infect. Immun.* **60:**3932–3936.

7. **Barnett, L. A., J. J. Ferretti, and M. W. Cunningham.** 1992. A 60-kDa acute rheumatic fever associated antigen of *Streptococcus pyogenes*, p. 216–218. *In* G. Orefici (ed.),

New Perspectives on Streptococci and Streptococcal Infections. Proceedings of the XI Lancefield International Symposium. (*Zbl. Bakt. Suppl.* **22.**) Gustav Fischer Verlag, New York, N.Y.

8. **Beachey, E. H., and G. H. Stollerman.** 1973. Mediation of cytotoxic effects of streptococcal M protein by non-type-specific antibody in human sera. *J. Clin. Invest.* **52:**2563–2570.

9. **Bessen, D., K. F. Jones, and V. A. Fischetti.** 1989. Evidence for two distinct classes of streptococcal M protein and their relationship to rheumatic fever. *J. Exp. Med.* **169:**269–283.

10. **Bessen, D. E., and V. A. Fischetti.** 1990. Differentiation between two biologically distinct classes of group A streptococci by limited substitutions of amino acids within the shared region of M protein-like molecules. *J. Exp. Med.* **172:**1757–1764.

11. **Bessen, D. E., L. G. Veasy, H. R. Hill, N. H. Augustine, and V. A. Fischetti.** 1995. Serologic evidence for a class I group A streptococcal infection among rheumatic fever patients. *J. Infect. Dis.* **172:**1608–1611.

12. **Bisno, A. L.** 1995. Non-suppurative poststreptococcal sequelae: rheumatic fever and glomerulonephritis, p. 1799–1810. *In* G. L. Mandell, J. E. Bennett, and R. Dolin (ed.), *Principles and Practice of Infectious Diseases*, vol. 2. Churchill Livingstone, New York, N.Y.

13. **Bisno, A. L., J. W. Wood, J. Lawson, S. Roy, E. H. Beachey, and G. H. Stallerman.** 1978. Antigens in urine of patients with glomerulonephritis and in normal human serum which cross-react with group A streptococci: identification and partial characterization. *J. Lab. Clin. Med.* **91:**500–513.

14. **Bronze, M. S., E. H. Beachey, and J. B. Dale.** 1988. Protective and heart-cross-reactive epitopes located within the NH2 terminus of type 19 streptococcal M protein. *J. Exp. Med.* **167:**1849–1859.

15. **Bronze, M. S., E. H. Beachey, J. M. Seyer, and J. B. Dale.** 1987. Protective and heart-cross-reactive epitopes of type 19 streptococcal M protein. *Trans. Assoc. Am. Physicians* **100:**80–84.

16. **Bronze, M. S., and J. B. Dale.** 1993. Epitopes of streptococcal M proteins that evoke antibodies that cross-react with human brain. *J. Immunol.* **151:**2820–2828.

17. **Carroll, P., D. Stafford, R. S. Schwartz, and B. D. Stollar.** 1985. Murine monoclonal anti-DNA autoantibodies bind to endogenous bacteria. *J. Immunol.* **135:**1086–1090.

18. **Cohen, I. R.** 1991. Autoimmunity to chaperonins in the pathogenesis of arthritis and diabetes. *Annu. Rev. Immunol.* **9:**567–589.

19. **Cohen, I. R., and D. B. Young.** 1991. Autoimmunity, microbial immunity, and the immunological homunculus. *Immunol. Today* **12:**105–110.

20. **Cromartie, W., J. Craddock, J. H. Schwab, S. Anderle, and C. Yang.** 1977. Arthritis in rats after systemic injection of streptococcal cells or cell walls. *J. Exp. Med.* **146:**1585–1602.

21. **Cromartie, W. J., and J. G. Craddock.** 1966. Rheumatic like cardiac lesions in mice. *Science* **154:**285–287.

22. **Cunningham, M. W.** 1996. Streptococci and rheumatic fever, p. 13–66. *In* N. R. Rose and H. Friedman (ed.), *Microorganisms and Autoimmune Disease.* Plenum Publishing Corp., New York, N.Y.

23. **Cunningham, M. W.** 1998. Molecular mimicry in auto-immunity and infection, p. 85–99. *In* D. J. LeBlanc, M. S. Lantz, and L. M. Switalski (ed.), *Microbial Pathogenesis: Current and Emerging Issues. Second Annual Indiana Conference.* Indiana University School of Dentistry, Indianapolis, Ind.

24. **Cunningham, M. W.** Pathogenesis of group A streptococci. Submitted for publication.

25. **Cunningham, M. W., S. M. Antone, J. M. Gulizia, and C. J. Gauntt.** 1992. Common epitopes shared between streptococcal M protein and viruses may be a link to autoimmunity, p. 534–536. *In* G. Orefici (ed.), *New Perspectives on Streptococci and Streptococcal Infections. Proceedings of the XI Lancefield International Symposium.* (*Zbl. Bakt. Suppl.* **22.**) Gustav Fischer Verlag, New York, N.Y.

26. **Cunningham, M. W., S. M. Antone, J. M. Gulizia, B. A. McManus, and C. J. Gauntt.** 1993. Alpha-helical coiled-coil molecules: a role in autoimmunity against the heart. *Clin. Immunol. Immunopathol.* **68:**118–123.

27. **Cunningham, M. W., S. M. Antone, J. M. Gulizia, B. M. McManus, V. A. Fischetti, and C. J. Gauntt.** 1992. Cytotoxic and viral neutralizing antibodies cross-react with streptococcal M protein, enteroviruses, and human cardiac myosin. *Proc. Natl. Acad. Sci. USA* **89:**1320–1324.

28. **Cunningham, M. W., S. M. Antone, M. Smart, R. Liu, and S. Kosanke.** 1997. Molecular analysis of human cardiac myosin-cross-reactive B- and T-cell epitopes of the group A streptococcal M5 protein. *Infect. Immun.* **65:** 3913–3923.

29. **Cunningham, M. W., N. K. Hall, K. K. Krisher, and A. M. Spanier.** 1985. A study of monoclonal antibodies against streptococci and myosin. *J. Immunol.* **136:**293–298.

30. **Cunningham, M. W., N. K. Hall, K. K. Krisher, and A. M. Spanier.** 1986. A study of anti-group A streptococcal monoclonal antibodies cross-reactive with myosin. *J. Immunol.* **136:**293–298.

31. **Cunningham, M. W., K. Krisher, and D. C. Graves.** 1984. Murine monoclonal antibodies reactive with human heart and group A streptococcal membrane antigens. *Infect. Immun.* **46:**34–41.

32. **Cunningham, M. W., J. M. McCormack, P. G. Fenderson, M. K. Ho, E. H. Beachey, and J. B. Dale.** 1989. Human and murine antibodies cross-reactive with streptococcal M protein and myosin recognize the sequence GLN-LYS-SER-LYS-GLN in M protein. *J. Immunol.* **143:** 2677–2683.

34. **Cunningham, M. W., J. M. McCormack, L. R. Talaber, J. B. Harley, E. M. Ayoub, R. S. Muneer, L. T. Chun, and D. V. Reddy.** 1988. Human monoclonal antibodies reactive with antigens of the group A *Streptococcus* and human heart. *J. Immunol.* **141:**2760–2766.

35. **Cunningham, M. W., and A. Quinn.** 1997. Immunological cross-reactivity between the class I epitope of streptococcal M protein and myosin, p. 887–892. *In* T. Horaud, A. Bouvet, R. Leclercq, H. de Montclos, and M. Sicard (ed.), *Streptococci and the Host.* Plenum Press, London, U.K.

36. **Cunningham, M. W., and R. A. Swerlick.** 1986. Polyspecificity of antistreptococcal murine monoclonal antibodies and their implications in autoimmunity. *J. Exp. Med.* **164:**998–1012.

38. **Dajani, A. S.** 1992. Guidelines for the diagnosis of rheumatic fever (Jones criteria, 1992 update). *JAMA* **268:** 2069–2073.

39. **Dale, J. B., and E. H. Beachey.** 1985. Epitopes of streptococcal M proteins shared with cardiac myosin. *J. Exp. Med.* **162:**583–591.

40. **Dale, J. B., and E. H. Beachey.** 1985. Multiple, heart-cross-reactive epitopes of streptococcal M proteins. *J. Exp. Med.* **161:**113–122.

41. **Dale, J. B., and E. H. Beachey.** 1986. Sequence of myosin-cross-reactive epitopes of streptococcal M protein. *J. Exp. Med.* **164:**1785–1790.

42. **Dale, J. B., and E. H. Beachey.** 1987. Human cytotoxic T lymphocytes evoked by group A streptococcal M proteins. *J. Exp. Med.* **166:**1825–1835.

43. **Degnan, B., J. Taylor, and J. A. Goodacre.** 1997. *Streptococcus pyogenes* type 5 M protein is an antigen, not a superantigen for human T cells. *Hum. Immunol.* **53:** 206–215.

44. **Dell, A., S. M. Antone, C. J. Gauntt, C. A. Crossley, W. A. Clark, and M. W. Cunningham.** 1991. Autoimmune determinants of rheumatic carditis: localization of epitopes in human cardiac myosin. *Eur. Heart J.* **12:** 158–162.

45. **Dudding, B. A., and E. M. Ayoub.** 1968. Persistence of streptococcal group A antibody in patients with rheumatic valvular disease. *J. Exp. Med.* **128:**1081–1098.

46. **Fenderson, P. G., V. A. Fischetti, and M. W. Cunningham.** 1989. Tropomyosin shares immunologic epitopes with group A streptococcal M proteins. *J. Immunol.* **142:**2475–2481.

47. **Fleischer, B., K.-H. Schmidt, D. Gerlach, and W. Kohler.** 1992. Separation of T-cell-stimulating activity from streptococcal M protein. *Infect. Immun.* **60:**1767–1770.

48. **Fujinami, R. S., and M. B. A. Oldstone.** 1985. Amino acid homology between the encephalitogenic site of myelin basic protein and virus: mechanism of autoimmunity. *Science* **230:**1043–1045.

49. **Fujinami, R. S., M. B. A. Oldstone, Z. Wroblewska, M. E. Frankel, and H. Koprowski.** 1983. Molecular mimicry in virus infections: cross reaction of measles virus phosphoprotein of herpes simplex virus protein with human intermediate filaments. *Proc. Natl. Acad. Sci. USA* **80:**2346–2350.

50. **Galvin, J. E., M. E. Hemric, K. E. Ward, and M. W. Cunningham.** Cytotoxic monoclonal antibody from rheumatic carditis recognizes human endothelium: implications in rheumatic heart disease. Submitted for publication.

51. **Gauntt, C., S. Tracy, N. Chapman, H. Wood, P. Kolbeck, A. Karaganis, C. Winfrey, and M. Cunningham.** 1995. Coxsackievirus-induced chronic myocarditis in murine models. *Eur. Heart J.* **16**(Suppl.)**:**56–58.

52. **Goldstein, I., B. Halpern, and L. Robert.** 1967. Immunological relationship between streptococcus A polysaccharide and the structural glycoproteins of heart valve. *Nature* **213:**44–47.

53. **Goroncy-Bermes, P., J. B. Dale, E. H. Beachey, and W. Opferkuch.** 1987. Monoclonal antibody to renal glomeruli cross-reacts with streptococcal M protein. *Infect. Immun.* **55:**2416–2419.

54. **Guilherme, L., E. Cunha-Neto, V. Coelho, R. Snitcowsky, P. M. A. Pomerantzeff, R. V. Assis, F. Pedra, J.**

Neumann, A. Goldberg, M. E. Patarroyo, F. Pileggi, and J. Kalil. 1995. Human heart-filtrating T cell clones from rheumatic heart disease patients recognize both streptococcal and cardiac proteins. *Circulation* **92:**415–420.

55. **Huber, S., J. Polgar, A. Moraska, M. Cunningham, P. Schwimmbeck, and P. Schultheiss.** 1993. T lymphocyte responses in CVB3-induced murine myocarditis. *Scand J. Infect. Dis. Suppl.* **88:**67–78.

56. **Huber, S. A., and M. W. Cunningham.** 1996. Streptococcal M protein peptide with similarity to myosin induces CD4+ T cell-dependent myocarditis in MRL/++ mice and induces partial tolerance against coxsackieviral myocarditis. *J. Immunol.* **156:**3528–3534.

57. **Huber, S. A., L. P. Job, and J. F. Woodruff.** 1980. Lysis of infected myofibers by coxsackievirus B3 immune lymphocytes. *Am. J. Pathol.* **98:**681–694.

58. **Huber, S. A., and P. A. Lodge.** 1986. Coxsackievirus B-3 myocarditis: identification of different pathogenic mechanisms in DBA/2 and BALB/c mice. *Am. J. Pathol.* **122:**284–291.

59. **Huber, S. A., A. Moraska, and M. Cunningham.** 1994. Alterations in major histocompatibility complex association of myocarditis induced by coxsackievirus B3 mutants selected with monoclonal antibodies to group A streptococci. *Proc. Natl. Acad. Sci. USA* **91:**5543–5547.

60. **Hutto, J. H., and E. M. Ayoub.** 1980. Cytotoxicity of lymphocytes from patients with rheumatic carditis to cardiac cells *in vitro*, p. 733–738. *In* S. E. Read and J. B. Zabriskie (ed.), *Streptococcal Diseases and the Immune Response.* Academic Press, Inc., New York, N.Y.

61. **Jones, K., S. S. Whitehead, M. W. Cunningham, and V. A. Fischetti.** Reactivity of rheumatic fever and scarlet fever patient sera with group A streptococcal M protein, myosin, and tropomyosin: a retrospective study. Submitted for publication.

62. **Jones, K. F., and V. A. Fischetti.** 1988. The importance of the location of antibody binding on the M6 protein for opsonization and phagocytosis of group A M6 streptococci. *J. Exp. Med.* **167:**1114–1123.

63. **Jones, K. F., S. A. Khan, B. W. Erickson, S. K. Hollingshead, J. R. Scott, and V. A. Fischetti.** 1986. Immunochemical localization and amino acid sequences of cross-reactive epitopes within the group A streptococcal M6 protein. *J. Exp. Med.* **164:**1226–1238.

64. **Jones, T. D.** 1944. The diagnosis of rheumatic fever. *JAMA* **126:**481–484.

65. **Kaplan, M. H.** 1963. Immunologic relation of streptococcal and tissue antigens. I. Properties of an antigen in certain strains of group A streptococci exhibiting an immunologic cross reaction with human heart tissue. *J. Immunol.* **90:**595–606.

66. **Kaplan, M. H., R. Bolande, L. Rakita, and J. Blair.** 1964. Presence of bound immunoglobulins and complement in the myocardium in acute rheumatic fever. Association with cardiac failure. *N. Engl. J. Med.* **271:**637–645.

67. **Kaplan, M. H., and M. Meyeserian.** 1962. Immunologic cross-reaction between group-A streptococcal cells and human heart tissue. *Lancet* **i:**706–710.

69. **Kaplan, M. H., and M. L. Suchy.** 1964. Immunologic relation of streptococcal and tissue antigens. II. Cross reactions of antisera to mammalian heart tissue with a cell wall constituent of certain strains of group A streptococci. *J. Exp. Med.* **119:**643–650.

70. **Kaplan, M. H., and K. H. Svec.** 1964. Immunologic relation of streptococcal and tissue antigens. III. Presence in human sera of streptococcal antibody cross reactive with heart tissue. Association with streptococcal infection, rheumatic fever, and glomerulonephritis. *J. Exp. Med.* **119:**651–666.

71. **Kefalides, N. A., N. Ohno, C. B. Wilson, H. Fillit, J. Zabriskie, and J. Rosenbloom.** 1993. Identification of antigenic epitopes in type IV collagen by use of synthetic peptides. *Kidney Int.* **43:**94–100.

72. **Kefalides, N. A., M. T. Pegg, N. Ohno, T. Poon-King, J. Zabriskie, and H. Fillit.** 1986. Antibodies to basement membrane collagen and to laminin are present in sera from patients with poststreptococcal glomerulonephritis. *J. Exp. Med.* **163:**588–602.

73. **Kil, K. S., M. W. Cunningham, and L. A. Barnett.** 1994. Cloning and sequence analysis of a gene encoding a 67-kilodalton myosin-cross-reactive antigen of *Streptococcus pyogenes* reveals its similarity to class II major histocompatibility antigens. *Infect. Immun.* **62:**2440–2449.

74. **Kraus, W., and E. H. Beachey.** 1988. Renal autoimmune epitope of group A streptococci specified by M protein tetrapeptide Ile-Arg-Leu-Arg. *Proc. Natl. Acad. Sci. USA* **85:**4516–4520.

75. **Kraus, W., J. B. Dale, and E. H. Beachey.** 1990. Identification of an epitope of type 1 streptococcal M protein that is shared with a 43-kDa protein of human myocardium and renal glomeruli. *J. Immunol.* **145:**4089–4093.

76. **Kraus, W., J. M. Seyer, and E. H. Beachey.** 1989. Vimentin-cross-reactive epitope of type 12 streptococcal M protein. *Infect. Immun.* **57:**2457–2461.

77. **Krisher, K., and M. W. Cunningham.** 1985. Myosin: a link between streptococci and heart. *Science* **227:**413–415.

78. **Lafer, E. M., J. Rauch, C. Andrezejewski, Jr., D. Mudd, B. Furie, R. S. Schwartz, and B. D. Stollar.** 1981. Polyspecific monoclonal lupus autoantibodies reactive with both polynucleotides and phospholipids. *J. Exp. Med.* **153:**897–909.

79. **Lange, C. F.** 1969. Chemistry of cross-reactive fragments of streptococcal cell membrane and human glomerular basement membrane. *Transplant. Proc.* **1:**959–963.

80. **Lange, C. F.** 1994. Localization of [C¹⁴] labeled anti-streptococcal cell membrane monoclonal antibodies (anti-SCM mAb) in mice. *Autoimmunity* **19:**179–191.

81. **Lindberg, L. H., and K. L. Vosti.** 1969. Elution of glomerular bound antibodies in experimental streptococcal glomerulonephritis. *Science* **166:**1032–1033.

82. **Lyampert, I. M., L. V. Beletskaya, N. A. Borodiyuk, E. V. Gnezditskaya, I. I. Rassokhina, and T. A. Danilova.** 1976. A cross-reactive antigen of thymus and skin epithelial cells common with the polysaccharide of group A streptococci. *Immunology* **31:**47–55.

83. **Lyampert, I. M., N. A. Borodiyuk, and G. A. Ugryumova.** 1968. The reaction of heart and other organ extracts with the sera of animals immunized with group A streptococci. *Immunology* **15:**845–854.

84. **Lyampert, I. M., O. I. Vvedenskaya, and T. A. Danilova.** 1966. Study on streptococcus group A antigens common with heart tissue elements. *Immunology* **11:**313–320.

85. **Manjula, B. N., A. S. Acharya, S. M. Mische, T. Fairwell, and V. A. Fischetti.** 1984. The complete amino acid sequence of a biologically active 197-residue fragment of

M protein isolated from type 5 group A streptococci. *J. Biol. Chem.* **259:**3686–3693.

86. **Manjula, B. N., and V. A. Fischetti.** 1980. Tropomyosin-like seven residue periodicity in three immunologically distinct streptococcal M proteins and its implications for the antiphagocytic property of the molecule. *J. Exp. Med.* **151:**695–708.

87. **Manjula, B. N., and V. A. Fischetti.** 1986. Sequence homology of group A streptococcal Pep M5 protein with other coiled-coil proteins. *Biochem. Biophys. Res. Commun.* **140:**684–690.

88. **Manjula, B. N., B. L. Trus, and V. A. Fischetti.** 1985. Presence of two distinct regions in the coiled-coil structure of the streptococcal Pep M5 protein: relationship to mammalian coiled-coil proteins and implications to its biological properties. *Proc. Natl. Acad. Sci. USA* **82:**1064–1068.

89. **Markowitz, A. S., and C. F. Lange.** 1964. Streptococcal related glomerulonephritis. I. Isolation, immunocytochemistry and comparative chemistry of soluble fractions from type 12 nephritogenic streptococci and human glomeruli. *J. Immunol.* **92:**565.

90. **McCarty, M.** 1956. Variation in the group specific carbohydrates of group A streptococci. II. Studies on the chemical basis for serological specificity of the carbohydrates. *J. Exp. Med.* **104:**629–643.

91. **McCarty, M.** 1964. Missing links in the streptococcal chain leading to rheumatic fever: The Duckett Jones Memorial Lecture. *Circulation* **29:**488–493.

92. **Mertens, N. M. J., J. E. Galvin, E. E. Adderson, and M. W. Cunningham.** Molecular analysis of cross-reactive anti-streptococcal/anti-myosin mouse monoclonal antibodies: nucleotide sequences for cross-reactivity. Submitted for publication.

93. **Michael, A. F., K. N. Drummond, R. A. Good, and R. L. Vernier.** 1966. Acute poststreptococcal glomerulonephritis: immune deposit disease. *J. Clin. Invest.* **45:**237–248.

94. **Miller, L. C., E. D. Gray, E. H. Beachey, and M. A. Kehoe.** 1988. Antigenic variation among group A streptococcal M proteins: nucleotide sequence of the serotype 5 M protein gene and its relationship with genes encoding types 6 and 24 M proteins. *J. Biol. Chem.* **263:**5668–5673.

95. **Mori, K., N. Kamakawaji, and T. Sasazuki.** 1996. Persistent elevation of immunoglobulin G titer against the C region of recombinant group A streptococcal M protein in patients with rheumatic fever. *Pediatr. Res.* **55:**502–506.

96. **Neu, N., N. R. Rose, K. W. Beisel, A. Herskowitz, G. Gurri-Glass, and S. W. Craig.** 1987. Cardiac myosin induces myocarditis in genetically predisposed mice. *J. Immunol.* **139:**3630–3636.

97. **Pruksakorn, S., B. Currie, E. Brandt, P. C., S. Hunsakunachai, A. Manmontri, J. H. Robinson, M. A. Kehoe, A. Galbraith, and M. F. Good.** 1994. Identification of T cell autoepitopes that cross-react with the C-terminal segment of the M protein of group A streptococci. *Int. Immunol.* **6:**1235–1244.

98. **Pruksakorn, S., A. Galbraith, R. A. Houghten, and M. F. Good.** 1992. Conserved T and B cell epitopes on the M protein of group A streptococci. Induction of bactericidal antibodies. *J. Immunol.* **149:**2729–2735.

99. **Putterman, C., and B. Diamond.** 1998. Immunization with a peptide surrogate for double stranded DNA (dsDNA) induces autoantibody production and renal immunoglobulin deposition. *J. Exp. Med.* **188:**29–38.

100. **Quinn, A., T. M. Shinnick, and M. W. Cunningham.** 1996. Anti-Hsp 65 antibodies recognize M proteins of group A streptococci. *Infect. Immun.* **64:**818–824.

101. **Quinn, A., K. Ward, V. Fischetti, M. Hemric, and M. W. Cunningham.** 1998. Immunological relationship between the class I epitope of streptococcal M protein and myosin. *Infect. Immun.* **66:**4418–4424.

102. **Robinson, J. H., M. C. Atherton, J. A. Goodacre, M. Pinkney, H. Weightman, and M. A. Kehoe.** 1991. Mapping T-cell epitopes in group A streptococcal type 5 M protein. *Infect. Immun.* **59:**4324–4331.

103. **Robinson, J. H., M. C. Case, and M. A. Kehoe.** 1993. Characterization of a conserved helper T-cell epitope from group A streptococcal M proteins. *Infect. Immun.* **61:**1062–1068.

104. **Sandson, J., D. Hamerman, and R. Janis.** 1968. Immunologic and chemical similarities between the streptococcus and human connective tissue. *Trans. Assoc. Am. Physicians* **81:**249–257.

105. **Sargent, S. J., E. H. Beachey, C. E. Corbett, and J. B. Dale.** 1987. Sequence of protective epitopes of streptococcal M proteins shared with cardiac sarcolemmal membranes. *J. Immunol.* **139:**1285–1290.

106. **Schett, G., Q. Xu, A. Amberger, R. Van der Zee, H. Recheis, J. Willeit, and G. Wick.** 1995. Autoantibodies against heat shock protein 60 mediate endothelial cytotoxicity. *J. Clin. Invest.* **96:**2569–2577.

107. **Schmidt, K.-H., D. Gerlach, L. Wollweber, W. Reichardt, K. Mann, J.-H. Ozegowski, and B. Fleischer.** 1995. Mitogenicity of M5 protein extracted from *Streptococcus pyogenes* cells is due to streptococcal pyrogenic exotoxin C and mitogenic factor MF. *Infect. Immun.* **63:**4569–4575.

108. **Schwab, J. H.** 1962. Analysis of the experimental lesion of connective tissue produced by a complex of C polysaccharide from group A streptococci. I. In vivo reaction between tissue and toxin. *J. Exp. Med.* **116:**17–28.

109. **Schwab, J. H.** 1964. Analysis of the experimental lesion of connective tissue produced by a complex of C polysaccharide from group A streptococci. II. Influence of age and hypersensitivity. *J. Exp. Med.* **119:**401–408.

110. **Schwab, J. H.** 1965. Biological properties of streptococcal cell wall particles. I. Determinants of the chronic nodular lesion of connective tissue. *J. Bacteriol.* **90:**1405–1411.

111. **Schwab, J. H., J. Allen, S. Anderle, F. Dalldorf, R. Eisenberg, and W. J. Cromartie.** 1982. Relationship of complement to experimental arthritis induced in rats with streptococcal cell walls. *Immunology* **46:**83–88.

112. **Schwimmbeck, P. L., and M. B. A. Oldstone.** 1989. Klebsiella pneumoniae and HLA B27-associated diseases of Reiter's syndrome and ankylosing spondylitis. *Curr. Top. Microbiol. Immunol.* **45:**45–56.

113. **Shikhman, A. R., and M. W. Cunningham.** 1994. Immunological mimicry between N-acetyl-beta-D-glucosamine and cytokeratin peptides. Evidence for a microbially driven anti-keratin antibody response. *J. Immunol.* **152:**4375–4387.

114. **Shikhman, A. R., and M. W. Cunningham.** 1997. Trick and treat: toward peptide mimic vaccines. *Nat. Biotechnol.* **15:**512–513.

115. **Shikhman, A. R., N. S. Greenspan, and M. W. Cunningham.** 1993. A subset of mouse monoclonal antibodies cross-reactive with cytoskeletal proteins and group A streptococcal M proteins recognizes N-acetyl-beta-D-glucosamine. *J. Immunol.* **151:**3902–3913.

116. **Shikhman, A. R., N. S. Greenspan, and M. W. Cunningham.** 1994. Cytokeratin peptide SFGSGFGGGY mimics N-acetyl-beta-D-glucosamine in reaction with antibodies and lectins, and induces in vivo anti-carbohydrate antibody response. *J. Immunol.* **153:**5593–5606.

117. **Swerlick, R. A., M. W. Cunningham, and N. K. Hall.** 1986. Monoclonal antibodies cross-reactive with group A streptococci and normal and psoriatic human skin. *J. Invest. Dermatol.* **87:**367–371.

118. **Tomai, M., J. A. Aileon, M. E. Dockter, G. Majumbar, D. G. Spinella, and M. Kotb.** 1991. T cell receptor V gene usage by human T cells stimulated with the superantigen streptococcal M protein. *J. Exp. Med.* **174:**285–288.

119. **Tomai, M., M. Kotb, G. Majumdar, and E. H. Beachey.** 1990. Superantigenicity of streptococcal M protein. *J. Exp. Med.* **172:**359–362.

120. **Tomai, M., P. M. Schlievert, and M. Kotb.** 1992. Distinct T cell receptor Vβ gene usage by human T lymphocytes stimulated with the streptococcal pyrogenic exotoxins and M protein. *Infect. Immun.* **60:**701–705.

121. **van de Rijn, I., J. B. Zabriskie, and M. McCarty.** 1977. Group A streptococcal antigens cross-reactive with myocardium: purification of heart reactive antibody and isolation and characterization of the streptococcal antigen. *J. Exp. Med.* **146:**579–599.

122. **Wang, B., P. M. Schlievert, A. O. Gaber, and M. Kotb.** 1993. Localization of an immunologically functional region of the streptococcal superantigen pepsin-extracted fragment of type 5 M protein. *J. Immunol.* **151:**1419–1429.

123. **Wedemayer, G. J., P. A. Patten, L. H. Wang, P. G. Schultz, and R. C. Stevens.** 1997. Structural insights into the evolution of an antibody combining site. *Science* **276:**1665–1669.

124. **Widdowson, J. P.** 1980. The M-associated protein antigens of group A streptococci, p. 125–147. *In* S. E. Read and J. B. Zabriskie (ed.), *Streptococcal Diseases and the Immune Response.* Academic Press, Inc., New York, N.Y.

125. **Widdowson, J. P., W. R. Maxted, D. L. Grant, and A. M. Pinney.** 1971. The relationship between M antigen and opacity factor in group A streptococci. *J. Gen. Microbiol.* **65:**69–80.

126. **Widdowson, J. P., W. R. Maxted, and A. M. Pinney.** 1976. An M-associated protein antigen (MAP) of group A streptococci. *J. Hyg.* **69:**553–564.

127. **Wu, X., B. Liu, P. L. Van der Merwe, N. N. Kalis, S. M. Berney, and D. C. Young.** 1998. Myosin-reactive autoantibodies in rheumatic carditis and normal fetus. *Clin. Immunol. Immunopathol.* **87:**184–192.

128. **Zabriskie, J. B.** 1967. Mimetic relationships between group A streptococci and mammalian tissues. *Adv. Immunol.* **7:**147–188.

129. **Zabriskie, J. B., and E. H. Freimer.** 1966. An immunological relationship between the group A streptococcus and mammalian muscle. *J. Exp. Med.* **124:**661–678.

130. **Zabriskie, J. B., K. C. Hsu, and B. C. Seegal.** 1970. Heart-reactive antibody associated with rheumatic fever: characterization and diagnostic significance. *Clin. Exp. Immunol.* **7:**147–159.

Extracellular Matrix Interactions with Gram-Positive Pathogens

GURSHARAN S. CHHATWAL AND KLAUS T. PREISSNER

8

Adherence to and invasion of eukaryotic cells are the main strategies used by pathogenic bacteria for colonization, evasion of immune defenses, survival, and causing disease in the mammalian hosts. Most gram-negative bacteria make use of pili to achieve adherence and invasion. Since gram-positive pathogens do not possess pili, they express specific cell surface components called adhesins that mediate their adherence to host tissues, thereby facilitating not only colonization but also invasion (37). Most of these adhesins function by recognizing and binding to various components of the host extracellular matrix (ECM) (44). ECM consists of many diverse structures and complex macromolecules that maintain the bulk of tissues and provide them with tensile strength and elasticity (39). In addition, ECM affects the cellular physiology of the organism and is critical for adhesion, migration, proliferation, and differentiation of many cell types. ECM not only serves as structural support for cells but also provides support for infiltrating pathogens to colonize and invade, particularly under conditions of injury and trauma. Cell surface adhesive components of the host are often recognized by pathogenic bacteria in a tissue- or cell-specific manner. Pathogens also contact host tissue fluids that contain a variety of adhesive components. Microbial binding may lead to structural and/or functional alterations of host proteins and to activation of cellular mechanisms that influence tissue and cell invasion of pathogens.

The interactions of bacteria with ECM, therefore, represent important pathogenicity mechanisms. The identification and characterization of host ECM molecules and complementary bacterial adhesins would contribute not only to a better understanding of the molecular aspects of pathogenesis but also to the design of novel strategies to control and manage infectious diseases. In this chapter we describe the interaction of ECM components with gram-positive pathogens. The first section describes the structure and function of ECM, and the second deals with the interactions of gram-positive pathogens with ECM components and their biological consequences. These interactions are summarized in Table 1.

EXTRACELLULAR MATRIX

The major structural components of the eukaryotic ECM are collagens that form different types of interstitial or basement membrane networks (64). The interstitial ECMs are mainly built up by fibril-forming collagens (types I, II, and III), whereas basement membranes contain a two-dimensional collagen type IV network. Other components, such as collagenous and noncollagenous glycoproteins, elastin, proteoglycans, hyaluronan, growth factors, and proteases, become associated with ECM, giving rise to their specialized structure and function at various locations in the body (25). The main portion of interstitial ECM is produced and deposited by imbedded connective tissue cells such as fibroblasts in the dermis, smooth muscle cells and fibroblasts in the vasculature, and osteoblasts and chondroblasts in bone and cartilage. In addition, melanocytes, fat, skeletal muscle, nerve and epithelial cells, and circulating blood cells, such as macrophages, granulocytes, and lymphocytes, participate in the synthesis and secretion of ECM material and determine the specific character of each tissue or organ. During embryonic development, the ECM is constantly rearranged, thereby regulating morphogenesis in an active and dynamic fashion. Likewise, the time-dependent modification and rebuilding of ECM at various locations in the body is essential for inflammation and wound healing. At these sites, the provisional ECM may provide bacterial entry and colonization.

Structure and Function of ECM Molecules

More than 25 different collagens have been defined on a molecular basis. Most collagens are the fibril-forming types I through III, produced and assembled during wound healing processes (50). The fibril-forming collagens, which also include types V and XI, are distinct from the non-fibril-forming collagens that form networks, such as type IV and type X collagens and the type VI collagen microfibrils. Other collagen types are mostly expressed in a tissue- or cell-specific manner and contribute to the appearance and mechanical properties of a given ECM.

Network-forming collagens, predominantly different isoforms of type IV, determine the structure of basement mem-

Gram-Positive Pathogens, ed. by V. A. Fischetti et al.
© 2000 American Society for Microbiology, Washington, D.C.

TABLE 1 Properties of ECM components involved in interaction with gram-positive pathogens

Parameter	Collagen type I, II	Collagen type IV	Elastin	Laminin	Fibrinogen	Fibronectin	Vitronectin	Thrombospondin
Mol. mass (kDa)	Multimeric fibers	Multimeric network	Multimeric network	900	350	440	78	420
Immobilized form	Yes	Yes	Yes	Yes	Yes	Yes	Yes	Yes
Soluble form	No	No	No	No	Yes	Yes	Yes	Yes
Plasma concentration (mg/ml)					2–4	0.2–0.4	0.2–0.4	Approx. 0.02
Major bacteria binding domain	A domain		30-kDa N-terminal		D fragment	29-kDa N-terminal domain/type I module	Hemopexin-like repeats, heparin-binding site	?
Interacting major gram-positive pathogen	Staphylococcus aureus	Streptococcus pyogenes	S. aureus, coagulase-negative staphylococci	S. aureus, Streptococcus sanguis	S. aureus, A, B, C, and G streptococci	S. aureus, coagulase-negative staphylococci, A, B, C, and G streptococci, Streptococcus pneumoniae, S. sanguis	S. aureus, coagulase-negative staphylococci, A, C, and G streptococci	S. aureus, A, C, and G streptococci

branes that are a specialized ECM. An intercalated laminin network that can also exist independently is linked to type IV collagen via nidogen/entactin. Heparan sulfate proteoglycans, usually perlecan, and adhesive glycoproteins are embedded into these supramolecular arrays (64). These components are in direct contact with epithelial, endothelial, striated muscle, fat, and nerve cells, which produce the individual components and deposit them in a polarized fashion toward the cellular basolateral side. Basement membranes serve quite distinct functions in the body, such as linking epithelia to the underlying interstitial ECM, covering skeletal myotubes as basal lamina, or serving as a filtration barrier in the glomerulus.

Laminins

Laminins are major constituents of basement membranes and are the first ECM proteins to be produced during embryogenesis. Because of their specific interactions with type IV collagen, proteoglycans, and other ECM components, as well as with several cell types, laminins play an important structural and functional role within basement membranes. Laminin, with a molecular mass of 900 kDa, is composed of three different chains, α, β, and γ, which are linked to each other by disulfide bridges. Owing to variations in chain composition, at least 10 different isoforms of laminin exist. Laminins are in close contact with a variety of cell types, including muscle, adipocytes, neurons, and endothelial and epithelial cells. Laminins are involved in promotion of cell proliferation, attachment, and chemotaxis, as well as neurite outgrowth and enhancement of angiogenesis.

Elastin

Another important molecule of ECM is elastin, which is the main component of elastic fiber. These fibers are responsible for the elasticity of lung, skin, and other tissues, particularly blood vessels. Elastin is a highly cross-linked polymer of the nonproteolytically modified, hydrophobic precursor tropoelastin. During fibrogenesis, secreted tropoelastin becomes stabilized into insoluble elastin by intermolecular cross-linking of lysines into desmosine and isodesmosine cross-links through the copper-dependent enzyme lysyl oxidase (26). The rubberlike mechanical properties of elastin result from repetitive hydrophobic domains of tropoelastin.

Adhesive Glycoproteins

Adhesive glycoproteins constitute important components of ECM. Upon vessel wall injury, particularly at sites of wound healing, initial adhesion of platelets and subsequent aggregation are dependent on adhesive glycoproteins, such as fibronectin and vitronectin, that are present in the subendothelial cell matrix and stored inside platelets and are secreted during this initial phase of hemostatic plug formation. Adhesive glycoproteins not only promote attachment via their Arg-Gly-Asp (RGD)-containing epitope (48) but also perform multiple functions by interacting with other ligands, such as heparan sulfate, collagens, or mediators of humoral defense mechanisms. These proteins also exist as soluble forms, which may differ from those in the subendothelium and α-granules of platelets owing to alternative splicing, differences in the state of polymerization, different conformations, or the transition into a self-aggregating molecule. The partitioning of these proteins between humoral and cell surface or ECM phases, together with inducible receptor sites on platelets or inflammatory cells, indicates that these molecules and responding cells are present in a "preactivated" state under quiescent conditions and become "activated" into functionally relevant forms once natural defense mechanisms are initiated or during vascular remodeling (48). Thus, these adhesion molecules not only provide versatile molecular links mediating adhesive processes and responsive reactions at localized sites but are also of major importance for the initial adherence phase of pathogens.

Fibronectin

Fibronectin is a ubiquitous adhesive protein that is essential for the adhesion of almost all types of cells. It is abundant in the circulation and at various extracellular sites. The characteristic form of the molecule in solution is a dimer generated by disulfide bridging at the carboxy terminus of two similar subunits, each with a molecular mass of about 220 kDa (22). In addition to approximately 30 intrachain disulfide bonds, two free sulhydryl groups per subunit are involved in the formation of high-molecular-weight polymers of fibronectin, which are predominantly found in extracellular tissues. The heterogeneity observed in fibronectin molecules isolated from plasma or tissue is due to variation in both the amino acid sequence and post-translational modifications. Although only one gene has been identified, variations in the carbohydrate content or structure, phosphorylation, sulfation, and acetylation are responsible for additional heterogeneity.

The sensitivity of fibronectin to proteolytic degradation has been used to identify the structure-function relationships of independent domains. The 30-kDa amino-terminal fragment contains the major acceptor site for factor XIIIa-mediated cross-linking and also bears the binding sites for heparin, fibrin, and bacteria, including *Staphylococcus aureus* (38). The well-known property of fibronectin to bind collagen or gelatin and to complement C1q is contained within the adjacent 40-kDa fragment, while the central portion of the molecule has no well-defined binding functions. The versatile integrin recognition sequence RGD was first recognized in the type III-11 repeat of fibronectin (46) and has been found in a large number of adhesive and nonadhesive proteins. Additional heparin-binding and fibrin-binding domains are located within the carboxy-terminal portion of the fibronectin molecule. Together with the RGD-containing cell attachment site, the heparin-binding domain is crucial for the establishment of stable focal adhesions, as has been demonstrated in fragment complementation assays. Synthesis and deposition of fibronectin as a self-associating fibrillar array into the growing ECM of adhesive cells occur by an active process, and the accumulation of fibronectin into an insoluble form is potentiated by disulfide bridge formation, as well as covalent cross-linking by transglutaminase/factor XIIIa.

Vitronectin

Another adhesive glycoprotein, vitronectin (47), is found in the circulation as a single-chain polypeptide with a molecular mass of 78 kDa and becomes associated with various ECM sites, particularly during tissue or vascular remodeling processes. Several immunofluorescent and histochemical studies suggest the deposition of vitronectin in a fibrillar pattern in loose connective tissue, in association with dermal elastic fibers in skin and with renal tissue (49). Moreover, the accumulation of terminal complement complexes along elastic fibers later in life, the association of vitronec-

tin with keratin bodies during keratinocyte programmed cell death (apoptosis), and colocalization of vitronectin with deposits of the terminal complement complex in kidney tissue from patients with glomerulonephritis suggest a role for vitronectin in preventing tissue damage in proximity to local complement activation. Although the mechanism of deposition of exogenous vitronectin alone or in association with other proteins into different tissues remains unclear, it may occur in a fashion similar to that for fibronectin; thus, both proteins are prominent candidates for promoting adherence of bacteria at various accessible ECM sites. The interaction of vitronectin with glycosaminoglycans or different types of native collagens and basement membrane-associated osteonectin and the cross-linking of vitronectin by transglutaminase/factor XIIIa are reactions that are likely to occur in the ECM in vivo as well (27, 54).

Fibrinogen

The adhesive glycoprotein fibrinogen is a major plasma glycoprotein that serves as the predominant macromolecular substrate for thrombin in the blood-clotting cascade. The primary structure of fibrinogen (18) together with biophysical studies (16) indicate that the 350-kDa molecule is composed of two identical sets of three polypeptide chains, $A\alpha$, $B\beta$, and γ, which are disulfide-bridged and organized in an antiparallel fashion. Upon selective and specific proteolytic attack by α-thrombin, two pairs of fibrinopeptides, A and B, are sequentially released, and the appearance of these peptides in the circulation is an indicator of thrombin activity in vivo. Covalent stabilization of the forming fibrin clot is mediated by transglutaminase/factor XIIIa-dependent cross-linking between γ and α chains. Together with invading cells and aggregating platelets, the fibrin clot constitutes the majority of the initial provisional ECM network for the wound healing process (14).

Fibrinogen is a multifunctional protein capable of binding to collagen, fibronectin, components of the fibrinolytic system, and a variety of eukaryotic cells, as well as to bacteria. In particular, while the fibrin clot is being organized, it may already serve as a cofactor surface for tissue plasminogen activator-dependent plasminogen formation, ultimately leading to clot lysis. Cell surface receptors for fibrinogen that belong to the family of integrins have been identified on mammalian cells; the platelet integrin α_{IIb}-β_3 (GP IIb/IIIa) is principally required for platelet aggregation. Integrin α_M-β_2 (complement receptor 3) on phagocytes may also recognize the ligand during wound healing and defense when phagocytic clearance of fibrin(ogen)-associated clot or cell fragments is required. The recognition of the distal end of the γ-chain of fibrinogen and of two RGD-containing sites and additional epitopes by these integrins indicates that fibrinogen may serve as a bridging component between surface receptors on different cells or other extracellular sites once they become exposed.

Thrombospondin

Thrombospondin, a multifunctional glycoprotein of 420 kDa, is stored in platelet granules and is secreted upon stimulation (28). The released thrombospondin becomes incorporated into fibrin clots. Thrombospondin belongs to a family of structurally unrelated members of the tenascin protein family, including osteonectin and osteopontin. These proteins promote divergent cellular functions (3).

Although they can associate with structural elements, such as collagen fibrils or basement membranes, they do not contribute to the structural integrity of connective tissues and are termed matricellular proteins (3). Rather, owing to highly regulated biosynthesis by various hormones and cytokines and during development, nerve regeneration, and vascular remodeling related to angiogenesis, these proteins become sequestered in the ECM and are therefore available for subsequent recruitment to different cell surface receptors, including integrins. In particular, thrombospondin plays a role in binding and modulation of cytokines and proteases. The level of thrombospondin in the circulation and other body fluids produced by vascular cells or secreted by activated platelets and its distribution in tissues appears to vary in correlation with different pathological states. Specifically, thrombospondin promotes cell attachment and spreading via αv-integrins in the vascular system but may also lead to destabilization of focal adhesions (40).

INTERACTION OF GRAM-POSITIVE PATHOGENS WITH ECM MOLECULES

Binding of Collagen to Gram-Positive Pathogens

Many staphylococcal and streptococcal species interact specifically with collagenous proteins. It has been speculated that these interactions might be important in the pathogenesis of various diseases, such as osteomyelitis and infective arthritis. Switalski et al. (59) isolated a surface protein of 135 kDa from S. aureus and tentatively identified it as a collagen receptor. Patti et al. (45) cloned and sequenced a gene, cna, encoding a collagen-binding protein from S. aureus. This collagen-binding protein exhibits all the main features of the surface protein of gram-positive bacteria. The ligand-binding domain of S. aureus is localized in the A region of collagen (45). So far, only a single gene has been identified for staphylococcal collagen-binding protein, although proteins of different molecular sizes have been reported from different strains, probably owing to strain-to-strain variation in the number of repeats (58). The collagen-binding protein from S. aureus is involved in the adherence of bacteria to cartilage. The mediation of adherence to cartilage is specific to collagen, because synthetic beads coated with collagen mediate the adherence to cartilage, whereas the beads coated with fibronectin show no adherence (58).

The binding to collagen by different streptococcal species has also been reported. Group A streptococci, isolated from patients with acute glomerulonephritis, specifically interact with type IV collagen. The binding is not dependent on the expression of M protein but involves another protein component, as shown by its sensitivity toward proteases. Animal pathogenic streptococci also interact with collagen type II. Streptococcus mutans interacts with collagen type I via a 16-kDa binding protein with high affinity, as shown by adherence of bacteria to a collagen-coated surface. Furthermore, collagen mediates adherence of S. mutans to dentin in the oral cavity, thereby playing a role in the pathogenesis of root surface caries (56). In infectious endocarditis, which is most commonly caused by streptococci and which is characterized by the formation of septic masses of platelets on the surfaces of heart valves, collagenlike platelet aggregation-associated protein of Streptococcus sanguis and direct interactions with host ECM

collagens enhance platelet accumulation and subsequent bacterial colonization. The binding of collagen to antigen I/II family of polypeptides of oral streptococci has also been reported (32). This interaction facilitates the invasion of bacteria into root dentinal tubes, which further underlines the importance of collagen-bacterial interactions.

Binding of Laminin to Gram-Positive Bacteria

Exposure of laminin to pathogenic bacteria is most frequently seen in damaged or inflamed tissues. Switalski et al. (60) described the binding of Streptococcus pyogenes to laminin. The binding was specific, and the binding component was a high-molecular-weight (10^6) cell surface protein. The binding of laminin was also reported in a Streptococcus gordonii strain isolated from a patient with infective endocarditis (55). The binding component was a 145-kDa cell wall protein that was regulated by the presence of laminin. The sera from patients with infective carditis contained antibodies against this protein, whereas no significant recognition was seen with sera from patients with valvulopathies, suggesting an increased expression of laminin-binding protein during infective endocarditis. Binding of laminin has also been reported with several oral and endocarditis strains of viridans streptococci (57). Most strains isolated from patients with endocarditis expressed laminin-binding protein, whereas only a few strains isolated from the oral cavity expressed this protein, indicating that laminin binding might be an important factor in the pathogenesis of viridans endocarditis.

Binding of Elastin to Gram-Positive Bacteria

The specific binding of elastin was described for a number of S. aureus strains (43). The binding component was identified as 25-kDa staphylococcal surface protein capable of binding elastin with high affinity. The bacterial binding domain was mapped to a 30-kDa N-terminal fragment of elastin. The staphylococcal elastin-binding protein has structural and functional features different from those of the mammalian elastin receptor. Although the function of elastin-binding protein in pathogenesis is not known, it has been postulated that this protein might contribute to infection in organs containing elastin, such as lungs, skin, and blood vessels.

Fibronectin, a Matrix Protein Most Sought After by Gram-Positive Pathogens

Fibronectin not only acts as a substrate for the adhesion of eukaryotic cells but also serves as the prototype of adhesion proteins that bind specifically to microorganisms (6, 71). Because it binds to over 16 bacterial species, the interactions with gram-positive bacteria have been characterized extensively. In particular, most S. aureus isolates, as well as various streptococcal strains, specifically bind and adhere to fibronectin, mediated by several different bacterial adhesins (17, 61, 62) with highly homologous recognition motifs for the adhesion protein. Two genes, fnbpA and fnbpB, encoding the fibronectin-binding protein of S. aureus have been described. The gene products show typical features of gram-positive surface proteins, and the binding domain was localized to a 38-amino-acid repeated sequence of these proteins.

A number of fibronectin-binding proteins have also been identified in S. pyogenes. Talay et al. (61, 62) identified the fibronectin-binding protein SfbI of group A streptococci that mediates the adherence of bacteria with epithelial cells. Hanski and Caparon (17) identified the fibronectin-binding protein F1 encoded by the gene *prtF*. The deduced amino acid sequences of SfbI and protein F1 indicated that both proteins were identical. SfbI protein consists of 638 amino acids and comprises five structurally distinct domains. The N-terminal signal peptide is followed by an aromatic domain and four proline-rich repeats, which are flanked by nonrepetitive spacer sequences. A second repeat region, distinct from the proline repeats, is located in the C-terminal part of the protein. Fibronectin binding was located to the C-terminal repeat region as well as to the nonrepetitive spacer sequence. Functional analysis showed that the repeat sequence represents the major adhesin of group A streptococci. The functional domains of SfbI were highly conserved in isolates belonging to different serotypes (63). Unlike M protein, SfbI protein did not show any cross-reactivity with host proteins (67). Besides its role in adherence, SfbI protein is also involved in the invasion of epithelial cells by group A streptococci. Synthetic beads coated with SfbI protein were readily internalized by epithelial cells, indicating that SfbI alone is sufficient for invasion (36, 37). SfbI protein, therefore, is the first defined invasin of group A streptococci. The cellular receptors responsible for protein F1 (SfbI)-mediated invasion have been identified as integrins capable of binding fibronectin (41). Besides its involvement in adherence and invasion, SfbI protein has also been shown to be a strong mucosal adjuvant (35). Mice immunized intranasally with ovalbumin coupled to SfbI evoked a substantially higher immunoglobulin A response in lung lavage than did mice immunized with ovalbumin alone. SfbI protein also evoked a protective immune response in mice. Animals immunized with SfbI protein were protected against the lethal challenge of group A streptococci belonging to different serotypes (15). These results underline the importance of SfbI protein in vaccine development. Kreikemeyer et al. (24) and Rakonjac et al. (51) showed that serum opacity factors of group A streptococci (SfbII and SOF22) also interacted with fibronectin specifically. The binding of fibronectin was located in the C-terminal repeat region of the molecule, which was not involved in the serum opacity factor activity. Since SOF is mainly a secreted protein, the role of its interaction with fibronectin is not yet clear. In in vivo experiments, however, synthetic beads coated with the repeat region of SOF showed significant adherence to epithelial cells (unpublished data). Another fibronectin-binding protein, Sffbp12, was described in group A streptococci with no homology to SfbI and SfbII protein (53). Sffbp12 shares a high degree of homology with fibronectin- and fibrinogen-binding proteins from *Streptococcus dysgalactiae* and *S. aureus*, respectively.

Animal pathogenic streptococci also interact with fibronectin, leading to their adherence to host epithelial cells (66). *S. dysgalactiae*, a cattle pathogen, expresses two different binding proteins, FnbA and FnbB (31). Their binding domain is located in the repeat region of both proteins, and the repeats of FnbA and FnbB do not show any sequence homology. Despite the difference in sequence of repeats for various fibronectin-binding proteins, they were capable of cross-inhibiting the binding of fibronectin to *S. aureus*, *S. dysgalactiae*, and *S. pyogenes* (23). The repeat motif of SfbI protein, therefore, conforms to the consensus sequence previously reported for FnbA, FnbB, and FnbpA.

Fibronectin binding has also been observed with coagulase-negative staphylococci (68). Unlike other gram-positive cocci, coagulase-negative staphylococci interact only with immobilized fibronectin, which allows them to colonize artificial devices such as intraocular lenses, prosthetic cardiac valves, vascular grafts, prosthetic joints, and intravascular catheters. Fibronectin incorporated in fibrin thrombi is also the cause of infection at sites of blood clots or damaged tissue (10). Some strains of *Streptococcus pneumoniae* also interact with immobilized fibronectin, but the pathogenic role of this interaction is not yet elucidated (70).

Binding of Vitronectin to Gram-Positive Bacteria

Vitronectin has equivalent effects on cellular adhesion and bacterial binding as are demonstrated for fibronectin, but owing to the deposition of vitronectin in the periphery, specific interactions with bacteria are likely to occur in damaged or altered tissues. Specific interaction of vitronectin with various strains of staphylococci and groups A, C, and G streptococci has been described (9). The binding of vitronectin to group G streptococci, but not to groups A and C streptococci, is inhibited by heparin, indicating the diversity of vitronectin-binding proteins among streptococci. Like fibronectin binding, the binding to vitronectin mediates the adherence of gram-positive bacteria to host cells (13, 66). The vitronectin-binding proteins therefore represent additional adhesins of gram-positive bacteria. The adhesin of *S. aureus* responsible for vitronectin interaction has striking similarity to a heparan sulfate-binding protein from the same staphylococcal strain, suggesting that the adhesion protein undergoes multiple interactions with different bacterial surface recognition sites (29).

Distant sequence homology exists between vitronectin and the heme-binding plasma protein hemopexin, whose ligand-binding properties strongly depend on the conformational flexibility of the protein. In addition to RGD- and heparin-dependent interactions with gram-positive bacteria, hemopexin-type repeats in vitronectin, as well as in hemopexin itself, have been identified as primary binding sites for group A streptococci (30).

Binding of Fibrinogen to Gram-Positive Bacteria

Together with invading cells and aggregating platelets, the fibrin clot represents the majority of initial provisional ECM network for sealing a wound site. At sites of trauma, fibrinogen therefore serves as a substrate for bacterial adhesion. Various fibrinogen-binding proteins mediating bacterial colonization in wounds or catheters have been described, of which those expressed by *S. aureus* are the best characterized (34). In particular, "clumping factor" serves to recognize the carboxy-terminal portion of fibrinogen γ-chain in a manner analogous to $\alpha_{IIb}\beta_3$-integrin binding. Moreover, the homology between metal ion-dependent adhesion sites of integrin subunits α_{IIb}, α_M, or clumping factor and an integrinlike protein from *Candida albicans* (21) indicates common mechanisms of fibrinogen binding in mammalian cells, lower eukaryotes, and prokaryotes. Staphylocoagulase serves as an additional fibrinogen-binding factor that is not involved in bacterial clumping but, owing to prothrombin binding and conversion, serves to promote fibrin formation or bacterial attachment onto fibrinogen-coated surfaces (12). In group A streptococci, M protein is a cell surface structure principally responsible for binding fibrinogen, and this binding contributes to the known antiopsonic property of M protein (72). Fibrinogen binding to streptococci of groups C and G also leads to inhibition of complement fixation and subsequent phagocytosis, indicating an important role of this interaction (7, 65). Furthermore, bacterial colonization was reduced in a mouse mastitis model by vaccination with *S. aureus* fibrinogen-binding proteins (33), providing a new concept for antimicrobial therapy.

The adherence of *S. aureus* to endothelial cells is mediated predominantly by fibrinogen as bridging molecule leading to acute endovascular infections (4). Although platelet aggregation via a fibrinogen-dependent mechanism can be induced by this pathogen, this process is independent of the aforementioned principal $\alpha_{IIb}\beta_3$-integrin-binding interactions with fibrinogen (1).

In addition to staphylococcal or streptococcal surface proteins that interact with ECM components, proteins released by these bacterial species can directly influence fibrin formation or dissolution. While staphylocoagulase binds and activates prothrombin to thrombin, staphylokinase and streptokinase interact stoichiometrically with plasminogen, resulting in plasmin formation, whereby the former fibrinolytic agent acts in a fibrin-specific manner. These strategies apparently allow effective fixation and subsequent penetration of these bacteria into wound areas.

Binding of Thrombospondin to Gram-Positive Bacteria

The high-affinity binding of thrombospondin has been described for *S. aureus* strains (19). This interaction mediates the adherence of staphylococci to tissues via activated platelets during inflammation or infection. A large number of protein A-negative and -positive staphylococcus isolates adhered to thrombospondin-coated synthetic disks, indicating that adherence is significantly promoted as a function of absorbed thrombospondin. Specific binding of thrombospondin has also been observed with groups A, C, and G streptococci (4a). Like *S. aureus*, the binding component in streptococci is resistant to trypsin. The exact biological function of streptococcal-thrombospondin interaction has yet to be elucidated.

Degradation of ECM by Gram-Positive Pathogens

Invasion of bacteria in tissues requires the degradation of matrix proteins. This is accomplished by various bacterial enzymes as well as by host-derived plasmin (2). Tissue-type plasminogen activators produced by many eukaryotic cells, particularly under inflammatory conditions, are responsible for plasmin formation. A number of bacteria, such as groups A, C, and G streptococci, express surface receptors for plasmin(ogen) that facilitate pericellular proteolysis and invasion (42, 52). A plasmin-binding protein, PAM, has been isolated from group A streptococci. This protein belongs to the M protein family and binds to the kringle domain of plasminogen. This binding was blocked by a lysine analog, indicating that lysine residues in the M-like protein participate in the interaction (73). The interaction of streptococci with plasmin or plasminogen might contribute to streptococcal invasion. It was found that plasmin bound to streptococci could be protected from inactivation by the plasmin inhibitor α_2-antiplasmin. The structural similarity of neutrophil-derived polypeptide "defensins" to plasminogen kringle motifs (20) suggests that their antimicrobial activity can be related to interference with plasmin formation, thereby preventing the spreading of

infection. As an acute-phase reactant in host defense, a circulating broad-spectrum proteinase inhibitor, α_2-macroglobulin, serves to eliminate complex proteinases via receptor-mediated endocytosis and may do so with microbial enzymes as well (11). Interestingly, various strains of streptococci exhibit specific binding to α_2-macroglobulin (8). A protein of 78 kDa was purified from group A streptococci that interacted specifically with native α_2-macroglobulin. Although this protein possessed no proteolytic activity, its interaction with α_2-macroglobulin led to a change in conformation similar to that obtained by α_2-macroglobulin-protease complexes (5). Through their interaction with α_2-macroglobulin, streptococci might gain access to host tissues, possibly via the α_2-macroglobulin receptor (69).

CONCLUSIONS

Gram-positive pathogens are still a major cause of serious human diseases. The rapid emergence of resistance to antibiotics in various gram-positive cocci, and the fact that no vaccine is available against most of these organisms, makes gram-positive pathogens a major health hazard. A better understanding of the molecular mechanism of gram-positive infections is a prerequisite for development of effective vaccines and for designing novel strategies for the prevention and treatment of these infections. Adherence and invasion are the important disease-causing mechanisms of gram-positive pathogens. Since interaction of these pathogens with the components of host ECM is involved both in adherence and in invasion, the underlying mechanisms of this interaction are of utmost importance for designing novel therapeutic strategies. The ECM-binding bacterial proteins (adhesins) can be used as vaccine candidates in an antiadhesin vaccine. These strategies are, however, complicated by the fact that not all adhesin genes are present in every strain and that environmental factors can affect expression of these genes. Because each microorganism can employ multiple mechanisms of adhesion to initiate infection, effective antiadhesion drugs may contain cocktails of various inhibitors.

A classic example of the potential use of bacterial cell surface components interacting with ECM is shown with fibronectin-binding protein SfbI from *S. pyogenes*. SfbI protein acts both as adhesin and invasin of group A streptococci. Mice immunized with this protein not only show reduced pharyngeal colonization but are also protected against lethal streptococcal challenge. Since SfbI is highly conserved in its functional domain in streptococcal isolates from different geographical regions, it does not cross-react with the host proteins, is highly immunogenic, and represents a promising vaccine candidate. Besides its role in adhesion and invasion, SfbI protein is a strong mucosal adjuvant, and because of its nontoxic nature it has an advantage over the toxin-based adjuvants. Immunization with *S. aureus* fibrinogen-binding proteins led to reduced colonization in a mouse mastitis model. Synthetic peptides representing the fibronectin-binding domain of *S. aureus* adhesins as well as antibodies against the fibronectin-binding domain of *S. aureus* were successful in blocking bacterial colonization on implanted foreign material. These examples clearly demonstrate the potential use of components involved in the interaction of ECM and gram-positive pathogens and justify further research in this field.

REFERENCES

1. **Bayer, A. S., P. M. Sullam, M. Ramos, C. Li, A. L. Cheung, and M. R. Yeaman.** 1995. *Staphylococcus aureus* induces platelet aggregation via a fibrinogen-dependent mechanism which is independent of principal platelet glycoprotein IIb/IIIa fibrinogen-binding domains. *Infect. Immun.* **63:**3634–3641.

2. **Border, C. C., R. Lottenberg, G. O. von Mering, K. H. Johnston, and M. D. P. Boyle.** 1991. Isolation of a prokaryotic plasmin receptor. Relationship to a plasminogen activator produced by the same micro-organism. *J. Biol. Chem.* **266:**4922–4928.

3. **Bornstein, P.** 1995. Diversity of function is inherent in matricellular proteins: an appraisal of thrombospondin 1. *J. Cell Biol.* **130:**503–506.

4. **Cheung, A. L., M. Krishnan, E. A. Jaffe, and V. A. Fischetti.** 1991. Fibrinogen acts as a bridging molecule in the adherence of *Staphylococcus aureus* to cultured human endothelial cells. *J. Clin. Invest.* **87:**2236–2245.

4a. **Chhatwal, G. S.** Unpublished data.

5. **Chhatwal, G. S., G. Albohn, and H. Blobel.** 1987. Novel complex formed between a nonproteolytic cell wall protein of group A streptococci and α2-macroglobulin. *J. Bacteriol.* **169:**3691–3695.

6. **Chhatwal, G. S., and H. Blobel.** 1987. Heterogeneity of fibronectin reactivity among streptococci as revealed by binding of fibronectin fragments. *Comp. Immunol. Microbiol. Infect. Dis.* **10:**99–108.

7. **Chhatwal, G. S., I. S. Dutra, and H. Blobel.** 1985. Fibrinogen binding inhibits the fixation of the third component of human complement on surface of groups A, B, C, and G streptococci. *Microbiol. Immunol.* **29:**973–980.

8. **Chhatwal, G. S., H. P. Müller, and H. Blobel.** 1983. Characterization of binding of human α_2-macroglobulin to group G streptococci. *Infect. Immun.* **41:**959–964.

9. **Chhatwal, G. S., K. T. Preissner, G. Müller-Berghaus, and H. Blobel.** 1987. Specific binding of the human S protein (vitronectin) to streptococci, *Staphylococcus aureus*, and *Escherichia coli*. *Infect. Immun.* **55:**1878–1883.

10. **Chhatwal, G. S., P. Valentin-Weigand, and K. N. Timmis.** 1990. Bacterial infection of wounds: fibronectin-mediated adherence of group A and C streptococci to fibrin thrombi in vitro. *Infect. Immun.* **58:**3015–3019.

11. **Chu, C. T., and S. V. Pizzo.** 1994. Alpha-2-macroglobulin, complement, and biologic defense: antigens, growth factors, microbial proteases, and receptor ligation. *Lab. Invest.* **71:**792–812.

12. **Dickinson, R. B., J. A. Nagel, D. McDevitt, T. J. Foster, R. A. Proctor, and S. L. Cooper.** 1995. Quantitative comparison of clumping factor- and coagulase-mediated *Staphylococcus aureus* adhesion to surface-bound fibrinogen under flow. *Infect. Immun.* **63:**3143–3150.

13. **Filippsen, L. F., P. Valentin-Weigand, H. Blobel, K. T. Preissner, and G. S. Chhatwal.** 1990. Role of complement S protein (vitronectin) in adherence of *Streptococcus dysgalactiae* to bovine epithelial cells. *Am. J. Vet. Res.* **51:**861–865.

14. **Gailit, J., and R. A. F. Clark.** 1994. Wound repair in the context of extracellular matrix. *Curr. Opin. Cell Biol.* **6:**717–725.

15. **Guzmán, C. A., S. R. Talay, G. Molinari, E. Medina, and G. S. Chhatwal.** 1999. Protective immune response against *Streptococcus pyogenes* in mice after intranasal vaccination with the fibronectin binding protein SfbI. *J. Infect. Dis.* **179:**901–906.

16. **Hall, C. E., and H. S. Slayter.** 1959. The fibrinogen molecule: its size, shape, and mode of polymerization. *J. Biophys. Biochem. Cytol.* **5:**11–15.

17. **Hanski, E., and M. Caparon.** 1992. Protein F, a fibronectin-binding protein, is an adhesin of the group A *Streptococcus pyogenes*. *Proc. Natl. Acad. Sci. USA* **89:**6172–6176.

18. **Henschen, A., F. Lottspeich, E. Töpfer-Petersen, and R. Warbinek.** 1979. Primary structure of fibrinogen. *Thromb. Haemostas.* **41:**662–670.

19. **Herrmann, M., S. J. Suchard, L. A. Boxer, F. A. Waldvogel, and D. P. Lew.** 1991. Thrombospondin binds to *Staphylococcus aureus* and promotes staphylococcal adherence to surfaces. *Infect. Immun.* **59:**279–288.

20. **Higazi, A. A. R., I. I. Barghouti, and R. Abumuch.** 1995. Identification of an inhibitor of tissue-type plasminogen activator-mediated fibrinolysis in human neutrophils—a role for defensin. *J. Biol. Chem.* **270:**9472–9477.

21. **Hostetter, M. K.** 1996. An integrin-like protein in *Candida albicans*: implications for pathogenesis. *Trends Microbiol.* **4:**242–246.

22. **Hynes, R. O., and K. M. Yamada.** 1982. Fibronectins: multifunctional modular glycoproteins. *J. Cell Biol.* **95:**369–377.

23. **Joh, J. J., K. House-Pompeo, J. M. Patti, S. Gurusiddappa, and M. Höök.** 1994. Fibronectin receptors from Gram-positive bacteria: comparison of active sites. *Biochemistry* **33:**6086–6092.

24. **Kreikemeyer, B., S. R. Talay, and G. S. Chhatwal.** 1995. Characterization of a novel fibronectin-binding surface protein in group A streptococci. *Mol. Microbiol.* **17:**137–145.

25. **Kreis, T., and R. Vale.** 1993. *Guidebook to the Extracellular Matrix and Adhesion Proteins.* Oxford University Press, Oxford, U.K.

26. **Labat-Robert, J., M. Bihari-Varga, and L. Robert.** 1990. Extracellular matrix. *FEBS Lett.* **268:**386–393.

27. **Lane, D. A., A. M. Flynn, G. Pejler, U. Lindahl, J. Choay, and K. T. Preissner.** 1987. Structural requirements for the neutralization of heparin-like saccharides by complement S protein/vitronectin. *J. Biol. Chem.* **262:**16343–16349.

28. **Lawler, J. W., and R. O. Hynes.** 1987. Structure organization of thrombospondin molecule. *Semin. Thromb. Hemostasis.* **13:**245–254.

29. **Liang, O. D., M. Maccarana, J. I. Flock, M. Paulsson, K. T. Preissner, and T. Wadström.** 1993. Multiple interactions between human vitronectin and *Staphylococcus aureus*. *Biochim. Biophys. Acta* **374:**1–7.

30. **Liang, O. D., K. T. Preissner, and G. S. Chhatwal.** 1997. The hemopexin-type repeats of human vitronectin are recognized by *Streptococcus pyogenes*. *Biochem. Biophys. Res. Commun.* **234:**445–449.

31. **Lindgren, P.-E., P. Speziale, M. McGavin, H.-J. Monstein, M. Höök, L. Visai, T. Kostiainen, S. Bozzini, and M. Lindberg.** 1992. Cloning and expression of two different genes from *Streptococcus dysgalactiae* encoding fibronectin receptors. *J. Biol. Chem.* **267:**1924–1931.

32. **Love, R. M., M. D. McMillan, and H. F. Jenkinson.** 1997. Invasion of dentinal tubules by oral streptococci is associated with collagen recognition mediated by the antigen I/II family of polypeptides. *Infect. Immun.* **65:** 5157–5164.

33. **Mamo, W., M. Boden, and J. I. Flock.** 1994. Vaccination with *Staphylococcus aureus* fibrinogen binding protein (FgBPs) reduces colonization of *S. aureus* in a mouse mastitis model. *FEMS Immun. Med. Microbiol.* **10:**47–53.

34. **McDevitt, D., T. Nanavaty, K. House-Pompeo, E. Bell, N. Turner, L. McIntire, T. Foster, and M. Höök.** 1997. Characterization of the interaction between the *Staphylococcus aureus* clumping factor (ClfA) and fibrinogen. *Eur. J. Biochem.* **247:**416–424.

35. **Medina, E., S. R. Talay, G. S. Chhatwal, and C. A. Guzmán.** 1998. Fibronectin-binding protein I of *Streptococcus pyogenes* is a promising adjuvant for antigens delivered by mucosal route. *Eur. J. Immunol.* **28:**1069–1077.

36. **Molinari, G., and G. S. Chhatwal.** 1998. Invasion and survival of *Streptococcus pyogenes* in eukaryotic cells correlates with the source of the clinical isolates. *J. Infect. Dis.* **177:**1600–1607.

37. **Molinari, G., S. R. Talay, P. Valentin-Weigand, M. Rohde, and G. S. Chhatwal.** 1997. The fibronectin-binding protein of *Streptococcus pyogenes*, SfbI, is involved in the internalization of group A streptococci by epithelial cells. *Infect. Immun.* **65:**1357–1363.

38. **Mosher, D. F., and R. A. Proctor.** 1980. Binding of factor XIIIa-mediated cross-linking of a 27-kilodalton fragment of fibronectin to *Staphylococcus aureus*. *Science* **209:**927–929.

39. **Mosher, D. F., J. Sottile, C. Wu, and J. A. McDonald.** 1992. Assembly of extracellular matrix. *Curr. Opin. Cell Biol.* **4:**810–818.

40. **Murphy-Ullrich, J. E., S. Gurusiddappa, W. A. Frazier, and M. Höök.** 1993. Heparin-binding peptides from thrombospondin 1 and 2 contain focal adhesion-labilizing activity. *J. Biol. Chem.* **268:**26784–26789.

41. **Ozeri, V., I. Rosenshine, D. F. Mosher, R. Fässler, and E. Hanski.** 1998. Roles of integrins and fibronectin in the entry of *Streptococcus pyogenes* into cells via protein F1. *Mol. Microbiol.* **30:**625–637.

42. **Pancholi, V., and V. A. Fischetti.** 1998. Alpha-enolase, a novel strong plasmin(ogen) binding protein on the surface of pathogenic streptococci. *J. Biol. Chem.* **273:** 14503–14515.

43. **Park, P. W., D. D. Roberts, L. E. Grosso, W. C. Parks, J. Rosenbloom, W. R. Abrams, and R. P. Mecham.** 1991. Binding of elastin to *Staphylococcus aureus*. *J. Biol. Chem.* **266:**23399–23406.

44. **Patti, J. M., B. L. Allen, M. J. McGavin, and M. Höök.** 1994. MSCRAMM-mediated adherence of microorganisms to host tissue. *Annu. Rev. Microbiol.* **48:**585–617.

45. **Patti, J. M., J. O. Boles, and M. Höök.** 1993. Identification and biochemical characterization of the ligand binding domain of the collagen adhesin from *Staphylococcus aureus*. *Biochemistry* **32:**11428–11435.

46. **Pierschbacher, M. D., and E. Ruoslahti.** 1984. Cell attachment activity of fibronectin can be duplicated by small synthetic fragments of the molecule. *Nature* **309:** 30–33.

47. **Preissner, K. T.** 1991. Structure and biological role of vitronectin. *Annu. Rev. Cell Biol.* **7:**275–310.

48. **Preissner, K. T., A. E. May, K. D. Wohn, M. Germer, and S. M. Kanse.** 1997. Molecular crosstalk between adhesion receptors and proteolytic cascades in vascular remodeling. *Thromb. Haemostas.* **78:**88–95.

49. **Preissner, K. T., and B. Pötzsch.** 1995. Vessel wall-dependent metabolic pathways of the adhesive proteins, von-Willebrand-factor and vitronectin. *Histol. Histopathol.* **10:**239–251.

50. **Raghow, R.** 1994. The role of extracellular matrix in post-inflammatory wound healing and fibrosis. *FASEB J.* **8:** 823–831.

51. **Rakonjac, J. V., J. C. Robbins, and V. A. Fischetti.** 1995. DNA sequence of the serum opacity factor of group A streptococci: identification of a fibronectin-binding repeat domain. *Infect. Immun.* **63:**622–631.

52. **Ringdahl, U., M. Svensson, A. C. Wistedt, T. Renn, R. Kellner, W. Müller-Esterl, and U. Sjöbring.** 1998. Molecular co-operation between protein PAM and streptokinase for plasmin acquisition by *Streptococcus pyogenes*. *J. Biol. Chem.* **273:**6424–6430.

53. **Rocha, C. L., and V. A. Fischetti.** 1997. Identification and characterization of a new protein from *Streptococcus pyogenes* having homology with fibronectin and fibrinogen binding proteins. *Adv. Exp. Med. Biol.* **418:**737–739.

54. **Rosenblatt, S., J. A. Bassuk, C. E. Alpers, E. H. Sage, R. Timpl, and K. T. Preissner.** 1997. Differential modulation of cell adhesion by interaction between adhesive and counteradhesive proteins: characterization of the binding of vitronectin to osteonectin (BM40, SPARC). *Biochem. J.* **324:**311–319.

55. **Sommer, P., C. Gleyzal, S. Guerret, J. Etienne, and J.-A. Grimaud.** 1992. Induction of a putative laminin-binding protein of *Streptococcus gordonii* in human infective endocarditis. *Infect. Immun.* **60:**360–365.

56. **Switalski, L. M., W. G. Butcher, P. C. Caufield, and M. S. Lantz.** 1993. Collagen mediates adhesion of *Streptococcus mutans* to human dentin. *Infect. Immun.* **61:** 4119–4125.

57. **Switalski, L. M., H. Murchinson, R. Timpl, R. Curtiss III, and M. Höök.** 1987. Binding of laminin to oral and endocarditis strains of viridans streptococci. *J. Bacteriol.* **169:**1095–1101.

58. **Switalski, L. M., J. M. Patti, W. Butcher, A. G. Gristina, P. Speziale, and M. Höök.** 1993. A collagen receptor in *Staphylococcus aureus* strains isolated from patients with septic arthritis mediates adhesion to cartilage. *Mol. Microbiol.* **7:**99–107.

59. **Switalski, L. M., P. Speziale, and M. Höök.** 1989. Isolation and characterization of a putative collagen receptor from *Staphylococcus aureus* strain Cowan 1. *J. Biol. Chem.* **264:**21080–21086.

60. **Switalski, L. M., P. Speziale, M. Höök, T. Wadström, and R. Timpl.** 1984. Binding of *Streptococcus pyogenes* to laminin. *J. Biol. Chem.* **259:**3734–3738.

61. **Talay, S. R., E. Ehrenfeld, G. S. Chhatwal, and K. N. Timmis.** 1991. Expression of the fibronectin-binding components of *Streptococcus pyogenes* in *Escherichia coli* demonstrates that they are proteins. *Mol. Microbiol.* **5:**1727–1734.

62. **Talay, S. R., P. Valentin-Weigand, P. G. Jerlström, K. N. Timmis, and G. S. Chhatwal.** 1992. Fibronectin-binding protein of *Streptococcus pyogenes*: sequence of the

binding domain involved in adherence of streptococci to epithelial cells. *Infect. Immun.* **60:**3837–3844.

63. **Talay, S. R., P. Valentin-Weigand, K. N. Timmis, and G. S. Chhatwal.** 1994. Domain structure and conserved epitopes of Sfb protein, the fibronectin-binding adhesin of *Streptococcus pyogenes*. *Mol. Microbiol.* **13:**531–539.

64. **Timpl, R.** 1989. Structure and biological activity of basement membrane proteins. *Eur. J. Biochem.* **180:**487–502.

65. **Traore, M. Y., P. Valentin-Weigand, G. S. Chhatwal, and H. Blobel.** 1991. Inhibitory effects of fibrinogen on phagocytic killing of streptococcal isolates from humans, cattle and horses. *Vet. Microbiol.* **28:**295–302.

66. **Valentin-Weigand, P., J. Grulich-Henn, G. S. Chhatwal, G. Müller-Berghaus, H. Blobel, and K. T. Preissner.** 1988. Mediation of adherence of streptococci to human endothelial cells by complement S-protein (vitronectin). *Infect. Immun.* **56:**2851–2855.

67. **Valentin-Weigand, P., S. R. Talay, A. Kaufhold, K. N. Timmis, and G. S. Chhatwal.** 1994. The fibronectin binding domain of the Sfb protein adhesin of *Streptococcus pyogenes* occurs in many group A streptococci and does not cross-react with heart myosin. *Microb. Pathog.* **17:**111–120.

68. **Valentin-Weigand, P., K. N. Timmis, and G. S. Chhatwal.** 1993. Role of fibronectin in staphylococcal coloni-zation of fibrin thrombi and plastic surfaces. *J. Med. Microbiol.* **38:**90–95.

69. **Valentin-Weigand, P., M. Y. Traore, H. Blobel, and G. S. Chhatwal.** 1990. Role of α2-macroglobulin in phagocytosis of group A and C streptococci. *FEMS Microbiol. Lett.* **70:**321–324.

70. **van der Flier, M., N. Chhun, T. M. Wizemann, J. Min, J. B. McCarthy, and E. I. Tuomanen.** 1995. Adherence of *Streptococcus pneumoniae* to immobilized fibronectin. *Infect. Immun.* **63:**4317–4322.

71. **Vercellotti, G. M., D. Lussenhop, P. K. Peterson, L. T. Furcht, J. B. McCarthy, H. S. Jacob, and C. F. Moldow.** 1984. Bacterial adherence to fibronectin and endothelial cells: a possible mechanism for bacterial tissue tropism. *J. Lab. Clin. Med.* **103:**34–43.

72. **Whitnack, E., and E. H. Beachy.** 1985. Inhibition of complement-mediated opsonization and phagocytosis of *Streptococcus pyogenes* by D fragments of fibrinogen and fibrin bound to cell surface M protein. *J. Exp. Med.* **162:**1983–1997.

73. **Wiestedt, A. C., U. Ringdahl, W. Müller-Esterl, and U. Sjöbring.** 1995. Identification of a plasminogen-binding motif in PAM, a bacterial surface protein. *Mol. Microbiol.* **18:**569–578.

Streptococci-Mediated Host Cell Signaling

VIJAYKUMAR PANCHOLI

9

To be successful pathogens, bacteria must adhere to, colonize, and invade the target tissue. For successful colonization and infection, bacteria usually need to sequentially engage (adhere) to surface-bound adhesins with complementary receptors on target cells. In this regard, group A streptococci show a remarkable tissue tropism for the human pharynx and the skin, causing a variety of diseases. The surface of group A streptococci presents an array of proteins that perform a variety of functions (40). As more information is available in the last decade on this aspect, it is noteworthy that most, if not all, surface proteins are multifunctional (40) (see chapter 2, this volume). In a given strain, one surface protein may perform several functions, or several proteins may participate to perform one particular function. Thus, the interaction of group A streptococci with their host cell is more complex than what was previously imagined. Besides this complexity, group A streptococci are of special concern because of their ability to cause autoimmune diseases such as rheumatic fever and acute poststreptococcal glomerulonephritis (76, 77). Reemergence of rheumatic fever cases and group A streptococcal invasive and fatal infections in the absence of any specific virulent clone suggests that host factors play a critical role in the final outcome of a variety of streptococcal diseases (8, 18, 52). Despite several years of intensive investigations into streptococcal pathogenesis, it has remained enigmatic as to why pharyngeal infection, and not other infections such as skin infections, is a prerequisite for rheumatic fever and rheumatic heart diseases (78).

It is becoming more apparent that microorganisms, using their surface proteins, interact with host cell receptor molecules and regulate intracellular signaling pathways to induce their own adherence, colonization, and internalization. While numerous studies have laid a sound foundation for the understanding of the cross-talk between gram-negative bacteria and host cells, thereby unraveling their basic virulence mechanisms (see reviews in references 4, 10, 11, 25, 33, 36, and 79), there is a distinct lack of information on the cell biology of group A streptococci-infected cells. Unlike gram-negative bacteria, the field of cellular microbiology in relation to infections due to gram-

positive cocci in general, and to pathogenic streptococci in particular, has remained essentially unexplored. Because tissue-tropic streptococcal diseases are as common and as severe in their clinical manifestations as those caused by their gram-negative counterparts and gram-positive *Listeria* sp. (5, 24), there is strong reason to believe that the gram-positive cocci do participate in cross-talk during bacterium-host cell interactions. Some recent reports show direct evidence of streptococci-mediated signaling events in the host cell that supports this notion (6, 57, 66, 75).

The purpose of this chapter is to review the available literature that has provided direct as well as indirect evidence that the initial interactions between pathogenic streptococci and their specific host cells result in the exchange of messages between them, which in turn determines the final outcome of the disease. Because this literature has only recently been available, only for group A streptococci and pneumococci among all streptococcal species, they are the focus of this chapter. It is likely, however, that other species of streptococci may display a similar capacity to evoke signaling events in their target cells.

While there are many missing links in our understanding of the precise signal transduction pathways evoked by these organisms in their respective host cells, the available literature suggests a common paradigm (25). Pathogenic gram-positive cocci through their surface protein(s) interact with specific receptors on target cells and induce a series of biochemical signals. These signals, which are characterized by the induction of phosphokinase enzymes and phosphorylation of several intracellular proteins, ultimately target the nucleus and lead to either generalized or specific gene activation. Activation of some of these genes may result in the modulation of interleukins or other cytokines, which may then initiate a proinflammatory response. These induced signals and resulting products could have several effects on the invasion of bacteria. For example, these induced signals may modulate cytoskeletal structure and/or specific host cell receptors or may destroy adjoining cells and disrupt natural protective barriers that could facilitate bacterial entry (Fig. 1).

Gram-Positive Pathogens, ed. by V. A. Fischetti et al.
© 2000 American Society for Microbiology, Washington, D.C.

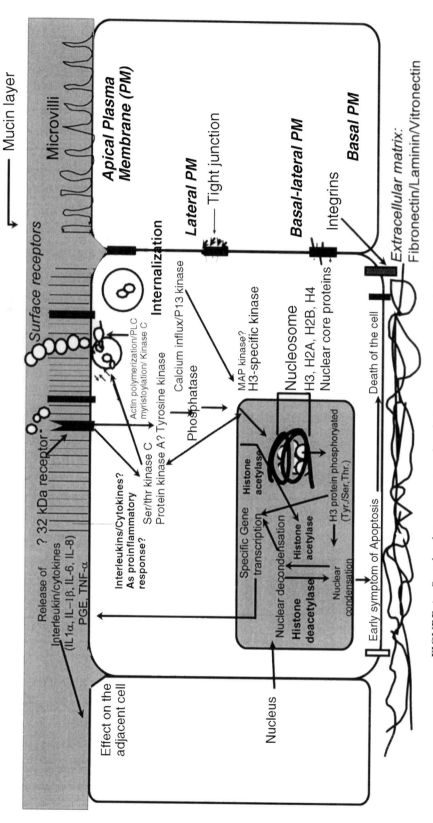

FIGURE 1 Postulated streptococci-mediated signaling events and their implications. The cartoon illustrates possible interrelations between intracellular signaling events evoked by streptococci in the polarized target cells, such as human pharyngeal cells. The illustrated possible pathways are hypothesized on the basis of available specific reports on streptococci-mediated signaling events (6, 57, 66, 75) and established signal transduction pathways in eukaryotic cells.

INITIAL INTERACTION OF GRAM-POSITIVE BACTERIAL SURFACE PROTEINS WITH MUCIN

The human respiratory epithelium is covered by a viscous layer of mucus, situated on the top of the cilia and resulting in a continuous flow toward the pharynx. By virtue of this anatomical disposition, all mucosal pathogens first interact with mucin, the main component of respiratory mucus, before they adhere to a specific host cell receptor on the epithelial cell. Mucins are a highly complex class of glycoproteins found in both secretory and bound forms and expressed by several genes (29, 69). The structural diversity among the mucin family also increases owing to the posttranslational O-glycosylation and sulfation that contribute to more than 50% of the molecular mass of the mucin molecule (69). Although mucin may serve as an initial protective barrier against invading pathogens, overproduction of mucin has been shown to exacerbate the pathological symptoms in diseases such as bacterial pneumonia and cystic fibrosis (22, 41, 65). While several reports have described the role of the interaction of mucin with gram-negative bacteria in the pathogenesis of cystic fibrosis and similarly with oral streptococci in oral diseases (20, 22, 41), limited information is available on the interaction of mucin with pathogenic gram-positive bacteria and its implication in the disease process (12, 19, 48, 68). Recently, we have shown that group A streptococci specifically bind to salivary/tracheobronchial mucin (64). The M protein and a 39-kDa cell wall-associated surface protein were found to mediate this property (64). The binding of mucin to group A streptococci was also found to increase streptococcal adherence to cultured human pharyngeal cells (64). How pathogens find their way through the mucin barrier to their specific target receptor on epithelial cells is unknown. The role of gram-positive mucolytic enzymes such as streptococcal neuraminidase (19) or other extracellular proteases (7, 45) in facilitating bacterial migration through the viscous mucin layer to target them to the epithelial surface-specific receptor before invasion requires further investigation. In recent years, considerable advances have been made toward our understanding of the structure and function of mucin glycoproteins and their physiological niche (29, 71). The activation of mucin genes via certain signal transduction pathways as a result of bacterial interactions has become more evident from a recent report (21). That study found that certain types of mucins such as MUC2 and MUC5AC are overexpressed in human bronchial explant and epithelial cell cultures in response to interactions with culture supernatants of both gram-negative bacteria (e.g., Pseudomonas aeruginosa) and gram-positive bacteria (Staphylococcus aureus, Streptococcus pyogenes, Staphylococcus epidermidis). This expression is found to be tyrosine-phosphorylation-dependent (21), indicating that a specific type of mucus production may be the result of the induction of a specific intracellular signaling event. Although respiratory cell-bound mucins, such as MUC1, are thought to be involved in cell adhesion (31, 43, 72), and their cytoplasmic domains have been shown to play a role in cytoskeletal rearrangements in certain disease conditions (26), it is not clear at present whether salivary mucin or cell-bound mucin serves as an adhesin or a functional receptor for bacterial surface proteins.

BINDING OF STREPTOCOCCAL ADHESINS TO HOST CELL RECEPTORS AND INTRACELLULAR SIGNALING EVENTS IN TARGET CELLS

Since several reports have conclusively proved in the case of gram-negative bacteria that cell-cell interactions via specific receptor-adhesins elicit a variety of specific signaling events that ultimately determine the disease outcome (4, 10, 11, 25, 33, 36, 79), it is suspected that streptococcal adherence cannot simply be a static process but may result in specific intracellular responses. For example, the C-repeat of the streptococcal M protein has been shown to bind CD46, a membrane-bound complement regulatory protein related to factor H that is proposed to serve as a cellular receptor for S. pyogenes on human keratinocytes (53). It is also suggested that this streptococcal-host cell contact may initiate proinflammatory response from keratinocytes during infection (75).

The evidence that streptococci upon contact with their target cells such as pharyngeal cells may regulate intracellular signaling events was first suggested when a novel multifunctional protein, streptococcal surface dehydrogenase (SDH), was identified on the surface of streptococci (55). SDH, which shows binding activity to a variety of mammalian proteins such as fibronectin, myosin, and actin, is also an ADP-ribosylating enzyme (56). Since ADP-ribosylation plays an important role in intracellular signaling events and is the mechanism by which many bacterial toxins show their deleterious effect on target cells (49), SDH, and hence streptococci, may play a significant role in regulating signaling events in target cells. The direct evidence that streptococci and SDH, upon their contact with pharyngeal cells, regulate their intracellular signaling was illustrated in a recent study in which the Detroit 562 pharyngeal cell line was used as a working in vitro infection model (57). That study showed that SDH binds selectively to a 30- to 32-kDa membrane protein of human pharyngeal cells. Since streptococcal adherence to pharyngeal cells can be inhibited in a dose-dependent manner in the presence of purified SDH (unpublished data), it was suggested that SDH may be an adhesin for the 30- to 32-kDa receptor on pharyngeal cells (57).

Although the structural identity of this receptor is not yet known, purified SDH and whole streptococci, when they interact with intact pharyngeal cells, activate both tyrosine and serine/threonine kinases in these cells (57). This activation results in induced serine/threonine and tyrosine phosphorylation of several cellular proteins, the most noticeable of which are those of 180-, 100-, 30-, 22-, 17-, and 8-kDa proteins (Fig. 2). The SDH-mediated induced phosphorylation of certain proteins (180 and 100 kDa), and more notably the de novo phosphorylation of the 17-kDa pharyngeal protein, was inhibited only in the presence of genistein, which specifically inhibits the catalytic action of protein tyrosine kinase (57) (Fig. 2A). Reactivity of the 17-kDa pharyngeal cell protein with specific antibodies against phosphotyrosine and phosphoserine indicates double phosphorylation of this protein (Fig. 2A). Intact streptococci or purified SDH are unable to induce phosphorylation of these cellular proteins or the de novo phosphorylation of the 17-kDa protein in the absence of cytoplasmic components of pharyngeal cells (Fig. 2B). This provides an additional proof that streptococci- or SDH-mediated interactions occur, possibly through the 30- to

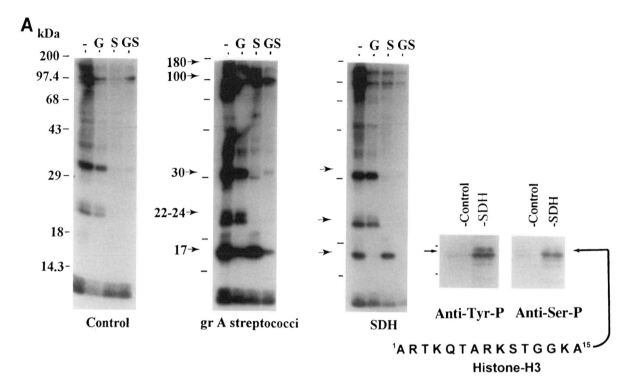

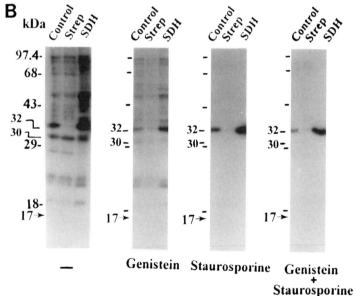

FIGURE 2 SDH/group A streptococci-mediated protein phosphorylation in intact pharyngeal cells. These experiments were carried out as described previously (57) in the presence of phosphokinase inhibitors genistein (G), staurosporine (S), or both (GS), and the profile of ^{32}P-labeled phosphorylated proteins in the membrane/particulate fraction was determined after sodium dodecyl sulfate-polyacrylamide gel electrophoresis and Western blotting, followed by autoradiography. (A) ^{32}P-labeled proteins in the membrane/particulate fraction after phosphorylation of intact pharyngeal cells. Note de novo phosphorylation of the 17-kDa protein, which is specifically inhibited in the presence of genistein. An accompanying Western blot shows the reaction of the 17-kDa phosphorylated protein with antiphosphotyrosine and antiphosphoserine monoclonal antibodies. The N-terminal amino acid sequence analysis of the 17-kDa protein revealed it to be histone H3 protein. (B) Phosphorylation of isolated membrane/particulate fraction-associated proteins of the pharyngeal cells in the absence of the cytoplasmic fraction. Note the absence of de novo phosphorylation of the 17-kDa protein as shown in A.

32-kDa putative surface receptor. This, in turn, evokes a specific biochemical signal that requires the cytoplasmic milieu for intracellular signal transduction, ultimately resulting in the induced or de novo phosphorylation of cellular proteins. Since the primary target of group A streptococci is the human pharynx, human pharyngeal cell lines such as Detroit 562 and FaDu have proved to be ideal in vitro infection models for these streptococcal studies (57). Evidence also suggests that intact streptococci are able to induce phosphorylation of pharyngeal cell proteins to a significantly greater extent than that by purified SDH, suggesting that, in addition to SDH, other streptococcal surface proteins may have the capacity to induce phosphorylation resulting in an enhanced signaling response (57).

In the case of *Streptococcus pneumoniae*, cell wall components have been shown to activate intracellular signaling events in astrocytes; however, the specific receptor on these cells is not known (66). In other studies, it was shown that the phosphorylcholine associated with cell wall teichoic acid plays a critical role in pneumococcal binding to endothelial or lung cells (16), possibly through the platelet-activating factor (PAF) receptor (16). Interestingly, this receptor is expressed only during the activated stage of the cell (16). Pneumococcal binding to resting cells is mediated through GalNAcβ1-4Gal and GalNAcβ1-3Gal receptors present on the cells, and their binding to de novo expressed PAF receptors and subsequent invasion of the cell is mediated through N-acetylglucosamine residues present on them (16, 28). The PAF receptor is known to participate in a variety of intracellular signaling events. However, surprisingly, pneumococcal adherence to and invasion of the target cell mediated by the PAF receptor occurs without evoking any intracellular signal transduction. Further dissection of biochemical events associated with pneumococcal adherence and invasion may provide a novel insight into the mechanism underlying pneumococcal tissue tropism and infection (15).

Another line of evidence that pneumococci are capable of inducing signal transduction was recently reported in target cells such as microglia and astrocytes (66). Microglial cells and astrocytes, which are found at the site of inflammation and which secrete various proinflammatory mediators, seem to play a critical role in the pneumococci-mediated inflammation and central nervous system immune responses (74). The study carried out on these cells showed that, like the lipopolysaccharide of gram-negative bacteria, pneumococcal cell wall (PCW) components are capable of inducing CD14-dependent signaling events in primary cultures of rat astrocytes and the human astrocytoma cell line U373MG (66). This study also showed that PCW initiates tyrosine phosphorylation and the subsequent activation of mitogen-activated proteins erk-1/erk-2 and p38 in astrocytes in a dose-dependent manner in the presence of CD14. Since the mitogen-activated protein kinase pathway ultimately results in the induction of transcription activation factor (67), this study provides an important insight into the mechanism of PCW-mediated secretion of proinflammatory mediators during the disease process. Cauwels et al. (6) have concluded that pneumococci-mediated cytokine production is dependent on several factors, including cell type, type of stimulus, and the cytokines measured. Together, these studies indicate that PCW and lipoteichoic acid act synergistically to initiate signaling events in various cell types through the CD14 receptor and can be demonstrated only in immature (but not in permanent) human mono-cytes (6), whereas the intact pneumococci do not use CD14 as a receptor to initiate signaling events. It seems likely that pneumococci, like group A streptococci (14, 57), use more than one receptor to activate target cells to mediate their own internalization (6).

ROLE OF INTEGRINS IN STREPTOCOCCI-MEDIATED SIGNALING EVENTS

Integrins are large $\alpha\beta$ heterodimeric membrane proteins found on the surface of a wide variety of mammalian cells (35, 36). These integrins seem to be involved in promoting adhesion to extracellular matrix proteins such as fibronectin, collagen, and laminin. Several fibronectin-binding proteins on the surface of group A streptococci have been described and include protein-F/Sfb1 (30, 47), protein F2 (39), serum opacity factor (61), protein H (27), streptococcal glyceraldehyde-3-phosphate dehydrogenase (SDH) (55), Fbp54 (13), and M1 (14). Sfb1 and protein F are found, at least in part, to mediate streptococcal invasion of HEp2 cells (47). Similarly, the M1 protein has recently been found to efficiently bind to the human lung epithelial cell lines, A549, in the presence of fibronectin via $\alpha5\beta1$ integrin (14). This integrin-mediated interaction seems to play a critical role in streptococcal invasion. While recognizing the importance of $\alpha5\beta1$ integrin-mediated streptococcal invasion of epithelial cells, that study emphasizes that the ability of fibronectin to mediate invasion varies considerably, depending on the source of the preparation (13, 14). Similar to *Yersinia* sp., group A streptococcal invasion is reported to be mediated by laminin (37), which recognizes $\alpha1\beta1$, $\alpha2\beta1$, $\alpha3\beta1$, $\alpha6\beta1$, or $\alpha7\beta1$ integrins on the cell surface (35, 36, 73). Thus, streptococci may use one or more integrins, depending on the cell type, as receptors for invasion and may subsequently evoke an integrin-specific signaling pathway. Bacterial invasion of a target epithelial cell is a dynamic process that involves active participation of the cytoskeletal structure (4, 9, 79). It seems reasonable to predict that binding to integrins allows streptococci to communicate with the cytoskeletal structure of the cell and regulates its entry into the cell. The studies showing the inhibition of streptococcal invasion of Detroit pharyngeal cells in the presence of the tyrosine and serine/threonine kinase inhibitors genistein and staurosporine (57) indicate that rearrangement of the cytoskeletal structure, which facilitates bacterial entry, is substantially regulated by the activation of these intracellular signaling enzymes (57). Although no direct proof is available for integrin-mediated signaling events during streptococcal interaction with epithelial cells, the ability of integrins to link extracellular matrix protein to the intracellular cytoskeleton proteins via their cytoplasmic domains has been shown to evoke the induction of protein kinase 3, IP3, and the induction of calcium influx (34, 36).

In all published reports, the fibronectin-binding property of group A streptococci is believed to have a primary role in initial adherence to, colonization of, and subsequent invasion of respiratory epithelial cells (14, 38, 47, 54). This view, which is based solely on in vitro unpolarized tissue culture cells, however, needs to be reconsidered since streptococci target polarized pharyngeal cells. Establishment of cell polarity is fundamental to differentiation and diversity of function in most cells. Integrin receptors, which are evenly distributed on transformed tissue culture cells, are, in fact, located on the basal rather than the apical side of the polarized cell (3, 23, 62). Streptococci, after entry into the host, are expected to interact with specific receptor(s) located on the apical surface before invasion. Hence, it is likely that during initial colonization and invasion of the intact epithelia, streptococci do not utilize their fibronectin-binding property since fibronectin is located on the basal side (Fig. 1). Once streptococci reach the extracellular matrix, their adherence to one or more of its components, such as fibronectin, may initiate signaling events in the cell via a specific class of integrins. These signaling events may, in turn, afford streptococci the ability to further invade these cells after the initial entry. It is reasonable to speculate that integrin-mediated signaling events play a critical role in streptococcal dissemination in tissues.

TARGETING THE NUCLEUS, AN ULTIMATE FUNCTION OF THE MEMBRANE RECEPTOR-ORIGINATED SIGNAL TRANSDUCTION

Cumulative efforts from many laboratories over the past several years have established that, within minutes after the activation of several membrane-signaling molecules, there occurs a sequential stimulation of several protein kinases, collectively known as the mitogen-activated protein kinase signaling cascade (67). This, in turn, eventually converges and activates several regulatory molecules in the cytoplasm and the nucleus (67). These regulatory molecules, collectively known as the STAT molecule (signal transducer in the cytoplasm and activator of transcription in the nucleus), are phosphorylated on a single tyrosine residue and then participate in the activation of distinct sets of target genes (17, 59). It has also become increasingly apparent that the status of gene transcription directly reflects the structure of chromatin (32, 80). The chromatin or the nucleosome is a highly ordered structure involving DNA supercoil stabilized by octameric core histone proteins H2A, H2B, H3, and H4 (44). Acetylation/deacetylation and phosphorylation/dephosphorylation of histone proteins play an important role in relaxation or condensation of the chromatin structure. In the past 3 years, it has become increasingly evident that acetylation/deacetylation and phosphorylation/dephosphorylation of core histone proteins play a crucial role in the activation and repression of gene transcription, which is important for growth and development (1, 2, 14, 32, 46, 63, 70). More recently, it has been confirmed that the enzymes (acetylase and deacetylase) that are involved in the modification of core histone proteins and hence the condensation/relaxation of chromatin are, in fact, transcriptional modulators (see reviews in references 32, 60, and 80). The report of streptococci/SDH-mediated de novo phosphorylation of the 17-kDa pharyngeal cell protein and its identification as histone H3 is the first to describe the possible role of histone phosphorylation in bacteria-mediated signaling in intact target cells (57) (Fig. 2A). Electron microscopy of SDH-treated pharyngeal cells showing distinct nuclear chromatin condensation supports the view that this effect is the direct result of the de novo phosphorylation of histone H3, although other possibilities may explain the condensation of the chromatin structure (57). It is likely that streptococci-mediated induction of protein phosphorylation in pharyngeal cells may be due in part to the upstream signaling events that target the nucleus. The implications of these findings can be viewed in terms of the ability of SDH/streptococci to induce gene transcription in the tar-

get cells. Hence, to understand the role of streptococci-mediated signaling events in pathogenesis, the relationship between the streptococci/SDH-mediated histone H3 phosphorylation and other serine/threonine and tyrosine phosphorylation events needs to be further investigated. Although histone phosphorylation has not yet been reported in other gram-positive bacteria-mediated host signaling events, the fact that pneumococci and their cell wall components activate certain mitogen-activated protein kinases such as erk-1/erk-2 and p38 (66) indicates that this bacteria-cell interaction may evoke a response in the nuclei of their target cells.

PROINFLAMMATORY RESPONSE IN HOST CELLS AFTER THEIR INTERACTION WITH STREPTOCOCCI

Proinflammatory response from the host cells in terms of the de novo or increased secretion of certain cytokines as a result of the bacterium-host cell interaction provides indirect evidence of not only signal transduction events, but also the activation of specific genes and subsequent nuclear responses. While this aspect is described in more detail in the other sections of this volume, the purpose of describing it in this chapter is to reveal the relationship between host cell signaling events and the secretion of certain cytokines that have the potential to evoke inflammatory responses.

There is substantial in vitro and in vivo evidence to confirm that, like lipopolysaccharide (81), pathogenic gram-positive cocci, their secretory products, and cell wall components are capable of inducing proinflammatory mediators such as interleukins and tumor necrosis factor alpha in a CD14-dependent manner (15, 50, 58, 66, 75). The role of staphylococcal and streptococcal exotoxins, which function as superantigens in the induction of mitogenic and inflammatory responses, is well established and is characterized by the release of a variety of chemokines and cytokines from peripheral blood monocytes (58). More evidence is available that confirms the ability of target epithelial cells or blood monocytes to release certain interleukins and tumor necrosis factor as a result of their interaction with intact gram-positive bacteria in a CD14-independent manner (6, 75). The examination by Wang et al. (75) of keratinocyte responses to adherent versus nonadherent group A streptococci has revealed a distinct pattern of expression of several proinflammatory molecules. That study reported further that both adherent and nonadherent streptococci are capable of inducing interleukin (IL)-6 and prostaglandin E_2; however, only adherent streptococci are capable of inducing IL1-α, IL-1β, and IL-8. Similarly, only the adherent streptococci are capable of damaging the keratinocytes. Since the production of cytokines and the damage to the keratinocytes are considerably reduced in streptolysin O-deficient streptococcal mutants, Wang et al. conclude that, to induce a proinflammatory response, streptococci also utilize some of their secretory products (75). In view of the fact that the primary target cell for group A streptococci is the pharyngeal cell, it is pertinent to explore the mechanism by which streptococci are able to evoke such an inflammatory response by using an in vitro pharyngeal cell model, preferably in a primary pharyngotonsillar organ culture. The variations in cytokine production, as described before, once again indicate that streptococci and their secretory products evoke a variety of signal transduction pathways in their target cells.

Identifying these pathways may provide important insights into the initial clinical picture of pharyngitis and the streptococcal ability to invade pharyngeal and other target cells.

CONCLUDING REMARKS

We have used the direct and indirect evidence of the consequences of streptococcal interactions with target epithelial cells to formulate our notion that streptococcal adherence is not merely a static process but is, in fact, highly regulated, dynamic, and complex. The skeleton of this dynamic process is the intracellular signaling events that ultimately determine the cells' fate and subsequently that of adjacent cells (Fig. 1). While together these events allow streptococci to disseminate in tissues, the biochemical changes in affected cells determine the clinical picture of streptococcal disease. Thus, changes in intracellular signaling events as a result of streptococcal interactions play a critical role in the disease process.

Compared with the amount of knowledge accumulated with regard to the capacity of gram-negative bacteria to evoke intracellular signaling, the knowledge gained so far for gram-positive bacteria-mediated host cell signaling is encouraging yet very limited. The mechanism of pathogenesis of streptococcal diseases is much more complex than that in many gram-negative infections since a variety of multifunctional proteins on the surface of streptococci have been shown to play a role in the disease process. This complexity is enhanced by the fact that the expression of some of these proteins is type specific or strain specific. It is not known at present how these proteins, whether individually or in concert, interact with host cells and ultimately evoke intracellular signaling. We may, therefore, need to decipher distinct signal transduction pathways induced as a result of these interactions to accurately define the pathogenesis of streptococcal diseases. The information on cellular microbiology of group A streptococci and pharyngeal cells may also unravel novel and important signaling pathways that link external cues to nuclear transcription in target cells. This information can also be utilized to design appropriate and novel therapeutic agents to intervene in the disease process as has been done for many septic shock and inflammatory diseases (42, 51).

REFERENCES

1. **Ajiro, K., and T. Nishimoto.** 1985. Specific site of histone H3 phosphorylation related to the maintenance of premature chromosome condensation. *J. Biol. Chem.* **260:** 15379–15381.
2. **Barratt, M. J., C. A. Hazzalin, E. Cano, and L. C. Mahadevan.** 1994. Mitogen-stimulated phosphorylation of histone H3 is targeted to a small hyperacetylation-sensitive fraction. *Proc. Natl. Acad. Sci. USA* **91:**4781–4785.
3. **Bissell, M. J.** 1993. Introduction: form and function in the epithelia. *Semin. Cell Biol.* **4:**157–159.
4. **Bliska, J. B., J. E. Galan, and S. Falkow.** 1993. Signal transduction in the mammalian cell during bacterial attachment and entry. *Cell* **73:**903–920.
5. **Bone, R. C.** 1994. Gram-positive organisms and sepsis. *Arch. Intern. Med.* **154:**26–34.
6. **Cauwels, A., E. Wan, M. Leismann, and E. Tuomanen.** 1997. Coexistence of CD14-dependent and independent pathways for stimulation of human monocytes by gram-positive bacteria. *Infect. Immun.* **65:**3255–3260.
7. **Chen, C. C., and P. P. Cleary.** 1989. Cloning and expression of the streptococcal C5a peptidase gene in *Escherichia coli:* linkage to the type 12 M protein gene. *Infect. Immun.* **57:**1740–1745.
8. **Cleary, P. P., E. L. Kaplan, J. P. Handley, A. Wlazlo, M. H. Kim, A. R. Hauser, and P. M. Schlievert.** 1992. Clonal basis for resurgence of serious *Streptococcus pyogenes* disease in the 1980s. *Lancet* **339:**518–521.
9. **Cossart, P.** 1997. Subversion of the mammalian cell cytoskeleton by invasive bacteria. *J. Clin. Invest.* **99:**2307–2311.
10. **Cossart, P., P. Boquet, S. Normark, and R. Rappuoli.** 1996. Cellular microbiology emerging. *Science* **271:**315–316.
11. **Cotter, P. A., and J. F. Miller.** 1996. Triggering bacterial virulence. *Science* **273:**1183–1184.
12. **Courtney, H. S., and D. L. Hasty.** 1991. Aggregation of group A streptococci by human saliva and effect of saliva on streptococcal adherence to host cells. *Infect. Immun.* **59:**1661–1666.
13. **Courtney, H. S., Y. Li, J. B. Dale, and D. L. Hasty.** 1994. Cloning, sequencing, and expression of a fibronectin/fibrinogen-binding protein from group A streptococci. *Infect. Immun.* **62:**3937–3946.
14. **Cue, D., P. E. Dombek, H. Lam, and P. P. Cleary.** 1998. *Streptococcus pyogenes* serotype M1 encodes multiple pathways for entry into human epithelial cells. *Infect. Immun.* **66:**4593–4601.
15. **Cundell, D. R., C. Gerard, I. Idanpaan-Heikkila, E. I. Tuomanen, and N. P. Gerard.** 1996. PAF receptor anchors *Streptococcus pneumoniae* to activated human endothelial cells. *Adv. Exp. Med. Biol.* **416:**89–94.
16. **Cundell, D. R., N. P. Gerard, C. Gerard, I. Indanpaan-Heikkila, and E. I. Tuomanen.** 1995. *Streptococcus pneumoniae* anchor to activated human cells by the receptor for platelet-activating factor. *Nature* **377:**435–438.
17. **Darnell, J. E.** 1998. STATs and gene regulation. *Science* **277:**1630–1635.
18. **Davies, H. D., A. McGeer, B. Schwartz, K. Green, D. Cann, A. E. Simor, and D. E. Low.** 1996. Invasive group A streptococcal infections in Ontario, Canada. *N. Engl. J. Med.* **335:**547–554.
19. **Davis, L., M. M. Baig, and E. M. Ayoub.** 1979. Properties of extracellular neuraminidase produced by group A streptococcus. *Infect. Immun.* **24:**780–786.
20. **Demuth, D. R., E. E. Golub, and D. Malamud.** 1990. Streptococcal-host interactions. *J. Biol. Chem.* **265:** 7120–7126.
21. **Dohrman, A., S. Miyata, M. Gallup, J.-D. Li, C. Chapelin, A. Coste, E. Escudier, J. Nadel, and A. Bashir.** 1998. Mucin gene (MUC 2 and MUC 5AC) upregulation by gram-positive and gram-negative bacteria. *Biochim. Biophys. Acta* **1406:**251–259.
22. **Doring, G., H. Obernessen, K. Botzenhart, B. Flehmig, N. Hoiby, and A. Hofmann.** 1983. Proteases of *Pseudomonas aeruginosa* in patients with cystic fibrosis. *J. Infect. Dis.* **147:**744–750.
23. **Drubin, D. G., and W. J. Nelson.** 1996. Origins of polarity. *Cell* **84:**335–344.

24. **Durand, M. L., S. B. Calderwood, D. J. Weber, S. I. Miller, V. S. Caviness, Jr., and M. M. Swartz.** 1993. Acute bacterial meningitis in adults: a review of 493 episodes. *N. Engl. J. Med.* **328:**21–28.

25. **Finlay, B. B., and P. Cossart.** 1997. Exploitation of mammalian host cell functions by bacterial pathogens. *Science* **276:**718–725.

26. **Forstner, G.** 1995. Signal transduction, packaging and secretion of mucins. *Annu. Rev. Physiol.* **57:**585–605.

27. **Frick, I.-M., K. L. Crossin, G. M. Edelman, and L. Bjorck.** 1995. Protein Hβa bacterial surface protein with affinity for both immunoglobulin and fibronectin type III domains. *EMBO J.* **14:**1674–1679.

28. **Garcia, R. C., D. R. Cundell, E. I. Tuomanen, L. F. Kolakowski, C. Gerard, and N. P. Gerard.** 1995. The role of N-glycosylation for functional expression of the human platelet-activating factor receptor. Glycosylation is required for efficient membrane trafficking. *J. Biol. Chem.* **270:**25178–25184.

29. **Gendler, S. J., and A. P. Spicer.** 1995. Epithelial mucin genes. *Annu. Rev. Physiol.* **57:**607–634.

30. **Hanski, E., and M. Caparon.** 1992. Protein F, a fibronection-binding protein, is an adhesin of the group A streptococcus, *Streptococcus pyogenes. Proc. Natl. Acad. Sci. USA* **89:**6172–6176.

31. **Hilkens, J., M. J. L. Ligtengerg, H. L. Vos, and S. V. Litvinov.** 1992. Cell membrane-associated mucins and their adhesion-modulating property. *Trends Biochem. Sci.* **17:**359–363.

32. **Hopkin, K.** 1997. Spools, switches, or scaffolds: how might histones regulate transcription? *J. NIH Res.* **9:**34–37.

33. **Hultgren, S. J., S. Abraham, M. Caparon, P. Falk, J. W. St. Geme III, and S. Normack.** 1993. Pilus and nonpilus bacterial adhesions: assembly and function in cell regulation. *Cell* **73:**887–901.

34. **Humphries, M. J.** 1994. Mechanisms of ligand binding by integrins. *Biochem. Soc. Trans.* **22:**275–282.

35. **Hynes, R. O.** 1992. Integrins: versatility, modulation, and signalling in cell adhesion. *Cell* **69:**11–25.

36. **Iseberg, R. R.** 1991. Discrimination between intracellular uptake and surface adhesion of bacterial pathogens. *Science* **252:**934–938.

37. **Iseberg, R. R., and G. T. V. Nhieu.** 1994. Binding and internalization of microorganisms by integrin receptors. *Trends Microbiol.* **2:**10–14.

38. **Jadoun, J., E. Burnstein, E. Hanski, and S. Sela.** 1996. Proteins M and F are required for efficient invasion of group A streptococci into cultured epithelial cells, abstr. L19. XIII Lancefield International Symposium on Streptococci and Streptococcal Diseases, Paris, France.

39. **Jaffe, J., S. Natanson-Yaron, M. G. Caparon, and E. Hanski.** 1996. Protein F2, a novel fibronectin-binding protein from *Streptococcus pyogenes*, possesses two binding domains. *Mol. Microbiol.* **21:**373–384.

40. **Kehoe, M. A.** 1994. Cell-wall-associated proteins in gram-positive bacteria, p. 217–261. *In* J.-M. Ghuysen and R. Hakenbeck (ed.), *Bacterial Cell Wall.* Elsevier Science Publishing, Inc., New York, N.Y.

41. **Klinger, J., B. Tandler, C. Kiedtke, and T. Boat.** 1984. Proteinases of *Pseudomonas aeruginosa* evoke mucins release by tracheal epithelium. *J. Clin. Invest.* **74:**1669–1678.

42. **Levitzki, A., and A. Gazit.** 1995. Tyrosine kinase inhibition: an approach to drug development. *Science* **267:**1782–1788.

43. **Livinov, S. V., and J. Hilkens.** 1993. The epithelial sialomucin, episialin, is sialylated during recycling. *J. Biol. Chem.* **268:**21364–21371.

44. **Luger, K., A. W. Mader, R. K. Richmond, D. F. Sargent, and T. J. Richmond.** 1997. Crystal structure of the nucleosome core particle at 2.8A resolution. *Nature* **389:**251–260.

45. **Lukomski, S., E. H. Burns, Jr., P. R. Wyde, A. Podbielski, J. Rurangirwa, D. K. Moore-Poveda, and J. M. Musser.** 1998. Genetic inactivation of an extracellular cysteine protease (SpeB) expressed by *Streptococcus pyogenes* decreases resistance to phagocytosis and dissemination to organs. *Infect. Immun.* **66:**771–776.

46. **Mahadevan, L. C., A. C. Willis, and M. J. Barratt.** 1991. Rapid histone H3 phosphorylation in response to growth factors, phorbol esters, okadaic acid, and protein synthesis inhibitors. *Cell* **65:**775–783.

47. **Molinari, G., S. R. Talay, P. Valentin-Weigand, M. Rohde, and G. S. Chhatwal.** 1998. The fibronectin-binding protein of *Streptococcus pyogenes*, SfbI, is involved in the internalization of group A streptococci by epithelial cells. *Infect. Immun.* **65:**1357–1363.

48. **Mosquera, J. A., V. N. Katiyar, J. Coello, and B. Rodriguez-Iturbe.** 1985. Neuraminidase production by streptococci from patients with glomerulonephritis. *J. Infect. Dis.* **151:**259–263.

49. **Moss, J., and M. Vaughan.** 1988. ADP-ribosylation of guanyl nucleotide binding proteins by bacterial toxins. *Adv. Enzymol.* **61:**303–379.

50. **Muller-Alouf, H., J. E. Alouf, D. Gerlach, J.-H. Ozegowski, C. Fitting, and J.-M. Cavaillon.** 1994. Comparative study of cytokine release by human peripheral blood mononuclear cells stimulated with *Streptococcus pyogenes* superantigenic erythrogenic toxins, heat-killed streptococci and lipopolysaccharide. *Infect. Immun.* **62:**4915–4921.

51. **Natanson, C., W. D. Hoffman, A. F. Suffredini, P. Q. Eichacker, and R. L. Danner.** 1994. Selected treatment strategies for septic shock based on proposed mechanisms of pathogenesis. *Ann. Intern. Med.* **120:**771–783.

52. **Norgren, M., A. Norrby, and S. E. Holm.** 1992. Genetic diversity in T1M1 group A streptococci in relation to clinical outcome of infection. *J. Infect. Dis.* **166:**1014–1020.

53. **Okada, N., M. K. Liszewski, J. P. Atkinson, and M. Caparon.** 1995. Membrane cofactor protein (CD46) is a keratinocyte receptor for the M protein of group A streptococcus. *Proc. Natl. Acad. Sci. USA* **92:**2489–2493.

54. **Okada, N., M. Watarai, V. Ozeri, E. Hanski, M. Caparon, and C. Sasakawa.** 1998. A matrix form of fibronectin mediates enhanced binding of *Streptococcus pyogenes* to host tissue. *J. Biol. Chem.* **272:**26978–26984.

55. **Pancholi, V., and V. A. Fischetti.** 1992. A major surface protein on group A streptococci is a glyceraldehyde-3-phosphate dehydrogenase with multiple binding activity. *J. Exp. Med.* **176:**415–426.

56. **Pancholi, V., and V. A. Fischetti.** 1993. Glyceraldehyde-3-phosphate dehydrogenase on the surface of group

A streptococci is also an ADP-ribosylating enzyme. *Proc. Natl. Acad. Sci. USA* **90:**8154–8158.

57. **Pancholi, V., and V. A. Fischetti.** 1997. Regulation of the phosphorylation of human pharyngeal cell proteins by group A streptococcal surface dehydrogenase (SDH): signal transduction between streptococci and pharyngeal cells. *J. Exp. Med.* **186:**1633–1643.

58. **Parsonnet, J.** 1989. Mediators in the pathogenesis of toxic shock syndrome: overview. *J. Infect. Dis.* **11**(Suppl.1):5263–5269.

59. **Pellegrini, S., and I. Dusanter-Fourt.** 1997. The structure, regulation and function of the Janus kinases (JAKs) and the signal transducers and activators of transcription (STATs). *Eur. J. Biochem.* **248:**615–633.

60. **Pennisi, E.** 1997. Opening the way to gene activity. *Science* **275:**155–157.

61. **Rakonjac, J. V., J. C. Robbins, and V. A. Fischetti.** 1995. DNA sequence of the serum opacity factor of group A streptococci: identification of a fibronectin-binding repeat domain. *Infect. Immun.* **63:**622–631.

62. **Rodriguez-Boulan, E., and W. J. Nelson.** 1989. Morphogenesis of the polarized epithelial cell phenotype. *Science* **245:**718–725.

63. **Roth, S. Y., and C. D. Allis.** 1996. Histone acetylation. *Cell* **87:**5–8.

64. **Ryan, P. R., V. Pancholi, and V. A. Fischetti.** 1998. Binding of group A streptococci to mucin and its implications in the colonization process, abstr. B-4, p. 56. *Abstr. 98th Gen. Meet. Am. Soc. Microbiol. 1998.* American Society for Microbiology, Washington, D.C.

65. **Sajjan, S. U., and J. F. Forstner.** 1998. Identification of the mucin-binding adhesin of *Pseudomonas cepacia* isolated from patients with cystic fibrosis. *Infect. Immun.* **60:**1434–1440.

66. **Schumann, R. R., D. Pfeil, D. Freyer, W. Buerger, N. Lamping, C. J. Kirschning, U. B. Goebel, and J. R. Weber.** 1998. Lipopolysaccharide and pneumococcal cell wall components activate the mitogen activated protein kinases (MAPK) erk-1, erk-2, and p38 in astrocytes. *Glia* **22:**295–305.

67. **Seger, R., and E. G. Krebs.** 1995. The MAPK signaling cascade. *FASEB J.* **9:**726–735.

68. **Shuter, J., V. B. Hatcher, and F. D. Lowy.** 1998. *Staphylococcus aureus* binding to human nasal mucin. *Infect. Immun.* **64:**310–318.

69. **Strous, G. J., and J. Dekker.** 1992. Mucin-type glycoproteins. *Crit. Rev. Biochem. Mol. Biol.* **27:**57–92.

70. **Sweet, M. T., G. Carlson, R. G. Cook, D. Nelson, and C. D. Allis.** 1997. Phosphorylation of linker histones by a protein kinase A-like activity in mitotic nuclei. *J. Biol. Chem.* **272:**916–923.

71. **Tabak, L. A.** 1995. In defense of the oral cavity: structure, biosynthesis, and function of salivary mucins. *Annu. Rev. Physiol.* **57:**547–564.

72. **Taylor-Papadimitriou, J., and O. J. Finn.** 1997. Biology, biochemistry and immunology of carcinoma-associated mucins. *Immunol. Today* **18:**105–107.

73. **Timpl, R., and J. C. Brown.** 1998. The laminins. *Matrix Biol.* **14:**275–281.

74. **Tuomanen, E., R. Austrian, and H. R. Masure.** 1995. Pathogenesis of pneumococcal infection. *N. Engl. J. Med.* **332:**1280–1284.

75. **Wang, B., N. Ruiz, A. Pentland, and M. Caparon.** 1997. Keratinocyte proinflammatory responses to adherent and nonadherent group A streptococci. *Infect. Immun.* **65:** 2119–2126.

76. **Wannamaker, L. W.** 1970. Differences between streptococcal infections of the throat and of the skin. *N. Engl. J. Med.* **282:**23–31.

77. **Wannamaker, L. W.** 1970. Differences between streptococcal infections of the throat and of the skin (second of two parts). *N. Engl. J. Med.* **282:**78–85.

78. **Wannamaker, L. W.** 1973. The chains that link the throat to the heart. *Circulation* **48:**9–18.

79. **Wick, M. J., J. L. Madara, B. N. Fields, and S. J. Normark.** 1991. Molecular cross talk between epithelial cells and pathogenic microorganisms. *Cell* **67:**651–659.

80. **Wolffe, A. P., J. Wong, and D. Pruss.** 1997. Activators and repressors: making use of chromatin to regulate transcription. *Genes Cells* **2:**291–302.

81. **Wright, S. D., R. A. Ramos, P. S. Tobias, R. J. Ulevitch, and J. C. Mathison.** 1990. CD14, a receptor for complexes of lipopolysaccharide (LPS) and LPS binding protein. *Science* **249:**1431–1433.

Vaccine Approaches To Protect against Group A Streptococcal Pharyngitis

VINCENT A. FISCHETTI

10

Streptococcus pyogenes (Lancefield group A) are human pathogens responsible for a wide range of diseases, the most common of which are nasopharyngeal infections and impetigo. More than 25 million cases of group A streptococcal infections occur each year in the United States, at a cost of over $1 billion to the public, in addition to losses in productivity. In recent years, an increase in streptococcal toxic shock and invasive infections (particularly necrotizing fasciitis) has been reported with certain strains of group A streptococci, resulting in rapid fatalities in up to 30% of the cases (42). Since there have been no reports of penicillin resistance in group A streptococci, streptococci-related diseases can be successfully treated with this antibiotic; however, with erythromycin, the second drug of choice for these bacteria, resistance is currently observed. Before treatment, group A streptococcal infections are usually associated with fever, significant discomfort, and generalized lethargy. About 3% of individuals with untreated or inadequately treated streptococcal pharyngitis develop rheumatic fever and rheumatic heart disease. While in the United States outbreaks of rheumatic fever are usually sporadic and infrequent, affecting only local areas of the country (46), in developing countries as many as 1% of school-age children are estimated to have rheumatic heart disease (16). Because of this, and the concern that penicillin-resistant strains may appear, there is a strong incentive to develop a safe and effective vaccine against group A streptococcal pharyngitis.

Although more prevalent during winter months, at any given time up to 30% of asymptomatic humans carry group A streptococci in their pharynx. Except under unusual circumstances and one serotype, there is no animal reservoir for these organisms. Thus, the eradication or significant reduction in the carriage of group A streptococci in the human pharynx would have a profound effect on the dissemination of this organism in the population and on the initiation of streptococcal disease. Even in the case of the highly invasive strains of streptococci, there is no evidence that they have their origins in other than pharyngeal sources.

Lancefield (33) showed more than 50 years ago that the surface M protein would be a prime candidate for a vaccine to protect against streptococcal infection. However, we have learned since that time that protection incurred by serum immunoglobulin G (IgG) to the M molecule is type specific, and antibodies directed to limited portions of the type-specific region may induce protection against only a limited number of strains within an M type. More recent studies have revealed that broad protection against streptococcal infection may be achieved by the induction of a local secretory response to exposed conserved sequences found within the M molecule. The approach prevents streptococcal colonization of the upper respiratory mucosa in a mouse model.

It has been known for decades that patients with rheumatic fever have serum IgG to human smooth muscle tissue at levels three to four times that found in normal serum. While these antibodies have also been found to react with streptococcal cell wall, membrane, and M protein N-terminal determinants, their role, if any, in the pathogenesis of rheumatic fever has not yet been proved. The fact that rheumatic fever and further cardiac damage occur around the time of the streptococcal infection, and not between infections when cross-reactive antibodies are still elevated (45), argues against the direct involvement of these antibodies in the disease process. Despite this, and until proved otherwise, it will be important to minimize the induction of cross-reactive antibodies in streptococcal vaccine preparations. This chapter concentrates on the progress to date toward the development of a vaccine to protect against streptococcal nasopharyngeal infection.

M PROTEIN STRUCTURE AND FUNCTION

Protective immunity to group A streptococcal infection is achieved through antibodies directed to the M protein (18), a major virulence factor present on the surface of all clinical isolates. M protein is a coiled-coil fibrillar protein composed of three major segments of tandem repeat sequences that extends nearly 60 nm from the surface of the streptococcal cell wall (18) (Fig. 1). The A- and B-repeats located within the N-terminal half are antigenically variable among the >100 known streptococcal types, with the

Gram-Positive Pathogens, ed. by V. A. Fischetti et al.

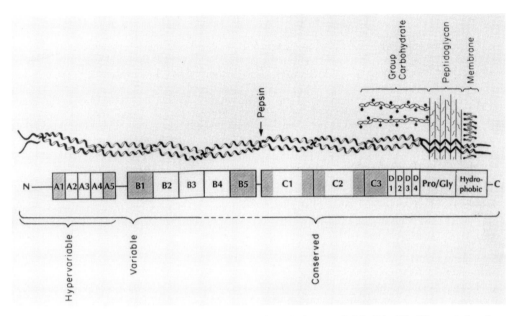

FIGURE 1 Proposed model of the M protein from M6 strain D471 (21, 25). The coiled-coil rod region extends about 60 nm from the cell wall with a short nonhelical domain at the NH_2 terminus. The Pro/Gly-rich region of the molecule is found within the peptidoglycan (36). The membrane-spanning segment is composed of predominantly hydrophobic amino acids, and a short charged tail extends into the cytoplasm. Data suggest that the membrane anchor may be cleaved shortly after synthesis (36). The A-, B-, and C-repeat regions are indicated along with those segments containing conserved, variable, and hypervariable epitopes among heterologous M serotypes. Pepsin designates the position of a pepsin-susceptible site near the center of the molecule.

N-terminal nonrepetitive region and A-repeats exhibiting hypervariability. The more C-terminal C-repeats, the majority of which are surface exposed, contain epitopes that are highly conserved among the identified M proteins (30). Because of its antigenically variable N-terminal region, the M protein provides the basis for the Lancefield serological typing scheme for group A streptococci (18) and a more contemporary molecular approach to typing group A streptococci (4).

The M protein is considered the major virulence determinant because of its ability to prevent phagocytosis when present on the streptococcal surface and thus, by this definition, all clinical isolates express M protein. This function may in part be attributed to the specific binding of complement factor H to both the conserved C-repeat domain and the fibrinogen bound to the B-repeats (26), preventing the deposition of C3b on the streptococcal surface. It is proposed that when the streptococcus contacts serum, the factor H bound to the M molecule inhibits or reverses the formation of C3b,Bb complexes and helps to convert C3b to its inactive form (iC3b) on the bacterial surface, preventing C3b-dependent phagocytosis. Studies have shown that antibodies directed to the B- and C-repeat regions of the M protein are unable to promote phagocytosis. This may be the result of the ability of factor H to also control the binding of C3b to the Fc receptors on these antibodies, resulting in inefficient phagocytosis. Antibodies directed to the hypervariable N-terminal region are opsonic, perhaps because they cannot be controlled by the factor H bound to the B- and C-repeat regions. Thus, it appears that the streptococcus has devised a method to protect its conserved region from being used against itself

by binding factor H to regulate the potentially opsonic antibodies that bind to these regions.

TYPE-SPECIFIC PROTECTION

The M protein has been a prime vaccine candidate to prevent group A streptococcal infections since Lancefield showed clearly that M protein-specific human and animal antibodies have the capacity to opsonize streptococci in preparation for phagocytic clearance. In general, serum IgG directed to the hypervariable NH_2-terminal portion of M protein leads to complement fixation and phagocytosis by polymorphonuclear leukocytes of streptococci of the homologous serotype (28). Even antibodies directed to whole group A streptococci will only allow phagocytosis of strains of the same M type in a phagocytic assay, suggesting that besides the M protein, no other streptococcal antigen is able to induce antibodies to override the antiphagocytic property of the M protein. Fox (22) has reviewed the early attempts in M protein vaccine development. Since the early 1970s few human trials have been realized. This is partially based on problems with hypersensitivity reactions found with the acid-extracted M protein preparations of the time and the fact that only type-specific protection was observed. In addition, repeated attempts to separate heterologous protein contaminants from the type-specific determinants proved unsuccessful. Except for one investigation, all streptococcal vaccine development since the early 1970s was based on animal studies in which the analysis of the immune response to M protein preparations was performed with and without adjuvants and in combination with other antigens.

In 1979 Beachey et al. (3) used pepsin-extracted M24 protein (PepM24, the N-terminal half of the native M24 molecule) (Fig. 1) to immunize human volunteers. Unlike earlier acid-extracted products (22), this highly purified fragment was found to be free of non-type-specific reactivity and did not induce delayed-type hypersensitivity tests in the skin of the 37 adult volunteers. Immunization with alum-precipitated PepM24 protein led to the development of type-specific opsonic antibodies in 10 of 12 volunteers, none of whom developed heart-reactive antibodies as determined by immunofluorescence. These studies clearly indicated that M protein vaccines free of sensitizing antigens could be produced but further emphasized the type specificity of the immune response.

Using these studies as a starting point, Beachey and co-workers (2) began to develop a type-specific epitope-based vaccine strategy to protect against streptococcal disease. It was soon learned that the complete PepM24 fragment was not necessary, but peptides representing the first 20 or so N-terminal amino acids of the M24 protein also evoked type-specific opsonic antibodies to M24 streptococci. Experiments with synthetic peptides of the N-terminus of M1, M5, M6, and M19 proteins resulted in the same conclusions. When a hybrid peptide was chemically synthesized representing the N-terminal sequence of both the M24 and M5 proteins and injected into animals with complete Freund's adjuvant, opsonic antibodies were produced to both M5 and M24 streptococci (1). Opsonic antibodies to three M proteins were obtained when the N-terminal sequences of three M protein sequences (M5-M6-M24) were synthesized in tandem and injected into rabbits. In a recent study, Dale et al. (12) used recombinant technology to prepare a tandem oligonucleotide array representing the N-terminal sequence of four serotypes (M24-M5-M6-M19). When the recombinant tetravalent fusion protein was purified and used to immunize rabbits, antibodies were raised against all four M proteins with substantial variation in both the enzyme-linked immunosorbent assay (ELISA) titer and opsonic activity to the four respective streptococcal types. These studies were subsequently repeated using an octavalent construct (14); however, in these studies, opsonic antibodies were produced against six of the eight serotypes used in the vaccine construct. As with the other studies using N-terminal sequences as a vaccine, none of the rabbits developed tissue cross-reactive antibodies as measured by immunofluorescence of cardiac tissue. These studies confirm and extend the earlier experiments of Beachey and colleagues and reveal that such an approach may be useful for the prevention of infection by specific streptococcal serotypes.

An important factor to consider in the development of a type-specific epitope-based vaccine is the potential of the streptococcus to generate new M serotypes by changing the amino-terminal portion of the M protein. For example, in the type 6 M protein of strain D471, type-specific opsonizing antibodies are directed against epitopes located both at the amino-terminal end (residues 1 through 21) and within the A-repeat block (28), which begins at amino acid 27 and continues to residue 96 in the 441-amino-acid M molecule (24). High-frequency, intragenic recombinational events within the A-repeat block can lead to a significant loss in the opsonizing ability of monospecific antibodies directed to this region (29). For instance, an opsonic antibody generated to the D471 parental strain showed some or no opsonizing activity to size-variant derivatives of this strain or other M6 streptococci isolated from patients. In

addition, Harbaugh et al. (23) showed that sequence variation also occurs within the nonrepeating N-terminal end of the M protein sequence from strains of the same serotype and that these changes ultimately affect the binding of opsonic antibodies. More recently, Penney et al. (37) showed a significant degree of sequence variation within the N-terminal hypervariable region of several strains identified serologically as M1. In all, these findings strongly indicate that opsonic antibodies induced by M protein N-terminal sequences from a given vaccine strain may prove to be ineffective or weakly effective against other strains of the same serotype. Therefore, a type-specific vaccine necessary to protect against a streptococcal infection would require a multivalent antigen corresponding to stable immunodeterminants on serotypes that together account for the majority of the nasopharyngeal isolates prevalent within the population at a given time. Thus far, this region has not been identified.

MUCOSAL VACCINE FOR NON-TYPE-SPECIFIC PROTECTION

At its peak incidence, 50% of children between the ages of 5 and 7 years suffer from streptococcal infection each year. Furthermore, the siblings of a child with a streptococcal pharyngitis are five times more likely to acquire the organism than is one of the parents. This decreased occurrence of streptococcal pharyngitis in adults might be explained by a nonspecific age-related host factor resulting in a decreased susceptibility to streptococci. Alternatively, protective antibodies directed to antigens common to a large number of group A streptococcal serotypes might arise as a consequence of multiple infections or exposures experienced during childhood. This could result in an elevated response to conserved M protein epitopes. This latter hypothesis is partly supported by our earlier studies on the immune response to the M protein in which it was found that the B-repeat domain (Fig. 1) was clearly immunodominant. When rabbits were immunized with the whole M protein molecule, the first detectable antibodies were directed to the B-repeat region and rose steadily with time. It was only after repeated M protein immunizations that antibodies were produced against the hypervariable A- and conserved C-repeat regions.

Early human trials by Fox and colleagues (15, 38) strongly suggested that mucosal vaccination with M protein was protective. Using highly purified acid-extracted M protein (which contains type-specific and conserved region fragments of M protein), volunteers were immunized either intranasally or subcutaneously. When they were challenged orally with virulent streptococci, the intranasally immunized volunteers displayed lower rates of both nasopharyngeal colonization and clinical illness compared with placebo controls. Volunteers immunized subcutaneously displayed only a reduction in clinical illness and showed no reduction in nasopharyngeal colonization.

Unlike antibodies to the N-terminal region of the M molecule, it was clear from our own studies that antibodies directed to the exposed C-repeat region were not opsonic. Because of this, experiments were performed to explore whether mucosal antibodies directed only to the conserved region of M protein could be responsible for protection against streptococcal infection. Taking advantage of the pepsin site in the center of the M molecule (separating the variable and conserved regions) (Fig. 1), the recombinant M6 protein was cleaved, and the N- and C-terminal fragments were separated by sodium dodecyl sulfate-polyacrylamide gel electrophoresis and Western blotted. When the blots were reacted with different adult human sera, all adults tested had antibodies to the C-terminal conserved region while, as expected, only sera that were opsonic for the M6 organisms reacted with the N-terminal variable region. Similar studies performed with M protein isolated from five different common serotypes (M3, M5, M6, M24, M29) revealed that sera from 10 of 17 adults tested did not have N-terminal-specific antibodies to these M types, while only two sera reacted with two serotypes and the remaining five sera reacted with only one serotype. However, all sera tested reacted to the C-terminal fragment of the M molecule. Similar results were seen when salivary IgA from adults and children were tested in ELISA against the N- and C-terminal halves of the M6 molecule (unpublished data). In all, this is further evidence that the relative resistance of adults to streptococcal pharyngitis is clearly not due to the presence of type-specific antibodies to multiple M serotypes of streptococci, but may perhaps be due to the presence of antibodies to conserved determinants.

From these findings we reasoned that an immune response to the conserved region of the M molecule might afford protection by inducing a mucosal response to prevent streptococcal colonization and ultimate infection. In view of the evidence that the conserved C-repeat epitopes of the M molecule are immunologically exposed on the streptococcal surface (30), it should be possible to generate mucosal antibodies that are reactive to the majority of streptococcal types, using only the conserved region antigen for immunization.

PASSIVE PROTECTION

Secretory IgA (sIgA) is able to protect mucosal surfaces from infection by pathogenic microorganisms despite the fact that its effector function differs from those of serum-derived immunoglobulins. When streptococci are administered intranasally to mice, they are able to cause death by first colonizing and then invading the mucosal barrier, resulting in dissemination of the organism to systemic sites. Using this model we first examined if sIgA, delivered directly to the mucosa, plays a role in protecting against streptococcal infection. Live streptococci were mixed with affinity-purified M protein-specific sIgA or IgG antibodies and administered intranasally to the animals (6). The results clearly showed that the anti-M-protein sIgA protected the mice against streptococcal infection and death, whereas the opsonic serum IgG administered in the same way was without effect. This indicated to us that sIgA can protect at the mucosa and may preclude the need for opsonic IgG in preventing streptococcal infection. This study was also one of the first to compare purified, antigen-specific sIgA and serum IgG for passive protection at a mucosal site.

In another laboratory, passive protection against streptococcal pharyngeal colonization was also shown by the oral administration of purified lipoteichoic acid (LTA) but not deacylated LTA before oral challenge in mice (11). The addition of anti-LTA by the same route also protected mice from oral streptococcal challenge. While several in vitro studies showed the importance of M protein and LTA in streptococcal adherence, these in vivo studies together with those presented above suggest that both M protein and LTA may play a key role in the colonization of

the mouse pharyngeal mucosa. However, it is uncertain whether this is also true in humans.

ACTIVE IMMUNIZATION WITH CONSERVED REGION PEPTIDES

To determine whether a local mucosal response directed to the conserved exposed epitopes of M protein can influence the course of mucosal colonization by group A streptococci, peptides corresponding to these regions were used as immunogens in a mouse model (7). Overlapping synthetic peptides of the conserved region of the M6 protein were covalently linked to the mucosal adjuvant cholera toxin B subunit (CTB) and administered intranasally to the mice in three weekly doses. Thirty days later, animals were challenged intranasally with live streptococci (either homologous M6 or heterologous M14), and pharyngeal colonization by the challenge organism was monitored by throat swabs for the presence of streptococci for 10 to 15 days. Mice immunized with the peptide-CTB complex showed a significant reduction in colonization with either M6 or M14 streptococci compared with mice receiving CTB alone (5) (Fig. 2). Thus, despite the fact that conserved-region peptides were unable to evoke an opsonic antibody response (28), these peptides have the capacity to induce a local immune response capable of influencing the colonization of group A streptococci at the nasopharyngeal mucosa in this model system. These findings were the first to demonstrate protection against a heterologous serotype of group A streptococci with a vaccine consisting of the widely shared C-repeat region of the M6 protein.

Confirmation of these findings was later published independently using a different streptococcal serotype as the immunizing and challenge strains (8). In a separate study, Pruksakorn et al. (40), using different criteria for streptococcal opsonization than those previously published (32), found that, when a peptide derived from the conserved region of the M protein was used to immunize mice, it induced antibodies capable of opsonizing type 5 streptococci and streptococci isolated from Aboriginal and Thai patients with rheumatic fever. These findings are in sharp contrast to those from the studies of Jones and Fischetti (28), who showed that antibodies to the conserved region of M protein are not opsonic. However, since the peptide reported by Pruksakorn et al. (40) is similar to one of the peptides used by Bessen and Fischetti (5) in their mucosal protection studies (see above), the induction of serum IgG during mucosal immunization may offer added protection against streptococcal infection.

VACCINIA VIRUS AS A VECTOR

To further verify the validity of using the M protein conserved region as a streptococcal vaccine, experiments were repeated in a vaccinia virus vector system. In these studies, the gene coding for the complete conserved region of the M6 molecule (from the pepsin site to the C terminus) was cloned and expressed in vaccinia virus producing the recombinant VV::M6 virus (19). Tissue culture cells infected with this virus were found to produce the conserved region of the M6 molecule. Animals immunized intranasally with only a single dose of recombinant virus were significantly protected from heterologous streptococcal challenge compared with animals immunized with wild-type virus (Fig. 3). When the extent of colonization was examined in those

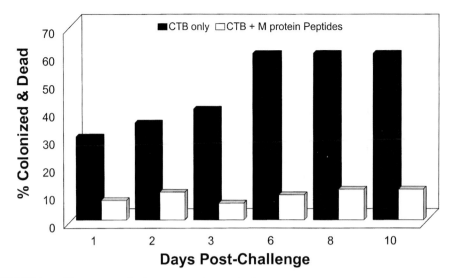

FIGURE 2 The extent of colonization of mice challenged with group A streptococci after oral immunization with M-protein conserved-region M6 peptides linked to CTB. Throats of orally immunized mice were swabbed each day after challenge with M14 streptococci, and the specimens were plated on blood plates to determine the extent of colonization compared with that of mice vaccinated with CTB only. Plates showing group A streptococci were scored as positive.

animals immunized with wild-type or the VV::M6 recombinant, the VV::M6-immunized animals showed a marked reduction in overall colonization, indicating that mucosal immunization reduced the bacterial load on the mucosa in these animals. Animals immunized intradermally with the VV::M6 virus and challenged intranasally showed no protection.

The approaches described above proved that induction of a local immune response was critical for protection against streptococcal colonization and that the protection was not dependent upon an opsonic response. However, in the event that the streptococcus was successful in penetrating the mucosa and establishing an infection, only then would type-specific antibodies be necessary to eradicate the organism. This idea may perhaps explain why adults sporadically develop a streptococcal pharyngitis. The success of these strategies not only forms the basis of a broadly protective vaccine for the prevention of streptococcal pharyngitis but may also offer insights for the development of other vaccines. For instance, a vaccine candidate previously shown to be ineffective by the parenteral route may prove to be successful by simply changing the site of immunization. Furthermore, these results emphasize the fact that in some cases antigens need to be presented to the

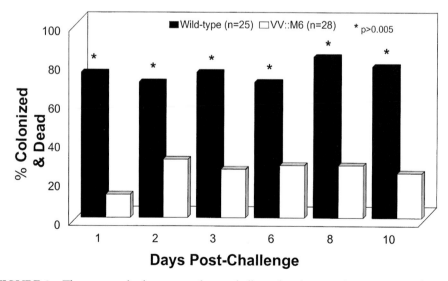

FIGURE 3 The extent of colonization of mice challenged with group A streptococci after oral immunization with recombinant vaccinia virus containing the gene for the whole conserved region of the M6 protein. Throats of orally immunized mice were swabbed each day after challenge with M14 streptococci, and the specimens were plated on blood plates to determine the extent of colonization compared with that of mice vaccinated with wild-type vaccinia only. Plates showing group A streptococci were scored as positive.

immune system in a specific fashion to ultimately induce a protective response.

THE USE OF GRAM-POSITIVE COMMENSALS AS VACCINE VECTORS

The CTB-linked peptide and vaccinia systems, although successful for protection against streptococcal infection, are not ideal. The CTB system requires large quantities of purified peptide and a linkage protocol that allows no more that two peptides per CTB molecule to enable the proper binding of CTB to GM_1 ganglioside (5). Although possible to achieve, these requirements make this type of vaccine relatively expensive to produce, even if recombinant technology were used to prepare the fusion molecules. Given the fact that such a vaccine ultimately must be administered in developing countries, the cost would likely be prohibitive. The vaccinia virus vector, on the other hand, is inexpensive but unlikely to gain approval from the Food and Drug Administration since oral/intranasal administration, which could result in serious complications, is required for effectiveness. Because of these limitations, we set out to develop a mucosal vaccine delivery system that is safe, effective, and inexpensive.

While several nonliving systems of delivering antigen to mucosal sites have been developed (7, 9, 17, 34), live vectors may afford a better and more natural response without the need to reimmunize to gain higher antibody titers. In most instances, live antigen delivery vectors are derived from bacteria (usually gram-negative) (43) or viruses (44) that are normally considered mammalian pathogens. Perhaps this is due in part to our better understanding of these organisms, making genetic manipulations easier. Usually these organisms have been extensively engineered to reduce their pathogenicity yet maintain certain invasive qualities (e.g., to invade the M cells of the gut mucosa) to induce a mucosal immune response. To circumvent some of the safety and environmental issues inherent in the wide-scale dissemination of engineered pathogens, we developed nonpathogenic gram-positive bacteria as vaccine vectors (39). In this system, foreign antigens are displayed on the surface of gram-positive human commensal organisms that colonize the niche invaded by the pathogen (oral, intestinal, vaginal). Colonization generates both an enhanced local IgA response to the foreign antigen and systemic IgG and T-cell responses. Unlike many other live bacterial systems, in which the foreign antigen is either retained in the cytoplasm, translocated to the periplasm, or in some cases secreted, the gram-positive vector anchors the foreign antigen to the cell for surface display (39). Since the cell wall peptidoglycan of the gram-positive cell is a natural adjuvant, an enhanced response is obtained when the engineered organisms are processed for antibody induction.

Our ability to accomplish this is based on the recent discovery that surface molecules from gram-positive bacteria that anchor via their C termini (of which >65 have now been sequenced) have a highly conserved C-terminal region that is responsible for cell attachment (20, 35) (see chapter 2, this volume).

A NEW GENERATION STREPTOCOCCAL VACCINE

Based on successful mouse studies examining the immune response to a recombinant antigen expressed on the surface of *Streptococcus gordonii*, a recombinant *S. gordonii* was engineered that comprised the C-terminal half of the M protein containing the exposed conserved region of the molecule. This segment was similar to that used successfully in the vaccinia virus experiments (see above) (19). *S. gordonii*, expressing this segment of the M protein on its surface, successfully colonized for up to 12 weeks all of the 10 rabbits immunized. During this time, the animals raised a salivary IgA (Fig. 4) and serum IgG (Fig. 5) response to the intact M protein. The amount of M protein-specific sIgA was up to 5% of the total IgA in the saliva of these animals. The IgA and IgG induced by this method did not cross-react with human heart tissue, as determined by immunofluorescence assay, or with tropomyosin, myosin, or vimentin by Western blot. Studies are currently in progress to determine protective effects of this delivery system in a mouse model.

The commensal delivery system would be ideal for developing countries, if proved to be effective. Because it is a live vector, it would be easy to administer and not likely to require additional doses. Also, since gram-positive bacteria are stable for long periods in the lyophilized state, a cold chain would not be required. Early studies show that, when reintroduced into the human oral cavity, *S. gordonii* is capable of persisting for over 2 years and is transmitted to other members of the family. For a developing country this factor could be ideal, since rarely is the whole population able to be immunized. However, it remains to be determined if this approach will induce a protective immune response in humans.

NON-M-PROTEIN APPROACHES TO PROTECT AGAINST STREPTOCOCCAL INFECTION

Other approaches have been developed recently that may be an alternative to using M protein as a vaccine candidate (13). Cleary and coworkers (10) have identified a group A streptococcal protease that specifically cleaves the human serum chemotaxin C5a, preventing its binding to polymorphonuclear neutrophils. This cleavage of C5a has been shown to reduce the influx of inflammatory cells at the site of a streptococcal infection. Expanding this finding, they demonstrated that delivery of a defective form of the streptococcal C5a peptidase molecule intranasally to mice showed promise in protecting against challenge against heterologous M serotypes (27). In these studies, immunized animals cleared the challenge streptococci from the throat more rapidly than did control animals.

Using a different strategy, Musser and his group have been working on the group A streptococcal cysteine protease known as streptococcal pyrogenic exotoxin B (SpeB). The gene for this protein is found in virtually all strains of group A streptococci, and in most cases its product is secreted from these organisms. Since cysteine proteases have been implicated in bacterial pathogenicity, Kapur et al. (31) found that a protease-negative SpeB mutant lost nearly all of its ability to cause death in mice when compared with wild-type organisms. They showed that mice passively immunized with rabbit IgG to cysteine protease exhibited a longer time to death than did control animals. Active immunization with cysteine protease gave the same result. Thus, while this vaccine approach prolonged the time to death, it did not prevent death.

N-Acetyl-glucosamine, a polysaccharide component of the streptococcal cell wall, is the group-specific antigen for

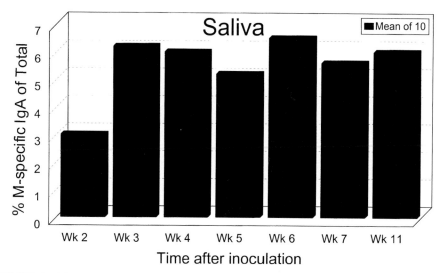

FIGURE 4 M protein-specific salivary IgA in rabbits colonized with *S. gordonii* expressing the conserved region on the cell surface. Salivary samples were taken after pylocarpine induction and tested in ELISA against the M protein.

group A streptococci. Since most people infected by streptococci will develop anti-*N*-acetyl-glucosamine antibodies, Zabriskie and colleagues attempted to determine if these antibodies have any effect on protecting against streptococcal infection. Using a modified in vitro phagocytic assay and a low bacterial inoculum, they found that anticarbohydrate antibodies specific for the *N*-acetyl-glucosamine were phagocytic. Comparing human and rabbit opsonic sera, they concluded that high titers of anti-*N*-acetyl-glucosamine-specific antibodies were effective in opsonization and phagocytosis of streptococci (41).

SUMMARY AND CONCLUSION

It would be naive to believe that a streptococcal infection is not a highly complex process, and perhaps no one single approach will control or prevent all aspects of infection by these organisms. While we may be able to prevent infection by using a mucosal approach, these types of vaccines are not usually sterilizing, and organisms introduced at high doses may break through and cause sporadic infections. This is more akin to how we believe most adults naturally become more resistant to streptococcal infections. One of the benefits of a successful mucosal vaccination scheme would be the reduction of streptococcal colonization in general, thus reducing the total number of these pathogens in the population. Since the main reservoir for group A streptococci for most streptococci-related illnesses is the human nasopharynx, reducing the carriage of these organisms by only 30% would have a profound impact on the dissemination of streptococci in the environment and thus a significant reduction in streptococcal disease in general.

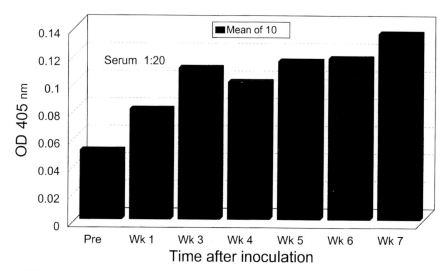

FIGURE 5 M protein-specific serum IgG in rabbits colonized with *S. gordonii* expressing the conserved region on the cell surface. Blood samples were taken at weekly intervals and tested in ELISA against the M protein.

Supported in part by USPHS Grant AI11822 and a grant from SIGA Pharmaceuticals.

REFERENCES

1. **Beachey, E. H., H. Gras-Masse, A. Tarter, M. Jolivet, F. Audibert, L. Chedid, and J. M. Seyer.** 1986. Opsonic antibodies evoked by hybrid peptide copies of types 5 and 24 streptococcal M proteins synthesized in tandem. *J. Exp. Med.* **163:**1451–1458.

2. **Beachey, E. H., J. M. Seyer, J. B. Dale, W. A. Simpson, and A. H. Kang.** 1981. Type-specific protective immunity evoked by synthetic peptide of *Streptococcus pyogenes* M protein. *Nature* **292:**457–459.

3. **Beachey, E. H., G. H. Stollerman, R. H. Johnson, I. Ofek, and A. L. Bisno.** 1979. Human immune response to immunization with a structurally defined polypeptide fragment of streptococcal M protein. *J. Exp. Med.* **150:** 862–877.

4. **Beall, B., R. Facklam, and T. Thompson.** 1995. Sequencing *emm*-specific polymerase chain reaction products for routine and accurate typing of group A streptococci. *J. Clin. Microbiol.* **34:**953–958.

5. **Bessen, D., and V. A. Fischetti.** 1988. Influence of intranasal immunization with synthetic peptides corresponding to conserved epitopes of M protein on mucosal colonization by group A streptococci. *Infect. Immun.* **56:** 2666–2672.

6. **Bessen, D., and V. A. Fischetti.** 1988. Passive acquired mucosal immunity to group A streptococci by secretory immunoglobulin A. *J. Exp. Med.* **167:**1945–1950.

7. **Bessen, D., and V. A. Fischetti.** 1990. Synthetic peptide vaccine against mucosal colonization by group A streptococci. I. Protection against a heterologous M serotype with shared C repeat region epitopes. *J. Immunol.* **145:** 1251–1256.

8. **Bronze, M. S., D. S. McKinsey, E. H. Beachey, and J. B. Dale.** 1988. Protective immunity evoked by locally administered group A streptococcal vaccines in mice. *J. Immunol.* **141:**2767–2770.

9. **Challacombe, S. J., D. Rahman, H. Jeffery, S. S. Davis, and D. T. O'Hagan.** 1992. Enhanced secretory IgA and systemic IgG antibody responses after oral immunization with biodegradable microparticles containing antigen. *Immunology* **76:**164–168.

10. **Cleary, P. P., U. Prahbu, J. B. Dale, D. E. Wexler, and J. Handley.** 1992. Streptococcal C5a peptidase is a highly specific endopeptidase. *Infect. Immun.* **60:**5219–5223.

11. **Dale, J. B., R. W. Baird, H. S. Courtney, D. L. Hasty, and M. S. Bronze.** 1994. Passive protection of mice against group A streptococcal pharyngeal infection by lipoteichoic acid. *J. Infect. Dis.* **169:**319–323.

12. **Dale, J. B., E. Y. Chiang, and J. W. Lederer.** 1993. Recombinant tetravalent group A streptococcal M protein vaccine. *J. Immunol.* **151:**2188–2194.

13. **Dale, J. B., P. P. Cleary, V. A. Fischetti, J. M. Musser, and J. B. Zabriskie.** 1997. Group A and group B streptococcal vaccine development. A round table presentation. *Adv. Exp. Med. Biol.* **418:**863–868.

14. **Dale, J. B., M. Simmons, E. C. Chiang, and E. Y. Chiang.** 1996. Recombinant, octavalent group A streptoccal M protein vaccine. *Vaccine* **14:**944–948.

15. **D'Alessandri, R., G. Plotkin, R. M. Kluge, M. K. Wittner, E. N. Fox, A. Dorfman, and R. H. Waldman.** 1978. Protective studies with group A streptococcal M protein vaccine. III. Challenge of volunteers after systemic or intranasal immunization with type 3 or type 12 group A streptococcus. *J. Infect. Dis.* **138:**712–718.

16. **Dodu, S. R. A., and S. Bothig.** 1989. Rheumatic fever and rheumatic heart disease in developing countries. *World Health Forum* **10:**203–212.

17. **Eldridge, J. H., J. K. Staas, J. A. Meulbroek, J. R. McGhee, T. R. Tice, and R. M. Gilley.** 1991. Biodegradable microspheres as a vaccine delivery system. *Mol. Immunol.* **28:**287–294.

18. **Fischetti, V. A.** 1989. Streptococcal M protein: molecular design and biological behavior. *Clin. Microbiol. Rev.* **2:** 285–314.

19. **Fischetti, V. A., W. M. Hodges, and D. E. Hruby.** 1989. Protection against streptococcal pharyngeal colonization with a vaccinia:M protein recombinant. *Science* **244:** 1487–1490.

20. **Fischetti, V. A., V. Pancholi, and O. Schneewind.** 1990. Conservation of a hexapeptide sequence in the anchor region of surface proteins of Gram-positive cocci. *Mol. Microbiol.* **4:**1603–1605.

21. **Fischetti, V. A., D. A. D. Parry, B. L. Trus, S. K. Hollingshead, J. R. Scott, and B. N. Manjula.** 1988. Conformational characteristics of the complete sequence of group A streptococcal M6 protein. *Proteins* **3:**60–69.

22. **Fox, E. N.** 1974. M proteins of group A streptococci. *Bacteriol. Rev.* **38:**57–86.

23. **Harbaugh, M. P., A. Podbielski, S. Hugl, and P. P. Cleary.** 1993. Nucleotide substitutions and small-scale insertion produce size and antigenic variation in group A streptococcal M1 protein. *Mol. Microbiol.* **8:**981–991.

24. **Hollingshead, S. K., V. A. Fischetti, and J. R. Scott.** 1986. Complete nucleotide sequence of type 6 M protein of the Group A streptococcus. *J. Biol. Chem.* **261:**1677–1686.

25. **Hollingshead, S. K., V. A. Fischetti, and J. R. Scott.** 1986. Complete nucleotide sequence of type 6 M protein of the group A streptococcus: repetitive structure and membrane anchor. *J. Biol. Chem.* **261:**1677–1686.

26. **Horstmann, R. K., H. J. Sievertsen, M. Leippe, and V. A. Fischetti.** 1992. Role of fibrinogen in complement inhibition by streptococcal M protein. *Infect. Immun.* **60:**5036–5041.

27. **Ji, Y., B. Carlson, A. Kondagunta, and P. P. Cleary.** 1997. Intranasal immunization with C5a peptidase prevents nasopharyngeal colonization of mice by the group A *Streptococcus*. *Infect. Immun.* **65:**2080–2087.

28. **Jones, K. F., and V. A. Fischetti.** 1988. The importance of the location of antibody binding on the M6 protein for opsonization and phagocytosis of group A M6 streptococci. *J. Exp. Med.* **167:**1114–1123.

29. **Jones, K. F., S. K. Hollingshead, J. R. Scott, and V. A. Fischetti.** 1988. Spontaneous M6 protein size mutants of group A streptococci display variation in antigenic and opsonogenic epitopes. *Proc. Natl. Acad. Sci. USA* **85:** 8271–8275.

30. **Jones, K. F., B. N. Manjula, K. H. Johnston, S. K. Hollingshead, J. R. Scott, and V. A. Fischetti.** 1985. Location of variable and conserved epitopes among the multiple

serotypes of streptococcal M protein. *J. Exp. Med.* **161:** 623–628.

31. **Kapur, V., J. T. Maffei, R. S. Greer, L. L. Li, G. J. Adams, and J. M. Musser.** 1994. Vaccination with streptococcal extracellular cysteine protease (interleukin-1 beta convertase) protects mice against challenge with heterologous group A streptococci. *Microb. Pathog.* **16:**443–450.

32. **Lancefield, R. C.** 1959. Persistence of type specific antibodies in man following infection with group A streptococci. *J. Exp. Med.* **110:**271–292.

33. **Lancefield, R. C.** 1962. Current knowledge of the type specific M antigens of group A streptococci. *J. Immunol.* **89:**307–313.

34. **McKenzie, S. J., and J. F. Halsey.** 1984. Cholera toxin B subunit as a carrier protein to stimulate a mucosal immune response. *J. Immunol.* **133:**1818–1824.

35. **Medaglini, D., G. Pozzi, T. P. King, and V. A. Fischetti.** 1995. Mucosal and systemic immune responses to a recombinant protein expressed on the surface of the oral commensal bacterium *Streptococcus gordonii* after oral colonization. *Proc. Natl. Acad. Sci. USA* **92:**6868–6872.

36. **Pancholi, V., and V. A. Fischetti.** 1988. Isolation and characterization of the cell-associated region of group A streptococcal M6 protein. *J. Bacteriol.* **170:**2618–2624.

37. **Penney, T. J., D. R. Martin, L. C. Williams, S. A. de Malmanche, and P. L. Bergquist.** 1995. A single *emm* gene-specific oligonucleotide probe does not recognise all members of the *Streptococcus pyogenes* M type 1. *FEMS Microbiol. Lett.* **130:**145–150.

38. **Polly, S. M., R. H. Waldman, P. High, M. K. Wittner, A. Dorfman, and E. N. Fox.** 1975. Protective studies with a group A streptococcal M protein vaccine. II. Challenge of volunteers after local immunization in the upper respiratory tract. *J. Infect. Dis.* **131:**217–224.

39. **Pozzi, G., M. Contorni, M. R. Oggioni, R. Manganelli, M. Tommasino, F. Cavalieri, and V. A. Fischetti.** 1992. Delivery and expression of a heterologous antigen on the surface of streptococci. *Infect. Immun.* **60:**1902–1907.

40. **Pruksakorn, S., B. Currie, E. Brandt, D. Martin, A. Galbraith, C. Phornphutkul, S. Hunsakunachal, A. Manmontri, and M. F. Good.** 1994. Towards a vaccine for rheumatic fever: identification of a conserved target epitope on M protein of group A streptococci. *Lancet* **344:** 639–642.

41. **Salvadori, L. G., M. S. Blake, M. McCarty, J. Y. Tai, and J. B. Zabriskie.** 1995. Group A streptococcus-liposome ELISA antibody titers to group A polysaccharide and opsonophagocytic capability to the antibodies. *J. Infect. Dis.* **171:**593–600.

42. **Stevens, D. L.** 1992. Invasive group A streptococcus infections. *Clin. Infect. Dis.* **14:**2–13.

43. **Tacket, C. O., D. M. Hone, R. Curtiss III, S. M. Kelly, G. Losonsky, L. Guers, A. M. Harris, R. Edelman, and M. M. Levine.** 1992. Comparison of the safety and immunogenicity of *delta-aroC delta-aroD* and *delta-cya delta-crp Salmonella typhi* strains in adult volunteers. *Infect. Immun.* **60:**536–541.

44. **Tartaglia, J., and E. Paoletti.** 1990. Live recombinant viral vaccines, p. 125–151. *In* M. H. V. van Regenmorterl and A. R. Neurath (ed.), *Immunochemistry of Viruses*, vol. II. Elsevier Science Publishing, Inc., New York, N.Y.

45. **van de Rijn, I., J. B. Zabriskie, and M. McCarty.** 1977. Group A streptococcal antigens cross-reactive with myocardium. Purification of heart-reactive antibody and isolation and characterization of the streptococcal antigen. *J. Exp. Med.* **146:**579–599.

46. **Veasy, L. G., S. E. Wiedmeier, G. S. Orsmond, H. D. Ruttenberg, M. M. Boucek, S. J. Roth, and V. F. Tait.** 1987. Resurgence of acute rheumatic fever in the intermountain area of the United States. *N. Engl. J. Med.* **316:** 421–427.

The Bacteriophages of Group A Streptococci

W. MICHAEL McSHAN

11

Bacteriophages (or phages), the ubiquitous viruses infecting almost all known species of bacteria, were discovered early in the twentieth century, apparently independently, by Twort in Britain and d'Hérelle in France (1). The ensuing years saw numerous studies in the new phenomenon of bacterial viruses, finding that phages could be isolated from water and soil as well as from the exterior and interior surfaces of humans and animals. With few exceptions, phages could be isolated that infected virtually all pathogenic and nonpathogenic bacteria. In those preantibiotic days, the idea of viruses that could kill many human pathogens attracted considerable attention in the scientific community, and much of the first work on the phages of the genus *Streptococcus* involved attempts to use these agents therapeutically to cure streptococci-associated diseases (reviewed by Evans [23]). The study of bacteriophages of streptococci and the related genera has maintained a consistent level of interest by the scientific community for a number of reasons, both medical and industrial. This chapter focuses on the influence that the phages of *Streptococcus pyogenes* (group A streptococci; GAS), both lytic and lysogenic, have on the biology and dissemination of virulence factors of this important gram-positive pathogen.

EARLY STUDIES

Evans, in a series of reports from the 1930s and 1940s, presented the first systematic studies on the bacteriophages of GAS (23–27). She initially investigated the possibility of phages in preventing or lessening the severity of streptococcal infections (23). Treatment of mice or rabbits with a virulent strain of GAS and a bacteriophage isolated from sludge did not show any protective effects for the infected animals, and in some cases, rabbits treated with antistreptococcal phages succumbed to an unusually violent course of disease. It is not known whether the strain of GAS used in these experiments harbored any endogenous temperate phages or the properties of the infecting phage. It is possible that the increase in virulence observed was caused by phage selection eliminating a less virulent subpopulation of the streptococci or conversion of the streptococci to a more virulent form, either by introduction of a toxin gene or by an increase in the expression of a somatic virulence gene like M protein, as was later observed with phage CS112 (72). Further, culture of many strains of GAS in blood or in the presence of human serum results in increased hyaluronic acid capsule production, which, as will be discussed, is an effective block to infection by the known lytic streptococcal bacteriophages (57).

One of the earliest facts to come to light concerning GAS bacteriophages was that the lysogenic state was very common (24), an observation that has been confirmed by numerous subsequent studies (9, 38, 46, 47, 81, 86). Another key observation made during the early days of GAS streptococcal investigations was that sterile culture filtrates from "scarlatina" (erythrogenic toxin A-producing) strains could cause toxigenic conversion of non-toxin-producing GAS (8, 32). The association of bacteriophages with this phenomenon was not made at that time since the discovery of lysogeny would not occur until several decades later. The link between toxigenic conversion of GAS and temperate phages was finally made by Zabriskie in 1964 (89) and is now recognized as an important factor for virulence in these bacteria.

LYTIC BACTERIOPHAGES
General Characteristics

Bacteriophages may typically be grouped by their life cycle into two categories: the lytic phages and the temperate (lysogenic) phages. The lytic bacteriophages infect their specific host bacterium, replicate their genome and assemble new virions, and then rupture the host to release the newly formed phages. Lytic phages do not enter into any extended relationship with their host cell following infection, and they do not directly alter the genotype of the host by the introduction of a novel gene, for example. However, lytic phages can play important roles in the shaping of the biology of their GAS hosts, either through elimination of the phage-susceptible members of a population consisting of more than one strain, selection for the rare

Gram-Positive Pathogens, ed. by V. A. Fischetti et al.
© 2000 American Society for Microbiology, Washington, D.C.

phage-resistant variants in a mostly homogeneous population, or by being the vectors of genetic exchange through generalized transduction.

Bacteriophage A25

Bacteriophage A25, the best-studied lytic phage of GAS, was originally isolated from Paris sewage by N. A. Boulgakov and described by Maxted (57, 58). Phage A25 was one of a series of phages isolated together; its name is derived from the letter of the Lancefield group of its propagating strain combined with the type number (M25). Because this bacteriophage is able to mediate generalized transduction in GAS (49), a number of studies have addressed various aspects of its biology, and although much is still unknown about A25, it remains the best-characterized virulent GAS phage. Phage A25 is also referred to in the literature as phage 12204, its designation by the American Type Culture Collection (ATCC 12204). The role of A25 in mediating generalized transduction is discussed in the section dedicated to that topic.

Only a few electron micrograph studies have been done on the bacteriophages of GAS, and those that have been studied, both lytic and temperate, have a similar structure with isometric heads and long noncontractile tails (Bradley's class B) (53, 70, 90). By the more recent classification scheme, these virions would belong to the large family of *Siphoviridae* (1). One virulent phage similar to phage A25 was examined in detail and shown to have an isometric, octahedral head measuring 58 to 60 nm across, with a long flexible tail that measures 180 to 190 nm in length and 10 nm in diameter (53, 90). The tail of this phage is composed of 8-nm circular subunits and terminates in a transverse plate with a single projecting spike that is about 20 nm long (70, 90). Within the capsid is contained the linear, double-stranded DNA genome of 34.6 kb in length (65). Whether the genome circularizes within the cell or its mode of replication is unknown. The physical map of phage A25, redrawn from Pomrenke and Ferretti (65), is shown in Fig. 1.

Fischetti and Zabriskie (30) found that phage A25 will not adsorb to mechanically disrupted GAS or isolated GAS cell walls but will adsorb to intact, living GAS. In contrast, Cleary and coworkers (14) showed that heat-killed streptococci could adsorb A25 almost as well as living cells, attributing the discrepancy with the other group's finding to differences in growth and suspending mediums as well as in the number of PFU used in the assay. The receptor for A25 is the peptidoglycan of GAS, and treatment of the cells with muralytic enzymes such as lysozyme or the group C streptococcus phage C1 lysin destroyed the receptor activity. Additionally, phage A25 will also adsorb to streptococci from groups C, G, O, and K. Group G, in addition to absorbing phage A25, can actually become infected, and with passage of the phage through group G organisms, the phage can be adapted to efficient propagation (14). Wannamaker et al. (80) demonstrated that A25 was able to

infect and produce plaques on 34 of 71 group C streptococcal strains (48%); the frequency of plaquing by the virulent group C phage, phage C1, on the same strains was 70%. The other GAS-virulent phages, A6, A12, and A27, were also able to infect these group C strains at frequencies ranging from 34 to 47%. Since the group C-specific carbohydrate and peptidoglycan are present in all members of that group, there must be surface molecules that either block or enhance access to those molecules, modulating the binding of phages A25 and C1.

The hyaluronic acid capsule of GAS provides an effective barrier against infection by many lytic bacteriophages, preventing attachment of the phage to the bacterial surface (57). In natural infections, the state of encapsulation by GAS varies according to the growth phase of the cells, with transcription of the hyaluronic acid synthesis (*has*) operon dropping to very low levels once stationary phase is reached (20, 48). It was observed by Maxted (57) that exogenous hyaluronidase must be added to achieve efficient infection of GAS by lytic phages like A25 in vitro. Temperate streptococcal phages, by contrast, usually encode a hyaluronidase and thus do not require the exogenous addition of this enzyme (45). One report observed trace amounts of hyaluronidase associated with phage A25 virions (2.7×10^{-16} units/PFU [4]), but whether this small amount of activity was due to a phage-specified protein or a streptococcal hyaluronidase was not determined.

One-step growth experiments showed that phage A25 has an average burst size of approximately 30 PFU per cell (52). While the cell culture temperature does not alter the burst size, the latent period during which phage replication occurs is strongly influenced, increasing from 39 min at 37°C to more than 100 min at 26°C. Nothing is specifically known about the replication of the A25 genome; however, one might predict some similarity to the replication of phage T1 of *Escherichia coli*. Like phage A25, T1 is a lytic transducing phage with a linear genome. While neither T4 nor T7 replicates its DNA by circular intermediates, the molecular details of the replication of the A25 genome are unknown.

Although exogenous hyaluronidase is required for infection of GAS by lytic phages such as A25, release of the newly synthesized virions is mediated by phage-encoded lysins that disrupt the thick, gram-positive bacterial cell wall. Hill and Wannamaker (35) isolated a phage-associated lysin, not associated with the viral particle, from cultures of *S. pyogenes* K56 that had lysed following infection with phage A25. In addition to its lytic activity against GAS, it also showed activity against streptococci from groups C, G, and H. In contrast to the better characterized bacteriophage C1 lysin (see below), the A25 lysin does not degrade isolated GAS cell walls, and heat-killed or chloramphenicol-treated intact cells do not detectably lyse. Such results argue that the A25 lysin is not a muramidase like the phage C1 lysin. Interestingly, chloroform-killed GAS lysed at a higher rate than live cells. Cardiolipin is a

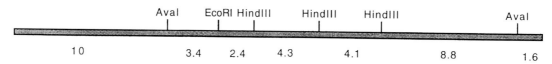

FIGURE 1 Physical map of the bacteriophage A25 genome. The size of the A25 genome is estimated to be 34.6 kb by restriction endonuclease mapping; the size of each fragment is given below in kilobase pairs. Redrawn from Pomrenke and Ferretti (65).

chloroform-soluble component of membranes that is known to be an inhibitor of autolysins, and the addition of cardiolipin to K56 cells protected them from lysis by the A25 lysin. Thus, although no autolysin has yet been identified in GAS, it is possible that the A25 lysin may function in a manner similar to the autolysin family of enzymes (35).

Lytic Phages and Streptococcal Virulence

While the toxin-carrying temperate phages of GAS have been long associated with increasing the virulence of the lysogen by toxigenic conversion, lytic phages may also play a role in selecting populations of streptococci better adapted to resisting host defenses. Maxted (58) found that exposure of many strains of GAS from a variety of M-types selected for the appearance of mucoid variants that were resistant to subsequent lytic phage infection. These mucoid variants had greatly increased hyaluronic acid production (often undetectable in the parental strain) and in some cases also increased M protein expression. While some of the GAS strains examined by Maxted were able to grow in normal human blood, those strains that ordinarily did not grow gave phage-surviving progeny that were able to survive in the presence of phagocytes. Parallel studies of mouse virulence produced similar results, with mucoid phage survivors of nonmucoid, avirulent strains acquiring the ability to survive and grow in mice (58). Evans (23) had previously reported that when GAS-infected mice were treated with bacteriophage in an effort to rescue them, the phage-treated mice died slightly more rapidly. While it is difficult to extrapolate from mouse studies to normal human infections, these observations may illustrate how the interaction of natural phage and GAS populations could lead to better-adapted or more virulent strains of these bacteria.

An important question that remained unanswered by the studies of Maxted concerned the appearance of the mucoid, phage-resistant progeny. Although the standard fluctuation test of Luria and Delbrück (50) was applied by Maxted to determine if random mutations in the population were sufficient to explain the appearance of the mucoid colonies, the results were inconclusive (58). In light of the later findings that temperate phages (72) and multigene activators like the *mga* gene product (13) can influence the expression of M protein and the hyaluronic acid capsule, it is conceivable that other factors, perhaps in addition to spontaneous mutations, influence the final population phenotype. Indeed, it is possible that A25 is a virulent mutant of a temperate phage that no longer is able to enter the lysogenic cycle but still can recombine with endogenous prophages in host GAS to generate novel bacteriophages that can mediate phenotypic changes in their host in a manner similar to phage SP24 (discussed below). This notion is strengthened by the observation that a temperate phage from group G streptococci, GT-234, is serologically related to phage A25 (17). The findings by Behnke and Malke (2, 3) that some streptococcal prophages prevent A25 from replicating in their host strain is similar to the prevention of phage superinfection by homoimmune temperate phages. Alternatively, the inhibition of A25 replication could be similar to the inhibition of phage T1 by *E. coli* lysogenic for lambda resulting from either the N or Q gene products, depending upon the stage of T1 replication (10, 33). Possibly more than one mechanism of inhibition of A25 by the different lysogens is in operation; some prophages apparently inhibit A25 by initiating premature lysis of the superinfected cell, while others appear to prevent some step of A25 adsorption or replication (3).

Bacteriophage C1 and the Phage-Associated Lysin

Although bacteriophage C1 of *Streptococcus equi* and *Streptococcus equisimilis* (group C streptococci; GCS) does not infect GAS, it deserves mentioning because it produces a lysin that degrades not only the cell walls of GCS but also the walls of GAS and group G streptococci. One of the interesting findings made by Evans was the lysis of GAS when cocultured with GCS infected with what is now known to be phage C1 (26). GAS were resistant to infection by this lytic phage, but they were rapidly lysed if exposed to the products released after the infection and lysis of GCS. This lysis, as was later discovered, was due to the action of a strong muramidase specified by this phage that attacked common cell wall structures shared by the two groups of streptococci.

Phage C1 is a small phage with an octahedral head with a diameter of 40 nm, and in contrast to most streptococcal phages, it lacks the usual long tail, having six to eight tail fibers with terminal knobs attached to a central core (90). The genome of phage C1 is a linear, double-stranded DNA molecule of about 17,000 bp in length (65). Whether the ends of the linear phage DNA are covalently linked to a peptide as in the similar, well-characterized bacteriophage ϕ29 of *Bacillus subtilis* is unknown. The latent period for phage C1 is about 20 min, with a burst size of 100 mature virions per infected cell (28, 31, 44). The host receptor for phage C1 is the group C carbohydrate, and the phage will bind to either intact group C cells or isolated group C cell walls (30). GAS lack this carbohydrate and thus are resistant to infection by phage C1.

At the conclusion of the replicative cycle, a phage-encoded lysin is produced that degrades the GCS cell wall, leading to rupture of the bacteria and release of the newly formed phage particles. The lysin is remarkable in that it has a molecular mass of over 100,000 (29, 69), twice the molecular weight of typical phage lysins (71) and requiring about 15% of the coding potential of the phage genome. The C1 lysin is an amidase that degrades the alanine-muramic acid linkage of the streptococcal peptidoglycan (90). The peptidoglycans of GAS share this structure with GCS and are thus also sensitive to the action of the lysin, even though they are not hosts for phage infection. A number of studies have used the C1 lysin, either purified or as a crude preparation, to facilitate the preparation of cell membrane-associated molecules or DNA from GAS (15, 84).

TEMPERATE BACTERIOPHAGES

General Characteristics

Temperate bacteriophages differ fundamentally from the lytic phages in that infection of the host bacterium can lead to two different outcomes. Sometimes phage replication proceeds immediately as in the lytic phages, leading to the lysis of the host and release of new virus particles. However, in response to favorable biological conditions, temperate phages can alternatively integrate a copy of their genome into a specific target site in the host chromosome, leading to the formation of a prophage. The integrated phage genome remains integrated in the bacterial DNA, being replicated with the host genome and passed onto daughter cells. The cells harboring the prophage are immune from superinfection by similar phages by the action of the repressor protein specified by the prophage (66). If cellular conditions become unfavorable or if the host cell is exposed to certain inducing agents such as UV light or mitomycin C, the integrated phage excises its genome from the host chromosome, enters the lytic cell of phage replication, and releases newly synthesized virions after rupture of the host.

Morphologically, the temperate phages of GAS resemble the lytic phages, having a distinct head and a long, noncontractile tail (45, 64, 78). The phages that have been characterized have linear or circular genomes that range from 32 to 42 kb in length (34, 74, 88). Burst sizes are usually low, in the range of 10 to 30 PFU per cell, although occasional strains produce burst sizes around 100 PFU per cell (44, 85). The frequency of lysogeny in GAS appears to be very high by the most recent estimates (>90%) (38, 86, 87), although the most extensive surveys have been done using phage-specific probes for DNA hybridization and thus do not distinguish between complete, viable prophages and defective or partial phage genomes. Therefore, the frequency of strains that can be induced to release viable phage is uncertain. The most striking and medically important feature of many temperate GAS phage genomes is inclusion of toxin genes that can alter the virulence of the host bacterium. While the role played by lytic streptococcal phages in pathogenesis is probably indirect, acting as vehicles of genetic exchange through generalized transduction, lysogeny by GAS bacteriophages can directly enhance the pathogenic potential of the host streptococcus through toxigenic conversion.

Toxigenic conversion is the introduction, following the establishment of lysogeny, of a temperate phage-encoded gene into its host bacterium that functions as a virulence factor. Phage-associated virulence factors have been observed to occur in a number of species of bacteria, including *Corynebacterium diphtheriae*, *Clostridium botulinum*, *Staphylococcus aureus*, *E. coli*, *Vibrio cholerae*, and *S. pyogenes* (6). Interestingly, the conversion of nontoxigenic GAS by cultural filtrates from toxigenic strains to scarlet fever toxin production may have been the earliest reported example of this phenomenon (8, 32), although the actual link to a bacteriophage was not made until many years later (89). These toxin genes are often positioned on the phage genome next to the genetic elements involved with site-specific integration of the phage into the bacterial chromosome, suggesting that such toxin genes may have originally become associated with the phage chromosome as a result of aberrant excision of the phage following induction that resulted in the acquisition of a bacterial gene. The fact that these toxin genes appear to play no role in the replication of the phage, being secreted proteins with signal peptides, strengthens the notion that such elements were acquired at some point late in the phage's evolutionary history (6). The actual origins of phage-associated toxin genes are difficult to trace, since the original genetic material may have originated from a different bacterial species and become established in the currently known system only after the appearance of a host-range variant of a phage. Some molecular evidence suggests that such events have indeed contributed to the dissemination of phage-associated toxin genes; for example, the *speA* gene of *S. pyogenes* and the enterotoxins B and C1 of *S. aureus* show a significant degree of homology and thus may share a common origin (83).

The erythrogenic toxins (pyrogenic exotoxins) were originally defined by Dick and Dick (21, 22) as substances that produced the characteristic erythematous skin reaction typical of scarlet fever. Three distinct members of this class of toxins have been described, named the A, B, and C toxins. The B toxin, encoded by *speB*, has been shown to be a protease encoded by the streptococcal genome, and it, as well as some more recently identified toxins whose links to bacteriophages have not been established, will not be considered further in this chapter. However, the genes for the A and C toxins, *speA* and *speC*, respectively, are elements in the genomes of temperate streptococcal phages.

The frequency of *speA* and *speC* in GAS, especially in relation to strains isolated from cases of scarlet fever or rheumatic fever, was examined by Yu and Ferretti (85–87). Since neither of these genes has been found except in association with a temperate phage, this survey provides an informative picture of lysogeny as related to human disease. A large selection of GAS strains were examined that were isolated from a variety of clinical diseases and M-types; these isolates were divided into three groups: strains associated with scarlet fever, strains associated with rheumatic fever, and strains associated with other streptococcal diseases (e.g., impetigo or cellulitis) but not scarlet fever or rheumatic fever (general strains). Overall, *speA* was found to be present in almost 30% of all strains surveyed, while *speC* was found in 50% of the strains. However, when each group of strains was examined, *speA* was present in only 15% of the general strains but in 50% of both the scarlet fever and rheumatic fever strains. The frequency of *speC* in all three groups was about 50%. Of the total number of strains examined (512), 30% carried neither *speA* nor *speC*. Strains carrying both toxins were rare, with only 5% of the general strains and 9% of the total strains having both *speA* and *speC*.

The possession of a hyaluronidase gene differentiates many temperate GAS phages from the lytic GAS phages. Kjems (44) observed that hyaluronidase was released by a lysogenic strain of GAS during phage multiplication, with at least some of the activity being associated with the isolated phage particles (45). About 33% of the hyaluronidase activity found in a lysate of Kjem's phage 22/22 could be sedimented by centrifugation with the intact phage particles. Of the remaining two-thirds enzymatic activity in the lysate, about half was found associated with phage fragments (including many free tails) and the rest as "free" hyaluronidase (45). Lysates of bacteriophage 12/12 (45) precipitated with polyethylene glycol and purified by isopycnic centrifugation also separated the hyaluronidase activity into two fractions, the main one being intact phages and the secondary, lighter fraction being unattached phage tails (64). A type 49 lysogen also had 25% of the hyaluronidase activity associated with the intact phage particles, but, in contrast to the other phages studied, most of the non-phage-associated enzyme was considered to be free and not associated with tail fragments (4). These varying amounts of free and phage-associated enzyme observed with different phages may reflect differences in phage tail structural stability or transcriptional activity of the hyaluronidase gene during phage replication. Specific phage antisera against a temperate phage from types 1, 12, and 22 GAS strains were able to neutralize hyaluronidase activity in lysates from the homologous strain. More than one type of phage-associated hyaluronidase can be identified serologically, and these phage enzymes are distinct from the bacterial cell-associated streptococcal hyaluronidase (45,

46). So far, two phage hyaluronidase genes have been cloned, sequenced, and analyzed in detail: *hylP*, isolated from a type 49 nephritic GAS strain lysogenic for phage H4489A (37), and *hylP2*, isolated from a type 22 GAS containing a phage that is probably defective (38). These two hyaluronidase genes are closely related, having 66.5% identity and 77.5% similarity at the nucleotide sequence level (38). The *hylP* gene specifies a region in the hyaluronidase protein that contains a series of 10 Gly-X-Y amino acid triplets, a structure reminiscent of collagen. Interestingly, this region is absent in *hylP2* and is obviously dispensable for hyaluronidase activity (38). However, the presence of a collagenlike motif could potentially lead to the induction of antibodies that cross-react with host tissues, as is seen in the polyarthritis that sometimes accompanies rheumatic fever. Further allelic variants of these phage hyaluronidase genes, possibly generated by recombination, have been identified (56).

T12 and the *speA*-Carrying Bacteriophages

Bacteriophage T12 is the prototypic phage that carries the erythrogenic toxin A gene (*speA*) in its chromosome. The transmissibility of the erythrogenic toxin A from a toxigenic strain to a cocultured nontoxigenic strain by culture filtrates was reported as early as 1926 by Cantacuzene and Boncieu (8) and 1927 by Frobisher and Brown (32). However, the connection between this transfer of toxigenicity and a bacteriophage vector was not made until 1964 when Zabriskie (89) demonstrated that transmission of the erythrogenic toxin was linked to induction and transfer of a lysogenic phage, phage T12. By using prolonged irradiation, he was able to obtain a derivative of the erythrogenic toxin A-producing GAS lysogen T25/41 that was cured of phage and no longer toxigenic. Zabriskie then demonstrated that this derivative strain, T25$_3$, could be restored to toxin production by infecting it with phages induced from scarlatina toxin strains. Although their exact role was still undetermined, it was clear that phages were involved in the production of type A erythrogenic toxin (89).

The first physical map of bacteriophage T12 was constructed by Johnson and Schlievert (39), showing that the phage chromosome was about 36 kb in length. Submolar fragments in restriction endonuclease digests of the phage genome led them to conclude that the genome of T12 was circularly permuted and terminally redundant. They further suggested from the mapping data that the T12 genome was replicated as a concatamer and packaged by a headful mechanism. The headful packaging mechanism was first described in bacteriophage T4 of *E. coli*: concatameric phage DNA is packaged starting at a specific site (*pac*) and proceeding by subsequent headfuls. Since the amount of DNA inserted in a headful is slightly more than one genome length into the phage head, the result is phage DNA molecules that are terminally redundant and circularly permuted. The presence of submolar and heterogeneous fragments in phage DNA preparations is a consequence of this type of packaging (77).

Final proof of bacteriophage association of this toxin gene came with the cloning of *speA* from the phage T12 genome and expression as a recombinant molecule independently by two laboratories (40, 82). Weeks and Ferretti (83) subsequently reported the complete DNA sequence of *speA*, showing that the gene was 753 bp in length and encoded a peptide with a molecular weight of 29,244 and a 30-amino-acid signal peptide, resulting in a molecular weight of 25,787 for the secreted protein. The carboxy terminus of SpeA was found to have extensive homology with

the carboxy-terminal regions of *S. aureus* enterotoxins B and C1, suggesting some common evolutionary history linking these toxins (83). These results were confirmed independently by Johnson et al. (42).

The physical map of phage T12 prepared by Johnson and Schlievert (39) was further examined in subsequent studies by Yu (85) and Yu and Ferretti (88). They found that the submolar restriction endonuclease fragments that were interpreted as evidence that phage T12 isolated from induction of *S. pyogenes* T25$_3$ was circularly permuted and terminally redundant actually were derived from a second phage residing in the T25$_3$ genome and coinduced with phage T12. The presence of this second phage genome was not initially detected in phage genomic DNA preparations because its chromosome does not resolve from phage T12 DNA by ordinary agarose gel electrophoresis, and separation of the two genomes can only be accomplished by reverse-field agarose gel electrophoresis (85, 88). This second phage is apparently defective in some key function and can only be rescued by some other phage such as T12 coinduced from strain T25$_3$ (85, 88). Whether the helper phage function provided by T12 is excision, replication, or encapsidation is unknown. The cryptic phage has a linear genome that is 36,000 bp in length. After separation of the cryptic phage DNA from phage T12 DNA, a corrected physical map of phage T12 was constructed (Fig. 2).

The complete DNA sequence of T12 was determined as well as a preliminary description of its gene structure and expression (60; unpublished results). T12 has a 35,066-bp genome (percent G+C = 38.5%) with divergent transcriptional directions for the regulatory genes and the structural genes similar to the recently described lactococcal phage r1t (79). The percent G+C of the region containing *speA* differs from the rest of the phage genome, being only 30%, and suggests that the toxin gene originated from a different source than the other phage DNA. Forty-nine open reading frames have been identified by function (*speA* and the integrase) or by homology to previously described phage genes. In general, little sequence homology exists with the lambdoid family of phages, but many regions of the T12 genome suggest a strong genetic relationship with the phages of *Lactobacillus lactis* and those of other gram-positive organisms.

Two other *speA*-carrying bacteriophages have been identified, phages 49 and 270 (Fig. 2). Both of these phages have distinct physical maps that distinguish them from phage T12: phage 270 has a genome of 32,000 bp, and phage 49 has a genome of 40,000 bp. As with T12, both phages have circular genomes. The only chromosomal region with significant homology, as determined by DNA hybridization, between the three phages is *speA* and its flanking regions (85, 88). Phage 270 does share the same integrase gene and attachment site with phage T12, whereas those elements in phage 49 are different and still uncharacterized (60, 62). Both phages 49 and 270 are capable of providing the necessary helper function to induce the cryptic phage from GAS T25$_3$ (85, 88).

Yu (85) and Yu and Ferretti (88) constructed DNA hybridization probes for regions unique to phage T12 or to phage 49 (Fig. 2). In a survey of 300 GAS strains, they found that 45% of the total strains contained T12-specific sequences, while 14% of the total strains had phage 49-associated sequences. Among the strains that were *speA*$^+$, 79% contained phage T12 and 35% contained phage 49 sequences. However, 95% of the strains that contained phage 49 sequences also contained *speA*, while only 66% of the strains that hybridized to the T12 probe also con-

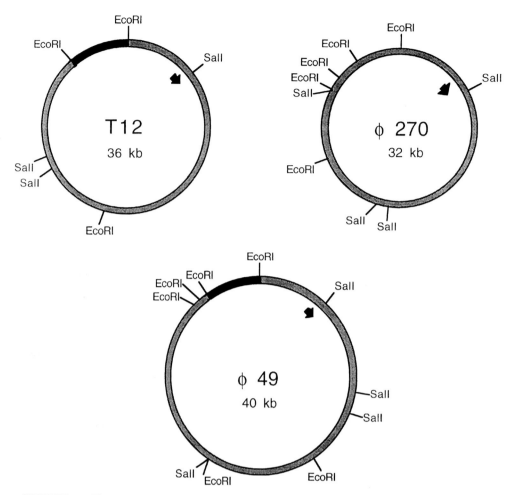

FIGURE 2 Physical maps of the genomes of bacteriophages T12, 270, and 49. The physical maps of the three *speA*-containing bacteriophages are shown with the restriction cleavage sites for *Eco*RI and *Sal*I. The location of *speA* on each phage genome is indicated by the arrow. The restriction fragments used as the phage T12- and phage 49-specific probe are shaded black. Redrawn from Yu and Ferretti (88).

tained *speA*. Thus, while phage T12 was more widely distributed in the survey population, the association between *speA* and phage 49 is greater than that of phage T12 and *speA*. It is noteworthy that a number of strains hybridized to either the T12- or phage 49-specific probe but did not contain *speA*. Such phages may serve as a reservoir for genetic recombination, allowing expansion or shift of host range for toxigenic phages.

The bacterial surface receptors for phage T12, phage 49, or any other toxigenic phage have not yet been identified, but some evidence suggests that these molecules may be present on some strains and not others, the receptors possibly being related to M-type. For example, M1 strains in the survey by Yu and Ferretti (88) had only sequences associated with T12 carriage, never those associated with phage 49, and *speA* in general was found with highest frequency in M1, M3, M49, T1, and T3/13 strains. However, 89% of the M3 strains that were *speA*⁺ showed evidence for either double lysogeny or the presence of some recombinant phage, hybridizing to both probes specific for phages T12 and 49; the remaining 11% hybridized only to the phage 49 probe (85, 88).

Temperate bacteriophages recombine with their host bacterial chromosome at a specific sequence as an integral part of their biological cycle; the molecular mechanisms that control this event in the best-characterized of the viruses, bacteriophage lambda of *E. coli*, have been described (66). These phages, upon infection of the bacterial cell, produce an integrase protein (Int) that mediates recombination between the bacterial chromosome at a specific site (bacterial attachment site, or *attB*) and a corresponding region on the phage chromosome (phage attachment site, or *attP*). There is a "core" sequence of 15 to 100 bp that is identical in *attB* and *attP* and serves as the site of recombination between the two genomes (7). Integration of the phage requires only *attP*, *attB*, Int, and a host-encoded protein, integration host factor; however, excision of the integrated phage from the bacterial chromosome requires two additional proteins, the phage-encoded excisionase (Xis) and the host protein Fis. The integrase gene and bacterial and phage attachment sites of T12 have been identified and characterized (62).

Bacteriophage T12 integrates by site-specific recombination into what was initially identified as a gene for a serine tRNA (attB), completing the downstream half of the gene by a 96-base duplication in the phage genome (62). As more sequence has been collected in this region as a result of the S. pyogenes genome sequencing project, it appears likely that the attachment site is in the 3' end of a longer molecule, a tmRNA (named for its dual tRNA- and mRNA-like character; it is also referred to as a 10sA RNA) (84a). The 3' end of tmRNA genes serve as att sites for phages or cryptic prophages in several species, including E. coli, V. cholerae, and Dichelobacter nodosus. This duplication allows transcription of the genetic element to continue unimpeded. The anticodon loop of the tRNA-like region is where the duplication begins between the phage T12 and S. pyogenes genomes and probably is the site of crossover between the phage and bacterial chromosomes. The integrase gene (int) and attachment region of φ270 are identical to phage T12, and DNA hybridization confirms that this phage integrates into the same site as T12 (62). By contrast, φ49 recombines with the streptococcal genome at a different location and contains no sequences that hybridize with T12 int (60). Not all bacteriophages that share attB with T12 are speA+; φ436 recombines with the bacterial chromosome at the T12 attB site and has the same integrase gene but does not contain speA (59). This sequence that serves as the T12 attachment site is conserved in a number of streptococcal species, including S. agalactiae, S. mutans, S. salivarius, S. downei, and S. gordonii (60). The widespread distribution of a phage attachment site in related species has interesting implications about the possible expansion of host range beyond the original host. Additionally, the construction of plasmid vectors containing the phage integrative functions can provide useful tools for the site-specific introduction of cloned sequences into GAS (61).

Phage CS112 and Phage SF370 (speC-Associated Phages)

The first association between pyrogenic exotoxin C (speC) and a bacteriophage was made by Colon-Whitt et al. (19) after inducing the speC+ S. pyogenes C203U to release phage that subsequently could convert a nontoxigenic strain to SpeC production. Phage CS112, the prototypic speC+ temperate bacteriophage, was later identified by Johnson and coworkers (41) in a similar fashion, inducing a phage from S. pyogenes CS112 that was capable of transferring speC production to strain K56. Phage CS112 has a circular genome of 40,800 bp that appears to be terminally redundant and circularly permuted, and the gene for speC is located near the phage attachment site, as was found to be true in the speA+ phages (Fig. 3) (34, 41). Goshorn and Schlievert (34) found that many strains of GAS harboring speC could not be induced to release active phages, although these strains often had associated phage CS112 DNA sequences. Some strains also contained the phage CS112-specific sequences but lacked the speC gene, suggesting that the recombination between phage CS112 and endogenous and possibly defective phages may have generated the observed noninducible speC phages. Strains of GAS with the highest frequency of speC carriage are M2, M4, M6, M12, T2, T4, T6, and T28 strains. The only serotype that frequently contains both speA and speC genes is the M3 serotype, which, interestingly, was also the serotype that showed a high frequency of hybridization to both the phage T12- and phage 49-specific probes (85–87).

The bacterial attachment site for phage CS112 remains unidentified, as does the length of DNA duplication between the bacterial and phage genomes. However, another speC+ bacteriophage has been identified in the genome of the GAS SF370, the strain of S. pyogenes being used for genomic sequencing at the University of Oklahoma. Although the analysis of this phage (phage SF370) is incomplete pending the completion of the genome project, it appears that this phage will share several characteristics with phage CS112, including a similar physical map (unpublished results). Preliminary results have revealed that the bacterial attachment site for phage SF370 is distinct from the one used by phage T12 and does not appear to interrupt any genes or contain a lengthy duplication with the phage genome (9a). Further analysis and comparison of phage CS112 and SF370 should provide valuable information about their biology and influence on speC distribution.

Bacteriophage SP24

The M protein is the major cell surface antigen of S. pyogenes and is responsible for the resistance to phagocytosis by human polymorphonuclear leukocytes by M+ streptococci. The expression of M protein is phenotypically variable, with some strains giving rise to M- colonial variants with some frequency. This spontaneous variation in M protein expression was related to the presence of an extrachromosomal element by Cleary and coworkers, who demonstrated that growth of GAS under conditions that typically cure bacteria of plasmids or phage resulted in high numbers of M- variants (11). Subsequently, a temperate bacteriophage, SP24, was isolated from an M12 strain known to vary in M protein expression. When phage SP24 was introduced into an M- strain derived from an M76 parent (CS112), the resulting lysogen expressed type 76 M protein (72). Careful serological analysis revealed that CS112 actually produced trace amounts of the M76 protein, and introduction of SP24 caused the M protein gene (emm) expression to return to levels similar to that of its parental strain.

Initially, introduction of phage SP24 resulted in lysogens with unstable expression of emm, with a high frequency of M+ lysogens reverting to M- phenotype. Stabilization of emm expression was achieved by sequential incubation of M+ lysogens in human blood; the recovery of stable M76-expressing streptococci following this treatment was not the result of merely selecting for parental revertants, since passage in human blood of unlysogenized CS112 recovered no M+ derivatives (72). Isolation and analysis of the phage genomes induced from these stable lysogens revealed that phage SP24 had undergone a rearrangement with an endogenous temperate phage present in S. pyogenes CS112, resulting in the substitution of a distinct 2.5 kb in the phage SP24 genome (Fig. 4) (73). Spanier and Cleary (75) further determined that this recombinant phage, SP272, was able to adsorb more rapidly to CS112 than the wild-type phage CS24 virions, possibly giving phage SP272 a selective advantage over its parent during multiple rounds of replication. The surface receptor for the phage is unknown. Integration of phage SP24 precedes the appearance of the M+ phenotype but is not sufficient to ensure conversion since lysogens may be isolated that are M- (75). The exact relation of this genome rearrangement in the bacteriophage to the stabilization of emm expression remains to be determined.

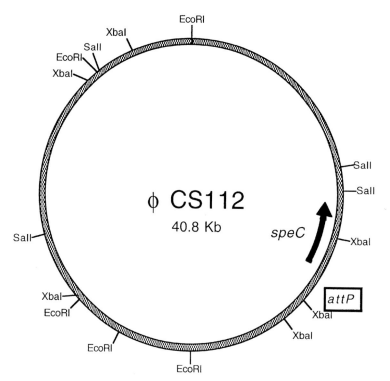

FIGURE 3 The physical map of the bacteriophage CS112 genome. The circularized form of the temperate phage CS112 genome is shown with the restriction endonuclease cleavage sites for *Eco*RI, *Xba*I, and *Sal*I. The location of the gene for pyrogenic exotoxin C (*speC*) is indicated by the arrow. The probable phage attachment site for site-specific recombination into the bacterial chromosome is indicated by *attP*. Redrawn from Goshorn and Schlievert (34).

The phage SP24 genome, isolated from intact virions, is a linear genome 42 kb in length (Fig. 4). Similar to the well-characterized phages P1, P22, and T1, phage SP24 appears to be terminally redundant and circularly permuted. After entry into the streptococcal cell, the redundant terminal ends allow circularization of the phage chromosome, a prerequisite for subsequent integration into the bacterial chromosome. Since the terminal redundancy involves about 7.9% of the genome, the circular form of the phage DNA is 39.8 kb in length (Fig. 4) (74).

TRANSDUCTION

The three common methods of genetic exchange within a species of bacteria are transformation, conjugation, and transduction. In GAS, neither transformation nor conjugation has been observed to occur naturally, although electrotransformation has become an important laboratory tool. However, transduction occurs in GAS and is mediated by both lytic and temperate bacteriophages. An important observation of infections caused by GAS is the periodic shift in severity and in the predominant clinical syndromes (67, 68). The genetic basis for these phenotypic shifts in natural populations of GAS has been investigated in several recent studies; the results have shown that considerable allelic variation exists at the DNA level in both bacteria- and phage-associated genes as well as in linkage between specific genes and clinical syndromes (5, 9, 43, 63, 86), more than would be expected to result from accumulated random mutations between genetically isolated individuals. Since transduction is the only known natural

means of genetic exchange, it is likely that this mechanism and its associated bacteriophages play an important role in the genetic shifts seen in GAS.

Transduction among GAS was first reported by Leonard et al. in 1968 (49). They identified five phages (two temperate and three lytic) that were able to transduce streptomycin resistance from a type 6 donor strain (56188) at varying frequencies to *S. pyogenes* strain K56 (M-type 12). Of these phages that mediated transduction, phage A25 transduced at the highest frequency (10^{-6} transducing units per PFU). Transfer of streptomycin resistance from an A25 lysate of a resistant host to the sensitive K56 donor strain was insensitive to deoxyribonuclease treatment of the lysate but could be prevented by phage-specific antisera. Both observations supported transduction as the mechanism of transfer, and subsequent work confirmed the ability of A25 (and several related phages) to mediate generalized transduction.

The presence of the hyaluronic capsule of GAS prevents infection by phage A25 and must be removed for transduction to occur. Removal is easily accomplished by the addition of hyaluronidase to the culture; however, the removal of the capsule exposes potential transductants to subsequent infection by any free A25 phage, and even if the input ratio of phage to host is kept below one, the overall recovery of transductants will be diminished. Two methods have been successfully employed to protect the transductants from reinfection and lysis. The first employs specific A25 antiphage serum to block any unabsorbed or progeny phage from infecting the transductants resulting from the initial absorption. Using this method, a 1,000-

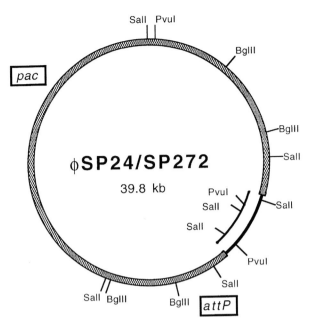

FIGURE 4 Physical map of the bacteriophage SP24/SP272 genome. The circularized form of the temperate phage SP24 genome is shown with the restriction endonuclease cleavage sites for BglII, PvuI, and SalI. Introduction of phage SP24 into GAS strain CS112 selects for recombinants that have a region of their chromosome (indicated with the thinner line) replaced by a homologous region from an endogenous prophage. This recombinant phage, SP272, is associated with stable expression of the M76 protein in CS112, a strain that normally produces only trace amounts of M protein. The divergent region in SP272 is shown with its relevant restriction sites interior to the SP24 map. The regions involved in packaging (pac) and integration (att) are indicated. The figure is redrawn from Spanier and Cleary (73–75).

fold increase in the *ery-2000* erythromycin-resistance gene was obtained (54). Equivalent results were achieved by using temperature-sensitive mutants of phage A25 to infect decapsulated K56. Malke (52) isolated two such mutations, ts-1 and ts-2, that were capable of plaquing on GAS K56 at 29.5°C but not at 37.5°C; the increasing effectiveness of transduction frequency was ts+ < ts−1 ≅ ts−2 < ts1−2 (52). No enhancement in transduction frequency was obtained by combining treatment with antiphage serum and the use of the temperature-sensitive mutant phages.

The efficiency of A25-mediated transduction also may be increased by UV irradiation of the transducing lysates before adsorption to the recipient streptococci (16, 51). Malke found that as UV fluence to the phage A25 lysate increases, the number of transductants increases, reaching the maximum value when a 90 to 99% reduction in PFU is achieved (54). At this level of phage inactivation, the frequency of transductants increased to 4×10^{-5} to 5×10^{-5} per initial PFU, resulting in approximately a 10-fold increase in transduction frequency. Colón and coworkers (16) also observed the same enhancement of transduction in UV-treated lysates, not only with phage A25 but also with the other lytic phages, A5 and A6. Interestingly, no enhancement in transduction frequency was observed when two temperate transducing phages were UV-treated. Twenty-six GAS recipient strains were tested for transduc-

tion by A25, A5, and A6 before and after UV irradiation of the phages. Thirteen of these strains, including M-types 6, 8, 12, 13, 19, and nontypables, were efficiently transduced only after UV treatment (16).

GAS strain T25$_3$, the laboratory host strain for the erythrogenic toxin A-carrying phage T12, can serve as donor for A25 transduction; however, T25$_3$-derived lysates produce 6 to 8 times fewer transducing particles that those derived from K56. This observed difference is not due to a diminished phage burst size, since A25 grown on T25$_3$ produces approximately the same number of PFU per cell as propagation on strain K56. Interestingly, the transducing lysate from a streptomycin derivative of T25$_3$ was able to transfer this antibiotic resistance to a homologous strain (T25$_3$ wild type) at a lower frequency than to a heterologous strain (SM60) (55).

Temperate GAS bacteriophages are also able to mediate the transfer of antibiotic resistance by transduction. Ubukata and coworkers (78) observed the transfer of resistance to tetracycline, chloramphenicol, macrolides, lincomycin, and clindamycin by induced temperate phages from T12 strains. Antibiotic resistance was transferred either alone or in combination, suggesting that the donor strain may have been harboring a genetic element conferring multiple drug resistance, such as the transposon Tn916. The transfer of erythromycin and streptomycin resistance by temperate phages induced by mitomycin C treatment was also observed by Hyder and Streitfeld (36). Malke (53) had previously observed that treatment of the GAS culture with nitrosoguanidine or another mutagen was necessary for the transfer of high-level erythromycin resistance when using a lytic phage to create the transducing lysate. However, transduction of high-level erythromycin resistance was mediated by the induced temperate phages without further treatment. However, that study did not address the question of whether the mitomycin C (a known mutagen) used to induce the prophages might be functioning in the same manner as the nitrosoguanidine in the earlier studies.

The transfer of genetic material from one streptococcal group to another by transduction is another potential means of the dissemination of genes that could affect virulence. Bacteriophages isolated from streptococcal groups A, E, and G were isolated that could lyse streptococci from serogroups A, C, G, H, and L, and in some cases, these lytic phages were also capable of propagation in one or more of the other serogroups (18). The same study also showed that phage A25 as well as two other lytic GAS phages could also transduce a number of group G strains to streptomycin resistance. The same workers subsequently isolated a temperate phage from a group G streptococcus (phage GT-234) that was capable of propagating on streptococci from serogroups A and C (17). Phage GT-234, which is serologically related to and morphologically indistinguishable from the lytic GAS phage A25, was able to transduce strains of serogroups A, C, and G to streptomycin resistance. Wannamaker et al. (80) demonstrated that genetic information (streptomycin resistance) could be transferred into group A strains by a transducing phage from GCS. This phage was a temperate phage induced from a clinical isolate of a GCS, and although most of the temperate phages induced from other strains of GCS did not produce plaques on GAS, this phage was unusual in that it was able to infect 40% of the GAS strains tested.

Horizontal transfer of genetic material in GAS has been shown by numerous studies to be important in the dissemination of genes in natural populations (5, 9, 43, 63, 86).

The impact of this genetic flow is readily appreciated when shifts in the prevalence and severity of streptococcal diseases occur (12, 76). The role of transduction and bacteriophages in horizontal transfer is assumed, yet no modern studies have directly addressed the question. The majority of the studies into streptococcal transduction were done before the advent of many of the current techniques of molecular biology, and so the time may have come to re-examine this phenomenon. A better understanding of streptococcal transduction may prove key to understanding the flow of genetic information in natural populations of GAS.

REFERENCES

1. **Ackermann, H.-W., and M. S. DuBow.** 1987. *Viruses of Prokaryotes: General Properties of Bacteriophages*, vol. 1. CRC Press, Boca Raton, Fla.

2. **Behnke, D., and H. Malke.** 1978. Bacteriophage interference in *Streptococcus pyogenes*. I. Characterization of prophage-host systems interfering with the virulent phage A25. *Virology* **85:**118–128.

3. **Behnke, D., and H. Malke.** 1978. Bacteriophage interference in *Streptococcus pyogenes*. II. A25 mutants resistant to prophage-mediated interference. *Virology* **85:**129–136.

4. **Benchetrit, L. C., E. D. Gray, and L. W. Wannamaker.** 1977. Hyaluronidase activity of bacteriophages of group A streptococci. *Infect. Immun.* **15:**527–532.

5. **Bessen, D. E., and S. K. Hollingshead.** 1994. Allelic polymorphism of *emm* loci provides evidence for horizontal gene spread in group A streptococci. *Proc. Natl. Acad. Sci. USA* **91:**3280–3284.

6. **Bishai, W. R., and J. R. Murphy.** 1988. Bacteriophage gene products that cause human disease, p. 683–724. *In* R. Calendar (ed.), *The Bacteriophages*, vol. 2. Plenum Press, New York, N.Y.

7. **Campbell, A. M.** 1992. Chromosomal insertion sites for phages and plasmids. *J. Bacteriol.* **174:**7495–7499.

8. **Cantacuzene, J., and O. Boncieu.** 1926. Modifications subies pare des streptococques d'origine non-scarlatineuse au contact des produits scarlatineux filtrés. *C.R. Acad. Sci.* **182:**1185.

9. **Chaussee, M. S., J. Liu, D. L. Stevens, and J. J. Ferretti.** 1996. Genetic and phenotypic diversity among isolates of *Streptococcus pyogenes* from invasive infections. *J. Infect. Dis.* **173:**901–908.

9a. **Chen, J., and McShan, W. M.** Unpublished observations.

10. **Christensen, J. R., M. C. Gawron, and J. Halpern.** 1978. Exclusion of bacteriophage T1 by bacteriophage lambda. I. Early exclusion requires lambda N gene product and host factors involved in N gene expression. *J. Virol.* **25:**527–534.

11. **Cleary, P. P., Z. Johnson, and L. Wannamaker.** 1975. Genetic instability of M protein and serum opacity factor of group A streptococci: evidence suggesting extrachromosomal control. *Infect. Immun.* **12:**109–118.

12. **Cleary, P. P., E. L. Kaplan, J. P. Handley, A. Wlazlo, M. H. Kim, A. R. Hauser, and P. M. Schlievert.** 1992. Clonal basis for resurgence of serious *Streptococcus pyogenes* disease in the 1980s. *Lancet* **339:**518–521.

13. **Cleary, P. P., L. McLandsborough, L. Ikeda, D. Cue, J. Krawczak, and H. Lam.** 1998. High-frequency intracellular infection and erythrogenic toxin A expression un-

14. **Cleary, P. P., L. W. Wannamaker, M. Fisher, and N. Laible.** 1977. Studies of the receptor for phage A25 in group A streptococci: the role of peptidoglycan in reversible adsorption. *J. Exp. Med.* **145:**578–593.

15. **Cohen, J. O., H. Gross, and W. K. Harrell.** 1977. Immunogenicity and characteristics of M protein released by phage-associated lysin from group-A streptococci types 1 and 23. *J. Med. Microbiol.* **10:**179–194.

16. **Colon, A. E., R. M. Cole, and C. G. Leonard.** 1970. Transduction in group A streptococci by ultraviolet-irradiated bacteriophages. *Can. J. Microbiol.* **16:**201–202.

17. **Colon, A. E., R. M. Cole, and C. G. Leonard.** 1971. Lysis and lysogenization of groups A, C, and G streptococci by a transducing bacteriophage induced from a group G streptococcus. *J. Virol.* **8:**103–110.

18. **Colon, A. E., R. M. Cole, and C. G. Leonard.** 1972. Intergroup lysis and transduction by streptococcal bacteriophages. *J. Virol.* **9:**551–553.

19. **Colon-Whitt, A., R. S. Whitt, and R. M. Cole.** 1979. Production of an erythrogenic toxin (streptococcal pyrogenic exotoxin) by a non-lysogenised group-A streptococcus, p. 64–65. *In* M. T. Parker (ed.), *Pathogenic Streptococci.* Reedbooks Ltd., Chertsey, U.K.

20. **Crater, D. L., and I. van de Rijn.** 1995. Hyaluronic acid synthesis operon (*has*) expression in group A streptococci. *J. Biol. Chem.* **270:**18452–18458.

21. **Dick, G. F., and G. H. Dick.** 1924. The etiology of scarlet fever. *JAMA* **82:**301–302.

22. **Dick, G. F., and G. H. Dick.** 1924. A skin test for susceptibility to scarlet fever. *JAMA* **82:**265–266.

23. **Evans, A. C.** 1933. Inactivation of antistreptococcus bacteriophage by animal fluids. *Public Health Rep.* **48:**411–426.

24. **Evans, A. C.** 1934. The prevalence of streptococcus bacteriophage. *Science* **80:**40–41.

25. **Evans, A. C.** 1934. Streptococcus bacteriophage: a study of four serological types. *Public Health Rep.* **49:**1386–1401.

26. **Evans, A. C.** 1940. The potency of nascent streptococcus bacteriophage B. *J. Bacteriol.* **39:**597–604.

27. **Evans, A. C., and E. M. Stockrider.** 1942. Another serologic type of streptococcic bacteriophage. *J. Bacteriol.* **42:**211–214.

28. **Fischetti, V. A., B. Barron, and J. B. Zabriskie.** 1968. Studies on streptococcal bacteriophages. I. Burst size and intracellular growth of group A and group C streptococcal bacteriophages. *J. Exp. Med.* **127:**475–488.

29. **Fischetti, V. A., E. C. Gotschlich, and A. W. Bernheimer.** 1971. Purification and physical properties of group C streptococcal phage-associated lysin. *J. Exp. Med.* **133:**1105–1117.

30. **Fischetti, V. A., and J. B. Zabriskie.** 1968. Studies on streptococcal bacteriophages. II. Adsorption studies on group A and group C streptococcal bacteriophages. *J. Exp. Med.* **127:**489–505.

31. **Friend, P. L., and H. D. Slade.** 1966. Characteristics of group A streptococcal bacteriophages. *J. Bacteriol.* **92:**148–154.

32. **Frobisher, M., and J. H. Brown.** 1927. Transmissible toxicogenicity of streptococci. *Bull. Johns Hopkins Hosp.* **41:**167–173.

33. **Gawron, M. C., J. R. Christensen, and T. M. Shoemaker.** 1980. Exclusion of bacteriophage T1 by bacteriophage lambda. II. Synthesis of T1-specific macromolecules under N-mediated excluding conditions. *J. Virol.* **35:**93–104.

34. **Goshorn, S. C., and P. M. Schlievert.** 1989. Bacteriophage association of streptococcal pyrogenic exotoxin type C. *J. Bacteriol.* **171:**3068–3073.

35. **Hill, J. E., and L. W. Wannamaker.** 1981. Identification of a lysin associated with a bacteriophage (A25) virulent for group A streptococci. *J. Bacteriol.* **145:**696–703.

36. **Hyder, S. L., and M. M. Streitfeld.** 1978. Transfer of erythromycin resistance from clinically isolated lysogenic strains of *Streptococcus pyogenes* via their endogenous phage. *J. Infect. Dis.* **138:**281–286.

37. **Hynes, W. L., and J. J. Ferretti.** 1989. Sequence analysis and expression in *Escherichia coli* of the hyaluronidase gene of *Streptococcus pyogenes* bacteriophage H4489A. *Infect. Immun.* **57:**533–539.

38. **Hynes, W. L., L. Hancock, and J. J. Ferretti.** 1995. Analysis of a second bacteriophage hyaluronidase gene from *Streptococcus pyogenes*: Evidence for a third hyaluronidase involved in extracellular enzymatic activity. *Infect. Immun.* **63:**3015–3020.

39. **Johnson, L. P., and P. M. Schlievert.** 1983. A physical map of the group A streptococcal pyrogenic exotoxin bacteriophage T12 genome. *Mol. Gen. Genet.* **189:**251–255.

40. **Johnson, L. P., and P. M. Schlievert.** 1984. Group A streptococcal phage T12 carries the structural gene for pyrogenic exotoxin type A. *Mol. Gen. Genet.* **194:**52–56.

41. **Johnson, L. P., P. M. Schlievert, and D. W. Watson.** 1980. Transfer of group A streptococcal pyrogenic exotoxin production to nontoxigenic strains of lysogenic conversion. *Infect. Immun.* **28:**254–257.

42. **Johnson, L. P., M. A. Tomai, and P. M. Schlievert.** 1986. Bacteriophage involvement in group A streptococcal pyrogenic exotoxin A production. *J. Bacteriol.* **166:**623–627.

43. **Kehoe, M. A., V. Kapur, A. M. Whatmore, and J. M. Musser.** 1996. Horizontal gene transfer among group A streptococci: implications for pathogenesis and epidemiology. *Trends Microbiol.* **4:**436–443.

44. **Kjems, E.** 1958. Studies on streptococcal bacteriophages. 2. Adsorption, lysogenization, and one-step growth experiments. *Acta Pathol. Microbiol. Scand.* **42:**56–66.

45. **Kjems, E.** 1958. Studies on streptococcal bacteriophages. 3. Hyaluronidase produced by the streptococcal phage-host cell system. *Acta Pathol. Microbiol. Scand.* **44:**429–439.

46. **Kjems, E.** 1960. Studies on streptococcal bacteriophages. 5. Serological investigation of phages isolated from 91 strains of group A haemolytic streptococci. *Acta Pathol. Microbiol. Scand.* **49:**205–212.

47. **Krause, R. M.** 1957. Studies on bacteriophages of hemolytic streptococci. I. Factors influencing the interaction of phage and susceptible host cell. *J. Exp. Med.* **106:**365–383.

48. **Leonard, B. A., M. Woischnik, and A. Podbielski.** 1998. Production of stabilized virulence factor-negative variants by group A streptococci during stationary phase. *Infect. Immun.* **66:**3841–3847.

49. **Leonard, C. G., A. E. Colón, and R. M. Cole.** 1968. Transduction in group A streptococcus. *Biochem. Biophys. Res. Commun.* **30:**130–135.

50. **Luria, S. E., and M. Delbrück.** 1948. Mutations of bacteria from virus sensitivity to virus resistance. *Genetics* **28:**491–511.

51. **Malke, H.** 1969. Transduction of *Streptococcus pyogenes* K 56. *Microbiol. Genet. Bull.* **31:**23.

52. **Malke, H.** 1969. Transduction of *Streptococcus pyogenes* K 56 by temperature-sensitive mutants of the transducing phage A25. *Z. Naturforsch. Teil B* **24:**1556–1561.

53. **Malke, H.** 1970. Characteristics of transducing group A streptococcal bacteriophages A 5 and A 25. *Arch. Gesamte Virusforsch.* **29:**44–49.

54. **Malke, H.** 1972. Transduction in group A streptococci. In L. W. Wannamaker and J. M. Matsen (ed.), *Streptococci and Streptococcal Diseases.* Academic Press, Inc., New York, N.Y.

55. **Malke, H.** 1973. Phage A25-mediated transfer induction of a prophage in *Streptococcus pyogenes. Mol. Gen. Genet.* **125:**251–264.

56. **Marciel, A. M., V. Kapur, and J. M. Musser.** 1997. Molecular population genetic analysis of a *Streptococcus pyogenes* bacteriophage-encoded hyaluronidase gene: recombination contributes to allelic variation. *Microb. Pathog.* **22:**209–217.

57. **Maxted, W. R.** 1952. Enhancement of streptococcal bacteriophage lysis by hyaluronidase. *Nature* **170:**1020–1021.

58. **Maxted, W. R.** 1955. The influence of bacteriophage on *Streptococcus pyogenes. J. Gen. Microbiol.* **12:**484–495.

59. **McShan, W. M., and J. J. Ferretti.** 1997. Genetic diversity in temperate bacteriophages of *Streptococcus pyogenes*: identification of a second attachment site for phages carrying the erythrogenic toxin A gene. *J. Bacteriol.* **179:**6509–6511.

60. **McShan, W. M., and J. J. Ferretti.** 1997. Genetic studies of erythrogenic toxin carrying temperate bacteriophages of *Streptococcus pyogenes*, p. 971–973. In T. Horaud, A. Bouvet, R. Leclercq, H. D. Montclos, and M. Sicard (ed.), *Streptococci and the Host.* Plenum Publishing Corp., New York, N.Y.

61. **McShan, W. M., R. E. McLaughlin, A. Nordstrand, and J. J. Ferretti.** 1998. Vectors containing streptococcal bacteriophage integrases for site-specific gene insertion. *Methods Cell Sci.* **20:**51–57.

62. **McShan, W. M., Y.-F. Tang, and J. J. Ferretti.** 1997. Bacteriophage T12 of *Streptococcus pyogenes* integrates into the gene encoding a serine tRNA. *Mol. Microbiol.* **23:**719–728.

63. **Musser, J. M., V. Kapur, J. Szeto, X. Pan, D. S. Swanson, and D. R. Martin.** 1995. Genetic diversity and relationships among serotype M1 strains of *Streptococcus pyogenes. Dev. Biol. Stand.* **85:**209–213.

64. **Niemann, H., A. Birch-Andersen, E. Kjems, B. Mansa, and S. Stirm.** 1976. Streptococcal bacteriophage 12/12-borne hyaluronidase and its characterization as a lyase (EC 4.2.99.1) by means of streptococcal hyaluronic acid and purified bacteriophage suspensions. *Acta Pathol. Microbiol. Scand. Sect. B* **84:**145–153.

65. **Pomrenke, M. E., and J. J. Ferretti.** 1989. Physical maps of the streptococcal bacteriophage A25 and C1 genomes. *J. Basic Microbiol.* **29:**395–398.

66. **Ptashne, M.** 1992. *A Genetic Switch: Phage Lambda and Higher Organisms*, 2nd ed. Cell Press, Blackwell Scientific Publications, Inc., Cambridge, Mass.

67. **Quinn, R. W.** 1982. Epidemiology of group A streptococcal infections—their changing frequency and severity. *Yale J. Biol. Med.* **55:**265–270.

68. **Quinn, R. W.** 1989. Comprehensive review of morbidity and mortality trends for rheumatic fever, streptococcal disease, and scarlet fever: the decline of rheumatic fever. *Rev. Infect. Dis.* **11:**928–953.

69. **Raina, J. L.** 1981. Purification of *Streptococcus* group C bacteriophage lysin. *J. Bacteriol.* **145:**661–663.

70. **Read, S. E., and R. W. Reed.** 1972. Electron microscopy of the replicative events of A25 bacteriophages in group A streptococci. *Can. J. Microbiol.* **18:**93–96.

71. **Sable, S., and S. Lortal.** 1995. The lysins of bacteriophages infecting lactic acid bacteria. *Appl. Microbiol. Biotechnol.* **43:**1–6.

72. **Spanier, J. G., and P. P. Cleary.** 1980. Bacteriophage control of antiphagocytic determinants in group A streptococci. *J. Exp. Med.* **152:**1393–1406.

73. **Spanier, J. G., and P. P. Cleary.** 1983. A DNA substitution in the group A streptococcal bacteriophage SP24. *Virology* **130:**514–522.

74. **Spanier, J. G., and P. P. Cleary.** 1983. A restriction map and analysis of the terminal redundancy in the group A streptococcal bacteriophage SP24. *Virology* **130:**502–513.

75. **Spanier, J. G., and P. P. Cleary.** 1985. Integration of bacteriophage SP24 into the chromosome of group A streptococci. *J. Bacteriol.* **164:**600–604.

76. **Stevens, D. L., M. H. Tanner, J. Winship, R. Swarts, K. M. Ries, P. M. Schlievert, and E. Kaplan.** 1989. Severe group A streptococcal infections associated with a toxic shock-like syndrome and scarlet fever toxin A. *N. Engl. J. Med.* **321:**1–7.

77. **Streisinger, G., R. S. Edgar, and G. H. Denhardt.** 1964. Chromosome structure in phage T4: circularity of the linkage map. *Proc. Natl. Acad. Sci. USA* **51:**775–779.

78. **Ubukata, K., M. Konno, and R. Fujii.** 1975. Transduction of drug resistance to tetracycline, chloramphenicol, macrolides, lincomycin and clindamycin with phages induced from *Streptococcus pyogenes*. *J. Antibiotics* **28:**681–688.

79. **van Sinderen, D., H. Karsens, J. Kok, P. Terpstra, M. H. J. Ruiters, G. Venema, and A. Nauta.** 1996. Sequence analysis and molecular characterization of the temperate lactococcal bacteriophage r1t. *Mol. Microbiol.* **19:**1343–1355.

80. **Wannamaker, L. W., S. Almquist, and S. Skjold.** 1973. Intergroup phage reactions and transduction between group C and group A streptococci. *J. Exp. Med.* **137:**1338–1353.

81. **Wannamaker, L. W., S. Skjold, and W. R. Maxted.** 1970. Characterization of bacteriophages from nephritogenic group A streptococci. *J. Infect. Dis.* **121:**407–418.

82. **Weeks, C. R., and J. J. Ferretti.** 1984. The gene for type A streptococcal exotoxin (erythrogenic toxin) is located in bacteriophage T12. *Infect. Immun.* **46:**531–536.

83. **Weeks, C. R., and J. J. Ferretti.** 1986. Nucleotide sequence of the type A streptococcal exotoxin gene from *Streptococcus pyogenes* bacteriophage T12. *Infect. Immun.* **52:**144–150.

84. **Wheeler, J., J. Holland, J. M. Terry, and J. D. Blainey.** 1980. Production of group C streptococcus phage-associated lysin and the preparation of *Streptococcus pyogenes* protoplast membranes. *J. Gen. Microbiol.* **120:**27–33.

84a. **Williams, K.** Personal communication.

85. **Yu, C.-E.** 1990. Ph.D. dissertation. University of Oklahoma, Oklahoma City.

86. **Yu, C.-E., and J. J. Ferretti.** 1989. Molecular epidemiologic analysis of the type A streptococcal exotoxin (erythrogenic toxin) gene (*speA*) in clinical *Streptococcus pyogenes* strains. *Infect. Immun.* **57:**3715–3719.

87. **Yu, C. E., and J. J. Ferretti.** 1991. Frequency of the erythrogenic toxin B and C genes (*speB* and *speC*) among clinical isolates of group A streptococci. *Infect. Immun.* **59:**211–215.

88. **Yu, C. E., and J. J. Ferretti.** 1991. Molecular characterization of new group A streptococcal bacteriophages containing the gene for streptococcal erythrogenic toxin A (*speA*). *Mol. Gen. Genet.* **231:**161–168.

89. **Zabriskie, J. B.** 1964. The role of temperate bacteriophage in the production of erythrogenic toxin by group A streptococci. *J. Exp. Med.* **119:**761–779.

90. **Zabriskie, J. B., S. E. Read, and V. A. Fischetti.** 1972. Lysogeny in streptococci, p. 99–118. *In* L. W. Wannamaker and J. M. Matsen (ed.), *Streptococci and Streptococcal Diseases: Recognition, Understanding, and Management*. Academic Press, Inc., New York, N.Y.

Molecular Epidemiology, Ecology, and Evolution of Group A Streptococci

DEBRA E. BESSEN AND SUSAN K. HOLLINGSHEAD

12

Within a bacterial species, there are often strains that differ from others in important biological properties. This is certainly the case for *Streptococcus pyogenes* (group A streptococci; GAS), whose members can cause a wide variety of human diseases, yet there does not appear to exist a single omnipotent strain. To better understand strain differences and their relevance to human disease, stable markers have been identified within organisms of this species. Epidemiological markers are useful for investigating outbreaks of disease, and they can also provide a reference point for deciphering the genetic organization of a bacterial population. The epidemiology of a microbial disease is often, in large part, a reflection of the ecology and evolution of the causative agent.

SEROLOGICAL MARKERS

During the 1920s, Rebecca Lancefield began work aimed at understanding the basis for protective immunity to GAS infection. Antibodies raised to extractable surface antigens, called M proteins, led to opsonophagocytosis of the strain from which the M protein was derived (27). However, the antibodies directed to the M protein of one organism often failed to protect against many other GAS isolates tested. A serological typing scheme arose through the development of antibodies directed to M proteins of different isolates. More than 80 distinct M-types have now been identified, and protective immunity to GAS infection is type specific.

More recent analysis of M proteins demonstrates that they form hairlike fibrils that extend about 60 nm from the surface of the bacterial cell. The M and M-like proteins are extremely complex in both structure and function. The determinants of serological type lie at the amino-termini (i.e., distal fibril tips) of many, but not all, M or M-like proteins. A review of M protein structure is presented in chapter 2 of this volume.

M protein serotyping overlaps with a second typing scheme that can be used for a major subgroup of GAS: those that produce opacity factor (OF). OF is a protein, either bound to the cell surface or secreted, that displays apolipoproteinase and fibronectin-binding activities (39). OF-typing is based on neutralization of apolipoproteinase activity by specific antibody, and it is highly correlative with M-type. The genes encoding M protein (*emm*) and OF (*sof*) map far apart on the GAS chromosome; however, both genes are transcriptionally regulated by the product of the *mga* gene (34). The existence of two independent molecules of high antigenic heterogeneity, which are tightly coupled in their association with typing sera, yet whose genes are physically unlinked, is seemingly paradoxical. Perhaps the selective pressure of host immunity acts to maintain nonoverlapping combinations of specific *emm* and *sof* alleles (16). A protective role for antibody directed to OF remains to be proved.

With the M-typing scheme in place, subsequent decades of epidemiological studies revealed that there are serotypes of GAS that have a strong tendency to cause throat infection but not skin infection, and, similarly, there are other serotypes often seen with impetigo but much less often with pharyngitis (5, 44). This observation gave rise to the concept of distinct throat and skin types. Similarly, only certain M-types were frequently observed in association with outbreaks of rheumatic fever (rheumatogenic types), and acute glomerulonephritis could be attributed to a small subset of M-types (nephritogenic types) as well (38). Thus, the M serotypic markers identify organisms as having a strong tendency to cause some disease, but not all disease, suggesting that there exists a degree of specialization among strains belonging to this species.

ECOLOGY AND PRIMARY INFECTION

The group A streptococcus is a free-living organism; however, its ecological niche appears to be quite narrow—its only known natural reservoir is the human. The primary sites for colonization of GAS involve two tissues: the nasal and oropharyngeal mucosal epithelium of the upper respiratory tract and the superficial layers of the epidermis. Whether the surface of unbroken skin allows GAS to colonize and reproduce, or if tiny breaks in the stratum corneum are required for gaining a foothold, is controversial.

Gram-Positive Pathogens, ed. by V. A. Fischetti et al.
© 2000 American Society for Microbiology, Washington, D.C.

military, child daycare) display deletion and duplication of sequence repeats within their *emm* genes. A loss of antigenic epitopes that are targets of protective antibody is one consequence of these genetic changes, perhaps by inducing conformational alterations that affect the amino-terminal end (24).

In contrast to the portion of *emm* genes encoding the surface-exposed part of the M protein molecule, the 3′ end is relatively conserved among different M-serotypes. A comparison of the nucleotide sequences of the 3′ ends of *emm* and *emm*-like genes, encoding the cell-associated regions of the mature M and M-like proteins, indicates that they constitute four major phylogenetic lineages; presumably, each lineage arose through duplication and subsequent divergence of an ancestral *emm* gene (Fig. 2) (20, 21). The sequence divergence observed at the 3′ ends is largely contained within the peptidoglycan-spanning domain coding region, which is flanked by highly conserved sequences both upstream and downstream, corresponding to the putative group A carbohydrate- and cytoplasmic membrane-spanning domains, respectively. The four principal peptidoglycan-spanning domain forms differ not only in amino acid sequence, but also in their lengths, ranging from 36 to 55 amino acid residues. The four phylogenetic lineages each represent a subfamily (SF) of *emm* genes, designated SF1 through SF4. Conceivably, distinct peptidoglycan structures are required to accommodate the assembly of each of the *emm* SF gene products on the cell surface.

GENOTYPIC MARKERS RELATED TO ECOLOGY

The content and relative chromosomal arrangements of *emm* SF genes are found to exist in only five basic patterns (A through E), with very few exceptions (Fig. 1). Pattern A, D, and E strains are the most common. Combining evolutionary genetic analysis with an epidemiologically well-defined collection of GAS derived largely from the United States, Trinidad, and Europe, it can be shown that *emm* pattern A–C organisms have a strong tendency to be isolated from the nasopharynx, as opposed to impetigo lesions, whereas the majority of pattern D strains are impetigo associated (4). Pattern E strains represent a third group, in which tissue site preference is less clear-cut. Interestingly, nearly all pattern E strains are OF producers, and, conversely, nearly all OF producers are pattern E, suggesting that their distinction as a unique population extends beyond their *emm* chromosomal arrangement. Similar observations have been made in tropical Australia for isolates derived from impetiginous lesions, which yield PCR products corresponding to the entire *emm* chromosomal region of the longer 7-kb form (15), presumably representing patterns D and/or E.

The *emm* patterns can be considered genetic markers for the principal tissue reservoir of a given strain or clone. In Connecticut, a recent population-based survey of invasive disease isolates showed that <2% were pattern D whereas 70% were patterns A–C, implicating the throat as the principal reservoir for transmission of clones causing invasive disease in this population (13). This makes sense because in this geographic region, GAS pharyngitis is highly prevalent whereas impetigo is uncommon.

Whether or not the *emm* chromosomal region represents a locus that is directly responsible for ecological differences is not yet known. The idea that there exist M-types associated with strong tissue site preferences raises several interesting issues related to genetic exchange and evolutionary divergence.

GENE TRANSFER AND EVOLUTION

Evolution is a two-step process: genetic change followed by natural selection. Accumulated point mutations that give rise to divergent alleles can be used to discern phylogenetic lineages within a bacterial species. Spontaneous mutations occur at a low frequency, whereas highly mutable loci, such as the coding region for the surface-exposed portion of *emm* gene products, can undergo change at much higher frequencies. Horizontal gene exchange between a donor and recipient cell provides an opportunity for the bacterium to acquire extensive genetic change through a single event. Thus, gene transfer is a major force in bacterial evolution. We are now only at the beginning stages of gaining knowledge on the evolution of GAS and understanding how it shapes both short- and long-term epidemiological trends. Nevertheless, it is useful to consider some of the key variables that effect genetic change and natural selection.

Horizontal gene transfer is a form of sexual reproduction used by bacteria, and it can be constrained by physical and/or biological barriers. The vehicles for horizontal gene movements include plasmids, conjugative transposons, bacteriophage, and naked DNA. The frequency of transfer events varies widely among different bacterial species and also differs within a bacterial species in accordance with the mechanism (conjugation, transformation, transduction), among several other factors. Through recombination, either homologous or site-specific, new genetic material can be incorporated into the bacterial chromosome and become stably inherited. There is ample evidence for the widespread presence of temperate bacteriophage among GAS, and a role for plasmids has been implicated in at least one form of antibiotic resistance. GAS are not known to be naturally competent for transformation by exogenous fragments of DNA, although the possibility cannot be ruled out.

One key question is how much genetic exchange takes place among GAS organisms. The population genetic structure of a bacterial species is a measure of the degree of linkage disequilibrium among chromosomal genes, or, in other words, the nonrandom association of alleles. The chromosomal loci used for measurements of population genetic structure encode housekeeping functions (e.g., lactate dehydrogenase) and are assumed to be selectively neutral. Linkage disequilibrium arises because genetic recombination has not been frequent enough to break up associations among the alleles at different loci. In a freely recombining bacterial population, the associations between alleles will be random and linkage equilibrium is attained (i.e., panmixia). By assessing the degree of linkage among the alleles of multiple neutral loci, one can estimate the genetic distance between the many isolates of a bacterial species. Methods used for making such measurements include multilocus enzyme electrophoresis (MLEE) and, more recently, multilocus sequence typing. The data can be subjected to statistical treatment, as in the index of association (I_A) measurement for linkage disequilibrium (33); an I_A value for GAS has been reported for a subset of strains. Comparisons between genetic distances as defined by MLEE and allelic variants of several virulence loci demonstrate that

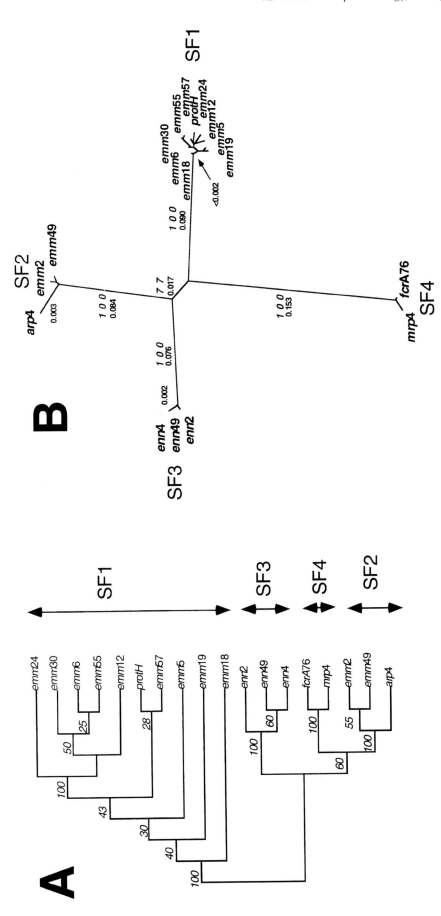

FIGURE 2 Phylograms of the 3' region of 18 *emm* family genes (20). DNA sequences (GenBank) are: *emm6* (M11338), *emm12* (U02480), *emm24* (M19031), *emm18* (S82224), *emm57* gene (X60959), *emmL2.1* and *emmL2.2* (X61276), *enn4* (Z11602), and *mrp4* (M87831). Alignment of the 3' region (about 432 bp) and references for the above sequences were previously published (20). (A) The consensus tree generated by exhaustive search under maximum parsimony with the aid of the program Phylip. (B) A tree constructed from the same alignment by the neighbor-joining method in the program PAUP4.0B (Swofford). Italic numbers for each tree indicate the percentage of time each branch was joined together under bootstrap analysis (1,000 replicates performed). Roman numbers on the tree indicate branch lengths as calculated by PAUP. SF1 through SF4 represent subfamilies 1 through 4, designated as those clusters that have 100% bootstrap support in both trees.

horizontal gene transfer does occur (reviewed in reference 26).

The existence of ecologically distinct populations of GAS may have important implications for the population genetic structure of this species. Horizontal gene exchange is most likely to occur when the donor and recipient cells occupy the same space at the same time. In rare instances, a single focus of infection can become simultaneously infected with a second GAS strain at either the throat or skin. Since there is a distinct population of GAS having a strong tendency to infect the throat, and a second population with a strong predilection for the skin, it would not be unexpected if the frequency of gene transfer is significantly higher within ecological populations than it is between populations. The geographic and seasonal differences in the peak incidences of pharyngitis and impetigo throughout the world may act to create even further distance between strains having a strong preference for one tissue over the other.

The ability of a new genetic variant to survive is tied to its fitness. If there exist a few loci that are responsible for ecological divergence, and one of those loci is altered via a recombinational event involving a second, ecologically distinct population, then the recombinant may be of lowered fitness and become quickly lost because it is not well adapted to the ecological niche of either population. For GAS, pattern A–C and pattern D strains may represent two ecological populations. Pattern E strains may be a third ecological population, and since many pattern E strains are readily found at both the throat and skin, they would not necessarily experience reproductive isolation from pattern A–C or D strains. Even if the frequency of recombinational exchange within an ecological population were equal to the between-population recombinational rate and, as a result, divergence among neutral loci approached zero, the extent of linkage disequilibrium among adaptive loci can still remain high. Thus, there is the potential for an uncoupling of neutral and adaptive divergence (10). One can ask which allelic combinations among GAS virulence genes display levels of linkage disequilibrium that far exceed the values for selectively neutral loci.

Horizontal genetic events leading to the emergence of a new clone are favored by conditions of high bacterial density. A period of high prevalence of human infection is paralleled by a proportionally higher density bacterial population, conditions that will, in turn, further favor horizontal genetic exchange between bacterial cells. A condition of highly endemic GAS impetigo, with multiple clones contributing to the disease burden, favors coinfection of a single skin lesion by multiple clones (7). This, in turn, should increase the probability of genetic exchange between different clones.

Because asymptomatic carriage of GAS at the throat can be of long duration, the chances of a second GAS strain establishing cocolonization in a carrier host are more favorable than those for a rapidly cleared infection. Carriage rates are highest in the younger school-aged children. The physiological state of the bacterial cells is likely to influence the chances of successful horizontal exchange. Experimental transformation of GAS by electroporation is enhanced by enzymatic removal of the hyaluronic acid capsule and alterations in peptidoglycan structure. Many bacteriophage of GAS produce hyaluronidase. From studies on OF-producing strains of GAS, it has been proposed that stationary phase and the accompanying high-density and low-nutrient conditions, whereby the transcription of genes encoding for M protein and capsular biosynthesis are decreased, parallel the state of bacteria during asymptomatic carriage; oligopeptide permease has been implicated as having a role in controlling this GAS phenotype (29). Perhaps of relevance is the finding that in a closely related Enterococci sp., oligopeptide permease plays an important role in intercellular signaling and conjugal transfer of genetic information (28). However, it is also possible that the biology of the carrier state varies for different strains. Using an M-type 3 strain, a serotype that is typically pattern A–C and throat-derived, an essential role for both M protein and capsule is demonstrated in persistent pharyngeal colonization of baboons (2). The mechanics of genetic transfer among GAS, and the ecological conditions that are most likely to promote the optimal physiological state for the donor and recipient cells, is an area of inquiry that remains to be more thoroughly explored.

For a bacterial species whose world is largely confined to the human population, the selective pressures that shape its genetic structure are intrinsic to the human condition. In the absence of a GAS vaccine, all protective immunity is acquired through prior infection. Bacteriocinlike inhibitory substances produced by Streptococcus salivarius, a commensal of the human oral cavity, are toxic to at least some strains of GAS, and thus the composition of the normal flora may selectively influence GAS survival. Antibiotic resistance in GAS has been described (41); however, it does not yet appear to be the potent driving force that it is for the global spread of select clones of the related bacterial species, Streptococcus pneumoniae. Nonetheless, a single genetic transfer event can change all that.

Human behavior is yet another factor in the evolution of microorganisms. Within the past 50 years, antibiotic use has become influential in the natural selection of many bacteria that reproduce in humans and farm animals. The GAS infection that triggers symptoms of sore throat and high fever is quickly eradicated in most of the developed world. But do variants of GAS exist that are of lowered virulence because of a diminished capacity to cause overt clinical symptoms and, if so, does the low virulence phenotype provide a selective advantage that allows for their emergence as highly prevalent clones? The basic reproductive rate (i.e., fitness) of GAS is directly proportional to its successful transmission to new hosts (30). Only if the symptoms of acute pharyngitis specifically enhance the transmission of GAS to a new host (i.e., virulence and transmission are positively coupled), will the advantage of evading near-certain death by antibiotics be counterbalanced by a decrease in transmissibility. In several recent large outbreaks of rheumatic fever in communities across the United States, a substantial proportion of cases were associated with clinically inapparent GAS infection that went unnoticed and untreated in the weeks prior to acute autoimmune attack.

TRANSMISSION DYNAMICS

From the evolutionary standpoint, invasive disease is a dead end for the infecting GAS organism because the severely ill patient becomes immobilized and, thus, opportunities for its transmission to new hosts are diminished. If virulence is too high and the primary host is killed off quickly, leaving insufficient time for transmission to a secondary host, then the bacterium will die along with the host. The measure for virulence is a rate function, and one must also consider the number of infected hosts who experience only mild to moderate morbidity, or no illness at all.

During a two-and-one-half-month period in early 1995, there was a confined outbreak of seven cases of severe invasive disease attributed to a single clone of M-type 3 GAS; six of the seven patients resided in one of two communities in southeastern Minnesota, and five had underlying medical conditions or predisposing factors (9). The M3 clone accounted for 78% of the GAS detected by screening elementary school children for pharyngeal carriage (92% participated); in all, 25% of the children harbored the M3 clone in the absence of any illness that would have otherwise prevented them from attending school. Of the cases of GAS-associated sore throat identified in community clinics, 28% of GAS isolates were identical to the M3 clone. The findings on asymptomatic carriage suggest that the M3 clone is highly prevalent and highly transmissible (i.e., has high fitness), but it is of low virulence in children. The M3 clone can give rise to acute pharyngitis (a state of intermediate virulence); however, the majority of pharyngitis cases are associated with GAS strains that are not nearly as well represented among the cases of invasive disease and carriage. Is the M3 clone best described as a "super killer" or as a "super transmitter"? Are the high levels of virulence that it can attain strictly dependent on host deficiencies? Is its transmission largely independent of pharyngitis?

In general terms, it is reasonable to assume that the rate of transmission of a microbial agent is directly proportional to the number of susceptible hosts (30). Host susceptibility to GAS infection appears to depend largely on the absence of antibodies specific for the infecting M-type (2, 27). Thus, the maintenance of a given GAS M-type in a community will be inversely proportional to herd immunity; if herd immunity rises above a certain threshold, that M-type may become altogether lost. Another parameter that affects transmission rate is the within-host density of the bacterial population: the higher it is, the better the chances are for transmission to a greater number of secondary hosts, at least up to a certain threshold (i.e., the primary host is not immobilized). It is also possible that the bacterium's tissue site limitation can be overridden if the infecting dose attains a critically high level. For example, it is not unusual for GAS impetigo to give rise to secondary infection at the upper respiratory tract at approximately 2 or 3 weeks after the initial onset of skin involvement (5).

There may also be qualitative, biological properties of the bacterium that can enhance its rate of transmission. For example, it has been proposed that expression of the fibronectin-binding protein, protein F, facilitates colonization during the earliest stage of GAS throat infection (43). In numerous epidemics of GAS pharyngitis, during which there was a high rate of person-to-person transmission, the causative agent was often found to be highly enriched for hyaluronic acid capsule (22, 45). Overall, we have incomplete knowledge of the biological requirements that must be met by the GAS for its successful journey into its next human incubator.

REFERENCES

1. **Anthony, B. F., E. L. Kaplan, L. W. Wannamaker, and S. S. Chapman.** 1976. The dynamics of streptococcal infections in a defined population of children: serotypes associated with skin and respiratory infections. *Am. J. Epidemiol.* **104:**652–666.

2. **Ashbaugh, C. D., M. H. Shearer, R. C. Kennedy, G. C. White, and M. R. Wessels.** 1998. M protein and hyaluronic acid capsule are required for group A streptococcal colonization of the baboon pharynx, abstr. B53A. *Program Abstr. 38th Intersci. Conf. Antimicrob. Agents Chemother.* American Society for Microbiology, Washington, D.C.

3. **Bessen, D. E., M. W. Izzo, E. J. McCabe, and C. M. Sotir.** 1997. Two-domain motif for IgG-binding activity by group A streptococcal *emm* gene products. *Gene* **196:** 75–82.

4. **Bessen, D. E., C. M. Sotir, T. L. Readdy, and S. K. Hollingshead.** 1996. Genetic correlates of throat and skin isolates of group A streptococci. *J. Infect. Dis.* **173:**896–900.

5. **Bisno, A. L.** 1995. *Streptococcus pyogenes*, p. 1786–1799. *In* G. L. Mandell, R. G. Douglas, and J. E. Bennett (ed.), *Principles and Practice of Infectious Diseases*, 4th ed. John Wiley & Sons, Inc., New York, N.Y.

6. **Breese, B. B., and C. B. Hall.** 1978. *Beta Hemolytic Streptococcal Diseases*. Houghton Mifflin, Boston, Mass.

7. **Carapetis, J., D. Gardiner, B. Currie, and J. D. Mathews.** 1995. Multiple strains of *Streptococcus pyogenes* in skin sores of Aboriginal Australians. *J. Clin. Microbiol.* **33:** 1471–1472.

8. **Cleary, P. P., L. McLandsborough, L. Ikeda, D. Cue, J. Krawczak, and H. Lam.** 1998. High-frequency intracellular infection and erythrogenic toxin A expression undergo phase variation in M1 group A streptococci. *Mol. Microbiol.* **28:**157–167.

9. **Cockerill, F. R., K. L. MacDonald, R. L. Thompson, F. Roberson, P. C. Kohner, J. Besser-Wiek, J. M. Manahan, J. M. Musser, P. M. Schlievert, J. Talbot, B. Frankfort, J. M. Steckelberg, W. R. Wilson, and M. T. Osterholm.** 1997. An outbreak of invasive group A streptococcal disease associated with high carriage rates of the invasive clone among school-aged children. *JAMA* **277:**38–43.

10. **Cohan, F. M.** 1996. The role of genetic exchange in bacterial evolution. *ASM News* **62:**631–636.

11. **Davies, H. D., A. McGeer, B. Schwartz, K. Green, D. Cann, A. E. Simor, and D. E. Low.** 1996. Invasive group A streptococcal infections in Ontario, Canada. *N. Engl. J. Med.* **335:**547–554.

12. **Facklam, R., B. Beall, A. Efstratiou, V. Fischetti, E. Kaplan, P. Kriz, M. Lovgren, D. Martin, B. Schwartz, A. Totolian, D. Bessen, S. Hollingshead, F. Rubin, J. Scott, and G. Tyrrell.** 1999. Report on an international workshop: demonstration of *emm* typing and validation of provisional M-types of group A streptococci. *Emerg. Infect. Dis.* **5:**247–253.

13. **Fiorentino, T. R., B. Beall, P. Mshar, and D. E. Bessen.** 1997. A genetic-based evaluation of principal tissue reservoir for group A streptococci isolated from normally sterile sites. *J. Infect. Dis.* **176:**177–182.

14. **Fischetti, V. A., M. Jarymowycz, K. F. Jones, and J. R. Scott.** 1986. Streptococcal M protein size mutants occur at high frequency within a single strain. *J. Exp. Med.* **164:** 971–980.

15. **Gardiner, D. L., A. M. Goodfellow, D. R. Martin, and K. S. Sriprakash.** 1998. Group A streptococcal Vir types are M-protein gene (*emm*) sequence type specific. *J. Clin. Microbiol.* **36:**902–907.

16. Gupta, S., M. C. J. Maiden, I. M. Feavers, S. Nee, R. M. May, and R. M. Anderson. 1996. The maintenance of strain structure in populations of recombining infectious agents. *Nat. Med.* **2:**437–442.

17. Haanes, E. J., and P. P. Cleary. 1989. Identification of a divergent M protein gene and an M protein related gene family in serotype 49 *Streptococcus pyogenes. J. Bacteriol.* **171:**6397–6408.

18. Harbaugh, M. P., A. Podbielski, S. Hugl, and P. P. Cleary. 1993. Nucleotide substitutions and small-scale insertion produce size and antigenic variation in group A streptococcal M1 protein. *Mol. Microbiol.* **8:**981–991.

19. Hollingshead, S. K., V. A. Fischetti, and J. R. Scott. 1987. Size variation in group A streptococcal M protein is generated by homologous recombination between intragenic repeats. *Mol. Gen. Genet.* **207:**196–203.

20. Hollingshead, S. K., T. Readdy, J. Arnold, and D. E. Bessen. 1994. Molecular evolution of a multi-gene family in group A streptococci. *Mol. Biol. Evol.* **11:**208–219.

21. Hollingshead, S. K., T. L. Readdy, D. L. Yung, and D. E. Bessen. 1993. Structural heterogeneity of the *emm* gene cluster in group A streptococci. *Mol. Microbiol.* **8:** 707–717.

22. Johnson, D. R., D. L. Stevens, and E. L. Kaplan. 1992. Epidemiological analysis of group A streptococcal serotypes associated with severe systemic infections, rheumatic fever, or uncomplicated pharyngitis. *J. Infect. Dis.* **166:** 374–382.

23. Johnsson, E., A. Thern, B. Dahlback, L. O. Heden, M. Wikstrom, and G. Lindahl. 1996. A highly variable region in members of the streptococcal M protein family binds the human complement regulator C4BP. *J. Immunol.* **157:**3021–3029.

24. Jones, K. F., S. K. Hollingshead, J. R. Scott, and V. A. Fischetti. 1988. Spontaneous M6 protein size mutants of group A streptococci display variation in antigenic and opsonogenic epitopes. *Proc. Natl. Acad. Sci. USA* **85:** 8271–8275.

25. Kaplan, E. L. 1980. The group A streptococcal upper respiratory tract carrier state: an enigma. *J. Pediatr.* **97:** 337–345.

26. Kehoe, M. A., V. Kapur, A. M. Whatmore, and J. M. Musser. 1996. Horizontal gene transfer among group A streptococci: implications for pathogenesis and epidemiology. *Trends Microbiol.* **4:**436–443.

27. Lancefield, R. C. 1962. Current knowledge of the type specific M antigens of group A streptococci. *J. Immunol.* **89:**307–313.

28. Leonard, B. A. B., A. Podbielski, P. J. Hedberg, and G. M. Dunny. 1996. *Enterococcus faecalis* pheromone binding protein, PrgZ, recruits a chromosomal oligopeptide permease system to import sex pheromone cCF10 for induction of conjugation. *Proc. Natl. Acad. Sci. USA* **93:**260–264.

29. Leonard, B. A. B., M. Woischnik, and A. Podbielski. 1998. Production of stabilized virulence factor-negative variants by group A streptococci during stationary phase. *Infect. Immun.* **66:**3841–3847.

30. Levin, B. R. 1996. The evolution and maintenance of virulence in microparasites. *Emerg. Infect. Dis.* **2:**93–102.

31. Leyden, J. J., R. Stewart, and A. M. Kligman. 1980. Experimental infections with group A streptococci in humans. *J. Invest. Dermatol.* **75:**196–201.

32. Martin, D. R., and K. S. Sriprakash. 1996. Epidemiology of group A streptococcal disease in Australia and New Zealand. *Recent Adv. Microbiol.* **4:**1–40.

33. Maynard Smith, J., N. H. Smith, M. O'Rourke, and B. G. Spratt. 1993. How clonal are bacteria? *Proc. Natl. Acad. Sci. USA* **90:**4384–4388.

34. McLandsborough, L. A., and P. P. Cleary. 1995. Insertional inactivation of *virR* in *Streptococcus pyogenes* M49 demonstrates that VirR functions as a positive regulator of ScpA, FcRA, OF, and M protein. *FEMS Microbiol. Lett.* **128:**45–52.

35. Moxon, E. R., P. B. Rainey, M. A. Nowak, and R. E. Lenski. 1994. Adaptive evolution of highly mutable loci in pathogenic bacteria. *Curr. Biol.* **4:**24–33.

36. Musser, J. M., V. Kapur, J. Szeto, X. Pan, D. S. Swanson, and D. M. Martin. 1995. Genetic diversity and relationships among *Streptococcus pyogenes* strains expressing serotype M1 protein: recent intercontinental spread of a subclone causing episodes of human disease. *Infect. Immun.* **63:**994–1003.

37. Podbielski, A. 1993. Three different types of organization of the vir regulon in group A streptococci. *Mol. Gen. Genet.* **237:**287–300.

38. Potter, E. V., M. Svartman, I. Mohammed, R. Cox, T. Poon-King, and D. P. Earle. 1978. Tropical acute rheumatic fever and associated streptococcal infections compared with concurrent acute glomerulonephritis. *J. Pediatr.* **92:**325–333.

39. Rakonjac, J. V., J. C. Robbins, and V. A. Fischetti. 1995. DNA sequence of the serum opacity factor of group A streptococci: identification of a fibronectin-binding repeat domain. *Infect. Immun.* **63:**622–631.

40. Relf, W. A., D. R. Martin, and K. S. Sriprakash. 1994. Antigenic diversity within a family of M proteins from group A streptococci: evidence for the role of frameshift and compensatory mutations. *Gene* **144:**25–30.

41. Seppala, H., A. Nissinen, H. Jarvinen, S. Huovinen, T. Henriksson, E. Herva, S. E. Holm, M. Jahkola, M.-L. Katila, T. Klaukka, S. Kontiainen, O. Liimatainen, S. Oinonen, L. Passi-Metsomaa, and P. Huovinen. 1992. Resistance to erythromycin in group A streptococci. *N. Engl. J. Med.* **326:**292–297.

42. Stenberg, L., P. W. O'Toole, J. Mestecky, and G. Lindahl. 1994. Molecular characterization of protein Sir, a streptococcal cell surface protein that binds both immunoglobulin A and immunoglobulin G. *J. Biol. Chem.* **269:** 13458–13464.

43. VanHeyningen, T., T. Fogg, D. Yates, E. Hanski, and M. Caparon. 1993. Adherence and fibronectin-binding are environmentally regulated in the group A streptococcus. *Mol. Microbiol.* **9:**1213–1222.

44. Wannamaker, L. W. 1970. Differences between streptococcal infections of the throat and of the skin. *N. Engl. J. Med.* **282:**23–31.

45. Wessels, M. R., and M. S. Bronze. 1994. Critical role of the group A streptococcal capsule in pharyngeal colonization and infection in mice. *Proc. Natl. Acad. Sci. USA* **91:**12238–12242.

46. Whatmore, A. M., V. Kapur, D. J. Sullivan, J. M. Musser, and M. A. Kehoe. 1994. Non-congruent relationships between variation in *emm* gene sequences and the population genetic structure of group A streptococci. *Mol. Microbiol.* **14:**619–631.

B. Group B Streptococci

Pathogenic Mechanisms and Virulence Factors of Group B Streptococci

VICTOR NIZET AND CRAIG E. RUBENS

13

Group B streptococci (GBS) are bacteria well adapted to asymptomatic colonization of human adults but a potentially devastating pathogen to certain susceptible infants. Because the newborn is quantitatively and qualitatively deficient in host defenses, including phagocytes, complement, and specific antibody, an environment exists in which a variety of potential GBS virulence factors are unveiled. The complex interactions between the bacterium and the newborn host that lead to disease manifestation can be divided into several important categories (Table 1). This chapter reviews GBS pathogenic mechanisms involved in adherence to epithelial surfaces, cellular invasion of epithelial and endothelial barriers, direct injury to host tissues, avoidance of immunologic clearance, and induction of the sepsis syndrome. Special attention is focused on recent molecular genetic discoveries that have led to the identification of a number of specific GBS determinants implicated in virulence to the newborn host.

ADHERENCE TO EPITHELIAL SURFACES

Essential to acquisition of the organism by the newborn, the first step in the pathogenesis of GBS disease is asymptomatic colonization of the female genital tract. In comparison to other microorganisms, GBS bind very efficiently to human vaginal cells, with maximal adherence at the acidic pH characteristic of vaginal mucosa (90). GBS also adhere effectively to human cells from a variety of fetal, infant, and adult tissues, including placental membranes, buccal or pharyngeal mucosa, and alveolar epithelium and endothelium, and each is potentially relevant to vertical transmission and production of invasive disease in the infant. It follows that either (i) the specific component of the host cell surface to which GBS attaches is widely distributed among human tissues, or (ii) GBS can adhere to multiple cell surface components.

Both ionic and hydrophobic interactions contribute to GBS host cell adherence. Epithelial cell binding is inhibited by pretreatment of the GBS with a variety of proteases, preincubation of the cells with hydrophobic GBS surface proteins, or preincubation of the GBS with antibodies to these proteins (70, 84). The GBS polysaccharide capsule itself attenuates the adherence potential of the organism (70). Potential mechanisms include steric hindrance of a high-affinity adhesin-receptor interaction or a decrease in repulsive forces generated between negatively charged sialic acid residues on the GBS capsule and the surface of the host cell.

A potential nonproteinaceous ligand involved in GBS adherence is lipotechoic acid (LTA), an amphiphilic glycolipid polymer that extends through the cell wall and is known to mediate host cell attachment by other gram-positive pathogens. Low-affinity binding of GBS to human buccal epithelial cells is competitively inhibited by preincubation with purified LTA (45). GBS with high levels of LTA bind more efficiently to buccal cells of fetal rather than adult origin, suggesting a potential link to the age-restricted susceptibility to GBS disease. Enzymatic treatment of the fetal buccal epithelial cells with trypsin or periodate abolishes LTA binding, indicating the presence of a glycoprotein receptor(s) on the fetal cells absent from the adult cells. Interestingly, topical administration of LTA or glycerol phosphate, a subunit component of LTA putatively involved in binding, may decrease GBS vaginal colonization of pregnant mice (8).

Like other pathogens, GBS appear to attach to host extracellular matrix proteins such as fibronectin, fibrinogen, and laminin. While GBS bind to fibronectin immobilized on polystyrene, they do not bind fibronectin in its soluble form (71). GBS adherence to fibronectin is therefore a low-avidity interaction, requiring the close proximity of multiple fibronectin molecules and GBS adhesins present in a solid-phase model. Adherence of GBS to laminin involves the gene *lmb*, which encodes a homolog of the streptococcal Lra1 adhesin family (68). Targeted mutagenesis of the *lmb* locus results in diminished adherence of GBS to immobilized laminin. Finally, analysis of a random plasmid integrational mutant library of GBS has identified a genetic locus involved in human fibrinogen binding (69). These genes share significant homology with several gram-positive two-component regulator systems. The association of this locus with the fibrinogen-binding properties

Gram-Positive Pathogens, ed. by V. A. Fischetti et al.
© 2000 American Society for Microbiology, Washington, D.C.

TABLE 1 Virulence attributes of group B streptococci implicated in neonatal infection

Pathogenic categories	Specific mechanisms	GBS determinants
Adherence to epithelial surfaces	Colonization of vaginal mucosa	Undefined protein adhesins
	Attachment to respiratory epithelium	Lipotechoic acid
	Fibronectin binding	
	Fibrinogen binding	
	Laminin binding	
Cellular invasion of epithelial and endothelial barriers	Chorionic invasion/transcytosis	Undefined protein invasins
	Lung epithelial invasion	
	Lung endothelial invasion	
	Brain endothelial invasion/transcytosis	
Direct injury to host tissues	Injury to placental membranes	β-Hemolysin[a]
	Injury to lung epithelial cells	Proteases
	Injury to lung endothelial cells	Collagenase
		CAMP factor
		Hyaluronate lyase
		Lipotechoic acid
Avoidance of immunologic clearance	Resistance to opsonophagocytosis	Polysaccharide capsule[a]
	Inhibition of complement deposition	C5a peptidase[a]
	Impairment of neutrophil recruitment	Surface C protein antigen[a]
	Nonimmune antibody binding	CAMP factor
	Resistance to intracellular killing	Hyaluronate lyase
Induction of the sepsis syndrome	Cytokine release (IL-1, TNF-α)	Cell wall peptidoglycan
	Nitric oxide synthase induction	β-hemolysin

[a]Role in virulence demonstrated by animal studies with isogenic, factor-deficient mutants.

of GBS was confirmed by targeted knockout of the putative response regulator. The precise contributions of any extracellular matrix protein-binding phenotypes to GBS pathogenesis have not yet been elucidated.

INVASION OF EPITHELIAL AND ENDOTHELIAL CELLS

GBS have been shown to penetrate and survive within several human cell types—a phenomenon that has been termed cellular invasion. This phenotypic property may account for the ability of GBS to traverse a number of host cellular obstacles, including the placental membranes, the alveoli of the infant lung, and the neonatal blood-brain barrier. The pathogenic consequence of bacterial passage through the respective barriers may be significant: amnionitis and overwhelming fetal exposure; bacteremia and systemic spread; entry into the central nervous system and meningitis. The electron micrographs in Fig. 1 show ultrastructural evidence of GBS cellular invasion from a series of in vivo and in vitro studies. In all cell types examined, intracellular GBS are found within membrane-bound vacuoles, suggesting that the organism somehow elicits its own endocytotic uptake. This virulence attribute of GBS is reminiscent of several enteric bacterial pathogens for which invasion of gut epithelial cells is a primary step in establishment of systemic infection.

It has long been suspected that GBS can penetrate into the amniotic cavity through intact placental membranes, because fulminant early-onset disease develops in some infants delivered by cesarean section with no identifiable obstetric risk factors. Rapid death and advanced lung inflammatory changes on autopsy of such patients strongly imply that the onset of infection occurred in utero. Experimentally, GBS are seen to migrate through freshly isolated chorioamniotic membranes, binding to the maternal surface by 2 h and appearing on the fetal surface within 8 h of inoculation (15). The nature of GBS placental penetration has been dissected further in studies using primary chorion and amnion cell cultures (87). GBS are highly invasive for chorion cells and are capable of transcytosing through intact chorion cell monolayers without disruption of intracellular tight junctions. However, although GBS adhere to amnion cells, they fail to invade these cells under a variety of assay conditions. The amnion may thus constitute a formidable host barrier against infection of the fetus, and GBS penetration into the amniotic cavity may require direct or indirect injury to amnion cells (see below) rather than cellular invasion and transcytosis (87).

Following aspiration of infected amniotic fluid, the initial focus of GBS infection is the newborn lung. Indeed, pneumonia with marked respiratory distress is a hallmark of early-onset infection. To disseminate from the alveolar space and gain access to the systemic circulation, GBS must traverse three host barriers: the alveolar epithelium, the pulmonary interstitium, and the pulmonary endothelium. In vivo evidence for GBS invasion of these host cells comes from a primate model in which early-onset pneu-

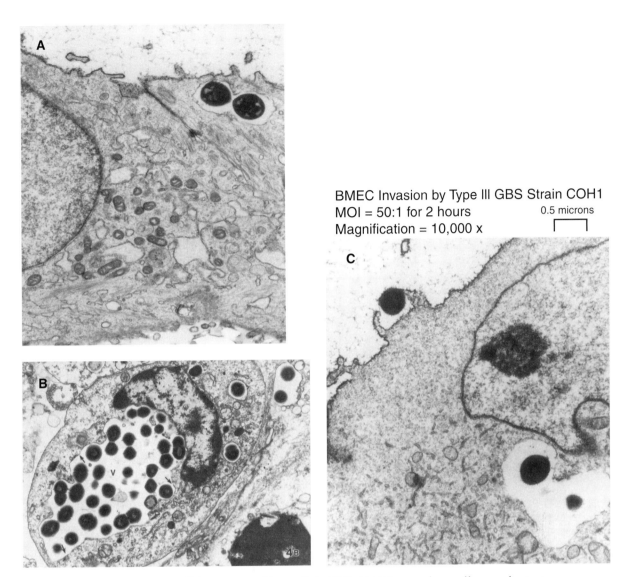

FIGURE 1 Intracellular invasion of host cells by GBS. (A) Primary cultures of human chorionic epithelial cells. (B) Respiratory epithelial cells of an infant macaque following infection by intra-amniotic inoculation. (C) Immortalized cultures of human brain microvascular endothelial cells. In each case, intracellular GBS are found within membrane-bound vacuoles (v). Host cytoskeletal changes in C suggest that the organism elicits its own endocytotic uptake.

monia and septicemia are established following intra-amniotic inoculation of the organism. Electron microscopy of lung tissue from the infant macaques demonstrated GBS within membrane-bound vacuoles of type I and II alveolar epithelial cells, interstitial fibroblasts, and capillary endothelial cells (16, 58). Subsequent tissue culture studies demonstrated morphologically comparable GBS invasion of human lung epithelial cells (59) and human umbilical vein or piglet lung endothelial cells (16).

The phenomenon of cellular invasion by GBS has been further elucidated by means of a quantitative tissue culture assay (59). This assay quantifies intracellular GBS from host cell lysates after noninvading bacteria have been killed by antibiotics that are active only extracellularly. GBS isolates of capsular serotypes Ia, Ib, Ia/c, II, and III all invade alveolar epithelial cells, although a small degree of strain variation exists in the magnitude of invasion (27).

GBS invade more efficiently into pulmonary microvascular endothelial cells than endothelial cells of pulmonary artery origin (20), suggesting a tropism for particular endothelial surfaces. To produce meningitis, GBS circulating in the bloodstream must subsequently penetrate human brain microvascular endothelial cells (BMEC), the single-cell layer that constitutes the blood-brain barrier. GBS invasion and transcytosis of intact, polar tissue culture monolayers of human BMEC were recently demonstrated (48). Serotype III strains, which account for the majority of central nervous system isolates, invaded BMEC more efficiently than did strains from other common GBS serotypes.

Time course studies in epithelial and endothelial cells indicate that GBS survive intracellularly for up to 24 h after invasion, without appreciable bacterial replication (48, 59). Addition of specific inhibitors demonstrates that active bacterial DNA, RNA, and protein synthesis is necessary for invasion and that the host endocytotic mechanism involves actin microfilaments (48, 59, 87). Invasion of BMEC appears to require microtubular cytoskeletal elements and active host protein synthesis as well (48). Uptake of GBS is associated with induction of protein-kinase signal transduction pathways in the epithelial cell (75). These pathways may be dependent upon calmodulin, because entry of GBS into epithelial cells is inhibited in a dose-dependent manner when the extracellular calcium concentration is reduced.

The polysaccharide capsule itself does not appear to be essential for GBS cellular invasion. Rather, an isogenic unencapsulated mutant invades epithelial and endothelial cells more efficiently than the type III clinical isolate from which it was derived (16, 27, 48). As discussed above, the GBS capsule decreases adherence of the organism to alveolar epithelium (70), presumably through steric interference of certain receptor-ligand interactions. Should initial adherence to the epithelial cell be a requirement for subsequent invasion, inhibition of invasion by the presence of capsule would be anticipated. Likewise, capsule may mask other "invasin" molecules on the GBS surface that promote epithelial cell uptake independent of adherence. Finally, one must consider the fact that the polysaccharide capsule confers an important survival advantage to GBS through inhibition of macrophage and neutrophil phagocytosis. Should alveolar epithelial cells share rudimentary aspects of their endocytotic uptake mechanism with these professional phagocytes, capsule inhibition of invasion may be a by-product of the stronger selective pressure placed on GBS to avoid immunologic clearance.

Direct experimental evidence is accumulating to indicate that cellular invasion is a crucial component in the pathogenesis of neonatal GBS disease. When tested in the tissue culture model, GBS isolates from the blood of infected neonates are significantly more invasive for respiratory epithelia than are isolates from vaginal carriers or colonized neonates without clinical symptoms (74). These results indicate that in vitro invasion of human epithelial cell monolayers is a marker for the ability of a GBS to produce invasive disease in vivo. Furthermore, a GBS transposon mutant has been created that is deficient in the ability to invade alveolar epithelial cells in tissue culture (34). In contrast to the parent strain, this mutant fails to produce bacteremia when administered by aerosolization to newborn rats. This finding is consistent with the hypothesis that cellular invasion is required for GBS to bypass the pulmonary epithelial barrier.

DIRECT INJURY TO HOST CELLULAR BARRIERS

Localized bacterial proliferation with associated injury to host tissues is apparent in pathologic specimens from patients with GBS disease, in particular the placenta (chorioamnionitis), lung (pneumonia), and brain (meningitis). GBS products, in particular a β-hemolysin, have been shown to be directly cytotoxic to human cells and thus represent potential virulence determinants of the organism. Disruption of epithelial and endothelial cell barriers through direct cell injury could theoretically facilitate placental penetration and systemic spread of the organism in the infected infant. Note that intracellular invasion and transcytosis by GBS, described in the previous section, can occur without significant injury to host cells in vitro. The relative contribution of intracellular invasion versus direct tissue injury to GBS pathogenesis remains to be determined—it is likely they act in concert to produce the clinical spectrum of invasive disease. In fulminant early-onset disease and pneumonia, autopsy findings of alveolar hemorrhage and proteinaceous exudate attest to significant pulmonary epithelial and endothelial cell destruction, through which GBS would have direct access to the circulation. By contrast, in late-onset GBS disease, a subacute clinical presentation without pneumonia is the rule. Intracellular invasion and transcytosis of the alveolar lining by colonizing bacteria may represent the principal mechanism of bacterial spread to the bloodstream. Subsequent invasion and transcytosis of BMEC by circulating GBS could result in meningitis, a frequent clinical complication.

The great majority (>98%) of GBS clinical isolates demonstrate beta-hemolysis when plated on sheep blood agar. The GBS hemolysin(s) itself has yet to be isolated, owing largely to the fact that its activity is unstable. High-molecular-weight carrier molecules such as albumin are required to preserve hemolysin activity in GBS culture supernatants (39). A role for β-hemolysin in lung epithelial and endothelial injury has been demonstrated by the development of GBS transposon mutants expressing either a nonhemolytic (NH) or hyperhemolytic (HH) phenotype (17, 46). When monolayers of human alveolar epithelial cells are exposed to log-phase GBS or stabilized hemolysin extracts of GBS cultures, cellular injury can be measured by lactate dehydrogenase release or trypan blue nuclear staining. Whereas NH strains produce no detectable injury beyond baseline (medium alone), hemolysin-producing strains induce injury to lung epithelial and endothelial cells in direct correlation to their ability to lyse sheep erythrocytes (17, 46). Electron microscopy studies of the injured lung epithelial cells reveal global loss of microvillus architecture, disruption of cytoplasmic and nuclear membranes, and marked swelling of the cytoplasm and organelles (46)—findings that suggest that β-hemolysin acts as a pore-forming cytolysin. Radiolabeled rubidium (^{86}Rb$^+$) and hemoglobin exhibit identical efflux kinetics from sheep erythrocytes exposed to GBS β-hemolysin, an indication that the toxin produces membrane lesions of large size (39).

In vitro, GBS injury to cultured lung epithelial and endothelial cells can be documented at bacterial concentrations of 10^6 (most hemolytic clinical isolate) to 10^8 (least hemolytic clinical isolate) GBS per ml (17, 46). When GBS pneumonia is induced in newborn primates, bacterial density reaches 10^9 to 10^{11}/g of lung tissue (58), indicating that an ample local reservoir exists for β-hemolysin production and host cell injury. GBS β-hemolysin expression is associated with dose-dependent increases in albumin transit across polar endothelial cell monolayers, suggesting it may contribute to pulmonary vascular leakage (17). Protein efflux from pulmonary vessels into the terminal air spaces is consistent with the clinical pattern of alveolar congestion and hyaline membrane formation seen in early-onset pneumonia. The extent of alveolar epithelial cell injury produced by HH strains is reduced in a stepwise fashion when dipalmitoyl phosphatidylcholine (DPPC), the major component of human surfactant, is added in concentrations corresponding to the physiologic increase in alveolar fluid DPPC during the third trimester of pregnancy (46). This finding provides a theoretical rationale for the increased incidence of severe GBS pneumonia in premature, surfactant-deficient neonates.

Animal virulence studies suggest a role for β-hemolysin expression in GBS disease acquired via the respiratory tract. Using intranasal inoculation, a comparison of chemically derived NH and HH GBS mutants shows that increased beta-hemolytic activity is associated with a decreased 50% lethal dose (LD_{50}) and an earlier time to death for a given inoculum (80). When the bacterial inoculum is delivered directly into the rat lung by transthoracic puncture, an NH GBS transposon mutant exhibits a 1,000-fold increased LD_{50} compared with wild type (47).

The molecular basis of GBS β-hemolysin expression has yet to be defined. In genetic studies, a recombinant plasmid containing GBS chromosomal DNA conferred a beta-hemolytic phenotype on host *Escherichia coli* and allowed identification of a potential GBS β-hemolysin gene encoding a protein of 230 amino acids (6). However, when an intragenic fragment of this gene sequence is used to create targeted knockouts in two wild-type GBS strains through integrational mutagenesis, hemolytic activity is unaffected, indicating the gene does not encode the major GBS β-hemolysin determinant (47). Analysis of a random plasmid integrational mutant library of GBS for nonhemolytic isolates has identified an accessory genetic locus involved in β-hemolysin expression (67). Three open reading frames (ORFs) encoding proteins with significant homologies to prokaryotic multidrug-resistance ABC transporters were sequenced. Targeted knockout of two of these ORFs reproduced the nonhemolytic phenotype, confirming that an ABC transporter-type function is required for the hemolytic activity of GBS.

Mechanisms by which GBS may promote placental membrane rupture and premature delivery are being examined. Isolated chorioamniotic membranes exposed to GBS have decreased tensile strength, elasticity, and work to rupture (63). Peptide fragments released from these membranes suggest the organism is producing one or more proteases that degrade the placental tissue. GBS proteolytic activity has been identified in culture supernatants, but no correlation exists between the proteolytic activity of a given strain and its virulence in a mouse model (10). Recently, a cell-associated collagenase activity of GBS was postulated. Antibodies raised against collagenase from *Clostridium histolyticum* cross-react with cell-associated proteins produced by GBS and inhibit GBS hydrolysis of a synthetic peptide collagen analog (28). Disruption of collagen fibrils could theoretically play a role in GBS penetration of the chorioamnion and premature rupture of membranes. However, when the gene for this enzyme (*pepB*) was cloned, sequenced, and expressed, it was found incapable of solubilizing a film of reconstituted rat tail collagen (36). Rather, it appears to be a zinc metallopeptidase capable of degrading a variety of small bioactive peptides (e.g., bradykinin, neurotensin).

GBS may induce placental membrane rupture indirectly by alteration of host cell processes. For example, the presence of GBS within the lower uterine cavity or cervix activates the maternal decidua cell peroxidase-H_2O_2-halide system, which could promote oxygen radical-induced damage to adjacent fetal membranes (62). Filtered extracts of GBS modify the arachidonic acid metabolism of cultured human amnion cells, favoring production of prostaglandin E_2 (2). High local concentrations of this compound are known to stimulate the onset of normal labor and may also be a mechanism for initiation of premature labor. GBS stimulate macrophage inflammatory protein-1a and interleukin-8 production from human chorion cells; these chemokines are important mediators signaling migration of inflammatory cells and may also contribute to the pathogenesis of infection-associated preterm labor (9).

GBS secrete a protein that degrades hyaluronic acid, an important component of the extracellular matrix in higher organisms. The GBS hyaluronate lyase gene has been cloned and expressed in *E. coli* and shares 50.7% amino acid identity with the pneumococcal hyaluronidase (37). Hyaluronate lyase is expressed in increased levels by type III GBS isolates, and strains obtained from neonates with bloodstream infections produced higher levels of the enzyme than strains from asymptomatically colonized infants or adults with noninvasive disease. The biological role of the hyaluronate lyase in GBS pathogenesis remains uncertain. Theoretically, breakdown of hyaluronic acid in the extracellular matrix could facilitate tissue spread by the organism.

Finally, the CAMP phenomenon refers to synergistic hemolytic zones produced by colonies of GBS streaked adjacent to colonies of *Staphylococcus aureus* on sheep blood agar. GBS CAMP factor is an extracellular protein of 23.5 kDa that further destabilizes and lyses erythrocyte membranes pretreated with beta-toxin, a staphylococcal sphingomyelinase. Human erythrocytes, like sheep erythrocytes, have not been observed to undergo hemolysis with CAMP factor alone. The CAMP factor gene (*cfb*) has been cloned and expressed in *E. coli*, and the recombinant protein elicits antibodies that inhibit the CAMP phenomenon (53). The limited evidence for direct toxicity comes from experiments in which partially purified CAMP factor preparation produces mortality in rabbits when injected intravenously (65). Finally, LTA purified from GBS is cytotoxic for a variety of human cell monolayers in tissue culture, including human embryonic brain cells and human embryonic amnion cells (44). The potential contribution of soluble LTA to GBS-induced tissue injury is uncertain, however, since little LTA is released from the cytoplasmic membrane under normal growth conditions.

AVOIDANCE OF IMMUNE CLEARANCE

Upon penetration of GBS into the lung tissue or bloodstream of the newborn infant, an immunologic response is recruited to clear the organism. The central elements of this response are host phagocytic cells of the neutrophil and, to a lesser extent, monocyte-macrophage cell lines. However, as is the case with most other pathogenic bacteria, effective phagocytosis of GBS by neutrophils and macrophages requires opsonization. Without the participation of specific antibodies and serum complement, phagocytosis of GBS is dramatically reduced. The predilection of certain neonates to suffer invasive GBS disease may thus reflect quantitative or qualitative deficits in (i) phagocytic cell function, (ii) serotype-specific anti-GBS immunoglobulin (Ig), and/or (iii) the classical and alternative complement pathways. In addition to these host factors, GBS possess a number of unique virulence attributes that interfere with effective opsonophagocytosis, chief among them the type-specific polysaccharide capsule.

GBS associated with human disease are almost invariably encapsulated, belonging to one of the nine recognized capsule serotypes: Ia, Ib, and II through VIII. With minor exceptions, the various GBS capsular polysaccharide antigens are composed of the same four component monosaccharides: glucose, galactose, N-acetylglucosamine, and sialic acid (type VI lacks N-acetylglucosamine, and type VIII contains rhamnose). However, serotype-specific epitopes of each polysaccharide are created by differences in the arrangement of component sugars into a unique repeating unit (82). Hyperimmune rabbit antisera directed against a given type-specific polysaccharide antigen provide passive protection to mice from lethal challenge with virulent strains from the homologous but not heterologous serotypes. A low level of human maternal anticapsular IgG is a major risk factor for development of invasive GBS infections in the neonate (1).

The biochemistry and immunology of GBS capsular polysaccharide have been studied most thoroughly in serotype III organisms. The native type III capsular polysaccharide is a high-molecular-weight polymer composed of more than 100 repeating pentasaccharide units. Each pentasaccharide unit contains a trisaccharide backbone of galactose, glucose, and N-acetylglucosamine with a side chain of galactose and a terminal sialic acid moiety (82). Sialic acid is known to be a critical element in the epitope of type III GBS polysaccharide capsule that confers protective immunity. After treatment with sialidase, the altered capsular polysaccharide fails to elicit protective antibodies against GBS infection. Moreover, protective antibodies derived from native type III capsule do not bind to the altered (asialo) capsule backbone structure (31). Human infants who possess antibodies that react only to the desialylated capsule remain at high risk for invasive disease.

A correlation between the sialic acid component of type III GBS capsule and animal virulence was first noted in studies employing chemical modification or spontaneous but genetically uncharacterized mutants. Organisms treated with sialidase are opsonized more effectively by complement through the alternative pathway and are consequently more readily phagocytosed by human neutrophils in vitro (11). Sialidase treatment of type III GBS resulted in diminished lethality of the organism upon intravenous administration to neonatal rats (64). Serial subculture of a wild-type GBS strain in the presence of type III specific antiserum allowed identification of mutants that lacked the terminal sialic acid component of the polysaccharide capsule. These mutants possessed a 1,000-fold greater LD_{50} following tail vein injection in mice (88).

Direct proof for the role of type III GBS capsule in virulence was provided by the construction of isogenic capsule-deficient mutants by means of Tn916 (or Tn916ΔE) mutagenesis (60). Libraries of GBS:Tn916 transconjugates were screened by immunoblot analysis for alterations in capsule expression. Two major types of type III GBS capsule mutants have been identified by this method. The first mutant phenotype completely lacked evidence of capsular material by immune electron microscopy and failed to react with antisera to type III GBS or to type 14 pneumococcus (which recognizes the asialo-core structure of type III capsule) (60, 83). The second mutant phenotype reacts only with pneumococcal type 14 antisera and has been shown by structural carbohydrate chemistry to lack specifically the terminal sialic acid residues of the native type III capsule (81).

Interference of effective C3 deposition by sialylated polysaccharide capsule appears to be an important virulence mechanism of GBS. Situated at the convergence of the classical and alternative complement pathways, deposition of C3 on the bacterial surface, with subsequent cleavage and degradation to opsonically active fragment C3b, is a pivotal element in host defense against invasive infections. C3 deposition and degradation occur on the sur-

face of GBS representing a variety of serotypes. However, the extent of C3 deposition by the alternative pathway is inversely related to the size and density of the polysaccharide capsule present on the surface of type Ib and type III GBS strains (40, 66). Isogenic type III mutant strains expressing a sialic acid-deficient capsule, or lacking capsular polysaccharide altogether, bind 8- to 16-fold greater amounts of C3 than does the parent strain (40). Moreover, C3 fragments bound to the acapsular mutant are predominantly in the active form, C3b, whereas the inactive form, C3bi, is predominantly bound to the surface of the parent strain. In comparison to the parent strains, the isogenic unencapsulated or asialo type III capsule mutants were susceptible to opsonophagocytosis in the presence of complement and peripheral blood neutrophils (40, 60, 83).

The type III GBS capsule mutants were also significantly less virulent in animal models of GBS infection. In a model of neonatal GBS pneumonia and bacteremia, neonatal rats have been inoculated with either the parent strain or an acapsular mutant by intratracheal injection. In animals that received the acapsular mutant, fewer GBS were recovered per gram of lung, more bacteria were associated with resident alveolar macrophages, and the animals became significantly less bacteremic than animals that received the parent strain (41). Subcutaneous injection of the acapsular or asialo mutants in neonatal rats resulted in similar LD_{50} values that were at least 100-fold greater than those obtained with the parent strain (60, 83). Together these data provided compelling evidence that the capsule protects the organism from phagocytic clearance during the initial pulmonary phase and later bacteremic phase of early-onset GBS infection.

The transposon insertion sites for both the asialo- and noncapsular GBS mutants map to the same 20-kb region of type III chromosome (33). Southern blot analyses with genomic DNA samples from other clinical isolates indicated that the entire capsule gene region is highly conserved among several GBS strains representing a variety of serotypes. Nucleotide sequence analysis of this region has revealed several contiguous ORFs that share significant homology to complex capsule and exopolysaccharide synthesis genes of other organisms (Fig. 2). The capsule genes are organized according to their role in synthesis: regulation, chain length, export, repeating unit synthesis, and sialic acid synthesis/activation. This arrangement has also been observed for the type Ia capsule genes, although the entire operon has yet to be completely characterized. The genes involved in sialic acid metabolism (cpsN, O, E, and F) are unique to GBS among gram-positive organisms, sharing similarity and function with sialic acid genes from E. coli K1 (25). Indeed, the low G+C content of the sialic acid genes from E. coli K1 may suggest that the origin of these genes arose from gram-positive bacteria. The other genes share significant homology to polysaccharide genes from type 14 capsule genes from Streptococcus pneumoniae and other organisms (Fig. 2). The common arrangement of the polysaccharide synthesis genes and their significant homology suggest important conservation of the gene structure among gram-positive bacteria that produce complex polysaccharides on their surface.

Two divergently transcribed ORFs, cpsX and cpsY, separated by a common regulatory region, were identified upstream of the cpsA–D genes involved in polysaccharide capsule biosynthesis in GBS (32, 89). These genes are implicated in the regulation of capsule expression in GBS, since the CpsX protein shares sequence similarities with LytR of Bacillus subtilis, an attenuator of transcription,

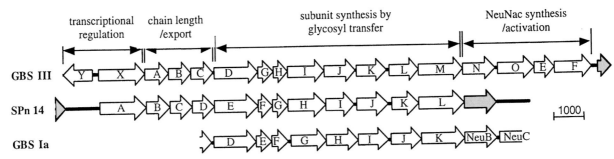

FIGURE 2 Cps gene loci for serotype III GBS (GBS III) compared with *S. pneumoniae* type 14 (Spn 14) and partial operon for GBS serotype Ia (GBS Ia). At the top, loci responsible for specific functions during biosynthesis are indicated, based on homology and genetic organization of complex polysaccharide synthesis loci found in many gram-positive bacteria.

while CpsY has similarity to genes belonging to the LysR family of transcriptional regulators. Genes of the LysR family are known to autoregulate their own expression as well as regulate the expression of divergently transcribed genes. Northern blot and primer extension analysis showed that *cpsC* was transcribed coordinately with the upstream ORFs *cpsB*, *cpsA*, and *cpsX* (89). The nature of the gene products encoded by these genes is unknown, but operon structures identical to *cpsXABC* have recently been identified in the capsule gene clusters of types 19F and 14 pneumococci, *Lactococcus lactis*, and *Streptococcus thermophilus*. Plasmid integrational mutagenesis has been employed to create targeted knockouts of the *cpsX*, *cpsA*, *cpsB*, and *cpsC* ORFs. In each case, the resultant GBS mutant failed to produce any immunologically detectable capsular polysaccharide or its precursors (89). These data have demonstrated that the capsule genes are expressed as a single large transcript starting at a promoter just 5′ of the *cpsX* gene and ending 3′ of the *cpsF* gene.

GBS are known to regulate the degree of capsular polysaccharide expression with cell growth rate (50). Organisms passed on solid growth media become less encapsulated (phase shifting) than when they are passed in animals (23). Mouse passage of various serotypes of GBS is followed by increases in sialylated capsule content that correlate with increased virulence (49). Subpopulations of heavily or poorly encapsulated GBS can be subcultured from strains isolated from infected infants. It is interesting to speculate that GBS may regulate capsule expression in response to the host environment at different stages in the pathogenic process. Unencapsulated mutants are more adherent and significantly more invasive for respiratory epithelium than is the parent strain (27, 70). These findings suggest that the capsule hinders interaction with the epithelial cell surface and/or partially masks important bacterial surface ligands that recognize receptors on the epithelial surface. In mucosal colonization and the early stages of newborn infection, a dynamic balance may exist between producing enough capsule to avoid immune clearance but not so much as to prevent bacterial-epithelial interaction. Once having invaded tissues or circulating in the bloodstream, an up-regulation of GBS capsule expression may then be favored as a means of preventing rapid opsonophagocytic clearance. In vitro, it appears that only brief periods of fast growth are required for up-regulation of *cps* gene-specific mRNAs and expression of high levels of cell-associated capsule (57).

The c antigen of GBS is a protein complex consisting of two distinct components: α, which is sensitive to the protease trypsin, and β, which is trypsin resistant. Overall, one or both components of the c protein are found in approximately 60% of GBS clinical isolates, but notably less than 1% of type III strains. The α component is expressed as a series of proteins that demonstrate a laddering pattern on immunoblots and vary greatly in molecular mass (14 to 145 kDa) among GBS strains (14). Large, identical tandem repeat units make up 74% of the DNA sequence in the α antigen gene and are likely to play a role in generating the size diversity observed in α antigen expression (43). The β component of c protein is typically expressed as a single 130-kDa protein capable of binding human IgA. Determination of the nucleotide sequence of the complete β antigen gene shows that it encodes a polypeptide typical of other gram-positive cell wall proteins (29).

Increasing evidence suggests that c protein may contribute to GBS virulence. Type II strains possessing both the α and β components of c protein produce significantly higher mortality in the infant rat model than type II strains lacking c protein (13). In vitro studies showed that type II strains without c protein were more easily killed than c protein-positive strains in an opsonophagocytic bactericidal assay (51). This limitation in phagocytosis of α- and β-containing type II strains is not due to differential uptake by the neutrophils, since the organisms are efficiently internalized, but rather to an apparent defect in intracellular killing. Protection against phagocytosis may reflect non-immune binding of IgA by the β component, which appears to interfere sterically with deposition of opsonically active complement protein C3 on the GBS surface (29). Inactivation of the α antigen gene (*bca*) by allelic replacement resulted in a five- to seven-fold attenuation in lethality in the mouse model when compared with the isogenic wild-type strain (35). Finally, antibodies to the α component of c protein did not passively protect neonatal mice from lethal challenge with the *bca* gene knockout mutant, suggesting the antigen is a target for protective immunity (35). Deletion in the repeat region of the α component of c protein enhances the pathogenicity of GBS in immune mice by (i) loss of a protective (conformational) epitope(s) and (ii) loss of antibody due to a decrease in antigen size relative to cell wall components and/or capsular polysaccharide (22).

In addition to opsonization, the serum complement system contributes to host defense through generation of soluble chemotactic factors that promote neutrophil mobilization, in particular C5a. A hallmark feature of severe early-onset GBS disease is the poor influx of host neutrophils into sites of tissue infection (58). GBS contribute to poor neutrophil mobilization by production of an enzyme that specifically cleaves and inactivates human C5a (26). GBS inactivate C5a by proteolytic cleavage between histidine-67 and lysine-68 near the C terminus (3). This finding is consistent with the known critical role of the C terminus of C5a in activation of human neutrophils as demonstrated by site-directed mutagenesis.

The normally cell-associated GBS C5a peptidase activity has been purified to apparent homogeneity by chromatographic methods, revealing an enzyme with a molecular mass of approximately 120,000 Da (5). C5a peptidase appears to be a serine esterase based on sensitivity to the inhibitor di-isopropyl fluorophosphate (5). The enzymatic activity of soluble C5a peptidase is completely neutralized by serum from normal human adults, in large part owing to naturally occurring IgG antibodies (26). IgG also neutralizes C5a peptidase on the surface of a capsule-deficient GBS mutant but fails to neutralize the enzyme on the surface of the intact encapsulated type III parent strain (26). This suggests that the GBS capsular polysaccharide serves to protect the cell-associated C5a peptidase from inactivation by naturally occurring antibodies. GBS expression of C5a peptidase reduces neutrophil recruitment to the lungs of C5-deficient mice reconstituted with human C5a (4).

Impairment of respiratory burst function may be an especially significant risk factor for infection with GBS, which is, surprisingly, 10-fold more resistant to killing by oxygen metabolites than is the catalase-producing *S. aureus* (86). The protection of GBS against oxidative damage is correlated in part with a greater endogenous content of the oxygen-metabolite scavenger glutathione. GBS can persist intracellularly in macrophages for 24 to 48 h, in part owing to the ability of the organism to impair host cell protein kinase C-dependent signal transduction required for microbicidal activity (7).

CAMP factor is released from log-phase GBS cultures into the medium and is capable of binding weakly to the Fc region of human IgG and IgM (30). The complete amino acid sequence of CAMP factor has been determined, and partial sequence homology exists with the Fc-binding regions of *S. aureus* protein A (61). In the mouse model, data suggest a role for CAMP factor in resistance to immune clearance. Mice treated with a sublethal dose of GBS but then injected with purified CAMP factor develop septicemia and die (12). However, neither CAMP factor alone nor various peptide fragments of the CAMP factor coadministered with the bacteria result in similar mortality. Interference of normal immune system function by GBS hyaluronate lyase is also possible, since human macrophages, neutrophils, and lymphocytes express the CD44 receptor that binds specifically to hyaluronan. Lymphocyte adherence to endothelial cells is inhibited by bovine testicular and fungal hyaluronidases, and presumably would also be sensitive to the GBS enzyme (54).

INDUCTION OF THE SEPSIS SYNDROME

If failures in epithelial barrier function and immunologic clearance allow GBS to establish bacteremia in the neonate, development of the sepsis syndrome and, in many cases, profound septic shock may be the consequence. Severe early-onset GBS disease is clinically indistinguishable from septic shock associated with gram-negative endotoxemia. Findings include systemic hypotension, persistent

pulmonary hypertension, tissue hypoxemia and acidosis, temperature instability, disseminated intravascular coagulation, neutropenia, and, ultimately, multiple-organ system failure. Because infusion of GBS produces similar pathophysiologic changes in neonatal animal models of sepsis, several investigations have begun to elucidate the patterns in which GBS activate host inflammatory mediators to induce sepsis syndrome and circulatory shock.

Animal models in which GBS are infused intravenously demonstrate a biphasic host inflammatory response (21). The acute phase (<1 h) is manifested by increased pulmonary artery pressure and decreased arterial oxygenation and is associated with a rise in serum levels of thromboxane B_2 (TxB_2), a stable metabolite marker of the pulmonary vasoconstrictor thromboxane A_2 (TxA_2). Pulmonary hypertension and hypoxemia persist through the late phase (2 to 4 h), in which a progressive pattern of systemic hypotension, decreased cardiac output, and metabolic acidosis develops together with hematologic abnormalities and organ system dysfunction. Inflammatory markers of the late phase include increases in serum TxB_2, tumor necrosis factor alpha (TNF-α), and 6-keto-prostaglandin F1α (a stable metabolite of prostacyclin). Experiments employing specific antagonists confirm the importance of these compounds in producing the hemodynamic alterations of GBS sepsis and demonstrate the involvement of additional mediators in the host inflammatory cascade.

As a known stimulator of both the cyclo-oxygenase and lipo-oxygenase pathways, interleukin-1 (IL-1) may occupy a proximal position in the cytokine cascade of septic shock. Treatment with an IL-1 receptor antagonist improves cardiac output and mean arterial pressure and increases the length of survival in piglets receiving a continuous infusion of GBS (77). In mice, GBS induce a Th1-like cytokine response (IL-2, gamma interferon, IL-12) in the absence of cytokines important in B-cell help (IL-4, IL-5, IL-10) (56). This pattern of response may allow GBS to evade antibody production important for clearance. The cytokine IL-12 may also play a particularly important role in systemic GBS infection. IL-12 elevation is seen 12 to 72 h after GBS challenge in the neonatal rat. Pretreatment with a monoclonal antibody against IL-12 results in greater mortality and levels of bacteremia, whereas therapeutic administration of IL-12 results in lower mortality and bloodstream CFU (38).

Some ambiguity exists as to the precise role played by TNF-α in GBS septicemia of the newborn. One can frequently detect TNF-α in the blood, urine, or cerebrospinal fluid of infants with invasive GBS disease (85). Human mixed mononuclear cell cultures exposed to GBS release TNF-α in a dose- and time-dependent manner; moreover, neonatal monocytes exhibit a larger TNF-α response than adult cells (85). GBS infusion in piglets is associated with TNF-α release during the late phase of hemodynamic response, but the TNF-α inhibitor pentoxiphylline has only modest effects on the ongoing pulmonary hypertension, hypoxemia, and systemic hypotension (18). Marked improvement in these hemodynamic parameters is seen when pentoxiphylline treatment is combined with indomethicin inhibition of TxB_2 and prostacyclin synthesis (73). Serum TNF-α levels in the mouse and rat also rise after challenge with GBS; however, administration of polyclonal or monoclonal anti-TNF-α antibody does not affect overall mortality in these models (72, 73).

In contrast to gram-negative pathogens and endotoxin, the specific nature of the GBS component(s) that triggers the host cytokine cascade is less well understood. Cell wall preparations of GBS cause TNF-α release from human monocytes in a manner requiring CD14 and complement receptors type 3 and 4 (42). The group B polysaccharide and peptidoglycan appear to be significantly greater stimulators of TNF-α release from monocytes than LTA or type-specific capsular polysaccharide (76). IL-1 and IL-6 release is also stimulated by soluble GBS cell wall antigens (79).

Studies using isogenic type III GBS mutants lacking polysaccharide capsule have shown that the presence of capsule has no effect on production of TNF-α by human mononuclear cells in vitro (85) and does not change the degree of pulmonary hypertension observed in vivo (19). In contrast, capsule-deficient mutants of a type Ib GBS strain actually produce a greater degree of pulmonary hypertension than the parent strain in the piglet model (52). The latter finding implies that the type-specific Ib capsular polysaccharide may actually cloak the GBS cell wall component responsible for triggering the early phase of the host inflammatory response. This property of GBS capsule may be an important virulence attribute, allowing the organism to multiply and spread beyond a pulmonary focus before adequate host clearance mechanisms are recruited.

Incubation of piglet mesenteric arteries with heat-killed GBS produces a marked hyporesponsiveness to noradrenaline, the endothelial cell-derived vasoconstrictor ET-1, and the synthetic TxA$_2$ analog U46619. This hyporesponsiveness appears to result from enhanced release of nitric oxide (NO), suggesting a role for nitric oxide synthetase (iNOS) induction in the systemic hypotension in GBS sepsis (78). GBS-treated pulmonary arteries also exhibited NO-mediated hyporesponsiveness to noradrenaline and ET-1 but responded normally to U46619 (78). Absence of the TxA$_2$-induced component of NO-mediated pulmonary hyporesponsiveness might help explain the coexistence of pulmonary hypertension with systemic hypotension during GBS sepsis syndrome. Studies employing isogenic HH and NH mutants demonstrate that the GBS β-hemolysin is involved in transcriptional induction of iNOS in murine macrophages in a dose- and time-dependent manner (55). Intravenous challenge of rabbits with HH mutants resulted in significantly higher mortality, higher median NO serum levels, and greater organ injury and disseminated intravascular coagulation when compared with wild-type or NH mutants (55). These findings corroborate those of earlier studies in which administration of filter-purified GBS β-hemolysin extracts to rabbits or rats produced dose-dependent hypotensive changes and a limited number of deaths due to shock, findings not seen with streptolysin S from *Streptococcus pyogenes* (24).

CONCLUSIONS

GBS infection represents a complex interaction between the bacterial pathogen and the susceptible newborn. Human infants exhibit several well-documented deficiencies in host defense mechanisms, creating an environment in which a variety of potential GBS virulence factors are revealed. Chief among these is the sialylated polysaccharide surface capsule, whose role in pathogenesis has been established by the creation of specific isogenic mutants defective in capsule expression. These mutants are more susceptible to opsonophagocytosis and exhibit decreased virulence in animal models. The transposon insertion sites identified a

specific chromosomal locus containing numerous genes involved in capsule biosynthesis. The α component of the c protein antigen has been cloned and sequenced, and targeted mutagenesis demonstrates that the gene product contributes to virulence and is also a target of protective immunity. Another virulence attribute, the C5a peptidase, has been characterized genetically and biochemically and demonstrated to impair neutrophil recruitment both in vitro and in vivo. Studies using isogenic mutants with altered β-hemolysin expression indicate the GBS hemolysin contributes to lung epithelial and endothelial cell injury in early-onset pneumonia and may also activate NO production important in the etiology of hypotensive shock. To date, however, neither the hemolysin molecule nor its structural gene has been positively identified.

The pathogen-host relationship between GBS and human newborns becomes increasingly complex when one considers that bacterial attributes that promote virulence at a given stage in pathogenesis may attenuate virulence at other stages. For example, presence of the antiphagocytic polysaccharide capsule has an inhibitory effect on GBS invasion of host epithelial and endothelial cells. However, some capsule must be synthesized on the mucosal surface to prevent clearance by resident macrophages. Furthermore, by cloaking proinflammatory cell wall components, encapsulated GBS strains are associated with less pulmonary hypertension and meningeal irritation than are unencapsulated mutants. These observations suggest that GBS may need to up- or down-regulate expression of capsular polysaccharide and other potential virulence factors at specific stages of the infectious process.

Recent advances in molecular genetic techniques applicable to gram-positive bacteria should allow (i) more in-depth molecular analyses of these and other potential GBS virulence factors (e.g., hyaluronate lyase, CAMP factor), (ii) identification of novel virulence factors involved in complex pathogenic processes such as host cell adherence and invasion, and (iii) construction of specific GBS mutants for virulence testing in relevant animal models of GBS disease. These findings will promote our understanding of the complex processes involved during maternal and newborn infections with perinatal pathogens. A fundamental understanding of GBS pathogenesis at the molecular level will provide invaluable information toward the rational design of therapeutic and immunoprophylactic strategies to combat this foremost of neonatal infections.

REFERENCES

1. Baker, C. J., and D. L. Kasper. 1976. Correlation of maternal antibody deficiency with susceptibility to neonatal group B streptococcal infection. *N. Engl. J. Med.* **294:** 753–756.

2. Bennett, P. R., M. P. Rose, L. Myatt, and M. G. Elder. 1987. Preterm labor: stimulation of arachidonic acid metabolism in human amnion cells by bacterial products. *Am. J. Obstet. Gynecol.* **156:**649–655.

3. Bohnsack, J. F., K. W. Mollison, A. M. Buko, J. C. Ashworth, and H. R. Hill. 1991. Group B streptococci inactivate complement component C5a by enzymic cleavage at the C-terminus. *Biochem. J.* **273:**635–640.

4. Bohnsack, J. F., K. Widjaja, S. Ghazizadeh, C. E. Rubens, D. R. Hillyard, C. J. Parker, K. H. Albertine, and H. R. Hill. 1997. A role for C5 and C5a-ase in the acute

neutrophil response to group B streptococcal infections. *J. Infect. Dis.* **175**:847–855.

5. **Bohnsack, J. F., X. N. Zhou, P. A. Williams, P. P. Cleary, C. J. Parker, and H. R. Hill.** 1991. Purification of the proteinase from group B streptococci that inactivates human C5a. *Biochim. Biophys. Acta.* **1079**:222–228.

6. **Conrads, G., A. Podbielski, and R. Lutticken.** 1991. Molecular cloning and nucleotide sequence of the group B streptococcal hemolysin. *Zentralbl. Bakteriol.* **275**:179–184.

7. **Cornacchione, P., L. Scaringi, K. Fettucciari, E. Rosati, R. Sabatini, G. Orefici, C. von Hunolstein, A. Modesti, A. Modica, F. Minelli, and P. Marconi.** 1998. Group B streptococci persist inside macrophages. *Immunology* **93**:86–95.

8. **Cox, F., L. Taylor, E. K. Eskew, and S. J. Mattingly.** 1993. Prevention of group B streptococcal colonization and bacteremia in neonatal mice with topical vaginal inhibitors. *J. Infect. Dis.* **167**:1118–1122.

9. **Dudley, D. J., S. S. Edwin, J. Van Wagoner, N. H. Augustine, H. R. Hill, and M. D. Mitchell.** 1997. Regulation of decidual cell chemokine production by group B streptococci and purified bacterial cell wall components. *Am. J. Obstet. Gynecol.* **177**:666–672.

10. **Durham, D. L., S. J. Mattingly, T. I. Doran, T. W. Milligan, and D. C. Straus.** 1981. Correlation between the production of extracellular substances by type III group B streptococcal strains and virulence in a mouse model. *Infect. Immun.* **34**:448–454.

11. **Edwards, M. S., D. L. Kasper, H. J. Jennings, C. J. Baker, and A. Nicholson-Weller.** 1982. Capsular sialic acid prevents activation of the alternative complement pathway by type III, group B streptococci. *J. Immunol.* **128**:1278–1283.

12. **Fehrenbach, F. J., D. Jurgens, J. Ruhlmann, B. Sterzik, and M. Ozel.** 1988. Role of CAMP-factor (protein B) in virulence, p. 351–7. *In* F. J. Feherenbach (ed.), *Bacterial Protein Toxins.* Gustav Fischer, Stuttgart, Germany.

13. **Ferrieri, P.** 1988. Surface-localized protein antigens of group B streptococci. *Rev. Infect. Dis.* **10**(Suppl 2): S363–S366.

14. **Flores, A. E., and P. Ferrieri.** 1996. Molecular diversity among the trypsin resistant surface proteins of group B streptococci. *Zentralbl. Bakteriol.* **285**:44–51.

15. **Galask, R. P., M. W. Varner, C. R. Petzold, and S. L. Wilbur.** 1984. Bacterial attachment to the chorioamniotic membranes. *Am. J. Obstet. Gynecol.* **148**:915–928.

16. **Gibson, R. L., M. K. Lee, C. Soderland, E. Y. Chi, and C. E. Rubens.** 1993. Group B streptococci invade endothelial cells: type III capsular polysaccharide attenuates invasion. *Infect. Immun.* **61**:478–485.

17. **Gibson, R. L., V. Nizet, and C. E. Rubens.** 1999. Group B streptococcal β-hemolysin promotes injury of lung microvascular endothelial cells. *Pediatr. Res.* **45**:626–634.

18. **Gibson, R. L., G. J. Redding, W. R. Henderson, and W. E. Truog.** 1991. Group B streptococcus induces tumor necrosis factor in neonatal piglets. Effect of the tumor necrosis factor inhibitor pentoxifylline on hemodynamics and gas exchange. *Am. Rev. Respir. Dis.* **143**:598–604.

19. **Gibson, R. L., G. J. Redding, W. E. Truog, W. R. Henderson, and C. E. Rubens.** 1989. Isogenic group B streptococci devoid of capsular polysaccharide or beta-hemolysin: pulmonary hemodynamic and gas

exchange effects during bacteremia in piglets. *Pediatr. Res.* **26**:241–245.

20. **Gibson, R. L., C. Soderland, W. R. Henderson, Jr., E. Y. Chi, and C. E. Rubens.** 1995. Group B streptococci (GBS) injure lung endothelium in vitro: GBS invasion and GBS-induced eicosanoid production is greater with microvascular than with pulmonary artery cells. *Infect. Immun.* **63**:271–279.

21. **Gibson, R. L., W. E. Truog, W. R. Henderson, Jr., and G. J. Redding.** 1992. Group B streptococcal sepsis in piglets: effect of combined pentoxifylline and indomethacin pretreatment. *Pediatr. Res.* **31**:222–227.

22. **Gravekamp, C., B. Rosner, and L. C. Madoff.** 1998. Deletion of repeats in the alpha C protein enhances the pathogenicity of group B streptococci in immune mice. *Infect. Immun.* **66**:4347–4354.

23. **Gray, B. M., and D. G. Pritchard.** 1992. Phase variation in the pathogenesis of group B streptococcal infections, p. 452–454. *In* G. Orefici (ed.), *New Perspectives on Streptococci and Streptococcal Infections.* Gustav Fischer Verlag, New York, N.Y.

24. **Griffiths, B. B., and H. Rhee.** 1992. Effects of haemolysins of groups A and B streptococci on cardiovascular system. *Microbios* **69**:17–27.

25. **Haft, R. F., M. R. Wessels, M. F. Mebane, N. Conaty, and C. E. Rubens.** 1996. Characterization of cpsF and its product CMP-N-acetylneuraminic acid synthetase, a group B streptococcal enzyme that can function in K1 capsular polysaccharide biosynthesis in *Escherichia coli. Mol. Microbiol.* **19**:555–563.

26. **Hill, H. R., J. F. Bohnsack, E. Z. Morris, N. H. Augustine, C. J. Parker, P. P. Cleary, and J. T. Wu.** 1988. Group B streptococci inhibit the chemotactic activity of the fifth component of complement. *J. Immunol.* **141**:3551–3556.

27. **Hulse, M. L., S. Smith, E. Y. Chi, A. Pham, and C. E. Rubens.** 1993. Effect of type III group B streptococcal capsular polysaccharide on invasion of respiratory epithelial cells. *Infect. Immun.* **61**:4835–4841.

28. **Jackson, R. J., M. L. Dao, and D. V. Lim.** 1994. Cell-associated collagenolytic activity by group B streptococci. *Infect. Immun.* **62**:5647–5651.

29. **Jerlstrom, P. G., S. R. Talay, P. Valentin-Weigand, K. N. Timmis, and G. S. Chhatwal.** 1996. Identification of an immunoglobulin A binding motif located in the beta-antigen of the c protein complex of group B streptococci. *Infect. Immun.* **64**:2787–2793.

30. **Jurgens, D., B. Sterzik, and F. J. Fehrenbach.** 1987. Unspecific binding of group B streptococcal cocytolysin (CAMP factor) to immunoglobulins and its possible role in pathogenicity. *J. Exp. Med.* **165**:720–732.

31. **Kasper, D. L., C. J. Baker, R. S. Baltimore, J. H. Crabb, G. Schiffman, and H. J. Jennings.** 1979. Immunodeterminant specificity of human immunity to type III group B streptococcus. *J. Exp. Med.* **149**:327–339.

32. **Koskiniemi, S., M. Sellin, and M. Norgren.** 1998. Identification of two genes, cpsX and cpsY, with putative regulatory function on capsule expression in group B streptococci. *FEMS Immunol. Med. Microbiol.* **21**:159–168.

33. **Kuypers, J. M., L. M. Heggen, and C. E. Rubens.** 1989. Molecular analysis of a region of the group B streptococcus chromosome involved in type III capsule expression. *Infect. Immun.* **57**:3058–3065.

34. La Penta, D., P. Framson, V. Nizet, and C. Rubens. 1997. Epithelial cell invasion by group B streptococci is important to virulence. *Adv. Exp. Med. Biol.* **418:**631–634.

35. Li, J., D. L. Kasper, F. M. Ausubel, B. Rosner, and J. L. Michel. 1997. Inactivation of the alpha C protein antigen gene, bca, by a novel shuttle/suicide vector results in attenuation of virulence and immunity in group B Streptococcus. *Proc. Natl. Acad. Sci. USA* **94:**13251–13256.

36. Lin, B., W. F. Averett, J. Novak, W. W. Chatham, S. K. Hollingshead, J. E. Coligan, M. L. Egan, and D. G. Pritchard. 1996. Characterization of PepB, a group B streptococcal oligopeptidase. *Infect. Immun.* **64:**3401–3406.

37. Lin, B., S. K. Hollingshead, J. E. Coligan, M. L. Egan, J. R. Baker, and D. G. Pritchard. 1994. Cloning and expression of the gene for group B streptococcal hyaluronate lyase. *J. Biol. Chem.* **269:**30113–30116.

38. Mancuso, G., V. Cusumano, F. Genovese, M. Gambuzza, C. Beninati, and G. Teti. 1997. Role of interleukin 12 in experimental neonatal sepsis caused by group B streptococci. *Infect. Immun.* **65:**3731–3735.

39. Marchlewicz, B. A., and J. L. Duncan. 1981. Lysis of erythrocytes by a hemolysin produced by a group B *Streptococcus* sp. *Infect. Immun.* **34:**787–794.

40. Marques, M. B., D. L. Kasper, M. K. Pangburn, and M. R. Wessels. 1992. Prevention of C3 deposition by capsular polysaccharide is a virulence mechanism of type III group B streptococci. *Infect. Immun.* **60:**3986–3993.

41. Martin, T. R., J. T. Ruzinski, C. E. Rubens, E. Y. Chi, and C. B. Wilson. 1992. The effect of type-specific polysaccharide capsule on the clearance of group B streptococci from the lungs of infant and adult rats. *J. Infect. Dis.* **165:**306–314.

42. Medvedev, A. E., T. Flo, R. R. Ingalls, D. T. Golenbock, G. Teti, S. N. Vogel, and T. Espevik. 1998. Involvement of CD14 and complement receptors CR3 and CR4 in nuclear factor-kappaB activation and TNF production induced by lipopolysaccharide and group B streptococcal cell walls. *J. Immunol.* **160:**4535–4542.

43. Michel, J. L., L. C. Madoff, K. Olson, D. E. Kling, D. L. Kasper, and F. M. Ausubel. 1992. Large, identical, tandem repeating units in the C protein alpha antigen gene, bca, of group B streptococci. *Proc. Natl. Acad. Sci. USA* **89:**10060–10064.

44. Miyazaki, S., O. Leon, and C. Panos. 1988. Adherence of *Streptococcus agalactiae* to synchronously growing human cell monolayers without lipoteichoic acid involvement. *Infect. Immun.* **56:**505–512.

45. Nealon, T. J., and S. J. Mattingly. 1985. Kinetic and chemical analyses of the biologic significance of lipoteichoic acids in mediating adherence of serotype III group B streptococci. *Infect. Immun.* **50:**107–115.

46. Nizet, V., R. L. Gibson, E. Y. Chi, P. E. Framson, M. Hulse, and C. E. Rubens. 1996. Group B streptococcal beta-hemolysin expression is associated with injury of lung epithelial cells. *Infect. Immun.* **64:**3818–3826.

47. Nizet, V., R. L. Gibson, and C. E. Rubens. 1997. The role of group B streptococci beta-hemolysin expression in newborn lung injury. *Adv. Exp. Med. Biol.* **418:**627–630.

48. Nizet, V., K. S. Kim, M. Stins, M. Jonas, E. Y. Chi, D. Nguyen, and C. E. Rubens. 1997. Invasion of brain microvascular endothelial cells by group B streptococci. *Infect. Immun.* **65:**5074–5081.

49. Orefici, G., S. Recchia, and L. Galante. 1988. Possible virulence marker for *Streptococcus agalactiae* (Lancefield Group B). *Eur. J. Clin. Microbiol. Infect. Dis.* **7:**302–305.

50. Paoletti, L. C., R. A. Ross, and K. D. Johnson. 1996. Cell growth rate regulates expression of group B Streptococcus type III capsular polysaccharide. *Infect. Immun.* **64:**1220–1226.

51. Payne, N. R., Y. K. Kim, and P. Ferrieri. 1987. Effect of differences in antibody and complement requirements on phagocytic uptake and intracellular killing of "c" protein-positive and -negative strains of type II group B streptococci. *Infect. Immun.* **55:**1243–1251.

52. Philips, J. B. D., J. X. Li, B. M. Gray, D. G. Pritchard, and J. R. Oliver. 1992. Role of capsule in pulmonary hypertension induced by group B streptococcus. *Pediatr. Res.* **31**(4 Pt. 1):386–390.

53. Podbielski, A., O. Blankenstein, and R. Lutticken. 1994. Molecular characterization of the cfb gene encoding group B streptococcal CAMP-factor. *Med. Microbiol. Immunol.* **183:**239–256.

54. Pritchard, D. G., B. Lin, T. R. Willingham, and J. R. Baker. 1994. Characterization of the group B streptococcal hyaluronate lyase. *Arch. Biochem. Biophys.* **315:**431–437.

55. Ring, A., V. Nizet, and J. L. Shenep. 1998. Group B streptococcal beta-hemolysin induces NO synthase in macrophages in vitro and multiorgan injury and death in vivo. Presented at "Septic Shock Caused by Gram-Positive Bacteria," Vibo Valentia, Italy.

56. Rosati, E., K. Fettucciari, L. Scaringi, P. Cornacchione, R. Sabatini, L. Mezzasoma, R. Rossi, and P. Marconi. 1998. Cytokine response to group B streptococcus infection in mice. *Scand. J. Immunol.* **47:**314–323.

57. Ross, R. A., H. H. Yim, C. E. Rubens, and L. C. Paoletti. Expression of type III polysaccharide by group B *Streptococcus* during adaptation to changing growth rate environments. Submitted for publication.

58. Rubens, C. E., H. V. Raff, J. C. Jackson, E. Y. Chi, J. T. Bielitzki, and S. L. Hillier. 1991. Pathophysiology and histopathology of group B streptococcal sepsis in *Macaca nemestrina* primates induced after intraamniotic inoculation: evidence for bacterial cellular invasion. *J. Infect. Dis.* **164:**320–330.

59. Rubens, C. E., S. Smith, M. Hulse, E. Y. Chi, and G. van Belle. 1992. Respiratory epithelial cell invasion by group B streptococci. *Infect. Immun.* **60:**5157–5163.

60. Rubens, C. E., M. R. Wessels, L. M. Heggen, and D. L. Kasper. 1987. Transposon mutagenesis of type III group B Streptococcus: correlation of capsule expression with virulence. *Proc. Natl. Acad. Sci. USA* **84:**7208–7212.

61. Ruhlmann, J., B. Wittmann-Liebold, D. Jurgens, and F. J. Fehrenbach. 1988. Complete amino acid sequence of protein B. *FEBS Lett.* **235:**262–266.

62. Sbarra, A. J., G. B. Thomas, C. L. Cetrulo, C. Shakr, A. Chaudhury, and B. Paul. 1987. Effect of bacterial growth on the bursting pressure of fetal membranes in vitro. *Obstet. Gynecol.* **70:**107–110.

63. Schoonmaker, J. N., D. W. Lawellin, B. Lunt, and J. A. McGregor. 1989. Bacteria and inflammatory cells reduce chorioamniotic membrane integrity and tensile strength. *Obstet. Gynecol.* **74:**590–596.

64. Shigeoka, A. O., N. S. Rote, J. I. Santos, and H. R. Hill. 1983. Assessment of the virulence factors of group

B streptococci: correlation with sialic acid content. *J. Infect. Dis.* **147:**857–863.

65. Shyur, S. D., H. V. Raff, J. F. Bohnsack, D. K. Kelsey, and H. R. Hill. 1992. Comparison of the opsonic and complement triggering activity of human monoclonal IgG1 and IgM antibody against group B streptococci. *J. Immunol.* **148:**1879–1884.

66. Smith, C. L., D. G. Pritchard, and B. M. Gray. 1991. Role of polysaccharide capsule in C3 deposition by type Ib group B streptococci (GBS), abstr. 450. *Program Abstr. 31st Intersci. Conf. Antimicrob. Agents Chemother.* American Society for Microbiology, Washington, D.C.

67. Spellerberg, B., B. Pohl, G. Hause, S. Martin, J. Weber-Heynemann, and R. Lütticken. 1999. Identification of genetic determinants for the hemolytic activity of *Streptococcus agalactiae* by ISS1 transposition. *J. Bacteriol.* **181:**3212–3219.

68. Spellerberg, B., E. Rozdzinski, S. Martin, J. Weber-Heynemann, N. Schmitzler, R. Lütticken, and A. Podbielski. 1999. Lmb, a protein with similarities to the Lra1 adhesin family, mediates attachment of *Streptococcus agalactiae* to human laminin. *Infect. Immun.* **67:**871–878.

69. Spellerberg, B., E. Rozdzinski, S. Martin, and R. Lutticken. 1998. Two-component regulatory system controls fibrinogen binding of Group B streptococci. *ASM Conference on Streptococcal Genetics, Vichy, France.* American Society for Microbiology, Washington, D.C.

70. Tamura, G. S., J. M. Kuypers, S. Smith, H. Raff, and C. E. Rubens. 1994. Adherence of group B streptococci to cultured epithelial cells: roles of environmental factors and bacterial surface components. *Infect. Immun.* **62:**2450–2458.

71. Tamura, G. S., and C. E. Rubens. 1995. Group B streptococci adhere to a variant of fibronectin attached to a solid phase. *Mol. Microbiol.* **15:**581–589.

72. Teti, G., G. Mancuso, and F. Tomasello. 1993. Cytokine appearance and effects of anti-tumor necrosis factor alpha antibodies in a neonatal rat model of group B streptococcal infection. *Infect. Immun.* **61:**227–235.

73. Teti, G., G. Mancuso, F. Tomasello, and M. S. Chiofalo. 1992. Production of tumor necrosis factor-alpha and interleukin-6 in mice infected with group B streptococci. *Circ. Shock* **38:**138–144.

74. Valentin-Weigand, P., and G. S. Chhatwal. 1995. Correlation of epithelial cell invasiveness of group B streptococci with clinical source of isolation. *Microb. Pathog.* **19:**83–91.

75. Valentin-Weigand, P., H. Jungnitz, A. Zock, M. Rohde, and G. S. Chhatwal. 1997. Characterization of group B streptococcal invasion in HEp-2 epithelial cells. *FEMS Microbiol. Lett.* **147:**69–74.

76. Vallejo, J. G., C. J. Baker, and M. S. Edwards. 1996. Roles of the bacterial cell wall and capsule in induction of tumor necrosis factor alpha by type III group B streptococci. *Infect. Immun.* **64:**5042–5046.

77. Vallette, J. D., Jr., R. N. Goldberg, C. Suguihara, T. Del Moral, O. Martinez, J. Lin, R. C. Thompson, and E. Bancalari. 1995. Effect of an interleukin-1 receptor antagonist on the hemodynamic manifestations of group B streptococcal sepsis. *Pediatr Res.* **38:**704–708.

78. Villamor, E., F. Perez Vizcaino, J. Tamargo, and M. Moro. 1996. Effects of group B Streptococcus on the responses to U46619, endothelin-1, and noradrenaline in isolated pulmonary and mesenteric arteries of piglets. *Pediatr. Res.* **40:**827–833.

79. von Hunolstein, C., A. Totolian, G. Alfarone, G. Mancuso, V. Cusumano, G. Teti, and G. Orefici. 1997. Soluble antigens from group B streptococci induce cytokine production in human blood cultures. *Infect. Immun.* **65:**4017–4021.

80. Wennerstrom, D. E., J. C. Tsaihong, and J. T. Crawford. 1985. Evaluation of the role of hemolysin and pigment in the pathogenesis of early onset group B streptococcal infection, p. 155–6. *In* Y. Kimura, S. Kotami, and Y. Shiokowa (ed.), *Recent Advances in Streptococci and Streptococcal Diseases.* Reedbooks, Bracknell, U.K.

81. Wessels, M. R., V. J. Benedi, D. L. Kasper, L. M. Heggen, and C. E. Rubens. 1991. The type III capsule and virulence of group B *Streptococcus*, p. 219–23. *In* G. M. Dunny, P. P. Cleary, and L. L. McKay (ed.), *Genetics and Molecular Biology of Streptococci, Lactococci, and Enterococci.* American Society for Microbiology, Washington, D.C.

82. Wessels, M. R., V. Pozsgay, D. L. Kasper, and H. J. Jennings. 1987. Structure and immunochemistry of an oligosaccharide repeating unit of the capsular polysaccharide of type III group B Streptococcus. A revised structure for the type III group B streptococcal polysaccharide antigen. *J. Biol. Chem.* **262:**8262–8267.

83. Wessels, M. R., C. E. Rubens, V. J. Benedi, and D. L. Kasper. 1989. Definition of a bacterial virulence factor: sialylation of the group B streptococcal capsule. *Proc. Natl. Acad. Sci. USA* **86:**8983–8987.

84. Wibawan, I. T., C. Lammler, and F. H. Pasaribu. 1992. Role of hydrophobic surface proteins in mediating adherence of group B streptococci to epithelial cells. *J. Gen. Microbiol.* **138**(Pt 6):1237–1242.

85. Williams, P. A., J. F. Bohnsack, N. H. Augustine, W. K. Drummond, C. E. Rubens, and H. R. Hill. 1993. Production of tumor necrosis factor by human cells in vitro and in vivo, induced by group B streptococci. *J. Pediatr.* **123:**292–300.

86. Wilson, C. B., and W. M. Weaver. 1985. Comparative susceptibility of group B streptococci and *Staphylococcus aureus* to killing by oxygen metabolites. *J. Infect. Dis.* **152:**323–329.

87. Winram, S. B., M. Jonas, E. Chi, and C. E. Rubens. 1998. Characterization of group B streptococcal invasion of human chorion and amnion epithelial cells in vitro. *Infect. Immun.* **66:**4932–4941.

88. Yeung, M. K., and S. J. Mattingly. 1983. Isolation and characterization of type III group B streptococcal mutants defective in biosynthesis of the type-specific antigen. *Infect. Immun.* **42:**141–151.

89. Yim, H. H., A. Nittayarin, and C. E. Rubens. 1997. Analysis of the capsule synthesis locus, a virulence factor in group B streptococci. *Adv. Exp. Med. Biol.* **418:**995–997.

90. Zawaneh, S. M., E. M. Ayoub, H. Baer, A. C. Cruz, and W. N. Spellacy. 1979. Factors influencing adherence of group B streptococci to human vaginal epithelial cells. *Infect. Immun.* **26:**441–447.

Surface Structures of Group B *Streptococcus* Important in Human Immunity

LAWRENCE C. PAOLETTI, LAWRENCE C. MADOFF, AND DENNIS L. KASPER

14

With their perspicacity in interpreting results obtained with a handful of immunological techniques that would be considered insensitive by today's standards, Rebecca Lancefield and coworkers in the 1930s set the stage for decades of research on *Streptococcus agalactiae*, also known as group B *Streptococcus* (GBS). In contrast to group A *Streptococcus*, which was isolated mainly from humans, the first strains of GBS studied by Lancefield were isolated from cows with mastitis or from "normal milk." At that time, GBS was a well-known cause of bovine mastitis and a concern mainly of the dairy industry (51).

The first report of GBS in humans appeared in 1938 (20). However, not until the 1960s, when several reports appeared in the clinical literature on the presence of hemolytic GBS among human newborns with sepsis and meningitis, did this pathogen receive serious attention from researchers (14, 17, 29). By the 1970s, GBS had eclipsed *Escherichia coli* as the most common bacterial cause of sepsis among newborns, with an incidence of early-onset disease of 2.9 per 1,000 births and a case-fatality rate of 46% (2). Despite variability among surveillance sites (reviewed in reference 92), rates of early-onset GBS disease remained high throughout the 1970s and 1980s, prompting the Institute of Medicine to declare in 1985 that the prevention of perinatal GBS disease was a national health priority (30).

Today, colonization with GBS—a risk factor for neonatal illness—has been reported in pregnant and nonpregnant adults and the elderly, not only in the United States (3, 92) but also worldwide (94). This situation accentuates the need to complete vaccine development in this country while considering a global vaccine initiative against this life-threatening pathogen. This chapter highlights critical advances in our understanding of the role(s) of GBS surface antigens (namely, the group B carbohydrate, the type-specific capsular polysaccharides [CPSs], and proteins) in immunity and of their application as components of experimental vaccines.

CARBOHYDRATE ANTIGENS OF GBS

Group B Carbohydrate

The group B carbohydrate is an antigen common to all strains and serotypes of GBS. Positioned proximal to the cell wall (97), the group B carbohydrate is composed of rhamnose, galactose, N-acetylglucosamine, and glucitol. These sugars form four different oligosaccharide units (76) linked by phosphodiester bonds to create a complex and highly branched tetra-antennary structure (Fig. 1). The terminal position and abundance of rhamnose suggested that this sugar would constitute or be part of an immunodominant epitope; this hypothesis was proved by inhibition studies using a number of derivative oligosaccharides with polyclonal and monoclonal antibodies to the group B carbohydrate (74).

Lancefield et al. (55) were the first to demonstrate the inability of antibody to the group B carbohydrate to protect mice from lethal challenge with viable GBS. Indeed, the presence of maternal antibody to the group B carbohydrate was associated with poor neonatal outcome despite good agreement between levels of specific IgG in maternal and cord sera (1). This result confirmed in humans what was already known in animals: that the group B carbohydrate was not important to natural immunity.

Still, investigators wondered whether a group B carbohydrate vaccine would be effective in eliciting functional antibody if the carbohydrate's immunogenicity were enhanced. Rainard (85) demonstrated an improved immune response in cows when the group B carbohydrate was coupled to ovalbumin. Antibody induced to this conjugate vaccine was of the immunoglobulin G (IgG) class, was specific for the major sugars of the group B carbohydrate, and was opsonically active against an unencapsulated GBS strain of bovine origin (85). That high-titered IgG specific for the group B carbohydrate was less active against highly encapsulated GBS strains of human origin was discerned in studies of a group B carbohydrate-tetanus toxoid conjugate vaccine (70). Adsorption studies using rabbit anti-

Gram-Positive Pathogens, ed. by V. A. Fischetti et al.
© 2000 American Society for Microbiology, Washington, D.C.

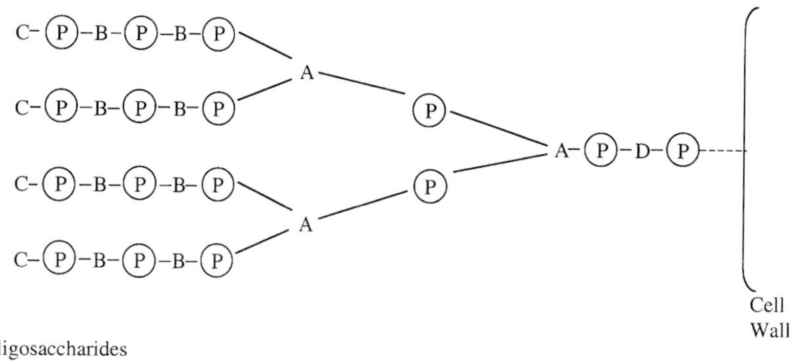

Oligosaccharides

A: α-L-Rha$p$-(1→3)- α-D-Gal$p$-(1→3)-ß-D-Glc$p$ NAc-(1→4)- α-L-Rha$p$-(1→2)-
[α-L-Rha$p$ (1→3) α-D- Gal$p$-(1→3)-ß-D-Glc$p$ NAc-(1→4)-]- α-L-Rha$p$-(1→2)-
α-L-Rha$p$-(1→1')-D Glucitol-(3'-1)- α-L-Rha$p$.

B: α-L-Rha$p$-(1→2)-[α-L-Rha$p$-(1→3) α-D- Gal$p$-(1→3)-ß-D-Glc$p$ NAc-(1→4)-]-
α-L-Rha$p$ (1→2)- α-L-Rha$p$-(1→1')-D-Glucitol-(3'-1)- α-L-Rha$p$.

C: α-L-Rha$p$-(1→2)- α-L-Rha$p$-(1→2)- α-L-Rha$p$-(1→1')-D-Glucitol-(3'-1)- α-L-
Rha$p$.

D: α-L-Rha$p$-(1→3)- α-D-Gal$p$-(1→3)-ß-D-Glc$p$ NAc-(1→3)- α-L-Rha$p$-(1→3)- α-
L-Rha$p$ (1→3)- α-L-Rha$p$-(1→3)-ß-L-Rha$p$-(1→4)-D-GlcNAc.

Ⓟ = phosphate

FIGURE 1 Proposed structure of the group B carbohydrate antigen, as modified from reference 74. The dashed line indicates that the linkage between the group B carbohydrate and the cell wall has not been discerned. Ⓟ, phosphate.

sera to this vaccine with GBS strains of human origin that were highly encapsulated or with an isogenic mutant strain that lacked CPS showed CPS interference with specific antibody binding (70); this finding confirmed the inability of group B-specific antibody to bind to the group B antigen on highly encapsulated GBS cells. Not surprisingly, antibody to bovine isolates that possess low levels of CPS may be effective in combating GBS disease in the cow, whereas the same antibody would not be effective against more highly encapsulated GBS of human origin. The ability of group B antibody to bind to poorly encapsulated GBS deserves further attention, especially since the expression of CPS by a GBS strain (of human origin) has been shown to be regulated by the cell's rate of growth (82), whereas the expression of the group B carbohydrate is not influenced by growth rate (87a).

Chemistry of Group B Streptococcal CPSs

With few exceptions, all strains of GBS isolated from humans are encapsulated (Fig. 2A) and can be classified on the basis of serology and CPS structure. Nine distinct GBS serotypes have thus far been identified: Ia, Ib, II, III, IV, V, VI, VII, and VIII. In the past, serotypes Ia, Ib, II, and III were equally prevalent in normal vaginal carriage and early-onset sepsis (i.e., that developing at less than 7 days

of age) (3). However, type V strains have recently emerged as an important cause of GBS infection (87), and strains of types VI and VIII have become prevalent among Japanese women (47). The chair models in Fig. 3 highlight the heterogeneity among the structures of GBS CPSs—this despite the fact that each repeating unit contains four of only five possible sugars (glucose, galactose, N-acetylglucosamine, rhamnose, and sialic acid). Below is a description of the arrangement of the saccharides in the repeating unit of each CPS.

Types Ia and Ib

Lancefield (52) was the first to notice serological cross-reactions between strains bearing Ia and Ib antigens, although she did not know what cellular structure contained the differing epitopes. Nuclear magnetic resonance analysis, performed 45 years after Lancefield's observation, showed that the native CPSs of GBS types Ia and Ib indeed are structurally similar, differing only in the linkage of the side chain galactose to N-acetylglucosamine (31); the type Ia CPS has a β-(1→4) linkage, and the type Ib CPS has a β-(1→3) linkage in this position. These linkages are critical to the immunospecificity of these CPSs. The repeating unit of Ia and Ib CPSs is a pentasaccharide with a disaccharide backbone and a trisaccharide side chain. Like all

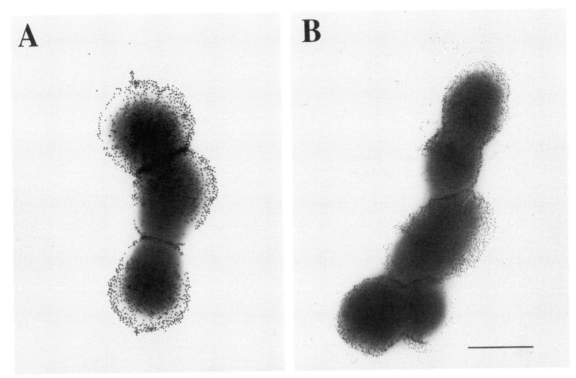

FIGURE 2 Immunogold-labeled electron micrographs showing surface localization of type Ia CPS (A) and the alpha C protein (B) of GBS strain A909. For CPS type Ia staining, rabbit antiserum to the CPS, followed by 20-nm-diameter-gold-labeled protein A, were used. For alpha C protein, rabbit antiserum to the purified one-repeat alpha C protein followed by 15-nm-diameter-gold-labeled protein A were used. Bar, 500 nm. Reprinted from *Infection and Immunity* (24), with permission.

GBS CPSs, each possesses an α-$(2{\rightarrow}3)$-linked sialic acid as a terminal side-chain saccharide.

Type II

The native type II CPS repeating unit structure is composed of galactose, glucose, N-acetylglucosamine, and sialic acid in a 3:2:1:1 molar ratio (34). This repeating unit structure has two side chains—a sialic acid residue and a galactose residue—linked separately to the pentasaccharide backbone.

Type III

The type III CPS has a trisaccharide backbone of glucose, N-acetylglucosamine, and galactose and a disaccharide side chain of sialic acid and galactose linked β-$(1{\rightarrow}4)$ to the backbone N-acetylglucosamine (101). Although sialic acid per se is not an immunodominant saccharide, the negatively charged carboxylate group on this sugar influences the conformation and thus the immunodeterminant epitopes of the entire polymer (33, 102). Removal of the terminal sialic acid from the repeating unit of the type III CPS with mild acid or neuraminidase results in a CPS with chemical and immunological identity to the pneumococcal type 14 CPS (39)—a finding that has enhanced our understanding of the roles of sialic acid in GBS immunity, as discussed further below.

Type IV

The GBS type IV CPS contains galactose, glucose, N-acetylglucosamine, and sialic acid in a 2:2:1:1 molar ratio

(99). The presence of sialic acid does not appear to be a critical immunodeterminant for type IV CPS, as it is for other GBS CPSs (99).

Type V

The repeating unit structure of serotype V CPS is composed of seven sugars. The CPS has a trisaccharide backbone consisting of glucose, galactose, and glucose, as well as two side chains. One side chain consists of glucose linked β-$(1{\rightarrow}3)$ to the backbone galactose. The other side chain is a trisaccharide of N-acetylglucosamine, galactose, and sialic acid. This side chain is linked β-$(1{\rightarrow}6)$ to a backbone glucose and terminates with sialic acid (100).

Type VI

The type VI CPS repeating unit comprises galactose, glucose, and sialic acid in a 2:2:1 molar ratio (44a, 96). Type VI was the first GBS CPS described that lacked N-acetylglucosamine.

Type VII

The GBS type VII repeating unit contains a trisaccharide backbone and a trisaccharide side chain with a molar ratio of 2:2:1:1 for galactose, glucose, N-acetylglucosamine, and sialic acid (43).

Type VIII

Recently elucidated, the structure of GBS CPS type VIII is the most unusual and simplest structure of all GBS serotypes. Like type VI CPS, the type VIII CPS lacks N-

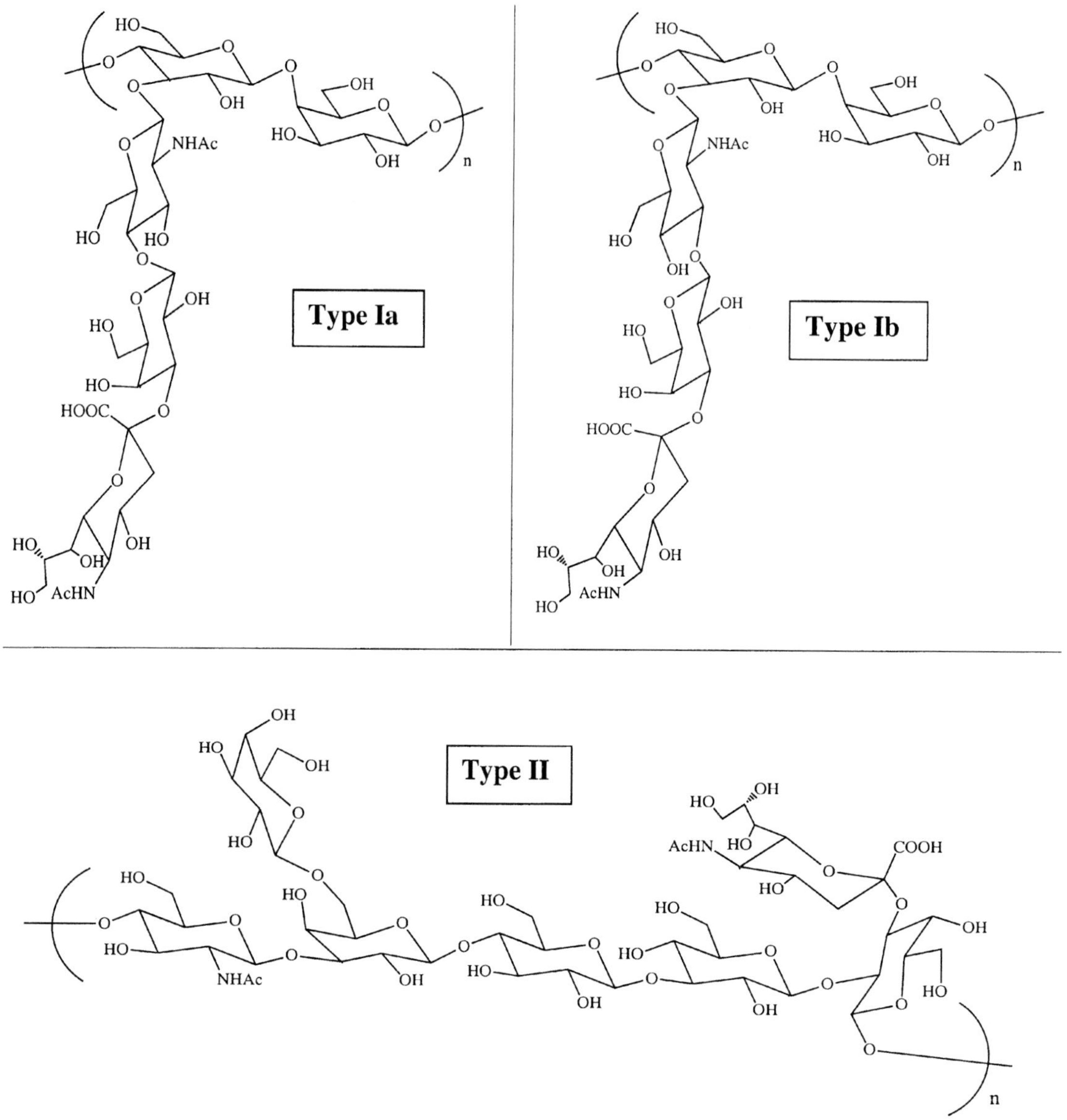

FIGURE 3 Chair models of the repeating units of GBS capsular polysaccharides.

acetylglucosamine; yet unlike any other GBS CPS this tetrasaccharide contains rhamnose, a sugar previously associated exclusively with the group B carbohydrate. The type VIII repeating unit contains glucose, galactose, rhamnose, and sialic acid in 1:1:1:1 molar ratio, with the sialic acid as the monosaccharide side chain (44).

Role of Sialic Acid as a Structural Feature and a Virulence Factor

Hot HCl or cold trichloroacetic acid (TCA) extraction methods used by Lancefield and Freimer (54) resulted in GBS antigens that were incomplete immunochemically. These investigators surmised that the missing component contributed to the serological specificity and negative charge of the CPS antigen. That GBS type-specific CPSs contain sialic acid was cited as a personal communication to Lancefield by Liu in 1972—a discovery that verified sialic acid indeed "represents the missing antigenic component of the carbohydrate in the TCA antigen" (53). The

acidic nature of the TCA extraction procedure resulted in the purification of an incomplete antigen because the α-(2→3) ketosidic linkage between sialic acid and galactose—a bond that is universal among all GBS type-specific CPSs—is acid labile.

Structural analysis of GBS type III CPS showed that the desialylated or core structure was identical to that of *Streptococcus pneumoniae* type 14 CPS. This serendipitous observation helped to define the importance of sialic acid on type III CPS in the formation of a conformationally dependent epitope (101, 102). Specifically, carboxylate groups on sialic acid residues form intramolecular hydrogen bonds with the galactose residues in the backbone of the type III CPS (33). Reduction of the carboxylate groups to hydroxymethyl groups destroyed the CPS antigenicity with antisera to the native type III CPS, while oxidation of the carboxylate groups with sodium *meta*-periodate to remove carbons 8 and 9 did not affect antigenicity. The finding that GBS CPS retains its antigenicity upon oxidation with

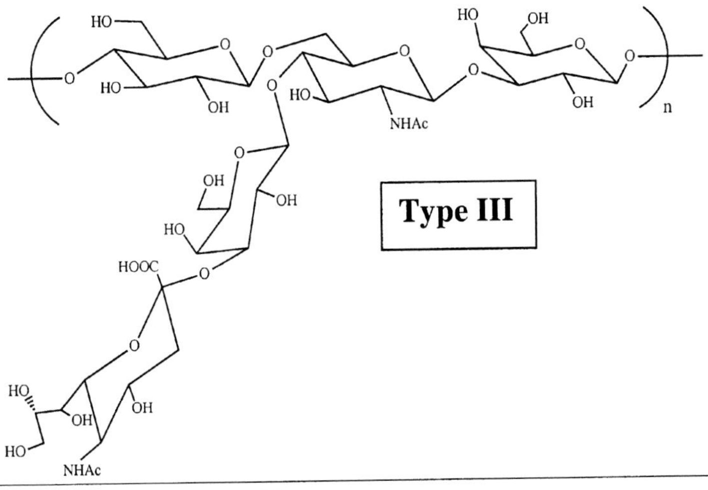

FIGURE 3 (Continued)

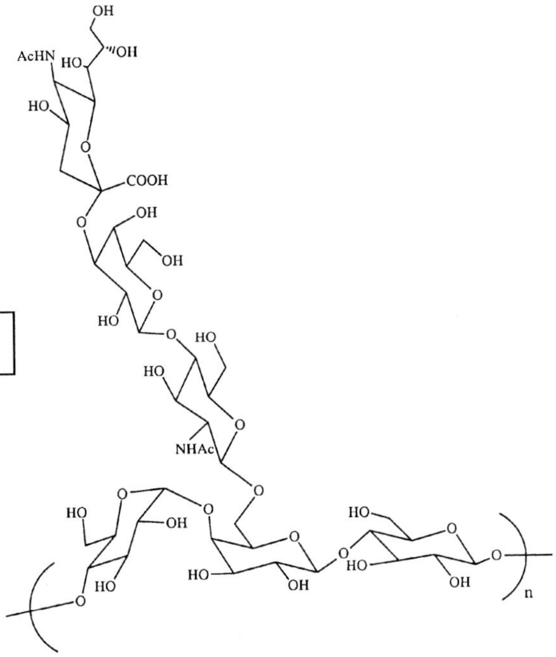

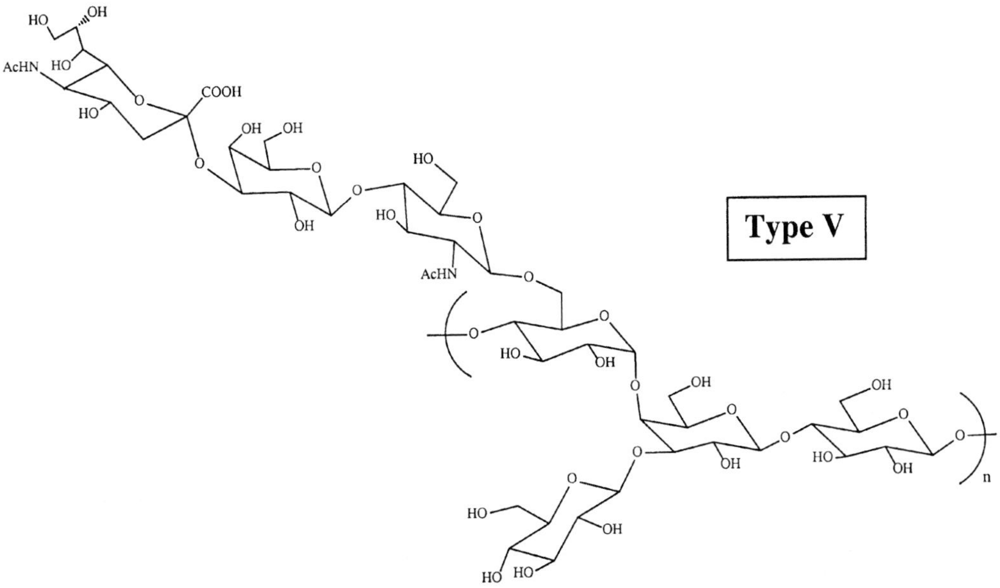

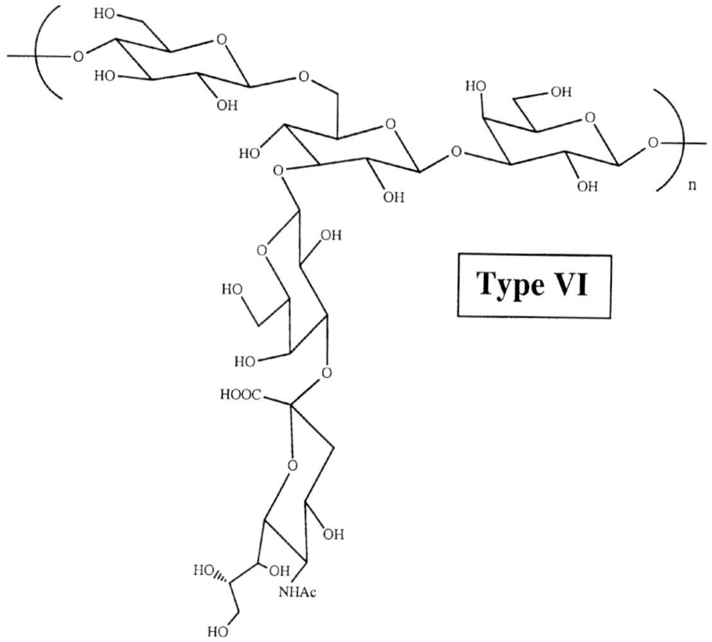

Type VI

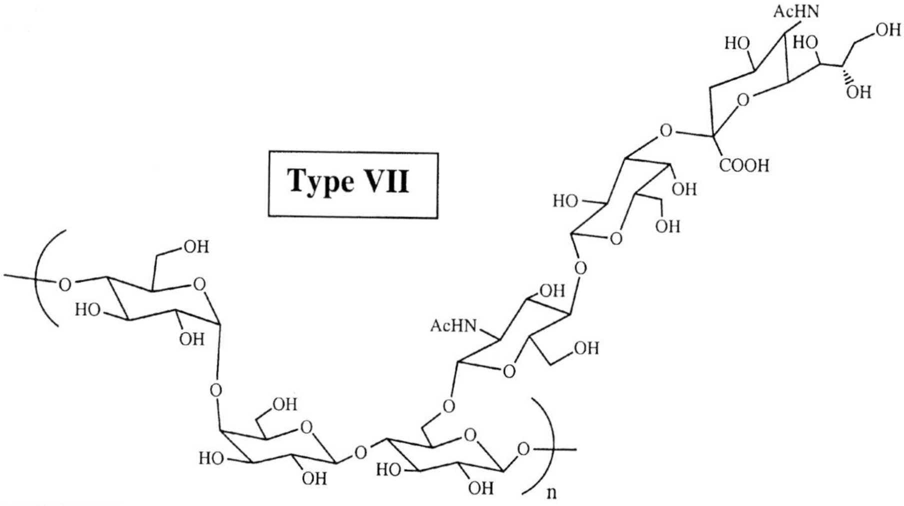

Type VII

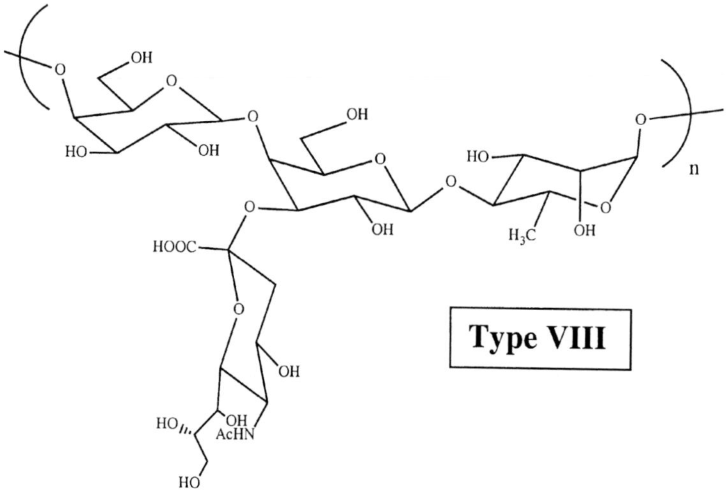

Type VIII

FIGURE 3   (Continued)

sodium periodate became important in the design of a coupling strategy to produce GBS conjugate vaccines. The requirement for sialic acid to impart antigenicity on GBS CPSs is not universal, however. Whereas the sialic acid residue on type Ia CPS is required for complete antigenicity, the binding of specific antisera to the structurally related type Ib does not depend on sialic acid (91a).

The presence of sialic acid on GBS creates a surface that does not activate the alternative pathway of complement (16). In contrast, GBS organisms grown in the presence of neuraminidase—a condition that results in a lower degree of surface sialylation—support complement activation in the presence or absence of specific antibody. Marques et al. (69) reported lower amounts of C3 deposition on GBS organisms that possessed a sialylated as opposed to a desialylated CPS. In all, these findings suggested a critical role for sialic acid in evading the host's natural immune mechanisms and thus in modulating GBS virulence.

The role of the complement pathway in the immune response to sialylated antigen (GBS type III CPS) and desialylated antigen (S. pneumoniae type 14 CPS) was examined in untreated mice and in mice treated with cobra venom factor to cause complement depletion (68). Complement depletion did not affect the immune response to the sialylated antigen, but this condition markedly reduced the specific antibody response to desialylated CPS. These results demonstrated the importance of the CPS structure in complement activation and in the generation of a specific antibody response (68).

The role of sialic acid as a virulence factor of GBS was confirmed in studies with a transposon mutant of type III strain COH 31r/s that lacked the ability to synthesize sialylated CPS (106). GBS strain COH 31-21 produced an asialo type III CPS that bound to pneumococcal type 14 CPS but not native type III CPS antisera. The virulence of the asialo mutant was 2 orders of magnitude lower than that of the wild-type parent strain and equal to that of another transposon mutant, COH 31-15, that completely lacked CPS (106).

Genetics of CPS Antigens

The transposon mutants described above were the first used to map the genes important in GBS CPS expression (46). Seven genes have since been identified (cpsA through cpsG) in type III GBS, with functions assigned or suggested by sequence homology (88). Three specific regions (88) designate genes involved in monosaccharide synthesis and activation (region I), oligosaccharide subunit synthesis (region II), and transport and polymerization of oligosaccharide subunits (region III), respectively. The product of cpsF, located in region I, has been determined to be a CMP-sialic acid synthetase, an enzyme that functions in the activation of sialic acid during the formation of type III CPS (25). The gene cpsD, located in region II, encodes a galactosyl transferase (89), and cpsA, cpsB, and cpsC are proposed to be associated with oligosaccharide polymerization. Recently, cpsX and cpsY have been identified upstream from cpsA and found to share sequence homologies with genes involved in CPS regulation in other gram-positive bacteria (45). Indeed, that 10 additional open reading frames in the CPS gene regions of GBS have been identified (109) underscores the complex process of CPS biosynthesis.

PROTEIN ANTIGENS OF GBS

While the importance of surface proteins in immunity to Streptococcus pyogenes has long been appreciated, their role in immunity to GBS was recognized only in the 1970s, when Lancefield et al. (55) described the cross-serotype protection afforded by rabbit antiserum to protease-sensitive epitopes. The intervening years have witnessed considerable research in this area. This work has led to a new appreciation of the diversity of GBS surface proteins and an increased awareness of their roles in immunity and pathogenicity. It is now apparent that most GBS strains express antigenic surface proteins (Table 1). Much of the interest in these proteins arises from a desire to identify protective antigens for inclusion in GBS vaccines.

Tandem Repeat-Containing Proteins of GBS

Alpha C Protein

Wilkinson and Eagon (108) described a protein extract obtained from a serotype Ic strain of GBS. On the basis of this description, Lancefield recognized that the previously designated serotype Ic in fact comprised the Ia polysaccharide and the Ibc protein. In seminal experiments, Lancefield and coworkers (55) demonstrated that antibodies to the protein component protected mice from lethal challenge with either Ia or Ib capsular serotype GBS strains that bore the Ibc proteins. Even in their initial work, Wilkinson and Eagon (108) recognized that the protein extract from the Ic GBS strain contained two moieties, one sensitive to digestion with trypsin, the other resistant. Bevanger and Maeland (12) described the trypsin-resistant component as the alpha antigen of the Ibc protein and the trypsin-sensitive component as the beta antigen. The term C proteins was coined to clarify the distinction between the proteins and the carbohydrate antigens; the presence of the C protein(s) was denoted by the letter following the capsular serotype (e.g., Ia/c) (28). Later work by a number of investigators identified the alpha and beta components as distinct proteins, independently expressed in different strains, and encoded by separate genes (62, 65, 72, 90, 91). Generally present in strains of Ia and Ib capsular serotype and frequently present in type II strains, these antigens are occasionally found in GBS of all serotypes. The trypsin-resistant antigen is now referred to as the alpha C protein (Fig. 2B) and the trypsin-sensitive antigen as the beta C protein. In one survey, these proteins were found in 59% of strains in a collection of 785 clinical isolates from the University of Minnesota (38). All Ib strains expressed C proteins, and 84% expressed both alpha and beta; 91% of Ia strains and 81% of type II strains expressed the alpha C protein. In a similar survey, Bevanger (10) found that 52% of human isolates of GBS expressed the alpha C protein and 25% expressed the beta C protein.

Monoclonal antibodies have been used to identify the alpha C protein and to define it as a mouse-protective epitope in GBS (65). In Western blot analysis, monoclonal antibody reacted with the vast majority of strains of GBS type Ia (all of 14 strains), type Ib (11 of 15 strains), and type II (all of 12 strains). Western blots of surface extracts from GBS strains also demonstrated an unusual phenotypic characteristic of this protein: a pattern of regularly spaced bands with a periodicity of approximately 8 kDa. Also evident on Western blots was size heterogeneity of the expressed protein from strain to strain, with the size of the largest band varying from <70 kDa to >160 kDa in a panel of 37 isolates. An interesting observation was that the protein sizes did not appear to be randomly distributed but clustered around 100 kDa, with relatively few isolates expressing either a very large or a very small alpha C protein.

TABLE 1 Characteristics of group B streptococcal protein antigens

Designation	Trypsin susceptibility	Laddering pattern	Antisera protective	Gene	Associated GBS serotype(s)
C proteins					
Alpha	Resistant	Yes	Yes	bca	Ia, Ib, II
Epsilon variant	?	Yes	?		Ia
Beta	Sensitive	No	Yes	bac	Ib
Rib protein	Resistant	Yes	Yes	rib	III
R proteins (R1, R2, R3, R4)	Resistant	?	Yes-type II No-type III	?	II, III
Type V alphalike protein	Resistant	Yes	Yes	?	V

?, unknown.

The gene encoding the alpha C protein was cloned from the prototype Ia/c strain A909 from the Lancefield collection (73). This gene, *bca*, consists of 3,063 nucleotides and encodes several identifiable regions. A signal sequence with homology to other gram-positive surface proteins was 56 amino acids in length (93). This sequence was followed by an amino-terminal domain of 170 amino acids with a predicted molecular mass of 18.8 kDa. The most striking feature of the gene was a series of nine long tandem repeats that had 246 nucleotides each (encoding 82 amino acids, or 7.9 kDa) and that were identical at the nucleotide level. The repeat region composed >75% of the mature protein, and the correspondence between the size of the repeat region and the distance between the bands on the Western blot suggested that the repeats gave rise to the ladder pattern seen on Western blots. Antibodies to both the repeat region and the amino-terminal domain have been shown to be protective in an animal model and to mediate opsonophagocytic killing in vitro (42). The carboxyl terminus showed homology to those of other gram-positive surface proteins, including an LPXTG domain (identified as a peptidoglycan-binding motif), a hydrophobic membrane-spanning region, and a charged tail thought to be involved in orienting and anchoring the protein to the cell wall (73).

A size discrepancy between the alpha C proteins of GBS isolates from mothers and those from their neonates suggested that a loss of repeats within the protein gene may occur during colonization or infection with GBS (63). This phenomenon was confirmed experimentally in an immune mouse model. While spontaneous deletions within the alpha C protein occur at a low rate (approximately 10^{-4}), passage through mice given antiserum to the alpha C protein gave rise to deletion rates approaching 50%. Moreover, the mutants bound poorly to the antiserum and were more poorly opsonized for phagocytic killing than the wild-type strain. In vitro experiments showed that purified alpha C proteins containing nine or 16 repeats bound with high affinity to antibodies to wild-type (nine-repeat) alpha C protein, whereas one- and two-repeat alpha C proteins bound with lower overall affinity (22). These differences were not due simply to a reduced number of antibody-binding sites but appeared to reflect a conformational dependence of the protein on repeat number. In contrast, antibodies to the one-repeat protein appeared to bind to all of the different constructs with similar overall affinity. Antibodies to the single- or double-repeat proteins showed greater protective efficacy in vivo and tended to allow fewer deletion mutants to escape. Further studies demonstrated that immunogenicity in mice was inversely related to repeat number (23): nine- and 16-repeat alpha C proteins evoked lower titers of antibody, particularly to the amino-terminal domain, than did the one- and two-repeat variants. These findings suggested that modulation of repeat number may be a mechanism for interaction with the host's immune system.

While many surface proteins, particularly those containing repeating regions, have been shown to interact with human host substances, no such binding has been defined for the alpha C protein. However, a genetically constructed null mutant of the alpha C protein in GBS strain A909 exhibited virulence that was modestly attenuated (by <1 $\log_{10}$) relative to that of the wild type in neonatal mice (58). The decreased mortality evident in the first 24 h after infection suggested a role in the early stages of pathogenesis. A single-repeat alpha C protein mutant in the same strain, derived by passage through immune mice, showed equivalent pathogenicity in naive mice but displayed enhanced lethality (>2 $\log_{10}$) in mice first immunized against the protein (24).

Rib Protein

While the alpha C protein is present mainly in GBS strains of types Ia, Ib, and II, a phenotypically similar but immunologically distinct protein, designated Rib, has been discovered in a type III GBS strain (93). This protein, present in 31 of 33 clinical type III isolates studied, showed a laddering phenotype on sodium dodecyl sulfate-polyacrylamide gel electrophoresis (SDS-PAGE), resistance to trypsin digestion, and variability in molecular size among clinical isolates. Antiserum to the protein protected between 62% and 90% of mice from challenge with GBS but did not cross-protect the animals against infection by alpha-positive GBS. Amino-terminal sequencing of the protein demonstrated homology to the alpha C protein with identity in 6 of 12 residues. When cloned and sequenced, the gene encoding this protein showed structural as well as sequence similarity to the alpha C protein (98). Like *bca*, the *rib* gene contains a signal sequence, an amino-terminal region, a series of tandem repeats, and a carboxy-terminal membrane anchor region (Fig. 4). Thus, a family of genetically similar, tandem repeat-containing proteins was identified in GBS.

Type V Alphalike Protein

During the 1980s, the previously obscure GBS serotype V emerged as an important pathogen in the United States. A novel protein identified in many type V strains was phenotypically similar to the alpha C and Rib proteins (48). Antiserum to the purified protein recognized a homologous protein (of variable size) in 25 of 41 clinical isolates. This antiserum protected 78% of neonatal mice from GBS challenge. Thus, the protein exhibited the characteristic features of this family of proteins: (i) protective efficacy against infection with a homologous GBS strain in a mouse model; (ii) laddering phenotype on SDS-PAGE with variable molecular size among clinical isolates; and (iii) resistance to trypsin digestion. In addition, this protein displayed amino acid sequence similarity both to the alpha C protein (17 of 18 evaluable residues) and to the Rib protein (5 of 10 residues). The gene encoding this protein was cloned and expressed in *E. coli* (49).

R Proteins

R proteins, also identified by their resistance to trypsin digestion, were initially recognized in *S. pyogenes* by Lancefield and Perlmann (56). Wilkinson (107) identified four immunologically distinct R proteins (R1, R2, R3, and R4) occurring in various combinations in streptococci of groups A, B, and C. Flores and Ferrieri (19) found R proteins of subtypes R1 and R4 in 37% (49 of 131) of GBS clinical isolates. The proteins were present predominantly in type II and III strains. Linden (60) showed that rabbit antiserum to an affinity-purified R protein from type III GBS protected mice against invasive infection by type II R-positive strains but not against that by type III R-positive strains. The presence of R protein antibodies in human serum correlated with the absence of invasive infection by R protein-expressing GBS strains in human neonates (61).

To date, no R protein has been well characterized either biochemically or genetically. However, the R proteins appear to share two important phenotypic characteristics with the members of the tandem repeat-containing family of GBS surface proteins: trypsin resistance and protective capacity. Bevanger et al. (11) isolated a protein from a serotype III strain, designated it an R-associated protein, and found that it was biochemically and immunologically highly similar to both R4 and Rib. Similarly, Lachenauer and Madoff (48) showed that the alphalike protein from a GBS type V strain was reactive with antiserum to R1. Further characterization of the R proteins will probably reveal that they are members of the same family as the other tandem repeat-containing GBS proteins and that the genes encoding them possess a similar structure.

Beta C Protein

The beta C protein was initially recognized as the trypsin-susceptible component of the C protein (108). Subsequent work has demonstrated its expression on the surface of approximately 10 to 25% of GBS strains, with predominance in Ib strains (10, 38, 62). Russell-Jones et al. (91) recognized that this antigen was a 130-kDa protein and that it bound specifically to human IgA. This binding occurs via the Fc portion of the IgA heavy chain and is of high affinity, with an affinity constant of 3.5×10^8 M^{-1} (59). The affinity is considerably higher for serum than for secretory IgA. The gene for the beta C protein was cloned by several groups (15, 27, 36, 72), and antibody to the cloned gene product was demonstrated to have protective capacity against infection in mice (72). Two complete nucleotide sequences for the gene (*bac*) have been published and differ minimally; one is 1,134 amino acids in length (27) and the other 1,164 amino acids (36). Other than a single amino acid difference, the two sequences differ only in the number of short XPZ repeats found in the carboxy-terminal half of the protein. Both contain consensus gram-positive signal sequences and carboxy-terminal membrane anchor regions that have extensive homology to other gram-positive surface proteins (including the GBS tandem-repeating proteins discussed above). Both groups of researchers studying the beta C protein detected IgA-binding capacity in the amino-terminal half of the molecule; however, Jerlstrom et al. (36) detected IgA binding within two separate nonoverlapping subclones, whereas Heden et al. (27) detected only a single IgA-binding region. Jerlstrom and colleagues (37) later localized IgA binding to a 73-amino-acid region containing a MLKKIE sequence and found that this sequence was required for IgA binding. These observations suggest that only one IgA-binding region exists (as is the case with other immunoglobulin-binding proteins) and that Jerlstrom's earlier finding may have resulted from the use of polyclonal IgA, which reacted in an immune fashion with a portion of the protein not involved in IgA binding.

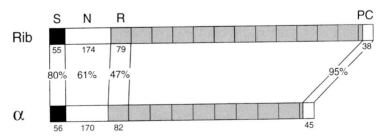

FIGURE 4 Comparison of the Rib and alpha C proteins. Overall structure of Rib from strain BM110 and alpha C protein from strain A909 and degree of amino acid residue identity between different regions of the proteins. S, signal peptide; N, NH₂-terminal region; R, one repeat; P, partial repeat; C, COOH-terminal region. The number of amino acids in each region is indicated. The Rib protein has 12 repeats of 79 amino acids, and the alpha C protein has nine repeats of 82 amino acids. Reprinted from the *Journal of Biological Chemistry* (98), with permission.

Antibodies elicited to the beta C protein appear to be highly protective (21, 64, 72). Immunization of mice with this protein protected their neonatal pups from invasive GBS infection at >90% efficacy, and antibodies facilitated bacterial opsonization for phagocytic killing by polymorphonuclear leukocytes (64). One group has found that antibodies to the beta C protein—unlike those to any other GBS antigen—are actually bactericidal in vitro, i.e., are able to kill GBS in the absence of polymorphonuclear leukocytes (21). The beta C protein has also served as an effective carrier protein in conjugate vaccines. A glycoconjugate containing equal quantities of beta C protein and type III CPS linked by reductive amination was broadly protective against invasive GBS infection in neonatal mice challenged with either type III GBS or a type Ia/C beta-positive strain (67). Rabbit antiserum to the conjugate vaccine was opsonic against multiple strains containing either type III CPS or beta-C protein. Similar conjugates with beta C protein have also been made with types Ia, II, and III CPSs and have elicited antibody titers comparable to those elicited by CPS-tetanus toxoid (TT) conjugates (75).

Other Surface Proteins

Several other surface proteins have been described in GBS. Brady et al. (13) recognized gamma and delta C proteins when polyclonal antiserum to the prototype C-protein strain A909 was absorbed with another alpha C protein-positive strain. Residual activity against surface proteins was recognized in several strains. Gamma protein reactivity was found only in alpha-positive strains. Michel and others (71) identified a protein in some alpha-positive GBS strains that reacted with monoclonal antibody to alpha C protein but varied in restriction digest pattern at the genetic level and in morphology on SDS-PAGE. Partial nucleotide sequencing of the gene encoding this alpha C protein variant showed sequence divergence from *bca* in the amino-terminal domain, and the variant was designated epsilon. It appears likely that the gamma protein reactivity identified by Brady and colleagues was, in fact, due to differences between alpha and epsilon and was seen when antiserum raised to the prototype alpha C protein was absorbed with an epsilon-positive GBS strain (12a). The reactivity was probably a marker for antigenic epitopes present in the original alpha C protein but not in the epsilon variant. The delta C protein detected in many type III GBS isolates has not been further characterized.

Many bovine GBS isolates express X protein. This protein has not been well characterized; however, in one study, an immunoblot of the protein showed multiple laddering bands that suggested homology with the tandem repeat-containing proteins (57). Immunization with an X protein elicited opsonic antibodies in cows (86).

The glutamine synthetase of GBS, a 52-kDa protein that complemented this enzymatic function in *E. coli*, was found in various serotypes of GBS (95). Evidence suggesting surface expression of this protein included the finding of amino-terminal signal sequence and carboxy-terminal membrane anchor homology with other surface proteins and the presence of antibodies to the protein in serum raised to whole bacteria. Pancholi and Fischetti (77) described a 45-kDa plasminogen-binding protein—SEN, or streptococcal surface enolase—on the surface of many streptococcal species, including GBS, and hypothesized a role for this protein in virulence. It has not been determined whether either of these proteins elicits protective antibody.

GBS VACCINES
CPS Vaccines

Clinical data and experimental observations support an important role for CPS-specific antibodies in the prevention or control of GBS disease. These data include the following: (i) GBS produces type-specific CPSs that are targets of protective immunity (39, 52). (ii) Low-level type-specific maternal antibody is a risk factor for disease in infants; conversely, high specific maternal IgG is associated with protection in the neonate (6). (iii) Type-specific antibodies are opsonically active in vitro (9) and are protective in animal models of infection (67). (iv) IgG to the type-specific CPSs can cross the placenta (8), a result that is critical to the concept of a maternal vaccine against neonatal GBS disease (4).

Native GBS types Ia, Ib, II, and III CPSs have been tested as vaccines in adults. The first clinical trial with GBS vaccine enrolled 33 healthy adults to evaluate the safety and immunogenicity of type III CPS extracted by two methods (4). Although both preparations were well tolerated, the EDTA extraction method produced a type III CPS that was more immunogenic than that obtained by TCA extraction. By the 1980s, more than 300 healthy adults, including 40 third-trimester pregnant women, were safely vaccinated with type III CPS or other GBS CPS vaccines at CPS doses ranging from 10 to 150 μg (5, 8). Although well tolerated, these antigens failed to induce a strong specific antibody response in the target population, i.e., subjects with low levels of preexisting CPS-specific antibody. However, type-specific antibody levels among responders (i.e., mainly those with preexisting specific antibody levels >2 μg/ml) increased sharply 4 weeks after CPS vaccination. High levels of CPS-specific antibody persisted for at least 2 years, and declining levels were restored to their peak upon revaccination with CPS. The class of human antibody among responders to GBS type III CPS (5) and to GBS type II CPS (18) was predominantly IgG. The response to CPS antigens differed with the serotype tested; native type II CPS was the most immunogenic (88% rate of response) and type Ia CPS the least (40%) (5). GBS CPS vaccines induced antibodies that were active against homologous GBS serotypes in opsonophagocytic assays and were passively protective in animal studies (5). By the late 1980s, preclinical and clinical results showed that, although the CPS structures were critical vaccine components, they were not sufficiently immunogenic by themselves to be effective GBS vaccines. Researchers then began to focus on the development of GBS CPS vaccines whose immunogenicity was enhanced by methods that preserved their native antigenic structures.

Animal models are useful and often vital in the development of new vaccines. A mouse model of maternal vaccination-neonatal GBS disease (Fig. 5) has been used extensively to evaluate the efficacy of actively administered GBS vaccines or of passively administered type-specific immune sera (67). This model, which focuses primarily on neonatal survival, has also been adapted to test the therapeutic potential of human antisera to GBS vaccines (81).

Polysaccharide-Protein Conjugate Vaccines

Conjugation technology has yielded an abundance of materials for use as reagents in experimental immunology or as potential vaccines. Without doubt, the clinical success of conjugate vaccines against *Haemophilus influenzae* type b

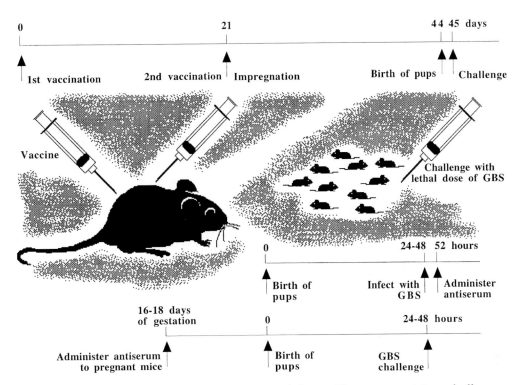

FIGURE 5 Mouse models of group B streptococcal disease. The immunogenicity and efficacy of GBS vaccines has been evaluated with the maternal vaccination-neonatal mouse model by actively vaccinating female mice, mating these mice, and subsequently challenging their offspring with GBS (top timeline). In a passive vaccination-protection model (bottom timeline), newborn pups born to dams that received immune sera during pregnancy are challenged with GBS as a means of measuring the functional capacity of IgG. The therapeutic potential of GBS vaccine-induced serum (middle timeline) has been ascertained by inoculating naive pups with GBS 4 h before administration of immune sera. In all three models, pups are infected or challenged with GBS within 24 to 48 h of birth, and survival is assessed 48 h after challenge.

infections spurred interest in adapting this technology to improve the immunogenicity of other poorly immunogenic carbohydrates, including those of GBS.

The first GBS CPS conjugate vaccines were prepared in 1990 by three different coupling strategies (Table 2). Native type III CPS was coupled to tetanus toxoid (III-TT) using adipic acid dihydrazide as a spacer molecule (50) or directly via aldehydes formed on a selected number of sialic acid residues (103). In a different approach, coupling of a type III oligosaccharide of 14 pentasaccharide repeating units to TT using a synthetic 6-C spacer molecule resulted in a single-site attachment of the reducing end of the oligosaccharide (79). The chemical methods used to create these and other bacterial conjugate vaccines have been reviewed in detail by Jennings and Sood (35). Despite differences in coupling chemistries, CPS size, purification methods, and the adjuvant used in the development and administration of these vaccines, all three conjugates were better immunogens in laboratory animals than was uncoupled type III CPS. Moreover, antibodies elicited by these GBS conjugate vaccines were of the IgG class and were functionally active when tested in vitro and in vivo.

With the coupling chemistry well in hand, progress was then made in developing conjugate vaccines against GBS serotypes other than type III (83, 105). Immunity in mice to multiple GBS serotypes was demonstrated in studies of a mixture of Ia-TT, Ib-TT, II-TT, and III-TT conjugates

administered as a single tetravalent vaccine (84). The lack of interference of one serotype with another in the tetravalent vaccine opened the way to the development of a multivalent GBS vaccine for use in humans. However, clinical observations of an increased prevalence of serotype V dictated its inclusion in a multivalent vaccine (26). Preclinical studies of a serotype V-TT conjugate vaccine have paved the way for this innovation (104).

To broaden coverage with a single vaccine construct, the beta C protein of GBS (discussed in detail above) has been used both as a carrier for type III CPS and as an immunogen in a conjugate vaccine eliciting protective antibodies to GBS strains of serotype III as well as to nontype III GBS strains that contain this protein (67). Although the results with this vaccine proved that broader coverage could be attained with a relevant carrier protein, the alpha C protein of GBS (see above) would be a better carrier for GBS CPSs since it is present on a larger number of GBS serotypes (66). A GBS type III-alpha C protein conjugate vaccine could potentially provide coverage against up to 90% of disease-causing strains of GBS.

Because protective antibodies to type III GBS CPS appear to recognize a conformational epitope, all conjugate vaccines for clinical use were prepared using a coupling method known to preserve both the native repeating unit structure and the conformation of the CPS (33). This conjugation method (32) is essentially a two-step process: (i)

TABLE 2 Characteristics of group B streptococcal conjugate vaccines

Capsular serotype(s)	Carrier protein(s)	Spacer	Coupling Method	Linkage	Study	Reference
III	TT	ADH[a]	CR[b]	Multisite, random	Preclinical	50
III oligosaccharide[c]	TT[d]	6-C[e]	RA[f]	Single, end-linked	Preclinical	79
III	TT	None	RA	Multisite, direct[g]	Preclinical	103
III oligosaccharide[h]	TT	6-C	RA	Single, end-linked	Preclinical	80
II	TT	None	RA	Multisite, direct	Preclinical	83
Ia	TT	None	RA	Multisite, direct	Preclinical	105
III	β-C[i]	None	RA	Multisite, direct	Preclinical	67
III	α-C[j]	None	RA	Multisite, direct	Preclinical	66
Ib	TT	None	RA	Multisite, direct	Preclinical	84
V	TT	None	RA	Multisite, direct	Preclinical	104
III	TT	None	RA	Multisite, direct	Clinical	40
Ia, Ib	TT	None	RA	Multisite, direct	Preclinical/clinical	7
Ia, II, III	TT and β-C	None	RA	Multisite, direct	Preclinical	75
II	TT	None	RA	Multisite, direct	Clinical	78
V	TT	None	RA	Multisite, direct	Clinical	78
V	CRM_{197}[k]	None	RA	Multisite, direct	Clinical	78

[a] Adipic acid dihydrazide.
[b] Carbodiimide reduction.
[c] 14 pentasaccharide repeating units.
[d] Monomeric tetanus toxoid.
[e] 6-Aminohexyl-1-β-D-galactopyranoside.
[f] Reductive amination using sodium cyanoborohydride.
[g] Via aldehydes formed on a selected number of sialic acid residues.
[h] 6 and 25 pentasaccharide repeating units.
[i] GBS beta C protein.
[j] GBS alpha C protein.
[k] Cross-reactive material, a diphtheria mutant toxoid isolated from *Clostridium diphtheriae*.

GBS CPS is subjected to mild oxidation with sodium periodate, which creates free aldehyde groups on a limited number of sialic acid residues on the CPS. (ii) The newly created aldehyde groups on the CPS serve as sites for irreversible coupling to free amino groups present on carrier proteins by a process called reductive amination. The end result is a conjugate vaccine that contains CPS covalently linked directly to the protein. The requirement for a covalent linkage between the CPS and carrier protein was necessary to the formation of efficacious GBS vaccines (Table 3).

TABLE 3 Effect of covalent attachment of polysaccharide to protein on the efficacy in mice of group B streptococcal polysaccharide conjugate vaccines[a]

Vaccine	No. of pups surviving/ no. challenged	Percent survival
III-TT	27/29	93
III+TT	0/33	0
III CPS	0/22	0
TT	0/10	0

[a] Female outbred mice received a priming dose of vaccine emulsified with complete Freund's adjuvant and were mated 2 weeks later. A booster dose of vaccine was administered with incomplete Freund's adjuvant 3 weeks after the priming dose. Newborn pups (< 36 h old) were challenged with an ordinarily lethal dose of GBS type III strain M781, and survival was assessed 48 h later. All groups received 1 µg per dose of type III capsular polysaccharide, either uncoupled (III CPS) or covalently coupled to tetanus toxoid (III-TT) or admixed with tetanus toxoid (III+TT).

Conjugate vaccines for use in phase 1 and phase 2 clinical trials have been individually prepared by the reductive amination method with GBS CPSs Ia, Ib, II, III, and V. Each CPS was individually coupled to TT, and a second type V conjugate vaccine was prepared with the mutant diphtheria toxoid CRM_{197} (78). Four uncoupled GBS CPS preparations (types Ia, Ib, II, and III) were used as control vaccines in these clinical trials.

Clinical Trials with III-TT Vaccine

The initial clinical trials with the second-generation (i.e., conjugate) GBS vaccines were designed to compare the safety and immunogenicity of conjugated and uncoupled type III CPS and to determine an optimal immunogenic dose (40). Both III-TT (at any CPS dose: 58, 14.5, 3.6 µg) and III CPS were well tolerated, with minimal reactogenicity reported among the 90 female recipients. The geometric mean concentration (GMC) of III CPS-specific IgG among the 30 recipients of the 58-µg dose (as CPS) of III-TT rose from 0.09 µg/ml before immunization to 4.89 µg/ml 2 weeks thereafter; during this same interval, the GMC of specific IgG among the 30 recipients of III CPS increased from 0.21 µg/ml to only 1.30 µg/ml ($P < 0.05$). The relative immunogenicity of III-TT appeared to be dose dependent, with a peak GMC of type-specific IgG of 2.72 µg/ml in the 14.5-µg group and 1.10 µg/ml in the 3.6-µg group (40). Moreover, antibodies elicited by immunization with III-TT recognized a conformationally dependent epitope of the CPS, promoted in vitro opsonophagocytosis and killing of GBS, and were functional in vivo in a neonatal mouse protection model (40).

Clinical Trials with Ia-TT and Ib-TT Vaccines

The Ia-TT and Ib-TT conjugate vaccines were also well tolerated and were more immunogenic than uncoupled homologous CPS in nonpregnant, healthy women (7). The GMC of type Ia-specific IgG among recipients of a 60-μg dose (as CPS) of Ia-TT rose from 0.5 μg/ml to a peak of 26.2 μg/ml 8 weeks after vaccination, whereas during the same interval antibody levels among recipients of Ia CPS rose from 0.2 to only 2.4 μg/ml. Levels of type Ia CPS-specific IgG remained high in recipients of the Ia-TT vaccine at 1 and 2 years after vaccination, a result demonstrating the durability of the response. The response to Ia-TT vaccines was dose dependent: 60-μg and 15-μg doses elicited higher antibody levels than did a 3.75-μg dose (7).

Like the other GBS conjugate vaccines, Ib-TT vaccine was safer and more immunogenic than uncoupled homologous CPS (7). The prevaccination GMC of type-specific antibody among recipients of Ib-TT was 0.4 μg/ml, a level that rose to a peak of 14.2 μg/ml 4 weeks after vaccination with a single 63-μg dose (as CPS). In contrast, among recipients of uncoupled type Ib CPS, the prevaccination type-specific antibody GMC of 0.4 μg/ml rose to only 4.4 μg/ml 4 weeks after vaccination. The response to Ib-TT vaccine was dose dependent, with the highest (63-μg) and intermediate (15.75-μg) doses resulting in the greatest magnitude of antibody response (7).

After the success of the Ia-TT, Ib-TT, and III-TT clinical trials, conjugate vaccines for types II and V—the other two GBS serotypes responsible for most disease in the United States—were prepared. These vaccines are currently in clinical trials (78). Preliminary results show a better IgG response to CPS after vaccination with the conjugate than after vaccination with homologous CPS.

SUMMARY

The seminal findings by Rebecca Lancefield and coworkers on the role of carbohydrate and protein antigens in GBS immunity have led a generation of researchers in immunity not only to a better understanding of these antigens in immunity but also toward the development of effective vaccines. Today, GBS remains a major bacterial cause of disease among neonates and a common cause of bovine mastitis (41). This organism is increasingly common in the nonpregnant population in the United States (26), especially among the elderly (recently reviewed in reference 92). However, as the twentieth century ends, we have the knowledge necessary to produce and formulate effective vaccines against GBS disease.

We thank Pamela McInnes at the National Institutes of Health for her unwavering support in the development of group B streptococcal vaccines. We also thank Ying Wang for assistance with the chair models of the capsular polysaccharides, Claudia Gravekamp for providing electron micrographs of GBS, Thomas DiCesare for illustration of the mouse model, and Julie McCoy for editorial assistance. L.C.P. is a recipient of the Milton Fund Award. Most of the GBS research conducted at the Channing Laboratory and cited in this review has been funded through the National Institute of Allergy and Infectious Diseases of the NIH.

REFERENCES

1. **Anthony, B. F., N. F. Concepcion, and K. F. Concepcion.** 1985. Human antibody to the group-specific polysaccharide of group B *Streptococcus. J. Infect. Dis.* **151:**221–226.

2. **Baker, C. J., and F. F. Barrett.** 1973. Transmission of group B streptococci among parturient women and their neonates. *J. Pediatr.* **83:**919–925.

3. **Baker, C. J., and M. S. Edwards.** 1995. Group B streptococcal infections, p. 980–1054. *In* J. S. Remington and J. O. Klein (ed.), *Infectious Diseases of the Fetus and Newborn Infant,* 4th ed. The W. B. Saunders Co., Philadelphia, Pa.

4. **Baker, C. J., M. S. Edwards, and D. L. Kasper.** 1978. Immunogenicity of polysaccharides from type III group B *Streptococcus. J. Clin. Invest.* **61:**1107–1110.

5. **Baker, C. J., and D. L. Kasper.** 1985. Group B streptococcal vaccines. *Rev. Infect. Dis.* **7:**458–467.

6. **Baker, C. J., D. L. Kasper, I. B. Tager, A. Paredes, S. Alpert, W. M. McCormack, and D. Goroff.** 1977. Quantitative determination of antibody to capsular polysaccharide in infection with type III strains of group B *Streptococcus. J. Clin. Invest.* **59:**810–818.

7. **Baker, C. J., L. C. Paoletti, M. R. Wessels, H.-K. Guttormsen, M. A. Rench, M. E. Hickman, and D. L. Kasper.** 1999. Safety and immunogenicity of capsular polysaccharide-tetanus toxoid conjugate vaccines for group B streptococcal types Ia and Ib. *J. Infect. Dis.* **179:**142–150.

8. **Baker, C. J., M. A. Rench, M. S. Edwards, R. J. Carpenter, B. M. Hays, and D. L. Kasper.** 1988. Immunization of pregnant women with a polysaccharide vaccine of group B *Streptococcus. N. Engl. J. Med.* **319:**1180–1220.

9. **Baltimore, R. S., D. L. Kasper, C. J. Baker, and D. K. Goroff.** 1977. Antigenic specificity of opsonophagocytic antibodies in rabbit anti-sera to group B streptococci. *J. Immunol.* **118:**673–678.

10. **Bevanger, L.** 1983. Ibc proteins as serotype markers of group B streptococci. *Acta Pathol. Microbiol. Immunol. Scand. Sect. B* **91:**231–234.

11. **Bevanger, L., A. I. Kvam, and J. A. Maeland.** 1995. A *Streptococcus agalactiae* R protein analysed by polyclonal and monoclonal antibodies. *APMIS* **103:**731–736.

12. **Bevanger, L., and J. A. Maeland.** 1979. Complete and incomplete Ibc protein fraction in group B streptococci. *Acta Pathol. Microbiol. Immunol. Scand. Sect. B* **87:**51–54.

12a.**Brady, L. J.** Personal communication and unpublished data.

13. **Brady, L. J., U. D. Daphtary, E. M. Ayoub, and M. D. P. Boyle.** 1988. Two novel antigens associated with group B streptococci identified by a rapid two-stage radioimmunoassay. *Infect. Immun.* **158:**965–973.

14. **Butter, M. N. W., and C. E. de Moor.** 1967. *Streptococcus agalactiae* as a cause of meningitis in the newborn, and of bacteraemia in adults. Differentiation of human and animal varieties. *J. Microbiol. Serol.* **33:**439–450.

15. **Cleat, P. H., and K. N. Timmis.** 1987. Cloning and expression in *Escherichia coli* of the Ibc protein genes of group B streptococci: binding of human immunoglobulin A to the beta antigen. *Infect. Immun.* **55:**1151–1155.

16. **Edwards, M. S., W. A. Nicholson, C. J. Baker, and D. L. Kasper.** 1980. The role of specific antibody in alternative complement pathway-mediated opsonophagocytosis of type III, group B *Streptococcus. J. Exp. Med.* **151:**1275–1287.

17. **Eickhoff, T. C., J. O. Klein, A. K. Daly, D. Ingall, and M. Finland.** 1964. Neonatal sepsis and other infections

due to group B beta-hemolytic streptococci. *N. Engl. J. Med.* **271:**1221–1228.

18. **Eisenstein, T. K., B. J. De Cueninck, D. Resavy, G. D. Shockman, R. B. Carey, and R. M. Swenson.** 1983. Quantitative determination in human sera of vaccine-induced antibody to type-specific polysaccharides of group B streptococci using an enzyme-linked immunosorbent assay. *J. Infect. Dis.* **147:**847–856.

19. **Flores, A. E., and P. Ferrieri.** 1989. Molecular species of R-protein antigens produced by clinical isolates of group B streptococci. *J. Clin. Microbiol.* **27:**1050–1054.

20. **Fry, R. M.** 1938. Fatal infections by haemolytic streptococcus group B. *Lancet* **i:**199–201.

21. **Fusco, P. C., J. W. Perry, S. M. Liang, M. S. Blake, F. Michon, and J. Y. Tai.** 1997. Bactericidal activity elicited by the beta C protein of group B streptococci contrasted with capsular polysaccharides. *Adv. Exp. Med. Biol.* **418:** 841–845.

22. **Gravekamp, C., D. S. Horensky, J. L. Michel, and L. C. Madoff.** 1996. Variation in repeat number within the alpha C protein of group B streptococci alters antigenicity and protective epitopes. *Infect. Immun.* **64:** 3576–3583.

23. **Gravekamp, C., D. L. Kasper, J. L. Michel, D. E. Kling, V. Carey, and L. C. Madoff.** 1997. Immunogenicity and protective efficacy of the alpha C protein of group B streptococci are inversely related to the number of repeats. *Infect. Immun.* **65:**5216–5221.

24. **Gravekamp, C., B. Rosner, and L. C. Madoff.** 1998. Deletion of repeats in the alpha C protein enhances the pathogenicity of group B streptococci in immune mice. *Infect. Immun.* **66:**4347–4354.

25. **Haft, R. F., M. R. Wessels, M. F. Mebane, N. Conaty, and C. E. Rubens.** 1996. Characterization of *cpsF* and its product CMP-*N*-acetylneuraminic acid synthetase, a group B streptococcal enzyme that can function in K1 capsular polysaccharide biosynthesis in *Escherichia coli*. *Mol. Microbiol.* **19:**555–563.

26. **Harrison, L. H., J. A. Elliott, D. M. Dwyer, J. P. Libonati, P. Ferrieri, L. Billmann, and A. Schuchat.** 1998. Serotype distribution of invasive group B streptococcal isolates in Maryland: implications for vaccine formulation. *J. Infect. Dis.* **177:**998–1002.

27. **Heden, L. O., E. Frithz, and G. Lindahl.** 1991. Molecular characterization of an IgA receptor from group B streptococci: sequence of the gene, identification of a proline-rich region with unique structure and isolation of N-terminal fragments with IgA-binding capacity. *Eur. J. Immunol.* **21:**1481–1490.

28. **Henricksen, J. P., J. Ferrieri, J. Jelinkova, W. Koehler, and W. R. Maxted.** 1984. Nomenclature of antigens of group B streptococci. *Int. J. Syst. Bacteriol.* **34:**500.

29. **Hood, M., A. Janney, and G. Dameron.** 1961. Beta hemolytic *Streptococcus* group B associated with problems of the perinatal period. *Am. J. Obstet. Gynecol.* **82:**809–818.

30. **Institutes of Medicine, Division of Health Promotion and Disease Prevention.** 1985. Appendix P: prospects for immunizing against *Streptococcus* group B, p. 424–438. *In Diseases of Importance in the United States*, vol. I. *New Vaccine Development: Establishing Priorities*. National Academy Press, Washington, D.C.

31. **Jennings, H. J., E. Katzenellenbogen, C. Lugowski, and D. L. Kasper.** 1983. Structure of native polysaccharide antigens of type Ia and type Ib group B *Streptococcus*. *Biochemistry* **22:**1258–1264.

32. **Jennings, H. J., C. Lugowski, and F. E. Ashton.** 1984. Conjugation of meningococcal lipopolysaccharide R-type oligosaccharides to tetanus toxoid as a route to a potential vaccine against group B *Neisseria meningitidis*. *Infect. Immun.* **43:**407–412.

33. **Jennings, H. J., C. Lugowski, and D. L. Kasper.** 1981. Conformational aspects critical to the immunospecificity of the type III group B streptococcal polysaccharide. *Biochemistry* **20:**4511–4518.

34. **Jennings, H. J., K. G. Rosell, E. Katzenellenbogen, and D. L. Kasper.** 1983. Structural determination of the capsular polysaccharide antigen of type II group B *Streptococcus*. *J. Biol. Chem.* **258:**1793–1798.

35. **Jennings, H. J., and R. K. Sood.** 1994. Synthetic glycoconjugates as human vaccines, p. 325–361. *In* Y. C. Lee and R. T. Lee (ed.), *Neoglycoconjugates: Preparation and Applications*. Academic Press, Inc., New York, N.Y.

36. **Jerlstrom, P. G., G. S. Chhatwal, and K. N. Timmis.** 1991. The IgA-binding beta antigen of the C protein complex of group B streptococci: sequence determination of its gene and detection of two binding regions. *Mol. Microbiol.* **5:**843–849.

37. **Jerlstrom, P. G., S. R. Talay, P. Valentin-Weigand, K. N. Timmis, and G. S. Chhatwal.** 1996. Identification of an immunoglobulin A binding motif located in the beta-antigen of the c protein complex of group B streptococci. *Infect. Immun.* **64:**2787–2793.

38. **Johnson, D. R., and P. Ferrieri.** 1984. Group B streptococcal Ibc protein antigen: distribution of two determinants in wild-type strains of common serotypes. *J. Clin. Microbiol.* **19:**506–510.

39. **Kasper, D. L., C. J. Baker, R. S. Baltimore, J. H. Crabb, G. Schiffman, and H. J. Jennings.** 1979. Immunodeterminant specificity of human immunity to type III group B *Streptococcus*. *J. Exp. Med.* **149:**327–339.

40. **Kasper, D. L., L. C. Paoletti, M. R. Wessels, H. K. Guttormsen, V. J. Carey, H. J. Jennings, and C. J. Baker.** 1996. Immune response to type III group B streptococcal polysaccharide-tetanus toxoid conjugate vaccine. *J. Clin. Invest.* **98:**2308–2314.

41. **Keefe, G. P.** 1997. *Streptococcus agalactiae* mastitis: a review. *Can. Vet. J.* **38:**429–437.

42. **Kling, D. E., C. Gravekamp, L. C. Madoff, and J. L. Michel.** 1997. Characterization of two distinct opsonic and protective epitopes within the alpha C protein of the group B *Streptococcus*. *Infect. Immun.* **65:**1462–1467.

43. **Kogan, G., J. R. Brisson, D. L. Kasper, C. von Hunolstein, G. Orefici, and H. J. Jennings.** 1995. Structural elucidation of the novel type VII group B *Streptococcus* capsular polysaccharide by high resolution NMR spectroscopy. *Carbohydr. Res.* **277:**1–9.

44. **Kogan, G., D. Uhrin, J. R. Brisson, L. C. Paoletti, A. E. Blodgett, D. L. Kasper, and H. J. Jennings.** 1996. Structural and immunochemical characterization of the type VIII group B *Streptococcus* capsular polysaccharide. *J. Biol. Chem.* **271:**8786–8790.

44a.**Kogan, G., D. Uhrin, J.-R. Brisson, L. C. Paoletti, D. L. Kasper, C. von Hunolstein, G. Orefici, and H. J. Jennings.** 1994. Structure of the type VI group B *Streptococcus* capsular polysaccharide determined by high resolution NMR spectroscopy. *J. Carbohydr. Chem.* **13:** 1071–1078.

45. Koskiniemi, S., M. Sellin, and M. Norgren. 1998. Identification of two genes, cpsX and cpsY, with putative regulatory function on capsule expression in group B streptococci. FEMS Immunol. Med. Microbiol. 21:159–168.

46. Kuypers, J. M., L. M. Heggen, and C. E. Rubens. 1989. Molecular analysis of a region of the group B Streptococcus chromosome involved in type III capsule expression. Infect. Immun. 57:3058–3065.

47. Lachenauer, C. S., D. L. Kasper, J. Shimada, Y. Iciman, H. Ohtsuka, M. Kaku, L. C. Paoletti, and L. C. Madoff. Serotypes VI and VIII predominate among group B streptococci isolated from pregnant Japanese women. J. Infect. Dis. 179:1030–1033.

48. Lachenauer, C. S., and L. C. Madoff. 1996. A protective surface protein from type V group B streptococci shares N-terminal sequence homology with the alpha C protein. Infect. Immun. 64:4255–4260.

49. Lachenauer, C. S., and L. C. Madoff. 1997. Cloning and expression in Escherichia coli of a protective surface protein from type V group B streptococci. Adv. Exp. Med. Biol. 418:615–618.

50. Lagergard, T., J. Shiloach, J. B. Robbins, and R. Schneerson. 1990. Synthesis and immunological properties of conjugates composed of group B Streptococcus type III capsular polysaccharide covalently bound to tetanus toxoid. Infect. Immun. 58:687–694.

51. Lancefield, R. C. 1934. A serological differentiation of specific types of bovine hemolytic streptococci (group B). J. Exp. Med. 59:441–458.

52. Lancefield, R. C. 1938. Two serological types of group B hemolytic streptococci with related, but not identical, type-specific substances. J. Exp. Med. 67:25–40.

53. Lancefield, R. C. 1972. Cellular antigens of group B streptococci, p. 57–65. In L. W. Wannamaker and J. M. Matsen (ed.), Streptococci and Streptococcal Diseases, Recognition, Understanding and Management. Academic Press, Inc., New York, N.Y.

54. Lancefield, R. C., and E. H. Freimer. 1966. Type-specific polysaccharide antigens of group B streptococci. J. Hyg. Camb. 64:191–203.

55. Lancefield, R. C., M. McCarty, and W. N. Everly. 1975. Multiple mouse-protective antibodies directed against group B streptococci. Special reference to antibodies effective against protein antigens. J. Exp. Med. 142:165–179.

56. Lancefield, R. C., and G. E. Perlmann. 1952. Preparation and properties of a protein (R antigen) occurring in streptococci of group A, type 28 and in certain streptococci of other serological groups. J. Exp. Med. 96:83–97.

57. Lautrou, Y., P. Rainard, B. Poutrel, M. S. Zygmunt, A. Venien, and J. Dufrenoy. 1991. Purification of the protein X of Streptococcus agalactiae with a monoclonal antibody. FEMS Microbiol. Lett. 64:141–145.

58. Li, J., D. L. Kasper, F. M. Ausubel, B. Rosner, and J. L. Michel. 1997. Inactivation of the alpha C protein antigen gene, bca, by a novel shuttle/suicide vector results in attenuation of virulence and immunity in group B Streptococcus. Proc. Natl. Acad. Sci. USA 94:13251–13256.

59. Lindahl, G., B. Akerstrom, J. P. Vaerman, and L. Stenberg. 1990. Characterization of an IgA receptor from group B streptococci: specificity for serum IgA. Eur. J. Immunol. 20:2241–2247.

60. Linden, V. 1983. Mouse-protective effect of rabbit anti-R-protein antibodies against group B streptococci type II carrying R-protein. Lack of effect on type III carrying R-protein. Acta Pathol. Microbiol. Immunol. Scand. Sect. B 91:145–151.

61. Linden, V., K. K. Christensen, and P. Christensen. 1983. Correlation between low levels of maternal IgG antibodies to R protein and neonatal septicemia with group B streptococci carrying R protein. Int. Arch. Allergy Appl. Immunol. 71:168–172.

62. Madoff, L. C., S. Hori, J. L. Michel, C. J. Baker, and D. L. Kasper. 1991. Phenotypic diversity in the alpha C protein of group B streptococcus. Infect. Immun. 59:2638–2644.

63. Madoff, L. C., J. L. Michel, E. W. Gong, D. E. Kling, and D. L. Kasper. 1996. Group B streptococci escape host immunity by deletion of tandem repeat elements of the alpha C protein. Proc. Natl. Acad. Sci. USA 93:4131–4136.

64. Madoff, L. C., J. L. Michel, E. W. Gong, A. K. Rodewald, and D. L. Kasper. 1992. Protection of neonatal mice from group B streptococcal infection by maternal immunization with beta C protein. Infect. Immun. 60:4989–4994.

65. Madoff, L. C., J. L. Michel, and D. L. Kasper. 1991. A monoclonal antibody identifies a protective C-protein alpha-antigen epitope in group B streptococcus. Infect. Immun. 59:204–210.

66. Madoff, L. C., L. C. Paoletti, J. L. Michel, E. W. Gong, and D. L. Kasper. 1994. Synthesis of a type III polysaccharide-recombinant alpha C protein conjugate vaccine for prevention of group B streptococcal infection. Clin. Infect. Dis. 19:602.

67. Madoff, L. C., L. C. Paoletti, J. Y. Tai, and D. L. Kasper. 1994. Maternal immunization of mice with group B streptococcal type III polysaccharide-beta C protein conjugate elicits protective antibody to multiple serotypes. J. Clin. Invest. 94:286–292.

68. Markham, R. B., W. A. Nicholson, G. Schiffman, and D. L. Kasper. 1982. The presence of sialic acid on two related bacterial polysaccharides determines the site of the primary immune response and the effect of complement depletion on the response in mice. J. Immunol. 128:2731–2733.

69. Marques, M. B., D. L. Kasper, M. K. Pangburn, and M. R. Wessels. 1992. Prevention of C3 deposition by capsular polysaccharide is a virulence mechanism of type III group B streptococci. Infect. Immun. 60:3986–3993.

70. Marques, M. B., D. L. Kasper, A. Shroff, F. Michon, H. J. Jennings, and M. R. Wessels. 1994. Functional activity of antibodies to the group B polysaccharide of group B streptococci elicited by a polysaccharide-protein conjugate vaccine. Infect. Immun. 62:1593–1599.

71. Michel, J. L., B. D. Beseth, L. C. Madoff, S. K. Olken, D. L. Kasper, and F. M. Ausubel. 1994. Genotypic diversity and evidence for two distinct classes of the C protein alpha antigen of group B Streptococcus, p. 331–332. In A. Totolian (ed.), Pathogenic Streptococci: Present and Future. Lancer Publications, St. Petersburg, Russia.

72. Michel, J. L., L. C. Madoff, D. E. Kling, D. L. Kasper, and F. M. Ausubel. 1991. Cloned alpha and beta C-protein antigens of group B streptococci elicit protective immunity. Infect. Immun. 59:2023–2028.

73. Michel, J. L., L. C. Madoff, K. Olson, D. E. Kling, D. L. Kasper, and F. M. Ausubel. 1992. Large identical tandem repeating units in the C protein alpha antigen gene, bca, of group B streptococci. Proc. Natl. Acad. Sci. USA 89:10060–10065.

74. Michon, F., R. Chalifour, R. Feldman, M. Wessels, D. L. Kasper, A. Gamian, V. Pozsgay, and H. J. Jennings. 1991. The alpha-L-(1→2)-trirhamnopyranoside epitope on the group-specific polysaccharide of group B streptococci. Infect. Immun. 59:1690–1696.

75. Michon, F., P. C. Fusco, A. J. D'Ambra, M. Laude-Sharp, K. Long-Rowe, M. S. Blake, and J. Y. Tai. 1997. Combination conjugate vaccines against multiple serotypes of group B streptococci. Adv. Exp. Med. Biol. 418: 847–850.

76. Michon, F., E. Katzenellenbogen, D. L. Kasper, and H. J. Jennings. 1987. Structure of the complex group-specific polysaccharide of group B Streptococcus. Biochemistry 26:476–486.

77. Pancholi, V., and V. A. Fischetti. 1998. Alpha-enolase, a novel strong plasmin(ogen) binding protein on the surface of pathogenic streptococci. J. Biol.Chem. 273: 14503–14515.

78. Paoletti, L. C., C. J. Baker, and D. L. Kasper. 1998. Neonatal group B streptococcal disease: progress towards a multivalent maternal vaccine, abstr. P16, p. 43. Abstr. First Annual Conference on Vaccine Research. National Foundation for Infectious Diseases, Washington, D.C.

79. Paoletti, L. C., D. L. Kasper, F. Michon, J. DiFabio, K. Holme, H. J. Jennings, and M. R. Wessels. 1990. An oligosaccharide-tetanus toxoid conjugate vaccine against type III group B Streptococcus. J. Biol Chem. 265:18278–18283.

80. Paoletti, L. C., D. L. Kasper, F. Michon, J. DiFabio, H. J. Jennings, T. D. Tosteson, and M. R. Wessels. 1992. Effects of chain length on the immunogenicity in rabbits of group B Streptococcus type III oligosaccharide-tetanus toxoid conjugates. J. Clin. Invest. 89:203–209.

81. Paoletti, L. C., J. Pinel, A. K. Rodewald, and D. L. Kasper. 1997. Therapeutic potential of human antisera to group B streptococcal glycoconjugate vaccines in neonatal mice. J. Infect. Dis. 175:1237–1239.

82. Paoletti, L. C., R. A. Ross, and K. D. Johnson. 1996. Cell growth rate regulates expression of group B Streptococcus type III capsular polysaccharide. Infect. Immun. 64: 1220–1226.

83. Paoletti, L. C., M. R. Wessels, F. Michon, J. DiFabio, H. J. Jennings, and D. L. Kasper. 1992. Group B Streptococcus type II polysaccharide-tetanus toxoid conjugate vaccine. Infect. Immun. 60:4009–4014.

84. Paoletti, L. C., M. R. Wessels, A. K. Rodewald, A. A. Shroff, H. J. Jennings, and D. L. Kasper. 1994. Neonatal mouse protection against infection with multiple group B streptococcal (GBS) serotypes by maternal immunization with a tetravalent GBS polysaccharide-tetanus toxoid conjugate vaccine. Infect. Immun. 62:3236–3243.

85. Rainard, P. 1992. Isotype antibody response in cows to Streptococcus agalactiae group B polysaccharide-ovalbumin conjugate. J. Clin. Microbiol. 30:1856–1862.

86. Rainard, P., Y. Lautrou, P. Sarradin, and B. Poutrel. 1991. Protein X of Streptococcus agalactiae induces opsonic antibodies in cows. J. Clin. Microbiol. 29:1842–1846.

87. Rench, M. A., and C. J. Baker. 1993. Neonatal sepsis caused by a new group B streptococcal serotype. J. Pediatr. 122:638–640.

87a.Ross, R. A., L. C. Madoff, and L. C. Paoletti. Regulation of cell component production by growth rate in group B Streptococcus. J. Bacteriol., in press.

88. Rubens, C. E., R. F. Haft, and M. R. Wessels. 1995. Characterization of the capsular polysaccharide genes of group B streptococci. Dev. Biol. Stand. 85:237–244.

89. Rubens, C. E., L. M. Heggen, R. F. Haft, and M. R. Wessels. 1993. Identification of cpsD, a gene essential for type III capsule expression in group B streptococci. Mol. Microbiol. 8:843–855.

90. Russell-Jones, G. J., and E. C. Gotschlich. 1984. Identification of protein antigens of group B streptococci, with special reference to the Ibc antigens. J. Exp. Med. 160:1476–1484.

91. Russell-Jones, G. J., E. C. Gotschlich, and M. S. Blake. 1984. A surface receptor specific for human IgA on group B streptococci possessing the Ibc protein antigen. J. Exp. Med. 160:1467–1475.

91a.Schifferle, R. E., H. J. Jennings, M. R. Wessels, E. Katzenellenbogen, R. Roy, and D. L. Kasper. 1985. Immunochemical analysis of the types Ia and Ib group B streptococcal polysaccharides. J. Immunol. 135:4164–4170.

92. Schuchat, A. 1998. Epidemiology of group B streptococcal disease in the United States: shifting paradigms. Clin. Microbiol. Rev. 11:497–513.

93. Stalhammar-Carlemalm, M., L. Stenberg, and G. Lindahl. 1993. Protein rib: a novel group B streptococcal cell surface protein that confers protective immunity and is expressed by most strains causing invasive infections. J. Exp. Med. 177:1593–1603.

94. Stoll, B. J., and A. Schuchat. 1998. Maternal carriage of group B streptococci in developing countries. Pediatr. Infect. Dis. J. 17:499–503.

95. Suvorov, A. N., A. E. Flores, and P. Ferrieri. 1997. Cloning of the glutamine synthetase gene from group B streptococci. Infect. Immun. 65:191–196.

96. von Hunolstein, C., S. D'Ascenzi, B. Wagner, J. Jelínková, G. Alfarone, S. Recchia, M. Wagner, and G. Orefici. 1993. Immunochemistry of capsular type polysaccharide and virulence properties of type VI Streptococcus agalactiae (group B streptococci). Infect. Immun. 61:1272–1280.

97. Wagner, M., B. Wagner, and V. R. Kubin. 1980. Immunoelectron microscopic study of the location of group-specific and type-specific polysaccharide antigens on isolated walls of group B streptococci. J. Gen. Microbiol. 120:369–376.

98. Wastfelt, M., M. Stalhammar-Carlemalm, A. M. Delisse, T. Cabezon, and G. Lindahl. 1996. Identification of a family of streptococcal surface proteins with extremely repetitive structure. J. Biol. Chem. 271:18892–18897.

99. Wessels, M. R., W. J. Benedí, H. J. Jennings, F. Michon, J. L. DiFabio, and D. L. Kasper. 1989. Isolation and characterization of type IV group B Streptococcus capsular polysaccharide. Infect. Immun. 57:1089–1094.

100. Wessels, M. R., J. L. DiFabio, V. J. Benedí, D. L. Kasper, F. Michon, J. R. Brisson, J. Jelínková, and H. J. Jennings. 1991. Structural determination and immuno-

chemical characterization of the type V group B streptococcus capsular polysaccharide. *J. Biol. Chem.* **266:** 6714–6719.

101. **Wessels, M. R., V. Pozsgay, D. L. Kasper, and H. J. Jennings.** 1987. Structure and immunochemistry of an oligosaccharide repeating unit of the capsular polysaccharide of type III group B Streptococcus. A revised structure for the type III group B streptococcal polysaccharide antigen. *J. Biol. Chem.* **262:**8262–8267.

102. **Wessels, M. R., A. Munoz, and D. L. Kasper.** 1987. A model of high-affinity antibody binding to type III group B *Streptococcus* capsular polysaccharide. *Proc. Natl. Acad. Sci. USA* **84:**9170–9174.

103. **Wessels, M. R., L. C. Paoletti, D. L. Kasper, J. L. DiFabio, F. Michon, K. Holme, and H. J. Jennings.** 1990. Immunogenicity in animals of a polysaccharide-protein conjugate vaccine against type III group B *Streptococcus. J. Clin. Invest.* **86:**1428–1433.

104. **Wessels, M. R., L. C. Paoletti, J. Pinel, and D. L. Kasper.** 1995. Immunogenicity and protective activity in animals of a type V group B streptococcal polysaccharide-tetanus toxoid conjugate vaccine. *J. Infect. Dis.* **171:**879–884.

105. **Wessels, M. R., L. C. Paoletti, A. K. Rodewald, F. Michon, J. DiFabio, H. J. Jennings, and D. L. Kasper.** 1993. Stimulation of protective antibodies against type Ia and Ib group B streptococci by a type Ia polysaccharide-tetanus toxoid conjugate vaccine. *Infect. Immun.* **61:**4760–4766.

106. **Wessels, M. R., C. E. Rubens, V. J. Benedí, and D. L. Kasper.** 1989. Definition of a bacterial virulence factor: sialylation of the group B streptococcal capsule. *Proc. Natl. Acad. Sci. USA* **86:**8983–8987.

107. **Wilkinson, H. W.** 1972. Comparison of streptococcal R antigens. *Appl. Microbiol.* **24:**669–670.

108. **Wilkinson, H. W., and R. G. Eagon.** 1971. Type-specific antigens of group B type Ic streptococci. *Infect. Immun.* **4:**596–604.

109. **Yim, H. H.** 1998. Regulation of capsular polysaccharide production in group B *Streptococcus,* abstr. 27, p. 36. ASM *Conference on Streptococcal Genetics, Vichy, France.* American Society for Microbiology, Washington, D.C.

Epidemiology of Group B Streptococcal Infections

SHARON BALTER, CYNTHIA G. WHITNEY, AND ANNE SCHUCHAT

15

Group B *Streptococcus* (GBS), or *Streptococcus agalactiae*, was first classified in the 1930s by Lancefield and Hare (40) in their studies on the serologic differentiation of streptococci. Lancefield's studies established that most puerperal infections of the day were due to group A streptococci, although they identified GBS from the vaginal cultures of asymptomatic women. GBS was better known at the time as the cause of bovine mastitis. GBS was first described as a human pathogen in a 1938 report of three women with fatal puerperal sepsis due to GBS (30). However, it was only rarely reported as a human pathogen until the 1960s, when it was increasingly seen as an adult pathogen and emerged as the leading cause of neonatal sepsis (25). Before this time, sepsis in newborns had been principally caused by *Escherichia coli* (25). The reasons for the emergence of GBS disease are unclear.

GBS DISEASE IN NONPREGNANT ADULTS

Incidence

Although GBS is commonly thought of as a cause of disease in neonates and pregnant women, it causes substantial morbidity and mortality among nonpregnant adults (Fig. 1) and appears to be increasing in incidence in that population. Incidence rates vary by area; a study in metropolitan Atlanta in 1992–1993 found the incidence of invasive disease among nonpregnant adults to be 5.9 cases per 100,000 population, a 37% increase from 1989–1990 (14). A similar study in Maryland in 1992–1993 found the incidence among nonpregnant adults to be 6.5/100,000 (37). Reported case-fatality rates range from 16% (68) to 70% (31); recent reports suggest a case-fatality rate of about 30% (26). The overall rate of invasive GBS disease is slightly lower for women, about 3.3/100,000, compared with 4.9/100,000 for men, but women account for more than half the cases among the elderly (26).

Of nine known GBS serotypes, the most common serotypes reported among adults are types III and V. Serotype V was first reported in 1985 and initially appeared to be a rare cause of infection (14). However, a recent population-based study found it to be the most common serotype causing disease in nonpregnant adults and the second most common serotype in pregnant women (37). The emergence of serotype V appears to be a relatively recent phenomenon. Although serotype V isolates would previously have been identified as nontypeable, studies done before serotype V was identified reported few nontypeable isolates; in the mid-1970s, only 4% of adults had nontypeable strains, whereas serotype V now accounts for 29% of adult cases (37).

Syndromes

The most common syndromes caused by GBS in adults are skin, soft tissue, and bone infections (26, 59). These infections are often complications of chronic diabetes or decubitus ulcers (26). Cellulitis, foot ulcers, and abscesses are the most common manifestations, but other infections, including necrotizing fasciitis, have occasionally been reported (24).

Another common presentation of GBS infection in adults is bacteremia without any identified focus. In one study, about a quarter of such patients had underlying hepatic or renal failure (26), although malignancy and diabetes also were common underlying conditions. Patients with indwelling catheters are at higher risk for GBS bacteremia (38). Polymicrobial bacteremia, often with *Staphylococcus aureus*, is identified in 26 to 30% of patients with GBS bacteremia (26, 38).

GBS can also cause pneumonia (26, 59). Pneumonia is a particularly severe form of GBS disease and has a high mortality rate (38, 63). In one small case series, all seven patients with GBS pneumonia died (63). Patients with GBS pneumonia appear to have more neurologic disease, dementia, or tracheoesophageal fistulas, suggesting that GBS pneumonia may be related to aspiration of the organism (26, 63). In the series of seven patients, six had severe underlying neurologic conditions, and one had cancer of the esophagus with tracheoesophageal fistula. Chest radiographs showed bilateral or lobar infiltrates, and although infections were polymicrobial, GBS was the predominant organism (63).

Gram-Positive Pathogens, ed. by V. A. Fischetti et al.
© 2000 American Society for Microbiology, Washington, D.C.

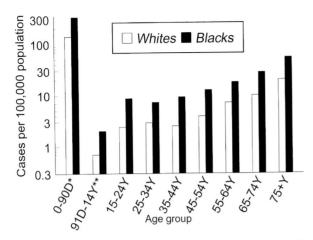

FIGURE 1 Incidence of invasive GBS disease by race and age, selected U.S. counties. *, D, age in days; **, Y, age in years. Reprinted from reference 68.

The epidemiology of GBS endocarditis has undergone a shift since the preantibiotic era. Previously, endocarditis was one of the most common syndromes caused by GBS infection, but GBS endocarditis is much rarer today (59). In the 1930s and 1940s, GBS endocarditis was predominantly an acute disease of pregnant women and most often affected the mitral valve (41, 59). A comparison of a series of patients with endocarditis from 1938 to 1945 with a series after 1962 found that in the latter series, about half the patients were men, and, while the mitral valve remained the most common site of infection, aortic and tricuspid valve disease was also seen (41). Cases of GBS endocarditis in the latter series were both acute and subacute. Large, friable vegetations are a common feature of the disease, often leading to embolization and requiring surgical intervention (41).

Less common GBS infections such as arthritis, urinary tract infections, meningitis, and peritonitis also occur. GBS arthritis is most commonly monoarticular, although it can be polyarticular (59). Diabetes, osteoarthritis, and underlying joint disease, including prosthetic joints, are common predisposing factors. In a case series, patients with GBS arthritis required both antibiotic therapy and drainage for recovery and, despite therapy, only about 50% completely recovered (24). GBS urinary tract infections occur in nonpregnant adults as well as pregnant women (23, 41). In men, GBS urinary tract infections appear to be associated with prostatic infections (23), although they may also occur with stones or other abnormalities of the urinary tract and in patients with diabetes (26, 41).

Risk Factors

Adults with chronic diseases are at higher risk for GBS infection. Conditions that increase the risk of GBS disease include diabetes, malignancy, liver disease, neurologic impairment, renal failure, other forms of immune impairment such as human immunodeficiency virus infection, and steroid use (23, 26, 38, 41, 59, 63). In one study, patients with any type of cancer had a 16-fold-higher risk of developing invasive GBS disease than did healthy adults (59), although another study found that this increased risk declined with age and that the elderly with cancer were at no greater risk than the elderly without cancer (26). A third study, which used multivariate analysis, found that

only cancer of the breast independently increased the risk of invasive GBS disease.

Diabetes is a commonly reported condition among patients with GBS disease (23, 26, 38, 41, 59). In a population-based study of risk factors, patients with diabetes had 10.5 times the risk of the general population (59). The precise mechanism underlying this increased risk is unclear. Diabetes increases the risk for GBS disease for both insulin-dependent and non-insulin-dependent patients across all age groups (26). Persons with diabetes do not have an increased rate of nose, throat, skin, or rectal GBS colonization compared with those without diabetes (18), but some evidence indicates that abnormalities in immune function, such as neutrophil phagocytosis or intracellular killing, may contribute to the increased risk for diabetic patients (62).

In addition to underlying disease, other risk factors for invasive GBS disease in adults include advanced age and black race (26, 59). Although the disease occurs in all age groups, in-hospital mortality is higher among older patients with GBS infections (26, 38). The mean age of GBS patients in several population-based studies was about 65 years (26, 38, 59). The explanation for the higher risk of GBS disease among black persons is unclear, although black race may be a surrogate for other factors. In one study, race was not an independent risk factor after controlling for chronic medical conditions (38). In a population-based study in metropolitan Atlanta, the incidence of invasive infection among blacks was twice as high as that among whites, but most of the disparity was associated with residence in the urban center; whites residing in the same area also had an increased risk compared with whites in other communities (26). The higher prevalence of diabetes and renal disease may also contribute to higher rates of GBS disease in this population (26).

GBS infection may be nosocomially acquired. One study found that among patients acquiring the infection in the hospital, GBS was independently associated with congestive heart failure, dementia, seizure disorder, and placement of a central venous line (38). Clustering of invasive GBS disease among hospitalized patients is rarely reported, suggesting that the underlying illness of the hospitalized population or invasive procedures introducing endogenous flora into patients predisposes them to disease, rather than the disease being transmitted from patient to patient. Because the gastrointestinal tract is thought to be a reservoir for GBS (5), there has been concern that sigmoidoscopy and colonoscopy can lead to disease postprocedure. Arthritis and pacemaker wire infections (24) due to GBS following sigmoidoscopy or colonoscopy have been reported.

GBS IN PREGNANT WOMEN AND NEWBORNS

Infections in Pregnant Women

GBS causes substantial morbidity in pregnant women. Of 307 cases of invasive GBS disease in adults identified in a population-based active surveillance study, 11% were in pregnant women (68). In one study, the attack rate of GBS infections in pregnant women was approximately 2/1,000 deliveries (48). GBS causes a variety of perinatal infections in pregnant women, including both symptomatic and asymptomatic bacteriuria, endometritis, amnionitis, meningitis, pyelonephritis, and postpartum wound infections (48).

It has also been suggested that GBS urinary tract infections or urinary tract, rectal, or genital colonization in pregnant women may lead to late-term abortions and preterm and low-birth-weight infants. A study comparing 150 women who presented with signs of threatened abortion with 100 women of similar gestational age demonstrated that women who aborted were more likely to be colonized with GBS than those who did not (21). The association was seen for both cervical colonization and urinary colonization. Smaller studies have both confirmed (47) and disputed (45) the association between GBS in the urinary tract and preterm delivery. In a large study done by the Vaginal Infections and Prematurity Group, 13,914 women were screened for GBS colonization during pregnancy. This study showed an increase in the risk of low-birth-weight and premature infants among women who were heavily colonized with GBS. Women with light colonization had pregnancy outcomes similar to those of the uncolonized women (52).

Sites of Maternal Colonization

Many pregnant women are colonized with GBS but are asymptomatic. However, GBS colonization in pregnant women is important because of the risk for transmission to their newborns. Newborns born to colonized mothers are more likely to develop GBS infection (6). Women may be colonized at multiple sites, including the rectum, vagina, cervix, and throat (7, 53). Rectal carriage is more common than vaginal carriage (7, 22). Cervical colonization is relatively less common, and throat colonization is even rarer (7, 22, 53). The Vaginal Infections and Prematurity Study, the largest of its kind, found the overall carriage rate among 13,914 pregnant women to be 21%, although only vaginal cultures were collected in this study (52).

The ability to detect GBS colonization depends both on the sites cultured and on the medium used. Using selective broth medium will result in higher recovery of GBS; one study demonstrated 50% improvement in the recovery of GBS with the use of selective broth medium compared with standard blood agar (50). In this study, which evaluated the impact of culture site, 94 women were cultured both vaginally and rectally; 29 women were positive for GBS. Of those, 3 (10%) were positive on vaginal culture only, 12 (41.4%) were positive on rectal culture only, and 14 (48.3%) were positive on both rectal and vaginal culture (50).

Duration of Maternal Colonization

GBS carriage may be chronic, intermittent, or transient (15, 22). In a large study of pregnant women who were screened for GBS once during a prenatal visit from the late first trimester through the third trimester, 67% of women who were GBS carriers and 8.5% of women who were not colonized with GBS at their prenatal visit were culture positive for GBS at delivery (16). The predictive value of prenatal culture for predicting colonization at delivery increased with shorter intervals between prenatal sampling and delivery. All (16 of 16) carriers who delivered within 5 weeks of their prenatal culture remained positive at the time of delivery. In a similar study of 754 women who were sequentially cultured for rectal and vaginal colonization with GBS in the second and third trimesters, 31% were carriers of GBS in the second trimester and 28% were carriers of GBS in the third trimester; only 17% were positive in both trimesters. Serotyping of the isolates indicated

that persistence of the same type was most common in anorectal-vaginal carriers and least likely in women who were vaginal carriers only; anorectal-vaginal carriers rarely acquired a new type. Of women who initially did not carry GBS, 16% acquired GBS between the second and third trimesters (22). Another study of 826 women at 35 to 36 weeks' gestation showed that the sensitivity of late antenatal cultures for identifying colonization status at delivery was 87% (67).

Risk Factors for Maternal Carriage

Efforts have been made to characterize groups with higher rates of GBS carriage. As part of the Vaginal Infections and Prematurity Study, 8,049 pregnant women underwent vaginal and cervical cultures for GBS colonization at 23 to 26 weeks' gestation (53). In a multivariate analysis, Caribbean Hispanics and blacks were more likely to carry GBS than were whites and Mexican Hispanics. The same study also found that GBS colonization was associated with increasing age, lower parity, and fewer years of education; current smoking was associated with a decreased risk of colonization. Colonization was only associated with "extreme" increases in sexual activity, defined as those women who had both frequent intercourse and multiple partners. Although carriage was associated with concurrent colonization with *Candida* spp., GBS carriage was not associated with concurrent infection with sexually transmitted pathogens, including *Neisseria gonorrhoeae*, *Trichomonas vaginalis*, or *Chlamydia trachomatis* (53). A lack of association between GBS colonization and sexually transmitted diseases has been observed in other studies (11). A study of nonpregnant college women showed that women with any sexual experience were more likely to be colonized than were those without sexual experience. That study also suggested that intrauterine devices may be associated with carriage (11). It did not demonstrate an association between GBS colonization and oral contraceptives, although this has been reported (56).

Infant Colonization

Infants born to GBS-colonized women are more likely to be colonized with GBS. In a study of 802 women in which investigators obtained endocervical and vaginal cultures at the 36th week of pregnancy and again at delivery, cultures were also obtained from infants' anterior nares, external ear canals, and the base of their umbilicus within 3 h of birth. Of infants born to mothers who had positive cultures at delivery, 37% were colonized with GBS at delivery and 49% were colonized by the time they were discharged. In contrast, of the 689 infants born to culture-negative mothers, only 1% acquired GBS colonization. All but one of the 31 GBS-colonized infants in this study were colonized with the same serotype as their mother (29).

Although infant colonization is presumed to occur in the uterus or birth canal, even if a mother is colonized in the rectum, and not the vagina, there is still a 17% rate of vertical transmission of GBS to the newborn (6). GBS can cross intact membranes, and neonatal acquisition from the mother can occur in newborns delivered by cesarean section. In a longitudinal study of vaginally delivered infants whose mothers were colonized with GBS, 61% acquired GBS colonization; of infants who were delivered by cesarean section to mothers who were colonized with GBS, 40% acquired GBS colonization with the same serotype as the mother. Infants whose mothers were heavily colonized were more likely to become colonized than those whose mothers were only lightly colonized (95% versus 31%) (6).

Infections in Newborns

GBS is a leading cause of neonatal bacterial disease in the United States. The case-fatality rate of GBS in newborns is about 4.0%: 4.5% for early-onset disease and 2.4% for late-onset disease (20). Before the use of prevention measures became widespread in the early and mid-1990s, GBS caused an estimated 7,600 cases of serious illness and 310 deaths among U.S. infants aged ≤90 days (68). Between 1993 and 1995, the overall annual incidence of early-onset GBS disease declined from 1.7 cases per 1,000 liveborn infants to 1.3 cases per 1,000 liveborn infants, a 24% decline (20).

Infections in newborns commonly present as bacteremia, meningitis, or pneumonia. There are two distinct syndromes: early-onset disease (EOD), which appears in the first week of life, usually within the first 24 h, and late-onset disease (LOD), which occurs on or after 7 days of age (10). Infants with EOD most commonly have sepsis or pneumonia; meningitis and bone and soft tissue infections can also occur. EOD is due to ascending infection from the genital tract or colonization of the infant during delivery. Amniotic infection can occur even when the membranes are intact (21), although both premature and prolonged rupture of membranes increase the risk of transmission. Case-fatality rates for EOD are estimated at 4.5% (20).

Late-onset disease is more likely to present with meningitis than is EOD, although LOD also can present as bone and soft tissue infections, urinary tract infections, or pneumonia (65). Although the disease is often thought of as occurring between the first 7 days and 12 weeks of life, in one study at a tertiary care center, 20% of cases occurred in infants older than 3 months of age (65). Infants with LOD are less likely to be severely ill upon presentation, and their deliveries are less likely to be characterized by predisposing obstetric complications (9).

Among patients with EOD, serotypes III and Ia are most commonly identified, although serotypes V and II are also seen. The majority of patients with LOD have serotype III, although serotype Ia is also fairly common (37).

Risk Factors for Early-Onset Neonatal Disease

Maternal carriage is clearly a risk factor for neonatal GBS disease (15). Infants born to heavily colonized mothers are more likely to develop invasive disease than are infants born to mothers who are lightly colonized or not colonized at all (34, 42). Without preventive measures, about 1 to 2% of neonates born to mothers with rectovaginal colonization will develop early-onset GBS disease (15).

Although the majority of cases of GBS disease occur in full term-infants, prematurity and low birth weight are risk factors for the development of invasive neonatal GBS disease (58). While premature infants may be more susceptible to invasive disease, it is also possible that prematurity and low birth weight are the result of GBS colonization in the mother (44, 52).

Maternal antibodies to the capsular polysaccharides of GBS protect against GBS disease. Antibodies are passively transferred from a mother to her fetus during pregnancy, offering protection against invasive disease. One study found lower antibody concentrations in women whose infants developed GBS disease than in colonized women whose infants did not (12), while another found that women whose infants developed GBS disease had lower antibody titers than the general population (35). The concentration of serum antibody against GBS increases with increasing age (4); this could explain in part why young maternal age is a risk factor for invasive neonatal GBS disease (57, 58).

Obstetric factors that have been associated with early-onset GBS disease include intrapartum fever (15, 57), rupture of membranes before labor (57), urinary tract infection during pregnancy (43, 57), prolonged labor (57), and prolonged rupture of membranes (15, 27). It has also been suggested that multiple gestation may be a risk factor for EOD, although this observation may be due to the fact that such infants are more likely to be low birth weight and premature (49).

Other risk factors for EOD include black race (57, 58) and history of previous miscarriage (58). One study identified internal monitoring for more than 12 h as a risk factor for GBS disease (66), although this was not an independent risk factor in other studies (57).

Risk Factors for Late-Onset Disease

Risk factors for late-onset GBS disease are less well understood than those for EOD, although obstetric factors do not appear to be as important. A cohort study in metropolitan Atlanta found that black infants had 35 times the rate of LOD than infants of other races. The only other risk factor found in that study was young maternal age (58). Another report suggested that breast feeding by mothers with GBS mastitis may play a role in transmission of GBS, but this route of infection probably accounts for a minority of cases (13).

Although vertical transmission accounts for the colonization of most infants, nosocomial or horizontal transmission of GBS may account for some LOD. Anthony et al. (6) found evidence of nosocomial spread of colonization in 2 of 10 cohorts of infants studied. In one cohort, an infant acquired GBS colonization of the same serotype as the infant's mother; five other infants born on the same or next day and cared for in the same half of the nursery acquired GBS colonization of the same serotype, although their mothers had consistently negative cultures. Outbreaks of GBS disease in nurseries have rarely been reported (13).

TREATMENT

Group B streptococci are uniformly susceptible to penicillin in vitro, and penicillin G is the treatment of choice for established cases of GBS. The organism is generally susceptible in vitro to ampicillin, vancomycin, teicoplanin, imipenem, and first-, second-, and third-generation cephalosporins, although the activity varies with each agent (24). A recent study found resistance to erythromycin in 7.4% of isolates and resistance to clindamycin in 3.4% of isolates (28). For newborns with presumptive GBS sepsis, the initial treatment should be penicillin G or ampicillin plus an aminoglycoside. Once GBS has been identified, penicillin G alone can be given.

For infants with GBS meningitis, the recommended dosage of penicillin G is 250,000 to 450,000 U/kg/day given intravenously in three divided doses for infants 7 days or younger; infants older than 7 days should receive 450,000 U/kg/per day given intravenously in four divided doses. Ampicillin can also be used; the dose is 200 mg/

kg/day given intravenously in three divided doses for infants 7 days or younger, and 300 mg/kg/day in four to six divided doses for infants over 7 days of age (2). For uncomplicated meningitis, 14 to 21 days of therapy is recommended, although longer treatment may be necessary in complicated cases (2). Infants with bacteremia of undefined focus require treatment for at least 10 days. For osteomyelitis, treatment is required for 4 weeks or more (2).

In adults, penicillin G is the treatment of choice. Ten days of treatment is recommended for bacteremia, soft tissue infections, pneumonia, and pyelonephritis, and 14 to 21 days is recommended for meningitis. For osteomyelitis, 3 to 4 weeks of therapy is required. Endocarditis patients should receive 4 to 6 weeks of therapy, and surgical debridement or valve replacement may be necessary (24).

PREVENTION

Strategies Studied for Neonatal Prevention

Many early-onset neonatal infections are preventable with currently available methods, namely, chemoprophylaxis. Several approaches to chemoprophylaxis have been studied. One approach was postnatal penicillin prophylaxis administered to newborns by intramuscular injection immediately after birth. Only one randomized trial was done, in which blood cultures were collected from all newborns before chemoprophylaxis; no differences were observed among the treated and untreated groups of low-birth-weight infants in this study (51). The mortality rates were also similar in both groups. Another study in a single hospital found a reduction in early-onset GBS infections, but there was no difference in mortality between the penicillin-treated group and the control group, and an increase in mortality associated with penicillin-resistant infections was initially identified in the treatment group, although it was not sustained (61). A long-term observational study in the same hospital did find lower rates of EOD among groups in which postnatal prophylaxis was used (0.25 and 0.63 per 1,000 births) versus those in which it was not (1.19, 1.59, and 1.95 per 1,000) (60). Most of the infants who got sick despite postnatal penicillin did so in the first 4 h after birth. Postnatal prophylaxis may not be effective against early-onset GBS infections that are acquired in utero well before birth.

Another approach to chemoprophylaxis has been to treat GBS carriers with oral antibiotics before labor and delivery. However, even when antibiotics effectively reduced colonization, colonization recurred (32, 36). In one study, 40 women in the study group and their husbands were treated simultaneously with oral penicillin and erythromycin for 12 to 14 days (32). At delivery, however, 67% of the women remained colonized, and there was no difference between the treatment group and 19 untreated women who served as controls. The only study in which oral chemoprophylaxis of carriers was effective in reducing infant colonization involved screening at 38 weeks and treating with oral penicillin (or erythromycin in case of allergy) until delivery, a regimen that would exclude preterm deliveries and could result in complications from prolonged antibiotic use (46). The efficacy of this regimen has not been tested against invasive disease.

The most successful approach to prevention has been the use of antibiotics during labor or after the onset of membrane rupture for women who are known carriers or who have risk factors for GBS disease in their infant. Several studies indicated that intrapartum chemoprophylaxis of colonized women reduced both neonatal colonization and EOD (19). A study that randomized only those carriers who had risk factors for neonatal disease (premature labor or prolonged rupture of membranes) demonstrated a significant decrease in both neonatal colonization and disease among those who received intrapartum ampicillin (17), and a study of 30,197 women who were screened for GBS at 32 weeks' gestation and treated with intrapartum penicillin showed 16 EOD infections in the study group compared with 27 in the control group of 26,195 women (1 per 1,000 births versus 5 per 1,000 births; $P = 0.04$) (33).

Prevention strategies for neonatal GBS disease can be effective. In a study that linked prevention policies to the number of EOD cases, hospitals with policies for GBS prevention had fewer cases of EOD (64). The decline in cases between 1993 and 1995 (20) was likely due to adoption of various prevention strategies by many clinicians.

An ideal method for identification of women who should receive chemoprophylaxis would be through the use of an appropriate rapid test at the time of delivery. Although rapid screening tests are available, none are sufficiently accurate for routine detection of GBS carriers at the time of delivery. A recent comparison of three immunoassays versus culture in selective broth media found that, although sensitivity of the immunoassays ranged from 36 to 100% for heavily colonized women, the sensitivity was only 12 to 37% for lightly colonized women (8). Until a more sensitive rapid test is available, prenatal cultures are the only method for adequate detection of GBS carriers. The accuracy of prenatal screening cultures is improved if they are collected late in pregnancy. In a study of 826 women, the sensitivity of antenatal cultures collected between 35 and 36 weeks was 87% and the specificity was 96%, while among patients with cultures collected 6 or more weeks before delivery, the sensitivity was only 43% and the specificity was 85% (67). Because carriage of GBS can be transient and recurrent, the best time to collect prenatal cultures has been a concern. Although culturing late in pregnancy ensures the reliability of the findings at delivery, women who deliver preterm and are therefore at a higher risk of transmission of neonatal disease are missed.

Current Recommendations for Prevention of Early-Onset Neonatal Disease

Building on earlier prevention research and on practical concerns, the Centers for Disease Control and Prevention, the American College of Obstetricians and Gynecologists, and the American Academy of Pediatrics issued consensus prevention guidelines in 1996 and early 1997 (1, 3, 19). The statements recommend using one of two strategies: a screening-based approach or a risk-based approach. In the screening-based approach, all women are screened at 35 to 37 weeks' gestation for rectovaginal colonization with GBS (Fig. 2). Intrapartum chemoprophylaxis is offered to all women who have at least one of the following: GBS bacteriuria during pregnancy, delivery of a previous infant with GBS disease, or positive prenatal culture. If results of prenatal culture are not known at delivery, intrapartum chemoprophylaxis should be given if the patient either is <37 weeks' gestation, has duration of membrane rupture ≥18 h, or has a temperature of ≥100.4°F (≥38°C). To ensure accurate results, both vaginal and rectal swabs should be collected and incubated in a single selective culture broth.

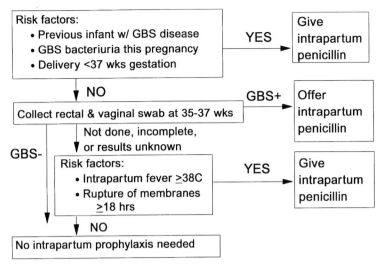

FIGURE 2 Prevention strategy using a screening-based approach. Reprinted from reference 19.

In the risk-based approach, intrapartum prophylaxis is given if one or more of the following risk factors are present: expected delivery at <37 weeks' gestation, duration of membrane rupture ≥18 h, temperature ≥100.4°F (≥38°C), history of previous infant with GBS disease, or GBS urinary tract infection (Fig. 3). Prenatal screening cultures are not routinely collected with this strategy.

For both strategies, the recommended intrapartum regimen is penicillin, 5-mU intravenous load and then 2.5 mU every 4 h until delivery, or, alternatively, ampicillin, 2-g load and then 1 g intravenously every 4 h until delivery. For penicillin-allergic women, clindamycin, 900 mg intravenously every 8 h until delivery, or erythromycin, 500 mg intravenously every 6 h until delivery, may be used. Intrapartum chemoprophylaxis should begin as soon as possible during labor for GBS-positive women and as soon as risk factors are identified for those women whose GBS status is unknown.

A retrospective case review of 245 cases of early-onset GBS disease estimated that the screening-based approach could have prevented 78% of cases, whereas a risk-based approach would have prevented 41% of cases (54). In ad-

dition, a model assessing 19 different prevention strategies also found these methods to be cost saving (55).

THE FUTURE

GBS disease is a major concern for pregnant women and their newborns and a growing problem among nonpregnant adults with chronic diseases. Although most neonatal GBS disease can be prevented through intrapartum prophylaxis, currently available strategies are not ideal. Logistic aspects can be cumbersome for strategies that rely on prenatal screening; strategies that involve identification of risk factors present at delivery may not allow adequate time for administration of chemoprophylaxis. In addition, currently recommended strategies involve giving intrapartum antimicrobial agents to approximately 20 to 25% of pregnant women, which may result in adverse reactions and possible emergence of antimicrobial-resistant organisms.

Development of a vaccine against GBS may provide a better long-term solution than chemoprophylaxis. A vaccine could be given to women during pregnancy or to

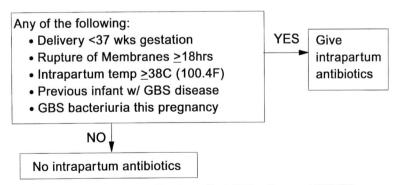

FIGURE 3 Prevention strategy using a risk-based approach. Reprinted from reference 19.

women who may soon become pregnant; transfer of antibodies across the placenta late in pregnancy would confer protection to the newborn. In addition, a vaccine could be effective against LOD and disease in pregnant women and nonpregnant adults. A vaccine for GBS is in development (39). Until it is available, however, more widespread adoption of recommended prevention measures for neonatal GBS is crucial.

REFERENCES

1. **American Academy of Pediatrics.** 1997. Revised guidelines for prevention of early-onset group B streptococcal (GBS) infection. *Pediatrics* **99**:489–495.

2. **American Academy of Pediatrics, Committee on Infectious Diseases, G. Peter, C. B. Hall, N. A. Halsey, S. M. Marcy, and L. K. Pickering.** 1997. *1997 Red Book: Report of the Committee on Infectious Diseases*, 24th ed., p. 494–501. American Academy of Pediatrics, Elk Grove Village, Ill.

3. **American College of Obstetricians and Gynecologists, Committee on Obstetric Practice.** 1996. *Prevention of Early-Onset Group B Streptococcal Disease in Newborns*, no. 173, p. 1–7. American College of Obstetricians and Gynecologists, Washington, D.C.

4. **Anthony, B. F., I. E. Concepcion, N. F. Concepcion, C. M. Vadheim, and J. Tiwari.** 1994. Relation between maternal age and serum concentration of IgG anitbody to type III group B streptococci. *J. Infect. Dis.* **170**:717–719.

5. **Anthony, B. F., R. Eisenstadt, J. Carter, S. K. Kwang, and C. J. Hobel.** 1981. Genital and intestinal carriage of group B streptococci during pregnancy. *J. Infect. Dis.* **143**:761–766.

6. **Anthony, B. F., D. M. Okada, and C. J. Hobel.** 1979. Epidemiology of the group B *Streptococcus*: maternal and nosocomial sources for infant acquisitions. *J. Pediatr.* **95**:431–436.

7. **Badri, M. S., S. Zawaneh, A. C. Cruz, G. Mantilla, H. Baer, W. N. Spellacy, and E. M. Ayoub.** 1977. Rectal colonization with group B *Streptococcus*: relation to vaginal colonization of pregnant women. *J. Infect. Dis.* **135**:308–312.

8. **Baker, C. J.** 1996. Inadequacy of rapid immunoassays for intrapartum detection of group B streptococcal carriers. *Obstet. Gynecol.* **88**:51–55.

9. **Baker, C. J., F. F. Barrett, R. C. Gordon, and M. D. Yow.** 1973. Suppurative meningitis due to streptococci of Lancefield group B: a study of 33 infants. *J. Pediatr.* **82**:724–729.

10. **Baker, C. J., and M. S. Edwards.** 1995. Group B streptococcal infections, p. 980–1054. *In* J. S. Remington and O. J. Klein (ed.), *Infectious Disease of the Fetus and Newborn Infant*, 4th ed. The W. B. Saunders Co., Philiadelphia, Pa.

11. **Baker, C. J., D. K. Goroff, S. Alpert, V. A. Crockett, S. H. Zinner, J. R. Evrard, B. Rosner, and W. M. McCormack.** 1977. Vaginal colonization with group B streptococcus: a study of college women. *J. Infect. Dis.* **135**:392–397.

12. **Baker, C. J., D. L. Kasper, I. B. Tager, A. Paredes, S. Alpert, W. M. McCormack, and D. Goroff.** 1977. Quantitative determination of antibody to capsular polysaccha-

ride in infection with type III strains of group B *Streptococcus. J. Clin. Invest.* **59**:810–818.

13. **Band, J. D., H. W. Clegg, P. S. Hayes, R. R. Facklam, J. Stringer, and R. E. Dixon.** 1981. Transmission of group B streptococci traced by use of multiple epidemiologic markers. *Am. J. Dis. Child.* **135**:355–358.

14. **Blumberg, H. M., D. S. Stephens, M. Modansky, M. Erwin, J. Elliot, R. R. Facklam, A. Schuchat, W. Baughman, and M. M. Farley.** 1996. Invasive group B streptococcal disease: the emergence of serotype V. *J. Infect. Dis.* **173**:365–373.

15. **Boyer, K. M., C. A. Gadzala, L. I. Burd, D. E. Fisher, J. B. Paton, and S. P. Gotoff.** 1983. Selective intrapartum chemoprophylaxis of neonatal group B streptococcal early-onset disease. I. Epidemiologic rationale. *J. Infect. Dis.* **148**:795–801.

16. **Boyer, K. M., C. A. Gadzala, P. D. Kelly, L. I. Burd, and S. P. Gotoff.** 1983. Selective intrapartum chemoprophylaxis of neonatal group B streptococcal early-onset disease. II. Predictive value of prenatal cultures. *J. Infect. Dis.* **148**:802–809.

17. **Boyer, K. M., and S. P. Gotoff.** 1986. Prevention of early-onset neonatal group B streptococcal disease with selective intrapartum chemoprophylaxis. *N. Engl. J. Med.* **314**:1665–1669.

18. **Casey, J. I., S. Maturlo, J. Albin, and S. C. Edberg.** 1982. Comparison of carriage rates of group B *Streptococcus* in diabetic and nondiabetic persons. *Am. J. Epidemiol.* **116**:704–708.

19. **Centers for Disease Control and Prevention.** 1996. Prevention of perinatal group B streptococcal disease: a public health perspective. *Morbid. Mortal. Weekly Rep.* **45**:1–24.

20. **Centers for Disease Control and Prevention.** 1997. Decreasing incidence of perinatal group B streptococcal disease—United States, 1993–1995. *Morbid. Mortal. Weekly Rep.* **46**:473–477.

21. **Daugaard, H. O., A. C. Thomsen, U. Henriques, and A. Ostergaard.** 1988. Group B streptococci in the lower urogenital tract and late abortions. *Am. J. Obstet. Gynecol.* **158**:28–31.

22. **Dillon, H. C., E. Gray, M. A. Pass, and B. M. Gray.** 1982. Anorectal and vaginal carriage of group B streptococci during pregnancy. *J. Infect. Dis.* **145**:794–799.

23. **Duma, R. J., A. N. Weinberg, T. F. Medrek, and L. J. Kunz.** 1969. Streptococcal infections: a bacteriologic and clinical study of streptococcal bacteremia. *Medicine* **48**:87–107.

24. **Edwards, M. S., and C. J. Baker.** 1995. *Streptococcus agalactiae* (group B *Streptococcus*), p. 1835–1845. *In* G. L. Mandell, J. E. Bennett, and R. Dolin (ed.), *Principles and Practice of Infectious Diseases*, 4th ed. Churchill Livingstone, Ltd., New York, N.Y.

25. **Eickhoff, T. C., J. O. Klein, A. K. Daly, D. Ingall, and M. Finland.** 1964. Neonatal sepsis and other infections due to group B beta-hemolytic streptococci. *N. Engl. J. Med.* **271**:1221–1228.

26. **Farley, M. M., R. C. Harvey, T. Stull, J. D. Smith, A. Schuchat, J. D. Wegner, and D. S. Stephens.** 1993. A population-based assessment of invasive disease due to group B *Streptococcus* in nonpregnant adults. *N. Engl. J. Med.* **328**:1807–1811.

27. Faxelius, G., K. Bremme, K. Kvist-Christensen, P. Christensen, and S. Ringertz. 1988. Neonatal septicemia due to group B streptococci—perinatal risk factors and outcome of subsequent pregnancies. *J. Perinat. Med.* **16:** 423–430.

28. Fernandez, M., M. Hickman, and C. J. Baker. 1998. Antimicrobial susceptibilities of group B streptococci isolated between 1992 and 1996 from patients with bacteremia or meningitis. *Antimicrob. Agents Chemother.* **42:**1517–1519.

29. Ferrieri, P., P. P. Cleary, and A. E. Seeds. 1976. Epidemiology of group B streptococcal carriage in pregnant women and newborn infants. *J. Med. Microbiol.* **10:**114.

30. Fry, R. M. 1938. Fatal infections by haemolytic streptococcus group B. *Lancet* **i:**199–201.

31. Gallagher, P. G., and C. Watanakunakorn. 1985. Group B streptococcal bacteremia in a community teaching hospital. *Am. J. Med.* **78:**795–800.

32. Gardner, S. E., M. D. Yow, L. J. Leads, P. K. Thompson, E. O. Mason, and D. J. Clark. 1979. Failure of penicillin to eradicate group B streptococcal colonization in the pregnant woman: a couple study. *Am. J. Obstet. Gynecol.* **135:**1062–1065.

33. Garland, S. M., and J. R. Fliegner. 1991. Group B streptococcus and neonatal infections: the case for intrapartum chemoprophylaxis. *Aust. N.Z. Obstet. Gynaecol.* **31:** 119–122.

34. Gerards, L. J., B. P. Cats, and J. A. A. Hoogkamp-Korstanje. 1985. Early neonatal group B streptococcal disease: degree of colonisation as an important determinant. *J. Infect.* **11:**119–124.

35. Grubb, R., K. K. Christensen, P. Christensen, and V. Linden. 1982. Association between maternal Gm allotype and neonatal septicemia with group B streptococci. *Immunogenetics* **9:**143–147.

36. Hall, R. T., W. Barnes, W. Krishnan, D. J. Harris, P. G. Rhodes, J. Fayez, and G. L. Miller. 1976. Antibiotic treatment of parturient women colonized with group B streptococci. *Am. J. Obstet. Gynecol.* **124:**630–634.

37. Harrison, L. H., J. Elliott, D. M. Dwyer, J. P. Libonati, P. Ferrieri, L. Billmann, A. Schuchat, and the Maryland Emerging Infections Program. 1998. Serotype distribution of invasive group B streptococcal isolates in Maryland: implications for vaccine formulation. *J. Infect. Dis.* **177:**998–1002.

38. Jackson, L. A., R. Hilsdon, M. M. Farley, L. H. Harrison, A. L. Reingold, B. D. Plikaytis, J. D. Wenger, and A. Schuchat. 1995. Risk factors for group B streptococcal disease in adults. *Ann. Intern. Med.* **123:**415–420.

39. Kasper, D. L., L. C. Paoletti, M. R. Wessels, H. K. Guttormsen, V. J. Carey, H. J. Jennings, and C. J. Baker. 1996. Immune response to type III group B streptococcal polysaccharide-tetanus toxoid conjugate vaccine. *J. Clin. Invest.* **98:**2308–2314.

40. Lancefield, R. C., and R. Hare. 1935. The serological differentiation of pathogenic and non-pathogenic strains of hemolytic streptococci from parturient women. *J. Exp. Med.* **61:**335–349.

41. Lerner, P. I., K. V. Gopalakrishna, E. Wolinsky, M. C. McHenry, J. S. Tan, and M. Rosenthal. 1977. Group B *Streptococcus* (*S. agalactiae*) bacteremia in adults: analysis of 32 cases and review of the literature. *Medicine* **56:** 457–473.

42. Lim, D. V., K. S. Kanarek, and M. E. Peterson. 1982. Magnitude of colonization and sepsis by group B streptococci in newborn infants. *Curr. Microbiol.* **7:**99–101.

43. Liston, T. E., R. E. Harris, S. Foshee, and D. M. Null. 1979. Relationship of neonatal pneumonia to maternal urinary and neonatal isolates of group B streptococci. *South. Med. J.* **72:**1410–1412.

44. McDonald, H., R. Vigneswaran, and J. A. O'Loughlin. 1989. Group B streptococcal colonization and preterm labor. *Aust. N.Z. J. Obstet. Gynaecol.* **29:**291–293.

45. McKenzie, H., M. L. Donnet, P. N. Howie, N. B. Patel, and D. T. Benvie. 1994. Risk of preterm delivery in pregnant women with group B streptococcal urinary infections or urinary antibodies to group B streptococcal and *E. coli* antigens. *Br. J. Obstet. Gynaecol.* **101:**107–113.

46. Merenstein, G., W. A. Todd, G. Brown, C. C. Yost, and T. Luzier. 1980. Group B β-hemolytic *Streptococcus*: randomized controlled treatment study at term. *Obstet. Gynecol.* **55:**315–318.

47. Moller, M., A. C. Thomsen, K. Borch, K. Dinesen, and M. Zdravkovic. 1984. Rupture of fetal membranes and premature delivery associated with group B streptococci in urine of pregnant women. *Lancet* **ii:**69–70.

48. Pass, M. A., B. M. Gray, and H. C. Dillon. 1982. Puerperal and perinatal infections with group B streptococci. *Am. J. Obstet. Gynecol.* **143:**147–152.

49. Pass, M. A., S. Khare, and H. C. Dillion. 1980. Twin pregnancies: incidence of group B streptococcal colonization and disease. *J. Pediatr.* **97:**635–637.

50. Philipson, E. H., D. A. Palermino, and A. Robinson. 1995. Enhanced antenatal detection of group B streptococcus colonization. *Obstet. Gynecol.* **85:**437–439.

51. Pyati, S. P., R. S. Pildes, N. M. Jacobs, R. S. Ramamurthy, T. F. Yeh, D. S. Raval, L. D. Lilien, P. Amma, and W. I. Metzger. 1983. Penicillin in infants weighing two kilograms or less with early-onset group B streptococcal disease. *N. Engl. J. Med.* **308:**1383–1388.

52. Regan, J. A., M. A. Klebanoff, R. P. Nugent, D. A. Eshenbach, W. C. Blackwelder, Y. Lou, R. S. Gibbs, P. J. Rettig, D. H. Martin, and R. Edelman. 1996. Colonization with group B streptococci in pregnancy and adverse outcome. *Am. J. Obstet. Gynecol.* **174:**1354–1360.

53. Regan, J. A., M. A. Klebanoff, R. P. Nugent, and Vaginal Infections and Prematurity Study Group. 1991. The epidemiology of group B streptococcal colonization in pregnancy. *Obstet. Gynecol.* **77:**604–610.

54. Rosenstein, N. E., A. Schuchat, and Neonatal Group B Streptococcal Disease Study Group. 1997. Opportunities for prevention of perinatal group B streptococcal disease: a multistate surveillance analysis. *Obstet. Gynecol.* **90:** 901–906.

55. Rouse, D. J., R. L. Goldenberg, S. P. Cliver, G. R. Cutter, S. T. Mennemeyer, and C. A. Fargason. 1994. Strategies for the prevention of early-onset neonatal group B streptococcal sepsis: a decision analysis. *Obstet. Gynecol.* **83:**483–494.

56. Schauf, V., and V. Hlaing. 1976. Group B streptococcal colonization in pregnancy. *Obstet. Gynecol.* **47:**719–721.

57. Schuchat, A., K. Deaver-Robinson, B. D. Plikaytis, K. Zangwill, J. Mohle-Boetani, and J. D. Wenger. 1994. Multistate case-control study of maternal risk factors for

neonatal group B streptococcal disease. *Pediatr. Infect. Dis. J.* **13**:623–629.

58. **Schuchat, A., M. J. Oxtoby, S. L. Cochi, A. K. Sikes, A. Hightowe, B. Plikaytis, and C. V. Broome.** 1990. Population-based risk factors for neonatal group B streptococcal disease: results of a cohort study in metropolitan Atlanta. *J. Infect. Dis.* **162**:672–677.

59. **Schwartz, B., A. Schuchat, M. J. Oxtoby, S. L. Cochi, A. Hightower, and C. V. Broome.** 1991. Invasive group B streptococcal disease in adults. *JAMA* **266**:1112–1114.

60. **Siegel, J. D., and N. B. Cushion.** 1996. Prevention of early-onset group B streptococcal disease: another look at single-dose penicillin at birth. *Obstet. Gynecol.* **87**:692–698.

61. **Siegel, J. D., G. H. J. McCracken, N. Threldkeld, B. M. DePasse, and C. R. Rosenfeld.** 1982. Single-dose penicillin prophylaxis of neonatal group B streptococcal disease. *Lancet* **i**:1426–1430.

62. **Tan, J. S., J. L. Anderson, C. Watanakunakorn, and J. P. Phair.** 1975. Neutrophil dysfunction in diabetes mellitus. *J. Lab. Clin. Med.* **85**:26–33.

63. **Verghese, A., S. L. Berk, L. J. Boelen, and J. K. Smith.** 1982. Group B streptococcal pneumonia in the elderly. *Arch. Intern. Med.* **142**:1642–1645.

64. **Whitney, C. G., B. D. Plikaytis, W. S. Gozansky, J. D. Wenger, and A. Schuchat.** 1996. Prevention practices for perinatal group B streptococcal disease: a multistate surveillance analysis. *Obstet. Gynecol.* **89**:28–32.

65. **Yagupsky, P., M. A. Menegus, and K. R. Powel.** 1991. The changing spectrum of group B streptococcal disease in infants: an eleven-year experience in a tertiary care hospital. *Pediatr. Infect. Dis. J.* **10**:801–808.

66. **Yancey, M. K., P. Duff, P. Kubilis, P. Clark, and B. H. Frentzen.** 1996. Risk factors for neonatal sepsis. *Obstet. Gynecol.* **87**:188–194.

67. **Yancey, M. K., A. Schuchat, L. K. Brown, V. L. Ventura, and G. R. Markenson.** 1996. The accuracy of late antenatal screening cultures in predicting genital group B streptococcal colonization at delivery. *Obstet. Gynecol.* **88**:811–815.

68. **Zangwill, K. M., A. Schuchat, and J. D. Wenger.** 1992. Group B streptococcal disease in the United States, 1990: report from a multistate active surveillance system. *Morbid. Mortal. Weekly Rep. CDC Surveill. Summ.* **41**:25–32.

Genetics and Pathogenicity Factors of Group C and G Streptococci

HORST MALKE

16

Die Systeme

Prächtig habt ihr gebaut. Du lieber Himmel! Wie treibt man, nun er so königlich erst wohnet, den Irrthum heraus!

Systems

Splendidly did you construct your sublime philosophical systems. Heaven! how shall we eject errors that live in such style.

J. W. von Goethe and F. von Schiller, Xenions

The Lancefield group C (GCS) and group G (GGS) streptococci carry the immunodeterminant residues *N*-acetylgalactosamine and rhamnose, respectively, on the oligosaccharide side chains of their cell wall carbohydrate antigens. These organisms are distributed in both humans and animals. They are isolated as opportunistic commensals from the skin, nose, throat, vagina, and gastrointestinal tract but may also be associated with clinically important infections of these sites and with hospital outbreaks. Serious diseases caused by GCS and GGS often resemble those due to group A streptococci (GAS) and include septicemia, pharyngitis, cellulitis, otitis media, septic arthritis, meningitis, infective endocarditis, and multiple organ abscesses (19). The genetic relationships of GCS and GGS strains are complicated and incompletely resolved, owing to the diversity of species within the serogroups or the present uncertainties in assigning species names, in particular, to the human GGS strains. The current classification (Table 1) relies on habitats, pathogenicity properties, physiological characteristics, and relationships of informational macromolecules, with serological grouping being useful for differentiating infraspecific biotypes. Before the advent of recombinant DNA technology, GCS or GGS had rarely been subjected to genetic studies. However, studies of the host range of bacteriophages isolated from GAS, GCS, and GGS, and of their transducing potential, provided evidence for intergroup phage reactions and intergroup transduction between strains belonging to different Lancefield groups, thus amending the original notion of the strict group specificity of streptococcal phage-host interactions (93, 108). The application of recombinant DNA techniques has advanced our understanding of GCS and GGS in diverse areas, and this chapter concentrates on the structure and function of genes and proteins studied at the molecular level in recent years.

MOLECULAR TAXONOMIC APPROACHES TO THE CLASSIFICATION OF GCS AND GGS

On the basis of 16S rRNA comparative sequence analysis, GCS and GGS fall into two species groups, the pyogenic and the anginosus group, the latter also known as the "*Streptococcus milleri*" group (45). Chromosomal DNA-DNA hybridization studies combined with biochemical tests indicate that the pyogenic streptococci previously named *Streptococcus dysgalactiae* and "*Streptococcus equisimilis*," together with those belonging to the large-colony-forming GGS and serogroup L strains, exhibit high levels of DNA sequence identity and, on the basis of the commonly accepted species level hybridization (≥70%), constitute a single species, *S. dysgalactiae* (23). Whole-cell protein electrophoresis revealed two subpopulations within this species, leading to differentiation between *S. dysgalactiae* subsp. *equisimilis* comprising GCS and GGS strains of human origin and *S. dysgalactiae* subsp. *dysgalactiae* that harbors animal GCS and serogroup L strains (107). GGS isolated from animals always appear to qualify as *Streptococcus canis*. On grounds of DNA sequence similarity, "*Streptococcus zooepidemicus*" is closely related to *Streptococcus equi*, warranting a similar subdivision of this species into the *S. equi* subsp. *equi* and *S. equi* subsp. *zooepidemicus* (23). Recent epidemiological studies have increasingly relied on genotypic classification methods such as restriction endonuclease digestion of total cell DNA followed by conventional (64) or pulsed-field gel electrophoresis (PFGE) analysis of macrorestriction fragment patterns (5), analysis of restriction fragment length polymorphism of rDNA (ribotyping) (87), randomly amplified polymorphic DNA (RAPD) fingerprinting (6, 32), multilocus enzyme electrophoresis (MLEE) (7), and typing based on comparative sequence analyses of the fast-evolving nonfunctional 16S-23S rRNA intergenic spacer region (11, 26). Taken together, the results obtained, by and large, corroborate the species classification given in Table 1 and, moreover, testify

Gram-Positive Pathogens, ed. by V. A. Fischetti et al.

TABLE 1 Classification of GCS and GGS[a]

Taxon	Lancefield group	Host	Specific distinguishing reactions
S. dysgalactiae subsp. *equisimilis*	C, G	Humans	Streptokinase activity on human plasminogen; C5a peptidase production; large-colony-forming; GGS bind IgG Fc fragment
S. dysgalactiae subsp. *dysgalactiae*	C, L	Various animals	No streptokinase activity on human plasminogen; no C5a peptidase production
S. equi subsp. *equi*	C	Horses	Ribose and sorbitol fermentation negative
S. equi subsp. *zooepidemicus*	C	Various animals, humans	Ribose and sorbitol fermentation positive
S. canis	G	Dogs, cows, cats	Beta-hemolytic; α- and β-D-galactosidase production
"*S. milleri*"	C, G, F, A	Humans	Beta-hemolytic; small-colony-forming; GGS do not bind IgG Fc fragment

[a] Adapted from references 12, 13, 19, 23, and 107.

to the utility of these methods for subspecific typing. For example, RAPD fingerprinting has proved useful in epidemiological investigations designed to delineate new and persistent *S. dysgalactiae* strains that caused intramammary infections in dairy cows (74). Comparison of the discriminatory power of different genomic typing methods revealed that RAPD excels MLEE for GGS (and GAS) typing (6) and that PFGE is even more efficacious than MLEE and RAPD, as indicated by its potential to identify 93 distinct types among a total of 41 GAS strains and 58 GCS or GGS strains, with no types common to strains of different species (5). Although a comprehensive battery of species-specific gene probes for identification of GCS and GGS has yet to be developed, first attempts to approach this goal have focused on oligonucleotides designed from 16S and 23S rRNA sequences (98). Alternatively, future differential genome analysis applied to GCS and GGS might identify genes that are responsible for species-specific features.

CELL SURFACE-ASSOCIATED PROTEINS

One hallmark of the gram-positive pathogens is the synthesis of specific cell wall-associated proteins that enable them to interact in various ways with proteins present in the body fluids or extracellular tissue matrix of their mammalian hosts. Such interactions may facilitate colonization, lead to molecular host mimicry, or interfere with various host defenses against invasion.

M and M-like Proteins

The major bona fide virulence determinant of GAS, M protein, is generally accepted to comprise polymorphic, fibrillar surface-exposed polypeptides that share common structural and functional features, among them a high propensity to form an alpha helix with a periodicity indicative of a coiled-coil structure, the capability of inhibiting phagocytosis by human neutrophils, and the potential to elicit opsonic antibodies that provide type-specific protection. Several studies conducted mainly in the 1980s provided highly suggestive evidence based on serological cross-

reactions with typing sera or monoclonal antibodies raised to M protein from GAS, phagocytosis inhibition, and DNA hybridization analysis that M proteins or M-like proteins are also encoded and expressed by GCS and GGS (24). Cloning and sequence determination of the corresponding genes afforded conclusive evidence in support of this notion. The first M protein gene, *emmG1*, to be isolated from a human GGS strain encodes a product with structural features characteristic of GAS class I M proteins that are epidemiologically associated with rheumatogenic, opacity factor-negative serotypes (14). As expected, structural similarity between MG1 and the GAS class I M proteins was significantly greater in the C-terminal regions (containing the anchor domains and the C- and D-repeats) than in the N-terminal portions (containing the A- and B-repeats) of the mature proteins (86–94% versus 31–35% amino acid identities), an observation consistent with a difference in antigenicity between MG1 and the GAS M types. An *emmG1* probe encoding the variable region of MG1 detected homologous DNA in seven additional M-positive GGS strains but also revealed at least four distinct *emm* alleles harbored by these strains. The essential aspects of these findings were confirmed and expanded by others (88, 91, 92, 95) using a greater and epidemiologically unrelated number of GGS isolates that also included strains recovered from animal sources. It is noteworthy that none of the DNA specimens isolated from GGS strains originating from animals hybridized with appropriate *emm* gene probes from GAS whereas all DNAs from human GGS strains (large-colony-forming isolates) did so (88, 92). Probes designed to differentiate between GAS M class I and M class II genes detected only GGS *emm* genes related to GAS M class I genes (88). The complete sequence of an additional GGS *emm* gene (*emmLG593*), together with the sequences of the 5′ ends of 30 additional GGS *emm* genes, revealed six distinct genetic types that, in paired comparisons, shared <95% sequence identity in the *emm* gene region coding for the mature N termini (88). This observation adds further support to the genetic heterogeneity of the M proteins of GGS and, if consistent with future supplementary opsonophagocytotic studies, may be used for the development of GGS M typing schemes for epidemiological purposes. Most interestingly, in the latter studies (88) as well as those conducted independently by others (91, 95), certain GGS *emm* genes were identified, defined portions of which exhibit exceedingly high levels of sequence identity to the variable regions of well-established GAS class I M protein genes, most notably *emm12* and *emm57*. Inasmuch as the results of total genomic DNA-DNA hybridization, chromosomal restriction endonuclease profiling, and MLEE all indicate that GGS are highly divergent from GAS in overall genomic character and, hence, have no recent GAS ancestors, the most likely explanation for the high local homology of certain *emm* genes in the two serogroups is intergroup horizontal genetic exchange followed by homologous recombination events (91). Although several scenarios can be envisaged to suggest possible modes of DNA transfer, the actual transfer mechanisms are entirely left in the dark.

In GCS isolated from horses, three M or M-like proteins have been characterized at the molecular level: SeM (nearly identical to FgBP) (70) produced by *S. equi* subsp. *equi*, and SzP and SzPSe produced by *S. equi* subsp. *zooepidemicus* (101, 102). SzP and SzPSe are structurally only marginally related to SeM. They are antigenically cross-reactive and are coded by allelic variants of the same hypervariable gene. Functionally, however, all three proteins resemble M proteins of GAS in being capable of inhibiting phagocytosis, eliciting serum opsonic and protective responses, and binding (equine) fibrinogen (70, 90, 101). The complete sequences now available for these proteins are consistent with earlier serological and other observations indicating that *S. equi* subsp. *equi* is antigenically conserved, whereas equine isolates of *S. equi* subsp. *zooepidemicus* exhibit great antigenic variation (44). These observations, together with data derived from comparative MLEE, suggest that *S. equi* subsp. *zooepidemicus* is the ancestral species from which a clone now named *S. equi* subsp. *equi* has emanated (44). In comparison to the SzP protein family, SeM appears to be the stronger opsonizing antigen in small laboratory animals (101) and, therefore, holds some promise with respect to the development of a vaccine efficacious against equine strangles (70, 90). Of note, in the regions external to the cell wall, the SeM and SzP family proteins share no obvious sequence homology with GAS M proteins (101). However, M antigens related to the GAS class I M proteins and to those of the human GGS strains have also been detected on GCS isolated from humans (9, 81). In this connection, the FAI protein found in a GCS isolate from a horse (99) deserves special attention. FAI consists of overlapping functional modules responsible for fibrinogen, immunoglobulin G (IgG), and albumin binding. Its three C-repeat copies share 40 to 50% sequence identity with the B-repeats of GAS M type 49 and the C-repeats of M12 but exhibit no sequence similarity to any portions of SeM or the SzP proteins. Since GAS are usually not distributed in hosts other than primates, it seems doubtful whether the strain producing FAI is a genuine horse strain and whether the *fai* gene results from intergroup DNA exchange (101).

IgG-Binding Proteins

The most thoroughly studied cell surface protein produced by many human GGS and GCS strains is protein G, which binds the heavy chains of IgG Fc and Fab with an affinity comparable to that of antigen. Conventional thinking implies a contribution to adherence, inhibition of phagocytosis, and host mimicry in the role that protein G might have in pathogenesis; however, strong evidence in favor of this notion is missing. Although the structure of protein G is reminiscent of staphylococcal protein A, the two polypeptides share, except for their C termini, little sequence identity (21, 37). The protein G genes (*spg*) of distinct *S. dysgalactiae* strains may differ in their actual nucleotide sequence and size. However, allowing for processes like duplication of internal repeats, and recombination and replication slippages involving repeats (21, 75), different *spg* alleles exhibit a very high level of sequence identity, making protein G a unique type among the diverse group of gram-positive Ig-binding proteins. Compared to protein A, protein G binds not only to human, rabbit, mouse, and guinea pig IgG but also to IgG from cow, horse, sheep, and goat, which protein A does not bind. In addition to showing wider IgG species reactivity, protein G, unlike protein A, also binds human subclass IgG3. Finally, protein G's reactivity is strictly limited to IgG (protein A also binds IgM) (37). Because of these properties, protein G has become one of the most versatile tools in immunochemistry. Dissection of the molecule by enzymatic and genetic means followed by functional analysis of defined fragments confirmed early observations indicating that protein G is bifunctional in binding both IgG and serum albumin and led to the identification of the regions responsible for these activities. Thus, IgG Fc binding resides in the C-terminal portion of the molecule and is mediated by two or more repeated domains, whereas the albumin-binding activity is a separate function of repeated domains in the N-terminal half (2, 37, 72). Fragmentation of human serum albumin located its single protein G-binding site to disulfide loops 6 to 8, with a 5.5-kDa pepsin fragment showing complete inhibition of the interaction between protein G and the full-length albumin (22). The three-dimensional structures of the 56-residue B1 and 57-residue B2 IgG Fc-binding domains of protein G from GGS strain GX7809 (21) have been determined by nuclear magnetic resonance (NMR) spectroscopy (35) and X-ray diffraction analysis (1), respectively. The two approaches yielded essentially identical overall structures characterized by a four-stranded β sheet and a helix lying diagonally across the sheet and connecting the outer two β strands, β2 and β3. The IgG-binding interface of this structure is formed by negative charges distributed around the helix and β3 (34). The entire folding motif is very compact and shows high thermal stability owing to extensive hydrogen bonding and tight packing of its buried hydrophobic core. Taking advantage of these properties, the folding thermodynamics and mechanism of the B1 segment have been studied (89) and an M13 phage-B1 domain display system developed to pinpoint primary structure requirements for thermodynamic stability of the B1 domain (76). Interestingly, the IgG-binding domains also bind, in a nonimmune manner, heavy-chain constant-region sequences of the Fab portion of IgG but use disparate parts of their structure to bind the heavy chains of Fc and Fab. The crystal structure of the complex between Fab and the third IgG-binding domain of protein G shows that the outer β2 strand of the molecule aligns in an antiparallel fashion with the last β strand of the constant heavy-chain domain of Fab, CH1 (18). In this manner, the β sheet structure is extended into protein G. Since the CH1 domain is highly conserved in the IgG Fab fragments from diverse species, these findings neatly explain the broad specificity of the protein G-IgG interaction. Moreover, provided that backbone-backbone hydrogen bonding between the interacting β strands is preserved, this arrangement allows Fab binding in the presence of considerable primary structure variability.

In recent years, IgG-binding proteins have been found that show interesting functional differences from the prototype protein G molecules. The *mag* gene cloned from a *S. dysgalactiae* strain codes for a polypeptide with three discrete domains, the C-terminal and central ones being homologous to the IgG and albumin binding regions, respectively, of protein G, whereas the unique N-terminal domain binds α_2-macroglobulin, a plasma proteinase inhibitor (41). The *mig* gene isolated from a case of bovine mastitis specifies a protein with binding activity to IgG, which resides in five C-terminal repeats, and to the α_2-macroglobulin-proteinase complex. The latter activity is located in the N-terminal portion of the Mig protein, which exhibits only weak sequence similarity to the α_2-macroglobulin-binding domain of the Mag protein. Furthermore, protein Mig has no albumin-binding activity (43). Finally, the Zag protein, isolated from an *S. zooepidemicus* strain, shows a molecular architecture that is similar to that of the Mag protein, but its N-terminal α_2-macroglobulin-binding domain represents a third unique type of this module with little sequence similarity to the corresponding regions of either the Mag or Mig proteins. However, inhibition assays showed that all three proteins compete for the same or overlapping sites in the α_2-macroglobulin molecule (42). The pathogenetic role of these proteins remains to be established; however, evidence has been published showing that proteinase-complexed α_2-macroglobulin bound to animal GCS inhibits phagocytosis in vitro (106).

Fibronectin-Binding Proteins

It was not until the beginning of the 1990s that surface proteins of GAS, GCS, or GGS that bind to fibronectin were characterized at the molecular level. The primary function of this adhesive matrix protein is to mediate substrate adhesion of eukaryotic cells via integrin receptors that bind to central parts of the molecule containing the RGD sequence motif. Among the best-studied GCS and GGS proteins responsible for fibronectin binding are those encoded by the *S. dysgalactiae* genes *fnbA*, *fnbB* (cloned from the same strain) (52, 54), and *gfbA* (47), the *S. equisimilis fnb* gene (53), and the *S. zooepidemicus fnz* gene of the animal Z5 strain, which also produces the Zag protein (see above) (55). The mature proteins appear to occur in two size classes of about 60 and 120 kDa, but they all consist of C-terminal repetitive domains that are close to the cell wall-spanning regions and mediate binding to the 29-kDa N-terminal segment of fibronectin. Comparison between the C-terminal repeats reveals a consensus sequence consisting of two adjacent glycins surrounded by highly acidic tripeptides that are required for binding activity, as indicated by the inhibitory effect of synthetic peptides that mimic individual repeats and show a high level of cross-reactivity (40, 65). In addition to these interactions, the sequence motif LAGESGET separating the first and second repeat of the Fnz protein has been identified as an important part of a fibronectin-binding domain that does not bind the 29-kDa fibronectin fragment (55). The fibronectin-binding proteins F1 and F2 of GAS also have two domains, each responsible for binding. The F1 domain N-terminal to the repetitive domain contains the LAGES-GET motif (39). The F2 protein, which is homologous to

FnbB and Fnb, contains a nonrepeated domain, designated UFBD, which lies upstream of the repeat region and exhibits fibronectin binding independently of the repeat domain (39). UFBD displays a high degree of sequence similarity to FnbB and Fnb regions N-terminal to their repetitive domains, strongly suggesting that these two proteins also use more than one domain for ligand binding. The same is true of FnbA, in which a previously unidentified region, Au, upstream of the repeats is capable of independent fibrinogen binding (94). Interestingly, a monoclonal antibody raised to FnbA enhances, rather than inhibits, fibronectin binding to protein fragments or synthetic peptides containing Au and recognizes only the Au-fibrinogen complex (94). Circular dichroism analysis suggested that the ligand binding site containing Au has little secondary structure, but on binding fibronectin is induced to form a predominantly β sheet structure, which is seen by the antibody (38). The infected host might respond to such ligand-induced conformational changes with antibodies directed against the complex, thus stabilizing, rather than blocking, tissue adherence of the pathogen (94). That fibronectin-binding proteins do contribute to adherence is shown, e.g., by the correlation between the presence of the *gfbA* gene (in 36% of GGS strains tested) and the ability of such cells to adhere to human skin fibroblasts, as well as by the transformability of a nonadherent strain to the adherence phenotype upon introduction of *gfbA* (47).

Plasmin(ogen)-Binding Proteins

Besides GAS, certain strains of human GCS and GGS as well as bovine GGS have been found to bind native human plasminogen and, with lesser affinity, an N-terminal plasminogen fragment containing kringles 1 to 3 (105). The bovine GGS isolates were additionally able to bind efficiently to miniplasminogen (B chain plus kringle 5). Correspondingly, Scatchard analysis revealed a two-phase interaction with the bovine strains, yielding equilibrium dissociation constants (K_d values) of about 20 and 400 nM for binding to plasminogen and miniplasminogen, respectively. This suggested that bovine GGS strains display two different binding substances that interact with different portions of plasminogen and show more than 10-fold-different affinities (105). However, in these experiments, the nature of the proteins responsible for binding remained unknown. Sequence information is available, however, for two related M-like proteins from human GCS and a GGS strain capable of binding plasminogen, fibrinogen, and serum albumin to distinct sites of these proteins (4).

The surprising discovery in GAS that the glycolytic enzyme glyceraldehyde-3-phosphate dehydrogenase (GAPDH), which possesses none of the hallmarks of grampositive cell surface proteins, can be localized to the cell wall to function as a major binding protein for a variety of mammalian proteins, including plasmin (56, 77), prompted the isolation and functional analysis of the corresponding gene (*gapC*) in the human group C strain H46A (30). In this strain, *gapC* is an essential gene that occurs in single copy and is abundantly transcribed (Fig. 1). Its product shares 95% sequence identity with *Streptococcus pyogenes* GAPDH. Moreover, the number and position of lysine residues shown to be involved in plasmin(ogen) binding (30, 56) are identical in the two proteins. The binding parameters for human plasminogen and plasmin of purified GapC protein were determined by real-time biospecific interaction analysis, with results indicating that the protein

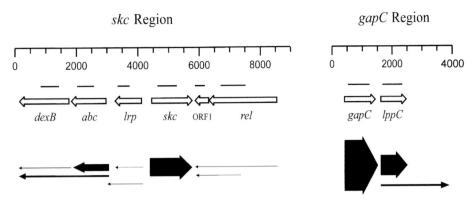

FIGURE 1 Organization of the streptokinase (*skc*) and GADPH (*gapC*) gene regions in the *S. equisimilis* H46A chromosome. Open arrows represent the genes (*dexB*, α-glucosidase; *abc*, ABC transporter; *lrp*, leucine-rich protein of unknown function; ORF1, open reading frame of unknown function; *rel*, bifunctional (p)ppGpp synthetase and (p)ppGppase; *lppC*, cytoplasmic membrane lipoprotein acid phosphatase) and their orientation. Black bars above the genes show the probes used for quantitative Northern hybridizations. Transcripts are represented by black arrows, the thickness of which reflects relative mRNA abundance.

binds plasmin and plasminogen with about 10-fold-different affinities (K_d values, 25 nM and 220 nM, respectively). Thus, the zymogen and active enzyme appear to possess low-affinity binding sites for the *gapC* gene product, a result at variance with that ascribed to GAS GAPDH, whose plasmin binding activity was reported to be 3 orders of magnitude higher (10). This conflict, which has not been satisfactorily resolved, may have to do with the new observation that, on reexamination, the GAS strain involved turned out to display other plasmin-binding substances on the surface in addition to GAPDH (17). A candidate substance could well be the glycolytic enzyme α-enolase, which has recently been identified as a novel ubiquitous streptococcal surface protein with strong plasmin(ogen)-binding capacity (78). Given the 2-μM concentration of plasminogen in plasma, even the low K_d value for the association of this protein with GapC would suggest that, in the vasculature, the great majority of the *S. equisimilis* H46A plasminogen binding sites are occupied, even if GapC is the only plasminogen-binding protein. A streptokinase-negative mutant of H46A, isolated to eliminate endogenous plasminogen-activating capacity, has indeed been shown to bind plasminogen, as indicated by the generation of cell surface-associated plasmin activity following exposure to exogenous plasminogen activators (61). A novel aspect characterizing the interaction of plasmin(ogen) with surface GAPDHs relates to the finding that limited prior treatment with plasmin greatly increases the subsequent plasminogen-binding capacity of these proteins (30). Presumably, the underlying mechanism involves the exposure of new plasminogen binding sites on GAPDH by plasmin, which preferentially catalyzes the hydrolysis of α-amino-substituted lysine (and arginine) esters. In the infected host, streptococci capable of transporting GAPDH (and α-enolase) to the cell wall and secreting streptokinase would thus appear to have evolved a remarkable system to generate and amplify cell surface protease activity. Host plasminogen bound with low affinity can be converted to the active enzyme directly on the cell surface. The cell-bound plasmin may serve to generate additional plasminogen binding sites on GAPDH to start new cycles of

plasminogen binding and activation, resulting in amplification of the tissue-invasive potential of the pathogen. Although there exist strains producing proteins unrelated to GAPDH that display high-affinity plasmin(ogen) binding sites (4), the special role of GAPDH (and possibly α-enolase) is due to its general occurrence on the surface of the pyogenic streptococci (77, 78). A major challenge to what is presently understood of the plasminogen-binding and -activating capacity of pathogenic streptococci relates to the identification of possible factors involved in the modulation of cell surface protease activity that may be necessary to avoid destructive proteolysis of surface structures important for colonization and invasion. It is interesting to note that a screen designed to identify *Escherichia coli* genes expressed preferentially by natural isolates when incubated in nutrient-poor aquatic medium identified a streptococcal *gapC* homolog (85% sequence homology), suggesting acquisition of the gene by horizontal transfer from a gram-positive organism and possible involvement in the preparation of cells for the colonization of a new host (20).

Complement Fragment C5a Peptidase

Complement fragment C5a functions as a chemotaxin that mediates phagocyte recruitment to the site of infection. GAS and human GGS have been found to produce a highly specific cell surface endopeptidase that cleaves C5a, thereby limiting the activity of this chemotactic signal (13). In Southern hybridization experiments, a probe specific for the GAS C5a peptidase gene *scpA* detected homologous *scpG* sequences in all human GGS strains tested but failed to detect the gene in GGS isolated from animals. The *scpA* and *scpG* genes appear to be very similar, and their products cross-react serologically (13). However, *scpG* shows different linkage relationships, because, in the GGS genome, it is not located 3′ adjacent to *emm* or *emm*-like genes, as is the case for *scpA* in the *mga* regulon of GAS (13, 95). Furthermore, the expression of *scpG* does not appear to be positively controlled by *mga* since this regulatory protein seems to have no homolog in GGS (95). As already discussed above, loss of synteny of the component genes of the *mga* regulon in GGS may be taken to indicate recent acquisition of individual genes by horizontal gene transfer.

CYTOPLASMIC MEMBRANE-ASSOCIATED ENZYMES

Hyaluronan Synthase

The hyaluronan (hyaluronic acid) capsule of the pyogenic streptococci has an important role in resistance to phagocytosis and in virulence. In GAS, the three genes required for hyaluronan synthesis, *hasA*, *hasB*, and *hasC*, constitute an operon (16). An authentic *hasA* homolog encoding hyaluronan synthase, which catalyzes the alternate addition of UDP-acetylglucosamine and UDP-glucuronic acid to form the capsule polymer, has been cloned from *S. equisimilis* (48). The GCS enzyme (seHAS) sequence is 72% identical to that of its homolog from GAS (spHAS) and, accordingly, the two proteins cross-react serologically. Furthermore, the predicted membrane topology of seHAS is identical to that of spHAS; both proteins are processive enzymes, and the size distributions of the hyaluronan chains they synthesize are similar. However, seHAS shows a 2-fold-faster chain elongation rate than spHAS, making it the most active hyaluronan synthase described thus far (48). These findings leave no doubt that a putative GCS hyaluronan synthase described previously by others (49) is of mistaken identity. Although a 6.2-kb restriction fragment from an encapsulated GCS strain was shown to hybridize with a *hasA*-specific probe and is potentially large enough to accommodate *hasB* and *hasC* also (16), definitive proof for the existence and functioning of this operon in GCS or GGS has not been reported.

Cytoplasmic Membrane Lipoprotein Acid Phosphatase

Attempts to elucidate the linkage relationships of the *S. equisimilis* H46A *gapC* gene resulted in the discovery of a novel streptococcal gene 3' adjacent to, and co-oriented with, *gapC* (31) (Fig. 1). This gene, designated *lppC*, encodes a membrane lipoprotein, as indicated by the presence of a lipoprotein-specific signal sequence cleavage and lipidation site (VTG⎮C), by globomycin sensitivity of signal sequence processing, and by the tight association of the protein with the streptococcal cytoplasmic membrane or the outer membrane of *E. coli* when heterologously expressed in this organism. Southern, Northern, and Western analyses revealed that *lppC* has homologs in GAS and is transcribed independently of *gapC* as monocistronic mRNA from a σ^{70}-like consensus promoter. Database searching identified additional *lppC* homologs in pathogenic species, including *Flavobacterium meningosepticum*, *Haemophilus influenzae*, *Helicobacter pylori*, and *Enterobacter cloacae* (57). Among those, of particular interest is the *hel* gene of *H. influenzae*, which encodes the major outer membrane antigen *e* (P4) of this organism. Protein *e* (P4), which exhibits 58% sequence similarity to LppC, has been reported to be involved in the uptake of hemin as a source of porphyrin, an essential nutrient for *H. influenzae* when grown in aerobic conditions (84). However, unlike the *hel* gene, *lppC* failed to complement *hemA* mutants of *E. coli* for growth on hemin. Moreover, *S. equisimilis* could not grow on hemin in iron-limited medium, indicating that *lppC* is not involved in hemin utilization (31). Rather, bi-

ochemical, serological, and genetic evidence has been provided to show that the LppC protein functions as an acid phosphatase (57). This finding establishes a novel function for streptococcal lipoproteins, suggests a function for LppC homologs in other species, and requires reevaluation of the primary role of *e* (P4). However, the physiological or possibly pathophysiological role of LppC has yet to be elucidated. In this connection, it is interesting to note that acid phosphatases from several pathogenic species have been recognized as virulence factors that support intracellular survival by inhibiting the respiratory burst (85).

EXTRACELLULAR PROTEINS

Streptokinase

The first streptokinase gene (*skc*) to be cloned (58) and sequenced (62) originates from the most potent and commercially exploited producer of this plasminogen activator, *S. equisimilis* H46A. Later, a streptokinase gene (*skg*) from a human GGS strain was also characterized (109), and the two encoded proteins were found to share 98% sequence identity. Subsequent research has appreciably expanded our knowledge of this important streptococcal protein in various areas, and part of this work has been reviewed previously (63). Here I focus on recent studies done to elucidate the expression control of *skn* genes and the streptokinase-plasminogen structure-function relationships.

Expression Control of *skn* Genes

Since gene regulation is influenced by the spatial association of genes on the chromosome, it is appropriate to point out that the streptokinase genes show homology of synteny in the genomes of GAS and of human GCS and GGS (28, 68) (Fig. 1). Conservation of linkage appears also to be true of an equine group C streptokinase gene that hybridizes with *skc* (28) but encodes a streptokinase that preferentially activates horse plasminogen and antigenically differs from the proteins produced by human strains (71a). The transcription pattern of the genes in the *skc* region shows that *skc* is transcribed most abundantly (Fig. 1). A comparison of *skn* expression levels between H46A and two GAS strains (NZ131 and A374) reveals significant differences, with the relative ratios of skc_{H46A}, ska_{NZ131}, and ska_{A374} expression amounting to, respectively, 1:0.30:0.25 at the protein level and 1:0.42:0.32 at the mRNA level (27). The different expression levels cannot be accounted for by different mRNA decay rates, as *skc* and *ska* mRNAs have similar chemical half-lives on the order of 2 to 3 min (62a). As expected from their orientation relative to that of the neighboring genes (Fig. 1), *skc* and *ska* have been shown to be transcribed as 1.3-kb monocistronic mRNA. Transcription is terminated by a hypersymmetrical intrinsic terminator with bidirectional activity shared by the oppositely oriented *skc* and *rel-orf1* transcription units (96, 97). Analysis of the *skc* promoter by S1 nuclease mapping and *lacZ* reporter gene transcriptional fusions introduced in single-copy genomic form into the H46A chromosome or in multiple plasmid-form copies into *E. coli* revealed the existence of two overlapping σ^{70}-like consensus promoters (29). Under normal experimental conditions, the strength of the upstream P1 promoter greatly exceeds that of P2, as judged by comparing the abundance of the two *skc* transcripts in either host and assessing reporter gene activity in the wild type and a single-base pair P1 promoter down mutant. Most important, deletions covering the 202-bp region (USR) that separates the promoters of the divergently transcribed *skc* and *lrp* genes greatly decrease the *skc* core promoter strength in *S. equisimilis* but not in *E. coli*. The USR is rich in AT tracts, and computer modeling as well as circular permutation analysis showed that it contains a segment of intrinsically bent DNA, with the bend center located at position −100 with respect to the major P1-directed transcription initiation site (36). The pivotal role of the bending locus as a required element for full promoter strength is supported by deletions showing that DNA sequences upstream of, or between, the bending locus and the −35 core promoter region do not substantially stimulate *skc* transcription. In addition, the stimulatory effect of the bent DNA is slightly sensitive to a 5-bp deletion immediately downstream of the bend center, which would appear to result in spatial misorientation of the bend locus (27) (Fig. 2). These deletion data also discount the possibility that factor-independent contacts of the RNA polymerase α-subunits to the −40 to −60 promoter upstream region, as shown for some *E. coli* promoters, are an important determinant of *skn* promoter activity. To define further the contribution of the individual sequence elements to *skc* promoter strength, the core promoters were dissected by alteration of their −10 hexamers and the resulting constructs fused in the presence or absence of the USR to a luciferase reporter gene to allow real-time expression data to be determined in the homologous streptococcal system and in *E. coli* (33). In *S. equisimilis*, the reporter gene activity increased in the following order of the involved sequence elements: P2 ≈ P2+USR < P1 < P1+P2 < P1+USR < P1+P2+USR. This shows that P1 and P2 alone are extremely weak, but when combined they act in a mutual contributory manner. In the USR-less arrangement, only the combined core promoters have measurable, albeit low, activity, and the USR stimulates only P1 and the combination of P1+P2. In contrast, in *E. coli*, virtually the entire promoter activity derives from the P1 promoter, and the USR is dispensable for full promoter activity. The distinctive feature of P1 appears to be its extended −10

region, with a TG motif located one base upstream of the −10 hexamer. In *E. coli*, such promoters have been shown to function activator-independently, even in the absence of a recognizable −35 region. Recently, region 2.5 of σ^{70}, previously not identified as interacting with promoters, has been implicated in contacting the TG motif (3). It is interesting to note that the corresponding region of σ^{42}, the streptococcal homolog of σ^{70}, is highly conserved and, in particular, does carry the residues (H455 and E458) implicated in recognition of the TG extension by σ^{70} (Fig. 3). Nevertheless, as shown by the above findings, the streptococcal system is much more stringent with respect to supplementary structural requirements for full *skn* promoter activity and displays strain-specific *skn* expression levels. Considering the latter, it is pertinent to point out that, although the intergenic regions between the *skn* and *lrp* genes of GCS and GAS are homologous, there exist sequence differences that are highest in the regions containing the bending locus. However, allele swap experiments in which single-copy promoter-*lacZ* fusions were exchanged between the above-mentioned GCS and GAS strains indicated that sequence differences in the wild-type regions relevant for transcription initiation cannot account for the differential *skn* expression levels. Thus, expression of the GAS constructs was up-regulated in the GCS genetic background and, conversely, expression of the GCS construct was down-regulated in the GAS background. This shows that the host genetic background dictates *skn* expression levels, suggesting the existence of a *trans*-acting factor with strain-specific activity that may contact the bending locus, thereby stimulating *skn* expression by more than 1 order of magnitude (27). Further work is required to support the existence of this factor and to determine its identity. Furthermore, the significance of P2 remains enigmatic. Its existence was postulated by the detection of distinct minor transcripts with identical 5′ ends in both the homologous host and *E. coli* (29). To date, no environmental conditions have been ascertained that might increase its activity. P2 does not possess the conserved T in the −10 hexamer,

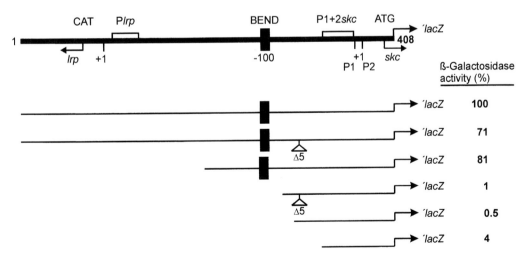

FIGURE 2 Identification of the bending locus (BEND) upstream of the streptokinase core promoter (P1+2*skc*) as the *cis*-acting element required for full *skc* expression. The scheme shows the relevant sequence features of the *skc-lrp* intergenic region and the position of deletions used to measure reporter gene (*lacZ*) activity of transcriptional fusions inserted as single copies via *dexB* into the *S. equisimilis* H46A chromosome. Transcription initiation sites (+1) and the start codons of *skc* and *lrp* are indicated. See text for more details.

Region 2.4   Region 2.5   Region 3

FIGURE 3 Sequence comparison of *E. coli* (3) and *S. pyogenes* σ factors (1a) between region 2.4 and the start of region 3. Identical and similar amino acids are marked by asterisks and dots, respectively. The residues identified in *E. coli* as contacting the TG extension of the −10 promoter hexamer are indicated by open boxes.

which, in *E. coli*, is contacted by specific amino acids (Q437 and T440) in region 2.4 of σ[70] (3). These amino acids (Q193 and T196) are also conserved in the corresponding σ[42] region of *S. pyogenes* (Fig. 3).

Structure-Function Relationships of Streptokinase

Streptokinase (SK) has at least three functions in the process of plasminogen (Pg) activation: (i) high affinity Pg binding to form a 1:1 stoichiometric Pg:SK complex; (ii) interacting with Pg to structure the active site, i.e., forming the virgin enzyme with amidolytic activity; and (iii) binding and processing of substrate Pg molecules, resulting in cleavage of the activating R561-V562 peptide bond. Before the recent solution of the crystal structure of SK complexed with microplasmin (μPm, i.e., Pg residues 542 to 791) (107a), various other biophysical methods, like nuclear magnetic resonance spectroscopy, circular dichroism, dynamic light scattering, or differential scanning calorimetry, have been employed to derive the overall domain organization of the protein (2a, 15, 69, 80, 100, 110, and references therein). In addition, structure-function relationships were studied by using defined fragments or specific mutant forms of the molecule (14a, 71, 83, 86, 112, 113, and references therein), including specific point mutations that impair the generation of a functional activator complex (50, 51). Furthermore, the functional significance of defined SK regions has been supported by anti-SK monoclonal antibodies that have binding specificities to these sequences and inhibit the formation of the SK-Pg activator complex (79). Figure 4 attempts to integrate the available information into a structural and functional map of the streptokinase molecule.

The solution structural studies have presented evidence that streptokinase is an elongated globular molecule that consists of at least three independent folding domains, designated A (approximately residues 1 to 145), B (residues 146 to 290), and C (residues 291 to 380), corresponding to the α, β, and γ domains recognized by X-ray crystallography. Fragments containing these autonomous domains adopt tertiary structures in solution that are closely related to those of the native molecule. Both terminal regions of the polypeptide chain are largely unstructured, explaining early observations showing that streptokinase muteins lacking 15 N-terminal or 41 C-terminal residues, or that truncated streptokinases fused to foreign proteins continue to activate plasminogen (46, 59, 60). In the binary complex, SK interacts with μPm most extensively through the α and γ domains, the former contacting, among others, the catalytic triad residues H and D, and the latter binding near the activation cleavage site (107a). However, the crystallographic model cannot rule out the possibility that the β domain interacts with the kringles of Pg in the Pg:SK activator complex, as previously proposed in an activation scheme based on functional studies in solution (113). Docking of substrate μPg to the μPm:SK activator complex such that the activation bond of Pg is positioned in the catalytic site of the resultant ternary complex reveals extensive contacts not only between the substrate molecule and μPm but also between the substrate and all three domains of SK on the rim, the extent of the latter interactions being in the descending domain order α, β, and γ (107a). Concerning the contact activation mechanism, the proposal derived from functional studies with truncated SK peptides (113) and the hypothesis resulting from the μPm:SK X-ray structure (107a) meet in assuming that the SK C or γ domain induces a conformational change in the active center of Pg to yield the virgin enzyme. The functional studies, capitalizing on including the Pg kringles in the interaction studies, further suggest that the A (or α) domain interacts with the kringle domain of a second Pg molecule. In the resultant ternary complex, the region near the activating peptide bond of the substrate molecule is restructured and brought in close proximity to the active center to facilitate the conversion to plasmin. In this scheme, SK peptide S60-K334 is essential for minimal activator capacity. This is greatly augmented by SK peptide V13-K59, which contains the N-terminal Pg binding site and functions to enhance substrate recognition. Taken together, these studies show that, except for the very ends of the polypeptide chain, most of the SK sequence is essential for efficient functionality, suggesting that any attempts at decreasing the size of the protein to reduce antigenicity may detrimentally affect Pg activation.

Streptolysins and Streptodornase

Based on the well-known observation that a positive antistreptolysin O test is not specific for recent GAS disease, successful attempts were made to clone and sequence the streptolysin O (*slo*) genes of GCS (*S. equisimilis*) and GGS (*S. canis*) (73). The *slo* gene proved to be highly conserved in the three serogroups, with sequence identity of the protein exceeding 98%. Expectedly, the *slo* gene was not detected in *S. zooepidemicus* (73) and *S. equi* (25). However, a streptolysin S-like activity has been characterized in the latter subspecies and shown to be subject to transposon inactivation at a single locus (25). The long-desired recent cloning and sequencing of a GAS gene (*sag*) shown to be associated with streptolysin S production (8) will certainly inspire further investigations into this unique and pathogenetically important streptococcal toxin, which has been difficult to characterize genetically.

In contrast to the highly conserved *slo* gene, the streptodornase gene *sdc*, first cloned from *S. equisimilis* H46A (111), encodes a DNase that exhibits only 48% sequence identity to a type D DNase encoded by the *sdaD* gene later isolated from a GAS strain (82). However, establishing the intergroup sequence relationships of the different streptococcal DNases must await the determination of the enzyme types that occur in GCS and GGS as well as the molecular characterization of the three remaining types from GAS.

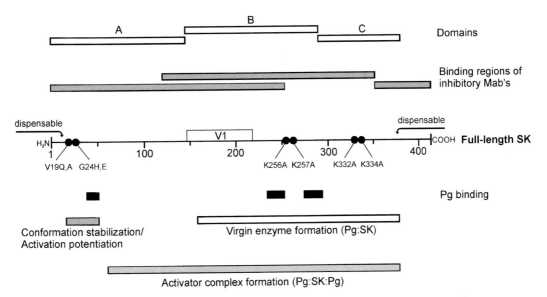

FIGURE 4 Structural and functional map of streptokinase (SK) produced by *S. equisimilis* H46A, based on data referenced in the text. The mature 414-amino-acid polypeptide is presented in the middle of the scheme, with the position of the major variable region, V1, indicated by an open box. Amino acid substitutions impairing plasminogen (Pg) activation are indicated by black circles. Autonomous structural domains (A, B, C) and regions important for the indicated functional activities as well as binding regions of inhibitory monoclonal antibodies (Mab's) are delineated by bars above and below the polypeptide chain.

Also, although implicated as potential virulence factors, the streptococcal DNases have, thus far, not directly been shown to be involved in pathogenicity.

THE STRINGENT AND RELAXED RESPONSES OF *S. EQUISIMILIS*

Sequence analysis of *S. equisimilis* chromosomal DNA 3' adjacent to the *skc* gene (68) led to the unexpected discovery of the 739-codon *rel* gene (Fig. 1), the first homolog from a gram-positive organism of the paralogous *E. coli relA* and *spoT* genes to be cloned, sequenced, and functionally characterized in some detail (66, 67). In *E. coli*, these genes are involved in the metabolism of guanosine 5',3'-polyphosphates [(p)ppGpp] that accumulate in response to starvation for amino acids or glucose exhaustion. The accumulated (p)ppGpp mediates a broad pleiotropic response hallmarked by a rapid shutdown of stable RNA (rRNA and tRNA) synthesis (the stringent response) and, in addition, by an increase or decrease of the synthesis of about half of the cellular proteins. As revealed by in vivo complementation experiments with appropriate *E. coli* mutants and verified by in vitro enzyme activity assays, the *S. equisimilis rel* gene, designated rel_{Seq} here, encodes a strong (p)ppGpp 3'-pyrophosphohydrolase activity [(p)ppGppase] and a weaker ribosome-independent (p)ppGpp synthetic activity resulting from its ATP:GTP 3'-pyrophosphoyltransferase capacity (66). In vivo, the net effect of the opposing Rel_{Seq} activities favors (p)ppGpp degradation, resulting in the failure of rel_{Seq} to complement the absence of *relA* and, hence, in its functional likening to *spoT*. Structural support for this inference provided the detection of immunological cross-reactivity of Rel_{Seq} with polyclonal antibodies to SpoT but not RelA and, conversely, of SpoT but not RelA with antibodies to Rel_{Seq}. However, despite these clear ob-servations, disruption of the chromosomal rel_{Seq} gene at codon 216 by insertion mutagenesis abolished the (p)ppGpp accumulation response following amino acid starvation of wild-type *S. equisimilis*, a phenotype characteristic of *E. coli relA* but not *spoT* mutants. Further functional characterization of rel_{Seq} showed that amino acid deprivation of *S. equisimilis* triggers a rel_{Seq}-dependent stringent response characterized by rapid (p)ppGpp accumulation at the expense of GTP and abrupt cessation of net RNA accumulation in the wild type but not in the insertion mutant, which displays a relaxed response (67). Thus, one of the most interesting results of this work is the finding that the Rel_{Seq} protein is bifunctional and shows activities that reside separately in the paralogous SpoT and RelA proteins from *E. coli*. This establishes an important variation of the *E. coli* paradigm for regulating (p)ppGpp levels and presumably the expression of genes controlled by (p)ppGpp. Evidence available so far, including that provided by the *S. pyogenes* genome sequencing project (1a), indicates that there is a unique streptococcal *rel* gene, unaccompanied by a sister *rel/spo* paralog. However, it is also clear that our present knowledge of rel_{Seq} function scratches just the surface of what might be in store concerning the importance of this gene. Largely neglected to be studied in gram-positive pathogens, the stringent response will certainly reveal many regulatory features necessary for a comprehensive understanding of the biology of these organisms.

CONCLUSIONS AND PROSPECTS

It is apparent from the preceding account that molecular biological investigations into GCS and GGS have preferentially sought to identify and characterize genes and proteins putatively involved in pathogenicity. In many cases, this work has been aided by knowledge obtained previously from studies of the counterparts of such determinants in GAS. This strategy led to the realization that GCS and GGS share many of the cell surface-associated as well as extracellular pathogenicity factors with GAS. Among the latter, a notable exception is the GAS pyrogenic exotoxins, which appear not to be produced by GCS or GGS. Given the possibility of intergroup phage reactions and the fact that most of these exotoxins are phage encoded, the apparent failure of GCS and GGS to become phage converted remains enigmatic. With few exceptions, the analysis of genetic linkage relationships has rarely been pursued in GCS and GGS, and the same is true regarding investigations into gene regulation. The little knowledge available suggests that both homology of synteny and drastic rearrangement of gene order will be observed in future work aimed at the development of a comparative genomics of GAS, GCS, and GGS. The elucidation of the phylogenetic relationships of these organisms may be complicated given the clear indications of horizontal gene transfer following speciation of the species to be compared. An additional complication is exemplified by the rearrangement in GGS of the genes of the GAS *mga* regulon. Although appearing to function together in pathogenesis and, hence, being closely linked in *S. pyogenes* to prevent separation by recombination, these genes may become dissociated from their common regulatory gene, *mga*, during transfer. On the other hand, this may be considered a great asset as it opens up the unique possibility of studying the evolution of regulation of coadapted genes in closely related species.

I apologize that space constraints sometimes obliged me to cite reviews or other secondary literature referring to primary research papers relevant to this chapter. Work by me and my coworkers mentioned here was supported by grants from the German Research Association; the Fonds of the Chemical Industry; and the Thuringian Ministry of Science, Research and Arts.

NOTE ADDED IN PROOF

After the submission of this manuscript, several studies pertinent to this chapter have appeared in print. A new gene, *sfs*, encoding a fibronectin-binding protein has been found in numerous strains of both subspecies of *Streptococcus equi*. The SFS protein, which shows no similarity to previously described fibronectin-binding proteins, inhibits the binding of fibronectin to collagen (54a). The 56-kDa GCS protein erroneously described as hyaluronan synthase (49) was reexamined and found to function as an extracellular kinase that becomes autophosphorylated in the presence of ATP, and then stimulates hyaluronan synthesis and capsule shedding from the surface of GCS (70a). The hyaluronan synthases seHAS and spHAS were overexpressed as His_6-tagged membrane proteins in *Escherichia coli*, purified to homogeneity, and characterized as lipid-dependent enzymes, the activity of which is strongly stimulated by cardiolipin and phosphatidylserine (103, 104). Deletion of Ile1 from mature streptokinase was demonstrated to block the transition of the initially inactive plasminogen-streptokinase complex to its active form, a rearrangement that is rate limiting at low temperature. This observation reinforces an earlier proposal that the N-terminus of streptokinase forms a salt bridge with Asp-740 of plasminogen, thus triggering the conformational change leading to the virgin enzyme (107b). This hypothesis contrasts with salt bridge formation of Lys-698 with Asp-740, a proposal based on the streptokinase-microplasmin crystal structure (107a).

REFERENCES

1. **Achari, A., S. P. Hale, A. J. Howard, G. M. Clore, A. M. Gronenborn, K. D. Hardman, and M. Whitlow.** 1992. 1.67-A X-ray structure of the B2 immunoglobulin-binding domain of streptococcal protein G and comparison to the NMR structure of the B1 domain. *Biochemistry* **31:**10449–10457.

1a. **Advanced Center for Genome Technology. University of Oklahoma.** 1998. *Streptococcus pyogenes* Genome Sequencing Project. http://www.genome.ou.edu/strep.html

2. **Akerström, B., E. Nielsen, and L. Björck.** 1987. Definition of IgG- and albumin-binding regions of streptococcal protein G. *J. Biol. Chem.* **262:**13388–13391.

2a. **Azuaga, A. I., N. D. Woodruff, F. Conejero-Lara, V. F. Cox, R. A. G. Smith, and C. M. Dobson.** 1999. Expression and characterization of the intact N-terminal domain of streptokinase. *Protein Sci.* **8:**443–446.

3. **Barne, K. A., J. A. Bown, S. J. W. Busby, and S. D. Minchin.** 1997. Region 2.5 of the *Escherichia coli* RNA polymerase σ^{70} subunit is responsible for the recognition of the "extended −10" motif at promoters. *EMBO J.* **16:**4034–4040.

4. **Ben Nasr, A., A. Wistedt, U. Ringdahl, and U. Sjöbring.** 1994. Streptokinase activates plasminogen bound to human group C and G streptococci through M-like proteins. *Eur. J. Biochem.* **222:**267–276.

5. **Bert, F., C. Branger, and N. Lambert-Zechovsky.** 1997. Pulsed-field gel electrophoresis is more discriminating than multilocus enzyme electrophoresis and random amplified polymorphic DNA analysis for typing pyogenic streptococci. *Curr. Microbiol.* **34:**226–229.

6. **Bert, F., B. Picard, C. Branger, and N. Lambert-Zechovsky.** 1996. Analysis of genetic relationships among strains of groups A, C and G streptococci by random amplified polymorphic DNA analysis. *J. Med. Microbiol.* **45:**278–284.

7. **Bert, F., B. Picard, N. Lambert-Zechovsky, and P. Goullet.** 1995. Identification and typing of pyogenic streptococci by enzyme electrophoretic polymorphism. *J. Med. Microbiol.* **42:**442–451.

8. **Betschel, S. D., S. M. Borgia, N. L. Barg, D. E. Low, and J. C. S. de Azavedo.** 1998. Reduced virulence of group A streptococcal Tn*916* mutants that do not produce streptolysin S. *Infect. Immun.* **66:**1671–1679.

9. **Bisno, A. L., C. M. Collins, and J. C. Turner.** 1997. M proteins of group C streptococci isolated from patients with acute pharyngitis, p. 745–748. *In* T. Horaud, M. Sicard, A. Bouvet, R. Leclerq, and H. DeMonclos (ed.), *Streptococci and the Host.* Plenum Press, New York, N.Y.

10. **Broeseker, T. A., M. D. Boyle, and R. Lottenberg.** 1988. Characterization of the interaction of human plasmin with its specific receptor on a group A streptococcus. *Microb. Pathog.* **5:**19–27.

11. **Chanter, N., N. Collin, N. Holmes, M. Binns, and J. Mumford.** 1997. Characterization of the Lancefield group C streptococcus 16S-23S RNA gene intergenic spacer and its potential for identification and sub-specific typing. *Epidemiol. Infect.* **118:**125–135.

12. **Cimolai, N., and A. C. H. Cheong.** 1992. Differentiation of species in human β-haemolytic group G streptococci using immunoglobulin Fc fragment receptor. *J. Clin. Pathol.* **45:**232–234.

13. **Cleary, P. P., J. Peterson, C. Chen, and C. Nelson.** 1991. Virulent human strains of group G streptococci express a C5a peptidase enzyme similar to that produced by group A streptococci. *Infect. Immun.* **59:**2305–2310.

14. Collins, C. M., A. Kimura, and A. L. Bisno. 1992. Group G streptococcal M protein exhibits structural features analogous to those of class I M protein of group A streptococci. *Infect. Immun.* 60:3689–3696.

14a. Conejero-Lara, F., J. Parrado, A. I. Azuaga, C. M. Dobson, and C. P. Ponting. 1998. Analysis of the interactions between streptokinase domains and human plasminogen. *Protein Sci.* 7:2190–2199.

15. Conejero-Lara, F., J. Parrado, A. I. Azuaga, R. A. G. Smith, C. P. Ponting, and C. M. Dobson. 1996. Thermal stability of the three domains of streptokinase studied by circular dichroism and nuclear magnetic resonance. *Protein Sci.* 5:2583–2591.

16. Crater, D. L., and I. van de Rijn. 1995. Hyaluronic acid synthesis operon (*has*) expression in group A streptococci. *J. Biol. Chem.* 270:18452–18458.

17. D'Costa, S. S., H. Wang, D. W. Metzger, and M. D. P. Boyle. 1997. Group A streptococcal isolate 64/14 expresses surface plasmin-binding structures in addition to Plr. *Res. Microbiol.* 148:559–572.

18. Derrick, J. P., and D. B. Wigley. 1992. Crystal structure of a streptococcal protein G domain bound to an Fab fragment. *Nature* 359:752–754.

19. Efstratiou, A. 1997. Pyogenic streptococci of Lancefield groups C and G as pathogens in man. *J. Appl. Microbiol. Symp. Suppl.* 83:72S–79S.

20. Espinosa-Urgel, M., and R. Kolter. 1998. *Escherichia coli* genes expressed preferentially in an aquatic environment. *Mol. Microbiol.* 28:325–332.

21. Fahnestock, S. R., P. Alexander, J. Nagle, and D. Filpula. 1986. Gene for an immunoglobulin-binding protein from a group G streptococcus. *J. Bacteriol.* 167:870–880.

22. Falkenberg, C., L. Björck, and B. Akerström. 1992. Localization of the binding site for streptococcal protein G on human serum albumin. Identification of a 5.5-kilodalton protein G binding albumin fragment. *Biochemistry* 31:1451–1457.

23. Farrow, J. A. E., and M. D. Collins. 1984. Taxonomic studies on streptococci of serological groups C, G and L and possibly related taxa. *Syst. Appl. Microbiol.* 5:483–493.

24. Fischetti, V. A. 1989. Streptococcal M protein: molecular design and biological behavior. *Clin. Microbiol. Rev.* 2:285–314.

25. Flanagan, J., N. Collin, J. Timoney, T. Mitchell, J. A. Mumford, and N. Chanter. 1998. Characterization of the haemolytic activity of *Streptococcus equi*. *Microb. Pathog.* 24:211–221.

26. Forsman, P., A. Tilsala-Timisjärvi, and T. Alatossava. 1997. Identification of staphylococcal and streptococcal causes of bovine mastitis using 16S-23S rRNA spacer regions. *Microbiology* 143:3491–3500.

27. Frank, C., K. Steiner, A. Bruckmann, and H. Malke. 1998. Streptokinase gene regulation, abstr. 2C–33. ASM Conference on Streptococcal Genetics, Vichy, France. American Society for Microbiology, Washington, D.C.

28. Frank, C., K. Steiner, and H. Malke. 1995. Conservation of the organization of the streptokinase gene region among pathogenic streptococci. *Med. Microbiol. Immunol.* 184:139–146.

29. Gase, K., T. Ellinger, and H. Malke. 1995. Complex transcriptional control of the streptokinase gene of *Streptococcus equisimilis* H46A. *Mol. Gen. Genet.* 247:749–758.

30. Gase, K., A. Gase, H. Schirmer, and H. Malke. 1996. Cloning, sequencing and functional overexpression of the *Streptococcus equisimilis* H46A *gapC* gene encoding a glyceraldehyde-3-phosphate dehydrogenase that also functions as a plasmin(ogen)-binding protein: purification and biochemical characterization of the protein. *Eur. J. Biochem.* 239:42–51.

31. Gase, K., G. Liu, A. Bruckmann, K. Steiner, J. Ozegowski, and H. Malke. 1997. The *lppC* gene of *Streptococcus equisimilis* encodes a lipoprotein that is homologous to the e (P4) outer membrane protein from *Haemophilus influenzae*. *Med. Microbiol. Immunol.* 186:63–73.

32. Gillespie, B. E., B. M. Jayarao, and S. P. Oliver. 1997. Identification of *Streptococcus* species by randomly amplified polymorphic deoxyribonucleic acid fingerprinting. *J. Dairy Sci.* 80:471–476.

33. Gräfe, S., T. Ellinger, and H. Malke. 1996. Structural dissection and functional analysis of the complex promoter of the streptokinase gene from *Streptococcus equisimilis* H46A. *Med. Microbiol. Immunol.* 185:11–17.

34. Gronenborn, A. M., and G. M. Clore. 1993. Identification of the contact surface of a streptococcal protein G domain complexed with a human Fc fragment. *J. Mol. Biol.* 233:331–335.

35. Gronenborn, A. M., D. R. Filpula, N. Z. Essig, A. Achari, M. Whitlow, P. T. Wingfield, and G. M. Clore. 1991. A novel, highly stable fold of the immunoglobulin binding domain of streptococcal protein G. *Science* 253:657–660.

36. Gross, S., K. Gase, and H. Malke. 1996. Localization of the sequence-determined DNA bending center upstream of the streptokinase gene *skc*. *Arch. Microbiol.* 166:116–121.

37. Guss, B., M. Eliasson, A. Olsson, M. Uhlén, A. K. Frej, H. Jörnvall, J. I. Flock, and M. Lindberg. 1986. Structure of the IgG-binding regions of streptococcal protein G. *EMBO J.* 5:1567–1575.

38. House-Pompeo, K., Y. Xu, D. Joh, P. Speziale, and M. Höök. 1996. Conformational changes in the fibronectin binding MSCRAMMs are induced by ligand binding. *J. Biol. Chem.* 271:1379–1384.

39. Jaffe, J., S. Natanson-Yaron, M. G. Caparon, and E. Hanski. 1996. Protein F2, a novel fibronectin-binding protein from *Streptococcus pyogenes*, possesses two binding domains. *Mol. Microbiol.* 21:373–384.

40. Joh, H. J., K. House-Pompeo, J. M. Patti, S. Gurusiddappa, and M. Höök. 1994. Fibronectin receptors from gram-positive bacteria: comparison of active sites. *Biochemistry* 33:6086–6092.

41. Jonsson, H., L. Frykberg, L. Rantamäki, and B. Guss. 1994. MAG, a novel plasma protein receptor from *Streptococcus dysgalactiae*. *Gene* 143:85–89.

42. Jonsson, H., H. Lindmark, and B. Guss. 1995. A protein G-related cell surface protein in *Streptococcus zooepidemicus*. *Infect. Immun.* 63:2968–2975.

43. Jonsson, H., and H. P. Müller. 1994. The type-III Fc receptor from *Streptococcus dysgalactiae* is also an α2-macroglobulin receptor. *Eur. J. Biochem.* 220:819–826.

44. Jorm, L. R., D. N. Love, G. D. Bailey, G. M. McKay, and D. A. Briscoe. 1994. Genetic structure of populations of β-haemolytic Lancefield group C streptococci from horses and their association with disease. *Res. Vet. Sci.* 57:292–299.

45. **Kawamura, Y., X. G. Hou, F. Sultana, H. Miura, and T. Ezaki.** 1995. Determination of 16S rRNA sequences of *Streptococcus mitis* and *Streptococcus gordonii* and phylogenetic relationships among members of the genus *Streptococcus. Int. J. Syst. Bacteriol.* **45:**406–408.

46. **Klessen, C., K. H. Schmidt, J. J. Ferretti, and H. Malke.** 1988. Tripartite streptokinase gene fusion vectors for gram-positive and gram-negative procaryotes. *Mol. Gen. Genet.* **212:**295–300.

47. **Kline, J. B., S. Xu, A. L. Bisno, and C. M. Collins.** 1996. Identification of a fibronectin-binding protein (GfbA) in pathogenic group G streptococci. *Infect. Immun.* **64:**2122–2129.

48. **Kumari, K., and P. H. Weigel.** 1997. Molecular cloning, expression, and characterization of the authentic hyaluronan synthase from group C *Streptococcus equisimilis. J. Biol. Chem.* **272:**32539–32546.

49. **Lansing, M., S. Lellig, A. Mausolf, I. Martini, F. Crescenzi, M. O'Regan, and P. Prehm.** 1993. Hyaluronate synthase: cloning and sequencing of the gene from *Streptococcus* sp. *Biochem. J.* **289:**179–184.

50. **Lee, A. Y., S. T. Jeong, I. C. Kim, and S. M. Byun.** 1997. Identification of the functional importance of valine-19 residue in streptokinase by N-terminal deletion and site-directed mutagenesis. *Biochem. Mol. Biol. Int.* **41:** 199–207.

51. **Lin, L. F., S. Oeun, A. Houng, and G. L. Reed.** 1996. Mutation of lysines in a plasminogen binding region of streptokinase identifies residues important for generating a functional activator complex. *Biochemistry* **35:**16879–16885.

52. **Lindgren, P. E., M. J. McGavin, C. Signäs, B. Guss, S. Gurusiddappa, M. Höök, and M. Lindberg.** 1993. Two different genes coding for fibronectin-binding proteins from *Streptococcus dysgalactiae:* the complete nucleotide sequences and characterization of the binding domains. *Eur. J. Biochem.* **214:**819–827.

53. **Lindgren, P. E., C. Signäs, L. Rantamäki, and M. Lindberg.** 1994. A fibronectin-binding protein from *Streptococcus equisimilis:* characterization of the gene and identification of the binding domain. *Vet. Microbiol.* **41:** 235–247.

54. **Lindgren, P. E., P. Speziale, M. McGavin, H. J. Monstein, M. Höök, L. Visai, T. Kostiainen, S. Bozzini, and M. Lindberg.** 1992. Cloning and expression of two different genes from *Streptococcus dysgalactiae* encoding fibronectin receptors. *J. Biol. Chem.* **267:**1924–1931.

54a.**Lindmark, H., and B. Guss.** 1999. SFS, a novel fibronectin-binding protein from *Streptococcus equi,* inhibits the binding between fibronectin and collagen. *Infect. Immun.* **67:**2383–2388.

55. **Lindmark, H., K. Jacobsson, L. Frykberg, and B. Guss.** 1996. Fibronectin-binding protein of *Streptococcus equi* subsp. *zooepidemicus. Infect. Immun.* **64:**3993–3999.

56. **Lottenberg, R., C. C. Broder, M. D. Boyle, S. J. Kain, B. L. Schroeder, and R. Curtiss III.** 1992. Cloning, sequence analysis, and expression in *Escherichia coli* of a streptococcal plasmin receptor. *J. Bacteriol.* **174:**5204–5210.

57. **Malke, H.** 1998. Cytoplasmic membrane lipoprotein LppC of *Streptococcus equisimilis* functions as an acid phosphatase. *Appl. Environ. Microbiol.* **64:**2439–2442.

58. **Malke, H., and J. J. Ferretti.** 1984. Streptokinase: cloning, expression, and excretion by *Escherichia coli. Proc. Natl. Acad. Sci. USA* **81:**3557–3561.

59. **Malke, H., and J. J. Ferretti.** 1991. Expression and properties of hybrid streptokinases extended by N-terminal plasminogen kringle domains, p. 184–189. *In* G. M. Dunny, P. P. Cleary, and L. L. McKay (ed.), *Genetics and Molecular Biology of Streptococci, Lactococci, and Enterococci.* American Society for Microbiology, Washington, D.C.

60. **Malke, H., D. Lorenz, and J. J. Ferretti.** 1987. Streptokinase: expression of altered forms, p. 143–149. *In* J. J. Ferretti and R. Curtiss III (ed.), *Streptococcal Genetics.* American Society for Microbiology, Washington, D.C.

61. **Malke, H., U. Mechold, K. Gase, and G. Gerlach.** 1994. Inactivation of the streptokinase gene prevents *Streptococcus equisimilis* H46A from acquiring cell-associated plasmin activity in the presence of plasminogen. *FEMS Microbiol. Lett.* **116:**107–112.

62. **Malke, H., B. Roe, and J. J. Ferretti.** 1985. Nucleotide sequence of the streptokinase gene from *Streptococcus equisimilis* H46A. *Gene* **34:**357–362.

62a.**Malke, H., K. Steiner, K. Gase, and C. Frank.** Expression and regulation of the streptokinase gene. *Methods,* in press.

63. **Malke, H., K. Steiner, K. Gase, U. Mechold, and T. Ellinger.** 1995. The streptokinase gene: allelic variation, genomic environment, and expression control. *Dev. Biol. Stand.* **85:**183–193.

64. **Martin, N. J., E. L. Kaplan, M. A. Gerber, M. A. Menegus, M. Randolph, K. Bell, and P. P. Cleary.** 1990. Comparison of epidemic and endemic group G streptococci by restriction enzyme analysis. *J. Clin. Microbiol.* **28:** 1881–1886.

65. **McGavin, M., S. Gurusiddappa, P. E. Lindgren, M. Lindberg, G. Raucci, and M. Höök.** 1993. Fibronectin receptors from *Streptococcus dysgalactiae* and *Staphylococcus aureus. J. Biol. Chem.* **268:**23946–23953.

66. **Mechold, U., M. Cashel, K. Steiner, D. Gentry, and H. Malke.** 1996. Functional analysis of a *relA/spoT* gene homolog from *Streptococcus equisimilis. J. Bacteriol.* **178:** 1401–1411.

67. **Mechold, U., and H. Malke.** 1997. Characterization of the stringent and relaxed responses of *Streptococcus equisimilis. J. Bacteriol.* **179:**2658–2667.

68. **Mechold, U., K. Steiner, S. Vettermann, and H. Malke.** 1993. Genetic organization of the streptokinase region of the *Streptococcus equisimilis* H46A chromosome. *Mol. Gen. Genet.* **241:**129–140.

69. **Medved, L. V., D. A. Solovjov, and K. C. Ingham.** 1996. Domain structure, stability and interactions in streptokinase. *Eur. J. Biochem.* **239:**333–339.

70. **Meehan, M., P. Nowlan, and P. Owen.** 1998. Affinity purification and characterization of a fibrinogen-binding protein complex which protects mice against lethal challenge with *Streptococcus equi* subsp. *equi. Microbiology* **144:** 993–1003.

70a.**Nickel, V., S. Prehm, M. Lansing, A. Mausolf, A. Podbielski, J. Deutscher, and P. Prehm.** 1998. An ectoprotein kinase of group C streptococci binds hyaluronan and regulates capsule formation. *J. Biol. Chem.* **273:**23668–23673.

71. Nihalani, D., R. Kumar, K. Rajagopal, and G. Sahni. 1998. Role of the amino-terminal region of streptokinase in the generation of a fully functional plasminogen activator complex probed with synthetic peptides. Protein Sci. 7:637–648.

71a.Nowicki, S. T., D. Minning-Wenz, K. H. Johnston, and R. Lottenberg. 1994. Characterization of a novel streptokinase produced by Streptococcus equisimilis of non-human origin. Thromb. Haemost. 72:595–603.

72. Nygren, P. A., C. Ljungquist, H. Thromborg, K. Nustad, and M. Uhlén. 1990. Species-dependent binding of serum albumins to the streptococcal receptor protein G. Eur. J. Biochem. 193:143–148.

73. Okumura, K., A. Hara, T. Tanaka, I. Nishiguchi, W. Minamide, H. Igarashi, and T. Yutsudo. 1994. Cloning and sequencing the streptolysin O genes of group C and group G streptococci. DNA Sequence 4:325–328.

74. Oliver, S. P., B. E. Gillespie, and B. M. Jayarao. 1998. Detection of new and persistent Streptococcus uberis and Streptococcus dysgalactiae intramammary infections by polymerase chain reaction-based DNA fingerprinting. FEMS Microbiol. Lett. 160:69–73.

75. Olsson, A., M. Eliasson, B. Guss, B. Nilsson, U. Hellman, M. Lindberg, and M. Uhlén. 1987. Structure and evolution of the repetitive gene encoding streptococcal protein G. Eur. J. Biochem. 168:319–324.

76. O'Neil, K. T., R. H. Hoess, D. P. Raleigh, and W. F. DeGrado. 1995. Thermodynamic genetics of the folding of the B1 immunoglobulin-binding domain from streptococcal protein G. Proteins 21:11–21.

77. Pancholi, V., and V. A. Fischetti. 1992. A major surface protein on group A streptococci is a glyceraldehyde-3-phosphate-dehydrogenase with multiple binding activity. J. Exp. Med. 176:415–426.

78. Pancholi, V., and V. A. Fischetti. 1998. α-Enolase, a novel strong plasmin(ogen) binding protein on the surface of pathogenic streptococci. J. Biol. Chem. 273:14503–14515.

79. Parhami-Seren, B., T. Keel, and G. L. Reed. 1996. Structural characterization of immunodominant regions of streptokinase recognized by murine monoclonal antibodies. Hybridoma 15:169–176.

80. Parrado, J., F. Conejero-Lara, R. A. G. Smith, J. M. Marshall, C. P. Ponting, and C. M. Dobson. 1996. The domain organization of streptokinase: nuclear magnetic resonance, circular dichroism, and functional characterization of proteolytic fragments. Protein Sci. 5:693–704.

81. Podbielski, A., B. Melzer, and R. Lütticken. 1991. Application of the polymerase chain reaction to study the M protein(-like) gene family in beta-hemolytic streptococci. Med. Microbiol. Immunol. 180:213–227.

82. Podbielski, A., I. Zagres, A. Flosdorff, and J. Weber-Heynemann. 1996. Molecular characterization of a major serotype M49 group A streptococcal DNase gene (sdaD). Infect. Immun. 64:5349–5356.

83. Reed, G. L., L. F. Lin, B. Parhami-Seren, and P. Kussie. 1995. Identification of a plasminogen binding region in streptokinase that is necessary for the creation of a functional streptokinase-plasminogen activator complex. Biochemistry 34:10266–10271.

84. Reidl, J., and J. J. Mekalanos. 1996. Lipoprotein e (P4) is essential for hemin uptake by Haemophilus influenzae. J. Exp. Med. 183:621–629.

85. Reilly, T. J., G. S. Baron, F. E. Nano, and M. S. Kuhlenschmidt. 1996. Characterization and sequencing of a respiratory burst-inhibiting acid phosphatase from Francisella tularensis. J. Biol. Chem. 271:10973–10983.

86. Rodríguez, P., P. Fuentes, M. Barro, J. G. Alvarez, E. Munoz, D. Collen, and H. R. Lijnen. 1995. Structural domains of streptokinase involved in the interaction with plasminogen. Eur. J. Biochem. 229:83–90.

87. Schnitzler, N., G. Haase, A. Podbielski, A. Kaufhold, C. Lämmler, and R. Lütticken. 1997. Human isolates of large colony-forming β-hemolytic group G streptococci form a distinct clade upon 16S rRNA gene analysis, p. 363–365. In T. Horaud, A. Bouvet, R. Leclercq, H. de Montclos, and M. Sicard (ed.), Streptococci and the Host. Plenum Press, New York, N.Y.

88. Schnitzler, N., A. Podbielski, G. Baumgarten, M. Mignon, and A. Kaufhold. 1995. M or M-like protein gene polymorphisms in human group G streptococci. J. Clin. Microbiol. 33:356–363.

89. Sheinerman, F. B., and C. L. Brooks. 1998. Calculations on folding of segment B1 of streptococcal protein G. J. Mol. Biol. 278:439–456.

90. Sheoran, A. S., B. T. Sponseller, M. A Holmes, and J. F. Timoney. 1997. Serum and mucosal antibody isotype responses to M-like protein (SeM) of Streptococcus equi in convalescent and vaccinated horses. Vet. Immunol. Immunopathol. 59:239–251.

91. Simpson, W. J., J. M. Musser, and P. P. Cleary. 1992. Evidence consistent with horizontal transfer of the gene (emm12) encoding serotype M12 protein between group A and group G pathogenic streptococci. Infect. Immun. 60:1890–1893.

92. Simpson, W. J., J. C. Robbins, and P. P. Cleary. 1987. Evidence for group A-related M protein genes in human but not animal-associated group G streptococcal pathogens. Microb. Pathog. 3:339–350.

93. Skjold, S. A., H. Malke, and L. W. Wannamaker. 1979. Transduction of plasmid-mediated erythromycin resistance between group-A and -G streptococci, p. 274–275. In M. T. Parker (ed.), Pathogenic Streptococci. Reedbooks Ltd., Chertsey, Surrey, England.

94. Speziale, P., D. Joh, L. Visai, S. Bozzini, K. House-Pompeo, M. Lindberg, and M. Höök. 1996. A monoclonal antibody enhances ligand binding of fibronectin MSCRAMM (adhesin) from Streptococcus dysgalactiae. J. Biol. Chem. 271:1371–1378.

95. Sriprakash, K. S., and J. Hartas. 1996. Lateral genetic transfers between group A and G streptococci for M-like genes are ongoing. Microb. Pathog. 20:275–285.

96. Steiner, K., and H. Malke. 1995. Transcription termination of the streptokinase gene of Streptococcus equisimilis H46A: bidirectionality and efficiency in homologous and heterologous hosts. Mol. Gen. Genet. 246:374–380.

97. Steiner, K., and H. Malke. 1997. Primary structure requirements for in vivo activity and bidirectional function of the transcription terminator shared by the oppositely oriented skc/rel-orf1 genes of Streptococcus equisimilis. Mol. Gen. Genet. 255:611–618.

98. Sultana, F., Y. Kawamura, X. G. Hou, S. E. Shu, and T. Ezaki. 1998. Determination of 23S rRNA sequences from members of the genus Streptococcus and characterization of genetically distinct organisms previously identified as

members of the *Streptococcus anginosus* group. *FEMS Microbiol. Lett.* **158**:223–230.

99. **Talay, S. R., M. P. Grammel, and G. S. Chhatwal.** 1996. Structure of a group C streptococcal protein that binds to fibrinogen, albumin and immunoglobulin G via overlapping modules. *Biochem. J.* **315**:577–582.

100. **Teuten, A. J., R. W. Broadhurst, R. A. G. Smith, and C. M. Dobson.** 1993. Characterization of structural and folding properties of streptokinase by n.m.r. spectroscopy. *Biochem. J.* **290**:313–319.

101. **Timoney, J. F., S. C. Artiushin, and J. S. Boschwitz.** 1997. Comparison of the sequences and functions of *Streptococcus equi* M-like proteins SeM and SzPSe. *Infect. Immun.* **65**:3600–3605.

102. **Timoney, J. F., J. Walker, M. Zhou, and J. Ding.** 1995. Cloning and sequence analysis of a protective M-like protein gene from *Streptococcus equi* subsp. *zooepidemicus*. *Infect. Immun.* **63**:1440–1445.

103. **Tlapak-Simmons, V. L., B. A. Baggenstoss, T. Clyne, and P. H. Weigel.** 1999. Purification and lipid dependence of the recombinant hyaluronan synthases from *Streptococcus pyogenes* and *Streptococcus equisimilis*. *J. Biol. Chem.* **274**:4239–4245.

104. **Tlapak-Simmons, V. L., B. A. Baggenstoss, K. Kumari, C. Heldermon, and P. H. Weigel.** 1999. Kinetic characterization of the recombinant hyaluronan synthases from *Streptococcus pyogenes* and *Streptococcus equisimilis*. *J. Biol. Chem.* **274**:4246–4253.

105. **Ullberg, M., I. Karlsson, B. Wiman, and G. Kronvall.** 1992. Two types of receptors for human plasminogen on group G streptococci. *Acta Pathol. Microbiol. Immunol. Scand.* **100**:21–28.

106. **Valentin-Weigand, P., M. Y. Traore, H. Blobel, and G. S. Chhatwal.** 1990. Role of α2-macroglobulin in phagocytosis of group A and C streptococci. *FEMS Microbiol. Lett.* **70**:321–324.

107. **Vandamme, P., B. Pot, E. Falsen, K. Kersters, and L. A. Devriese.** 1996. Taxonomic study of Lancefield streptococcal groups C, G, and L (*Streptococcus dysgalactiae*) and proposal of *S. dysgalactiae* subsp. *equisimilis* subsp. nov. *Int. J. Syst. Bacteriol.* **46**:774–781.

107a. **Wang, X., X. Lin, J.A. Loy, J. Tang, and X. C. Zhang.** 1998. Crystal structure of the catalytic domain of human plasmin complexed with streptokinase. *Science* **281**:1662–1665.

107b. **Wang, S., G. L. Reed, and L. Hedstrom.** 1999. Deletion of Ile1 changes the mechanism of streptokinase: evidence for the molecular sexuality hypothesis. *Biochemistry* **38**:5232–5240.

108. **Wannamaker, L. W., S. Almquist, and S. Skjold.** 1973. Intergroup phage reactions and transduction between group C and group A streptococci. *J. Exp. Med.* **137**:1338–1353.

109. **Walter, F., M. Siegel, and H. Malke.** 1989. Nucleotide sequence of the streptokinase gene from a group G-*Streptococcus*. *Nucleic Acids Res.* **17**:1262.

110. **Welfle, K., R. Misselwitz, A. Schaup, D. Gerlach, and H. Welfle.** 1997. Conformation and stability of streptokinases from nephritogenic and nonnephritogenic strains of streptococci. *Proteins* **27**:26–35.

111. **Wolinowska, R., P. Ceglowski, J. Kok, and G. Venema.** 1991. Isolation, sequence and expression in *Escherichia coli*, *Bacillus subtilis* and *Lactococcus lactis* of the DNase (streptodornase)-encoding gene from *Streptococcus equisimilis* H46A. *Gene* **106**:115–119.

112. **Young, K. C., G. Y. Shi, Y. F. Chang, B. I. Chang, L. C. Chang, M. D. Lai, W. J. Chuang, and H. L. Wu.** 1995. Interaction of streptokinase and plasminogen. *J. Biol. Chem.* **270**:29601–29606.

113. **Young, K. C., G. Y. Shi, D. H. Wu, L. C. Chang, B. I. Chang, C. P. Ou, and H. L. Wu.** 1998. Plasminogen activation by streptokinase via a unique mechanism. *J. Biol. Chem.* **273**:3110–3116.

Pathogenicity Factors in Group C and G Streptococci

GURSHARAN S. CHHATWAL AND SUSANNE R. TALAY

17

Group C and G streptococci constitute a heterogeneous complex of streptococcal species that either reside as apathogenic commensals in humans and animals or act as causative agents of severe infection and organ damage associated with high mortality rates. In this chapter we give a short overview of the various streptococcal species, the diseases they cause, and the pathogenicity factors identified during recent years that contribute to the virulence behavior of group C and G streptococci.

Lancefield group C streptococci can be divided into two morphologic groups: large- and small-colony-forming. The small-colony phenotype is expressed by the "*Streptococcus milleri*" group, a heterogeneous complex of streptococci that has been taxonomically classified on a molecular level and is currently divided into the three species *Streptococcus anginosus*, *Streptococcus constellatus*, and *Streptococcus intermedius* (79). Members of these species are recognized as common commensal organisms of the human oral cavity, gastrointestinal tract, and genitourinary tract (56). However, the clinical significance of this group is increasingly enhanced by the role these organisms play in serious pyogenic infections (52). Traditionally recognized as important animal and human pathogens are the large-colony-forming group C streptococci. These streptococci belong either to the species *Streptococcus equi*, which can be divided into the two subspecies *S. equi* subsp. *equi* and *S. equi* subsp. *zooepidemicus*, or to the species *Streptococcus dysgalactiae*, which can be divided into *S. dysgalactiae* subsp. *equisimilis* and *S. dysgalactiae* subsp. *dysgalactiae*. (For convenience, the latter are hereafter referred to as *S. equisimilis* and *S. dysgalactiae*.) Infections caused by large-colony-forming group C streptococci in humans are mostly due to *S. equisimilis* and, to a lesser extent, *S. zooepidemicus*. *S. equisimilis* can cause bacteremia, cellulitis, peritonitis, septic arthritis, pneumonia, endocarditis, and, as recently shown, acute pharyngitis (8, 71). *S. zooepidemicus* can cause bacteremia, meningitis, septic arthritis, endocarditis, and pneumonia (2, 71). Generally, large-colony-forming group C streptococcal bacteremia in humans is either community acquired or found to be associated with exposure to animals or their products. Although occurring less frequently than bacte-

remia caused by streptococcal serotypes A and B, it is characterized by high mortality rates (up to 25%) as well as other major sequelae that reflect the potentially high level of virulence of these organisms.

In domestic animals the species *S. equi* and *S. dysgalactiae* as well as the respective subspecies are of clinical relevance. *S. equisimilis* can cause pneumonia, arthritis, septicemia, and abscesses. *S. dysgalactiae* is mainly responsible for bovine mastitis. *S. equi* subsp. *equi*, a pathogen primarily restricted to horses and donkeys, causes strangles, a highly contagious respiratory illness characterized by abscesses in the mandibular and pharyngeal lymph nodes (9). This organism is not considered to be part of the normal flora because of its close association with disease. In contrast, *S. equi* subsp. *zooepidemicus* is an opportunistic commensal that can cause rhinopharyngitis, pneumonia, metritis, neonatal septicemia, and wound infections in horses (9, 78) as well as disease in other domestic animals such as cattle, sheep, pigs, and chicken (1). As mentioned above, it is also responsible for a range of zoonotic infections in humans.

The beta-hemolytic group G *Streptococcus* was first identified by Lancefield and Hare (37) in 1935 and can be serotypically distinguished from other streptococci by its type-specific polysaccharide antigen. Initially it had not been considered to be pathogenic, since the organism was found as part of the normal flora of the skin, pharynx, gastrointestinal tract, and vagina. During recent years, however, it has been shown to be an important pathogen that may cause life-threatening human infections and can also cause disease in animals. Human pathogenic group G streptococcal strains are taxonomically part of the *S. dysgalactiae* complex, in which they form a distinct clade (57, 75). The characteristics of group G streptococci isolated from humans are similar to those of *Streptococcus pyogenes*, the serotype group A streptococcus (22, 65, 75). The pathogen is commonly isolated from the throat, skin lesions, and the bloodstream, depending on the kind of infection. Group G streptococci can cause epidemic pharyngitis, meningitis, puerperal sepsis, peritonitis, cellulitis, arthritis, wound infection, septicemia, infective endocarditis, and

Gram-Positive Pathogens, ed. by V. A. Fischetti et al.
© 2000 American Society for Microbiology, Washington, D.C.

glomerulonephritis (27, 29, 64). Group G streptococci also reside in domestic animals such as cattle, sheep, cats, and dogs that either constitute healthy carriers of the bacterium or develop diseases such as bovine mastitis.

In the following section, we discuss factors that enable group C and G streptococci to infect their hosts and cause disease. Such factors include adhesive structures that initiate the infection process, antiphagocytic factors that enable the bacterium to evade the host's immune system, factors that are potentially involved in spreading in tissues, and factors that specifically bind, degrade, or damage host components.

MECHANISMS FOR ADHERENCE

Adhesion of microorganisms to host tissues represents a critical phase in the development of infection, and a successful principle of bacterial tissue targeting is binding to components of the extracellular matrix (13, 51). The high-molecular-weight glycoprotein fibronectin is one of the matrix compounds involved in the cell-targeting mechanisms of gram-positive cocci. Fibronectin itself is responsible for substrate adhesion of eukaryotic cells via specific cell surface factors of the integrin family. It also specifically interacts with other matrix components, such as collagen, fibrin, and sulfated glycosaminoglycans, that designate this molecule to fulfill a multifunctional role within the extracellular network (see reference 31 for review). Being present in the extracellular matrix of most tissues, as well as in plasma and other body fluids in its soluble form, fibronectin represents an exquisite target for bacteria to exploit the cell attachment properties of this molecule by linking the pathogen to specific target cells. Epithelial cells of the upper respiratory tract of humans are bathed in secretions containing fibronectin in its soluble form. Once bound to the bacterial surface, it will enable the pathogen to attach and subsequently colonize the primary site of infection. During recent years, several genes coding for fibronectin-binding proteins have been identified and characterized from group C and G streptococci, as shown in Table 1 (7, 39–42). The proteins expressed by the distinct species share a common architecture, reflecting the close functional relationship between these molecules. Presence of putative signal sequence at the N terminus as well as wall- and membrane-spanning regions at the C terminus characterize the proteins as membrane-anchored surface proteins. The binding domains for fibronectin are located within the C-terminal part of the proteins and are composed of 3 to 5 repetitive units that consist of 35 to 37 amino acid residues. *S. dysgalactiae* possesses two genes, *fnbA* and *fnbB*, coding for fibronectin-binding proteins (41). Since only low sequence similarity is exhibited by the two genes, it was assumed that they did not evolve via gene duplication but evolved independently. However, FnBA does not seem to be expressed under standard conditions. Functional studies

on the repetitive binding units employing either synthetic peptides or chemically modified peptide fragments revealed a conserved core sequence (45). This core element, defined as ED(T/S)(X9,10)GG(X3,4)(I/V)DF, contains the essential di- and tripeptides GG and (I/V)DF that are required for functional activity. Cross-inhibition studies on the fibronectin-binding activity of different bacterial species indicate that structural motifs from different fibronectin receptors recognize a common or similar site within the N-terminal fragment of fibronectin (45). In addition, mediation of adherence is more efficient with bacteria recognizing the N-terminal region of the fibronectin molecule than with those interacting with the C-terminal region (14, 72). Another region of the fibronectin molecule is recognized by a nonrepetitive stretch of amino acid residues in FNZ from *S. zooepidemicus* (42), a fibronectin receptor that shows low sequence similarity with the proteins characterized from *S. dysgalactiae* and *S. equisimilis*. This region, which is also present in SfbI protein of *S. pyogenes* (67), acts independently from the repeat region in binding to fibronectin (59). Interestingly, this region is also present in GfbA, the recently identified fibronectin-binding protein of group G streptococci (46). However, direct interaction between GfbA and fibronectin has not yet demonstrated any strong correlation between the adhesive properties of the group G streptococcal isolates and the presence of the *gfbA* gene. Furthermore, heterologous expression of the GfbA protein enabled the bacterium to adhere to human skin fibroblasts (7). It is very likely that the fibronectin-binding proteins of group C and G streptococci identified so far are indeed adhesins like their homologs in *S. pyogenes*. However, further work will be needed not only to gain substantial knowledge of the adherence process mediated by fibronectin, but also to elaborate whether these molecules can mediate cell invasion, as recently shown in the case of SfbI/protein F of *S. pyogenes* (32, 46).

Fibronectin-mediated attachment is not restricted to the mucosa but is also found in injured tissues. Many of the fibronectin-binding bacteria, including streptococci, are common wound pathogens. Any tissue damage will provoke exposure of the host's fibronectin molecule, which is then no longer hidden behind the first line of defense, the epithelial cell barrier. It is very likely that these bacteria gain access to the cells as well as the underlying tissue via fibronectin that either constitutes a part of the extracellular matrix of the traumatic tissue or is present in de novo-formed fibrin thrombi. Involvement of fibronectin in the adherence mechanism of group C streptococci was shown by mimicking this biological situation in an in vitro model (18). In this model *S. equi* and *S. dysgalactiae* adhered to fibrin thrombi via fibronectin incorporated into the fibrin matrix, as shown by substantially reduced adherence when fibrin thrombi were generated using fibronectin-depleted plasma (18).

TABLE 1 Fibronectin-binding proteins of group C and G streptococci

Organism	Protein	Size in aa	No. of repeats	References
Streptococcus dysgalactiae	FnBA	1,092	3	39, 41
Streptococcus dysgalactiae	FnBB	1,118	3	39, 41
Streptococcus equi subsp. *zooepidemicus*	FNZ	597	5	42
Streptococcus equisimilis	FnB	1,169	3	40
Group G *Streptococcus*	GfbA	580	3	7

An alternative way for group C and G streptococci to adhere to host cells is via binding to vitronectin and collagen. Vitronectin is a multifunctional serum protein that affects the humoral immune system by binding to and inhibiting the complement membrane attack complex (see reference 54 for review). It is also a major matrix-associated adhesive glycoprotein that regulates blood coagulation. Collagen, a matrix protein that influences the structure of tissues and is also involved in cell attachment, proliferation, and cell differentiation, can be divided into at least 14 types. Both group C and G streptococci specifically interact with vitronectin (17). The binding of vitronectin to group G streptococci but not to group C streptococci was influenced by heparin, indicating that binding components of group C and G streptococci interact with different domains of vitronectin (17). Vitronectin binding, however, mediates the adherence of both group C and G streptococci to epithelial cells as well as endothelial cells (24, 73). Although the role of vitronectin in adherence of C and G streptococci is well established, the underlying mechanisms and the bacterial factors involved are not yet known. It has been shown that group A streptococci bind vitronectin via its hemopexinlike domain (38), and it would be interesting to see if group C and G streptococci also use this mechanism for their interaction with vitronectin. A collagen type II-binding protein isolated from *S. zooepidemicus* that has the potential to mediate attachment to collagen-rich tissues has been described (76). However, to date none of the group G or C streptococcal factors have been characterized on the molecular level and thus require further investigation.

ANTIPHAGOCYTIC FACTORS

One feature of pathogenic streptococci that strongly contributes to their virulence is the ability to resist phagocytosis. The streptococcal M protein, first identified and characterized in group A *Streptococcus* (see reference 25 for review), is one of the major antiphagocytic factors of streptococci. It is a fibrillar surface-exposed molecule that deters opsonization of the organism by the alternative complement pathway and thus evades the host's nonspecific immune defense mechanism. As a consequence, bacteria carrying the M protein are capable of surviving and multiplying in blood. The effect of not being phagocytosed by polymorphonuclear leukocytes is based on strongly reduced deposition of C3 to the bacterial surface, which could be restored by addition of M-type-specific antibodies. Specific interaction of M protein with the host proteins fibrinogen and factor H was suggested to be responsible for affecting the alternative complement pathway (15, 30).

As in group A streptococci, human-pathogenic strains of group G and C streptococci carry M proteins. Protein MG1, the first group G streptococcal M-like protein characterized on the molecular level (21), exhibits typical structural and biological features of M proteins, such as coiled-coil structure and the ability to generate type-specific opsonizing antibodies. Protein MG1 shares highly homologous sequences with the C-terminal repeat region of class I M proteins, which are frequently associated with rheumatic fever. M proteins of group G streptococci are also responsible for conferring resistance to phagocytosis (10). Studies on recurrently infected mice, however, showed that the animals neither acquired M-specific protective immunity against the cellulitis-causing group G *Streptococcus* (5) nor had opsonic antibodies, an effect of the disease that is in contrast to the results observed with group A streptococci. Extended epidemiological characterization of *emm* genes in human-pathogenic group G streptococci revealed that they all exhibited features of class I *emm* genes, whereas animal pathogenic strains failed to contain homologous sequences (58). The high similarity of the *emm* genes from both species gave support to the hypothesis that in group G streptococci the region containing a class I *emm* gene has been acquired through horizontal transfer (61) and that mosaiclike polymorphism in the flanking regions is the result of subsequent rearrangements. It is thus not surprising that homologous M protein sequences can also be demonstrated in group C streptococci of human origin (53). Recent characterization of the *emm* genes of acute pharyngitis-associated isolates of *S. equisimilis* also revealed the presence of *emm* genes highly homologous to those found in group G streptococci (4). This finding underlines the potential clinical relevance of certain M-type group C streptococci of human origin.

A different situation is found in animal-pathogenic group C streptococci. The well-characterized equine pathogens *S. equi* and *S. zooepidemicus* display M proteins that are distinct from those of group A or human-pathogenic group C and G streptococci and thus appear only distantly related (69, 70). Although they share common features, such as the ability to bind fibrinogen, prevent phagocytosis (6), act as protective antigens, and display a mainly alpha-helical fimbrillar structure, other characteristics such as A- and B-repeat regions as well as primary sequence homology are missing. The 58-kDa M-like protein SeM of *S. equi* and the 40-kDa protein SzPSe are distinct from each other, but both bind fibrinogen. SeM was postulated to be conserved and to act as a protective antigen against *S. equi* but not *S. zooepidemicus* infection. The data collected on the equine streptococcal M-like proteins add substantial evidence to the assumption that *S. equi* is a conserved clonal pathogen derived from the genetically diverse *S. zooepidemicus* (69). Furthermore, studies on the M-like proteins of the different species and their relationship not only revealed the spreading of factors via horizontal gene transfer but also showed that distinct hosts may constitute a strong barrier for interspecies gene transfer. Human strains of group G streptococci express a C5a peptidase enzyme similar to that produced in group A streptococci. C5a peptidase represents an additional factor contributing to the antiphagocytic properties of pathogenic streptococci by specifically destroying the host chemotaxin C5a, leading to a limited recruitment of polymorphonuclear leukocytes (19).

The different behavior of human- and animal-pathogenic group C and G streptococci with regard to their antiphagocytic factors might also be explained by the presence of non-M-like proteins capable of binding fibrinogen (12). Fibrinogen binding masks the bacterial surface, leading to reduced C3 fixation and subsequent phagocytosis irrespective of the nature of the binding component (15). Except for the group C streptococcal FAI protein, which binds to fibrinogen, albumin, and immunoglobulin G (IgG) via overlapping modules (66), the non-M-like fibrinogen-binding proteins have not been characterized.

PROTEIN G AND FUNCTIONALLY RELATED PROTEINS

Streptococcal protein G, a surface molecule associated with the majority of group C and G streptococcal isolates of human origin, has separate and repetitively arranged binding sites for several ligand proteins (23, 28, 50, 63). It represents a structurally and functionally well-characterized type III IgG Fc receptor interacting with a wide species range of immunoglobulins, as well as human serum albumin, kininogen, and α_2-macroglobulin. Protein G exhibits a modular structure in which the binding sites for IgG are located within the C-terminal repeat region (28, 63). The central A/B-repeat region constitutes the binding domain for serum albumin (63), and the N-terminal E region is responsible for interacting with the native form of α_2-macroglobulin (47). α_2-Macroglobulin is a large plasma glycoprotein that can act as a broad-spectrum proteinase inhibitor undergoing conformational switching during the inhibitory process. It interacts specifically with different streptococcal species (16) and forms a novel complex differing from the α_2-macroglobulin-protease complex (11). In contrast to human pathogenic strains of group G streptococci that exclusively bind to the native (slow) form of α_2-macroglobulin via protein G, animal-derived isolates of bovine and equine origin bind the proteinase-complexed (fast) form of the molecule. Two protein G-related proteins from a mastitis-causing *S. dysgalactiae* strain, proteins MIG (35) and MAG (33), as well as protein ZAG (34) from *S. zooepidemicus*, were characterized on the molecular level and shown to be responsible for binding the fast form of α_2-macroglobulin. Like protein G, proteins MAG and ZAG also exhibited serum albumin and type III Fc receptor activity, whereas protein MIG lacked albumin-binding activity. In contrast, protein DG12, a protein G analog that was isolated from a group G streptococcus of bovine origin, exhibited albumin binding but lacked α_2-macroglobulin- and IgG-binding properties (62). Similar to fibrinogen, binding of α_2-macroglobulin can influence phagocytosis of *S. equi* and *S. dysgalactiae*, as shown in experiments in which the fast form of α_2-macroglobulin was capable of inhibiting phagocytosis of group C streptococci by polymorphonuclear leukocytes (74).

Binding of group C and G streptococci to serum albumin and the Fc portion of the IgG molecule—achieved either by distinct proteins or by combining binding domains on one molecule—is a widely found principle (66). Although little is known about the role of these binding proteins in the infection process and their contribution to virulence, they represent a clear example of convergent evolution emphasizing selective advantages for the pathogen. Some of the speculative functions discussed are alternative adherence mechanisms, molecular mimicry, sequestering of IgG, and sensory systems (20).

ENZYMES AND TOXINS

In this section, we discuss factors that appear to be involved in tissue spreading or are able to cause multiple effects in the host. After successfully adhering to the target tissue and evading the host's phagocytic attacks, spreading of pathogenic streptococci in tissues is regarded to be an important step for the onset and development of an invasive disease. One of the factors that are probably involved in this process is streptokinase, a protein found in A, G, and C streptococci (48). The virulence potential of this protein lies in its ability to interact with plasminogen to form a streptokinase-plasminogen complex, thereby generating plasmin, a key enzyme in the fibrinolytic system that is able to break down tissue barriers. M proteins co-

ordinately interact with the secreted streptokinase by binding either fibrinogen (77) or plasminogen (68). For group A streptococci, two distinct pathways are postulated for acquisition of plasminlike activity by the bacterium. The first involves the enzymatic generation of plasmin and direct capture of the active enzyme, which is, however, still inhibitable by its natural inhibitor α_2-antiplasmin. The second involves fibrinogen and is characterized by its resistance to natural regulators (43). Examination of the streptokinase genes of human group C and G streptococci revealed allelic variation of the latter (68). Some of the identified alleles were previously found in group A streptococci, whereas others were unique for the tested strains. Although believed to have implications for the virulence potential of the organism, the biological significance of the variation found in the middle of the protein is still unclear. However, some studies do not show an association between streptokinase variants and disease manifestations such as acute poststreptococcal glomerulonephritis.

Streptolysins were recognized early as the product of various serotype streptococci. Streptolysin O (SLO) is the prototype of a family of thiol-activated cytolysins produced by the genus *Streptococcus* as well as by other gram-positive bacteria, including *Bacillus*, *Clostridium*, and *Listeria* species. Like other members of this toxin family, SLO can reversibly be inactivated by oxidation and is inhibited by cholesterol, which constitutes its natural receptor on the eukaryotic cell. The cytotoxic activity is based on the ability of SLO to form large oligomeric hydrophilic transmembrane pores via a two-step mechanism. Monomeric water-soluble toxin molecules first bind to cholesterol in the lipid bilayer. Lateral diffusion of the monomers in the membrane then leads to aggregation of the monomers, resulting in the formation of supramolecular arc- and ring-shaped structures constituting the pores through which passive flux of ions and macromolecules takes place (3). Besides its general cytotoxic effects, SLO is of considerable clinical interest because of the cardiotoxic effects that are thought to contribute to the induction of rheumatic heart disease (36). Genes coding for SLO of group C and G streptococci are almost identical to that of group A streptococci (49) and have been shown to act synergistically with streptococcal cysteine protease in producing lung injuries in a rat model (60) as well as with an adhesin in modulating signaling responses of keratinocytes during in vitro infection (55). The animal-pathogenic species *S. equi* lacks the production of SLO but appears to secrete a streptolysin S-like toxin (26). Streptolysin S belongs to a distinct group of hemolytic toxins that are characterized by their resistance to oxidation and sensitivity to trypan blue (44). In contrast to SLO, streptolysin S is not immunogenic, is difficult to purify and to clone, and thus has not been characterized on the molecular level. Although insertional inactivation of *S. pyogenes* streptolysin S production was shown to correlate with inactivation of a riboflavin salvage pathway and was therefore proposed to be involved in nutrient acquisition in the host (43), its function in group C and G streptococci is unknown.

CONCLUSIONS

Streptococci of serotype G and C represent a heterogeneous but exquisitely host-adapted range of streptococcal isolates that span the variety of professional commensals, opportunistic pathogens, and exclusive pathogens for many

TABLE 2 Pathogenicity factors of group C and G streptococci

Streptococcal serotype	Bacterial factor	Host factor involved	Function
C, G	Fibronectin-binding proteins	Soluble fibronectin	Adherence
C, G	Vitronectin-binding proteins	Vitronectin	Adherence
C	Collagen-binding protein	Collagen	Adherence
C, G	M and M-like proteins	Fibrinogen, factor H, IgG, plasmin(ogen)	Antiphagocytic, tissue spreading, adherence (?)
G	C5a peptidase	Cleavage of C5a	Antiphagocytic, limited recruitment of polymorphonuclear leukocytes
C, G	Streptokinase	Plasmin activation (M protein as cofactor)	Spreading factor
C, G	Protein G	Albumin, α_2-macroglobulin, IgG Fc, IgG Fab, kininogen	Antiphagocytic? Adherence? Sensory factor?
C, G	Streptolysin O	Cell membranes	Toxic to cells, multiple effects
C	FAI protein	Fibrinogen, albumin, IgG	Antiphagocytic? Sensory factor?

mammalian species, including humans (Table 2). Analyzing the bacterial factors that specifically interact with components of the host has not only allowed insight into the biochemical principles as well as some functional strategies of these streptococci but has also revealed interesting evolutionary aspects, including convergent development, horizontal spread, and module shuffling. However, many open questions concerning the complex network of bacterial factors, their synergistic functions, and their general contribution to virulence still remain to be resolved.

We thank K. Mummenbrauer and H. Brink for their help during manuscript preparation.

REFERENCES

1. **Barnham, M., G. Cole, A. Efstratiou, J. R. Tagg, and S. A. Skjold.** 1987. Characterization of *Streptococcus zooepidemicus* (Lancefield group C) from human and selected animal infections. *Epidemiol. Infect.* **98:**171–182.

2. **Barnham, M., J. Kerby, R. S. Chandler, and M. R. Millar.** 1989. Group C streptococci in human infection: a study of 308 isolates with clinical correlations. *Epidemiol. Infect.* **102:**379–390.

3. **Bhakdi, S., J. Tranum Jensen, and A. Sziegoleit.** 1985. Mechanism of membrane damage by streptolysin-O. *Infect. Immun.* **47:**52–60.

4. **Bisno, A., C. M. Collins, and J. Turner.** 1996. M proteins of group C streptococci isolated from patients with acute pharyngitis. *J. Clin. Microbiol.* **34:**2511–2515.

5. **Bisno, A., and J. M. Gaviria.** 1997. Murine model of recurrent group G streptococcal cellulitis: no evidence of protective immunity. *Infect. Immun.* **65:**4926–4930.

6. **Boschwitz, J. S., and J. F. Timoney.** 1994. Inhibition of C3 deposition on *Streptococcus equi* subsp. *equi* by M protein: a mechanism for survival in equine blood. *Infect. Immun.* **62:**3515–3520.

7. **Bradford Kline, J., S. Xu, A. L. Bisno, and C. M. Collins.** 1996. Identification of a fibronectin-binding protein (GfbA) in pathogenic group G streptococci. *Infect. Immun.* **64:**2122–2129.

8. **Bradley, S. F., J. J. Gordon, D. D. Baumgartner, W. A. Marasco, and C. A. Kauffman.** 1991. Group C streptococcal bacteremia: analysis of 88 cases. *Rev. Infect. Dis.* **13:**270–280.

9. **Bryans, J. T., and B. O. Moore.** 1972. Group C streptococcal infections of the horse, p. 327–338. *In* L. W. Wannamaker and J. M. Matsen (ed.), *Streptococci and Streptococcal Diseases: Recognition, Understanding and Management.* Academic Press, Inc., New York, N.Y.

10. **Campo, R. E., D. R. Schultz, and A. Bisno.** 1995. M proteins of group G streptococci: mechanisms of resistance to phagocytosis. *J. Infect. Dis.* **171:**601–606.

11. **Chhatwal, G. S., G. Albohn, and H. Blobel.** 1987. Novel complex formed between a nonproteolytic cell wall protein of group A streptococci and α_2-macroglobulin. *J. Bacteriol.* **169:**3691–3695.

12. **Chhatwal, G. S., and H. Blobel.** 1986. Binding of human fibrinogen to streptococci and its role in streptococcal pathogenicity, p. 239–249. *In* D. Lane, A. Henschen, and K. Jasani (ed.), *Fibrinogen, Fibrin Formation and Fibrinolysis,* Walter de Gruyter and Co., Berlin, Germany.

13. **Chhatwal, G. S., and H. Blobel.** 1986. Binding of host plasma proteins to streptococci and their role in streptococcal pathogenicity. *IRCS Med. Sci.* **14:**1–3.

14. **Chhatwal, G. S., and H. Blobel.** 1987. Heterogeneity of fibronectin reactivity among streptococci as revealed by binding of fibronectin fragments. *Comp. Immunol. Microbiol. Infect. Dis.* **10:**99–108.

15. **Chhatwal, G. S., I. S. Dutra, and H. Blobel.** 1985. Fibrinogen binding inhibits the fixation of the third component of human complement on surface of groups A, B, C, and G streptococci. *Microbiol. Immunol.* **29:**973–980.

16. **Chhatwal, G. S., H. P. Müller, and H. Blobel.** 1983. Characterization of human α_2-macroglobulin to group G streptococci. *Infect. Immun.* **41:**959–964.

17. **Chhatwal, G. S., K. T. Preissner, G. Müller-Berghaus, and H. Blobel.** 1987. Specific binding of the human S protein (vitronectin) to streptococci, *Staphylococcus aureus,* and *Escherichia coli.* *Infect. Immun.* **55:**1878–1883.

18. Chhatwal, G. S., P. Valentin Weigand, and K. N. Timmis. 1990. Bactreial infection of wounds: fibronectin-mediated adherence of Group A and C streptococci to fibrin thrombi in vitro. *Infect. Immun.* **58**:3015–3019.

19. Cleary, P., J. Peterson, C. Chen, and C. Nelson. 1991. Virulent human strains of group G streptococci express a C5a peptidase enzyme similar to that produced by group A streptococci. *Infect. Immun.* **59**:2305–2310.

20. Cleary, P., and D. Retnoningrum. 1994. Group A streptococcal immunoglobulin-binding proteins: adhesins, molecular mimicry, or sensory proteins? *Trends Microbiol.* **2**:131–136.

21. Collins, C. M., A. Kimura, and A. Bisno. 1992. Group G streptococcal M protein exhibits structural features analogous to those of class I M protein of group A streptococci. *Infect. Immun.* **60**:3689–3696.

22. Deibel, R. H., and H. W. Seedey. 1974. *Streptococcus*, p. 490–509. *In* R. E. Buchanan and N. E. Gibbons (ed.), *Bergey's Manual of Determinative Bacteriology*, 8th ed. The Williams & Wilkins Co., Baltimore, Md.

23. Fahnestock, S. R., P. Alexander, J. Nagle, and D. Filipula. 1986. Gene for an immunoglobulin-binding protein from a group G *Streptococcus*. *J. Bacteriol.* **167**:870–880.

24. Filippsen, L. F., P. Valentin-Weigand, H. Blobel, K. T. Preissner, and G. S. Chhatwal. 1990. Role of complement S protein (vitronectin) in adherence of *Streptococcus dysgalactiae* to bovine epithelial cells. *Am. J. Vet. Res.* **51**:861–865.

25. Fischetti, V. A. 1989. Streptococcal M protein: molecular design and biological behavior. *Clin. Microbiol. Rev.* **2**:285–314.

26. Flanagan, J., N. Collin, J. Timoney, T. Mitchell, J. A. Mumford, and N. Chanter. 1998. Characterization of the haemolytic activity of *Streptococcus equi*. *Microb. Pathog.* **24**:211–221.

27. Gnann, J. W., B. M. Gray, F. M. Griffin, and W. E. Dismukes. 1987. Acute glomerulonephritis following group G streptococcal infections. *J. Infect. Dis.* **156**:411–412.

28. Guss, B., M. Eliasson, A. Olsson, M. Uhlén, A. C. Frej, H. Jörnvall, J. I. Flock, and M. Lindberg. 1986. Structure of the IgG-binding regions of streptococcal protein G. *EMBO J.* **5**:1567–1575.

29. Hill, H. R., G. G. Cladwell, E. Wilson, D. Hager, and R. A. Zimmermann. 1969. Epidemic of pharyngitis due to streptococci of Lancefield group-G. *Lancet* **ii**:371–374.

30. Horstmann, R. D., H. J. Sievertsen, J. Knobloch, and V. A. Fischetti. 1988. Antiphagocytic activity of streptococcal M protein: selective binding of complement control protein factor H. *Proc. Natl. Acad. Sci. USA* **85**:1657–1661.

31. Hynes, R. O., and K. M. Yamada. 1982. Fibronectins: multifunctional modular glycoproteins. *J. Cell Biol.* **95**:369–377.

32. Jadoun, J., V. Ozeri, E. Burnstein, E. Skutelsky, E. Hanski, and S. Sela. 1998. Protein F1 is required for efficient entry of *Streptococcus pyogenes* into epithelial cells. *J. Infect. Dis.* **178**:147–158.

33. Jonsson, H., L. Frykberg, L. Rantamäki, and B. Guss. 1994. MAG, a novel plasma protein receptor from *Streptococcus dysgalactiae*. *Gene* **143**:85–89.

34. Jonsson, H., H. Lindmark, and B. Guss. 1995. A protein G-related cell surface protein in *Streptococcus zooepidemicus*. *Infect. Immun.* **63**:2968–2975.

35. Jonsson, H., and H. P. Müller. 1994. The type-II Fc receptor from *Streptococcus dysgalactiae* is also an $\alpha2$ macroglobulin receptor. *Eur. J. Biochem.* **220**:819–826.

36. Kellner, A., A. W. Bernheimer, A. S. Carlson, and E. B. Freeman. 1956. Loss of myocardial contractibility induced in isolated mammalian hearts by streptolysin O. *J. Exp. Med.* **104**:361–373.

37. Lancefield, R. C., and R. Hare. 1935. The serological differentiation of pathogenic and nonpathogenic streptococci from parturient women. *J. Exp. Med.* **61**:335–349.

38. Liang, O. D., K. T. Preissner, and G. S. Chhatwal. 1997. The hemopexin-type repeats of human vitronectin are recognized by *Streptococcus pyogenes*. *Biochem. Biophys. Res. Commun.* **234**:445–449.

39. Lindgren, P. E., M. J. McGavin, C. Signäs, B. Guss, S. Gurusiddappa, M. Höök, and M. Lindberg. 1993. Two different genes coding for fibronectin-binding proteins from *Streptococcus dysgalactiae*. The complete nucleotide sequences and characterization of the binding domains. *Eur. J. Biochem.* **214**:819–827.

40. Lindgren, P. E., C. Signäs, L. Rantamaki, and M. Lindberg. 1994. A fibronectin-binding protein from *Streptococcus equisimilis*: characterization of the gene and identification of the binding domain. *Vet. Microbiol.* **41**:235–247.

41. Lindgren, P. E., P. Speziale, M. McGavin, H. J. Monstein, M. Höök, L. Visai, T. Kostiainen, S. Bozzini, and M. Lindberg. 1993. Cloning and expression of two different genes from *Streptococcus dysgalactiae* encoding fibronectin receptors. *J. Biol. Chem.* **267**:1924–1931.

42. Lindmark, H., K. Jacobsson, L. Frykberg, and B. Guss. 1996. Fibronectin-binding protein of *Streptococcus equi* subsp. *zooepidemicus*. *Infect. Immun.* **64**:3993–3999.

43. Liu, S., S. Sela, G. Cohen, J. Jadoun, A. Cheung, and I. Ofek. 1997. Insertional inactivation of streptolysin S expression is associated with altered riboflavin metabolism in *Streptococcus pyogenes*. *Microb. Pathog.* **22**:227–234.

44. Loridan, C., and J. E. Alouf. 1986. Purification of RNA core induced streptolysin S and isolation and haemolytic characteristics of the carrier-free toxin. *J. Gen. Microb.* **132**:307–315.

45. McGavin, M. J., S. Gurusiddappa, P. E. Lindgren, and M. Lindberg. 1993. Fibronectin receptors from *Streptococcus dysgalactiae* and *Staphylococcus aureus*. Involvement of conserved residues in ligand binding. *J. Biol. Chem.* **268**:23946–23953.

46. Molinari, G., S. R. Talay, P. Valentin-Weigand, M. Rohde, and G. S. Chhatwal. 1997. The fibronectin-binding protein of *Streptococcus pyogenes*, SfbI, is involved in the internalization of group A streptococci by epithelial cells. *Infect. Immun.* **65**:1357–1363.

47. Müller, H. P., and L. K. Rantamäki. 1995. Binding of native $\alpha2$ macroglobulin to human group G streptococci. *Infect. Immun.* **63**:2833–2839.

48. Nasr, B., A. Wistedt, U. Ringdahl, and U. Sjöbring. 1994. Streptokinase activates plasminogen bound to human group C and G streptococci through M-like proteins. *Eur. J. Biochem.* **222**:267–276.

49. Okumura, K., A. Hara, T. Tanaka, I. Nishiguchi, W. Minamide, H. Igarashi, and T. Yutsudo. 1994. Cloning and sequencing the streptolysin O (SLO) genes of group C and G streptococci. *DNA Sequence* **4**:325–328.

50. Olsson, A., M. Eliasson, B. Guss, B. Nilsson, U. Hellmann, M. Lindberg, and M. Uhlén. 1987. Structure and evolution of the repetitive gene encoding streptococcal protein G. *Eur. J. Biochem.* **168:**319–324.

51. Patti, J. M., and M. Höök. 1994. Microbial adhesins recognizing extracellular matrix macromolecules. *Curr. Opin. Cell Biol.* **6:**752–758.

52. Piscitelli, S., J. Shwed, P. Schreckenberger, and L. Danziger. 1992. *Streptococcus milleri* group: renewed interest in an elusive pathogen. *Eur. J. Clin. Microbiol. Infect. Dis.* **11:**491–498.

53. Podbielski, A., M. Mignon, J. Weber-Heynemann, N. Schnitzler, R. Ltticken, and A. Kaufhold. 1994. Characterization of groups C (GCS) and G (GGS) streptococcal M protein (*emm*) genes, p. 234–236. *In* A. Totolian (ed.), *Pathogenic Streptococci: Present and Future.* Lancer Publications, St. Petersburg, Russia.

54. Preissner, K. T. 1991. Structure and biological role of vitronectin. *Annu. Rev. Cell Biol.* **7:**275–310.

55. Ruiz, N., B. Wang, A. Pentland, and M. Caparon. 1998. Streptolysin O and adherence synergistically modulate proinflammatory responses of keratinocytes to group A streptococci. *Mol. Microbiol.* **27:**337–346.

56. Ruoff, K. L. 1988. *Streptococcus anginosus* ("*S. milleri*"): the unrecognized pathogen. *Clin. Microbiol. Rev.* **1:**102–108.

57. Schnitzler, N., G. Haase, A. Podbielski, A. Kaufhold, C. Lämmler, and R. Lütticken. 1997. Human isolates of large colony forming β hemolytic group G streptococci form a distinct clade upon 16S rRNA gene analysis, p. 363–365. *In* T. Horaud (ed.), *Streptococci and the Host.* Plenum Press, New York, N.Y.

58. Schnitzler, N., A. Podbielski, G. Baumgarten, M. Mignon, and A. Kaufhold. 1995. M or M-like gene polymorphisms in human group G streptococci. *J. Clin. Microbiol.* **33:**356–363.

59. Sela, S., A. Aviv, A. Tovi, I. Burstein, M. G. Cparon, and E. Hanski. 1993. Protein F: an adhesin of *Streptococcus pyogenes* binds fibronectin via two distinct domains. *Mol. Microbiol.* **19:**1049–1055.

60. Shanley, T. P., D. Schrier, V. Kapur, M. Kehoe, J. M. Musser, and P. A. Ward. 1996. Streptococcal cysteine protease augments lung injury induced by products of group A streptococci. *Infect. Immun.* **64:**870–877.

61. Simpson, W. J., J. M. Musser, and P. Cleary. 1992. Evidence consistent with horizontal transfer of the gene (*emm12*) encoding serotype M12 protein between group A and group G pathogenic streptococci. *Infect. Immun.* **60:**1890–1893.

62. Sjöbring, U. 1992. Isolation and molecular characterization of a novel albumin-binding protein from group G streptococci. *Infect. Immun.* **60:**3601–3608.

63. Sjöbring, U., L. Björck, and W. Kastern. 1991. Streptococcal protein G. Gene structure and protein binding properties. *J. Biol. Chem.* **266:**399–405.

64. Smyth, E. G., A. P. Pallelett, and R. N. Davidson. 1988. Group G streptococcal endocarditis: two case reports, review of the literature and recommendations for treatment. *J. Infect.* **16:**169–176.

65. Sriprakash, K. S., and J. Hartas. 1997. Genetic mosaic upstream of *scpG* in human group G streptococci contains sequences from group A streptococcal virulence regulon. *Adv. Exp. Med. Biol.* **418:**749–751.

66. Talay, S. R., M. Grammel, and G. S. Chhatwal. 1996. Structure of a group C streptococcal protein that binds to fibrinogen, albumin and immunoglobulin G via overlapping modules. *Biochem. J.* **315:**577–582.

67. Talay, S. R., P. Valentin-Weigand, P. G. Jerlström, G. S. Chhatwal, and K. N. Timmis. 1992. Fibronectin-binding protein of *Streptococcus pyogenes*: sequence of the binding domain involved in adherence of streptococci to epithelial cells. *Infect. Immun.* **60:**3837–3844.

68. Tewodros, W., I. Karlsson, and G. Kronvall. 1996. Allelic variation of the streptokinase gene in beta-hemolytic streptococci group C and G isolates of human origin. *FEMS Immunol. Med. Microbiol.* **13:**29–34.

69. Timoney, J. F., S. C. Artiushin, and J. S. Boschwitz. 1997. Comparison of the sequences and functions of *Streptococcus equi* M-like proteins SeM and SzPSe. *Infect. Immun.* **65:**3600–3605.

70. Timoney, J. F., J. Walker, M. Zhou, and J. Ding. 1995. Cloning and sequence analysis of a protective M-like protein gene from *Streptococcus equi* subsp. *zooepidemicus*. *Infect. Immun.* **63:**1440–1445.

71. Turner, J. C., F. G. Hayden, M. C. Lobo, C. E. Ramirez, and D. Murren. 1997. Epidemiologic evidence for Lancefield group C beta-hemolytic streptococci as a cause of exudative pharyngitis in college students. *J. Clin. Microbiol.* **35:**1–4.

72. Valentin-Weigand, P., G. S. Chhatwal, and H. Blobel. 1988. Adherence of streptococcal isolates from cattle and horses to their respective host epithelial cells. *Am. J. Vet. Res.* **49:**1485–1488.

73. Valentin-Weigand, P., J. Grulich-Henn, G. S. Chhatwal, G. Müller-Berghaus, H. Blobel, and K. T. Preissner. 1988. Mediation of adherence of streptococci to human endothelial cells by complement S protein (vitronectin). *Infect. Immun.* **56:**2851–2855.

74. Valentin-Weigand, P. M. Y. Traore, H. Blobel, and G. S. Chhatwal. 1990. Role of α_2-macroglobulin in phagocytosis of group A and C streptococci. *FEMS Microbiol. Lett.* **70:**321–324.

75. Van Damme, P., B. Pot, E. Falsen, K. Kersters, and L. A. DeVriese. 1996. Taxonomic study of Lancefield streptococcal Groups C, G, and L (*Streptococcus dysgalactiae*) and proposal of *S. dysgalactiae* subsp. *equisimilis* subsp. Nov. *Int. J. Syst. Bacteriol.* **46:**774–781.

76. Visai, L., S. Bozzini, G. Raucci, A. Toniolo, and P. Speziale. 1995. Isolation and characterization of a novel collagen-binding protein from *Streptococcus pyogenes* strain 6414. *J. Biol. Chem.* **270:**347–353.

77. Wang, H., R. Lottenberg, and D. P. Boyle. 1995. A role for fibrinogen in the streptokinase dependent acquisition of plasmin(ogen) by group A streptococci. *J. Infect. Dis.* **171:**85–92.

78. Welsh, R. D. 1984. The significance of *Streptococcus zooepidemicus* in the horse. *Equine Practice* **6:**6–16.

79. Whiley, R. A., and D. Beighton. 1991. Emended descriptions and recognition of *Streptococcus constellatus*, *Streptococcus intermedius*, and *Streptococcus anginosus* as distinct species. *Int. J. Syst. Bacteriol.* **41:**1–5.

Group C and Group G Streptococcal Infections: Epidemiologic and Clinical Aspects

HAROLD R. OSTER AND ALAN L. BISNO

18

Streptococci of groups C and G are associated with infections of humans and animals. These organisms are found as commensals in the throat, skin, and occasionally genitourinary tract, and the epidemiologic and clinical patterns of human disease reflect this ecologic niche.

GROUP C STREPTOCOCCI

Taxonomy

Streptococci bearing the group C antigen comprise several species. These species can be initially divided by colony morphology. In this chapter, we are concerned primarily with the species that form large colonies (≥ 0.5 mm): *Streptococcus dysgalactiae* and *Streptococcus equi*. *Bergey's Manual of Determinative Bacteriology* (31) currently classifies as subspecies of *S. equi* the following group C streptococci: *S. equi* subsp. *equi*, *S. equi* subsp. *equisimilis*, and *S. equi* subsp. *zooepidemicus*. Based on DNA homology, others group these organisms differently, placing *S. equi* subsp. *equisimilis* in a *S. dysgalactiae* complex (47, 54) or listing all large-colony group C streptococci under *S. dysgalactiae* (13). In this discussion, we use the scheme in *Bergey's Manual* and for convenience refer to the subspecies as *S. equi*, *S. equisimilis*, and *S. zooepidemicus*. Streptococci bearing the group C antigen that form small or minute colonies (<0.5 mm) are often designated as members of the "*Streptococcus milleri*" (*Streptococcus anginosus*) group (15). Organisms of this group may belong to Lancefield groups A, C, G, or F, or may not be groupable (31). As many as 75% of clinical isolates bearing the group C antigen are actually of the "*S. milleri*" group (44).

Epidemiology

The small-colony forms often colonize the upper respiratory tract and the gastrointestinal tract. These microorganisms cause a variety of diseases, including endocarditis, pneumonia, pleural empyema, and abscesses of subcutaneous tissues, brain, and intra-abdominal organs (36, 48).

Animal Infections

Large-colony group C streptococci (henceforth, group C streptococci) are pathogenic in animals. *S. equi* is occasionally found in the upper respiratory tract of normal horses and is the causative agent of equine strangles. This acute, contagious, and deadly respiratory disease has led to explosive epidemics in horse stables and has serious potential economic consequences for horse fanciers (11). *S. zooepidemicus* is a cause of infection in a variety of animal species, including cows, rabbits, and swine. *S. equisimilis* can be a commensal of many animals and causes animal disease similar to that caused by *S. zooepidemicus*. *S. dysgalactiae* is a cause of mastitis in cows and various infections in lambs (24, 28).

Human Infections

Studies of group C streptococcal infections in humans are complicated by taxonomic uncertainty. Many case series include the small-colony forms, while others do not distinguish between the various taxa within the group. Most cases of human infection in which the taxon is identified are caused by *S. equisimilis*, followed by *S. zooepidemicus*. The epidemiology of group C streptococcal infections generally correlates with the source of the organism in nature. *S. equisimilis* is a common commensal of humans (45). For this reason, transmission, when it occurs, is more likely to be from person to person, and most cases are sporadic in nature, rather than associated with common source outbreaks. When outbreaks do occur, they are generally associated with close personal contact or perhaps with environmental contamination. By contrast, *S. zooepidemicus* is generally associated with exposure to animals or to common source outbreaks, especially consumption of contaminated dairy products. In the following sections we describe the characteristics of the more common human infections—those due to *S. equisimilis* and *S. zooepidemicus*—as well as of the few reported cases due to *S. equi* and *S. dysgalactiae*.

Pharyngitis

Clusters of pharyngitis cases due to *S. zooepidemicus* are related to a common source, usually consumption of unpasteurized dairy products. In two such outbreaks, poststreptococcal acute glomerulonephritis ensued. Duca et al.

(21) described 85 patients with pharyngitis due to S. *zoo-epidemicus* following the ingestion of improperly pasteurized milk. Eighty-seven percent of the patients were adults. Approximately one-third of the patients developed acute glomerulonephritis, generally in the second or third week of illness. In a smaller outbreak, five members of a family developed an upper respiratory infection related to S. *zooepidemicus* after consuming unpasteurized milk (5). Three of the five family members subsequently developed glomerulonephritis, which was confirmed in one case by renal biopsy.

The role of group C streptococci in sporadic cases of pharyngitis is still somewhat controversial. S. *equisimilis* is by far the most frequent group C streptococcus isolated from patients with pharyngitis. Streptococci of this species are also often cultured, however, from the throats of healthy individuals. There are a few large studies examining the relationship of S. *equisimilis* to acute, sporadic pharyngitis. Turner et al. (53) studied students reporting to a college health service with acute pharyngitis and compared them with controls without infectious problems. Group C streptococci were cultured at a higher rate from those with pharyngitis than from the control group. Patients with positive cultures for group C streptococci were more likely to have features suggestive of a bacterial infection, such as exudative tonsillitis and anterior cervical lymphadenopathy, than were those with negative cultures. Furthermore, these group C strains resisted phagocytosis in human blood and contained genomic DNA encoding an M protein similar in structure to that of group A streptococci, providing further evidence of possible human virulence (8). In a later study, these authors described 265 students with exudative pharyngitis and compared them with 75 patients with rhinovirus infection and 162 students with noninfectious problems. S. *equisimilis* was isolated significantly more frequently from patients with exudative pharyngitis than from either control group (52). Twenty-two cases of pharyngitis from which group C streptococci were isolated occurred during the fall of 1974 in a school for boys with learning disabilities (6). Although it is likely in this epidemiologic setting that the infecting strains were S. *equisimilis*, they were unfortunately not speciated.

Infections of Skin and Soft Tissue

Infection due to group C streptococci may complicate ulcers associated with diabetes mellitus, immobility (45), or venous and lymphatic compromise of any cause. Recurrent cellulitis may occur, for example, in the saphenous venectomy limb of patients who have undergone coronary artery bypass grafting (2) or in the extremities of individuals who have had axillary, pelvic, or femoral node dissection for cancer. Soft tissue abscesses and even necrotizing fasciitis can occur as well, usually following puncture wounds or other trauma (15, 45). Most cases of skin and soft tissue infection with group C streptococci in which the taxon is identified are due to S. *equisimilis*. In one large series with 102 isolates from superficial skin and wound infections, all but one were S. *equisimilis* (4). In contrast to S. *equisimilis*, the rare cases of skin and soft tissue infection due to S. *zooepidemicus* and S. *equi* usually involve exposure to animals. One case of cellulitis with bacteremia due to S. *zooepidemicus* was reported in a renal transplant patient who was exposed to horses at a show (35). A case of severe facial cellulitis due to S. *equi* was reported in another man who also had equine exposure (11).

Other Localized Deep Tissue Infections

As with other streptococcal pathogens, group C streptococci have been associated with infection of most body sites. These reports usually involve one or a few cases and have been described elsewhere (24, 28, 45). Ortel et al. (39) reviewed 10 patients with group C streptococcal arthritis previously reported; two cases were caused by S. *equisimilis*, two by S. *zooepidemicus*, and one by "S. *milleri*"; the remainder were unspecified. Only one patient, infected with S. *zooepidemicus*, had animal exposure, while one-half had preexisting rheumatologic disease. None of the patients had severe underlying illness.

Maternal and Neonatal Infections

Group C streptococci can be found in the normal female genitourinary tract, but their presence often indicates infection. There have been at least two outbreaks of puerperal fever caused by S. *equisimilis*. Thirty-three confirmed cases in England were caused by a single strain of S. *equisimilis*. Clinical features included fever and signs of perineal infection. Sources of infection were postulated by the authors to be environmental, because the organism was cultured from toilet seats and bath plug holes. The organism was, however, also cultured from the throats of many of the nursing staff (51). In another outbreak, also in England, though 4 years later, seven women developed puerperal fever due to S. *equisimilis*. Interestingly, the isolates were identical to the strain responsible for the first outbreak. Though S. *equisimilis* was not isolated from the environment, it was speculated that transmission may have occurred through common use of a toilet seat (27). These epidemiologic and microbiologic data suggesting transmission by fomites must be interpreted with caution. The role of environmental contamination versus nosocomial person-to-person transmission in such outbreaks remains to be determined.

Neonatal group C streptococcal infection is rare. In one case, meningitis due to S. *equisimilis* developed in an infant whose mother was being treated for chorioamnionitis at the time of delivery (25). In another case, a preterm infant developed meningitis due to S. *dysgalactiae*; the source of infection was not determined, as the mother was not ill and the organism could not be cultured from her (41).

Bacteremia and Serious Invasive Disease

Individual case reports document instances of meningitis (33), pneumonia (56), and various other invasive infections caused by group C streptococci. More informative epidemiologically are series describing large numbers of cases. Bradley et al. (9) reviewed 88 cases of bacteremia reported in the literature. Twenty-one of these patients reported exposure to animals or animal products, and as expected, most had bacteremia due to S. *zooepidemicus*. Ten patients had consumed unpasteurized milk, four patients were farmers, one was a butcher, and several had other contact with animals. In the same series, 24 patients with definite or probable endocarditis were described. Of these, five cases were due to S. *zooepidemicus*, four were due to S. *equisimilis*, and the remainder were unspecified. Animal exposure was noted only in patients with infection due to S. *zooepidemicus* or unspeciated organisms. Underlying cardiac disease was seen in 60% of the patients in whom adequate information was available.

Carmeli et al. (16) reported 10 cases of group C streptococcal bacteremia in Israel and reviewed several other case series (Table 1). In this review, which encompassed

TABLE 1 Group C streptococcal bacteremia: comparison of case reports and cases reported in population-based studies[a]

	Case reports	Ohio (Ref. 45)	Hong Kong (Ref. 59)	North Yorkshire, England (Ref. 3)	Madrid, Spain (Ref. 7)	Israel (Ref. 16)	Population studies (All)
Total no. of patients	80	23	11	5	10	10	59
Underlying diseases (no.)	53	23	7	3	10	9	52[b]
Malignant neoplasm	14	5	2	1	2	4	14
Alcohol/cirrhosis	4	3	2	1	1	3	10[b]
Heart disease	15	6	2	1	2	3	14
Diabetes mellitus	4	3	0	0	0	2	8
Injecting drug abuse	3	1	0	0	2	0	3
Exposure to animal (no.)	23	0	0	0	0	0	0[b]
Streptococcus species (no.)	39	2	11	3	10	0	26
S. equisimilis	20	2	0	3	5	0	10
S. zooepidemicus	16	0	11	0	1	0	12
S. equi	2	0	0	0	2	0	2
S. dysgalactiae	1	0	0	0	2	0	2
Clinical syndromes (no.)							
Endovascular	28	3	2	1	2	0	8[b]
Primary bacteremia	19	3	1	0	4	4	12
Central nervous system infection	11	1	0	0	1	1	3
Pneumonia	4	2	2	1	2	1	8
Skin infection	4	5	4	1	1	4	15[b]
Other	16	9	2	2	0	0	13
Morbidity (no.)	17	6	1			2	9
Mortality (no.)	18	6	2	3	4	2	17

[a] Reproduced from reference 16, with permission.
[b] Statistically significant difference ($P < 0.05$) between case reports and population-based studies.

Bradley's, some patients were infected with each of the species of group C and had primary bacteremia, or bacteremia secondary to pharyngitis, epiglottitis, pericarditis, pneumonia, skin and soft tissue infection, endocarditis, or infected aneurysm. Of note is the high prevalence of serious underlying diseases in the patients with bacteremia, greater than 70% overall. Such illnesses were found to be more common in population-based studies than in compilations of case reports. Malignancy, cardiovascular disease, and immunosuppression were most frequent. Human immunodeficiency virus infection and injectable drug use were also seen. Again, exposure to animals or animal products was generally restricted to patients with *S. zooepidemicus*. Edwards et al. (23) described such an outbreak in West Yorkshire. Over a 5-day period, 11 patients presented with bacteremia due to *S. zooepidemicus*. Presentations included primary septicemia, endocarditis, infected aneurysm, and meningitis. All 11 patients had consumed unpasteurized milk from the same source. Yuen et al. (59) reported 11 cases of *S. zooepidemicus* bacteremia with sepsis over a 4-year period in Hong Kong. The patients had a variety of presenting syndromes, and 55% had a serious underlying illness. None of the patients reported exposure to animals or animal products. After further investigation, it was felt that the infections were acquired from ingestion of undercooked pork. This practice is common in that population of Hong Kong. Furthermore, condemned septicemic pigs were found to be infected with *S. zooepidemicus* strains whose DNA fingerprints were identical to the human isolates.

GROUP G STREPTOCOCCI

Taxonomy

The taxonomy of group G streptococci is as subject to controversy as that of group C streptococci. Like group C streptococci, streptococci of group G can be subdivided by colony size. *Bergey's Manual* (31) designates all strains producing large colonies (>0.5 mm) as *Streptococcus canis*. Some authors use the term *S. canis* to refer to only animal strains of large-colony group G streptococci (24). Farrow and Collins (26) used DNA hybridization techniques and biochemical reactions to show that large-colony strains are closely related to *S. dysgalactiae* of group C, suggesting that they are the same species. Others have gone farther, proposing the term *S. dysgalactiae* subsp. *equisimilis* for all large-colony group C and group G streptococci (54). For simplicity, and to reflect the unique epidemiology of the taxon, we use the term group G streptococci to refer to all human strains of large-colony group G streptococci. Certain similarities between these large-colony forms and group A streptococci are evident. For example, many strains of group G streptococci isolated from serious human infections possess M proteins analogous in structure and function to those of group A streptococci (14, 18). Small-colony streptococci (<0.5 mm) bearing the group G antigen belong to the "*S. milleri*" group, previously discussed in this chapter.

Epidemiology

Animal Infections

The animal strains of group G streptococci can be found as commensals of domestic animals, including dogs and cattle. They can also cause a variety of animal infections, including lymphadenitis, dermatitis, abscesses, arthritis, and mastitis (19, 28).

Human Infections

Large-colony group G streptococci (henceforth, group G streptococci) frequently colonize the throats of healthy persons (42). As detailed below, however, these microorganisms have been implicated in a wide variety of mild to life-threatening infections of throat, skin, and deep tissues.

Pharyngitis

Group G streptococci have clearly been linked to outbreaks of pharyngitis. Many of these outbreaks have been related to a common source, usually a food product. In one such outbreak during a single week in 1968, 176 students at a college were evaluated for pharyngitis (30). The attack rate in the student body was 31%. Signs and symptoms were similar to those characteristic of group A streptococcal pharyngitis, suggesting an etiologic role for the organism. Epidemiologic investigation linked the outbreak to contaminated egg salad. In another common source outbreak, 72 persons who attended a convention developed pharyngitis, with group G streptococci isolated from most who had cultures performed (50). All of the patients had consumed chicken salad prepared by a single cook whose throat culture was positive for the organism.

An epidemic of group G streptococcal pharyngitis involving 68 students occurred over a 1-week period at a North Carolina college (34). Because no common food source could be identified, the author concluded that the mode of spread was most likely person to person. The very sharp epidemic curve and brief duration of the outbreak suggest, however, that contamination of a common food vehicle is more likely. In support of this conclusion are the facts that all students interviewed had eaten in the campus cafeteria in the week preceding illness and that one student with a positive culture was a food handler.

As with group C streptococci, the role of group G streptococci in endemically occurring pharyngitis remains somewhat controversial. Gerber et al. (29) studied 221 patients who presented to a pediatric office with clinical findings suggestive of streptococcal pharyngitis over a 6-month period. Group A streptococci were isolated from 41% and group G streptococci from 25% of throat cultures. The latter were mainly isolated during a single 8-week period. In previous years in this pediatric practice, only 1 to 2% of throat cultures had been positive for group G streptococci. Initially, 75% of group G streptococci were misidentified as group A streptococci on the basis of bacitracin susceptibility. Evidence was quite strong that the group G streptococcus was a cause of pharyngitis. There were no significant differences between patients with cultures positive for groups G and A streptococci with respect to clinical findings at presentation, duration of illness, or the percentage with a significant increase in antistreptolysin O titer (44% and 39%, respectively).

Skin and Soft Tissue Infections

Group G streptococci not infrequently cause skin and soft tissue infections, and the skin is often the portal of entry for serious invasive disease and bacteremia. These infections can manifest as pyoderma, cellulitis, erysipelas, surgical wound infections, and abscesses (10, 20, 38). Like group C streptococci, group G infections are often associated with venous stasis and lymphedema (2). Chronic ulcers of various etiologies place the patient at increased risk as well. Group G streptococci were, for example, isolated more than twice as often as group A streptococci from patients admitted to a Swedish dermatological ward (38), and more than half of the 34 group G streptococcal skin infections were related to chronic leg ulcers (38). Injectable-drug users seem to be at increased risk for cellulitis and skin abscesses due to group G streptococci, and the skin is the usual source of bacteremia in such patients (see below) (20). Burn patients are also at risk for skin infections with this organism; such individuals accounted for 8% of cutaneous group G streptococci infections in one series (10). Group G streptococci have been linked to skin graft infection with subsequent loss of the graft (43). The majority of patients with serious group G streptococcal skin and soft tissue infections have underlying diseases. These are most commonly malignancy, cardiovascular disease, alcoholism, and diabetes mellitus (40, 55).

Infectious Arthritis

Numerous cases of infectious arthritis due to group G streptococci have been reported. Serious medical illnesses and previous joint disease were common features in these patients. In five cases of group G streptococcal infectious arthritis from the UCLA hospital system, all patients had prior joint disease and two had infected prostheses (37). In a recent series of seven patients, only one patient had no underlying systemic or rheumatologic illness (46). The remaining six patients had prior trauma, surgery, or inflammation at the affected joint. Four patients also had underlying medical conditions, including diabetes mellitus, alcoholism, and cardiovascular disease. In a review of 50 previously reported cases of group G streptococcal arthritis, more than one-third of patients had chronic joint disease, while just under half of the patients had one of four underlying conditions: malignancy, alcoholism, diabetes mellitus, or injectable-drug use (12). Osteomyelitis also occurs but is reported less frequently than infectious arthritis. In these cases, there is also often a significant underlying disease (10, 55).

Neonatal Infections

Colonization of neonates with group G streptococci seems to be a common finding. In one study, cultures were taken from the nose and umbilicus of more than 3,000 neonates over a 1-year period at the New York Hospital (22). The monthly incidence of positive cultures for group G streptococci ranged from 41 to 76%. Seven cases of neonatal sepsis due to this organism were diagnosed over the same time period. Five of the seven cases occurred in the setting of complications of pregnancy or childbirth. In a larger review, encompassing this series, premature or prolonged rupture of the amniotic membranes was the most common risk factor associated with group G streptococcal infection (17). As in the previous series, most cases were related to obstetrical complications.

Bacteremia and Serious Invasive Disease

Group G streptococci can cause invasive disease, although bacteremia due to this organism is an uncommon occurrence. Auckenthaler et al. (1) reported 38 patients who were bacteremic with group G streptococci at the Mayo Clinic-affiliated hospitals, representing 0.25% of all patients with positive blood cultures over a 10-year period. Seventy percent of the patients acquired the infection in the community, and the skin was the portal of entry in approximately three-quarters of the patients. Most of the hospital-acquired bacteremias involved a postoperative wound or a transcutaneous procedure. The patients tended to be older, with most being in the sixth to eighth decades. Many patients had venous insufficiency, lymphedema, or another cause of chronic lower extremity edema. In a review from Boston University, 29 patients with group G streptococcal bacteremia were identified over a 3-year period (58). The median age of the affected patients was 68 years, and one-half had a skin infection as the primary source of the bacteremia. In another series, six cases of bacteremia occurred in injectable-drug users (20). The portal of entry for these patients was the skin. All of the infected patients had injected drugs for at least 10 years.

As is the case with skin and soft tissue infections, a common theme in the literature regarding group G streptococcal bacteremia is the high incidence of serious underlying disease, especially malignancy (1, 32, 40, 55, 58). In the Mayo Clinic series (1), for example, approximately one-half of the patients had an underlying malignancy, equally divided between hematologic and solid tumors. Many other patients reported in the various series have suffered from alcoholism, diabetes mellitus, or neurologic disease.

Pleuropulmonary infection is occasionally the initial source in group G streptococcal bacteremia. In one review of such infections, seven patients with bacteremic pleuropulmonary infections were noted (55). Again, serious underlying disease was the rule; six of the patients had lymphoma or diabetes mellitus or chronically used injectable drugs or alcohol.

Endocarditis

Endocarditis due to group G streptococci is uncommon. Like bacteremia, the disease tends to occur in older patients with serious underlying conditions. Preexisting valvular disease is noted in about one-half of all patients. In a review of 40 cases (49), the average age was 56 years and the overall mortality was 36%. Underlying disease was present in about one-half of the patients; six patients had a malignancy, six were diabetic, four were alcoholics, and three were injectable-drug users. Also, one-half of the patients had known preexisting valvular disease, with mitral regurgitation being the most common abnormality. Three cases occurred in patients with prosthetic valves. In a series of seven cases not included in the above review, the average age of the patients was 72 years, and only one patient was younger than 60 (57). Underlying medical conditions and/or preexisting valvular disease were noted in most cases.

SUMMARY

Although most research on beta-hemolytic streptococci has focused on group A organisms, streptococci of serogroups C and G are being increasingly recognized as important human pathogens. Small-colony strains (<0.5 mm) are generally classified as members of the "S. milleri" group and are associated with a variety of pyogenic infections. The large-colony (≥0.5 mm) strains are similar in morphology and often in clinical expression to *Streptococcus pyogenes*. Most human group C streptococcal infections are caused by person-to-person transmission of *S. equisimilis*, but infections due to *S. zooepidemicus* (and, rarely, to *S. equi*) are zoonoses. Transmission of these latter species occurs by animal contact or by contamination of food products and has been associated with the development of poststreptococcal glomerulonephritis. Both group G and C streptococci cause infections of throat and skin and soft tissues. Moreover, strains of both serogroups invade the bloodstream and disseminate widely to many deep tissue sites, including endocardium. Life-threatening invasive infections due to streptococci of groups C and G occur most frequently in patients with severe underlying medical diseases.

Both groups C and G streptococci are highly susceptible to penicillin in vitro. Issues in therapy are discussed elsewhere (28).

REFERENCES

1. **Auckenthaler, R., P. E. Hermans, and J. A. Washington II.** 1983. Group G streptococcal bacteremia: clinical study and review of the literature. *Rev. Infect. Dis.* **5:**196–204.

2. **Baddour, L. M., and A. L. Bisno.** 1985. Non-group A beta-hemolytic streptococcal cellulitis: association with venous and lymphatic compromise. *Am. J. Med.* **79:**155–159.

3. **Barnham, M.** 1989. Invasive streptococcal infections in the era before the acquired immune deficiency syndrome: a 10 years' compilation of patients with streptococcal bacteraemia in North Yorkshire. *J. Infect.* **18:**231–248.

4. **Barnham, M., J. Kerby, R. S. Chandler, and M. R. Millar.** 1989. Group C streptococci in human infection: a study of 308 isolates with clinical correlations. *Epidemiol. Infect.* **102:**379–390.

5. **Barnham, M., T. J. Thornton, and K. Lange.** 1983. Nephritis caused by *Streptococcus zooepidemicus* (Lancefield group C). *Lancet* **i:**945–948.

6. **Benjamin, J. T., and V. A. J. Perriello.** 1976. Pharyngitis due to group C hemolytic streptococci in children. *J. Pediatr.* **89:**254–256.

7. **Berenguer, J., I. Sampedro, E. Cercenado, J. Baraia, M. Rodriquez-Creixems, and E. Bouza.** 1992. Group C beta-hemolytic streptococcal bacteremia. *Diagn. Microbiol. Infect. Dis.* **15:**151–155.

8. **Bisno, A. L., C. M. Collins, and J. C. Turner.** 1996. M proteins of group C streptococci isolated from patients with acute pharyngitis. *J. Clin. Microbiol.* **34:**2511–2515.

9. **Bradley, S. F., J. J. Gordon, D. D. Baumgartner, W. A. Marasco, and C. A. Kauffman.** 1991. Group C streptococcal bacteremia: analysis of 88 cases. *Rev. Infect. Dis.* **13:**270–280.

10. **Brahmadathan, K. N., and G. Koshi.** 1989. Importance of group G streptococci in human pyogenic infections. *J. Trop. Med. Hyg.* **92:**35–38.

11. **Breiman, R. F., and F. J. Silverblatt.** 1986. Systemic *Streptococcus equi* infection in a horse handler—a case of human strangles. *West. J. Med.* **145:**385–386.

12. **Bronze, M. S., S. Whitby, and D. R. Schaberg.** 1997. Group G streptococcal arthritis: case report and review of the literature. *Am. J. Med. Sci.* **313:**239–243.

13. **Bruckner, D. A., and P. Colonna.** 1997. Nomenclature for aerobic and facultative bacteria. *Clin. Infect. Dis.* **25:**1–10.

14. **Campo, R. E., D. R. Schultz, and A. L. Bisno.** 1995. M-proteins of group G streptococci: mechanisms of resistance to phagocytosis. *J. Infect. Dis.* **171:**601–606.

15. **Carmeli, Y., and K. L. Ruoff.** 1995. Report of cases of and taxonomic considerations for large-colony-forming Lancefield group C streptococcal bacteremia. *J. Clin. Microbiol.* **33:**2114–2117.

16. **Carmeli, Y., J. M. Schapiro, D. Neeman, A. M. Yinnon, and M. Alkan.** 1995. Streptococcal group C bacteremia: survey in Israel and analytical review. *Arch. Intern. Med.* **155:**1170–1176.

17. **Carstensen, H., C. Pers, and O. Pryds.** 1988. Group G streptococcal neonatal septicaemia: two case reports and a brief review of the literature. *Scand. J. Infect. Dis.* **20:**407–410.

18. **Collins, C. M., A. Kimura, and A. L. Bisno.** 1992. Group G streptococcal M protein exhibits structural features analogous to class I M protein of group A streptococci. *Infect. Immun.* **60:**3689–3696.

19. **Corning, B. F., J. C. Murphy, and J. G. Fox.** 1991. Group G streptococcal lymphadenitis in rats. *J. Clin. Microbiol.* **29:**2720–2723.

20. **Craven, D. E., A. I. Rixinger, A. L. Bisno, T. A. Goularte, and W. R. McCabe.** 1986. Bacteremia caused by group G streptococci in parenteral drug abusers: epidemiological and clinical aspects. *J. Infect. Dis.* **153:**988–992.

21. **Duca, E., G. Teodorovici, C. Radu, A. Vita, P. Talasman-Niculescu, E. Bernescu, C. Feldi, and V. Rosca.** 1969. A new nephritogenic streptococcus. *J. Hyg.* **67:**691–698.

22. **Dyson, A. E., and S. E. Read.** 1981. Group G streptococcal colonization and sepsis in neonates. *J. Pediatr.* **99:**944–947.

23. **Edwards, A. T., M. Roulson, and M. J. Ironside.** 1988. A milkborne outbreak of serious infection due to *Streptococcus zooepidemicus* (Lancefield group C). *Epidemiol. Infect.* **101:**43–51.

24. **Efstratiou, A., G. Colman, G. Hahn, J. F. Timoney, J. M. Boeufgras, and D. Monget.** 1994. Biochemical differences among human and animal streptococci of Lancefield group C or group G. *J. Med. Microbiol.* **41:**145–148.

25. **Faix, R. G., E. I. Soskolne, and R. E. Schumacher.** 1997. Group C streptococcal infection in a term newborn infant. *J. Perinatol.* **17:**79–82.

26. **Farrow, J. A. E., and M. D. Collins.** 1984. Taxonomic studies on streptococci of serological groups C, G and L and possibly related taxa. *Syst. Appl. Microbiol.* **5:**483–493.

27. **Galloway, A., I. Noel, A. Efstratiou, E. Saint, and D. R. White.** 1994. An outbreak of group C streptococcal infection in a maternity unit. *J. Hosp. Infect.* **28:**31–37.

28. **Gaviria, J. M., and A. L. Bisno.** Group C and G streptococci. *In* E. L. Kaplan and D. L. Stevens. (ed.), *Streptococcal Infections.* Oxford Press, New York, N.Y., in press.

29. **Gerber, M. A., M. F. Randolph, N. J. Martin, M. F. Rizkallah, P. P. Cleary, E. L. Kaplan, and E. M. Ayoub.** 1991. Community-wide outbreak of group G streptococcal pharyngitis. *Pediatrics* **87:**598–603.

30. **Hill, H. R., G. G. Caldwell, E. Wilson, D. Hager, and R. A. Zimmerman.** 1969. Epidemic of pharyngitis due to streptococci of Lancefield group G. *Lancet* **ii:**371–374.

31. **Holt, J. G., N. R. Krieg, P. H. A. Sneath, J. T. Staley, and S. T. Williams.** 1994. Gram-positive cocci, p. 527–558. *In Bergey's Manual of Determinative Bacteriology,* 9th ed. The Williams & Wilkins Co., Baltimore, Md.

32. Liu, C. E., T. N. Jang, F. D. Wang, L. S. Wang, and C. Y. Liu. 1995. Invasive group G streptococcal infections: a review of 37 cases. *Chung Hua I Hsueh Tsa Chih* **56:**173–178.

33. Low, D. E., M. R. Young, and G. K. M. Harding. 1980. Group C streptococcal meningitis in an adult. Probable acquisition from a horse. *Arch. Intern. Med.* **140:**977–978.

34. McCue, J. D. 1982. Group G streptococcal pharyngitis: analysis of an outbreak at a college. *JAMA* **248:**1333–1336.

35. McKeage, M. J., M. W. Humble, and R. B. I. Morrison. 1990. *Streptococcus zooepidemicus* cellulitis and bacteremia in a renal transplant recipient. *Aust. N.Z. J. Med.* **20:**177–178.

36. Molina, J. M., Leport, C., Bure, A., Wolff, M., Michon, C., and Vilde, J. L. 1991. Clinical and bacterial features of infections caused by *Streptococcus milleri. Scand. J. Infect. Dis.* **23:**659–666.

37. Nakata, M. M., J. H. Silvers, and L. George. 1983. Group G streptococcal arthritis. *Arch. Intern. Med.* **143:**1328–1330.

38. Nohlgard, C., A. Bjorklind, and H. Hammar. 1992. Group G streptococcal infections on a dermatological ward. *Acta Derm. Venereol.* **72:**128–130.

39. Ortel, T. L., J. Kallianos, and H. A. Gallis. 1990. Group C streptococcal arthritis: case report and review. *Rev. Infect. Dis.* **12:**829–837.

40. Packe, G. E., D. F. Smith, T. M. S. Reid, and C. C. Smith. 1991. Group G streptococcal bacteremia—a review of thirteen cases in Grampian. *Scot. Med. J.* **36:**42–44.

41. Quinn, R. J. M., A. F. Hallett, P. C. Appelbaum, and R. C. Cooper. 1978. Meningitis caused by *Streptococcus dysgalactiae* in a preterm infant. *Am. J. Clin. Pathol.* **70:**948–950.

42. Reid, H. F. M., D. C. J. Bassett, T. Poon-King, J. B. Zabriskie, and S. E. Read. 1985. Group G streptococci in healthy school-children and in patients with glomerulonephritis in Trinidad. *J. Hyg.* **94:**61–68.

43. Rider, M. A., and J. C. McGregor. 1994. Group G streptococcus—an emerging cause of graft loss? *Br. J. Plast. Surg.* **47:**346–348.

44. Ruoff, K. L., L. J. Kunz, and M. J. Ferraro. 1985. Occurrence of *Streptococcus milleri* among beta-hemolytic streptococci isolated from clinical specimens. *J. Clin. Microbiol.* **22:**149–151.

45. Salata, R. A., P. I. Lerner, D. M. Shlaes, K. V. Gopalakrishna, and E. Wolinsky. 1989. Infections due to Lancefield group C streptococci. *Medicine* **68:**225–239.

46. Schattner, A., and K. L. Vosti. 1998. Bacterial arthritis due to beta-hemolytic streptococci of serogroups A, B, C,

F, and G. Analysis of 23 cases and a review of the literature. *Medicine* **77:**122–139.

47. Schnitzler, N., G. Haase, A. Podbielski, A. Kaufhold, C. Lammler, and R. Lutticken. 1997. Human isolates of large colony-forming beta hemolytic group G streptococci form a distinct clade upon 16S rRNA gene analysis. *Adv. Exp. Med. Biol.* **418:**363–365.

48. Singh, K. P., A. Morris, S. D. Lang, D. M. MacCulloch, and D. A. Bremner. 1988. Clinically significant *Streptococcus anginosus (Streptococcus milleri)* infections: a review of 186 cases. *N.Z. Med. J.* **101:**813–816.

49. Smyth, E. G., A. P. Pallett, and R. N. Davidson. 1988. Group G streptococcal endocarditis: two cases reports, a review of the literature and recommendations for treatment. *J. Infect.* **16:**169–176.

50. Stryker, W. S., D. W. Fraser, and R. R. Facklam. 1982. Foodborne outbreak of group G streptococcal pharyngitis. *Am. J. Epidemiol.* **116:**533–540.

51. Teare, E. L., R. D. Smithson, A. Efstratiou, W. R. Devenish, and N. D. Noah. 1989. An outbreak of puerperal fever caused by group C streptococci. *J. Hosp. Infect.* **13:**337–347.

52. Turner, J. C., F. G. Hayden, M. C. Lobo, C. E. Ramirez, and D. Murren. 1997. Epidemiologic evidence for Lancefield group C beta-hemolytic streptococci as a cause of exudative pharyngitis in college students. *J. Clin. Microbiol.* **35:**1–4.

53. Turner, J. C., G. F. Hayden, D. Kiselica, J. Lohr, C. F. Fishburne, and D. Murren. 1990. Association of group C beta-hemolytic streptococci with endemic pharyngitis among college students. *JAMA* **264:**2644–2647.

54. Vandamme, P., B. Pot, E. Falsen, K. Kersters, and L. A. Devriese. 1996. Taxonomic study of Lancefield streptococcal groups C, G, and L (*Streptococcus dysgalactiae*) and proposal of S. dysgalactiae subsp. equisimilis subsp. nov. *Int. J. Syst. Bacteriol.* **46:**774–781.

55. Vartian, C., P. I. Lerner, D. M. Shlaes, and K. V. Gopalakrishna. 1985. Infections due to Lancefield group G streptococci. *Medicine* **64:**75–88.

56. Vartian, C. V. 1991. Bacteremic pneumonia due to group C streptococci. *Rev. Infect. Dis.* **13:**1029–1030.

57. Venezio, F. R., R. M. Gullberg, G. O. Westenfelder, J. P. Phair, and F. V. Cook. 1986. Group G streptococcal endocarditis and bacteremia. *Am. J. Med.* **81:**29–34.

58. Watsky, K. L., N. Kollisch, and P. Densen. 1985. Group G streptococcal bacteremia: the clinical experience at Boston University Medical Center and a critical review of the literature. *Arch. Intern. Med.* **145:**58–61.

59. Yuen, K. Y., W. H. Seto, C. H. Choi, W. Ng, S. W. Ho, and P. Y. Chau. 1990. *Streptococcus zooepidemicus* (Lancefield group C) septicaemia in Hong Kong. *J. Infect.* **21:**241–250.

The Cell Wall of *Streptococcus pneumoniae*

ALEXANDER TOMASZ AND WERNER FISCHER

19

In this chapter we provide a brief overview and update of structural and functional aspects of the pneumococcal cell wall. The two major components—the teichoic acid and the peptidoglycan—are discussed separately with some obvious and unavoidable overlaps, mostly when anatomical and functional aspects and cell biology are discussed. Historically, studies of the pneumococcal cell wall were motivated by such unique features as the presence of choline in the teichoic acids and the pleomorphic changes that accompany removal or alteration of these residues; changes in peptidoglycan composition in penicillin resistance; an easily modulated autolytic system; and the demonstration of host-related (e.g., inflammatory) activities of cell wall components.

THE ANATOMY OF PNEUMOCOCCAL CELL WALL

The overwhelming majority of "natural" isolates of pneumococci express on their outermost surface one of the 90 chemically different capsular polysaccharides (see chapter 20, this volume). Under these diverse structures lies the cell wall, which, as far as the resolution of currently used analytical techniques can tell, is much more uniform. This seems to be true both for the teichoic acid (39) and for the peptidoglycan (36): the two macromolecular components that make up the bulk of the cell wall in roughly equal (milligram to milligram) proportions (28). Deviations from this species-specific peptidoglycan structure were observed among naturally occurring variants (36) and in laboratory constructs (37).

The cell wall of *Streptococcus pneumoniae* R36A appears in electron microscopic thin sections prepared by the method of Kellenberger (44) (osmium tetraoxide and glutaraldehyde fixation followed by uranylacetate and lead citrate staining) as a band of uniform width composed of two electron-dense lines (3 to 4 nm each) enclosing a wider low-density layer (6 to 8 nm). The distribution of teichoic acid chains appears to be uniform within this wall layer (38), and this is presumably also true for the peptidoglycan. Some anatomically differentiated areas may be identified

through electron microscopy. These are (i) the equatorial areas where cell wall growth becomes centripetal (formation of septum or crosswall) and which represent the "growth zones" of the entire wall (see below); (ii) in dividing cells a circumferential thickening ("hump") of the cell wall appears at the place of the incipient septa. (iii) Parallel, or perhaps just before the beginning of the formation of septum, the hump appears to be split at the center, and the two halves begin to "move" on the cell surface symmetrically to the left and to the right of the ingrowing septum, coupled to the growth and eventual division of the cell into two daughter cells. Owing to the conservative mode of replication of the pneumococcal cell wall, these two half-humps are morphological age markers: they divide the cell wall of each pneumococcal cell into two hemispheres that differ in age by one cell generation. (iv) Separation of daughter cells at the end of cell division may be inhibited in pneumococci by several means (19, 33, 42). Under these conditions one can observe by electron microscopy a thin bridge of cell wall material connecting neighboring bacteria to one another within the long chains of cells.

Little is known about the mechanism and control of these phenomena except that both cell wall synthetic and hydrolytic enzymes must participate.

STRUCTURE OF PEPTIDOGLYCAN

Purified cell walls of the nonencapsulated strain R6St (streptomycin resistant) of *S. pneumoniae* were hydrolyzed by the pneumococcal amidase (*N*-acetyl muramic acid L-alanine amidase; the product of the *lytA* gene) under conditions in which this enzyme can quantitatively release the peptide units of cell wall muropeptides. Next, the family of peptides were separated by high-performance liquid chromatography (HPLC), similar but not identical to the analytical system introduced by Glauner et al. (16) for the separation of muropeptides from the cell wall of *Escherichia coli*. Size fractionation, determination of amino acid composition and NH$_2$ termini, partial sequencing of the peptides generated by HPLC, and analysis by time-of-flight mass spectrometry have allowed the identification of a surprisingly large number of monomeric, dimeric, and trimeric peptides (14) showing a diversity comparable to that seen among the muropeptide species identified in *E. coli*. In the structural assignments it was assumed that the amino acids within the stem peptides had the usual, alternating sequence of L and D amino acids beginning with L-alanine in position 1, followed by a D-isoglutamine and then by L-lysine. The carboxy terminus in the stem peptide is occupied by two consecutive D-alanine residues; however, such muropeptides with intact pentapeptide residue are rare in pneumococci cultivated under normal growth conditions. Extension of this analytical technique to cell walls of clinical isolates from a large variety of isolation sites and dates and expressing a variety of capsules has led to the proposition that the wall peptide composition of *S. pneumoniae* grown in the commonly used semisynthetic media and harvested in the late exponential phase is constant and characteristic of the species (36). The most abundant monomer of the pneumococcal peptidoglycan was a tripeptide, the most frequent dimer a directly cross-linked tritetrapeptide. Interestingly, the representation of carboxy-terminal alanine was extremely rare, suggesting the presence of powerful D,D and D,L carboxypeptidases. This was confirmed by testing the wall peptide composition in a PBP3-defective mutant (34) and in the laboratory strain R36A growing in the presence of subinhibitory concentrations of clavulanic acid, a selective inhibitor of PBP3 in this bacterium (35). The peptidoglycan produced under these conditions showed accumulation of peptide species that retained carboxy-terminal D-alanine residues. Pneumococci grown in the presence of clavulanate also showed abnormal physiological properties: premature induction of stationary phase autolysis, hypersensitivity to lysozyme, and reduced MICs for deoxycholate and penicillin.

An interesting feature of the peptide network was the presence of both directly and indirectly cross-linked components. In the latter, alanyl-serine or alanyl-alanine dipeptides formed the cross-link. In terms of cross-linking mode, the pneumococcal cell wall may be classified as either A1α or A3α (14), depending on which dimer one chooses. A massive distortion of peptidoglycan composition in the direction of a preponderance of indirectly cross-linked components was demonstrated in some penicillin-resistant clinical isolates (13).

A method that uses muramidase digestion of the pneumococcal cell wall followed by HPLC separation of muropeptides, through an adaptation of the method of Glauner et al., was described recently (18).

PEPTIDOGLYCAN COMPOSITION AND PENICILLIN RESISTANCE

The first series of highly penicillin-resistant clinical isolates examined by the HPLC method were from South Africa. It was in these isolates that the mechanism of resistance, namely, reduction in antibiotic "affinity" of penicillin-binding proteins (PBPs), was identified for the first time (50). It was also in genetic crosses with these isolates used as DNA donors that the stepwise nature of penicillin resistance (i.e., the sequential reduction in the penicillin affinity of several high-molecular-weight PBPs in parallel with the gradually increasing penicillin MIC) was recognized (50).

Analysis of the penicillin-resistant South African clinical isolates revealed that they produced cell walls of a radically different composition from the one seen in the penicillin-susceptible and nonencapsulated laboratory isolate (13). When we extended the HPLC analysis to the walls of several penicillin-susceptible clinical isolates and several resistant strains (all but one from South Africa), we confirmed the striking shift toward indirectly cross-linked wall peptide composition in the resistant isolates. A link between resistance to penicillin and abnormal wall composition was also suggested by the analysis of genetic crosses: a shift toward the distorted wall composition of the resistant DNA donor was observed in genetic transformants above certain MICs. It was suggested that the anomalous wall peptide composition reflected the altered substrate preference of the penicillin-resistant PBPs from the linear to the branched wall peptide precursors.

However, examination of a larger number of penicillin-resistant pneumococci from diverse geographic sources made the correlation between low-affinity PBPs, cell wall composition, and penicillin resistance less clear. Abnormal wall peptide composition was shown to be a property of the particular penicillin-resistant pneumococcal genetic lineage rather than an obligatory correlate of resistance itself. No cell wall muropeptide alterations could be detected in highly penicillin-resistant transformants generated by heterologous cross with resistant *Streptococcus mitis* as the DNA donor (18). Also, the linkage between the penicillin-resistant phenotype and abnormal wall composition was lost if the first round of genetic transformation was followed by a secondary backcross into the same susceptible recipient (32). Nevertheless, several of these secondary transformants began to show defective growth in drug-free medium, somewhat similar to the physiological abnormalities seen in highly penicillin-resistant laboratory step mutants, which, like clinical isolates, had low-affinity PBPs and cell walls of abnormal cross-linking mode (37). The observations suggest that in some strains or constructs the performance of penicillin-resistant PBPs in cell wall synthesis may be impaired or suboptimal. For these PBPs to function efficiently both in the presence and in the absence of the antibiotic, resistant bacteria may also have to acquire some additional compensatory factors, possibly via recombination with heterologous sources.

CELL WALL COMPOSITION AND AUTOLYSIS

Pneumococci have an auxotrophic requirement for choline that may also, however, be satisfied by other aminoalcohols such as ethanolamine (42). One of the several striking alterations observed in such ethanolamine-grown pneumococci was that they became resistant to autolysis-inducing agents and conditions such as treatment with detergents, hypertonic media, penicillin, and other inhibitors of cell wall synthesis or extended incubation in the stationary phase of growth. Cell walls isolated from ethanolamine-grown bacteria (which contained this nonmethylated amino alcohol in place of the normal choline residues in the wall teichoic acids) were completely resistant to the hydrolytic action of the pneumococcal amidase when incubated with this enzyme in vitro. However, after chemical conversion of the ethanolamine residues to choline (by in vitro treatment of the walls with methyl iodide), the same cell walls became perfect substrates for enzymatic hydrolysis

(20). These observations initiated several interesting lines of investigation into the mechanism of autolysis and into the physiological role of choline residues in the cell wall and cell membrane teichoic acids (see the section on teichoic acids).

ACTIVATION OF THE AUTOLYTIC AMIDASE IN VITRO AND IN VIVO

Several lines of investigations have clearly established that the major enzyme involved with autolytic phenomena in pneumococci is the gene product of *lytA*, an enzyme catalyzing the hydrolysis of the amide bond between the *N*-acetylmuramic acid and L-alanine residues in the pneumococcal peptidoglycan. During in vitro hydrolysis of walls, there is no evidence that the amidase would preferentially hydrolyze amide bonds in the structurally different muropeptides. This is in contrast to findings made with pneumococci in which autolysis was induced in vivo by adding penicillin to the growth medium. A culture of pneumococci was biosynthetically labeled with radioactive lysine in its cell wall peptides, and the molecular size of the cell wall pieces solubilized and released to the medium during penicillin lysis was determined. Even when more than 70% of the culture was lysed, the cell wall was not released as free peptides (as would happen during in vitro hydrolysis) but rather as huge sheets of material in which the muropeptide network was still covalently linked to the teichoic acids (12). These observations indicate that in the intact wall in vivo in some strategically located muropeptide units are attacked preferentially by the autolytic amidase. These muropeptides may be involved with the cross-linking of parallel sheets of cell wall.

GROWTH ZONE AND CELL WALL SEGREGATION

Similar to other streptococci, pneumococci also incorporate new cell wall units into the preexisting wall material at a single growth zone located at the cellular equator. Both peptidoglycan and teichoic acid units enter the pneumococcal surface at this growth zone, which could be visualized by exploiting the unique selectivity of the pneumococcal amidase for choline-containing segments of the cell wall. Pneumococci were grown in ethanolamine-containing medium, in which they are completely resistant to the autolytic amidase. Upon addition of trace amounts of radiolabeled choline to such a culture, the bacteria immediately shifted to the utilization of choline so that the nascent wall units that began to incorporate into the cell surface produced a cell wall that was susceptible to hydrolysis by exogenous amidase added to the medium. It was possible to show by electron microscopy that this enzymatic microsurgery has selectively removed only a thin, equatorially located band of cell wall, thus identifying the anatomical site of the wall growth zone (24).

The inhibition of cell separation in ethanolamine-grown pneumococci has allowed the design of experiments to test the mode of inheritance of pneumococcal cell walls. Pneumococci labeled in their wall by tritiated choline were shifted to an ethanolamine-containing medium in which the bacteria continued to grow in the form of chains of cells, i.e., "linear clones" in which the distribution of radioactively labeled bacteria could be followed (by autora-diography) as a function of cell generations in the ethanolamine-containing medium. Since the teichoic acid choline does not exhibit turnover during growth, the localization of radioactively labeled cells within the chains of bacteria could provide an insight into the mode of wall segregation. The finding was that the radioactive label remained in large clusters in association with cells that were located either at the tips or at the center of chains. The results demonstrate the conservation of large hemispherical segments of the cell wall that are passed on intact to daughter cells during cell division (4).

TEICHOIC ACID AND LIPOTEICHOIC ACID

Pneumococci are unique because their teichoic acid (TA) and lipoteichoic acid (LTA) possess identical repeat and chain structures, whereas in other gram-positive bacteria TAs and LTAs are structurally and biosynthetically distinct entities (8). TAs are cell wall components and phosphodiester-linked to peptidoglycan; LTAs are components of the plasma membrane, being hydrophobically anchored to the outer membrane layer through their fatty acids (Fig. 1).

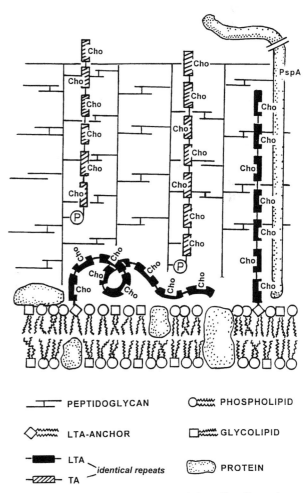

FIGURE 1 Diagrammatic sketch of the cell wall membrane complex of pneumococci. Pneumococcal surface protein (PspA) is shown as an example of a surface protein with a choline-binding signature.

In 1930, long before LTAs and TAs had been discovered and defined, pneumococcal TA was described as pneumococcal C polysaccharide by Tillet and coworkers (41). Thirteen years later, pneumococcal LTA was isolated by Goebel and his colleagues (17) and named lipocarbohydrate or pneumococcal F antigen owing to its fatty acid content and immunological properties. In these early studies, a structural relationship between C polysaccharide and lipocarbohydrate was suggested, and in contrast to the various strain-specific capsular polysaccharides, lipocarbohydrate and C polysaccharide were considered pneumococcal common antigens. This was recently confirmed by serological methods that showed that all 83 known capsular types of *S. pneumoniae* possess C polysaccharide and F antigen (39). The two polymers differ immunologically, as Forssman antigenicity is associated with the LTA (1, 5). As shown by immunoelectron microscopy, C polysaccharide is uniformly distributed on both the inside and outside of the cell walls, and LTA is located on the surface of the cytoplasmic membrane (38).

Choline, the surface signature of pneumococci, was detected as a component of TA and LTA by Tomasz (42) and Briles and Tomasz (5), respectively. The complex structures of TA and LTA could not be unraveled before modern analytical techniques became available (see review by Fischer [9]). In 1980, Jennings and coworkers (21) published the first complete structure of pneumococcal TA. The structure of pneumococcal LTA was clarified in 1992 (1), and subsequent reinvestigation of the TA, isolated from the same strain from which the LTA had been isolated, revealed that both polymers possess identical chain structures (10).

STRUCTURE OF LTA AND TA

As shown in Fig. 2, the chains of LTA and TA are made up of identical repeats that consist of D-glucose, the rare positively charged amino sugar 2-acetamido-4-amino-2,4,6-trideoxy-D-galactose (AATGal), two *N*-acetyl-D-galactosaminyl residues, phosphocholine, and ribitol 5-phosphate. The repeats are joined together by phosphodiester bonds between O5 of the ribitol and O6 of the glucosyl residue of adjacent repeats. The functionally important phosphocholine residues are phosphodiester-linked to O6 of the *N*-acetyl-D-galactosaminyl residues. The number of phosphocholine residues per repeat is strain specific (Table 1): in LTA and TA of strain R6, the majority of repeats carry two phosphocholine residues, whereas most of the repeats in strain Rx1 are substituted with one.

The chain of LTA is phosphodiester-linked to O6 of the terminal sugar residue of D-Glc*p*(β1-4)-D-AATGal*p*(β1-3)-D-Glc*p*(α1-3)acyl$_2$Gro, a unique glycolipid that contains the positively charged AATGal residue intercalated between two D-glucopyranosyl residues (Fig. 2). Whereas in other gram-positive bacteria the chain of LTA is linked to a membrane glycolipid, the lipid anchor of pneumococcal LTA does not occur in the free state in the membrane, where Gal(α1-2)Glc(α1-3)acyl$_2$Gro is the major glycolipid (9). From the molar ratio of intrachain phosphate to glycerol, an average of 6 to 7 repeats per chain is calculated for both strains (Table 1).

Microheterogeneity of LTA became apparent by hydrophobic interaction chromatography and mass spectral analysis (1). The chain of LTA may vary in length between two and eight repeats, with the range differing from sample to sample. On sodium dodecyl sulfate-polyacrylamide gel electrophoresis, LTA yields a ladderlike pattern of up to six bands, each differing from the next by one repeat. Species with one phosphocholine per repeat are distinguished from species with two by higher mobility of the individual bands.

Pneumococcal TA is linked to the peptidoglycan by a phosphodiester to O6 of some of the MurNAc-residues, which is demonstrated by the release of MurN-6-P on HCl hydrolysis of cell walls (25) and of muropeptides (for structure, see Fig. 3). Although the release of MurNAc-6-P indicates the presence of a linkage unit, which in other gram-positive bacteria connects the TA chain and muramic acid by a phosphodiester bond, the typical components of already known linkage units, namely, glycerophosphate and *N*-acetyl-mannosamine, could not be detected either in the hydrolysate of pneumococcal cell walls or in that of TA-containing muropeptides. Since no other sugar was found either, an acid-degradable sugar like AATGal is suggested as a component of the linkage unit (9).

In LTA and TA the AATGal residues are of conformational importance because the positively charged amino groups interact electrostatically or by hydrogen bonding with the negatively charged phosphate groups on the adjacent glucosyl residues (9, 23). The strength of this interaction compels the positively charged amino group on C-4 from the axial into the equatorial position, which results in conformational mobility between the common $^4C_1$ chair conformation and the energetically less favored $^5S_1$ screw conformation. Owing to the presence of AATGal in each repeat, the conformation of the whole chain is affected.

CELL WALL COMPOSITION AND DISTRIBUTION OF TEICHOIC ACID IN PEPTIDOGLYCAN

The cell wall components of strain Rx1 are shown in Table 2. Insight into the distribution of TA in the peptidoglycan network has been obtained by cleavage of pneumococcal cell walls with muramidases, followed by separation of TA-substituted and nonsubstituted muropeptides on molecular sieve columns (15). As judged from the distribution between the high- and low-molecular-weight fractions of peptide markers (glutamic acid, radiolabeled lysine), 30 to 40% of the muropeptides are present in the TA-containing fraction. Cleavage with muramidases of different substrate specificity led to the characterization of the TA-carrying muropeptides depicted in Fig. 3. A structure similar to that of muropeptide IV was proposed by Garcia-Bustos and Tomasz (15). As can be seen, in muropeptides containing two and three disaccharide moieties, only one of them carries a TA chain. The ratio MurNAc-6-P/MurNAc, determined in whole peptidoglycan, indicates that the TA-substituted muramic acid fraction varies between 15 and 19% (28, 48) but may in some preparations approach 30% (10b).

PHASE VARIATIONS IN COLONIAL MORPHOLOGY

Pneumococci can undergo spontaneous reversible phase variation between transparent and opaque phenotypes at high frequency (for further details, see chapter 22). These phenotypes are distinguished by their colony morphology, when viewed with oblique transmitted light on transparent agar surfaces (46). The transparent phenotype is associated with lower amounts of capsular polysaccharide, a higher

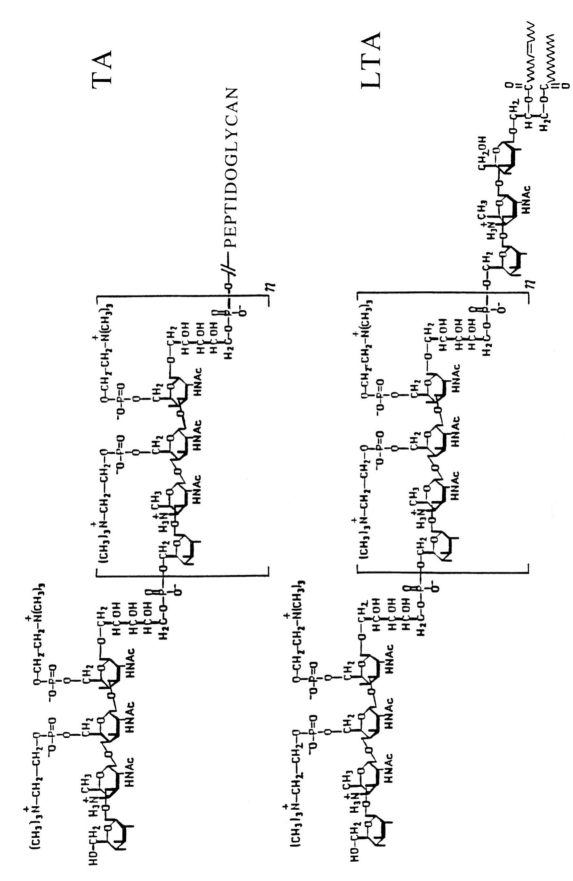

FIGURE 2 Pneumococcal TA and LTA. Depicted are the LTA and TA of strain R6 in which most of the repeats carry two phosphocholine residues (1, 10). In strain Rx1, the majority of the repeats carry one phosphocholine residue, which is attached to the non-ribitol-linked *N*-acetylgalactosaminyl residue (10a).

TABLE 1 Composition of the LTA and TA from *S. pneumoniae* Rx1 and R6

| Component | Molar ratio to intrachain phosphate (total phosphate − choline-linked phosphate) | | | |
| | Rx1 | | R6 | |
	TA	LTA	TA	LTA
Total phosphate	2.17	2.13	2.86	2.78
Choline	1.17	1.13	1.86	1.78
Intrachain phosphate	1.00	1.00	1.00	1.00
Glucose	1.19	1.10	0.89	1.01
Galactosamine	2.13	2.20	2.00	1.98
Ribitol[a]	1.04	1.02	1.00	
Quinovosamine[b]	+	+	+	+
Glycerol[c]		0.15		0.16

[a] Ribitol + 2,5-anhydroribitol.
[b] Quinovosamine is indicative of 2-acetamido-4-amino-2,4,6-trideoxy-galactose (see text).
[c] Glycerol was released from the glyceroglycolipid anchor of LTA (see text).

cellular content of TA, and accordingly higher amounts of choline in cell walls. For the opaque phenotype, the reverse is true: it has greater amounts of capsular polysaccharide and a lower content of teichoic acids (22). On Western blot analysis, the LTAs extracted from both variants show a similar ladderlike pattern, suggesting identical chain structures and average chain length. In addition, transparent variants express greater amounts of choline-binding protein A (CbpA), while opaque phenotypes have greater amounts of pneumococcal surface protein A (PspA) (31). A genetic locus conferring opacity has been identified, but its function remains to be clarified.

TABLE 2 Components of purified cell walls from *S. pneumoniae* Rx1

Component	Concn. (μmol/mg [dry wt])
Teichoic acid	
Total phosphate	1.11
Intrachain phosphate[a]	0.56
Galactosamine	1.06
Glucose	0.63
Ribitol[b]	0.64
Choline	0.55
Peptidoglycan	
Muramic acid	0.34
Muramic acid-6-phosphate	0.07
Glucosamine	0.30
Glutamic acid	0.52
Serine	0.07
Alanine	0.85
Lysine	0.51

[a] For definition, see Table 1.
[b] Ribitol + 2,5-anhydroribitol.

In an animal model of systemic infection following intraperitoneal inoculation of mice, the opaque phenotype was significantly more virulent than the transparent (22). In contrast to opaque isolates, transparent variants are capable of colonizing the nasopharynx in an infant rat model of carriage (46), whereby CbpA plays an essential role (31). Transparent variants also display enhanced ability to adhere to type II pneumocytes and vascular endothelial cells. In addition, through their surface-exposed phosphocholine,

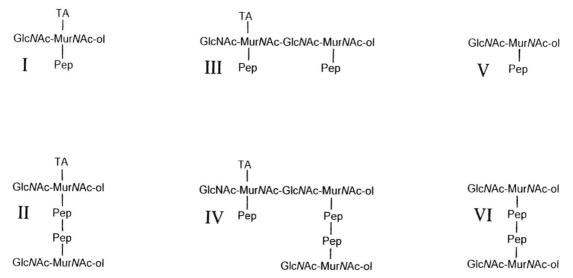

FIGURE 3 Teichoic acid-containing muropeptides. Muropeptides I and II were released by a muramidase that cleaves the glycosidic bonds of TA-substituted and nonsubstituted MurNAc-residues; III and IV were released by a muramidase that cleaves the glycosidic bonds of nonsubstituted MurNAc-residues only. The TA-free muropeptides V and VI were released by both enzymes. After enzymatic hydrolysis, cleavage points were marked by reduction with borohydride. Pep, stem peptides.

transparent variants bind more effectively to the receptor of platelet-activating factor, which is expressed on the surface of cytokine-activated lung and vascular endothelial cells. Binding to the platelet-activating factor receptor enhances pneumococcal adherence and invasion of host cells (6).

THE NUTRITIONAL REQUIREMENT OF CHOLINE

Pneumococci are not able to synthesize the choline required for the synthesis of LTA and TA. Moreover, choline is an essential growth factor but can be substituted in this function by nutritional ethanolamine (30). Although ethanolaminephosphate is incorporated into LTA and TA at the same positions as phosphocholine (10a), it is not able to replace phosphocholine functionally. Ethanolamine-grown cells lack transformability (42), and there are several proteins in pneumococci that specifically recognize and bind to phosphocholine residues. These proteins contain distinct N-terminal domains that are responsible for their specific biological activities, whereas the C-terminal domains are highly homologous and contain 6 to 10 choline-recognizing repeats of 20 amino acids each. The question of whether there are differences in the choline-binding domains between pneumococcal strains that contain one or two phosphocholines per repeat has not yet been addressed.

Autolysins

N-Acetylmuramyl-L-alanine-amidase, the major autolytic enzyme (LytA) of pneumococci, was the first reported protein to contain a C-terminal choline-binding domain of six choline-recognizing repeats (for review, see Lopez et al. [26]). LytA may be involved with separation of daughter cells during cell division and is responsible for stationary-phase penicillin- and deoxycholate-induced cell lysis (42). Binding of the amidase to phosphocholine on TA is a prerequisite for peptidoglycan hydrolysis (20, 26). Phosphocholine on TA or LTA is also required for the conversion of the inactive E form into the active C form (2, 26, 45). Pneumococci possess a second minor autolysin, characterized as endo-β-1,4-N-acetylglucosaminidase, which also requires TA-linked phosphocholine residues for activity (11).

Since in ethanolamine-grown cells both autolysins are inactive, ethanolamine-grown cells form long chains and are resistant to all kinds of autolysis. However, the requirement of LytA for built-in phosphocholine residues is no longer seen when solubilized cell wall polymers are degraded to muropeptides (15) or when the TA is removed from peptidoglycan (33). These observations may be interpreted to indicate that TA prevents the access of autolysin to its cell wall substrate and that this effect is overcome by binding of the enzyme to the phosphocholine residues. The activity of LytA must be strictly regulated in growing cells. At the present time, however, the cellular location of LytA, the mechanism that keeps the enzyme inactive, and the mode of triggering its activity are still a matter of debate (48, 49).

Pneumococcal Surface Protein A

Another well-characterized protein that contains a C-terminal domain of 10 choline-recognizing repeats is pneumococcal surface protein A (PspA) (3). In contrast to surface proteins of other gram-positive bacteria, PspA is anchored to the outer layer of the plasma membrane with choline-mediated interaction between membrane-associated LTA and its C-terminal repeat region (49). Accordingly, in ethanolamine-grown cells, or in mutants that lack phosphocholine on LTA, PspA is no longer retained and is lost into the surrounding medium (48, 49). The function of PspA is unknown. Yet it is important in pneumococcal virulence, is the immunodeterminant protein antigen on *S. pneumoniae* cells, and is capable of eliciting protective immunity in animal models of pneumococcal infection (3, 27). Two recent observations are particularly noteworthy: (i) intranasal immunization of mice with PspA prevented not only colonization of the nasopharynx but also subsequent invasive disease (47); and (ii) mice were also protected from fatal infections by oral immunization with an avirulent live recombinant *Salmonella* strain producing PspA (29). PspC is another surface protein whose choline-binding domain is 90% homologous to the choline-binding domain of PspA (3).

Other Choline-Binding Proteins

Recently, in pneumococcal lysates at least eight novel choline-binding proteins were identified by use of a choline affinity matrix (31). Four of these proteins were located at the bacterial surface, and five reacted with human reconvalescent sera. Antisera to these proteins passively protected mice challenged with a lethal dose of pneumococci in the peritoneum. CbpA, the predominant component of this mixture, was characterized as an adhesin and a determinant of virulence. The deduced sequence from the corresponding gene showed a distinct N-terminal region and a C-terminal domain consisting of 10 repeated choline-binding domains, nearly identical to PspA. CbpA is involved in adherence of pneumococci to cytokine-activated human cells and plays an important role in nasopharyngeal colonization.

CHOLINE-INDEPENDENT STRAINS

A choline-independent strain, R6Cho⁻, was recovered from a heterologous cross with DNA from *Streptococcus oralis* (33). *S. oralis* also incorporates phosphocholine into its cell wall TA but, unlike pneumococci, does not have a nutritional requirement for choline. Another choline-independent strain, JY2190, is a mutant that was generated by serial passage of strain Rx1 in chemically defined medium containing decreasing concentrations of ethanolamine with each passage (9, 48). None of the two strains had acquired the ability to synthesize choline or ethanolamine, because phosphorylated aminoalcohols could not be detected on TA and LTA. Despite the absence of phosphocholine, there was no alteration either in the structure of LTA and TA or in cell wall composition, including the stem peptide profile. Only the phosphate content of cell walls was reduced, consistent with the absence of phosphocholine. In vivo, the lack of active autolysins became apparent by impaired cell separation at the end of cell division and by resistance against stationary phase and penicillin-induced lysis. Owing to the absence of choline from LTA, PspA was lost into the surrounding medium, whereas despite choline-free TA, the amidase was retained on the cells. Both choline-independent strains retained the capacity to incorporate nutritional choline into teichoic acids: when grown in the presence of choline, PspA was retained, cells were separated normally and became penicillin sensitive, and phosphocholine was discovered on LTA and TA.

The metabolic background to dispense with the nutritional requirement for choline is not yet understood. As outlined recently (9, 48), a hint may come from in vitro experiments that provided evidence that the synthesis of peptidoglycan, which is the basis of cell growth, was inhibited by choline deprivation (7). If one assumes that pneumococcal TA, like the TAs of other gram-positive bacteria, is synthesized in linkage to polyprenol phosphate and that only completed phosphocholine- or phospho-ethanolamine-substituted TA is transferred to peptidoglycan, polyprenol-linked TA lacking these substituents would not be transferred but would trap polyprenol phosphate, rendering it unavailable for peptidoglycan synthesis. In this hypothesis, the nutritional requirement for choline resides in a recognition site of the transferase for phosphoaminoalcohols on TA. Accordingly, the mutation in JY2190 may render the activity of the transferase independent of this regulation. Strain R6Cho⁻ may have acquired by the heterologous cross the TA transferase from *S. oralis*, which is apparently not under this kind of control.

HOST-RELATED BIOLOGICAL ACTIVITIES OF PNEUMOCOCCAL CELL WALL COMPONENTS

The capacity of pneumococcal cell walls (as well as other gram-positive wall preparations) to induce the production of preinflammatory cytokines (tumor necrosis factor α, interleukin [IL]-6, IL-8, and IL-1) from peripheral mononuclear phagocytes, apparently through the CD-14 pathway, has been reported by several laboratories. While the specific wall component(s) responsible has not yet been identified, both intact pneumococcal cell wall peptidoglycan and disaccharide peptides carry cytokine-inducing activity. Pneumococcal wall and wall subcomponents were shown to induce slow-wave sleep in rabbits and inflammation and other signs of meningitis when introduced into the rabbit or rat models of disease. Additional evidence for the multitude of interactions between the cell wall of pneumococci and host-derived factors is the attachment of complement and C polysaccharide-reactive protein (CRP) to the choline teichoic acids. CRP, part of the host's innate defense system, is in humans a major acute-phase reactant. It contains a phosphocholine recognition site, and Ca^{2+}-mediated binding of phosphocholine to this site activates the classical pathway of the complement cascade and generates host defense-related complement fragments (e.g., opsonins) (40). Administration of human CRP protected mice against fatal pneumococcal infections, presumably by binding to the surface-exposed phosphocholine. That CRP is actually an important host defense molecule in pneumococcal disease became evident when human CRP transgenic mice were doped with lipopolysaccharide and then challenged with an intraperitoneally administered lethal dose of pneumococci (40). Host-related biological activities of pneumococcal cell wall components have been reviewed recently (43).

REFERENCES

1. **Behr, T., W. Fischer, J. Peter-Katalinic, and H. Egge.** 1992. The structure of pneumococcal lipoteichoic acid. Improved preparation, chemical and mass spectrometric studies. *Eur. J. Biochem.* **207:**1063–1075.

2. **Briese, T., and R. Hakenbeck.** 1985. Interaction of the pneumococcal amidase with lipoteichoic acid and choline. *Eur. J. Biochem.* **146:**417–427.

3. **Briles, D. E., S. K. Hollingshead, E. Swiatlo, A. Brooks-Walter, A. Szalai, A. Virolainen, L. S. McDaniel, K. A. Benton, P. White, K. Prellner, A. Hermansson, P. C. Aerts, H. van Dijk, and M. J. Crain.** 1997. PspA and PspC: their potential for use as pneumococcal vaccines. *Microb. Drug Resist.* **3:**401–408.

4. **Briles, E. B., and A. Tomasz.** 1970. Radioautographic evidence for equatorial wall growth in a gram positive bacterium: segregation of choline ³H-labeled teichoic acid. *J. Cell Biol.* **47:**786–790.

5. **Briles, E. B., and A. Tomasz.** 1973. Pneumococcal Forssman antigen. A choline-containing LTA. *J. Biol. Chem.* **248:**6394–6397.

6. **Cundell, D. R., N. P. Gerard, C. Gerard, I. Idanpaan-Helklia, and E. I. Tuomanen.** 1995. *Streptococcus pneumoniae* anchor to activated human cells by the receptor for platelet-activating factor. *Nature* **377:**435–438.

7. **Fischer, H., and A. Tomasz.** 1985. Peptidoglycan cross-linking and teichoic acid attachment in *Streptococcus pneumoniae*. *J. Bacteriol.* **163:**46–54.

8. **Fischer, W.** 1990. Bacterial phosphoglycolipids and lipoteichoic acids, p. 123–234. *In* M. Kates (ed.), *Handbook of Lipid Research*, vol. 6. *Glycolipids, Phosphoglycolipids, and Sulfoglycolipids*. Plenum Press, New York, N.Y.

9. **Fischer, W.** 1997. Pneumococcal lipoteichoic and teichoic acid. *Microb. Drug Resist.* **3:**309–325.

10. **Fischer, W., T. Behr, R. Hartmann, J. Peter-Katalinic, and H. Egge.** 1993. Teichoic acid and lipoteichoic acid of *Streptococcus pneumoniae* possess identical chain structures. A reinvestigation of teichoic acid (C polysaccharide). *Eur. J. Biochem.* **215:**851–857.

10a.**Fischer, W., R. Hartmann, and G. Pohlentz.** Unpublished data.

10b.**Fischer, W., K. Leopold, and C. Emilius.** Unpublished results.

11. **Garcia, P., J. L. Garcia, E. Garcia, and R. Lopez.** 1989. Purification and characterization of the autolysin glycosidase of *Streptococcus pneumoniae*. *Biochem. Biophys. Res. Commun.* **158:**251–256.

12. **Garcia-Bustos, J., and A. Tomasz.** 1989. Mechanism of pneumococcal cell wall degradation in vitro and in vivo. *J. Bacteriol.* **171:**114–119.

13. **Garcia-Bustos, J., and A. Tomasz.** 1990. A biological price of antibiotic resistance: major changes in the peptidoglycan structure of penicillin-resistant pneumococci. *Proc. Natl. Acad. Sci. USA* **87:**5414–5419.

14. **Garcia-Bustos, J. F., B. T. Chait, and A. Tomasz.** 1987. Structure of the peptide network of pneumococcal peptidoglycan. *J. Biol. Chem.* **262:**15400–15405.

15. **Garcia-Bustos, J. F., and A. Tomasz.** 1987. Teichoic acid-containing muropeptides from *Streptococcus pneumoniae* as substrates for the pneumococcal autolysin. *J. Bacteriol.* **169:**447–453.

16. **Glauner, B., J.-V. Holtje, and U. Schwarz.** 1988. The composition of the murein of *Escherichia coli*. *J. Biol. Chem.* **263:**10888–10095.

17. Goebel, W. F., T. Shedlovsky, G. I. Lavin, and M. H. Adams. 1943. The heterophil antigen of *Pneumococcus*. *J. Biol. Chem.* **148:**1–15.

18. Hakenbeck, R., A. Konig, I. Kern, M. van der Linden, W. Keck, D. Billot-Klein, R. Legrand, B. Schoot, and L. Gutmann. 1998. Acquisition of five high-M_r penicillin-binding protein variants during transfer of high-level β-lactam resistance from *Streptococcus mitis* to *Streptococcus pneumoniae*. *J. Bacteriol.* **180:**1831–1840.

19. Holtje, J.-V., and A. Tomasz. 1975. Lipoteichoic acid: a specific inhibitor of autolysin activity in pneumococcus. *Proc. Natl. Acad. Sci. USA* **72:**1690–1694.

20. Holtje, J.-V., and A. Tomasz. 1975. Specific recognition of choline residues in the cell wall teichoic acid by the N-acetylmuramyl-L-alanine amidase of pneumococcus. *J. Biol. Chem.* **250:**6072–6076.

21. Jennings, H. J., C. Lugowski, and N. M. Young. 1980. Structure of the complex polysaccharide C-substance from *Streptococcus pneumoniae* type 1. *Biochemistry* **19:**4712–4719.

22. Kim, J. O., and J. N. Weiser. 1998. Association of intrastrain phase variation in quantity of capsular polysaccharide and teichoic acid with the virulence of *Streptococcus pneumoniae*. *J. Infect. Dis.* **177:**368–377.

23. Klein, R. A., R. Hartmann, H. Egge, T. Behr, and W. Fischer. 1996. The aqueous solution structure of a lipoteichoic acid from *Streptococcus pneumoniae* strain R6 containing 2,4-diamino-2,4,6-trideoxy-galactose: evidence for conformational mobility of the galactopyranose ring. *Carbohydr. Res.* **281:**79–98.

24. Laitinen, H., and A. Tomasz. 1990. Changes in composition of peptidoglycan during maturation of the cell wall in pneumococci. *J. Bacteriol.* **172:**5961–5967.

25. Liu, T.-Y., and E. C. Gotschlich. 1967. Muramic acid phosphate as a component of the muropeptide of Gram-positive bacteria. *J. Biol. Chem.* **242:**471–476.

26. Lopez, R., E. García, P. García, and J. L. García. 1997. The pneumococcal cell wall degrading enzymes: a modular design to create new lysins? *Microb. Drug Resist.* **3:**199–211.

27. McDaniel, L. S., J. S. Sheffield, P. Delucchi, and D. E. Briles. 1991. PspA, a surface protein of *Streptococcus pneumoniae*, is capable of eliciting protection against pneumococci of more than one capsular type. *Infect. Immun.* **59:**222–228.

28. Mosser, J. L., and A. Tomasz. 1970. Choline-containing teichoic acid as a structural component of pneumococcal cell wall and its role in sensitivity to lysis by an autolytic enzyme. *J. Biol. Chem.* **245:**287–298.

29. Nayak, A. R., S. A. Tinge, R. C. Tart, L. S. McDaniel, D. E. Briles, and R. Curtiss III. 1998. A live recombinant avirulent oral *Salmonella* vaccine expressing pneumococcal surface protein A induces protective responses against *Streptococcus pneumoniae*. *Infect. Immun.* **66:**3744–3751.

30. Rane, L., and Y. Subbarow. 1940. Nutritional requirements of the pneumococcus. 1. Growth factors for types I, II, V, VII, VIII. *J. Bacteriol.* **40:**695–704.

31. Rosenow, C., P. Ryan, J. N. Weiser, S. Johnson, P. Fontan, A. Ortqvist, and H. R. Masure. 1997. Contribution of novel choline-binding proteins to adherence, colonization and immunogenicity of *Streptococcus pneumoniae*. *Mol. Microbiol.* **25:**819–829.

32. Severin, A., A. M. S. Figueiredo, and A. Tomasz. 1996. Separation of abnormal cell wall composition from penicillin resistance through genetic transformation of *Streptococcus pneumoniae*. *J. Bacteriol.* **178:**1788–1792.

33. Severin, A., D. Horne, and A. Tomasz. 1997. Autolysis and cell wall degradation in a choline-independent strain of *Streptococcus pneumoniae*. *Microb. Drug Resist.* **3:**391–400.

34. Severin, A., C. Schuster, R. Hakenbeck, and A. Tomasz. 1992. Altered murein composition in a DD-carboxypeptidase mutant of *Streptococcus pneumoniae*. *J. Bacteriol.* **174:**5152–5155.

35. Severin, A., E. Severina, and A. Tomasz. 1997. Abnormal physiological properties and altered cell wall composition in *Streptococcus pneumoniae* grown in the presence of clavulanic acid. *Antimicrob. Agents Chemother.* **41:**504–510.

36. Severin, A., and A. Tomasz. 1996. Naturally occurring peptidoglycan variants of *Streptococcus pneumoniae*. *J. Bacteriol.* **178:**168–174.

37. Severin, A., M. V. Vaz Pato, A. M. Sa Figueiredo, and A. Tomasz. 1995. Drastic changes in the peptidoglycan composition of penicillin resistant laboratory mutants of *Streptococcus pneumoniae*. *FEMS Microbiol. Lett.* **130:**31–35.

38. Sørensen, U. B. S., J. Blom, A. Birch-Andersen, and J. Henrichsen. 1988. Ultrastructural localization of capsules, cell wall polysaccharide, cell wall proteins, and F antigen in pneumococci. *Infect. Immun.* **56:**1890–1896.

39. Sørensen, U. B. S., and J. Henrichsen. 1987. Cross reaction between pneumococci and other streptococci due to C polysaccharide and F antigen. *J. Clin. Microbiol.* **25:**1854–1859.

40. Szalai, A. J., A. Agrawal, T. J. Greenhough, and J. E. Volanakis. 1997. C-Reactive protein. Structural biology, gene expression, and host defense function. *Immunol. Res.* **16:**127–136.

41. Tillet, W. S., W. F. Goebel, and O. T. Avery. 1930. Chemical and immunological properties of a species-specific carbohydrate of pneumococci. *J. Exp. Med.* **52:**895–900.

42. Tomasz, A. 1968. Biological consequences of the replacement of choline by ethanolamine in the cell wall of pneumococcus: chain formation, loss of transformability, and loss of autolysis. *Proc. Natl. Acad. Sci. USA* **59:**86–93.

43. Tomasz, A. 1999. The pneumococcal cell wall. *In* A. Tomasz (ed.), *Streptococcus pneumoniae: Molecular Biology and Mechanisms of Disease—Update for the 1990s*, in press. Mary Ann Liebert, Inc., New York, N.Y.

44. Tomasz, A., J. D. Jamieson, and E. Ottolenghi. 1964. The fine structure of *Diplococcus pneumoniae*. *J. Cell Biol.* **22:**453–467.

45. Tomasz, A., and M. Westphal. 1971. Abnormal autolytic enzyme in a pneumococcus with altered teichoic acid composition. *Proc. Natl. Acad. Sci. USA* **68:**2627–2630.

46. Weiser, J. N., R. Austrian, P. K. Sreenivasan, and H. R. Masure. 1994. Phase variation in pneumococcal opacity: relationship between colonial morphology and nasopharyngeal colonization. *Infect. Immun.* **62:**2582–2589.

47. Wu, H. Y., M. H. Nahm, Y. Guo, M. W. Russell, and D. E. Briles. 1997. Internasal immunization with PspA

(Pneumococcal surface protein A) can prevent carriage and infection with *Streptococcus pneumoniae*. *J. Infect. Dis.* **175:**839–846.

48. **Yother, J., K. Leopold, J. White, and W. Fischer.** 1998. Generation and properties of a *Streptococcus pneumoniae* mutant which does not require choline or analogs for growth. *J. Bacteriol.* **180:**2093–2101.

49. **Yother, J., and J. M. White.** 1994. Novel surface attachment mechanism of the *Streptococcus pneumoniae* protein PspA. *J. Bacteriol.* **176:**2962–2985.

50. **Zighelboim, S., and A. Tomasz.** 1980. Penicillin-binding proteins of the multiply antibiotic-resistant South African strains of *Streptococcus pneumoniae*. *Antimicrob. Agents Chemother.* **17:**434–442.

Streptococcus pneumoniae Capsular Polysaccharide

JAMES C. PATON AND JUDY K. MORONA

20

BACKGROUND

The presence of what is now recognized as the polysaccharide capsule on the surface of *Streptococcus pneumoniae* was noted by Pasteur in the first published description of the organism in 1880, and since that time it has been the direct or indirect focus of intensive investigation (reviewed by Austrian [4, 5]). Studies during the first three decades of the 20th century demonstrated the existence of multiple capsular types of *S. pneumoniae* and the fact that antibodies to the capsule conferred type-specific protection against challenge in laboratory animals. The capsular material itself was isolated by Dochez and Avery in 1917 (21), but the fact that it was immunogenic led them to believe that this "soluble substance of the pneumococcus" was proteinaceous in nature. It was not until 1925 that Avery and colleagues (8, 10) demonstrated that the pneumococcal capsule consisted of polysaccharide, the first nonprotein antigen to be recognized.

The capsule forms the outermost layer of all fresh clinical isolates of *S. pneumoniae*. It is approximately 200 to 400 nm thick (59) and, with the exception of type 3, appears to be covalently attached to the outer surface of the cell wall peptidoglycan (60). A total of 90 structurally and serologically distinct capsular polysaccharide (CPS) types have been recognized to date (27), and the chemical structures of the repeat units for many of these have been determined (reviewed by van Dam et al. [64]). The simplest CPS types are linear polymers with repeat units comprising two or more monosaccharides. The more complicated structural types are branched polysaccharides with repeat unit backbones composed of one to six monosaccharides plus additional side chains. Two nomenclature systems for CPS serotypes have been developed, but the Danish system, which combines antigenically cross-reacting types into groups, is now preferred to the American system, which lists serotypes in chronological order of discovery.

The polysaccharide capsule is considered to be a sine qua non of pneumococcal virulence (5). As mentioned above, all fresh clinical isolates of *S. pneumoniae* are encapsulated, and spontaneous nonencapsulated derivatives of such strains are almost completely avirulent. Indeed, as early as 1931, Avery and Dubos (7) demonstrated that enzymic depolymerization of the CPS of a type 3 pneumococcus increased its 50% lethal dose (LD_{50}) more than 10^5-fold. More recently, a similar effect on virulence of type 3 *S. pneumoniae* was achieved by transposon mutagenesis of a gene essential for capsule production (67). The clear morphological distinction between encapsulated ("smooth") and nonencapsulated ("rough") pneumococci, as well as the massive difference in virulence, facilitated early studies on the phenomenon of "capsular transformation." This was first demonstrated by Griffith (25), who found that a proportion of mice injected with a mixture of live rough and killed smooth pneumococci died. Smooth pneumococci expressing the same capsular serotype as the killed smooth strain were isolated from the blood of these mice. The transformation reaction was subsequently performed in vitro and led to the seminal discovery that the "transforming principle," the carrier of genetic information, was in fact DNA (9).

ROLE IN VIRULENCE

The precise manner in which the pneumococcal capsule contributes to virulence is not fully understood, although it is known to have strong antiphagocytic properties in nonimmune hosts. The majority of CPS serotypes are highly charged at physiological pH, and this may directly interfere with interactions with phagocytes (39). Pneumococcal cell wall teichoic acid (often referred to as C polysaccharide) is capable of activating the alternative complement pathway. In addition, antibodies to this and other cell surface constituents (e.g., surface proteins), which are found in most adult sera, may result in activation of the classical complement pathway, as does interaction of the teichoic acid with C-reactive protein. However, the capsule forms an inert shield, which appears to prevent either the Fc region of immunoglobulin G or iC3b fixed to deeper cell surface structures from interacting with receptors on phagocytic cells (52, 70). Pneumococci belonging to different CPS serotypes vary in their capacity to resist phagocytosis in vitro and also in their ability to elicit a humoral immune response (64), which no doubt accounts in large part for the fact that certain types are far more commonly associated with human disease (5). Otherwise isogenic pneumococci expressing different CPS serotypes, generated by in vitro or in vivo transformation, also exhibit marked differences in virulence for mice (31, 53). However, other factors contribute to some extent, as Kelly et al. (31) demonstrated that virulence is also influenced by the recipient strain. The difference in virulence between pneumococcal serotypes is clearly a function of the biological properties of the CPS itself and is not simply related to the thickness of capsule. For example, type 3 and type 37 pneumococci both produce very thick capsules, but only the former is of high virulence for humans or laboratory animals (5). Notwithstanding this, analyses of a series of mutants producing different amounts of CPS have demonstrated that within a given strain and serotype, virulence of *S. pneumoniae* is directly related to capsular thickness (41).

CPS-BASED VACCINES

In the immune host, binding of specific antibody to the CPS results in opsonization and rapid clearance of the invading pneumococci. For this reason a polyvalent pneumococcal CPS vaccine was licensed in the late 1970s. The vaccine provides serotype-specific protection, and the current formulation contains CPS purified from 23 of the 90 recognized types. However, the distribution of *S. pneumoniae* serotypes varies both temporally and geographically. The current vaccine covers approximately 85 to 90% of disease-causing serotypes in the United States or Europe, but in parts of Asia, coverage is <60% (39). Furthermore, serotype prevalence data are scanty for many developing countries, and so vaccine coverage is uncertain. The CPS vaccine is undoubtedly protective in healthy adults (against invasive infections caused by those serotypes included in the vaccine), but the efficacy is much lower in other groups at high risk of pneumococcal infection, such as the elderly; patients with underlying pulmonary, cardiac or renal disease; and immunocompromised patients. Efficacy is poorest in young children, for whom the existing formulation has little or no demonstrable clinical benefit. Polysaccharides are T-cell-independent antigens and so are poorly immunogenic in children under 5 years. This is particularly so for the five pneumococcal CPS types that most commonly cause invasive disease in children (22).

Conjugation of CPS antigens to protein carriers converts them into T-cell-dependent antigens, which are far more immunogenic in young children. Indeed, phase I/II clinical trials have shown that pneumococcal CPS-protein conjugate vaccines are highly immunogenic in human infants and induce immunological memory (57). Several large phase III trials are in progress. However, any protection will still be serotype specific, and the number of types covered by the conjugate formulations has been reduced from 23 to at most 11. Thus, although these vaccines are likely to provide markedly improved protection compared with nonconjugated purified CPS vaccines, it will be directed against a much more limited serotype range.

GENES ENCODING BIOSYNTHESIS OF CPS

Although the chemical structures have been determined for a number of CPS types (64), until recently compara-

tively little was known of the genes encoding CPS biosynthesis, and even less is known about how these are regulated or how the pneumococcus acquired the capacity to synthesize such a vast array of CPS serotypes. CPS production requires a complex pathway, including synthesis of the component monosaccharides, activation of each to a nucleotide precursor, coordinated transfer of each sugar, in sequence, to the repeating oligosaccharide, and subsequent polymerization, export, and attachment to the cell surface. The frequency of transformation of pneumococci from one CPS type to another observed during the classical studies of Griffith, Avery, and others suggested that at least those genes encoding the serotype-specific components of the CPS biosynthetic machinery were closely linked on the chromosome. Additional genetic and biochemical studies by Austrian, Bernheimer, and colleagues (6, 12) demonstrated that during transformation, CPS biosynthesis genes are transferred as a cassette. In the vast majority of cases, incorporation of the donor *cps* locus into the recipient cell chromosome was accompanied by loss of the original *cps* locus, consistent with genetic recombination between homologous sequences flanking the serotype-specific *cps* genes. However, transformation of a nonencapsulated derivative of a type 3 strain with DNA from a type 1 strain occasionally resulted in encapsulated transformants expressing both type 1 and type 3 CPS. Such binary encapsulated strains contained the defective type 3 *cps* locus as well as the functional type 1 locus. Binary encapsulation could also be achieved by transformation of the type 3 mutant with DNA from other serotypes whose CPS contained glucuronic acid (GlcA), a component of type 3 CPS. The recipient strains used in these studies had mutations in the gene encoding UDP-glucose (UDP-Glc) dehydrogenase, which is required for synthesis of UDP-GlcA. Thus, in the binary transformants, UDP-GlcA required for synthesis of type 3 CPS could be supplied by the UDP-Glc dehydrogenase encoded by a second *cps* locus. The frequency with which DNA from certain binary strains transformed unrelated pneumococci to the binary phenotype suggested a degree of linkage between the two *cps* loci, and possibly a tandem arrangement in some cases (12).

During the last 5 years, genes encoding CPS biosynthesis in several *S. pneumoniae* serotypes have been cloned and sequenced, and data are now available for types 1, 2, 3, 4, 14, 19A, 19B, 19C, 19F, 23F, and 33F (1, 19, 26, 29, 29a, 34, 36, 40, 44–48, 51, 56). The type 3 locus (designated *cps3* [20] or *cap3* [24]) and the type 19F locus (designated *cps19f* [26]) were the first to be completely sequenced and were shown to be located at the same position in the chromosome, between *dexB* and *aliA* (1, 19, 26, 45). However, there are major differences between these loci, suggesting distinct mechanisms of CPS biosynthesis in these strains.

Type 3 CPS Biosynthesis Genes

Type 3 CPS has a simple disaccharide repeat unit comprising Glc and GlcA. The *cps3* (*cap3*) locus contains only three intact genes, which are transcribed as a single unit (1, 19). The first gene (*cps3D* or *cap3A*) encodes the UDP-Glc dehydrogenase required for the synthesis of UDP-GlcA (2). The second gene (*cps3S* or *cap3B*) encodes the type 3 synthase, a processive β-glycosyltransferase that links the alternating Glc and GlcA moieties via distinct glycosidic bonds (1, 19). There is a significant degree of amino acid sequence similarity between Cps3S/Cap3B and other bacterial polysaccharide synthases, including HasA, which synthesizes the hyaluronic acid capsule of group A streptococci. These synthases have a common predicted architecture, with four transmembrane domains and a large central cytoplasmic domain. This latter region is believed to contain two distinct catalytic sites capable of forming the two different glycosidic linkages (30). Interestingly, transformation with plasmids carrying *cps3S/cap3B* alone is sufficient to direct synthesis of type 3 CPS in *Escherichia coli* or in *S. pneumoniae* serotypes 1, 2, 5, or 8, all of which contain UDP-GlcA. In smooth heterologous *S. pneumoniae* hosts, expression of *cps3S/cap3B* resulted in binary encapsulated strains producing type 3 as well as the original CPS type. In the *E. coli* transformants, a significant proportion of the type 3 CPS appeared in the periplasm, indicating that additional type-specific genes are not required for transport of CPS across the cell membrane (3). Indeed, there are no precedents for existence of dedicated export systems for any bacterial polysaccharides synthesized by such processive transferases. The C-terminal hydrophobic domains of these proteins have been predicted to form a pore or channel in the membrane, through which the growing polysaccharide chain is extruded as it is synthesized (30). The final complete gene in the *cps3/cap3* locus (*cps3U* or *cap3C*) encodes a Glc-1-phosphate uridylyltransferase. The product of this enzyme (UDP-Glc) is present in all pneumococci, regardless of CPS serotype, indicating that a functional homolog of *cps3U/cap3C* is located elsewhere in the chromosome. This is consistent with the finding that insertion-duplication mutagenesis of this gene does not abrogate type 3 CPS biosynthesis (1, 19). An additional open reading frame (ORF), designated *cps3M*, is located 3′ to *cps3U/cap3C* and has homology to genes encoding phosphoglucomutases (19). However, it is truncated at its C terminus and may not encode a functional enzyme; insertion-duplication mutagenesis of *cps3M* also has no impact on encapsulation (14, 19).

Type 19F CPS Biosynthesis Genes

Type 19F CPS is more complex than type 3, with a biological repeat unit consisting of →2)-α-L-Rha*p*-(1-PO$_4^-$→ 4)-β-D-Man*p*NAc-(1→4)-α-D-Glc*p*-(1→ (45). The *cps19f* locus is also much more complex than *cps3/cap3*, which is suggestive of a more elaborate biosynthetic mechanism. It is 15 kb long and consists of 15 genes, designated *cps19fA* to *cps19fO*, that appear to be arranged as a single transcriptional unit (Fig. 1) (26, 45). Insertion-duplication mutants in 13 of the 15 genes were constructed in a smooth-type 19F strain, all of which resulted in a rough (unencapsulated) phenotype. The contribution of individual genes could not be determined using these mutants owing to the possible polar effects of the mutagenesis procedure. Nevertheless, these studies did confirm that each gene is part of an operon essential for CPS production. In contrast, insertion-duplication mutagenesis of regions immediately 5′ or 3′ to *cps19f* had no impact on encapsulation (26, 45). Comparison with sequence databases and complementation analysis has allowed assignment of functions for most of the *cps19f* gene products, as well as a putative biosynthetic pathway for type 19F CPS. Cps19fA, the product of the first gene of the operon, has homology with LytR of *Bacillus subtilis* (a transcriptional attenuator of the autolytic genes *lytABC*) and may therefore have a regulatory function. Cps19fB has homology to proteins encoded by other gram-positive polysaccharide biosynthesis loci, but its function is currently unknown. Cps19fC and Cps19fD have homology with the first and last thirds of ExoP of *Rhizobium*, which is involved in exopolysaccharide chain length

regulation and possibly export. Cps19fC also has homology with conserved motifs of Cld/Rol proteins (O-antigen chain length regulators) from several gram-negative bacteria (54). All of these proteins have two membrane-spanning hydrophobic domains; the N and C termini are located in the cytoplasm, while the central portion is exposed on the external side of the cell membrane. Cps19fD is cytoplasmic, and like the C-terminal portion of ExoP, contains a putative ATP binding domain (26). Thus, Cps19fC and Cps19fD probably function together in chain length regulation and possibly also export of type 19F CPS in an analogous fashion to ExoP (55). Cps19fE is almost identical to a *S. pneumoniae* type 14 *cps* gene (*cps14E*) that has been shown experimentally to encode a glucosyl transferase, catalyzing the first step in polysaccharide biosynthesis, namely, transfer of Glc-1-phosphate to a lipid carrier (possibly undecaprenyl phosphate) in the cell membrane (33, 34). Cps19fF has homology to TagA of *B. subtilis* and so is likely to be an N-acetyl-mannosamine transferase. Cps19fG has homology to LicD of *Haemophilus influenzae*, the function of which is unknown. It also has similarity to the central portion of Mnn4, which is required for addition of mannosylphosphate to N-linked oligosaccharides in *Saccharomyces cerevisiae* (53a). Cps19fH has homology with rhamnosyl transferases, and it is possible that it functions with Cps19fG to form the phosphodiester linkage between rhamnose and N-acetyl-mannosamine in the type 19F CPS repeat unit. Cps19fI is very hydrophobic, with 12 potential membrane-spanning domains. It is probably the polysaccharide polymerase, as it has limited similarity with O-antigen polymerases (Rfc proteins) of *E. coli* and *Shigella*. Cps19fJ is also very hydrophobic and is probably the oligosaccharide repeat unit transporter, as it has similarity with RfbX proteins, the *E. coli/Shigella/Salmonella* O-antigen repeat unit transporters. It also has homology to the CapF protein of *Staphylococcus aureus*. Cps19fK is a UDP-N-acetylglucosamine-2-epimerase and has a high degree of homology with the *E. coli* RffE protein, which is a component of the pathway for biosynthesis of the enterobacterial common antigen. Indeed, cloned *cps19fK* is capable of complementing an *rffE* mutation, when expressed in *E. coli* (45). The last four genes of the *cps19f* locus encode enzymes required for dTDP-rhamnose biosynthesis. Cps19fL is a glucose-1-phosphate thymidylyl transferase; Cps19fM is a dTDP-4-keto-6-deoxyglucose-3,5-epimerase; Cps19fN is a dTDP-glucose-4,6-dehydratase; and Cps19fO is a dTDP-L-rhamnose synthase. These proteins have a very high degree of homology (up to 70% amino acid identity for Cps19fL) with the dTDP-rhamnose pathway enzymes of *Shigella flexneri* (RfbBDAC). Indeed, a cloned DNA fragment containing *cps19fL-O* is capable of complementing a *rfbBDAC*-deleted *S. flexneri rfb* operon in an *E. coli* K-12 background (45).

Based on the above, one can hypothesize that biosynthesis of type 19F CPS occurs by a mechanism analogous to that proposed for Rol/Cld- and Rfc-dependent O-antigen assembly in *Salmonella enterica* serogroups B and E (69). The initial step, catalyzed by Cps19fE, involves transfer of Glc-1-phosphate to a lipid carrier on the cytoplasmic face of the cell membrane. Cps19fE has several large hydrophobic domains in its N-terminal portion, which would anchor it to the membrane and facilitate interaction with the lipid carrier (26). Cps19fF, G, and H then catalyze the sequential transfer of the other component monosaccharide precursors (synthesized in the cytoplasm by Cps19fK, L, M, N, and O) to form the trisaccharide repeat unit. These

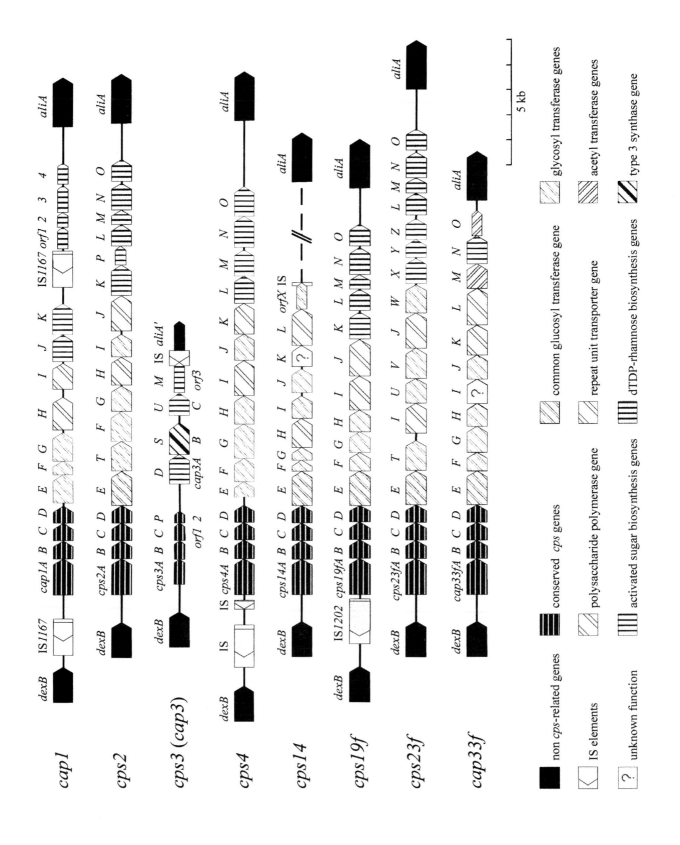

lipid-linked repeat units are then translocated from the cytoplasmic to the extracellular side of the cell membrane and polymerized in a blockwise fashion, extending the polysaccharide at the reducing terminus. These two steps are catalyzed by Cps19fJ and I, respectively, both of which are integral membrane proteins. The process of translocation and polymerization may be closely linked, and the two proteins may form a complex in the membrane. In *Salmonella* and *Shigella* O-antigen assembly, Rol/Cld has been proposed to regulate chain length by modulating interaction between the lipid-linked nascent O antigen and either the polymerase Rfc, or RfaL, a ligase responsible for transfer of O antigen to the lipid A core molecule (50, 69). Cps19fC and D may perform a similar function in *S. pneumoniae* type 19F. Pneumococcal CPS is believed to be covalently linked to the cell wall peptidoglycan (60), but the precise nature of this linkage and the enzyme responsible are unknown.

CPS Biosynthesis Genes in Other Serotypes

As mentioned above, sequence data for a number of other pneumococcal *cps* loci have recently become available. For most of these, functions of the gene products have been proposed on the basis of amino acid sequence similarities with known proteins. However, Kolkman et al. (33, 34, 36) have biochemically characterized the four glycosyltransferases encoded by genes in the type 14 *cps* locus. Also, a gene in the type 1 locus (*cap1K*) has been proved to encode a UDP-Glc dehydrogenase by complementation analysis (51). Although the size of each *cps* locus varies (from 11 to 18 genes), the organization closely resembles that of *cps19f*, as shown in Fig. 1, and this is suggestive of a common mechanism of biosynthesis. All *cps* loci are located between *dexB* and *aliA* in the pneumococcal chromosome. Available sequence data beyond the 3′ end of the type 14 locus do not include *aliA* (36), but we have shown by PCR that this gene is indeed located about 4 kb downstream of *cps14L* (47). There is a highly conserved consensus σ⁷⁰ promoter site immediately upstream of the first gene, and all appear to be transcribed as single units. The transcriptional start site has also been determined by primer extension analysis of the type 1 locus (51).

Homologs of *cps19fA-D* comprise the 5′ portion of each *cps* locus sequenced to date, suggesting an important common function in CPS biosynthesis. The only exception to this is the type 3 locus. Although a corresponding region with extensive sequence homology to *cps19fA-D* is located between *dexB* and the *cps3DSM/cap3ABC* genes referred to above, there are several deletions and frame-shift mutations, including loss of the conserved promoter sequence. Only one of the four ORFs remains intact, and the region

does not appear to be transcribed (1, 19). This, of course, is consistent with the distinct mechanisms of biosynthesis for type 3 and type 19F CPS described above. Interestingly, homologs of *cps19fA-D* are also present in the CPS biosynthesis loci of other members of the genus *Streptococcus*. Staphylococcal capsule loci, however, contain homologs of *cps19fB-D* (albeit in a different gene order), but not *cps19fA*.

The central portion of each *cps* locus encodes the glycosyltransferases responsible for assembly of the oligosaccharide repeat units, the polysaccharide polymerase, and the repeat unit transporter. Sequence comparison and hybridization studies have shown that the putative polymerase and transporter genes are highly specific, and cross-hybridization only occurs between serotypes with near-identical repeat unit backbones. Similarly, the glycosyltransferases exhibit a high degree of substrate specificity and form distinct glycosidic linkages. Accordingly, only a few of the transferases examined to date are common to unrelated serotypes. In all such cases, the type of glycosidic linkage and the sugar residues involved are identical. For example, a high stringency homolog of *cps19fF* is present in types 9N and 9V, and the type 19F, 9N, and 9V CPS repeat units all contain ManNAc with a $\beta(1{\rightarrow}4)$ linkage to Glc (45). The most widely distributed transferase is Cps19fE and its homologs, which add Glc to the lipid carrier, a common first step in biosynthesis of a large number of CPS serotypes. Glucosyl transferase activity has been demonstrated in nearly all serotypes tested that contain Glc in their CPS, except of course type 3 (35). Probes based on *cps19fE* (or the virtually identical *cps14E*) hybridized at high stringency with roughly half of such strains (26, 35). However, a second allele with approximately 70% nucleotide and amino acid sequence identity is present in all Glc-containing serotypes tested that did not hybridize to *cps19fE* (47).

The 3′ regions of the various *cps* loci encode enzymes for synthesis of activated monosaccharide precursors. The exception to this is type 14, but its CPS is composed of Glc, Gal, and GlcNAc, and the activated forms of these sugars are ubiquitous in pneumococci. Activated monosaccharide synthesis genes would be expected to be more widely distributed among different *S. pneumoniae* serotypes than genes encoding specific glycosyltransferases. Even so, there is substantial deviation among functional homologs. For example, the UDP-Glc dehydrogenases encoded by *cap1K*, *cps2K*, and *cps3D/cap3A* exhibit 60 to 90% amino acid sequence identity to each other. Also, the UDP-N-acetylglucosamine-2-epimerases encoded by *cps19fK* and *cps4L* are only 66% identical. However, high-stringency homologs of the four genes at the 3′ terminus of the *cps19f*

FIGURE 1 Organization of the *cps* loci from various *S. pneumoniae* serotypes. The organizations are based on published data for types 1 (51), 2 (29), 3 (1, 19), 14 (34, 36), 19F (26, 45), 23F (44, 56), and 33F (40). For type 4, the organization is based on analysis of contigs in the partial genome sequence for *S. pneumoniae* type 4 released by the Institute for Genomic Research (29a). Gene and locus designations are as published; ambiguities result from independent studies of the type 3 locus and so both designations are included. ORFs within the DNA sequence are indicated by large boxed arrows. Highly conserved ORFs, or those encoding proteins belonging to a particular functional group, are identified as shown in the legend below the figure. Assignment of an ORF to a given function-related group has been based on the published information for each locus, as well as on additional database comparisons for some of the ORFs. The narrow boxed arrows represent cryptic ORFs not required for CPS biosynthesis in the respective serotype. The broken line for *cps14* represents unsequenced DNA between *orfX* and *aliA*.

locus, which encode synthesis of dTDP-rhamnose from Glc-1-phosphate, have been shown to be present in all of 10 serotypes tested that were known to contain rhamnose in their CPS (45). Sequence data are available for two of these other cps loci (types 2 and 23F), and in each case the last four genes at the 3′ end have 91 to 99% nucleotide sequence identity to cps19fL-O (16, 29, 44, 56). Interestingly, Muñoz et al. (51) have reported that all type 1 pneumococci examined also contain high-stringency homologs of cps19fL-O, even though this serotype does not contain rhamnose in its CPS. Sequence analysis demonstrated that these genes (designated orf1-4) are located in the same position relative to aliA as they are in the rhamnose-containing serotypes (Fig. 1). However, orf1-4 are separated from the 3′ end of the cap1 locus by an insertion element IS1167; there is no evidence that they are transcribed, and orf4 is also truncated. Thus, it appears that these cryptic rhamnose genes are nonfunctional, and their presence suggests that the ancestor of type 1 pneumococci may have been a serotype containing rhamnose (51). Apparently nonfunctional genes are also present at the 3′ termini of some other cps loci. The truncated putative phosphoglucomutase gene cps3M of type 3 has already been mentioned (19). Another gene (orfX), the predicted product of which is 37% identical to Cps23fW, a putative glycerol-2-phosphate transferase, is also found at the 3′ end of the type 14 locus. This activity is not required for synthesis of type 14 CPS, and insertional mutagenesis of orfX has no effect on encapsulation. Also, the transcription termination signal for cps14 is located within orfX, and there is no evidence of an alternative promoter site immediately upstream (36). Interestingly, glycerol 2-phosphate is found in the CPS of the structurally and genetically related type 15A, and it is possible that the presence of the cryptic gene in type 14 may reflect a common ancestry. Similarly, a nonfunctional, frame-shift mutated gene (cap33fO) is located at the 3′ end of the type 33F locus, but an intact copy, which probably encodes an acetyl transferase, is present in type 33A (40).

Comparison of cps Loci in Serogroup 19

The Danish pneumococcal serotype nomenclature system groups CPS serotypes that cross-react immunologically, presumably as a consequence of structural similarities resulting in shared epitopes. If such related types share a common ancestor, analysis of the cps loci might be expected to provide insights into the mechanism of generation of capsular diversity. Group 19 comprises serotypes 19F, 19A, 19B, and 19C, and the structure of their CPS repeat units is shown in Fig. 2A. Types 19F and 19A are very closely related; they have an identical trisaccharide repeat unit and differ only in the nature of the glycosidic linkage formed during polymerization. Both these types are important causes of human disease, particularly in young children. Types 19B and 19C are only occasionally associated with human disease; they have more complicated CPS repeat units than types 19F and 19A, but differ from each other only by one additional Glc side chain in type 19C. The cps loci of all four members of group 19 have now been characterized (26, 45–48); the organization of each locus, and the degree of nucleotide sequence homology with cps19f, is shown in Fig. 2B. Thirteen genes (cps19A–H and K–O) are conserved in all four serogroup members. Nearly all of the common genes from types 19F, 19B, and 19C are >95% identical to each other, whereas those from the type 19A cps locus are more divergent. Al-

though the two loci are identical in terms of the number and arrangement of the genes present, sequence homology between individual cps19a and cps19f genes varies from 70 to 99% (at both the DNA and the deduced amino acid sequence level). This sequence divergence is surprising given that, theoretically, only an alteration in cps19fI, such that an α (1→3) rather than an α (1→2) linkage is formed during polymerization of the repeat units, would be required to change a type 19F pneumococcus into type 19A. This suggests either that the type 19F and 19A loci diverged in the distant evolutionary past or that their component genes originated from different sources. The latter alternative is supported by the fact that the G+C contents of most of the genes of the cps19a locus are consistently higher (up to 6.5%) than that for the respective cps19f gene (Fig. 3). Southern hybridization analysis using the cps19fC and D genes as probes had previously indicated that while high-stringency homologs for these genes were present in about half of all serotypes tested, including types 19B and 19C, there was negligible hybridization with the other types, including type 19A (45). We have since shown that all those types that did not react with cps19fC and D probes hybridize at high stringency with cps19aC and D probes (47). Sequence analysis of the cpsC genes from a total of 13 S. pneumoniae serotypes confirms the existence of two distinct classes (designated I and II). Nucleotide sequence identity between class I and class II genes is approximately 70 to 80%, but genes within a class exhibit >95% identity (47). Thus, the marked difference between the genes at the 5′ end of the type 19A cps locus and homologs in other members of group 19 is almost certainly a consequence of recombination between portions of class I and class II loci, rather than divergence over a long period of time.

High-stringency homologs of cps19fI and J, which encode the type 19F polysaccharide polymerase and repeat unit transporter, respectively, are not found in the type 19B cps locus. In type 19B this region of the cps locus (between cps19bH and cps19bK) contains five genes (Fig. 2B) encoding an unrelated polymerase and repeat unit transporter, as well as two additional putative glycosyltransferases and an unknown protein that might possibly be involved in synthesis of an activated ribose precursor (46). Transformation studies indicated that these five genes encode all of the functions required to convert a type 19F pneumococcus to type 19B. Interestingly, however, this region did not encode an additional ManNAc transferase for addition of the fourth sugar in the repeat unit backbone, and so we hypothesized that the transferase encoded by cps19bF is capable of adding ManNAc at two different positions (46).

The type 19C cps locus contains 19 genes, and at 21 kb it is the largest pneumococcal capsule gene cluster characterized to date. It is virtually identical to the type 19B locus (the 18 common genes exhibit >95% nucleotide sequence identity) and differs only in the insertion of a putative glucosyl transferase gene (cps19cS) between cps19cK and cps19cL. The presence of this gene accounts for the additional Glc side chain in the otherwise identical repeat unit structures (Fig. 2A).

Analysis of the G+C content of individual genes in the group 19 cps loci (Fig. 3) indicates a mosaic structure, with loci made up of blocks of genes that may have originated from different species. For example, the four type 19F genes encoding dTDP-rhamnose biosynthesis (cps19fLMNO) have G+C contents ranging from 41.5 to 42.3%, which is significantly higher than that of flanking sequences and for

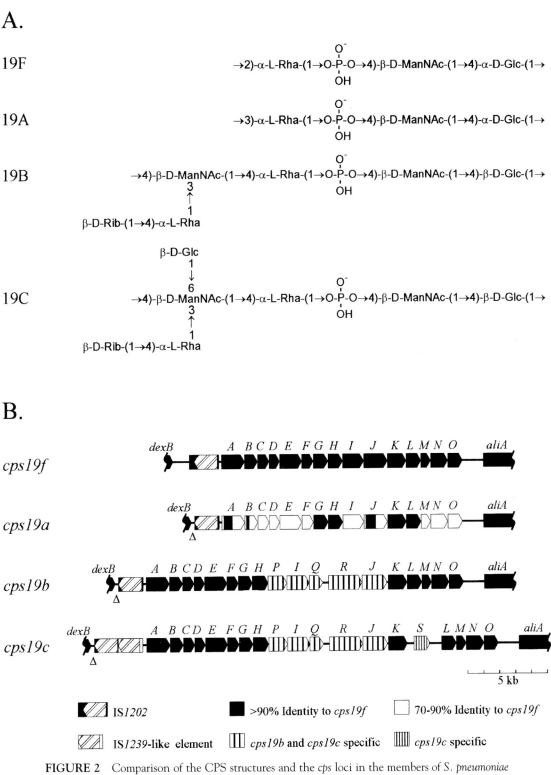

FIGURE 2 Comparison of the CPS structures and the *cps* loci in the members of *S. pneumoniae* serogroup 19. (A) Structures of the biological repeat units for CPS from *S. pneumoniae* types 19F, 19A, 19B, and 19C. These are based on published chemical repeat unit structures (13, 64), adjusting for the fact that Glc is the first sugar of the biological repeat unit. (B) Organization of the group 19 *cps* loci. For each gene (or part thereof), the approximate degree of nucleotide sequence identity with the respective type 19F gene is indicated. For type 19C, the region *cps19cPIQRJ* is >95% identical to the respective type 19B region. The presence of DNA deletions in the *dexB*-IS*1202* region is indicated by Δ.

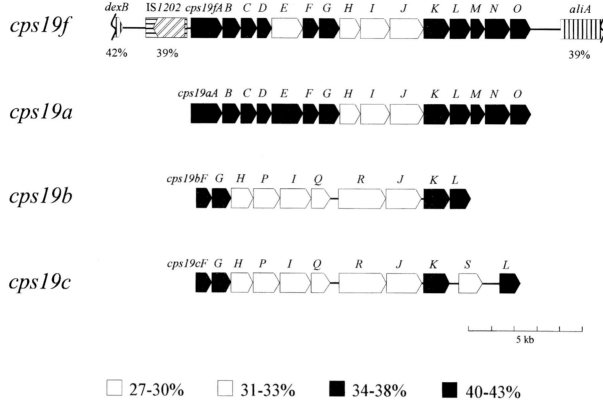

FIGURE 3 G+C content of the group 19 *cps* loci. Individual *cps* genes are shaded in accordance with their approximate G+C content. For flanking genes, the percent G+C is shown on the map. Portions of the *cps19b* and *cps19c* loci are not shown, as they are >95% identical to *cps19f*.

pneumococcal genes in general. Conversely, the block of five genes that converted a type 19F pneumococcus to type 19B (*cps19bPIQRJ*) have G+C contents of 27.2 to 29.7%, which are significantly lower than that of flanking regions. Variability in G+C content of individual genes, or clusters of genes, has been noted in other pneumococcal *cps* loci (36, 40, 56) and is a common feature of such loci in other bacterial species (58).

A final feature worthy of mention is the presence of an insertion sequence designated IS*1202* immediately upstream of the *cps19f* promoter. This 1,747-bp insertion element is unusual in that it is flanked by a very long direct repeat target sequence (27 bp) (26, 43). IS*1202* sequences are found at the same position in representatives of all four members of group 19. However, the *dexB*-IS*1202* regions in types 19A, 19B, and 19C contain an identical series of deletions of 58, 742, and 321 bp. This removes most of the DNA between *dexB* and the stop codon of the putative IS*1202* transposase. The type 19C strain that we sequenced also contains a second insertion sequence, the transposase of which has deduced amino acid sequence homology to IS*1239* from *Streptococcus pyogenes*. This element has integrated into the right inverted repeat sequence of IS*1202* (48).

From the above, one can hypothesize that type 19F may be the ancestor of the other members of serogroup 19. An early event in subsequent evolution may have involved the deletion mutations in the *dexB*-IS*1202* region. A recombination event in which the *cps19fIJ* region was replaced by *cps19bPIQRJ* may then have led to the emergence of

type 19B. An additional recombination incorporating *cps19cS* would then be required to create type 19C; insertion of the IS*1239*-related element may have occurred either before or after this event. Type 19A presumably also deviated from the type 19F ancestor, and evolution of the distinct CPS type may have involved a number of recombination events, some of which involved DNA fragments from a class II locus. It should be emphasized, however, that such a model is highly speculative, and *cps* sequence data from additional representatives of each of the members of group 19 are required for a definitive analysis.

CAPSULAR TRANSFORMATION IN VIVO

There is growing evidence that the phenomenon of capsular transformation first observed by Griffith (25) is a common phenomenon in vivo. Application of modern molecular typing techniques has resulted in the detection of otherwise genetically indistinguishable pneumococci expressing different capsular types. This has been particularly evident in clonal groups that are resistant to multiple antibiotics. Indeed, derivatives of a highly successful, multiply resistant type 23F clone (which originated in Spain) expressing types 3, 9N, 14, 19A, and 19F capsules have been isolated (11, 15, 17, 53). The sequences of the regions from *dexB-cpsB* and from *cpsL-aliA* have recently been determined for the parental multiresistant type 23F strain and eight otherwise identical type 19F clinical isolates. Examination of polymorphisms in the conserved regions of the two *cps* loci indicate that in each case, the 5′ recombination occurred upstream of *dexB*. In six of the eight type 19F strains, the 3′ crossover point was downstream of *aliA*. However, in the other two, a recombination crossover point between the introduced type 19F sequences and the type 23F chromosome was identified; this was in *cpsM* in one strain and *cpsN* in the other (16). Thus, capsule switching involves exchange of very large DNA fragments, ranging from at least 15 kb to over 22.5 kb. The existence of multiple crossover points as well as additional minor polymorphisms within the type 19F-derived *cps* genes also indicated that the eight multiply resistant type 19F strains that were studied arose as a consequence of a minimum of four independent transformation events involving different type 19F donors. It therefore appears that these capsule switching events may be relatively common among pneumococci in nature (16). Multiple serotypes of *S. pneumoniae* are frequently carried concurrently in the human nasopharynx (5), providing ample opportunity for exchange of DNA between types. In addition to enhancing the spread of drug resistance among diverse capsular types, these exchanges may also provide a mechanism for evasion of serotype-specific host immune defenses, such as those resulting from immunization with pneumococcal CPS-protein conjugate vaccines, which provide cover against a limited range of serotypes.

Complete or partial insertion elements have been located adjacent to 9 of the 11 *cps* loci sequenced to date, and in type 1 the locus is flanked at both the 5′ and 3′ ends by IS*1167*. These *cps*-flanking regions appear to be common targets for insertion elements, and this has led to the suggestion that they may play a role in horizontal transfer of *cps* genes (36, 51). There are precedents for this in other bacteria; the *H. influenzae* type b capsule genes, for example, are located on a 17-kb compound transposon (38). However, the identification of crossover points within the *cps* loci of the type 19F derivatives of the multiresistant type 23F *S. pneumoniae* clone referred to above confirms that at least in these cases, capsular exchange occurred as a consequence of homologous recombination rather than by transposition. Nevertheless, Muñoz et al. (51) have demonstrated that IS*1167* sequences flanking part of the *cap1* locus cloned in a plasmid could direct ectopic integration of these genes into copies of IS*1167* located elsewhere in the pneumococcal chromosome, resulting in genetically binary strains.

REGULATION OF CPS PRODUCTION

Colonization of the nasopharyngeal mucosa is an essential first step in the pathogenesis of pneumococcal disease and is presumed to involve interaction between pneumococcal adhesins and specific receptors on host epithelial cells. In a proportion of cases, asymptomatic carriage progresses to invasive disease, and although the events involved are poorly understood, it is clearly a watershed in the bacteria-host relationship. In recent years evidence has emerged that this transition involves a major switch in expression of important virulence determinants, as the pneumococcus adapts to the altered microenvironment (63). Maximal expression of capsule is clearly essential for systemic virulence, but the degree of exposure of other important pneumococcal surface structures, such as the adhesins, may also be influenced by capsular thickness. Nonencapsulated pneumococci exhibit higher adherence to human respira-

tory epithelial (A549) cells in vitro than otherwise isogenic derivatives expressing either type 3 or type 19F capsules (62). Thus, the very feature (encapsulation) that is absolutely essential for systemic virulence of *S. pneumoniae* could be disadvantageous during the colonization phase. Pneumococci have recently been shown to undergo a bidirectional phase variation between two distinct colonial morphologies, described as opaque and transparent. The transparent phenotype exhibits increased in vitro adherence to buccal epithelial cells and cytokine-activated A549 cells relative to opaque variants of the same strain, as well as an enhanced capacity to colonize the nasopharynx of infant rats (18, 68). On the other hand, the opaque form is associated with massively increased virulence in animal models of systemic disease, and this correlates with increased production of CPS relative to cell wall teichoic acid, compared with the transparent phenotype (32). Phase variation also correlated with alteration in levels of several surface proteins, but the molecular mechanism involved has yet to be elucidated.

Clearly, the capacity to regulate CPS production, at either the transcriptional or translational level, is important for the survival of the pneumococcus in different host environments. Transcripts of many polysaccharide loci from gram-negative bacteria have large leader sequences, which form a series of stem-loop structures (42). One of these contains the conserved JUMPstart sequence (28), which is part of two cis-acting sequences known as *ops* (for operon polarity suppressor). Interaction of these sequences with RfaH (a transcription antiterminator) and the RNA polymerase prevents the formation of other stem-loops in the leader sequence, which would otherwise result in premature termination (42). However, transcription of pneumococcal *cps* genes cannot be regulated in the same manner. Primer extension analysis has confirmed that the conserved σ^{70} promoter site upstream of all *S. pneumoniae cps* loci (except type 3) is functional, and the transcriptional start site is only about 20 nucleotides upstream of the initiation codon of the first *cps* gene (51). This leader sequence is too short for formation of any stem-loop structures.

The presence of a 115-bp repeated element has been noted upstream of the *cps* promoter in *S. pneumoniae* types 1, 3, 14, and 19F. This element appears to be specific to pneumococci, and copies have been found in the vicinity of other genes believed to be associated with virulence (36, 51). These sequences have no obvious function, although Kolkman et al. (36) have suggested that they might have a regulatory function for coordinately controlled expression of virulence-related genes. However, examination of sequence data for *S. pneumoniae* type 4 released by the Institute for Genomic Research indicates that there are at least 40 copies of sequences with >80% identity to the element in the genome. In type 19F, the copy near the *cps* locus is actually separated from the *cps* promoter by IS*1202*, and so it is difficult to imagine its involvement in a global regulatory mechanism. Moreover, insertion of the mutagenesis vector pVA891 into IS*1202*, which places the 115-bp element more than 8 kb upstream of the promoter, is known not to affect type 19F CPS production (43). As previously mentioned, in type 19F, IS*1202* is inserted immediately 5′ to the −35 sequence of the *cps* promoter and so the above result, in all probability, precludes the involvement of any upstream region in transcriptional regulation of *cps19f*. It is possible, however, that the level of transcription of *cps* loci varies from strain to strain. Although the −10 and −35 sequences themselves are highly conserved, Llull et al. (40) have noted minor variations in flanking sequences. In type 37, for example, the −10 and −35 sequences are separated by 16 nucleotides rather than 17, and this correlated with a markedly lower promoter strength. A 4-nucleotide deletion immediately 5′ to the −35 sequence in type 33F, relative to type 1, also correlated with a slight reduction in promoter strength (40). The extent to which such differences impact on the level of encapsulation is uncertain, but it is curious in the light of these findings that type 37, which had the weakest *cps* promoter, produces one of the thickest capsules of all *S. pneumoniae* serotypes (5).

At present the mechanism of regulation of CPS production in *S. pneumoniae* is not understood. The only product of the *cps* loci likely to be involved is CpsA, which resembles a *B. subtilis* transcriptional attenuator. We have recently constructed a derivative of *S. pneumoniae* type 19F with an in-frame deletion mutation in *cps19fA*. This mutant still produces type 19F CPS, as judged by quellung reaction. Also, colonies growing on blood agar appeared smooth, but they were much smaller than the parental type 19F strain, despite the lack of any obvious difference in growth rate (49). Additional phenotypic characterization of this mutant, including transcriptional studies of *cps19f* and quantitation of total CPS production, are in progress. In group B streptococci, an additional gene located upstream of the *cps19fA* homolog, but transcribed in the opposite orientation, encodes a protein (CpsY) with similarity to the LysR family of DNA-binding transcriptional regulators (37). However, no such gene is adjacent to any of the pneumococcal *cps* loci examined to date. In *Lactococcus lactis*, the plasmid-encoded exopolysaccharide biosynthesis locus *eps* exhibits some organizational similarities to pneumococcal *cps* loci. However, a gene encoding a protein with similarity to LytR is found at the end of the *eps* locus rather than at the beginning, and on the opposite DNA strand. Also, two additional genes are found at the 5′ end of the locus (between the promoter and homologs of *cps19fC, D,* and *B*), one of which encodes a protein (EpsR) with similarity to DNA-binding regulatory proteins (65), but it has no similarity to any pneumococcal gene products. Thus, it appears that distinct transcriptional regulatory mechanisms may exist for polysaccharide biosynthesis loci in different gram-positive species.

It is also possible that genes located elsewhere on the chromosome may influence CPS production in *S. pneumoniae*. Watson et al. (66) suggested that a region 3′ to the pneumococcal *lytA* (autolysin) gene might be involved in regulation of capsular expression, as this appeared to be the site of Tn916 insertion in a type 3 mutant that had lost the capacity to produce CPS. However, García et al. (23) subsequently demonstrated that deletion of this region of the chromosome had no impact on encapsulation. An additional possibility is that regulation of CPS production is indirect and mediated by availability of precursors or cofactors. For example, the mechanism of synthesis of the pneumococcal cell wall teichoic acid (C polysaccharide) is likely to be similar to CPS. Accordingly, if teichoic acid synthesis is subject to direct regulation, this may influence the availability of common precursors (e.g., UDP-Glc or the lipid carrier) for CPS biosynthesis. Such a phenomenon would be consistent with the observation that phase variation in *S. pneumoniae* has opposite effects on the total amounts of CPS and teichoic acid (32). Saturation of a common lipid carrier pool with the disaccharide precursor for peptidoglycan synthesis has previously been proposed

to explain blockade of exopolysaccharide production in *Streptococcus thermophilus* with a mutation in *pbp2b*, one of its penicillin-binding protein genes (61).

CONCLUDING REMARKS

In this chapter we have attempted to summarize the current state of knowledge on the CPS of *S. pneumoniae*, with particular reference to the genes encoding biosynthesis of this most important of all pneumococcal surface antigens. Notwithstanding the insights gained from the elegant classical genetic studies on CPS production carried out before the early 1970s, research in recent years has been revolutionized by the advent of modern methods for gene cloning, DNA amplification, and sequence analysis. The complete nucleotide sequences for the *cps* loci from 11 of the 90 known serotypes are now available, with several additional loci currently under investigation. The functions of many of the individual genes in these loci await confirmation by conventional biochemical and genetic analysis. Nevertheless, access to the enormous body of information now available on sequence databases, combined with knowledge of the chemical structures for many of the CPS repeat units, has enabled accurate predictions of function for a significant proportion of these genes. It has also been possible to predict the mechanisms of CPS biosynthesis in pneumococci by analogy with those operating in gram-negative bacteria. The existence of two distinct mechanisms for CPS biosynthesis in *S. pneumoniae* has already been recognized. Type 3 CPS is polymerized directly from activated monosaccharide precursors by a processive transferase. However, in all other types investigated to date, CPS repeat units appear to be assembled on the cytoplasmic side of the membrane attached to a lipid carrier and then translocated across the membrane, polymerized, and presumably attached to the cell wall. Much remains to be learned about the precise molecular events involved in both of these processes, and about how CPS production in pneumococci is regulated. Detailed biochemical and mutational analysis is also required to elucidate the precise functions of the four genes at the 5′ end of the *cps* loci, which presumably encode important common steps in polysaccharide biosynthesis in pneumococci, as well as in other gram-positive genera. Given the importance of capsules to the virulence of *S. pneumoniae* and several other gram-positive pathogens, such conserved components of the CPS biosynthesis machinery may prove to be useful targets for novel antimicrobial strategies.

We sincerely thank Gianni Pozzi, Alexander Tomasz, Ernesto García, and Rubens López for communicating sequence data and/or manuscripts describing the cps loci of serotypes 2, 23F, and 33F prior to publication. We are also grateful to Renato Morona for many helpful discussions. Work in our laboratory is supported by a grant from the National Health and Medical Research Council of Australia.

REFERENCES

1. **Arrecubieta, C., E. García, and R. López.** 1995. Sequence and transcriptional analysis of a DNA region involved in the production of capsular polysaccharide in *Streptococcus pneumoniae* type 3. Gene **167:**1–7.

2. **Arrecubieta, C., R. López, and E. Gárcia.** 1994. Molecular characterization of *cap3A*, a gene from the operon required for the synthesis of the capsule of *Streptococcus pneumoniae* type 3: sequencing of mutations responsible

for the unencapsulated phenotype and localization of the capsular cluster on the pneumococcal chromosome. *J. Bacteriol.* **176:**6375–6383.

3. **Arrecubieta, C., R. López, and E. García.** 1996. Type 3-specific synthase of *Streptococcus pneumoniae* (Cap3B) directs type 3 polysaccharide biosynthesis in *Escherichia coli* and in pneumococcal strains of different serotypes. *J. Exp. Med.* **184:**449–455.

4. **Austrian, R.** 1981. Pneumococcus: the first one hundred years. *Rev. Infect. Dis.* **3:**183–189.

5. **Austrian, R.** 1981. Some observations on the pneumococcus and on the current status of pneumococcal disease and its prevention. *Rev. Infect. Dis.* **3**(Suppl.):S1–S17.

6. **Austrian, R., H. P. Bernheimer, E. E. B. Smith, and G. T. Mills.** 1959. Simultaneous production of two capsular polysaccharides by pneumococcus. II. The genetic and biochemical bases of binary capsulation. *J. Exp. Med.* **110:**585–602.

7. **Avery, O. T., and R. Dubos.** 1931. The protective action of a specific enzyme against type III pneumococcus infections in mice. *J. Exp. Med.* **54:**73–89.

8. **Avery, O. T., and M. Heidelberger.** 1925. Immunological relationships of cell constituents of pneumococcus. *J. Exp. Med.* **42:**367–376.

9. **Avery, O. T., C. M. MacLeod, and M. McCarty.** 1944. Studies on the chemical nature of the substance inducing transformation of pneumococcal types. Induction of transformation by a desoxyribonucleic acid fraction isolated from pneumococcus type III. *J. Exp. Med.* **79:**137–158.

10. **Avery, O. T., and, H. J. Morgan.** 1925. Immunological reactions of the isolated carbohydrate and protein of pneumococcus. *J. Exp. Med.* **42:**347–353.

11. **Barnes, D. M., S. Whittier, P. H. Gilligan, S. Soares, A. Tomasz, and F. W. Henderson.** 1995. Transmission of multidrug-resistant serotype 23F *Streptococcus pneumoniae* in group day care: evidence suggesting capsular transformation of the resistant strain *in vivo*. *J. Infect. Dis.* **171:**890–896.

12. **Bernheimer, H. P., I. E. Wermundsen, and R. Austrian.** 1967. Qualitative differences in the behavior of pneumococcal deoxyribonucleic acids transforming to the same capsular type. *J. Bacteriol.* **93:**320–333.

13. **Beynon, L. M., J. C. Richards, M. B. Perry, and P. J. Kniskern.** 1991. Antigenic and structural relationships within group 19 *Streptococcus pneumoniae*: chemical characterization of the specific capsular polysaccharides of type 19B and 19C. *Can. J. Chem.* **70:**218–232.

14. **Caimano, M. J., G. G. Hardy, and J. Yother.** 1998. Capsule genetics in *Streptococcus pneumoniae* and a possible role for transposition in the generation of the type 3 locus. *Microb. Drug Resist.* **4:**11–23.

15. **Coffey, T. J., C. G. Dowson, M. Daniels, J. Zhou, C. Martin, B. G. Spratt, and J. M. Musser.** 1991. Horizontal gene transfer of multiple penicillin-binding protein genes and capsular biosynthetic genes in natural populations of *Streptococcus pneumoniae*. *Mol. Microbiol.* **5:**2255–2260.

16. **Coffey, T. J., M. C. Enright, M. Daniels, J. K. Morona, R. Morona, W. Hryniewicz, J. C. Paton, and B. G. Spratt.** 1998. Recombinational exchanges at the capsular polysaccharide biosynthetic locus lead to frequent serotype changes among natural isolates of *Streptococcus pneumoniae*. *Mol. Microbiol.* **27:**73–83.

17. **Coffey, T. J., M. C. Enright, M. Daniels, P. Wilkinson, S. Berron, A. Fenoll, and B. G. Spratt.** 1998. Serotype 19A variants of the Spanish serotype 23F multiresistant clone of *Streptococcus pneumoniae*. *Microb. Drug Resist.* **4:**51–55.

18. **Cundell, D. R., J. N. Weiser, J. Shen, A. Young, and E. I. Tuomanen.** 1995. Relationship between colonial morphology and adherence of *Streptococcus pneumoniae*. *Infect. Immun.* **63:**757–761.

19. **Dillard, J. P., M. W. Vandersea, and J. Yother.** 1995. Characterization of the cassette containing genes for type 3 capsular polysaccharide biosynthesis in *Streptococcus pneumoniae*. *J. Exp. Med.* **181:**973–983.

20. **Dillard, J. P., and J. Yother.** 1994. Genetic and molecular characterization of capsular polysaccharide biosynthesis in *Streptococcus pneumoniae* type 3. *Mol. Microbiol.* **12:**959–972.

21. **Dochez, A. R., and O. T. Avery.** 1917. The elaboration of specific soluble substance by pneumococcus during growth. *J. Exp. Med.* **26:**477–493.

22. **Douglas, R. M., J. C. Paton, S. J. Duncan, and D. Hansman.** 1983. Antibody response to pneumococcal vaccination in children younger than five years of age. *J. Infect. Dis.* **148:**131–137.

23. **García, E., C. Arrecubieta, and R. López.** 1996. The *lytA* gene and the region located downstream of this gene are not involved in the formation of the type 3 capsular polysaccharide of *Streptococcus pneumoniae*. *Curr. Microbiol.* **33:**133–135.

24. **García, E., P. García, and R. López.** 1993. Cloning and sequencing of a gene involved in the synthesis of the capsular polysaccharide of *Streptococcus pneumoniae* type 3. *Mol. Gen. Genet.* **239:**188–195.

25. **Griffith, F.** 1928. The significance of pneumococcal types. *J. Hyg.* **27:**113–159.

26. **Guidolin, A., J. K. Morona, R. Morona, D. Hansman, and J. C. Paton.** 1994. Nucleotide sequence of an operon essential for capsular polysaccharide biosynthesis in *Streptococcus pneumoniae* type 19F. *Infect. Immun.* **62:**5384–5396.

27. **Henrichsen, J.** 1995. Six newly recognized types of *Streptococcus pneumoniae*. *J. Clin. Microbiol.* **33:**2759–2762.

28. **Hobbs, M., and P. R. Reeves.** 1994. The JUMPstart sequence: a 39 bp element common to several polysaccharide gene clusters. *Mol. Microbiol.* **12:**855–856.

29. **Iannelli, F., B. J. Pearce, and G. Pozzi.** 1999. The type 2 capsule locus of *Streptococcus pneumoniae*. *J. Bacteriol.* **181:**2652–2654.

29a. **Institute for Genomic Research:** http://www.tigr.org/pub/data/s_pneumoniae/

30. **Keenleyside, W. J., and C. Whitfield.** 1996. A novel pathway for O-polysaccharide biosynthesis in *Salmonella enterica* serovar Borreze. *J. Biol. Chem.* **271:**28581–28592.

31. **Kelly, T., J. P. Dillard, and J. Yother.** 1994. Effect of genetic switching of capsular type on virulence of *Streptococcus pneumoniae*. *Infect. Immun.* **62:**1813–1819.

32. **Kim, J. O., and J. N. Weiser.** 1998. Association of intrastrain phase variation in quantity of capsular polysaccharide and teichoic acid with the virulence of *Streptococcus pneumoniae*. *J. Infect. Dis.* **177:**368–377.

33. **Kolkman, M. A. B., D. A. Morrison, B. A. M. van der Zeijst, and P. J. M. Nuijten.** 1996. The capsule polysac-

charide synthesis locus of *Streptococcus pneumoniae* serotype 14: identification of the glycosyltransferase gene *cps14E*. *J. Bacteriol.* **178**:3736–3741.

34. **Kolkman, M. A. B., B. A. M. van der Zeijst, and P. J. M. Nuijten.** 1997. Functional analysis of glycosyltransferases encoded by the capsular polysaccharide locus of *Streptococcus pneumoniae* serotype 14. *J. Biol. Chem.* **272:** 19502–19508.

35. **Kolkman, M. A. B., B. A. M. van der Zeijst, and P. J. M. Nuijten.** 1998. Diversity of capsular polysaccharide synthesis gene clusters in *Streptococcus pneumoniae*. *J. Biochem.* **123**:937–945.

36. **Kolkman, M. A. B., W. Wakarchuk, P. J. M. Nuijten, and B. A. M. van der Zeijst.** 1997. Capsular polysaccharide synthesis in *Streptococcus pneumoniae* serotype 14: molecular analysis of the complete *cps* locus and identification of genes encoding glycosyltransferases required for the biosynthesis of the tetrasaccharide subunit. *Mol. Microbiol.* **26**:197–208.

37. **Koskiniemi, S., M. Sellin, and M. Norgren.** 1998. Identification of two genes, *cpsX* and *cpsY*, with putative regulatory function on capsule expression in group B streptococci. *FEMS Immunol. Med. Microbiol.* **21**:159–168.

38. **Kroll, J. S., B. M. Loynds, and E. R. Moxon.** 1991. The *Haemophilus influenzae* capsulation gene cluster: a compound transposon. *Mol. Microbiol.* **5**:1549–1560.

39. **Lee, C.-J., S. D. Banks, and J. P. Li.** 1991. Virulence, immunity and vaccine related to *Streptococcus pneumoniae*. *Crit. Rev. Microbiol.* **18**:89–114.

40. **Llull, D., R. López, E. García, and R. Muñoz.** 1998. Molecular structure of the gene cluster responsible for the synthesis of the polysaccharide capsule of *Streptococcus pneumoniae* type 33F. *Biochim. Biophys. Acta* **1449**:217–224.

41. **MacLeod, C. M., and M. R. Krauss.** 1950. Relation of virulence of pneumococcal strains for mice to the quantity of capsular polysaccharide formed in vitro. *J. Exp. Med.* **92**:1–9.

42. **Marolda, C. L., and M. A. Valvano.** 1998. Promoter region of the *Escherichia coli* O7-specific lipopolysaccharide gene cluster: structural and functional characterization of an upstream untranslated mRNA sequence. *J. Bacteriol.* **180**:3070–3079.

43. **Morona, J. K., A. Guidolin, R. Morona, D. Hansman, and J. C. Paton.** 1994. Isolation, characterization and nucleotide sequence of IS*1202*, an insertion sequence of *Streptococcus pneumoniae*. *J. Bacteriol.* **176**:4437–4443.

44. **Morona, J. K., D. C. Miller, T. J. Coffey, C. J. Vindurampulle, B. G. Spratt, R. Morona, and J. C. Paton.** 1999. Molecular and genetic characterization of the capsule biosynthesis locus of *Streptococcus pneumoniae* type 23F. *Microbiology* **145**:781–789.

45. **Morona, J. K., R. Morona, and J. C. Paton.** 1997. Characterization of the locus encoding the *Streptococcus pneumoniae* type 19F capsular polysaccharide biosynthetic pathway. *Mol. Microbiol.* **23**:751–763.

46. **Morona, J. K., R. Morona, and J. C. Paton.** 1997. Molecular and genetic characterization of the capsule biosynthesis locus of *Streptococcus pneumoniae* type 19B. *J. Bacteriol.* **179**:4953–4958.

47. **Morona, J. K., R. Morona, and J. C. Paton.** 1999. Analysis of the 5′ portion of the type 19A capsule locus identifies two classes of *cpsC*, *cpsD*, and *cpsE* genes in *Streptococcus pneumoniae*. *J. Bacteriol.* **181**:3599–3605.

48. **Morona, J. K., R. Morona, and J. C. Paton.** Comparative genetics of capsular polysaccharide biosynthesis in *Streptococcus pneumoniae* types belonging to serogroup 19. *J. Bacteriol.* **181**, in press.

49. **Morona, J. K., R. Morona, and J. C. Paton.** Unpublished observations.

50. **Morona, R., L. van den Bosch, and P. A. Manning.** 1995. Molecular, genetic, and topological characterization of O-antigen chain length regulation in *Shigella flexneri*. *J. Bacteriol.* **177**:1059–1068.

51. **Muñoz, R., M. Mollerach, R. López, and E. García.** 1997. Molecular organization of the genes required for the synthesis of type 1 capsular polysaccharide of *Streptococcus pneumoniae*: formation of binary encapsulated pneumococci and identification of cryptic dTDP-rhamnose biosynthesis genes. *Mol. Microbiol.* **25**:79–92.

52. **Musher, D. M.** 1992. Infections caused by *Streptococcus pneumoniae*: clinical spectrum, pathogenesis, immunity and treatment. *Clin. Infect. Dis.* **14**:801–807.

53. **Nesin, M., M. Ramirez, and A. Tomasz.** 1998. Capsular transformation of a multidrug-resistant *Streptococcus pneumoniae* in vivo. *J. Infect. Dis.* **177**:707–713.

53a. **Odani, T., Y. Shimma, A. Tanaka, and Y. Jigami.** 1996. Cloning and analysis of the *MNN4* gene required for phosphorylation of *N*-linked oligosaccharides in *Saccharomyces cerevisiae*. *Glycobiology* **6**:805–810.

54. **Paton, J. C., J. K. Morona, and R. Morona.** 1997. Characterization of the capsular polysaccharide biosynthesis locus of *Streptococcus pneumoniae* type 19F. *Microb. Drug Resist.* **3**:89–99.

55. **Paulsen, I. T., A. M. Beness, and M. H. Saier, Jr.** 1997. Computer-based analyses of the protein constituents of transport systems catalysing export of complex carbohydrates in bacteria. *Microbiology* **143**:2685–2699.

56. **Ramirez, M., and A. Tomasz.** 1998. Molecular characterization of the complete 23F capsular polysaccharide locus of *Streptococcus pneumoniae*. *J. Bacteriol.* **180**:5273–5278.

57. **Rennels, M. B., K. M. Edwards, H. L. Keyserling, K. S. Reisinger, D. A. Hogerman, D. V. Madore, I. Chang, P. R. Paradiso, F. J. Malinoski, and A. Kimura.** 1998. Safety and immunogenicity of heptavalent pneumococcal vaccine conjugated to CRM197 in United States infants. *Pediatrics* **101**:604–611.

58. **Roberts, I. S.** 1996. The biochemistry and genetics of capsular polysaccharide production in bacteria. *Annu. Rev. Microbiol.* **50**:285–315.

59. **Sorensen, U. B. S., J. Blom, A. Birch-Andersen, and J. Henrichsen.** 1988. Ultrastructural localization of capsules, cell wall polysaccharide, cell wall proteins, and F antigen in pneumococci. *Infect. Immun.* **56**:1890–1896.

60. **Sorensen, U. B. S., J. Henrichsen, H.-C. Chen, and S. C. Szu.** 1990. Covalent linkage between the capsular polysaccharide and the cell wall peptidoglycan of *Streptococcus pneumoniae* revealed by immunochemical methods. *Microb. Pathog.* **8**:325–334.

61. **Stingele, F., and B. Mollet.** 1996. Disruption of the gene encoding penicillin-binding protein 2b (*pbp2b*) causes al-

tered cell morphology and cease in exopolysaccharide production in *Streptococcus thermophilus* Sfi6. *Mol. Microbiol.* **22:**357–366.

62. **Talbot, U., A. W. Paton, and J. C. Paton.** 1996. Uptake of *Streptococcus pneumoniae* by respiratory epithelial cells. *Infect. Immun.* **64:**3772–3777.

63. **Tuomanen, E. I., and H. R. Masure.** 1997. Molecular and cellular biology of pneumococcal infection. *Microb. Drug Resist.* **3:**297–308.

64. **van Dam, J. E. G., A. Fleer, and H. Snippe.** 1990. Immunogenicity and immunochemistry of *Streptococcus pneumoniae* capsular polysaccharides. *Antonie van Leeuwenhoek* **58:**1–47.

65. **van Kranenburg, R., J. D. Marugg, I. I. van Swam, N. J. Willem, and W. M. de Vos.** 1997. Molecular characterization of the plasmid-encoded *eps* gene cluster essential for exopolysaccharide biosynthesis in *Lactococcus lactis.* *Mol. Microbiol.* **24:**387–397.

66. **Watson, D. A., V. Kapur, D. M. Musher, J. W. Jacobson, and J. M. Musser.** 1995. Identification, cloning and sequencing of DNA essential for encapsulation of *Streptococcus pneumoniae.* *Curr. Microbiol.* **31:**251–259.

67. **Watson, D. A., and D. M. Musher.** 1990. Interruption of capsule production in *Streptococcus pneumoniae* serotype 3 by insertion of transposon Tn916. *Infect. Immun.* **58:**3135–3138.

68. **Weiser, J. N., R. Austrian, P. K. Sreenivasan, and H. R. Masure.** 1994. Phase variation in pneumococcal opacity: relationship between colonial morphology and nasopharyngeal colonization. *Infect. Immun.* **62:**2582–2589.

69. **Whitfield, C.** 1995. Biosynthesis of lipopolysaccharide O antigens. *Trends Microbiol.* **3:**178–185.

70. **Winkelstein, J. A.** 1981. The role of complement in the host's defense against *Streptococcus pneumoniae.* *Rev. Infect. Dis.* **3:**289–298.

Streptococcus pneumoniae: Invasion and Inflammation

KHOOSHEH GOSINK AND ELAINE TUOMANEN

21

Streptococcus pneumoniae is the leading cause of invasive bacterial disease in the very young and the elderly (44). Thus, it is currently the single most significant infection, be it gram positive or gram negative. A model of pathogenesis must explain that *S. pneumoniae* can behave as a transient commensal, colonizing the nasopharynx of 40% of healthy adults and children with no adverse effects (2). Children carry this pathogen in the nasopharynx asymptomatically for about 4 to 6 weeks, often several serotypes at a time (22). Having then acquired serotype-specific immunity, they go on to acquire a new serotype every 2 months or so, presumably up to 90 different times. On this backdrop of community spread arise such local and systemic infections as otitis media, pneumonia, sepsis, and meningitis. By the age of 5 years, the majority of children in the United States will have experienced at least one case of pneumococcal otitis media. It is estimated that 25% of all community-acquired pneumonia is due to pneumococcus (12). While the incidence of meningitis is far less (1 in 100,000 per annum), pneumococcal meningitis has a 25% mortality rate, which is higher than that of other meningeal pathogens (44). In addition, 50% of the survivors sustain permanent neurological sequelae. With increasing evidence of antibiotic-resistant organisms, the lack of a broadly effective vaccine, and the ability of pneumococcus to transfer genes for resistance, capsule, and virulence via transformation, it is imperative to develop an understanding of the mechanism by which pneumococcus causes disease.

The pneumococcus is unique among gram-positive pathogens, demonstrating a pattern of disease shared only by gram-negative *Haemophilus influenzae* and *Neisseria meningitidis*. These three pathogens share carriage in the nasopharynx and targeting of lung, ear, blood, and brain for infection. They also share physiology rare in other pathogens, including natural transformation of DNA and autolysis in stationary phase. Comparison of the complete genome sequences of pneumococcus (25a) and *H. influenzae* indicates that very different gene products perform similar functions in most, but not all, cases. There are no type III secretion systems or outer membrane proteins of the gram-negative pathogens in pneumococcus. There are also relatively few classical gram-positive surface proteins (anchored by an LPXTGE motif). The only exotoxin, pneumolysin, is trapped within the cytoplasm. This chapter reviews the surface of the pneumococcus with special reference to host-bacterial interactions, the pathogenesis of pneumococcal infection, and the induction of the inflammatory response.

PNEUMOCOCCAL SURFACE COMPONENTS

Capsule

Pneumococcus is enveloped by a polysaccharide capsule. Once outside the mucosal surface, encapsulated strains are $>10^5$ times more virulent than unencapsulated strains. The capsule interferes with phagocytosis by leukocytes, not by virtue of its amount but rather by its chemical composition (2). There are 90 chemically different capsular serotypes, explaining why acquisition of this bacterium is as common as the common cold. Aside from generating serotype-specific immunity, the capsule is relatively inert. Infusion of capsular polysaccharides into the lung or the subarachnoid space does not cause an inflammatory response. The capsule also does not interfere with the ability of the underlying components, such as the cell wall and surface proteins, to interact with the host defense system (60).

Cell Wall

In contrast to capsular polysaccharides, the cell wall is a potent inducer of inflammation (51, 56). In fact, challenge with cell wall components can re-create many of the symptoms of pneumonia, otitis media, and meningitis in experimental models. The pneumococcal cell wall is roughly six layers thick and is composed of peptidoglycan with teichoic acid attached to approximately every third *N*-acetylmuramic acid residue (Fig. 1). Lipoteichoic acid is another important glycopolymer of the cell surface and is chemically identical to the teichoic acid but is attached to the cell membrane by a lipid moiety. The peptidoglycan is synthesized by a set of cell surface enzymes that bind the

Gram-Positive Pathogens, ed. by V. A. Fischetti et al.
© 2000 American Society for Microbiology, Washington, D.C.

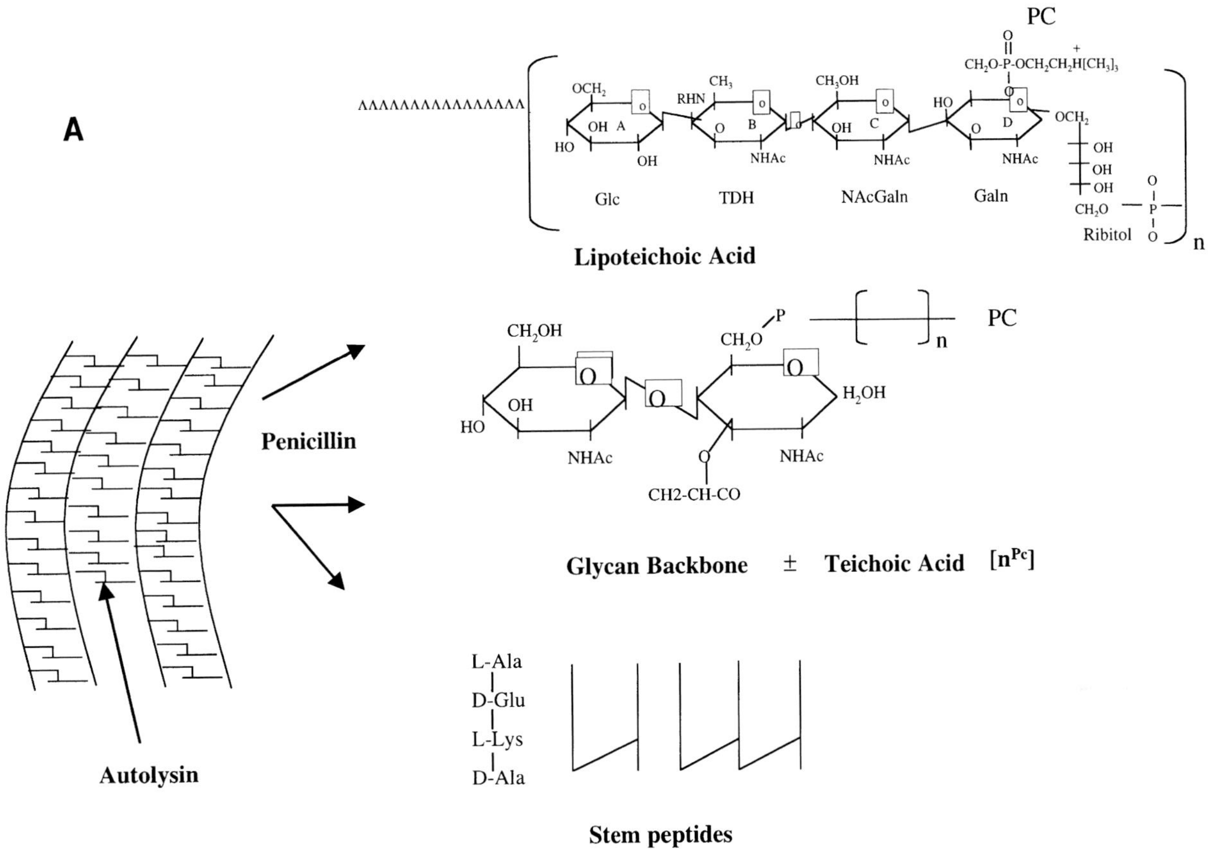

FIGURE 1 Structure of the pneumococcal cell wall and its relationship to inflammation. (A) Penicillin induces cell wall degradation by the autolysin releasing cell wall fragments such as lipoteichoic acid, glycan polymers with and without teichoic acid, and small stem peptides. All teichoicated species contain phosphorylcholine (PC), a key component increasing inflammatory activity. (B) (Next page) All of these components interact with a variety of human cells, which in turn produce inflammatory mediators. Particularly important in this response are platelet-activating factor (PAF) and interleukin-1 (IL-1). These mediators combine to produce the symptomatology of pneumococcal infection, including changes in blood flow, fluid balance in the tissue, and leukocytosis. Glc, glucose; TDH, trideoxyhexose; NAcGaln, *N*-acetylgalactosamine; Galn, galactosamine; L-Ala, L-alanine; D-Glu, D-glucose; L-Lys, L-lysine; TNF, tumor necrosis factor; MIP, macrophage inflammatory protein; NO, nitric oxide; PGE_2, prostaglandin E_2; IC, intracranial.

antibiotic penicillin by virtue of its structural similarity to the cell wall constituent D-alanyl-D-alanine (penicillin-binding proteins). These enzymes are responsible for constructing the scaffolding of the wall from intracellular disaccharide pentapeptides, and changes in their affinity for substrate are the molecular mechanism for penicillin resistance (28). The genes for penicillin-binding proteins are distributed widely on the chromosome, and sequence changes accompanying penicillin resistance are believed to be imported as cassettes from other streptococci by natural transformation (18).

Both the teichoic acid and the lipoteichoic acid contain phosphorylcholine that has been recognized as an important element in the biology of pneumococcus (49). Two choline residues are covalently added to each carbohydrate repeat (66). The genetic locus for this process has been identified, and mutations eliminating the ability to add choline to the surface appear to be crippling or lethal for the pneumococcus (67). The phosphorylcholine is a key molecule for invasion and acts both as an adhesin and as a docking site for proteins referred to as choline-binding proteins. It has been shown that other respiratory pathogens such as *Haemophilus*, *Pseudomonas*, *Neisseria*, and *Mycoplasma* spp. also have phosphorylcholine decorating their surface lipopolysaccharides (59). This may indicate the interaction of a diverse group of respiratory pathogens with a host molecule through the choline moiety.

Pneumococcal Proteins

Most gram-positive bacteria harbor both exotoxins and cell surface-associated virulence determinants. Pneumococci can produce three hemolysins but do not release a barrage of exotoxins during growth (Table 1). The most potent toxin, pneumolysin, is a classical thiol-activated cytotoxin (30) (Fig. 2). Pneumolysin is stored intracellularly and is released upon lysis of pneumococci by autolysin. This protein, toxic to nearly every eukaryotic cell, forms oligomers resulting in transmembrane pores that ultimately lead to cell lysis. Pneumolysin can also stimulate the production of inflammatory cytokines, inhibit beating of the epithelial

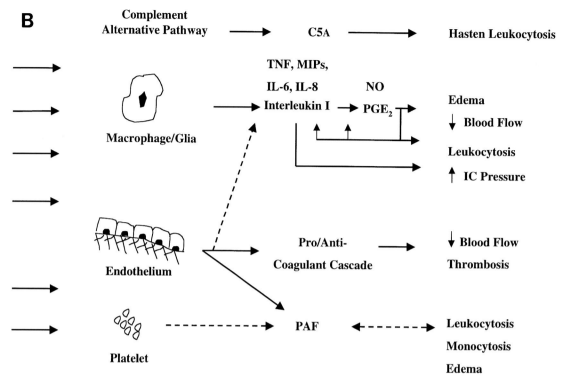

FIGURE 1   (*Continued*)

cell cilia, inhibit lymphocyte proliferation, decrease the bactericidal activity of neutrophils, and activate complement (6, 41). A second hemolysin activity has been described but has not yet been identified at the genetic level (10). Finally, pneumococci also produce hydrogen peroxide in amounts greater than those produced by human leukocytes. This small molecule is a potent hemolysin (46).

From genomic analysis for signal sequences and genetic analysis using *phoA* fusion technology to identify exported proteins (35), it is estimated that pneumococci harbor >500 surface proteins. *S. pneumoniae* has various mechanisms by which these proteins are attached to the surface. One family of proteins, perhaps as many as 15, uses a well-characterized LPXTGE motif that is a cleavage site as well as an anchor for covalent attachment to the cell wall (43). One example of a gene in this family is *zmpB*, which encodes a zinc metalloprotease and has been shown to play a role in virulence (32). Recent evidence indicates that ZmpB is required for the processing and export of a number of proteins to the surface of the pneumococcus. In ZmpB-deficient strains, autolysin and CbpA are found intracellularly but fail to be exported to the cell surface. The absence of CbpA on the cell surface results in reduced adhesion and intranasal colonization. An immunoglobulin A (IgA) protease is also a member of the LPXTGE family of cell wall adducts (57).

A second family of surface proteins (lipoproteins) contains an LXXC motif in the N terminus that serves as a cleavage site and a covalent binding site for palmitic acid. There are on the order of 20 proteins containing this motif in the pneumococcal genome. A number of these proteins have been implicated in pathogenesis. Mutations in two peptide permease genes, *amiA* and *plpA*, resulted in decreased binding to glycoconjugates present on resting endothelial and lung cells (15). Two neuraminadases, NanA and NanB, attach to the cell wall by the LXXC motif (34). These enzymes appear to be important in cleaving sialic acid from the host cell surface during the process of colonization and inflammation (27).

As has been found in other streptococci and staphylococci, glycolytic enzymes are present on the pneumococcal surface (see chapter 2, this volume). The mechanism of the association with the surface remains unknown. The *spxB* gene encodes a pyruvate oxidase that decarboxylates pyruvate to produce acetyl phosphate plus H_2O_2 and CO_2 (46). Mutation in this gene results in reduced adherence to all cell types and greatly attenuates virulence. A surface-associated enolase has been suggested to participate in pneumococcal binding to fibronectin and the vascular cell adhesion molecule of endothelia (62).

The most unique group of proteins on the pneumococcal surface are the choline-binding proteins (CBPs) (Fig. 3). The 12 CBPs noncovalently bind to the choline moiety of the cell wall, thereby reversibly clipping proteins with diverse functions onto the bacterial surface. This is a novel

TABLE 1   Characteristics of pneumococcal hemolysins

Hemolysin	Hemolysis type	Expressed aerobically
H_2O_2	α	Yes
Pneumolysin	β	No
2nd hemolysin	β	No

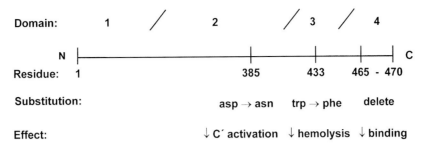

FIGURE 2 Domain structure of pneumolysin. Pneumolysin has three functionally separate domains: one activating complement, one causing hemolysis, and the other binding to cholesterol. Site-specific mutations alter these properties individually. Compiled from references 30 and 4.

mechanism of attachment for virulence determinants. CBPs all share a common C-terminal choline-binding domain that consists of 2 to 10 repeats of 20 amino acids (Fig. 4) (19). This construct is found in surface proteins of two other bacteria: clostridia and *Streptococcus mutans*. The N termini of the CBPs are distinct, indicating diverse functions of the CBPs. This family of proteins includes several proteins that play a role in pneumococcal infection: PspA, a well-defined protective antigen; LytA, an autolysin; and CbpA, an adhesin.

PspA is a surface protein that has been shown to be structurally and antigenically variable (29). PspA is a 65-kDa protein with 10 choline-binding repeats (Fig. 4). The N terminus is highly charged and is predicted to be predominantly α-helix (64). It has as yet no defined function; however, immunization with antibodies against PspA or purified truncated PspA confer protection in mice challenged with pneumococcal strains of various capsular se-

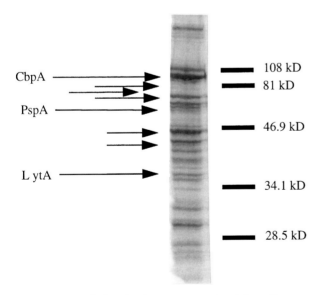

FIGURE 3 Choline-binding proteins eluted from *S. pneumoniae* serotype 4. The CBPs were eluted by soaking bacterial pellets in saline containing 2% choline for 20 min (48). A concentrated, dialyzed supernatant was separated by polyacrylamide gel electrophoresis. Arrows indicate the position of proteins containing repeats consistent with choline-binding domains. Coomassie stain. Molecular size markers are indicated on right.

rotypes (48). Insertion-duplication mutants of PspA have reduced virulence, indicating that PspA is necessary for full virulence in mice.

LytA is the autolytic enzyme that is responsible for pneumococcal lysis in stationary phase as well as in the presence of antibiotics (50). This protein has two functional domains: a C-terminal domain with six choline-binding repeats that anchor the protein on the cell wall, and an N-terminal domain that provides amidase catalytic function (19). An autolysin-negative mutant has variably reduced virulence as compared to wild type. This may arise from the lack of release of cell wall and pneumolysin, resulting in a greatly reduced inflammatory response (53).

CbpA has recently been identified as a major pneumococcal adhesion (40). It is a 110-kDa protein with eight choline-binding repeats that are nearly identical to PspA (Fig. 4). The N terminus (amino acids 1 to 433) consists of two repeats, each containing three α-helices (Fig. 5). A *cbpA*-deficient mutant is not only defective in colonization of the nasopharynx in an infant rat model but also fails to bind to activated type II human lung cells as well as endothelial cells (40). In the absence of CbpA, pneumococci fail to enter and traverse an in vitro blood-brain barrier, indicating a critical role for this protein in invasion (38). CbpA has also been described to bind to secretory IgA and the third component of complement (23, 45) (Fig. 5). It is not known if these various binding properties arise from an ability of CbpA to serve as a lectin and bind a common carbohydrate on these targets or whether there are multiple target-binding domains in the molecule.

In addition to these three well-characterized choline-binding proteins, the recent sequencing of the pneumococcal genome has revealed the existence of nine new CBPs (Fig. 3). These proteins all have between two and 10 choline-binding repeats in the C-terminal region. The roles of these CBPs may be restricted to virulence since their absence does not affect transformation or autolysis (21a).

In summary, the pneumococcal surface is characterized by a thick cell wall that has choline as an important modification. Figure 6 depicts, by immunoelectron microscopy and as a schematic, the cell surface density of free choline and an example of the distribution of a CBP, LytA, on the surface. Clipping of the CBPs onto and off the choline can be achieved rapidly by various regulatory mechanisms, allowing a repopulation of the surface and a "morphing" of the function of the surface. This variability is balanced by the more permanent fixation of a smaller array of other surface proteins similar to those more widely distributed in

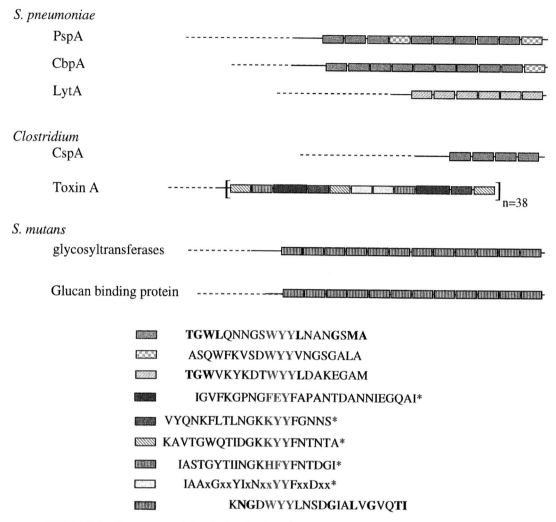

FIGURE 4 Comparison of the choline-binding domains and glucan-binding domains of streptococci and clostridia. The sequences provided are examples of a repeat unit and do not indicate a true consensus sequence. Asterisk indicates a consensus sequence as described by Wren (63). Bold letters indicate the residues conserved within the family of repeats, and the blue highlighted residues identify a common motif among all the repeats. Compiled from references 21, 39, 40, 42, and 65.

gram-positive bacteria in general (LPXTGE and LXXC motifs). Hypervirulent clones are believed to exist among pneumococci, particularly in the population of antibiotic-resistant strains. Whereas previously there was a general impression that a single "body" underlay the variety of 90 different capsules, it is now more reasonable to hypothesize that at least several different basic pneumococcal bodies exist. Differences between these bodies are likely to be manifested in the molecular constituents of the cell surface. Surface proteins are subject to phase variation (see chapter 21, this volume) and to other regulatory elements, such as peptide permeases (15) and over a dozen two-component signal transduction systems. For instance, reversible phase variation of colony morphology from opaque to transparent changes the amounts of capsular material, choline in the cell wall, and amounts of CBPs PspA, CbpA, and LytA (58). The sensor regulators appear to respond to small peptides released by the bacteria in a cell density-dependent fashion (24) (see chapter 23, this volume). These multiple regulatory circuits allow the pneumococcus to progress through cell density-dependent changes in physiology: transformation at low density, adherence and invasion in midlogarithmic phase, and autolysis in stationary phase (14).

ADHERENCE AND INVASION

The receptors for pneumococci in the nasopharynx have not been definitively determined. Several carbohydrates inhibit adherence to the Detroit cell line, suggesting a role at this site. This is particularly true for sialic acid (3). Loss of eukaryotic sialic acid by the action of pneumococcal neuraminidase accompanies the advance of pneumococci up the eustachian tube to the middle ear (27). From the nasopharynx the pneumococci also spread by aerosol down

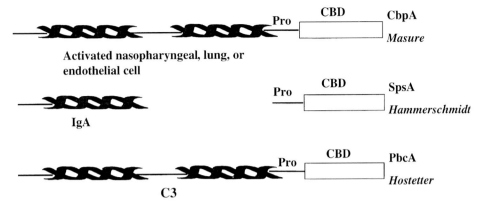

FIGURE 5 Comparison of the three published sequences and functions for CbpA homologs. CbpA was described as an adhesin with lectin-binding activity for activated human cells (40). SpsA was described as a protein that can bind but not cleave secretory IgA (23). *spsA* sequence is missing a substantial portion of the middle of the gene. PbcA was described as a protein that can bind but not cleave C3 of the complement cascade and promotes pneumococcal binding to activated cells (45). Sequences indicated in parallel are >95% identical. Coils represent α-helix. Pro, proline-rich region; CBD, choline-binding domain.

to the lower respiratory tract. At this site, many pulmonary pathogens appear to recognize the surface glycoconjugate N-acetyl-D-galactosamine linked either β1,3 or β1,4 to galactose (26). The pneumococcal ligands for sialic acid or lactosamine have not been identified, and there are no apparent differences in the abilities to bind to these receptors between invasive and noninvasive isolates or variants (16). This implies that this level of binding is fairly universal and consistent with asymptomatic colonization.

In contrast to colonization, invasion appears to be a more stringently selected capability. It remains to be determined exactly where invasion occurs. Pneumococci can traverse the respiratory epithelium in vivo and in vitro, but the route taken from the respiratory epithelium to the bloodstream may be via lymphatics or direct invasion of endothelial cells. Pneumococcus is a low-efficiency invader: ~0.2% of the inoculum invades, as compared with classically invasive bacteria such as *Salmonella* and *Shigella* spp. in which greater than 2 to 3% of the inoculum invades host cells (13). Invasion is considered a multistage process initiated by adherence. A key step in the transition from adherence to invasion is activation of host cells, resulting in expression of new receptors. Activated cells synthesize glycoconjugates bearing sialic acid and lacto-N-neotetraose. These determinants and others may be displayed on several eukaryotic proteins that are recognized by pneumococci, such as platelet-activating factor (PAF) receptor (13), secretory IgA (23), and the third component of complement (45). A critical adhesive ligand on the pneumococcus that binds all of these targets is CbpA. A direct interaction of the phosphorylcholine of the cell wall with the PAF receptor is also described, consistent with the ability of this receptor to recognize choline in the natural ligand PAF (13).

A two-chamber cell culture system (5) using a monolayer of A549 lung cells or human brain microvascular endothelial cells has demonstrated the kinetics and molecular components of the pathway for transcytosis of cells by pneumococci (38) (Fig. 7). The bacteria enter vacuoles that transit through the cell over several hours. Once in a vacuole, the number of viable pneumococci is decreased as

a result of intracellular death, transcytosis, or recycling back to the original point of entry. Seventy percent of the bacteria are internalized via a pathway dependent on the receptor for PAF. Pneumococcal transmigration does not require de novo DNA, RNA, or protein synthesis. Although both opaque and transparent variants can be internalized, transcytosis is restricted to the transparent variants and requires CbpA. The few opaque variants that are internalized die intracellularly. Unlike PAF, the binding of pneumococcus to the PAF receptor does not result in the activation of a G-protein-mediated signal transduction pathway. It is known that PAF receptor is rapidly internalized in response to the binding of its ligand PAF, and upon release of PAF is recycled back to the apical side. This recycling pathway appears to be hijacked by pneumococcus, thereby increasing the efficiency of invasion by re-presenting transcytosis-competent bacteria to the apical surface. It has been suggested that the continued association of pneumococcus with the PAF receptor in the vacuole may somehow impair the recycling pathway and lead to transcytosis (38). It is also possible that binding of pneumococcus induces an as yet unidentified signal transduction pathway.

INFLAMMATION

The pneumococcus has served as a prototype for understanding the molecular mechanism of the inflammatory response to gram-positive bacteria (52). The arsenal of potentially inflammatory components includes a set of intracellular cytotoxins, cell wall components (peptidoglycan and teichoic acid), and secreted proteins. The presence of pneumococci in a tissue in and of itself is not sufficient to cause an inflammatory response, even when introduced into a sterile site such as the lung or subarachnoid space (61). In such healthy tissues, it requires challenge with ~100,000 bacteria per ml to trigger an inflammatory response (53). In contrast, the pneumococcus becomes invasive and inflammation ensues with as few as 10 bacteria if a preceding proinflammatory signal is supplied (25). This

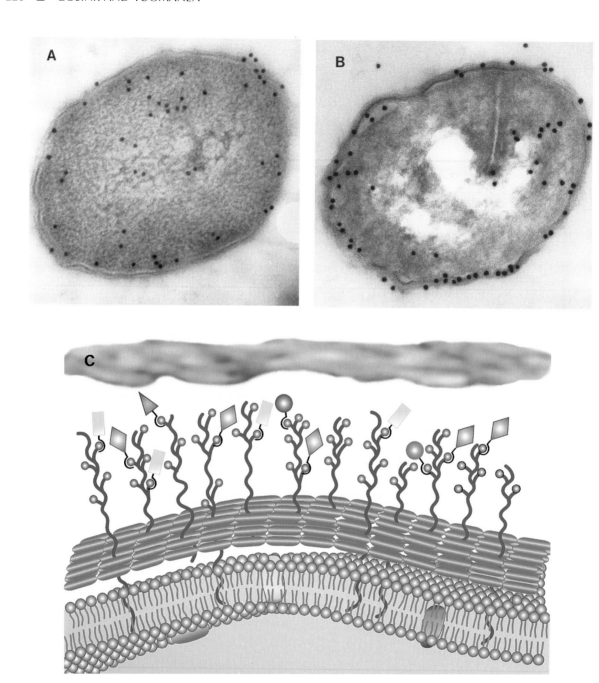

FIGURE 6 Immunohistochemical and schematic depiction of the choline biology of the pneumococcal surface. Immunogold labeling of pneumococci with (A) TEPC-15 antibody recognizing free choline and (B) antiautolysin antibody. These two images contrast free (A) versus CBP-bound (B) choline. (C) Schematic view of the capsule (blue), cell wall (green), and membrane (red). The teichoic and lipoteichoic acids are indicated as dark blue lines bearing choline (circles). A proportion of these are capped by choline-binding proteins. Figure courtesy of K. G. Murti, St. Jude Electron Microscopy Core Facility.

signal is a cytokine in experimental systems or an antecedent viral infection in clinical situations.

Adherence and entry of the bacteria into epithelia, endothelia, and leukocytes trigger several host response pathways: the chemokine/cytokine cascade, the coagulation cascade, and the complement cascade (Fig. 1). The cell wall directly activates the alternative pathway of the complement cascade, which in turn hastens the accumulation of leukocytes (60). As a counterattack, the bacterial surface protein CbpA binds C3 without cleavage, and the bacteria also degrade C3, suggesting mechanisms to avoid fixing complement to the cell wall (1, 45). Activation of the coagulation cascade results in the production of a procoagulant state on human cell surfaces leading to enhanced formation of thrombosis, a characteristic of sites of pneumococcal infection (20). Contact with pneumococci also

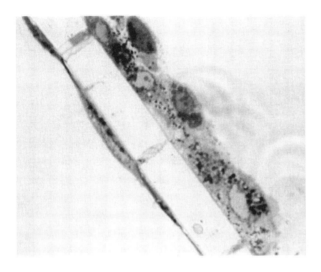

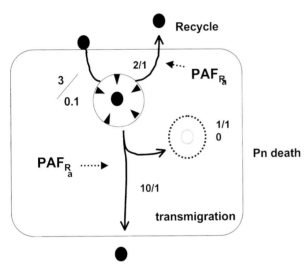

FIGURE 7 Microscopic and schematic depictions of pneumococcal invasion. (Top) Pneumococci (dark blue gram positive) bind to and are internalized into type II A549 lung cells on top of a microporous filter. Upon exiting the base of the lung cell, the bacteria pass through the filter pores and invade and transcytose across primary human endothelial cells under the filter. Figure courtesy of C. Rosenow, Rockefeller University. (Bottom) The schematic illustrates the differences in the transcytosis process between opaque and transparent phase variants (transparent/opaque). Three fates are: (i) entry and recycling to the apical surface, favored for the transparent bacteria and inhibitable by PAF receptor antagonist (PAF Ra); (ii) entry and death within the vacuole, favored for opaque bacteria; (iii) entry and transmigration across the cell, overwhelmingly favored by transparent bacteria. Pn, pneumococci. Figure modified from Ring et al. (38).

elicits the production of several cytokines, including interleukin-1β (IL-1β) from human cells (37). The activation of these cytokines occurs by several different pathways. One involves the binding of peptidoglycan to CD14, a cell surface receptor known to initiate the inflammatory response for endotoxin (36). However, non-CD14-dependent stimulation of cytokine production also occurs, and the mechanism is as yet unknown (11). One possibility in-

volves the interaction of the choline of the pneumococcus with the receptor for the chemokine, PAF. The result of induction of these signaling cascades is the activation of NF-κB and induction of tumor necrosis factor and IL-1, IL-6, and IL-8 (47). This results in altered vascular permeability and creation of a serous exudate.

The generation of chemokines, such as PAF and C5a, and cytokines, particularly IL-1β and IL-8, directs the recruitment of leukocytes by classical and unusual means. The leukocyte adhesion molecule CD18 is responsible for the recruitment of leukocytes to all body sites in response to pneumococci (55). However, in the lung approximately half of the leukocytes are recruited by a CD18-independent pathway that is not well understood (9, 17). In the absence of antibody, the acute-phase reactant, C-reactive protein, binds directly to the choline of the cell wall and serves to opsonize the bacteria for phagocytosis (31). The appearance of anticapsular antibodies initiates resolution of the infection in that phagocytosis becomes efficient. The production of an IgA protease and the ability of CbpA to bind to secretory IgA serve to promote bacterial survival in this setting (23, 57).

The inflammatory response can cause considerable tissue damage. In addition, as pneumococci begin to lyse in response to host defense molecules and antimicrobial agents, they release cell wall, pneumolysin, and other components that lead to greater inflammation and cytotoxic effects (53). Pneumolysin, hydrogen peroxide, and perhaps a second hemolysin (Table 1) not only kill cells but also strongly induce the production of nitric oxide (7). This potent vasoactive mediator is a major agent operative in septic shock. Thus, the death of pneumococci itself produces a reactive burst in the inflammatory response, causing more damage to the host, as is particularly well illustrated in meningitis. In laboratory animals and human patients, pneumococcal variants that do not lyse in response to penicillin show reduced tissue damage (54). Down-modulation of the host inflammatory response during the early phases of antibiotic therapy has proved successful as an adjunctive therapy to reduce sequelae in meningitis (33). Although many of the players in cell wall-induced inflammation are also involved in host response to endotoxin, there are differences, and some of these differences may be important in the development of agents that will improve the outcome of sepsis caused by both gram-positive and gram-negative bacteria.

In addition to direct damage by cytotoxins and bystander effects during an exuberant host response, human cells can undergo apoptosis in response to pneumococci. This is especially evident during meningitis, when neurons of the hippocampus are killed by this mechanism (68). New studies suggest that significant protection from neuronal injury can be achieved by inhibition of apoptosis by caspase inhibitors (8). The molecular mechanism of this event remains to be determined.

CONCLUSION

The availability of the pneumococcal genome provides a great tool for the understanding and discovery of factors involved in virulence, regulation of virulence, and modulation of disease progression. These in turn can provide a better understanding of the processes of inflammation and invasion, suggesting new antibiotics and vaccine candidates to better control the progression of pneumococcal disease or to prevent it altogether. New aspects of pathogenesis already brought to light by the study of the pneumococcus include the prominent role of choline in the surface interactions of pulmonary pathogens, perhaps indicating a relationship to the choline biology of PAF and its receptor so prominent in the lung. The existence of a noncovalent coat anchored on a bacterial surface that can be stripped and replaced is a novel problem in extracellular architecture and its regulation. The invasion and recycling of pneumococci in and out of human cells suggests a dynamic invasive capability for a pathogen not suited for intracellular survival. Finally, the pneumococcus serves to mark inflammatory pathways that remain to be discovered, including a non-CD14 inflammatory network and a non-CD18 mechanism of leukocyte recruitment. It still remains unclear why people infected with gram-positive bacteria get sick and die.

REFERENCES

1. **Angel, C., M. Rusek, and M. Hostetter.** 1994. Degradation of C3 by *Streptococcus pneumoniae. J. Infect. Dis.* **170:**600–608.

2. **Austrian, R.** 1986. Some aspects of the pneumococcal carrier state. *J. Antimicrob. Chemother.* **18**(Suppl A)**:**35–45.

3. **Barthelson, R., A. Mobasseri, D. Zopf, and P. Simon.** 1998. Adherence of *Streptococcus pneumoniae* to respiratory epithelial cells is inhibited by sialylated oligosaccharides. *Infect. Immun.* **66:**1439–1444.

4. **Berry, A., J. Alexander, T. Mitchell, P. Andrew, D. Hansman, and J. Paton.** 1995. Effect of defined point mutations in the pneumolysin gene on the virulence of *Streptococcus pneumoniae. Infect. Immun.* **63:**1969–1974.

5. **Birkness, K., B. Swisher, E. White, E. Long, E. Ewing, and F. Quinn.** 1995. A tissue culture bilayer model to study the passage of *Neisseria meningitidis. Infect. Immun.* **63:**402–409.

6. **Boulnois, G. J., J. C. Paton, T. J. Mitchell, and P. W. Andrew.** 1991. Structure and function of pneumolysin, the multifunctional, thiol-activated toxin of *Streptococcus pneumoniae. Mol. Microbiol.* **5:**2611–2616.

7. **Braun, J. S., R. Novak, G. Gao, P. J. Murray, and J. Shenep.** 1999. Pneumolysin, a protein toxin of *Streptococcus pneumoniae*, induces nitric oxide production from macrophages. *Infect. Immun.* **67:**3753–3756.

8. **Braun, J., R. Novak, S. Bodmer, J. Cleveland, and E. Tuomanen.** 1999. Neuroprotection by a caspase inhibitor in acute bacterial meningitis. *Nat. Med.* **5:**298–302.

9. **Cabellos, C., D. E. MacIntyre, M. Forrest, M. Burroughs, S. Prasad, and E. Tuomanen.** 1992. Differing roles of platelet-activating factor during inflammation of the lung and subarachnoid space. *J. Clin. Invest.* **90:**612–618.

10. **Canvin, J., J. Paton, G. Boulnois, P. Andrew, and T. Mitchell.** 1997. *Streptococcus pneumoniae* produces a second hemolysin that is distinct from pneumolysin. *Microb. Pathog.* **22:**129–132.

11. **Cauwels, A., E. Wan, M. Leismann, and E. Tuomanen.** 1997. Coexistence of CD14-dependent and independent pathways for stimulation of human monocytes by gram-positive bacteria. *Infect. Immun.* **65:**3255–3260.

12. **Centers for Disease Control and Prevention.** 1995. Pneumonia and influenza death rates: United States, 1979–1994. *Morbid. Mortal. Weekly Rep.* **44:**535–536.

13. **Cundell, D., N. Gerard, C. Gerard, I. Idanpaan-Heikkila, and E. Tuomanen.** 1995. *Streptococcus pneumoniae* anchors to activated eukaryotic cells by the receptor for platelet activating factor. *Nature* **377:**435–438.

14. **Cundell, D., H. Masure, and E. Tuomanen.** 1995. The molecular basis of pneumococcal infection: an hypothesis. *Clin. Infect. Dis.* **21:**S204–S212.

15. **Cundell, D., B. Pearce, J. Sandros, A. Naughton, and H. Masure.** 1995. Peptide permeases from *Streptococcus pneumoniae* affect adherence to eucaryotic cells. *Infect. Immun.* **63:**2493–2498.

16. **Cundell, D., and E. Tuomanen.** 1994. Receptor specificity of adherence of *Streptococcus pneumoniae* to human type II pneumocytes and vascular endothelial cells in vitro. *Microb. Pathog.* **17:**361–374.

17. **Doerschuk, C. M., R. K. Winn, H. O. Coxson, and J. M. Harlan.** 1990. CD18-dependent and -independent mechanisms of neutrophil emigration in the pulmonary and systemic microcirculation of rabbits. *J. Immunol.* **144:** 2327–2333.

18. **Dowson, C., B. Barcus, S. King, P. Pickerill, A. Whatmore, and M. Yeo.** 1997. Horizontal gene transfer and the evolution of resistance and virulence determinants in *Streptococcus. J. Appl. Microbiol.* **83:**42S–51S.

19. **Garcia, J., R. Sanchez-Beato, F. Medrano, and R. Lopez.** 1998. Versatility of choline-binding domain. *Microb. Drug Resist.* **4:**25–36.

20. **Geelen, S., C. Bhattacharyya, and E. Tuomanen.** 1992. Induction of procoagulant activity on human endothelial cells by *Streptococcus pneumoniae. Infect. Immun.* **60:** 4179–4183.

21. **Giffard, P., and N. Jacques.** 1994. Definition of a fundamental repeating unit in streptococcal glycosyltransferase glucan-binding regions and related sequences. *J. Dent. Res.* **73:**1133–1141.

21a.**Gosink, K., and H. R. Masure.** Unpublished data.

22. **Gray, B., and H. Dillon.** 1986. Clinical and epidemiologic studies of pneumococcal infection in children. *Pediatr. Infect. Dis.* **5:**201–207.

23. **Hammerschmidt, S., S. Talay, P. Brandtzaeg, and G. Chhatwal.** 1997. SpsA, a novel pneumococcal surface protein with specific binding to secretory immunoglobulin A and secretory component. *Mol. Microbiol.* **25:**1113–1124.

24. **Havarstein, L., P. Gaustad, I. Nes, and D. Morrison.** 1996. Identification of the streptococcal competence pheromone receptor. *Mol. Microbiol.* **21:**965–971.

25. **Idanpaan-Heikkila, I., P. Simon, C. Cahill, K. Sokol, and E. Tuomanen.** 1997. Oligosaccharides interfere with the establishment and progression of experimental pneumococcal pneumonia. *J. Infect. Dis.* **176:**704–712.

25a.**Institute for Genomic Research.** http://www.tigr.org/ tdb/mdb/mdb.html

26. **Krivan, H. C., D. D. Roberts, and V. Ginsburg.** 1988. Many pulmonary pathogenic bacteria bind specifically to the carbohydrate sequence GalNacB1-4Gal found in some glycolipids. *Proc. Natl. Acad. Sci. USA* **85:**6157–6161.

27. Linder, T., R. Dandiles, D. Lime, and T. DeMaria. 1994. Effect of intranasal inoculation of *Streptococcus pneumoniae* on the structure of the surface carbohydrates of the chinchilla eustachian tube and middle ear mucosa. *Microb. Pathog.* **16:**435–441.

28. Marton, A., M. Gulyas, R. Munoz, and A. Tomasz. 1991. Extremely high incidence of antibiotic resistance in clinical isolates of *Streptococcus pneumoniae* in Hungary. *J. Infect. Dis.* **163:**542–548.

29. McDaniel, L. S., J. S. Sheffield, E. Swiatlo, J. Yother, M. J. Crain, and D. E. Briles. 1992. Molecular localization of variable and conserved regions of pspA and identification of additional pspA homologous sequences in *Streptococcus pneumoniae*. *Microb. Pathog.* **13:**261–269.

30. Mitchell, T., and P. Andrew. 1997. Biological properties of pneumolysin. *Microb. Drug Resist.* **3:**19–26.

31. Mold, C., K. Edwards, and H. Gewura. 1982. Binding of C-reactive protein to bacteria. *Infect. Immun.* **38:**392–395.

32. Novak, R., K. Gosink, and H. R. Masure. Regulation of surface proteins by an extracellular metalloproteases. Submitted for publication.

33. Odio, C. M., I. Faingezicht, M. Paris, M. Nassar, A. Baltodano, J. Rogers, X. Saez-Llorens, K. D. Olsen, and G. H. McCracken, Jr. 1991. The beneficial effects of early dexamethasone administration in infants and children with bacterial meningitis. *N. Engl. J. Med.* **324:**1525–1531.

34. Paton, J., A. Berry, and R. Lock. 1997. Molecular analysis of putative pneumococcal virulence proteins. *Microb. Drug Resist.* **3:**1–10.

35. Pearce, B., Y. Yin, and H. Masure. 1993. Genetic identification of exported proteins in *Streptococcus pneumoniae*. *Mol. Microbiol.* **9:**1037–1050.

36. Pugin, J., D. Heumann, A. Tomasz, V. Kravchenki, Y. Akamatsu, M. Nishijima, M. Lauser, P. Tobias, and R. Ulevitch. 1994. CD14 is a pattern recognition receptor. *Immunity* **1:**509–516.

37. Riesenfeld-Orn, I., S. Wolpe, J. F. Garcia-Bustos, M. K. Hoffman, and E. Tuomanen. 1989. Production of interleukin-1 but not tumor necrosis factor by human monocytes stimulated with pneumococcal cell surface components. *Infect. Immun.* **57:**1890–1893.

38. Ring, A., J. Weiser, and E. Tuomanen. 1998. Pneumococcal trafficking across the blood-brain barrier. *J. Clin. Invest.* **102:**1–14.

39. Ronda, C., J. Garcia, E. Garcia, J. Sanchez-Puelles, and R. Lopez. 1987. Biological role of the pneumococcal amidase: cloning of the lytA gene. *Eur. J. Biochem.* **164:**621–624.

40. Rosenow, C., P. Ryan, J. Weiser, S. Johnson, P. Fontan, A. Ortqvist, and H. Masure. 1997. Contribution of a novel choline binding protein to adherence, colonization, and immunogenicity of *Streptococcus pneumoniae*. *Mol. Microbiol.* **25:**819–829.

41. Rubins, J., D. Charboneau, C. Fasching, A. Berry, J. Paton, J. Alexander, P. Andrew, T. Mitchell, and E. Janoff. 1996. Distinct role for pneumolysin's cytotoxic and complement activities in the pathogenesis of pneumococcal pneumonia. *Am. J. Respir. Crit. Care Med.* **153:**1339–1346.

42. Sanchez-Beato, A., C. Ronda, and J. Garcia. 1995. Tracking the evolution of the bacterial choline binding domain: molecular characterization of the *Clostridium* cspA gene. *J. Bacteriol.* **177:**1098–1103.

43. Schneewind, O., A. Fowler, and K. Faull. 1995. Structure of the cell wall anchor of surface proteins in *Staphylococcus aureus*. *Science* **268:**103–106.

44. Schuchat, A., K. Robinson, J. Wenger, L. Harrison, M. Farley, A. Reingold, L. Lefkowitz, and B. Perkins. 1997. Bacterial meningitis in the United States. *N. Engl. J. Med.* **337:**970–976.

45. Smith, B., and M. Hostetter. 1998. Characterization of a pneumococcal surface protein that binds complement protein C3 and its role in adhesion, abstr. D-122, p. 233. *Abstr. 98th Gen. Meet. Am. Soc. Microbiol. 1998.* American Society for Microbiology, Washington, D.C.

46. Spellberg, B., D. Cundell, J. Sandros, B. Pearce, I. Idänpään-Heikkilä, C. Rosenow, and H. Masure. 1996. Pyruvate oxidase as a determinant of virulence in *Streptococcus pneumoniae*. *Mol. Microbiol.* **19:**803–813.

47. Spellberg, B., C. Rosenow, W. Sha, and E. Tuomanen. 1996. Pneumococcal cell wall activates NF-κB in human monocytes. *Microb. Pathog.* **20:**309–317.

48. Talkington, D. F., D. L. Crimmins, D. C. Voellinger, J. Yother, and D. E. Briles. 1991. A 43-kilodalton pneumococcal surface protein, PspA: isolation, protective abilities, and structural analysis of the amino-terminal sequence. *Infect. Immun.* **59:**1285–1289.

49. Tomasz, A. 1967. Choline in the cell wall of a bacterium: novel type of polymer-linked choline in pneumococcus. *Science* **157:**694–697.

50. Tomasz, A., A. Albino, and E. Zanati. 1970. Multiple antibiotic resistance in a bacterium with suppressed autolytic system. *Nature* **227:**138–140.

51. Tomasz, A., and K. Saukkonen. 1989. The nature of cell wall-derived inflammatory components of pneumococci. *Pediatr. Infect. Dis. J.* **8:**902–903.

52. Tuomanen, E., R. Austrian, and H. Masure. 1995. The pathogenesis of pneumococcal infection. *N. Engl. J. Med.* **332:**1280–1284.

53. Tuomanen, E., H. Liu, B. Hengstler, O. Zak, and A. Tomasz. 1985. The induction of meningeal inflammation by components of the pneumococcal cell wall. *J. Infect. Dis.* **151:**859–868.

54. Tuomanen, E., H. Pollack, A. Parkinson, M. Davidson, R. Facklam, R. Rich, and O. Zak. 1988. Microbiological and clinical significance of a new property of defective lysis in clinical strains of pneumococci. *J. Infect. Dis.* **158:**36–43.

55. Tuomanen, E., K. Saukkonen, S. Sande, C. Cioffe, and S. D. Wright. 1989. Reduction of inflammation, tissue damage, and mortality in bacterial meningitis in rabbits treated with monoclonal antibodies against adhesion-promoting receptors of leukocytes. *J. Exp. Med.* **170:**959–969.

56. Tuomanen, E. I., A. Tomasz, B. Hengstler, and O. Zak. 1985. The relative role of bacterial cell wall and capsule in the induction of inflammation in pneumococcal meningitis. *J. Infect. Dis.* **151:**535–540.

57. Wani, J., J. Gilbert, A. Plaut, and J. Weiser. 1996. Identification, cloning, and sequencing of the immunoglobulin A1 protease gene of *Streptococcus pneumoniae*. *Infect. Immun.* **64:**2240–2245.

58. Weiser, J., R. Austrian, P. Sreenivasan, and H. Masure. 1994. Phase variation in pneumococcal opacity: relation-

ship between colonial morphology and nasopharyngeal colonization. *Infect. Immun.* **62:**2582–2589.

59. **Weiser, J., M. Shchepetov, and S. Chong.** 1997. Decoration of lipopolysaccharide with phosphorylcholine: a phase-variable characteristic of *Haemophilus influenzae. Infect. Immun.* **65:**943–950.

60. **Winkelstein, J., and A. Tomasz.** 1978. Activation of the alternative complement pathway by pneumococcal cell wall teichoic acid. *J. Immunol.* **120:**174–178.

61. **Winternitz, M., and A. Hirschfelder.** 1913. Studies upon experimental pneumonia in rabbits. *J. Exp. Med.* **17:** 657–665.

62. **Wizemann, T., J. Min, J. McCarthy, and E. Tuomanen.** 1996. Adherence of *Streptococcus pneumoniae* to VCAM-1 and the HepII region of fibronectin, abstr. B-358, p. 216. *Abstr. 96th Gen. Meet. Am. Soc. Microbiol. 1996.* American Society for Microbiology, Washington, D.C.

63. **Wren, B.** 1991. A family of clostridal and streptococcal ligand-binding proteins with conserved C-terminal repeat sequences. *Mol. Microbiol.* **5:**797–803.

64. **Yother, J., and D. E. Briles.** 1992. Structural properties and evolutionary relationships of PspA, a surface protein of *Streptococcus pneumoniae*, as revealed by sequence analysis. *J. Bacteriol.* **174:**601–609.

65. **Yother, J., G. L. Handsome, and D. E. Briles.** 1992. Truncated forms of PspA that are secreted from *Streptococcus pneumoniae* and their use in functional studies and cloning of the *pspA* gene. *J. Bacteriol.* **174:**610–618.

66. **Yother, J., K. Leopold, J. White, and W. Fischer.** 1998. Generation and properties of a *Streptococcus pneumoniae* mutant which does not require choline or analogs for growth. *J. Bacteriol.* **180:**2093–2101.

67. **Zhang, J., I. Idanpaan-Heikkila, W. Fischer, and E. Tuomanen.** The pneumococcal *lic D2* is involved in phosphorylcholine metabolism. Submitted for publication.

68. **Zysk, G., W. Bruck, J. Gerber, Y. Bruck, H. Prange, and R. Nau.** 1996. Anti-inflammatory treatment influences neuronal apoptosis cell death in the dentate gyrus in experimental pneumococcal meningitis. *J. Neuropathol. Exp. Neurol.* **55:**722–728.

Phase Variation of *Streptococcus pneumoniae*

JEFFREY N. WEISER

22

Streptococcus pneumoniae undergoes spontaneous, reversible phenotypic variation, or phase variation, which is readily visualized as differences in colony morphology. Each isolate, therefore, is a heterogeneous population of organisms that differ in multiple characteristics. Cell surface features that vary in amount in association with colony opacity include capsular polysaccharide and the choline-containing teichoic acid. In addition, the distribution of a family of proteins noncovalently anchored to the cell surface by binding to choline differs in opaque and transparent pneumococci. Opaque variants, which express more capsular polysaccharide and less teichoic acid, are more virulent in animal models of sepsis but colonize the nasopharynx poorly. In contrast, transparent variants, which have less capsular polysaccharide and more teichoic acid, colonize the nasopharynx in animal models more efficiently but are avirulent. This suggests that phase variation generates a mixed population that may allow for selection of organisms in vivo with characteristics permissive for either carriage or systemic infection.

PHENOTYPIC VARIATION IN GRAM-POSITIVE BACTERIA

The spontaneous, reversible, on-and-off switching or phase variation of virulence determinants is a well-recognized property of many gram-negative pathogens. This ability allows for the selection of variants with optimal characteristics for an individual host or host environment and appears to be particularly common among the nonenteric pathogens such as *Haemophilus influenzae*, *Neisseria gonorrhoeae*, and *Neisseria meningitidis* (7, 8, 28, 36). In contrast, the enteric pathogens seem to depend more on sensing of the environment followed by programmed alterations in gene expression. There have been occasional reports of phase variation in gram-positive bacteria, but the molecular mechanisms involved and the precise role in host-pathogen interaction have not been well established (2, 10, 16, 19, 20).

We recently described phase variation in *S. pneumoniae*, the pneumococcus, and characterized its relationship to colonization and the pathogenesis of infection (35). The focus of this laboratory has been the identification of variably expressed cell surface components as a means of gaining insight into the pathogenesis of pneumococcal disease at a molecular level. *S. pneumoniae* is highly proficient at colonization of its human host. Despite its narrow host range, it is capable of considerable flexibility, as demonstrated by the ability of different strains to synthesize a vast repertoire of at least 90 unique capsular polysaccharides. The pneumococcus has, in addition, the capacity to thrive in a number of host environments, including the bloodstream and the mucosal surface of the nasopharynx. As is the case for other respiratory tract pathogens that frequently cause invasive infection, the ability of the pneumococcus to adapt to these varied environments requires changes in the expression of specific cell surface molecules (Table 1).

CHARACTERISTICS OF PHENOTYPIC VARIATION IN THE PNEUMOCOCCUS

As is sometimes the case for other bacterial species, phenotypic variation in the pneumococcus can be appreciated by detailed examination of colony morphology (35). Since a colony is an array of closely approximated organisms, differences in their physical characteristics may affect the packing of organisms within the colony. In some cases these differences may alter the passage of light through the colony, resulting in altered colony appearance. When viewed with oblique, transmitted light and magnification on transparent medium, it is possible to observe opaque and transparent colony forms in colonies derived from the same isolate (Fig. 1). Variation in colony morphology appears to be common to all strains, although it is more readily appreciated in isolates of certain serotypes, possibly because some capsule types may act to obscure phenotypic differences. Opacity variation is not apparent on nontranslucent medium, such as blood agar, which probably accounts for why these differences had not previously been described.

Gram-Positive Pathogens, ed. by V. A. Fischetti et al.
© 2000 American Society for Microbiology, Washington, D.C.

TABLE 1 Summary of differences in characteristics of phase variants of *Streptococcus pneumoniae*

Characteristic	Colony morphology	
	Opaque	Transparent
Physiology	Dome-shaped colonies	Umbilicated colonies
	Larger colonies	More rapid autolysis
		Higher efficiency of natural transformation
Cell surface		
Protein	Increased amounts of PspA	Increased amounts of CbpA
		Increased amounts of autolysin (LytA)
Nonprotein	Increased content of capsular polysaccharide	Increased amounts of phosphorylcholine
		Increased content of C-polysaccharide (teichoic acid)
Pathogenesis	Decreased binding of serum C-reactive protein	Increased adherence to activated type II pneumocytes
	Decreased opsonophagocytosis	Increased invasion and transcytosis in human brain microvascular endothelial cells
	Increased virulence (animal model of sepsis)	More efficient colonization (animal model of carriage)

There is spontaneous, reversible variation between colony phenotypes. It is possible to detect sectored colonies resulting from phase variation during the clonal expansion of a single organism as it forms a colony. The frequency of switching is highly variable from isolate to isolate (range 10^{-3} to 10^{-6}/generation) and appears to be independent of in vitro growth conditions, including pH, temperature,

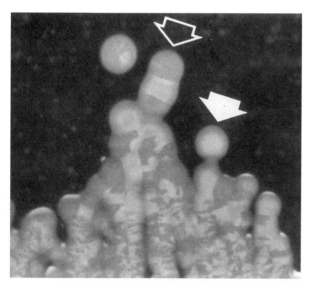

FIGURE 1 Colonies of a type 18C clinical isolate of *S. pneumoniae* showing phenotypic variation between opaque (solid arrow) and transparent (open arrow) colony forms when viewed with oblique, transmitted illumination on a transparent surface. Magnification, ×240.

and osmolarity. Under standard culture conditions, pneumococcal isolates are highly heterogeneous populations. It is possible, however, to separate many strains into nearly uniform populations of opaque and transparent and, in some cases, intermediate forms for direct comparison.

CORRELATION BETWEEN OPACITY VARIATION AND PNEUMOCOCCAL INFECTION

Animal models were used to determine whether differences in colony morphology correlated with a difference in the ability of the pneumococcus to colonize and infect a host. The relative ability of opaque and transparent variants to colonize the nasopharynx, the initial step in the pathogenesis of pneumococcal disease, was assessed in an infant rat model (35). This is a convenient model for obtaining washes of the nasopharynx to determine the quantity as well as the phenotype of colonizing organisms. Following a single intranasal inoculum, the number of organisms in the nasopharynx expands rapidly and the pups remain heavily colonized for at least several weeks. Pneumococcal carriage by infant rats, furthermore, does not appear to be limited to certain serotypes. When equal inocula (10^3 CFU) of relatively uniform populations of opaque or transparent variants of the same strain were compared in this model, only the transparent organisms were able to establish dense and stable colonization of the mucosal surface of the nasopharynx. After challenge with a large inoculum (10^7 CFU) of an opaque variant, the pups gradually became colonized, but by the end of 7 days the opaque variant had been cleared and there was heavy colonization with transparent forms. This suggested that since such a large inoculum contained a small number of transparent variants, these were selected for from among the heterogeneous inoculum.

Evidence that the transparent phenotype is selected for during nasopharyngeal colonization left in question the biological role of the opaque phenotype. This was addressed by using intraperitoneal rather than intranasal inoculation in order to bypass the requirements of colonization (12). Because infant rats are not highly susceptible to invasive pneumococcal infection, it was necessary to perform these experiments in an adult mouse model of sepsis. This required the use of mouse virulent serotypes in which the opaque and transparent forms are easily distinguished. Equal intraperitoneal inocula of relatively uniform populations of opaque or transparent variants of the same strain (10^7 CFU) were compared in this model. All mice receiving opaque organisms died of sepsis. In contrast, of the few mice that expired following inoculation with transparent organisms, splenic cultures revealed only organisms that had reverted to a more opaque phenotype. This suggests that during invasive infection, there is a strong selection for organisms with the opaque phenotype.

Based on the animal experiments, it appears that the pneumococcus phase varies between at least two forms: one adapted for nasopharyngeal colonization and the other for events following colonization. The animal experiments, therefore, demonstrated the relevance of opacity variation to the pathogenesis of pneumococcal infection.

GENETIC BASIS OF OPACITY VARIATION

Two approaches, genetic and biochemical, were taken to define the bacterial characteristics that change in association with colony opacity that may contribute to the differences observed in the animal models. The genetic approach relied on the ability to transform the pneumococcus at high efficiency (25). Chromosomal DNA from an opaque isolate was used to transform a transparent recipient so as to acquire the opaque phenotype. With "opaque" DNA from some strains, the frequency of obtaining opaque colonies was far higher than that in controls with "transparent" DNA, in which opaque colonies were seen only as a result of background phase variation. A chromosomal library in bacteriophage lambda constructed with DNA from one such strain was then screened to identify a single clone able to transform the transparent recipient to the opaque phenotype. Analysis of this opacity locus revealed two genes, *glpD* and *glpF*, with homology to the glycerol regulon genes in other bacteria. Following these genes was a sequence with homology to repetitive pneumococcal intergenic elements, BOX A and C, upstream of an open reading frame that could encode a 126-amino-acid protein with no homology to current entries in sequence databases (15). Although the analysis of this complex region is in progress, several observations are of potential relevance. The stem-loop-forming element, BOX A-C, was not present in the same locus of the recipient strain. Introduction of these elements during transformation affected colonial morphology, possibly by altering expression of the 126-amino-acid open reading frame, a putative regulatory gene, downstream from the BOX element. Mutagenesis confirmed that this element increases the frequency of variation in opacity. Strains lacking the BOX A-C element appear to phase vary at lower rates, and DNA with transforming activity contains this element.

When the BOX element is incorporated into a transparent recipient, the increased frequency of variation makes it possible to detect higher numbers of opaque colonies.

RELATIONSHIP BETWEEN OPACITY VARIATION AND CARBOHYDRATE-CONTAINING STRUCTURES

The biochemical comparison of pneumococcal variants has focused on two major cell surface structures, the capsular polysaccharide and C polysaccharide, or cell wall teichoic acid. The capsule is a well-recognized virulence determinant, and relatively small differences in the amount of capsular polysaccharide have been shown to have a significant effect on the ability of the pathogen to cause infection (14). Opacity variation is present in unencapsulated mutants, which suggests that differences in the amount or composition of the capsule are not the sole factor responsible for phenotypic variation. Variants of the same strain are indistinguishable using type-specific antiserum in a quellung reaction, suggesting that their capsules are antigenically related. Immunoelectron microscopy demonstrated a larger and denser zone of immunoreactive capsular material in opaque pneumococci (11). A capture enzyme-linked immunosorbent assay (ELISA) with type-specific monoclonal antibodies (MAbs) was then used to quantify amounts of capsular polysaccharide in organisms of each phenotype (12). This procedure showed that opaque pneumococci have 1.2- to 22-fold more capsular polysaccharide than the related transparent variants.

Differences in the other carbohydrate structure on the cell surface, the teichoic acids, were analyzed by taking advantage of the unique structure of these molecules, which contain choline in the form of covalently bound choline phosphate or phosphorylcholine (ChoP) (1, 6). Choline, which is obtained exclusively from the nutrient medium, is essential for growth of both transparent and opaque variants. It is estimated that approximately 90% of the choline incorporated into cells is localized to the cell wall teichoic acid, with the remainder in the lipoteichoic acid (pneumococcal Forssman antigen), which is anchored in the plasma membrane (31). The teichoic acids consist of polymers that contain from two to eight identical repeating units linked by ribitol phosphate, with each unit containing two phosphorylcholine residues (see chapter 21, this volume).

The quantity of teichoic acid per cell was compared in opaque and transparent variants by measuring the incorporation of [^{3}H]choline from the culture medium (12). After growth to the same density, transparent variants incorporated 3.9- to 8.7-fold more of the radiolabel per cell than did opaque variants of the same strain. The correlation between choline incorporation and colony morphology was confirmed by showing that the level of incorporation in a spontaneous revertant of an opaque to a transparent variant was essentially equivalent to that of the transparent form of the same strain. Cells grown in the presence of [^{3}H]choline were fractionated to determine the cellular location of the increased choline incorporated into organisms with the transparent phenotype. The majority of the label was found in the sodium dodecyl sulfate-insoluble cell fraction, which includes the crude cell wall and the cell wall-associated teichoic acid. These differences in the amount of cellular ChoP were also demonstrated by a quantitative ELISA assay using a natural murine immunoglobulin A MAb, TEPC-15, that reacts specifically to phosphorylcholine and has been shown to bind to phosphorylcholine on the pneumococcus (13). Finally, opaque and transparent phase variants were compared by flow cytometry using the MAb TEPC-15. Although both phenotypes reacted with the MAb, the 5.6-fold greater intensity of reactivity for organisms of the transparent form suggests that there is more surface-exposed phosphorylcholine associated with this phenotype.

These findings would suggest that either the structure of the teichoic acid or the amount of teichoic acid per cell is subject to variation, since the only significant reservoir of cellular choline is the teichoic acids. Differences in teichoic acid structure could account for these observations if the amount of choline per teichoic acid chain varied owing to either differences in the average number of repeating units per chain or the number of phosphorylcholine residues per repeating unit. Either of these differences in structure would be expected to affect the size distribution of teichoic acid chains. Western analysis using the MAb TEPC-15 was used to visualize lipoteichoic acid (sodium dodecyl sulfate-soluble fraction). The lipoteichoic acid showed a ladderlike array of doublet bands, which represents chains with differing numbers of repeating units. When samples from equivalent numbers of opaque and transparent organisms were compared, there was no appreciable difference in the migration of each chain or in the average chain length. The intensity of reactivity for each band, however, was greater for the transparent variant. This observation suggests that the higher amount of choline in transparent pneumococci may be a result of increased numbers of teichoic acid residues per cell rather than variation in structure or average number of repeating units per chain. Since the cell wall-associated teichoic acid is covalently attached to muramic acid residues in the peptidoglycan and only a fraction of the muramic acid residues appear to be linked to a teichoic acid chain, the differences in the amount of teichoic acid observed in this study could be a consequence of differing proportions of muramic acid residues having an attached teichoic acid chain (5).

VARIATION IN THE EXPRESSION OF CELL SURFACE PROTEINS

The relationship of several previously identified cell surface proteins to opacity variation has been examined. The role of cell lysis or autolysis in this phenomenon was considered because of several observations: (i) Opaque colonies are dome shaped whereas transparent colonies are umbilicated after equivalent incubation conditions. (ii) Electron micrographs of organisms grown under identical conditions revealed that there was breakdown of the cell wall structures only in the transparent cells (35). (iii) Opaque organisms undergo spontaneous lysis more slowly and are more resistant to the detergent deoxycholate (25). Autolysis occurs once the bacteria reach stationary phase through the enzymatic degradation of the cell wall primarily by the major murein amidase, LytA (autolysin). Differences in rates of autolysis could result from variations in the peptidoglycan substrate or the expression of the amidase. The hydrolysis of purified cell walls of opaque and transparent organisms as determined by high-pressure liquid chromatography analysis of stem peptides released by treatment with amidase was indistinguishable (34). This result made it un-

likely that the peptidoglycan was responsible for the observed differences in autolysis. The expression of autolysin in opaque and transparent variants was compared using antiserum to LytA (provided by R. Lopez). Low levels of autolysis in opaque variants correlated with decreased levels of immunoreactive LytA on colony immunoblots and Western analysis. Mutants in which the lytA gene has been interrupted have an altered colony morphology but are still capable of displaying phenotypic variation. This indicated that LytA is only one factor contributing to opacity variation. The question of whether LytA, which is present in higher amounts on the cell surface of transparent variants, has a role in the more efficient colonization by this phenotype was examined in the infant rat model of pneumococcal carriage. LytA⁻ mutants in encapsulated strains (provided by J. Paton) were indistinguishable when compared with LytA⁺ parent strains in their ability to colonize the infant rat nasopharynx.

The hypothesis that other cell surface proteins are expressed in higher amounts in transparent variants, like LytA, and contribute to colonization was examined as follows. LytA is known to anchor to the cell by binding to ChoP on the lipoteichoic acid. Antiserum to pneumococcal proteins that adhered to a choline column (provided by R. Masure) was used to compare the differential expression of choline-binding proteins. In addition to LytA, this antiserum recognized at least two other proteins that could be eluted from cells by incubation in high concentrations of choline. One of these proteins was the pneumococcal surface protein PspA, which was present in greater quantities in opaque variants (12). Although the function of PspA is unknown, it has been shown to contribute to the virulence of the pneumococcus in mice (17). The other choline-binding protein, designated CbpA, was present in higher amounts in transparent variants. Mutagenesis of the gene encoding this protein eliminated its expression and resulted in diminished adherence to type II pneumocytes and a more than 2-log decrease in the number of organisms colonizing the infant rat nasopharynx (24). CbpA, therefore, may contribute to the enhanced ability of transparent pneumococci to colonize the mucosal surface of the nasopharynx. Two other reports have suggested additional functions, the binding to secretory immunoglobulin A (SpsA) and to complement component C3 (PbcA), to what appears to be the identical pneumococcal protein (9, 27). More recently analysis of the pneumococcal genome sequence has revealed a number of additional genes whose translation products have sequences related to the choline-binding domain (38). Thus, the family of choline-binding proteins may be more extensive, and differences in their expression on the cell surface seem likely considering the variation in the ChoP anchor.

ROLE OF PHOSPHORYLCHOLINE IN THE BIOLOGY OF S. PNEUMONIAE

Since teichoic acid is a major component of the cell surface, differences in amounts of ChoP that distinguish opaque and transparent phase variants might have numerous effects on the cell. Evidence that the cell surface expression of multiple choline-binding proteins varies in association with variation of the ChoP anchor has been discussed. In addition, ChoP on the pneumococcus may interact directly with host molecules. Choline, which is generally not present in prokaryotes, is a major component of eukaryotic membrane lipids as phosphatidylcholine and is found on many host structures.

Pneumococci have been shown to adhere to human buccal epithelial cells, type II pneumocytes, human lung epithelial cells, as well as vascular endothelial cells. For each of these cell types, transparent organisms bind in greater numbers than the related opaque organisms (4). This result correlates with the enhanced ability of the transparent variant to colonize the nasopharynx. The adherence of transparent, but not opaque, pneumococci is augmented by stimulation of resting human cells with the cytokines interleukin-1 or tumor necrosis factor. Adherence to stimulated cells correlates with the ability of transparent variants to bind to cells transfected with the receptor for platelet-activating factor (rPAF) (3). Inflammatory cytokines activate the expression of the rPAF, which has been identified on many host tissues. It has been proposed that the pneumococcus interacts with the rPAF by structural mimicry of PAF. Since both the cell surface of the pneumococcus and PAF contain ChoP, this structure may be crucial to the binding of the bacteria to this host cell target, an interaction that is inhibited by rPAF antagonists and high concentrations of exogenous choline. This interaction has also been shown to require the presence of choline on the organism, since pneumococci grown in the presence of ethanolamine in lieu of choline are poorly adherent. It is possible that the enhanced adherence to activated cells by transparent organisms is due to the increased expression of ChoP associated with this phenotype. An alternative explanation is that differences in adherence between phenotypes are due to changes in the cell surface expression of choline-binding proteins rather than a direct result of differences in content of ChoP. Phase variation in the display of the amount of ChoP and subsequent differences in choline-binding proteins may allow the organism to switch from an adherent to a nonadherent state. This may be critical to the organism's ability to exist both in the nasopharynx, where attachment to cells may be beneficial for prolonged carriage, and in the bloodstream, where adherence to cells may confer a disadvantage for survival.

Interaction with host cells may also be a factor in the ability of the pneumococcus to breach the blood-brain barrier, as occurs in the pathogenesis of meningitis. This process has been modeled using human brain microvascular endothelial cells in culture (21). Pneumococci are able to invade these cells and transcytose to the basal surface through a mechanism that involves the rPAF. The process of invasion is inhibited by the presence of capsular polysaccharide and is approximately sixfold more efficient for transparent variants that have smaller capsules.

CONTRIBUTION OF PHENOTYPIC VARIATION TO HOST CLEARANCE MECHANISMS

The expression of elevated amounts of ChoP associated with the transparent phenotype may, in addition, promote clearance of the organism by the host. Human serum contains abundant natural antibody to ChoP, and a serum protein, C-reactive protein (CRP), recognizes this structure and has been shown to promote the in vivo clearance of the pneumococcus (18, 29, 30). CRP, an acute-phase reactant whose levels may rise 1,000-fold during inflammation, is capable of activating the classical pathway of

complement by binding to C1q (32). Only transparent pneumococci appear to bind significant amounts of human CRP (11). These factors may lead to a selection of pneumococci with less cell surface ChoP, i.e., of the opaque phenotype, once an organism has bypassed the mucosal barrier.

The contribution of the capsular polysaccharide to the pathogenesis of pneumococcal infection has been ascribed to its antiphagocytic properties. The possibility that the increased virulence of opaque pneumococci could be due to the antiphagocytic effect of the higher amount of capsular polysaccharide associated with this phenotype was addressed in an opsonophagocytosis assay (22, 23). For variants from an individual isolate, the amount of immune serum required for opsonophagocytosis by HL-60 cells in culture was proportional to the quantity of capsular polysaccharide. The average titer of serum required to achieve 50% killing was significantly greater for opaque compared with transparent variants. This suggests that the ability of opaque organisms to cause sepsis may be a result of diminished clearance by the combined activity of antibody, CRR, complement, and phagocytes.

RELATIONSHIP TO OTHER PATHOGENS OF THE HUMAN RESPIRATORY TRACT

Phase variation in the expression of ChoP is not unique to the pneumococcus, although this structure is distinctly uncommon in prokaryotes. Recently, this laboratory has reported that another pathogen of the human respiratory tract, *H. influenzae*, a gram-negative bacterium that has a life cycle with many similarities to that of *S. pneumoniae*, decorates its cell surface with ChoP (37). In the case of *H. influenzae*, ChoP is found on the lipopolysaccharide (36). In fact, MAb TEPC-15 cross-reacts with ChoP on the surface glycolipids of both *H. influenzae* (lipopolysaccharide) and *S. pneumoniae* (lipoteichoic acid). *H. influenzae* also acquires choline from the growth medium, although, unlike for the pneumococcus, it is not a nutritional requirement. The expression of ChoP on the lipopolysaccharide requires a previously identified chromosomal locus containing four genes, *licA–D*. Phase variation is mediated by a molecular switch based on multiple tandem repeats of the sequence 5′-CAAT′-3′ within the open reading frame of *licA*. The gene product of *licA* has homology to eukaryotic choline kinases, suggesting that the bacterial pathway for choline incorporation has common features with that of eukaryotes.

Using degenerate primers based on the active site of choline kinases, a *licA* homolog was isolated from *S. pneumoniae* (26). Extension of the sequence revealed homologs of *licB* and *licC* from *H. influenzae*, suggesting that the pathway for choline incorporation is similar in the pneumococcus. The only other bacteria in which a *licA* homolog has been described are members of the genus *Mycoplasma*. Since *Mycoplasma*, *Haemophilus*, and *Streptococcus* all share the ability to infect the human respiratory tract, there may be a common mechanism involving ChoP in their pathogenesis. More recently, the TEPC-15 binding has been detected on pili of *N. meningitidis* and a 43-kDa membrane protein in *Pseudomonas aeruginosa* (33). The ChoP epitope on this protein of unknown function in *P. aeruginosa* undergoes variation in expression based on growth temperature. Since pili in *N. meningitidis* also undergo phase variation, it appears that at least four major pathogens that may infect the human respiratory tract have the ChoP epitope on their cell surface and display phase variation in either the presence or amount of this unusual prokaryotic structure. *H. influenzae* decorated with ChoP were efficiently killed by CRP in the presence of complement (36). This finding confirmed the role of this innate host defense mechanism in targeting organisms containing cell surface ChoP and provides a possible explanation for why bacteria must down-regulate its expression.

REFERENCES

1. Behr, T., W. Fischer, J. Peter-Katalinic, and H. Egge. 1992. The structure of pneumococcal lipoteichoic acid. *Eur. J. Biochem.* **207:**1063–1075.

2. Christensen, G., L. Baddour, B. Madison, J. Parisi, S. Abraham, D. Hasty, J. Lowrance, J. Josephs, and W. Simpson. 1990. Colonial morphology of staphylococci on Memphis agar: phase variation of slime production, resistance to beta-lactam antibiotics, and virulence. *J. Infect. Dis.* **161:**1153–1169.

3. Cundell, D. R., N. P. Gerard, C. Gerard, I. Idanpaan-Heikkila, and E. I. Tuomanen. 1995. *Streptococcus pneumoniae* anchor to activated human cells by the receptor for platelet-activating factor. *Nature* **377:**435–438.

4. Cundell, D. R., J. N. Weiser, J. Shen, A. Young, and E. I. Tuomanen. 1995. Relationship between colonial morphology and adherence of *Streptococcus pneumoniae*. *Infect. Immun.* **63:**757–761.

5. Fischer, H., and A. Tomasz. 1985. Peptidoglycan cross-linking and teichoic acid attachment in *Streptococcus pneumoniae*. *J. Bacteriol.* **163:**46–54.

6. Fischer, W., T. Behr, R. Hartmann, J. Peter-Katalinic, and H. Egge. 1993. Teichoic acid and lipoteichoic acid of *Streptococcus pneumoniae* possess identical chain structures. A reinvestigation of teichoid acid (C polysaccharide). *Eur. J. Biochem.* **215:**851–857.

7. Gotschlich, E. C. 1994. Genetic locus for the biosynthesis of the variable portion of *Neisseria gonorrhoeae* lipooligosaccharide. *J. Exp. Med.* **180:**2181–2190.

8. Hammerschmidt, S., A. Muller, H. Sillmann, M. Muhlenhoff, R. Borrow, A. Fox, J. van Putten, W. D. Zollinger, R. Gerady-Schahn, and M. Frosch. 1996. Capsule phase variation in *Neisseria meningitidis* serogroup B by slipped-strand mispairing in the polysialyltransferase gene (siaD): correlation with bacterial invasion and the outbreak of meningococcal disease. *Mol. Microbiol.* **20:**1211–1220.

9. Hammerschmidt, S., S. R. Talay, P. Brandtzaeg, and G. S. Chhatwal. 1997. SpsA, a novel pneumococcal surface protein with specific binding to secretory immunoglobulin A and secretory component. *Mol. Microbiol.* **25:**1113–1124.

10. Jones, G., D. Clewell, L. Charles, and M. Vickerman. 1996. Multiple phase variation in haemolytic, adhesive and antigenic properties of *Streptococcus gordonii*. *Microbiology* **142:**181–189.

11. Kim, J., J. H. Wani, U. B. S. Sørensen, J. Blom, and J. N. Weiser. 1998. Characterization of phenotypic variants of *Streptococcus pneumoniae*, abstr. B-2, p. 56. *Abstr. 98th Gen. Meet. Am. Soc. Microbiol. 1998.* American Society for Microbiology. Washington, D.C.

12. Kim, J., and J. Weiser. 1998. Association of intrastrain phase variation in quantity of capsular polysaccharide and teichoic acid with the virulence of *Streptococcus pneumoniae*. *J. Infect. Dis.* **177:**368–377.

13. Leon, M. A., and N. M. Young. 1971. Specificity for phosphorylcholine of six murine myeloma proteins reactive with pneumococcus C polysaccharide and beta-lipoprotein. *Biochemistry* **10:**1424–1429.

14. MacLeod, C. M., and M. R. Krauss. 1950. Relation of virulence of pneumococcal strains for mice to the quantity of capsular polysaccharide formed in vitro. *J. Exp. Med.* **92:**1–9.

15. Martin, B., O. Humbert, M. Camara, E. Guenzi, J. Walker, T. Mitchell, P. Andrew, M. Prudhomme, G. Alloing, R. Hakenbeck, D. A. Morrison, G. J. Boulnois, and J. Claverys. 1992. A highly conserved repeated DNA element located in the chromosome of *Streptococcus pneumoniae*. *Nucleic Acids Res.* **20:**3479–3483.

16. McCarty, M. 1966. The nature of the opaque colony variation in group A streptococci. *J. Hyg.* **64:**185–190.

17. McDaniel, L. S., J. Yother, M. Vijayakumar, L. McGarry, W. R. Guild, and D. E. Briles. 1987. Use of insertional inactivation to facilitate studies of biological properties of pneumococcal surface protein A (PspA). *J. Exp. Med.* **165:**381–394.

18. Mold, C., S. Nakayama, T. Holzer, H. Gewurz, and T. Du Clos. 1981. C-reactive protein is protective against *Streptococcus pneumoniae* infection in mice. *J. Exp. Med.* **154:**1703–1708.

19. Pincus, S. H., R. L. Cole, E. Kamanga-Sollo, and S. H. Fischer. 1993. Interaction of group B streptococcal opacity variants with the host defense system. *Infect. Immun.* **91:**3761–3768.

20. Pincus, S. H., R. L. Cole, M. R. Wessels, M. D. Corwin, E. Kamanga-Sollo, S. F. Hayes, W. Cieplak, and J. Swanson. 1992. Group B streptococcal variants. *J. Bacteriol.* **174:**3739–3749.

21. Ring, A., J. N. Weiser, and E. I. Tuomanen. 1998. Pneumococcal penetration of the blood-brain barrier: molecular analysis of a novel re-entry path. *J. Clin. Invest.* **102:**347–360.

22. Romero-Steiner, S., M. Carvalho, S. Barnardi, J. Kim, J. Weiser, and G. Carlone. 1998. Decreased opsonophagocytic activity in the pneumococcal opaque phenotype is associated with higher capsular polysaccharide concentration. *Program Abstr. 38th Intersci. Conf. Antimicrob. Agents Chemother.* American Society for Microbiology, Washington, D.C.

23. Romero-Steiner, S., D. Libutti, L. B. Pais, J. Dykes, P. Anderson, J. C. Whitin, H. L. Keyserling, and G. M. Carlone. 1997. Standardization of an opsonophagocytic assay for the measurement of functional antibody activity against *Streptococcus pneumoniae* using differentiated HL-60 cells. *Clin. Diag. Lab. Immunol.* **4:**415–422.

24. Rosenow, C., P. Ryan, J. N. Weiser, S. Johnson, P. Fontan, A. Ortqvist, and H. R. Masure. 1997. Contribution of novel choline-binding proteins to adherence, colonization and immunogenicity of *Streptococcus pneumoniae*. *Mol. Microbiol.* **25:**819–829.

25. Saluja, S. K., and J. N. Weiser. 1995. The genetic basis of colony opacity in *Streptococcus pneumoniae*: evidence for the effect of box elements on the frequency of phenotypic variation. *Mol. Microbiol.* **16:**215–227.

26. Shchepetov, M., and J. Weiser. 1998. Identification of a locus involved in teichoic acid biosynthesis and choline incorporation in *Streptococcus pneumoniae*, abstr. B-7, p. 57. *Abstr. 98th Gen. Meet. Am. Soc. Microbiol. 1998.* American Society for Microbiology, Washington, D.C.

27. Smith, B., Q. Cheng, and M. Hostetter. 1998. Characterization of a pneumococcal surface protein that binds complement protein C3 and its role in adhesion, abstr. D-122, p. 233. *Abstr. 98th Gen. Meet. Am. Soc. Microbiol. 1998.* American Society for Microbiology, Washington, D.C.

28. Stern, A., and T. F. Meyer. 1987. Common mechanism controlling phase and antigenic variation in pathogenic Neisseria. *Mol. Microbiol.* **1:**5–12.

29. Szalai, A. J., A. Agrawal, T. J. Greenhough, and J. E. Volanakis. 1997. C-reactive protein. *Immunol Res.* **16:**127–136.

30. Szalai, A. J., D. E. Briles, and J. E. Volanakis. 1996. Role of complement in C-reactive-protein-mediated protection of mice from *Streptococcus pneumoniae. Infect. Immun.* **64:**4850–4853.

31. Tomasz, A. 1981. Surface components of *Streptococcus pneumoniae. Rev. Infect. Dis.* **3:**190–210.

32. Volanakis, J. E., and M. H. Kaplan. 1974. Interaction of C-reactive protein complexes with the complement system. II. Consumption of guinea pig complement by CRP complexes. Requirement for human C1q. *J. Immunol.* **113:**9–17.

33. Weiser, J., J. Goldberg, N. Pan, L. Wilson, and M. Virji. 1998. The phosphorylcholine epitope undergoes phase variation on a 43 kD protein in *Pseudomonas aeruginosa* and on pili of pathogenic *Neisseria. Infect. Immun.* **66:**4263–4267.

34. Weiser, J., Z. Markiewicz, E. Tuomanen, and J. Wani. 1996. Relationship between phase variation in colony morphology, intrastrain variation in cell wall physiology and nasopharyngeal colonization by *Streptococcus pneumoniae. Infect. Immun.* **64:**2240–2245.

35. Weiser, J. N., R. Austrian, P. K. Sreenivasan, and H. R. Masure. 1994. Phase variation in pneumococcal opacity: relationship between colonial morphology and nasopharyngeal colonization. *Infect. Immun.* **62:**2582–2589.

36. Weiser, J. N., N. Pan, K. L. McGowan, D. Musher, A. Martin, and J. C. Richards. 1998. Phosphorylcholine on the lipopolysaccharide of *Haemophilus influenzae* contributes to persistence in the respiratory tract and sensitivity to serum killing mediated by C-reactive protein. *J. Exp. Med.* **187:**631–640.

37. Weiser, J. N., M. Shchepetov, and S. T. H. Chong. 1997. Decoration of lipopolysaccharide with phosphorylcholine; a phase-variable characteristic of *Haemophilus influenzae. Infect. Immun.* **65:**943–950.

38. Yother, J., and J. M. White. 1994. Novel surface attachment mechanism of the *Streptococcus pneumoniae* protein PspA. *J. Bacteriol.* **176:**2976–2985.

Genetics of *Streptococcus pneumoniae*

JANET YOTHER

23

The origins of genetics in *Streptococcus pneumoniae* can be traced to studies that began in the late 1800s with the isolation of nonencapsulated variants and ultimately led to the discovery of bacterial gene transfer by Griffith in 1928 and the identification of DNA as the genetic material by Avery, MacLeod, and McCarty in 1944 (4, 20, 71). Transformation was the hallmark of these and many other studies analyzing the transfer of capsular polysaccharide biosynthetic genes, as well as nutritional and antibiotic markers. Today, it remains the single most important technique in the genetic analysis of *S. pneumoniae*. Many of the *S. pneumoniae* strains commonly used in genetic studies are descended from isolates obtained by Avery in 1916 and first described in 1928 by Dawson (62, 66). In fact, all of the nonencapsulated isolates used to study the transformation process are derived from a single type 2 clinical isolate and its nonencapsulated variants used in the 1944 studies of Avery et al. Table 1 provides the lineage for a number of these strains, along with properties pertinent to this chapter. Recently, physical and genetic mapping studies have provided more detailed information regarding the *S. pneumoniae* chromosome (Fig. 1), and a preliminary, partial genomic sequence is now available from the Institute for Genomic Research (TIGR) (26b). This chapter highlights much of the current information regarding *S. pneumoniae* genetics, and the reader is referred to references within those cited for further details of the elegant earlier studies that laid the foundations for our present knowledge.

GENE TRANSFER

Both transformation and conjugation have been described in *S. pneumoniae*. Transformation serves as the primary, and perhaps sole, means of transferring chromosomal genes. Conjugation occurs with plasmids that are capable of self-transfer or mobilization and with conjugative transposons that are integrated into the chromosome. The latter mediate their own transfer but do not cotransfer linked chromosomal markers. Conjugative transposons and chromosomal genes transferred by transformation are the major mechanisms known to be involved in the spread of anti-biotic resistance in *S. pneumoniae*. Generalized transduction has not been observed to occur, although a process termed pseudotransduction, which involves properties of both transduction and transformation, has been described for one pneumococcal bacteriophage.

Transformation

Natural transformation in *S. pneumoniae* can be divided into several stages: (i) the induction of competence, (ii) binding of double-stranded DNA to the cell surface, (iii) digestion of bound DNA into discrete double-stranded fragments, (iv) uptake of single-stranded linear DNA, (v) binding of the internalized DNA by single-stranded binding proteins, and (vi) recombination of homologous DNA into the chromosome by a single-strand replacement mechanism. Successful transformation of plasmids requires the uptake of two complementary, overlapping DNA strands. Essentially any DNA can be taken up, as no specific recognition sequences are required. DNA that is not homologous to the pneumococcal chromosome or that cannot replicate independently is degraded and becomes part of the nucleotide precursor pool.

Induction of Competence

S. pneumoniae becomes competent for transformation for only a brief period during the early exponential phase of growth (68). The specific point at which competence is induced is determined by both the cell density and the culturing conditions. Maximal transformation efficiencies are usually observed in a rich medium containing calcium and serum or albumin, at a starting pH of about 7.0 to 7.2. Under optimum culture conditions, transformation of a given marker from a donor chromosomal DNA preparation can be observed in 1% or more of the recipients. Alterations in starting pH affect both the timing and efficiency of transformation (11). A low pH tends to prevent the development of competence, whereas higher pH values result in multiple waves of competence that occur over a broader range of cell densities and rarely reach the peak efficiency. Under optimum conditions, cotransformation of two unlinked chromosomal markers occurs near the fre-

Gram-Positive Pathogens, ed. by V. A. Fischetti et al.

TABLE 1 Derivation and properties of commonly used *S. pneumoniae* strains

Strain	Capsule[a]	Derivation and relevant properties[b]
D39	Type 2⁺	Clinical isolate (1928); virulent; pDP1⁺; *comC1*
R36	Type 2⁻	D39 passaged 36× in anti-type 2 serum (1944); does not revert to Cps⁺ (see R36A, below); Hex⁺; pDP1⁺
R36N	Type 2⁻	Colony variant of R36 (1944); nontransformable; pDP1⁺
R36NC	Type 2⁻	Transformable derivative of R36N (1947); *Dpn*I; Hex⁺; pDP1⁺
R36A	Type 2⁻	Colony variant of R36 (1944); deletions demonstrated in type 2 capsule locus (reference 26a); Hex⁺; pDP1⁻
R6	Type 2⁻	Single-colony isolate of R36A (1962); *Dpn*I; Hex⁺; pDP1⁻
R6x	Type 2⁻	R6 transformed to the Rx Hex⁻ phenotype (1973)
Rx	Type 3*	R36A transformed to type 3 encapsulation (1949) and subsequently isolated as a spontaneous "nonencapsulated" derivative (1959) (but see Rx1, below); Hex⁻ (*hexB*⁻); pDP1⁻
Rx1	Type 3*	Highly transformable derivative of Rx; contains the type 3 *cps* locus and, consequently, a 5′ deletion of *ali* (*plpA*); produces small amounts of type 3 capsule (~20% of that observed with most type 3 strains); Cps phenotype due to point mutation in *cps3D*; virulence is not restored by transformation to normal type 3 encapsulation (unlike transformation of D39 to type 3); *Dpn*0; Hex⁻ (*hexB*⁻); pDP1⁻; *comC1*
A66	Type 3⁺	Clinical isolate (1928); virulent; donor in 1944 transformation studies of Avery et al.; *Dpn*I; *comC2*
TIGR[c]	Type 4⁺	Clinical isolate, virulent; used for genome sequencing; *Dpn*I; *comC2*

[a]The *cps* genotype is given, and the symbols indicate whether the strain is encapsulated (+), nonencapsulated (−), or intermediate (*). Of the type 3* strains, only Rx1 has been demonstrated to produce capsular polysaccharide (13a, 15).

[b]Properties are listed for strains in which they have been demonstrated, and may be the same for progenitors. Lineages and original references are described for most strains in references 62 and 66. Numbers in parentheses indicate date of original description. Except as noted, all strains are avirulent. *Dpn* phenotypes are from reference 41. The *Dpn*0 phenotype in Rx1 is apparently due to a mutation in the *Dpn*I-encoding locus. Hex phenotypes are from references 52 and 66. *comC* alleles are from reference 51 or analysis of the *S. pneumoniae* genome sequence (26b). All D39 derivatives are expected to carry the *comC1* allele.

[c]The designation of the strain used by TIGR for sequencing is not known.

quency expected for two independent events, indicating that essentially all cells in the culture become competent. Cotransformation of unlinked markers during subsequent, postpeak waves of competence occurs at a frequency higher than expected, indicating that only a fraction of these cells are actually capable of being transformed. These differing outcomes make it possible to manipulate the culture conditions, depending on the desired outcome of the transformation. In addition, the high transformation frequencies observed under optimum conditions allow for the introduction of DNA for which there is no selectable marker but for which a suitable screen exists. Indeed, the use of isolated, homogeneous DNA fragments, such as that obtained from cloned DNA, has been observed to result in transformation frequencies as high as 50% (15, 64).

The development of competence is mediated by the competence factor (68), a small peptide that induces expression of genes encoding transformation-related proteins. Different *S. pneumoniae* isolates contain one of two alleles of *comC*, the gene that encodes the competence-stimulating peptide (CSP) (51). The 41-amino-acid translation products from *comC1* and *comC2* each contain an identical 24-amino-acid N-terminal leader sequence that possesses a Gly-Gly processing site common to peptide bacteriocins. The remaining 17 amino acids, which represent the mature, active peptides of CSP-1 and CSP-2, differ at eight positions. Sequence analysis of encapsulated strains representing multiple capsular serotypes has shown that each has either a *comC1* or *comC2* allele and that the

respective alleles are highly conserved. In general, however, transformation of encapsulated isolates is not readily observed under the growth conditions described above. In many cases, the failure to transform is related to the presence of the capsule, which impedes the induction of competence (15, 72). Noncompetent cultures, as well as many encapsulated strains, have been induced to competence by the addition of exogenous competence factor isolated from competent cells of the nonencapsulated R6 strain, by supernatants of competent cultures of the Rx1 strain, and by the addition of synthetic CSP-1 or CSP-2, depending on the allele present in the recipient strain (51, 68, 72). The conditions for inducing competence are less strict under these circumstances than for natural induction, and cultures at essentially any cell density can be induced. The optimum CSP concentration for inducing competence is strain dependent. In addition, because competence is a transient state, the time of exposure to the competence factor prior to DNA addition is critical, usually ranging from 10 to 20 min, and is also strain dependent. Although it has not yet been identified, there is evidence for a macromolecular inhibitor that causes the cessation of the competent phase (68).

The identification of the *comC1* and *comC2* alleles and their respective peptides has made possible the transformation of a large number of encapsulated isolates. The efficiencies of transformation can vary widely, however, and there remain many strains (approximately 50% of those tested) that have not been transformed. In addition, some

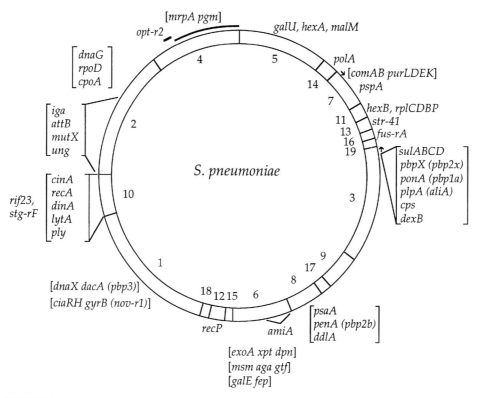

FIGURE 1 *S. pneumoniae* physical and genetic map. The map shows SmaI fragments and is based on the original data of Gasc et al. (19) for the R6 chromosome. Additional gene locations were obtained through a combined analysis of this map, a second R6 physical and genetic map (34), published map locations of specific genes, and homology searches of the TIGR type 4 *S. pneumoniae* genome sequence (26b). Bracketed genes are closely linked, in the order shown. Where known, their location and orientation on the chromosome are indicated by an arrow. The relative positions of genes localized to the same fragment but contained in different brackets or separated by a comma are not known. Alternative gene names are shown in parentheses. The size of the genome has been estimated to be from 2,034 kb (34) to 2,272 kb (19). *lytA* was originally mapped to SmaI fragment number 9 (19) but was later reported to be on fragment 10 (40). *stg-rF* and *rif-23*, which are linked to *lytA*, are therefore also shown on fragment 10. Additional cell wall biosynthetic genes are located adjacent to *penA* and *ddlA* on fragment 8. *dnaX* was reported to be on fragment 15 (34) but is adjacent to *dacA* in the type 4 genome sequence and is therefore shown on fragment 1. It is possible that the genome arrangements may prove different among strains.

strains transform more efficiently when competence is induced using supernatants from competent Rx1 cultures than when using the appropriate synthetic CSP. These observations suggest that other factors are also important in the induction of competence and the ability of recipients to be transformed.

The induction of competence leads to a transient restriction of most cellular protein synthesis, with high-level expression of about 14 proteins expected to be involved in transformation (38). Competence develops through a quorum-sensing mechanism in which CSP serves as the pheromone signal. Recent genetic studies have characterized some of the events and genetic loci involved in detecting and responding to the CSP signal (summarized in Fig. 2). Export of CSP is mediated by an ABC transporter encoded by the *comAB* locus, which is not closely linked to *comC* (25). *comC* is the first gene in an operon with *comD* and *comE*, which encode the sensor and response regulator components, respectively, of a two-component signal transduction system (46). In *Streptococcus gordonii*, which contains homologs of ComCDE and undergoes com-

petence in a manner similar to *S. pneumoniae*, ComD has been shown to be the CSP receptor (24). It likely serves the same function in *S. pneumoniae*. Expression of *comCDE* is enhanced approximately 40-fold upon addition of exogenous CSP. Maximal expression occurs within 5 min and returns to near basal levels by 20 min. Loss of *comE* abolishes competence induction and *comCDE* expression. An activated ComE may be involved in regulating the expression of *comCDE* and other competence-related genes. ComE apparently binds upstream of *comC*, and a number of loci (*rec, cil, cel, cgl*, and *coi*) known to be induced by CSP and required for transformation are preceded by a common sequence that may function as a competence-specific promoter (3, 9, 46, 47). It is not yet known whether ComE acts, either directly or indirectly, to regulate expression of these loci.

Other genes or loci necessary for specific steps in transformation have been described and are discussed in the appropriate sections below. Mutant analyses have also identified several additional loci whose precise points of involvement are not entirely clear. Adc mutants transform

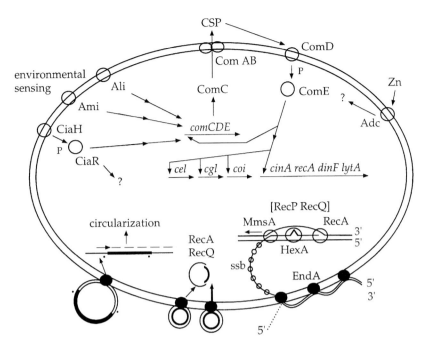

FIGURE 2 Transformation. The model is based on information discussed in the text. The putative regulation of gene expression may be either direct or indirect. Phosphorylation of response regulator molecules by their cognate histidine kinase sensors is indicated by P. Uptake and integration of linear DNA is shown in the lower right portion, with monomer and dimer plasmid uptake shown in the central and left lower portions, respectively. Plasmid uptake will occur by the same mechanism as linear DNA uptake. For monomers, complementary overlapping strands will pair, and DNA synthesis will complete the double-stranded circular molecule. For the dimer, a small fragment of the complementary strand can serve as a primer for DNA synthesis, and circularization can occur as described in the text. Dots on the dimer molecule indicate homologous sites.

poorly unless Zn is added, and the addition of CSP in the absence of Zn does not induce competence. Therefore, the effect is likely downstream of induction and may be related to transmission of the signal from ComD to ComE (16). Point mutations in *ciaH* result in no CSP being exported and a failure to respond to exogenously added CSP. In contrast, insertion mutations in either *ciaH* or *ciaR* do not affect competence. CiaH and CiaR are the sensor and response regulator components, respectively, of a two-component signal transduction system with homology to the PhoB/R system of *Escherichia coli* (as well as other sensor/response pairs). The point mutations apparently result in constitutive phosphorylation of CiaH and, consequently, constitutive activity of CiaR. Thus, activated CiaR appears to repress competence expression. It is not known whether the failure of the constitutive mutants to respond to exogenous CSP results from CiaR repression of gene expression at loci other than *comCDE*, or if ComD and ComE simply are not present owing to complete repression at this locus (22). Oligopeptide permeases may also be involved in sensing the environment. Mutations in the *ami* and *ali* (also referred to as *plp*) loci, and especially those in more than one locus, affect the spontaneous development of competence but do not alter the response to exogenous CSP induction (3, 45).

DNA Binding and Uptake

Double-stranded DNA can bind to competent cells at multiple locations (50 binding sites have been estimated), where it is subjected to single-strand nicks. A constitutively expressed membrane-associated endonuclease (EndA) converts the nicks into double-strand cuts, resulting in fragments with a median length of about 7 kb (53). Nucleolytic digestion of one strand occurs in the 5' to 3' direction, while the other strand is taken into the cell in the 3' to 5' direction (36). Binding and subsequent uptake of a single large molecule may occur at multiple independent sites. Hence, even though digestion results in fragments of relatively small size, linkage of markers hundreds of kilobases apart can be detected using carefully prepared DNA. The average small size also does not preclude the uptake and integration of large DNA fragments, or of large heterologous regions flanked by sequences homologous to the recipient chromosome, as evidenced by the exchange of genetic cassettes encoding the capsular polysaccharide biosynthetic enzymes (reviewed in reference 71).

It has been proposed that EndA is part of a DNA uptake complex (53). Although the specific components of such a complex have not been identified, several genes induced by CSP and required for efficient transformation encode predicted products homologous to proteins involved in DNA binding, uptake, and the assembly of DNA transport machinery. These genes are contained in the loci referred to as *cel* or *cilE*, *cgl* or *cilD*, and *cilB* (9, 47).

Chromosomal Integration

Once inside the cell, the DNA is contained in an eclipse complex, where it is protected from cellular nucleases by interaction with a 19.5-kDa single-strand binding protein and no longer has transforming activity (70). Homologous

other species of streptococci present in the oral cavity (67). Temperate phage appear to be present in the majority of pneumococcal isolates and occur in a wide variety of capsular serotypes (6). However, as shown using isogenic derivatives of encapsulated and nonencapsulated strains, the presence of the capsule prevents infection by virulent ω phages under laboratory conditions (7).

Properties of some of the better-characterized pneumococcal phage are shown in Table 2. Some important similarities between many of the phages have been noted (reviewed in reference 18). Cp-1 and HB-3, as well as their related phages, contain a terminal protein covalently linked to the 5′ end of each DNA strand. In the case of Cp-1, and possibly the other phage, the terminal protein serves as a primer in initiation of DNA replication for the linear molecule. The mechanism is like that found in adenovirus and *Bacillus subtilis* phage φ29, with which Cp-1 has structural similarity. As in these other viruses, the DNA of Cp-1 and its related phage contains inverted terminal repeat structures. A number of the pneumococcal phages are also related by the structural organization of their lytic enzymes. Cp-1, Cp-9, Dp-1, and HB-3 require choline in the pneumococcal cell wall for adsorption and cell lysis. The lytic enzyme in each phage contains a C-terminal choline-binding domain that is also present in the pneumococcal autolysin and numerous other choline-binding proteins of S. pneumoniae. The N-terminal region of HB-3 encodes an amidase that is virtually identical to that of the host autolysin, whereas the same region in Cp-1 and Cp-9 encodes a lysozyme. In Dp-1, a hydrolase similar to that found in the *Lactococcus lactis* phage BK5-T is encoded. The lysozyme found in Cp-1 and Cp-9 is also encoded by the N-terminal region of the Cp-7 lytic enzyme, but the C-terminal region of this protein does not encode a choline-binding domain, and hence this phage does not require choline-containing cell walls for infection. These enzymes represent examples of modular evolution.

THE *Dpn* RESTRICTION SYSTEM

Most S. pneumoniae isolates contain the genes necessary for expression of either the *Dpn*I or *Dpn*II restriction system.

The genes for each of these systems are located at the same site in the respective chromosomes, thus forming genetic cassettes that can be exchanged as units between strains (27). The *Dpn*I endonuclease (DpnC) cleaves double-stranded DNA containing the methylated sequence 5′-GmeATC-3′, whereas the *Dpn*II endonuclease (DpnB) cleaves unmethylated DNA of the same sequence. In addition to their respective endonucleases, the *Dpn*I gene cassette encodes a second protein (DpnD) of unknown function, while the *Dpn*II cassette encodes the adenine methyltransferases, DpnA and DpnM, that are necessary for methylating the host DNA to prevent restriction (55). DpnM acts only on double-stranded DNA, whereas DpnA can methylate both double- and single-stranded DNA (10).

The major function of the *Dpn* system is to protect the cell against entry of phage DNA. Neither of the *Dpn* endonucleases attacks single-stranded DNA or hemimethylated double-stranded DNA. Thus, natural transformation of chromosomal DNA is not affected. In contrast, plasmid transformation is affected to some extent, with an approximate 50% reduction observed for transfers between strains of differing *Dpn* types as compared with transfers between strains of the same type (10). The lack of complete restriction of methylated plasmids by *Dpn*I cells apparently results from the mechanism of plasmid uptake, which can give rise to large segments of hemimethylated DNA following DNA synthesis to restore both strands of the incoming plasmid. Complete restriction of nonmethylated plasmids by the *Dpn*II system is prevented by the action of DpnA, which methylates incoming single-stranded DNA. In the absence of such methylation, restriction of the newly restored plasmid that is not methylated on either strand apparently occurs faster than methylation of the double-stranded DNA by DpnM. Single-stranded, unmethylated plasmid DNA introduced by conjugation into *Dpn*II cells is also severely restricted, presumably because synthesis of the complementary strand by enzymes transferred with the DNA occurs more rapidly than protective methylation of the single strand by DpnA (10). The introduction of plasmids by electroporation, or other methods that involve the uptake of double-stranded DNA, is also strongly affected by the *Dpn* system (29).

TABLE 2 Properties of some S. pneumoniae bacteriophage[a]

Phage	Structure	DNA	% GC[b]	Burst size	N-terminal[c]	C-terminal[d]
ω8 (virulent)	Octahedral hexagonal head (60 nm) Flexible tail (180 × 10 nm) Tail fiber (90 nm)	50 kb, ds, linear	49	100		
Cp-1 (virulent)	Irregular hexagonal head (60 × 45 nm) Tail (20 × 15 nm) Neck appendages Head fibers	19.3 kb, ds, linear terminal protein	42	10	Lysozyme	Choline
Dp-1 (virulent)	Polyhedral head (67 nm) Tail (155 nm) Lipid envelope	57 kb, ds, linear	27	100	Amidase	Choline
HB-3 (temperate)	Head (65 nm) Tail (156 × 1 nm)	40 kb, ds, linear terminal protein			Amidase	Choline

[a] Information is from citations in the text and references therein.
[b] The GC content of the S. pneumoniae chromosome is approximately 39%.
[c] Activity of the N-terminal domain of the lytic enzyme.
[d] Substrate (choline) bound by the C-terminal domain of the lytic enzyme.

PROMOTERS

The existence of extended −10 sites in promoter sequences was first noted in the *Dpn*II locus and was subsequently found to be relatively common in *S. pneumoniae*, in contrast to the relatively rare occurrence of such sequences in *E. coli*. The promoter of the *Dpn*II operon contains an extended −10 site (TATGGTATAAT) but does not have a −35 site. Analysis of 35 additional known or putative *S. pneumoniae* promoters revealed that 60% had the complete extended −10 consensus sequence (TNTGNTATAAT) and 71% had at least the TG motif of the extension. None lacked a −35 site, and those lacking a −10 extension tended to have −35 sequences close to the *E. coli* consensus (55). Numerous *S. pneumoniae* sequences exhibit promoter activity in *E. coli* but not in *S. pneumoniae*, apparently as a result of the AT richness of the *S. pneumoniae* DNA (approximately 39% GC) and of more stringent sequence recognition requirements of the *S. pneumoniae* RNA polymerase. Such requirements could be fulfilled by the extended −10 sequences. Excessive promoter activity of *S. pneumoniae* DNA in *E. coli* has been used to explain difficulties encountered in cloning pneumococcal sequences. While it is clear that difficulties exist, sequences with promoter activity strong enough to destabilize cloning vectors appear to be rare in the *S. pneumoniae* chromosome (14). The expression of toxic products, the use of large fragments or high-copy-number plasmids, and sequence-specific effects unrelated to promoter activity appear to be more likely causes.

INSERTION SEQUENCES AND REPETITIVE ELEMENTS

Four distinct insertion sequences have been identified in the chromosomes of *S. pneumoniae* isolates. Their properties are summarized in Table 3. The BOX elements, many of which are located near numerous genes involved in transformation and virulence, represent another group of pneumococcal repetitive sequences (31). Three subunits of 59, 45, and 50 bp (box A, box B, and box C, respectively) compose the element, although not all are present in every copy. As many as 25 copies have been detected in some strains. Although the role of the BOX elements is not known, their location has prompted the suggestion that they may be involved in coordinate regulation of competence and virulence gene expression.

Recent analyses of the capsule loci of *S. pneumoniae* have found them to be replete with insertion elements, repetitive sequences, and partial gene fragments (71). Copies of IS1167 flank the type 1 genes, a partial copy is located downstream of the type 14 locus, and a derivative comprising only an internal fragment (*tnpA*) with about 50% homology is found within the type 3 locus. A partial transposase sequence is also located upstream of the type 14 locus, and a partial sequence with homology to the insertion element-like H-repeat sequences of *E. coli* is present in the type 3 locus. IS1202 is located upstream of the type 19F locus, and its recognition sequence is present in a similar site in the type 3 locus. In addition, a highly conserved 115-bp sequence found upstream of the genes encoding hyaluronidase, neuraminidases A and B, and penicillin-binding protein 3 is present upstream of the type 1, 3, 14, and 19F capsule loci. The capsule loci are not unlike the chromosome as a whole, in which numerous repetitive elements and transposases are apparent both experimentally, as noted above, and through inspection of the *S. pneumoniae* genome sequence.

THE GENERATION AND ANALYSIS OF MUTANTS

The ability of *S. pneumoniae* to be naturally transformed makes possible a number of techniques that are not always easily performed in other bacteria. That same property, however, demands that special care be taken in the conclusions drawn from mutagenesis experiments. As noted above, the frequency of spontaneous mutations increases during transformation with homologous DNA, and mutations other than those desired are often observed to occur. In addition, a large amount of DNA that is both linked and unlinked to the gene of interest can be incorporated during chromosomal transformations. Consequently, before drawing conclusions regarding the effect of the intended mutation, it is essential either to observe a consistent phenotype among multiple, independently isolated mutants or to specifically repair the known mutation and observe reversion to the parental phenotype. In the case of chromosomal transformations between nonisogenic strains, mutants are backcrossed several times to the parent strain to reduce the amount of extraneous donor DNA.

The most frequently used mutagenesis technique in *S. pneumoniae* involves insertion-duplication. In this procedure, a fragment of *S. pneumoniae* DNA is cloned into a plasmid that is able to replicate in *E. coli* but not *S. pneumoniae*. The clone can thus be propagated in *E. coli* and used to transform *S. pneumoniae*, where homologous recombination into the chromosome at the target site is necessary to select for an antibiotic resistance encoded by the plasmid vector. Target sequences of less than 200 bp can be used successfully, but larger fragments increase the frequency of transformation, owing in part to the higher likelihood of digestion of the surface-bound DNA within the cloned fragment (see Plasmid Transformation, above). As noted above, the use of RecA⁺ *E. coli* strains can also be used to enhance transformation frequencies, if recombination within the cloned fragment is not expected to be

TABLE 3 Insertion sequences found in *S. pneumoniae*

Element[a]	Size (bp)	Terminal inverted repeats	Target (bp)	Copies	%GC	Distribution[b]	Reference
IS1202	1,747	23 bp, imperfect	27	1–5	39.5	Mainly type 19	37
IS1167	1,435	24 bp, imperfect	8	3–12	38	Multiple, 11/22	73
IS1381	846	20 bp, imperfect	7	5–7	38	Multiple, 8/8	56
IS1515	871	12 bp, perfect	3	2–13	39	Type 1, 17/17; others, 3/20	42

[a] Each element contains an open reading frame or frames with homology to transposase sequences.
[b] The capsular serotypes of strains in which the element occurs; numbers indicate the fraction of tested strains containing the element.

an issue. Transformation and subsequent integration into the *S. pneumoniae* chromosome will result in gene disruption if the cloned fragment is an internal portion of a gene, or restoration of the entire gene if the fragment overlaps one end of the gene (Fig. 3). In the latter case, a mutant phenotype should not be observed, unless the target sequence is part of an operon. Thus, a convenient and essential control when using an internal fragment for gene disruption is to use a 3′-overlapping fragment that restores the complete gene and should, in the absence of polar effects, have the parental phenotype. Integration of the clone results in a duplication of the target sequence. Consequently, precise excision of insertions occurs at low frequency, and resolved plasmids can be recovered by transformation into *E. coli*. Unlike many clones containing large *S. pneumoniae* DNA fragments, those containing fragments of the size necessary for generating insertion-duplication mutations are generally stable in *E. coli*. Hence, libraries can be used to mutagenize the *S. pneumoniae* chromosome, and unknown target fragments can easily be cloned, as can surrounding DNA following digestion, ligation, and transformation to *E. coli*, using the integration vector for replication. Despite the ability of the insertions to resolve, they are relatively stable, and resolution usually occurs at a frequency of <1/10⁵. Thus, reversion is not generally apparent in the absence of a strong selection for the revertant phenotype.

Insertion-duplication using vectors containing reporter genes, including *cat* and *lacZ*, has been used successfully to analyze gene expression in *S. pneumoniae* (2, 9, 15, 47). In the case of the latter, high endogenous levels of β-galactosidase make it necessary to isolate LacZ mutants for expression studies. Integration with vectors permitting PhoA translational fusions is used to study exported proteins (45). Most of the vectors used for insertion-duplication, as well as most replicating vectors, encode erythromycin resistance for selection in both *S. pneumoniae* and *E. coli* (2, 9, 15, 47, 72). Other commonly used antibiotic markers are chloramphenicol, kanamycin, spectinomycin, and tetracycline (47, 64, 65). Plasmids carrying β-lactamase-encoding genes, whether expressed or not, should not be used because naturally occurring resistance due to β-lactamase has not yet appeared in *S. pneumoniae*.

Transposon mutagenesis has been used with limited success in *S. pneumoniae*, but a promising approach involves the recently described mariner in vitro mutagenesis technique (1). Here, linear DNA is mutagenized in vitro using the *Himar*1 transposase and a minitransposon containing an antibiotic resistance. Gaps generated at the insertion site are repaired, and the linear fragments are transformed to *S. pneumoniae* with selection for the antibiotic marker. Site specificity is minimal, and saturation mutagenesis appears possible.

Mapping of point or other small chromosomal mutations utilizes cotransformation with known selectable markers for linkage analyses. As noted earlier, linkage can be observed with markers that are hundreds of kilobases apart owing to binding and uptake of a single DNA molecule at multiple sites on the cell surface. More precise mapping, even in the absence of a linked marker, can be achieved using restriction enzyme-digested chromosomal DNA and demonstrating that transfer of the mutant or linked phenotype occurs with smaller fragments. Electrophoretic separation of the restricted DNA allows individual fractions, either purified or contained in melted slices from a low-gelling temperature agarose gel, to be transformed directly to *S. pneumoniae*. The small number of fragments contained in the fraction can then usually be cloned and tested individually for the ability to transfer the phenotype. As noted earlier, cloned fragments normally yield high

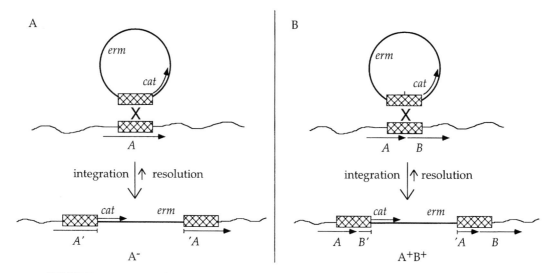

FIGURE 3 Insertion-duplication mutagenesis and restoration. (A) The effect of using an internal gene fragment to direct insertion of a nonreplicating plasmid into the chromosome. Duplication of the target fragment occurs, and the gene is disrupted by the plasmid insertion, resulting in an insertion-duplication mutation. (B) The target fragment overlaps the ends of two genes. Insertion results in duplication of the target fragment, but both genes are completely reconstructed and the result is an insertion-duplication restoration. Both genes should be functional, unless they form part of an operon, in which case the plasmid insertion would be polar on the downstream gene. The figure shows a selectable erythromycin-resistance gene (*erm*) and a promoterless chloramphenicol-resistance reporter gene (*cat*), as discussed in reference 15.

transformation frequencies owing to the large concentration of homogeneous DNA. They can be transformed either as linear restriction fragments or as the intact clone without selection for the antibiotic marker. In the latter case, the double crossover event to introduce or repair the mutation in the recipient chromosome occurs about 500-fold more frequently than the single crossover event necessary to integrate the plasmid and cause gene disruption. Fragments of less than 200 bp can be used for mapping, but efficiency increases with increasing fragment size. PCR fragments can similarly be used as a source of homogeneous DNA for transformation mapping.

I am indebted to Walter Guild, Sandy Lacks, Don Morrison, and Alex Tomasz, whose encouragement and freely shared knowledge helped turn a brief adventure in pneumococcal genetics into what has surely become a lifelong endeavor. Work in my laboratory is supported by Public Health Service grants AI28457 and GM53017 from the National Institutes of Health.

REFERENCES

1. **Akerley, B. J., E. J. Rubin, A. Camilli, D. J. Lampe, H. M. Robertson, and J. J. Mekalanos.** 1998. Systematic identification of essential genes by *in vitro mariner* mutagenesis. *Proc. Natl. Acad. Sci. USA* **95:**8927–8932.

2. **Alloing, G., C. Granadel, D. A. Morrison, and J. P. Claverys.** 1996. Competence pheromone, oligopeptide permease, and induction of competence in *Streptococcus pneumoniae. Mol. Microbiol.* **21:**471–478.

3. **Alloing, G., B. Martin, C. Granadel, and J. P. Claverys.** 1998. Development of competence in *Streptococcus pneumoniae*: pheromone autoinduction and control of quorum sensing by the oligopeptide permease. *Mol. Microbiol.* **29:**75–83.

4. **Avery, O. T., C. M. MacLeod, and M. McCarty.** 1944. Studies on the chemical nature of the substance inducing transformation of pneumococcal types. Induction of transformation by a desoxyribonucleic acid fraction isolated from pneumococcus Type III. *J. Exp. Med.* **79:**137–158.

5. **Ayoubi, P., A. O. Kilic, and M. N. Vijayakumar.** 1991. Tn5253, the pneumococcal Ω(*cat tet*) BM6001 element, is a composite structure of two conjugative transposons, Tn5251 and Tn5252. *J. Bacteriol.* **173:**1617–1622.

6. **Bernheimer, H. P.** 1979. Lysogenic pneumococci and their bacteriophages. *J. Bacteriol.* **138:**618–624.

7. **Bernheimer, H. P., and J. G. Tiraby.** 1976. Inhibition of phage infection by pneumococcus capsule. *Virology* **73:**308–309.

8. **Buu-Hoi, A., and T. Horodniceanu.** 1980. Conjugative transfer of multiple antibiotic resistance markers in *Streptococcus pneumoniae. J. Bacteriol.* **143:**313–320.

9. **Campbell, E. A., S. Y. Choi, and H. R. Masure.** 1998. A competence regulon in *Streptococcus pneumoniae* revealed by genomic analysis. *Mol. Microbiol.* **27:**929–939.

10. **Cerritelli, S., S. S. Springhorn, and S. A. Lacks.** 1989. DpnA, a methylase for single-strand DNA in the *Dpn*II restriction system, and its biological function. *Proc. Natl. Acad. Sci. USA* **86:**9223–9227.

11. **Chen, J. D., and D. A. Morrison.** 1987. Modulation of competence for genetic transformation in *Streptococcus pneumoniae. J. Gen. Microbiol.* **133:**1959–1967.

12. **Claverys, J. P., and S. A. Lacks.** 1986. Heteroduplex deoxyribonucleic acid base mismatch repair in bacteria. *Microbiol. Rev.* **50:**133–165.

13. **Courvalin, P., and C. Carlier.** 1986. Transposable multiple antibiotic resistance in *Streptococcus pneumoniae. Mol. Gen. Genet.* **205:**291–297.

13a. **Dillard, J. P., M. W. Vandersea, and J. Yother.** 1995. Characterization of the cassette containing genes for type 3 capsular polysaccharide biosynthesis in *Streptococcus pneumoniae. J. Exp. Med.* **181:**973–983.

14. **Dillard, J. P., and J. Yother.** 1991. Analysis of *Streptococcus pneumoniae* sequences cloned into *Escherichia coli*: effect of promoter strength and transcription terminators. *J. Bacteriol.* **173:**5105–5109.

15. **Dillard, J. P., and J. Yother.** 1994. Genetic and molecular characterization of capsular polysaccharide biosynthesis in *Streptococcus pneumoniae* type 3. *Mol. Microbiol.* **12:**959–972.

16. **Dintilhac, A., G. Alloing, C. Granadel, and J. P. Claverys.** 1997. Competence and virulence of *Streptococcus pneumoniae*: Adc and PsaA mutants exhibit a requirement for Zn and Mn resulting from inactivation of putative ABC metal permeases. *Mol. Microbiol.* **25:**727–739.

17. **Engel, H. W., N. Soedirman, J. A. Rost, W. J. van Leeuwen, and J. D. van Embden.** 1980. Transferability of macrolide, lincomycin, and streptogramin resistances between group A, B, and D streptococci, *Streptococcus pneumoniae*, and *Staphylococcus aureus. J. Bacteriol.* **142:**407–413.

18. **Garcia, P., A. C. Martin, and R. Lopez.** 1997. Bacteriophages of *Streptococcus pneumoniae*: a molecular approach. *Microb. Drug Resist.* **3:**165–176.

19. **Gasc, A. M., L. Kauc, P. Barraille, M. Sicard, and S. Goodgal.** 1991. Gene localization, size, and physical map of the chromosome of *Streptococcus pneumoniae. J. Bacteriol.* **173:**7361–7367.

20. **Griffith, F.** 1928. The significance of pneumococcal types. *J. Hyg.* **27:**113–159.

21. **Grist, R. W., and L. O. Butler.** 1983. Effect of transforming DNA on growth and frequency of mutation of *Streptococcus pneumoniae. J. Bacteriol.* **153:**153–162.

22. **Guenzi, E., A. M. Gasc, M. A. Sicard, and R. Hakenbeck.** 1994. A two-component signal-transducing system is involved in competence and penicillin susceptibility in laboratory mutants of *Streptococcus pneumoniae. Mol. Microbiol.* **12:**505–515.

23. **Hall, M., and S. Matson.** 1999. The *Escherichia coli* MutL protein physically interacts with MutH and stimulates the MutH-associated endonuclease activity. *J. Biol. Chem.* **274:**1306–1312.

24. **Havarstein, L. S., P. Gaustad, I. F. Nes, and D. A. Morrison.** 1996. Identification of the streptococcal competence-pheromone receptor. *Mol. Microbiol.* **21:**863–869.

25. **Hui, F. M., L. Zhou, and D. A. Morrison.** 1995. Competence for genetic transformation in *Streptococcus pneumoniae*: organization of a regulatory locus with homology to two lactococcin A secretion genes. *Gene* **153:**25–31.

26. **Humbert, O., M. Prudhomme, R. Hakenbeck, C. G. Dowson, and J. P. Claverys.** 1995. Homeologous recombination and mismatch repair during transformation in *Streptococcus pneumoniae*: saturation of the Hex mismatch repair system. *Proc. Natl. Acad. Sci. USA* **92:**9052–9056.

26a. **Ianelli, F., B. J. Pearce, and G. Pozzi.** 1999. The type 2 capsule locus of *Streptococcus pneumoniae. J. Bacteriol.* **181:**2652–2654.

26b. **Institute for Genomic Research.** http://www.tigr.org and http://www.ncbi.nlm.nih.gov

27. **Lacks, S. A., B. M. Mannarelli, S. S. Springhorn, and B. Greenberg.** 1986. Genetic basis of the complementary DpnI and DpnII restriction systems of S. pneumoniae: an intercellular cassette mechanism. *Cell* **46:**993–1000.

28. **Lefrancois, J., M. M. Samrakandi, and A. M. Sicard.** 1998. Electrotransformation and natural transformation of *Streptococcus pneumoniae*: requirement of DNA processing for recombination. *Microbiology* **144:**3061–3068.

29. **Lefrancois, J., and A. M. Sicard.** 1997. Electrotransformation of *Streptococcus pneumoniae*: evidence for restriction of DNA on entry. *Microbiology* **143:**523–526.

30. **Martin, B., P. Garcia, M.-P. Castanië, and J.-P. Claverys.** 1995. The recA gene of *Streptococcus pneumoniae* is part of a competence-induced operon and controls lysogenic induction. *Mol. Microbiol.* **15:**367–379.

31. **Martin, B., O. Humbert, M. Camara, E. Guenzi, J. Walker, T. Mitchell, P. Andrew, M. Prudhomme, G. Alloing, R. Hakenbeck, D. A. Morrison, G. J. Boulnois, and J.-P. Claverys.** 1992. A highly conserved repeated DNA element located in the chromosome of *Streptococcus pneumoniae*. *Nucleic Acids Res.* **20:**3479–3483.

32. **Martin, B., G. J. Sharples, O. Humbert, R. G. Lloyd, and J. P. Claverys.** 1996. The mmsA locus of *Streptococcus pneumoniae* encodes a RecG-like protein involved in DNA repair and in three-strand recombination. *Mol. Microbiol.* **19:**1035–1045.

33. **Masure, H., B. Pearce, H. Shio, and B. Spellerberg.** 1998. Membrane targeting of RecA during genetic transformation. *Mol. Microbiol.* **27:**845–852.

34. **Matsushima, P., and R. Baltz.** 1996. Physical mapping of partial DNA sequences from the *Streptococcus pneumoniae* R6 genome, abstr. H-63, p. 494. *Abstr. 96th Gen. Meet. Am. Soc. Microbiol. 1996.* American Society for Microbiology, Washington, D.C.

35. **McDonnell, M., C. Ronda-Lain, and A. Tomasz.** 1975. "Diplophage": a bacteriophage of *Diplococcus pneumoniae*. *Virology* **63:**577–582.

36. **Mejean, V., and J. P. Claverys.** 1993. DNA processing during entry in transformation of *Streptococcus pneumoniae*. *J. Biol. Chem.* **268:**5594–5599.

37. **Morona, J. K., A. Guidolin, R. Morona, D. Hansman, and J. C. Paton.** 1994. Isolation, characterization, and nucleotide sequence of IS1202, an insertion sequence of *Streptococcus pneumoniae*. *J. Bacteriol.* **176:**4437–4443.

38. **Morrison, D. A., and M. F. Baker.** 1979. Competence for genetic transformation in pneumococcus depends on synthesis of a small set of proteins. *Nature* **282:**215–217.

39. **Morrison, D. A., S. A. Lacks, W. R. Guild, and J. M. Hageman.** 1983. Isolation and characterization of three new classes of transformation-deficient mutants of *Streptococcus pneumoniae* that are defective in DNA transport and genetic recombination. *J. Bacteriol.* **156:**281–290.

40. **Mortier-Barriere, I., A. de Saizieu, J. P. Claverys, and B. Martin.** 1998. Competence-specific induction of recA is required for full recombination proficiency during transformation in *Streptococcus pneumoniae*. *Mol. Microbiol.* **27:**159–170.

41. **Muckerman, C. C., S. S. Springhorn, B. Greenberg, and S. A. Lacks.** 1982. Transformation of restriction endonuclease phenotype in *Streptococcus pneumoniae*. *J. Bacteriol.* **152:**183–190.

42. **Munoz, R., R. Lopez, and E. Garcia.** 1998. Characterization of IS1515, a functional insertion sequence in *Streptococcus pneumoniae*. *J. Bacteriol.* **180:**1381–1388.

43. **Ottolenghi, E., and R. Hotchkiss.** 1960. Appearance of genetic transforming activity in pneumococcal cultures. *Science* **132:**1257–1258.

44. **Pearce, B., A. Naughton, E. Campbell, and H. Masure.** 1995. The rec locus, a competence-induced operon in *Streptococcus pneumoniae*. *J. Bacteriol.* **177:**86–93.

45. **Pearce, B. J., A. M. Naughton, and H. R. Masure.** 1994. Peptide permeases modulate transformation in *Streptococcus pneumoniae*. *Mol. Microbiol.* **12:**881–892.

46. **Pestova, E. V., L. S. Havarstein, and D. A. Morrison.** 1996. Regulation of competence for genetic transformation in *Streptococcus pneumoniae* by an auto-induced peptide pheromone and a two-component regulatory system. *Mol. Microbiol.* **21:**853–862.

47. **Pestova, E. V., and D. A. Morrison.** 1998. Isolation and characterization of three *Streptococcus pneumoniae* transformation-specific loci by use of a lacZ reporter insertion vector. *J. Bacteriol.* **180:**2701–2710.

48. **Porter, R. D., and W. R. Guild.** 1976. Characterization of some pneumococcal bacteriophages. *J. Virol.* **19:**659–667.

49. **Porter, R. D., N. B. Shoemaker, G. Rampe, and W. R. Guild.** 1979. Bacteriophage-associated gene transfer in pneumococcus: transduction or pseudotransduction? *J. Bacteriol.* **137:**556–567.

50. **Poyart-Salmeron, C., P. Trieu-Cuot, C. Carlier, and P. Courvalin.** 1990. The integration-excision system of the conjugative transposon Tn1545 is structurally and functionally related to those of lambdoid phages. *Mol. Microbiol.* **4:**1513–1521.

51. **Pozzi, G., L. Masala, F. Iannelli, R. Manganelli, L. S. Havarstein, L. Piccoli, D. Simon, and D. A. Morrison.** 1996. Competence for genetic transformation in encapsulated strains of *Streptococcus pneumoniae*: two allelic variants of the peptide pheromone. *J. Bacteriol.* **178:**6087–6090.

52. **Prudhomme, M., B. Martin, V. Mejean, and J. P. Claverys.** 1989. Nucleotide sequence of the *Streptococcus pneumoniae* hexB mismatch repair gene: homology of HexB to MutL of *Salmonella typhimurium* and to PMS1 of *Saccharomyces cerevisiae*. *J. Bacteriol.* **171:**5332–5338.

53. **Puyet, A., B. Greenberg, and S. A. Lacks.** 1990. Genetic and structural characterization of endA. A membrane-bound nuclease required for transformation of *Streptococcus pneumoniae*. *J. Mol. Biol.* **213:**727–738.

54. **Reizer, J., A. Reizer, A. Bairoch, and M. H. Saier, Jr.** 1993. A diverse transketolase family that includes the RecP protein of *Streptococcus pneumoniae*, a protein implicated in genetic recombination. *Res. Microbiol.* **144:**341–347.

55. **Sabelnikov, A. G., B. Greenberg, and S. A. Lacks.** 1995. An extended −10 promoter alone directs transcription of the DpnII operon of *Streptococcus pneumoniae*. *J. Mol. Biol.* **250:**144–155.

56. **Sanchez-Beato, A. R., E. Garcia, R. Lopez, and J. L. Garcia.** 1997. Identification and characterization of IS1381, a new insertion sequence in *Streptococcus pneumoniae*. *J. Bacteriol.* **179:**2459–2463.

57. **Saunders, C. W., and W. R. Guild.** 1981. Monomer plasmid DNA transforms *Streptococcus pneumoniae*. *Mol. Gen. Genet.* **181:**57–62.

58. Schuster, C., M. van der Linden, and R. Hakenbeck. 1998. Small cryptic plasmids of *Streptococcus pneumoniae* belong to the pC194/pUB110 family of rolling circle plasmids. *FEMS Microbiol Lett.* **164:**427–431.

59. Scott, J., and G. Churchward. 1995. Conjugative transposition. *Annu. Rev. Microbiol.* **49:**367–397.

60. Shoemaker, N. B., M. D. Smith, and W. R. Guild. 1979. Organization and transfer of heterologous chloramphenicol and tetracycline resistance genes in pneumococcus. *J. Bacteriol.* **139:**432–441.

61. Sibold, C., Z. Markiewicz, C. Latorre, and R. Hakenbeck. 1991. Novel plasmids in clinical strains of *Streptococcus pneumoniae*. *FEMS Microbiol Lett.* **61:**91–95.

62. Smith, M. D., and W. R. Guild. 1979. A plasmid in *Streptococcus pneumoniae*. *J. Bacteriol.* **137:**735–739.

63. Smith, M. D., N. B. Shoemaker, V. Burdett, and W. R. Guild. 1980. Transfer of plasmids by conjugation in *Streptococcus pneumoniae*. *Plasmid* **3:**70–79.

64. Stassi, D. L., P. Lopez, M. Espinosa, and S. A. Lacks. 1981. Cloning of chromosomal genes in *Streptococcus pneumoniae*. *Proc. Natl. Acad. Sci. USA* **78:**7028–7032.

65. Tao, L., D. J. LeBlanc, and J. J. Ferretti. 1992. Novel streptococcal-integration vectors for gene cloning and inactivation. *Gene* **120:**105–110.

66. Tiraby, G., M. S. Fox, and H. Bernheimer. 1975. Marker discrimination in deoxyribonucleic acid-mediated transformation of various *Pneumococcus* strains. *J. Bacteriol.* **121:**608–618.

67. Tiraby, J. G., E. Tiraby, and M. S. Fox. 1975. Pneumococcal bacteriophages. *Virology* **68:**566–569.

68. Tomasz, A., and R. D. Hotchkiss. 1964. Regulation of the transformability of pneumococcal cultures by macromolecular cell products. *Proc. Natl. Acad. Sci. USA* **51:**480–487.

69. Vijayakumar, M. N., and S. Ayalew. 1993. Nucleotide sequence analysis of the termini and chromosomal locus involved in site-specific integration of the streptococcal conjugative transposon Tn5252. *J. Bacteriol.* **175:**2713–2719.

70. Vijayakumar, M. N., and D. A. Morrison. 1983. Fate of DNA in eclipse complex during genetic transformation in *Streptococcus pneumoniae*. *J. Bacteriol.* **156:**644–648.

71. Yother, J. 1999. Common themes in the genetics of streptococcal capsular polysaccharides, p. 161–184. *In* J. B. Goldberg (ed.), *Genetics of Bacterial Polysaccharides*. CRC Press, Inc., Boca Raton, Fla.

72. Yother, J., L. S. McDaniel, and D. E. Briles. 1986. Transformation of encapsulated *Streptococcus pneumoniae*. *J. Bacteriol.* **168:**1463–1465.

73. Zhou, L., F. M. Hui, and D. A. Morrison. 1995. Characterization of IS*1167*, a new insertion sequence in *Streptococcus pneumoniae*. *Plasmid* **33:**127–138.

Pneumococcal Vaccines

D. E. BRILES, J. C. PATON, E. SWIATLO, AND M. H. NAHM

24

With the discovery of the pneumococcus in 1881 it became apparent that this gram-positive pathogen was a major cause of serious and often fatal pneumonia (70). It is also a major cause of meningitis and otitis media in children. Pneumococci are the largest cause of community-acquired pneumonia in the developed world. In the United States they cause over 10,000 deaths annually, primarily among the elderly. In the preantibiotic era the pneumococcus was also a major cause of death among younger adults. In the developing world pneumococci are an important cause of childhood deaths due to bacterial respiratory infection following viral disease. Around the world, such infections kill an estimated one million children annually. Within recent years about one-third to one-half of pneumococci recovered from humans have become at least partially resistant to penicillin, and penicillin-resistant strains are frequently also resistant to other common antibiotics. The rise of antibiotic resistance among pneumococci has already complicated treatment, especially of meningitis, and threatens to greatly increase the morbidity and mortality caused by pneumococci unless new control measures are developed.

It has long been recognized that the best management of most infectious disease is prevention. Vaccines offer the prospect of a highly cost-effective means of preventing morbidity and mortality caused by pneumococci. This chapter provides a concise summary of issues critical to the development and application of pneumococcal vaccines. There are several relatively recent reviews that address this topic in more detail (14, 15, 46, 64). In the preantibiotic era, vaccination attempts utilized whole killed pneumococci injected parenterally. Although such vaccines were sometimes protective in humans, they were also highly reactogenic. These killed vaccines were mainly used to elicit antibody in animals for passive treatment of infected humans (24). With the realization that much of the protective antibody was directed against the type-specific capsular polysaccharides (PS), most subsequent vaccine attempts focused on the use of mixtures of the isolated polysaccharides to elicit protection (54). More recent developments have utilized protein-polysaccharide conjugates to increase the immunogenicity of the polysaccharides, es-

pecially in children (8, 52, 64). The wide diversity of pneumococcal PS types, the inability of young children to make adequate responses to most soluble polysaccharides, and the complexity (and anticipated high expense) of conjugate vaccines have led to studies of the possibility of using pneumococcal proteins as vaccines.

PNEUMOCOCCAL CAPSULAR POLYSACCHARIDE VACCINE

Soon after the pneumococcal capsule was identified as a major immunogen in the 1930s, subsequent vaccine development focused on purified PS. Pneumococci are now known to possess one of at least 90 different PS types, although pneumococci of less than 30 types cause the vast majority of human disease. Early preparations were successfully used in nursing homes and military recruits during World War II. It had become apparent that immunization with pneumococcal polysaccharide was effective in preventing bacteremic pneumococcal pneumonia, and two hexavalent polysaccharide vaccines were commercially introduced. However, in the late 1940s antibiotics were becoming widely available and the perceived impact of pneumococcal infections had greatly diminished. The two vaccines were withdrawn from the market for lack of demand.

Despite the widespread use of antibiotics and the increasing sophistication of hospital intensive care technology, morbidity and mortality from invasive pneumococcal infections remained high. It was clear that antibiotics alone would not be sufficient to diminish the impact of pneumococcal disease. Renewed interest in pneumococcal vaccines led to further clinical trials of polysaccharide vaccines in adults at high risk for invasive infections. Two large trials, in South Africa and New Guinea, confirmed the efficacy of multivalent polysaccharide vaccines in preventing invasive pneumococcal disease in adults. Based largely on the results of these trials, a 14-valent polysaccharide vaccine was licensed in the United States in 1977. The formulation was expanded to include 23 PS types in 1983,

Gram-Positive Pathogens, ed. by V. A. Fischetti et al.
© 2000 American Society for Microbiology, Washington, D.C.

and this is the currently approved pneumococcal vaccine in the United States.

Following the introduction of the pneumococcal polysaccharide vaccine, clinical trials continued, with mixed results. Some large prospective studies have raised doubts as to the effectiveness of the polysaccharide vaccine (43, 65). Because of ethical problems in denying at-risk subjects an approved and recommended vaccine, large randomized trials of pneumococcal vaccine are difficult to design. Many small case-control trials have shown variable effectiveness of the vaccine, and the current polysaccharide formulation is generally considered to have an efficacy of 60 to 80% in preventing bacteremic pneumococcal infection.

One frequently postulated reason for the incomplete protection afforded by the polysaccharide vaccine is the poor immunogenicity of PS in elderly adults. Response to vaccination varies by age and presence of underlying chronic medical conditions. Certain PS types in the 23-valent vaccine, notably types 6B, 9V, 19F, and 23F, induce relatively weak antibody responses that diminish to prevaccination levels in a short time (62). In addition to quantity, a significant minority of elderly adults have impaired functional antibody responses to certain serotypes, as measured by in vitro avidity and opsonophagocytosis assays (58).

POLYSACCHARIDE-PROTEIN CONJUGATE VACCINES

Children less than 2 years of age have the highest incidence of invasive pneumococcal disease and present a uniquely difficult group to immunize. Young children respond poorly to polysaccharide antigens and are particularly susceptible to infections with encapsulated bacteria. The reasons for this poor immune response are complex and not fully known, but it is clear that children less than 2 years of age do not have a fully developed capacity to respond to many T-cell-independent polysaccharide antigens. Polysaccharides bind and cross-link antigen receptors on the surface of B cells and directly activate B cells without significant involvement from $CD4^+$ T cells. This is not to imply that T cells cannot modulate a polysaccharide antibody response. Antibody responses to polysaccharide antigens in adults can be amplified in an antigen-nonspecific manner by $CD4^+$ T cells; however, the same in vitro observations can be made when neonatal T cells are used with adult B cells. The immunological deficit in young children is generally thought to be immaturity of B cells. Bacterial PS can activate complement via the alternative pathway and generate C3 split products, including C3d. Activation of B cells by polysaccharides is probably dependent on an additional interaction between C3d and complement receptor 2, its cognate ligand. Neonatal B cells express low levels of complement receptor 2, and this may represent the primary defect that makes young children unresponsive to polysaccharide antigens (53). Whatever the underlying mechanisms, children less than 2 years of age do not respond to most of the common pediatric pneumococcal serotypes, and alternative vaccination strategies are needed to protect this susceptible population.

The poor immunogenicity of polysaccharides in the very young and elderly has spurred the development of pneumococcal vaccines that involve T cell responses. Antibody responses to protein antigens are intact in the first year of life and in the elderly. Also, protein antigens induce memory T cells and antibody responses that can be boosted by repeated immunization. An effective means to induce T-dependent responses to polysaccharides is covalent coupling of polysaccharide epitopes to a carrier protein. This strategy has proved highly successful in development of a protein-polysaccharide vaccine for *Haemophilus influenzae* type b. The exact nature of the recruitment of T cells by conjugate vaccines remains debatable; however, a prevailing model suggests that B cells internalize the polysaccharide-protein conjugates, process the conjugate, and present peptide fragments to peptide-specific T cells. Effective conjugate vaccines require physical attachment of polysaccharides to proteins—mixtures of peptides and polysaccharides will not induce T-dependent polysaccharide responses (22). Moreover, B cells primed with conjugates can be subsequently activated (boosted) by exposure to polysaccharides alone and presumably by natural infection (25). These observations have led to the development of pneumococcal conjugate vaccines, and early study results indicate that the conjugate vaccines elicit protective antibodies in children younger than 2 years of age.

The distribution of pneumococcal serotypes in pediatric infections is more narrowly restricted than that of adult infections. Conjugate vaccines are relatively complicated and expensive to manufacture but may be feasible in protecting children in the developed world against pneumococcal infection because of this restricted distribution. Many clinical trials using pneumococcal polysaccharides conjugated to tetanus or diphtheria toxoids, or meningococcal outer membrane proteins, have demonstrated the safety and immunogenicity of pneumococcal conjugate vaccines in very young children (30). In a manner similar to that seen with the *H. influenzae* type b conjugate vaccine, immunization of children with PS vaccine reduced carriage of pneumococcal serotypes included in the vaccine (20, 42). Finally, large-scale clinical trials completed in the United States have demonstrated that conjugate vaccines can greatly reduce the incidence of invasive pneumococcal infection in young children, and these vaccines should soon become available for clinical uses (8).

Conjugate vaccines have also been tested in elderly adults and are safe and immunogenic (51, 63). These studies found that conjugate vaccines were not more immunogenic than the current polysaccharide vaccine. Some of the newer conjugate vaccines may be more effective than the current polysaccharide vaccine in adults. However, the pneumococcal serotypes responsible for invasive infections in adults are more diverse than those in children, and the formulation of conjugate vaccines to cover this larger variety of serotypes may prove to be a developmental and financial hurdle that may be difficult to overcome.

Alternative methods to enhance the immunogenicity of polysaccharides have been investigated, but their development is still in the early stages. Mucosal delivery of serotype 23F polysaccharide with the B subunit of cholera toxin has been shown to induce specific IgA at mucosal surfaces. In contrast, incorporation of type 23F polysaccharide in liposomes has not been shown to enhance immunogenicity in the same system (68). Mucosal immunization with 6B polysaccharide conjugated to tetanus toxoid has been shown to reduce nasopharyngeal carriage in mice (71). Future refinements in delivery of polysaccharide antigens and immunomodulatory compounds will continue to make the pneumococcal capsule a prime target for vaccine-induced immune responses. (See chapter 20, this volume, for more details on pneumococcal capsules.)

PROTEIN VACCINES

Overview

Polyvalent pneumococcal vaccines based on purified PS have been available for two decades, but their clinical efficacy has been limited by poor immunogenicity in high-risk groups (particularly young children). As discussed above, the problem of poor vaccine immunogenicity in children is being addressed by conjugation of the PS to protein carriers, thereby converting the PS from T-cell-independent to T-cell-dependent antigens. However, serotype coverage will be more limited, as it is unlikely that more than 11 serotypes will be included in such conjugate formulations. Nasopharyngeal colonization with *Streptococcus pneumoniae* is thought to be a prerequisite for invasive disease. There are some data from the conjugate vaccine trials indicating that, although carriage of vaccine types was reduced, the vacated niche was promptly occupied by nonvaccine serotypes known to cause invasive disease in humans (42). In addition, the cost of the conjugate vaccines is likely to be high; thus, their use may be low in third world countries where the need for effective pediatric vaccines is the greatest. In view of this, much recent attention has focused on the possibility of developing vaccines based on pneumococcal protein antigens common to all serotypes (46). Such proteins, being T-cell-dependent antigens, are likely to be highly immunogenic in human infants and able to elicit immunological memory. The pneumococcal proteins also have potential as carriers for the PS in the conjugate vaccines.

The discovery of cross-reactive proteins was preceded by studies demonstrating that antibodies directed against the phosphocholine epitopes of teichoic and lipoteichoic acids could be protective against pneumococcal infections in mice (11, 13, 26). This observation made it clear that the capsule was not the only pneumococcal antigen that could elicit a protective response. Since antibodies to the phosphocholine epitopes of teichoic acids are not made in young children, this discovery by itself did not provide a means of developing an efficacious noncapsular pneumococcal vaccine. Subsequent studies have identified several protein antigens that are immunogenic in young children and that have also been shown to elicit protective immune responses in mice. These antigens include the surface protein PspA; autolysin, an enzyme on the pneumococcal cell wall; and pneumolysin, a cytoplasmic protein that is released when pneumococci are autolysed. These three proteins and several others are described below.

PspA

Pneumococcal surface protein A (PspA) is produced by all pneumococci (19). The protein is important for virulence (37) and has been shown to interfere with activation of C3 (12). Although serologically variable when examined with monoclonal antibody, PspA is highly cross-reactive when examined with polyclonal sera (19). Monoclonal and polyclonal antibodies to PspA can passively protect mice from otherwise fatal bacteremia and sepsis caused by pneumococci (34, 35). Subcutaneous and intraperitoneal immunization with PspA has been able to protect mice against fatal infections, and this protection has been observed to be highly cross-protective regardless of PspA type (16, 35, 67). Intranasal immunization with PspA has also been shown to protect against nasopharyngeal carriage in an adult mouse carriage model, raising the possibility that vaccines could be developed to prevent carriage and transmission of pneumococci (71).

A number of pneumococcal proteins, including PspA, are able to bind choline and share similar, but usually not indistinguishable, choline-binding domains near the C-terminal ends of the proteins (10, 36, 55, 72, 73). The N-terminal end of PspA is composed of a largely coiled-coil α-helical sequence that is responsible for most of the cross-protective immunity elicited by PspA. Paradoxically, this region also contains most of the serologic variability within the PspA molecule (33).

The serum of virtually all adults, and most children over 7 months of age, contains detectable antibody to PspA (69). The levels of antibody are higher in adults than in children. It is thus possible that natural antibodies to PspA may contribute more to the immunity to pneumococcal infection in adults than in young children. It is hoped that immunization to elicit high levels of antibodies to PspA may be able to protect young children as well as adults from infections with pneumococci.

PspC is a protein that is highly similar to PspA in its overall domain structure and in its proline-rich and choline-binding domains (12). The gene for this protein was originally identified through its close similarity to PspA (36). Antibodies to PspC are able to protect against pneumococcal infection, apparently through cross-reactivity with PspA (17). This protein has also been independently discovered by others based on its ability to bind secretory IgA (SpsA) and choline (CbpA), and it has potential roles in colonization and adherence (23, 55). The full potential of this protein as a vaccine has yet to be determined.

Pneumolysin

All pneumococci produce pneumolysin, a potent 53-kDa thiol-activated pore-forming cytolysin, which can attack any cell that has cholesterol in its plasma membrane (44). It is a bifunctional toxin and, in addition to its cytotoxic properties, is capable of directly activating the classical complement pathway in the absence of specific antibody, with a concomitant reduction in serum opsonic activity (50). This latter property is mediated by its capacity to bind the Fc region of human immunoglobulin G (IgG) (39).

Structure-function analysis of pneumolysin has indicated that a domain toward the C terminus of the toxin (amino acids 427 to 437), which includes a unique cysteine residue, is critical for cytotoxicity (9). This cysteine motif is highly conserved among other members of the thiol-activated cytolysin family. Several single-amino-acid substitutions within this region (and other regions involved in cell binding and pore formation) reduce the cytotoxicity of pneumolysin by up to 99.9%. A separate region, which has a degree of amino acid homology with human C-reactive protein, is responsible for IgG binding and complement activation, and a mutation in this domain (Asp385→Asn) interferes with both properties (39).

In vitro studies have demonstrated that highly purified pneumolysin has a variety of detrimental effects on cells and tissues, which undoubtedly contribute to the pathogenesis of disease (reviewed by Paton et al. [44] and Paton [45]). Also, complete inactivation of the pneumolysin gene in either a type 2 or type 3 pneumococcus has been shown to reduce virulence for mice challenged by both the intranasal and intraperitoneal routes, as judged by median survival time and 50% lethal dose (LD_{50}) (6, 7). Compared with the wild-type strain, intranasal challenge with the pneumolysin-negative pneumococci resulted in a less severe inflammatory response, a reduced rate of multiplication within the lung, a reduced capacity to injure the alveolar-capillary barrier, and a delayed onset of bacteremia (18, 57). Pneumolysin also plays a critical role after bacteremia has developed, by preventing the generation of inflammation-based immunity that would otherwise be able to keep bacteremia in check (3). S. pneumoniae derivatives, in which the wild-type pneumolysin gene was replaced by mutated genes encoding toxins with point mutations affecting either or both of the cytotoxic and complement activation properties (4), have also been used to confirm distinct roles for both toxin activities in the pathogenesis of pneumococcal pneumonia (1, 56).

Collectively, the above studies establish the importance of pneumolysin in the pathogenesis of pneumococcal disease and identify it as a target for vaccination. It has been known for many years that immunization with purified pneumolysin protects mice against challenge with highly virulent pneumococci (48). Potential problems of toxicity have been overcome by site-directed mutagenesis of regions of the toxin essential for cytotoxicity and complement activation, as described above. Genes encoding these recombinant pneumolysin toxoids (pneumolysoids) have also been inserted into *Escherichia coli* expression vectors, permitting large-scale production of antigens at low cost (49). Sequence analysis of pneumolysin genes from a wide range of *S. pneumoniae* serotypes has confirmed that there is very little variation in primary amino acid sequence (>99% identity) (32, 40) and so a single vaccine antigen should provide coverage against all pneumococci regardless of serotype. Indeed, immunization with pneumolysoid provides a significant degree of protection against at least nine different serotypes of *S. pneumoniae* (2). It should be emphasized, however, that toxin-neutralizing antibodies elicited by immunization with pneumolysoid would not be expected to have opsonophagocytic activity. Thus, the protection elicited by a pneumolysin-containing vaccine may be enhanced by incorporation of protection-eliciting surface antigens such as PS or PspA.

Pneumolysin has also shown promise as a carrier for the otherwise poorly immunogenic PS in experimental conjugate vaccine formulations. Immunization of mice with pneumolysoid conjugated to type 19F PS elicited a strong and boostable antibody response to both protein and PS moieties and provided infant mice with a high degree of protection against challenge with *S pneumoniae* (29, 49). Similar results have been reported for conjugates of native pneumolysin with type 18C PS (27). A recent comparison of tetravalent pneumolysoid-PS or tetanus toxoid-PS conjugate vaccines (incorporating capsular polysaccharide types 6B, 14, 19F, and 23F) demonstrated that pneumolysoid was at least as good a carrier protein as tetanus toxoid and, in the case of type 23F, superior (38). Clearly, such antigens have the potential to evoke a significant anti-PS response, as well as an anti-virulence-protein response, thereby conferring comprehensive protection against pneumococcal disease in humans.

Pneumococcal Surface Adhesin A (PsaA)

PsaA is a 37-kDa lipoprotein produced by all pneumococci (59) that was initially thought to be an adhesin (61). *psaA*-negative pneumococci are virtually avirulent for mice and exhibit a reduced capacity to adhere in vitro to human type II pneumocytes (5). The *psaA* gene is part of a complex operon encoding the components of an ABC transport sys-tem. Dintilhac et al. (21) have proposed that the *psaA* operon encodes a manganese transporter and that *S. pneumoniae* requires Mn^{2+} for growth. The three-dimensional structure of PsaA has recently been determined; interestingly, the metal binding site was found to be occupied by Zn^{2+}, rather than Mn^{2+}, suggesting the possibility of dual specificity (28). The dimensions of PsaA (approximately 7 nm at its longest axis) are such that any protein anchored to the cell membrane via a lipid moiety will not be exposed on the outer surface of the pneumococcus. Thus, PsaA is unlikely to be an adhesin per se, and the virtual avirulence of *psaA*-negative pneumococci can be explained either by a requirement for Mn^{2+} (or Zn^{2+}) as a cofactor or for regulation of expression of other virulence determinants (e.g., adhesins), or by growth retardation due to an inability to scavenge these metal ions in vivo. Recently, it has been reported that *psaA*-negative pneumococci do not express certain choline-binding proteins on their surface, including autolysin (LytA) and CbpA (41); the latter protein has previously been shown to be an adhesin with specificity for receptors on cytokine-activated epithelial and endothelial cells (55).

The profound effect of mutagenesis of *psaA* on virulence suggested that PsaA might be a suitable vaccination target. Concerns were raised as to the possibility of antigenic variation, given the unexpected degree of sequence divergence between the *psaA* gene from the type 2 *S. pneumoniae* strain D39 (5) and that originally reported for the homolog from the closely related strain R36A (61). However, the sequence of *psaA* from a type 6B pneumococcus was subsequently shown to be almost identical to that from D39 (60). Furthermore, restriction fragment length polymorphism analysis of PCR-amplified *psaA* from 80 *S. pneumoniae* strains representing 23 PS serotypes indicated that the gene is in fact highly conserved, suggesting that the original sequence reported to be from R36A may be incorrect (61). To date, the only published study of the protective immunogenicity of PsaA is that of Talkington et al. (66), who demonstrated protection of mice against intravenous challenge with type 3 *S. pneumoniae*. This protection is presumably due to in vivo blockade of metal ion transport and would necessitate penetration of antibody through the capsule and cell wall layers such that it can bind to PsaA attached to the outer face of the plasma membrane. The efficiency of this may well be influenced by thickness of the capsule, which is believed to be upregulated during invasive infection. It will therefore be of interest to assess the protective efficacy of immunization with PsaA in animal models of systemic disease, as well as nasopharyngeal colonization. A further issue that warrants consideration is the possibility that inhibition of PsaA function by exogenous antibody may also interfere with expression of autolysin and so induce a state of penicillin tolerance (41).

Other Pneumococcal Proteins

Several other putative pneumococcal virulence proteins, including the major pneumococcal autolysin (LytA), two neuraminidases (NanA and NanB), and hyaluronidase, have been examined for vaccine potential (reviewed by Paton et al. [47]). Of these, the most effective was LytA, which elicited a level of protection similar to that achieved by immunization with pneumolysin. This protective effect appeared to be largely attributable to blockade of the release of pneumolysin; the toxin is located in the cytoplasm of growing pneumococci and is released into the external medium only when the cells undergo autolysis. Immunization with a combination of pneumolysin and LytA did not increase the degree of protection over that obtained using either antigen alone. Also, immunization with LytA provided no protection against challenge with high doses of pneumolysin-negative pneumococci (31). Immunization of mice with either of the two neuraminidases has been shown to confer statistically significant protection, but in both cases protection was poorer than that achieved by immunization with pneumolysin, and there was no additive protective effect. However, the efficacy of immunization with both NanA and NanB has yet to be determined. Immunization of mice with purified pneumococcal hyaluronidase, on the other hand, does not confer demonstrable protection against challenge (47).

The Case for a Multivalent Pneumococcal Protein Vaccine

Of the pneumococcal proteins proposed as vaccine candidates to date, pneumolysin, PspA, and PsaA have all been shown to contribute to the virulence of *S. pneumoniae* and to be produced by virtually all clinical isolates. Moreover, for pneumolysin and PsaA, there appears to be no significant antigenic variation from isolate to isolate, and PspA contains cross-protective epitopes despite variation in the N-terminal region. The likely role of PspC as a mediator of nasopharyngeal colonization justifies serious examination of its vaccine potential as well. Furthermore, analysis of open reading frames in the pneumococcal genome sequence may identify additional proteins worthy of consideration. If the various proteins function at different stages of the pathogenic process, there is a strong possibility that immunization with a combination of antigens will provide additive protection. Moreover, there may be differences in the relative protective capacities of the individual antigens against particular *S. pneumoniae* strains. Thus, a combined pneumococcal protein vaccine may elicit a higher degree of protection against a wider variety of strains than any single antigen.

REFERENCES

1. Alexander, J. E., A. M. Berry, J. C. Paton, J. B. Rubins, P. W. Andrew, and T. J. Mitchell. 1998. Amino acid changes affecting the activity of pneumolysin alter the behavior of pneumococci in pneumonia. *Microb. Pathog.* **24:** 167–174.

2. Alexander, J. E., R. A. Lock, C. C. A. M. Peeters, J. T. Poolman, P. W. Andrew, T. J. Mitchell, D. Hansman, and J. C. Paton. 1994. Immunization of mice with pneumolysin toxoid confers a significant degree of protection against at least nine serotypes of *Streptococcus pneumoniae*. *Infect. Immun.* **62:**5683–5688.

3. Benton, K. A., M. P. Everson, and D. E. Briles. 1995. A pneumolysin-negative mutant of *Streptococcus pneumoniae* causes chronic bacteremia rather than acute sepsis in mice. *Infect. Immun.* **63:**448–455.

4. Benton, K. A., J. C. Paton, and D. E. Briles. 1997. The hemolytic and complement-activating properties of pneumolysin do not contribute individually to virulence in a pneumococcal bacteremia model. *Microb. Pathog.* **23:** 201–209.

5. Berry, A. M., and J. C. Paton. 1996. Sequence heterogeneity of PsaA, a 37-kilodalton putative adhesin essential

6. Berry, A. M., J. C. Paton, and D. Hansman. 1992. Effect of insertional inactivation of the genes encoding pneumolysin and autolysin on the virulence of *Streptococcus pneumoniae* type 3. *Microb. Pathog.* **12:**87–93.

7. Berry, A. M., J. Yother, D. E. Briles, D. Hansman, and J. C. Paton. 1989. Reduced virulence of a defined pneumolysin-negative mutant of *Streptococcus pneumoniae*. *Infect. Immun.* **57:**2037–2042.

8. Black, S., H. Shinefield, P. Ray, L. Edwi, B. Fireman, T. K. P. V. S. Group, R. Auystrian, G. Siber, J. Hackell, K. Robert, and I. Chang. 1998. Efficacy of heptavalent conjugate pneumococcal vaccine (Wyeth Lederle) in 37,000 infants and children: results of the northern California Kaiser Permanente efficacy trial, abstr. LB-9, p. 23. *Program Abstr. 38th Intersci. Conf. Antimicrob. Agents Chemother.* American Society for Microbiology, Washington, D.C.

9. Boulnois, G. J., J. C. Paton, T. J. Mitchell, and P. W. Andrew. 1991. Structure and function of pneumolysin, the multifunctional, thiol-activated toxin of *Streptococcus pneumoniae*. *Mol. Microbiol.* **5:**2611–2616.

10. Briese, T., and R. Hakenbeck. 1985. Interaction of the pneumococcal amidase with lipoteichoic acid and choline. *Eur. J. Biochem.* **146:**417–427.

11. Briles, D. E., C. Forman, J. C. Horowitz, J. E. Volanakis, W. H. Benjamin, Jr., L. S. McDaniel, J. Eldridge, and J. Brooks. 1989. Antipneumococcal effects of C-reactive protein and monoclonal antibodies to pneumococcal cell wall and capsular antigens. *Infect. Immun.* **57:**1457–1464.

12. Briles, D. E., S. K. Hollingshead, E. Swiatlo, A. Brooks-Walter, A. Szalai, A. Virolainen, L. S. McDaniel, K. A. Benton, P. White, K. Prellner, A. Hermansson, P. C. Aerts, H. Van Dijk, and M. J. Crain. 1997. PspA and PspC: their potential for use as pneumococcal vaccines. *Microb. Drug Resist.* **3:**401–408.

13. Briles, D. E., M. Nahm, K. Schroer, J. Davie, P. Baker, J. Kearney, and R. Barletta. 1981. Antiphosphocholine antibodies found in normal mouse serum are protective against intravenous infection with type 3 *Streptococcus pneumoniae*. *J. Exp. Med.* **153:**694–705.

14. Briles, D. E., J. C. Paton, M. H. Nahm, and E. Swiatlo (ed.). *Immunity to Streptococcus pneumoniae*. Lippincott Williams & Wilkins, New York, N.Y., in press.

15. Briles, D. E., E. Swiatlo, and K. Edwards. Vaccine strategies for *Streptococcus pneumoniae*. *In* D. L. Stevens (ed.), *Streptococci*. Oxford University Press, New York, N.Y., in press.

16. Briles, D. E., R. C. Tart, E. Swiatlo, J. P. Dillard, P. Smith, K. A. Benton, B. A. Ralph, A. Brooks-Walter, M. J. Crain, S. K. Hollingshead, and L. S. McDaniel. 1998. Pneumococcal diversity: considerations for new vaccine strategies with an emphasis on pneumococcal surface protein A (PspA). *Clin. Microbiol. Rev.* **11:**645–657.

17. Brooks-Walter, A., R. C. Tart, D. E. Briles, and S. K. Hollingshead. The *pspC* gene of *Streptococcus pneumoniae* encodes a polymorphic protein PspC, which is capable of eliciting protective immunity to pneumococcal bacteremia and sepsis. Submitted for publication.

18. Canvin, J. R., A. P. Marvin, M. Sivakumaran, J. C. Paton, G. J. Boulonois, P. W. Andrew, and T. J. Mitchell.

1995. The role of pneumolysin and autolysin in the pathology of pneumoniae and septicemia in mice infected with a type 2 pneumococcus. *J. Infect. Dis.* **172**:119–123.

19. **Crain, M. J., W. D. Waltman, J. S. Turner, J. Yother, D. E. Talkington, L. M. McDaniel, B. M. Gray, and D. E. Briles.** 1990. Pneumococcal surface protein A (PspA) is serologically highly variable and is expressed by all clinically important capsular serotypes of *Streptococcus pneumoniae. Infect. Immun.* **58**:3293–3299.

20. **Dagan, R., R. Melamed, M. Muallem, L. Piglansky, D. Greenburg, O. Abramson, P. M. Mendalman, N. Bohidar, and P. Yagupsky.** 1996. Reduction of nasopharyngeal carriage of pneumococci during the second year of life by a heptavalent conjugate pneumococcal vaccine. *J. Infect. Dis.* **174**:1271–1278.

21. **Dintilhac, A., G. Alloing, C. Granadel, and J.-P. Claverys.** 1997. Competence and virulence of *S. pneumoniae*: Adc and PsaA mutants exhibit a requirement for Zn and Mn resulting from inactivation of metal permeases. *Mol. Microbiol.* **25**:727–739.

22. **Eskola, J., H. Kayhty, H. Peltola, V. Karanko, P. H. Makela, J. Samuleson, and L. K. Gordon.** 1985. Antibody levels achieved in infants by course of *Haemophilus influenzae* type B polysaccharide/diphtheria toxoid conjugate vaccine. *Lancet* **i**:1184–1186.

23. **Hammerschmidt, S., S. R. Talay, P. Brandtzaeg, and G. S. Chhatwal.** 1997. SpsA, a novel pneumococcal surface protein with specific binding to secretory immunoglobulin A and secretory component. *Mol. Microbiol.* **25**:1113–1124.

24. **Heffron, R.** 1939. *Pneumonia.* The Commonwealth Fund, New York, N.Y.

25. **Insel, R. A., and P. W. Anderson.** 1986. Oligosaccharide-protein conjugate vaccines induce and prime for oligoclonal IgG antibody responses to the *Haemophilus influenzae* b capsular polysaccharide in human infants. *J. Exp. Med.* **163**:262–269.

26. **Kenny, J. J., G. Guelde, R. T. Fisher, and D. L. Longo.** 1993. Induction of phosphocholine-specific antibodies in X-linked immune deficient mice: in vivo protection against a *Streptococcus pneumoniae* challenge. *Int. Immunol.* **6**:561–568.

27. **Kuo, J., M. Douglas, H. K. Ree, and A. A. Lindberg.** 1995. Characterization of a recombinant pneumolysin and its use as a protein carrier for pneumococcal type 18C conjugate vaccine. *Infect. Immun.* **63**:2706–2713.

28. **Lawrence, M. C., P. A. Pilling, A. D. Ogunniyi, A. M. Berry, and J. C. Paton.** Crystal structure of pneumococcal surface antigen PsaA. Submitted for publication.

29. **Lee, C.-J., R. A. Lock, P. W. Andrew, T. J. Mitchell, D. Hansman, and J. C. Paton.** 1994. Protection in infant mice from challenge with *Streptococcus pneumoniae* type 19F by immunization with a type 19F polysaccharide-pneumolysoid conjugate. *Vaccine* **12**:875–877.

30. **Levine, M., G. C. Woodrow, J. B. Kaper, and G. S. Cobon.** 1997. *Conjugate Vaccines against Streptococcus pneumoniae,* 2nd ed. Marcel Dekker, Inc., New York, N.Y.

31. **Lock, R. A., D. Hansman, and J. C. Paton.** 1992. Comparative efficacy of autolysin and pneumolysin as immunogens protecting mice against infection by *Streptococcus pneumoniae. Microb. Pathog.* **12**:137–143.

32. **Lock, R. A., Q. Q. Zhang, A. M. Berry, and J. C. Paton.** 1996. Sequence variation in the *Streptococcus pneumoniae* pneumolysin gene affecting haemolytic activity and electrophoretic mobility of the toxin. *Microb. Pathog.* **21**: 71–83.

33. **McDaniel, L. S., B. A. Ralph, D. O. McDaniel, and D. E. Briles.** 1994. Localization of protection-eliciting epitopes on PspA of *Streptococcus pneumoniae* between amino acid residues 192 and 260. *Microb. Pathog.* **17**:323–337.

34. **McDaniel, L. S., G. Scott, J. F. Kearney, and D. E. Briles.** 1984. Monoclonal antibodies against protease sensitive pneumococcal antigens can protect mice from fatal infection with *Streptococcus pneumoniae. J. Exp. Med.* **160**:386–397.

35. **McDaniel, L. S., J. S. Sheffield, P. Delucchi, and D. E. Briles.** 1991. PspA, a surface protein of *Streptococcus pneumoniae*, is capable of eliciting protection against pneumococci of more than one capsular type. *Infect. Immun.* **59**:222–228.

36. **McDaniel, L. S., J. S. Sheffield, E. Swiatlo, J. Yother, M. J. Crain, and D. E. Briles.** 1992. Molecular localization of variable and conserved regions of *pspA*, and identification of additional *pspA* homologous sequences in *Streptococcus pneumoniae. Microb. Pathog.* **13**:261–269.

37. **McDaniel, L. S., J. Yother, M. Vijayakumar, L. McGarry, W. R. Guild, and D. E. Briles.** 1987. Use of insertional inactivation to facilitate studies of biological properties of pneumococcal surface protein A (PspA). *J. Exp. Med.* **165**:381–394.

38. **Michon, F., P. C. Fusco, C. A. S. A. Minetti, M. Laude-Sharp, S. Moore, D. P. Remeta, I. Heron, and M. S. Blake.** 1998. Multivalent pneumococcal capsular polysaccharide conjugate vaccines employing genetically detoxified pneumolysin as a carrier protein. *Vaccine* **16**:1732–1741.

39. **Mitchell, T. J., P. W. Andrew, F. K. Saunders, A. N. Smith, and G. J. Boulnois.** 1991. Complement activation and antibody binding by pneumolysin via a region of the toxin homologous to a human acute-phase protein. *Mol. Microbiol.* **5**:1883–1888.

40. **Mitchell, T. J., F. Mendez, J. C. Paton, P. W. Andrew, and G. J. Boulnois.** 1990. Comparison of pneumolysin genes and proteins from *Streptococcus pneumoniae* types 1 and 2. *Nucleic Acids Res.* **18**:4010.

41. **Novak, R., E. Charpentier, J. S. Braun, and E. I. Toumanen.** 1998. Penicillin tolerance genes in *Streptococcus pneumoniae*: the ABC type Mn permease complex, abstr. 1D-26, p. 62–63. *Abstr. ASM Conference on Streptococcal Genetics, Vichy, France.* American Society for Microbiology, Washington, D.C.

42. **Obaro, S. K., R. A. Adegbola, W. A. S. Banya, and B. M. Greenwood.** 1996. Carriage of pneumococci after pneumococcal vaccination. *Lancet* **348**:271–272.

43. **Ortqvist, A., J. Hedlund, L.-A. Burman, E. Elbel, M. Hofer, M. Leinonen, I. Lindbald, B. Sundelof, and M. Kalin.** 1998. Randomised trial of 23-valent pneumococcal capsular polysaccharide vaccine in prevention of pneumonia in middle-aged and elderly people. Swedish pneumococcal vaccine group study. *Lancet* **351**:399–403.

44. **Paton, J., P. Andrew, G. Boulnois, and T. Mitchell.** 1993. Molecular analysis of the pathogenicity of *Streptococcus pneumoniae*: the role of pneumococcal proteins. *Annu. Rev. Microbiol.* **47**:89–115.

45. **Paton, J. C.** 1996. The contribution of pneumolysin to the pathogenicity of *Streptococcus pneumoniae*. *Trends Microbiol.* **4:**103–106.

46. **Paton, J. C.** 1998. Novel pneumococcal surface proteins: role in virulence and vaccine potential. *Trends Microbiol.* **6:**85–87; discussion 87–88.

47. **Paton, J. C., A. M. Berry, and R. A. Lock.** 1997. Molecular analysis of putative pneumococcal virulence proteins. *Microb. Drug. Resist.* **3:**3–10.

48. **Paton, J. C., R. A. Lock, and D. C. Hansman.** 1983. Effect of immunization with pneumolysin on survival time of mice challenged with *Streptococcus pneumoniae*. *Infect. Immun.* **40:**548–552.

49. **Paton, J. C., R. A. Lock, C.-J. Lee, J. P. Li, A. M. Berry, T. J. Mitchell, P. W. Andrew, D. Hansman, and G. J. Bulnois.** 1991. Purification and immunogenicity of genetically obtained pneumolysin toxoids and their conjugation to *Streptococcus pneumoniae* type 19F polysaccharide. *Infect. Immun.* **59:**2297–2304.

50. **Paton, J. C., B. Rowan-Kelly, and A. Ferrante.** 1984. Activation of human complement by the pneumococcal toxin pneumolysin. *Infect. Immun.* **43:**1085–1087.

51. **Powers, D. C., E. L. Anderson, K. Lottenbach, and C. M. Mink.** 1996. Reactogenicity and immunogenicity of a protein-conjugated pneumococcal oligosaccharide vaccine in older adults. *J. Infect. Dis.* **173:**1014–1018.

52. **Rennels, M. B., K. M. Edwards, H. L. Keyserling, K. S. Reisinger, D. A. Hogerman, D. V. Madore, I. Chang, P. R. Paradiso, F. J. Malinoski, and A. Kimura.** 1998. Safety and immunogenicity of heptavalent pneumococcal vaccine conjugated to CRM197 in United States infants. *Pediatrics* **101:**604–611.

53. **Rijkers, G. T., E. A. M. Sanders, M. A. Breukels, and B. J. M. Zegers.** 1996. Responsiveness of infants to capsular polysaccharides: implications for vaccine development. *Rev. Med. Microbiol.* **7:**3–12.

54. **Robbins, J. B., R. Austrian, C.-J. Lee, S. C. Rastogi, G. Schiffman, J. Henrichsen, P. H. Makela, C. V. Broome, R. R. Facklam, R. H. Tiesjema, and J. C. Parke, Jr.** 1983. Considerations for formulating the second-generation pneumococcal capsular polysaccharide vaccine with emphasis on the cross-reactive types within groups. *J. Infect. Dis.* **148:**1136–1159.

55. **Rosenow, C., P. Ryan, J. N. Weiser, S. Johnson, P. Fontan, A. Ortqvist, and H. R. Masure.** 1997. Contribution of novel choline-binding proteins to adherence, colonization and immunogenicity of *Streptococcus pneumoniae*. *Mol. Microbiol.* **25:**819–829.

56. **Rubins, J. B., D. Charboneau, C. Fashing, A. M. Berry, J. C. Paton, J. E. Alexander, P. W. Andrew, T. J. Mitchell, and E. N. Janoff.** 1996. Distinct roles for pneumolysin's cytotoxic and complement activities in pathogenesis of pneumococcal pneumonia. *Am. J. Respir. Crit. Care Med.* **153:**1339–1346.

57. **Rubins, J. B., D. Charboneau, J. C. Paton, T. J. Mitchell, and P. W. Andrew.** 1995. Dual function of pneumolysin in the early pathogenesis of purine pneumococcal pneumonia. *J. Clin. Invest.* **95:**142–150.

58. **Rubins, J. B., A. K. G. Puri, D. Carboneau, R. MacDonald, N. Opstad, and E. N. Janoff.** 1998. Magnitude, duration, quality, and function of pneumococcal vaccine responses in elderly adults. *J. Infect. Dis.* **178:**431–440.

59. **Russell, H., J. A. Tharpe, D. E. Wells, E. H. White, and J. Johnson.** 1990. Monoclonal antibody recognizing a species-specific protein from *Streptococcus pneumoniae*. *J. Clin. Microbiol.* **28:**2191–2195.

60. **Sampson, J. S., Z. Furlow, A. M. Whitney, D. Williams, R. Facklam, and G. M. Carlone.** 1997. Limited diversity of *Streptococcus pneumoniae pspA* among pneumococcal vaccine serotypes. *Infect. Immun.* **65:**1967–1971.

61. **Sampson, J. S., S. P. O'Connor, A. R. Stinson, J. A. Tharpe, and H. Russell.** 1994. Cloning and nucleotide sequence analysis of *psaA*, the *Streptococcus pneumoniae* gene encoding a 37-kilodalton protein homologous to previously reported *Streptococcus* sp. adhesins. *Infect. Immun.* **62:**319–324.

62. **Sankilampi, U., P. O. Honkanen, A. Bloigu, and M. Leinonen.** 1997. Persistence of antibodies to pneumococcal capsular polysaccharide vaccine in the elderly. *J. Infect. Dis.* **176:**1100–1104.

63. **Shelly, M. A., H. Jacoby, G. J. Riley, B. T. Graves, M. Pichichero, and J. J. Treanor.** 1997. Comparison of pneumococcal polysaccharide and CRM_{197} conjugated pneumococcal oligosaccharide vaccines in young and elderly adults. *Infect. Immun.* **65:**242–247.

64. **Siber, G. R.** 1994. Pneumococcal disease: prospects for a new generation of vaccines. *Science* **265:**1385–1387.

65. **Simberkoff, M. S., A. P. Cross, M. Al-Ibrahim, A. L. Baltch, P. J. Geiseler, J. Nadler, A. S. Richmond, R. P. Smith, G. Schiffman, D. S. Shepard, and J. P. VanEeckhout.** 1986. Efficacy of pneumococcal vaccine in high-risk patients. *N. Engl. J. Med.* **315:**1318–1327.

66. **Talkington, D. F., B. G. Brown, J. A. Tharpe, A. Koening, and H. Russell.** 1996. Protection of mice against fatal pneumococcal challenge by immunization with pneumococcal surface adhesion A (PsaA). *Microb. Pathog.* **21:**17–22.

67. **Tart, R. C., L. S. McDaniel, B. A. Ralph, and D. E. Briles.** 1996. Truncated *Streptococcus pneumoniae* PspA molecules elicit cross-protective immunity against pneumococcal challenge in mice. *J. Infect. Dis.* **173:**380–386.

68. **Van Cott, J. L., T. Kobayashi, M. Yamamoto, S. Pillai, J. R. McGhee, and H. Kiyono.** 1996. Induction of pneumococcal polysaccharide-specific mucosal immune responses by oral immunization. *Vaccine* **14:**392–398.

69. **Virolainen, A., W. Russell, S. Rapola, D. E. Briles, and H. Kayhty.** 1996. Human antibodies to pneumococcal surface protein A (PspA), abstr. G-38, p. 150. *Program Abstr. 36th Intersci. Conf. Antimicrob. Agents Chemother.* American Society for Microbiology, Washington, D.C.

70. **White, B.** 1938. *The Biology of Pneumococcus*. The Commonwealth Fund, New York, N.Y.

71. **Wu, H.-Y., M. Nahm, Y. Guo, M. Russell, and D. E. Briles.** 1997. Intranasal immunization of mice with PspA (pneumococcal surface protein A) can prevent intranasal carriage and infection with *Streptococcus pneumoniae*. *J. Infect. Dis.* **175:**893–846.

72. **Yother, J., and D. E. Briles.** 1992. Structural properties and evolutionary relationships of PspA, a surface protein of *Streptococcus pneumoniae*, as revealed by sequence analysis. *J. Bacteriol.* **174:**601–609.

73. **Yother, J., and J. M. White.** 1994. Novel surface attachment mechanism for the *Streptococcus pneumoniae* protein PspA. *J. Bacteriol.* **176:**2976–2985.

Pathogenicity of Enterococci

LYNN E. HANCOCK AND MICHAEL S. GILMORE

25

INTRODUCTION

Enterococcal pathogenicity was initially addressed at the end of the 19th century by MacCallum and Hastings (46), who isolated an organism from a case of acute endocarditis and designated it *Micrococcus zymogenes* based on its fermentative properties. The organism was shown to be resistant to desiccation, heating to 60°C, and several antiseptics, including carbolic acid and chloroform (46). It was also found to be lethal when injected intraperitoneally in white mice and was capable of producing endocarditis in a canine model (46). A century later, enterococci are prominent among nosocomial pathogens, ranking second only to *Escherichia coli* in total nosocomial infections, accounting for more than 12% of all cases (56).

Infections caused by the genus *Enterococcus* (most notably *Enterococcus faecalis*, which accounts for ~80% of all infections) include urinary tract infections, bacteremia, intra-abdominal infections, and endocarditis (33, 53). The problem of nosocomial enterococcal infection is compounded by multiple antibiotic resistance. A comparison of outcomes for patients with bacteremia due to vancomycin-resistant *Enterococcus faecium* or vancomycin-susceptible *E. faecium* found a median length of stay of 46 days after the first episode of bacteremia in the group of patients with vancomycin-resistant *E. faecium*, compared with 19 days for patients infected by a susceptible strain (43). The presence of vancomycin-resistant enterococci in the bloodstream has also been associated with increased mortality (14). Patients with enterococcal bacteremia were observed to be twice as likely to die (37% versus 16%) when the infecting isolate was resistant to vancomycin (14). However, more recent studies indicate that vancomycin resistance is not a major predictor for clinical outcome (45, 48). Whether more recent findings describe cases that benefit from further evolution in the treatment of vancomycin-resistant enterococci requires further analysis. As vancomycin frequently represents the last available therapeutic agent for multiply antibiotic-resistant enterococci, the rapid increase in vancomycin resistance (33) indicates that enterococcal infection will pose an increasing therapeutic challenge.

Resistance trends among *E. faecalis* and *E. faecium* were recently reviewed (33). Data compiled for approximately 15,000 isolates over a 3-year period (1995 to 1997) showed that while resistance to ampicillin and vancomycin is relatively uncommon among *E. faecalis* isolates (<2%), *E. faecium* showed a general trend toward increasing resistance to both ampicillin (83%) and vancomycin (52%). Resistance alone, however, does not explain the pervasiveness of enterococci in nosocomial infections. *E. faecalis*, while remaining sensitive to vancomycin and ampicillin, continues to be the most frequently encountered enterococcal isolate, accounting for 79% of enterococcal infections (33). Reasons for the disparity in the number of infections caused by *E. faecalis* and *E. faecium* are not well known. One explanation for the overrepresentation of *E. faecalis* among clinical isolates may simply relate to natural abundance. Several studies indicate that *E. faecalis* is more abundant in the human gastrointestinal tract (3, 57). In one such study (57), the numbers of *E. faecalis* were on average 100-fold higher than those of *E. faecium*. An alternative explanation is that the preponderance of *E. faecalis* infections is attributable to enhanced virulence, a prospect for which there is some evidence (33, 38).

This chapter focuses on the mechanisms by which enterococci cause human disease. Because of space considerations, only representative works are cited. Comprehensive citations may be found in several reviews of enterococcal pathogenesis (33, 38, 40, 52).

Biology and Epidemiology

The classification of enterococci as group D streptococci dates back to the scheme established by Rebecca Lancefield in the early 1930s (42). In 1984, enterococci were given formal genus status after DNA-DNA and DNA-RNA hybridization studies demonstrated a more distant relationship with the streptococci (64).

Enterococci are generally considered commensals of the gastrointestinal tract of a variety of organisms, including humans, and are morphologically indistinguishable from other streptococci. They are found in a number of environments, probably because of dissemination in animal ex-

Gram-Positive Pathogens, ed. by V. A. Fischetti et al.
© 2000 American Society for Microbiology, Washington, D.C.

crement and environmental persistence. Several intrinsic features of *Enterococcus* may allow members of this genus to survive for extended periods of time, leading to its persistence and nosocomial spread. *E. faecalis* is able to grow in 6.5% NaCl, at temperatures ranging from 10 to 45°C, and can survive a treatment of 30 min at 60°C. Additionally, the organism grows in the presence of 40% bile salts and over a broad pH range (53). The earliest descriptions of the organism noted that it was "hardy and tenacious of life" (46).

Several studies have examined the nature of the environmental ruggedness of the organism (16, 18). *E. faecalis* was observed to adapt to the presence of lethal levels of bile salts and detergents, such as sodium dodecyl sulfate, when first cultured at sublethal levels for as little as 5 s (16). The ability of enterococci to adapt and persist in the presence of detergents may allow them to survive inadequate cleaning regimens, contributing to their persistence in the hospital. Numerous epidemiologic studies have shown that enterococci can be transmitted from person to person in the hospital (8, 38). This transmission typically occurs via the hands of health care workers or on clinical instruments, such as ear-probe thermometers (61).

Antibiotic Resistance

The intrinsic ruggedness of enterococci also confers an unusual level of tolerance to several classes of antibiotics, including aminoglycosides, β-lactams, and quinolones. For example, the resistance of enterococci to aminoglycosides results from the ability of enterococci to block the uptake of the drug at the cell wall (29, 51). Consequently, aminoglycosides are only effective against enterococci when used in combination with cell wall-active antibiotics. This combination treatment modality has been compromised, however, by the rapid spread of high-level aminoglycoside resistance among enterococci ($>2,000$ $\mu g/ml$) (30). Although the mechanism of high-level resistance was determined to be the result of a bifunctional enzyme (15), the molecular basis for the intrinsic resistance of enterococci to low levels of aminoglycosides remains to be determined. Further discussion of the genetic and physiologic basis for antibiotic resistance can be found elsewhere in this volume (chapters 26 and 63).

PATHOGENIC MECHANISMS

To infect, enterococci must first be able to colonize, primarily at mucosal surfaces. From the site of colonization, the organism must then evade the host clearance and ultimately produce pathologic changes in the host, either through direct toxic activity or indirectly by inducing an inflammatory response (40).

Colonization and Translocation

Enterococci normally colonize the gastrointestinal tract of humans. They are found in relative abundance in human feces (10^5 to 10^7 organisms per g) (57). A close association is likely to exist between enterococci and its host, or the organism would be eliminated owing to normal intestinal motility (38). Studies in progress are examining the specific binding of enterococci to intestinal epithelium (63). Many infection-derived enterococcal isolates were found to be clonal, indicating nosocomial transmission. Moreover, a number of studies have documented patient colonization following hospital admission and have shown that coloni-

zation with multiply resistant strains is a predisposing factor for subsequent infection (33, 55). To colonize the lower bowel, enterococci must survive transit through the low pH of the stomach. Several studies have examined the acid tolerance of *E. faecalis* (17, 71). Flahaut et al. (17) demonstrated that exposure of *E. faecalis* to a sublethal pH (pH 4.8) for 15 to 30 min protected the organism from a normally lethal challenge at pH 3.2. Suzuki and colleagues (71) have shown that an *E. faecalis* mutant defective in F_1-F_0 H^+-ATPase activity was unable to grow at pH <6. The H^+-ATPase is used to regulate the cytoplasmic pH of *E. faecalis* by proton extrusion. This enzyme has been shown to be activated at low pH (71). From these studies, it is apparent that enterococci possess the ability to withstand the low gastric pH, which would facilitate colonization. This attribute may be critical in the ability of multidrug-resistant enterococcal strains to colonize the intestinal tract and cause hospital ward outbreaks. Whether infection-derived enterococcal isolates show enhanced acid tolerance has yet to be determined.

Therapy with antibiotics possessing little antienterococcal activity is a key predisposing factor leading to enterococcal colonization and infection (8). Studies in mice with antibiotic-induced intestinal *E. faecalis* overgrowth demonstrated that organisms can adhere to epithelial surfaces of the ileum, cecum, and colon (74, 75). These same studies showed that enterococci possess the ability to translocate from the intestinal lumen to the mesenteric lymph nodes, liver, and spleen (74, 75). As prior antibiotic therapy appears to be a predisposing factor for enterococcal infection, antibiotic-induced intestinal overgrowth by *E. faecalis*, followed by translocation of the organism into the circulation, may offer one explanation for bacteremias of unknown etiology (23).

The mechanisms responsible for enterococcal translocation are not clearly defined. One hypothesis is that enterococci are phagocytosed by tissue macrophages or intestinal epithelial cells and are transported across the intestinal wall to the underlying lymphatic system. Failure to kill the phagocytosed organisms could then lead to systemic spread (76). Olmsted et al. (59) examined the role of the plasmid-encoded surface protein, aggregation substance, in the ability of *E. faecalis* to be internalized by cultured intestinal epithelial (HT-29) cells. The presence of aggregation substance significantly augmented *E. faecalis* internalization by HT-29 cells. However, in contrast to the 1 order of magnitude in enhanced uptake efficiency conferred by aggregation substance, a difference of 3 orders of magnitude was observed between various enterococcal strains tested. This indicated that additional unknown features of this species play major roles as determinants of internalization efficiency.

The vast majority (87 of 91) of *E. faecalis* isolates produce superoxide (O_2^-), whereas *E. faecium* isolates (5 of 13) do so less frequently (32). When clinical and commensal isolates of *E. faecalis* were compared for O_2^- production, strains associated with bacteremia produced O_2^- in vitro at a rate 60% higher than did stool isolates. What role, if any, O_2^- production plays in enterococcal pathogenesis, however, is not well defined. Perhaps enterococcal strains capable of producing O_2^- are better adapted physiologically to utilize limited resources in the intestinal environment, leading to overgrowth of the organism. Alternatively, O_2^- production may enhance niche control in proximity to the intestinal epithelium. The membrane-damaging effects of oxygen radicals may then potentiate the ability of the organism to translocate across a weakened epithelial barrier. The enrichment of this trait among *E. faecalis* infection-derived isolates presents intriguing questions and warrants further study.

One of the enigmas of nosocomial enterococcal infection not easily explained is the ready colonization of an ecological niche already occupied by members of the same species. As noted, antibiotics lacking substantial antienterococcal activity (i.e., antibiotics that do not deleteriously affect indigenous enterococci) are important predisposing factors for infection. These infections are frequently caused by multiply resistant enterococcal isolates that have been exogenously acquired and appear to have out-competed indigenous enterococci in the absence of direct selection. The numbers of exogenously acquired, multiply resistant enterococci would be expected to be minuscule compared with the numbers of indigenous enterococci that are present and well positioned to occupy any niche suitable for enterococcal colonization. The fact that exogenous, multiply resistant, nosocomially transmitted enterococci efficiently colonize the gastrointestinal tract suggests that they may not compete directly for the same niche as indigenous strains. A novel surface protein capable of enabling enterococci to colonize a new area of the gastrointestinal tract, an area perhaps less endowed with immune clearance mechanisms, may explain the ability of these outbreak strains to efficiently colonize and cause disease. Another mechanism that could render colonization by exogenous strains noncompetitive would be an ability to utilize nutrients in a new or enhanced way. For instance, an enhanced ability to utilize mammalian intestinal mucin as an energy source may enable outbreak strains to colonize deeper layers of the mucosa in more intimate association with intestinal epithelial cells. Few studies have comprehensively examined and compared the physiology of commensal strains and enterococcal strains that caused multiple infections, and so at present, these prospects remain completely speculative. However, a new surface protein of novel structure has been described that is enriched among endocarditis (40%) and bacteremia isolates (29%) but is rare among fecal isolates (<3%) (66, 66a). This protein shares homology with other surface proteins of gram-positive bacteria (C-alpha and Rib proteins from group B streptococci and M protein from group A streptococci). The exact function of this protein in the biology of *E. faecalis* is not currently known, but its enrichment in clinical isolates predicts an important role.

Bacteremia

Nosocomial surveillance data for the period October 1986 to April 1997 list enterococci as the third most common cause of nosocomial bacteremia, accounting for 12.8% of all isolates (56). The translocation of enterococci across an intact intestinal epithelial barrier is thought to lead to many bacteremias with no identifiable source (38, 76). Other identifiable sources for enterococcal bacteremia include intravenous lines, abscesses, and urinary tract infections (38). The risk factors for mortality associated with enterococcal bacteremia include severity of illness, patient age, and use of broad-spectrum antibiotics such as third-generation cephalosporins or metronidazole (68). Huycke et al. (34) showed that patients infected with hemolytic, gentamicin-resistant *E. faecalis* strains had a fivefold-increased risk for death within 3 weeks compared with patients infected with nonhemolytic, gentamicin-susceptible strains. Moreover, mode of treatment was not associated

with outcome, discounting the contribution of aminoglycoside resistance to this enhanced lethality of infection. In a more recent study, Caballero-Granado et al. (7) analyzed the clinical outcome, including mortality, for bacteremia caused by *Enterococcus* spp. with or without high-level gentamicin resistance. Mortality associated with high-level gentamicin resistance (29%) was not significantly different from that of gentamicin-susceptible strains (28%). In addition, these workers found no significant difference in the length of hospitalization after acquisition of enterococcal bacteremia. Taken together, these studies suggest that high-level aminoglycoside resistance does not affect clinical outcome and that the presence of the *E. faecalis* cytolysin (hemolysin) may enhance the severity of the infection. A number of well-controlled independent animal studies confirm the toxicity of the enterococcal cytolysin. Cytolysin significantly lowers the 50% lethal dose (LD_{50}) of the infecting strain for mice (13, 35, 50). As discussed below, cytolysin also contributes to the acute toxicity of lupine endocarditis and endophthalmitis models (35, 39).

Urinary Tract

Enterococci have been estimated to account for 110,000 urinary tract infections (UTI) annually in the United States (33). A few studies have been aimed at understanding the interaction of enterococci with uroepithelial tissue (25, 41, 73). Kreft et al. (41) showed a potential role for the plasmid-encoded aggregation substance in the adhesion of enterococci to renal epithelial cells. *E. faecalis* cells harboring the pheromone-responsive plasmid pAD1, or various isogenic derivatives, were better able to bind to the cultured pig renal tubular cell line LLC-PK than were plasmid-free cells. Kreft et al. also showed that a synthetic peptide containing the fibronectin motif Arg-Gly-Asp-Ser could inhibit binding. This structural motif mediates the interaction between fibronectin and eukaryotic surface receptors of the integrin family (20).

Guzman and coworkers (25) analyzed strains of *E. faecalis* isolated from either UTI or endocarditis for their ability to adhere to urinary tract epithelial cells and the Girardi heart cell line. UTI isolates adhered to the urinary tract epithelial cells in vitro, whereas strains from endocarditis adhered efficiently to the Girardi heart cell line. A key observation from these experiments was that growth in pooled human serum enhanced the binding of UTI isolates to the Girardi heart cell line (eightfold increase). The authors noted that the serum-dependent alterations to cell adhesion were lost by several subcultures in brain heart infusion broth (25). In a later study, *E. faecalis* adherence was found to be mediated by carbohydrate antigens present on the cell surface (26). Thus, the nature of the interaction of enterococci with uroepithelial tissue appears to be quite complex, involving surface adhesins of protein and/or carbohydrate nature.

Endocarditis

Of the diverse infections caused by enterococci, infective endocarditis is one of the most therapeutically challenging (49). Enterococci are the third leading cause of infective endocarditis, accounting for 5 to 20% of cases of native valve endocarditis, and 6 to 7% of prosthetic valve endocarditis (49). As noted above, enterococci cultured in serum exhibit enhanced binding to Girardi heart cells. This interaction is inhibited by periodate treatment of the bacterial cell as well as competitive inhibition of binding, by prior incubation of the target cells with specific sugar residues, including D-galactose and L-fucose (26). This suggests that a carbohydrate antigen mediates the adherence of enterococci to cultured heart cells that were derived from the right auricular appendage (Girardi heart).

The presence of the pheromone-responsive plasmid pAD1 enhances vegetation formation in enterococcal endocarditis (9). By comparing endocarditis caused by isogenic mutants in either cytolysin (hemolysin) production or aggregation substance, which are encoded on pAD1, it was observed that the presence of cytolysin contributed to overall lethality (6 of 11 animals killed, compared with 2 of 13 killed by the noncytolytic mutant; $P < 0.01$), whereas the presence of aggregation substance led to a twofold increase in mean vegetation weight. It was noted, however, that all strains tested were able to cause endocarditis, even the plasmid-free controls. These data suggest that the virulence traits encoded by auxiliary genetic elements can enhance the pathogenicity of the organism but may not be essential in establishing infection.

Serum from a patient with *E. faecalis* endocarditis was used to identify an *E. faecalis* antigen selectively expressed in serum but not in broth culture (44). This protein antigen, designated EfaA, had a predicted molecular weight of 34,768. Database homology searches revealed extensive sequence similarity with several streptococcal adhesins. The authors hypothesized that this surface antigen might function as an important adhesin in endocarditis, but there are no published data to support this.

Endophthalmitis

Colonization of host tissue may play a role in the pathogenesis of endophthalmitis. Enterococci are among the most destructive agents that cause this postoperative complication of cataract surgery (27, 39). Experiments designed to determine whether aggregation substance targeted *E. faecalis* to alternate anatomical structures within the eye showed that enterococci attach to membranous structures in the vitreous, but that such adherence is not dependent on the presence of aggregation substance (37). In summary, the preponderance of data indicate that *E. faecalis* adhesion to host tissues is complex and involves multiple adhesins, including surface carbohydrates as well as proteins.

Immune Evasion

For enterococci to maintain an infection, they must successfully evade both specific and nonspecific host defense mechanisms. Other gram-positive pathogens possess attributes that allow them to survive in the host despite powerful nonspecific host defenses mediated primarily by professional phagocytes, i.e., neutrophils, monocytes, and macrophages. These factors include antiphagocytic polysaccharide capsules, antiphagocytic surface proteins, such as the group A streptococcal M protein, and various secreted toxins with direct toxicity for phagocytic cells.

Studies designed to characterize the host response to enterococcal infection have been conducted (2, 19, 28, 58). Harvey et al. (28) concluded that "neutrophil mediated killing of enterococci was largely a function of complement with antibody playing a less essential but potentially important role." Arduino et al. (2) reached similar conclusions with regard to the role of complement in the clearance of enterococci by polymorphonuclear cells. The latter study also tested several defined virulence traits, which included gelatinase, cytolysin, and aggregation substance of *E. faecalis* and found no significant correlation between a given trait and resistance to phagocytosis. However, these phagocytosis assays were performed under con-

ditions unlikely to support cytolysin expression (72), and production of the other traits in this environment was similarly not controlled. Although cytolysin is not made at appreciable levels in standard laboratory media, it can be detected when organisms are cultured in serum (unpublished data), so it may be of value to reexamine the roles of these factors under more physiologic conditions.

In the course of studying neutrophil-mediated phagocytosis, Arduino et al. (1) identified a strain of *E. faecium* that exhibited increased resistance to phagocytosis. They were able to detect electron-dense clumps adjacent to the cell wall, consistent with the presence of capsular material. It was noted, however, that similar electron-dense clumps surrounded the cell wall of strains of *E. faecalis* that appeared to be sensitive to polymorphonuclear cell-mediated clearance. The nature of the material providing protection to phagocytosis for the *E. faecium* strain was shown to be protease resistant and periodate sensitive, implicating a carbohydrate. Further characterization of this material has not been reported. In a recent case report, Bottone and colleagues (6) isolated three highly mucoid-encapsulated strains of *E. faecalis* from patients with UTI. Whether these encapsulated strains are more resistant to phagocytic killing by neutrophils remains to be demonstrated.

Additional evidence for the importance of antienterococcal antibodies in promoting clearance by opsonophagocytic killing was recently reported by Gaglani et al. (19). These authors compared killing efficacy by neutrophils using normal human serum, hypogammaglobulinemic serum, or normal human serum adsorbed with the homologous bacterial strain. Their findings suggest that normal human serum has sufficient antienterococcal antibodies to promote >90% reduction of the bacterial inoculum with serum concentrations as low as 0.5%. Hypogammaglobulinemic serum also promoted >90% reduction of the bacterial inoculum, but only at serum concentrations above 5%. Adsorption with the homologous strain significantly reduced bactericidal activity of the normal human serum. These findings indicate that the normal human host possesses antibodies to enterococci that aid in opsonophagocytic killing, which might be expected for a normal bowel organism in constant association with the host. The protective role of antienterococcal antibodies should be examined to determine whether such protection extends from the strain level to either genus or species. Work in progress is examining these issues (60, 67).

In 1992, Maekawa et al. (47) proposed a new serologic typing scheme for *E. faecalis*, in which *E. faecalis* isolates were classified into 21 distinct serotypes, but four serotypes accounted for 72% of all typable strains, indicating that certain serotypes are more prevalent among clinical isolates. The exact nature of the serotype antigens is not yet known; it could be protein (such as M protein in the case of group A streptococci), or a polysaccharide (as is the case for group B streptococci and *Streptococcus pneumoniae*).

More recently, it has been observed that enterococci possess the ability to survive within professional phagocytes (21, 70). Preliminary evidence (70) indicates an important role for aggregation substance in the adherence, entry, and survival in macrophages. Further studies into the basis for the survival of enterococci within phagocytes will add to our understanding of how enterococci escape immune surveillance.

Pathologic Tissue Damage

Following adhesion to host cell surfaces, and evasion of the host immune response, the last step in the pathogenesis of infection is the production of pathologic changes in the host. Such changes can be induced by the host inflammatory cascade or by direct tissue damage as a result of secreted toxins or proteases. Both mechanisms have been observed in studies of E. faecalis pathogenesis.

Indirect Tissue Damage

Enterococcal lipoteichoic acid (LTA), also known as the group D streptococcal antigen, has been implicated in a variety of biological processes (24–26, 54). Some properties ascribed to LTA include modulating the host immune response as well as mediating the adherence of enterococci to host cells. Bhakdi et al. (4) found the LTA from enterococci to be as inflammatory as lipopolysaccharide of gram-negative bacteria. LTA may also contribute to the ability of enterococci to exchange and rapidly disseminate genetic determinants, a subject that has been investigated extensively (12) (see chapter 26, this volume).

The role of LTA and aggregation substance (AS) in cardiac infections was recently examined (65). Strains of E. faecalis defective in AS and the enterococcal binding substance (EBS), which is at least partially derived from LTA, did not induce clinical signs of illness when injected intraventricularly at levels of 10^8 CFU/ml. However, EBS$^+$AS$^-$ or EBS$^-$AS$^+$ strains induced signs of illness and pericardial inflammation. All rabbits injected with the EBS$^+$AS$^+$ strain developed illness and died. Surprisingly little inflammation was observed in rabbits injected with the EBS$^+$AS$^+$ strain despite the lethality observed. The authors state that such observations are consistent with the presence of a superantigen. The presence of LTA (EBS) and AS together may mediate effects on the host immune response that differ from those seen when either component acts alone. These results warrant further investigation, as one of the strains tested possessed multiple transposon insertions, and the culture fluid extract used to demonstrate superantigen behavior was crude and of unknown physiologic relevance in terms of concentration. Nevertheless, the hypothesis that AS may bind and display LTA in a conformation that promotes an enhanced inflammatory response is intriguing.

Direct Tissue Damage

The enterococcal cytolysin and the zinc metalloprotease (gelatinase) are secreted factors well suited to contribute to disease severity (9, 13, 22, 35, 39). The role of the enterococcal cytolysin in disease pathogenesis has been well established in several independent laboratories and models (9, 35, 39, 50). The cytolysin is enriched in clinical isolates and occurs at a frequency of 45 to 60% (31, 34, 36). The cytolysin is a unique bacterial toxin that is distantly related to lantibiotic bacteriocins, a family of small, posttranslationally modified antimicrobial peptides (5). The cytolysin possesses both toxin and bacteriocin activities and may provide several levels of selective advantage for E. faecalis strains expressing this trait. We observed that the presence of the cytolytic phenotype promotes the appearance of enterococci in the blood when compared with a noncytolytic isogenic mutant in a mouse model of septicemia (10). A substantial body of data supports the role for cytolysin in enterococcal infection, both in humans (34) and in animal models (9, 35, 39).

The most direct and quantitative evidence for pathologic damage attributable to the cytolysin was obtained using a rabbit model of endophthalmitis (39). This model was selected because of the natural aberrations in the intraocular immune response, allowing a robust infection to be established with as few as 10 organisms. This limited response provides the offending bacterium an opportunity to adapt to in vivo growth conditions and environmental cues. Moreover, highly sensitive and quantitative measurements of the evolution of disease can be made. A role for the cytolysin in tissue pathology was unambiguously demonstrated both by a reduction in retinal function as measured by electroretinography (B-wave response), and complete destruction of retinal architecture by 24 h postinfection. The contribution of cytolysin to the severity of disease has also been observed in animal models of systemic disease (35) and endocarditis (9). These findings conclusively demonstrate the importance of the E. faecalis cytolysin as a major virulence factor in E. faecalis infections. However, the cytolytic phenotype is a variable trait typically encoded on pheromone-responsive plasmids (38). In about one-half of E. faecalis infections, the cytolysin is absent, emphasizing the importance of other traits in pathogenicity (11, 31, 34, 36).

The enterococcal gelatinase may also play a measurable role in systemic disease (13), as well as in a caries model using germ-free rats (22). Dupont et al. (13) showed a reduced LD_{50} for mice injected with gelatinase-producing (Gel$^+$) strains. These studies were limited, however, in that the basis for comparison was a Gel$^-$ strain that was generated by chemical mutagenesis, and therefore other uncharacterized traits may have contributed to the observed attenuation.

Using germ-free rats, Gold et al. (22) showed that a proteolytic (Gel$^+$) strain exhibited cariogenic activity, whereas three nonproteolytic strains exhibited little cariogenicity. There is some indication that clinical isolates may be enriched for this trait, as greater than 50% of isolates from both endocarditis and other clinical sources exhibited gelatinase activity, whereas only 27% of community fecal isolates possessed this trait (11). In this study, highly virulent lineages that caused multiple infections were discounted by study design, skewing the correlation between phenotype and disease incidence (11).

Although demonstration of the involvement of gelatinase in tissue pathology and virulence requires the comparison of isogenic mutants, in vitro targets of this enzyme do provide clues for a potential role (69). The gene for gelatinase, gelE, encodes an enzyme that shares significant homology with neutral proteinases from Bacillus species and elastase from Pseudomonas aeruginosa (69). The proposed targets of this enzyme include gelatin, casein, hemoglobin, and other bioactive peptides, including the E. faecalis sex pheromones. The observation that enterococcal pheromones are potent chemoattractants (62), and that gelatinase can cleave these and other bioactive peptides, indicates that gelatinase at least has the potential to modulate the host response to enterococcal infection.

CONCLUSIONS AND FUTURE PERSPECTIVES

Enterococci are well adapted for survival and persistence in a variety of adverse environments, including sites of infection and inanimate hospital surfaces. The rapid emergence of antimicrobial resistance among enterococci undoubtedly also contributes to their emergence as prominent nosocomial pathogens, making them among the most difficult to treat. An understanding of the molecular pathogenesis of enterococcal infection, however, is in its infancy, largely because rogue commensals do not fit the mold of what has emerged as the modern conception of pathogenesis. These organisms are not highly toxigenic, highly invasive, or highly infectious by most measures. They do, nevertheless, cause a substantial amount of human disease, highlighting the inadequacy of existing pathogenesis precepts. Additional study on the pathogenesis of enterococcal infection is urgently needed to reverse the trend toward larger numbers of refractory infections caused by these organisms, for which we have limited understanding.

We thank Viswanathan Shankar, Mark Huycke, Wolfgang Haas, Pravina Srinivas, Phil Coburn, Brett Shepard, Michelle Callegan, and Mary Booth for helpful discussions and comments. Portions of the work described were supported by PHS grants AI41108 and EY08289.

REFERENCES

1. **Arduino, R. C., K. Jacques-Palaz, B. E. Murray, and R. M. Rakita.** 1994. Resistance of *Enterococcus faecium* to neutrophil-mediated phagocytosis. *Infect. Immun.* **62:** 5587–5594.

2. **Arduino, R. C., B. E. Murray, and R. M. Rakita.** 1994. Roles of antibodies and complement in phagocytic killing of enterococci. *Infect. Immun.* **62:**987–993.

3. **Benno, Y., K. Susuki, K. Susuki, K. Narisawa, W. R. Bruce, and T. Mitsuoka.** 1986. Comparison of the fecal microflora in rural Japanese and urban Canadians. *Microbiol. Immunol.* **30:**521–532.

4. **Bhakdi, S., T. Klonisch, P. Nuber, and W. Fischer.** 1991. Stimulation of monokine production by lipoteichoic acids. *Infect. Immun.* **59:**4693–4697.

5. **Booth, M. C., C. P. Bogie, H. G. Sahl, R. J. Siezen, K. L. Hatter, and M. S. Gilmore.** 1996. Structural analysis and proteolytic activation of *Enterococcus faecalis* cytolysin, a novel lantibiotic. *Mol. Microbiol.* **21:**1175–1184.

6. **Bottone, E. J., L. Patel, P. Patel, and T. Robin.** 1998. Mucoid encapsulated *Enterococus faecalis*: an emerging morphotype isolated from patients with urinary tract infections. *Diagn. Microbiol. Infect. Dis.* **31:**429–430.

7. **Caballero-Granado, F. J., J. M. Cisneros, R. Luque, M. Torres-Tortosa, F. Gamboa, F. Diez, J. L. Villanueva, R. Perez-Cano, J. Pasquau, D. Merino, A. Menchero, D. Mora, M. A. Lopez-Ruz, and A. Vergara.** 1998. Comparative study of bacteremias caused by *Enterococcus* spp. with and without high-level resistance to gentamicin. *J. Clin. Microbiol.* **36:**520–525.

8. **Chenoweth, C., and D. Schaberg.** 1990. The epidemiology of enterococci. *Eur. J. Clin. Microbiol. Infect. Dis.* **9:**80–89.

9. **Chow, J. W., L. A. Thal, M. B. Perri, J. A. Vazquez, S. M. Donabedian, D. B. Clewell, and M. J. Zervos.** 1993. Plasmid-associated hemolysin and aggregation substance production contribute to virulence in experimental enterococcal endocarditis. *Antimicrob. Agents Chemother.* **37:** 2474–2477.

10. **Coburn, P. S., H.-G. Sahl, R. J. Siezen, and M. S. Gilmore.** 1997. The *Enterococcus faecalis* cytolysin: a novel toxin with roots in the lantibiotic family, p. 409–410. *Abstr. Eighth European Workshop on Bacterial Protein Toxins, 1997.* Gustav Fischer, Jena, Germany.

11. **Coque, T. M., J. E. Patterson, J. M. Steckelberg, and B. E. Murray.** 1995. Incidence of hemolysin, gelatinase, and aggregation substance among enterococci isolated from patients with endocarditis and other infections and from feces of hospitalized and community-based persons. *J. Infect. Dis.* **171:**1223–1229.

12. **Dunny, G. M., B. A. Leonard, and P. J. Hedberg.** 1995. Pheromone-inducible conjugation in *Enterococcus faecalis*: interbacterial and host-parasite chemical communication. *J. Bacteriol.* **177:**871–876.

13. **Dupont, H., P. Montravers, J. Mohler, and C. Carbon.** 1998. Disparate findings on the role of virulence factors of *Enterococcus faecalis* in mouse and rat models of peritonitis. *Infect. Immun.* **66:**2570–2575.

14. **Edmond, M. B., J. F. Ober, J. D. Dawson, D. L. Weinbaum, and R. P. Wenzel.** 1996. Vancomycin-resistant enterococcal bacteremia: natural history and attributable mortality. *Clin. Infect. Dis.* **23:**1234–1239.

15. **Ferretti, J. J., K. S. Gilmore, and P. Courvalin.** 1986. Nucleotide sequence analysis of the gene specifying the bifunctional 6′-aminoglycoside acetyltransferase 2″-aminoglycoside phosphotransferase enzyme in *Streptococcus faecalis* and identification and cloning of gene regions specifying the two activities. *J. Bacteriol.* **167:**631–638.

16. **Flahaut, S., J. Frere, P. Boutibonnes, and Y. Auffray.** 1996. Comparison of the bile salts and sodium dodecyl sulfate stress responses in *Enterococcus faecalis*. *Appl. Environ. Microbiol.* **62:**2416–2420.

17. **Flahaut, S., A. Hartke, J. Giard, and Y. Auffray.** 1997. Alkaline stress response in *Enterococcus faecalis*: adaptation, cross-protection, and changes in protein synthesis. *Appl. Environ. Microbiol.* **63:**812–814.

18. **Flahaut, S., A. Hartke, J. C. Giard, A. Benachour, P. Boutibonnes, and Y. Auffray.** 1996. Relationship between stress response toward bile salts, acid and heat treatment in *Enterococcus faecalis*. *FEMS Microbiol. Lett.* **138:** 49–54.

19. **Gaglani, M. J., C. J. Baker, and M. S. Edwards.** 1997. Contribution of antibody to neutrophil-mediated killing of *Enterococcus faecalis*. *J. Clin. Immunol.* **17:**478–484.

20. **Galli, D., F. Lottspeich, and R. Wirth.** 1990. Sequence analysis of *Enterococcus faecalis* aggregation substance encoded by the sex pheromone plasmid pAD1. *Mol. Microbiol.* **4:**895–904.

21. **Gentry-Weeks, C. R., J. M. Keith, A. Pikis, and R. Karkhoff-Schweizer.** 1998. Survival of *Enterococcus faecalis* in mouse peritoneal macrophages and J774A.1 monocyte-macrophage cells, abstr. B-14, p. 58. *Abstr. 98th Gen. Meet. Am. Soc. Microbiol. 1998.* American Society for Microbiology, Washington, D.C.

22. **Gold, O. G., H. V. Jordan, and J. van Houte.** 1975. The prevalence of enterococci in the human mouth and their pathogenicity in animal models. *Arch. Oral Biol.* **20:** 473–477.

23. **Graninger, W., and R. Ragette.** 1992. Nosocomial bacteremia due to *Enterococcus faecalis* without endocarditis. *Clin. Infect. Dis.* **15:**49–57.

24. **Guzman, C. A., C. Pruzzo, and L. Calegari.** 1991. *Enterococcus faecalis*: specific and non-specific interactions with human polymorphonuclear leukocytes. *FEMS Microbiol. Lett.* **68:**157–162.

25. **Guzman, C. A., C. Pruzzo, G. LiPira, and L. Calegari.** 1989. Role of adherence in pathogenesis of *Enterococcus faecalis* urinary tract infection and endocarditis. *Infect. Immun.* **57:**1834–1838.

26. **Guzman, C. A., C. Pruzzo, M. Plate, M. C. Guardati, and L. Calegari.** 1991. Serum dependent expression of

Enterococcus faecalis adhesins involved in the colonization of heart cells. *Microb. Pathog.* **11**:399–409.

27. **Han, D. P., S. R. Wisniewski, L. A. Wilson, M. Barza, A. K. Vine, B. H. Doft, and S. F. Kelsey.** 1996. Spectrum and susceptibilities of microbiologic isolates in the Endophthalmitis Vitrectomy Study. *Am. J. Ophthalmol.* **122**:1–17.

28. **Harvey, B. S., C. J. Baker, and M. S. Edwards.** 1992. Contributions of complement and immunoglobulin to neutrophil-mediated killing of enterococci. *Infect. Immun.* **60**:3635–3640.

29. **Hewitt, W. L., S. J. Seligman, and R. A. Deigh.** 1966. Kinetics of the synergism of penicillin-streptomycin and penicillin-kanamycin for enterococci and its relationship to L-phase variants. *J. Lab. Clin. Med.* **67**:792–807.

30. **Horodniceanu, T., L. Bougueleret, N. El-Solh, G. Bieth, and F. Delbos.** 1979. High-level, plasmid-borne resistance to gentamicin in *Streptococcus faecalis* subsp. *zymogenes*. *Antimicrob. Agents Chemother.* **16**:686–689.

31. **Huycke, M. M., and M. S. Gilmore.** 1995. Frequency of aggregation substance and cytolysin genes among enterococcal endocarditis isolates. *Plasmid* **34**:152–156.

32. **Huycke, M. M., W. Joyce, and M. F. Wack.** 1996. Augmented production of extracellular superoxide by blood isolates of *Enterococcus faecalis*. *J. Infect. Dis.* **173**:743–746.

33. **Huycke, M. M., D. F. Sahm, and M. S. Gilmore.** 1998. Multiple-drug resistant enterococci: the nature of the problem and an agenda for the future. *Emerg. Infect. Dis.* **4**:239–249.

34. **Huycke, M. M., C. A. Spiegel, and M. S. Gilmore.** 1991. Bacteremia caused by hemolytic, high-level gentamicin-resistant *Enterococcus faecalis*. *Antimicrob. Agents Chemother.* **35**:1626–1634.

35. **Ike, Y., H. Hashimoto, and D. B. Clewell.** 1984. Hemolysin of *Streptococcus faecalis* subspecies *zymogenes* contributes to virulence in mice. *Infect. Immun.* **45**:528–530.

36. **Ike, Y., H. Hashimoto, and D. B. Clewell.** 1987. High incidence of hemolysin production by *Enterococcus (Streptococcus) faecalis* strains associated with human parenteral infections. *J. Clin. Microbiol.* **25**:1524–1528.

37. **Jett, B. D., R. V. Atkuri, and M. S. Gilmore.** 1998. *Enterococcus faecalis* localization in experimental endophthalmitis: role of plasmid-encoded aggregation substance. *Infect. Immun.* **66**:843–848.

38. **Jett, B. D., M. M. Huycke, and M. S. Gilmore.** 1994. Virulence of enterococci. *Clin. Microbiol. Rev.* **7**:462–478.

39. **Jett, B. D., H. G. Jensen, R. E. Nordquist, and M. S. Gilmore.** 1992. Contribution of the pAD1-encoded cytolysin to the severity of experimental *Enterococcus faecalis* endophthalmitis. *Infect. Immun.* **60**:2445–2452.

40. **Johnson, A. P.** 1994. The pathogenicity of enterococci. *J. Antimicrob. Chemother.* **33**:1083–1089.

41. **Kreft, B., R. Marre, U. Schramm, and R. Wirth.** 1992. Aggregation substance of *Enterococcus faecalis* mediates adhesion to cultured renal tubular cells. *Infect. Immun.* **60**:25–30.

42. **Lancefield, R. C.** 1933. A serological differentiation of human and other groups of hemolytic streptococci. *J. Exp. Med.* **57**:571–595.

43. **Linden, P. K., A. W. Pasculie, R. Manez, D. J. Kramer, J. J. Fung, A. D. Pinna, and S. Kusne.** 1995. Differences

in outcomes for patients with bacteremia due to vancomycin-resistant *Enterococcus faecium* or vancomycin-susceptible *E. faecium*. *Clin. Infect. Dis.* **22**:663–670.

44. **Lowe, A. M., P. A. Lambert, and A. W. Smith.** 1995. Cloning of an *Enterococcus faecalis* endocarditis antigen: homology with adhesins from some oral streptococci. *Infect. Immun.* **63**:703–706.

45. **Lucas, G. M., N. Lechtzin, W. Puryear, L. L. Yau, C. W. Flexner, and R. D. Moore.** 1998. Vancomycin-resistant and vancomycin-susceptible enterococcal bacteremia: comparison of clinical features and outcomes. *Clin. Infect. Dis.* **26**:1127–1133.

46. **MacCallum, W. G., and T. W. Hastings.** 1899. A case of acute endocarditis caused by *Micrococcus zymogenes* (Nov. Spec.), with a description of the microorganism. *J. Exp. Med.* **4**:521–534.

47. **Maekawa, S., M. Yoshioka, and Y. Kumamoto.** 1992. Proposal of a new scheme for the serological typing of *Enterococcus faecalis* strains. *Microbiol. Immunol.* **36**:671–681.

48. **Mainous, M. R., P. A. Lipsett, and M. O'Brien.** 1997. Enterococcal bacteremia in the surgical intensive care unit. Does vancomycin resistance affect mortality? The Johns Hopkins SICU Study Group. *Arch. Surg.* **132**:76–81.

49. **Megran, D. W.** 1992. Enterococcal endocarditis. *Clin. Infect. Dis.* **15**:63–71.

50. **Miyazaki, S., A. Ohno, I. Kobayashi, T. Uji, K. Yamaguchi, and S. Goto.** 1993. Cytotoxic effect of hemolytic culture supernatant from *Enterococcus faecalis* on mouse polymorphonuclear neutrophils and macrophages. *Microbiol. Immunol.* **37**:265–270.

51. **Moellering, R. C., and A. N. Weinberg.** 1971. Studies on the antibiotic synergism against enterococci: II. Effects of various antibiotics on the uptake of ^{14}C-labeled streptomycin by enterococci. *J. Clin. Invest.* **50**:2580–2584.

52. **Moellering, R. C. J.** 1992. Emergence of *Enterococcus* as a significant pathogen. *Clin. Infect. Dis.* **14**:1173–1178.

53. **Moellering, R. C. J.** 1995. *Enterococcus* species, *Streptococcus bovis*, and *Leuconostac* species, p. 1826–1835. *In* G. L. Mandell, J. E. Bennett, and R. Dolin (ed.), *Principles and Practices of Infectious Diseases*, 4th ed. Churchill Livingston, New York, N.Y.

54. **Montravers, P., J. Mohler, L. Saint Julien, and C. Carbon.** 1997. Evidence of the proinflammatory role of *Enterococcus faecalis* in polymicrobial peritonitis in rats. *Infect. Immun.* **65**:144–149.

55. **Murray, B. E.** 1998. Diversity among multidrug-resistant enterococci. *Emerg. Infect. Dis.* **4**:37–47.

56. **National Nosocomial Infections Surveillance.** 1997. National nosocomial infections surveillance (NNIS) report, data summary from October 1986-April 1997, issued May 1997. *Am. J. Infect. Control* **25**:477–487.

57. **Noble, C. J.** 1978. Carriage of group D streptococci in the human bowel. *J. Clin. Pathol.* **31**:1182–1186.

58. **Novak, R. M., T. J. Holzer, and C. R. Libertin.** 1993. Human neutrophil oxidative response and phagocytic killing of clinical and laboratory strains of *Enterococcus faecalis*. *Diagn. Microbiol. Infect. Dis.* **17**:1–6.

59. **Olmsted, S. B., G. M. Dunny, S. L. Erlandsen, and C. L. Wells.** 1994. A plasmid-encoded surface protein on *Enterococcus faecalis* augments its internalization by cul-

tured intestinal epithelial cells. *J. Infect. Dis.* **170:**1549–1556.

60. **Ortiz, A., J. Sarwar, R. Naso, and A. Fattom.** 1998. Antibodies to an *Enterococcus* type 1 capsular polysaccharide clear bacteremia and organ seeding in mice challenged with a sub-lethal dose of *Enterococcus faecalis* bacteria, abstr. B-25, p. 59. *Abstr. 98th Gen. Meet. Am. Soc. Microbiol. 1998.* American Society for Microbiology, Washington, D.C.

61. **Porwancher, R., A. Sheth, S. Remphrey, E. Taylor, C. Hinkle, and M. Zervos.** 1997. Epidemiological study of hospital-acquired infection with vancomycin-resistant *Enterococcus faecium*: possible transmission by an electronic ear-probe thermometer. *Infect. Control Hosp. Epidemiol.* **18:**771–773.

62. **Sannomiya, P. A., R. A. Craig, D. B. Clewell, A. Suzuki, M. Fujino, G. O. Till, and W. A. Marasco.** 1990. Characterization of a class of nonformylated *Enterococcus faecalis*-derived neutrophil chemotactic peptides: the sex pheromones. *Proc. Natl. Acad. Sci. USA* **87:**66–70.

63. **Sartingen, S., B. Leonard, R. Marre, and E. Rodzinski.** 1998. Aggregation substance promotes the adherence and internalization of *Enterococcus faecalis* by intestinal epithelial cells, abstr. B-79, p. 79. *Abstr. 98th Gen. Meet. Am. Soc. Microbiol. 1998.* American Society for Microbiology, Washington, D.C.

64. **Schleifer, K. H., and R. Kilpper-Balz.** 1984. Transfer of *Streptococcus faecalis* and *Streptococcus faecium* to the genus *Enterococcus* nom. rev. as *Enterococcus faecalis* comb. nov. and *Enterococcus faecium* comb. nov. *Int. J. Syst. Bacteriol.* **34:**31–34.

65. **Schlievert, P. M., P. J. Gahr, A. P. Assimacopoulos, M. M. Dinges, J. A. Stoehr, J. W. Harmala, H. Hirt, and G. M. Dunny.** 1998. Aggregation and binding substances enhance pathogenicity in rabbit models of *Enterococcus faecalis* endocarditis. *Infect. Immun.* **66:**218–223.

66. **Shankar, V., A. S. Baghdayan, and M. S. Gilmore.** 1998. Esp, a novel surface protein enriched among clinical isolates of *Enterococcus faecalis*, abstr. B-17, p. 58. *Abstr. 98th Gen. Meet. Am. Soc. Microbiol. 1998.* American Society for Microbiology, Washington, D.C.

66a. **Shankar, V., A. S. Baghdayan, M. M. Huycke, G. Lindahl, and M. S. Gilmore.** 1999. Infection-derived *Enterococcus faecalis* strains are enriched in *esp*, a gene encoding a novel surface protein. *Infect. Immun.* **67:**193–200.

67. **Sood, R. K., M. Poth, S. Shepherd, A. Patel, R. Naso, and A. Fattom.** 1998. Capsular serotyping of *Enterococcus faecalis*: isolation, characterization, and immunogenicity of capsular polysaccharide isolated from *E. faecalis* type 1, abstr. E-19, p. 238. *Abstr. 98th Gen. Meet. Am. Soc. Microbiol. 1998.* American Society for Microbiology, Washington, D.C.

68. **Stroud, L., J. Edwards, L. Danzing, D. Culver, and R. Gaynes.** 1996. Risk factors for mortality associated with enterococcal bloodstream infections. *Infect. Control Hosp. Epidemiol.* **17:**576–580.

69. **Su, Y. A., M. C. Sulavik, P. He, K. K. Makinen, P. Makinen, S. Fiedler, R. Wirth, and D. B. Clewell.** 1991. Nucleotide sequence of the gelatinase gene (*gelE*) from *Enterococcus faecalis* subsp. *liquefaciens*. *Infect. Immun.* **59:**415–420.

70. **Suessmuth, S., M. Susa, S. Sartingen, R. Marre, and E. Rodzinski.** 1998. *Enterococcus faecalis*: adherence, entry and survival in macrophages is promoted by aggregation substance (AS), abstr. B-84, p. 69. *Abstr. 98th Gen. Meet. Am. Soc. Microbiol. 1998.* American Society for Microbiology, Washington, D.C.

71. **Suzuki, T., C. Shibata, A. Yamaguchi, K. Igarashi, and H. Kobayashi.** 1993. Complementation of an *Enterococcus hirae* (*Streptococcus faecalis*) mutant in the alpha subunit of the H$^+$-ATPase by cloned genes from the same and different species. *Mol. Microbiol.* **9:**111–118.

72. **Todd, E. W.** 1934. A comparative serological study of streptolysins derived from human and from animal infections, with notes on pneumococcal haemolysin, tetanolysin, and staphylococcus toxin. *J. Pathol. Bacteriol.* **39:**299–321.

73. **Tsuchimori, N. R., R. Hayashi, A. Shino, T. Yamazaki, and K. Okonogi.** 1994. *Enterococcus faecalis* aggravates pyelonephritis caused by *Pseudomonas aeruginosa* in experimental ascending mixed urinary tract infection in mice. *Infect. Immun.* **62:**4534–4541.

74. **Wells, C. L., and S. L. Erlandsen.** 1991. Localization of translocating *Escherichia coli*, *Proteus mirabilis*, and *Enterococcus faecalis* within cecal and colonic tissues of monoassociated mice. *Infect. Immun.* **59:**4693–4697.

75. **Wells, C. L., R. P. Jechorek, and S. L. Erlandsen.** 1990. Evidence for the translocation of *Enterococcus faecalis* across the mouse intestinal tract. *J. Infect. Dis.* **162:**82–90.

76. **Wells, C. L., M. A. Maddaus, and R. L. Simmons.** 1988. Proposed mechanisms for the translocation of intestinal bacteria. *Rev. Infect. Dis.* **10:**958–979.

Enterococcal Genetics

KEITH E. WEAVER

26

The majority of interest in enterococcal genetics has been generated in response to three landmark discoveries: (i) identification of the first conjugative plasmids whose transfer systems are induced by an identifiable signal, (ii) identification of the first transposons capable of intercellular transposition, and (iii) the evolution of vancomycin resistance. Since even the genes for vancomycin resistance are generally located on plasmids and transposons, most efforts to date to understand genetic processes in enterococci have focused on extrachromosomal elements. With the recent release by the Institute for Genomic Research (TIGR) of the partial genome sequence of *Enterococcus faecalis*, this situation will undoubtedly change rapidly. I begin this chapter with some preliminary analysis of the *E. faecalis* chromosome sequence, with particular emphasis on homologs of genes known to play a role in genetic processes in other bacteria. This is followed by sections dealing with each of the landmark systems noted above.

GENES INVOLVED IN CHROMOSOMAL REPLICATION, INHERITANCE, AND RECOMBINATION

A preliminary analysis of the *E. faecalis* chromosome sequence, with particular emphasis on the genes involved in replication, inheritance, and recombination, was performed using the partial sequences made available by TIGR on the Internet (35a). Specifically, *E. faecalis* homologs for the *dnaA*, *soj/spoOJ*, *minCDE/divIVA*, *mukB*, and *ruvABC* genes were sought and their genetic context determined. The *E. faecalis recA* and *ftsZ* genes were previously identified (17, 54) and are mentioned briefly below.

In a comparison of genes surrounding the replication origins of *Bacillus subtilis* and *Pseudomonas putida*, Ogasawara and Yoshikawa (49) concluded that gene order in this region was highly conserved among eubacteria. Examination of the *E. faecalis* genome sequence identified a strong *dnaA* homolog flanked by eight conserved *dnaA* boxes (60) (Fig. 1) that likely functions as an origin of chromosomal replication. This conclusion is supported by the identification of *dnaN*, *recF*, *gyrB*, and *gyrA* homologs in the ex-

pected order downstream of *dnaA*. However, downstream of *gyrA* were a number of genes located upstream of *dnaA* in *B. subtilis* (Fig. 2A). Unfortunately, the sequence upstream of *E. faecalis dnaA* was not present on the same contig. In *B. subtilis*, the *thdF*, *gidA*, *gidB*, *soj*, and *spoOJ* genes, among others, are closely linked upstream of *dnaA* (Fig. 2B). Genes homologous to *soj* (55%) and *spoOJ* (71%), implicated in *B. subtilis* chromosomal segregation (18, 71), were identified on another *E. faecalis* contig. As in *B. subtilis*, these genes were flanked by the *gidB* and *yyaF* genes, but some intervening genes were absent in *E. faecalis*. The 5′ end of the *gidB* gene was not located on the contig and could not be identified in the released sequence, so it is impossible to say if the gene order upstream of *gidB* is conserved. A homolog of the *gidA* gene (71%), which is located upstream of *gidB* in *B. subtilis*, *P. putida*, and *Escherichia coli*, was located on a separate *E. faecalis* contig. As in *B. subtilis*, a *thdF* homolog (68%) was located immediately upstream of *gidA*, but the surrounding genes were not related to genes located in the *B. subtilis* origin region. This may represent a major departure from the conservation usually observed around the chromosomal origin, but it is still possible that a second copy of *gidA* and *thdF* exists upstream of *E. faecalis gidB*. No homolog of the putative *E. coli* chromosome segregation gene *mukB* was identified (47), but *mukB* is also absent from the *B. subtilis* genome (39).

The *minCDE* genes of *E. coli* are involved in suppressing polar septation and ensuring proper medial septum placement. *B. subtilis* carries homologs of the *minCD* genes but not the *minE* topological specificity factor. The *divIVA* gene appears to perform the role of *minE* in *B. subtilis* and is also involved in the switch to polar septation during sporulation (7). Since *E. faecalis* cells are cocci and apparently have no poles, it is perhaps not surprising that no strong *min* homologs were identified within the *E. faecalis* sequence. However, a gene with 54% homology to *B. subtilis divIVA* was identified. It is located immediately upstream of the isoleucyl tRNA synthetase gene (*ileS*) as in *B. subtilis* (Fig. 2C). Three genes separate the *E. faecalis divIVA* gene from the key cell division gene *ftsZ*. Two of these genes, *ylmE*

Gram-Positive Pathogens, ed. by V. A. Fischetti et al.
© 2000 American Society for Microbiology, Washington, D.C.

FIGURE 1 Organization of the putative origin of chromosomal replication of *E. faecalis*. (A) Gene order in the origin region. Gene designations were assigned on the basis of homology to gene products identified in database searches. Levels of homology to known proteins were as follows: DnaA, 54% identical to *B. subtilis* DnaA; DnaN, 52% identical to *B. subtilis* DnaN; RecF, 58% identical to *Staphylococcus aureus* RecF; GyrB, 78% identical to *Streptococcus pneumoniae* GyrB; GyrA, 64% identical to *B. subtilis* GyrA; Ssb, 52% identical to *B. subtilis* Ssb; YybT, 38% identical to *B. subtilis* YybT; DnaC, 64% identical to *B. subtilis* DnaC; PurA, 75% identical to *B. subtilis* PurA. (B) Organization of suspected *dnaA* boxes surrounding the *dnaA* gene. Orientation of the boxes is shown by the direction of the arrows relative to the consensus *dnaA* box sequence previously defined (60). Size of boxes and distance between boxes, but not the relative length of the *dnaA* gene, are shown to scale.

and *ylmF*, bear strong resemblance to hypothetical coding regions upstream of the *B. subtilis divIVA* gene. *B. subtilis divIVA* is also linked to the *ftsZ* gene cluster, but 12 genes of various function separate *ftsZ* from *divIVA* in this organism. The conservation of gene order within the *ftsZ* gene cluster between *E. coli*, *B. subtilis*, *Staphylococcus aureus*, and *E. faecalis* has been previously reported (54).

The isolation of an *E. faecalis recA* homolog was previously reported by Dybvig et al. (17). The *recA* homolog was also found in the TIGR sequence but was located on a contig too small to identify neighboring genes. Homologs of the branch migration helicase genes *ruvA* and *ruvB* (70) were identified in the *E. faecalis* sequence. The gene products were 44% and 66% homologous to *B. subtilis* RuvA and RuvB, respectively, but the adjacent genes were not similar to those linked to the *B. subtilis ruvAB* genes. Surprisingly, genes homologous to pAD1 *traA* (25%) and pCF10 *prgZ* (42%) were closely linked downstream and upstream of *ruvAB*, respectively (Fig. 2D). TraA and PrgZ are involved in sensing and transducing the conjugative pheromone signal for these *E. faecalis* plasmids (see below). Although the *E. faecalis* genome sequence probably contains some plasmid sequences, it is unlikely that the *ruvAB* homologs and the associated *traA* and *traC* genes are located on a pheromone-responsive plasmid, since genes surrounding them show similarity to genes located on the *E. coli* and *B. subtilis* chromosome and not plasmid genes. It will be interesting to determine what role, if any, these chromosomal homologs of plasmid-encoded genes play in the biology of the host. No homolog of the Holliday junction resolvase *ruvC* (70) was found in either the *B. subtilis* or *E. faecalis* sequence.

It should be noted that these analyses are based on an incomplete *E. faecalis* sequence. No doubt the gene order and organization of these important genes will be further refined as more sequence data become available.

PHEROMONE-RESPONSIVE PLASMIDS

The enterococci play host to a wide variety of plasmids and may provide a reservoir for dissemination of genetic material via conjugation to other gram-positive organisms.

Three classes of plasmids commonly found in enterococci have been studied in detail. Two of these, the rolling-circle replicating plasmids and the theta replicating pAMβ1 plasmids, are not restricted to the enterococci (i.e., have a broad host range) and have been reviewed extensively elsewhere (9, 37). The third class, the pheromone-responsive plasmids, replicate primarily in *E. faecalis* and remain one of only two plasmid systems in which a specific signal for induction of conjugation has been identified (the quorum-sensing system of *Agrobacterium* Ti plasmids is the other [24]). The reader is referred to several extensive reviews for more detailed information concerning pheromone plasmid systems (12, 15, 16, 66, 73).

Conjugation in cells containing pheromone-responsive plasmids is normally repressed in the absence of an appropriate recipient. When cocultivated with or exposed to culture filtrates (CF) of plasmid-free cells, plasmid-containing cells respond by producing a plasmid-encoded surface adhesin called aggregation substance (AS). AS binds to a ligand called binding substance that is present on the surface of both donor and recipient cells, leading to the formation of macroscopic cellular aggregates that provide the contact required for plasmid transfer. Lipoteichoic acid appears to be at least one component of binding substance. Donor cells containing a particular pheromone plasmid will not aggregate in response to CF from cells containing the same plasmid but will respond to CF from cells containing different pheromone plasmids, indicating a degree of specificity in the signal-sensing apparatus. The pheromone signals specific for four plasmids have been purified from inducing CF. All are small, hydrophobic peptides, seven to eight amino acids long, containing a single hydroxy amino acid. Synthetic peptides of the appropriate sequence function as effective inducing signals for their cognate plasmids. Therefore, unlike the peptide quorum-sensing systems of some other gram-positive bacteria (38), no posttranslational processing is required for activation of the *E. faecalis* pheromones.

Once a cell obtains a copy of a pheromone plasmid, CF from that strain is no longer capable of inducing an aggregation response in cells containing that plasmid. This plasmid-specific pheromone "shutdown" is accomplished by

A

Bs	*oriC-dnaA-dnaN-yyaA-recF-yyaB-gyrB-gyrA-rrnO*
Ef	*oriC-dnaA-dnaN- -recF- -gyrB-gyrA-ssb- -yybT-rpII- -dnaC- -purA*
Bs	*oriC- 14 genes -ssb-37 genes-yybT-rpII-5 genes-dnaC-yycE-purA*

B

Ef	*-gidB- -soj-spoOJ-orf- -yyaF*
Bs	*oriC-rpmH-rnpA-spoIIIJ-jag-thdF-gidA-gidB-yyaA-soj-spoOJ-yyaC-yyaD-yyaE-yyaF*
Ef	*-orf-thdF-gidA-gutD-orf-gutM-pthB*

C

Bs	*ftsA-ftsZ-8 genes-ylmE-ylmF-ylmG-ylmH-divIVA-ileS-*
Ef	*ftsA-ftsZ- -ylmE-ylmF- orf -divIVA-ileS-*

D

Ef	*Bs ycdI-Bs yceA-Bs purR-Bs tms-Bs yfkN-traC-ruvA-ruvB-Ec yhjC-Ec ygjK-traA*

FIGURE 2 Comparison of gene order in selected regions of the E. *faecalis* and B. *subtilis* chromosome. Gene order for B. *subtilis* genes is depicted as in reference 39. Gene order for E. *faecalis* genes was determined from sequence released by TIGR on the Internet (35a). Bs, B. *subtilis*; Ef, E. *faecalis*. Gaps are introduced in some regions to align related genes. Empty gaps indicate that the genes on each side are contiguous on the indicated chromosome. If multiple genes separate the two listed genes, the number of genes is shown. The designation *orf* indicates genes with no matching sequence in the B. *subtilis* chromosome. (A) The E. *faecalis* chromosomal origin of replication and genes downstream of *dnaA*. Figure 1 shows a more detailed map of this region. The top *Bs* sequence represents the analogous region in the B. *subtilis* chromosome. The bottom *Bs* sequence represents the gene order on the opposite side of the origin. The *Ef* sequence was derived from TIGR contig 6232. Note that gene order between the *Ef ori* and *gyrA* is well conserved with *Bs* but that many *Ef* genes downstream of *gyrA* are located on the opposite side of the *Bs ori*. Detail of the first 14 genes adjacent to the *Bs ori* on the opposite side from *dnaA* is shown in B. (B) E. *faecalis* genes related to B. *subtilis* genes on the opposite side of the *ori* from *dnaA*. The *Ef gidB* gene was localized on TIGR contig 6366; the *gidA* gene was on contig 6273. Not shown on the *gidB* fragment are two genes located downstream of *yyaF* with only limited homology to known genes. The genes downstream of *gidA* encode products 50 to 60% identical to proteins involved in sugar metabolism in organisms other than B. *subtilis*. No related genes are located near *gidA* in B. *subtilis*. (C) E. *faecalis* genes in the *ftsZ-divIVA* region. The *divIVA* gene was located on E. *faecalis* TIGR contig 6309. A gene encoding a product homologous to glucose-6-phosphate-1-dehydrogenase was located downstream of *Ef divIVA* and encoded on the opposite strand. No similar gene was linked to *Bs divIVA*. (D) E. *faecalis ruvAB*-containing fragment. The E. *faecalis ruvAB* genes were located on TIGR contig 6209. Surrounding *ruvAB* were genes whose products were ≈50% homologous to hypothetical proteins encoded on the B. *subtilis* (prefaced Bs) and *Escherichia coli* (prefaced Ec) chromosomes. These genes were not localized near *ruvAB* in their respective genomes. The products of the *traC* and *traA* genes were 25% and 42% homologous to their pAD1 and pCF10 homologs, respectively.

the interplay of at least two plasmid-encoded determinants, an integral membrane protein that interferes with the processing and/or secretion of the pheromone and an inhibitor peptide that acts as a competitive inhibitor of the pheromone. The inhibitor peptides are biochemically similar to the pheromones, and each specifically inhibits only the pheromone signal for the plasmid on which it is encoded. The relative importance of these two systems in pheromone shutdown may vary among the pheromone plasmids. For example, in pAD1-containing cells only inhibitor is secreted, indicating that the membrane protein, TraB, effectively limits pheromone secretion. In contrast, in pCF10-containing cells, pCF10-specific pheromone is still secreted despite the synthesis of the TraB homolog PrgY, but is precisely balanced by the production of inhibitor.

Eighteen distinct sex pheromone plasmids have been described, ranging in size from 37 to 91 kb (34). A region of ≈18 kb encoding a functional replicon, a regulatory region for pheromone sensing and response, and surface exclusion and AS genes has been mapped for three pheromone plasmids: pAD1, pCF10, and pPD1. Although significant differences exist in the details of the sequence and functional mechanisms of the genes, the general order of functional units is highly conserved. A generalized map is shown in Fig. 3. Other comparisons have been published previously (16, 23, 34). The "Kill" segment is nearly 100% conserved at the DNA sequence level. It includes three small open reading frames (ORFs) of unknown function and a determinant of ≈400 bp designated *par*. *par* is essential for stable inheritance of the pAD1 replicon and ap-

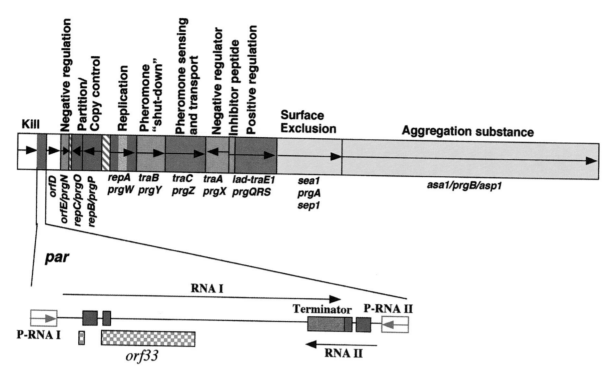

FIGURE 3 Generalized map of an 18-kb segment of three pheromone plasmids, pAD1, pCF10, and pPD1. Determinants colored blue are involved in replication, copy control, and stable inheritance. Hatched blue boxes are sequence repeats involved in plasmid maintenance. The light blue box indicates the poorly conserved internal repeats present within all initiator proteins of the pAD1 class. Red and green colored boxes are genes involved in negative and positive regulation of the pheromone response, respectively. The order of the pheromone shutdown and pheromone-sensing and transport determinants is reversed in pCF10 and pPD1. The yellow boxes are structural genes involved in surface exclusion and aggregation. White boxes indicate genes with no known function. Arrows indicate the direction of transcription of each gene. Gene names for pAD1 and pCF10 genes are shown below the map, pAD1 first. Except for the inhibitor, positive regulation, surface exclusion, and AS genes, pPD1 uses the pAD1 designations. pPD1 designations for surface exclusion, and AS genes are shown last. The mechanism of positive regulation for pPD1 has not been determined in detail, so gene names are not shown. The pPD1 inhibitor gene is *ipd*. Sequence downstream of *repB/prgP* has not been determined for pPD1. The insert shows the organization of the *par* region, required for stable inheritance of the replicon. The green boxes indicate positions of the promoters (P-RNAI and P-RNAII) for the convergently transcribed RNAs. Purple boxes indicate the regions encoding complementary sequences in each RNA. These sequences are probably important for RNAI-RNAII interaction. The checkerboard box designated *orf33* shows the coding region and associated ribosome binding site for a 33-amino-acid ORF believed to encode the toxic component of the *par* postsegregational killing system.

pears to function via a postsegregational killing mechanism (25, 68, 69). A detailed map of *par* is shown in Fig. 3. *par* encodes two convergently transcribed RNAs, RNAI and RNAII, initiating from each end of *par* and terminating at a common bidirectional rho-independent terminator. RNAI is the toxic component of the system and encodes a 33-amino-acid peptide that may be the toxin. Evidence from electron microscopy and flow cytometry indicates that the toxin target is involved in either chromosomal separation or cell division (unpublished observations). RNAII appears to suppress toxin function by an unusual antisense mechanism. Although the RNAI and RNAII genes overlap only at the terminator, each reads in opposite directions across a pair of direct repeats (purple boxes in Fig. 3).

Therefore, RNAII contains regions of complementarity with both the 3′ and 5′ ends of RNAI. Furthermore, the RNAI repeats are positioned such that interaction with RNAII would sequester the ribosome binding site of *orf33*, presumably interfering with its translation. As in other postsegregational killing systems, the *par* antidote (RNAII) is less stable than the toxin (RNAI). This differential stability is essential for activation of the toxin in plasmid-free segregants.

Immediately to the right of *par* in pAD1 are two genes, *orfD* and *orfE*, with unknown functions. Neither has a detectable effect on plasmid maintenance, but they have not been tested for effects on conjugation functions. In pCF10 *orfD* is absent and an *orfE* homolog, designated *prgN*, is located adjacent to *par*. *prgN* has been implicated in negative control of the pheromone response in pCF10. Since *prgN* and *orfE* are 86% homologous, *orfE* might be expected to perform a similar function in pAD1.

Downstream from *orfE* is a segment of DNA, shown in blue in Fig. 3, sufficient for replication and copy control in *E. faecalis* (67). This "basic replicon" consists of three ORFs (solid blue) and a variety of sequence elements (hatched and lighter blue sections) that are involved in replicon function. *repB* and its pCF10 and pPD1 homologs belong to the ParA/SopA family of plasmid partition and chromosome segregation proteins (44, 71). The ParA/SopA proteins are usually cotranscribed with a second group of conserved proteins (ParB/SopB), and the two function together to segregate replicon copies to daughter cells. *repC* and the similarly positioned pCF10 *prgO* gene could be functional homologs of ParB/SopB, but they show no sequence homology with each other or with other *parB/sopB* genes. Transposon insertions in both *repB* and *repC* have been shown to affect plasmid stability, and mutations in *repB* also affect plasmid copy number. The effects of *repB* insertions must be interpreted with caution, however, since they may be polar on *repC*. In addition, it is possible that the copy-up phenotype results from a secondary mutation that partially compensates for severe segregational instability.

Upstream of the *repB* homologs of all three plasmids are a series of DNA sequence repeats; related repeats are present downstream of the *repC* homologs of pAD1 and pCF10 (hatched boxes). The organization and sequence of these repeats are different in each plasmid. The simplest organization is observed in pCF10. Six tandem repeats with a consensus of ATATATXXX are located upstream of *prgP*, with a seventh repeat separated from the last tandem repeat by about 25 bp and positioned such that protein binding could regulate *prgP* transcription. Downstream of *prgO* are two sets of four tandem repeats separated by 10 bp. In pAD1 the repeat consensus sequence is TAGTARRR, and the number of repeats upstream of *repB* is greatly expanded compared with pCF10. A total of 26 repeats are present upstream of *repB*. Twenty-five repeats are in two clusters of 13 and 12 tandem repeats separated by 76 bp. The 26th repeat is separated from the last tandem repeat by 18 bp and is again positioned in a way that suggests it may be involved in regulation of *repB* transcription. Three tandem repeats are located downstream of *repC*. The situation in pPD1 is more complex. Although larger direct and inverted repeats were originally reported upstream of the *repB* homolog (23), work with pAD1 and pCF10 suggested that these large repeats might consist of sets of 8- or 9-bp repeats. Indeed, examination of the repeats revealed the presence of 13 repeats with a consensus sequence of ATTTACACA. Unlike pAD1 and pCF10, these repeats are scattered in small clusters throughout the *repA-repB* intergenic region and are present in both direct and inverted orientations. Again, some of the repeats are positioned such that they could regulate *repB* transcription. The arrangement of any potential *repC* downstream repeats in pPD1 is not known because this region has yet to be sequenced. The arrangement of the *repBC* genes and their associated repeats is reminiscent of the *parAB* systems and their associated partition sites (72), although in these systems the upstream and downstream repeats have distinct sequences for binding ParA for autoregulation and ParB for partition, respectively. It is therefore tempting to propose

that a RepBC complex binds the repeats for both partition and autoregulation of transcription. Deletion of the *repC* downstream repeats in pAD1 resulted in a nonfunctional replicon (67), but it is unclear whether this is due to the disruption of an essential replication function or production of a replicon so unstable it cannot be maintained in the host. Spontaneous expansion of the pAD1 *repB* upstream repeats has also been associated with loss of regulation of the conjugative response and constitutive aggregation (30). The molecular basis of this phenomenon is unknown but may not be directly related to the normal function of the repeats.

The final ORF of the basic replicon, *repA/prgW*, is essential for replication and probably represents the replication initiator. Homologs of these proteins have been identified on four nonpheromone plasmids of varying sizes in *Bacillus natto* (63), *S. aureus* (5, 26), and *Lactobacillus helveticus* (62). In three of these plasmids, the *repA* homolog was shown to be sufficient for replication, and in two plasmids the replication origin was localized within the initiator coding region. Comparison of the sequence and pattern of conservation of these putative replication initiators is instructive. Although each gene contains a series of repeats near the center of the protein coding sequence (light blue shading in Fig. 3), the DNA sequence of the repeats is the least conserved portion of the gene. It is this repeat region that has been implicated as the origin of replication. The N terminus is conserved in the derived amino acid sequence of each initiator protein. The C terminus is even more highly conserved than the N terminus among the pheromone plasmid proteins but is poorly conserved when compared with the nonpheromone plasmid initiators. These results suggest that the pheromone plasmid replicons may be part of a larger replicon family that is widely dispersed among gram-positive bacteria. This conclusion has several interesting evolutionary implications. First, *repBC* and their associated repeats, absent in the nonpheromone plasmids, would appear to be evolutionary add-ons and are probably not essential for replication. However, selected pheromone plasmids may have become so dependent on their stabilization functions that they are effectively essential for maintenance. Second, the pattern of homology suggests that the N terminus of the initiators serves an essential role in initiator function, such as strand separation or interaction with host components of the replication machinery. The C-terminal conservation of pheromone plasmid initiators suggests that this region may perform some unique role essential for pheromone plasmid function. The central repeat regions probably represent the replication origins, which are bound specifically by their cognate initiators. Finally, the fact that these replicons are dispersed among a variety of gram-positive bacteria suggests either that the ancestral replicon predated the divergence of the host organisms, or that it had a host range much broader than is currently observed with pheromone plasmids. Interestingly, it has been observed that PrgW binds pheromone and that pheromone is required for plasmid maintenance. Furthermore, provision of pheromone in a host not normally permissive for plasmid replication, *Lactococcus lactis*, allowed plasmid establishment (15). Perhaps the evolving pheromone plasmids became dependent upon pheromone for replication, as well as conjugation, effectively restricting their host range. Also of interest is the observation that the pAD1 origin of conjugative transfer is located within the *repA* repeat region (1). The colocalization of transfer and replication origins is unique among conjugative plasmids and suggests that RepA and its homologs on other pheromone plasmids may play a very special role in both replication and conjugation.

Downstream of *repA* is a 5- to 6-kb region required for regulation of the conjugative pheromone response (red and green boxes in Fig. 3). All three pheromone plasmids encode determinants within this region with similar functions but widely varying degrees of homology and, in some cases, mechanisms of action. These include (i) a gene encoding an integral membrane protein involved in pheromone shutdown (*traB*/*prgY*); (ii) a gene encoding a protein required for pheromone binding and maximum pheromone sensitivity (*traC*/ *prgZ*); (iii) a gene encoding a negative regulatory protein (*traA*/*prgX*); (iv) a gene encoding the inhibitor peptide; (v) and a positive regulatory region including DNA, RNA, and protein elements. The *traB*/*prgY* and *traC*/*prgZ* genes are probably cotranscribed in both pAD1 and pCF10 even though the order of the genes is reversed. Although both pAD1 TraB and pCF10 PrgY are essential negative regulators of the pheromone response, their effects on pheromone secretion are different. Expression of TraB alone results in undetectable levels of extracellular pheromone, while PrgY expression has no detectable effect on extracellular levels but appears to eliminate a membrane-bound pool of pheromone, thereby protecting the cell from autoinduction. The molecular basis for this difference is unknown. TraC and PrgZ are homologous to the OppA subunit of oligopeptide permeases (Opp) and are postulated to interact directly with the host Opp system to facilitate import of external pheromone. While these proteins increase the sensitivity of plasmid-containing cells to pheromone, they are not essential for an effective pheromone response, suggesting that some pheromone may diffuse directly across the membrane or is imported nonspecifically by the host Opp system. It is possible that TraC/PrgY is essential in nature, where pheromone levels may be much lower than those commonly used to induce a response in the laboratory.

Adjacent to the *traBC* homologs and transcribed in the opposite direction is a key negative regulator of the pheromone response, *traA* in pAD1 and pPD1 and *prgX* in pCF10. Although the *traA*/*prgX* genes are relatively poorly conserved, null mutations in each lead to complete derepression of AS and conjugation. However, certain mutations in the 3′ end of pAD1 *traA* incompletely derepress the system and lose pheromone responsiveness. This phenotype implicates the C terminus of TraA as important for signal transduction, and further work has proved that TraA from both pAD1 and pPD1 binds cognate pheromone (22, 46). In the case of pAD1, the binding site was localized to the C terminus. In pPD1 it was demonstrated that TraA binds pheromone with greater affinity and specificity than the external pheromone-binding protein, TraC, suggesting that TraA is actually the key specificity determinant. These results support a model in which TraC facilitates pheromone binding by the host Opp system, which then passes the peptide into the cytoplasm where it interacts with and inactivates or otherwise alters TraA function. pAD1 TraA has been shown to act in *trans* while pCF10 PrgX appears to act preferentially in *cis* (32).

Upstream of *traA*/*prgX* and divergently transcribed from it are the genes for the respective inhibitor peptides. The inhibitor genes are translated as precursors resembling signal peptides and then processed to mature inhibitor peptides. Transcription from the inhibitor peptide promoter continues downstream into a region critical for control of the pheromone response. However, the mechanism by which regulation occurs is quite different in pAD1 and pCF10 (for detailed models see references 4 and 12). In pAD1, a pair of transcriptional terminators, TTS1 and TTS2, are located between the inhibitor gene, *iad*, and a gene encoding a *trans*-acting positive regulatory protein, *traE1*. In the absence of pheromone, transcription from the *iad* promoter is terminated at the TTS terminators; pheromone induction results in readthrough of the terminators and transcription of *traE1*. TraE1 up-regulates its own transcription from a promoter located within TTS2 and also induces transcription of the structural genes required for conjugation. TraA is known to bind upstream of *iad* and negatively regulate its promoter. It may also participate in antitermination at the TTS terminators, although a mechanism for this function has not been described. Another putative negative regulatory gene, *traD*, was identified between *iad* and TTS1 and is transcribed from the opposite strand (14). It was originally proposed that a 23-amino-acid peptide encoded on *traD* might enhance termination at TTS1/TTS2. However, mutations affecting translation of the peptide had no effect on regulation, indicating that the transcript is the functional product of *traD* (14a), perhaps acting as an antisense regulator.

As in pAD1, the key event in pCF10 positive regulation of the pheromone response is extension of the transcript for the inhibitor peptide, *prgQ*. But in this case, extension appears to be facilitated by pheromone-enhanced translation of downstream proteins that stabilize the longer RNA, rather than by factor-mediated antitermination. Pheromone enhances translation by direct ribosome binding. The pCF10 positive effector is not a *trans*-acting transcriptional activator but the extended RNA itself, designated Q_L. A protein designated PrgS, encoded on Q_L, facilitates binding of Q_L to the ribosome. Only Q_L-bound ribosomes efficiently translate the gene for AS, *prgB*. Antitermination at a site downstream of *prgS* is also required to allow transcription from the *prgQ* promoter through *prgB*. The role of PrgX in this regulatory scheme is currently unclear. The pCF10 regulatory model includes several unique mechanisms that have not been previously described. It will be interesting to learn the molecular details of these mechanisms.

Downstream of the regulatory region are genes for a surface exclusion protein, which reduces transfer between cells containing related plasmids, and AS (34). These genes have >90% identity at the nucleotide sequence level between the three pheromone plasmids, and homologs have been identified by Southern blotting on all pheromone plasmids but pAM373. Each gene is frequently preceded by a small ORF of unknown function. Although genes extending more than 20 kb downstream of the AS gene have been shown to be pheromone inducible (12), little is known about their function. Sequence determination 7.5 kb and 2.7 kb downstream of *asa1* and *asp1*, the AS genes of pAD1 and pPD1, respectively, revealed no sequence or organizational similarity (34). Southern blotting experiments indicated that the genes downstream of *asp1* were fairly restricted in occurrence, found in only one other pheromone plasmid, while the genes downstream of *asa1* were more widespread, occurring even in pPD1 at a different location. Undoubtedly, further examination of these genes will reveal how the AS is assembled on the cell surface, how the conjugation pore is formed, and how the DNA is processed.

Finally, it should be noted that the pheromone plasmids carry a number of genes of interest in addition to those involved in replication, stability, and conjugation. These include the cytolysin (27) and UV resistance genes (50) of pAD1, the bacteriocin gene of pPD1 (65), and a variety of antibiotic-resistance genes, including the first β-lactamase identified in the family *Streptococcaceae* (45) and vancomycin resistance (29).

CONJUGATIVE TRANSPOSONS

As in most other bacterial cells, a variety of mobile genetic elements have been identified in the *E. faecalis* genome. In addition to the common classes of transposons (insertion sequences, composite transposons, and Tn3-type transposons), a new class of transposons, the conjugative transposons, was first identified and described in *E. faecalis*. It is now known that these elements are widespread in grampositive bacteria, and an evolutionarily distinct lineage of conjugative transposons from the gram-negative genus *Bacteroides* has also been studied extensively. The recent discovery of a conjugative transposon in *Salmonella senftenberg* (35) suggests that these types of elements may be more widespread than originally believed and, because of their broad host range, may play a very important role in eubacterial evolution. This discussion is limited to the Tn916-like transposons. The reader is referred to several comprehensive reviews for information on other systems and more detail on the molecular mechanisms of transposition (13, 55, 59, 61).

The current model for conjugative transposition of the Tn916-Tn1545 family is outlined in Fig. 4. Unlike most transposons, integrated conjugative transposons are not flanked by direct repeats of target sequence. Preferred insertion sites tend to be AT-rich and possess intrinsic curvature, but the DNA sequence is not conserved and flanking sequences on each end are heterologous. Excision is carried out by a transposon-encoded enzyme of the λ integrase family. Tn916 integrase (Int) produces staggered cuts at each end of the integrated transposon, flush with the transposon end on one strand and 6 nucleotides into the host sequence on the other. The excised transposon contains 5' overhangs of host-derived sequence on each end; these overhangs are referred to as coupling sequences. The transposon is circularized at the coupling sequences. Since the joint is derived from chromosomal DNA at each end that is not complementary, a 6-bp heteroduplex is formed in the circular intermediate at the coupling sequences. The circular intermediate does not replicate, so the heteroduplex sequence may be maintained until transfer. However, recent evidence indicates that one strand of the heteroduplex is preferentially repaired in *E. faecalis* (42); the left coupling sequence is preferentially retained with the transposon. Transposon DNA is apparently processed for transfer in a manner similar to that in conjugative plasmids, since a *cis*-acting sequence, similar structurally and functionally to plasmid-encoded *oriT*s, has been identified (36). Thus, DNA transfer would occur via formation of a single-strand nick and transfer of the nicked strand into the recipient. The mechanism by which this strand is transported across the donor and recipient cell envelopes is not known, but it does not appear to require cell fusion. How or whether the strand is circularized and its complement synthesized is also unknown. Integration is believed to occur by a reversal of the excision process.

Staggered cuts are made at each end of the coupling sequence and in the chosen target. The coupling sequence and target single-stranded extensions are then "annealed" and ligated to complete integration. Since target selection is not sequence specific, transposon insertion results in heteroduplex formation at each end of the transposon. The heteroduplexes are then resolved after the first round of chromosomal replication or by repair processes.

The complete sequence of Tn916 has been determined and the function of many genes has been identified by Tn5 insertion (21). A physical and functional map is shown in Fig. 5A. Thus far, most work to determine the mechanism of action of specific gene products has focused on Int and Xis. As in λ integration, Int is required for both integration and excision, whereas Xis stimulates excision (57). Int has two DNA-binding domains. The N-terminal domain (Int-N) binds to direct repeat sequences near each end of Tn916 (Fig. 5B). The C-terminal domain (Int-C) binds to the transposon termini and the coupling sequences as well as to target sequences. Int-C contains the conserved active site residues involved in the nicking-closing activities of the Int protein family. Covalent binding of Int-C to the 3' phosphate group of cleaved DNA, presumably via the conserved Tyr residue, has been demonstrated (64), suggesting that Tn916 Int functions via the same mechanism as λ-Int and Flp. The fact that Int must bind and cleave at the coupling sequences, which vary from insert to insert, may explain the wide range of conjugation frequencies (from $<10^{-9}$ to $>10^{-4}$ per donor cell) observed using different donor strains. The Xis protein binds between the Int-C and Int-N binding sites on the left end and just beyond the Int-N binding site on the right end (58). Nuclease protection experiments indicate that, like λ-Xis, Tn916 Xis binds as a dimer, bends the DNA, and may wrap the DNA around itself. Whether Xis facilitates Int function simply by bending the DNA or by protein-protein interaction, or both, is not clear. Host factors may also participate in Tn916 excision.

Tn916 does not mobilize plasmids into which it has integrated, nor does it mobilize chromosomal genes adjacent to its integration site during intercellular transposition. This suggests that some mechanism suppresses utilization of the *oriT* site until after the transposon has excised from its host genome and circularized. Two different models have been proposed to link excision and circularization to expression of conjugation functions; these will be referred to as the TraA (13) and Tet-inducible models (11). The TraA model proposes that the key positive regulator of conjugation is the product of *orf5*, a putative 9.2-kDa protein called TraA. TraA is proposed to activate its own transcription, transcription of the *xis/int* promoter, and transcription from a promoter located upstream of *orf24* proposed to drive cotranscription of all conjugation genes. TraA may bind to conserved sequences located upstream of all three of these promoters. When integrated within the host chromosome, the *traA* promoter is proposed to function at a low level that occasionally results in sufficient TraA production to induce excision. After excision, transcription of *traA* is enhanced by several strong outward-reading promoters located near the right end of the transposon that become fused to the transposon left end upon circularization. The increased production of TraA results in high-level expression of conjugation functions and transposon transfer. Transcription across the fusion point in the circular intermediate also represses Int and Xis production either by transcriptional interference or by an

teicoplanin. Indeed, a single amino acid change in $VanS_B$ allows VanB to respond to teicoplanin (3). The exact signal to which VanS responds is unknown but may differ in the VanA and VanB systems. Recent results suggest that $VanS_A$ may respond to the accumulation of a peptidoglycan precursor, while $VanS_B$ interacts directly with the antibiotic (3a).

In addition to the signaling system, four other determinants are conserved between VanA and VanB: (i) VanH, an enzyme that reduces pyruvate to D-Lac, providing the substrate for depsipeptide synthesis; (ii) VanA/B, an enzyme related to D-Ala-D-Ala ligases with altered substrate specificity, preferentially producing D-Ala-D-Lac precursors; (iii) VanX, a D,D-dipeptidase that hydrolyzes D-Ala-D-Ala dipeptides produced by the host ligase that would compete with D-Ala-D-Lac for incorporation into peptidoglycan precursors; and (iv) VanY, a D,D-carboxypeptidase that hydrolyzes peptidoglycan precursors containing a terminal D-Ala. The VanA cluster encodes an additional gene, vanZ, which confers low-level resistance to teicoplanin by an unknown mechanism. The VanB cluster encodes a gene, vanW, unrelated to any VanA gene and of unknown function. Although the gene order in the two systems differs, the strong conservation, particularly in the vanHAX genes, suggests they have a common origin. Recent evidence suggests that the VanA and VanB ligases may have originated in the glycopeptide-producing actinomycetes (43). Another ligase gene, designated vanD, that was recently identified in E. faecium is 69% homologous to vanA and vanB (51). Homologs of the vanR, vanS, vanY, vanH, and vanX genes were also found linked to vanD (10a).

Both the VanA and VanB gene clusters are frequently, if not always, associated with mobile genetic elements. VanA is located on a Tn3-type transposon designated Tn1546 and is frequently found on conjugative plasmids. In one E. faecium outbreak, a unique mechanism of facilitating horizontal transfer of VanA was observed (31). In this case, a nonconjugative Tn1546-bearing plasmid was observed to integrate into a pheromone plasmid via a mechanism involving a series of insertion sequences located outside of Tn1546. Conjugative transfer was enhanced by the fact that plasmid fusion occurred within a homolog of prgX, a key negative regulator of the pheromone response in pCF10-like plasmids (see above). The resulting insertional inactivation of prgX derepressed conjugation function, and transfer occurred in the absence of pheromone. VanB has been localized on a variety of mobile elements. In some strains, conjugative transfer of VanB has been associated with the movement of large (90 to 250 kb), chromosomally located elements that may be complex conjugative transposons (19). VanB was further localized on one of these elements to a 64-kb composite transposon, Tn1547, capable of transposition to plasmids under laboratory conditions. In other cases, VanB resistance was associated with conjugative plasmids (74), although the identification of similar genes on a variety of plasmids suggested a transposon location for these genes as well. Recently, a VanB gene cluster was localized on a Tn916-like element, designated Tn5382 (10). Although independent mobility of this element was not demonstrated, it appears to be associated with a larger, 130- to 160-kb chromosomally located conjugative element. Interestingly, the insertion site of Tn5382 was immediately downstream of a pbp5 gene associated with high-level ampicillin resistance, explaining the close association of these two resistance markers among clinical E. faecium isolates.

VanC-type resistance is a low-level intrinsic resistance to vancomycin commonly found in enterococcal species other than E. faecalis and E. faecium. Like the vanA and vanB genes, vanC encodes an enzyme with homology to D-Ala-D-Ala ligases. However, the VanC ligase produces D-Ala-D-Ser dipeptides rather than D-Ala-D-Lac depsipeptides. Peptidoglycan precursors terminating in D-Ser have a reduced affinity for vancomycin. vanC genes are chromosomally located and have not been associated with mobile elements.

The fact that some genes for glycopeptide resistance reside on mobile genetic elements has increased concerns that resistance might be transferred to more dangerous gram-positive pathogens. Indeed, transfer to S. aureus, Streptococcus pyogenes, and various Listeria species has been observed under laboratory conditions (6, 8, 48). Among clinical isolates, however, vanA and vanB have been found outside the enterococci in only Oerskovia turbata, Arcanobacterium haemolyticum, Streptococcus bovis, and Bacillus circulans (41, 52, 53). Thus far the few moderately vancomycin-resistant S. aureus clinical isolates appear to resist vancomycin action via a different mechanism (33). Perhaps expression of vanA/vanB is incompatible with the pathogenic mechanisms or survival of S. aureus in its host, or perhaps sufficient opportunities for transfer have not yet occurred in nature. Only time and continued vigilance will tell.

CONCLUDING REMARKS

Because of their uniqueness and accessibility, most work on the genetics of enterococci has focused on mobile genetic elements and their role in disseminating antibiotic resistance. Even so, there is still much work to be done. Most work on the pheromone-responsive plasmids has focused on the signal-sensing and regulatory networks, with work on the replication and maintenance functions receiving some attention. Except for the extensive work done on AS and the recent localization of the origin of transfer, almost nothing is known about the genes involved in mating pair formation or DNA processing. Comparative studies with pheromone plasmids other than the three examined so far would be useful in tracing the evolution of these unique plasmids. A detailed study of pAM373, which appears to be evolutionarily distinct from the other pheromone plasmids, is badly needed. Despite a great deal of effort, the chromosomal genes responsible for pheromone elaboration have yet to be identified. Now that the E. faecalis genome sequence is available, this problem will hopefully be resolved and efforts can be focused on determining how the pheromones are processed and secreted. Indeed, prospective genes for the pCF10, pPD1, and pOB1 specific pheromones have been located and are currently under investigation (14b). The sequences appear to be embedded in lipoprotein signal peptides, as was previously observed for a pheromonelike activity produced by the S. aureus conjugative plasmid pSK41 (20). Recently, a chromosomal gene that appears to facilitate pheromone processing or secretion was also identified (1a). As in the pheromone plasmids, very little is known about mating pair formation or DNA processing in the conjugative transposons, and many details of the mechanism regulating expression of the transfer genes still need to be worked out. With the recent identification of a functional origin of transfer and the development of regulatory models making testable predictions, mechanisms are now available to address these questions. Antimicrobial resistance remains a significant problem in the enterococci and, hopefully, a greater understanding of the variety of mobile genetic elements responsible for the dissemination of the genes responsible will help control this problem. In addition, a greater knowledge of the chromosomal genes required for essential host functions may provide novel targets for future generations of antibiotics.

NOTE ADDED IN PROOF

Recent DNA sequence data downstream of the published pPD1 *repB* sequence (23) revealed the presence of a gene analogous to the pAD1 and pCF10 *repC*/*prgO* genes. Within eight bases of the termination codon of this gene was an extended inverted repeat containing two copies of the repeat sequence identified between the pPD1 *repA* and *repB* genes. Two more direct repeats were identified further downstream. A sequencing error was also identified within the published pPD1 *repB* sequence that resulted in premature termination of the RepB product. Correction of this sequence results in a predicted *repB* product clearly related to pAD1 and pCF10 RepB/PrgP and containing all of the motifs typical of the ParA/SopA partition proteins. Thus, the organization of the pPD1 *repBC* region conforms to that observed in the other pheromone plasmids. Interestingly, two more *repBC* pairs have been identified on plasmids present in other low-G+C gram-positive bacteria. The first pair was identified on the conjugative plasmid pAW63 from *Bacillus thuringiensis* (71a). The published sequence identifies 18 complete and partial, direct and inverted repeats with the consensus sequence AAAGATAC upstream of *rep63B*. Further examination of the sequence downstream of ORF6, the *repC* analog, revealed the presence of another pair of repeats in inverted orientation (1b). In the pAW63 replicon, however, the *repBC* genes and repeats are associated with a replication initiator of the pAMβ1 family, unrelated to the pheromone plasmid initiators. The second *repBC* pair was identified during a computer search of GenBank for homologs of pAD1 *repC*. This pair is located on the *Lactobacillus reuteri* plasmid pTE15 and contains the only protein with significant homology to another RepC (32% homology and 58% similarity to pAD1 RepC) (accession number AF036766). Examination of the sequence upstream of the *repB* homolog revealed a complex series of two different repeats interspersed with one another in direct and inverted orientation. The consensus sequence of these repeats is GTAGTAA and GTTATATATT. No similar repeats were detected downstream of *repC*. Unlike the other plasmids, no apparent replication initiator protein was located upstream of the repeats, and the relationship of this plasmid to the pheromone plasmids is unknown. Nevertheless, these results suggest that the *repBC* genes, with their associated repeats, are organized in an independent module that can interact with unrelated initiator proteins to facilitate the maintenance of plasmids in gram-positive bacteria.

REFERENCES

1. **An, F. Y., and D. B. Clewell.** 1997. The origin of transfer (*oriT*) of the enterococcal, pheromone-responding, cytolysin plasmid pAD1 is located within the *repA* determinant. *Plasmid* **37**:87–94.

1a. **An, F. Y., and D. B. Clewell.** Personal communication.

1b. **Andrup, L.** Personal communication.

2. **Arthur, M., and P. Courvalin.** 1993. Genetics and mechanisms of glycopeptide resistance in enterococci. *Antimicrob. Agents Chemother.* **37**:1563–1571.

3. **Baptista, M., F. Depardieu, P. Reynolds, P. Courvalin, and M. Arthur.** 1997. Mutations leading to increased levels of resistance to glycopeptide antibiotics in VanB-type enterococci. *Mol. Microbiol.* **25**:93–105.

3a. **Baptista, M., P. Rodrigues, F. Depardieu, P. Courvalin, and M. Arthur.** 1999. Single-cell analysis of glycopeptide resistance gene expression in teicoplanin-resistant mutants of a VanB-type *Enterococcus faecalis*. *Mol. Microbiol.* **32**:17–28.

4. **Bensing, B. A., D. A. Manias, and G. M. Dunny.** 1997. Pheromone cCF10 and plasmid pCF10-encoded regulatory molecules act post-transcriptionally to activate expression of downstream conjugation functions. *Mol. Microbiol.* **24**:285–294.

5. **Berg, T., N. Firth, S. Apisiridej, A. Hettiaratchi, A. Leelaporn, and R. A. Skurray.** 1998. Complete nucleotide sequence of pSK41: evolution of staphylococcal conjugative multiresistance plasmids. *J. Bacteriol.* **180**:4350–4359.

6. **Biavasco, F., E. Giovanetti, A. Miele, C. Vignaroli, B. Faccinelli, and P. E. Varaldo.** 1996. In vitro conjugative transfer of VanA vancomycin resistance between clinical enterococci and listeriae of different species. *Eur. J. Clin. Microbiol. Infect. Dis.* **15**:50–59.

7. **Bouché, J.-P., and S. Pichoff.** 1998. On the birth and fate of bacterial division sites. *Mol. Microbiol.* **29**:19–26.

8. **Brisson-Noel, A., S. Dutka-Malen, C. Molinas, R. Leclercq, and P. Courvalin.** 1990. Cloning and heterospecific expression of the resistance determinant *vanA* encoding high-level resistance to glycopeptides in *Enterococcus faecium*. *Antimicrob. Agents Chemother.* **34**:924–927.

9. **Bruand, C., E. L. Chatelier, S. D. Ehrlich, and L. Jannière.** 1993. A fourth class of theta-replication plasmids: the pAMβ1-family from Gram positive bacteria. *Proc. Natl. Acad. Sci. USA* **90**:11668–11672.

10. **Carias, L. L., S. D. Rudin, C. J. Donsky, and L. B. Rice.** 1998. Genetic linkage and cotransfer of a novel, *vanB*-containing transposon (Tn5382) and a low-affinity penicillin-binding protein 5 gene in a clinical vancomycin-resistant *Enterococcus faecium* isolate. *J. Bacteriol.* **180**:4426–4434.

10a. **Casadewall, B., and P. Courvalin.** 1999. Characterization of the *vanD* glycopeptide resistance gene cluster from *Enterococcus faecium* BM4339. *J. Bacteriol.* **181**:3644–3648.

11. **Celli, J., and P. Trieu-Cuot.** 1998. Circularization of Tn916 is required for expression of the transposon-encoded transfer functions: characterization of long tetracycline-inducible transcripts reading through the attachment site. *Mol. Microbiol.* **28**:103–117.

12. **Clewell, D. B.** 1993. Bacterial sex pheromone-induced plasmid transfer. *Cell* **73**:9–12.

13. **Clewell, D. B., S. E. Flannagan, and D. D. Jaworski.** 1995. Unconstrained bacterial promiscuity: the Tn916-Tn1545 family of conjugative transposons. *Trends Microbiol.* **3**:229–236.

14. **de Freire Bastos, M. C., K. Tanimoto, and D. B. Clewell.** 1997. Regulation of transfer of the *Enterococcus faecalis* pheromone-responding plasmid pAD1: temperature-sensitive transfer mutants and identification of a new regulatory determinant, *traD*. *J. Bacteriol.* **179**:3250–3259.

14a. **de Freire Bastos, M. C., H. Tomita, K. Tanimoto, and D. B. Clewell.** 1998. Regulation of the *Enterococcus faecalis* pAD1-related sex pheromone response: analyses of *traD* expression and its role in controlling conjugation functions. *Mol. Microbiol.* **30**:381–392.

14b.**Dunny, G., and D. Clewell.** Personal communication.

15. **Dunny, G. M., and B. A. B. Leonard.** 1997. Cell-cell communication in Gram-positive bacteria. *Annu. Rev. Microbiol.* **51**:527–564.

16. **Dunny, G. M., B. A. B. Leonard, and P. J. Hedberg.** 1995. Pheromone-inducible conjugation in *Enterococcus faecalis*: interbacterial and host-parasite chemical communication. *J. Bacteriol.* **177**:871–876.

17. **Dybvig, K., S. K. Hollingshead, D. G. Heath, D. B. Clewell, F. Sun, and A. Woodard.** 1992. Degenerate oligonucleotide primers for enzymatic amplification of *recA* sequences from gram-positive bacteria and mycoplasmas. *J. Bacteriol.* **174**:2729–2732.

18. **Errington, J.** 1998. Dramatic new view of bacterial chromosome segregation. *ASM News* **64**:210–217.

19. **Evers, S., J. R. Quintiliani, and P. Courvalin.** 1996. Genetics of glycopeptide resistance. *Microb. Drug Resist.* **2**:219–223.

20. **Firth, N., P. D. Fink, L. Johnson, and R. A. Skurray.** 1994. A lipoprotein signal peptide encoded by the staphylococcal conjugative plasmid pSK41 exhibits an activity resembling that of *Enterococcus faecalis* pheromone cAD1. *J. Bacteriol.* **176**:5871–5873.

21. **Flannagan, S. E., L. A. Zitzow, Y. A. Su, and D. B. Clewell.** 1994. Nucleotide sequence of the 18 kb conjugative transposon Tn916 from *Enterococcus faecalis*. *Plasmid* **32**:350–354.

22. **Fujimoto, S., and D. B. Clewell.** 1998. Regulation of the pAD1 sex pheromone response of *Enterococcus faecalis* by direct interaction between the cAD1 peptide mating signal and the negatively regulating, DNA-binding TraA protein. *Proc. Natl. Acad. Sci. USA* **95**:6430–6435.

23. **Fujimoto, S., H. Tomita, E. Wakamatsu, K. Tanimoto, and Y. Ike.** 1995. Physical mapping of the conjugative bacteriocin plasmid pPD1 of *Enterococcus faecalis* and identification of the determinant related to the pheromone response. *J. Bacteriol.* **177**:5574–5581.

24. **Fuqua, W. C., and S. C. Winans.** 1994. A LuxR-LuxI type regulatory system activates *Agrobacterium* Ti plasmid conjugal transfer in the presence of a plant tumor metabolite. *J. Bacteriol.* **176**:2796–2806.

25. **Gerdes, K., A. P. Gultyaev, T. Franch, K. Pedersen, and N. D. Mikkelsen.** 1997. Antisense RNA-regulated programmed cell death. *Annu. Rev. Genet.* **31**:1–31.

26. **Gering, M., F. Götz, and R. Brückner.** 1996. Sequence and analysis of the replication region of the *Staphylococcus xylosis* plasmid pSX267. *Gene* **182**:117–122.

27. **Gilmore, M. S., R. A. Segarra, M. C. Booth, C. P. Bogie, L. R. Hall, and D. B. Clewell.** 1994. Genetic structure of the *Enterococcus faecalis* plasmid pAD1-encoded cytolytic toxin system and its relationship to lantibiotic determinants. *J. Bacteriol.* **176**:7335–7344.

28. **Haldimann, A., S. L. Fisher, L. L. Daniels, C. T. Walsh, and B. L. Wanner.** 1997. Transcriptional regulation of the *Enterococcus faecium* BM4147 vancomycin resistance gene cluster by the VanS-VanR two-component regulatory system in *Escherichia coli* K-12. *J. Bacteriol.* **179**:5903–5913.

29. **Handwerger, S., M. J. Pucci, and A. Kolokathis.** 1990. Vancomycin resistance is encoded on a pheromone response plasmid in *Enterococcus faecium* 228. *Antimicrob. Agents Chemother.* **34**:358–360.

30. **Heath, D. G., F. Y. An, K. E. Weaver, and D. B. Clewell.** 1995. Phase variation of conjugation functions of *En-*terococcus faecalis* plasmid pAD1 relates to changes in number of direct repeat (iteron) sequences. *J. Bacteriol.* **177**:5453–5459.

31. **Heaton, M. P., L. F. Discotto, M. J. Pucci, and S. Handwerger.** 1996. Mobilization of vancomycin resistance by transposon-mediated fusion of a VanA plasmid with an *Enterococcus faecium* sex pheromone-response plasmid. *Gene* **171**:9–17.

32. **Hedberg, P. J., B. A. B. Leonard, R. E. Ruhfel, and G. M. Dunny.** 1996. Identification and characterization of the genes of *Enterococcus faecalis* plasmid pCF10 involved in replication and in negative control of pheromone inducible conjugation. *Plasmid* **35**:46–57.

33. **Hiramatsu, K., H. Hanaki, T. Ino, K. Yabuta, T. Oguri, and F. C. Tenover.** 1997. Methicillin-resistant *Staphylococcus aureus* clinical strain with reduced vancomycin susceptibility. *J. Antimicrob. Chemother.* **40**:135–136.

34. **Hirt, H., R. Wirth, and A. Muscholl.** 1996. Comparative analysis of 18 sex pheromone plasmids from *Enterococcus faecalis*: detection of a new insertion element on pPD1 and implications for the evolution of this plasmid family. *Mol. Gen. Genet.* **252**:640–647.

35. **Hochhut, B., K. Jahreis, J. W. Lengeler, and K. Schmid.** 1997. CTnscr94, a conjugative transposon found in Enterobacteria. *J. Bacteriol.* **179**:2097–2102.

35a.**Institute for Genomic Research.** ftp://ftp.tigr.org/pub/data/e_faecalis

36. **Jaworski, D. D., and D. B. Clewell.** 1995. A functional origin of transfer (*oriT*) on the conjugative transposon Tn916. *J. Bacteriol.* **177**:6644–6651.

37. **Khan, S. A.** 1997. Rolling-circle replication of bacterial plasmids. *Microbiol. Mol. Biol. Rev.* **61**:442–455.

38. **Kleerebezem, M., L. E. N. Quadri, O. P. Kuipers, and W. M. de Vos.** 1997. Quorum sensing by peptide pheromones and two-component signal-transduction systems in Gram-positive bacteria. *Mol. Microbiol.* **24**:895–904.

39. **Kunst, F., et al.** 1997. The complete genome of the Gram-positive bacterium *Bacillus subtilis*. *Nature* **390**:249–256.

40. **Leclercq, R., and P. Courvalin.** 1997. Resistance to glycopeptides in Enterococci. *Clin. Infect. Dis.* **24**:545–556.

41. **Ligozzi, M., G. L. Cascio, and R. Fontana.** 1998. *vanA* gene cluster in a vancomycin-resistant clinical isolate of *Bacillus circulans*. *Antimicrob. Agents Chemother.* **42**:2055–2059.

42. **Manganelli, R., S. Rucci, and G. Pozzi.** 1997. The joint of Tn916 circular intermediates is a homoduplex in *Enterococcus faecalis*. *Plasmid* **38**:71–78.

43. **Marshall, C. G., G. Broadhead, B. K. Leskiw, and G. D. Wright.** 1997. D-Ala-D-Ala ligases from glycopeptide antibiotic-producing organisms are highly homologous to the enterococcal vancomycin-resistance ligases VanA and VanB. *Proc. Natl. Acad. Sci. USA* **94**:6480–6483.

44. **Motallebi-Veshareh, M., D. A. Rouch, and C. M. Thomas.** 1990. A family of ATPases involved in active partitioning of diverse bacterial plasmids. *Mol. Microbiol.* **4**:1455–1463.

45. **Murray, B. E., F. An, and D. B. Clewell.** 1988. Plasmids and pheromone response of the β-lactamase producer *Streptococcus* (*Enterococcus*) *faecalis* HH22. *J. Antimicrob. Chemother.* **32**:547–551.

46. **Nakayama, J., Y. Takanami, T. Horii, S. Sakuda, and A. Suzuki.** 1997. Molecular mechanism of peptide-specific pheromone signaling in *Enterococcus faecalis*: function of pheromone receptor TraA and pheromone binding protein TraC encoded by pPD1. *J. Bacteriol.* **180:**449–456.

47. **Niki, H., R. Imamura, M. Kitaoka, K. Yamanaka, T. Ogura, and S. Hiraga.** 1992. *E. coli* MukB protein involved in chromosome partition forms a homodimer with a rod-and-hinge structure having DNA binding and ATP/GTP binding activities. *EMBO J.* **11:**5101–5110.

48. **Nolbe, W. C., Z. Virani, and R. G. A. Cree.** 1992. Co-transfer of vancomycin and other resistance genes from *Enterococcus faecalis* NCTC 12201 to *Staphylococcus aureus*. *FEMS Microbiol. Lett.* **93:**159–168.

49. **Ogasawara, N., and H. Yoshikawa.** 1992. Genes and their organization in the replication origin region of the bacterial chromosome. *Mol. Microbiol.* **6:**629–634.

50. **Ozawa, Y., K. Tanimoto, S. Fujimoto, H. Tomita, and Y. Ike.** 1997. Cloning and genetic analysis of the UV resistance determinant (*uvr*) on the *Enterococcus faecalis* pheromone-responsive conjugative plasmid pAD1. *J. Bacteriol.* **179:**7468–7475.

51. **Perichon, B., P. Reynolds, and P. Courvalin.** 1997. VanD-type glycopeptide-resistant *Enterococcus faecium* BM4339. *Antimicrob. Agents Chemother.* **41:**2016–2018.

52. **Power, G. M., Y. H. Abdullah, H. G. Talsania, W. Spice, S. Aathithan, and G. L. French.** 1995. *vanA* genes in vancomycin-resistant clinical isolates of *Oerskovia turbata* and *Arcanobacterium* (*Corynebacterium*) *haemolyticum*. *J. Antimicrob. Chemother.* **36:**595–606.

53. **Poyart, C., C. Pierre, G. Quesne, B. Pron, P. Berche, and P. Trieu-Cuot.** 1997. Emergence of vancomycin resistance in the genus Streptococcus: characterization of a *vanB* transferable determinant in *Streptococcus bovis*. *Antimicrob. Agents Chemother.* **41:**24–29.

54. **Pucci, M. J., J. A. Thanassi, L. F. Discotto, R. E. Kessler, and T. J. Dougherty.** 1997. Identification and characterization of cell wall–cell division gene clusters in pathogenic gram-positive cocci. *J. Bacteriol.* **179:**5632–5635.

55. **Rice, L. B.** 1998. Tn*916* family conjugative transposons and dissemination of antimicrobial resistance determinants. *Antimicrob. Agents Chemother.* **42:**1871–1877.

56. **Rice, L. B., and L. L. Carias.** 1997. Transfer of Tn*5385*, a composite, multiresistance chromosomal element from *Enterococcus faecalis*. *J. Bacteriol.* **180:**714–721.

57. **Rudy, C., K. L. Taylor, D. Hinerfeld, J. R. Scott, and G. Churchward.** 1997. Excision of a conjugative transposon *in vitro* by the Int and Xis proteins of Tn*916*. *Nucleic Acids Res.* **25:**4061–4066.

58. **Rudy, C. K., J. R. Scott, and G. Churchward.** 1997. DNA binding by the Xis protein of the conjugative transposon Tn*916*. *J. Bacteriol.* **179:**2567–2572.

59. **Salyers, A. A., N. B. Shoemaker, A. M. Stevens, and L.-Y. Li.** 1995. Conjugative transposons: an unusual and diverse set of integrated gene transfer elements. *Microbiol. Rev.* **59:**579–590.

60. **Schaper, S., and W. Messer.** 1995. Interaction of the initiator protein DnaA of *Escherichia coli* with its DNA target. *J. Biol. Chem.* **270:**17622–17626.

61. **Scott, J. R., and G. G. Churchward.** 1995. Conjugative transposition. *Annu. Rev. Microbiol.* **49:**367–397.

62. **Takiguchi, R., H. Hashiba, K. Aoyama, and S. Ishii.** 1989. Complete nucleotide sequence and characterization of a cryptic plasmid from *Lactobacillus helveticus subsp. jugurti*. *Appl. Environ. Microbiol.* **55:**1653–1655.

63. **Tanaka, T., and M. Ogura.** 1998. A novel *Bacillus natto* plasmid pLS32 capable of replication in *Bacillus subtilis*. *FEBS Lett.* **422:**243–246.

64. **Taylor, K. L., and G. Churchward.** 1997. Specific DNA cleavage mediated by the integrase of conjugative transposon Tn*916*. *J. Bacteriol.* **179:**1117–1125.

65. **Tomita, H., S. Fujimoto, K. Tanimoto, and Y. Ike.** 1997. Cloning and genetic and sequence analysis of the bacteriocin 21 determinant encoded on the *Enterococcus faecalis* pheromone-responsive conjugative plasmid pPD1. *J. Bacteriol.* **179:**7843–7855.

66. **Weaver, K. E.** 1995. pAD1 replication and maintenance, p. 89–98. *In* J. J. Ferretti, M. S. Gilmore, T. R. Klaenhammer, and F. Brown (ed.), *Genetics of Streptococci, Enterococci, and Lactococci*, vol. 85. S. Karger, Basel.

67. **Weaver, K. E., D. B. Clewell, and F. An.** 1993. Identification, characterization, and nucleotide sequence of a region of *Enterococcus faecalis* pheromone-responsive plasmid pAD1 capable of autonomous replication. *J. Bacteriol.* **175:**1900–1909.

68. **Weaver, K. E., K. D. Jensen, A. Colwell, and S. Sriram.** 1996. Functional analysis of the *Enterococcus faecalis* plasmid pAD1-encoded stability determinant *par*. *Mol. Microbiol.* **20:**53–63.

69. **Weaver, K. E., K. D. Walz, and M. S. Heine.** Isolation of a derivative of *Escherichia coli-Enterococcus faecalis* shuttle vector pAM401 temperature sensitive for maintenance in *E. faecalis*, and its use in evaluating the mechanism of pAD1 *par*-dependent stabilization. *Plasmid* **40:**225–232.

70. **West, S. C.** 1998. Processing of recombination intermediates by the RuvABC proteins. *Annu. Rev. Genet.* **31:**213–244.

71. **Wheeler, R. T., and L. Shapiro.** 1997. Bacterial chromosome segregation: is there a mitotic apparatus? *Cell* **88:**577–579.

71a. **Wikes, A. L. Smidt, O. A. Økstad, A.-B. Kolstø, J. Mahillon, and L. Andrup.** 1999. Replication mechanism and sequence analysis of the replicon of pAM63, a conjugative plasmid from *Bacillus thuringiensis*. *J. Bacteriol.* **181:**3193–3200.

72. **Williams, D. R., and C. M. Thomas.** 1992. Active partitioning of bacterial plasmids. *J. Gen. Microbiol.* **138:**1–16.

73. **Wirth, R.** 1994. The sex pheromone system of *Enterococcus faecalis*. More than just a plasmid collection mechanism? *Eur. J. Biochem.* **222:**235–246.

74. **Woodford, N., D. Morrison, A. P. Johnson, A. C. Bateman, J. G. M. Hastings, T. S. J. Elliott, and B. Cookson.** 1995. Plasmid-mediated *vanB* glycopeptide resistance in enterococci. *Microb. Drug Resist.* **1:**235–240.

Pathogenesis of Oral Streptococci

R. R. B. RUSSELL

27

LIFE IN THE MOUTH

The oral cavity represents an environment that is warm, moist, and well provided with nutrients in the form of a steady supply of host-derived macromolecules, with the added bonus of several meals a day supplemented by whatever intermittent snacks a particular individual may fancy. Within the mouth, there is a range of habitats differing in such properties as the supply of oxygen, nutrient flow, pH, and the nature of substratum—hard (teeth) or soft (mucosal tissues). Small wonder then, that the oral cavity supports a rich and abundant microflora, though any colonizing bacterium must have evolved properties to overcome some of the hazards encountered, such as fluctuations in the composition of nutrient supply, local oxygen availability, shear forces due to saliva flow and mastication, and a range of host defense mechanisms.

Species of streptococci are well represented among the bacteria found in the oral cavity, which harbors several hundred different identified species of bacteria—a number that is probably a great underestimate of the true complexity of the flora, since there remain many taxa of uncertain status and many microscopically observable microbes that have not yet been isolated in laboratory culture. Some of these streptococci seem to be ubiquitous among the human populations studied, while others have a rather more restricted distribution. When they have been sought, identical or closely related streptococci have also been found in a wide variety of animal species, so streptococci are clearly part of the normal commensal flora of mammals; the purpose of this chapter is to consider the problems that arise when this commensal relationship breaks down and the oral streptococci become opportunistic pathogens.

THE TAXING PROBLEM OF TAXONOMY

The question of how to classify the oral streptococci has been tackled since the start of the 20th century, but even today one can still find textbooks in which all are lumped together under the descriptive term "viridans streptococci," which refers to the greening reaction, or alpha-hemolysis, produced on blood agar. Such a classification based on he-

molysis is not, however, truly interchangeable with "oral streptococci" since some of these may show beta-hemolysis while the alpha-hemolytic phenotype is not restricted to species of streptococci found in the oral cavity. In a comprehensive review of the history of the developments in classification of the oral streptococci, Whiley and Beighton (59) list 18 currently recognized species and describe the changes in taxonomic position and nomenclature that have taken place. A clearer picture gradually emerges as new technical approaches have been introduced—biotyping (4), serotyping, DNA-DNA hybridization, and, most recently, ribosomal RNA sequencing, all leading to shifts in our understanding. There is no reason to think that this process is completed, with two new species being proposed in 1998 and ongoing arguments about nomenclature. The relationships between a wide range of species determined by rRNA gene sequence analysis by Kawamura et al. (27) are shown in Fig. 1, while Table 1 lists the species regularly isolated from the human oral cavity.

The mitis group contains the largest number of named species of oral streptococci. rRNA data show that *Streptococcus pneumoniae* also lies in this group (27) though its habitat is considered to be the nasopharynx rather than the oral cavity. *S. pneumoniae* is closely related to *Streptococcus mitis* and *Streptococcus oralis*, and there is now a substantial body of evidence showing that there is extensive exchange of genetic information between the species. This was first reported for a penicillin-binding protein when genes in penicillin-resistant isolates of *S. pneumoniae* were shown to have a mosaic structure, with segments clearly identical to genes of *S. oralis* (8, 12). The commensal oral streptococci thus offer a pool of genetic material, which can undergo gene shuffling with an important pathogen and lead to the emergence of resistant strains (43). Such shuffling occurs not just for antibiotic-resistance genes under the pressure of antibiotic selection but has been demonstrated in genes for immunoglobulin A (IgA) protease and the *comCDE* loci required for competence (22, 42). There is thus reason to think that there may be extensive mixing of other genetic traits within the mitis group, since all are naturally competent and share a common response

Gram-Positive Pathogens, ed. by V. A. Fischetti et al.
© 2000 American Society for Microbiology, Washington, D.C.

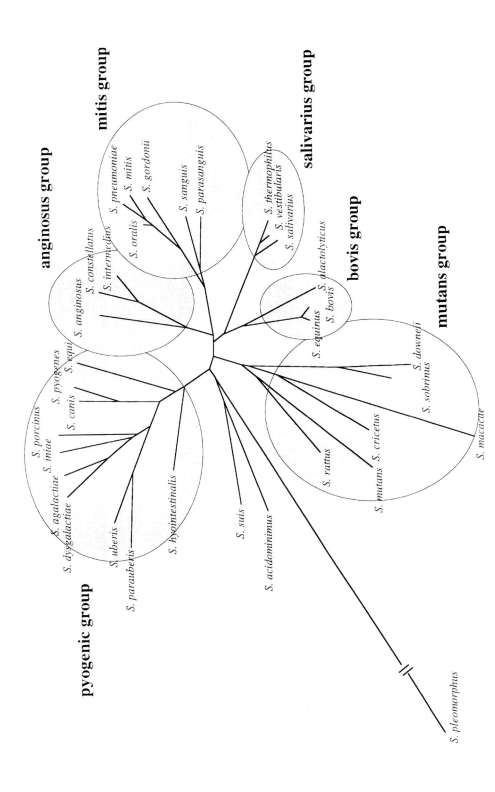

FIGURE 1 Phylogenetic relationships among 34 *Streptococcus* species by 16S rRNA gene sequence analysis. Reproduced from Kawamura et al. (27) with permission. These species found in the human mouth are listed in Table 1.

TABLE 1. Streptococci commonly found in the human mouth

Group	Species	Occurrence	Virulence properties
Mitis group	S. mitis S. oralis S. gordonii S. sanguis S. parasanguis S. crista S. infantis S. peroris	Most common group, pioneer species in plaque colonization, most frequent cause of endocarditis	Survival in bloodstream; attachment to heart valves; induction of tissue damage (inflammation?)
Mutans group	S. mutans S. sobrinus	Mainly found in plaque, associated with dental caries	Colonization of plaque; acidogenesis; aciduricity; extracellular glucan synthesis
Salivarius group	S. salivarius S. vestibularis	Colonize principally mucosal surfaces	Rarely pathogenic
Anginosus group	S. anginosus S. constellatus S. intermedius	Found in gingival crevice, associated with abscesses	Survival in anaerobic conditions; resistance to host defenses; tissue damage by degradative enzymes

mechanism to a competence-stimulating pheromone (22). This raises the question as to whether the currently defined species can be separated by clear boundaries, or whether they represent a continuum, with many mosaic isolates displaying a mixture of properties now regarded as characteristic of individual species (41). This is particularly so for the *S. mitis/S. oralis/S. pneumoniae* group, into which the newly described *Streptococcus infantis* and *Streptococcus peroris* may also fit (28). It is possible that the existence of a high level of natural competence and consequent ready exchange of genes between strains sharing a common pherotype (22) may also explain the observed heterogeneity within the anginosus group (61). In contrast, species boundaries appear to be clearly delineated in the mutans and salivarius groups, which have not been shown to have the *comCDE* operon (22).

WHAT'S IN A NAME?

The properties of *Streptococcus gordonii* and *Streptococcus sanguis* are considered by Jenkinson in chapter 29 of this volume. In the ongoing process of species classification, it has recently been suggested that *S. sanguis*, *Streptococcus parasanguis*, and *Streptococcus crista* should be changed to *S. sanguinis*, *S. parasanguinis*, and *S. cristatus*, respectively, because the latter names are correct in Latin grammar (55). While the argument is reasonable, an alternative view has been put forward (14) pointing out that such changes may be undesirable because of the confusion they would introduce to the literature. In the long run, it is only usage by the scientific community (and editorial policy) that determines what terms stand the test of time, but it should be noted that there is no microbiological, as opposed to linguistic, basis for a change.

ACQUISITION OF ORAL STREPTOCOCCI

The oral cavity is sterile in utero, but although during birth the neonate is exposed to all the complex microflora of the birth canal, these organisms fail to colonize, illustrating the highly selective environment of the mouth. Oral flora is

rapidly established soon after birth, with streptococci being numerically dominant, particularly *S. salivarius*, *S. mitis*, and *S. oralis* (19, 40), which colonize the mucosal surfaces and dorsum of the tongue. From where do these infecting streptococci come? There is little evidence for the source of the species mentioned, but the transmission of *S. mutans* has been studied in some detail. *S. mutans* preferentially colonizes hard surfaces in the mouth and hence its appearance is delayed until eruption of the first teeth, with the majority of infants acquiring it during a "window of infectivity" around a median age of 26 months (7). There is a correlation between the levels of mutans streptococci in the mouths of infants and in their mothers, and fingerprinting methods have been used to demonstrate that the most common source of infection is the mother (34). While an infant may initially carry only one clonal line of *S. mutans* (identified by ribotyping or restriction fragment length polymorphism), by adulthood he or she may have up to seven different clonotypes though these seem to be remarkably stable once established (1, 31, 32). In marked contrast, there is a great degree of diversity in the population structure of *S. mitis* in any one individual, and there is evidence of rapid turnover of dominant clonal types, with any one type being resident for only a few weeks (18, 23, 44). It seems probable that the diversity of types of *S. mitis* is due to its natural competence and the ready exchange of genetic information between strains. In other words, new clonotypes emerge by recombination between residents rather than by superinfection with new strains. This free mixing and shuffling of DNA is also seen in the *Neisseria* spp., which similarly colonize mucosal surfaces, and it has been proposed that it may represent an evolved mechanism for generating antigenic diversity that helps the bacteria to evade the host immune defenses. While this remains to be tested, it is interesting to note that the mutans streptococci, which seem to be genetically stable, are also successful colonizers of the oral cavity.

MECHANISM OF COLONIZATION

The immediate challenge for any bacterium entering the mouth is to avoid being spat out again or swallowed. Early work by Gibbons established both that oral streptococci can adhere to surfaces in the mouth and that they exhibit a specificity of adhesion that may at least partly explain why they colonize particular sites (20, 25, 30). Most work has been done on the formation of dental plaque, which complies with the model for formation of a biofilm: a cleaned tooth surface rapidly becomes coated with a conditioning layer of host-derived and bacterial macromolecules to which particular pioneer species attach, with streptococci (*S. mitis*, *S. oralis*, and *S. sanguis*) being predominant. Plaque then develops by a combination of growth of attached pioneers to form microcolonies and the accretion of further bacteria by interspecies linking. This coaggregation again involves specific binding between bacterial surface adhesins, usually proteins, and receptors, which are generally carbohydrates. There is now an extensive literature on the specificity of these interactions, and elegant proposals have been put forward to explain how a network of binding interactions between different species could lead to the stabilization of a multispecies biofilm. The molecular basis of these interbacterial binding reactions and bacterial binding to host salivary or tissue macromolecules has recently been reviewed (26, 30, 62). However, despite the detailed laboratory knowledge, there remains scant direct evidence of the significance of the contribution of coaggregation to colonization or pathogenesis. One study, in which enamel chips coated with coaggregating and noncoaggregating streptococci were placed in the mouth, found no difference in the subsequent accumulation of species predicted to bind to the target organisms (51). Similarly, the significance of binding host macromolecules is unclear, though there are reports that binding of laminin acts as a signal affecting expression of *S. gordonii* surface proteins (53) and that the ability to bind collagen may be essential for the invasion of dentine tubules by *S. mutans* (see chapter 29, this volume). The collagen binding is associated with the presence of a particular protein called antigen I/II, and experiments with mutants lacking I/II also suggest that it contributes to dental caries, identifying I/II as a virulence determinant (26).

A driving force for the identification of streptococcal binding interactions has been the search for suitable targets for interventions that would moderate plaque formation or the progression of disease (notably dental caries). Thus, it has been proposed that bacterial control could be achieved by provision of an excess of soluble adhesin or receptor molecules, perhaps as a mouthwash, which could disrupt or prevent the establishment of streptococci at their normal attachment site (47). An alternative approach is to exploit the immune system to develop a vaccine approach to control.

IMMUNOLOGICAL PROCESSES IN THE MOUTH

Antibodies to the oral streptococci can be detected from an early age, though the pattern of response against different antigens shows great variability. Thus, studies using Western blot analysis to characterize the response are often difficult to interpret because of cross-reaction between antigens of different species. Despite numerous studies, however, there has emerged no convincing evidence that the immune response controls the developing microflora. This remains one of the great puzzles, not just for the oral cavity but for other permanently colonized mucosal sites—how is

the balance between commensal bacteria and host response maintained in equilibrium? Recent attention has been focused on the contribution of the major secretory antibody, IgA, and of the IgA1 proteases that cleave it (29, 37). Such a protease is produced by *S. sanguis*, *S. mitis*, and *S. oralis*, three species that are among the most successful pioneer species in plaque formation. In the context of the mixed flora of plaque, however, the protease produced by these species may benefit neighboring species too (56), though, again, this effect has yet to be demonstrated in vivo. If IgA does influence the oral microflora, it is clearly unable to eliminate streptococci, and the interplay between the immune system and the oral streptococci must be very subtle. On the other hand, there is now substantial evidence that a strongly enhanced response induced by immunization or by passive application of antibody (36) can influence the ability of streptococci to colonize and/or cause disease, and the topic of a vaccine against dental caries, based upon *S. mutans* antigens such as I/II, has been reviewed (21, 48).

METABOLISM OF DENTAL PLAQUE

All streptococci in the oral cavity are faced with a fluctuating environment, but most is known about those species found in supragingival dental plaque. There, depending on the site in the mouth and the stage of development of plaque, wide variation is experienced in availability of nutrients, oxygen, and pH (38). The streptococci, as facultative anaerobes, are of central importance in the metabolism of plaque, not only because they are the organisms that are well equipped to survive in the fluctuating conditions but because they facilitate the survival of other, strictly anaerobic species (6). The various species of streptococci differ both in their ability to generate acid from dietary carbohydrate and in their acid tolerance. A consequence of this is that frequent exposure to carbohydrate and the consequent fall in pH, to values that may go as low as pH 4.0 in carious lesions, serve to enrich the population of aciduric species and particularly the mutans streptococci (5, 38).

The generation of acid by streptococci occurs in the presence of a broad range of dietary carbohydrates, but sucrose shows a particular association with dental caries. This is in part explained by the widespread use of sucrose in foodstuffs and particularly snack foods, but sucrose is also the sole substrate for extracellular fructosyltransferases and glucosyltransferases of streptococci. The polymers (fructans and glucans) produced by these enzymes are important to overall plaque economy by serving as energy store, altering the permeability of plaque, and stabilizing the plaque biofilm by bacteria-bacteria and bacteria-surface adhesion (9). These aspects are considered in more detail by Kuramitsu in chapter 28 of this volume.

ORAL STREPTOCOCCI AS PATHOGENS

The oral streptococci may be viewed as opportunistic pathogens, their opportunities occurring under two broad groups of conditions: (i) when environmental changes result in an ecological shift that favors an overgrowth of particular species of streptococci, and (ii) when they escape their normal habitat and establish at another site, usually as a consequence of some breakdown in normal host defenses.

DENTAL CARIES

The disease of dental caries manifests as localized dissolution of the enamel surface of the tooth, which may advance to reach the underlying dentine and even penetrate as far as the tooth pulp. Caries lesions most commonly occur in pits and fissures, or in the area of contact between adjacent teeth; i.e., at stagnation sites where plaque has the chance to accumulate undisturbed. Decay is most frequently found in the occlusal (biting) surfaces of molar teeth but under certain circumstances may show a different distribution, for example, in the rapidly progressing "nursing bottle caries" of infants put to sleep with their incisor teeth effectively bathed in a sugar-rich solution from the feeding bottle, dummy, or even the mother's breast. The central pathogenic process of caries, in which plaque bacteria utilize dietary carbohydrate to generate the acid that demineralizes enamel, has been apparent for over a century, but the complexity of the plaque microflora rendered identification of causative organisms a difficult task. Debate raged between proponents of the specific plaque hypothesis, which proposed that particular species were responsible, and the nonspecific plaque hypothesis, which advanced the view that all plaque, regardless of its composition, was bad. The most recent variant, advanced by Marsh and Bradshaw (38), is the ecological plaque hypothesis, which applies our understanding of the properties of the various plaque species to explain the shifts in the bacterial population under the environmental stress of frequent doses of sucrose or other fermentable carbohydrate. Thus, the drops in pH following sugar exposure favor the bacteria that survive and multiply well under these conditions. If these species are themselves good acid producers, there is a ratchet effect whereby the increased number of bacteria produce even more acid at the next sugar exposure, so there is a shift in the composition of the plaque population.

Of all the species of bacteria found in dental plaque, only the mutans streptococci have been shown to have a clear association with the initiation of dental caries. Cross-sectional and longitudinal studies, carried out principally in the United States and Scandinavia, show that increased levels of mutans streptococci are strong predictors of current or future decay (35). Since S. mutans is the predominant species, being found in the vast majority of individuals, it seems clear that it is the principal etiological agent. However, most studies have not distinguished between S. mutans and Streptococcus sobrinus, so, although the latter is much less frequent, its contribution to caries in individual cases is unclear. While S. mutans is now generally accepted as being of prime importance, it is also recognized that caries can occur even when this species cannot be detected. This indicates that there are other bacteria, or combinations of bacterial species, that may fill the same ecological niche as S. mutans and hence generate the same conditions in plaque that lead to sufficient acid production to cause decay. According to the ecological plaque hypothesis, it will be the niche that is of importance rather than the particular species, which happens to fit the unique physiological profile of that niche at any particular time. Not surprisingly, bacteria that are isolated from carious lesions and that share properties with S. mutans are also mostly streptococci, belonging to a variety of species, though the features of these isolates of non-mutans streptococci that make them more acidogenic and aciduric than is normally characteristic of their species remain to be discovered (49).

The known association between diet and dental caries should lead to caution in extrapolating results from one human population to another. Surveys in a number of African countries have revealed that high levels of mutans streptococci can be found with no correspondingly high incidence of dental caries (57). This underscores the importance of diet in creating the local conditions that give the oral streptococci the opportunity to become pathogens.

From the above, it is apparent that the full complement of biological properties of mutans streptococci (and other streptococci that on occasion cause caries) contribute to pathogenicity, ranging from initial adherence and establishment in plaque through to multiplication in competition with other plaque bacteria, carbohydrate fermentation, and acid tolerance. In the absence of a cariogenic dietary challenge, mutans streptococci are harmless commensals in balance with other plaque bacteria and the host defenses. The most straightforward way to avoid disturbance of this balance, which may lead to disease, is therefore dietary control. For alternative preventative or therapeutic strategies directed against S. mutans or other plaque bacteria, we are provided with any number of possible targets, since many cellular components are essential for successful colonization and growth of mutans streptococci in the mouth (47).

PERIODONTAL DISEASE

Periodontal diseases affect the soft tissues and bone supporting the tooth and are acknowledged to have a microbial etiology, though the nature of the host response plays a powerful role in the occurrence and severity of disease. Streptococci are prominent members of the plaque flora above the gingival margin but subgingivally are displaced by gram-negative anaerobic rods. Investigations into the microbiology of periodontal disease have indicated that high numbers of streptococci at a site are generally an indicator of health, though a recent report found that Streptococcus constellatus was found at higher prevalence in patients with refractory periodontitis (10). Conditions in the periodontal pocket are alkaline as a result of proteolysis of blood and tissue components, so it is not surprising that species such as S. mutans are not favored. Why S. constellatus may flourish in disease, and not the other members of the anginosus group that are also normally present in the gingival crevice, is not clear.

ABSCESSES

The pulp of the tooth and the periapical region around the root are normally sterile, but when bacteria gain access, e.g., owing to extensive caries or to trauma to the tooth, they may multiply within the confined space to produce an abscess. A wide range of species can be isolated from dental abscesses, but individual abscesses generally harbor only three or four species (33). Among the species isolated, streptococci are the most common, particularly S. anginosus, which in one study was found in 33% of acute dentoalveolar abscesses (17). All three species of the anginosus group can be isolated from the oral cavity, particularly in the gingival crevice around the margins of the teeth. However, in contrast to the other oral streptococci, which are rarely found in other body sites, they also appear to be normal inhabitants of the gastrointestinal tract and vagina. The three species are also frequently isolated from purulent infections of internal organs, including the brain, liver, lungs, and spleen, with some indication that each shows a predilection for particular sites (60). Most abscesses are mixed infections, and it has been proposed that the facultatively anaerobic streptococci are important in the early stages of abscess formation, establishing local conditions suitable for subsequent invasion by strict anaerobes (50). Further development of the abscesses may depend upon synergistic interactions in which certain species benefit others by inactivation of host defenses or a number of species form a nutritional "consortium" for the efficient sequential degradation of host macromolecules (particularly glycoproteins). In this regard, the production by the anginosus group of a spectrum of degradative enzymes allows them to multiply in a location where they are remote from any diet and must obtain nutrients by breakdown of host tissue. Enzymes produced include glycosidases, nucleases, hyaluronidase, sialidase, and chondroitin sulfate depolymerase (24). There has been a report of a novel cytotoxin, intermedilysin, produced by *Streptococcus intermedius* (39). This cytotoxin shows some homology to pneumolysin but has a different spectrum of action. This is the only report of such a cytotoxin produced by one of the oral streptococci.

SYSTEMIC INFECTIONS

Because the anginosus streptococci are naturally found at several body sites, there is no a priori case to consider the oral cavity the source of infection for abscesses elsewhere in the body. There is the strong possibility that streptococci from a periapical abscess would gain entry to the bloodstream, but there have been no definitive reports, rather than circumstantial evidence, linking an oral infection to a distant abscess. For other oral streptococci, however, there is a much stronger case for believing in an oral source for disseminated infections. Electrophoretic fingerprinting techniques have demonstrated a correspondence between oral isolates and isolates from blood (45), infected prosthetic joint replacements (3), and heart valve lesions (16). Even very mild manipulations in the mouth can result in transfer of bacteria into the bloodstream—one study in children found a transient bacteremia in 38.5% of episodes of toothbrushing (46)—while procedures such as dental extraction (especially if there is an abscess underlying the extracted tooth) will produce a massive shower of organisms, including streptococci, into the circulation. In healthy individuals this poses little problem, but the consequences can be severe if the defense mechanisms are defective, as, for example, in patients with neutropenic cancer. *S. mitis* has been reported to be the species most frequently isolated, with a disturbingly high incidence of antibiotic-resistant strains (2).

The systemic disease with which oral streptococci are most clearly associated is infective endocarditis (sometimes referred to as subacute bacterial endocarditis). Approximately half of all cases of infective endocarditis are attributable to streptococci, the most common species identified being *S. sanguis*, *S. oralis*, and *S. gordonii* (11). The properties of these species are considered further by Jenkinson in chapter 29 of this volume.

VIRULENCE FACTORS

The oral streptococci are normal commensals of the human mouth and as such play a beneficial role in colonization resistance, excluding potentially pathogenic species. Under environmental stress such as imposition of a high-sugar diet or changes in saliva flow, homeostasis is disrupted and a microbial population capable of the sustained acid attack that results in dental caries becomes established. The mutans streptococci are considered the principal etiological agents of caries, and we now know a good deal about their properties. However, the challenges for the future are not just to explore the molecular pathogenesis of *S. mutans* but to understand the complex interactions in the plaque ecosystem, the balance between harmless and harmful species, and the means by which both these types of bacteria maintain a balance with the protective factors of the host. Dental caries is a consequence of acid production by the universal process of glycolysis. It seems clear that in a similar way the anginosus group causes tissue damage in abscesses simply by being present in increased numbers and carrying out their normal metabolic activity, degrading macromolecules to derive a source of carbon and nitrogen. These metabolic processes thus all can contribute to the virulence of the streptococci and are amenable to study in laboratory model systems such as chemostats. Much less is known about how the oral streptococci interact with the cells of an infected host, and we are only beginning to gather information on how they can induce or suppress cellular functions. It is now clear, however, that strong cellular responses can be elicited by streptococcal surface components (13, 15, 52, 54, 58), and further study of these responses should yield new insights into pathogenic mechanisms in abscess formation and valve damage in endocarditis.

REFERENCES

1. **Alaluusua, S., S. J. Alaluusua, J. Karjalainen, M. Saarela, T. Holttinen, M. Kallio, P. Höältta, H. Torkko, P. Relander, and S. Asikainen.** The demonstration by ribotyping of the stability of oral *Streptococcus mutans* infection over 5 to 7 years in children. *Arch. Oral Biol.* **39:**467–471.

2. **Alcaide, F., J. Carratala, J. Liñares, F. Gudiol, and R. Martin.** 1996. In vitro activities of eight macrolide antibiotics and RP-59500 (quinpristin-dalfopristin) against viridans group streptococci isolated from blood of neutropenic cancer patients. *Antimicrob. Agents Chemother.* **40:**2117–2120.

3. **Bartzokas, C. A., R. Johnson, M. Jane, M. V. Martin, P. K. Pearce, and Y. Saw.** 1994. Relation between mouth and haematogenous infection in total joint replacements. *Br. Med. J.* **309:**506–508.

4. **Beighton, D., J. M. Hardie, and R. A. Whiley.** 1991. A scheme for the identification of viridans streptococci. *J. Med. Microbiol.* **35:**367–372.

5. **Bowden, G. H., and I. R. Hamilton.** 1998. Survival of oral bacteria. *Crit. Rev. Oral Biol. Med.* **9:**54–85.

6. **Bradshaw, D. J., P. D. Marsh, G. K. Watson, and C. Allison.** 1997. Oral anaerobes cannot survive oxygen stress without interacting with facultative/aerobic species as a microbial community. *Lett. Appl. Microbiol.* **25:**385–387.

7. **Caufield, P. W., G. R. Cutter, and A. P. Dasanayake.** 1993. Initial acquisition of mutans streptococci by infants: evidence for a discrete window of infectivity. *J. Dent. Res.* **72:**37–45.

8. **Coffey, T. J., C. G. Dowson, M. Daniels, and B. G. Spratt.** 1993. Horizontal spread of an altered penicillin-binding protein 2B gene between *Streptococcus pneumoniae* and *Streptococcus oralis*. *FEMS Microbiol. Lett.* **110:**335–339.

9. **Colby, S. M., and R. R. B. Russell.** 1997. Sugar metabolism by mutans streptococci. *J. Appl. Bacteriol. Symp. Suppl.* **83:**80S–88S.

10. **Colombo, A. P., A. D. Haffajee, F. E. Dewhirst, B. J. Paster, C. M. Smith, M. A. Cugini, and S. S. Sockransky.** 1998. Clinical and microbiological features of refractory periodontitis. *J. Clin. Periodontol.* **25:**169–180.

11. **Douglas, C. W. I., J. Heath, K. K. Hampton, and F. E. Preston.** 1993. Identity of viridans streptococci isolated from cases of infective endocarditis. *J. Med. Microbiol.* **39:**179–182.

12. **Dowson, C. G., T. J. Coffey, C. Kell, and R. A. Whiley.** 1993. Evolution of penicillin resistance in *Streptococcus pneumoniae*; the role of *Streptococcus mitis* in the formation of a low affinity PBP2B in *S. pneumoniae*. *Mol. Microbiol.* **9:**635–643.

13. **Elting, L. S., G. P. Bodey, and B. H. Keefe.** 1992. Septicemia and shock syndrome due to viridans streptococci: a case-control study of predisposing factors. *Clin. Infect. Dis.* **14:**1201–1207.

14. **Euzéby, J. P.** 1998. Proposal to amend rule 61 of the international code of nomenclature of bacteria (1990 revision). *Int. J. Syst. Bacteriol.* **48:**611–612.

15. **Ferreira, P., A. Brás, D. Tavares, M. Vilanova, A. Ribeiro, A. Videira, and M. Arala-Chaves.** 1997. Purification, and biochemical and biological characterization of an immunosuppressive and lymphocyte mitogenic protein secreted by *Streptococcus sobrinus*. *Int. Immunol.* **9:**1735–1743.

16. **Fiehn, N.-E., E. Gutschik, T. Larsen, and J. M. Bangsborg.** Identity of streptococcal blood isolates and oral isolates from two patients with infective endocarditis. *J. Clin. Microbiol.* **33:**1399–1401.

17. **Fisher, L. E., and R. R. B. Russell.** 1993. The isolation and characterisation of milleri group streptococci from dental periapical abscesses. *J. Dent. Res.* **72:**1191–1193.

18. **Fitzsimmons, S., M. Evans, C. Pearce, M. J. Sheridan, R. Wientzen, G. Bowden, and M. F. Cole.** 1996. Clonal diversity of *Streptococcus mitis* biovar 1 isolates from the oral cavity of human neonates. *Clin. Diagn. Lab. Immunol.* **3:**517–522.

19. **Frandsen, E. V. G., V. Pedrazzoli, and M. Kilian.** 1991. Ecology of viridans streptococci in the oral cavity and pharynx. *Oral Microbiol. Immunol.* **6:**129–133.

20. **Gibbons, R. J.** 1984. Microbial ecology. Adherent interactions which may affect microbial ecology in the mouth. *J. Dent. Res.* **63:**378–385.

21. **Hajishengallis, G., and S. M. Michalek.** 1999. Current status of a mucosal vaccine against dental caries. *Oral Microbiol. Immunol.* **14:**1–20.

22. **Håvarstein, L. S., R. Hakenbeck, and P. Gaustad.** 1997. Natural competence in the genus *Streptococcus*: evidence that streptococci can change pherotype by interspecies recombinational exchanges. *J. Bacteriol.* **179:**6589–6594.

23. **Hohwy, J., and M. Kilian.** 1995. Clonal diversity of the *Streptococcus mitis* biovar 1 population in the human oral cavity and pharynx. *Oral Microbiol. Immunol.* **10:**19–25.

24. **Homer, K. H., L. Denbow, R. A. Whiley, and D. Beighton.** 1993. Chondroitin sulphate depolymerase and hyaluronidase activities of viridans streptococci determined by a sensitive spectrophotometric method. *J. Clin. Microbiol.* **31:**1648–1651.

25. **Hsu, S. D., J. O. Cisar, A. L. Sandberg, and M. Kilian.** 1994. Adhesive properties of viridans streptococcal species. *Microb. Ecol. Health Dis.* **7:**125–137.

26. **Jenkinson, H. F., and R. J. Lamont.** 1997. Streptococcal adhesion and colonization. *Crit. Rev. Oral Biol. Med.* **8:**175–200.

27. **Kawamura, Y., X.-G. Hou, F. Sultana, H. Miura, and T. Ezaki.** 1995. Determination of 16S rRNA sequences of *Streptococcus mitis* and *Streptococcus gordonii* and phylogenetic relationships among members of the genus *Streptococcus*. *Int. J. Syst. Bacteriol.* **45:**406–408.

28. **Kawamura, Y., X.-G. Hou, Y. Todome, F. Sultana, K. Hirose, S.-E. Shu, T. Ezaki, and H. Ohkuni.** 1998. *Streptococcus peroris* sp. nov. and *Streptococcus infantis* sp. nov.: new members of the *Streptococcus mitis* group, isolated from human clinical specimens. *Int. J. Syst. Bacteriol.* **48:**921–927.

29. **Kilian, M., J. Reinholdt, H. Lomholt, K. Poulsen, and E. V. G. Frandsen.** 1996. Biological significance of IgA1 proteases in bacterial colonization and pathogenesis: critical evaluation of experimental evidence. *APMIS* **104:**321–338.

30. **Kolenbrander, P. E., and J. London.** 1993. Adhere today, here tomorrow: oral bacterial adherence. *J. Bacteriol.* **175:**3247–3252.

31. **Kozai, K., D. S. Wang, H. J. Sandham, and H. I. Phillips.** 1991. Changes in strains of mutans streptococci induced by treatment with chlorhexidine varnish. *J. Dent. Res.* **70:**1252–1257.

32. **Kulkarni, G. V., K. H. Chan, and H. J. Sandham.** 1989. An investigation into the use of restriction endonuclease analysis for the study of transmission of mutans streptococci. *J. Dent. Res.* **68:**1155–1161.

33. **Lewis, M. A. O., T. W. MacFarlane, and D. A. McGowan.** 1990. A microbiological and clinical review of the acute dentoalveolar abscess. *Br. J. Oral Maxillofac. Surg.* **28:**359-366.

34. **Li, Y., and P. W. Caufield.** 1995. The fidelity of initial acquisition of mutans streptococci by infants from their mothers. *J. Dent. Res.* **74:**681–685.

35. **Loesche, W. J.** 1986. Role of *Streptococcus mutans* in human dental decay. *Microbiol. Rev.* **50:**353–380.

36. **Ma, J. K.-C., B. Y. Hikmat, K. Wycoff, N. D. Vine, D. Chargelegue, L. Yu, M. B. Hein, and T. Lehner.** 1998. Characterization of a recombinant plant monoclonal secretory antibody and preventive immunotherapy in humans. *Nat. Med.* **4:**601–606.

37. **Marcotte, H., and M. C. Lavoie.** 1998. Oral microbial ecology and the role of salivary immunoglobulin A. *Microbiol. Mol. Biol. Rev.* **62:**71–109.

38. **Marsh, P. D., and D. J. Bradshaw.** 1997. Physiological approaches to the control of oral biofilms. *Adv. Dent. Res.* **11:**176–185.

39. **Nagamune, H., C. Ohnishi, A. Katsuura, K. Fushitani, R. A. Whiley, A. Tsuji, and Y. Matsuda.** 1996. Intermedilysin, a novel cytotoxin specific for human cells, secreted by *Streptococcus intermedius* UNS46 isolated from a human liver abscess. *Infect. Immun.* **64:**3093–3100.

40. Pearce, C., G. H. Bowden, M. Evans, S. P. Fitzsimmons, J. Johnson, M. J. Sheridan, R. Wirntzen, and M. F. Cole. Identification of pioneer viridans streptococci in the oral cavity of human neonates. *J. Med. Microbiol.* **42:**67–72.

41. Poulsen, K., and M. Kilian. 1998. Genetic relationships between *Streptococcus pneumoniae*, *Streptococcus mitis*, and *Streptococcus oralis*, abstr. 2D-05, p. 81. ASM *Conference on Streptococcal Genetics, Vichy, France.* American Society for Microbiology, Washington, D.C.

42. Poulsen, K., J. Reinholdt, C. Jespersgaard, K. Boye, T. A. Brown, M. Hauge, and M. Kilian. 1998. A comprehensive genetic study of streptococcal immunoglobulin A1 proteases: evidence for recombination within and between species. *Infect. Immun.* **66:**181–190.

43. Reichmann, P., A. König, J. Liñares, F. Alcaide, F. C. Tenover, L. McDougal, S. Swidsinski, and R. Hakenbeck. 1997. A global gene pool for high-level cephalosporin resistance in commensal *Streptococcus* species and *Streptococcus pneumoniae*. *J. Infect. Dis.* **176:**1001–1012.

44. Reinholdt, J., and M. Kilian. 1998. Clonal diversity and turnover of oral streptococci in humans, abstr. 2D-02, p. 81. ASM *Conference on Streptococcal Genetics, Vichy, France.* American Society for Microbiology, Washington, D.C.

45. Richard, P., G. Amador Del Valle, P. Moreau, N. Milpied, M. P. Felice, T. Daeschler, J. L. Harousseau, and H. Richet. 1995. Viridans streptococcal bacteraemia in patients with neutropenia. *Lancet* **345:**1607–1609.

46. Roberts, G. J., H. S. Holzel, M. R. J. Sury, N. A. Simmons, P. Gardner, and P. Longhurst. 1997. Dental bacteremia in children. *Pediatr. Cardiol.* **18:**24–27.

47. Russell, R. R. B. 1994. Control of specific plaque bacteria. *Adv. Dent. Res.* **8:**285–290.

48. Russell, R. R. B., and N. W. Johnson. 1987. The prospects for vaccination against dental caries. *Br. Dent. J.* **162:**29–34.

49. Sansone, C., J. van Houte, K. Joshipura, R. Kent, and H. C. Margolis. 1993. The association of mutans streptococci and non-mutans streptococci capable of acidogenesis at a low pH with dental caries on enamel and root surfaces. *J. Dent. Res.* **72:**508–516.

50. Shinzato, T., and A. Saito. 1994. A mechanism of pathogenicity of "*Streptococcus milleri* group" in pulmonary infection: synergy with an anaerobe. *J. Med. Microbiol.* **40:**118–123.

51. Skopek, R. J., W. F. Liljemark, C. G. Bloomquist, and J. D. Rudney. 1993. Dental plaque development on defined streptococcal surfaces. *Oral Microbiol. Immunol.* **8:**16–23.

52. Soares, R., P. Ferreira, M. M. G. Santarem, M. Teixeira da Silva, and M. Arala-Chaves. 1990. Low T- and B-cell reactivity is an apparently paradoxical request for murine immunoprotection against *Streptococcus mutans*. *Scand. J. Immunol.* **31:**361–366.

53. Sommer, P., C. Gleyzal, S. Guerret, J. Etienne, and J. A. Grimaud. 1992. Induction of a putative laminin-binding protein of *Streptococcus gordonii* in human infective endocarditis. *Infect. Immun.* **60:**360–365.

54. Stinson, M. W., R. McLaughlin, S. H. Choi, Z. E. Juarez, and J. Barnard. 1998. Streptococcal histone-like protein: primary structure of *hlpA* and protein binding to lipoteichoic acid and epithelial cells. *Infect. Immun.* **66:**259–265.

55. Trüper, H. G., and L. de' Clari. 1997. Necessary correction of specific epithets formed as substantives (nouns) "in apposition." *Int. J. Syst. Bacteriol.* **47:**908–909.

56. Tyler, B. M., and M. F. Cole. 1998. Effect of IgA1 protease on the ability of secretory IgA1 antibodies to inhibit the adherence of *Streptococcus mutans*. *Microbiol. Immunol.* **42:**503–508.

57. van Palenstein Helderman, W. H., M. I. N. Matee, J. S. van der Hoeven, and F. H. M. Mikx. 1996. Cariogenicity depends more on diet than the prevailing mutans streptococcal species. *J. Dent. Res.* **75:**535–545.

58. Vernier, A., M. Diab, M. Soell, G. Haan-Archipoff, A. Beretz, D. Wachsmann, and J.-P. Klein. 1996. Cytokine production by human epithelial and endothelial cells following exposure to oral viridans streptococci involves lectin interactions between bacteria and cell surface receptors. *Infect. Immun.* **64:**3016–3022.

59. Whiley, R. A., and D. Beighton. 1998. Current classification of the oral streptococci. *Oral Microbiol. Immunol.* **13:**196–216.

60. Whiley, R. A., D. Beighton, T. G. Winstanley, H. Y. Fraser, and J. M. Hardie. 1992. *Streptococcus intermedius*, *Streptococcus constellatus*, and *Streptococcus anginosus* (the *Streptococcus milleri* group): association with different body sites and clinical infections. *J. Clin. Microbiol.* **30:**243–244.

61. Whiley, R. A., L. M. C. Hall, J. M. Hardie, and D. Beighton. 1997. Genotypic and phenotypic diversity within *Streptococcus anginosus*. *Int. J. Syst. Bacteriol.* **47:**645–650.

62. Whittaker, C. J., C. M. Klier, and P. E. Kolenbrander. 1996. Mechanisms of adhesion by oral bacteria. *Annu. Rev. Microbiol.* **50:**513–552.

Streptococcus mutans: Molecular Genetic Analysis

HOWARD K. KURAMITSU

28

Dental caries (tooth decay) have plagued humans since the dawn of civilization and still constitute one of the most common human infectious diseases. Although not life threatening and currently decreasing in severity in more advanced countries, billions of dollars are still spent annually in the United States to treat this disease. It has been recognized for almost a century that caries result from the accumulation of oral bacteria in the form of gelatinous masses termed dental plaque. These organisms ferment a variety of dietary carbohydrates to produce acids, most importantly lactic acid, that demineralize the inorganic matrix of teeth to produce decay. Of the hundreds of species that inhabit the human oral cavity, *Streptococcus mutans*, initially isolated in 1924, has been primarily implicated in this disease (24, 38, 60). A variety of oral streptococcal species originally classified as *S. mutans* strains are now recognized as separate species and are collectively referred to as mutans streptococci. Where appropriate, I will use the proper species designations for these strains.

In an attempt to identify the virulence factors associated with *S. mutans*, genetic approaches have been crucial (7, 34, 50). In this chapter I briefly discuss the genetic approaches utilized before the advent of gene cloning and focus primarily on the cloning of genes from *S. mutans* along with the parallel development of genetic tools for their exploitation in analyzing the virulence of these organisms. Finally, the potential utilization of genome sequence information currently being generated for these organisms is discussed, as well as some of the remaining unresolved issues regarding cariogenicity. Space limitations restrict this discussion to genes for which strong in vitro and in vivo evidence exists for relevance to virulence.

IDENTIFICATION OF THE VIRULENCE FACTORS OF *S. MUTANS* BEFORE THE ERA OF GENE CLONING

The recognition that human oral bacterial isolates related to *S. mutans* were the most active in promoting dental decay in rodent model systems prompted a comparison of the phenotypic properties of these organisms with the other, more numerous, noncariogenic oral streptococci (24). Such approaches suggested that *S. mutans* strains could be distinguished from these latter organisms primarily on the basis of three unique properties: their ability to produce large amounts of water-insoluble glucan polymers from sucrose, their capacity to rapidly ferment sucrose to lactic acid, and their strong aciduricity (38).

Since rodent model systems for examining the cariogenicity of bacteria have been available for more than 30 years (60), it was possible to examine a specific property of a mutans streptococcus for its role in cariogenicity if appropriate mutants were available. Freedman et al. (16) utilized both spontaneous and chemically induced mutants of *Streptococcus sobrinus* 6715 in this manner. These results confirmed the earlier suggestions that insoluble glucan synthesis as well as intracellular polysaccharide storage are important virulence factors in these organisms. However, before the advent of gene cloning, it was difficult to isolate mutants of the mutans streptococci that were altered in a variety of potential virulence traits because of limitations in mutant detection strategies.

INITIAL DEVELOPMENT OF MOLECULAR GENETIC APPROACHES FOR EXAMINING THE VIRULENCE OF *S. MUTANS*

The initial success in expressing heterologous bacterial genes in *Escherichia coli* prompted attempts at similar expression of genes from *S. mutans*, initially in the laboratory of Curtiss (11). Utilizing a complementation strategy, an aspartic-semialdehyde dehydrogenase gene was isolated from strain UAB62 in *E. coli* utilizing plasmid vectors (32). Subsequently, this laboratory isolated several genes suggested to play important roles in cariogenicity (11). The first successful isolation of a glucosyltransferase (*gtf*) gene involved in sucrose-dependent colonization was carried out by Gilpin et al. (21) with a lambda phage cloning system.

One of the advantages of utilizing a molecular genetic approach for examining the virulence of a microorganism is the ability to construct defined mutants with cloned genes. This became possible for several strains of *S. mutans*,

Gram-Positive Pathogens, ed. by V. A. Fischetti et al.
© 2000 American Society for Microbiology, Washington, D.C.

but not *S. sobrinus*, with the demonstration of natural transformation in some strains of *S. mutans* (48). Following insertional inactivation of cloned *S. mutans* genes, principally with erythromycin-resistance markers, it was possible to construct stable defined mutants defective in the targeted genes after transformation and a Campbell-like double crossover integration event. Using this strategy, a large collection of defined *S. mutans* mutants is now currently available for testing in both in vitro and in vivo model systems (34).

During this same period when gene cloning and transformation systems were being developed for *S. mutans*, *E. coli*-streptococcal shuttle plasmid and suicide vectors were successfully constructed (12, 41). These vectors could be utilized for the reintroduction of cloned genes into transformable *S. mutans* strains as well as allowing for the introduction of transposons into these strains for mutant isolation. Therefore, by the mid-1980s most of the tools required for genetic analysis of *S. mutans* virulence were in place.

UTILIZATION OF CLONED GENES TO CHARACTERIZE THE VIRULENCE PROPERTIES OF *S. MUTANS*

Colonization Properties

Several approaches suggested that mutans streptococci colonize tooth surfaces via a two-stage process: (i) initial sucrose-independent attachment followed by (ii) sucrose-dependent accumulation (59). An examination of the properties of *spaA* mutants of *S. sobrinus* 6715 suggested that the SpaA protein is an important virulence factor in initial attachment to tooth surfaces (11). Interestingly, homologous proteins have also been identified on the cell surfaces of other mutans streptococci (35, 47) as well as on *Streptococcus sanguis* strains (13). The genes coding for these proteins have been isolated and sequenced, and potential adhesin as well as cell wall-anchoring domains have been identified (6).

To directly evaluate the role of the homologous cell surface proteins in tooth colonization (the *S. mutans* proteins have been named P1, antigen I/II, Pac, SpaP, and B in different laboratories), mutants defective in the *spaP* gene have been constructed (36). One of these mutants was demonstrated to be markedly attenuated in attaching to saliva-coated hydroxylapatite in vitro, which is consistent with an important role in sucrose-independent attachment. However, when implanted into sucrose-fed rats, the mutant was as cariogenic as the parental organism (4). It was concluded from these studies that these results did not necessarily obviate a role for the SpaP protein in colonization but suggested that *S. mutans* displays multiple mechanisms of tooth attachment. Therefore, these molecular genetic approaches did not allow for clear delineation of a significant role for these proteins in tooth colonization.

By contrast, the utilization of cloned *gtf* genes has clearly demonstrated the importance of these genes in cariogenicity in rodent model systems (34). Following the initial genetic observation that at least two distinct *gtf* genes were present in *S. mutans* strains (21), it was demonstrated that three *gtf* genes are present on the chromosome of most strains of *S. mutans* (34). Two of these genes, *gtfB* and *gtfC*, are tandemly localized on the chromosomes, are expressed from their own promoters, and code for proteins,

previously called GTF-I and GTF-SI, respectively, that synthesize predominantly water-insoluble glucans. The third gene, *gtfD*, expresses an enzyme that is responsible for primer-dependent water-soluble glucan formation. That each of these genes is important for smooth-surface caries formation in rodents was demonstrated following the implantation of mutants defective in each gene (63). Similar general conclusions regarding the importance of the *gtf* genes in cariogenicity were also reached in a different rat model system (45). Homologous genes coding for GTF activity have also been identified in several other oral streptococci (20).

Another specific property associated with the mutans streptococcal strains related to glucan metabolism was their ability to be agglutinated in the presence of high-molecular-weight dextrans (19). This property was suggested to aid in the colonization of teeth by these organisms in the presence of dietary sucrose, since the glucans synthesized from the disaccharide could play a role in the aggregation of microorganisms to form dental plaque (24). However, a specific role in cariogenesis has not yet been established for glucan-mediated aggregation.

It was suggested that since the Gtfs could bind glucans as an acceptor, these enzymes or their proteolytic peptide fragments might act as glucan-binding proteins (44). Moreover, at least three distinct proteins with glucan-binding properties have now been identified in the mutans streptococci (1, 54, 58). A mutant defective in one of these genes, *gbpA*, has been constructed using the isolated gene and implanted in a rat caries model system (27). Surprisingly, this mutant induced more caries than the parental strain and also displayed increased recombination between the *gtfB* and *gtfC* genes. However, the molecular basis for these observations has yet to be determined. Therefore, the role of the glucan-binding proteins in the cariogenicity of the mutans streptococci is still uncertain.

Carbohydrate Metabolism

Earlier biochemical analyses had suggested that *S. mutans* strains could ferment a variety of sugars to lactic acid (24). Evidence for the presence of phosphoenolpyruvate-dependent sugar phosphotransferase transport systems (PEP-PTS) in these organisms was also indicated (57). Since a strong correlation between the levels of sucrose in the diet and caries incidence was suggested (37), much attention has been focused on the metabolism of this disaccharide. Metabolic studies have demonstrated that the majority of sucrose metabolized by the mutans streptococci is transported into the cells for fermentation rather than serving as substrates for the synthesis of extracellular polysaccharides (60). Evidence for both PEP-PTS and non-PTS transport of sucrose into the cells was provided following biochemical analysis (57).

The genes encoding the EII^{scr} and sucrose-6-phosphate hydrolase proteins, *scrA* and *scrB*, of the sucrose-PTS system were isolated from a strain of *S. mutans* (39, 53) as well as from strain 6715 of *S. sobrinus* (10). For each organism, both genes are present on opposite DNA strands and appear to be expressed from individual promoters (30). Immediately downstream from the *scrB* gene another gene, *scrR*, which appears to function as a regulator of the expression of the two genes, has been identified (30). However, the mechanism by which the ScrR protein regulates the expression of the genes involved in sucrose transport remains to be determined. It was also suggested (49) that the trehalose-PTS system of *S. mutans* may also act as a low-affinity transport system for sucrose in these organisms. In addition, the more recent isolation and characterization of the genes of the *msm* operon suggested that this operon may play a role in non-PTS-dependent sucrose transport (51). Since sucrose-grown cells appear to express the highest levels of ScrA and ScrB (30, 55) whereas the sucrose-PTS activity is repressed in the presence of high sucrose concentrations (15), it is likely that an additional rate-limiting factor of the sucrose PTS is regulated by sucrose.

One of the earliest genetic studies attempting to define the virulence properties of mutans streptococci utilized chemically induced mutants of *S. mutans* that were defective in intracellular polysaccharide storage (16). The implantation of such mutants in rats suggested that this property was an important virulence factor which could explain how cariogenic plaque might continually produce acids during periods when teeth were most susceptible to decay (periods of low salivary flow when food was also absent). More recently, the *S. mutans* genes encoding the enzymes required for intracellular glycogen storage have been isolated and a defined mutant defective in this property constructed from an inactivated gene (26). As predicted from the earlier studies, intracellular polysaccharide storage-defective mutants were attenuated in caries formation.

Since plaque dextrans and fructans could also serve as reserve sources of carbohydrates, it has been suggested that dextranases as well as fructanases could be important virulence factors (24). However, the utilization of monospecific *ftf* and *fruA* mutants defective in fructan synthesis and hydrolysis, respectively, has suggested a plausible role for *fruA* but not *ftf* in a program-fed rat caries model (8). Moreover, monospecific dextranase mutants have not yet been evaluated in animal models as a fructanase mutant has.

Aciduricity of *S. mutans* as a Virulence Factor

Comparative biochemical analysis of oral streptococci suggested that the relative aciduricity of *S. mutans* strains could serve as an important cariogenic property (34). The demonstration of relatively high levels of membrane-associated, H^+-translocating ATPase activity in *S. mutans* as well as more acidic pH optima for glycolytic activities relative to several other oral bacterial suggested a molecular basis for this property (2). In addition, subsequent approaches have identified additional factors that might also contribute to the aciduricity of *S. mutans*. The utilization of a Tn916 mutagenesis strategy identified the *dgk* gene expressing diacylglycerol kinase activity as a critical factor in the aciduricity of strain GS5 (64). Nevertheless, a specific role for the *dgk* gene in aciduricity was not suggested, since inactivation of this gene also affected the ability of strain GS5 to grow at elevated temperatures and at high osmolarity. However, the molecular basis for such an alteration in the stress responses of *S. mutans* has not yet been determined.

NOVEL GENETIC APPROACHES FOR EXAMINING THE VIRULENCE OF *S. MUTANS*

Transposon Mutagenesis

Transposon mutagenesis has been utilized to identify virulence-related genes as well as to construct gene fusions for regulation studies. Pioneering studies primarily from Clew-

ell's laboratory (18) suggested that the transposon Tn916 might be useful in this regard for oral streptococci. Initial attempts with this transposon and naturally transformable *S. mutans* strains suggested the feasibility of such an approach (9). Genes involved in aciduricity (64) as well as in bacteriocin synthesis (9) have been now been identified using Tn916 mutagenesis. Despite the fact that insertion of the transposon into the chromosome of target organisms is not entirely random (18), the screening of sufficient numbers of Tn916 insertions has proved to be essential in this regard. Utilizing suicide vectors as a means of introducing Tn916 into highly transformable *S. mutans* strains, one of the major limiting factors in this approach has been the development of suitable screening procedures for identifying specific mutant phenotypes.

An alternative transposon mutagenesis strategy based upon derivatives of the gram-positive transposon Tn917 has been recently introduced to identify several genes in the poorly transformable *S. mutans* JH1005 (23). This strategy was developed initially for use with strains with low rates of transformation. Therefore, the rare introduction of Tn917 harbored on a temperature-sensitive replicating plasmid into poorly transformable *S. mutans* strains would allow for sufficient insertion events into the host chromosome following transposition and elimination of the plasmid. This approach has been successful in identifying several different classes of genes (23). More recently, this approach has been utilized to generate *lacZ* fusions in *S. mutans* (29).

Reporter Systems for Investigating Gene Regulation in *S. mutans*

Following the identification and characterization of a number of genes that appear to be important in cariogenicity of *S. mutans* strains, it was of great interest to investigate the regulation of expression of these genes. Despite the relative difficulty of extracting RNA from *S. mutans*, procedures have now been developed to successfully utilize Northern blotting as well as reverse transcription-PCR analyses with these organisms (30, 33). These approaches have generally determined that, as in other bacteria, gene regulation appears to occur primarily at the level of transcription in *S. mutans*.

To evaluate the regulation of expression of a gene, gene fusion technology has proved to be extremely useful. Initially, *lacZ* translational fusions were constructed in *S. mutans* GS5 to examine the regulation of expression of the *scrA* gene (55). This strategy has more recently been utilized to examine the regulation of expression of the *scrB* gene (30).

An alternative strategy to utilize gene fusion technology in *S. mutans* has been the construction of chloramphenicol resistance (Cm^r) transcriptional fusions (31). Work with these constructs suggests that the expression of the *gtf* genes may be regulated by the attachment of the organisms to solid surfaces.

In Vivo Expression Technology Systems for *S. mutans*

Since it is very likely that a microorganism may express genes in vivo that are not normally expressed under laboratory conditions, the ability to identify such genes should be relevant to virulence. Several systems based upon the in vivo expression technology strategy (42) have been developed primarily for gram-negative pathogens. In addition, a similar system has recently been developed for streptococci (28) and should also be useful in *S. mutans*. Likewise, the expression of a specific *S. mutans* gene in vivo has now been demonstrated with the Cm^r fusion approach (22).

Utilization of the *S. mutans* Genome Sequence

As this chapter is being written, the sequencing of the genome of *S. mutans* UA159 is in progress. The availability of the nucleotide and deduced amino acid sequences of the genes constituting the genome will have a profound effect on the genetic analysis of these organisms. In many cases, this will obviate the requirement for the development of novel strategies to identify clones containing a specific gene. For example, to isolate a specific *S. mutans* gene it will only be necessary to have available the sequence of a portion of a homologous gene or motif from another organism. This sequence could then be utilized to identify the corresponding open reading frame in the *S. mutans* genome via computer homology programs. Specific oligonucleotide primers can then be synthesized to amplify the gene for direct isolation by PCR. These amplified fragments could also be used for convenient gene inactivation to examine the phenotype of mutants altered in the target gene.

Software programs to identify genes in protein databases with specific functions (e.g., membrane proteins) are also currently under development (5). Likewise, genes for cell-associated proteins containing cell wall-binding motifs or extracellular proteins containing signal sequences could also be identified with similar programs. The genes could then be isolated following PCR amplification, and their respective roles could be determined following gene inactivation or the more recently developed antisense strategy (52). In addition, the encoded proteins could be expressed and purified, and antibodies could be prepared for immunoelectron microscopic localization of the protein in *S. mutans*. As homologous motifs for various classes of proteins are recognized, it is likely that computer programs for identifying these sequences will become increasingly useful in searching the genome databases.

It is also likely that the *S. mutans* genome sequence will reveal novel genes that could play unanticipated roles in the virulence of these organisms. This is suggested by the observation that a significant fraction of the open reading frames identified in the bacterial genomes sequenced up to now have no obvious homology with genes from other organisms. The determination of the role of these genes in virulence will require novel approaches and should constitute a challenge equivalent to the designing of novel clone detection strategies in the pre-genome sequencing era. The ordering of genes on the *S. mutans* chromosome may also aid in the identification of "pathogenicity islands," which have been demonstrated for other pathogenic bacteria (46).

Finally, a comparison of the *S. mutans* genome sequence with those from other microorganisms will aid in evaluating the evolution of these oral bacteria. Since genome sequences of the *Streptococcus pneumoniae* and *Streptococcus pyogenes* chromosomes are now available, it will be of interest to determine if patterns of evolution can be discerned from such comparison of these streptococci.

SOME UNRESOLVED ISSUES REGARDING THE CARIOGENICITY OF *S. MUTANS*

Role of Glucans in Tooth Colonization

It is now recognized that many of the major virulence properties of *S. mutans* have been identified (3, 34, 38, 60).

However, as pointed out in these reviews, a number of significant questions still remain to be answered. Since the role of sucrose in the development of human dental caries has been recognized for many years, it is not surprising that sucrose metabolism has been a primary focus of caries research. Nevertheless, it is still not clear what constitutes the glucan products that are involved in *S. mutans* colonization of teeth. Although it is recognized that it is the α-1,3-rich water-insoluble glucan fraction that is primarily responsible for this property, the precise chemical characterization of the "adherent" glucans still remains elusive. Furthermore, the respective roles of the multiple Gtfs expressed by these organisms in synthesizing the adherent glucans is still not clear. Recent evidence suggests that the GtfC enzyme of *S. mutans* strains appears to be especially important in this regard (17, 61), but the molecular basis for this has not yet been established. In addition, evidence for the presence of four Gtfs in some of the mutans streptococci (25) raises additional questions regarding how they interact in synthesizing the colonization-promoting glucans. Perhaps the utilization of novel Gtfs, constructed as a result of site-directed mutagenesis (56), that synthesize unique glucans may prove useful in answering some of these questions.

Basic questions regarding the mechanism of action of the Gtfs also still remain to be resolved (43). Models involving single and dual sucrose-binding sites on the enzymes have been proposed, along with proposals that glucose is added to the reducing, nonreducing, or both ends of the acceptor glucan molecules. It is likely that these questions will ultimately be resolved by a combination of site-directed mutagenesis coupled with future structural studies of these enzymes.

It has also been suggested that the glucans synthesized by *S. mutans* may play additional roles in cariogenesis other than serving as a means of enhancing the attachment of the organisms to tooth surfaces (62). This is further suggested by more recent results with rodent model systems that indicate that mutants defective in the *gtfD* gene are attenuated in smooth surface caries formation despite the fact that such mutants display wild-type levels of sucrose-dependent colonization of smooth surfaces, at least in vitro (63). Thus, the glucans may serve as permeability barriers within the dental plaque biofilm, which could affect the transport of sugars into plaque and/or the diffusion of acids from plaque. It will be of interest to utilize *gtf* mutant strains as well as strains genetically engineered to synthesize unique glucans to test this hypothesis using novel plaque permeability models in animal systems.

It has also been suggested that the dextranases expressed by *S. mutans* may play a significant role in modulating the structure of the glucans synthesized by these organisms (14). Chemically induced mutants of *S. sobrinus* defective in dextranase activity are fully virulent in monoinfected rats but are less virulent when implanted into conventional rats (16). However, the molecular basis for this difference still remains to be determined.

Sucrose-Independent Colonization Mechanisms

As described earlier, although in vitro analysis suggested an important role for the SpaP protein in colonization, rat model implantation did not clearly show such involvement (4). Therefore, there may be multiple mechanisms (ionic interactions, binding of the cells to glucans present in pellicle, hydrophobic interactions between the cells and the pellicle proteins, nonspecific trapping during mastication) that could obscure the role of the SpaP protein in attachment (4). Clearly, the relative importance of these attachment mechanisms still remains to be determined.

Role of Strain Differences in Caries Formation

Preliminary experiments have suggested that there are differences in cariogenic potential between different *S. mutans* isolates when tested in rats (34). However, it is not known if these results can be extrapolated to human dental caries. Until recently, it has been difficult to differentiate between distinct strains of these organisms. However, the development of more recent molecular approaches (restriction fragment length polymorphism, plasmid analysis, ribotyping, bacteriocin typing) has made it possible to readily discriminate between different strains of *S. mutans* (3, 34). Such differences could result from chromosomal rearrangements, as indicated by the observation that deletion of the genes for the *msm* operon has occurred in some strains (37). In addition, the presence of insertion elements in some, but not all, strains of *S. mutans* has been documented (40) and could also influence gene rearrangements on the chromosomes of these organisms. Whether or not such alterations can result in strains with different cariogenic potential in the human oral cavity still remains to be determined.

Properties of *S. mutans* in Dental Plaque

Since it is clear that the phenotype of a bacterium expressed in laboratory culture may not entirely mimic the properties expressed by the same organism in vivo, it will be important to examine gene expression in *S. mutans* under conditions that occur within dental plaque. The utilization of chemostat and biofilm model systems with strains containing reporter constructs represents an initial attempt to accomplish this goal (3, 7, 14). These studies have demonstrated that the genes coding for enzymes involved in exopolymer synthesis can be regulated by environmental changes (pH, nutrient availability) that might be expected to occur in the oral cavity.

It is also very likely that the presence of other microorganisms as well as host factors can readily influence the properties of oral bacteria (24). However, very little information is currently available regarding the influence of plaque bacteria or host factors (antibodies or other antibacterial factors present in saliva) on the virulence properties of *S. mutans* in vivo.

PROSPECTS FOR THE FUTURE

Although the introduction of molecular genetic approaches in deciphering the cariogenicity of *S. mutans* has significantly increased our understanding of the virulence properties of these organisms, it is not yet clear if this increased knowledge can be translated into more effective anticaries therapies. Ideally, this information could be used to develop novel anticaries strategies (vaccines, antimicrobials) as realized or proposed for other pathogenic bacteria. Perhaps our enhanced understanding of the physiology of *S. mutans* that the genome sequencing approach promises may prove beneficial in this regard. It will be of great interest to determine if such strategies will have a significant impact in further reducing the incidence of dental caries, as current nongenetic approaches have (improved dental hygiene, fluoridation).

This review is dedicated to the memory of a good friend and former colleague, Dennis Perry, who was instrumental in developing a gene transfer system for S. mutans. I also thank Robert Burne for critically reviewing the manuscript as well as several colleagues for communicating unpublished material. Work in my laboratory was supported by NIH grant DE03258. Space limitations prevented the citation of a number of publications related to S. mutans genetics, and their absence is not intended to imply the lack of relevance.

REFERENCES

1. **Banas, J. A., R. R. B. Russell, and J. J. Ferretti.** 1990. Sequence analysis of the gene for the glucan-binding protein of *Streptococcus mutans* Ingbritt. *Infect. Immun.* **58:** 667–673.

2. **Bender, G. R., S. V. Sutton, and R. E. Marquis.** 1986. Acid tolerance, proton permeabilities, and membrane ATPases of oral streptococci. *Infect. Immun.* **53:**331–338.

3. **Bowden, G. H. W., and I. R. Hamilton.** 1998. Surivival of oral bacteria. *Crit. Rev. Oral Biol. Med.* **9:**54–85.

4. **Bowen, W. H., K. M. Schilling, E. Giertsen, S. Pearson, S. F. Lee, A. S. Bleiweis, and D. Beeman.** 1991. Role of a cell surface-associated protein in adherence and dental caries. *Infect. Immun.* **59:**4606–4609.

5. **Boyd, D., C. Schierle, and J. Beckwith.** 1998. How many membrane proteins are there? *Protein Sci.* **7:**201–205.

6. **Brady, L. J., D. A. Piacentini, P. J. Crowley, P. C. F. Oyston, and A. S. Bleiweis.** 1992. Differentiation of salivary agglutinin-mediated adherence and aggregation of mutans streptococci by use of monoclonal antibodies against the major surface adhesin P1. *Infect. Immun.* **60:** 1008–1017.

7. **Burne, R. A.** 1998. Oral streptococci... products of their environments. *J. Dent. Res.* **77:**445–452.

8. **Burne, R. A., Y.-Y. M. Chen, D. L. Wexler, H. Kuramitsu, and W. H. Bowen.** 1996. Cariogenicity of *Streptococcus mutans* strains with defects in fructan metabolism assessed in a program-fed specific-pathogen-free rat model. *J. Dent. Res.* **75:**1572–1577.

9. **Caufield, P. W., G. R. Shah, and S. K. Hollingshead.** 1990. Use of transposon Tn916 to inactivate and isolate a mutacin-associated gene from *Streptococcus mutans*. *Infect. Immun.* **58:**4126–4135.

10. **Chen, Y.-Y. M., and D. L. LeBlanc.** 1997. Genetic analysis of scrA and scrB from *Streptococcus sobrinus* 6715. *Infect. Immun.* **60:**3739–3746.

11. **Curtiss, R., III.** 1985. Genetic analysis of *Streptococcus mutans* virulence. *Curr. Top. Microbiol. Immunol.* **118:** 253–277.

12. **Dao, M. L., and J. J. Ferretti.** 1985. *Streptococcus-Escherichia coli* shuttle vector pSA3 and its use in cloning of streptococcal genes. *Appl. Environ. Microbiol.* **49:**115–119.

13. **Demuth, D. R., C. A. Davis, A. M. Comer, R. J. Lamont, R. S. Leboy, and D. Malamud.** 1988. Cloning and expression of *Streptococcus sanguis* surface antigen that interacts with a human salivary agglutinin. *Infect. Immun.* **56:**2484–2490.

14. **Ellen, R. P., and R. A. Burne.** 1996. Conceptual advances in research on the adhesion of bacteria to surfaces, p. 201–247. *In* M. Fletcher (ed.), *Bacterial Adhesion: Molecular and Ecological Diversity*. Wiley-Liss Inc., New York, N.Y.

15. **Ellwood, D. C., and I. R. Hamilton.** 1982. Properties of *Streptococcus mutans* Ingbritt growing on limiting sucrose in a chemostat: repression of phosphoenolpyruvate phosphotransferase transport system. *Infect. Immun.* **36:**576–581.

16. **Freedman, M. L., J. M. Tanzer, and A. L. Coykendall.** 1981. The use of genetic variants in the study of dental caries, p. 247–269. *In* J. M. Tanzer (ed.), *Animal Models in Cariology.* Information Retrieval Inc., Washington, D.C.

17. **Fujiwara, T., M. Tamesada, Z. Bian, S. Kawabata, S. Kimura, and S. Hamada.** 1996. Deletion and reintroduction of glucosyltransferase genes of *Streptococcus mutans* and role of their gene products in sucrose dependent cellular adherence. *Microb. Pathog.* **20:**225–233.

18. **Gawron-Burke, C., and D. B. Clewell.** 1984. Regeneration of insertionally inactivated streptococcal DNA fragments after excession of transposon Tn*916* in *Escherichia coli*: strategy for targeting and cloning genes from grampositive bacteria. *J. Bacteriol.* **159:**214–221.

19. **Gibbons, R. J., and R. J. Fitzgerald.** 1969. Dextran-induced agglutination of *Streptococcus mutans* and its potential role in the formation of microbial dental plaques. *J. Bacteriol.* **98:**341–346.

20. **Giffard, P. M., and N. A. Jacques.** 1994. Definition of a fundamental repeating unit in streptococcal glucosyltransferase glucan-binding regions and related sequences. *J. Dent. Res.* **73:**1133–1141.

21. **Gilpin, M. L., R. R. B. Russell, and P. Morrissey.** 1985. Cloning and expression of two *Streptococcus mutans* glucosyltransferases in *Escherichia coli* K-12. *Infect. Immun.* **49:**414–416.

22. **Grey, W. T., R. Curtiss III, and M. C. Hudson.** 1997. Expression of the *Streptococcus mutans* fructosyltransferase gene within a mammalian host. *Infect. Immun.* **65:**2488–2490.

23. **Gutierrez, J. A., P. J. Crowley, D. P. Brown, J. D. Hillman, P. Youngman, and A. S. Bleiweis.** 1996. Insertional mutagenesis and recovery of interrupted genes of *Streptococcus mutans* by using transposon Tn*917*: preliminary characterization of mutants displaying acid sensitivity and nutrient requirements. *J. Bacteriol.* **174:**4166–4175.

24. **Hamada, S., and H. D. Slade.** 1980. Biology, immunology, and cariogenicity of *Streptococcus mutans*. *Microbiol. Rev.* **44:**331–384.

25. **Hanada, N., Y. Yamashita, Y. Shibata, S. Sato, T. Katayama, T. Takehara, and M. Inoue.** 1991. Cloning of a *Streptococcus sobrinus gtf* gene that encodes a glucosyltransferase which produces a high-molecular-weight water-soluble glucan. *Infect. Immun.* **59:**3434–3438.

26. **Harris, G. S., S. M. Michalek, and R. Curtiss III.** 1992. Cloning a locus involved in *Streptococcus mutans* intracellular polysaccharide accumulation and virulence testing of an intracellular polysaccharide-deficient mutant. *Infect. Immun.* **60:**3175–3185.

27. **Hazlett, K. R. O., S. M. Michalek, and J. A. Banas.** 1998. Inactivation of the *gbpA* gene of *Streptococcus mutans* increases virulence and promotes in vivo accumulation of recombination between the glucosyltransferase B and C genes. *Infect. Immun.* **66:**2180–2185.

28. **Herzberg, M. C., M. W. Meyer, A. Kilic, and L. Tao.** 1997. Host-pathogen interactions in bacterial endocarditis: streptococcal virulence in the host. *Adv. Dent. Res.* **11:**69–74.

29. **Hillman, J. D.** Personal communication.

30. **Hiratsuka, K., B. Wang, Y. Sato, and H. Kuramitsu.** 1998. Regulation of sucrose-6-phosphate hydrolase activity in *Streptococcus mutans*: characterization of the *scrR* gene. *Infect. Immun.* **66:**3736–3743.

31. **Hudson, M. C., and R. Curtiss III.** 1990. Regulation of expression of *Streptococcus mutans* genes important to virulence. *Infect. Immun.* **58:**464–470.

32. **Jagusztyn-Krynicka, E. K., M. Smorawinska, and R. Curtiss III.** 1982. Expression of a *Streptococcus mutans* aspartic-semialdehyde dehyrogenase gene cloned into plasmid pBR322. *J. Gen. Microbiol.* **128:**1135–1145.

33. **Jayaraman, G. C., J. E. Penders, and R. A. Burne.** 1997. Transcriptional analysis of the *Streptococcus mutans hrcA*, *grpE*, and *dnaK* genes and regulation of the expression in response to heat shock and environmental acidification. *Mol. Microbiol.* **25:**329–341.

34. **Kuramitsu, H. K.** 1993. Virulence factors of mutans streptococci: role of molecular genetics. *Crit. Rev. Oral Biol. Med.* **4:**159–176.

35. **Lee, S. F., A. Progulske-Fox, and A. S. Bleiweis.** 1988. Molecular cloning and expression of a *Streptococcus mutans* major surface protein antigen P1 (I/II) in *Escherichia coli*. *Infect. Immun.* **56:**2114–2119.

36. **Lee, S. F., A. Progulske-Fox, G. W. Erdos, D. A. Piacenti, G. Y. Arakawa, P. J. Crowley, and A. S. Bleiweis.** 1989. Construction and characterization of isogenic mutants of *Streptococcus mutans* deficient in major surface protein antigen P1 (I/II). *Infect. Immun.* **57:**3306–3313.

37. **Lewis, C. R., and R. R. B. Russell.** 1997. Chromosomal deletions in *Streptococcus mutans*. *Adv. Exp. Med. Biol.* **418:**677–679.

38. **Loesche, W. J.** 1986. Role of *Streptococcus mutans* in human dental decay. *Microbiol. Rev.* **50:**353–380.

39. **Lunsford, R. D., and F. L. Macrina.** 1986. Molecular cloning and characterization of *scrB*, the structural gene for the *Streptococcus mutans* phosphoenolpyruvate-dependent sucrose phosphotransferase system sucrose-6-phosphate hydrolase. *J. Bacteriol.* **166:**426–434.

40. **Macrina, F. L., K. R. Jones, C. A. Alpert, B. M. Chassy, and S. M. Michalek.** 1991. Repeated DNA sequences involved in mutations affecting transport of sucrose into *Streptococcus mutans* V403 via the phosphoenolpyruvate phosphotransferase system. *Infect. Immun.* **59:**1535–1543.

41. **Macrina, F. L., J. A. Tobian, K. R. Jones, R. P. Evans, and D. B. Clewell.** 1982. A cloning vector able to replicate in *Escherichia coli* and *Streptococcus sanguis*. *Gene* **19:**345–353.

42. **Mahan, M. J., J. W. Tobias, J. M. Slauch, P. C. Hanna, R. J. Collier, and J. J. Mekalanos.** 1995. Antibody-based selection for bacterial genes that are specifically induced during infection of a host. *Proc. Natl. Acad. Sci. USA* **92:**669–673.

43. **Mooser, G.** 1992. Glycosidases and glycosyltransferases. *The Enzymes* **20:**187–231.

44. **Mooser, G., and C. Wong.** 1988. Isolation of a glucan-binding domain of glucosyltransferase (1,6-α-glucan synthase) from *Streptococcus sobrinus*. *Infect. Immun.* **56:**880–884.

45. **Munro, C., S. M. Michalek, and F. L. Macrina.** 1991. Cariogenicity of *Streptococcus mutans* V403 glucosyltransferase and fructosyltransferase mutants constructed by allelic exchange. *Infect. Immun.* **59:**2316–2323.

46. **Ochman, H., F. C. Sonicini, F. Solomon, and E. A. Groisman.** 1996. Identification of a pathogenicity island required for *Salmonella* survival in host cells. *Proc. Natl. Acad. Sci. USA* **93:**7800–7804.

47. **Okahashi, N., C. Sasakawa, M. Yoshikawa, S. Hamada, and T. Koga.** 1989. Cloning of a surface protein antigen from serotype c *Streptococcus mutans*. *Mol. Microbiol.* **3:**221–228.

48. **Perry, D., and H. K. Kuramitsu.** 1981. Genetic transformation of *Streptococcus mutans*. *Infect. Immun.* **32:**1295–1297.

49. **Poy, F., and G. R. Jacobson.** 1990. Evidence that a low affinity sucrose phosphotransferse activity in *Streptococcus mutans* GS-5 is a high affinity trehalose uptake system. *Infect. Immun.* **58:**1479–1480.

50. **Russell, R. R. B.** 1994. The application of molecular genetics to the microbiology of dental caries. *Caries Res.* **28:**69–82.

51. **Russell, R. R. B., J. Aduse-Opoku, I. C. Sutcliffe, L. Tao, and J. J. Ferretti.** 1992. A binding protein-dependent transport system in *Streptococcus mutans* responsible for multiple sugar metabolism. *J. Biol. Chem.* **267:**4631–4637.

52. **Sato, T., J. Wu, and H. Kuramitsu.** 1998. The *sgp* gene regulates stress responses of *Streptococcus mutans*: utilization of an antisense RNA strategy to investigate essential gene function. *FEMS Microbiol. Lett.* **159:**241–245.

53. **Sato, Y., F. Poy, G. R. Jacobson, and H. K. Kuramitsu.** 1989. Characterization and sequence analysis of the *scrA* gene encoding enzyme II^{scr} of the *Streptococcus mutans* phosphoenolpyruvate-dependent sucrose phosphotransferase system. *J. Bacteriol.* **171:**263–271.

54. **Sato, Y., Y. Yamamoto, and H. Kizaki.** 1997. Cloning and sequence analysis of the *gbpC* gene encoding a novel glucan-binding protein of *Streptococcus mutans*. *Infect. Immun.* **65:**668–675.

55. **Sato, Y., Y. Yamamoto, R. Suzuki, H. Kizaki, and H. K. Kuramitsu.** 1991. Construction of *scrA:lacZ* fusions to investigate regulation of the sucrose PTS of *Streptococcus mutans*. *FEMS Microbiol. Lett.* **79:**339–346.

56. **Shimamura, A., Y. J. Nakano, H. Mukasa, and H. K. Kuramitsu.** 1994. Identification of amino acid residues in *Streptococcus mutans* glucosyltransferases influencing the structure of the glucan product. *J. Bacteriol.* **176:**4845–4850.

57. **Slee, A. M., and J. M. Tanzer.** 1979. Phosphoenolpyruvate-dependent sucrose phosphotransferase activity in five serotypes of *Streptococcus mutans*. *Infect. Immun.* **26:**783–786.

58. **Smith, D. J., H. Akita, W. F. King, and M. A. Taubman.** 1994. Purification and antigenicity of a novel glucan-binding protein of *Streptococcus mutans*. *Infect. Immun.* **65:**668–675.

59. **Staat, R. H., S. D. Langley, and R. J. Doyle.** 1980. *Streptococcus mutans* adherence: presumptive evidence for protein-mediated attachment followed by glucan-dependent cellular accumulation. *Infect. Immun.* **27:**675–681.

60. **Tanzer, J. M.** 1992. Microbiology of dental caries, p. 377–424. *In* J. Slots and M. A. Taubman (ed.), *Contemporary Oral Microbiology and Immunology.* Mosby Year Book, St. Louis, Mo.

61. **Tsumori, H., and H. Kuramitsu.** 1997. The role of the *Streptococcus mutans* glucosyltransferases in the sucrose-dependent attachment to smooth surfaces: essential role of the GtfC enzyme. *Oral Microbiol. Immunol.* **12:**274–280.

62. **van Houte, J., J. Russo, and K. S. Prostak.** 1989. Increased pH-lowering ability of *Streptococcus mutans* cell mass associated with extracellular glucan-rich matrix material and the mechanism involved. *J. Dent. Res.* **68:**457–459.

63. **Yamashita, Y., W. H. Bowen, R. A. Burne, and H. K. Kuramitsu.** 1993. Role of *Streptococcus mutans* gtf genes in caries induction in the specific-pathogen-free rat model. *Infect. Immun.* **61:**3811–3817.

64. **Yamashita, Y., T. Takehara, and H. K. Kuramitsu.** 1993. Molecular characterization of a *Streptococcus mutans* mutant altered in environmental stress responses. *J. Bacteriol.* **175:**6220–6228.

Genetics of *Streptococcus sanguis*

HOWARD F. JENKINSON

29

For many years it was convenient to refer to the non-mutans group oral streptococci, which did not ferment mannitol, as sanguis group streptococci. These bacteria make up the highest proportion of streptococci found within the human oral cavity, colonizing the hard (dental) and soft (epithelial) tissues, and, unlike the mutans group organisms, they generally cause little or no dental decay in laboratory animals. However, a reappraisal of sanguis group bacterial classification by Kilian et al. (22) resulted in the designation of four oral streptococcal species: *S. gordonii*, *S. mitis*, *S. oralis*, and *S. sanguis*. *S. gordonii* is phenotypically very similar to *S. sanguis*, and most strains of these species are Lancefield group H antigen positive. Unlike *S. gordonii*, *S. sanguis* produces immunoglobulin A1 (IgA1) protease and does not bind α-amylase. *S. oralis* and *S. mitis* are group H antigen (Blackburn) negative (22). *S. oralis* produces IgA1 protease and neuraminidase and is distinguished biochemically from *S. gordonii* and *S. sanguis* by its inability to hydrolyze L-arginine. *S. mitis* is the least clearly defined species. The currently accepted collective designation for these bacteria, which now includes three additional species, is the mitis group comprising *Streptococcus crista*, *S. gordonii*, *S. mitis*, *S. oralis*, *S. parasanguis*, *S. pneumoniae*, and *S. sanguis* (48). Species within this group demonstrate 95% 16S rRNA sequence homology (21); the most closely related organisms, *S. mitis*, *S. oralis*, and *S. pneumoniae*, show 99% 16S rRNA sequence homology, although they exhibit only 60% overall genomic DNA homology (21).

The mitis group bacteria are commensals of the human oral cavity and nasopharynx. *S. mitis* and *S. oralis* are major pioneer species colonizing the oral mucosa in neonates. Production of IgA1 protease, hexosaminidases, and neuraminidase by these organisms may provide a colonization advantage under exposure of secretory IgA antibodies in mother's milk and permit growth of bacteria on sugars released from host glycoconjugates. Following tooth eruption, the oral flora become more highly complex, and there are increases in the isolation frequencies of all streptococcal species, especially *S. gordonii* and *S. sanguis*. These organisms are found at most adult oral sites and have a high affinity for binding to salivary glycoprotein pellicles as well as for binding to other oral bacteria (reviewed in reference 19). Hence, streptococci are believed to form the foundation layers on teeth and other oral surfaces to which other organisms bind and therefore are highly significant in the development of oral bacterial communities. Since, for the most part, we live in commensal harmony with these oral streptococci, there is an underlying notion that they are relatively nonpathogenic. However, *S. gordonii* and *S. sanguis* are among the most common bacterial pathogens associated with an increasingly prevalent and serious endovascular condition, infective endocarditis. Historically, the characteristics of *S. sanguis*-like organisms that have been best studied are competence for DNA-mediated transformation and adhesion to saliva-coated surfaces. Consequently, the main focus of this chapter is to consider the molecular mechanisms and genetic control of processes associated with these characteristics.

COMPETENCE DEVELOPMENT AND DNA-MEDIATED TRANSFORMATION

Competence Pheromones

S. gordonii and *S. pneumoniae* (pneumococcus), and possibly all mitis group organisms (see below), are naturally transformable and attain a special developmental state (competence) at some point in their growth cycle during which DNA may be taken up. Development of competence in *S. gordonii* (formerly *S. sanguis*) has long been known to depend upon the production of an extracellular competence factor that is heat resistant and trypsin sensitive and is produced in a defined window of growth in batch culture, usually early exponential phase (25). Major advances in understanding of the competence and transformation processes, especially in *S. pneumoniae*, have come about following purification of the pneumococcal competence factor (designated competence-stimulating peptide or pheromone, CSP) (12). Based upon the sequence of natural competence factor, a 17-amino-acid-residue peptide (Csp1) was synthesized and shown to stimulate competence de-

Gram-Positive Pathogens, ed. by V. A. Fischetti et al.
© 2000 American Society for Microbiology, Washington, D.C.

velopment by *S. pneumoniae* cells, more or less irrespective of their growth phase, thus confirming competence factor as an unmodified oligopeptide pheromone (12). The gene encoding CSP (*comC*) in pneumococcus and *S. gordonii* is located immediately upstream of the *comD* and *comE* genes encoding, respectively, the sensor histidine kinase and response regulator protein of a two-component signal transduction system (14). The ComD protein is the receptor for the *comC* gene product and senses extracellular levels of CSP. Subsequently it has been shown that the *comCDE* locus, which is flanked by two conserved tRNA genes, is also present in *S. mitis*, *S. oralis*, *S. sanguis*, and a number of other oral streptococcal species (15).

All the streptococcal CSPs identified to date, with the apparent exception of an *S. sanguis* peptide (GenBank accession no. AJ000872), belong to a class of peptides, including bacteriocins, that are synthesized as precursors containing within the N-terminal leader peptide a conserved double-glycine processing site (13). A dedicated ATP-binding cassette (ABC)-type transporter is responsible for proteolytic cleavage of the leader and translocation of the C-terminal peptide out across the cytoplasmic membrane. Three *comC* alleles have been found so far in *S. pneumoniae*, each encoding 17-amino-acid-residue peptides processed from 41-mer prepeptides (38). Three *comC* alleles have also been found in *S. gordonii*; two of these, designated *comC1* and *comC2*, encode 19-amino-acid-residue CSPs that contain only five identical residues, but which are processed from 50-amino-acid-residue prepeptides that carry identical leader sequences (14). The ComC1 and ComC2 pheromones will efficiently stimulate competence only in strains of the corresponding pherotype; that is, the ComD receptor will respond only to the cognate CSP (14). Sequencing of genes encoding CSP and ComD receptor from different streptococcal species indicates that mosaic structures have arisen as a result of recombination between allelic variants (15). The spectrum of oral streptococcal pherotypes could therefore be immensely broad. Three CSPs for *S. mitis* have been identified, and these are most similar to the pneumococcal CSPs. On the other hand, ComC precursor sequences (accessed from GenBank) obtained for *S. crista*, *S. gordonii*, *S. oralis*, and *S. sanguis* are quite distinct.

The ComC precursor is believed to be processed and secreted by an ABC transporter and accessory protein encoded by the *comA* and *comB* genes. ComA proteins from *S. pneumoniae* and *S. gordonii* are 85% identical and have significant homology to the bacteriocin transporters, and *comA* mutants are transformation deficient (27, 28). Downstream of *comB* in *S. gordonii* is a region designated *comX* that complements transformation deficiency in *S. gordonii* Wicky (28). This region has the potential to encode three different peptides, one of which, designated ComX, is a basic 52-amino-acid-residue peptide that might be required for competence independently of CSP (28). Thus it is possible that CSP is not the only signaling molecule involved, and indeed there is evidence that other environmentally responsive pathways impact on competence development (see below).

Genetic and Environmental Regulation of Competence Development

The CSP product of the *comC* gene is proposed to be a central signaling molecule that mediates quorum sensing in cultures of *S. gordonii* and *S. pneumoniae*. Experimental ev-

idence suggests that the mechanisms involved in competence development and transformation are fundamentally similar in these two species of streptococci. Accordingly, a model depicting the genetic pathways involved in the regulation of competence development in *S. gordonii* may be proposed (Fig. 1), based on data obtained also for pneumococcus. The *comC* gene product is probably processed and secreted to the external medium by the ComAB transporter. The sensor domain of ComD responds to a threshold concentration of CSP by autophosphorylating and subsequently activating (by phosphorylation) the response regulatory protein ComE (Fig. 1). Activated ComE (ComE-P) binds upstream of *comC* and regulates transcription of *comCDE* to synchronize development of competence within the culture (Fig. 1). ComE-P acts also to induce transcription of, or activate, other regulatory proteins, resulting in the expression of "later" competence genes such as *cipA*, *comYABCD*, and *recA* (27). In this model, then, ComE is the regulatory kingpin of the transformation process.

Although this simplified control circuit accounts for how the competent state is (auto)induced, a more complex regulatory network may be envisaged because a number of additional intrinsic and environmental factors affect transformation. For example, once competence is induced it decays quite rapidly in pneumococcus (but less rapidly in *S. gordonii* Challis), and previously competent cells are refractory to further CSP induction for at least one cell generation (27). Also, an ABC-type binding protein-dependent oligopeptide permease designated Hpp, that preferentially transports hexa- and heptapeptides (17), is involved in competence development by *S. gordonii*. Mutations in the peptide-binding protein genes of the homologous Ami locus in pneumococcus alter the timing of competence development (1). To account for these observations, additional regulatory circuits must operate to provide temporal controls on the competence pathway. One possibility is that levels of ComE-P are under control of an intracellular ComE-specific phosphatase activity (1), with dephosphorylation shutting off *comCDE* transcription as well as expression of other competence-induced loci (CIL) (Fig. 1). In this scenario, it could be postulated that levels of peptides, or of a specific competence-associated peptide (processed ComX?), sensed via the Hpp system, could inhibit ComE-phosphatase activity (Fig. 1). The result of this would be to provide an additional level of control on the ComE regulatory circuit and a means whereby alternative environmental signals might be integrated into the developmental system. A regulatory hierarchy responsive to more than one environmental peptide signal operates in the control of competence development and sporulation initiation in *Bacillus subtilis* (24).

The activity of the Hpp oligopeptide transporter also influences expression of the *S. gordonii cshA* gene, which encodes a cell wall-linked surface protein adhesin (31, 32). Transcript levels from the *cshA* gene normally increase in late exponential growth phase and appear to be sensitive to an extracellular signal sensed via Hpp (32) (Fig. 1). It may be of considerable ecological significance, then, that the activities of this oligopeptide transporter can modulate both adhesion and competence, considering that oral streptococci grow within the human host principally in biofilms. Under these conditions the regulated expression of bacterial cell surface adhesins would be important for the development and maintenance of a community responsive to intercellular peptide signals.

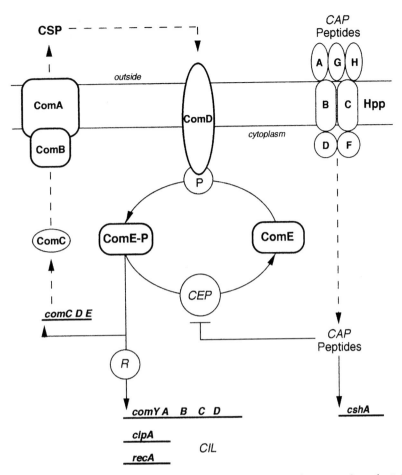

FIGURE 1 Schematic model for genetic control of competence induction in *S. gordonii* showing possible regulatory networks influencing the central autoinduction pathway. Extracellular CSP is sensed by ComD, with concomitant phosphorylation of ComE, leading to regulation of transcription of *comCDE* (autoregulation) and transcriptional induction via regulatory protein (*R*) of genes at competence-induced loci (*CIL*). Levels of phosphorylated ComE (ComE-P) may be modulated by the activity of a specific phosphatase (*CEP*), which in turn could be inhibited by a competence-associated peptide (*CAP*). Expression of *cshA* adhesin gene is also sensitive to "peptide control" through functioning of the Hpp oligopeptide transporter. See text for further explanation.

DNA Uptake and Recombination

When cells are competent, double-stranded DNA binds to a cell surface receptor complex; one strand is nicked and then enters the cell, becoming bound to an "eclipse" protein (27). The DNA-protein complex then synapses with the chromosome if sufficient base homology is present. Alternatively, if the input DNA carries an origin of replication, the DNA may become established extrachromosomally. Only some of the genes and proteins involved in the DNA recognition, uptake, and recombination processes have been characterized. A competence-specific complex comprising two DNA binding proteins and a nickase has been extracted from the surface of *S. gordonii* cells (27), while the *comY(ABCD)* operon, homologous to *comG(ABCD)* in *B. subtilis*, encodes polypeptides thought to mediate DNA uptake (29). The *cipA* gene encodes a protein necessary for chromosomal integration of DNA and for plasmid reassembly (37).

Natural transformation is a mechanism whereby related species acquire genetic diversity. Growth and survival of recombinants in vivo are determined by host-

environmental selective pressures such as the availability of nutrients and sites for adhesion, and the level and specificity of immune defenses. Genes for many streptococcal cell surface proteins contain amino acid repeat blocks (19), and it is probable that the DNA sequences encoding these are preferred sites for recombination, thus generating diversity in substrate-binding functions and antigenicity (27). In the laboratory, transformation has been utilized successfully in many genetic studies of oral streptococci. To generate mutants, gene replacement techniques in particular have been employed. In the most preferred method, linear DNA comprising a selectable marker inserted within a cloned fragment of chromosomal DNA is transformed into the host strain, and the marker is integrated into the chromosome by homologous recombination (double crossover), resulting in allelic replacement of the target site. This is an efficient mutagenesis technique in streptococci that are naturally competent (27). Site-directed mutagenesis of streptococcal genes may also be achieved with intact, nonreplicating suicide vectors carrying a selectable marker adjacent to a cloned region of chromosomal homology.

Plasmid integration and duplication of the chromosomal target are accomplished by a "Campbell-like" recombination event between homologous DNA (18). Integration of a selectable marker into *S. pneumoniae* and *S. gordonii* by double-crossover homologous recombination is favored over insertion-duplication (27). Replicative plasmid transformation of *S. gordonii* is highly efficient (46) and has been utilized for complementation analysis, expression of heterologous genes, and the isolation of selectable markers such as TetM (46), which has been used widely in streptococcal genetics. While allelic replacement with linear transforming DNA can be readily achieved in naturally transformable *S. gordonii*, it appears to be accomplished in electrocompetent *S. parasanguis* only by transforming with circular double-stranded plasmid DNA (8). It remains to be determined if this is an unusual feature of *S. parasanguis* transformation or if normal processes occurring during pheromone-induced transformation are bypassed in electro-transformation.

ADHESION

Multiple Adhesive Interactions

Insertional mutagenesis methods (as described above) in combination with functional analyses of purified native, or recombinant, streptococcal surface proteins have revealed much information about the molecular basis of adhesion (19). The *S. sanguis*-like organisms, being predominant colonizers of salivary pellicle, carry a diverse range of adhesins for salivary molecules. They also express adhesins that bind host cells as well as other microbial cells (11). Multiple adhesins for salivary molecules, host cells and tissues, and other bacterial cells can be theorized to confer numerous advantages to the streptococci. Adhesins of differing specificities will result in more avid binding, increase the probability of an individual cell engaging an oral cavity receptor, and allow binding to the wide range of receptors present at different body sites (19). Interbacterial adhesion is especially relevant in the complex microbiota of the oral cavity. Many of the later bacterial arrivals in the development of plaque communities include pathogenic gram-negative species such as *Porphyromonas gingivalis* and *Treponema denticola* that bind directly to streptococci (19, 49). Interbacterial binding is mediated in the main by protein-carbohydrate recognitions (49), and the best characterized of these interactions are those occurring between different streptococcal species and between streptococci and another major component of dental plaque, *Actinomyces naeslundii*. Strains of most mitis group streptococci that exhibit lactose-sensitive binding to *A. naeslundii* produce antigenically diverse linear cell wall phosphopolysaccharides containing the hostlike recognition motifs GalNAcβ1→3Gal or Galβ1→3GalNAc. These motifs are recognized by the fimbrial lectins of *A. naeslundii* (4) and by GalNAc-sensitive lectins present on other streptococcal cells, *Haemophilus parainfluenzae* and *Prevotella loescheii* (49). Strains of *S. gordonii* and *S. sanguis* that produce these cell wall polysaccharides appear to be negative for the Lancefield group H antigen (Blackburn), which is lipoteichoic acid (40).

Surface Protein Adhesins

Cells of *S. gordonii*, *S parasanguis*, and *S. sanguis* show a range of surface structures. A majority of *S. sanguis* strains carry peritrichous fibrils of approximate lengths of 50 to 70 nm, while some isolates produce tufts of fibrils and some strains carry fimbriae (9, 10). It has proved difficult to demonstrate a definite association between production of surface fibrils and adhesion. However, recent work has shown that mutations in *S. gordonii cshA* (31) or *S. parasanguis fap1* (50) genes affect production of fibrils or fimbriae, respectively, in these species. CshA is a 259-kDa cell wall-associated protein in *S. gordonii* and was first identified following gene inactivation experiments as conferring the property of cell surface hydrophobicity (31, 33). The polypeptide comprises 2,508 amino acid (aa) residues, and the region encompassing residues 879 to 2,417 contains 13 repeat blocks of 101 aa residues that are implicated in determining hydrophobicity. The nonrepetitive N-terminal region spanning 836 aa residues appears to mediate adhesion of *S. gordonii* cells to immobilized fibronectin and *A. naeslundii* cells (30). Isogenic *cshA* mutants of *S. gordonii* are deficient in binding these receptors but are unaffected in adhesion to salivary pellicle; they show reduced hydrophobicity and are unable to colonize the oral cavity of mice (33). Fap1 is a 200-kDa surface protein in *S. parasanguis* with sequence regions rich in threonine and serine and no similarity to CshA. Isogenic *fap1* mutants lack fimbriae and are deficient in binding salivary pellicle but are unaffected in surface hydrophobicity. On the growing list of oral streptococcal secreted proteins that are involved in adhesion, colonization, and virulence (Table 1), CshA and Fap1 are the most likely candidates for cell surface structures.

Perhaps the best characterized of the oral streptococcal adhesins are those protein members of the antigen I/II family. The genes encoding these polypeptides were first described in *Streptococcus mutans*, but similar genes have now been detected in most indigenous oral streptococcal species (19). *S. gordonii* is unusual in expressing two antigen I/II polypeptides, designated SspA and SspB (Table 1), that are the products of tandemly arranged and independently expressed chromosomal genes (5). Both these polypeptides bind salivary agglutinin glycoprotein, found in parotid saliva, in a lectinlike reaction (5); bind to type I collagen (26); mediate adhesion of cells to *A. naeslundii* (5); and bind *P. gingivalis* (2). The multiple binding properties are attributed to the presence of more than one functional domain with unique amino acid sequence and binding specificity. The antigen I/II polypeptides may be capable of distinguishing between immobilized or soluble forms of salivary glycoproteins, thus allowing streptococci to colonize oral tissues coated with salivary components despite the presence of excess soluble forms of the components in saliva.

Sialic acid (NeuNAc)-binding adhesins are present on most strains of *S. gordonii* and *S. sanguis*. The antigen I/II polypeptides SspA and SspB contribute, at least in part, to NeuNAc-sensitive binding of *S. gordonii* cells to salivary agglutinin glycoprotein and to NeuNAc-sensitive agglutination by *S. gordonii* cells of human erythrocytes (5). However, the major Ca^{2+}-dependent lectin activity for α2-3-linked sialic acid-containing receptors in *S. gordonii* is associated with surface fibrillar structures (45). These were shown to contain a unique antigen (Hs), composed of protein and wheat germ agglutinin-reactive carbohydrate of >200 kDa, that mediates hemagglutination of human erythrocytes (45). Potential receptors for this Hs antigen within the host include salivary mucins, secretory IgA1, and leukosialin on polymorphonuclear leukocytes.

S. gordonii and several other mitis group species (48) are able to avidly bind the most abundant salivary enzyme, α-amylase. In *S. gordonii* this is mediated by a cell surface 20-kDa polypeptide (AbpA) (Table 1), the expression of

TABLE 1 Secreted polypeptides associated with properties of adhesion, colonization, and virulence in S. gordonii, S. parasanguis, and S. sanguis

Polypeptide	Molecular mass (kDa)[a]	Surface retention[b]	Properties or function	Reference
S. gordonii				
CshA	259	WAP	Fibrillar determinant of cell surface hydrophobicity; adhesion to fibronectin, A. naeslundii, S. oralis, Candida albicans	19, 30, 31, 33
CshB	245	WAP	Cooperative in expression of CshA	30, 33
SspA	171	WAP	Binds salivary glycoproteins, collagen, A. naeslundii, P. gingivalis	2, 5, 19, 26
SspB	160	WAP	Binds similar substrates to SspA	2, 5, 19, 26
AbpA	20	WAP, RP	Binds α-amylase	39
GtfG	153	RP	Extracellular glucan synthesis	20, 47
HppA	76	MAL	Oligopeptide uptake, competence; binds A. naeslundii, serum factors	17, 18
S. parasanguis				
Fap1	200	WAP	Fimbrial protein; adhesion to salivary pellicle	50
FimA	35	MAL	Putative metal ion-binding; adhesion to fibrin and salivary pellicle	3, 9, 23
S. sanguis				
Iga	200	RP	Human IgA1 protease	36
PAAP[c]	150	WAP	Binds platelet surface glycoproteins; platelet aggregation	6, 7

[a] Molecular mass values calculated from inferred amino acid sequences of mature polypeptides (where known) or estimated from sodium dodecyl sulfate-polyacrylamide gel electrophoresis.

[b] WAP, wall-anchored protein; RP, released protein; MAL, membrane-associated lipoprotein.

[c] PAAP, platelet aggregation-associated protein.

which is confined to the region of the cell division septum on dividing cells, or to polar zones on single cells (39). Amylase binding might be a means by which a surface enzymic activity is acquired by bacteria for generation of oligosaccharides for uptake and metabolism. In addition, α-amylase present in salivary pellicle could serve as a receptor for adhesion. Adhesive and nutritional functions of surface proteins are probably closely linked. Adhesins may act as "capture" proteins to provide substrates for degradation by cell surface enzyme complexes, while the solute-binding components of nutrient uptake systems may function as adhesins (9, 17, 23).

Many oral streptococci produce glucosyltransferases that hydrolyze sucrose and polymerize the glucose into glucans (see chapter 27, this volume). *S. gordonii* carries a single *gtfG* gene encoding an enzyme that synthesizes a mixed α-1,3- and α-1,6-linked glucan that may promote adhesion, aggregation, and accumulation of bacterial cells on surfaces. Glucosyltransferase activity undergoes reversible phase variation, and variants expressing only 30% or less of wild-type glucosyltransferase activity do not accumulate in significant numbers within glucans on surfaces. Other phenotypes, such as adhesion to salivary pellicle, binding to A. naeslundii, and β-hemolysin production, are also subject to phase variation (20). Therefore, generation of phase variants in vivo might aid dispersal of cells to colonize different sites. A positive gene regulatory protein, designated Rgg, that controls *gtfG* expression has been identified in *S. gordonii* and in other mitis group streptococci (47); however, there is no evidence for direct involvement of this regulatory gene in phase variation.

VIRULENCE-ASSOCIATED FACTORS

The IgA1 metalloproteinases produced by *S. sanguis*, *S. oralis*, *S. mitis*, and *S. pneumoniae* are specific endopeptidases that cleave the Pro-227–Thr-228 peptide bond in the hinge region of the heavy chain of human IgA1, including the secretory form (S-IgA1). By cleaving IgA1, these enzymes interfere with mucosal immune protective functions. All *iga* genes sequenced have the potential to encode proteins of approximate molecular mass of 200 kDa (Table 1) containing the Zn-binding motif HEMTH.(X)$_{19}$.E and a typical gram-positive bacterial cell wall anchor motif LPxTG located unusually close to the N terminus (36). Variations in number and sequence of tandemly repeated elements present within the N-terminal regions, arising through intragenic homologous recombination, may contribute to antigenic diversity and therefore to immune evasion.

Upon gaining access to the bloodstream, *S. sanguis*-like streptococci can infect the heart valves and endocardium. Streptococcal virulence in endocarditis involves adherence to extracellular matrix proteins and interactions of bacterial cells with platelets. *S. sanguis* is capable of inducing platelet aggregation in a multistep process (6). A class I adhesin mediates initial attachment of bacteria to platelets, and aggregation is then induced by an antigenically distinct class II adhesin designated platelet aggregation-associated protein (PAAP) (Table 1). The platelet-interactive domain of PAAP has been identified as the heptapeptide PGEQGPK shared by the platelet-binding domains of collagen types I and III (6). The outcome of these activities is the irreversible cross-linking of platelets and plasma proteins into a thrombus, a major pathogenic determinant in endocarditis and occlusive vascular disease. The presence of environmental collagen alters the expression and conformation of PAAP produced by *S. sanguis* (7): this may, in turn, promote continued accumulation of platelets into *S. sanguis*-induced endocardial vegetations. Collagen sensing by *S. gordonii* involves up-regulation of surface protein expression (26, 43) and imparts a developmental growth

phages makes use of a phage-specified integrase that catalyzes insertion of the phage at a specific bacterial target, which is often localized at or near a tRNA gene. Using elements of lactococcal bacteriophage TP901-1, a site-specific integrative vector was designed to obtain chromosomal single-copy integration (7). This system should allow the stable insertion of foreign genes for stable expression and can also be used to study expression of genes in single copy under various growth conditions.

Suppressor Strains

Nonsense suppressor strains are classically used in *E. coli* genetics to analyze the phenotypes of point mutations. Nonsense suppressor strains of *L. lactis* were isolated, and the nonsense suppressor genes were cloned. Plasmids carrying the suppressor genes could suppress an otherwise lethal nonsense mutation in the cell. This property was exploited to construct and establish a food-grade plasmid (i.e., having no foreign DNA) containing the suppressor genes in a suppressible purine auxotrophic strain; this plasmid is stable in a milk medium that cannot sustain growth of a purine auxotroph (19).

Regulated and Optimized Promoters

Regulated promoters constitute a powerful tool to examine the roles of a particular gene (18, 20). Studies in lactococci have undergone a significant leap through in-depth dissection and application of the highly regulated nisin biosynthesis pathway. (Nisin is a bacteriocin encoded by a conjugative transposon.) The promoter for the nisin biosynthesis gene, *nisA*, is regulated by *nisR* and *nisK* gene products. In the absence of nisin, no promoter activity is detected, as measured by the presence or absence of gene products. Addition of sublethal amounts of nisin results in a strong induction of promoter activity (18). This system has rapidly been adopted for studies requiring controlled gene expression and has been shown to be functional in other gram-positive bacteria (18, 25).

Phage promoters are often highly regulated by phage-encoded repressors. This feature can be exploited to generate inducible promoter systems. In one study, similarities between the *L. lactis* r1t phage repressor and the *E. coli* bacteriophage λ C1 repressor, for which a thermosensitive variant exists, led to the engineering of a thermosensitive variant of the lactococcal repressor (61). This controlled system may potentially be used in bacteria other than *L. lactis*.

The dissection of metabolic and stress response pathways has also been fruitful in the design of regulated promoters (see reference 46 for review). The identification and use of pH-, salt-, or cold-regulated promoters may have particular advantages when induced expression at different times of fermentation is desired (e.g., when salt is added at a step in cheese production) (9, 72, 74).

Constitutive expression of promoters at fixed levels can be valuable for quantitative physiological studies or for fine-tuning of gene expression in biotechnology. A set of synthetic promoters that differ by the sequence and length of spacers between the consensus sequences was designed, and they have a range of constitutive activity that differs by a factor of 6,000 (40).

Cell Lysis Systems

Controlled cell lysis is a potentially powerful means of arresting cellular and metabolic activity; in fermentation, it may additionally result in a coordinated release of enzymes that could accelerate product maturation. The host autolysin, AcmA (8), or bacteriophage-encoded lysins and holins (which allow lysin release) are good candidates for this purpose. This application is potentially useful in controlling cell growth in fermented dairy products, as well as for enzyme release (74). Expression of lysin and holin by a nisin-induced promoter does indeed appear to accelerate cheese ripening (16). Note that this approach has obvious applications to growth control of bacterial pathogens in food products.

The cell envelope is an important barrier protecting the cell from stress situations. Cell wall damage via autolytic enzymes can render cells more sensitive to environmental conditions. It was found that bacteria that have undergone even partial cell wall damage are permeable to small labeled probes used in standard in situ hybridization methods, while undamaged cells are not (4). The method developed for studies in lactococci to estimate the cell wall state can be used for various gram-positive bacteria (4).

Protein Export Reporter

Protein export reporters can be precious tools in determining the properties and roles of surface and extracellular proteins, notably in environmental sensing and cell communication. In *L. lactis*, very few genome-encoded exported proteins have been studied, possibly because fusion reporters adapted from *E. coli* are commonly used (e.g., see reference 67) but may be poorly expressed in gram-positive bacteria (69). Recently, a reporter for protein export in gram-positive bacteria has been perfected and applied to *L. lactis* (52, 69). The reporter, the nuclease of *S. aureus*, is a stable, very well characterized protein that is active when present as an amino- or carboxy-terminal fusion to other peptides and is faithful in reporting export events and in determining membrane protein topology (69). It has already been used in *L. lactis* to follow expression of exported proteins under various environmental conditions. The major advantages of using the nuclease over previously described export reporters are that it rapidly assumes its conformation and so avoids degradation, and as few as ~300 nuclease molecules per cell can be detected in colony assays (69).

METABOLISM

Metabolic products derived from lactococcal fermentation are of great interest to cheese manufacturers, as they contribute substantially to the taste of foods that we eat. A major focus of lactococcal research is therefore in the metabolic pathways and their manipulation to alter the yields of particular products. However, basic metabolic functions may have far-reaching effects, and metabolic manipulations can result in dramatic changes in bacterial growth characteristics and cell survival.

Carbon Metabolism

From a simplistic viewpoint, lactococci seem to use sugars to provide energy and amino acids to synthesize proteins (Fig. 2). (The dairy lactococci have multiple nutritional requirements for amino acids and vitamins, probably resulting from their adaptation to a life in milk.) Lactose is the major sugar source in milk, and through its uptake and degradation lactococci generate energy in glycolysis. In contrast, casein, the major protein component in milk, is degraded to provide the major carbon source for anabolism.

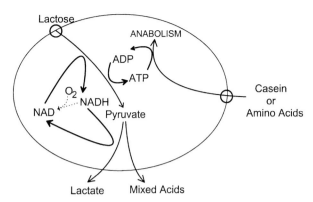

FIGURE 2 Carbon metabolism in *L. lactis*. Fermentation of sugar results in ATP production, which in turn is used for anabolism. During anaerobic conditions and rapid sugar flux, all sugar is converted to lactate (homolactic fermentation). When the sugar flux is slower, or the oxygen concentration high, mixed acid fermentation is observed. The latter two conditions are characterized by lower NADH-NAD ratios than those found during homolactic fermentations. Note that the major part of the carbon for anabolism is derived from amino acids (or casein) supplied in the medium.

The flow of carbon for energy production may therefore be almost separated from the flow of carbon for anabolism in these bacteria, making them ideally suited for metabolic studies.

Sugars

All species belonging to the genus *Lactococcus* produce acid from glucose, fructose, mannose, and *N*-acetylglucosamine. The species *L. lactis* used for dairy fermentation is known for its mainly homolactic fermentation of lactose and other sugars. *L. lactis* subsp. *lactis* strains are more versatile than *L. lactis* subsp. *cremoris* in their use of diverse sugar sources, including maltose, ribose, and trehalose (79). The sugar may be transported by a plasmid-encoded phosphotransferase system in dairy strains or (albeit at a slower rate) by a permease in nondairy strains (11).

Fermentation of carbohydrate may be shifted from homolactic (lactate production) to mixed acid fermentation (acetic acid, formic acid, CO_2, and ethanol produced in addition to lactate). Two very different sets of conditions accompany this change: (i) altered redox state created by increased aeration during growth, or (ii) a reduced flow of the sugar used for energy production.

(i) Oxygen appears to be involved in maintaining the NADH-NAD ratio, which itself seems to regulate the switch between homolactic and mixed acid fermentation (11); aerobic conditions result in oxidation of NADH to NAD (catalyzed by NADH oxidases), thereby reducing the NADH-NAD ratio in the cell. Lactate dehydrogenase is active at high NADH-NAD ratios (i.e., low oxygen), while glyceraldehyde-3-phosphate dehydrogenase (GAPDH) is inhibited. Increasing the amounts of NADH oxidase from a nisin-inducible promoter was shown to decrease the NADH-NAD ratio and to decrease the in vivo activity of lactate dehydrogenase. The increased pool size of pyruvate was in this case directed to acetoin and the flavor compound diacetyl (54). Other types of engineering (e.g., through mutations that block specific pathways [1, 17]) can also result in altered flavor properties in lactococcal fer-

mentation. Note that lactate dehydrogenase is essential in *Streptococcus mutans* but not in *L. lactis*, possibly because *S. mutans* lacks the alternative pathways for reoxidation of NADH that are present in *L. lactis*.

(ii) Sugar flow may be decreased if galactose is used as an energy source or when lactose is transported by a permease (29). All sugar-carbon utilization produces pyruvate regardless of the growth conditions, and the pathway used for further conversion determines if the fermentation is homolactic or mixed acid. Decreased sugar flow thus favors activity of enzymes giving rise to mixed fermentation products. In contrast, when carbon fluxes are high, GAPDH is a bottleneck in glycolysis, resulting in high pools of intermediates upstream of GAPDH. These pools inhibit pyruvate formate-lyase in one of the fermentation pathways from pyruvate, resulting in homolactic fermentation under anaerobic conditions.

The genetic organization of enzymes involved in sugar utilization may reveal regulation at the transcriptional level. *L. lactis* appears to coordinate the expression of three genes involved in key but distinct steps in fermentation—*pfk*, *pyk*, and *ldh* encoding phosphofructokinase, pyruvate kinase, and lactate dehydrogenase, respectively—by having them present in one operon (called *las*). The existence of this operon may prevent unwanted accumulation of glycolytic intermediates. Available sequence information for *S. pyogenes* and *S. pneumoniae* indicates that the *ldh* gene is not within this operon, suggesting that a common regulation of the three genes may be unique to lactococci.

Amino Acids

In a milk medium, lactococci derive amino acids from casein via hydrolysis by the extracellular protease PrtP, transport of some of the generated peptides, and further degradation by a multitude of intracellular peptidases (see reference 48 for review). The amino acids that are readily available in a milk medium are used directly as amino acid building blocks and also as a general carbon supply for other forms of anabolism in lactococci. Proteases are plasmid encoded, further indicating the importance of plasmid-carried features for optimal growth in milk.

Peptidases are of key importance for amino acid utilization. There are at least 14 peptidases with different specificities. These intracellular enzymes play important roles in bacterial survival. Although no particular phenotype is observed when these mutants are examined in laboratory media, lactococci grow poorly or die in milk fermentation conditions.

Dairy lactococci differ from plant lactococci in that they require several amino acids for growth. Surprisingly, however, strains of both origins appear to have the necessary genes for biosynthesis. The active biosynthetic operons are regulated by specific RNA structures that are affected by amino acid availability (see reference 46 for review). However, dairy lactococci require Ile, Leu, Val, and His and sometimes Arg, Met, Pro, and/or Glu (79). Sequencing of genes encoding the biosynthetic pathways for the branched-chain amino acids (*leu* and *ilv*) and *his* revealed that the amino acid requirements in dairy strains resulted from multiple mutations in the structural genes but not from deletions of these genes. This may suggest that mutations accumulated as an economic measure in strains maintained in a dairy environment.

Nucleotide Metabolism

Pathways of nucleotide biosynthesis have been extensively characterized in numerous organisms, including lactococci, and the pathways involved seem to be highly similar. Nevertheless, studies of the genes involved, their organization, and regulation have been rewarding in lactococci. Mutants in nucleotide metabolism may display many different phenotypes, since nucleotides are not only substrates for DNA and RNA polymerases but are also substrates or allosteric effectors for many enzymes, and furthermore constitute parts of different coenzymes. Modulation of nucleotide pools in the cell is also influenced by the presence of exogenous nucleobases or nucleosides in the medium. Therefore, knowledge of the pathways of uptake and utilization of these compounds (the so-called salvage pathways) is important in order to interpret or predict the bacterial response to changes in the medium and to increased intracellular degradation of nucleic acids.

As seen above for operons involved in carbon metabolism, gene organization in lactococci has unique characteristics. For example, the genes encoding the pyrimidine biosynthesis pathway leading to the formation of UMP are organized in at least four different operons in *L. lactis*, and in a single operon in all other gram-positive bacteria. Similarly, purine biosynthesis genes involved in the first 10 steps leading to IMP production are located in at least three separate operons in *L. lactis*, and in a single operon in *B. subtilis*.

Pyrimidine biosynthesis is regulated by attenuation and antitermination by PyrR. The PyrR protein binds to PyrR "boxes" present on the 5' ends of the untranslated mRNAs on three of the identified pyrimidine biosynthetic operons (59). PyrR binding blocks the formation of an antiterminator structure, resulting in transcriptional termination, similarly to what is observed in *B. subtilis* (56). Mutation of the *pyrR* gene (present on one of the *L. lactis* operons) results in increased levels of the pyrimidine biosynthetic enzymes (59a).

Purine biosynthesis in *L. lactis* is positively controlled by PurR, and *purR* mutants are purine auxotrophs. Activation of transcription at the *purC* and *purD* operons occurs by activating bound PurR. Present data suggest that the PurR activating effector is PRPP (5-phosphoribosyl-1-pyrophosphate). PRPP levels are reduced when purine is added, as they are used to synthesize purine mononucleotides; thus, addition of purine may indirectly result in decreased gene expression (44). A Pur box consensus sequence was identified by mutational analysis (43). It is interesting that the very homologous PurR in *B. subtilis* works as a repressor protein.

The ability to utilize exogenous nucleobases or nucleosides present in the medium or formed from intracellular degradation of nucleic acids via the salvage pathways can reflect to what extent cells can survive under various stress or growth conditions. This capacity seems to differ between microorganisms. The nucleobases uracil, guanine, and adenine are taken up and converted to nucleotide monophosphates, while cytosine cannot be utilized by lactococci (58). Furthermore, all nucleosides except cytidine can be degraded to the corresponding nucleobase. The pyrimidine nucleosides may also be directly converted to nucleotides since the corresponding nucleoside kinases (*udk* and *tdk*) have been shown to be functional in lactococci. It was recently shown that low nucleotide pool sizes may serve as internal stress signals that provoke expression of stress response genes in *L. lactis* (see below) (24, 70, 71). The different responses to addition of exogenous nucleotide precursors emphasize the importance of the salvage pathways.

RECOMBINATION FUNCTIONS

The genomes of lactococci are dynamic. Genome comparisons by pulse-field gels reveal large genomic rearrangements affecting as much as 40% of the genome (14, 15). What provokes genome rearrangements? The environment may provoke DNA damage or provide other stress signals that trigger the cell to modify its genome. Studies have shown that phage infection can provoke profound alterations in genome organization (15). Stress conditions may provoke DNA damage, which can in turn increase the frequency of rearrangements. In milk and particularly under industrial conditions, stresses may be frequent, thus favoring genetic rearrangements and endogenous prophage activation. *L. lactis* itself is equipped for genetic rearrangements; some lactococci encode chromosome mobilization functions that allow large DNA segments to be transferred between species. Insertion elements, which are abundant in *L. lactis* (6), can be responsible for rearrangements and in some cases have been shown to be highly conserved between lactococci and thermophilic streptococci.

Homologous recombination (HR) functions also play a key role in DNA rearrangement and transfer and are also important in maintaining genome integrity in the case of chromosome breakage or a replication fork block. A good knowledge of HR is needed to understand population variation among lactococci and is also an indispensable tool for genetic studies.

In *E. coli*, at least 30 different genes are involved in HR (see reference 47 for review). Genetic studies (21, 22, 26) and genome sequencing (6, 83a) confirm the presence of many of the same genes in *L. lactis*. Two key enzymes, RecA (22) and the functional analog of RecBCD, called RexAB (26), have been particularly studied. Analyses of the roles of both functions have led to surprising findings.

RecA

RecA is a highly conserved protein that is needed for the majority of recombination events and also regulates expression of DNA repair proteins; a LexA protein (a repressor that is cleaved by RecA) is present in *L. lactis* by sequence analysis (6). In addition to its role in HR, the *recA* gene was shown to have a regulatory role in lactococcal stress response, as growth of a *recA* mutant is both heat and oxygen sensitive (23) (see section on stress below).

RexAB

Unprotected double-stranded DNA (dsDNA) ends are not usually present in cells. Such ends could be provided by incoming DNA (e.g., in DNA transformation, conjugation, transduction, or phage infection) or by a break in the chromosome. Should this DNA be degraded or repaired? In *E. coli*, the HR machinery leaves the decision to the RecBCD enzyme, an exonuclease and helicase (see reference 47 for review). RecBCD distinguishes its own DNA from foreign DNA by the presence of a frequently occurring sequence on the *E. coli* genome called Chi (5'-GCTGGTGG-3'); it degrades dsDNA until it recognizes Chi, at which point its exonuclease activity is modulated; residual helicase activity unwinds dsDNA and thereby stimulates recombination. Despite extensive studies in *E. coli*, the existence of similar interactions in other organisms remains undetermined. Tests performed in lactococci have cleared up the questions of ubiquity of dsDNA end repair mechanisms. As in *E. coli*, an exonuclease/helicase (RexAB) processes dsDNA ends in *L. lactis*, and its exonuclease activity is blocked by a short DNA sequence (5'GCGCGTG-3'; see reference 5). However, despite this functional equivalence, RexAB shows nearly no sequence similarity with RecBCD, and the Chi sites are different (5, 26). Streptococcal sequence analyses reveal that they have RexAB-like rather than RecBCD-like enzymes, but it is likely that the Chi sites they recognize are different. RecBCD-like enzymes are reported to have a role in natural competence in *B. subtilis* (33). Mutational analyses of the RexAB homologs will help determine whether the enzyme is needed during competence in streptococci.

Plasmid- or Phage-Encoded HR Functions

Remarkably, three key DNA recombination and/or repair functions have been discovered on plasmids or phages. (i) A gene encoding a RecA-like protein was found on a plasmid conferring phage resistance (by an abortive infection mechanism); the identified ORF represents the only known example of a RecA-like protein in which key amino acids needed for RecA function are not conserved. Although no activity was attributed to a *recA*-like gene, the chromosomal *recA* copy is needed for the abortive infection mechanism to function (30). (ii) RuvC is known to mediate cleavage of Holliday junction recombination intermediates in *E. coli*. A gene encoding a RuvC-like nuclease was characterized from an *L. lactis* phage; despite good sequence conservation, it appeared that the phage protein recognizes replication forks, while RuvC does not (3). Interestingly, analogs of the *E. coli* RuvC protein do not seem to be present on *L. lactis* or *B. subtilis* genomes (6, 60a). (iii) UmuCD is needed for error-prone DNA repair in *E. coli*. A functional UmuC-like protein is plasmid encoded in *L. lactis* (27).

Introns

A novel group II intron was discovered in *L. lactis* and found to be involved in conjugative transfer (as reported for both chromosomal and plasmid transfer; 12, 77). Characterization of the intron shows that the encoded protein has reverse transcriptase and RNA maturase activities and, in combination with intron RNA, also has DNA endonuclease activity; these functions mediate homing of the introns into new targets by a RecA-independent mechanism (12).

STRESS RESISTANCE AND LIFESTYLE OF *L. LACTIS*

The varied environments faced by microorganisms suggest that they are often faced with different types of stress conditions, including oxidation, temperature shifts, and acidification. Survival under stress also seems to contribute to virulence of pathogens, as mutants affected in stress response genes (e.g., σ^s, *recA*, *hflB*) show attenuated virulence. When the environment is artificial, e.g., in industrial fermentation, stress conditions may be particularly harsh and the cell may not have the machinery to survive. During industrial processes, bacteria may be exposed to osmotic and oxidative stress (in starter preparation), low pH (by lactic acid production during growth), and temperature variations (40°C during cheddar cheese cooking and 8 to 16°C during cheese ripening). These conditions can each result in DNA, protein, and/or lipid damage and can also produce other cellular changes, all of which can eventually lead to cell death. Although studies confirm that *L. lactis* is equipped with stress response mechanisms (Table 3), its survival is conditional. Lactococci seem to survive well when conditioned by gradual and intermediate stress conditions. However, exposure to rapid environmental changes can be lethal.

For example, actively growing lactococci are exquisitely sensitive to a rapid drop in pH. Exposure of cells to pH 4 for 3 h can result in approximately 10,000-fold drop in viability. However, if cells are first exposed for 15 min to an intermediate pH (pH 5) and then submitted to the pH 4 condition, complete survival is observed, indicating that the intermediate pH 5 exposure gives the cells time to adapt to harsher conditions (35, 71). It is observed that exposure to mild stress conditions (such as low pH, salt, heat, cold, UV rays, or an oxidative stress) results in the induction of numerous genes (9, 34, 42, 65, 66, 72). For example, a nonlethal heat shock results in a 2- to 100-fold induction of HS-specific mRNA transcripts (2).

Among the stress response genes identified in lactococci (see reference 71 for review) are those that encode conserved chaperones (DnaK, DnaJ, GrpE, and GroESL) or proteases (HflB, ClpP, and Lon [Lon was identified by sequence homologies]). These stress response proteins are strongly interactive; e.g., specificities of proteases may be directed by the presence of particular chaperone proteins. These proteins are best known for their roles in degradation and repair of misfolded or unfinished proteins and in basic cell processes, such as DNA replication and secretion. But they also have important regulatory roles in stress response (described below). Many other proteins are implicated in stress response and may be induced by a specific stress condition. For example, two-component (sensor-regulator) systems are able to detect and react to specific environmental signals (64). Also, expression of GadB and C, putatively involved in glutamate transport by an antiporter, is induced by salt and acid; glutamate transport presumably involves efflux of H⁺, thereby maintaining intracellular pH (72). Expression of superoxide dismutase, needed for removal of toxic radicals, is acid inducible (73).

Stress response and its regulation may differ according to the microorganism and the conditions it must withstand. For example:

(i) σ^{32}, the sigma factor that regulates heat shock response in *E. coli* (38), seems to be absent in *L. lactis* and other gram-positive bacteria. Alternative σ factors (other than the vegetative σ factor) have been identified in gram-positive bacteria, including *S. aureus*, *Listeria monocytogenes*, and *B. subtilis*, where they are implicated in stress response and, where relevant, virulence. The existence of alternative σ factors in *L. lactis* remains to be confirmed.

(ii) One means of heat shock response regulation in *B. subtilis* is based on a palindromic structure called CIRCE, present upstream of chaperone genes (i.e., the *groELS* operon and the *dnaK* operon encoding *hrcA*, *grpE*, *dnaK*, and *dnaJ*) and a repressor called *hrcA* (75). Both CIRCE and *hrcA* are present in *L. lactis* and appear be involved in negative regulation of HS genes; the lactococcal HrcA homolog was shown to negatively regulate a CIRCE-controlled gene in *E. coli* (11a). However, concerning regulation, *dnaK*, which has no effect on HrcA levels or activity in *B. subtilis*, appears to be needed for *hrcA* activity in *L. lactis* (45).

(iii) The lactococcal *recA* gene product is involved in response to DNA damage and oxidative stress and, sur-

TABLE 3 Stress response mutants of *L. lactis*

Mutant	Stress sensitivity or resistance	Reference
Stress-sensitive		
recA	DNA-damaging agents, oxygen, high temperature	23
sodA	Low pH, oxygen	73
dnaK	High temperature	45
hflB	High salt, high temperature, cold	63
clpP	High temperature, protein-damaging agent	28
Stress-resistant[a]		
guaA	Low pH,[b] oxygen,[b,c] high temperature,[b,c] starvation[b,c]	24, 70, 71
relA	Low pH,[b] oxygen,[b] high temperature,[b] starvation[b]	70, 71
pstS	Low pH,[b] oxygen,[b,c] high temperature,[c] starvation[c]	24, 70, 71
deoB	Low pH,[b] oxygen,[c] high temperature,[c] starvation[c]	24, 70, 71
hpt	Low pH,[b] oxygen,[c] high temperature,[c] starvation[b,c]	24, 70, 71
tktA	Oxygen,[c] high temperature,[c] starvation[c]	24
pnpA	Oxygen,[c] high temperature[c]	24

[a]Mutants were selected by transposition mutagenesis for improved resistance of (i) the wild-type MG1363 strain to a combination of sublethal stress conditions that together were lethal, or (ii) the *recA* MG1363 strain to heat stress (the *recA* mutation attributes thermosensitivity; 23).

[b]Determined in a wild-type MG1363 background.

[c]Determined in a *recA* MG1363 background.

prisingly, to heat stress (23). Investigation of the latter phenotype revealed that RecA is likely to be involved in regulation of HS proteins; inactivation of the *recA* gene results in decreased amounts of CIRCE-regulated chaperone proteins, while levels of HS protease HflB (a negative regulator of the HS response in *E. coli* that degrades σ^{32}; 38, 63) are increased at all temperatures. It was proposed that in *L. lactis* RecA controls levels of the HflB protease, which in turn may regulate chaperone proteins by governing stability of a putative positive regulator. The identification of the factors involved in this putative regulatory pathway remains to be clarified. Note that *recA* was identified as a virulence locus in *S. aureus*, in signature-tagged mutagenesis tests (60). The observation in *L. lactis* that the *recA* mutant is stress sensitive may explain that finding.

(iv) Proteases seem to play an important role in stress response regulation (38). In *L. lactis*, inactivation of the ClpP protease resulted in the stabilization of several proteins, thus revealing potential targets of the protease (28).

(v) Additional regulatory factors, affecting metabolic flux, seem to govern stress response in *L. lactis* (24, 70, 71) (see below) and have not been reported to have such effects in other bacteria.

Like other microorganisms, *L. lactis* shows cross protection; i.e., exposure to one stress condition can also protect the cell from other stress conditions (34, 36, 42, 65) (Table 4). For example, (i) UV irradiation, which leads to the induction of numerous proteins, also allows resistance to various stresses such as heat, acid, or hydrogen peroxide; (ii) starvation enhances global stress resistance; and (iii) exposure to a moderate acid stress results in improved resistance to acid, ethanol, heat, NaCl, or H_2O_2. Although cross protection is not always the case, it appears that numerous genes can help lactococci resist more than a single stress condition (Table 4). This is not surprising, as different stress conditions may affect the cell in the same way and give rise to the same type of cellular damage.

One powerful way to identify genes involved in stress resistance is by genetic selection. Transposon insertional

mutagenesis was used to isolate stress-resistant strains of lactococci (24, 70, 71) (Table 3). A combination of high temperature (37°C) and low pH (below 6) (i.e., similar to conditions in the stomach) is lethal for lactococci (and possibly for other organisms), although each condition alone is nonlethal (70, 71). For the *L. lactis recA* strain, high temperature alone is enough to kill the strain. Thirty *L. lactis* mutants that survived lethal stress conditions were identified. These mutant strains were classified according to their ability to survive challenge by heat, oxidation, low pH, or starvation. Some mutants were specifically resistant to the stress used for the selection; for example, *glnP* is specifically acid resistant after low pH plus high temperature selection (70), and a putative redox factor is specifically temperature resistant after selection for high temperature resistance in the *recA* mutant (24). However, some of the mutants (*deoB*, *guaA*, *tktA*, *hpt*, and *pstS*) were obtained independently in both selections and were multi-stress resistant (Table 3), suggesting their roles in general stress response. It appears that the key effects of these mutants are in maintaining intracellular metabolic pools: *deoB*, *guaA*, *tktA*, *hpt*, and *relA* affect guanosine-

TABLE 4 Cross protection by a single stress in *L. lactis*

Adaptation[a]	Cross protection				
	HS	Acid	EtOH	H_2O_2	NaCl
Acid	++	++	++	++	++
DNA damage	++	++	++	+	?
Oxidative (H_2O_2)	++	−	?	++	?
Heat shock	++	++	?	?	?
Osmotic (NaCl)	?	−	?	?	++
Carbon starvation	++	++	++	++	+

[a]Cells were first adapted to nonlethal stress conditions, then challenged by the same or different stress conditions. HS, heat shock; EtOH, ethanol. See references 34, 36, 65, and 71.

phosphate pools, while *pstS* affects phosphate pools. Low intracellular levels of these metabolite pools in the mutants may constitute a starvation signal to induce a stress response. These mutant strains also show better long-term survival than their nonmutated parents. It thus appears that a general stress response is induced in *L. lactis* when intracellular guanosine-phosphate and phosphate pools are low. These metabolites may thus be specific intracellular stress sensors that regulate a general stress response of the cell. Although the above studies reveal a physiological state that provokes stress response, the regulatory genes remain to be identified.

Stress-resistant lactococci have several potential uses. First, such strains are potentially valuable in dairy fermentation. Their greater resistance to stress may overcome the survival variability seen in conventional strains. The specifically acid-resistant strains may provide resistance to extreme acid pH conditions or may be better at maintaining a neutral internal pH. Stress-resistant strains may survive longer in fermentation and may also be more resistant to harsh storage conditions (like freezing and lyophilization). Second, as such strains may survive better in the harsh environments encountered in the gut, they may be attractive for probiotic uses. Third, LAB are potentially valuable candidates for production of molecules with medical or biotechnological uses; lactococci are nontoxic and can be used to express and export proteins or other molecules of interest, either for industrial production or in the gut. LAB with improved survival properties may have advantages over conventional strains.

LACTOCOCCAL DEFENSE AGAINST ITS ENVIRONMENT

L. lactis has several means of protecting itself against a hostile environment. As mentioned above, stress response pathways seem to be particularly overlapping, and thus a single stress condition may induce an active response against multiple stress conditions. Other lactococcal functions that are more specific may be active in combating stress. For example, toxic products such as bile, quaternary compounds, and antibiotics may be actively pumped out of the cell by specialized transport functions. Competing bacteria and phage may be eliminated by the production of lactic acid (which may inhibit growth and also be toxic) or by the production of hydrogen peroxide (under aerobic conditions, through the action of superoxide dismutase or cellular oxidases; 17). Also, bacterial "warfare" may be realized by bacteriocins that have widely different host spectra (see reference 62 for review). In industrial settings, phage are a major menace that can affect fermentations; abortive infection mechanisms can block various steps in the phage multiplication cycle (see reference 13 for review). One surprising observation, discussed below, concerns a multidrug-resistance pump similar to that found in humans.

Bacteria can better tolerate their environment if they eliminate toxic substances. Among the numerous transport systems that shuttle metabolites in and out of the cell, some mediate drug expulsion and consequently can confer drug tolerance (see reference 81 for review). In *L. lactis*, one multidrug pump having specificity for a wide range of amphiphilic, cationic drugs (including antibiotics, quaternary ammonium compounds, aromatic dyes, and phosphonium ions) is LmrA. The *lmrA* gene encodes an efflux pump that is responsible for the export of toxic molecules such as ethidium in exchange for H^+ influx. Surprisingly, LmrA (590 amino acids) is very similar to the human multidrug-resistance *p*-glycoprotein, thus raising questions about the origins of the pump. Judging from sequence analyses, an LmrA dimer would be the functional equivalent of the *p*-glycoprotein (LmrA is 32% identical to half of the *p*-glycoprotein, particularly within known functional domains). Remarkably, LmrA is functional in eukaryotic cells and is able to replace *p*-glycoprotein defects, thus making *L. lactis* an excellent model to study drug extrusion (81). Note that sequence comparisons predict an LmrA homolog in *S. pneumoniae* (an ORF with ~30% identity over 539 amino acids is present).

CONCLUSIONS

L. lactis is perhaps the microorganism most eaten by humans. It is a member of a family composed of pathogens (e.g., *S. pneumoniae*, *S. pyogenes*), commensal microorganisms (e.g., *Streptococcus gordonii*, *S. mutans*), and food microorganisms (e.g., *S. thermophilus* and *L. lactis*). Studies of *L. lactis* may allow us to understand how little pathogens and nonpathogens differ. As a bacterium that acidifies its own medium, *L. lactis* may have a high capacity for stress resistance when preadapted; stress-resistant mutants with constitutive stress resistance can be selected. As a food microorganism, *L. lactis* may come into close contact with other bacteria in both the food environment and the gut. As a nontoxic bacterium that secretes relatively few proteins in quantity, *L. lactis* may also be an organism of choice for oral vaccine design (see reference 83 for review) and particularly for the extracellular expression of antigens and enzymes for biotechnological uses (52, 68, 69). The development of surface display systems in LAB will be potentially useful in the development of oral vaccines based on the nontoxic LAB. As an organism that is present on plants, in milk, in dairy products, and in the gut, *L. lactis* may be the organism of choice for studies on the influence of environmental stress on evolution.

We thank our colleagues H. Ingmer, P. Ruhdal Jensen, M. Kilstrup, J. Kok, W. Konings, E. Maguin, J. Martinussen, B. Poolman, F. Rallu, and A. Sorokin for communicating their work to the authors. We are also grateful to our laboratory colleagues M. El Karoui, S. Kulakauskas, P. Langella, Y. Le Loir, J.-C. Piard, I. Poquet, and S. Sourice, for their insights on lactococci that were incorporated into this chapter.

REFERENCES

1. **Arnau, J., F. Jørgensen, S. M. Madsen, A. Vrang, and H. Israelsen.** 1997. Cloning, expression and characterization of the *Lactococcus lactis pfl* gene, encoding pyruvate formate-lyase. *J. Bacteriol.* **179:**5884–5891.

2. **Arnau, J., K. I. Sørensen, K. F. Appel, F. K. Vogensen, and K. Hammer.** 1996. Analysis of heat shock gene expression in *Lactococcus lactis* MG1363. *Microbiology* **142:**1685–1691.

3. **Bidnenko, E., S. D. Ehrlich, and M. C. Chopin.** 1998. *Lactococcus lactis* phage operon coding for an endonuclease homologous to RuvC. *Mol. Microbiol.* **28:**823–834.

4. **Bidnenko, E., C. Mercier, J. Tremblay, P. Tailliez, and S. Kulakauskas.** 1998. Estimation of the state of the bacterial cell wall by fluorescent *in situ* hybridization. *Appl. Environ. Microbiol.* **64:**3059–3062.

5. **Biswas, I., E. Maguin, S. D. Ehrlich, and A. Gruss.** 1995. A 7-base-pair sequence protects DNA from exonucleolytic degradation in *Lactococcus lactis. Proc. Natl. Acad. Sci. USA* **92:**2244–2248.

6. **Bolotin, A., S. Mauger, K. Malarme, A. Sorokin, and S. D. Ehrlich.** 1998. *Lactococcus lactis* IL1403 diagnostic genomics, abstr. O3. *ASM Conference on Streptococcal Genetics, Vichy, France.* American Society for Microbiology, Washington, D.C.

7. **Brøndsted, L., and K. Hammer.** 1999. Use of the integration elements encoded by the temperate lactococcal bacteriophage TP901-1 to obtain chromosomal single copy transcriptional fusions in *Lactococcus lactis. Appl. Environ. Microbiol.* **65:**752–758.

8. **Buist, G., H. Karsens, A. Nauta, D. van Sinderen, G. Venema, and J. Kok.** 1997. Autolysis of *Lactococcus lactis* caused by induced overproduction of its major autolysin, AcmA. *Appl. Environ. Microbiol.* **63:**2722–2728.

8a. **Center for Microbial Ecology. Michigan State University.** http://www.cme.msu.edu/RDP/

9. **Chapot-Chartier, M. P., C. Schouler, A. S. Lepeuple, J. C. Gripon, and M. C. Chopin.** 1997. Characterization of cspB, a cold-shock-inducible gene from *Lactococcus lactis*, and evidence for a family of genes homologous to the *Escherichia coli* cspA major cold shock gene. *J. Bacteriol.* **179:**5589–5593.

10. **Coakley, M., G. Fitzgerald, and R. P. Ros.** 1997. Application and evaluation of the phage resistance- and bacteriocin-encoding plasmid pMRC01 for the improvement of dairy starter cultures. *Appl. Environ. Microbiol.* **63:**1434–1440.

11. **Cocaign-Bousquet, M., C. Garrigues, P. Loubiere, and N. D. Lindley.** 1996. Physiology of pyruvate metabolism in *Lactococcus lactis. Antonie van Leeuwenhoek* **70:**253–267.

11a. **Cochu, A., A. Gruss, and P. Duwat.** Negative regulation of chaperone genes by an HrcA-like element in *Lactococcus lactis.* Submitted for publication.

12. **Cousineau, B., D. Smith, S. Lawrence-Cavanagh, J. E. Mueller, J. Yang, D. Mills, D. Manias, G. Dunny, A. M. Lambowitz, and M. Belfort.** 1998. Retrohoming of a bacterial group II intron: mobility via complete reverse splicing, independent of homologous DNA recombination. *Cell* **94:**451–462.

13. **Daly, C., G. F. Fitzgerald, and R. Davis.** 1996. Biotechnology of lactic acid bacteria with special reference to bacteriophage resistance. *Antonie van Leeuwenhoek* **70:**99–110.

14. **Daveran-Mingot, M. L., N. Campo, P. Ritzenthaler, and P. Le Bourgeois.** 1998. A natural large chromosomal inversion in *Lactococcus lactis* is mediated by homologous recombination between two insertion sequences. *J. Bacteriol.* **180:**4834–4842.

15. **Davidson, B. E., N. Kordias, M. Dobos, and A. J. Hillier.** 1996. Genomic organization of lactic acid bacteria. *Antonie van Leeuwenhoek* **70:**161–183.

16. **de Ruyter, P. G., O. P. Kuipers, W. C. Meijer, and W. M. de Vos.** 1997. Food-grade controlled lysis of *Lactococcus lactis* for accelerated cheese ripening. *Nat. Biotechnol.* **15:**976–979.

17. **de Vos, W. M.** 1996. Metabolic engineering of sugar catabolism in lactic acid bacteria. *Antonie van Leeuwenhoek* **70:**223–242.

18. **de Vos, W. M., M. Kleerebezem, and O. P. Kuipers.** 1997. Expression systems for industrial Gram-positive bacteria with low guanine and cytosine content. *Curr. Opin. Biotechnol.* **8:**547–553.

19. **Dickely, F., D. Nilsson, E. B. Hansen, and E. Johansen.** 1995. Isolation of *Lactococcus lactis* nonsense suppressors and construction of a food-grade cloning vector. *Mol. Microbiol.* **15:**839–847.

20. **Djordjevic, G. M., and T. R. Klaenhammer.** 1998. Inducible gene expression systems in *Lactococcus lactis. Mol. Biotechnol.* **9:**127–139.

21. **Duwat, P., A. Cochu, S. D. Ehrlich, and A. Gruss.** 1997. Characterization of *Lactococcus lactis* UV-sensitive mutants obtained by IS*1* transposition. *J. Bacteriol.* **179:**4473–4479.

22. **Duwat, P., S. D. Ehrlich, and A. Gruss.** 1992. Use of degenerate primers for polymerase chain reaction cloning and sequencing of the *Lactococcus lactis* subsp. *lactis* recA gene. *Appl. Environ. Microbiol.* **58:**2674–2678.

23. **Duwat, P., S. D. Ehrlich, and A. Gruss.** 1995. The recA gene of *Lactococcus lactis*: characterization and involvement in oxidative and thermal stress. *Mol. Microbiol.* **17:**1121–1131.

24. **Duwat, P., S. D. Ehrlich, and A. Gruss.** 1998. Effects of metabolic flux on stress response pathways in *Lactococcus lactis. Mol. Microbiol.* **31:**845–858.

25. **Eichenbaum, Z., M. J. Federle, D. Marra, W. M. de Vos, O. P. Kuipers, M. Kleerebezem, and J. R. Scott.** 1998. Use of the lactococcal nisA promoter to regulate gene expression in gram-positive bacteria: comparison of induction level and promoter strength. *Appl. Environ. Microbiol.* **64:**2763–2769.

26. **El Karoui, M., D. Ehrlich, and A. Gruss.** 1998. Identification of the lactococcal exonuclease/recombinase and its modulation by the putative Chi sequence. *Proc. Natl. Acad. Sci. USA* **95:**626–631.

27. **Frankiel, H.** 1995. Thesis. Institut National de la Recherche Agronomique, Jouy en Josas, France.

28. **Frees, D., and H. Ingmer.** 1999. ClpP participates in degradation of misfolded proteins in *Lactococcus lactis. Mol. Microbiol.* **31:**79–88.

29. **Garrigues, C., P. Loubiere, N. D. Lindley, and M. Cocaign-Bousquet.** 1997. Control of the shift from homolactic acid to mixed-acid fermentation in *Lactococcus lactis*: predominant role of the NADH/NAD+ ratio. *J. Bacteriol.* **179:**5282–5287.

30. **Garvey, P., A. Rince, C. Hill, and G. F. Fitzgerald.** 1997. Identification of a recA homolog (recALP) on the conjugative lactococcal phage resistance plasmid pNP40: evidence of a role for chromosomally encoded recAL in abortive infection. *Appl. Environ. Microbiol.* **63:**1244–1251.

31. **Godon, J. J., K. Jury, C. A. Shearman, and M. J. Gasson.** 1994. The *Lactococcus lactis* sex-factor aggregation gene cluA. *Mol. Microbiol.* **12:**655–663.

31a. **Gruss, A.** Unpublished data.

32. **Guedon, G., F. Bourgoin, M. Pebay, Y. Roussel, C. Colmin, J. M. Simonet, and B. Decaris.** 1995. Characterization and distribution of two insertion sequences, IS*1191* and iso-IS*981*, in *Streptococcus thermophilus*: does intergeneric transfer of insertion sequences occur in lactic acid bacteria co-cultures? *Mol. Microbiol.* **16:**69–78.

33. Haijema, B. J., M. Noback, A. Hesseling, J. Kooistra, G. Venema, and R. Meima. 1996. Replacement of the lysine residue in the consensus ATP-binding sequence of the AddA subunit of AddAB drastically affects chromosomal recombination in transformation and transduction of *Bacillus subtilis*. *Mol. Microbiol.* **21:**989–999.

34. Hartke, A., S. Bouche, X. Gansel, P. Boutibonnes, and Y. Auffray. 1994. Starvation-induced stress resistance in *Lactococcus lactis* subsp. *lactis* IL1403. *Appl. Environ. Microbiol.* **60:**3474–3478.

35. Hartke, A., S. Bouche, J. C. Giard, A. Benachour, P. Boutibonnes, and Y. Auffray. 1996. The lactic acid stress response of *Lactococcus lactis* subsp. *lactis*. *Curr. Microbiol.* **33:**194–199.

36. Hartke, A., S. Bouche, J. M. Laplace, A. Benachour, P. Boutibonnes, and Y. Auffray. 1995. UV-inducible proteins and UV-induced cross protection against acid, ethanol, H_2O_2, or heat treatments in *Lactococcus lactis* subsp. *lactis*. *Arch. Microbiol.* **163:**329–336.

37. Havarstein, L. S., G. Coomaraswamy, and D. A. Morrison. 1995. An unmodified heptadecapeptide pheromone induces competence for genetic transformation in *Streptococcus pneumoniae*. *Proc. Natl. Acad. Sci. USA* **92:** 11140–11144.

38. Herman, C., D. Thévenet, R. D'Ari, and P. Bouloc. 1995. Degradation of σ^{32}, the heat shock regulator in *Escherichia coli*, is governed by HflB. *Proc. Natl. Acad. Sci. USA* **92:**3516–3520.

38a.Institut National de la Recherche Agronomique (INRA). http://spock.jouy.inra.fr

39. Jannière, J., A. Gruss, and S. D. Ehrlich. 1993. Plasmids from gram-positive bacteria, p. 625–644. *In* J. A. Hoch, A. L. Sonensheim, and R. Losick (ed.), *Bacillus subtilis and Other Gram-Positive Bacteria: Biochemistry, Physiology, and Molecular Genetics.* American Society for Microbiology, Washington, D.C.

40. Jensen, P. R., and K. Hammer. 1998. The sequence of spacers between the consensus sequences modulates the strength of prokaryotic promoters. *Appl. Environ. Microbiol.* **64:**82–87.

41. Khan, S. A. 1997. Rolling-circle replication of bacterial plasmids. *Microbiol. Mol. Biol. Rev.* **61:**442–455.

42. Kilstrup, M., S. Jacobsen, K. Hammer, and F. K. Vogensen. 1997. Induction of heat shock proteins DnaK, GroEL, and GroES by salt stress in *Lactococcus lactis*. *Appl. Environ. Microbiol.* **63:**1826–1837.

43. Kilstrup, M., S. G. Jessing, S. B. Wichmand-Jørgensen, M. Madsen, and D. Nilsson. 1998. Activation control of *pur* gene expression in *Lactococcus lactis*: proposal for a consensus activator binding sequence based on deletion analysis and site-directed mutagenesis of *purC* and *purD* promoter regions. *J. Bacteriol.* **180:**3900–3906.

44. Kilstrup, M., and J. Martinussen. 1998. A transcriptional activator, homologous to the *Bacillus subtilis* PurR repressor, is required for expression of purine biosynthetic genes in *Lactococcus lactis*. *J. Bacteriol.* **180:**3907–3916.

45. Koch, B., M. Kilstrup, F. K. Vogensen, and K. Hammer. 1998. Induced levels of heat shock proteins in a *dnaK* mutant of *Lactococcus lactis*. *J. Bacteriol.* **180:**3873–3881.

46. Kok, J. 1996. Inducible gene expression and environmentally regulated genes in lactic acid bacteria. *Antonie van Leeuwenhoek* **70:**129–145.

47. Kowalczykowski, S. C., D. A. Dixon, A. K. Eggleston, S. D. Lauder, and W. M. Rehrauer. 1994. Biochemistry of homologous recombination in *Escherichia coli*. *Microbiol. Rev.* **58:**401–465.

48. Kunji, E. R. S., I. Mierau, A. Hagting, B. Poolman, and W. N. Konings. 1996. The proteolytic systems of lactic acid bacteria. *Antonie van Leeuwenhoek* **70:**187–221.

49. Langella, P., Y. Le Loir, S. D. Ehrlich, and A. Gruss. 1993. Efficient plasmid mobilization by pIP501 in *Lactococcus lactis* subsp. *lactis*. *J. Bacteriol.* **175:**5806–5813.

50. Le Bourgeois, P., M. Lautier, L. van den Berghe, M. J. Gasson, and P. Ritzenthaler. 1995. Physical and genetic map of the *Lactococcus lactis* subsp. *cremoris* MG1363 chromosome: comparison with that of *Lactococcus lactis* subsp. *lactis* IL 1403 reveals a large genome inversion. *J. Bacteriol.* **177:**2840–2850.

51. Leenhouts, K., G. Buist, A. Bolhuis, A. ten Berge, J. Kiel, I. Mierau, M. Dabrowska, G. Venema, and J. Kok. 1996. A general system for generating unlabelled gene replacements in bacterial chromosomes. *Mol. Gen. Genet.* **253:**217–224.

52. Le Loir, Y., A. Gruss, S. D. Ehrlich, and P. Langella. 1998. A nine-residue synthetic propeptide enhances secretion efficiency of heterologous proteins in *Lactococcus lactis*. *J. Bacteriol.* **180:**1895–1903.

53. Liu, C. Q., N. Khunajakr, L. G. Chia, Y. M. Deng, P. Charoenchai, and N. W. Dunn. 1997. Genetic analysis of regions involved in replication and cadmium resistance of the plasmid pND302 from *Lactococcus lactis*. *Plasmid* **38:**79–90.

54. Lopez de Felipe, F., M. Kleerebezem, W. M. de Vos, and J. Hugenholtz. 1998. Cofactor engineering: a novel approach to metabolic engineering in *Lactococcus lactis* by controlled expression of NADH oxidase. *J. Bacteriol.* **180:**3804–3808.

55. Lopez de Felipe, F., C. Magni, D. de Mendoza, and P. Lopez. 1995. Citrate utilization gene cluster of the *Lactococcus lactis* biovar diacetylactis: organization and regulation of expression. *Mol. Gen. Genet.* **246:**590–599.

56. Lu, Y., and R. L. Switzer. 1996. Evidence that the *Bacillus subtilis* pyrimidine regulatory protein PyrR acts by binding to *pyr* mRNA at three sites in vivo. *J. Bacteriol.* **178:**5806–5809.

57. Maguin, E., H. Prëvost, S. D. Ehrlich, and A. Gruss. 1996. Efficient insertional mutagenesis in lactococci and other gram-positive bacteria. *J. Bacteriol.* **178:**931–935.

58. Martinussen, J., and K. Hammer. 1995. Powerful methods to establish chromosomal markers in *Lactococcus lactis*: an analysis of pyrimidine salvage pathway mutants obtained by positive selections. *Microbiology* **141:**1883–1890.

59. Martinussen, J., and K. Hammer. 1998. The *carB* gene encoding the large subunit of carbamoylphosphate synthetase from *Lactococcus lactis* is transcribed monocistronically. *J. Bacteriol.* **180:**4380–4386.

59a.Martinussen, J., and K. Hammer. Unpublished data.

60. Mei, J. M., F. Nourbakhsh, C. W. Ford, and D. W. Holden. 1997. Identification of *Staphylococcus aureus* virulence genes in a murine model of bacteraemia using signature-tagged mutagenesis. *Mol. Microbiol.* **26:**399–407.

60a.MICADO. http://locus.jouy.inra.fr/cgi-bin/genmic/madbase/progs/madbase.operl

60b. **National Center for Biotechnology Information. National Institutes of Health.** http://www.ncbi.nlm.nih.gov/BLAST/unfinishedgenome.html

61. **Nauta, A., B. van den Burg, H. Karsens, G. Venema, and J. Kok.** 1997. Design of thermolabile bacteriophage repressor mutants by comparative molecular modeling. *Nat. Biotechnol.* **15:**980–983.

62. **Nes, I. F., D. B. Diep, L. S. Havarstein, M. B. Brurberg, V. Eijsink, and H. Holo.** 1996. Biosynthesis of bacteriocins in lactic acid bacteria. *Antonie van Leeuwenhoek* **70:** 113–128.

63. **Nilsson, D., A. A. Lauridsen, T. Tomoyasu, and T. Ogura.** 1994. A *Lactococcus lactis* gene encodes a membrane protein with putative ATPase activity that is homologous to the essential *Escherichia coli ftsH* gene product. *Microbiology* **140:**2601–2610.

64. **O'Connell-Motherway, M., G. F. Fitzgerald, and D. van Sinderen.** 1997. Cloning and sequence analysis of putative histidine protein kinases isolated from *Lactococcus lactis* MG1363. *Appl. Environ. Microbiol.* **63:**2454–2459.

65. **O'Sullivan, E., and S. Condon.** 1997. Intracellular pH is a major factor in the induction of tolerance to acid and other stresses in *Lactococcus lactis*. *Appl. Environ. Microbiol.* **63:**4210–4215.

66. **Panoff, J. M., S. Legrand, B. Thammavongs, and P. Boutibonnes.** 1994. The cold shock response of *Lactococcus lactis* subsp. *lactis*. *Curr. Microbiol.* **29:**213–216.

67. **Perez-Martinez, G., J. Kok, G. Venema, J. M. van Dijl, H. Smith, and S. Bron.** 1992. Protein export elements from *Lactococcus lactis*. *Mol. Gen. Genet.* **234:**401–411.

68. **Piard, J. C., I. Hautefort, V. A. Fischetti, S. D. Ehrlich, M. Fons, and A. Gruss.** 1997. Cell wall anchoring of the *Streptococcus pyogenes* M6 protein in various lactic acid bacteria. *J. Bacteriol.* **179:**3068–3072.

69. **Poquet, I., S. D. Ehrlich, and A. Gruss.** 1998. An export-specific reporter designed for gram-positive bacteria: application to *Lactococcus lactis*. *J. Bacteriol.* **180:** 1904–1912.

70. **Rallu, R., A. Gruss, S. D. Ehrlich, and E. Maguin.** Pleiotropic acid-resistant mutants of *Lactococcus lactis*. Submitted for publication.

71. **Rallu, F., A. Gruss, and E. Maguin.** 1996. *Lactococcus lactis* and stress. *Antonie van Leeuwenhoek* **70:**243–251.

72. **Sanders, J. W., K. Leenhouts, J. Burghoorn, J. R. Brands, G. Venema, G., and J. Kok.** 1998. A chloride-inducible acid resistance mechanism in *Lactococcus lactis* and its regulation. *Mol. Microbiol.* **27:**299–310.

73. **Sanders, J. W., K. J. Leenhouts, A. J. Haandrikman, G. Venema, and J. Kok.** 1995. Stress response in *Lactococcus lactis*: cloning, expression analysis, and mutation of the lactococcal superoxide dismutase gene. *J. Bacteriol.* **177:** 5254–5260.

74. **Sanders, J. W., G. Venema, and J. Kok.** 1997. A chloride-inducible gene expression cassette and its use in induced lysis of *Lactococcus lactis*. *Appl. Environ. Microbiol.* **63:**4877–4882.

75. **Schulz, A., and W. Schumann.** 1996. *hrcA*, the first gene of the *Bacillus subtilis dnaK* operon encodes a negative regulator of class I heat shock genes. *J. Bacteriol.* **178:** 1088–1093.

76. **Seegers, J. F., S. Bron, C. M. Franke, G. Venema, and R. Kiewiet.** 1994. The majority of lactococcal plasmids carry a highly related replicon. *Microbiology* **140:**1291–1300.

77. **Shearman, C., J. J. Godon, and M. Gasson.** 1996. Splicing of a group II intron in a functional transfer gene of *Lactococcus lactis*. *Mol. Microbiol.* **21:**45–53.

78. **Stiles, M. E., and W. H. Holzapfel.** 1997. Lactic acid bacteria of foods and their current taxonomy. *Int. J. Food Microbiol.* **36:**1–29.

79. **Teuber, M.** 1995. The genus *Lactococcus*, p. 173–234. *In* B. J. B. Wood and W. H. Holzapfel (ed.), *The Genera of Lactic Acid Bacteria*. Blackie Academic and Professional, Glasgow, U.K.

80. **van Kranenburg, R., and W. M. de Vos.** 1998. Characterization of multiple regions involved in replication and mobilization of plasmid pNZ4000 coding for exopolysaccharide production in *Lactococcus lactis*. *J. Bacteriol.* **180:**5285–5290.

81. **van Veen, H. W., and W. N. Konings.** 1997. Multidrug transporters from bacteria to man: similarities in structure and function. *Semin. Cancer Biol.* **8:**183–191.

82. **Vaughan, E. E., and W. M. de Vos.** 1995. Identification and characterization of the insertion element IS*1070* from *Leuconostoc lactis* NZ6009. *Gene* **155:**95–100.

83. **Wells, J. M., K. Robinson, L. M. Chamberlain, K. M. Schofield, and R. W. Le Page.** 1996. Lactic acid bacteria as vaccine delivery vehicles. *Antonie van Leeuwenhoek* **70:** 317–330.

THE STAPHYLOCOCCUS

SECTION EDITOR: Richard P. Novick

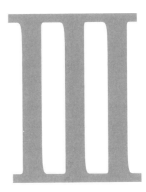

THIS SECTION IS A COMPREHENSIVE SURVEY OF THE BIOLOGY OF staphylococci, including various coagulase-negative species as well as *S. aureus*, focusing on recent developments. Starting with the genome of *S. aureus*, which is virtually complete at this date, this section goes on to cover basic information on genetics and metabolism, areas that have received, at best, spotty coverage, owing to the much more complete story for related gram-positive species such as *Bacillus subtilis*, which has recently been covered in a comprehensive ASM monograph by that name. Indeed, the basic metabolic processes are largely conserved among these and other gram-positive species. The overall emphasis is therefore on areas that are especially relevant to staphylococcal pathogenesis, including plasmids and transposons, epidemiology and typing, toxins and other exoproteins and their activities, adhesins, capsules, pathogenicity and its regulation, antibiotic resistance, and experimental infections, plus a strong emphasis on non-*aureus* species. As with the other sections of the book, the literature coverage is not encyclopedic, but includes key recent publications that can serve as entrée into the areas covered.

TABLE 1 Characteristics of the genus *Staphylococcus* in comparison with other genera classified as members of the family *Micrococcaceae*

Characteristic	Genus			
	Staphylococcus	*Micrococcus*	*Planococcus*	*Stomatococcus*
GC content of DNA (mol% G + C)	30–35	70–75	40–51	56–60
Cell wall composition (more than 2 mol of glycin per mol of glutaminic acid in peptidoglycan)	+	−	−	−
Type of fructose-1,6-diphosphate aldolase	I	II	ND[a]	II
Cytochrome *c*	−	+	ND	+
Characteristics used in bacteriological diagnostics				
Sensitivity to lysostaphin	+	−	−	−
Sensitivity to furazolidone	−	+	ND	ND

their incorrect identification as a coagulase-negative staphylococcal species.

It is hoped that genotyping methods such as those involving PCR analysis of specific characteristics such as rRNA genes (21), *femA*, nuclease, or coagulase (1), or more general methods such as PCR analysis of repetitive sequences, will soon replace phenotypic methods entirely.

EPIDEMIOLOGICAL TYPING

General Aspects

Outbreaks of *S. aureus* infections are in most cases due to the clonal expansion of a particular strain. Coagulase-negative staphylococci from colonization and infections in a particular hospital unit are mostly polyclonal; epidemic clonal spread has rarely been described. Typing in epidemiology has the goal of discriminating the epidemic clone by characteristics that differ from those of epidemiologically unrelated strains. The ideal typing characteristic should be stable within the epidemic strain and sufficiently diverse within the species population. Diversity results from random, nonlethal genetic events, which are assayed by the typing system (e.g., neutral mutations influencing the charge of an enzyme of primary metabolism and thus its motility in multilocus enzyme electrophoresis; deletions or insertions influencing the distance between restriction sites and thus also restriction endonuclease cleavage patterns). Genetic diversity among strains possessing clinical significance with regard to pathogenicity or antibiotic resistance as selection factors will be limited when a particular clonal group that is optimally adapted to the hospital situation has been disseminated. The development of more sensitive molecular techniques during the past 10 years has led to an increased ability to detect subsets of evolutionary lineages among related groups of bacterial strains.

Requirements for Typing Systems

Characteristics used for typing should be demonstrable in nearly all of the isolates within a species in order to guarantee a high typeability. Results of typing have to be reproducible in repeated laboratory testing of the same isolate as well as during the course of an epidemic and between different laboratories. A particular typing method should generate a sufficient number of types (discriminatory power), and the frequency of the most frequent type should be less than 5% of the number of isolates of the species population investigated.

Phenotypic Typing Methods for *S. aureus*

The oldest phenotypic typing method is serotyping. Although typeability and reproducibility seemed to have been

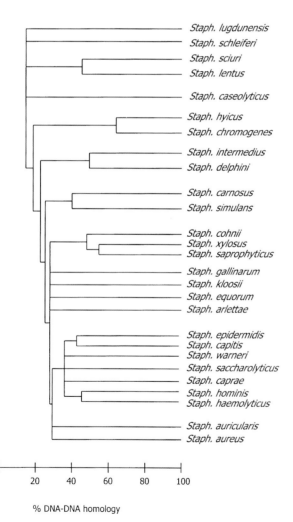

FIGURE 1 Phylogenetic relations among staphylococcal species as deduced from DNA-DNA hybridization (21).

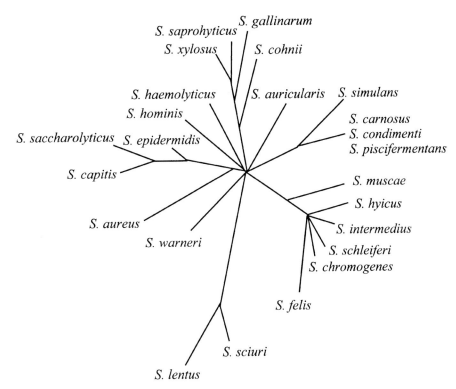

FIGURE 2 Phylogenetic relationship of staphylococci reflected by the 16S rRNA-based tree. Reprinted from reference 18.

fairly good, it was later replaced by phage typing, which has a much higher discriminatory power. In phage typing, isolates are discriminated by their susceptibility or resistance to lysis by each member of a set of typing phages. In general terms the specificity of phage patterns is based on receptor specificity in adsorption, DNA restriction, superinfection immunity of lysogenic cells, and further mechanisms of postadsorption interference. The phage-typing system of S. aureus was established empirically. Receptor specificity is obviously not a factor in phage typing of S. aureus. Later studies revealed that DNA restriction and modification is probably the most common mechanism of variation (24), but superinfection immunity is also a determinant of typing patterns.

Ideally, each unrelated strain in a population should exhibit a specific phage pattern. For a number of strains this is actually found, and phage patterns correspond to specific molecular typing patterns; e.g., most S. aureus strains that produce toxic shock syndrome toxin, exfoliative toxin, or borderline methicillin have specific phage patterns; strains of types 94, 96, or 95 also represent specific genotypes (19, 31).

Most strains of S. aureus, however, are lysogenic; prophages influence phage patterns by superinfection immunity and DNA restriction; and variations in lysogeny occur frequently. Because of this variability, phage typing is of very limited use, especially for multiply resistant strains. Another problem is nontypeability, which requires expanding the bacteriophage panel.

Bacteriocinogenicity typing, which is based on the capacity to form bacteriocins active against different indicator strains, was also tried in S. aureus and was abandoned because of unsatisfactory reproducibility and discriminatory

power. This is not surprising since bacteriocins are very often plasmid determined and have broad host ranges.

As with other bacterial pathogens, exoprotein profiles have been used as an epidemiological marker for typing coagulase-negative staphylococci and S. aureus. In S. aureus this approach is impaired by unsatisfactory resolution of the various proteins over a broad molecular mass range in polyacrylamide gel electrophoresis. This was overcome by using immunoblots (14). One has to remember, however, that genes for exoproteins can be located on accessory genetic elements (e.g., on prophages, as known for SEA and SAK) and that lysogenization can influence the expression of other exoproteins, as is the case for geh and hlb.

Besides resistance to antibiotics, clinical isolates of S. aureus often exhibit resistance to heavy metal ions and different cationic substances. These characteristics have also been used for typing (resistotyping). Most of them are encoded by plasmids and therefore may be lost or newly acquired, which compromises the stability of typing results—a problem also encountered with typing by plasmid profiles (see below).

Genotypic Typing Methods

In general terms, genotypic typing is a means of sampling the genome but must focus on regions that have a desirable degree of variability. How one decides on the appropriate degree of variability is clearly the key question in attempting to develop typing schemes. As a general rule for epidemiology, several levels of discrimination would seem to be required. The most sensitive level that would be required for epidemiology would generate a single pattern for a set of obviously clonal strains from a specific outbreak, different from patterns for epidemiologically unrelated strains. A second level would be able to differentiate from unrelated strains the widely disseminated progeny of such a clone, and a third level would be used to characterize strains common to particular clinical conditions—which might or might not have a definable common genotype.

In the following sections, the various genotyping methods applied to *S. aureus* are described in historical order.

Plasmid Profile Analysis

Plasmid profile analysis became applicable after the introduction of microliter techniques for plasmid isolation (8). Staphylococcal strains, especially coagulase-negative staphylococci, often differ in the number and size of their plasmids, easily detectable in agarose gels. The discriminatory power is low but can be improved by determining restriction endonuclease cleavage patterns of the plasmids. The value of plasmid profile analysis is limited, however, since plasmids can be spontaneously lost or newly acquired. The plasmids of hospital strains can be related, as is the case, for example, for conjugative plasmids carrying aminoglycoside-resistance determinants from the United States or Australia. This method is still in use for a quick comparison of isolates of coagulase-negative staphylococci from several blood cultures derived from the same patients to exclude contaminants.

Multilocus Enzyme Electrophoresis

Multilocus enzyme electrophoresis reflects indirectly genotypic polymorphisms by differences in the electrophoretic mobilities of individual, soluble enzymes of primary metabolism. Differences in electrophoretic mobility are due to mutations affecting the charge of an enzyme. This method is rather laborious and therefore not broadly used in epidemiology. It also has the drawback that mutations not affecting the charge of enzymes are missed; therefore, it is not a true genotyping method. For studies of population genetics, however, it has been extremely useful, e.g., for showing clonal relatedness of *S. aureus* producing toxic shock syndrome toxin-1; clonal relatedness of *S. aureus* exhibiting phage pattern 95; and clonal diversity of MRSA (15, 16).

Restriction Fragment Length Polymorphism

Restriction fragment length polymorphism typing is based on the randomness of the distribution of restriction endonuclease cleavage sites on the bacterial genome, which is reflected by fragment length. Large fragments generated with infrequently cleaving enzymes are resolved by pulsed-field gel electrophoresis. Differences between strains in the size and number of fragments derived from digestion of genomic DNA can be due to quite different genetic events, such as point mutations that change the endonuclease recognition site, insertions, deletions, inversions, or transpositions. Thus, a single point mutation leading to loss or gain of one particular *Sma*I site results in a three-fragment difference between the mutant and the parental strain. A three-fragment difference is also seen after an insertion or deletion of a transposable element or a prophage containing one recognition site (Fig. 3), whereas translocation or inversion of DNA containing one recognition site would cause a four-fragment difference.

Small fragments derived from digestion with frequently cutting enzymes cannot be sufficiently separated by electrophoresis. A particular local DNA sequence with variable cleavage sites can readily be resolved by hybridizing specific probes for this sequence to whole cellular DNA digests.

The most common application of this is ribotyping, since *S. aureus* possesses up to eight copies of the rRNA operon, having length polymorphisms for the rRNA gene spacers. There are also polymorphisms with regard to the location of restriction endonuclease cleavage sites on rRNA gene spacers and on DNA sequences neighboring the rRNA gene operons. In comparative trials, although ribotyping was found to be less discriminative than *Sma*I macrorestriction patterns (28), it was successfully used for the discrimination of MRSA from different continents (9) and can also be used for speciation.

Southern blot hybridization with probes based on chromosomal accessory genetic elements has also been used effectively for strain discrimination. One example is methicillin resistance, based on acquisition of the *mecA* gene, which is located on "*mec*-associated DNA," a DNA element of about 50 kb, often including Tn554. The *mecA* gene contains a cleavage site for the *Cla*I enzyme. Hybridization of probes for *mecA* and for Tn554 to *Cla*I digests of genomic DNA of MRSA detects polymorphisms within *mec*-associated DNA and neighboring regions; it revealed a probable ancestral relatedness of "classical" MRSA from worldwide sources (11).

Field inversion gel electrophoresis and pulsed-field gel electrophoresis allow a sufficient separation of fragments with comparably high molecular masses (macrorestriction patterns). As the conditions for lysis of the staphylococcal cells in agarose blocks, deproteinization of DNA, cleavage, and electrophoresis are well standardized, the results are highly reproducible. For *S. aureus* the enzyme *Sma*I has been most useful, yielding 8 to 20 fragments ranging from 8 to 800 kb in size.

For application to epidemiological tracing, staphylococcal isolates are regarded as different strains if their *Sma*I macrorestriction patterns differ in four or more bands (26). In other words, the occurrence of two or more genetic events influencing the distribution of restriction sites is taken to indicate an epidemiological difference. This criterion is rather arbitrary and could lead to the wrong conclusion: for example, translocation or inversion involving DNA containing one or more sites will affect four or more fragments. Although *Sma*I macrorestriction patterns are rather stable during outbreaks of nosocomial infections with MRSA, genomic rearrangements have been observed along with the dissemination of particular MRSA clones beyond a particular site or region (32).

Typing by PCR

PCR is able to produce millions of copies from a particular DNA stretch within a short time. Polymorphisms detected by PCR can be based on genetic events taking place between the location of primer binding sequences, thus leading to different lengths of amplimers; also, length polymorphisms in repetitive DNA sequences can easily be detected by PCR (Fig. 4).

Arbitrarily Primed PCR

A prerequisite to the use of PCR for typing is the availability of DNA sequences for the whole bacterial chromosome or at least for particular genes. Until the DNA sequence of the *S. aureus* chromosome has been analyzed for more repetitive sequences useful for PCR typing, arbitrary primers are used as genetic markers, as has been done with other organisms (30). The use of single short primers of arbitrary nucleotide sequence leads to randomly amplified polymorphic DNA (RAPD).

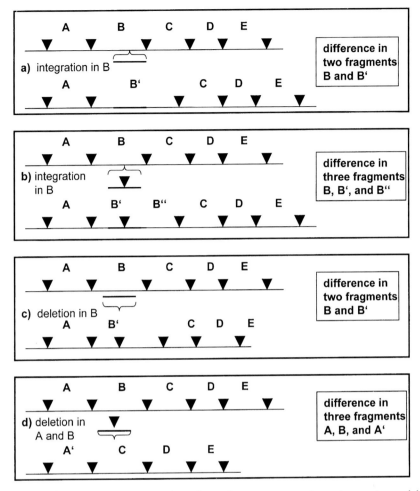

FIGURE 3 Changes in restriction endonuclease cleavage patterns due to integration or deletion of DNA sequences. Brackets indicate restriction endonuclease cleavage site.

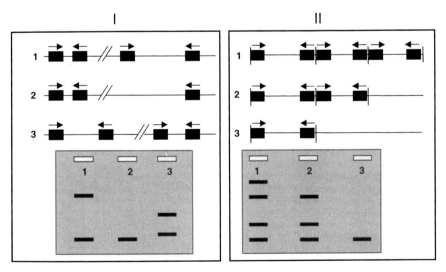

FIGURE 4 Examples of PCR typing. (Left) Variations in the location of primer binding sites on the bacterial genome. (Right) Variations in number and size of repetitive sequences and corresponding PCR patterns.

RAPD primers consist of 8 to 10 nucleotides, and RAPD PCR is performed at comparably low annealing temperatures. Thus, these primers also recognize sequences not completely homologous to them. This can lead to problems of reproducibility, even in repetition of the same typing experiment—especially when more than one method of template DNA preparation or more than one batch of the polymerase is used (see also reference 28) and, most important, when the procedure is performed by different laboratories. Among a variety of primers, ERIC-2 and BG2 have been especially useful (27), particularly for typing MRSA, for which a concordant clonal delineation in comparison with SmaI macrorestriction analysis has been described (25).

PCR of Repetitive Sequences

All bacterial genomes harbor repetitive sequences that can be used for epidemiological typing. In most cases these motifs occur throughout the entire chromosome but rarely within genes. For S. aureus, repeats like the enterobacterial repetitive intergenic consensus (ERIC) in Enterobacteriaceae have not yet been described. Consequently, the ERIC-2 primers have been used for S. aureus but at a rather low annealing temperature as with RAPD. This situation has been improved by the use of intragenic repetitive regions that have been found in the coagulase (coa) and protein A (spa) genes (7, 23). Although the polymorphism in the coa gene can lead to amplimers of different length and with different AluI cleavage sites, the coa polymorphism has not been as discriminative as SmaI macrorestriction patterns (23). In contrast, the spa gene shows such a high degree of polymorphism that it tends to be more sensitive than necessary; that is, it generates more than one pattern within a set of obviously clonal strains.

PCR of rRNA Gene Spacer Sequences

With the exception of a few bacterial species, rRNA operons are present in 2 to 11 copies per prokaryotic chromosome. Because of homologous sequences of the rRNA operon, intrachromosomal recombinational events can lead to polymorphisms. Thus, rRNA gene spacers can differ with regard to sequences and length in the same chromosome and between strains. Length variations can be assessed by PCR, with primers binding to the nonvariable sequences of the 16S and 23S rRNA genes. In S. aureus this PCR is much less discriminatory than SmaI macrorestriction patterns (3, 33). However, isolates obviously belonging to the same clonal groups, e.g., certain strains producing toxic shock syndrome toxin or strains with borderline resistance to methicillin or strains exhibiting phage pattern 95, exhibit common rRNA gene spacer patterns. Thus, these patterns are more useful for a grouping of strains in population analysis than for direct epidemiological analysis of local outbreaks (3).

PCR for DNA Stretches Flanked by Insertion Elements

IS257 seems to be widely disseminated and has a significant role in the evolution of multiple-resistance plasmids; IS256 is associated with Tn4001, carrying genes for gentamicin resistance and efflux of cationic compounds. Despite the mobility of IS257, chromosomal insertions are stable within the time frame of most epidemiological analyses. PCR for length polymorphism of inter-IS256 sequences has been found to be useful for typing multiresistant MRSA (5). Its value for typing aminoglycoside-sensitive S. aureus strains among MRSA may be limited. The frequent association of IS256 with plasmids is problematic because of plasmid acquisition or loss.

PCR for DNA Stretches Flanked by Other Known Sequences

The S. aureus chromosome contains the attachment region of Tn916 at several locations. Nucleotide sequences neighboring ribosomal binding sites are obviously rather conserved (17). If in the right orientation, PCR with primer sequences for both known sequences (tar916-shida PCR) amplifies fragments of different lengths and can be applied for typing S. aureus (4). The availability of the sequence of the S. aureus chromosome now offers new options for PCR typing based on length polymorphism of stretches flanked by previously unknown repetitive sequences. Once such new forms of PCR reveal sufficient reproducibility and discriminatory power, typing by PCR that uses low annealing temperatures will be phased out.

Combined Use of Different Molecular Typing Methods

In addition to outbreak analysis, epidemiological surveillance of S. aureus must be able to track interhospital dissemination and the evolution of multiresistant strains. Strains exhibiting one particular macrorestriction or PCR pattern may not necessarily be identical or related to a common ancestor. A strong match of the results of several typing techniques based on different genomic polymorphisms, however, indicates clonal relatedness. This approach has been, above all, applied to the epidemiology of MRSA. For example, extraregional dissemination of a particular MRSA clone in Spain and Portugal was clearly demonstrated by SmaI macrorestriction analysis and patterns of mecA and Tn554 hybridization to ClaI digests and Tn554 to SmaI digests (20). The clonal group of the European phage pattern 77 MRSA, which has been disseminated in western and central Europe and probably originated from the Iberian peninsula, exhibits a characteristic SmaI macrorestriction pattern. It can be subdifferentiated by hybridizing IS256 to SmaI-digested DNA (13).

We have investigated the presence of the mecA gene coding region in strains of clonal groups of S. aureus with phage pattern 95, phage pattern 94,96, and phage pattern 29 that were until recently sensitive to methicillin. Close clustering of S. aureus within these clonal groups by SmaI macrorestriction patterns was confirmed by three different PCR patterns (rRNA gene spacers, ERIC-2, and tar916-shida [33]).

The application to typing of strain-specific DNA probes derived from different RAPD segments has recently been demonstrated. This approach is based on cloning and sequencing of RAPD amplimers followed by selection of strain-specific probes for hybridization to EcoRI fragments. Used for typing a set of selected 59 S. aureus strains, this technique was as discriminative as SmaI macrorestriction patterns (29).

Another typing attempt used subtractive hybridization to screen for and identify conserved DNA sequences associated with epidemic MRSA. In an Australian clonal MRSA population, a specific series of six tandem repeats with no significant nucleotide sequence homologies to previously described protein coding sequences was identified (6), confirming the clonality of these strains. The use of

TABLE 2 Comparison of genotyping systems applied to *S. aureus*

Typing systems	Reproducibility	Discriminatory power
Multilocus enzyme electrophoresis	High	Good
Restriction fragment length polymorphisms		
*Sma*I-macrorestriction patterns	High	Excellent
Ribotyping	Good	Moderate
Hybridization of *mecA* and Tn*554* probes to genomic digests	High	Good
PCR-detected polymorphisms		
rRNA gene spacer	High	Moderate
RAPD (ERIC-2 and BG2 primers)	Moderate (some PCR products inconsistently seen)	Good
tar916-shida	Moderate (some PCR products inconsistently seen)	Good
Inter-IS*256*	Moderate (some PCR products inconsistently seen)	High for gentamicin-resistant isolates
coa repetitive regions	High	Moderate
spa repetitive regions	High	High but overdiscrimination
Plasmid profiles	High	Weak

these sequences for hybridization to genomic digests will probably be useful for typing MRSA. A comparison of the various genotyping methods with respect to reproducibility and discriminatory power is shown in Table 2.

Interpretation of Typing Results and Data Analysis

Ideally, outbreak isolates represent the recent clonal progeny of one particular precursor and therefore exhibit unique genotypical patterns that are clearly different from those of epidemiologically unrelated strains. One has to remember, however, that genotypical patterns of outbreak strains can change owing to genomic rearrangements and

TABLE 3 Criteria for the interpretation of *Sma*I macrorestriction patterns

Microbiological and epidemiological interpretation	No. of genetic events leading to pattern differences	No. of fragment differences in comparison to the index strain
Indistinguishable; isolate belongs to the outbreak clonal population	0	0
Closely related; isolate probably belongs to the outbreak clonal population	1	1–3
Possibly related; isolate represents a subclone of the outbreak population	2	4–6
Unrelated; isolate does not belong to the outbreak population	3	≥7

that in case of a wide dissemination of one particular epidemic clone, secondary outbreaks may not necessarily be related. General principles already established for the interpretation of molecular typing data (21) are based on fragment or amplimer differences (Table 3). The application of these criteria to interpretation of macrorestriction patterns is rather unequivocal. It is hoped that criteria for interpretation of PCR patterns might emerge from collaborative studies.

Large numbers of macrorestriction (or amplimer) patterns cannot be compared by eye. Computer-aided systems for image processing, storage of a densitogram for each lane of a gel or of molecular mass patterns of the fragments in each lane, and cluster analysis have been developed (2). Currently available commercial systems are mainly based on density curves obtained from fluorometric analysis of gels by means of Pearson's coefficient. A much more reliable alternative is fragment matching based on molecular mass (2), since this could control for differences in conditions, especially between laboratories.

REFERENCES

1. **Brakstad, O. G., J. A. Maeland, and Y. Tveten.** 1993. Multiplex polymerase chain reaction for detection of genes for *Staphylococcus aureus* thermonuclease and methicillin resistance and correlation of oxacillin resistance. *Acta Pathol. Microbiol. Scand.* **101**:681–688.
2. **Claus, H., C. Cuny, B. Pasemann, and W. Witte.** 1998. A database system for fragment patterns of genomic DNA of *Staphylococcus aureus*. *Zbl. Bakteriol.* **287**:105–116.
3. **Cuny, C., H. Claus, and W. Witte.** 1996. Discrimination of *S. aureus* by PCR for r-RNA gene spacer size polymorphisms and comparison to *Sma*I macrorestriction patterns. *Zentralbl. Bakteriol.* **283**:466–476.
4. **Cuny, C., and W. Witte.** 1996. Typing of *Staphylococcus aureus* by PCR for DNA sequences flanked by transposon Tn*916* target region and ribosomal binding site. *J. Clin. Microbiol.* **34**:1502–1505.

5. Deplano, A. M., O. Vancechonutte, G. Verschraegen, and M. J. Struelens. 1997. Typing of *Staphylococcus aureus* and *Staphylococcus epidermidis* strains by PCR analysis of inter IS256 spacer length polymorphism. *J. Clin. Microbiol.* **35**:2580–2587.

6. El-Adhami, W. A., P. R. Stewart, and K. I. Matthai. 1997. The isolation and cloning of chromosomal DNA specific for a clonal population of *Staphylococcus aureus* by substractive hybridization. *J. Med. Microbiol.* **46**:987–997.

7. Frenay, H. M. E., J. P. G. Theelen, L. M. Schouls, M. J. E. Vandenbroucke-Grauls, J. Verhoef, W. J. Van Leeuwen, and F. P. Mooi. 1994. Discrimination of epidemic and nonepidemic methicillin-resistant *Staphylococcus aureus* strains in the basis of protein A gene polymorphism. *J. Clin. Microbiol.* **32**:846–847.

8. Goering, R. V., and E. A. Ruff. 1993. Comparative analysis of conjugative plasmids mediating gentamicin resistance in *Staphylococcus aureus*. *Antimicrob. Agents Chemother.* **24**:450–452.

9. Hiramatsu, K. I. 1995. Molecular evolution of MRSA. *Microbiol. Immunol.* **39**:531–543.

10. Kloos, W. E., and D. W. Lambe, Jr. 1991. *Staphylococcus*, p. 222–237. *In* A. Balows, W. J. Hausler, Jr., K. L. Herrmann, H. D. Isenberg, and H. J. Shadomy (ed.), *Manual of Clinical Microbiology*, 5th ed. American Society for Microbiology, Washington, D.C.

11. Kreiswirth, B., J. Kornblum, R. D. Arbeit, W. Eisner, J. N. Maslow, A. McGee, D. E. Low, and R. P. Novick. 1993. Evidence for a clonal origin of methicillin resistance in *Staphylococcus aureus*. *Science* **259**:227–230.

12. Ludwig, W., and H. H. Schleifer. 1994. Bacterial phylogeny based on 16S and 23S rRNA sequence analysis. *FEMS Microbiol. Rev.* **15**:155–173.

13. Morvan, L. A., S. Arbeit, C. Dodard, and N. El Solh. 1997. Contribution of a typing method based on IS256 probing of SmaI-digested cellular DNA to discrimination of European phage type 77 methicillin-resistant *Staphylococcus aureus* strains. *J. Clin. Microbiol.* **35**:1415–1423.

14. Mulligan, M., R. Y. Kwok, D. M. Citron, J. F. John, and P. B. Smith. 1988. Immunoblots, antimicrobial resistance, and bacteriophage typing of oxacillin-resistant *Staphylococcus aureus*. *J. Clin. Microbiol.* **26**:2395–2401.

15. Musser, J. M. 1996. Molecular population genetic analysis of emerged bacterial pathogens: selected insights. *Emerg. Infect. Dis.* **2**:1–17.

16. Musser, J., and R. K. Selander. 1990. Genetic analysis of natural populations of *Staphylococcus aureus*, p. 59–67. *In* R. P. Novick (ed.), *Molecular Biology of the Staphylococci*. VCH Publishers, New York, N.Y.

17. Novick, R. P. 1990. The *Staphylococcus aureus* as a molecular genetic system, p. 7–37. *In* R. P. Novick (ed.), *Molecular Biology of the Staphylococci*. VCH Publishers, New York, N.Y.

18. Probst, A., C. Hertel, L. Richter, L. Wassill, W. Ludwig, and W. Hammes. 1998. *Staphylococcus condimenti* sp. nov., from soy sauce mash, and *Staphylococcus carnosus* (Schleifer and Fischer, 1982) subsp. *utilis* subsp. nov. *Int. J. Syst. Bacteriol.* **48**:651–658.

19. Rosdahl, W. T., W. Witte, J. M. Musser, and J. O. Jarlow. 1994. *Staphylococcus aureus* of type 95, spread of a single clone. *Epidemiol. Infect.* **113**:463–470.

20. Santos-Sanches, I., M. A. De Sousa, S. I. Calheiros, L. Felicito, I. Pedra, and H. De Lencastre. 1995. Multidrug-resistant Iberian epidemic clone of methicillin-resistant *Staphylococcus aureus* endemic in a hospital in Northern Portugal. *Microb. Drug Resist.* **1**:299–306.

21. Saruta, K., T. Matsunaga, M. Kono, S. Hoshina, S. Ikawa, O. Sakai, and K. Machida. 1997. Rapid identification and typing of *Staphylococcus aureus* by nested PCR amplified ribosomal DNA spacer region. *FEMS Microbiol. Lett.* **146**:271–278.

22. Schleifer, K. H., and R. M. Kroppenstedt. 1990. Chemical and molecular classification of staphylococci. *J. Appl. Bacteriol. Symp. Suppl.* **1990**:9S–45S.

23. Schwarzkopf, A., and H. Karch. 1994. Genetic variation in *Staphylococcus aureus* coagulase genes: potential and limit for use as epidemiological marker. *J. Clin. Microbiol.* **32**:2407–2412.

24. Stobberingh, E. E., and K. C. Winkler. 1977. Restriction deficient mutants of *Staphylococcus aureus*. *J. Gen. Microbiol.* **90**:359–367.

25. Struelens, M., R. Bax, A. Deplano, V. G. Quint, and A. Van Belkum. 1993. Concordant clonal delineation of methicillin-resistant *Staphylococcus aureus* by macrorestriction analysis and polymerase chain reaction genome fingerprinting. *J. Clin. Microbiol.* **31**:1964–1970.

26. Tenover, F. C., R. D. Arbeit, and R. V. Goering. 1997. How to select and interpret molecular strain typing methods for epidemiological studies of bacterial infections: a review for healthcare epidemiologists. *Infect. Control Hosp. Epidemiol.* **18**:426–439.

27. Van Belkum, A. 1994. DNA fingerprinting of medically important microorganisms by use of PCR. *Clin. Microbiol. Rev.* **7**:174–184.

28. Van Belkum, A., R. Bax, and G. Prevost. 1994. Comparison of four genotyping assays for epidemiological study of methicillin-resistant *Staphylococcus aureus*. *Eur. J. Clin. Microbiol. Infect. Dis.* **13**:420–424.

29. Van Leeuwen, W., M. Simons, J. Schijs, H. Verbrugh, and A. Van Belkum. 1996. On the nature and use of randomly amplified DNA from *Staphylococcus aureus*. *J. Clin. Microbiol.* **34**:2770–2777.

30. Versalovic, J., T. Koeuth, and J. R. Lupski. 1991. Distribution of repetitive DNA sequences in eubacteria and application to fingerprinting of bacterial genomes. *Nucleic Acids Res.* **19**:6823–6831.

31. Witte, W., C. Cuny, and H. Claus. 1993. Clonal relatedness of *Staphylococcus aureus* strains from infections in humans as deduced from genomic DNA fragment patterns. *Med. Microbiol. Lett.* **2**:72–79.

32. Witte, W., C. Cuny, O. Zimmermann, R. Rüchel, M. Höpken, R. Fischer, and J. Wagner. 1994. Stability of genomic DNA fragment patterns in methicillin-resistant *Staphylococcus aureus* during the course of intra- and interhospital spread. *Eur. J. Epidemiol.* **10**:743–748.

33. Witte, W., M. Kresken, C. Braulke, and C. Cuny. 1997. Increasing incidence and widespread dissemination of methicillin-resistant *Staphylococcus aureus* (MRSA) in hospitals in central Europe, with special reference to German hospitals. *Clin. Microbiol. Infect.* **3**:414–422.

Genetic and Physical Map of the Chromosome of *Staphylococcus aureus* 8325

JOHN J. IANDOLO

32

The genome map of *Staphylococcus aureus* owes much of its development to the efforts of Peter A. Pattee and colleagues at Iowa State University (13, 25). By focusing on the *S. aureus* phage group III strain NCTC 8325, his group originated studies to generate and map mutant strains. The map has continued to evolve, and recently an effort was undertaken by my group to sequence and annotate the entire genome of strain 8325. A second strain (COL) is being sequenced by the Institute for Genomic Research.

Mutants used to analyze the genetic organization of the chromosome were constructed using a combination of methods that included UV light, various chemical mutagens, and the transposons Tn551, Tn917, Tn4001, and Tn916 (15, 22, 24). The extreme fastidiousness of this organism presented a problem regarding the availability of nutritional markers for analysis that is still of concern today. Genetic markers were analyzed by transduction and by transformation albeit at low frequencies of transfer. Moreover, the inability to move genetic markers at high frequency and the lack of selectable phenotypes caused many genetic loci to be marked by transposition insertion (Ω) sites. Some of these sites are located in known genes and were used to identify phenotypic changes induced by insertional mutation, while others are silent and simply serve to note insertion sites on the genome. This led to the construction of three linkage groups consisting of a mix of knock-out insertions, silent insertions, and nutritional auxotrophies (26).

Transduction was the first mechanism exploited for genetic analysis of *S. aureus* and is the main tool for fine-structure mapping. However, genetic arrays investigated by this method are limited in size because the average headful of phage DNA of staphylococcal phages is about 40 to 45 kb (33) (ca. 1.0 to 1.5% of the chromosome). The report of Lindberg and coworkers (19) of transformation in *S. aureus* NCTC 8325, coupled with improvements in chromosomal DNA isolation procedures, allowed the analysis of larger fragments of chromosomal DNA. It was found that, to be cotransducible, two genes had to be at least 20% cotransformable; this led to the conclusion that transformation could transfer up to about 200 kb of DNA. This

in turn led to an extension of the linkage groups to cover most of the genome but still left three unclosed gaps in the circular chromosome. The relationship of these linkage groups and generation of a circular genetic map were made possible through the development of a computer-assisted protoplast fusion technique (31). Transformation analysis was then utilized to confirm the linkage data generated by the protoplast fusion analysis (32). With the advent of genome-sized DNA electrophoretic technology, the three chromosomal gaps were closed, and the Pattee group was the first to construct a circular physical map and correlate fragments with known genetic markers.

A major advance in the construction of the maps has occurred as a result of the deposition of large quantities of sequence information in the GenBank database and through our efforts to sequence the entire *S. aureus* genome. Although this project is not complete, it has allowed us to clarify some ambiguities and to add additional loci on the map. It has been our policy to use these periodic updates in the linkage map and the deposition of the genome sequence on our web site (1) to make this information available to all staphylococcal researchers.

THE PHYSICAL MAP

There is, as expected within the genus *Staphylococcus*, a great deal of similarity among physical maps of the phage groups (13). Restriction fragment length polymorphisms have been found and for the most part can be rationalized as the result of lysogeny with many different phages (30). Virtually all staphylococci are lysogenized at least singly, and many are multiply lysogenized (6). This sometimes makes electrophoretic data of genome digests particularly difficult to interpret correctly.

However, fragmentation data using four restriction endonucleases have led to the construction of a reliable physical map of the phage group III strain NCTC 8325 genome. The restriction map data were correlated with genetic markers by hybridization and by genetic crosses. The result of these analyses is presented in Fig. 1. Many more markers are present on this version of the genetic map as a result

Gram-Positive Pathogens, ed. by V. A. Fischetti et al.
© 2000 American Society for Microbiology, Washington, D.C.

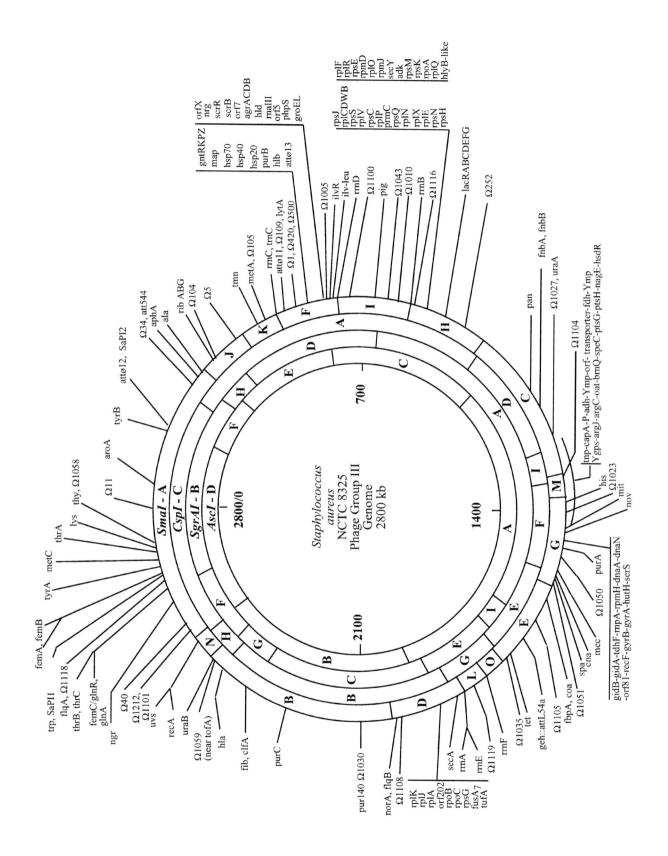

of analysis of genome sequencing data. These additional loci were found to be linked to known genes and could be unambiguously placed on the map. Other markers (notably adhesin molecules) have been shown by hybridization to be associated with various SmaI fragments (29) but have not been mapped to specific loci. A complete list of known, mapped, unmapped, and hypothetical genes and transposon insertion sites is presented in Table 1. In addition to mapped genes, Table 1 contains many of the genes that have been localized to a particular SmaI fragment but not mapped relative to other markers. Furthermore, some of the equivalent genetic positions of genes or genetic elements that do not occur naturally in the 8325 genome are indicated. They are known to transfer naturally and may be present or absent in strains whose genetic pedigree is not known or not clearly defined. Since there are no allelic counterparts in strains that do not contain these elements, they are regarded as accessory genetic material and possibly provide a selective advantage in certain environmental situations. Nevertheless, they are described below and listed in Table 1 for completeness.

Many of the housekeeping functions identified in *S. aureus* are highly conserved and similar in organization to other bacteria. In particular, ribosomal protein operons (SmaI-H and D), ribosomal RNA operons (various SmaI-junction fragments), and genes around the origin of replication (SmaI-G) are consistent with the gene order found in the *Bacillus subtilis* (1a) and *Escherichia coli* (2) genomes. The genome also carries a variable number of mobile genetic elements, including insertion sequences, transposons, bacteriophages, and foreign DNA inserts of uncharacterized type. The additional DNA sequences are often responsible for virulence properties, such as enterotoxin production, production of toxic shock syndrome toxin, or resistance to methicillin. These variable modular elements of staphylococcal genomes are summarized below.

Insertion Sequences and Transposons

Insertion sequences and transposons have been identified in a variety of strains of *S. aureus*. Insertion sequence IS256 is a component of Tn4001, and multiple copies of this element have been detected on the chromosomes of certain clinical *S. aureus* strains (7). IS431 (IS257) is a component of Tn4003 and is found on a number of antibiotic-resistance-encoding and conjugative plasmids of *S. aureus* (3, 27). This insertion sequence is also found in the methicillin-resistance region of the chromosome. IS1181 was initially identified on an *S. aureus* plasmid, and copies of this element are present in the 8325 genome (11). Transposons identified in *S. aureus* include those utilized in the Pattee studies to map the chromosome. These are Tn551, encoding macrolide-lincosamide-streptogramin B resistances, and Tn4001, encoding erythromycin, gentamicin, neomycin, and kanamycin resistance (21). Additional transposons include Tn4003, encoding trimethoprim resistance, and Tn4002 and Tn4201, encoding β-lactamase (12, 17). Site-specific transposons include Tn552 (encoding β-lactamase), Tn554 (encoding resistance to erythromycin and spectinomycin), and Tn3582 (which is closely related to Tn554) (23, 28, 34). These transposable elements have preferred primary insertion sites in the chromosome. For Tn553 this consists of an integrated copy of *bin-res* and for Tn554 a unique site, *att554* (Fig. 1). The latter transposon can insert at low frequency at secondary sites if the primary site is deleted (23). These are discussed in more detail in chapter 33, this volume.

Prophages

S. aureus 8325 is known to carry three temperate bacteriophages. Bacteriophages φ11 and φ13 are located in SmaI fragment F while φ12 has been mapped to SmaI fragment A. In addition to altering contour-clamped homogeneous

FIGURE 1 Physical and genetic map of the genome of the *S. aureus* phage group III strain NCTC 8325. Information concerning the individual markers is summarized in Table 1. The order of the majority of the markers shown is based on genetic analysis (see references 13, 24–26), DNA hybridization of SmaI, CspI, SgrAI, and AscI restriction endonuclease digestions of the chromosome, and direct analysis of genome sequence. The fragments were resolved by pulsed-field agarose gel electrophoresis and hybridized with appropriate probes. SmaI fragment O was identified and mapped by Wada et al. (35). Of the markers shown, *lys, thy, Ω11, Ω402, aphA, attφ11, Ω420, Ω500, purB, bla, ilvR, leu, pig, uraA, his, mit, nov, dnaD, purA, fus, pur–140, tofA, ngr, thrC, flqA, Ω401, tyrA,* and *metC* have not been physically mapped; their map locations are based only on multifactorial transformation analyses. The genomes of the prophages φ12 and φ13 contain an internal SgrAI restriction endonuclease recognition site. As a result, φ12 confers an additional restriction site in SgrAI–B and φ13 in SgrAI–D (13). In addition, each of the ribosomal RNA operons (*rrnA–F*) contains an internal SmaI recognition site that defines the junctions between SmaI L/D, H/I, K/F, I/F, and L/O. The orientation of the markers in parentheses is not known relative to the remainder of the map. The entire chromosome is about 2,800 kb, and the individual fragments are drawn to scale. Approximately 125 to 150 kb of DNA represents the three known prophages (φ11, φ12, φ13) that lysogenize the NCTC 8325 genome. There are several small DNA fragments that have not yet been placed on the physical map. These include SgrAI-G, -J, and -K representing about 300 kb, at least two CspI fragments representing about 100 kb, and one SmaI fragment representing about 10 kb.

Note: The mutant strains used in generating the map were maintained as part of the collection of P. A. Pattee. They include derivatives generated by his research group as well as contributions from staphylococcal researchers around the world. Upon his retirement, I acquired the collection and merged it with my own. The strains are freely available to investigators. Requests for strains can be sent to the Staphylococcal Genetic Stock Center (SGSC) at my address at the front of this volume or directly by e-mail (John-Iandolo@ouhsc.edu). I also invite colleagues to consider deposition of marked strains and clones into the collection for distribution (either full or restricted) from a central source. Information about the staphylococcal genome sequencing projects can be found in reference 1 (NCTC 8325) and reference 13a (COL).

TABLE 1 Genetic markers of *S. aureus* 8325

Gene symbol[a]	Phenotype	Map location[b]	Reference(s)
adh[c]	Alcohol dehydrogenase	GMC	13, 25
adk	Adenosine kinase	H	13, 25
agr[d]	Accessory gene regulator of several exoproteins and toxins (*agrACBD*)	F	13, 25
ala[d]	L-Alanine requirement	A	13, 24, 25
aphA	Aminoglycoside phosphotransferase (APH3-III)	A	13
argJ	Amino acid acetyltransferase/deacetylase	GMC	Unpublished
argC	Arginine requirement	GMC	Unpublished
attφ11	Prophage φ11 integration site	F	13, 25
attφ12	Prophage φ12 integration site	A	13, 25
attφ13	Prophage φ13 integration site (in *hlb*)	F	9, 13, 25
attφL54a	Prophage φL54a integration site (in *geh*)	E	13, 18, 25
att554	Primary integration site for Tn554	A	13, 25
bla (*pen*)	β-Lactamase production	F	13, 25
brnQ	Transport system 1 for Leu, Ile, Val	GMC	Unpublished
capA–P	Capsule genes	GMC	Unpublished
coa	Coagulase	E	13, 25
clfA	Clumping factor	B	13, 25, 29
cna	Collagen adhesin	G	13, 25, 29
dnaA	DNA biosynthesis, initiation	G	13, 25
dnaN	DNA polymerase III holoenzyme, β subunit	G	Unpublished
ebpS	Elastin-binding protein	A	13, 25, 29
ermB	Impaired erythromycin resistance by Tn551	In Tn*551*	13, 25
fdh	Formic dehydrogenase	GMC	Unpublished
femAB[d] (*Ω2OO3*)	Factor essential for expression of methicillin resistance; see *Ω2004*	A	13, 25
femC	Reduction of methicillin resistance levels	A	13, 25
fib	Fibrinogen-binding protein	B	13, 25, 29
fnbA	Fibronectin-binding protein A	C	13, 25, 29
fnbB	Fibronectin-binding protein B	C	13, 25, 29
flqA	Fluoroquinolone resistance	A	13, 25
flqB	Quinolone resistance	D	13, 25
*fusA*₇	Fusidic acid resistance	D	13, 25, 31
geh[e]	Glycerol ester hydrolase (lipase)	E	13, 18, 25
gidA[c]	Glucose-inhibited division protein	G	Unpublished
gidB[c]	Glucose-inhibited division protein	G	Unpublished
glnA	Glutamine synthetase	A	13, 25
glnR	Glutamine synthetase repressor	A	13, 25
gnt	Gluconate utilization (*gntRKPZ*)	F	13, 25
groEL[c]	Chaperone protein	F	Unpublished
gyrA	DNA gyrase; ciprofloxacin resistance	G	13, 25
gyrB	DNA gyrase	G	13, 25
his	L-Histidine requirement (*hisEABCDG*)	G or M	13, 25
hla	α-Toxin structural gene	B	13, 25
hlb[e]	β-Toxin	F	9, 13, 25
hld	Delta toxin, 5′ end of RNAIII	F	13, 25
hlg	Gamma toxin (*hlgABC*)	C	13, 25
hlyB-like[c]	ABC transporter	H	Unpublished
hsdR	Host restriction, endonuclease R	GMC	
hsp20	Heat shock protein	F	13, 25
hsp40	Heat shock protein	F	13, 25
hsp70	Heat shock protein	F	13, 25
hutH	Histidase	G	
ilv-leu[d]	L-Isoleucine, L-valine, L-leucine requirement (*ilvABCD-leuABCD*)	F	13, 25

(Continued on next page)

TABLE 1 (*Continued*)

Gene symbol[a]	Phenotype	Map location[b]	Reference(s)
ilvR	Resistance to D–leucine	F	13, 25
lacR	Lactose repressor	H	13
lacA	Galactose-6-phosphate isomerase subunit	H	13
lacB	Galactose-6-phosphate isomerase subunit	H	13
lacC	Tagatose-6-phosphate kinase	H	13
lacD	Tagatose-1,6-bisphosphate aldolase	H	13
lacE	Enzyme IIB of lactose phosphotransferase transport system	H	13
lacF	Enzyme IIA of lactose phosphotransferase transport system	H	13
lacG	Phosphobetagalactosidase	H	13
lip[d]	Lipoic acid requirement	A	13, 25
lys[d]	L-Lysine requirement (*lysOABFG*)	A	13, 25
lytA	N-Acetylmuramyl-L-alanine amidase (peptidoglycan hydrolase)	F (in ϕ11)	13, 25
map	Matrix adhesin protein	C	13, 25, 29
mecA	Methicillin resistance; *pbp2a*	G	13, 25
metA	L-Methionine requirement	K	13, 22, 25
MetC[d]	L-Methionine requirement; β-cystathionase	A	13, 25
mit	Enhanced sensitivity to mitomycin C, nitrosoguanidine, and UV light	G	13, 25
mdr	Multidrug resistance transporter	D	13, 25
nagE[c]	N-Acetylglucosamine-specific enzyme II of phosphotransferase transport system	GMC	Unpublished
ngr	Apurinic endonuclease deficiency	A	13, 25
norA	Hydrophilic quinolone resistance	D	13, 25
nov	Novobiocin resistance	G	13, 25
nrg	Unknown function	F	
nuc	Staphylococcal nuclease		13, 25
oat[c]	Ornithine aminotransferase	GMC	Unpublished
orf[c]	Hypothetical transporter protein	GMC	Unpublished
orf5[c]	Linked hypothetical protein	F	Unpublished
orf7[c]	*agr*-linked hypothetical protein	F	Unpublished
orf81[c]	Replication-related protein	G	Unpublished
orf202	Hypothetical protein	D	Unpublished
orfX[c]	Hypothetical protein	F	Unpublished
pan[d]	Pantothenate requirement	C	13, 25
pbp4	Penicillin-binding protein 4	D	13, 25
phpS[c]	Unknown function	F	Unpublished
pig	Golden-yellow pigment deficiency	I	13, 25
prmC[c]	50S ribosomal protein methylase	H	Unpublished
ptsG[c]	Glucose-specific enzyme II of phosphotransferase transport system	GMC	Unpublished
ptsH[c]	PEP phosphotransferase, Hpr	GMC	Unpublished
purA	Adenine requirement	G	13, 25, 26
purB[d]	Adenine + guanine requirement	F	13, 25, 26
purC[d]	Purine requirement	B	13, 25, 32
purD[d]	Guanine requirement	B	13, 25
pur-140	Purine requirement	B	13, 25
recA	Homologous recombination deficiency	N	13, 25
recF	Recombination	G	
rib[d]	Riboflavin requirement (*ribABG*)	J	13, 24, 25
rif	Rifampin resistance	D	13, 25
rnaIII[c]	RNAIII, regulatory molecule for exoprotein synthesis	F	Unpublished
rnpA[c]	RNase P	G	Unpublished
rplA[c]	50S ribosomal protein	D	Unpublished
rplB[c]	50S ribosomal protein	H	Unpublished

(*Continued on next page*)

TABLE 1 Genetic markers of *S. aureus* 8325 (*Continued*)

Gene symbol[a]	Phenotype	Map location[b]	Reference(s)
rplC[c]	50S ribosomal protein	H	Unpublished
rplD[c]	50S ribosomal protein	H	Unpublished
rplE[c]	50S ribosomal protein	H	Unpublished
rplF[c]	50S ribosomal protein	H	Unpublished
rplJ[c]	50S ribosomal protein	D	Unpublished
rplK[c]	50S ribosomal protein	D	Unpublished
rplN[c]	50S ribosomal protein	H	Unpublished
rplO[c]	50S ribosomal protein	H	Unpublished
rplP[c]	50S ribosomal protein	H	Unpublished
rplQ[c]	50S ribosomal protein	H	Unpublished
rplR[c]	50S ribosomal protein	H	Unpublished
rplV[c]	50S ribosomal protein	H	Unpublished
rplW[c]	50S ribosomal protein	H	Unpublished
rplX[c]	50S ribosomal protein	H	Unpublished
rpmD[c]	50S ribosomal protein	H	Unpublished
rpmH[c]	50S ribosomal protein	G	Unpublished
rpmJ[c]	50S ribosomal protein	H	Unpublished
rpoA[c]	RNA polymerase α subunit	H	Unpublished
rpoB	RNA polymerase β subunit	D	13, 25
rpoC[c]	RNA polymerase β' subunit	D	Unpublished
rpsC[c]	30S ribosomal protein	H	Unpublished
rpsE[c]	30S ribosomal protein	H	Unpublished
rpsG[c]	30S ribosomal protein	D	Unpublished
rpsH[c]	30S ribosomal protein	H	Unpublished
rpsJ[c]	30S ribosomal protein	H	Unpublished
rpsK[c]	30S ribosomal protein	H	Unpublished
rpsM[c]	30S ribosomal protein	H	Unpublished
rpsN[c]	30S ribosomal protein	H	Unpublished
rpsQ[c]	30S ribosomal protein	H	Unpublished
rpsS[c]	30S ribosomal protein	H	Unpublished
rrnA	rRNA operon	D/L junction	13, 25
rrnB	rRNA operon	H/I junction	13, 25
rrnC	rRNA operon	K/F junction	13, 25
rrnD	rRNA operon	F/I junction	13, 25
rrnE	rRNA operon	D/L junction	13, 25
rrnF	rRNA operon	L/O junction	13, 25
sak[f]	Staphylokinase production	F (in ϕ13)	10, 13, 25
secA	Secretion of proteins	D	13
serS	Seryl tRNA synthetase	G	
spa	Staphylococcal protein A	G	13, 25
sarA	Exoprotein regulatory locus	D	13, 25
sarB	Exoprotein regulatory locus	D	13, 25
scrB[c]	Sucrose permease	F	Unpublished
scrR[c]	Sucrose permease regulatory gene	F	Unpublished
secY	*Secretion of proteins*	H	*13*
speC	Ornithine decarboxylase	GMC	13
tagD	Glycerol-3-phosphate cytidyltransferase	D	13
tdhF	Thiophene/furan oxidation protein	G	
tet	Tetracycline resistance	E	13, 25
thrA[d]	L-Lysine + L-methionine + L-threonine requirement; failure to convert aspartate to aspartic β-semialdehyde	A	13, 24, 25

(*Continued on next page*)

TABLE 1 (*Continued*)

Gene symbol[a]	Phenotype	Map location[b]	Reference(s)
thrB[d]	L-Threonine requirement; homoserine kinase deficiency	A	13, 24, 25
thrC[d]	L-Threonine requirement; threonine synthetase deficiency	A	13, 25
thy	Thymine requirement	A	13, 25, 26
tmn	Tetracycline and minocycline resistance	K	13, 25
tnp	Transposase	GMC	
tofA	Temperature-sensitive osmotically remedial cell wall synthesis; D-glutamate addition defect	B	13, 25
trnC	tRNA genes (27 tandem genes)	K/F junction	13, 25
trp[d]	L-Tryptophan requirement (trpABFCDE)	A	13, 25
tufA[c]	Elongation factor Tu	D	
tyrA[d]	L-Tyrosine requirement	A	13, 24, 25
tyrB[d]	L-Tyrosine requirement	A	13, 24, 25
uraA	Uracil requirement	C	13, 25, 26
uraB[d]	Uracil requirement	B	13, 24, 25
uvr	Enhanced sensitivity to UV light (recA allele)	N[g]	13, 25
Ymp[c]	Hypothetical membrane protein	GMC	Unpublished
Ymp[c]	Membrane protein	GMC	Unpublished
Ygps[c]	Hypothetical glycerol phosphate dehydrogenase	GMC	Unpublished
SaPI1	Pathogenicity island prototype; 15.2-kb element containing the toxic shock syndrome toxin gene. Located at ca. 2,650 kb on the map.	A	20
SaPI2	Pathogenicity island prototype; 15.2 kb-element containing the toxic shock syndrome toxin gene. Located at c. 180 kb on the map.	A	20
Ω–Ω100	Silent insertions of Tn551	Various	13, 25
Ω1000–Ω1099	Silent insertions of Tn551	Various	13, 25
Ω401 and Ω402	Original insertion sites for toxic shock syndrome toxin mobile element. Now identified as pathogenicity islands SaPI1 and SaPI2 (see above).	A	8
Ω420 (Ω42)	Insertion site of pI258	F	13, 25
Ω500 (Ω50)	Insertion site of pI258	F	13, 25
Ω1100–Ω1199	Silent insertions of Tn916	Various	13, 25
Ω101–Ω120; Ω1200–Ω1299	Silent insertions of Tn4001	Various	13, 25
Ω2004	Insertion site of Tn551 that impairs Mec; may affect penicillin-binding proteins; see femA	A	13, 25

[a] Gene designations in parentheses are old designations.

[b] Refers to specific *SmaI* restriction fragments in Fig. 1. Genes identified on a specific fragment but not shown in Fig. 1 have only been mapped by pulsed-field gel electrophoresis and DNA hybridization. Other genes (see Fig. 1 legend) have only been mapped by multifactorial transformation analyses.

[c] Genes whose identity has been ascertained from genome sequence information and whose location was identified by relation to known markers.

[d] Phenotype for which insertional inactivation with Tn551, Tn917, and/or Tn4001 is known.

[e] Controlled by negative phage conversion.

[f] Controlled by positive phage conversion.

[g] Chromosomal location of *uvr* is based on physical and genetic map data for Ω1073 and Ω1074, which exhibit greater than 50% cotransduction with *uvr* and *recA*.

electric field (CHEF) gel profiles, staphylococcal phages are responsible for both positive and negative phenotypic conversion. Certain prophages have been found to carry the determinants for enterotoxin A production, staphylokinase, or certain transposons (5, 10, 16). Negative phage conversion has also been documented. Insertion of bacteriophage φ13 into the genome of strain 8325 results in loss of β-toxin production, while φL54a, whose *att* site is in the lipase gene (*geh*), inactivates glycerol ester hydrolase production (9, 10, 18).

Variable Genetic Elements

DNA sequences can be found in the chromosomes of certain staphylococcal isolates for which there is no equiva-

lent allelic sequence in other strains. They represent important components of the *S. aureus* genome and are discussed in detail in chapter 33, this volume. They include pathogenicity islands such as those carrying *tst* and *seb*, resistance islands such as that carrying *mecA*, and uncharacterized elements such as those containing determinants of capsule production and collagen-binding proteins (4, 8, 13, 14, 17, 20). Additional such elements are likely to be found as new clinical strains are compared with the sequenced 8325 genome. Along with prophages, they represent serious complications for the interpretation of macrorestriction (CHEF) patterns (see chapter 31, this volume) and may also affect phage-typing patterns.

REFERENCES

1. **Advanced Center for Genome Technology. University of Oklahoma.** http://www.genome.ou.edu

1a. **Anagnostopoulos, C., P. J. Piggot, and J. A. Hoch.** 1993. The genetic map of *Bacillus subtilis*, p. 425–462. *In* A. L. Sonenshein, J. A. Hoch, and R. Losick (ed.), *Bacillus subtilis and Other Gram-Positive Bacteria: Biochemistry, Physiology, and Molecular Genetics.* American Society for Microbiology, Washington, D.C.

2. **Bachmann, B. J.** 1987. Linkage map of *Escherichia coli* K-12, edition 7, p. 807–876. *In* F. C. Neidhardt, J. L. Ingraham, B. Low, B. Magasanik, M. Schaechter, and H. E. Umbarger (ed.), *Escherichia coli and Salmonella typhimurium: Cellular and Molecular Biology.* American Society for Microbiology, Washington, D.C.

3. **Barberis-Maino, L., B. Berger-Bächi, H. Weber, W. D. Beck, and F. H. Kayser.** 1987. IS431, a staphylococcal insertion sequence-like element related to IS26 from *Proteus vulgaris. Gene* **59:**107–113.

4. **Berger-Bächi, B., A. Strässle, J. G. Gustafson, and F. H. Kayser.** 1992. Mapping and characterization of multiple chromosomal factors involved in methicillin resistance in *Staphylococcus aureus. Antimicrob. Agents Chemother.* **36:**1367–1373.

5. **Betley, M. J., and J. J. Mekalanos.** 1988. Nucleotide sequence of the type A staphylococcal enterotoxin gene. *J. Bacteriol.* **170:**34–41.

6. **Blair, J. E., and R. E. O. Williams.** 1961. Phage typing of staphylococci. *Bull. W. H. O.* **24:**771–784.

7. **Byrne, M. E., D. A. Rouch, and R. A. Skurray.** 1989. Nucleotide sequence analysis of IS256 from the *Staphylococcus aureus* gentamicin-tobramycin-kanamycin-resistance transposon Tn4001. *Gene* **81:**361–367.

8. **Chu, M. C., B. N. Kreiswirth, P. A. Pattee, R. P. Novick, M. E. Melish, and J. F. James.** 1988. Association of toxic shock toxin-1 determinant with a heterologous insertion at multiple loci in the *Staphylococcus aureus* chromosome. *Infect. Immun.* **56:**2702–2708.

9. **Coleman, D. C., J. P. Arbuthnott, H. M. Pomeroy, and T. H. Birkbeck.** 1986. Cloning and expression in *Escherichia coli* and *Staphylococcus aureus* of the beta-lysin determinant from *Staphylococcus aureus*: evidence that bacteriophage conversion of beta-lysin activity is caused by insertional inactivation of the beta-lysin determinant. *Microb. Pathog.* **1:**549–564.

10. **Coleman, D. C., D. S. Sullivan, R. J. Russell, J. P. Arbuthnott, B. F. Carey, and H. M. Pomeroy.** 1989. *Staphylococcus aureus* bacteriophages mediating the simultaneous lysogenic conversion of β-lysin, staphylokinase and enterotoxin A: molecular mechanism of triple conversion. *J. Gen. Microbiol.* **135:**1679–1697.

11. **Derbise, A., K. G. H. Dyke, and N. El Solh.** 1994. Isolation and characterization of IS1181, an insertion sequence from *Staphylococcus aureus. Plasmid* **31:**252–264.

12. **Gillespie, M. T., B. R. Lyon, and R. A. Skurray.** 1988. Structural and evolutionary relationships of β-lactamase transposons from *Staphylococcus aureus. J. Gen. Microbiol.* **134:**2857–2866.

13. **Iandolo, J. J., J. P. Bannantine, and G. C. Stewart.** 1997. Genetic and physical map of the chromosome of *Staphylococcus aureus*, p. 39–54. *In* G. L. Archer and K. Crosley (ed.), *Staphylococci and Staphylococcal Diseases.* Churchill-Livingstone, New York, N.Y.

13a. **Institute for Genomic Research.** http://www.tigr.org

14. **Johns, M. B., Jr., and S. A. Khan.** 1988. Staphylococcal enterotoxin B gene is associated with a discrete genetic element. *J. Bacteriol.* **170:**4033–4039.

15. **Jones, J. M., S. C. Yost, and P. A. Pattee.** 1987. Transfer of the conjugal tetracycline resistance transposon Tn916 from *Streptococcus faecalis* to *Staphylococcus aureus* and identification of some insertion sites in the staphylococcal chromosome. *J. Bacteriol.* **169:**2121–2131.

16. **Kondo, I., S. Itoh, and Y. Yoshizawa.** 1981. Staphylococcal phages mediating the lysogenic conversion of staphylokinase. *Zentralbl. Bakteriol.* **10**(Suppl.)**:**357.

17. **Lee, C. Y.** 1995. Association of staphylococcal type 1 capsule genes with a discrete genetic element. *Gene* **167:**115–119.

18. **Lee, C. Y., and J. J. Iandolo.** 1986. Lysogenic conversion of staphylococcal lipase is caused by insertion of the bacteriophage L54a genome into the lipase structural gene. *J. Bacteriol.* **166:**385–391.

19. **Lindberg, M., J.-E. Sjostrom, and T. Johansson.** 1972. Transformation of chromosomal and plasmid characters in *Staphylococcus aureus. J. Bacteriol.* **109:**844–847.

20. **Lindsay, J. A., A. Ruzin, H. F. Ross, N. Kurepina, and R. P. Novick.** 1998. The gene for toxic shock is carried by a family of mobile pathogenicity islands in *Staphylococcus aureus. Mol. Microbiol.* **29:**527–543.

21. **Lyon, B. R., J. W. May, and R. A. Skurray.** 1984. Tn4001: a gentamicin and kanamycin resistance transposon in *Staphylococcus aureus. Mol. Gen. Genet.* **193:**554–556.

22. **Mahairas, G. G., B. R. Lyon, R. A. Skurray, and P. A. Pattee.** 1989. Genetic analysis of *Staphylococcus aureus* with Tn4001. *J. Bacteriol.* **171:**3968–3972.

23. **Murphy, E., S. Philips, I. Edelman, and R. P. Novick.** 1981. Tn554: isolation and characterization of plasmid insertions. *Plasmid* **5:**292–305.

24. **Pattee, P. A.** 1981. Distribution of Tn551 insertion sites responsible for auxotrophy on the *Staphylococcus aureus* chromosome. *J. Bacteriol.* **145:**479–488.

25. **Pattee, P. A.** 1993. The genetic map of *Staphylococcus aureus*, p. 489–496. *In* A. L. Sonenshein, J. A. Hoch, and R. Losick (ed.), *Bacillus subtilis and Other Gram-Positive Bacteria: Biochemistry, Physiology, and Molecular Genetics.* American Society for Microbiology, Washington, D.C.

26. **Pattee, P. A., and D. S. Neveln.** 1975. Transformation analysis of three linkage groups in *Staphylococcus aureus. J. Bacteriol.* **124:**201–211.

27. **Rouch, D. A., L. J. Messerotti, L. S. L. Loo, C. A. Jackson, and R. A. Skurray.** 1989. Trimethoprim resistance transposon Tn4003 from *Staphylococcus aureus* encodes genes for a dihydrofolate reductase and thymidylate synthetase flanked by three copies of IS257. *Mol. Microbiol.* **3:**161–175.

28. **Rowland, S. J., and K. G. Dyke.** 1989. Characterization of the staphylococcal β-lactamase transposon Tn552. *EMBO J.* **8:**2761–2773.

29. **Smeltzer, M. S., A. F. Gillaspy, F. L. Pratt, Jr., M. D. Thames, and J. J. Iandolo.** 1997. Prevalence and chro-

mosomal map location of *Staphylococcus aureus* adhesin genes. *Gene* **196:**249–259.

30. **Smeltzer, M. S., M. E. Hart, and J. J. Iandolo.** 1994. The effect of lysogeny on the genomic organization of *Staphylococcus aureus. Gene* **138:**51–57.

31. **Stahl, M. L., and P. A. Pattee.** 1983. Computer-assisted chromosome mapping by protoplast fusion in *Staphylococcus aureus. J. Bacteriol.* **154:**395–405.

32. **Stahl, M. L., and P. A. Pattee.** 1983. Confirmation of protoplast fusion-derived linkages in *Staphylococcus aureus* by transformation with protoplast DNA. *J. Bacteriol.* **154:** 406–412.

33. **Stewart, P. R., H. G. Waldron, J. S. Lee, and P. R. Matthews.** 1985. Molecular relationships among serogroup B bacteriophages of *Staphylococcus aureus. J. Virol.* **55:**111–116.

34. **Townsend, D. E., S. Bolton, N. Ashdown, D. I. Annear, and W. B. Grubb.** 1986. Conjugative staphylococcal plasmids carrying hitch-hiking transposons similar to Tn*554:* intra- and interspecies dissemination of erythromycin resistance. *Aust. J. Exp. Biol. Med. Sci.* **64:**367–379.

35. **Wada, A., H. Ohta, K. Kulthanen, and K. Hiramatsu.** 1993. Molecular cloning and mapping of 16S-23S rRNA gene complexes of *Staphylococcus aureus. J. Bacteriol.* **175:** 7483–7487.

Genetics: Accessory Elements and Genetic Exchange

NEVILLE FIRTH AND RONALD A. SKURRAY

33

As has been the case for other bacterial genera, studies into the genetic basis of staphylococcal pathogenicity and, in particular, antimicrobial resistance have revealed in some strains the presence of determinants not normally represented in the genome. Although not fundamental requirements for survival per se, such accessory elements invariably encode functions required to meet the demands of a particular environmental niche. It is clear that the acquisition, maintenance, and dissemination of accessory elements have been central to the ongoing success of staphylococci as pathogens. Staphylococci represent a salient illustration of the adaptability afforded to microorganisms by access to additional functions through gene transfer mechanisms.

MECHANISMS OF GENETIC EXCHANGE

Although DNA can be introduced into staphylococci in the laboratory via each of the three traditional bacterial gene transfer mechanisms of transformation, transduction, and conjugation, the latter two are believed to be the most significant mediators of natural genetic exchange; transformation is very inefficient, has a curious cofactor requirement that can be satisfied by a component of phage 55C, and is thought to be limited by extracellular nucleases and/or restriction systems encoded by staphylococci (58, 65). In addition to their own transfer, staphylococcal conjugative plasmids (see below) also facilitate the transmission of other nonconjugative plasmids, either by mobilization, if the other plasmid encodes a specialized relaxation system (71), or more generally via cointegrate formation, and potentially subsequent resolution, in a process termed conduction (9, 59). A novel but poorly understood mechanism of genetic exchange, named mixed-culture transfer or phage-mediated conjugation, has also been identified in staphylococci (52, 58). Although phage, or perhaps components of phage, play a role in mixed-culture transfer, the process has been shown to be mechanistically distinct from transduction. There have been several reports of subinhibitory levels of some antibiotics enhancing the efficiency of DNA transfer between staphylococci; however, the basis of this stimulation remains unclear (1, 11, 25).

Identical or nearly identical accessory genes, elements, and plasmids have been detected in different staphylococcal species and other bacterial genera, such as enterococci and streptococci (30, 64, 69). Such observations suggest that, directly and/or indirectly, the gene transfer mechanisms operating in staphylococci facilitate not only intraspecific transfer, but also interspecific and intergeneric exchange, and hence access to an extended and shared reservoir of determinants (30, 69).

STAPHYLOCOCCAL PLASMIDS

One or more plasmids are usually found in clinical isolates of *Staphylococcus aureus* and coagulase-negative staphylococci (65). Most staphylococcal plasmids can be categorized into one of three main classes based on physical/genetic organization and functional characteristics (64, 69), although a newly described group of plasmids, the pSK639 family (56), should probably be considered a fourth class. A number of other plasmids that do not appear to fall within these classes await further investigation. Fifteen plasmid incompatibility groups (Inc1-15) have so far been identified in staphylococci (64, 90); however, plasmid classification is increasingly being inferred from DNA sequence data, particularly of replication regions. Whereas some staphylococcal plasmids are phenotypically cryptic, most that have been described encode antimicrobial resistance determinants; additionally, other clinically significant properties, such as enterotoxin production, have been attributed to several plasmids (2, 12). Although our knowledge is skewed by the medical bias of staphylococcal research, it is likely that plasmids are prevalent in all staphylococcal species.

Small Rolling-Circle (RC) Plasmids

The most thoroughly characterized staphylococcal plasmids are those that utilize asymmetric rolling-circle (RC) replication via a single-stranded DNA (ssDNA) intermediate (for a detailed review of RC plasmid replication, see ref-

Gram-Positive Pathogens, ed. by V. A. Fischetti et al.
© 2000 American Society for Microbiology, Washington, D.C.

erence 50). Probably reflecting constraints imposed by this replication strategy, RC (sometimes alternatively called ssDNA) plasmids are normally less than 5 kb in size and are rarely found to contain transposable elements (64). This restriction in RC plasmid size probably results from increased production of high-molecular-weight DNA by larger plasmids. Thought to represent a defect in the termination step of RC replication, accumulation of high-molecular-weight DNA impairs cell growth, leading to counterselection against host cells and hence the plasmid itself (40). RC plasmids are usually maintained at 10 to 60 copies per cell and are phenotypically cryptic or carry only a single resistance gene, although there are examples of the carriage of two determinants (65).

Staphylococcal RC plasmids can be subdivided into four families, exemplified by pT181, pC194, pE194, and pSN2, based on replication region sequence similarity (64). Representative RC plasmids are shown in Fig. 1. These families contain plasmids from different staphylococcal species and other bacterial genera, such as *Bacillus*, *Lactobacillus*, and *Streptococcus*, attesting to the horizontal transmission of RC plasmids (40). Some of these plasmids encode a mobilization system (e.g., *mobABC* on pC221; Fig. 1). A locus consisting of a gene, *pre*, and site, RS$_A$, originally identified as a site-specific recombination function on plasmids such as pT181 and pE194 (Fig. 1) (64), may function as a second type of mobilization system, possibly involving conduction (35, 37, 70a). RC plasmids, including those in different families, often share highly similar DNA segments, such that they appear to be mosaic structures consisting of dis-

crete functional cassettes encoding replication, resistance, and sometimes recombination and/or mobilization functions (64, 72). Exchange of these segments is thought to be promoted by the replication strategy of RC plasmids and their extended regions of sequence similarity (36). Site-specific recombination functions such as *pre*, which favor cointegrate formation rather than multimer resolution (64), may also be involved.

pSK639 Family Plasmids

A group of plasmids related to a prototype, pSK639, has been identified in strains of *Staphylococcus epidermidis* (56). In size and characteristics, these plasmids can be considered to fall between the RC and multiresistance groups of staphylococcal plasmids. Like RC plasmids, pSK639 is relatively small (8 kb) and confers only trimethoprim resistance, mediated by a trimethoprim-insensitive dihydrofolate reductase encoded by the *dfrA* gene (Fig. 2) (3, 77). Moreover, pSK639 family plasmids possess mobilization regions closely related to those of the pT181 family RC plasmids pC221 (Fig. 1), pC223, and pS194 (3). However, pSK639-type plasmids are proposed to utilize a replication region related to that of the theta-replicating *Lactococcus lactis* plasmid pWV02 (3, 40). This family is also distinguished from the RC plasmids by the carriage of insertion sequence (IS) elements, in the form of multiple copies of IS257 (56). Indeed, there are examples of these elements mediating the cointegrative fusion of a pSK639-type plasmid and an RC plasmid, such as pT181 in the case of pSK818 (Fig. 2), to form hybrid multiresistance plasmids (56).

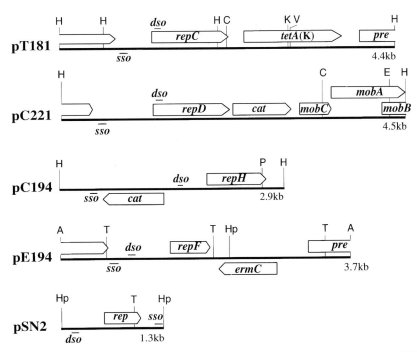

FIGURE 1 Maps of the RC plasmid family prototypes, pT181, pC194, pE194, and pSN2, and the mobilizable pT181 family plasmid, pC221 (69); see text for additional references. Plasmid sizes are shown on the right. Genes (arrowed boxes) and loci encoding the following functions are indicated: *cat*, chloramphenicol resistance; *dso*, double-stranded origin of DNA replication; *ermC*, erythromycin resistance; *mobA/mobB/mobC*, plasmid mobilization; *pre*, plasmid recombination/mobilization; *rep/repC/repD/repF/repH*, initiation of plasmid replication; *sso*, single-stranded origin of DNA replication; *tetA*(K), tetracycline resistance. Restriction sites shown: A, *AluI*; C, *ClaI*; E, *EcoRI*; H, *HindIII*; Hp, *HpaII*; K, *KpnI*; P, *PvuII*; T, *TaqI*; V, *EcoRV*.

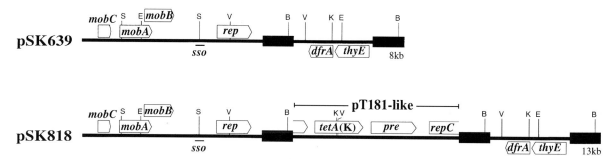

FIGURE 2 Maps of representative pSK639 family plasmids, pSK639 and pSK818 (3, 56). The locations of IS257 copies are marked by solid boxes, and the position of the cointegrated pT181-like plasmid within pSK818 is indicated. Plasmid sizes are shown on the right. Genes (arrowed boxes) and loci encoding the following functions are indicated: *dfrA*, trimethoprim resistance; *mobA/mobB/mobC*, plasmid mobilization; *pre*, plasmid recombination/mobilization; *rep/repC*, initiation of plasmid replication; *sso*, single-stranded origin of DNA replication; *tetA*(K), tetracycline resistance; *thyE*, thymidylate synthetase. Restriction sites shown: B, *Bgl*II; E, *Eco*RI; K, *Kpn*I; S, *Sal*I; V, *Eco*RV.

To date, autonomous pSK639 family plasmids have been identified only in *S. epidermidis*. However, it is now known that the composite transposonlike structure designated Tn4003 (77) and related elements evident on a number of *S. aureus* multiresistance plasmids (see below) actually represent a cointegrated copy of a pSK639-like plasmid (30, 56). The pSK639 family has recently been extended by the characterization of two new members, pIP1629 and pIP1630, also from *S. epidermidis*, that confer resistance to type A streptogramins and, in the case of pIP1630, trimethoprim resistance (10).

Multiresistance Plasmids

Staphylococcal multiresistance plasmids are usually 15 to 40 kb in size and exist at approximately 5 copies per cell (65). These plasmids typically encode several antimicrobial resistance determinants, often in association with transposable elements (30, 69). Because of their size, these plasmids are presumed to replicate via the theta mode (40), a notion supported by limited experimental (82) and sequence data (27, 31). Two groups of multiresistance plasmids have been recognized, the β–lactamase/heavy metal resistance plasmids and the pSK1 family.

S. aureus strains from the 1960s and 1970s commonly contained plasmids that conferred resistance to β-lactam antibiotics and heavy metals or other inorganic ions. These plasmids characteristically carried a Tn552-like β-lactamase-encoding transposon or a derivative thereof (24, 78) and frequently possessed a composite structure designated Tn4004, conferring resistance to mercuric ions, and operons encoding resistance to arsenical and/or cadmium ions (Fig. 3) (58, 80). Additionally, some β-lactamase/heavy metal resistance plasmids carry the transposons Tn4001 or Tn551, conferring resistance to aminoglycosides (34, 58) and macrolide-lincosamide-streptogramin B (MLS) antibiotics (66), respectively, and/or a *qacA* or *qacB* gene mediating multidrug resistance to antiseptics and disinfectants (Fig. 3) (58).

Based on structural similarities, five families (α, β, γ, δ, and orphan) of β-lactamase/heavy metal resistance plasmids have been recognized (80). The α and γ families are closely related, and plasmids that appear to represent hybrids of members from different families have also been isolated (Fig. 3) (34, 80). The replication initiation regions

from two γ family β-lactamase/heavy metal resistance plasmids, pSX267 from *Staphylococcus xylosus* (31) and pI9789::Tn552 (Fig. 3) from *S. aureus* (19a), have been characterized and found to possess a high degree of identity at the nucleotide level. The deduced replication initiation proteins of these plasmids exhibit amino acid sequence similarity to those from enterococcal pheromone-response plasmids, such as pAD1, and *Lactobacillus* plasmids, such as pLJ1.

Plasmids related to the prototype pSK1 (Fig. 4) were first identified in epidemic *S. aureus* and coagulase-negative staphylococcal strains isolated in Australia during the 1980s and later in isolates from Europe (69). In addition to a *qacA* gene mediating multidrug resistance to antiseptics/disinfectants carried ubiquitously by pSK1 family plasmids (76, 86), members of this family variously encode Tn4001, which confers resistance to gentamicin and other aminoglycosides (75), a Tn552-like β-lactamase transposon, Tn4002 (33), and/or a composite structure designated Tn4003 that mediates trimethoprim resistance (Fig. 4) (77). Tn4003 is now thought to represent a vestige of a cointegrated pSK639-like plasmid (see above) (30, 56). The complete nucleotide sequence of pSK1 has been determined, revealing that its replication initiation region is related to those of the β-lactamase/heavy metal resistance plasmids pSX267 and pI9789::Tn552 (27). Antimicrobial resistance genes and associated elements account for slightly less than half of pSK1, the remainder corresponding to a 15-kb DNA segment conserved on all known members of the pSK1 family (84). In addition to the replication region, this conserved segment contains a number of open reading frames, the functions of which remain to be established.

Conjugative Plasmids

The largest known staphylococcal multiresistance plasmids (30 to 60 kb) are those that encode their own conjugative transfer (65, 69). Only one family of conjugative plasmids has been examined in detail and includes *S. aureus* plasmids such as pSK41 (13, 29), pGO1 (61), and pJE1 (Fig. 5) (25). Members of this plasmid family have been identified in association with gentamicin resistance in some strains since the mid-1970s and with mupirocin resistance more recently (59, 69). Structurally related conjugative

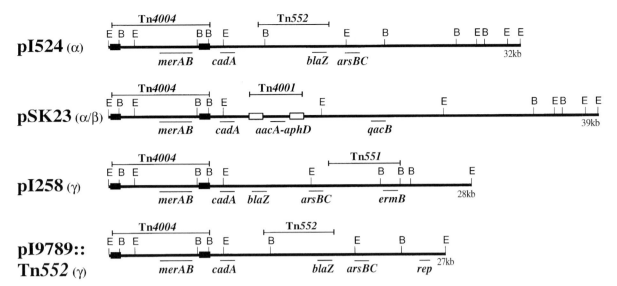

FIGURE 3 Maps of representative β-lactamase/heavy metal multiresistance plasmids pI524, pSK23, pI258, and pI9789::Tn552 (69); see text for additional references. The family to which each plasmid belongs is shown in parentheses on the left, and plasmid sizes are indicated on the right. The positions of transposons are shown above each map. Open and solid boxes denote copies of IS256 and IS257, respectively. Loci encoding the following functions are indicated beneath the maps: *aacA-aphD*, aminoglycoside resistance; *arsBC*, arsenic resistance; *blaZ*, penicillin resistance; *cadA*, cadmium resistance; *ermB*, erythromycin resistance; *merAB*, mercury resistance; *qacB*, multidrug resistance to antiseptics and disinfectants; *rep*, initiation of plasmid replication. Restriction sites shown: B, *Bgl*II; E, *Eco*RI.

plasmids have been found in coagulase-negative staphylococci (7, 48), and interspecific transfer has been demonstrated (5, 47). Conjugative transfer of pSK41 family plasmids occurs only on solid surfaces and does so at comparatively low frequencies, in the range of 10^{-5} to 10^{-7} transconjugants per donor cell (59).

The nucleotide sequence of pSK41 has been completed (13). The pSK41 replication initiation region is related to those of the staphylococcal multiresistance plasmids pSX267, pI9789::Tn552, and pSK1 and in turn to plasmids, including pAD1 and pLJ1, from other genera (see above) (13, 27). Conjugation by pSK41 family plasmids is mediated by a transfer system composed of approximately 15 genes (13, 17, 29, 61, 81); two of these are adjacent to the

origin of transfer, *oriT*, whereas the remainder are located within a 14-kb transfer-associated region, *tra*, which is flanked by copies of IS257 (Fig. 5). Despite the evolutionary relationship between the replication region of pSK41 and that of transmissible enterococcal pheromone-response plasmids like pAD1, the pSK41-type conjugation system is instead related to those of the *L. lactis* plasmid pMRC01 (23) and the broad-host-range plasmid pIP501 (13), originally identified in *Streptococcus agalactiae*. This relatedness is manifested by similarity in *tra* product amino acid sequences and genetic organization (13, 23, 27a). Furthermore, several of the products from these gram-positive conjugation systems possess similarity to proteins from gram-negative transfer systems, such as the *Escherichia coli*

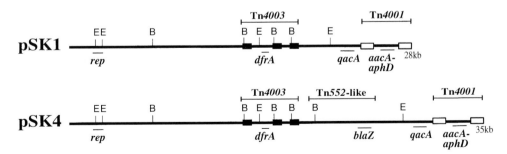

FIGURE 4 Maps of representative pSK1 family multiresistance plasmids, pSK1 and pSK4 (69); see text for additional references. Plasmid sizes are indicated on the right. The positions of transposons are indicated above each map. Open and solid boxes denote copies of IS256 and IS257, respectively. Loci encoding the following functions are indicated beneath the maps: *aacA-aphD*, aminoglycoside resistance; *blaZ*, penicillin resistance; *dfrA*, trimethoprim resistance; *qacA*, multidrug resistance to antiseptics and disinfectants; *rep*, initiation of plasmid replication. Restriction sites shown: B, *Bgl*II; E, *Eco*RI.

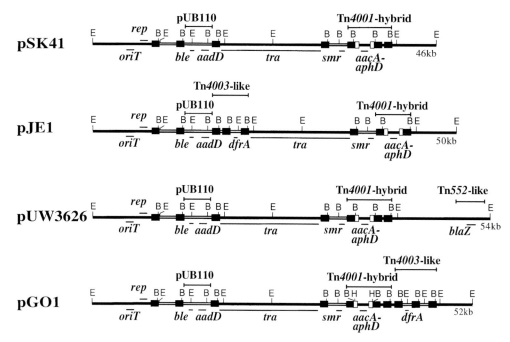

FIGURE 5 Maps of representative conjugative multiresistance plasmids pSK41, pJE1, pUW3626, and pGO1 (69); see text for additional references. Plasmid sizes are indicated on the right. The positions of transposons and an integrated copy of the RC plasmid pUB110 are indicated above each map; the latter and other small cointegrated plasmids are denoted by double lines. Solid and open boxes denote copies of IS*257* and truncated copies of IS*256*, respectively. Loci encoding the following functions are indicated beneath the maps: *aadD/aacA-aphD*, aminoglycoside resistances; *blaZ*, penicillin resistance; *ble*, bleomycin resistance; *dfrA*, trimethoprim resistance; *oriT*, origin of conjugative DNA transfer; *rep*, initiation of plasmid replication; *smr*, multidrug resistance to antiseptics and disinfectants; *tra*, conjugative transfer. Restriction sites shown: B, *BglII*; E, *EcoRI*. The position of the *BglII* site within IS*257* indicates the orientation of the element.

F plasmid conjugation system and the *Agrobacterium tumefaciens* Ti and Tra DNA transfer systems (29, 61). This suggests that pSK41/pIP501-like conjugation systems belong to a superfamily of macromolecular transport mechanisms (28). Although the details of staphylococcal conjugation are yet to be elucidated, the basic process is likely to resemble that of the more extensively characterized gram-negative counterparts, viz., transfer of a single strand of plasmid DNA via direct cell-cell contact. Mobilization of relaxed coresident plasmids (71) and the demonstrated nicking of pGO1 *oriT* (17) are consistent with this notion. However, unlike the related gram-negative DNA transfer systems, no pilus-like structure seems to be associated with staphylococcal conjugation, perhaps reflecting the distinction in the cell envelope organizations and accounting for the solid surface requirement for conjugative transfer.

Multiple copies of IS*257* are a feature of pSK41-like plasmids. For example, pGO1 contains nine directly repeated copies and a single copy in the opposite orientation (Fig. 5). Several of these elements delimit DNA segments that carry resistance genes, and it is now known that in most instances these correspond to cointegrated copies of smaller plasmids (13, 83). Such determinants include the aminoglycoside adenyltransferase gene, *aadD*, and the bleomycin resistance gene, *ble*, on a copy of pUB110 and the small multidrug resistance determinant, *smr* (formerly *qacC/D*), on another RC plasmid (Fig. 5). A third cointegrated RC plasmid, to which no phenotype has yet been ascribed, is also present on most of these conjugative plasmids. Plasmids such as pJE1, and probably pGO1, also contain a Tn*4003*-like structure now known to represent a fourth integrated plasmid, evidently a pSK639-like trimethoprim-resistance plasmid (see above). In each case so far examined, the replication functions of the smaller cointegrated plasmids appear to have been inactivated either by insertion of IS*257* during fusion of the two replicons or by subsequent deletions adjacent to this element (13). Similar IS*257*-associated events are presumably responsible for the Tn*4001*-IS*257* hybrid structures found on most pSK41 family plasmids, thereby rendering them transpositionally defective (15). Analysis of the pSK41 sequence suggests that this family of plasmids has arisen by accretion of resistance determinants by preexisting conjugative plasmids, rather than via the en bloc incorporation of a conjugation system into a resistance plasmid (13).

TRANSPOSONS AND INSERTION SEQUENCES

Transposable elements detected in staphylococci are listed in Table 1 together with relevant features and associations. For many, designation as a transposable element is based on possession of diagnostic characteristics, such as terminal inverted repeats, flanking insertion sequences and/or target

TABLE 1 Staphylococcal insertion sequences and transposons[a]

Element	Size (kb)	Associated resistance(s)/ other phenotype(s)[b]	Relevant gene(s)	TIR (bp)/ flanking IS[c]	Target duplication (bp)	Location(s)[d]
IS256	1.3	Gentamicin/kanamycin/ tobramycin	aacA-aphA	26	8	C, P
IS257[e]	0.8	Antiseptics/disinfectants	smr	27	8	P
		Bleomycin	ble			C, P
		Gentamicin/kanamycin/ tobramycin	aacA-aphA			C, P
		Kanamycin/neomycin/ paromomycin/tobramycin	aadD			C, P
		Mercury	merA, merB			C, P
		Mupirocin	mupA			P
		Tetracycline	tetA(K)			C, P
		Trimethoprim	dfrA			P
		Virginiamicin	vgb			P
		Preprolysostaphin	lss			P
		Lysostaphin immunity factor	lif			P
IS1181	2.0	Kanamycin/neomycin	aphA-3	23	8	C
		Streptomycin	aadE			C
IS1182	1.9	Kanamycin/neomycin	aphA-3	33	8	C
		Streptomycin	aadE			C
IS1272	1.9	Methicillin	mecA	16	Unknown	C
IS1293	1.3	Preprolysostaphin	lss	26	Unknown	P
		Lysostaphin immunity factor	lif			P
Tn551	5.3	Macrolides/lincosamides/ streptogramin B (MLS)	ermB	40	5	P
Tn552[f]	6.1	Penicillins	blaZ	116	6/7	C, P
Tn554[g]	6.7	Macrolides/lincosamides/ streptogramin B (MLS)	ermA	Absent	Absent	C, P
		Spectinomycin	spc			
Tn3854	4.5	Kanamycin/neomycin streptomycin	Unknown	Unknown	Unknown	P
Tn4001[h]	4.7	Gentamicin/kanamycin/ tobramycin	aacA-aphD	IS256 (I)[i]	8	C, P
Tn4003	4.7	Trimethoprim	dfrA	IS257 (D)[j]	8	P
Tn4004	7.8	Mercury	merA, merB	IS257 (D)	8	P
Tn4291	7.5	Methicillin	mecA	Unknown	Unknown	C

(Continued on next page)

TABLE 1 (Continued)

Element	Size (kb)	Associated resistance(s)/ other phenotype(s)[b]	Relevant gene(s)	TIR (bp)/ flanking IS[c]	Target duplication (bp)	Location(s)[d]
Tn5404	16	Kanamycin/neomycin Streptomycin	aphA-3 aadE	116	6	C
Tn5405	12	Kanamycin/neomycin Streptomycin	aphA-3 aadE	IS1182 (I)	8	C

[a] See Paulsen et al. (69), Firth and Skurray (30), and the text for references.
[b] In the case of insertion sequences (IS), association is based on probable involvement in the acquisition, dissemination, and/or expression of the resistance/phenotype.
[c] TIR, terminal inverted repeat.
[d] C, chromosome; P, plasmid.
[e] IS257 is also known as IS431.
[f] Tn3852, Tn4002, and Tn4201 are likely to be similar or identical to Tn552.
[g] Tn3853 is likely to be similar or identical to Tn554.
[h] Tn3851 and Tn4031 are likely to be similar or identical to Tn4001.
[i] (I), inverted orientation.
[j] (D), direct orientation.

duplications, presence of an open reading frame encoding a transposase homolog, and/or detection in varied genetic contexts, rather than formal demonstration of mobility in a recA recombination-defective host. As outlined for specific elements below, in addition to facilitating the translocation of genes between replicons, IS elements are increasingly being shown to play more subtle roles in phenotypic expression and genome evolution.

Inverted copies of IS256 flank the composite transposon Tn4001 (Fig. 3 and 4) (75). Tn4001 and elements related to it are thought to be largely responsible for the emergence of linked resistance to the aminoglycosides gentamicin, tobramycin, and kanamycin in staphylococci, having been found on the chromosome and multiresistance plasmids (Fig. 3–5) in clinical S. aureus and coagulase-negative staphylococcal isolates from around the world (58, 69). Indeed, some of the earliest reported gentamicin-resistant strains, dating from 1975, have been shown to carry a Tn4001-like structure on the chromosome (91). Resistance is mediated by the gene aacA-aphD, which encodes a bifunctional enzyme possessing both aminoglycoside acetyltransferase and phosphotransferase activities (75). Transposons closely related to Tn4001 are also evident in enterococci (43) and streptococci (44). In enterococci, IS256 and related IS elements are also associated with elements encoding resistance to vancomycin (73) and erythromycin (74).

Roles in gene expression and regulation have been attributed to IS256. It has been suggested that a hybrid promoter partially specified by IS256 directs transcription of the aacA–aphD resistance gene of Tn4001 (75). The potential of IS256 to form such hybrid promoters has been demonstrated, and a role for this phenomenon in methicillin heteroresistance has been proposed (60). IS256 is also implicated in the control of virulence of biofilm-forming S. epidermidis, since alternating insertion and excision of this element is thought to be responsible for phase variation in polysaccharide intercellular adhesin synthesis (93). Abolition of biofilm production has been correlated with independent insertions of IS256 into several structural genes of the operon that mediates polysaccharide intercellular adhesin production (see chapter 46, this volume); as expected for IS256, these insertions generated 8-bp target duplications. Analysis of biofilm-forming revertants that were isolated revealed precise excision of the IS256 copy together with one copy of the duplicated target sequence, such that the wild-type sequence was restored.

IS257 has been found in diverse genetic contexts in staphylococci, variously associated with a number of determinants (Table 1). The prevalence of resistance genes flanked by copies of this element resulted in the designation of several such composite structures as transposons, viz., Tn4003 and Tn4004. Although the possibility that these structures can behave as legitimate transposons cannot be excluded, such an organization is probably a consequence of the transposition mechanism of IS257. This IS element is now thought to undergo nonresolved replicative transposition (30, 83). The expected outcome of such an event involving two replicons is cointegration, so that the replicons become fused with a directly repeated copy of IS257 at each junction, as has been experimentally demonstrated by Needham et al. (63). Many IS257-flanked segments represent plasmids cointegrated into larger plasmids or the chromosome (30, 83). Frequently observed deletion events adjacent to IS257 may result from intramolecular transposition of IS257. However, replicon fusions and sequence deletions have also been shown to have resulted from homologous recombination between preexisting copies of IS257 (13).

The implications of the activities of IS257 are best illustrated by the pSK41 family plasmids (Fig. 5). It would seem that IS257-mediated cointegration has enabled these conjugative plasmids to collect functions, in the form of smaller plasmids, as they move horizontally through bacterial populations. Insertions and flanking deletions mediated by this element provide a mechanism for the inactivation or removal of deleterious sequences, such as redundant replication functions. Similarly, deletion of transposition functions has immobilized the Tn4001-IS257 hybrid elements on these plasmids (Fig. 5), thereby enhancing the maintenance of the plasmids under aminoglycoside selection.

A central role for IS257 has been suggested in the evolution of staphylococcal trimethoprim resistance, including the probable capture of the resistance gene dfrA from the chromosome of an S. epidermidis strain (30, 83). Furthermore, an IS257-hybrid promoter is responsible for transcription of this determinant (55, 83). In some plasmids, the high-level trimethoprim resistance typically conferred by this determinant has been moderated as a consequence of IS257-associated flanking deletions. Analogous IS257-hybrid promoters are likely to play a role in the expression of at least some other genes associated with this element (83).

IS1272 is thought to have contributed to the evolution of methicillin resistance in S. aureus and coagulase-negative staphylococci (6, 8, 42). A remnant of IS1272 is evident adjacent to a deletion within the methicillin resistance (mec) region of some strains (see below). This deletion removed mecI and part of mecR1, regulatory genes that control transcription of the divergent mecA gene, thereby resulting in higher levels of methicillin resistance. Multiple intact copies of IS1272 have been found on the chromosomes of S. aureus and coagulase-negative staphylococcal strains, and it is particularly prevalent in Staphylococcus haemolyticus (8).

Three Tn3-type transposons have been detected in staphylococci, viz., Tn551, Tn552, and Tn5404 (30, 69). The MLS transposon Tn551 is closely related to Tn917 from Enterococcus faecalis but confers constitutive rather than inducible MLS resistance (65). Tn551 was originally identified on the β-lactamase/heavy metal resistance plasmid pI258 (Fig. 3) but has subsequently been shown to transpose to numerous sites in other large plasmids and the chromosome. Despite a tendency to "hot-spot," it has proved an invaluable tool for mapping and mutagenesis studies (45, 65, 68).

In contrast to Tn551, the β-lactamase transposon Tn552 seems to be restricted to a very limited set of insertion sites on the chromosome and plasmids (69). In multiresistance and conjugative plasmids, the insertion sites of Tn552 and Tn552-like elements are located within an intergenic region upstream of a gene encoding a probable recombinase, which is homologous to the resolvase carried by these transposons (13, 70). The basis of this site specificity is unclear, since no obvious nucleotide sequence similarity could be discerned between the insertion region on the pSK41 family conjugative plasmid pUW3626 (Fig. 5) and the near-identical insertion sites on pI9789::Tn552 and pSK4, which are members of the β-lactamase/heavy metal resistance and pSK1 family plasmids, respectively (Fig. 3 and 4) (13). Despite such insertional specificity, Tn552-type transposons are thought to represent the source of all staphylococcal β-lactamase genes (24). However, in many plasmids, only remnants of such transposons remain, presumably as a consequence of multiple insertion events and/or rearrangements promoted by the recombinase systems present on these elements and plasmids.

Tn5404 appears to have resulted from the transposition of a chromosomal element to a plasmid in a clinical S. aureus strain (22). This transposon shares an invertible segment with an adjacent copy of Tn552, such that an entire copy of one or the other of these elements is generated depending on the orientation of the segment (22). Another invertible segment within Tn5404 represents a composite structure designated Tn5405 (22), which is bounded by copies of IS1182, an element related to IS1272 (see above) (21). Comparison of this pair of IS1182 elements with an independent chromosomal copy revealed that the presumptive transposase gene of the former contains a point mutation (21). Moreover, one of the IS1182 elements flanking Tn5405 is interrupted by a copy of IS1181, which is both prevalent and active in S. aureus (16, 21).

The central region of Tn5405 has been detected in the absence of flanking IS1182 elements. In addition to aadE and aphA-3 aminoglycoside-resistance genes, such regions sometimes encode a functional sat4 streptothricin-resistance determinant and appear to have been disseminated among other genera, including enterococci and Campylobacter sp. (20). Elucidation of any relationship between Tn5405 and Tn3854, which confers the same resistance phenotype, awaits characterization of the latter.

An incomplete vestige of an element related to IS1181 has been detected on the plasmid pACK1, from Staphylococcus simulans biovar staphylolyticus ATCC 1362. This element, designated IS1293, is located at one end of a segment that contains the genes encoding preprolysostaphin (lss) and lysostaphin immunity factor (lif) (87). It is likely that the lss-lif segment of this plasmid, which possesses a truncated remnant of IS257 at its other end, has been acquired through horizontal transfer (87).

The MLS- and spectinomycin-resistance transposon, Tn554, is an unusual transposable element that lacks terminal inverted repeats and generates no target sequence duplications upon insertion (62). Tn554 transposes at an efficiency approaching 100% into a unique site in the S. aureus and S. epidermidis chromosomes, termed att554, with a preference for one orientation. However, natural isolates have been identified that carry this or related elements at different secondary sites, including some located in mec regions (see below); a Tn554-like element, Tn3853, has also been detected on a conjugative plasmid from S. epidermidis, pWG25 (88). Such secondary insertions can be generated, albeit at much lower frequency, if att554 is absent or occupied by a preexisting copy of Tn554.

In staphylococci, the reassortment of resistance genes is largely limited to the activities of IS elements and transposons, since no mechanisms related to the integrons of gram-negative bacteria (39) have been detected. Rather, resistance gene clusters in staphylococci have commonly been assembled via sequential cointegrative capture of small resistance plasmids mediated by IS257, a process that resembles, particularly in its outcome, the accretion of resistance genes within integrons.

PROPHAGES

Most clinical S. aureus isolates are lysogenic, commonly containing multiple prophages (65). Infection by temperate phages has been found to influence the staphylococcal host in several ways. Lysogeny and other factors, including restriction-modification systems, modulate the susceptibility of a strain to phage infection. This phenomenon has formed the basis of a phage-typing system, employing an international reference set of phages, that has been used for epidemiological and evolutionary analysis of S. aureus strains; this methodology is increasingly being superseded by more contemporary molecular approaches (4) (see chapter 31, this volume).

Temperate phages have been shown, or are suspected, to carry several virulence-associated functions, thereby possessing the potential to induce lysogenic conversion of infected hosts. These include the genes for enterotoxin A (sea) (14), staphylokinase (sak) (51), and probably enterotoxin E (see) (19), as well as a silent enterotoxin variant (*sezA*) (85). In contrast, lysogenization has been shown to abolish production of β-toxin (β-hemolysin) and lipase (glycerol ester hydrolase) owing to the presence of phage attachment sites within their respective structural genes, *hlb* and *geh*, on the *S. aureus* chromosome (18, 54).

OTHER ACCESSORY ELEMENTS

In staphylococci, a number of accessory functions have been ascribed to discrete segments of DNA present in the chromosome of some strains. However, in several cases the nature of the segments or the mechanisms by which they have been incorporated remain to be elucidated.

The 14.6-kb operon responsible for type-1 capsular polysaccharide biosynthesis (see chapter 37, this volume) is contained within a chromosomal segment of approximately 34 kb that is unique to type 1 *S. aureus* strains (53). Nucleotide sequencing demonstrated the presence of enterotoxin- and transposaselike genes at one end of this *cap1* element but did not reveal any repeated sequences at the junctions of the structure (67); mobility of the element has not been demonstrated.

The gene encoding collagen adhesin, *cna*, is carried on a DNA segment that interrupts a sequence that is contiguous in Cna⁻ *S. aureus* strains. Nucleotide sequencing has demonstrated that *cna* is the only gene within this element, which is approximately 4 kb in length, and did not reveal any recognizable repeat sequences at its termini (32).

Production of enterotoxin B has been associated with a discrete element of at least 27 kb that encodes the structural gene, *seb* (formerly *entB*) (49). However, *seb* has also been associated with a penicillinase plasmid (2), raising the possibility that this element may correspond to an integrated plasmid (49).

The family of accessory elements that carry the gene encoding toxic shock syndrome toxin-1, *tst*, conforms to the pathogenicity island paradigm (38, 57). One such element, *S. aureus* pathogenicity island 1 (SaPI1), has been completely sequenced. SaPI1 is 15.2 kb in length and contains a number of open reading frames in addition to *tst*, including putative genes for an integrase and a second superantigen toxin (57). Specific interactions between SaPIs and particular bacteriophage result in high-frequency transduction. SaPI1 is transduced at high frequency only by phage 80α and has been shown to insert into a specific chromosomal site via a *recA*-independent mechanism. In contrast, a distinct but related element, SaPI2, from a prototypical menstrual toxic shock syndrome strain, is located at a different chromosomal location and is efficiently transduced only by phage 80 (57). In the case of SaPI1, excision and replication, but not integration, are dependent on the presence of 80α. Integration and excision of SaPI1 are proposed to occur via classical Campbell recombination since the circular intermediate contains a copy of the duplicated 17-bp sequence that flanks the integrated form (57).

Most methicillin-resistant *S. aureus* isolates contain a DNA segment not present in methicillin-sensitive strains. Although generically termed the *mec* region, there is considerable variation in both composition and size (20 to 60 kb) of *mec* regions from different strains (Fig. 6) (42). Methicillin resistance is mediated, at least in part, by the *mecA* gene, which encodes the 76-kDa low-affinity penicillin-binding protein PBP2′ (see chapter 63, this volume). Two regulatory loci, *mecI* and *mecR1*, are commonly found upstream of, and transcribed divergently to, *mecA*

(however, see IS*1272*, above). The *mec* region appears to act as a chromosomal hot spot for the insertion of additional antimicrobial resistance determinants, often in association with transposable elements. These include Tn*554* or related elements encoding resistance to erythromycin and spectinomycin, or cadmium, or IS*257*-flanked segments conferring resistance to mercury, tetracycline, and/or aminoglycosides and bleomycin; the latter two segments are known to be cointegrated copies of the plasmids pT181 and pUB110, respectively (Fig. 6) (30, 83).

Consistent with the notion that *mec* regions represent horizontally transferable segments, closely related segments have been found in coagulase-negative staphylococcal strains. Indeed, the ubiquitous carriage of a *mecA* homolog by *Staphylococcus sciuri* has led to the suggestion that this or a closely related coagulase-negative staphylococcus represents the origin of the *mec* determinants found in other species (92). Current evidence suggests that *mec* regions have been introduced into *S. aureus* on a limited number of occasions (42).

Nucleotide sequencing has revealed the presence of terminal inverted repeats at the ends of the *mec* region and confirmed an identical insertion site in different *S. aureus* strains (Fig. 6) (41), consistent with previous chromosome mapping; a common chromosomal location is similarly observed in *S. epidermidis* and *S. haemolyticus* (26). The *mec* region has been proposed to represent a chromosomal cassette (45a). The element previously provisionally designated Tn*4291* (Table 1) (89) presumably corresponds to such a *mec* region cassette. Two recombinase genes identified in the 52-kb *mec* region of the Japanese *S. aureus* isolate N315 (Fig. 6) have been shown to mediate the excision and circularization of this cassette structure. These genes have also been shown to mediate site- and orientation-specific insertion into the chromosome of a plasmid carrying them and a *mec* region attachment sequence (46). Although it is possible that homologous recombination has also played a role in the integration of the *mec* region in some strains (26), the observations above indicate that the *mec* region can behave as a site-specific mobile genetic element. The size and features of the *mec* region suggest that it could be considered a "resistance island" (38).

PERSPECTIVES

Gene transfer mechanisms, together with accessory elements such as plasmids, transposable elements, prophages, and pathogenicity and resistance islands, serve as catalysts for rapid microbial adaptation through quantum evolution, by providing access to a reservoir of functions shared by different types of organisms. As outlined above, the functions associated with accessory elements have a major influence on the potential of staphylococci to survive and cause disease. It is likely that ongoing whole genome sequencing projects will reveal new accessory elements, both chromosomal and plasmid encoded, and supplement our understanding of the roles played by those elements identified previously. However, by definition, accessory elements are only present in subsets of strains; results of the analysis of specific strains, no matter how thoughtfully chosen, are therefore unlikely to be representative of the full range of elements exploited by staphylococci. The obvious significance of the accessory elements to the success of staphylococci demands a broad focus of study in terms of species and strains, and the differences between them, to

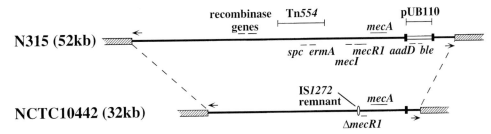

FIGURE 6 Maps of the *mec* regions in the chromosomes of the *S. aureus* strains N315 and NCTC10442 (41, 42). The sizes of the proposed *mec* cassettes are shown in parentheses. The positions of Tn554 and a cointegrated copy of the RC plasmid pUB110 on N315 are indicated; the plasmid is denoted by double lines. The locations of two recombinase genes identified in the N315 *mec* region are also shown. Solid boxes denote copies of IS257. On NCTC10442, the open oval indicates the position of an IS1272 remnant located adjacent to a deletion that removed *mecI* and left part of *mecR1* (ΔmecR1). Arrowed lines denote terminal inverted repeats. Non-*mec* region DNA is shown hatched. Loci encoding the following functions are indicated: *aadD*, aminoglycoside resistance; *ble*, bleomycin resistance; *ermA*, MLS resistance; *mecA*, methicillin resistance; *mecI/mecR1*, regulation of methicillin resistance; *spc*, spectinomycin resistance.

achieve a comprehensive understanding of the pathogens that belong to this genus.

We wish to thank our colleagues who provided manuscripts and data before publication, and Melissa Brown and Ian Paulsen for helpful comments on the manuscript. Work in the laboratory of R. A. S. on staphylococcal genetics is supported by grants from the National Health and Medical Research Council (Australia).

REFERENCES

1. **Al-Masaudi, S. B., M. J. Day, and A. D. Russell.** 1991. Effect of some antibiotics and biocides on plasmid transfer in *Staphylococcus aureus*. *J. Appl. Bacteriol.* **71:**239–243.

2. **Altboum, Z., I. Hertman, and S. Sarid.** 1985. Penicillinase plasmid-linked genetic determinants for enterotoxins B and C1 production in *Staphylococcus aureus*. *Infect. Immun.* **47:**514–521.

3. **Apisiridej, S., A. Leelaporn, C. D. Scaramuzzi, R. A. Skurray, and N. Firth.** 1997. Molecular analysis of a mobilizable theta-mode trimethoprim resistance plasmid from coagulase-negative staphylococci. *Plasmid* **38:**13–24.

4. **Arbeit, R. D.** 1997. Laboratory procedures for epidemiological analysis, p. 253–286. *In* K. B. Crossley and G. L. Archer (ed.), *The Staphylococci in Human Disease.* Churchill Livingstone, New York, N.Y.

5. **Archer, G. L., D. R. Dietrick, and J. L. Johnston.** 1985. Molecular epidemiology of transmissible gentamicin resistance among coagulase-negative staphylococci in a cardiac surgery unit. *J. Infect. Dis.* **151:**243–251.

6. **Archer, G. L., D. M. Niemeyer, J. A. Thanassi, and M. J. Pucci.** 1994. Dissemination among staphylococci of DNA sequences associated with methicillin resistance. *Antimicrob. Agents Chemother.* **38:**447–454.

7. **Archer, G. L., and J. Scott.** 1991. Conjugative transfer genes in staphylococcal isolates from the United States. *Antimicrob. Agents Chemother.* **35:**2500–2504.

8. **Archer, G. L., J. A. Thanassi, D. M. Niemeyer, and M. J. Pucci.** 1996. Characterization of IS1272, an insertion sequence-like element from *Staphylococcus haemolyticus*. *Antimicrob. Agents Chemother.* **40:**924–929.

9. **Archer, G. L., and W. D. Thomas, Jr.** 1990. Conjugative transfer of antimicrobial resistance genes between staph-

ylococci, p. 115–122. *In* R. P. Novick (ed.), *Molecular Biology of the Staphylococci.* VCH Publishers, New York, N.Y.

10. **Aubert, S., K. G. H. Dyke, and N. El Solh.** 1998. Analysis of two *Staphylococcus epidermidis* plasmids coding for resistance to streptogramin A. *Plasmid* **40:**238–242.

11. **Barr, V., K. Barr, M. R. Millar, and R. W. Lacey.** 1986. Beta-lactam antibiotics increase the frequency of plasmid transfer in *Staphylococcus aureus*. *J. Antimicrob. Chemother.* **17:**409–413.

12. **Bayles, K. W., and J. J. Iandolo.** 1989. Genetic and molecular analyses of the gene encoding staphylococcal enterotoxin D. *J. Bacteriol.* **171:**4799–4806.

13. **Berg, T., N. Firth, S. Apisiridej, A. Hettiaratchi, A. Leelaporn, and R. A. Skurray.** 1998. Complete nucleotide sequence of pSK41: evolution of staphylococcal conjugative plasmids. *J. Bacteriol.* **180:**4350–4359.

14. **Betley, M. J., and J. J. Mekalanos.** 1985. Staphylococcal enterotoxin A is encoded by phage. *Science* **229:**185–187.

15. **Byrne, M. E., M. T. Gillespie, and R. A. Skurray.** 1990. Molecular analysis of a gentamicin resistance transposon-like element on plasmids isolated from North American *Staphylococcus aureus* strains. *Antimicrob. Agents Chemother.* **34:**2106–2113.

16. **Chesneau, O., R. Lailler, A. Derbise, and N. El Solh.** 1999. Transposition of IS1181 in the genomes of *Staphylococcus* and *Listeria*. *FEMS Microbiol. Lett.* **177:**93–100.

17. **Climo, M. W., V. K. Sharma, and G. L. Archer.** 1996. Identification and characterization of the origin of conjugative transfer (*oriT*) and a gene (*nes*) encoding a single-stranded endonuclease on the staphylococcal plasmid pGO1. *J. Bacteriol.* **178:**4975–4983.

18. **Coleman, D., J. Knights, R. Russell, D. Shanley, T. H. Birkbeck, G. Dougan, and I. Charles.** 1991. Insertional inactivation of the *Staphylococcus aureus* beta-toxin by bacteriophage phi 13 occurs by site- and orientation-specific integration of the phi 13 genome. *Mol. Microbiol.* **5:**933–939.

19. **Couch, J. L., M. T. Soltis, and M. J. Betley.** 1988. Cloning and nucleotide sequence of the type E staphylococcal enterotoxin gene. *J. Bacteriol.* **170:**2954–2960.

19a.**Curnock, S., and K. G. H. Dyke.** Unpublished data.

20. Derbise, A., G. de Cespedes, and N. El Solh. 1997. Nucleotide sequence of the *Staphylococcus aureus* transposon, Tn5405, carrying aminoglycosides resistance genes. *J. Basic Microbiol.* 37:379–384.

21. Derbise, A., K. G. Dyke, and N. El Solh. 1996. Characterization of a *Staphylococcus aureus* transposon, Tn5405, located within Tn5404 and carrying the aminoglycoside resistance genes, *aphA-3* and *aadE*. *Plasmid* 35:174–188.

22. Derbise, A., K. G. H. Dyke, and N. El Solh. 1995. Rearrangements in the staphylococcal β-lactamase-encoding plasmid, pIP1066, including a DNA inversion that generates two alternative transposons. *Mol. Microbiol.* 17:769–779.

23. Dougherty, B. A., C. Hill, J. F. Weidman, D. R. Richardson, J. C. Venter, and R. P. Ross. 1998. Sequence and analysis of the 60 kb conjugative, bacteriocin-producing plasmid pMRC01 from *Lactococcus lactis* DPC3147. *Mol. Microbiol.* 29:1029–1038.

24. Dyke, K., and P. Gregory. 1997. Resistance mediated by β-lactamase, p. 139–157. *In* K. B. Crossley and G. L. Archer (ed.), *The Staphylococci in Human Disease*. Churchill Livingstone, New York, N.Y.

25. Evans, J., and K. G. H. Dyke. 1988. Characterization of the conjugation system associated with the *Staphylococcus aureus* plasmid pJE1. *J. Gen. Microbiol.* 134:1–8.

26. Fey, P. D., M. W. Climo, and G. L. Archer. 1998. Determination of the chromosomal relationship between *mecA* and *gyrA* in methicillin-resistant coagulase-negative staphylococci. *Antimicrob. Agents Chemother.* 42:306–312.

27. Firth, N., S. Apisiridej, S. Curnock, K. G. H. Dyke, and R. A. Skurray. Unpublished data.

27a. Firth, N., T. Berg, and R. A. Skurray. 1999. Evolution of conjugative plasmids from Gram-positive bacteria. *Mol. Microbiol.* 31:1598–1600.

28. Firth, N., K. Ippen-Ihler, and R. A. Skurray. 1996. Structure and function of the F factor and mechanism of conjugation, p. 2377–2401. *In* F. C. Neidhardt, R. Curtiss III, J. L. Ingraham, E. C. C. Lin, K. B. Low, Jr., B. Magasanik, W. S. Reznikoff, M. Riley, M. Schaechter, and H. E. Umbarger (ed.), *Escherichia coli and Salmonella: Cellular and Molecular Biology*, 2nd ed. ASM Press, Washington D.C.

29. Firth, N., K. P. Ridgway, M. E. Byrne, P. D. Fink, L. Johnson, I. T. Paulsen, and R. A. Skurray. 1993. Analysis of a transfer region from the staphylococcal conjugative plasmid pSK41. *Gene* 136:13–25.

30. Firth, N., and R. A. Skurray. 1998. Mobile elements in the evolution and spread of multiple-drug resistance in staphylococci. *Drug Resist. Updates* 1:49–58.

31. Gering, M., F. Götz, and R. Bruckner. 1996. Sequence and analysis of the replication region of the *Staphylococcus xylosus* plasmid pSX267. *Gene* 182:117–122.

32. Gillaspy, A. F., J. M. Patti, F. L. Pratt, Jr., J. J. Iandolo, and M. S. Smeltzer. 1997. The *Staphylococcus aureus* collagen adhesin-encoding gene (*cna*) is within a discrete genetic element. *Gene* 196:239–248.

33. Gillespie, M. T., B. R. Lyon, and R. A. Skurray. 1988. Structural and evolutionary relationships of β-lactamase transposons from *Staphylococcus aureus*. *J. Gen. Microbiol.* 134:2857–2866.

34. Gillespie, M. T., and R. A. Skurray. 1986. Plasmids in multiresistant *Staphylococcus aureus*. *Microbiol. Sci.* 3:53–58.

35. Grohmann, E., E. L. Zechner, and M. Espinosa. 1997. Determination of specific DNA strand discontinuities with nucleotide resolution in exponentially growing bacteria harboring rolling circle-replicating plasmids. *FEMS Microbiol. Lett.* 152:363–369.

36. Gruss, A., and S. D. Ehrlich. 1989. The family of highly interrelated single-stranded deoxyribonucleic acid plasmids. *Microbiol. Rev.* 53:231–241.

37. Guzmán, L. M., and M. Espinosa. 1997. The mobilization protein, MobM, of the streptococcal plasmid pMV158 specifically cleaves supercoiled DNA at the plasmid *oriT*. *J. Mol. Biol.* 266:688–702.

38. Hacker, J., G. Blum-Oehler, I. Muhldorfer, and H. Tschape. 1997. Pathogenicity islands of virulent bacteria: structure, function and impact on microbial evolution. *Mol. Microbiol.* 23:1089–1097.

39. Hall, R. M., and C. M. Collis. 1995. Mobile gene cassettes and integrons: capture and spread of genes by site-specific recombination. *Mol. Microbiol.* 15:593–600.

40. Helinski, D. R., A. E. Toukdarian, and R. P. Novick. 1996. Replication control and other stable maintenance mechanisms of plasmids, p. 2295–2324. *In* F. C. Neidhardt, R. Curtiss III, J. L. Ingraham, E. C. C. Lin, K. B. Low, Jr., B. Magasanik, W. S. Reznikoff, M. Riley, M. Schaechter, and H. E. Umbarger (ed.), *Escherichia coli and Salmonella: Cellular and Molecular Biology*. ASM Press, Washington, D.C.

41. Hiramatsu, K. 1995. Molecular evolution of MRSA. *Microbiol. Immunol.* 39:531–543.

42. Hiramatsu, K., N. Kondo, and I. Teruyo. 1996. Genetic basis for molecular epidemiology of MRSA. *J. Infect. Chemother.* 2:117–129.

43. Hodel-Christian, S. L., and B. E. Murray. 1992. Comparison of the gentamicin resistance transposon Tn5281 with regions encoding gentamicin resistance in *Enterococcus faecalis* isolates from diverse geographic locations. *Antimicrob. Agents Chemother.* 36:2259–2264.

44. Horaud, T., G. de Cespédès, and P. Trieu-Cuot. 1996. Chromosomal gentamicin resistance transposon Tn3706 in *Streptococcus agalactiae* B128. *Antimicrob. Agents Chemother.* 40:1085–1090.

45. Iandolo, J. J., J. P. Bannantine, and G. C. Stewart. 1997. Genetic and physical map of the chromosome of *Staphylococcus aureus*, p. 39–53. *In* K. B. Crossley and G. L. Archer (ed.), *The Staphylococci in Human Disease*. Churchill Livingstone, New York, N.Y.

45a. Ito, T., Y. Katayama, and K. Hiramatsu. 1999. Cloning and nucleotide sequence determination of the entire *mec* DNA of pre-methicillin-resistant *Staphylococcus aureus* N315. *Antimicrob. Agents Chemother.* 43:1449–1458.

46. Ito, T., Y. Katayama, and K. Hiramatsu. Unpublished data.

47. Jaffe, H. W., H. M. Sweeney, C. Nathan, R. A. Weinstein, S. A. Kabins, and S. Cohen. 1980. Identity and interspecific transfer of gentamicin-resistance plasmids in *Staphylococcus aureus* and *Staphylococcus epidermidis*. *J. Infect. Dis.* 141:738–747.

48. Jaffe, H. W., H. M. Sweeney, R. A. Weinstein, S. A. Kabins, C. Nathan, and S. Cohen. 1982. Structural and phenotypic varieties of gentamicin resistance plasmids in hospital strains of *Staphylococcus aureus* and coagulase-negative staphylococci. *Antimicrob. Agents Chemother.* 21:773–779.

49. **Johns, M. B., Jr., and S. A. Khan.** 1988. Staphylococcal enterotoxin B gene is associated with a discrete genetic element. *J. Bacteriol.* **170:**4033–4039.

50. **Khan, S. A.** 1997. Rolling-circle replication of bacterial plasmids. *Microbiol. Mol. Biol. Rev.* **61:**442–455.

51. **Kondo, I., and K. Fujise.** 1977. Serotype B staphylococcal bacteriophage singly converting staphylokinase. *Infect. Immun.* **18:**266–272.

52. **Lacey, R. W.** 1980. Evidence for two mechanisms of plasmid transfer in mixed cultures of *Staphylococcus aureus*. *J. Gen. Microbiol.* **119:**423–435.

53. **Lee, C. Y.** 1995. Association of staphylococcal type-1 capsule-encoding genes with a discrete genetic element. *Gene* **167:**115–119.

54. **Lee, C. Y., and J. J. Iandolo.** 1986. Lysogenic conversion of staphylococcal lipase is caused by insertion of the bacteriophage L54a genome into the lipase structural gene. *J. Bacteriol.* **166:**385–391.

55. **Leelaporn, A., N. Firth, M. E. Byrne, E. Roper, and R. A. Skurray.** 1994. Possible role of insertion sequence IS257 in dissemination and expression of high- and low-level trimethoprim resistance in staphylococci. *Antimicrob. Agents Chemother.* **38:**2238–2244.

56. **Leelaporn, A., N. Firth, I. T. Paulsen, and R. A. Skurray.** 1996. IS257-mediated cointegration in the evolution of a family of staphylococcal trimethoprim resistance plasmids. *J. Bacteriol.* **178:**6070–6073.

57. **Lindsay, J. A., A. Ruzin, H. F. Ross, N. Kurepina, and R. P. Novick.** 1998. The gene for toxic shock toxin is carried by a family of mobile pathogenicity islands in *Staphylococcus aureus*. *Mol. Microbiol.* **29:**527–545.

58. **Lyon, B. R., and R. Skurray.** 1987. Antimicrobial resistance of *Staphylococcus aureus*: genetic basis. *Microbiol. Rev.* **51:**88–134.

59. **Macrina, F. L., and G. L. Archer.** 1993. Conjugation and broad host range plasmids in streptococci and staphylococci, p. 313–329. *In* D. B. Clewell (ed.), *Bacterial Conjugation.* Plenum Press, New York, N.Y.

60. **Maki, H., and K. Murakami.** 1997. Formation of potent hybrid promoters of the mutant *llm* gene by IS256 transposition in methicillin-resistant *Staphylococcus aureus*. *J. Bacteriol.* **179:**6944–6948.

61. **Morton, T. M., D. M. Eaton, J. L. Johnston, and G. L. Archer.** 1993. DNA sequence and units of transcription of the conjugative transfer gene complex (*trs*) of *Staphylococcus aureus* plasmid pGO1. *J. Bacteriol.* **175:**4436–4447.

62. **Murphy, E.** 1990. Properties of the site-specific transposable element Tn554, p. 123–135. *In* R. P. Novick (ed.), *Molecular Biology of the Staphylococci.* VCH Publishers, New York, N.Y.

63. **Needham, C., W. C. Noble, and K. G. H. Dyke.** 1995. The staphylococcal insertion sequence IS257 is active. *Plasmid* **34:**198–205.

64. **Novick, R. P.** 1989. Staphylococcal plasmids and their replication. *Annu. Rev. Microbiol.* **43:**537–565.

65. **Novick, R. P.** 1990. The *Staphylococcus* as a molecular genetic system, p. 1–37. *In* R. P. Novick (ed.), *Molecular Biology of the Staphylococci.* VCH Publishers, New York, N.Y.

66. **Novick, R. P., I. Edelman, M. D. Schwesinger, A. D. Gruss, E. C. Swanson, and P. A. Pattee.** 1979. Genetic translocation in *Staphylococcus aureus*. *Proc. Natl. Acad. Sci. USA* **76:**400–404.

67. **Ouyang, S. O., and C. Y. Lee.** Unpublished data.

68. **Pattee, P. A., H.-C. Lee, and J. P. Bannantine.** 1990. Genetic and physical mapping of the chromosome of *Staphylococcus aureus*, p. 41–58. *In* R. P. Novick (ed.), *Molecular Biology of the Staphylococci.* VCH Publishers, New York, N.Y.

69. **Paulsen, I. T., N. Firth, and R. A. Skurray.** 1997. Resistance to antimicrobial agents other than β-lactams, p. 175–212. *In* K. B. Crossley and G. L. Archer (ed.), *The Staphylococci in Human Disease.* Churchill Livingstone, New York, N.Y.

70. **Paulsen, I. T., M. T. Gillespie, T. G. Littlejohn, O. Hanvivatvong, S. J. Rowland, K. G. H. Dyke, and R. A. Skurray.** 1994. Characterisation of *sin*, a potential recombinase-encoding gene from *Staphylococcus aureus*. *Gene* **141:**109–114.

70a. **Projan, S. J.** Unpublished data.

71. **Projan, S. J., and G. L. Archer.** 1989. Mobilization of the relaxable *Staphylococcus aureus* plasmid pC221 by the conjugative plasmid pGO1 involves three pC221 loci. *J. Bacteriol.* **171:**1841–1845.

72. **Projan, S. J., and R. Novick.** 1988. Comparative analysis of five related staphylococcal plasmids. *Plasmid* **19:**203–221.

73. **Quintiliani, R., Jr., and P. Courvalin.** 1996. Characterization of Tn1547, a composite transposon flanked by the IS16 and IS256-like elements, that confers vancomycin resistance in *Enterococcus faecalis* BM4281. *Gene* **172:**1–8.

74. **Rice, L. B., L. L. Carias, and S. H. Marshall.** 1995. Tn5384, a composite enterococcal mobile element conferring resistance to erythromycin and gentamicin whose ends are directly repeated copies of IS256. *Antimicrob. Agents Chemother.* **39:**1147–1153.

75. **Rouch, D. A., M. E. Byrne, Y. C. Kong, and R. A. Skurray.** 1987. The *aacA-aphD* gentamicin and kanamycin resistance determinant of Tn4001 from *Staphylococcus aureus*: expression and nucleotide sequence analysis. *J. Gen. Microbiol.* **133:**3039–3052.

76. **Rouch, D. A., D. S. Cram, D. DiBerardino, T. G. Littlejohn, and R. A. Skurray.** 1990. Efflux-mediated antiseptic resistance gene *qacA* from *Staphylococcus aureus*: common ancestry with tetracycline- and sugar-transport proteins. *Mol. Microbiol.* **4:**2051–2062.

77. **Rouch, D. A., L. J. Messerotti, L. S. Loo, C. A. Jackson, and R. A. Skurray.** 1989. Trimethoprim resistance transposon Tn4003 from *Staphylococcus aureus* encodes genes for a dihydrofolate reductase and thymidylate synthetase flanked by three copies of IS257. *Mol. Microbiol.* **3:**161–175.

78. **Rowland, S., and K. G. H. Dyke.** 1990. Tn552, a novel transposable element from *Staphylococcus aureus*. *Mol. Microbiol.* **4:**961–975.

80. **Shalita, Z., E. Murphy, and R. P. Novick.** 1980. Penicillinase plasmids of *Staphylococcus aureus*: structural and evolutionary relationships. *Plasmid* **3:**291–311.

81. **Sharma, V. K., J. L. Johnston, T. M. Morton, and G. L. Archer.** 1994. Transcriptional regulation by TrsN of conjugative transfer genes on staphylococcal plasmid pGO1. *J. Bacteriol.* **176:**3445–3454.

82. **Sheehy, R. J., and R. P. Novick.** 1975. Studies on plasmid replication. V. Replicative intermediates. *J. Mol. Biol.* **93:**237–253.

83. **Skurray, R. A., and N. Firth.** 1997. Molecular evolution of multiply-antibiotic-resistant staphylococci. *Ciba Found. Symp.* **207:**167–183.

84. **Skurray, R. A., D. A. Rouch, B. R. Lyon, M. T. Gillespie, J. M. Tennent, M. E. Byrne, L. J. Messerotti, and J. W. May.** 1988. Multiresistant *Staphylococcus aureus:* genetics and evolution of epidemic Australian strains. *J. Antimicrob. Chemother.* **21:**19–38.

85. **Soltis, M. T., J. J. Mekalanos, and M. J. Betley.** 1990. Identification of a bacteriophage containing a silent staphylococcal variant enterotoxin gene (*sezA*$^+$). *Infect. Immun.* **58:**1614–1619.

86. **Tennent, J. M., B. R. Lyon, M. Midgley, I. G. Jones, A. S. Purewal, and R. A. Skurray.** 1989. Physical and biochemical characterization of the *qacA* gene encoding antiseptic and disinfectant resistance in *Staphylococcus aureus. J. Gen. Microbiol.* **135:**1–10.

87. **Thumm, G., and F. Gotz.** 1997. Studies on prolysostaphin processing and characterization of the lysostaphin immunity factor (Lif) of *Staphylococcus simulans* biovar *staphylolyticus. Mol. Microbiol.* **23:**1251–1265.

88. **Townsend, D. E., S. Bolton, N. Ashdown, D. I. Annear, and W. B. Grubb.** 1986. Conjugative, staphylococcal plasmids carrying hitch-hiking transposons similar to Tn*554*: intra- and interspecies dissemination of erythromycin resistance. *Aust. J. Exp. Biol. Med. Sci.* **64:**367–379.

89. **Trees, D. L., and J. J. Iandolo.** 1988. Identification of a *Staphylococcus aureus* transposon (Tn*4291*) that carries the methicillin resistance gene(s). *J. Bacteriol.* **170:**149–154.

90. **Udo, E. E., and W. B. Grubb.** 1991. A new incompatibility group plasmid in *Staphylococcus aureus.* FEMS *Microbiol. Lett.* **62:**33–36.

91. **Wright, C. L., M. E. Byrne, N. Firth, and R. A. Skurray.** 1998. A retrospective molecular analysis of gentamicin resistance in *Staphylococcus aureus* strains from UK hospitals. *J. Med. Microbiol.* **47:**173–178.

92. **Wu, S., C. Piscitelli, H. de Lencastre, and A. Tomasz.** 1996. Tracking the evolutionary origin of the methicillin resistance gene: cloning and sequencing of a homologue of *mecA* from a methicillin susceptible strain of *Staphylococcus sciuri. Microb. Drug Resist.* **2:**435–441.

93. **Ziebuhr, W., V. Krimmer, S. Rachid, I. Losner, F. Götz, and J. Hacker.** 1999. A novel mechanism of phase variation of virulence in *Staphylococcus epidermidis:* evidence for control of the polysaccharide intercellular adhesion synthesis by alternating insertion and excision of the insertion sequence element IS*256. Mol. Microbiol.* **32:**345–356.

Carbohydrate Catabolism: Pathways and Regulation

REINHOLD BRÜCKNER AND JOANNIS BASSIAS

34

The central pathways of carbon metabolism are conserved in virtually all organisms, but details of specific biosynthetic and degradative pathways vary considerably between bacteria, plants, and animals. In addition, metabolic capacities differ widely among bacteria. As a consequence, generalization and even comparative approaches do not appear to be suitable to completely understand the physiology of a certain bacterium. For that end, biosynthetic potential and nutritional requirements have to be defined for each bacterial strain individually. On the other hand, the comparison of related organisms, e.g., AT-rich gram-positive bacteria, may reveal common principles in physiology and regulation. A prominent example would perhaps be carbon catabolite repression, which is executed in AT-rich gram-positive organisms by a different mechanism than in enteric bacteria (20, 41).

In *Staphylococcus aureus* and other staphylococcal species, relatively little is known at the molecular level about carbohydrate utilization, biosynthetic pathways, and nutritional requirements. The limited knowledge on sugar utilization systems is especially surprising, since *S. aureus* was the first gram-positive bacterium in which the phosphotransferase system (PTS), a complex enzyme system involved in sugar uptake and regulation of cellular metabolism, was described. Numerous biochemical and physiological studies, initiated before gene isolation techniques became available, have not been pursued and extended to the gene level. Therefore, information on transcriptional regulation of genes involved in catabolic or biosynthetic pathways is fragmentary. In other microorganisms, especially in *Escherichia coli* and *Bacillus subtilis*, these studies have led to fundamental concepts of molecular biology and gene regulation. Considering the need of a pathogen to multiply in the host organism to cause disease, there is still good reason to investigate cellular metabolism in the staphylococci. Since earlier work on sugar utilization and other aspects of intermediary metabolism has already been summarized (3, 23), we concentrate here on the latest development in staphylococcal carbohydrate metabolism.

GLYCOLYTIC PATHWAYS

Glucose catabolism in *S. aureus* predominantly proceeds through the Embden-Meyerhof-Parnas (EMP) pathway (Fig. 1), but the pentose phosphate cycle is also operative. There is no evidence for the Entner-Doudoroff pathway. Depending on the growth conditions, about 85% of glucose is catabolized via the EMP pathway. The major end product of anaerobic glucose metabolism is lactate, while under aerobic growth conditions acetate and CO_2 are predominant (3). Glucose enters the EMP pathway as glucose 6-phosphate, which is produced either directly by PTS-mediated transport or by the activity of a glucose kinase. Various other carbohydrates, hexoses, hexitols, or disaccharides such as fructose, mannitol, sucrose, or maltose are fed into the EMP pathway by the activity of peripheral sugar-specific enzymes, such as PTS permeases, hydrolases, kinases, and dehydrogenases.

THE PHOSPHOTRANSFERASE SYSTEM

The phosphoenolpyruvate-carbohydrate PTS, which is widely distributed among eubacteria, catalyzes the concomitant uptake and phosphorylation of a great variety of carbohydrates (31). The system consists of sugar-specific PTS permeases, often referred to as enzymes II (EII), and two proteins, enzyme I (EI) and histidine-containing protein (HPr), that participate in the phosphorylation of all PTS-transported carbohydrates and, therefore, have been termed the general PTS proteins. The specific permeases are composed of up to four protein domains (EIIA, B, C, D), at least one of which is membrane bound. These protein domains may be detached or fused as a single polypeptide chain. On the other hand, the general PTS proteins are soluble, cytoplasmic proteins. The phosphoryl-transfer chain starts with EI and phosphoenolpyruvate and proceeds via the phosphocarrier protein HPr to the EIIA and EIIB domains of the PTS permeases. Subsequently, the incoming sugar is phosphorylated by EIIB (Fig. 1).

As mentioned above, *S. aureus* was the first gram-positive species in which PTS activity was demonstrated. The uptake of glucose, mannose, mannitol, glucosamine, N-acetylglucosamine, sucrose, lactose, galactose, and β-glucosides is reported to be PTS dependent (23). Glucose entry may not completely rely on PTS activity, since HPr

Gram-Positive Pathogens, ed. by V. A. Fischetti et al.
© 2000 American Society for Microbiology, Washington, D.C.

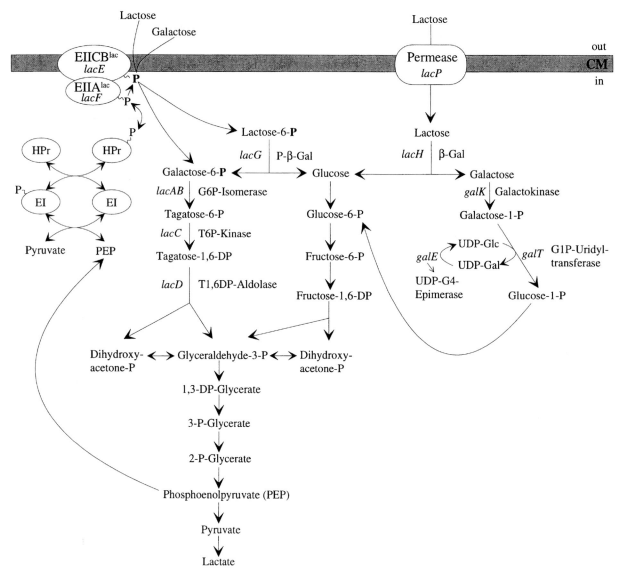

FIGURE 1 Alternative lactose catabolic pathways in staphylococci. Transport of lactose and galactose and their catabolism are shown. In *S. aureus*, lactose and galactose are transported by the PTS. Internalized lactose 6-phosphate is hydrolyzed by a phospho-β-galactosidase to galactose 6-phosphate and glucose. Galactose 6-phosphate is catabolized through the tagatose 6-phosphate pathway. This pathway most likely exists in staphylococci exhibiting lactose PTS activity. In *S. xylosus* and probably other staphylococcal species that do not possess a lactose PTS, a permease is responsible for the transport of lactose. Galactose uptake has not been studied in these species. Nonphosphorylated lactose is hydrolyzed by a β-galactosidase to yield glucose and galactose. Galactose is likely catabolized through the Leloir pathway. In both staphylococcal species, glucose 6-phosphate, produced by a glucose kinase, enters the EMP pathway, the main glycolytic pathway in staphylococci. Only the galactoside-specific genes and their encoded products are mentioned in the pathways. Abbreviations: CM, cytoplasmic membrane; EI, enzyme I; EIIA^lac, lactose-specific enzyme IIA; EIICB^lac, lactose-specific enzyme IICB; HPr, histidine-containing protein; β-Gal, β-galactosidase; P-β-Gal, phospho-β-galactosidase; G6P-Isomerase, galactose-6-phosphate isomerase; G1P-Uridyltransferase, galactose-1-phosphate uridyltransferase; T6P-Kinase, tagatose-6-phosphate kinase; T1,6DP-Aldolase, tagatose-1,6-diphosphate aldolase; UDP-Gal, UDP-galactose; UDP-Glc, UDP-glucose; UDP-G4-Epimerase, UDP-galactose-4 epimerase; PEP, phosphoenolpyruvate; P, phosphate; DP, diphosphate.

or EI mutant strains still ferment glucose, although at a reduced rate (37). The disaccharide maltose is mainly taken up independently of PTS, but additional minor PTS-mediated transport cannot be ruled out (37). On the other hand, utilization of glycerol clearly does not involve PTS components. Several other carbohydrates, such as cellobiose, raffinose, xylose, and arabinose, do not serve as carbon sources for *S. aureus*. Metabolism of others, such as trehalose, has not been investigated. The utilization of the pentose xylose, a trait found in only a few staphylococcal species (23), does not involve the PTS system in *Staphylococcus xylosus* (43, 44).

At the molecular level, information on genes encoding the general PTS proteins and the lactose PTS is available from *S. aureus*. The primary structure of *S. aureus* HPr was initially determined by sequencing the purified protein (2). Subsequently, the corresponding DNA was also sequenced (25). The HPr gene, *ptsH*, is followed by *ptsI*, the gene encoding the general PTS protein enzyme I. The genetic organization, *ptsHI*, is identical to that found in *Staphylococcus carnosus* (12, 24) and encountered in the majority of bacteria possessing PTS. It appears that the two genes are transcribed from a promoter in front of *ptsH*. The short spacing of only 2 bp between *ptsH* and *ptsI* and the apparent lack of a Shine-Dalgarno sequence even suggest that *ptsH* and *ptsI* translation may be coupled. As inferred from biochemical data on PTS activity, expression of the *ptsHI* operon is constitutive, which seems to be reasonable, since the enzymes are needed for the uptake of a variety of carbohydrates.

LACTOSE AND GALACTOSE UTILIZATION

Metabolism of lactose and galactose in *S. aureus* is initiated by PTS-dependent uptake and phosphorylation of the sugars, yielding lactose 6-phosphate and galactose 6-phosphate, respectively. Glucose and galactose 6-phosphate are produced from intracellular lactose 6-phosphate by a phospho-β-galactosidase. Glucose is metabolized via the EMP pathway, whereas galactose 6-phosphate is degraded via the tagatose 6-phosphate pathway (Fig. 1), which was first discovered in *S. aureus*. The corresponding genes are arranged in a heptacistronic operon and were designated *lacABCDFEG*, since the locus was originally defined by mutations abolishing lactose utilization. The genes *lacFE* encode the galactoside-specific components of the PTS, enzyme IIA and enzyme IICB (4). The terminal *lacG* gene encodes the phospho-β-galactosidase (5). The *lacABCD* cluster specifies the enzymes of the tagatose 6-phosphate pathway, with *lacAB* encoding galactose-6-phosphate isomerase, probably forming a heteromultimeric complex, *lacC* for tagatose-6-phosphate kinase and *lacD* for tagatose-1,6-diphosphate aldolase (39). Upstream from the *lacABCDFEG* operon and arranged in the same orientation is the *lacR* gene, encoding the repressor of the lactose operon (29). LacR resembles several lactose repressors from gram-positive bacteria (9) that belong to the DeoR regulator family, named after the repressor of the deoxyribonucleotide operon in *E. coli* (47).

Addition of lactose or galactose to the growth medium leads to induction of the *lac* genes, with galactose 6-phosphate being the intracellular inducer. Apart from this specific regulation, the lactose operon of *S. aureus* is also subject to global carbon catabolite repression, but the nature of the regulatory mechanism remains unclear (29). It is conceivable that the described catabolite repression of the *lac* operon of *S. aureus* is actually caused by glucose-mediated inducer exclusion (37, 40).

The lactose metabolism pathway described for *S. aureus* is not universal for all staphylococci. In *S. xylosus*, for example, lactose is taken up in nonphosphorylated form by a lactose permease, which shows significant similarity to members of the GPH protein family that transport galactosides, pentoses, and hexuronides (30). Transporters of this family were found in a variety of bacteria, including enteric and lactic acid bacteria. The lactose permease is encoded by *lacP*, which forms a bicistronic operon with the following *lacH* gene, specifying a β-galactosidase (1). After lactose hydrolysis, glucose is phosphorylated by a glucose kinase and catabolized through the EMP pathway and galactose likely by the Leloir pathway (Fig. 1) (27). The genes encoding the enzymes for the Leloir pathway have recently been cloned from *S. carnosus* (26). The *lacR* gene, upstream and in the opposite orientation from the *lacPH* operon of *S. xylosus*, encodes an activator belonging to the AraC/XylS family (17). In contrast to *S. aureus*, only lactose, but not galactose, in the culture medium induced the *lacPH* operon (1).

GLUCOSE UTILIZATION

Apart from the genes responsible for galactoside transport and catabolism, no other carbohydrate uptake systems have been described in *S. aureus*. It is therefore of interest to consider genes from other staphylococcal species, whose counterparts are likely to exist in *S. aureus*. In *S. carnosus*, two genes located right next to each other, *glcA* and *glcB*, have been cloned, both of which complemented an *E. coli* mutant strain deficient in glucose uptake (8). The GlcA/B proteins were highly similar to each other (69% identity) and to glucose-specific enzyme II proteins from *B. subtilis* or *E. coli*. The two *S. carnosus* PTS permeases had fused EII domains in the order EIICBA. Immediately upstream of *glcA*, a gene, *glcT*, was detected whose deduced amino acid sequence shows a high degree of similarity to bacterial regulators involved in antitermination (45). Interestingly, the activity of these regulators is controlled by PTS-mediated phosphorylation. A putative transcriptional terminator partially overlapping an inverted repeat, that could be the target site for the antiterminator protein GlcT, is found in the *glcT-glcA* intergenic region. This organization resembles the *glcT-ptsG* region of *B. subtilis*, encoding the GlcT antiterminator protein and the glucose-specific enzyme II, respectively (46). Therefore, *glcA* expression in *S. carnosus* is most likely controlled by antitermination. An incomplete open reading frame, whose deduced protein shares 83% identical residues with the carboxy-terminal third of GlcA of *S. carnosus*, is present in the preliminary *S. aureus* NCTC 8325 genome sequence database (38). Therefore, *S. aureus* probably possesses at least one gene encoding an enzyme II specific for glucose. It will be interesting to see whether the second gene, *glcB*, and the putative antiterminator gene, *glcT*, are also present.

In addition to PTS-dependent glucose transport, *S. aureus* is apparently able to take up glucose by PTS-independent routes. One such system has been identified in *S. xylosus*, but the mechanism by which glucose enters the cell has not been elucidated (13). A glucose kinase, which would be needed for catabolism of nonphosphorylated glucose, is present in S. xylosus as well as in S. aureus (38, 49).

MANNITOL UTILIZATION

The mannitol phosphotransferase system has been described for S. carnosus. The system consists of an EIICB enzyme, encoded by mtlA, and of EIIA, encoded by mtlF, which together form the mannitol-specific PTS permease. The mtlA and mtlF genes are clustered on the S. carnosus chromosome with mtlD, the gene for mannitol-1-phosphate dehydrogenase, which produces fructose 6-phosphate. The gene mtlA is about 2 kb away from mtlF and mtlD (14, 15). The nucleotide sequence of this intervening region has not been determined. S. aureus possesses an EIIA protein specific for mannitol with the same apparent molecular mass as the S. carnosus enzyme and virtually identical amino acids at the amino terminus and around the phosphorylation site (35). Therefore, the mtl systems in both organisms appear to be very similar.

SUCROSE UTILIZATION

The sucrose PTS permease, analyzed in S. xylosus and encoded by scrA, is composed of fused EIIBC domains (48). The EIIA domain, which is essential for PTS-mediated sucrose uptake, has not been identified in S. xylosus. Based on the analysis of sucrose utilization in E. coli and B. subtilis, it appears questionable whether a sucrose-specific EIIA protein exists. In both organisms, the EIIA domain specific for glucose, either as a separate enzyme or fused to the IICB domains of the glucose permease, serves as phosphoryl donor for the sucrose-specific EII enzymes. The GlcA/B proteins mentioned above would thus be good candidates for being involved in sucrose uptake.

Internalized sucrose 6-phosphate is cleaved by sucrose-phosphate hydrolase, the gene product of scrB, to yield glucose 6-phosphate and fructose (7). Fructose is subsequently phosphorylated by a fructokinase encoded by scrK (6). The genes scrB and scrK are probably cotranscribed. However, the sucrose permease gene scrA is not located in close vicinity to the scrBK genes. Expression of scrA as well as scrB is induced by sucrose in the medium. Regulation is dependent on the LacI/GalR-type repressor ScrR, whose gene scrR is found upstream of scrB (18). An scr gene cluster, scrR, scrB, and scrK, is also present in S. aureus (38). It is located next to agrA, the gene encoding the response regulator of the accessory gene regulator system (agr) (28). The gene for the sucrose permease has not yet been detected.

MALTOSE UTILIZATION

Maltose utilization in S. xylosus is dependent on an α-glucosidase or maltase, whose gene, malA, is the second gene of the malRA operon, of which malR encodes a regulator belonging to the LacI/GalR family (10). The same enzymatic activity as mediated by MalA in S. xylosus has been characterized in S. aureus, and recently the corresponding gene appeared in the preliminary database of the S. aureus genome sequence (38). Therefore, both staphylococcal species seem to cleave maltose by this enzyme. However, the mechanism of maltose transport remains to be elucidated. In any case, at least one glucose moiety has to be phosphorylated by the glucose kinase for complete maltose catabolism.

XYLOSE UTILIZATION

Although xylose utilization is scarce among staphylococci, the pathway in S. xylosus will be discussed, since it represents the only molecular information on staphylococcal pentose utilization. Three S. xylosus genes, xylA, xylB, and xylR, involved in xylose catabolism have been identified (43, 44). They encode the xylose isomerase XylA, the xylulose kinase XylB, and the regulator XylR. XylA and XylB generate xylulose 5-phosphate, which enters the pentose phosphate cycle. Genes encoding proteins for xylose transport have not yet been identified, but it is established that xylose enters the cells in nonmodified form by a PTS-independent route. Transcriptional analysis yielded the finding that xylAB is transcribed from one promoter in front of xylA, while xylR is transcribed independently. Transcription initiated at the xylA promoter is inducible by the addition of xylose to the growth medium, whereas xylR expression is constitutive with respect to xylose. Regulation of xylAB expression is dependent on a functional XylR protein, which acts as a repressor.

GLUCOSE-MEDIATED REGULATION

The availability of glucose leads to regulatory processes often referred to as glucose effect or carbon catabolite repression (40, 41). In S. aureus, numerous publications describe the influence of glucose on a variety of cellular processes, like utilization of alternative carbon sources, production of extracellular enzymes, activity of glycolytic enzymes, or cytochrome content. Especially the production of potential virulence factors attracted considerable attention. However, the mechanism by which glucose exerts its regulatory effect has not been elucidated so far. The analysis of catabolite repression of inducible systems is quite often complicated by the inability of the inducer to enter the cells in appreciable amounts, when glucose is also present in the medium. In addition to this process, referred to as inducer exclusion (40), inducer expulsion, the rapid removal of internalized inducer, has been described for a number of gram-positive bacteria (40). While inducer exclusion is found in a wide variety of bacteria, inducer expulsion is encountered less frequently. The latter does not appear to be operative in S. aureus (40). The molecular mechanisms leading to inducer exclusion in gram-positive bacteria are not yet completely understood.

In gram-positive bacteria, one form of catabolite repression relies on a transcriptional regulator, termed catabolite control protein A (CcpA) (19), and a protein kinase, HPr kinase, that phosphorylates the PTS phosphocarrier HPr in an ATP-dependent manner. According to the current model, HPr phosphorylated at the serine residue at position 46 binds to CcpA, which thereupon binds to catabolite repression operators designated cre (catabolite responsive element) (21). However, the recent identification of NADP as a corepressor for Bacillus CcpA suggests a more complex regulatory circuit (22). The existence of HPr kinase activity in S. aureus was demonstrated some time ago (37), and more recently the gene, hprK, was identified (6). It is encoded downstream of uvrA, encoding subunit A of excinuclease ABC, and followed by lgt, encoding diacylglyceryl transferase (32). The same genetic organization was also found in *Staphylococcus epidermidis* and *S. xylosus* (6). In *B. subtilis*, in which the first *hprK* gene was identified, inactivation of the gene resulted in the loss of catabolite repression (16, 36). Inactivation studies in the staphylococci are under way.

The gene for CcpA awaits identification in *S. aureus*, but its involvement in catabolite repression was demonstrated in *S. xylosus* (11). In this species, a *cre* operator located in the *malRA* promoter region is responsible for CcpA-mediated regulation (11). The *S. aureus* gene *pckA*, encoding phosphoenolpyruvate carboxykinase, an essential enzyme for gluconeogenesis, is apparently subject to the same type of control (42). A putative *cre* operator is located around the start site of the *pckA* gene, and promoter probe studies revealed a 22-fold repression of the promoter activity in the presence of glucose. Therefore, CcpA-mediated catabolite repression is very likely to exist in *S. aureus*.

On the other hand, there is also evidence for glucose-mediated repression of gene expression that does not depend directly on CcpA. It has been reported that expression of RNAII and RNAIII in the *agr* locus is subject to glucose control at low pH (34). In addition, enterotoxin type C mRNA is strongly reduced in the presence of glucose even in an *agr*-negative strain (33). In both cases, *cre* operators are not apparent in the respective promoter regions. Thus, glucose control could be exerted by a CcpA-independent mechanism or by factors controlled by CcpA. In any case, it will be interesting to determine the nature of glucose regulation in these systems.

RESPONSE TO GLUCOSE LIMITATION

Bacteria quite often encounter nutrient-limiting conditions in their natural environment. To survive prolonged periods of nutrient deprivation, many bacteria have developed starvation-survival strategies for persistence. When *S. aureus* cultures are starved for glucose, phosphate, amino acids, or multiple nutrients, a dramatic loss of viability of up to 99.9% within the first 2 days is observed (50). After this initial drop in viability, the remaining cells developed a long-term starvation-survival state under glucose- or multiple-nutrient limitations, whereas amino acid or phosphate starvation led to a further loss of viability and eventually to nonculturability. Long-term survival of the cells was associated with changes in cell morphology and increased resistance to acid shock and oxidative stress and was dependent on cell wall and protein synthesis. It appears, therefore, that glucose- or multiple-nutrients starvation induces differential protein synthesis and, most likely, also differential gene expression. It will be interesting to determine which genes and proteins are induced during the starvation period, enabling *S. aureus* to survive under nutrient-limiting conditions.

We thank B. Krismer, F. Götz, and N. Schnell for communicating results before publication, and S. Dilsen for help in literature searches. Work in Tübingen was sponsored by the European Union Biotech Program (BIOC2-CT92-0137) and the Deutsche Forschungsgemeinschaft (BR947/3-1).

REFERENCES

1. **Bassias, J., and R. Brückner.** 1998. Regulation of lactose utilization genes in *Staphylococcus xylosus. J. Bacteriol.* **180:**2273–2279.

2. **Beyreuther, K., H. Raufuss, O. Schrecker, and W. Hengstenberg.** 1977. The phosphoenolpyruvate-dependent phosphotransferase system of *Staphylococcus aureus.* 1. Amino-acid sequence of the phosphocarrier protein HPr. *Eur. J. Biochem.* **75:**275–286.

3. **Blumenthal, H. J.** 1972. Glucose catabolism in staphylococci, p. 111–135. *In* J. O. Cohen (ed.), *The Staphylococci.* Wiley Interscience, New York, N.Y.

4. **Breidt, F., Jr., W. Hengstenberg, U. Finkeldei, and G. C. Stewart.** 1987. Identification of the genes for the lactose-specific components of the phosphotransferase system in the *lac* operon of *Staphylococcus aureus. J. Biol. Chem.* **262:**16444–16449.

5. **Breidt, F., Jr., and G. C. Stewart.** 1986. Cloning and expression of the phospho-beta-galactosidase gene of *Staphylococcus aureus* in *Escherichia coli. J. Bacteriol.* **166:**1061–1066.

6. **Brückner, R.** Unpublished results.

7. **Brückner, R., E. Wagner, and F. Götz.** 1993. Characterization of a sucrase gene from *Staphylococcus xylosus. J. Bacteriol.* **175:**851–857.

8. **Christiansen, I., and W. Hengstenberg.** 1996. Cloning and sequencing of two genes from *Staphylococcus carnosus* coding for glucose-specific PTS and their expression in *Escherichia coli* K-12. *Mol. Gen. Genet.* **250:**375–379.

9. **de Vos, W. M., and E. E. Vaughan.** 1994. Genetics of lactose utilization in lactic acid bacteria. *FEMS Microbiol. Rev.* **15:**217–237.

10. **Egeter, O., and R. Brückner.** 1995. Characterization of a genetic locus essential for maltose-maltotriose utilization in *Staphylococcus xylosus. J. Bacteriol.* **177:**2408–2415.

11. **Egeter, O., and R. Brückner.** 1996. Catabolite repression mediated by the catabolite control protein CcpA in *Staphylococcus xylosus. Mol. Microbiol.* **21:**739–749.

12. **Eisermann, R., R. Fischer, U. Kessler, A. Neubauer, and W. Hengstenberg.** 1991. Staphylococcal phosphoenolpyruvate-dependent phosphotransferase system. Purification and protein sequencing of the *Staphylococcus carnosus* histidine-containing protein, and cloning and DNA sequencing of the *ptsH* gene. *Eur. J. Biochem.* **197:**9–14.

13. **Fiegler, H., J. Bassias, I. Jankovic, and R. Brückner.** 1999. Identification of a gene in *Staphylococcus xylosus* encoding a novel glucose uptake protein. *J. Bacteriol.* **181:**4929–4936.

14. **Fischer, R., R. Eisermann, B. Reiche, and W. Hengstenberg.** 1989. Cloning, sequencing and overexpression of the mannitol-specific enzyme-III-encoding gene of *Staphylococcus carnosus. Gene* **82:**249–257.

15. **Fischer, R., and W. Hengstenberg.** 1992. Mannitol-specific enzyme II of the phosphoenolpyruvate-dependent phosphotransferase system of *Staphylococcus carnosus. Eur. J. Biochem.* **204:**963–969.

16. **Galinier, A., M. Kravanja, R. Engelmann, W. Hengstenberg, M. C. Kilhoffer, J. Deutscher, and J. Haiech.** 1998. New protein kinase and protein phosphatase families mediate signal transduction in bacterial catabolite repression. *Proc. Natl. Acad. Sci. USA* **95:**1823–1828.

17. **Gallegos, M.-T., R. Schleif, A. Bairoch, K. Hofmann, and J. L. Ramos.** 1997. AraC/XylS family of transcriptional regulators. *Microbiol. Mol. Biol. Rev.* **61:**393–410.

18. **Gering, M., and R. Brückner.** 1996. Transcriptional regulation of the sucrase gene of *Staphylococcus xylosus* by the repressor ScrR. *J. Bacteriol.* **178:**462–469.

19. **Henkin, T. M.** 1996. The role of the CcpA transcriptional regulator in carbon metabolism in *Bacillus subtilis*. *FEMS Microbiol. Lett.* **135:**9–15.

20. **Hueck, C. J., and W. Hillen.** 1995. Catabolite repression in *Bacillus subtilis*: a global regulatory mechanism for the gram-positive bacteria? *Mol. Microbiol.* **15:**395–401.

21. **Hueck, C. J., W. Hillen, and M. H. Saier, Jr.** 1994. Analysis of a *cis*-active sequence mediating catabolite repression in gram-positive bacteria. *Res. Microbiol.* **145:**503–518.

22. **Kim, J. H., M. I. Voskuil, and G. H. Chambliss.** 1998. NADP, corepressor for the *Bacillus* catabolite control protein CcpA. *Proc. Natl. Acad. Sci. USA* **95:**9590–9595.

23. **Kloos, W. E., K.-H. Schleifer, and F. Götz.** 1991. The genus *Staphylococcus*, p. 1369–1420. *In* A. Balows, H. G. Trüper, M. Dwarkin, W. Harder, and K. H. Schleifer (ed.), *The Procaryotes*. Springer-Verlag, Heidelberg, Germany.

24. **Kohlbrecher, D., R. Eisermann, and W. Hengstenberg.** 1992. Staphylococcal phophoenolpyruvate-dependent phosphotransferase system: molecular cloning and nucleotide sequence of the *Staphylococcus carnosus ptsI* gene and expression and complementation studies of the gene product. *J. Bacteriol.* **174:**2208–2214.

25. **Kravanja, M., R. Engelmann, I. Christiansen, and W. Hengstenberg.** GenBank X93205.

26. **Krismer, B., and F. Götz.** Personal communication.

27. **Maxwell, E. S., K. Kurahashi, and H. M. Kalckar.** 1962. Enzymes of the Leloir pathway. *Methods Enzymol.* **5:**174–189.

28. **Novick, R. P., S. J. Projan, J. Kornblum, H. F. Ross, G. Ji, B. Kreiswirth, F. Vandenesch, and S. Moghazeh.** 1995. The *agr* P2 operon: an autocatalytic sensory transduction system in *Staphylococcus aureus*. *Mol. Gen. Genet.* **248:**446–458.

29. **Oskouian, B., and G. C. Stewart.** 1990. Repression and catabolite repression of the lactose operon of *Staphylococcus aureus*. *J. Bacteriol.* **172:**3804–3812.

30. **Poolman, B., J. Knol, C. van der Does, P. J. F. Henderson, W.-J. Liang, G. Leblanc, T. Pourcher, and I. Mus-Veteau.** 1996. Cation and sugar selectivity determinants in a novel family of transport proteins. *Mol. Microbiol.* **19:**911–922.

31. **Postma, P. W., J. W. Lengeler, and G. R. Jacobson.** 1993. Phosphoenolpyruvate: carbohydrate phosphotransferase systems of bacteria. *Microbiol. Rev.* **57:**543–594.

32. **Qi, H. Y., K. Sankaran, K. Gan, and H. C. Wu.** 1995. Structure-function relationship of bacterial prolipoprotein diacylglyceryl transferase: functionally significant conserved regions. *J. Bacteriol.* **177:**6820–6824.

33. **Regassa, L. B., J. L. Couch, and M. J. Betley.** 1991. Steady-state staphylococcal enterotoxin type C mRNA is affected by a product of the accessory gene regulator (*agr*) and by glucose. *Infect. Immun.* **59:**955–962.

34. **Regassa, L. B., R. P. Novick, and M. J. Betley.** 1992. Glucose and nonmaintained pH decrease expression of the accessory gene regulator (*agr*) in *Staphylococcus aureus*. *Infect. Immun.* **60:**3381–3388.

35. **Reiche, B., R. Frank, J. Deutscher, N. Meyer, and W. Hengstenberg.** 1988. Staphylococcal phosphoenolpyruvate-dependent phosphotransferase system: purification and characterization of the mannitol-specific enzyme III^{mtl} of *Staphylococcus aureus* and *Staphylococcus carnosus* and

homology with the enzyme II^{mtl} of *Escherichia coli*. *Biochemistry* **27:**6512–6516.

36. **Reizer, J., C. Hoischen, F. Titgemeyer, C. Rivolta, R. Rabus, J. Stülke, D. Karamata, M. H. Saier, Jr., and W. Hillen.** 1998. A novel protein kinase that controls carbon catabolite repression in bacteria. *Mol. Microbiol.* **27:**1157–1169.

37. **Reizer, J., S. L. Sutrina, M. H. Saier, Jr., G. C. Stewart, A. Peterkofsky, and P. Reddy.** 1989. Mechanistic and physiological consequences of HPr(ser) phosphorylation on the activities of the phosphoenolpyruvate:sugar phosphotransferase system in gram-positive bacteria: studies with site-specific mutants of HPr. *EMBO J.* **8:**2111–2120.

38. **Roe, B. A., A. Dorman, F. Z. Najar, S. Clifton, and J. Iandolo.** *Staphylococcus aureus* Genome Sequencing Project. http://www.genome.ou.edu

39. **Rosey, E. L., B. Oskouian, and G. C. Stewart.** 1991. Lactose metabolism by *Staphylococcus aureus*: characterization of *lacABCD*, the structural genes of the tagatose 6-phosphate pathway. *J. Bacteriol.* **173:**5992–5998.

40. **Saier, M. H., Jr., S. Chauvaux, G. M. Cook, J. Deutscher, I. T. Paulsen, J. Reizer, and J.-J. Ye.** 1996. Catabolite repression and inducer control in Gram-positive bacteria. *Microbiology* **142:**217–230.

41. **Saier, M. H., Jr., S. Chauvaux, J. Deutscher, J. Reizer, and J.-J. Ye.** 1995. Protein phosphorylation and regulation of carbon metabolism in gram-negative versus gram-positive bacteria. *Trends Biochem. Sci.* **20:**267–271.

42. **Scovill, W. H., H. J. Schreier, and K. W. Bayles.** 1996. Identification and characterization of the *pckA* gene from *Staphylococcus aureus*. *J. Bacteriol.* **178:**3362–3364.

43. **Sizemore, C., E. Buchner, T. Rygus, C. Witke, F. Götz, and W. Hillen.** 1991. Organization, promoter analysis and transcriptional regulation of the *Staphylococcus xylosus* xylose utilization operon. *Mol. Gen. Genet.* **227:**377–384.

44. **Sizemore, C., B. Wieland, F. Götz, and W. Hillen.** 1992. Regulation of *Staphylococcus xylosus* xylose utilization genes at the molecular level. *J. Bacteriol.* **174:**3042–3048.

45. **Stülke, J., M. Arnaud, G. Rapoport, and I. Martin-Verstraete.** 1998. PRD—a protein domain involved in PTS-dependent induction and carbon catabolite repression of catabolic operons in bacteria. *Mol. Microbiol.* **28:**865–874.

46. **Stülke, J., I. Martin-Verstraete, M. Zagorec, M. Rose, A. Klier, and G. Rapoport.** 1997. Induction of the *Bacillus subtilis ptsGHI* operon by glucose is controlled by a novel antiterminator, GlcT. *Mol. Microbiol.* **25:**65–78.

47. **Valentin-Hansen, P., P. Hojrup, and S. Short.** 1985. The primary structure of the DeoR repressor from *Escherichia coli* K-12. *Nucleic Acids Res.* **13:**5927–5936.

48. **Wagner, E., F. Götz, and R. Brückner.** 1993. Cloning and characterization of the *scrA* gene encoding the sucrose-specific enzyme II of the phosphotransferase system from *Staphylococcus xylosus*. *Mol. Gen. Genet.* **241:**33–41.

49. **Wagner, E., S. Marcandier, O. Egeter, J. Deutscher, F. Götz, and R. Brückner.** 1995. Glucose kinase-dependent catabolite repression in *Staphylococcus xylosus*. *J. Bacteriol.* **177:**6144–6152.

50. **Watson, S. P., M. O. Clements, and S. J. Foster.** 1998. Characterization of the starvation-survival response of *Staphylococcus aureus*. *J. Bacteriol.* **180:**1750–1758.

Respiration and Small-Colony Variants of *Staphylococcus aureus*

RICHARD A. PROCTOR

35

STAPHYLOCOCCAL RESPIRATION

Staphylococcus aureus respiration and carbohydrate metabolism follow patterns very similar to those in *Escherichia coli* (1), with glycolysis in staphylococci following the nearly universal patterns found in all biological systems. Glycolysis results in the formation of some ATP. However, under aerobic conditions, much more ATP is formed when the pyruvate formed from the breakdown of glucose or fructose enters the citric acid cycle where it is oxidized to CO_2 and H_2O. The oxidation of these substrates requires that electrons flow through the electron transport chain. The reducing equivalents (electrons plus hydrogen ions) result from the oxidation of carbohydrates, amino acids, and lipids with the simultaneous reduction of NAD and FAD to NADH and $FADH_2$, respectively. Electrons flow from NADH and $FADH_2$ to menaquinone and cytochromes (Fig. 1) as these substrates are oxidized. The synthesis of ATP associated with electron transport occurs in the large transmembrane F_0F_1 ATPase (15, 16). In *S. aureus*, there is only one quinone, menaquinone, whereas in *E. coli*, menaquinone is used for aerobic electron transport, while fumarate and ubiquinone are involved in anaerobic metabolism (4, 5, 9). Glucose and fructose follow the glycolytic pathway to produce ATP when electron transport is interrupted. In contrast, uptake of mannitol, xylose, lactose, sucrose, maltose, and glycerol requires ATP and complex regulation via the phosphoenolpyruvate transferase system (41), and this is likely to be disrupted in respiratory-defective variants because these substrates cannot be used for growth by these bacteria.

Respiratory-defective mutants of *S. aureus* have been associated with unusually persistent, recurrent, and antibiotic-resistant infections (24, 33–38, 47, 49). These respiratory variants have been found to have defects in hemin and menaquinone biosynthesis, resulting in defective electron transport (Fig. 1). This leads to slow growth; hence, they are classified as small-colony variants (SCVs). While mutations in any essential gene can cause slow growth and small colonies, this chapter considers a subset of these mutants that were first observed in patients who had unusually persistent infections and were later found to have defects in oxidative metabolism and electron transport. The SCV acronym as used here refers only to slow-growing mutants defective in electron transport. These electron-transport variants produce reduced amounts of lytic toxins and are able to persist within host cells (2, 33–38).

HISTORY OF SMALL-COLONY VARIANTS

The first report of SCVs was in 1910 and concerned *Salmonella typhi* (13). Initially, these were called gonidial mutants, a term that was abbreviated to G forms or colonies (20). The idea that G forms were part of the natural growth cycle of bacteria arose because they were frequently found among many bacterial genera (33). Further work proved that they were not part of the growth cycle when it was demonstrated that many of these variants readily reverted to the large-colony form and that their small colony size related to more exacting metabolic requirements (reviewed in references 33–38). Subsequent studies revealed that a variety of host or environmental factors could select for mutants with this phenotype. Studies dating from the 1930s and 1940s reported decreased respiratory rates, restricted carbohydrate utilization, low ATP levels, reduced fermentation rates, and diminished dye reduction in SCVs from multiple bacterial species (33–38). It was also observed by a number of early investigators that SCVs produced no pigment and were auxotrophic for hemin or menadione. All of the phenotypic changes that were reported for these SCVs can be related to an interruption of electron transport.

MULTIPLE PHENOTYPIC CHANGES ASSOCIATED WITH INTERRUPTED ELECTRON TRANSPORT

A comparison of staphylococcal SCVs with parental wild type revealed a highly characteristic phenotype. A large number of phenotypic changes are apparent: small, nonpigmented, nonhemolytic colonies that fail to ferment mannitol and are resistant to aminoglycoside antibiotics. All aspects of this dramatic phenotype can be attributed

Gram-Positive Pathogens, ed. by V. A. Fischetti et al.
© 2000 American Society for Microbiology, Washington, D.C.

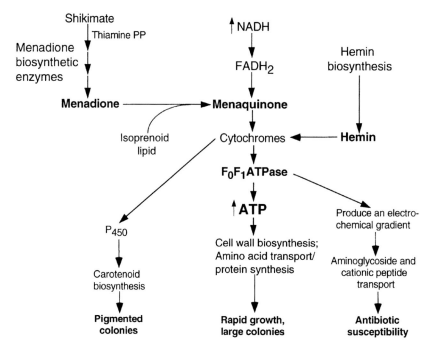

FIGURE 1 Relationship between electron transport and the small-colony-variant phenotype in *S. aureus*.

to the loss of one or another of the enzymes critical to the biosynthesis of electron transport chain components (Fig. 1).

The initial interest in electron transport arose because many of the clinical and laboratory isolates of *S. aureus* SCVs were found to be hemin or menadione auxotrophs (reviewed in references 33–38). Failure to produce hemin blocks the synthesis of cytochromes. As this is often the result of mutations in genes controlling prophyrin biosynthesis, the SCV phenotype can be reversed by giving hemin or its precursors. *S. aureus* is easier to study than *E. coli* because it is permeable to hemin (23). Hemin auxotrophy may also result from the inability to insert ferric iron into the pyrrole structure used for cytochrome biosynthesis, but interestingly, such auxotrophs can insert ferrous iron into catalase, suggesting that the defect is in the ferrochelatase (26). Similarly, mutations in the menadione biosynthetic genes result in a loss of menaquinone. Adding hemin or menadione completely reverses the SCV phenotype in clinical isolates, laboratory-generated strains, and a site-directed *hemB* mutant (2, 35, 46, 48). Anaerobic growth also reduplicates the SCV phenotype in *S. aureus* (2, 37), which is not surprising because menaquinone is not synthesized under anaerobic conditions (17). In contrast, *E. coli* has two quinones: menaquione is produced under anaerobic conditions, whereas ubiquinone is used when oxygen is present (4). Consequently, anaerobic growth in *E. coli* does not reduplicate the SCV phenotype in *E. coli*.

Slow growth may occur owing to several factors when electron transport is inoperative. One direct effect of interrupting electron transport is reduced synthesis of teichoic acid, because the biosynthesis of phosphatidyl glycerol, a teichoic acid precursor, requires electron transport (19, 26, 28). Also, ATP is required for many cellular functions, such as uptake of amino acids and carbohydrates and cell wall biosynthesis; hence, organisms interrupted in electron transport have a lower capacity to produce ATP

and show slower growth, resulting in small colonies. We found that the interdivision time for *S. aureus* 6850, a clinical isolate (parent strain), was 20 min versus 180 min for *S. aureus* JB-1, a menadione auxotrophic SCV, when grown in tryptic soy broth (2). Hale et al. (20) found similar growth rates with a *S. aureus* parent versus SCV at 20 min versus 120 min, respectively, in nutrient broth. In contrast, a hemin auxotrophic SCV had a 131- to 173-min generation time versus its parent's time of 53 min (28). While there are clearly differences in growth rates of several respiratory-defective strains of *S. aureus*, all of these variants produce small colonies and fulfill the definition of an SCV, i.e., an organism that forms colonies that are 10-fold smaller than its parent.

Electron transport is also required for carotenoid biosynthesis. Electrons are shuttled through the electron transport system to the P_{450} system, which is linked to pigment formation (22). Hence, SCVs that are defective in electron transport form nonpigmented colonies.

The failure of SCVs to ferment many carbohydrates, such as mannitol or xylose, can also be related to defects in electron transport (33–38). While glucose and fructose can be used to produce ATP via the glycolytic pathway, the metabolism of other sugars requires an intact citric acid cycle for their efficient utilization. Interruption of electron transport results in the accumulation of NADH, which down-regulates the citric acid cycle, thereby blocking the fermentation of mannitol and other carbohydrates.

Resistance of SCVs to aminoglycosides and other antibacterial substances is a further consequence of defects in electron transport (3, 29, and reviewed in reference 34). A transmembrane potential (negative on the inside) is formed as electrons flow through the F_0F_1 ATPase. This charge gradient enhances the accumulation of aminoglycosides that are positively charged and associate with the negatively charged *S. aureus* membrane when a sufficiently large electrochemical gradient is present (12, 18, 28). Consequently, menadione and hemin auxotrophs are readily recovered when *S. aureus* is exposed to gentamicin at relatively high concentrations (4 to 8 times the MIC) (2, 29, 34) because they are resistant to the drug owing to decreased uptake. Similarly, some lantibiotics and host cationic proteins are most active when the bacterial membrane has a strongly negative transmembrane potential (27, 34). Thus the development of electron transport-defective *S. aureus* SCVs in the absence of antibiotic pressure might be explained by selection for resistance to positively charged antibacterial host peptides or proteins (27, 33–35, 46). Only those compounds requiring a large membrane electrochemical gradient for uptake will be affected, because 60 to 70% of the normal membrane potential in *S. aureus* can be established in the absence of electron transport through ATP production via the glycolytic pathway when glucose or fructose is present (6). Glucose allows the F_0F_1 ATPase to break down ATP as protons are transported out of the cell, thereby producing a membrane potential (15, 16).

Other early observations about the SCV phenotype, such as a decreased ability to reduce dyes and to utilize oxygen by *S. aureus* SCVs, are also consistent with decreased electron transport (33–38). Thus, most of the features reported for clinical and laboratory-generated *S. aureus* SCVs can be explained by the model shown in Fig. 1.

OTHER CONSIDERATIONS ABOUT ELECTRON TRANSPORT AND SCV

Some *S. aureus* SCVs have primary defects in unsaturated fatty acid biosynthesis, and this also may cause defects in electron transport (19, 26, 44). Menaquinone is formed from menadione plus a repeating isoprenoid tail (usually eight repeating units in *S. aureus*) (19, 44). This lipid tail is required for association with the bacterial membrane, and the cytochromes subsequently nucleate around the quinone to form the electron transport chain (17). A similar situation was seen with some *Salmonella typhimurium* SVCs that formed normal colonies (i.e., grew rapidly) when the medium was supplemented with oleic acid (50). Hence, some *S. aureus* SCVs may be electron transport deficient because they cannot synthesize unsaturated fatty acids.

Inhibition or mutation of the membrane complex that synthesizes ATP and creates the membrane potential, the F_0F_1 ATPase, would also be expected to produce SCVs. Thus, mutations in the genes involved in the biosynthesis of the F_0F_1 ATPase in several bacterial species result in the formation of SCVs. By using Tn551 mutagenesis in the presence of hemin, menadione, and gentamicin, we isolated *S. aureus* mutants that show a typical SCV phenotype with reduced ATP production on minimal medium, increased resistance to gentamicin, but no auxotrophy for menadione or hemin (unpublished data). Menadione and hemin auxotrophs were eliminated since they are susceptible to gentamicin under these conditions. These *S. aureus* mutants resemble F_0F_1 ATPase mutants in other species showing increased oxygen utilization. High oxygen use is due to increased activity of dehydrogenases and type b cytochromes, which remove reducing equivalents formed during glycolysis (21). We have also found that bacterial electron transport inhibitors Z69 and Z90 are able to reproduce the SCV phenotype (unpublished data). These observations give further support for the model in Fig. 1.

Stable *S. aureus* SCVs with metabolic characteristics suggestive of electron transport variants have been recovered by growth in medium containing lithium chloride (51). While other salts (e.g., barium nitrite) selected for SCVs, these strains were very unstable. This is of some interest as lithium chloride is known to interfere with the F_0F_1 ATP-like ATPase in *Ilyobacter tartaricus* (30). Perhaps, lithium chloride damages organisms with an intact electron transport chain, hence, selecting for SCVs.

S. aureus subsp. *anaerobius*, an animal pathogen, is commonly isolated from abscesses in sheep, forms small colonies, and is deficient in porphyrin synthesis, but its characteristics are different from those of *S. aureus* SCVs (11). This species is unable to grow in an aerobic environment, fails to produce catalase when supplemented with hemin, and produces a violet pigment. This organism has been isolated only once from humans (39), but its distinctive characteristics should readily separate it from *S. aureus* SCVs.

WELL-CHARACTERIZED HEMIN- OR MENAQUINONE-DEFICIENT *S. AUREUS* STRAINS

S. aureus JB-1 is a menadione auxotrophic SCV that was selected with gentamicin from its parent, *S. aureus* 6850, an organism isolated from a patient with bacteremia and multiple metastatic sites of infection (2). *S. aureus* JB-1 produces less ATP, transmembrane potential, pigment, alpha-toxin, and coagulase; fails to ferment mannitol; uses less oxygen than its parent strain; is more resistant than its parent to aminoglycosides and to lantibiotics that require a large membrane potential for uptake; is able to survive on biomaterial surfaces despite massive concentrations of antibiotics; and is able to persist within cultured endothelial cells (2, 7, 27, 35). This complete SCV phenotype is reversed by the addition of menadione or *o*-succinylbenzoate but not by an earlier menadione biosynthetic precursor, chorismate (2), suggesting that it has a mutation in the menadione pathway prior to the *o*-succinylbenzoate intermediate.

Some hemin auxotrophs selected by resistance to gentamicin have also been characterized in detail (2, 28). These fail to ferment lactose and mannitol; have a uniform ability to grow on glucose, but not on oxidizable but nonfermentable substrates such as succinate, α-ketoglutarate, lactate, or malate; are negative for pigment formation, DNase activity, beta-hemolysis, catalase, benzidine reduction (suggesting the absence of cytochromes), tetrazolium dye reduction (an assay for electron transport activity); show little O_2 utilization during exponential phase growth; accumulate lactate; have increased pyruvate kinase and phosphohexose isomerase (suggesting the use of glycolysis for growth); have decreased uptake of tritiated tobramycin but are susceptible to novobiocin and chloramphenicol; and persist within cultured endothelial cells (2, 48). The SCV phenotype is reversed by hemin and in some cases hemin precursors such as Δ-aminolevulinic acid. Some of these SCVs are gram negative and coagulase negative (28), but this is not uniformly found (2, 25, 33–38).

Standard transformational and linkage analysis of known mutations (auxotrophies) of the hemin biosynthetic pathway allowed mapping of the hemin biosynthetic pathway in *S. aureus* in 1968 (45). These mutants were selected by resistance to kanamycin, and they grew as small colonies. More recently, it was found that mutations in this pathway cause the dramatic SCV phenotype (2, 26) as do transposon mutations (23). Nevertheless, the fact that single point mutations could have such a profoundly pleiotropic effect was not generally accepted until one of the *hem* genes, *hemB*, encoding Δ-aminolevulinic acid dehydratase, had been cloned, sequenced, and insertionally inactivated (48). The *S. aureus hemB* knockout mutant had the entirely predictable and complete SCV phenotype (2, 23) as described above, which was complemented by the cloned *hemB* and reversed by the addition of hemin to the culture medium. (The *S. aureus hemB* is quite closely related to the corresponding gene from other species [*Bacillus subtilis*, 62% identity; *E. coli*, 55%; mouse, 54%; pea, 55%; and spinach, 54%] and can complement the *E. coli hemB* mutant [23, 48].) In addition, this *hemB* knockout mutant persisted within cultured endothelial cells because it enters these cells, but it does not lyse them owing to decreased alpha-toxin production (48). Both the protein and mRNA for alpha-toxin are reduced. Complementation of the mutant in *trans* or addition of hemin reversed the phenotype and allowed the organism to lyse the endothelial cells.

CLINICAL STUDIES OF SCV

In recent studies, *S. aureus* SCVs have been shown to produce persistent, recurrent, and antibiotic-resistant infections (24, 37, 47). Patients infected with SCVs often have a history of long-term treatment with antibiotics given either parenterally (especially aminoglycosides) or locally in the form of antibiotic impregnated beads. They may have prolonged disease-free intervals (as long as 53 years), but during recurrences they show unusual resistance to apparently active antibiotic therapy (24, 37, 47). The organism may be misidentified as viridans streptococci or coagulase-negative staphylococci (25, 37). These studies are consistent with laboratory models and clinical anecdotes. For example, persistence of SCVs occurred in the kidneys of mice despite increased susceptibility to complement (49), and one of the first reported persistent and recurrent human SCV cases occurred in a patient treated with penicillin and streptomycin whose *S. aureus* aortic valve endocarditis had a late recurrence with the same strain (32). The ability of SCVs to persist within host cells helps to shield them from some antibiotics (2). Taken together, SCVs present a challenge for treatment because the intracellular location, slow growth, and decreased antibiotic uptake reduce the efficacy of antibiotic treatment (2, 8, 10, 14, 28, 37).

The clinical microbiologist must be alerted to look for SCVs, because their slow growth means that they may either be missed entirely or may be overgrown by more rapidly dividing strains or revertants. Furthermore, they are easily misidentified because of their unusual colony morphology and biochemical profile (24, 25, 37). Missing an SCV in a mixed culture may have serious consequences because it is more antibiotic resistant than the parent strain, leading to a major reporting error if it is not identified. Also, failure to identify the SCVs as a variant subpopulation may cause the clinician to believe that a new, rather than recurrent, infection is present, thereby hiding an antibiotic failure. Care should be used when selecting media, as some contain high concentrations of menadione and/or hemin (e.g., brain heart infusion and Schaedler's broths), whereas other media will allow the expression of the menadione and hemin auxotrophic SCV phenotype (tryptic soy and Mueller Hinton broths are low in hemin and menadione).

MENADIONE AND HEMIN AUXOTROPHS PREDOMINATE AMONG CLINICAL ISOLATES

While many different mutations might produce slow growth, menadione and hemin auxotrophs predominate among clinical *S. aureus* SCV isolates (33–38). There may be several reasons for this. First, the mutation must result in a variant that has some survival advantage. Mutations causing defects in electron transport produce variants that are more resistant to positively charged antimicrobials and that are able to persist within host cells owing to reduced toxin production (discussed below). Second, the mutation must not produce a defect that can be readily reversed by compounds naturally present in the host. Both menadione and hemin are at low concentrations in mammalian cells, whereas other substances, e.g., tryptophan, would be supplied by the intracellular milieu. Because *S. aureus* can survive anaerobically, electron transport is not absolutely essential for survival. Taken together, these conditions favor the selection of mutants in menadione and hemin biosynthesis.

INSTABILITY OF THE SCV PHENOTYPE: A REQUIREMENT FOR REVERSION TO THE VIRULENT PHENOTYPE

One of the earlier observations concerning SCVs from clinical specimens was that the phenotype was often unstable. For descriptive purposes, we consider strains "highly unstable" when they lose their SCV phenotype on the first passage (common), "stable" when the phenotype is maintained through 10 passes on solid medium (unusual), and "highly stable" when it can be repeatedly passed (rare). The instability of SCVs has made studying these variants difficult.

Slow-growing variants are often seen following penicillin treatment; however, these strains have been proved to be quite unstable (43). In contrast, streptomycin-selected SCVs are stable and have characteristics very similar to those of the now well-characterized SCVs that are defective in electron transport (3). The precise defects in penicillin-selected SCVs have not been defined, but the aminoglycoside-resistant SCVs are due to mutations in genes relating to the biosynthesis of components used in the electron transport chain. Because the genes for cytochromes, menaquinone, and the F_0F_1 ATPase have not yet been cloned in *S. aureus*, the mechanism behind the instability is unknown. One could postulate instability mechanisms such as a hot spot for point mutations, mismatching of repeated sequences in the promoter regions, and others. Alternatively, there could be strong selection for invertible elements that facilitate the production of these variants, because resistance to antibiotics and ability to persist inside host cells are very favorable to bacterial survival.

ELECTRON TRANSPORT, RESPIRATION, AND TOXIN PRODUCTION IN *S. AUREUS*

Connections between toxin production and environmental conditions have been noted in *S. aureus* for many years. For example, anaerobic growth results in markedly decreased hemolysis and results in the typical SCV phenotype (2). In addition, exogenous glucose, nonacidic pH, or Mg^{2+} at >10 to 12 μg/ml results in decreased carotenoid pigment, alpha-toxin, and toxic shock syndrome toxin production (31, 40). A consistent reduction in alpha-toxin production is seen in clinical *S. aureus* SCVs (33–38) as well as in *hemB* mutants (23, 48). This occurs even though a *hemB* mutant is able to grow over 7 orders of magnitude and enter into stationary phase, a situation in which *agr* is normally strongly activated and toxins are produced (48). These data suggest a link between respiratory metabolism and toxin production. In support of this concept, we have found that two electron transport inhibitors (Z69 and Z90) at sub-MICs produce typical SCVs in *S. aureus* that are unable to hemolyze rabbit erythrocytes or lyse cultured endothelial cells (unpublished data). Characterizing the signals in the pathways that link metabolism and toxin production may provide opportunities for modifying signaling with drugs to prevent disease, as has been suggested by work with glycerol monolaurate wherein *agr* and the toxins it regulates are decreased (42).

REFERENCES

1. **Baldwin, J. E., and H. Krebs.** 1981. The evolution of metabolic cycles. *Nature* **291:**381–382.
2. **Balwit, J. M., P. van Langevelde, J. M. Vann, and R. A. Proctor.** 1994. Gentamicin-resistant, menadione and hemin auxotrophic *Staphylococcus aureus* persist within cultured endothelial cells. *J. Infect. Dis.* **170:**1033–1037.
3. **Barbour, R. G. H.** 1950. Small colony variants ("G" forms) produced by *Staphylococcus pyogenes* during the development of resistance to streptomycin. *Aust. J. Exp. Biol. Med Sci.* **28:**411–421.
4. **Bentley, R., and R. Meganathan.** 1982. Biosynthesis of vitamin K (menaquinone) in bacteria. *Microbiol. Rev.* **46:**241–280.
5. **Bishop, D. H. L., K. D. Pandya, and H. K. King.** 1962. Ubiquinone and vitamin K in bacteria. *Biochem. J.* **83:**606–614.
6. **Chinn, B. D.** 1936. Characteristics of small colony variants with special reference to *Shigella paradysenteriae sonne. J. Infect Dis.* **59:**137–157.
7. **Chuard, C., P. E. Vaudaux, R. A. Proctor, and D. P. Lew.** 1997. Decreased susceptibility to antibiotic killing of small colony variants of *Staphylococcus aureus* in fluid phase and on fibronectin-coated surfaces. *J. Antimicrob. Chemother.* **39:**603–608.
8. **Clerch, B., E. Rivera, and M. Llagostera.** 1996. Identification of a pKM101 region which confers a slow growth rate and interferes with susceptibility to quinonolone in *Escherichia coli* AB1157. *J. Bacteriol.* **178:**5568–5572.
9. **Collins, M. D., and D. Jones.** 1981. Distribution of isoprenoid quinone structure types in bacteria and their taxonomic implications. *Microbiol. Rev.* **45:**316–354.
10. **Darouiche, R. O., and R. J. Hamill.** 1994. Antibiotic penetration of and bactericidal activity within endothelial cells. *Antimicrob. Agents Chemother.* **38:**1059–1064.
11. **de la Fuente, R., K. H. Schleifer, F. Götz, and H.-P. Köst.** 1986. Accumulation of porphyrins and pyrrole pigments by *Staphylococcus aureus* ssp. *anaerobius* and its aerobic mutant. *FEMS Microbiol. Lett.* **45:**183–188.
12. **Eisenberg, E. S., L. J. Mandel, H. R. Kaback, and M. H. Miller.** 1984. Quantitative association between electrical potential across the cytoplasmic membrane and early gentamicin uptake and killing in *Staphylococcus aureus. J. Bacteriol.* **157:**863–867.
13. **Eisenberg, P.** 1910. Untersuchungen über die variablen Typhusstamm (*Bacterium typhi mutabile*), sowie über eine eigentumliche hemmende Wirkung des gewöhnlichen Agar, verursacht durch Autoklavierung, abstr. 56, p. 208. *Zbl. Bakteriol. Abstr.*, vol. 1.
14. **Eng, R. H. K., F. T. Padberg, S. M. Smith, E. N. Tan, and C. E. Cherubin.** 1991. Bactericidal effects of antibiotics on slowly and nongrowing bacteria. *Antimicrob. Agents Chemother.* **35:**1824–1828.
15. **Fillingame, R. H.** 1997. Coupling H^+ transport and ATP synthesis in F_0F_1 ATP synthases: glimpses of interacting parts in a dynamic molecular machine. *J. Exp. Biol.* **200:**217–224.
16. **Fillingame, R. H., P. C. Jones, W. Jiang, F. I. Valiyaveetil, and O. Y. Dmitriev.** 1998. Subunit organization and structure in the F_0 sector of *Escherichia coli* F_0F_1 ATP synthase. *Biochim. Biophys. Acta* **1365:**135–142.
17. **Frerman, F. E., and D. C. White.** 1967. Membrane lipid changes during formation of a functional electron transport system in *Staphylococcus aureus. J. Bacteriol.* **94:**1868–1874.
18. **Gilman, S., and V. Saunders.** 1986. Accumulation of gentamicin by *Staphylococcus aureus*: the role of transmembrane electrical potential. *J. Antimicrob. Chemother.* **17:**37–44.
19. **Goldenbaum, P. E., and D. C. White.** 1974. Role of lipid in the formation and function of the respiratory system of *Staphylococcus aureus. Ann. N.Y. Acad. Sci.* **236:**115–123.
20. **Hale, J. H.** 1947. Studies on staphylococcal mutation: characteristics of the "G" (gonidial) variant and factors concerned in its production. *Br. J. Exp. Pathol.* **28:**202–210.
21. **Jensen, P. R., and O. Michelsen.** 1992. Carbon and energy metabolism of *atp* mutants of *Escherichia coli. J. Bacteriol.* **174:**7635–7641.
22. **Joyce, G. H., and D. C. White.** 1971. Effect of benso(a)pyrene and piperonyl butoxide on formation of respiratory system, phospholipids, and carotenoids of *Staphylococcus aureus. J. Bacteriol.* **106:**403–411.
23. **Kafala, B., and A. Sasarman.** 1994. Cloning and sequence analysis of the *hemB* gene of *Staphylococcus aureus. Can. J. Microbiol.* **40:**651–657.
24. **Kahl, B., R. A. Proctor, A. Schulze-Everding, M. Herrmann, H. G. Koch, I. Harms, and G. Peters.** 1998. Persistent infection with small colony variant strains of *Staphylococcus aureus* in patients with cystic fibrosis. *J. Infect. Dis.* **177:**1023–1029.
25. **Kahl, B., C. von Eiff, M. Herrmann, G. Peters, and R. A. Proctor.** 1996. Staphylococcal small colony variants present a challenge to clinicians and clinical microbiologists. *Antimicrob. Infect. Dis. Newsl.* **15:**59–63.
26. **Kaplan, M. W., and W. E. Dye.** 1976. Growth requirements of some small-colony-forming variants of *Staphylococcus aureus. J. Clin. Microbiol.* **4:**343–348.
27. **Koo, S. P., A. S. Bayer, H.-G. Sahl, R. A. Proctor, and M. R. Yeaman.** 1996. Staphylocidal action of thrombininduced platelet microbicidal protein (tPMP) is not solely dependent on transmembrane potential ($\Delta\Psi$). *Infect. Immun.* **64:**1070–1074.
28. **Lewis, L. A., K. Li, M. Bharosay, M. Cannella, V. Jorgenson, R. Thomas, D. Pena, M. Velez, B. Pereira, and A. Sassine.** 1990. Characterization of gentamicin-resistant

respiratory-deficient (Res⁻) variant strains of *Staphylococcus aureus*. *Microbiol. Immunol.* **34:**587–605.

29. **Musher, D. M., R. E. Baughn, G. B. Templeton, and J. N. Minuth.** 1977. Emergence of variant forms of *Staphylococcus aureus* after exposure to gentamicin and infectivity of the variants in experimental animals. *J. Infect. Dis.* **136:**360–369.

30. **Neumann, S., U. Matthey, G. Kaim, and P. Dimroth.** 1998. Purification and properties of the F_0F_1 ATPase of *Ilyobacter tartaricus*, a sodium ion pump. *J. Bacteriol.* **180:**3312–3316.

31. **Novick, R.** 1993. Staphylococcus, p. 17–33. *In* A. L. Sonenshein, J. A. Hoch, and R. Losick (ed.). *Bacillus subtilis and Other Gram-Positive Bacteria.* American Society for Microbiology, Washington, D.C.

32. **Nydahl, B. C., and W. L. Hall.** 1965. The treatment of staphylococcal infection with nafcillin with a discussion of staphylococcal nephritis. *Ann. Intern. Med.* **63:**27–43.

33. **Proctor, R. A.** 1994. Microbial pathogenic factors: small colony variants, p. 77–90. *In* A. L. Bisno and F. A. Waldvogel (ed.), *Infections Associated with Indwelling Medical Devices,* 2nd ed. American Society for Microbiology, Washington, D.C.

34. **Proctor, R. A.** 1998. Bacterial energetics and antimicrobial resistance. *Drug Resist. Updates* **1:**227–235.

35. **Proctor, R. A., J. M. Balwit, and O. Vesga.** 1994. Variant subpopulations of *Staphylococcus aureus* can cause persistent and recurrent infections. *Infect. Agents Dis.* **3:**302–312.

36. **Proctor, R. A., and G. Peters.** 1998. Small colony variants in staphylococcal infections: diagnostic and therapeutic implications. *Clin. Infect. Dis.* **27:**419–423.

37. **Proctor, R. A., P. van Langevelde, M. Kristjansson, J. N. Maslow, and R. D. Arbeit.** 1995. Persistent and relapsing infections associated with small colony variants of *Staphylococcus aureus. Clin. Infect. Dis.* **20:**95–102.

38. **Proctor, R. A., O. Vesga, M. F. Otten, S.-P. Koo, M. R. Yeaman, H.-G. Sahl, and A. S. Bayer.** 1996. *Staphylococcus aureus* small colony variants cause persistent and resistant infections. *Chemotherapy* (Basel) **42**(Suppl. 2):47–52.

39. **Ray, S. C., R. Schulick, D. Flayhart, and J. Dick.** 1997. First report of infection with *Staphylococcus aureus* subspecies *anaerobius*, abstr. 567. 35th Infect. Dis. Soc. Am. Meeting.

40. **Regassa, L. B., R. P. Novick, and M. J. Betley.** 1992. Glucose and nonmaintained pH decrease expression of the accessory gene regulator (*agr*) in *Staphylococcus aureus. Infect. Immun.* **60:**3381–3388.

41. **Reizere, J., M. H. Saier, Jr., J. Deutscher, F. Grenier, J. Thompson, and W. Hengstenberg.** 1988. The phosphoenolpyruvate:sugar phosphotransferase system in gram-positive bacteria: properties, mechanism, and regulation. *Crit. Rev. Microbiol.* **15:**297–338.

42. **Schlievert, P. M., J. R. Dringer, M. H. Kim, S. J. Projan, and R. P. Novick.** 1992. Effect of glycerol monolaurate on bacterial growth and toxin production. *Antimicrob. Agents Chemother.* **36:**626–631.

43. **Schnitzer, R. J., L. J. Camagni, and M. Buck.** 1943. Resistance of small colony variants (G forms) of a staphylococcus toward the bacteriostatic activity of penicillin. *Proc. Soc. Exp. Biol. Med.* **53:**75–89.

44. **Taber, H. W.** 1993. Respiratory chains, p. 199–212. *In* A. L. Sonenshein, J. A. Hoch, and R. Losick (ed.), *Bacillus subtilis and Other Gram-Positive Bacteria.* American Society for Microbiology, Washington, D.C.

45. **Tien, W., and D. C. White.** 1968. Linear sequential arrangement of genes for the biosynthetic pathway of heme in *Staphylococcus aureus. Proc. Natl. Acad. Sci. USA* **61:**1392–1398.

46. **Vesga, O., J. M. Vann, D. Brar, and R. A. Proctor.** 1996. *Staphylococcus aureus* small colony variants are induced by the endothelial cell intracellular milieu. *J. Infect. Dis.* **173:**739–742.

47. **von Eiff, C., D. Bettin, R. A. Proctor, B. Rolauffs, N. Lindner, W. Winkelmann, and G. Peters.** 1997. Recovery of small colony variants of *Staphylococcus aureus* following gentamicin bead placement for osteomyelitis. *Clin. Infect. Dis.* **25:**1250–1251.

48. **von Eiff, C., C. Heilmann, R. A. Proctor, C. Woltz, G. Peters, and F. Götz.** 1997. A site-directed *Staphylococcus aureus hemB* mutant is a small-colony variant which persists intracellularly. *J. Bacteriol.* **179:**4706–4712.

49. **Wise, R. I., and W. W. Spink.** 1954. The influence of antibiotics on the origin of small colonies (G variants) of *Micrococcus pyogenes* var. *aureus. J. Clin. Invest.* **33:**1611–1622.

50. **Xu, K., J. Delling, and T. Elliott.** 1992. The genes required for heme synthesis in *Salmonella typhimurium* include those encoding alternative functions for aerobic and anaerobic coproporphyrinogen oxidation. *J. Bacteriol.* **174:**3953–3963.

51. **Youmans, G. P., and E. Delves.** 1942. The effect of inorganic salts on the production of small colony variants by *Staphylococcus aureus. J. Bacteriol.* **44:**127–136.

The Staphylococcal Cell Wall

ALEXANDER TOMASZ

36

INTRODUCTION

Historical Overview

Ever since the recognition in the late 1940s to early 1950s of the cell wall as a unique anatomical component of all eubacterial cells, *Staphylococcus aureus* has often served as the gram-positive model in wall-related studies. One of the first demonstrations that bacterial cell walls can be isolated as physical entities with the size and shape of the whole bacterium was with *S. aureus*. It was in *S. aureus* that the UDP-linked amino sugar-containing wall precursor peptides were discovered, providing the first insights into the unique pathway of cell wall biosynthesis. The history of interest in the staphylococcal cell wall also reflects the history of success and failure of the antibiotic era vis-à-vis staphylococci as human pathogens. The clue that eventually led to the discovery of penicillin (and later to the autolytic enzymes) was provided by the lysis of staphylococcal colonies in the vicinity of a mold contaminant on an agar plate in Fleming's laboratory. Elucidation of the mode of action of several important antibiotics in the 1960s and 1970s has been intimately linked to studies on the biosynthesis of staphylococcal cell walls since the beginning of the antibiotic era. This includes the mode of action of penicillin and other β-lactam antibiotics as specific inhibitors (acylating agents) directed against the active site of transpeptidases—key enzymes in the assembly of bacterial cell walls. It was mainly from such studies in *S. aureus*, in parallel with studies done in *Escherichia coli*, that by the early 1980s a coherent picture emerged concerning steps in the biosynthetic pathway that lead to the formation of the lipid-linked disaccharide pentapeptide. With some structural variations, this pentapeptide was shown to be the universal building block of cell wall peptidoglycan in both gram-positive and gram-negative bacteria. For summaries of various aspects of studies on the cell walls of staphylococci up to the late 1980s, see references 18 and 21a.

In our time, the resurgence of interest in various aspects of the staphylococcal cell wall again seems in good part related to recent events of the antibiotic era, namely, the appearance and extensive spread of methicillin-resistant *S. aureus* (MRSA) and the shift during the last decade in the etiology of human infectious diseases from gram-negative toward gram-positive species, particularly staphylococci. Thus, many recent studies on the staphylococcal cell wall are linked to studies on the mechanism of antibiotic resistance (primarily resistance to methicillin and, most recently, to vancomycin) and to some aspects of staphylococcal virulence.

The Changing Image of Cell Walls

During the earlier eras, biochemists and microbiologists working with cell walls viewed them as more or less inert exoskeletons essential to withstand the turgor pressure of cytoplasm. However, these views are changing rapidly with the recognition of the complexity of chemical structure and biosynthetic pathways, the large number of genetic determinants, and the intimate involvement of cell walls with host-related functions of staphylococci. Indeed, it is perhaps the image of staphylococcal cell walls that has undergone the most substantial change in our era.

The rapid accumulation of new information is making it clear that, eventually, all the morphological, biochemical, and genetic information about staphylococcal cell walls will have to be integrated into a coherent cell biological model. This model would explain how the flow of precursor molecules and their polymerization into the preexisting cell wall leads to the formation of a multilayered supermolecular cell wall "sacculus" that has the same size and shape as the particular bacterium; how the nascent, innermost layer of this envelope "matures" while moving outward toward the cell surface; how and why and through what signals and by what catalysts the outermost layers are shed into the medium during wall turnover; and how the spatial and temporal accuracy of septation is controlled to allow the formation of two bacterial cells with the same genetic and biochemical composition. This, of course, is true not only for staphylococci but for all gram-positive and gram-negative bacteria as well. Clearly, in a very real sense, the ultimate integration of all metabolic activities of a bacterial cell must occur on the level of cell wall.

Gram-Positive Pathogens, ed. by V. A. Fischetti et al.
© 2000 American Society for Microbiology, Washington, D.C.

In addition to this revival of interest in microbial cell biology, two approaches that have been making the greatest impact on discoveries in this field are the introduction of high-resolution analytical techniques (high-pressure liquid chromatography [HPLC] and mass spectrometry) and the increasing application of molecular genetic approaches made possible primarily by the recognition that antibiotic resistance provides selectable phenotypes to identify genes of complex functions in cell wall metabolism (12). A coherent overview and comprehensive coverage of data in a period of such rapid expansion in a field would be difficult. Therefore, the purpose of this chapter would be better served by pointing out some of the most recent findings and emerging trends.

After a brief review of the anatomy of staphylococcal cell walls, the new information is covered under four headings: High-Resolution Analysis of the Staphylococcal Peptidoglycan; "Methicillin-Conditional" Mutants in Cell Wall Metabolism; Penicillin-Binding Proteins; and Complex Functions of Cell Walls.

Anatomy of *S. aureus* Cell Wall

Most clinical isolates of *S. aureus* express, on their outermost surface, one of the 11 chemically different capsular polysaccharides that have been identified so far (34). The chemical subunit structures of these important and often antiphagocytic carbohydrate polymers have been elucidated, and rapid advances have been made in the identification of genetic determinants and organization of capsular loci (42) (for more detailed coverage, see chapter 37, this volume). Underneath these somewhat variant surface layers is the staphylococcal cell wall. In electron-microscopic thin sections followed by heavy metal staining, the cell wall appears as a triple layer of two electron-dense lines. Unlike in streptococci, cell divisions in *S. aureus* occur in three consecutive division planes, each at right angles to one another, and proper orientation of cell wall septa must involve a complex and superbly controlled mechanism. Cyclic morphological alterations of cell walls during growth, division, and separation of daughter cells, under conditions of normal growth, and also during exposure to a variety of antimicrobial agents have been documented in the elegant electron microscopy studies of Giesbrecht and colleagues (19). Basic compositional/structural features of the *S. aureus* cell wall were established in earlier studies. It is composed of a highly cross-linked A3α-type peptidoglycan with pentaglycine oligopeptide units connecting the ϵ-amino group of the lysine component of one muropeptide to the penultimate D-alanine of another. This peptidoglycan, together with ribitol-type teichoic acid chains (attached to the 6-hydroxyl groups of some of the N-acetylmuramic acid residues of the glycan chain), surrounds the *S. aureus* cell in the form of a multilayered envelope (Fig. 1).

HIGH-RESOLUTION ANALYSIS OF THE STAPHYLOCOCCAL PEPTIDOGLYCAN

Progress in the high-resolution chemistry of the staphylococcal cell wall came from the introduction of HPLC and mass spectrometric methods (56) for the analysis of the peptidoglycan. First, studies with gel-permeation HPLC established the presence of muropeptide oligomers of lengths extending to nine units and beyond (45). This was followed by the adaptation of the reverse-phase HPLC system

(originally developed for the analysis of *E. coli* cell walls) in combination with mass spectrometry for the analysis of strains of *S. aureus* (7). The primary motivation in these studies was to better understand the mechanism of methicillin resistance in the highly and homogeneously resistant MRSA strain COL and its large number of transposon mutants, but the results obtained are valid for both resistant and susceptible strains of *S. aureus*. Together with studies on the cell walls of *Streptococcus pneumoniae* (see chapter 19, this volume), analysis of the peptidoglycan of the methicillin-resistant *S. aureus* provided the first high-resolution view of the complexity of cell wall structures in gram-positive bacteria.

Enzymatic hydrolysis (with the M1 muramidase) of the staphylococcal peptidoglycan (isolated from purified cell walls by removal of the teichoic acid chains with hydrofluoric acid) was followed by borohydride reduction and separation of the muropeptide components on a reverse-phase HPLC column. This method resolved the hydrolysate to more than 21 distinct UV-absorbing peaks plus a "hump" of unresolved material of longer retention times that made up more than 50% of the muropeptides.

Analysis of muropeptide components obtained after digestion with muramidase, or with the combination of muramidase plus lysostaphin (a bacteriolytic endopeptidase that attacks the pentaglycine bridge), for amino acid and amino sugar composition and for molecular mass (fast atom bombardment mass spectrometry), revealed the main structural principle staphylococci use in building the peptidoglycan. The major monomeric building block is a disaccharide pentapeptide carrying D-isoglutamine in position 2, an intact D-alanyl-D-alanine carboxy terminus, and a pentaglycine substituent attached to the ϵ-amino group of the lysine residue. These monomeric units make up about 6% of all muropeptides. Another 20% consist of dimers in which the pentaglycine substituent of one muropeptide unit is cross-linked to the penultimate D-alanine of a neighboring one. About 40% consist of higher oligomers containing three to nine muropeptide units generated by the same cross-linking principle, and still higher oligomers account for an additional 15 to 25% of the muropeptide units.

The retention time of muropeptides on the reverse-phase column was shown to be related to the logarithm of the number of disaccharide-peptide-pentaglycyl units per muropeptide species. When the whole peptidoglycan was hydrolyzed by a combination of muramidase digestion (which breaks bonds in the glycan chain) and lysostaphin treatment (which breaks peptide cross-links), the number of muropeptide components separable on the HPLC column was reduced to essentially three simple monomers, carrying fragments of the oligoglycine cross bridge. Thus, the "hump" of unresolved components appears to represent higher oligomers of the same basic muropeptide unit.

During polymerization of muropeptide precursors to the mature peptidoglycan, the carboxy-terminal D-alanine of the "donor" muropeptides is lost in each transpeptidation step, and, therefore, each cross-linked muropeptide contains only one intact D-alanyl-D-alanine residue (that belonging to the original "acceptor" molecule in the cross-link), independent of the length of the oligomer. This fact appears to be of importance in some glycopeptide-resistant laboratory mutants of *S. aureus* (see later).

In addition to the major monomeric unit and its cross-linked multimeric derivatives, the staphylococcal peptidoglycan also contains structural variants in which

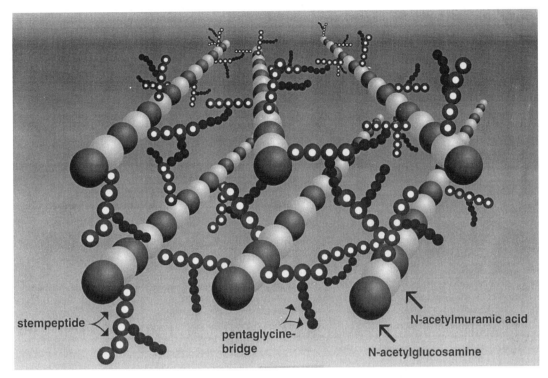

FIGURE 1 Three-dimensional structure of staphylococcal peptidoglycan. The straight lines of large bowls represent the sugar moieties of the peptidoglycan. Each globe in these lines symbolizes an amino sugar, N-acetylglucosamine (black globe), or N-acetylmuramic acid (white globe). Stempeptides, branching from N-acetylmuramic acid, are characterized by small dark globes with a white center. The connecting interpeptide bridges (pentaglycines) between the stempeptides are shown as small black globes. Schematic drawing by Peter Giesbrecht, Thomas Kersten, Heiner Maidhof, and Jörg Wecke, Robert Koch-Institute, Berlin, Germany; reproduced with permission.

the pentapeptide monomer carries only one glycine unit or none on the ϵ-amino group of the lysine residue. Monomers in which a single alanine or one alanine plus four glycines replace the pentaglycine bridge can also be detected, and all these monomers occur in various combinations among dimeric muropeptides as well.

Stability of Muropeptide Composition

Analysis of peptidoglycan prepared from a large number of S. aureus isolates of different clonal types (as defined by multilocus enzyme analysis), susceptible or resistant to methicillin (either homogeneous or heteroresistant), contemporary as well as preantibiotic era S. aureus isolates, all showed virtually identical HPLC muropeptide patterns, provided that the cells were grown in the same medium and in a "balanced" state of growth and harvested in the late exponential phase of growth (37a). Selective Tn551 inactivation of the mecA gene did not cause any detectable change in muropeptide composition (8). These data suggest that, as in findings with other bacteria, the muropeptide composition of S. aureus cell walls is specific for the species.

Variations in Peptidoglycan Composition

Effect of Antibiotics

Growth of the highly methicillin-resistant S. aureus strain COL in a wide range of subinhibitory concentrations of methicillin caused striking changes in muropeptide composition: the representation of trimeric plus higher oligo-

meric components was drastically reduced (from a combined representation of close to 50% to less than 10%) while the proportion of monomeric and dimeric components increased (from about 15% up to about 50%). This major compositional change was virtually complete at methicillin concentrations of about 5 μg/ml (i.e., at antibiotic concentrations that would fully acylate the four normal penicillin-binding proteins [PBPs] in membrane preparations) and did not change further even in the presence of 750 μg/ml antibiotic in the medium (the MIC for methicillin in this strain was 1,600 μg/ml). It was suggested that the abruptness of this change in peptidoglycan composition may represent the switching over from the normal wall biosynthetic system (the normal PBPs) to another one (PBP2A, encoded by the methicillin-resistance gene mecA) capable of functioning in the presence of high concentrations of methicillin (8). The anomalous composition of the peptidoglycan produced under these conditions would then reflect the limited capacity of PBP2A for cross-linking more than single monomeric muropeptides. Nevertheless, similar albeit much less abrupt compositional changes were also observed when methicillin-susceptible staphylococci or an isogenic derivative of strain COL (with inactivated mecA) were grown in sub-MIC concentrations of methicillin.

Effect of Growth Phase and Media Composition

Secondary modification of the staphylococcal peptidoglycan (O acetylation of muramic acid residues and changes in the N acetylation of the amino sugars) has been described earlier, as a function of the culture's growth phase or during treatment with chloramphenicol. A recent study followed up earlier observations on the effect of exogenous glycine on the structure of staphylococcal peptidoglycan. Growth of the highly methicillin-resistant MRSA strain COL in the presence of high concentrations of glycine (between 0.06 and 0.25 M) resulted in a dramatic distortion of the HPLC profile of muropeptides: there was extensive reduction in cross-linkage, and major monomeric, dimeric, trimeric, and some higher oligomeric muropeptide components were gradually replaced (parallel with the increasing glycine concentration) by novel components that were similar to the normal muropeptides except that their carboxy terminal D-alanine residues were replaced by glycine. Such glycine-terminating muropeptides appear to be poor substrates for transpeptidases since oligomers higher than pentameric compounds were rare in the peptidoglycan of cells grown in high concentrations of glycine (9).

Effect of Cefotaxime Resistance

Laboratory mutants selected for increased cefotaxime resistance contain an unusual muropeptide dimer in their peptidoglycan. The molar proportion of this dimer increases in parallel with the increasing resistance level. Analytical work with HPLC, MALDI (matrix assisted laser desorption) mass spectrometry, and sequencing showed that the composition of this unusual compound was similar to that of the major muropeptide dimer except for one thing: in the new compound, the two monomeric components appear to be cross-linked not by one but by two oligoglycine units generating a macrocyclic (14-member) peptide ring structure that included the pentaglycine cross bridges of both muropeptide monomers. These doubly cross-linked muropeptides may be considered products of an abnormal, two-step transpeptidase reaction in which each monomeric component has served both as "acceptor" and "donor." These doubly cross-linked muropeptides appear to be the signature of an abnormally acting transpeptidase in the mutants (3).

Effect of Vancomycin and Teicoplanin Resistance

Mutants with high MICs for these two glycopeptide antibiotics were isolated from the highly methicillin-resistant S. aureus strain COL. The peptidoglycan of the mutants showed a grossly distorted muropeptide composition: there was a large increase in the molar proportion of the major monomeric muropeptide carrying the intact D-alanyl-D-alanine carboxy termini and a major reduction in the representation of dimeric and oligomeric components. It was suggested that some of the unusual properties of these mutants and the mechanism of resistance are related to the enrichment of the cell walls in muropeptide monomers. Mutants growing in the presence of vancomycin or teicoplanin were able to remove the antibiotics from the medium during growth, and the antibacterial agents could subsequently be recovered in biologically active form from the cell wall fraction of the bacteria. Muropeptides terminating in the D-alanyl-D-alanine residues are known to form the binding sites for glycopeptide antibiotics. The mechanism of resistance may then be related to trapping the glycopeptides in the mature layer of the peptidoglycan enriched for muropeptide monomers, thus preventing the antibiotic molecules from reaching sites of wall biosynthesis at the plasma membrane (43, 44).

"METHICILLIN-CONDITIONAL" MUTANTS IN CELL WALL METABOLISM

Transposon mutants of methicillin-resistant S. aureus (later named fem mutants or auxiliary mutants) were originally isolated to clarify some complex genetic features of staphylococcal methicillin resistance. The first Tn551 mutant (later named femA) was isolated by Berger-Bächi in 1983 (2), even before the basic genetic mechanism of methicillin resistance was recognized. Subsequently, additional transposon mutants with reduced methicillin resistance were isolated by Kornblum et al. (29) and Murakami and Tomasz (36). In several of these, the inserts were in genes other than femA, increasing the number of fem mutants from one (femA) to four (femA, femB, femC, femD). Unexpectedly, practically all staphylococcal transposon mutants selected for reduced resistance have retained an intact mecA (methicillin-resistance gene), and expression of mecA leading to its protein product PBP2A also remained unaffected despite the massive reduction in the methicillin MIC of the mutants. Biochemical analysis of the first fem mutants indicated that the inserts were in genes involved in staphylococcal cell wall synthesis (Table 1).

The identification of cell wall defects and reduced methicillin MIC in the first few fem mutants suggested that methicillin resistance of MRSA may provide a relatively easily selectable phenotype for the identification of genes in cell wall synthesis and metabolism (12). As a working model, it was proposed that the methicillin MIC in such "methicillin-conditional" mutants may primarily reflect the success of the mecA gene product (PBP2A). PBP2A is positioned close to the end of a long and complex metabolic pathway in catalyzing peptidoglycan incorporation in the presence of antibiotics (β-lactams) that can competitively inhibit this function. In this model, quantitative fluctuation of the methicillin MIC was supposed to serve as a sensitive gauge to register errors, perturbations, or interference with any step in wall synthesis that precedes (or in any way assists) the terminal wall synthetic reaction catalyzed by PBP2A. Such errors may include the production of structurally incorrect cell wall precursors that may not compete successfully with the antibiotic for the active site of PBP2A (11); or such incorrectly structured muropeptides may not be able to perform some as yet unidentified effector function in wall synthesis (analogous perhaps to regulatory functions of cell wall metabolites in the control of β-lactamase expression; see reference 25). Abnormally structured stem peptides that are incorporated into the cell wall may create localized structural defects or may serve as signals for murein hydrolases. It was assumed that any of these defects may cause a decrease in the MIC for methicillin.

For these reasons, and to maximize sensitivity of this "MIC gauge," we chose the highly and homogeneously resistant MRSA strain COL as the common background of a transposon library that was systematically screened for all insertional mutants in which the methicillin MIC was reduced by at least a factor of 10 (12). In earlier Tn551 transposition experiments, the level of selection was either ill defined or arbitrarily set for maximal change.

The complexity of cell wall structure and biosynthesis suggests that the number of genes involved will be very large. Our model assumes that many of these genes may be identified through the methicillin-conditional mutant method. Observations that have appeared in the literature have so far borne out these expectations. Tn551 MRSA

TABLE 1 Genes involved in staphylococcal methicillin resistance

Gene	Relevant function	Site of Tn551 (or Tn918) insertion	Reduction in methicillin resistance from MIC = 1600 μg/ml	Reference
glmM	Synthesis of GlucNac-1-PO$_4$	In open reading frame	1.5	27, 54
murE	Addition of lysine to muropeptide	3 nulceotides from end carboxy terminus	25.0	31
femA and femB	Synthesis of pentaglycine cross-links	At carboxy terminus	1.5	30, 36
femC	Regulation of glutamine synthetase	In promoter	3–6	20
Sigma B	Stress response	In open reading frame	25.0	54
PBP2	Major wall synthetic enzyme	150-nucleotide deletion at carboxy terminus	12.0	40
fmt	PBP-like protein	At carboxy terminus	64.0	28
llm	Lipid-linked wall precursor?	33 amino acid residues to C terminus	12.5	32

mutants selected for reduced MIC have led to insights into hitherto unknown steps in staphylococcal wall precursor biosynthesis (femA/B, gluM, murE, llm), have identified a unique pbp-like gene (fmt), and have led to the implication of cooperative functioning of the "normal" PBP2 and PBP2A; several of the Tn551 mutants also caused extensive reduction of vancomycin and/or teicoplanin resistance. Next, some relevant properties of these mutants are summarized.

femA/B

femA and femB were shown to be two closely related but distinct genes that are part of an operon (46). femA and femB mutants produced cell walls in which the normal length of oligoglycine cross bridges was reduced. The composition of mutant walls suggested that the synthesis of the pentaglycine bridges occurred in several steps: the block in femA mutants would be just past the addition of the first glycyl unit, while in femB mutants, the enrichment of the cell wall in muropeptides carrying a triglycyl substituent pointed to a block past the addition of the third glycine unit (10). In the original femA and femB mutants, the transposon insert was located in the promoter region (femA) or close to the end of the gene (femB), thus allowing partial expression in both cases. In the early femA mutants the large increase in monoglycyl peptides was also accompanied by an increase in muropeptides in which one of the glycine residues was replaced by serine or alanine. Interestingly, the gene epr from Staphylococcus capitis, with homology to femA and femB, caused the addition of serine residues to the pentapeptide bridges when introduced into wild-type S. aureus on a multicopy plasmid (48). Recent work in which the entire femA/B operon was eliminated by allelic replacement confirmed and clarified earlier findings: the femA/B null mutant produced peptidoglycan in which most muropeptide subunits carried only monoglycyl substituents; complementation of the null mutant with either femA or femA/B caused modification in the peptido-

glycan that was consistent with the involvement of femA with the addition of the second and third glycine residues to the cross bridge and femB with the addition of the fourth and fifth. There was no change in the proportion of the linear pentapeptide carrying no glycine residues in any of these mutants, implying that addition of the first glycine to the muropeptide unit is under the control of another gene (femX) as yet to be identified. The femA/B null mutants were fully viable but showed morphological abnormalities, decreased growth rate, and radically reduced methicillin resistance (46).

The nature of the catalytic activities of FemA and FemB proteins (each about 49 kDa) is not known. The glycine substituents composing the interpeptide cross bridges are thought to originate in a form linked to tRNA molecules. However, femA and femB show no homology to the staphylococcal tRNA synthetase genes. Staphylococci contain three distinct glycyl tRNA species, each different from the glycyl tRNA participating in normal protein synthesis. It has been suggested that the three unique tRNA species correspond to the three steps in the formation of the pentaglycine chain. Step one involves the attachment of the first glycine to the ϵ-amino group of the muropeptide lysine residue, and steps 3 and 4 each involve the extension of the bridge by two additional glycine units. The hypothetical femX and fem genes A and B seem to be involved in this sequence at the first, second, and fourth stage, respectively. Attachment of the oligoglycine bridges to the nascent muropeptides most likely occurs in the lipid phase of biosynthesis, since cell wall precursors with the corresponding structures were not detected in the wall precursor fraction of staphylococci extractable by water solvents.

femC

Different types of alterations in muropeptide patterns were observed in additional Tn551 mutants, again selected for reduced methicillin resistance. In one of these mutants (femC), the HPLC pattern showed a decrease in the proportion of cross-linked peptides, an increase in the major monomeric component, and the appearance of clusters of multiple muropeptide bands, each eluting with slightly altered retention times but close to the position of major muropeptide peaks. Mass spectrometric analysis showed that the anomalous multiple peaks differed from the nearest "normal" muropeptides by integer multiples of the mass unit 1, which is the difference between the molecular mass of an NH_2 and an OH group. The anomalous muropeptides were shown to contain glutamic acid units with free α-carboxyl group rather than the normal isoglutamine residues [37]. Cloning and sequencing of *femC* showed that the insert was in *glnR*, a gene regulating the transcription (and/or activity) of glutamine synthetase *glnA*, causing a glutamine deficiency in the bacteria that apparently led to the formation of nonamidated muropeptides [20].

The properties of the *femC* mutant indicate that the amino donor in the amidation of the α-carboxyl group of the D-glutamic acid residues is glutamine. The enzyme catalyzing the addition of D-glutamic acid to the muropeptide precursor has been purified from *S. aureus*, but the nature of the enzyme catalyzing the amidation reaction is not known, nor is it clear at what stage of muropeptide synthesis this reaction occurs. Amidation may occur on several levels since muramyl dipeptides containing either D-isoglutamine or free D-glutamic acid were detected both in the peptidoglycan and in the water-extractable precursor fraction of another auxiliary mutant, defective in *murE* (see later).

femD (glmM)

The only abnormalities noticed in the HPLC profile of muropeptides in the Tn*551* mutant RUSA315 (originally also named *femD* and located in the *Sma*I-I fragment of the chromosome) were the deficits in muropeptide 1 (a linear pentapeptide with free [unsubstituted] ϵ-amino group on the lysine residue) and muropeptide 8 (muropeptide with a single alanine substituent on the lysine residue). Cloning and sequencing followed by purification of the protein identified *femD* as the staphylococcal *glmM* encoding a phosphoglucosamine mutase, an enzyme essential for the conversion of glucosamine 6-phosphate to glucosamine 1-phosphate. Glucosamine 1-phosphate is an obligatory substrate in the enzymatic reactions that lead to the production of UDP-*N*-acetylglucosamine, a key cell wall metabolite. The Tn*551* insert in mutant RUSA315 was in the structural gene of *glmM*, yet the mutant was able to grow (although with reduced growth rate), suggesting that in staphylococci, in contrast to *E. coli*, an alternative pathway for the synthesis of acetylglucosamine-1-PO_4 may exist. The staphylococcal *glmM* gene introduced into an *E. coli* mutant with inactivated *glmM* could fully express and replace the missing function in the mutant [27]. The staphylococcal *glmM* shows a considerable degree of homology to two genes in the urease operon: *ureD* of *Mycobacterium leprosi* and *ureC* in *Helicobacter* sp. [54]. The significance of this similarity is not clear at the present time. Sequencing upstream of the *gluM* in *S. aureus* has identified an operon that includes the structural gene for arginase [55a].

The Staphylococcal murE Gene

Inactivation of a gene located on the *Sma*I fragment B caused extensive reduction in methicillin resistance. HPLC profiles of the peptidoglycan showed the appearance of unusual muropeptides eluting with short retention times from the column. These were identified as disaccharide dipep-tides, terminating either in free D-glutamic acid or in D-isoglutamine. The same peptides linked to UDP-*N*-acetylmuramic acid were also detected in the wall precursor fraction of the mutants. Sequencing has identified the gene as the *murE* of *S. aureus*. The gene encoded a deduced two-domain protein structure similar to the one seen in the *murE* ligase of *E. coli*: the C-terminal domain showed properties of substrate binding, and the N-terminal domain showed properties common to ATP-binding portions of all amino acid ligases [31]. The Tn*551* was inserted three nucleotides from the end of the gene, which must be involved with the addition of lysine residues to the growing muropeptide chain.

The *llm* Gene

The *llm* gene was identified in a new Tn*551* mutant with increased autolysis rates associated with decreased methicillin resistance. *llm* encodes a lipophilic protein of 351 amino acid residues with sequence similarities to the *mraY* genes, and it may represent the first staphylococcal gene encoding the lipid-linked muropeptide precursor [32].

The *fmt* Gene

The *fmt* gene was identified in a Tn*551* mutant of strain COL-located on the *Sma*I fragment B of the chromosome. The putative 397-amino-acid-residue protein showed a limited sequence similarity to the carboxypeptidase R61, a hydropathy pattern similar to that seen in PBPs 1, 2A, and 4 of *S. aureus*, and two of the three conserved motifs (SXXK and S[Y]XN) characteristic of the penicillin-binding domains of PBPs [28].

The Staphylococcal *pbp2* as an Auxiliary Gene

A recent finding of considerable conceptual importance, both for the mechanism of methicillin resistance and for the functioning of PBPs in cell wall assembly, was the finding that one of the mutants—RUSA130—in the large Tn*551* library selected for reduced methicillin resistance had the insert in *pbp2* [40]. In current models, catalysis of cell wall assembly in MRSA growing in the presence of antibiotics is by the low-affinity PBP2A, since the complement of normal staphylococcal PBPs would be inactivated by the antibiotic. Yet, genetic backcrosses and successful complementation of the Tn*551* mutant with the intact *pbp2* made it unambiguously clear that the defective gene in RUSA130 was the genetic determinant of the normal staphylococcal PBP2, implying that some kind of functioning of PBP2 remains essential for cell wall synthesis in an MRSA strain even in the presence of high concentrations of methicillin. It is conceivable that the function involved is a penicillin-insensitive transglycosylase activity that may be present in this bimodular PBP. This finding may be the first evidence to suggest that in staphylococci, synthesis of the cell wall may be catalyzed by a multienzyme complex similar to the one postulated for *E. coli* [24].

PENICILLIN-BINDING PROTEINS

Tentative roles for the four staphylococcal PBPs were originally assigned on the basis of morphological/biochemical effects of β-lactams that showed more or less selective binding to individual PBPs [17]. Recent genetic work using allelic replacement has clearly established the essential nature of PBP1 [53]. The structural gene of PBP2 was cloned and sequenced (21, 35), and critically located point mutations in it were associated with methicillin resistance in both laboratory mutants and some clinical isolates that did not carry the *mecA* gene but showed low-level β-lactam resistance (51). A transposon insertional mutant in the *pbp2* gene was shown to cause major reduction in methicillin resistance in the *mecA*-carrying MRSA strain COL (40). Involvement of PBP2 with vancomycin resistance was suggested by the apparent overproduction of the protein in a staphylococcal mutant (6), and the reduced β-lactam affinity of PBP2 was noted in cefotaxime-resistant mutants (3). The structural gene of PBP4 was cloned and sequenced (15, 22), and an extensive deletion in the promoter region was identified in β-lactam-resistant laboratory isolates that showed an increase in peptidoglycan cross-linking (23). These findings confirm earlier proposals that PBP4 can act as a secondary transpeptidase. The possibility that expression of the PBP4 determinant is controlled via the upstream ABC transporter (which may be part of an operon) is being considered (14).

The first division and cell wall (DCW) cluster of cell division and cell wall-related genes (41) was described for *S. aureus*. As in the *E. coli* DCW cluster, it contains a PBP gene (*pbpA* for PBP1) at one end, followed by some determinants of muropeptide biosynthesis (*mreY* and *murD*) and then by cell division-related genes (*div1B*, *ftsA*, *ftsZ*). However, in contrast to *E. coli*, the genes are not overlapping, and most of the muropeptide synthesis genes are absent. One of these (*murE*) was recently identified at a distance from the DCW (31).

COMPLEX FUNCTIONS OF CELL WALLS

The capacity of staphylococcal cell walls, peptidoglycan, and teichoic acid to induce the production of proinflammatory cytokines has been repeatedly documented (50). Two global regulatory genes—*agr* and *sar*—that control the production and release of several staphylococcal virulence-related factors were shown to influence staphylococcal autolysis (induced by penicillin or Triton X-100) in an opposing fashion (16). The carboxy-terminal end of staphylococcal surface proteins (such as protein A) was shown to be covalently bound to the oligoglycine cross bridge of

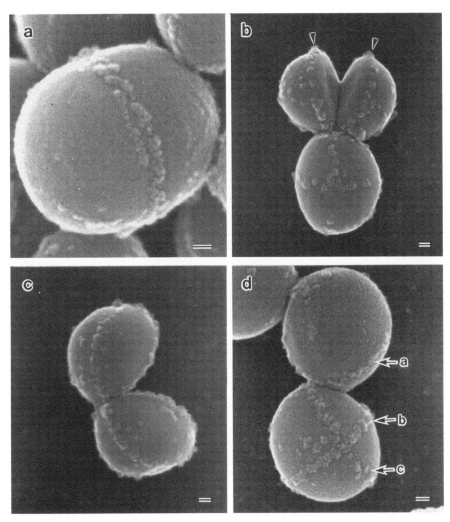

FIGURE 2 Localization of *atl* gene products on the cell surface of *S. aureus* during the division cycle as determined by scanning electron microscopy. Panels a to d show the immunogold-labeling patterns on cells at various stages of the cell cycle. Bar, 100 nm. Reproduced from Yamada et al. (57).

peptidoglycan muropeptide monomers, through a threonine residue in the protein (generated after cleaving of the wall sorting signal LPXTG), which is then linked to the amino-terminal glycine residue of the muropeptide (52). The structure of these cross bridges is under the control of the *femA/B* genes. Therefore, the already pleomorphic *femA/B* mutant (showing defective wall structure, reduced methicillin and lysostaphin resistance, and slow wall turnover) may also be affected in virulence.

The active involvement of the cell wall in complex functions is also being recognized in studies on the transport (release and uptake) of large molecules (13). Increased susceptibility of *femA/B* null mutants to some antibacterial agents other than β-lactams may be related to the increased porosity of the peptidoglycan (30). Increased extractability of proteins was noted in a highly teicoplanin-resistant staphylococcal mutant with greatly decreased peptidoglycan cross-linking. Both resistance and extractability (wall porosity?) were abolished in a Tn551 mutant selected for reduced resistance to teicoplanin (44).

Murein Hydrolases in Cell Division and Antibiotic Resistance

The cell biological functions of staphylococcal murein hydrolases and their regulation are being intensively investigated, and these studies benefited greatly from the introduction of sodium dodecyl sulfate-polyacrylamide gel electrophoresis for profiling bacteriolytic enzymes and from genetic work on the cloning and sequencing of murein hydrolases (26, 33). Regulatory loci affecting autolysis as well as morphology were identified (4, 5). Oshida and colleagues (38) identified the autolysin gene *atl* encoding a bifunctional protein of a deduced molecular size of 137,381 Da. The protein was shown to contain an amidase as well as an endo-β-N-acetylglucosaminidase domain that undergo proteolytic processing at the cell surface. The availability of detailed sequence information on *atl* has led to some extremely interesting observations. The still "double-headed" *atl* gene product containing the fused amidase and glucosaminidase proteins was found to be localized by immunogold labeling on the plasma membrane as a circumferential ring at the future cell divisional site, and this localization precedes the appearance of centripetally growing cell wall (57) (Fig. 2). The findings suggest that this enzyme, known to be involved with cell separation, may function in conjunction with some staphylococcal *fts*-type protein in targeting the site of future cell division. The molecular "signature" that directs the autolysin gene to its site of action at the cellular equator was shown (by site-directed mutagenesis) to reside within the three repeat elements of the *atl* gene (1).

Alteration of autolytic properties (reduction or increase in rates) has been consistently seen in staphylococcal mutants isolated for changed β-lactam resistance. Stepwise increases in autolytic rates (and cell wall turnover rates) in parallel with the increase in resistance level were demonstrated in cefotaxime-resistant laboratory mutants (3). Reduced rates of wall turnover (and autolysis) were observed in *femA*, *femB*, and *femC* mutants. The reverse situation—major reduction (and heterogeneity) in methicillin resistance—was observed in several, but not all transductants generated by crossing the inactivated *atl* into the highly methicillin-resistant strain COL (39).

SUMMARY

Through these findings the staphylococcal cell wall is beginning to emerge not only as a critically important exoskeleton responsible for the structural integrity of the bacterium, but also as an organelle of multiple functions: it is intimately involved with interactions with the system of innate immunity of the host; provides covalent attachment sites for a number of host-related surface proteins; and occupies center stage in resistance to β-lactam and glycopeptide antibiotics. Some observations also begin to throw light on how the control of degradative and synthetic activities and morphogenetic principles come together in the assembly and replication of cell walls.

I thank P. Giesbrecht and M. Sugai for allowing the reproduction of their illustrations in this review.

REFERENCES

1. **Baba, T., and O. Schneewind.** 1998. Targeting of muralytic enzymes to the cell division site of gram-positive bacteria: repeat domains direct autolysin to the equatorial surface ring of *Staphylococcus aureus*. *EMBO J.* **17**:4639–4646.

2. **Berger-Bächi, B.** 1983. Insertional inactivation of staphylococcal methicillin resistance by Tn551. *J. Bacteriol.* **154**:479–487.

3. **Boneca, I. G., N. Xu, D. A. Gage, B. L. M. de Jonge, and A. Tomasz.** 1997. Structural characterization of an abnormally cross-linked muropeptide dimer that is accumulated in the peptidoglycan of methicillin- and cefotaxime-resistant mutants of *Staphylococcus aureus*. *J. Biol. Chem.* **272**:29053–29059.

4. **Brunskill, E. W., and K. W. Bayles.** 1996. Identification of LytSR-regulated genes from *Staphylococcus aureus*. *J. Bacteriol.* **178**:5810–5812.

5. **Brunskill, E. W., B. L. de Jonge, and K. W. Bayles.** 1997. The *Staphylococcus aureus scdA* gene: a novel locus that affects cell division and morphogenesis. *Microbiology* **143**:2877–2882.

6. **Daum, R. S., S. Gupta, R. Sabbagh, and W. M. Milewski.** 1992. Characterization of *Staphylococcus aureus* isolates with decreased susceptibility to vancomycin and teicoplanin: isolation and purification of a constitutively produced protein associated with decreased susceptibility. *J. Infect. Dis.* **166**:1066–1072.

7. **De Jonge, B. L. M., Y.-S. Chang, D. Gage, and A. Tomasz.** 1992. Peptidoglycan composition in heterogeneous Tn551 mutants of a methicillin-resistant *Staphylococcus aureus* strain. *J. Biol. Chem.* **267**:11255–11259.

8. **De Jonge, B. L. M., Y.-S. Chang, D. Gage, and A. Tomasz.** 1992. Peptidoglycan composition of a highly methicillin-resistant *Staphylococcus aureus* strain: the role of penicillin binding protein 2A. *J. Biol. Chem.* **267**:11248–11254.

9. **De Jonge, B. L. M., Y.-S. Chang, N. Xu, and D. Gage.** 1996. Effect of exogenous glycine on peptidoglycan composition and resistance in a methicillin-resistant *Staphylococcus aureus* strain. *Antimicrob. Agents Chemother.* **40**:1498–1503.

10. **De Jonge, B. L. M., T. Sidow, Y.-S. Chang, H. Labischinski, B. Berger-Bächi, D. A. Gage, and A. Tomasz.** 1993. Altered muropeptide composition in *Staphylococcus aureus* strains with an inactivated *femA* locus. *J. Bacteriol.* **175**:2779–2782.

11. **De Lencastre, H., B. L. M. de Jonge, P. R. Matthews, and A. Tomasz.** 1994. Molecular aspects of methicillin

resistance in *Staphylococcus aureus*. *J. Antimicrob. Chemother.* **33**:7–24.

12. **De Lencastre, H., and A. Tomasz.** 1994. Reassessment of the number of auxiliary genes essential for expression of high-level methicillin resistance in *Staphylococcus aureus*. *Antimicrob. Agents Chemother.* **38**:2590–2598.

13. **Dijkstra, A. J., and W. Keck.** 1996. Peptidoglycan as a barrier to transenvelope transport. *J. Bacteriol.* **178:** 5555–5562.

14. **Domanski, T. L., and K. W. Bayles.** 1995. Analysis of *Staphylococcus aureus* genes encoding penicillin-binding protein 4 and an ABC-type transporter. *Gene* **167:**111–113.

15. **Domanski, T. L., B. L. M. de Jonge, and K. W. Bayles.** 1997. Transcription analysis of the *Staphylococcus aureus* gene encoding penicillin-binding protein 4. *J. Bacteriol.* **179:**2651–2657.

16. **Fujimoto, D. F., and K. W. Bayles.** 1998. Opposing roles of the *Staphylococcus aureus* virulence regulators, Agr and Sar, in Triton X-100- and penicillin-induced autolysis. *J. Bacteriol.* **180:**3724–3726.

17. **Georgopapadakou, N. H., B. A. Dix, and Y. R. Mauriz.** 1986. Possible physiological functions of penicillin-binding proteins in *Staphylococcus aureus*. *Antimicrob. Agents Chemother.* **29:**333–336.

18. **Ghuysen, J.-M., and R. Hakenbeck (ed.).** 1994. *Bacterial Cell Wall.* Elsevier Science B.V., The Netherlands.

19. **Giesbrecht, P., T. Kersten, H. Maidhof, and J. Weeke.** The staphylococcal cell wall: morphogenesis and fatal variations in the presence of penicillin. *Microbiol. Mol. Biol. Rev.*, in press.

20. **Gustafson, J., A. Strässle, H. Hächler, F. H. Kayser, and B. Berger-Bächi.** 1994. The *femC* locus of *Staphylococcus aureus* required for methicillin resistance includes the glutamine synthetase operon. *J. Bacteriol.* **176:**1460–1467.

21. **Hackbarth, C. J., T. Kocagoz, S. Kocagoz, and H. F. Chambers.** 1995. Point mutations in *Staphylococcus aureus* PBP2 gene affect penicillin-binding kinetics and are associated with resistance. *Antimicrob. Agents Chemother.* **39:**103–106.

21a.**Hakenbeck, R., J.-V. Höltje, and H. Labischinski (ed.).** 1983. *The Target of Penicillin.* Walter de Gruyter, Berlin, Germany.

22. **Henze, U. U., and B. Berger-Bächi.** 1995. *Staphylococcus aureus* penicillin-binding protein 4 and intrinsic β-lactam resistance. *Antimicrob. Agents Chemother.* **39:**2415–2422.

23. **Henze, U. U., and B. Berger-Bächi.** 1996. Penicillin-binding protein 4 overproduction increases β-lactam resistance in *Staphylococcus aureus*. *Antimicrob. Agents Chemother.* **40:**2121–2125.

24. **Höltje, J.-V.** 1996. Molecular interplay of murein synthases and murein hydrolases in *Escherichia coli*. *Microb. Drug Resist.* **2:**99–103.

25. **Jacobs, C., J.-M. Frère, and S. Normark.** 1997. Cytosolic intermediates for cell wall biosynthesis and degradation control inducible β-lactam resistance in gram-negative bacteria. *Cell* **88:**823–832.

26. **Jayaswal, R. K., Y. I. Lee, and B. J. Wilkinson.** 1990. Cloning and expression of a *Staphylococcus aureus* gene encoding a peptidoglycan hydrolase activity. *J. Bacteriol.* **172:**5783–5788.

27. **Jolly, L., S. Wu, J. van Heijenoort, H. de Lencastre, D. Mengin-Lecreulx, and A. Tomasz.** 1997. The *femR315*

gene from *Staphylococcus aureus*, the interruption of which results in reduced methicillin resistance, encodes a phosphoglucosamine mutase. *J. Bacteriol.* **179:**5321–5325.

28. **Komatsuzawa, H., M. Sugai, K. Ohta, T. Fujiwara, S. Nakashima, J. Suzuki, C. Y. Lee, and H. Suginaka.** 1997. Cloning and characterization of the *fmt* gene which affects the methicillin resistance level and autolysis in the presence of Triton X-100 in methicillin-resistant *Staphylococcus aureus*. *Antimicrob. Agents Chemother.* **41:**2355–2361.

29. **Kornblum, J., B. J. Hartman, R. P. Novick, and A. Tomasz.** 1986. Conversion of a homogeneously methicillin resistant strain of *Staphylococcus aureus* to heterogeneous resistance by Tn551-mediated insertional inactivation. *Eur. J. Clin. Microbiol.* **5:**714–718.

30. **Ling, B., and B. Berger-Bächi.** 1998. Increased overall antibiotic susceptibility in *Staphylococcus aureus femAB* null mutants. *Antimicrob. Agents Chemother.* **42:**936–938.

31. **Ludovice, A. M., S. Wu, and H. de Lencastre.** 1998. Molecular cloning and DNA sequencing of the *Staphylococcus aureus* UDP-N-acetylmuramyl tripeptide synthetase (*murE*) gene, essential for the optimal expression of methicillin resistance. *Microb. Drug Resist.* **4:**85–90.

32. **Maki, H., T. Yamaguchi, and K. Murakami.** 1994. Cloning and characterization of a gene affecting the methicillin resistance level and the autolysis rate in *Staphylococcus aureus*. *J. Bacteriol.* **176:**4993–5000.

33. **Mani, N., P. Tobin, and R. K. Jayaswal.** 1993. Isolation and characterization of autolysis-defective mutants of *Staphylococcus aureus* created by Tn917-lacZ mutagenesis. *J. Bacteriol.* **175:**1493–1499.

34. **Moreau, M., J. C. Richards, J. M. Fournier, R. A. Byrd, W. W. Karakawa, and W. F. Vann.** 1990. Structure of the type-5 capsular polysaccharide of *Staphylococcus aureus*. *Carbohydr. Res.* **201:**285–297.

35. **Murakami, K., T. Fujimura, and M. Doi.** 1994. Nucleotide sequence of the structural gene for the penicillin-binding protein 2 of *Staphylococcus aureus* and the presence of a homologous gene in other staphylococci. *FEMS Microbiol. Lett.* **117:**131–136.

36. **Murakami, K., and A. Tomasz.** 1989. Involvement of multiple genetic determinants in high-level methicillin-resistance in *Staphylococcus aureus*. *J. Bacteriol.* **171:**874–879.

37. **Ornelas-Soares, A., H. de Lencastre, B. de Jonge, D. Gage, Y.-S. Chang, and A. Tomasz.** 1993. The peptidoglycan composition of a *Staphylococcus aureus* mutant selected for reduced methicillin resistance. *J. Biol. Chem.* **268:**26268–26272.

37a.**Ornelas-Soares, A., H. de Lencastre, and A. Tomasz.** Unpublished observations.

38. **Oshida, T., M. Sugai, H. Komatsuzawa, Y.-M. Hong, H. Suginaka, and A. Tomasz.** 1995. A *Staphylococcus aureus* autolysin that has an *N*-acetylmuramoyl-L-alanine amidase domain and an endo-β-N-acetylglucosaminidase domain: cloning, sequence analysis and characterization. *Proc. Natl. Acad. Sci. USA* **92:**285–289.

39. **Oshida, T., and A. Tomasz.** 1992. Isolation and characterization of a Tn551-autolysis mutant of *Staphylococcus aureus*. *J. Bacteriol.* **174:**4952–4959.

40. **Pinho, M., A. M. Ludovice, S. Wu, and H. de Lencastre.** 1997. Massive reduction in methicillin resistance by transposon inactivation of the normal PBP2 in a methicillin

resistant strain of *Staphylococcus aureus*. *Microb. Drug Resist.* **3:**409–413.

41. **Pucci, M. J., J. A. Thanassi, L. F. Discotto, R. E. Kessler, and T. J. Dougherty.** 1998. Identification and characterization of cell wall-cell division gene clusters in pathogenic gram-positive cocci. *J. Bacteriol.* **179:**5632–5635.

42. **Sau, S., N. Bhasin, E. R. Wann, J. C. Lee, T. J. Foster, and C. Y. Lee.** 1997. The *Staphylococcus aureus* allelic genetic loci for serotype 5 and 8 capsule expression contain the type-specific genes flanked by common genes. *Microbiology* **143:**2395–2405.

43. **Sieradzki, K., and A. Tomasz.** 1997. Inhibition of cell wall turnover and autolysis by vancomycin in a highly vancomycin-resistant mutant of *Staphylococcus aureus*. *J. Bacteriol.* **179:**2557–2566.

44. **Sieradzki, K., and A. Tomasz.** 1998. Suppression of glycopeptide resistance in a highly teicoplanin resistant mutant of *Staphylococcus aureus* by transposon inactivation of genes involved in cell wall synthesis. *Microb. Drug Resist.* **4:**159–168.

45. **Snowden, M. A., and H. R. Perkins.** 1990. Peptidoglycan cross-linking in *Staphylococcus aureus*: an apparent random polymerisation process. *Eur. J. Biochem.* **191:**373–377.

46. **Strandén, A. M., K. Ehlert, H. Labischinski, and B. Berger-Bächi.** 1997. Cell wall monoglycine cross bridges and methicillin hypersusceptibility in a *femAB* null mutant of methicillin-resistant *Staphylococcus aureus*. *J. Bacteriol.* **179:**9–16.

47. **Strauss, A., G. Thumm, and F. Götz.** 1998. Influence of *lif*, the lysostaphin immunity factor, on acceptors of surface proteins and cell wall sorting efficiency in *Staphylococcus carnosus*. *J. Bacteriol.* **180:**4960–4962.

48. **Sugai, M., T. Fujiwara, K. Ohta, H. Komatsuzawa, M. Ohara, and H. Suginaka.** 1997. *epr*, which encodes glycylglycine endopeptidase resistance, is homologous to *femAB* and affects serine content of peptidoglycan cross bridges in *Staphylococcus capitis* and *Staphylococcus aureus*. *J. Bacteriol.* **179:**4311–4318.

50. **Timmerman, C. P., E. Mattson, L. Martinez-Martinez, L. de Graaf, J. A. G. van Strijp, V. H. Verbrugh, J. Verhoef, and A. Fleer.** 1993. Induction of release of tumor necrosis factor from human monocytes by staphylococci and staphylococcal peptidoglycans. *Infect. Immun.* **61:**4167–4172.

51. **Tomasz, A., H. B. Drugeon, H. M. de Lencastre, D. Jabes, L. McDougall, and J. Bille.** 1989. New mechanism for methicillin resistance in *Staphylococcus aureus*: clinical isolates that lack the PBP 2a gene and contain normal penicillin-binding proteins with modified penicillin-binding capacity. *Antimicrob. Agents Chemother.* **33:**1869–1874.

52. **Ton-That, H., K. F. Faull, and O. Schneewind.** 1997. Anchor structure of staphylococcal surface proteins. A branched peptide that links the carboxyl terminus of proteins to the cell wall. *J. Biol. Chem.* **272:**22285–22292.

53. **Wada, A., and H. Watanabe.** 1998. Penicillin-binding protein 1 of *Staphylococcus aureus* is essential for growth. *J. Bacteriol.* **180:**2759–2765.

54. **Wu, S., H. de Lencastre, A. Sali, and A. Tomasz.** 1996. A phosphoglucomutase-like gene essential for the optimal expression of methicillin resistance in *Staphylococcus aureus*: Molecular cloning and DNA sequencing. *Microb. Drug Resist.* **2:**277–286.

55. **Wu, S., H. de Lencastre, and A. Tomasz.** 1996. Sigma-B, a putative operon encoding alternate sigma factor of *Staphylococcus aureus* RNA polymerase: molecular cloning and DNA sequencing. *J. Bacteriol.* **178:** 6036–6042.

55a. **Wu, S. W., and H. de Lencastre.** Transcriptional analysis of the *glmM* gene cluster in *Staphylococcus aureus*. Submitted for publication.

56. **Xu, N., Z.-H. Huang, B. L. M. de Jonge, and D. A. Gage.** 1997. Structural characterization of peptidoglycan muropeptides by matrix-assisted laser desorption ionization mass spectrometry and postsource decay analysis. *Anal. Biochem.* **248:**7–14.

57. **Yamada, S., M. Sugai, H. Komatsuzawa, S. Nakashima, T. Oshida, A. Matsumoto, and H. Suginaka.** 1996. An autolysin ring associated with cell separation of *Staphylococcus aureus*. *J. Bacteriol.* **178:**1565–1571.

Staphylococcal Capsule

CHIA Y. LEE AND JEAN C. LEE

37

Many eubacteria produce capsular polysaccharides on their cell surface. The capsule may play an important role in the interaction of bacteria with their immediate environment. For many bacterial pathogens, capsules are important virulence factors that allow the microbe to evade the host immune system. More than 90% of clinical isolates of *Staphylococcus aureus* produce capsular polysaccharides, which have been classified into 11 serotypes. These capsules can also be divided into two distinct groups on the basis of colony morphology. Mucoid-type capsules include the serotype 1 and 2 capsules; strains producing these capsules are heavily encapsulated and are mucoid on solid medium. Microcapsules include the remaining serotype 3 to 11 capsules; strains with these capsules have a thin capsular layer and form nonmucoid colonies on solid medium. As discussed below, this grouping is significant with respect not only to morphology but also to genetic criteria and virulence properties. Because mucoid strains are easily recognized, they were used as the prototype in early studies of staphylococcal capsules. Serotype 2 strain Smith Diffuse was the first encapsulated isolate of *S. aureus* to be characterized. Its biological properties are similar to those of serotype 1 strain M. An excellent review by Wilkinson (48) summarizes the early studies that focused on these mucoid-type *S. aureus* strains. Because of the prevalence of serotypes 5 and 8 among clinical isolates (2, 3, 14, 16, 35), recent studies have focused on the biology of these microencapsulated nonmucoid strains. In this chapter, we review recent developments in the genetics, regulation, and virulence properties associated with *S. aureus* capsules.

STRUCTURE

Capsular polysaccharides of serotypes 1, 2, 5, and 8 have been purified, and their biochemical structures have been determined (11, 12, 27, 29). The trisaccharide repeating units of type 5 and type 8 capsules are very similar. They share the same sugar composition and differ only in the linkages between the amino sugars and the position of O acetylation. The repeating units of these four serotypes are shown below:

Type 1: →4)-α-D-GalNAcAp(1→4)-α-D-GalNAcAp(1→3)-α-D-FucNAcp(1-
(A taurine residue is linked by an amide bond to every fourth D-GalNAcAp residue.)

Type 2: →4)-β-D-GlcNAcAp(1→4)-β-D-GlcN(*N*-acetylalanyl)AcAp-(1-

Type 5: →4)-3-O-Ac-β-D-ManNAcAp(1→4)-α-L-FucNAcp(1→3)-β-D-FucNAcp(1-

Type 8: →3)-4-O-Ac-β-D-ManNAcAp(1→3)-α-L-FucNAcp(1→3)-β-D-FucNAcp(1-

(Abbreviations: GalNAcA, *N*-acetylgalactosaminuronic acid; FucNAc, *N*-acetylfucosamine; GlcNAcA, *N*-acetylglucosaminuronic acid; ManNAcA, *N*-acetylmannosaminuronic acid; O-Ac, O-acetyl.)

GENETICS

Only in recent years has a molecular approach to the study of *S. aureus* capsule expression been taken. The genetic determinant of the serotype 1 capsule produced by strain M was the first to be cloned and sequenced (18, 24). A genomic library from strain M was screened for clones that conferred the mucoid phenotype to chemically derived capsule mutants (18). The *cap1* gene cluster, located in the *S. aureus* chromosome, contains 13 genes designated *cap1A* through *cap1M*. Molecular characterization, mutagenesis, and transcriptional analysis of the *cap1* locus showed that the *cap1* genes were transcribed in the same orientation into an ~14.6-kb transcript (24, 31). Several internal promoters within the *cap1* operon were identified by genetic complementation of chemically induced and allele-specific mutants (31). As shown by gene fusions with *Pseudomonas xylE* as the reporter gene, the internal promoters are much weaker than the primary promoter located upstream of the first gene, *cap1A*. Nevertheless, the internal promoters are biologically functional (31).

More recently, the *cap5* and *cap8* gene clusters required for the production of type 5 and type 8 capsules, respectively, were cloned. The *cap5* gene cluster was targeted by transposon mutagenesis (23, 38), whereas the *cap8* gene cluster was identified by screening of a genomic library of serotype 8 strain Becker with *cap1* gene probes under low-stringency conditions (39). Further analysis revealed that the *cap5* and *cap8* loci each contain 16 closely linked genes, *cap5(8)A* through *cap5(8)P*, transcribed in one orientation. Twelve of the 16 genes composing the two gene clusters are nearly identical, whereas the remaining four genes in each gene cluster are type specific (38). As shown in Fig. 1, the type-specific genes are located in the central region flanked by the common genes. Sequence comparison revealed moderate homology (61 to 71% amino acid identity) of *cap5(8)A-D* with *cap1A-D*, respectively (38, 39). This limited homology apparently allowed the cloning of the *cap8* gene cluster by hybridization.

The *cap8* genes have been subjected to detailed molecular characterization (41). Like the *cap1* genes, the 16 *cap8* genes are transcribed by a primary promoter into a large transcript (~17 kb). Gene fusion studies showed numerous weak internal promoters within the *cap8* operon that are biologically active (33). Although the *cap1* and *cap8* operons are similarly transcribed by a primary promoter located at the beginning of the operon, the *cap1A* promoter is about 60-fold stronger than the *cap8A* promoter in their respective genetic backgrounds (31, 41). The relative strengths of these two primary promoters may explain why the type 1 capsule is produced in greater abundance than the type 8 capsule.

The entire *cap1* operon was shown to be specific to type 1 strains by Southern hybridization under high-stringency conditions. In fact, the 14.6-kb *cap1* gene cluster is part of a discrete genetic element 33.3 to 35.8 kb in length (19). Recent sequencing of the junctions revealed that the element contains a transposase-like gene and an enterotoxin-like gene near one end (32). However, no repeated sequences were found at the junctions. It remains to be determined whether the *cap1* genes reside on a mobile genetic element.

In contrast, the presence of common genes flanking type-specific genes indicates that the two *cap5* and *cap8* loci are allelic. The finding of these common genes in strains of several different capsule serotypes (including serotypes 1 and 2) and in nontypeable strains suggests that all strains of *S. aureus* contain an allelic microcapsule-related locus (23, 39). These results also suggest that mucoid type 1 strain M must contain two capsule gene clusters. In fact, the second capsule locus from strain M has been cloned but not yet characterized (39). It is interesting that both *cap5(8)* and *cap1* loci mapped closely on the same 175-kb *Sma*I-G fragment of the physical map of strain NCTC8325 (32, 38).

REGULATION

Type 1 capsule expression is only modestly affected by environmental factors (22). An analysis of the strain M *cap1* primary promoter showed that deletions upstream of the −35 region had no effect on promoter activity as measured by an *xylE* reporter gene (31). The fact that the *cap1* promoter apparently requires no upstream *cis*-acting element for activity suggests that the genes for mucoid-type capsules are constitutively expressed. Nonetheless, the production of mucoid-type capsules has been shown to be unstable both in vitro and in vivo. In strain M, nonmucoid colonies arise at a frequency of 10^{-4} at 37°C and 10^{-2} to 3.8×10^{-1} at 43°C (18); in type 2 strain Smith Diffuse, the rate is 10^{-2} at 37 or 44°C (43). Loss of mucoid phenotype in strain M was due not to rearrangement but rather to random mutations within the genes in the *cap1* locus (25). Switching between mucoid and nonmucoid phenotypes has been shown to occur upon animal passage. When nonmucoid mutants of strain M were injected intraperitoneally into mice, mucoid colonies were recovered from the peritoneal washings of mice that succumbed to challenge (42). In contrast, loss of the mucoid phenotype was observed in a sublethal renal abscess infection. Organisms isolated from mouse kidneys early in the infection (before day 10) were all mucoid. However, by day 24, the majority of colonies cultured from the kidneys of infected animals were no longer mucoid, even though they were otherwise phenotypically identical to the mucoid challenge strain (19a). Thus, mucoid-type capsules, though not regulated at the transcriptional level, may be regulated by mutation and reversion in the structural genes required for capsule production.

In contrast to the production of mucoid-type capsules, that of microcapsules is influenced by environmental factors. *S. aureus* cultivated on agar plates or in vivo (in the rabbit endocarditis infection model) expressed >300-fold more type 8 capsule than did broth-grown *S. aureus* (22).

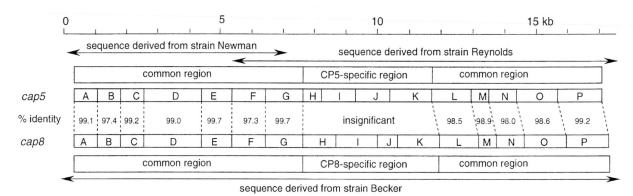

FIGURE 1 Comparison of *cap5* and *cap8* gene clusters. The *cap5* sequence is derived from strains Newman and Reynolds and the *cap8* sequence from strain Becker. Gene designations are shown in boxes. Percent identity indicates the amino acid identity of the deduced proteins between the two clusters. Both gene clusters are transcribed from left to right.

Likewise, 4- to 8-fold more capsule was produced by staphylococci grown in an iron-limited, defined medium than by cells grown in an iron-replete medium (22). Production of type 5 capsule was enhanced under conditions of high oxygen tension but reduced under alkaline growth conditions or in the presence of yeast extract (8, 45). In the presence of carbon dioxide, type 5 capsule production was reduced (13). The effect of carbon dioxide on type 8 capsule production has been shown to be dependent on the *S. aureus* strain as well as on the growth medium (8a, 19a). Isolates of *S. aureus* from bovine mastitis have been shown to produce increased levels of capsular polysaccharide in the presence of milk (26, 46). Recently, type 5 and type 8 capsules have been shown to be positively regulated by *agr*, a quorum-sensing global regulator that simultaneously regulates many gene products in *S. aureus* (7, 40). In high-density cell populations, induced RNAIII (the effector of the *agr* response) up-regulates many extracellular virulence factors while down-regulating many cell wall-associated gene products (36). The fact that microcapsules, unlike other cell wall-associated proteins, are positively regulated by *agr* suggests that microcapsules may be important at a different stage of the pathogenic process.

Studies at the molecular level on the regulation of *S. aureus* microcapsule production have just begun. Sequencing of the type 5 and type 8 capsule gene clusters has revealed several inverted and direct repeats upstream of the primary promoter. One of these repeats, a 10-bp inverted repeat located 14 bp upstream of the *cap8* promoter, is required for promoter activity (33). A chromosomal mutation within the 10-bp repeat reduced capsule production to an undetectable level. A DNA fragment containing the inverted repeat served as a protein-binding site in a gel mobility shift experiment. These results suggest that a positive regulator binding to the inverted repeat is necessary for type 8 capsule production. Preliminary data suggest that the positive regulatory protein has a molecular mass of ~80 kDa, and N-terminal sequencing suggests that it is a novel protein (33). It is possible that certain environmental effects on capsule act through this putative regulator in controlling expression of the *cap8* operon.

BIOSYNTHESIS

Sequence comparison of staphylococcal capsule genes with proteins of known functions has led to the assignment of putative functions to most of the 16 *cap5(8)* genes (38). Although the biosynthesis of staphylococcal capsules is not well understood, the predicted functions of several of the genes have been confirmed by a combination of genetic and biochemical approaches. The trisaccharide repeating unit of the type 5 capsular polysaccharide contains an O-acetyl group on the third carbon of the ManNAcA component. On the basis of its homology to various acetyltransferase genes, the *cap5H* gene was proposed to function in ManNAcA O acetylation. This prediction was supported by the fact that a mutant with a Tn918 insertion at the *cap5H* gene produced wild-type levels of O-deacetylated type 5 capsule, as determined by nuclear magnetic resonance and immunological methods (5). Similarly, *cap8J* is most likely the O acetylation gene in the *cap8* gene cluster; this conclusion is based on sequence homology and on preliminary experiments showing that a monoclonal antibody specific to O-acetylated type 8 capsule failed to react with a *cap8J*-specific mutant (40). The *cap5H* and *cap8J*

genes are located in the type-specific region in their respective operons, a fact consistent with O acetylation of the type 5 and 8 capsules at different positions of ManNAcA.

The functions of two *cap5* genes, *cap5O* and *cap5P*, have been elucidated by genetic experiments in which the two genes complemented their *Escherichia coli* homologs, the *rffE* and *rffD* genes, respectively (17). The *rffE* gene encodes UDP-N-acetylglucosamine-2-epimerase, and the *rffD* gene encodes UDP-N-acetylmannosamine dehydrogenase. Both genes are required for the biosynthesis of ManNAcA, one of the three sugar residues of the *E. coli* enterobacterial common antigen (37). Since ManNAcA is also a component of *S. aureus* type 5 and type 8 capsules, the *cap5O* and *cap5P* genes (and the nearly identical *cap8O* and *cap8P* genes) are likely to have functions equivalent to those of the *rffE* and *rffD* genes, respectively, for the synthesis of UDP-ManNAcA in *S. aureus*. Recent studies have confirmed the function of *cap5P* by demonstrating UDP-GlcNAc-2-epimerase activity in vitro from the cloned *S. aureus* gene product (6).

The *S. aureus cap8B* gene is homologous to several proteins thought to be involved in the chain-length determination of polysaccharides (38). A *cap8B* mutant produces the same amount of capsule as the wild type, but with lower molecular weight (40). This result suggests that *cap8B* is involved in chain-length determination (as is the nearly identical *cap5B*).

ROLE IN VIRULENCE

The role of staphylococcal capsules in virulence has been studied extensively. Early studies indicated that the mucoid-type capsules were important antiphagocytic virulence factors that masked C3b deposited on the bacterial cell wall, preventing its recognition by receptors on phagocytic cells (34). That the amount of capsule produced is important for virulence has been confirmed by comparing 50% lethal dose (LD_{50}) values of transposon-induced mutants of a type 1 strain (20). The virulence properties of the type 1 capsule were also confirmed by a mouse lethality study using a genetic knockout mutant (24). The critical role of mucoid-type capsules in staphylococcal virulence is therefore well established.

The correlation of microcapsules with virulence is not so clear. An early study reported that microencapsulated strains of *S. aureus* resisted phagocytic killing and that antibodies to the capsules induced type-specific phagocytosis in vitro (15). This claim was later disputed by studies showing that strains with type 5 and type 8 capsules were no more virulent than their capsule-negative mutants generated by chemical or transposon mutagenesis (1, 49). Moreover, the type 5 capsule was shown to attenuate bacterial virulence in a rat model of catheter-induced endocarditis. Capsule-specific antibodies did not protect rats against endocarditis when the animals were challenged intravenously (4, 28). However, Fattom et al. (9) showed that immunization with the type 5 and type 8 capsules coupled to protein protected mice from a lethal dose of *S. aureus* administered intraperitoneally and from a sublethal dose that resulted in disseminated infection. Protection conferred by antibodies to the conjugate vaccine was later confirmed in a rat catheter-induced endocarditis model in which the animals were challenged intraperitoneally rather than intravenously (21). The intraperitoneal challenge route allows for a gradual generation of bacteremia and organ seeding, and that is perhaps more clinically relevant than the events following intravenous challenge (9).

Direct evidence that microcapsules are important virulence factors comes from recent animal studies. Nilsson et al. (30) showed that mice inoculated with *S. aureus* expressing a type 5 capsule had a higher frequency of arthritis and a more severe form of the disease than did animals inoculated with nonencapsulated mutant strains. Furthermore, macrophages were better able to ingest and kill nonencapsulated *S. aureus* mutants than the parental strain. Similarly, Thakker et al. (47) demonstrated that a serotype 5 strain of *S. aureus* was cleared less readily from the bloodstream than were capsule-deficient mutants. The parental strain was susceptible to phagocytic killing only in the presence of capsular antibodies and complement. In contrast, the capsular mutants were opsonized for phagocytosis by nonimmune serum with complement activity. Taken together, the results of these studies indicate that the capsule is important in staphylococcal virulence, although this conclusion is clearly dependent on the animal model of infection tested. One study suggested that microcapsules contribute to virulence by acting as adhesins (44); however, these results require independent confirmation.

CONCLUSION

Studies on staphylococcal capsules, in general, are lagging behind those on capsules of other bacterial pathogens. Although recent molecular genetic studies have significantly advanced our knowledge of staphylococcal capsule biology, much remains to be explored. For example, we have just begun to investigate how capsules are synthesized and how they are regulated in response to various environmental conditions. These studies are pivotal for elucidating the role of the capsule in the pathogenesis of staphylococcal diseases. Recently, microcapsules have been used as targets in vaccine development (10). This treatment may be an important adjunct in combating diseases caused by *S. aureus*, especially given the advent of multidrug-resistant clinical isolates. Our understanding of the biology of capsules will be invaluable to the rational designing and development of a staphylococcal polysaccharide vaccine. Furthermore, the mechanism whereby the serotype 5 and 8 capsules act as virulence factors has not been investigated. Current knowledge indicates that the virulence properties of these microcapsules are animal model dependent and therefore disease specific. Further studies on virulence properties of microencapsulated strains are therefore warranted. Since the foundation for molecular studies has been laid, we should witness major advancement in the area of staphylococcal capsule research in the next few years.

This work was supported by NIH grants AI37027 (to C. Y. L.) and AI29040 (to J. C. L.). We thank Ouyang Shu, Subrata Sau, Navnnet Bhasin, and Kevin Kiser for their significant contributions to this work.

REFERENCES

1. **Albus, A., R. D. Arbeit, and J. C. Lee.** 1991. Virulence of *Staphylococcus aureus* mutants altered in type 5 capsule production. *Infect. Immun.* **59:**1008–1014.

2. **Albus, A., J. M. Fournier, C. Wolz, A. Boutonnier, M. Ranke, N. Hoiby, H. Hochkeppel, and G. Doring.** 1988. *Staphylococcus aureus* capsular types and antibody response to lung infection in patients with cystic fibrosis. *J. Clin. Microbiol.* **26:**2205–2209.

3. **Arbeit, R. D., W. W. Karakawa, W. F. Vann, and J. B. Robbins.** 1984. Predominance of two newly described capsular polysaccharide types among clinical isolates of *Staphylococcus aureus. Diagn. Microbiol. Infect. Dis.* **2:**85–91.

4. **Baddour, L. M., C. Lowrance, A. Albus, J. H. Lowrance, S. K. Anderson, and J. C. Lee.** 1992. *Staphylococcus aureus* microcapsule expression attenuates bacterial virulence in a rat model of experimental endocarditis. *J. Infect. Dis.* **165:**749–753.

5. **Bhasin, N., A. Albus, F. Michon, P. J. Livolsi, J.-S. Park, and J. C. Lee.** 1998. Identification of a gene essential for O-acetylation of the *Staphylococcus aureus* type 5 capsular polysaccharide. *Mol. Microbiol.* **27:**9–21.

6. **Bhasin, N., K. B. Kiser, L. Deng, and J. C. Lee.** 1998. Purification and biochemical characterization of UDP-GlcNAc-2-epimerase encoded by the *cap5P* gene of *Staphylococcus aureus* type 5 strain Reynolds, abstr. B-415. *Abstr. 98th Gen. Meet. Am. Soc. Microbiol. 1998.* American Society for Microbiology, Washington, D.C.

7. **Dassy, B., T. Hogan, T. J. Foster, and J. M. Fournier.** 1993. Involvement of the accessory gene regulator (*agr*) in expression of type 5 capsular polysaccharide by *Staphylococcus aureus. J. Gen. Microbiol.* **139:**1301–1306.

8. **Dassy, B., W. T. Stringfellow, M. Lieb, and J. M. Fournier.** 1991. Production of type 5 capsular polysaccharide by *Staphylococcus aureus* grown in a semi-synthetic medium. *J. Gen. Microbiol.* **137:**155–1162.

8a.**Doring, G.** Personal communication.

9. **Fattom, A., J. Sarwar, A. Ortiz, and R. Naso.** 1996. A *Staphylococcus aureus* capsular polysaccharide (CP) vaccine and CP-specific antibodies protect mice against bacterial challenge. *Infect. Immun.* **64:**1659–1665.

10. **Fattom, A., R. Schneerson, D. C. Watson, W. W. Karakawa, D. Fitzgerald, I. Pastan, X. Li, J. Shiloach, D. A. Bryla, and J. Robbins.** 1993. Laboratory and clinical evaluation of conjugate vaccines composed of *Staphylococcus aureus* type 5 and type 8 capsular polysaccharides bound to *Pseudomonas aeruginosa* recombinant exoprotein A. *Infect. Immun.* **61:**1023–1032.

11. **Fournier, J. M., W. F. Vann, and W. W. Karakawa.** 1984. Purification and characterization of *Staphylococcus aureus* type 8 capsular polysaccharide. *Infect. Immun.* **45:**87–93.

12. **Hanessian, S., and T. H. Haskell.** 1964. Structural studies on staphylococcal polysaccharide antigens. *J. Biol. Chem.* **239:**2758–2764.

13. **Herbert, S., D. Worlitzsch, B. Dassy, A. Boutonnier, J. M. Fournier, G. Bellon, A. Dalhoff, and G. Doring.** 1997. Regulation of *Staphylococcus aureus* capsular polysaccharide type 5: CO_2 inhibition in vitro and in vivo. *J. Infect. Dis.* **176:**431–438.

14. **Hochkeppel, H. K., D. G. Braun, W. Vischer, A. Imm, S. Sutter, U. Staeubli, R. Guggenheim, E. L. Kaplan, A. Boutonnier, and J. M. Fournier.** 1987. Serotyping and electron microscopy studies of *Staphylococcus aureus* clinical isolates with monoclonal antibodies to capsular polysaccharide type 5 and type 8. *J. Clin. Microbiol.* **25:**526–530.

15. **Karakawa, W. W., A. Sutton, R. Schneerson, A. Karpas, and W. F. Vann.** 1988. Capsular antibodies induce type-specific phagocytosis of *Staphylococcus aureus* by human

polymorphonuclear leukocytes. *Infect. Immun.* **56:**1090–1095.

16. **Karakawa, W. W., and W. F. Vann.** 1982. Capsular polysaccharides of *S. aureus. Semin. Infect. Dis.* **4:**285–293.

17. **Kiser, K. B., and J. C. Lee.** 1998. *Staphylococcus aureus cap5O* and *cap5P* genes functionally complement mutations affecting enterobacterial common-antigen biosynthesis in *Escherichia coli. J. Bacteriol.* **180:**403–406.

18. **Lee, C. Y.** 1992. Cloning of genes affecting capsule expression in *Staphylococcus aureus* strain M. *Mol. Microbiol.* **6:**1515–1522.

19. **Lee, C. Y.** 1995. Association of staphylococcal type-1 capsule-encoding genes with a discrete genetic element. *Gene* **167:**115–119.

19a.**Lee, J. C.** Unpublished data.

20. **Lee, J. C., M. J. Betley, C. A. Hopkins, N. E. Perez, and G. B. Pier.** 1987. Virulence studies, in mice, of transposon-induced mutants of *Staphylococcus aureus* differing in capsule size. *J. Infect. Dis.* **156:**741–750.

21. **Lee, J. C., J.-S. Park, S. E. Shepherd, V. Carey, and A. Fattom.** 1997. Protective efficacy of antibodies to the *Staphylococcus aureus* capsular polysaccharides in a modified rat model of endocarditis. *Infect. Immun.* **65:**4146–4151.

22. **Lee, J. C., S. Takeda, P. Livolsi, and L. C. Paoletti.** 1993. Effects of in vitro and in vivo growth conditions on expression of type 8 capsular polysaccharide by *Staphylococcus aureus. Infect. Immun.* **61:**1853–1858.

23. **Lee, J. C., S. Xu, A. Albus, and P. J. Livolsi.** 1994. Genetic analysis of type 5 capsular polysaccharide expression by *Staphylococcus aureus. J. Bacteriol.* **176:**4883–4889.

24. **Lin, W. S., T. Cunneen, and C. Y. Lee.** 1994. Sequence analysis and molecular characterization of genes required for the biosynthesis of type 1 capsular polysaccharide in *Staphylococcus aureus. J. Bacteriol.* **176:**7005–7016.

25. **Lin, W. S., and C. Y. Lee.** 1996. Instability of type 1 capsule production in *Staphylococcus aureus*, abstr. B-236. *Abstr. 96th Gen. Meet. Am. Soc. Microbiol. 1996.* American Society for Microbiology, Washington, D.C.

26. **Mamo, W., F. Rozgonyi, S. Hjertén, and T. Wadström.** 1987. Effect of milk on surface properties of *Staphylococcus aureus* from bovine mastitis. *FEMS Microbiol. Lett.* **48:**195–200.

27. **Moreau, M., J. C. Richards, J. M. Fournier, R. A. Byrd, W. W. Karakawa, and W. F. Vann.** 1990. Structure of the type-5 capsular polysaccharide of *Staphylococcus aureus. Carbohydr. Res.* **201:**285–297.

28. **Nemeth, J., and J. C. Lee.** 1995. Antibodies to capsular polysaccharides are not protective against experimental *Staphylococcus aureus* endocarditis. *Infect. Immun.* **63:**375–380.

29. **Murthy, S. V. K. N., M. A. Melly, T. M. Harris, C. G. Hellerqvist, and J. H. Hash.** 1983. The repeating sequence of the capsular polysaccharide of *Staphylococcus aureus. Carbohydr. Res.* **117:**113–123.

30. **Nilsson, I.-M., J. C. Lee, T. Bremell, C. Rydén, and A. Tarkowski.** 1997. The role of staphylococcal polysaccharide microcapsule expression in septicemia and septic arthritis. *Infect. Immun.* **65:**4216–4221.

31. **Ouyang, S., and C. Y. Lee.** 1997. Transcriptional analysis of type 1 capsule genes in *Staphylococcus aureus. Mol. Microbiol.* **23:**473–482.

32. **Ouyang, S., and C. Y. Lee.** Unpublished data.

33. **Ouyang, S., S. Sau, and C. Y. Lee.** 1998. Analysis of the promoter for the expression of type 8 capsular polysaccharide in *Staphylococcus aureus*, abstr. B-416. *Abstr. 98th Gen. Meet. Am. Soc. Microbiol. 1998.* American Society for Microbiology, Washington, D.C.

34. **Peterson, P. K., B. J. Wilkinson, Y. Kim, D. Schmeling, and P. G. Quie.** 1978. Influence of encapsulation on staphylococcal opsonization and phagocytosis by human polymorphonuclear leukocytes. *Infect. Immun.* **19:**943–949.

35. **Poutrel, B., A. Boutonnier, L. Sutra, and J. M. Fournier.** 1988. Prevalence of capsular polysaccharide types 5 and 8 among *Staphylococcus aureus* isolates from cow, goat, and ewe milk. *J. Clin. Microbiol.* **26:**38–40.

36. **Projan, S. J., and R. P. Novick.** 1997. The molecular basis of pathogenicity, p. 55–81. *In* K. B. Crossley and G. L. Archer (ed.), *The Staphylococci in Human Disease.* Churchill Livingstone, New York, N.Y.

37. **Rick, P. D., and R. P. Silver.** 1996. Enterobacterial common antigen and capsular polysaccharides, p. 104–122. *In* F. C. Neidhardt, R. Curtiss III, J. L. Ingraham, E. C. C. Lin, K. B. Low, Jr., B. Magasanik, W. S. Reznikoff, M. Riley, M. Schaechter, and H. E. Umbarger (ed.), *Escherichia coli and Salmonella: Cellular and Molecular Biology,* 2nd ed. ASM Press, Washington, D.C.

38. **Sau, S., N. Bhasin, E. R. Wann, J. C. Lee, T. J. Foster, and C. Y. Lee.** 1997. The *Staphylococcus aureus* allelic genetic loci for serotype 5 and 8 capsule expression contain the type-specific genes flanked by common genes. *Microbiology* **143:**2395–2405.

39. **Sau, S., and C. Y. Lee.** 1996. Cloning of type 8 capsule genes and analysis of gene clusters for the production of different capsular polysaccharides in *Staphylococcus aureus. J. Bacteriol.* **178:**2118–2126.

40. **Sau S., and C. Y. Lee.** Unpublished data.

41. **Sau, S., J. Sun, and C. Y. Lee.** 1997. Molecular characterization and transcriptional analysis of type 8 genes in *Staphylococcus aureus. J. Bacteriol.* **179:**1614–1621.

42. **Scott, A. C.** 1969. A capsulate *Staphylococcus aureus. J. Med. Microbiol.* **2:**253–260.

43. **Smith, R. M., J. T. Parisi, L. Vidal, and J. N. Baldwin.** 1977. Nature of the genetic determinant controlling encapsulation in *Staphylococcus aureus* Smith. *Infect. Immun.* **17:**231–234.

44. **Soell, M., M. Diab, G. Haan-Archipoff, A. Beretz, C. Herbelin, B. Poutrel, and J.-P. Klein.** 1995. Capsular polysaccharide types 5 and 8 of *Staphylococcus aureus* bind specifically to human epithelial (KB) cells, endothelial cells, and monocytes and induce release of cytokines. *Infect. Immun.* **63:**1380–1386.

45. **Stringfellow, W. T., B. Dassy, M. Lieb, and J. M. Fournier.** 1991. *Staphylococcus aureus* growth and type 5 capsular polysaccharide production in synthetic media. *Appl. Environ. Microbiol.* **57:**618–621.

46. **Sutra, L., C. Mendolia, P. Rainard, and B. Poutrel.** 1990. Phagocytosis of mastitis isolates of *Staphylococcus aureus* and expression of type 5 capsular polysaccharide are influenced by growth in the presence of milk. *J. Clin. Microbiol.* **28:**2253–2258.

47. **Thakker, M., J. S. Park, V. Carey, and J. C. Lee.** *Staphylococcus aureus* serotype 5 capsular polysaccharide is an-

tiphagocytic and enhances bacterial virulence in a murine bacteremia model. *Infect. Immun.* **66:**5183–5189.

48. **Wilkinson, B. J.** 1983. Staphylococcal capsules and slime, p. 481–523. *In* C. S. G. Easmon and G. Adlam (ed.), *Staphylococci and Staphylococcal Infections.* Academic Press, Inc., New York, N.Y.

49. **Xu, S., R. D. Arbeit, and J. C. Lee.** 1992. Phagocytic killing of encapsulated and microencapsulated *Staphylococcus aureus* by human polymorphonuclear leukocytes. *Infect. Immun.* **60:**1358–1362.

Staphylococcus aureus Exotoxins

GREGORY A. BOHACH AND TIMOTHY J. FOSTER

38

Staphylococcus aureus exotoxins fall into two general groups: (i) membrane-active agents and (ii) toxins with superantigen (SAg) activity. The former group plays a contributory role in pathogenesis, and no distinct clinical syndrome results directly from their activity. *S. aureus* SAgs include the pyrogenic toxin (PT) family and the exfoliative toxins (ETs). These toxins share a set of immunomodulatory activities as a result of their SAg function that promote the ability of most PTs to induce toxic shock syndrome (TSS). Some SAgs such as the enterotoxins and ETs have additional activities that endow them with the ability to induce other diseases, such as staphylococcal food poisoning and scalded skin syndrome.

MEMBRANE-ACTIVE AGENTS

Alpha-Toxin

Alpha-toxin is a cytolytic pore-forming toxin and is one of the most potent bacterial toxins known (reviewed in reference 8). Alpha-toxin is especially toxic for rabbits (50% lethal dose [LD_{50}] = 1.3 μg), and rabbit erythrocytes are also very sensitive to lysis by this toxin. Human erythrocytes are approximately 1,000-fold less sensitive. It is suspected that two cell-binding mechanisms exist. Cells such as rabbit erythrocytes are suspected to have high-affinity binding sites, while other cells such as human erythrocytes have only low-affinity sites (33). The low-affinity binding mechanism is responsible for alpha-toxin damage to protein-free liposomes at high toxin concentrations. The identity of the high-affinity binding site is unknown but is probably a membrane protein.

Alpha-toxin is secreted as a monomer of 293 residues. Upon binding to the membrane, the monomer oligomerizes to form a ring-shaped pore (Fig. 1). Although originally interpreted to be a hexamer, structural studies have provided evidence for both hexameric and heptameric rings. Only part of the toxin penetrates the bilayer while the bulk of it remains on the surface. The pore is a water-filled channel. A model for toxin assembly (Fig. 2) was deduced from mutagenesis and chemical modification experiments (72).

Toxin at various stages of assembly differs in protease susceptibility. The amino latch region (residues 1 to 20) is susceptible to protease in forms $\alpha 1$, $\alpha 1^*$, and $\alpha 7^*$ but is resistant in the heptameric pore of $\alpha 7$, suggesting that folding of the latch is part of the last stage of toxin-pore assembly. A membrane-associated glycine-rich region of largely β structure (residues 110 to 148) is protease sensitive only in the monomer $\alpha 1$, suggesting that it is hidden from protease attack, even though membrane penetration is the last stage. Evidence in formulating the model also came from cysteine-scanning mutagenesis. Single cysteine substitutions were derivatized with a polarity-sensitive fluorescence probe and reacted with liposomes. Residues in the glycine-rich region, particularly residues 126 to 140, showed a marked shift in fluorescence, indicating that they are embedded in a hydrophobic lipid environment.

The crystal structure of an alpha-toxin heptamer is mushroom-shaped and forms a channel ranging in diameter from 14 to 46 Å (65). The lower half of a 14-stranded antiparallel β-barrel penetrates and spans the lipid bilayer to form the pore. Each subunit contributes two amphipathic β-strands encompassing residues 118 to 140, with residue 128 forming the turning point at the membrane face. At the top of the cap the channel diameter is 28 Å and is lined by the N-terminal amino latches. The channel widens to 46 Å before narrowing again to 15 Å at the junction of the cap and stem. The stem pore varies from 14 to 24 Å depending on side chains in the lumen. Within the stem are two bands of hydrophobic residues. Both ends are defined by rings of acidic and basic residues. At the top of the cap domain, the amino latch makes extensive contacts with its clockwise-related neighbor. In the body of the cap, adjacent protomers make contact though β-sandwiches.

Most *S. aureus* isolates possess *hla*, the structural gene for alpha-toxin, but are variable in its expression or in the amount of toxin expressed. TSS isolates are often nonhemolytic and less inflammatory and often have a nonsense mutation in *hla*. Work with mutants in animal models indicated that strains expressing alpha-toxin are more virulent than isogenic derivatives (29). Although adults usually

Gram-Positive Pathogens, ed. by V. A. Fischetti et al.
© 2000 American Society for Microbiology, Washington, D.C.

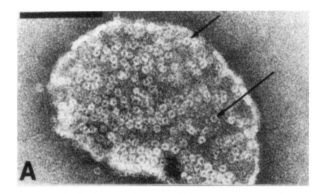

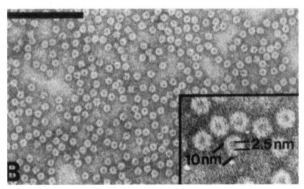

FIGURE 1 Membrane pore formation by alpha-toxin. (A) Rabbit erythrocyte membrane fragment negatively stained following lysis with alpha-toxin. Arrows designate representative ring-shaped structures (10 nm) on the membrane. (B) Ring-shaped alpha-toxin multimers isolated in detergent solution. (Inset) The rings are magnified so that the internal channel (2.5 nm) and ring perimeter (10 nm) are clearly visible.

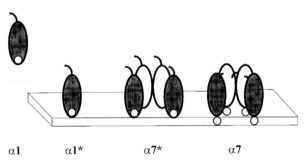

α1   α1*   α7*   α7

FIGURE 2 Model for alpha-toxin assembly based on crystallographic data and structure-function experiments. The model depicts the formation of a heptameric ring (α7 and α7*). A cross section of the ring revealing only four monomers is shown so that the proposed structural alterations are visible. In this model, alpha-toxin is expressed and secreted as a monomer (α1). α1, bound to the target membrane (designated α1*), promotes assembly of the heptamer (α7). In the final stage of assembly, β-sheets in each monomer (depicted as small circles) insert into the membrane forming a channel, and the N-terminal latches contact adjacent monomers rendering them resistant to proteolysis.

contain antibody, high-affinity binding sites may allow the toxin to affect susceptible cells at very low concentrations. In vitro and in vivo, alpha-toxin is hemolytic, cytotoxic, dermonecrotic, and lethal. Several cell types, including erythrocytes, mononuclear immune cells, epithelial and endothelial cells, and platelets, can be killed by alpha-toxin. Cell death directly from membrane damage has been attributed to impaired osmoregulation, cation and small molecule influx/efflux, and apoptosis (39). Sublethal quantities of alpha-toxin can allow Ca^{2+} influx, activating phospholipases and arachidonic acid metabolism. Prostaglandins and/or leukotrienes can be generated, resulting in vasoactive effects that augment the direct lethality of alpha-toxin for endothelial cells. Additionally, proinflammatory cytokines and procoagulatory compounds released from activated monocytes and platelets, respectively, contribute to effects of alpha-toxin systemically, particularly for the cardiovascular system and the lungs.

Beta-Toxin

In 1935, Glenny and Stevens differentiated beta-toxin from alpha-toxin by antibody neutralization and showed that it lysed sheep, but not rabbit, erythrocytes (reviewed in reference 5). Beta-toxin is also unique in its ability to induce hot-cold lysis. Incubation of sensitive erythrocytes with toxin at 37°C results in little or no lysis. However, lysis ensues if treated erythrocytes are chilled below 10°C. On the basis of this activity, it has been concluded that beta-toxin production is more frequent with animal isolates (~88%) than with human isolates (11 to 45%). We have noted, however, that the use of antisera is a more sensitive means of detecting expression of the toxin. Thus, many human isolates that do not exhibit a hot-cold reaction express beta-toxin that is detectable by immunoblotting. The S. aureus beta-toxin gene hlb (58) carries the attachment site for serological group F-converting phages (15). In lysogens, the gene is disrupted by prophage integration. Beta-toxin is very basic and is often the most abundant protein in culture supernatants, facilitating its purification from S. aureus and Staphylococcus intermedius. Based on sequencing data, S. aureus beta-toxin is a 33,742-Da protein with two cysteines that are likely to form a disulfide linkage required for activity. The S. intermedius toxin is similar but not identical to S. aureus beta-toxin (21). Both proteins are related to sphingomyelinase of Bacillus cereus and to a similar enzyme produced by Leptospira interrogans. Structure-function studies on staphylococcal beta-toxin are limited, but it is likely that its mode of action mimics the B. cereus enzyme, which has been better characterized (71).

Beta-toxin displays species-dependent activity. Sheep, cow, and goat erythrocytes are most sensitive. Human erythrocytes are intermediate in sensitivity, whereas murine and canine erythrocytes are resistant. Beta-toxin is a neutral sphingomyelinase, and the degree of erythrocyte sensitivity depends on membrane sphingomyelin content. It acts as a type C phosphatase, hydrolyzing sphingomyelin to phosphorylcholine and ceramide (Fig. 3A). Beta-toxin-induced hemolysis and sphingomyelin hydrolysis require divalent cations. Mg^{2+} is most effective, although either Co^{2+} or Mn^{2+} can enhance activity. Ca^{2+} and Zn^{2+} are inhibitory. Beta-toxin enzyme activity occurs maximally at or around 37°C but is negligible at 4°C. This suggests that hot-cold lysis has two stages—sphingomyelin hydrolysis followed by physical disruption of the membrane at lower temperatures. Unlike lesions induced by pore-forming toxins, beta-toxin causes invaginations of selected regions of

A

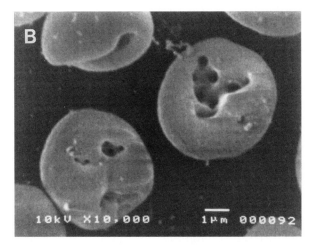

PHOSPHORYLCHOLINE

CERAMIDE

SPHINGOMYELIN

FIGURE 3 Properties of staphylococcal beta-toxin. (A) Sphingomyelin chemical formula showing beta-toxin cleavage site resulting in generation of phosphorylcholine and ceramide. (B) Scanning electron micrograph showing lesions in human erythrocyte membranes caused by beta-toxin after shifting the temperature to 4°C.

the membrane (Fig. 3B). Sphingomyelin is located in membrane outer leaflet patches. Cohesive forces are apparently sufficient to hold the ceramide hydrolysis product in position in the membrane. On cooling, a phase separation occurs, with condensation of ceramide into pools and collapse of the bilayer (43).

Results of early toxicity studies with beta-toxin should be interpreted with caution since toxin preparations were often contaminated with alpha-toxin. Subcutaneous administration of purified beta-toxin causes erythema in rabbits, and, in general, beta-toxin is at least 10- to 160-fold less toxic than alpha-toxin in animal models. The use of isogenic *S. aureus* strains has shown that beta-toxin contributes to pathogenesis in a murine mastitis model (29) and that it has a small role to play in ocular keratitis (51). Beta-toxin is leukotoxic to a variety of cells. Leukocytes are also susceptible to hot-cold lysis, although sublytic concentrations or incubation at physiological temperatures results in abnormal function. Beta-toxin inhibits monocyte migration and stimulates release of interleukin 1β (IL-1β), IL-6 receptor, and soluble CD14 from human monocytes

(74). The effects on other leukocytes are less pronounced. Although a number of investigations have failed to demonstrate hot-cold lysis of neutrophils, impaired chemotaxis and Fc binding were reported. Free ceramide is a potent second messenger and is involved in a number of cascade reactions leading to kinase and phosphatase induction, thereby promoting apoptosis. Although beta-toxin induces apoptosis in human leukemic cell lines and murine fibrosarcoma cell lines (38), Walev et al. (74) could not demonstrate a role for ceramide in the effects of beta-toxin on monocytes.

Delta-Toxin

In 1947, Williams and Harper (76) proposed the existence of delta-toxin as the fourth cytolytic *S. aureus* toxin. Delta-toxin is unique since it is small, heat stable, has surfactive properties, and is lytic toward many types of membranes from most animal species, including those on erythrocytes, other cells, organelles, and even bacterial protoplasts (reviewed in reference 5). Also, cytoplasmic leakage and lysis of cells exposed to delta-toxin occur without a demonstrable lag, similar to treatment with Triton X-100 and other detergents (62).

Delta-toxin, a 26-residue peptide similar to bee venom mellitin, is encoded by the *hld* gene located near the 5' end of the Agr RNAIII transcript. It is produced by nearly all *S. aureus* isolates and by a high percentage of other staphylococcal species. At least two variants of delta-toxin exist; toxins expressed by human and canine strains of *S. aureus* are only 62% identical and are immunologically distinct. Molecular modeling suggested that delta-toxin is a helical amphipathic peptide with its hydrophobic and hydrophilic residues on opposing sides of the helix. High-resolution nuclear magnetic resonance studies have yielded similar structures, but differences exist between solution and membrane-bound toxin (20, 69). In solution, residues 2 to 20 form a stable helix, whereas the C-terminal residues are more flexible. Bound to lipid micelles, residues 5 to 23 form an extended helix. The exact location and orientation of the peptide in relation to the membrane are still unclear. It is likely that the toxin inserts at least partly into the lipid bilayer, disordering lipid chain dynamics. It has been suggested that the mode of lysis involves formation of channels in membranes composed of aggregates of six molecules of delta-toxin (46).

Since delta-toxin exhibits activity toward a broad spectrum of cells, it is potentially cytotoxic for tissues and could have an adverse effect on leukocytes. Although delta-toxin induces dermonecrosis when administered intradermally into the skin of rabbits, extremely high concentrations are required to cause lethality in laboratory animals (77). It is only minimally immunogenic but is inhibited by binding to proteins, cholesterol, and phospholipids in serum.

Gamma-Toxin, Leukocidin, and Other Bicomponent Toxins

Gamma-toxin, leukocidin, and other bicomponent toxins are a family of proteins encoded by the *hlg* and *luk-PV* loci (Fig. 4). All of the toxins in this family contain two synergistically acting proteins: one S component (LukS-PV, HlgA, or HlgC) and one F component (LukF-PV or HlgB), designated on the basis of their mobility (slow or fast) in ion-exchange chromatography. The prototype bicomponent toxins are the closely related Panton-Valentine leukocidin (PVL) and gamma-toxin. The PVL S and F

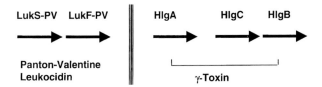

FIGURE 4 Organization of bicomponent toxin genes in a strain harboring both the *hlg* and *luk-PV* loci. Any S and F component may combine to generate a unique bicomponent toxin. The two prototype bicomponent toxins, Panton-Valentine leukocidin (PVL) and gamma-toxin, are composed of LukS-PV+LukF-PV and HlgA+HlgB, respectively.

components are LukS-PV and LukF-PV (56). Gamma-toxin likewise contains S and F components designated HlgA and HlgB, respectively.

In most *S. aureus* strains, the *hlg* locus encodes three polypeptides. Two (HlgA and HlgC) are S components and are related to LukS-PV, encoded by the *luk-PV* locus (Fig. 5). HlgB is similar to LukF-PV (18). More than 99% of clinical isolates carry the *hlg* locus, while only 2% of clinical isolates carry the *luk* genes and express PVL. The latter isolates, which also contain *hlg*, produce all three S components and both F components. Any S component can combine with each of the F proteins, leading to formation of six toxin combinations in PVL-producing strains, or to two combinations (HlgB+HlgA or HlgB+HlgC) in strains harboring only the *hlg* locus. Only PVL and gamma-toxin have been specifically named. Other bicomponent toxins are designated by listing their two components. A novel variant of bicomponent toxin (LukE+LukD) expressed by *S. aureus* Newman has been reported (31).

PVL can stimulate and lyse neutrophils and macrophages. Other normal cells are not affected by PVL, which is nonhemolytic (55). Gamma-toxin is strongly hemolytic and 90-fold less leukotoxic. Like PVL, the LukF-PV component paired with either HlgC or HlgA promotes leukotoxic activity but poor hemolytic activity. HlgB+HlgC is weakly hemolytic but is leukotoxic, albeit less than PVL. PVL is active against human and rabbit leukocytes but not murine, ovine, or guinea pig cells. Of 11 species of animals tested, erythrocytes from rabbits were the only cells lysed efficiently by gamma-toxin. Human cells were lysed but were 1,000-fold less sensitive than rabbit cells (4). Despite this, additional toxicity may be displayed in vivo. Animal studies have shown that mice, rabbits, and guinea pigs are all sensitive to gamma-toxin.

Synergistic function involves sequential binding of the F and S components. The action of gamma-toxin on erythrocytes involves initial binding of HlgB (F) followed by HlgA (S) (52) and subsequent generation of a pore. Ring-shaped structures (2.1 to 2.4 nm in diameter) purified from erythrocytes consist of a complex (150 to 250 kDa) of HlgA and HlgB in a 1:1 ratio (40, 67). Similarly, the two components of the PVL also bind sequentially to human neutrophils, although most reports indicate that the 32-kDa S component binds first (16). Although the stoichiometry of S- and F-related subunits on interactions with human neutrophils is not clear at present, it is believed that PVL forms pores. The number of toxin molecules bound per cell and the pore diameter are affected by divalent cations, especially Ca^{2+}. At physiological Ca^{2+} levels, the pores are permeable to small divalent ions but not to ethidium bromide. Pores formed in the absence of

Ca^{2+} are larger (at least 0.78 nm in diameter) and allow the passage of ethidium bromide (26). In vitro studies have shown an affinity of S components of both PVL and gamma-toxin for GM1 gangliosides (50, 53), although the molecular nature of the receptor has not been conclusively determined.

Initial exposure of neutrophils to PVL in vitro results in swelling and rounding of the cells and their nuclei. Cell lysis eventually occurs and is preceded by degranulation and nuclear rupture (50). One of the major effects of PVL is a rapid accumulation of intracellular free Ca^{2+}, possibly by activation of an endogenous Ca^{2+} channel (17). PVL has been associated with certain types of cutaneous infections; antibodies provide some protection and are associated with less severe disease (78).

TOXINS WITH SUPERANTIGEN ACTIVITY
Staphylococcal Pyrogenic Toxin Superantigens

Classical PTs include staphylococcal enterotoxins (SEs), TSS toxin-1 (TSST-1), and streptococcal pyrogenic exotoxins (9). Recently, additional SEs, types G through J, have been identified (49, 79). SEs are classified largely on antigenic differences (Table 1), and nine major SE types are known to be expressed. Minor molecular variants exist for SEC and TSST-1. The PTs may be expressed by animal isolates as well as human strains (23).

The SEs are best known for causing staphylococcal food poisoning, which results from ingestion of food contaminated with toxin. Staphylococcal food poisoning is the leading cause of food-borne microbial intoxication worldwide and is usually linked to improper storage of food (34). Most cases are self-limiting and have a mean duration of 23.6 h. Patients typically present with emesis after a short incubation period (mean of 4.4 h). Other symptoms, including nausea, diarrhea, and abdominal pain or cramping, are also common. The SEs are stable in the gastrointestinal tract and indirectly stimulate the emetic reflex center. Although the vagus nerve is involved, molecular events involved are unclear (24). There is mounting evidence that mast cell activation occurs and that inflammatory mediators and neuropeptide substance P are released upon SE activity in the gastrointestinal tract and elsewhere (2, 11, 63). TSST-1 is not implicated in staphylococcal food poisoning.

TSS is an acute systemic illness. The criteria for defining TSS cases were established in 1981 (60). Patients present with hypotension, fever, rash, and desquamation during convalescence and have involvement of at least three additional organ systems. Some investigators have proposed implementing a revised case definition to include patients with less severe disease, such as those whose syndrome is attenuated as a result of early treatment (54). Staphylococcal TSS may manifest in either of two general forms, menstrual or nonmenstrual. Menstrual TSS occurs in women whose vaginal/cervical mucosa are colonized by TSST-1-producing *S. aureus*. Tampon use is a risk factor in menstrual TSS, and a correlation between tampon absorbency and risk of developing TSS has been established. Nonmenstrual TSS may result from *S. aureus* infection elsewhere in the body. Either SEs or TSST-1 may mediate the nonmenstrual form.

Shared biological properties attributed to the staphylococcal pyrogenic toxin superantigens (PTSAgs) include induction of fever, hypotension, T-cell proliferation, im-

A

```
LukS-PV (49775)  MVKKRLLAATLSLGIITPIATSFHE-SKADNNIENIGDGA--EVVKRTEDTSSDKWGVTQNIQFDFVKDKKYNKDALILK  (77)
HlgC    (P83)    *L*NKI**T***VSLLA*L*NPLL*NA**A*DT*D**K*NDV*II*****KT*N*************************** (80)
HlgC    (5R)     *L*NKI*TT***VSLLA*L*NPLL*NA**A*DT*D**K*SDI*II*****KT*N*************************** (80)
HlgC    (MRSA)   *L*NKI**T***VSLLA*L*NPLL*NA**A*DT*D**K*SDI*II*****KT*N****************T********** (80)
HlgA    (5R)     *I*NKI*T***AV*L*A*L*NP*I*I***E*K**D**Q*****II***Q*IT*KRLAI*******************VV* (80)
HlgA    (49775)  *I*NKI*T***AV*L*A*L*NP*I*I***E*K**D**Q*****II***Q*IT*KRLAI*******************VV* (80)
HlgA    (MRSA)   *I*NKI*T***AV*L*A*L*NP*I*I***E*K**D**Q*****II***Q*IT*KRLAI*******************VV* (80)

LukS-PV (49775)  MQGFINSKTTYYNYKNTDHIKAMRWPFQYNIGLKTNDPNVDLINYLPKNKIDSVNVSQTLGYNIGGNFNSGPSTGGNGSF (157)
HlgC    (P83)    *****S*R*******N***S***********KY*S**********E*T************Q*A**L****** (160)
HlgC    (5R)     *****S*R*******K*N*V***********KY*S**********E*T************Q*A**L****** (160)
HlgC    (MRSA)   *****S*R*******K*N*V***********KY*S**********E*T************Q*A**L****** (160)
HlgA    (5R)     *****S*R***SDL*KYPY**R*I******S***K*S**********AD***K********Q*A**I**S*** (160)
HlgA    (49775)  *****S*R***SDL*KYPY**R*I******S***K*S**********AD***K********Q*A**I**S*** (160)
HlgA    (MRSA)   *****S*R***SDL*KYPY**R*I******S***K*S**********AD***K********Q*A**I**S*** (160)

LukS-PV (49775)  NYSKTISYNQQNYISEVEHQNSKSVQWGIKANSFITSLGKMSGHDPNLFVGYKPYSQNPRDYFVPDNELPPLVHSGFNPS (237)
HlgC    (P83)    ****S***T****V****Q******L**V*****A*ES*QK*AF*SD******H*KD*******S******Q****** (240)
HlgC    (5R)     ****S***T****V****Q******L**V*****A*ES*QK*AF*SD******H*KD*******S******Q****** (240)
HlgC    (MRSA)   ****S***T****V****Q******L**V*****A*ES*QK*AF*SD******H*KD*******S******Q****** (240)
HlgA    (5R)     **********K**VT****S****G*K**V*****V*PN*QV*AY*QY**AQ-D*TGPAA*******Q****IQ***** (239)
HlgA    (49775)  **********K**VT****S****G*K**V*****V*PN*QV*AY*QY**AQ-D*TGPAA*******Q****IQ***** (239)
HlgA    (MRSA)   **********K**VT****S****G*K**V*****V*PN*QV*AY*QY**AQ-N*TGPAA*******Q****IQ***** (239)

LukS-PV (49775)  FIASVSHEKGSGDTSEFEITYGRNMDVTHA--TRRTTHYGNSYLEGSRIHKAFVNRNYTVKYEVNWKTHEIKVKGHN (312)
HlgC    (P83)    ***T*******S**************IK*S*******D*H*V*N**K****************Q* (317)
HlgC    (5r)     ***T*******S**************IK*S*******D*H*V*N******************Q* (317)
HlgC    (MRSA)   ***T*******S**************IK*S*******D*H*V*N*****************Q* (317)
HlgA    (5R)     **TTL***R*K**K***********A*Y*YV**HR-------*AVD*K*D**K***V***********V*I*SITPK (311)
HlgA    (49775)  **TTL****K*K***********A*Y*YV**PR-------*AVD*K*D**K***V***********V*I*SITPK (311)
HlgA    (MRSA)   **TTL***R*K**K***********A*Y*YV**HR-------*AVD*K*D**K***V***********V*I*SITPK (311)
```

B

```
LukF-PV (49775)  MK--KIVKSREVTSIALLLLSNTLDAAQHITPVSEKKVDDKITLYKTTATSDSDKLKISQILTFNFIKDKSYDKDTLILKAA (80)
HlgB    (P83)    *NMN*L***SVA**M********AN*EGK****V*******V*********A****F*****************V*** (82)
HlgB    (5R)     **MN*L***SVA**M*****G*AN*EGK****V*******V*********A****F*****************V***T (82)
HlgB    (MRSA)   **MN*L***SVA**M*****G*AN*EGK****V*******V*********A****F*****************V***T (82)

LukF-PV (49775)  GNIYSGYTKPNPKDTISSQFYWGSKYNISINSDSNDSVNVVDYAPKNQNEEFQVQQTVGYSYGGDINISNGLSGGGNGSKSF (162)
HlgB    (P83)    ***N****ER*****YDF*KI***A***V**S*Q*******N*L**TF****S*********L**NTA* (164)
HlgB    (5R)     ***N**FV****N*YDF*KL***A***V**S*Q*******A*********N*L**TF****S*******L**NTA* (164)
HlgB    (MRSA)   ***N**FV****N*YDF*KL***A***V**S*Q*******A*********N*L**TF****S*******L**NTA* (164)

LukF-PV (49775)  SETINYKQESYRTSLDKRTNFKKIGWDVEAHKIMNNGWGPYGRDSYHSTYGNEMFLGSRQSNLNAGQNFLEYHKMPVLSRGN (244)
HlgB    (P83)    ************T*SRN**Y*NV**G*************F*P*****L**AG***SAY****IAQ*Q**L***S* (246)
HlgB    (5R)     ************T*SRN**Y*NV**G*************F*P*****L**AG***SAY****IAQ*Q**L***S* (246)
HlgB    (MRSA)   ************-*SRN**Y*NV**G*********-*******F*P*****L**AG***SAY****IAQ*Q**L***S* (244)

LukF-PV (49775)  FNPEFIGVLSRKQNAAKKSKITVTYQSEMDRYTNFWINFNWIGNNYKDHIRATHTSIYEVDWENHTVKLIDTQSKEKNPMS (325)
HlgB    (P83)    ****LS***HR*DG**********R***L*QIR*NG*Y*A**NFKTR*FK*T**I*****K***L**KET*N*K (325)
HlgB    (5R)     *****LS***HR*DG**********R***L*QIR*NG*Y*A**NFKTR*FK*T**I*****K***L**KET*N*K (325)
HlgB    (MRSA)   ***R*LS***HR*D***********R***L*QIR*NG*Y*A**NFKTR*FK*T**I*****K***L**KET*N*K (323)
```

FIGURE 5 Sequence alignment showing relatedness of *luk* and *hlg* S and F gene products from various *S. aureus* isolates (in parentheses on the left), obtained from reference 56 and GenBank (accession numbers D42143, L01055, X64389, S53213, and X81586). Residue numbers are indicated on the right. (A) S components; (B) F components.

munosuppression, enhanced susceptibility to endotoxin shock, and induction of cytokines (9). In 1989, key reports (30, 61, 75) helped to elucidate the mechanisms by which PTSAgs function. They interact in a unique way with the immune system: (i) they cause polyclonal proliferation of T cells bearing certain T-cell receptor (TCR) Vβ elements; (ii) they bind specifically to major histocompatibility complex (MHC) class II (MHCII) on antigen-presenting cells (APCs), and this binding is required for maximum T-cell stimulation; and (iii) this interaction results in deletion of stimulated Vβ-expressing T cells. Molecules with these properties are termed SAgs, and the PTs are prototypic SAgs.

SAgs bind unprocessed to MHCII at a region distinct from the peptide binding groove (37, 41). Since the MHCII-bound toxin stimulates T cells based on their Vβ sequences rather than antigen specificity, many T cells (up to 30%) are activated. This type of cellular stimulation, plus elevated cytokine levels, adversely affect the host. Combined, they cause fever, lethal shock, immunosuppression, and several clinical syndromes, such as TSS and possibly certain autoimmune diseases. The major cytokines induced initially include IL-1, tumor necrosis factor alpha and beta, gamma interferon, and IL-2. The role of these cytokines and secondary proinflammatory mediators in the pathogenesis of TSS has been reviewed in detail (47). It is proposed that *S. aureus* benefits from immunosuppression induced by the PTSAgs. Immunosuppression may be demonstrated in vitro and in patients. For example, recurrent TSS has been attributed to a failure to generate neutralizing antitoxin antibodies. Systemic exposure to TSST-1 or SEs causes a decline in function and levels of certain lymphocyte populations (75). Apoptosis mediated by Fas/Fas ligand is one mechanism by which SAgs delete certain

TABLE 1 General properties of currently known PTSAgs[a]

Toxin[b]	Molecular mass (Da)	pI	Human Vβ specificity
SEA	27,100	6.8–7.3	1.1, 5.3, 6.3, 6.4, 6.9, 7.3, 7.4, 9.1, 18
SEB	28,336	8.6	3, 12, 13.2, 14, 15, 17, 20
SEC1	27,496	8.6	3, 12, 13.2, 14, 15, 17, 20
SEC2	27,531	7.0	3, 12, 13.1, 13.2, 14, 17, 20
SEC3$_{FRI909}$	27,588	8.0	3, 12, 13.1, 13.2, 14, 17, 20
SEC3$_{FRI913}$	27,563	8.1	3, 12, 13.1, 13.2, 14, 17, 20
SEC$_{bovine}$	27,618	7.6	3, 12, 13.2, 14, 15, 17, 20
SEC$_{ovine}$	27,517	7.6	3, 12, 13.2, 14, 15, 17, 20
SEC$_{canine}$	27,600	7.0	3, 12, 13.2, 14, 15, 17, 20
SEE	26,425	8.5	5.1, 6.3, 6.4, 6.9, 8.1, 18
SEG	27,042	ND[c]	ND
SEH	25,145	5.7	ND
SEI	24,928	ND	ND
SEJ	ND[d]	ND	ND
TSST-1	22,049	7.2	2
TSST$_{ovine}$	22,095	8.5	ND

[a] Data compiled from references 47 and 79.
[b] Mature secreted form.
[c] ND, Not determined or reported.
[d] Based on nucleotide sequencing, the intracellular translation product is a 31,210-Da precursor. The N terminus or size of the extracellular form was not reported.

populations. CD8$^+$ cells suppress CD4$^+$ cell responses by inducing apoptosis of CD4$^+$ cells through ligation with Fas. In addition, SAg-activated T cells can undergo anergy in which they fail to proliferate or secrete IL-2 (28, 44). High doses of SAg also promote B-cell apoptosis and the down-regulation of immunoglobulin-secreting B cells (66).

Crystal structures have been reported for several PTSAgs (reviewed in reference 47). They are all, including TSST-1, compact ellipsoidal proteins folded into two domains composed of mixed α/β structures (Fig. 6A). The smaller domain of the SEs, domain B, contains most of the N-terminal half of the protein, excluding the N terminus. This domain is similar in topology to the well-known oligonucleotide/oligosaccharide binding motif, although there is at present no evidence that binding to carbohydrates or nucleic acids occurs or is required for function. The top of domain B of SEs also contains the disulfide loop, a conserved feature of all known SEs. The larger domain A has a β-grasp motif and is composed of C-terminal residues, plus the N terminus extending over the top. The interface between the two domains is delineated by a short shallow cavity at the top of the molecule and by a large groove extending all the way along the back of the molecule. Despite lacking sequence homology with the SEs, TSST-1 shares a similar domain structural organization. Four key SE features are absent in TSST-1: (i) the SE N terminus, which folds back over domain A; (ii) an α-helix in the loop at the base of domain B; (iii) a second α-helix positioned in the groove between both domains; and (iv) the disulfide loop. SEs A, D, and E require Zn^{2+} to stimulate T cells most efficiently (68). In SEA, Zn^{2+} is bound to the external portion of domain A, near the N terminus through residues H187, H225, and D227. Although not apparently required for activity, the SECs bind a Zn^{2+} atom at the base of the groove on the back side of the molecule (10). This binding site, including an HEXXH motif, resembles thermolysin, the prototype of one class of metalloenzymes. SED has a second Zn^{2+} binding site in a position similar to that of SECs. The role of this second site in these toxins may be to stabilize a local structural configuration (68).

Each PTSAg has a different affinity for, and recognizes a particular repertoire of, MHCII molecules. The toxins have evolved a variety of modes for binding to this receptor (13). An SEB:HLA-DR1 crystal structure (37) showed that SEB interacts with the HLA-DR1 α1-chain at a concave surface on the receptor adjacent to but outside of the peptide-binding groove. This interaction requires 19 SEB residues on the edge and top of domain B, in and near the cysteine loop. Binding to the receptor in this manner orients domain A away from the receptor α-chain. The crystal structure of HLA-DR1 complexed with TSST-1 (41) is similar to the SEB:HLA-DR1 structure. However, TSST-1 extends further over the top of the receptor, contacting the bound peptide and the β-chain. Binding of other SEs to MHCII is more complex. SEA has a low-affinity MHCII binding site that overlaps that of SEB/TSST-1, plus a high-affinity site on the outside of domain A near the N terminus (36). The high-affinity site presumably involves coordination of Zn^{2+} through three toxin residues and residue 81 of the MHCII β-chain. The existence of two MHCII binding sites per toxin suggests that two MHCII molecules could be cross-linked (70). Similar modes of action, including formation of homodimers by some toxins, have been proposed for other SEs with high-affinity Zn^{2+} binding sites on the outer face of domain A (SED, SEE, and SEH) (3, 68).

PTSAgs interact with a defined TCR repertoire (Table 1), determined by sequences of TCR Vβ and by toxin residues in the shallow cavity at the top of the molecule (19). A crystal structure of SEC3 complexed with part of the murine TCR β-chain (25) revealed that the toxin binds residues in the CDR1, CDR2, and HV4 loops of Vβ. Superimposing crystal structures from (i) the Vβ:SEC3 com-

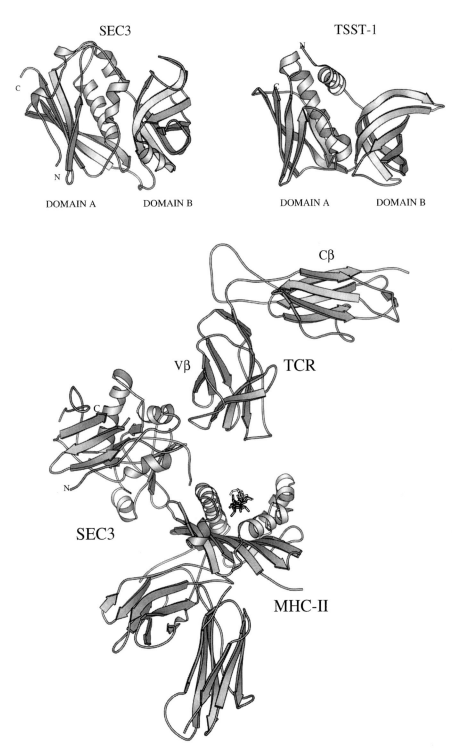

FIGURE 6 Structural properties and receptor interactions of PTSAgs. (A) A structural comparison of SEC3 and TSST-1. Ribbon diagrams shown are based on crystal structures published for the two toxins (10, 22). The two structures are oriented so that the TCR binding cavity in each is located at the top and the cysteine loop, unique to the SEs, is on the upper-right corner of SEC3. Both toxins possess a similar domain organization and an overall topology despite having several important differences, as discussed in the text. (B) A model of the trimolecular complex with SEB or SEC bound to TCR and MHCII (adapted from data in references 25 and 37). In this model, SAgs orient the two receptors away from each other, inducing an aberrant mechanism of T-cell activation. Note that antigenic peptide associated with MHCII is positioned away from the TCR binding site.

plex, (ii) the SEB:HLA-DR1 complex, and (iii) a TCR Vα homodimer led to a trimolecular complex model representing SAg binding to the T cell and APC (Fig. 6B). In this model the PTSAg wedges between the TCR and the MHCII. This orients the peptide binding cleft away from the TCR in a configuration greatly different from typical antigen presentation. The overall affinity of the entire complex determines the effectiveness of the stimulation (42). Binding by toxins with low affinity for the TCR can be compensated for by stronger binding to MHCII, and vice versa.

The roles of SAg activity and T-cell proliferation in staphylococcal food poisoning and TSS have been the subject of considerable investigative attention, and some evidence exists that SAg function alone is not significant to induce these illnesses. For example, studies with various SEs have generated mutants that dissociate T-cell proliferation from emesis. Substitution of residue 25 in the SEA TCR binding cavity dramatically reduces T-cell proliferation with no demonstrable effect on emesis (32). Similar conclusions were drawn from studies with SEB and SEC (1, 35). Although the disulfide bond and cysteine loop have been proposed to be involved in emesis, the contents of the loops are quite variable. Therefore, it is unlikely that the cysteines themselves or the loop residues play a key direct role in emesis. It is more likely that the disulfide bond maintains a conformation and orientation of other residues crucial for emesis (10, 35). It is suspected that a highly conserved stretch of residues in the SE small domain, immediately downstream from the disulfide bond, may mediate emesis and overlap or affect a site required for T-cell proliferation (32, 35).

It is generally agreed that the massive production of cytokines coinciding with T-cell proliferation contributes significantly to PTSAg-induced lethal shock, although work with TSST-1 site-specific mutants has suggested that induction of lethality is more complex than originally suspected. The cause of reduced or absent in vivo toxicity for most TSST-1 mutants can be traced to their defect in T-cell proliferative ability. Generally, the defect is attributed to an amino acid alteration that interferes with their ability to bind to either the TCR or MHCII. However, there is substantial evidence to suggest that an additional property of PTSAgs, separate from their T-cell proliferative ability, is sufficient to induce lethality. One interesting mutant, Q136A, retains the ability to induce T-cell proliferation but is devoid of lethal function in vivo (22). Residue 136 in TSST-1 is located in a largely buried site in the toxin molecule. The crystal structure of the Q136A mutant revealed that substitution of glutamine with alanine at position 136 caused a dramatic change in the conformation of the β7-β9 loop covering the back of the central α-helix (Fig. 6A).

Exfoliative (Epidermolytic) Toxins

The ETs have been conclusively implicated in staphylococcal scalded skin syndrome (SSSS). As defined by Melish and Glasgow (45), SSSS includes a spectrum of staphylococcal illnesses in patients, predominantly children, characterized by formation of bullae or skin blisters and a potential for widespread peeling. According to their designation, SSSS includes Ritter's disease, toxic epidermal necrosis, bullous impetigo, and some cases of erythema. Widespread skin peeling occurs in patients lacking ET antibodies, but localized lesions, as in bullous impetigo, develop if antibody is present in the serum. Melish and Glasgow also found that inoculating SSSS isolates in new-born mice generated a positive Nikolsky's sign and sterile skin lesions similar to those in humans with SSSS. They proposed that a soluble toxin, identified shortly thereafter, was responsible for effects in the mouse model. This model is still standard methodology for assessing the epidermolytic effects of ETs.

S. aureus isolates, especially phage group II, can express two antigenic forms of ETs, designated ETA and ETB (reviewed in reference 6). ETA is encoded by a chromosomal gene (eta), while the etb structural gene is on a plasmid. The two proteins share greater than 40% identity over 242 and 246 residues in ETA and ETB, respectively. In addition, both proteins are approximately 25% identical to the S. aureus serine (V8) protease; although they possess the serine protease catalytic triad of H72, D120, and S195, they are not detectably proteolytic (57). Crystal structures have been solved for ETA (12, 73). Structurally, it belongs to a family of serine proteases of which chymotrypsin is the prototype, containing two perpendicular β-barrel domains and C-terminal α-helix (Fig. 7). ETA possesses a unique N-terminal 15-residue α-helix that affects the position of loop D adjacent to the catalytic site. In one conformation, the peptide bond between residues P192 and G193 is flipped 180 degrees compared with other serine proteases, so that substrates would not have access to the active site. The putative latent protease might be activated by binding to a cellular receptor via the N-terminal helix, shifting loop D and opening the active site.

Lesions in SSSS in humans and mice are characterized by separation of stratum granulosa cells, causing intraepidermal skin peeling. With sufficient doses, ETs can induce effects in as little as 10 min. The physical separation of the skin coincides with desmosome degeneration along a plane that eventually marks the cleavage plane. Histological observations preceding desmosome degradation include gap formations between cells and distended intercellular spaces. Neither the mechanism of skin splitting by ETs nor the target of the toxins is known. Experimental evidence suggests that ET protease functions are essential. Mutant toxins with alterations of residues in the catalytic triad are not epidermolytic (55, 59). The ETs lack detectable protease activity but do have esterase activity (7). Although they bind to keratohyalin granules in cells of the stratum granulosum (64), they also bind to granules in cells of animals that are not susceptible to the toxins. Also, intact cells are impervious to the toxin. Several studies suggest that the toxins act extracellularly and are not cytotoxic for the cells in the epidermis.

The ETs also function as SAgs, inducing T-cell proliferation via a mechanism that requires APCs and occurs in a Vβ-dependent manner. Compared with PTSAgs, the ETs are 100-fold less potent in inducing T-cell proliferation and are less toxic in rabbits (48). Initial reports (14) suggested that ETs cause stimulation of human T cells bearing Vβ2. In contrast, Fleischer and Bailey (27) suggested that the ETs were not SAgs but that the putative activity resulted from SAg contamination of the ET preparations. They were able to purify commercial ETA such that fractions with the ability to stimulate human Vβ2 T cells were separated from the toxin. Monday et al. (48) clarified this issue and confirmed that both ETs are, in fact, SAgs. However, highly purified and recombinant ETs expressed in either S. aureus or Escherichia coli selectively expanded human T cells expressing Vβs 3, 12, 13.2, 14, 15, and 17, but not Vβ2. It is not clear whether SAg activity contributes to SSSS, since skin-peeling ability and T-cell proliferation are separable by mutagenesis. One ETA mutant with an altered

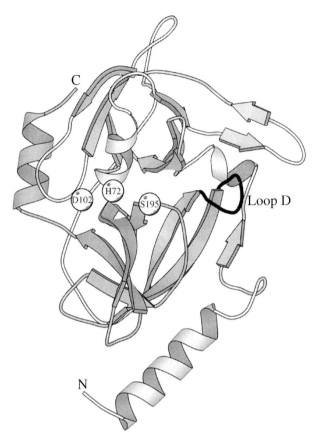

FIGURE 7 Ribbon diagram of the ETA crystal structure showing important functional features. Similar to other chymotrypsinlike proteases, ETA has two β-barrel domains and a C-terminal α-helix. The N-terminal domain, which includes a highly charged α-helix, is unique and is suspected to be involved in receptor binding. The positions of residues H72, D102, and S195, comprising the putative catalytic triad, are superimposable with the analogous residues of α-thrombin. D164 in loop D controls access of substrate to the protease active site by hydrogen bonding to G193. This causes the P192-G193 peptide bond to flip 180 degrees compared with that seen in other serine proteases and may explain the lack of demonstrable proteolytic activity in vitro. Binding of the N-terminal α-helix to its receptor has been proposed to cause a shift in the position of loop D and thereby the P192-G193 peptide bond, allowing access to the active site in vivo (73).

active-site serine (S195C) is unable to induce skin peeling despite inducing T-cell proliferation similarly to the native toxin (73).

We thank S. Bhakdi, D. Boehm, K. Dziewanowska, Y. Piemont, D. Ohlendorf, P. Schlievert, and G. Vath for helpful discussions, review of the manuscript, and in some cases providing figures. C. Deobald and M. Marshall assisted in preparation of the manuscript and figures. This work was supported by grants from the U.S. Public Health Service (AI28401) and the United Dairymen of Idaho (to G. A. B.) and from The Wellcome Trust, The Health Research Board of Ireland, BioResearch Ireland, and Inhibitex, Inc. (to T. J. F.).

REFERENCES

1. **Alber, G., D. K. Hammer, and B. Fleischer.** 1990. Relationship between enterotoxic- and T lymphocyte-stimulating activity of staphylococcal enterotoxin B. *J. Immunol.* **144:**4501–4506.

2. **Alber, G., P. H. Scheuber, B. Reck, B. Sailer-Kramer, A. Hartmann, and D. K. Hammer.** 1989. Role of substance P in immediate-type skin reactions induced by staphylococcal enterotoxin B in unsensitized monkeys. *J. Allergy Clin. Immunol.* **84:**880–885.

3. **Al-Daccak, R., K. Mehindate, F. Damdoumi, P. Etongue-Mayer, H. Nilsson, P. Antonsson, M. Sundstrom, M. Dohlsten, R. P. Sekaly, and W. Mourad.** 1998. Staphylococcal enterotoxin D is a promiscuous superantigen offering multiple modes of interactions with the MHC class II receptors. *J. Immunol.* **160:**225–232.

4. **Alouf, J. E.** 1977. Cell membranes and cytolytic bacterial toxins, p. 220–270. *In* P. Cuatrecasas (ed.), *Receptors and Recognition*, series B, vol. 1. *The Specificity and Action of Animal, Bacterial and Plant Toxins.* Chapman and Hall Ltd., London, U.K.

5. **Arbuthnott, J. P.** 1982. Bacterial cytolysins (membrane-damaging toxins), p. 107–129. *In* P. Cohen and S. van Heyningen (ed.), *Molecular Action of Toxins and Viruses.* Elsevier Biomedical Press, Amsterdam, The Netherlands.

6. **Bailey, C. J., B. P. Lockhart, M. B. Redpath, and T. P. Smith.** 1995. The epidermolytic (exfoliative) toxins of *Staphylococcus aureus. Med. Microbiol. Immunol.* **184:**53–61.

7. **Bailey, C. J., and M. B. Redpath.** 1992. The esterolytic activity of epidermolytic toxins. *Biochem. J.* **284:**177–180.

8. **Bhakdi, S., and J. Tranum-Jensen.** 1991. Alpha-toxin of *Staphylococcus aureus. Microbiol. Rev.* **55:**733–751.

9. **Bohach, G. A., D. J. Fast, R. D. Nelson, and P. M. Schlievert.** 1990. Staphylococcal and streptococcal pyrogenic toxins involved in toxic shock syndrome and related illnesses. *Crit. Rev. Microbiol.* **17:**251–272.

10. **Bohach, G. A., L. M. Jablonski, C. F. Deobald, Y. I. Chi, and C. V. Stauffacher.** 1995. Functional domains of staphylococcal enterotoxins, p. 339–356. *In* M. Ecklund, J. L. Richard, and K. Mise (ed.), *Molecular Approaches to Food Safety: Issues Involving Toxic Microorganisms.* Alaken, Inc., Fort Collins, Colo.

11. **Boyle, T., V. Lancaster, R. Hunt, P. Gemski, and M. Jett.** 1994. Method for simultaneous isolation and quantitation of platelet activating factor and multiple arachidonate metabolites from small samples: analysis of effects of *Staphylococcus aureus* enterotoxin B in mice. *Anal. Biochem.* **216:**373–382.

12. **Cavarelli, J., G. Prevost, W. Bourguet, L. Moulinier, B. Chevrier, B. Delagoutte, A. Bilwes, L. Mourey, S. Rifai, Y. Piemont, and D. Moras.** 1997. The structure of *Staphylococcus aureus* epidermolytic toxin A, an atypic serine protease, at 1.7 Å resolution. *Structure* **5:**813–824.

13. **Chintagumpala, M. M., J. A. Mollick, and R. R. Rich.** 1991. Staphylococcal toxins bind to different sites on HLA-DR. *J. Immunol.* **147:**3876–3881.

14. **Choi, Y., B. Kotzin, L. Herron, J. Callahan, P. Marrack, and J. Kappler.** 1989. Interaction of *Staphylococcus aureus* toxin superantigens with human T cells. *Proc. Natl. Acad. Sci. USA* **86:**8941–8945.

15. **Coleman, D. C., J. P. Arbuthnott, H. M. Pomeroy, and T. H. Birkbeck.** 1986. Cloning and expression in *Escherichia coli* and *Staphylococcus aureus* of the beta-lysin determinant from *Staphylococcus aureus*: evidence that bacteriophage conversion of beta-lysin activity is caused

by insertional inactivation of the beta-lysin determinant. *Microb. Pathog.* **1:**549–564.

16. **Colin, D. A., I. Mazurier, S. Sire, and V. Finck-Barbancon.** 1994. Interaction of the two components of leukocidin from *Staphylococcus aureus* with human polymorphonuclear leukocyte membranes: sequential binding and subsequent activation. *Infect. Immun.* **62:**3184–3188.

17. **Colin, D. A., O. Meunier, L. Staali, H. Monteil, and G. Prevost.** 1996. Action mode of two components pore-forming leucotoxins from *Staphylococcus aureus*. *Med. Microbiol. Immunol.* **185:**107–114.

18. **Cooney, J., Z. Kienle, T. J. Foster, and P. W. O'Toole.** 1993. The gamma-hemolysin locus of *Staphylococcus aureus* comprises three linked genes, two of which are identical to the genes for the F and S components of leukocidin. *Infect. Immun.* **61:**678–771.

19. **Deringer, J. R., R. J. Ely, C. V. Stauffacher, and G. A. Bohach.** 1996. Subtype-specific interactions of type C staphylococcal enterotoxins with the T-cell receptor. *Mol. Microbiol.* **22:**523–534.

20. **Dufourc, E. J., J. Dufourcq, T. H. Birkbeck, and J. H. Freer.** 1990. δ-Haemolysin from *Staphylococcus aureus* and model membranes. A solid-state ^{2}H-NMR and ^{31}P-NMR study. *Eur. J. Biochem.* **187:**581–587.

21. **Dziewanowska, K., V. E. Edwards, J. R. Deringer, G. A. Bohach, and D. J. Guerra.** 1996. Comparison of the beta-toxins from *Staphylococcus aureus* and *Staphylococcus intermedius*. *Arch. Biochem. Biophys.* **335:**102–108.

22. **Earhart, C. A., D. T. Mitchell, D. L. Murray, D. M. Pinheiro, M. Matsumura, P. M. Schlievert, and D. H. Ohlendorf.** Structures of five mutants of toxic shock syndrome toxin-1 with reduced biological activity. *Biochemistry* **37:**7194–7202.

23. **Edwards, V. M., J. R. Deringer, S. D. Callantine, C. F. Deobald, P. H. Berger, V. Kapur, C. V. Stauffacher, and G. A. Bohach.** 1997. Characterization of the canine type C enterotoxin produced by *Staphylococcus intermedius* pyoderma isolates. *Infect. Immun.* **65:**2346–2352.

24. **Elwell, M. R., C. T. Liu, R. O. Spertzel, and W. R. Beisel.** 1975. Mechanisms of oral staphylococcal enterotoxin B-induced emesis in the monkey. *Proc. Soc. Exp. Biol. Med.* **148:**424–427.

25. **Fields, B. A., E. L. Malchiodi, H. Li, X. Ysern, C. V. Stauffacher, P. M. Schlievert, K. Karjalainen, and R. A. Mariuzza.** 1996. Crystal structure of a T-cell receptor beta-chain complexed with a superantigen. *Nature* **384:**188–192.

26. **Finck-Barbancon, V., G. Duportail, O. Meunier, and D. A. Colin.** 1993. Pore formation by two-component leukocidin from *Staphylococcus aureus* within the membrane of human polymorphonuclear leukocytes. *Biochim. Biophys. Acta* **1182:**275–282.

27. **Fleischer, B., and C. J. Bailey.** 1992. Recombinant epidermolytic (exfoliative) toxin A of *Staphylococcus aureus* is not a superantigen. *Med. Microbiol. Immunol.* **180:**273–279.

28. **Florquin, S., and L. Aaldering.** 1997. Superantigens: a tool to gain new insight into cellular immunity. *Res. Immunol.* **148:**373–386.

29. **Foster, T. J., M. O'Reilly, P. Phonimdaeng, J. Cooney, A. H. Patel, and A. J. Bramley.** 1990. Genetic studies of virulence factors of *Staphylococcus aureus*. Properties of coagulase and gamma-toxin and the role of alpha-toxin,

beta-toxin and protein A in the pathogenesis of *S. aureus* infections, p. 403–417. *In* R. P. Novick (ed.), *Molecular Biology of the Staphylococci*, VCH Publishers, Cambridge, U.K.

30. **Fraser, J. D.** 1989. High-affinity binding of staphylococcal enterotoxins A and B to HLA-DR. *Nature* **339:**221–223.

31. **Gravet, A., D. A. Colin, R. Keller, H. Giradot, H. Monteil, and G. Prevost.** 1998. Characterization of a novel structural member, LukE-LukD, of the bi-component staphylococcal leucotoxins family. *FEBS Lett.* **436:**202–208.

32. **Harris, T. O., and M. J. Betley.** 1995. Biological activities of staphylococcal enterotoxin type A mutants with N-terminal substitutions. *Infect. Immun.* **63:**2133–2140.

33. **Hildebrand, A., M. Roth, and S. Bhakdi.** 1991. *Staphylococcus aureus* alpha-toxin: dual mechanisms of binding to target cells. *J. Biol. Chem.* **266:**17195–17200.

34. **Holmberg, S. D., and P. A. Blake.** 1984. Staphylococcal food poisoning in the United States. New facts and old misconceptions. *JAMA* **251:**487–489.

35. **Hovde, C. J., J. C. Marr, M. L. Hoffmann, S. P. Hackett, Y. I. Chi, K. K. Crum, D. L. Stevens, C. V. Stauffacher, and G. A. Bohach.** 1994. Investigation of the role of the disulphide bond in the activity and structure of staphylococcal enterotoxin C1. *Mol. Microbiol.* **13:**897–909.

36. **Hudson, K. R., R. E. Tiedemann, R. G. Urban, S. C. Lowe, J. L. Strominger, and J. D. Fraser.** 1995. Staphylococcal enterotoxin A has two cooperative binding sites on major histocompatibility complex class II. *J. Exp. Med.* **182:**711–720.

37. **Jardetsky, T. S., J. H. Brown, J. C. Gorga, L. J. Stern, R. G. Urban, Y. I. Chi, C. V. Stauffacher, J. L. Strominger, and D. C. Wiley.** 1994. Three-dimensional structure of a human class II histocompatibility molecule complexed with superantigen. *Nature* **368:**711–718.

38. **Jarvis, W. D., R. N. Kolesnick, F. A. Fornari, R. S. Traylor, D. A. Gewirtz, and S. Grant.** 1994. Induction of apoptotic DNA damage and cell death by activation of the sphingomyelin pathway. *Proc. Natl. Acad. Sci. USA* **91:**73–77.

39. **Jonas, D., I. Walev, T. Berger, M. Liebetrau, M. Palmer, and S. Bhakdi.** 1994. Novel path to apoptosis: small transmembrane pores created by staphylococcal alpha-toxin in T lymphocytes evoke internucleosomal DNA degradation. *Infect. Immun.* **62:**1304–1312.

40. **Kaneko, J., O. Toshiko, T. Tomita, and Y. Kamio.** 1997. Sequential binding of staphylococcal γ-hemolysin to human erythrocytes and complex formation of the hemolysin on the cell surface. *Biosci. Biotech. Biochem.* **61:**846–851.

41. **Kim, J., R. G. Urban, J. L. Strominger, and D. C. Wiley.** 1994. Toxic shock syndrome toxin-1 complexed with a class II major histocompatibility molecule HLA-DR1. *Science* **266:**1870–1874.

42. **Leder, L., A. Llera, P. M. Lavoie, M. I. Lebedeva, H. Li, R. P. Sekaly, G. A. Bohach, P. J. Gahr, P. M. Schlievert, K. Karjalainen, and R. A. Mariuzza.** 1998. A mutational analysis of the binding of staphylococcal enterotoxins B and C3 to the T cell receptor beta chain and major histocompatibility complex class II. *J. Exp. Med.* **187:**823–833.

43. **Low, D. K. R., and J. H. Freer.** 1977. Biological effects of highly purified β-lysin (sphingomyelinase C) from *Staphylococcus aureus*. *FEMS Microbiol. Lett.* **2:**133–138.

44. Mahlknecht, U., M. Herter, M. K. Hoffmann, D. Niethammer, and G. E. Dannecker. 1996. The toxic shock syndrome toxin-1 induces anergy in human T cells in vivo. *Hum. Immunol.* **45:**42–45.

45. Melish, M. E., and L. A. Glasgow. 1970. The staphylococcal scalded skin syndrome: development of an experimental model. *N. Engl. J. Med.* **282:**1114–1119.

46. Mellor, I. R., D. H. Thomas, and M. S. P. Sansom. 1988. Properties of ion channels formed by *Staphylococcus aureus* δ-toxin. *Biochim. Biophys. Acta* **942:**280–294.

47. Monday, S. R., and G. A. Bohach. Genetic, structural, biological, pathophysiological and clinical aspects of *Staphylococcus aureus* enterotoxins and toxic shock syndrome toxin-1. *In* J. E. Alouf and J. H. Freer (ed.), *Sourcebook of Bacterial Protein Toxins*, 2nd ed., in press. Academic Press, Inc., London, U.K.

48. Monday, S. R., G. M. Vath, W. A. Ferens, C. Deobald, J. V. Rago, P. J. Gahr, D. Monie, J. J. Iandolo, S. K. Chapes, W. C. Davis, D. H. Ohlendorf, P. M. Schlievert, and G. A. Bohach. Unique superantigen activity of staphylococcal exfoliative toxins. *J. Immunol.* **181:**4550–4559.

49. Munson, S. H., M. T. Tremaine, M. J. Betley, and R. A. Welch. 1998. Identification and characterization of staphylococcal enterotoxin types G and I from *Staphylococcus aureus*. *Infect. Immun.* **66:**3337–3348.

50. Noda, M., and I. Kato. 1991. Leukocidal toxins, p. 243–251. *In* J. E. Alouf and J. H. Freer (ed.), *Sourcebook of Bacterial Protein Toxins*. Academic Press, Inc., London, U.K.

51. O'Callaghan, R. J., M. C. Callegan, J. M. Moreau, L. C. Green, T. J. Foster, O. M. Hartford, L. S. Engel, and J. M. Hill. 1997. Specific roles of alpha-toxin and beta-toxins during *Staphylococcus* corneal infection. *Infect. Immun.* **65:**1571–1578.

52. Ozawa, T., J. Kaneko, and Y. Kamio. 1995. Essential binding of LukF of staphylococcal γ-hemolysin followed by the binding of HγII for the hemolysis of human erythrocytes. *Biosci. Biotech. Biochem.* **59:**1181–1183.

53. Ozawa, T., J. Kaneko, H. Narija, K. Izaki, and Y. Kamio. 1994. Inactivation of the γ-hemolysin HγII component by addition of monoganglioside G_{M1} to human erythrocyte. *Biosci. Biotech. Biochem.* **58:**602–605.

54. Parsonnet, J. 1998. Case definition of staphylococcal TSS: a proposed revision incorporating laboratory findings, p. 15. *In* F. Arbuthnott and B. Furman (ed.), *Proceedings of the European Conference on Toxic Shock Syndrome*. The Royal Society of Medicine Limited, London, U.K.

55. Prevost, G., P. Coupie, P. Prevost, S. Gayet, P. Petiau, B. Cribier, H. Monteil, and Y. Piemont. 1995. Epidemiological data on *Staphylococcus aureus* strains producing synergohymenotropic toxins. *J. Med. Microbiol.* **42:**237–245.

56. Prevost, G., B. Cribier, P. Coupie, P. Petiau, G. Supersac, V. Finck-Barbancon, H. Monteil, and Y. Piemont. 1995. Panton-Valentine leukocidin and gamma-hemolysin from *Staphylococcus aureus* ATCC 49775 are encoded by distinct genetic loci and have difference biological activities. *Infect. Immun.* **63:**4121–4129.

57. Prevost, G., S. Rifai, M. L. Chaix, S. Meyer, and Y. Piemont. 1992. Is the His72, Asp120, Ser195 constitutive of the catalytic site of staphylococcal exfoliative toxin A?, p. 488–489. *In* B. Witholt (ed.), *Bacterial Protein Toxins*. Gustav Fischer, Stuttgart, Germany.

58. Projan, S. J., J. Kornblum, B. Kreiswirth, S. L. Moghazeh, W. Eisner, and R. P. Novick. 1989. Nucleotide sequence: the β-hemolysin gene of *Staphylococcus aureus*. *Nucleic Acid Res.* **17:**3305.

59. Redpath, M. B., T. J. Foster, and C. J. Bailey. 1991. The role of the serine protease active site in the mode of action of epidermolytic toxin of *Staphylococcus aureus*. *FEMS Microbiol. Lett.* **81:**151–156.

60. Reingold, A. L., N. T. Hargrett, K. N. Shands, B. B. Dan, G. P. Schmid, B. Y. Strickland, and C. V. Broome. 1982. Toxic shock syndrome surveillance in the United States, 1980 to 1981. *Ann. Intern. Med.* **96:**875–880.

61. Rich, R. R., J. A. Mollick, and R. G. Cook. 1989. Superantigens: interaction of staphylococcal enterotoxins with MHC class II molecules. *Trans. Am. Clin. Climatol. Assoc.* **101:**195–204.

62. Rogalsky, M. 1979. Nonenteric toxins of *Staphylococcus aureus*. *Microbiol. Rev.* **43:**320–360.

63. Scheuber, P. H., C. Denzlinger, C. D. Wilker, G. Beck, D. Keppler, and D. K. Hammer. 1987. Staphylococcal enterotoxin B as a nonimmunological mast cell stimulus in primates: the role of endogenous cysteinyl leukotrienes. *Int. Arch. Allergy Appl. Immunol.* **82:**289–291.

64. Smith, T. P., D. A. John, and C. J. Bailey. 1987. The binding of epidermolytic toxin from *Staphylococcus aureus* to mouse epidermal tissue. *Histochem. J.* **19:**137–149.

65. Song, L., M. R. Hobaugh, C. Shustak, S. Cheley, H. Bayley, and J. E. Gouaux. 1996. Structure of the staphylococcal α-hemolysin, a heptameric transmembrane pore. *Science* **274:**1859–1865.

66. Stohl, W., J. E. Elliott, D. H. Lynch, and P. A. Kiener. 1998. CD95 (Fas)-based, superantigen-dependent, CD4+ T cell-mediated down-regulation of human in vitro immunoglobulin responses. *J. Immunol.* **160:**5231–5238.

67. Sugawara, N., T. Tomita, and Y. Kamio. 1997. Assembly of γ-hemolysin into a pore-forming ring-shaped complex on the surface of human erythrocytes. *FEBS Lett.* **410:**333–337.

68. Sundstrom, M., L. Abrahmsen, P. Antonsson, K. Mehindate, W. Mourad, and M. Dohlsten. 1996. The crystal structure of staphylococcal enterotoxin type D reveals Zn2+-mediated homodimerization. *EMBO J.* **15:**6832–6840.

69. Tappin, M. J., A. Pastore, R. S. Norton, J. H. Freer, and I. D. Campbell. 1988. High-resolution ^{1}H NMR study of the solution structure of δ-hemolysin. *Biochemistry* **27:**1643–1647.

70. Tiedemann, R. E., and J. D. Fraser. 1996. Cross-linking of MHC class II molecules by staphylococcal enterotoxin A is essential for antigen-presenting cell and T cell activation. *J. Immunol.* **157:**3958–3966.

71. Tomita, T., Y. Ueda, H. Tamura, R. Taguchi, and H. Ikezawa. 1993. The role of acidic amino-acid residues in catalytic and adsorptive sites of *Bacillus cereus* sphingomyelinase. *Biochim. Biophys. Acta* **1203:**85–92.

72. Valeva, A., A. Weisser, B. Walker, M. Kehoe, H. Bayley, S. Bhakdi, and M. Palmer. 1996. Molecular architecture of a toxin pore: a 15 residue sequence lines the transmembrane channel of staphylococcal α-toxin. *EMBO J.* **15:**1857–1864.

73. **Vath, G. M., C. A. Earhart, J. V. Rago, M. H. Kim, G. A. Bohach, P. M. Schlievert, and D. H. Ohlendorf.** 1997. The structure of the superantigen exfoliative toxin A suggests a novel regulation as a serine protease. *Biochemistry* **36:**1559–1566.

74. **Walev, I., U. Weller, S. Strauch, T. Foster, and S. Bhakdi.** 1996. Selective killing of human monocytes and cytokine release provoked by sphingomyelinase (beta toxin) of *Staphylococcus aureus. Infect. Immun.* **64:**2974–2979.

75. **White, J., A. Herman, A. M. Pullen, R. Kubo, J. W. Kappler, and P. Marrack.** 1989. The V beta-specific superantigen staphylococcal enterotoxin B: stimulation of mature T cells and clonal deletion in neonatal mice. *Cell* 56:27–35.

76. **Williams, R. E. O., and G. H. Harper.** 1947. Staphylococcal haemolysins on sheep blood agar with evidence for a fourth haemolysin. *J. Pathol. Bacteriol.* **59:**69–78.

77. **Wiseman, G. M.** 1975. The hemolysins of *Staphylococcus aureus. Bacteriol. Rev.* **39:**317–344.

78. **Woodin, A. M.** 1970. Staphylococcal leukocidin, p. 327–355. *In* T. C. Montie, S. Kadis, and S. J. Ajl (ed.), *Microbial Toxins.* Academic Press, Inc., New York, N.Y.

79. **Zhang, S., J. J. Iandolo, and G. C. Stewart.** 1998. The enterotoxin D plasmid of *Staphylococcus aureus* encodes a second enterotoxin determinant (*sej*). *FEMS Microbiol. Lett.* **168:**227–233.

Extracellular Enzymes

STAFFAN ARVIDSON

39

Staphylococcus aureus produces a large number of extracellular proteins that have enzymatic activity or that can activate host enzymes. In a review published in 1983, 15 enzymes and enzyme activators were described (1). Since that time a few more enzymes have been added to the list (Table 1), and new information on the structure and function of many exoenzymes has been added, mainly as a result of cloning and sequencing of their respective genes.

In this chapter I describe the basic characteristics of the enzymes secreted by *S. aureus* and what is known about their respective genes, synthesis, and biological functions. Not all enzymes are described; some are omitted because they are described elsewhere in this volume, others because of a lack of significant new information, and others simply because of limited space.

COAGULASE

Coagulase production is the principal criterion used in the clinical microbiology laboratory for the identification of *S. aureus*. Although a few strains of *S. aureus* do not produce detectable amounts of coagulase, all strains seem to possess a coagulase gene (*coa*) (50).

Coagulase binds with human prothrombin in a 1:1 molar ratio to form a complex named staphylothrombin, which can convert fibrinogen to fibrin (21). Unlike the physiological activation of thrombin, formation of staphylothrombin does not involve proteolytic cleavage of prothrombin. The mechanism of this nonproteolytic activation of prothrombin is so far unknown.

Coagulase also binds to fibrinogen (4). A fraction of coagulase, which is associated with the bacterial cell surface, was therefore thought to be responsible for the clumping of bacterial cells when mixed with plasma. However, clumping factor is a distinct fibrinogen-binding protein that can promote binding of bacteria to solid-phase fibrinogen, in contrast to coagulase, which binds only soluble fibrinogen (31). Coagulase does not possess a cell wall-anchoring sequence (LPXTG) but seems to be associated with the bacterial cells by other means. A fibrinogen-binding protein (FbpA) belonging to the coagulase family,

with a unique sequence, LPXSITG, in the middle of the molecule, has been characterized (9). Although this potential anchoring signal is not followed by a typical membrane-spanning region and a charged tail, it is speculated that the LPXSITG motif could participate in anchoring this new type of coagulase to the cell wall.

Eight serotypes of coagulase have been identified by neutralization tests. The primary structures of coagulases belonging to serotypes I, II, and III have been compared (37). Four distinct segments of coagulase can be recognized: (i) a typical signal peptide of 26 amino acid residues; (ii) a highly variable N-terminal region of 150 to 270 amino acid residues, which are only 50% identical between the serotypes; (iii) a central region with over 90% identical residues; and (iv) a C-terminal region composed of 5, 6, or 8 tandem repeats of 27 amino acid residues. The prothrombin-binding region of coagulase is in the variable N terminus, while fibrinogen binds to the C-terminal repeat region (31).

Although the 81-bp tandem repeats encoding the C-terminal region of coagulase are well conserved, individual repeats differ in the presence or absence of *Alu*I and *Cfo*I restriction endonuclease sites. Based on the restriction fragment length polymorphism of the 3′ end of the coagulase gene, it has been possible to discriminate between isolates of *S. aureus* in epidemiological studies (16, 17).

Production of coagulase is negatively regulated by *agr*. Consistent with the regulatory model, production of coagulase in wild-type bacteria is maximum during early exponential growth, while in *agr* mutants production of coagulase continues throughout growth. However, the levels of *coa*-specific transcript and extracellular coagulase activity were much lower in the *agr* mutant than in the wild type, indicating that *agr* also has a positive effect on *coa* expression (25). Transcription of *coa* is also affected by the *sae* locus. In an *agr:sae* double mutant, *coa*-specific mRNA levels were markedly reduced compared with the *agr* single mutant, suggesting that *sae* is a positive regulator (15).

The role of coagulase in staphylococcal pathogenesis is unclear. One possible function could be that fibrin clotting around a focal infection protects the bacteria from various

Gram-Positive Pathogens, ed. by V. A. Fischetti et al.
© 2000 American Society for Microbiology, Washington, D.C.

A third putative thiol protease seems to be produced from a gene (*sasp* ORF2) transcribed together with the Glu-specific V8 protease (18, 34). The COOH-terminal 174-amino-acid residues of this predicted 393-amino-acid protein shows 47% identity to mature staphopain (Fig. 1). The presence of a typical N-terminal signal peptide and signal peptidase cleavage site suggests that this is an extracellular thiol protease. The N-terminal half of the protein, which consists of 34% charged amino acids, showed no obvious similarity to any known proteins and may represent a propeptide that is cleaved off after secretion. Nothing is known about the enzymatic activity of this putative protease.

Metalloprotease

The amino acid sequence and the three-dimensional structure of *S. aureus* metalloprotease (aureolysin, protease III) have been determined (2). This extracellular enzyme is a 33-kDa, calcium-binding, zinc-requiring endopeptidase with optimum activity at neutral pH. While zinc appears to be required for the catalytic activity, calcium ions seem to stabilize the protein against proteolytic degradation. Like other neutral metalloproteases, it cleaves peptide bonds on the N-terminal side of bulky hydrophobic amino acid residues. The metalloprotease appears to be involved in processing of V8 protease zymogen (13). Its three-dimensional structure is very similar to that of thermolysin from *Bacillus thermoproteolyticus*, neutral protease from *Bacillus cereus*, and elastase from *Pseudomonas aeruginosa* (2). All the amino acid residues involved in substrate binding and the catalytic activity are conserved among these proteases. However, some differences were seen in the active-site cleft, which could explain the lack of elastase activity of aureolysin as compared with the other enzymes. The amino acid sequence of *S. aureus* metalloprotease is 79% identical to that of the extracellular mature form (300 amino acid residues) of *Staphylococcus epidermidis* metalloprotease (SepA). SepA is synthesized as a preproenzyme of 507 amino acids (49). It is therefore reasonable to believe that the *S. aureus* protease is also synthesized as a preproenzyme, although the corresponding gene has not yet been identified.

LIPASE

Glycerol Ester Hydrolase

True lipases are glycerol ester hydrolases that degrade water-insoluble long-chain triacylglycerols at the lipid-water interface. Interaction with the substrate interface leads to an increased enzymatic activity, known as interfacial activation. Most lipases are also active against acyl *p*-nitrophenylesters, Tweens (polyoxyethylenesorbitan), and sometimes phospholipids. Related enzymes hydrolyzing preferentially water-soluble glycerol esters are also called lipases but should properly be referred to as short-chain glycerol ester hydrolases, or esterases. Both types of enzyme belong to the family of serine esterases, possessing a catalytic triad consisting of a serine, a histidine, and either glutamic acid or aspartic acid.

Both types of enzyme are produced by *S. aureus*. Esterase activity toward water-soluble glycerol esters is produced by nearly all strains, while true lipases are less common. True lipase is produced by strains TEN5 and PS54 (24, 41). This enzyme hydrolyzes long-chain triacylglycerols as well as water-soluble triacylglycerols and Tweens. It has a pH optimum for activity at 8.0 to 8.5 and is stimulated by calcium ions. TEN5 lipase has an apparent molecular mass of 44 to 46 kDa, while the predicted product of the lipase gene (*geh*) has a molecular mass of 76.4 kDa. Lipase is synthesized as preproenzymes with a typical signal peptide of 37 amino acids followed by a propeptide of 258 amino acid residues, which is cleaved by an extracellular protease after secretion. The prolipase and the mature lipase show the same enzymatic activity. The propeptide, which is highly hydrophilic, has been suggested to stimulate secretion of the more hydrophobic lipase (41).

Production of lipase by strain PS54 is subject to negative lysogenic conversion by bacteriophage L54a, which can insert in the lipase coding sequence (26). The attachment site is located at the 3′ end of the gene, close to the active-site histidine.

The "lipase" produced by strain NTCT8530 cleaves preferentially short-chain triacylglycerols, while most long-chain lipids are not hydrolyzed at all (44). Maximum activity is toward butyryl esters at pH 6.5. It does not show interfacial activation and should therefore not be classified as a true lipase. For clarity, this enzyme will henceforth be referred to as staphylococcal butyryl esterase. The gene encoding butyryl esterase was originally denoted as *lip* (Table 1) and later as *geh* (36). As these names have already been used for other *S. aureus* genes (26), *beh* (butyryl ester hydrolase) would be more appropriate. In Table 1 the butyryl esterase gene is called *geh*NTCT8530.

Like the true lipase, the butyryl esterase is synthesized as a preproenzyme with a predicted molecular mass of about

```
Staphopain:   YNEQYVNKLENFKIRETQGNNGWCAGYTMSALLNATYNTNKYHAEAVMRF 50
              QY N L+NFKIRE Q +N WCAG++M+ALLNAT NT+ Y+A  +MR
sasp ORF 2:   DQVQYENTLKNFKIREQQFDNSWCAGFSMAALLNATKNTDTYNAHDIMRT 241


Staphopain:   LHPNLQGQQFQFTGLTPREMIYFGQTQGRSPQLLNRMTTYNEVDNLTKNN 100
              L+P + Q      P +MI +G++QGR    + +Y +VD LTK+N
sasp ORF 2:   LYPEVSEQDLPNCATFPNQMIEYGKSQGRDIHYQEGVPSYEQVDQLTKDN 291


Staphopain:   KGIAILGSRV-ESRNGMHAGHAMAVVGNAKLNNGQEVIIIWNPWDNGFMT 150
              GI IL   V ++ N  H GHA+AVVGNAK+N+ QE +I WNPWD
sasp ORF 2:   VGIMILAQSVSQNPNDPHLGHALAVVGNAKIND-QEKLIYWNPWDTELSI 341


Staphopain:   QDAKNNVIPVSNGDHYQWYSSIYGY 174
              QDA ++++ +S   Y WY S+ GY
sasp ORF2:    QDADSSLLHLSFNRDYNWYGSMIGY 365
```

FIGURE 1 Comparison of the amino acid sequences of the open reading frame downstream of *sasp* (*sasp* ORF 2) with that of staphopain. Consensus amino acid residues around the putative active-site cysteine (C), histidine (H), and asparagine (N) are underlined.

75 kDa. Mature enzyme has a molecular mass of 44 to 48 kDa. The proregion of the butyryl esterase is only 14% identical to that of the lipase. However, the hydropathy and flexibility plots of the proregions are nearly superimposable, suggesting a structural rather than a catalytic role of the propeptide. Gene fusion experiments indicate that the proregion is important for the secretion of butyryl esterase or other heterologous proteins (27). The mature esterase is 54% identical to the true lipase, with the highest homology around the active-site residues, Ser-412, Asp-603, and His-645, while it is 83% identical to lipase (GehSE1) from *S. epidermidis*, which is a typical butyryl esterase (45).

The function of lipases is not fully understood, but they may be important for bacterial nutrition. A role in virulence has also been suggested, based on the observation that staphylococcal lipase impairs granulocyte function (41). Free fatty acids, resulting from lipase activity, are also known to impair the immune system. On the other hand, long-chain fatty acids are bacteriocidal and would therefore seem to interfere with pathogenicity. However, fatty acids are detoxified by fatty acid-modifying enzyme, which is secreted by most lipase-producing strains of *S. aureus* (see below).

It has been reported that expression of lipase is positively regulated by *agr* and negatively regulated by *sar* (8). However, since lipase activity was assayed using substrates that do not discriminate between the true lipase and the butyryl esterase, it is not known which enzyme is regulated.

PI-Phospholipase C

Two different phospholipases C are produced by *S. aureus*. One is the hemolytic sphingomyelinase referred to as β-hemolysin, which is described in chapter 38, this volume. The other is a phosphatidylinositol-specific (PI) phospholipase, PI-PLC, which was identified in *S. aureus* culture supernatants over 35 years ago. Because it did not show cytotoxicity, it was long overlooked as a potential virulence factor. However, PI-PLC can degrade membrane-associated inositol phospholipids and release glycan-PI–anchored cell surface proteins, thereby interfering with important eukaryotic cell functions (30). Over 150 proteins are known to be bound to the cell surface via glycan-PI. Among these are several hydrolytic ectoenzymes, adhesion molecules, and various receptors and surface antigens. Inositol phospholipids are also involved in many signal transduction processes. Thus, PI-PLC has the potential to compromise host cell functions in a number of ways that could contribute to staphylococcal pathogenicity.

The gene encoding PI-PLC has been cloned and sequenced (Table 1), and the gene product has been characterized (12). The *plc* gene encodes a mature protein with a calculated molecular mass of 34,107 Da, which is in good agreement with the apparent molecular mass of purified PI-PLC (32 kDa). Northern blotting revealed a *plc*-specific mRNA of approximately 1 kb, indicating that the transcript is monocistronic. A typical signal peptide of 26 amino acid residues was removed from the N terminus during secretion. The mature enzyme showed high specificity for PI with optimum activity at pH 5.5 to 6.0. Although PI-PLC from *S. aureus* does not contain any cysteine residues, enzyme activity is inhibited by low concentrations of thiol-reactive agents such as $HgCl_2$ and *p*-chloromercuribenzoate. The reason for this is not understood.

All fresh clinical isolates of *S. aureus* were found to produce PI-PLC, although the amount of enzyme produced

varied 30-fold between strains (12). Interestingly, when frozen stock cultures of these strains were tested several months later, most of the strains produced substantially less PI-PLC. Specific mRNA could only be detected in those strains producing high amounts of PI-PLC, suggesting that the wide range of PI-PLC production between stains might reflect differences at the transcriptional level. Analysis of PI-PLC production in isogenic pairs of wild-type and *agr* mutant strains revealed 80% reduced levels of PI-PLC in the mutants as compared with the parental strains, indicating that expression of *plc* is positively regulated by an *agr*-dependent mechanism. Different levels of *agr* activity could thus explain the differences in PI-PLC production between clinical isolates, as has been shown for other *agr*-regulated genes.

Fatty Acid-Modifying Enzyme (FAME)

Staphylococcal abscesses contain long-chain free fatty acids, and other neutral lipids, that are bacteriocidal to *S. aureus*. FAME, which is found in culture supernatants of about 80% of *S. aureus* strains, can inactivate these bacteriocidal lipids by catalyzing the esterification of these lipids to alcohols, preferably cholesterol (20, 29). Bacteriocidal lipids are released from glycerides in the abscess, presumably by the action of staphylococcal lipase. This is consistent with the observation that most strains that produce lipase also produce FAME. It has been shown that FAME-producing strains of *S. aureus* are more virulent in a murine model (35).

Saturated fatty acids with 15 to 19 carbons are most efficiently esterified by FAME. The enzyme has a pH optimum between 5.5 and 6.0 and a temperature optimum of about 40°C. Enzyme activity is inhibited by tri- and diglycerides with unsaturated fatty acid chains. FAME has also been identified in culture supernatants from *S. epidermidis* (5).

Production of FAME was markedly reduced in *agr* mutant strains as compared with the corresponding parental strains (6). Consistent with positive regulation of FAME by *agr*, maximum production of FAME appeared during the postexponential phase of growth in wild-type strains. Production of FAME was also reduced in a *sar* mutant strain, which is consistent with the reduced *agr* activity observed in *sar* mutants.

HYALURONATE LYASE

Hyaluronic acid is a ubiquitous component of the extracellular matrix of vertebrates. Extracellular enzymes that could hydrolyze hyaluronic acid were therefore among the first enzymes to be implicated in bacterial pathogenesis. Hyaluronic acid is a linear polysaccharide composed of repeating units of D-glucuronic acid(1-β-3)N-acetyl-D-glucosamine(1-β-4). Three classes of hyaluronidases are recognized based on their different mechanisms of action. Hyaluronidases of bacterial origin, including that from *S. aureus*, which degrade hyaluronic acid by a β-elimination mechanism yielding disaccharides that contain glucuronosyl residues with a double bond, are named hyaluronate lyases.

The reading frame of staphylococcal hyaluronate lyase, *hysA* (Table 1), encodes a protein of 807 amino acid residues with a calculated molecular mass of 92 kDa (14). An N-terminal signal sequence of 40 amino acids ending with a typical signal peptidase cleavage site suggests that the molecular mass of the secreted enzyme would be 87.4 kDa. This is close to the reported value of 84 kDa.

The amino acid sequence of staphylococcal hyaluronate lyase shares homology with the hyaluronate lyases from *Streptococcus pneumoniae*, *Streptococcus agalactiae*, and *Propionobacterium acne* (46). The greatest similarity is seen with the pneumococcal enzyme (36% identical). A histidine residue (His-479) in hyaluronate lyase from *S. agalactiae*, which is essential for catalytic activity, is present in a region that is highly conserved in the enzymes from the four different species (28). Several other regions, rich in basic residues that are thought to be involved in hyaluronate binding, are also conserved.

Earlier work has shown that staphylococcal hyaluronate lyase activity is inhibited by thiol-reactive agents, indicating that free SH groups are essential (1). The presence of two cysteine residues in the staphylococcal enzyme is consistent with these observations. However, the homologous hyaluronate lyases from *S. pneumoniae* and *S. agalactiae* lack cysteine residues, suggesting that cysteines are not part of the active site.

Unlike most other extracellular enzymes from *S. aureus*, hyaluronate lyase is produced only during the exponential phase of growth, suggesting that it is not regulated by *agr* (48).

NUCLEASE

Thermostable nuclease (TNase, also called staphylococcal nuclease [Snase] or nuclease A) is produced by nearly all strains of *S. aureus* and has been used as a diagnostic criterion for this species. However, highly related thermonucleases are also produced by *Staphylococcus hyicus* (68% identity to *S. aureus* TNase) and *Staphylococcus intermedius* (45% identity) (7). TNase hydrolyzes single- or double-stranded DNA and RNA at the 5′ position of phosphodiester bonds by a calcium-dependent mechanism. It is one of the most extensively studied enzymes in terms of protein structure and catalytic properties (19, 51). Since the enzyme can refold spontaneously after thermal unfolding, it has been widely used as a model to study protein-folding mechanisms (43). The mature extracellular form of the enzyme consists of 149 amino acids (molecular mass 16.8 kDa) and has a globular form. Amino acid residues Asp-21 and Asp-40 are involved in binding a single Ca^{2+}, while Arg-35, Glu-43, and Arg-87, together with Ca^{2+}, are involved in substrate binding and catalysis.

The nuclease gene (*nuc*) encodes a protein of 228 amino acid residues (Table 1). An unusually long signal peptide (60 amino acids), containing two highly hydrophobic stretches of approximately the same length, separated by a region containing three basic residues, is cleaved off during secretion both in *S. aureus* and in heterologous systems (i.e., *B. subtilis* and *Corynebacterium glutamicum*). The secreted form of the enzyme (nuclease B) has a 19-residue N-terminal propeptide, which is cleaved by an extracellular protease. The propeptide appears to function as a secretion enhancer (47). The propeptide did not stimulate secretion when it was attached to another protein, indicating that the propeptide acts as a specific secretion enhancer for the nuclease. The propeptides of other staphylococcal exoenzymes may have a similar effect (27).

Investigations have been supported by grant 4513 from the Swedish Medical Research Council.

REFERENCES

1. **Arvidson, S.** 1983. Extracellular enzymes from *Staphylococcus aureus*, p. 745–808. *In* C. S. F. Easmon and C. Adlam (ed.), *Staphylococci and Staphylococcal Infections*, vol. 2. Academic Press, Inc., London, U.K.

2. **Banbula, A., J. Potempa, J. Travis, C. Fernandes-Catalan, K. Mann, R. Huber, W. Bode, and F. J. Medrano.** 1998. Amino-acid sequence and three-dimensional structure of the *Staphylococcus aureus* metalloproteinase at 1.72 Å resolution. *Structure* **6:**1185–1192.

3. **Birktoft, J. J., and K. Breddam.** 1994. Glutamyl endopeptidases. *Methods Enzymol.* **244:**114–126.

4. **Bodén, M. K., and J.-I. Flock.** 1989. Fibrinogen-binding protein/clumping factor from *Staphylococcus aureus*. *Infect. Immun.* **57:**2358–2363.

5. **Chamberlain, N. R., and S. A. Brueggemann.** 1997. Characterisation and expression of fatty acid modifying enzyme produced by *Staphylococcus epidermidis*. *J. Med. Microbiol.* **46:**693–697.

6. **Chamberlain, N. R., and B. Imanoel.** 1996. Genetic regulation of fatty acid modifying enzyme from *Staphylococcus aureus*. *J. Med. Microbiol.* **44:**125–129.

7. **Chesneau, O., and N. El Solh.** 1994. Primary structure and biological features of a thermostable nuclease isolated from *Staphylococcus hyicus*. *Gene* **145:**41–47.

8. **Cheung, A. L., J. M. Koomey, C. A. Butler, S. J. Projan, and V. A. Fischetti.** 1992. Regulation of exoprotein expression in *Staphylococcus aureus* by a locus (*sar*) distinct from *agr*. *Proc. Natl. Acad. Sci. USA* **89:**6462–6466.

9. **Cheung, A. L., S. J. Projan, R. E. Edelstein, and V. A. Fischetti.** 1995. Cloning, expression, and nucleotide sequence of a *Staphylococcus aureus* gene (*fbpA*) encoding a fibrinogen-binding protein. *Infect. Immun.* **63:**1914–1920.

10. **Coleman, D. C., D. J. Sullivan, R. J. Russel, J. P. Arbuthnott, B. F. Carey, and H. M. Pomeroy.** 1989. *Staphylococcus aureus* bacteriophages mediating simultaneous lysogenic conversion of beta-lysin, staphylokinase and enterotoxin A: molecular mechanism of triple conversion. *J. Gen. Microbiol.* **135:**1679–1697.

11. **Collen, D.** 1998. Staphylokinase: a potent, uniquely fibrin-selective thrombolytic agent. *Nat. Med.* **4:**279–284.

12. **Daugherty, S., and M. G. Low.** 1993. Cloning, expression, and mutagenesis of phosphatidylinositol-specific phospholipase C from *Staphylococcus aureus*: a potential staphylococcal virulence factor. *Infect. Immun.* **61:**5078–5089.

13. **Drapeau, G. P.** 1978. Role of a metalloprotease in activation of the precursor of staphylococcal protease. *J. Bacteriol.* **136:**607–613.

14. **Farrell, A. M., D. Taylor, and K. T. Holland.** 1995. Cloning, nucleotide sequence determination and expression of the *Staphylococcus aureus* hyaluronate lyase gene. *FEMS Microbiol. Lett.* **130:**81–85.

15. **Giraudo, A. T., A. L. Cheung, and R. Nagel.** 1997. The *sae* locus of *Staphylococcus aureus* controls exoprotein synthesis at the transcriptional level. *Arch. Microbiol.* **168:**53–58.

16. **Goh, S. H., S. K. Byrne, J. L. Zhang, and A. W. Chow.** 1992. Molecular typing of *Staphylococcus aureus* on the basis of coagulase gene polymorphism. *J. Clin. Microbiol.* **30:**1642–1645.

17. **Hookey, J. V., J. F. Richardson, and B. D. Cookson.** 1998. Molecular typing of *Staphylococcus aureus* based

PCR restriction fragment polymorphism and DNA sequence analysis of the coagulase gene. *J. Clin. Microbiol.* **36**:1083–1089.

18. **Hufnagel, W.** Personal communication.

19. **Hynes, T. R., and R. O. Fox.** 1991. The crystal structure of staphylococcal nuclease refined at 1.7 Å resolution. *Proteins* **10**:92–105.

20. **Kapral, F. A., S. Smith, and D. Lal.** 1992. The esterification of fatty acids by *Staphylococcus aureus* fatty acid modifying enzyme (FAME) and its inhibition by glycerides. *J. Med. Microbiol.* **37**:235–237.

21. **Kawabata, S., T. Morita, S. Iwanaga, and H. Igarashi.** 1985. Enzymatic properties of staphylothrombin, an active molecular complex formed between staphylocoagulase and human prothrombin. *J. Biochem.* **98**:1603–1614.

22. **Kiess, M., B. Hofmann, S. Weissflog, and D. Schomburg.** 1998. *SWISS-PROT* database. Accession number P81297.

23. **Kondo, I., and K. Fujise.** 1977. Serotype B staphylococcal bacteriophage singly converting staphylokinase. *Infect. Immun.* **18**:266–272.

24. **Kotting, J., H. Eibl, and F. J. Fehrenbach.** 1988. Substrate specificity of *Staphylococcus aureus* (TEN5) lipase with isomeric oleyl-sn-glycerol esters as substrate. *Chem. Phys. Lipids.* **47**:117–122.

25. **Lebeau, C., F. Vandenesch, T. Greenland, R. P. Novick, and J. Etienne.** 1994. Coagulase expression in *Staphylococcus aureus* is positively and negatively modulated by an *agr*-dependent mechanism. *J. Bacteriol.* **176**:5534–5536.

26. **Lee, C. Y., and J. J. Iandolo.** 1986. Lysogenic conversion of staphylococcal lipase is caused by insertion of the bacteriophage L54a genome into the lipase structural gene. *J. Bacteriol.* **166**:385–391.

27. **Liebl, W., and F. Götz.** 1986. Studies on lipase directed export of β-lactamase in *Staphylococcus carnosus*. *Mol. Gen. Genet.* **204**:166–173.

28. **Lin, B., W. F. Averett, and D. G. Pritchard.** 1997. Identification of a histidine residue essential for enzymatic activity of group B streptococcal hyaluronate lyase. *Biochem. Biophys. Res. Commun.* **231**:379–382.

29. **Long, J. P., J. Hart, W. Albers, and F. A. Kapral.** 1992. The production of fatty acid modifying enzyme (FAME) and lipase by various staphylococcal species. *J. Med. Microbiol.* **37**:232–234.

30. **Marques, M. B., P. F. Weller, J. Parsonnet, B. J. Ransil, and A. Nicholson-Weller.** 1989. Phosphatidylinositol-specific phospholipase C, a possible virulence factor for *Staphylococcus aureus*. *J. Clin. Microbiol.* **27**:2451–2454.

31. **McDevitt, D., P. Vaudaux, and T. J. Foster.** 1992. Genetic evidence that bound coagulase of *Staphylococcus aureus* is not clumping factor. *Infect. Immun.* **60**:1514–1523.

32. **McGavin, M. J., C. Zahradka, K. Rice, and J. E. Scott.** 1997. Modification of the *Staphylococcus aureus* fibronectin binding phenotype by V8 protease. *Infect. Immun.* **65**: 2621–2628.

33. **Moravia, P., E. Morfeldt, and S. Arvidson.** Unpublished results.

34. **Morfeldt, E., and S. Arvidson.** Unpublished results.

35. **Mortensen, J. E., T. R. Shryock, and F. A. Kapral.** 1992. Modification of bactericidal fatty acids by an enzyme of *Staphylococcus aureus*. *J. Med. Microbiol.* **36**:293–298.

36. **Nikoleit, K., R. Rosenstein, H. M. Verheij, and F. Götz.** 1995. Comparative biochemical and molecular analysis of the *Staphylococcus hyicus*, *Staphylococcus aureus* and a hybrid lipase. *Eur. J. Biochem.* **228**:732–738.

37. **Phonimdaeng, P., M. O'Reilly, P. Nowlan, A. J. Brame, and T. J. Foster.** 1990. The coagulase of *Staphylococcus aureus* 8325-4. Sequence analysis and virulence of site-specific coagulase-deficient mutants. *Mol. Microbiol.* **4**: 393–404.

38. **Potempa, J., A. Dubin, G. Korzus, and J. Travis.** 1988. Degradation of elastin by cysteine proteinase from *Staphylococcus aureus*. *J. Biol. Chem.* **263**:2664–2667.

39. **Potempa, J., W. Watorek, and J. Travis.** 1986. The inactivation of human α1-protease inhibitor by proteases from *Staphylococcus aureus*. *J. Biol. Chem.* **261**:14330–14334.

40. **Rabijn, A., H. L. DeBondt, and C. Deranter.** 1997. Three-dimensional structure of staphylokinase, a plasminogen activator with therapeutic potential. *Nat. Struct. Biol.* **4**:357–360.

41. **Rollof, J., and S. Normark.** 1992. In vivo processing of *Staphylococcus aureus* lipase. *J. Bacteriol.* **174**:1844–1847.

42. **Sawai, T., K. Tomono, K. Yanagihara, Y. Yamamoto, M. Kaku, Y. Hirakata, H. Koga, T. Tashiro, and S. Kohno.** 1997. Role of coagulase in a murine model of hematogenous pulmonary infection induced by intravenous injection of *Staphylococcus aureus* enmeshed in agar beads. *Infect. Immun.* **65**:466–471.

43. **Shortle, D., Y. Wang, J. R. Gillespie, and J. O. Wrabl.** 1996. Protein folding for realists: a timeless phenomenon. *Protein Sci.* **5**:991–1000.

44. **Simons, J.-W., H. Adams, R. C. Cox, N. Dekker, F. Götz, A. J. Slotboom, and H. M. Verheij.** 1996. The lipase from *Staphylococcus aureus*: expression in *Escherichia coli*, large-scale purification and comparison of substrate specificity to *Staphylococcus hyicus* lipase. *Eur. J. Biochem.* **242**:760–769.

45. **Simons, J.-W., M. D. VanKampen, S. Riel, F. Götz, M. R. Egmond, and H. M. Verheij.** 1998. Cloning, purification and characterisation of the lipase from *Staphylococcus epidermidis*. *Eur. J. Biochem.* **253**:675–683.

46. **Steiner, B., S. Romero-Steiner, D. Cruce, and R. George.** 1997. Cloning and sequencing of the hyaluronate lyase gene from *Propionibacterium acnes*. *Can. J. Microbiol.* **43**:315–321.

47. **Suciu, D., and M. Inouye.** 1996. The 19-residue propeptide of staphylococcal nuclease has a profound secretion-enhancing ability in *Escherichia coli*. *Mol. Microbiol.* **21**:181–195.

48. **Taylor, D., and K. T. Holland.** 1991. Differential regulation of toxic shock syndrome toxin-1 and hyaluronate lyase production by *Staphylococcus aureus*. *Zentralbl. Bakteriol. Mikrobiol. Hyg. Suppl.* **21**:209–212.

49. **Teufel, P., and F. Götz.** 1993. Characterization of an extracellular metalloprotease with elastase activity from *Staphylococcus epidermidis*. *J. Bacteriol.* **175**:4218–4224.

50. **Vandenesch, F., C. Lebeau, M. Bes, D. McDevitt, T. Greenland, R. P. Novick, and J. Etienne.** 1994. Coagulase deficiency in clinical isolates of *Staphylococcus aureus* involves both transcriptional and post-transcriptional defects. *J. Med. Microbiol.* **40**:344–349.

51. **Weber, D. J., A. G. Gittis, G. P. Mullen, C. Abeygunawardana, E. E. Lattman, and A. S. Mildvan.** 1992. NMR docking of a substrate into the X-ray structure of staphylococcal nuclease. *Proteins* **13**:275–287.

Staphylococcal Surface Proteins

MAGNUS HÖÖK AND TIMOTHY J. FOSTER

40

In communicating with its environment, *Staphylococcus aureus*, like other bacteria, uses surface proteins. Surface proteins are responsible for the adherence of bacteria to supporting structures and participate in the import of nutrients into the cell and the export of different cellular products. Changes in the environment surrounding the bacteria might be sensed through transmembrane protein systems and result in intracellular signaling. Also, proteins or protein segments that ultimately are to be secreted by the bacteria to influence the environment may for a shorter or longer period of time be associated with or linked to components of the cell wall and act as cell surface proteins.

Studies of surface proteins in *S. aureus* have, in the recent past, focused on the analysis of proteins that appear to mediate adherence of bacteria to various extracellular matrix components. Members of this subfamily of adhesins, which have been called MSCRAMMs (microbial surface components recognizing adhesive matrix molecules) (38, 40), appear to be anchored to the cell wall. These proteins are presented in Fig. 1. They include protein A, the prototype of cell wall anchor proteins that is best known for its ability to bind the Fc domain of immunoglobulin G (IgG) but recently has also been shown to interact with the von Willebrand factor (9); a collagen-binding MSCRAMM, CNA (41); two fibronectin-binding MSCRAMMs, FnbpA and FnbpB (17); two fibrinogen-binding MSCRAMMs, ClfA (24) and ClfB (31); and the recently characterized Sdr (SD repeat) subfamily of proteins (18) that so far includes three members and for which ligands have not yet been identified. This chapter deals primarily with the structure and function of the cell wall-associated proteins.

ANCHORING OF PROTEINS TO THE CELL WALL

The mechanisms involved in secretion and anchoring of cell wall-associated proteins are incompletely understood, although the principal steps have been revealed (see reference 51 and references therein). The cell wall-associated proteins contain a fairly long signal peptide at the N

terminus, which allows them to be secreted through the Sec pathway. Furthermore, the C-terminal domain contains the sequence motif LPXTG, followed by a stretch of hydrophobic amino acid residues, allowing the polypeptide chain to transverse the membrane, and ending with a handful of positively charged residues, which presumably are located on the cytoplasmic side of the membrane. It is believed that as the proteins are secreted, the hydrophobic C-terminal tail remains in the membrane and anchors the protein at the cell surface. The LPXTG motif is subsequently recognized by a hypothetical enzyme called "sortase" with transamidase activity, which catalyzes the cleavage of the protein and the formation of a covalent bond between the threonine residue in the protein and glycine residues in the pentaglycine cross bridge between adjacent muropeptide units in the cell wall.

CELL WALL-ASSOCIATED PROTEINS

Protein A

Many of the cell wall-associated proteins on *S. aureus* act as MSCRAMM and mediate adherence of the bacteria to extracellular matrix components. However, protein A (SpA), the first staphylococcal surface protein to be characterized, is primarily known for its ability to bind the Fc region of mammalian IgG. As is obvious from the cartoon model of SpA in Fig. 1, the structural organization of this molecule is somewhat different from that of other characterized cell wall-associated proteins. The N-terminal signal sequence is followed by tandem repeats of five homologous IgG-binding domains labeled E, D, A, B, and C. Each of the IgG-binding domains is composed of a ~60-amino-acid-residue unit that forms three α-helices as shown by nuclear magnetic resonance (47). The structure of the subdomain B in complex with the constant domain (Fc) of IgG subclass 1 has been solved by x-ray analyses of a co-crystal (5). The binding surface between the two molecules appeared to involve nine amino acid residues in the IgG fragment and 11 amino acid residues in the B domain (7).

Gram-Positive Pathogens, ed. by V. A. Fischetti et al.
© 2000 American Society for Microbiology, Washington, D.C.

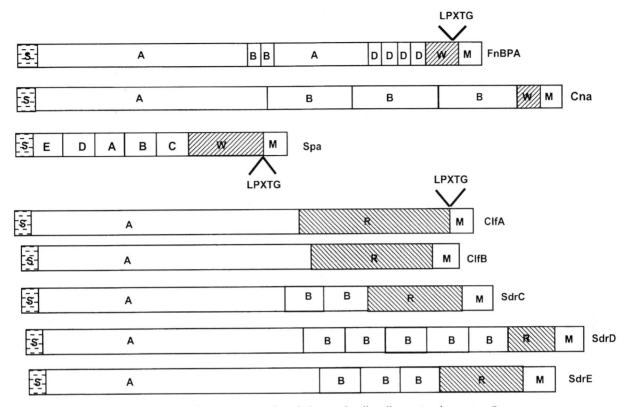

FIGURE 1 Structural organization of staphylococcal cell wall-associated proteins. S represents signal sequences, R represents the Ser-Asp dipeptide repeats, W represents cell wall-spanning segments, and M represents membrane-spanning regions and positively charged residues. For further details, see the text.

The binding characteristics and specificity of the SpA IgG interaction have been analyzed in great detail (21) and will not be discussed here.

The importance of SpA in staphylococcal infections is somewhat unclear. It is possible that, by binding IgG, SpA could interfere with the phagocytosis of opsonized bacteria via receptors on the host cell. Active SpA fragments released from the bacteria in this respect would be more effective than SpA anchored to the staphylococcal cell wall. In fact, SpA can be detected in bacterial culture media, although the mechanism of release is unclear. The recent observation that SpA can mediate bacterial adherence to von Willebrand factor (9), an extracellular matrix protein important in normal hemostasis, suggests that SpA may play additional roles in the infectious process.

Fibronectin-Binding MSCRAMMs

Binding of fibronectin to *S. aureus* was first reported by P. Kuusela in 1978 (20). Two structurally similar proteins, FnbpA (45) and FnbpB (17), were later demonstrated to be responsible for this binding activity. Both proteins contain a ~500-amino-acid N-terminal stretch of unique sequence and unknown function. In FnbpA, this stretch is interrupted by a 35-amino-acid segment (called B) of unknown function that is repeated once. The fibronectin-binding activity has been localized to a ~50-amino-acid unit repeated four times and partially a fifth time. This region is found just outside the cell wall-associated domain. Several streptococcal species also contain fibronectin-binding MSCRAMMs with a similar structural organiza-

tion (15, 22, 44, 50). Synthetic peptides mimicking the different repeat units bind fibronectin and effectively inhibit bacterial binding of the matrix protein (15). Earlier work demonstrated that *S. aureus* binds primarily to the N-terminal domain of fibronectin (29). This domain is composed of five copies of a type I folding motif, and this motif, in pairs, seems to form the binding sites recognized by the MSCRAMMs (12). Recent studies suggest that the repeat units contain several discrete binding motifs that interact with different sites in fibronectin (14). The MSCRAMM-fibronectin interaction involves structural rearrangements with immunological consequences. In the MSCRAMM, the domain containing the ligand-binding repeat units appears to have an unordered structure that acquires a defined conformation on binding to fibronectin (11, 42). This conformational change is accompanied by the formation of neo epitopes that can be demonstrated both by monoclonal antibodies (46) and by antibodies isolated from patients recovering from staphylococcal infections (4). The antibodies recognizing the neo epitopes do not interfere with the MSCRAMM-fibronectin interaction but rather stabilize the MSCRAMM-ligand complex and appear to promote fibronectin binding.

The Fnbps mediate bacterial adherence to fibronectin-containing substrates. Fibronectin is found in the extracellular matrix of most tissues and also in a soluble form in many body fluids. Following implantation, a foreign object such as a catheter or other biomaterial rapidly becomes coated with host proteins such as fibronectin and fibrinogen. Bacteria recognize and adhere to this protein coat

(52). In addition, the Fnbps also appear to play a major role in the invasion of host cells. The MSCRAMMs recruit and bind soluble fibronectin, which subsequently is recognized by integrins on the host cell, resulting in phagocytosis of the host protein-coated bacteria. This mechanism was initially demonstrated for fibronectin-binding streptococci (28, 35) and recently also for *S. aureus* (13).

The Collagen-Binding MSCRAMM

Some strains of *S. aureus* express a collagen-binding MSCRAMM called CNA. The presence of this MSCRAMM is sufficient and necessary for bacteria to adhere to collagenous tissues such as cartilage (48). The structural organization of CNA is similar to that of most *S. aureus* MSCRAMMs (Fig. 1). Following the N-terminal signal peptide is a so-called A domain that consists of ~500 amino acid residues of unique sequence. This domain is followed by region B with a tandemly repeated structure. The repeating unit in the B region is composed of 180 amino acid residues and occurs in one to four copies in *cna* from different strains (6). The C-terminal domain is composed of a typical cell wall-associated region, including the LPXTG motif, a hydrophobic transmembrane region, and a short cytoplasmic region composed primarily of positively charged residues. The collagen-binding activity has been located in a 190-amino-acid segment within the A domain (38). The crystal structure of this subdomain has been solved at 1.8 Å resolution (49). The polypeptide folds like a "jelly roll" in two β-sheets connected by a short α-helix. A trench traverses one of the β-sheets, and molecular modeling suggests that this trench can accommodate a collagen triple helix. The collagen-binding function of the trench was further demonstrated by site-specific mutagenesis, in which changes in single amino acid residues forming the walls of the trench resulted in mutated proteins with dramatically reduced collagen-binding activities (49).

Analyses of a recombinant version of the intact A domain suggest that this domain is folded into an elongated structure rather than a sphere. This might suggest that the A domain is a mosaic protein composed of several subdomains that are independently folded and have different functions (43).

The collagen-binding MSCRAMM is a virulence factor in a mouse model of septic arthritis, as shown by infecting animals with pairs of strains differing only with respect to the presence or absence of an active *cna* gene (39). The number of mice that developed arthritis was reduced by 30 to 40% in animals infected with CNA-negative strains compared with those infected with staphylococci expressing CNA. These results show that, although CNA contributes to the development of septic arthritis, other bacterial components can compensate, at least to some degree, for a lack of CNA. In view of this observation, it was surprising to find that vaccination of mice with a recombinant form of the CNA A domain provided effective protection in a mouse sepsis model against challenge with a CNA-expressing strain (32). The protective effect is immunoglobulin mediated, since antibodies raised in rats conferred protection in mice when used in a passive immunization experiment. These results demonstrate that MSCRAMMs should be considered targets in immunopreventive and therapeutic strategies to combat staphylococcal infections.

Fibrinogen-Binding MSCRAMMs

S. aureus has long been known to clump in the presence of blood plasma (30). Fibrinogen was shown to be the plasma component that induces clumping (10), and two fibrinogen-binding MSCRAMMs present on *S. aureus* cells were found to mediate clumping. These MSCRAMMs, called ClfA (24) and ClfB (31), are structurally related and are composed of an N-terminal signal sequence, a 500-amino-acid A domain, an R domain consisting of serine-aspartic acid dipeptide repeats followed by a short cell wall-associated domain, a transmembrane, and a cytoplasmic domain. The fibrinogen-binding sites are located to the A domains (25, 31), whereas the R domains, which can be of variable length (23), are needed for the presentation of the proteins at the bacterial cell surface (8). The ligand-binding site in ClfA has been further mapped to a subfragment of the A domain corresponding to residues 220 to 559 (25). Although the structural organization of ClfA and ClfB is similar, the amino acid sequences of the ligand-binding A domains are only 27% identical. Furthermore, ClfA and ClfB bind to different sites in fibrinogen. ClfA recognizes the C terminus of the γ-chain (26), whereas ClfB binds to the α- and β-chains (31). ClfA and ClfB are also differentially expressed. ClfB is produced only in early logarithmic growth and is removed from the bacterial cell surface at later growth stages. ClfA, on the other hand, seems to be produced throughout the growth cycle (31).

The interaction of ClfA with fibrinogen has been studied in some detail. A 17-amino-acid synthetic peptide corresponding to the C terminus of the fibrinogen γ-chains binds to a recombinant form of the A domain of ClfA with a K_d of 10^{-6} M in an interaction that is inhibited by the presence of Ca^{2+} ions (34). The A domain contains a motif reminiscent of a Ca^{2+}-binding EF-hand, and site-specific mutations in this motif resulted in a protein with lower affinity for the γ-chain peptide but less sensitivity to Ca^{2+} ions (34). ClfA exhibits fibrinogen-binding characteristics similar to those of the platelet integrin $\alpha IIb\beta 3$ (34). Both cell surface proteins bind to the same site in fibrinogen in an interaction that is affected by Ca^{2+}. In fact, a recombinant form of the ClfA A domain is a very effective inhibitor of fibrinogen-dependent platelet aggregation and a potentially useful antithrombotic agent (26). This design is probably not accidental but represents a bacterial counterweapon against a host defense mechanism. Platelet aggregation results in the degranulation of alpha granules and the release of, among other things, antimicrobial peptides (53). Inhibition of platelet aggregation would prevent the exposure of staphylococci to these potentially harmful peptides.

ClfA is a virulence factor in a rat endocarditis model, as shown by a lower infection rate by the *clfA* mutant compared with the *clfA*$^+$ wild-type strain (52). ClfA and ClfB can also promote bacterial adherence to host-conditioned biomaterial, particularly material which has been in contact with the host for a short time (<6 h) (31, 52).

Sdr Proteins

In a Southern blot, it was recently shown that *S. aureus* can contain at least three additional genes with R domain-encoding segments (18). The corresponding proteins were called SdrC, SdrD, and SdrE for SD repeat-containing proteins C, D, and E, using a terminology by which ClfA and ClfB could be called SdrA and SdrB, respectively. The structural organization of SdrC, SdrD, and SdrE proteins is similar to that of ClfA and ClfB except that a B domain composed of a 110-amino-acid unit repeated two to five times is inserted between the A domain and the R domain. The B units contain high-affinity Ca^{2+} binding sites that strictly conform to an EF-hand. Removal of Ca^{2+} from this site results in a dramatic conformational change in the protein, resulting in a transition from a globular form to an elongated structure (19). The A domains of the Sdr proteins, including the ClfA and ClfB proteins, are of similar size (~500 amino acid residues), and there is a 20 to 30% sequence identity between the different protein species. It appears likely that these A domains can engage in interactions with host proteins, although ligands for the SdrC, SdrD, and SdrE proteins have not yet been reported. Recently, a gene encoding a fibrinogen-binding Sdr protein was found in *Staphylococcus epidermidis* (33).

NONCOVALENTLY ANCHORED SURFACE PROTEINS

Extraction of staphylococcal cells with 1 M LiCl results in the release from the cells of a major surface protein that is capable of interacting with a variety of proteins and peptides, including many extracellular matrix proteins of the host (27). Sequence analysis of the corresponding gene (16) revealed that the encoded protein was largely composed of six repeats of a 110-amino-acid motif with a central portion composed of a subdomain with high homology to the peptide-binding groove of the β-chain of major histocompatibility complex class II (MHCII) molecules. The extracted staphylococcal protein was therefore called Map (MHCII analogous protein). Map contains an N-terminal signal peptide and is probably secreted by the Sec system but does not have features characteristic of cell wall-anchored proteins and is therefore probably not covalently linked to the cell wall. Recent studies suggest that Map can act as a transplantable substrate for bacterial adherence and interacts with specific cell wall-anchored proteins (19a).

Lysostaphin is a secreted preproenzyme produced by "*Staphylococcus staphylolyticus*" that is capable of hydrolyzing pentaglycine cross bridges in peptidoglycans of, for example, *S. aureus*. Since the producing organism is "resistant," the enzyme may function to kill competing staphylococci in mixed bacterial populations. To perform this function effectively, lysostaphin has, in addition to its enzymatic activity, a targeting domain located in the C terminus that allows the protein to bind to cell surface components of *S. aureus* (3). This mechanism provides an alternative to covalent cell anchoring for locating proteins to the surface of staphylococcal cells.

Other surface proteins that are bound noncovalently to the cell wall are autolysins (1, 2) and the elastin-binding protein EbpS (36).

It is likely that in the near future we will have sequence information on all surface proteins present on *S. aureus*. The challenge to the research community is then to determine the role of the individual proteins in the disease process.

REFERENCES

1. **Baba, T., and O. Schneewind.** 1996. Target cell specificity of a bacteriocin molecule: a C-terminal signal directs lysostaphin to the cell wall of *Staphylococcus aureus*. *EMBO J.* **15**:4789–4797.

2. **Baba, T., and O. Schneewind.** 1998. Instruments of microbial warfare: bactericin synthesis, toxicity and immunity. *Trends Microbiol.* **6**:66–71.

3. **Baba, T., and O. Schneewind.** 1998. Targeting of muralytic enzymes to the cell division site of Gram-positive baceria: repeat domains directed autolysin to the equational surface ring of *Staphylococcus aureus*. *EMBO J.* **17**:4639–4646.

4. **Casolini, F., L. Visai, D. Joh, G. G. Conaldi, A. Toniolo, M. Höök, and P. Speziale.** 1998. Antibody response to fibronectin-binding MSCRAMM in patients diagnosed with *Staphylococcus aureus* infections. *Infect. Immun.* **66:** 5433–5442.

5. **Deisenhofer, J.** 1982. Crystallographic refinement and atomic models of a human Fc fragment and its complex with fragment B of Protein A from *Staphylococcus aureus* at 2.9- and 2.8-Å resolution. *Biochemistry* **20**:2361–2370.

6. **Gillaspy, A. F., J. M. Patti, F. L. Pratt, Jr., J. J. Iandolo, and M. S. Smeltzer.** 1997. The *Staphylococcus aureus* collagen adhesin-encoding gene (*cna*) is within a discrete genetic element. *Gene* **196**:239–248.

7. **Gouda, H., M. Shiraiski, H. Takahashi, K. Kalo, H. Torigoe, Y. Arala, and I. Shimada.** 1998. NMR study of the interaction between the B domain of staphylococcal Protein A and the Fc portion of immunoglobulin G. *Biochemistry* **37**:129–136.

8. **Hartford, O., P. Francois, P. Vaudaux, and T. J. Foster.** 1997. The dipeptide repeat region of the fibrinogen-binding protein (clumping factor) is required for functional expression of the fibrinogen-binding domain on the Staphylococcal surface. *Mol. Microbiol.* **25**:1065–1076.

9. **Hartlieb, J., N. Koehler, R. Dickinson, S. Chhatwal, J. J. Sixma, T. Foster, G. Peters, B. Kehrel, and M. Hermann.** 1998. Binding mechanisms of *Staphylococcus aureus* to von Willebrand Factor: protein A revisited, abstr. B-080. *Program Abstr. 38th Intersci. Conf. Antimicrob. Agents Chemother.* American Society for Microbiology, Washington, D.C.

10. **Hawiger, J. S., D. K. Hammond, S. Timmons, and A. Z. Budzynski.** 1978. Interaction of human fibrinogen with staphylococci: presence of a binding region on normal and abnormal fibrinogen interacting with staphylococcal clumping factor. *Blood* **51**:799–812.

11. **House-Pompeo, J., Y. Xu, D. Joh, P. Speziale, and M. Höök.** 1996. Conformational changes in the fibronectin binding MSCRAMM are induced by ligand binding. *J. Biol. Chem.* **271**:1379–1384.

12. **Huff, S., Y. V. Matusuka, M. J. McGavin, and K. C. Ingham.** 1994. Interaction of N-terminal fragments of fibronectin with synthetic and recombinant D motifs from its binding protein of *Staphylococcus aureus* studied using fluorescence anisotropy. *J. Biol. Chem.* **269**:15563–15570.

13. **Joh, D., T. Fowler, E. Wann, S. Johansson, and M. Höök.** MSCRAMMs and integrins participate in staphylococcal host cell invasion. Unpublished data.

14. **Joh, D., P. Speziale, S. Gurusiddappa, J. Manor, and M. Höök.** 1998. Multiple specificities of the staphylococcal and streptococcal fibronectin-binding microbial surface components recognizing adhesive matrix molecules. *Eur. J. Biochem.* **258**:897–905.

15. **Joh, H. J., K. House-Pompeo, J. M. Patti, S. Gurusiddappa, and M. Höök.** 1994. Fibronectin receptors from Gram-positive bacteria: comparison of active sites. *Biochemistry* **33**:6086–6092.

16. **Jönsson, K., D. McDevitt, M. H. McGavin, J. M. Patti, and M. Höök.** 1995. *Staphylococcus aureus* expresses a ma-

jor histocompatibility complex class II analog. *J. Biol. Chem.* **270:**21457–21460.

17. Jönsson, K., C. Signäs, H.-P. Müller, and M. Lindberg. 1991. Two different genes encode fibronectin binding proteins in *Staphylococcus aureus*—the complete nucleotide sequence and characterization of the 2nd gene. *Eur. J. Biochem.* **202:**1041–1048.

18. Josefsson, E., K. W. McCrea, D. Ni Eidhin, D. O'Connell, J. Cox, M. Höök, and T. J. Foster. 1998. Three new members of the serine-aspartate repeat protein multigene family of *Staphylococcus aureus*. *Microbiology* **144:**3387–3395.

19. Josefsson, E., D. O'Connell, T. J. Foster, I. Durussel, and J. A. Cox. 1998. The binding of calcium to the B-repeat segment of SdrD, a cell surface protein of *Staphylococcus aureus*. *J. Biol. Chem.* **273:**31145–31152.

19a. Kreikemeyer, B., D. McDevitt, and M. Höök. 1 Unpublished results.

20. Kuusela, P. 1978. Fibronectin binds to *Staphylococcus aureus*. *Nature* **276:**718–720.

21. Langone, J. J. 1982. Protein A of *Staphylococcus aureus* and related immunoglobulin receptors produced by streptococci and pneumococci. *Adv. Immunol.* **32:**157–252.

22. Lindgren, P.-E., M. J. McGavin, C. Signäs, B. Guss, S. Gurusiddappa, M. Höök, and M. Lindberg. 1993. Two different genes coding for fibronectin-binding proteins from *Streptococcus dysgalactiae*. *Eur. J. Biochem.* **214:**819–827.

23. McDevitt, D., and T. J. Foster. 1995. Variation in the size of the repeat region of the fibrinogen receptor (clumping factor) of *Staphylococcus aureus* strains. *Microbiology* **141:**937–943.

24. McDevitt, D., P. Francois, P. Vaudaux, and T. J. Foster. 1994. Molecular characterization of the clumping factor (fibrinogen receptor) of *Staphylococcus aureus*. *Mol. Microbiol.* **11:**237–248.

25. McDevitt D., P. Francois, P. Vaudaux, and T. J. Foster. 1995. Identification of the ligand binding domain of the surface-located fibrinogen receptor (clumping factor) of *Staphylococcus aureus*. *Mol. Microbiol.* **16:**895–907.

26. McDevitt, D., T. Nanavaty, K. House-Pompeo, E. Bell, N. Turner, L. McIntire, T. J. Foster, and M. Höök. 1997. Characterization of the interaction between the *Staphylococcus aureus* clumping factor (ClfA) and fibrinogen. *Eur. J. Biochem.* **247:**416–424.

27. McGavin, M. H., D. Krajewska-Pietrasik, C. Rydén, and M. Höök. 1993. Identification of a *Staphylococcus aureus* extracellular matrix-binding protein with broad specificity. *Infect. Immun.* **61:**2479–2485.

28. Molinari, G., S. R. Talay, P. Valentin-Weigand, M. Rohde, and G. S. Chhatwal. 1997. The fibronectin binding protein of *Streptococcus pyogenes* Sfb7 is involved in the internalization of group A streptococci by epithelial cells. *Infect. Immun.* **65:**1357–1363.

29. Mosher, D. F., and R. A. Proctor. 1980. Binding and factor XIII$_a$-mediated cross-linking of a 27 kilodalton fragment of fibronectin to *Staphylococcus aureus*. *Science* **209:**927–929.

30. Much, H. 1908. Uber eine Varstafe des Fibrinfermentes in Kulturen von *Staphylokokkus aureus*. *Biochem. Z.* **14:**253–263.

31. Ni Eidhin, D., S. Perkins, P. Francois, P. Vaudaux, M. Höök, and T. J. Foster. 1998. Clumping factor B (ClfB) a new surface-located fibrinogen-binding adhesin of *Staphylococcus aureus*. *Mol. Microbiol.* **32:**245–257.

32. Nilsson, I. M., J. M. Patti, T. Bremell, M. Höök, and A. Tarkowski. 1998. Vaccination with a recombinant fragment of the collagen adhesin provides protection against *Staphylococcus aureus*-mediated septic death. *J. Clin.Invest.* **101:**2640–2649.

33. Nilsson, M., L. Feykberg, J.-I. Flock, L. Pei, M. Lindberg, and B. Guss. 1998. A fibrinogen-binding protein of *Staphylococcus epidermidis*. *Infect. Immun.* **66:**2666–2673.

34. O'Connell, D. P., T. Nanavaty, D. McDevitt, S. Gurusiddappa, M. Höök, and T. J. Foster. 1998. The fibronectin-binding MSCRAMM (clumping factor) of *Staphylococcus aureus* has an integrin-like Ca2+-dependent inhibitory site. *J. Biol. Chem.* **273:**6821–6829.

35. Ozeri, V., I. Rosenshine, D. F. Mosher, R. Fässler, and E. Hanski. 1998. Roles of integrins and fibronectin in the entry of *Streptococcus pyogenes* into cells via protein F1. *Mol. Microbiol.* **30:**625–637.

36. Park, P. W., T. J. Broekelmann, B. R. Mecham, and R. P. Mecham. 1999. Characterization of the elastin binding domain in the cell-surface 25 kDa elastin-binding protein of *Staphylococcus aureus* (BbpS). *J. Biol. Chem.* **274:**2845–2850.

37. Patti, J. M., B. L. Allen, M. J. McGavin, and M. Höök. 1994. MSCRAMMs mediate adherence of microorganisms to host tissues. *Annu. Rev. Microbiol.* **48:**585–617.

38. Patti, J. M., J. O. Boles, and M. Höök. 1993. Identification and biochemical characterization of the ligand binding domain of the collagen adhesin from *Staphylococcus aureus*. *Biochemistry* **32:**11428–11435.

39. Patti, J. M., T. Bremell, D. Krajewska-Pietrasik, A. Abdelnour, A. Tarkowski, C. Rydén, and M. Höök. 1994. The *Staphylococcus aureus* collagen adhesin is a virulence determinant in experimental septic arthritis. *Infect. Immun.* **62:**152–161.

40. Patti, J. M., and M. Höök. 1994. Microbial adhesins recognizing extracellular matrix macromolecules. *Curr. Opin. Cell Biol.* **6:**752–758.

41. Patti, J. M., K. Jönsson, B. Guss, L. W. Switalski, K. Wiberg, M. Lindberg, and M. Höök. 1992. Molecular characterization and expression of a gene encoding a *Staphylococcus aureus* collagen adhesin. *J. Biol. Chem.* **267:**4766–4772.

42. Penkett, C. J., C. Redfield, J. A. Jones, I. Dodd, J. Hubbard, R. A. G. Smith, L. J. Smith, and C. M. Dobson. 1998. Structural and dynamical characterization of biologically active unfolded fibronectin-binding protein from *Staphylococcus aureus*. *Biochemistry* **37:**17054–17067.

43. Rich, R. L., B. Demeler, K. Ashby, C. C. S. Deivanayagam, J. W. Petrich, J. M. Patti, V. L. N. Sthanam, and M. Höök. 1998. Domain structure of the *Staphylococcus aureus* collagen adhesin. *Biochemistry* **37:**15423–15433.

44. Sela, S., A. Aviv, A. Tovi, I. Burstein, M. G. Caparon, and E. Hanski. 1993. Protein F: an adhesin of *Streptococcus pyogenes* binds fibronectin via two distinct domains. *Mol. Microbiol.* **10:**1049–1055.

45. Signäs, C., G. Raucci, K. Jönsson, P.-E. Lindgren, G. Anatharamaiah, M. Höök, and M. Lindberg. 1989. Nucleotide sequence of the gene for a fibronectin-binding protein from *Staphylococcus aureus* and its use in the synthesis of biologically active peptides. *Proc. Natl. Acad. Sci. USA* **86:**697–703.

46. Speziale, P., D. Joh, L. Visai, S. Bozzini, K. House-Pompeo, M. Lindberg, and M. Höök. 1996. A monoclonal antibody enhances ligand binding of a fibronectin MSCRAMM (adhesin) from *Streptococcus dysgalactiae*. *J. Biol. Chem.* **271:**1371–1378.

47. Starovasnik, M. A., N. J. Skelton, M. P. O'Connell, R. F. Kelley, D. Reilly, and W. J. Fairbrother. 1996. Solution structure of the E-domain of staphylococcal protein A. *Biochemistry* **35:**1558–1569.

48. Switalski, L. M., J. M. Patti, W. Butcher, A. G. Gristina, P. Speziale, and M. Höök. 1993. A collagen receptor on *Staphylococcus aureus* strains isolated from patients with septic arthritis mediates adhesion to cartilage. *Mol. Microbiol.* **7:**99–107.

49. Symersky, J., J. M. Patti, M. Carson, K. House-Pompeo, M. Teale, D. Moore, L. Jin, A. Schneider, L. J. DeLucas, M. Höök, and V. L. N. Sthanam. 1997. Structure of the collagen-binding domain from a *Staphylococcus aureus* adhesin. *Nat. Struct. Biol.* **4:**833–838.

50. Talay, S. R., P. Valentin-Weigand, K. N. Timmis, and G. Chhatwal. 1994. Domain structure and conserved epitopes of Sfb protein, the fibronectin binding adhesin of *Streptococcus pyogenes*. *Mol. Microbiol.* **13:**531–539.

51. Ton-That, H., H. Tabischinski, B. Berger-Bächi, and O. Schneewind. 1998. Anchor structure of staphylococcal surface proteins III role of the FemA, FemB and FemX factors in anchoring surface proteins to the bacterial cell wall. *J. Biol. Chem.* **273:**29143–29149.

52. Vaudaux, P. E., P. Francois, R. A. Proctor, D. McDevitt, T. J. Foster, R. M. Albrecht, D. P. Lew, H. Wabers, and S. L. Cooper. 1995. Use of adhesion-defective mutants of *Staphylococcus aureus* to define the role of specific plasma proteins in promoting bacterial adhesion to canine arteriovenous shunts. *Infect. Immun.* **63:**585–590.

53. Yeaman, M. R., S. M. Puentes, D. C. Norman, and A. S. Bayer. 1992. Partial purification and staphylocidal activity of thrombin-induced platelet microbicidal protein. *Infect. Immun.* **60:**1202–1209.

Pathogenicity Factors and Their Regulation

RICHARD P. NOVICK

41

Staphylococcus aureus is an extraordinarily versatile pathogen, causing a wide spectrum of infections, which can be viewed as three general types—(i) superficial lesions such as small skin abscesses and wound infections; (ii) systemic and life-threatening conditions such as osteomyelitis, tropical myositis, endocarditis, pneumonia, and septicemia; and (iii) toxinoses such as toxic shock, toxic epidermal necrolysis, and food poisoning. As is typical of gram-positive bacteria, *S. aureus* depends for its pathogenicity largely on a set of extracellular and cell wall-associated proteins, and, accordingly, staphylococcal pathogenicity is multifactorial. In most staphylococcal disease conditions, with the exception of the toxinoses, no single factor is responsible, and it has generally been difficult to define precisely the role of any given factor. This point is addressed at some length in a review by Projan and Novick (65).

Pathogenicity factors may be viewed as a subset of a large class of accessory gene products, products that are not required for growth and cell division but are advantageous in particular environments. Many of these products are encoded by accessory genetic elements such as plasmids, transposons, prophages, and pathogenicity islands. *S. aureus* and other staphylococci, which are facultative pathogens, are able to adapt to at least three distinct types of environment. They can live freely outside of any host, they can exist as external colonizers or commensals, and they can live within the tissues of a host organism, causing disease. It stands to reason that such organisms will have developed overlapping subsets of accessory genes, dedicated to and regulated by the concomitant environmental exigencies. In this light, the disease state is seen as simply one of several states of existence for the organism and the relevant gene set, the virulence genes, as simply representing an adaptation to that type of existence.

Traditionally, bacterial pathogenicity or virulence factors are products whose role in the disease process is either clearly demonstrable, e.g., toxins, or more or less obvious on the basis of biological properties, e.g., enzymes that degrade tissue components. As these genes are actively expressed in laboratory cultures, they and their products have been relatively easy to identify and study (Table 1). In *S.*

aureus, most belong to an extensive regulon, controlled by *agr*, which down-regulates surface proteins during exponential phase in laboratory cultures and up-regulates secreted proteins postexponentially. The presence within a global virulence regulon of genes encoding catabolic enzymes, such as proteases, nucleases, and lipases, which can be regarded as nutritional factors as well as virulence factors, becomes understandable as representing an overlap between the environmental and pathogenicity gene subsets. The coordinate postexponential expression of similar catabolic enzymes by soil organisms such as *Bacillus subtilis*, which are certainly not pathogens, could indicate that a catabolic enzyme regulon has been borrowed from soil bacteria by gram-positive pathogens for incorporation into their virulence regulons.

Although the *agr* regulon includes most of the genes that belong to the "traditional" virulence gene set, it is far from the whole story for *S. aureus*. Several other important regulatory genes have been identified, and certain environmental signals have long been known to influence virulence gene expression. The putative regulatory genes through which these signals act are as yet unknown. Additionally, signature-tagged mutagenesis (STM) and in vitro expression technology (IVET) have revealed numerous loci in which transposon insertion interferes with growth in vivo but not in vitro, leading to severe attenuation of a model infection (55). Identification of the genes involved, which requires ruling out polar effects, and an understanding of the roles of such genes, may lead to important insights into the infective process and must certainly lead to the identification of additional, hitherto unknown regulatory factors.

In this chapter, available information on the identification and regulation of pathogenicity genes in staphylococci is summarized, and an attempt is made to develop a comprehensive view of the way in which the organism adapts to the host environment.

PATHOGENICITY FACTORS

For present purposes, there are two sets of pathogenicity factors—those listed in Table 1 and, provisionally, those

Gram-Positive Pathogens, ed. by V. A. Fischetti et al.
© 2000 American Society for Microbiology, Washington, D.C.

TABLE 1 Pathogenicity factors and regulatory genes

Genes[a]	Regulatory locus			Environmental factors[b]										Reference
	agr	sarA	sae	NaCl	Pxp	Clin	Nov	Gluc	Low pH	Low O₂	EDTA, EGTA	GML	Trp[2c]	
agr	↑↑	↑		↑	↑	↓		↓		↓		↑		62
sarA	↑			↑	↑			↓			↑			16
hla	↑15×	↑3×		↓↓50×	↑↑	↓	↑	↓		↓	↑	→	→	16, 18, 25a, 35, 67
hlb	↑				↑	↓	↑		↓	↓	↑	→	→	
hld	↑↑	↑		↑	↑	↓		↓	↓			→	→	
hlg					↑							→	→	
pvl					↑									
spr(v8)	↑↑	↓1,000×		↑	↑			↑						18, 25a
3′ spr	↑	→			↑									
Metalloprotease	↑	↓↓		↓↓	↑									18, 25a
tst	↑	↑2×		↓↓	↑	↓	↑	→	↓	↓	↑	→		16, 79
entA					↑							↑		
entB[d]	↑3×	↑3×		→	↑	↓		↓	↓	↓		→		3, 18, 25a
entC	↑			↓↓	↑	↓		↓	↓	↓		→		3, 76
eta, etb	↑↑			eta ↓		↓	↓							
geh	↑	↑		↑	↑	→								35
Murein hydrolase (32 kDa)	↑	↓↓			↑									30
Murein hydrolase (>50 kDa)	↓↓				↑									32
clfA	↑↑	↑		↑										
FAME	↑				↑									
hal	↑													
spa	↓	↓20×	↓3–5×	↓↓	↓	↑	↑			↓10×	↑	→		16, 35
fbp		↑			↑	↓								
fnbA	↓	↑		↑	↑	↓						→		71
coa	↓↓		↑↑		↑	↓								35
nuc	↑				↑									35

[a] Gene abbreviations: *agr*, accessory gene regulator; *clfa*, clumping factor A; *coa*, coagulase; *ent*, enterotoxin; *eta*, exfoliative toxin; *etb*, exfoliative toxin B; *fbp*, fibrinogen-binding protein; *geh*, glycerol ester hydrolase; *hal*, hyaluronidase; *hla*, α-hemolysin; *hlb*, β-hemolysin; *hld*, δ-hemolysin; *hlg*, γ-hemolysin; *nuc*, staphylococcal nuclease; *pvl*, Panton-Valentin leukocidin; *sar*, staphylococcal accessory regulator; *sae*, staphylococcal accessory gene expression; *spa*, protein A; *spr*, serine protease; *tst*, toxic shock syndrome toxin-1.

[b] Abbreviations: Clin, clindamycin (certain other antibiotics that inhibit protein synthesis have similar effects); Gluc, glucose; GML, glycerol monolaurate; Nov, novobiocin or other gyrase inhibitor; Pxp, postexponential phase.

[c] Trp, effects of limitation of an essential amino acid.

[d] *entB* is representative of other enterotoxin genes (*entC*, *entD*, *entE*) that are regulated similarly.

identified by STM and IVET, many of which are of unknown function. The standard list (Table 1), which has been compiled on the basis of obvious roles in pathogenesis or membership in the *agr* regulon, is essentially a list of most known exoproteins. The list includes secreted proteins, such as cytotoxins, immunotoxins, and enzymes; surface proteins, including protein A (SpA); and the MSCRAMMs (microbial surface components recognizing adhesive matrix molecules), such as fibrinogen-binding protein, fibronectin-binding protein, collagen-binding protein, and others. The secreted proteins are assumed to enable the organism to attack the local cellular and structural elements of diverse tissues and organs, and the MSCRAMMs enable the bacteria to adhere to tissue components and to resist phagocytosis and other host defenses (4, 65) (see chapters 39 and 40, this volume). Agr also appears to down-regulate one autolysin (27) and two or more penicillin-binding proteins (64). However, the significance of this for virulence is unknown.

For heuristic purposes, it is not considered useful to include in the list of pathogenicity determinants all genes that are required for the establishment and/or maintenance of an infection; although random mutagenesis with in vivo screening will identify many such genes, most of them are standard housekeeping genes, which may prove attractive as novel therapeutic targets but provide little if any enlightenment on the causation of disease. Whether structural components of the cell envelope, which have an important role in septic shock, should be included is problematic; although these components are encoded by housekeeping genes, their role in disease, in addition to enabling growth, is to induce an excessive and detrimental host response. Since regulation of these components vis-à-vis the disease process is unknown, they are not considered here.

Most, if not all, of the *S. aureus* virulence factors are variably expressed by one or another of the coagulase-negative species, as shown in Table 2. For any of these species, most strains express none of these factors; a few strains express one or a few of them. It is interesting that many strains of *Staphylococcus warneri* and *Staphylococcus lugdunensis* express either β-hemolysin or staphylokinase (69), as is the case with *S. aureus*, in which staphylokinase is carried by a converting phage that inserts into the β-hemolysin structural gene. Detailed information on the activities and properties of the various pathogenicity factors is provided in chapters 38 and 39.

REGULATORY MODALITIES

Information on regulation exists only for the factors listed in Table 1. These factors are expressed in vitro by *S. aureus* according to a carefully orchestrated temporal program (Fig. 1); for those recently identified by STM (55) and IVET (51), it is known only that they are preferentially expressed in vivo. The in vitro expression program schematized in Fig. 1 is assumed, in the absence of any direct data, to be relevant to the development of an infection in vivo, which presumably requires that the various proteins be produced at appropriate times and in appropriate amounts. Very early in the exponential phase, the synthesis of surface proteins, including protein A (SpA), coagulase (Coa), fibronectin-binding proteins (FnBPA and FnBPB), and others, is rapidly induced. Later in the exponential phase, synthesis of these proteins is sharply down-regulated at the transcriptional level. During the postexponential

phase, transcription of a large set of genes encoding extracellular pathogenicity factors, including cytotoxins, hemolysins, superantigens, and tissue-degrading enzymes, is up-regulated. Synthesis of these factors continues for a relatively short time and is then switched off as cells enter stationary phase (3a). It is not known whether specific regulatory factors are responsible for this downshift. Similar in vitro temporal programs have been described for other bacterial pathogens, e.g., *Bordetella pertussis* (37, 70), which activate their virulence program in response to specific signals, such as increased temperature, that are characteristic of the host tissue environment.

In vitro studies have identified several regulatory elements, of which the most thoroughly characterized and probably the most important is *agr* (accessory gene regulation). The importance of the *agr* system is supported by in vivo studies showing that *agr* mutants are greatly attenuated in virulence in each of several animal models tested, including mouse mammary infections (26), arthritis (1), and subcutaneous abscesses (7), and rabbit endocarditis (16) (see chapters 44). Most recently, *agr* mutants have been found to be defective for the induction of apoptosis in infected mammary epithelial cells, implying the existence of an *agr*-up-regulated apoptosis-inducing factor(s) (83). Results such as these obviously support the inference that the above list of products of *agr*-up-regulated genes includes many that are frank pathogenicity factors, but the reciprocal down-regulation of surface proteins by *agr* is also relevant. Although *agr* mutants are greatly attenuated, they are not avirulent; *agr*, *spa* mutants, however, are entirely avirulent in the mouse mammary infection model (26), which is particularly noteworthy because protein A is overproduced in *agr* mutants and may compensate in some unknown way for the absence of the other factors.

The temporal program of exoprotein gene expression shown in Fig. 1 implies the existence of several regulatory elements in addition to *agr*. Fragmentary observations point to five of these: (i) *sar* is required for the optimal function of *agr* promoters P2 and P3 and is a transcription factor for several genes independently of *agr*; (ii) *sae* (known only as a Tn*551* insertion [34]) is required for the expression of *coa* and may thus correspond to an early-exponential-phase signal; (iii) AgrA may also be involved in *spa* up-regulation at this stage, although it is clearly not required; (iv) an RNA structure modifier, such as the H-NS-like protein SptA (86), is likely to act on RNAIII during mid-exponential phase; (v) an independent timing signal, which may represent a change in the nutritional milieu, is necessary for the postexponential up-regulation of exoprotein expression. Two additional regulatory mutations have been identified by Tn*551* insertions: 1E3 (20) and Xpr (77). 1E3 appears to be a suppressor of a partial *agr* defect in the strain ISP479, in which it was isolated, and Xpr is an *agr* mutation associated with a high frequency of chromosomal rearrangements.

A stationary-phase σ-factor, σ^B, homologous to σ^B of *B. subtilis* and σ^S of *Escherichia coli*, has recently been identified in *S. aureus* (22, 84) and shown to be required for the expression of one of the σ^B transcripts (84) and of genes involved in pigment synthesis, defense against oxidative stress, and other functions that have at least an ancillary role in virulence. The only other σ^B promoter thus far identified in *S. aureus* is the *sar* P3 promoter (53), so SarA could be involved in the regulation of σ^B-responsive genes. An important observation is that strain 8325-4, the standard laboratory strain used throughout the world for

TABLE 2 Pathogenicity factors produced by non-aureus staphylococci[a]

Organism	Xhl	Ahl	Bhl	Lip	Dhl	FAME	Ghl	TSS	SEA	FgBP	SEB	SEC	SED	SEE	ET	Coa	SAK	Cap	Nuc	Las	Spa	FnBP
S. aureus	+	+	+	+	+	+	+	+	+	+	+	+	+	+	+	+	+	+	+	+	+	+
S. intermedius					+		+	+	+			+				+						
S. hyicus				+	+										+		+	+		+		
S. epidermidis	+			+	+	+		+		+							+			+		
S. xylosus								+				+					+					
S. sciuri								+						+			+					
S. lugdunensis			+		+					+												
S. warneri										+								+				
S. schleiferi			+																			+
S. haemolyticus	+									+								+				+
S. saprophyticus										+												
S. hominis					+					+							+					
S. simulans										+				+				+				
S. equorum														+								
S. lentus														+			+					
S. capitis														+						+		
S. chromogenes																				+		

[a] Many of these proteins have been identified by their activities only and therefore need not be the same as the corresponding S. aureus protein. Abbreviations: Ahl, α-hemolysin; Bhl, β-hemolysin; Cap, type 5 or 8 capsule; Coa, coagulase; Dhl, δ-hemolysin; ET, exfoliative toxin; FAME, fatty acid monoestterifying enzyme; FgBP, fibrinogen-binding protein; FnBP, fibronectin-binding protein; Ghl, γ-hemolysin; Las, elastase; Lip, lipase; Nuc, staphylococcal nuclease; SAK, staphylokinase; SEA, enterotoxin A; SEB, enterotoxin B; SEC, enterotoxin C; SED, enterotoxin D; SEE, enterotoxin E; Spa, protein A; TSS, toxic shock syndrome toxin-1; Xhl, uncharacterized hemolysin.

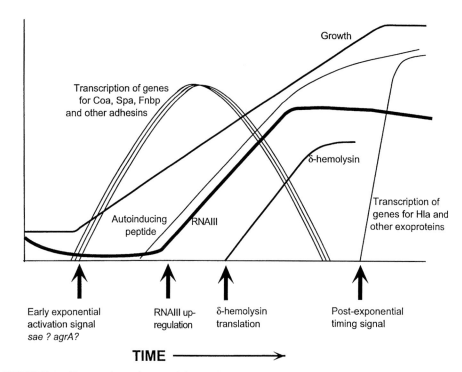

FIGURE 1 Temporal regulation of the virulence response in *S. aureus*. Regulatory events in the virulence response are shown in relation to a standard in vitro growth curve. At the beginning of exponential phase, two or more signals activate synthesis of protein A, fibronectin-binding protein, other surface proteins, and coagulase. During exponential phase, the autoinducing peptide accumulates, eventually reaching a concentration sufficient to activate synthesis of RNAIII, which immediately down-regulates transcription of the surface protein and coagulase genes. δ-Hemolysin, encoded by RNAIII, is not translated until approximately 1 h later. As the cells enter postexponential phase, RNAIII plus an unknown timing signal combine to up-regulate transcription of the hemolysins, exotoxins, and other secreted proteins.

many of the studies on *S. aureus* pathogenesis and regulation, is defective for *rsbU*, a gene required for the activation of σ^B (46). This means that many of the results described here and elsewhere have been obtained in the absence of σ^B function.

The *plaC* gene, initially identified as a chromosomal gene increasing the copy number of plasmid pT181 (39), encodes a sigma factor that is used by the *agr* P2 promoter (8, 39a) as well as by the pT181 countertranscript promoter. As PlaC is similar to σ^{70} of *B. subtilis*, it is probably the standard vegetative sigma factor of *S. aureus*; if so, induction of *agr* expression would not be controlled by the switching of sigma factors.

Regulatory Interactions

Considering the *agr* regulation system as a linear pathway of sequentially acting functions, one may imagine four types of regulatory elements—those acting upstream of *agr*, *agr* itself, those acting downstream, and those acting outside of *agr*, that is, epistatic to it. At present, it is clear that *sar* acts within the *agr* pathway, by affecting *agr* transcription, for all of the target genes and is epistatic to it for some, whereas *sae* may act in the *agr* pathway at a point downstream of the *agr* locus for some target genes but is clearly epistatic to it for others. It is very likely that there are other, as yet unidentified regulatory elements within such a scheme.

In Table 1 is a summary of the known genes and environmental conditions that affect the expression of pathogenicity factors by *S. aureus* as defined above. These are described briefly in the following sections. An important proviso for most of the observations described here and elsewhere is that they are based largely on one or a very few strains, and there is reason to believe that there may be profound strain differences in regulatory modalities, such as the above-mentioned variation in σ^B expression.

REGULATORY ELEMENTS

The *agr* System

The idea that staphylococcal virulence factors are coordinately regulated was originally conceived on the basis of mutations that jointly affected their expression (12, 85) and was confirmed by the isolation of a chromosomal Tn*551* insertion that had this phenotype (52, 68). Cloning of the chromosomal sequences surrounding this insertion (50, 63) led to the identification of a global regulatory locus, *agr*, that reciprocally regulates the level of expression of surface proteins and secreted proteins. *agr* is a complex locus that consists of two divergent transcription units, driven by promoters P2 and P3 (Fig. 2). The P2 operon contains four genes, *agrB*, *D*, *C*, and *A*, all of which are required for transcriptional activation of the *agr* system

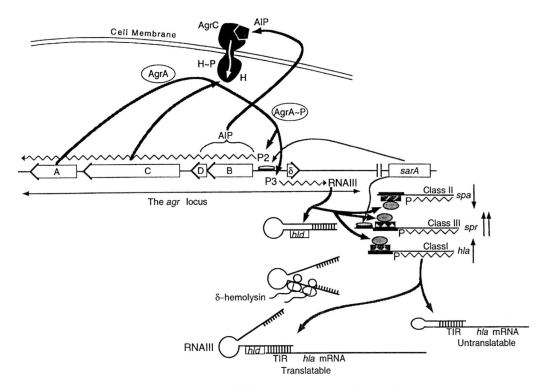

FIGURE 2 The *agr* regulatory circuit. Wavy arrows represent the two divergent *agr* transcripts, RNAII and RNAIII, generated from promoters P2 and P3, respectively. RNAII determines four gene products, AgrA, B, C, and D. AgrA and C are the response regulator and receptor-histidine protein kinase components of a classical signal transduction pathway, and AgrB and D combine to generate an autoinducing peptide (AIP), which is the activating ligand for AgrC. Activated AgrA, plus an accessory transcription factor, SarA, up-regulates both P2 and P3. The latter transcript, RNAIII, is the effector of *agr*-specific regulation and also determines δ-hemolysin. The AIP is shown binding to the extracellular domain of AgrC and inducing a signal (white arrow) that causes autophosphorylation of the conserved histidine in the cytoplasmic domain. This phosphate is transferred to AgrA, giving AgrA~P, which, in conjunction with SarA, activates P2 and P3, setting in motion the autocatalytic *agr* activation circuit. RNAIII immediately blocks transcription of class II exoprotein genes (*spa* and other surface protein genes) and is predicted to fold into an untranslatable configuration. An unknown cellular factor then causes unfolding and permits translation of *hld* and, at the same time, permits the interaction with *hla* mRNA that allows translation of α-hemolysin. RNAIII up-regulates transcription of the class I and III exoprotein genes (class III represents those that are down-regulated by SarA) and down-regulates those of class II. It is not clear whether the putative unfolding of RNAIII is required for transcriptional activation of these genes.

(60). The primary function of this four-gene unit is to activate the two major *agr* promoters, P2 and P3 (45, 50, 63), significantly aided by a second regulatory product, SarA (20). The actual effector of *agr*-dependent exoprotein gene regulation, however, is the P3 transcript RNAIII (40, 45, 61), which acts primarily at the level of transcription. *agr* is conserved throughout the staphylococci with interesting variations, especially in the *B-D-C* region.

The Mechanism of *agr* Activation

AgrC is the receptor and AgrA the presumed response regulator of a two-component signal transduction pathway. This pathway is autoinduced by a posttranslationally modified octapeptide that is processed from within the 46-residue *agrD* product (41). Note that the P2 operon shows two levels of autocatalysis since it encodes its own autoactive response regulator, AgrA, as well as the autoinducing peptide (AIP), the activator for its own signal receptor (Fig. 2).

AgrB, a 26-kDa transmembrane protein (42a), determines the specificity of processing (42) and is probably responsible for secretion of the peptide as well as for the modification, a cyclic thiolactone bond between an internal cysteine and the C-terminal carboxyl (42, 54, 62). Structure-function studies of this peptide have shown that linear variants are inactive and therefore that the ring structure is essential. Most changes in the primary amino acid sequence cause the peptide to become an inhibitor rather than an activator of the *agr* response. This is also the consequence of replacing the thiolactone bond with a lactone or an amide, indicating that the thiolactone bond is essential for *agr* activation and that the ring structure must be present for biological activity. Together, these results suggest that activation may involve a covalent bond between the peptide and its receptor, and that inhibition by variant peptides may involve a different, noncompetitive interaction, perhaps through allosteric changes in the receptor (54). Peptides are responsible for the autoinduction of bacteriocin and lantibiotic syntheses by lactobacilli and other gram-positive bacteria (44, 59) and for competence autoinduction in *Streptococcus pneumoniae* (33) and *B. subtilis* (75). All of these serve as ligands for receptors in typical signal transduction pathways; however, none contains an internal thioester ring, nor has any other naturally occurring peptide with such a ring been identified.

Binding of the AIP to AgrC, a protein with four or five transmembrane α-helices, induces autophosphorylation of the latter on a conserved cytoplasmic histidine residue (49), and it is assumed that the phosphate group is then transferred to AgrA. AgrA is a 34-kDa cytoplasmic protein (63) with the several sequence motifs characteristic of bacterial response regulators. It is presumably activated by transphosphorylation of its conserved aspartate residue by AgrC~P. Although the primary function of AgrA is activation of the two major *agr* promoters (50), purified AgrA does not detectably bind to the intergenic region between the P2 and P3 transcription initiation sites (57), either because only the activated form binds or because some other factor, possibly SarA, is jointly required (19). Suggestive evidence has recently been obtained that AgrA may participate in the activation of *coa* transcription (47).

It is interesting that signal receptors in gram-positive bacteria typically have multiple transmembrane helices (5, 9, 23, 34, 38, 67, 76) and are activated by peptides, whereas those of gram-negative bacteria generally have one, rarely two, transmembrane helices and are activated by a variety of signals other than peptides. It is also notable that autoinducers in gram-negative bacteria are *N*-acylhomoserine lactones rather than peptides, and, with the exception of the *Vibrio harveyi* autoinducer (13), these act intracellularly rather than via signal transduction. A partial explanation for this difference is that peptides over a certain size do not penetrate the outer membrane of gram-negative bacteria, whereas homoserine lactones are freely diffusible.

The P3 Operon

Autoinduction of the P2 operon results in the concomitant activation of transcription from the *agr* P3 promoter, producing the 514 P3 transcript RNAIII, the effector of *agr*-specific target gene regulation (61). The P3 operon encodes one translated product, the 26-amino-acid δ-hemolysin peptide, which has no overall role in the regulatory function of the *agr* locus (40, 61), though a nonsense mutation in codon 3 of δ-hemolysin results in the failure to activate transcription of *spr*, the gene for serine protease (40). This may be a consequence of the effects on RNAIII secondary structure rather than any direct regulatory role for the δ-hemolysin peptide. RNAIII analogs encoded by other staphylococci vary considerably in sequence but are predicted to have similar overall secondary structure. In Fig. 3 is an alignment of RNAIII sequences from several species (61, 77, 79). With the exception of *S. lugdunensis* (26), all encode δ-hemolysin (11, 77).

The primary regulatory function of RNAIII is at the level of transcription (61), probably mediated through one or more regulatory proteins. Thus, we have observed a strong promoter-specific increase in *hla* transcription and a similarly strong inhibition of *spa* transcription immediately following addition of erythromycin or chloramphenicol to a growing culture of either an *agr*⁺ or an *agr*-null strain (68a). These effects are consistent with effects of RNAIII on the translation or function of labile regulatory proteins. Other RNAs interacting with regulatory proteins have recently been described (3, 74). RNAIII deletions (61),

trans-specific RNAIIIs, and interspecies RNAIII hybrids (11, 77) have given different levels of expression for several exoproteins, suggesting that the molecule contains subdomains with separable functions, perhaps representing sites of interaction with different regulatory proteins. It is certainly likely that RNAIIIs of non-aureus staphylococci have a regulatory role in their native species; to date, however, their activity has been demonstrated only in *S. aureus* (11, 77). Thus, in an RNAIII-deficient mutant, RNAIII-*sl* (from *S. lugdunensis*) partially restored the *agr* phenotype: at the mRNA level, *hla* (α-hemolysin) expression was restored, *hlb* (β-hemolysin) expression was partially restored but lipase, *spr* (serine protease), and nuclease expression were not restored, and *spa* (protein A) was not downregulated (13). On the other hand, RNAIII-*se* (from *S. epidermidis*), RNAIII-*sw* (from *S. warneri*), and RNAIII-*ss* (from *S. simulans*), when introduced into *S. aureus* WA400, completely repressed *spa* transcription but stimulated transcription of *hla* and serine protease (*spr*) genes less efficiently than RNAIII-*sa* (78). Further, the 5′ end of the RNAIII molecule may be involved in translational regulation (see below), whereas the 3′ end seems to be important for both the transcriptional activation and repression functions of the molecule (13, 64). Thus, a chimeric construct (RNAIII-*sl-sa*), consisting of the 5′ end of RNAIII-*sl* including the promoter fused to the 3′ end of RNAIII-*sa*, had greater activity in restoration of *agr* regulation of β-hemolysin, protein A, and, to a lesser extent, lipase and serine protease in *S. aureus* than did unmodified RNAIII-*sl* (13). Tests with *S. epidermidis*-*S. aureus* RNAIII hybrids showed that both the 5′ and 3′ halves of RNAIII are involved in the transcriptional regulation of *hla* and *spr* (80).

Translational Regulation

RNAIII contains a series of nucleotides in its 5′ region complementary to the leader region of *hla* and another series of nucleotides in its 3′ region complementary to the 5′ region of *spa*. Deletion of the 5′ end, containing the sequences complementary to the α-hemolysin mRNA leader, abolished translation of α-hemolysin without affecting transcription of the *hla* gene (61), and deletion of the 3′ end, containing the sequences complementary to the *spa* mRNA leader, eliminated RNAIII-specific inhibition of Spa translation (unpublished data). The requirement of the 5′ end of RNAIII for α-hemolysin translation suggested that the *hla* mRNA leader must have a secondary structure that precludes translation and that pairing with the complementary sequences in RNAIII must prevent this inhibitory secondary structure from forming so as to permit translation. This hypothesis has been confirmed by Morfeldt et al. (56), who demonstrated the formation of a specific complex between RNAIII and the *hla* leader that preempted the formation of an inhibitory pairing in the *hla* leader that blocked the translation initiation signals. In the computer-predicted secondary structure of RNAIII, however, the sequences that would unblock *hla* translation are paired and would be unavailable for complementary pairing with the *hla* leader. This pairing would also be expected to block δ-hemolysin translation, and so we suspect that if RNAIII assumed this configuration, it would have to be unfolded in order to permit δ-hemolysin translation as well as to interact with the *hla* leader. We have observed that δ-hemolysin is not translated until about 1 h after the synthesis of RNAIII, consistent with a requirement for a posttranslational change in its conformation (6). Moreover, deletion of the 3′ half of RNAIII, which would eliminate

```
Consensus    ACAGTT  G   AAATAT   A TTA G A T TT A    A     AA ATGTTTTAAT  A
SepiRNAIII   TTACAGTTGAGTACTAAATATTGCTATTTACGAAATTTTAATCTTTAGATGAAAAAATCATGTTTTAATAGACT
6390RNAIII   ------------------------------------------------------------------------
SwarRNAIII   TTACAGTTTAGTATAAAATATTGCAAGTTAGGAATTTTTTAATTCAAAGGGTTATAACCATGTTTTAATAGACT
SlugRNAIII   ------------------------------------------------------------------------
SsimRNAIII   AGACAGTTATGATTGAAATATCATAAGTTAGGGATTATTTAACATTAAATTTAGAAATGATGTTTTAATGAAGA

Consensus    TATCACAGA GATGTGAT GAAA   AG T A   TT    AA TA       A
SepiRNAIII   CATATCACAGATGATGTGATTGAAAGATAGTTGAAAAATTTGCTTAATCTAGTCGAGTGAATGTTAAATTCATT
6390RNAIII   --GATCACAGA-GATGTGATGGAAA-ATAGTTGATGAGT--------------------TGTTTAATT--TT
SwarRNAIII   CATATCACAGA-GATGTGATGGAAATAGAGTTGAAAATT-TGCTTAATCTACTTATTAGAATTATTTTTCTAAG
SlugRNAIII   --TATCACAGA-GATGTGATAGATATT--GATAAAGCTTTTTGAAAAC------------------------
SsimRNAIII   TGTATCACAGA-GATGTGATAAGAATTCAGGTAACACATTACATAAACATATCGTGTTTGAAGTCAATGTGGTT

Consensus             C T A     A  A TG   TT  ATGGCA C   ATC T TTC A  TCGGTGA
SepiRNAIII   CGTATCCATTACCTTAATT--C-GAAAGGAGTGAAGTTATGATGGCAGCAGATATCAT-TTCTACAATCGGTGAT
6390RNAIII   AAGAATTTTTTATCTTAATT--AAGGAAGGAGTGA--TTTCAATGGCACAAGATATCAT-TTCAACAATCGGTGAC
SwarRNAIII   TAATTCTAGTTCCATTACTTGAAGGAAGGAGTGA--TTTCAATGGCAGCAGATATCAT-TTCAACAATCGGTGAT
SlugRNAIII   ------------------------TAATTGG----ATTGTAGCTACGATAAGCTT-TTCAATTTTC------
SsimRNAIII   ACCTTACTTCATCTTAATATTTCAAAGAAATTGG----TTTATGGCATCGTTAATATTCTTCGTAGGTC------

Consensus    TTAGTAA AT ATTAT   AC GT AA A T C  A  AA              T   A G      T
SepiRNAIII   TTAGTAAAATGGATTATCGATACAGTTAATAAATTCAAAAAAATAATT------------TTTGAATGAGTCTATT
6390RNAIII   TTAGTAAAATGGATTATCGACACAGTGAACAAATTCACTAAAAAA--------------TAAGATGAATAATTA
SwarRNAIII   TTAGTAAAATTAATCATTAACACTGTGAAAAAATTCCAAAAATAATTTTTTCTAGTTAGTTAAGATGAGTATTTA
SlugRNAIII   ----CAACAATGATTGTAGTAATTGTT----GGTGCTGTAT----------------------------ATT
SsimRNAIII   ----TAACATGCCTTAGACTAACTGAT----CATCCTATATTGAAAAG------------AAAGGGGGCAATA

Consensus                                                           ATA   T
SepiRNAIII   G------------------------------------------------------------------------
6390RNAIII   A------------------------------------------------------------------------
SwarRNAIII   GGATTTACAATAATTCAAATGGAAGGAGTGATTTCAATGACTGCAGATATCATTTCAACAATTGGTGATTTTGT
SlugRNAIII   CTCA-------------------------------------ATAATATATG----------------
SsimRNAIII   CACAT-------------------------------GGCAGGCGATATCGTAGGCACTATCGGTGAATTCGT

Consensus                            T.A.T A        T  T TAACT TGTTTTCTTCGT
SepiRNAIII   ---------------------------TAACT---------------T--TTGTAACTTTGTTTTCTTCGT
6390RNAIII   ---------------------------TTACT--------------TTCATTGTAAATTTGTTATCTTCGT
SwarRNAIII   AAAATGGATTTTAGATACAGTAAAAAAATTCACTAAATAATAGTAGAAATTCATTGTAACTTTGTTTTCTTCGT
SlugRNAIII   ---------------------------TTATTCAT---------------TCTTAACTATGTTTTCTTCGT
SsimRNAIII   GAAA-----------------------TTAATCATCGA----------AACAGTTCAAAAATTCACTCAA

Consensus    A A  T  A TA GT T G  TAAGC CATCCCAACTTAAT AT    ATG AAAATT G
SepiRNAIII   ATA---ATTAATACTATTAG---TGAGTTGTTGAGC-CATCCCAACTTAATAATTTACTAATATAAACTAAGCA
6390RNAIII   A-------TAGTACTAAAAGT-ATGAGTTATTAAGC-CATCCCAACTTAATAA-----CCATGTAAAATTAGCA
SwarRNAIII   A-------------TAAAAG--GTGAGTTATTAAGC-CATCCCAACTTAATAA------CTTGCAACATTAGTA
SlugRNAIII   TAA--ATTGATTAATTATTGT---------TAAGC-CATCCCAACTTAATGATAATTG-ATG---AATTG---
SsimRNAIII   AAATAATTCCTTATTTATTGTCCTTTGAGTATAAGCACTAAATTGTTTAATCCTTATAGCAGG---AATTGGTG

Consensus     TG   A CATT   T  T      TAG TTTCCT GG CTCA G TA TAT    T A   A C T
SepiRNAIII   AGTGA-GAAGCATTTGCTAGTA--ACTGTAG-TTTCCTTGGACTCAGTGTTACGTATTATTCTTAGCTACCTTA
6390RNAIII   AGTGA-GTAACATTTGCTAGTA--GAGTTAG-TTTCCTTGGACTCAGTGCTATGTATTTTTCTTAATTATCAT-
SwarRNAIII   GGTGACCAAACATTTACTAATATGACTGTAGTTTTCCTTGGACTCAGTGTTACGTATTATTCTTAGCTACCTTT
SlugRNAIII   ------CTAACATTGAGTTGTG--------------TAGGGCTCA---------ATGTAGATAATTACGC-T-
SsimRNAIII   TTTGTCCTTGCATTGATTCCTG--------------CATCTCTCATA--TATGTATATCGGAAATGAAGCGT-
```

FIGURE 3 Comparison of RNAIII gene sequences. The five available RNAIII sequences were aligned with gaps introduced as needed to display the several striking conserved motifs. "Consensus" indicates at least four matches out of five; in places where only three sequences are present, all three must match. δ-Hemolysin coding regions are indicated in bold italics, and potentially translatable regions upstream of and in frame with them are indicated in italics. Note that *S. warneri* RNAIII contains two copies of the δ-hemolysin coding frame and that the predicted *Staphylococcus simulans* δ-hemolysin is 53 amino acids in length—almost exactly twice the length of the others. Sepi, *Staphylococcus epidermidis*; Swar, *S. warneri*; Slug, *S. lugdunensis*; Ssim, *S. simulans*; 6390 is a standard *S. aureus* group I strain.

```
Consensus    AT G TA     TTCT    A GTAA  TATCGT  A AACATT AATTTATCA      A     G TAAAT
SepiRNAIII   AATAGGTAATTATTTCT-AGCATGTAAGCTATCGT-AAACAACATTCAATTTATCATGTTAAAT--AGATAAAT
6390RNAIII   TATAGATAATTATTTCT-AGCATGTAAGCTATCGT-AAACAACA-TCGATTTATCA---TTATT--TGATAAAT
SwarRNAIII   AATTAGGTATT-TTTCT-AACATGTAAGCTATCGTAAAACAACATTCAATTTATCATTATAAAAAGAGATAAAT
SlugRNAIII   --TAGCTA--CTTAACTGTAGAAGGTAATTATTGT------AC---TAGTGTTACT-----CTAGTCAGAAGA-
SsimRNAIII   --TCGGTATGCATATCATTGTAGTTAAATTATCGT-TAGTAACATTAAATTTATCA-----AATATCGGTAAAT

Consensus    TCACTAAA TTTTTTCATAATTAATAACATCCCCA AAAAATAGATTG AAAAATAACTGT AAAACATTCCCT
SepiRNAIII   TCACTAAAATTTTTTCATAATTAATAACATCCCCA-AAAAATAGATTG-AAAAATAACTGT-AAAACATTCCCT
6390RNAIII   ----AAAATTTTTTTCATAATTAATAACATCCCC--AAAAATAGATTGAAAAAATAACTGT-AAAACATTCCCT
SwarRNAIII   TCACTAAAATTTTTTCATAATTAATAACATCCCCAAAAAAATAGATTGAAAAAATAACTGTAAAAACATTCCCT
SlugRNAIII   -----------TTTTCATAATTAATAACATCCCCA--AATTATTATTG--AAAATAACTGTAAAAACATTCCCT
SsimRNAIII   TCACTAAATTTTTTTCATAATTAATAACATCCCCA--AAAATAGATAG-AAAAATAACTGT--AAACATTCCCT

Consensus    -TAATAATAAGT A    A G CGTGAG C CTCCCAA CTCACGG
SepiRNAIII   -TAATAATAAGTTATC--AAGCCGTGAGTCTCTCCCAAGCTCACGGCT------
6390RNAIII   -TAATAATAAGTA-----TGGTCGTGAGCCCCTCCCAAGCTCGCGGCCTTTTTT
SwarRNAIII   -TAATAATAAGTTATCAAAAGCCGTGAGCCCCTCCCAAGCTCACGGT-------
SlugRNAIII   -TAATAAGCGAAAA----AAGTCGTGAGTCACTCCCAAACTCACGACTTTTA--
SsimRNAIII   TGAATAATAAGTAA----AACCCATGAGTTCCTCCCAAACTCATGGGT------
```

FIGURE 3 *Continued.*

the putative inhibitory intramolecular pairing, eliminated this delay. Presumably, such a change in conformation would also be required for the activation of δ-hemolysin translation by RNAIII. The agent of this putative post-translational change in RNAIII conformation, however, is not known. One possibility is translation of the 18 translatable codons 5′ to *hld*, the δ-hemolysin coding sequence. It has recently been observed that a small H-NS analog, StaA, in *E. coli* affects RNA secondary structure and is synthesized for a short period during the exponential phase of growth (86). Such a protein could be responsible for modifying RNAIII conformation so as to permit its translation. In Fig. 2 is a scheme modeling this rather complex regulatory circuit. However, a recently proposed secondary structure for RNAIII is not consistent with this concept (11a).

Sequence Variation in the *agr* Locus

In Fig. 4 is presented a comparison of the *agrB, C,* and *D* sequences from several *S. aureus* strains and from *S. lugdunensis* (42, 79) and *S. epidermidis* (80). The boxed region in AgrD contains the AIP sequences. Those of *S. aureus* groups I and II, of *S. lugdunensis*, and of one *S. epidermidis* strain, SeI, have been determined directly (44, 65); the others are inferred. Note that the central cysteine is the only absolutely conserved residue. The N-terminal third of AgrB and the C-terminal half of AgrC are highly conserved; between these conserved regions is a highly variable region encompassing the C-terminal two-thirds of AgrB, all of AgrD, and the N-terminal half of AgrC (42). Note that the dividing line between the variable and conserved regions is very sharp, consistent with a cassette-switching or other site-specific recombinational exchange mechanism, and that this sequence organization is conserved in the two non-aureus staphylococci thus far analyzed. Conserved motifs within the divergent region probably represent conserved functions. It seems likely that a remarkable evolutionary mechanism underlies this divergence, since the three genes must have diverged in concert to maintain the specificity of their processing and receptor-binding activities. This mechanism could involve some sort of hypervariability-generating mechanism in which large numbers of random mutations occur, with selection for the functional combinations.

The *sar* Locus

sar was initially identified as a Tn*917* insertion that decreased the expression of several of the exoprotein genes belonging to the *agr* regulon (15) (Table 1). The *sar* locus contains three overlapping transcription units, 0.56, 0.8, and 1.2 kb, with a common 3′ end, that have confusingly been designated *sarA, C,* and *B*, respectively (Fig. 5)—this despite the presence of but a single gene, also (and properly) referred to as *sarA*, that is included in all three transcripts and encodes SarA, a 14.7-kDa DNA-binding protein distantly related to VirF of *Shigella flexneri*. Two other small open reading frames are present, upstream of *sarA*, whose translation has never been demonstrated. SarA up-regulates transcription of the two *agr* operons approximately 5- to 10-fold by binding to the intergenic region (14a, 38, 60). SarA also affects the expression of certain exoprotein genes directly, independently of its affects on *agr*, and is thus best viewed as a general transcription factor, one of whose targets is the *agr* locus. Deletions in the region upstream of *sarA* eliminate the ability of cell-free extracts to gel-shift the *agr* target (35) and of SarA to regulate the expression of some of the exoprotein genes (17). These effects are probably due to a decrease in SarA production. SarA footprints a 26-nucleotide AT-rich sequence in the intergenic region between the *agr* P2 and P3 promoters (21). This sequence element was previously identified by Projan (64a) in the 5′ regions of several of the exoprotein genes as well as in the *agr* locus. It is proposed here to designate this sequence the "Sar box" (see Fig. 6); however, its significance is uncertain, as SarA has recently been shown to bind AT-rich DNA nonspecifically (76a).

A *sar* homolog has been identified in *S. epidermidis* that complements a *sar* mutant of *S. aureus*. The *S. epidermidis* gene lacks the two small open reading frames present in the latter (25), supporting the idea that their putative translation products are not important in *sar* function. No regulatory function has yet been demonstrated for *sar* in non-aureus staphylococci, most probably because no mutants that could be tested have yet been isolated.

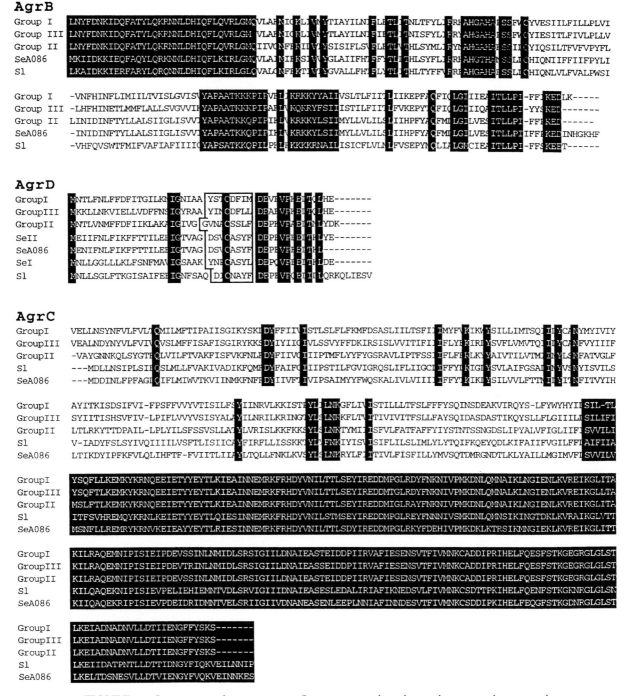

FIGURE 4 Comparison of *agr* sequences. Sequences are aligned according to similarities, and regions of high similarity are shaded. Within these regions, nonidentical amino acid residues are not distinguished. Sequences were determined by the Novick and Arvidson laboratories, with the exception of the *S. epidermidis* sequences, which were determined by Van Wamel et al. (81) and by Otto et al. (62). The boxed region in the AgrD sequences represents the respective AIPs. Those of *S. aureus* groups I and II and of *S. epidermidis* A086 have verified by amino acid sequencing and/or peptide synthesis. The others have been predicted from the corresponding nucleotide sequences.

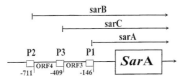

FIGURE 5 The *sar* locus. Sar B, C, and A represent three overlapping transcripts with promoters P2, P3, and P1, respectively, shown as small open boxes with numbers representing distances from the *sarA* start codon. Figure kindly provided by Ambrose Cheung.

sae

sae may be a third gene regulating exoprotein synthesis in *S. aureus*, although it has been characterized thus far only as a chromosomal Tn*551* insertion in ISP479 (31, 32). Available information on *sae* is very limited, as it has not been made available to the scientific community. The *sae* mutation affects the expression of many staphylococcal exoprotein genes, with the notable exception of protein A, serine protease, and lipase, which are unaffected. Most important, the mutation has no effect on the expression of *sar* or *agr*, suggesting that it acts downstream of or epistatically to these genes in the exoprotein expression pathway. Thus, there may be at least one required regulatory product downstream from or outside of the *agr* pathway.

Environmental Factors

Several environmental signals affect the expression of target genes in the *agr* regulon. Thus, 1 M NaCl blocks RNAIII transcription (35a) and also blocks transcription of many *agr*-up-regulated genes such as *hla* and *spr* (14), as well as of *agr*-down-regulated genes such as *spa* and *fnp* (14). These effects appear to be independent of RNAIII expression. It has long been known that expression of many of the exoprotein genes is catabolite repressed, which could account for the observation that a postexponential signal, in addition to RNAIII, is required for the up-regulation of the exoprotein genes (78). This signal could represent a nutritional change in the medium, such as the exhaustion of glucose. Perhaps these virulence factors are catabolite repressed because their basic function is nutritional—thus,

they would not be needed in a nutritionally rich (glucose-containing) environment. A notable exception is serine protease—staphylococci require several amino acids and these, of course, cannot be produced from glucose but can be obtained by the hydrolysis of external proteins. Catabolite repression is thus consistent with a primary nutritional role for the toxic exoproteins.

It has also been observed that subinhibitory concentrations of antibiotics that inhibit protein synthesis (30), surfactants such as glycerol monolaurate (66), lowering of pH or O$_2$ tension, or limitation of essential amino acids (48) have a general inhibitory effect on exoprotein gene expression. The inhibitory effects of these environmental factors, many of which act independently of *agr*, suggest that the cell has a specific regulatory modality that selectively switches off expression of accessory genes, such as virulence factors and other exoproteins, when the organism is stressed by inhibition of protein synthesis. Consistent with this is the observation that respiration-defective mutants (small-colony variants) also show a global reduction in exoprotein synthesis (see chapter 35, this volume). The inclusion of 1 M NaCl in this list is puzzling, since it is rather harmless for staphylococci. Its effects may involve depolarization of the cellular membrane, since they are reversed by glycine betaine (82), which reestablishes the electrochemical gradient, and so may or may not be part of the hypothetical general stress response. It is also possible that some or all of these effects are mediated through changes in DNA supercoiling (72) or through the acquisition or loss of specific σ factors by RNA polymerase. Such observations suggest that there are additional unidentified regulatory genes through which these environmental stimuli must act.

Divalent cations (3a, 14, 25a) have also been observed to affect the expression of many of these same virulence genes (Table 1), and β-lactam antibiotics seem to have an effect opposite to that of the protein synthesis inhibitors, namely, stimulating expression of some of the exoprotein genes, possibly owing to cross-talk between signaling systems.

Bacterial Interference Caused by the *agr* Autoinduction System

A survey of various staphylococci, including diverse *S. aureus* strains and representatives of other staphylococcal species, revealed, remarkably, that the *agr* autoinducing peptides from some strains could inhibit the *agr* response in other strains. On the basis of this activity, *S. aureus* strains could be divided into four major groups such that within any one group, each strain produced a peptide that could activate the *agr* response in the other members of that group, whereas between groups the autoinducers were mutually inhibitory (24a, 42, 77a). Genetic analysis revealed that the inhibitory and activating activities of any strain were always encoded by a single gene, *agrD*, suggesting that a single peptide was responsible for both activities. This was confirmed by purification of the autoinducer from several strains; in each case, a single peptide species was obtained that activated the *agr* response by the strain that produced it and others in its group and inhibited the response in strains of the other two groups (42). Finally, the above-mentioned thioester-containing synthetic autoinducers had the same activating and inhibiting activities as the culture supernatants containing the native peptides (54).

```
agr locus
  RNAIII +1(1572)    82 nt  Sar box  72 nt   RNAII +1(1721)


  agr       ATTTGTATTTAATATTTTAACATAAA*    19/26
  spa       CTTTAAATTTAATTATAAATATAGAT*    19/26
  entB,C    ATTTTCTTTTAATATTTTTTTAATTG*    24/26
  tst       ATTTTTAATTAATATATATTTAAACA     20/26
  hlb       ATTTATTATTAATATTTATTTAATTG     21/26
  hla       ATTTTTTATTAATAGTTAATTAATTG*    25/26
  FnBP      CTTTGTATGCAATATATATGTGAGTT*    17/26


  Consensus ATTTnTATTTAATATTTATTTAAnTn
```

FIGURE 6 The Sar box. The *agr* locus is represented diagrammatically to indicate the location of the Sar box in relation to the start points of the two major *agr* transcripts, RNAII and RNAIII. Sequences marked with an asterisk have been shown to bind SarA, and, for *agr*, *spa*, and *hla*, their deletion has been shown to eliminate SarA regulation (14a, 23). Numbers in parentheses represent the sequence coordinates of these start points (65, 66). For *hla*, the distance between the Sar box and the transcriptional start is 33 nucleotides. See Table 1 for proteins corresponding to genes.

Three observations suggest that the peptide-determined autoinduction specificity may be biologically relevant. First, *S. aureus* 502A was used in the late 1960s to block umbilical stump colonization of newborn infants by the then rampant *S. aureus* 80/81 complex and was also used successfully to treat furunculosis by displacing the causative strains (73). In our hands, 80/81 strains usually belong to autoinducer group I, whereas 502A belongs to group II, and its autoinducer is a highly potent inhibitor of *agr* activation in strains of groups I and III. A possible implication is that the 502A peptide was responsible for the observed interference; if so, one might infer that this interference involves inhibition of the expression of *agr*-regulated factors. An obvious problem with this hypothesis is that *agr* downregulates most or all of the known surface proteins, many of which are adhesins and are generally regarded as colonization factors. However, the actual factors that are responsible for interference are not known, nor has it been demonstrated that the known adhesins correspond to colonization factors in vivo.

A second observation is that there are significant correlations between *agr* groupings and clinical syndromes. Thus, the vast majority of toxic shock strains of *S. aureus*, including those responsible for staphylococcal scarlet fever, belong to autoinducer group III; most of the strains producing exfoliative toxin (ETA or ETB) belong to *agr* group IV; and most endocarditis strains belong to groups I and II (24a, 49a). The *agr* group III toxic shock syndrome strains have many other features in common, including phage type (2), electrotype (58), and weak expression of most extracellular toxic products (71), suggesting that they may represent a clone. The same is probably true of the *agr* group IV ETA strains, which all belong to phage group II and are indistinguishable by pulsed-field gel electrophoresis (45a).

A third observation is that the synthetic group II thiolactone peptide greatly attenuated an experimental infection by a group I strain (Fig. 7) (54).

It appears, therefore that the *agr* specificity grouping represents the first subdivision of *S. aureus* (and probably other species) based on the fundamental biology of the organism. Only surveys of a large number of additional strains will reveal whether this is a valid impression.

In Table 1 are summarized the known genes and environmental factors influencing the expression of staphylococcal virulence factors. Included is a list of some of the in vivo-expressed genes. Of particular interest is the *sae* mutation, whose effects are independent of *agr*. This mutation seems to interfere with a regulatory paradigm rather different from that of *agr*, in that a different set of exoprotein genes is blocked, including *coa*, and genes that are reciprocally regulated by *agr* do not show this reciprocality. Mutations such as this indicate clearly that we still have a great deal to learn about virulence gene regulation in *S. aureus*.

Staphylococcal Pathogenesis

Aside from the toxinoses (toxic shock syndrome, scalded skin syndrome, and food poisoning), whose pathogenesis is understandable on the basis of the mechanism of action of the respective toxins, the pathogenesis of classical staphylococcal infections, such as abscesses, endocarditis, pneumonia, and osteomyelitis, is highly complex and involves several more or less distinct processes after the organism has gained access to a tissue compartment: adherence to matrix molecules that may be in the intercellular space or on cell surfaces, mediated by the MSCRAMMs (see chapter 40, this volume); attraction of polymorphonuclear leukocytes by chemotactic factors and destruction of these and other cellular elements by locally acting toxins (chapter 38); activation of fibrinogen clotting and then dissolution; degradation of intercellular matrix by proteases and other exoenzymes (chapter 39); degradation of complement and/or immunoglobulins and, in some situations, invasion of epithelial or endothelial cells (chapter 42), followed by the induction of apoptosis in the former (10, 83). Synthesis of most of the virulence factors involved in this process is regulated by *agr*, *sar*, *sae*, and combinations of these genes. Superimposed on these bacterial activities is a very robust inflammatory response, including the activation of cellular signaling pathways leading to the production of cytokines and other proinflammatory mediators. The local inflammatory response is usually sufficient to contain the organism, leading to an abscess—a pocket of semiliquid necrotic tissue, degraded tissue components, dead polymorphonuclear leukocytes and bacteria, surrounded by a capsule composed of fibrin and infiltrating inflammatory cells—but when the innate immune response is inadequate, the bacteria may escape the primary lesion and disseminate via the bloodstream. This may result in overwhelming sepsis, with shock caused by the teichoic acid-induced activation of cellular signaling pathways, leading to the overproduction of tumor necrosis factor alpha (36). Variations in the expression of different virulence factors may be responsible for strain-dependent tissue tropisms. A major unknown in this scenario is the role of the recently identified virulence factors encoded by IVET and STM-identified genes. Although staphylococcal lesions obviously develop according to a regular temporal progression of events, a major unknown is whether this progression has any resemblance to that seen in laboratory cultures, as diagrammed in Fig. 1, and whether it is mediated by any of the regulatory factors identified in vitro. The only relevant information at present is a recent finding that one of the three *sar* promoters, P2, which is silent in vitro, is selectively activated during experimental endocarditis (18). Since *S. aureus* strains differ widely in the patterns of expression of the various virulence factors, it has long been thought that there is significant strain specificity for different types of infections (that is, aside from the frank toxinoses). Available data on this point remain too fragmentary to warrant any description here.

Relation of the Autoinduction-Density-Sensing Paradigm to Staphylococcal Pathogenesis

It is widely accepted that genes regulated by autoinducers are not expressed until a critical population density is reached, at which point, but not earlier, their expression becomes adaptive. The critical threshold population density would then coincide with a critical autoinducer concentration. Autoinducing circuits of this type are considered to represent cell density-sensing mechanisms and thus to represent a means of intercell communication in a population of bacteria (28, 43). This type of density sensing could be highly useful in the context of an abscess, in which the up-regulated genes would be activated when the organisms become crowded and nutritionally deprived. This is parallel in an interesting way to the activation of *Vibrio fischeri* bioluminescence (24, 29), which is the prototypical density-sensing autoinduction system. Here, the bioluminescence response, which occurs in the light organs of various deep-sea fishes and other animals, does not occur

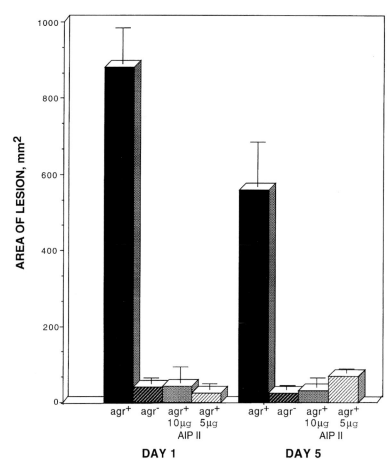

FIGURE 7 Effect of an inhibitory AIP on a mouse subcutaneous abscess. Hairless mice were injected subcutaneously with ~10⁸ organisms of RN6390B (agr⁺), with or without the group II AIP as indicated, or with RN6911 (agr⁻). Lesions were measured daily, and areas were calculated by approximating an elliptical shape.

until the bacteria have become sufficiently dense to generate the amount of light required by the host organism.

The application of this principle to *S. aureus* presents some interesting problems. On the one hand, the density-dependent activation of pathogenicity determinants in the well-developed abscess pocket would seem to be clearly adaptive. The bacteria have been walled off by the host defense system, are crowded, and are under nutritional stress. The autoinduction of secreted pathogenicity factors could enable the bacteria to express these factors at high levels while there is still time, enabling them to break out of the pocket and disseminate to initiate new foci of infection. Although this rationale could explain why *S. aureus* may use autoinduction in localized infection, it is not easily applicable to nonlocalized infections such as septicemia and endocarditis. It may be relevant in this context that the group III and IV strains are largely responsible for toxic shock and for toxic epidermal necrolysis, respectively, toxinoses that can be produced by small numbers of organisms in the absence of a frankly purulent lesion. It is also important to point out, in conclusion, that although the known in vitro regulatory systems governing the production of staphylococcal pathogenicity factors represent a very elegant and self-consistent view of how gene functions may be organized for efficient expression, direct data on whether this paradigm operates in vivo have yet to be ob-

tained. Additionally, it is difficult to imagine that specific environmental signals, such as those listed in Table 1, do not have a major role. Only by direct studies of gene expression and regulation in vivo will it be possible to develop a full understanding of how the staphylococcal genome is organized for optimal functioning in the hostile environment of the host tissues.

REFERENCES

1. **Abdelinour, A., S. Arvidson, T. Bremell, C. Ryden, and A. Tarkowski.** 1993. The accessory gene regulator (*agr*) controls *Staphylococcus aureus* virulence in a murine arthritis model. *Infect. Immun.* **61:**3879–3885.

2. **Altemeier, W. A., S. A. Lewis, P. M. Schlievert, M. S. Bergdoll, H. S. Bjornson, J. L. Staneck, and B. A. Crass.** 1982. *Staphylococcus aureus* associated with toxic shock syndrome. *Annu. Intern. Med.* **96:**978–982.

3. **Altuvia, S., A. Zhang, L. Argaman, A. Tiwari, and G. Storz.** 1998. The *Escherichia coli* OxyS regulatory RNA represses fhlA translation by blocking ribosome binding. *EMBO J.* **17:**6069–6075.

3a. **Arvidson, S.** Personal communication.

4. **Arvidson, S. O.** 1983. Extracellular enzymes from *Staphylococcus aureus*, p. 745–808. *In* C. S. F. Easmon and C.

Adlam (ed.): *Staphylococci and Staphyloccocal Infections*, vol. 2. Academic Press, London, U.K.

5. **Axelsson, L., and A. Holck.** 1995. The genes involved in production of and immunity to sakacin A, a bacteriocin from *Lactobacillus sake* Lb706. *J. Bacteriol.* **177:**2125–2137.

6. **Balaban, N., and R. P. Novick.** 1995. Translation of RNAIII, the *Staphylococcus aureus agr* regulatory RNA molecule, can be activated by a 3′-end deletion. *FEMS Microbiol. Lett.* **133:**155–161.

7. **Barg, N., C. Bunce, L. Wheeler, G. Reed, and J. Musser.** 1992. Murine model of cutaneous infection with gram-positive cocci. *Infect. Immun.* **60:**2636–2640.

8. **Basheer, R., and S. Iordanescu.** 1991. The *Staphylococcus aureus* chromosomal gene plaC, identified by mutations amplifying plasmid pT181, encodes a sigma factor. *Nucleic Acids Res.* **19:**4921–4924.

9. **Ba-Thein, W., M. Lyristis, K. Ohtani, I. T. Nisbet, H. Hayashi, J. I. Rood, and T. Shimizu.** 1996. The virR/virS locus regulates the transcription of genes encoding extracellular toxin production in *Clostridium perfringens. J. Bacteriol.* **178:**2514–2520.

10. **Bayles, K. W., C. A. Wesson, L. E. Liou, L. K. Fox, G. A. Bohach, and W. R. Trumble.** 1998. Intracellular *Staphylococcus aureus* escapes the endosome and induces apoptosis in epithelial cells. *Infect. Immun.* **66:**336–342.

11. **Benito, Y., G. Lina, T. Greenland, J. Etienne, and F. Vandenesch.** 1998. Trans-complementation of a *Staphylococcus aureus agr* mutant by *Staphylococcus lugdunensis agr* RNAIII. *J. Bacteriol.* **180:**5780–5783.

11a. **Benito, Y., and F. Vandenesch.** Personal communication.

12. **Bjorklind, A., and S. Arvidson.** 1980. Mutants of *Staphylococcus aureus* affected in the regulation of exoprotein synthesis. *FEMS Microbiol. Lett.* **7:**203–206.

13. **Cao, J. G., Z. Y. Wei, and E. A. Meighen.** 1995. The lux autoinducer-receptor interaction in *Vibrio harveyi*: binding. *Biochem. J.* **312:**439–444.

14. **Chan, P. F., and S. J. Foster.** 1998. The role of environmental factors in the regulation of virulence-determinant expression in *Staphylococcus aureus* 8325-4. *Microbiology* **144:**2469–2479.

14a. **Cheung, A.** Personal communication.

15. **Cheung, A. L., J. M. Coomey, C. A. Butler, S. J. Projan, and V. A. Fischetti.** 1992. Regulation of exoprotein expression in *Staphylococcus aureus* by a locus (*sar*) distinct from *agr. Proc. Natl. Acad. Sci. USA* **89:**6462–6466.

16. **Cheung, A. L., K. J. Eberhardt, E. Chung, M. R. Yeaman, P. M. Sullam, M. Ramos, and A. S. Bayer.** 1994. Diminished virulence of sar⁻/agr⁻ mutant of *Staphylococcus aureus* in the rabbit model of endocarditis. *J. Clin. Invest.* **94:**1815–1822.

17. **Cheung, A. L., K. Eberhardt, and J. H. Heinrichs.** 1997. Regulation of protein A synthesis by the *sar* and *agr* loci of *Staphylococcus aureus. Infect. Immun.* **65:**2243–2249.

18. **Cheung, A. L., C. N. Nast, and A. S. Bayer.** 1998. Selective activation of *sar* promoters with the use of green fluorescent protein transcriptional fusions as the detection system in the rabbit endocarditis model. *Infect. Immun.* **66:**5988–5993.

19. **Cheung, A. L., and S. J. Projan.** 1994. Cloning and sequencing of sarA of *Staphylococcus aureus*, a gene required for the expression of *agr. J. Bacteriol.* **176:**4168–4172.

20. **Cheung, A. L., C. Wolz, M. R. Yeaman, and A. S. Bayer.** 1995. Insertional inactivation of a chromosomal locus that modulates expression of potential virulence determinants in *Staphylococcus aureus. J. Bacteriol.* **177:**3220–3226.

21. **Chien, Y., and A. L. Cheung.** 1998. Molecular interactions between two global regulators, sar and agr, in *Staphylococcus aureus. J. Biol Chem.* **273:**2645–2652.

22. **Deora, R., T. Tseng, and T. K. Misra.** 1997. Alternative transcription factor sigmaSB of *Staphylococcus aureus*: characterization and role in transcription of the global regulatory locus *sar. J. Bacteriol.* **179:**6355–6359.

23. **Diep, D. B., L. S. Havarstein, J. Nisssen-Meyer, and I. F. Nes.** 1994. The gene encoding plantaricin A, a bacteriocin from *Lactobacillus plantarum* C11, is located on the same transcription unit as *agr*-like regulatory system. *Appl. Environ. Microbiol.* **60:**160–166.

24. **Eberhard, A.** 1972. Inhibition and activation of bacterial luciferase synthesis. *J. Bacteriol.* **109:**1101–1105.

24a. **Figueiredo, A. M. S., and R. P. Novick.** Unpublished data.

25. **Fluckiger, U., C. Wolz, and A. L. Cheung.** 1998. Characterization of a *sar* homolog of *Staphylococcus epidermidis. Infect. Immun.* **66:**2871–2878.

25a. **Foster, S.** Personal communication.

26. **Foster, T. J., M. O'Reilly, P. Phonimdaeng, J. Cooney, A. H. Patel, and A. J. Bramley.** 1990. Genetic studies of virulence factors of *Staphylococcus aureus*. Properties of coagulase and gamma-toxin, alpha-toxin, beta-toxin and protein A in the pathogenesis of *S. aureus* infections, p. 403–420. In R. P. Novick (ed.), *Molecular Biology of the Staphylococci*. VCH Publishers, New York, N.Y.

27. **Fujimoto, D. F., and K. W. Bayles.** 1998. Opposing roles of the *Staphylococcus aureus* virulence regulators, Agr and Sar, in Triton X-100- and penicillin-induced autolysis. *J. Bacteriol.* **180:**3724–3726.

28. **Fuqua, C., S. C. Winans, and E. P. Greenberg.** 1996. Census and consensus in bacterial ecosystems: the LuxR-LuxI family of quorum-sensing transcriptional regulators. *Annu. Rev. Microbiol.* **50:**727–751.

29. **Fuqua, W. C., and S. C. Winans.** 1994. A LuxR-LuxI type regulatory system activates *Agrobacterium* Ti plasmid conjugal transfer in the presence of a plant tumor metabolite. *J. Bacteriol.* **176:**2796–2806.

30. **Gemmel, C. G., and A. M. A. Shibl.** 1976. The control of toxin and enzyme biosynthesis in staphylococci by antibiotics, p. 657–664. In J. Jeljaszewicz (ed.), *Staphylococci and Staphylococcal Diseases*. Gustav Fischer Verlag, Stuttgart, Germany.

31. **Giraudo, A., C. Raspanti, A. Calzolari, and R. Nagel.** 1994. Characterization of a Tn551-mutant of *Staphylococcus aureus* defective in the production of several exoproteins. *Can. J. Microbiol.* **8:**677–681.

32. **Giraudo, A. T., A. L. Cheung, and R. Nagel.** 1997. The sae locus of *Staphylococcus aureus* controls exoprotein synthesis at the transcriptional level. *Arch. Microbiol.* **168:**53–58.

33. **Håvarstein, L. S., G. Coomaraswamy, and D. A. Morrison.** 1995. An unmodified heptadecapeptide pheromone induces competence for genetic transformation in *Streptococcus pneumoniae. Proc. Natl. Acad. Sci. USA* **92:**11140–11144.

34. **Håvarstein, L. S., P. Gaustad, I. F. Nes, and D. A. Morrison.** 1996. Identification of the streptococcal competence-pheromone receptor. *Mol. Microbiol.* **21:**863–869.

35. **Heinrichs, J. H., M. G. Bayer, and A. L. Cheung.** 1996. Characterization of the *sar* locus and its interaction with *agr* in *Staphylococcus aureus*. *J. Bacteriol.* **178:**418–423.

35a. **Herbert, S., and R. P. Novick.** Unpublished data.

36. **Heumann, D., C. Barras, A. Severin, M. P. Glauser, and A. Tomasz.** 1994. Gram-positive cell walls stimulate synthesis of tumor necrosis factor. *Infect. Immun.* **62:**2715–2721.

37. **Huh, Y. J., and A. A. Weiss.** 1991. A 23-kilodalton protein, distinct from BvgA, expressed by virulent *Bordetella pertussis* binds to the promoter region of *vir*-regulated toxin genes. *Infect. Immun.* **59:**2389–2395.

38. **Huhne, K., L. Axelsson, A. Holck, and L. Krockel.** 1996. Analysis of the sakacin P gene cluster from *Lactobacillus sake* Lb674 and its expression in sakacin-negative Lb. sake strains. *Microbiology* **142:**1437–1448.

39. **Iordanescu, S.** 1989. *Staphylococcus aureus* chromosomal mutation plaC1 amplifies plasmid pT181 by depressing synthesis of its negative-effector countertranscripts. *J. Bacteriol.* **171:**4831–4835.

39a. **Iordanescu, S.** Personal communication.

40. **Janzon, L., and S. Arvidson.** 1990. The role of the delta-lysin gene (*hld*) in the regulation of virulence genes by the accessory gene regulator (*agr*) in *Staphylococcus aureus*. *EMBO J.* **9:**1391–1399.

41. **Ji, G., R. Beavis, and R. Novick.** 1995. Cell density control of staphylococcal virulence mediated by an octapeptide pheromone. *Proc. Natl. Acad. Sci. USA* **92:**12055–12059.

42. **Ji, G., R. Beavis, and R. P. Novick.** 1997. Bacterial interference caused by autoinducing peptide variants. *Science* **276:**2027–2030.

42a. **Ji, G., and R. P. Novick.** Unpublished data.

43. **Kleerebezem, M., L. E. Quadri, O. P. Kuipers, and W. M. de Vos.** 1997. Quorum sensing by peptide pheromones and two-component signal-transduction systems in Gram-positive bacteria. *Mol. Microbiol.* **24:**895–904.

44. **Klein, C., C. Kaletta, and K. D. Entian.** 1993. Biosynthesis of the lantibiotic subtilin is regulated by a histidine kinase/response regulator system. *Appl. Environ. Microbiol.* **59:**296–303.

45. **Kornblum, J., B. Kreiswirth, S. J. Projan, H. Ross, and R. P. Novick.** 1990. *agr*: a polycistronic locus regulating exoprotein synthesis in *Staphylococcus aureus*, p. 373–402. In R. P. Novick (ed.), *Molecular Biology of the Staphylococci*. VCH Publishers, New York, N.Y.

45a. **Kornblum, J., and R. P. Novick.** Unpublished data.

46. **Kullik, I., and P. Giachino.** 1997. The alternative sigma factor sigmaB in *Staphylococcus aureus*: regulation of the sigB operon in response to growth phase and heat shock. *Arch. Microbiol.* **167:**151–159.

47. **Lebeau, C., F. Vandenesch, T. Greeland, R. P. Novick, and J. Etienne.** 1994. Coagulase expression in *Staphylococcus aureus* is positively and negatively modulated by an *agr*-dependent mechanism. *J. Bacteriol.* **176:**5534–5536.

48. **Leboeuf-Trudeau, T., J. de'Repentigny, R. M. Frenette, and S. Sonea.** 1969. Tryptophan metabolism and toxin formation in *S. aureus* Wood 46 strain. *Can. J. Microbiol.* **15:**1–7.

49. **Lina, G., S. Jarraud, G. Ji, T. Greenland, A. Pedraza, J. Etienne, R. P. Novick, and F. Vandenesch.** 1998. Transmembrane topology and histidine protein kinase activity of AgrC, the *agr* signal receptor in *Staphylococcus aureus*. *Mol. Microbiol.* **28:**655–662.

49a. **Lina, G., F. Vandenesch, and J. Etienne.** Personal communication.

50. **Lofdahl, S., E. Morfeldt, L. Janzon, and S. Arvidson.** 1988. Cloning of a chromosomal locus (*exp*) which regulates the expression of several exoprotein genes in *Staphylococcus aureus*. *Mol. Gen. Genet.* **211:**435–440.

51. **Lowe, A. M., D. T. Beattie, and R. L. Deresiewicz.** 1998. Identification of novel staphylococcal virulence genes by in vivo expression technology. *Mol. Microbiol.* **27:**967–976.

52. **Mallonee, D. H., B. A. Glatz, and P. Pattee.** 1982. Chromosomal mapping of a gene affecting enterotoxin A production in *Staphylococcus aureus*. *Appl. Environ. Microbiol.* **43:**397–402.

53. **Manna, A. C., M. G. Bayer, and A. L. Cheung.** 1998. Transcriptional analysis of different promoters in the sar locus in *Staphylococcus aureus*. *J. Bacteriol.* **180:**3828–3836.

54. **Mayville, P., G. Ji, R. Beavis, H.-M. Yang, M. Goger, R. P. Novick, and T. W. Muir.** 1999. Structure-activity analysis of synthetic autoinducing thiolactone peptides from *Staphylococcus aureus* responsible for virulence. *Proc. Natl. Acad. Sci. USA* **96:**1218–1223.

55. **Mei, J. M., F. Nourbakhsh, C. W. Ford, and D. W. Holden.** 1997. Identification of *Staphylococcus aureus* virulence genes in a murine model of bacteraemia using signature-tagged mutagenesis. *Mol. Microbiol.* **26:**399–407.

56. **Morfeldt, E., D. Taylor, A. von Gabain, and S. Arvidson.** 1995. Activation of alpha-toxin translation in *Staphylococcus aureus* by the trans-encoded antisense RNA, RNAIII. *EMBO J.* **14:**4569–4577.

57. **Morfeldt, E., K. Tegmark, and S. Arvidson.** 1996. Transcriptional control of the *agr*-dependent virulence gene regulator, RNAIII, in *Staphylococcus aureus*. *Mol. Microbiol.* **21:**1227–1237.

58. **Musser, J. M., P. M. Schlievert, A. W. Chow, P. Ewan, B. N. Kreiswirth, V. T. Rosdahl, A. S. Naidu, W. Witte, and R. K. Selander.** 1990. A single clone of *Staphylococcus aureus* causes the majority of cases of toxic shock syndrome. *Proc. Natl. Acad. Sci. USA* **87:**225–229.

59. **Nes, I. F., L. S. Havarstein, and H. Holo.** 1995. Genetics of non-lantibiotic bacteriocins. *Dev. Biol. Stand.* **85:**645–651.

60. **Novick, R. P., S. Projan, J. Kornblum, H. Ross, B. Kreiswirth, and S. Moghazeh.** 1995. The *agr* P-2 operon: an autocatalytic sensory transduction system in *Staphylococcus aureus*. *Mol. Gen. Genet.* **248:**446–458.

61. **Novick, R. P., H. F. Ross, S. J. Projan, J. Kornblum, B. Kreiswirth, and S. Moghazeh.** 1993. Synthesis of staphylococcal virulence factors is controlled by a regulatory RNA molecule. *EMBO J.* **12:**3967–3975.

62. **Otto, M., R. Sussmuth, G. Jung, and F. Gotz.** 1998. Structure of the pheromone peptide of the *Staphylococcus epidermidis* agr system. *FEBS Lett.* **424:**89–94.

63. **Peng, H.-L., R. P. Novick, B. Kreiswirth, J. Kornblum, and P. Schlievert.** 1988. Cloning, characterization and se-

quencing of an accessory gene regulator (*agr*) in *Staphylococcus aureus*. *J. Bacteriol.* **179:**4365–4372.

64. **Piriz Duran, S., F. H. Kayser, and B. Berger-Bächi.** 1996. Impact of sar and agr on methicillin resistance in *Staphylococcus aureus*. *FEMS Microbiol. Lett.* **141:**255–260.

64a. **Projan, S.** Personal communication.

65. **Projan, S., and R. Novick.** 1997. The molecular basis of virulence, p. 55–81. *In* G. Archer and K. Crossley (ed.), *Staphylococci in Human Disease*. Churchill Livingstone, New York, N.Y.

66. **Projan, S. J., S. Brown-Skrobot, P. Schlievert, S. L. Moghazeh, and R. P. Novick.** 1994. Glycerol monolaurate inhibits the production of β-lactamase, toxic shock syndrome toxin-1 and other staphylococcal exoproteins by interfering with signal transduction. *J. Bacteriol.* **176:**4204–4209.

67. **Quadri, L. E., M. Kleerebezem, O. P. Kuipers, W. M. de Vos, K. L. Roy, J. C. Vederas, and M. E. Stiles.** 1997. Characterization of a locus from *Carnobacterium piscicola* LV17B involved in bacteriocin production and immunity: evidence for global inducer-mediated transcriptional regulation. *J. Bacteriol.* **179:**6163–6171.

68. **Recsei, P., B. Kreiswirth, M. O'Reilly, P. Schlievert, A. Gruss, and R. Novick.** 1986. Regulation of exoprotein gene expression by *agr*. *Mol. Gen. Genet.* **202:**58–61.

68a. **Ross, H., and R. P. Novick.** Unpublished data.

69. **Sawicka-Grzelak, A., A. Szymanowska, A. Mlynarczyk, and G. Mlynarczyk.** 1993. [Production of staphylokinase and hemolysin by coagulase-negative staphylococcus]. *Med. Dosw. Mikrobiol.* **45:**7–10.

70. **Scarlato, V., B. Arico, A. Prugnola, and R. Rappuoli.** 1991. Sequential activation and environmental regulation of virulence genes in *Bordetella pertussis*. *EMBO J.* **10:**3971–3975.

71. **Schlievert, P., M. Osterholm, J. Kelly, and R. Nishimura.** 1982. Toxin and enzyme characterization of *Staphylococcus aureus* isolates from patients with and without toxic-shock syndrome. *Ann. Intern. Med.* **96:**937–940.

72. **Sheehan, B. J., T. J. Foster, C. J. Dorman, S. Park, and G. S. Stewart.** 1992. Osmotic and growth-phase dependent regulation of the eta gene of *Staphylococcus aureus*: a role for DNA supercoiling. *Mol. Gen. Genet.* **232:**49–57.

73. **Shinefield, H. R., J. C. Ribble, and M. Boris.** 1971. Bacterial interference between strains of *Staphylococcus aureus*, 1960–1971. *Am. J. Dis. Child.* **121:**148–152.

74. **Sledjeski, D. D., A. Gupta, and S. Gottesman.** 1996. The small RNA, DsrA, is essential for the low temperature expression of RpoS during exponential growth in *Escherichia coli*. *EMBO J.* **15:**3993–4000.

75. **Solomon, J., B. Lazazzera, and A. Grossman.** 1996. Purification and characterization of an extracellular peptide factor that affects two different developmental pathways in *Bacillus subtilis*. *Genes Dev.* **10:**2014–2024.

76. **Solomon, J., R. Magnuson, A. Sruvastavam, and A. Grossman.** 1995. Convergent sensing pathways mediate response to two extracellular competence factors in *Bacillus subtilis*. *Genes Dev.* **9:**547–558.

76a. **Tegmark, K., and S. Arvidson.** Personal communication.

77. **Tegmark, K., E. Morfeldt, and S. Arvidson.** 1998. Regulation of *agr*-dependent virulence genes in *Staphylococcus aureus* by RNAIII from coagulase-negative staphylococci. *J. Bacteriol.* **180:**3181–3186.

77a. **Vandenesch, F.** Personal communication.

78. **Vandenesch, F., J. Kornblum, and R. P. Novick.** 1991. A temporal signal, independent of *agr*, is required for *hla* but not for *spa* transcription in *Staphylococcus aureus*. *J. Bacteriol.* **173:**6313–6320.

79. **Vandenesch, F., S. Projan, B. Kreiswirth, J. Etienne, and R. P. Novick.** 1993. Agr-related sequences in *Staphylococcus lugdunensis*. *FEMS Microbiol. Lett.* **111:**115–122.

80. **Van Wamel, W. J., G. van Rossum, J. Verhoef, C. M. Vandenbroucke-Grauls, and A. C. Fluit.** 1998. Cloning and characterization of an accessory gene regulator (agr)-like locus from *Staphylococcus epidermidis*. *FEMS Microbiol. Lett.* **163:**1–9.

81. **Van Wamel, W. J. B., A. G. A. Welten, J. Verhoef and A. C. Fluit.** 1997. Diversity in the accessory gene regulator (*agr*) locus of *Staphylococcus epidermidis*, abstr. B-33, p. 34. *Abstr. 97th Gen. Meet. Am. Soc. Microbiol. 1997*. American Society for Microbiology, Washington, D.C.

82. **Vijaranakul, U., M. J. Nadakavukaren, D. O. Bayles, B. J. Wilkinson, and R. K. Jayaswal.** 1997. Characterization of an NaCl-sensitive *Staphylococcus aureus* mutant and rescue of the NaCl-sensitive phenotype by glycine betaine but not by other compatible solutes. *Appl. Environ. Microbiol.* **63:**1889–1897.

83. **Wesson, C. A., L. E. Liou, K. M. Todd, G. A. Bohach, W. R. Trumble, and K. W. Bayles.** 1998. *Staphylococcus aureus agr* and *sar* global regulators influence internalization and induction of apoptosis. *Infect. Immun.* **66:**5238–5243.

84. **Wu, S., H. de Lencastre, and A. Tomasz.** 1996. Sigma-B, a putative operon encoding alternate sigma factor of *Staphylococcus aureus* RNA polymerase: molecular cloning and DNA sequencing. *J. Bacteriol.* **178:**6036–6042.

85. **Yoshikawa, M., F. Matsuda, M. Naka, E. Murofushi, and Y. Tsunematsu.** 1974. Pleiotropic alteration of activities of several toxins and enzymes in mutants of *Staphylococcus aureus*. *J. Bacteriol.* **119:**117–122.

86. **Zhang, A., V. Derbyshire, J. L. Salvo, and M. Belfort.** 1995. *Escherichia coli* protein StpA stimulates self-splicing by promoting RNA assembly in vitro. *RNA* **1:**783–793.

Staphylococcus aureus–Eukaryotic Cell Interactions

FRANKLIN D. LOWY

42

Staphylococcus aureus causes a diversity of diseases that range from minor skin and soft tissue infections to life-threatening systemic infections, including endocarditis and sepsis. Clinically, endocarditis is characterized by local cardiac valvular destruction and metastatic infections, while sepsis is characterized by vasculitis, coagulopathy, and multiorgan dysfunction. Endovascular involvement is a critical component of these different life-threatening diseases. Staphylococci must interact with the endothelium in order to establish endovascular infections or to traverse the endovascular space and create tissue-based abscesses (20, 23). In vitro studies demonstrate that staphylococci colonize and invade endothelial cells and, as a consequence of this process, cause significant cellular alterations. These cellular events are likely to contribute to the pathogenesis of staphylococcal disease. This chapter focuses on *S. aureus*–endothelial cell interactions as a model of staphylococcal invasion of eukaryotic cells. Staphylococcal invasion of endothelial cells has been the most extensively studied and is among the most important events in the pathogenesis of invasive systemic disease. When relevant, reference will also be made to staphylococcal interaction with other cell types.

ADHERENCE

Staphylococcal adherence to endothelial cells is the critical first step in the invasion process. The apparent tropism of staphylococci for endothelium likely accounts for the selective internalization of this pathogen. Staphylococci adhere to endovascular tissue and endothelial cells grown in tissue culture more avidly than do other bacterial species. This was first demonstrated by Gould et al. (16) using human and canine cardiac valvular tissue. Subsequent studies confirmed these observations using human and porcine endothelial cells grown in tissue culture (21, 25, 35). In general, the bacterial species most commonly associated with acute bacterial endocarditis were also the most adherent (Fig. 1) (21, 25). *S. aureus* adherence to endothelial cells is saturable in a dose- and time-dependent manner, suggesting a specific adhesin-receptor interaction (32). Ad-

herence appears to be mediated by bacterial-endothelial cell, protein-protein interactions. Several candidate surface proteins that bind endothelial cells have been identified, although a functional role, i.e., the ability to inhibit staphylococcal adherence to endothelial cells, has not been demonstrated (8, 14, 31). Protein A, one of the major surface proteins of *S. aureus*, did not inhibit adherence (31). The potential role of the structurally related family of staphylococcal surface proteins designated MSCRAMMs (microbial surface components recognizing adhesive matrix molecules) is also unclear. These proteins facilitate *S. aureus* colonization of different tissues by binding to extracellular matrix components such as collagen or fibrinogen (26). Their ability to mediate staphylococcal adherence to eukaryotic cells is unknown.

Little is known about endothelial cell receptors for staphylococci. A 50-kDa endothelial cell membrane protein has been partially characterized. This protein competitively inhibited staphylococcal adherence to endothelial cells in vitro (32). Campbell and Johnson (8) have also identified several candidate proteins using ^{125}I-labeled endothelial cell surface proteins.

Serum components may function as bridging ligands for this interaction, by serving as receptors for both staphylococci and endothelial cells. Cheung et al. (11) showed that adherence of *S. aureus* to human endothelial cells was increased in the presence of fibrinogen and other blood constituents. Fibronectin has also been hypothesized to play a similar role (17).

Variation in growth conditions alters adherence of staphylococci to endothelial cells. Cheung et al. (10) demonstrated that endothelial cells grown in the presence of tumor necrosis factor alpha are more susceptible to staphylococcal infection. Blumberg et al. (7) showed that the absence of acidic fibroblast growth factor in the medium increased adherence while the addition of endotoxin and interleukin-1 (IL-1) was without effect.

Microenvironmental changes, such as alteration of the matrix underlying endothelial cells, also alters staphylococcal adherence. Alston et al. (2) demonstrated that the extracellular matrix elaborated by *S. aureus*-infected

Gram-Positive Pathogens, ed. by V. A. Fischetti et al.
© 2000 American Society for Microbiology, Washington, D.C.

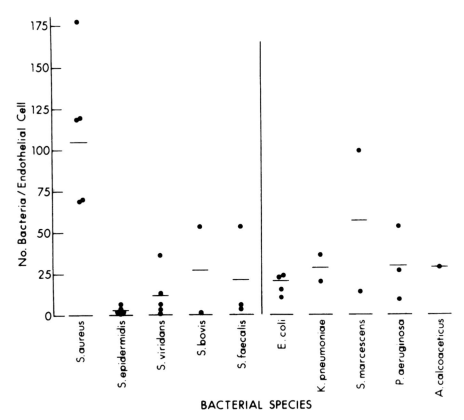

FIGURE 1 Adherence of 32 clinical blood culture isolates to human umbilical vein endothelial cells after a 2-h incubation at 37°C. The average initial bacterial inoculum was 2.2×10^8 CFU/ ml. Reprinted from *Infection and Immunity* (25).

endothelial cells signaled phenotypic changes in newly seeded endothelial cells. These cells became more susceptible to subsequent infection. This altered susceptibility appeared to be due to a reduction in the sulfation of the matrix heparan sulfate proteoglycans that resulted in a similar reduction in sulfation of the cellular heparan sulfate proteoglycans. Reducing cellular heparan sulfate proteoglycans by growth of the endothelial cells in the presence of chlorate produced similar results (2).

These studies suggest that subtle alterations in cellular growth conditions produce significant changes in the susceptibility of endothelial cells to infection. These changes may not produce detectable pathologic changes in vivo but may nevertheless significantly increase the possibility of infection at a particular site.

Bacterial growth conditions also influence staphylococcal adherence to endothelial cells. Staphylococci harvested during exponential growth phase are more adherent than those harvested during stationary phase, suggesting that the proteins mediating bacterial adherence are under the control of global regulatory genes such as *agr* or *sar* (9, 27, 31) (see chapter 41, this volume).

INVASION

Ogawa et al. (25) first demonstrated that following adherence, staphylococci are endocytosed into membrane-bound vacuoles in an endothelial cell-mediated process (see Fig. 2). Subsequent studies demonstrated that endocytosis is blocked by cytochalasin B, an inhibitor of phagocytosis.

Viable bacteria are not required for this process. UV-killed, but not heat-killed, bacteria are internalized, suggesting that critical heat-sensitive proteins are required (18, 24, 33).

Following internalization, there is fusion of the phagosome with cellular lysosomes and limited, if any, intracellular bacterial replication. Endothelial cells are ineffective phagocytes, and staphylococci appear capable of prolonged intracellular survival (see chapter 35). Bacteria may be released into the medium, spread to the underlying extracellular matrix, or cause progressive cellular damage, ultimately destroying the cell (18, 24, 33). Vann and Proctor (34) used isogenic alpha-toxin-producing and -nonproducing strains to demonstrate that cellular damage was alpha-toxin mediated.

Internalization of staphylococci by mammary cells and osteoblasts has also been reported. The internalization process in primary or immortalized mammary cell lines, as well as in osteoblasts, is similar to that in endothelial cells. The internalization process appears dependent on initial bacterial contact followed by internalization in a eukaryotic cell-initiated process (1, 12, 19).

INTRACELLULAR SURVIVAL OF *S. AUREUS*

The intracellular environment provides a potential sanctuary, protecting staphylococci from host defense mechanisms as well as from the bactericidal effects of antibiotics. Vesga et al. (36) demonstrated that the intraendothelial cell milieu selects for slow-growing, respiration-defective

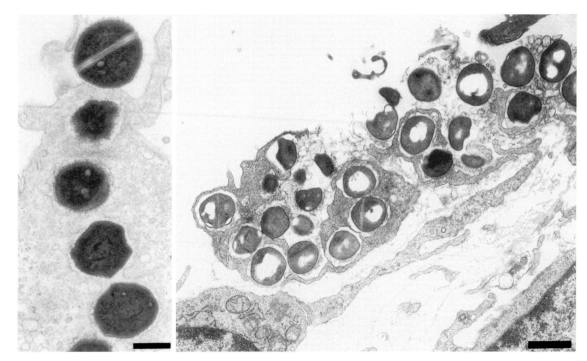

FIGURE 2   Demonstration of endothelial cell phagocytosis of *S. aureus* in vitro. (Left) Staphylococci incubated with human umbilical vein endothelial cells in tissue culture (30 min). The bacteria are phagocytized, enclosed within a membrane-bound vacuole, and transported into the cell. Bar, 0.5 μm. (Right) Section of rabbit aorta incubated with staphylococci (bacteria incubated with tissue for 30 min, then replaced with medium for a 5.5-h incubation). The endothelial cell contains a large number of bacteria enclosed within vacuoles. The cell has ruptured, releasing bacteria into the medium. Bar, 1.0 μm. Reprinted from the *Journal of Ultrastructural and Molecular Structural Research* (24) with the permission of Academic Press.

mutants that are phenotypically small-colony variants (see chapter 35). In contrast to wild-type strains, these mutants, often menadione-deficient, are capable of prolonged intracellular survival. The intracellular environment may also protect these mutants from the lethal effects of antimicrobial agents. One feature of these variants that appears to facilitate their intracellular survival is their inability to produce alpha-toxin (3, 36). These variants are phenotypically reversed by hemin or menadione and recover their virulence (3). This phenomenon provides a potential explanation for the well-known ability of staphylococci to cause persistent or recurrent infections. Bacteria may survive intracellularly and cause recurrent disease on withdrawal of antibiotics (28).

CONSEQUENCES OF STAPHYLOCOCCAL INVASION OF ENDOTHELIAL CELLS IN VITRO

A variety of cellular changes occur as a result of staphylococcal invasion (Fig. 3). Fc receptors, adhesion molecules, and tissue factor are expressed on the cell surface, and cytokines are released into the medium following bacterial internalization (5, 6, 13, 37, 38). Staphylococcal adherence to the cell surface is not sufficient to initiate these changes.

Fc receptors are expressed within 5 min of exposure to staphylococci and do not require de novo protein synthesis, suggesting that they exist preassembled in the cytosol or are unmasked after cell injury (6). The nature of cellular injury necessary for Fc receptor expression is nonspecific and has been reported following cytomegaloviral infection, trypsin treatment, or ingestion of latex beads (30).

In contrast to Fc receptor expression, release of cytokines by *S. aureus*-infected endothelial cells requires several hours, suggesting the need for protein synthesis. Staphylococcal internalization is required for cytokine release, but it is not sufficient. Ingestion of UV-killed bacteria induces cytokine expression, while ingestion of latex beads does not. Different strains of staphylococci have different abilities to induce expression (37). This observation may be one reason for differences in the severity of clinical presentations induced by different staphylococcal strains (22).

Detection of cytokine message and protein following *S. aureus* infection varies for the different cytokines and is also different from the response induced by endotoxin. IL-6 message was detected 3 h following infection. IL-6 protein was first noted at 12 h. IL-1β mRNA was also detected at 12 h, while IL-1 protein was initially seen at 24 h (37). IL-8 message was detected within 1 h. Secreted IL-8 protein was first noted at 6 h. Expression of all the cytokines persisted for 72 h. The IL-8-containing supernatant from infected endothelial cells had functional activity, attracting polymorphonuclear leukocytes across an endothelial cell monolayer in an in vitro transmigration assay (38). The cytokine colony-stimulating factor-1 was not released following infection, suggesting that a limited set of cytokines was expressed (37).

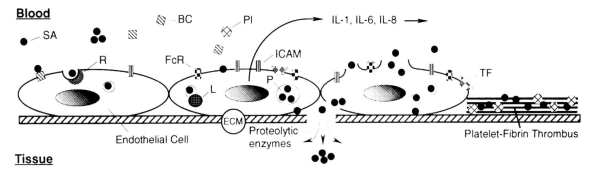

FIGURE 3 Model for *S. aureus* invasion of endovascular tissue. The sequence of events progresses from left to right. Circulating staphylococci (SA) adhere to endothelial cells directly via adhesin-receptor (R) interactions or by bridging ligands that include blood constituents (BC) such as fibrinogen (11, 32). Modifications of the endothelium resulting from microenvironmental changes (such as alterations of the extracellular matrix [ECM]) can signal changes in cellular susceptibility to infection (2). Staphylococci may also adhere to sites of endovascular damage where platelet (PI)-fibrin thrombi have attached following endothelial cell injury (20). Staphylococci are phagocytized in an endothelial cell-mediated process. The bacteria are internalized into phagosomes (P) and may undergo phagolysosomal (L) fusion (24). Staphylococci elaborate proteolytic enzymes that facilitate their spread to adjoining tissues and release into the bloodstream. Tissue factor (TF) is expressed by infected endothelial cells, facilitating fibrin deposition and progression of vegetation formation (13). Once in the adjoining subepithelial tissues, the bacteria elicit an inflammatory response that results in abscess formation. This sequence of events contributes to the establishment of metastatic foci of infection as well as the pathogenesis of endocarditis when cardiac endothelium is involved (23). Phagocytosis also results in expression of Fc receptors (FcR) and adhesion molecules (e.g., ICAM), as well as the release of IL-1, -6, and -8 by endothelial cells. As a result, leukocytes adhere to endothelial cells and diapedese to sites of infection (5, 6).

S. aureus-infected endothelial cells become more adhesive for monocytes and granulocytes in vitro. This effect is staphylococcal concentration dependent. Granulocyte adhesion was mediated by CD11/CD18 integrins, while monocytes were bound via CD11a/CD18 and CD49d/CD29 (VLA-4) integrins. Endothelial cells expressed intercellular adhesion molecule-1 (ICAM-1, CD54), vascular cell adhesion molecule-1 (CD106), and major histocompatibility complex class 1 protein following infection. In contrast, expression of P-selectin (CD62P), E-selectin (CD62E), PECAM-1 (CD31), and ICAM-2 (CD102) was unchanged (5). This process is important in attracting leukocytes to sites of inflammation.

APOPTOSIS IN MAMMARY EPITHELIAL CELLS

As noted above, staphylococcal internalization by bovine mammary cells has also been reported. Bayles et al. (4) reported that following entry, some bacteria escape from the membrane-bound vacuoles and induce cellular apoptosis. Similar effects were recently reported by Menzies and Kourteva (24a) in endothelial cells infected by *S. aureus*. Bayles et al. (4) hypothesized that staphylococcal beta-toxins facilitate the release of bacteria from vacuoles, increasing levels of ceramide, an inducer of apoptosis. The potential clinical relevance of this observation is as yet uncertain. Staphylococci may evade the bactericidal effects of both professional and nonprofessional phagocytes by escaping into the cytoplasm. In addition, they remain protected from the host immune response, fostering intracellular survival (4).

CLINICOPATHOLOGIC CORRELATIONS

S. aureus is classically considered an extracellular pathogen. Its ability to invade eukaryotic cells remains primarily an in vitro observation that has not been well documented in laboratory animal models of infection or in pathological specimens. Its relevance in the pathogenesis of staphylococcal disease is therefore not established. Other extracellular pathogens, including *Neisseria gonorrhoeae* and *Haemophilus influenzae*, have also been shown in in vitro studies to invade eukaryotic cells in a similar fashion, suggesting that a common mechanism may exist (15, 29).

Despite these reservations, the studies performed to date suggest a sequence of events that explain the ability of *S. aureus* to invade endovascular tissue and cause persistent infections (Fig. 3). Staphylococci bind endothelium via multiple, specific adhesin-receptor, protein-protein interactions, enabling the bacteria to effectively colonize different tissues under a variety of clinical settings. There are inducible as well as constitutive cellular receptors that facilitate *S. aureus* adherence to endothelial cells. Expression of these receptors may be modified by microenvironmental changes that result from injury or ischemia to local tissues. The ability to invade and survive within endothelial and other epithelial cells provides a potential site for prolonged bacterial survival that can then serve as a nidus for recurrent infections. In addition, the changes resulting from internalization of staphylococci precipitate a series of cellular changes that contribute to the pathogenesis of life-threatening diseases. These events include the ability of *S. aureus* to cause (i) endocarditis by adhering to and invading damaged or undamaged valvular tissue and causing progressive local cardiovascular damage; (ii) metastatic infections by traversing endothelial or other epithelial cell barriers, elaborating proteolytic enzymes and spreading to adjoining tissues; or (iii) a sepsis syndrome similar to that produced by endotoxin resulting from endothelial cell injury with the expression of endothelial cell receptors and the release of cellular cytokines (18, 20, 23).

AREAS FOR FUTURE STUDY

The studies performed to date on the process of invasion have been primarily descriptive and have not yet provided an understanding of the responsible mechanisms. Little is known about the proteins that mediate adherence and invasion of endothelial cells or the process that initiates endothelial cell-mediated phagocytosis. The signaling pathways responsible for initiation of the cellular changes and the release of cytokines are also unexplored. The clinical relevance of endothelial cell invasion and its consequences has not been validated in studies with laboratory animal models of infection. Similar studies with other eukaryotic cells are even more limited. It is therefore apparent that much remains to be examined in this potentially important area of staphylococcal pathogenesis.

I gratefully acknowledge the thoughtful review of this chapter by Victor B. Hatcher. Work was supported in part by a grant-in-aid from the American Heart Association and by grants (DA09656 and DA11868) from the National Institute on Drug Abuse.

REFERENCES

1. **Almeida, R. A, K. R. Matthews, E. Cifrian, A. J. Guidry, and S. P. Oliver.** 1996. *Staphylococcus aureus* invasion of bovine mammary epithelial cells. *J. Dairy Sci.* **79:**1021–1026.

2. **Alston, W. K., D. A. Elliott, M. E. Epstein, V. B. Hatcher, M. Tang, and F. D. Lowy.** 1997. Extracellular matrix heparan sulfate modulates endothelial cell susceptibility to *Staphylococcus aureus.* *J. Cell. Physiol.* **173:**102–109.

3. **Balwit, J. M., P. van Langevelde, J. M. Vann, and R. A. Proctor.** 1994. Gentamicin-resistant menadione and hemin auxotrophic *Staphylococcus aureus* persist within cultured endothelial cells. *J. Infect. Dis.* **179:**1033–1037.

4. **Bayles, K. W., C. A. Wesson, L. E. Liou, L. K. Fox, G. A. Bohach, and W. R. Trumble.** 1998. Intracellular *Staphylococcus aureus* escapes the endosome and induces apoptosis in epithelial cells. *Infect. Immun.* **66:**336–342.

5. **Beekhuizen, H., J. S. van de Gevel, B. Olsson, I. J. van Benten, and R. van Furth.** 1997. Infection of human vascular endothelial cells with *Staphylococcus aureus* induces hyperadhesiveness for human monocytes and granulocytes. *J. Immunol.* **158:**774–782.

6. **Bengualid, V., V. B. Hatcher, B. Diamond, E. A. Blumberg, and F. D. Lowy.** 1990. *Staphylococcus aureus* infection of human endothelial cells potentiates Fc receptor expression. *J. Immunol.* **145:**4279–4283.

7. **Blumberg E. A., V. B. Hatcher, and Lowy F. D.** 1988. Acidic fibroblast growth factor modulates *Staphylococcus aureus* adherence to human endothelial cells. *Infect. Immun.* **56:**1470–1474.

8. **Campbell, K. M., and C. M. Johnson.** 1990. Identification of *Staphylococcus aureus* binding proteins on isolated porcine cardiac valve cells. *J. Lab. Clin. Med.* **115:**217–223.

9. **Cheung, A. L., J. M. Koomey, C. A. Butler, S. J. Projan, and V. A. Fischetti.** 1992. Regulation of exoprotein expression in *Staphylococcus aureus* by a locus (*sar*) distinct from *agr.* *Proc. Natl. Acad. Sci. USA* **89:**6462–6466.

10. **Cheung A. L., J. M. Koomey, S. Lee, E. A. Jaffe, and V. A. Fischetti.** 1991. Recombinant human tumor necrosis factor alpha promotes adherence of *Staphylococcus aureus* to cultured human endothelial cells. *Infect. Immun.* **59:**3827–3831.

11. **Cheung A. L., M. Krishnan, E. A. Jaffe, and V. A. Fischetti.** 1991. Fibrinogen acts as a bridging molecule in the adherence of *Staphylococcus aureus* to cultured human endothelial cells. *J. Clin. Invest.* **87:**2236–2245.

12. **Craven, N., and J. C. Anderson.** 1979. The location of *Staphylococcus aureus* in experimental chronic mastitis in the mouse and the effect on the action of sodium cloxacillin. *Br. J. Exp. Pathol.* **60:**453–459.

13. **Drake, T. A., and M. Pang.** 1988. *Staphylococcus aureus* induces tissue factor expression in cultured human cardiac valve endothelium. *J. Infect. Dis.* **157:**749–756.

14. **Elliott, D., A. P. Andersen, L. Hanau, V. B. Hatcher, and F. D. Lowy.** 1997. Identification of a *S. aureus* adhesin that mediates adherence to human endothelial cells, abstr. 87. 35th Infectious Diseases Society of America Meeting, San Francisco, Calif.

15. **Geme, J. W., III, and S. Falkow.** 1990. *Haemophilus influenzae* adheres to and enters cultured human epithelial cells. *Infect. Immun.* **58:**4036–4044.

16. **Gould, K., C. H. Ramirez-Ronda, R. K. Holmes, and J. P. Sanford.** 1975. Adherence of bacteria to heart valves *in vitro. J. Clin. Invest.* **56:**1364–1370.

17. **Hamill, R. J.** 1987. Role of fibronectin in infective endocarditis. *J. Infect. Dis.* **9:**S360–S371.

18. **Hamill, R. J., J. M. Vann, and R. A. Proctor.** 1986. Phagocytosis of *Staphylococcus aureus* by cultured bovine aortic endothelial cells: model for postadherence events in endovascular infections. *Infect. Immun.* **54:**833–836.

19. **Hudson, M. C., W. K. Ramp, N. C. Nicholson, A. S. Williams, and M. T. Nousiainen.** 1995. Internalization of *Staphylococcus aureus* by cultured osteoblasts. *Microb. Pathog.* **19:**409–419.

20. **Ing, M. B., L. M. Baddour, and A. S. Bayer.** 1997. Bacteremia and infective endocarditis: pathogenesis, diagnosis, and complications, p. 331–354. *In* K. B. Crossley and G. L. Archer (ed.), *The Staphylococci in Human Disease.* Churchill Livingstone, New York, N.Y.

21. **Johnson, C. M., G. A. Hancock, and G. D. Goulin.** 1988. Specific binding of *Staphylococcus aureus* to cultured porcine cardiac valvular endothelial cells. *J. Lab. Clin. Med.* **112:**16–22.

22. **Kessler, C. M., E. Nussbaum, and C. U. Tuazon.** 1991. Disseminated intravascular coagulation associated with *Staphylococcus aureus* septicemia is mediated by peptidoglycan-induced platelet aggregation. *J. Infect. Dis.* **164:**101–107.

23. **Lowy, F. D.** 1998. *Staphylococcus aureus* infections. *N. Engl. J. Med.* **339:**520–532.

24. **Lowy, F. D., J. Fant, L. Higgins, S. Ogawa, and V. Hatcher.** 1988. *Staphylococcus aureus*–human endothelial cell interactions. *J. Ultrastruct. Mol. Struct. Res.* **98:**137–146.

24a. **Menzies, B. E., and I. Kourteva.** 1998. Internalization of *Staphylococcus aureus* by endothelial cells induces apoptosis. *Infect. Immun.* **66:**5994–5998.

25. **Ogawa, S. K., E. R. Yurberg, V. B. Hatcher, M. A. Levitt, and F. D. Lowy.** 1985. Bacterial adherence in endocarditis: adherence to human endothelial cells. *Infect. Immun.* **50:**218–224.

26. **Patti, J. M., B. L. Allen, M. J. McGavin, and M. Höök.** 1994. MSCRAMM-mediated adherence of microorganisms to host tissues. *Annu. Rev. Microbiol.* **48:**585–617.

27. **Peng, H. L., R. P. Novick, B. Kreiswirth, J. Kornblum, and P. Schlievert.** 1988. Cloning, characterization, and sequencing of an accessory gene regulator (*agr*) in *Staphylococcus aureus. J. Bacteriol.* **170:**4365–4372.

28. **Proctor, R. A., P. Van Langevelde, M. Kristjannsson, J. N. Maslow, and R. B. Arbeit.** 1995. Persistent and relapsing infections associated with small-colony variants of *Staphylococcus aureus. Clin. Infect. Dis.* **20:**95–102.

29. **Richardson, W. P., and J. C. Sadoff.** 1988. Induced engulfment of *Neisseria gonorrhoeae* by tissue culture cells. *Infect. Immun.* **56:**2512–2514.

30. **Ryan, U. S.** 1986. Phagocytic properties of endothelial cells, vol. 3, p. 33–49. *In* U. S. Ryan (ed.), *Endothelial Cells.* CRC Press, Inc., Boca Raton, Fla.

31. **Tompkins, D. C., L. J. Blackwell, V. B. Hatcher, D. A. Elliott, C. O'Hagan-Sotsky, and F. D. Lowy.** 1992. *Staphylococcus aureus* proteins that bind to human endothelial cells. *Infect. Immun.* **60:**965–969.

32. **Tompkins, D. C., V. B. Hatcher, D. Patel, G. A. Orr, L. L. Higgins, and F. D. Lowy.** 1990. A human endothelial cell membrane protein that binds *Staphylococcus aureus* in vitro. *J. Clin. Invest.* **85:**1248–1254.

33. **Vann, J. M., and R. A. Proctor.** 1987. Ingestion of *Staphylococcus aureus* by bovine endothelial cells results in time- and inoculum-dependent damage to endothelial cell monolayers. *Infect. Immun.* **55:**2155–2163.

34. **Vann, J. M., and R. A. Proctor.** 1988. Cytotoxic effects of ingested *Staphylococcus aureus* on bovine endothelial cells: role of *S. aureus* α-hemolysin. *Microb. Pathog.* **4:**443–453.

35. **Vercellotti, G. M., D. Lussenhop, P. K. Peterson, L. T. Furcht, J. B. McCarthy, H. S. Jacob, and C. F. Moldow.** 1984. Bacterial adherence to fibronectin and endothelial cells: a possible mechanism for bacterial tissue tropism. *J. Lab. Clin. Med.* **103:**34–43.

36. **Vesga, O., M. C. Groeschel, M. F. Otten, D. W. Brar, J. V. Vann, and R. A. Proctor.** 1996. *Staphylococcus aureus* small colony variants are induced by the endothelial cell intracellular milieu. *J. Infect. Dis.* **173:**739–742.

37. **Yao, L., V. Bengualid, F. D. Lowy, J. Gibbons, V. B. Hatcher, and J. W. Berman.** 1995. Internalization of *Staphylococcus aureus* by endothelial cells induces cytokine gene expression. *Infect. Immun.* **63:**1835–1839.

38. **Yao, L., F. D. Lowy, and J. W. Berman.** 1996. Interleukin-8 gene expression in *Staphylococcus aureus*-infected endothelial cells. *Infect. Immun.* **64:**3407–3409.

The Epidemiology of *Staphylococcus* Infections

FRED C. TENOVER AND ROBERT P. GAYNES

43

The staphylococci are a remarkably diverse group of organisms that cause a variety of diseases ranging from innocuous urinary tract infections to often fatal forms of endocarditis. Staphylococci have been recognized as serious pathogens for over a century (29), yet despite large-scale efforts to halt their spread, particularly in hospitals, they remain the most common cause of community- and nosocomially acquired bacteremia. Of the 2 million patients who acquire a nosocomial infection annually in the United States, approximately 260,000 will have an infection associated with *Staphylococcus aureus* (12, 34). The disease spectrum of the staphylococci includes abscesses, bacteremia, central nervous system infections, endocarditis, osteomyelitis, pneumonia, urinary tract infections, and a host of syndromes caused by exotoxins, including bullous impetigo, food poisoning, scalded skin syndrome, and toxic shock syndrome. Remarkably, in addition to being the leading cause of bacteremia in the United States, staphylococci also are among the four most common causes of food-borne illness, surpassing even campylobacteriosis and listeriosis (1). Herein, some of the common illnesses caused by staphylococci, particularly *S. aureus*, are described.

COMMUNITY-ACQUIRED INFECTIONS
Colonization and Infection of Skin and Soft Tissues

Staphylococci colonize a sizable portion of the human population. Colonization affords organisms, such as *S. aureus*, the opportunity to gain access to skin sites, which when infected can serve as a source for more serious diseases, such as bacteremia, endocarditis, or toxemias (e.g., toxic shock syndrome). More commonly, however, breaches of skin sites result in either abscesses or furuncles (commonly known as boils) that can progress to carbuncles (more serious, deep-seated infections of several hair follicles). Approximately 20% of the population is stably colonized with *S. aureus*, and as many as 30 to 50% of the population may show transient colonization of the nares, axilla, perineum, or vagina (13, 24, 26, 31). Diabetics, intravenous (i.v.) drug

users, patients on dialysis, and patients with acquired immunodeficiency syndrome have higher rates of *S. aureus* colonization (14, 36, 47, 54, 55). Hospitalized patients and health care workers are also at higher risk of becoming colonized for extended periods of time (13).

Colonization, particularly of the nares by *S. aureus*, leads to hand carriage, and from the hands, the organisms are frequently spread to other areas of the body. Thus, staphylococci often follow a nose to hands to wound route of infection (13, 55). Decreasing colonization, such as through the use of nasal ointments containing mupirocin, can aid in reducing both nosocomial and familial spread of staphylococci that can cause diseases such as recurrent folliculitis and furuncles (21, 24, 25, 54). Normally, large numbers of staphylococci are required to cause infection, but disruption of skin barriers, as seen in i.v. drug users or in diabetics, increases the risk of staphylococcal infections. Dialysis patients are at particularly high risk for infection (54, 55).

Skin Infections

Approximately one-half of all skin infections are caused by *S. aureus* (11, 44). Infections include folliculitis, cellulitis, furuncles, carbuncles, hydradenitis suppurtiva, mastitis, pyodermas, and pyomyositis. Impetigo, which involves release of epidermolytic toxins, can range from mild, recurrent infections to the more serious bullous impetigo, characterized by blisters that continually break and become infected, to the potentially life-threatening scalded skin syndrome (29, 44). Chronic *S. aureus* skin and soft tissue infections may be suggestive of a host factor disorder such as chronic granulomatous disease, an X-linked genetic disorder with frequent pyogenic infections. While cellulitis is usually treated with antimicrobial chemotherapy, the major mode of treatment for folliculitis and local abscesses is surgical drainage rather than antimicrobial agents.

Bacteremia and Endocarditis

Virtually any *S. aureus* infection can lead to bacteremia. *S. aureus* causes about 11 to 38% of community-acquired bacteremia (7, 24, 52). Mortality from *S. aureus* bacteremia

Gram-Positive Pathogens, ed. by V. A. Fischetti et al.
© 2000 American Society for Microbiology, Washington, D.C.

ranges from 11 to 48%, a figure that has increased steadily for a number of years (32). In a recent review of data on adult bacteremia from three hospitals by Weinstein et al. (52), S. aureus was the most common cause of clinically significant bacteremia (18.9%) and coagulase-negative staphylococci (CoNS) (9.2%) were the third most common cause (when organisms judged to be probable contaminants were excluded from analysis) (51). These results are similar to data from the Mayo Clinic (7), where S. aureus was the most common cause of bacteremia (18.4%) and CoNS were the third most common cause (10.4%), with both organism groups showing a significant increase in frequency over the preceding 5-year period. In Weinstein's study, 50.6% of S. aureus infections were community acquired. The majority of S. aureus and CoNS infections were related to indwelling lines or intravascular devices. However, the proportion of cases of S. aureus bacteremia was significantly lower in neutropenic patients, owing in part to the increase in gram-negative bacteremia arising from gastrointestinal sources. Among all blood-borne pathogens, CoNS showed the lowest associated mortality (5.5%), while associated mortality for S. aureus was 11.9%.

Approximately 10 to 40% of community-acquired cases of S. aureus bacteremia progress to endocarditis (8, 23, 32). This figure is higher in i.v. drug users, often because they are heavily colonized with S. aureus and have frequent breaches of skin barriers, and is lower in patients with nosocomial bacteremia. Most patients have no apparent focus for the bacteremia preceding endocarditis. S. aureus is unusual in that it can cause infectious endocarditis on a normal, native heart valve. S. aureus infectious endocarditis can present as right-sided endocarditis, primarily in i.v. drug users; left-sided native valve endocarditis; or prosthetic valve endocarditis (PVE) (18, 22). Left-sided infectious endocarditis is usually more severe, occurring typically in older patients, and is associated with more complications, including heart failure (51%) and extracranial embolization, when compared with right-sided infectious endocarditis. Embolization, which more commonly occurs in left-sided endocarditis, can involve many organ systems, including kidney, bone, spleen, and major blood vessels (23, 32), and may lead to deep tissue infections, including life-threatening brain abscesses.

Only 2 to 13% of cases of CoNS bacteremia progress to endocarditis, although CoNS are the most common cause of PVE (18, 22). PVE is divided into early PVE, i.e., those cases occurring within 60 days of valve replacement, and late PVE, which includes those infections occurring later than 60 days after valve replacement (22). Thirty to 67% of early PVE cases are caused by CoNS, compared with only 20 to 28% of late-onset PVE cases. Late PVE caused by CoNS is usually community acquired, while early onset is nosocomial. In many cases of late PVE there is no obvious source. Most late-onset PVE cases are caused by methicillin-resistant strains of *Staphylococcus epidermidis* (23), which are common in both community and hospitalized patients.

Osteomyelitis

Staphylococcal osteomyelitis is classified as either acute or chronic (24, 50). Acute hematogenous osteomyelitis is usually a disease of children, primarily neonates, in whom it affects the long bones of the lower extremity (39, 50). The disease can be cured if detected early and treated with intensive antimicrobial therapy. Frequent therapeutic failures are seen, particularly if therapy is discontinued in less than

4 weeks (39). In adults, S. aureus bacteremia rarely leads to osteomyelitis of long bones; instead, vertebral bodies are more commonly affected. Chronic osteomyelitis, which can complicate open fractures or penetrating wounds, is usually an infection of at least 6 months' duration, is more commonly seen in adults, and is often refractory to cure with antimicrobial therapy. Chronic osteomyelitis can also develop after an initial, but failed, attempt at antimicrobial therapy. Although fluoroquinolones have shown excellent activity in vitro against gram-negative pathogens, none of the current quinolones provide optimal staphylococcal coverage even when combined with other agents such as rifampin (39, 50).

Toxin-Mediated Diseases

Several staphylococcal diseases are mediated by toxins, including impetigo (see above), food poisoning, and toxic shock syndrome. Staphylococcal food poisoning is a result of ingesting one of several staphylococcal enterotoxins, the most ubiquitous of which is enterotoxin A (20, 53). Since this is a true toxemia, viable staphylococci are not always present in the contaminated vehicle. Staphylococcal food poisoning is characterized by nausea, vomiting, headache (which can be severe), and, less commonly, diarrhea (1, 20, 53). Time to onset of symptoms after consuming contaminated food averages 4.4 h (20). Since it is usually a self-limiting disease, rarely if ever leading to systemic infection, and reporting by state health departments is passive, the true incidence of staphylococcal food poisoning is unknown (1).

The disease mediated by toxic shock syndrome toxin-1 (TSST-1) was first described in 1978 (46). Toxic shock syndrome has been associated with the use of highly absorbent tampons (42) but has been described in association with many types of S. aureus infections, some of which can be quite minor (9, 38). Toxic shock syndrome consists of fever above 102°F; hypotension; rash, usually followed by desquamation of the skin, especially on the palms and soles; hyperemia of mucous membranes; and involvement of multiple organ systems, as evidenced by diarrhea, thrombocytopenia, cardiopulmonary dysfunction, or a variety of other symptoms (9, 42). TSST-1 is a potent superantigen eliciting a variety of cytokines that, along with tumor necrosis factor, contribute to the severity of illness (41, 48). A single clone of S. aureus causes the majority of TSST illness (33). Toxin-mediated diseases caused by other staphylococcal toxins, including scalded skin syndrome, are known to be infrequent in the United States, but there are few data available on the prevalence of such diseases in other parts of the world.

NOSOCOMIAL INFECTIONS

Staphylococci are among the most common causes of hospital-acquired infections, including bacteremia, surgical site infections, pneumonia, and urinary tract infections (12, 34, 35). Outbreaks of nosocomial infections with S. aureus occur in patients across the spectrum of ages, from adults in tertiary care centers to infants in hospital nurseries. Outbreaks in neonates usually result in skin infections and bacteremia, although more serious diseases, such as osteomyelitis and meningitis, can occur (17). Transmission of S. aureus strains from health care workers to patients have resulted in mediastinitis (16), nosocomial toxic shock syndrome (27), and the spread of antimicrobial-resistant strains within an institution (10). One particularly virulent strain of S. aureus was shown to have spread from the index patient, who introduced the strain into Canada from an Indian hospital, to multiple patients in medical centers across Canada (40). Table 1 shows data on the frequency of pathogens associated with nosocomial infections from the National Nosocomial Infections Surveillance (NNIS) System, Centers for Disease Control and Prevention (Atlanta, Ga.). S. aureus and Escherichia coli were the most commonly isolated nosocomial pathogens for all infections reported to NNIS by U.S. hospitals using a hospital-wide surveillance approach from 1990 through 1996 (Table 1). S. aureus was rarely isolated from nosocomially acquired urinary tract infections but is common at other sites and may account for as many as 260,000 nosocomial infections in the United States each year. The overall frequency of S. aureus as a pathogen associated with nosocomial infections has not changed in the past decade. CoNS were isolated almost twice as often as S. aureus from nosocomial bloodstream infections. The increasing frequency of CoNS bacteremia is due to the increasing use of intravascular devices and an increased recognition of CoNS as significant pathogens (30). However, S. aureus remains the most common cause of surgical site infections.

Bloodstream Infections

Primary bloodstream infections are defined as infections in which no other primary site can be discerned (3). Diagnosis is usually based on positive blood cultures drawn from one or more venipuncture sites or from intravascular catheters. S. aureus was the second most frequent pathogen associated with bloodstream infections in the NNIS system (34) and a frequent cause of nosocomial bacteremia in other countries as well (21, 43, 49). As noted in the study of Weinstein et al. (52), these infections are often related to the presence of an intravascular device. Bloodstream infections that are associated with these devices are serious and more often occur in patients who have had (i) central i.v. lines that were urgently or emergently placed, (ii) lines that have been in place for at least 72 h, and (iii) lines that are in femoral veins rather than subclavian veins. Removal of the purportedly infected catheter, followed by bacteriologic culture using semiquantitative methods, can be helpful for diagnosis and a guide for antimicrobial chemotherapy.

Many medical and surgical procedures in which a device is left in place can lead to a device-associated infection with or without bloodstream infection. Staphylococci are often associated with this type of infection (7, 49, 52). In situations where an implantable device is left in place, CoNS are the most frequent cause of infections (3, 30, 52).

Pneumonia

S. aureus is an unusual cause of community-acquired pneumonia but is a common etiologic agent of pneumonia in the hospital setting, frequently as a consequence of influenza (29). Typically, S. aureus pneumonia develops in elderly patients approximately 4 to 14 days following the onset of influenza. Although such patients initially begin to improve clinically from influenza, they become increasingly ill with symptoms of high fever, productive sputum, and often respiratory failure. This postinfluenza pneumonia, commonly associated with S. aureus and Streptococcus pneumoniae, usually responds to antimicrobial agents if treated early but can be fatal.

Risk factors for S. aureus pneumonia are similar to those for nosocomial pneumonia in general and include recent surgery, particularly thoracic or high abdominal surgery; chronic obstructive lung disease; ventilator therapy; older age; and immunosuppressive therapy (29). Mortality is approximately 15 to 20%. Therapy must be aggressive and usually consists of a β-lactamase-resistant penicillin, such as nafcillin, or vancomycin for several weeks.

Hematogenous spread of S. aureus to the lungs during bloodstream infection, especially as a consequence of right-sided endocarditis (32), produces a different clinical and radiologic appearance than the pneumonia described above. The physical and X-ray findings do not usually show signs of consolidation. Rather, X rays show multiple, discrete pulmonary infiltrates that sometimes cavitate in a few days, representing metastatic foci of infection. Therapy consists of a β-lactamase-resistant penicillin, such as nafcillin, or vancomycin for several weeks. Identifying the source of the bloodstream infection and appropriate antimicrobial chemotherapy are essential for cure.

S. aureus remains the most common cause of empyema, i.e., infections in the pleural space. Acute empyema usually results from direct extension from a pneumonia or lung abscess or from hematogenous seeding (29). Chest pain, fever, shortness of breath, and signs of pleural effusion are nearly always present. Thoracocentesis is required for diagnosis, but therapy nearly always requires chest tube placement and, frequently, surgery since loculations are common complications of S. aureus infections.

TABLE 1 Percentage distribution of the five most common nosocomial pathogens isolated from major infection sites, 1990 through 1996[a]

Pathogen	All sites (n = 101,821 isolates)	UTI (n = 35,079 isolates)	SSI (n = 17,671 isolates)	BSI (n = 14,424 isolates)	PNEU (n = 13,433 isolates)	Other (n = 21,214 isolates)
S. aureus	13	2	20	16	19	18
E. coli	12	24	8	5	4	4
Coagulase-negative staphylococci	11	4	14	31	2	14
Enterococcus spp.	10	16	12	9	2	5
Pseudomonas aeruginosa	9	11	8	3	17	7

[a]Hospital-wide component, National Nosocomial Infections Surveillance System, Centers for Disease Control and Prevention. UTI, urinary tract infection; SSI, surgical site infection; BSI, blood stream infection; PNEU, pneumonia. Data from reference 33.

Surgical Site Infections

Surgical site infections (SSIs) constituted approximately 15% of the infections reported to the NNIS system by hospitals that collected hospital-wide surveillance data (12). SSIs are a major infection control concern since they are associated with serious morbidity, mortality, and high health care cost. The risk of SSI is related to a number of factors. Among the most important are the operative procedure performed; the degree of microbiologic contamination of the operative field; duration of operation; and the intrinsic susceptibility of the patient, due to advanced age, malnutrition, trauma, loss of skin integrity (e.g., burns), or the presence of an infection at a distant site. Typically occurring 7 to 10 days following the operative procedure, SSIs have a wide range of onset times, i.e., from 2 to 30 days. The time of onset may be strongly influenced by the administration of the appropriate antimicrobial agents prophylactically. The dose, timing, and selection of the prophylactic agents of choice have been reviewed elsewhere (25).

S. *aureus* is the most common cause of SSIs. This is largely due to its common presence on the skin or mucous membranes of patients, although other factors tend to influence the development of an SSI, as noted above. The use of prosthetic materials has a strong influence on the development of an SSI with S. *aureus*. Elek and Conen (11) have shown that a subcutaneous injection of $>10^5$ CFU of S. *aureus* was easily controlled by host defense mechanisms, whereas 3×10^2 of the same organism led to infection in the presence of prosthetic material such as a suture. The development of a glycocalix by most strains of CoNS and many strains of S. *aureus* is thought to protect these organisms from host defenses on the surface of prosthetic material (29). For effective treatment, removal of the foreign material is usually necessary, followed by antimicrobial therapy (25).

INFECTION CONTROL CONSIDERATIONS

Isolation Practices

Transmission of infection requires three elements: a source of infecting microorganisms, a susceptible host, and a means of transmission for the microorganism (15). For S. *aureus* infections, the human sources may be the patient, health care personnel, or, occasionally, outside visitors. These sources may be either infected or colonized with S. *aureus*. The role of the environment in transmitting staphylococcal infections remains controversial, although contaminated products, such as anesthetics, can clearly transmit infections (28). The host factors related to infection with S. *aureus* depend on the site of infection, the patient's immunologic status, and the infectious dose. Transmission may be by direct or indirect contact with the source or by contact with a common vehicle. Nosocomial outbreaks of food poisoning by toxin-producing S. *aureus* are rare and are an uncommon mode of transmission for S. *aureus* in health care settings.

Contact transmission is the most important and frequent mode of transmission for S. *aureus* (5, 15). Direct contact transmission involves contact of body surface to body surface and physical transfer of S. *aureus* to the host from an infected or colonized person (13). For example, a surgical resident, known to be colonized in his nares with methicillin-resistant S. *aureus* (MRSA), transmitted staphylococcal infections to six patients during two distinct outbreaks (16). The organisms from the patients were shown to be the same strain as that harbored by the resident by bacteriophage typing and antibiogram analysis. Several attempts to eradicate the MRSA from the resident's nares with various antimicrobial agent combinations were initially unsuccessful. However, mupirocin was eventually successful in eradicating the organism. A similar case of transmission of S. *aureus* from a neurosurgeon, who was a disseminator of a TSST-1-producing strain of S. *aureus* to two patients who developed symptoms of toxic shock syndrome following surgical procedures, was reported by Kreiswirth et al. (27). DNA probe typing was used to confirm the similarity of the three isolates from the neurosurgeon and the two patients. The nosocomial infections occurred 5 years apart.

Indirect contact transmission involves contact of a susceptible host with a contaminated intermediate object, usually inanimate, such as gloves that are not changed between patients. In the health care setting, indirect contact transmission, usually on the gloves or hands of health care personnel, is often the dominant mode of transmission for S. *aureus*. Hand washing remains the single most important measure to reduce the risk of transmitting microorganisms from one person to another (12, 15). In addition to hand washing, gloves play an important role in reducing the risks of transmission of S. *aureus* in health care settings. However, wearing gloves does not eliminate the need for hand washing, because gloves may have small, inapparent defects or may be torn during use.

Containment measures for patients with infections with S. *aureus* usually require standard precautions, which include hand washing after touching body fluids or contaminated items, whether or not gloves are worn; clean, nonsterile glove use before touching mucous membranes or body fluids; clean, nonsterile gown use to protect skin and prevent soiling of clothing during procedures and patient care activities that are likely to generate splashes of body fluids; and handling patient care equipment, environmental surfaces, and linens in a manner that prevents skin and mucous membrane exposures and avoids transfer of microorganisms to other patients or the environment. In general, private rooms are not needed if patients can undertake the usual personal hygiene measures and are unlikely to contaminate the environment (15). If a patient has an SSI or draining abscess with S. *aureus*, contact isolation may be needed. Contact isolation requires a private room, routine gown use, and dedicated patient care equipment for the duration of illness. Nurseries pose special problems for containment of MRSA infections. In one institution, continual transmission of an endemic strain of MRSA over a 3-year period was interrupted only when the institution applied an antibacterial solution of three dyes to the umbilical stumps of infants on the unit to reduce colonization with MRSA, increased the number of nurses on the unit to alleviate understaffing, and added a dedicated infection control nurse to tighten infection control efforts (17). The few cases of MRSA that occurred after these interventions were shown by pulsed-field gel electrophoresis typing to be caused by strains different from the endemic MRSA strain that had plagued the institution for over 3 years (Fig. 1). Indeed, the endemic strain had been eradicated from the medical center.

Antimicrobial Resistance in the Hospital Setting

Several previous reports suggest that the prevalence of S. *aureus* strains resistant to methicillin, oxacillin, or nafcillin

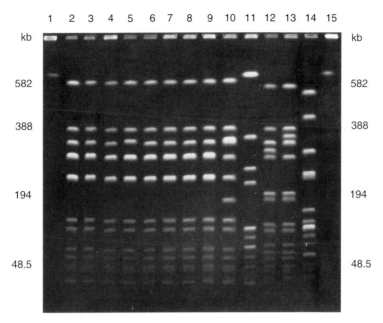

FIGURE 1 Pulsed-field gel electrophoresis patterns of SmaI-digested chromosomal DNA from 12 *S. aureus* isolates from a hospital undergoing a long-term outbreak of MRSA in a neonatal ICU. Lanes 1 and 15, molecular size standards; lanes 2 to 9, MRSA isolates obtained from patients before the interventions; lanes 10 to 13, strains of MRSA isolated after the interventions, showing eradication of the endemic MRSA strain; lane 14, an unrelated MRSA isolate from a patient during the endemic period (see reference 17).

has increased in the United States and abroad (2, 4, 21, 37, 43, 55) and that such strains have the propensity to cause outbreaks (2, 10, 17, 40). A review of 50,574 *S. aureus* nosocomial infections reported to the NNIS system from 1987 to 1997 showed that 29% were resistant to the semisynthetic β-lactam agents oxacillin, nafcillin, or methicillin. The percent MRSA among all hospitals rose from 14.3% in 1987 to 39.7% in 1997; however, the rate of increase depended upon the hospital's bed-size category (≤500 beds or >500 beds) and the patient's intensive care unit (ICU) status (34). Overall, for the decade 1987 to 1997, the percent MRSA differed between *S. aureus* isolated from patients in an ICU and patients in noncritical care wards (non-ICU), 32.0% and 26.3%, respectively (*P* < 0.001). When we examine the data by hospital bed size and by year, the percent MRSA among both ICU and non-ICU patients in hospitals with >500 beds rose steadily until 1993, then leveled off at 37.6% for ICU patients and 38.6% for non-ICU patients; in 1996, the percentage began to rise again for both ICU patients (42.2% in 1997) and non-ICU patients (42.4% in 1997) (Fig. 2). In smaller hospitals (≤500 beds), the rate of increase of *S. aureus* isolated from ICUs increased steadily to 35.2% in 1997, whereas the *S. aureus* isolated from non-ICUs had a somewhat slower rate of increase, to 27.3% in 1997 (Fig. 3).

Although overt resistance to vancomycin has not been documented in clinical isolates, recent reports have demonstrated clinical infections with *S. aureus* isolates intermediate to vancomycin (MICs = 8 μg/ml) (6, 19, 45). The development of vancomycin resistance among *S. aureus* isolates would dramatically impact our ability to control the spread of MRSA in health care settings (4, 6).

Treatment of the *S. aureus* Carrier State

Staphylococcal infection and carriage occur commonly in humans. Fortunately, health care personnel serve as reser-

voirs (and disseminators) of *S. aureus* much less commonly than do patients (5). Carriage of *S. aureus* is most common in the anterior nares, although other skin sites may be involved. The frequency of carriage among health care personnel ranges between 20 and 90%, but fewer than 10% of health care workers are ever linked to disseminating *S. aureus*, probably because they are colonized with low numbers of organisms. Carriage of *S. aureus* in the nares has been shown to correspond to hand carriage, and persons with skin lesions caused by the organism are more likely than asymptomatic nasal carriers to disseminate the bacteria (54, 55). Culture surveys of health care personnel can detect carriers but do not indicate who among them have disseminated *S. aureus*. Thus, such surveys are not cost-effective and may subject personnel with positive cultures

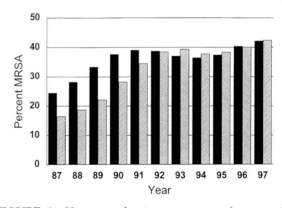

FIGURE 2 Histogram showing percentage of nosocomial MRSA infections in ICUs (black bars) versus non-ICUs (gray bars) in hospitals with >500 beds.

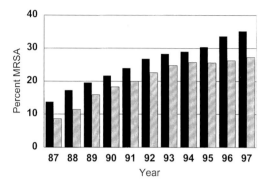

FIGURE 3 Histogram showing percentage of nosocomial MRSA infections in ICUs (black bars) versus non-ICUs (gray bars) in hospitals with ≤500 beds.

to unnecessary treatment and removal from duty. An epidemiologic investigation of clusters of *S. aureus* infections is needed to identify personnel who are disseminators of *S. aureus*. Such implicated personnel can then be removed from clinical duties until carriage has been eradicated.

Several antimicrobial regimens have been used to eradicate carriage of *S. aureus*. These include oral agents, i.e., ciprofloxacin, trimethoprim-sulfamethoxazole, and rifampin, or topical agents such as mupirocin (2, 21, 54, 55). Unfortunately, antimicrobial resistance to all these agents has emerged, underscoring the need for judicious antimicrobial treatment of the carrier state and avoidance of the indiscriminate culture survey and broad-scale treatment of culture-positive personnel.

Restriction from patient care activities or food handling is indicated for personnel who have draining skin lesions with *S. aureus* until they have received appropriate therapy. No work restrictions are needed for colonized personnel unless they have been epidemiologically implicated in *S. aureus* transmission in a facility.

REFERENCES

1. **Armstrong, G. L., J. Hollingsworth, and J. G. Morris, Jr.** 1998. Bacterial foodborne disease, p. 109–138. *In* A. S. Evan and P. S. Brachman (ed.), *Bacterial Infections of Humans. Epidemiology and Control*, 3rd ed., Plenum Medical Book Company, New York, N.Y.

2. **Aubry-Damon, H., P. Legrand, C. Brun-Buisson, A. Astier, C. J. Soussy, and R. Leclercq.** 1997. Reemergence of gentamicin-susceptible strains of methicillin-resistant *Staphylococcus aureus*: roles of an infection control program and changes in aminoglycoside use. *Clin. Infect. Dis.* **25:**647–653.

3. **Banerjee, S., G. Emori, D. H. Culver, R. P. Gaynes, W. J. Martone, T. G. Emori, T. C. Horan, J. R. Edwards, W. R. Jarvis, J. S. Tolson, T. S. Henderson, J. M. Hughes, and the National Nosocomial Infections Surveillance (NNIS) System.** 1991. Trends in nosocomial bloodstream infections in the United States, 1980–89. *Am. J. Med.* **91**(Suppl. 3B)**:**86S–89S.

4. **Barrett, S. P., R. V. Mummery, and B. Chattopadhyay.** 1998. Trying to control MRSA causes more problems than it solves. *J. Hosp. Infect.* **39:**85–93.

5. **Bolyard, E. A., O. C. Tablan, and W. W. Williams.** 1998. Guideline for infection control in health care personnel, 1998. *Am. J. Infect. Control* **26:**289–354.

6. **Centers for Disease Control and Prevention.** 1997. *Staphylococcus aureus* with reduced susceptibility to vancomycin—United States, 1997. *Morbid. Mortal. Weekly Rep.* **46:**765–766.

7. **Cockerill, F. R., III, J. G. Hughes, E. A. Vetter, R. A. Mueller, A. L. Weaver, D. M. Ilstrup, J. E. Rosenblatt, and W. R. Wilson.** 1997. Analysis of 281,797 consecutive blood cultures performed over an eight-year period: trends in microorganisms isolated and the value of anaerobic culture of blood. *Clin. Infect. Dis.* **24:**403–418.

8. **Dajani, A. S., K. A. Taubert, W. Wilson, A. F. Bolger, A. Bayer, P. Ferrieri, M. H. Gewitz, S. T. Shulman, S. Nouri, J. W. Newburger, C. Hutto, T. J. Pallasch, T. W. Gage, M. E. Levison, G. Peter, and G. Zuccaro, Jr.** 1997. Prevention of bacterial endocarditis: recommendations by the American Heart Association. *Clin. Infect. Dis.* **25:**1448–1458.

9. **Davis, J. P., P. Joan Chesney, P. J. Wand, M. LaVenture, and the Investigation and Laboratory Team.** 1980. Toxic-shock syndrome. *N. Engl. J. Med.* **303:**1429–1435.

10. **Dominguez, M. A., H. de Lancaster, J. Linares, and A. Tomasz.** 1994. Spread of a dominant methicillin-resistant *Staphylococcus aureus* (MRSA) clone during an outbreak of MRSA disease in a Spanish hospital. *J. Clin. Microbiol.* **32:**2081–2087.

11. **Elek, S. D., and P. E. Conen.** 1957. The virulence of *Staphylococcus aureus* for man. A study of the problems of wound infection. *Br. J. Exp. Pathol.* **38:**573–577.

12. **Emori, T. G., and R. P. Gaynes.** 1993. An overview of nosocomial infections including the role for the microbiology laboratory. *Clin. Microbiol. Rev.* **6:**428–442.

13. **Fekety, F. R., Jr.** 1964. The epidemiology and prevention of staphylococcal infection. *Medicine (Baltimore)* **43:**593–618.

14. **Frank, U., F. D. Daschner, G. Schulgen, and J. Mills.** 1997. Incidence and epidemiology of nosocomial infections in patients infected with human immunodeficiency virus. *Clin. Infect. Dis.* **25:**318–320.

15. **Garner, J. S.** 1996. Guideline for isolation precautions in hospitals. *Infect. Control Hosp. Epidemiol.* **17:**53–80.

16. **Gaynes, R., R. Marosok, J. Mowry-Hanley, C. Laughlin, K. Foley, C. Friedman, and M. Kirsh.** 1991. Mediastinitis following coronary artery bypass surgery: a 3 year review. *J. Infect. Dis.* **163:**117–121.

17. **Haley, R. W., N. B. Cushion, F. C. Tenover, T. L. Bannerman, D. Dryer, J. Ross, P. J. Sanchez, and J. D. Siegel.** 1995. Eradication of endemic methicillin-resistant *Staphylococcus aureus* infections from a neonatal intensive care unit. *J. Infect. Dis.* **171:**614–624.

18. **Harris, S. L.** 1992. Definitions and demographic characteristics, p. 1–18. *In* D. Kaye (ed.). *Infective Endocarditis*, 2nd ed. Raven Press, Ltd., New York, N.Y.

19. **Hiramatsu, K., H. Hanaki, T. Ino, K. Yabuta, T. Oguri, and F. C. Tenover.** 1997. Methicillin-resistant *Staphylococcus aureus* clinical strain with reduced vancomycin susceptibility. *J. Antimicrob. Chemother.* **40:**135–136.

20. **Holmberg, S. D., and P. A. Blake.** 1984. Staphylococcal food poisoning in the United States. New facts and old misconceptions. *JAMA* **251:**487–489.

21. **Irish, D., I. Eltringham, A. Teall, H. Pickett, H. Farelly, S. Reith, and B. Cookson.** 1998. Control of an outbreak of an epidemic methicillin-resistant *Staphylococcus aureus* also resistant to mupirocin. *J. Hosp. Infect.* **39:**19–26.

22. **Ivert, T. S. A., W. E. Dismukes, C. G. Cobbs, E. H. Blackstone, J. W. Kirklin, and L. A. Bergdahl.** 1984. Prosthetic valve endocarditis. *Circulation* **69:**223–232.

23. **Karchmer, A. W.** 1992. Staphylococcal endocarditis, p. 225–249. *In* D. Kaye (ed.), *Infective Endocarditis*, 2nd ed. Raven Press, Ltd., New York, N.Y.

24. **Kauffman, C. A., and S. F. Bradley.** 1997. Epidemiology of community-acquired infection, p. 287–308. *In* K. B. Crossley and G. L. Archer (ed.), *The Staphylococci in Human Disease.* Churchill Livingstone, New York, N.Y.

25. **Kernodle, D. S., and A. B. Kaiser.** 1996. Postoperative infections and antimicrobial prophylaxis, p. 2742–2756. *In* G. L. Mandell, J. E. Bennett, and R. Dolin (ed.), *Principles and Practice of Infectious Diseases*, 4th ed. Churchill Livingstone, New York, N.Y.

26. **Kluytmans, J., A. van Belkum, and H. Verbrugh.** 1997. Nasal carriage of *Staphylococcus aureus*: epidemiology, underlying mechanisms, and associated risks. *Clin. Microbiol. Rev.* **10:**505–520.

27. **Kreiswirth, B. N., G. R. Kravitz, P. M. Schlievert, and R. P. Novick.** 1986. Nosocomial transmission of a strain of *Staphylococcus aureus* causing toxic shock syndrome. *Ann. Intern. Med.* **105:**704–707.

28. **Kuehnert, M. J., R. M. Webb, E. M. Jochimsen, G. A. Hancock, M. J. Arduino, S. Hand, M. Currier, and W. R. Jarvis.** 1997. *Staphylococcus aureus* bloodstream infections among patients undergoing electroconvulsive therapy traced to breaks in infection control and possible extrinsic contamination by propofol. *Anesth. Analg.* **85:**420–425.

29. **Lowy, F. D.** 1998. *Staphylococcus aureus* infections. *N. Engl. J. Med.* **339:**520–532.

30. **Martin, M. A., M. A. Pfaller, and R. P. Wenzel.** 1989. Coagulase-negative staphylococcal bacteremia. Mortality and hospital stay. *Ann. Intern. Med.* **110:**9–16.

31. **Martin, R. R., V. Buttram, P. Besch, J. J. Kirkland, and G. P. Petty.** 1982. Nasal and vaginal *Staphylococcus aureus* in young women: quantitative studies. *Ann. Intern. Med.* **96**(Pt. 2)**:**951–953.

32. **Mortara, L. A., and A. S. Bayer.** 1993. *Staphylococcus aureus* bacteremia and endocarditis. *Infect. Dis. Clin. North Am.* **7:**53–68.

33. **Musser, J. M., P. M. Schlievert, A. W. Chow, P. Ewan, B. N. Kreiswirth, V. T. Rosdahl, A. S. Naidu, W. Witte, and R. K. Selander.** 1990. A single clone of *Staphylococcus aureus* causes the majority of cases of toxic shock syndrome. *Proc. Natl. Acad. Sci. USA* **87:**225–229.

34. **National Nosocomial Infections Surveillance System.** 1996. National Nosocomial Infections Surveillance (NNIS) report, data summary from October 1986–April 1996, issued May 1996. *Am. J. Infect. Control* **24:**380–388.

35. **Náwas, T., A. Hawwari, E. Hendrix, J. Hebden, R. Edelman, M. Martin, W. Campbell, R. Naso, R. Schwalbe, and A. I. Fattom.** 1998. Phenotypic and genotypic characterization of nosocomial *Staphylococcus aureus* isolates from trauma patients. *J. Clin. Microbiol.* **36:**414–420.

36. **Noble, W. C., H. A. Valkenburg, and C. H. I. Wolters.** 1967. Carriage of *Staphylococcus aureus* in random samples of a normal population. *J. Hyg. (London)* **65:**567–573.

37. **Panlilio, A. L., D. H. Culver, R. P. Gaynes, S. Banerjee, T. S. Henderson, J. S. Tolson, and W. J. Martone.** 1992. Methicillin-resistant *Staphylococcus aureus* in U.S. hospitals, 1975–1991. *Infect. Control Hosp. Epidemiol.* **13:**582–586.

38. **Parsonnet, J.** 1996. Nonmenstrual toxic shock syndrome: new insights into diagnosis, pathogenesis, and treatment, p. 1–20. *In* J. S. Remington and M. N. Swartz (ed.), *Current Clinical Topics in Infectious Diseases.* Blackwell Science, Cambridge, U.K.

39. **Rissing, J. P.** 1997. Antimicrobial therapy for chronic osteomyelitis in adults: role of the quinolones. *Clin. Infect. Dis.* **25:**1327–1333.

40. **Roman, R. S., J. Smith, M. Walker, S. Byrne, K. Ramotar, B. Dycjk, A. Kabani, and L. E. Nicolle.** 1997. Rapid geographic spread of a methicillin-resistant *Staphylococcus aureus* strain. *Clin. Infect. Dis.* **25:**698–705.

41. **Schlievert, P. M., K. N. Shands, B. B. Dan, G. P. Schmid, and R. D. Nishimura.** 1981. Identification and characterization of an exotoxin from *Staphylococcus aureus* associated with toxic shock syndrome. *J. Infect. Dis.* **143:**509–516.

42. **Shands, K. N., G. P. Schmid, B. B. Dan, D. Blum, R. J. Guidotti, N. T. Hargrett, R. L. Anderson, D. L. Hill, C. V. Broome, J. D. Band, and D. W. Fraser.** 1980. Toxic-shock syndrome in menstruating women: association with tampon use and *Staphylococcus aureus* and clinical features in 52 cases. *N. Engl. J. Med.* **303:**1436–1442.

43. **Struelens, M. J., R. Mertens, and the Groupement pour le Dépistage, l'Etude et la Prévention des Infections Hospitalières.** 1994. National survey of methicillin-resistant *Staphylococcus aureus*. *Eur. J. Clin. Microbiol. Infect. Dis.* **13:**56–63.

44. **Swartz, M., and A. N. Weinberg.** 1987. Infections due to gram-positive bacteria, p. 2100–2121. *In* T. B. Fitzpatrick, K. A. Arndt, and W. H. Clark (ed.), *Dermatology in General Medicine*, 3rd ed. McGraw-Hill, New York, N.Y.

45. **Tenover, F. C., M. V. Lancaster, B. C. Hill, C. D. Steward, S. A. Stocker, G. A. Hancock, C. M. O'Hara, S. A. McAllister, N. C. Clark, and K. Hiramatsu.** 1998. Characterization of staphylococci with reduced susceptibilities to vancomycin and other glycopeptides. *J. Clin. Microbiol.* **36:**1020–1027.

46. **Todd, J., M. Fishaut, F. Kapral, and T. Welch.** 1978. Toxic-shock syndrome associated with phage-group-I staphylococci. *Lancet* **2:**1116–1118.

47. **Tuazon, C. U., and J. N. Sheagren.** 1974. Increased rate of carriage of *Staphylococcus aureus* among narcotic addicts. *J. Infect. Dis.* **129:**725–727.

48. **Uchiyama, T., X. J. Yan, K. Imanishi, and J. Yagi.** 1994. Bacterial superantigens—mechanism of T cell activation by the superantigens and their role in the pathogenesis of infectious diseases. *Microbiol. Immunol.* **38:**245–256.

49. **Vallúes, J., C. León, and F. Alvarez-Lerma.** 1997. Nosocomial bacteremia in critically ill patients: a multicenter study evaluating epidemiology and prognosis. *Clin. Infect. Dis.* **24:**387–395.

50. **Waldvogel, F. A., and H. Vasey.** 1980. Osteomyelitis: the past decade. *N. Engl. J. Med.* **303:**360–363.

51. **Weinstein, M. P.** 1996. Current blood culture methods and systems: clinical concepts, technology, and interpretation of results. *Clin. Infect. Dis.* **23:**40–46.

52. **Weinstein, M. P., M. L. Towns, S. M. Quartey, S. Mirrett, L. G. Reimer, G. Parmigiani, and L. B. Reller.** 1997. The clinical significance of positive blood cultures in the 1990s: a prospective comprehensive evaluation of the microbiology, epidemiology, and outcome of bacteremia and fungemia in adults. *Clin. Infect. Dis.* **24:**584–602.

53. **Wieneke, A. A., D. Roberts, and R. J. Gilbert.** 1993. Staphylococcal food poisoning in the United Kingdom, 1969–1990. *Epidemiol. Infect.* **110:**519–531.

54. **Yu, V. L., A. Goetz, M. Wagnener, P. B. Smith, J. D. Rihs, J. Hanchett, and J. J. Zuravleff.** 1986. *Staphylococcus aureus* nasal carriage and infection in patients on hemodialysis: efficacy of antibiotic prophylaxis. *N. Engl. J. Med.* **315:**91–96.

55. **Zimakoff, J., F. B. Pedersen, L. Bergen, J. Baagø-Nielsen, B. Daldorph, F. Espersens, B. G. Hansen, N. Hoiby, O. B. Jepsen, P. Joffe, H. J. Kolmos, M. Klausen, K. Kristoffersen, J. Ladefoged, S. Olesen-Larsen, V. T. Rosdahl, J. Scheibel, B. Storm, and P. Tofte-Jensen.** 1996. *Staphylococcus aureus* carriage and infections among patients in four haemo- and peritoneal-dialysis centers in Denmark. *J. Hosp. Infect.* **33:**289–300.

Animal Models of Experimental *Staphylococcus aureus* Infection

L. VINCENT COLLINS AND ANDREJ TARKOWSKI

44

Staphylococcus aureus infections remain a permanent threat to humankind, being associated with high morbidity and mortality. Staphylococci can give rise to a diverse spectrum of diseases ranging from cutaneous infections to life-threatening conditions such as sepsis, endocarditis, and arthritis. The increasing prevalence of immunocompromised subjects and the appearance of methicillin-resistant staphylococci should prompt researchers to reach a better understanding of the host-bacterium relationship, a prerequisite for better therapeutic and preventive measures. Studies of pathogenetic mechanisms in human *S. aureus* infections have met with shortcomings, most of them related to uncertainty as to the exact time of the onset of the disease as well as the ethical problems related to manipulations of the host immune system and/or the invading bacterium. Use of laboratory animals should overcome most of these problems. In this respect the mouse is the most versatile animal to use owing to (i) the development of spontaneous staphylococcal infections (13), (ii) the availability of many inbred and genetically well-characterized mouse strains, (iii) the availability of large numbers of strains either lacking a certain gene(s) of potential interest (so-called gene knockout mice) or having overexpression of a certain gene(s) (so-called transgeneic mice), and (iv) last, but not least, the existing knowledge on the murine immune system. However, some diseases cannot be replicated or monitored easily in mice, and in these cases it has been necessary to develop alternative animal models.

In many instances it is clear that subjects displaying skin defects (e.g., due to eczema), damaged joints (e.g., due to rheumatoid arthritis), or defective heart valves (e.g., due to rheumatic fever) will be at risk of acquiring *S. aureus* infection. Also, patients with installed prostheses or indwelling catheters will be more prone to develop staphylococcal infections originating at these sites. Animal models to mimic these situations are particularly difficult to obtain.

The aim of this chapter is to discuss experimental models of *S. aureus* infections, including toxic shock, sepsis, endocarditis, colonization of joints and bones, mastitis, eye and skin infections, and septic arthritis. We have excluded those studies carried out primarily to test antibiotic therapies and have instead concentrated on models that provide an insight into the pathogenesis of *S. aureus*. As an example of a model for studying staphylococcal disease, we present the murine model of septic arthritis and sepsis and discuss how models such as this might be used to formulate treatment and prophylaxis regimens.

ANIMAL MODELS OF STAPHYLOCOCCAL DISEASE

The various disease syndromes associated with *S. aureus* infections and some proposed animal models are listed in Table 1.

Toxic Shock and Sepsis

Cases of toxic shock syndrome (TSS) due to the TSST-1 toxin of *S. aureus* are predominantly linked to tampon use in women. TSS acquired in this way is difficult to reproduce in animals, although some success has been achieved with a simulation of tampon use in rabbits (49). Attempts have also been made to induce TSS or a TSS-like syndrome by other means in laboratory animals (39, 58), but a definitive model for this disease has not been forthcoming.

Septicemia caused by *S. aureus* is usually the result of hematogenous infection from surgical intervention or from a wound site. Animal models of sepsis in which the bacteria are injected intravenously, e.g., in chickens (20) or mice (77), closely mimic the hematogenous spread of disease seen in humans and have proved useful in defining the components of the host immune system involved in resistance to colonization and septic shock.

Endocarditis

Animal models of endocarditis typically entail catheterization via the right carotid artery into the left ventricle, resulting in traumatization of the aortic valve. The formation of a mature vegetation on the valve ensues, comprising platelets, fibrin, inflammatory cells, and depositions

TABLE 1 Animal models used to study disease syndromes associated with *S. aureus* infection

Disease	Animal model	Reference
Toxic shock and sepsis	Rabbit	44
	Baboon	36, 53
	Chicken	20
	Mouse	69
Endocarditis	Rat	57
	Guinea pig	43
	Rabbit	56
Osteomyelitis	Dog	22, 23, 62
	Chicken	24, 25, 60
	Rat	30, 52
	Sheep	35
	Rabbit	68
	Guinea pig	50
Mastitis	Mouse	17, 27, 41, 42
	Rabbit	4
	Cow	65
	Sheep	5, 66
Wound infection	Rat	49
	Guinea pig	10
Skin abscess	Mouse	15
Eye infection	Guinea pig	21
	Rat	7
	Rabbit	9
Septic arthritis	Rabbit	40, 54, 59
	Rat	12
	Mouse	13, 14

of fibrinogen and fibronectin that serve as sites of adhesion for *S. aureus* (62). A rat model of endocarditis has been developed (31) along with a number of other models, most notably in the guinea pig (48) and the rabbit (61).

In the rat model of endocarditis, it was shown that immunization with a fibronectin fusion protein afforded protection against intravenous (i.v.) challenge with *S. aureus* (63). A similar rat model was used to demonstrate that passive immunization with antibodies raised against *S. aureus* type 5 capsular polysaccharide was protective against serotype 5-induced endocarditis (43).

The pivotal role played by certain staphylococcal virulence factors in endocarditis has also been elucidated through use of the rat model. These factors include the fibronectin adhesin (40), the capsular polysaccharide (6), clumping factor (50), and the collagen adhesin (33). In addition, in the rabbit model it was demonstrated that *agr*- and *sar*-regulated factors are important in bacterial colonization of valves (18).

Osteomyelitis

Colonization of the bones and joints by *S. aureus* may occur as a result of either hematogenous infection or follow-

ing local trauma. The virulence factors involved in survival in blood, homing and attachment to joint tissues and bone, induction of inflammation, and subsequent joint destruction are complex, but an understanding of some of these processes is emerging from pathological observations of the disease in animal models that resemble those encountered in humans. Hematogenous, *S. aureus*-induced osteomyelitis in either the acute or chronic form can be simulated in animal models, e.g., in dogs (22, 23), chickens (24, 25, 67), and rats (32), with the associated inflammatory reactions and bone changes. Chronic osteomyelitis, on the other hand, can be induced by direct injection of *S. aureus* into the tibial marrow cavity in sheep (38), rats (57), and dogs (69).

Posttraumatic osteomyelitis following instillation of *S. aureus* into fractured leg bones has been studied in rabbits (76) and guinea pigs (55). These animals provide useful models for the development of therapeutic interventions in human bone infection following trauma.

Conventional therapies for acute hematogenous osteomyelitis involve surgical drainage and antibiotic therapy. New therapies are being investigated using biodegradable implants impregnated with antibiotics in rabbit (16, 19) and rat (36) models to treat experimental *S. aureus*-induced osteomyelitis.

Mastitis

Mastitis due to *S. aureus* in ruminants is an economically important disease against which there is as yet no reliable efficacious vaccine. The mouse model of intramammary inoculation has been used extensively in studying this disease and represents a reproducible and cheap simulation of the disease in dairy cattle (17). Various attempts have been made in the past to design vaccines directed against the bacterial surface polysaccharides and proteins (adhesins) and against the toxins of *S. aureus* and to test their efficacy in bovine, mouse, and rabbit models (26, 68). Mamo and colleagues (47) utilized the mouse model to demonstrate that immunization with a fragment of the fibronectin-binding protein reduced the incidence of severe mastitis. In a related study, mice vaccinated with the fibrinogen-binding protein showed a reduced incidence and severity of mastitis (46). Rabbits vaccinated with a toxoid derivative were protected against the lethal gangrenous form of mastitis, but there was no protection against abscess formation (4). Adjuvant whole-cell–toxoid combinations have also shown promise in the protection of heifers (74) and ewes (73) from clinical mastitis, as has a killed whole-cell, toxoid, liposome-enclosed exopolysaccharide preparation in ewes (5). More recently, a live attenuated *S. aureus* strain was used to provide protection in mice when inoculated locally during late pregnancy or early lactation (28). It is clear, therefore, that the mouse model is a valuable tool for elucidating the role of specific antigens in mastitis and for testing potential vaccine preparations.

Skin and Wound Infections

Staphylococcal infections of the skin can be minor or life threatening, depending on the integrity of the skin surface and the invasiveness of the bacterial strain. Traumatization of the skin surface is often used to investigate *S. aureus* wound infectivity. As examples, a rat model was used to show that a fibrinogen-binding protein of *S. aureus* played a key role in wound infection (54), and an invasive burn wound infection model in guinea pigs was used to study the role of host factors in combating infection of burns

(10). Alternatively, bacteria are injected subcutaneously into mice along with microbeads or other foreign material, and lesions are measured after a given time period (15). This method allows comparison of the ability of different bacterial isolates to induce dermonecrosis or abscess formation.

Eye Infections

A model of staphylococcal keratitis in guinea pigs has been described (21), and a similar model in rats was used to determine the effects of antibiotic and corticosteroid treatments on the progression of S. aureus infection of the cornea (7). Radial incision of the rabbit conjunctiva followed by inoculation of S. aureus resulted in purulent conjunctivitis and a model for testing antimicrobial treatments in this disease (9).

Septic Arthritis

Well-defined human studies of infectious arthritis are almost nonexistent, as the time of induction of infection is often unknown. Furthermore, adequate tissue biopsies are rarely obtained. Experimental models of arthritis employing intra-articular injection in rabbits (45, 59, 66) have overcome some of these difficulties. However, in cases of human septic arthritis, staphylococci typically reach joints through hematogenic spreading rather than by local inoculation. Thus, the intra-articular route of inoculation bypasses the early, and potentially critical, stages of pathogenesis. During these stages the bacteria would conceivably be required to adapt to the environment within the host, to survive bactericidal components in the blood, to home to bone and synovial tissue, and to penetrate these structures to reach the joint cavity. A model of hematogenically transmitted joint and bone infection was clearly desirable. A spontaneous hematogenous arthritis in mice (13) and rats (12) was detected and described, proving that rodents may occasionally be natural hosts for staphylococci. The murine infectious model provides the undisputed advantage of using well-defined and inbred strains of mice—a prerequisite for analyzing the genetic and immunological background of the host's susceptibility. In this model approximately 90% of the mice develop septic arthritis, with disease initiation usually 24 h after i.v. inoculation of S. aureus. The characteristics of the murine model closely mirror those seen in human septic arthritis, especially showing a very high frequency and degree of periarticular bone erosivity (11). For this reason the murine model has proved extremely useful in elucidating the role of several virulence factors of S. aureus in septic arthritis, which include collagen adhesin (56), bone sialoprotein (14), Agr- and Sar-regulated components (1, 51), and the polysaccharide capsule (52). The model of septic arthritis and sepsis has also been valuable in underscoring the importance of crucial components of the host immune system, e.g., neutrophils (72) (Fig. 1), T lymphocytes (2) (Fig. 2), B lymphocytes (78) (Fig. 3), and major histocompatibility complex class II expression (3) in the pathogenesis of septic arthritis.

Kidney Abscess and Nephritis

Hematogenously acquired S. aureus infection resulting in arthritis, as described above, is typically accompanied by chronic renal abscesses. Acute suppurative infections following i.v. injection of S. aureus have been reported in mice (29, 30) and rabbits (27). In the mouse model, infection leads to the formation of multiple cortical and med-

ullar abscesses (30, 64, 65). Histopathological examination of the focal and diffuse infiltrates in the kidney cortex and medulla reveals the presence of both phagocytic cells and CD4$^+$ and CD8$^+$ lymphocytes (71). The TSST-1 (44, 71) and capsule (42) have been implicated as important virulence factors in this syndrome. A combination of corticosteroids and antibiotics was shown to be more efficacious in treating septic murine nephritis than were antibiotics alone, suggesting that T-lymphocyte infiltration plays a significant role in this disease (70).

GENE KNOCKOUT MOUSE MODELS

Cytokines play a critically important role in the pathogenesis of S. aureus infection, and the modulation of specific cytokines is attracting substantial interest as a means of treating disease. For instance, it has been observed that both gamma interferon (IFN-γ) mRNA and protein expression are clearly increased during the course of S. aureus infection in mice (79). Inactivation of the receptor for IFN-γ by gene knockout technology removes the functional IFN-γ receptor in mice. When these mice were inoculated with S. aureus, more frequent and severe arthritis compared with wild-type littermates was noted. Also, sepsis-triggered mortality in the IFN-γ receptor knockout mice was increased in early (but not late) stages of the infection (81). In contrast, supplementation of normal mice with extrinsic IFN-γ significantly decreased staphylococcal sepsis-triggered mortality on one hand, but enhanced the development of arthritis on the other hand (80). Interleukin-4 (IL-4) is another cytokine produced by T lymphocytes. Mice not able to produce this compound are protected against septic arthritis and sepsis-triggered mortality. Our data suggest that this protection is due to the role of IL-4 as an inhibitory factor in phagocytosis, decreasing clearance of bacteria during ongoing infection (35). Yet another cytokine produced predominantly by monocytes/macrophages is tumor necrosis factor. This cytokine is known to be one of the major causative factors involved in degradation of cartilage and subchondral bone in aseptic arthritis, e.g., rheumatoid arthritis. In the case of septic arthritis, tumor necrosis factor increases joint destruction but ameliorates the severity of sepsis (34). Much more remains to be learned about the activities and concentrations of cytokines and chemokines, the synergy between different cytokines, and the responding cells in the tissues of experimentally infected animals at different stages of the disease process. In the future it may be possible, using the various animal models, to improve the outcome of S. aureus infection by treatments with either anti-inflammatory cytokines (e.g., IL-10, IL-4, and/or transforming growth factor β) or antagonists to proinflammatory cytokines (e.g., IL-1 receptor antagonist) or by the introduction of antagonists to NF-κB, an intracellular messenger mediating the activation of many proinflammatory cytokines.

DEVELOPMENT OF NEW THERAPIES USING ANIMAL MODELS

Anti-Staphylococcal Therapies

The emergence of antibiotic-resistant staphylococci and in particular the methicillin-resistant S. aureus (MRSA) strains has stimulated a resurgence in the development of

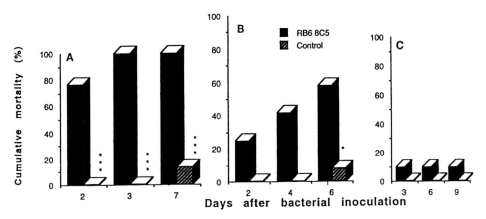

FIGURE 1 Mortality of mice treated with monoclonal anti-granulocyte antibody (RB6-8C5) or an equivalent amount of control rat immunoglobulin G, 2 h before an i.v. injection of 3×10^7 CFU (A), 8×10^6 CFU (B), or 1.5×10^6 CFU (C) of *S. aureus*. *, $P < 0.005$; ***, $P < 0.001$. Reprinted from *Infection and Immunity* (72) with permission of the publisher.

new antibiotics with anti-staphylococcal activities and alternatives to classical antibacterial therapies. Notable examples of the latter approach are recent studies using peptides produced by bacteria to modulate staphylococcal virulence (8, 37).

Vaccination Strategies Using Animal Models

An ideal vaccine candidate should induce responses that prevent bacterial adherence, promote opsonophagocytic killing by leukocytes, and neutralize secreted toxic proteins as suggested by Lee (41). Such a vaccine has thus far not been developed. Infection with *S. aureus* itself does not per se provide any protection against subsequent infections. Indeed, serological studies have, in general, failed to reveal a correlation between bacterial antibody titers and protective capacity. In recent times three distinct approaches to achieve immunization have been assayed. The first involves

vaccination with staphylococcal polysaccharides alone or conjugated to a protein carrier. Results from experimental trials show that this approach clearly decreases the severity of infection in models of *S. aureus* peritonitis, endocarditis, bacteremia, and renal abscess formation (41). One potential obstacle in these immunizations is the multitude of different capsular polysaccharide serotypes ($n = 1$) that can be used by various staphylococcal strains. An encouraging note, however, is that two polysaccharide serotypes (numbers 5 and 8) constitute about 75% of all disease-inducing staphylococcal strains.

A second possibility that has been explored is to trigger protective immunity against staphylococcal superantigens. To induce immunity using native superantigens is not advisable since these molecules will trigger severe inflammation, potentially leading to septic shock. An alternative used by Woody and colleagues (75) is to employ entero-

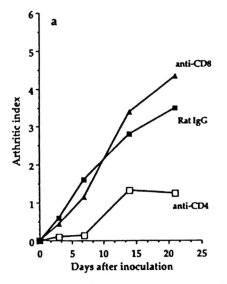

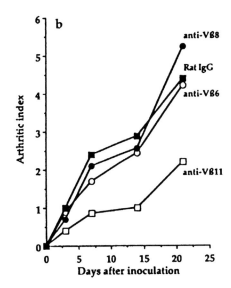

FIGURE 2 Effects of in vivo depletion of CD4$^+$, CD8$^+$, Vβ6$^+$, Vβ8$^+$, and Vβ11$^+$ T lymphocytes on the course of *S. aureus* arthritis triggered by TSST-1-producing bacteria. Mice were inoculated i.v. with *S. aureus*, and depletion was achieved using specific monoclonal antibodies. Reprinted from the *European Journal of Immunology* (2) with permission of the publisher.

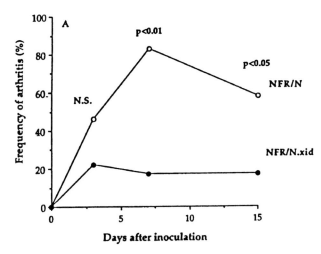

FIGURE 3 Frequency of arthritis in intact NFR/N and B-cell-deficient NFR/N.xid mice after an i.v. injection of *S. aureus*. N.S., not significant. Reprinted from the *Journal of Immunology* (78) with permission of the publisher.

toxins devoid of their superantigenic properties but still expressing major antigenic determinants. Vaccination of mice with staphylococcal enterotoxin B mutated in a hydrophobic loop dominating the interface with major histocompatibility complex class II locus did not trigger an inflammatory response but induced immunity. Antibodies to this molecule were able to protect a majority of mice against lethal staphylococcal enterotoxin B challenge (75). However, even here the high number of different toxins that may be produced by various virulent staphylococcal strains makes the task of developing a universal staphylococcal vaccine difficult.

S. aureus binds to numerous components of the extracellular matrix, including collagens, laminin, bone sialoprotein, fibronectin, fibrinogen, and vitronectin. Antibodies directed to staphylococcal adhesins and thus interfering with bacterial binding to the host tissue might prevent successful colonization. Indeed, experiments involving immunization with fibronectin-binding protein provided partial protection against experimental endocarditis and mastitis in rodents (reviewed in reference 41). Our laboratory has demonstrated that immunization of mice with recombinant collagen binding adhesin significantly alleviates the outcome of severe, life-threatening *S. aureus* sepsis (Fig. 4) (53). Here, vaccination-induced antibodies displayed protective properties, as proved by passive transfer experiments. Since collagen adhesin is expressed on the majority of staphylococci involved in invasive infections, it is another promising vaccine candidate that should merit evaluation in clinical trials.

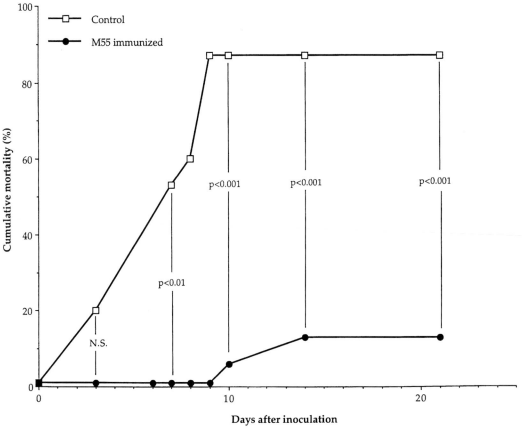

FIGURE 4 Cumulative mortality of mice vaccinated with the recombinant collagen-binding adhesin M55 or bovine serum albumin and subsequently challenged i.v. with a collagen adhesin-expressing heterologous *S. aureus* strain. N.S., not significant. Reprinted from the *Journal of Clinical Investigation* (53) with permission of the publisher.

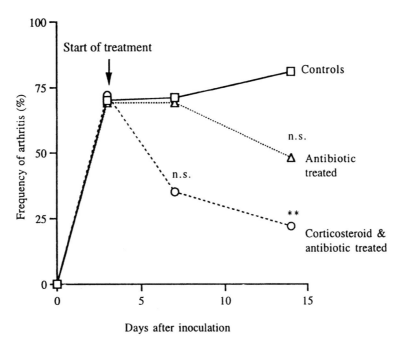

FIGURE 5 Effect of combined corticosteroid and antibiotic treatment versus antibiotic treatment alone on the frequency of *S. aureus* arthritis. **, $P < 0.01$; n.s., not significant. Reprinted from *Arthritis and Rheumatism* (60) with permission of the publisher.

Treatment of Ongoing Infection

Antibiotics are and will continue to be the standard treatment for systemic staphylococcal infections. However, despite adequate use of antibiotics and consequent eradication of staphylococci, continuous tissue destruction may occur. This can be clearly seen in the case of septic arthritis and is mediated by the exaggerated activation of the host immune response (see above). In addition, septic shock caused by *S. aureus* will not successfully respond to antibiotic treatment alone. Thus, there is room for improvement with respect to treatment of ongoing *S. aureus* infections. We have recently shown that concomitant corticosteroid and antibiotic treatment will, despite the inhibitory role of steroids on neutrophil function, clearly down-regulate the severity of septic arthritis (Fig. 5) and nephritis (60, 70).

FINAL COMMENTS

During the last decade, the use of experimental models of staphylococcal infections has clarified the involvement of several bacterial virulence factors as well as many hematopoietic cell types and their products in the pathogenesis of infection. Much still remains obscure, for example, how bacterial virulence factors act in vivo and the many details of the host-bacterium interplay. More information is required about the risk factors for acquiring staphylococcal infection to properly select subjects for vaccination procedures, and new modalities should be developed to treat ongoing infections with combinations of anti-inflammatory agents, passive immunization, and antibiotics to minimize the risk of sequels. Animal models that mimic the etiology, progression, and pathology of the disease in the natural host are a crucial component in the development of therapeutic strategies.

REFERENCES

1. **Abdelnour, A., S. Arvidson, T. Bremell, C. Ryden, and A. Tarkowski.** 1993. The accessory gene regulator (*agr*) controls *Staphylococcus aureus* virulence in a murine arthritis model. *Infect. Immun.* **61:**3879–3885.

2. **Abdelnour, A., T. Bremell, R. Holmdahl, and A. Tarkowski.** 1994. Clonal expansion of T lymphocytes causes arthritis and mortality in mice infected with toxic shock syndrome toxin-1-producing staphylococci. *Eur. J. Immunol.* **24:**1161–1166.

3. **Abdelnour, A., Y. X. Zhao, R. Holmdahl, and A. Tarkowski.** 1997. Major histocompatibility complex class II region confers susceptibility to *Staphylococcus aureus* arthritis. *Scand. J. Immunol.* **45:**301–307.

4. **Adlam, C., P. D. Ward, A. C. McCartney, J. P. Arbuthnott, and C. M. Thorley.** 1977. Effect of immunization with highly purified alpha- and beta-toxins on staphylococcal mastitis in rabbits. *Infect. Immun.* **17:**259–266.

5. **Amorena, B., R. Baselga, and I. Albizu.** 1994. Use of liposome-immunopotentiated exopolysaccharide as a component of an ovine mastitis staphylococcal vaccine. *Vaccine* **12:**243–249.

6. **Baddour, L. M., C. Lowrance, A. Albus, J. H. Lowrance, S. K. Anderson, and J. C. Lee.** 1992. *Staphylococcus aureus* microcapsule expression attenuates bacterial virulence in a rat model of experimental endocarditis. *J. Infect. Dis.* **165:**749–753.

7. **Badenoch, P. R., G. J. Hay, P. J. McDonald, and D. J. Coster.** 1985. A rat model of bacterial keratitis. Effect of antibiotics and corticosteroid. *Arch. Ophthalmol.* **103:**718–722.

8. **Balaban, N., T. Goldkorn, R. T. Nhan, L. B. Dang, S. Scott, R. M. Ridgley, A. Rasooly, S. C. Wright, J. W. Larrick, R. Rasooly, and J. R. Carlson.** 1998. Autoin-

ducer of virulence as a target for vaccine and therapy against *Staphylococcus aureus. Science* **280:**438–440.

9. **Behrens-Baumann, W., and T. Begall.** 1993. [Reproducible model of a bacterial conjunctivitis.] *Ophthalmologica* **206:**69–75.

10. **Bjornson, A. B., H. S. Bjornson, N. A. Lincoln, and W. A. Altemeier.** 1984. Relative roles of burn injury, wound colonization, and wound infection in induction of alterations of complement function in a guinea pig model of burn injury. *J. Trauma* **24:**106–115.

11. **Bremell, T., A. Abdelnour, and A. Tarkowski.** 1992. Histopathological and serological progression of experimental *Staphylococcus aureus* arthritis. *Infect. Immun.* **60:** 2976–2985.

12. **Bremell, T., S. Lange, R. Holmdahl, C. Ryden, G. K. Hansson, and A. Tarkowski.** 1994. Immunopathological features of rat *Staphylococcus aureus* arthritis. *Infect. Immun.* **62:**2334–2344.

13. **Bremell, T., S. Lange, L. Svensson, E. Jennische, K. Grondahl, H. Carlsten, and A. Tarkowski.** 1990. Outbreak of spontaneous staphylococcal arthritis and osteitis in mice. *Arthritis Rheum.* **33:**1739–1744.

14. **Bremell, T., S. Lange, A. Yacoub, C. Ryden, and A. Tarkowski.** 1991. Experimental *Staphylococcus aureus* arthritis in mice. *Infect. Immun.* **59:**2615–2623.

15. **Bunce, C., L. Wheeler, G. Reed, J. Musser, and N. Barg.** 1992. Murine model of cutaneous infection with grampositive cocci. *Infect. Immun.* **60:**2636–2640.

16. **Calhoun, J. H., and J. T. Mader.** 1997. Treatment of osteomyelitis with a biodegradable antibiotic implant. *Clin. Orthoped.* 206–214.

17. **Chandler, R. L.** 1970. Experimental bacterial mastitis in the mouse. *J. Med. Microbiol.* **3:**273–282.

18. **Cheung, A. L., K. J. Eberhardt, E. Chung, M. R. Yeaman, P. M. Sullam, M. Ramos, and A. S. Bayer.** 1994. Diminished virulence of a sar⁻/agr⁻ mutant of *Staphylococcus aureus* in the rabbit model of endocarditis. *J. Clin. Invest.* **94:**1815–1822.

19. **Dahners, L. E., and C. H. Funderburk.** 1987. Gentamicin-loaded plaster of Paris as a treatment of experimental osteomyelitis in rabbits. *Clin. Orthoped.* 278–282.

20. **Daum, R. S., W. H. Davis, K. B. Farris, R. J. Campeau, D. M. Mulvihill, and S. M. Shane.** 1990. A model of *Staphylococcus aureus* bacteremia, septic arthritis, and osteomyelitis in chickens. *J. Orthoped. Res.* **8:**804–813.

21. **Davis, S. D., L. D. Sarff, and R. A. Hyndiuk.** 1978. Staphylococcal keratitis. Experimental model in guinea pigs. *Arch. Ophthalmol.* **96:**2114–2116.

22. **Deysine, M., H. D. Isenberg, and G. Steiner.** 1983. Chronic haematogenous osteomyelitis; studies on an experimental model. *Int. Orthoped.* **7:**69–78.

23. **Deysine, M., E. Rosario, and H. D. Isenberg.** 1976. Acute hematogenous osteomyelitis: an experimental model. *Surgery* **79:**97–99.

24. **Emslie, K. R., and S. Nade.** 1983. Acute hematogenous staphylococcal osteomyelitis. A description of the natural history in an avian model. *Am. J. Pathol.* **110:**333–345.

25. **Emslie, K. R., N. R. Ozanne, and S. M. Nade.** 1983. Acute haematogenous osteomyelitis: an experimental model. *J. Pathol.* **141:**157–167.

26. **Foster, T. J.** 1991. Potential for vaccination against infections caused by *Staphylococcus aureus. Vaccine* **9:**221–227.

27. **Freedman, L. R.** 1960. Experimental pyelonephritis. VI. Observations on susceptibility of the rabbit kidney to infection by a virulent strain of *Staphylococcus aureus. Yale J. Biol. Med.* **32:**272–279.

28. **Gomez, M. I., V. E. Garcia, M. M. Gherardi, M. C. Cerquetti, and D. O. Sordelli.** 1998. Intramammary immunization with live-attenuated *Staphylococcus aureus* protects mice from experimental mastitis. *FEMS Immunol. Med. Microbiol.* **20:**21–27.

29. **Gorrill, R. H.** 1951. Experimental staphylococcal infections in mice. *Br. J. Exp. Pathol.* **32:**151–155.

30. **Gorrill, R. H.** 1958. The establishment of staphylococcal abscesses in the mouse kidney. *Br. J. Exp. Pathol.* **39:**203–212.

31. **Gutschik, E.** 1983. Experimental staphylococcal endocarditis: an overview. *Scand. J. Infect. Dis. Suppl.* **41:**87–94.

32. **Hienz, S. A., H. Sakamoto, J. I. Flock, A. C. Morner, F. P. Reinholt, A. Heimdahl, and C. E. Nord.** 1995. Development and characterization of a new model of hematogenous osteomyelitis in the rat. *J. Infect. Dis.* **171:** 1230–1236.

33. **Hienz, S. A., T. Schennings, A. Heimdahl, and J. I. Flock.** 1996. Collagen binding of *Staphylococcus aureus* is a virulence factor in experimental endocarditis. *J. Infect. Dis.* **174:**83–88.

34. **Hultgren, O., H. P. Eugster, J. D. Sedgwick, H. Korner, and A. Tarkowski.** 1998. TNF/lymphotoxin-alpha double-mutant mice resist septic arthritis but display increased mortality in response to *Staphylococcus aureus. J. Immunol.* **161:**5937–5942.

35. **Hultgren, O., M. Kopf, and A. Tarkowski.** 1998. *Staphylococcus aureus*-induced septic arthritis and septic death is decreased in IL-4-deficient mice: role of IL-4 as promoter for bacterial growth. *J. Immunol.* **160:**5082–5087.

36. **Itokazu, M., T. Ohno, T. Tanemori, E. Wada, N. Kato, and K. Watanabe.** 1997. Antibiotic-loaded hydroxyapatite blocks in the treatment of experimental osteomyelitis in rats. *J. Med. Microbiol.* **46:**779–783.

37. **Ji, G., R. C. Beavis, and R. P. Novick.** 1995. Cell density control of staphylococcal virulence mediated by an octapeptide pheromone. *Proc. Natl. Acad. Sci. USA* **92:** 12055–12059.

38. **Kaarsemaker, S., G. H. Walenkamp, and A. E. van de Bogaard.** 1997. New model for chronic osteomyelitis with *Staphylococcus aureus* in sheep. *Clin. Orthoped.* 246–252.

39. **Kohrman, K. A., J. J. Kirkland, and P. J. Danneman.** 1989. Response of various animal species to experimental infection with different strains of *Staphylococcus aureus. Rev. Infect. Dis.* **11**(Suppl. 1):S231–S236. (Discussion, **11**[Suppl. 1]:S236–S237.)

40. **Kuypers, J. M., and R. A. Proctor.** 1989. Reduced adherence to traumatized rat heart valves by a low-fibronectin-binding mutant of *Staphylococcus aureus. Infect. Immun.* **57:**2306–2312.

41. **Lee, J. C.** 1996. The prospects for developing a vaccine against *Staphylococcus aureus. Trends Microbiol.* **4:**162–166.

42. **Lee, J. C., M. J. Betley, C. A. Hopkins, N. E. Perez, and G. B. Pier.** 1987. Virulence studies, in mice, of transposon-induced mutants of *Staphylococcus aureus* differing in capsule size. *J. Infect. Dis.* **156:**741–750.

43. **Lee, J. C., J. S. Park, S. E. Shepherd, V. Carey, and A. Fattom.** 1997. Protective efficacy of antibodies to the

Staphylococcus aureus type 5 capsular polysaccharide in a modified model of endocarditis in rats. *Infect. Immun.* **65:** 4146–4151.

44. **Lee, J. C., N. E. Perez, and C. A. Hopkins.** 1989. Production of toxic shock syndrome toxin 1 in a mouse model of *Staphylococcus aureus* abscess formation. *Rev. Infect. Dis.* **11**(Suppl. 1):S254–S259.

45. **Linhart, W. E., S. Spendel, G. Weber, and S. Zadravec.** 1990. Septic arthritis—an experimental animal model useful in free oxygen radical research. *Z. Versuchstierkd.* **33:**65–71.

46. **Mamo, W., M. Boden, and J. I. Flock.** 1994. Vaccination with *Staphylococcus aureus* fibrinogen binding proteins (FgBPs) reduces colonisation of *S. aureus* in a mouse mastitis model. *FEMS Immunol. Med. Microbiol.* **10:**47–53.

47. **Mamo, W., P. Jonsson, J. I. Flock, M. Lindberg, H. P. Muller, T. Wadstrom, and L. Nelson.** 1994. Vaccination against *Staphylococcus aureus* mastitis: immunological response of mice vaccinated with fibronectin-binding protein (FnBP-A) to challenge with *S. aureus*. *Vaccine* **12:** 988–992.

48. **Maurin, M., H. Lepidi, B. La Scola, M. Feuerstein, M. Andre, J. F. Pellissier, and D. Raoult.** 1997. Guinea pig model for *Staphylococcus aureus* native valve endocarditis. *Antimicrob. Agents Chemother.* **41:**1815–1817.

49. **Melish, M. E., S. Murata, C. Fukunaga, K. Frogner, and C. McKissick.** 1989. Vaginal tampon model for toxic shock syndrome. *Rev. Infect. Dis.* **11**(Suppl. 1):S238–S246. (Discussion, **11**[Suppl. 1]:S246–S247.)

50. **Moreillon, P., J. M. Entenza, P. Francioli, D. McDevitt, T. J. Foster, P. Francois, and P. Vaudaux.** 1995. Role of *Staphylococcus aureus* coagulase and clumping factor in pathogenesis of experimental endocarditis. *Infect. Immun.* **63:**4738–4743.

51. **Nilsson, I. M., T. Bremell, C. Ryden, A. L. Cheung, and A. Tarkowski.** 1996. Role of the staphylococcal accessory gene regulator (*sar*) in septic arthritis. *Infect. Immun.* **64:**4438–4443.

52. **Nilsson, I. M., J. C. Lee, T. Bremell, C. Ryden, and A. Tarkowski.** 1997. The role of staphylococcal polysaccharide microcapsule expression in septicemia and septic arthritis. *Infect. Immun.* **65:**4216–4221.

53. **Nilsson, I. M., J. M. Patti, T. Bremell, M. Hook, and A. Tarkowski.** 1998. Vaccination with a recombinant fragment of collagen adhesin provides protection against *Staphylococcus aureus*-mediated septic death. *J. Clin. Invest.* **101:**2640–2649.

54. **Palma, M., S. Nozohoor, T. Schennings, A. Heimdahl, and J. I. Flock.** 1996. Lack of the extracellular 19-kilodalton fibrinogen-binding protein from *Staphylococcus aureus* decreases virulence in experimental wound infection. *Infect. Immun.* **64:**5284–5289.

55. **Passl, R., C. Muller, C. C. Zielinski, and M. M. Eibl.** 1984. A model of experimental post-traumatic osteomyelitis in guinea pigs. *J. Trauma* **24:**323–326.

56. **Patti, J. M., T. Bremell, D. Krajewska-Pietrasik, A. Abdelnour, A. Tarkowski, C. Ryden, and M. Hook.** 1994. The *Staphylococcus aureus* collagen adhesin is a virulence determinant in experimental septic arthritis. *Infect. Immun.* **62:**152–161.

57. **Power, M. E., M. E. Olson, P. A. Domingue, and J. W. Costerton.** 1990. A rat model of *Staphylococcus aureus* chronic osteomyelitis that provides a suitable system for studying the human infection. *J. Med. Microbiol.* **33:** 189–198.

58. **Quimby, F., and H. T. Nguyen.** 1985. Animal studies of toxic shock syndrome. *Crit. Rev. Microbiol.* **12:**1–44.

59. **Riegels-Nielson, P., N. Frimodt-Moller, and J. S. Jensen.** 1987. Rabbit model of septic arthritis. *Acta Orthoped. Scand.* **58:**14–19.

60. **Sakiniene, E., T. Bremell, and A. Tarkowski.** 1996. Addition of corticosteroids to antibiotic treatment ameliorates the course of experimental *Staphylococcus aureus* arthritis. *Arthritis Rheum.* **39:**1596–1605.

61. **Sande, M. A.** 1981. Evaluation of antimicrobial agents in the rabbit model of endocarditis. *Rev. Infect. Dis.* **3**(Suppl.):S240–S249.

62. **Santoro, J., and M. E. Levison.** 1978. Rat model of experimental endocarditis. *Infect. Immun.* **19:**915–918.

63. **Schennings, T., A. Heimdahl, K. Coster, and J. I. Flock.** 1993. Immunization with fibronectin binding protein from *Staphylococcus aureus* protects against experimental endocarditis in rats. *Microb. Pathog.* **15:**227–236.

64. **Smith, I. M., A. P. Wilson, E. C. Hazard, W. K. Hummer, and M. E. Dewey.** 1960. Death from staphylococci in mice. *J. Infect. Dis.* **107:**369–378.

65. **Smith, J. M., and R. J. Dubos.** 1956. The behavior of virulent and avirulent staphylococci in the tissues of normal mice. *J. Exp. Med.* **103:**87–108.

66. **Smith, R. L., G. Kajiyama, and D. J. Schurman.** 1997. Staphylococcal septic arthritis: antibiotic and nonsteroidal anti-inflammatory drug treatment in a rabbit model. *J. Orthoped. Res.* **15:**919–926.

67. **Speers, D. J., and S. M. Nade.** 1985. Ultrastructural studies of adherence of *Staphylococcus aureus* in experimental acute hematogenous osteomyelitis. *Infect. Immun.* **49:** 443–446.

68. **Sutra, L., and B. Poutrel.** 1994. Virulence factors involved in the pathogenesis of bovine intramammary infections due to *Staphylococcus aureus*. *J. Med. Microbiol.* **40:**79–89.

69. **Varshney, A. C., H. Singh, R. S. Gupta, and S. P. Singh.** 1989. Experimental model of staphylococcal osteomyelitis in dogs. *Indian J. Exp. Biol.* **27:**816–819.

70. **Verba, V., E. Sakiniene, and A. Tarkowski.** 1997. Beneficial effect of glucocorticoids on the course of haematogenously acquired *Staphylococcus aureus* nephritis. *Scand. J. Immunol.* **45:**282–286.

71. **Verba, V., and A. Tarkowski.** 1996. Participation of V beta 4(+)-, V beta 7(+)-, and V beta 11(+)-T lymphocytes in haematogenously acquired *Staphylococcus aureus* nephritis. *Scand. J. Immunol.* **44:**261–266.

72. **Verdrengh, M., and A. Tarkowski.** 1997. Role of neutrophils in experimental septicemia and septic arthritis induced by *Staphylococcus aureus*. *Infect. Immun.* **65:**2517–2521.

73. **Watson, D. L.** 1988. Vaccination against experimental staphylococcal mastitis in ewes. *Res. Vet. Sci.* **45:**16–21.

74. **Watson, D. L.** 1992. Vaccination against experimental staphylococcal mastitis in dairy heifers. *Res. Vet. Sci.* **53:**346–353.

75. **Woody, M. A., T. Krakauer, R. G. Ulrich, and B. G. Stiles.** 1998. Differential immune responses to staphylococcal enterotoxin B mutations in a hydrophobic loop

dominating the interface with major histocompatibility complex class II receptors. *J. Infect. Dis.* **177:**1013–1022.

76. **Worlock, P., R. Slack, L. Harvey, and R. Mawhinney.** 1988. An experimental model of post-traumatic osteomyelitis in rabbits. *Br. J. Exp. Pathol.* **69:**235–244.

77. **Yao, L., J. W. Berman, S. M. Factor, and F. D. Lowy.** 1997. Correlation of histopathologic and bacteriologic changes with cytokine expression in an experimental murine model of bacteremic *Staphylococcus aureus* infection. *Infect. Immun.* **65:**3889–3895.

78. **Zhao, Y. X., A. Abdelnour, R. Holmdahl, and A. Tarkowski.** 1995. Mice with the xid B cell defect are less susceptible to developing *Staphylococcus aureus*-induced arthritis. *J. Immunol.* **155:**2067–2076.

79. **Zhao, Y. X., A. Ljungdahl, T. Olsson, and A. Tarkowski.** 1996. In situ hybridization analysis of synovial and systemic cytokine messenger RNA expression in superantigen-mediated *Staphylococcus aureus* arthritis. *Arthritis Rheum.* **39:**959–967.

80. **Zhao, Y. X., I. M. Nilsson, and A. Tarkowski.** 1998. The dual role of interferon-gamma in experimental *Staphylococcus aureus* septicaemia versus arthritis. *Immunology* **93:**80–85.

81. **Zhao, Y. X., and A. Tarkowski.** 1995. Impact of interferon-gamma receptor deficiency on experimental *Staphylococcus aureus* septicemia and arthritis. *J. Immunol.* **155:** 5736–5742.

Cellular and Extracellular Defenses against Staphylococcal Infections

JERROLD WEISS, ARNOLD S. BAYER, AND MICHAEL YEAMAN

45

Staphylococci are prominent members of the normal flora of humans that can also produce a wide variety of invasive infections (48). These include infections of wounds, prostheses and other foreign bodies, bones, endocardium, the bloodstream, and further metastatic sites. Breaches of the skin and mucosal barriers greatly increase the likelihood of invasive staphylococcal infections, affirming the importance of these peripheral barriers in maintaining a normally asymptomatic host-bacterial relationship. A cardinal sign of staphylococcal invasion, especially that of *Staphylococcus aureus*, is the acute inflammatory host response in which the influx of polymorphonuclear leukocytes (PMN) is a very prominent feature. This reflects the major role of these professional phagocytes in first-line defense against invading bacteria (15). Deficiencies in the mobilization or function of PMN are associated with increased susceptibility to infection by many extracellular bacterial pathogens, including staphylococci. However, recent evidence has revived interest in secretion-based extracellular defenses against staphylococci and other gram-positive bacteria (71, 72, 83). Hallmarks of many staphylococcal infections (e.g., *S. aureus* abscesses, bacterial vegetations at injured endothelium, *Staphylococcus epidermidis* foreign body infections) include deposition of bacteria in physical states that are likely refractory to phagocytosis (34, 66). Thus, extracellular antibacterial action may be much more important in normal host defense against staphylococci than has been generally appreciated.

ANTISTAPHYLOCOCCAL ACTION OF PHAGOCYTES (PMN)

In general, the action of PMN at extravascular sites of infection requires a highly regulated series of PMN responses resulting in the directed migration of PMN from blood to sites of infection, sequestration of bacterial prey, and intracellular cytotoxic action (15). A gradient of bacteria- and host-derived chemotaxins and other mediators trigger complementary changes in the surface properties of the PMN and endothelium and in PMN contractility, leading to sequential adherence and transendothelial migration of PMN

in the immediate vicinity of underlying infectious sites. How invading staphylococci provoke acute inflammation is still not precisely known but almost certainly includes the proinflammatory action of derivatives or fragments of the peptidoglycan matrix of the cell wall and other cell wall-associated bacterial products such as lipoteichoic acid (10, 28, 41). PMN at inflammatory sites may have increased phagocytic and cytotoxic capacities reflecting pleiotropic effects of bacterial proinflammatory products as well as of host chemokines and cytokines (19). In mouse models of staphylococcal infection, introduced genetic deficiencies in either endothelial adhesion molecules (e.g., ICAM-1; 68) or host cytokines (e.g., tumor necrosis factor alpha; 30) may increase susceptibility to infection or decrease outcome severity. These seemingly paradoxical findings are consistent with the belief that normally protective acute inflammatory host responses to staphylococcal infection may also have harmful effects depending on specific features of the infection (e.g., size of the bacterial inoculum).

Maximum efficiency of intraphagocytic destruction of staphylococci requires fluid-phase opsonins to facilitate delivery of bacteria to phagocytes and subsequent uptake. Plasma-derived and extracellular matrix-associated proteins (e.g., fibronectin) that can interact simultaneously with cell wall-associated bacterial proteins and surface sites on PMN may promote bacterial clearance. However, the most potent host opsonins are cell wall- and capsule-directed antibodies and fragments of C3 derived from complement activation (35, 53, 67). Cell walls of nonencapsulated bacteria can promote complement activation via the alternative pathway, leading to deposition of C3b(I) and targeting of bacteria to complement receptors (CR3, CR4) on phagocytes. Antibodies to staphylococcal cell walls are normally present in plasma, reflecting the frequent exposure to staphylococci. These antibodies are both directly opsonic via Fc$_\gamma$ receptors on phagocytes and complement activating and likely contribute to the high levels of "native" resistance to invasive staphylococcal infection (42). Bacteremic isolates of *S. aureus* are usually encapsulated. The presence of a capsular layer surrounding the bacterial cell wall can reduce the efficiency of antibody and complement inter-

Gram-Positive Pathogens, ed. by V. A. Fischetti et al.
© 2000 American Society for Microbiology, Washington, D.C.

action with the cell wall and, hence, the opsonic activity of normal plasma (63, 67). In certain models of infection, type-specific anticapsular antibodies are protective, presumably by promoting bacterial clearance (43). Similarly, the capsular slime layer of coagulase-negative *S. epidermidis* likely impedes phagocytic clearance of bacteria adherent to foreign bodies; antibodies to these surface constituents may be opsonic as well as antiadherent.

Uptake by professional phagocytes of staphylococci and most other bacterial prey triggers activation of a cellular oxidative response (respiratory burst) and fusion of cytoplasmic granules with the phagocytic vacuole (15) (Fig. 1). Both events, together with gradual vacuolar acidification, act to create an intensely noxious microenvironment in which the ingested bacterium resides. The extent to which PMN and mononuclear phagocytes (monocytes and macrophages) differ in their bactericidal and digestive capacities toward staphylococci has not been carefully examined. However, the granules of PMN appear to be enriched in substances that can exert bactericidal effects either in concert with the respiratory burst (e.g., myeloperoxidase [MPO]) or independent of oxygen (e.g., defensins, bactericidal/permeability-increasing protein [BPI], cathelicidins) (18). Conversely, mononuclear phagocytes (especially macrophages) appear to be better equipped to digest bacteria and bacterial remnants (36, 73). These properties are consistent with the primary and secondary roles of PMN and mononuclear phagocytes, respectively, in host responses to invading bacteria. Antibacterial compounds within the granules of PMN include peptides and proteins that can directly kill staphylococci (e.g., phospholipase A2 [PLA2], defensins, cathelicidins, cathepsin G) (18). However, in vitro killing of staphylococci by both PMN and mononuclear phagocytes is reduced, especially at higher bacterial loads, when phagocytosis is not accompanied by activation of the respiratory burst (57, 75). Moreover, in chronic granulomatous disease (CGD), a disorder characterized by recurrent, potentially life-threatening infections in which host phagocytes fail to mount a respiratory burst during phagocytosis, the major microbial pathogen is *S. aureus*; infections with *S. epidermidis* are also common (15). Genetically engineered mice lacking essential components of the respiratory burst oxidase (see below) also display reduced host resistance to staphylococci in vitro and in vivo (32, 54). Thus, an essential role of the phagocyte oxidative response in host defense against staphylococci is well established. Experiments in mice (49) suggest an important role for nitric oxide (NO) in antistaphylococcal

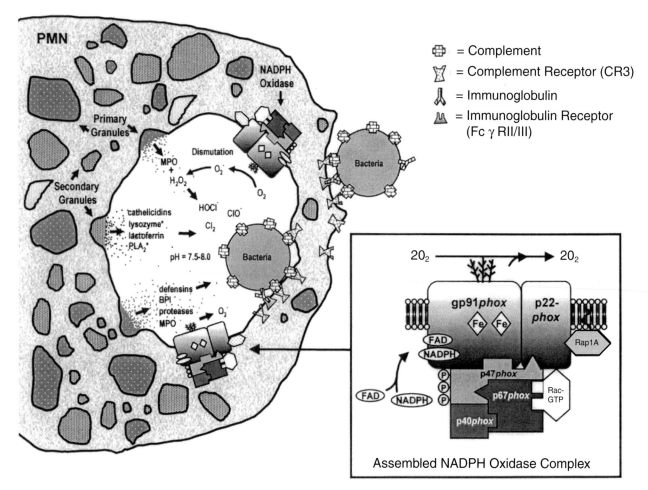

FIGURE 1 Schematic representation of PMN engaged in phagocytosis: attachment and internalization of bacterial prey into the phagocytic vacuole, fusion of cytoplasmic granules with the phagosome to deliver antibacterial peptides and proteins, and mobilization of assembled NADPH oxidase within the phagolysosome.

defenses, but the cellular origin of NO and downstream reactive nitrogen metabolites at sites of infection and their role (e.g., cytotoxicity versus signaling) and prominence in human responses to infection are not yet firmly established (25). The extent to which O_2-independent antistaphylococcal agents within and outside PMN further contribute to normal host resistance to staphylococci requires further study. In patients with CGD, prophylactic gamma interferon therapy has significantly reduced the incidence and consequences of infection without remarkable changes in phagocyte oxidase activity, indicating the likely importance of other host determinants of resistance to staphylococci (78).

Invading bacteria reaching the bloodstream are promptly met by PMN (and monocytes) and cleared from the circulation. However, at sites where endothelial monolayers are not fully intact, bacteria may adhere and become less susceptible to phagocytic clearance. At high multiplicities of infection in vitro, S. aureus can invade cultured vascular endothelial cells (5, 82). The fate of intraendothelial staphylococci is uncertain. Bacterial invasion induces up-regulation of endothelial cell adhesion molecules for attachment of monocytes and PMN (5) that could play a role in clearance of infected cells and disruption of endothelial integrity. Apoptotic responses of the infected endothelium have also been described (50). Staphylococcal infections in CGD are overwhelmingly of extravascular nature (15), consistent with the normal clearance function of phagocytes in this disease but also raising the possibility that other mechanisms of host defense against intravascular infections are operative.

ANTISTAPHYLOCOCCAL ACTION OF PLATELETS

Gram-positive pathogens are, by far, the most frequent causes of endovascular infections, including infective endocarditis, vascular catheter sepsis, intravascular graft infections, and infections associated with hemodialysis shunts (31). In these clinical settings, S. aureus, S. epidermidis, and other coagulase-negative staphylococci as well as the viridans group of streptococci and the enterococci are the major etiologic agents (31). A common feature in these clinical syndromes is microbial colonization of damaged endothelium. At such sites, platelets represent the initial and most abundant blood cells to arrive (83). Traditionally, it has been thought that platelets promote the development of endovascular infections by providing an adhesive surface upon which bacteremic organisms can dock to initiate infection. In contrast, recent evidence suggests that platelets serve an antimicrobial role (Fig. 2) (83). Platelets can internalize bacteria and express on their surface a number of molecules similar to those involved in neutrophil-based host defenses (e.g., complement receptors, Fc_γ receptors, and selectins). Likely to be of greater importance, however, especially toward bacteria attached to surfaces, is the ability of platelets to release an array of antimicrobial (poly)peptides in response to physiologic stimuli present at sites of endovascular damage (83). Platelet products released during degranulation include proteins directly cytotoxic toward gram-positive bacteria (e.g., platelet microbicidal proteins [PMPs] [83] and group IIA PLA2 [40, 71]) and also molecules chemotactic for PMN, further enhancing the local mobilization of host defenses at endovascular sites of infection (83). Under these conditions,

locally secreted antistaphylococcal agents may include products of PMN, platelets, and/or the endothelium and include both cytotoxic proteins and downstream metabolites of oxidative and NO metabolism (Fig. 2C). Rabbits rendered profoundly and selectively thrombocytopenic after induction of infection are impaired in their ability to clear viridans streptococci from damaged endocardium (60). Moreover, strains of viridans streptococci and S. aureus that are relatively resistant to platelet-derived antibacterial proteins (e.g., thrombin-induced PMP-1 [tPMP-1]; see below) are more virulent in animal models of infectious endocarditis owing mainly to increased bacterial proliferation after adherence to sterile vegetations (9, 12).

ANTISTAPHYLOCOCCAL ACTIVITY OF BODY FLUIDS

The complement system confers upon normal plasma potent cytotoxic activity against a broad range of nonencapsulated gram-negative bacteria. In contrast, with the exception of exquisitely lysozyme-sensitive soil bacteria, gram-positive bacteria, including staphylococci, streptococci, and enterococci, are generally resistant to normal plasma and grow readily in this body fluid (17, 71). However, serum from certain animal species (e.g., rats and rabbits but not humans) is potently bactericidal toward these and many other gram-positive bacteria (17, 71). The bactericidal activity of rabbit and rat sera toward S. aureus and several other gram-positive bacteria is mainly attributable to group IIA PLA2 released from platelets during in vitro clotting (17, 47, 72). This enzyme is naturally mobilized from a variety of intravascular and extravascular sources during inflammation and thereby confers potent extracellular bactericidal activity toward many gram-positive bacteria as part of the body's response to bacterial invasion. This has been demonstrated both in sterile peritoneal inflammatory exudates in rabbits induced by intraperitoneal administration of glycogen (72) and in plasma of baboons following intravenous bacterial challenge (71). In tears, where exposure to bacterial invaders is virtually constant, PLA2 levels are constitutively high and again appear to account for much of the antistaphylococcal activity of this body fluid (56). Recent studies have revealed a remarkable array of secretory (poly)peptides produced by a variety of specialized epithelia that, as in more primitive organisms, provide an additional early shield against microbial invasion (22).

MOLECULAR CHARACTERISTICS OF ANTISTAPHYLOCOCCAL HOST DEFENSE SYSTEMS

Oxygen-Dependent Systems in PMN (Phagocytes)

Respiratory Burst Oxidase

Activation of professional phagocytes, as normally occurs during phagocytosis, triggers the assembly of a multicomponent enzyme complex that catalyzes the transfer of reducing equivalents from NADPH to molecular oxygen, yielding superoxide anion (Fig. 1) (1). Essential components of the respiratory burst NADPH oxidase include both transmembrane and cytosolic proteins. The former is

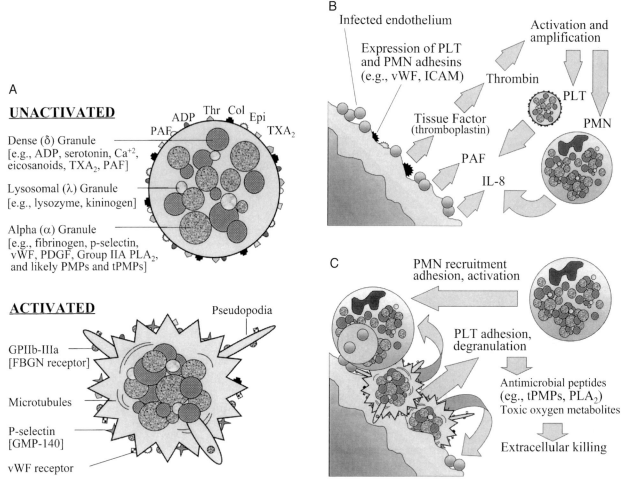

A

UNACTIVATED

ADP   Thr Col Epi
PAF   TXA₂

Dense (δ) Granule
[e.g., ADP, serotonin, Ca⁺², eicosanoids, TXA₂, PAF]

Lysosomal (λ) Granule
[e.g., lysozyme, kininogen]

Alpha (α) Granule
[e.g., fibrinogen, p-selectin, vWF, PDGF, Group IIA PLA₂, and likely PMPs and tPMPs]

ACTIVATED

Pseudopodia

GPIIb-IIIa
[FBGN receptor]

Microtubules

P-selectin
[GMP-140]

vWF receptor

B

Infected endothelium

Expression of PLT and PMN adhesins (e.g., vWF, ICAM)

Activation and amplification

Thrombin

Tissue Factor (thromboplastin)

PLT

PMN

PAF

IL-8

C

PMN recruitment adhesion, activation

PLT adhesion, degranulation

Antimicrobial peptides (eg., tPMPs, PLA₂)
Toxic oxygen metabolites

Extracellular killing

FIGURE 2 (A) Schematic representation of platelets before and after activation. Note changes upon cell activation in cell architecture and surface receptors that facilitate platelet adherence and degranulation. Also note the presence of antimicrobial peptides and proteins (e.g., platelet microbicidal proteins [PMPs], thrombin-induced PMPs [tPMPs], and group IIA PLA2) believed to be stored in the alpha granule. PLT, platelets. (B) Cross-talk between vascular endothelium, platelets (PLT), and PMN in response to localized infection. Infected endothelium expresses products that either directly (e.g., platelet activating factor [PAF] and interleukin-8 [IL-8]) or indirectly (tissue factor) trigger platelet and PMN recruitment and activation and up-regulate surface receptors for (activated) platelets and PMN. (C) Recruitment, adherence, and activation of platelets and PMN (and endothelium) in juxtaposition to adherent bacteria. Degranulation and activation of respiratory burst in PMN and NO production by endothelium (not shown) lead to localized extracellular mobilization of antimicrobial peptides and proteins and toxic oxygen and nitrogen metabolites. Note that products of platelet degranulation include proteins known (e.g., platelet factor IV) or believed (e.g., tPMPs, PMPs, PLA2) to recruit and/or potentiate antimicrobial functions of PMN. Thus, secreted products may provide mechanisms for extracellular killing of (adherent) bacteria that are refractory to phagocytosis and for enhanced uptake and intracellular destruction of bacteria still susceptible to phagocytosis.

an unusual b-type cytochrome (b_{558}) consisting of a 91-kDa glycoprotein (gp91-$phox$) and a 22-kDa subunit (p22-$phox$). Cytosolic components of the oxidase include p47-$phox$, p67-$phox$, p40-$phox$, and Rac1 or Rac2, low-molecular-weight G proteins. gp91-$phox$ has the intrinsic molecular features needed to reduce oxygen univalently via NADPH, but the other components are necessary for intracellular stability (p22-$phox$) and optimal oxidase activity. Thus, acquired gene defects in either the X-linked gp91-$phox$ or in

autosomal genes encoding p22-, p47-, or p67-$phox$ confer a CGD phenotype, although disease severity can vary markedly (14, 15).

Before cell activation, most cytochrome b_{558} resides within secondary granules, with additional subpools in secretory vesicles and in the plasma membrane (42). During phagocytosis, cytochrome b_{558} is mobilized from these sites to the forming phagocytic vacuole or phagolysosome, where it serves as a docking site for the cytosolic oxidase components. Multiple sites of contact between the cytochrome and the soluble oxidase components have been implicated, including SH3 domains in p47- and p67-*phox* and proline-rich targets of SH3 domains in p22-, p47-, and p67-*phox*. The importance of these sites in oxidase assembly and function have been most dramatically illustrated in a CGD patient who had a homozygous point mutation in p22-*phox*. The mutation (P156Q) disrupted a proline-rich region in p22-*phox* normally involved in interactions with an SH3 domain of p47-*phox* during oxidase activation (15).

Full assembly of the respiratory burst oxidase appears to occur exclusively at membrane sites contiguous with the attached or internalized target bacterium. No assembly or oxidase activity is detected at remaining secondary granules (52). The possible roles of the submembranous actin-based cytoskeleton and/or in situ alterations of cytochrome b_{558} conformation in specifying the sites of oxidase assembly and activation within the cell are under study. Based on the predicted topology of gp91-*phox*, O_2^- is released either within the lumen of the phagosome during phagocytosis or across the extracytoplasmic leaflet of the plasma membrane when particle attachment and cell activation are not coupled to particle uptake.

Bactericidal Reactive Oxidants

Which products of the oxidative response are responsible for bacterial killing and how are still unclear (25). Most O_2^- formed by the active oxidase is dismutated to H_2O_2. Neither O_2^- nor H_2O_2, at the levels produced, is sufficient to directly account for O_2-dependent killing of staphylococci. A variety of secondary oxidants derived from O_2^- or H_2O_2 with potent bactericidal (cytotoxic) properties have been described, including hydroxyl radical, singlet oxygen, hypochlorous acid, *N*-chloramines, and peroxynitrite (25). Some products (e.g., hydroxyl radical, peroxynitrite) are very short-lived and have a very short range of action, whereas others (e.g., *N*-chloramines) are more long-lived and hence potentially able to act at a distance. Together, these products may target a wide range of cellular constituents, including DNA, lipids, protein thiols, and Fe/S centers, accounting for their cytotoxicity against virtually all microbes but potentially host cells as well. A protective role of these oxidants in host defense is achieved by coupling oxidase activation to microbial recognition and constraining the half-life of the oxidase and of the oxidants generated. The prominence and function of specific oxidants, either within phagolysosomes or at interfaces containing adherent bacteria, remain uncertain. However, chlorination of ingested particles, including *S. aureus*, has been demonstrated and likely reflects the action of HOCl. Levels of bacterial chlorination during killing of ingested *S. aureus* by PMN and of bacteria exposed to reagent HOCl are similar, suggesting that enough HOCl is produced during phagocytosis to be responsible for bacterial killing. Killing of *S. aureus* by PMN in vitro is substantially MPO dependent (24), consistent with the catalytic role of MPO in formation of HOCl from H_2O_2 and Cl^-. MPO (and HOCl) can also react with many other anions, including nitrite. The latter may provide a mechanism whereby NO and 2° reactive nitrogen oxides (25) are produced in settings (e.g., human phagocytes) where endogenous NO synthase activity may be limiting. However, in macrophages where MPO is limiting (15) or at infectious sites where other host cells expressing NO synthase are prominent (e.g., endothelium), other oxidants, including peroxynitrite derived from reaction of O_2^- and NO (4), may play a greater role. This may account, in part, for the much greater susceptibility to staphylococcal infection in CGD than in MPO deficiency.

Oxygen-Independent Systems

A wide variety of mammalian (poly)peptides has been described that can kill staphylococci and many other gram-positive bacteria under arbitrary in vitro conditions (18). The ability of these proteins, alone or in concert with other host defense systems, to exert antibacterial effects under conditions that much more closely simulate natural infectious sites requires much more study. We focus our description below on those agents whose mechanism of action against *S. aureus* has been most extensively studied.

Group IIA PLA2

Mammals, including humans, produce many different proteins with PLA2 activity, including a subset of low-M_r (13,000 to 18,000) enzymes (11, 64). Humans express at least four different isoforms of the low-M_r PLA2 (64). Hallmarks of these proteins include a highly compact three-dimensional structure stabilized by 6 to 8 disulfides and a Ca^{2+}-dependent catalytic machinery. Whereas the overall structure and catalytic properties of these enzymes are closely similar, their sites of expression and mobilization and apparent biological functions differ greatly. Thus, within granules of phagocytes and platelets and in inflammatory fluids, the most prominent isoform is the group IIA PLA2 (33, 40, 79). This isoform alone expresses potent antibacterial activity against gram-negative bacteria in concert with other host defense systems and independently against *S. aureus* and many other gram-positive bacteria (71, 72, 79). The 90% lethal dose (LD_{90}) of group IIA PLA2 toward *S. aureus* is ca. 1 nM in artificial medium (e.g., tissue culture fluid) and in natural biological fluids. In contrast to the properties of many other antibacterial peptides and proteins, the group IIA PLA2 is fully active in media containing physiological levels of monovalent and divalent cations. Many other species of gram-positive bacteria are highly sensitive to group IIA PLA2, including *S. epidermidis*, alpha- and beta-hemolytic streptococci (e.g., viridans, pneumococci, groups A and B), enterococci (*E. faecalis* and *E. faecium*), and *Listeria monocytogenes* (27, 56, 72). In contrast, host cells are highly resistant to group IIA PLA2, at least under normal conditions (40, 51). In the venom of many snakes, homologs of mammalian group IIA PLA2 are present that have potent neuro-, cardio-, or myotoxic effects against mammalian targets but are inactive against *S. aureus* even at high (100 μM) concentrations (72). Work is ongoing to better define the structural characteristics of these related PLA2s that account for the specific cytotoxic properties of snake venom PLA2 toward mammalian targets and of group IIA PLA2 against bacteria.

Bacterial killing by PLA2 involves several stepwise interactions with the bacterial envelope, including binding, penetration of the cell wall, degradation of membrane phospholipids, and activation of autolysins (20). Whereas group IIA PLA2 is potently active against intact gram-positive bacteria such as *S. aureus*, protoplasts fully denuded of cell wall are susceptible to all low-M_r PLA2s (46). Thus, the selective ability of group IIA PLA2 to attack gram-positive bacteria reflects the selective ability of this subset of enzymes to bind and penetrate the cell wall to gain access to phospholipids in the cell membrane. These initial steps depend on the highly cationic properties of the group IIA PLA2 (net charge in different species ranges from +12 to +17), a property unique to this isoform. Mutations that either reduce the surface basicity of the enzyme or disrupt the catalytic machinery greatly reduce or eliminate antibacterial activity, reflecting the importance of both the cationic and catalytic properties in the potent antibacterial activity of group IIA PLA2 (46, 72).

The ability of bound PLA2 to penetrate the cell wall to attack phospholipids in the cell membrane is diminished in nongrowing bacteria and enhanced in bacteria treated with β-lactams, even at subinhibitory doses (20). The binding of the cationic PLA2 to anionic moieties in the cell wall (e.g., glycerol-phosphate polymer of lipoteichoic acid) may cause displacement and local activation of autolysins and thus promote penetration of PLA2. Envelope sites of new cell wall synthesis and remodeling or of β-lactam-induced reduced cross-linking likely represent the sites most permissive for PLA2 penetration (20). Bacterial killing follows degradation of virtually all phospholipids present in the bacterial membrane before PLA2 treatment but apparently also requires diffuse activation of autolysins. Autolysin-deficient bacteria can tolerate massive phospholipid degradation, implying impressive reparative abilities of PLA2-treated bacteria, before autolysin activation leads to irreversible cell envelope alterations (20).

In PMN-rich acute inflammatory exudates, levels of group IIA PLA2 are far higher in the inflammatory fluid than within the phagocytes, suggesting that antibacterial action by PLA2 may be mainly exerted in the extracellular fluid (79). However, PLA2 bound to extracellular bacteria can be carried piggyback into the phagocytic vacuole during phagocytosis, supplementing limiting amounts of PLA2 derived from the granules and increasing intracellular bacterial phospholipolysis and destruction (74, 80). At least some bacteria resistant to phagocytosis remain sensitive to extracellular PLA2. Encapsulated and nonencapsulated S. aureus are equally sensitive to the PLA2, although bacteria surrounded by much larger capsular layers (e.g., type III pneumococci) seem much more resistant (71, 72). Bacterial clumping induced by cross-linking of bacteria to each other through multivalent interactions with specific plasma proteins (e.g., fibrinogen) or to surfaces coated with plasma or matrix proteins is a hallmark of S. aureus and may be a progenitor to abscess and biofilm formation. Clumped bacteria that are likely fully refractory to phagocytosis are still (somewhat) sensitive to extracellular PLA2, depending on the extent of clumping (16). Also depending on the extent of clumping, proteases, such as plasmin, can penetrate and dissolve protein cross-links holding bacterial clumps together and thereby increase bacterial sensitivity to PLA2 action. Thus, the extracellular mobilization of group IIA PLA2 during inflammation, together with local activation of proteolytic activity, could provide a mechanism by which the host can control proliferation and survival of S. aureus even after bacterial clumping.

PMPs

Stimulation of rabbit or human platelets with thrombin, as naturally occurs at sites of endovascular damage, leads to extracellular accumulation of potent antimicrobial activity against gram-positive pathogens (83). Several antimicrobial (poly)peptides are released from platelets treated with thrombin. They were originally termed thrombodefensins since they were believed to represent the platelet version of neutrophil defensins (8). However, the major antibac-

terial substances toward gram-positive bacteria that are present in these secretions, the so-called PMPs and group IIA PLA2, are unrelated to defensins.

The PMPs within rabbit platelets compose an array of seven antimicrobial peptides (87). These peptides are all cationic, heat-stable, and range in molecular mass from 6.0 to 9.0 kDa, thus distinct in size from either defensins or group IIA PLA2. Other structural features that distinguish PMPs from defensins include the prominence of lysine residues and the presence of two to four, rather than six, cystines in the PMPs (26). Of the seven PMPs, five are isolated by acid extraction of washed platelets (PMPs 1 to 5), whereas two other peptides (tPMP-1 and tPMP-2) are released following thrombin stimulation. The latter two peptides differ by only a single amino acid (glycine in tPMP-1 and arginine in tPMP-2); preliminary sequence data for PMPs 1 to 5 suggest that they also derive from distinct genes. Human platelets contain antimicrobial peptides analogous to PMPs 1 to 5 and the tPMPs (61). These human PMPs have been termed thrombocidins. They contain the chemokine C-X-C motif and appear to be derived from known platelet proteins (platelet basic protein, neutrophil-activating peptide II, and connective tissue-activating peptide III). The presence of the C-X-C chemokine motif strongly suggests that, in addition to their direct antimicrobial properties, these peptides play an indirect host defense function via their chemoattractant properties for inflammatory cells (e.g., PMN).

All seven PMPs are active against S. aureus and Candida albicans in nanomolar concentrations. PMPs 1 to 5 are more active at acid pH, whereas tPMP-1 and -2 are more active at neutral pH. Of these peptides, the antimicrobial action of tPMP-1 has been studied most intensively. Except for the enterococci, tPMP-1 is microbicidal against most of the clinically relevant gram-positive pathogens tested to date, including S. aureus, S. epidermidis, the viridans streptococci, and also Candida sp. (85, 87). In vitro studies of tPMP-1 susceptibility of bacteremic isolates of S. aureus, S. epidermidis, and the viridans streptococci from patients with either infectious endocarditis-related or -unrelated infections revealed that isolates from patients with infectious endocarditis tended to be substantially more resistant in vitro to tPMP-1 than were isolates not associated with infectious endocarditis (81). A recent follow-up study of 60 bacteremic isolates from a single medical center (Duke University, Durham, N.C.) also revealed that strains from patients with infectious endocarditis or vascular catheter infections were significantly more resistant in vitro to tPMP-1 than were isolates arising from soft tissue abscesses (2). These results suggest that tPMP-1 resistance provides the invading microbe with a survival advantage at sites of endovascular damage.

Within 5 to 15 min of exposure of S. aureus to tPMP-1, perturbations of the cell membrane are manifested by ultrastructural analysis, suggesting that, like other endogenous antimicrobial peptides, tPMP-1 might target the microbial cell membrane as part of its lethal pathway (84). In support of this hypothesis, auxotrophs of S. aureus that exhibit defects in their ability to generate a normal transmembrane electric potential ($\Delta\psi$) are more resistant to killing by tPMP-1 than are their parental counterparts with normal $\Delta\psi$. Dietary supplementation (e.g., menadione) of the $\Delta\psi$-deficient mutants to meet their auxotrophic requirements restores the $\Delta\psi$ to near parental levels and reconstitutes tPMP-1 susceptibility (84). The ability of tPMP-1 to increase bacterial membrane permeability to propidium iodide is also diminished in the $\Delta\psi$ mutants and restored by addition of menadione (84). Studies by Koo et al. (37) have suggested that the overall proton motive force ($\Delta\psi$ and ΔpH) is important in tPMP-1-induced microbicidal action. The dependence of tPMP-1 on $\Delta\psi$ for its microbicidal action further distinguishes this peptide from neutrophil defensins, which appear to kill *S. aureus* in a $\Delta\psi$-independent manner. Addition of tPMP-1 to model planar lipid bilayers causes membrane permeabilization and disruption of the bilayer in a voltage-dependent manner with maximal effects induced at a trans-negative voltage orientation relative to the site of addition of the peptide (39). Well-defined, voltage-gated pores formed by defensins in similar model membranes are not formed by tPMP-1. The staphylocidal activity of tPMP-1 is substantially reduced under in vitro conditions at which microbial membrane energetics are reduced (e.g., stationary phase, low temperature) (38). Finally, cytoplasmic membrane fluidity of tPMP-1-resistant strains of *S. aureus* obtained either by serial passage, transposon mutagenesis, or plasmid carriage is markedly different from that of their genetically related tPMP-1-susceptible counterparts.

Although initial effects of tPMP-1 target the bacterial cell membrane, the discrepancy between early membrane effects of tPMP-1 (which occur within minutes of exposure) and the delayed microbicidal effects (which require 1 to 2 h of treatment) suggests that other targets are involved in the lethal pathway of tPMP-1. Pretreatment of tPMP-1-susceptible strains with antibiotics that block either 50S ribosome-dependent protein synthesis or DNA gyrase subunit B actions completely inhibit tPMP-1-induced microbicidal effects. These studies support the concept that tPMP-1-induced lethality involves intracellular targets as well as membrane perturbations.

In addition to its direct microbicidal effects against gram-positive pathogens, tPMP-1 also causes prolonged postexposure growth-inhibitory effects against staphylococci, similar to the effects of cell wall-active antibiotics such as oxacillin and vancomycin (86). In contrast to the microbicidal effects of tPMP-1, these growth-inhibitory properties of the peptide are observed in both tPMP-1-susceptible and -resistant strains. Simultaneous exposure of staphylococci to tPMP-1 and antibiotics such as oxacillin produces synergistic bactericidal effects. In addition, preexposure of *S. aureus* or *Candida* sp. to tPMP-1 significantly reduces the capacity of these organisms to bind to platelets in vitro, an effect that can be further magnified by exposure to other antimicrobial agents (88).

The contribution of PMPs to extracellular host defenses against adherent bacteria depends on mechanisms that induce release of these antibacterial peptides. To date, three physiologically relevant stimuli have been shown to prompt the release of one or more PMPs in vitro. Each of these stimuli is associated with either endothelial damage or microbial colonization of such sites. As noted above, thrombin is a potent stimulus for the release of tPMP-1 and tPMP-2 from rabbit platelets (87). *S. aureus*-induced rabbit platelet aggregation also results in secretion from platelets of substances with biologic activity and electrophoretic properties compatible with PMPs. Moreover, in the presence of staphylococcal alpha-toxin, a major secretory virulence factor of *S. aureus*, rabbit platelets are lysed coincident with the release of PMP-1 and PMP-2 (3).

Defensins

Defensins are a family of highly abundant 3.5- to 4.5-kDa cationic peptides (44). Their concentration within the primary granules of PMN is extraordinarily high. Defensins

are also present in many mucosal secretions and in certain macrophages. Two closely related subfamilies of peptides have been identified, so-called α- and β-defensins, with distinct spacing and arrangement of three disulfides that stabilize an otherwise similar triple-stranded antiparallel β-sheet with cationic/hydrophobic (amphiphilic) character (90). Both α- and β-defensins have been identified in primary granules of PMN and in mucosal secretions, but α-defensins are generally in myeloid cells whereas β-defensins are often of epithelial origin. In humans, four α-defensins, HNP (human neutrophil defensin) 1 to 4, are found in neutrophils, two different α-defensins (HNP-5 and -6) are in small intestinal Paneth cells, and human β-defensin-1 is highly expressed in kidney, pancreas, and, to a lesser extent, other epithelial cells (89). Mucosal β-defensins are induced by inflammatory stimuli (13, 58) and appear to play a role in bactericidal activity of human lung fluid (23). All known human defensin genes are encoded within a cluster of genes on chromosome 8p23. Defensins and defensinlike cysteine-rich peptides are widely expressed ancient components of antimicrobial host defenses.

The α-defensins are synthesized as ca. 95-residue noncytotoxic preforms (21). Efficient sorting and cleavage to generate the mature cytotoxic peptides present in the granules depends on the propiece. In addition, the net charge properties of the anionic propiece and cationic mature peptide have coevolved to maintain the charge neutrality of the proform despite wide variation in the net (+) charge of the mature defensin. This charge neutralization within the nascent polypeptide is thought to protect the producing cell (e.g., PMN) from the possible toxicity of the mature peptide during biogenesis and targeting of defensins for storage in granules (65). Abundant acidic glycosaminoglycans may serve a similar function in the granules (18).

Defensins are broadly cytotoxic toward bacteria, especially gram-positive bacteria, fungi, metazoan parasites, and mammalian cells (18, 44). Hence, the cytotoxicity of defensins lacks great intrinsic specificity. However, storage of these peptides within granules and delivery to phagolysosomes (Fig. 1) should direct the peptides' cytotoxicity more selectively toward the ingested microbial prey. Independent microbicidal effects are expressed in vitro at micromolar concentrations under hypotonic conditions, but much higher concentrations of defensins are required when physiological concentrations of salts are present. Thus, secreted (i.e., diluted) defensins are much less likely to exert cytotoxic effects unless retained within a microcompartment. Secreted defensins may also be trapped within inactive complexes by other host proteins in plasma and other body fluids, further reducing the likelihood of extracellular cytotoxic action (18). However, at least certain defensins can act in synergy with other host defense agents, which may permit these peptides to contribute to antibacterial host defense at concentrations that have no independent antibacterial activity (45).

The more highly cationic (type 1) α-defensins (e.g., rabbit defensins NP-1 and NP-2) do not require target pathogens to be metabolically active to produce a microbicidal effect. In contrast, the less cationic type 2 α-defensins (e.g., HNP-1 and -2) are maximally active against metabolically active targets. These features have important implications in antimicrobial host defense, since infecting microbes may vary in their envelope charge and/or metabolic activities, particularly within distinct infection sites and different microenvironments of the host. For example, persistent *S. aureus* small-colony variants exhibit reduced oxygen consumption, ATP production, and membrane bioenergetics (55). Thus, an array of defensins exist that are likely suited to act against a broad spectrum of microbes, including those of differing metabolic status.

Defensins target and disrupt the bacterial cytoplasmic membrane (59). Staphylococcal membranes exhibit blebbing and distortion after exposure to rabbit PMN granule extracts rich in defensins, group IIA PLA2, and other cationic proteins (70). Initial interactions are likely driven by electrostatic attraction between the cationic defensins and anionic surface lipids. Insertion of bound defensins into model membranes is driven by transmembrane potential and involves the formation of multimeric pores (77). Differences in effects on model membranes between particular defensin species appear to correlate with the propensity of the defensin to form stable dimers in the context of lipid environments, simulating the bacterial cytoplasmic membrane (29). However, these differences do not obviously correlate with differences in cytotoxic activity.

Several other effects of defensins have been described in vitro that are independent of their cytotoxic activities but could contribute to a role in host responses to infection and injury. These include chemotactic activity toward monocytes (62) and T cells (7), antagonism of ACTH action at its receptor, stimulation of epithelial cell growth, and secretion of interleukin-8 and inhibition of fibrinolysis, the latter suggesting a mechanism by which PMN could maintain a fibrin barrier at sites of infection (18). Apparently subcytotoxic doses of HNP-3 have protective effects in experimental infections (including S. aureus) in mice, possibly by promoting leukocyte accumulation at sites of infection (76).

Other Agents

Recent advances in the study of host defenses have revealed a remarkable array of antibiotic peptides and proteins present not only within granules of professional phagocytes but also along many epithelial layers lining the internal and external surfaces of the body. Lysozyme, the first mammalian antibacterial protein discovered, is generally limited in its independent antibacterial action to nonpathogenic gram-positive bacteria. High-level resistance to lysozyme, exhibited by most gram-positive pathogens, reflects extensive cross-linking and other modifications of the peptidoglycan matrix, which impedes access of lysozyme to its substrate (GlcNAc-MurNAc) in the bacterial cell wall (18).

SUMMARY AND PERSPECTIVES

There has been great progress in the molecular definition and characterization of mechanisms of host defense. In some cases, this has led to a molecular understanding of specific host defense defects (e.g., CGD) that are associated with greatly increased susceptibility to staphylococcal infections. The ability of knockout mice to substantially reproduce the phenotype of CGD offers hope that similar approaches will be useful in testing the role of other host defense systems when analogous human deficiencies are either unknown or nonexistent. Mice rendered devoid of PMN elastase activity by this experimental approach exhibit an unexpected vulnerability to gram-negative bacterial infections but not to staphylococcal infections (6), further supporting the belief inferred from in vitro studies that the ability to defend against a diverse array of micro-

organisms is in part effected by the existence of a multiplicity of defense systems. However, the power of these approaches should not be overstated. In many instances, experimental infections in mice provide poor models of human infections, perhaps, in part, reflecting significant differences in the arsenal of mouse and human defenses and the handling by these different animal species of important bacterial products (18). In the case of elastase function in host defense, in vitro studies have not yielded compelling evidence of antibacterial function, leaving unsettled whether the effects of elastase in mice in vivo are direct or indirect. Even in CGD, the actual means by which normal phagocytes utilize reactive oxidants to kill ingested staphylococci remains unknown. Moreover, in the specific context of staphylococcal infections in which extracellular bacteria can occupy niches (e.g., abscesses, biofilms) that are likely refractory to phagocytes, a better understanding of the action of extracellular host defenses in these settings is needed. As the problem of antibiotic resistance among staphylococci continues to grow (69), the study of endogenous antibiotics that normally function in the extracellular environment of the host may also lead to the identification and development of novel therapeutic compounds.

REFERENCES

1. Babior, B. 1999. Review: NADPH oxidase: an update. *Blood* **93:**1464–1476.

2. Bayer, A. S., D. Cheng, M. R. Yeaman, G. R. Corey, R. S. McClelland, L. J. Harrel, and V. G. Fowler, Jr. 1998. In vitro resistance to thrombin-induced microbicidal protein among clinical bacteremic isolates of *Staphylococcus aureus* correlates with an endovascular infectious source. *Antimicrob. Agents Chemother.* **42:**3169–3172.

3. Bayer, A. S., M. D. Ramos, B. E. Menzies, M. R. Yeaman, A. Shen, and A. L. Cheung. 1997. Hyperproduction of alpha-toxin by *Staphylococcus aureus* results in paradoxically reduced virulence in experimental endocarditis—host defense role for platelet microbicidal proteins. *Infect. Immun.* **65:**4652–4660.

4. Beckman, J. S., and W. H. Koppenol. 1996. Nitric oxide, superoxide, and peroxynitrite: the good, the bad, and ugly. *Am. J. Physiol.* **271:**C1424–C1437.

5. Beekhuizen, H., J. S. van de Gevel, B. Olsson, I. J. van Benten, and R. van Furth. 1997. Infection of human vascular endothelial cells with *Staphylococcus aureus* induces hyperadhesiveness for human monocytes and granulocytes. *J. Immunol.* **158:**774–782.

6. Belaaouaj, A., R. McCarthy, M. Baumann, Z. Gao, T. Ley, S. Abraham, and S. Shapiro. 1998. Mice lacking neutrophil elastase reveal impaired host defense against gram-negative bacterial sepsis. *Nat. Med.* **4:**615–618.

7. Chertov, O., D. F. Michiel, L. Xu, J. M. Wang, K. Tani, W. J. Murphy, D. L. Longo, D. D. Taub, and J. J. Oppenheim. 1996. Identification of defensin-1, defensin-2 and CAP37/azurocidin as T-cell chemoattractant proteins released from interleukin-8-stimulated neutrophils. *J. Biol. Chem.* **271:**2935–2940.

8. Dankert, J. 1988. Role of platelets in early pathogenesis of viridans group streptococcal endocarditis: a study of thrombodefensins. Ph.D. thesis. University of Groningen, The Netherlands.

9. **Dankert, J., J. van der Werff, S. A. J. Zaat, W. Joldersma, D. Klein, and J. Hess.** 1995. Involvement of bactericidal factors from thrombin-stimulated platelets in clearance of adherent viridans streptococci in experimental infective endocarditis in rabbits. *Infect. Immun.* **63:**663–671.

10. **De Kimpe, S. J., M. Kengatharan, C. Thiemermann, and J. R. Vane.** 1995. The cell wall components peptidoglycan and lipoteichoic acid from *Staphylococcus aureus* act in synergy to cause shock and multiple organ failure. *Proc. Natl. Acad. Sci. USA* **92:**10359–10363.

11. **Dennis, E. A.** 1997. The growing phospholipase A2 superfamily of signal transduction enzymes. *Trends Biochem.* **22:**1–2.

12. **Dhawan, V. G., A. S. Bayer, and M. R. Yeaman.** 1998. Influence of in vitro susceptibility to thrombin-induced platelet microbicidal protein on the progression of experimental *Staphylococcus aureus* endocarditis. *Infect. Immun.* **66:**3476–3479.

13. **Diamond, G., J. P. Russell, and C. L. Bevins.** 1996. Inducible expression of an antibiotic peptide gene in lipopolysaccharide-challenged tracheal epithelial cells. *Proc. Natl. Acad. Sci. USA* **93:**5156–5160.

14. **Dinauer, M.** 1993. The respiratory burst oxidase and the molecular genetics of chronic granulomatous disease. *Crit. Rev. Clin. Lab. Sci.* **30:**329–369.

15. **Dinauer, M. C., W. M. Nauseef, and P. E. Newburger.** 1999. Inherited disorders of phagocyte killing. *In* A. L. Beaudet, W. S. Sly, D. Vallee, B. Vogelstein, and B. Childs (ed.), *The Metabolic Basis of Inherited Disease.* 8th ed. McGraw-Hill Book Co., New York, N.Y.

16. **Dominiecki, M. E., and J. Weiss.** 1999. Antibacterial action of extracellular mammalian group IIA phospholipase A2 against grossly clumped *Staphylococcus aureus.* *Infect. Immun.* **67:**2299–2305.

17. **Donaldson, D. M., and J. G. Tew.** 1977. Beta-lysin of platelet origin. *Bacteriol. Rev.* **41:**501–513.

18. **Elsbach, P., J. Weiss, and O. Levy.** 1999. Oxygen-independent antimicrobial systems of phagocytes, p. 801–817. *In* J. I. Gallin, R. Snyderman, and C. Nathan (ed.), *Inflammation. Basic Principles and Clinical Correlates,* 3rd ed. Lippincott-Raven, New York, N.Y.

19. **Ferrante, A., A. J. Martin, E. J. Bates, D. H. B. Goh, D. P. Harvey, D. Parsons, D. A. Rathjen, G. Russ, and J.-M. Dayer.** 1993. Killing of *Staphylococcus aureus* by tumor necrosis factor-α-activated neutrophils. *J. Immunol.* **151:**4821–4828.

20. **Foreman-Wykert, A. K., Y. Weinrauch, P. Elsbach, and J. Weiss.** 1999. Cell-wall determinants of the bactericidal action of group IIA phospholipase A2 against Gram-positive bacteria. *J. Clin. Invest.* **103:**715–721.

21. **Ganz, T.** 1994. Biosynthesis of defensins and other antimicrobial peptides. *Ciba Found. Symp.* **186:**62–71.

22. **Ganz, T., and J. Weiss.** 1997. Antimicrobial peptides of phagocytes and epithelia. *Semin. Hematol.* **34:**343–354.

23. **Goldman, M. J., G. M. Anderson, E. D. Stolzenberg, U. P. Kari, M. Zasloff, and J. M. Wilson.** 1997. Human beta-defensin-1 is a salt-sensitive antibiotic in lung that is inactivated in cystic fibrosis. *Cell* **88:**553–560.

24. **Hampton, M. B., A. J. Kettle, and C. C. Winterbourn.** 1996. Involvement of superoxide and myeloperoxidase in oxygen-dependent killing of *Staphylococcus aureus* by neutrophils. *Infect. Immun.* **64:**3512–3517.

25. **Hampton, M. B., A. J. Kettle, and C. C. Winterbourn.** 1998. Inside the neutrophil phagosome: oxidants, myeloperoxidase, and bacterial killing. *Blood* **92:**3007–3017.

26. **Harwig, S. S., T. Ganz, and R. I. Lehrer.** 1994. Neutrophil defensins. Purification, characterization and antimicrobial testing. *Methods Enzymol.* **236:**160–176.

27. **Harwig, S. S., L. Tan, X. D. Qu, Y. Cho, P. B. Eisenhauer, and R. I. Lehrer.** 1995. Bactericidal properties of a murine intestinal phospholipase A2. *J. Clin. Invest.* **95:**603–610.

28. **Heumann, D., C. Barras, A. Severin, M. P. Glauser, and A. Tomasz.** 1994. Gram-positive cell walls stimulate synthesis of tumor necrosis factor alpha and interleukin-6 by human monocytes. *Infect. Immun.* **62:**2715–2721.

29. **Hristova, K., M. E. Selsted, and S. H. White.** 1996. Interactions of monomeric rabbit neutrophil defensin with bilayers: comparison with dimeric human defensin HNP-2. *Biochemistry* **35:**11888–11894.

30. **Hultgren, O., H.-P. Eugster, J. D. Sedgwick, H. Körner, and A. Tarkowski.** 1998. TNF/lymphotoxin-α double mutant mice resist septic arthritis but display increased mortality in response to *Staphylococcus aureus.* *J. Immunol.* **161:**5937–5942.

31. **Ing, M. B., L. Baddour, and A. S. Bayer.** 1997. Staphylococcal bacteremia and infective endocarditis—pathogenesis, diagnosis and complications. *In* G. Archer and K. Crossley (ed.), *Staphylococci and Staphylococcal Diseases.* Churchill-Livingstone Publishers, New York, N.Y.

32. **Jackson, S. H., J. I. Gallin, and S. M. Holland.** 1995. The p47phox mouse knock-out model of chronic granulomatous disease. *J. Exp. Med.* **182:**751–758.

33. **Kallajoki, M., and T. J. Nevalainen.** 1997. Expression of Group II phospholipase A2 in human tissues, p. 8–16. *In* W. Uhl, T. J. Nevalainen, and M. W. Buchler (ed.), *Phospholipase A2: Basic and Clinical Aspects in Inflammatory Diseases,* vol. 24. S. Karger, Basel, Switzerland.

34. **Kapral, F. A.** 1966. Clumping of *Staphylococcus aureus* in the peritoneal cavity of mice. *J. Bacteriol.* **92:**1188–1195.

35. **Karakawa, W. W., A. Sutton, R. Schneerson, A. Karpas, and W. F. Vann.** 1988. Capsular antibodies induce type-specific phagocytosis of capsulated *Staphylococcus aureus* by human polymorphonuclear leukocytes. *Infect. Immun.* **56:**1090–1095.

36. **Katz, S. S., Y. Weinrauch, R. S. Munford, P. Elsbach, and J. Weiss.** Lipopolysaccharide deacylation following extracellular or intracellular killing of *Escherichia coli* by rabbit inflammatory exudates. Submitted for publication.

37. **Koo, S.-P., A. S. Bayer, R. A. Proctor, H.-G. Sahl, and M. R. Yeaman.** 1996. Staphylocidal action of platelet microbicidal protein is not solely dependent on intact transmembrane potential. *Infect. Immun.* **60:**1070–1074.

38. **Koo, S.-P., M. R. Yeaman, and A. S. Bayer.** 1996. Staphylocidal action of platelet microbicidal protein is modified by microenvironment and target cell growth phase. *Infect. Immun.* **64:**3758–3764.

39. **Koo, S.-P., M. R. Yeaman, C. C. Nast, and A. S. Bayer.** 1997. The bacterial cell membrane is a principal target for the staphylocidal action of thrombin-induced platelet microbicidal protein. *Infect. Immun.* **65:**4795–4800.

40. **Kudo, I., M. Murakami, S. Hara, and K. Inoue.** 1993. Mammalian non-pancreatic phospholipases A2. *Biochim. Biophys. Acta* **1171:**217–231.

41. Kusonoki, T., E. Hailman, T. S. C. Juan, H. S. Lichenstein, and S. D. Wright. 1995. Molecules from *Staphylococcus aureus* that bind CD14 and stimulate innate immune responses. *J. Exp. Med.* **182:**1673–1682.

42. Lee, J. C. 1996. The prospects for developing a vaccine against *Staphylococcus aureus*. *Trends Microbiol.* **4:**162–166.

43. Lee, J. C., J.-S. Park, S. E. Shepherd, V. Carey, and A. Fattom. 1997. Protective efficacy of antibodies to the *Staphylococcus aureus* type 5 capsular polysaccharide in a modified model of endocarditis in rats. *Infect. Immun.* **65:**4146–4151.

44. Lehrer, R. I., and T. Ganz. 1996. Endogenous vertebrate antibiotics: defensins, protegrins, and other cysteine-rich antimicrobial peptides. *Ann. N.Y. Acad. Sci.* **797:**228–239.

45. Levy, O., C. E. Ooi, J. Weiss, R. I. Lehrer, and P. Elsbach. 1994. Individual and synergistic effects of rabbit granulocyte proteins on *Escherichia coli*. *J. Clin. Invest.* **94:**672–682.

46. Liang, N. S., R. Koturi, L. M. Madsen, M. Gelb, and J. Weiss. Structural properties of target and effector that determine the selective ability of Group IIA phospholipase A2 to act against Gram-positive bacteria. Submitted for publication.

47. Liang, N. S., J. C. Lee, and J. Weiss. Mobilization of rat Group IIA phospholipase A2 in response to sub-lethal *Staphylococcus aureus* infection. Submitted for publication.

48. Lowy, F. D. 1998. Medical progress: *Staphylococcus aureus* infections. *N. Engl. J. Med.* **339:**520–532.

49. McInnes, I. B., B. Leung, X. Q. Wei, C. C. Gemmell, and F. Y. Liew. 1998. Septic arthritis following *Staphylococcus aureus* infection in mice lacking inducible nitric oxide synthase. *J. Immunol.* **160:**308–315.

50. Menzies, B. E., and I. Kourteva. 1998. Internalization of *Staphylococcus aureus* by endothelial cells induces apoptosis. *Infect. Immun.* **66:**5994–5998.

51. Murakami, M., Y. Nakatani, and I. Kudo. 1996. Type II secretory phospholipase A2 associated with cell surfaces via C-terminal heparin-binding lysine residues augments stimulus-initiated delayed prostaglandin generation. *J. Biol. Chem.* **271:**30041–30051.

52. Nauseef, W., B. Volpp, S. McCormick, K. Leidal, and R. Clark. 1991. Assembly of the neutrophil respiratory burst oxidase. *J. Biol. Chem.* **266:**5911–5917.

53. Peterson, P. K., B. J. Wilkinson, Y. Kim, D. Schmeling, S. D. Douglas, P. G. Quie, and J. Verhoef. 1978. The key role of peptidoglycan in the opsonization of *Staphylococcus aureus*. *J. Clin. Invest.* **61:**597–609.

54. Pollock, J. D., D. A. Williams, M. A. Gifford, L. L. Li, X. Du, J. Fisherman, S. H. Orkin, C. M. Doerschuk, and M. C. Dinauer. 1995. Mouse model of X-linked chronic granulomatous disease, an inherited defect in phagocyte superoxide production. *Nat. Genet.* **9:**202–209.

55. Proctor, R. A., J. M. Balwit, and O. Vesga. 1994. Variant subpopulations of *Staphylococcus aureus* as a cause of persistent infections. *Infect. Agents Dis.* **3:**302–312.

56. Qu, X. D., and R. I. Lehrer. 1998. Secretory phospholipase A2 is the principal bactericide for staphylococci and other gram-positive bacteria in human tears. *Infect. Immun.* **66:**2791–2797.

57. Quie, P. G., J. G. White, B. Holmes, and R. A. Good. 1967. In vitro bactericidal capacity of human polymorphonuclear leukocytes: diminished activity in chronic granulomatous disease of childhood. *J. Clin. Invest.* **46:**668–679.

58. Schonwetter, B. S., E. D. Stolzenberg, and M. A. Zasloff. 1995. Epithelial antibiotics induced at sites of inflammation. *Science* **267:**1645–1648.

59. Shimoda, M., K. Ohki, Y. Shimamota, and O. Kohashi. 1995. Morphology of defensin-treated *Staphylococcus aureus*. *Infect. Immun.* **63:**2886–2891.

60. Sullam, P. M., U. Frank, M. G. Tauber, M. Yeaman, A. Bayer, and H. F. Chambers. 1993. Effect of thrombocytopenia on the early course of streptococcal endocarditis. *J. Infect. Dis.* **168:**910–914.

61. Tang, Y. Q., M. R. Yeaman, and M. E. Selsted. 1995. Purification, characterization and antimicrobial properties of peptides released from thrombin-induced human platelets. *Blood* **86:**910A.

62. Territo, M. C., T. Ganz, M. E. Selsted, and R. Lehrer. 1989. Monocyte-chemotactic activity of defensins from human neutrophils. *J. Clin. Invest.* **84:**2017–2020.

63. Thakker, M., J.-S. Park, V. Carey, and J. C. Lee. 1998. *Staphylococcus aureus* serotype 5 capsular polysaccharide is antiphagocytic and enhances bacterial virulence in a murine bacteremia model. *Infect. Immun.* **66:**5183–5189.

64. Tischfield, J. A. 1997. A reassessment of the low molecular weight phospholipase A2 gene family in mammals. *J. Biol. Chem.* **272:**17247–17250.

65. Valore, E. V., E. Martin, S. S. Harwig, and T. Ganz. 1996. Intramolecular inhibition of human defensin HNP-1 by its propiece. *J. Clin. Invest.* **97:**1624–1629.

66. Vaudaux, P. E., G. Zulian, E. Huggler, and F. A. Waldvogel. 1985. Attachment of *Staphylococcus aureus* to polymethymethacrylate increases its resistance to phagocytosis in foreign body infection. *Infect. Immun.* **50:**472–477.

67. Verbrugh, H. A., P. K. Peterson, B. Y. Nguyen, S. P. Sisson, and Y. Kim. 1982. Opsonization of encapsulated *Staphylococcus aureus*: the role of specific antibody and complement. *J. Immunol.* **129:**1681–1687.

68. Verdrengh, M., T. A. Springer, J.-C. Gutierrez, and A. Tarkowski. 1996. A role of intercellular adhesion molecule 1 in pathogenesis of staphylococcal arthritis and in host defense against staphylococcal bacteremia. *Infect. Immun.* **64:**2804–2807.

69. Waldvogel, F. A. 1999. New resistance in *Staphylococcus aureus*. *N. Engl. J. Med.* **340:**556–557.

70. Walton, E. 1978. The preparation, properties and action on *Staphylococcus aureus* of purified fractions from the cationic proteins of rabbit polymorphonuclear leukocytes. *Br. J. Exp. Pathol.* **59:**416–431.

71. Weinrauch, Y., C. Abad, N. S. Liang, S. F. Lowry, and J. Weiss. 1998. Mobilization of potent plasma bactericidal activity during systemic bacterial challenge: role of group IIA phospholipase A2. *J. Clin. Invest.* **102:**633–638.

72. Weinrauch, Y., P. Elsbach, L. M. Madsen, A. Foreman, and J. Weiss. 1996. The potent anti-*Staphylococcus aureus* activity of a sterile rabbit inflammatory fluid is due to a 14-kD phospholipase A2. *J. Clin. Invest.* **97:**250–257.

73. Weinrauch, Y., S. S. Katz, R. S. Munford, P. Elsbach, and J. Weiss. 1999. Deacylation of purified lipopolysaccharide by cellular and extracellular components of a ster-

ile rabbit peritoneal inflammatory exudate. *Infect. Immun.* **67:**3376–3382.

74. **Weiss, J., M. Inada, P. Elsbach, and R. M. Crowl.** 1994. Structural determinants of the action against *Escherichia coli* of a human inflammatory fluid phospholipase A2 in concert with polymorphonuclear leukocytes. *J. Biol. Chem.* **269:**26331–26337.

75. **Weiss, J., M. Victor, O. Stendahl, and P. Elsbach.** 1982. Killing of Gram-negative bacteria by polymorphonuclear leukocytes: the role of an O_2-independent bactericidal system. *J. Clin. Invest.* **69:**959–970.

76. **Welling, M. M., P. S. Hiemstra, M. T. van den Barselaar, A. Paulusma-Annema, P. H. Nibbering, E. K. J. Pauwels, and W. Calame.** 1998. Antibacterial activity of human neutrophil defensins in experimental infections in mice is accompanied by increased leukocyte accumulation. *J. Clin. Invest.* **102:**1583–1590.

77. **White, S. H., W. C. Wimley, and M. E. Selsted.** 1995. Structure, function, and membrane integration of defensins. *Curr. Opin. Struct. Biol.* **5:**521–527.

78. **Woodman, R., R. Erickson, J. Rae, H. Jaffe, and J. Curnutte.** 1992. Prolonged recombinant interferon-γ therapy in chronic granulomatous disease: evidence against enhanced neutrophil oxidase activity. *Blood* **79:**1558–1562.

79. **Wright, G., C. E. Ooi, J. Weiss, and P. Elsbach.** 1990. Purification of a cellular (granulocyte) and an extracellular (serum) phospholipase A2 that participate in the destruction of *Escherichia coli* in a rabbit inflammatory exudate. *J. Biol. Chem.* **265:**6675–6681.

80. **Wright, G. C., J. Weiss, K.-S. Kim, H. Verheij, and P. Elsbach.** 1990. Bacterial phospholipid hydrolysis enhances the destruction of *Escherichia coli* ingested by rabbit neutrophils. Role of cellular and extracellular phospholipases. *J. Clin. Invest.* **85:**1925–1935.

81. **Wu, T., M. R. Yeaman, and A. S. Bayer.** 1994. Resistance to platelet microbicidal protein in vitro among bacteremic staphylococcal and viridans streptococcal isolates correlates with an endocarditis source. *Antimicrob. Agents Chemother.* **38:**729–732.

82. **Yao, L., V. Bengualid, F. D. Lowy, J. J. Gibbons, V. B. Hatcher, and J. W. Berman.** 1995. Internalization of *Staphylococcus aureus* by endothelial cells induces cytokine gene expression. *Infect. Immun.* **63:**1835–1839.

83. **Yeaman, M. R.** 1997. The role of platelets in antimicrobial host defense. *Clin. Infect. Dis.* **25:**951–970.

84. **Yeaman, M. R., A. S. Bayer, S.-P. Koo, W. Foss, and P. M. Sullam.** 1998. Platelet microbicidal proteins and neutrophil defensin disrupt the *Staphylococcus aureus* cytoplasmic membrane by distinct mechanisms of action. *J. Clin. Invest.* **101:**178–187.

85. **Yeaman, M. R., A. S. Ibrahim, J. E. Edwards, A. S. Bayer, and M. A. Ghannoum.** 1993. Thrombin-induced platelet microbicidal protein is fungicidal in vitro. *Antimicrob. Agents Chemother.* **37:**546–553.

86. **Yeaman, M. R., D. C. Norman, and A. S. Bayer.** 1992. Platelet microbicidal protein enhances the bactericidal and post-antibiotic effects in *Staphylococcus aureus*. *Antimicrob. Agents Chemother* **36:**1665–1660.

87. **Yeaman, M. R., A. Shen, Y. Tang, A. S. Bayer, and M. A. Selsted.** 1997. Isolation and antimicrobial activity of microbicidal proteins from rabbit platelets. *Infect. Immun.* **65:**1023–1031.

88. **Yeaman, M. R., P. R. Sullam, P. F. Dazin, and A. S. Bayer.** 1994. Platelet microbicidal protein alone and in combination with antibiotics reduces adherence of *Staphylococcus aureus* to platelets in vitro. *Infect. Immun.* **62:**3416–3423.

89. **Zhao, C., I. Wang, and R. I. Lehrer.** 1996. Widespread expression of beta-defensin hBD-1 in human secretory glands and epithelial cells. *FEBS Lett.* **396:**319–322.

90. **Zimmermann, G. R., P. Legault, M. E. Selsted, and A. Pardi.** 1995. Solution structure of bovine neutrophil beta-defensin-12: the peptide fold of the beta-defensin is identical to that of the classical defensins. *Biochemistry* **34:**13663–13671.

Biology and Pathogenicity of *Staphylococcus epidermidis*

CHRISTINE HEILMANN AND GEORG PETERS

46

Coagulase-negative staphylococci (CoNS) form a group of more than 30 defined species. They share many basic biological properties with the much smaller group of coagulase-positive staphylococci typified by *Staphylococcus aureus*, the predominant species. However, CoNS show substantial differences with *S. aureus*, especially with respect to their ecology and their pathogenic potential. The normal habitat of CoNS are skin and mucous membranes of humans and animals. Here they represent a major part of the normal aerobic flora. Only 12 species are normally found in specimens of human origin, of which some occur only rarely. Diagnostically, CoNS are separated from *S. aureus* by their inability to produce free coagulase. They can further be differentiated on the basis of their novobiocin susceptibility: novobiocin-resistant CoNS resemble the *Staphylococcus saprophyticus* group, and novobiocin-susceptible CoNS mainly resemble the *Staphylococcus epidermidis* group (Table 1). By far the most frequently isolated species, from normal flora as well as from clinical specimens, is *S. epidermidis*.

This chapter deals with the current knowledge about *S. epidermidis*. It especially focuses on the pathogenicity of *S. epidermidis*, the underlying biological properties, and how these contrast with those of *S. aureus*. The basic biology is covered in other chapters, but a brief overview of the disease spectrum is given to place the studies on pathogenesis in perspective. The last part of the chapter deals with one ecological aspect, lantibiotics, which are potentially important for bacterial interference on skin and mucous membranes.

SPECTRUM OF DISEASE

CoNS, particularly *S. epidermidis*, are among the most frequently isolated microorganisms in the clinical microbiology laboratory. The vast majority of infections attributed to CoNS are nosocomial infections. Data from the National Nosocomial Infections Surveillance (Centers for Disease Control and Prevention, Atlanta, Ga.) show that CoNS are among the five most commonly reported pathogens in hospitals and the most frequently reported isolate in nosocomial bloodstream infections. There is a striking difference in the spectrum of clinical presentation of diseases caused by staphylococci: *S. aureus* causes a broad variety of pyogenic infections as well as toxin-mediated diseases in the normal host, and novobiocin-resistant CoNS, particularly *S. saprophyticus*, are found in urinary tract infections. In contrast, *S. epidermidis* rarely causes pyogenic infections in the normal host, except for natural valve endocarditis, and there is little to suggest any role in toxin-mediated diseases. However, when the host is compromised, *S. epidermidis* may even be the predominant cause of infection (45).

One such group is intravenous drug (heroin) abusers who develop right-sided endocarditis. It is hypothesized that endothelial microlesions caused by heroin microcrystals and repeated bacteremia with high inocula of *S. epidermidis* due to unsterile injection procedures are important causative factors. A second group is immunocompromised patients: *S. epidermidis* is the leading cause of septicemia, with the onset later than 48 h in premature newborns. Depending on the gestational age, the opsonophagocytosis system of premature newborns is not well enough developed to handle even bacteria with low pathogenic potential such as *S. epidermidis*. Patients with aplasia (neutropenia) after cytostatic and/or immunosuppressive therapy are also highly susceptible to *S. epidermidis* septicemia owing to the low number of functioning polymorphonuclear neutrophils (PMNs). The sources of infection in all these patients are skin and mucous membranes, and the port of entry is very often an intravascular catheter. A third group, probably the most important, is patients with foreign bodies such as indwelling catheters or implanted polymer devices of various materials (e.g., polyethylene, polyurethane, silicon rubber), increasingly used in diagnostic or therapeutic procedures. Infection is the major complication associated with the use of such devices, and overall, CoNS, mainly *S. epidermidis*, are the most frequently isolated microorganisms in these infections. Depending on the kind of device and its insertion site, different infection syndromes generate a variety of clinical presentations (Table 2) (45).

The clinical picture of *S. epidermidis* infections markedly differs from that of *S. aureus* infections. Normally, there are

Gram-Positive Pathogens, ed. by V. A. Fischetti et al.
© 2000 American Society for Microbiology, Washington, D.C.

TABLE 1 Coagulase-negative staphylococci (CoNS) found in human specimens

Novobiocin-susceptible CoNS
   S. epidermidis group
      S. epidermidis
      S. haemolyticus
      S. hominis
      S. capitis
      S. warneri

   S. simulans
   S. auricularis

   S. lugdunensis
   S. schleiferi

Novobiocin-resistant CoNS
   S. saprophyticus
   S. cohnii
   S. xylosus

no fulminant signs of infection, and the clinical course is more subacute or even chronic. Accordingly, making the diagnosis of *S. epidermidis* infection is often difficult. Therapy is especially problematic in foreign body (polymer)-associated infections: despite the use of appropriate antibiotics with proven in vitro efficacy and despite a generally functioning host response, it is often not possible to eradicate the focus on the infected device. Thus, removal of the device and subsequent renewal become necessary (45).

Therefore, *S. epidermidis* infection contrasts with *S. aureus* infection in that *S. epidermidis* requires an especially predisposed host. Only in such a circumstance can *S. epidermidis* change from a commensal or saprophytic organism

TABLE 2 Foreign body (polymer)-associated *S. epidermidis* infections[a]

Septicemia/endocarditis
   Intravascular catheters
   Vascular prostheses
   Pacemaker leads
   Defibrillator systems
   Prosthetic heart valves
   Left ventricular assist devices

Peritonitis
   Ventriculo-peritoneal CSF shunts
   CAPD catheter systems

Ventriculitis
   Internal and external CSF shunts

Chronic polymer-associated syndromes[b]
   Prosthetic joint (hip) loosening
   Fibrous capsular contracture syndrome after mammary augmentation with silicone prostheses
   Late-onset endophthalmitis after implantation of artificial intraocular lenses following cataract surgery

  [a]CAPD, continuous ambulatory peritoneal dialysis; CSF, cerebrospinal fluid.
  [b]At least in some instances, the role of *S. epidermidis* is very probable, but more data are needed.

in the human cutaneous or mucocutaneous ecosystem to a pathogen.

BIOFILM FORMATION

The most important step in the pathogenesis of *S. epidermidis* foreign body-associated infectious diseases is the colonization of the polymer surface by forming multilayered cell clusters, which are embedded in an amorphous extracellular material (35). Infection of the polymer likely occurs by inoculation with only a few bacteria from the patient's skin or mucous membranes during implantation of the device. The colonizing bacteria together with the extracellular material, which is mainly composed of cell wall teichoic acids (17), and host products are referred to as biofilm. The presence of large adherent biofilms on explanted intravascular catheters has been demonstrated by scanning electron microscopy (Fig. 1) (34).

Biofilm formation proceeds in two steps: rapid attachment of the bacteria to the surface is followed by a prolonged accumulation phase that involves cell proliferation and intercellular adhesion. In recent years, significant progress has been made in defining the molecular mechanisms involved in biofilm formation, which are summarized in Fig. 2.

Initial Attachment

Microbial adherence to biomaterials depends on the cell surface characteristics of the bacteria and on the nature of the polymer material. Factors involved include physicochemical forces such as charge, van der Waal's forces, and hydrophobic interactions. Cell surface hydrophobicity and initial adherence have been attributed to bacterial surface-associated proteins. With the aid of monoclonal antibodies that efficiently block adherence, the antigenically related staphylococcal surface proteins SSP-1 and SSP-2 (280 kDa and 250 kDa, respectively) have been identified as fimbrialike polymers involved in *S. epidermidis* 354 adherence to polystyrene (44), but the genes and protein sequences are not yet available. Recently, the gene encoding the surface-associated autolysin AtlE of *S. epidermidis*

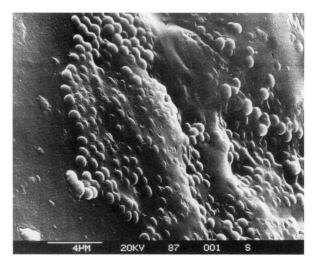

FIGURE 1 Scanning electron micrograph of an early stage of biofilm formation by *S. epidermidis* on a polyethylene surface. Reprinted from reference 35, with permission.

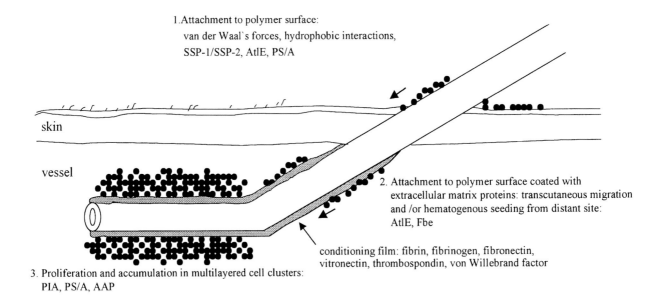

1. Attachment to polymer surface:

van der Waal`s forces, hydrophobic interactions,

SSP-1/SSP-2, AtlE, PS/A

skin

vessel

2. Attachment to polymer surface coated with
extracellular matrix proteins: transcutaneous migration
and /or hematogenous seeding from distant site:
AtlE, Fbe

conditioning film: fibrin, fibrinogen, fibronectin,
vitronectin, thrombospondin, von Willebrand factor

3. Proliferation and accumulation in multilayered cell clusters:
PIA, PS/A, AAP

FIGURE 2　Model of different phases of S. epidermidis biofilm formation on a prosthetic polymer device and bacterial factors involved. SSP-1/SSP-2, staphylococcal surface proteins; AtlE, autolysin; PS/A, polysaccharide/adhesin; Fbe, fibrinogen-binding protein; PIA, polysaccharide intercellular adhesin; AAP, accumulation-associated protein.

O-47, which mediates primary attachment of bacterial cells to a polymer surface, has been cloned and sequenced (12). The 148-kDa AtlE shows high similarity to the S. aureus autolysin Atl (61% identical amino acids) and is proteolytically cleaved into two bacteriolytically active domains, a 60-kDa amidase and a 52-kDa glucosaminidase. In the central part of the protein, there are three repetitive sequences, possibly involved in the adhesive function.

Aside from proteins, a polysaccharide structure, capsular polysaccharide/adhesin (PS/A), has been associated with initial adherence and slime production (30). Tn917 mutants deficient in PS/A and initial adherence do not cause endocarditis in a rabbit model, in contrast to the isogenic parent. Furthermore, immunization with PS/A results in protection against infection.

Interaction of *S. epidermidis* with Extracellular Matrix Proteins

While the direct interaction between bacteria and naked polymer surfaces plays a crucial role in the early stages of the adherence process in vitro and probably also in vivo, additional factors may be important in later stages of adherence in vivo, because implanted material rapidly becomes coated with plasma and extracellular matrix proteins such as fibronectin, fibrinogen, vitronectin, thrombospondin, and von Willebrand factor. Some of these host factors could serve as specific receptors for colonizing bacteria. Indeed, adherence of clinical coagulase-positive and -negative staphylococcal isolates is significantly promoted by surface-bound fibronectin in comparison with surface-bound albumin. Adherence of all S. aureus strains tested is markedly promoted by immobilized fibrinogen, while adherence of S. epidermidis strains to immobilized fibrinogen was found to vary significantly among different strains (31).

Various host factor-binding proteins from S. aureus have already been cloned and sequenced, among them fibronectin-binding proteins and fibrinogen-binding proteins. Relatively few data on host factor-binding proteins of CoNS

are available. However, recently a fibrinogen-binding protein (Fbe) in S. epidermidis was cloned and sequenced (31). Sequence comparison revealed that the 119-kDa Fbe shows similarity to the cell wall-bound fibrinogen receptor (clumping factor, ClfA) of S. aureus, which has been shown to contribute to virulence. The overall organization of Fbe corresponds to that of other surface-associated proteins, including an LPXTG motif and the characteristic membrane-spanning region.

For the autolysin AtlE from S. epidermidis mediating primary attachment to a polystyrene surface (see above), vitronectin-binding activity was also found (12). There is preliminary evidence for an in vivo role of AtlE: in a central venous catheter–associated infection model, only 50% of rats challenged with an atlE-negative Tn917 mutant developed an infection versus 80% of rats challenged with the isogenic wild-type strain. Recently, another adhesin/autolysin (Aas), which exhibits significant homology to Atl and AtlE, was cloned and sequenced from S. saprophyticus (14). Aas binds to fibronectin and to sheep erythrocytes, leading to hemagglutination. Thus, these autolysins may represent a novel class of staphylococcal adhesins.

Accumulation Process

After succeeding in primary attachment to the polymer surface, bacteria proliferate and accumulate in multilayered cell clusters, which requires intercellular adhesion. Transposon mutants that are not able to accumulate in multilayered cell clusters lack a specific polysaccharide antigen referred to as polysaccharide intercellular adhesin (PIA). Purification and structural analysis of PIA revealed that it consists of a major polysaccharide I (>80%) and a minor polysaccharide II (<20%). Polysaccharide I is a linear β-1,6-linked glucosaminoglycan mainly composed of at least 130 2-deoxy-2-amino-D-glucopyranosyl residues, of which 80 to 85% are N-acetylated. Polysaccharide II is structurally related to polysaccharide I but has a lower content of non-N-acetylated glucosaminyl residues and contains a small amount of phosphate and ester-linked succinyl residues. Thus, PIA represents a so far unique structure (22).

The genes (*icaABC*) that mediate cell clustering and PIA synthesis in *S. epidermidis* have been cloned and sequenced (13). Recently, an additional open reading frame (*icaD*) located between *icaA* and *icaB* and overlapping both genes was identified, and the function of the respective gene products in PIA synthesis was analyzed (8). Evidence has been provided that the proposed N-acetylglucosaminyltransferase activity is carried out by IcaA. However, IcaA alone exhibited only low transferase activity. Coexpression of *icaA* encoding the catalytic enzyme together with *icaD* led to a significant increase in activity and to synthesis of N-acetylglucosamine oligomers with a maximal length of 20 residues. Only in the presence of *icaC* does IcaAD catalyze the synthesis of long-chain oligomers reacting with PIA-specific antiserum. In a mouse model of foreign body infection, a PIA-negative mutant was shown to be significantly less virulent than the isogenic wild-type strain; therefore, PIA represents an important pathogenicity factor. Furthermore, PIA mediates hemagglutination (5). The hemagglutinin of *S. epidermidis* has been demonstrated to be associated with biofilm production and to be a polysaccharide structure. A study investigating the pathogenic properties of strains obtained from polymer-associated septicemic disease compared with saprophytic skin and mucosal isolates demonstrated a strong correlation of biofilm formation and presence of the *ica* gene cluster essentially associated with disease isolates (46).

Formerly, another antigenic marker of slime production and accumulation designated as slime-associated antigen (SAA) was identified. Changes in the purification procedure have now shown that the composition of SAA differs from that originally described and that SAA mainly consists of N-acetylglucosamine. Hence, it has been concluded that SAA and PIA may be the same antigenic structure (1).

Most recently, it has been reported that PS/A production is also determined by the *ica* gene cluster and that PS/A is chemically related to PIA (25). Both antigens are characterized by the common β-1,6-linked polyglucosamine backbone, but PS/A can be distinguished from PIA by molecular size (>250,000 kDa PS/A versus ~28,000 Da PIA) and the presence of succinate groups on the majority of the amino groups of the glucosamine residues. However, the authors did not present detailed structural analysis data.

Another factor seems to be essential for accumulation and biofilm formation in *S. epidermidis*. A 140-kDa extracellular protein, AAP (accumulation-associated protein), missing in the accumulation-negative mutant M7 was shown to be essential for accumulative growth in certain *S. epidermidis* strains on polymer surfaces (16, 36). An antiserum raised against the purified protein inhibited accumulation of the wild-type strain RP62A up to 98%, whereas the preimmune serum did not. Biochemical and functional properties clearly differentiate AAP from other factors known to be involved in biofilm formation. The function of AAP in the accumulation process is speculated to be the anchoring of PIA to the cell surface, because the mutant M7 produces PIA that is only loosely attached to the cell surface in contrast to the wild type (21a).

OTHER POTENTIAL VIRULENCE FACTORS
Extracellular Enzymes and Toxins

The establishment of an infection and the survival of the bacteria in the host depend on the ability to invade host tissues and to evade host defense systems, respectively. For this, staphylococci, in particular *S. aureus*, have developed multiple mechanisms, including production of a variety of extracellular proteins and enzymes, such as protein A, lipases, proteases, esterases, phospholipases, fatty-acid modifying enzymes, as well as production of hemolysins, and toxins with superantigenic properties such as enterotoxins, exfoliative toxins, and toxic shock syndrome toxin-1 (TSST-1). Additionally, proteases may play a role in proteolytic inactivation of host defense mechanisms such as antibodies and platelet microbicidal proteins as well as in destruction of tissue proteins causing increased invasiveness.

In *S. epidermidis*, an extracellular metalloprotease with elastase activity has been detected, and its gene has been cloned and sequenced (41). Previously, an elastase from *S. epidermidis* that degrades human secretory immunoglobulin A, immunoglobulin M, serum albumin, fibrinogen, and fibronectin has been identified as a cysteine protease and thus assumed to be a virulence factor (39); however, the corresponding gene has not yet been cloned. An extracellular serine protease is involved in epidermin processing (7) (see below). The genes of two lipases from *S. epidermidis* exhibiting a high degree of similarity (97.8% identical amino acids) have been cloned and sequenced, and they have been proposed to be involved in skin colonization (4, 38). The characterization and expression of fatty-acid modifying enzyme in *S. epidermidis* have been described (2).

In contrast to *S. aureus*, which produces all of the above-mentioned toxins in a strain-dependent manner, *S. epidermidis* is much less toxigenic. *S. epidermidis* can produce delta-toxin, which differs from the *S. aureus* delta-toxin in only three amino acids (26). The delta-toxin is encoded by *hld*, which is a component of the regulatory *agr* system (see below) and acts by formation of pores in the membrane, leading to the lysis of erythrocytes and other mammalian cells. Reports on unusual *S. epidermidis* strains producing enterotoxin C or TSST-1 (23) are controversial.

Factors Involved in Inflammatory Reaction and Host Defense

A serious consequence of *S. epidermidis* polymer-associated infection is septicemia. In the pathophysiology of inflammatory events in septicemia, the production of cytokines such as tumor necrosis factor alpha (TNF-α), interleukin-1β (IL-1β), and IL-6 is thought to play a major role. Peptidoglycan and teichoic acid, cell wall components purified from an *S. epidermidis* strain, stimulate human monocytes to release TNF-α, IL-1, and IL-6 in a concentration-dependent manner. Further studies revealed that human serum strongly increases peptidoglycan-induced TNF-α release by human monocytes (24).

Another feature of extracellular products of *S. epidermidis* is the interference with several neutrophil functions. Although, the extracellular slime substance itself has been found to induce a significant chemotactic response in human PMNs (18), it decreases the phagocytic activity of murine peritoneal macrophages in a dose-dependent fashion (37). Moreover, when human PMNs were preincubated with increasing amounts of slime, the responsiveness to known chemotactic stimuli such as FMLP (N-formyl-methionyl-leucyl-phenylalanine) and zymosan-activated serum was inhibited. In addition, preincubation of PMNs with slime stimulates PMN degranulation, especially after previous treatment with cytochalasin B. This effect was dose dependent and particularly pronounced with lactoferrin.

This may lead to the waste of antibacterial cellular products after contact with slime, which results, together with the reduced chemotactic responsiveness, in a decreased ability for intracellular killing. Indeed, in a surface opsonophagocytosis model, there was significantly less killing of the accumulation-positive strain *S. epidermidis* RP62A by human PMNs than of its accumulation-negative mutant M7 (17a).

Extracellular slime produced by *S. epidermidis* has also been shown to reduce the blastogenic response of human peripheral mononuclear cells to T-cell mitogens (phytohemagglutinin and streptococcal blastogen A) in a dose-dependent manner (9). This was indicated by a decrease in the incorporation rate of [³H]thymidine and in the number of blastic cells in the cell cultures. The underlying mechanisms are still unclear, as is the biological relevance.

Iron Acquisition Inside the Host

The common prerequisite for all bacterial pathogens to establish an infection is the ability to proliferate within the mammalian host. As do all bacteria, the staphylococci require iron for their growth; however, the free iron concentration (10^{-18} M) in the extracellular body fluids, owing to the presence of high-affinity iron-binding glycoproteins such as transferrin or lactotransferrin, is much too low to support staphylococcal growth.

The mechanisms by which the staphylococci acquire iron from transferrin are not well understood. In general, there are two known mechanisms for iron acquisition. One mechanism involves the synthesis and secretion of low-molecular-mass iron chelators (siderophores), which remove iron from transferrin. The siderophore-iron complexes are then taken up by specific bacterial transport systems. The siderophores staphyloferrin A and B (481 and 448 Da, respectively), first isolated from *Staphylococcus hyicus*, were also found to be produced by *S. epidermidis* under iron-restricted conditions (21). The second mechanism by which bacteria assimilate iron depends on direct contact between the host transferrin and a bacterial surface receptor.

Both *S. epidermidis* and *S. aureus* express a number of iron-repressible cell wall- and cytoplasmic membrane-associated proteins when isolated during infection in humans as well as when grown in vivo in laboratory animal infections (29). These include a 42-kDa cell wall protein that functions as a receptor for human transferrin (28) and a 32-kDa cytoplasmic membrane-associated lipoprotein (3). Human transferrin is an approximately 80-kDa monomeric protein with two distinguishable homologous domains, termed the N and the C lobes, each of which contains an iron-binding site. Recent results demonstrate that the staphylococci efficiently remove iron from human transferrin sequentially from the N lobe and then from the C lobe via a receptor-mediated process that seems to be energy dependent (28). Cloning and sequencing of the DNA region encoding the 32-kDa cytoplasmic membrane-associated lipoprotein of *S. epidermidis* revealed that the corresponding gene (*sitC*) is part of a translationally coupled, iron-regulated operon (*sitABC*) that encodes an ABC-type transporter (3). It is speculated that this novel ABC transporter is involved in either siderophore- or transferrin-mediated iron uptake in *S. epidermidis*.

REGULATION OF VIRULENCE FACTORS

Fur-like Protein

Because of the low availability of iron in the host, many bacterial pathogens use low iron concentrations as a signal to activate certain virulence factors, including toxins, adhesins, and invasins. The corresponding genes and genes involved in the biosynthesis and transport of siderophores are often regulated by Fur (ferric uptake regulator), which has been studied extensively in gram-negative bacteria. Fur is a DNA-binding repressor protein that binds, only in the presence of iron, to a consensus sequence termed the Fur box located within the promoter region of the target genes. When iron levels are low, Fur does not bind, and genes are transcribed. Recently, a gene (*fur*) for a Fur-like protein was identified in *S. epidermidis* (10). Within the -35 promoter region of the *fur* gene, a sequence motif was detected with low similarity to Fur boxes. Although the *S. epidermidis* Fur protein is unable to complement an *Escherichia coli fur* mutant, the Fur protein of *E. coli* binds to the Fur box of the *S. epidermidis fur* promoter region. The role of the Fur regulatory mechanism in the expression of virulence factors in *S. epidermidis* has not yet been determined.

DtxR Homolog

An alternative iron-dependent repressor in gram-positive bacteria is DtxR (diphtheria toxin repressor), first identified as a repressor of diphtheria toxin synthesis in *Corynebacterium diphtheriae*. Although it has a similar function, DtxR shares no homology with Fur and belongs to a newly identified family of iron-dependent repressors. DtxR homologs have been found in several bacterial species, where they regulate genes encoding iron transport systems, heme oxygenase, and virulence determinants.

Most recently, a DtxR homolog, designated SirR (staphylococcal iron regulator repressor), was identified in *S. epidermidis* by sequence analysis of the DNA region located upstream of the *sitABC* operon (15) (see above). Within the *sitABC* promoter/operator region, a palindromic sequence referred to as the Sir box was found. The Sir box overlaps the transcriptional start site of *sitABC* and shows high homology to the DtxR operator consensus sequence. DNA mobility shift assays confirmed that SirR binds to the Sir box only in the presence of metal ions such as Fe^{2+} and Mn^{2+}. Southern hybridization experiments revealed that there are at least five Sir boxes in the *S. epidermidis* genome and at least three in the *S. aureus* genome, suggesting that SirR controls the expression of multiple target genes. Because an additional Sir box within the *sirR* operator is missing, SirR does not seem to be autoregulated like the *S. epidermidis* Fur-like protein.

The *agr* (Accessory Gene Regulator) Locus

An *agr* homolog in *S. epidermidis* has been identified and sequenced (32, 43). DNA sequence analysis revealed a pronounced similarity between the *S. epidermidis* and *S. aureus agr* systems (see chapter 41, this volume). The extracellular signaling molecule produced by a typical *S. epidermidis* strain is a cyclic octapeptide (DSVCASYF) that is encoded by *agrD* and contains a thiolester linkage between the central cysteine and the C-terminal carboxyl group (32). This octapeptide exhibits activity at nanomolar concentrations. Sequence comparison revealed no striking similarity between the signaling peptides of *S. epidermidis*, *S. aureus*, or *Staphylococcus lugdunensis* (hepta-, octa-, or nonapeptides) except for the central cysteine and its distance to the C terminus. Therefore, these conserved structural features are thought to be necessary for thiolactone formation. The AgrD proteins of *S. epidermidis* and *S. aureus* show evident similarity in the region located C terminal of the signaling peptides, suggesting that this region represents a structural element important for the modifying reaction probably mediated by AgrB. AgrB shows an overall identity of 51.3% between both species. The *S. epidermidis* histidine kinase AgrC shares 50.5% identical amino acids with the *S. aureus* AgrC, with pronounced similarity in the C-terminal portion and low similarity in the N-terminal portion. These sequence data are also in agreement with the *S. lugdunensis* data, leading to the hypothesis that the N-terminal part of AgrC represents the region binding the signaling peptides that differ in sequence, while the C-terminal part interacts with the highly conserved response regulator AgrA (87.3% identity between *S. epidermidis* and *S. aureus*). In addition, an RNAIII homolog in *S. epidermidis* (560 nucleotides) was shown to regulate virulence gene expression in *S. aureus* (40). The *S. epidermidis* RNAIII had the ability to completely repress transcription of protein A and the ability to activate transcription of the alpha-toxin (*hla*) and serine protease (*ssp*) genes in an RNAIII-deficient *S. aureus* mutant. However, the stimulatory effect was reduced compared with that of *S. aureus* RNAIII. Especially the first 50 nucleotides and last 150 nucleotides of RNAIII were found to be highly similar in *S. epidermidis* and *S. aureus*. Construction and analysis of *S. epidermidis*-*S. aureus* RNAIII hybrids showed that both the 5′ and 3′ halves of the RNA molecule are important for the regulatory function.

The *sar* (Staphylococcal Accessory Regulator) Locus

In *S. aureus*, another global regulator, *sar*, also controls exoprotein synthesis by modulating the expression of *agr* (see chapter 41, this volume). The *sar* locus in *S. aureus* contains a major open reading frame, *sarA*, preceded by two smaller open reading frames. DNA mobility shift assays demonstrated that the *sar* gene products bind to an *agr* P2 promoter fragment, probably leading to activation of transcription of RNAII and subsequently RNAIII. A *sar* homolog of *S. epidermidis* has been cloned and sequenced (6), and it was revealed that the SarA protein of *S. epidermidis* is nearly identical (84%) to SarA of *S. aureus*. In contrast, the *sarA* flanking DNA sequence shows only 50% identity between both strains, and the two smaller open reading frames are absent in *S. epidermidis*. Remarkably, an *S. epidermidis sar* fragment including *sarA* and the upstream flanking region interacts with an *agr* promoter fragment of *S. aureus*. Moreover, functional analysis confirmed that the *S. epidermidis sar* homolog was able to restore alpha-toxin production in an *S. aureus sar* mutant (6). Because most of the typical virulence determinants of *S. aureus* are missing in *S. epidermidis*, it has to be clarified which genes are under the control of *agr* and *sar* in *S. epidermidis* and other CoNS. Possible candidates include genes encoding proteases, delta-toxin, lipases, autolysin, fibrinogen-binding protein, AAP, PS/A, and PIA production.

LANTIBIOTICS

Another biological property that distinguishes *S. epidermidis* from *S. aureus* is the production of bacteriocins called lantibiotics, which are active against gram-positive bacteria. Their production may play a substantial role in bacterial interference on skin and mucous membranes and thus create an ecological niche for *S. epidermidis*. Lantibiotics are antibiotic peptides that contain the rare thioether amino acids lanthionine and/or methyllanthionine. Type A lantibiotics act by inducing the formation of pores in the cytoplasmic membrane. *S. epidermidis* and other gram-positive bacteria such as *Bacillus subtilis* and lactobacilli

produce lantibiotics. Among those produced by *S. epidermidis* are the well-characterized epidermin and Pep5, as well as the newly identified epilancin K7 and epicidin 280. In general, these peptides are gene encoded and posttranslationally modified. The genes involved are organized in biosynthetic gene clusters located on plasmids. Nine genes implicated in epidermin production are encoded on the 54-kb plasmid pTü32. They include *epiA* as the structural gene for the epidermin precursor peptide and *epiB*, *epiC*, *epiD*, and *epiP*, which are involved in posttranslational modification. The flavoprotein EpiD catalyzes the oxidative decarboxylation of the C-terminal cysteine residue of the precursor peptide (20). EpiB and EpiC are assumed to catalyze the dehydration of serine and threonine residues and the formation of thioether bonds (19). EpiP is an extracellular serine protease that processes the 52-amino-acid epidermin precursor peptide into the mature 22-amino-acid peptide antibiotic (7). Bacteria producing bacteriocins, including lantibiotics, are always immune to their own bacteriocin. The immunity against epidermin and increased epidermin production is mediated by *epiF*, *epiE*, and *epiG*. These genes encode the proposed components of an ABC transporter system (33). EpiQ is a regulator that activates the transcription of *epiABCD* and *epiFEG*.

The biosynthesis of Pep5 is carried out by the *pep-TIAPBC* gene cluster located on the 20-kb plasmid pED503 (27). Pep5 biosynthesis involves *pepA* as the structural gene; the genes *pepB* and *pepC* encode putative modification enzymes; *pepP* encodes an intracellular leader peptidase that cleaves off the N-terminal leader sequence; *pepT* encodes a translocator of the ABC-type transporter, which exports the mature lantibiotic; and *pepI* is implicated in conferring immunity. Most recently, the novel lantibiotic epicidin 280, whose amino acid composition exhibits 75% similarity to Pep5, has been described and its biosynthetic gene cluster analyzed (11).

The structural gene *elkA* of epilancin K7 has been cloned and sequenced (42). Preceding *elkA*, a gene, *elkT*, encoding a proposed translocator protein has been detected that possibly mediates the export of epilancin K7. The gene *elkP* is located downstream of *elkA* and encodes a putative leader peptidase essential for processing.

FUTURE ASPECTS

Our knowledge of the biology and pathogenicity of CoNS, particularly *S. epidermidis*, has significantly increased in recent years. But we are still far away from a sufficient understanding of this versatile microorganism. We still know very little about the biology of *S. epidermidis* in its normal habitat, the cutaneous/mucocutaneous flora. Increased research in this area is obviously necessary to gain more insight into the complex balance mechanisms between bacterium and human host. This will help us to better understand when and how *S. epidermidis* can change from saprophyte to pathogen. Improvement in the armamentarium of molecular methods will enable us to analyze not only the genome but also the proteome of *S. epidermidis* in the future. But this has to be complemented by further research on the functional level, including the development of good animal models.

REFERENCES

1. **Baldassarri, L., G. Donnelli, A. Gelosia, M. C. Voglino, A. W. Simpson, and G. D. Christensen.** 1996. Purification and characterization of the staphylococcal

slime-associated antigen and its occurrence among *Staphylococcus epidermidis* clinical isolates. *Infect. Immun.* **64:** 3410–3415.

2. **Chamberlain, N. R., and S. A. Brueggemann.** 1997. Characterisation and expression of fatty acid modifying enzyme produced by *Staphylococcus epidermidis. J. Med. Microbiol.* **46:**693–697.

3. **Cockayne, A., P. J. Hill, N. B. Powell, K. Bishop, C. Sims, and P. Williams.** 1998. Molecular cloning of a 32-kilodalton lipoprotein component of a novel iron-regulated *Staphylococcus epidermidis* ABC transporter. *Infect. Immun.* **66:**3767–3774.

4. **Farrell, A. M., T. J. Foster, and K. T. Holland.** 1993. Molecular analysis and expression of the lipase of *Staphylococcus epidermidis. J. Gen. Microbiol.* **139:**267–277.

5. **Fey, P. D., J. S. Ulphani, C. Heilmann, F. Götz, D. Mack, and M. E. Rupp.** 1998. Polysaccharide intercellular adhesin (PIA) mediates hemagglutination (HA) in *Staphylococcus epidermidis,* abstr. B-40. *Abstr. 98th Gen. Meet. Am. Soc. Microbiol. 1998.* American Society for Microbiology, Washington, D.C.

6. **Fluckiger, U., C. Wolz, and A. L. Cheung.** 1998. Characterization of a *sar* homolog of *Staphylococcus epidermidis. Infect. Immun.* **66:**2871–2878.

7. **Geissler, S., F. Götz, and T. Kupke.** 1996. Serine protease EpiP from *Staphylococcus epidermidis* catalyzes the processing of the epidermin precursor peptide. *J. Bacteriol.* **178:** 284–288.

8. **Gerke, C., A. Kraft, R. Süssmuth, O. Schweitzer, and F. Götz.** 1998. Characterization of the N-acetylglucosaminyltransferase activity involved in the biosynthesis of the *Staphylococcus epidermidis* polysaccharide intercellular adhesin. *J. Biol. Chem.* **273:**18586–18593.

9. **Gray, E. D., G. Peters, M. Verstegen, and W. E. Regelmann.** 1984. Effect of extracellular slime substance from *Staphylococcus epidermidis* on the human cellular immune response. *Lancet* **i:**365–367.

10. **Heidrich, C., K. Hantke, G. Bierbaum, and H. G. Sahl.** 1996. Identification and analysis of a gene encoding a Fur-like protein of *Staphylococcus epidermidis. FEMS Microbiol. Lett.* **140:**253–259.

11. **Heidrich, C., U. Pag, M. Josten, J. Metzger, R. W. Jack, G. Bierbaum, G. Jung, and H. G. Sahl.** 1998. Isolation, characterization, and heterologous expression of the novel lantibiotic epicidin 280 and analysis of its biosynthetic gene cluster. *Appl. Environ. Microbiol.* **64:**3140–3146.

12. **Heilmann, C., M. Hussain, G. Peters, and F. Götz.** 1997. Evidence for autolysin-mediated primary attachment of *Staphylococcus epidermidis* to a polystyrene surface. *Mol. Microbiol.* **24:**1013–1024.

13. **Heilmann, C., O. Schweitzer, C. Gerke, N. Vanittanakom, D. Mack, and F. Götz.** 1996. Molecular basis of intercellular adhesion in the biofilm-forming *Staphylococcus epidermidis. Mol. Microbiol.* **20:**1083–1091.

14. **Hell, W., H.-G. W. Meyer, and G. Gatermann.** 1998. Cloning of *aas,* a gene encoding a *Staphylococcus saprophyticus* surface protein with adhesive and autolytic properties. *Mol. Microbiol.* **29:**871–881.

15. **Hill, P. J., A. Cockayne, P. Landers, J. A. Morrissey, C. M. Sims, and P. Williams.** 1998. SirR, a novel iron-dependent repressor in *Staphylococcus epidermidis. Infect. Immun.* **66:**4123–4129.

16. **Hussain, M., M. Herrmann, C. von Eiff, F. Perdreau-Remington, and G. Peters.** 1997. A 140-kilodalton extracellular protein is essential for the accumulation of *Staphylococcus epidermidis* strains on surfaces. *Infect. Immun.* **65:**519–524.

17. **Hussain, M., M. H. Wilcox, and P. J. White.** 1993. The slime of coagulase-negative staphylococci: biochemistry and relation to adherence. *FEMS Microbiol. Rev.* **10:** 191–207.

17a.**Johnson, G.** Personal communication.

18. **Johnson, G. M., D. A. Lee, W. E. Regelmann, and G. Peters.** 1986. Interference with granulocyte function by *Staphylococcus epidermidis* slime. *Infect. Immun.* **54:**13–20.

19. **Kupke, T., and F. Götz.** 1996. Expression, purification, and characterization of EpiC, an enzyme involved in the biosynthesis of the lantibiotic epidermin, and sequence analysis of *Staphylococcus epidermidis epiC* mutants. *J. Bacteriol.* **178:**1335–1340.

20. **Kupke, T., C. Kempter, G. Jung, and F. Götz.** 1995. Oxidative decarboxylation of peptides catalyzed by flavoprotein EpiD. Determination of substrate specificity using peptide libraries and neutral loss mass spectrometry. *J. Biol. Chem.* **270:**11282–11289.

21. **Lindsay, J. A., and T. V. Riley.** 1994. Staphylococcal iron requirements, siderophore production, and iron-regulated protein expression. *Infect. Immun.* **62:**2309–2314.

21a.**Mack, D.** Personal communication.

22. **Mack, D., W. Fischer, A. Krokotsch, K. Leopold, R. Hartmann, H. Egge, and R. Laufs.** 1996. The intercellular adhesin involved in biofilm accumulation of *Staphylococcus epidermidis* is a linear beta-1,6-linked glucosaminoglycan: purification and structural analysis. *J. Bacteriol.* **178:**175–183.

23. **Marin, M. E., M. C. de la Rosa, and I. Cornejo.** 1992. Enterotoxigenicity of *Staphylococcus* strains isolated from Spanish dry-cured hams. *Appl. Environ. Microbiol.* **58:** 1067–1069.

24. **Mattsson, E., J. Rollof, J. Verhoef, H. van Dijk, and A. Fleer.** 1994. Serum-induced potentiation of tumor necrosis factor alpha production by human monocytes in response to staphylococcal peptidoglycan: involvement of different serum factors. *Infect. Immun.* **62:**3837–3843.

25. **McKenney, D., J. Hübner, E. Muller, Y. Wang, D. Goldmann, and G. B. Pier.** 1998. The *ica* locus of *Staphylococcus epidermidis* encodes production of the capsular polysaccharide/adhesin. *Infect. Immun.* **66:**4711–4720.

26. **McKevitt, A. I., G. L. Bjornson, C. A. Mauracher, and D. W. Scheifele.** 1990. Amino acid sequence of a delta-like toxin from *Staphylococcus epidermidis. Infect. Immun.* **58:**1473–1475.

27. **Meyer, C., G. Bierbaum, C. Heidrich, M. Reis, J. Suling, M. I. Iglesias-Wind, C. Kempter, E. Molitor, and H. G. Sahl.** 1995. Nucleotide sequence of the lantibiotic Pep5 biosynthetic gene cluster and functional analysis of PepP and PepC. Evidence for a role of PepC in thioether formation. *Eur. J. Biochem.* **232:**478–489.

28. **Modun, B., R. W. Evans, C. L. Joannou, and P. Williams.** 1998. Receptor-mediated recognition and uptake of iron from human transferrin by *Staphylococcus aureus* and *Staphylococcus epidermidis. Infect. Immun.* **66:**3591–3596.

29. **Modun, B. J., A. Cockayne, R. Finch, and P. Williams.** 1998. The *Staphylococcus aureus* and *Staphylococcus epider-*

midis transferrin-binding proteins are expressed *in vivo* during infection. *Microbiology* **144**:1005–1012.

30. **Muller, E., J. Hübner, N. Gutierrez, S. Takeda, D. A. Goldmann, and G. B. Pier.** 1993. Isolation and characterization of transposon mutants of *Staphylococcus epidermidis* deficient in capsular polysaccharide/adhesin and slime. *Infect. Immun.* **61**:551–558.

31. **Nilsson, M., L. Frykberg, J. I. Flock, L. Pei, M. Lindberg, and B. Guss.** 1998. A fibrinogen-binding protein of *Staphylococcus epidermidis. Infect. Immun.* **66**:2666–2673.

32. **Otto, M., R. Süssmuth, G. Jung, and F. Götz.** 1998. Structure of the pheromone peptide of the *Staphylococcus epidermidis agr* system. *FEBS Lett.* **424**:89–94.

33. **Peschel, A., and F. Götz.** 1996. Analysis of the *Staphylococcus epidermidis* genes *epiF*, *-E*, and *-G* involved in epidermin immunity. *J. Bacteriol.* **178**:531–536.

34. **Peters, G., R. Locci, and G. Pulverer.** 1981. Microbial colonization of prosthetic devices. II. Scanning electron microscopy of naturally infected intravenous catheters. *Zentralbl. Bakteriol. Mikrobiol. Hyg. B* **173**:293–299.

35. **Peters, G., R. Locci, and G. Pulverer.** 1982. Adherence and growth of coagulase-negative staphylococci on surfaces of intravenous catheters. *J. Infect. Dis.* **146**:479–482.

36. **Schumacher-Perdreau, F., C. Heilmann, G. Peters, F. Götz, and G. Pulverer.** 1994. Comparative analysis of a biofilm-forming *Staphylococcus epidermidis* strain and its adhesion-positive, accumulation-negative mutant M7. *FEMS Microbiol. Lett.* **117**:71–78.

37. **Shiau, A. L., and C. L. Wu.** 1998. The inhibitory effect of *Staphylococcus epidermidis* slime on the phagocytosis of murine peritoneal macrophages is interferon-independent. *Microbiol. Immunol.* **42**:33–40.

38. **Simons, J. W., M. D. van Kampen, S. Riel, F. Götz, M. R. Egmond, and H. M. Verheij.** 1998. Cloning, purification and characterisation of the lipase from *Staphy-*

lococcus epidermidis—comparison of the substrate selectivity with those of other microbial lipases. *Eur. J. Biochem.* **253**:675–683.

39. **Sloot, N., M. Thomas, R. Marre, and S. Gatermann.** 1992. Purification and characterisation of elastase from *Staphylococcus epidermidis. J. Med. Microbiol.* **37**:201–205.

40. **Tegmark, K., E. Morfeldt, and S. Arvidson.** 1998. Regulation of *agr*-dependent virulence genes in *Staphylococcus aureus* by RNAIII from coagulase-negative staphylococci. *J. Bacteriol.* **180**:3181–3186.

41. **Teufel, P., and F. Götz.** 1993. Characterization of an extracellular metalloprotease with elastase activity from *Staphylococcus epidermidis. J. Bacteriol.* **175**:4218–4224.

42. **Van de Kamp, M., H. W. van den Hooven, R. N. Konings, G. Bierbaum, H. G. Sahl, O. P. Kuipers, R. J. Siezen, W. M. de Vos, C. W. Hilbers, and F. J. van de Ven.** 1995. Elucidation of the primary structure of the lantibiotic epilancin K7 from *Staphylococcus epidermidis* K7. Cloning and characterisation of the epilancin-K7-encoding gene and NMR analysis of mature epilancin K7. *Eur. J. Biochem.* **230**:587–600.

43. **Van Wamel, W. J., G. van Rossum, J. Verhoef, C. M. Vandenbroucke-Grauls, and A. C. Fluit.** 1998. Cloning and characterization of an accessory gene regulator (*agr*)-like locus from *Staphylococcus epidermidis. FEMS Microbiol. Lett.* **163**:1–9.

44. **Veenstra, G. J., F. F. Cremers, H. van Dijk, and A. Fleer.** 1996. Ultrastructural organization and regulation of a biomaterial adhesin of *Staphylococcus epidermidis. J. Bacteriol.* **178**:537–541.

45. **Von Eiff, C., C. Heilmann, and G. Peters.** 1998. *Staphylococcus epidermidis*: why is it so successful? *Clin. Microbiol. Infect.* **4**:297–300.

46. **Ziebuhr, W., C. Heilmann, F. Götz, P. Meyer, K. Wilms, E. Straube, and J. Hacker.** 1997. Detection of the intercellular adhesion gene cluster (*ica*) and phase variation in *Staphylococcus epidermidis* blood culture strains and mucosal isolates. *Infect. Immun.* **65**:890–896.

Biology and Pathogenicity of Staphylococci Other than *Staphylococcus aureus* and *Staphylococcus epidermidis*

GERARD LINA, JEROME ETIENNE, AND FRANÇOIS VANDENESCH

47

Staphylococci other than *Staphylococcus aureus* and *Staphylococcus epidermidis* correspond to a vast group of more than 40 species and subspecies, all of which are frequently (though erroneously, since some of them are phenotypically coagulase positive) referred to as coagulase-negative staphylococci (CNS). These bacteria are commensals on animal and human skin, and some of them have become pathogens by virtue of medical progress both in human and veterinary medicine. Until the mid-1970s identification of non-*aureus* species of staphylococci was technically difficult and their taxonomy was ill-defined, but since 1975 advances in staphylococcal systematics have enabled the description of a number of staphylococcal species. These include species whose pathogenic potential has been clearly demonstrated in humans (e.g., *S. haemolyticus*, *S. saprophyticus*, *S. lugdunensis*, *S. schleiferi*, *S. warneri*, and *S. caprae*) or in animals (e.g., *S. intermedius*, *S. hyicus*, and *S. simulans*) (45). Many new species and subspecies of CNS have now been described; their biology and pathogenic potential are only beginning to be defined. However, it is becoming more and more apparent that various staphylococcal species express proteins that were until recently thought to belong exclusively to *S. aureus*. This review will not extensively discuss the biology of all known species of staphylococci but will focus on species of CNS selected for their pathogenic potential and those for which recent information has had a significant effect on the general knowledge of staphylococcal biology.

STAPHYLOCOCCUS LUGDUNENSIS

S. lugdunensis was originally described in 1988. Isolates of this species can be misidentified as *S. aureus* on the basis of clumping factor production (58 to 79% positive) and from the expression (although weak) of thermostable DNase, two characteristics that are typical of, and often considered diagnostic of, *S. aureus*.

Diseases

S. lugdunensis is a human commensal, more pathogenic than most other species of *Staphylococcus* other than *S.*

aureus, causing primary infections of the human skin and postoperative wound infections, with a predominance of sites below the waist (77). Deep-seated infections often associated with bacteremia occur, including peritonitis, prosthetic infections, osteomyelitis, septic arthritis, and infective endocarditis (76, 77). Indeed, *S. lugdunensis* appears to be the cause of severe and destructive infective endocarditis; the outcome is usually more favorable in patients who undergo valve replacement (21). This suggests that any isolate of *S. lugdunensis* should be assumed to be a pathogen unless proved otherwise. *S. lugdunensis* has been associated with mastitis in goats.

Genomic Diversity and Natural Habitat

An unusual feature of *S. lugdunensis* is its highly genotypic conservation when studied with genomic markers such as plasmid profiling and restriction endonuclease analysis of chromosomal DNA (77). Pulsed-field gel electrophoresis (PFGE) confirmed the lack of biodiversity of this species when compared with *S. epidermidis*. It has been suggested that *S. lugdunensis* is a skin commensal. However, the predominance of sites of infection below the waist (76) suggests that its preferential site for colonization may be the perineum rather than the entire skin surface.

Antibiotic Susceptibility

Isolates of *S. lugdunensis* are usually highly susceptible to antistaphylococcal antibiotics: only about 25% of strains produce β-lactamase (76). Resistance to cadmium has been detected in almost 60% of *S. lugdunensis* isolates and found to be carried by a 3.16-kb plasmid (prototype plasmid pLUG10) that is a member of the pT181 group of class I staphylococcal plasmids (16). However, the cadmium resistance genes (*cadB cadX*) present on pLUG10 are usually found on plasmids of the pC194 group of class I plasmids such as pOX6 of *S. aureus*. This suggests that cadmium resistance in staphylococci may be conferred by homologous gene sets carried by class I plasmids of different origins.

Determinants of Pathogenicity

The synthesis of an extracellular slime substance or glycocalyx, a factor interfering with phagocyte function and

Gram-Positive Pathogens, ed. by V. A. Fischetti et al.

TABLE 1 Extracellular factors produced by *Staphylococcus* species other than *S. aureus* and *S. epidermidis*

Strain	Enzymes						
	Esterase (Lambe, 1990; activity [51])	Lipase			FAME (Long, 1992; activity [55])	Urease	
		Lambe, 1990; activity (51)	Long, 1992; activity (55)	Ayora, 1994; gene (6)		Schleifer, 1986; activity (65)	Gaterman, 1989; Jose, 1991; gene (28, 40)
S. capitis	NT[a]	NT	1/6	NT	4/6	<10%	NT
S. caprae	NT	NT	2/6	NT	4/6	>90%	NT
S. cohnii	NT	NT	5/6	NT	5/6	<10%	NT
S. haemolyticus	NT	NT	0/9	NT	0/9	<10%	NT
S. hominis	6/6	6/6	3/6	NT	3/6	>90%	NT
S. hyicus	NT	NT	NT	*shl*		11–89%	NT
S. intermedius	NT	NT	NT	NT	NT	>90%	NT
S. lugdunensis	6/6	6/6	0/10	NT	0/10	NT	NT
S. saprophyticus	NT	NT	9/9	NT	7/9	>90%	Cloned and sequenced
S. schleiferi	6/6	6/6	10/10	NT	8/10	NT	NT
S. simulans	NT	NT	1/10	NT	8/10	>90%	NT
S. warneri	6/6	6/6	3/5	NT	3/5	>90%	NT
S. xylosus	NT	NT	NT	NT	NT	>90%	Cloned and sequenced

[a]NT, not tested.

involved in the pathogenesis of foreign body infections, has been demonstrated in *S. lugdunensis*. The glycocalyx activates monocyte prostaglandin E_2 production, which in turn contributes to the inhibition of T-cell proliferation. This activation of monocytes also results in modulation of tumor necrosis factor alpha and nitric oxide production, two significant antimicrobial activities of macrophages (68). Production of other enzymes, including esterase, fatty acid-modifying enzyme (FAME), and lipase (50, 55), that may act as invasion factors has been reported (Table 1). *S. lugdunensis* binds to collagen, fibronectin, vitronectin, laminin, fibrinogen, thrombospondin, plasminogen, and human immunoglobulin G immobilized on latex beads (61, 62) (Table 2). However, *S. lugdunensis* showed only moderate attachment on fibronectin-coated polymethylmethacrylate coverslips, and none of 10 strains of *S. lugdunensis*

showed any significant attachment to fibrinogen (26a). *S. lugdunensis* induced abscess formation in a mouse model, and this was enhanced by the presence of a foreign body (51).

Production of synergistic hemolysin, protease, lipase, and esterase by *S. lugdunensis* has been variably observed (34, 51, 80) (Table 3). The synergisic hemolysin has been termed SLUSH; it consists of three very similar 43-residue peptides highly related to the three "gonococcal growth inhibitor" (GGI) peptides of *S. haemolyticus*, which also possesses synergistic hemolytic activity (22, 83). SLUSH may also exhibit some antibacterial activity in that crude supernatants from hemolytic, but not from nonhemolytic, *S. lugdunensis* strains show antibacterial activity on several staphylococcal species (22). Moreover, SLUSH has no homologies with the *S. aureus* delta-hemolysin and, unlike

TABLE 2 Binding of various *Staphylococcus* species to extracellular matrix proteins[a]

Species	Binding (no. of positive/total no. of strains tested) to extracellular matrix proteins:							
	Fn			Cn		Vn (PAA and I-PBA) (61, 62)	Lm (PAA and I-PBA) (61, 62)	Fg (PAA and I-PBA) (61, 62)
	(PAA) (57)	(PAA and I-PBA) (61, 62)	(PAA and PCR) (58)	(PAA) (57)	(PAA and I-PBA) (61, 62)			
S. capitis	3/6	0/2	1/2	6/6	2/2	0/2	1/2	NT
S. chromogenes	4/47	NT	NT	32/47	NT	NT	NT	NT
S. cohnii	5/14	2/2	NT	9/14	2/2	2/2	1/2	NT
S. haemoliticus	NT	1/1	15/16	NT	1/1	1/1	1/1	NT
S. hominis	10/16	2/2	6/6	10/16	2/2	2/2	2/2	NT
S. hyicus	37/171	0/2	6/7	127/171	1/2	0/2	0/2	NT
S. intermedius	NT	NT	1/2	NT	NT	NT	NT	NT
S. lugdunensis	NT	9/11*	3/3	NT	8/11	10/11*	11/11*	7/11*
S. saprophyticus	NT	0/2	1/1	NT	1/2	0/2	0/2	NT
S. simulans	4/18	0/2	NT	14/18	1/2	0/2	0/2	NT
S. warneri	1/4	1/2	3/5	2/4	2/2	0/2	1/2	NT
S. xylosus	5/19	1/2	NT	16/19	2/2	1/2	1/2	NT

[a]Fn, fibronectin; Cn, collagen; Vn, vitronectin; Lm, laminin; Fg, fibrinogen; PAA, particle agglutination assays; I-PBA, iodine-labeled protein binding assays; PCR, PCR amplification of *fbnB*-like genes using primers based on conserved sequences of *S. aureus fbnbA* and *fbnbB*; NT, not tested; *, low-affinity binding.

TABLE 4 Superantigenic toxins produced by *Staphylococcus* species other than *S. aureus* and *S. epidermidis*

Species	Enterotoxins (SEs), TSST-1, exfoliative or synergohymenotropic toxin detected by:											
	ELISA					Vernozy-Rozand, 1996, + Southern blot (81)	Gene sequence (Edwards, 1997 [25])	Emetic assay		Mitogenic assay (Edwards, 1997 [25])	Skin assay (piglet) (Andresen, 1998 [5])	Leukotoxicity on PMNs (Prevost, 1995 [63])
	Almazan, 1987 (4)	Bautista, 1988 (10)	Hirooka, 1988 (36)	Valle, 1990 (73)	Orden, 1992 (59)			Adesyuu, 1984 (1)	Edwards, 1997 (25)			
S. capitis	NT^a	NT	NT	NT	NT	1/11 SEE, *see*	NT	NT	NT	NT	NT	NT
S. caprae	NT		NT	1/18 SEA 2/18 SEB 1/18 SEC 1/18 SED 1/18 SEE	NT	NT	NT	NT	NT	NT	NT	NT
S. chromogenes	NT		NT	3/23 SEC	NT	0/1 SEE	NT	2/2 Not SEA-E	NT	NT	NT	NT
S. cohnii	NT	1/4 SEC	NT	0/6 SEA-E 0/3 SEA-E	NT	0/2 SEE	NT	NT	NT	NT	NT	NT
S. equorum	NT		NT		NT	2/24 SEE, *see*	NT	NT	NT	NT	NT	NT
S. haemolyticus	NT	1/6 SEA 1/6 SEB 3/6 SEC 4/6 SED	NT	1/64 SEA 3/64 SEB 15/64 SEC 4/64 SEE	NT	NT	NT	NT	NT	NT	NT	NT
S. hyicus	NT		NT	2/13 SEC	NT	NT	NT	3/3 Not SEA-E	NT	NT	12/60 ExhA 20/60 ExhB 11/60 ExhC; not similar to ETA-B	NT
S. intermedius	2/66 SEA 1/66 SEB 13/66 SEC 4/66 SED 1/66 SEE		6/73 SEC 7/73 SED 6/73 SEE 4/73-TSST-1	NT	NT	NT	sec canine	NT	SEC canine	SEC canine	NT	51/51 and gene *lukS-1/lukF-1* characterized
S. lentus	NT		NT	2/3 SEE	NT	1/7 SEE, *see*	NT NT NT	NT	NT	NT		
S. saprophyticus	NT		NT	1/13 SEB 3/13 SEC 2/13 SEE	NT	NT	NT	NT	NT	NT	NT	
S. sciuri	NT		NT	1/20 SEC 3/20 SEE	NT	0/1 SEE	NT	NT	NT	NT	NT	
S. simulans	NT		NT	1/45 SEA 2/45 SEB 5/45 SEC	NT	5/51 SEE, *see*	NT	NT	NT	NT	NT	
S. warneri	NT		NT		NT	0/6 SEE	NT	NT	NT	NT	NT	
S. xylosus	NT	4/5 SED	NT	4/23 SEC 2/23 SEE	2/3-TSST-1	1/24 SEE	NT	NT	NT	NT	NT	NT

^a NT, not tested

Genomic Diversity and Natural Habitat

S. schleiferi is a rather genetically homogeneous species since a very limited polymorphism is detected by PFGE or by plasmid analysis. However, ribotyping combined with PFGE appeared to be efficient in detecting intraspecific variations among *S. schleiferi* isolates (49). The ecological niches of this microorganism are at present not well known, except for its natural occurrence in the human preaxillary flora (20).

Antibiotic Susceptibility

S. schleiferi is usually susceptible to all antistaphylococcal antibiotics including penicillin. This is unusual in light of the nosocomial acquisition of infections attributed to the organism.

Determinants of Pathogenicity

S. schleiferi causes subcutaneous abscess formation in mice (51). Production of virulence factors such as glycocalyx, esterase, protease, β-hemolysin, FAME, and lipase has been described (Table 1) (34, 51, 55). In a recent unpublished study (26a), 9/11 strains of *S. schleiferi* showed a moderate to strong attachment on fibronectin- and fibrinogen-coated polymethylmethacrylate coverslips, equivalent to that of *S. aureus*. No fibrinogen-binding surface components homologous to the ClfA protein of *S. aureus* were identified by either Western immunoblotting or analysis of the ClfA antigen by fluorescence-activated cell sorting, and the putative fibrinogen adhesin of *S. schleiferi* is still unknown. In contrast, *S. schleiferi* expressed one fibronectin-binding adhesin homologous to the fibronectin-binding protein(s) of *S. aureus* according to four criteria: (i) adhesion of *S. schleiferi* to fibronectin-coated surfaces was blocked by a recombinant D-repeat peptide of *S. aureus* FnBPa; (ii) a high-molecular-weight protein band of ca. 200 kDa, which comigrated with fibronectin-binding proteins of *S. aureus*, was identified by fibronectin ligand affinity Western blotting; (iii) reverse transcriptase PCR of mRNAs of 10 of 11 strains of *S. schleiferi* using forward and reverse primers of a highly conserved region of *fnb* genes of *S. aureus* 8325-4 revealed a constant-size 331-bp fragment; (iv) sequencing of the reverse transcriptase PCR-amplified 331-bp fragment in two strains of *S. schleiferi* exhibiting the highest fibronectin-binding activity revealed very high homology with an *fnb* region of *S. aureus* controls (26a). Comparable results were obtained by S. J. Peacock and T. Foster (62a) in that Western ligand affinity blotting of cell wall-associated protein extracted from *S. schleiferi* isolates demonstrated a fibronectin-binding protein with an apparent molecular size of 200 kDa (for one strain) or 180 kDa (two clinical isolates). PCR analysis using primers designed to amplify the 345-bp fragment of *S. aureus* 8325-4 *fnbA* that encodes the D1-D3 binding domain of FnbPA gave a product of comparable size.

STAPHYLOCOCCUS CAPRAE

S. caprae (together with *S. gallinarum*) was described in 1983. Both these species were originally isolated from animals (65).

Diseases

S. caprae has been usually associated with goats, but since 1991 an increasing number of laboratories report isolating the organism from human clinical specimens of both community- and hospital-acquired infections (67). Among these, urinary tract infection, skin infection, bacteremia associated with intravenous access, endocarditis, and bone and joint postsurgical infection were reported.

Genomic Diversity and Natural Habitat

S. caprae was originally isolated from goat's milk; it is the most prevalent CNS in mastitis-free goat's milk and has not been isolated from milk of cows or sheep. *S. caprae* demonstrated considerable conservation in chromosome structure as indicated by PFGE analysis of *S. caprae* isolates from various geographical locations. However, human isolates of *S. caprae* were distinguished from goat isolates on the basis of their SmaI PFGE profile and to some extent on the basis of their cellular fatty acids (45) and their ribotypes (75). The normal habitat of *S. caprae* in humans remains to be discovered, but circumstantial evidence from its association with nosocomial infections suggests it is likely to be a commensal of the skin flora.

Antibiotic Susceptibility

S. caprae is usually susceptible to most antistaphylococcal antibiotics including penicillin, but all the strains tested to date (both of human and animal origin) have been uniformly resistant to fosfomycin. Resistance to methicillin has been detected, although infrequently in human isolates (67, 75).

Determinants of Pathogenicity

Most isolates of *S. caprae* express a synergistic hemolysin resembling the delta-lysin of *S. aureus*. Southern blot hybridization using *S. aureus* *hld* and *slush* probes revealed that *S. caprae* contains *hld* and *slush* homolog genes, both responsible for synergistic-hemolytic activity (23) (Table 3). Enterotoxin production by certain strains isolated from healthy goats has been detected by ELISA (73) (Table 4). This has not been confirmed by emetic or mitogenic assays. Production of enzymes, including lipase, FAME, and urease (50, 55), that may act as invasion factors has been reported.

STAPHYLOCOCCUS WARNERI

S. warneri was first described in 1975 and is usually found on human skin in small populations (45). A nearly subspecies group, represented by non-human *S. warneri*, is one of the major staphylococcal species found living on the skin and nasal membranes of various prosimians and monkeys.

Diseases

S. warneri has the usual characteristics of the CNS and can cause significant infection both in the community and in the hospital. *S. warneri* has been reported to cause bacteremia, infective endocarditis, cerebrospinal fluid shunt infection, subdural empyema, vertebral osteomyelitis, and urinary tract infection (15).

Determinants of Pathogenicity

S. warneri induces abscess formation in a mouse model with foreign body implant but less frequently than *S. schleiferi*, *S. lugdunensis*, and *S. epidermidis* (51). *S. warneri* binds to collagen and fibronectin immobilized on latex beads (57, 58, 62) (Table 2). Production of glycocalyx, lipase, protease, and esterase has been described (51). The lipase of

S. warneri is secreted as a protein with an apparent molecular mass of 90 kDa that is processed in the culture supernatant to a protein of 45 kDa. Purified lipase has a broad substrate specificity, and results of inhibition studies are consistent with the presence of a serine residue at the catalytic site (70). A synergistic hemolysin that is produced by the majority of the isolates has been attributed to *hld* homologs within RNAIII (23, 71) (Table 3). RNAIII from *S. warneri* has been cloned and sequenced. It has an estimated length of 684 nt, and its sequence shows a high degree of identity, especially in the first 50 and last 150 nt, with that of RNAIII-sa. Surprisingly, it contains two nonidentical copies of delta-like hemolysin, both 25 amino acids in length, that differ in seven and five residues, respectively, from *S. aureus* delta-hemolysin (71). However, the distribution of the charged residues suggests that the variants of delta-hemolysin produced by *S. warneri* can still form amphipatic alpha-helices (71). *S. warneri* RNAIII has the ability to regulate severereal *agr*-dependent genes of *S. aureus* (71). Enterotoxin production by certain strains isolated from healthy goats has been detected by ELISA (73) (Table 4). This has not been confirmed by emetic or mitogenic assays. Production of enzymes, including esterase, lipase, FAME, and urease (50, 55), that may act as invasion factors has been reported (Table 1).

STAPHYLOCOCCUS PASTEURI

S. pasteuri is a new species of CNS described in 1993 in human, animal, and food specimens (17). *S. pasteuri* strains are phenotypically similar to *S. warneri*, but a clear-cut distinction between these two species may be obtained by comparing their rRNA gene restriction patterns (17). A synergistic hemolysin that is produced by most of the isolates has been attributed to an *hld* homolog (23) (Table 3).

STAPHYLOCOCCUS SIMULANS

Described by Schleifer and Kloos in 1975, this species was named *simulans* for having some phenotypic similarities to *S. aureus* (65).

Diseases

S. simulans may be associated with a variety of animal infections including bovine and ovine mastitis, feline and canine pyodermas, and abscesses (27). *S. simulans* has been isolated as a rare cause of human infection including urinary tract infection, wound, bone, and joint infections, septicemia, and native valve endocarditis (38).

Genomic Diversity and Natural Habitat

This species is isolated from the skin and urethra of healthy humans, skin of other mammals, and food including goat's milk and cheese (45, 65, 81). In a study on nursing home residents, 35% of carriage sites screened were positive; the organism was most frequently carried in the perineum, but also in the nose and hairline (8).

Antibiotic Susceptibility

The susceptibility of *S. simulans* isolates to benzylpenicillin, methicillin, erythromycin, tetracycline, and kanamycin is variable. Some strains carry plasmids which encode beta-lactamase or ribosomal methylase genes which confer macrolide and lincosamide resistance (7, 65). Methicillin-resistant isolates have been reported (8).

Determinants of Pathogenicity

Binding of *S. simulans* to fibronectin and collagen immobilized on latex beads has been described (57, 62) (Table 2). Production of enzymes, including lipase, FAME, and urease (50, 55), that may act as invasion factors has been reported (Table 1). The synergistic hemolysis produced by *S. simulans* (35) has been attributed to an *hld* homolog within RNAIII that has been cloned and sequenced (71) (Table 3). The RNAIII of *S. simulans* has an estimated length of 573 nt, and its sequence shows a high degree of identity, especially in the first 50 and last 150 nt, with that of RNAIII-sa. The predicted 26-residue delta-hemolysin from *S. simulans* shows 19 of 26 amino acids identical to those of *S. aureus* (71). *S. simulans* RNAIII has the ability to regulate several *agr*-dependent genes of *S. aureus* (71). Enterotoxin production by certain strains isolated from goat's milk and cheese has been detected by ELISA; the results obtained were further confirmed by Southern blotting using two oligonucleotide probes specific for the *S. aureus* enterotoxin E gene (81) (Table 4). This has not been confirmed by emetic or mitogenic assays. *S. simulans* has been observed to produce capsular polysaccharide (14).

The lysostaphin (Lss)-producing strain designated *S. simulans* biovar *staphylolyticus* carries a large beta-lactamase plasmid encoding Lss and the Lss immunity factor (Lif). Lss is an extracellular glycylglycine endopeptidase that lyses other staphylococci by hydrolyzing thepolyglycine interpeptide bridges in their cell wall peptidoglycan. Lss is synthesized as a 493-amino-acid preprotein with a 36-amino-acid signal peptide, a propeptide of 211 amino acids, and a mature protein of 246 amino acids. Prolysostaphin is processed in the culture supernatant by an extracellular cysteine protease. Lif, which is encoded in the opposite direction, confers lysostaphin immunity by increasing the serine/glycine ratio of the interpeptide bridges. Lif shows similarity to FemA and FemB proteins, which are involved in the biosynthesis of the glycine interpeptide bridge of staphylococcal peptidoglycan (72). Lss was shown to be more effective in the treatment of experimental *S. aureus* endocarditis in rabbits than vancomycin and could become a novel therapeutic agent (19).

STAPHYLOCOCCUS CAPITIS

S. capitis subsp. *ureolyticus* is distinguished from *S. capitis* subsp. *capitis* by its urease activity (45). *S. capitis* has been reported to cause urinary tract infections, catheter-related bacteremia, cellulitis, cerebrospinal fluid shunt infection, and infective endocarditis (52). Although the scalp is the most usual habitat of *S. capitis* (45), it was not unequivocally identified as the portal of entry in these cases of endocarditis.

S. capitis binds to laminin, collagen, and fibronectin immobilized on latex beads (57, 58, 62) (Table 2). Production of enzymes, including lipase and FAME (50, 55), that may act as invasion factors has been reported (Table 1). A synergistic hemolysin that is produced by the majority of the isolates has been attributed to an *hld* homolog by Southern blot hybridization (Table 3) (23). The production of enteroxin by strains isolated from goat's milk and cheese has been reported (81). Enterotoxin was detected by ELISA and by Southern blotting using two oligonucleotide probes specific for the *S. aureus* enterotoxin E (81) (Table 4). Enterotoxin function, however, has not been confirmed by mitogenic or emetic assays.

S. capitis EPK1 produces a 35-kDa glycylglycine endopeptidase (ALE-1) which hydrolyzes interpeptide pentaglycine chains of the cell wall peptidoglycan of *S. aureus*. So far, *S. capitis* EPK1 is the only producer of ALE-1 among tested *S. capitis* strains (68a). Characterization of the enzyme activity and cloning of the plasmid-encoded *ale-1* gene revealed that ALE-1 is very similar to prolysostaphin produced by *S. simulans* biovar *staphylolyticus* (69). Protein homology search suggests that ALE-1 and Lss are members of a Zn^{2+} protease family with a 38-amino-acid-long motif, Tyr-X-His-X(11)-Val-X(12/20)-Gly-X(5-6)-His (69). The *epr* gene located upstream of and in the opposite orientation to *ale-1* confers resistance to ALE-1 and Lss. As observed with Lif (responsible for Lss immunity), the *epr* product may be involved in the addition of serine to the pentapeptide peptidoglycan precursor (69).

STAPHYLOCOCCUS HOMINIS

S. hominis is one of the major staphylococcal species inhabiting the skin of humans and is considered as an opportunistic pathogen of low virulence. A new subspecies, *S. hominis* subsp. *novobiosepticus*, isolated from human blood cultures, a wound, a breast abscess, and a catheter tip, has been recently described (46). *S. hominis* binds to vitronectin, laminin, collagen, and fibronectin immobilized on latex beads (57, 58, 62) (Table 2). A synergistic hemolysin that is produced by the majority of the isolates could not be attributed to *slush* or *hld* homologs by Southern blot hybridization (Table 3), possibly suggesting the existence of a third staphylococcal synergistic hemolysin (23). Production of enzymes, including esterase, lipase, FAME, and urease (50, 55), that may act as invasion factors has been reported (Table 1).

STAPHYLOCOCCUS XYLOSUS

S. xylosus has hosts in all mammalian orders (43) but has rarely been associated with human infections. It is used as starter culture in the production of dry sausage and fermented fish. *S. xylosus* binds to vitronectin, laminin, collagen, and fibronectin immobilized on latex beads (57, 62) (Table 2). A synergistic hemolysin that is produced by approximately half of the isolates has been attributed to a *slush* homolog (23) (Table 3). The production of enterotoxin and toxic shock syndrome toxin 1 (TSST-1) by strains isolated from goat's milk and cheese (81), sheep's milk (10), healthy goats (73), and sheep, goat, and cow mastitis (59) has been detected by ELISA. In one study the results obtained were further confirmed by Southern blotting using two oligonucleotide probes specific for the *S. aureus* enterotoxin E gene (81) (Table 4). Enterotoxin function has not been confirmed by mitogenic or emetic assays. The biology of nutrition and metabolism of *S. xylosus* has been the subject of increasing attention in recent years. The lactose utilization genes and the mechanism of catabolite repression in *S. xylosus* have been characterized (9). A genetic locus essential for maltose-maltotriose utilization in *S. xylosus* has been identified and its function has been characterized. The serine acetyltransferase gene (*cysE*) of *S. xylosus* was identified by transposon mutagenesis; it is surrounded by genes encoding glutamyl-tRNA synthetase (*gltX*) and cysteinyl-tRNA synthetase (*cysS*) in an organization identical to that found in *Bacillus subtilis* and *Bacillus stearothermophilus* (26). The sucrose-specific regulon for sucrose utilization by *S. xylosus* has been characterized. A gene encoding urease which may act as an invasion factor has been reported (Table 1) (40).

STAPHYLOCOCCUS INTERMEDIUS
Diseases and Habitat

S. intermedius is the most important non-*aureus* species that produces coagulase; it is, however, taxonomically distinct from *S. aureus*. It is the predominant coagulase-positive staphylococcus in the dog's mouth and in skin infections of dogs. It has also been found in a wide range of other animal species including pigeons, minks, cats, foxes, raccoons, gray squirrels, goats, and horses (12). *S. intermedius* has rarely been found in human beings, even among individuals with frequent animal exposure, but it is a common and potentially invasive zoonotic pathogen of canine-inflicted human wounds (53). It has been isolated in rare cases of non-canine-inflicted wounds, a case of infective endocarditis, and a case of catheter-related bacteremia (56). *S. intermedius* was also considered to be the etiologic agent in an outbreak of food intoxication involving butter-blend products that resulted in more than 265 cases in the western United States; all the isolates were reported to produce enterotoxin A (42). In fact, the true frequency of *S. intermedius* in non-canine-inflicted wounds remains unclear, as it can be confused with *S. aureus* in medical laboratory analysis on the basis of coagulase production.

Genomic Diversity

Genomic DNA fingerprinting by PFGE of *S. intermedius* isolates revealed a high degree of polymorphism (66). Conventional genomic DNA fingerprinting (DNA cut with *Bgl*II) suggested that isolates from healthy dogs and those from canine pyoderma belong to distinct clusters (2). Ribotyping did not confirm these results; isolates from cases of healthy dogs and canine pyoderma were indistinguishable. However, pigeon and equine strains showed a variety of ribotypes, including those of the canine isolates, suggesting the exchange of strains between animal species (47).

Determinants of Pathogenicity

Binding of *S. intermedius* to fibronectin immobilized on latex beads has been described (58) (Table 2). *S. intermedius* produces a 42-kDa immunoglobulin binding protein, specific for the Fc domain, that shows close functional and antigenic similarity to protein A of *S. aureus* (32). It is expressed both in cell wall-bound and secreted forms. *S. intermedius* produces a staphylocoagulase which resembles that of *S. aureus* in its rate and method of action on prothrombin but is antigenically distinct from the *S. aureus* coagulase (64). It has not been characterized at the molecular level. *S. intermedius* has been described as a beta- and delta-hemolysin producer, but only the beta-hemolysin has been characterized; the protein has been purified and its biochemical properties have been compared with those of *S. aureus* beta-hemolysin. Both toxins have similar enzymatic properties, belong to the class of neutral sphingomyelinases C, and have a high specificity for sphingomyelin with identical kinetic parameters. Despite these similarities, the sizes and amino acid compositions of the two toxins differ; there was no detectable cysteine residue in *S. intermedius* beta-hemolysin. The available N-terminal amino acid sequence of *S. intermedius* beta-hemolysin shows only 9 of 19 residues identical to *S. aureus* beta-hemolysin, confirming the lack of identity between the two toxins (24). A synergohymenotropic toxin (SHT) that is produced by *S. intermedius* isolates has been characterized. Like the Panton-Valentine leucocidin (PVL), it is made of two components (LukS-I and LukF-I) encoded in the *luk-I* operon and is leukotoxic on polymorphonuclear leukocytes from various species. It is weakly hemolytic on rabbit erythrocytes (63).

Production of enterotoxins A, C, D, and E and TSST-1 by strains of *S. intermedius* isolated from dog infections (4, 36) and from butter-blend products responsible for a food intoxication (42) has been reported. From these toxins, only enterotoxin C (SEC) has been characterized at the molecular level. Called SEC canine, it is a 239-amino-acid protein with >95% sequence identity with the SEC variants produced by *S. aureus* (SEC1, SEC2, SEC3). Purified SEC canine induces an emetic response in monkeys and induces proliferation of T cells in a V_β-dependent manner with the same profile as that induced by SEC1 (25) (Table 4). Production of urease that may act as an invasion factor has been reported (Table 1).

STAPHYLOCOCCUS COHNII

S. cohnii subsp. *urealyticum* and *S. cohnii* subsp. *cohnii* are recently designated subspecies of the species *cohnii* (48). These subspecies (formerly subspecies 1 and 2) differ by phenotypic and metabolic properties and in their host range: *S. cohnii* subsp. *urealyticum* has been isolated from both humans and other primates, whereas *S. cohnii* subsp. *cohnii* has been isolated only from humans (45). *S. cohnii* has been isolated from human urinary tract and wound infections, septic arthritis, and meningitis. Susceptibility to benzylpenicillin, erythromycin, tetracycline, and chloramphenicol is variable. Methicillin-resistant strains have been isolated. A staphylococcal plasmid carrying two novel genes, *vatC* and *vgbB*, encoding resistance to streptogramin A and B antibiotics has been recently described. *vatC* encodes a 212-amino-acid acetyltransferase that inactivates streptogramin A and exhibits 58.2 to 69.8% amino acid identity with the staphylococcal Vat and VatB protein and the *Enterococcus faecium* SatA protein. *vgbB* encodes a 295-amino-acid lactonase that inactivates streptogramin B and shows 67% amino acid identity with the staphylococcal Vgb lactonase (3).

S. cohnii binds to fibronectin, collagen, vitronectin, and laminin immobilized on latex beads (57, 62) (Table 2). A synergistic hemolysin that is produced by the majority of the isolates has been attributed to a *slush* homolog by Southern blot hybridization (Table 3) (23). The production of enteroxin by strains isolated from sheep's milk (10), detected by ELISA, has been reported (Table 4). Enterotoxin function has not been confirmed by mitogenic or emetic assays. Production of enzymes, including lipase and FAME (50, 55), that may act as invasion factors has been reported (Table 1).

STAPHYLOCOCCUS SCIURI

S. sciuri (*S. sciuri* subsp. *sciuri*, *S. sciuri* subsp. *carnaticus*, and *S. sciuri* subsp. *rodentium*) is commonly isolated from the skin of rodents and somewhat less frequently from the skin of ungulates, carnivora, and marsupials (44, 65). A homolog of the *S. aureus* methicillin resistance gene *mecA* was recently shown to be ubiquitous in 134 independent isolates of *S. sciuri*. Among these, isolates of *S. sciuri* subsp. *sciuri* and *S. sciuri* subsp. *carnaticus* showed only marginal, if any, resistance to methicillin (MIC, 0.75–6.0 μg/ml), while most isolates of *S. sciuri* subsp. *rodentium* expressed a heterogeneous methicillin resistance phenotype. Investigation of the genetic organization of the *mecA* region in the three subspecies revealed that *S. sciuri* strains can contain two different forms of *mecA*. One form is virtually identical to the *mecA* gene of *S. aureus* and is present in *S. sciuri* subsp. *rodentium*; it is proposed that these strains may have acquired a *mecA* gene from *S. aureus* or *S. epidermidis* (85). The second form, which can coexist with the first one in *S. sciuri* subsp. *rodentium*, shows somewhat less similarity (79.5%) to *mecA* of *S. aureus* and appears to be the predominant form in *S. sciuri* subspecies. Strains carrying only this form are suceptible to methicillin, suggesting that it is a silent gene that may be an evolutionary relative or precursor of the *mecA* gene of *S. aureus* (85).

Resistance to lincosamide and streptogramin A with susceptibility to erythromycin has been observed in some strains of *S. sciuri* subsp. *sciuri* from human origin and could not be attributed to any of the known genes conferring resistance to macrolide-lincosamide-streptogramin antibiotics (54a).

STAPHYLOCOCCUS HYICUS

S. hyicus (*S. hyicus* subsp. *hyicus* and *S. hyicus* subsp. *chromogenes*) is an opportunistic pathogen found in pigs and cattle (45, 65). *S. hyicus* binds to vitronectin, laminin, collagen, and fibronectin immobilized on latex beads (58, 62). *S. hyicus* strains isolated from pigs (but not from cows) produce a 42-kDa immunoglobulin binding protein, specific for the Fc domain, that shows close functional and antigenic similarity with protein A of *S. aureus* (32).

S. hyicus strains isolated from pigs affected with exudative epidermitis produce three antigenically distinct exfoliative toxins of approximately 30 kDa designated ExhA, ExhB, and ExhC, apparently not similar to *S. aureus* ETA and ETB (5). These toxins produce exfoliation in piglet and chicken skin assays. However, exfoliation was not observed using other animals (mouse, rat, guinea pig, hamster, dog, or cat).

S. hyicus subsp. *hyicus* produces a lipase termed SHL (Table 1) that is secreted as an 86-kDa proenzyme and is processed to the mature 46-kDa enzyme by the extracellular metalloprotease ShpII (6). Evidence has been presented showing that the pro-region of the lipase acts as an intramolecular chaperone which facilitates translocation of the native lipase and of a number of completely unrelated proteins fused to the propeptide. It was also observed that the pro-region protects the proteins from proteolytic degradation (31). SHL is exceptional in that it displays a high phospholipase activity, hydrolyzes neutral lipids, and has no chain length preference. Site-directed mutagenesis and domain exchange were used to determine that in the C-terminal domain it is Ser-356 which mainly determines phospholipase activity (74).

The production of enterotoxin by strains isolated from healthy goats (73) has been detected by ELISA. In another study, the production of an emetic toxin has been observed in a monkey emetic model (Table 4). However, the emetic toxin was not immunologically reactive with enterotoxins A through E (1). Production of urease which may act as an invasion factor has been reported (Table 1).

STAPHYLOCOCCUS CARNOSUS

S. carnosus is used as starter culture in the production of dry sausage and fermented fish. It is poorly pathogenic, and its natural habitat has not been determined. The physical and genetic map of S. carnosus chromosome has been determined (82), and specific vectors have been developed for gene cloning and expression of heterologous protein in this species, including a highly efficient surface display expression system. For instance, the cholera toxin B subunit (CTB) from Vibrio cholerae was properly expressed in this system; thus the cell surface display of heterologous receptors on S. carnosus could be considered as potential live bacterial vaccine delivery systems for administration by the mucosal route (54).

STAPHYLOCOCCUS GALLINARUM

S. gallinarum derives its name from its avian host, mainly the skin of poultry, but it may also be found in other birds (45, 65). S. gallinarum produces the bacteriocin gallidermin, which exhibits activity against propionibacteria. Like epidermin, Pep5, and epilancin K7 from S. epidermidis, gallidermin is a member of the class of lanthionine-containing peptide antibiotics also designated lantibiotics. Gallidermin is highly homologous to epidermin; it differs only in a Leu/Ile exchange in position 6. The biosynthesis of all of these lantibiotics proceeds from structural genes which code for prepeptides that are enzymatically modified to give the mature peptides. The additional genes coding for transporters, immunity functions, regulatory proteins, and the modification enzymes which catalyze the biosynthesis of the rare amino acids are found in gene clusters adjacent to the structural genes (60).

REFERENCES

1. **Adesiyun, A. A., S. R. Tatini, and D. G. Hoover.** 1984. Production of enterotoxin(s) by Staphylococcus hyicus. Vet. Microbiol. **9:**487–495.

2. **Allaker, R. P., N. Garrett, L. Kent, W. C. Noble, and D. H. Lloyd.** 1993. Characterisation of Staphylococcus intermedius isolates from canine pyoderma and from healthy carriers by SDS-PAGE of exoproteins, immunoblotting and restriction endonuclease digest analysis. J. Med. Microbiol. **39:**429–433.

3. **Allignet, J., N. Liassine, and N. El Solh.** 1998. Characterization of a staphylococcal plasmid related to pUB110 and carrying two novel genes, vatC and vgbB, encoding resistance to streptogramins A and B and similar antibiotics. Antimicrob. Agents Chemother. **42:**1794–1798.

4. **Almazan, J., R. de la Fuente, E. Gomez-Lucia, and G. Suarez.** 1987. Enterotoxin production by strains of Staphylococcus intermedius and Staphylococcus aureus isolated from dog infections. Zentralbl. Bakteriol. Mikrobiol. Hyg. A **264:**29–32.

5. **Andresen, L. O.** 1998. Differentiation and distribution of three types of exfoliative toxin produced by Staphylococcus hyicus from pigs with exudative epidermitis. FEMS Immunol. Med. Microbiol. **20:**301–310.

6. **Ayora, S., P. E. Lindgren, and F. Gotz.** 1994. Biochemical properties of a novel metalloprotease from Staphylococcus hyicus subsp. hyicus involved in extracellular lipase processing. J. Bacteriol. **176:**3218–3223.

7. **Barcs, I., and L. Janosi.** 1992. Plasmids encoding for erythromycin ribosomal methylase of Staphylococcus epidermidis and Staphylococcus simulans. Acta Microbiol. Hung. **39:**85–92.

8. **Barnham, M., R. Horton, J. M. P. Smith, J. Richardson, R. R. Marples, and S. Reith.** 1996. Methicillin-resistant Staphylococcus simulans masquerading as MRSA in a nursing home. J. Hosp. Infect. **34:**331–337.

9. **Bassias, J., and R. Bruckner.** 1998. Regulation of lactose utilization genes in Staphylococcus xylosus. J. Bacteriol. **180:**2273–2279.

10. **Bautista, L., P. Gaya, M. Medina, and M. Nunez.** 1988. A quantitative study of enterotoxin production by sheep milk staphylococci. Appl. Environ. Microbiol. **54:**566–569.

11. **Benito, Y., G. Lina, T. Greenland, J. Etienne, and F. Vandenesch.** 1998. trans-Complementation of Staphylococcus aureus agr mutant by S. lugdunensis agr-RNAIII. J. Bacteriol. **180:**5780–5783.

12. **Biberstein, E. L., S. Jang, and D. C. Hirsh.** 1984. Species distribution of coagulase-positive staphylococci in animals. J. Clin. Microbiol. **19:**610–615.

13. **Billot-Klein, D., L. Gutmann, D. Bryant, D. Bell, J. van Heijenoort, J. Grewal, and D. M. Shlaes.** 1996. Peptidoglycan synthesis and structure in Staphylococcus haemolyticus expressing increasing levels of resistance to glycopeptide antibiotics. J. Bacteriol. **178:**4696–4703.

14. **Burriel, A. R.** 1998. In vivo presence of capsular polysaccharide in coagulase-negative staphylococci of ovine origin. New Microbiol. **21:**49–54.

15. **Buttery, J. P., M. Easton, S. R. Pearson, and G. G. Hogg.** 1997. Pediatric bacteremia due to Staphylococcus warneri: microbiological, epidemiological, and clinical features. J. Clin. Microbiol. **35:**2174–2177.

16. **Chaouni, L., T. Greenland, J. Etienne, and F. Vandenesch.** 1996. Nucleic acid sequence and affiliation of pLUG10, a novel cadmium resistance plasmid from Staphylococcus lugdunensis. Plasmid **36:**1–8.

17. **Chesneau, O., A. Morvan, F. Grimont, H. Labischinski, and N. El Solh.** 1993. Staphylococcus pasteuri sp. nov., isolated from human, animal, and food specimens. Int. J. Syst. Bacteriol. **43:**237–244.

18. **Cintas, L. M., P. Casaus, H. Holo, P. E. Hernandez, I. F. Nes, and L. S. Havarstein.** 1998. Enterocins L50A and L50B, two novel bacteriocins from Enterococcus faecium L50, are related to staphylococcal hemolysins. J. Bacteriol. **180:**1988–1994.

19. **Climo, M. W., R. L. Patron, B. P. Goldstein, and G. L. Archer.** 1998. Lysostaphin treatment of experimental methicillin-resistant Staphylococcus aureus aortic valve endocarditis. Antimicrob. Agents Chemother. **42:**1355–1360.

20. **Dacosta, A., H. Lelièvre, G. Kirkorian, M. Célard, P. Chevalier, F. Vandenesch, J. Etienne, and P. Touboul.** 1998. Role of the preaxillary flora in pacemaker infections. A prospective study. Circulation **97:**1791–1795.

21. **De Hondt, G., M. Ieven, C. Vandermersch, and J. Colaert.** 1997. Destructive endocarditis caused by Staphylococcus lugdunensis. Case report and review of the literature. Acta Clin. Belg. **52:**27–30.

22. **Donvito, B., J. Etienne, L. Denoroy, T. Greenland, Y. Benito, and F. Vandenesch.** 1997. Synergistic hemolytic activity of *Staphylococcus lugdunensis* is mediated by three peptides encoded by a non-*agr* genetic locus. *Infect. Immun.* **65:**95–100.

23. **Donvito, B., J. Etienne, T. Greenland, C. Mouren, V. Delorme, and F. Vandenesch.** 1997. Distribution of the synergistic haemolysin genes *hld* and *slush* with respect to *agr* in human staphylococci. *FEMS Microbiol. Lett.* **151:** 139–144.

24. **Dziewanowska, K., V. M. Edwards, J. R. Deringer, G. A. Bohach, and D. J. Guerra.** 1996. Comparison of the beta-toxins from *Staphylococcus aureus* and *Staphylococcus intermedius*. *Arch. Biochem. Biophys.* **335:**102–108.

25. **Edwards, V. M., J. R. Deringer, S. D. Callantine, C. F. Deobald, P. H. Berger, V. Kapur, C. V. Stauffacher, and G. A. Bohach.** 1997. Characterization of the canine type C enterotoxin produced by *Staphylococcus intermedius* pyoderma isolates. *Infect. Immun.* **65:**2346–2352.

26. **Fiegler, H., and R. Bruckner.** 1997. Identification of the serine acetyltransferase gene of *Staphylococcus xylosus*. *FEMS Microbiol. Lett.* **148:**181–187.

26a.**François, P., and P. Vaudaux.** Personal communication.

27. **Fthenakis, G. C., R. R. Marples, J. F. Richardson, and J. E. Jones.** 1994. Some properties of coagulase-negative staphylococci isolated from cases of ovine mastitis. *Epidemiol. Infect.* **112:**171–176.

28. **Gatermann, S., and R. Marre.** 1989. Cloning and expression of *Staphylococcus saprophyticus* urease gene sequences in *Staphylococcus carnosus* and contribution of the enzyme to virulence. *Infect. Immun.* **57:**2998–3002.

29. **Gatermann, S. G., and K. B. Crossley.** 1997. Urinary tract infections, p. 493–508. *In* K. B. Crossley and G. L. Archer (ed.), *The Staphylococci in Human Disease.* Churchill Livingstone Inc., New York, N.Y.

30. **Giovanetti, E., F. Biavasco, A. Pugnaloni, R. Lupidi, G. Biagini, and P. E. Varaldo.** 1996. An electron microscopic study of clinical and laboratory-derived strains of teicoplanin-resistant *Staphylococcus haemolyticus*. *Microb. Drug Resist.* **2:**239–243.

31. **Gotz, F., H. M. Verheij, and R. Rosenstein.** 1998. Staphylococcal lipases: molecular characterisation, secretion, and processing. *Chem. Phys. Lipids* **93:**15–25.

32. **Greene, R. T., and C. Lammler.** 1992. Isolation and characterization of immunoglobulin binding proteins from *Staphylococcus intermedius* and *Staphylococcus hyicus*. *Zentralbl. Veterinaermed. B* **39:**519–525.

33. **Hajek, V., H. Meugnier, M. Bes, Y. Brun, F. Fiedler, Z. Chmela, Y. Lasne, J. Fleurette, and J. Freney.** 1996. *Staphylococcus saprophyticus* subsp. *bovis* subsp. nov., isolated from bovine nostrils. *Int. J. Syst. Bacteriol.* **46:** 792–796.

34. **Hébert, G. A.** 1990. Hemolysins and other characteristics that help differentiate and biotype *Staphylococcus lugdunensis* and *Staphylococcus schleiferi*. *J. Clin. Microbiol.* **28:** 2425–2431.

35. **Hébert, G. A., C. G. Crowder, G. A. Hancock, W. R. Jarvis, and C. Thornsberry.** 1988. Characteristics of coagulase-negative staphylococci that help differentiate these species and other members of the family Micrococcaceae. *J. Clin. Microbiol.* **26:**1939–1949.

36. **Hirooka, E. Y., E. E. Muller, J. C. Freitas, E. Vicente, Y. Yoshimoto, and M. S. Bergdoll.** 1988. Enterotoxigenicity of *Staphylococcus intermedius* of canine origin. *Int. J. Food Microbiol.* **7:**185–191.

37. **Igimi, S., E. Takahashi, and T. Mitsuoka.** 1990. *Staphylococcus schleiferi* subsp. *coagulans* subsp. nov., isolated from the external auditory meatus of dogs with external ear otitis. *Int. J. Syst. Bacteriol.* **40:**409–411.

38. **Jansen, B., F. Schumacher-Perdreau, G. Peters, G. Reinhold, and J. Schonemann.** 1992. Native valve endocarditis caused by *Staphylococcus simulans*. *Eur. J. Clin. Microbiol. Infect. Dis.* **11:**268–269. (Letter.)

39. **Ji, G., R. Beavis, and R. P. Novick.** 1997. Bacterial interference caused by autoinducing peptide variants. *Science* **276:**2027–2030.

40. **Jose, J., S. Christians, R. Rosenstein, F. Gotz, and H. Kaltwasser.** 1991. Cloning and expression of various staphylococcal genes encoding urease in *Staphylococcus carnosus*. *FEMS Microbiol Lett.* **64:**277–281.

41. **Kazembe, P., A. E. Simor, A. E. Swarney, L. G. Yap, B. Kreiswirth, J. Ng, and D. E. Low.** 1993. A study of the epidemiology of an endemic strain of *Staphylococcus haemolyticus* (TOR-35) in a neonatal intensive care unit. *Scand. J. Infect. Dis.* **25:**507–513.

42. **Khambarty, F. M., R. W. Bennett, and D. B. Shah.** 1994. Application of pulsed-field gel electrophoresis to the epidemiological characterization of *Staphylococcus intermedius* implicated in a food-related outbreak. *Epid. Infect.* **113:** 75–81.

43. **Kloos, W. E.** 1986. Ecology of human skin, p. 37–50. *In* P. A. Mardh and K. H. Schleifer (ed.), *Coagulase-Negative Staphylococci*. Almqvist & Wiksell International, Stockholm, Sweden.

44. **Kloos, W. E., D. N. Ballard, J. A. Webster, R. J. Hubner, A. Tomasz, I. Couto, G. L. Sloan, H. P. Dehart, F. Fiedler, K. Schubert, H. de Lencastre, I. S. Sanches, H. E. Heath, P. A. Leblanc, and A. Ljungh.** 1997. Ribotype delineation and description of *Staphylococcus sciuri* subspecies and their potential as reservoirs of methicillin resistance and staphylolytic enzyme genes. *Int. J. Syst. Bacteriol.* **47:**313–323.

45. **Kloos, W. E., and T. L. Bannerman.** 1994. Update on clinical significance of coagulase-negative staphylococci. *Clin. Microbiol. Rev.* **7:**117–140.

46. **Kloos, W. E., C. G. George, J. S. Olgiate, P. L. Van, M. L. McKinnon, B. L. Zimmer, E. Muller, M. P. Weinstein, and S. Mirrett.** 1998. *Staphylococcus hominis* subsp. *novobiosepticus* subsp. nov., a novel trehalose- and N-acetyl-D-glucosamine-negative, novobiocin- and multiple-antibiotic-resistant subspecies isolated from human blood cultures. *Int. J. Syst. Bacteriol.* **3:**799–812.

47. **Kloos, W. E., and J. F. Wolfshohl.** 1979. Evidence of desoxyribonucleotide sequence divergence between staphylococci living on human and other primate skin. *Curr. Microbiol.* **3:**167–172.

48. **Kloos, W. E., and J. F. Wolfshohl.** 1991. *Staphylococcus cohnii* subspecies: *Staphylococcus cohnii* subsp. *cohnii* subsp. nov. and *Staphylococcus cohnii* subsp. *urealyticum* subsp. nov. *Int. J. Syst. Bacteriol.* **41:**284–289.

49. **Kluytmans, J., H. Berg, P. Steegh, F. Vandenesch, J. Etienne, and A. van Belkum.** 1998. Outbreak of *Staphylococcus schleiferi* wound infections: strain characterization by randomly amplified polymorphic DNA analysis, PCR

ribotyping, conventional ribotyping, and pulsed-field gel electrophoresis. *J. Clin. Microbiol.* **36:**2214–2219.

50. **Krzeminski, Z., and A. Raczynska.** 1990. Elastolytic activity of staphylococci isolated from human oral cavity. *Med. Dosw. Mikrobiol.* **42:**1–4.

51. **Lambe, D. W., K. P. Fergusson, J. L. Keplinger, C. G. Gemmell, and J. H. Kalbfleisch.** 1990. Pathogenicity of *Staphylococcus lugdunensis, Staphylococcus schleiferi,* and three other coagulase-negative staphylococci in a mouse model and possible virulence factors. *Can. J. Microbiol.* **36:**455–463.

52. **Latorre, M., P. M. Rojo, R. Franco, and R. Cisterna.** 1993. Endocarditis due to *Staphylococcus capitis* subspecies *ureolyticus. Clin. Infect. Dis.* **16:**343–344.

53. **Lee, J.** 1994. *Staphylococcus intermedius* isolated from dog-bite wounds. *J. Infect.* **29:**105–118.

54. **Liljeqvist, S., P. Samuelson, M. Hansson, T. N. Nguyen, H. Binz, and S. Stahl.** 1997. Surface display of the cholera toxin B subunit on *Staphylococcus xylosus* and *Staphylococcus carnosus. Appl. Environ. Microbiol.* **63:**2481–2488.

54a.**Lina, G., A. Quaglia, M. E. Reverdy, R. Leclercq, F. Vandenesch, and J. Etienne.** 1999. Distribution of genes encoding resistance to macrolides, lincosamides, and streptogramins among staphylococci. *Antimicrob. Agents Chemother.* **43:**1062–1066.

55. **Long, J. P., J. Hart, W. Albers, and F. A. Kapral.** 1992. The production of fatty acid modifying enzyme (FAME) and lipase by various staphylococcal species. *J. Med. Microbiol.* **37:**232–234.

56. **Mahoudeau, I., X. Delabranche, G. Prevost, H. Monteil, and Y. Piemont.** 1997. Frequency of isolation of *Staphylococcus intermedius* from humans. *J. Clin. Microbiol.* **35:** 2153–2154.

57. **Miedzobrodzki, J., A. S. Naidu, J. L. Watts, P. Ciborowski, K. Palm, and T. Wadstrom.** 1989. Effect of milk on fibronectin and collagen type I binding to *Staphylococcus aureus* and coagulase-negative staphylococci isolated from bovine mastitis. *J. Clin. Microbiol.* **27:**540–544.

58. **Minhas, T., H. A. Ludlam, M. Wilks, and S. Tabaqchali.** 1995. Detection by PCR and analysis of the distribution of a fibronectin- binding protein gene (*fbn*) among staphylococcal isolates. *J. Med. Microbiol.* **42:**96–101.

58a.**Novick, R.** Personal communication.

59. **Orden, J. A., J. Goyache, J. Hernandez, A. Domenech, G. Suarez, and E. Gomez-Lucia.** 1992. Production of staphylococcal enterotoxins and TSST-1 by coagulase negative staphylococci isolated from ruminant mastitis. *Zentralbl. Veterinaermed.* B **39:**144–148.

60. **Ottenwalder, B., T. Kupke, S. Brecht, V. Gnau, J. Metzger, G. Jung, and F. Gotz.** 1995. Isolation and characterization of genetically engineered gallidermin and epidermin analogs. *Appl. Environ. Microbiol.* **61:**3894–3903.

61. **Paulsson, M., A. Ljungh, and T. Wadstrom.** 1992. Rapid identification of fibronectin, vitronectin, laminin, and collagen cell surface binding proteins on coagulase-negative staphylococci by particle agglutination assays. *J. Clin. Microbiol.* **30:**2006–2012.

62. **Paulsson, M., C. Petersson, and A. Ljungh.** 1993. Serum and tissue protein binding and cell surface properties of *Staphylococcus lugdunensis. J. Med. Microbiol.* **38:**96–102.

62a.**Peacock, S. J., G. Lina, J. Etienne, and T. Foster.** 1999. *Staphylococcus schleiferi* subsp. *schleiferi* expresses a fibronectin-binding protein. *Infect. Immun.* **67:**4272–4275.

63. **Prevost, G., T. Bouakham, Y. Piemont, and H. Monteil.** 1995. Characterisation of a synergohymenotropic toxin produced by *Staphylococcus intermedius. FEBS Lett.* **376:** 135–140.

64. **Raus, J., and D. N. Love.** 1990. Comparison of the staphylocoagulase activities of *Staphylococcus aureus* and *Staphylococcus intermedius* on Chromozym-TH. *J. Clin. Microbiol.* **28:**207–210.

65. **Schleifer, K. H.** 1986. Gram-positive cocci, p. 999–1103. *In* P. H. A. Sneath, N. S. Mair, M. E. Sharpe, and J. G. Holt (ed.), *Bergey's Manual of Systematic Bacteriology.* The Williams & Wilkins Co., Baltimore, Md.

66. **Shimizu, A., H. A. Berkhoff, W. E. Kloos, C. G. George, and D. N. Ballard.** 1996. Genomic DNA fingerprinting, using pulsed-field gel electrophoresis, of *Staphylococcus intermedius* isolated from dogs. *Am. J. Vet. Res.* **57:**1458–1462.

67. **Shuttleworth, R., R. J. Behme, A. McNabb, and W. D. Colby.** 1997. Human isolates of *Staphylococcus caprae:* association with bone and joint infections. *J. Clin. Microbiol.* **35:**2537–2541.

68. **Stout, R. D., Y. Li, A. R. Miller, and D. W. J. R. Lambe.** 1994. Staphylococcal glycocalyx activates macrophage prtostaglandin E2 and interleukin 1 production and modulates tumor necrosis factor alpha and nitric oxide production. *Infect. Immun.* **62:**4160–4166.

68a.**Sugai, M.** Personal communication.

69. **Sugai, M., T. Fujiwara, K. Ohta, H. Komatsuzawa, M. Ohara, and H. Suginaka.** 1997. *epr,* which encodes glycylglycine endopeptidase resistance, is homologous to *femAB* and affects serine content of peptidoglycan cross bridges in *Staphylococcus capitis* and *Staphylococcus aureus. J. Bacteriol.* **179:**4311–4318.

70. **Talon, R., N. Dublet, M. C. Montel, and M. Cantonnet.** 1995. Purification and characterization of extracellular *Staphylococcus warneri* lipase. *Curr. Microbiol.* **30:**11–16.

71. **Tegmark, K., E. Morfeldt, and S. Arvidson.** 1998. Regulation of *agr*-dependent virulence genes in *Staphylococcus aureus* by RNAIII from coagulase-negative staphylococci. *J. Bacteriol.* **180:**3181–3186.

72. **Thumm, G., and F. Gotz.** 1997. Studies on prolysostaphin processing and characterization of the lysostaphin immunity factor (Lif) of *Staphylococcus simulans* biovar *staphylolyticus. Mol. Microbiol.* **23:**1251–1265.

73. **Valle, J., E. Gomez-Lucia, S. Piriz, J. Goyache, J. A. Orden, and S. Vadillo.** 1990. Enterotoxin production by staphylococci isolated from healthy goats. *Appl. Environ. Microbiol.* **56:**1323–1326.

74. **Van Kampen, M., J. W. Simons, N. Dekker, M. R. Egmond, and H. M. Verheij.** 1998. The phospholipase activity of *Staphylococcus hyicus* lipase strongly depends on a single Ser to Val mutation. *Chem. Phys. Lipids* **93:**39–45.

75. **Vandenesch, F., S. J. Eykyn, M. Bes, H. Meugnier, J. Fleurette, and J. Etienne.** 1995. Identification and ribotypes of *Staphylococcus caprae* isolates isolated as human pathogens and from goat milk. *J. Clin. Microbiol.* **33:** 888–892.

76. **Vandenesch, F., S. J. Eykyn, J. Etienne, and J. Lemozy.** 1995. Skin and post-surgical wound infections due to *Staphylococcus lugdunensis. Clin. Microbiol. Infect.* **1:**73–74.

77. **Vandenesch, F., S. J. Eykyn, and J. Etienne.** 1995. Infections caused by newly-described species of coagulase-negative staphylococci. *Rev. Med. Microbiol.* **6:**94–100.

78. **Vandenesch, F., C. Lebeau, M. Bes, G. Lina, B. Lina, T. Greenland, Y. Benito, Y. Brun, J. Fleurette, and J. Etienne.** 1994. Clotting activity in *Staphylococcus schleiferi* subspecies from human patients. *J. Clin. Microbiol.* **32:**388–392.

79. **Vandenesch, F., S. Projan, B. Kreiswirth, J. Etienne, and R. P. Novick.** 1993. *agr* related sequences in *Staphylococcus lugdunensis.* FEMS *Microbiol. Lett.* **111:**115–122.

80. **Vandenesch, F., M. J. Storrs, F. Poitevin-Later, J. Etienne, P. Courvalin, and J. Fleurette.** 1991. Delta-like haemolysin produced by *Staphylococcus lugdunensis.* FEMS *Microbiol. Lett.* **78:**65–68.

81. **Vernozy-Rozand, C., C. Mazuy, G. Prevost, C. Lapeyre, M. Bes, Y. Brun, and J. Fleurette.** 1996. Enterotoxin production by coagulase-negative staphylococci isolated from goats' milk and cheese. *Int. J. Food Microbiol.* **30:**271–280.

82. **Wagner, E., J. Doskar, and F. Gotz.** 1998. Physical and genetic map of the genome of *Staphylococcus carnosus* TM300. *Microbiology* **144:**509–517.

83. **Watson, D. C., M. Yaguchi, J. G. Bisaillon, R. Beaudet, and R. Morosoli.** 1988. The amino acid sequence of a gonococcal growth inhibitor from *Staphylococcus haemolyticus.* *Biochem. J.* **252:**87–93.

84. **Weiss, K., D. Rouleau, and M. Laverdiere.** 1996. Cystitis due to vancomycin-intermediate *Staphylococcus saprophyticus.* *J. Antimicrob. Chemother.* **37:**1039–1040.

85. **Wu, S., H. de Lencastre, and A. Tomasz.** 1998. Genetic organization of the *mecA* region in methicillin-susceptible and methicillin-resistant strains of *Staphylococcus sciuri.* *J. Bacteriol.* **180:**236–242.

Antibiotic Resistance in the Staphylococci

STEVEN J. PROJAN

48

This chapter summarizes specific resistance mechanisms found in the staphylococci. Several other chapters in this section discuss resistance to various specific classes of antimicrobial agents across the gram-positive spectrum.

The importance of antimicrobial chemotherapy for the treatment of staphylococcal infections cannot be overstated. Before the widespread use of antimicrobials, *Staphylococcus aureus* bacteremia was fatal approximately 90% of the time (51). While the therapeutic use of penicillin G in the early 1940s greatly reduced mortality, the first resistant strains were described almost immediately (17). Today we have at our disposal a large number of nominally efficacious antimicrobial agents that are active against the staphylococci. Unfortunately, we are also confronted with a dazzling array of resistance determinants and mutant strains compromising the utility of all but one of the classes of antimicrobial agents against one of the most virulent of pathogenic organisms.

THE CONSEQUENCE OF ANTIMICROBIAL RESISTANCE IN STAPHYLOCOCCI

What are the consequences of resistance with respect to the staphylococci? First and foremost, resistance effectively reduces the number of therapeutic options available to treat staphylococcal infections. While actual reports of clinical failures of antimicrobial chemotherapy for staphylococcal infections are rare, they are increasing in frequency. In addition, there are data that suggest a far greater risk to the patient infected with a resistant organism as opposed to a patient infected with a susceptible strain of staphylococci. In at least one report, the attributable mortality for patients infected with strains of methicillin-resistant *S. aureus* (MRSA) was 10-fold higher than that for patients infected with methicillin-susceptible *S. aureus* strains (35). However, it should be pointed out that some infections, even those caused by susceptible staphylococci, do not respond well to antimicrobial chemotherapy; among these are osteomyelitis and bacterial endocarditis, which often require surgical intervention (63).

RESISTANCE AND THE SOURCE OF THE INFECTION

When discussing resistance among the staphylococci, it is important to draw a distinction between community-acquired and hospital-acquired (nosocomial) infections. Almost uniformly, resistance rates among community isolates are significantly lower than resistance rates for nosocomial isolates. Ascertaining the source of an infection can, therefore, have important implications in choosing a course of antimicrobial chemotherapy. The fact that hospital isolates appear to have higher rates of resistance implies that these isolates are epidemiologically distinct from community isolates, and the molecular biology of these strains certainly bears that hypothesis out. This also implies that nosocomial isolates represent microflora that are resident in the hospital and are either transferred from patient to patient via health care workers or are transferred from colonized health care workers to patients. As surgical site infection is one of the nosocomial infections often attributed to staphylococci, one may wonder how, before the advent of antimicrobial chemotherapy, such infections were prevented, given our current reliance on antibiotic prophylaxis. One practice was to perform the surgery in a mist of carbolic acid (phenol), which, while not well tolerated by many participating in such surgeries, did serve to prevent a large number of infections. The modern lesson to be drawn from this is that we now use antibiotics as a substitute for good hygiene and sound surgical practice.

FACTORS THAT HAVE LED TO THE EMERGENCE AND DISSEMINATION OF RESISTANCE

Discussions on the root cause(s) of bacterial drug resistance often generate more heat than light. These discussions are complicated by a lack of supporting data for many of the hypotheses brought forward to explain why we observe resistance. However, it is clear that, as stated by Levy, "Given sufficient time and drug use, antibiotic resistance will emerge" (29).

Gram-Positive Pathogens, ed. by V. A. Fischetti et al.
© 2000 American Society for Microbiology, Washington, D.C.

Antimicrobial resistance among the staphylococci is not a phenomenon limited to antibiotics. Strains resistant to arsenicals and mercury were identified well before what is now known as the antibiotic era, but for the purposes of this chapter the discussion will be limited to agents currently used clinically to treat staphylococcal infections (Table 1).

MECHANISMS OF RESISTANCE

It is now dogma that resistance falls into one of three mechanistic classes: (i) prevention of accumulation within the bacterial cells usually via efflux of the agent out of the bacterial cell by either dedicated or general efflux pumps (34), (ii) alteration of the molecular target of the antibiotic, or (iii) inactivation of the antibiotic. In fact, even for the above list there are some resistance genes that do not fall neatly into any of these three categories (e.g., resistance to the semisynthetic tetracycline minocycline via ribosome protection), and, in the case of some if not most classes of drugs, more than one resistance mechanism is at play.

GENETICS OF RESISTANCE

From the genetic point of view, resistance falls into one of two classes: mutation of a bacterial gene or acquisition of a dedicated resistance gene from some other organism by some form of genetic exchange (transduction, conjugation, or transformation). In general, resistance via mutation is

TABLE 1 Agents used to treat staphylococcal infections

Agent	Molecular target
Cell wall biosynthesis inhibitors	
Beta-lactams (ampicillin)	Penicillin-binding proteins (transpeptidation)
Glycopeptides (vancomycin, teicoplanin)	MurNac pentapeptide (transglycosylation)
Transcription inhibitors	
Rifampin	RNA polymerase
DNA inhibitors	
Fluoroquinolones (ciprofloxacin)	DNA gyrase; topoisomerase IV
Protein synthesis inhibitors	
Aminoglycosides (gentamicin)	Ribosome
Tetracyclines (minocycline, doxycycline)	
Chloramphenicol	
Macrolides-lincosamides-streptogramin B (erythromycin)	
Streptogramin A	
Pseudomonic acid (mupirocin)	
Fusidic acid	
Essential small-molecule biosynthesis inhibitors	
Sulfonamides	Dihydropterate synthetase
Trimethoprim	Dihydrofolate reductase

an alteration of the target site of the antibiotic, although increased expression of either the target (and titration of the antimicrobial agent) or of a nonspecific efflux pump also can be the result of a mutation. Acquired resistance determinants are, by and large, dedicated to narrow classes of compounds but can run the gamut of mechanisms and are usually inducible (meaning that the gene encoding the resistance is expressed by the host bacterium when it is exposed to the drug, at concentrations insufficient to inhibit bacterial growth). Dissemination of these acquired resistance determinants varies in frequency, depending on the type of genetic element carrying the resistance determinant.

As pointed out below (and in chapter 62), the presence in a strain of the methicillin-resistance determinant is not by itself sufficient for the bacterium to be phenotypically methicillin resistant. Apparently, a series of other genetic events must take place for methicillin resistance to be manifested. Although this case may be unique, this implies that acquisition of a resistance determinant by itself may not be sufficient for the expression of resistance.

BETA-LACTAM RESISTANCE

In the history of antimicrobial chemotherapy, the most useful antistaphylococcal agents have been the beta-lactam antibiotics, the prototype of which is penicillin. These agents, which include several structural classes (e.g., penams, penems, cephamycins, cephalosporins, carbapenems, and monobactams), all contain one common structural feature, the beta-lactam ring. This four-membered ring structure can be thought of as the first peptide mimetic in that the target of these drugs is the transpeptidase domain of the cell wall biosynthetic enzymes referred to as penicillin-binding proteins. Binding of the antibiotic blocks peptidoglycan biosynthesis and often results in the subsequent induction of autolysins, leading to lysis of the bacterial cells. As a rule, the beta-lactams are far more effective in killing growing (dividing) bacteria.

Resistance

Not long after the introduction of penicillin into clinical practice, the first beta-lactamase-producing staphylococci were described. The rapid dissemination of strains carrying genes encoding the beta-lactamases, enzymes that hydrolyze the beta-lactam ring and therefore inactivate the antibiotic, had the effect of almost ending the antibiotic era before it got started. Today over 90% of clinical isolates of S. aureus produce beta-lactamases.

However, semisynthetic penicillins were developed (e.g., methicillin and oxacillin) that could be hydrolyzed by the staphylococcal beta-lactamases only very poorly and were, for many years, extremely effective at controlling most staphylococcal infections. In fact, it was not until the advent of MRSA that antibiotic resistance among the staphylococci again became a public health concern. Subsequent to the discovery of the beta-lactamases in S. aureus, it was found that the genes encoding their production are all very similar, with only minor differences in amino acid sequence and substrate specificity. Despite these minor differences it was possible to classify beta-lactamase by serology (60). The genes encoding both the beta-lactamase structural gene and the two regulatory genes are found on a transposable element (Tn552) which itself is often found on a plasmid (46).

"Methicillin" Resistance

The term MRSA is something of a misnomer because these strains can often be resistant to all clinically available beta-lactam antibiotics, rendering these once-potent agents ineffective in treating staphylococcal infections. Recently, there have been several reports of novel beta-lactams that have good levels of activity against MRSA. However, none of these are in clinical development as of yet. Interestingly, methicillin is no longer used therapeutically, and many researchers who have long used the agent in their work are finding it difficult to obtain supplies for their studies; often, oxacillin is used instead. The genetic basis of this resistance to all beta-lactam antibiotics is the *mecA* gene, which encodes a penicillin-binding protein called PBP2a, which has a low affinity for beta-lactams but is still capable of transpeptidation. However, the mere presence of the *mecA* gene is not sufficient for high-level methicillin resistance. One of the most interesting phenomena among certain strains carrying the *mecA* gene is a high degree of heterotypic expression of resistance to methicillin. This heterotypy is revealed by population analyses of a genetically homogeneous culture of *S. aureus*. The observation is that a large proportion of the bacteria in the culture are apparently susceptible to the antibiotic (typically, oxacillin is used), and a relatively small proportion is resistant to concentrations as high as 800 μg/ml. The apparently resistant colonies, however, maintain the same heterotypic phenotype when recultured and similarly analyzed (58).

While part of this phenomenon may be associated with expression of the *mecA* gene, which apparently requires the genetic inactivation of the BlaI cognate MecI repressor for expression of PBP2a (33), mutations in genes that are still poorly defined can apparently convert this resistance to homogeneous high level, as described in chapter 62. Clearly, the phenomenon of heterotypic expression has complicated susceptibility testing.

GLYCOPEPTIDES

Because MRSA strains are uniformly resistant to beta-lactams and are often resistant to most if not all of the agents discussed below, the last remaining line of defense is the glycopeptide antibiotics vancomycin and teicoplanin. These antibiotics are also cell wall biosynthesis inhibitors and block polymerization by binding to the terminal D-Ala-D-Ala of the Mur-Nac-pentapeptide.

Resistance

While to date resistance to the glycopeptide antibiotics (vancomycin and teicoplanin) in the staphylococci has been limited to *Staphylococcus haemolyticus* (which is considered to be a special case not applicable to other staphylococci), a number of clinical isolates of *S. aureus* with reduced susceptibility to the glycopeptides have been described. These strains, called GISA (for glycopeptide intermediate-susceptible *S. aureus*), have arisen in a very few patients who have been subject to long-term antimicrobial chemotherapy using vancomycin. While the mechanism of this resistance is not yet clear, it is likely to be the result of several stepwise mutations yielding incrementally higher MICs. Similar mutants arise in laboratory strains, also subject to long-term selection.

However, it should be noted that the therapeutic use of teicoplanin is somewhat controversial. It has not been approved for use in the United States, and it has been sug-gested that teicoplanin, when used clinically, may actually select for vancomycin-resistant staphylococci similar to the GISA strains described above (50).

RIFAMPIN

Rifampin, a member of the rifamycin class of antibiotics, inhibits transcription by attacking the beta-subunit of RNA polymerase (61). Because RNA polymerase is required both for transcription and in the initiation of DNA replication, it is effectively a bifunctional antimicrobial and as such is especially potent in inhibiting the growth of most bacteria, including the staphylococci.

Resistance

Despite the fact that rifampin is one of the most potent antistaphylococcal agents in vitro, with MICs of 0.03 μg/ml and lower, resistance readily arises by mutation in the *rpoB* gene (which encodes the beta-subunit of RNA polymerase), resulting in a lower affinity for the antibiotic (8). Therefore, when rifampin is used therapeutically, it is often used in combination with other antibiotics (e.g., fusidic acid).

FLUOROQUINOLONES

The fluoroquinolone antimicrobials (which include ciprofloxacin, levofloxacin, and trovafloxacin) are one of the few classes of antibacterial agents that are not based on a natural product. Rather, these drugs are derivatives of nalidixic acid and have found wide utility owing to their broad spectrum of activity and oral bioavailability. The fluoroquinolones specifically target the bacterial type II topoisomerases, DNA gyrase and topoisomerase IV. These enzymes have double-strand breaking-and-joining activity, with DNA gyrase being responsible for introducing negative superhelical turns and topoisomerase IV for decatenation of the two daughter chromosomes on termination of replication. Treatment of bacteria with fluoroquinolones results in lethal accumulation of double-strand breaks. As with the beta-lactams, the fluoroquinolones are more active against dividing bacteria. Among the gram-positive bacteria, topoisomerase IV has been demonstrated to be the primary target of the fluoroquinolones, while DNA gyrase has been shown to be the primary target in gram-negative bacteria.

Resistance

Resistance to the fluoroquinolones is a result of a series of stepwise mutations in the genes encoding DNA gyrase (*gyrA* and *gyrB*) and topoisomerase IV (*grlA* and *grlB*). In addition, mutations that result in the up-regulation of an efflux pump (NorA) have been identified both in vitro and among clinically resistant isolates (24). It has been hypothesized that fluoroquinolones that would inhibit the activity of both DNA gyrase and topoisomerase IV equally would not be subject to the kind of bootstrapping of resistance that we currently observe (64). It has been suggested that natural products are poor starting points for the development of new antibiotics, because the producing organisms (which are usually bacteria) are immune to the action of the antibiotics they produce, and the genetic information for that immunity will eventually be transferred into the target pathogenic bacteria after environmental selection with the antibiotic (57). In fact, this is undoubtedly the case in tetracycline-producing strains (52), which harbor genes encoding tetracycline-specific efflux proteins clearly related to the efflux genes found in pathogenic bacteria (16). However, as we have seen with the fluoroquinolones, resistance can also readily arise de novo by mutation to a wholly synthetic class of agents as well as to natural antibiotics such as rifampin.

AMINOGLYCOSIDES

The aminoglycoside class of antibiotics, as exemplified by streptomycin, is second in quantity only to the penicillins in therapeutic use. Drugs in this class include amikacin, gentamicin, netilmicin, and tobramycin. In general, they are protein synthesis inhibitors that bind irreversibly to the 30S subunit of the bacterial ribosome. The result of this binding is that aminoacyl-tRNAs are apparently unable to bind productively to the acceptor site, preventing elongation of the peptide chain. Mutations in the streptomycin-binding site can result in less effective inhibition but can also result in loss of fidelity in protein synthesis, with the insertion of incorrect amino acid residues into the elongating peptide. However, some of the aminoglycosides other than streptomycin may bind to multiple sites on the 30S subunit and are, therefore, less prone to mutations that alter the target of these antibiotics (47). The aminoglycosides are normally described as bactericidal, which may be related to irreversible binding to their target. Aminoglycosides are also thought to be bactericidal because they are thought to have membrane-disruption properties and to promote their own uptake into bacterial cells, which may contribute to their lethality. Spectinomycin, which lacks an amino sugar, has the same molecular target (the 30S ribosomal subunit) but binds reversibly and is bacteriostatic. The membrane-penetrating activity of the aminoglycosides may also explain why aminoglycosides can synergize with other antimicrobials (especially the beta-lactams).

Resistance

In general, resistance to the aminoglycosides commonly used therapeutically is via enzymatic inactivation of the antibiotics (reviewed by Shaw et al. [49]). The plethora of aminoglycoside-modifying enzymes found in bacteria probably derived from both producing organisms and mutation of cellular housekeeping genes (37). In the case of the staphylococci, the modifying enzymes appear to be of exogenous origin, as they are found on large and small plasmids and transposons. Perhaps the most common aminoglycoside-resistance gene is the aac(6')-aph(2″) gene found on Tn4001 (19). This gene actually encodes a single protein with two functional (and separable) domains having different enzymatic activities, aac(6') having an acetylating activity and aph(2″) a phosphorylating activity. Tn4001 and related transposons are often found associated with high-molecular-weight conjugative plasmids (56) but can also be found on the staphylococcal chromosome (19). Somewhat less frequently found is ant(4')-Ia (encoding an adenylylase) followed by aph(3')-IIIa, a phosphorylase. In addition, these genes (and the enzymes that they encode) are often found in combination (31). The streptomycin-resistance gene on plasmid pS194, based on similarity to better-defined enzymes, is likely to be an adenylyltransferase (39).

Another potential resistance strategy is the formation of small-colony (or electron transport) variants. The role of small-colony variants play in actual infections is still a subject of conjecture, but they can be isolated in vitro by selection for aminoglycoside resistance (see chapter 35).

TETRACYCLINES

The tetracyclines are a class of structurally related compounds that include minocycline and doxycycline and are characterized by four interlocking six-carbon rings. These antibiotics inhibit protein synthesis by binding to the 30S subunit of the ribosome and block the entry of aminoacyl-tRNAs into the acceptor site. In general, the tetracyclines are bacteriostatic for the staphylococci (38) despite a mechanism of action similar to that of the aminoglycosides. This may be a function of the relative affinities of these drugs for their targets; i.e., aminoglycosides bind more avidly than tetracyclines.

Resistance

There are two tetracycline-resistance genes that predominate among staphylococci. The tet(K) gene, almost exclusively found on pT181 and related plasmids, encodes a tetracycline efflux pump that is somewhat unusual in that it has 14 membrane-spanning alpha-helices as opposed to the far more common 12 membrane-spanning alpha-helices seen widely in the major facilitator superfamily of transport proteins (36). While Tet(K) is relatively efficient in removing tetracycline from cells, it is not very effective in eliminating either doxycycline or minocycline. Far less frequently found in staphylococci is tet(L), which encodes an efflux pump similar in structure to that of tet(K). A third tetracycline-resistance gene, tet(M), occurring in S. aureus, is virtually identical to the tet(M) gene originally identified on the transposons Tn916 and Tn1545 in enterococci. However, in S. aureus tet(M) is apparently found at a constant chromosomal location and lacks the ability to transpose (32). While pT181 plasmids are ubiquitous, the tet(M) determinant is probably more clinically significant in that it encodes resistance to all tetracyclines, including minocycline and doxycycline. The expression of both tet(K) and tet(M) has been shown to be induced by a subinhibitory concentration of the tetracyclines. While the mechanism of induction of tet(K) has not been well defined, the tet(M) gene is induced by a ribosome-stalling mechanism, relieving transcriptional attenuation of the nascent mRNA (11).

CHLORAMPHENICOL

Chloramphenicol inhibits protein synthesis by blocking peptide bond formation. The antibiotic binds to the 50S ribosome subunit and interferes with the binding of the aminoacyl moiety of the aminoacyl-tRNA (38). Despite its potency and the fact that relatively few clinical isolates are resistant to chloramphenicol, it is rarely used in the treatment of infected patients. The reason for this is that approximately 1 in 50,000 patients treated with chloramphenicol develops aplastic anemia, which is invariably fatal. However, with the emergence of glycopeptide-resistant staphylococci, leaving only novel therapies and agents of questionable efficacy, chloramphenicol treatment may represent a viable therapeutic option for life-threatening infections.

Resistance

To date, resistance to chloramphenicol in the staphylococci is associated solely with inactivation by chloramphenicol acetyltransferases encoded by *cat* genes. The *cat* genes encoding these enzymes in the staphylococci have been found on a diversity of small plasmids (e.g., pC194, pC221, pC223, pUB112). Despite the diversity of the replicons with which they are associated, these *cat* genes are all clearly related to, and are also similar to, genes encoding chloramphenicol acetyl transferases in other genera. The staphylococcal genes appear to be uniformly inducible, with the version found on pUB112 the most intensely studied. The primary mechanism of induction appears to be translational attenuation (15), which is a common theme among chloramphenicol-resistance genes (30).

MACROLIDES-LINCOSAMIDES-STREPTOGRAMIN B (MLS)

The MLS protein synthesis inhibitors are related both structurally and mechanistically. They function by binding to the 50S ribosomal subunit and inhibiting the peptidyl transferase reaction (38). These agents are bacteriostatic for the staphylococci unless combined with agents such as streptogramin A (see below).

Resistance

The predominant form of resistance to these agents is methylation of the 23S rRNA encoded by three related genes (*ermA*, *ermB*, and *ermC*). This alteration totally eliminates the binding of the drug to its target. While all three of these determinants have been associated with mobile genetic elements, perhaps the most widespread of these is the *ermC* gene, which has been found mostly on small plasmids such as pE194 (25) or pE5 (40). The *ermC* gene nominally confers inducible resistance to macrolides (as does *ermA*), such as erythromycin, but not to lincosamides, such as clindamycin. This means that macrolides (at low concentrations) induce the expression of resistance to any MLS antibiotic, whereas lincosamides (like clindamycin) do not (14). Therefore, a strain harboring a plasmid carrying a "wild-type" *ermC* gene would be resistant to erythromycin but clindamycin susceptible. However, deletions in the mRNA leader sequence upstream from the *ermC* coding sequence render the expression of resistance constitutive. Therefore, it is not surprising that, as the use of clindamycin has become more widespread, constitutively resistant strains have become commonplace (55). This is actually an important observation. It has long been presumed that constitutive expression of any resistance determinant would result in an unacceptable metabolic burden to the bacterium, resulting in a competitive disadvantage and subsequent counterselection. In fact, the ability of constitutively resistant strains to be maintained even in the absence of apparent selection demonstrates that, for at least this example, the metabolic burden is not always unacceptable. This also implies that merely removing selection (by restricting the use of certain antibiotics) will not necessarily result in a significant decrease in the proportion of resistant isolates.

Recently, a gene encoding efflux pumps specific for these agents has been reported. MsrA is a macrolide-specific ABC (ATP-binding cassette) transporter encoded by *msrA* (43, 44). Inactivation has been described for a streptogramin B compound, virginiamycin B, by the product of the *vgb* gene, a hydrolase (4). There has also been a report of inactivation (62) by a single clinical isolate of *S. aureus*, but the responsible gene or protein has not been described as yet.

STREPTOGRAMIN COMBINATIONS

Streptogramin combinations, essentially natural product antimicrobials, such as virginiamycin and pristinamycin, consist of a combination of at least a streptogramin A- and a streptogramin B-type antibacterial (59). The streptogramin A compounds are bacteriostatic protein synthesis inhibitors that bind to the peptidyl transferase domain of the 50S ribosomal subunit and prevent the binding of the 3'-termini of aminoacyl-tRNA and peptidyl-tRNAs. The streptogramin B compounds also bind to the 50S ribosomal subunit (perhaps at subunit L4), inhibit the peptidyl transferase reaction (38), and are also bacteriostatic for staphylococci. However, the combination of a streptogramin A and a streptogramin B is often bactericidal (8), and a combination of semisynthetic derivatives of the pristinamycin components (referred to as quinupristin and dalfopristin individually and as Synercid collectively) has recently been introduced for the treatment of infections with gram-positive bacteria (27).

Resistance to Streptogramin A and B Combinations

Resistance to agents of the streptogramin B type is discussed above. With respect to the A-type compounds, two forms of resistance have been noted: inactivation by acetylation and efflux. Several structurally related "virginiamycin acetyl transferase" genes have been described by El Solh and colleagues (3). These now include at least four genes (*vatA*, *vatB*, *vatC*, and *sat4*) found in the staphylococci. To date, a two-gene complex encoding a putative ABC transporter (*vga* and *vgaB*) has also been described (2), and a plasmid, pIP630, has been identified carrying *vga*, *vatA*, and *vgb* (encoding a streptogramin B hydrolase described above).

As indicated above, the streptogramin combinations are considered bactericidal; however, activity toward a strain carrying resistance to either the A or B class is bacteriostatic, and it is considered that the critical factor in resistance to the combination is the presence of a gene encoding an acetyl transferase (1).

FUSIDIC ACID

Fusidic acid, which is steroidal in structure, inhibits protein synthesis by binding to elongation factor G (EF-G), although it had been suggested that, given the excess in EF-G within bacterial cells, fusidic acid may act by other mechanisms such as blocking EF-Tu–aminoacyl-tRNA complex formation (13). However, as mutations in the gene encoding EF-G in *S. aureus* lead to resistance, it is likely that EF-G is, indeed, the target of fusidic acid. It should be noted that, while fusidic acid is not approved for use in the United States, it has been used effectively (mainly as a topical agent) and in combination with other agents, such as rifampin, for the treatment of MRSA infections.

Resistance

Resistance to fusidic acid is mainly due to mutations in the gene encoding EF-G, the *fusA* gene. These mutations putatively decrease fusidic acid binding to its target. Plasmid-encoded fusidic acid resistance (*fusB*) has also been reported and is presumably mediated by an efflux mechanism (16).

SULFONAMIDES AND TRIMETHOPRIM

Mechanism of Action

Sulfonamides and trimethoprim are generally used in combination (e.g., Bactrim, cotrimoxazole). The sulfonamides are wholly synthetic analogs of *p*-aminobenzoic acid and are competitive inhibitors of dihydropteroate synthetase, an enzyme in the biosynthetic pathway for dihydrofolate. Dihydrofolate, a substrate of dihydrofolate reductase (DHFR), is reduced to tetrahydrofolate, which is a necessary cofactor for the biosynthesis of the amino acids glycine and methionine and of purines and pyrimidines. The toxic effect is probably due mainly to the depletion of thymidylate, resulting in "thymineless death." Because trimethoprim, a nucleoside analog, is an inhibitor of bacterial DHFRs, the combination of sulfonamides and trimethoprim blocks two sequential steps in the pathway of several essential bacterial metabolites. Eukaryotic cells are generally not susceptible to trimethoprim because it has much lower affinity for eukaryotic versions of DHFRs (22).

Resistance

Sulfonamide resistance in the staphylococci, which arose soon after the introduction of the sulfa drugs, is chromosomally encoded (by the *sulA* gene) and is attributed to the overproduction of *p*-aminobenzoate (28). While overproduction of the bacterial DHFR can provide low-level resistance to trimethoprim, in general high-level resistance is provided by a DHFR that has low affinity for trimethoprim. Genes encoding this enzyme are often found on large conjugative plasmids like pGO1 or pSK1 (5, 45) and, together with a gene encoding a thymidylate synthase (*thyE*), are flanked by two copies of the insertion sequence IS257.

MUPIROCIN

Formerly known as pseudomonic acid, mupirocin blocks protein synthesis by competitively inhibiting isoleucyl tRNA (26). Mupirocin (formulated as Bactroban) has come into wide use as a topical agent for the treatment of gram-positive infections and more recently has been employed successfully to treat nasal carriers of MRSA, especially those in chronic care settings (e.g., nursing homes) and hospital staff (42).

Resistance

Low-level resistance to mupirocin is likely due to mutations in the chromosomally encoded isoleucyl tRNA synthetase, while high-level (and clinically relevant) resistance is clearly due to an acquired gene (*mupA*) encoding an isoleucyl tRNA synthetase with reduced affinity for mupirocin (18, 23). To date, the *mupA* gene has been found associated with both small and large plasmids but always flanked by two copies of IS257. The origin of the *mupA* gene has yet to be traced.

COMPOUNDS IN DEVELOPMENT

While the increase in multiply resistant strains of staphylococci has stimulated research directed toward new classes of antistaphylococcal drugs and improved versions of the antibiotics we currently employ, there are relatively few promising new drugs on the horizon. One new class, the wholly synthetic oxazolidinones, are bacteriostatic protein synthesis inhibitors and have performed well in preliminary clinical trials (54). Also in various stages of clinical development are new versions of glycopeptides, macrolides, cephalosporins, fluoroquinolones and tetracyclines. Some of these updated versions have excellent potency against resistant staphylococci both in vitro and in vivo. Also, some compounds previously set aside are once again being tested clinically for antistaphylococcal activity. These include the lipoteichoic acid synthesis inhibitor daptomycin (9); everninomicin, which inhibits protein synthesis; and the cell wall biosynthesis inhibitor ramoplanin (7). Interestingly, the endopeptidase lysostaphin has proved extremely effective (better than any other agent in clinical use or development) as an antistaphylococcal agent in an endocarditis model (12) but has drawn little interest among pharmaceutical companies, probably owing to its narrow spectrum of activity and an observed rapid emergence of resistance.

REFERENCES

1. **Allignet, J., and N. El Solh.** 1995. Diversity among the gram-positive acetyltransferases inactivating streptogramin A and structurally related compounds and characterization of a new staphylococcal determinant, *vatB*. *Antimicrob. Agents Chemother.* **39**:2027–2036.

2. **Allignet, J., and N. El Solh.** 1997. Characterization of a new staphylococcal gene, *vgaB*, encoding a putative ABC transporter conferring resistance to streptogramin A and related compounds. *Gene* **202**:133–138.

3. **Allignet, J., N. Liassine, and N. El Solh.** 1998. Characterization of a staphylococcal plasmid related to pUB110 and carrying two novel genes, *vatC* and *vgbB*, encoding resistance to streptogramins A and B and similar antibiotics. *Antimicrob. Agents Chemother.* **42**:1794–1798.

4. **Allignet, J., V. Loncle, P. Mazodier, and N. El Solh.** 1988. Nucleotide sequence of a staphylococcal plasmid gene, *vgb*, encoding a hydrolase inactivating the B components of virginiamycin-like antibiotics. *Plasmid* **20**:271–275.

5. **Archer, G. L., J. P. Coughter, and J. L. Johnston.** 1986. Plasmid-encoded trimethoprim resistance in staphylococci. *Antimicrob. Agents Chemother.* **29**:733–740.

6. **Aubry-Damon, H., C. J. Soussy, and P. Courvalin.** 1998. Characterization of mutations in the *rpoB* gene that confer rifampin resistance in *Staphylococcus aureus*. *Antimicrob. Agents Chemother.* **42**:2590–2594.

7. **Billot-Klein, D., D. Shlaes, D. Bryant, D. Bell, R. Legrand, L. Gutmann, and J. van Heijenoort.** 1997. Presence of UDP-N-acetylmuramyl-hexapeptides and -heptapeptides in enterococci and staphylococci after treatment with ramoplanin, tunicamycin, or vancomycin. *J. Bacteriol.* **179**:4684–4688.

8. **Bouanchaud, D. H.** 1997. In-vitro and in-vivo antibacterial activity of quinupristin/dalfopristin. *J. Antimicrob. Chemother.* **39**(Suppl. A):15–21.

9. Canepari, P., and M. Boaretti. 1996. Lipoteichoic acid as a target for antimicrobial action. *Microb. Drug Resist.* 2:85–89.

10. Chopra, I. 1976. Mechanisms of resistance to fusidic acid in *Staphylococcus aureus. J. Gen. Microbiol.* 96:229–238.

11. Clewell, D. B., S. E. Flannagan, and D. D. Jaworski. 1995. Unconstrained bacterial promiscuity: the Tn*916*-Tn*1545* family of conjugative transposons. *Trends Microbiol.* 3:229–236.

12. Climo, M. W., R. L. Patron, B. P. Goldstein, and G. L. Archer. 1998. Lysostaphin treatment of experimental methicillin-resistant *Staphylococcus aureus* aortic valve endocarditis. *Antimicrob. Agents Chemother.* 42:1355–1360.

13. Cundliffe, E. 1972. The mode of action of fusidic acid. *Biochem. Biophys. Res. Commun.* 46:1794–1801.

14. Denoya, C. D., D. H. Bechhofer, and D. Dubnau. 1986. Translational autoregulation of *ermC* 23S rRNA methyltransferase expression in *Bacillus subtilis. J. Bacteriol.* 168:1133–1141.

15. Dick, T., and H. Matzura. 1990. Chloramphenicol-induced translational activation of *cat* messenger RNA in vitro. *J. Mol. Biol.* 212:661–668.

16. Doyle, D., K. J. McDowall, M. J. Butler, and I. S. Hunter. 1991. Characterization of an oxytetracycline-resistance gene, *otrA*, of *Streptomyces rimosus. Mol. Microbiol.* 5:2923–2933.

17. Fleming, A. 1942. In vitro tests of penicillin potency. *Lancet* i:732.

18. Gilbart, J., C. R. Perry, and B. Slocombe. 1993. High-level mupirocin resistance *in Staphylococcus aureus*: evidence for two distinct isoleucyl-tRNA synthetases. *Antimicrob. Agents Chemother.* 37:32–38.

19. Gillespie, M. T., B. R. Lyon, L. J. Messerotti, and R. A. Skurray. 1987. Chromosome- and plasmid-mediated gentamicin resistance in *Staphylococcus aureus* encoded by Tn*4001. J. Med. Microbiol.* 24:139–144.

20. Gregory, P. D., R. A. Lewis, S. P. Curnock, and K. G. Dyke. 1997. Studies of the repressor (BlaI) of beta-lactamase synthesis in *Staphylococcus aureus. Mol. Microbiol.* 24:1025–1037.

21. Guay, G. G., S. A. Khan, and D. M. Rothstein. 1993. The *tet*(K) gene of plasmid pT181 of *Staphylococcus aureus* encodes an efflux protein that contains 14 transmembrane helices. *Plasmid* 30:163–166.

22. Hitchings, G. H. 1973. Mechanism of action of trimethoprim-sulfamethoxazole I. *J. Infect. Dis.* 128:S433.

23. Hodgson, J. E., S. P. Curnock, K. G. H. Dyke, R. Morris, D. R. Sylvester, and M. S. Gross. 1994. Molecular characterization of the gene encoding high-level mupirocin resistance in *Staphylococcus aureus* J2870. *Antimicrob. Agents Chemother.* 38:1205–1208.

24. Hooper, D. C. 1995. Quinolone mode of action. *Drugs* 49(Suppl. 2):10–15.

25. Horinouchi, S., and B. Weisblum. 1982. Nucleotide sequence and functional map of pE194, a plasmid that specifies inducible resistance to macrolide, lincosamide, and streptogramin type B antibodies. *J. Bacteriol.* 150:804–814.

26. Hughes, J., and G. Mellows. 1978. Inhibition of isoleucyl-transfer ribonucleic acid synthetase in *Escherichia coli* by pseudomonic acid. *Biochem. J.* 176:305–318.

27. Jones, R. N., C. H. Ballow, D. J. Biedenbach, J. A. Deinhart, and J. J. Schentag. 1998. Antimicrobial activity of quinupristin-dalfopristin (RP 59500, Synercid) tested against over 28,000 recent clinical isolates from 200 medical centers in the United States and Canada. *Diagn. Microbiol. Infect. Dis.* 31:437–451.

28. Landy, M., N. W. Larkum, E. J. Oswald, and P. Streighoff. 1943. Increased synthesis of p-aminobenzoic acid associated with the development of sulfonamide resistance in *Staphylococcus aureus. Science* 97:265.

29. Levy, S. B. 1998. Multidrug resistance—a sign of the times. *N. Engl. J. Med.* 338:1376–1378.

30. Lovett, P. S. 1996. Translation attenuation regulation of chloramphenicol resistance in bacteria—a review. *Gene* 179:157–162.

31. Miller, G. H., F. J. Sabatelli, L. Naples, R. S. Hare, and K. J. Shaw. 1995. The most frequently occurring aminoglycoside resistance mechanisms—combined results of surveys in eight regions of the world. The Aminoglycoside Resistance Study Groups. *J. Chemother.* 7(Suppl. 2):17–30.

32. Nesin, M., P. Svec, J. R. Lupski, G. N. Godson, B. Kreiswirth, J. Kornblum, and S. J. Projan. 1990. Cloning and nucleotide sequence of a chromosomally encoded tetracycline resistance determinant, *tet*A(M), from a pathogenic, methicillin-resistant strain of *Staphylococcus aureus. Antimicrob. Agents Chemother.* 34:2273–2276.

33. Niemeyer, D. M., M. J. Pucci, J. A. Thanassi, V. K. Sharma, and G. L. Archer. 1996. Role of *mecA* transcriptional regulation in the phenotypic expression of methicillin resistance in *Staphylococcus aureus. J. Bacteriol.* 178:5464–5471.

34. Nikaido, H. 1994. Prevention of drug access to bacterial targets: permeability barriers and active efflux. *Science* 264:382–388.

35. O'Kane, G. M., T. Gottlieb, and R. Bradbury. 1998. Staphylococcal bacteraemia: the hospital or the home? A review of *Staphylococcus aureus* bacteraemia at Concord Hospital in 1993. *Aust. N.Z. J. Med.* 28:23–27.

36. Pao, S. S., I. T. Paulsen, and M. H. Saier, Jr. 1998. Major facilitator superfamily. *Microbiol. Mol. Biol. Rev.* 62:1–34.

37. Paradise, M. R., G. Cook, R. K. Poole, and P. N. Rather. 1998. Mutations in *aarE*, the *ubiA* homolog of *Providencia stuartii*, result in high-level aminoglycoside resistance and reduced expression of the chromosomal aminoglycoside 2'-N-acetyltransferase. *Antimicrob. Agents Chemother.* 42:959–962.

38. Pestka, S. 1971. Inhibitors of ribosome functions. *Annu. Rev. Microbiol.* 25:487–562.

39. Projan, S. J., S. Moghazeh, and R. P. Novick. 1988. Nucleotide sequence of pS194, a streptomycin-resistance plasmid from *Staphylococcus aureus. Nucleic Acids Res.* 16:2179–2187.

40. Projan, S. J., M. Monod, C. S. Narayanan, and D. Dubnau. 1987. Replication properties of pIM13, a naturally occurring plasmid found in *Bacillus subtilis*, and of its close relative pE5, a plasmid native to *Staphylococcus aureus. J. Bacteriol.* 169:5131–5139.

41. Rao, G. G. 1998. Risk factors for the spread of antibiotic-resistant bacteria. *Drugs* 55:323–330.

42. Raz, R., D. Miron, R. Colodner, Z. Staler, Z. Samara, and Y. Keness. 1996. A 1-year trial of nasal mupirocin in

the prevention of recurrent staphylococcal nasal colonization and skin infection. *Arch. Intern. Med.* **156:**1109–1112.

43. **Ross, J. I., E. A. Eady, J. H. Cove, and S. Baumberg.** 1996. Minimal functional system required for expression of erythromycin resistance by *msrA* in *Staphylococcus aureus* RN4220. *Gene* **183:**143–148.

44. **Ross, J. I., E. A. Eady, J. H. Cove, W. J. Cunliffe, S. Baumberg, and J. C. Wootton.** 1990. Inducible erythromycin resistance in staphylococci is encoded by a member of the ATP-binding transport super-gene family. *Mol. Microbiol.* **4:**1207–1214.

45. **Rouch, D. A., L. J. Messerotti, L. S. Loo, C. A. Jackson, and R. A. Skurray.** 1989. Trimethoprim resistance transposon Tn4003 from *Staphylococcus aureus* encodes genes for a dihydrofolate reductase and thymidylate synthetase flanked by three copies of IS257. *Mol. Microbiol.* **3:**161–175.

46. **Rowland, S. J., and K. G. Dyke.** 1990. Tn552, a novel transposable element from *Staphylococcus aureus. Mol. Microbiol.* **4:**961–975.

47. **Schlessinger, D., and G. Medoff.** 1975. Streptomycin, dehidrostreptomycin amd the gentamicins, p. 535–549. *In* J. W. Corcoran and F. E. Hahn (ed.), *Antibiotics*, vol. 3. Springer-Verlag, New York, N.Y.

48. **Schwarz, S., P. D. Gregory, C. Werckenthin, S. Curnock, and K. G. Dyke.** 1996. A novel plasmid from *Staphylococcus epidermidis* specifying resistance to kanamycin, neomycin and tetracycline. *J. Med. Microbiol.* **45:**57–63.

49. **Shaw, K. J., P. N. Rather, R. S. Hare, and G. H. Miller.** 1993. Molecular genetics of aminoglycoside resistance genes and familial relationships of the aminoglycoside-modifying enzymes. *Microbiol. Rev.* **57:**138–163.

50. **Shlaes, D. M., and J. H. Shlaes.** 1995. Teicoplanin selects for *Staphylococcus aureus* that is resistant to vancomycin. *Clin. Infect. Dis.* **20:**1071–1073.

51. **Smith, I. M., and A. B. Vickers.** 1960. Natural history of 338 treated and untreated patients with staphylococcal septicaemia. *Lancet* **i:**1318.

52. **Stone, M. J., and D. H. Williams.** 1992. On the evolution of functional secondary metabolites (natural products). *Mol. Microbiol.* **6:**29–34.

53. **Su, Y. A., P. He, and D. B. Clewell.** 1992. Characterization of the *tet*(M) determinant of Tn916: evidence for regulation by transcription attenuation. *Antimicrob. Agents Chemother.* **36:**769–778.

54. **Swaney, S. M., H. Aoki, M. C. Ganoza, and D. L. Shinabarger.** 1998. The oxazolidinone linezolid inhibits initiation of protein synthesis in bacteria. *Antimicrob. Agents Chemother.* **42:**3251–3255.

55. **Thakker-Varia, S., W. D. Jenssen, L. Moon-McDermott, W. P. Weinstein, and D. T. Dubin.** 1987. Molecular epidemiology of macrolides-lincosamides-streptogramin B resistance in *Staphylococcus aureus* and coagulase-negative staphylococci. *Antimicrob. Agents Chemother.* **31:**735–743.

56. **Thomas, W. D., Jr., and G. L. Archer.** 1989. Mobility of gentamicin resistance genes from staphylococci isolated in the United States: identification of Tn4031, a gentamicin resistance transposon from *Staphylococcus epidermidis. Antimicrob. Agents Chemother.* **33:**1335–1341.

57. **Tomasz, A.** 1994. Multiple-antibiotic-resistant pathogenic bacteria. A report on the Rockefeller University Workshop. *N. Engl. J. Med.* **330:**1247–1251.

58. **Tomasz, A., S. Nachman, and H. Leaf.** 1991. Stable classes of phenotypic expression in methicillin-resistant clinical isolates of staphylococci. *Antimicrob. Agents Chemother.* **35:**124–129.

59. **Vazquez, D.** 1975. The Streptogramin family of antibiotics, p. 521–534. *In* J. W. Corcoran and F. E. Hahn (ed.), *Antibiotics*, vol. 3. Springer-Verlag, New York, N.Y.

60. **Voladri, R. K., M. K. Tummuru, and D. S. Kernodle.** 1996. Structure-function relationships among wild-type variants of *Staphylococcus aureus* beta-lactamase: importance of amino acids 128 and 216. *J. Bacteriol.* **178:**7248–7253.

61. **Wehrli, W., and M. Staehelin.** 1971. Actions of the rifamycins. *Bacteriol. Rev.* **35:**290–309.

62. **Wondrack, L., M. Massa, B. V. Yang, and J. Sutcliffe.** 1996. Clinical strain of *Staphylococcus aureus* inactivates and causes efflux of macrolides. *Antimicrob. Agents Chemother.* **40:**992–998.

63. **Yu, V. L., G. D. Fang, T. F. Keys, A. A. Harris, L. O. Gentry, P. C. Fuchs, N. M. Wagener, and E. S. Wong.** 1994. Prosthetic valve endocarditis: superiority of surgical valve replacement versus medical therapy only. *Ann. Thoracic Surg.* **58:**1073–1077.

64. **Zhao, X., C. Xu, J. Domagala, and K. Drlica.** 1997. DNA topoisomerase targets of the fluoroquinolones: a strategy for avoiding bacterial resistance. *Proc. Natl. Acad. Sci. USA* **94:**13991–13996.

THE LISTERIA

SECTION EDITOR: Daniel A. Portnoy

L ISTERIOSIS AND *LISTERIA* MONOCYTOGENES are of interest to a broad range of investigators, including food microbiologists, clinicians, immunologists, medical microbiologists, and even cell biologists. Thus, while *L. monocytogenes* remains one of the leading causes of mortality from food-borne infections in the United States, it has emerged as an excellent model system to study basic aspects of bacterial pathogenesis and cell-mediated immunity.

L. monocytogenes is particularly interesting as a food-borne pathogen in that it is ubiquitous in nature, has been isolated from numerous species of animals, and has a temperature range for growth of 4°C to 44°C. The reader with particular interest in food microbiology is directed to the recently revised and expanded edition of *Listeria, Listeriosis, and Food Safety*, edited by Elliot T. Ryser and Elmer H. Martin and published by Marcel Dekker, Inc. (1999).

L. monocytogenes has now been used by immunologists for four decades to study cell-mediated immunity. Indeed, much of our current understanding of cell-mediated immunity has been derived from the murine model of listeriosis. It may be true that more is known about the murine response to *L. monocytogenes* than any other bacterial pathogen; however, it should be noted that the vast majority of these studies have used an intravenous or intraperitoneal route of inoculation and the relevance of these studies to the natural, oral route of infection is not always clear. This is an area of future growth and development.

During the past 10 years, *L. monocytogenes* has emerged as a model system to study basic aspects of intracellular pathogenesis. Many new genetic techniques have been developed, ranging from transposon mutagenesis to allelic exchange and the development of reporter systems and, most recently, the discovery of transducing phages. We have learned about the capacity of *L. monocytogenes* to enter cells, escape from a vacuole, grow in the mammalian cell cytosol, exploit a host system of actin-based motility, and spread from cell to cell, all without contact with the humoral immune system. The mechanism of *L. monocytogenes* movement has become a model system with which to dissect basic mechanisms of actin-based motility. The *L. monocytogenes* model is described in some detail in the chapter on cytoskeleton in the college and graduate textbook *Molecular Biology of the Cell*, 3rd edition, by Bruce Alberts et al. (Garland Publishing, New York, N.Y., 1994).

Epidemiology and Clinical Manifestations of *Listeria monocytogenes* Infection

WALTER F. SCHLECH III

49

EPIDEMIOLOGY

Listeria monocytogenes is a gram-positive motile facultative anaerobe that inhabits a broad ecologic niche (56). With selective media it can be readily isolated from soil, water, and vegetation, including raw produce designated for human consumption (17), without further processing. Contamination of meat and vegetables is on the surface and is relatively common, with up to 15% of these foods harboring the organism. In addition, the organism is a transient inhabitant of both animal and human gastrointestinal tracts (21, 29). This site is the likely source for the organism in invasive listeriosis when it occurs.

The organism is psychrophilic and enjoys a competitive advantage against other gram-positive and gram-negative microorganisms in cold environments, such as refrigerators. It may also be amplified in spoiled food products, particularly when spoilage leads to increased alkalinity. Feeding of spoiled silage with a high pH has resulted in epidemics of listeriosis in sheep or cattle (28).

Several large food-borne outbreaks of listeriosis in humans have parallels to epidemic listeriosis in animals. These outbreaks, which have been attributed to coleslaw (41), unpasteurized cheeses (6, 24), pasteurized milk (11), and several meat products (15, 30), have established that human infection by *L. monocytogenes* has a food-borne origin. Many other food products have been implicated in both epidemic and sporadic disease (Table 1). Uncertainty exists as to the reason that outbreaks of listeriosis occur in human populations, although the 50% infective dose (ID_{50}) in sporadic disease is probably high. Enhancement of organism-specific virulence factors may also play a role in epidemic disease, although all isolates of *L. monocytogenes* have the constitutive ability to produce all the virulence factors characteristic of the species. Serotype 4b and MEE type 1 organisms are overrepresented in epidemic disease in humans, suggesting that these characteristics could be markers for other virulence factors that are yet undetermined.

Recent evidence has suggested that most sporadic cases of listeriosis are also food borne. These reports remained anecdotal until large case-controlled studies of sporadic disease implicated food products, including cold meats, turkey franks, and delicatessen-type foods, as vehicles for development of sporadic invasive listeriosis in humans (42, 45).

Our current understanding of the epidemiology of human listeriosis suggests that the organism is a common contaminant of food products, perhaps as much as 15% of foods, and that ingestion of small numbers of *L. monocytogenes* probably occurs on a daily basis in human populations. Amplification of the organism in biofilms or on food products undergoing processing but not pasteurization and kept at cold temperatures allows overgrowth of *L. monocytogenes* in contrast to inhibition of other organisms. Subsequent ingestion of large numbers of the organism may overwhelm local host-defense systems in the gastrointestinal tract and reticuloendothelial systems of the liver and spleen with subsequent development of invasive disease. Other organism-specific or host-specific risk factors that may play a role in sporadic or epidemic listeriosis are incompletely defined (44). The annual rate of sporadic listeriosis in Europe and North America is usually <1/100,000 population per year (13, 38). Active surveillance carried out in hospital microbiology laboratories in specific geographic regions in the United States has confirmed data obtained from passive reporting systems (7). Sporadic listeriosis appears to be more common in the spring and summer months. This could be explained by seasonal variations in the types of food products eaten by human populations, with higher-risk products eaten in the warmer months. In addition, data suggest that preexisting damage to the gastrointestinal mucosa by other microorganisms such as those that are associated with viral gastroenteritis may allow translocation of *L. monocytogenes* from the gastrointestinal tract with subsequent development of invasive disease (46). These viral pathogens often have seasonal patterns that overlap with those of invasive listeriosis.

Demographic data from surveillance studies indirectly reveal several host-specific risk factors for invasive listeriosis. Infection is most commonly seen in the first 30 days of life or in patients older than 60. In the first instance, the fetus is infected during maternal sepsis with *L. mono-*

Gram-Positive Pathogens, ed. by V. A. Fischetti et al.
© 2000 American Society for Microbiology, Washington, D.C.

TABLE 1 Some foods implicated in published reports of food-borne listeriosis

Coleslaw (cabbage)	Strawberries
Pasteurized whole milk	Nectarines
Chocolate milk	Patés
Mexican-style cheese	Goat cheese
Soft cheeses (different types)	Salted mushrooms
Delicatessen foods	Lettuce
Ice cream	Alfalfa tablets
Salami	Pork tongue in aspic
Shrimp salad	"Rillettes"
Uncooked hotdogs	Rice salad
Turkey franks	Undercooked chicken
Fresh cream	Blueberries
Baby corn	

cytogenes or from perivaginal and perianal colonization of the mother by transition through the birth canal. Host defense against listeriosis is impaired in those infants with underdeveloped macrophage and cell-mediated immune function, and invasive listeriosis is more likely to occur if colonization of the liver, respiratory tract, or gastrointestinal tract has occurred. A unique outbreak of neonatal listeriosis in Costa Rica has been described: the vehicle was L. monocytogenes-contaminated mineral oil used to clean infants after delivery from healthy mothers with cross contaminations of shared mineral oil (43). The index case was infected through the traditional route of maternal-fetal transmission. The increased risk of invasive listeriosis in older patients reflects the increasing incidence of immunosuppressive conditions, such as solid tumors and hemologic malignancy, in this age group. Control of early infection in humans and in animal models is highly dependent on an intact gastrointestinal mucosa and effective macrophage function in the liver, spleen, and peritoneum following bacterial translocation from the gastrointestinal tract. Both these protective events can be impaired by the primary disease or by chemotherapy or radiation-induced damage. In addition, treatment of malignancy and the use of immunosuppressive agents with a specific effect on cell-mediated immune function, such as corticosteroids or cyclosporin A (40), predispose to invasive infection by diminishing L. monocytogenes-specific host responses that occur after the initial phase of infection.

The cell-mediated immune response to L. monocytogenes is normally impaired in pregnant women (51) and, accompanied by the decreased gastrointestinal motility (55) seen in pregnancy, may predispose to invasive listeriosis and subsequent transplacental infection of the infant. This results in "early-onset" listeriosis characterized by the delivery of an often premature and severely ill infant. Spontaneous recovery of the mother from Listeria sepsis normally occurs following delivery of the infant.

In "late-onset" listeriosis, the infant is infected through maternal gastrointestinal carriage of L. monocytogenes without sepsis, and the infant is infected during transition through a colonized birth canal. In these cases, clinical disease in the infant develops 7 to 14 days later. Direct cutaneous invasion is unlikely, and it is believed that aspiration of the organism into the respiratory tract or swallowing of the organism by the infant may occur during the incubation period.

Several large outbreaks of a febrile gastroenteritis syndrome (8, 37, 39) have further highlighted the importance of L. monocytogenes as a food-borne pathogen. In these outbreaks, with an average incubation period of approximately 24 h, attack rates were much higher than those reported for outbreaks of invasive listeriosis. The reported vehicles for these more typical food-borne infections have included shrimp (37), rice salad (39), and chocolate milk (8). In the latter case, the chocolate milk was heavily contaminated ($>10^9$ CFU/ml of L. monocytogenes), suggesting that the high attack rate was related to the dose ingested and not to enhanced intrinsic virulence of the particular infecting strain of L. monocytogenes.

While a predisposition to invasive listeriosis is seen in patients with malignancy or organ transplant, recent data suggest that specific impairment of the immune system caused by human immunodeficiency virus (HIV) infection is an important factor in sporadic listeriosis (20). Several studies report that attack rates for invasive listeriosis in HIV-positive patients are 500- to 1,000-fold greater than that in the general population. In California, where active surveillance of hospital laboratories for listeriosis is carried out, a reduction in invasive listeriosis cases in HIV infection has been brought about by widely promulgated dietary recommendations to prevent food-borne illness and by the use of prophylaxis for Pneumocystis carinii pneumonia, primarily trimethoprim-sulfamethoxazole, to which L. monocytogenes is susceptible (10). A recent reduction in the overall incidence of listeriosis in non-HIV-positive patients can also be attributed to distribution of dietary recommendations to populations at risk, including pregnant women, patients with malignancies, and organ transplant recipients (53). Perhaps more important, the decreased incidence of listeriosis may also be due to the promotion of guidelines to promote universal awareness of the problem in the food-processing industry, which has undertaken Hazard Analysis at Critical Control Points (HACCP) programs to reduce contamination of foods with L. monocytogenes as well as with other food-borne pathogens such as Salmonella sp., Campylobacter sp., and Escherichia coli (3). These activities have provided increased protection in the face of increased public demand for fresh, unprocessed food products that may not have been cooked or pasteurized and that by definition present a greater degree of risk for food-borne illness.

In addition to HACCP programs, regulatory agencies have aggressively pursued the control of L. monocytogenes contamination of food. The U.S. Food and Drug Administration has instituted a zero-tolerance policy for L. monocytogenes in its industry sampling programs (54). Other countries have less stringent guidelines and have adopted less stringent guidelines, allowing a small amount of contamination ($<10^2$ CFU/g) in order to strike a balance between protection of public health and needless condemnation of otherwise edible food products. While invasive listeriosis may be more common in some countries in Europe than in the United States, it is not clear whether these differences can be attributed to less stringent standards in Europe that allow more L. monocytogenes in the food supply.

CLINICAL DISEASE DUE TO L. MONOCYTOGENES

A wide variety of clinical syndromes have been associated with L. monocytogenes infection in both animals and humans (Table 2). The earliest descriptions of *L. monocytogenes* sepsis were described in an epizootic affecting South African rodents (35) and in laboratory colonies of rabbits (33). One distinguishing characteristic of infection in rabbits was the production of monocytosis in blood, which suggested the species name monocytogenes. A monocytosis-producing antigen has been described as a virulence factor of *L. monocytogenes* (47), but monocytosis in the peripheral blood is not a characteristic of invasive infection in humans.

Many wild and domesticated animals are subject to invasive listeriosis. Animals acquire the organism from the environment through grazing, amplified by fecal contamination of soil and vegetation. Specific syndromes with parallels in human disease have been recognized in animals. In New Zealand in the 1930s, Gill (14) described circling disease, a rhombencephalitis of sheep that may affect flocks fed spoiled silage. *L. monocytogenes* has also been implicated as a cause of abortion and prematurity in ruminants. Intravenous and oral models of *L. monocytogenes* infection in rodents can duplicate the illness seen in the natural state in animals, including maternal sepsis and abortion (23).

The clinical syndromes associated with listeriosis in humans have been more recently described. Nyfeldt (34) described *L. monocytogenes* as a cause of infectious mononucleosis in 1929, but this was subsequently discounted. Neonatal listeriosis was initially described in postwar Europe in premature septic newborns in East Germany (36). This description of early-onset listeriosis was followed by reports of neonatal meningitis (late-onset listerosis) occurring somewhat later in the postpartum period. *L. monocytogenes* as a cause of meningitis in neonates is third to group B streptococci and *E. coli* in the developed world (52). In less developed countries, gram-negative meningitis with *E. coli* or *Salmonella* species is more common, but *Listeria* meningitis still occurs.

Early-onset neonatal listeriosis has characteristic clinical features, including prematurity, sepsis at birth, fever, a diffuse maculopapular cutaneous eruption, and evidence of significant hepatic involvement with jaundice (9). The mortality rate of early-onset listeriosis, even with treatment, is very high, and stillbirth is also common in this setting. Autopsy findings in cases of early-onset listeriosis show significant chorioamnionitis in placental remnants and granulomas in multiple organs, particularly the liver and spleen, in infected infants. The original descriptions

from East Germany characterized the entire syndrome as "granulomatosis infantiseptica."

The mothers of these septic infants may be asymptomatic but commonly have flu-like or pyelonephritis symptoms before the early onset of labor, and their blood cultures are frequently positive for *L. monocytogenes*. Symptoms in the mother include fever, chills, and malaise, which resolve spontaneously following delivery of the infected infant and placenta. Anecdotal case reports suggest that early treatment of the mother who has *Listeria* sepsis can prevent transplacental infection or treat the fetus in utero with subsequent delivery of a normal uninfected infant (19). Unfortunately, this usually only happens when a community-based outbreak of *L. monocytogenes* has been identified and physicians are aware of the problem in a particular geographic region through public health alerts.

Late-onset neonatal meningitis due to *L. monocytogenes* has the typical features of the same syndrome caused by other organisms in this setting, including fever, irritability, bulging fontanelles, and meningismus (22). These symptoms usually develop 1 to 2 weeks following delivery. The mother has usually had an uncomplicated pregnancy, delivery, and postpartum course with no signs of sepsis. The clinical syndrome usually dictates a lumbar puncture, and the cerebrospinal fluid (CSF) in 50% of the cases will reveal the organism by Gram's stain. CSF cultures are usually positive, although the organism may be isolated simultaneously or only from the blood in some cases. The CSF shows other characteristics of bacterial meningitis, including a high polymorphonuclear leukocyte count, elevated protein, and low glucose with a decrease in the CSF-serum glucose ratio.

ADULT MENINGOENCEPHALITIS

L. monocytogenes is an uncommon cause of bacterial meningitis in adults. There are two clinical presentations. The first is a typical subacute bacterial meningitis characterized by fever, headache, and neck stiffness (49). Because the organism is not commonly seen on Gram's stain of CSF, and because the cell counts are lower than in other forms of bacterial meningitis, an initial diagnosis of viral meningitis is commonly made before culture of the organism from CSF or blood. The onset of the syndrome can occur over several days, unlike meningococcal or pneumococcal meningitis, which have a more abrupt onset. During epidemics of food-borne listeriosis, *Listeria* meningitis can occur in apparently healthy individuals of all ages. In sporadic disease, patients more commonly have obvious defects in cell-mediated immune function that predispose them to listeriosis.

The second form of central nervous system listeriosis in adults is a rhombencephalitis that has features characteristic of the same illness in animals described as circling disease (1). Fever, headache, nausea, and vomiting occur early, with signs of meningeal irritation less commonly present. Subsequently, patients develop multiple cranial nerve abnormalities accompanied by cerebellar dysfunction, including ataxia. Fever may not be present in up to 15% of patients, which makes the diagnosis more difficult and more suggestive of other noninfectious disorders. CSF pleocytosis may be minimal, and the organism is rarely seen on Gram's stain. The diagnosis is established by culture of CSF or blood. Magnetic resonance imaging is the best diagnostic study and frequently demonstrates typical multiple microabscesses of the cerebellum and diencephalon (Fig.

TABLE 2 Major clinical syndromes associated with *Listeria monocytogenes*

Neonatal meningitis	Hepatitis
Meningoencephalitis in adults	Liver abscess
Rhombencephalitis	Cutaneous infections (in animal workers)
Bacteremia in infants or adults	Endophthalmitis
Native or prosthetic valve endocarditis	Febrile gastroenteritis
Spontaneous bacterial peritonitis	CAPD[a] peritonitis
Pneumonia	Septic arthritis
Osteomyelitis	

[a]Continuous ambulatory peritoneal dialysis.

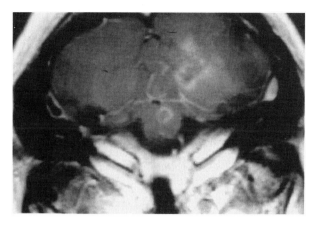

FIGURE 1 Magnetic resonance imaging of the cerebellum and midbrain of a 72-year-old, previously healthy man. Spinal fluid and blood cultures demonstrated *L. monocytogenes*. Well-formed microabscesses typical of listeriosis can be seen within the brain parenchyma.

1). The mortality rate in this condition approaches 50%, and despite treatment, residual morbidity, including permanent cranial nerve palsies and ataxia, may persist.

LISTERIA SEPSIS

Listeria sepsis, or bacteremia without central nervous system involvement, represents one-third of adult cases of invasive listeriosis. The symptoms are nonspecific but usually include fever and chills. As noted above, in pregnant women *Listeria* sepsis often masquerades as pyelonephritis or "flu." The diagnosis is often established in retrospect following delivery of an infected infant. In nonpregnant adults, *Listeria* sepsis almost always occurs in patients with malignancy, organ transplant, or other immunocompromised states (16). In these settings, the presentation is also nonspecific and mimics sepsis with other gram-positive and gram-negative pathogens. The mortality rate in these series is 30%.

OTHER CLINICAL SYNDROMES

Cutaneous Listeriosis

Cutaneous listeriosis is an occupational hazard of veterinary workers exposed to infected amniotic fluid or placental remnants that are removed from the birth canal of animals (31). Occasionally, cutaneous infection, including conjunctivitis, has been seen in laboratory workers. Cutaneous listeriosis is characterized by low-grade fever and multiple papulopustular lesions of the skin from which the organism can be isolated. Its appearance is similar to the rash seen in infants with early-onset disseminated listeriosis. In adults, the condition may resolve spontaneously without treatment, but the infection itself should be entirely preventable with appropriate gloving and other protective wear.

Bacterial Endocarditis Caused by *L. monocytogenes*

Bacterial endocarditis presumably follows transient bacteremia from a gastrointestinal source, followed by establish-

ment of endovascular infection on an abnormal heart valve. *L. monocytogenes* is an uncommon cause of native valve endocarditis, and over 50% of cases that have been described involve prosthetic valves (50). Infection with *L. monocytogenes* is usually found as part of the late prosthetic valve endocarditis syndrome. Diagnostic criteria for *Listeria* endocarditis include presence of a prosthetic valve with or without vegetation and continuous bacteremia with *L. monocytogenes*. Septic emboli and abscess formations in other organs are relatively frequent. In native valve endocarditis, *L. monocytogenes* can sometimes follow previous episodes of streptococcal bacterial endocarditis or other valvular heart disease. Patients with malignancy, diabetes, steroid therapy, and renal transplantation have been described with *Listeria* endocarditis. Their presentation is nonspecific for *L. monocytogenes* and includes prolonged fever, chills, and ultimately signs of congestive heart failure. Septic embolization occurs in two-thirds of patients, and aortic and mitral valve involvement are most common. *L. monocytogenes* can also cause arterial infections that involve prosthetic abdominal and aortic grafts or native abdominal aortic aneurysms (12). The mortality of this condition approaches 33%, and surgical replacement appears to favorably affect the outcome.

Hepatitis and Liver Abscess Due to *L. monocytogenes*

L. monocytogenes has been described as a cause of acute hepatitis in several case reports (4). It occurs as acute onset of fever and jaundice accompanied by positive blood cultures for *L. monocytogenes*. The diagnosis is usually unsuspected. Severe disease with death has been described, and autopsy or liver biopsy generally reveals microabscesses and occasionally granulomas similar to those seen in severe neonatal disease. Predisposing factors include cirrhosis, although *Listeria* hepatitis can occur in a normal host.

Solitary and multiple liver abscesses with fever are also described (5). Bacteremia occurs in one-half of these patients. Predisposing factors include diabetes mellitus, transplantation, cirrhosis, and alcoholism. Aspiration of the abscess demonstrates the organism. The mortality rate is 50%, and postmortem examination often reveals abscesses in other organs as well. Patients with multiple abscesses appear to do worse than those with solitary abscess, despite appropriate treatment.

Listeria Peritonitis

L. monocytogenes can also cause isolated episodes of peritonitis (48). It is most commonly seen in patients undergoing continuous ambulatory peritoneal dialysis (CAPD) and the organisms are isolated from the dialysate or from blood culture. The organisms presumably cause infection through translocation from the gastrointestinal tract in patients who have ingested the organism with food. This complication is extremely rare and represents <1% of all cases of CAPD peritonitis. *L. monocytogenes* can also cause spontaneous bacterial peritonitis in patients with hepatic cirrhosis from alcohol or hemochromatosis. The mortality rate is low, and laboratory and clinical features are typical of spontaneous bacterial peritonitis due to other organisms. The organism can rarely cause this disease in patients who have undergone liver transplantation.

Musculoskeletal Infection

L. monocytogenes is a very uncommon cause of osteomyelitis (27). Reports of *Listeria* osteomyelitis emphasize the role of diabetes mellitus or leukemia as predisposing factors, particularly when long-term corticosteroids are administered. Relapses have been described despite effective antibiotic therapy. Septic arthritis due to *L. monocytogenes* appears to be more common than bone infection, and rheumatoid arthritis appears to be a frequent associated condition. Infection may follow joint injection with corticosteroids in the latter disease. Infection has been described in prosthetic hips and knees as well as in native joints. With prolonged antibiotic therapy, medical treatment alone, as opposed to removal of the prosthetic joint, may be successful. Deaths are rare and are normally due to the underlying disease.

Gastroenteritis

A febrile gastroenteritis syndrome has been described for listeriosis (8, 37, 39). Gastrointestinal prodromal symptoms, such as diarrhea or abdominal pain, have been common in large outbreaks of food-borne adult listeriosis, but sepsis and meningoencephalitis have been the usual presenting syndromes. Population-based attack rates for invasive listeriosis have been low in this setting. In *Listeria* gastroenteritis, a more typical food-borne illness occurs with high attack rates among the individuals exposed to the vehicle of infection. Most patients are well before development of the infection. While bacteremia has occurred in some patients, primary symptoms are diarrhea, fever, fatigue, chills, and myalgias occurring 24 h following exposure. This incubation period is considerably shorter than the 3- to 4-week period for more usual forms of invasive listeriosis.

Pregnant women appear to be more likely to have sepsis in these outbreaks. Isolation of *L. monocytogenes* with selective media from stool has been rare, but serologic tests have been used to help define the extent of the outbreaks. In the three outbreaks reported to date, rice salad, shrimp salad, and chocolate milk have been the reported vehicles, and high colony counts of the organism (up to 10^9 CFU/g) appear to be present in the contaminated food. Invasive listeriosis can also be a result of loss of gastrointestinal integrity due to other gastrointestinal tract pathogens such as *Shigella* sp. (25) or to presumed viral gastroenteritis (46). This may account for some sporadic and epidemic cases of *Listeria* sepsis and meningoencephalitis.

Diagnosis of Listeriosis

Diagnosis of all forms of *L. monocytogenes* infection depends on isolation of the organism from a normally sterile site, usually blood or cerebrospinal fluid. Gram's stain of specimens of sterile spinal fluid, peritoneal fluid, or joint aspirates occasionally, but unpredictably, reveals gram-positive coccobacilli, characteristic of *L. monocytogenes*. In some forms of central nervous system infection, particularly rhombencephalitis, several samples may have to be obtained in order to isolate the organism.

In situations where antibiotic therapy has already been administered, isolation of the organism from a nonsterile site may support a diagnosis of listeriosis. In pregnant women, stool or vaginal cultures may be positive when selective media for *L. monocytogenes* are used for culture. In febrile gastroenteritis syndromes, where traditional pathogens have not been isolated using standard media, culture of the stool with selective media for *L. monocytogenes* may also demonstrate the organism.

Serologic tests, until recently, have not been useful in defining infection (2). In an outbreak of febrile gastroenteritis, retrospective analysis of sera from most patients has demonstrated high levels of anti-listeriolysin A antibody (8). However, serologic tests are not likely to be useful for the acute diagnosis of listeriosis. Whether direct detection of PCR products from the *L. monocytogenes* "pathogenicity island" will help in early diagnosis of *Listeria* meningitis remains an experimental hypothesis.

Treatment of Listeriosis

L. monocytogenes remains susceptible to most β-lactam antibiotics, with the exception of cephalosporins, to which the organism is always resistant (18). Because newer cephalosporins are commonly used for the treatment of nonspecific sepsis syndromes or for the empiric treatment of bacterial meningitis, specific therapy for listeriosis may be delayed for some patients. When listeriosis is a likely diagnosis, the use of ampicillin or, in penicillin-allergic patients, vancomycin will provide empiric coverage for *L. monocytogenes* until the diagnosis is made by culture.

A combination of ampicillin and gentamicin is the current therapy of choice for all forms of listeriosis (26). Ampicillin is not bactericidal for *L. monocytogenes*, and in vitro and in vivo data suggest that an additive or synergistic effect with gentamicin may improve outcome. No randomized controlled clinical trials of therapy in humans have been carried out, however.

Trimethoprim-sulfamethoxazole, with or without the addition of rifampin, is an alternative treatment regimen that has been recommended. In one retrospective study, the combination of amoxicillin and cotrimoxazole was found to be more effective than ampicillin and gentamicin (32).

Trimethoprim-sulfamethoxazole has also been used as a prophylactic agent against a number of microorganisms, including *P. carinii*, in patients with HIV infection and patients undergoing chemotherapy for leukemia or lymphoma. This drug would be effective in protecting against *L. monocytogenes* infections. The use of prophylaxis has been temporally associated with a decrease in the incidence of listeriosis in these compromised hosts in combination with dietary guidelines that have been issued for these patients in recent years (10).

The duration of treatment for invasive listeriosis has not been studied. Relapses appear to be uncommon, and 2 to 3 weeks of therapy with ampicillin and gentamicin is sufficient for most forms of listeriosis. Rhombencephalitis with abscess formation in the central nervous system may require more prolonged therapy, but data are not available that support treatment beyond 4 weeks (26).

REFERENCES

1. **Armstrong, R. W., and P. C. Fung.** 1993. Brainstem encephalitis due to *Listeria monocytogenes*: case report and review. *Clin. Infect. Dis.* **16:**689–702.

2. **Berche, P., K. A. Reich, M. Bonnichon, J. L. Beretti, C. Geoffroy, J. Raveneau, P. Cossart, J. L. Gaillard, P. Geslin, H. Kreis, and M. Veron.** 1990. Detection of anti-listeriolysin O for serodiagnosis of human listeriosis. *Lancet* **335:**624–627.

3. **Bernard, D., and W. Sveum.** 1994. Industry perspectives on *Listeria monocytogenes* in foods: manufacturing and processing. *Dairy Food Environ Sanit.* **14:**140–143.

4. **Bourgeois, N., F. Jacobs, M. L. Tavares, F. Rickaert, C. Deprez, C. Liesnard, F. Moonens, J. Van de Stadt, M.**

Gelin, and M. Adler. 1993. *Listeria monocytogenes* hepatitis in a liver transplant recipient: a case report and review of the literature. *J. Hepatol.* **18:**284–289.

5. Braun, T. I., D. Travis, R. R. Dee, and R. E. Nieman. 1993. Liver abscess due to *Listeria monocytogenes*: case report and review. *Clin. Infect. Dis.* **17:**267–269.

6. Bula, C. J., J. Bille, and M. P. Glauser. 1995. An epidemic of food borne listeriosis in Western Switzerland: description of 57 cases involving adults. *Clin. Infect. Dis.* **20:**66–72.

7. Centers for Disease Control. 1992. Food-borne listeriosis—United States, 1988–1990. *Morbid. Mortal. Weekly Rep.* **41:**251, 257–258.

8. Dalton, C. B., C. C. Austin, J. Sobel, P. S. Hayes, W. F. Bibb, L. M. Graves, B. Swaminathan, M. E. Proctor, and P. M. Griffin. 1997. An outbreak of gastroenteritis and fever due to *Listeria monocytogenes* in milk. *N. Engl. J. Med.* **336:**100–105.

9. Evans, J. R., A. C. Allen, D. A. Stinson, R. Bortolussi, and L. J. Peddle. 1985. Perinatal listeriosis: report of an outbreak. *Pediatr. Infect. Dis.* **4:**237–241.

10. Ewert, D. P., L. Lieb, P. S. Hayes, M. W. Reeves, and L. Mascola. 1995. *Listeria monocytogenes* infection and serotype distribution among HIV-infected persons in Los Angeles County, 1985-1992. *J. Acquired Immune Defic. Syndr. Hum. Retrovirol.* **8:**461–465.

11. Fleming, D. W., S. L. Cochi, K. L. MacDonald, J. Brondum, P. S. Hayes, B. D. Plikaytis, M. B. Holmes, A. Auduria, C. V. Broome, and A. L. Reingold. 1985. Pasteurized milk as a vehicle of infection in an outbreak of listerosis. *N. Engl. J. Med.* **312:**404–407.

12. Gautoar, A. R., L. A. Cone, D. R. Woodard, R. J. Mahler, R. D. Lynch, and D. H. Stoltzman. 1992. Arterial infections due to *Listeria monocytogenes*: report of four cases and review of world literature. *Clin. Infect. Dis.* **14:**23–28.

13. Gellin, B. J., C. V. Broome, W. F. Bibb, R. E. Weaver, S. Gaventa, L. Mascola, and the Listeriosis Study Group. 1991. Epidemiology of listeriosis in the United States—1986. *Am. J. Epidemiol.* **133:**392–401.

14. Gill, D. A. 1937. Ovine bacterial encephalitis (circling disease) and the bacterial genus Listerella. *Aust. Vet. J.* **13:**46–56.

15. Goulet, V., A. Lapoutre, P. Rocourt, A. Courtieu, P. Dehaumont, and P. Veit. 1993. Epidemie de Listeriose en France—bilin final et resultat de l'enquète epidemiologique. *Bull. Epidemiol. Hebdom.* **4:**13–14.

16. Goulet, V., and P. Marchetti. 1996. Listeriosis in 225 non-pregnant patients in 1992. Clinical aspects and outcome in relationship to predisposing conditions. *Scand. J. Infect. Dis.* **28:**367–374.

17. Graves, L. M., B. Swaminathan, G. W. Ajello, G. D. Malcolm, R. E. Weaver, R. Ransom, K. Dever, B. D. Plikaytis, A. Schuchat, J. D. Wenger, R. W. Pinner, C. V. Broome, and the Listeria Study Group. 1992. Comparison of three selective enrichment methods for the isolation of *Listeria monocytogenes* from naturally contaminated foods. *J. Food Prot.* **55:**952–959.

18. Hof, H., T. Nichterlein, and M. Kretschmar. 1997. Management of listeriosis. *Clin. Microbiol. Rev.* **10:**345–357.

19. Hume, O. S. 1976. Maternal *Listeria monocytogenes* septicemia with sparing of the fetus. *Obstet. Gynecol.* **48:**33S–34S.

20. Jurado, R. L., M. M. Farley, E. Pereira, R. C. Harvey, A. Schuchat, J. D. Wenger, and D. S. Stephens. 1993. Increased risk of meningitis and bacteremia due to *Listeria monocytogenes* in patients with human immunodeficiency virus infection. *Clin. Infect. Dis.* **17:**224–227.

21. Kampelmacher, E. H., and L. M. Van Noorle-Jansen. 1969. Isolation of *Listeria monocytogenes* from feces of clinically healthy humans and animals. *Zentralbl. Bakteriol. Paris Abt. I Orig.* **211:**353–359.

22. Kessler, S. L., and A. S. Dajani. 1990. Listeria meningitis in infants and children. *Pediatr. Infect. Dis. J.* **9:**61–63.

23. Lanmerding, A. M., K. A. Glass, A. Gendron-Fitzpatrick, and M. P. Doyle. 1992. Determination of virulence of different stains of *Listeria monocytogenes* and *Listeria innocua* by oral inoculation of pregnant mice. *Appl. Bacteriol. Microbiol.* **58:**3991–4000.

24. Linnan, M. J., L. Mascola, X. D. Lou, V. Goulet, S. May, G. Salminen, D. W. Hird, M. L. Yonekura, P. Hayes, R. Weaver, A. Audurier, B. D. Plikaytis, S. L. Fannin, A. Kleks, and C. V. Broome. 1988. Epidemic listeriosis associated with Mexican-style cheese. *N. Engl. J. Med.* **319:**823–828.

25. Lorber, B. 1991. Listeriosis following shigellosis. *Rev. Infect. Dis.* **13:**865–866.

26. Lorber, B. 1997. Listeriosis. *Clin. Infect. Dis.* **24:**1–9.

27. Louthrenoo, W., and H. R. Schumacher, Jr. 1990. *Listeria monocytogenes* osteomyelitis complicating leukemia: report and literature review of Listeria osteoarticular infections. *J. Rheumatol.* **17:**107–110.

28. Low, J. C., and C. P. Renton. 1985. Septicemia, encephalitis and abortions in a housed flock of sheep caused by *Listeria monocytogenes* type 1/2. *Vet. Rec.* **114:**147–150.

29. Mascola, L., F. Sorvillo, V. Goulet, B. Hall, R. Weaver, and M. J. Linnan. 1992. Fecal carriage of *Listeria monocytogenes*—observations during a community-wide, common-source outbreak. *Clin. Infect. Dis.* **15:**557–558.

30. McLauchlin, J., S. M. Hall, S. K. Velani, and R. J. Gilbert. 1991. Human listerosis and pate: a possible association. *Br. Med. J.* **303:**773–775.

31. McLauchlin, J., and J. C. Low. 1994. Primary cutaneous listeriosis in adults: an occupational disease in veterinarians and farmers. *Vet. Rec.* **135:**615–617.

32. Merle-Melet, M., L. Dossou-Glete, P. Maurer, P. Meyer, A. Lozniewski, O. Kuntzburger, M. Weber, and A. Gerard. 1996. Is amoxicillin-cotrimoxazole the most appropriate antibiotic regimen for Listeria meningoencephalitis. Review of 22 cases and the literature. *J. Infect.* **33:**79–85.

33. Murray, E. G. D., R. A. Webb, and M. B. R. Swann. 1926. A disease of rabbits characterized by large mononuclear leucocytosis, caused by a hitherto undescribed bacillus: Bacterium monocytogenes. *J. Pathol. Bacteriol.* **29:**407–439.

34. Nyfeldt, A. 1929. Etiologie de la mononuclease infectieuse. *C. R. Soc. Biol. (Paris)* **101:**590–592.

35. Pirie, J. H. H. 1927. A new disease of veld rodents, "Tiger River Disease." *Publ. S. Afr. Inst. Med. Res.* **3:**163–186.

36. Potel, J. 1952. Zur granulomatosis infantiseptica. *Zentralbl. Bakteriol. Parasitenkd. Abt. I Orig.* **158:**329–331.

37. Riedo, F. X., R. W. Pinner, M. L. Tosca, M. L. Cartter, M. L. Graves, N. W. Reeves, R. E. Weaver, B. D. Plikaytis, and C. V. Broome. 1994. A point-source food

borne listeriosis outbreak: documented incubation period and possible mild illness. *J. Infect Dis.* **170:**693–696.

38. **Rocourt, J., C. H. Jacquet, and J. Bille.** 1997. Human listeriosis, 1991-1992, p. 1–42. *In Food Safety Issues Division of Food and Nutrition.* World Health Organization, Geneva, Switzerland.

39. **Salamina, G., E. Dalle Donne, A. Niccolini, G. Poda, D. Cesaroni, M. Brici, M. Maldini, R. Fini, A. Schuchat, B. Swaminathan, W. Bibb, J. Rocourt, N. Binkin, and S. Salmaso.** 1996. A foodborne outbreak of gastroenteritis due to *Listeria monocytogenes. Epidemiol. Infect.* **117:**429–436.

40. **Schlech, W. F., III.** 1993. An animal model of foodborne *Listeria monocytogenes* virulence: effect of alterations in local and systemic immunity on invasive infection. *Clin. Invest. Med.* **16:**219–225.

41. **Schlech, W. F., III, P. M. Lavigne, R. A. Bortolussi, A. C. Allen, E. V. Haldane, A. J. Wort, A. W. Hightower, S. E. Johnson, S. H. King, E. S. Nicholls, and C. V. Broome.** 1983. Epidemic listeriosis—evidence for transmission by food. *N. Engl. J. Med.* **308:**203–206.

42. **Schuchat, A., K. A. Deaver, J. D. Wenger, B. D. Plikaytis, L. Mascola, R. W. Pinner, A. L. Reingold, C. V. Broome, and the Listeriosis Study Group.** 1992. Role of foods in sporadic listeriosis. I. Case-control study of dietary risk factors. *JAMA* **267:**2041–2045.

43. **Schuchat, A., A. C. Lizano, C. V. Broome, B. Swaminathan, C. Kim, and K. Winn.** 1991. Outbreak of neonatal listerosis associated with mineral oil. *Pediatr. Infect. Dis.* **10:**183–189.

44. **Schuchat, A., B. Swaminathan, and C. V. Broome.** 1991. Epidemiology of listeriosis. *Clin. Microbiol. Rev.* **4:**169–183.

45. **Schwartz, B., C. V. Broome, G. R. Brown, A. W. Hightower, C. A. Ciesielski, S. Gaventa, B. G. Gellin, L. Mascola, and the Listeriosis Study Group.** 1988. Association of sporadic listeriosis with consumption of uncooked hotdogs and undercooked chicken. *Lancet* **ii:**779–782.

46. **Schwartz, B., D. Hexter, C. V. Broome, A. W. Hightower, R. V. Hirschhorn, J. D. Porter, P. S. Hayes, W. F.** Bibb, B. Lorber, and D. G. Faris. 1989. Investigation of an outbreak of listeriosis; new hypothesis for the etiology of epidemic *Listeria monocytogenes* infections. *J. Infect. Dis.* **159:**680–685.

47. **Shuin, D. T., and S. B. Galsworthy.** 1982. Stimulation of monocyte production by an endogenous mediator induced by a component from *Listeria monocytogenes. Immunology* **46:**343–351.

48. **Sivalingam, J. J., P. Martin, H. S. Fraimow, J. C. Yarze, and L. S. Friedman.** 1992. *Listeria monocytogenes* peritonitis: case report and literature review. *Am. J. Gastroenterol.* **87:**1839–1845.

49. **Skogberg, K. J., J. Syrjanen, M. Jahkola, O. V. Reinkonen, J. Paavonen, J. Ahonen, S. Kontiainen, P. Ruutu, and V. Valtonen.** 1992. Clinical presentation and outcome of listeriosis in patients with and without immunosupressive therapy. *Clin. Infect. Dis.* **14:**815–821.

50. **Spyrou, N., M. Anderson, and R. Foale.** 1997. Listeria endocarditis: current management and patient outcome—world literature review. *Heart* **4:**380–383.

51. **Sridama, V., F. Pacni, S.-L. Wang, A. Moawad, M. Reilly, and L. J. Degroot.** 1982. Decreased levels of helper T-cells: a possible cause of immunodeficiency in pregnancy. *N. Engl. J. Med.* **307:**352–356.

52. **Synnott, M. B., D. L. Morse, and S. M. Hall.** 1994. Neonatal meningitis in England and Wales: a review of routine national data. *Arch. Dis. Child.* **71:**F75–80.

53. **Tappero, J. W., A. Schuchat, K. A. Deaver, L. Mascola, and J. D. Wenger for the Listeriosis Study Group.** 1995. Reduction in the incidence of human listeriosis in the United States: effectiveness of prevention efforts? *JAMA* **273:**1118–1122.

54. **Thompson, P., P. A. Salisbury, C. Adams, and D. L. Archer.** 1990. U.S. food legislation. *Lancet* **336:**1557–1559.

55. **Wald, A., D. H. Vanthiel, L. Hoechstetter, J. S. Gavaler, K. M. Egler, R. Verm, L. Scott, and R. Lester.** 1982. Effect of pregnancy on gastrointestinal transit. *Dig. Dis. Sci.* **27:**1015–1018.

56. **Welshimer, H. J., and J. Donker-Voet.** 1971. *Listeria monocytogenes* in nature. *Appl. Microbiol.* **21:**516–519.

Immune and Inflammatory Responses to *Listeria monocytogenes* Infection

ALYCE FINELLI AND ERIC G. PAMER

50

Listeria monocytogenes has been used as a model pathogen for investigations of mammalian immunity to bacterial infection for nearly four decades (74). Early studies by Mackaness (45) demonstrated that murine immunity to *L. monocytogenes* is mediated by cells and not by serum components, a concept referred to as cell-mediated immunity. Since these studies, *L. monocytogenes* has become a standard pathogen for investigations of cell-mediated immunity and also nonspecific inflammatory responses to bacterial infection. The cumulative studies of murine infection with *L. monocytogenes* arguably have provided the most complete and sophisticated picture of mammalian immune responses to a bacterial pathogen.

L. MONOCYTOGENES PATHOGENESIS INFLUENCES THE IMMUNE RESPONSE

L. monocytogenes is a gram-positive bacterium. As such, it does not contain lipopolysaccharide and does not cause, upon mammalian infection, septic shock as would be seen following infection with a gram-negative bacterium. *L. monocytogenes* does, however, contain within its cell membrane and wall other components, such as lipoteichoic acids, that can be bound by macrophage receptors and that may play an important role in modulating early inflammatory responses in the mammalian hosts. The immune response to intracellular pathogens is influenced by the pathogen's localization and activity within the infected host cell. Full protective immunity is obtained only by infection with live *L. monocytogenes* capable of intracellular survival (7). *L. monocytogenes* enters cells but rapidly escapes the phagosome by secreting listeriolysin (see chapter 53, this volume), a membranolytic protein that enables bacteria to enter the cytosol of the infected cell (55). Although this strategy allows *L. monocytogenes* to escape from the hostile environment of phagolysosomes, it subjects this pathogen to surveillance by the major histocompatibility complex (MHC) class I antigen-processing pathway. The MHC class I antigen-processing pathway degrades cytosolic proteins into peptides, which are transported into the endoplasmic reticulum and bound by newly synthesized MHC class I molecules (52). Complexes of MHC class I and cytosolically generated peptides traffic to the cell surface, where cytolytic T cells detect MHC class I molecules presenting foreign peptides. This antigen-processing pathway, which plays a major role in combating viral infection, also plays a major role in the defense against *L. monocytogenes* infection (53). The consequences of a particular microbial pathogenesis strategy for the resulting immune response are demonstrated very clearly by *L. monocytogenes* infection.

MURINE INFECTION WITH *L. MONOCYTOGENES*

The mouse model of *L. monocytogenes* infection was developed nearly 40 years ago with the discovery that infection with a low inoculum is readily cleared from mice and results in high resistance to subsequent infection with much higher doses of bacteria (74). Subsequent studies have characterized the complex network of inflammatory and immune responses elicited by *L. monocytogenes* infection. In the following sections, the innate inflammatory responses and the T-lymphocyte-mediated immune responses to *L. monocytogenes* infection are discussed.

INNATE CELLULAR INFLAMMATORY RESPONSES TO *L. MONOCYTOGENES* INFECTION

Although the murine immune response to *L. monocytogenes* infection has become a paradigm for T-cell-mediated immune responses to intracellular pathogens, work in the last 10 years has demonstrated the important, indeed critical, role of non-antigen-specific innate inflammation in containing *L. monocytogenes* infection. This crucial role of nonspecific immunity was first suggested by experiments with SCID mice: these mice, which lack T- and B-cell-mediated immunity, are able to control *L. monocytogenes* infection, although they are unable to clear infection (2). The roles of various inflammatory cells and inflammatory cytokines are discussed in the following sections.

Macrophage-Mediated Clearance of *L. monocytogenes*

Initial defense against systemic infection by *L. monocytogenes* relies upon splenic and hepatic macrophages. Following intravenous injection, the great majority of *L. monocytogenes* cells are taken up by these tissue macrophages and killed, accounting for the rapid decrease in viable bacteria during the first few hours of infection (49). Recent studies suggest that the macrophage scavenger receptor plays a role in the uptake and killing of *L. monocytogenes* during the early stages of infection (68). The critical role that macrophages play in the presentation of *L. monocytogenes* antigens to T lymphocytes is discussed in later sections of this chapter.

Neutrophil-Mediated Killing of *L. monocytogenes*

Neutrophils play a key role in defense against infection by *L. monocytogenes*. In the first few days following *L. monocytogenes* infection, histologic analysis of spleen and liver shows the presence of many neutrophils. Furthermore, immune-mediated resistance to *L. monocytogenes* infection correlates with enhanced accumulation of neutrophils at sites of infection, suggesting that neutrophils play a major role in defense against infection (18). The protective role of neutrophils in the defense against *L. monocytogenes* infection was demonstrated with antibodies that block migration of neutrophils and monocytes to sites of inflammation (60). Blocking neutrophil migration greatly exacerbated *L. monocytogenes* infection in the liver, suggesting that neutrophils lyse infected hepatocytes and kill *L. monocytogenes* to curtail infection (13). Further, studies with neutrophil-depleting monoclonal antibodies showed that neutrophils play a larger role in containing *L. monocytogenes* hepatic infection than splenic infection (14). The first few days following infection appear to be the time when neutrophils are critical in the defense against *L. monocytogenes* infection. Mice lacking reduced nicotinamide adenine dinucleotide phosphate oxidase activity have increased susceptibility to *L. monocytogenes* infection, underscoring the importance of the role of neutrophils and the oxidative burst in early defenses against this pathogen (22). T-lymphocyte-mediated immunity, however, does not depend upon neutrophil-mediated inflammation and killing of bacteria, since splenocytes from *L. monocytogenes*-infected, neutrophil-depleted mice confer specific immunity upon transfer to naive mice (16). Interestingly, transfer of immune splenocytes to neutrophil-depleted recipient mice does not result in protection, indicating that immune T cells require neutrophils to abolish *L. monocytogenes* infection (1).

The Role of Natural Killer Cells in Immunity to *L. monocytogenes*

Natural killer (NK) cells are cytolytic cells that also produce inflammatory cytokines, such as gamma interferon (IFN-γ). Target cell recognition by NK cells is complex and can involve the recognition of glycosylation changes of surface glycoproteins and also of the absence of surface MHC class I molecules, as occurs during some viral infections. NK cells are also involved in the defense against *L. monocytogenes* infection. Lymph nodes draining tissues infected with *L. monocytogenes* contain large numbers of IFN-γ-secreting cells, which are predominantly NK cells (23,

70). It has been demonstrated that depletion of NK cells from mice with specific antibodies results in exacerbation of *L. monocytogenes* infection (23). It is likely that interleukin-12 production by *L. monocytogenes*-infected macrophages promotes NK cell expansion during infection. However, it is unclear how these NK cells are recruited to sites of infection, and what the NK cell targets are at sites of infection. Although it is not known for certain whether NK cells can specifically recognize *L. monocytogenes*-infected cells, it has been proposed that changes in surface glycosylation induced by *L. monocytogenes* infection may target infected cells for lysis by NK cells (76). It has been shown that heat-killed *L. monocytogenes* bacteria can induce IFN-γ secretion by NK cells in the presence of tumor necrosis factor (78). Defense against *L. monocytogenes* infection is highly dependent upon IFN-γ, as demonstrated by the fact that mice lacking the receptor for IFN-γ or the gene for IFN-γ succumb to sublethal doses of *L. monocytogenes* (28, 34). The finding that MHC class I-restricted cytolytic T lymphocytes need not express IFN-γ to confer protective immunity (28) suggests that the IFN-γ produced by NK cells is critical for early defense against *L. monocytogenes* infection.

CYTOKINES AND SOLUBLE MEDIATORS OF DEFENSE AGAINST *L. MONOCYTOGENES*

Much of the innate immune response to pathogens such as *L. monocytogenes* depends upon the production of inflammatory cytokines by infected cells and responding effector cells such as NK cells. In the following sections, we discuss the roles of some of the mediators in the response to *L. monocytogenes* infection. The cytokine milieu generated during the initial inflammatory response to *L. monocytogenes* infection is likely to play a large role in determining the quality and magnitude of the specific immune response.

IFN-γ

IFN-γ is a cytokine that activates macrophages to become more effective microbial killers through generation of reactive oxygen and nitrogen metabolites. In addition, IFN-γ increases the capacity of many cell types to process and present antigens to T lymphocytes. Antibodies that neutralize IFN-γ dramatically worsen *L. monocytogenes* infection. Indeed, mice in which the gene for IFN-γ or its receptor is disrupted succumb to even low doses of *L. monocytogenes* (28, 34). Multiple cell types, including NK cells, CD8+ T cells, and CD4+ T cells, secrete IFN-γ during the course of *L. monocytogenes* infection. It is likely, however, that IFN-γ secreted during the early stages of infection, before the expansion of antigen-specific T lymphocytes, plays a major role in ensuring the survival of the host.

Tumor Necrosis Factor Alpha

Tumor necrosis factor alpha (TNF-α) is induced by lipopolysaccharide of gram-negative bacteria and mediates septic shock in animals. In the case of murine infection with *L. monocytogenes*, however, TNF-α is a cytokine that is essential for survival. Administration of TNF-α blocking antibodies to infected mice dramatically worsened infection (29). Furthermore, in mice with a disruption in the TNF-α receptor gene, *L. monocytogenes* infection was uniformly lethal (54, 61). In other studies, it was found that TNF-α plays a role in the activation of macrophages in SCID mice (3). Taken together, these studies indicate that TNF-α is a critical factor in the defense against *L. monocytogenes*.

Interleukin-1

Interleukin-1 (IL-1) is an inflammatory cytokine that mediates many of the systemic symptoms associated with severe infection. IL-1 is expressed in the livers and spleens of mice as early as 1 day after *L. monocytogenes* infection. The amount of IL-1 that is found directly correlates with the severity of infection (30). Administration of antibodies specific for the type I IL-1 receptor to infected mice resulted in 100- to 1,000-fold higher numbers of *L. monocytogenes* bacteria in the livers and spleens than in those of untreated mice. Blocking IL-1 inhibited the influx of neutrophils into the peritoneal cavity following infection, suggesting that IL-1 plays a role in neutrophil recruitment (58). Furthermore, blocking IL-1 diminished macrophage sensitivity to IFN-γ and the expression of MHC class II molecules by macrophages following *L. monocytogenes* infection, suggesting that IL-1 also mediates its protective effect by facilitating macrophage activation (59). IL-1, in combination with IL-12, has also been shown to enhance $\gamma\delta$-T-cell proliferation and expression of IFN-γ (67). It is possible that IL-1α and IL-1β may have overlapping roles in the defense against *L. monocytogenes* infection, since mice in which the gene for IL-1β was disrupted were shown to have similar resistance to *L. monocytogenes* infection as wild-type mice (79).

IL-12

IL-12 enhances NK cell proliferation and CD8$^+$-T-cell expansion and promotes differentiation of CD4$^+$ T cells into TH1 (IFN-γ- and IL-2-secreting) cells rather than TH2 (IL-4-secreting) cells (8). Exposure of macrophages to heat-killed *L. monocytogenes* cells elicits IL-12 secretion, an effect that may be attributable to lipoteichoic acids in the bacterial cell wall (12). Administration of IL-12 to *L. monocytogenes*-infected mice enhanced bacterial clearance (33). Conversely, blocking IL-12 with specific antibody exacerbated *L. monocytogenes* infection (77). In SCID mice, IL-12 blocking antibodies also worsened *L. monocytogenes* infection; this deleterious effect could be reversed by concomitant administration of IFN-γ (72). Thus, IL-12 production after *L. monocytogenes* infection may be essential for generating early, IFN-γ-mediated responses. While immunization of mice with dead *L. monocytogenes* bacteria generally does not prime a protective immune response, a recent study found that IL-12-treated mice immunized with heat-killed *L. monocytogenes* bacteria developed protective immunity (47).

IL-6

IL-6 also plays a protective role in the immune response to *L. monocytogenes* infection. As found with IL-1, the levels of IL-6 found in mice during *L. monocytogenes* infection directly reflect the severity of infection (43). Mice that received exogenous IL-6 shortly before infection had 10-fold fewer bacteria in the spleen and liver 2 and 4 days later (44). The protective effect of IL-6 may be mediated by neutrophils, as suggested by experiments with immunodeficient mice lacking B and T cells (20). Given these findings, it is not surprising that mice with disruption of the IL-6 gene are more susceptible to *L. monocytogenes* infection, with more extensive liver infection and fewer circulating neutrophils than control mice (20, 38).

IL-4 and IL-10

CD4$^+$-T-cell responses can be subdivided into those that support T-cell-mediated immunity (TH1 responses) and those that support humoral immunity (TH2 responses). Cytokines that promote T-cell-mediated immunity enhance the clearance of *L. monocytogenes*, while cytokines that promote humoral immunity appear to worsen *L. monocytogenes* infections. For example, in mice injected with antibodies that block IL-4, a cytokine produced by TH2 T lymphocytes, the number of *L. monocytogenes* bacteria in the liver and spleen is reduced by a factor of up to 100 after infection (27). Similar findings were noted when IL-10 was studied in SCID mice (71). Injection of mice with immune complexes, which promotes IL-10 secretion by macrophages, increased their susceptibility to *L. monocytogenes* infection. Exacerbation of infection by immune complexes, however, could be prevented by coadministration of antibodies that blocked IL-10. Mice with a genetic disruption of the IL-10 gene have increased resistance to *L. monocytogenes* (19).

Signal Transduction and Gene Activation Required for *L. monocytogenes* Specific Immunity

The complex inflammatory responses to *L. monocytogenes* infection require the transmission of signals from the surface of effector cells to the nucleus and the subsequent activation of genes encoding the mediators of the inflammatory response. As might be expected, signal transducer and activator of transcription (STAT) proteins play key roles in the defense against *L. monocytogenes*. Specifically, STAT1 and STAT4 are involved in IFN-γ- and IL-12-mediated responses, respectively (36, 46). At the nuclear level, NF-κB is a transcription factor that regulates the expression of IL-1β, IL-2, IL-6, IFN-β, and TNF-α. Mice in which the gene for the p50 subunit of NF-κB has been disrupted are more susceptible to *L. monocytogenes* infection (64). Similarly, mice lacking NF-IL6, which regulates the expression of IL-6, IL-1α, IL-8, TNF-α, granulocyte colony-stimulating factor, and nitric oxide synthase, are also more susceptible to *L. monocytogenes* infection (69). Mice deficient in the IFN consensus sequence-binding protein or IFN regulatory factor 2 also manifest an impaired ability to clear *L. monocytogenes* infections, while mice lacking IFN regulatory factor 1, impairing their ability to produce nitric oxide, clear infection normally (24). These studies are beginning to outline the network of signaling pathways and gene activation events involved in the defense against intracellular bacterial infection.

Nitric Oxide-Mediated Clearance of *L. monocytogenes*

Nitric oxide plays an important role in the defense against microbial pathogens. Nitric oxide can either act directly or can react with hydrogen peroxide to form peroxynitrite, which enhances killing by the oxidative burst. Nitric oxide production participates in macrophage-mediated killing of *L. monocytogenes*. Inhibition of nitric oxide synthase with aminoguanidine decreases the in vitro ability of macrophages to kill intracellular *L. monocytogenes* and, if administered in vivo, dramatically increases the mortality of normal and SCID mice infected with *L. monocytogenes* (4).

Chemokine-Mediated Recruitment of Effector Cells During *L. monocytogenes* Infection

Chemokines and their receptors are integral for the coordinated movement of immune and inflammatory cells to sites of infection. The role of individual chemokines and their receptors in the response to *L. monocytogenes* infection is just beginning to be deciphered. Mice deficient for CCR2, the receptor for monocyte chemoattractant protein-1, are highly susceptible to *L. monocytogenes* and are unable to clear infections (39). A somewhat different result was obtained with mice lacking the receptor for IL-8, a chemokine specific for neutrophils. These mice had enhanced resistance to early infection but had a predilection for developing chronic infections (17). It is likely that the murine model of *L. monocytogenes* infection will be useful for characterizing the roles of the rapidly increasing numbers of chemokines and chemokine receptors in mediating cellular immune responses.

T LYMPHOCYTE RESPONSES TO *L. MONOCYTOGENES*

γδ T Cells in the Early Defense Against *L. monocytogenes*

T lymphocytes can express either γδ- or αβ-T-cell receptors. The function and specificity of the αβ T cells are well understood in general, and their role in the defense against *L. monocytogenes* infection is described later in this chapter. On the other hand, γδ T cells constitute a diverse set of lymphocytes with an array of different effector functions and with a range of specificities that is only partially understood. Because of their prevalence within mucosal and epidermal tissues, γδ T cells are thought to participate in early defenses against infections at these sites. In keeping with this notion, γδ T lymphocytes appear to contribute to early defenses against *L. monocytogenes* infection. For example, *L. monocytogenes* infection of the mouse peritoneum results in an early influx of γδ T cells, which is followed by the appearance of αβ T cells (50). The magnitude of the γδ-T-cell response in the spleen and liver appears to correlate directly with the severity of infection (5). Antibody-mediated depletion of γδ T cells from infected mice results in an exacerbation of *L. monocytogenes* infection, with roughly 100-fold more bacteria in mouse spleens 5 days after infection (32). However, γδ-T-cell-depleted mice eventually clear the infection, in contrast to mice depleted of αβ T cells. Studies in mice that have been genetically engineered to lack either αβ or γδ T cells or both αβ and γδ T cells demonstrate that early *L. monocytogenes* infection is controlled to a much greater extent by γδ T cells than by αβ T cells (48). In both αβ- and γδ-T-cell-knockout mice, however, sublethal doses of *L. monocytogenes* are eventually cleared. In contrast, mice lacking both γδ and αβ T cells develop chronic infections with *L. monocytogenes*. As with NK cells, the specificity of γδ T cells responding to *L. monocytogenes* infection has not been determined. Although γδ-T-cell expansion may be stimulated by bacterial antigens, there is also evidence that IL-1 and IL-12, which are produced by *L. monocytogenes*-infected macrophages, can drive expansion of γδ T cells (67).

αβ T Cells in the Clearance of *L. monocytogenes* Infection

Complete clearance of *L. monocytogenes* infection and long-term immunity are mediated by αβ T lymphocytes. αβ T cells can interact with MHC class II molecules, in the case of CD4+ T cells, or MHC class I molecules, in the case of CD8+ T cells. MHC class II molecules usually present peptides that are derived from exogenous proteins that are degraded in an endosomal compartment (15). MHC class I molecules, on the other hand, generally present peptides derived from the degradation of proteins in the cytosol of the antigen-presenting cell (52). The role of CD8+ and CD4+ T cells in the defense against *L. monocytogenes* infection was first investigated by transferring these T-cell subsets from *L. monocytogenes*-immunized mice to naive recipients. These experiments demonstrated that CD8+ cytolytic T lymphocytes (CTLs) protect naive mice from lethal infection (37). Further experiments dissecting the relative roles of CD4+ or CD8+ T cells were performed in mice that were devoid of either the MHC class I or class II antigen-processing pathway (40). These studies demonstrated that mice lacking MHC class II antigen presentation and CD4+ T cells have a slightly prolonged course of infection following *L. monocytogenes* infection. Mice that lacked MHC class I antigen presentation and CD8+ T cells, however, suffered more prolonged infections but were still eventually able to clear the infection. It is interesting to note that recovered mice lacking CD8+ T cells have a lower level of immunity than do immune mice lacking CD4+ T cells. These studies, therefore, demonstrated that neither CD4+ nor CD8+ T cells alone are essential for clearance of infection or immunity. However, both CD4+ cells and, probably to a greater extent, CD8+ cells do contribute to protective immunity against *L. monocytogenes* infection.

CD8+ CTLs were believed to mediate resistance to *L. monocytogenes* infection not only by lysing infected cells, but also by secreting IFN-γ in the infected microenvironment, thus activating local macrophages. Surprisingly, CTLs derived from mice in which the gene for IFN-γ has been disrupted are capable of providing protection against *L. monocytogenes*, indicating that production of this cytokine is not critical for CTL-mediated immunity, but rather the cytolytic function of the CTL is the crucial factor (28). The importance of cell lysis by CTLs was demonstrated in mice in which the gene for perforin was disrupted (35). These perforin knockout mice have diminished resistance to primary infection and reduced immunity following sublethal infection. Furthermore, adoptive transfer of perforin-deficient immune splenocytes conferred approximately 100-fold less protection than did control immune splenocytes. Thus, lysis of *L. monocytogenes*-infected cells appears to play a significant role in the ability of CTLs to confer protection.

ANTIGEN SPECIFICITY OF T LYMPHOCYTES

MHC Class II-Restricted T Cells

Before entering the cytosol of infected cells, *L. monocytogenes* traverses the endosomal compartment, thereby exposing itself to the MHC class II antigen-processing pathway (56). The first *L. monocytogenes* protein identified as a T-cell-detected antigen was listeriolysin, the essential virulence factor enabling bacterial escape into the cytosol (6). Listeriolysin contains several epitopes that are presented by MHC class II molecules to CD4+ T lymphocytes (62). Another antigen that is detected by *L. monocytogenes*-specific CD4+ T cells was identified by expression cloning (63). Most recently, the secreted protein p60, a murein hydrolase, was shown to be presented by MHC class II molecules to CD4+ T cells (25). CD4+ T lymphocytes from immunized mice respond to many *L. monocytogenes* proteins separated by two-dimensional electrophoresis, indicating that many different proteins feed into the MHC class II antigen-processing pathway during the course of infection (21). Because only a small number of antigens have been identified, it is unclear whether secreted, cell wall-associated, or intrabacterial proteins form the major targets of *L. monocytogenes*-specific CD4$^+$ T lymphocytes.

MHC Class Ia-Restricted CTLs

L. monocytogenes enters the cytosolic compartment of infected cells and secretes proteins that are degraded by host proteasomes, generating peptides that feed into the MHC class I antigen-processing pathway (53). Studies with BALB/c mice have shown that at least three secreted proteins of *L. monocytogenes* are presented to CD8$^+$ T cells by the conventional MHC class I antigen-processing pathway (10, 53). Specifically, listeriolysin, the major virulence factor, p60, and mpl, a secreted metalloprotease, are degraded in the cytosol, and the generated peptide fragments are presented by the murine H2-K^d MHC class I molecule. Quantitative studies of these peptides in infected macrophages indicate that each is processed and presented with distinct kinetics. Surprisingly, the magnitude of the T-cell response to each of these peptides does not correlate with the amount of peptide presented on the surface of infected macrophages (11, 75). Moreover, the largest CTL response occurs to a peptide that derives from a protein that is nearly undetectable in infected cells.

The kinetics of the T-cell response to *L. monocytogenes* infection have recently been investigated. Although CTLs respond to different peptides that derive from distinct antigens, the time period of CTL expansion following infection is similar for all T cells, regardless of their peptide specificity (11). This finding suggests that the inflammatory reaction to *L. monocytogenes* infection, rather than the presence of peptide, determines the kinetics of T-cell expansion. The important interactions between the innate inflammatory responses and the specific T-cell responses are increasingly being appreciated (73).

MHC Class Ib-Restricted CTLs

MHC class Ib molecules, also referred to as nonclassical MHC molecules, are encoded within the MHC and share many characteristics with MHC class Ia molecules, including their general structure and association with β2-microglobulin (42, 65). In one important respect, however, these two families of molecules differ: while MHC class Ia molecules are highly polymorphic, MHC class Ib molecules are highly conserved; i.e., genetically diverse individuals are likely to express different MHC class Ia molecules but similar MHC class Ib molecules. One of the most interesting MHC class Ib molecules in mice is H2-M3. This molecule specifically binds peptides that contain N-formyl methionine at the amino terminus and thus has the ability to selectively bind peptides that are of bacterial origin (51). Indeed, following infection of mice with *L. monocytogenes*, H2-M3 presents bacterially derived peptides to CD8$^+$ CTLs. Three different *L. monocytogenes*-derived peptides that are presented to CTLs by H2-M3 have been identified (26, 41, 57). In addition to H2-M3, there is evidence suggesting that another MHC class Ib molecule, Qa-1b, may also present *L. monocytogenes*-derived peptides to CTLs (9).

ROLE OF BACTERIAL PROTEIN SECRETION IN T-CELL-MEDIATED IMMUNITY

The majority of proteins synthesized by bacteria are retained within the bacterial cytosol or cell wall, while a minority are secreted into the bacterial environment. The secreted proteins, however, would be anticipated to have the greatest access to the antigen-processing pathways, particularly the MHC class I antigen pathway. Therefore, it is not very surprising that all of the *L. monocytogenes*-derived, MHC class I-restricted epitopes either derive from secreted proteins or are peptides released by bacteria (see above). Studies of *L. monocytogenes* antigens expressed by *Salmonella typhimurium* demonstrated that protein secretion is essential for the priming of protective immunity (31). This work was recently extended when it was shown that immunity was protective only against bacteria that secreted a chimeric, CTL epitope-containing protein and not against *L. monocytogenes* bacteria that retained the epitope intrabacterially (66).

CONCLUDING REMARKS

The bacterial pathogen *L. monocytogenes* has proved to be an invaluable tool for studies of immune responses to infection. As new technologies are developed, studies of *L. monocytogenes* infection will enable deeper dissection of host defenses against infection. Undoubtedly, future studies of experimental listeriosis will continue to contribute to our understanding of cell-mediated immune responses.

REFERENCES

1. **Appelberg, R., A. G. Castro, and M. T. Silva.** 1994. Neutrophils as effector cells of T-cell-mediated, acquired immunity in murine listeriosis. *Immunology* **83:**302–307.

2. **Bancroft, G. J., R. D. Schreiber, and E. R. Ununue.** 1991. Natural immunity: a T-cell-independent pathway of macrophage activation defined in the SCID mouse. *Immunol. Rev.* **124:**5–24.

3. **Bancroft, G. J., K. C. Sheehan, R. D. Schreiber, and E. R. Unanue.** 1989. Tumor necrosis factor is involved in the T cell-independent pathway of macrophage activation in SCID mice. *J. Immunol.* **143:**127–130.

4. **Beckerman, K. P., H. W. Rogers, J. A. Corbett, R. D. Schreiber, M. L. McDaniel, and E. R. Unanue.** 1993. Release of nitric oxide during the T cell-independent pathway of macrophage activation. *J. Immunol.* **150:** 888–895.

5. **Belles, C., A. K. Kuhl, A. J. Donoghue, Y. Sano, R. L. O'Brien, W. Born, K. Bottomly, and S. R. Carding.** 1996. Bias in the gamma delta T cell response to *Listeria monocytogenes*. V delta 6.3+ cells are a major component of the gamma delta T cell response to *Listeria monocytogenes*. *J. Immunol.* **156:**4280–4289.

6. **Berche, P., J. L. Gaillard, C. Geoffroy, and J. E. Alouf.** 1987. T cell recognition of listeriolysin O is induced during infection with *Listeria monocytogenes*. *J. Immunol.* **139:** 3813–3821.

7. **Berche, P., J. L. Gaillard, and P. J. Sansonetti.** 1987. Intracellular growth of *Listeria monocytogenes* as a prerequisite for *in vivo* induction of T cell-mediated immunity. *J. Immunol.* **138:**2266–2271.

8. **Biron, C. A., and R. T. Gazzinelli.** 1995. Effects of IL-12 on immune responses to microbial infections: a key mediator in regulating disease outcome. *Curr. Opin. Immunol.* **7:**485–496.

9. **Bouwer, H. G., M. S. Seaman, J. Forman, and D. J. Hinrichs.** 1997. MHC class Ib restricted cells contribute

to antilisterial immunity: evidence for Qa-1b as a key restricting element for Listeria-specific CTLs. *J. Immunol.* **159:**2795–2801.

10. **Busch, D. H., A. G. A. Bouwer, D. Hinrichs, and E. G. Pamer.** 1997. A nonamer peptide derived from *Listeria monocytogenes* metalloprotease is presented to cytolytic T lymphocytes. *Infect. Immun.* **65:**5326–5329.

11. **Busch, D. H., I. M. Pilip, S. Vijh, and E. G. Pamer.** 1998. Coordinate regulation of complex T cell populations responding to bacterial infection. *Immunity* **8:**353–362.

12. **Cleveland, M. G., J. D. Gorham, T. L. Murphy, E. Tuomanen, and K. M. Murphy.** 1996. Lipoteichoic acid preparations of gram-positive bacteria induce interleukin-12 through a CD14-dependent pathway. *Infect. Immun.* **64:**1906–1912.

13. **Conlan, J. W., and R. J. North.** 1991. Neutrophil-mediated dissolution of infected host cells as a defense strategy against a facultative intracellular bacterium. *J. Exp. Med.* **174:**741–744.

14. **Conlan, J. W., and R. J. North.** 1994. Neutrophils are essential for early anti-Listeria defense in the liver, but not in the spleen or peritoneal cavity, as revealed by a granulocyte-depleting monoclonal antibody. *J. Exp. Med.* **179:**259–268.

15. **Cresswell, P.** 1994. Assembly, transport and function of MHC class II molecules. *Annu. Rev. Immunol.* **12:**259–293.

16. **Czuprynski, C. J., J. F. Brown, N. Maroushek, R. D. Wagner, and H. Steinberg.** 1994. Administration of anti-granulocyte mAb RB6-8C5 impairs the resistance of mice to *Listeria monocytogenes* infection. *J. Immunol.* **152:**1836–1846.

17. **Czuprynski, C. J., J. F. Brown, H. Sternberg, and D. Carroll.** 1998. Mice lacking the murine interleukin-8 receptor homologue demonstrate paradoxical responses to acute and chronic experimental infection with *Listeria monocytogenes*. *Microb. Pathog.* **24:**17–23.

18. **Czuprynski, C. J., P. M. Henson, and P. A. Campbell.** 1985. Enhanced accumulation of inflammatory neutrophils and macrophages mediated by transfer of T cells from mice immunized with *Listeria monocytogenes*. *J. Immunol.* **134:**3449–3454.

19. **Dai, W. J., G. Kohler, and F. Brombacher.** 1997. Both innate and acquired immunity to *Listeria monocytogenes* infection are increased in IL-10 deficient mice. *J. Immunol.* **158:**2259–2268.

20. **Dalrymple, S. A., L. A. Lucian, R. Slattery, T. McNeil, D. M. Aud, S. Fuchino, F. Lee, and R. Murray.** 1995. Interleukin-6-deficient mice are highly susceptible to *Listeria monocytogenes* infection: correlation with inefficient neutrophilia. *Infect. Immun.* **63:**2262–2268.

21. **Daugelat, S., C. H. Ladel, B. Schoel, and S. H. E. Kaufmann.** 1994. Antigen-specific T-cell responses during primary and secondary *Listeria monocytogenes* infection. *Infect. Immun.* **62:**1881–1888.

22. **Dinauer, M. C., M. B. Deck, and E. R. Unanue.** 1997. Mice lacking reduced nicotinamide adenine dinucleotide phosphate oxidase activity show increased susceptibility to early infection with *Listeria monocytogenes*. *J. Immunol.* **158:**5581–5583.

23. **Dunn, P. L., and R. J. North.** 1991. Early gamma interferon production by natural killer cells is important in

defense against murine listeriosis. *Infect. Immun.* **59:**2892–2900.

24. **Fehr, T., G. Schoedon, B. Odermatt, T. Holtschke, M. Schneemann, M. F. Bachmann, T. W. Mak, I. Horak, and R. M. Zinkernagel.** 1997. Crucial role of interferon consensus sequence binding protein, but neither of interferon regulatory factor 1 nor of nitric oxide synthesis for protection against murine listeriosis. *J. Exp. Med.* **185:**921–931.

25. **Geginat, G., M. Lalic, M. Kretschmar, W. Goebel, H. Hof, D. Palm, and A. Bubert.** 1998. Th1 cells specific for a secreted protein of *Listeria monocytogenes* are protective *in vivo*. *J. Immunol.* **160:**6046–6055.

26. **Gulden, P. H., P. Fischer, N. E. Sherman, W. Wang, V. H. Engelhard, J. Shabanowitz, D. H. Hunt, and E. G. Pamer.** 1996. A *Listeria monocytogenes* pentapeptide is presented to cytolytic T lymphocytes by the H2-M3 MHC class Ib molecule. *Immunity* **5:**73–79.

27. **Haak-Frendscho, M., J. F. Brown, Y. Iizawa, R. D. Wagner, and C. J. Czuprynski.** 1992. Administration of anti-IL-4 monoclonal antibody 11B11 increases the resistance of mice to *Listeria monocytogenes* infection. *J. Immunol.* **148:**3978–3985.

28. **Harty, J. T., and M. J. Bevan.** 1995. Specific immunity to *Listeria monocytogenes* in the absence of INF gamma. *Immunity* **3:**109–118.

29. **Havell, E.** 1989. Evidence that tumor necrosis factor has an important role in antibacterial resistance. *J. Immunol.* **143:**2894–2899.

30. **Havell, E. A., L. L. Moldawer, D. Helfgott, P. L. Kilian, and P. G. Sehgal.** 1992. Type I IL-1 receptor blockade exacerbates murine listeriosis. *J. Immunol.* **148:**1486–1492.

31. **Hess, J., I. Gentschev, D. Miko, M. Welzel, C. Ladel, W. Goebel, and S. H. E. Kaufmann.** 1996. Superior efficacy of secreted over somatic antigen display in recombinant Salmonella vaccine induced protection against listeriosis. *Proc. Natl. Acad. Sci. USA* **93:**1458–1463.

32. **Hiromatsu, K., Y. Yoshikai, G. Matsuzaki, S. Ohga, K. Muramori, K. Matsumoto, J. A. Bluestone, and K. Nomoto.** 1992. A protective role of γ/δ T cells in primary infection with *Listeria monocytogenes* in mice. *J. Exp. Med.* **175:**49–56.

33. **Hsieh, C. S., S. E. Macatonia, C. S. Tripp, S. F. Wolf, A. O'Garra, and K. M. Murphy.** 1993. Development of TH1 CD4+ T cells through IL-12 produced by Listeria-induced macrophages. *Science* **260:**547–549.

34. **Huang, S., W. Hendriks, A. Althage, S. Hemmi, H. Bluethmann, R. Kamijo, J. Vilcek, R. M. Zinkernagel, and M. Aguet.** 1993. Immune response in mice that lack the interferon-γ receptor. *Science* **259:**1742–1745.

35. **Kagi, D., B. Ledermann, K. Burki, H. Hengartner, and R. F. Zinkernagel.** 1994. CD8 T cell-mediated protection against an intracellular bacterium by perforin-dependent cytotoxicity. *Eur. J. Immunol.* **24:**3068–3072.

36. **Kaplan, M. H., Y. L. Sun, T. Hoey, and M. J. Grusby.** 1996. Impaired IL-12 responses and enhanced development of Th2 cells in Stat4-deficient mice. *Nature* **382:**174–177.

37. **Kaufmann, S. H. E., E. Hug, and G. DeLibero.** 1986. *Listeria monocytogenes* reactive T lymphocyte clones with cytolytic activity against infected target cells. *J. Exp. Med.* **164:**363–368.

38. Kopf, M., H. Baumann, G. Freer, M. Freudenberg, M. Lamers, T. Kishimoto, R. Zinkernagel, H. Bluethmann, and G. Kohler. 1994. Impaired immune and acute-phase responses in interleukin-6-deficient mice. *Nature* **368:** 339–342.

39. Kurihara, T., G. Warr, J. Loy, and R. Bravo. 1997. Defects in macrophage recruitment and host defense in mice lacking the CCR2 chemokine receptor. *J. Exp. Med.* **186:** 1757–1762.

40. Ladel, C. H., I. E. F. Flesch, J. Arnoldi, and S. H. E. Kaufmann. 1994. Studies with MHC-deficient knock-out mice reveal impact of both MHC I- and MHC II-dependent T cell responses on *Listeria monocytogenes* infection. *J. Immunol.* **153:**3116–3122.

41. Lenz, L. L., B. Dere, and M. J. Bevan. 1996. Identification of an H2-M3-restricted Listeria epitope: implications for antigen presentation by M3. *Immunity* **5:**63–72.

42. Lindahl, K. F., D. E. Byers, V. Dabhi, R. Hovik, E. P. Jones, G. P. Smith, C.-R. Wang, H. Xiao, and M. Yoshino. 1997. H2-M3, a full service class Ib histocompatibility antigen. *Annu. Rev. Immunol.* **15:**851–879.

43. Liu, Z., R. J. Simpson, and C. Cheers. 1992. Recombinant interleukin-6 protects mice against experimental infection. *Infect. Immun.* **60:**4402–4406.

44. Liu, Z., R. J. Simpson, and C. Cheers. 1994. Role of IL-6 in activation of T cells for acquired cellular resistance to *Listeria monocytogenes. J. Immunol.* **152:**5375–5380.

45. Mackaness, G. B. 1962. Cellular resistance to infection. *J. Exp. Med.* **116:**381–406.

46. Meraz, M. A., J. M. White, K. C. Sheehan, E. A. Bach, S. J. Rodig, A. S. Dighe, D. H. Kaplan, J. K. Riley, A. C. Greenlund, D. Campbell, K. Carver-Moore, R. N. DuBois, R. Clark, M. Aguet, and R. D. Schreiber. 1996. Targeted disruption of the Stat1 gene in mice reveals unexpected physiologic specificity in the JAK-STAT signaling pathway. *Cell* **84:**431–442.

47. Miller, M. A., M. J. Skeen, and H. K. Ziegler. 1995. Nonviable bacterial antigens administered with IL-12 generate antigen-specific T cell responses and protective immunity against *Listeria monocytogenes. J. Immunol.* **155:**4817–4828.

48. Mombaerts, P., J. Arnoldi, F. Russ, S. Tonegawa, and S. H. Kaufmann. 1993. Different roles of alpha beta and gamma delta T cells in immunity against an intracellular bacterial pathogen. *Nature* **365:**53–56.

49. North, R. J., P. L. Dunn, and J. W. Conlan. 1997. Murine listeriosis as a model of antimicrobial defense. *Immunol. Rev.* **158:**26–36.

50. Ohga, S., Y. Yoshikai, Y. Takeda, K. Hiromatsu, and K. Nomoto. 1990. Sequential appearance of gamma/delta- and alpha/beta-bearing T cells in the peritoneal cavity during an i.p. infection with *Listeria monocytogenes. Eur. J. Immunol.* **20:**533–538.

51. Pamer, E. G., M. J. Bevan, and K. F. Lindahl. 1993. Do nonclassical, class Ib MHC molecules present bacterial antigens to T cells? *Trends Microbiol.* **1:**35–38.

52. Pamer, E. G., and P. Cresswell. 1998. Mechanisms of MHC class I-restricted antigen processing. *Annu. Rev. Immunol.* **16:**323–358.

53. Pamer, E. G., A. J. A. M. Sijts, M. S. Villanueva, D. H. Busch, and S. Vijh. 1997. MHC class I antigen processing of *Listeria monocytogenes* proteins: implications for dominant and subdominant CTL responses. *Immunol. Rev.* **158:**129–136.

54. Pfeifer, K., T. Matsuyama, T. M. Kundig, A. Wakeham, K. Kenjihara, A. Shahinian, K. Wiegmann, P. S. Ohashi, M. Kronke, and T. W. Mak. 1993. Mice deficient for the 55 kd tumor necrosis factor receptor are resistant to endotoxic shock, yet succumb to *L. monocytogenes* infection. *Cell* **73:**457–467.

55. Portnoy, D. A., P. S. Jacks, and D. J. Hinrichs. 1988. Role of hemolysin for the intracellular growth of *Listeria monocytogenes. J. Exp. Med.* **167:**1459–1471.

56. Portnoy, D. A., R. D. Schreiber, P. Connelly, and L. G. Tilney. 1989. Gamma interferon limits access of *Listeria monocytogenes* to the macrophage cytoplasm. *J. Exp. Med.* **170:**2141–2146.

57. Princiotta, M. F., L. L. Lenz, M. J. Bevan, and U. D. Staerz. 1998. H2-M3 restricted presentation of a Listeria-derived leader peptide. *J. Exp. Med.* **187:**1711–1720.

58. Rogers, H. W., K. C. F. Sheehan, L. M. Brunt, S. K. Dower, E. R. Unanue, and R. D. Schreiber. 1992. Interleukin 1 participates in the development of anti-Listeria responses in normal and SCID mice. *Proc. Natl. Acad. Sci. USA* **89:**1011–1015.

59. Rogers, H. W., C. S. Tripp, R. D. Schreiber, and E. R. Unanue. 1994. Endogenous IL-1 is required for neutrophil recruitment and macrophage activation during murine Listeriosis. *J. Immunol.* **153:**2093–2101.

60. Rosen, H., S. Gordon, and R. J. North. 1989. Exacerbation of murine listeriosis by a monoclonal antibody specific for the type 3 complement receptor of myelomonocytic cells. *J. Exp. Med.* **170:**27–37.

61. Rothe, J., W. Lesslauer, H. Lotscher, Y. Lang, P. Koebel, F. Kontgen, A. Althage, R. Zinkernagel, M. Steinmetz, and H. Bluethmann. 1993. Mice lacking the tumour necrosis factor receptor 1 are resistant to TNF-mediated toxicity but highly susceptible to infection by *Listeria monocytogenes. Nature* **364:**798–802.

62. Safley, S. A., C. W. Cluff, N. E. Marshall, and H. K. Ziegler. 1991. Role of listeriolysin-O (LLO) in the T lymphocyte response to infection with *Listeria monocytogenes.* Identification of T cell epitopes of LLO. *J. Immunol.* **146:** 3604–3616.

63. Sanderson, S., D. J. Campbell, and N. Shastri. 1995. Identification of a CD4+ T cell-stimulating antigen of pathogenic bacteria by expression cloning. *J. Exp. Med.* **182:**1751–1757.

64. Sha, W. C., H. C. Liou, E. I. Tuomanen, and D. Baltimore. 1995. Targeted disruption of the p50 subunit of NF-κB leads to multifocal defects in immune responses. *Cell* **80:**321–330.

65. Shawar, S. M., J. M. Vyas, J. R. Rodgers, and R. R. Rich. 1994. Antigen presentation by major histocompatibility complex class I-B molecules. *Annu. Rev. Immunol.* **12:**839–880.

66. Shen, H., J. F. Miller, X. Fan, D. Kolwyck, R. Ahmed, and J. T. Harty. 1998. Compartmentalization of bacterial antigens: differential effects on priming of CD8 T cells and protective immunity. *Cell* **92:**535–545.

67. Skeen, M. J., and H. K. Ziegler. 1995. Activation of γδ T cells for production of IFN-γ is mediated by bacteria via macrophage-derived cytokines IL-1 and IL-12. *J. Immunol.* **154:**5832–5841.

68. Suzuki, H., Y. Kurihara, M. Takeya, N. Kamada, M. Kataoka, K. Jishage, O. Ueda, H. Sakaguchi, T. Higashi, T. Suzuki, Y. Takashima, Y. Kawabe, O. Cynshi, Y. Wada, M. Honda, H. Kurihara, H. Aburatani, T. Doi, A. Matsumoto, S. Azuma, T. Noda, Y. Toyoda, H. Itakura, Y. Yazaki, S. Horiuchi, K. Takahashi, J. K. Kruijt, T. J. C. van Berkel, U. P. Steinbrecher, S. Ishibashi, N. Maeda, S. Gordon, and T. Kodama. 1997. A role for macrophage scavenger receptors in atherosclerosis and susceptibility to infection. *Nature* **386:**292–296.

69. Tanaka, T., S. Akira, K. Yoshida, M. Umemoto, Y. Yoneda, N. Shirafuji, H. Fujiwara, S. Suematsu, N. Yoshida, and T. Kishimoto. 1995. Targeted disruption of the NF-IL6 gene discloses its essential role in bacteria killing and tumor cytotoxicity by macrophages. *Cell* **80:**353–361.

70. Teixeira, H. C., and S. H. E. Kaufmann. 1994. Role of NK1.1+ cells in experimental Listeriosis. *J. Immunol.* **152:**1873–1882.

71. Tripp, C. S., K. P. Beckerman, and E. R. Unanue. 1995. Immune complexes inhibit antimicrobial responses through interleukin-10 production. *J. Clin. Invest.* **95:**1628–1634.

72. Tripp, C. S., M. K. Gately, J. Hakimi, P. Ling, and E. R. Unanue. 1994. Neutralization of IL-12 decreases resistance to Listeria in SCID and C.B-17 mice. Reversal by IFN-gamma. *J. Immunol.* **152:**1883–1887.

73. Unanue, E. R. 1997. Studies in listeriosis show the strong symbiosis between the innate cellular system and the T-cell response. *Immunol. Rev.* **158:**11–25.

74. Unanue, E. R. 1997. Why listeriosis? A perspective on cellular immunity to infection. *Immunol. Rev.* **158:**5–11.

75. Vijh, S., and E. G. Pamer. 1997. Immunodominant and subdominant CTL responses to *Listeria monocytogenes* infection. *J. Immunol.* **158:**3366–3371.

76. Villanueva, M. S., C. J. M. Beckers, and E. G. Pamer. 1994. Infection with *Listeria monocytogenes* impairs sialic acid addition to host cell glycoproteins. *J. Exp. Med.* **62:**1881–1888.

77. Wagner, R. D., H. Steinberg, J. F. Brown, and C. J. Czuprynski. 1994. Recombinant interleukin-12 enhances resistance of mice to *Listeria monocytogenes* infection. *Microb. Pathog.* **17:**175–186.

78. Wherry, J. C., R. D. Schreiber, and E. R. Unanue. 1991. Regulation of IFN-production by natural killer cells in SCID mice: role of TNF and bacterial stimuli. *Infect. Immun.* **59:**1709–1718.

79. Zheng, H., D. Fletcher, W. Kozak, M. Jiang, K. J. Hofmann, C. A. Conn, D. Soszynski, C. Grabiec, M. E. Trumbauer, A. Shaw, M. J. Kostura, K. Stevens, H. Rosen, R. J. North, H. Y. Chen, M. J. Tocci, M. J. Kluger, and L. H. T. Van der Ploeg. 1995. Resistance to fever induction and impaired acute-phase response in interleukin-1β-deficient mice. *Immunity* **3:**9–19.

Genetic Tools for Use with *Listeria monocytogenes*

NANCY E. FREITAG

51

An increasing number of genetic tools have become available in recent years for the molecular characterization of *Listeria monocytogenes*. Plasmid vectors, reporter genes, systems designed for transposon mutagenesis, heterologous expression systems, and, most recently, transducing phage have all greatly advanced the experimental capacity to generate, characterize, and complement mutations within *L. monocytogenes* and to define the functional roles of gene products. This chapter is a brief description of some of the genetic tools currently available for use with *L. monocytogenes*. Key references are given throughout to provide sources for expanded details on plasmid constructions, assay conditions, and other technical aspects. The variety of genetic tools described on the following pages are meant to be representative of the resources available to those interested in *L. monocytogenes* genetics and should not be considered a complete and exhaustive list of all available plasmid vectors, transposons, and so on.

PHYSICAL MAPS OF THE *L. MONOCYTOGENES* CHROMOSOME

L. monocytogenes has a genome size of approximately 3,000 kb. Physical maps have been generated for two different serotypes. Pulsed-field gel electrophoresis has been used to establish a circular physical map of *L. monocytogenes* LO28 (serotype 1/2c) (36). The locations of several genes, including the *hly* locus and *inlAB*, have been mapped to defined intervals, as well as six putative rRNA operons. A physical map has also been generated for *L. monocytogenes* Scott A (serotype 4b) (25).

An *L. monocytogenes* genome project is currently being coordinated by Pascale Cossart (Pasteur Institute, Paris, France). The project, under an EC contract, began Sept. 1, 1998, and is expected to be completed in less than 2 years. The genome sequence of strain EGD-e (formerly known as EGD) (34) will be released in progress to the public on a future Web site (http://www.pasteur.fr/units/gmp/Gmp_projects.html) whenever contigs reach 100 kb with an estimated error rate less than one error per 5,000 nucleotides (15).

PLASMID VECTORS

Plasmid vectors are critical tools for many aspects of bacterial genetics. The ability to complement chromosomal gene mutations, to deliver transposable elements, to monitor gene expression via reporter-gene fusions, and to introduce specific mutations into the *L. monocytogenes* chromosome via plasmid integration has tremendously advanced molecular characterization of *L. monocytogenes* pathogenesis. A number of plasmid vectors developed for use in other gram-positive bacteria have been used successfully in *L. monocytogenes*, and several of these have been improved or modified since their original constructions. Table 1 lists a sample of these plasmids and a brief summary of their utility for *L. monocytogenes* genetic analyses. The plasmids listed have been divided into three groups. Group I includes those plasmids that have been used for complementation studies of *L. monocytogenes* gene mutations and for general gene expression; these plasmids contain both gram-negative and gram-positive origins of replication. Group II features plasmids used for the delivery of transposons into *L. monocytogenes*, primarily derivatives of the transposon Tn917. Group III includes plasmid vectors used for allelic exchange with *L. monocytogenes* chromosomal sequences and for the construction of chromosomal and plasmid reporter-gene fusions. Group III includes plasmid vectors with gram-positive temperature-sensitive origins of replication that facilitate integration and excision of plasmid vectors into and out of the *L. monocytogenes* chromosome, suicide vectors for generation of chromosomal insertion mutations, and autonomously replicating shuttle vectors designed for the construction of reporter gene fusions in plasmids. Several of the plasmids listed in Table 1 are described in more detail later in this section.

Introduction of Plasmid DNA into *L. monocytogenes*

Introduction of plasmid DNA into *L. monocytogenes* can be accomplished by transformation of *L. monocytogenes* protoplasts (10, 63) or by electroporation (42). Electroporation provides a simple and rapid method, and trans-

Gram-Positive Pathogens, ed. by V. A. Fischetti et al.
© 2000 American Society for Microbiology, Washington, D.C.

TABLE 1 Representative *L. monocytogenes* plasmid vectors

Plasmid	Size (kb)	Origin of rep. gram−/ gram+[a]	Origin of transfer[b]	Description	Reference
Group I: for general cloning and gene expression					
pAM401	10.4	Yes/yes	No	Shuttle vector, Cm[c]	66
pMK3	7.2	Yes/yes	No	Shuttle vector, Kn[r]	56
pMK4	5.6	Yes/yes	No	Shuttle vector, Cm[r]	56
pAT18	6.6	Yes/yes	Yes	Shuttle vector, Em[r]	60
pCON-2	7.6	Yes/yes	Yes	Shuttle vector, Cm[r]	A. Milenbachs and P. Youngman
pCON2-101	10.1	Yes/yes	Yes	Shuttle vector, Cm[r], low copy number in *E. coli*	A. Milenbachs and P. Youngman
pSPAC	5.7	Yes/yes	No	Expression vector, places gene under control of SPAC promoter, Cm[r]	69
Group II: for delivery of transposons					
pAM118	30.1	Yes/yes	Yes	Shuttle vector with Tn916, Em[r], Tet[r]	22
pLTV1	20.6	Yes/yes (ts)[d]	No	Shuttle vector with Tn917-lac, Em[r], Cm[r], Tet[r]	10
pLTV3	22.1	Yes/yes (ts)	No	Shuttle vector with Tn917-lac, Em[r], Cm[r], Tet[r], Ble[r]	10
pTV32-OK	14.1	Yes (ts)/yes (ts)	No	Shuttle vector with Tn917-lac, pWVO1 ts origin, Em[r]	J. Behari and P. Youngman
pTV32-OKT	14.8	Yes (ts)/yes (ts)	Yes	Shuttle vector with Tn917-lac, pWVO1 ts origin, Em[r]	J. Behari and P. Youngman
Group III: for allelic exchange, chromosomal gene insertion, reporter gene fusions					
pLSV1	6.3	Yes/yes (ts)	No	ts integrational vector, Em[r]	67
pKSV7	6.9	Yes/yes (ts)	No	ts integrational vector, Cm[r]	54
pCON-1	7.6	Yes/yes (ts)	Yes	ts integrational vector, Cm[r]	T. Foulger and P. Youngman
pHS-LV	10.8	Yes/yes (ts)	No	ts integrational vector for insertion of genes within *orf*Z, Em[r]	52
pNF579	7.6	Yes/yes (ts)	No	ts integrational vector for creating transcriptional *gfp* fusions, Cm[r]	N. Freitag
pNF580	8.1	Yes/yes (ts)	No	ts integrational vector for creating transcriptional *gus* fusions, Cm[r]	N. Freitag
pLCR	12	Yes/yes (ts)	No	ts integrational vector for creating transcriptional *lacZ* and *cat* fusions, Kn[r]	40
pAUL-A	9.2	Yes/yes (ts)	No	ts integrational vector for insertional mutagenesis and directional cloning, Em[r]	48
pTEX5235	4.8	Yes/no	Yes	Suicide vector for gene disruptions, Sp[r]	58
pTCV-lac	12	Yes/yes	Yes	Shuttle vector for construction of *lacZ* transcriptional fusions, low copy number, Em[r]	47
pSB292	7.1	Yes/yes	No	Shuttle vector for construction of *luxAB* transcriptional fusions, Kn[r], Cm[r]	43

[a] Presence of plasmid origin of replication function in gram-negative/gram-positive bacteria.

[b] Presence or absence of origin of transfer, allowing plasmid conjugation.

[c] Antibiotic resistance markers for drug selection in *L. monocytogenes*: Cm, chloramphenicol; Kn, kanamycin; Em, erythromycin; Sp, spectinomycin; Ble, bleomycin; Tet, tetracycline

[d] ts, temperature-sensitive plasmid origin of replication.

formants are generally recovered following 1 to 2 days' growth on selective media. The efficiency of transformation via electroporation can be increased by growing *L. monocytogenes* in the presence of low amounts of penicillin to interfere with cell wall synthesis (42); however, transformation frequencies even after penicillin treatment are generally not very high (usually 100 to 1,000 colonies per μg of DNA in our hands, although efficiencies as high as 4×10^6 colonies per μg of DNA have been reported for some strains [42]).

Recently, methods have been reported that improve the transfer of plasmid DNA into *L. monocytogenes* via conjugation (62). Plasmid vectors have been constructed that can be transferred directly from an *Escherichia coli* donor to *L. monocytogenes* at frequencies as high as 10^{-3} transconjugants per donor CFU. This improved efficiency of plasmid introduction should add to the repertoire of genetic techniques thus far available. For example, it may now be possible to easily generate knockout mutations in *L. monocytogenes* target genes by introduction via conjugation of suicide vectors (vectors with no gram-positive origins of replication) that insert into chromosomal genes following homologous recombination. A vector recently described by Teng et al. (58) derived from pAT18 (60), for the generation of insertion mutants in *Enterococcus* species, can be expected to fulfill this function in *L. monocytogenes* (plasmid pTEX5235 in Table 1). Suicide vectors have been successfully introduced into *L. monocytogenes* via electroporation, but the frequency of isolation of insertion mutants using this method is low and requires at least 1 kb of homologous DNA for plasmid integration into the chromosome (4). Improved suicide vectors might also be designed to facilitate the cloning of DNA sequences adjacent to chromosomal regions of interest. In addition, conjugal transfer of plasmid DNA into *L. monocytogenes* should facilitate the generation of *L. monocytogenes* plasmid libraries in *E. coli* and introduction of those libraries with reasonable efficiency into *L. monocytogenes* for complementation of unknown or uncharacterized genomic mutations.

Plasmid Vectors for Gene Expression and for Complementation Studies

A variety of shuttle plasmids with broad host ranges that are capable of replication in both *E. coli* and *L. monocytogenes* have been described (Table 1). For the expression of gene products via their own promoters or for gene complementation experiments, plasmid vectors such as pMK4 (56) and pAM401 (66) have been routinely used. These plasmids contain multiple cloning sites and antibiotic-resistance genes that are selectable in both gram-positive and gram-negative bacteria. Recently constructed and improved shuttle vectors include those containing origins of transfer to allow conjugation directly from *E. coli* hosts into *L. monocytogenes*, and shuttle vectors that are maintained at low copy numbers in *E. coli* to facilitate the characterization of *L. monocytogenes* genes that may be toxic at high copy. Such vectors include the multifunctional pCON-2 and pCON2-101 plasmids (constructed by Andrea Milenbachs in Philip Youngman's lab), shown in Fig. 1. pCON-2 and pCON2-101 both contain antibiotic-resistance genes selectable in gram-negative and gram-positive bacteria, multiple cloning sites within regions capable of alpha complementation (68), origins for conjugative transfer (65), and origins for plasmid replication in gram-negative and gram-positive bacteria. pCON2-101 has two additional advantages: (i) a pSC101-derived origin that maintains the

plasmid at low copy number in *E. coli*; and (ii) a f1 origin derived from filamentous bacteriophage for the generation of single-stranded DNA templates.

The inducible SPAC promoter, developed for use in *Bacillus subtilis*, has also been successfully used for controlling gene expression in *L. monocytogenes* (19). The SPAC expression system, developed by Yansura and Henner (69), is based on the concept of using the *E. coli lac* repressor to control the transcription of recombinant genes in *B. subtilis*. A promoterless copy of the gene of interest is inserted downstream from a *lac* operator-regulated promoter in the presence of the *lacI* gene, which is under the control of a constitutive gram-positive promoter. Confirmation of the utility of SPAC as a method for controlling gene expression in *L. monocytogenes* was carried out by placing the green fluorescent protein (GFP) gene of *Aequorea victoria* (12) under the control of the SPAC promoter on plasmid pSPAC and by introducing the construct into *L. monocytogenes*. Expression of GFP was found to be dependent upon the addition of IPTG (isopropyl-β-D-thiogalactopyranoside) to the medium (19); thus, the SPAC system is suitable and useful for situations in which it is desirable to control the expression of a gene of interest in *L. monocytogenes*.

Plasmid Vectors for Allelic Exchange of *L. monocytogenes* Chromosomal DNA

A powerful method allowing detailed examination of *L. monocytogenes* genetic elements is the ability to introduce specific mutations into the *L. monocytogenes* chromosome through the process of allelic exchange. First described in *L. monocytogenes* by Michel et al. (37), the replacement of wild-type chromosomal sequences with defined nucleotide substitutions, deletions, or with a reporter gene construct has since been simplified by the development and use of shuttle plasmids carrying gram-positive, temperature-sensitive origins of replication. The method, described by Camilli et al. (11), is depicted in Fig. 2 and proceeds as follows: plasmids are introduced into *L. monocytogenes* at temperatures permissive for plasmid replication, and transformants are isolated on selective media. Plasmid-containing strains are then grown in the presence of drug selection at temperatures nonpermissive for plasmid replication; these are conditions that enrich for bacteria containing integrations of plasmid DNA into homologous regions of the *L. monocytogenes* chromosome (Fig. 2A and B). After selection of plasmid-integrant colonies, the plasmid is then excised from the chromosome and lost by growing the bacteria at temperatures permissive for plasmid replication in the absence of drug selection (Fig. 2C and D). Bacterial colonies are then screened for drug sensitivity and for the presence of the chromosomal mutation. If the plasmid construct contains homologous chromosomal sequences of approximately equal size flanking the mutation, and if the mutation confers no growth defect or advantage upon bacterial colonies, then bacteria that contain the mutation of interest within their chromosomal sequences generally represent about 50% of the recovered population.

An integrational vector developed by Smith and Youngman (54) has several features that make it very useful for the process of allelic exchange in *L. monocytogenes*. Plasmid pKSV7 contains gram-positive, temperature-sensitive replication functions derived from pE194ts; the copy number of this plasmid at 32°C in *B. subtilis* is about 5 (23, 64, 70). pKSV7 also carries a ColE1 origin of replication, chloramphenicol and β-lactamase resistance genes for antibi-

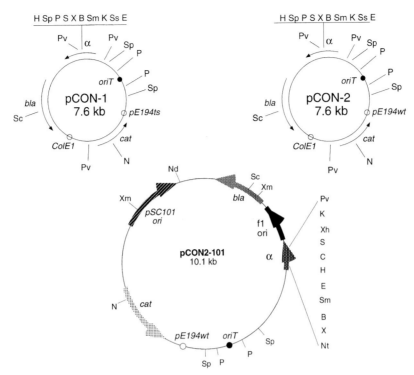

FIGURE 1 New plasmid vectors for gene expression/complementation and chromosomal integration in *L. monocytogenes*. ColE1 rep, replication functions derived from pBR322 (6); pSC101 ori, replication functions derived from pSC101 (55); f1 ori, single-stranded phage origin of replication derived from Promega phagemid pGEM-3Zr (+); oriT, origin for conjugative transfer derived from plasmid pCTC1 (65); pE194wt ori, replication functions derived from plasmid pE194 (28); pE194ts, temperature-sensitive replication functions derived from pE194ts (64); bla, a β-lactamase gene derived from pBR322; cat, a chloramphenicol resistance gene derived from pC194 (28); α, a fragment of the *E. coli lac* operon capable of alpha complementation (see reference 68); B, *Bam*HI recognition sites; C, *Cla*I sites; E, *Eco*RI sites; H, *Hin*dIII sites; K, *Kpn*I sites; N, *Nco*I sites; Nd, *Nde*I sites; Nt, *Not*I sites; P, *Pst*I sites; Pv, *Pvu*II sites; S, *Sal*I sites; Sc, *Sca*I sites; Sm, *Sma*I sites; Sp, *Sph*I sites; Ss, *Ssp*I sites; X, *Xba*I sites; Xh, *Xho*I sites; Xm, *Xmn*I sites.

otic selection, and a cluster of multiple cloning sites within *lacZ* sequences for insert screening by alpha complementation. pKSV7 has been successfully used for the generation of single- and multiple-base substitutions, deletions, and insertions within the *L. monocytogenes* chromosome (for examples, see references 11, 20, and 31). A similar vector, pCON-1, has been constructed by Tracey Foulger in the Youngman laboratory. pCON-1 contains the added advantage of an origin of transfer element and is shown in Fig. 1.

A vector has also been constructed for inserting genes of interest into a nonessential region of the *L. monocytogenes* chromosome. Vector pHS-LV was designed to introduce genes or sequences by allelic exchange into the *L. monocytogenes orfZ* region, an open reading frame of unknown function located at the 3′ end of the *prfA* regulon. This vector has been used extensively for the construction of *L. monocytogenes* strains expressing foreign antigens (30, 51, 52).

TRANSPOSONS

Transposable elements provide an effective means of creating random insertional mutations in bacterial populations. Several different transposons have been used in the generation of libraries of *L. monocytogenes* insertion mutants. Tn*1545* and Tn*916* are conjugative transposons that have been introduced successfully into *L. monocytogenes*; however, the use of these two transposons has been restricted by both the relatively low frequencies of transposition observed and the limitations imposed on the randomness of insertion by the requirements for sequence homology between both ends of the elements and sequences surrounding the sites of integration (14, 49). The Tn*3*-like transposon Tn*917* has proved more useful and has been used successfully in many gram-positive bacteria, including *L. monocytogenes*. Tn*917* generates extremely stable insertional mutations with a relatively high degree of randomness. Two modified forms of the Tn*917* transposon were constructed by Camilli et al. (10) and have been used in the generation of large-scale libraries of *L. monocytogenes* mutants (1, 10, 57). These modified Tn*917* elements have several advantages: (i) the transposons are carried on vectors with temperature-sensitive origins of replication that simplify the recovery of chromosomal insertions; (ii) they carry a promoterless copy of *E. coli lacZ* for the generation of transcriptional fusions; and (iii) they contain several features (a ColE1 replicon, an *E. coli*-selectable antibiotic-resistance gene, and a polylinker cloning site) that facilitate recovery in *E. coli* of chromosomal DNA adjacent to the sites of insertion.

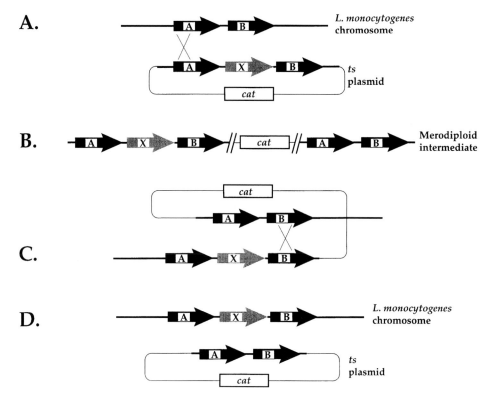

FIGURE 2 Schematic diagram for construction of chromosomal mutations in *L. monocytogenes*. The method depicted here and described in the text is suitable for the introduction of insertions, deletions, or single or multiple nucleotide substitutions within the *L. monocytogenes* chromosome. (A) Chromosomal integration of the temperature-sensitive plasmid vector by homologous recombination between plasmid-encoded genes A and B and the chromosomal alleles. The designated crossover points are arbitrary and can occur on either side of gene X. (B) Growth of bacterial cultures in the presence of chloramphenicol at temperatures nonpermissive for plasmid replication selects for the merodiploid intermediates that result from plasmid integration into the chromosome. (C) Merodiploid intermediate strains are then passed for several generations without selective pressure at temperatures permissive for plasmid replication. Spontaneous excision of the integrated plasmid from the chromosome occurs. (D) Excised plasmids are cured at temperatures nonpermissive for plasmid replication in the absence of drug selection.

J. Behari in P. Youngman's lab has recently developed new vectors for delivery of Tn917*lac* (2). These vectors include pTV32-OK, shown in Fig. 3, and an *oriT*-containing derivative of pTV32-OK, pTV32OK-T, for efficient delivery of the vector via conjugation. pTV32-OK and pTV32OK-T are substantially smaller in size than the previously described Tn917-delivery vectors. Both plasmids take advantage of the *repA-ts* broad-host-range origin of replication derived from plasmid pWVO1 (24) and the selectable antibiotic-resistance gene *aph3A*, which expresses kanamycin resistance in both *E. coli* and gram-positive hosts (61).

Tn10-derived transposons that are active in *B. subtilis* have been described (44). These transposons may be suitable for *L. monocytogenes* mutagenesis, but such use of these vectors has not yet been described.

REPORTER GENES

Several reporter genes developed for use in other systems have proved useful for monitoring *L. monocytogenes* transcriptional gene regulation. Transcriptional fusions to reporter genes such as *lacZ*, *gus*, *gfp*, *lux*, and *cat* have all been constructed in *L. monocytogenes* and have been used successfully to monitor patterns of bacterial gene expression (Table 1). The advantages and disadvantages of some of these reporter systems are discussed briefly below.

Fusions to β-Galactosidase

lacZ fusions have been generated either by transposon insertion (Tn917*lac* elements) (10, 35) or via integrative plasmids (40). β-Galactosidase activity can be measured for both extracellular and intracellularly grown bacteria (35), and the use of 4-methyl-umbelliferyl-β-D-galactopyranoside (MUG) as a substrate enables a higher degree of sensitivity than is possible with the colorimetric assay using o-nitro-phenyl-β-D-galactopyranoside (ONPG) (71). MUG assays should be carried out in the presence of 0.1% Triton X-100 to enhance bacterial permeability for the MUG substrate (33, 39). Intracellular assays of bacterial β-galactosidase activity should also control for any endogenous β-galactosidase activity present in the infected cell lines. One method to reduce the background contribution of endogenous mammalian β-galactosidase activity

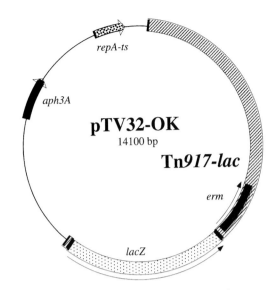

FIGURE 3 New vector for delivery of Tn*917lac* transposable element. *repA-ts*, broad-host-range, temperature-sensitive origin of replication derived from plasmid pWVO1(24); *aphA3*, selectable antibiotic-resistance gene that expresses kanamycin resistance in both *E. coli* and gram-positive hosts (61); *erm*, an inducible erythromycin-resistance gene naturally associated with Tn*917* (59); *lacZ*, a promoterless copy of the *E. coli lacZ* gene, substituted with a ribosome-binding site derived from *B. subtilis* gene *spoVG* (see reference 72).

is to adjust the pH of the assay conditions to approximately 8.0 (33, 39).

Fusions to β-Glucuronidase

gusA, the gene encoding β-glucuronidase, has been used as a reporter gene in a variety of prokaryotic systems. Recently, transcriptional fusions to *gus* have been constructed and introduced into *L. monocytogenes* (3, 18, 38). The use of *gus* as a reporter gene has at least two advantages over *lacZ*: (i) there is very little endogenous β-glucuronidase activity in *L. monocytogenes*, and (ii) *gus* fusions are far superior for assessing expression phenotypes on solid media (using 5-bromo-4-chloro-3-indolyl glucuronide [X-Gluc] as an indicator). Colorimetric assays (29) are easily performed, and fluorometric assay substrates are also commercially available and can be used as described above for the MUG assay of β-galactosidase activity (71), with the substitution of 4-methyl-umbelliferyl-β-D-glucuronide for 4-methyl-umbelliferyl-β-D-galactopyranoside. A sample vector for the generation of *L. monocytogenes* transcriptional gene fusions to *gus* is shown in Fig. 4 (pNF580) (see also Table 1).

Fusions to *gfp*

GFP of *A. victoria* (12) is a useful reporter gene system for the examination of gene expression in that it requires no cofactors for fluorescent activity and can be used in living systems as well as in fixed samples. It is possible to monitor the timing of *L. monocytogenes* gene expression for individual bacteria located within infected animal cells by using single-copy, chromosomally located transcriptional fusions to *gfp* (19). GFP fluorescence can be detected for bacteria located either in the cytoplasm or within vacuoles of in-

fected host cells following fixation and microscopic examination (Fig. 5). A sample vector for the generation of *L. monocytogenes* transcriptional gene fusions to *gfp* is shown in Fig. 4 (pNF579) (see also Table 1). Constitutive expression of GFP within *L. monocytogenes* may also provide a means of distinguishing different *L. monocytogenes* strains within an infected host cell. Dietrich et al. (17) have used plasmid-borne fusions to *gfp* to monitor *L. monocytogenes* promoter activity as well as the efficiency of plasmid delivery to infected cells by an attenuated suicide strain of *L. monocytogenes*.

Fusions to *lux* and *cat*

Transcriptional fusions to plasmid-borne *luxAB*-encoded luciferase were used by Park and Kroll (41) to monitor *L. monocytogenes* gene expression in a variety of growth media (pSB292) (Table 1). Luciferase activity was assessed by using a luminometer and measuring the peak of light produced.

Moors et al. (40) have developed a plasmid vector that allows transcriptional fusions to both *cat* (encoding chloramphenicol transacetylase) and *lacZ* to be constructed in single copy in the *L. monocytogenes* chromosome (Fig. 6). This vector, pLCR, enables both extracellular and intracellular measurement of target gene expression in *L. monocytogenes*.

Many other genes have potential uses as reporter systems in *L. monocytogenes*. Bubert et al. (7) have constructed transcriptional gene fusions to the *L. monocytogenes iap* gene that encodes an extracellular protein with a murein hydrolase activity necessary for septum separation. The *iap* gene product can be detected by immunoblotting and enzyme-linked immunosorbent assay, and bacteria that express target promoter-*iap* fusions filament and form long chains. Another gene of potential interest and use in *L. monocytogenes* is the gene encoding the *Staphylococcus aureus*-secreted nuclease (Nuc) (45). Nuc has the potential of becoming an export-specific reporter system for general use in gram-positive bacteria, similar to the gram-negative reporters PhoA and BlaM. Nuc activity requires an extracellular location, and Nuc⁺ colonies can be identified on plates overlaid with toluidine blue-DNA-agar. The Nuc system may facilitate the identification and characterization of gene products exported under a variety of conditions in *L. monocytogenes*.

HETEROLOGOUS EXPRESSION SYSTEMS FOR *L. MONOCYTOGENES* GENE PRODUCTS

In certain circumstances, it may be beneficial to use a heterologous expression system to monitor protein function or promoter activity. The naturally competent and well-studied bacterium *B. subtilis* has several advantages to recommend it as a heterologous expression system host for *L. monocytogenes* gene products. In addition to being a fellow gram-positive bacterium, a variety of genetic tools have been developed over the years to aid in the rapid generation of reporter gene fusions and inducible promoter constructs. *B. subtilis* has been used as a host system to demonstrate the ability of the *L. monocytogenes* listeriolysin O protein to promote lysis of phagosomal membranes in the absence of any additional *L. monocytogenes* proteins (5), to aid in the identification of the *L. monocytogenes plcA*

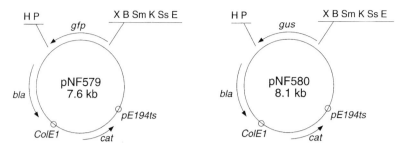

FIGURE 4 New vectors designed for the construction of *gfp* and *gus* transcriptional gene fusions. *gfp*, a promoterless copy of the *A. victoria gfp* gene (12), substituted with a ribosome-binding site derived from SD1 of *ermC* (16); *gus*, a promoterless copy of the *E. coli gusA* gene (PCR-amplified from plasmid pMLK100 [32], substituted with a ribosome-binding site derived from SD1 of *ermC* [16]). Restriction sites and other features are as described in the legend of Fig. 1.

gene product (9), to demonstrate the ability of PrfA to activate *hly* expression in the absence of any additional *L. monocytogenes* proteins (21), and to measure the relative levels of PrfA-dependent activation of target promoter sequences (50). The *B. subtilis* expression systems used in each of these cases took advantage of readily available plasmid vectors and bacteriophage for rapid generation of the constructs of interest.

L. MONOCYTOGENES BACTERIOPHAGE

Perhaps the most important genetic tool recently developed for *L. monocytogenes* genetic analysis has been the isolation and use of generalized transducing bacteriophages. Bacteriophage have been used for years as a method of typing *L. monocytogenes* strains, and an extensive collection of bacteriophages have been assembled for this purpose. Until recently, however, little utilization was made of these bacteriophages in terms of advancing their potential as tools for *L. monocytogenes* genetics. The availability of

transducing bacteriophage has several advantages for analysis of bacteria. They can be used to verify that a specific transposon insertion is associated with a mutant phenotype; they can facilitate the rapid introduction of selectable mutations into new strain backgrounds; they can be used to move point mutations closely associated with selectable markers into new strains; and they are useful for localized mutagenesis and fine structure genetic mapping.

David Hodgson (Department of Biological Sciences, University of Warwick, U.K.) has organized a collection of *L. monocytogenes* bacteriophages and tested them for transduction ability (26). A number of bacteriophages were found to be capable of transduction for *L. monocytogenes* serotype 1/2a and 4b strains. These bacteriophages are in the process of being further characterized, but at least two have been put into current use as genetic tools for *L. monocytogenes*. Brief descriptions of these bacteriophages are provided below.

P35

A generalized transducing phage, P35 was isolated from silage and grows on the 1/2a serotype strains SLCC5764, 10403S, LO28, and EGD (26). It is a temperate, citrate-sensitive, chloroform-sensitive bacteriophage that only grows at room temperature. Examination of purified phage

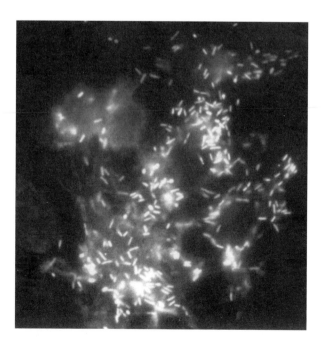

FIGURE 5 *L. monocytogenes* expressing GFP from a chromosomal *actA-gfp* transcriptional gene fusion during growth in the mouse macrophagelike tissue culture cell line J774 (53).

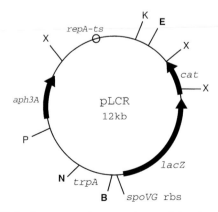

FIGURE 6 Integrational vector designed for the construction of *lacZ* and *cat* transcriptional gene fusions. pLCR contains a promoterless *Bacillus pumilus cat-86* and *E. coli lacZ* gene with an upstream *spoVG* ribosome-binding site and BamHI site (40). *trpA*, transcriptional terminator (13). Restriction sites and other features are as described in other figure legends.

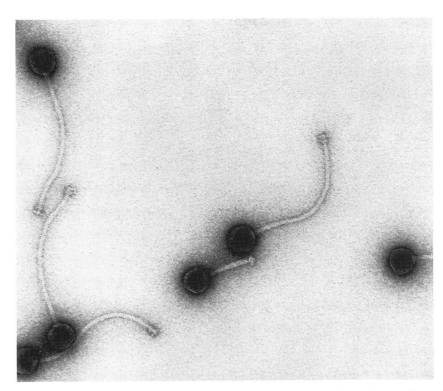

FIGURE 7   Electron micrograph of *L. monocytogenes* generalized transducing bacteriophages P35 and φ10403S. P35 is the short-tailed bacteriophage shown near the center of the micrograph; φ10403S is the long-tailed species. Micrograph courtesy of R. Calendar and R. Inman.

reveals a large capsid with about 38 kb of DNA and a tail of modest length (Fig. 7) (27). Although P35 can exist transiently in a carrier state, stable lysogenization has not been observed (8). It is capable of low-frequency transduction and interstrain transduction between 10403, EGD, and LO28 and has been shown to infect 75% of all serotype 1/2 strains tested (26).

U153

A second phage with very high transduction frequency, U153, was isolated by D. Hodgson from *L. monocytogenes* CUSI153. It grows on SLCC5764 and 10403S but not on LO28 or EGD. It has a genome of about 40 kb, and on the basis of five molecules examined so far, the genome is not unique (8). U153 has a long, snaky tail. It is capable of high-frequency transduction and interstrain transduction between 10403, NCTC7973, and SLCC5764 (26).

R. Calendar and colleagues have found that *L. monocytogenes* 10403S carries a bacteriophage, φ10403S, that looks very similar to U153 (8) (Fig. 7). Both bacteriophages have the same attachment specificity, since lysogenization by U153 can cure φ10403S. φ10403S and U153 have viral interference properties: both can interfere with a bacteriophage isolated from strain EGD, whereas the bacteriophage from EGD cannot interfere with φ10403S and U153. This interference property probably explains why so few *L. monocytogenes* phages grow on 10403S (8).

P35 and U153 have been used successfully by several labs (46) to generate new strain constructions and to verify the association between transposon insertions and mutant phenotypes. The availability of transducing phage opens enormous potential for the further characterization and genetic manipulation of *L. monocytogenes*. D. Hodgson is willing to send out any of the *L. monocytogenes* bacteriophages

as requested. In addition, anyone who is using an unusual strain of *L. monocytogenes* may send it to him, and he will test it with his collection of transducing bacteriophage to see which are capable of growth on that strain.

SUMMARY

The expanding number of tools available for the detailed characterization and genetic manipulation of *L. monocytogenes* increases the utility of this organism as a model system for the molecular study of an intracellular pathogen. Innovative efforts by many investigators have accelerated our ability to understand multiple aspects of bacterial pathogenesis in terms of gene expression, signal transduction, and protein function outside and inside infected host cells. The progress has been impressive, especially when one considers that little was known about bacterial components of *L. monocytogenes* infection as recently as just over a decade ago.

I thank A. Milenbachs, J. Behari, P. Youngman, M. Moors, D. Higgins, L. Shetron-Rama, A. Benson, P. Cossart, D. Portnoy, R. Calendar, R. Inman, and D. Hodgson and colleagues in their laboratories for many helpful discussions and for the communication of data before publication. Work was supported by Public Health Service grant AI41816 from the National Institutes of Health.

REFERENCES

1. **Annous, B. A., L. A. Becker, D. O. Bayles, D. R. Labeda, and B. J. Wilkinson.** 1997. Critical role of anteiso-C15:0 fatty acid in the growth of *Listeria monocytogenes* at low temperatures. *Appl. Environ. Microbiol.* **63:**3887–3894.

2. **Behari, J., and P. Youngman.** Personal communication.

3. **Behari, J., and P. Youngman.** 1998. Regulation of *hly* expression in *Listeria monocytogenes* by carbon sources and pH occurs through separate mechanisms mediated by PrfA. *Infect. Immun.* **66:**3635–3642.

4. **Benson, A.** Personal communication.

5. **Bielecki, J., P. Youngman, P. Connelly, and D. A. Portnoy.** 1990. *Bacillus subtilis* expressing a haemolysin gene from *Listeria monocytogenes* can grow in mammalian cells. *Nature* **345:**175–176.

6. **Bolivar, R., R. L. Rodriquez, P. J. Greene, M. C. Betlach, H. L. Heynecker, H. W. Boyer, J. H. Crosa, and S. Falkow.** 1977. Construction and characterization of new cloning vehicles. II. A multipurpose cloning system. *Gene* **2:**95–113.

7. **Bubert, A., H. Kestler, M. Gotz, R. Bockmann, and W. Goebel.** 1997. The *Listeria monocytogenes iap* gene as an indicator gene for the study of PrfA-dependent regulation. *Mol. Gen. Genet.* **256:**54–62.

8. **Calendar, R.** Personal communication.

9. **Camilli, A., H. Goldfine, and D. A. Portnoy.** 1991. *Listeria monocytogenes* mutants lacking phosphatidylinositol-specific phospholipase C are avirulent. *J. Exp. Med.* **173:**751–754.

10. **Camilli, A., D. A. Portnoy, and P. Youngman.** 1990. Insertional mutagenesis of *Listeria monocytogenes* with a novel Tn917 derivative that allows direct cloning of DNA flanking transposon insertions. *J. Bacteriol.* **172:**3738–3744.

11. **Camilli, A., L. G. Tilney, and D. A. Portnoy.** 1993. Dual roles of *plcA* in *Listeria monocytogenes* pathogenesis. *Mol. Microbiol.* **8:**143–157.

12. **Chalfie, M., Y. Tu, G. Euskirchen, W. W. Ward, and D. C. Prasher.** 1994. Green fluorescent protein as a marker for gene expression. *Science* **263:**802–805.

13. **Christie, G. E., P. J. Farnham, and T. Platt.** 1981. Synthetic sites for transcription termination and a functional comparison with tryptophan operon termination sites *in vitro*. *Proc. Natl. Acad. Sci. USA* **78:**4180–4184.

14. **Clewell, D. B., S. E. Flannagan, Y. Ike, J. M. Jones, and C. Gawron-Burke.** 1988. Sequence analysis of termini of conjugative transposon Tn916. *J. Bacteriol.* **170:**3046–3052.

15. **Cossart, P.** Personal communication.

16. **Denoya, C. D., D. H. Bechhofer, and D. Dubnau.** 1986. Translational autoregulation of *ermC* 23S rRNA methyltransferase expression in *Bacillus subtilis*. *J. Bacteriol.* **168:**1133–1141.

17. **Dietrich, G., A. Bubert, I. Gentschev, Z. Sokolovic, A. Simm, A. Catic, S. H. E. Kaufmann, J. Hess, A. Szalay, and W. Goebel.** 1998. Delivery of antigen-encoding plasmid DNA into the cytosol of macrophages by attenuated suicide *Listeria monocytogenes*. *Nat. Biotechnol.* **16:**181–185.

18. **Freitag, N. E.** Unpublished data.

19. **Freitag, N. E., and K. E. Jacobs.** 1999. Examination of *Listeria monocytogenes* intracellular gene expression with the green fluorescent protein of *Aequorea victoria*. *Infect. Immun.* **67:**1844–1852.

20. **Freitag, N. E., and D. A. Portnoy.** 1994. Dual promoters of the *Listeria monocytogenes prfA* transcriptional activator

appear essential in vitro but are redundant *in vivo*. *Mol. Microbiol.* **12:**845–853.

21. **Freitag, N. E., P. Youngman, and D. A. Portnoy.** 1992. Transcriptional activation of the *Listeria monocytogenes* hemolysin gene in *Bacillus subtilis*. *J. Bacteriol.* **174:**1293–1298.

22. **Gawron-Burke, C., and D. B. Clewell.** 1984. Regeneration of insertionally inactivated streptococcal DNA fragments after excision of transposon Tn916 in *Escherichia coli*: strategy for targeting and cloning of genes from gram-positive bacteria. *J. Bacteriol.* **159:**214–221.

23. **Gryczan, T. J., J. Hahn, S. Contente, and D. Dubnau.** 1982. Replication and incompatibility properties of plasmid pE194 in *Bacillus subtilis*. *J. Bacteriol.* **152:**722–735.

24. **Gutierrez, J. A., P. J. Crowley, D. P. Brown, J. D. Hillman, P. Youngman, and A. S. Bleiweis.** 1996. Insertional mutagenesis and recovery of interrupted genes of *Streptococcus mutans* by using transposon Tn917: preliminary characterization of mutants displaying acid sensitivity and nutritional requirements. *J. Bacteriol.* **178:**4166–4175.

25. **He, W., and J. B. Luchansky.** 1997. Construction of the temperature-sensitive vectors pLUCH80 and pLUCH88 for delivery of Tn917::NotI/SmaI and use of these vectors to derive a circular map of *Listeria monocytogenes* Scott A, a serotype 4b isolate. *Appl. Environ. Microbiol.* **63:**3480–3487.

26. **Hodgson, D. A.** Personal communication.

27. **Hodgson, D. A., R. Calendar, and R. B. Inman.** Personal communication.

28. **Iordanescu, S.** 1976. Three distinct plasmids originating in the same *Staphylococcus aureus* strain. *Arch. Roum. Pathol. Exp. Microbiol.* **35:**111–118.

29. **Jefferson, R. A., S. M. Burgess, and D. Hirsh.** 1986. β-Glucuronidase from *Escherichia coli* as a gene-fusion marker. *Proc. Natl. Acad. Sci. USA* **83:**8447–8451.

30. **Jensen, E. R., R. Selvakumar, H. Shen, R. Ahmed, F. O. Wettstein, and J. F. Miller.** 1997. Recombinant *Listeria monocytogenes* vaccination eliminates papillomavirus-induced tumors and prevents papilloma formation from viral DNA. *J. Virol.* **71:**8467–8474.

31. **Jones, S., and D. A. Portnoy.** 1994. Characterization of *Listeria monocytogenes* pathogenesis in a strain expressing perfringolysin O in place of listeriolysin O. *Infect. Immun.* **62:**5608–5613.

32. **Karow, M., and P. J. Piggot.** 1995. Construction of *gusA* transcriptional fusion vectors for *Bacillus subtilis* and their utilization for studies of spore formation. *Gene* **163:**69–74.

33. **Klarsfeld, A. D., P. L. Goossens, and P. Cossart.** 1994. Five *Listeria monocytogenes* genes preferentially expressed in infected mammalian cells: *plcA*, *purH*, *purD*, *pyrE* and an arginine ABC transporter gene, *arpJ*. *Mol. Microbiol.* **13:**585–597.

34. **Leimeister-Wachter, M., and T. Chakraborty.** 1989. Detection of listeriolysin, the thiol-dependent hemolysin in *Listeria monocytogenes*. *Infect. Immun.* **57:**2350–2357.

35. **Mengaud, J., S. Dramsi, E. Gouin, J. A. Vazquez-Boland, G. Milon, and P. Cossart.** 1991. Pleiotropic control of *Listeria monocytogenes* virulence factors by a gene that is autoregulated. *Mol. Microbiol.* **5:**2273–2283.

36. **Michel, E., and P. Cossart.** 1992. Physical map of the *Listeria monocytogenes* chromosome. *J. Bacteriol.* **174:**7098–7103.

37. **Michel, E., K. A. Reich, R. Favier, P. Berche, and P. Cossart.** 1990. Attenuated mutants of the intracellular bacterium *Listeria monocytogenes* obtained by single amino acid substitutions in listeriolysin O. *Mol. Microbiol.* **4:** 2167–2178.

38. **Milenbachs, A. A., and P. Youngman.** Personal communication.

39. **Moors, M.** Personal communication.

40. **Moors, M. A., B. Levitt, P. Youngman, and D. A. Portnoy.** 1999. Expression of listeriolysin O and ActA by intracellular and extracellular *Listeria monocytogenes*. *Infect. Immun.* **67:**131–139.

41. **Park, S. F., and R. G. Kroll.** 1993. Expression of listeriolysin and phosphatidylinositol-specific phospholipase C is repressed by the plant-derived molecule cellobiose in *Listeria monocytogenes*. *Mol. Microbiol.* **8:**653–661.

42. **Park, S. F., and G. S. A. B. Stewart.** 1990. High-efficiency transformation of *Listeria monocytogenes* by electroporation of penicillin-treated cells. *Gene* **94:**129–132.

43. **Park, S. F., G. S. A. B. Stewart, and R. G. Kroll.** 1992. The use of bacterial luciferase for monitoring the environmental regulation of expression of genes encoding virulence factors in *Listeria monocytogenes*. *J. Gen. Microbiol.* **138:**2619–2627.

44. **Petit, M.-A., C. Bruand, L. Janniere, and S. D. Ehrlich.** 1990. Tn*10*-derived transposons active in *Bacillus subtilis*. *J. Bacteriol.* **172:**6736–6740.

45. **Poquet, I., S. D. Ehrlich, and A. Gruss.** 1998. An export-specific reporter designed for gram-positive bacteria: application to *Lactococcus lactis*. *J. Bacteriol.* **180:**1904–1912.

46. **Portnoy, D. A., D. A. Hodgson, and P. Youngman.** Personal communication.

47. **Poyart, C., and P. Trieu-Cuot.** 1997. A broad-host-range mobilizable shuttle vector for the construction of transcriptional fusions to β-galactosidase in gram-positive bacteria. *FEMS Microbiol. Lett.* **156:**193–198.

48. **Schaferkordt, S., and T. Chakraborty.** 1995. Vector plasmid for insertional mutagenesis and directional cloning in *Listeria* spp. *BioTechniques* **19:**720–725.

49. **Scott, J. R., P. A. Kirchman, and M. E. Caparon.** 1988. An intermediate in transposition of the conjugative transposon Tn*916*. *Proc. Natl. Acad. Sci. USA* **85:**4809–4813.

50. **Sheehan, B., A. Klarsfeld, T. Msadek, and P. Cossart.** 1995. Differential activation of virulence gene expression by PrfA, the *Listeria monocytogenes* virulence regulator. *J. Bacteriol.* **177:**6469–6476.

51. **Shen, H., J. F. Miller, X. Fan, D. Kolwyck, R. Ahmed, and J. T. Harty.** 1998. Compartmentalization of bacterial antigens: differential effects on priming of CD8 T cells and protective immunity. *Cell* **92:**535–545.

52. **Shen, H., M. K. Slifka, M. Matloubian, E. R. Jensen, R. Ahmed, and J. F. Miller.** 1995. Recombinant *Listeria monocytogenes* as a live vaccine vehicle for the induction of protective anti-viral cell-mediated immunity. *Proc. Natl. Acad. Sci. USA* **92:**3987–3991.

53. **Shetron-Rama, L., and N. E. Freitag.** Unpublished data.

54. **Smith, K., and P. Youngman.** 1992. Use of a new integrational vector to investigate compartment-specific ex-

55. **Stoker, N. G., N. F. Fairweather, and B. G. Spratt.** 1982. Versatile low-copy-number plasmid vectors for cloning in *Escherichia coli*. *Gene* **18:**335–341.

56. **Sullivan, M., R. E. Yasbin, and F. E. Young.** 1984. New shuttle vectors for *B. subtilis* and *E. coli* which allow rapid detection of inserted fragments. *Gene* **29:**21–26.

57. **Sun, A. N., A. Camilli, and D. A. Portnoy.** 1990. Isolation of *Listeria monocytogenes* small-plaque mutants defective for intracellular growth and cell-to-cell spread. *Infect. Immun.* **58:**3770–3778.

58. **Teng, R., B. E. Murray, and G. M. Weinstock.** 1998. Conjugal transfer of plasmid DNA from *Escherichia coli* to Enterococci: a method to make insertion mutations. *Plasmid* **39:**182–186.

59. **Tomich, P. K., F. Y. An, and D. B. Clewell.** 1980. Properties of erythromycin-inducible transposon Tn*917*. *J. Bacteriol.* **141:**1366–1374.

60. **Trieu-Cuot, P., C. Carlier, C. Poyart-Salmeron, and P. Courvalin.** 1991. Shuttle vectors containing a multiple cloning site and a *lacZ* α gene for conjugal transfer of DNA from *Escherichia coli* to gram-positive bacteria. *Gene* **102:**99–104.

61. **Trieu-Cuot, P., and P. Courvalin.** 1983. Nucleotide sequence of the *Streptococcus faecalis* plasmid gene encoding the 3′5″-aminoglycoside phosphotransferase type III. *Gene* **23:**331–341.

62. **Trieu-Cuot, P., E. Derlot, and P. Courvalin.** 1993. Enhanced conjugative transfer of plasmid DNA from *Escherichia coli* to *Staphylococcus aureus* and *Listeria monocytogenes*. *FEMS Microbiol. Lett.* **109:**19–24.

63. **Vicente, M. F., J. C. Perez-Diaz, and F. Baquero.** 1987. A protoplast transformation system for *Listeria* sp. *Plasmid* **18:**89–92.

64. **Villafane, R., D. H. Bechhofer, C. S. Narayanan, and D. Dubnau.** 1987. Replication control genes of plasmid pE194. *J. Bacteriol.* **169:**4822–4829.

65. **Williams, D. R., D. I. Young, and M. Young.** 1990. Conjugative plasmid transfer from *Escherichia coli* to *Clostridium acetobutylicum*. *J. Gen. Microbiol.* **136:**819–826.

66. **Wirth, R., F. Y. An, and D. B. Clewell.** 1986. Highly efficient protoplast transformation system for *Streptococcus faecalis* and a new *Escherichia coli-S. faecalis* shuttle vector. *J. Bacteriol.* **165:**831–836.

67. **Wuenscher, M. D., S. Kohler, W. Goebel, and T. Chakraborty.** 1991. Gene disruption by plasmid integration in *Listeria monocytogenes*: insertional inactivation of the listeriolysin determinant *lisA*. *Mol. Gen. Genet.* **228:**177–182.

68. **Yanisch-Perron, C., J. Vieira, and J. Messing.** 1985. Improved M13 phage cloning vectors and host strains: nucleotide sequences of the M13mp18 and pUC19 vectors. *Gene* **33:**103–119.

69. **Yansura, D. G., and D. J. Henner.** 1984. Use of the *Escherichia coli* lac repressor and operator to control gene expression in *Bacillus subtilis*. *Proc. Natl. Acad. Sci. USA* **81:**439–443.

70. **Youngman, P.** 1990. Use of transposons and integrational vectors for mutagenesis and construction of gene fusions in *Bacillus* species, p. 221–266. *In* C. R. Harwood and

pression of the *Bacillus subtilis* spoIIM gene. *Biochimie* **74:** 705–711.

S. M. Cutting (ed.), *Molecular Biology Methods for Bacillus*. John Wiley & Sons, Chichester, U.K.

71. **Youngman, P.** 1987. Plasmid vectors for recovering and exploiting Tn*917* transpositions in *Bacillus* and other Gram-positives, p. 79–103. *In* K. C. Hardy (ed.), *Plasmids: A Practical Approach*. IRL Press, Oxford, U.K.

72. **Youngman, P., J. B. Perkinds, and K. Sandman.** 1985. Use of Tn*917*-mediated transcriptional fusions to *lacZ* and *cat-86* for the identification and study of regulated genes in *Bacillus subtilis*, p. 47–54. *In* J. A. Hoch and P. Setlow (ed.), *Molecular Biology of Microbial Differentiation*. American Society for Microbiology, Washington, D.C.

Regulation of Virulence Genes in Pathogenic *Listeria* spp.

WERNER GOEBEL, JÜRGEN KREFT, AND REGINE BÖCKMANN

52

Regulation of virulence genes occurs in many pathogenic bacteria, e.g., *Bordetella pertussis*, *Vibrio cholerae*, *Shigella flexneri*, *Salmonella typhimurium*, *Yersinia enterocolitica*, and the gram-positive *Staphylococcus aureus*, in a fashion that allows the coordinate and differential expression of the virulence factors at the right time during infection. Physicochemical parameters which change when the pathogen enters a host, such as temperature, pH value, deficiency in essential metal ions, especially iron, or nutrient starvation, may act as signals influencing the differential expression of these genes.

The gram-positive, rod-shaped, nonsporulating, and facultative intracellular bacterium *Listeria monocytogenes* is able to cause systemic infections in humans and animals (24, 38). This ability requires internalization of the bacteria into mammalian host cells, intracellular survival, and spreading into neighboring host cells in order to move from the primary site of infection, which is normally the intestine, to the peripheral organs, i.e., liver and spleen and then further into the peripheral blood circuit and eventually to the brain. During a successful infection, *L. monocytogenes* may thus encounter and invade a number of host cells and tissues. This in turn may demand the precisely timed expression of a variety of virulence factors.

THE VIRULENCE GENES OF *L. MONOCYTOGENES* AND *LISTERIA IVANOVII*

In the past years, several virulence genes have been identified in *L. monocytogenes* and the related animal pathogen *L. ivanovii*. The functions of the virulence factors were characterized in part, and it was shown that they were involved either in adherence and uptake by the host cell or in intracellular replication and cell-to-cell spreading of the listeriae. Genes essential for intracellular replication and intra- and intercellular mobility are clustered in a chromosomal region that is present at the same location in all *L. monocytogenes* and *L. ivanovii* isolates tested and even in the avirulent *Listeria seeligeri* (albeit in a largely inactivated form) between the two housekeeping genes, *ldh*

and *prs*, encoding lactate dehydrogenase and phosphoribosyl-pyrophosphate synthetase, respectively (25, 32, 33, 44, 53). The other category of virulence genes includes those encoding the large and small internalins, the functions of which are only understood in part (15, 18, 22, 45). *L. monocytogenes* and *L. ivanovii* seem to differ widely with respect to these *inl* genes and their localization on the chromosome (15, 19, 36). Another protein with implications in virulence is the extracellular protein p60, which has peptidoglycan hydrolase activity (58) but seems also to enhance adherence and uptake of *L. monocytogenes* into some host cells (31). Other proteins possibly involved in virulence are catalase and superoxide dismutase (4), the stress protease ClpC (51), and a specific uptake system for phosphorylated sugars (47).

REGULATION OF THE LISTERIAL VIRULENCE GENES BY ENVIRONMENTAL PARAMETERS

A number of studies have shown that the expression of the listerial virulence genes is specifically influenced by temperature (34), pH (13), carbon sources (13, 40), and various stress conditions (2, 3, 52). Below 20°C, where *L. monocytogenes* is still capable of proficient growth, none of the PrfA-regulated virulence genes (see below) seem to be expressed, although they are readily transcribed at 37°C and at least some of them, e.g., *hly* and *actA*, are induced even more at heat shock temperature (54). Repression of *hly* by cellobiose and the phenolic beta-glucoside arbutin has been described (6, 42, 43), but recent data show a similar inhibitory effect on *hly* expression by several mono- and disaccharides, including glucose. This suggests a more general catabolic effect due to the metabolism of these sugars (40). Starvation conditions, e.g., incubation in minimal essential medium (MEM), or incubation in brain heart infusion medium (BHI) containing charcoal, induces most genes of the *prfA* gene cluster, some of the *inl* genes, and the uptake of phosphorylated sugars (3, 18, 47, 49). There is also evidence that iron limitation may influence the expression of some virulence genes (1, 11, 12). Expression of the p60 protein seems to be regulated on the translational level (28), while transcription of the p60-encoding *iap* gene occurs constitutively at all growth temperatures. The intracellular environment of mammalian host cells also exhibits a stimulatory effect on the expression of some of the PrfA-regulated genes. This rather complex regulation pattern of the listerial virulence genes under intracellular conditions is discussed later in this chapter.

THE PLEIOTROPIC REGULATORY FACTOR PrfA

The only regulatory factor molecularly characterized up to now that is crucial for the expression of most virulence genes is the positive regulatory factor A, PrfA, encoded by the first gene of the common virulence gene cluster. This cluster, comprising a total of six genes (*prfA*, *plcA*, *hly*, *mpl*, *actA*, and *plcB*), is therefore often referred to as the PrfA virulence gene cluster or the PrfA regulon (5, 30, 44, 53). Since this gene cluster also occurs in *L. ivanovii* and *L. seeligeri* (see above) it is not surprising that both of these *Listeria* species also carry intact *prfA* genes. These *prfA* genes and their products share extended sequence homology (about 80% identity on the DNA and protein levels) with their *L. monocytogenes* counterpart (Fig. 1). PrfA is necessary for the expression of all genes of the PrfA gene cluster and all known genes encoding small internalins in *L. monocytogenes* and *L. ivanovii*, while expression of the genes for the large internalins (so far identified only in *L. monocytogenes*) is either partially dependent on PrfA, i.e., *inlA* and *inlB* (37), or even independent, i.e., *inlG*, *inlH*, and *inlE* (45). PrfA also seems to activate the transcription of genes involved in the uptake of hexose phosphates in *L. monocytogenes* (47). Interestingly, the functionally homologous *uhp* genes in *Escherichia coli* are positively regulated by Crp(cAMP), which, as discussed below, is closely related to PrfA. There is preliminary evidence for a negative function of PrfA in the regulation of some genes in *L. monocytogenes*, e.g., *clpC* (50) and the gene for a 64-kDa protein (54), as well as in *L. ivanovii* (33).

PrfA IS A MEMBER OF THE Crp/Fnr FAMILY OF TRANSCRIPTION ACTIVATORS

Based on its amino acid sequence, PrfA is clearly related to the Crp/Fnr family of transcription activators (5, 30, 33, 53). The best-characterized members of this family of regulatory proteins are Crp (or Cap) and Fnr of *E. coli*. Crp regulates expression of genes involved in catabolite functions and is the major regulator in catabolite repression (29); Fnr regulates the cellular response to anaerobic growth conditions in *E. coli* (55). Other members of this family, such as FixK of *Rhizobium meliloti* and NtcA of *Synechococcus* spp., are involved in the regulation of nitrogen metabolism. A direct regulatory role in the expression of virulence genes has been shown for HlyX of *Actinobacillus pleuropneumoniae* and Crp of *Xanthomonas campestris*. Loss of Crp also leads to a significant attenuation of virulence in *Salmonella* and *Shigella* spp.

Most members of the Crp/Fnr family have been found in gram-negative bacteria, and, with the Flp protein of *Lactobacillus casei* (26), PrfA is one of only two well-characterized representatives of this family in gram-positive bacteria.

Extensive molecular studies have been carried out on cyclic AMP (cAMP)-binding factor Crp (Cap), and since PrfA shares extended sequential and structural similarity with Crp, a short overview of the most essential features of Crp might be useful for the understanding of PrfA.

MOLECULAR AND FUNCTIONAL ASPECTS OF Crp (Cap)

Crp, the catabolite activator protein (therefore also named Cap), is crucial in *E. coli* for the utilization of many different carbon sources in the absence of glucose. Crp forms in solution a homodimer with a molecular mass of 45 kDa. The monomeric Crp (209 amino acids) consists of an N-terminal domain (amino acids 1 to 129) and a C-terminal domain (amino acids 139 to 209), which are connected by a short hinge region. The N-terminal domain participates

```
                                                                         #
L.monocytogenes EGD/LO28  MNAQAEEFKK YLETNGIKPK QFHKKELIFN QWDPQEYCIF LYDGITKLTS  50
L.ivanovii ATCC 19119     -D---AD--N F--------- K-K--DI--- --------V- -----A---N
L.seeligeri SLCC 3954     -STK--D--E ---S---Q-- Y-R---F--S --E-N----- -HE-VA----
                              #          #       #
                          ISENGTIMNL QYYKGAFVIM SGFIDTETSV GYYNLEVISE QATAYVIKIN 100
                          -----S---- -------I-- --SL--GKPL ---------K T-----L--S
                          ---S-D-L-- -------I-- T------K-L -------V-E --A--I---S
                                                                          *
                          ELKELLSKNL THFFYVFQTL QKQVSYSLAK FNDFSINGKL GSICGQLLIL 150
                          D--N-V-S-I -QL--II-A- ---------- -----V---- ----------
                          D----V--D- KQL--IID-- ---------- -----S---- ------F---

                          TYVYGKETPD GIKITLDNLT MHELGYSSGI AHSSAVSRII SKLKQEKVIV 200
                          -----EKN-- ----A----- -Q-------- --C------L ----K----E
                          A----E---N ------EK-- -Q---C---- ---------- ------N--E
                                              |--  helix-turn-helix  --|

                          YKNSCFYVQN LDYLKRYAPK LDEWFYLACP ATWGKLN               237
                          --H-S----- -----KI--- ---------- S--E-F-
                          --D-Y--IK- IA---KI--- I--------H NS-D-F-
```

FIGURE 1 Comparison of the amino acid sequences of the PrfA proteins from *L. monocytogenes*, *L. ivanovii*, and *L. seeligeri*. Pound sign (#) denotes conserved glycines at the turns of the *β*-roll structure; asterisk (*) indicates the position of the G145S substitution leading to a constitutively active PrfA (48, 56). Sequences are from references 32, 33, 35, and 39.

in cAMP binding and homodimerization. The region of amino acids 19 to 99 consists of short, antiparallel β-sheets (β-roll structure) that are separated by conserved glycine residues. Adjacent to this domain is a region termed helix C, which is involved in dimer formation and is also essential for cAMP binding. Detailed studies have shown that cAMP is bound to the N-terminal domain by ionic and hydrogen bonds via the amino acids G71, E72, R82, S83, R123, T127, and S128 (57). The C-terminal domain carries between amino acids 168 and 191 the helixE-turn-helixF motif essential for DNA binding. The two helices interact with DNA by polar side chains and positive-charged residues of certain amino acids. Activation of transcription from Crp-cAMP-dependent promoters requires contact with the RNA polymerase involving the Crp activation regions AR1 (amino acids 156 to 164), AR2 (amino acids 19, 21, and 101), and AR3 (amino acids 52 to 58) (41, 59). Figure 2 summarizes the functionally important features of Crp.

In the absence of cAMP, Crp binds nonspecifically to DNA. Complex formation of Crp with cAMP probably influences the orientation of helix F, thereby enhancing binding of the Crp-cAMP complex to the specific target sequence. Glucose limitation leads to a cellular increase in the level of cAMP. Two molecules of cAMP bind to homodimeric Crp, thereby causing a conformational change. The Crp-cAMP complex recognizes and binds to a sym-metric consensus sequence, TGTGA-N6-TCACA, which is located at varying positions (between −40 and −200 bp) upstream of the transcriptional start sites of Crp-regulated genes. Based on the location of the Crp-binding site, two well-defined classes of Crp-dependent promoters are distinguished (8, 9, 17): class I promoters carry the Crp-binding site in variable positions upstream of the −10 promoter box (in *lacP1*, a typical class I promoter, it is located at position −61.5). The interaction with the C-terminal domain of the alpha subunit (alpha-CTD) of RNA polymerase occurs in this case via a single AR1 region of the Crp dimer, which forms an exposed loop structure. Class II promoters, e.g., *galP1*, carry the Crp-binding sequence at position −41.5, thus overlapping the −35 promoter box. In this case, both subunits of Crp are in contact with RNA polymerase. The transcriptional activation of class II promoters is more complex than that of class I promoters and requires the interaction of alpha-CTD with AR1 of the promoter-distal subunit and of the N-terminal domain of the alpha subunit (alpha-NTD) of RNA polymerase with AR2 of the promoter-proximal Crp subunit. It is this interaction that catalyzes the transition of the closed RNA polymerase-promoter complex into the transcription-competent open complex. There is an additional interaction of the sigma subunit of RNA polymerase with AR3, at least under in vitro conditions. Another class of promoters, typified by the *malK* promoter, needs an addi-

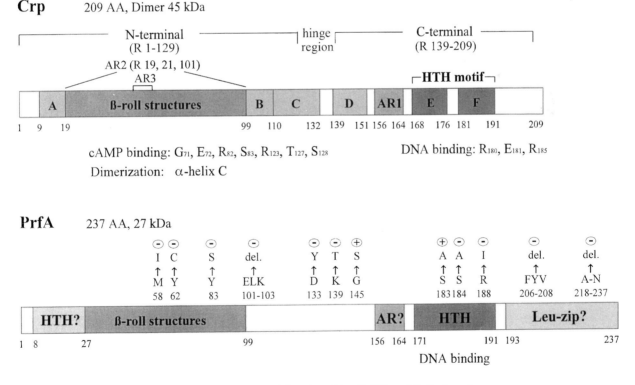

FIGURE 2 Schematic comparison of Crp from *E. coli* (57) and PrfA from *L. monocytogenes*. The functionally important features of Crp, e.g., the N-terminal β-roll structures, the C-terminal helix-turn-helix (HTH) motif, and the activating regions AR1, AR2, and AR3 are indicated, as well as amino acids known to be involved in cAMP and DNA binding. Capital letters (A to D) denote alpha-helical regions. For PrfA the predicted N-terminal HTH motif and the putative leucine-zipper (Leu-zip) structure are marked. Amino acid substitutions in PrfA that cause a decreased (−) or increased (+) activity are shown above.

tional protein for transcriptional activation (29). The Crp-binding sequence is in this case further upstream than it is in class I or class II promoters, and its position varies.

SIMILARITIES BETWEEN PrfA AND Crp

Although the direct comparison of the amino acid sequences of Crp (207 amino acids) and PrfA (237 amino acids) shows only 20% identity, there are significant sequential and structural similarities between these two regulatory proteins, particularly in some of the functional regions of Crp described above. These similarities are outlined in Fig 2. The N-terminal domain of PrfA also exhibits the short β-sheets interrupted by glycine residues (β-roll structure). The amino acid residues essential for the binding of cAMP to Crp and for dimerization are less well conserved in PrfA, but interestingly, the exchange of G145 into serine leads to a mutant PrfA that causes a derepressed phenotype of most PrfA-regulated virulence genes (see below). There are comparable mutations in helix D of Crp (G145S and A144T) that lead to a cAMP-independent active conformation of Crp (23). These data support the view that PrfA may also be converted to a transcriptional active form by binding of a cofactor.

There are also sequential similarities in PrfA to the activation region AR1 of Crp (53). The C-terminal domain of PrfA contains a helix-turn-helix (HTH) motif (amino acids 171 to 191) with substantial sequence homology to that of Crp. Moreover, Sheehan et al. (52) have provided direct evidence by site-specific mutagenesis that S183 and S184 are essential for binding of PrfA to DNA.

Unlike Crp, PrfA contains an additional putative HTH motif (amino acids 8 to 27) and an extended C terminus with a putative leucine zipper motif (33). Both of these additional structures are absent in any of the other members of the Crp/Fnr family. The functions of these two regions of PrfA are not yet known.

ALTERATIONS IN DEFINED POSITIONS OF PrfA LEAD TO FUNCTIONAL CHANGES

As previously noted (49), *L. monocytogenes* isolates differ in the expression of virulence genes when grown in standard media like BHI. This difference can be most easily recognized by the hemolytic phenotype (dependent on the level of expression of the *hly* gene) or the lecithinase activity (dependent on the expression of *plcB*) when these strains are grown on blood agar or egg yolk plates, respectively. While most wild-type strains show low expression of both phenotypes (type 1 strains), others show high activity (type 2 strains). There is a third type of strain exhibiting intermediate activity (type 3). Induction of both hemolytic and lecithinase activity is observed with type 1 (and type 3), but not with type 2 strains after the shift of BHI-grown cultures into MEM or by the addition of activated charcoal to BHI (49). Specific amino acid changes in PrfA, e.g., G145S and C229Y, correlate with the transition of type 1 to type 2 strains. A mutation leading to a G145S substitution in PrfA of a type 1 strain leads to a highly activated PrfA protein (48). As already mentioned, this amino acid is located in a region of PrfA corresponding to helix D of Crp. Mutations in Crp affecting helix D lead to a conformation that exhibits a cAMP-independent transcriptional activation. The significance of the C229Y exchange in PrfA in (at least some) type 2 strains remains

unclear at the moment. The correlation of type 2 with the altered PrfA protein is further supported by the observation that transfer of the *prfA* gene from a type 2 to a type 1 strain shifts the hemolytic phenotype to type 2 (2, 48).

Using an assay that selects for mutations in PrfA that decrease its transcriptional activation, we have obtained several point mutations (M58I, Y62C, Y83S) in the β-roll structure (amino acids 27 to 99), in the hinge region connecting the N-terminal domain to the C-terminal domain (D133Y, K139T), and in the HTH DNA-binding domain of PrfA (R188I) and several short in-frame deletions (Fig. 2). Some of these mutations were obtained several times in independent experiments; e.g., an R188I exchange in the HTH motif was obtained six times. As shown by Sheehan et al. (52), an S184A exchange also leads to decreased binding of PrfA to its target sequence and reduced expression of virulence genes, while the S183A exchange increases binding of PrfA to this site and causes enhanced expression of virulence genes. Experimental evidence suggests that cAMP most probably is not an effector for PrfA; amino acids essential for cAMP binding in Crp that are still somewhat conserved in PrfA may simply represent relics of a common ancestral precursor of the two regulatory proteins. As discussed later, there is, however, some evidence for the involvement of a low-molecular-weight effector for PrfA distinct from cAMP, and one may speculate that the altered "cAMP pocket" of PrfA (as compared with the cAMP pocket of Crp) may represent its binding site. A deletion removing most of the PrfA-specific leucine-zipperlike C terminus of PrfA (which is entirely absent in Crp and the other members of this family) also abolishes the activity of PrfA (33).

REGULATION OF PrfA SYNTHESIS

The transcription of the *prfA* gene and hence the cellular level of PrfA is subject to a complex regulation circuit. Two promoters, *prfAP1* and *prfAP2*, are positioned 113 and 30 bp, respectively, in front of the *prfA* start codon. Transcription from these promoters leads to monocistronic *prfA* transcripts of 0.9 and 0.8 kb, respectively. PrfA protein seems to negatively influence this transcription, since the amount of the monomeric *prfA* transcripts is significantly enhanced in the absence of functional PrfA. Moreover, a deletion in the putative −35 box of *prfAP2* induces transcription of *prfA* from *prfAP1* (20, 21). As previously described (33), there is a degenerated PrfA-binding site located at position −37 from the transcriptional start site, i.e., shifted by about a one-half helical turn relative to the start site. PrfA binding to this "PrfA box" may cause the repression of *prfA* transcription. It has indeed been shown that Crp can act as a transcriptional repressor of class II promoters (see above) when it binds to a displaced position, thereby blocking binding of RNA polymerase to the promoter (29). In addition to the monocistronic transcription, *prfA* is transcribed as bicistronic mRNA together with *plcA*, from a promoter located in front of *plcA*, which is activated by PrfA (39). This positive autoregulation of *prfA* transcription seems to be essential for the proper expression of the virulence genes, since the interruption of the *plcA-prfA* operon by transposon insertions into the intergenic region leads to a significant drop in PrfA-dependent gene expression (10). A similar decrease in virulence gene expression is also observed by the deletion of the two promoters in front of *prfA*. This has led to a model for the regulation of *prfA* expression that assumes that *prfA* is first transcribed from one (or both) of its own promoters. This monocistronic transcript generates enough PrfA protein to activate the PrfA-dependent promoter in front of *plcA*. The bicistronic *plcA-prfA* transcript leads to an increased level of cellular PrfA. Excessive synthesis of PrfA is prohibited by negative feedback exerted by binding of PrfA to the putative PrfA box of *prfAP2*. Interestingly, in *L. seeligeri*, the *plcA* and *prfA* genes are interrupted by a large insertion; consequently, *prfA* is very weakly expressed (32). Furthermore, there is circumstantial evidence that the amount of PrfA protein may, in addition, be controlled on the translational level since under some conditions it is not directly proportional to the mRNA level (46). Further work is needed to completely elucidate this complex regulatory network.

THE PrfA-BINDING SITE (PrfA BOX) AND ITS INTERACTION WITH PrfA

All known PrfA-regulated promoters of *L. monocytogenes*, *L. ivanovii*, and *L. seeligeri* possess a conserved symmetric sequence (PrfA box) at around position −40 from the transcriptional start site. This arrangement of the PrfA box relative to the −10 promoter box is similar to that of the Crp-dependent class II promoters (see above). The consensus sequence of the 14-bp PrfA box is TTAACA-NN-TGTTAA. This sequence is perfectly maintained in the common PrfA box of the *plcA* and *hly* promoters. The corresponding PrfA boxes of the promoters for *mpl*, *actA*, and *inlC* exhibit a 1-bp change each, while the PrfA box of the *inlA-inlB* promoter is even more degenerated by 2-bp substitutions (2, 16, 37). In *L. ivanovii*, the common PrfA box of the *plcA/ilo* and that of the *actA* and *i-inlE* promoter also show perfect symmetry, while all other identified PrfA boxes of *L. ivanovii* again show at least a 1-bp change (Fig. 3). The symmetry of the PrfA-binding site suggests that this regulatory protein, like other members of the Crp/Fnr family, may bind as homodimer. There is indeed some experimental evidence supporting this assumption (14).

By electrophoretic mobility shift studies and DNase I footprinting, it has been shown (1, 14) that purified PrfA protein binds to oligonucleotides containing essentially the 14-bp PrfA box, forming a defined complex (CIII). As expected, the efficiency of PrfA binding depends on the "quality" of the PrfA box; e.g., binding is most efficient to the PrfA box of the *hly* promoter and weakest to that of the *inlA* promoter (14). Furthermore, it has been shown (56) that the activated PrfA with the G145S exchange binds more efficiently to the PrfA box than does the corresponding wild-type PrfA protein, suggesting that the strength of PrfA binding to its target sequence may be directly correlated with the expression of the corresponding PrfA-dependent gene.

Electrophoretic mobility shift experiments with PrfA-containing cell-free extracts of *L. monocytogenes* and the short PrfA-box target sequences do not yield CIII complexes, but specific slow-migrating complexes (CI and CII) are obtained with longer DNA probes containing the entire promoter sequence, including the PrfA box. Binding of purified PrfA protein to the longer target DNA results again in the formation of a faster-migrating complex (CIII), which can be, however, converted into CI by the addition of *L. monocytogenes* extracts devoid of PrfA protein. Conversion is highest if the extract is derived from listeriae

```
L. monocytogenes EGD

              PrfA-box                    -10

hly      TTAACA TT TGTTAA    - 23bp -    TAGAAT

plcA     TTAACA AA TGTTAA    - 22bp -    TAAGAT

mpl      TTAACA AA TGTaAA    - 22bp -    TATAAT

actA     TTAACA AA TGTTAg    - 21bp -    GATATT

inlC     TTAACg CT TGTTAA    - 22bp -    TAACAT

inlA     aTAACA TA aGTTAA    - 21bp -    TATTAT


L. ivanovii ATCC19119

ilo      TCTTTAACA TT TGTTAAAGA  - 23bp -   TACAAT

plcA     TCTTTAACA AA TGTTAAAGA  - 21bp -   TAAGAT

mpl        TTTAACA AA TGTcAAA    - 22bp -   TATAAT

actA      ATTAACA AA TGTTAAT     - 21bp -   TATTCT

i-inlC     TTAACg CT TGTTAA      - 22bp -   TAACAT

i-inlD     TTAACt TT TGTTAt      - 22bp -   TATTAT

i-inlE     TTAACA TT TGTTAA      - 22bp -   TATGAT

i-inlF     TTAACt TT TGTTAt      - 22bp -   TAGAAT
```

FIGURE 3 Sequence comparison of PrfA-regulated virulence gene promoters from *L. monocytogenes* and *L. ivanovii*. Sequences are from published data or from our own unpublished results.

kept under "inducing" conditions, such as incubation of the *L. monocytogenes* culture in MEM or exposure to heat shock temperature (see above). The formation of CI and CII complexes is less pronounced with extracts from bacteria grown in rich medium or in MEM with an excess of iron (1).

A cell-free extract from a *prfA*-deletion mutant of *L. monocytogenes* has been fractionated on sucrose gradients. When added to an assay containing PrfA and the longer (see above) *hly* promoter probe, two fractions could be identified that affect CIII complex formation in two different ways. The top fraction enhances CIII formation significantly, suggesting that a low-molecular-weight component may convert the PrfA protein into a conformation that binds more efficiently to its target sequence. This observation is reminiscent of the effect that cAMP exerts on the binding of Crp to its target sequence. The other fraction, banding at around 3.5 S (corresponding to a protein of about 35 to 70 kDa), shifts the CIII complex in part to the CI complex (see above). The protein component(s) responsible for this activity has been termed Paf (for PrfA-activating factor). Recently, we showed that purified RNA polymerase from MEM-grown *L. monocytogenes* (lacking *prfA*) forms with the *hly* (and *actA*) promoter probe the CII complex, which is converted into the CI complex upon addition of PrfA protein. This result suggests that Paf copurifies with this RNA polymerase preparation or is even part of the RNA polymerase. This RNA polymerase preparation correctly initiates transcription from the *hly* and *actA* promoters in the presence of PrfA in vitro.

INTRACELLULAR EXPRESSION OF PrfA-REGULATED GENES

The results described in the other chapters in this section suggest that regulation of virulence genes mediated by PrfA may be complex and may involve environmental parameters as well as additional bacterial factors. So far, little is known about the in vivo regulation of listerial virulence genes. Undoubtedly, the proper regulation of the virulence genes by PrfA plays a decisive role in *L. monocytogenes* infection, since a *prfA*-deficient mutant of *L. monocytogenes* is entirely avirulent in the mouse model (39); furthermore, there is no advantage in infectivity by the in vitro "activated" virulence phenotype of type 3 *L. monocytogenes* strains compared with the "repressed" virulence phenotype of type 1 strains under in vivo conditions (27, 49). Virulence of type 1 strains in the mouse may be even higher than that of type 3 strains, suggesting that inducibility of PrfA may be crucial in vivo.

As a first step in understanding the in vivo regulation of virulence genes by PrfA, we have studied the expression of some PrfA-regulated genes in mammalian host cells by determining the level of their transcripts by a modified reverse transcription-PCR procedure and the activities of the corresponding promoters using *gfp* as reporter gene (7, 18). Interestingly, rather low levels of *prfA* expression are always observed in the phagosome and in the cytosol of phagocytic as well as nonphagocytic cells, although expression of some of the PrfA-dependent virulence genes is rather high. This suggests that activation of PrfA may play a more essential role inside host cells than the amount of PrfA. As expected, expression of *hly* and *plcA* is high in the phagosome, while *actA*, hardly expressed under extracellular conditions and in the phagosome of the type 1 strain *L. monocytogenes* EGD, is highly induced in the cytosol (2). On the other hand, there is preferential expression of *inlC* in the cytosol, occurring at a later time in infection than that of *actA*, while *inlA* and *inlB* are hardly expressed inside cells. Preliminary evidence indicates a variable expression of these virulence genes, depending on the type of host cells.

CONCLUSION AND OUTLOOK

The extensive studies on the molecular aspects of *L. monocytogenes* virulence have led to the identification of a variety of virulence genes involved in the various steps of the intracellular infection cycle. The differential regulation of most of these virulence genes is intimately connected with the pleiotropic regulatory protein PrfA. This transcriptional activator of the pathogenic *Listeria* species, *L. monocytogenes* and *L. ivanovii*, shares common properties with other members of the Crp/Fnr family to which it belongs, but it also possesses unique features. Like most regulators of this family, the PrfA protein most probably binds as homodimer to its symmetric target sequence and may enhance (or inhibit) binding of RNA polymerase to PrfA-dependent promoters. There is, however, growing evidence that PrfA is embedded in other regulatory circuits of *L. monocytogenes* and *L. ivanovii*, which may modulate the activity as well as the amount of PrfA. Much has to be learned about the external physicochemical and metabolic

parameters as well as the additional listerial factors influencing PrfA-dependent gene regulation under extracellular and particularly under intracellular conditions. The present data suggest that these additional parameters may affect gene regulation by PrfA in two ways: (i) by changing the conformation of PrfA, which in turn may influence binding of PrfA to the different target sites located in the corresponding promoters with positive (and possibly also negative) effects on the initiation of transcription of PrfA-regulated genes, and (ii) by inducing gene expression of *prfA*, thus increasing the amount of cellular PrfA, which again may have positive or negative effects on the transcription of PrfA-regulated genes. Finally, the participation of additional listerial protein factors that may interact with PrfA directly to modulate its binding to the target sites and/or to RNA polymerase, similar to the situation of Crp-dependent promoters that require additional factors, should also be considered. The recently described Paf may represent such a factor. The three-dimensional structure of PrfA and the complete identification of the PrfA regulon by the upcoming genome sequence of *L. monocytogenes* may have a strong impact on the further study of virulence gene regulation in *Listeria*. Undoubtedly, the precise knowledge of the mechanisms involved in the regulation of the listerial virulence genes will be crucial for the understanding of the pathogenesis of infections by pathogenic *Listeria* spp.

We are grateful to J. Bohne, A. Bubert, C. Dickneite, R. Lampidis, and Z. Sokolovic for communicating unpublished results and for helpful discussions. We thank J. Daniels for critical reading of the manuscript. Our work has been supported by grants from the Deutsche Forschungsgemeinschaft (SFB165-B4) and the Fonds der Chemischen Industrie.

REFERENCES

1. **Böckmann, R., C. Dickneite, B. Middendorf, W. Goebel, and Z. Sokolovic.** 1996. Specific binding of the *Listeria monocytogenes* transcriptional regulator protein PrfA to target sequences requires additional factor(s) and is influenced by iron. *Mol. Microbiol.* **22:**643–653.

2. **Bohne, J., H. Kestler, C. Uebele, Z. Sokolovic, and W. Goebel.** 1996. Differential regulation of the virulence genes of *Listeria monocytogenes. Mol. Microbiol.* **20:**1189–1198.

3. **Bohne, J., Z. Sokolovic, and W. Goebel.** 1994. Transcriptional regulation of *prfA* and PrfA-regulated virulence genes in *L. monocytogenes. Mol. Microbiol.* **11:**1141–1150.

4. **Brehm, K., S. Bernard, T. Nichterlein, and J. Kreft.** Contribution of catalase and superoxide dismutase to *Listeria monocytogenes* virulence. Submitted for publication.

5. **Brehm K., J. Kreft, M.-T. Ripio, and J.-A. Vázquez-Boland.** 1996. Regulation of virulence gene expression in pathogenic *Listeria. Microbiol. SEM* **12:**219–236.

6. **Brehm, K., M.-T. Ripio, J. Kreft, and J.-A. Vázquez-Boland.** The *bvr* locus of *Listeria monocytogenes* mediates virulence gene repression by β-glucosides. *J. Bacteriol.*, in press.

7. **Bubert, A., S.-K. Chun, L. Papatheodorou, A. Simm, W. Goebel, and Z. Sokolovic.** 1999. Differential virulence gene expression by *Listeria monocytogenes* growing within host cells. *Mol. Gen. Genet.* **261:**323–336.

8. **Busby, S., and R. H. Ebright.** 1994. Promoter structure, promoter recognition, and transcription activation in procaryotes. *Cell* **79:**743–746.

9. **Busby, S., and R. H. Ebright.** 1997. Transcription activation at Class II CAP-dependent promoters. *Mol. Microbiol.* **23**:853–859.

10. **Camilli, A., L. G. Tilney, and D. A. Portnoy.** 1993. Dual roles of *plcA* in *Listeria monocytogenes* pathogenesis. *Mol. Microbiol.* **8**:143–157.

11. **Conte, M. P., C. Longhi, M. Polidoro, G. Petrone, V. Buonfiglio, S. Di Santo, E. Papi, L. Seganti, P. Visca, and P. Valenti.** 1996. Iron availability affects entry of *Listeria monocytogenes* into enterocytelike cell line Caco-2. *Infect. Immun.* **64**:3925–3929.

12. **Cowart, R. E., and B. G. Foster.** 1981. The role of iron in the production of haemolysin by *Listeria monocytogenes*. *Curr. Microbiol.* **6**:287–290.

13. **Datta, A. R., and M. H. Kothary.** 1993. Effects of glucose, growth temperature, and pH on listeriolysin O production in *Listeria monocytogenes*. *Appl. Environ. Microbiol.* **59**:3495–3497.

14. **Dickneite, C., R. Böckmann, A. Spory, W. Goebel, and Z. Sokolovic.** 1998. Differential interaction of the transcription factor PrfA and the PrfA-activating factor (Paf) of *Listeria monocytogenes* with target sequences. *Mol. Microbiol.* **27**:915–928.

15. **Dramsi, S., P. Dehoux, M. Lebrun, P. L. Goossens, and P. Cossart.** 1997. Identification of four new members of the internalin multigene family of *Listeria monocytogenes* EGD. *Infect. Immun.* **65**:1615–1625.

16. **Dramsi, S., C. Kocks, C. Forestier, and P. Cossart.** 1993. Internalin-mediated invasion of epithelial cells by *Listeria monocytogenes* is regulated by the bacterial growth state, temperature and the pleiotropic activator PrfA. *Mol. Microbiol.* **9**:931–941.

17. **Ebright, R. H.** 1993. Transcription activation at Class I CAP-dependent promoters. *Mol. Microbiol.* **8**:797–802.

18. **Engelbrecht, F., S.-K. Chun, C. Ochs, J. Hess, F. Lottspeich, W. Goebel, and Z. Sokolovic.** 1996. A new PrfA-regulated gene of *Listeria monocytogenes* encoding a small, secreted protein which belongs to the family of internalins. *Mol. Microbiol.* **21**:823–837.

19. **Engelbrecht, F., C. Dickneite, R. Lampidis, M. Götz, U. DasGupta, and W. Goebel.** 1998. Sequence comparison of the chromosomal regions encompassing the internalin C genes (*inlC*) of *Listeria monocytogenes* and *Listeria ivanovii*. *Mol. Gen. Genet.* **257**:186–197.

20. **Freitag, N. E., and D. A. Portnoy.** 1994. Dual promoters of the *Listeria monocytogenes prfA* transcriptional activator appear essential *in vitro* but are redundant *in vivo*. *Mol. Microbiol.* **12**:845–853.

21. **Freitag, N. E., L. Rong, and D. A. Portnoy.** 1993. Regulation of the *prfA* transcriptional activator of *Listeria monocytogenes*: multiple promoter elements contribute to intracellular growth and cell-to-cell spread. *Infect. Immun.* **61**:2537–2544.

22. **Gaillard, J.-L., P. Berche, C. Frehel, E. Gouin, and P. Cossart.** 1991. Entry of *Listeria monocytogenes* into cells is mediated by internalin, a repeat protein reminiscent of surface antigens from gram-positive cocci. *Cell* **65**:1127–1141.

23. **Garges, S., and S. Adhya.** 1988. Cyclic AMP-induced conformational change of cyclic AMP receptor protein (CRP): intragenic suppressors of cyclic AMP-independent CRP mutations. *J. Bacteriol.* **170**:1417–1422.

24. **Gellin, B. G., and C. V. Broome.** 1989. Listeriosis. *JAMA* **261**:1313–1320.

25. **Gouin, E., J. Mengaud, and P. Cossart.** 1994. The virulence gene cluster of *Listeria monocytogenes* is also present in *Listeria ivanoii*, an animal pathogen, and *Listeria seeligeri*, a nonpathogenic species. *Infect. Immun.* **62**:3550–3553.

26. **Irvine, A. S., and J. R. Guest.** 1993. *Lactobacillus casei* contains a member of the CRP-FNR family. *Nucleic Acids Res.* **21**:753.

27. **Kathariou, S., J. Rocourt, H. Hof, and W. Goebel.** 1988. Levels of *Listeria monocytogenes* hemolysin are not directly proportional to virulence in experimental infections in mice. *Infect. Immun.* **56**:534–536.

28. **Köhler, S., A. Bubert, M. Vogel, and W. Goebel.** 1991. Expression of the *iap* gene coding for protein p60 of *Listeria monocytogenes* is controlled on the posttranscriptional level. *J. Bacteriol.* **173**:4668–4674.

29. **Kolb, A., S. Busby, H. Buc, S. Garges, and S. Adhya.** 1993. Transcriptional regulation by cAMP and its receptor protein. *Annu. Rev. Biochem.* **62**:749–795.

30. **Kreft, J., J. Bohne, R. Gross, H. Kestler, Z. Sokolovic, and W. Goebel.** 1995. Control of *Listeria monocytogenes* virulence by the transcriptional regulator PrfA, p. 129–142. *In* R. Rappuoli, V. Scarlato, and B. Aricó (ed.), *Signal Tranduction and Bacterial Virulence*. R. G. Landes Company, Austin, Tex.

31. **Kuhn, M., and W. Goebel.** 1989. Identification of an extracellular protein of *Listeria monocytogenes* possibly involved in intracellular uptake by mammalian cells. *Infect. Immun.* **57**:55–61.

32. **Lampidis, R., M. Emmerth, I. Karunasagar, and J. Kreft.** Unpublished data.

33. **Lampidis, R., R. Gross, Z. Sokolovic, W. Goebel, and J. Kreft.** 1994. The virulence regulator protein of *Listeria ivanovii* is highly homologous to PrfA from *Listeria monocytogenes* and both belong to the Crp-Fnr family of transcription regulators. *Mol. Microbiol.* **13**:141–151.

34. **Leimeister-Wächter, M., E. Domann, and T. Chakraborty.** 1992. The expression of virulence genes in *Listeria monocytogenes* is thermoregulated. *J. Bacteriol.* **174**:947–952.

35. **Leimeister-Wächter, M., C. Haffner, E. Domann, W. Goebel, and T. Chakraborty.** 1990. Identification of a gene that positively regulates expression of listeriolysin, the major virulence factor of *Listeria monocytogenes*. *Proc. Natl. Acad. Sci. USA* **87**:8336–8340.

36. **Lingnau, A., T. Chakraborty, K. Niebuhr, E. Domann, and J. Wehland.** 1996. Identification and purification of novel internalin-related proteins in *Listeria monocytogenes* and *Listeria ivanovii*. *Infect. Immun.* **64**:1002–1006.

37. **Lingnau, A., E. Domann, M. Hudel, M. Bock, T. Nichterlein, J. Wehland, T. Chakraborty.** 1995. Expression of *Listeria monocytogenes* EGD *inlA* and *inlB* genes, whose products mediate bacterial entry into tissue culture cell lines, by PrfA-dependent and -independent mechanisms. *Infect. Immun.* **64**:1002–1006.

38. **Lorber, B.** 1997. Listeriosis. *Clin. Infect. Dis.* **24**:1–11.

39. **Mengaud, J., S. Dramsi, E. Gouin, J. A. Vazquez-Boland, G. Milon, and P. Cossart.** 1991. Pleiotropic control of *Listeria monocytogenes* virulence factors by a gene that is autoregulated. *Mol. Microbiol.* **5**:2273–2283.

40. **Milenbachs, A. A., D. P. Brown, M. Moors, and P. Youngman.** 1997. Carbon-source regulation of virulence

gene expression in *Listeria monocytogenes*. *Mol. Microbiol.* **23:**1075–1085.

41. **Niu, W., Y. Kim, G. Tau, T. Heyduk, and R. Ebright.** 1996. Transcription activation at Class II CAP-dependent promoters: two interactions between CAP and RNA polymerase. *Cell* **87:**1123–1134.

42. **Park, S. F.** 1994. The repression of listeriolysin O expression in *Listeria monocytogenes* by the phenolic beta-D-glucoside, arbutin. *Lett. Appl. Microbiol.* **19:**258–260.

43. **Park, S. F., and R. G. Kroll.** 1993. Expression of listeriolysin and phosphatidyl-inositol-specific phospholipase C is repressed by the plant-derived molecule cellobiose in *Listeria monocytogenes*. *Mol. Microbiol.* **8:**653–661.

44. **Portnoy, D. A., T. Chakraborty, W. Goebel, and P. Cossart.** 1992. Molecular determinants of *Listeria monocytogenes* pathogenesis. *Infect. Immun.* **60:**1263–1267.

45. **Raffelsbauer, D., A. Bubert, F. Engelbrecht, J. Scheinpflug, A. Simm, J. Hess, S. H. E. Kaufmann, and W. Goebel.** 1999. The gene cluster *inlC2DE* of *Listeria monocytogenes* contains additional new internalin genes and is important for virulence in mice. *Mol. Gen. Genet.* **260:**144–158.

46. **Renzoni, A., A. Klarsfeld, S. Dramsi, and P. Cossart.** 1997. Evidence that PrfA, the pleiotropic activator of virulence genes in *Listeria monocytogenes*, can be present but inactive. *Infect. Immun.* **65:**1515–1518.

47. **Ripio, M.-T., K. Brehm, M. Lara, M. Suárez, and J.-A. Vázquez-Boland.** 1997. Glucose-1-phosphate utilization by *Listeria monocytogenes* is PrfA dependent and coordinately expressed with virulence factors. *J. Bacteriol.* **197:**7174–7180.

48. **Ripio, M.-T., G. Dominquez-Bernal, M. Lara, M. Suárez, and J.-A. Vázquez-Boland.** 1997. A Gly145Ser substitution in the transcriptional activator PrfA causes constitutive overexpression of virulence factors in *Listeria monocytogenes*. *J. Bacteriol.* **179:**1533–1540.

49. **Ripio, M.-T., G. Dominquez-Bernal, M. Suárez, K. Brehm, P. Berche, and J.-A. Vázquez-Boland.** 1996. Transcriptional activation of virulence genes in wild-type strains of *Listeria monocytogenes* in response to a change in the extracellular medium composition. *Res. Microbiol.* **147:**311–384.

50. **Ripio, M.-T., J.-A. Vázquez-Boland, Y. Vega, S. Nair, and P. Berche.** 1998. Evidence for expressional crosstalk between the central virulence regulator PrfA and the stress response mediator ClpC in *Listeria monocytogenes*. *FEMS Microbiol. Lett.* **158:**45–50.

51. **Rouquette, C., M.-T. Ripio, E. Pellegrini, J. M. Bolla, R. I. Tascon, J.-A. Vázquez-Boland, and P. Berche.** 1996. Identification of a ClpC ATPase required for stress tolerance and *in vivo* survival of *Listeria monocytogenes*. *Mol. Microbiol.* **21:**977–987.

52. **Sheehan, B., A. Klarsfeld, R. Ebright, and P. Cossart.** 1996. A single substitution in the putative helix-turn-helix motif of the pleiotropic activator PrfA attenuates *Listeria monocytogenes* virulence. *Mol. Microbiol.* **20:**785–797.

53. **Sheehan, B., C. Kocks, S. Dramsi, E. Gouin, A. D. Klarsfeld, J. Mengaud, and P. Cossart.** 1994. Molecular and genetic determinants of the *Listeria monocytogenes* infectious process. *Curr. Top. Microbiol. Immunol.* **192:**187–216.

54. **Sokolovic, Z., J. Riedel, M. Wuenscher, and W. Goebel.** 1993. Surface-associated, PrfA-regulated proteins of *Listeria monocytogenes* synthesized under stress conditions. *Mol. Microbiol.* **8:**219–227.

55. **Spiro, S., and J. R. Guest.** 1990. FNR and its role in oxygen-regulated gene expression in *Escherichia coli*. *FEMS Microbiol. Rev.* **75:**399–482.

56. **Vega, Y., C. Dickneite, M.-T. Ripio, R. Böckmann, B. González-Zorn, S. Novella, G. Dominquez-Bernal, W. Goebel, and J.-A. Vázquez-Boland.** 1998. Functional similarities between the *Listeria monocytogenes* virulence regulator PrfA and cyclic AMP receptor protein: the PrfA* (Gly145 Ser) mutation increases binding affinity for target DNA. *J. Bacteriol.* **180:**6655–6660.

57. **Weber, I. T., and T. A. Steitz.** 1987. Structure of a complex of catabolite gene activator protein and cyclic AMP refined at 2.5 Å resolution. *J. Mol. Biol.* **198:**311–326.

58. **Wuenscher, M. D., S. Köhler, A. Bubert, U. Gerike, and W. Goebel.** 1993. The *iap* gene of *Listeria monocytogenes* is essential for cell viability, and its gene product, p60, has bacteriolytic activity. *J. Bacteriol.* **175:**3491–3501.

59. **Zhou, Y., T. J. Merkel, and R. H. Ebright.** 1994. Characterization of the activating region of *Escherichia coli* catabolite gene activator protein (CAP) II. Role at class I and class II CAP-dependent promoters. *J. Mol. Biol.* **243:**603–610.

The Cell Biology of Invasion and Intracellular Growth by *Listeria monocytogenes*

PASCALE COSSART AND DANIEL A. PORTNOY

53

There is overwhelming histological, morphological, and immunological evidence that *Listeria monocytogenes* is a facultative intracellular pathogen that grows intracellularly both in animals and in tissue culture models of infection. Its capacity to survive within macrophages was noted very early on by Mackaness (39) and believed to be one of its key properties for virulence. It is now appreciated that *L. monocytogenes* grows in a variety of cell types both in vivo and in cell culture. Indeed, *L. monocytogenes* grows in all adherent cells examined. Nearly 10 years ago (46, 68), the stages describing the intracellular growth and cell-to-cell spread of *L. monocytogenes* were defined morphologically. During the past decade, the molecular mechanisms that control invasion, intracellular growth, and cell-to-cell spread have been elucidated. In this chapter, each intracellular stage depicted in Fig. 1 is covered in detail.

MORPHOLOGICAL DESCRIPTION OF THE ENTRY PROCESS INTO NONPHAGOCYTIC CELLS

The process of entry of *L. monocytogenes* into nonphagocytic cells has been examined by both transmission and scanning electron microscopy (Fig. 2) (43). Internalization of the bacteria starts by a close apposition of the plasma membrane on the bacterium, which becomes progressively enwrapped within the cell. This phenomenon is usually referred to as the "zipper" mechanism, in contrast to the "trigger" mechanism used by *Salmonella* and *Shigella* spp., a process that resembles macropinocytosis. The zipper mechanism for bacterial entry was first described in the case of invasin-mediated entry of *Yersinia* sp. (26). In that case, the surface protein invasin interacts with $\beta 1$ integrins to promote uptake of yersiniae. As described below, the same type of ligand receptor interaction occurs for *L. monocytogenes*. The situation is very different in *Salmonella* or *Shigella* spp., for which triggering of signaling events seems to be mediated by proteins directly translocated within the mammalian cell via a type III secretion system.

BACTERIAL PROTEINS INVOLVED IN ENTRY INTO TISSUE CULTURED CELLS

Internalin (InlA), a Surface Protein Linked to the Cell Wall via a LPXTG Motif

Internalin (InlA) is an 800-amino-acid protein that displays two regions of repeats, the first consisting of 15 successive leucine-rich repeats (LRRs) of 22 amino acids (11, 16) (see Fig. 3). LRRs are motifs with a constant periodicity of leucine residues, generally found in proteins involved in strong protein-protein interactions, although the LRRs are not necessarily part of the domain of interaction. The second repeat region is made of two 70-amino-acid residues and a third partial repeat of 49 amino acids. In addition, internalin has all the features of a protein that is targeted to and exposed on the bacterial surface, i.e., a signal peptide, and a C-terminal region containing a LPXTG peptide followed by a hydrophobic sequence and a few charged residues. This type of C terminus is found in more than 50 gram-positive bacterial surface proteins and allows covalent linkage of the protein to the peptidoglycan, after cleavage of the T-G link, as demonstrated for protein A of *Staphylococcus aureus* (60).

Several lines of evidence indicate that internalin is sufficient for entry in cells expressing its receptor (see below). Indeed, expression of *inlA*, the gene encoding internalin in *Listeria innocua* and also in the more distantly related gram-positive bacterium *Enterococcus faecalis*, confers invasiveness to these noninvasive species (16, 36). Moreover, latex beads coated with internalin are readily internalized (36).

The region of internalin essential for entry into mammalian cells is the LRRs, as first shown by inhibition of entry with monoclonal antibodies raised against internalin and recognizing the LRRs and then confirmed by a deletion analysis carried out on *inlA* (42). Moreover, *L. innocua* cells expressing the internalin LRRs and the interrepeat region fused to the C-terminal part of protein A are able to invade cells expressing the internalin receptor (see below) (36).

Internalin is expressed on the bacterial surface and can be found in varying amounts in the culture supernatants, an observation that raised the interesting possibility that

Gram-Positive Pathogens, ed. by V. A. Fischetti et al.
© 2000 American Society for Microbiology, Washington, D.C.

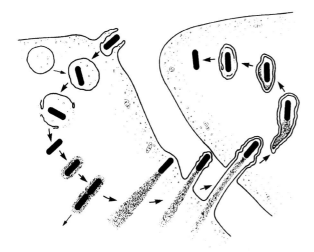

FIGURE 1   Model of the cell biology of *L. monocytogenes* infection. The stippled regions represent host actin filaments. Reprinted from Tilney and Portnoy (68), with permission from the *Journal of Cell Biology*.

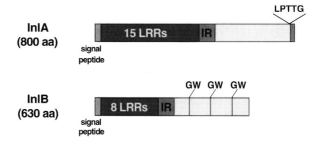

FIGURE 3   Schematic representation of InlA and InlB. LRR and IR indicate leucine-rich repeat and interrepeat regions, respectively. GW represents the GW repeats present in InlB. The LPTTG sequence indicated is involved in cell wall anchoring of InlA.

internalin should be released to elicit entry (12). Genetic evidence leading to truncation of the C-terminal domain of internalin or swapping of this domain with that of the surface protein ActA has now clearly established that the surface-associated form of internalin mediates entry (35).

A recent analysis of strains of *L. monocytogenes* of various origins (clinical strains or food isolates) has shown that internalin is not always expressed on the surface and can

be secreted as a truncated protein in the culture supernatant owing to mutations in *inlA*. This is the situation found in strain LO28, a strain that has been extensively used for genetic studies and that displays a fully virulent phenotype in the mouse model (28). There are several possible interpretations of these data that question either the role of internalin in vivo or the animal model used or both.

InlB, an Invasion Protein with GW Modules, a Novel Cell Wall Association Motif

InlB is a 630-amino-acid protein found both at the bacterial surface and to some extent in bacterial culture supernatants (10). It has a signal sequence and eight tandem

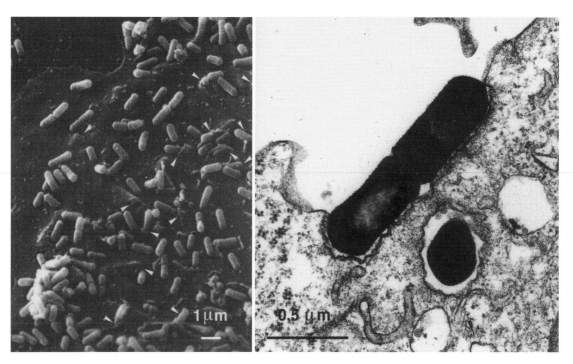

FIGURE 2   Electron micrographs of *L. monocytogenes* entering Caco-2 cells. (Left) Thin section of infected Caco-2 cells. (Right) Scanning electron micrograph. Reprinted from reference 43 with permission from *Cell*.

LRRs very similar to those of internalin (see Fig. 3). However, InlB does not contain the LPXTG motif or any hydrophobic region that would suggest a possible transmembrane region. Recent experiments indicate that the 232 C-terminal amino acids of InlB are necessary and sufficient to anchor InlB to the bacterial surface (3). This region contains three tandem repeats of approximately 80 amino acids beginning with the sequence GW, hence the name given to these repeats. Interestingly, similar repeats are found in a newly identified surface protein of *L. monocytogenes* named Ami. However, in Ami, there are eight GW modules and the protein is surface associated and cannot be detected in the culture supernatant. Conversely, when the C-terminal part of Ami is fused to InlB, InlB cannot be found in the culture supernatant. Similar GW repeats are present in the C-terminal part of lysostaphin. Lysostaphin is secreted by *Staphylococcus simulans* and associates with the cell wall of *S. aureus* even when exogenously added. Externally added InlB is also able to associate with *L. monocytogenes* and several other gram-positive bacteria. This external association leads to entry of a Δ*inlB* mutant and also promotes entry of the noninvasive species *Staphylococcus carnosus* and *L. innocua*, suggesting that InlB may interact with the cell wall after secretion or release from the bacterial surface, and that this interaction could contribute to invasion (3). In fact, latex beads coated with InlB enter efficiently into Vero cells, HEp2 cells, and HeLa cells, demonstrating that InlB is sufficient for entry (4, 48). How InlB associates with the cell surface is currently under investigation. Whether this loose association of InlB onto the bacterial surface plays a role in early signaling events is also the subject of current research.

Internalin and InlB are two members of the internalin multigene family characterized by the presence of LRRs similar to those in internalin. This family contains seven members. InlC (or IrpA) was identified as an abundant secreted protein whose presence depends on PrfA, the transcription regulator of all known virulence genes (8, 14). The other genes (*inlC2*, *inlD*, *inlE*, and *inlF*) were identified by screening DNA libraries of strain EGD at low stringency, using *inlA* as a probe (11). InlC is secreted and accordingly does not display the LPXTG motif or the GW modules. The other proteins have the LPXTG motif. Their expression has been analyzed at the transcription level. The genes are expressed, albeit much less efficiently than *inlA* and *inlB* under the same conditions. It had been anticipated that these *inl* genes could also play a role in entry. Simple and multiple deletions have shown that this does not seem to be the case (8, 11, 14). It should be noted that on the basis of hybridization studies at low stringency, all *Listeria* species seem to have *inl* genes (16). Their structure and their function have not yet been investigated.

Other Proteins Involved in Entry

It has been suggested that P60 and ActA could also participate in invasion (1, 32, 33). The noninvasive phenotype of P60 mutants seems to be an indirect consequence of the murein hydrolase activity of P60 (mutants form very long chains) (73). The role of ActA in invasion is less clear. Its role in invasion is due to its putative capacity to interact with proteoglycans. Attachment and subsequent entry of *L. monocytogenes* into CHO epithelial cells, as well as in IC-21 macrophages, are impaired by treatment of bacteria with heparin or heparan sulfate. Treatment of cells with heparinase III inhibits adhesion and invasion. Mutant CHO cells deficient in heparan sulfate synthesis are less permissive than wild-type CHO cells. Taken together, these data and other indirect evidence suggest that heparan sulfate plays a role in invasion (1). In addition, the similarity between a sequence in the N-terminal part of ActA and a peptide present in the circumsporozoite protein of *Plasmodium falciparum* and known to interact with heparan sulfate suggests that the impaired capacity of an ActA mutant to invade mammalian cells might be due to a lack of interaction with heparan sulfate. These observations are somewhat intriguing and deserve further investigation.

BACTERIAL-HOST CELL INTERACTIONS

Internalin and InlB, Two Invasion Proteins with Different Specificities

The genes encoding internalin and InlB form a bicistronic operon that is completely silent in the noninvasive transposon mutants first identified by Gaillard et al. (16). Such mutants are noninvasive for a large number of cell lines. In contrast, mutations affecting only *inlA* or only *inlB* have different phenotypes; internalin promotes entry into human intestinal epithelial cell lines, such as Caco-2 cells and Lovo cells, and InlB elicits entry into HeLa cells, Vero cells, certain hepatocytes, and some other cell lines (10, 25).

E-Cadherin, the Receptor for Internalin

The internalin receptor on the human epithelial cell line Caco-2 was identified using affinity chromatography (43). It is the cell adhesion molecule E-cadherin, a 110-kDa transmembrane glycoprotein that normally mediates calcium-dependent cell-cell adhesion, through homophilic interactions between extracellular domains. Cadherins are tissue specific, with E-cadherin specifically expressed in epithelial cells, N-cadherin in neuronal cells, and so on. These proteins play a critical role in cell sorting during development and maintenance of tissue cohesion and architecture during adult life. In polarized epithelial cells, E-cadherin is mainly expressed at the adherens junctions and on the basolateral face of the cell. Integrity of the intracytoplasmic domain of cadherins is required for optimal intercellular adhesion. This domain interacts with proteins named catenins, which in turn interact with the actin cytoskeleton, highlighting the importance of the cytoskeleton in maintaining adhesion of adjacent epithelial cells.

A set of transfected cell lines was used to demonstrate that the internalin–E-cadherin interaction promotes not only specific binding but also entry of *L. monocytogenes* (43). Both *L. monocytogenes* and *L. innocua* expressing internalin enter efficiently into cells expressing the chicken E-cadherin (L-CAM), in contrast to cells expressing N-cadherin or no cadherin. Similar results have also recently been obtained with internalin-coated beads (36), definitively establishing that E-cadherin acts as a receptor for internalin and promotes entry. How and where the LRR region of internalin interacts with E-cadherin is unknown. Preliminary experiments indicate that the cytoplasmic domain of E-cadherin is required for entry of the bacteria but not adhesion (unpublished results).

Given the basolateral localization of E-cadherin in polarized epithelial cells, these data suggest that *L. monocytogenes* would not enter enterocytes by their apical face. This hypothesis is in agreement with the observation that *L. monocytogenes* preferentially invades Caco-2 cell islets by their basolateral face (18).

The InlB receptor has not been identified. Indirect evidence indicates that it is not E-cadherin.

HOST CELL SIGNALING DURING INVASION

Tyrosine kinase inhibitors (65, 71) or actin polymerization inhibitors such as cytochalasin D (17) inhibit entry but not adhesion of *L. monocytogenes*, suggesting that these inhibitors do not affect the bacteria-cell receptor interaction but more specifically affect steps downstream from the initial interaction and require tyrosine kinase activity and an intact actin cytoskeleton. The kinases and the protein directly involved in actin rearrangements have not been identified. Recent evidence indicates that the phosphoinositide (PI) 3-kinase P85/P110, a signaling protein implicated in actin polymerization in response to receptor stimulation and tyrosine phosphorylation, is involved in entry of *L. monocytogenes* in mammalian cells (25). Thus, one possible implication of tyrosine phosphorylation in entry would be in the bacterial activation of this lipid kinase.

In resting cells, PI 3-kinase is a cytosolic heterodimeric (P85/P110) protein. Upon receptor stimulation, it translocates from the cytosol to a membrane-associated tyrosine-phosphorylated protein, e.g., an activated tyrosine kinase receptor or a tyrosine-phosphorylated adapter protein. Migration to the plasma membrane stimulates activity of PI 3-kinase by placing the enzyme in a compartment where its substrates are located and also by inducing conformational changes after protein-protein interactions. PI 3-kinase phosphorylates the D3 position of the inositol ring of PI, PIP, and PIP2, giving rise to PI3P, PI3,4P2, and PI3,4,5P3. These last two phosphoinositides are virtually absent in resting cells, and their levels increase dramatically upon stimulation. PI3,4P2 and PI3,4,5P3 are not substrates for known phospholipases and appear to act as second messengers by interacting with various kinases such as the serine/threonine kinase Akt.

Pretreatment of various mammalian cells with wortmannin, a fungal metabolite that specifically inhibits PI 3-kinase, impairs *L. monocytogenes* entry (25). The effect is maximal in Vero cells, in which entry is InlB dependent. In addition, expression of a dominant-negative form of P85 (DP85) also abrogates *L. monocytogenes* entry, establishing that the P85/P110 PI 3-kinase is required for entry (25). Measurements of the cellular levels of phosphoinositides reveal that entry of *L. monocytogenes* stimulates synthesis of both PI3,4P2 and PI3,4,5P3 in Vero cells. This synthesis is inhibited by pretreating cells with wortmannin, but not by pretreatment with cytochalasin D, suggesting that cytoskeletal rearrangements may occur downstream from the PI 3-kinase stimulation during bacterial invasion. Interestingly, PI 3-kinase stimulation is inhibited by genistein, an inhibitor of tyrosine kinases. Accordingly, shortly after infection, PI 3-kinase associates with at least one tyrosine-phosphorylated protein, and it is possible that this association stimulates PI 3-kinase activation.

A Δ*inlB* mutant that still adheres to Vero cells does not efficiently stimulate PI 3-kinase and does not stimulate association of PI 3-kinase with tyrosine-phosphorylated proteins. Thus, InlB seems to play a critical role in the stimulation of P85/P110.

How stimulation of PI 3-kinase affects bacterial invasion is not known, but one possible mechanism is by controlling actin polymerization or reorganization of the actin cytoskeleton. PI3,4P2 and PI3,4,5P3 are able to uncap barbed ends of actin filaments in permeabilized platelets, suggesting a simple means by which stimulation of PI 3-kinase activity by adherent bacteria could drive local cytoskeletal changes needed for entry. It is also possible that a high concentration of phosphoinositides may affect the local curvature of the lipid bilayer facilitating bacterial internalization. The generation of phospholipids in the membrane may also attract proteins that specifically bind to these phospholipids but are normally absent, e.g., proteins with plekstrin homology (PH) domains. PI 3-kinase itself can interact with other signaling proteins that regulate organization of the actin cytoskeleton, such as pp125FAK or Rho-GTPases, and such interactions may play a role in entry. PI 3-kinase has also been implicated in endocytic processes, and it is possible that some of the components used for endocytosis are also employed by *L. monocytogenes* to gain entry into host cells.

Intriguingly, the internalin–E-cadherin-mediated entry in Caco-2 cells, which is also affected by wortmannin, does not stimulate PI 3-kinase activity. In these cells, the basal levels of PI3,4P2 and PI3,4,5P3 do not increase significantly upon entry. In fact, these lipids are already present at high concentration in uninfected Caco-2 cells. Thus, in Caco-2 cells, the internalin–E-cadherin pathway may exploit a preactivated PI 3-kinase.

The future challenges are to identify the upstream tyrosine-phosphorylated proteins and the downstream targets of PI 3-kinase products. Preliminary experiments indicate that one of the phosphorylated proteins is c-Cbl (24). A recent study in HeLa cells in which entry is InlB dependent (4) demonstrated a role for MEK1 mitogen-activated protein (MAP) kinase kinase/ERK2 in entry (66). Treatment with wortmannin does not affect ERK2 activation. How the PI 3-kinase pathway and the MEK1/ERK2 pathway contribute to entry in HeLa cells deserves investigation (66).

RELEVANCE OF INTERNALIN AND InlB

Two important issues are currently a matter of debate: (i) What is the real contribution of internalin and/or InlB for entry in various cell lines? (ii) What is the relevance of internalin and InlB in vivo?

Three main types of tissue cultured cells have been examined for the role of internalin and InlB: epithelial cells, hepatocytes, and endothelial cells of various origins. Internalin-coated beads enter epithelial cells that express E-cadherin or transfected fibroblasts that express E-cadherin (36). Thus, all cells that express E-cadherin are expected to have an internalin-mediated pathway of entry. InlB-coated beads are able to enter Vero cells, HeLa cells, or HEp2 cells, demonstrating that the InlB pathway is also sufficient for entry. This pathway seems to be more widely used than the internalin pathway, as demonstrated by the effect of *inlB* mutations on entry in many cell types. In principle, the contribution of both pathways for entry should require the presence of both the internalin and the InlB receptors. This question cannot rigorously be addressed before the identification of the InlB receptor.

Both InlA and InlB are necessary for efficient entry in Caco-2, HeLa, Henle 407, PtK2, A549, and HEp2 cells

(38). However, other data indicate that internalin promotes entry in Caco-2 cells or cells expressing E-cadherin (16, 36) but seems to play virtually no role for entry into Vero, CHO, HeLa, and HEp2 cells, where InlB is the major invasion factor (11, 25). Clearly, experimental conditions, strains, and other factors may explain some of the discrepancies.

In hepatocytes of either human or murine origin, there is converging evidence that InlB plays a major role for entry (11, 19). In endothelial cells, several diverging studies were published with some role for internalin (13) or no role for either internalin or InlB (23). However, the most recent report, using human umbilical vein endothelial cells (HUVEC), seems to provide evidence that InlB is the major contributor to entry (51).

In the mouse model, internalin mutants behave like wild type after infection by the oral route, suggesting that internalin does not play a major role in the crossing of the intestinal barrier, at least in the mouse model (10, 19). Identical results were obtained using the rat ileal loop model (57). InlB also does not seem to play a role in translocation through the gut in the mouse model. In contrast, it appears to play a role in the hepatic phase of the infection (10, 19, 21, 22). InlB mutants replicate less efficiently than wild type in the liver (10, 19, 21, 22). Whether InlB mediates entry in hepatocytes in vivo, or plays a role in intrahepatocyte survival, is still unclear.

INVASION: CONCLUDING REMARKS

L. monocytogenes can enter into mammalian cells by at least two pathways (Fig. 4). One of them, the internalin–E-cadherin pathway, can be compared to that of *Yersinia* sp., in which the bacterial factor invasin interacts with a

β1 integrin, a cell adhesion molecule known to be connected to the cytoskeleton. Although the molecules involved are different, some functional similarity exists between the two systems, such as the sensitivity to tyrosine-kinase inhibitors and to cytochalasin D. The other strategy for entry is InlB dependent. This pathway seems to share similarities with growth factor-mediated signaling pathways. However, the residual level of entry of a ΔinlAB mutant suggests that other mechanisms of entry may exist that remain to be discovered.

LISTERIOLYSIN O AND ESCAPE FROM A VACUOLE

Intracellular pathogens can be divided into those that reside within a host vacuole and those, like *L. monocytogenes*, that escape and grow directly in the host cytosol (45). This critical aspect of *L. monocytogenes* pathogenesis was not clear until electron microscopy studies were performed using tissue culture models of infection (46, 68). Before this time, most studies used primary cultures of mouse peritoneal macrophages in which the majority of bacteria are killed and hence restricted to the vacuolar compartment. In contrast, most cell lines lack bacteriocidal capacity and allow the majority of ingested bacteria access to the cytosol. Even bone marrow-derived macrophages allow approximately 50% of the bacteria into the cytosol (7, 61), whereas resident peritoneal macrophages kill approximately 90% of ingested bacteria (54). Activated macrophages are extremely effective at killing *L. monocytogenes*.

There is overwhelming evidence that the primary *L. monocytogenes* determinant responsible for escape from a vacuole and thus entrance into the cytosol is listeriolysin O (LLO), encoded by the *hly* gene. First, mutants lacking

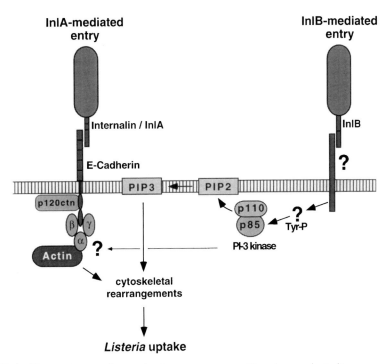

FIGURE 4 Two pathways of entry by *L. monocytogenes*. Question marks indicate steps that are not understood. PIP2 indicates the phospholipid phosphoinositol-4,5-bisphosphate (one of the substrates for PI 3-kinase), and PIP3 represents phosphoinositol-3,4,5-trisphosphate (one of the products of this kinase).

LLO are absolutely avirulent and fail to escape from a vacuole (53). Second, expression of LLO by either *Bacillus subtilis* or *L. innocua* promotes the capacity of these nonpathogenic bacteria to escape from a vacuole and to grow in the cytosol. Third, purified LLO encapsulated into pH-sensitive liposomes can mediate dissolution of a vacuole (37).

LLO is a member of a large family of pore-forming hemolysins secreted by gram-positive pathogens, including streptolysin O (SLO) secreted by *Streptococcus pyogenes* and perfringolysin O (PFO) secreted by *Clostridium perfringens* (64, 69). The crystal structure of PFO has recently been solved (58). All members of the family are thought to share a common mechanism of action that involves binding to membrane cholesterol followed by insertion, oligomerization of between 20 and 80 monomers, and pore formation. The major known difference between LLO and other members of the family is that LLO has an acidic pH optimum (20, 55).

SLO and PFO are thought to act extracellularly and promote pathogenesis by blocking chemotaxis of neutrophils (69). This raises the question of what, if anything, is unique to LLO. To address this question, PFO was cloned and expressed in *L. monocytogenes* under the control of the endogenous *hly* promoter. PFO was able to mediate vacuolar escape at approximately 50% of the efficiency of LLO. However, after a few bacteria divisions in the cytosol, the host cell became permeabilized and died (27). This led to the hypothesis that LLO has properties that render it nontoxic. First, it appears to have a short cytosolic half-life. Second, it has an acidic pH optimum that may make it less toxic in the cytosol. Third, although *hly* is transcribed at high levels in the cytosol (44), LLO may not be efficiently translated and/or secreted in the cytosol.

Beauregard et al. (2) developed a method using a pH-sensitive dye to measure the pH of *L. monocytogenes*-containing vacuoles in macrophages and to measure the time after internalization in which vacuoles were perforated. They showed that the average pH of vacuolar perforation was 5.9 and that an inhibitor of the vacuolar proton pump prevented vacuolar perforation. Furthermore, bacteria were able to cause perforation at times as short as 3 min. However, it should be noted that this study examined vacuolar perforation, not escape of the bacteria from the phagosome.

While it is clear that LLO is largely responsible for mediating escape from a vacuole in most cell lines, two *L. monocytogenes* phospholipases C (PLCs) also play a role. Mutants lacking both the broad-range PLC (PlcB) and a phosphatidylinositol-specific PLC (PlcA) escape from a macrophage vacuole at 50% of the efficiency of wild type (61). Surprisingly, PlcB can even replace LLO in some cell lines, such as the human epithelial cell lines HeLa and Henle 407 (40). While the latter observation may be a tissue culture artifact, it must be considered when formulating a model for the role of LLO in vacuolar escape, i.e., under some circumstances, LLO-minus *L. monocytogenes* cells can still escape from a vacuole.

The following model for vacuolar escape is consistent with the above data. Subsequent to internalization, LLO is synthesized, secreted, and binds to membrane cholesterol. Optimal pore formation occurs between pH 5.5 and 6.0, which is the pH of an early endosome. One consequence of pore formation is the elevation of vacuolar pH, which may cause the cessation of vacuolar maturation and thus

allow the *L. monocytogenes* PLCs, perhaps in concert with host PLCs, to mediate vacuolar dissolution.

ActA AND ACTIN-BASED MOTILITY

Upon escape from a vacuole and arrival to the host cell cytosol, *L. monocytogenes* begins rapid growth and becomes encapsulated by host actin filaments and other actin-binding proteins (46, 68). After a couple of bacterial cell divisions, the bacteria begin to move in the cytosol at measurable rates. Movement is entirely dependent on polymerization of host actin (59, 67). New filaments form at the rear end of moving bacteria and then depolymerize. Moving bacteria have a characteristic "tail" composed of actin filaments and actin-binding proteins. The length of the tail reflects the rate of bacterial movement.

The *L. monocytogenes* ActA protein is absolutely necessary for actin-based motility (5, 9, 29). Mutants lacking ActA escape normally from a vacuole, grow at wild-type growth rates in the cytosol, but grow as microcolonies and are unable to spread from cell to cell. Importantly, ActA mutants are 1,000-fold less virulent in the murine model of infection. There is also substantial evidence that ActA is not only necessary, but sufficient for actin-based motility (15, 31, 52, 62).

The ActA protein can be divided into three domains (34, 52, 63). The C-terminal portion of ActA encodes a transmembrane domain that anchors ActA to the bacterial surface. Truncation of this domain results in secretion of functional ActA into the growth medium. This is an important observation as it facilitates the purification and biochemical characterization of ActA. The central region of ActA contains four short proline-rich repeats (consensus sequence DFPPPPTDEEL). These proline-rich repeats act as binding sites for a host protein called VASP, which in turn binds to another host protein called profilin (50, 63). Both VASP and profilin localize at the interface of moving *L. monocytogenes* and its actin tail. The phenotype of mutants lacking the proline-rich repeats is that bacteria move at rates approximately one-third of wild type. This phenotype is consistent with a role for profilin in enhancing the rate of actin polymerization. The N-terminal domain of ActA is essential and sufficient for actin polymerization. Deletions of this region result in an ActA null phenotype. The role of this region in mediating actin polymerization is being actively studied. Purified ActA or an N-terminal fragment acts as an actin nucleator when mixed 1:1 with a complex of host proteins called the Arp2/3 complex (72). This is quite an exciting finding, as the Arp2/3 complex is being proposed as being critical for actin nucleation by host cells, yet by itself it is a weak nucleator. Thus, ActA seems to be fulfilling the function of an unknown host factor.

Immunofluorescence analysis of ActA has shown that it surrounds the bacteria, although there is a higher concentration at the rear end of moving bacteria (30, 49). However, VASP appears to be localized in an asymmetric distribution. Thus, ActA appears to be functionally asymmetric on the bacterial surface. One explanation for this functional asymmetry is that dimerization of ActA (47) is concentration dependent and occurs at one end of the bacteria.

CELL-TO-CELL SPREAD

ActA is clearly an essential determinant of *L. monocytogenes* pathogenesis, whose role is most likely to facilitate cell-to-cell spread. Thus, actin-based motility per se is not important, but rather the fact that it promotes the interaction of intracytosolic bacteria with host membrane. The actual host molecules that link the actin tail to the membrane are unknown, nor is it known if ActA interacts directly with the host membrane. Further, the surface molecules that mediate cell-cell interaction are not known, nor is it known if cell-to-cell spread is mediated by an active or passive process in the recipient cell.

Although nothing is yet known about the mechanism of cell-to-cell spread, it is known that the two bacterial phospholipases and metalloprotease (Mpl) play a role (41, 61, 70). Mutants lacking any one of these enzymes have modest defects upon cell-to-cell spread, as determined by plaque formation in monolayers of mouse L2 cells. However, mutants lacking both PLCs have a large defect in cell-to-cell spread and a 500-fold decrease in virulence. The broad-range PLC of *L. monocytogenes* (PlcB) is synthesized as an inactive precursor that can be proteolytically activated by the *L. monocytogenes* metalloprotease (Mpl) (56). However, mutants lacking PlcA and Mpl spread considerably better than a double mutant lacking PlcA and PlcB. This observation suggested that there might be an Mpl-independent pathway for PlcB activation. Indeed, in the absence of Mpl, PlcB is proteolytically activated during infection by a host cysteine protease (41). These results illustrate the complexity of assigning specific roles to each of these gene products. There appears to be an overlap between the role of the two PLCs as well as between Mpl and a host protease. Perhaps these different determinants have different relative importance depending on the cells or animal species infected.

CELL-TO-CELL SPREAD ALLOWS THE BACTERIA TO AVOID BOTH THE HUMORAL AND CELLULAR IMMUNE RESPONSE

As discussed in chapter 50 of this volume, acquired immunity to *L. monocytogenes* is largely mediated by the effector function of CD8[+] T cells. Antibody plays no measurable role in immunity. The cell biology of infection is consistent with the immunological perspective; i.e., once bacteria enter a cell, they need not be exposed to the humoral immune response. However, growth in the cytosol potentially exposes bacterial proteins to the major histocompatibility complex (MHC) class I pathway of antigen processing and presentation. Indeed, *L. monocytogenes* antigens presented in association with the MHC class I antigen H2-K[d] are derived from secreted bacterial proteins. Surprisingly, the dominant antigen recognized by immune CD8[+] T cells is derived from LLO (6). Cells infected with *L. monocytogenes* are recognized and lysed by immune CD8[+] T cells. Thus, the capacity of *L. monocytogenes* to spread cell to cell allows the bacteria to escape both the humoral and cellular immune response.

This chapter could not have been written without the work of many of our collaborators who over the years have worked on invasion. They are greatly appreciated. In addition, we thank L. Braun for critical reading and M. Lecuit for help with the figures.

REFERENCES

1. **Alvarez-Dominguez, C., J. A. Vazquez-Boland, E. Carrasco-Marin, P. Lopez-Mato, and F. Leyva-Cobian.** 1997. Host-cell heparan sulfate proteoglycans mediate attachment and entry of *Listeria monocytogenes*, and the listerial surface protein ActA is involved in heparan sulfate receptor recognition. *Infect. Immun.* **65:**78–88.

2. **Beauregard, K. E., K. Lee, R. J. Collier, and J. A. Swanson.** 1997. pH-dependent perforation of macrophage phagosomes by listeriolysin O from *Listeria monocytogenes*. *J. Exp. Med.* **186:**1159–1163.

3. **Braun, L., S. Dramsi, P. Dehoux, H. Bierne, G. Lindahl, and P. Cossart.** 1997. InlB: an invasion protein of *Listeria monocytogenes* with a novel type of surface association. *Mol. Microbiol.* **25:**285–294.

4. **Braun, L., H. Ohayon, and P. Cossart.** 1998. The InlB protein of *Listeria monocytogenes* is sufficient to promote entry into mammalian cells. *Mol. Microbiol.* **27:**1077–1087.

5. **Brundage, R. A., G. A. Smith, A. Camilli, J. A. Theriot, and D. A. Portnoy.** 1993. Expression and phosphorylation of the *Listeria monocytogenes* ActA protein in mammalian cells. *Proc. Natl. Acad. Sci. USA* **90:**11890–11894.

6. **Busch, D. H., I. M. Pilip, S. Vijh, and E. G. Pamer.** 1998. Coordinate regulation of complex T cell populations responding to bacterial infection. *Immunity* **8:**353–362.

7. **de Chastellier, C., and P. Berche.** 1994. Fate of *Listeria monocytogenes* in murine macrophages: evidence for simultaneous killing and survival of intracellular bacteria. *Infect. Immun.* **62:**543–553.

8. **Domann, E., S. Zechel, A. Lingnau, T. Hain, A. Darji, T. Nichterlein, J. Wehland, and T. Chakraborty.** 1997. Identification and characterization of a novel PrfA-regulated gene in *Listeria monocytogenes* whose product, IrpA, is highly homologous to internalin proteins, which contain leucine-rich repeats. *Infect. Immun.* **65:**101–109.

9. **Domann, E., J. Wehland, M. Rohde, S. Pistor, M. Hartl, W. Goebel, M. Leimeister-Wächter, M. Wuenscher, and T. Chakraborty.** 1992. A novel bacterial virulence gene in *Listeria monocytogenes* required for host cell microfilament interaction with homology to the proline-rich region of vinculin. *EMBO J.* **11:**1981–1990.

10. **Dramsi, S., I. Biswas, E. Maguin, L. Braun, P. Mastroeni, and P. Cossart.** 1995. Entry of *L. monocytogenes* into hepatocytes requires expression of InlB, a surface protein of the internalin multigene family. *Mol. Microbiol.* **16:**251–261.

11. **Dramsi, S., P. Dehoux, M. Lebrun, P. Goossens, and P. Cossart.** 1997. Identification of four new members of the internalin multigene family in *Listeria monocytogenes* strain EGD. *Infect. Immun.* **65:**1615–1625.

12. **Dramsi, S., C. Kocks, C. Forestier, and P. Cossart.** 1993. Internalin-mediated invasion of epithelial cells by *Listeria monocytogenes* is regulated by the bacterial growth state, temperature and the pleiotropic activator, *prfA*. *Mol. Microbiol.* **9:**931–941.

13. **Drevets, D. A., R. T. Sawyer, T. A. Potter, and P. A. Campbell.** 1995. *Listeria monocytogenes* infects human endothelial cells by two distinct mechanisms. *Infect. Immun.* **63:**4268–4276.

14. **Engelbrecht, F., S.-K. Chun, C. Ochs, J. Hess, F. Lottspeich, W. Goebel, and Z. Sokolovic.** 1996. A new PrfA-regulated gene of *Listeria monocytogenes* encoding a small, secreted protein which belongs to the family of internalins. *Mol. Microbiol.* **21:**823–837.

15. **Friederich, E., E. Gouin, R. Hellio, C. Kocks, P. Cossart, and D. Louvard.** 1995. Targeting of *Listeria monocytogenes* ActA protein to the plasma membrane as a tool to dissect both actin-based cell morphogenesis and ActA function. *EMBO J.* **14:**2731–2744.

16. **Gaillard, J. L., P. Berche, C. Frehel, E. Gouin, and P. Cossart.** 1991. Entry of *L. monocytogenes* into cells is mediated by Internalin, a repeat protein reminiscent of surface antigens from gram-positive cocci. *Cell* **65:**1127–1141.

17. **Gaillard, J. L., P. Berche, J. Mounier, S. Richard, and P. Sansonetti.** 1987. In vitro model of penetration and intracellular growth of *Listeria monocytogenes* in the human enterocyte-like cell line Caco-2. *Infect. Immun.* **55:**2822–2829.

18. **Gaillard, J. L., and B. B. Finlay.** 1996. Effect of cell polarization and differentiation on entry of *Listeria monocytogenes* into enterocyte-like Caco-2 cell line. *Infect. Immun.* **64:**1299–1308.

19. **Gaillard, J. L., F. Jaubert, and P. Berche.** 1996. The *inlAB* locus mediates the entry of *Listeria monocytogenes* into hepatocytes *in vivo. J. Exp. Med.* **183:**359–369.

20. **Geoffroy, C., J.-L. Gaillard, J. E. Alouf, and P. Berche.** 1987. Purification, characterization, and toxicity of the sulfhydryl-activated hemolysin listeriolysin O from *Listeria monocytogenes. Infect. Immun.* **55:**1641–1646.

21. **Gregory, S. H., A. J. Sagnimeni, and E. J. Wing.** 1996. Expression of the *inlAB* operon by *Listeria monocytogenes* is not required for entry into hepatic cells in vivo. *Infect. Immun.* **64:**3983–3986.

22. **Gregory, S. H., A. J. Sagnimeni, and E. J. Wing.** 1998. Internalin B promotes the replication of *Listeria monocytogenes* in mouse hepatocytes. *Infect. Immun.* **65:**5137–5141.

23. **Greiffenberg, L., Z. Sokolovic, H. J. Schnittler, A. Spory, R. Bockman, W. Goebel, and M. Kuhn.** 1997. *Listeria monocytogenes*-infected human umbilical vein endothelial cells—internalin-independent invasion, intracellular growth, movement and host responses. *FEMS Microbiol. Lett.* **157:**163–170.

24. **Ireton, K., and P. Cossart.** 1997. Host-pathogen interactions during entry and actin-based movement of *Listeria monocytogenes. Annu. Rev. Genet.* **31:**113–138.

25. **Ireton, K., B. Payrastre, H. Chap, W. Ogawa, H. Sakaue, M. Kasuga, and P. Cossart.** 1996. A role for phosphoinositide 3-kinase in bacterial invasion. *Science* **274:**780–782.

26. **Isberg, R. R., and G. Tran Van Nhieu.** 1994. Two mammalian cell internalization strategies used by pathogenic bacteria. *Annu. Rev. Genet.* **27:**395–422.

27. **Jones, S., and D. A. Portnoy.** 1994. Characterization of *Listeria monocytogenes* pathogenesis in a strain expressing perfringolysin O in place of listeriolysin O. *Infect. Immun.* **62:**5608–5613.

28. **Jonquieres, R., H. Bierne, J. Mengaud, and P. Cossart.** 1998. The *inlA* of *Listeria monocytogenes* LO28 harbors a nonsense mutation resulting in release of internalin. *Infect. Immun.* **66:**3420–3422.

29. **Kocks, C., E. Gouin, M. Tabouret, P. Berche, H. Ohayon, and P. Cossart.** 1992. *L. monocytogenes*-induced actin assembly requires the *actA* gene product, a surface protein. *Cell* **68:**521–531.

30. **Kocks, C., R. Hellio, P. Gounon, H. Ohayon, and P. Cossart.** 1993. Polarized distribution of *Listeria monocytogenes* surface protein ActA at the site of directional actin assembly. *J. Cell Sci.* **105:**699–710.

31. **Kocks, C., J. B. Marchand, E. Gouin, H. d'Hauteville, P. J. Sansonetti, M. F. Carlier, and P. Cossart.** 1995. The unrelated surface proteins ActA of *Listeria monocytogenes* and IcsA of *Shigella flexneri* are sufficient to confer actin-based motility to *L. innocua* and *E. coli* respectively. *Mol. Microbiol.* **18:**413–423.

32. **Kohler, S., A. Bubert, M. Vogel, and W. Goebel.** 1991. Expression of the *iap* gene coding for protein p60 of *Listeria monocytogenes* is controlled on the post-transcriptional level. *J. Bacteriol.* **173:**4668–4674.

33. **Kuhn, M., and W. Goebel.** 1989. Identification of an extracellular protein of *Listeria monocytogenes* possibly involved in intracellular uptake by mammalian cells. *Infect. Immun.* **57:**55–61.

34. **Lasa, I., V. David, E. Gouin, J.-B. Marchand, and P. Cossart.** 1995. The amino-terminal part of ActA is critical for the actin-based motility of *Listeria monocytogenes*; the central proline-rich region acts as a stimulator. *Mol. Microbiol.* **18:**425–436.

35. **Lebrun, M., J. Mengaud, H. Ohayon, F. Nato, and P. Cossart.** 1996. Internalin must be on the bacterial surface to mediate entry of *Listeria monocytogenes* into epithelial cells. *Infect. Immun.* **57:**55–61.

36. **Lecuit, M., H. Ohayon, L. Braun, J. Megaud, and P. Cossart.** 1997. Internalin of *Listeria monocytogenes* with an intact leucine-rich repeat region is sufficient to promote internalization. *Infect. Immun.* **65:**5309–5319.

37. **Lee, K. D., Y. K. Oh, D. A. Portnoy, and J. A. Swanson.** 1996. Delivery of macromolecules into cytosol using liposomes containing hemolysin from *Listeria monocytogenes. J. Biol. Chem.* **271:**7249–7252.

38. **Lingnau, A., E. Domann, M. Hudel, M. Bock, T. Nichterlein, J. Wehland, and T. Chakraborty.** 1995. Expression of the *Listeria monocytogenes* EGD *inlA* and *inlB* genes, whose products mediate bacterial entry into tissue culture cell lines, by PrfA-dependent and -independent mechanisms. *Infect. Immun.* **63:**3896–3903.

39. **Mackaness, G. B.** 1962. Cellular resistance to infection. *J. Exp. Med.* **116:**381–406.

40. **Marquis, H., V. Doshi, and D. A. Portnoy.** 1995. The broad-range phospholipase C and a metalloprotease mediate listeriolysin O-independent escape of *Listeria monocytogenes* from a primary vacuole in human epithelial cells. *Infect. Immun.* **63:**4531–4534.

41. **Marquis, H., H. Goldfine, and D. A. Portnoy.** 1997. Proteolytic pathways of activation and degradation of a bacterial phospholipase C during intracellular infection by *Listeria monocytogenes. J. Cell Biol.* **137:**1381–1392.

42. **Mengaud, J., M. Lecuit, M. Lebrun, F. Nato, J.-C. Mazie, and P. Cossart.** 1996. Antibodies to the leucine-rich repeat region of internalin block entry of *Listeria monocytogenes* into cells expressing E-cadherin. *Infect. Immun.* **64:**5430–5433.

43. **Mengaud, J., H. Ohayon, P. Gounon, R. M. Mege, and P. Cossart.** 1996. E-cadherin is the receptor for internalin, a surface protein required for entry of *Listeria monocytogenes* into epithelial cells. *Cell* **84:**923–932.

44. **Moors, M. A., B. Levitt, P. Youngman, and D. A. Portnoy.** 1999. Expression of listeriolysin O and ActA by in-

tracellular and extracellular *Listeria monocytogenes*. *Infect. Immun.* **67**:131–139.

45. Moulder, J. M. 1985. Comparative biology of intracellular parasitism. *Microbiol. Rev.* **49**:298–337.

46. Mounier, J., A. Ryter, M. Coquis-Rondon, and P. J. Sansonetti. 1990. Intracellular and cell-to-cell spread of *Listeria monocytogenes* involves interaction with F-actin in the enterocytelike cell line Caco-2. *Infect. Immun.* **58**: 1048–1058.

47. Mourrain, P., I. Lasa, A. Gautreau, A. Pugsley, and P. Cossart. 1997. ActA is a dimer. *Proc. Natl. Acad. Sci. USA* **94**:10034–10039.

48. Müller, S., T. Hain, P. Pashalidis, A. Lignau, E. Domann, T. Chakraborty, and J. Wehland. 1998. Purification of the *inlB* gene product of *Listeria monocytogenes* and demonstration of its biological activity. *Infect. Immun.* **66**: 3128–3133.

49. Niebuhr, K., T. Chakraborty, M. Rohde, T. Gazlig, B. Jansen, P. Kollner, and J. Wehland. 1993. Localization of the ActA polypeptide of *Listeria monocytogenes* in infected tissue culture cell lines: ActA is not associated with actin "comets." *Infect. Immun.* **61**:2793–2802.

50. Niebuhr, K., F. Ebel, R. Frank, M. Reinhard, E. Domann, U. D. Carl, U. Walter, F. B. Gertler, J. Wehland, and T. Chakraborty. 1997. A novel proline-rich motif present in ActA of *Listeria monocytogenes* and cytoskeletal proteins is the ligand for the EVH1 domain, a protein module present in the Ena/VASP family. *EMBO J.* **16**: 5433–5444.

51. Parida, S. K., E. Domann, M. Rohde, S. Muller, A. Darji, T. Hain, et al. 1998. Internalin B is essential for adhesion and mediates the invasion of *Listeria monocytogenes* into human endothelial cells. *Mol. Microbiol.* **28**: 81–93.

52. Pistor, S., T. Chakraborty, K. Niebuhr, E. Domann, and J. Wehland. 1994. The ActA protein of *Listeria monocytogenes* acts as a nucleator inducing reorganization of the actin cytoskeleton. *EMBO J.* **13**:758–763.

53. Portnoy, D. A., T. Chakraborty, W. Goebel, and P. Cossart. 1992. Molecular determinants of *Listeria monocytogenes* pathogenesis. *Infect. Immun.* **60**:1263–1267.

54. Portnoy, D. A., R. D. Schreiber, P. Connelly, and L. G. Tilney. 1989. Gamma interferon limits access of *Listeria monocytogenes* to the macrophage cytoplasm. *J. Exp. Med.* **170**:2141–2146.

55. Portnoy, D. A., R. K. Tweten, M. Kehoe, and J. Bielecki. 1992. Capacity of listeriolysin O, streptolysin O, and perfringolysin O to mediate growth of *Bacillus subtilis* within mammalian cells. *Infect. Immun.* **60**:2710–2717.

56. Poyart, C., E. Abachin, I. Razafimanantsoa, and P. Berche. 1993. The zinc metalloprotease of *Listeria monocytogenes* is required for maturation of phosphatidylcholine phospholipase C: direct evidence obtained by gene complementation. *Infect. Immun.* **61**:1576–1580.

57. Pron, B., C. Boumaila, F. Jaubert, S. Sarnacki, J.-P. Monnet, P. Berche, and J.-L. Gaillard. 1998. Comprehensive study of the intestinal stage of listeriosis in a rat ligated ileal loop system. *Infect. Immun.* **66**:747–755.

58. Rossjohn, J., S. C. Feil, W. J. McKinstry, R. K. Tweten, and M. W. Parker. 1997. Structure of a cholesterol-binding, thiol-activated cytolysin and a model of its membrane form. *Cell* **89**:685–692.

59. Sanger, J. M., J. W. Sanger, and F. S. Southwick. 1992. Host cell actin assembly is necessary and likely to provide the propulsive force for intracellular movement of *Listeria monocytogenes*. *Infect. Immun.* **60**:3609–3619.

60. Schneewind, O., A. Fowler, and K. F. Faull. 1995. Structure of the cell wall anchor of surface proteins in *Staphylococcus aureus*. *Science* **268**:103–106.

61. Smith, G. A., H. Marquis, S. Jones, N. C. Johnston, D. A. Portnoy, and H. Goldfine. 1995. The two distinct phospholipases C of *Listeria monocytogenes* have overlapping roles in escape from a vacuole and cell-to-cell spread. *Infect. Immun.* **63**:4231–4237.

62. Smith, G. A., D. A. Portnoy, and J. A. Theriot. 1995. Asymmetric distribution of the *Listeria monocytogenes* ActA protein is required and sufficient to direct actin-based motility. *Mol. Microbiol.* **17**:945–951.

63. Smith, G. A., J. A. Theriot, and D. A. Portnoy. 1996. The tandem repeat domain in the *Listeria monocytogenes* ActA protein controls the rate of actin-based motility, the percentage of moving bacteria, and the localization of vasodilator-stimulated phosphoprotein and profilin. *J. Cell Biol.* **135**:647–660.

64. Smyth, C. J., and J. L. Duncan. 1978. Thiol-activated (oxygen-labile) cytolysins, p. 129–183. *In* J. Jeljaszewicz and T. Wasstrom (ed.), *Bacterial Toxins and Cell Membranes*. Academic Press Inc., New York, N.Y.

65. Tang, P., I. Rosenshine, and B. B. Finlay. 1994. *Listeria monocytogenes*, an invasive bacterium, stimulates MAP kinase upon attachment to epithelial cells. *Mol. Biol. Cell* **5**:455–464.

66. Tang, P., C. Sutherland, M. R. Gold, and B. B. Finlay. 1998. *Listeria monocytogenes* invasion of epithelial cells requires the MEK/ERK2 mitogen-activated protein kinase pathway. *Infect. Immun.* **66**:1106–1112.

67. Theriot, J. A., T. J. Mitchison, L. G. Tilney, and D. A. Portnoy. 1992. The rate of actin-based motility of intracellular *Listeria monocytogenes* equals the rate of actin polymerization. *Nature* **357**:257–260.

68. Tilney, L. G., and D. A. Portnoy. 1989. Actin filaments and the growth, movement, and spread of the intracellular bacterial parasite, *Listeria monocytogenes*. *J. Cell Biol.* **109**:1597–1608.

69. Tweten, R. K. 1995. Pore-forming toxins in gram-positive bacteria, p. 207–229. *In* J. A. Roth, C. A. Bolin, K. A. Brogden, F. C. Minion, and M. J. Wannemuehler (ed.), *Virulence Mechanisms of Bacterial Pathogens*, 2nd ed. American Society for Microbiology, Washington, D.C.

70. Vazquez-Boland, J.-A., C. Kocks, S. Dramsi, H. Ohayon, C. Geoffroy, J. Mengaud, and P. Cossart. 1992. Nucleotide sequence of the lecithinase operon of *Listeria monocytogenes* and possible role of lecithinase in cell-to-cell spread. *Infect. Immun.* **60**:219–230.

71. Velge, P., E. Bottreau, B. Kaeffer, N. Yurdusev, P. Pardon, and N. Van Langendonck. 1994. Protein tyrosine kinase inhibitors block the entries of *Listeria monocytogenes* and *Listeria ivanovii* into epithelial cells. *Microb. Pathog.* **17**:37–50.

72. Welch, M. D., J. Rosenblatt, J. Skoble, D. A. Portnoy, and T. J. Mitchison. 1998. Interaction of human Arp2/3 complex and the *Listeria monocytogenes* ActA protein in actin filament nucleation. *Science* **281**:105–108.

73. Wuenscher, M., S. Kohler, A. Bubert, U. Gerike, and W. Goebel. 1993. The *iap* gene of *Listeria monocytogenes* is essential for cell viability and its gene product, p60, has bacteriolytic activity. *J. Bacteriol.* **175**:3491–3501.

SPORE-FORMING PATHOGENS AND GRAM-POSITIVE ACTINOMYCETALES

SECTION EDITOR: Julian I. Rood

ISEASES SUCH AS BOTULISM, tetanus, gas gangrene, anthrax, and diphtheria are classical diseases that were recognized well before the advent of the science of microbiology and what we now regard as modern medicine. These syndromes were known before the germ theory of disease had been verified and before the concept of spontaneous generation had been disproved. Their analysis provided some of the paradigm formative studies in microbiology: consider the demonstration by Koch in 1877 that *Bacillus anthracis* was the causative agent of anthrax and the identification of diphtheria toxin in 1888 by Roux and Yersin and tetanus toxin in 1889 by Kitasato. Despite the fact that these diseases have been studied for many years, there have been some very exciting discoveries in the past decade. The crystal structures of key toxins such as botulinum toxin, diphtheria toxin, anthrax toxin components, and *Clostridium perfringens* α-toxin have been elucidated. The precise enzymatic role and substrate of the botulinum and tetanus metalloprotease neurotoxins have been elucidated, and definitive evidence that α-toxin is the essential toxin in *C. perfringens*-mediated gas gangrene has been obtained. Only now are we determining the actual mode of action of the anthrax toxins and how their various components interact. These are exciting times indeed for research on some classical bacterial diseases.

This section also deals with several diseases whose etiology or importance has become apparent only relatively recently. The best example is pseudomembranous colitis, which is caused by *Clostridium difficile*. This organism produces two of the largest toxins that are known, toxins whose action as monoglucosyltransferases has only been elucidated in the past few years. These chapters also deal with what can be regarded as emerging infections such as norcardiasis and actinomycosis.

The chapters are essentially self-contained and are generally focused on a particular pathogen or pathogenic genus. Hence there are separate chapters on diphtheria, anthrax, *Nocardia*, and *Actinomyces*. The exception to this structure is the pathogenic clostridia. A general chapter on the genetics of these organisms is followed by three individual chapters that are centered on pathogenesis. These chapters are organized on the basis of the mechanism of pathogenesis rather than the bacterial species. Clostridial diseases all involve the elucidation of powerful protein toxins. Irrespective of which bacterium produces these toxins, they primarily affect either the nervous system or the gastrointestinal tract, or cause diseases that involve extensive tissue damage and necrosis. Therefore, it was considered that a more coherent understanding of the pathogenesis of these diseases would result from having separate chapters that discussed the neurotoxic, enterotoxic, and histotoxic clostridia.

The most severely affected parts of the world are central Asia and western areas of Africa. Endemic anthrax in major wildlife reserves in southern Africa threatens several endangered species. An "anthrax belt" exists from Turkey to Pakistan. In North America, anthrax is endemic in northern Alberta and the Northwest Territories. Sporadic outbreaks of anthrax occur within the United States; northwest Mississippi/southeast Arkansas and western Texas are considered areas of endemicity. The disease is also endemic throughout Mexico, Central America, and in many South American countries (53).

B. ANTHRACIS

Physiology, Morphology, and Taxonomy

B. anthracis is a facultative anaerobe and grows in most rich undefined media with a doubling time of approximately 30 min. The only absolute nutritional requirements for growth are methionine and thiamine. However, most strains grow poorly on glucose-salts medium containing only these amino acids. Commonly used minimal media contain nine or more amino acids. Uracil, adenine, guanine, and manganese stimulate growth of some strains. The optimal growth temperature for B. anthracis is 37°C, and cells cannot grow above 43°C (47). In most culture conditions, the rod-shaped cells form long bamboolike chains characteristic of the species, and sporulating cells carry elliptic, centrally located spores. In infected tissues, the chains are shorter than those formed during growth in vitro, and spores are generally not apparent (52). B. anthracis cells are capsulated in infected tissues and when grown in appropriate in vitro conditions (see below).

B. anthracis has been taxonomically aligned with Bacillus cereus, Bacillus thuringiensis, and Bacillus mycoides based on physiological, morphological, and DNA similarities. B. anthracis is most closely related to B. cereus, as evidenced by results of DNA hybridization experiments, the sequences of the genes encoding 16S and 23S rRNA, and the intergenic regions between the 16S and 23S rRNA genes and between gyrA and gyrB (1, 22). Distinguishing phenotypic characteristics of B. anthracis include lack of motility, weak hemolytic activity, sensitivity to gamma bacteriophage and penicillin, and the elaboration of the specific virulence factors, anthrax toxin and capsule.

Genetic variability within B. anthracis strains has been assessed using amplified fragment length polymorphism marker analysis and amplification of specific chromosomal and plasmid-associated loci by PCR. Molecular analysis of the B. anthracis genome has revealed the most genetically uniform bacterial species known. A variable DNA sequence designated VNTR (variable number of tandem repeats) is particularly useful as a discriminatory marker, and five allelic categories of B. anthracis have been assigned based on the presence of two to six copies of a 12-bp tandem repeat. There is a correlation between allelic category and geographic distribution, indicating that the VNTR marker can serve as an epidemiological tool. The VNTR lies within an open reading frame designated vrrA (variable region with repetitive sequence) that is predicted to encode a 30-kDa membrane-associated protein. However, no function has been associated with the putative gene product (25, 26).

Virulence Plasmids

Fully virulent strains of B. anthracis carry two large plasmids, pXO1 (182 kb) and pXO2 (93 kb). Characterized genes on plasmid pXO1 include the structural genes for the toxin proteins, cya (edema factor), lef (lethal factor), and pag (protective antigen); a regulatory gene, atxA; and a gene encoding a type I topoisomerase, topA. Plasmid pXO2-encoded genes include those required for synthesis of a poly-D-glutamic acid capsule, capB, capC, and capA; a gene associated with capsule depolymerization, dep; and a regulatory gene, acpA.

The complete DNA sequences of pXO1 and pXO2 are known (AF065404) (38, 39). The GC content of each plasmid is 33%, similar to that of the B. anthracis chromosome. Sequence annotation suggests that approximately 65% of the plasmid DNA represents coding regions. Together, pXO1 and pXO2 may encode as many as 200 proteins, yet little is known about the expression or function of these potential genes. The only phenotype associated with pXO2 is the ability to synthesize capsule. The presence of pXO1, on the other hand, is associated with a number of phenotypes. In addition to the inability to synthesize toxin, pXO1⁻ strains grow more poorly on certain minimal media, are more sensitive to some bacteriophages, produce less capsule material, and sporulate earlier and at a higher frequency than pXO1⁺ strains (46). Some of these phenotypes are associated with the pXO1-encoded regulatory gene atxA (see below).

pXO1 and pXO2 carry numerous open reading frames that appear to be related to mobile genetic elements. For example, pXO1 has at least 13 open reading frames that are related to insertion sequences or transposition genes, including three homologs of the IS231 sequence commonly associated with the crystal toxin genes of B. thuringiensis (39). The presence of sequences resembling mobile elements suggests the potential for structural diversity among plasmids from different isolates. However, inversion of a 40-kb region in pXO1 is the only report of an altered plasmid structure (47).

Studies of plasmid stability are limited. Spontaneous loss of pXO1 during growth in laboratory media is rare; however, pXO1-cured isolates can be obtained following growth at 43°C (46). Unlike pXO1, pXO2 is lost spontaneously at a relatively high frequency. pXO2⁺ strains occasionally yield rough colonies of noncapsulated cells when streaked to solid media and incubated in appropriate conditions. In one study, approximately one-half of the capsule-negative mutants isolated had lost pXO2. The remaining mutants retained the plasmid but lost the ability to synthesize the capsule (15). pXO2 curing can be induced by growing strains in the presence of novobiocin. Epidemiological investigations have revealed environmental isolates of B. anthracis that are pXO1⁺ pXO2⁻ in addition to strains harboring both virulence plasmids. The significance of plasmid pXO2-cured strains in environmental samples is subject to speculation (51).

Mechanisms of plasmid replication have not been investigated. The origin of replication for each plasmid has been mapped to distinct plasmid DNA fragments (43, 54). However, specific genes associated with plasmid replication have not been identified.

ANTHRAX TOXINS

Unique Variation of the A-B Model

B. anthracis secretes two toxins, lethal toxin (LeTx) and edema toxin (EdTx), that have distinct roles in anthrax disease. Each toxin is composed of two proteins, one with enzymatic activity specific for the toxin and one that functions in receptor binding and translocation of the toxin across the target cell membrane. Protective antigen (PA) (M_r 82,684) is the binding/translocating component of both toxins. The enzymatic component of LeTx is called lethal factor (LF) (M_r 90,237). The enzymatic component of EdTx is termed edema factor (EF) (M_r 89,840). Intradermal coinjection of PA and LF (LeTx) causes death, while coinjection of PA and EF (EdTx) produces edema. None of the three proteins has a toxic effect when administered alone. As its name implies, PA is immunogenic when injected into susceptible animals.

In recent years, the X-ray crystallographic structure of PA, identification of LF activities, and analysis of host responses to LeTx have complemented a wealth of biochemical information. Readers are referred to comprehensive reviews by Leppla (29) and Hanna (21).

PA-Mediated Toxin Entry into Host Cells

The current model for EdTx and LeTx entry is shown in Fig. 1. PA binds with high affinity to an unidentified receptor found on many cell types. Upon binding, PA is cleaved by furin or a furinlike protease, releasing a 20-kDa amino-terminal fragment (PA_{20}). The remaining 63-kDa receptor-bound fragment (PA_{63}) then oligomerizes to form a ring-shaped heptamer. The heptamer binds EF and LF competitively, and the entire complex is internalized and trafficked to the endosome. In the low-pH environment of the endosome, the PA_{63} heptamer undergoes a conformational change such that it converts from a "prepore" to a pore (channel) in the membrane. Subsequently, EF and LF associated with the heptamer translocate the membrane and contact their cytosolic targets (29, 40, 61).

X-ray crystallographic structures of PA and the water-soluble PA_{63} heptamer, investigations of PA_{63}-membrane interactions, and site-directed mutagenesis studies have revealed a detailed understanding of PA structure and function. The PA monomer is organized mainly into antiparallel β-sheets and has four domains, as shown in Fig. 2 (40). PA domain 1 (residues 1 to 249) contains the protease recognition site (Arg-164–Lys–Lys–Arg-167). Cleavage releases PA_{20}, leaving subdomain 1′ as the amino terminus of PA_{63}. Removal of PA_{20} is essential for EF and LF binding, and mutations that eliminate the cleavage site render PA nontoxic (29).

Domain 2 of PA (residues 250 to 487) contains a disordered amphipathic loop that is thought to play a role in insertion of the PA_{63} heptamer into the cell membrane. This large flexible loop projects outward from the side of each monomer within the soluble heptamer. It has been proposed that prepore-to-pore conversion involves a conformational rearrangement of the seven loops of the PA_{63} heptamer such that they combine to form a transmembrane 14-stranded β-barrel (40). Under low-pH conditions, the heptamer forms cation-selective channels in artificial and cell membranes (14, 37). Electrophysiological measurements on membrane-inserted heptamers containing chemically modified residues are consistent with the membrane-spanning β-barrel model (5).

PA domain 3 (residues 488 to 594) carries a hydrophobic patch that is thought to play a role in protein-protein interactions (40). Domain 4 at the carboxy terminus of PA (residues 595 to 735) is required for receptor binding. Short carboxy-terminal deletions of PA reduce or eliminate cell-binding activity. Receptor binding can also be blocked by certain monoclonal antibodies that react with regions within domain 4 (29).

Interaction of PA with EF and LF

EF and LF compete for the same high-affinity binding site on PA_{63}. Monoclonal antibody studies and computer mod-

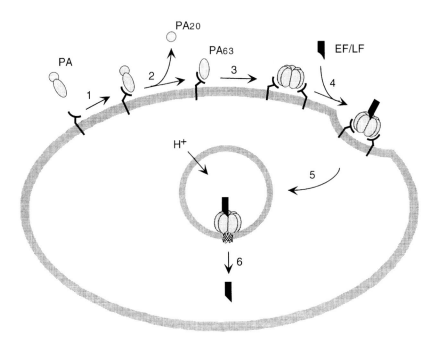

FIGURE 1 Model of anthrax toxin entry into eukaryotic cells. 1, binding of PA to its receptor; 2, proteolytic activation of PA and dissociation of PA_{20}; 3, self-association of monomeric PA_{63} to form the heptameric prepore; 4, binding of EF/LF to the prepore; 5, endocytosis of the receptor/PA_{63}/ligand complex; 6, pH-dependent insertion of PA_{63} and the translocation of the ligand. Reprinted from reference 61.

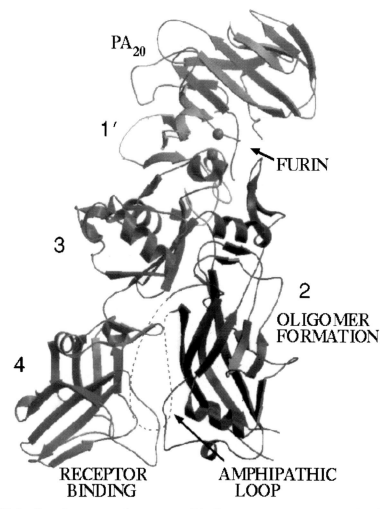

FIGURE 2 Crystal structure of monomeric PA. Domain 1 comprises PA$_{20}$ plus domain 1'. Domains 2 to 4 are as indicated. Two Ca^{2+} ions are shown as spheres. Figure courtesy of R. C. Liddington, University of Leicester.

eling of the PA$_{63}$ heptamer predict that this site is located at the amino-terminal region of PA$_{63}$ (subdomain 1b) and the adjacent domain 3 (29, 40). Biochemical data indicate that each PA$_{63}$ monomer binds one EF or LF, suggesting a maximum of seven LF or EF molecules bound to the ring structure (21).

The amino-terminal 250-amino-acid regions of EF and LF have similar amino acid sequences and hydrophilicity profiles, suggesting that these regions of the proteins interact with PA$_{63}$. Mutagenesis of the amino-terminal region of LF can result in mutants with reduced PA$_{63}$-binding ability. Furthermore, when the amino-terminal domain of LF is genetically fused to heterologous proteins, the fusion proteins are internalized by cells in a PA-dependent manner (29, 36).

LeTx

LeTx is the central effector of shock and death from systemic anthrax and is therefore considered to be the primary virulence factor of *B. anthracis*. Injection of LeTx into susceptible animals produces symptoms nearly identical to those of infection with *B. anthracis* (29). Moreover, the LF gene, *lef*, is required for virulence in a mouse model (41).

Although LeTx is lethal for a number of laboratory animals, the physiological effect of the toxin is most apparent in a Fischer 344 rat model. Following intravenous injection of large amounts of LeTx, the rats succumb to pulmonary edema within 40 min. In vitro, macrophages are uniquely susceptible to the toxin, despite the fact that numerous cell types can bind PA. At high doses, LeTx induces rapid total cytolysis of mouse peritoneal macrophages and macrophage cell lines within 90 min. Thus, macrophage lysis is commonly used as a sensitive and convenient bioassay for LF activity (29).

The mechanism by which LF induces cytolysis of macrophages is unclear. It has been proposed that cytolysis results from overexposure to cytokines or reactive oxygen intermediates. In vitro, sublytic amounts of LeTx induce release of interleukin 1 (IL-1) and tumor necrosis factor from macrophages, and higher levels of LeTx cause overproduction of reactive oxygen intermediates, including superoxide anions (21). Macrophages may be important mediators in toxin-induced shock and death. The elaboration of cytokines could account for the shock induced in systemic anthrax victims. Following toxin challenge, mice are protected from lethality by administration of anti-IL-1 or IL-1 receptor antagonists. Moreover, depletion of macrophages from mice render them insensitive to LeTx challenge (21).

Identification of a specific enzymatic activity and physiological substrate(s) for LF has been an intense area of investigation for many years. LF contains the characteristic zinc-binding motif of metalloproteases, and this motif appears to be essential for toxicity (29). A number of small synthetic peptides can be cleaved by LF in vitro. This activity is zinc dependent and adversely affected by mutations in a thermolysinlike active site of LF (20). LF has also been shown to cleave the amino terminus of mitogen-activated protein kinases (MAPKKs) Mek1 and Mek2. In vivo, cleavage of MAPKKs by LF inactivates the enzymes and inhibits the MAPK signal transduction pathway, a key signaling pathway that leads to activation of transcription factors in the nucleus and other effectors throughout the cell (10, 59).

Considering the fundamental role of the MAPK signaling pathway, it is likely that inhibition of MAPKK activities is important in anthrax pathogenesis. However, the unique sensitivity of macrophages to LF cannot be explained by this enzymatic activity or by other catalytic activities observed in vitro. In another recent study, cultured macrophages from inbred mouse strains were found to vary in their sensitivity to LF. The genetic polymorphism maps to a single gene, *Ltx1*, on mouse chromosome 11. The polymorphism in *Ltx1* influences intoxication events downstream of toxin entry into the cytosol, and it has been proposed that the polymorphism may represent genetic differences in a substrate for LF (42). This finding indicates an additional substrate for LF, as the genes that encode Mek1 and Mek2 map to other loci.

EdTx

The edema observed in cutaneous anthrax infections is attributed to the anthrax EdTx. Subcutaneous administration of EdTx gives rise to edema at the site of inoculation in laboratory animals (29). The contribution of EdTx to systemic anthrax is not clear. In a mouse model for systemic anthrax, strains deleted for the *cya* gene are attenuated only 10-fold (41).

EF is a calmodulin-dependent adenylyl cyclase. Treatment of some cell types with EdTx results in dramatic increases in cyclic AMP, in some cases rising 1,000-fold to reach 2,000 μmol/mg of cell protein. Routine assay for edema factor is performed by measuring in vitro adenylyl cyclase activity (29).

EF-catalyzed increases in cyclic AMP levels disrupt normal second-messenger signaling, potentially resulting in numerous effects on cellular metabolism. EdTx appears to disrupt antibacterial responses of phagocytes. Human neutrophils treated with the toxin have reduced phagocytic and oxidative burst abilities but increased chemotactic responses to N-formyl-Met-Leu-Phe (29). Cultured monocytes treated with EdTx express reduced levels of lipopolysaccharide-inducible tumor necrosis factor alpha but secrete elevated levels of IL-6 (21).

SURFACE STRUCTURES

Capsule

When *B. anthracis* is grown in appropriate conditions (see below), the outermost surface of vegetative cells is covered by a capsule. Encapsulated cells are strongly resistant to phagocytosis, establishing the capsule as an important virulence factor (31). The *B. anthracis* capsule is unusual in that it is a polypeptide composed of gamma-linked alpha-peptide chains of 50 to 100 D-glutamic acid residues (47).

Three pXO2-encoded genes, *capB*, *capC*, and *capA*, encoding proteins of 44, 16, and 46 kDa, respectively, are required for capsule synthesis. Results of hydropathicity analysis and localization of the gene products in *Escherichia coli* minicells indicate that all three proteins are membrane associated (30, 31). However, their functions with regard to capsule synthesis have not been elucidated. Another pXO2-encoded gene, *dep*, encodes a 51-kDa protein that catalyzes the hydrolysis of capsule material into lower-molecular-weight polymers. Although expression of this gene during infection has not been reported, it has been proposed that these low-molecular-weight peptides act to inhibit host defenses (56).

S-Layer

Between the capsule and the cell wall of *B. anthracis* lies a paracrystalline surface layer called the S-layer. The S-layer was first observed on noncapsulated cells, and its formation is independent of capsule synthesis (34). The S-layer is composed of two 94-kDa proteins, EA1 and Sap. EA1 (extractable antigen 1) was identified initially as a vegetative cell protein that reacted strongly with serum from animals vaccinated with a live noncapsulated *B. anthracis* strain (12). EA1 appears to constitute the main S-layer lattice and is the major cell-associated protein (13, 35). Sap was first identified as an abundant secreted protein in culture supernatants of noncapsulated strains. The Sap protein is less tightly associated with the S-layer than EA1 (11). Adherence of the capsule to the cell surface is not dependent upon synthesis of the S-layer proteins. A mutant strain, carrying deletions in both S-layer genes, is fully capsulated (34).

It is not known whether the *B. anthracis* S-layer has a specific role in virulence. S-layers are produced by many bacteria and are believed to function in molecular sieving, cell shape maintenance, and phage fixation. It has been proposed that S-layers produced by pathogenic bacteria affect binding to host cells and protect against complement-mediated killing. It is difficult to imagine such functions for the *B. anthracis* S-layer, because structural and immunological analyses indicate that cells are completely covered by the capsule (34). However, it is not clear whether all *B. anthracis* cells are capsulated at all stages of infection. It is conceivable that the S-layer may have a protective role in noncapsulated cells.

Exosporium

The surface of *B. anthracis* spores is of interest considering that infection usually begins with entry of spores rather than vegetative cells. Electron micrographs of *B. anthracis* indicate the presence of an exosporium, a loose-fitting, diffuse layer covering the spore (18). The chemical composition of the *B. anthracis* exosporium has not been examined. However, a similar structure covering *B. cereus* spores contains proteins, lipids, and carbohydrates (33). Electron micrographs of *B. anthracis* and *B. cereus* spores also show fine filaments of about 720 Å in length and 100 Å in diameter extending out from the exosporium (17). There has been some speculation that the filaments and/or other components of the exosporium may function in uptake or other interactions with host cells. Unfortunately, there has been little functional characterization of these structures.

VIRULENCE GENE EXPRESSION

Host Cues

Toxin and capsule synthesis by *B. anthracis* represents an intriguing example of coordinate expression of virulence genes in response to host-related cues. Expression of the toxin and capsule genes by *B. anthracis* during growth in certain media is enhanced in the presence of bicarbonate or under elevated (5% or greater) atmospheric CO_2. This CO_2 effect on toxin and capsule synthesis is specific and not simply due to the buffering capacity of dissolved bicarbonate during bacterial growth or to decreased oxygen levels (47). Elevated CO_2 is postulated to be a physiologically significant signal during anthrax infection. Concentrations of bicarbonate (about 20 mM) and CO_2 (about 40 mm Hg) in humans are similar to those that activate toxin production during growth in vitro (28).

CO_2-enhanced gene expression is at the level of transcription (3, 7, 27, 44, 58). In the case of the toxin genes, mutants harboring transcriptional *lacZ* and *cat* fusions have been used to monitor relative promoter activity. At late log phase, when toxin gene promoter activity and toxin protein yields are highest, expression of a *pag-lacZ* transcriptional fusion on pXO1 is induced 5- to 8-fold during growth in 5% CO_2 compared with growth in air (0.03% CO_2). Growth in 20% CO_2 increases the transcription up to 19-fold (27).

Growth temperature also affects expression of the toxin genes. When cells are grown in elevated CO_2, promoter activity of the three toxin genes is increased 4- to 6-fold during incubation at 37°C, compared with incubation at 28°C (44). Unlike CO_2, temperature does not appear to globally regulate virulence gene expression; no temperature effects on capsule gene expression have been reported.

Regulatory Genes

All of the known *B. anthracis* virulence genes and their associated regulatory elements are located on pXO1 and pXO2 (Fig. 3). The toxin genes, *cya*, *lef*, and *pag*, encoding EF, LF, and PA, respectively, are located noncontiguously within a 30-kb region of pXO1 (39). Genes associated with synthesis and depolymerization of the D-glutamic acid capsule, *capB*, *capC*, *capA*, and *dep*, are located contiguously and in the same direction of transcription on pXO2 (56). Transposon-mediated insertion mutagenesis of strains harboring single virulence plasmids resulted in identification

of two *trans*-acting positive regulators of virulence genes. The pXO1-encoded gene *atxA* (anthrax toxin activator) activates transcription of the three toxin genes (9, 55). The pXO2-encoded gene *acpA* activates transcription of *capB* (58). It is not known if the capsule genes are transcribed as an operon, and the effect(s) of the regulatory genes on steady-state levels of *capC*, *capA*, and *dep* mRNA(s) has not been determined.

The AtxA and AcpA proteins have amino acid sequence and functional similarities. AtxA and AcpA are of similar size (56 and 55 kDa, respectively), and the predicted amino acid sequences of the proteins are 28% identical and 51% similar throughout. The *atxA* gene cloned on a multicopy plasmid in a pXO2$^+$ strain can activate expression of *capB* in the presence or absence of *acpA*. However, *acpA* cloned on a multicopy plasmid in a pXO1$^+$ strain does not activate toxin gene expression (57). Strains carrying pXO1 and pXO2 produce more capsule than strains harboring only pXO2 (15). Enhanced capsule synthesis by pXO1$^+$ pXO2$^+$ strains is attributed to the presence of *atxA* on pXO1 (16).

The unidirectional crosstalk between *atxA* and *acpA* indicates that *acpA* has a higher specificity for target genes compared with *atxA*. Indeed, *atxA* appears to have a global effect on gene expression. Screens for additional *atxA*-dependent and CO_2-enhanced genes have revealed numerous loci on pXO1 that appear to contain promoters that are regulated coordinately with the toxin genes. Moreover, an *atxA*-null strain exhibits poor growth on certain minimal media, as does a pXO1-cured strain (23).

The *atxA* gene is essential for virulence of *B. anthracis* in a mouse model. Inoculation of mice with high doses of spores of the toxigenic noncapsulated Sterne strain causes a lethal disease (60). An *atxA*-null mutant of the Sterne strain is avirulent in mice. Moreover, the antibody response to all three toxin proteins is decreased significantly in *atxA*-null infected mice, compared with those infected with the parent strain. Therefore, *atxA* appears to regulate toxin gene expression during infection (9). The role of *acpA* in virulence has not been addressed experimentally.

The mechanism(s) by which the two regulatory proteins activate gene expression is not known. Although it has been suggested that the predicted amino acid sequences of AtxA and AcpA indicate helix-turn-helix DNA-binding motifs (58), specific DNA-binding ability has not been demonstrated for either regulator. Minimal upstream sequences required for activation of *pag* and *capB* have been determined (9, 57). Comparison of the promoter regions of these genes and other *atxA*-activated genes does not reveal obvious sequence similarities. However, the AT-rich nature of the genome (approximately 67% for the plasmids and the chromosome) makes such comparisons difficult.

The relationship between CO_2-enhanced virulence gene expression and regulatory gene function is not clear. The *atxA* gene is essential for transcription from the unique start sites of *cya* and *lef* and for transcription from the major start site, P1, of the *pag* gene. AtxA can activate toxin gene expression in air-grown cells, although the steady-state levels of *cya*, *lef*, and *pag* mRNA transcripts are significantly increased during growth in elevated CO_2 (9, 27). Expression of the *atxA* gene is unaffected by the CO_2 signal; steady-state levels of *atxA* mRNA and protein do not differ in pXO1$^+$ cells cultured in the two growth conditions (8). Thus, CO_2 may affect the function of AtxA or the function or expression of some other regulator. Conversely, expression of the *acpA* gene is increased in pXO2$^+$ cells

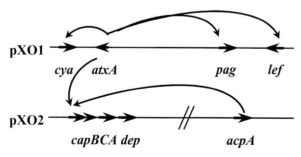

FIGURE 3 Model depicting *B. anthracis* virulence gene regulation.

grown in elevated CO_2 (58), suggesting that AcpA is limiting for capsule gene expression in cells grown in air.

trans-acting factors other than AtxA and AcpA may affect virulence gene expression. Transposon mutagenesis of a pXO1$^+$ strain resulted in a mutant that synthesizes all three toxin proteins in the absence of elevated CO_2. It has been proposed that the mutant lacks a repressor protein that binds to toxin gene promoter DNA in the absence of the CO_2 signal (55). However, the gene for the putative repressor has not been identified. In addition, overexpression of *atxA* appears to reduce *pag* expression, suggesting that some other factor may be titrated by excess AtxA. If an additional *trans*-acting factor exists, it cannot be titrated by adding more target promoter sequences. When the *pag* promoter was cloned on a multicopy plasmid and introduced into a pXO1$^+$ strain, no change in PA synthesis was observed (45).

Expression of the *pag* gene in *Bacillus subtilis* has led to some interesting observations. The *atxA* gene activates expression of a transcriptional *pag-lacZ* fusion when the genes are cloned on separate multicopy plasmids in *B. subtilis* (55). Expression of a *pag-lacZ* fusion is also enhanced in *abrB* and *ccpA* mutants of *B. subtilis*, implicating growth phase regulation and catabolic repression of *pag* in this species (2). It is not known whether homologs of these genes have similar functions in *B. anthracis*.

GENETIC TOOLS

Plasmids and Antibiotic-Resistance Markers Used in *B. anthracis*

Several plasmid vectors derived from *Staphylococcus aureus* plasmids pUB110, pC194, and pE194 and *B. cereus* plasmid pBC16 can replicate and express antibiotic resistance in *B. anthracis*. Most *B. anthracis* strains are sensitive to tetracycline, kanamycin (and neomycin), chloramphenicol, spectinomycin, and MLS (macrolide-lincosamide-streptogramin B) antibiotics, making selection for plasmid-borne markers encoding the appropriate resistance determinants feasible.

Electroporation

B. anthracis is not transformable by methods commonly used for other gram-positive organisms. However, plasmids can be electroporated into *B. anthracis* at frequencies up to 10^4 transformants per μg of DNA. Electroporation frequencies are highest when plasmids are isolated from DNA methyltransferase-deficient (*dam*) *E. coli* strains or from *B. subtilis* 168, a strain that is not known to possess adenine-methylating activities. These observations suggest that *B. anthracis* has a restriction system dependent on a specific pattern of adenine methylation for the recognition of foreign DNA (32). There are no reports of electroporation of linear DNA fragments into *B. anthracis*.

Transduction and Conjugation

Genetic transfer between *B. anthracis* and the closely related species *B. cereus* and *B. thuringiensis* can occur by transduction and conjugation. Bacteriophage CP-51 can transduce various plasmids among these species and is useful for transduction of chromosomal markers between *B. anthracis* strains (47). CP-51-mediated transduction of pXO2 from *B. anthracis* to *B. cereus* was instrumental in demonstrating that the plasmid was associated with capsule

synthesis (15). *B. thuringiensis* plasmids pXO11 and pXO12 are self-transmissible and transfer to *B. cereus*, *B. anthracis*, and other *B. thuringiensis* subspecies. *B. anthracis* strains harboring pXO12 can transfer pXO1 and pXO2 to cured *B. anthracis* strains by conduction involving the transposon Tn*4430*, which is found on the conjugative plasmid. Conjugative plasmid transfer between *B. subtilis* and *B. anthracis* has also been reported. *B. subtilis natto* plasmid pLS20 facilitates the transfer of the tetracycline-resistance plasmid pBC16 from *B. subtilis* to *B. anthracis* (47).

Plasmids can also be transferred directly from *E. coli* to *B. anthracis* using a recombinant conjugal system. Shuttle vectors containing replication origins and selectable markers functional in both species and the transfer origin of the IncP plasmid RK2 can be mobilized by self-transmissible IncP plasmids coresident in *E. coli* donors (48).

Gene Insertions and Replacements

It is possible to generate specific insertions and gene replacements in *B. anthracis* by homologous recombination. Integrative vectors, carrying *B. anthracis* DNA sequences and lacking a replication origin functional in *B. anthracis*, can be transferred from *E. coli* to *B. anthracis* with selection for plasmid-encoded markers using the RK2 mating system. This strategy was used to disrupt each of the anthrax toxin genes on pXO1, creating important strains for studies of toxin gene function (41).

Specific DNA insertions into the *B. anthracis* genome can also be achieved using pE194-derived plasmids, which are temperature sensitive for replication. Constructs carrying *B. anthracis* DNA sequences can be electroporated into *B. anthracis* with initial selection at 32°C, the permissive temperature. Mutants carrying the plasmid inserted into the *B. anthracis* genome can be isolated following growth at 43°C (6). Unfortunately, *B. anthracis* will not grow at temperatures exceeding 43°C, and incomplete curing of pE194-derived vectors at this temperature makes this system relatively inefficient.

Finally, the *E. coli*–*B. anthracis* shuttle vector pUTE29 can be used to generate gene replacements and insertions. Derivatives of pUTE29 can be electroporated into *B. anthracis* with selection for the plasmid-encoded tetracycline-resistance marker. However, the plasmid is readily lost from *B. anthracis* in the absence of selection. When DNA sequences flanking a *B. anthracis* gene are cloned flanking a selectable marker in pUTE29, the marker can be introduced into the *B. anthracis* genome by a double-crossover recombination event. This method has been particularly useful for construction of strains deleted for specific genes (9).

Random Mutagenesis

Point mutants of *B. anthracis* can be isolated following UV mutagenesis of spores. Rifampicin- and streptomycin-resistant mutants and numerous amino acid, purine, and pyrimidine auxotrophs have been obtained by this method (4). Insertion mutations can be generated using *Enterococcus faecalis* transposons Tn*916* and Tn*917*. Each method has disadvantages. Low electroporation and *E. coli*–*B. anthracis* mating frequencies make it difficult to screen genomic libraries for clones that complement point mutants. Tn*916* insertion mutants are unstable in the absence of selection for the tetracycline-resistance gene encoded by the transposon (24). Tn*917* insertion mutants are stable in the absence of selection. However, insertions can be accompanied by local deletions and rearrangements of *B. an*-

thracis DNA (27). In *B. anthracis* strains harboring either pXO1 or pXO2, Tn*917* transposes preferentially to the virulence plasmids relative to the chromosome for reasons that are not clear (23).

Measuring Gene Expression

Transcriptional fusions employing promoterless genes for *β*-galactosidase and chloramphenicol acetyltransferase have been used in *B. anthracis* to monitor gene expression (3, 44). Measurements of specific galactosidase activity associated with these strains must be interpreted with caution. Specific enzyme activity of *B. anthracis* cells producing high levels of *β*-galactosidase can be highly variable when cells are grown in liquid media. Autolysis of the cells can lead to release of inactive *β*-galactosidase into the culture medium (45). Methods for RNA extraction, Northern hybridization, primer extension, and RNase protection have been developed for *B. anthracis* (27, 49).

ANTHRAX VACCINES

The Sterne Strain

The current veterinary vaccine for anthrax is a spore preparation of an attenuated *B. anthracis* strain isolated by M. Sterne in 1937. The Sterne strain and its derivatives have been used worldwide for more than 50 years. This strain produces the anthrax toxin proteins but is noncapsulated owing to the absence of pXO2. Although the vaccine is effective, there are safety concerns. The strain has a low level of virulence in some animal species, and it can occasionally result in necrosis at the injection site (19).

PA

The human vaccine for anthrax is a noncellular vaccine composed primarily of PA. The vaccine is produced in the United States and the United Kingdom using similar methods. PA is obtained from a toxigenic noncapsulated strain by adsorbing culture supernatant to aluminum hydroxide or precipitating with alum (47). Efficacy of the vaccine has been established in studies of laboratory animals and in a field study of an exposed and susceptible human population. However, vaccination requires a series of doses and repeated boosters to maintain immunity. Rare adverse reactions have been reported. More important, in some laboratory animals, this vaccine does not provide protection against all virulent *B. anthracis* strains (19).

Results of animal studies indicate that the Sterne live spore vaccine is more protective than the human PA vaccine. Numerous explanations have been proposed for this observation. The antigenic structure of purified PA may differ from that of PA produced in vivo. If a critical sustained level of PA is required for maximum immunity, this may only be achieved by the live vaccine. *B. anthracis* antigens other than PA may supplement the immune response to PA. And finally, cellular immunity may contribute to protection, and the live vaccine would stimulate this to a higher degree than the PA vaccine (19).

FUTURE DIRECTIONS

Dramatic progress has been made regarding the structure and function of the anthrax toxin proteins, and advances are likely to continue at a rapid pace. Further investigations of LF activity will facilitate development of toxin inhibitors

useful in treatment of the disease. Current understanding of the mechanism of PA-mediated translocation across membranes has already spurred development of PA-mediated protein delivery systems. Such systems have potential therapeutic applications and could serve as novel vaccines. Identification of the cellular receptor for PA will be important for these studies and fill a gap in current understanding of toxin entry.

The molecular biology of *B. anthracis* has also advanced rapidly, allowing facile identification of the organism. Progress in this area, accompanied by information regarding the genomes of *B. cereus* and *B. thuringiensis*, is likely to fuel renewed interest in genetic exchange between these closely related species.

Much information has been gained regarding virulence gene regulation, but mechanisms for controlled expression of virulence genes are most likely complex. The functions of proteins encoded by regulatory genes are not known, and additional regulatory elements have been implicated but not identified. Patterns of gene expression in response to changing environmental stimuli during the course of infection are unknown. Genes required for the initial stages of infection such as attachment and germination (not necessarily in that order) remain to be identified.

The surface structure of *B. anthracis* has received renewed interest, leading to identification of the *B. anthracis* S-layer proteins. Surprisingly, the roles of the Cap proteins in capsule biosynthesis have not been pursued. Investigation of the surface structures of cells and spores may have implications for initial interactions between *B. anthracis* and the host organism.

Work in the author's laboratory was supported by Public Health Service grant AI33537 from the National Institutes of Health. I thank Anthony Costa, Alex Hoffmaster, and Edward Nikonowicz for critical reading of the manuscript.

REFERENCES

1. **Ash, C., J. A. Farrow, M. Dorsch, E. Stackebrandt, and M. D. Collins.** 1991. Comparative analysis of *Bacillus anthracis*, *Bacillus cereus*, and related species on the basis of reverse transcriptase sequencing of 16S rRNA. *Int. J. Syst. Bacteriol.* **41:**343–346.

2. **Baillie, L., A. Moir, and R. Manchee.** 1998. The expression of the protective antigen of *Bacillus anthracis* in *Bacillus subtilis*. *J. Appl. Microbiol.* **84:**741–746.

3. **Bartkus, J. M., and S. H. Leppla.** 1989. Transcriptional regulation of the protective antigen gene of *Bacillus anthracis*. *Infect. Immun.* **57:**2295–2300.

4. **Battisti, L., B. D. Green, and C. B. Thorne.** 1985. Mating system for transfer of plasmids among *Bacillus anthracis*, *Bacillus cereus*, and *Bacillus thuringiensis*. *J. Bacteriol.* **162:** 543–550.

5. **Benson, E. L., P. D. Huynh, A. Finkelstein, and R. J. Collier.** 1998. Identification of residues lining the anthrax protective antigen channel. *Biochemistry* **37:**3941–3948.

6. **Brown, D. P., L. Ganova-Raeva, B. D. Green, S. R. Wilkinson, M. Young, and P. Youngman.** 1994. Characterization of *spo0A* homologues in diverse *Bacillus* and *Clostridium* species identifies a probable DNA-binding domain. *Mol. Microbiol.* **14:**411–426.

7. **Cataldi, A., A. Fouet, and M. Mock.** 1992. Regulation of *pag* gene expression in *Bacillus anthracis*: use of a *pag*-

lacZ transcriptional fusion. *FEMS Microbiol. Lett.* **98:**89–94.

8. **Dai, Z., and T. M. Koehler.** 1997. Regulation of anthrax toxin activator gene (*atxA*) expression in *Bacillus anthracis*: temperature, not CO₂/bicarbonate, affects AtxA synthesis. *Infect. Immun.* **65:**2576–2582.

9. **Dai, Z., J.-C. Sirard, M. Mock, and T. M. Koehler.** 1995. The *atxA* gene product activates transcription of the anthrax toxin genes and is essential for virulence. *Mol. Microbiol.* **16:**1171–1181.

10. **Duesbery, N. S., C. P. Webb, S. H. Leppla, V. M. Gordon, K. R. Klimpel, T. D. Copeland, N. G. Ahn, M. K. Oskarsson, K. Fukasawa, K. D. Paull, and G. F. Vande Woude.** 1998. Proteolytic inactivation of MAP-kinase-kinase by anthrax lethal factor. *Science* **280:**734–737.

11. **Etienne-Toumelin, I., J. C. Sirard, E. Duflot, M. Mock, and A. Fouet.** 1995. Characterization of the *Bacillus anthracis* S-layer: cloning and sequencing of the structural gene. *J. Bacteriol.* **177:**614–620.

12. **Ezzell, J. W., Jr., and T. G. Abshire.** 1988. Immunological analysis of cell-associated antigens of *Bacillus anthracis*. *Infect. Immun.* **56:**349–356.

13. **Farchaus, J. W., W. J. Ribot, M. B. Downs, and J. W. Ezzell.** 1995. Purification and characterization of the major surface array protein from the avirulent *Bacillus anthracis* Delta Sterne-1. *J. Bacteriol.* **177:**2481–2489.

14. **Finkelstein, A.** 1994. The channel formed in planar lipid bilayers by the protective antigen component of anthrax toxin. *Toxicology* **87:**29–41.

15. **Green, B. D., L. Battisti, T. M. Koehler, and C. B. Thorne.** 1985. Demonstration of a capsule plasmid in *Bacillus anthracis*. *Infect. Immun.* **49:**291–297.

16. **Guignot, J., M. Mock, and A. Fouet.** 1997. AtxA activates the transcription of genes harbored by both *Bacillus anthracis* virulence plasmids. *FEMS Microbiol. Lett.* **147:**203–207.

17. **Hachisuka, Y., K. Kojima, and T. Sato.** 1966. Fine filaments on the outside of the exosporium of *Bacillus anthracis* spores. *J. Bacteriol.* **91:**2382–2384.

18. **Hachisuka, Y., S. Kozuka, and M. Tsujikawa.** 1984. Exosporia and appendages of spores of *Bacillus* species. *Microbiol. Immunol.* **28:**619–624.

19. **Hambleton, P., and P. C. Turnbull.** 1990. Anthrax vaccine development: a continuing story. *Adv. Biotechnol. Processes* **13:**105–122.

20. **Hammond, S. E., and P. C. Hanna.** 1998. Lethal factor active-site mutations affect catalytic activity in vitro. *Infect. Immun.* **66:**2374–2378.

21. **Hanna, P.** 1998. Anthrax pathogenesis and host response. *Curr. Top. Microbiol. Immunol.* **225:**13–35.

22. **Harrell, L. J., G. L. Andersen, and K. H. Wilson.** 1995. Genetic variability of *Bacillus anthracis* and related species. *J. Clin. Microbiol.* **33:**1847–1850.

23. **Hoffmaster, A. R., and T. M. Koehler.** 1997. The anthrax toxin activator gene *atxA* is associated with CO₂-enhanced non-toxin gene expression in *Bacillus anthracis*. *Infect. Immun.* **65:**3091–3099.

24. **Ivins, B. E., S. L. Welkos, G. B. Knudson, and S. F. Little.** 1990. Immunization against anthrax with aromatic compound-dependent (Aro⁻) mutants of *Bacillus anthracis* and with recombinant strains of *Bacillus subtilis* that pro-

duce anthrax protective antigen. *Infect. Immun.* **58:**303–308.

25. **Jackson, P. J., E. A. Walthers, A. S. Kalif, K. L. Richmond, D. M. Adair, K. K. Hill, C. R. Kuske, G. L. Andersen, K. H. Wilson, M. Hugh-Jones, and P. Keim.** 1997. Characterization of the variable-number tandem repeats in *vrrA* from different *Bacillus anthracis* isolates. *Appl. Environ. Microbiol.* **63:**1400–1405.

26. **Keim, P., A. Kalif, J. Schupp, K. Hill, S. E. Travis, K. Richmond, D. M. Adair, M. Hugh-Jones, C. R. Kuske, and P. Jackson.** 1997. Molecular evolution and diversity in *Bacillus anthracis* as detected by amplified fragment length polymorphism markers. *J. Bacteriol.* **179:**818–824.

27. **Koehler, T. M., Z. Dai, and M. Kaufman-Yarbray.** 1994. Regulation of the *Bacillus anthracis* protective antigen gene: CO₂ and a *trans*-acting element activate transcription from one of two promoters. *J. Bacteriol.* **176:**586–595.

28. **Lentner, C.** 1981. Units of measurement, body fluids, composition of the body, nutrition. *In Geigy Scientific Tables*, vol. 1. Ciba Geigy, Basel, Switzerland.

29. **Leppla, S. H.** 1995. Anthrax toxins, p. 543–572. *In* J. Moss, B. Iglewski, M. Vaughan, and A. T. Tu (ed.), *Bacterial Toxins and Virulence Factors in Disease.* Marcel Dekker, Inc., New York, N.Y.

30. **Makino, S., C. Sasakawa, I. Uchida, N. Terakado, and M. Yoshikawa.** 1988. Cloning and CO₂-dependent expression of the genetic region for encapsulation from *Bacillus anthracis.* *Mol. Microbiol.* **2:**371–376.

31. **Makino, S.-I., I. Uchida, N. Terakado, C. Sasakawa, and M. Yoshikawa.** 1989. Molecular characterization and protein analysis of the *cap* region, which is essential for encapsulation in *Bacillus anthracis.* *J. Bacteriol.* **171:**722–730.

32. **Marrero, R., and S. L. Welkos.** 1995. The transformation frequency of plasmids into *Bacillus anthracis* is affected by adenine methylation. *Gene* **152:**75–78.

33. **Matz, L. L., T. C. Beaman, and P. Gerhardt.** 1970. Chemical composition of exosporium from spores of *Bacillus cereus.* *J. Bacteriol.* **101:**196–201.

34. **Mesnage, S., E. Tosi-Couture, P. Gounon, M. Mock, and A. Fouet.** 1998. The capsule and S-layer: two independent and yet compatible macromolecular structures in *Bacillus anthracis.* *J. Bacteriol.* **180:**52–58.

35. **Mesnage, S., E. Tosi-Couture, M. Mock, P. Gounon, and A. Fouet.** 1997. Molecular characterization of the *Bacillus anthracis* main S-layer component: evidence that it is the major cell-associated antigen. *Mol. Microbiol.* **23:**1147–1155.

36. **Milne, J. C., S. R. Blanke, P. C. Hanna, and R. J. Collier.** 1995. Protective antigen-binding domain of anthrax lethal factor mediates translocation of a heterologous protein fused to its amino- or carboxy-terminus. *Mol. Microbiol.* **15:**661–666.

37. **Milne, J. C., and R. J. Collier.** 1993. pH-dependent permeabilization of the plasma membrane of mammalian cells by anthrax protective antigen. *Mol. Microbiol.* **10:**647–653.

38. **Okinaka, R.** 1999. Personal communication.

39. **Okinaka, R. T., K. Cloud, O. Hampton, A. Hoffmaster, K. K. Hill, P. Keim, T. M. Koehler, G. Lamke, S. Kumano, J. Mahillon, D. Manter, Y. Martinez, D. Ricke, R. Svensson, and P. J. Jackson.** Unpublished data.

40. Petosa, C., R. J. Collier, K. R. Klimpel, S. H. Leppla, and R. C. Liddington. 1997. Crystal structure of the anthrax toxin protective antigen. *Nature* **385:**833–838.

41. Pezard, C., P. Berche, and M. Mock. 1991. Contribution of individual toxin components to virulence of *Bacillus anthracis. Infect. Immun.* **59:**3472–3477.

42. Roberts, J. E., J. W. Watters, J. D. Ballard, and W. F. Dietrich. 1998. Ltx1, a mouse locus that influences the susceptibility of macrophages to cytolysis caused by intoxication with *Bacillus anthracis* lethal factor, maps to chromosome 11. *Mol. Microbiol.* **29:**581–591.

43. Robertson, D. L., T. S. Bragg, S. Simpson, R. Kaspar, W. Xie, and M. T. Tippetts. 1990. Mapping and characterization of *Bacillus anthracis* plasmids pXO1 and pXO2. *Salisbury Med. Bull.* **68**(Spec. Suppl.):55–58.

44. Sirard, J.-C., M. Mock, and A. Fouet. 1994. The three *Bacillus anthracis* toxin genes are coordinately regulated by bicarbonate and temperature. *J. Bacteriol.* **176:**5188–5192.

45. Sirard, J.-C., M. Mock, and A. Fouet. 1995. Molecular tools for the study of transcriptional regulation in *Bacillus anthracis. Res. Microbiol.* **146:**729–737.

46. Thorne, C. B. 1985. Genetics of *Bacillus anthracis*, p. 56–62. *In* L. Leive (ed.), *Microbiology.* American Society for Microbiology, Washington, D.C.

47. Thorne, C. B. 1993. *Bacillus anthracis*, p. 113–124. *In* A. L. Sonenshein, J. A. Hoch, and R. Losick (ed.), *Bacillus subtilis and Other Gram-Positive Bacteria: Biochemistry, Physiology, and Molecular Genetics.* American Society for Microbiology, Washington, D.C.

48. Trieu-Cuot, P., C. Carlier, C. Poyart-Salmeron, and P. Courvalin. 1991. Shuttle vectors containing a multiple cloning site and a lacZ alpha gene for conjugal transfer of DNA from *Escherichia coli* to gram-positive bacteria. *Gene* **102:**99–104.

49. Tsui, H.-C. T., A. J. Pease, T. M. Koehler, and M. E. Winkler. 1994. Detection and quantitation of RNA transcribed from bacterial chromosomes and plasmids, p. 179–204. *In* K. W. Adolph (ed.), *Molecular Microbiology Techniques*, Part A. Academic Press, Inc., San Diego, Calif.

50. Turnbull, P. C. 1991. Anthrax vaccines: past, present and future. *Vaccine* **9:**533–539.

51. Turnbull, P. C., R. A. Hutson, M. J. Ward, M. N. Jones, C. P. Quinn, N. J. Finnie, C. J. Duggleby, J. M. Kramer, and J. Melling. 1992. *Bacillus anthracis* but not always anthrax. *J. Appl. Bacteriol.* **72:**21–28.

52. Turnbull, P. C. B. 1991. *Bacillus*, p. 233–245. *In* S. Baron (ed.), *Medical Microbiology*, 4th ed. The University of Texas Medical Branch at Galveston.

53. Turnbull, P. C. B. (ed.). 1996. *Proceedings of the International Workshop on Anthrax*, vol. 87, Special Supplement. Salisbury Medical Society, Salisbury, U.K.

54. Uchida, I., K. Hashimoto, S. Makino, C. Sasakawa, M. Yoshikawa, and N. Terakado. 1987. Restriction map of a capsule plasmid of *Bacillus anthracis. Plasmid* **18:**178–181.

55. Uchida, I., J. M. Hornung, C. B. Thorne, K. R. Klimpel, and S. H. Leppla. 1993. Cloning and characterization of a gene whose product is a *trans*-activator of anthrax toxin synthesis. *J. Bacteriol.* **175:**5329–5338.

56. Uchida, I., S. Makino, C. Sasakawa, M. Yoshikawa, C. Sugimoto, and N. Terakado. 1993. Identification of a novel gene, *dep*, associated with depolymerization of the capsular polymer in *Bacillus anthracis. Mol. Microbiol.* **9:**487–496.

57. Uchida, I., S. Makino, T. Sekizaki, and N. Terakado. 1997. Cross-talk to the genes for *Bacillus anthracis* capsule synthesis by atxA, the gene encoding the *trans*-activator of anthrax toxin synthesis. *Mol. Microbiol.* **23:**1229–1240.

58. Vietri, N. J., R. Marrero, T. A. Hoover, and S. L. Welkos. 1995. Identification and characterization of a *trans*-activator involved in the regulation of encapsulation by *Bacillus anthracis. Gene* **152:**1–9.

59. Vitale, G., R. Pellizzari, C. Recchi, G. Napolitani, M. Mock, and C. Montecucco. 1998. Anthrax lethal factor cleaves the N-terminus of MAPKKs and induces tyrosine/threonine phosphorylation of MAPKs in cultured macrophages. *Biochem. Biophys. Res. Commun.* **248:**706–711.

60. Welkos, S. L., and A. M. Friedlander. 1988. Pathogenesis and genetic control of resistance to the Sterne strain of *Bacillus anthracis. Microb. Pathog.* **4:**53–69.

61. Wesche, J., J. L. Elliott, P. O. Falnes, S. Olsnes, and R. J. Collier. 1998. Characterization of membrane translocation by anthrax protective antigen. *Biochemistry* **37:**15737–15746.

Clostridial Genetics

DENA LYRAS AND JULIAN I. ROOD

55

The genus *Clostridium* consists of an extremely diverse group of primarily gram-positive bacteria that have traditionally been grouped together based on their anaerobic growth requirements and their ability to produce heat-resistant endospores. Consequently, the different members of the genus are very dissimilar and the genus lacks phylogenetic coherence. There are 120 different species within the genus, 35 of which can be considered capable of causing disease in humans or animals (44).

The most common feature of the pathogenic clostridia is that the cell and tissue damage that they cause primarily results from the production of potent extracellular toxins. Although somewhat artificial in that it crosses species boundaries, it is useful to divide the pathogenic clostridia into three major groups based upon their resultant disease pathology. These groups consist of the neurotoxic clostridia, which produce toxins that affect the nervous system; the enterotoxic clostridia, which produce toxins that affect the gastrointestinal tract; and the histotoxic clostridia, whose necrotic pathology results from the production of one or more toxins that affect the structural and functional integrity of host cells located at or near the site of infection. The division into these three groups is used in the clostridial chapters of this volume rather than a species-specific approach because it leads to a more unified understanding of the pathogenesis process. This chapter focuses on the genetics of the pathogenic clostridia, dealing exclusively with the major clostridial pathogens *Clostridium perfringens*, *Clostridium difficile*, *Clostridium botulinum*, and *Clostridium tetani*.

PHYLOGENETICS OF THE PATHOGENIC CLOSTRIDIA

The initial application of homology studies using 23S rRNA molecules led to the division of 56 clostridial species into four major groups (24). Groups I and II consisted of well-defined species with a low (24 to 32%) percent G + C content; group III consisted of low percent G + C species that did not fit into the other groups; and group IV consisted of high (41 to 45%) percent G+C organisms.

The major clostridial pathogens *C. perfringens* and *C. botulinum* belonged to group I, whereas *C. tetani* was a group II species. Subsequent studies involving sequence analysis of 16S rRNA molecules have confirmed these findings and have shown that, based on phylogenetic analysis, many members of the genus *Clostridium* are closely related to other bacteria (10, 44). The original group I organisms (24) are primarily located in 16S rRNA cluster I (Fig. 1), and it has been suggested that these organisms, which include *C. perfringens*, *Clostridium novyi*, *C. tetani*, and *C. botulinum*, are the true members of the genus *Clostridium*. Of the other major pathogens, *Clostridium septicum* and *Clostridium chauvoei* belong to cluster II, which may well remain within the genus *Clostridium* (44). However, *C. difficile* and *Clostridium sordellii* belong to cluster XI. These organisms should eventually be reclassified and designated as belonging to another genus. However, apart from the confusion that such changes in nomenclature will cause among diagnostic microbiologists and clinicians, there is another problem in that cluster I includes the type species of the genus *Sarcina*. As the older name would, according to taxonomic convention, have precedence, this is clearly an untenable situation (10, 44).

THE GENETICS OF *C. BOTULINUM*

Toxin-Encoding Bacteriophages and Plasmids of *C. botulinum*

There are seven distinct toxin types of *C. botulinum*, the causative agent of both human and animal botulism. These types are distinguished by their ability to produce antigenically distinct botulinum neurotoxins (BoNT). Phylogenetically, these isolates represent at least three quite distinct strains (Fig. 1), which in any other genus would be classified as separate species. However, because of the disease significance of these organisms, clinical considerations have been allowed to override phylogenetics and all of these isolates are still designated as *C. botulinum*.

The nonproteolytic *C. botulinum* types B, E, and F represent a distinct grouping of isolates that have almost 100%

Gram-Positive Pathogens, ed. by V. A. Fischetti et al.
© 2000 American Society for Microbiology, Washington, D.C.

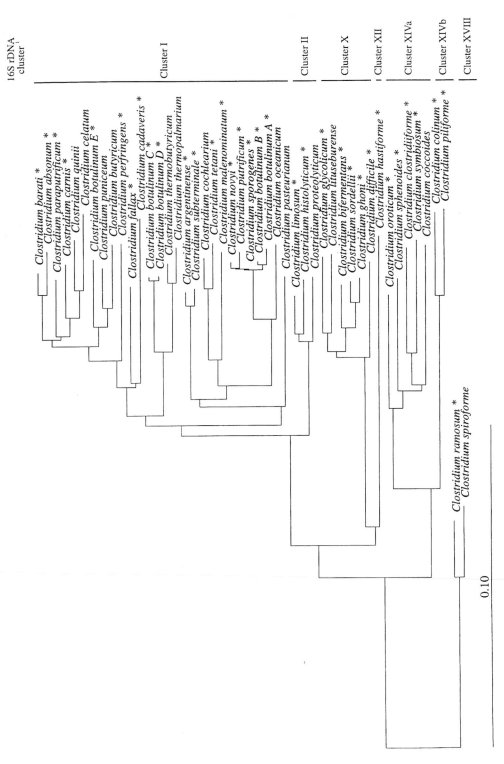

FIGURE 1 Phylogenetic relationships of the pathogenic clostridia. The scale bar indicates 10 base changes per 100 nucleotides. Reproduced from Stackebrandt and Rainey (44) with the kind permission of the authors and Academic Press.

16S rRNA sequence identity to toxigenic strains of *Clostridium butyricum* (10). By contrast, the proteolytic type A and F strains are virtually indistinguishable from *Clostridium putrificum* and *Clostridium sporogenes* by phylogenetic analysis. *C. botulinum* types C and D and *C. novyi* also are essentially identical species on phylogenetic grounds, whereas *C. botulinum* type G strains are quite distinct, being very closely related to *Clostridium subterminale* (10). Note that none of these *C. botulinum* toxin types are closely related to *C. tetani*.

This apparent phylogenetic chaos becomes more understandable when the location of the genes encoding the various botulinum neurotoxins is analyzed. The genes encoding the various BoNT proteins are located in a gene cluster that also includes the genes encoding a hemagglutinin protein and a nontoxic nonhemagglutinin protein as well as several other open reading frames and a putative regulatory gene (21). It has been known for some time that the genes encoding the BoNT/C and BoNT/D toxins, and the associated gene cluster, are located on pseudolysogenic bacteriophage that are not integrated into the chromosome but are maintained in an extrachromosomal state (21, 22). Curing of these bacteriophage leads to the concomitant loss of the ability to produce the BoNT/C or BoNT/D toxins and their associated proteins such as hemagglutinin protein and nontoxic nonhemagglutinin proteins, a property that is restored upon reinfection. In addition, type C and type D strains of *C. botulinum* that have been cured of their toxigenic bacteriophage are indistinguishable from type A strains of *C. novyi* that have been cured of a bacteriophage that carries the *C. novyi* alpha-toxin gene. Strains can be readily interconverted between *C. botulinum* types C and D and *C. novyi* type A by infection with the respective BoNT/C-, BoNT/D-, and *C. novyi* alpha-toxin-specific bacteriophages. This points out the phylogenetic inconsistency of the current classification system.

C. botulinum type C and D strains also produce the C3 exoenzyme, an extracellular ADP-ribosyltransferase that acts on the mammalian Rho protein. The toxigenic bacteriophage that carry the BoNT/C and BoNT/D genes also carry the C3 structural gene. This gene is located on an unstable 21.5-kb fragment that has flanking directly repeated 6-bp Tn554-like core motifs. Deletion of the 21.5-kb region leads to the loss of one of these repeats, but it is not known if deletion results from site-specific or homologous recombination events. These observations have prompted the suggestion, for which there is no direct experimental evidence, that the C3 gene may be located on a transposable genetic element (20).

Bacteriophages have also been identified in type A, B, E, and F strains of *C. botulinum*, but in these isolates they are integrated into the chromosome and are difficult to cure (21). There is no evidence that toxin production in these toxin types is bacteriophage encoded (22).

Plasmids have been reported in almost all *C. botulinum* toxin types, but with one exception these plasmids do not encode toxin genes (22). The major exception involves type G strains, in which it has been clearly shown that the BoNT/G gene is located on a 114-kb plasmid that also encodes the hemagglutinin and nontoxic nonhemagglutinin proteins and the ability to produce the bacteriocin boticin G. In addition, although there is no evidence that the BoNT/E gene is plasmid determined in *C. botulinum*, a BoNT/E homolog that appears to be encoded on a large plasmid has been identified in *C. butyricum* (19).

Genetic Manipulation of *C. botulinum*

Although methods for the genetic manipulation of *C. perfringens* are well established, it has taken some time for similar methods to be developed for the neurotoxic clostridia. The enterococcal tetracycline-resistance transposon Tn916 has been shown to encode its own conjugative transfer from *Enterococcus faecalis* to *C. botulinum* at low but acceptable frequencies. The transposon appears to insert at several different sites in the *C. botulinum* chromosome, and multiple insertions are common. Several auxotrophic mutants have been isolated. In addition, Tn916 mutagenesis of *C. botulinum* type A strain 62A has been used to isolate mutants defective in their ability to produce BoNT/A (23).

Pulsed-field gel electrophoresis of *Mlu*I-, *Sma*I-, and *Rsr*II-digested DNA has been used to determine the size of the genome of strain 62A, but the 4.04-Mb genome has not been mapped (27). Fourteen Tn916 insertion mutants of strain 62A were also mapped by pulsed-field gel electrophoresis, with 5 of 14 transconjugants containing single Tn916 insertions at different sites in the chromosome. Tn916-mediated deletions were also observed (27), and the Tn916-derived Tox⁻ mutant LNT01 has been shown to contain a deletion of ca. 70 kb in the toxin gene region (23).

A transformation method has been developed for the plasmid-free *C. botulinum* type A strain Hall A, but it is relatively inefficient, yielding a maximum of ca. 3.4×10^3 transformants per μg of DNA (47). The process involves the electroporation of polyethylene glycol-treated mid-log-phase cells with the streptococcal vector pGK12, which encodes both erythromycin and chloramphenicol resistance. Other workers have used the *E. faecalis* shuttle vector pAT19 to transform strain 62A and to study the regulation of BoNT/A expression. These genetic experiments have provided evidence that toxin gene expression is activated by the *botR*/A gene product (32).

A versatile conjugation system has now been developed for introducing cloned genes into *C. botulinum* (5). This system involves the use of the mobilizable *C. perfringens*–*Escherichia coli* shuttle vector pJIR1457 (30) and the *E. coli* donor strain S17-1. Conjugative transfer of pJIR1457 from *E. coli* to the *C. botulinum* Hall A and 62A strains occurs at high frequency, and the resultant transconjugants stably maintain the plasmid. Therefore, this process can be routinely used to introduce cloned genes back into *C. botulinum*. For example, a pJIR1457 derivative containing a gene encoding a derivative of the BoNT/A light chain has been transferred to the toxin mutant LNT01 by conjugation, and the resultant transconjugant has been shown to produce high levels of modified light chain (5).

THE GENETICS OF *C. tetani*

Toxin-Encoding Plasmids of *C. tetani*

The gene encoding the tetanus neurotoxin (TeNT) has been shown to be plasmid determined. Initial analysis of 21 toxigenic strains of *C. tetani* revealed that they all contained a large plasmid that was absent from cured nontoxigenic strains. These observations were confirmed by subsequent studies that led to the identification and mapping of the 75-kb plasmid pCL1, which carries the TeNT structural gene on a 16.5-kb *Eco*RI fragment (22).

Genetic Manipulation of *C. tetani*

Tn916 has also been shown to encode its own transfer from *E. faecalis* to *C. tetani* (45), with the transposon inserting at multiple sites in the recipient genome. In addition, the transconjugants were able to act as donors in subsequent matings with either *C. tetani* or *E. faecalis*, the former at frequencies as high as 3.9×10^{-4} transconjugants per donor cell. However, there have been no further studies on this promising transposon mutagenesis system in *C. tetani*.

Transformation methods have also been developed recently for *C. tetani* (33). Electroporation was used to introduce the regulatory gene *tetR* into *C. tetani* and to show that overexpression of *tetR* leads to increased transcription of the TeNT structural gene and overexpression of TeNT. The transformation process involved the preparation of electrocompetent *C. tetani* cells in an anaerobic glove chamber and the electroporation of these cells in a sealed cuvette outside the chamber. The vector was a derivative of the high-copy-number *E. faecalis* shuttle plasmid pAT19. The development of these methods represents the beginning of an exciting new phase for genetic studies in *C. tetani*.

THE GENETICS OF *C. PERFRINGENS*

Antibiotic Resistance in *C. perfringens*

Chloramphenicol Resistance

Chloramphenicol resistance in *C. perfringens* is not as common as erythromycin or tetracycline resistance and is mediated by the production of chloramphenicol acetyltransferase (CAT) enzymes (29). The *catP* chloramphenicol-resistance gene is encoded on transposons that are located on large conjugative tetracycline-resistance plasmids (1). Comparative analysis of the nucleotide sequence of *catP* showed that, apart from the *catD* gene from *C. difficile*, it does not have significant similarity to other *cat* genes. However, the deduced CatP monomer has significant similarity at the amino acid sequence level, including the carboxy terminus, which contains the active site. CatP is most closely related to CAT monomers from *C. difficile*, *Vibrio anguillarum*, and *Campylobacter coli*, all of which have a four-amino-acid deletion in comparison to other CAT determinants (29). Recently, the *catP* gene was identified in *Neisseria meningitidis* (16).

A different chloramphenicol-resistance gene, *catQ*, has also been detected in a single strain of *C. perfringens* (29). The CatQ monomer is most closely related to the CatB monomer from *C. butyricum* and does not contain the four-amino-acid deletion present in the other clostridial *cat* genes. Therefore, this determinant seems to have evolved independently of *catP*, although it is possible that they share a common ancestor (29).

Erythromycin Resistance

The study of erythromycin- or macrolide-lincosamide-streptogramin B (MLS) resistance in *C. perfringens* has predominantly focused on the ErmBP determinant, which belongs to the ErmB-ErmAM hybridization class (29). The *ermBP* gene is located on a 63-kb nonconjugative plasmid, pIP402, although it is not widespread in *C. perfringens*. The 738-bp *ermBP* gene is identical to the *erm* gene from the promiscuous *E. faecalis* plasmid pAMβ1 and has at least 98% nucleotide sequence identity to other *ermB/ermAM*

genes (4). In contrast to many of these genes, *ermBP* and the pAMβ1 gene are not preceded by a leader peptide sequence and are constitutively expressed (4). An open reading frame (ORF) of unknown function, ORF3, which is present downstream of the *erm* genes from Tn917, pAMβ1, pAM77, and pIP501, is also found downstream of *ermBP* (Fig. 2) (4).

The *ermBP* gene is located between two almost identical directly repeated sequences, DR1 (1,341 bp) and DR2 (1,340 bp) (Fig. 2), which may encode a protein, ORF298, that has similarity to chromosomal and plasmid partitioning proteins (4). In both DR1 and DR2 there are almost identical 47-bp palindromic sequences, *palA* and *palB* (Fig. 2) (4). Comparative analysis has shown that both the pAMβ1 and pIP501 *erm* regions have DR2 but have a 975-bp internal deletion in DR1, whereas other determinants have small portions of DR1 but none have complete copies (Fig. 2) (29). In each case the deletion endpoints are different, suggesting that they have arisen from separate deletion events (4).

Hybridization analyses have shown that a second *erm* gene, *ermQ*, is also found in *C. perfringens*. This determinant represents the most common erythromycin-resistance determinant in *C. perfringens* (29). The wider distribution of *ermQ* may reflect differences in the mechanisms by which the two determinants are disseminated (29).

Tetracycline Resistance

Tetracycline resistance is the most common resistance phenotype found in *C. perfringens*, and resistant strains usually carry at least two distinct resistance genes (29). All tetracycline-resistant *C. perfringens* strains carry the *tetA(P)* gene. In more than half of these strains a second, overlapping resistance gene, *tetB(P)*, is found downstream of *tetA(P)* (28). The Tet P determinant was isolated from the conjugative plasmid pCW3 and is found on all known tetracycline-resistance plasmids from *C. perfringens*. It can also be encoded on nonconjugative plasmids or on the chromosome. A *tet(M)*-like gene is found in most of the isolates that do not carry *tetB(P)*, but it is not associated with the *tetA(P)* gene (28). Despite the wide distribution of the *tetB(P)* and *tet(M)* genes, no isolates have been found that carry both genes (28).

The *tetA(P)* gene encodes a 46-kDa transmembrane protein, TetA(P), which mediates the active efflux of tetracycline from the cell (43). Hydrophobicity analysis suggests that TetA(P) represents a novel type of efflux protein in comparison to other tetracycline efflux proteins. It appears to have two major hydrophilic domains that are not centrally located, whereas the prototype tetracycline efflux proteins have two large related six-transmembrane domains that are separated by a large central cytoplasmic domain (43). Site-directed mutagenesis has shown that amino acids that are predicted to be located in or near transmembrane domains, namely, Glu-52, Glu-59, and Glu-89, are essential for the active efflux of tetracycline (26). Random mutagenesis experiments have led to the identification of other residues that are required for tetracycline efflux. These residues are primarily located in the cytoplasmic loops 2-3 and 4-5, and the putative periplasmic loop 7-8 (3). Isolates that do not carry *tetB(P)* carry the *tetA408(P)* gene, which is 37 bp shorter than *tetA(P)* (28). The TetA408(P) protein has four amino acid changes within its last transmembrane domain, but the transmembrane domains of the protein have not been significantly altered and the efflux phenotype is intact (28).

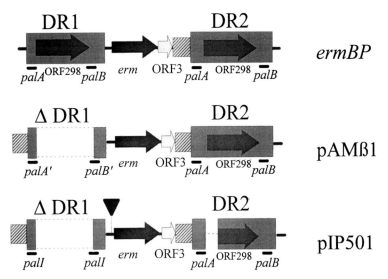

FIGURE 2 Comparison of the genetic organization of the *ermB/ermAM* gene regions. Regions of nucleotide sequence similarity are indicated by the same shading. The approximate extents of the deletions in the pAMβ1 and pIP501 DR1-like sequences (ΔDR1) are indicated by the dotted lines. The filled arrows indicate the individual ORFs and their direction of transcription. The approximate locations of the palindromic sequences (*palA, palB*, and *palI*) but not their sizes are indicated by the boldface lines below the filled boxes. The *palA'* and *palB'* sequences in pAMβ1 and the *palI* sequences in pIP501 represent the portions of the *ermBP*-derived *palA* and *palB* homologs that are present at the ends of the deletion in these DR1-like sequences. Leader peptide sequences are indicated by filled triangles. Modified from Lyras and Rood (29).

The *tetB*(P) gene encodes a putative 72.6-kDa protein that has significant amino acid sequence similarity to Tet M-like cytoplasmic tetracycline-resistance proteins and is therefore likely to function by a ribosomal protection mechanism (43). This gene encodes resistance to both tetracycline and minocycline, as do the *tet*(M)-like genes, whereas *tetA*(P) only confers resistance to tetracycline (28, 43).

Plasmids of *C. perfringens*

Many different types of plasmids have been found in C. *perfringens*, including plasmids that encode antibiotic resistance, bacteriocin production and immunity, and virulence factors or toxins. Many of these plasmids, especially those encoding toxins, have not been extensively studied.

C. *perfringens* is the only member of the clostridia in which conjugative antibiotic-resistance plasmids have been found. Two such plasmids have been examined in detail, the 47-kb tetracycline-resistance plasmid pCW3 and the 54-kb tetracycline- and chloramphenicol-resistance plasmid pIP401. Detailed restriction maps of both pCW3 and pIP401 have been constructed; comparison of these maps reveals that the two plasmids are very closely related. Further studies have shown that all conjugative tetracycline-resistance plasmids from C. *perfringens* are closely related to or identical to pCW3 (29).

Loss of chloramphenicol resistance often occurs during conjugative transfer of pIP401 and is associated with the loss of approximately 6 kb from the plasmid. Subsequent studies showed that this segment comprises the transposable element Tn4451. The pIP401 replicon can therefore be considered to be a pCW3 plasmid that contains Tn4451 (1). Nonconjugative antibiotic-resistance plasmids have also been identified in C. *perfringens*, the most notable be-

ing the 63-kb plasmid pIP402, which carries *ermBP* and therefore encodes MLS resistance (42).

The best-studied of all C. *perfringens* plasmids is the 10.2-kb bacteriocin plasmid pIP404, which has 10 ORFs (17). One of these genes, *bcn*, encodes the bacteriocin BCN5. Another gene, *uviA*, which is situated in an operon located downstream of *bcn*, encodes bacteriocin immunity. The origin of replication of pIP404 has been defined, copy-number control elements have been studied, and UV-inducible promoters responsible for the expression of *bcn* and the *uviAB* operon have been identified (42).

Recent studies have shown that a number of virulence factors or toxins of C. *perfringens* are located on plasmids. These studies involved digestion with the intron-encoded endonuclease I-*Ceu*I, which only cuts within rRNA operons found on the chromosome. Analysis of I-*Ceu*I digests of 16 C. *perfringens* isolates confirmed that the *plc* (alpha-toxin), *pfoA* (theta-toxin), *colA* (kappa-toxin), and *nagH* (sialidase) genes are chromosomally located, whereas the *etx* (epsilon-toxin), *cpb* (beta-toxin), *iap/ibp* (iota-toxin), *lam* (lambda-toxin), and urease genes are located on large extrachromosomal elements that do not contain I-*Ceu*I sites (25). The beta2-toxin gene, *cpb2*, which does not hybridize with *cpb*, is also found on a large plasmid (18). Another study has confirmed that the C. *perfringens* urease genes, *ureABC*, are found on large plasmids that also often encode *etx*, *iap/ibp*, or the enterotoxin gene *cpe* (14). Very little is known about any of the large toxin plasmids. Finally, it can be concluded that distinguishing C. *perfringens* type A isolates from type B to type E isolates clearly depends on the presence in the latter isolates of plasmids that confer the ability to produce beta-, epsilon-, or iota-toxin.

The genetic location of the *cpe* gene is more variable than that of the other toxin genes. In C. *perfringens* strains isolated from food poisoning outbreaks, the *cpe* gene is chromosomally determined, whereas in enterotoxin-producing isolates of animal origin and isolates associated with non-food-borne gastrointestinal infections, *cpe* is plasmid determined (9, 11).

Transposons and Insertion Sequences of *C. perfringens*

The most well-characterized transposon from *C. perfringens* is Tn4451, which carries the *catP* gene. This transposon is lost at high frequency from recombinant plasmids in *E. coli* and is also excised from its host plasmid pIP401 during conjugative transfer in *C. perfringens*. Transposition of Tn4451 has been demonstrated in *E. coli* but not in the original host *C. perfringens* (29). The complete nucleotide sequence has been determined, and the 6,338-bp transposon has been found to encode six genes (Fig. 3) (2). Similar transposons, Tn4452 and Tn4453, have been identified in *C. perfringens* and *C. difficile*, respectively.

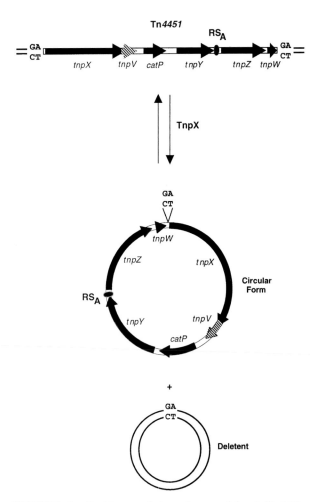

FIGURE 3 Excision and insertion model of Tn4451. Tn4451 encodes six genes, as shown by the arrows, and is flanked by directly repeated GA dinucleotides. The TnpX-mediated excision of Tn4451 leads to the formation of the circular form of the transposon, which carries one of the dinucleotides at the joint, and a deletion plasmid, in which one dinucleotide remains at the deletion site. Modified from Crellin and Rood (12).

Apart from *catP*, the two genes encoded by Tn4451 that have been studied in most detail are *tnpX* and *tnpZ*. The *tnpX* gene is required for the spontaneous excision of the element in both *E. coli* and *C. perfringens*, since a derivative of Tn4451 with an internal deletion in *tnpX* is stable in both organisms. The provision of *tnpX* in *trans* restores the ability of the transposon to excise. Hybridization and PCR studies showed that the TnpX-mediated excision of Tn4451 results in the formation of a circular form of the transposon. This behavior is similar to that of the conjugative transposons Tn916 and Tn1545, where the first step in transposition involves the formation of a nonreplicating circular intermediate. It has therefore been proposed that the circular form of Tn4451 represents the first step in transposition of this element.

Sequence analysis indicated that the site-specific recombinase TnpX has a conserved resolvase/invertase domain (2) that we have shown by site-directed mutagenesis to be essential for the excision of Tn4451 (12). Analysis of Tn4451 target sites also showed that target-site specificity is evident during the transposition process. The Tn4451 target sequence was found to resemble the junction of the circular form, with insertion occurring at a GA dinucleotide and Tn4451 insertions being flanked by directly repeated GA dinucleotides that are also found at the joint of the circular form where the left and right termini of Tn4451 are fused (12). Based on these results, a model for the excision and insertion of Tn4451 was proposed whereby the resolvase/invertase domain of TnpX catalyzes the formation of a 2-bp staggered nick on either side of the GA dinucleotide located at the ends of Tn4451, at the junction of the circular form and at the insertion site (Fig. 3). Analysis of Tn4451 derivatives with altered GA dinucleotides at the left and right ends provided the experimental evidence to support this model (12).

The *tnpZ* gene encodes a protein that has amino acid sequence similarity to a group of plasmid mobilization and recombination proteins that compose the Mob/Pre family. These proteins interact with an upstream palindromic sequence, the RS_A site, resulting in plasmid nicking, which is the first step in DNA mobilization. There is an RS_A-like sequence upstream of *tnpZ*, and further studies have shown that, in the presence of a chromosomally integrated copy of the broad-host-range IncP plasmid RP4 in *E. coli*, TnpZ promotes plasmid mobilization in *cis* and functions in *trans* to allow the mobilization of a coresident plasmid carrying an RS_A site (13). TnpZ also facilitates the conjugative transfer of plasmids from *E. coli* to *C. perfringens*. Site-directed mutagenesis of the RS_A site results in a significantly reduced mobilization frequency, confirming that it is required for the TnpZ-mediated mobilization process. TnpZ is unique in that it is the only known example of a Mob/Pre protein associated with a transposable genetic element (13). Furthermore, Tn4451 is the first element that is mobilizable, but not self-transmissible, to be identified from a gram-positive bacterium. The presence of mobilizable transposons may enable nonconjugative plasmids to be mobilized in the presence of other transfer-proficient factors. This raises the interesting possibility that TnpZ may play a role in the transmission, and hence widespread dissemination, of other resistance genes. Tn4451 therefore seems to be a highly evolved element that has acquired some properties of both conjugative and nonconjugative transposons (13, 29).

A number of other insertion sequences and transposons have been identified in *C. perfringens*, namely, IS1151,

IS*1469*, IS*1470*, and Tn*5565*. However, studies on these elements are not as advanced as those on Tn*4451*. The interesting feature of these elements is that they seem to be associated with toxin genes, and, although there is no direct experimental evidence, they have been implicated in the genetic transfer or mobilization of these virulence genes.

IS*1151* is a 1,696-bp element that is located 96 bp upstream of the plasmid-encoded epsilon-toxin gene, *etx*, in a C. perfringens type D strain. An IS*1151*-like element is located near *etx* in all type B and D isolates. IS*1151* contains two terminal 23-bp regions that constitute a perfect inverted repeat and has sequence similarity to the IS*231* family of insertion elements from *Bacillus thuringiensis*, among others (29). It was postulated that IS*1151* may be derived from IS*231* and that, like IS*231* in B. thuringiensis, IS*1151* may be involved in virulence gene transfer or mobilization in C. perfringens. Located upstream of IS*1151* is an ORF with significant similarity to the transposase-encoding *tnpA* gene from the Tn*3* family of transposons, which includes the cryptic Tn*4430* element from B. thuringiensis. IS*231* preferentially inserts at the ends of Tn*4430* in B. thuringiensis. The presence of IS*1151* and a region with similarity to Tn*4430* suggests that a similar situation may exist in C. perfringens (41).

The *cpe* gene may be chromosomal or plasmid determined and in one strain was found to be located on the same plasmid as the *etx* gene. Sequence analysis has shown that *cpe* is closely associated with a number of insertion sequences and in some chromosomal strains may be located on a transposon (7). All isolates that produce enterotoxin contain a 789-bp IS*200*-like element, IS*1469*, located 1.2 kb upstream of the *cpe* gene. In the chromosomal *cpe* strain NCTC8239, immediately upstream of IS*1469* is another insertion sequence, IS*1470*A, which belongs to the IS*30* family (Fig. 4). A second copy of IS*1470*, IS*1470*B, is found 1.1 kb downstream of *cpe* in this strain (Fig. 4). This structure appears to be that of a compound transposon, designated Tn*5565*, which contains an internal copy of IS*1469* and the *cpe* gene (6, 7). In strains isolated from food poisoning outbreaks, *cpe* is chromosomal and appears to be located on the same restriction fragment, suggesting that it is located on the same compound transposon (9). There is no evidence at this time that this putative transposon is capable of transposition. However, recent studies using PCR analysis have provided evidence that this element may excise to produce a circular form, which may represent a transposition intermediate (6).

In enterotoxin-producing strains of animal origin, the *cpe* gene is plasmid determined (9, 11). In these strains, *cpe* is associated with IS*1469* as before, but not with IS*1470*. These plasmids also carry IS*1151*, which seems to be confined to isolates that have plasmid-borne *cpe* genes (11). In one such isolate, F3686, IS*1151* is located 260 bp downstream of the *cpe* gene, whereas in another isolate, 945P, IS*1151* is linked to the *etx* gene but not to *cpe*, although both genes are carried on the same plasmid (11). Recent studies have shown that *cpe* is also plasmid encoded in C. perfringens strains associated with non-food-borne gastrointestinal infections (9).

Genetic Manipulation of *C. perfringens*

C. perfringens serves as the paradigm species for clostridial genetics. Methods for the genetic manipulation of C. perfringens are now well established, making this organism very amenable to genetic analysis. There are several well-characterized shuttle plasmids that can reliably and reproducibly be introduced into C. perfringens cells by transformation or conjugation. Methods for transposon mutagenesis and homologous recombination are also well established (40).

Although a number of C. perfringens–E. coli shuttle vectors have been constructed (Table 1), the most widely used shuttle vectors are derivatives of the plasmid pJIR418 (30, 40). This plasmid, which has been completely sequenced, contains the origins of replication of the E. coli plasmid pUC18 and the C. perfringens plasmid pIP404 together with the multiple cloning site and *lacZ′* gene from pUC18 (Table 1). It also contains the *ermBP* and *catP* genes, which confer erythromycin and chloramphenicol resistance, respectively, in both C. perfringens and E. coli. Detection of recombinants can be achieved by X-Gal (5-bromo-4-chloro-3-indolyl-β-D-galactopyranoside) screening in E. coli or by insertional inactivation of the *ermBP* or *catP* genes in either E. coli or C. perfringens.

Two derivatives of pJIR418, pJIR750 and pJIR751, have been constructed by deletion of the *ermBP* and *catP* genes, respectively, thereby resulting in sequenced vectors that carry a single antibiotic-resistance gene (Table 1). Recent studies have involved the construction of derivatives of pJIR750 and pJIR751, designated pJIR1456 and pJIR1457, respectively, which can be mobilized into C. perfringens via conjugation from an appropriate E. coli donor (30).

A promoter probe plasmid based on pJIR418 has also been constructed for use in C. perfringens. This plasmid, pPSV (Table 1), was constructed by deletion of the *catP* gene from pJIR418 followed by the addition of a promoterless *catP* gene downstream of the multiple cloning site (34). The *catP* gene has also been used as a reporter system in the promoter probe shuttle vector pTCATT (Table

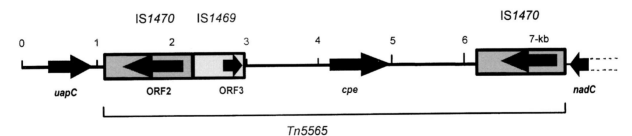

FIGURE 4 Genetic organization of the putative enterotoxin transposon Tn*5565*. The organization of the chromosomal *cpe* gene region is shown. ORFs are indicated by the arrows. The insertion elements are shown by the gray boxes. The scale is shown in kilobases. Modified from Rood (41).

TABLE 1 Properties of C. perfringens–E. coli shuttle plasmids

Shuttle plasmid[a]	Size (kb)	C. perfringens		E. coli	
		Replicon	Selection[b]	Replicon	Selection[b]
pHR106	7.9	pJU122	Cm	pSL100	ApCm
pJIR418	7.4	pIP404	EmCm	pUC18	EmCmXG
pJIR750	6.6	pIP404	Cm	pUC18	CmXG
pJIR751	6.0	pIP404	Em	pUC18	EmXG
pJIR1456[c]	6.9	pIP404	Cm	pUC18	CmXG
pJIR1457[c]	6.2	pIP404	Em	pUC18	EmXG
pPSV	6.9	pIP404	Em	pUC18	EmXG
pTCATT	6.4	pIP404	Em	pUC18	Em

[a]For details and sources, see references 30 and 40. Based on Table 5.1 from reference 40.
[b]Ap, Cm, Em, resistance to ampicillin, chloramphenicol, and erythromycin, respectively. XG, screening for β-galactosidase production on X-Gal medium.
[c]These plasmids carry the RP4 oriT region and can therefore be mobilized, in the presence of plasmid RP4, to recipient cells (30).

1), which contains a promoterless catP gene flanked by transcriptional terminators. However, PCR-mediated regeneration of the 5′ terminus of the catP gene is required during the construction of promoter fusions in pTCATT, limiting its usefulness (8). Other studies have attempted to develop reporter systems for use in C. perfringens. One system involved the construction of a plasmid that contains the luxAB genes from Vibrio fischeri under the control of the plc promoter. Luciferase activity and bioluminescence were obtained from this plasmid in C. perfringens (39). In another study, reporter plasmids using the E. coli gusA gene cloned into pJIR750 were constructed, and β-glucuronidase production was used successfully to monitor the sporulation-specific regulation of the cpe gene in C. perfringens (46).

The most reliable and reproducible way of introducing shuttle plasmids into C. perfringens is by using a transformation method that utilizes electroporation technology. The efficiency of transformation is very dependent on the host strain. A reasonable transformation frequency (3 × 10^5 transformants per µg DNA) was obtained when the C. perfringens isolate strain 13 was used as the transformation recipient. Consequently, strain 13 and derivatives of this strain are now the most widely used strains in studies of C. perfringens genetics (40).

The studies aimed toward introducing shuttle plasmids into C. perfringens showed that some strains cannot be transformed by electroporation (40). Recently, a new method for the introduction of DNA into C. perfringens has been developed that may overcome this limitation. This involved the construction of the shuttle plasmids pJIR1456 and pJIR1457, which carry the origin of transfer, or oriT, region from the broad-host-range conjugative plasmid RP4 (30). These new shuttle plasmids can be mobilized from the E. coli donor S17-1, which carries RP4 chromosomally, to a C. perfringens recipient at high efficiency. They can also be introduced into C. perfringens using the conventional electroporation process. The development of this system may allow C. perfringens strains that cannot be transformed via electroporation to be genetically manipulated (30).

Methods of transposon mutagenesis in C. perfringens have been developed, but these methods are not very efficient. The most successful mutagenesis methods have utilized the conjugative tetracycline-resistance transposon Tn916 (40). In these studies, C. perfringens cells were transformed with a suicide plasmid carrying Tn916, and transformants were selected on medium containing tetracycline. Hybridization analysis showed that Tn916 had inserted at different sites on the chromosome. However, multiple insertion events were common, as were deletion events. Since the C. perfringens transposons Tn4451 and Tn4452 have not been shown to transpose in this organism, there is still a need for more efficient transposon mutagenesis methods.

Homologous recombination and allelic exchange methods have also been used successfully for genetic analysis in C. perfringens (40). Homologous recombination may involve either single- or double-crossover events. Single-crossover events have been used to integrate suicide plasmids into the plc and virR genes, enabling an analysis of the function of these genes to be performed. Allelic exchange, achieved through a double-crossover event, has also been used to disrupt the chromosomally located plc and pfoA genes, and the plasmid-borne tnpX gene of Tn4451, again enabling functional analysis of these genes (40). Since two crossovers are required, allelic exchange is a less efficient process than obtaining mutants by a single-crossover event. However, the disadvantage of single-crossover events is that wild-type genes can be regenerated from the mutants by further homologous recombination events. This is not a problem with allelic exchange, which results in inherently stable mutants. It is worth noting that the efficiency by which mutants are obtained by allelic exchange depends on the size of the homologous regions that flank the gene of interest, and it is recommended that there be at least 1 kb of flanking DNA present on either side of the gene to be replaced (40).

THE GENETICS OF C. DIFFICILE
Antibiotic Resistance in C. difficile

Chloramphenicol Resistance

Early studies on chloramphenicol resistance in C. difficile W1 involved cloning of the catD gene, which is very closely related to the catP gene from C. perfringens. In contrast to catP, catD is present in multiple copies on the C. difficile chromosome (29). Recent studies have shown that both copies are associated with transposons, designated Tn4453a and Tn4453b. These elements are very similar to

Tn4451 (31). It is therefore concluded that catD and catP have evolved relatively recently from a common ancestor (29).

Erythromycin Resistance

Early studies of erythromycin resistance in C. difficile demonstrated that the conjugative transfer of MLS resistance was common, unlike the situation in C. perfringens. Analysis of C. difficile 630 indicated that transfer appears to involve a chromosomal determinant, designated Tn5398. The erythromycin-resistance gene carried on Tn5398 also belongs to the ermB/ermAM class and is designated ermBZ (29). Recent studies have shown that this determinant consists of two ermBZ genes, ermBZ1 and ermBZ2, with ORF3 present downstream of each ermBZ gene. The two ermBZ genes are separated by a 1.34-kb direct repeat sequence that is almost identical to direct repeats associated with the ErmB/AM class of MLS-resistance determinants from C. perfringens and E. faecalis. Both ermBZ genes are also flanked by variants of the 1.34-kb direct repeat sequence (15). The duplication and genetic arrangement of the two ermBZ genes is novel for a member of the ErmB/AM class of MLS-resistance determinants. Overall, the extent of the similarity between the erm determinants suggests that they are derived from a common progenitor. However, it appears that the sequences that flank the erm genes have diverged extensively even though the genes are very highly conserved. This conclusion is supported by the fact that many of these determinants are located on conjugative plasmids or transposons that are easily disseminated (29).

Tetracycline Resistance

Studies on tetracycline-resistance determinants from C. difficile are not as advanced as those from C. perfringens. A tet(M) gene has been identified in the conjugative C. difficile strains 630 and 662. The gene from strain 630 is transferable at a low frequency and is constitutively expressed. Genetic studies have shown that it is located on a conjugative transposon, designated Tn5397 (35).

Transposons of *C. difficile*

Four transposons from C. difficile have been described, specifically, the conjugative transposons Tn5397 and Tn5398 and the mobilizable transposons Tn4453a and Tn4453b. The interesting feature of these elements is that they, or the resistance determinants which they encode, have significant homology to elements from other bacteria.

The conjugative transposon Tn5397 encodes a tet(M)-like tetracycline-resistance determinant and can be transferred to Bacillus subtilis and back to C. difficile. Tn5397 integrates into the C. difficile chromosome at two specific sites, whereas it integrates into B. subtilis at various sites. The entire transposon has been cloned, and a recombinant plasmid carrying Tn5397 has been used to transform B. subtilis to tetracycline resistance. The resulting transformants could transfer Tn5397 to C. difficile by filter mating (29). Physical and genetic analysis indicates that Tn5397 is related to the Tn916/Tn1545 family of conjugative transposons. Tn5397 and Tn916 are virtually identical in regions that are known to be required for intercellular transposition of Tn916 but are unrelated in regions involved in integration and excision from the target site. Also, the ends of the two elements are different (35). Sequencing of the Tn5397 region located upstream of tet(M) indicated that the element contains a group II intron that

is inserted into a gene that, in Tn916, is required for intercellular transposition. Tn5397 retains its ability to conjugate, despite having this gene disrupted by the intron, which provides indirect evidence that the intron in Tn5397 is functional. Furthermore, by comparison with other group II introns, the intron in Tn5397 contains all the necessary information for mobility (35).

The C. difficile isolate that carries Tn5397 also carries a second conjugative transposon, Tn5398, which encodes erythromycin (MLS) resistance and which is transferred independently of Tn5397. This element is transferred at low frequency to C. difficile CD37 and also to Staphylococcus aureus and B. subtilis. Transconjugants derived from the latter species can act as donors in matings using strain CD37 as the recipient. Hybridization studies revealed that, like Tn5397 and Tn916, Tn5398 integrates in a site-specific manner in C. difficile but randomly in B. subtilis (29, 37).

The chloramphenicol resistance gene catD from C. difficile W1 is encoded on the transposons Tn4453a and Tn4453b. These elements are structurally and functionally related to Tn4451 from C. perfringens, but the three elements are distinguishable by restriction mapping (31). As with Tn4451, Tn4453a, and Tn4453b are excised precisely from recombinant plasmids, generating a circular form. This process is mediated by Tn4453-encoded tnpX genes. The joint of the circular form is also very similar to that of Tn4451. These results suggest that the Tn4453-encoded TnpX proteins bind to comparable target sequences and function in a manner similar to that of TnpX from Tn4451 (31). Tn4453a and Tn4453b are mobilizable transposons, since, like Tn4451, they can be transferred to suitable recipient cells in the presence of RP4. This mobilization system may play a significant role in the dissemination of Tn4451- and Tn4453-like transposons to different bacterial genera and species (31). It is therefore worth noting that there is a cat gene in N. meningitidis that is identical to catP and almost identical to catD (16). The finding that almost identical transposons are found in C. perfringens and C. difficile suggests that genetic exchange may have occurred directly between these two organisms or through an intermediate bacterial host.

Genetic Manipulation of *C. difficile*

The development of methods by which C. difficile can be genetically manipulated has not progressed significantly. In early studies, Tn916 or derivatives of this transposon were transferred from B. subtilis into C. difficile CD37 by filter mating. Tn916 is the only element not originally from C. difficile to have been stably introduced into this organism. The C. difficile transconjugants each contained one copy of the transposon integrated into the same position in the genome. This site-specific insertion is also observed with Tn5397 and Tn5398. Tn916 is of limited value as a cloning vector because of its large size and the fact that it is not an autonomously replicating element. However, the site-specific insertion of Tn916 means that genes can reproducibly be introduced into a defined site in the C. difficile genome (29).

The limitations of using Tn916 as a cloning vector have been partly addressed. A plasmid that carries a segment of Tn916 has been constructed and introduced into a B. subtilis strain containing a chromosomal copy of Tn916, where it integrated into this element by homologous recombination. The recombinant transposon was then transferred by conjugation to C. difficile, where it was then inserted into the chromosome (36). This method was used successfully to introduce a fragment of the *C. difficile* toxin B gene back into the appropriate *C. difficile* strain. Although this approach could be used for other general cloning purposes in *C. difficile*, it is very cumbersome and has not been used on a regular basis (36).

Finally, pulsed-field gel electrophoresis, using the restriction enzymes *Sac*II and *Nru*I, has been employed to generate a physical map of the 4.4-Mb chromosome of the toxigenic *C. difficile* strain ATCC 43594 (38). This was the first step toward gaining an understanding of the genome organization of *C. difficile*. Currently, the entire sequence of the genome of strain 630 is being determined. Further details are available; see reference 42a.

We apologize to the many colleagues whose original work could not be cited. Owing to strict limitations in the number of permissible references, we have cited review articles wherever possible. The work carried out in this laboratory was generously supported by grants from the Australian National Health and Medical Research Council.

REFERENCES

1. **Abraham, L. J., and J. I. Rood.** 1987. Identification of Tn*4451* and Tn*4452*, chloramphenicol resistance transposons from *Clostridium perfringens. J. Bacteriol.* **169:**1579–1584.

2. **Bannam, T. L., P. K. Crellin, and J. I. Rood.** 1995. Molecular genetics of the chloramphenicol-resistance transposon Tn*4451* from *Clostridium perfringens:* the TnpX site-specific recombinase excises a circular transposon molecule. *Mol. Microbiol.* **16:**535–551.

3. **Bannam, T. L., and J. I. Rood.** Identification of structural and functional domains of the tetracycline efflux protein TetA(P) from *Clostridium perfringens. Microbiology,* in press.

4. **Berryman, D. I., and J. I. Rood.** 1995. The closely related *ermB*/AM genes from *Clostridium perfringens, Enterococcus faecalis* (pAMβ1), and *Streptococcus agalactiae* (pIP501) are flanked by variants of a directly repeated sequence. *Antimicrob. Agents Chemother.* **30:**1830–1834.

5. **Bradshaw, M., M. Goodnough, and E. Johnson.** 1998. Conjugative transfer of the *E. coli*-C. *perfringens* shuttle vector pJIR1457 to *Clostridium botulinum* type A strains. *Plasmid* **40:**233–237.

6. **Brynestead, S., and P. E. Granum.** 1999. Evidence that Tn*5565*, which includes the enterotoxin gene in *Clostridium perfringens*, can have a circular form which may be a transposition intermediate. *FEMS Microbiol. Lett.* **170:**281–286.

7. **Brynestad, S., B. Synstad, and P. E. Granum.** 1997. The *Clostridium perfringens* entertotoxin gene is on a transposable genetic element in type A human food poisoning strains. *Microbiology* **143:**2109–2115.

8. **Bullifent, H. L., A. Moir, and R. W. Titball.** 1995. The construction of a reporter system and use for the investigation of *Clostridium perfringens* gene expression. *FEMS Microbiol. Lett.* **131:**99–105.

9. **Collie, R. E., and B. A. McClane.** 1998. Evidence that the enterotoxin gene can be episomal in *Clostridium perfringens* isolates associated with nonfoodborne human gastrointestinal disease. *J. Clin. Microbiol.* **36:**30–36.

10. **Collins, M. D., P. A. Lawson, A. Willems, J. J. Cordoba, J. Fernandez-Garayzabal, P. Garcia, J. Cai, H. Hippe, and J. A. Farrow.** 1994. The phylogeny of the genus *Clos-*

11. **Cornillot, E., B. Saint-Joanis, G. Daube, S.-I. Katayama, P. E. Granum, B. Canard, and S. T. Cole.** 1995. The enterotoxin gene (*cpe*) of *Clostridium perfringens* can be chromosomal or plasmid-borne. *Mol. Microbiol.* **15:**639–647.

12. **Crellin, P. K., and J. I. Rood.** 1997. The resolvase/invertase domain of the site-specific recombinase TnpX is functional and recognizes a target sequence that resembles the junction of the circular form of the *Clostridium perfringens* transposon Tn*4451. J. Bacteriol.* **179:**5148–5156.

13. **Crellin, P. K., and J. I. Rood.** 1998. Tn*4451* from *Clostridium perfringens* is a mobilizable transposon that encodes the functional Mob protein, TnpZ. *Mol. Microbiol.* **27:**631–642.

14. **Dupuy, B., G. Daube, M. R. Popoff, and S. T. Cole.** 1997. *Clostridium perfringens* urease genes are plasmid borne. *Infect. Immun.* **65:**2313–2320.

15. **Farrow, K. A., D. Lyras, and J. I. Rood.** Unpublished results.

16. **Galimand, M., G. Gerbaud, M. Guibourdenche, J.-Y. Riou, and P. Courvalin.** 1998. High-level chloramphenicol resistance in *Neisseria meningitidis. N. Engl. J. Med.* **339:**868–874.

17. **Garnier, T., and S. T. Cole.** 1988. Complete nucleotide sequence and genetic organization of the bacteriocinogenic plasmid, pIP404, from *Clostridium perfringens. Plasmid* **19:**134–150.

18. **Gibert, M., C. Jolivet-Renaud, and M. R. Popoff.** 1997. Beta2 toxin, a novel toxin produced by *Clostridium perfringens. Gene* **203:**65–73.

19. **Hauser, D., M. Gibert, P. Boquet, and M. R. Popoff.** 1992. Plasmid localization of a type E botulinal neurotoxin gene homologue in toxigenic *Clostridium butyricum* strains, and absence of this gene in non-toxigenic *C. butyricum* strains. *FEMS Microbiol. Lett.* **99:**251–256.

20. **Hauser, D., M. Gibert, M. W. Eklund, P. Boquet, and M. R. Popoff.** 1993. Comparative analysis of C3 and botulinal neurotoxin genes and their environment in *Clostridium botulinum* types C and D. *J. Bacteriol.* **175:**7260–7268.

21. **Henderson, I., T. Davis, M. Elmore, and N. Minton.** 1997. The genetic basis of toxin production in *Clostridium botulinum* and *Clostridium tetani*, p. 261–294. *In* J. Rood, B. McClane, J. Songer, and R. Titball (ed.), *The Clostridia: Molecular Biology and Pathogenesis.* Academic Press, Inc., London, U.K.

22. **Johnson, E.** 1997. Extrachromosomal virulence determinants in the clostridia, p. 35–48. *In* J. Rood, B. McClane, J. Songer, and R. Titball (ed.), *The Clostridia: Molecular Biology and Pathogenesis.* Academic Press, Inc., London, U.K.

23. **Johnson, E., W. J. Lin, Y. Zhou, and M. Bradshaw.** 1997. Characterization of neurotoxin mutants in *Clostridium botulinum* type A. *Clin. Infect. Dis.* **25**(Suppl. 2)**:**S168–S170.

24. **Johnson, J. L., and B. S. Francis.** 1975. Taxonomy of the clostridia: ribosomal acid homologies among the species. *J. Gen. Microbiol.* **88:**229–244.

25. **Katayama, S., B. Dupuy, G. Daube, B. China, and S. T. Cole.** 1996. Genome mapping of *Clostridium perfringens* strains with I-*Ceu*I shows many virulence genes to be plasmid-borne. *Mol. Gen. Genet.* **251:**720–726.

26. **Kennan, R. M., L. M. McMurry, S. B. Levy, and J. I. Rood.** 1997. Glutamate residues located within putative transmembrane helices are essential for TetA(P)-mediated tetracycline efflux. *J. Bacteriol.* **179:**7011–7015.

27. **Lin, W.-J., and E. A. Johnson.** 1995. Genome analysis of *Clostridium botulinum* type A by pulsed-field gel electrophoresis. *Appl. Environ. Microbiol.* **61:**4441–4447.

28. **Lyras, D., and J. I. Rood.** 1996. Genetic organization and distribution of tetracycline resistance determinants in *Clostridium perfringens. Antimicrob. Agents Chemother.* **40:**2500–2504.

29. **Lyras, D., and J. I. Rood.** 1997. Transposable genetic elements and antibiotic resistance determinants from *Clostridium perfringens* and *Clostridium difficile,* p. 73–92. *In* J. I. Rood, B. A. McClane, J. G. Songer, and R. W. Titball (ed.), *The Clostridia: Molecular Biology and Pathogenesis.* Academic Press, Inc., London, U.K.

30. **Lyras, D., and J. Rood.** 1998. Conjugative transfer of RP4-*oriT* shuttle vectors from *Escherichia coli* to *Clostridium perfringens. Plasmid* **39:**160–164.

31. **Lyras, D., C. Storie, A. S. Huggins, P. K. Crellin, T. L. Bannam, and J. I. Rood.** 1998. Chloramphenicol resistance in *Clostridium difficile* is encoded on Tn*4453* transposons that are closely related to Tn*4451* from *Clostridium perfringens. Antimicrob. Agents Chemother.* **42:**1563–1567.

32. **Marvaud, J., M. Gibert, K. Inoue, Y. Fujinaga, K. Oguma, and M. Popoff.** 1998. *botR/A* is a positive regulator of botulinum neurotoxin and associated non-toxin protein genes in *Clostridium botulinum* A. *Mol. Microbiol.* **29:**1009–10018.

33. **Marvaud, J.-C., U. Eisel, T. Binz, H. Niemann, and M. R. Popoff.** 1998. TetR is a positive regulator of the tetanus toxin gene in *Clostridium tetani* and is homologous to BotR. *Infect. Immun.* **66:**5698–5702.

34. **Matsushita, C., O. Matsushita, M. Koyama, and A. Okabe.** 1994. A *Clostridium perfringens* vector for the selection of promoters. *Plasmid* **31:**317–319.

35. **Mullany, P., M. Pallen, M. Wilks, J. Stephen, and S. Tabaqchali.** 1996. A group II intron in a conjugative transposon from the gram-positive bacterium, *Clostridium difficile. Gene* **174:**145–150.

36. **Mullany, P., M. Wilks, L. Puckey, and S. Tabaqchali.** 1994. Gene cloning in *Clostridium difficile* using Tn*916* as a shuttle conjugative transposon. *Plasmid* **31:**320–323.

37. **Mullany, P., M. Wilks, and S. Tabaqchali.** 1995. Transfer of macrolide-lincosamide-streptogramin B (MLS) resistance in *Clostridium difficile* is linked to a gene homologous with toxin A and is mediated by a conjugative transposon, Tn*5398. J. Antimicrob. Chemother.* **35:**305–315.

38. **Norwood, D. A., Jr., and J. A. Sands.** 1997. Physical map of the *Clostridium difficile* chromosome. *Gene* **201:**159–168.

39. **Phillips-Jones, M. K.** 1993. Bioluminescence (*lux*) expression in the anaerobe *Clostridium perfringens. FEMS Microbiol. Lett.* **106:**265–270.

40. **Rood, J. I.** 1997. Genetic analysis in *C. perfringens,* p. 65–72. *In* J. I. Rood, B. A. McClane, J. G. Songer, and R. W. Titball (ed.), *The Clostridia: Molecular Biology and Pathogenesis.* Academic Press, Inc., London, U.K.

41. **Rood, J. I.** 1998. Virulence genes of *Clostridium perfringens. Annu. Rev. Microbiol.* **52:**333–360.

42. **Rood, J. I., and S. T. Cole.** 1991. Molecular genetics and pathogenesis of *Clostridium perfringens. Microbiol. Rev.* **55:**621–648.

42a. **The Sanger Centre.** http://www.sanger.ac.uk/Projects/C_difficile/

43. **Sloan, J., L. M. McMurry, D. Lyras, S. B. Levy, and J. I. Rood.** 1994. The *Clostridium perfringens* TetP determinant comprises two overlapping genes: *tetA*(P) which mediates active tetracycline efflux and *tetB*(P) which is related to the ribosomal protection family of tetracycline resistance determinants. *Mol. Microbiol.* **11:**403–415.

44. **Stackebrandt, E., and F. A. Rainey.** 1997. Phylogenetic relationships, p. 3–19. *In* J. I. Rood, B. A. McClane, J. G. Songer, and R. W. Titball (ed.), *The Clostridia: Molecular Biology and Pathogenesis.* Academic Press, Inc., London, U.K.

45. **Volk, W. A., B. Bizzini, K. R. Jones, and F. L. Macrina.** 1988. Inter- and intrageneric transfer of Tn*916* between *Streptococcus faecalis* and *Clostridium tetani. Plasmid* **19:**255–259.

46. **Zhao, Y., and S. B. Melville.** 1998. Identification and characterization of sporulation-dependent promoters upstream of the enterotoxin gene (*cpe*) of *Clostridium perfringens. J. Bacteriol.* **180:**136–142.

47. **Zhou, Y., and E. Johnson.** 1993. Genetic transformation of *Clostridium botulinum* Hall A by electroporation. *Biotechnol. Lett.* **15:**121–126.

Neurotoxigenic Clostridia

ERIC A. JOHNSON

56

Neurotoxigenic clostridia are those species that produce neurotoxins affecting the nervous system of humans and animals (14, 18, 27, 31). The classic neurotoxigenic diseases are botulism and tetanus, which are caused by the action of extremely toxic neurotoxins (NTs) produced by *Clostridium botulinum* and *Clostridium tetani*, respectively. Botulinum neurotoxins (BoNTs) and tetanus neurotoxin (TeNT) are the most potent toxins known (14, 18, 28, 33), primarily owing to their extraordinary neurospecificity and to their catalytic cleavage of neuronal substrates at exceedingly low concentrations of NT ($\leq 10^{-12}$ M) (27, 28). Recently, other strains of clostridia (*Clostridium baratii*, *Clostridium butyricum* [14]) that produce BoNTs have been recognized, indicating that the genes encoding BoNTs and associated proteins of the toxin complexes can be transferred to nonpathogenic clostridia. During the past two decades, toxigenic clostridia have been isolated with increasing frequency from human specimens and have attracted considerable interest as etiologic agents of disease (14, 31). The control of neurotoxic clostridial diseases in humans and animals presents considerable challenges to physicians and veterinarians.

The neurotoxic clostridia *C. tetani* and *C. botulinum* (18, 36, 40, 41) were isolated more than 100 years ago and were demonstrated to cause true toxemias, acting solely through the action of their NTs. The interest in BoNT has increased greatly in recent years because of its remarkable and extraordinarily effective utility as a pharmacological agent for treatment of neuronal diseases (16, 32). Interest in BoNT has also been incited by awareness of its possible implementation in bioterrorism (9). The biochemistry, structure, genetics, and the action of BoNTs and TeNT have become important tools in cell biology and have stimulated much interest in their basic nature and their actions on the human nervous system (1, 4, 29, 30). Several contemporary books and extensive reviews describe the taxonomy, physiology, genetics, and diseases caused by the toxigenic clostridia as well as the biochemistry, immunology, pharmacology, and medical uses of the NTs (1, 27, 30, 31, 35, 36). This chapter describes the properties of C. *botulinum* and C. *tetani* with an emphasis on pathogenesis and new findings on the organisms and their NTs.

CLINICAL ASPECTS OF BOTULISM AND TETANUS

Botulism

Botulism is a rare but often severe disease caused by the extremely potent BoNTs produced by C. *botulinum* and certain other clostridia (10, 14, 18, 36). Botulism is a true toxemia and was historically recognized as a deadly form of food poisoning occurring through the consumption of contaminated foods (40). BoNTs can also cause disease by its entry into the circulation from wounds or from the intestinal tract of susceptible infants and adults colonized by BoNT-producing clostridia (36). Although there is little historical record before the 19th century, botulism was suspected to occur in ancient cultures, and certain dietary laws and food processing methods probably evolved as a result of the disease (36). BoNTs in contaminated foods were noted in the 1700s to cause muscle paralysis and suffocation, with a high fatality rate exceeding 50%. The bacterial and toxigenic etiology was discovered in the 1890s by Van Ermengem in a remarkable series of experiments (40). His descriptions of the disease and properties of the toxin have been aptly reviewed (31, 36), and the principles he established still form the foundation of knowledge for prevention of food-borne botulism. Besides classical food botulism, wound botulism was discovered in 1943 (14), and infection and colonization of susceptible infant and adult gastrointestinal tracts was recognized in the 1970s and 1980s (14, 18, 36). All four types of botulism are quite rare in their incidence and may consequently be misdiagnosed as more common paralytic diseases such as Guillain-Barré syndrome, myasthenia gravis, tick paralysis, stroke, or nervous system infections. Infant botulism is the most prevalent form of botulism in the United States, but food-borne botulism is the most common form of the disease in most other regions of the world. C. *botulinum* is not as adept as C. *tetani* in colonizing wounds, and wound botulism is extremely rare compared with tetanus. However, since 1991 the numbers of cases of wound botulism have increased markedly, primarily in intravenous drug users in California (24).

Gram-Positive Pathogens, ed. by V. A. Fischetti et al.
© 2000 American Society for Microbiology, Washington, D.C.

BoNT-producing clostridia synthesize seven serotypes of BoNTs, designated A through G. The toxins occur in complexes with nontoxic proteins (32, 37), which impart stability to the labile BoNTs. The BoNTs can be neutralized by specific antitoxins obtained by immunization of animals with toxoids. Six independent evolutionary lineages of BoNT-producing clostridia have been proposed (10), emphasizing the ability of the toxin genes to be transferred among clostridia and the potential for discovery of new neurotoxic clostridial species in the future.

Irrespective of the source of the toxin (food, wound, or infant or adult intestinal infection), the neuroparalytic symptoms are similar (9, 36). The hallmark clinical symptomatology of botulism is a bilateral and descending weakening and paralysis of skeletal muscles. The incubation time for onset of symptoms varies with the type of botulism, the serotype of BoNT, and the quantity of BoNT that enters the circulation. Wound botulism usually has a relatively long incubation period of 4 to 14 days. Classical food-borne botulism occurs following the consumption of food contaminated with preformed BoNT, and the vast majority of cases are caused by types A, B, and E, and rarely by type F. The most severe and long-lasting food-borne botulism occurs with type A. The incubation time of food-borne botulism varies with the BoNT serotype and quantity of toxin ingested. In type A cases, the onset time is usually 12 to 36 h following consumption of the toxic food. The incubation period can be as short as 2 h when high quantities of toxin are ingested (9), or as long as several days to weeks with type B or E BoNTs and ingestion of low quantities of toxin (18).

The site of action of all serotypes of BoNT is the presynaptic terminals of motor neurons, where, after internalization into the presynaptic nerve terminal, the NT causes a blockade of the release of acetylcholine and regional flaccid paralysis of muscles innervated by the intoxicated nerves (9, 35). Botulinum toxin probably also triggers or inhibits physiological processes other than cholinergic neurotransmitter release, such as autonomic nerve function, but these processes are only beginning to be elucidated. In most cases of botulism, cranial nerves are first affected, particularly those innervating the eyes, and the first symptoms are blurred and double vision, dilated pupils, and drooping eyelids (Fig. 1). The eyes sluggishly respond to light in a darkened room. These abnormalities of cranial innervated muscles are followed by a descending and progressive paralysis characterized by difficulty in swallowing, weakness of the neck, dry mouth, and problems in speaking. Weakness of the upper limbs is common, and in severe cases respiratory muscle weakness generally ensues, requiring mechanical ventilation to prevent fatality by suffocation. BoNT can probably paralyze every striated muscle of the body, but there is apparently no interference with nervous activity of the central nervous system at in vivo concentrations (18, 27). The patient's hearing remains normal, consciousness is not lost, and the victim is cognizant of the progression of the disease. Certain episodes of botulism have been reported that affect autonomic and sensory functions. A patient's awareness of loss of muscle activity can lead to considerable emotional distress such as anxiety and depression.

The fatality rate from food-borne botulism has decreased from 50% to ≤9% in recent years (9). Recovery from botulism is prolonged, requiring weeks to months, but is usually complete and patients regain full normal function. Symptoms of fatigue, dry mouth, and blurred vision

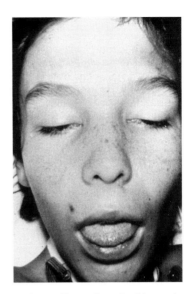

FIGURE 1 A boy suffering from food-borne botulism showing the prominent effects of BoNT on the eyes and the descending paralysis to other regions of the face. Photograph courtesy of Charles L. Hatheway (deceased), Centers for Disease Control and Prevention, Atlanta, Ga.

can persist for months, and recovery of autonomic abilities may take longer than neuromuscular functions. Antibodies do not generally develop in patients who have experienced food-borne botulism but can develop in infants or adults with intestinal botulism who are exposed to toxin for continual periods of time. Antibodies have been reported in individuals who have been repeatedly injected with type A botulinum toxin for therapeutic purposes (16, 33).

Electromyography can aid in detecting defective neuromuscular transmission and botulism (9), and methods to differentially diagnose botulism from other neurological diseases have been described (9). The definitive diagnosis of botulism depends on the detection of BoNT in the patient's serum, feces, and/or in food that was consumed before onset of the disease (14). Currently, the only reliable assay for BoNT is the mouse bioassay together with neutralization of mouse toxicity with type-specific antitoxins (one source is the Centers for Disease Control and Prevention [CDC], Atlanta, Ga.). Enzyme-linked immunosorbent assays (ELISAs) have also been used to detect NT antigen. Since the catalytic activities of BoNTs and TeNT on neuronal substrates have been discovered in recent years, it is possible that assays measuring BoNT endopeptidase activity could partially replace or augment the mouse assay.

No specific antidote is currently available for preventing botulism or reversing paralysis once it ensues. However, early administration of antibodies can decrease the severity of the disease and slow its progression, resulting in a shorter hospital stay and more rapid recovery (9, 14, 18). Complete recovery probably requires the formation of new functional neuromuscular junctions with resumption of neurotransmission. The reason that a single exposure to exceedingly low concentrations ($\leq 10^{-12}$ M) of BoNT can cause such a long-lasting disease is currently not understood and is a very intriguing phenomenon.

Botulinum toxin has been considered a potential biological warfare agent that could be administered in aerosols,

TABLE 1 Properties of botulinum (BoNT) and tetanus (TeNT) neurotoxins

Toxin	Gene location[a]	Representative species affected	Specific toxicity (10^8)[b]	Light-chain substrate	Peptide bond cleaved[c]
BoNT/A	C	Humans, chickens	1.05–1.86	SNAP-25	Gln-Arg
BoNT/B	C	Humans, cattle, horses	0.98–1.14	VAMP	Gln-Phe
BoNT/C$_1$	B	Birds, cattle, dogs, minks	0.88	Syntaxin SNAP-25	Lys-Ala Arg-Ala
BoNT/D	B	Cattle	1.60	VAMP	Lys-Leu
BoNT/E	C	Humans, fish, aquatic birds	0.21–0.25	SNAP-25	Arg-Ile
BoNT/F	C	Humans	0.16–0.40	VAMP	Gln-Lys
BoNT/G	P	Unknown	0.1–0.3	VAMP	Ala-Ala
TeNT	P	Humans, cattle, horses, sheep, dogs, chickens, other animals		VAMP	Gln-Phe

[a]Gene location: C, chromosome; B, bacteriophage; P, plasmid. For putative chromosomal locations, this location is inferred from PCR amplification of "chromosomal" DNA preparations, except for type A, in which toxin gene mutations have been mapped to the chromosome (see references 5 and 18).

[b]Specific toxicity refers to toxins activated by trypsinization when necessary for maximum toxicity. Toxicities are per milligram of protein. Most of the reported data are from Sugiyama (37). Experiments in our laboratory and previous studies (32, 37) have shown that oral doses of types A or B toxin complexes in mice are 10 to 1,000 times greater than the intraperitoneal or intravenous lethal dose depending on the size of the complex, diluent, and other factors.

[c]The specific peptide bond cleaved is shown. However, the clostridial neurotoxins require a minimum peptide length of >20 amino acid residues and a characteristic substrate tertiary structure for catalytic activity.

teriophages, or plasmids (11) (Table 1). The genes encoding BoNT/C and BoNT/D are associated with bacteriophages, and toxin production by *C. botulinum* types C and D strains is notoriously unstable (11, 18). The genes encoding the *C. botulinum* type G toxin gene cluster is located on a large plasmid, and type G toxin is produced in very low quantities (11, 18).

The genes encoding the botulinum toxin complexes occur as clusters and consist of two operons that are transcribed in opposite directions (Fig. 3). One operon contains the genes for nontoxic nonhemagglutinin protein (NTNH) and BoNT, and the other contains the gene for the hemagglutinin protein. Between the two operons is a gene (*botR/A*, *orf21*, or *orfX*) that encodes a 21-kDa protein that has been demonstrated to positively regulate expression of NTNH and BoNT (23). In a *C. botulinum* type A strain constructed in our laboratory that is deleted for the toxin gene cluster, we have observed increased expression of a recombinant gene construct containing the BoNT light (L)-chain gene fused to the NTNH gene promoter

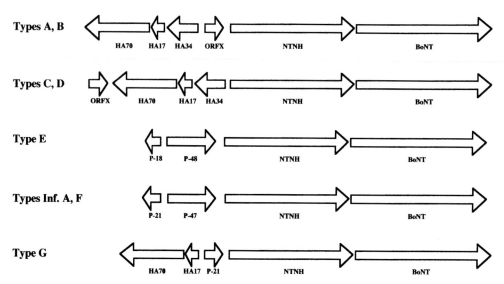

FIGURE 3 Arrangement of the genes in the neurotoxin gene clusters of *C. botulinum*. Diagram courtesy of Marite Bradshaw, University of Wisconsin.

and containing the gene encoding BotR. Since L chain is expressed at much higher levels by this recombinant plasmid than in wild-type strains, it is likely that regulatory factors that negatively affect BoNT expression are also present in *C. botulinum* type A, but these have not been elucidated. The production and proteolytic processing of BoNT are influenced by nitrogen nutrition (14, 18, 33), suggesting the involvement of global regulatory mechanisms in BoNT expression and proteolytic activation. Sequencing of the regions flanking the BoNT gene cluster has shown that there exists considerable polymorphism in these regions and that there are differences in location of some genes involved in BoNT production, such as those involved in cell lysis (5). These data indicate that transcriptional regulatory proteins control expression of BoNT, but additional research is needed to further elucidate the molecular mechanisms of BoNT gene regulation.

In addition to the primary gene cluster, silent chimeric neurotoxin type B gene sequences have been identified in type A *C. botulinum* strains at a surprisingly high frequency (10, 14, 18). Nucleotide sequence analysis showed that silent BoNT/B genes usually have high homology to the gene from authentic proteolytic type B *C. botulinum*. However, the silent genes contain stop signals and deletions that prevent functional expression.

With the exception of *C. botulinum* types C and D, strains of *C. botulinum* generally only produce one neurotoxin type (14, 18, 33, 36). There are reports, however, of strains that produce two toxin types, one of which is produced in much higher titers than the other, such as Af, Bf, and Ba (18, 36), where the major toxin produced is designated by the uppercase letter. Recent studies have demonstrated that the *C. botulinum* neurotoxin genes occur in a cluster that is highly polymorphic or mosaic in its genetic structure (5, 10, 14). In summary, the genes for proteins of the toxin complexes are linked in clusters and may be associated with unstable genetic elements such as bacteriophages or plasmids.

C. tetani produces two virulence factors, the highly toxic TeNT and a hemolysin commonly termed tetanolysin. TeNT is produced after active growth, does not have a signal sequence, is not actively secreted, and is released upon cell lysis (33). It was shown at Harvard during development of TeNT vaccine that production of TeNT is dependent on a histidine-containing peptide in the medium, suggesting that nitrogen regulation is involved in production (33), as has been observed in *C. botulinum* (14, 17). In addition to nutritional regulation that probably occurs by global regulatory mechanisms, specific regulatory proteins closely linked to the TeNT gene also regulate toxin production. A TetR-encoding gene in *C. tetani* with (50 to 65%) identity to *botR* was recently shown to be located immediately upstream from the TeNT structural gene. The gene encoded a 21.5-kDa protein that possessed features of a DNA-binding protein, including a helix-turn-helix motif and a basic pI (9.53) (22). It showed homology with other putative regulatory genes in *Clostridium* spp., including ~30% identity with *uviA* from *C. perfringens* and *txeR* from *Clostridium difficile*. Overexpression of TetR in *C. tetani* increased TeNT expression and the level of the corresponding mRNA, which indicated that TetR is a positive transcriptional regulator of the TeNT gene. Chemically defined media have been developed that support the growth of *C. tetani*, but these media give low toxin titers and lead to strain degeneration (33). Although TeNT is produced mainly after growth has

ceased, it is not associated with sporulation (33). Highly toxigenic strains of both *C. tetani* and *C. botulinum* (type A) tend to sporulate poorly. These observations suggest that sporulation and toxin synthesis are governed by opposing regulatory mechanisms.

BoNT AND TeNT STRUCTURE

"Crystalline" botulinum toxin was first precipitated from culture fluid in 1908 by Somer and later purified by other researchers (32, 33, 37). Researchers in the United States and Japan demonstrated that the large (900-kDa) crystalline type A molecule could be separated into toxic and nontoxic moieties (see references 32, 33, and 37 for reviews). Later, using centrifugation techniques or chromatography on ion exchange resins, the NT component was isolated (37).

BoNTs are proteins of about 150 kDa that naturally exist as components of progenitor toxic complexes: as the M complex (ca. 300 kDa) consisting of BoNT associated with an NTNH protein of 120 to 140 kDa, or as the L and LL complexes (about 450 and 900 kDa, respectively) in which the M complex associates with a hemagglutinin protein(s) (32, 37). The nontoxic proteins in the complexes have been demonstrated to provide protection during manipulations and during passage through the gastrointestinal tract (Table 1). TeNT is produced as a single peptide of ca. 150 kDa and is not known to form complexes with nontoxic proteins.

BoNTs and TeNT are produced as single-chain molecules of ca. 150 kDa that achieve their characteristic high toxicities (Table 1) by posttranslational proteolytic cleavage to form a di-chain molecule composed of an L chain (~50 kDa) and a heavy (H) chain (~100 kDa) linked by a disulfide bond (35, 37). BoNTs and TeNT consist of three basic functional domains (Fig. 4): (i) L chain, the catalytic domain that has endopeptidase activity on neuronal substrates; (ii) H_N, the translocation domain residing in the N-terminal region of the H chain; and (iii) H_C, the receptor-binding domain located in the C-terminal region of the H chain. The gene and amino acid sequences, structure, and pharmacology of TeNT and BoNTs have been thoroughly reviewed (10, 27, 29, 31).

TeNT and BoNTs comprise a unique group of zinc proteases with certain unusual properties compared with other metalloproteases. The amino acid sequences of all seven BoNTs and TeNT have been deduced from the corresponding genes. The overall amino acid identity is about 40% for the eight NTs (29), and TeNT and BoNT have striking regions of homology, particularly in the residues defining the catalytic active site, in the translocation domain, and in the two cysteine residues forming the disulfide bond connecting the H chain and the L chain (28, 29). The least degree of homology is in the carboxyl region of the H chain, which is involved in neurospecific binding. Various other regions have low similarities of amino acid sequence (29). The high homologies of amino acid sequences from physiologically and genetically distinct clostridia indicate that a single ancestral gene was dispersed by lateral transfer among the clostridia.

Solving the crystal structure of BoNT/A and BoNT/A complex (6, 20) proved to be a daunting endeavor and has led to considerable insight into the structure of BoNT. After the isolation of diffracting crystals and preliminary X-ray analysis were reported in 1991, the crystal structure of

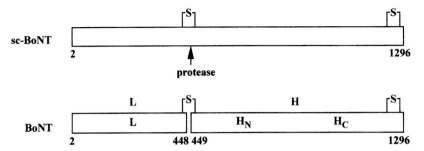

FIGURE 4 Proteolytic cleavage sites and basic domain structure of BoNT/A. After proteolytic activation, the NT consisting of an L chain and an H chain can be defined to three basic domains: (i) L chain, catalytic domain; (ii) H_N, translocation domain; and (iii) H_C, receptor-binding domain. Diagram courtesy of Marite Bradshaw, University of Wisconsin.

the entire 1,285-amino-acid di-chain structure was determined at 3.3 Å resolution (20) (Fig. 5). The overall shape of BoNT/A is rectangular with dimensions of ca. 45 Å × 105 Å × 130 Å, and the molecule shows a linear arrangement of the three functional domains with no contact between the catalytic and binding domains. The three functional domains have separated and distinct structures, with the exception of an unusual loop that encircles the perimeter of the catalytic domain (20). The finding of this loop was unexpected and presents a puzzle regarding catalysis since it covers the catalytic active site. It is possible that the site opens on contact with the substrate or when

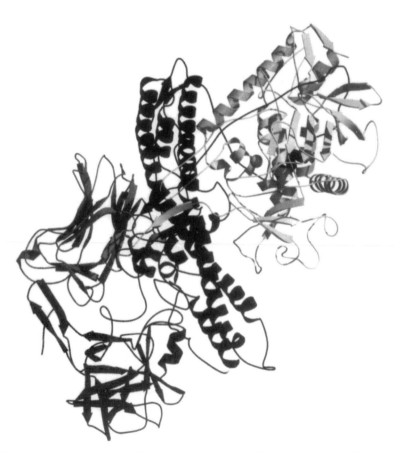

FIGURE 5 Diagram of the three-dimensional structure of BoNT/A. The catalytic domain is located in the upper right, the translocation domain in the center, the N-terminal binding subdomain at the left, and the C-terminal binding domain at the lower left. The catalytic zinc is represented as a ball. The overall structure is 45 Å × 105 Å × 130 Å. Drawing courtesy of Ray C. Stevens, University of California, Berkeley.

internalized within neurons, but the mechanism of catalysis will require further studies. The structure revealed that the ganglioside-binding C-terminal subdomain has structural homology with proteins known to interact with sugars such as the H_C fragment of TeNT, serum amyloid P, sialidase, various lectins, and the cryia and insecticidal-δ-endotoxin, which bind glycoproteins and create leakage channels in membranes (20). The proteins with the most homology to the catalytic domain are thermolysin and leishmanolysin. The translocation domain was distinct in structure from bacterial pore-forming toxins and showed more resemblance to coiled-coil viral proteins such as HIV-1, gp41/GCN4, influenza hemagglutinin, and the Moloney murine leukemia virus transmembrane fragment (20). BoNT appears to consist of a hybrid of varied structural motifs that evolved by combination of functional subunits to generate a highly toxic pathogenic molecule.

Burkhard et al. (6) also successfully determined the three-dimensional structure of the 900-kDa botulinum type A complex to 15 Å resolution in an effort to understand the biological significance of the large complex. The complex was triangular in shape, had an estimated radius of about 110 Å, and possessed six distinct cylindrical lobes. The BoNT component appeared to be located in the center of the complex, where it possibly was protected from the external environment.

Although the complete three-dimensional structure of TeNT has not yet been solved at high resolution, Umland et al. (39) have determined the structure of the receptor-binding fragment H_C of TeNT, resolved to 2.7 Å. It contained two closely associated domains, including a variation of the beta-trefoil motif in the C-terminal domain and a putative ganglioside-binding site.

MECHANISMS OF INHIBITION OF NEUROTRANSMISSION BY BoNT AND TeNT

Poisoning of nerves by TeNT and BoNTs involves a multistep process (4, 12, 13, 28, 35), including binding to gangliosides on presynaptic nerve cells, internalization by receptor-mediated endocytosis and/or synaptic vesicle reuptake, induction of channel formation, and internalization of the L catalytic domain into the presynaptic neuronal cytosol. The postulated model for trafficking of the L chain into the nerve cytosol follows models elucidated for protein toxins with intracellular targets such as diphtheria toxin, *Pseudomonas aeruginosa* exotoxin A, ricin, and certain lipid-containing viruses. Binding occurs by the H_C region of the H chain, probably initially to polysialogangliosides [(TeNt: highest affinity to G_{T1b}, G_{D1b}; lower affinity to G_{M1} and G_{D1a}); (BoNT: varying affinities to G_{D1a}, G_{T1b}, and others depending on serotype)] (12, 28). It is unlikely that polysialogangliosides are the sole receptors of BoNTs and TeNT, since the affinities ($\sim 10^{-8}$ M) do not provide sufficient binding needed for in vivo NT intoxication (28). In clinical botulism and tetanus, the peripheral concentrations of BoNT and TeNT are in the subpicomolar range ($\leq 10^{-12}$ M). Experiments indicate that protein receptors are also involved (28), and although candidate proteins have been suggested, they have not been conclusively identified and characterized for the various serotypes of BoNT and for TeNT (12, 28).

After being incorporated within endosomes, acidification of the endosome environment has been postulated to induce pore formation and entry of the L chain into the

nerve cytosol following the model for diphtheria toxin (see chapter 59, this volume). Once internalized, the zinc-dependent catalytic domain of the seven serotypes of BoNTs interact with and cleave neuronal soluble *N*-ethylmaleimide-sensitive fusion protein receptor (SNARE) proteins involved in neurotransmitter vesicle transport to the membrane, vesicle-membrane fusion, and calcium-induced exocytosis (28). The SNAREs syntaxin and SNAP-25 are associated with the cytoplasmic face of presynaptic nerve terminals, and VAMP is inserted into the synaptic vesicle membrane (Fig. 6). The three SNAREs, VAMP, SNAP-25, and syntaxin, possess a distinct three-dimensional motif (the SNARE motif) that is required for specific proteolysis by the NTs (13, 28). This nonapeptide motif occurs not only in mammals but also in *Drosophila melanogaster*, *Torpedo mermorata*, and *Arabidopsis thaliana*, but the NTs act poorly on substrates from certain of these organisms, probably owing to mutations within or outside the SNARE motif. BoNTs and TeNT are unusual among zinc-dependent proteases in requiring a substrate of at least 16 amino acids in length, probably because the NTs recognize the shape of the substrate and not specific peptide bonds.

The use of modern molecular biology methods to investigate exocytosis, combined with electrophysiology, has led to tremendous advances in our understanding of the processes of neurotransmission and its inhibition by BoNTs and TeNT. In vitro, the SNAREs SNAP-25, VAMP, and syntaxin assemble into a stable ternary complex (13), and disassembly requires an ATPase *N*-ethylmaleimide-sensitive fusion (NSF) protein together with SNAPs (soluble NSF-attachment proteins). These SNAREs are most susceptible to cleavage when they are not assembled in the tight ternary complex. In vivo, both fast and slow phases of membrane fusion and neurotransmitter secretion are blocked by all but one of the different BoNTs (BoNT/A) or TeNT, suggesting that the SNAREs integral for secretion must exist both in a ternary complex and free within the neuron. Although much progress has been made in understanding the mechanisms by which clostridial NTs inhibit neurotransmission, many gaps exist in our knowledge of the mechanisms of the BoNTs and TeNT. The mechanism of receptor-mediated endocytosis and trafficking of the toxins to the nerve cytosol has not been solved. Besides being involved in exocytosis, SNAP-25 and other neuronal proteins perform several functions in the nerve, including axonal growth, maturation, and synaptogenesis (13, 28). Exocytosis is a complex process that undoubtedly involves many molecular interactions, and the role of NTs in altering these processes is unclear at this time. Much additional research is needed to understand their mechanisms and to validate the in vitro models using in vivo systems.

OTHER FACTORS INFLUENCING PATHOGENESIS OF *C. TETANI* AND *C. BOTULINUM*

Although BoNTs and TeNT are the primary determinants of virulence in neurotoxigenic clostridia, wound or intestinal infections also require additional virulence mechanisms. *C. tetani* is adept at colonizing wounds while *C. botulinum* does so poorly, whereas *C. botulinum* but not *C. tetani* colonizes and causes disease in the gastrointestinal tract of infants and susceptible adults. The factors that govern the differential colonization of wounds by *C. tetani* and *C. botulinum* have not been established. It is possible that the hemolysin of *C. tetani* promotes infection by enabling

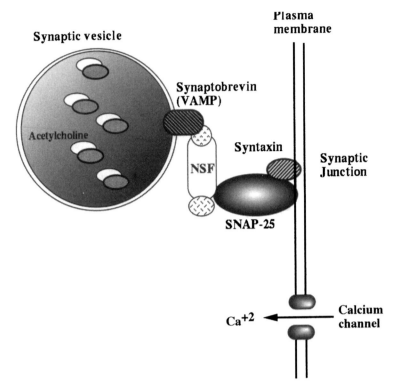

FIGURE 6 Diagram of proteins involved in trafficking synaptic vesicles containing acetylcholine to the neuronal plasma membrane. The SNARE proteins synaptobrevin (VAMP), syntaxin, and SNAP-25 form a ternary complex that is resistant to cleavage by botulinum toxin. In their free forms, the SNARE proteins are readily cleaved by BoNTs and TeNT at specific sites (see Table 1). N-ethylmaleimide fusion protein (NSF) is required for complex integrity and contact of vesicle with target membrane. Release of neurotransmitters from vesicles fused to the presynaptic membrane is triggered by influx of calcium. BoNTs and TeNT prevent exocytosis of neurotransmitter by interfering with trafficking of synaptic vesicles and possibly membrane fusion and exocytotic release of neurotransmitter by specific cleavage of SNARE proteins. Diagram courtesy of Michael Goodnough, University of Wisconsin.

the organism to acquire iron and other nutrients required for clostridial growth, but other virulence factors may be involved as well.

Sugiyama and colleagues (reviewed in reference 36) employed rodent animal models to study factors governing colonization of the intestinal tract by *C. botulinum*. The age dependence observed in humans was also duplicated in mice and rats. BoNT was produced in the colon after 2 to 3 days in 7- to 13-day-old mice challenged with *C. botulinum* type A spores. Mice 8 to 10 days old were most susceptible to this nonsymptomatic colonization. The average CD_{50} (50% colonizing dose) of *C. botulinum* type A spores was less than 1,000 spores in 9-day-old mice. The age-dependent susceptibility of mice to intestinal colonization by *C. botulinum* appears to be the counterpart of the clinical infection occurring in human infants less than 12 months of age. The age-related susceptibility in mice was not due to chronological age itself, since adult germfree mice are highly susceptible. BoNT was produced mostly in the large bowel, with slightly more toxin in the cecum than in the colon.

The role of indigenous flora in resistance was supported when germfree mice acquired a complex gut flora when exposed to normal mice in their cages. Peroral challenges with 10^5 spores were ineffective because of protection by the normal gut flora. However, the normally resistant adult mouse became susceptible when fed a kanamycin-erythromycin mixture that changes the nature of the gut flora. Pretreatment with metronidazole also sensitized mice to infection, suggesting that obligate anaerobes contributed to resistance of colonization. Interestingly, *C. perfringens* and *C. difficile* were identified in the germfree model as protecting against challenge by 10^5 spores. Two *Lactobacillus*, one *Bacteroides*, four fusiform, and one spirochete strain did not prevent colonization of germfree mice with 50 type A spores but did reduce the mortality rate.

Smith and Sugiyama (36) postulated that asymptomatic colonization of humans may occur more frequently than is currently believed, and overt botulism may occur only when BoNT reaches a site in the gastrointestinal tract from which it is actively absorbed. BoNT has been detected in the stools of healthy human adults, and colonization has been documented in adults whose gastrointestinal flora are altered by antibiotic therapy (18, 36). Intestinal lesions caused by *C. botulinum* have been observed in horses and other animals. The colonization of the intestinal tract by *C. tetani* has recently been postulated to cause neurological and developmental disabilities in humans, including autism

(3). Inhibition of neurotransmitter release by TeNT could have a much greater impact on central nervous system function and development than previously recognized.

SAFETY IN WORKING WITH BoNTs AND TeNT

BoNTs and TeNT are extremely toxic molecules and are considered the most potent poisons known (33, 34). Toxicity of BoNT/A has been estimated as 0.2 ng/kg of body weight in humans, and the lethal dose is 1 μg or less. Because the consequences of an accidental intoxication are so severe, safety must be a primary concern of scientists studying these toxins. The CDC recommends biosafety level 3 primary containment and personnel precautions for facilities making large quantities of the NTs. All personnel who work in the laboratory should be immunized with a pentavalent (A-E) toxoid available from the CDC. A biosafety manual should be posted in the laboratory and should contain the proper emergency phone numbers and procedures for emergency response, spill control, and decontamination. When performing steps in which aerosols may be created, special precautions need to be taken. A class II or III biological safety cabinet or respiratory protection should be used. The use of needles and syringes for bioassays requires extreme caution.

Beginning in 1997, C. botulinum cultures and toxins were included in a group of select agents whose transfer has been controlled by the CDC. To transfer these agents, both the person sending and the person receiving them must be registered with the CDC and exchange the appropriate approval forms.

CONCLUSIONS AND PERSPECTIVES

Remarkable advances have been achieved during the past decade in elucidating the biochemistry, structure, and pharmacological mechanisms of BoNTs and TeNT. The discovery that the light chains of the clostridial NT family cleave neuronal substrates involved in transmitter vesicle transport and exocytosis in the presynaptic nerve terminal has contributed greatly to an understanding of the mechanisms of clostridial NTs. These advances have certainly contributed to the remarkable success of botulinum toxin as a pharmacological agent for the treatment of myriad hyperactive muscle diseases. Much information has also been gained regarding the genes encoding the toxins, and we are beginning to determine their mechanism of regulation and release from the cell. Paradoxically, lagging behind is knowledge of pathogenesis by the neurotoxic clostridia, such as the virulence mechanisms required for infection of the gastrointestinal tract of susceptible infants and adults by C. botulinum, factors affecting the propensity of C. tetani compared with C. botulinum to infect wounds, and the role of other microorganisms and host defense mechanisms in controlling infection and neurotoxin activity. To prevent further human illness and deaths by neurotoxic clostridia, antidotes are urgently needed that can reverse the detrimental affects of BoNTs and TeNT once they are bound and internalized in nerves. Further knowledge of the properties of the neurotoxigenic clostridia and their NTs will also lead to improved pharmaceuticals for prevention as well as treatment of human disease using these most potent toxins.

Research in the author's laboratory has been supported by the NIH, USDA, University of Wisconsin, and industry sponsors of the Food Research Institute, University of Wisconsin-Madison.

REFERENCES

1. Aktories, K. (ed.). 1997. *Bacterial Toxins. Tools in Cell Biology and Pharmacology. A Laboratory Companion.* Chapman and Hall, London, U.K.

2. Bleck, T. P. 1989. Clinical aspects of tetanus, p. 379–398. *In* L. L. Simpson (ed.), *Botulinum Toxin and Tetanus Toxin.* Academic Press, Inc., San Diego, Calif.

3. Bolte, E. R. 1998. Autism and *Clostridium tetani*. *Med. Hypoth.* **51:**133–144.

4. Boquet, P., P. Munro, C. Fiorentini, and I. Just. 1998. Toxins from anaerobic bacteria: specificity and molecular mechanisms of action. *Curr. Opin. Microbiol.* **1:**66–74.

5. Bradshaw, M., M. C. Goodnough, and E. A. Johnson. 1998. Conjugative transfer of the *Escherichia coli*-*Clostridium perfringens* shuttle vector pJIR1457 to *Clostridium botulinum* type A strains. *Plasmid* **40:**233–237.

6. Burkhard, F., F. Chen, G. M. Kuziemko, and R. C. Stevens. 1997. Electron density projection map of the botulinum neurotoxin 900-kilodalton complex by electron crystallography. *J. Struct. Biol.* **120:**78–84.

7. Cato, E. P., W. L. George, and S. M. Finegold. 1986. Genus *Clostridium*, p. 1141–1200. *In* P. H. A. Sneath, N. S. Mair, M. E. Sharpe, and J. G. Holt (ed.), *Bergey's Manual of Systematic Bacteriology*, vol. 2. The Williams & Wilkins Co., Baltimore, Md.

8. Centers for Disease Control and Prevention. 1998. *Media for Isolation, Characterization, and Identification of Obligately Anaerobic Bacteria.* U.S. Department of Health and Human Services, Public Health Service, Centers for Disease Control and Prevention, Atlanta, Ga.

9. Cherington, M. 1998. Clinical spectrum of botulism. *Muscle & Nerve* **21:**701–710.

10. Collins, M. D., and A. K. East. 1998. Phylogeny and taxonomy of the food-borne pathogen *Clostridium botulinum* and its neurotoxins. *J. Appl. Microbiol.* **84:**5–17.

11. Eklund, M. W., F. T. Poysky, and W. H. Habig. 1989. Bacteriophages and plasmids in *Clostridium botulinum* and *Clostridium tetani* and their relationship to production of toxins, p. 25–51. *In* L. L. Simpson (ed.), *Botulinum Toxin and Tetanus Toxin.* Academic Press, Inc., San Diego, Calif.

12. Halpern, J. L., and E. A. Neale. 1995. Neurospecific binding, internalization, and retrograde axonal transport, p. 221–242. *In* C. Montecucco (ed.), *Clostridial Neurotoxins.* Springer-Verlag, Berlin, Germany.

13. Hanson, P. I., J. E. Heuser, and R. Jahn. 1997. Neurotransmitter release—four years of SNARE complexes. *Curr. Opin. Neurobiol.* **7:**310–315.

14. Hatheway, C. L., and E. A. Johnson. 1998. *Clostridium*: the spore-bearing anaerobes, p. 731–782. *In* L. Collier, A. Balows, and M. Sussman (ed.), *Topley & Wilson's Microbiology and Microbial Infections*, 9th ed., vol. 2. *Systematic Bacteriology.* Arnold, London, U.K.

15. Holdeman, L. V., E. P. Cato, and W. E. C. Moore. 1979. *Anaerobe Laboratory Manual*, 4th ed. Virginia Polytechic Institute and State University, Blacksburg.

16. Jankovic, J., and M. Hallett (ed.). 1994. *Therapy with Botulinum Toxin.* Marcel Dekker, Inc., New York, N.Y.

17. **Johnson, E. A.** 1999. Anaerobic fermentations, p. 139–150. *In* A. L. Demain and J. E. Davies (ed.), *Manual of Industrial Microbiology and Biotechnology,* 2nd ed. ASM Press, Washington, D.C.

18. **Johnson, E. A., and M. C. Goodnough.** 1998. Botulism, p. 723–741. *In* L. Collier, A. Balows, and M. Sussman (ed.), *Topley & Wilson's Microbiology and Microbial Infections,* 9th ed., vol. 2. *Systematic Bacteriology.* Arnold, London, U.K.

19. **Johnson, J. L., and B. S. Francis.** 1975. Taxonomy of the clostridia : ribosomal ribonucleic acid homologies among the species. *J. Gen. Microbiol.* **88:**229–244.

20. **Lacy, D. B., W. Tepp, A. C. Cohen, B. R. DasGupta, and R. C. Stevens.** 1998. Crystal structure of botulinum neurotoxin type A and implications for toxicity. *Nat. Struct. Biol.* **5:**898–902.

21. **Lambert, H. P. (ed.).** 1991. *Infections of the Central Nervous System.* B. C. Decker, Inc., Philadelphia, Pa.

22. **Marvaud, J.-C., U. Eisel, T. Binz, H. Niemann, and M. R. Popoff.** 1998. TetR is a positive regulator of the tetanus toxin gene in *Clostridium tetani* and is homologous to *botR. Infect. Immun.* **66:**5698–5702.

23. **Marvaud, J. C., M. Gibert, K. Inoue, Y. Fujinaga, K. Oguma, and M. R. Popoff.** 1998. botR/A is a positive regulator of botulinum neurotoxin and associated nontoxin protein genes in *Clostridium botulinum* A. *Mol. Microbiol.* **29:**1009–1018.

24. **Maselli, R. A., W. Ellis, R. N. Mandler, F. Sheikh, G. Senton, S. Knox, H. Salari-Namin, M. Agius, R. L. Wollman, and D. P. Richman.** 1997. Cluster of wound botulism in California: clinical, electrophysiological, and pathologic study. *Muscle & Nerve* **20:**1284–1295.

25. **Middlebrook, J. E., and J. E. Brown.** 1995. Immunodiagnosis and immunotherapy of tetanus and botulinum neurotoxins, p. 89–123. *In* C. Montecucco (ed.), *Clostridial Neurotoxins.* Springer-Verlag, Berlin, Germany.

26. **Miyamoto, O., J. Minami, T. Toyoshima, T. Nakamura, T. Masada, S. Nagao, T. Negi, T. Itano, and A. Okabe.** 1998. Neurotoxicity of *Clostridium perfringens* epsilon-toxin for the rat hippocampus via the glutamatergic system. *Infect. Immun.* **66:**2501–2508.

27. **Montecucco, C. (ed.)** 1995. *Clostridial Neurotoxins.* Springer-Verlag, Berlin, Germany.

28. **Montecucco, C., and G. Schiavo.** 1995. Structure and function of tetanus and botulinum neurotoxins. *Q. Rev. Biophys.* **28:**423–472.

29. **Niemann, H.** 1991. Molecular biology of the clostridial neurotoxins, p. 303–348. *In* J. H. Alouf and J. H. Freer (ed.), *A Sourcebook of Bacterial Protein Toxins.* Academic Press, Inc., London, U.K.

30. **Rappouli, R., and C. Montecucco (ed.).** 1997. *Guidebook to Protein Toxins and Their Use in Cell Biology.* Oxford University Press, Oxford, U.K.

31. **Rood, J. I., B. A. McClane, G. G. Songer, and R. W. Titball (ed.).** 1997. *The Clostridia: Molecular Biology and Pathogenesis.* Academic Press, Inc., San Diego, Calif.

32. **Sakaguchi, G.** 1983. *Clostridium botulinum* toxins. *Pharmacol. Ther.* **19:**165–194.

33. **Schantz, E. J., and E. A. Johnson.** 1992. Properties and use of botulinum toxin and other microbial neurotoxins in medicine. *Microbiol. Rev.* **56:**80–99.

34. **Shone, C. C., and H. S. Tranter.** 1995. Growth of clostridia and preparation of their neurotoxins, p. 143–160. *In* C. Montecucco (ed.), *Clostridial Neurotoxins.* Springer-Verlag, Berlin, Germany.

35. **Simpson, L. L. (ed.).** 1989. *Botulinum Neurotoxin and Tetanus Toxin.* Academic Press, Inc., San Diego, Calif.

36. **Smith, L. D. S., and H. Sugiyama.** 1988. *Botulism. The Organism, Its Toxins, the Disease.* Charles C Thomas, Springield, Ill.

37. **Sugiyama, H.** 1980. *Clostridium botulinum* neurotoxin. *Microbiol. Rev.* **44:**419–448.

38. **Udwadia, F. E.** 1994. *Tetanus.* Oxford University Press, Oxford, U.K.

39. **Umland, T. C., L. M. Wingert, S. Swaminathan, W. F. Furey, J. J. Schmidt, and M. Sax.** 1997. Structure of the receptor binding fragment HC of tetanus neurotoxin. *Nat. Struct. Biol.* **4:**788–792.

40. **Van Ermengem, E.** 1979. Classics in infectious disease. A new anaerobic bacillus and its relation to botulism. *Rev. Infect. Dis.* **1:**701–719. (Originally published in 1897 as: Ueber einen neuen anaeroben Bacillus und seine Beziehungen zum Botulismus. *Z. Hyg. Infektionskr.* **26:**1–56.)

41. **Willis, A. T.** 1969. *Clostridia of Wound Infection.* Butterworths, London, U.K.

Enterotoxic Clostridia: *Clostridium perfringens* Type A and *Clostridium difficile*

BRUCE A. McCLANE, DAVID M. LYERLY, J. SCOTT MONCRIEF, AND TRACY D. WILKINS

57

The clostridia are gram-positive, anaerobic, spore-forming rods, several species of which cause gastrointestinal disease in humans or domestic animals. The involvement of some clostridia in gastrointestinal disease is not surprising given the anaerobic nature of these bacteria and their ability to produce toxins, some of which are active on the gastrointestinal tract. In this chapter, we discuss two enterotoxin-producing clostridia that rank among the most important enteric pathogens of humans, i.e., *Clostridium difficile* and enterotoxin-positive type A strains of *Clostridium perfringens*.

ENTEROTOXIN-PRODUCING *C. PERFRINGENS* TYPE A

The arsenal of *C. perfringens* currently contains at least 14 different toxins; however, a single isolate of this bacterium never expresses all 14 of these toxins. Based upon this observation, a commonly used classification scheme (38, 39) assigns *C. perfringens* isolates to five toxin types (A through E), depending upon their ability to produce the four major lethal toxins (Table 1).

The major lethal toxins are not the only biomedically important *C. perfringens* toxins; some *C. perfringens* isolates, mostly belonging to type A, express *C. perfringens* enterotoxin (CPE). Although CPE-positive *C. perfringens* type A strains represent <5% of the global *C. perfringens* population (38, 39), these bacteria are very important human gastrointestinal pathogens, causing *C. perfringens* type A food poisoning, as well as non-food-borne human gastrointestinal diseases such as sporadic diarrhea (diarrhea in the absence of food poisoning or antibiotic therapy) and antibiotic-associated diarrhea (AAD). Although not discussed further in this chapter, these bacteria are also an important cause of diarrhea in domestic animals (47).

C. perfringens Type A Food Poisoning

Epidemiology

C. perfringens type A food poisoning currently ranks as the second most common food-borne illness in the United States, where over 600,000 cases are estimated to occur each year, at annual costs exceeding $120 million (38, 39). The prevalence of this food poisoning stems, in large part, from several attributes of *C. perfringens* (38, 39). First, the ubiquitous presence of *C. perfringens* in soil, animal and human feces, river sediments, etc., provides this bacterium with ample opportunity to contaminate foods. Second, the very short (<10 min) doubling time, and relative aerotolerance, of *C. perfringens* permits rapid growth in foods, which is necessary for reaching the large bacterial load (>10^6 vegetative cells per g of food) required to initiate *C. perfringens* type A food poisoning. Finally, the ability of *C. perfringens* to survive in incompletely cooked foods is facilitated by the heat tolerance of this bacterium's vegetative cells, which have an optimal growth temperature of 43 to 45°C but can grow at temperatures up to at least 50°C, and the even greater heat resistance of its spores.

Recognized outbreaks of *C. perfringens* type A food poisoning are usually very large, averaging about 100 cases (38, 39). The large size of these outbreaks stems, in part, from the association of *C. perfringens* type A food poisoning with institutional settings. Institutions are predisposed for the development of this form of food poisoning because they often prepare foods in advance and then store these foods before serving, which, as described below, can facilitate the development of *C. perfringens* type A food poisoning. However, the large size of most recognized *C. perfringens* type A food poisoning outbreaks is also somewhat misleading, since outbreaks of this illness are typically overlooked by public health officials unless many people are affected.

The most important contributing factor to outbreaks of *C. perfringens* type A food poisoning is holding foods under improper storage conditions (38, 39); because of the relative heat tolerance of vegetative *C. perfringens* cells, foods should either be stored under refrigeration or at temperatures above 70°C to prevent bacterial growth. Incomplete cooking, which may allow *C. perfringens* endospores (or vegetative cells) to survive cooking and then germinate in the improperly cooked food, is the second most common contributing factor to *C. perfringens* type A food poisoning

Gram-Positive Pathogens, ed. by V. A. Fischetti et al.

ecules such as amino acids. Several lines of evidence (32, 38, 39) which indicate that large complex formation directly causes the onset of these CPE-induced small molecule membrane permeability alterations include (i) kinetic studies, which show a close correlation between large complex formation and the onset of membrane permeability changes; (ii) studies with CPE deletion fragments, which demonstrate a direct correlation between the amount of large complex present in a CPE-treated cell and the extent of CPE-induced permeability alterations suffered by that cell; and (iii) the observation that blockage of CPE cytotoxicity at 4°C appears to result directly from an inhibition of large complex formation, since all earlier steps in CPE action occur at this low temperature. Once CPE-induced small molecule permeability changes are initiated, the mammalian cell rapidly dies from lysis owing to a breakdown in the cell's colloid osmotic equilibrium or to metabolic disturbances. For example, CPE completely inhibits macromolecular synthesis, at least in part by causing the loss of cytoplasmic ions and precursors through the CPE-permeabilized plasma membrane.

Recent studies (50) have shown that membranes offer CPE substantial protection from pronase digestion when the toxin is sequestered in large complex. This membrane-mediated protection is consistent with the possibility that a portion of the CPE molecule becomes inserted into the lipid bilayer of the plasma membrane when the toxin is localized in large complex. If true, this CPE insertion event, which may also involve one or more membrane proteins, may correspond to the formation of a pore that allows small molecules to pass freely through the plasma membrane. This CPE insertion/pore formation hypothesis receives support from previous studies (38, 39) showing that sonication of isolated CPE into artificial lipid bilayers results in channel formation. Interestingly, even if CPE insertion does occur as a consequence of large complex formation, both pronase challenge and antibody probe studies (31, 50) indicate that at least a small portion of this enterotoxin species still remains on the membrane surface.

How does the CPE protein mediate this action? The presence of receptor-binding activity in the C-terminal half of the enterotoxin was initially suggested by studies using chemically cleaved native CPE; this tentative assignment was then unambiguously confirmed when a recombinant $CPE_{171-319}$ fragment was shown to exhibit similar binding characteristics as native CPE (38, 39). Subsequent subcloning experiments then further localized receptor-binding activity to the 30 C-terminal amino acids. Recent studies (32) using recombinant C-terminal CPE-deletion fragments demonstrated that removing the five C-terminal amino acids from CPE eliminates both receptor-binding and cytotoxic activity. Besides confirming the importance of the extreme C-terminal region of CPE for receptor-binding activity, this result also provides strong evidence that a second, independent receptor-binding domain does not exist elsewhere in the CPE molecule.

While C-terminal CPE fragments such as $CPE_{171-319}$ bind well to receptors, they completely lack cytotoxicity (32). Since this result indicates that sequences in the N-terminal half of CPE are necessary for cytotoxic activity, CPE, like most bacterial toxins, apparently segregates its binding and toxicity domains. Although some amino acids in the N-terminal half of CPE are clearly required for cytotoxicity, the presence of some N-terminal amino acids in native CPE partially inhibits biological activity (38, 39).

For example, limited trypsin or chymotrypsin digestion, which removes the first 25 or 36 N-terminal amino acids, respectively, from the toxin, increases CPE activity two- to threefold. It has been suggested that these intestinal proteases might similarly activate CPE in the intestine during infection.

Trypsin-activated CPE exhibits the same binding properties as native CPE, a result that is consistent with the binding activity being located at the extreme C terminus. Therefore, removing N-terminal CPE sequences must facilitate some postbinding step in CPE action. A recent study (32) identified the nature of this step by showing that two recombinant N-terminal CPE-deletion fragments (CPE_{37-319} and CPE_{45-319}) that exhibit the same two- to threefold activation of cytotoxic activity as trypsin- or chymotrypsin-activated native CPE, have similar binding properties as native CPE but form 2 to 3 times more large complex than native CPE. Therefore, removing up to the first 45 N-terminal amino acid residues from native CPE apparently enhances toxicity by promoting large complex formation. This finding provides additional evidence for the importance of large complex formation for CPE cytotoxicity, a contention that receives even more support from the characterization of another N-terminal CPE deletion fragment (CPE_{53-319}); this polypeptide exhibits similar binding characteristics as native CPE but forms no large complex and is completely nontoxic.

As mentioned earlier, CPE is expressed only during sporulation, where it often accounts for 15% or more of the total protein inside the sporulating C. perfringens cell (38, 39). This abundant CPE expression does not result from a gene dosage effect, since each cpe-positive C. perfringens cell apparently carries only a single copy of the cpe gene. Northern blot analyses and studies with cpe reporter constructs (19, 42) indicate that CPE expression is regulated at the transcriptional level, with cpe transcription starting soon after the onset of sporulation. Studies using translational fusions containing nested deletions of the DNA region upstream of the cpe open reading frame (ORF) fused to a gusA reporter gene have recently identified three cpe promoter elements that can drive sporulation-associated CPE expression (51). Several observations (19, 38, 39, 51) suggest that the cpe gene is transcribed as a monocistronic message: (i) Northern blot analyses have detected a single 1.2-kb cpe message in sporulating cultures of cpe-positive isolates; (ii) all three of the recently identified cpe promoters lie immediately upstream of the cpe ORF; and (iii) a putative loop structure has been identified 36 bp downstream of the 3′ end of the cpe ORF; since this downstream loop is followed by an oligo-T tract, it may function as a rho-independent transcription terminator.

Interestingly, the DNA sequences located immediately upstream of the putative transcriptional start sites of the three cpe promoters exhibit significant homology to Bacillus subtilis consensus sequences for RNA polymerase containing SigE or SigK (51). Furthermore, SigE and SigK homologs have now reportedly been identified in C. perfringens (51). Collectively, these findings suggest that SigE and SigK may play a role in initiating CPE synthesis during sporulation. Since these two sporulation-dependent sigma factors should be present in most or all C. perfringens isolates, this hypothesis receives support from recent studies (19) demonstrating that several naturally cpe-negative C. perfringens isolates express CPE, in a sporulation-associated manner, after being transformed with a low-copy-number cpe-containing plasmid. Some preliminary evidence suggests that additional factors besides SigE and SigK may also play a role in regulating CPE expression. Specifically, putative Hpr binding sequences have been identified upstream and downstream of *cpe*, and preliminary data suggest that *C. perfringens* may carry an *hpr* homolog (13). Since Hpr is a transition state regulator involved in the repression of some genes in *B. subtilis*, an Hpr homolog may repress CPE synthesis during vegetative growth of *C. perfringens*.

The ORF of the *cpe* gene appears to be highly conserved in *cpe*-positive type A isolates. Not a single base pair difference was detected (16) in the *cpe* ORFs from eight different type A *C. perfringens* isolates, which originated from diverse sources, including food poisoning, veterinary diarrhea, and non-food-borne human gastrointestinal diseases. Some type C and D isolates also express a serologically identical CPE (38, 39), but the *cpe* gene present in these type C and D isolates has not yet been sequenced.

Interestingly, recent studies (5) have established that most (or all) *C. perfringens* type E isolates carry *cpe* sequences that, although 90% homologous to the type A *cpe* ORF, are silent because they consistently lack the *cpe* initiation codon, promoters, and ribosome-binding site. They also contain nine nonsense and two frameshift mutations. These silent *cpe* sequences are highly conserved among type E isolates, although the isolates themselves do not share any apparent clonal relationship. Furthermore, these type E *cpe* sequences are always immediately adjacent to iota-toxin genes on a plasmid. Collectively, these results suggest that the progenitor type E isolate originated when a DNA element containing iota-toxin genes was acquired from another bacterium, such as *Clostridium spiroforme*, which is known to carry iota-toxin-like genes. After this interspecies transfer, a recombinational or insertional event presumably occurred between the foreign DNA carrying the iota-toxin genes and a *cpe*-containing plasmid already present in the recipient *C. perfringens* cell. This insertion or recombination event resulted in loss of the *cpe* promoter; once CPE expression was disrupted, other mutations then accumulated in the *cpe* ORF. The highly conserved nature of the *cpe* sequences in a number of apparently unrelated type E isolates strongly suggests that the plasmid carrying these silent *cpe* sequences and the functional iota-toxin genes has recently been mobilized from the progenitor type E isolate to several different *cpe*-negative *C. perfringens* type A isolates, thereby converting them to type E isolates. If correct, this scenario for the origin of type E isolates represents some of the best evidence to date for the horizontal transfer of virulence genes in the clostridia.

The single copy of the *cpe* gene present in CPE-positive *C. perfringens* type A isolates can have either a chromosomal or a plasmid location (16–18, 29). Interestingly, there is a relationship between *cpe* location and certain CPE-associated diseases, with most food poisoning isolates carrying a chromosomal *cpe* gene, whereas most (or all) human non-food-borne disease isolates, and most veterinary disease isolates, carry a plasmid-borne *cpe* gene (16, 17).

Genome mapping studies (38, 39) indicate that the *cpe* gene is located in a highly variable region of the chromosome in the food poisoning isolate NCTC 8798, suggesting that this gene may be present on an integrated mobile genetic element. This hypothesis receives further support from studies (14) showing that IS*1470* sequences are located upstream and downstream of *cpe* in NCTC 8239, a food poisoning strain carrying a chromosomal *cpe* gene.

Based upon this information, and additional sequencing data, it has been suggested (14) that, in food poisoning strains such as NCTC 8239, *cpe* is located on a 6.3-kb compound transposon that has inserted into the chromosome between two housekeeping genes, *nadC*, which potentially encodes a quinolate phosphoribosyltransferase and *uapC*, which encodes a putative purine permease. While the *cpe*-positive strains carrying a plasmid-borne *cpe* gene have not been as thoroughly characterized as chromosomal *cpe* strains, it is known that these strains consistently carry another insertion element, IS*1151*, which is always located downstream of the *cpe* gene. In addition, an IS*1469* element is located immediately upstream of both the plasmid-borne and chromosomal *cpe* genes; in chromosomal *cpe* isolates, this IS*1469* sequence lies within the putative compound transposon. While these associations between insertion elements and *cpe* suggest that *cpe* may be disseminated by transposable elements, this hypothesis remains to be experimentally proved.

Immunology

Protective immune responses do not contribute substantially to the rapid resolution of *C. perfringens* type A food poisoning (38, 39). Instead, symptoms of this illness are thought to quickly abate, in large part, because diarrhea flushes the gastrointestinal tract, thus removing unbound CPE and CPE-producing *C. perfringens* cells from the intestines. Most Americans have detectable levels of CPE antibodies in their serum, but there is no evidence that prior exposure to *C. perfringens* type A food poisoning provides any significant long-term protection against future acquisition of this illness.

Treatment and Prevention

Treatment of *C. perfringens* type A food poisoning is symptomatic. Because antimicrobials are not used to treat this food-borne illness, there are few reliable data available regarding the antibiotic-resistance patterns of *C. perfringens* food poisoning isolates.

No vaccine is currently licensed for the prevention of *C. perfringens* type A food poisoning. However, epitope mapping studies (38, 39) have demonstrated the presence of a neutralizing linear epitope in the 30 C-terminal amino acids of CPE. Since C-terminal CPE fragments are not cytotoxic, these C-terminal CPE fragments hold promise as potential candidates for the development of a CPE vaccine. This hypothesis receives support from studies (38, 39) showing that mice developed high titers of CPE-neutralizing serum antibodies when they were immunized with a conjugate containing the 30 C-terminal CPE amino acids coupled to a thyroglobulin carrier. To obtain a strong CPE-neutralizing IgA response in the intestines, further studies are needed to develop antigen delivery systems for these C-terminal CPE fragments. If successful, a CPE vaccine might potentially be used in such high-risk populations as nursing home residents or, perhaps, in veterinary medicine to protect domestic animals from CPE-associated disease.

CPE-Associated Non-Food-Borne Human Gastrointestinal Disease

In the past 10 to 15 years, CPE-producing *C. perfringens* has been identified as a major cause of non-food-borne human gastrointestinal diseases, with surveys (9, 15, 17) suggesting that CPE-producing *C. perfringens* is responsible for

~10 to 15% of all cases of AAD, which is a frequency approaching that of C. difficile-induced AAD, and 5 to 20% of all cases of sporadic diarrhea. Some intriguing recent evidence (15) suggests that an AAD patient can become sick from a coinfection involving both C. difficile and CPE-producing C. perfringens.

The antibiotic sensitivity of CPE-positive C. perfringens AAD isolates remains unclear, but some preliminary information (9, 15) suggests that treatment with several different antibiotics (e.g., penicillins) can predispose patients to C. perfringens AAD. Preliminary epidemiological data also suggest (9, 15) that, like C. difficile-induced AAD, C. perfringens AAD can be acquired from the environment; i.e., this illness is a true infection. Presumably, AAD then develops when the acquired CPE-positive C. perfringens isolates proliferate in the gut of patients whose gastrointestinal microflora have been disturbed by antibiotics or other factors. Most individuals suffering from C. perfringens AAD are compromised (e.g., transplant patients, burn patients, AIDS patients), but it is unclear whether these compromising conditions predispose patients to this gastrointestinal disease or if this association simply reflects higher rates of antibiotic usage in these compromised patient populations.

The pathogenesis of sporadic diarrhea from CPE-positive C. perfringens is also poorly understood at present. The only predisposing factor identified to date for this illness is age; i.e., C. perfringens sporadic diarrhea is more common in the elderly (15).

As mentioned earlier, C. perfringens isolates associated with non-food-borne human gastrointestinal disease consistently carry a plasmid-borne cpe gene, which distinguishes them from food poisoning isolates carrying a chromosomal cpe gene (17). These genotypic differences between CPE-associated non-food-borne human gastrointestinal disease isolates and food poisoning isolates are very interesting, since the symptoms of CPE-associated non-food-borne human gastrointestinal disease are typically more severe and long lasting (non-food-borne gastrointestinal illnesses can last for several weeks) than the symptoms of C. perfringens type A food poisoning. However, a recent study (16) detected no obvious phenotypic differences between food poisoning and non-food-borne gastrointestinal disease isolates, indicating that the relative contributions of host and bacterial factors to the pathogenesis of CPE-associated non-food-borne human gastrointestinal illnesses will clearly require further study.

C. DIFFICILE

C. difficile is an opportunistic pathogen that causes nosocomial diarrhea and colitis after the normal gastrointestinal flora has been altered, most typically by antibiotics (3, 4). This organism causes a spectrum of gastrointestinal diseases of the colon ranging from milder AAD to fulminating pseudomembranous colitis, which can be lethal if not treated. C. difficile causes essentially all cases of pseudomembranous colitis but only 25% or fewer of AAD cases; perhaps 15% of AAD cases are due to CPE-positive C. perfringens type A, with the remaining cases of AAD being currently undiagnosed. These cases could be due to infection by other clostridia, Salmonella sp., Escherichia coli, Candida albicans, or Staphylococcus aureus, or could result from unconjugated bile acids reaching the colon.

C. difficile-mediated disease develops from the production of two toxins, A and B, which are also referred to as the enterotoxin and cytotoxin, respectively. Clostridium sordellii, which is primarily a wound pathogen, produces toxin HT, or hemorrhagic toxin, and toxin LT, or lethal toxin, which immunologically cross-react with these C. difficile toxins; toxins A and HT are both enterotoxic as well as cross-reactive, while toxins B and LT are cross-reactive. These four toxins, along with alpha-toxin from Clostridium noyvi, compose a new class of large clostridial cytotoxins whose cytotoxic, enterotoxic, and lethal activity is due to their effects on the cytoskeleton, causing cell rounding and, eventually, cell death. The level of cytotoxic activity varies among these toxins, with toxin B being the most lethal; it is 100 to 1,000 times more lethal than toxin A against most cell lines. Toxins A, HT, and LT are comparable in their cytotoxic activity. These four toxins are all lethal when injected into animals but vary greatly in their enterotoxic activity. Toxin A binds to specific carbohydrates on the mucosa of the colon and causes extensive damage, whereas toxin B binds to cells that are not surrounded by a carbohydrate layer and thus is toxic for cells exposed by the action of toxin A. There is some evidence that toxins A and B act synergistically during the onset and development of disease (36). Some strains of C. difficile produce slight variants of these toxins that are more like C. sordellii toxins. In addition, some C. difficile strains produce toxin B but not toxin A (48); toxin B from these strains may have more enterotoxic activity than toxin B from A+/B+ strains.

Bacteriology and Diagnosis

The ability of C. difficile to produce subterminal spores makes this organism difficult to eradicate in hospitals. Unlike the expression of CPE, there is no direct correlation between sporulation and toxin production in C. difficile. Sporulation of C. difficile can be stimulated by sodium cholate and sodium taurocholate. This property can be used to improve recovery of the organism from clinical specimens.

C. difficile fluoresces yellow when grown on solid media, especially media containing blood. It also produces isocaproic acid, a volatile fatty acid produced during fermentation, and p-cresol, which is produced during tyrosine metabolism (6). Although isocaproic acid production is unusual, it cannot be used for definitive identification since this fatty acid is also produced by other clostridia, including Clostridium bifermentans, C. sordellii, and Clostridium sporogenes. In the clinical laboratory, isolates are often identified by their characteristic "horse dung" odor.

Laboratory tests used as diagnostic aids for C. difficile disease include bacterial culture, latex agglutination, tissue culture, and ELISAs (11, 30, 36). Bacterial culture is difficult and inconsistent owing to variations in anaerobiosis, specimen collection, and media. Cycloserine-cefoxitin-fructose agar (CCFA) is the selective medium most often used for isolation of C. difficile. This medium is an egg yolk agar base containing cycloserine (500 μg/ml) and cefoxitin (16 μg/ml), along with 5% egg yolk suspension (23). Morphologically, the organism grows on CCFA as flat colonies that have a yellow fluorescence. These colonies are best viewed after 72-h incubation. Broth media can be made selective for C. difficile by the addition of cycloserine and cefoxitin.

Isolation of C. difficile is not sufficient for diagnosis because ~20% of C. difficile isolates do not produce toxins and, therefore, are not pathogenic. Additional testing by tissue culture or ELISA must be performed to demonstrate that the isolates can produce toxin. Latex agglutination tests detect glutamate dehydrogenase, which is a nontoxic metabolic enzyme produced by all strains of C. difficile (35). Some other anaerobes, including C. sporogenes, certain types of C. botulinum, and Peptostreptococcus anaerobius, produce glutamate dehydrogenase enzymes that cross-react with C. difficile glutamate dehydrogenase and thus produce false-positive reactions. As with bacterial culture, additional testing must be done to confirm the presence of C. difficile toxin in specimens testing positive by latex agglutination.

The presence of C. difficile toxin in stools can be detected by tissue culture assay or by ELISA. The tissue culture test is the more sensitive assay when performed properly, although some of the newer ELISAs are approaching similar levels of sensitivity. The detection of toxin by tissue culture is less controlled than that with ELISAs, and its accuracy varies greatly between hospital laboratories. Tissue culture also requires 24 to 48 h to obtain results, and the toxin must remain active in the stool for detection. ELISAs are a more controlled test, with a rapid turnaround time (<3 h), and can detect inactive toxin that remains serologically active. Recently, several immunochromatography dipstick tests have become available commercially. Although these tests offer results in <30 min, unpublished results presented at recent national meetings suggest that the sensitivity of these formats may be less than that of the ELISAs.

A common misconception about C. difficile-induced diarrhea is that C. difficile strains are resistant to the inciting antibiotic. In fact, most C. difficile strains are susceptible to a wide variety of antibiotics associated with the onset of disease, including penicillin, ampicillin, erythromycin, tetracycline, chloramphenicol, and clindamycin. C. difficile-induced gastrointestinal disease often starts only after the antibiotic has been discontinued and antibiotic levels have dropped in the intestine. Some C. difficile strains have been identified that are relatively resistant to chloramphenicol, rifampin, erythromycin, clindamycin, and tetracycline; resistance to some of these antibiotics is conferred by chromosomally carried determinants, and resistance to tetracycline may result from a resistance determinant carried by a conjugative transposon. However, since antibiotic levels in the colon are often much higher than those in the bloodstream, this resistance must be interpreted cautiously. For practical purposes, all human C. difficile strains are susceptible to vancomycin and metronidazole, the antibiotics most commonly used to treat C. difficile-induced gastrointestinal disease; however, there have been reports of metronidazole-resistant C. difficile in horses (37).

Epidemiology

C. difficile is the major cause of nosocomial diarrhea in industrialized countries (34). Outbreaks of this disease are increasingly common in hospitals and health care facilities, particularly Veterans Administration hospitals and large medical centers with large numbers of older susceptible patients. The increased incidence of C. difficile-induced gastrointestinal disease results, in part, from increased recognition of this illness, with most health care facilities now testing for the presence of C. difficile toxins in stools. Of utmost importance to the health care professional is knowledge of how this organism is transmitted and the role that health care workers play in this transmission. In a study on the acquisition and spread of C. difficile that was carried out over an 11-month period in an endemic setting (41), it was shown that (i) patient-to-patient transmission

is common; (ii) not all people who become infected develop clinical symptoms; and (iii) health care workers often transmit the organism. The number of asymptomatic carriers in these settings may exceed the number of persons who develop disease; these asymptomatic carriers are an epidemiologic concern since they may spread C. difficile to more susceptible patients.

The persistence of spores in the hospital makes this disease especially difficult to control. Spores are present throughout the rooms of patients, and their persistence is directly associated with increased risk of infection and disease. Spores can also be isolated from the hands, clothes, and shoes of health care workers, and typing studies have identified isolates that are the same as those from patients. Exposure to patients is not considered a threat to health care workers, although there is one reported incident in which healthy workers developed diarrhea following exposure to a patient with C. difficile disease (20).

Up to 50% of asymptomatic infants (up to the age of several months) carry toxigenic C. difficile in their normal flora (36). The mechanism responsible for their protection from disease is not known, but some evidence indicates that infants tend to be colonized with serogroups that are not typically associated with adult disease and that often produce lower levels of toxin in vitro, suggesting that infants tend to be colonized with weakly toxigenic isolates. However, there are some instances in which asymptomatic infants have as much toxin in their colon as adults suffering from diarrhea and pseudomembranous colitis. Many infants also carry serogroup F strains, which behave phenotypically, at least in vitro, as A−/B+ isolates. Additional studies are needed to examine the toxin genes of these isolates at the molecular level.

The role of infants as a reservoir for the organism is still unclear, and currently no precautionary measures are used to limit the potential spread of the organism from nurseries to other hospital areas. The resistance of infants or further identification of host specificity may represent a key to designing new treatments for C. difficile disease. The resistance afforded by age is not limited to infants, since most affected patients with disease are over the age of 50; younger patients are much more likely to have toxin in their colons but remain asymptomatic.

Patients can harbor multiple C. difficile strains at one time and can have both toxigenic and nontoxigenic isolates simultaneously present in their colon (36). Relapses are a common problem, especially in those patients remaining in hospitals for extended stays. In most instances, relapse results from the initial inciting strain, not from infection with a new strain. Most relapses probably occur from incomplete eradication of the organism following treatment with vancomycin or metronidazole, or from reacquiring the strain from spores present in the hospital room (34).

Pathogenesis

C. difficile cannot compete with the normal flora of adult humans or other animals and only grows to high numbers when the flora have been eliminated or drastically altered by antibiotics or chemotherapy. When the antibiotic declines below levels inhibitory to C. difficile, the ingested spores germinate and vegetative cells grow to high density in the colon. Whether or not the organism attaches to the colonic wall is unknown.

As the organism grows, it releases the toxins due to autolysis; thus, like CPE, neither toxin A nor toxin B appears to be secreted. The toxins then enter the host cell via receptor-mediated endocytosis and generalized pinocytosis and exert their molecular action. The initial diarrhea induced by C. difficile is caused by the toxins directly injuring the intestinal mucosa; however, the action of these toxins alone cannot explain the massive tissue destruction that occurs in pseudomembranous colitis. Much of this damage results from extensive inflammation due to infiltrating leukocytes. The role of inflammation in this process is substantiated by the presence of necrotic tissue and dead leukocytes in the pseudomembranes that develop in untreated disease. There are also elevated levels of prostaglandin E_2 and leukotriene B_4, both of which are associated with inflammation. Toxin A also has chemotactic properties, which most likely exacerbates the disease (45).

The hamster has been used as the animal model of choice (44) for studying the pathogenic mechanisms of C. difficile because hamsters are more susceptible than rats or mice and because treatment with antibiotics predisposes the hamster, like humans, to C. difficile disease. For example, when hamsters are given an approximately 3-mg dose of clindamycin intragastrically, they become infected with spores from contaminated bedding and then proceed to develop diarrhea and lethargy, with death following within several days. Furthermore, strains isolated from human cases of C. difficile-induced gastrointestinal disease are as pathogenic for hamsters as hamster isolates and, as for humans, disease in hamsters is caused only by toxin-producing strains (3, 7, 36).

Virulence Factors

Immunization of hamsters against toxins A and B is completely protective, indicating that the symptoms of C. difficile disease are mediated by these two toxins. Immunization of hamsters with only toxin A results in some protection against diarrhea, and mortality is reduced. Interestingly, C. difficile strains that produce the same amount of toxin can vary in virulence for hamsters, suggesting that other, still unrecognized, virulence factors also contribute to disease, perhaps by facilitating C. difficile growth in vivo. Some C. difficile isolates produce a thin capsule (40 to 80 nm in thickness) whose synthesis appears to be regulated by the level of glucose in the environment (6, 7). However, no association has been demonstrated between this capsule and either virulence or toxin production. Some isolates also produce fimbriae (8), but little is known about their production and, again, no correlation has been demonstrated between these fimbriae and virulence.

Toxins A and B are 308,000 M_r and 269,000 M_r, respectively, which makes them the largest known bacterial toxins (21). Both toxins are produced as single polypeptides and then released following autolysis of the bacterial cell. The toxin structural genes, toxA and toxB, have extensive homology; at the amino acid level they are 49% identical and 63% similar when conservative substitutions are considered. Both toxins have a complex contiguous series of repeating units at the C terminus. These units compose one-third of each toxin and consist of small repeating units within larger units. In toxin A, the repeating units bind to galactose-containing residues; the repeating units of toxin B may also be involved in binding, but the receptor has not been identified. The repeating subunits of toxins A and B have extensive similarity to the glucosyltransferases of Streptococcus mutans and Streptococcus sobrinus (GtfB, GtfC, and GtfI), which bind to carbohydrates (49). Most of the antibody elicited by immunization with

native toxin A reacts with the repeating units of this toxin; two sites (starting at amino acids 2097 and 2355) in this repeating region appear to be especially immunodominant, and these regions react with monoclonal antibodies that neutralize the enterotoxic activity of toxin A and inhibit toxin A binding to carbohydrates.

Upstream of the repeating units are several conserved features, including four cysteine residues located in almost identical positions, a central hydrophobic region, and a putative nucleotide-binding region (Fig. 2B). The N-terminal portions of toxins A and B, which include all of these regions, exhibit about 50% homology. The hydrophobic region, which is composed of a stretch of ~50 amino acids, may represent a membrane-spanning region that is involved in toxin uptake and intracellular processing. Altering the conserved cysteines, deleting the internal hydrophobic region, or removing the repeating units of toxin B by site-directed or deletion mutagenesis results in a 90% loss of cytotoxic activity. Modifying the histidine residue located in the potential nucleotide-binding site to glutamine causes a 99% loss in activity, suggesting a key role for this region in the toxic activity of toxin B (2).

The chromosomal genes encoding toxins A and B are located in a 19.6-kb pathogenicity islet, referred to as the toxigenicity element, which is chromosomally located and contains five distinct genes, designated txeR, toxB, txe2, toxA, and txe3 (Fig. 2A). The txeR, toxB, txe2, and toxA genes are transcribed in the same direction, but txe3 is read in the opposite direction. The toxB gene is located ~1 kb upstream of the toxA gene. The proteins encoded by txeR, txe2, and txe3 are not directly involved in the toxicity of toxins A or B, as evident from the fact that the toxA and toxB genes, when cloned independently, encode fully active toxins A and B (21, 25).

The boundaries of the toxigenicity element carrying the C. difficile toxin genes have been defined using a series of strains that vary by 5 orders of magnitude in their in vitro toxigenicity. Interestingly, the boundaries in both weak and strong toxin-producing strains are highly conserved, and there are no obvious discrepancies in the size of the inserts that could explain the broad range of toxin expression among C. difficile isolates. In nontoxigenic strains, a small chromosomal element, which was found to be 127 bp in one study (23a) but 115 bp in a second study (10), occupies the same chromosomal element as the toxigenicity element. There are no insertion sequences flanking the toxigenic element that might provide a mechanism for genetic exchange. However, there are inverted repeat regions within the small insertion region of nontoxigenic strains (10).

Most toxigenic strains of C. difficile produce both toxin A and B; nontoxigenic strains, which are also nonpathogenic, do not carry the toxin genes. Toxin production occurs during the stationary phase, under conditions (such as growth in dialysis tubing) that limit the growth of the organism. Toxin production is repressed by glucose (12), and when the organism is grown in rich medium in freestanding culture, very little toxin is produced.

There is remarkable DNA sequence conservation in the pathogenicity islet of both weakly and highly toxigenic C. difficile strains. Furthermore, the level of toxin-specific mRNA present in a C. difficile isolate directly correlates with the amount of toxin produced by that isolate, implying that the wide variation in toxin expression involves transcriptional regulation. The individual transcripts for txeR, toxB, txe2, and toxA are produced during stationary

(A) PATHOGENICITY ISLET ENCODING TOXINS A AND B

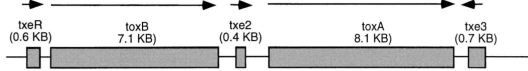

(B) PRIMARY STRUCTURE

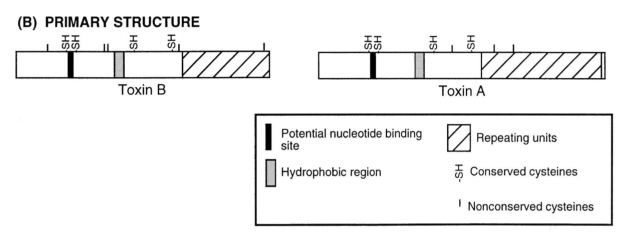

FIGURE 2 (A) The pathogenicity islet carrying the *toxA* and *toxB* genes. The entire DNA fragment composing the pathogenicity islet is approximately 19.6 kb. (B) Structural features conserved between toxins A and B.

phase, whereas *txe3* is transcribed during exponential growth. In addition to these smaller transcripts, a much larger (17.5 kb) transcript has been observed that apparently carries the information from *txe1*, *toxB*, *txe2*, and *toxA*. This large transcript is present at very low levels and probably results from incomplete transcriptional termination during high-level expression. Most transcription of *toxA* and *toxB* probably initiates from the individual promoters for these two genes; both promoters are very strong when the *txeR* gene product is present. The *txeR* gene, which encodes a small (22 kDa) protein containing a high number of lysine residues in its C terminus, positively regulates toxin expression (43). Experiments have shown that *E. coli* supplied with *txeR* in *trans* increased its expression of reporter sequences fused to either the *toxA* or *toxB* promoter. Whether or not *txeR* is part of a two-component regulatory system remains to be determined. It encodes a protein with a low level of homology to extracellular function sigma factors, which act as general stress response sigma factors.

Toxins A and B are monoglucosyltransferases (1) that cleave UDP-glucose and transfer glucose to Rho proteins, which are members of a subfamily of small GTP-binding proteins involved in the regulation of the cellular cytoskeleton and other cell functions, such as cell adhesion, microfilament organization, and nuclear signaling. Toxins A and B modify several Rho subtypes (RhoA, RhoB, and RhoC) as well as the functionally related Rac and Cdc42 proteins. The acceptor amino acid for enzymatic action by toxin A or B is residue Thr-37 in Rho. The monoglucosylation of Thr-37 in Rho represents a novel mechanism that offers molecular biologists new tools for studying the

cytoskeletal machinery of mammalian cells. The enzymatic region of toxins A and B resides in their N-terminal portion, since a recombinant fragment composing the 63-kDa N-terminal region of these toxins retains enzymatic activity (24). More detailed analysis (1) suggests that the region centered around residues 516 to 542 of these toxins is especially critical for activity. The effect of these toxins on the cytoskeletal system, through Rho modification, may cause detachment of the epithelial cells lining the mucosa, resulting in leakage of serous fluid into the intestines, which is then followed by diarrhea and the onset of colitis.

Toxin A binds to Galα1-3Galβ1-4GlcNAc and other galactose-containing residues (36). Toxin A is a more potent cytotoxin for tissue culture lines such as F9 teratocarcinoma mouse cells and HT-29 cells, both of which have large amounts of this trisaccharide on their surface. Humans do not produce this trisaccharide, but toxin A also binds to the carbohydrate antigens designated I, X, and Y (the latter two also referred to as Lewis X, Le^x, and Lewis Y, Le^y), which are produced in abundance by human intestinal epithelial cells. One theory for the resistance of infants to *C. difficile* disease is that infants lack toxin receptors; however, toxin A binds to human infant intestinal mucosa, which is known to contain the carbohydrate antigens that bind toxin A. In addition to uptake by receptor-mediated mechanisms, both toxins probably enter cells via generalized pinocytosis.

Immunology

The important contribution of inflammation to *C. difficile* disease has already been described. With regard to protective immune effects, little is known about the specific immunity that develops to *C. difficile* disease, although there is some evidence that convalescent patients develop mucosal and systemic antibodies against toxin A during infection (26). Relapses, which in the United States occur in nearly 20% of patients (36), are a major problem and represent a serious and debilitating form of the disease. Multiple relapses can occur, probably because the patient's intestinal tract is more susceptible after the initial damage and the fact that *C. difficile* spores often remain in the patient's environment.

Therapy and Prevention

Since the onset of *C. difficile* disease in hospitalized patients can be triggered by any agent that disrupts the normal intestinal microflora, the disease can often be treated simply by stopping the inciting agent. If illness continues, vancomycin or metronidazole can be administered orally (4, 45). Vancomycin is poorly absorbed, so high gastrointestinal concentrations can be achieved; however, there are concerns that overuse of vancomycin for treating non-life-threatening cases of *C. difficile* disease might contribute to the development of vancomycin resistance among commensal and pathogenic gastrointestinal flora. Although metronidazole is readily absorbed, it is therapeutically useful for treating *C. difficile* disease, particularly for the milder AAD cases. Metronidazole is also less expensive than vancomycin.

Alternative therapeutic approaches now being evaluated include the use of probiotics. For example, in a 23-month study, the incidence of AAD decreased from 31% in patients receiving a placebo to 9% in patients receiving *Saccharomyces boulardii*, a nonpathogenic yeast that remains viable in the human intestinal tract (40). Another probiotic, a human isolate of *Lactobacillus* sp. named strain GG, has also been given to humans; patients suffering relapses of *C. difficile* disease responded favorably following daily doses of 10^{10} *Lactobacillus* cells over a period of 7 to 10 days. Patients who received fecal enemas (consisting of freshly passed stool from the patient's spouse or child) were cured within 24 h. Administration of bacterial mixtures containing *Streptococcus faecalis*, *Clostridium innocuum*, *Clostridium ramosum*, *Bacteroides ovatus*, *Bacteroides vulgatus*, *Bacteroides thetaiotaomicron*, *E. coli*, *Clostridium bifermentans*, and *Peptostreptococcus productus* also reportedly caused a return to normal bowel habits within 24 h (36). Although probiotic therapy holds promise, it should be emphasized that these approaches have only been evaluated with low numbers of patients, so it remains unclear which specific approach should be used. However, probiotic therapy is appealing since this approach is aimed at restoration of the normal intestinal flora, which offers the most effective means of inhibiting the growth of *C. difficile*.

Antibodies against toxins A and B can neutralize all of their toxin activities, and immunization of laboratory animals against these toxins is completely protective against disease (36). These observations suggest that passive immunotherapy, such as oral administration of bovine milk antibodies against toxins A and B, or giving human immune globulin, holds promise for treating *C. difficile* disease. Furthermore, passive immunotherapy may be valuable for prevention, as well as therapy, since this approach protects against disease in laboratory animals (36). Passive immunotherapy should be of potential benefit to patients at risk, especially those undergoing prophylactic antibiotic therapy and/or those situated in endemic settings such as nursing homes or Veterans Administration hospitals. The

administration of antibodies also shows promise as an alternative to vancomycin or metronidazole therapy for patients suffering milder forms of disease and as adjunctive therapy in combination with other treatments.

Certain basic precautions should be taken to help control outbreaks of *C. difficile* disease. In endemic settings, disposable vinyl gloves should be used, and proper washing procedures need to be implemented (34, 36). Proper disinfection should be performed and equipment should be cleaned. Patient isolation may be appropriate, and cleaning patients' rooms with agents such as phosphate-buffered hypochlorite (1,600 ppm) may reduce transmission. In most instances, the incidence of disease can be reduced simply by educating health care workers about the disease and how it is spread. Antibiotic usage patterns should also be monitored and physicians alerted when endemics are first detected.

NOTE ADDED IN PROOF

The importance of CPE for the gastrointestinal pathogenesis of *C. perfringens* isolates causing both *C. perfringens* type A food poisoning and CPE-associated non-food-borne gastrointestinal illnesses has recently received important new support from rabbit ileal loop studies conducted with *cpe* knock-out mutants (M. R. Sarker, R. J. Carman, and B. A. McClane, *Mol. Microbiol.*, in press).

The RVP1 homolog proteins that act as CPE receptors have recently been renamed claudins-3 and -4 (K. Morita, M. Furuse, K. Fujimoto, and S. Tsukita, *Proc. Natl. Acad. Sci. USA* **96:** 511–516, 1999).

Because of page limits, this chapter has cited published reviews for older studies. We thank Eva Wieckowski for preparing Fig. 1. Some of the research cited for C. perfringens enterotoxin was supported by Public Health Service grant AI 19844-15 from the National Institute of Allergy and Infectious Diseases and by USDA grant 9802822 from the Ensuring Food Safety Research Program (to B.A.M.). Some of the research cited for the molecular biology of C. difficile toxins A and B was supported by Public Health Service grant AI15749 from the National Institute of Allergy and Infectious Diseases (to T.D.W. and D.M.L.).

REFERENCES

1. **Aktories, K., and I. Just.** 1995. Monoglucosylation of low-molecular-mass GTP-binding Rho proteins by clostridial cytotoxins. *Trends Cell Biol.* **5:**441–443.

2. **Barroso, L. A., J. S. Moncrief, D. M. Lyerly, and T. D. Wilkins.** 1994. Mutagenesis of the *Clostridium difficile* toxin B genes and effect on cytotoxic activity. *Microb. Pathog.* **16:**297–303.

3. **Bartlett, J. G.** 1994. *Clostridium difficile*: history of its role as an enteric pathogen and the current state of knowledge about the organism. *Clin. Infect. Dis.* **18:**S265–S272.

4. **Bartlett, J. G.** 1995. Antibiotic-associated diarrhea, p. 893–904. *In* M. J. Blaser, P. D. Smith, J. I. Ravdin, H. B. Greenberg, and R. L. Guerrant (ed.), *Infections of the Gastrointestinal Tract.* Raven Press, New York, N.Y.

5. **Billington, S. J., E. U. Wieckowski, M. R. Sarker, D. Bueschel, J. G. Songer, and B. A. McClane.** 1998. *Clostridium perfringens* type E animal isolates with highly conserved, silent enterotoxin gene sequences. *Infect. Immun.* **66:**4531–4536.

6. **Bongaerts, G. P. A., and D. M. Lyerly.** 1997. Role of bacterial metabolism and physiology in the pathogenesis of *Clostridium difficile* disease. *Microb. Pathog.* **22:**253–256.

7. **Borriello, S. P.** 1990. Pathogenesis of *Clostridium difficile* of the gut. *J. Med. Microbiol.* **33**:207–215.

8. **Borriello, S. P., H. A. Davies, and F. E. Barclay.** 1988. Detection of fimbriae amongst strains of *Clostridium difficile*. *FEMS Microbiol. Lett.* **49**:65–67.

9. **Borriello, S. P., H. E. Lawson, F. E. Barclay, and A. R. Welch.** 1987. *Clostridium perfringens* enterotoxin-associated diarrhoea. *In* S. P. Borriello (ed.), *Recent Advances in Anaerobic Microbiology*. Martinus Nijhoff, Boston, Mass.

10. **Braun, V., T. Hundsberger, P. Leukel, M. Sauerborn, and C. von Eichel-Streiber.** 1996. Definition of the single integration site of the pathogenicity locus in *Clostridium difficile*. *Gene* **181**:29–38.

11. **Brazier, J. S.** 1998. The diagnosis of *Clostridium difficile*-associated disease. *J. Antimicrob. Chemother.* **41**:29–40.

12. **Bruno, D., and A. L. Sonenshein.** 1998. Regulated transcription of *Clostridium difficile* toxin genes. *Mol. Microbiol.* **27**:107–120.

13. **Brynestad, S., L. A. Iwanejko, G. Stewart, and P. E. Granum.** 1994. A complex array of Hpr sequences proximal to the enterotoxin gene in *Clostridium perfringens* type A. *Microbiology* **140**:97–104.

14. **Brynestad, S., B. Synstad, and P. E. Granum.** 1997. The *Clostridium perfringens* enterotoxin gene is on a transposable element in type A human food poisoning strains. *Microbiology* **143**:2109–2115.

15. **Carman, R. J.** 1997. *Clostridium perfringens* in spontaneous and antibiotic-associated diarrhoea of man and other animals. *Rev. Med. Microbiol.* **8**(Suppl. 1):S43–S46.

16. **Collie, R. E., J. F. Kokai-Kun, and B. A. McClane.** 1998. Phenotypic characterization of enterotoxigenic *Clostridium perfringens* isolates from non-foodborne human gastrointestinal diseases. *Anaerobe* **4**:69–79.

17. **Collie, R. E., and B. A. McClane.** 1998. Evidence that the enterotoxin gene can be episomal in *Clostridium perfringens* isolates associated with non-foodborne human gastrointestinal diseases. *J. Clin. Microbiol.* **36**:30–36.

18. **Cornillot E., B. Saint-Joanis, G. Daube, S. Katayama, P. E. Granum, B. Carnard, and S. T. Cole.** (1995) The enterotoxin gene (*cpe*) of *Clostridium perfringens* can be chromosomal or plasmid-borne. *Mol. Microbiol.* **15**:639–647.

19. **Czeczulin, J. R., R. E. Collie, and B. A. McClane.** 1996. Regulated expression of *Clostridium perfringens* enterotoxin in naturally *cpe*-negative type A, B, and C isolates of *C. perfringens*. *Infect. Immun.* **64**:3301–3309.

20. **Delmee, M.** 1989. *Clostridium difficile* infection in health-care workers. *Lancet* **ii**:1095.

21. **Dove, C. H., S. Z. Wang, S. B. Price, C. J. Phelps, D. M. Lyerly, T. D. Wilkins, and J. L. Johnson.** 1990. Molecular characterization of the *Clostridium difficile* toxin A gene. *Infect. Immun.* **58**:480–488.

23. **George, W. L., V. L. Sutter, D. Citron, and S. M. Finegold.** 1979. Selective and differential medium for isolation of *Clostridium difficile*. *J. Clin. Microbiol.* **9**:214–219.

23a. **Hammond, G. A., and J. L. Johnson.** 1995. The toxigenic element of *Clostridium difficile* strain VPI 10463. *Microb. Pathogen.* **19**:203–213.

24. **Hofmann, F., C. Busch, U. Prepens, I. Just, and K. Aktories.** 1997. Localization of the glucosyltransferase activity of *Clostridium difficile* toxin B to the N-terminal part of the holotoxin. *J. Biol. Chem.* **272**:11074–11078.

25. **Johnson, J. L., C. Phelps, L. Barroso, M. D. Roberts, D. M Lyerly, and T. D. Wilkins.** 1990. Cloning and expression of the toxin B gene of *Clostridium difficile*. *Curr. Microbiol.* **20**:397–401.

26. **Johnson, S., W. D. Sypura, D. N. Gerding, S. L. Ewing, and E. N. Janoff.** 1995. Selective neutralization of a bacterial enterotoxin by serum immunoglobulin A in response to mucosal disease. *Infect. Immun.* **63**:3166–3173.

27. **Katahira, J., N. Inoue, Y. Horiguchi, M. Matsuda, and N. Sugimoto.** 1997. Molecular cloning and functional characterization of the receptor for *Clostridium perfringens* enterotoxin. *J. Cell Biol.* **136**:1239–1247.

28. **Katahira, J., H. Sugiyama, N. Inoue, Y. Horiguchi, M. Matsuda, and N. Sugimoto.** 1997. *Clostridium perfringens* enterotoxin utilizes two structurally related membrane proteins as functional receptors *in vivo*. *J. Biol. Chem.* **272**:26652–26658.

29. **Katayama S. I., B. Dupuy, G. Daube, B. China, and S. T. Cole.** (1996) Genome mapping of *Clostridium perfringens* strains with *I-Ceu* I shows many virulence genes to be plasmid-borne. *Mol. Gen. Genet.* **251**:720–726.

30. **Knoop, F. C., M. Owen, and I. C. Crocker.** 1993. *Clostridium difficile*: clinical disease and diagnosis. *Clin. Microbiol. Rev.* **6**:251–265.

31. **Kokai-Kun, J. F., and B. A. McClane.** 1996. Evidence that a region(s) of the *Clostridium perfringens* enterotoxin molecule remains exposed on the external surface of the mammalian plasma membrane when the toxin is sequestered in small or large complexes. *Infect. Immun.* **64**:1020–1025.

32. **Kokai-Kun, J. F., and B. A. McClane.** 1997. Deletion analysis of the *Clostridium perfringens* enterotoxin. *Infect. Immun.* **65**:1014–1022.

33. **Kraukauer, T., B. Fleischer, D. L. Stevens, B. A. McClane, and B. G. Stiles.** 1997. *Clostridium perfringens* enterotoxin lacks superantigenic activity but induces interleukin-6 response from human peripheral blood mononuclear cells. *Infect. Immun.* **65**:3485–3488.

34. **Lyerly, D. M.** 1993. Epidemiology of *Clostridium difficile* disease. *Clin. Microbiol. Newsl.* **15**:49–52.

35. **Lyerly, D. M., L. A. Barroso, and T. D. Wilkins.** 1991. Identification of the latex-reactive protein of *Clostridium difficile* as glutamate dehydrogenase. *J. Clin. Microbiol.* **29**:2639–2642.

36. **Lyerly, D. M., and T. D. Wilkins.** 1995. *Clostridium difficile*, p. 867–891. *In* M. J. Blaser, P. D. Smith, J. I. Ravdin, H. B. Greenberg, and R. L. Guerrant (ed.), *Infections of the Gastrointestinal Tract*. Raven Press, New York, N.Y.

37. **Magdesian, K. G., J. E. Madigan, D. C. Hirsh, S. S. Jang, Y. J. Tang, T. E. Carpenter, L. M. Hansen, and J. Silva, Jr.** 1997. *Clostridium difficile* and horses: a review. *Rev. Med. Microbiol.* **8**:S46–S48.

38. **McClane, B. A.** 1997. *Clostridium perfringens*, p. 192–234. *In* M. P. Doyle, L. R. Beuchat, and T. J. Montville (ed.), *Food Microbiology: Fundamentals and Frontiers*. ASM Press, Washington, D.C.

39. **McClane, B. A.** The action, genetics and synthesis of *Clostridium perfringens* enterotoxin. *In* J. W. Cary, M. A. Stein, and D. Bhatnagar (ed.), *Microbial Foodborne Diseases: Mechanisms of Pathogenesis and Toxin Synthesis*. Technomic Press, Lancaster, Pa., in press.

40. **McFarland, L. V., and G. W. Elmer.** 1995. Biotherapeutic agents: past, present and future. *Microecol. Ther.* **23:**46–73.

41. **McFarland, L. V., C. M. Surawicz, and W. E. Stamm.** 1990. Risk factors for *Clostridium difficile* carriage and *C. difficile*-associated diarrhea in a cohort of hospitalized patients. *J. Infect. Dis.* **162:**678–684.

42. **Melville, S. B., R. G. Labbe, and A. L. Sonenshein.** 1994. Expression from the *Clostridium* perfringens *cpe* promoter in *C. perfringens* and *Bacillus subtilis. Infect. Immun.* **62:**5550–5558.

43. **Moncrief, J. S., L. A. Barroso, and T. D. Wilkins.** 1997. Positive regulation of *Clostridium difficile* toxins. *Infect. Immun.* **65:**1105–1108.

44. **Onderdonk, A. B.** 1988. Role of the hamster model of antibiotic-associated colitis in defining the etiology of the disease, p. 115–125. *In* R. D. Rolfe and S. M. Finegold (ed.), *Clostridium difficile: Its Role in Intestinal Disease.* Academic Press, Inc., New York, N.Y.

45. **Pothoulakis, C., I. Castagliuolo, C. P. Kelly, and J. T. Lamont.** 1993. *Clostridium difficile*-associated diarrhea and colitis: pathogenesis and therapy. *Int. J. Antimicrob. Agents* **3:**17–32.

46. **Shih, N. J., and R. G. Labbe.** 1996. Sporulation-promoting ability of *Clostridium perfringens* culture fluids. *Appl. Environ. Microbiol.* **62:**1441–1443.

47. **Songer, J. G.** 1996. Clostridial enteric diseases of domestic animals. *Clin. Microbiol. Rev.* **9:**216–234.

48. **Torres, J. F.** 1991. Purification and characterization of toxin B from a strain of *Clostridium difficile* that does not produce toxin A. *J. Med. Microbiol.* **35:**40–44.

49. **von Eichel-Streiber, C., M. Sauerborn, and H. K. Kuramitsu.** 1992. Evidence for a modular structure of the homologous repetitive C-terminal carbohydrate-binding sites of *Clostridium difficile* toxins and *Streptococcus mutans* glucosyltransferases. *J. Bacteriol.* **174:**6707–6710.

50. **Wieckowski, E. U., J. F. Kokai-Kun, and B. A. McClane.** 1998. Characterization of membrane-associated *Clostridium perfringens* enterotoxin following pronase treatment. *Infect. Immun.* **66:**5897–5905.

51. **Zhao, Y., and S. B. Melville.** 1998. Identification and characterization of sporulation-dependent promoters upstream of the enterotoxin gene (*cpe*) of *Clostridium perfringens. J. Bacteriol.* **180:**136–142.

Histotoxic Clostridia

DENNIS L. STEVENS AND JULIAN I. ROOD

58

Histotoxic clostridial infection is a general term coined over a century ago that referred to gas gangrene and malignant edema in humans and blackleg in cattle (Table 1). In the last half of the 20th century, novel histotoxic infections have been described, such as necrotic enteritis, neutropenic enterocolitis, and spontaneous gas gangrene—all of which occur exclusively in humans—and abomasal ulceration in cattle (Table 1). These infections are rapidly progressive, are associated with gas in tissue, and manifest impressive tissue destruction, shock, and frequently death (reviewed in reference 41).

GENERAL CHARACTERISTICS OF HISTOTOXIC CLOSTRIDIA

Although the histotoxic clostridia are classified as gram-positive, spore-forming, anaerobic bacilli, not all of them are definitely so. Specifically, these bacteria readily lose the crystal violet stain in vitro and invariably appear as gram-negative rods after overnight culture. Similarly, in vivo, smears made of gangrenous lesions invariably demonstrate predominantly gram-negative rods. Further, not all of the histotoxic clostridia are strict anaerobes. For example, *Clostridium histolyticum* is aerotolerant and will form colonies on freshly prepared blood agar medium incubated aerobically. *Clostridium septicum* is less aerotolerant but will grow slowly at ambient oxygen tensions as well. In contrast, *Clostridium perfringens* and *Clostridium novyi* require anaerobic conditions, the latter being inhibited by oxygen tensions of greater than 0.05% (39).

THE MICROBIOLOGICAL NICHE OF HISTOTOXIC CLOSTRIDIA

The main habitats of all of the histotoxic clostridia are the soil and intestinal contents of humans and animals. *C. perfringens* is the most widespread of the histotoxic clostridia, with the quantity of organisms in soil being proportional to the degree and duration of animal husbandry in the region. For example, *C. perfringens* can reach as high as 5×10^5 per g of soil in the fertile valleys of Europe. The number of *C. perfringens* organisms carried by healthy humans has varied from 100 per g of stool among Swedish subjects to as high as 1×10^9 per g of stool among some Japanese subjects (30). These concepts are important, since the source of bacteria in histotoxic infections of humans is invariably related to either endogenous infection from a bowel source or contamination of deep wounds with soil. In contrast to all the other histotoxic clostridia in nature, *Clostridium chauvoei* resides largely in the gut of ruminants; hence, blackleg in cattle is invariably an infection caused by endogenous flora.

VIRULENCE FACTORS OF THE HISTOTOXIC CLOSTRIDIA

Our current understanding of the potent toxins produced by histotoxic clostridia is based upon studies done between World Wars I and II when gas gangrene was a major complication of battlefield injuries. Investigators of this period designated the major lethal toxins of these bacteria with Greek letters; the letter alpha was always used to designate the most potent toxin. A marvelous review of these data can be found in the monograph by Smith (38). Over the ensuing 50 years, modern technology has provided a greater understanding of mechanisms of action of some of these factors (Table 2).

STRUCTURE/FUNCTION OF THE MAJOR HISTOTOXIC CLOSTRIDIAL EXOTOXINS

The major *C. perfringens* extracellular toxins implicated in gas gangrene are alpha-toxin and theta-toxin. Alpha-toxin is a lethal, hemolytic toxin that has both phospholipase C and sphingomyelinase activities. The *C. perfringens* alpha-toxin has an N-terminal domain with sequence similarity to phospholipase C enzymes from other bacteria and a smaller β-sandwich C-terminal domain that is responsible for calcium-dependent membrane binding and hemolysis. Both domains are required for toxicity. Analysis of the alpha-toxin structure by X-ray crystallography has confirmed the two-domain structure, with the larger α-helical

Gram-Positive Pathogens, ed. by V. A. Fischetti et al.
© 2000 American Society for Microbiology, Washington, D.C.

TABLE 1 Histotoxic clostridial infections

Type of infection	Species infected	Organism
Gas gangrene 　　Traumatic	Humans	
		Clostridium perfringens type A
		Clostridium septicum
		Clostridium histolyticum
		Clostridium sordellii
		Clostridium novyi
		Clostridium fallax
Spontaneous	Humans	*Clostridium septicum*
Malignant edema	Humans	*Clostridium septicum*
Blackleg	Cattle	*Clostridium chauvoei*
Necrotic enteritis	Humans	*Clostridium perfringens* type C
Neutropenic enterocolitis	Humans	*Clostridium septicum*
Abomasal ulceration	Cattle	*Clostridium perfringens* type A

N-terminal domain containing the active enzymatic site. The C-terminal C2-like domain has strong structural analogy to eukaryotic phospholipid and/or calcium-binding C2 domains, such as has been found in intracellular second messenger proteins and human arachidonate 5'-lipoxygenase (52). The tyrosine residues of the alpha-toxin C2-like domain interact with the calcium/phosphatidylcholine complex of eukaryotic cell membranes, leading to binding of the toxin to the membrane surface (18) and the generation of intracellular messengers, such as diacylglycerol and ceramide (13).

The theta-toxin from *C. perfringens* is a member of the thiol-activated cytolysin family, now termed cholesterol-dependent cytolysins. These pore-forming toxins contain a conserved ECTGLAWEWWR motif. Modification of the invariant cysteine residue leads to loss of hemolytic activity; however, this residue can be replaced with alanine without loss of function. The crystal structure of theta-toxin has also been determined (32). The protein has an unusual elongated shape with 40% β-sheet structure and four discontinuous domains. The conserved C-terminal tryptophan-rich motif described above is located in an elongated loop near the tip of domain 4, the membrane insertion domain. This motif appears to be close to the putative cholesterol-binding domain. Based on the crystal structure, a model for the insertion of theta-toxin into the membrane has been proposed (34). This model involves the oligomerization of the theta-toxin monomers upon contact with the cell membrane, followed by insertion of the oligomer into the membrane via the formation of a hydrophobic structure involving the binding of cholesterol to the protruding domain 4, which can then span the cell membrane. Insertion of the oligomers into the membrane results in the formation of a membrane pore and leads to cell lysis. Although theta-toxin can lyse mammalian cells, genetic studies have shown that theta-toxin, by itself, is

TABLE 2 Major virulence factors of the histotoxic clostridia

Clinical infection	Organism	Virulence factor	Mechanism of action
Traumatic gas gangrene	*Clostridium perfringens* type A	Alpha-toxin	Phospholipase C
		Theta-toxin	Cytolysin
		Kappa-toxin	Collagenase
		Mu-toxin	Hyaluronidase
		Nu antigen	Deoxyribonuclease
Traumatic gas gangrene	*Clostridium histolyticum*	Alpha-toxin	Hemolytic, cytotoxic, lethal, antigenically related to alpha-toxin of *Clostridium septicum*
		Beta-toxin	Collagenase
		Gamma-toxin	Thiol-activated protease
		Delta-toxin	Elastase
		Epsilon-toxin	Thiol-activated cytolysin
Traumatic gas gangrene	*Clostridium novyi*	Alpha-toxin	Dermonecrotic; causes gelatinous edema
		Gamma-toxin	Phospholipase C
		Delta-toxin	Thiol-activated cytolysin
Traumatic and spontaneous gas gangrene and neutropenic enterocolitis in humans	*Clostridium septicum*	Alpha-toxin	Cytotoxic, lethal, hemolytic, antigenically related to alpha-toxin of *Clostridium histolyticum*
		Delta-toxin	Thiol-activated cytolysin
Gas gangrene and malignant edema	*Clostridium sordellii*	Alpha-toxin	Phospholipase C
		Beta-toxin	"Edema-producing factor"
		Gamma-toxin	Protease
		?	Phospholipase A
		?	Lysolecithinase
		?	Thiol-activated cytolysin
Enteritis necroticans	*Clostridium perfringens* type C	Beta-toxin	Cytolytic for intestinal villi
Abomasal ulceration in cattle	*Clostridium perfringens* type A	Beta-toxin	
Blackleg in cattle	*Clostridium chauvoei*	Alpha-toxin	Lethal, necrotizing, lethal
		Beta-toxin	Deoxyribonuclease
		Gamma-toxin	Hyaluronidase
		Delta-toxin	Thiol-activated cytolysin

not essential in causing mortality since an insertionally inactivated chromosomal structural gene (*pfoA*) mutant is still lethal in the mouse myonecrosis model (3). However, these same genetic studies as well as passive immunization studies in mice (10) suggest that theta-toxin does contribute to pathogenesis by enhancing the morbidity associated with gas gangrene, likely by its ability to modulate the inflammatory response to infection (10, 51) (see below).

The other major toxin that has been extensively studied is the alpha-toxin from *C. septicum*. While its importance in pathogenesis has been demonstrated (5), its effects on host systems have not been determined, in part because it is not yet possible to genetically manipulate *C. septicum*. The toxin is secreted as an inactive prototoxin that is cleaved near the C terminus by eukaryotic proteases such as trypsin or membrane-bound furin to form the active toxin (17). The toxin then oligomerizes on the membrane and inserts, forming a membrane pore and resulting in colloid-osmotic lysis (33). Crystallographic analysis of the related toxin, aerolysin (6), indicates that the proteolytic cleavage site is located in an exposed loop that also appears to be present in alpha-toxin.

GENETIC REGULATION OF TOXIN PRODUCTION IN *C. PERFRINGENS*

Global Regulation of Toxin Production

One of the most common mechanisms for regulating the expression of bacterial virulence genes involves two-component signal transduction. In this process an environmental or chemical signal, which may be specific for the human host, interacts with a sensor histidine kinase located in the bacterial cell membrane. The resultant autophosphorylation of the kinase enables it to act as a phosphodonor and leads to the phosphorylation of a cytoplasmic protein, the cognate response regulator. The activated response regulator usually interacts with specific virulence genes to activate or repress their transcription (24).

Little is known about which *C. perfringens* genes are differentially expressed in vivo or in vitro, although it is clear that all of the major extracellular toxins are produced when the cells are grown in artificial medium. However, there is convincing experimental evidence that toxin production in *C. perfringens* is regulated by a two-component signal transduction system that consists of a sensor histidine kinase, VirS, and a response regulator, VirR (27, 35, 37). These genes are located in an operon (8, 27) on the *C. perfringens* chromosome, at a site quite distinct from the regions carrying the various toxin structural genes (15). Mutation of either the *virS* or *virR* genes leads to the complete loss of the ability to produce theta-toxin and a reduction in the amount of alpha-toxin, collagenase (kappa-toxin), sialidase, and protease that is produced. Complementation of the mutants with the respective cloned wild-type *virS* or *virR* genes restores the wild-type toxin phenotype (27, 35). In addition, virulence studies carried out with the mouse myonecrosis model have shown that a Tn916-derived *virS* mutant has significantly reduced virulence, which is consistent with its reduced toxin production. These effects on virulence could also be reversed by complementation with a functional *virS* gene (27).

Analysis of mRNA transcripts from the wild-type, mutant, and complemented derivatives has confirmed that regulation occurs at the level of transcription (8). These studies have also shown that the VirS/R system differentially activates the transcription of the various toxin genes. The theta-toxin structural gene *pfoA* is expressed from a major promoter whose expression is totally dependent upon the presence of a functional *virRS* operon (8). Subsequent studies have shown that the VirR protein binds to directly repeated sequences located immediately upstream of the *pfoA* promoter (14). The alpha-toxin structural gene *plc* is also expressed from a single promoter, but transcription from this promoter is only partially VirR dependent: there is a basal level of VirR-independent transcription. By contrast, transcription of the *colA* gene, which encodes the extracellular collagenase, is transcribed from two separate promoters, one of which is VirR dependent and one of which is VirR independent (8). These data explain why theta-toxin production is totally dependent on the VirS/R system, whereas mutation of either *virS* or *virR* only partially reduces the production of alpha-toxin or collagenase.

Although some knowledge has been gained regarding the effect of the VirS/R system on its target structural genes, almost nothing is known about the signal that triggers the regulatory cascade. It is possible that a small, intercellular signaling molecule produced by *C. perfringens*, termed substance A, may function as a quorum sensor controlling the expression of the VirS/R regulon (reviewed in reference 37). This hypothesis suggests that once the concentration of *C. perfringens* cells reaches a certain density in host tissues, the level of substance A becomes sufficient to bind to VirS, activate the regulatory cascade, and increase toxin production. The net effect of the increased expression of the cytolysins, hydrolytic enzymes, and toxins such as theta-toxin, alpha-toxin, and collagenase could be increased to tissue destruction with the concomitant liberation of essential nutrients.

The Role of Upstream Sequences in the Regulation of Alpha-Toxin Production

The production of alpha-toxin also appears to be regulated by a mechanism that involves repeated AT-rich regions located upstream of the *plc* gene (37). Immediately upstream of the −35 box of the *plc* promoter are three directly repeated d(A)$_{5-6}$ sequences, with a 10- to 11-bp periodicity, located within a 31-bp AT-rich (30 of 31 nucleotides) region that confers significant DNA bending, suggesting that DNA topology plays a role in the regulation of alpha-toxin production (53). Deletion of the poly(A) tracts progressively decreases the amount of DNA curvature that is observed in vitro, as well as decreasing the levels of *plc*-specific mRNA observed in *C. perfringens*. In addition, disruption of the periodicity of the repeats by the insertion of bases between the repeats has similar effects. These results indicate that the AT-repeats are important for optimal *plc* transcription and alpha-toxin production (28). It has been postulated that the poly(A) tracts, by conferring a temperature-dependent DNA bending on the region upstream of the *plc* promoter, increase *plc* transcription and alpha-toxin production under lower-temperature saprophytic conditions, such as those experienced when the organism encounters decaying animal material in the environment (28). Increasing the supply of nutrients to the bacterial population again appears to be the major force driving the regulatory process.

Other studies have involved the comparison of a *C. perfringens* type A strain and a type C strain that produces 30 times less alpha-toxin (26). Gel mobility shift assays carried out with crude extracts from both strains suggested that the type A strain but not the type C strain contained a protein that was able to bind a 376-bp fragment that was internal to the *plc* gene. Although it was suggested that this putative protein was involved in the regulation of alpha-toxin production, it has not yet been purified or identified. Comparison of other type A and B strains of *C. perfringens* also indicates that strain-dependent regulatory systems are important in the control of alpha-toxin production (12).

The Putative Role of the *pfoR* Gene in the Regulation of Theta-Toxin Production

The production of theta-toxin may be controlled not only by the VirR/S regulon but also by another gene, *pfoR*, located 591 bp upstream of *pfoA* (36). The putative PfoR protein has motifs commonly found in regulatory proteins, including a helix-turn-helix motif. On multicopy plasmids in *Escherichia coli*, deletion of *pfoR* or removal of the helix-turn-helix motif decreases relative theta-toxin expression about 20-fold. Based on these results it was concluded that PfoR was an activator of *pfoA* expression (36). Since all of these studies were carried out in *E. coli*, confirmation of this hypothesis is dependent upon the isolation and characterization of a *C. perfringens pfoR* mutant. The way in which the VirS/R system interacts with the putative PfoR-dependent regulation of *pfoA* is not known. VirR does not appear to act via PfoR since, as already discussed, it binds directly to the *pfoA* promoter region. RNA slot blots showed that the low-level *pfoR*-specific transcript is still expressed in a *virS* mutant but that increased *pfoR* expression is observed when a multicopy *virRS* operon is introduced (8). These results suggest that there is some interaction between VirR and the putative *pfoR* activation system.

THE PATHOGENESIS OF HISTOTOXIC INFECTIONS

The events leading to full-blown histotoxic infection in humans or animals occur in three distinct stages. Stage 1 can be initiated by either bacteria or spores, although the elemental composition and oxidative reduction potential of the tissue is a critical factor in either event. In traumatic gas gangrene, this is most critical for *C. perfringens* and *C. novyi*. It is important to recognize that for these two organisms trauma must be extensive. Indeed 40 to 60% of cases of gas gangrene caused by these organisms are associated with trauma sufficient to interrupt the blood supply to large muscle groups. Interruption of blood flow rapidly results in a drop in oxidation reduction potential from +170 to +50 mV. At neutral pH the latter value is sufficient for growth of most anaerobic bacteria, including the clostridia. Anaerobic conversion of muscle glycogen into lactic acid by endogenous muscle enzymes increases hydrogen ion concentration, allowing clostridia to proliferate at oxidation reduction potentials higher than +50 mV (38). Hypoxic muscle tissue releases endogenous lysosomal enzymes converting muscle protein into peptides and amino acids that are also vital to clostridial growth and toxin production. Release of these factors is particularly critical for *C. perfringens*, which requires over 20 amino acids and vitamins for growth (39). In contrast, the more aerotolerant *C. septicum* can infect muscle and tissue with normal oxygen tensions, and hence it can cause either nontraumatic or spontaneous gas gangrene. In addition, the nutritional requirements for *C. septicum* are less demanding in terms of amino acids, although vitamins such as thiamine, riboflavin, pyridoxine niacin, biotin, and cobalamin are required.

In either case, once the organism begins to proliferate (stage 2), the pH and redox potential decline further, providing ideal conditions for growth and toxin production. Factors known to increase production of alpha-toxin of *C. perfringens* include (i) peptides containing glycine and branched-chain amino acids such as valine, or tyrosine-containing peptides of low molecular weight; (ii) a carbohydrate source of starch, dextran or fructose, but not glucose; (iii) pH below 7.0; and (iv) total electrolyte concentration of 0.1 to 0.15 M with a ratio of sodium to potassium of 2:1. Interestingly, injection of recombinant alpha-toxin results in a major drop in local pH as measured by nuclear magnetic resonance imaging (42).

Stage 3 is the consequence of the elaboration of potent extracellular toxins that cause local and regional tissue destruction and, upon reaching the systemic circulation, cause organ failure, circulatory collapse, and death. Of the plethora of clostridial exotoxins known, the alpha-, theta-, and kappa-toxins from *C. perfringens* have been the best characterized.

Two independent studies have shown that the alpha-toxin is an essential toxin in the disease process. First, vaccination with a purified recombinant protein consisting of the C-terminal alpha-toxin domain (amino acids 247 to 370) has been shown to protect mice from experimental *C. perfringens* infection (56). Second, an alpha-toxin (*plc*) mutant constructed by allelic exchange has been shown to be avirulent in a mouse myonecrosis model (3). Complementation of the chromosomal mutation with a recombinant plasmid carrying a wild-type *plc* gene restores the ability to cause disease, providing clear genetic evidence for the essential role of alpha-toxin in the disease process.

HISTOTOXIC CLOSTRIDIA AND THE HUMAN IMMUNE SYSTEM: THE INITIAL ENCOUNTER

Toxin production has been documented in vivo by demonstrating the progressive appearance of both theta- and alpha-toxins at the site of the experimental infection by 4 h (30). Still, continued toxin production and extension of infection are dependent upon the outcome of the initial interaction of locally proliferating bacteria with the immune system. It is clear that *C. perfringens* is capable of activation of the complement cascade with the generation of serum-derived chemotactic factors and opsonins via the alternative complement cascade (49). This suggests that *C. perfringens* is opsonized and ingested by human polymorphonuclear leukocytes (PMN) in the presence of nonimmune serum. Migration of PMN to the site of clostridial proliferation is driven by the generation of bacteria-induced serum-derived chemotactic factors (C3a and C5a) (49). This response is likely amplified in vivo by the generation of the potent chemokine interleukin-8 (IL-8), induced by bacterial exotoxins such as alpha-toxin (11). This initial and rapid response by the host arrests bacterial proliferation in the vast majority of patients with wounds contaminated with *C. perfringens*, since progression of infection to full-blown clinical gas gangrene develops in only 0.03 to 5.2% of traumatic open wounds despite the fact that 3.8 to 39% of these wounds are contaminated with histotoxic clostridial species such as *C. perfringens* (1).

Thus, clinical evidence suggests that host factors dominate in the initial encounter and overcome the constitutive and inducible arsenal of toxins produced by these microbes. Clinical factors within the wound that tip the balance in favor of bacterial proliferation include the number of bacteria, the presence of foreign material such as soil, and, importantly, interruption of arterial blood supply.

Clostridial exotoxin production in situ dramatically affects the next stage of the host response. As shown by Bryant et al. (10) in experimental models, PMN influx (chemotaxis and diapedesis) is intact and luxurious where killed C. perfringens are injected into thigh muscles of mice (Table 3). In wild-type infection, both experimentally (10) and in human cases (40), a conspicuous absence of leukocytes at the site of C. perfringens proliferation is the hallmark of gas gangrene. The anti-inflammatory effect (suppression of PMN influx) by both theta-toxin and alpha-toxin in vivo has been demonstrated in studies utilizing either purified or recombinant toxins (10, 51) (Table 3) or isogenic mutants lacking theta-toxin or alpha-toxin (3, 51). The mechanism of attenuation of this host response can be inferred from in vitro studies demonstrating direct concentration-dependent effects of theta-toxin upon PMN viability (49), directed and random migration (49), ability to generate superoxide anion (10, 49), reduced ability to ingest opsonized zymosan (10) or killed C. perfringens (42), and heterologous desensitization of chemotactic factor receptors (10, 49). The effects of alpha-toxin on phagocyte function are no less remarkable. Alpha-toxin primes resting PMN respiratory burst in response to opsonized zymosan with resultant maximal generation of oxygen radical production (9, 49). Morphologically, PMN treated with alpha-toxin assume a cytokineplast form, with multiple daughter cells connected by a cytoplasmic bridge (9, 45). Premature priming of cells derails anaerobic glycolysis in favor of hexose monophosphate shunt activity, subverting generation of ATP necessary for the cytoskeletal rearrangements involved in chemotaxis and ingestion.

PROPAGATION OF TISSUE HYPOXIA BEYOND THE SITE OF INITIAL TRAUMA

Recent experimental studies by Bryant and Stevens and colleagues (10, 51), as well as the classical clinical observations by McNee and Dunn (29), demonstrate not only an absence of PMN at the site of active infection (see

above), but also accumulation or entrapment of PMN within the postcapillary venules. In vitro studies suggest that theta-toxin and alpha-toxin up-regulate endothelial adherence molecules such as P-selectin (13), E-selectin, and ICAM (11). Additionally, theta-toxin up-regulates the important neutrophil adherence molecule CD11b/CD18 (10). Theta- and alpha-toxins are also capable of rapidly inducing endothelial synthesis of the lipid autocoid platelet-activating factor (PAF) but by different mechanisms. PAF synthesis induced by theta-toxin is likely the consequence of increased cytosolic calcium concentrations due to increased influx or mobilization of intracellular stores, with rapid activation of endogenous phospholipase A_2 leading to hydrolysis of phosphatidylcholine with release of arachidonic acid and incorporation of acetyl-coenzyme A into the Sn-2 position of lysophosphatidylcholine, yielding PAF (55). Alpha-toxin activates endothelial cells by direct formation of the intracellular messengers ceramide and diacylglycerol, the latter being a known regulator of protein kinase C (13). In either case, PAF, a potent endothelial receptor for platelets and PMN, is produced. Coincubation of PMN and endothelial cells in the presence of theta-toxin demonstrates tight adherence of PMN to endothelium via a CD11b/CD18 mechanism (10). Such adherence reaches sufficient magnitude to attenuate blood flow while further damaging endothelial cells by neutrophil-dependent mechanisms such as degranulation (release of hydrolytic enzymes) and oxygen radical production (51).

In the absence of a host response sufficient to limit bacterial growth, clostridial proliferation is rapid, leading to the local accumulation of toxin. Higher in situ concentrations of C. perfringens toxins further inhibit PMN influx and reach concentrations sufficient to cause cytotoxicity to emigrated cells. Absorption of toxin in postcapillary venules likely activates endothelium lining these vessels, resulting in propagation of the previously described endothelial cell-PMN interactions to more proximal tissue.

PATHOGENESIS OF SHOCK AND ORGAN FAILURE

Cardiovascular collapse and end organ failure occur late in the course of gas gangrene caused by histotoxic clostridia such as C. perfringens and C. septicum. "Tachycardia with feebleness of the pulse has followed the onset of pain and characteristically, has been out of proportion to the degree of elevation of the temperature" (1). This description offers a striking contrast to the early manifestations of septic shock caused by gram-negative bacteria such as E. coli, in which a hyperdynamic picture is most common even in the face of low blood pressures. For example, a rapid heart rate is generally associated with high cardiac outputs and a bounding pulse. In gas gangrene, at the onset of tachycardia, the blood pressure is normal but then drops precipitously. Although these descriptions from the 1970s are useful, the subsequent clinical literature has not offered more modern measurements of cardiovascular parameters in humans with clostridial gas gangrene. Experimental studies in animals do provide some useful insights into the dynamics of cardiovascular dysfunction induced by C. perfringens toxins. Initial studies demonstrated that purified as well as recombinant theta- and alpha-toxins were lethal to mice by intravenous injection, with alpha-toxin having the greatest potency.

In an awake rabbit model, recombinant theta-toxin, but not recombinant alpha-toxin, caused a significant increase

TABLE 3 Anti-inflammatory effects of C. perfringens alpha- and theta-toxins[a]

Material injected	PMN influx		Tissue destruction (myonecrosis)	
	Early	Late	Early	Late
Viable, wild-type C. perfringens	−	−	+	+ + + +
Killed, washed C. perfringens	+ + +	+ + + +	−	−
Killed C. perfringens + alpha-toxin	−	−	+ +	+ + +
Killed C. perfringens + theta-toxin	−	+	+ +	+ + +

[a]Adapted from reference 51.

in heart rate that was apparent within 60 min of toxin infusion (2). Interestingly, theta-toxin did not cause a significant drop in mean arterial pressure (Fig. 1). In contrast, crude toxin preparations induced a dramatic drop in mean arterial blood pressure within 2 h after the infusion of toxin (Fig. 1). Alpha-toxin also caused a significant reduction in blood pressure (Fig. 1), but the onset of alpha-toxin-induced hypotension was 1 h later (i.e., at 180 min post-toxin infusion) than that caused by the crude toxin (120 min post-toxin infusion). These data suggest additive or even synergistic interactions of toxins such as alpha- and theta-toxins. Both recombinant alpha-toxin and crude toxin caused significant drops in cardiac index as early as 90 min after infusion of the respective toxins (2) (Fig. 2).

MOLECULAR MECHANISMS OF TOXIN-INDUCED SHOCK

As reflected by the increased mortality in the rabbits receiving recombinant alpha-toxin or crude toxin, a greater and more rapid reduction in cardiac index was also measured in these groups compared with those receiving recombinant theta-toxin or normal saline (2) (Fig. 2).

These data suggest that one mechanism of cardiovascular dysfunction induced by alpha-toxin is related to direct myocardial toxicity. Indeed, a direct and dose-dependent reduction in myocardial contractility (dF/dt) occurred in isolated atrial strips bathed with recombinant alpha-toxin (2). Alpha-toxin has also been shown to affect the inotropic cardiac response in isolated embryonic chick heart preparations (31). However, alpha-toxin may

also contribute indirectly to shock by stimulating production of endogenous mediators such as tumor necrosis factor (TNF) (Fig. 3, 4) (43) and PAF (13).

Theta-toxin reduces systemic vascular resistance and markedly increases cardiac output (2, 50). This afterload reduction occurs, undoubtedly, through induction of endogenous mediators that cause relaxation of blood vessel wall tension such as PAF and prostaglandin I_2 (55). Reduced vascular tone develops rapidly and, to maintain adequate tissue perfusion, a compensatory host response is required to either increase cardiac output or rapidly expand the intravascular blood volume. In contrast, patients with gram-negative sepsis compensate for hypotension by markedly increasing cardiac output; however, this adaptive mechanism may not be possible in C. perfringens-induced shock owing to direct suppression of myocardial contractility by alpha-toxin (50). Theta-toxin, like other thiol-activated cytolysins, likely contributes to septic shock through additional indirect routes, including augmented release of TNF, IL-1, and IL-6 (19, 25). The roles of TNF, IL-1, and IL-6, as well as the potent endogenous vasodilators, bradykinin and nitric oxide, in shock associated with C. perfringens gas gangrene have not been elucidated.

TREATMENT OF C. PERFRINGENS GAS GANGRENE

Aggressive debridement of devitalized tissue, as well as rapid repair of compromised vascular supply, and prophylactic antibiotics greatly reduce the frequency of gas gan-

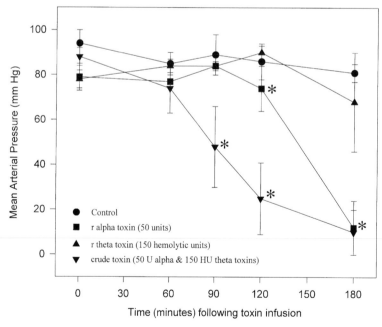

FIGURE 1 Effects of clostridial exotoxins on mean arterial pressure. Rabbits with stable baseline vital signs were given an intravenous infusion of normal saline, crude toxin preparation, recombinant alpha-toxin, or recombinant theta-toxin. Mean arterial pressures were monitored continuously via a catheter placed in the carotid artery. Each data point represents the mean arterial pressure (± standard error) determined by using six animals, with triplicate determinations for each time point. Asterisks indicate values significantly different from control values (P < 0.05) using Student's test. ●, Control (normal saline); ■, recombinant alpha-toxin; ▲, recombinant theta-toxin; ▼, crude toxin. Reprinted, with permission, from reference 2.

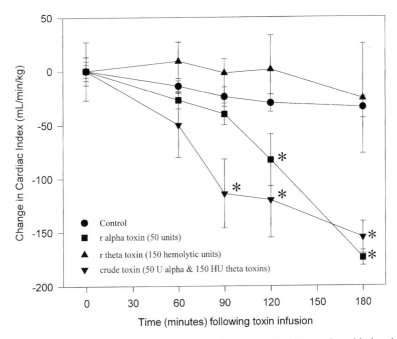

FIGURE 2 Effects of clostridial exotoxins on cardiac index. Rabbits with stable baseline vital signs were given an intravenous infusion of normal saline, crude toxin preparation, recombinant alpha-toxin, or theta-toxin. Cardiac index was measured over a 3-h period using a thermodilution technique. Each data point represents the mean cardiac index ($\pm$ standard error) determined by using six animals, with triplicate determinations for each time point. Asterisks indicate values significantly different from control values ($P < 0.05$). ●, Control (normal saline); ■, recombinant alpha-toxin; ▲, recombinant theta-toxin; ▼, crude toxin. Reprinted, with permission, from (2).

grene in contaminated deep wounds (7, 16, 40 41). Intramuscular epinephrine, prolonged application of tourniquets, and surgical closure of traumatic wounds should be avoided. Patients presenting with gas gangrene of an extremity have a better prognosis than those with truncal or intra-abdominal gas gangrene, largely because it is dif-

ficult to adequately debride such lesions (7, 16, 20). In addition, patients with associated bacteremia and intravascular hemolysis have the greatest likelihood of progressing to shock and death. Patients who are in shock at the time diagnosis is made have the highest mortality (20).

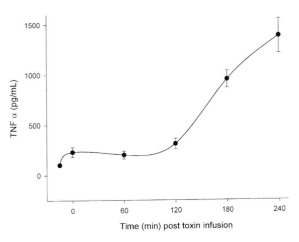

FIGURE 3 In vivo TNF-α production is induced by *C. perfringens* alpha-toxin. Serum samples were obtained from rabbits infused with recombinant alpha-toxin. TNF-α was measured by enzyme-linked immunosorbent assay. Each data point represents the mean TNF-α production ($\pm$ standard error) on duplicate serum samples run in triplicate. Reprinted, with permission, from (43).

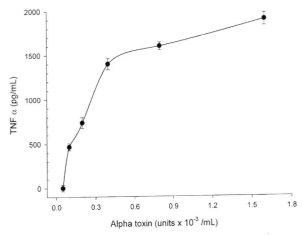

FIGURE 4 Alpha-toxin-induced TNF-α production by human mononuclear cells. TNF-α was measured in supernatant fluid from 10^6 human mononuclear cells stimulated with recombinant alpha-toxin. Each data point represents the mean TNF-α production ($\pm$ standard error) on samples collected at 24 h and assayed in duplicate using a commercial enzyme-linked immunosorbent assay. Reprinted, with permission, from (43).

SUPPRESSION OF TOXIN SYNTHESIS BY SPECIFIC CLASSES OF ANTIBIOTICS CORRELATES WITH IN VIVO EFFICACY

Based strictly on in vitro susceptibility data, most textbooks state that penicillin is the drug of choice (7, 16). However, experimental studies in mice suggest that clindamycin has the greatest efficacy and penicillin the least (46, 47). Other agents with greater efficacy than penicillin include erythromycin, rifampin, tetracycline, chloramphenicol, and metronidazole (46, 47). Slightly greater survival has been observed in animals receiving both clindamycin and penicillin; in contrast, antagonism was observed with penicillin plus metronidazole (46). Because some strains (2 to 5%) are resistant to clindamycin, a combination of penicillin and clindamycin is warranted. Based on his experimental studies and his vast clinical experience with gas gangrene, William Altemeier also recommended a protein synthesis inhibitor, tetracycline, plus penicillin (1). Thus, given an absence of efficacy data from a clinical trial in humans, the best treatment would appear to be a combination of antibiotics such as clindamycin and penicillin.

The failure of penicillin in experimental clostridial myonecrosis may be related to continued toxin production by filamentous forms of the organism induced by this cell wall-active agent (48). In contrast, the efficacy of clindamycin and tetracycline may be related to their ability to rapidly inhibit toxin synthesis (48).

The use of hyperbaric oxygen (HBO) is controversial, although some nonrandomized studies have reported excellent results with HBO therapy when combined with antibiotics and surgical debridement (4, 20, 21). Some experimental studies demonstrated that HBO alone was an effective treatment if the inoculum was small and treatment was begun immediately (23). In contrast, other studies have demonstrated that HBO was only of slight benefit when combined with penicillin (44). In these studies, however, survival was greater with clindamycin alone than with either HBO alone, penicillin alone, or HBO plus penicillin (44). The benefit of HBO, at least theoretically, is to inhibit bacterial growth (22), preserve marginally perfused tissue, and inhibit toxin production (54). Interestingly, Altemeier did not use HBO in patients and was able to realize a mortality rate of less than 15% using only surgical debridement and antibiotics (tetracycline plus penicillin) (1).

Therapeutic strategies directed against toxin expression in vivo, such as neutralization with specific antitoxin antibody or inhibition of toxin synthesis, may be valuable adjuncts to traditional antimicrobial regimens. Currently, antitoxin is no longer available. Future strategies may target endogenous proadhesive molecules such that toxin-induced vascular leukostasis and resultant tissue injury are attenuated.

REFERENCES

1. Altemeier, W. A., and W. D. Fullen. 1971. Prevention and treatment of gas gangrene. JAMA 217:806–813.

2. Asmuth, D. A., R. D. Olson, S. P. Hackett, A. E. Bryant, R. K. Tweten, J. Y. Tso, T. Zollman, and D. L. Stevens. 1995. Effects of Clostridium perfringens recombinant and crude phospholipase C and theta toxins on rabbit hemodynamic parameters. J. Infect. Dis. 172:1317–1323.

3. Awad, M. M., A. E. Bryant, D. L. Stevens, and J. I. Rood. 1995. Virulence studies on chromosomal α-toxin and θ-toxin mutants constructed by allelic exchange provide genetic evidence for the essential role of α-toxin in Clostridium perfringens-mediated gas gangrene. Mol. Microbiol. 15:191–202.

4. Bakker, D. J. 1988. Clostridial myonecrosis, p. 153–172. In J. C. Davis and T. K. Hunt (ed.), Problem Wounds: The Role of Oxygen. Elsevier Science Publishing, Inc., New York, N.Y.

5. Ballard, J., A. Bryant, D. Stevens, and R. K. Tweten. 1992. Purification and characterization of the lethal toxin (alpha-toxin) of Clostridium septicum. Infect. Immun. 60:784–790.

6. Ballard, J., B. A. Crabtree, J. Roe, and R. K. Tweten. 1995. The primary structure of Clostridium septicum alpha-toxin exhibits similarity with that of Aeromonas hydrophila aerolysin. Infect. Immun. 63:340–344.

7. Bartlett, J. G. 1990. Gas gangrene (other clostridium-associated diseases), p. 1851–1860. In G. L. Mandell, R. G. Douglas, and J. E. Bennett (ed.), Principles and Practice of Infectious Diseases. Churchill Livingstone, Ltd., New York, N.Y.

8. Ba-Thein, W., M. Lyristis, K. Ohtani, I. T. Nisbet, H. Hayashi, J. I. Rood, and T. Shimizu. 1996. The virR/virS locus regulates the transcription of genes encoding extracellular toxin production in Clostridium perfringens. J. Bacteriol. 178:2514–2520.

9. Bryant, A., D. Stevens, and J. Tso. 1991. Effects of alpha toxin from Clostridium perfringens (Cp) on PMNL, abstr. B-311, p. 77. Abstr. 91st Gen. Meet. Am. Soc. Microbiol. 1991. American Society for Microbiology, Washington, D.C.

10. Bryant, A. E., R. Bergstrom, G. A. Zimmerman, J. L. Salyer, H. R. Hill, R. K. Tweten, H. Sato, and D. L. Stevens. 1993. Clostridium perfringens invasiveness is enhanced by effects of theta toxin upon PMNL structure and function: the roles of leukocytotoxicity and expression of CD11/CD18 adherence glycoprotein. FEMS Immunol. Med. Microbiol. 7:321–336.

11. Bryant, A. E., and D. L. Stevens. 1996. Phospholipase C and perfringolysin O from Clostridium perfringens upregulate endothelial cell-leukocyte adherence molecule 1 and intercellular leukocyte adherence molecule 1 expression and induce interleukin-8 synthesis in cultured human umbilical vein endothelial cells. Infect. Immun. 64:358–362.

12. Bullifent, H. L., A. Moir, M. M. Awad, P. T. Scott, and R. W. Titball. 1996. The level of expression of α-toxin by different strains of Clostridium perfringens is dependent upon differences in promoter structure and genetic background. Anaerobe 2:365–371.

13. Bunting, M., D. E. Lorant, A. E. Bryant, G. A. Zimmerman, T. M. McIntyre, D. L. Stevens, and S. M. Prescott. 1997. Alpha toxin from Clostridium perfringens induces proinflammatory changes in endothelial cells. J. Clin. Invest. 100:565–574.

14. Cheung, J. K., and J. I. Rood. Unpublished observations.

15. Cole, S. T., and B. Canard. 1997. Structure, organization and evolution of the genome of Clostridium perfringens, p. 49–63. In J. I. Rood, B. A. McClane, J. G. Songer, and R. W. Titball (ed.), The Clostridia: Molecular Biology and Pathogenesis. Academic Press, Inc., London, U.K.

16. Gorbach, S. L. 1992. Clostridium perfringens and other clostridia, p. 1587–1596. In S. L. Gorbach, J. G. Bartlett,

and N. R. Blacklow (ed.), *Infectious Diseases*. The W.B. Saunders Co., Philadelphia, Pa.

17. **Gordon, V., R. Benz, K. Fujii, S. Leppla, and R. Tweten.** 1997. *Clostridium septicum* alpha-toxin is proteolytically activated by furin. *Infect. Immun.* **65:**4130–4134.

18. **Guillouard, I., P. M. Alzari, B. Saliou, and S. T. Cole.** 1997. The carboxy-terminal C2-like domain of the α-toxin from *Clostridium perfringens* mediates calcium-dependent membrane recognition. *Mol. Microbiol.* **26:**867–876.

19. **Hackett, S. P., and D. L. Stevens.** 1992. Streptococcal toxic shock syndrome: synthesis of tumor necrosis factor and interleukin-1 by monocytes stimulated with pyrogenic exotoxin A and streptolysin O. *J. Infect. Dis.* **165:**879–885.

20. **Hart, G. B., R. C. Lamb, and M. B. Strauss.** 1983. Gas gangrene. I. A collective review. *J. Trauma* **23:**991–1000.

21. **Heimbach, R. D., I. Boerema, W. H. Brummelkamp, and W. G. Wolfe.** 1977. Current therapy of gas gangrene, p. 153–176. *In* J. C. Davis and T. K. Hunt (ed.), *Hyperbaric Oxygen Therapy*. Undersea Medical Society, Bethesda, Md.

22. **Hill, G. B., and S. Osterhout.** 1972. Experimental effects of hyperbaric oxygen on selected clostridial species. I. In vitro studies. *J. Infect. Dis.* **125:**17–25.

23. **Hill, G. B., and S. Osterhout.** 1972. Experimental effects of hyperbaric oxygen on selected clostridial species. II. In vivo studies on mice. *J. Infect. Dis.* **125:**26–35.

24. **Hoch, J. A., and T. J. Silhavy (ed.).** 1995. *Two-Component Signal Transduction*. American Society for Microbiology, Washington, D.C.

25. **Houldsworth, S., P. W. Andrew, and T. J. Mitchell.** 1994. Pneumolysin stimulates production of tumor necrosis factor alpha and interleukin-1 beta by human mononuclear phagocytes. *Infect. Immun.* **62:**1501–1503.

26. **Katayama, S. I., O. Matsushita, J. Minami, S. Mizobuchi, and A. Okabe.** 1993. Comparison of the alpha-toxin genes of *Clostridium perfringens* type A and C strains: evidence for extragenic regulation of transcription. *Infect. Immun.* **61:**457–463.

27. **Lyristis, M., A. E. Bryant, J. Sloan, M. M. Awad, I. T. Nisbet, D. L. Stevens, and J. I. Rood.** 1994. Identification and molecular analysis of a locus that regulates extracellular toxin production in *Clostridium perfringens*. *Mol. Microbiol.* **12:**761–777.

28. **Matsushita, C., O. Matsushita, S. Katayama, J. Minami, K. Takai, and A. Okabe.** 1996. An upstream activating sequence containing curved DNA involved in activation of the *Clostridium perfringens plc* promoter. *Microbiology* **142:**2561–2566.

29. **McNee, J. W., and J. S. Dunn.** 1917. The method of spread of gas gangrene into living muscle. *Br. Med. J.* **1:**727–729.

30. **Noyes, H. E., W. L. Pritchard, F. B. Brinkley, and J. A. Mendelson.** 1964. Analyses of wound exudates for clostridial toxins. *J. Bacteriol.* **87:**623–629.

31. **Regal, J. F., and F. E. Shigeman.** 1980. The effect of phospholipase C on the responsiveness of cardiac receptors. I. Inhibition of the adrenergic inotropic response. *J. Pharmacol. Exp. Ther.* **214:**282–290.

32. **Rossjohn, J., S. C. Feil, W. J. McKinstry, R. K. Tweten, and M. W. Parker.** 1997. Structure of a cholesterol-

binding, thiol-activated cytolysin and a model of its membrane form. *Cell* **88:**685–692.

33. **Sellman, B. R., B. L. Kagan, and R. K. Tweten.** 1997. Generation of a membrane-bound, oligomerized pre-pore complex is necessary for pore formation by *Clostridium septicum* alpha toxin. *Mol. Microbiol.* **23:**551–558.

34. **Shepard, L. A., A. P. Heuck, B. D. Hamman, J. Rossjohn, M. W. Parker, K. R. Ryan, A. E. Johnson, and R. K. Tweten.** 1998. Identification of a membrane-spanning domain of the thiol-activated pore-forming toxin *Clostridium perfringens* perfringolysin O: an alpha-helical to beta-sheet transition identified by fluorescence spectroscopy. *Biochemistry* **37:**14563–14574.

35. **Shimizu, T., W. Ba-Thein, M. Tamaki, and H. Hayashi.** 1994. The *virR* gene, a member of a class of two-component response regulators, regulates the production of perfringolysin O, collagenase and hemagglutinin in *Clostridium perfringens*. *J. Bacteriol.* **176:**1616–1623.

36. **Shimizu, T., A. Okabe, J. Minami, and H. Hayashi.** 1991. An upstream regulatory sequence stimulates expression of the perfringolysin O gene of *Clostridium perfringens*. *Infect. Immun.* **59:**137–142.

37. **Shimizu, T., A. Okabe, and J. I. Rood.** 1997. Regulation of toxin production in *C. perfringens*, p. 451–470. *In* J. I. Rood, B. A. McClane, J. G. Songer, and R. W. Titball (ed.), *The Clostridia: Molecular Biology and Pathogenesis*. Academic Press, Inc., London, U.K.

38. **Smith, L. D. S.** 1975. Clostridial wound infections, p. 321–324. *In* L. D. S. Smith (ed.), *The Pathogenic Anaerobic Bacteria*. Charles C Thomas, Springfield, Ill.

39. **Smith, L. D. S.** 1975. Clostridium, p. 109–114. *In* L. D. S. Smith (ed.), *The Pathogenic Anaerobic Bacteria*. Charles C Thomas, Springfield, Ill.

40. **Stevens, D. L.** 1995. Clostridial infections, p. 13.1–13.9. *In* D. L. Stevens and G. L. Mandell (ed.), *Atlas of Infectious Diseases*. Churchill Livingstone, Ltd., Philadelphia, Pa.

41. **Stevens, D. L.** 1996. Clostridial myonecrosis and other clostridial diseases, p. 2090–2093. *In* J. C. Bennett and F. Plum (ed.), *Cecil Textbook of Medicine*. The W. B. Saunders Co., Philadelphia, Pa.

42. **Stevens, D. L.** Unpublished observations.

43. **Stevens, D. L., and A. E. Bryant.** 1997. Pathogenesis of *Clostridium perfringens* infection: mechanisms and mediators of shock. *Clin. Infect. Dis.* **25:**S160–S164.

44. **Stevens, D. L., A. E. Bryant, K. Adams, and J. T. Mader.** 1993. Evaluation of hyperbaric oxygen therapy for treatment of experimental *Clostridium perfringens* infection. *Clin. Infect. Dis.* **17:**231–237.

45. **Stevens, D. L., A. E. Gibbons, and R. A. Bergstrom.** 1989. Ultrastructural changes in human granulocytes induced by purified exotoxins from *Clostridium perfringens*, abstr. J-17, p. 244. *Abstr. 89th Annu. Meet. Am. Soc. Microbiol. 1989*. American Society for Microbiology, Washington, D.C.

46. **Stevens, D. L., B. M. Laine, and J. E. Mitten.** 1987. Comparison of single and combination antimicrobial agents for prevention of experimental gas gangrene caused by *Clostridium perfringens*. *Antimicrob. Agents Chemother.* **31:**312–316.

47. **Stevens, D. L., K. A. Maier, B. M. Laine, and J. E. Mitten.** 1987. Comparison of clindamycin, rifampin, tetracycline, metronidazole, and penicillin for efficacy in pre-

vention of experimental gas gangrene due to *Clostridium perfringens. J. Infect. Dis.* **155:**220–228.

48. **Stevens, D. L., K. A. Maier, and J. E. Mitten.** 1987. Effect of antibiotics on toxin production and viability of *Clostridium perfringens. Antimicrob. Agents Chemother.* **31:** 213–218.

49. **Stevens, D. L., J. Mitten, and C. Henry.** 1987. Effects of alpha and theta toxins from *Clostridium perfringens* on human polymorphonuclear leukocytes. *J. Infect. Dis.* **156:** 324–333.

50. **Stevens, D. L., B. E. Troyer, D. T. Merrick, J. E. Mitten, and R. D. Olson.** 1988. Lethal effects and cardiovascular effects of purified alpha- and theta-toxins from *Clostridium perfringens. J. Infect. Dis.* **157:**272–279.

51. **Stevens, D. L., R. K. Tweten, M. M. Awad, J. I. Rood, and A. E. Bryant.** 1997. Clostridial gas gangrene: evidence that alpha and theta toxins differentially modulate the immune response and induce acute tissue necrosis. *J. Infect. Dis.* **176:**189–195.

52. **Titball, R. W., D. L. Leslie, S. Harvey, and D. Kelly.** 1991. Hemolytic and sphingomyelinase activities of *Clostridium perfringens* alpha-toxin are dependent on a domain homologous to that of an enzyme from the human arachidonic acid pathway. *Infect. Immun.* **59:**1872–1874.

53. **Toyonaga, T., O. Matsushita, S.-I. Katayama, J. Minami, and A. Okabe.** 1992. Role of the upstream regulon containing an intrinsic DNA curvature in the negative regulation of the phospholipase C gene of *Clostridium perfringens. Microbiol. Immunol.* **36:**603–613.

54. **van Unnik, A. J. M.** 1965. Inhibition of toxin production in *Clostridium perfringens* in vitro by hyperbaric oxygen. *Antonie van Leeuwenhoek* **31:**181–186.

55. **Whatley, R. E., G. A. Zimmerman, D. L. Stevens, C. J. Parker, T. M. McIntyre, and S. M. Prescott.** 1989. The regulation of platelet activating factor production in endothelial cells—the role of calcium and protein kinase C. *J. Biol. Chem.* **264:**6325–6333.

56. **Williamson, E. D., and R. W. Titball.** 1993. A genetically engineered vaccine against alpha-toxin of *Clostridium perfringens* protects against experimental gas gangrene. *Vaccine* **11:**1253–1258.

Corynebacterium diphtheriae: Iron-Mediated Activation of DtxR and Regulation of Diphtheria Toxin Expression

JOHN F. LOVE AND JOHN R. MURPHY

59

One of the problems revealed by the breakdown of the former Soviet Union was the decreased proportion of the Soviet public vaccinated against diphtheria toxin, the primary virulence factor produced by toxigenic strains of *Corynebacterium diphtheriae*. Although the diphtheria toxoid vaccine has been available since the 1920s and is widely used in most industrialized countries, in 1990 only 68% of Russian children had received appropriate vaccination (30). In contrast, in the United States the vaccination rates against diphtheria during the same period were 90 to 95% (2). The result, possibly exacerbated by falling living standards and crumbling medical infrastructure, was an epidemic of diphtheria in Russia, which quickly spread into Ukraine and other neighboring countries. The incidence of clinical diphtheria escalated each year, and by 1994 there were nearly 40,000 cases in Russia alone, resulting in over 1,000 deaths (20). Although vaccination programs have begun to slow the advance of the epidemic, diphtheria remains a serious health issue for many countries in the region.

The interplay between *C. diphtheriae* and its environment, governing toxin production and human disease, is relatively well documented (42). Since the mechanism of pathogenesis by *C. diphtheriae* is relatively uncomplicated, diphtheria lends itself well as a model for other toxin-mediated diseases. Remarkably, before a given strain of *C. diphtheriae* can cause clinical diphtheria, it first must be infected and lysogenized by a corynebacteriophage that carries the structural gene for diphtheria toxin, *tox*. While diphtheria toxin is the primary virulence factor expressed by *C. diphtheriae*, the control of *tox* expression is regulated by an iron-activated regulatory element, the diphtheria toxin repressor (DtxR), which is encoded on the genome of the bacterial host. The DtxR-mediated regulation of diphtheria *tox* expression has been well studied and has served as a paradigm for the control of iron-sensitive virulence gene expression.

C. diphtheriae is a gram-positive, nonmotile, club-shaped bacillus. Three distinct colony types (*mitis*, *intermedius*, and *gravis*) have been identified. Most strains of *C. diphtheriae* require the addition of nicotinic and pantothenic acid to culture medium for growth. Corynebacteria have been shown to have cell walls that contain arabinose, galactose, and mannose, and a toxic 6,6′-diester of trehalose that is composed of corynemycolic and corynemycolenic acids.

Clinical diphtheria is most commonly seen as an infection of the upper respiratory tract; however, cutaneous diphtheria has also been reported. In each instance, the site of infection is characterized by a pseudomembrane, a fibrin membrane at the base of an ulcerative lesion that forms as a result of the combined effects of bacterial colonization and growth, toxin production, necrosis of the underlying tissue, and the immune response of the host. The symptoms of clinical nasopharyngeal diphtheria range from mild pharyngitis with fever to potentially fatal airway obstruction due to the combined effects of the pseudomembrane and profound edema of the neck. Since diphtheria toxin is delivered to all organ systems by the circulation, life-threatening systemic complications may also be associated with loss of motor function, and congestive heart failure may develop as a result of the toxin's action on peripheral neurons and myocardium. While the pathogenesis of diphtheria is based upon the ability of a given strain to both colonize a given host and produce diphtheria toxin, to date very little is known about colonization factors that may be associated with virulence. In contrast, there is now a detailed understanding of the molecular biology and genetics of both the regulation of diphtheria toxin expression and the structure-function relationships and mode of action of the toxin.

IRON-SENSITIVE EXPRESSION OF VIRULENCE FACTORS IN *C. DIPHTHERIAE*

The diphtheria toxin structural gene *tox* has long been known to be expressed by *C. diphtheriae* under conditions of iron starvation; work by Pappenheimer and Johnson as early as 1936 (44) demonstrated the inhibitory effect of iron on toxin production. Like the expression of siderophores and heme oxygenase (53, 54), diphtheria toxin is produced only when iron becomes the growth rate-limiting substrate. The production of siderophores and heme oxy-

Gram-Positive Pathogens, ed. by V. A. Fischetti et al.
© 2000 American Society for Microbiology, Washington, D.C.

genase by *C. diphtheriae* is notably an adaptive response to low iron and results in the acquisition of this essential nutrient, whereas the derepression of the diphtheria *tox* gene appears to act indirectly by the induction of tissue damage. While the structural genes encoding siderophores and heme oxygenase are carried on the *C. diphtheriae* genome, the structural gene for diphtheria toxin is carried on the genomes of a family of closely related corynebacteriophages (7, 65). Interestingly, the regulation of expression of the *tox* gene, as well as the genes involved in iron acquisition and utilization, are under the control of the *C. diphtheriae*-encoded iron-activated repressor DtxR.

THE DIPHTHERIA TOXIN REPRESSOR DtxR

The discovery that the diphtheria *tox* gene was carried by corynebacteriophage β raised an interesting and important question: is the regulation of *tox* mediated by a viral or corynebacterial determinant? Work by Murphy et al. (35) first suggested that the corynephage *tox* gene was regulated by a corynebacterial protein and that Fe^{2+} by itself does not directly control the expression of diphtheria toxin. Further work by Kanei et al. (25) and Murphy et al. (36) with various *C. diphtheriae* and corynephage β mutants led Murphy and Bacha in 1979 (33) to propose a model for the regulation of *tox* expression in which a *cis*-acting viral element and a *trans*-acting *C. diphtheriae* product together control the expression of *tox* in an iron-dependent manner.

As shown in Fig. 1, the *cis*-acting viral element was found to be the *tox* operator, a 27-bp interrupted palindromic sequence (16, 25, 49) that overlaps the −10 region of the *tox* promoter (4). This upstream region from the diphtheria toxin structural gene has been shown to be the *tox* promoter/operator *toxPO*. In 1989, Fourel et al. (13) used DNase protection assays to show that an element or elements found in crude *C. diphtheriae* extracts binds specifically to the *tox* operator sequence. Mutant strains of *C. diphtheriae* that were *cis*-dominant for toxin production (68) were later found to carry mutations in the *tox* operator alone (26).

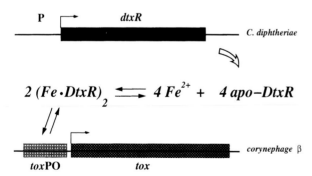

FIGURE 1 The expression of the diphtheria toxin structural gene *tox* is regulated by the Fe^{2+}-activated repressor DtxR. The *dtxR* gene is carried on the genome of *C. diphtheriae*, whereas *tox* is carried on the genome of toxigenic corynebacteriophage. In the presence of Fe^{2+}, apo-DtxR forms active dimers, (Fe · DtxR)₂, and two dimers bind to the diphtheria *tox* operator and repress *tox* expression. Under iron-limiting conditions, the ternary complex 2(Fe · DtxR)₂-*tox* operator dissociates and diphtheria toxin is expressed.

The *trans*-acting factor hypothesized by Murphy and Bacha in 1979 (33) was cloned by Boyd et al. (5) by screening genomic libraries of nontoxigenic, nonlysogenic *C. diphtheriae* in recombinant *Escherichia coli* that carried a transcriptional fusion between *toxPO* and *lacZ*. The factor found, named diphtheria toxin regulatory protein, or DtxR, was a 226-amino-acid protein. DtxR was shown to repress the expression of β-galactosidase from a *toxPO-lacZ* fusion in an iron-dependent fashion. The functional activity of DtxR in *C. diphtheriae* was subsequently demonstrated by studies performed by Schmitt and Holmes (54): expression of *dtxR* in C7hm723(βtox+), a strain of *C. diphtheriae* in which toxin expression is iron insensitive, resulted in conversion to the wild-type iron-sensitive phenotype.

DtxR: BIOCHEMICAL AND GENETIC ANALYSIS

In 1992, Tao et al. (58) used gel mobility shift assays with purified DtxR and ^{32}P-labeled *toxPO* as a probe to explore the transition metal ion activation of the repressor and its binding to the *tox* operator. This work demonstrated that activation of apo-DtxR and subsequent binding to the *tox* operator required a divalent transition metal ion. Since binding to the ^{32}P-labeled probe was blocked by the addition of either excess nonradioactive probe, anti-DtxR antisera, or the cation chelator 2,2′-dipyridyl, binding was clearly specific and dependent upon metal ion activation. Tao and Murphy (59) and Schmitt and Holmes (55) both showed that DtxR protects *toxPO* from DNase digestion following activation by one of a number of divalent cations (e.g., Co^{2+}, Fe^{2+}, Ni^{2+}, Cd^{2+}, or Mn^{2+}). Interestingly, Zn^{2+} was a weak activator of DtxR, and Cu^{2+} failed to activate repressor activity. This in vitro analysis of metal ion activation of apo-DtxR confirmed and extended earlier observations on the inhibitory effect of transition metal ions on the expression of diphtheria toxin by toxigenic strains of *C. diphtheriae* (17).

In 1994, Tao and Murphy (61) demonstrated by in vitro affinity enrichment and gel mobility-shift selection that the minimal nucleotide sequence necessary for DtxR binding is a 9-bp palindromic sequence separated by either a single C or G nucleotide as follows: 5′-T.AGGTTAGC/ GCTAACCT.A-3′. Since a family of related target sequences that bound DtxR with the same apparent affinity as the 27-bp *tox* operator was isolated, it was apparent that DtxR could function as a global regulatory element in the regulation of iron-sensitive genes in *C. diphtheriae*, as earlier postulated (54). Recently, a number of genes with upstream DtxR-binding sites were isolated and characterized (28, 53, 55, 57a). Several studies employed site-directed mutagenesis of DtxR to elucidate structure-function relationships within the repressor, especially the metal ion-binding and DNA-binding domains of DtxR. Since DtxR contains only a single cysteine residue, and disulfide bond-linked dimers of DtxR are inactive, Tao and Murphy (60) used saturation site-directed mutagenesis of Cys-102 to demonstrate that this residue plays a role in the coordination of Fe^{2+}. The substitution of Cys with all 20 amino acids, except Asp, resulted in the complete loss of repressor activity. DtxR(C102D) retained an iron-dependent active phenotype, albeit not to wild-type levels. Wang et al. (67) utilized bisulfite mutagenesis of *dtxR* to analyze the role of particular residues in DtxR function. These investigators also postulated that the ion-binding domain included Cys-102 and further that the DNA-binding domain (residues 1 to 52) formed a helix-turn-helix motif similar to the DNA-binding domains of many other regulatory proteins.

CRYSTALLOGRAPHIC ANALYSIS OF APO-, METAL ION-ACTIVATED, AND ACTIVATED DtxR COMPLEXED WITH THE DIPHTHERIA *tox* OPERATOR

DtxR was the first iron-activated DNA-binding repressor for which an X-ray crystal structure was solved. In 1995, both Qiu et al. (47) and Schiering et al. (52) published structures at similar resolution. Initially, both groups were able to assign coordinates to amino acids 1 through 136; however, they were unable to assign coordinates to the C-terminal 90-amino-acid residues owing to the high thermal factors of this portion of the repressor. More recently, Qiu et al. (46) reported the DtxR coordinates at 2.0 Å resolution, at which level the C-terminal domain was found to fold into an SH3-like structure. White et al. (69) have solved the structure of Ni^{2+}-activated DtxR(C102D) *tox*PO complex.

DtxR contains a total of eight α-helices, six of which are contained within the N-terminal two-thirds of the protein. Helices B and C and the three-amino-acid connecting loop between them (residues 27 to 50) constitute the helix-turn-helix motif and form the DNA-binding site. Surprisingly, activated DtxR(C102D) binds to *tox*O as two pairs of dimers (69). Each DtxR dimer binds to one of the palindromic sequences on almost opposite faces of the operator (Fig. 2). Knowledge of the structure of the ternary complex confirms and extends earlier observations that the DtxR footprint encompassed a region of 30 bp (59).

While the overall mechanism of target DNA recognition by DtxR is similar to that of other prokaryotic repressors, there are some noteworthy interactions. The α-helix C of DtxR is responsible for most interactions with DNA and inserts into the major groove of the double helix. Each helix-turn-helix motif makes a total of nine interactions with backbone phosphate groups: three from α-helix B, two from the turn, and four from α-helix C. The guanidinium group of Arg-60 binds in the minor groove of DNA, thereby making a bridge to additional phosphates. Owing to a structural rearrangement of DtxR upon binding to the *tox* operator, Thr-7 in α-helix A also interacts with a backbone phosphate. Interestingly, only a single amino acid, Gln-43, appears to associate directly with a DNA base. Again, observations made by crystallographic analysis confirm and extend earlier biochemical genetic analysis of the repressor (55, 67).

Metal Ion-Binding and Activation of Repressor Activity

Early observations made by saturation and equilibrium dialysis suggested that DtxR carried a single class of metal ion-binding site with an apparent K_d of 2×10^{-6} to 9×10^{-7} M (62, 67). Paradoxically, X-ray crystallographic analysis of metal ion-DtxR complexes clearly revealed two metal ions bound per monomer (47, 52). Using site-directed mutagenesis of each of the coordinating amino acid residues, Ding et al. (9) demonstrated that the primary metal ion-binding site involved in activation of DNA binding is composed of amino acids that form an octahedral coordination center. The primary metal ion-binding site required for DtxR(C102D) and DtxR activation is composed of the side chains of Met-10, Asp-102, Cys-102, Glu-105, His-106, and a water molecule (Fig. 3). This structural water molecule is also held in place by the main-chain carbonyl oxygen of Leu-4. In contrast, the binding of metal ion to the ancillary site does not appear to be of functional significance and may represent a crystallographic artifact.

The binding of activating metal ions by DtxR results in two distinct structural changes in the repressor. First, α-helices C in the dimeric structure move 2 Å closer together (52). This overall caliperlike or pincer movement in the structure places the two C helices in a position for more favorable helix-turn-helix interactions with the major groove in the *tox* operator (69). The second change in protein structure following ion binding is a helix-to-coil transition of the first six N-terminal amino acid residues (69). This change in structure appears to occur upon ion binding, rather than upon target DNA binding, as DtxR without *tox* operator demonstrated the transition upon addition of Ni^{2+} (52). In the coil form, the backbone carbonyl oxygen of Leu-4 becomes available for hydrogen bonding with the structural water molecule, which is one of the ligands for the primary metal ion-binding site. Since these events (i.e., ion binding and helix-to-coil transition) are codependent, it is likely that they occur simultaneously. Interestingly, the helix-to-coil transition has a second important consequence: the removal of an unfavorable stearic interaction between DtxR and DNA. Without the transition, the N terminus of the repressor would have an extended α-helix A and thereby inhibit binding to the *tox* operator.

Importantly, the activating metal ion does not form a bridge between DtxR and target operators. In contrast, the *trp* repressor, for which tryptophan is a corepressor, also binds to its DNA recognition site as two pairs of dimers. However, the *trp* repressor dimers interact with each other (27), and the corepressor tryptophan is bound at the DNA-protein interface (71). Although both DtxR and the *trp* repressor mediate repression through corepressors and bind to their respective operators as two dimers, the molecular mechanisms involved in repression are in fact very different.

DtxR Protein-Protein Interactions

The protein-protein interactions stabilizing DtxR dimers arise mostly from hydrophobic associations. Several nonpolar amino acids contained within α-helices D, E, and F and their intervening loops compose this hydrophobic network (52). This region contains amino acids 85 through 116 of DtxR. One internal hydrogen bond exists, between Glu-100 and Trp-104; mutation of either of these residues drastically reduces repressor activity (67). This network of hydrophobic interactions is likely responsible for the weak equilibrium between monomeric and dimeric apo-DtxR that was demonstrated by Tao et al. (62).

C-Terminal Domain of DtxR

The X-ray structure of the C-terminal domain of DtxR shows that this region of the repressor is composed of five antiparallel β-sheets and two short α-helices (46). Twigg et al. (64) and Wang et al. (66) have recently expressed the C-terminal SH3-like domain of DtxR, DtxR(Δ1-129), and determined its solution structure by heteronuclear nuclear magnetic resonance. These studies have confirmed the observations of Qiu et al. (46) that this region of DtxR

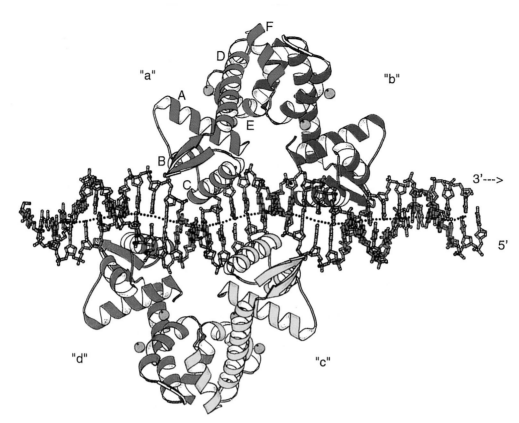

FIGURE 2 Structure of the 2[Ni. DtxR(C102)]$_2$-*tox* operator complex. Residues 3 to 120 in each DtxR(C102D) monomer are designated "a" to "d." Ribbons and arrows are used to designate α-helices (A to F) and β-strands in each monomer. The 33-bp DNA segment carries the 27-bp interrupted palindromic *tox* operator sequence. Adapted from White et al. (69).

folds into an SH3-like structure. Interestingly, Wang et al. (66) also report that, like eukaryotic SH3 domains of Hck, Src, and Itk, the C-terminal domain of DtxR may associate with a proline-rich peptide derived from the linker region between the N- and C-terminal domains of the repressor. The interactions between the C-terminal domain and the proline-rich linker region of DtxR (i.e., amino acids 125 to 139) possibly stabilize the monomeric form of the re-

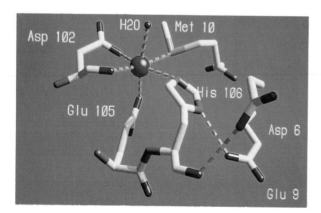

FIGURE 3 Partial X-ray structure of the primary metal ion-binding and activation site of DtxR(C102D). In this structure Ni^{2+} is coordinated by the side chains of Met-10, Asp-102, Glu-105, His-106, and a water molecule.

pressor. The binding of activating metal ions in the primary binding site would then induce a conformational change that would lead to C-terminal domain dissociation and formation of stable DtxR dimers.

dtxR ALLELES AND HOMOLOGS IN OTHER GRAM-POSITIVE PROKARYOTES

Boyd et al. (3) cloned and determined the nucleotide sequence of the *dtxR* alleles from the PW8, 1030, and C7hm723 strains of *C. diphtheriae*, and Nakao et al. (39, 40) determined the nucleotide sequence of the *dtxR* alleles from a number of *C. diphtheriae* clinical isolates from Russia and Ukraine. The comparison of the deduced DtxR amino acid sequences in these strains revealed a remarkable conservation in their respective N-terminal domains, whereas multiple changes were observed in their respective C-terminal domains.

Although DtxR from *C. diphtheriae* has been extensively studied, a number of closely related homologs from various gram-positive organisms have been identified. As shown in Fig. 4, there is extraordinary conservation of amino acid sequence within the N-terminal domains of DtxR and its homologs. IdeR (iron-dependent regulator) has been found to be the DtxR homolog in several species of *Mycobacterium*: *Mycobacterium tuberculosis* H37Rv, *Mycobacterium bovis* BCG, *Mycobacterium leprae*, *Mycobacterium avium*, and *Mycobacterium smegmatis* (10, 56). IdeR and DtxR are 78% identical and 90% homologous over their N-terminal 140

```
                1        10        20        30        40        50
       Bl DtxR  MKDLVDTTEM YLRTIYELEE EGIVPLRARI AERLEQSGPT VSQTVARMER
       Cd DtxR  MKDLVDTTEM YLRTIYELEE EGVTPLRARI AERLEQSGPT VSQTVARMER
       Ms IdeR  MNDLVDTTEM YLRTIYDLEE EGVVPLRARI AERLDQSGPT VSQTVSRMER
       Mt IdeR  MNELVDTTEM YLRTIYDLEE EGVTPLRARI AERLDQSGPT VSQTVSRMER
       Sl DesR  MSGLIDTTEM YLRTILELEE EGVVPMRARI AERLDQSGPT VSQTVARMER
       Sp DesR  MSGLIDTTEM YLRTILELEE EGVVPMRARI AERLDQSGPT VSQTVARMER

       CONSENSUS  M  L#DTTEM YLRTI @LEE EGV P RARI AERL@QSGPT VSQTV RMER


                        60        70        80        90        100
       Bl DtxR  DGLVHVSPDR SLEMTPEGRS LAIAVMRNDR LAERLLTDII GLDIHKVHDE
       Cd DtxR  DGLVVVASDR SLQMTPTGRT LATAVMRKHR LAERLLTDII GLDINKVHDE
       Ms IdeR  DGLLHVAGDR HLELTDKGRA LAVAVMRKHR LAERLLVDVI LPWEDGVHAE
       Mt IdeR  DGLLRVAGDR HLELTEKGRA LAIAVMRKHR LAERLLVDVI GLPWEEVHAE
       Sl DesR  DGLVSVAADR HLELTDEGRR LATRVMRKHR LAECLLVDVI GLEWEQVHAE
       Sp DesR  DGLVSVAPDR HLELTEEGRR LATRVMRKHR LAECLLVDVI GLEWEQVHAE

       CONSENSUS  DGL# V  DR  L  T  GR  LA  VMR@ R LAE LL D#I      VH E


                       110       120       130       140       150
       Bl DtxR  ACRWEHVMSD EVERRLVEVL DDVHRSPFGN PIPGLGEIGL DQADEPDSGV
       Cd DtxR  ACRWEHVMSD EVERRLVKVL KDVSRSPFGN PIPGLDELGV GNSDAAAPGT
       Ms IdeR  ACRWEHVMSE EVERRLVQVL ENPTTSPFGN PIPGLTELAV TPGVNTEDVS
       Mt IdeR  ACRWEHVMSE DVERRLVKVL NNPTTSPFGN PIPGLVELGV GPEPGADDAN
       Sl DesR  ACRWEHVMSE AVERRVLELL RHPTESPYGN PIPGLEELGE TDGADPFLDE
       Sp DesR  ACRWEHVMSE AVERRVLELL RHPTESPYGN PIPGLEELGE KDGADPFLDE

       CONSENSUS  ACRWEHVMS@ VERR## #L   SP GN PIPGL EL


                       160       170       180       190       200
       Bl DtxR  RAIDLPLGEN LKARIVQLNE ILQVDLEQFQ ALTDAGVEIG TEVDIINEQG
       Cd DtxR  RVIDAATSMP RKVRIVQINE IFQVETDQFT QLLDADIRVG SEVEIVDRDG
       Ms IdeR  LVRLTELPVG MPVAVVVRQL TEHVQGDTDL IGRLKEAGVV PNARVTVEAN
       Mt IdeR  LVRLTELPAG SPVAVVVRQL TEHVQGDIDL ITRLKDAGVV PNARVTVETT
       Sl DesR  GMVSLADLDP GQEGKTVVVR RIGEPIQTDA QLMYTLRRAG VQPGSVVSVT
       Sp DesR  GMVSLAELDP GAEGKTVVVR RIGEPIQTDA QLMYTLRRAG VQPGSVVSVT

       CONSENSUS


                       210       220
       Bl DtxR  RVVITHNGSS VELIDDLAHA VRVEKVEG
       Cd DtxR  HITLSHNGKD VELLDDLAHT IRIEEL
       Ms IdeR  NNGGVMIVIP GHEQVELPHH MAHAVKKKVE KVEKV
       Mt IdeR  PGGGVTIVIP GHENVTLPHE MAHAVKVEKV
       Sl DesR  ESAGGVLVGS GGEAAELEAD TASHVFVAKR
       Sp DesR  EAAGGGVLVG SSGEAAELET DVASHVFVAKP

       CONSENSUS
```

FIGURE 4 Comparison of the deduced amino acid sequence of *dtxR* alleles from *C. diphtheriae* (Cd) and *Brevibacterium lactofermentum* (Bl) with the homologous *ideR* alleles from *Mycobacterium tuberculosis* (Mt) and *Mycobacterium smegmatis* (Ms) and the *desR* alleles from *Streptomyces lividans* (Sl) and *Streptomyces pilosus* (Sp). The consensus sequence indicates those amino acid residues that are identical, as well as those that are conserved with respect to hydrophobic amino acids (#) and charged amino acids (@).

amino acids (10). *Streptomyces lividans* and *Streptomyces pilosus* have been found to carry DtxR homologs (19) that share 73% identity with DtxR in their respective N-terminal 139 amino acids; however, the relative identity drops to only 20% in the comparison of their C-terminal domains (19). In *Streptomyces* spp., the DtxR homologs have been designated DesR, or desferrioxamine operon repressors. Remarkably, in many cases these homologous iron-dependent regulatory proteins have been shown to be functional homologs of DtxR and to be capable of recognizing and regulating gene expression through the diphtheria *tox* operator (18, 41, 56).

The genome of *Treponema pallidum*, the spirochete that causes syphilis, carries the *troR* gene, which appears to be a truncated homolog of *dtxR*, since only the N-terminal 153 amino acids are present (22). Although the function of this homolog is largely unknown, the conservation of amino acids essential for DtxR function suggests that they have a similar role.

DIPHTHERIA TOXIN: BIOCHEMISTRY AND CYTOTOXICITY

Diphtheria toxin is the primary virulence factor expressed by toxigenic strains of *C. diphtheriae*. As discussed above, the structural gene encoding diphtheria toxin is carried by a family of closely related corynebacteriophage, the best studied of which is corynephage β (6). The diphtheria *tox* mRNA has been shown to be associated with membrane-bound polysomes, and the nascent toxin is cotranslationally secreted in precursor form (57). Following removal of the 19-amino-acid signal sequence, the mature form of the toxin is released into the surrounding medium.

Diphtheria toxin is a 535-amino-acid single-chain protein (58,342 molecular weight) with a 50% lethal dose (LD_{50}) in sensitive mammalian species of approximately 100 ng/kg of body weight (14, 16, 24, 49). Following mild digestion with trypsin, or other serine proteases, the toxin may be separated under denaturing conditions into two polypeptide chains: the N-terminal 193-amino-acid fragment A and the C-terminal 342-amino-acid fragment B (11, 15). Biochemical and genetic analysis has shown that the A fragment of diphtheria toxin is an enzyme that catalyzes the NAD^+-dependent ADP-ribosylation of eukaryotic elongation factor 2 (EF2) according to the following equation:

$$NAD^+ + EF2 \rightleftarrows ADPR\text{-}EF2 + nicotinamide + H^+$$

The B fragment of diphtheria toxin has been shown to facilitate the delivery of fragment A across the eukaryotic cell membrane to the cytosol and to carry the native receptor-binding domain. Choe et al. (8) first described the X-ray crystal structure of diphtheria toxin, which was subsequently corrected by Bennett et al. (1). The X-ray crystal structure analysis of diphtheria toxin clearly demonstrated that the toxin is a three-domain protein: the catalytic (C) domain (fragment A), the transmembrane (T) domain, and the receptor-binding (R) domain (Fig. 5).

The diphtheria toxin-mediated intoxication of sensitive eukaryotic cells has been shown to be a sequential process in which each structural domain plays an essential role. The intoxication process is initiated by the interaction of diphtheria toxin with its cell surface receptor. Naglich et al. (38) have shown that the diphtheria toxin receptor is the heparin-binding epidermal growth factor-like precursor (HB-EGF-LP). In addition, DAP-27 (the 27,000-molecular-weight monkey homolog of human CD9) has been shown to modulate the binding of diphtheria toxin to the cell surface and to enhance the sensitivity of given cells toward the toxin (6, 23, 31). Since an association between HB-EGF-LP and DAP-27 has not yet been detected, it is likely that the role of DAP-27 is to maintain large numbers of HB-EGF-LP on the cell surface.

Once bound to its cell surface receptor, diphtheria toxin is internalized by receptor-mediated endocytosis in coated pits (32). Either on the cell surface or in the early endocytic vesicle, the α-carbon backbone of diphtheria toxin is cleaved by the cellular protease furin after Arg-193, which is positioned in the sensitive loop between fragments A and B.

The diphtheria toxin T domain, which is composed of nine α-helices and their connecting loops, appears to play at least two important roles in the intoxication process (22). The first helical layer (i.e., helices 1 to 3) is amphipathic and appears to stabilize the toxin on the luminal surface of the endocytic vesicle membrane. As the lumen of the early endosome is acidified through the action of the vesicular ATPase, the decrease in pH causes a partial denaturation of the T domain and spontaneous insertion of helices 5 to 9 into the plane of the vesicle membrane, thereby forming a pore or channel through the membrane. This insertion event appears to begin with the third helix layer, channel-forming helices 8 and 9, which is then stabilized by the insertion of the second helix layer, composed of helixes 5, 6, and 7. As the channel is forming, an extended denatured C domain appears to be inserted into the nascent pore by a mechanism that may involve positioning of the carboxy-terminal end of the C domain by T-domain helix 1. The disulfide bond between the C and T domains is reduced either in the lumen of the endocytic vesicle (50) or as it emerges from the vesicle in the cytosol. Once the first few hydrophobic residues of the C domain reach the cytosol, it is possible that cellular chaparonins facilitate the entry process by sequentially binding and refolding the denatured polypeptide into an active conformation as it emerges from the lumen of the early endosome (29, 70a). The delivery of a single molecule of the C domain to the cytosol of a eukaryotic cell has been shown to be sufficient to bring about the death of the cell (70).

TURNING THE TABLES ON DIPHTHERIA TOXIN: FUSION PROTEIN TOXINS AND THE DEVELOPMENT OF A NEW CYTOTOXIC AGENT FOR THE TREATMENT OF CUTANEOUS T-CELL LYMPHOMA

Since the first step in the intoxication of eukaryotic cells is receptor binding, the use of protein engineering methods to "redesign" native diphtheria toxin by receptor-binding domain substitution has become attractive (34). A family of fusion protein toxins, each with a different targeting ligand, has been genetically constructed and characterized (37). One such fusion protein toxin, composed of the C and T domains of diphtheria toxin to which human interleukin-2 has been genetically fused ($DAB_{389}IL\text{-}2$), has been extensively evaluated in human clinical trials for the treatment of interleukin-2 receptor-positive leukemias and lymphomas (12).

Two randomized, double-blind studies were evaluated in a phase III clinical trial designed to rigorously test the po-

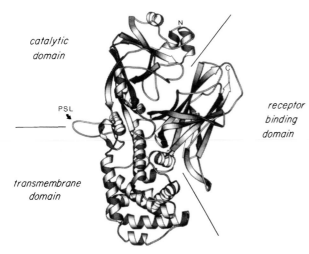

catalytic
domain

PSL

transmembrane
domain

receptor
binding
domain

N

C

FIGURE 5 Ribbon diagram of the X-ray crystal structure of native diphtheria toxin (1, 8). The relative positions of the catalytic, transmembrane, and receptor-binding domains are indicated. PSL, protease-sensitive loop; N, amino terminus; C, carboxy terminus.

tential efficacy of treating cutaneous T-cell lymphoma patients with DAB$_{389}$IL-2. The first study evaluated the intravenous administration of DAB$_{389}$IL-2 for up to eight courses of therapy at either 9 μg/kg/day or 18 μg/kg/day in patients with advanced refractory disease; the second arm of the trial evaluated DAB$_{389}$IL-2 in patients with less advanced disease at 9 μg/kg/day and 18 μg/kg/day and included a placebo control group.

Upon completion of the first study, analysis of the overall response rate in patients who met the inclusion criteria demonstrated that 30% of the total number of patients had a 50% or greater reduction in their tumor burden for at least 6 weeks following treatment with DAB$_{389}$IL-2. In this refractory patient population, 10% of the total number of patients who were evaluable had either a complete response (in the absence of histologic analysis) or a complete clinical response (documented by histology).

Consistent with earlier clinical studies, the most common adverse events experienced by this patient population were chills and fever, malaise, and nausea and vomiting. Less frequent adverse events included hypotension, edema, rash, and capillary leak syndrome. Unlike many other non-Hodgkin's lymphomas, cutaneous T-cell lymphoma is clearly a devastating malignancy in which patients suffer a substantial and often disfiguring disability during the course of their illness. The analysis of results from the first arm of the phase III trial has clearly demonstrated that DAB$_{389}$IL-2 therapy has the ability to offer substantial reduction in tumor burden and relief from constitutional symptoms in a large percentage of patients who are otherwise refractory. In June 1998, the U.S. Food and Drug Administration's Oncologic Drugs Advisory Committee unanimously recommended that DAB$_{389}$IL-2 (ONTAK) receive expedited approval by the agency for the treatment of refractory cutaneous T-cell lymphoma. In 1986, Murphy et al. (34) "envisioned that these chimeric molecules might serve as . . . targeted toxins for the treatment of human malignancies." DAB$_{389}$IL-2 is the first recombinant receptor-targeted biologic to achieve this status.

EPIDEMIOLOGY OF DIPHTHERIA

Outbreaks of clinical diphtheria almost always occur in individuals who have not become immunized and who have been exposed to a "carrier" (i.e., an individual who carries toxigenic C. diphtheriae in his or her nasopharyngeal flora). The infection is spread by droplet from person to person. While colonization of a susceptible individual with a toxigenic strain certainly plays a role in pathogenesis, Pappenheimer and Murphy (45) have also demonstrated that transmission of toxigenicity may occur by in situ lysogenic conversion of an autochthonous nontoxigenic strain of C. diphtheriae to toxigenicity.

In 1923, Ramon (48) demonstrated that the treatment of diphtheria toxin with formaldehyde resulted in the formation of a "toxoid" that could be used as an immunogen for vaccination against diphtheria. Remarkably, immunization with diphtheria toxoid results in the production of neutralizing antibodies that not only block native diphtheria toxin from binding its cell surface receptor but protects against clinical disease as well (42, 72). Mass immunizations of populations with diphtheria toxoid have led to a dramatic decrease in the incidence of diphtheria. One of the best-chronicled studies of the effects of mass immuni-

zation comes from Rumania. Before the large-scale immunization program began in 1958, only 60% of the population was immune to diphtheria toxin, as measured by Schick test reactivity; by 1979, that proportion had jumped to 97% (43, 51). Over the same period, diphtheria morbidity dropped from approximately 600 per 100,000 in 1958 to only 1 per 100,000 in 1972 (43, 51).

Saragea et al. (51) also observed a concomitant reduction in the relative proportion of toxigenic strains of C. diphtheriae that were isolated during that same period in Rumania. Between 1955 and 1966, 86% of C. diphtheriae clinical isolates were toxigenic; that number fell to 5% by 1977. Importantly, the overall prevalence of C. diphtheriae isolated from the general population did not change over this period. The results are clear: immunization with diphtheria toxoid eliminated a selective advantage for the spread of toxigenic C. diphtheriae among the population. However, nontoxigenic strains of C. diphtheriae continued to fill the ecologic niche.

Data from the recent diphtheria epidemic in Russia and the New Independent States teach a similar lesson (21). The relative number of Russian children vaccinated with diphtheria toxoid decreased from over 80% in 1980 to 68% in 1990 (30), a value that falls well below the 95% minimum that the World Health Organization suggests is necessary to prevent an epidemic. The increase in the relative number of individuals who were not vaccinated or in whom titers of anti-diphtheria toxin antibodies fell below protective levels has allowed toxigenic strains of C. diphtheriae to spread through the population, resulting in the current diphtheria epidemic.

Recent molecular epidemiologic analysis of toxigenic strains of C. diphtheriae isolated from this epidemic has provided further insight into the virulence of this pathogen. The tox and dtxR genes from 72 strains of C. diphtheriae that were recently isolated from Russia and Ukraine have been PCR amplified and sequenced. The tox gene is highly conserved: fewer than 6% of the strains carried a mutation in tox; importantly, none of those mutations coded for an amino acid change. This extraordinary conservation of amino acid sequence again suggests the importance of the structure-function relationships within diphtheria toxin. Sequencing of dtxR genes provided somewhat different results: only 20% of the strains were without a mutation in the diphtheria toxin repressor. While most mutations in dtxR did not give rise to amino acid substitutions, nine substitutions were found. Importantly, all nine residue substitutions were found in the C-terminal domain of DtxR (i.e., distal to residue 147) (39, 40). As discussed previously, the C-terminal region of DtxR has no known function and biochemically appears to be unnecessary for the three known activities of DtxR (ion binding, DNA binding, and protein-protein interaction).

The extreme conservation of the diphtheria toxin primary sequence suggests that the evolution of new tox alleles that would encode a toxin sufficiently different to avoid neutralization by anti-diphtheria toxoid antibodies is highly unlikely. This is good news and argues against fears that an evolved "super-diphtheria" could make vaccination with diphtheria toxoid obsolete.

The high conservation of the N-terminal half of DtxR underscores the importance of this iron-dependent repressor in the control of virulence gene expression. Mutations in DtxR that give rise to loss of repressor activity would allow constitutive expression of not just tox but all other iron-sensitive genes as well. Constitutive expression of siderophores could conceivably result in iron loading that might lead to DNA damage (63).

We thank Xiaochun Ding and Dagmar Ringe for providing the X-ray crystal structure of the DtxR primary metal ion-binding site (Fig. 3). J. R. M. is partially supported by Public Health Service grant AI21628 from the National Institute of Allergy and Infectious Diseases.

REFERENCES

1. **Bennett, M. J., S. Choe, and D. Eisenberg.** 1994. Domain swapping: entangling alliances between proteins. *Proc. Natl. Acad. Sci. USA* **91:**3127–3131.

2. **Bisgard, K. M., I. R. B. Hardy, T. Popovic, P. M. Strebel, M. Wharton, R. T. Chen, and S. C. Hadler.** 1998. Respiratory diphtheria in the United States, 1980 through 1995. *Am. J. Public Health* **88:**787–791.

3. **Boyd, J., K. Hall, and J. R. Murphy.** 1992. Characterization of *dtxR* alleles from *Corynebacterium diphtheriae* strains PW8, 1030, and C7hm723. *J. Bacteriol.* **174:**1268–1272.

4. **Boyd, J., and J. R. Murphy.** 1988. Analysis of the diphtheria *tox* promoter by site-directed mutagenesis. *J. Bacteriol.* **170:**5949–5952.

5. **Boyd, J., M. Oza, and J. R. Murphy.** 1990. Molecular cloning and DNA sequence analysis of an iron dependent diphtheria *tox* regulatory element (*dtxR*) from *Corynebacterium diphtheriae*. *Proc. Natl. Acad. Sci. USA* **87:**5968–5972.

6. **Brown, J. G., B. D. Almond, J. G. Naglich, and L. Eidels.** 1993. Hypersensitivity to diphtheria toxin by mouse cell expressing both diphtheria toxin receptor and CD9 antigen. *Proc. Natl. Acad. Sci. USA* **90:**8184–8188.

7. **Buck, G. A., R. E. Cross, T. P. Wong, J. Lorea, and N. Groman.** 1985. DNA relationships among some *tox*-bearing corynebacteriophages. *Infect. Immun.* **49:**679–684.

8. **Choe, S., M. J. Bennett, G. Fugii, P. M. G. Curmi, K. A. Kantardjieff, R. J. Collier, and D. Eisenberg.** 1992. The crystal structure of diphtheria toxin. *Nature* **357:**216–222.

9. **Ding, X., H. Zeng, N. Schiering, D. Ringe, and J. R. Murphy.** 1966. Identification of the primary metal ion-activation sites of the diphtheria *tox* repressor by X-ray crystallography and site-directed mutagenesis. *Nat. Struct. Biol.* **3:**382–387.

10. **Doukhan, L., M. Predich, G. Nair, O. Dussurget, I. Manic-Mulec, S. T. Cole, D. R. Smith, and I. Smith.** 1995. Genomic organization of the mycobacterial sigma gene cluster. *Gene* **165:**67–70.

11. **Drazin, R., J. Kandel, and R. J. Collier.** 1971. Structure and activity of diphtheria toxin. II. Attack by trypsin at a specific site within the intact toxin molecule. *J. Biol. Chem.* **246:**1504–1510.

12. **Foss, F. M., M. N. Saleh, J. G. Krueger, J. C. Nichols, and J. R. Murphy.** 1997. Diphtheria toxin fusion proteins. *Curr. Top. Microbiol. Immunol.* **234:**63–81.

13. **Fourel, G., A. Phalipon, and M. Kaczorek.** 1989. Evidence for direct regulation of diphtheria toxin gene transcription by an Fe2+-dependent DNA-binding repressor, DtoxR, in *Corynebacterium diphtheriae*. *Infect. Immun.* **57:**3221–3225.

14. **Gill, D. M.** 1982. Bacterial toxins: a table of lethal amounts. *Microbiol. Rev.* **46:**86–94.

15. **Gill, D. M., and A. M. Pappenheimer, Jr.** 1971. Structure-activity relationships in diphtheria toxin. *J. Biol. Chem.* **246:**1492–1495.

16. **Greenfield, L., M. J. Bjorn, G. Horn, D. Fong, G. A. Buck, R. J. Collier, and D. A. Kaplan.** 1983. Nucleotide sequence of the structural gene for diphtheria toxin carried by corynebacteriophage β. *Proc. Natl. Acad. Sci. USA* **80:**6853–6857.

17. **Groman, N. B., and K. Judge.** 1979. Effects of metal ions on diphtheria toxin production. *Infect. Immun.* **26:**1065–1070.

18. **Günter, K., C. Toupet, and T. Schupp.** 1993. Characterization of an iron-regulated promoter involved in desferrioxamine B synthesis in *Streptomyces pilosus*: repressor-binding site and homology to the diphtheria toxin gene promoter. *J. Bacteriol.* **175:**3295–3302.

19. **Günter-Seeboth, K., and T. Schupp.** 1995. Cloning and sequence analysis of the *Corynebacterium diphtheriae dtxR* homologue from *Streptomyces lividans* and S. *pilosus* encoding a putative iron repressor. *Gene* **166:**117–119.

20. **Hardham, J. M., L. V. Stamm, S. F. Porcella, J. G. Frye, N. Y. Barnes, J. K. Howell, S. L. Mueller, J. D. Radolf, G. M. Weinstock, and S. J. Norris.** 1997. Identification and transcriptional analysis of a *Treponema pallidum* operon encoding a putative ABC transport system, an iron-activated repressor protein homolog, and a glycolytic pathway enzyme homolog. *Gene* **197:**47–64.

21. **Hardy, I. R. B., S. Dittman, and R. W. Sutter.** 1996. Current situation and control strategies for resurgence of diphtheria in newly independent states of the former Soviet Union. *Lancet* **347:**1739–1744.

22. **Hu, H.-Y., P. D. Hunth, J. R. Murphy, and J. C. vanderSpek.** 1998. The effects of helix breaking mutations in the diphtheria toxin transmembrane domain helix layers of the fusion toxin DAB₃₈₉IL-2. *Protein Eng.* **11:**101–107.

23. **Iwamoto, R., H. Senoh, Y. Okada, and E. Mekada.** 1991. An antibody that inhibits the binding of diphtheria toxin to cells revealed the association of a 27-kDa membrane protein with the diphtheria toxin receptor. *J. Biol. Chem.* **266:**20463–20469.

24. **Kaczorek, M., F. Delpyyroux, N. Chenciner, R. E. Streek, J. R. Murphy, P. Boquet, and P. Tiollais.** 1983. Nucleotide sequence and expression in *Escherichia coli* of the CRM228 diphtheria toxin gene. *Science* **221:**855–858.

25. **Kanei, C., T. Uchida, and M. Yoneda.** 1977. Isolation from *Corynebacterium diphtheriae* C7(β) of bacterial mutants that produce toxin in medium containing excess iron. *Infect. Immun.* **18:**203–209.

26. **Krafft, A. E., S. P. Tai, C. Coker, and R. K. Holmes.** 1992. Transcription analysis and nucleotide sequence of *tox* promoter/operator mutants of corynebacteriophage beta. *Microb. Pathog.* **13:**85–92.

27. **Lawson, C. L., and J. Carey.** 1993. Tandem binding in crystals of a trp repressor/operator half site complex. *Nature* **366:**178–182.

28. **Lee, J. H., T. Wang, K. Ault, J. Liu, M. P. Schmitt, and R. K. Holmes.** 1997. Identification and characterization of three new promoter/operators from *Corynebacterium*

diphtheriae that are regulated by the diphtheria toxin repressor (DtxR) and iron. *Infect. Immun.* **65:**4273–4280.

29. **Lund, R. A.** 1995. The role of molecular chaparones in vivo. *Essays Biochem.* **29:**113–129.

30. **Maurice, J.** 1995. Russian chaos breeds diphtheria outbreak. *Science* **167:**1416–1417.

31. **Mitamura, T., R. Iwamoto, T. Umata, T. Yomo, I. Urabe, M. Tsuneoka, and E. Mekada.** 1992. The 27-kD diphtheria toxin receptor-associated protein (DRAP27) from Vero cells is the monkey homologue of human CD9 antigen: expression of DRAP27 elevates the number of diphtheria toxin receptors on toxin-sensitive cells. *J. Cell Biol.* **118:**1389–1399.

32. **Moya, M., A. Dautry-Versat, B. Goud, D. Louvard, and P. Boquet.** 1985. Inhibition of coated-pit formation in Hep2 cells blocks the cytotoxicity of diphtheria toxin but not ricin toxin. *J. Cell Biol.* **101:**548–559.

33. **Murphy, J. R., and P. Bacha.** 1979. Studies of the regulation of diphtheria toxin production, p. 181–186. *In* D. Schlessinger (ed.), *Microbiology—1979.* American Society for Microbiology, Washington, D.C.

34. **Murphy J. R., W. Bishai, M. Borowski, A. Miyanohara, J. Boyd, and S. Nagle.** 1986. Genetic construction, expression, and melanoma selective cytotoxicity of a diphtheria toxin α-melanocyte stimulating hormone fusion protein. *Proc. Natl. Acad. Sci. USA* **83:**8258–8262.

35. **Murphy, J. R., A. M. Pappenheimer, Jr., and S. Tayart de Borms.** 1974. Synthesis of diphtheria *tox* gene products in *Escherichia coli* extracts. *Proc. Natl. Acad. Sci. USA* **71:**11–15.

36. **Murphy, J. R., J. Skiver, and G. McBride.** 1976. Isolation and partial characterization of a corynebacteriophage *tox* operator constitutive-like mutant lysogen of *Corynebacterium diphtheriae. J. Virol.* **18:**235–244.

37. **Murphy, J. R., and J. C. vanderSpek.** 1995. Targeting diphtheria toxin to growth factor receptors. *Semin. Cancer Biol.* **6:**259–267.

38. **Naglich, J. G., J. E. Matherall, D. W. Russell, and L. Eidels.** 1992. Expression cloning of a diphtheria toxin receptor: identity with a heparin-binding EGF-like growth factor precursor. *Cell* **69:**1051–1061.

39. **Nakao, H., I. K. Mazurova, T. Glushkevich, and T. Popovic.** 1997. Analysis of heterogeneity of *Corynebacterium diphtheriae* toxin gene, *tox*, and its regulatory element, *dtxR*, by direct sequencing. *Res. Microbiol.* **148:**45–54.

40. **Nakao, H., J. M. Pruckler, I. K. Mazurova, O. V. Narvskaia, T. Glushkevich, V. F. Marijevski, A. N. Kravetz, B. I. Fields, I. K. Wachsmuth, and T. Popovic.** 1996. Heterogeneity of diphtheria toxin gene, *tox*, and its regulatory element, *dtxR*, in *Corynebacterium diphtheriae* strains causing epidemic diphtheria in Russia and Ukraine. *J. Clin. Microbiol.* **34:**1711–1716.

41. **Oguiza, J. A., X. Tao, A. T. Marcos, J. F. Martin, and J. R. Murphy.** 1995. Molecular cloning and characterization of the *Corynebacterium diphtheriae dtxR* homolog from *Brevibacterium lactofermentum. J. Bacteriol.* **177:**465–467.

42. **Pappenheimer, A. M., Jr.** 1977. Diphtheria toxin. *Annu. Rev. Biochem.* **46:**69–94.

43. **Pappenheimer, A. M., Jr.** 1980. Diphtheria: studies on the biology of an infectious disease, p. 45–73. *In The Harvey Lectures,* series 76. Academic Press, Inc., New York, N.Y.

44. **Pappenheimer, A. M., Jr., and S. J. Johnson.** 1936. Studies in diphtheria toxin production. I: The effects of iron and copper. *Br. J. Exp. Pathol.* **17:**335–341.

45. **Pappenheimer, A. M., Jr., and J. R. Murphy.** 1983. Studies on the molecular epidemiology of diphtheria. *Lancet* **ii:**923–926.

46. **Qiu, X., E. Pohl, R. K. Holmes, and W. G. J. Hol.** 1996. High-resolution structure of the diphtheria toxin repressor complexed with cobalt and manganese reveals an SH3-like third domain and suggests a possible role of phosphate as co-repressor. *Biochemistry* **35:**12292–12302.

47. **Qiu, X., C. L. Verlinde, S. Zhang, M. P. Schmitt, R. K. Holmes, and W. G. J. Hol.** 1995. Three-dimensional structure of the diphtheria toxin repressor in complex with divalent cation co-repressors. *Structure* **3:**87–100.

48. **Ramon, G.** 1923. Sur la concentration du serum antidiphtherique et l'isoletment de la antitoxine. *C.R. Soc. Biol.* **88:**167–168.

49. **Ratti, G., R. Rappuoli, and G. Giannini.** 1983. The complete nucleotide sequence of the gene coding for diphtheria toxin in the corynephage omega (*tox+*) genome. *Nucleic Acids Res.* **11:**6589–6595.

50. **Ryser, H.-J., R. Mandel, and F. Ghani.** 1991. Cell surface sulfhydryls are required for the cytotoxicity of diphtheria toxin but not ricin toxin in Chinese hamster ovary cells. *J. Biol. Chem.* **266:**18439–18442.

51. **Saragea, A., P. Maximescu, and E. Meitert.** 1979. Corynebacterium diphtheriae: microbiological methods used in clinical and epidemiological investigations. *Methods Microbiol.* **13:**61–176.

52. **Schiering, N., X. Tao, H. Zeng, J. R. Murphy, G. A. Petsko, and D. Ringe.** 1995. Structures of the apo- and the metal ion-activated forms of the diphtheria tox repressor from *Corynebacterium diphtheriae. Proc. Natl. Acad. Sci. USA* **92:**9843–9850.

53. **Schmitt, M. P.** 1997. Transcription of the *Corynebacterium diphtheriae* hmuO gene is regulated by iron and heme. *Infect. Immun.* **65:**4634–4641.

54. **Schmitt, M. P., and R. K. Holmes.** 1991. Characterization of a defective diphtheria toxin repressor (*dtxR*) allele and analysis of *dtxR* transcription in wild-type and mutant strains of *Corynebacterium diphtheriae. Infect. Immun.* **59:**3903–3908.

55. **Schmitt, M. P., and R. K. Holmes.** 1994. Cloning, sequence, and footprint analysis of two promoter/operators from *Corynebacterium diphtheriae* that are regulated by the diphtheria toxin repressor (DtxR) and iron. *J. Bacteriol.* **176:**1141–1149.

56. **Schmitt, M. P., M. Predich, L. Doukhan, I. Smith, and R. K. Holmes.** 1995. Characterization of an iron-dependent regulatory protein (IdeR) of *Mycobacterium tuberculosis* as a functional homolog of the diphtheria toxin repressor (DtxR) from *Corynebacterium diphtheriae. Infect. Immun.* **63:**4284–4289.

57. **Smith, W. P., P. C. Tai, J. R. Murphy, and B. D. Davis.** 1980. A precursor in the cotranslational secretion of diphtheria toxin. *J. Bacteriol.* **141:**184–189.

57a. **Sun, L., and J. R. Murphy.** Unpublished data.

58. **Tao, X., J. Boyd, and J. R. Murphy.** 1992. Specific binding of the diphtheria *tox* regulatory element DtxR to the *tox* operator requires divalent cations and a 9-base-pair interrupted palindromic sequence. *Proc. Natl. Acad. Sci. USA* **89:**5897–5901.

59. **Tao, X., and J. R. Murphy.** 1992. Binding of the metal-loregulatory protein DtxR to the diphtheria *tox* operator requires a divalent heavy metal ion and protects the palindromic sequence from DNase I digestion. *J. Biol. Chem.* **267:**21761–21764.

60. **Tao, X., and J. R. Murphy.** 1993. Cysteine-102 is positioned in the metal binding activation site of the *Corynebacterium diphtheriae* regulatory element DtxR. *Proc. Natl. Acad. Sci. USA* **90:**8524–8528.

61. **Tao, X., and J. R. Murphy.** 1994. Determination of the DtxR consensus binding site by *in vitro* affinity selection. *Proc. Natl. Acad. Sci. USA* **91:**9646–9650.

62. **Tao, X., H. Zeng, and J. R. Murphy.** 1995. Heavy metal ion activation of the diphtheria *tox* repressor (DtxR) results in the formation of stable homodimers. *Proc. Natl. Acad. Sci. USA* **92:**6803–6807.

63. **Touate, D., M. Jacques, B. Tardat, L. Bouchard, and S. Despied.** 1995. Lethal oxidative damage and mutagenesis are generated by iron in Δ*fur* mutants of *Escherichia coli*: protective role of superoxide dismutase. *J. Bacteriol.* **177:**2305–2314.

64. **Twigg, P. D., G. P. Wylie, G. Wang, D. L. D. Caspar, J. R. Murphy, and T. M. Logan.** 1999. Expression and assignment of the 1H, 15N, and 13C resonances of the C-terminal domain of the diphtheria toxin repressor. *J. Biomol. NMR* **13:**197–198.

65. **Uchida, T., D. M. Gill, and A. M. Pappenheimer, Jr.** 1971. Mutation in the structural gene for diphtheria toxin carried by temperate phage β. *Nat. New Biol.* **233:**8–11.

66. **Wang, G., G. P. Wylie, P. D. Twigg, D. L. D. Caspar, J. R. Murphy, and T. M. Logan.** 1999. Solution structure and peptide binding studies of the C-terminal Src homology 3-like domain of the diphtheria toxin repressor protein. *Proc. Natl. Acad. Sci. USA* **96:**6119–6124.

67. **Wang, Z., M. P. Schmitt, and R. K. Holmes.** 1994. Characterization of mutations that inactivate the diphtheria toxin repressor. *Infect. Immun.* **62:**1600–1608.

68. **Welkos, S. L., and R. K. Holmes.** 1981. Regulation of toxinogenesis in *Corynebacterium diphtheriae*. Mutations in bacteriophage β that alter the effect of iron on toxin production. *J. Virol.* **37:**936–945.

69. **White, A., X. Ding, J. R. Murphy, and D. Ringe.** 1998. Structure of metal ion-activated diphtheria toxin repressor/*tox* operator complex. *Nature* **394:**502–506.

70. **Yamaizumi, M., E. Mekada, T. Uchida, and Y. Okada.** 1978. One molecule of diphtheria toxin fragment A introduced into a cell can kill the cell. *Cell* **15:**245–250.

70a. **Zeng, H., and J. R. Murphy.** Unpublished data.

71. **Zhang, H., D. Zhao, M. Revington, W. Lee, X. Jia, C. Arrowsmith, and O. Jardetzky.** 1994. The solution structures of the *trp* repressor-operator DNA complex. *J. Mol. Biol.* **238:**592–614.

72. **Zucker, D. R., and J. R. Murphy.** 1984. Monoclonal antibody analysis of diphtheria toxin. I. Localization of epitopes and neutralization of cytotoxicity. *Mol. Immunol.* **21:**785–793.

Actinomyces: Surface Macromolecules and Bacteria-Host Interactions

MARIA K. YEUNG

60

Members of the genus *Actinomyces* are nonmotile, non-spore-forming, facultatively anaerobic, filamentous gram-positive bacteria. Differentiation of species within this genus has been difficult owing to the lack of reliable methods and markers. However, recent advances in recombinant DNA technology and knowledge of prokaryotic 16S rRNA sequences have allowed a precise phylogenic reorganization of more than 12 *Actinomyces* species and a determination of their intergeneric relationships with other actinomycetes genera consisting of bacteria with high G+C genomes (37). Of the many species of *Actinomyces*, *Actinomyces pyogenes*, formerly a *Corynebacterium* species, is relatively homogeneous. In contrast, strains of *Actinomyces viscosus* and *Actinomyces naeslundii* (in particular, those isolated from the oral cavity) are most heterogeneous, owing in part to many shared phenotypic characteristics and cross-reactive antigens among these bacteria. The use of various specific absorbed antibodies and biochemical tests has been instrumental in cataloging these bacteria into various serotypes and taxonomic groups or clusters (15, 16). More recently, *A. naeslundii* was further classified into two subspecies, designated genospecies 1 and 2, based on DNA homology studies and immunochemical analyses (24). Thus, human isolates of *A. naeslundii* previously designated *A. naeslundii* serotype I have been renamed *A. naeslundii* genospecies 1, while those human strains previously classified as *A. viscosus* serotypes II and III and *A. naeslundii* serotype II have been renamed *A. naeslundii* genospecies 2. All nonhuman isolates previously identified as *A. viscosus* remain members of the species *A. viscosus*.

The majority of *Actinomyces* species are indigenous flora found only in the oral environment. These bacteria, including strains of *A. naeslundii* and *A. viscosus*, predominate in both supra- and subgingival dental plaque (12, 28). Like other indigenous flora, certain strains of *Actinomyces* species are opportunistic pathogens and are implicated in gingivitis, periodontitis, and root caries (30, 33, 44). In addition, strains of *Actinomyces bovis*, *Actinomyces israelii*, and *A. viscosus* are prevalent causes of cervicofacial, abdominal, and thoracic actinomycosis in humans and domestic animals (45). The nonoral *A. pyogenes* is frequently isolated from domestic animals and is implicated in pyogenic infections (4).

PRIMARY COLONIZERS AND DENTAL BIOFILM FORMATION

Bacterial colonization is an important initial step in establishing a disease process, and those bacteria that colonize the oral cavity have evolved unique mechanisms affording them the ability to adhere to and persist in, or be removed from, the microenvironment. For example, in this unique habitat, several salivary glycoproteins, such as agglutinins or mucins, and the normal flushing action of saliva flow promote bacterial clearance. The formation of a microbial biofilm (dental plaque) on tooth and mucosal surfaces follows a definite sequence characterized by the presence of a microbiota rich in facultative anaerobic gram-positive bacterial species during the initial stages to one rich in anaerobic gram-negative bacteria at the later stages (12, 28). Among the greater than 300 bacterial taxa from plaque samples, those most frequently isolated belong to one of 22 genera, and of these genera, the genus *Actinomyces* is the most prevalent (30). These organisms, along with the viridans streptococci, are among the few gram-positive bacterial species that are the primary colonizers of the oral cavity. Numerous in vivo and in vitro studies have demonstrated the significance of these bacteria in the initiation and progression of plaque formation. High numbers of these bacteria are recovered within minutes from cleaned tooth surfaces. An increasing number of these organisms are recovered from enamel chips placed in the oral cavity of human subjects during the initial 24 h of plaque development (33). The deposition of these primary colonizers provides sites for attachment of other plaque bacteria, including many gram-negative bacteria found at the late stages of plaque formation (12, 25). In studies with various animal models, the failure of putative gram-negative periodontal pathogens to establish and maintain themselves in the oral cavity, in the absence of primary colonizers, strongly supports the role of these colonizers as "bridging" organisms. Many putative periodontal pathogens are

Gram-Positive Pathogens, ed. by V. A. Fischetti et al.
© 2000 American Society for Microbiology, Washington, D.C.

known to express proteases and other enzymes associated with cytokine induction and/or host tissue inflammation and destruction (46). Thus, these early colonizers are crucial to the maintenance of health and disease in the oral environment.

THE ORAL *ACTINOMYCES* SPECIES

Among the various oral *Actinomyces* species, early studies demonstrated that A. *naeslundii* and A. *viscosus* are predominant in plaque samples. These bacteria are found on both soft and hard tissue surfaces of the oral environment and persist from infancy to adulthood (12). However, a recent report suggests that A. *odontolyticus* is more prevalent in plaque samples harvested 2, 4, and 8 h following the initiation of plaque formation while A. *naeslundii* and A. *viscosus* are isolated in high numbers between 8 and 24 h (27). Significantly, *Actinomyces odontolyticus*, found only on the tongue surface, is not implicated in periodontal disease, whereas the proportions of A. *naeslundii* and A. *viscosus* are correlated with health and disease, consistent with findings established previously in numerous other studies. Additional in vivo examination of plaque samples from a large population would help define more precisely the order and distribution of various *Actinomyces* species recovered during the early phase of plaque development.

Intergeneric and intrageneric coaggregation are unique mechanisms that lead to accumulation and increase in bacterial cell mass and diversity within the microenvironment (5, 25). Whereas some organisms coaggregate with a few bacterial strains of several unrelated genera, others interact with only a particular bacterial strain. Thus, the initial flora composition (or change from the usual makeup thereof) significantly influences the organization of organisms attaching to the growing biofilm in subsequent plaque developmental stages. Oral *Actinomyces* species coaggregate with viridans streptococci and a few gram-negative bacteria that are putative periodontal pathogens. At least six different adherence mechanisms are characterized that mediate coaggregations among strains of *Actinomyces* and the oral streptococcal species (25). Many of these interactions involve protein-carbohydrate interactions that are inhibitable by the addition of mono- or disaccharides. Interestingly, lactose-sensitive coaggregation is most prevalent among these bacteria. Of the >80% of freshly isolated viridans streptococci that coaggregate with strains of A. *naeslundii* and A. *viscosus*, >85% are reversible in the presence of lactose (7, 25).

A few surface components have been identified from A. *naeslundii* and A. *viscosus* that mediate bacterial colonization to oral surfaces. Figure 1 illustrates the characteristics of some of the major components from A. *naeslundii* T14V, a representative human strain of A. *naeslundii* genospecies 2. Molecular studies of some of these components have been initiated recently to determine their contribution to the success of *Actinomyces* species as a primary colonizer of the oral environment. *Actinomyces* fimbriae are major cell surface components that are correlated with attachment to the tooth enamel and various eukaryotic cells, including mucosal epithelial cells, polymorphonuclear leukocytes, and erythrocytes (3, 5, 8, 10, 17, 41, 42), and coaggregation with certain strains of viridans streptococci (25). Whereas coaggregation among plaque bacteria results in growth and accumulation of bacterial biofilm, the ability to form bacterial aggregates may enhance the pathogenic properties of

these bacteria. For example, bacterial aggregates consisting of A. *israelii* alone or *Actinomyces* and oral streptococci are resistant to phagocytosis and killing by neutrophils in vitro and in vivo (14, 34). On the other hand, binding of *Actinomyces* strains to polymorphonuclear leukocytes results in the release of inflammatory mediators that precedes phagocytosis and bacterial killing (41). *Actinomyces* fimbriae also are involved in bacterial binding to secretory immunoglobulin A1 (39), an innate host defense mechanism promoting bacterial clearance. Nonfimbrial cell surface molecules also have been identified that play a direct or indirect role in bacterial adherence. These include sialidase (10), which cleaves terminal sialic acid residues from sialoglycoproteins, thereby unmasking receptors for *Actinomyces* type 2 fimbriae-mediated bacterial adherence (5), and other antigens that bind to selected strains of gram-negative bacterial species, including *Prevotella*, *Porphyromonas*, and the common fungal agent *Candida albicans* (19, 26, 32). The latter interactions involve the expression of protein or polysaccharide receptors on *Actinomyces* species. Thus, the tropism of *Actinomyces* strains toward various host tissue surfaces is dictated, in part, by the presence of specific ligand-receptor pairs available for interaction at any given time. This chapter describes our current knowledge concerning the *Actinomyces* surface molecules that are either directly or indirectly involved in bacterial adherence, with an emphasis on recent advances in molecular analyses of A. *naeslundii* fimbriae and sialidase. Strategies used by these bacteria to persist in the oral environment are also discussed.

ACTINOMYCES FIMBRIAE

Only a limited number of gram-positive bacterial species express cell surface fimbriae or fibrillalike macromolecules; they include strains of *Streptococcus* and *Corynebacterium* species (51). Fimbriae are detected from a few *Actinomyces* species. However, more information is available concerning fimbria from A. *naeslundii* and A. *viscosus*, owing in part to the fact that a high percentage of strains of these species express cell surface fimbriae. Most, if not all, A. *naeslundii* genospecies 2 isolates produce two distinct types of fimbria, designated 1 and 2, while at least one strain belonging to A. *naeslundii* genospecies 1 expresses only one fimbrial type (type 2) (5, 8). Moreover, in A. *naeslundii* genospecies 2 T14V, no gross inhibition of fimbriae production is detected when the bacteria are incubated either at 37°C (generation time of approximately 90 min) or at 42°C (generation time of approximately 2.5 h with a total cell yield reduced to 50 to 70% of that obtained at 37°C) in complex media (Fig. 2).

Actinomyces spp. express peritrichous fimbriae of approximately 3.5 to 4.5 nm in diameter that appear as long bundles or branches of individual appendages extending from the cell surface (13). Thus, these fimbriae are not rigid like type 1 fimbriae from *Escherichia coli* but resemble the E. *coli* K88 fimbriae. Isolated fimbriae from A. *naeslundii* strains T14V and WVU45 have been obtained, and specific polyclonal and monoclonal antifimbriae antibodies have been generated (5, 6). Little or no cross-immunoreactivity is detected between A. *naeslundii* T14V type 1 and type 2 fimbriae, and only weak cross-immunoreactivity is detected between antibodies directed against type 2 fimbriae from strains T14V and WVU45 (5). Studies using these antibodies in electron immunogold mi-

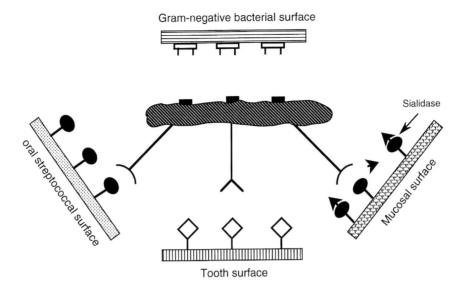

Gram-negative bacterial surface

oral streptococcal surface

Mucosal surface

Sialidase

Tooth surface

FIGURE 1 A model of *Actinomyces* cell surface macromolecules and host receptor interaction. The specifice bacteria-host interactions include type 1 fimbriae (⅄)-mediated adherence to salivary acidic PRPs (♀) that coat the tooth enamel (▦); a lectin activity associated with type 2 fimbriae (⅄) that bind to galactose- or *N*-actylgalactosamine-containing sugar moieties of polysaccharides (♀) on oral streptococcal cell walls (▭) or sialoglycoproteins on mucosal cell surface (▦), which is observed after removal of sialic acid residues by sialidase; and other surface receptors or ligands (▬) that bind to surface antigens on gram-negative bacteria (▤).

croscopy reveal that, in strain T14V, type 1 fimbriae are relatively shorter and closer to the cell surface than type 2 fimbriae, which extend farther away from the cell surface (13).

Fimbriae from *Actinomyces* strains are sheared from the cell surface by sonication, a routine procedure proved to be effective in the removal of fimbriae from most gram-negative bacteria. However, significantly higher yields are obtained by digesting the bacteria with lysozyme, which acts on cell wall peptidoglycan and facilitates further re-

lease of cell-associated molecules (51). Unlike fimbriae from the majority of gram-negative bacteria, *Actinomyces* fimbriae do not dissociate into monomeric subunits under conditions, including incubation in saturated guanidine hydrochloride or boiling at low pH (<2.0), that are effective in completely dissociating fimbriae of gram-negative bacteria. However, partial dissociation of fimbriae from *Actinomyces* strains is observed by boiling purified isolated fimbriae in a buffer containing sodium dodecyl sulfate (SDS) and disulfide bond-dissociating agents, such as 2-

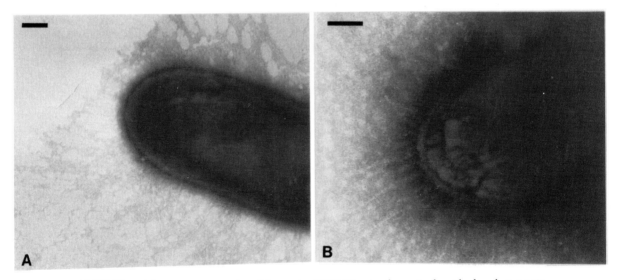

FIGURE 2 Electron micrographs of *A. naeslundii* T14V stained negatively with phosphotungstic acid (1%), showing the expression of cell surface fimbriae from bacteria grown in a complex medium at (A) 37°C and (B) 42°C. Bar, 100 nm.

mercaptoethanol or dithiothreitol (5, 51). A typical SDS-polyacrylamide gel electrophoresis (SDS-PAGE) profile consists of multiple protein bands that range from 35 to 200 kDa (52, 54, 56). Importantly, these proteins are immunostained with polyclonal and monoclonal antifimbriae antibodies, indicating that proteins of larger molecular masses represent multimeric forms of the structural fimbrial subunit (5, 6). To date, no effective methods have been identified that are capable of complete dissociation of *Actinomyces* fimbriae into monomeric subunits. In contrast, recent genetic analyses suggest that fimbriae from these organisms may be covalently linked to the cell wall peptidoglycan (51, 56), which constitutes a major difference between fimbriae from gram-positive bacteria and those from gram-negative bacteria. Further genetic and biochemical analyses are needed to verify this proposed hypothesis.

TYPE 1 FIMBRIAE

Actinomyces type 1 fimbriae mediate bacterial adherence to the tooth surface via interaction with salivary acidic proline-rich proteins (PRPs), which serve as the receptors (5, 17). PRPs are a series of complex nonglycosylated proteins that are most abundant (40% of total protein) in saliva and exhibit high affinity for hydroxyapatite surfaces. Type 1 fimbriae-mediated adherence also is observed, albeit to a lesser degree, to statherin, another salivary glycoprotein present at a lower concentration in the oral cavity. However, fimbriated *Actinomyces* strains do not adhere to soluble PRPs and conversely, no binding is observed between isolated fimbriae and PRPs immobilized on a solid surface (5, 8). Studies have indicated that the amino-terminal 30-amino-acid peptide, which is involved in binding of PRPs to hydroxyapatite, does not participate in specific interactions with strains that express type 1 fimbriae. A 44-amino-acid peptide at the carboxyl-terminal end of PRPs enhances adherence to fimbriated bacteria but is insufficient to abolish the observed binding activity (17), suggesting that the domain involved in ligand-receptor interaction is located elsewhere on the PRPs. On the other hand, the nature of the receptor-binding site associated with *Actinomyces* type 1 fimbriae remains to be elucidated. It is not known whether the binding sites are located on the fimbrial structural subunit or are products of a gene(s) distinct from that encoding the subunit gene. In this regard, several fimbriae-mediated adherence models have been identified based on genetic and biochemical analyses of fimbriae from gram-negative bacteria, including those from *E. coli*, *Pseudomonas* sp., and others (21, 23). In *E. coli* expressing type 1 fimbriae or pyelonephritis-associated pili, an adhesin distinct from the structural subunit mediates bacterial attachment to their respective host receptors (23). These structures are located at the tip and/or along the length of the structural fimbriae. In contrast, a specific domain at the carboxyl-terminal portion of the structural subunit is implicated in mediating binding of type IV pili expressed by *Pseudomonas* species to receptors on epithelial cells (21). Genetic analyses of *Actinomyces* fimbriae should provide insights into whether these models also apply to fimbriae-mediated adherence in gram-positive bacteria.

The *A. naeslundii* T14V type 1 fimbrial subunit gene (*fimP*) is the first of seven genes arranged in clusters on a 9.3-kb chromosomal DNA region initially cloned and expressed in *E. coli* (52). A protein of approximately 65 kDa from the recombinant clone was identified by Western blot analysis and immunostained with polyclonal and monoclonal anti-type 1 fimbriae antibodies. However, the protein predicted from the nucleotide sequence is approximately 54 kDa and has no sequence homology to other bacterial fimbrial or nonfimbrial proteins (53). Evidence that fimbriae from *Actinomyces* are composed of a major structural subunit is provided by the finding that the multiple protein bands separated on SDS-PAGE of purified fimbriae from *A. naeslundii* T14V are immunostained with antibody prepared for the protein expressed in an *E. coli* recombinant clone carrying *fimP* (52). Moreover, purified subunit from the recombinant clone lacks the 30-amino-acid signal peptide of the precursor subunit as predicted by the nucleotide sequence (53). Thus, not only is the *Actinomyces* fimbrial protein expressed in *E. coli*, but the precursor subunit also is processed by a leader sequence peptidase in this host. However, antibody prepared against the structural fimbrial subunit does not inhibit the adherence of bacteria to PRPs (6). Although this finding may indicate the absence of the specific antibody directed against the epitopes involved in adherence, the possibility remains that the adhesive component of *Actinomyces* type 1 fimbriae is encoded by a separate fimbria-associated gene.

The other six genes (*orf1* through *orf6*) adjacent to *fimP* on the *A. naeslundii* T14V chromosome were identified initially based on the nucleotide sequence (Fig. 3). These genes are transcribed in the same direction as *fimP* and are thought to be fimbria-associated genes by virtue of their close proximity to *fimP*. The codon preference of each encoded protein is consistent with that expected of *Actinomyces* strains. However, similar to the fimbrial structure subunit protein, none of these putative proteins share sequence homology to fimbria-associated proteins or any proteins of other bacteria. That these *orf* genes are fimbrial associated was established only recently, when genetic tools become available for *Actinomyces* species (51). These include the identification of a broad-host-range plasmid, pJRD215, that replicates autonomously in *Actinomyces* strains, the expression of two genes carried on this plasmid encoding resistance to kanamycin and streptomycin, and the development of transformation by electroporation in *Actinomyces* species. Moreover, the demonstration that site-specific mutations in the *Actinomyces* chromosome can be generated by homologous recombination with the aid of integration plasmids provides a strategy for the construction of mutants critical to the study of genes and their functions. Analysis of a series of mutants in which a DNA segment of each *orf* or *fimP* has been deleted and substituted with the kanamycin-resistance gene cassette reveals that the knockout mutants of *orf1*, *orf2*, or *fimP* do not express type 1 fimbriae. In contrast, knockout mutants of *orf3*, *orf4*, or *orf5* express only the 65-kDa monomeric subunit. None of these allelic replacement mutants is capable of adhering to PRPs (56). Thus, all six *orf* genes and *fimP* are required for the synthesis and function of *A. naeslundii* T14V type 1 fimbriae.

Whereas *fimP* encodes the major structural component of type 1 fimbriae, the biochemical and functional characteristics of the six fimbria-associated proteins remain to be delineated. It is likely that some of these proteins may be involved in regulation of *fimP* expression while some are structural components involved in the transport and assembly of fimbrial subunits. Analysis of the predicted protein sequences reveals certain unique features in these fimbria-associated proteins (Table 1). Of interest is the significant sequence homology at the carboxyl-terminal

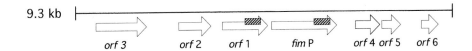

9.3 kb

orf 3   *orf 2*   *orf 1*   *fim* P   *orf* 4 *orf* 5   *orf* 6

FIGURE 3 Genomic map of the 9.3-kb region of *A. naeslundii* T14V chromosome showing the relative locations of the structural fimbrial subunit gene, *fimP*, and the six fimbria-associated genes, designated *orf1* through *orf6*. All the identified genes in the type 1 fimbrial gene cluster are transcribed in the same direction. Symbols: ——, *A. naeslundii* T14V chromosomal DNA; ⇨, fimbrial subunit or fimbria-associated gene; ▨, consensus membrane-anchoring motif of gram-positive bacterial surface antigens.

portion of the protein encoded by *ORF1* and that encoded by *fimP*. Moreover, the *orf1*-encoded protein expressed in *E. coli* is immunostained with anti-*A. naeslundii* T14V whole bacteria antibody, but not with the anti-type 1 fimbriae antibody (56). Thus, *orf1* shares epitopes common to certain cell surface antigens that are distinct from type 1 fimbriae of this bacterium. The significance of the latter

observation is that it is highly likely that *orf1* is expressed in *A. naeslundii* T14V and that it may be surface located. It also is possible that the ORF1 protein may play a role in mediating binding of type 1 fimbriae to PRPs. Further analysis of ORF1 and the other fimbria-associated proteins will provide insight into their contribution to the overall *A. naeslundii* T14V type 1 fimbriae biogenesis and function.

TABLE 1 *A. naeslundii* T14V fimbrial genes and gene products

Gene	Gene product	Molecular mass (Da)[a]	pI[b]	Potential transmembrane helices[c]	Motifs/remarks[d]
Type 1 fimbriae					
fimP	Structural subunit	56,899	5.2	aa 9–31 aa 504–524	Leader sequence at the amino terminus; cell wall sorting sequence and the membrane-anchoring motifs, LPXTG, at the carboxyl terminus
orf1	Accessory fimbrial protein	39,280	6.5	aa 344–364	Consensus gram-positive membrane-anchoring motif, LPXTG, at the carboxyl terminus; shares sequence homology with *fimP*-encoded protein
orf2	Accessory fimbrial protein	44,040	9.7	None detected	None
orf3	Accessory fimbrial protein	28,114	12.1	aa 20–39	None
orf4	Accessory fimbrial protein	30,878	9.8	aa 62–80 aa 154–173	Shares sequence homology with the fimbrial-associated protein encoded by *orf365* of the type 2 fimbrial gene cluster
orf5	Accessory fimbrial protein	17,797	4.3	aa 24–46	None
orf6	Accessory fimbrial protein	16,669	12.6	aa 1–17 aa 24–41 aa 66–84 aa 88–104	None
Type 2 fimbriae					
fimA	Structural subunit	56,037	6.11	aa 12–33 aa 501–522	Leader sequence at the amino terminus; cell wall sorting sequence and the membrane-anchoring motifs, LPXTG, at the carboxyl-terminus
orf365	Accessory fimbrial protein	39,438	5.05	aa 236–255	Shares sequence homology with the fimbrial-associated protein encoded by *orf4* of the type 1 fimbrial gene cluster

[a]Mass of predicted precursor protein derived from nucleotide sequence.
[b]pI values of predicted proteins derived from nucleotide sequence.
[c]Each amino acid (aa) domain listed has >90% probability of being a transmembrane helix.
[d]Special features and motifs identified by homology search against proteins in public databases.

Finally, it should be noted that, whereas at least seven genes have now been identified that constitute the *A. naeslundii* T14V type 1 fimbrial cluster, the possibility exists that additional genes located elsewhere on the genome may be involved in fimbrial biogenesis.

TYPE 2 FIMBRIAE

Actinomyces type 2 fimbriae mediate bacterial adherence (coaggregation) to oral streptococci and various mammalian cells, including erythrocytes, mucosal epithelial cells, and polymorphonuclear leukocytes that have been treated with sialidase (3, 5, 8, 10, 42). These interactions involve a lectin activity associated with the *Actinomyces* fimbriae and receptors consisting of either N-acetylgalactosamine or galactose that constitute parts of the oral streptococcal cell wall polysaccharide (5, 7, 8) or O-linked oligosaccharide of mammalian membrane glycoproteins or glycolipids (including neutraglycolipids and gangliosides) that are frequently linked to terminal sialic acid residues (3, 41, 42). Thus, in contrast to type 1 fimbriae-mediated adherence, which involves protein-protein interactions, type 2 fimbriae-mediated adherence is characterized by protein-carbohydrate interactions between ligand and receptor. The fimbriae-associated lectin recognizes receptor structures consisting of Gal-β1→3-GalNAc or GalNAc-β1→3-Gal, but not those containing Gal-β1→4-Gal or Gal-β1→4-GalNAc (5). Bacterial attachment is inhibited by galactose, N-acetylgalactosamine, methyl-β-D-galactoside, certain β-linked galactosides such as lactose, and certain plant lectins including peanut agglutinin and lectin from *Bauhinia purpurea* (3, 5, 42). Subtle differences of type 2 fimbriae-associated lectin specificity have been detected among various *Actinomyces* strains. Thus, some of the structures listed are more potent adherence inhibitors for some strains than others (5). As with type 1 fimbriae-mediated bacterial adherence, attachment to host surfaces involving type 2 fimbriae is observed only between immobilized receptors and fimbriated bacteria. No adherence is detected with soluble receptors from mammalian cells or oral streptococci, or with purified fimbriae. However, bacterial agglutination is detected with isolated fimbriae that had been aggregated previously by monoclonal anti-type 2 fimbriae antibodies and sialidase-treated erythrocytes (5). Thus, adherence mediated by these fimbriae depends on the cooperative effect of multivalent binding, which amplifies low-affinity binding sites.

Several cell wall polysaccharides that serve as receptors for type 2 fimbriae of *Actinomyces* spp. have been purified, and the fine structures have been determined from a few representative strains of oral streptococci (7). These polysaccharides consist of repeating hexasaccharide or heptasaccharide units linked by phosphodiester bonds. The immunogenic determinant is found at the reducing end while the fimbrial receptor structures consisting of terminal galactose or N-acetylgalactosamine are located at the nonreducing end, respectively, of each unit. Antibodies directed against these polysaccharides have been obtained and, with only one exception, no cross-immunoreactivity is detected among these antigens. Thus, the fimbriae-associated lectin interacts with receptors on surfaces that are structurally related but not identical. In contrast, only a limited number of receptors for type 2 fimbriae have been characterized from mammalian tissue surfaces. These include a 180-kDa salivary glycoprotein that coats epithelial

cells (1). Attachment of strains expressing type 2 fimbriae to the purified glycoprotein is observed only when the protein is adsorbed onto the buccal epithelium. The observed binding is pH dependent and lactose sensitive, and antiglycoprotein antibody inhibits adherence in a dose-dependent manner. It is likely that this salivary glycoprotein serves as a modulator in *Actinomyces* colonization onto buccal mucosa and/or other surfaces in the oral environment. Plant lectins exhibiting specificities similar to those of *Actinomyces* type 2 fimbriae are useful probes in the identification of receptors from mammalian cells, since isolated fimbriae from these bacteria do not bind to receptors. Partial purification of a 160-kDa sialoglycoprotein from the human KB epithelial cell surface is accomplished with the use of peanut agglutinin and the lectin from *B. purpurea* (3). Binding of *A. naeslundii* WVU45 to the partially purified protein immobilized on nitrocellulose is inhibited in the presence of wheat germ agglutinin, which interacts with sialic acid residues. Incubation of the 160-kDa protein with sialidase results in the appearance of a protein with a slightly retarded electrophoretic mobility of approximately 200 kDa. Binding of the latter protein band to *A. naeslundii* WVU45 is abolished by the addition of the peanut or *B. purpurea* lectin, thus providing support for its role as a receptor for *Actinomyces* type 2 fimbriae. More recently, a 130-kDa glycoprotein and two major glycolipid receptors were identified from human polymorphonuclear leukocytes with the aid of five different plant lectins that inhibit adherence of *Actinomyces* strains to polymorphonuclear leukocytes and subsequent bacterial killing (42). Further studies of these receptors should improve our understanding of their biochemical and immunological properties and their physiological role and expression in the respective mammalian cells. Whereas the structures and linkages of the oligosaccharides of those molecules that serve as receptors for type 2 fimbriae of *Actinomyces* strains are well defined, the nature of the receptor binding site(s) associated with type 2 fimbriae/lectin has not been elucidated. As with type 1 fimbriae-mediated adherence, the "adhesive" component(s) may be a part of the type 2 fimbrial subunit or, alternatively, the product of a gene distinct from that encoding the subunit protein.

Genetic analysis of *Actinomyces* type 2 fimbrial gene clusters from strains T14V and WVU45 have been initiated only recently (51, 54). A recombinant strain has been obtained that produces a protein of approximately 59 kDa that is immunostained with polyclonal and monoclonal antibodies directed against type 2 fimbriae from strain T14V (54). Sequence analyses and data from Edman degradation show that the type 2 fimbrial subunit gene, *fimA*, encodes a precursor subunit (calculated M_r of 56,037) with a leader peptide of 31 amino acid residues that is processed in the recombinant. As with *A. naeslundii* T14V type 1 fimbriae, no sequence homology is detected between the type 2 fimbrial subunit and other bacterial proteins. However, a 31% global sequence identity is detected between the type 1 and type 2 fimbrial subunits of strain T14V (54). Several regions of homology are found distributed throughout the entire protein sequence, each consisting of >8 amino acid residues and containing a conserved proline residue. A second gene, *orf365*, 3′ to *fimA* and encoding a protein of approximately 39 kDa, shares significant sequence homology to that encoded by *orf4*, a fimbria-associated gene 3′ to *fimP* in the *A. naeslundii* T14V type 1 fimbrial gene cluster (Fig. 3 and Table 1) (56). The possibility exists that these proteins may have similar functions in fimbriae synthesis. More significantly, this finding indicates that similarities exist in the organization of genes involved in the biogenesis of type 1 and type 2 fimbriae in *A. naeslundii* T14V. Isogenic mutants of *fimA* and *orf365* have been generated by substituting a deletion corresponding to an internal region of each gene with the kanamycin-resistance gene cassette. No fimbrial proteins are detected in the *fimA* knockout mutant, and only monomeric fimbrial subunits are expressed in the *orf365* knockout mutant. The subunits do not form aggregates or self-assemble into high-molecular-weight fimbriae (54). In addition, neither of these knockout mutants coaggregates with oral streptococci expressing cell wall polysaccharide receptors for type 2 fimbriae of *Actinomyces* species. Thus, bacterial adherence mediated by this fimbrial type also depends on the presence of fully assembled fimbriae. Further studies are needed to isolate and characterize additional genes required for the synthesis and function of type 2 fimbriae.

Only the structural fimbrial subunit gene of *A. naeslundii* WVU45 type 2 fimbriae has been cloned and sequenced to date (51). The structural subunit is approximately 59 kDa, as determined by Western blot analysis with anti-*A. naeslundii* WVU45 type 2 fimbriae antibody. The subunit precursor (calculated M_r of 56,574) also contains a leader sequence of 32 amino acids that is processed by peptidase in the *E. coli* recombinant clone. As expected, no sequence homology is detected between this fimbrial subunit and other bacterial proteins. In contrast, sequence homologies of 77% and 33% are noted between this fimbrial subunit and the subunits of *A. naeslundii* T14V type 2 and type 1 fimbriae, respectively (53, 54). Isolation of other fimbria-associated genes constituting the gene cluster of type 2 fimbriae in strain WVU45 will allow a comparison of the gene clusters involved in the biogenesis of type 2 fimbriae between WVU45 and T14V.

IMPACT OF GENETIC APPROACH ON THE STUDY OF *ACTINOMYCES* FIMBRIAE

Little is known about fimbriae from gram-positive bacteria. *A. naeslundii* T14V, which expresses two types of fimbriae that are immunologically and functionally distinct, is an ideal model for studies of gram-positive bacterial fimbriae. Although the genetic analysis of *Actinomyces* fimbriae is still in its infancy, results generated to date have added new insights concerning their structure and function in these bacteria. Analysis of the subunits of *A. naeslundii* T14V type 1 and type 2 and *A. naeslundii* WVU45 type 2 fimbriae reveals that their predicted secondary structure, hydropathy, and topology profiles are quite similar. Thus, the lack of cross-reactivity between type 1 and type 2 fimbriae indicates that the regions of significant sequence homology (53, 54) are unrelated to their antigenic determinants and suggests a common pattern of protein folding among the different fimbrial subunits. Studies that examine the distribution of fimbriae among various *Actinomyces* species indicate that the *A. naeslundii* T14V type 1 fimbrial gene (*fimP*) is conserved among many strains of *A. naeslundii* genospecies 2 of both human and nonhuman origin and also is detected in some strains of *A. bovis* (51). Similarly, conservation of the *A. naeslundii* T14V type 2 fimbrial subunit gene (*fimA*) is observed in many human and nonhuman strains of *A. naeslundii* genospecies 1 and 2 (36). The lack of sequence homology between *Actinomyces* fimbriae and other bacterial proteins indicates that fimbriae from

Actinomyces species are derived from a common ancestor distinct from that in other bacteria.

Analysis of the three *Actinomyces* fimbrial subunits reveals that, in general, they are larger than the majority of fimbrial subunits in gram-negative bacteria, which range between 20 and 30 kDa (21, 23). Similar to most gram-negative bacterial fimbriae, *Actinomyces* fimbriae are composed of a major structural subunit, and the precursors contain a signal peptide sequence for export to the cell surface (51). Fimbriae from *Actinomyces* spp. are predominantly hydrophilic, whereas those from most gram-negative bacteria are predominantly hydrophobic. Thus, monomeric fimbrial subunits from these gram-positive species do not form aggregates in solution under conditions that favor self-assembly. In addition, the *Actinomyces* fimbrial subunits are heat-modifiable proteins (52), a feature not previously noted in other bacterial fimbriae. All three *Actinomyces* fimbrial subunit precursors contain a consensus cell wall sorting sequence at the carboxyl terminus that is highly conserved among most, if not all, gram-positive bacterial surface proteins (43). The sorting sequence begins with a membrane-anchoring motif, LPXTG, followed by a domain rich in hydrophobic amino acids and a charged tail consisting of predominantly hydrophilic amino acid residues. Studies of the cell surface proteins of *Staphylococcus aureus* and other gram-positive bacteria indicate that the cell wall sorting sequence is required for covalent association of cell wall proteins via transpeptidation (31). The sorting sequence signals translocation and the subsequent retention of precursor cell surface proteins to the cytoplasmic membrane, where enzymatic cleavage between the threonine (T) and glycine (G) of the LPXTG membrane-anchoring motif then takes place. The released protein with a carboxyl threonine residue forms an amide bond with the pentaglycyl cross bridge and becomes covalently linked to the cell wall peptidoglycan. It is not known whether the cell wall sorting sequence is involved in fimbrial subunit polymerization and/or assembly in *Actinomyces* species. An earlier study indicates that the intact *A. naeslundii* T14V type 1 fimbrial subunit expressed in *E. coli* was associated with the inner membrane, while an amino-truncated subunit was secreted into the periplasm (52). A recent study demonstrates that an antibody to a 20-amino-acid peptide corresponding to the carboxyl terminus of the *A. naeslundii* T14V type 2 subunit precursor reacted strongly with the protein from a mutant expressing only the monomeric fimbrial subunits but did not react with assembled fimbriae isolated from this strain. Conversely, the antibody directed against isolated type 2 fimbriae did not react with the carboxyl-terminal peptide (54). These observations suggest that the cell wall sorting sequence participates in *Actinomyces* fimbrial biogenesis. Importantly, existing data suggest that the fimbrial precursor subunit of *Actinomyces* species is subjected to two posttranslational events, i.e., processing of the leader signal and the carboxyl-terminal cell wall sorting sequence at the N- and C-terminal ends of the protein during subunit assembly. Clearly, further studies are needed to elucidate the mechanism of fimbriae synthesis in these gram-positive bacteria.

The functional properties of *Actinomyces* fimbriae have been inferred by the observation that only those strains expressing type 1 fimbriae are able to bind to saliva-coated hydroxyapatite and that similar adherence is not obtained with strains, such as WVU45, that express only type 2 fimbriae or with spontaneous mutants that lack type 1 fimbriae (5). Likewise, that type 2 fimbriae mediate adherence to oral streptococci and mammalian cells is based on data that indicate that strain WVU45, and spontaneous mutants lacking type 1 fimbriae, are capable of binding to these various cell surfaces (3, 5). Indirect support for the different adherence properties of type 1 and type 2 fimbriae is provided by the observation that fluorescein-labeled anti-type 1 fimbriae antibody reacts with those *Actinomyces* strains in dental plaque expressing type 1 and type 2 fimbriae, while the labeled anti-type 2 fimbriae antibody reacts with strains on mucosal surface expressing only type 2 fimbriae. However, adherence inhibition has not been demonstrated using antifimbriae antibodies, which would substantiate the unique functions proposed for these fimbriae. Thus, the lack of adherence of fimbriae-deficient isogenic mutants to specific receptors (i.e., no binding between *fimP* or *fimA* knockout mutant to PRPs or oral streptococci, respectively) provides definitive evidence for the functional characteristics defined previously for each *Actinomyces* fimbrial type. Moreover, the lack of binding in mutants expressing only type 1 or type 2 fimbrial subunits to receptor-containing surfaces indicates that fimbriae-mediated adherence in *Actinomyces* is correlated with the presence of fully assembled fimbriae on the cell surface (54, 56). The mechanism of fimbriae-mediated adherence in *Actinomyces* will emerge from future studies focusing on the genetic and biochemical characteristics of these surface molecules. Such information will have significant impact on the design of strategies to interfere with bacteria-host interaction.

NONFIMBRIAL *ACTINOMYCES* CELL SURFACE ANTIGENS/RECEPTORS

Sialidase

Sialidases are ubiquitous among most pathogenic prokaryotic and eukaryotic organisms, including *Trypanosoma* species (35). Enzymatic cleavage of sialic acid residues from glycoproteins and/or glycolipids on mammalian cells exposes otherwise masked receptors that mediate adherence to and/or invasion of the host tissue. Sialidase activity is present in saliva, and many oral bacteria are known to produce sialidase (10, 29). Certain oral streptococci adhere to sialic acid-containing receptors, whereas others interact with receptors that are masked by sialic acid residues (7, 8). Thus, the accumulation of sialidase in the oral environment has a significant impact on the distribution of bacterial species in this microenvironment (11).

A high percentage (79% of fresh isolates) of *A. naeslundii* and *A. viscosus* strains express various levels of sialidase (10, 29, 48). Greater than 80% of the enzyme in the majority of *Actinomyces* strains is cell associated, while ~10% is found in the growth medium. Both the cell-associated and extracellular forms of the enzyme act on soluble substrates, but only the extracellular form of the enzyme hemagglutinates erythrocytes (10), suggesting that the location and/or conformation of the cell-associated enzyme may be inaccessible. However, in two *A. naeslundii* strains (T14V and DSM 43798), these two forms of the enzyme have identical enzymatic and biochemical properties (48, 55). Since binding of *Actinomyces* via type 2 fimbriae depends on removal of sialic acids from mammalian cell surface molecules (5), the production of sialidase by these bacteria may provide a concerted means to enhance bacterial adherence in vivo.

The sialidase gene from *A. naeslundii* strains T14V and DSM 43789 has been cloned and sequenced (22, 50, 55). Each predicted protein contains five 12-amino-acid sequences, referred to as the "Asp block," that are highly conserved among all prokaryotic and eukaryotic sialidases (38). The overall molecular size of the predicted proteins in these bacteria is comparable (M_r 92,871 and 113,000 for strains T14V and DSM 43798, respectively) and is significantly larger than that of other bacterial sialidases (38). The enzymes from these strains are predominantly hydrophobic and exhibit 89% sequence homology, including three regions of nearly 100% sequence identity (50). Unlike *Actinomyces* fimbrial genes, the sialidase gene shares an ancestor common to other prokaryotic and eukaryotic sialidases. Conservation of the T14V sialidase gene, *nanH*, also is observed in other human or nonhuman isolates of *Actinomyces* species (50).

The *A. naeslundii* T14V precursor sialidase contains two potential transmembrane helices; one corresponds to a typical leader peptide, and the other is a 33-amino-acid sequence at the carboxyl terminus. The latter sequence has characteristics that resemble the cell wall sorting sequence of gram-positive surface proteins except that this sequence begins with a pentapeptide, LSRTG, compared with the consensus motif LPXTG (43). Whereas the leader sequence is likely to be involved in the expression of extracellular sialidase, it is premature to speculate what mechanisms mediate the expression of the cell-associated sialidase in these bacteria. However, since only one copy of the *nanH* gene is detected in strain T14V, the cell-associated and extracellular forms of the enzyme must have been subjected to different posttranslational modifications. Two promoters arranged in tandem are identified upstream of the initiation codon of the *A. naeslundii* T14V *nanH* (50). It is not known whether all *Actinomyces* sialidase genes are controlled by dual promoters or, alternatively, whether their presence is correlated with the levels of sialidase activity among the "high" or "low" sialidase producers. Finally, it is unclear whether additional accessory genes may be required for the expression of sialidase in *Actinomyces* spp., although a *nanH* knockout mutant has no detectable cell-associated or extracellular sialidase activity (36). Examination of the regulation of sialidase expression in these bacteria under various environmental conditions should further our understanding of its contribution to oral microbial ecology.

Other Nonfimbrial Surface Structures

Nonfimbrial macromolecules also are expressed by *Actinomyces* species that mediate adherence to host tissues and other plaque bacteria, including those gram-negative bacteria implicated in specific forms of periodontal disease. In vivo studies that examine colonization by *Actinomyces* species reveal that implantation by these bacteria depends on the age, animal species, and diet of the hosts (2). Whereas colonization in certain outbred strains of mice is observed in strains expressing type 1 or type 2 fimbriae, colonization also is demonstrated with nonfimbriated *Actinomyces* strains, suggesting that other cell surface components unrelated to fimbriae are involved. Nonfimbrial surface components also may be involved in cytokine stimulation in mammalian monocytes, since there are no differences in the level of matrix metalloproteinase induction between wild-type strains of *A. naeslundii* T14V and isogenic *fimA*, *fimP*, or *nanH* knockout mutants (49). The identification and characterization of the molecules involved will provide insight into the complex nature of these in vivo interactions.

Coaggregation between selected strains of *Prevotella intermedia* and *Actinomyces* species, including A. *naeslundii*, A. *israelii*, and A. *odontolyticus*, is mediated by a protein or glycoprotein on the *Prevotella* cell surface and a carbohydrate or carbohydrate-containing molecule on the *Actinomyces* cell surface (32). However, no details are available with regard to the properties of the participating molecules from either coaggregation partner. On the other hand, several *Porphyromonas gingivalis* surface components, including fimbriae, a trypsin-like enzyme, and outer membrane proteins (18, 26, 40), have been implicated in adherence to strains of A. *naeslundii*. Moreover, evidence that a 40-kDa outer membrane protein from *P. gingivalis* 381 mediates adherence to A. *naeslundii* ATCC 19246 has been obtained from the finding that several monoclonal antibodies directed against this outer membrane are potent coaggregation inhibitors (40). *Actinomyces* species also adhere to C. *albicans*, a frequently isolated commensal fungal agent in immunocompromised patients. Preliminary data indicate that a glycoprotein or protein on selected C. *albicans* strains interacts with a carbohydrate or carbohydratelike molecule on strains of A. *naeslundii* and A. *viscosus* (19). Molecular study of nonfimbrial surface components from *Actinomyces* should broaden our knowledge of mechanisms that regulate interaction between these bacteria and late colonizers and other fungal agents in the oral environment.

EVASION OF HOST DEFENSE/IMMUNE MECHANISMS: TICKET TO BACTERIAL PERSISTENCE

Commensal organisms must develop strategies to permit establishment in the host environment and also to coexist in harmony with the host. The ideal bacteria-host interaction is one in which the bacteria exhibit immunity to host defensive mechanisms and, conversely, the host exhibits tolerance to the bacteria or fails to recognize them as foreign. *Actinomyces* species have developed and equipped themselves with faculties that allow them to survive many barriers and to persist as members of a predominant normal flora throughout the life span of the host. Strains of *Actinomyces* express multiple surface macromolecules, in addition to fimbriae, that interact with high specificity to both hard and soft tissue surfaces and to other organisms. Thus, firm attachment through multiple mechanisms is enhanced, and the probability of establishment in the host environment is increased. However, nonspecific host innate defensive components, such as secretory immunoglobulins, salivary proteins, and other molecules that mimic receptor structures on the bacterial cell surface, are present in the oral environment and are potent adherence inhibitors. To circumvent these obstacles, *Actinomyces* spp. express surface fimbriae that bind to receptors with low affinity but with high specificity (5, 8). Consequently, these bacteria bind only to receptors associated with surfaces, which in turn promotes colonization. Ironically, established bacteria are a sustained source of antigen for specific host immune responses whose function is to target "foreign" agents. In a mouse model, humoral response to *Actinomyces* spp. and fimbriae is genetically controlled (20). Significant variation in immunoglobulin G (IgG) response among mice with diverse genetic backgrounds is observed. Serum IgG from high responders, i.e., those mice capable of

mounting a high IgG response, is a better adherence inhibitor in vitro than IgG from low responders. In addition, the low-responder phenotype appears to dominate in this animal model. In the bovine model, *Actinomyces*-specific antibody titers in colostrum and milk were lower than those against other oral bacteria (47). However, since this study did not examine genetic control of immunoglobulin response in cows, it is not clear whether low responsiveness to *Actinomyces* spp. can be correlated with low antibody titer. In a recent longitudinal study of human humoral immunity, *Actinomyces* strains could not be isolated from infants before tooth eruption, but after birth, low levels of secretory IgA1 and IgA2 antibody that reacted with strains of A. *naeslundii* genospecies 1 and 2 were detected (9). In addition, while total secretory antibody levels and *Actinomyces* bacterial cell counts continued to increase from birth to 2 years of age, antibody specific for these bacteria declined during the same period, suggesting that *Actinomyces* sp. induces a limited secretory IgA response in humans. Taken together, while specific IgA or IgG antibody to *Actinomyces* spp. and fimbriae or nonfimbriae antigens is present in saliva and sera in humans and in animals immunized with these organisms (9, 20), and while antibody directed against whole bacteria can inhibit adherence to host receptors (20), insufficient amounts of specific antibody can confer protection against removal of the bacteria from the host. It appears, then, that *Actinomyces* spp. are capable of inducing low immune response and/or hyporesponsiveness, further ensuring survival in the host. Other strategies, such as phase variation and antigenic variation mechanisms, frequently observed in gram-negative bacteria, have been proposed as methods to evade host immune defenses by these gram-positive bacteria (9). However, further studies are needed to support this hypothesis. Clearly, recent advances in the development of genetic tools suitable for *Actinomyces* species will aid such studies, and invaluable information will emerge to improve our knowledge regarding the success of *Actinomyces* species as both a commensal and potential opportunistic pathogen in the host environment.

I thank L. N. Lee, University of Texas Health Science Center at San Antonio, for a critical review of this chapter and assistance in the preparation of the figures. Work was supported by a grant (DE11102) from the National Institute of Dental and Craniofacial Research.

REFERENCES

1. **Babu, J. P., M. K. Dabbous, and S. N. Abraham.** 1991. Isolation and characterization of a 180-kilodalton salivary glycoprotein which mediates the attachment of *Actinomyces naeslundii* to human buccal epithelial cells. *J. Periodontal Res.* **26:**97–106.
2. **Beem J. E., C. G. Hurley, W. E. Nesbitt, D. F. Croft, R. G. Marks, J. O. Cisar, and W. B. Clark.** 1996. Fimbrial-mediated colonization of murine teeth by *Actinomyces naeslundii*. *Oral Microbiol. Immunol.* **11:**259–265.
3. **Brennan, M. J., J. O. Cisar, and A. L. Sandberg.** 1986. A 160-kilodalton epithelial cell surface glycoprotein recognized by plant lectins that inhibit the adherence of *Actinomyces naeslundii*. *Infect. Immun.* **52:**840–845.
4. **Carter, G. R., and M. M. Chengappa.** 1991. *Essentials of Veterinary Bacteriology and Mycology*, 4th ed. Lea and Febiger, Philadelphia, Pa.
5. **Cisar, J. O.** 1986. Fimbrial lectins of the oral actinomyces, p. 183–196. *In* D. Mirrelman (ed.), *Microbial Lectins and*

Agglutinins: Properties and Biological Activity. John Wiley & Sons, Inc., New York, N.Y.

6. **Cisar, J. O., E. L. Barsumian, R. P. Siraganian, W. B. Clark, M. K. Yeung, S. D. Hsu, S. H. Curl, A. E. Vatter, and A. L. Sandberg.** 1991. Immunochemical and functional studies of *Actinomyces viscosus* T14V type 1 fimbriae with monoclonal and polyclonal antibodies directed against the fimbrial subunit. *J. Gen. Microbiol.* **137:** 1971–1979.

7. **Cisar, J. O., A. L. Sandberg, C. Abeygunawardana, G. P. Reddy, and C. A. Bush.** 1995. Lectin recognition of host-like saccharide motifs in streptococcal cell wall polysaccharides. *Glycobiology* **5:**655–662.

8. **Cisar, J. O., Y. Takashashi, R. S. Rhul, J. A. Donkersloot, and A. L. Sandberg.** 1997. Specific inhibitors of a bacterial adhesion: observations from the study of grampositive bacteria that initiate biofilm formation on the tooth surface. *Adv. Dent. Res.* **11:**168–175.

9. **Cole, M. F., S. Bryan, M. K. Evans, C. L. Pearce, M. J. Sheridan, P. A. Sura, R. Wientzen, and G. H. W. Bowden.** 1998. Humoral immunity to commensal oral bacteria in human infants: salivary antibodies reactive with *Actinomyces naeslundii* genospecies 1 and 2 during colonization. *Infect. Immun.* **66:**4283–4289.

10. **Costello, A. H., J. O. Cisar, P. E. Kolenbrander, and O. Gabriel.** 1979. Neuraminidase-dependent hemagglutination of human erythrocytes by human strains of *Actinomyces viscosus* and *Actinomyces naeslundii. Infect. Immun.* **26:**563–572.

11. **Davis G., and R. J. Gibbons.** 1990. Accessible sialic acid content of oral epithelial cells from healthy and gingivitis subjects. *J. Periodontal Res.* **25:**250–253.

12. **Ellen, R. P.** 1982. Oral colonization by gram-positive bacteria significant to periodontal disease, p. 98–111. *In* R. J. Genco and S. E. Mergenhagen (ed.), *Host-Parasite Interactions in Periodontal Diseases.* American Society for Microbiology, Washington, D.C.

13. **Ellen, R. P., I. A. Buivids, and J. R. Simardone.** 1989. *Actinomyces viscosus* fibril antigens detected by immunogold electron microscopy. *Infect. Immun.* **57:**1327–1331.

14. **Figdor, D., U. Sjogren, S. Sorlin, G. Sundqvist, and P. N. R. Nair.** 1992. Pathogenicity of *Actinomyces israelii* and *Arachnia propionica*: experimental infection in guinea pigs and phagocytosis and intracellular killing by human polymorphonuclear leukocytes in vitro. *Oral Microbiol. Immunol.* **7:**129–136.

15. **Fillery, E. D., G. H. Bowden, and J. M. Hardie.** 1978. A comparison of strains of bacteria designated *Actinomyces viscosus* and *Actinomyces naeslundii. Caries Res.* **12:**299–312.

16. **Firtel, M., and E. Fillery.** 1988. Distribution of antigenic determination between *Actinomyces viscosus* and *Actinomyces naeslundii. J. Dent. Res.* **67:**15–20.

17. **Gibbons, R. J., and D. I. Hay.** 1988. Human salivary acidic proline-rich proteins and statherin promote the attachment of *Actinomyces viscosus* LY7 to apatitic surfaces. *Infect. Immun.* **56:**439–445.

18. **Goulbourne, P. A., and R. P. Ellen.** 1991. Evidence that *Porphyromonas (Bacteroides) gingivalis* fimbriae function in adhesion to *Actinomyces viscosus. J. Bacteriol.* **173:**5266–5274.

19. **Grimaudo, N. J., W. E. Nesbitt, and W. B. Clark.** 1996. Coaggregation of *Candida albicans* with oral *Actinomyces* species. *Oral Microbiol. Immunol.* **11:**59–61.

20. **Haber, J., C. M. Grinnell, J. E. Beem, and W. B. Clark.** 1991. Genetic control of serum antibody responses of inbred mice to type 1 and type 2 fimbriae from *Actinomyces viscosus* T14V. *Infect. Immun.* **59:**2364–2369.

21. **Hahn, H. P.** 1997. The type-4 pilus is the major virulence-associated adhesin of *Pseudomonas aeruginosa*—a review. *Gene* **192:**99–108.

22. **Henningsen, M., P. Roggentin, and R. Schauer.** 1991. Cloning, sequencing and expression of the sialidase gene from *Actinomyces viscosus* DSM 43798. *Biol. Chem. Hoppe-Seyler* **372:**1065–1072.

23. **Hultgren, S. J., C. H. Jones, and S. Normark.** 1996. Bacterial adhesins and their assembly, p. 2730–2756. *In* F.C. Neidhardt, R. Curtiss III, J. L. Ingraham, E. C. C. Lin, K. B. Low, Jr., B. Magasanik, W. S. Reznikoff, M. Riley, M. Schaechter, and H. E. Umbarger (ed.), *Escherichia coli and Salmonella: Cellular and Molecular Biology.* American Society for Microbiology, Washington, D.C.

24. **Johnson, J. L., L. V. H. Moore, B. Kaneko, and W. E. C. Moore.** 1990. *Actinomyces georgiae* sp. nov., *Actinomyces gerencseriae* sp. nov., designation of two genospecies of *Actinomyces naeslundii*, and inclusion of *A. naeslundii* serotypes II and III and *Actinomyces viscosus* serotype II in *A. naeslundii* genospecies 2. *Int. J. Syst. Bacteriol.* **40:** 273–286.

25. **Kolenbrander, P. E.** 1991. Coaggregation: adherence in the human oral microbial ecosystem, p. 303–329. *In* M. Dworkin (ed.), *Microbial Cell-Cell Interactions.* American Society for Microbiology, Washington, D.C.

26. **Li, J., R. P. Ellen, C. I. Hoover, and J. R. Felton.** 1991. Association of proteases of *Porphyromonas (Bacteroides) gingivalis* with its adhesion to *Actinomyces viscosus. J. Dent. Res.* **70:**82–86.

27. **Liljemark, W. F., C. G. Bloomquist, C. L. Bandt, B. L. Pihlstrom, J. E. Hinrichs, and L. F. Wolff.** 1993. Comparison of the distribution of *Actinomyces* in dental plaque on inserted enamel and natural tooth surfaces in periodontal health and disease. *Oral Microbiol. Immunol.* **8:**5–15.

28. **Marsh, P. D., and M. V. Martin.** 1992. Dental plaque, p. 98–132. *In* P. D. Marsh and M. V. Martin (ed.), *Oral Microbiology.* Chapman & Hall, Ltd., London, U.K.

29. **Moncla, B. J., and P. Braham.** 1989. Detection of sialidase (neuraminidase) activity in *Actinomyces* species by using 2'-(4-methylumbelliferyl) α-D-N-acetylneuraminic acid in a filter paper spot test. *J. Clin. Microbiol.* **27:** 182–184.

30. **Moore, W. E. C., and L. V. H. Moore.** 1994. The bacteria of periodontal disease. *Periodontology 2000* **5:**66–77.

31. **Navarre, W. W., H. Ton-That, K. F. Faull, and O. Schneewind.** 1998. Anchor structure of staphylococcal surface proteins. II. COOH-terminal structure of muramidase and amidase-solubilized surface protein. *J. Biol. Chem.* **273:**29135–29142.

32. **Nesbitt, W. E., H. Fukushima, K.-P. Leung, and W. B. Clark.** 1993. Coaggregation of *Prevotella intermedia* with oral *Actinomyces* species. *Infect. Immun.* **61:**2011–2014.

33. **Nyvad, B., and M. Kilian.** 1987. Microbiology of the early colonization of human enamel and root surfaces in vivo. *Scand. J. Dent. Res.* **95:**369–380.

34. Ochiai, K., T. Kurita-Ochiai, Y. Kamino, and T. Ikeda. 1993. Effect of co-aggregation on the pathogenicity of oral bacteria. *J. Med. Microbiol.* **39:**183–190.

35. Pereira, M. E. A., J. S. Mejia, E. Ortega-Barria, D. Matzilevich, and R. P. Prioli. 1991. The *Trypanosoma cruzi* neuraminidase contains sequences similar to bacterial neuraminidase, YWTD repeats of the low density lipoprotein receptor, and type III modules of fibronectin. *J. Exp. Med.* **174:**179–191.

36. Ragsdale, P. A., and M. K. Yeung. Unpublished data.

37. Ramos, C. P., G. Foster, and M. D. Collins. 1997. Phylogenetic analysis of the genus *Actinomyces* based on 16S rRNA gene sequences: description of *Arcanobacterium phocae* sp. nov., *Arcanobacterium bernardiae* comb. nov., and *Arcanobacterium pyogenes* comb. nov. *Int. J. Syst. Bacteriol.* **47:**46–53.

38. Roggentin, P., B. Rothe, J. B. Kaper, J. Galen, L. Lawrisuk, E. R. Vimr, and R. Schauer. 1989. Conserved sequences in bacterial and viral sialidases. *Glycoconjugate J.* **6:**349–353.

39. Ruhl, S., A. L. Sandberg, M. F. Cole, and J. O. Cisar. 1996. Recognition of immunoglobulin A1 by oral *Actinomyces* and streptococcal lectins. *Infect. Immun.* **64:**5421–5424.

40. Saito, S., K. Hirastsuka, M. Hayakawa, H. Takiguchi, and Y. Abiko. 1997. Inhibition of a *Porphyromonas gingivalis* colonizing factor between *Actinomyces viscosus* ATCC 19246 by monoclonal antibodies against recombinant 40-kDa outer-membrane protein. *Gen. Pharmac.* **28:**675–680.

41. Sandberg, A. L., L. L. Mudrick, J. O. Cisar, J. A. Metcalf, and H. L. Malech. 1988. Stimulation of superoxide and lactoferrin release from polymorphonuclear leukocytes by the type 2 fimbrial lectin of *Actinomyces viscosus* T14V. *Infect. Immun.* **56:**267–269.

42. Sandberg, A. L., S. Ruhl, R. A. Joralmon, M. J. Brennan, M. J. Sutphin, and J. O. Cisar. 1995. Putative glycoprotein and glycolipid polymorphonuclear leukocyte receptors for the *Actinomyces naeslundii* WVU45 fimbrial lectin. *Infect. Immun.* **63:**2625–2631.

43. Schneewind, O., D. Mihaylova-Petkov, and P. Model. 1993. Cell wall sorting signals in surface proteins of Gram-positive bacteria. *EMBO J.* **12:**4803–4811.

44. Schüpbach, P., V. Osterwalder, and B. Guggenheim. 1996. Human root caries: microbiota of a limited number of root caries lesions. *Caries Res.* **30:**52–64.

45. Smego, R. A., Jr., and G. Foglia. 1998. Actinomycosis. *Clin. Infect. Dis.* **26:**1255–1263.

46. Socransky, S. S., and A. D. Haffajee. 1991. Microbial mechanisms in the pathogenesis of destructive periodontal diseases: a critical assessment. *J. Periodontal Res.* **26:**195–212.

47. Takahashi, N., G. Eisenhuth, I. Lee, N. Laible, S. Binion, and C. Schachtele. 1992. Immunoglobulins in milk from cows immunized with oral strains of *Actinomyces*, *Prevotella*, *Porphyromonas*, and *Fusobacterium*. *J. Dent. Res.* **71:**1509–1515.

48. Teufel, M., P. Roggentin, and R. Schauer. 1989. Properties of sialidase isolated from *Actinomyces viscosus* DSM 43798. *Biol. Chem. Hoppe-Seyler* **370:**435–443.

49. Wahl, L. A., and M. K. Yeung. Unpublished data.

50. Yeung, M. K. 1995. Complete nucleotide sequence of the *Actinomyces viscosus* T14V sialidase gene: presence of a conserved repeating sequence among strains of *Actinomyces* spp. *Infect. Immun.* **61:**109–116.

51. Yeung, M. K. 1999. Molecular analysis of *Actinomyces* genes. *Crit. Rev. Oral Biol. Med.* **10:**120–138.

52. Yeung, M. K., B. M. Chassy, and J. O. Cisar. 1987. Cloning and expression of a type 1 fimbrial subunit of *Actinomyces viscosus* T14V. *J. Bacteriol.* **169:**1678–1683.

53. Yeung, M. K., and J. O. Cisar. 1990. Sequence homology between the subunits of two immunologically and functionally distinct types of *Actinomyces* spp. *J. Bacteriol.* **172:**2462–2468.

54. Yeung, M. K., J. A. Donkersloot, J. O. Cisar, and P. A. Ragsdale. 1998. Identification of a gene involved in assembly of *Actinomyces naeslundii* T14V type 2 fimbriae. *Infect. Immun.* **66:**1482–1491.

55. Yeung, M. K., and S. R. Fernandez. 1991. Isolation of a neuraminidase gene from *Actinomyces viscosus* T14V. *Appl. Environ. Microbiol.* **37:**3062–3069.

56. Yeung, M. K., and P. A. Ragsdale. 1997. Synthesis and function of *Actinomyces naeslundii* T14V type 1 fimbriae require the expression of additional fimbria-associated genes. *Infect. Immun.* **65:**2629–2639.

The Pathogenesis of *Nocardia*

BLAINE L. BEAMAN

61

Nocardiae are gram-positive, partially acid-fast, filamentous bacteria that grow by apical extension, forming elongated cells with lateral branching (Fig. 1A). These filaments divide by fragmentation into shorter rods and coccoid cells. Therefore, all of the nocardial cells are branching filaments at log phase of growth, whereas stationary-phase nocardiae are short pleomorphic rods, coccobacilli, and cocci. The nocardiae are relatively slow-growing (doubling time >2 h), and they tend to form variable, hard, tenacious colonies with aerial filamentation (Fig. 1B). The nocardiae belong to the aerobic actinomycetes group of bacteria (Fig. 1); nevertheless, they are phylogenetically related to the corynebacteria, mycobacteria, rhodococci, and other mycolic acid-containing organisms. Most species of *Nocardia* have been recovered from soil, plant material, and water in most regions of the world. Therefore, it is generally believed that diseases caused by the nocardiae result from either a respiratory or a traumatic exposure to these sources. However, soil isolates of "pathogenic" species of *Nocardia* often exhibit either no or low virulence toward laboratory animals. On the other hand, clinical isolates of the same species are usually moderately to highly virulent (5). These observations suggest that some intermediate host may be involved in up-regulating and maintaining nocardial virulence in the environment. Protozoa such as amebae and small worms (e.g., nematodes) have been suggested as possible intermediate hosts that could maintain an evolutionary pressure on the preservation of invasion and virulence genes within many environmental pathogens (24, 47). This is an engaging hypothesis that may be supported by observations that *Nocardia* spp. reside naturally in dinoflagellates (51). Furthermore, additional pressure to maintain nocardial virulence in nature may occur as the result of survival and growth within the gut of various insects, as was reported in cockroaches (53).

HUMAN NOCARDIOSIS

Diseases in humans caused by nocardiae may be divided into at least six general categories based on the route of infection, site of disease, and subsequent pathological responses (5). There are no pathognomonic clinical manifestations for nocardiosis, and diseases caused by these organisms can mimic a wide variety of other conditions (5, 36, 37). Furthermore, the nocardiae in clinical samples may be quite variable in Gram-stained preparations, where they may appear as beaded filamentous cells, rods, cocci, "diphtheroids," variably gram-positive bacteria, and variably gram-negative organisms (Fig. 1C). The nocardiae are also variably acid-fast, so some strains may be mistaken for mycobacteria. Exposure of most humans to nocardiae probably occurs frequently, and based on experimental studies in animals, some of these exposures almost certainly lead to infection (27). Therefore, it is reasonable to assume that subclinical nocardial infections in humans may be relatively common. However, progression of nocardial infections to clinical disease is not recognized with high frequency, suggesting that both innate and acquired host defenses against nocardiae are quite effective (5). Numerous factors contribute to the progression from an inapparent infection to overt disease (5, 36, 37).

The incidence of human disease caused by nocardiae is not known; however, nocardiosis has been reported in most regions of the world (5, 15, 34, 36). Many of these investigators suggest that infections caused by *Nocardia* spp. are on the rise, and various studies suggest an incidence rate of disease varying from 0.3% to as much as 4.2% of all individuals with pulmonary diseases. On a global scale, these observations suggest that more than a million people develop clinical signs of nocardiosis each year. Unfortunately, most of these cases are misdiagnosed, and the correct diagnosis occurs either serendipitously or retrospectively (5, 15, 34, 36). The incidence of subclinical infection by the nocardiae is even less well documented. Hubble et al. (27), using serological analysis with antigens produced and secreted only by nocardiae during growth, reported that as many as 50% of the elderly people in Kansas were subclinically infected by nocardiae. At the same time, it was shown that approximately 20% of university students on a California campus had antibody to the secreted, diagnostic nocardial antigens. These studies suggest a very high incidence of exposure with possible subclinical infection by nocardiae (5, 27).

Gram-Positive Pathogens, ed. by V. A. Fischetti et al.
© 2000 American Society for Microbiology, Washington, D.C.

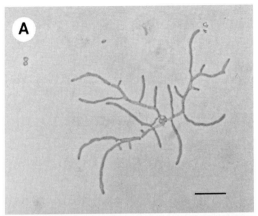

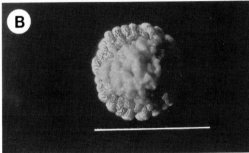

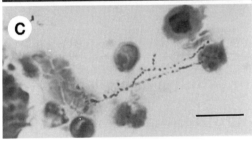

FIGURE 1 General characteristics of nocardiae. (A) Phase-contrast micrograph of *Nocardia* spp. grown on tryptone agar for 12 h. Note the typical branching, filamentous growth characterized as "nocardioform morphology." Bar, 10 μm. Reprinted from reference 8 with permission from the publisher. (B) Typical colonial morphology of *Nocardia asteroides* grown on glucose yeast extract agar at 37°C for 14 days. Bar, 1 cm. (C) Gram stain of *N. asteroides* in a smear from an abscess. Note the typical beaded appearance of the branching filaments. Bar, 15 μm.

Pulmonary Nocardiosis

It is generally thought that the nocardiae residing on dust particles become airborne and are then inhaled (5, 36, 37). Depending on the size of the particle, it can be deposited on the epithelial surface of either the upper or lower airways. The subsequent events occurring at this location depend upon the bacterial numbers, the growth state, and the virulence properties of individual nocardial cells present on the particle. If a virulent nocardial cell is present, then infection may ensue. Log-phase cells of nocardiae are typically much more virulent and invasive for animals and tissue culture cells than are stationary-phase organisms from the same culture (5). Indeed, log-phase cells of the most virulent strains of *Nocardia asteroides* behave as invasive, primary pathogens, whereas stationary-phase cells

act more like opportunistic pathogens requiring some host deficit for infection to occur (5). It is most likely that nocardiae aerosolized with dust particles would be dehydrated, starved, mixed with other microorganisms, and not in log phase of growth. Therefore, this encounter in an immunologically intact host should be abortive, but in an immunocompromised individual the nocardiae may survive and grow. Pulmonary encounters with "healthy" nocardiae in log phase of growth would be predicted to occur infrequently. These findings might provide some explanation for the observations that pulmonary nocardiosis is recognized most frequently in compromised hosts but that it can occasionally present as an aggressive disease in previously healthy individuals with no identifiable deficits in host defenses (5, 36, 37).

In the lung, nocardiae grow as facultative intracellular pathogens that induce predominantly a polymorphonuclear neutrophil (PMN) response with abscess formation, with or without calcification, although granulomas have been reported (5). The organisms may either remain localized within the bronchioles and alveoli, or they may invade aggressively through the lung parenchyma. The lesions may present as a localized, expanding infiltration; mild, self-limited infiltration; diffuse pneumonia; multinodular pulmonary infiltrates; lobar consolidation; interstitial infiltration; necrotizing pneumonia; indolent, progressive fibrosis; or empyema with pleural effusions (5, 14, 54). The clinical presentation of pulmonary nocardiosis is extremely variable, and it may be confused with many other pulmonary diseases. Misdiagnosis of pulmonary nocardiosis is often reported (5, 14, 54).

Extrapulmonary Nocardiosis

Primary nocardiosis outside of the lungs is probably the result of direct implantation from a contaminated source. The most frequently recognized site is in the eye, usually following some form of trauma (5, 46). Osteomyelitis, bone infections, and arthritis have been reported after various injuries (5), and endocarditis and pericarditis have been reported subsequent to surgical procedures (5).

Systemic Nocardiosis

Systemic nocardiosis is defined as a disease in which two or more locations in the body become infected by the same nocardial strain. Regardless of the site of the original primary infection, dissemination of the nocardiae to other sites may occur (5, 36). The skin, central nervous system, retinas of the eyes, and kidneys present frequent secondary foci; however, any region of the body can become involved following dissemination (5, 36).

Central Nervous System Nocardiosis

Approximately one-fourth of the cases of nocardiosis reported in the literature (excluding mycetomas) involve the central nervous system (CNS), especially the brain (5). Contrary to the general belief of many clinicians, brain infections caused by nocardiae may be primary, with no identifiable lesions elsewhere in previously healthy individuals (5). Indeed, more than 40% of these CNS infections occur in previously healthy individuals with no identifiable predisposing condition (5, 40). Therefore, the perception that CNS infection by nocardia occurs only in immunocompromised patients is not accurate (5, 40). Animal models show that intravenous (i.v.) inoculation with log-phase cells of most pathogenic *Nocardia* spp. results in preferential invasion of the brain (5). Taken together, these observations indicate that nocardiae are primary brain pathogens in humans and other mammals (5).

Nocardiae may induce a variable response in the brain depending upon the patient, the strain of nocardia, and the location of the infection (5, 40). Nocardial infections in the brain may be silent, with diagnosis made by accident (5); they may be insidious and difficult to diagnose (5); they may be acute with rapid progression (5); or they may be chronic with gradual progression, resulting in neurological deficits lasting from months to years (5). The types of lesions produced in the brain may be just as variable. Abscess formation is common, but there can be diffuse infiltration with no focal lesions as well as organized granulomata with giant cells (5).

Cutaneous, Subcutaneous, and Lymphocutaneous Nocardiosis

Many nonimmunocompromised individuals probably develop cutaneous and subcutaneous infections caused by all of the pathogenic species of *Nocardia* following traumatic exposure to contaminated soil (e.g., a rose thorn puncture). In most instances, these infections are self-limited and induce transitory cellulitis, pustules, or abscesses at the site of injury (5, 29, 30). However, in some individuals, these persist and progress to form a chronic, expanding cutaneous or subcutaneous lesion that requires medical attention (5, 29, 30). Occasionally, the infection extends into the lymphatics, affecting the regional lymph nodes. This form of lymphocutaneous nocardiosis may resemble sporotrichosis, which is usually caused by the fungus *Sporothrix schenckii*, and it is called sporotrichoid nocardiosis. Most cases of lymphocutaneous and sporotrichoid nocardiosis are caused by *Nocardia brasiliensis* (5, 45).

Mycetoma

A mycetoma is a chronic, progressive, pyogranulomatous disease that usually develops at the site of a localized injury such as a thorn prick (1, 5). Mycetomas are divided into two types, based on the specific etiology. Those caused by a variety of fungi are called eumycetomas; those caused by aerobic actinomycetes are referred to as actinomycetomas (1, 5). Three species of actinomycetes (*Actinomadura madurae*, *N. brasiliensis*, and *Streptomyces somaliensis*) account for most cases of actinomycetomas recognized throughout the tropical and subtropical regions of the world. Occasionally, other aerobic actinomycetes, including all species of nocardia, cause actinomycetomas (1, 5).

SPECIES OF *NOCARDIA* PATHOGENIC FOR ANIMALS

The nocardiae were first recognized as pathogens in animals such as cows and dogs (5); however, many species of *Nocardia* cause diseases in a wide variety of animals, including birds, fish, invertebrates, and mammals (5, 9). The first disease attributed to nocardia is bovine farcy, described by Nocard in 1888 (5, 9). At about the same time that Nocard recognized actinomycetes in bovine farcy, similar organisms were found to cause disease in dogs and in humans (5, 9). Indeed, in 1889 Trevisan named Nocard's bovine isolate *Nocardia farcinica*, and in 1896 Blanchard listed Eppinger's human isolate as *N. asteroides* (5, 9). Currently, 13 species of *Nocardia* have been published and validated according to the rules of the International Committee on Systematic Bacteriology (21): *N. asteroides*, *N. brasiliensis*, *Nocardia*

brevicatena, *Nocardia carnea*, *Nocardia crassostreae*, *N. farcinica*, *Nocardia flavorosea*, *Nocardia nova*, *Nocardia otitidiscaviarum*, *Nocardia pseudobrasiliensis*, *Nocardia seriolae*, *Nocardia transvalensis*, and *Nocardia vaccinii*. In addition, there are a variety of other species of *Nocardia* listed in journals that focus primarily on biotechnology and industrial microbiology. None of these strains appears to have medical relevance, and their relationship, if any, to the 13 species listed above is not known.

Nocardial infections are well described in the veterinary literature. In general, the same species of *Nocardia* that cause disease in humans (*N. asteroides*, *N. brasiliensis*, *N. farcinica*, *N. nova*, *N. otitidiscaviarum*, *N. pseudobrasiliensis*, and *N. transvalensis*) induce the equivalent disease in most other mammals, with *N. asteroides* representing the dominant pathogen (5, 9). *N. seriolae* induces abscesses in fish (5), *N. crassostreae* produces lesions in oysters (23), and *N. vaccinii* causes galls in blueberries (5). To my knowledge, these last three species of *Nocardia* have never been identified from human infections. *N. brevicatena* and *N. carnea* have been recovered occasionally from clinical material, but their roles in disease are not established, and the recently named *N. flavorosea* has not been associated with disease (17).

N. asteroides, *N. farcinica*, and *N. otitidiscaviarum* have caused significant outbreaks worldwide in dairy cattle, usually in the form of mastitis (5). In one published series from California, more than 1,600 cows at five dairies were involved. Many of these animals either died or had to be euthanized, thus representing a large economic loss (5). Interestingly, *N. brasiliensis* has not been reported to cause mastitis in dairy animals, even though it was found to cause mastitis in a human (39). None of the other nocardial species have been reported to cause infections in cows (5).

Since the initial reports on bovine farcy in the late 19th century, there have been hundreds of cases of nocardiosis described in animals. These include both domestic animals such as dogs, cats, sheep, pigs, goats, and horses (5, 9, 35) and wild animals such as raccoons, antelope, gazelles, whales, dolphins, porpoises, seals, deer, monkeys and other primates, armadillos, fox, mongoose (5, 9), water buffalo (43), a variety of birds (5, 9), crayfish (5, 9), various fish (5, 9), and oysters (23).

ANIMAL MODELS TO STUDY NOCARDIOSIS, PATHOGENESIS, AND HOST RESPONSES

A variety of animal species have been utilized as models to investigate nocardiosis. The first reported use of rabbits and guinea pigs as experimental models to study human disease caused by *N. asteroides* was in 1891 by Eppinger (5, 9). In 1902, MacCallum (5, 9) demonstrated that healthy dogs were quite susceptible to systemic infection by i.v. inoculation with *N. asteroides*. He suggested that dogs were ideal experimental models for studying human nocardiosis (5, 9). Over the next 70 years, many investigators attempted to establish mice and rats as the preferred experimental models for investigating nocardial pathogenesis. The results of these investigations were variable, often contradictory, and every conceivable outcome was reported (5, 9). These vacillating reports are the result of early misconceptions regarding the nature of nocardiae and the general lack of standardization of methods. Many other animals, such as monkeys, have been explored as models for investigating nocardial pathogenesis (5, 9). These studies have shown that a wide variety of animals are susceptible to nocardial infection (5, 9). Experimentally, the nocardiae induce similar types of diseases following inoculation into most mammals, including mice, which indicates a general lack of unique host responses (5, 9). Therefore, the responses observed in murine models probably have their counterparts in most other mammalian species, including humans. Regardless of the earlier erratic responses reported by different investigators utilizing mice as models for nocardiosis, the mouse remains the best, most versatile general model for investigating mechanisms of host-nocardial interactions (5, 9).

TISSUE CULTURES TO INVESTIGATE NOCARDIA-HOST INTERACTIONS

Nocardiae are facultatively intracellular pathogens that resist the microbicidal activities of PMN. They grow within mononuclear phagocytes from mice, rabbits, guinea pigs, and humans (5). Once inside macrophages, virulent strains of *N. asteroides* inhibit phagosome-lysosome fusion (Fig. 2A), block phagosomal acidification, and grow (5, 19). In contrast, phagosome-lysosome fusion is facilitated in activated macrophages in the presence of *Nocardia*-specific antibody and T lymphocytes from preimmunized animals (Fig. 2B); the intracellular bacteria are either inhibited or killed (19). Furthermore, virulent strains of *N. asteroides* actively invade a variety of host cells both in vitro (Fig. 3) and in vivo (Fig. 4). The invasion of these host cells appears to be both specific and multifactorial. The differential adherence to and specificity for invasion of cells by *N. asteroides* is growth stage dependent. In addition, different strains of nocardiae possess distinct surface-associated ligands that bind to either epithelial cells in the lung (Fig. 4C and D) or endothelial cells in the brain (Fig. 4A and B). A model strain, *N. asteroides* GUH-2, has all the attributes listed above, and it is neuroinvasive. Therefore, the mechanisms of nocardial-host interactions have been studied most extensively using this strain of *N. asteroides*. By comparing and contrasting a variety of human, veterinary, and environmental isolates of different species of *Nocardia* with *N. asteroides* GUH-2, a spectrum of virulence patterns and a progression of host interactions have been recognized (4, 5, 10, 11).

The in vitro interactions of nocardiae with monocytes, macrophages, and PMN have been studied extensively. Many of these investigations have focused on determining the mechanisms for the ability of *N. asteroides* to resist intracellular killing and then grow within these phagocytes (5). It was shown that virulent strains of *N. asteroides* secrete a superoxide dismutase (SOD) that becomes surface associated. This SOD, in combination with high levels of intracytoplasmic catalase, protects the nocardiae from the oxidative killing mechanisms of PMN, monocytes, and macrophages (5). Once phagocytized, these organisms appear to remain within phagosomes wherein they inhibit phagosome-lysosome fusion (Fig. 2). In addition, the nocardiae block or completely neutralize the acidification of the phagosome so that the intraphagosomal pH remains above pH 7 (5). At the same time, virulent strains of *N. asteroides* modulate lysosomal enzyme content, especially within macrophages (5). There is selective reduction in acid phosphatase activity, while lysozyme and esterase-neutral protease activities either increase or remain un-

changed (5). It was found that the strains of nocardiae that dramatically reduce acid phosphatase activity in macrophages also utilize this enzyme as a carbon source (5). Furthermore, when added to culture medium containing glutamate as the carbon source, acid phosphatase synergistically enhances nocardial growth. These effects are not observed with strains of *N. asteroides* that do not alter acid phosphatase activity in macrophages (5).

In addition to macrophages, monocytes, and PMN, the interactions of *N. asteroides* with a wide variety of both primary and established tissue culture cells have been reported (Fig. 3A through F). Primary cultures of astrocytes and microglia from newborn mice demonstrated differential interactions with *N. asteroides* GUH-2 (10). Stationary-phase cells had specific longitudinal adherence to the surface of astroglial cells that had a stellate morphology consistent with type II (Fig. 3F). In contrast, there was little or no adherence of nocardiae to the surface of the more prevalent astroglia that had the type I polygonal morphology (Fig. 3F) (10). In these preparations, the microglia rapidly phagocytized both filamentous and coccoid cells of nocardiae. However, only filamentous bacteria exhibiting apical adherence penetrated into the astroglia. Once inside, the nocardiae grew in astroglia, whereas growth in microglia was inhibited (10).

Comparative and differential interactions of log-phase and stationary-phase cells of *N. asteroides* GUH-2 were investigated in monocyte-macrophage cell lines J774A.1 and P388D1, bovine pulmonary artery endothelial cell line CPAE, rat glial cell line C6, and human astrocytoma cell lines CCF-STTG1 and U-373 MG (11). The interactions of log-phase and stationary-phase cells of *N. asteroides* GUH-2 in these six cell lines were different. There was a significant difference in uptake between log- and stationary-phase nocardiae in the phagocytic cell lines J774A.1 and P388D1. In cell line J774A.1, significantly more (>5-fold) log-phase cells of GUH-2 were internalized than stationary-phase organisms, and cytochalasin inhibited this uptake (>95% reduction) (11). Nevertheless, filaments of log-phase GUH-2 cells still adhered to the surface and penetrated into the cell in the absence of "phagocytosis" in cytochalasin-treated cells of J774A.2 (11). In contrast, human astrocytoma U-373 MG cells demonstrated little difference in effectiveness of internalization between log-phase and stationary-phase nocardiae, and cytochalasin did not block uptake (<10% reduction) (11). In all cell lines, only log-phase cells of *N. asteroides* GUH-2 exhibited a distinct filament tip-associated invasion process that appeared to be quite different from typical or classical phagocytosis (11). This filament tip-associated penetration did not appear to be affected by either cytochalasin or colchicine. On the other hand, stationary-phase cells of *N. asteroides* GUH-2 adhered longitudinally to the host cell surface with no evidence of apical penetration. These data suggest that there are multiple growth-stage-dependent surface components that are involved in the adherence to and internalization of *N. asteroides* (11).

DIFFERENTIAL ADHERENCE TO AND INVASION OF HOST CELLS IN VIVO AND IN VITRO

Host-pathogen interactions studied in tissue culture systems are artificial by nature, and these types of investiga-

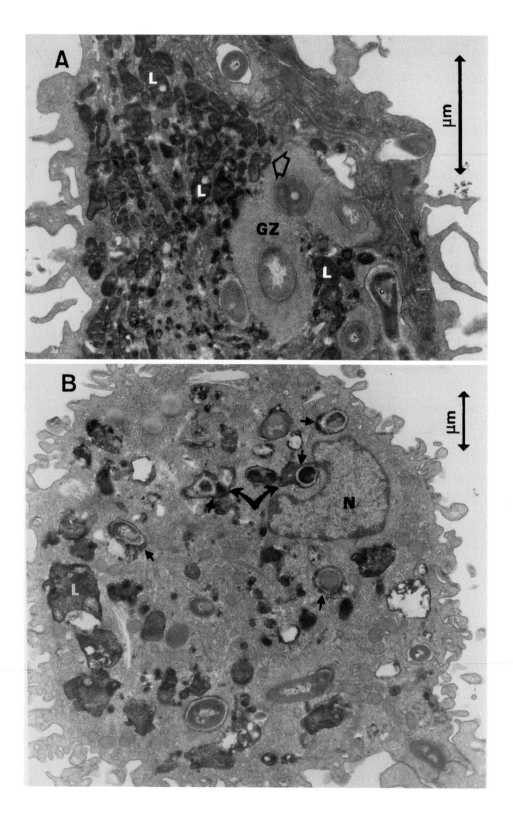

tions pose serious concerns about the validity of the conclusions drawn from the resultant data. Therefore, data obtained by in vitro studies must be validated in a relevant animal model system in vivo. The mouse may be utilized as a model for analyzing nocardial-host interactions, because responses to nocardiae in mice, rats, guinea pigs, rabbits, monkeys, dogs, and humans appear to be similar (5). Furthermore, data on nocardial interactions in mice indicate that nocardiae behave the same in host cells maintained in vitro (5, 6, 10, 11). For example, nocardial interactions on the pulmonary epithelium in mice appear to be the same as those with pulmonary epithelial cells grown in tissue culture. Therefore, nocardial interactions with specific types of host cells grown in vitro should be relevant for studying pathogenic mechanisms (5, 6, 10, 11).

Cells of *N. asteroides* GUH-2, in log phase of growth, adhere by way of the filament tip to the surface of both pulmonary epithelial cells (Clara cells) (Fig. 4C and D) and brain capillary endothelial cells in mice (Fig. 4A and B) (4, 6, 7). In contrast, stationary-phase organisms from the same culture do not adhere apically to either these Clara cells in the bronchioles or capillary endothelial cells in the brain (4, 6, 7). Unlike log-phase cells, stationary-phase organisms of *N. asteroides* GUH-2 adhere longitudinally to host cells, and in vitro they exhibit high specificity for primary type II astroglial cells (10). As noted above, only nocardial filaments attach apically to the surface of the host cells (Fig. 3A), and then the bacterium rapidly penetrates this surface to become internalized (Fig. 3B and D) (4, 6, 7). The same pattern of adherence to and invasion of epithelial, endothelial, and astroglial cell lines occurs in vitro (Fig. 3A through D) as observed in vivo (Fig. 4) following incubation with log-phase cells of *N. asteroides* GUH-2 (4, 6, 7, 10, 11).

The differential and selective adherence displayed by nocardiae both in vitro and in vivo suggests distinct multiple ligands for host cells on the nocardial surface (4, 6, 7, 10, 11). Furthermore, the expression of these components on the nocardial surface appears to be growth stage dependent (4, 6, 7, 10, 11). Utilizing different antibodies to block nocardial attachment supports the hypothesis that nocardiae possess growth-stage-dependent ligands that recognize surface receptors on different types of host cells (6). Some of these nocardial components appear to be specific proteins expressed only at the filament tip of nocardiae during active growth (43-kDa and 36-kDa proteins), whereas others may be glycolipids (trehalose dimycolates) incorporated into the entire cell envelope of the nocardiae

during all stages of growth (5, 6). A 43-kDa protein on the filament tip of log-phase cells of *N. asteroides* GUH-2 appears to be involved in apical attachment to and invasion of epithelial cells in the murine lung and HeLa cells in vitro (6). At the same time, a 36-kDa filament tip protein seems to be important in apical adherence to and penetration of capillary endothelial cells in the murine brain. Antibody against this 36-kDa protein blocks attachment and invasion in the brain but not the lung. Conversely, antibody specific for the 43-kDa tip protein blocks interactions in the lung but has no apparent effect on the interactions in the brain (6). At the same time, monoclonal antibody prepared against trehalose dimycolates from *N. asteroides* GUH-2 appears to decrease longitudinal adherence of both log- and stationary-phase nocardiae to many types of host cells (5, 6).

NOCARDIAL INVASION OF THE BRAIN

N. asteroides, *N. farcinica*, *N. nova*, *N. pseudobrasiliensis*, *N. transvalensis*, *N. brasiliensis*, and *N. otitidiscaviarum* cause CNS infections in humans (5, 36, 55) and animals (9). These infections are most frequently recognized as brain abscesses (5).

The interactions of *N. asteroides* in experimental CNS infections in animals have been investigated. Certain strains, exemplified by *N. asteroides* GUH-2, are selectively neuroinvasive (3–5, 7). These neuroinvasive nocardiae adhere to capillary endothelial cells within specific regions of the brain, wherein they invade through the endothelial cell into the brain parenchyma (Fig. 4A and B). During this invasion process, there may or may not be an ensuing inflammatory response. Furthermore, the integrity of the blood-brain barrier may remain intact (3–5, 7). We demonstrated that both mice and monkeys infected i.v. with neuroinvasive nocardiae develop either overt or cryptic CNS infection.

Even though it is well documented that nocardiae induce lesions in the human brain, the nature of the early interactions is not known. Based on animal models, the nocardiae probably reach the brain initially by way of the bloodstream, either by dissemination from a distant site (e.g., the lungs) or by direct inoculation following traumatic exposure (e.g., a puncture wound by a contaminated thorn or splinter). In addition to invasion of the brain, many neuroinvasive strains of *N. asteroides* also invade

---

FIGURE 2 Electron micrographs of phagosome-lysosome interactions in activated rabbit alveolar macrophages infected with *N. asteroides* GUH-2. The lysosomes were prelabeled with horseradish peroxidase for the purpose of ultrastructural and histochemical localization. (A) A section showing inhibition of phagosome-lysosome fusion and the intact appearance of the nocardiae. Many of the bacteria appear to be surrounded by a large granular zone (GZ) that prevents contact between the lysosome (L) and the phagosome (open arrow). The nocardiae were preincubated with 20% normal rabbit serum. Bar, 1.0 μm. Reprinted from reference 19 with permission from the publisher. (B) Phagosome-lysosome fusion and bacterial cellular damage are prominent (arrows) in the same preparation of macrophages shown in panel A except that these phagocytes were incubated with specifically primed lymph node lymphocytes and the bacteria were preincubated with sera and pulmonary lavage fluid from immunized rabbits. Note that the extensive bacterial damage and enhanced phagosome-lysosome fusion presented in this figure did not occur if any one of the components (primed lymphocytes, antibody, or pulmonary lavage fluid from immunized rabbits) was deleted. Bar, 1.0 μm. Reprinted from reference 19 with permission from the publisher.

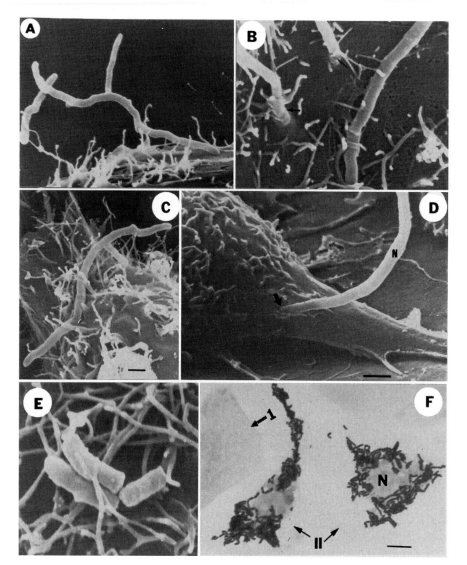

FIGURE 3 The comparative interactions of N. *asteroides* with different types of host cells grown in tissue culture. (A) Apical attachment of a log-phase nocardial filament to the surface of a HeLa cell. Note that this tip-associated adherence precedes penetration and invasion. Reprinted from reference 6 with permission from the publisher. (B) Apical penetration of the HeLa cell by three nocardial filaments (arrows). Reprinted from reference 6 with permission from the publisher. (C) Longitudinal adherence of a log-phase cell of N. *asteroides* to the surface of a HeLa cell. Note that only the bacterial filaments that attach by way of the filament tip appear to penetrate into the host cell. Bar, 1.0 μm. Reprinted from reference 6 with permission from the publisher. (D) Nocardial filament (N) tip invading through the surface (arrow) of an astrocytoma cell (CCF-STTG1) even after pretreatment of the tissue culture with cytochalasin. Bar, 1.0 μm. Reprinted from reference 11 with permission from the publisher. (E) Longitudinal association of stationary-phase bacilli to the microvilli of the HeLa cell. Bacteria attached in this manner do not invade the host cell. Reprinted from reference 6 with permission from the publisher. (F) Light micrograph of stationary-phase cells of N. *asteroides* GUH-2 adherent to the surface of type II astroglia (II). Bar, 10.0 μm. Note the arrangement of bacteria clustered around the nuclear region. Contrast with the adjacent cuboidal type I astroglia (1) with the total absence of adherent nocardiae.

bronchiolar epithelial cells (Fig. 4C and D). In mice, log-phase cells of neuroinvasive nocardiae injected i.v. invade neurons and reside intracellularly, surrounded by numerous layers of membrane of unknown origin (Fig. 5A) (3). The innermost layer of these membranes always appears to be tightly associated with the outermost layer of the nocardial cell wall (Fig. 5A, arrow). Exactly the same process is ob-served in the brain of monkeys following i.v. injection with the same nocardial strain (Fig. 5B, arrow) (3). Since this process occurs identically in mice and monkeys, it seems reasonable to suggest that the same events take place when nocardiae become blood-borne in humans. It is not known whether cryptic disease ensues following this invasion in humans.

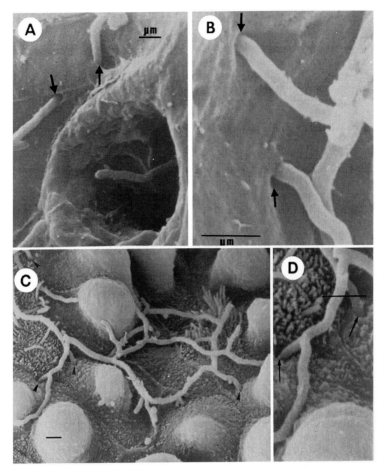

FIGURE 4 Scanning electron micrographs of differential attachment to and penetration of cells within the brain and lungs of mice by log-phase cells of N. *asteroides* GUH-2. (A) Apical penetration of capillary endothelial cells within an arteriole in the thalamus by nocardial filaments (arrows). Bar, 1.0 μm. Reprinted from reference 7 with permission from the publisher. (B) High-magnification view of two nocardial filaments penetrating into the endothelium of an arteriole in the region of the hypothalamus (arrows). Bar, 1.0 μm. Reprinted from reference 7 with permission from the publisher. (C) Nocardial interactions in the bronchiole of a C57BL/6 mouse 3 h after intranasal administration of log-phase cells of N. *asteroides* GUH-2. Note the association with nonciliated epithelial cells and apical penetration of Clara cells (arrows). Bar, 1.0 μm. (D) A high-magnification view of nocardial penetration into bronchiolar epithelial cells as shown in panel C. Bar, 1.0 μm.

Overt invasion by nocardiae results in lesions that vary pathologically from a diffuse cellular infiltration of PMN to either an abscess or a granuloma in mammals such as mice, monkeys, rabbits, and humans (5). On the other hand, cryptic invasion induces few or no inflammatory infiltrates, but instead induces subtle cellular changes, including neurodegeneration and axonal alterations, especially along the myelin sheath (3). In both mice and monkeys, the nocardiae invade neurons in the basal ganglia. Furthermore, intraneuronal and intraparenchymal inclusions are produced in the brain after a prolonged incubation period of several months (32). Indeed, cryptic invasion of the brain in mice following i.v. inoculation of log-phase cells of N. *asteroides* GUH-2 results in dopaminergic neurodegeneration in the substantia nigra, leading to a levodopa-responsive movement disorder that shares many features with parkinsonism (32). Thus, sublethal, intravenous inoculation of mice with log-phase cells of N. *aster-*

oides GUH-2 results in an infectious disease model of parkinsonism (32). The relationships, if any, of this process to the human disease are unknown.

THE GLYCOLIPIDS OF NOCARDIAE

Mycobacterium, Corynebacterium, Nocardia, Rhodococcus, Tsukamurella, and *Gordona* are bacterial genera that possess mycolic acids in their cell envelope (5). All of these genera contain species that may be pathogenic for humans; therefore, this group of organisms is collectively named mycolic acid-containing pathogens. Mycolic acids are unique α-branched, β-hydroxylated fatty acids that represent a major component of the cell wall, and they are associated with defining many of the basic biological properties of mycolic acid-containing pathogens (5). Relatively small amounts of mycolic acid reside freely within the envelope. Instead,

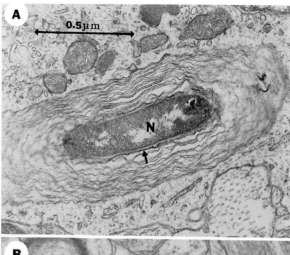

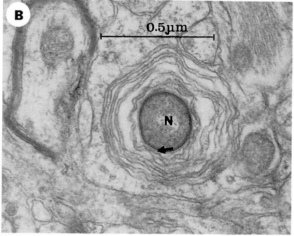

FIGURE 5 Comparative ultrastructure of *N. asteroides* GUH-2 growing within cells of the brain of a mouse and a monkey. Note that there is no inflammatory infiltration at the site of nocardial invasion in either animal. (A) A cell of *N. asteroides* GUH-2 growing within a neuron in the midbrain of a mouse 24 h after tail vein injection of a suspension of log-phase nocardiae. Note the numerous layers of membrane surrounding the ultrastructurally intact bacterium, with the innermost layer of membrane tightly adherent to the bacterial surface (arrow). N, nocardial filament. Bar, 0.5 μm. Reprinted from reference 3 with permission from the publisher. (B) A cross-section of a nocardial cell growing within the brain of a monkey 48 h after i.v. injection (leg vein) of a suspension of log-phase cells of *N. asteroides* GUH-2 as in panel A. Note the numerous layers of membrane, with the innermost layer adherent to the bacterial surface (arrow). Bar, 0.5 μm.

most of these long-chained fatty acids are covalently linked to a sugar moiety to form a glycolipid. Indeed, the bulk of the mycolic acids are bound to the arabinose portion of the arabinogalactan polymer, which, in turn, is covalently linked to the peptidoglycan of the cell wall (5). In addition, there are significant amounts of freely associated glycolipids consisting of mycolic acids covalently bound to other sugars, such as glucose and trehalose. The biological activities of the trehalose mycolates have received most of the research attention, and the literature on these compounds is extensive. Trehalose dimycolate may be the most biologically active of these glycolipids (2, 5, 25, 49).

In 1950, Bloch (25) isolated a compound from *Mycobacterium tuberculosis* that was believed to be responsible for both virulence and a characteristic serpentine pattern of growth. Because the virulence of *M. tuberculosis* for guinea pigs corresponded to the presence of the cordlike growth, and this growth was associated further with a substance that was soluble in organic solvents, this substance was named cord factor (5, 25). Bloch demonstrated that petroleum ether dissolved this substance from *M. tuberculosis* H37RV, and it imparted important biological activities to unrelated bacilli that were coated with this extract. Later studies demonstrated that this compound was composed of two mycolic acid molecules linked to trehalose to form trehalose 6,6'-dimycolate (TDM). Thus, TDM is synonymous with cord factor, and these two terms are currently used interchangeably by numerous investigators. Data suggest that TDM may actually impart serpentine cord formation in growing strains of virulent *M. tuberculosis*, even though earlier investigators expressed skepticism about this relationship (13). It is beyond the scope of this chapter to discuss the extensive literature on the structure and biology of TDMs (5, 25).

TDM has been shown to possess a wide variety of biological activities, depending upon the source of the TDM (thus different chemical structures) and the manner of preparation and presentation within various bioassays. The published effects of TDM on biological systems include the following: (i) TDM from *N. asteroides* in oil is lethal for mice (5); (ii) TDM targets and disrupts mitochondrial function (5, 25); (iii) TDM uncouples oxidative phosphorylation (5, 25); (iv) TDM induces apoptotic death in cells (42); (v) TDM inhibits phagosome-lysosome fusion in macrophages (16); (vi) TDM prevents calcium-dependent fusion in liposomal membranes (18); (vii) TDM is membrane interactive, altering hydration, rigidity, fluidity, and permeability in lipid bilayers (18); (viii) TDM induces granulomas in mice (5, 38); (ix) TDMs are potent immunoadjuvants (33); (x) TDMs stimulate γδ T cells and modulate cytokine production from lymphocytes (26); (xi) TDM has potent anticancer potential (5, 25, 38); (xii) TDM interferes with the coagulation system in blood (5, 25); (xiii) TDM is antigenic (31); (xiv) TDM can be used as a diagnostic antigen (20); (xv) TDM activates protein kinase C and induces tumor necrosis factor (48); (xvi) TDM can induce serpentine cord formation (13); and (xvii) TDM coupled to a protein carrier is an effective vaccine against virulent *M. tuberculosis* in mice (5, 25). Not all TDMs are equally active in the above-listed properties; indeed, the TDMs in the reports described above were obtained from different organisms grown under different culture conditions. Nevertheless, strains of nocardia that possess TDMs are more virulent than those that do not (28), and TDMs of nocardiae possess most of the same properties described above (5, 28). These observations support the hypothesis that the structures of the mycolic acid moieties are important for biological activities of the different TDMs. It is difficult to assess the role of chemical structure on a specific biological activity, because TDMs purified from most mycolic acid-containing bacteria are mixtures that have different sizes of mycolic acid side chains linked to trehalose (25). It is possible to synthesize artificial TDMs in the laboratory, but the biological relationships of these TDMs to the naturally occurring mixtures appear not to be the same.

It has been established clearly that TDM is a potent immunomodulator and immunoadjuvant, even though the mechanisms for these activities are not understood. Mycobacterial TDMs stimulate an increase of γδ T cells in human cord blood T lymphocytes (52), and TDM from *Mycobacterium bovis* preferentially increases the numbers of γδ T cells in the lungs of mice (44). Oswald et al. (41) reported that TDMs from *M. tuberculosis* induced interleukin-12 to mediate macrophage priming through induction of gamma interferon (41), and Beckman et al. (12) demonstrated that mycolic acids from the TDMs of *M. tuberculosis* are CD1b-restricted antigens (12). Also, mycobacterial TDMs cause an increase in natural killer, intermediate CD3, and γδ T cells during granuloma formation in the murine lungs (50).

The biological properties of some of these glycolipids in nocardiae and their possible relationship to nocardial pathogenesis have received a modicum of attention. It was first suggested in the early 1980s that the amount of trehalose dimycolate present in the cell wall of *N. asteroides* correlated directly with virulence (5). Thus, strains with large amounts of TDM were highly virulent, whereas strains with no detectable TDM were either avirulent or had very low virulence for mice (5, 28). In addition, small amounts of TDM from *N. asteroides* were lethal for mice following intraperitoneal injection of a mineral oil emulsion. Strains of *N. asteroides* that possessed TDM inhibited phagosome-lysosome fusion in macrophages, while strains that did not have TDM did not inhibit fusion (5, 28). Furthermore, TDMs purified from *N. asteroides* and coated onto yeast cells prevented phagosome-lysosome fusion when phagocytized by human monocytes. In contrast, these same yeast cells without TDM induced strong phagosome-lysosome fusion in monocytes (5, 18). Similar results were reported using TDMs from *M. tuberculosis*. Based on the above observations, it is probable that TDM plays a critical role in nocardial pathogenesis and host responsiveness to nocardial infection (5).

NOCARDIAL FACTORS IMPORTANT FOR PATHOGENESIS

The mechanisms of nocardial pathogenesis are incompletely understood. However, some of the factors contributing to pathogenesis and virulence of nocardiae have been determined (5). Both the conditions and phase of growth affect nocardial virulence (5), and nocardial strains capable of survival within macrophages are most virulent (5). The inhibition of phagosome-lysosome fusion is important for the intracellular survival and growth of nocardiae (5). This inhibition appears to be the result of surface glycolipids such as TDM (5, 18). Furthermore, there is a relationship between TDM content, inhibition of phagosome-lysosome fusion, and nocardial virulence for mice (5). Therefore, as described above, TDMs (cord factor) appear to be virulence factors (5, 28). Other cell wall components as well as the structure of individual mycolic acids in the envelope correlate with nocardial virulence (5). The cell walls of nocardiae contain powerful mitogens for both B cells and T cells. At least one of these is a T-cell-independent B-cell mitogen (5). The most virulent strains of nocardiae secrete SOD, and they have increased levels of intracytoplasmic catalase (5). Thus, both SOD and catalase are important virulence factors for nocardiae (5), since the level of nocardial resistance to the oxidative killing mechanisms of PMN and macrophages is important as a virulence determinant (5). Virulent strains of *N. asteroides* block or

neutralize phagosomal acidification in macrophages (5), and they decrease intralysosomal acid phosphatase and modulate lysosomal enzyme content in macrophages. These same strains utilize acid phosphatase as a sole carbon source. All of these activities are concordant with increased nocardial virulence, and they may be involved in pathogenesis (5). Cell-associated and secreted hemolysins appear to be associated with nocardial virulence, and they may be important for nocardial invasion of host cells (5).

Cell wall-defective nocardial cells (L forms) are involved in nocardial persistence, chronic infection, and cryptic invasion. L forms of certain strains of nocardiae are involved in pathogenesis (5). Some strains of *N. asteroides* are neuroinvasive whereas others are not. The strains of *N. asteroides* that invade both the lungs and brain of mice equally well are significantly more virulent than those strains that invade only the lungs (4). Nocardial L forms may be involved in cryptic invasion and persistence in the CNS (5).

Many of the neuroinvasive strains of *N. asteroides* selectively invade the basal ganglia, with specificity for the substantia nigra (3–5, 7, 32). Cryptic invasion of the basal ganglia results in neurodegeneration and selective destruction of dopaminergic neurons in the substantia nigra of mice, resulting in a parkinsonian movement disorder (5, 32). Furthermore, this selective adherence to and invasion of host tissues by log-phase cells of *N. asteroides* is mediated by filament tip-associated proteins. For example, a 43-kDa protein on the surface of the tip of log-phase cells of *N. asteroides* GUH-2 is involved in adherence to epithelial cells in the lung. At the same time, a 36-kDa tip-associated protein appears to be involved with adherence to capillary endothelial cells in the brain (6).

HOST FACTORS IMPORTANT FOR PROTECTION AGAINST NOCARDIAE

Many components of both the innate and acquired host responses are involved in protection from pathogenic species of *Nocardia* (5). First, PMN and macrophages are central to host resistance to these bacteria (5). However, PMN and macrophages alone do not kill the most virulent strains of *N. asteroides*. These professional phagocytes require the complete, intact machinery of the immune system to kill virulent nocardiae (5, 19). In addition, an immunologically specific subtype of T lymphocyte from immunized animals binds to the surface of log-phase cells of *N. asteroides* and induces lysis of these bacteria. This T-cell-mediated killing of nocardiae is not associated with natural killer cell activity and appears to be independent of antibody and complement (5). In contrast, B lymphocytes appear not to be required for host resistance to nocardiae; their role in protection of the host against nocardial infection is not clear (5). Immunosuppressive treatment with compounds such as cyclophosphamide, prednisone, or other substances that decrease cell-mediated responses significantly enhances host susceptibility to progressive, disseminated disease caused by even the least virulent strains of *N. asteroides* (5). Thus, T lymphocytes, activated macrophages, and cell-mediated immune responses appear to be most important for host resistance to nocardiae (5). However, such factors as the normal microflora of the animal as well as the γδ subset of T cells are important in the early stages of innate host resistance to nocardial infection (5, 22).

of protein kinase C by mycobacterial cord factor, trehalose 6-monomycolate, resulting in tumor necrosis factor-alpha release in mouse lung tissues. *Jpn. J. Cancer Res.* **86:**749–755.

49. **Syed, S. S., and R. L. Hunter, Jr.** 1997. Studies on the toxic effects of quartz and a mycobacterial glycolipid, trehalose 6,6′-dimycolate. *Ann. Clin. Lab. Sci.* **27:** 375–383.

50. **Tabata, A., K. Kaneda, H. Watanabe, T. Abo, and I. Yano.** 1996. Kinetics of organ-associated natural killer cells and intermediate CD3 cells during pulmonary and hepatic granulomatous inflammation induced by mycobacterial cord factor. *Microbiol. Immunol.* **40:**651–658.

51. **Tosteson, T. R., D. L. Ballantine, C. G. Tosteson, V. Hensley, and A. T. Bardales.** 1989. Associated bacterial flora, growth, and toxicity of cultured benthic dinoflagellates *Ostreopsis lenticularis* and *Gambierdiscus toxicus. Appl. Environ. Microbiol.* **55:**137–141.

52. **Tsuyuguchi, I., H. Kawasumi, C. Ueta, I. Yano, and S. Kishimoto.** 1991. Increase of T-cell receptor γ/δ-bearing T cells in cord blood of newborn babies obtained by in vitro stimulation with mycobacterial cord factor. *Infect. Immun.* **59:**3053–3059.

53. **Umunnabuike, A. C., and E. A. Irokanulo.** 1986. Isolation of *Campylobacter* subsp. *jejuni* from Oriental and American cockroaches caught in kitchens and poultry houses in Vom, Nigeria. *Int. J. Zoonoses* **13:**180–186.

54. **Uttamchandani, R. B., G. L. Daikos, R. R. Reyes, M. A. Fischl, G. M. Dickinson, E. Yamaguchi, and M. R. Kramer.** 1994. Nocardiosis in 30 patients with advanced human immunodeficiency virus infection: clinical features and outcome. *Clin. Infect. Dis.* **18:**348–353.

55. **Wallace, R. J., Jr., B. A. Brown, Z. Blacklock, R. Ulrich, K. Jost, J. M. Brown, M. M. McNeil, G. Onyi, V. A. Steingrube, and J. Gibson.** 1995. New *Nocardia* taxon among isolates of *Nocardia brasiliensis* associated with invasive disease. *J. Clin. Microbiol.* **33:**1528–1533.

ANTIBIOTIC AND HEAVY METAL RESISTANCE MECHANISMS

SECTION EDITOR: Richard P. Novick

GRAM-POSITIVE PATHOGENS HAVE ACQUIRED OR DEVELOPED much the same set of antibiotic resistances as have the gram-negatives; in many cases the same mechanisms are used by both groups, and homologous genes are involved. Significant exceptions exist in relation to antibiotics that do not penetrate the gram-negative outer membrane, such as glycopeptides, macrolides, and some β-lactams, to which these organisms are intrinsically resistant. In most cases, resistance genes have been developed in environmental contexts, representing defenses against other organisms or self-protection by antibiotic producers, and are imported by pathogens via mobile genetic elements. Several cases of mutational resistance have occurred in recent years, particularly to the gyrase inhibitors and to rifampin and other antimycobacterial drugs. A modest difference between the two groups is that gram-positive pathogens tend to possess multiple mono-resistance plasmids whereas gram-negative organisms more frequently collect multiple resistances on a single large plasmid.

This section covers resistance to the major classes of antibiotics, addressing frequency, mechanism of resistance, and mode of transfer, to provide a comprehensive view of the problem as it now exists. Antibiotic resistances that are more common in particular groups of organisms are discussed elsewhere in this volume, within the respective sections.

Mechanisms of Resistance to β-Lactam Antibiotics

DOUGLAS S. KERNODLE

62

The introduction of penicillin into clinical use in 1941 had a profound impact on the treatment of diseases caused by gram-positive pathogens (36). A variety of infections associated with serious morbidity became curable, including some that had been almost uniformly fatal, such as meningitis caused by *Streptococcus pneumoniae* and endocarditis caused by streptococci, staphylococci, and enterococci.

Within a short time, however, resistant bacteria that failed to respond to penicillin therapy were identified. In 1944 Kirby reported the production of a penicillin-inactivating enzyme by *Staphylococcus aureus*, and analysis of stored isolates from earlier decades established that resistant strains existed before the clinical use of penicillin (14, 15). Over the next couple of decades, β-lactamase-producing isolates of *S. aureus* became endemic in many hospitals as well as in the community, rendering penicillin ineffective in the treatment of most staphylococcal infections and prompting the development of penicillinase-resistant penicillins.

Shortly after methicillin was introduced in 1960, *S. aureus* isolates exhibiting resistance to methicillin were reported, apparently by an intrinsic mechanism not involving drug inactivation (3, 6). Similarly, *S. pneumoniae* isolates with intermediate resistance to penicillin were identified by 1967, and highly resistant pneumococcal isolates had appeared in South Africa by 1977 (36), again by a mechanism other than penicillin degradation. It has subsequently been shown that these examples of resistance involve changes in the membrane proteins that are active in peptidoglycan biosynthesis and are the normal targets of penicillin.

Antibiotic degradation by β-lactamase and alterations in penicillin-binding membrane proteins remain the major mechanisms by which gram-positive pathogens express resistance to β-lactams. Antibiotic tolerance in which the concentration of a β-lactam required to kill an isolate greatly exceeds the concentration needed to inhibit it (MBC-MIC ratio ≥32) may also be of clinical importance among some gram-positive species. Diminished β-lactam permeability through the outer cellular structures, which is a major mechanism of resistance to β-lactams among gram-negative species, possibly contributes to resistance expression in some higher gram-positive pathogens, such as nocardiae and mycobacteria, but does not appear to be a major mechanism of resistance among most gram-positive species.

MECHANISM OF ACTION OF β-LACTAMS

Gram-positive pathogens are characterized by a thick peptidoglycan layer, which is a relatively inelastic structure that confers shape on the organism and protects it against damage from differences in the osmotic pressure of its cytoplasm and its external environment. Peptidoglycan consists of alternating subunits of two amino sugars, N-acetylglucosamine and N-acetylmuramic acid. The N-acetylmuramic acid units are linked to peptide chains, many of which are interlinked to each other, providing cell wall stability.

Peptidoglycan biosynthesis is a dynamic process that requires the breaking of covalent bonds to enable the insertion of newly synthesized amino sugar cell wall subunits. To grow and divide, bacteria need autolytic enzymes (murein hydrolases with glycosidase, glucosaminidase, and amidase activities) and synthetic enzymes (glycosyltransferases, transpeptidases, and carboxypeptidases).

The major theory involving the mechanism of action of β-lactams concerns their structural similarity to the D-alanyl-D-alanine carboxy-terminal region of peptides involved in the cross-linking of peptidoglycan. Penicillins, cephalosporins, and other β-lactams acylate the active-site serine of cell transpeptidases, forming a stable acylenzyme that lacks catalytic activity (21, 24). Inhibition of peptidoglycan synthesis by covalent binding of β-lactams to cell wall synthetic enzymes may allow unopposed and ultimately suicidal autolytic activity. Cell lysis is not always seen, however, and in some species β-lactams appear to exert their antibacterial effect via other mechanisms. For example, benzylpenicillin induces a dose-dependent, rapid loss of total cellular RNA in group A streptococci without cell lysis, and the loss of RNA correlates with the loss of cell viability (44). Autolysis-dependent and autolysis-

Gram-Positive Pathogens, ed. by V. A. Fischetti et al.
© 2000 American Society for Microbiology, Washington, D.C.

independent mechanisms of killing have been described in pneumococci (45).

The penicillin-interactive enzymes involved in cell wall biosynthesis are localized on the outer face of the cytoplasmic membrane. Protein bands corresponding to these enzymes can be visualized on fluorograms following sodium dodecyl sulfate-polyacrylamide gel electrophoresis of bacterial membranes using radioactive penicillin for labeling, hence the name penicillin-binding proteins (PBPs) (42).

The strong antibacterial efficacy and low toxicity of β-lactams for eukaryotic cells have helped to make them the most highly developed class of antibacterial agents in clinical use. Historically, the most important have been penicillins and cephalosporins; however, in recent years other types have been introduced into clinical practice, including carbapenems, monobactams, and carbacephems.

β-LACTAMASE

β-Lactamases catalyze the inactivation of penicillin and other β-lactams by covalently binding to the β-lactam ring. This is essentially the same reaction that occurs when β-lactams bind to the active site of PBPs, with the major difference being that whereas the PBPs are inactivated because the reaction is not appreciably reversible, with β-lactamases there is subsequent hydrolysis of the covalent bond by a molecule of water such that the antibiotic is inactivated and there is regeneration of the active β-lactamase (Fig. 1).

β-Lactamases can be divided into four major groups based on their active site, molecular mass, and primary structure (21, 42). Class A enzymes have a serine active site, molecular mass of about 30 kDa, and usually have greater penicillinase than cephalosporinase activity. Class B contains the Zn^{2+} active-site metallo-β-lactamases, which usually exhibit a broad spectrum of activity, including the hydrolysis of carbapenems. They are found in *Bacillus cereus*, *Bacteroides* spp., and gram-negative aerobic species. Class C enzymes also have a serine active site but are larger than class A β-lactamases with a molecular mass of about 40 kDa and exhibit predominant cephalosporinase activity. Class D enzymes have a serine active site and exhibit oxacillinase activity. β-Lactamase classification schemes based on functional characteristics have also been described (5). Excepting the class B β-lactamase produced by *B. cereus*, all of the known β-lactamases produced by gram-positive species are class A enzymes (21).

The serine active site β-lactamases strongly resemble PBPs with several conserved primary structural elements that compose the active site, including the SXXK tetrad, the SDN triad, and the KTG triad (Table 1), as well as similarities in the three-dimensional positioning of α-helices and β-strands (21, 30). Some PBPs have detectable β-lactamase activity, including PBP 4 of *S. aureus* and peptide fragments of the pneumococcal PBPs (18, 22). The clinical relevance of this activity is unclear, and it does not appear to contribute much to resistance.

The ability of β-lactamase to hydrolyze a β-lactam can be expressed in simple kinetic terms: REH = k_{cat}/K_m, where REH is relative efficiency of hydrolysis, k_{cat} is the turnover number (i.e., the number of molecules of substrate degraded per second per molecule of enzyme), and K_m is the Michaelis constant, which reflects the affinity of the enzyme for its substrate. From this equation, it is apparent that there are two basic ways to alter drug structure in order

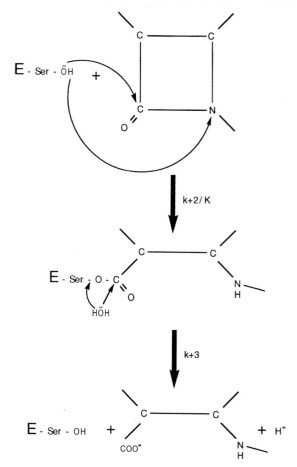

FIGURE 1 Interaction between the β-lactam ring and the active-site serine of penicillin-interactive enzymes. The chemical reaction is represented by the equation:

$$E + S \overset{K}{\rightleftharpoons} E \cdot S \overset{k+2}{\longrightarrow} E - S^* \underset{+H_2O}{\overset{k+3}{\longrightarrow}} E + P$$

where E + enzyme, S = β-lactam substrate, E · S = Michaelis complex, E − S* = acylenzyme, P = product, K = dissociation constant, and k+2 and k+3 are first-order rate constants, respectively, for the acylation step and hydrolysis of the acylenzyme. With PBPs, k+3 is low such that the acylenzyme is inert or hydrolyzed very slowly. With β-lactamases, k+3 is large and the antibiotic is rapidly hydrolyzed.

to reduce the REH: (i) by increasing the K_m, i.e., decreasing the binding affinity of the drug for the enzyme such that a higher concentration of the antibiotic is required to achieve half-maximal binding; and (ii) by reducing the k_{cat}, i.e., making the β-lactam intrinsically more resistant to enzymatic hydrolysis. A third strategy is to inhibit the enzyme's activity by the addition of a competing substance.

Each of these strategies has been used in the development of β-lactams active against β-lactamase-producing *S. aureus*. For example, via the addition of bulky side chains, the K_m values of the antistaphylococcal penicillins such as methicillin and oxacillin are much higher than the K_m of benzylpenicillin (Table 2). Because of their high K_m values,

TABLE 1 Highly conserved elements of penicillin-interactive proteins in *Staphylococcus*, *Streptococcus*, *Enterococcus*, and *Bacillus* species[a]

Proteins	SXXK tetrad	SDN triad	KTG triad
Class A β-lactamases	S*-X-X-K	S-D-N	K-T-G
		S-D-S	K-S-G
HMW class A PBPs	S*-X-X-K	S-R-N	K-T-G
		S-F-N	
		S-K-N	
HMW class B PBPs	S*-X-X-K	S-D-N	K-T-G
		S-C-N	K-S-G
		S-S-N	
HMW class C PBPs	S*-T-Y-K	Y-G-N	K-T-G
LMW class A PBPs	S*-X-X-K	S-A-N	K-T-G
		N-A-N	F-V-G
		S-G-N	
		S-S-N	
LMW class B PBPs	S*-X-X-K	Y-S-N	H-S-G
LMW class C PBPs	S*-X-X-K	S-N-N	K-T-G

*The active-site serine.

[a]The PBPs are classified according to the scheme of Ghuysen (24) as reported by Massova and Mobashery (42). HMW, high molecular weight; LMW, low molecular weight.

at clinically obtainable peak serum antibiotic levels of about 60 μg/ml there is minimal binding of methicillin or the isoxazolyl penicillins to *S. aureus* β-lactamase. Without binding, acylenzyme formation and hydrolysis cannot proceed, and therefore little or no antibiotic inactivation is observed. In contrast, some of the cephalosporins (e.g., cephalothin) have much lower k_{cat} values while having K_m values that are comparable to or even lower than that of benzylpenicillin. Although such agents form a Michaelis complex with β-lactamase at clinically relevant concentrations, drug inactivation proceeds slowly. Finally, clavulanic acid, sulbactam, and tazobactam bind covalently to the active site of staphylococcal β-lactamase and inactivate the enzyme. These agents have been combined with labile penicillins, e.g., ampicillin and amoxicillin, improving the antistaphylococcal activity of the latter. The similar structure of the penicillin-interactive regions of β-lactamases and PBPs (30, 42) creates a conundrum for new β-lactam development, as structural modifications need to be made that result not only in diminished inactivation of a compound by β-lactamase but simultaneously in retention of the ability to bind covalently to the target PBPs.

The resistance phenotype in β-lactamase-producing gram-positive bacteria differs from that observed with gram-negative species and is associated with an "inoculum effect" in which the MIC depends upon the number of

bacteria tested (56). That is, a large inoculum of β-lactamase-producing *S. aureus* exhibits much higher MICs than does a small inoculum (Table 3). This effect is related to the fact that with a large inoculum there is more preformed β-lactamase as well as more bacteria that can be induced to make β-lactamase than with the small inoculum. With gram-negative bacteria an inoculum effect is not generally observed, as the presence of an outer membrane and the sequestration of β-lactamase within the periplasmic space enable even single bacterial cells to exhibit high-level resistance.

EXAMPLES OF β-LACTAMASE-MEDIATED RESISTANCE IN GRAM-POSITIVE PATHOGENS

S. aureus

Around 90% of *S. aureus* isolates currently recovered from clinical specimens produce β-lactamase. The structural gene for β-lactamase, *blaZ*, and two regulatory genes, *blaI* and *blaR1*, usually reside on a plasmid, although a chromosomal location has been identified in some strains (14, 15). Some β-lactamase plasmids contain genes for resis-

TABLE 2 β-Lactam hydrolysis by *S. aureus* β-lactamase[a]

Antibiotic	k_{cat} (s^{-1})	K_m (μM)	REH[b]	Stability[c]
Benzylpenicillin	171	50	3.4	1
Methicillin	17	10,000	0.0017	2,000
Cephalothin	0.015	6.8	0.0022	1,545
Cefazolin	1.01	18.4	0.054	63

[a]Assays were performed using purified type A *S. aureus* β-lactamase (31, 72).

[b]Relative efficacy of hydrolysis (k_{cat}/K_m).

[c]Stability is expressed as a ratio of the stability of benzylpenicillin, which was set at 1.

TABLE 3 Effect of inoculum size upon the MIC of ampicillin of isogeneic *S. aureus* isolates

Isolate	MIC (μg/ml) for inoculum size (CFU)					
	10^1	10^2	10^3	10^4	10^5	10^6
VK7114	≤0.12	≤0.12	≤0.12	≤0.12	≤0.12	≤0.12
VK7115	0.25	1	2	4	32	≥128

[a]Modified from reference 34 with permission of the *Journal of Infectious Diseases*. VK7114 and VK7115 are isogeneic strains of *S. aureus* that differ only in the absence or presence, respectively, of a 1.1 kb *Hind*III *blaZ*-containing DNA fragment such that only VK7115 produces β-lactamase.

tance to other antibiotics, in particular, gentamicin and erythromycin.

Four variants of *S. aureus* β-lactamase termed types A, B, C, and D have been reported, initially by using serologic techniques and subsequently by detecting differences in the kinetics of hydrolysis of selected β-lactam substrates (14, 15, 33). The kinetic differences have been linked to amino acid substitutions at residue 128, close to the SDN loop (amino acids 130 to 132, Ambler numbering [1]) and residue 216, which is close to the β3 strand containing the KTG triad (Fig. 2). These sites flank the active-site cleft of the enzyme, and the substitutions modify, respectively, the k_{cat} and K_m values for various substrates (70). The prototypic A and C enzymes differ by six amino acids, type A and D by five amino acids, and type C and D by 11 amino acids (14). The type B enzyme, which is encoded by a chromosomal gene in some phage group II isolates, shows 15 or 16 amino acid differences with each of the type A, type C, and type D enzymes (69). Two additional novel *S. aureus* β-lactamases with unique substrate profiles have recently been identified and sequenced (31).

Most strains produce both extracellular and cell membrane-bound forms of β-lactamase, with the exoenzyme usually composing from 30 to 60% of the total β-lactamase activity. Following cleavage of the 24-residue leader peptide, the exoenzyme contains 257 amino acids with a high proportion of basic residues. Its highly basic nature (pI values of 9.7 to 10.1) may enable it to stay close to the bacterium via electrostatic interactions with anionic cell wall structures (72). The cell-associated form is six amino acids longer than the extracellular enzyme and is attached to fatty acids in the cell membrane via a glyceride-thioether bond (49). Some strains make only the cell-associated form and cannot make extracellular β-lactamase, a property that appears to be linked to amino acid substitutions in the portion of the enzyme corresponding to the normal cleavage site for proteolytic separation of the exoenzyme from the leader peptide (16).

The regulatory system for the induction of β-lactamase production in *S. aureus* involves derepression (14, 15). Under basal (uninduced) conditions a repressor binds to the promoter of *blaZ*, but with exposure to β-lactam antibiotics (induced) derepression occurs, typically producing a 30- to 100-fold increase in penicillinase activity. Two regulatory loci (*blaI* and *blaR1*) are usually directly upstream of *blaZ* on a plasmid (Fig. 3), whereas another putative regulatory locus (*blaR2*) has a chromosomal location (9, 14). *blaI* encodes the repressor. It has been cloned and expressed, and DNA footprinting studies suggest that the purified 126-amino-acid protein forms a homodimer that binds DNA in regions of dyad symmetry in the *blaR1-blaZ* intergenic region (14, 24a). *blaR1* encodes a high-molecular-weight (HMW) class C PBP of 585 amino acids and putative signal transducer-sensor properties. The C-terminal portion appears to be a penicillin-interactive protein based on its homology with the OXA-2 class D β-lactamase of gram-negative enteric bacilli (14).

Recently, Gregory et al. (24a) elucidated molecular aspects of how β-lactams induce β-lactamase production. During induction the repressor is converted to a smaller form, apparently by proteolytic removal of a 3-kDa fragment. The truncated repressor lacks the ability to bind to the *blaR1-blaZ* intergenic region, resulting in increased transcription of *blaZ*. The protein that mediates the conversion of the repressor has not yet been identified.

Several categories of isolates with mutations in the induction system have been identified (15). Magno-constitutive mutants produce large amounts of β-lactamase even in the absence of an inducer. Mutations in *blaI* that lead to impaired blocking of the *blaZ* operator can produce this phenotype. Alternatively, a defect in a chromosomal locus, called *blaR2*, has been implicated. Cohen and Sweeney (9) found that among mutants recovered following ethyl methanesulfonate-induced mutagenesis of *S. aureus*, the magno-constitutive phenotype was usually linked to the chromosome rather than to the β-lactamase plasmid. Little is known about *blaR2*, and the original *blaR2* mutants have been lost (14). It has been suggested that *blaR2* may encode an antirepressor or even a protease that converts the repressor to its non-DNA-binding form (14, 15, 24a). At least one magno-constitutive mutant, strain FAR4, in which the type D *S. aureus* β-lactamase was originally discovered, has been recovered from clinical specimens. Micro-constitutive mutants produce a small amount of β-lactamase and are little affected by inducers (15). Mutations in *blaR1* encoding the membrane transducing-sensor protein are believed to usually account for this phenotype. Meso-inducible mutants have reduced induction capability compared with most *S. aureus* strains. This phenotype appears to be attributable to a chromosomal defect, because wild-type plasmids assume the meso-inducible phenotype when transduced into a meso-inducible host (15).

The production of large amounts of β-lactamase has been associated with borderline susceptibility to the antistaphylococcal penicillins. The term borderline-susceptible (or borderline-resistant) *S. aureus* (BSSA) became popular

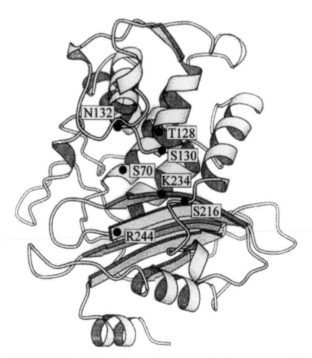

FIGURE 2 Ribbon figure of the type A β-lactamase of *S. aureus*. The active-site serine (S70) along with S130, N132, K234, and R244 catalyze the hydrolysis of β-lactam antibiotics. Amino acid substitutions close to the active-site cleft at residues 128 and 216 are responsible for the kinetic differences observed among the wild-type variants of *S. aureus* β-lactamase (69, 70).

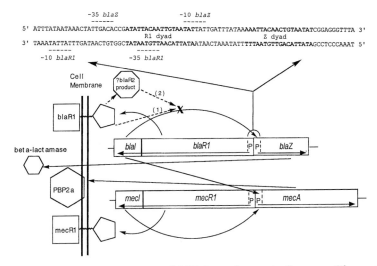

FIGURE 3 Regulation of β-lactamase and PBP 2a production in *S. aureus*. The production of β-lactamase, encoded by *blaZ*, is normally suppressed by the binding of the *blaI*-encoded repressor to the *blaR1-blaZ* intergenic region. This region includes two areas of dyad symmetry close to the −10 and −35 sequences for the *blaR1/blaI* and *blaZ* promoters (24a). Induction of β-lactamase is believed to be initiated by the binding of β-lactam to the transmembrane domain of a signal-transducing PBP encoded by *blaR1*, leading ultimately to repressor degradation with loss of its DNA-binding properties such that *blaZ* can be transcribed. Details of how the BlaR1-penicillin complex signals repressor degradation are unclear. Gregory et al. (24a) have suggested that the apparent degradation of the repressor could result from conformational changes to BlaR1 brought about by β-lactam binding that activate a protease in its cytoplasmic domain (mechanism 1). Alternatively, the putative chromosomal gene *blaR2* could encode a repressor-inactivating protease which the BlaR1-penicillin complex either activates or causes to be induced (mechanism 2). The production of the low-affinity PBP 2a, encoded by *mecA*, is regulated in a similar fashion to β-lactamase. There is homology between *mecI* and *blaI*, *mecR1* and *blaR1*, and the promoter and N-terminal portions of *blaZ* and *mecA*. This homology is strong enough that *blaI* can restore the normal inducible phenotype to *S. aureus* isolates, which produce large amounts of PBP 2a constitutively because of the absence of or a defect in the *mecI* locus.

after McDougal and Thornsberry's report in 1986 of *S. aureus* with a distinctive phenotype that included all of the following: (i) borderline MICs to oxacillin; (ii) lowering of oxacillin MICs into the clearly susceptible range by β-lactamase inhibitors; (iii) rapid hydrolysis of the chromogenic cephalosporin nitrocefin; and (iv) high MICs of benzylpenicillin (32, 43). Most of the isolates exhibiting all of these traits belong to phage group 94/96. They possess a 17.2-kb plasmid and make large amounts of the type A variant of *S. aureus* β-lactamase. When cured of their β-lactamase plasmid, they still exhibit an MIC slightly above that of other non-β-lactamase-producing strains by an unclear mechanism (2). BSSA have been associated with *S. aureus* wound infections in surgery that occur despite prophylaxis with cefazolin, a relatively β-lactamase-labile cephalosporin (32).

Massidda et al. (41) have suggested that BSSA strains produce a distinct methicillinase in addition to regular β-lactamase and speculate that it might be the *blaR1* product (i.e., the membrane transducer with homology to OXA-class D β-lactamases). However, the construction of an *S. aureus* strain with the BSSA phenotype by transforming a β-lactamase plasmid assembled from components, including *blaR1*, originating from non-BSSA isolates into a plasmid-free phage group 94/96 strain (34) is at variance with the hypothesis that the *blaR1* of BSSA possesses unique methicillinase properties. Furthermore, *S. aureus* strains that are not phage group 94/96 but that have been trans-

formed with the 17.2-plasmid from a BSSA strain do not exhibit the BSSA phenotype (2). It appears that the phage group 94/96 host background is critical to the expression of the BSSA phenotype, although the mechanism is unclear.

Terms such as borderline resistant, low-level resistant, and borderline susceptible have also been applied to other isolates of *S. aureus* that do not exhibit high-level β-lactamase production. Most such strains either (i) have alterations in their normal PBPs such that they have reduced binding affinity for β-lactams (66) or (ii) contain *mecA*, the gene encoding PBP 2a, but exhibit MICs around the breakpoint for designation as methicillin susceptible or methicillin resistant rather than the high MICs exhibited by most methicillin-resistant *S. aureus* (MRSA) strains (6). The latter group of strains could have genetic modifications of one or more of the *fem* genes (see below).

Coagulase-Negative Staphylococci

β-Lactamase is produced by most clinical isolates of coagulase-negative staphylococci. There is some evidence for β-lactamases with different substrate specificities, at least two of which are similar to the *S. aureus* enzymes (65).

Enterococci

β-Lactamase production appears to be a recently acquired trait in enterococci. It was first recognized in 1981 when an *Enterococcus faecalis* isolate was serendipitously observed to exhibit an inoculum effect to penicillin and subsequently has been identified in strains from several countries (47). Most of the described isolates make the type A variant of *S. aureus* β-lactamase, although a second β-lactamase exhibiting 10 amino acid differences from the prototypic type A sequence and nearly identical to a novel *S. aureus* β-lactamase has also been identified (67). Although some of the β-lactamase-producing strains contain intact β-lactamase-regulatory genes, i.e., *blaI* and *blaR1*, enterococci lack inducibility and produce small amounts of β-lactamase constitutively (67). In most β-lactamase-producing strains of enterococci, *blaZ* has been reported to reside on plasmid DNA, although chromosomal integration has occasionally been observed.

Bacillus species

B. cereus produces a class B zinc ion-dependent metallo-β-lactamase (21). *Bacillus* species including *B. cereus*, *Bacillus licheniformis*, *Bacillus subtilis*, and *Bacillus mycoides* make class A β-lactamases (42). Those from *B. licheniformis* 749/C and *B. cereus* 569/H exhibit about 40% amino sequence homology with *S. aureus* β-lactamase, suggesting a common ancestral origin for these genes (15). Furthermore, class A β-lactamase induction in *B. licheniformis* involves a regulatory mechanism similar to that described for *S. aureus*, and their *blaR1*-encoded proteins share strong homology (14).

Other Gram-Positive Pathogens

Nocardia spp. and some isolates of *Rhodococcus equi* make class A β-lactamases (19, 71). β-Lactamase production has not been observed in *Listeria monocytogenes*, *S. pneumoniae*, the viridans streptococci, or other streptococcal species. This includes the group A streptococci that have been extensively exposed to penicillin G via rheumatic fever prophylaxis and that often cohabit the pharynx with β-lactamase-producing *S. aureus*.

PENICILLIN-BINDING PROTEINS

Most bacterial species contain four to eight PBPs with molecular masses of 35 to 120 kDa (22, 26). By convention, the PBPs are numbered in order of diminishing molecular mass (i.e., the largest is PBP 1). Once the number of PBPs for a species is assigned, subsequently identified enzymes are numbered as derivatives of the previously established ones (22). In general, PBPs are species specific (e.g., *S. pneumoniae* PBPs are different from *S. aureus* PBPs, which are different from enterococcal PBPs). However, the PBPs of *S. pneumoniae* and viridans streptococcal species, including *Streptococcus mitis*, *Streptococcus oralis*, and *Streptococcus sanguis*, contain identical or closely related DNA sequences (26), and there is a high degree of homology of PBPs from different enterococcal species (20, 39).

A classification scheme that divides PBPs into six categories has been proposed by Ghuysen (24) (Table 4). HMW PBPs have a two-domain structure (24, 42). The PBP-binding C-terminal domain is involved in peptidoglycan transpeptidation. The N-terminal domain has been associated with transglycosylase activity in the HMW class A PBPs 1a and 1b of *Escherichia coli* (42), and there is recent evidence to suggest it has a similar function in PBP 1a of *S. pneumoniae* (11). However, in most HMW PBPs the function of the N-terminal domain remains undetermined (42). Furthermore, transglycosylases that are not

TABLE 4 Classification of PBPs and examples of each from gram-positive species[a]

PBP type	Examples
HMW class A	PBP 1a, *Streptococcus pneumoniae*
	PBP 1a, *Streptococcus oralis*
	PBP 2, *Staphylococcus aureus*
HMW class B	PBP 2b, *Streptococcus pneumoniae*
	PBP 2x, *Streptococcus pneumoniae*
	PBP 2a, *Staphylococcus aureus*
	(methicillin-resistant)
	PBP 5, *Enterococcus faecium*
	PBP 5, *Enterococcus faecalis*
HMW class C	BLAR1, *Bacillus licheniformis*
	BLAR1, *Staphylococcus aureus*
	MECR1, *Staphylococcus epidermidis*
LMW class A	PBP 4, *Staphylococcus aureus*
	Streptococcus pyogenes M49
	PBP 5, *Bacillus subtilis* 168
LMW class B	PBP 4*, *Bacillus subtilis* W168
LMW class C	*Bacillus subtilis* 168

[a] The PBPs are classified according to the scheme of Ghuysen (24) as reported by Massova and Mobashery (42). HMW, high molecular weight; LMW, low molecular weight.

PBPs have been identified in *S. aureus*, suggesting that the transglycosylase and transpeptidase activities may reside in different proteins in some gram-positive species (52). The HMW class B PBPs include several low-affinity PBPs that have been implicated in β-lactam resistance in gram-positive species and are discussed in more detail below. The HMW class C PBPs have been implicated in signal transduction associated with the induction of β-lactamase production in *B. licheniformis* and *S. aureus* and with the induction of PBP 2a production in *S. aureus* (42). Low-molecular-weight PBPs have single-domain structures and exhibit primarily carboxypeptidase activity (42). Some also exhibit transpeptidase activity. Not all of the PBPs are essential for viability, and the HMW PBPs are generally believed to be the lethal targets for β-lactam antibiotics.

In general, there is a direct relationship between the binding affinity of a β-lactam with vital PBPs and the inhibition of bacterial growth. For example, methicillin-susceptible strains of *S. aureus* (MSSA) produce four major PBPs with molecular masses of about 85, 81, 75, and 45 kDa, respectively referred to as PBPs 1, 2, 3, and 4 (6). The staphylocidal activity of a β-lactam correlates with its affinity for binding to PBP 1, which may be the primary transpeptidase, and PBP 3, which is involved in septation (7, 22). PBP 2 has been resolved into two components and is a transpeptidase functioning in stationary-phase cells (22), whereas PBP 4 exhibits carboxypeptidase activity. β-Lactams have different binding affinities for these PBPs (7, 23). For example, a 0.1 μg/ml solution of benzylpenicillin saturates *S. aureus* PBPs 1, 2, and 3, whereas a 100 μg/ml solution is required to saturate PBP 4. In contrast, cefoxitin binds well to PBP 4, with saturation at 0.5 μg/ml, but less well to PBPs 1, 2, and 3, with saturation at concentrations of 10, 2.0, and 10 μg/ml, respectively. β-Lactams and other cell wall-active agents work best when bacteria are growing rapidly, presumably because the autolysins are highly expressed during rapid growth. Cell wall-active agents are much less effective against bacteria that are in the postexponential growth phase or are surface adherent.

Because different gram-positive species exhibit heterogeneity in their PBPs, there is a wide range in the natural susceptibility of different species to penicillin and other β-lactams (36). Historically, group A streptococci and isolates of pneumococci from before 1967 have been exquisitely susceptible to benzylpenicillin, with MICs of about 0.01 μg/ml. Similarly, non-β-lactamase-producing strains of S. aureus generally exhibit MICs of benzylpenicillin of about 0.03 μg/ml. At the other extreme, penicillin-susceptible enterococci are naturally much less responsive to penicillin, with MIC values of about 2 μg/ml. Although L. monocytogenes is susceptible to benzylpenicillin and the aminopenicillins, cephalosporins have such low affinity for its essential PBP (PBP 3) that this species is considered to be intrinsically resistant to the cephalosporins. For a similar reason, enterococci are uniformly resistant to the cephalosporins.

The terms susceptible, intermediately resistant, and resistant are defined differently for different gram-positive species (Table 5). For example, with the pneumococci, even a small elevation in the MIC of ampicillin to the 0.12- to 1.0-μg/ml range indicates intermediate resistance, and isolates with MICs ≥2.0 μg/ml are considered resistant. In contrast, an enterococcus with an MIC of 8 μg/ml is still considered to be susceptible to ampicillin (48).

There are several PBP-related mechanisms by which gram-positive pathogens exhibit increased resistance to β-lactams. These include (i) point mutations in normal PBPs that reduce the binding affinity of β-lactams; (ii) overexpression of a normal PBP that has low affinity for β-lactams compared with the other normal PBPs; and (iii) acquisition of a gene encoding a new PBP from an entirely exogenous source or via recombination with a host PBP or β-lactamase gene.

EXAMPLES OF PBP-MEDIATED RESISTANCE IN GRAM-POSITIVE PATHOGENS

S. aureus

S. aureus exhibits resistance to β-lactams by all three of the PBP-related mechanisms described above. Point mutations in the penicillin-binding domains of S. aureus PBP 2 have been identified that result in low-level resistance to methicillin, and the other major PBPs probably can be similarly affected (6, 66). Most of the isolates exhibiting this type of resistance have been selected for in a laboratory setting, and it does not appear to be common among clin-

ical isolates despite the clinical use of β-lactams for nearly six decades. In addition, although PBP 4 in S. aureus is not essential for viability (22), its overexpression causes small but reproducible increases in β-lactam resistance (3).

From a clinical and epidemiologic perspective, clearly the most important form of PBP-mediated resistance to β-lactams involves the production of PBP 2a. MRSA contain 30 to 50 kb of additional DNA not found in methicillin-susceptible strains that is integrated into the chromosome near the pur-nov-his locus (3, 6) (see chapter 33, this volume). MSSA produce four major PBPs as discussed above, and MRSA also produce a fifth PBP of 78 kDa, called PBP 2a and encoded by a unique gene called mecA. PBP 2a has less binding affinity for most β-lactams than do the normal PBPs 1, 2, and 3. However, as is the case for the normal PBPs, there are differences in the binding affinity of different β-lactam agents. Some agents such as benzylpenicillin, ampicillin, and cefamandole bind relatively well, whereas the binding affinities of methicillin, oxacillin, and nafcillin are very poor (7). In the presence of methicillin, the cell wall of MRSA is altered. This results in a new peptidoglycan with fewer oligomeric muropeptides. This altered cell wall may be the synthetic product of PBP 2a, which is believed to take over the peptidoglycan transpeptidation normally performed by the methicillin-sensitive PBPs.

Similarly to β-lactamase, the production of PBP 2a in most S. aureus isolates is inducible. Furthermore, the genetic regulatory systems for the induction of β-lactamase and PBP 2a production are related (Fig. 3). It has been known since the early 1970s that the transduction of a mutant noninducible β-lactamase plasmid into an isolate of MRSA markedly suppresses its expression of resistance (8). It was subsequently shown that the regulatory genes for PBP 2a production, mecI and mecR1, are 61% and 34% homologous, respectively, with blaI and blaR1 (4), including their positioning upstream of the promoter for the respective structural genes, mecA and blaZ.

Exactly how PBP 2a evolved is unclear. Staphylococcus sciuri contains a mecA homolog for which the deduced amino acid sequence bears strong similarity to PBP 2a; however, it does not confer a methicillin-resistant phenotype (6). Also, the promoter and N-terminal portions of PBP 2a bear strong homology to S. aureus β-lactamase, and it has been suggested that PBP 2a might have originated from recombination of a gene encoding β-lactamase and a PBP (63). The homology between PBP 2a and S. aureus β-lactamase includes palindromic regions within the promoter that are candidates for the repressor-binding site (63). This combined with the strong homology between blaI and mecI appears to be the biological reason underlying the ability of blaI-blaR1 to restore inducibility to an isolate of MRSA either lacking or with a defective mecI-mecR1 locus (6). In the absence of a functional mecI-mecR1 or blaI-blaR1 regulatory element, magno-constitutive PBP 2a production occurs.

Since S. aureus β-lactamase and PBP 2a can utilize the same induction pathways, it has been suggested that the putative blaR2 locus reported to affect β-lactamase production is also involved in mecA regulation (3). Niemeyer et al. (50) have identified an MRSA strain that exhibits high basal mecA transcription unresponsive to mecI-mediated repression, suggesting that additional coregulatory factors are important. Whether this could represent the blaR2 locus is uncertain, but the analogy between magno-constitutive β-lactamase production unaffected by blaR1-blaI and

TABLE 5 MIC interpretive standards for gram-positive pathogens against ampicillin[a]

Organism	Ampicillin MIC (μg/ml) to be classified as:		
	Susceptible	Intermediate	Resistant
Staphylococci	≤ 0.12		≥ 0.25
Enterococci	≤ 8		≥ 16
S. pneumoniae	≤ 0.06	0.12–1	≥ 2
Other streptococci	≤ 0.12	0.25–2	≥ 4

[a] MIC interpretive standards as recommended by NCCLS (48).

high-level basal PBP 2a production unaffected by *mecR1-mecI* invites speculation. Such observations make it clear that there is much about the nature of the induction mechanism for β-lactamase and PBP 2a that remains unknown.

Most MRSA isolates exhibit heterogeneity in the expression of resistance, with a small fraction of the total population growing at a much higher antibiotic concentration than the majority (3). External factors such as temperature, osmolality, and light influence the proportion of the bacterial cell population that exhibits resistance (3, 6).

The expression of methicillin resistance is also influenced by chromosomal genes involved in cell wall biosynthesis, the *fem* (factors essential for the expression of methicillin-resistance) genes (3), as well as genes affecting autolysis (4). The reason that there is not as clear a relationship between the amount of PBP 2a and methicillin resistance among MRSA as there is between the amount of β-lactamase and benzylpenicillin resistance among MSSA probably reflects the effect of these other genes. Antibiotics that inhibit early steps in cell wall biosynthesis, including fosfomycin and cycloserine, cause a reduction in methicillin resistance as well as a change from homogeneous to heterogeneous expression of the methicillin-resistant phenotype (60). Selective inhibition of PBP 3 activity by cephradine has a similar effect. Another gene, *fmt* (factor that affects the methicillin-resistance level and autolysis in the presence of Triton X-100), that affects the expression of methicillin resistance and encodes a putative 397-amino-acid protein with homology to carboxypeptidases and β-lactamases has recently been identified (35).

Selected β-lactams are active against MRSA and methicillin-resistant coagulase-negative staphylococci in vitro at clinically achievable antibiotic concentrations. This activity has been correlated primarily with their binding affinity for PBP 2a (7). Furthermore, some of the agents that bind PBP 2a best are β-lactamase-labile (e.g., benzylpenicillin, ampicillin), and their activity is enhanced considerably by the addition of inhibitors of staphylococcal β-lactamase such as clavulanic acid, sulbactam, and tazobactam. However, not all of the in vitro observations regarding β-lactam activity against MRSA can be explained on the basis of PBP 2a binding affinity and β-lactamase stability. There is growing evidence to suggest that PBP 4 function may be important in MRSA by catalyzing cross-links that provide cell wall stability to the abnormal peptidoglycan synthesized by PBP 2a. Inhibition of PBP 4 results in poor cross-linkage of peptidoglycan that is more susceptible to autolysis (53). The combination of a β-lactam that binds relatively well to PBP 2a with one that binds to PBP 4 appears to be synergistic against MRSA. In addition, some new β-lactam agents currently in development appear to exhibit greater binding to PBP 2a.

Coagulase-Negative Staphylococci

Although less well studied, in general the mechanism of methicillin resistance in the coagulase-negative staphylococci generally involves a PBP 2a that is nearly identical to that seen in *S. aureus*. However, a few methicillin-resistant strains that lack *mecA* have been described (64).

Enterococci

Enterococci possess at least five and occasionally as many as nine PBPs (26). The major mechanism of enterococcal resistance to penicillins involves increased production of PBP 5, which exhibits a lower affinity for penicillin than the other PBPs (20). In general, penicillin resistance has been more strongly associated with isolates of *Enterococcus*

faecium than *E. faecalis* and has also been observed in other enterococcal species, including *Enterococcus raffinosus* and *Enterococcus gallinarum*. Much of the work involving β-lactam resistance in enterococci has been performed using a swine-derived isolate of *Enterococcus hirae* (ATCC 9790). This work has shown that among *E. hirae* mutants there is a direct correlation between the amount of PBP 5 produced and the MIC of penicillin. Furthermore, the penicillin MIC corresponds to the minimal concentration that saturates PBP 5. Conversely, reduced production or the absence of PBP 5 causes increased susceptibility to benzylpenicillin in mutants of *E. faecium*, yielding MICs similar to those of group A and other streptococcal species that do not produce a low-affinity PBP. Antiserum raised against PBP 5 of *E. hirae* is cross-reactive with PBPs of similar size in *E. faecium* and *E. faecalis* (39). PBP 5 of *E. hirae* exhibits 33% amino acid identity with PBP 2a of *S. aureus* as well as 24% and 25% identity, respectively, with PBPs 2x and 2b of *S. pneumoniae* (17, 26). Each of these is classified as a low-affinity HMW class B PBP (Table 4). PBP 5 overproduction derives from a deletion in the PBP 5 synthesis repressor gene (*psr*), which is located just upstream of the gene encoding PBP 5 (40). Peptidoglycan synthesized by PBP 5 differs from that produced normally by enterococci in that oligomers higher than trimers cannot be found in the PBP 5-synthesized product, although total cross-linking was comparable owing to increased dimer formation (61).

Recently, it has been shown that strains of *E. faecium* exhibiting very high resistance produce small amounts of PBP 5, which has been altered to exhibit even lower affinity for binding penicillin (55). Nucleotide point mutations resulting in amino acid substitutions close to the SDN triad within PBP 5 were demonstrated in these highly resistant strains.

S. pneumoniae

Penicillin-susceptible strains of *S. pneumoniae* possess five HMW PBPs (1a, 1b, 2x, 2a, and 2b) and one low-molecular-weight PBP (PBP 3) (26, 38), with different strains exhibiting considerable antigenic variation (36). PBPs 2x and 2b are the primary targets for penicillins (26). The intermediate penicillin-resistant phenotype can be derived in the laboratory by selection on antibiotic-containing agar and is attributed to point mutations in PBP 2x or PBP 2b (26).

The pattern found among most penicillin-resistant clinical isolates appears to be more complex than just a combination of point mutations of the pneumococcal PBP genes. These isolates exhibit a wide range of PBP profiles, and the large number of nucleotide changes suggests that the resistant PBPs originated by horizontal transfer of PBP genes or portions thereof from other streptococci. Homologs of the pneumococcal PBP 2x gene have been identified in *S. oralis* and *S. mitis* (59). These genes contain heterologous DNA sequence blocks with similar or identical junctions flanking them that serve as hot spots for recombination. Replacement of a portion of the PBP 2x gene of *S. pneumoniae* with the heterologous corresponding region of the gene from *S. oralis* or *S. mitis* can occur by crossover at the junctional area of homology, producing a mosaic PBP 2x that is slightly more resistant to cefotaxime despite the fact that both the donor strains and the recipient pneumococcal strain were cefotaxime susceptible (59). The development of β-lactam-resistant PBPs in *S. pneumoniae* probably involves a combination of the accumulation of point mutations in the PBP genes of commensal streptococcal species followed by the horizontal transfer and recombination of these PBP genes with their homologs in *S. pneumoniae*. Mosaic structures encoding low-affinity PBP variants have also been described in the PBP 2b and PBP 1a genes (12) and more recently in the PBP 2a and PBP 1b genes (27). Mosaic PBPs have been associated with alterations in the muropeptide structure in some penicillin-resistant isolates of pneumococci (57).

Whereas high-level resistance to penicillins generally involves changes in up to four of the five HMW PBPs (i.e., 1a, 2x, 2a, and 2b), resistance to the third-generation cephalosporins can be seen involving alterations in two PBPs (1a and 2x), as PBP 2b is not a natural target for the cephalosporins (26, 38, 54). PBP 2x and PBP 1a are closely linked on the chromosome, and high-level resistance to cefotaxime and intermediate resistance to penicillin have been transferred to a susceptible *S. pneumoniae* strain via a single round of transformation (46). Recently, specific amino acid substitutions have been shown to affect the magnitude of resistance to penicillin. Among strains already exhibiting altered *pbp2x* and *pbp2b* genes, a Thr to Ala or Ser substitution at residue 371 within the STMK active site (residues 370 to 373) of PBP 1a causes high-level penicillin resistance (62).

Selected capsular serotypes are overrepresented among penicillin-resistant strains including serotypes 6, 9, 14, 19, and 23. Often, all resistant strains within a region are of the same serotype and have the same PBP banding profile. However, when strains from different regions are analyzed, the picture is one of multiple clones of penicillin-resistant pneumococci with unique PBP profiles. These findings support the hypothesis that resistance has emerged in *S. pneumoniae* independently on multiple occasions, but there has been extensive dissemination of the highly resistant clones (54).

Viridans Streptococci

Historically, the viridans streptococci have been highly susceptible to β-lactam antibiotics, and most clinical isolates have remained susceptible. Over the past decade, however, cefotaxime- and benzylpenicillin-resistant isolates of *S. mitis* and *S. oralis* with low-affinity PBPs have been identified (54). One *S. mitis* strain with high-level resistance to both cefotaxime and benzylpenicillin (MICs >32 μg/ml) had alterations in all five HMW PBPs (1a, 1b, 2x, 2a, and 2b), and this resistance could be transferred to *S. pneumoniae* (27). There is also evidence that PBP 2b genes from penicillin-resistant *S. pneumoniae* have spread into *S. oralis* and *S. sanguis* (13). Interspecies and intraspecies gene transfer and recombination appear to be the major mechanism underlying β-lactam resistance for viridans streptococci as well as for *S. pneumoniae*.

Other Gram-Positive Pathogens

Clinical isolates of group A streptococci have remained highly susceptible to benzylpenicillin (MIC <0.01 μg/ml); however, intermediate resistance (MIC of 0.2 μg/ml) mediated by changes in PBPs 2 and 3 can be selected by antibiotic exposure in the laboratory (25). Group B streptococci are intrinsically about 10-fold less susceptible than group A streptococci to penicillin. Most clinical isolates have remained penicillin susceptible, although intermediate resistance (MICs of 0.25 to 2.0 μg/ml) has been described. Historically, *Bacillus anthracis* has been uniformly susceptible to benzylpenicillin (MICs of about 0.03 μg/ml) despite its exceedingly close relation to *B. cereus*, which is frequently resistant. However, in the past decade there have been scattered reports of isolates with reduced sus-

ceptibility, although the mechanism is not known (37). Altered PBP patterns have been reported in imipenem-resistant isolates of *R. equi* (51).

TOLERANCE

The term tolerance has been applied to bacteria that are inhibited but not killed by β-lactam antibiotics. It was first used in 1970 to describe laboratory strains of *S. pneumoniae* with a suppressed autolytic system that did not cause cell lysis, such that the strain lost viability at much lower rates than usual with exposure to penicillin at concentrations above the MIC (68).

Tolerance differs from other types of resistance mechanisms in that it does not involve an increase in the MIC of the drug compared with other isolates of the same species (68). In many clinical laboratories, it has been defined as an MBC-MIC ratio of ≥32; however, it has been noted that using this arbitrary breakpoint, the original pneumococcal isolate in which tolerance was reported would not have been considered tolerant (68). The use of different definitions of tolerance along with numerous technical factors that affect MBC test results has led to inconsistencies and confusion about the clinical relevance of tolerance in the medical literature (28, 58).

Genotypic tolerance originates from several mechanisms (28). Among tolerant mutants isolated in the laboratory, the reduced production of N-acetylmuramic acid-L-alanine amidase has been observed in *S. pneumoniae* and diminished muramidase production in *E. faecium*. In some strains of *B. subtilis*, tolerance involves the inactivation of autolysin by a protease. In group A streptococci, tolerance has been associated with changes in PBPs. Multiple other defects that could possibly lead to reduced killing of bacteria without affecting the MIC have been proposed (28).

In clinical isolates, tolerance appears to be more prevalent among gram-positive than gram-negative species (28). Groups A, B, C, and G streptococci exhibiting tolerance to penicillin have been recovered from clinical specimens (10, 25). In viridans streptococci, tolerance has been reported to adversely affect the response to antibiotic therapy in animal models of endocarditis (68). Clinical isolates of enterococci are now almost always tolerant to penicillin (58), in contrast to "penicillin-virgin" isolates of *E. faecalis* recovered from persons in the Solomon Islands in the 1960s (29). However, the latter group can be easily induced to develop tolerance when exposed to pulsed doses of benzylpenicillin in vitro, and it has been suggested that the frequent use of β-lactams in clinical practice might have contributed to the widespread prevalence of tolerant enterococci that is now seen (29). The prevalence of tolerance in *S. aureus* has been debated, and it has been more difficult to demonstrate the clinical relevance of tolerance with *S. aureus* than with other gram-positive species (58, 68).

Phenotypic tolerance refers to the ability of all bacteria to avoid the bactericidal activity of antibiotics at times when they are nongrowing (dormant) or growing only slowly. It may be a more important clinical problem than genotypic tolerance (68).

SPREAD OF RESISTANCE AND IMPLICATIONS FOR THE FUTURE

After nearly six decades of use, the effectiveness of β-lactams in infections caused by gram-positive pathogens has been seriously eroded by the development and spread of β-lactam resistance. Plasmid-encoded β-lactamase production is now widespread in *S. aureus* and has been transferred to enterococci, although it has not yet been reported in streptococcal species. The low-affinity PBP 2a of staphylococci came from an as yet undetermined source; however, clones expressing it have disseminated widely and are a common clinical problem. Enterococci typically demonstrate tolerance to the killing activity of penicillin, making its synergistic combination with an aminoglycoside necessary in the treatment of enterococcal endocarditis; however, the development of penicillin resistance by several mechanisms now threatens the effectiveness of the combination. The development of mosaic genes encoding low-affinity PBPs in *S. pneumoniae* and the viridans streptococci by a combination of point mutations and recombination of homologous portions of DNA between species is rapidly making β-lactams less useful in serious infections caused by these pathogens. These changes bear strong evidence of the power of selective antibiotic pressure in the clinical setting and the genetic versatility of gram-positive bacteria in adapting to survive. Success in containing the dissemination of β-lactam-resistant gram-positive pathogens will be difficult to achieve without strategies to control antibiotic utilization.

REFERENCES

1. Ambler, R. P., A. F. W. Coulson, J. M. Frère, J. M. Ghuysen, B. Joris, M. Forsman, R. C. Levesque, G. Tiraby, and S. G. Waley. 1991. A standard numbering scheme for the class A beta-lactamases. *Biochem. J.* **276:** 269–270.

2. Barg, N., H. Chambers, and D. Kernodle. 1991. Borderline susceptibility to antistaphylococcal penicillins is not conferred exclusively by the hyperproduction of β-lactamase. *Antimicrob. Agents Chemother.* **35:**1975–1979.

3. Berger-Bächi, B. 1997. Resistance not mediated by β-lactamase (methicillin resistance), p. 158–174. *In* K. B. Crossley and G. L. Archer (ed.), *The Staphylococci in Human Disease.* Churchill Livingstone, Ltd., New York, N.Y.

4. Brakstad, O. G., and J. A. Mœland. 1997. Mechanisms of methicillin resistance in staphylococci. *Acta Pathol. Microbiol. Immunol. Scand.* **105:**264–276.

5. Bush, K., G. A. Jacoby, and A. A. Medeiros. 1995. A functional classification scheme for β-lactamases and its correlation with molecular structure. *Antimicrob. Agents Chemother.* **39:**1211–1233.

6. Chambers, H. F. 1997. Methicillin resistance in staphylococci: molecular and biochemical basis and clinical implications. *Clin. Microbiol. Rev.* **10:**781–791.

7. Chambers, H. F., and M. Sachdeva. 1990. Binding of β-lactam antibiotics to penicillin-binding proteins in methicillin-resistant *Staphylococcus aureus. J. Infect. Dis.* **161:**1170–1176.

8. Cohen, S., C. J. Gibson, and H. M. Sweeney. 1972. Phenotypic suppression of methicillin resistance in *Staphylococcus aureus* by mutant noninducible penicillinase plasmids. *J. Bacteriol.* **112:**682–689.

9. Cohen, S., and H. M. Sweeney. 1968. Constitutive penicillinase formation in *Staphylococcus aureus* owing to a mutation unlinked to the penicillinase plasmid. *J. Bacteriol.* **95:**1368–1374.

10. Dagan, R., M. Ferne, M. Sheinis, M. Alkan, and E. Katzenelson. 1987. An epidemic of penicillin-tolerant Group A streptococcal pharyngitis in children living in a closed community: mass treatment with erythromycin. *J. Infect. Dis.* **156:**514–516.

11. Di Guilmi, A. M., N. Mouz, J. P. Andrieu, J. Hoskins, S. R. Jaskunas, J. Gagnon, O. Dideberg, and T. Vernet. 1998. Identification, purification, and characterization of transpeptidase and glycosyltransferase domains of *Streptococcus pneumoniae* penicillin-binding protein 1a. *J. Bacteriol.* **180:**5652–5659.

12. Dowson, C. G., T. J. Coffey, C. Kell, and R. A. Whitley. 1993. Evolution of penicillin resistance in *Streptococcus pneumoniae:* the role of *Streptococcus mitis* in the formation of a low affinity PBP 2B in *S. pneumoniae. Mol. Microbiol.* **9:**635–643.

13. Dowson, C. G., A. Hutchinson, N. Woodford, A. P. Johnson, R. C. George, and B. G. Spratt. 1990. Penicillin-resistant viridans streptococci have obtained altered penicillin-binding protein genes from penicillin-resistant strains of *Streptococcus pneumoniae. Proc. Natl. Acad. Sci. USA* **87:**5858–5862.

14. Dyke, K., and P. Gregory. 1997. Resistance to β-lactam antibiotics: resistance mediated by β-lactamases, p. 139–157. *In* K. B. Crossley and G. L. Archer (ed.), *The Staphylococci in Human Disease.* Churchill Livingstone, Ltd., New York, N.Y.

15. Dyke, K. G. H. 1979. β-Lactamases of *Staphylococcus aureus,* p. 291–310. *In* J. M. T. Hamilton-Miller and J. T. Smith (ed.), *Beta-Lactamases.* Academic Press, Inc., New York, N.Y.

16. East, A. K., S. P. Curnock, and K. G. H. Dyke. 1990. Change of a single amino acid in the leader peptide of a staphylococcal β-lactamase prevents the appearance of the enzyme in the medium. *FEMS Microbiol. Lett.* **69:** 249–254.

17. El Kharroubi, A., P. Jacques, G. Piras, J. Van Beeumen, J. Coyette, and J.-M. Ghuysen. 1991. The *Enterococcus hirae* R40 penicillin-binding protein 5 and the methicillin-resistant *Staphylococcus aureus* penicillin-binding protein 2' are similar. *Biochem. J.* **280:**463–469.

18. Ellerbrok, H., and R. Hakenbeck. 1988. Penicillin-degrading activities of peptides from pneumococcal penicillin-binding proteins. *Eur. J. Biochem.* **171:**219–224.

19. Fierer, J., P. Wolf, L. Seed, T. Gay, K. Noonan, and P. Haghighi. 1987. Non-pulmonary *Rhodococcus equi* infections in patients with acquired immune deficiency syndrome (AIDS). *J. Clin. Pathol.* **40:**556–558.

20. Fontana, R., M. Aldegheri, M. Ligozzi, H. Lopez, A. Sucari, and G. Satta. 1994. Overproduction of a low-affinity penicillin-binding protein and high-level ampicillin resistance in *Enterococcus faecium. Antimicrob. Agents Chemother.* **38:**1980–1983.

21. Frère, J-M. 1995. Beta-lactamases and bacterial resistance to antibiotics. *Mol. Microbiol.* **16:**385–395.

22. Georgopapadakou, N. H. 1993. Penicillin-binding proteins and bacterial resistance to β-lactams. *Antimicrob. Agents Chemother.* **37:**2045–2053.

23. Georgopapadakou, N. H., and F. Y. Liu. 1980. Binding of β-lactam antibiotics to penicillin-binding proteins of *Staphylococcus aureus* and *Streptococcus faecalis:* relation to antibacterial activity. *Antimicrob. Agents Chemother.* **18:** 834–836.

24. Ghuysen, J.-M. 1991. Serine β-lactamases and penicillin-binding proteins. *Annu. Rev. Microbiol.* **45:**37–67.

24a. Gregory, P. D., R. A. Lewis, S. P. Curnock, and K. G. H. Dyke. 1997. Studies of the repressor (BlaI) of β-lactamase synthesis in *Staphylococcus aureus. Mol. Microbiol.* **24:**1025–1037.

25. Gutmann, L., and A. Tomasz. 1982. Penicillin-resistant and penicillin-tolerant mutants of Group A streptococci. *Antimicrob. Agents Chemother.* **22:**128–136.

26. Hakenbeck, R., and J. Coyette. 1998. Resistant penicillin-binding proteins. *Cell. Mol. Life Sci.* **54:**332–340.

27. Hakenbeck, R., A. König, I. Kern, M. van der Linden, W. Keck, D. Billot-Klein, R. LeGrand, B. Schoot, and L. Gutmann. 1998. Acquisition of five high-M_r penicillin-binding protein variants during transfer of high-level β-lactam resistance from *Streptococcus mitis* to *Streptococcus pneumoniae. J. Bacteriol.* **180:**1831–1840.

28. Handwerger, S., and A. Tomasz. 1985. Antibiotic tolerance among clinical isolates of bacteria. *Rev. Infect. Dis.* **7:**368–386.

29. Hodges, T. L., S. Zighelboim-Daum, G. M. Eliopoulos, C. Wennersten, and R. C. Moellering, Jr. 1992. Antimicrobial susceptibility changes in *Enterococcus faecalis* following various penicillin exposure regimens. *Antimicrob. Agents Chemother.* **36:**121–125.

30. Joris, B., P. Ledent, O. Dideberg, E. Fonzé, J. Jamotte-Brasseur, J. A. Kelly, J. M. Ghuysen, and J. M. Frère. 1991. Comparison of the sequences of class A β-lactamases and of the secondary structure elements of penicillin-recognizing proteins. *Antimicrob. Agents Chemother.* **35:**2294–2301.

31. Kernodle, D. S. Unpublished data.

32. Kernodle, D. S., D. C. Classen, C. W. Stratton, and A. B. Kaiser. 1998. Association of borderline-susceptible strains of *Staphylococcus aureus* with surgical wound infections. *J. Clin. Microbiol.* **36:**219–222.

33. Kernodle, D. S., C. W. Stratton, L. W. McMurray, J. R. Chipley, and P. A. McGraw. 1989. Differentiation of beta-lactamase variants of *Staphylococcus aureus* by substrate hydrolysis profiles. *J. Infect. Dis.* **159:**103–108.

34. Kernodle, D. S., R. K. R. Voladri, and A. B. Kaiser. 1998. β-Lactamase production diminishes the prophylactic efficacy of ampicillin and cefazolin in a guinea pig model of *Staphylococcus aureus* wound infection. *J. Infect. Dis.* **177:**701–706.

35. Komatsuzawa, H., M. Sugai, K. Ohta, T. Fujiwara, S. Nakashima, J. Suzuki, C. Y. Lee, and H. Suginaka. 1997. Cloning and characterization of the *fmt* gene which affects the methicillin resistance level and autolysis in the presence of Triton X-100 in methicillin-resistant *Staphylococcus aureus. Antimicrob. Agents Chemother.* **41:**2355–2361.

36. Kucers, A. 1997. Penicillin G, p. 3–70. *In* A. Kucers, S. M. Crowe, M. L. Grayson, and J. F. Hoy (ed.), *The Use of Antibiotics: A Clinical Review of Antibacterial, Antifungal, and Antiviral Drugs,* 5th ed. Butterworth-Heinemann, Oxford, U.K.

37. Lalitha, M. K., and M. K. Thomas. 1997. Penicillin resistance in *Bacillus anthracis. Lancet* **349:**1522.

38. Leclercq, R. 1997. Antibiotic resistance in streptococci and enterococci. Where are we, where are we going? An opening lecture. *Adv. Exp. Med. Biol.* **418:**419–427.

39. Ligozzi, M., M. Aldegheri, S. C. Predari, and R. Fontana. 1991. Detection of penicillin-binding proteins immunologically related to penicillin-binding protein 5 of *Enterococcus hirae* ATCC 9790 in *Enterococcus faecium* and *Enterococcus faecalis. FEMS Microbiol. Lett.* **67:**335–339.

40. Ligozzi, M., F. Pittaluga, and R. Fontana. 1993. Identification of a genetic element (*psr*) which negatively controls expression of *Enterococcus hirae* penicillin-binding protein 5. *J. Bacteriol.* **175:**2046–2051.

41. Massidda, O., M. P. Montanari, and P. E. Varaldo. 1992. Evidence for a methicillin-hydrolysing β-lactamase in *Staphylococcus aureus* strains with borderline susceptibility to this drug. *FEMS Microbiol. Lett.* **92:**223–227.

42. Massova, I., and S. Mobashery. 1998. Kinship and diversification of bacterial penicillin-binding proteins and β-lactamases. *Antimicrob. Agents Chemother.* **42:**1–17.

43. McDougal, L. K., and C. Thornsberry. 1986. The role of β-lactamase in staphylococcal resistance to penicillinase-resistant penicillins and cephalosporins. *J. Clin. Microbiol.* **23:**832–839.

44. McDowell, T. D., and K. E. Reed. 1989. Mechanism of penicillin killing in the absence of bacterial lysis. *Antimicrob. Agents Chemother.* **33:**1680–1685.

45. Moreillon, P., Z. Markiewicz, S. Nachman, and A. Tomasz. 1990. Two bactericidal targets for penicillin in pneumococci: autolysis-dependent and autolysis-independent killing mechanisms. *Antimicrob. Agents Chemother.* **34:**33–39.

46. Muñoz, R., C. G. Dowson, M. Daniels, T. J. Coffey, C. Martin, R. Hakenbeck, and B. G. Spratt. 1992. Genetics of resistance to third-generation cephalosporins in clinical isolates of *Streptococcus pneumoniae. Mol. Microbiol.* **6:**2461–2465.

47. Murray, B. E. 1992. β-Lactamase-producing enterococci. *Antimicrob. Agents Chemother.* **36:**2355–2359.

48. National Committee for Clinical Laboratory Standards. 1998. *Performance Standards for Antimicrobial Susceptibility Testing; Eighth Informational Supplement.* NCCLS Approved Standard M100-S8. National Committee for Clinical Laboratory Standards, Wayne, Pa.

49. Nielsen, J. B. K., and J. O. Lampen. 1982. Membrane-bound penicillinases in Gram-positive bacteria. *J. Biol. Chem.* **257:**4490–4495.

50. Niemeyer, D. M., M. J. Pucci, J. A. Thanassi, V. K. Sharma, and G. L. Archer. 1996. Role of *mecA* transcriptional regulation in the phenotypic expression of methicillin resistance in *Staphylococcus aureus. J. Bacteriol.* **178:**5464–5471.

51. Nordmann, P., M. H. Nicolas, and L. Gutmann. 1993. Penicillin-binding proteins of *Rhodococcus equi*: potential role in resistance to imipenem. *Antimicrob. Agents Chemother.* **37:**1406–1409.

52. Park, W., and M. Matsuhashi. 1984. *Staphylococcus aureus* and *Micrococcus luteus* peptidoglycan transglycosylases that are not penicillin-binding proteins. *J. Bacteriol.* **157:**538–544.

53. Qoronfleh, M. W., and B. J. Wilkinson. 1986. Effect of growth of methicillin-resistant and -susceptible *Staphylococcus aureus* in the presence of β-lactams on peptidoglycan structure and susceptibility to lytic enzymes. *Antimicrob. Agents Chemother.* **29:**250–257.

54. Reichmann, P., A. König, J. Liñares, F. Alcaide, F. C. Tenover, L. McDougal, S. Swidsinski, and R. Haken-

beck. 1997. A global gene pool for high-level cephalosporin resistance in commensal *Streptococcus* species and *Streptococcus pneumoniae. J. Infect. Dis.* **176**:1001–1012.

55. **Rybkine, T., J. L. Mainardi, W. Sougakoff, E. Collatz, and L. Gutmann.** 1998. Penicillin-binding protein 5 sequence alterations in clinical isolates of *Enterococcus faecium* with different levels of beta-lactam resistance. *J. Infect. Dis.* **178**:159–163.

56. **Sabath, L. D., L. Garner, C. Wilcox, and M. Finland.** 1975. Effect of inoculum and of beta-lactamase on the anti-staphylococcal activity of thirteen penicillins and cephalosporins. *Antimicrob. Agents Chemother.* **8**:344–349.

57. **Severin, A., and A. Tomasz.** 1996. Naturally occurring peptidoglycan variants of *Streptococcus pneumoniae. J. Bacteriol.* **178**:168–174.

58. **Sherris, J. C.** 1986. Problems in in vitro determination of antibiotic tolerance in clinical isolates. *Antimicrob. Agents Chemother.* **30**:633–637.

59. **Sibold, C., J. Henrichsen, A. König, C. Martin, L. Chalkley, and R. Hakenbeck.** 1994. Mosaic *pbp*X genes of major clones of penicillin-resistant *Streptococcus pneumoniae* have evolved from *pbp*X genes of a penicillin-sensitive *Streptococcus oralis. Mol. Microbiol.* **12**:1013–1023.

60. **Sieradzki, K., and A. Tomasz.** 1997. Suppression of beta-lactam antibiotic resistance in a methicillin-resistant *Staphylococcus aureus* through synergic action of early cell wall inhibitors and some other antibiotics. *J. Antimicrob. Chemother.* **39**(Suppl. A):47–51.

61. **Signoretto, C., M. Boaretti, and P. Canepari.** 1998. Peptidoglycan synthesis by *Enterococcus faecalis* penicillin binding protein 5. *Arch. Microbiol.* **170**:185–190.

62. **Smith, A. M., and K. P. Klugman.** 1998. Alterations in PBP 1a essential for high-level penicillin resistance in *Streptococcus pneumoniae. Antimicrob. Agents Chemother.* **42**:1329–1333.

63. **Song, M. D., M. Wachi, M. Doi, F. Ishino, and M. Matsuhashi.** 1987. Evolution of an inducible penicillin-target protein in methicillin-resistant *Staphylococcus aureus* by gene fusion. *FEBS Lett.* **221**:167–171.

64. **Suzuki, E., K. Hiramatsu, and T. Yokota.** 1992. Survey of methicillin-resistant clinical strains of coagulase-negative staphylococci for *mecA* gene distribution. *Antimicrob. Agents Chemother.* **36**:429–434.

65. **Thore, M.** 1992. β-Lactamase substrate profiles of coagulase-negative skin staphylococci from orthopaedic inpatients and staff members. *J. Hosp. Infect.* **22**:229–240.

66. **Tomasz, A., H. B. Drugeon, H. M. de Lancastre, D. Jabes, L. McDougall, and J. Bille.** 1989. New mechanism for methicillin resistance in *Staphylococcus aureus*: clinical isolates that lack the PBP 2a gene and contain normal penicillin-binding proteins with modified penicillin-binding capacity. *Antimicrob. Agents Chemother.* **33**:1869–1874.

67. **Tomayko, J. F., K. K. Zscheck, K. V. Singh, and B. E. Murray.** 1996. Comparison of the β-lactamase gene cluster in clonally distinct strains of *Enterococcus faecalis. Antimicrob. Agents Chemother.* **40**:1170–1174.

68. **Toumanen, E., D. T. Durack, and A. Tomasz.** 1986. Antibiotic tolerance among clinical isolates of bacteria. *Antimicrob. Agents Chemother.* **30**:521–527.

69. **Voladri, R. K. R., and D. S. Kernodle.** 1998. Characterization of a chromosomal gene encoding type B β-lactamase of *Staphylococcus aureus. Antimicrob. Agents Chemother.* **42**:3163–3168.

70. **Voladri, R. K. R., M. K. R. Tummuru, and D. S. Kernodle.** 1996. Structure-function relationships among wild-type variants of *Staphylococcus aureus* β-lactamase: importance of amino acids 128 and 216. *J. Bacteriol.* **178**:7248–7253.

71. **Wallace, R. J., Jr., P. Vance, A. Weissfeld, and R. R. Martin.** 1978. β-Lactamase production and resistance to β-lactam antibiotics in *Nocardia. Antimicrob. Agents Chemother.* **14**:704–709.

72. **Zygmunt, D. J., C. W. Stratton, and D. S. Kernodle.** 1992. Characterization of four β-lactamases produced by *Staphylococcus aureus. Antimicrob. Agents Chemother.* **36**:440–445.

Resistance to Glycopeptides in Gram-Positive Pathogens

HENRY S. FRAIMOW AND PATRICE COURVALIN

63

For over 30 years, glycopeptide antibiotics have been main-stays of therapy for infections due to gram-positive pathogens. Initially introduced into clinical practice in the 1960s, usage of the glycopeptide vancomycin skyrocketed during the 1980s as one of the few antimicrobial agents effective against methicillin-resistant staphylococci. Intrinsic and at the time poorly understood resistance to vancomycin was observed in several generally nonpathogenic gram-positive bacterial genera such as *Leuconostoc*, *Pediococcus*, and *Lactobacillus*. However, the emergence of acquired glycopeptide resistance in clinically significant pathogens was not anticipated, and some suggested that such resistance would not occur. Glycopeptide-resistant enterococci were first isolated in Europe in 1986 (26, 44) and shortly thereafter in the United States, and over the past decade they have become a worldwide problem (30). Glycopeptide resistance in staphylococci was initially observed in coagulase-negative staphylococci, especially in *Staphylococcus haemolyticus* and less commonly *Staphylococcus epidermidis*, but during the past 2 years the medical community has been shaken by reports of clinical isolates of *Staphylococcus aureus* with reduced susceptibility to vancomycin (11, 24). Since glycopeptides have become the antibiotics of last resort for gram-positive infections, especially those due to methicillin-resistant *S. aureus* (MRSA), the emergence of glycopeptide resistance in this virulent pathogen is potentially catastrophic. Coincident with the emergence of glycopeptide resistance, there has been a tremendous explosion of information on the epidemiology, mechanisms, and genetics of glycopeptide resistance in both pathogenic and nonpathogenic gram-positive bacteria.

STRUCTURE AND MECHANISM OF ACTION OF GLYCOPEPTIDES

Glycopeptide antimicrobial agents are natural products that are produced by various species of soil-dwelling actinomycetes, such as *Amycolatopsis orientalis* and *Amycolatopsis coloradensis* (34, 47). Thus far over 200 different glycopeptides have been isolated. Two of these, vancomy-

cin and teicoplanin, have been successfully developed and approved for use in humans. Vancomycin is widely used throughout the United States, Europe, and other parts of the world, although it was only recently approved for use in Japan. Teicoplanin is available in Europe but not in the United States. Another agent, avoparcin, has been widely used in farming and in veterinary medicine in some parts of Europe. Several other glycopeptides, such as ristocetin, have been extensively investigated in vitro, but use in humans has been limited by toxicity. With the recent emergence of glycopeptide-resistant enterococci and staphylococci, there has been renewed interest by the pharmaceutical industry in developing modified semisynthetic glycopeptides with enhanced activity against resistant organisms.

Glycopeptide antibiotics are inhibitors of peptidoglycan synthesis. Unlike other cell wall synthesis inhibitors such as beta-lactams, glycopeptides do not bind directly to cell wall biosynthetic enzymes but instead interfere with peptidoglycan synthesis by complexing with carboxy-terminal alanine residues on peptidoglycan precursors (34). All glycopeptides are large, complex heptapeptide structures with molecular weights of 1,200 to 1,500 and consist of an aglycone core of six peptide residues linked to a triphenyl ether, with a variety of complex sugar groups attached to the heptapeptide core (47). Because of their large, rigid, hydrophobic structure, glycopeptides do not penetrate bacterial cellular membranes. They can only interact with gram-positive cell wall precursors after translocation of the latter on the undecaprenol lipid carrier onto the outer surface of the cytoplasmic membrane. The terminal D-alanyl–D-alanine (D-Ala–D-Ala) moiety of the UDP-*N*-acetyl muramyl pentapeptide precursors binds tightly within the groove in the aglycone portion of the glycopeptide molecule and blocks progression to the subsequent transglycosylation steps in peptidoglycan synthesis (Fig. 1 and 2). Complexing of glycopeptide to the D-Ala–D-Ala residues also interferes with the subsequent reactions catalyzed by D,D-transpeptidases and D,D-carboxypeptidases necessary for the anchoring of the peptidoglycan complex. Blockage of transglycosylation also leads to trapping of lipid

Gram-Positive Pathogens, ed. by V. A. Fischetti et al.
© 2000 American Society for Microbiology, Washington, D.C.

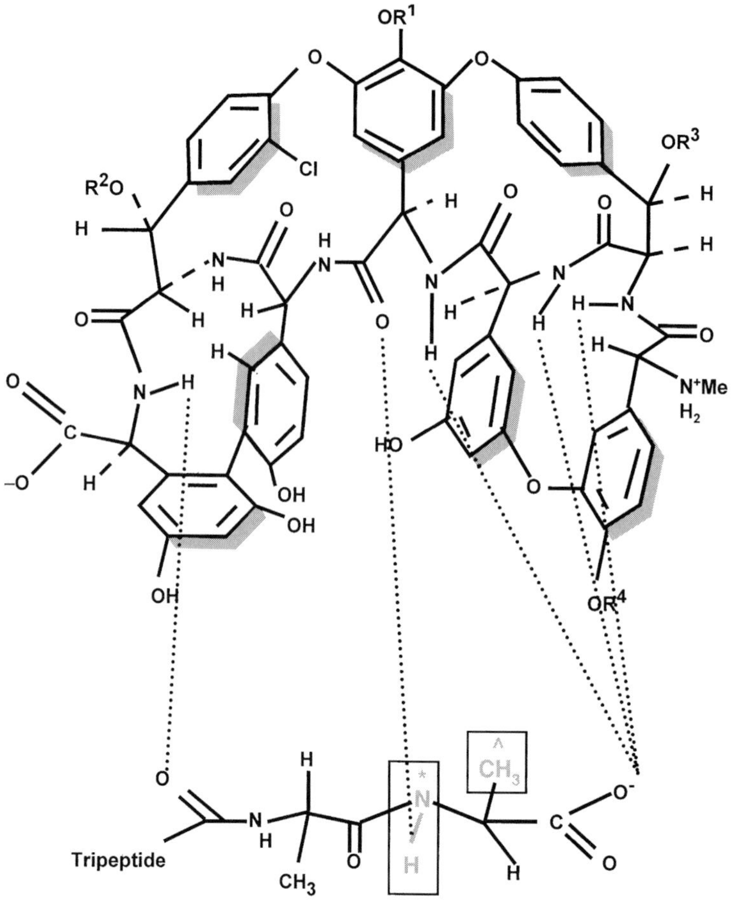

FIGURE 1 Structure of the glycopeptide:peptidyl-D-Ala–D-Ala complex. Binding of the glycopeptide to the D-Ala–D-Ala residue on peptidoglycan precursors involves five hydrogen bond interactions (indicated by the dashed lines). In D-Ala–D-Lac-terminating precursors, the NH of the amide group (*) is replaced by an oxygen in an ester linkage, eliminating the central hydrogen bond. In D-Ala–D-Ser-terminating precursors, the methyl side chain of the carboxy-terminal D-Ala (∧) is substituted by a hydroxymethyl (CH₂OH) side chain.

carrier and accumulation of peptidoglycan precursors in the cytoplasm. Some glycopeptides, especially teicoplanin, also bind to the cell membrane, but this is probably unimportant for their general mechanism of action. The glycopeptide target structures, pentapeptides terminating in D-Ala–D-Ala, are ubiquitous among most species of bacteria. The spectrum of activity of glycopeptides encompasses nearly all gram-positive cocci, including staphylococci, streptococci, and enterococci, and gram-positive rods, including *Bacillus* species and most *Corynebacterium* species and other diphtheroids. Glycopeptides also have activity against many gram-positive anaerobes, including *Clostridium difficile*.

OVERVIEW OF POTENTIAL MECHANISMS OF RESISTANCE TO GLYCOPEPTIDES

Because of their unique mechanism of action and ability to interfere with multiple critical reactions in peptidoglycan synthesis, acquired resistance to glycopeptides was deemed unlikely. Resistance to glycopeptides could theoretically occur by a variety of mechanisms, including deg-

radation of the antibiotic, exclusion of the antibiotic from the target site, or modification or elimination of the D-Ala–D-Ala target. Thus far, no naturally occurring glycopeptide-modifying or -degrading enzymes have been characterized from resistant organisms or from glycopeptide-producing strains. Gram-negative bacteria are generally resistant to glycopeptides owing to the presence of the outer membrane, which effectively prevents drug access to the developing cell wall. Certain outer-membrane-deficient *Escherichia coli* mutants have increased susceptibility to vancomycin.

Several gram-positive bacterial species are intrinsically highly resistant to vancomycin, including all species of *Leuconostoc*, *Pediococcus*, and *Erysipelothrix* and most species of *Lactobacillus*. Intrinsic resistance in these species is due to the presence of alternative cell wall synthesis intermediates, most commonly a depsipeptide that terminates in D-alanyl–D-lactate (D-Ala–D-Lac) rather than the D-Ala–D-Ala pentapeptide (7). The conformational change created by the ester rather than the amide bond in the depsipeptide significantly alters its interaction within the groove in the aglycone structure of the glycopeptide and

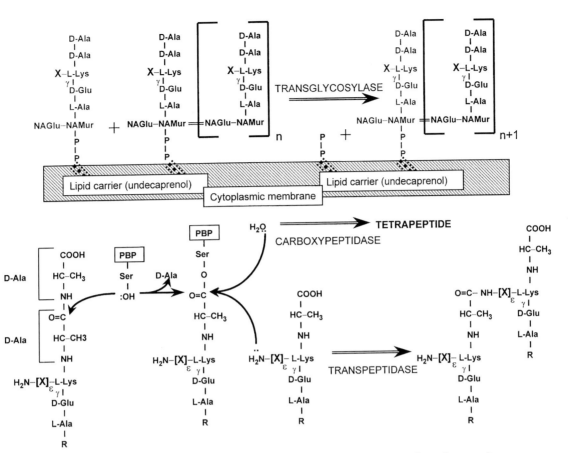

FIGURE 2 Transglycosylation and subsequent reactions in the cell wall synthesis pathway catalyzed by penicillin-binding proteins (PBPs). Binding of glycopeptide to the terminal D-Ala–D-Ala stem inhibits both the transglycosylation reaction and the subsequent carboxypeptidase and transpeptidase steps. In the initial step of the carboxypeptidase or transpeptidase reactions, the terminal D-Ala is cleaved during the initial binding of the active serine site of the PBPs to form the acyl-enzyme complex. The same acyl-enzyme intermediate would be produced if the terminal D-Ala residue were replaced by either D-Lac or D-Ser. Adapted from reference 4.

eliminates one hydrogen bond, resulting in binding affinity >1,000-fold less than that of the pentapeptide terminating in D-Ala–D-Ala (8, 46). Intrinsic low-level resistance can also be seen with alternative pentapeptides terminating in D-alanyl–D-serine (D-Ala–D-Ser), which are found in the uncommon motile enterococcal species *Enterococcus gallinarum*, *Enterococcus casseliflavus*, and *Enterococcus flavescens*. The terminal residue of the pentapeptide or depsipeptide is cleaved during subsequent cross-linking reactions; thus, there should be little consequence of employing this altered precursor structure on the ultimate composition and stability of the mature peptidoglycan, provided that the modified precursors can be processed by enzymes in the chromosomal cell wall synthesis pathways. Some intrinsically glycopeptide-resistant species possess multiple D,D ligases, allowing for the synthesis of both D-Ala–D-Ala and alternative precursors. Thus, although resistance may be intrinsic to these species, it is not necessarily constitutively expressed. The rationale for maintaining diverse cell wall synthesis pathways under natural conditions remains unclear.

Acquired resistance to glycopeptides results in development of resistance in strains of a genus previously considered to be uniformly susceptible; strains can acquire one or multiple different resistance mechanisms. Clinically important acquired resistance occurs predominantly in enterococci and staphylococci, although the resistance elements found in *Enterococcus faecalis* and *Enterococcus faecium* have also been found in other organisms. The mechanisms of acquired resistance to glycopeptides in enterococci have been well characterized and reflect changes in target analogous to those in intrinsically resistant organisms (4, 46). Acquired glycopeptide resistance in enterococci can result in decreased susceptibility to vancomycin alone or cross-resistance to both vancomycin and teicoplanin; resistant strains are generally referred to as glycopeptide-resistant enterococci (GRE). GRE possess one of several inducible, transferable gene clusters that not only are responsible for synthesis of alternative cell wall precursors terminating in D-Ala–D-Lac or D-Ala–D-Ser but simultaneously eliminate precursors terminating in D-Ala–D-Ala. The specifics of these enterococcal resistance elements are described in detail below. Glycopeptide resistance in staphylococci, although only incompletely understood, occurs by different mechanisms (22, 40). Resistance in staphylococci is selectable in individual strains in vitro without introduction of new genetic elements. Resistance does not appear to be mediated by alteration in the D-Ala–D-Ala glycopeptide-binding site but instead reflects more general alterations in cell wall composition that limit access of the glycopeptide to nascent cell wall precursors.

GLYCOPEPTIDE RESISTANCE IN ENTEROCOCCI: HISTORICAL AND EPIDEMIOLOGICAL FEATURES

The first well-characterized isolates of *E. faecalis* and *E. faecium* resistant to vancomycin were described in Europe and in the United States in the mid-1980s (26, 38, 43). Subsequently, a large outbreak was reported from a renal failure unit at the Dulwich Infirmary in England in 1986, focusing much broader attention on this problem (44). This prolonged outbreak consisted of numerous patients colonized or infected with isolates of *E. faecalis* and *E. faecium* with vancomycin MICs of >64 μg/ml. GRE have subsequently emerged during the late 1980s and 1990s as major nosocomial pathogens throughout Europe and the United States, although with significant variation in the rates of isolation and pattern of spread in different geographic regions. Data from the Centers for Disease Control and Prevention's National Nosocomial Infection Surveillance Project chart the emergence of GRE in the United States (Fig. 3) (30). Vancomycin resistance in U.S. hospitals steadily increased from 0.3% of nosocomial isolates in 1989 to 23.2% of intensive care unit and 15.4% of nonintensive care unit isolates in 1997. GRE have also been reported from countries throughout Europe, including Great Britain, France, Spain, Germany, Belgium, and Scandinavia. In European countries, the prevalence rates in hospitalized patients vary widely but in general are lower than those found in the United States (19). GRE have also been reported recently from Australia and Japan.

Numerous studies have specifically addressed the epidemiology of GRE in the hospital setting (30). The major reservoir for enterococci in humans is the lower gastrointestinal tract. Factors associated with the spread of GRE are those permitting the establishment of enterococci in the gastrointestinal tract of hospitalized patients as well as those increasing the "enterococcal burden" and facilitating dissemination of resistant organisms through person-to-person or environmental contact. The predominant risk factor for acquisition of GRE is exposure to antimicrobial agents. Enterococci are intrinsically resistant to antimicrobials routinely used in hospitalized patients, including cephalosporins, clindamycin, and other antianaerobic agents. Before the emergence of GRE, cephalosporin exposure was demonstrated to be a risk factor for increased levels of gastrointestinal colonization by enterococci and was implicated in the overall increasing incidence of enterococcal infections in hospitalized patients. In various studies, exposure to both parenteral and oral vancomycin, cephalosporins, metronidazole, and other antimicrobial agents is linked to acquisition of GRE. In addition, the overall "intensity" of antimicrobial exposure is important. Other risk factors include general debility, longer duration of hospital stay, and exposure to particular hospital units such as intensive care units, hematology-oncology units, liver transplant units, and renal failure units. These factors all reflect increased risk of antibiotic exposure as well as increased risk of exposure to GRE from sicker, more actively disseminating patients. Differences in the rates of colonization and infection in Europe and the United States may reflect differences in hospital antimicrobial usage patterns, especially glycopeptide usage.

Once colonized, patients may shed GRE from the gastrointestinal tract for over a year, even in the absence of continued antimicrobial pressure. Persistently colonized patients, especially unknown ones, represent an ongoing reservoir for dissemination of GRE when transferred to other hospitals, intermediate care facilities, or nursing homes. Prevalence of colonization may be up to 10 times higher than infection in the midst of an outbreak. Diarrhea and fecal incontinence markedly increase the risk of skin colonization and potential spread of organisms from persistently colonized patients (30). Enterococci are durable and are able to survive up to a week on inanimate surfaces in the hospital environment; they have been recovered from countertops, rectal thermometer probes, bed rails, telephones, television remote controls, and other surfaces in patient rooms as well as on the hands and stethoscopes of health care personnel.

PERCENTAGE OF ENTEROCOCCAL ISOLATES RESISTANT TO VANCOMYCIN, 1989-1997

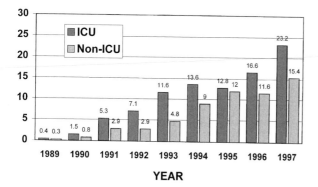

FIGURE 3 Percentage of nosocomial enterococci reported as resistant to vancomycin. Adapted from the National Nosocomial Infection Surveillance System, Hospital Infections Program, National Center for Infectious Diseases, Centers for Disease Control and Prevention (30).

VANCOMYCIN-RESISTANCE PHENOTYPES AND GENOTYPES

With the first appearance of clinical isolates of GRE, it was apparent that strains could be divided by specific characteristics into VanA, VanB, and VanC phenotypes (4). VanA strains include predominantly *E. faecium* and *E. faecalis* but occasionally other enterococcal species such as *Enterococcus avium*, *Enterococcus durans*, *Enterococcus hirae*, and *Enterococcus raffinosus*. They usually demonstrate inducible high-level resistance to both vancomycin and teicoplanin, with vancomycin MICs of ≥256 μg/ml. VanB strains are almost exclusively *E. faecium* and *E. faecalis*, demonstrate a range of vancomycin-resistance levels from as low as 4 μg/ml to >1,000 μg/ml, and are generally susceptible to teicoplanin. VanC strains are exclusively of the low-virulence, motile enterococcal species *E. casseliflavus*, *E. gallinarum*, and *E. flavescens* and are characterized by intrinsic low levels of resistance (MICs of 2 to 32 μg/ml). The genetic and molecular bases for these different patterns of resistance are discussed below. Currently, optimal characterization of strains is based on genotypic methods such as gene amplification by PCR and DNA hybridization for detecting specific resistance genes. Occasional isolates may demonstrate altered phenotypic profiles, such as *vanB*-genotype strains resistant to teicoplanin (21). Genotypic methods have also aided in localization of resistance elements to the host chromosome or to plasmids. Additional genotypes, including *vanD* and *vanE*, have also recently been recognized (12, 32). Genotypic classification of vancomycin-resistant enterococci is summarized in Table 1. As the genetic elements of the various enterococcal vancomycin-resistance clusters and their homologs in other species have been characterized, the designations for the resistance genes and proteins have become increasingly confusing and will require revision into a more unified nomenclature system.

NONHUMAN SOURCES FOR VANCOMYCIN-RESISTANT ENTEROCOCCI

There remains considerable speculation on the initial sources of human GRE isolates. Although GRE were initially isolated at similar times in Europe and the United States, studies show very different patterns of spread in the community setting (19, 30). Despite the rapid dissemination of GRE in U.S. hospitals, fueled by high levels of antimicrobial usage, rates of fecal colonization in healthy non-hospital volunteers remain low. Although both *vanA* and *vanB* genotypes are prevalent in the United States, the overall clonal diversity of isolates in U.S. hospitals is fairly low. In Europe, rates of colonization in healthy volunteers are much higher, and the wide clonal diversity of these community isolates is also reflected in the diversity of hospital isolates, although most hospital and community isolates are of the *vanA* genotype. In Belgium, fecal carriage rates in the community of up to 28% are reported from healthy volunteers given oral glycopeptide challenge, although most European studies demonstrate lower but still

significant colonization rates of 2 to 5% (19). GRE were recovered from sewage in several German cities and in England at times when hospital infection rates were low, and GRE have been recovered from other environmental sources, including animal feces, especially from poultry and pig farms. GRE have also been recovered from chicken carcasses and ground meat, suggesting a clear route of spread into the human food chain. GRE with indistinguishable pulsed-field gel electrophoresis patterns have been isolated from farm animals and humans, further supporting a role for animal reservoirs in human infection (1). Suspicion has focused on the widespread use of the glycopeptide avoparcin as a food additive in European countries as far back as the 1970s, predating the marked increases in vancomycin usage in humans during the 1980s. Rates of GRE may be higher in countries with heavier avoparcin usage, and individual farms where avoparcin was used have shown much higher rates of GRE recovery from animal feces (1). These observations support a model of widespread multiclonal emergence and dissemination of *vanA* genotype enterococcal strains in animals, with subsequent colonization of humans.

LABORATORY DETECTION OF GRE

The unexpected arrival of glycopeptide resistance in enterococci quickly demonstrated the inadequacy of available testing methods for the detection of GRE (42). VanA-type strains generally are readily detected by all commercial testing methods, including disc diffusion, E-test, and automated broth microdilution systems. However, low to intermediately resistant VanB-type strains were not well detected by either broth microdilution or diffusion methods, owing to the low levels of expression of resistance and the need for adequate incubation for inducing expression of resistance. The National Committee for Clinical Laboratory Standards in 1992 revised the criteria for vancomycin testing of enterococci by disc diffusion, changing the requirement for "fully susceptible" to a zone size of ≥17

TABLE 1 Genotypic and phenotypic characterization of glycopeptide-resistant enterococci

Resistance genotype	Predominant phenotype[a]	Mode of expression	Predominant location	Transferable elements	Alternate precursor	Species found in
vanA	Vanco ≥ 256 Teico ≥ 32	Inducible	Plasmid (chromosome)	Tn*1546* and related	D-Ala–D-Lac	E. faecium, E. faecalis, E. hirae, E. avium, E. durans, E. mundtii[b], B. circulans, O. turbata, A. haemolyticum
vanB	Vanco 4 to 1,000 Teico ≤ 1	Inducible	Chromosome (plasmid)	Tn*1547* Tn*5382*	D-Ala–D-Lac	E. faecium, E. faecalis, S. bovis
vanC1 *vanC2* *vanC3*	Vanco 2 to 32 Teico ≤ 1	Constitutive or inducible	Chromosome	?	D-Ala–D-Ser	E. gallinarum (vanC1) E. casseliflavus (vanC2) E. flavescens (vanC3)
vanD	Vanco 64 to 256 Teico 4 to 32	Constitutive or inducible	Chromosome	?	D-Ala–D-Lac	E. faecium
vanE	Vanco = 16 Teico = 0.5	?	Chromosome?	?	D-Ala–D-Ser	E. faecalis

[a] Expressed as MIC to vancomycin (Vanco) or teicoplanin (Teico) as μg/ml.
[b] Also in other enterococcal species including *vanC* genotype strains.

mm on Mueller-Hinton agar. There have also been improvements made in the automated test methods. However, some GRE will still be reported as susceptible by these methods. A commercial plate-based screening method using brain heart infusion agar with vancomycin (6 μg/ml) spotted with a 10^6-CFU inoculum is recommended as an adjunctive method for testing of equivocal or critical isolates. A number of genotypic methods for detection of *vanA*, *vanB*, and *vanC* genes have been described, and several are now in commercial development, including PCR, hybridization probes, and cycled probe reactions. Rapid identification of resistant strains is a critical component of public health recommendations for control of GRE, such as those proposed by the Hospital Infection Control Practices Advisory Committee (10).

OVERVIEW OF THE GENETIC AND MOLECULAR BASIS OF ENTEROCOCCAL GLYCOPEPTIDE RESISTANCE

The basis of all modes of glycopeptide resistance in enterococci involves alteration of composition of the terminal dipeptide in muramyl pentapeptide cell wall precursors to a structure with decreased binding affinity to glycopeptides (7) (see Fig. 4). This can be achieved by replacing the terminal D-Ala with either D-Lac or D-Ser. Replacement of the D-Ala–D-Ala pentapeptide terminus with a D-Ala–D-Lac ester linkage changes the conformation of the ligand, eliminates the central hydrogen bond critical to glycopeptide binding, and results in over 1,000-fold lower binding affinity. Replacement with D-Ala–D-Ser reduces binding affinity sevenfold (6). The terminal D-Lac or D-Ser residue would be cleaved during subsequent cross-linking reactions and should not alter the final composition of mature peptidoglycan. The bacterial enzyme that ligates D-Ala–D-Ala to tripeptide to form pentapeptide precursors readily incorporates D-Ala–D-X structures in addition to D-Ala–D-Ala. However, D-Lac-terminating depsipeptide precursors may be processed differently from pentapeptide precursors by penicillin-binding proteins (PBPs) during subsequent transpeptidation reactions (20).

The central enzyme required for glycopeptide resistance in enterococci is thus a cellular ligase of altered specificity, preferentially synthesizing D-Ala–D-Lac (VanA or VanB) or D-Ala–D-Ser (VanC) rather than D-Ala–D-Ala. This pathway also requires a mechanism for synthesis of the D-X (D-Lac or D-Ser) residue; for D-Lac this is a dehydrogenase (VanH) that generates D-Lac from pyruvate. Synthesis of exclusively D-Lac-terminating depsipeptide should yield a high-level glycopeptide resistance phenotype analogous to that of the intrinsically highly glycopeptide-resistant organisms. However, the enterococcal-resistance pathway also requires a mechanism for elimination of precursors terminating in D-Ala–D-Ala, since coproduction of both D-Ala- and D-Lac-terminating precursors does not yield a resistance phenotype (2). If sufficient D-Ala-terminating precursors are present, glycopeptides can still complex to these precursors bound to the lipid carrier, sequestering the lipid carrier and blocking transport of additional precursors across the cellular membrane. A resistance phenotype could theoretically occur by downregulation of the constitutively expressed D-Ala–D-Ala ligase, but such a mechanism has not been demonstrated. The other essential enzyme for expression of resistance is a D,D-dipeptidase (VanX) that hydrolyzes D-Ala–D-Ala,

creating a "futile cycle" of synthesis and subsequent hydrolysis of D-Ala–D-Ala dipeptide and markedly depleting D-Ala-terminating precursors (4, 36). Mature pentapeptide target for vancomycin can also be eliminated by D,D-carboxypeptidases (VanY) that cleave the terminal alanine from pentapeptide. The requirements for a dehydrogenase or related activity, ligase activity, and a D,D-dipeptidase are universal features of acquired enterococcal glycopeptide-resistance clusters as well as glycopeptide-resistance clusters of glycopeptide-producing organisms (4, 29).

THE *vanA* RESISTANCE CLUSTER

The first enterococcal resistance cluster to be characterized in detail was the *vanA* operon, conveying high-level resistance to both vancomycin and teicoplanin. The *vanA* cluster was initially found as part of Tn*1546*, a nonconjugative transposon of the Tn*3* superfamily of transposons (3). In most strains that have been studied in detail, the *vanA* gene cluster is incorporated in Tn*1546* or other highly conserved related elements and is usually found on self-transferable conjugative plasmids, accounting for the apparent ease of transfer of *vanA* resistance. In one instance, the *vanA* element was found on a plasmid in association with an enterococcal pheromone response gene, further facilitating horizontal dissemination. The *vanA*-containing element or portions of the gene cluster can also be found on the bacterial chromosome and can be associated with other complex transposable elements. The structure of Tn*1546* (Fig. 5) is composed of nine open reading frames (ORFs) flanked by 38-bp imperfect inverted repeats. ORF1 and ORF2 encode a transposase and a resolvase required for mobilization of the transposon. The remaining seven ORFs encode proteins specifically related to expression of glycopeptide resistance. The structure and function of these genes, their protein products, and the resulting cell wall precursor structures have been elegantly studied in vitro in functional assays as well as in laboratory enterococcal constructs carrying partial complements of the *vanA* gene cluster in *cis* or in *trans*. The *vanHAX* cluster comprises a single transcriptional element with a single promoter and includes the three genes essential for expression of glycopeptide resistance. The VanH protein is a dehydrogenase catalyzing production of D-Lac from pyruvate and is structurally similar to other cellular dehydrogenases (8). The product of *vanA* is a D,D ligase that preferentially synthesizes D-Ala–D-Lac, although small amounts of D-Ala–D-Ala can also be produced. The VanA ligase shares approximately 30 to 35% amino acid identity with other bacterial D-Ala–D-Ala (Ddl) ligases as well as with the D-Ala–D-Lac ligases of the intrinsically resistant *Leuconostoc*, *Pediococcus*, and *Lactobacillus* spp., but *vanA* is phylogenetically distinct from both of these groups (15). The *vanX* gene encodes a metallo-dipeptidase that preferentially hydrolyzes D-Ala–D-Ala.

Transcriptional activation of *vanHAX* is regulated by the VanRS two-component regulatory system. The *vanR* and *vanS* genes encode the signal sensor molecule VanS and the response regulator VanR, respectively (2). VanS responds to the presence of subinhibitory concentrations of glycopeptides in the cellular medium by autophosphorylation followed by phosphorylation of VanR, resulting in transcription of the glycopeptide-resistance genes. The N-terminal portion of the VanS protein consists of a membrane-associated domain that appears to function as

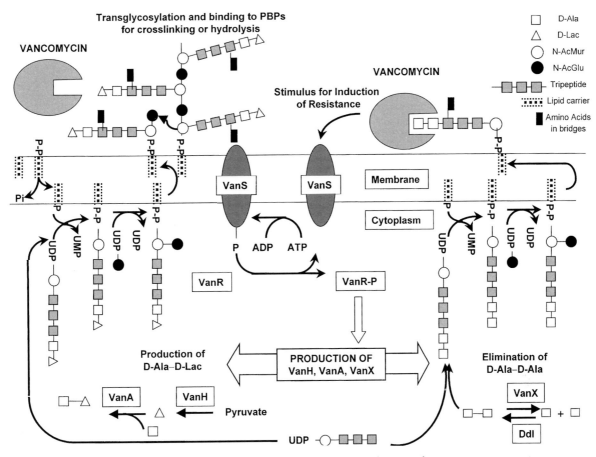

FIGURE 4 General scheme of peptidoglycan synthesis in glycopeptide-resistant enterococci. The enterococcal Ddl ligases produce D-Ala–D-Ala residues that are incorporated into peptidoglycan precursors. D-Ala–D-Ala-terminating precursors interact with glycopeptides, preventing transglycosylation and leading to the accumulation of cytoplasmic precursors. The VanRS signal transducing system responds directly or indirectly (via accumulation of cytoplasmic precursors) to the presence of glycopeptides, leading to induction of transcription of the *vanHAX* operon in the cytoplasm. Induction of these genes results in production of D-Lac–containing precursors as well as elimination of the pool of D-Ala–D-Ala precursors. Adapted from reference 4.

the sensor and demonstrates low homology to other known proteins. The C-terminal portion of the *vanS* gene includes multiple motifs that are highly conserved among other signal sensors of two-component regulatory systems and are involved in the autophosphorylation of a conserved histidine residue and transfer of this phosphate to the response regulator VanR following exposure to appropriate stimuli. In the phosphorylated state, VanR functions as a transcriptional activator of *vanHAX*. The specific molecular signal(s) for induction of the resistance cluster is unknown.

Other transglycosylation inhibitors with sites of action distinct from glycopeptides, including moenomycin, also cause induction of *vanA*, suggesting that the stimulus for VanS may be inhibition of transglycosylation and accumulation of cell wall precursors rather than the glycopeptide itself. Other structurally and mechanistically unrelated compounds may also induce expression of the *vanA* system. When inserted into *E. coli* strains, crosstalk can occur between the *vanRS* system and other two-component systems like *phoA*, suggesting that similar crosstalk may occur be-

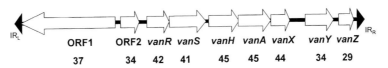

FIGURE 5 Structure of Tn*1546* (10,851 bp) carrying the *vanA* vancomycin-resistance gene cluster. The nine open reading frames are delineated by 38-bp imperfect inverted repeats and include genes with resolvase and transposase activity (ORF1 and ORF2) as well as genes involved in regulation (*vanR* and *vanS*), synthesis of D-Ala–D-Lac (*vanH* and *vanA*), and hydrolysis of D-Ala–D-Ala precursors (*vanX* and *vanY*). The percent G+C content of each gene is also shown.

tween *vanRS* and other enterococcal response regulators in vivo (41).

The remainder of the *vanA* complex includes two additional genes, *vanY* and *vanZ*. Unlike the *vanHAX* cluster, neither gene is obligatory for the expression of glycopeptide resistance. The VanY protein is a D,D-carboxypeptidase that cleaves terminal D-Ala from pentapeptide residues and can increase the level of glycopeptide resistance by further eliminating binding targets. The function of *vanZ* is poorly understood. VanZ has no known homologs but appears to be important in mediating increased resistance to teicoplanin.

The G+C content of the glycopeptide-resistance genes *vanRS* and *vanHAX* ranges from 41 to 45% and differs from those of the transposase and resolvase genes (37 and 34%) and the enterococcal chromosome (35%) (3). This suggests that these genes were initially acquired from one or multiple nonenterococcal sources and subsequently integrated into transposable elements before their acquisition by enterococci.

TRANSMISSION OF THE *vanA* CLUSTER

One important feature of the VanA resistance system is its ease of horizontal transfer to enterococci and other species. Conjugative transfer of VanA-type plasmids between enterococcal strains of numerous species occurs readily in the laboratory, and there are many examples of clinical outbreaks involving multiple strains and species of enterococci harboring the same VanA-type plasmids. This may also explain the clonal heterogeneity of VanA-type strains found in environmental isolates and population-based surveys of healthy outpatients. In early experiments using *vanA* strain BM4165, vancomycin resistance was transferable in vitro to *Streptococcus lactis*, *Streptococcus sanguis*, *Streptococcus pyogenes*, and *Listeria monocytogenes*, with varying degrees of expression of resistance in the recipients (4, 26). Successful transfer of the *vanA* operon to *S. aureus* has been achieved in vitro and in vivo using an indirect selection method, with expression of high-level resistance in the recipient strain and no apparent deleterious effects on growth; further attempts to introduce the *vanA* gene cluster into *S. aureus* have been discouraged (31). Despite successful laboratory transfer, there have not been any reported clinical strains of staphylococci of the VanA type, suggesting that there may be a biologic disadvantage resulting from incorporation of the *vanA* cluster into staphylococci in vivo. The *vanA* operon has also been found in other gram-positive species isolated from clinical material, including rare strains of *Oerskovia turbata*, *Arcanobacterium haemolyticum*, and *Bacillus circulans*. The *vanA*-containing elements and sequences of the *vanA* genes in the *Oerskovia* and *Arcanobacterium* strains are remarkably similar in structure to those found in Tn*1546*, and their appearance following outbreaks of vancomycin resistance in enterococci suggests that they were acquired from enterococci rather than the converse. In the *B. circulans* isolate, the structure of the vancomycin-resistance genes was highly homologous to that found in Tn*1546*, but resistance was carried on the chromosome in a distinct genetic element (27).

STRUCTURE AND FUNCTION OF THE *vanB* GENE COMPLEX

vanB genotype strains are characterized by variable levels of resistance to vancomycin and, with rare exceptions, sus-

ceptibility to teicoplanin. VanB is found almost exclusively in *E. faecium* and *E. faecalis*, but has also been found in a *Streptococcus bovis* strain. The structure of the *vanB* resistance cluster from the initial U.S. isolate, *E. faecalis* V583, is shown in Fig. 6 (16). The organization of the *vanB* cluster is similar to that of the *vanA* operon; however, there appears to be much more heterogeneity both in specific gene sequences within the cluster and in the overall organization of the resistance elements. The $VanR_B$ and $VanS_B$ regulatory elements of the *vanB* cluster have only low-level amino acid identity with $VanR_A$ and $VanS_A$ of the *vanA* cluster (34 and 23%, respectively), but they contain similar motifs that clearly distinguish them as a two-component regulatory system homologous to $VanR_AS_A$. There is very little comparability between the N-terminal "sensing" portions of $VanS_A$ and $VanS_B$, suggesting that they may respond to different signals. There are also functional differences between $VanS_A$ and $VanS_B$, in that $VanS_B$ does not respond to the presence of teicoplanin, although teicoplanin-inducible derivatives can be selected with defined mutations in both the N-terminal sensing and C-terminal kinase regions of $VanS_B$. Analogous to the *vanA* system, $VanR_BS_B$ controls transcriptional activation of the downstream $vanH_B$, *vanB*, and $vanX_B$ genes required for expression of vancomycin resistance. These genes are closely related to the D,D-dehydrogenase, D-Ala–D-Lac ligase, and D,D-dipeptidase of the *vanA* cluster, with amino acid identities of 67, 76, and 71%, respectively. Sequencing of *vanB* genes from numerous strains reveals heterogeneity of up to 5% between different isolates. This has implications for the development of reliable sequence-based methods for detection of VanB-type strains. In addition to the $vanH_BX_B$ genes, the *vanB* cluster also contains a D,D-carboxypeptidase, $VanY_B$, with low-level homology to $VanY_A$, as well as a novel gene, *vanW*, that is unique to the *vanB* cluster and of unknown function. There is no homolog of *vanZ* in the *vanB* cluster.

Higher levels of D,D-dipeptidase activity and the resulting higher ratio of D-Lac-containing depsipeptide precursors to pentapeptide precursors correlate with increased resistance to vancomycin in *vanB* strains, suggesting that $vanX_B$ expression and the efficiency of elimination of D-Ala–D-Ala is an important variable in the phenotypic diversity of VanB-type strains (5). Resistance to teicoplanin can be selected in laboratory and clinical isolates via constitutive expression of the $vanH_BBX_B$ complex, bypassing the need for a specific induction signal. In constitutive strains, discrete mutations are found in the N-terminal kinase portion of the $vanS_B$ gene, resulting in either altered or truncated $VanS_B$. Teicoplanin resistance can also occur through alterations in the induction specificity of the $vanS_B$ sensor, as described above.

VanB is found most commonly in the chromosome and can be transferred in the absence of plasmids by conjugation in association with transfer of large blocks of DNA of sizes of 90 to 250 kb. The first *vanB* transposable element to be described from *E. faecalis* V583 was Tn*1547*, a large composite transposon containing the *vanB* gene cluster flanked by insertion sequences related to the IS*256* family (33). Recently, VanB-type resistance in multiple *E. faecium* isolates from northeastern Ohio was found to be carried by a 27-kb element of transposonlike structure, designated Tn*5382*, with proposed excisase and integrase genes and terminal regions with high similarity to the broad-host-range conjugative transposon Tn*916* (9). Tn*5382* is also capable of transfer between enterococci as part of an even larger genetic element that also includes the *E. faecium*

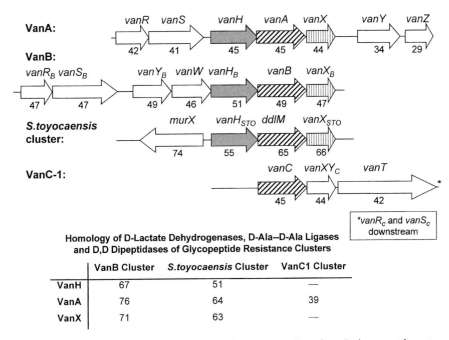

FIGURE 6   Comparison of the organization of the *vanA*, *vanB*, and *vanC* glycopeptide-resistance gene clusters and the glycopeptide-producing organism *S. toyocaensis*. G+C content of the respective genes is shown underneath. Also shown is the amino acid identity of the various resistance ligases, dehydrogenases, and dipeptidases (compared with VanA, VanH, and VanX).

gene for PBP5, which confers high-level resistance to ampicillin, perhaps accounting for the simultaneous appearance of both high-level ampicillin resistance and vancomycin resistance among many *E. faecium* strains. The *vanB* cluster has also occasionally been found on a plasmid.

ENTEROCOCCAL *vanC* GENOTYPES

VanC-type glycopeptide resistance is confined to several species of enterococci that are occasionally isolated from clinical material, especially fecal specimens, and are only rarely associated with human infection. These include *E. gallinarum*, *E. casseliflavus*, and *E. flavescens*. Most strains of *E. gallinarum* and *E. casseliflavus* are motile, and some *E. casseliflavus* strains produce a yellow pigment. *E. flavescens* is genetically very closely related to *E. casseliflavus*. These organisms demonstrate low-level vancomycin resistance with MICs of 2 to 32 μg/ml. VanC-type strains with very high vancomycin MICs have been described; however, such strains carried both *vanC* and *vanA* resistance genes. Resistance is an intrinsic feature of all these organisms and may be expressed either constitutively or inducibly. The mechanism of VanC-type resistance is due to production of pentapeptide terminating in D-Ser rather than D-Ala or D-Lac; vancomycin binds less avidly to D-Ala–D-Ser residues, but binding is not decreased to the extent occurring with precursors terminating in D-Lac (6). VanC-type strains remain susceptible to teicoplanin. The genes encoding the D-Ala–D-Ser ligases from *E. gallinarum* (*vanC-1*), *E. casseliflavus* (*vanC-2*), and *E. flavescens* (*vanC-3*) have been sequenced and functionally characterized in vitro (14). The VanC proteins constitute a family of closely related D-Ala–D-Ser ligases with high identity to each other but low-level similarity to either D-Ala–D-Ala or D-Ala–D-Lac ligases (15). Additional D-Ala–D-Ala specific ligases have also been identified in *E. gallinarum* and *E.*

casseliflavus; the presence of two different ligases would be required for the inducible expression of vancomycin resistance in VanC-type strains. The complete *vanC-1* operon from *E. gallinarum* has recently been sequenced and characterized. In addition to the VanC-1 D-Ala–D-Ser ligase, this operon includes the *vanT* and *vanXY*$_C$ genes. *vanT* encodes a membrane-bound serine racemase that catalyzes conversion of L-Ser to D-Ser. The product of *vanXY*$_C$ is a unique bifunctional D,D-dipeptidase/D,D-carboxypeptidase that provides activities similar to both VanX and VanY from the *vanA* cluster by eliminating the pool of D-Ala–D-Ala dipeptide and pentapeptide precursors. Located downstream from the clustered *vanC-1*, *vanT*, and *vanXY*$_C$ genes are reading frames whose predicted products have 50% and 42% amino acid identity with VanR and VanS, respectively, and constitute the presumed glycopeptide sensor VanR$_C$S$_C$ that determines inducibility for this resistance system (35).

OTHER ENTEROCOCCAL GLYCOPEPTIDE-RESISTANCE GENOTYPES

VanD-type resistance has recently been described in a single *E. faecium* strain from Europe and three strains from Massachusetts. *E. faecium* BM4339, the prototype strain, demonstrates moderate resistance to vancomycin with an MIC of 64 μg/ml but also low-level resistance to teicoplanin with an MIC of 4 μg/ml (32). The *vanD* cluster appears functionally similar to the *vanA* and *vanB* clusters. Strain BM4339 contains the *vanD* gene encoding a D-Ala–D-Lac ligase with 69% sequence identity to the VanA and VanB ligases. Genes encoding a dehydrogenase analogous to VanH and a D,D-dipeptidase similar to VanX have also been identified. A putative vancomycin sensor system has been found in association with the *vanD* cluster. Resistance in BM4339 appears to be constitutively expressed and located in the chromosome. In surveys of GRE strain collections, this genotype appears to be rare.

An additional enterococcal-resistance genotype, identified as *vanE*, has also been described (12). This mechanism has been found in a single isolate with a moderate level of resistance to vancomycin. The *vanE* resistance ligase appears to be a D-Ala–D-Ser ligase analogous to but genetically distinct from the *vanC* ligases. Thus, VanE appears to be the first example of an acquired VanC-like enterococcal-resistance mechanism utilizing a D-Ser-terminating precursor. Additional genes and cellular location of the *vanE* cluster are being determined.

GLYCOPEPTIDE-DEPENDENT ENTEROCOCCI

An unusual vancomycin-resistance phenotype in several clinical GRE isolates is vancomycin dependence (17). These strains grow poorly or not at all in the absence of vancomycin but grow well in vancomycin-supplemented media. Vancomycin-dependent enterococci have been isolated serendipitously from clinical specimens. Initial growth of organisms can occur in blood culture bottles or plates owing to carryover of low concentrations of vancomycin from clinical material, but organisms will then fail to grow on subculture without vancomycin supplementation. Most strains have been isolated from patients receiving long-term vancomycin therapy and are of the *vanA* or *vanB* genotype. The mechanism of vancomycin dependence is related to loss of function of the constitutive enterococcal Ddl ligase. Point mutations in *ddl* genes resulting in either truncated protein or alteration of essential amino acid residues have been demonstrated in clinical isolates and laboratory-derived dependent mutants. In a *ddl*-inactivated strain, cell wall synthesis is dependent on the D-Ala–D-Lac ligase and therefore can occur only after vancomycin induction and expression of the *vanA* or *vanB* gene cluster. Loss of the dependence phenotype can evolve either through further mutations leading to constitutive expression of the *vanA* or *vanB* operons or by mutations restoring D-Ala–D-Ala ligase activity (45). Although vancomycin-dependent strains have caused significant infections, including bacteremia, such isolates remain more a laboratory curiosity than a major clinical problem.

ORIGINS OF GLYCOPEPTIDE-RESISTANCE GENES

The lack of any documented high-level glycopeptide resistance in enterococci before 1986 and the significant differences in G+C content of the *vanA* and *vanB* glycopeptide-resistance clusters from other enterococcal genes strongly imply recent acquisition of these genes by enterococci from other organisms (4). However, the natural reservoir for these genes remains obscure. Potential reservoirs include intrinsically glycopeptide-resistant organisms such as *Leuconostoc*, *Lactobacillus*, or *Pediococcus* spp. or glycopeptide-producing organisms such as *Amycolatopsis orientalis*. None of these has been found to contain any of the enterococcal resistance ligases or any of the other genes from the *vanA* or *vanB* resistance clusters. Lactobacilli and the other intrinsically resistant organisms do contain D-Ala–D-Lac ligases functionally analogous to VanA and VanB, but by sequence analysis these genes are phylogenetically distinct from *vanA* or *vanB* (15). *Leuconostoc* and

Lactobacillus strains also contain homologs of the VanH and $VanH_B$ lactate dehydrogenases, but these genes are only distantly related (~30% amino acid homology) to the enterococcal dehydrogenases and are not found clustered with the D-Ala–D-Lac ligase genes. Recent studies in the glycopeptide-producing organisms *Streptomyces toyocaensis* and *A. orientalis* have revealed gene clusters with close homology to the dehydrogenase–ligase–D,D-dipeptidase complexes of GRE (28, 29) (Fig. 6). The *ddlM* gene product of *S. toyocaensis* and the *ddlN* gene product of *A. orientalis* demonstrate >60% amino acid sequence similarity with the deduced sequences of VanA, VanB, and VanD but much lower homology with other D-Ala–D-Lac or D-Ala–D-Ala ligases. Similarly, the VanH homologs in these strains have 51 to 55% amino acid identity with the VanH and $VanH_B$ gene products, and the VanX homologs are 61 to 63% identical with the enterococcal VanX proteins. The organization of the *vanH-ddl-vanX* cluster in the glycopeptide producers parallels that of the *vanA* and *vanB* cluster, with the three genes closely linked and under control of a single promoter (29). Using degenerate oligonucleotides, similar clusters have been found in other glycopeptide producers. However, no genes homologous to *vanY*, *vanW*, or *vanZ* and no vancomycin-sensing systems homologous to the *vanRS* and $vanR_BS_B$ systems have been found in the vancomycin-resistance clusters of these organisms. It is unlikely that the glycopeptide-producing actinomycetes are the immediate source of the resistance genes found in enterococci. The G+C content of the clusters from glycopeptide-producing organisms is much higher than that of the *vanA* or *vanB* clusters (>60% versus 40–44%), reflecting the high G+C content of actinomycetes in general. However, these organisms are potential reservoirs for a large family of glycopeptide-resistance genes that may have been incorporated by other soil organisms and ultimately found their way into enterococci.

Another homolog of the *vanA* and *vanB* cluster has been found in all strains of the soil-dwelling biopesticide *Bacillus popilliae* and in some strains of the closely related organism *Bacillus lentimorbus* (37). The organization of glycopeptide resistance in *B. popilliae* is phenotypically similar to *vanB* in that it is induced following exposure to vancomycin, with resulting high-level resistance to vancomycin but not teicoplanin. *B. popilliae* contains both a D-Ala–D-Ala ligase homologous to the product of the *ddl* gene of *Bacillus subtilis* and a second inducible ligase with 69 to 75% amino acid homology to VanA and VanB.

GLYCOPEPTIDE RESISTANCE IN *S. AUREUS*

Recent reports of clinical isolates of *S. aureus* with reduced susceptibility to glycopeptides from patients failing glycopeptide therapy have brought renewed attention to several observations regarding glycopeptide therapy for staphylococcal infections (24, 43). Investigators had previously described in vitro selection of *S. aureus* strains with reduced vancomycin susceptibility after serial passage on vancomycin (13, 40). MICs of these laboratory-selected resistant derivatives were as high as 24 μg/ml, although a more recent mutant characterized by Sieradzki and Tomasz (40) had an MIC of 100 μg/ml. Resistant mutants have been generated in vitro from both methicillin-susceptible and methicillin-resistant *S. aureus* (MRSA) strains. The mechanisms of increased resistance of these mutants were at first poorly understood, and their clinical relevance was not widely appreciated, especially in the absence of any documented clinical resistance. Clinicians have also noted that patients with significant *S. aureus* infections caused by apparently susceptible isolates have "failed" therapy with glycopeptides. Since resistance may be only one of multiple potential explanations for treatment failure, clinical failures were generally attributed to causes other than drug resistance, although many clinicians have considered vancomycin to be only a mediocre antistaphylococcal drug. In retrospect, some of these failures may have been due to acquired glycopeptide resistance similar to that documented in recent cases.

In discussing glycopeptide resistance in *S. aureus*, it is important to clarify the various terms used to define these strains (43). In the United States, where vancomycin is the only glycopeptide currently available, strains with vancomycin MICs of >4 and <32 μg/ml are considered intermediately resistant to vancomycin or are referred to as VISA strains; strains with MICs of ≥32 μg/ml would be vancomycin resistant (VRSA), although no such clinical isolates have been described thus far. The terms glycopeptide intermediately resistant *S. aureus* (GISA) and glycopeptide-resistant *S. aureus* (GRSA) are used to include isolates that are intermediately resistant or resistant to either teicoplanin or vancomycin. This term is more widely used in European countries where both vancomycin and teicoplanin are available. Although strains with increased resistance to vancomycin demonstrate increased resistance to teicoplanin, the converse may not be true, which may be important for the optimal detection of glycopeptide-resistant staphylococci. To add to the confusion, there are also differences in the MIC breakpoints used to define susceptibility, intermediate resistance, and complete resistance established by the various national laboratory standards committees. The method of testing and media used may also be relevant for defining strains with increased glycopeptide resistance.

Teicoplanin-resistant *S. aureus* strains were shown to emerge in patients receiving treatment with teicoplanin, but these observations initially caused little concern, as the teicoplanin-resistant strains appeared to remain susceptible to vancomycin (25). The first well-documented clinical VISA strain was isolated from Japan in 1996 (24). This strain, designated Mu50, was an MRSA isolate from an immunocompromised child previously treated with vancomycin and at the time failing vancomycin therapy; it was intermediately resistant (or decreasingly susceptible) to vancomycin, with an MIC of 8 μg/ml. This MIC remained stable during serial passage. Subsequent epidemiological studies from Japan have described large numbers of so-called heteroresistant *S. aureus* strains (23). These strains have reproducibly selectable subpopulations that will grow on brain heart infusion agar containing vancomycin at concentrations of 8 to 16 μg/ml. Such isolates composed up to 20% of MRSA in the hospital where Mu50 was isolated and 1.3% of MRSA from throughout Japan, although no other "true" resistant strains similar to Mu50 were found. Derivatives of heteroresistant strains with higher-level vancomycin resistance behaved like strain Mu50 in vitro. The resistant and heteroresistant Japanese strains were of the same IIA clonal type. Subsequent to this, there have been three well-documented reports of unrelated MRSA strains with vancomycin MICs of 8 μg/ml isolated in the United States during 1997 and 1998 from Michigan, New Jersey, and New York (43). All were associated with clinical failure of vancomycin therapy and occurred in pa-

tients with extensive prior vancomycin exposure for treatment of MRSA; in addition, all patients were receiving dialysis. Extensive epidemiological investigations revealed no secondary cases of infection or colonization in association with these three cases. There have recently been several other preliminary reports of similar strains in Europe and the United States. Guidelines for the appropriate isolation of patients infected with glycopeptide-resistant *S. aureus* strains have been published (11).

The mechanisms of resistance in *S. aureus* are distinct from those elaborated by enterococci. Although, as mentioned, the *vanA* gene cluster has been transferred to *S. aureus* in vitro, no genes analogous to any of the enterococcal vancomycin-resistance ligases have been found in clinical isolates of resistant staphylococci. Phenotypic features have been described that are common to many of the clinical and laboratory glycopeptide-resistant *S. aureus* strains (Table 2). These include slower growth rates and heterogeneous morphology when plated on vancomycin-containing medium, a markedly thickened cell wall, altered susceptibility to lysostaphin, and in some instances a paradoxical increased susceptibility to beta-lactams. In Mu50 there is evidence of increased activation of cell wall synthesis, including increased levels of PBP2 and PBP2′ and increased incorporation of *N*-acetylglucosamine, consistent with the observed thickening of the cell wall, although the degree of cross-linking within the thickened cell wall ap-

TABLE 2 Characteristics of clinical isolates of *S. aureus* with reduced glycopeptide susceptibility[a]

Clinical and epidemiological features of source patients and isolates

- Prolonged vancomycin exposure
- Chronic medical illness or hospitalization
- Indwelling devices
- End-stage renal disease
- Genetically identical glycopeptide-susceptible isolates found in source patients as well as other patients
- No secondary spread of resistant clones to other patients demonstrated

Susceptibility profiles of isolates

- Vancomycin MIC of 8–16 μg/ml
- Vancomycin susceptible by disc diffusion
- Homogeneous or heterogeneous expression of glycopeptide resistance
- All also MRSA (heterogeneous or homogeneous expression)
- Oxacillin MIC decreased in some clones with higher vancomycin MICs
- Potential synergy between vancomycin and oxacillin

Morphologic and other phenotypic characteristics of isolates

- Heterogeneous colony morphology on plates
- Slower growth rates of more resistant clones
- Multicellular aggregates in liquid media
- Thickened extracellular layer (cell wall?) by electron microscopy after growth in vancomycin
- Decrease in vancomycin concentration in growth medium with trapping of drug in extracellular matrix
- Increased amounts of some PBPs, especially PBP2 or PBP2′

[a]From references 11, 22, 23, and 43.

pears to be reduced (22). Other clinical and laboratory glycopeptide-resistant isolates also demonstrate increased amounts of PBP2 or other PBPs. Glycopeptide-resistant strains also appear to have an enhanced ability to absorb vancomycin from antibiotic-containing media, perhaps owing to binding of glycopeptide to a surplus of terminal D-Ala–D-Ala residues in the loosely adherent thickened cell wall. One proposed mechanism for resistance is that the increased "sponging" of vancomycin into the thickened cell wall effectively excludes vancomycin from accessing the actively metabolizing peptidoglycan adjacent to the membrane. The increased levels of PBPs may also contribute to resistance by competing more aggressively with vancomycin for binding to D-Ala–D-Ala terminal residues. Other observations initially proposed as having potential significance in glycopeptide-resistant staphylococci, such as the presence of a staphylococcal dehydrogenase homologous to VanH and the appearance of novel 35- to 40-kDa membrane proteins, have not been demonstrated to be important for glycopeptide resistance. Despite the intriguing phenomenological observations, the understanding of specific mechanisms or genetic events responsible for glycopeptide resistance remains incomplete. However, it is likely that resistance is related to a series of mutational or regulatory events rather than a single event. In addition, although glycopeptide resistance can be selected in vitro in both methicillin-susceptible and -resistant S. aureus, all clinical glycopeptide-resistant isolates thus far have been MRSA, suggesting that MRSA strains are more readily primed for evolution toward glycopeptide resistance.

Optimal laboratory methods for detection of glycopeptide-resistant staphylococci remain poorly defined (43). Disc diffusion methods failed to differentiate clinical VISA strains from susceptible strains, and broth microdilution methods may be suboptimal, particularly those with incubation times of less than 24 h. In practice, any S. aureus isolate with an MIC of 4 μg/ml or greater by any method should be further investigated. The use of commercial agar plates currently employed for screening GRE is also being evaluated. Testing with teicoplanin rather than vancomycin appears to discriminate more easily between susceptible and VISA strains and has also been suggested as an initial screening method. Optimization of laboratory methods is hindered by the relatively small number of isolates available for study and the lack of understanding of the clinical significance of heteroresistant strains that may be fairly common when actively screened for.

GLYCOPEPTIDE RESISTANCE IN COAGULASE-NEGATIVE STAPHYLOCOCCI

Glycopeptide resistance is more common in coagulase-negative staphylococci (CoNS) than in S. aureus, although these organisms do not present the same threat posed by glycopeptide-resistant S. aureus. Clinical isolates of glycopeptide-resistant CoNS were first described in 1987, and there have been numerous reports of similar isolates since that time (39). However, over the past decade the rate of CoNS with decreased susceptibility to glycopeptides has remained quite low—less than 1% in several recent surveys. It is not clear whether this reflects a true lack of emergence of resistance or inadequacy of testing methods. Most vancomycin resistance has been described in S. haemolyticus and S. epidermidis, with the highest prevalence in strains of S. haemolyticus (18). Unlike clinical glycopeptide-

resistant S. aureus isolates with MICs of only 8 μg/ml, MICs of vancomycin for some S. haemolyticus strains have been as high as 128 μg/ml. Most resistant isolates are recovered from patients receiving prolonged or repeated courses of glycopeptide therapy and have been particularly problematic in renal failure and oncology units and in neonatal intensive care units. Stepwise selection of vancomycin-resistant mutants of S. haemolyticus and S. epidermidis is easily accomplished in vitro. As in S. aureus, resistance to teicoplanin appears to develop more readily than resistance to vancomycin.

None of the glycopeptide-resistant CoNS strains studied to date has contained a VanA-type or VanB-type resistance mechanism or gene cluster. The resistance mechanisms in CoNS appear to be generally similar to those in S. aureus strains. Some of the laboratory-derived and clinical glycopeptide-resistant CoNS share phenotypic characteristics of resistant S. aureus strains, including thicker cell walls, smaller colony size with slower growth rates, increased levels of PBP2, and increased susceptibility to beta-lactams. Novel membrane proteins of varying sizes, generally from 35 to 40 kDa, have been detected in some isolates, but the significance of this is unclear. Analysis of cell wall precursor pools in one glycopeptide-resistant S. haemolyticus strain revealed small amounts of peptide terminating in D-Lac, although the low concentration of lactate-containing precursors seemed inadequate to account for the resistance observed in this strain. Alterations in the composition of the peptidoglycan peptide cross bridges have also been described in resistant strains. In CoNS, as in S. aureus, there is no evidence of incorporation of enterococcal or other novel glycopeptide-resistance genes to account for the observed glycopeptide resistance.

REFERENCES

1. **Aarestrup, F. M., P. Ahrens, M. Madsen, L. V. Pallesen, R. L. Poulsen, and H. Westh.** 1996. Glycopeptide susceptibility among Danish Enterococcus faecium and Enterococcus faecalis isolates of animal and human origin and PCR identification of genes within the VanA cluster. Antimicrob. Agents Chemother. **40:**1938–1940.

2. **Arthur, M., C. Molinas, and P. Courvalin.** 1992. The VanS-VanR two-component regulatory system controls synthesis of depsipeptide peptidoglycan precursors in Enterococcus faecium BM4147. J. Bacteriol. **174:**2582–2591.

3. **Arthur, M., C. Molinas, F. Depardieu, and P. Courvalin.** 1993. Characterization of Tn1546, a Tn3-related transposon conferring glycopeptide resistance by synthesis of depsipeptide peptidoglycan precursors in Enterococcus faecium BM4147. J. Bacteriol. **175:**117–127.

4. **Arthur, M., P. Reynolds, and P. Courvalin.** 1996. Glycopeptide resistance in enterococci. Trends Microbiol. **4:** 401–407.

5. **Baptista, M., F. Depardieu, P. E. Reynolds, P. Courvalin, and M. Arthur.** 1997. Mutations leading to increased levels of resistance to glycopeptide antibiotics in VanB-type enterococci. Mol. Microbiol. **25:**93–105.

6. **Billot-Klein, D., D. Blanot, L. Gutmann, and J. van Heijenoort.** 1994. Association constants for the binding of vancomycin and teicoplanin to N-acetyl-D-alanyl-D-alanine and N-acetyl-D-alanyl-D-serine. Biochem J. **304**(Pt 3):1021–1022.

7. **Billot-Klein, D., L. Gutmann, E. Collatz, and J. van Heijenoort.** 1992. Analysis of peptidoglycan precursors

in vancomycin-resistant enterococci. *Antimicrob. Agents Chemother.* **36**:1487–1490.

8. Bugg, T. D. H., G. D. Wright, S. Dutka-Malen, M. Arthur, P. Courvalin, and C. T. Walsh. 1991. Molecular basis for vancomycin resistance in *Enterococcus faecium* BM4147: biosynthesis of a depsipeptide peptidoglycan precursor by vancomycin resistance proteins VanH and VanA. *Biochemistry* **30**:10408–10415.

9. Carias, L. L., S. D. Rudin, C. J. Donskey, and L. B. Rice. 1998. Genetic linkage and co-transfer of a novel, *vanB*-containing transposon (Tn*5382*) and a low-affinity penicillin-binding protein 5 gene in a clinical vancomycin-resistant *Enterococcus faecium* isolate. *J. Bacteriol.* **180**: 4426–4434.

10. Centers for Disease Control and Prevention. 1995. Recommendations for preventing the spread of vancomycin resistance: recommendations of the Hospital Infection Control Practices Advisory Committee (HICPAC). *Morbid. Mortal. Weekly Rep. Rec. Rep.* **44**(12):1–20.

11. Centers for Disease Control and Prevention. 1997. Interim guidelines for prevention and control of staphylococcal infection associated with reduced susceptibility to vancomycin. *Morbid. Mortal. Weekly Rep.* **46**:626–628, 635–636.

12. Courvalin, P. 1998. The Van alphabet, abstr. S-37, p. 622. *Program Abstr. 38th Intersci. Conf. Antimicrob. Agents Chemother.* American Society for Microbiology, Washington, D.C.

13. Daum, R. S., S. Gupta, R. Sabbagh, and W. M. Milewski. 1992. Characterization of *Staphylococcus aureus* isolates with decreased susceptibility to vancomycin and teicoplanin: isolation and purification of a constitutively produced protein associated with decreased susceptibility. *J. Infect. Dis.* **166**:1066–1072.

14. Dutka-Malen, S., C. Molinas, M. Arthur, and P. Courvalin. 1992. Sequence of the *vanC* gene of *Enterococcus gallinarum* BM4174 encoding a D-alanine:D-alanine ligase-related protein necessary for vancomycin resistance. *Gene* **112**:53–58.

15. Evers, S., B. Casadewall, M. Charles, S. Dutka-Malen, M. Galimand, and P. Courvalin. 1996. Evolution of structure and substrate specificity in D-alanine:D-alanine ligases and related enzymes. *J. Mol. Evol.* **42**:706–712.

16. Evers, S., and P. Courvalin. 1996. Regulation of VanB-type vancomycin resistance gene expression by the VanS$_B$-VanR$_B$ two-component regulatory system in *Enterococcus faecalis* V583. *J. Bacteriol.* **178**:1302–1309.

17. Fraimow, H. S., D. L. Jungkind, D. W. Lander, D. R. Delso, and J. L. Dean. 1994. Urinary tract infection with an *Enterococcus faecalis* isolate that requires vancomycin for growth. *Ann. Intern. Med.* **121**:22–26.

18. Froggatt, J. W., J. L. Johnston, and G. L. Archer. Antimicrobial resistance in nosocomial isolates of *Staphylococcus haemolyticus*. *Antimicrob. Agents Chemother.* **33**: 460–466.

19. Goossens, H. J. 1998. Spread of vancomycin resistant enterococci: differences between the United States and Europe. *Infect. Control Hosp. Epidemiol.* **19**:546–551.

20. Gutmann, L., S. al-Obeid, D. Billot-Klein, M. L. Guerrier, and E. Collatz. 1994. Synergy and resistance to synergy between beta-lactam antibiotics and glycopeptiodes against glycopeptide-resistant strains of *Enterococcus faecium*. *Antimicrob. Agents Chemother.* **38**:824–829.

21. Hayden, M. K., G. M. Trenholme, J. E. Schultz, and D. F. Sahm. 1993. In vivo development of teicoplanin resistance in a VanB *Enterococcus faecium* isolate. *J. Infect. Dis.* **167**:1224–1227.

22. Hiramatsu, K. 1998. The emergence of *Staphylococcus aureus* with reduced susceptibility to vancomycin in Japan. *Am. J. Med.* **104**(5A):7S–10S.

23. Hiramatsu, K., N. Aritaka, H. Hanaki, S. Kawasaki, Y. Hosoda, S. Hori, Y. Fuckuchi, and I. Kobayashi. 1997. Vancomycin-resistant *Staphylococcus aureus*: dissemination of heterogeneously resistant strains in a Japanese hospital. *Lancet* **350**:1670–1673.

24. Hiramatsu, K., H. Hanaki, T. Ino, K. Yabuta, T. Oguri, and F. C. Tenover. 1997. Methicillin-resistant *Staphylococcus aureus* clinical strain with reduced vancomycin susceptibility. *J. Antimicrob. Chemother.* **40**:135–136.

25. Kaatz, G. W., S. M. Seo, N. J. Dorman, and S. A. Lerner. 1990. Emergence of teicoplanin resistance during therapy of *Staphylococcus aureus* endocarditis. *J. Infect. Dis.* **162**:103–108.

26. Leclercq, R., E. Derlot, J. Duval, and P. Courvalin. 1988. Plasmid-mediated vancomycin and teicoplanin resistance in *Enterococcus faecium*. *N. Engl. J. Med.* **319**: 157–161.

27. Ligozzi, M., G. Lo Cascio, and R. Fontana. 1998. *vanA* gene cluster in a vancomycin-resistant clinical isolate of *Bacillus circulans*. *Antimicrob. Agents Chemother.* **42**: 2055–2059.

28. Marshall, C. G., G. Broadhead, B. Leskiw, and G. D. Wright. 1997. D-Ala-D-Ala ligases from glycopeptide antibiotic-producing organisms are highly homologous to the enterococcal vancomycin-resistance ligases VanA and VanB. *Proc. Natl. Acad. Sci. USA* **94**:6480–6483.

29. Marshall, C. G., I. A. D. Lessard, I.-S. Park, and G. D. Wright. 1998. Glycopeptide antibiotic resistance genes in glycopeptide-producing organisms. *Antimicrob. Agents Chemother.* **42**:2215–2220.

30. Martone, W. J. 1998. Spread of vancomycin-resistant enterococci: why did it happen in the United States? *Infect. Control Hosp. Epidemiol.* **19**:539–545.

31. Noble, W. C., Z. Virani, and R. Cree. 1992. Co-transfer of vancomycin and other resistance genes from *Enterococcus faecalis* NCTC 12201 to *Staphylococcus aureus*. *FEMS Microbiol. Lett.* **93**:195–198.

32. Perichon, B., P. Reynolds, and P. Courvalin. 1997. VanD-type glycopeptide-resistant *Enterococcus faecium* BM4339. *Antimicrob. Agents Chemother.* **41**:2016–2018.

33. Quintiliani, R. J., and P. Courvalin. 1996. Characterization of Tn*1547*, a composite transposon flanked by the IS*16* and IS*256*-like elements, that confers vancomycin resistance in *Enterococcus faecalis* BM428. *Gene* **172**:1–8.

34. Reynolds, P. E. 1989. Structure, biochemistry, and mechanism of action of glycopeptide antibiotics. *Eur. J. Clin. Microbiol. Infect. Dis.* **8**:943–950.

35. Reynolds, P. E., C. A. Arias, and P. Courvalin. The *vanC* operon of *Enterococcus gallinarum* BM4174: sequence of *vanXY*$_C$ and characterization of the protein product as a D,D-dipeptidase/D,D-carboxypeptidase, abstr. C-85, p. 93. *Program Abstr. Intersci. Conf. Antimicrob. Agents Chemother.* American Society for Microbiology, Washington, D.C.

36. Reynolds, P. E., F. Depardieu, S. Dutka-Malen, M. Arthur, and P. Courvalin. 1994. Glycopeptide resistance

mediated by enterococcal transposon Tn*1546* requires production of VanX for hydrolysis of D-alanyl-D-alanine. *Mol. Microbiol.* **13:**1065–1070.

37. **Rippere, K., R. Patel, J. R. Uhl, K. E. Piper, J. M. Steckelberg, B. C. Kline, F. R. Cockerill, and A. A. Yousten.** 1998. DNA sequence resembling *vanA* and *vanB* in the vancomycin-resistant biopesticide *Bacillus popilliae. J. Infect. Dis.* **178:**584–588.

38. **Sahm, D., J. Kissinger, M. S. Gilmore, P. R. Murray, R. Mulder, J. Solliday, and B. Clarke.** 1989. In vitro susceptibility studies of vancomycin-resistant *Enterococcus faecalis. Antimicrob. Agents Chemother.* **33:**1588–1591.

39. **Schwabe, R. S., J. T. Stapleton, and P. H. Gilligan.** 1987. Emergence of vancomycin resistance in coagulase-negative staphylococci. *N. Engl. J. Med.* **316:**927–931.

40. **Sieradzki, K., and A. Tomasz.** 1997. Inhibition of cell wall turnover and autolysis by vancomycin in a highly vancomycin-resistant mutant of *Staphylococcus aureus. J. Bacteriol.* **179:**2557–2566.

41. **Silva, J. C., A. Haldimann, M. K. Prahalad, C. T. Walsh, and B. L. Wanner.** 1998. In vivo characterization of the type A and B vancomycin resistant enterococci (VRE) VanRS two-component systems in *Escherichia coli:* a nonpathogenic model for studying the VRE signal transduction pathways. *Proc. Natl. Acad. Sci. USA* **95:** 11951–11956.

42. **Swenson, J. M., N. C. Clark, M. J. Ferraro, D. F. Sahm, G. Doern, M. A. Pfaller, L. B. Reller, M. P. Weinstein, R. J. Zabransky, and F. C. Tenover.** 1994. Development of a standardized screening method for detection of vancomycin-resistant enterococci. *J. Clin. Microbiol.* **32:** 1700–1704.

43. **Tenover, F. C., M. V. Lancaster, B. C. Hill, C. D. Steward, S. A. Stocker, G. A. Hancock, C. M. O'Hara, S. K. McAllister, N. C. Clark, and K. Hiramatsu.** 1998. Characterization of staphylococci with reduced susceptibilities to vancomycin and other glycopeptides. *J. Clin. Microbiol.* **36:**1020–1027.

44. **Uttley, A. H., C. H. Collins, J. Naidoo, and R. C. George.** 1988. Vancomycin-resistant enterococci. *Lancet* **i:**57–58.

45. **Van Bambeke, F., M. Chauvel, P. E. Reynolds, H. S. Fraimow, and P. Courvalin.** 1999. Vancomycin-dependent *Enterococcus faecalis* clinical isolates and revertant mutants. *Antimicrob. Agents Chemother.* **43:**41–47.

46. **Walsh, C. T., S. L. Fisher, L.-S. Park, M. Prahalad, and Z. Wu.** 1996. Bacterial resistance to vancomycin: five genes and one missing hydrogen bond tell the story. *Chem. Biol.* **3:**21–28.

47. **Yao, R. C., and L. W. Crandall.** 1994. Glycopeptides: classification, occurrence and discovery, p. 1–27. *In* R. Nagarajan (ed.), *Glycopeptide Antibiotics.* Marcel Dekker, Inc., New York, N.Y.

Aminoglycoside Resistance in Gram-Positive Bacteria

KAREN J. SHAW AND GERARD D. WRIGHT

64

The aminoglycoside-aminocyclitol antibiotics are cationic molecules that find clinical use in the treatment of infections caused by gram-positive and gram-negative bacteria. The aminoglycosides are bacterially derived natural products, or semisynthetic derivatives of these, and consist of a six-member aminocyclitol ring (a cyclohexane ring substituted with hydroxyl and amino groups) to which aminosugars are linked. The diversity in these molecules arises from the characteristics of the aminocyclitol ring and the nature and number of substituent aminosugars (62). Consequently, there have been hundreds of different aminoglycosides identified since the initial discoveries of streptomycin and neomycin by Waksman and colleagues some 50 years ago (74, 90), and many of these have found clinical use.

The aminoglycosides are water-soluble basic molecules that have affinity for the bacterial ribosome; in particular, many of these molecules target the 16S rRNA (52, 91). The molecular structure of paromomycin bound to a synthetic 27-nucleotide RNA molecule that models the A-site target sequence has recently been reported (24). The physical interaction between the aminoglycoside antibiotic and the bacterial translational machinery has many consequences, including the misreading of mRNA transcripts, resulting in the production of membrane-damaging peptides (14, 15). The net effect is that many of these antibiotics are bactericidal; thus, they are highly effective and potent therapeutic agents.

AMINOGLYCOSIDE-RESISTANCE GENES AND ENZYME

Resistance to aminoglycoside antibiotics is manifested through both intrinsic and acquired mechanisms. Intrinsic resistance can be the result of decreased uptake of drug, point mutations in either rRNA or ribosomal proteins, increased expression of housekeeping genes such that aminoglycoside resistance occurs, or under the aegis of enzymes that methylate the target rRNA. Acquired aminoglycoside antibiotic resistance is thus far manifested only through covalent modification of the compounds and is the primary mechanism of resistance in clinical isolates. Chemical modification of aminoglycosides can occur either through N acetylation, O phosphorylation, or O adenylylation, with the net result of loss of strategic affinity for the target rRNA. Several of the transferases that catalyze these reactions have been cloned from gram-positive organisms (Table 1) and are detailed below.

The nomenclature of aminoglycoside-modifying enzymes includes a three-letter designation in capitals of the general mechanism (ANT = aminoglycoside nucleotidyl transferase; APH = aminoglycoside phosphotransferase; ACC = aminoglycoside acetyl transferase), a number in parentheses indicating the enzyme's target site, and a roman numeral indicating the specific enzyme itself, sometimes followed by a lowercase letter specifying the gene; e.g., ANT (4')-IIa is the second to be identified in the series of aminoglycoside adenylyl transferases that adenylate the aminoglycoside molecule at its 4' site.

Aminoglycoside O-Adenylyltransferases (ANT)

Two related streptomycin-resistance determinants have been cloned from Enterococcus faecalis [ant(6)-Ia] (57) and Bacillus subtilis [ant(6)-Ib] (56). The former is located on the enterococcal 80.7-kb conjugative R plasmid, pJH1, which also harbors the aminoglycoside kinase gene aph(3')-IIIa (see below), in addition to resistance determinants for tetracycline (tetL) and erythromycin (ermAM) (2). On the other hand, ant(6)-Ib is a B. subtilis chromosomal gene that has been characterized (36, 38), and the regiospecificity of adenyl transfer was confirmed by nuclear magnetic resonance (NMR) analysis (55) (Fig. 1). The plasmid-encoded ANT(6)-Ia enzyme from E. faecalis has been overexpressed and purified from Escherichia coli. The dimeric enzyme (2) is highly specific for streptomycin and demonstrates a Michaelis constant (K_m) for streptomycin of 100 μM and ATP of 102 μM (54).

Spectinomycin resistance in Staphylococcus aureus and E. faecalis is conferred by two distinct adenylylating enzymes encoded by ant(9)-Ia and ant(9)-Ib, respectively (Fig.

Gram-Positive Pathogens, ed. by V. A. Fischetti et al.
© 2000 American Society for Microbiology, Washington, D.C.

TABLE 1 Aminoglycoside-resistance determinants in gram-positive organisms that do not produce aminoglycoside antibiotics

Gene	Organism	Location	Phenotype[a]	Reference
Adenylyltransferases				
ANT(6)-Ia	*Enterococcus faecalis*	Plasmid	Strep	57
ANT(6)-Ib	*Bacillus subtilis*	Chromosome	Strep	56
ANT(9)-Ia	*Staphylococcus aureus*	Plasmid	Spec	53
ANT(9)-Ib	*Enterococcus faecalis*	Plasmid	Spec	39
ANT(4′)-Ia	*Staphylococcus aureus*	Plasmid	Kan, Tob, Amik, Neo, Dibek	43
ANT(4′)-Ia	*Bacillus brevis*	Chromosome	Kan, Tob, Amik, Neo	80
Phosphotransferases				
APH(3′)-IIIa	*Enterococcus faecalis, Staphylococcus aureus, Streptococcus pneumoniae*	Plasmid Tn*1545*	Kan, Amik, Isep, Neo, But, Ribos, Livid	86, 28, 5
APH(2″)-Ia[b]	*Enterococcus faecalis, Staphylococcus aureus*	Plasmid	Gent, Net, Siso, Kan, Tob, Amik, Isep, Neo, But, Ribos, Livid	22, 66
APH(2″)-Ic	*Enterococcus gallinarum*	Plasmid	Gent, Net, Kan, Tob	8
APH(2″)-Id	*Enterococcus casseliflavus*	Chromosome	Gent, Net, Kan, Tob	87
Acetyltransferases				
AAC(6′)-Ie[b]	*Enterococcus faecalis, Staphylococcus aureus*	Plasmid	Kan, Tob, Amik, Neo, Fort	22, 66
AAC(6′)-Ii	*Enterococcus faecium*	Chromosome	Kan, Tob, Amik, Neo	10

[a] Kan, kanamycin; Tob, tobramycin; Dibek, dibekacin; Amik, amikacin; Isep, isepamicin; Neo, neomycin; But, butirosin; Ribos, ribostamycin; Livid, lividomycin; Gent, gentamicin C; Net, netilmicin; Siso; sisomicin; Fort, fortimicin; Strep, streptomycin; Spec, spectinomycin.

[b] Part of a bifunctional enzyme, AAC(6′)-APH(2″).

2) (39, 53). The gene products are 36% identical (57% similar), and the proteins are of comparable size, approximately 39 kDa, yet they differ markedly in their predicted pIs: 8.04 for ANT(9)-Ia and 6.22 for ANT(9)-Ib. The precise catalytic mechanisms of these enzymes are unknown at present.

The best-studied adenylyltransferase enzyme is ANT(4′)-Ia from *S. aureus* (43), which also has been found in enterococci (6). This enzyme confers resistance to a number of aminoglycosides, including dibekacin and tobramycin but not gentamicin C (40). Resistance to the 4′-deoxyaminoglycoside dibekacin has been interpreted as evidence for 4″-adenyltransferase activity (Fig. 3) (71). ANT(4′)-Ia has been purified and characterized from *S. aureus* (40) and from *B. subtilis* bearing the staphylococcal plasmid pUB110 (67). Analysis of the product of

ANT(4′)-Ia–catalyzed adenylylation of tobramycin by NMR and mass spectrometry unambiguously identified the site of adenylylation to be the 4′-OH of kanamycin (40). A thermostable enzyme that differs from ANT(4′)-Ia by a Thr-130–Lys substitution has been purified from *Bacillus stearothermophilus* (43). This enzyme has been crystallized (33), and the structure of the apo-enzyme (68) and the kanamycin·AMPCPP ternary complex (60) was determined to 3.0 Å and 2.5 Å resolution, respectively.

The ANT(4′)-Ia enzyme is a homodimer with two active-site regions that lie at the dimer interface (Fig. 4). Each monomer contributes to the binding of both the aminoglycoside and the nucleotide substrates, so that only the dimer is catalytically competent. Ser-39 can form a hydrogen bond with the γ-phosphate of AMPCPP, and Asp-50 and Glu-192 are Mg^{2+} ligands. These residues are con-

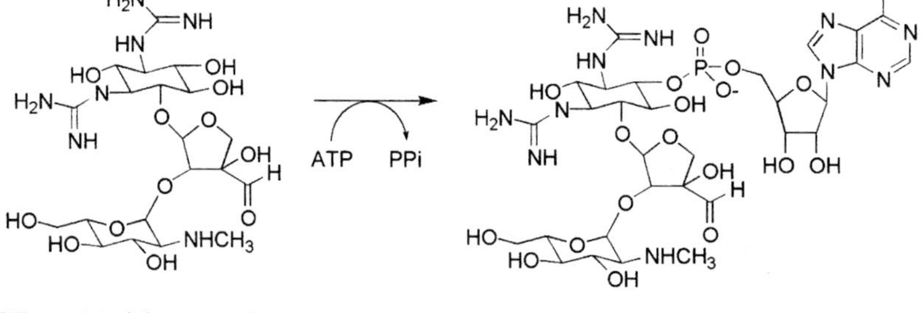

FIGURE 1 Modification of streptomycin by ANT(6). The structure of the adenylylated compound has been determined by NMR (55).

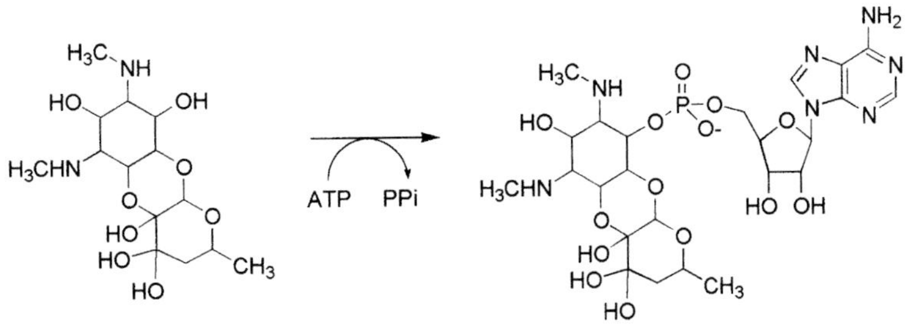

FIGURE 2 Proposed modification of spectinomycin by ANT(9).

served in most other ANTs and in other enzymes that utilize adenylyltransfer, such as DNA polymerase B (73). This similarity in function of conserved amino acids is paralleled by a general conservation in secondary structure around the nucleotide-binding region for these enzymes (30). The stereospecificity of adenylyltransfer by ANT(2″)-Ia from *Enterobacteriaceae* has been determined to go with inversion of configuration at the α-phosphorus, consistent with direct attack by the aminoglycoside 2″-OH on this center (89). By analogy, a similar direct-transfer mechanism has been proposed for ANT(4′)-Ia with an active-site Glu (Glu-145) acting as an active-site general base for the deprotonation of the AMP-accepting hydroxyl group (60). This mechanism has yet to be validated by additional studies, such as site-directed mutagenesis.

Another ANT(4′) enzyme is produced by *Bacillus brevis* (80). This enzyme was partially purified, and the native molecular mass was estimated to be 45 kDa. All four ribonucleoside triphosphates were utilized by this enzyme, as were dATP, dTTP, and dCTP; however, nucleoside diphosphates were not. As expected for this class of enzyme, ANT(4′) had an absolute requirement for divalent cations (Mn^{2+}, Zn^{2+}, Mg^{2+}). The resistance profile is similar to

that of the ANT(4′)-II enzyme isolated from *Pseudomonas aeruginosa* (77) since dibekacin resistance is not conferred, whereas ANT(4′)-Ia inactivates dibekacin. Failure to isolate aminoglycoside-sensitive isolates of *B. brevis* after treatment with acriflavin and ethidium bromide (plasmid-curing agents) suggests that the gene encoding this enzyme may be chromosomal. In support of this, the activity was detected in all strains of *B. brevis* tested but not in other species of *Bacillus*, suggesting the presence of a species-specific chromosomal gene.

Aminoglycoside O-Phosphotransferases (APH)

A common mechanism of aminoglycoside resistance is phosphorylation of hydroxyl groups, a reaction requiring ATP and a phosphotransferase (APH) catalyst. One of the most prevalent resistance phenotypes is resistance to kanamycin and neomycin, but not gentamicin or tobramycin. This is the hallmark of enzymes that transfer a phosphate group to the 3′-hydroxyl of kanamycin, APH(3′). There are at present seven phenotypic classes of APH(3′) (78), although only one such enzyme, APH(3′)-IIIa, is associated with gram-positive clinical isolates thus far. Among other gram-positive bacteria, APH(3′)-V isozymes are

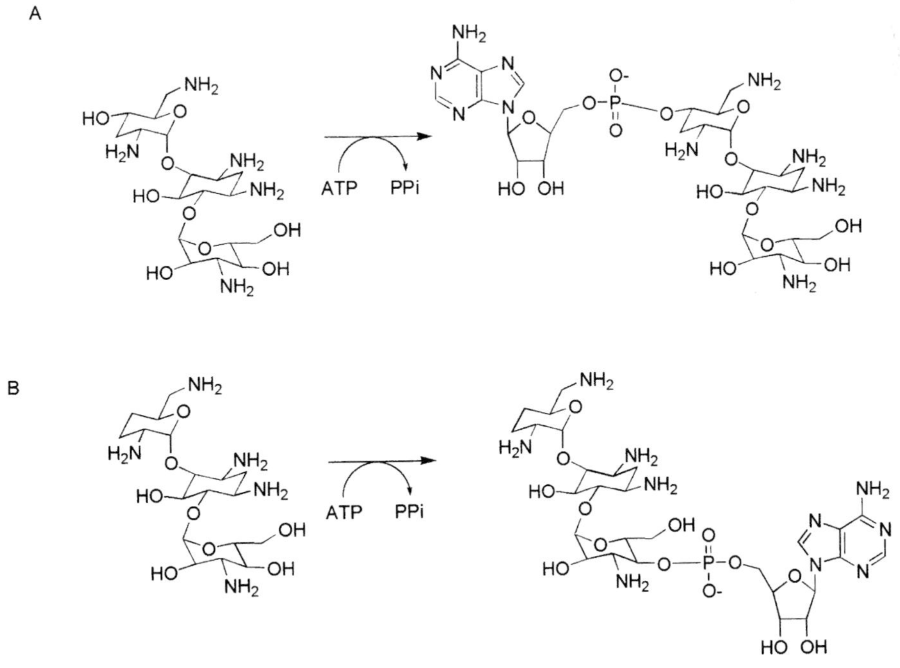

FIGURE 3 Modification of aminoglycosides by ANT(4′)-Ia. (A) Modification of the 4′-OH of tobramycin. (B) Modification of the 4″-OH of dibekacin.

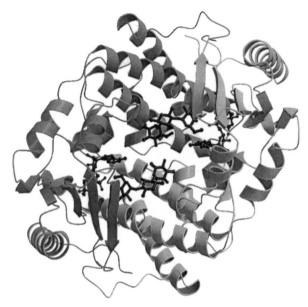

FIGURE 4 Three-dimensional structure of ANT(4')-Ia. Two views rotated by 90° of the dimeric enzyme bound to two molecules each of kanamycin A and AMPCPP (60).

found in aminoglycoside antibiotic-producing actinomycetes (32, 69, 70, 83), and APH(3')-IV has been cloned and characterized from the butirosin producer *Bacillus circulans* (72), where presumably these enzymes confer immunity to the antibiotics produced.

APH(3')-IIIa confers resistance to kanamycin, neomycin, paromomycin, and ribostamycin (Fig. 5). The N1-substituted aminoglycosides butirosin, amikacin, and isepamicin are substrates, although the MICs for the latter two are lower, commensurate with smaller k_{cat}/K_m values (46). APH(3')-IIIa also confers resistance to lividomycin

A, an aminoglycoside that lacks a 3'-hydroxyl group. This resistance has been attributed to phosphorylation at the 5″-hydroxyl group of the pentose (86), a prediction that has been validated by isolation and NMR characterization of phospho-lividomycin A (85). In addition, some 4,5-disubstituted 2-deoxystreptamine aminoglycosides such as neomycin can be diphosphorylated at positions 3' and 5″ (Fig. 5B). The gene encoding APH(3')-IIIa has been cloned from the *S. aureus* plasmid pSH2 (28), from plasmid pJH1 isolated from *E. faecalis* (86), and also from *Streptococcus pneumoniae*, where it is apparently chromosomally encoded (9). The APH(3')-IIIa enzyme has been overexpressed in *E. coli*, purified, and characterized extensively (44, 46–48, 84–86).

The APH(3')-IIIa enzyme is isolated either as a 31-kDa monomer or as a covalent dimer consisting of two monomer molecules linked head to tail through two disulfide bonds between Cys-19 and Cys-156. Both the monomer and dimer are kinetically indistinguishable in the steady state, unlike the example of ANT(4')-Ia, in which the active site lies at the dimer interface (31, 46). Steady-state kinetic analysis confirmed the broad-substrate specificity of the enzyme with K_m values between 7 and 35 μM for most aminoglycoside substrates and 198 μM and 245 μM, respectively, for isepamicin and amikacin, the poorest substrates tested (46, 49). As expected, tobramycin and dibekacin, both 3'-deoxyaminoglycosides, were not substrates but were competitive inhibitors of kanamycin phosphorylation and noncompetitive inhibitors of ATP utilization, indicating that these compounds bind to the aminoglycoside-binding region of the enzyme and form abortive, dead-end complexes with ATP (47). The k_{cat}, or turnover number, for APH(3')-IIIa represents the rate at saturating substrate concentration and was remarkably consistent, varying only twofold at 1.7 to 3.9 s^{-1} for all aminoglycoside substrates analyzed (46). On the other hand, k_{cat}/K_m, the rate at subsaturating antibiotic concentration, varied from 0.1 to 1.8 × 10^5 M^{-1} s^{-1} (46). Plotting MIC

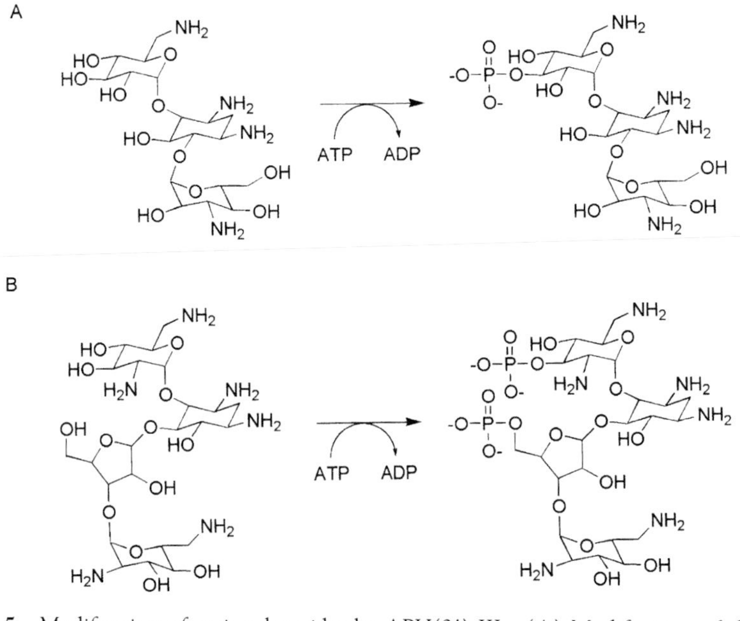

FIGURE 5 Modification of aminoglycosides by APH(3')-IIIa. (A) Modification of the 3'-OH of kanamycin A. (B) Modification of the 3' and 5″-hydroxyls of neomycin C.

versus both rates indicated a positive correlation with k_{cat}/K_m and not with k_{cat}. These results parallel those obtained with several other aminoglycoside-modifying enzymes (3, 16, 64) and are consistent with the hypothesis that these enzymes have evolved to modify aminoglycosides efficiently at low concentrations (below K_m) rather than at higher (saturating) concentrations, where the cell may already be overwhelmed by the action of the compounds. Thus, they likely represent bona fide detoxifying catalysts.

The values for k_{cat}/K_m determined for APH(3')-IIIa were on the order of 10^4 to 10^5 M^{-1} s^{-1}, significantly below the limiting rate of diffusion of small molecules in solution (approximately 10^8 to 10^9 M^{-1} s^{-1}). Enzymes that operate in this latter range are said to be "perfect catalysts" in that the specificity of the enzyme cannot be increased beyond this value. Antibiotic-resistance enzymes such as β-lactamases can operate close to this limit (25), as does APH(3')-Ia, the most prevalent APH(3') found in gram-negative clinical isolates (81). A survey of several synthetically prepared aminoglycosides in which amino groups were selectively replaced with hydrogen revealed that, unlike the enzymes of gram-negative bacteria, APH(3')-Ia and APH(3')-IIa (65), APH(3')-IIIa from gram-positive organisms is relatively tolerant to the loss of substrate amino groups (45). This can be interpreted by the proposal that, despite the fact that APH(3')-Ia has higher catalytic specificity for some aminoglycosides, as assessed by k_{cat}/K_m approaching diffusion control, APH(3')-IIIa displays a broader tolerance for different aminoglycoside substrates, which is "paid for" by a decrease in specificity.

The three-dimensional structure of APH(3')-IIIa has been determined to a resolution of 2.2 Å complexed with the product ADP (31). The most striking aspect of the structure is the dramatic overall three-dimensional similarity with Ser/Thr and Tyr kinases (Fig. 6) despite the very low (<10%) amino acid sequence homology between protein kinases and APHs. The enzyme is bilobal, with an N-terminal domain consisting largely of β-sheets and a C-terminal α-helical domain. The active site of the enzyme lies at the juncture of the two domains, and it is within this region that the few amino acid residues that are invariant between protein kinases and APH(3') are located. For example, Lys-44, which had been suggested to line the ATP-binding pocket, based on affinity labeling of APH(3')-IIIa (44), bridges the α- and β-phosphates of ADP in a fashion that is comparable to that seen in protein kinase structures. Similarly, Asn-195 is an Mg^{2+} ligand in protein and aminoglycoside kinases. Asp-190 is also conserved and located within the enzyme active site. The position of this residue is consistent with a role in deprotonation of the reactive hydroxyl group of the aminoglycoside substrate (either 3' or 5"). Consequently, mutagenesis to Ala significantly reduces enzyme activity (31). This is consistent with the proposed mechanism of phosphoryl transfer catalyzed by APH(3')-IIIa, in which direct attack of the aminoglycoside antibiotic at the γ-phosphate of ATP occurs rather than formation of a phospho-enzyme intermediate (84). However, the precise role of Asp-190 in phosphoryl transfer remains ambiguous at present.

The impressive structural similarity between protein kinases and APH(3')-IIIa is further supported by functional and inhibition studies. APH(3')-IIIa and the bifunctional AAC(6')-APH(2") (see below) have both been demonstrated to have low-level, selective protein serine kinase activity (12). Protein and peptide substrates such as myelin basic protein and peptides derived from myristolated alanine-rich C-kinase substrate (MARCKS) protein were substrates for both enzymes, with k_{cat}/K_m values on the order of 10^2 M^{-1} s^{-1}. Additionally, known inhibitors of protein kinases such as the isoquinoline sulfonamides (H- and CKI series) and the flavanoids quercetin and genistein were shown to inhibit both APH(3')-IIIa and only the aminoglycoside kinase activity of AAC(6')-APH(2") (13). As expected, inhibition with these compounds was competitive with ATP and noncompetitive with aminoglycoside substrates, consistent with their known binding modes in protein kinases. The common elements of structure, function, and inhibition suggest a common ancestry for protein and aminoglycoside kinases, perhaps evolved in an actinomycete antibiotic-producing organism in which Ser/Thr kinases have been recently cloned (94).

Four proteins have been identified that phosphorylate aminoglycosides in the (2") position; however, only three are observed in gram-positive bacteria. APH(2")-Ia is the carboxyl-terminal domain of the commonly occurring bifunctional protein AAC(6')-APH(2"). The phosphorylating domain has been shown to be separable from, and function independently of, the acetylating domain when the APH(2") domain is placed under the control of a heterologous promoter (22, 78). The phosphorylation of gentamicin (A, B, C1, C2, and C1a), sisomicin, amikacin, isepamicin, tobramycin, neomycin, paromomycin, lividomycin, and kanamycin (A and B) (among others) has been demonstrated to be catalyzed by the bifunctional enzyme (11, 41). The presence of the bifunctional enzyme not only encodes high-level resistance to clinically important aminoglycosides but also eliminates the synergy between aminoglycosides and cell wall-active agents, such as ampicillin or vancomycin. NMR characterization of the phosphorylated product of kanamycin A revealed 2"-phosphorylation, as predicted (Fig. 7A) (1). The enzyme also catalyzes the

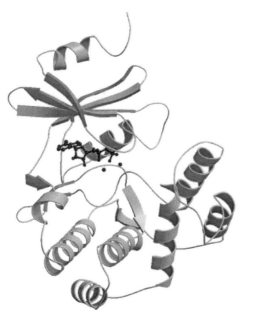

FIGURE 6 Three-dimensional structure of the APH(3')-IIIa·ADP complex. Access to the active site is through the cleft visible on the right (31).

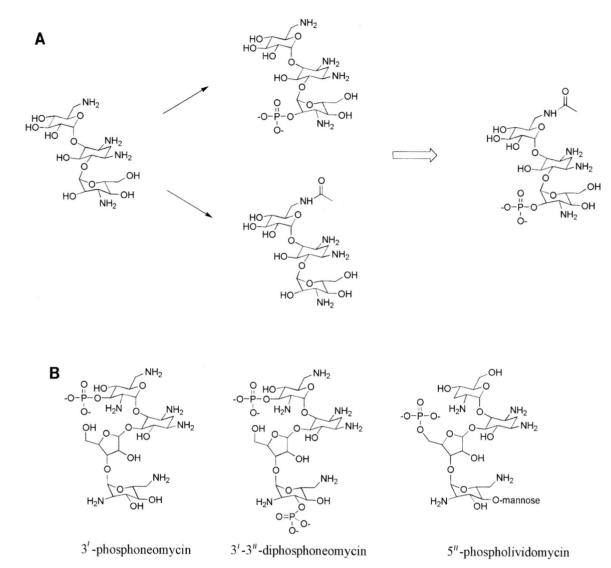

3′-phosphoneomycin   3′-3″-diphosphoneomycin   5″-phospholividomycin

FIGURE 7 Enzyme activity of the bifunctional enzyme AAC(6′)-APH(2″). A) Modification of kanamycin A results in 2″-phosphorylation and 6′-acetylation. Some bisubstituted product is also produced. (B) Modification of neomycin C results in 3′-phosphorylation and 3′-3″-diphosphorylation, and modification of lividomycin A results in 5″-phosphorylation.

phosphorylation of aminoglycosides with 4,5-disubstituted 2-deoxystreptamine rings such as neomycin, and here phosphorylation has been demonstrated to occur at the 3′, 3″, and 5″ positions (Fig. 7B) (11).

The gene that encodes APH(2″)-Ic was identified in and cloned from a plasmid carried by a gentamicin-resistant veterinary isolate of *Enterococcus gallinarum* (8). Significant increases in resistance to gentamicin, tobramycin, 6′-*N*-ethylnetilmicin, 5-episisomicin, kanamycin, and netilmicin were observed in a strain carrying the cloned gene; smaller increases in the MIC of dibekacin, amikacin, and neomycin were also observed (8). An intragenic probe for *aph(2″)-Ic* hybridized to 51 clinical isolates of enterococci, including 12 with intermediate levels of gentamicin resistance (MICs of 64 to 512 μg/ml). Positive hybridization was observed with chromosomal DNA of one *Enterococcus faecium* isolate and with plasmid DNA of one *E. faecalis*

isolate (8). Since the level of resistance to gentamicin that is observed in these two strains is only 256 μg/ml, this resistance gene may be missed by clinical laboratories that routinely use gentamicin levels at 500 to 2,000 μg/ml in screening for high-level aminoglycoside resistance (8). Although the presence of this gene compromises ampicillin-gentamicin synergy, synergy with netilmicin, amikacin, or isepamicin should be maintained (8).

The *aph(2″)-Id* gene was recently cloned from chromosomal DNA of a clinical isolate of *Enterococcus casseliflavus* that was highly resistant to gentamicin (MIC >2,000 μg/ml) (87). When this gene was cloned in a high-copy plasmid, significant increases in the MIC of gentamicin, dibekacin, tobramycin, 6′-*N*-ethylnetilmicin, and 5-episisomicin were observed in *E. coli* DH5α transformants (87). No changes in the MIC were observed for amikacin or neomycin. An intragenic probe hybridized to 17 of 118 enterococcal clinical isolates, whereas the common bifunctional *aac(6′)-Ie-aph(2″)-Ia* gene was not found in these 17 isolates.

Aminoglycoside N-acetyltransferases (AAC)

Acetyltransferases use acetyl coenzyme A to modify the 1, 2′, 3, or 6′ amino groups of aminoglycosides. Although over 25 genes have been identified that encode AAC(6′) enzymes, only two of them are found in gram-positive bacteria. The first is a bifunctional enzyme, AAC(6′)-APH(2″), that contains two separable domains, as noted above (22, 41, 78). The 5′ portion of the gene [*aac(6′)-Ie*] encodes resistance to kanamycin, tobramycin, netilmicin, amikacin, and fortimicin and shares sequence similarity to other AAC(6′) enzymes.

In a survey of aminoglycoside-resistant bacteria, 41.7% of aminoglycoside-resistant staphylococci used the AAC(6′)-APH(2′) mechanism. This mechanism was the most commonly occurring aminoglycoside-resistant mechanism in both staphylococci and *E. faecalis* (51).

The bifunctional AAC(6′)-APH(2″) mechanism was previously thought to be restricted to gram-positive organisms. However, Kettner et al. (34) have shown the occurrence of a similar mechanism in gram-negative bacteria, although fortimicin resistance is not conferred (61). The genes encoding this mechanism have been cloned and share considerable sequence homology with the sequences from gram-positive bacteria (61). However, in gram-negative bacteria, the two domains exist as separate genes, and the order of the genes is reversed [*aph(2″)-Ib* is proximal to *aac(6′)-Im*] (61).

The 57-kDa bifunctional AAC(6′)-APH(2″) has been purified from *S. aureus* and *Staphylococcus epidermidis* (88) and expressed in *E. coli* (1) and *B. subtilis* (11). The predicted acetylation of the 6′-amino group of kanamycin A has been demonstrated by NMR, and, in addition, some disubstituted product (2″-phospho + 6′-acetyl) has been recovered in reaction mixtures that included purified recombinant bifunctional enzyme, ATP, and acetyl coenzyme A (1). A bisubstituted product of arbekacin has also been recovered from extracts of *S. aureus* (35). The purified bifunctional enzyme can also catalyze O-acetyl transfer to the 6′-hydroxy aminoglycosides lividomycin and paromomycin (11). Thus, this enzyme confers the broadest substrate specificity known for AAC(6′) enzymes.

The second acetyltransferase observed in gram-positive bacteria is chromosomally encoded by *aac(6′)-Ii*, which encodes resistance to kanamycin, tobramycin, netilmicin, and amikacin. DNA hybridization using an *aac(6′)-Ii* probe demonstrated the species specificity of this gene: while all 26 strains of *E. faecium* tested hybridized to the probe, no hybridization was observed with any of the 44 other enterococcal strains, representing 13 different species (10). Insertional inactivation of *aac(6′)-Ii* abolished resistance to aminoglycosides that are substrates of the AAC(6′)-I enzyme (10).

The AAC(6′)-Ii protein shows significant sequence homology (>40%) to the AAC(6′)-Ia protein from *Citrobacter diversus* (82) and *Shigella sonnei* (23) as well as other AAC(6′) enzymes found in gram-negative bacteria (7, 29). The 20.7-kDa AAC(6′)-Ii enzyme has been overexpressed in *E. coli*, purified, and characterized (92). Like most other aminoglycoside-modifying enzymes, the purified AAC(6′)-Ii protein is isolated as a homodimer. Steady-state kinetic analysis demonstrated that, although many

aminoglycosides interact with the enzyme, acetylation occurred only with aminoglycosides possessing C-6′ amine and not a 6′-hydroxy group (92), differentiating it from AAC(6′)-APH(2″). Rather, aminoglycosides with 6′-hydroxyl groups were competitive inhibitors of kanamycin acetylation. Specificity constant (k_{cat}/K_m) values were relatively low compared with those of other aminoglycoside-modifying enzymes (10^4 M^{-1} s^{-1}). Furthermore, unlike plasmid-borne aminoglycoside-modifying enzymes that have been evaluated, MICs for the chromosomal AAC(6′)-Ii enzyme are correlated with k_{cat} (the rate at substrate saturation) and not k_{cat}/K_m (the rate at low substrate concentrations). These data suggest that this enzyme is not optimally evolved for aminoglycoside inactivation and supports the concept that the chromosomally encoded aminoglycoside enzymes may have alternative housekeeping roles in the cell [e.g., the AAC(2′)-Ia in *Providencia stuartii* has been implicated in peptidoglycan acetylation (59)]. Consistent with this hypothesis is the observation that purified AAC(6′)-Ii can acetylate some proteins in vitro (93). Furthermore, the recent determination of the three-dimensional structure of the enzyme has revealed a structural motif in common with yeast histone acetyltransferase (93).

Although enzymes that acetylate other amino groups of aminoglycosides have been identified, including AAC(1), AAC(2′), and AAC(3), these activities have thus far been detected only in gram-negative bacteria.

PREVALENCE AND DISTRIBUTION OF AMINOGLYCOSIDE-MODIFYING ENZYMES

Aminoglycoside-resistance mechanisms in gram-positive pathogens are primarily limited to the bifunctional AAC(6′)-APH(2″), ANT(4′)-Ia, and APH(3′)-IIIa. In a study of aminoglycoside usage and the distribution of resistance mechanisms, no correlation was observed between these two factors (50, 51). In this survey of staphylococcal isolates from a wide range of geographic areas, AAC(6′)-APH(2″) was the most common mechanism, occurring in 90% of resistant isolates. ANT(4′)-Ia was present in 44% of these isolates, and APH(3′)-IIIa was less common, occurring in only 15% of the resistant isolates. These resistance mechanisms were found either alone or in combination. In fact, previous studies have shown that 49% of resistant isolates had multiple resistance mechanisms (51). The incidence of these resistance mechanisms in *Staphylococcus* and *Enterococcus* spp. is comparable (79).

MODES OF TRANSFER

Enterococci and streptococci are generally naturally resistant to low levels of aminoglycosides, whereas staphylococci are not. High-level resistance is usually due to the acquisition of genes encoding aminoglycoside-modifying enzymes. The genes encoding aminoglycoside-modifying enzymes can be plasmid or transposon borne or chromosomally located (Table 1). The most commonly found plasmid-borne genes are the bifunctional *aac(6′)-aph(2″)*, *ant(4′)-Ia*, and *aph(3′)-IIIa*, which are found alone and in all possible combinations (51).

The *ant(4′)-Ia* gene is commonly found in staphylococci but has been detected in at least one strain of enterococcus (6, 58) and in *B. stearothermophilus* (43). This gene has been found both on small plasmids and on several large plasmids (20, 40). The *ant(4')-Ia* gene was cloned from plasmid pUB110, a small, 4.5-kb plasmid that also encodes resistance to bleomycin by an open reading frame that directly follows the *ant(4')-Ia* gene. Plasmid pUB110 is transferable to *B. stearothermophilus* (43) and has been shown to be completely stable under nonselective conditions in *B. subtilis* (63, 75). This plasmid is often found in clinical isolates of *S. aureus* and has been observed in strains containing only this plasmid, in strains also containing larger conjugative plasmids, integrated into larger conjugative plasmids, or integrated into the genome. The conjugative plasmid pSK41 has been shown to contain a complete copy of pUB110, flanked by IS257 elements (4). It is likely that pUB110 is integrated in other closely related conjugative plasmids, including pJE1, pSH7, pSH8, and pSH9 (4). It is possible that IS257 mediates pUB110 integration into other plasmids by transposition-cointegrate formation or by homologous recombination between IS257 elements on other plasmids or chromosomes. A chromosomal copy of pUB110, flanked by IS257 elements, has been observed adjacent to the *mecA* gene in methicillin-resistant strains of *S. aureus* (MRSA) (18, 76). The fortuitous location of the *ant(4')-Ia* gene on pUB110 and the subsequent insertion of this plasmid into larger conjugative plasmids may be responsible for the rapid dissemination of this gene.

The bifunctional *aac(6')-aph(2")* gene has been identified in both staphylococci and streptococci. The environment of the gene can be variable. It has been identified within Tn4001 present in pSK1 in *S. aureus* (66), *S. faecalis* plasmid pIP800 (22), and *S. epidermidis* TK406 (88). Restriction analysis suggests that the bifunctional gene is often found on structurally related conjugative plasmids that range in size from 20 to 57 kb (27, 42). These plasmids can also carry the *ant(4')-Ia* gene, the *aph(3')-IIIa* gene, and other antibiotic- and antiseptic-resistance determinants (42).

The *aac(6')-aph(2")* gene has also been localized to the chromosome of *S. aureus* isolates from France (21). Furthermore, clinical strains with insertions of Tn4001 (containing the bifunctional gene) into different chromosomal sites have been identified (26). Transposition of Tn4001 to both plasmid and chromosomal sites has been demonstrated in vitro (42), suggesting that this transposition may be at least partially responsible for the widespread distribution of the bifunctional gene.

The *ant(6)-Ia* and *aph(3')-IIIa* genes are also found in both staphylococcal and enterococcal species (58). They are linked within Tn5405, a staphylococcal composite transposon (17). This transposon carries three antibiotic-resistance genes in the order *ant(6)-Ia*, *sat4* (streptothricin resistance), and *aph(3')-IIIa*, in addition to three unknown open reading frames. The three antibiotic-resistance genes were also found in *S. aureus* plasmid pIP1718, which lacks the insertion elements flanking the sequences on the transposon. In a recent study (17), 50 epidemiologically unrelated staphylococci, collected between 1978 and 1995, were chosen because they hybridized with an *aph(3')-IIIa* probe. All 50 were shown to contain this conserved gene arrangement, although some rearrangements in the *sat4* gene were observed. The conserved gene arrangement is the likely explanation for the curious finding that all 38 streptomycin-susceptible strains of staphylococci and enterococci that hybridized with the *ant(6)-Ia* probe were also resistant to kanamycin by production of *aph(3')-IIIa* (58). It is likely that, although the *ant(6)-Ia* gene is somehow

defective in these strains, the linkage to *aph(3')-IIIa* is conserved.

This linked structure of genes was observed to be chromosomally located in some staphylococcal strains, whereas it was plasmid borne in others (17). The *E. faecalis* plasmid pJH1 also contains the three genes in this order, as does *Campylobacter coli* (17). These data suggest horizontal intergeneric transfer of a family of related mobile elements (17). The *aph(3')-IIIa* gene has also been cloned from Tn1545 from *S. pneumoniae* (5). This element also confers macrolide-lincosamide-streptogramin resistance (*ermA*) and resistance to tetracycline.

The *S. aureus* transposon Tn554 specifies resistance to spectinomycin [*ant(9)-Ia*] and macrolide-lincosamide-streptogramin B (*ermA*) (53). Tn554 transposes with a very high efficiency to a unique site in the *S. aureus* chromosome (*att554*) and with a lower efficiency to other secondary chromosomal or plasmid sites. It has never been found naturally on a plasmid (37). MRSA clinical isolates responsible for a 1987 outbreak of infections in New Jersey have been shown to contain Tn554 closely associated with the *mec* region (*att155*), a site that is absent in methicillin-susceptible isolates (19). This region was also shown to harbor *ant(4')-Ia* within an integrated copy of pUB110. In a survey of 3,398 MRSA isolates from more than 40 hospitals, over 90% contained Tn554 (spectinomycin resistance) (37). Many of these isolates were shown to harbor independent insertions of Tn554. The *mecA*/Tn554 hybridization patterns are useful epidemiological tools for following hospital outbreaks (37). The association of the *ant(9)-Ia* gene within Tn554 and in MRSA strains may provide a fortuitous vehicle for the spread of this gene, independent of aminoglycoside selection.

The *ant(9)-Ib* gene of *E. faecalis* was cloned from a strain that also carried determinants for erythromycin resistance and tetracycline resistance. These genes have been shown to be transferable by conjugation to plasmid-free strains (39). The transfer of the *ant(9)-Ib* gene and/or the gene encoding erythromycin resistance was always accompanied by the transfer of tetracycline resistance (39). The precise arrangement of these genes is unknown.

A probe specific for the *aph(2")-Ic* gene, recently cloned from a veterinary isolate of *E. gallinarum*, was used to test for the presence of this gene in 51 enterococcal clinical isolates. This probe hybridized to chromosomal DNA of one *E. faecium* isolate and to plasmid DNA of one *E. faecalis* isolate (8). The genetic environment of this gene in these isolates has not been further characterized. There is currently no evidence for the spread of the remaining chromosomal aminoglycoside genes *ant(6)-Ib*, *ant(4')-IIb*, *aph(2")-Id*, and *aac(6')-Ii* outside their species of origin.

CONCLUSIONS

Genes conferring resistance to aminoglycosides have been shown to be both chromosomally located and plasmid borne in gram-positive bacteria. A variety of mechanisms have been shown to disseminate these genes, either alone or in combination with other resistance determinants. The widespread nature of these resistance genes has greatly limited the clinical usefulness of aminoglycosides. In particular, the bifunctional AAC(6')-APH(2") determinant, which confers high-level resistance to virtually all important aminoglycosides, has had a devastating effect on aminoglycoside use. Reversal of this trend will require the identification of inhibitors that will block both enzyme activities found in this protein. Such compounds could revive the utility of aminoglycosides in the clinic.

REFERENCES

1. **Azucena, E., I. Grapsas, and S. Mobashery.** 1997. Properties of a bifunctional bacterial antibiotic resistance enzyme that catalyzes ATP-dependent 2″-phosphorylation and acetyl-CoA-dependent 6′-acetylation of aminoglycosides. *J. Am. Chem. Soc.* **119:**2317–2318.

2. **Banai, M., and D. J. LeBlanc.** 1983. Genetic, molecular, and functional analysis of *Streptococcus faecalis* R plasmid pJH1. *J. Bacteriol.* **155:**1094–1104.

3. **Bongaerts, G. P. A., and L. Molendijk.** 1984. Relation between aminoglycoside 2″-O-nucleotidyltransferase activity and aminoglycoside resistance. *Antimicrob. Agents Chemother.* **25:**234–237.

4. **Byrne, M. E., M. T. Gillespie, and R. A. Skurray.** 1991. 4′,4″ adenyltransferase activity on conjugative plasmids isolated from *Staphylococcus aureus* is encoded on an integrated copy of pUB110. *Plasmid* **25:**70–75.

5. **Caillaud, F., P. Trieu-Cuot, C. Carlier, and P. Courvalin.** 1987. Nucleotide sequence of the kanamycin resistance determinant of the pneumococcal transposon Tn*1545*: evolutionary relationships and transcriptional analysis of *aphA-3* genes. *Mol. Gen. Genet.* **207:**509–513.

6. **Carlier, C., and P. Courvalin.** 1990. Emergence of 4′,4″-aminoglycoside nucleotidyltransferase in enterococci. *Antimicrob. Agents Chemother.* **34:**1565–1569.

7. **Centron, D., and P. H. Roy.** 1998. Characterization of the 6′-N-aminoglycoside acetyltransferase gene *aac(6′)-Iq* from the integron of a natural multiresistance plasmid. *Antimicrob. Agents. Chemother.* **42:**1506–1508.

8. **Chow, J. W., M. J. Zervos, S. A. Lerner, L. A. Thal, S. M. Donabedian, D. D. Jaworski, S. Tsai, K. J. Shaw, and D. B. Clewell.** 1997. A novel gentamicin resistance gene in *Enterococcus. Antimicrob. Agents Chemother.* **41:**511–514.

9. **Collatz, E., C. Carlier, and P. Courvalin.** 1983. The chromosomal 3′,5″-aminoglycoside phosphotransferase in *Streptococcus pneumoniae* is closely related to its plasmid-coded homologs in *Streptococcus faecalis* and *Staphylococcus aureus. J. Bacteriol.* **156:**1373–1377.

10. **Costa, Y., M. Galimand, R. Leclercq, J. Duval, and P. Courvalin.** 1993. Characterization of the chromosomal *aac(6′)-Ii* gene specific for *Enterococcus faecium. Antimicrob. Agents Chemother.* **37:**1896–1903.

11. **Daigle, D. M., D. W. Hughes, and G. D. Wright.** 1999. Prodigious substrate specificity of AAC(6′)-APH(2″), an aminoglycoside antibiotic resistance determinant in enterococci and staphylococci. *Chem. Biol.* **6:**99–110.

12. **Daigle, D. M., G. A. McKay, P. R. Thompson, and G. D. Wright.** 1999. Aminoglycoside phosphotransferases required for antibiotic resistance are also serine protein kinases. *Chem. Biol.* **6:**11–18.

13. **Daigle, D. M., G. A. McKay, and G. D. Wright.** 1997. Inhibition of aminoglycoside antibiotic resistance enzymes by protein kinase inhibitors. *J. Biol. Chem.* **272:**24755–24758.

14. **Davis, B. D.** 1987. Mechanism of action of aminoglycosides. *Microbiol. Rev.* **51:**341–350.

15. **Davis, B. D., L. L. Chen, and P. C. Tai.** 1986. Misread protein creates membrane channels: an essential step in the bactericidal action of aminoglycosides. *Proc. Natl. Acad. Sci. USA* **83:**6164–6168.

16. **DeHertogh, D. A., and S. A. Lerner.** 1985. Correlation of aminoglycoside resistance with K_ms and V_{max}/K_m ratios of enzymatic modification of aminoglycosides by 2″-O-nucleotidyltransferase. *Antimicrob. Agents Chemother.* **27:**670–671.

17. **Derbise, A., S. Aubert, and N. El Solh.** 1997. Mapping the regions carrying the three contiguous antibiotic resistance genes *aadE*, *sat4*, and *aphA-3* in the genomes of staphylococci. *Antimicrob. Agents Chemother.* **41:**1024–1032.

18. **Dubin, D. T., and B. D. Davis.** 1961. The effect of streptomycin on potassium flux in *Escherichia coli. Biochim. Biophys. Acta* **52:**400–402.

19. **Dubin, D. T., P. R. Matthews, S. G. Chikramane, and P. R. Stewart.** 1991. Physical mapping of the *mec* region of an American methicillin-resistant *Staphylococcus aureus* strain. *Antimicrob. Agents Chemother.* **35:**1661–1665.

20. **El Solh, N., J. M. Fouace, Z. Shalita, D. H. Bouanchaud, R. P. Novick, and Y. A. Chabbert.** 1980. Epidemiological and structural studies of *Staphylococcus aureus* R plasmids mediating resistance to tobramycin and streptogramin. *Plasmid* **4:**117–120.

21. **El Solh, N., N. Moreau, and S. D. Ehrlich.** 1986. Molecular cloning and analysis of *Staphylococcus aureus* chromosomal aminoglycoside resistance genes. *Plasmid* **15:**53–58.

22. **Ferretti, J. J., K. S. Gilmore, and P. Courvalin.** 1986. Nucleotide sequence analysis of the gene specifying the bifunctional 6′-aminoglycoside acetyltransferase 2″-aminoglycoside phosphotransferase enzyme in *Streptococcus faecalis* and identification and cloning of gene regions specifying the two activities. *J. Bacteriol.* **167:**631–638.

23. **Fling, M. E., J. Kopf, and C. Richards.** 1985. GenBank accession no. M86913.

24. **Fourmy, D., M. I. Recht, S. C. Blanchard, and J. D. Puglisi.** 1996. Structure of the A site of *Escherichia coli* 16S ribosomal RNA complexed with an aminoglycoside antibiotic. *Science* **274:**1367–1371.

25. **Frère, J. M.** 1995. Beta-lactamases and bacterial resistance to antibiotics. *Mol. Microbiol.* **16:**385–395.

26. **Gillespie, M. T., B. R. Lyon, L. J. Messerotti, and R. A. Skurray.** 1987. Chromosome- and plasmid-mediated gentamicin resistance in *Staphylococcus aureus* encoded by Tn*4001. J. Med. Microbiol.* **24:**139–144.

27. **Goering, R. V., and E. A. Ruff.** 1983. Comparative analysis of conjugative plasmids mediating gentamicin resistance in *Staphylococcus aureus. Antimicrob. Agents Chemother.* **24:**450–452.

28. **Gray, G. S., and W. M. Fitch.** 1983. Evolution of antibiotic resistance genes: the DNA sequence of a kanamycin resistance gene from *Staphylococcus aureus. Mol. Biol. Evol.* **1:**57–66.

29. **Hannecart-Pokorni, E., F. Depuydt, L. de Wit, E. van Bossuyt, J. Content, and R. Vanhoof.** 1997. Characterization of the 6′-N-aminoglycoside acetyltransferase gene *aac(6′)-Im* [corrected] associated with a *sulI*-type integron. *Antimicrob. Agents Chemother.* **41:**314–318.

30. **Holm, L., and C. Sander.** 1995. DNA polymerase β belongs to an ancient nucleotidyltransferase superfamily. *Trends Biol. Chem.* **20:**345–347.

31. **Hon, W. C., G. A. McKay, P. R. Thompson, R. M. Sweet, D. S. C. Yang, G. D. Wright, and A. M. Berghuis.** 1997. Structure of an enzyme required for aminoglycoside resistance reveals homology to eukaryotic protein kinases. *Cell* **89:**887–895.

32. **Hoshiko, S., C. Nojiri, K. Matsunaga, K. Katsumata, E. Satoh, and K. Nagaoka.** 1988. Nucleotide sequence of the ribostamycin phosphotransferase gene and of its control region in *Streptomyces ribosidificus*. *Gene* **68:**285–296.

33. **Kanikula, A. M., H. H. Liao, J. Sakon, H. M. Holden, and I. Rayment.** 1992. Crystallization and preliminary crystallographic analysis of a thermostable mutant of kanamycin nucleotidyltransferase. *Arch. Biochem. Biophys.* **295:**1–4.

34. **Kettner, M., T. Macickova, and V. J. Krcmery.** 1991. Susceptibility of amikacin- and gentamicin-resistant clinical isolates of gram-negative bacteria to isepamicin, p. 273–275. *In* V. J. Krcmery, D. Adam, O. Balint, and E. Rubinstein (ed.), *Antimicrobial Chemotherapy in Immunocompromised Host*, vol. XL. International Congress of Chemotherapy, Berlin.

35. **Kondo, S., A. Tamura, S. Gomi, Y. Ikeda, T. Takeuchi, and S. Mitsuhashi.** 1993. Structures of enzymatically modified products of arbekacin by methicillin-resistant *Staphylococcus aureus*. *J. Antibiot.* **46:**310–315.

36. **Kono, M., K. Ohmiya, T. Kanda, N. Noguchi, and K. O'Hara.** 1987. Purification and characterization of chromosomal streptomycin adenyltransferase from derivatives of *Bacillus subtilis* Marburg 168. *FEMS Microbiol. Lett.* **40:**223–228.

37. **Kreiswirth, B., J. Kornblum, R. D. Arbeit, W. Eisner, J. N. Maslow, A. McGeer, D. E. Low, and R. P. Novick.** 1993. Evidence for a clonal origin of methicillin resistance in *Staphylococcus aureus*. *Science* **259:**227–230.

38. **Kunst, F., N. Ogasawara, I. Moszer, A. M. Albertini, G. Alloni, V. Azevedo, M. G. Bertero, P. Bessieres, A. Bolotin, S. Borchert, R. Borriss, L. Boursier, A. Brans, M. Braun, S. C. Brignell, S. Bron, S. Brouillet, C. V. Bruschi, B. Caldwell, V. Capuano, N. M. Carter, S. K. Choi, J. J. Codani, I. F. Connerton, and A. Danchin.** 1997. The complete genome sequence of the gram-positive bacterium *Bacillus subtilis*. *Nature* **390:**249–256.

39. **LeBlanc, D. J., L. N. Lee, and J. M. Inamine.** 1991. Cloning and nucleotide base sequence analysis of a spectinomycin adenyltransferase AAD(9) determinant from *Enterococccus faecalis*. *Antimicrob. Agents Chemother.* **35:**1804–1810.

40. **Le Goffic, F., A. Martel, M. L. Capmau, B. Baca, P. Goebel, H. Chardon, C. J. Soussy, J. Duval, and D. H. Bouanchaud.** 1976. New plasmid-mediated nucleotidylation of aminoglycoside antibiotics in *Staphlococcus aureus*. *Antimicrob. Agents Chemother.* **10:**258–264.

41. **Le Goffic, F., A. Martel, N. Moreau, M. L. Capmau, C. J. Soussy, and J. Duval.** 1977. 2″-O-Phosphorylation of gentamicin components by a *Staphylococcus aureus* strain carrying a plasmid. *Antimicrob. Agents Chemother.* **12:**26–30.

42. **Lyon, B. R., M. T. Gillespie, and R. A. Skurray.** 1987. Detection and characterization of IS256, an insertion sequence in *Staphylococcus aureus*. *J. Gen. Microbiol.* **133:**3031–3038.

43. **Matsumura, M., Y. Katakura, T. Imanaka, and S. Aiba.** 1984. Enzymatic and nucleotide sequence studies of a kanamycin-inactivating enzyme encoded by a plasmid from thermophilic bacilli in comparison with that encoded by plasmid pUB110. *J. Bacteriol.* **160:**413–420.

44. **McKay, G. A., R. A. Robinson, W. S. Lane, and G. D. Wright.** 1994. Active-site labeling of an aminoglycoside antibiotic phosphotransferase (APH(3′)-IIIa). *Biochemistry* **33:**14115–14120.

45. **McKay, G. A., J. Roestamadji, S. Mobashery, and G. D. Wright.** 1996. Recognition of aminoglycoside antibiotics by enterococcal-staphylococcal aminoglycoside 3′-phosphotransferase type IIIa: role of substrate amino groups. *Antimicrob. Agents Chemother.* **40:**2648–2650.

46. **McKay, G. A., P. R. Thompson, and G. D. Wright.** 1994. Broad spectrum aminoglycoside phosphotransferase type III from *Enterococcus*: overexpression, purification, and substrate specificity. *Biochemistry* **33:**6936–6944.

47. **McKay, G. A., and G. D. Wright.** 1995. Kinetic mechanism of aminoglycoside phosphotransferase type IIIa: evidence for a Theorell-Chance mechanism. *J. Biol. Chem.* **270:**24686–24692.

48. **McKay, G. A., and G. D. Wright.** 1996. Catalytic mechanism of enterococcal kanamycin kinase (APH(3′)-IIIa): viscosity, thio, and solvent isotope effects support a Theorell-Chance mechanism. *Biochemistry* **35:**8680–8685.

49. **McKay, G. A., and G. D. Wright.** 1998. Unpublished results.

50. **Miller, G. H., F. J. Sabatelli, R. S. Hare, Y. Glupczynski, P. Mackey, D. Shlaes, K. Shimizu, and K. J. Shaw.** 1997. The most frequent aminoglycoside resistance mechanisms—changes with time and geographic area: a reflection of aminoglycoside usage patterns? Aminoglycoside Resistance Study Groups. *Clin. Infect. Dis.* **24:**S46–S62.

51. **Miller, G. H., F. J. Sabatelli, L. Naples, R. S. Hare, and K. J. Shaw.** 1995. The most frequently occurring aminoglycoside resistance mechanisms—combined results of surveys in eight regions of the world. Aminoglycoside Resistance Study Groups. *J. Chemother.* **7**(Suppl. 2):17–30.

52. **Moazed, D., and H. F. Noller.** 1987. Interaction of antibiotics with functional sites in 16S ribosomal RNA. *Nature* **27:**389–394.

53. **Murphy, E.** 1985. Nucleotide sequence of a spectinomycin adenyltransferase AAD(9) determinant from *Staphylococcus aureus* and its relationship to AAD(3″)(9). *Mol. Gen. Genet.* **200:**33–39.

54. **O'Connor, T., and G. D. Wright.** 1998. Unpublished results.

55. **O'Hara, K., K. Ohmiya, and M. Kono.** 1988. Structure of adenylylated streptomycin synthesized enzymatically by *Bacillus subtilis*. *Antimicrob. Agents Chemother.* **32:**949–950.

56. **Ohmiya, K., T. Tanaka, N. Noguchi, K. O'Hara, and M. Kono.** 1989. Nucleotide sequence of the chromosomal gene coding for the aminoglycoside 6-adenylyltransferase from *Bacillus subtilis* Marburg 168. *Gene* **78:**377–378.

57. **Ounissi, H., and P. Courvalin.** 1987. Nucleotide sequences of streptococcal genes, p. 275. *In* J. J. Ferretti and R. Curtiss III (ed.), *Streptococcal Genetics*. American Society for Microbiology, Washington, D.C.

58. **Ounissi, H., E. Derlot, C. Carlier, and P. Courvalin.** 1990. Gene homogeneity for aminoglycoside-modifying enzymes in gram-positive cocci. *Antimicrob. Agents Chemother.* **34:**2164–2168.

59. Payie, K. G., P. N. Rather, and A. J. Clarke. 1995. Contribution of gentamicin 2'-N-acetyltransferase to the O acetylation of peptidoglycan in *Providencia stuartii*. *J. Bacteriol.* **177:**4303–4310.

60. Perdersen, L. C., M. M. Benning, and H. M. Holden. 1995. Structural investigation of the antibiotic and ATP-binding sites in kanamycin nucleotidyltransferase. *Biochemistry* **34:**13305–13311.

61. Petrin, J., R. Kuvelkar, H. Munayyer, M. Kettner, R. S. Hare, G. H. Miller, and K. J. Shaw. 1995. Presented at "Molecular Genetics of Bacteria and Phages," Cold Spring Harbor, N.Y.

62. Piepersberg, W. 1997. Molecular biology, biochemistry, and fermentation of aminoglycoside antibiotics, p. 81–163. *In* W. Strohl (ed.), *Biotechnology of Industrial Antibiotics*, 2nd ed. Marcel Dekker, Inc., New York, N.Y.

63. Polak, J., and R. P. Novick. 1982. Closely related plasmids from *Staphylococcus aureus* and soil bacilli. *Plasmid* **7:**152–162.

64. Radika, K., and D. B. Northrop. 1984. Correlation of antibiotic resistance with V_{max}/K_m ratio of enzymatic modification of aminoglycosides by kanamycin acetyltransferase. *Antimicrob. Agents Chemother.* **25:**479–482.

65. Roestamadji, J., I. Grapsas, and S. Mobashery. 1995. Loss of individual electrostatic interactions between aminoglycoside antibiotics and resistance enzymes as an effective means to overcoming bacterial drug resistance. *J. Am. Chem. Soc.* **117:**11060–11069.

66. Rouch, D. A., M. E. Byrne, Y. C. Kong, and R. A. Skurray. 1987. The *aacA-aphD* gentamicin and kanamycin resistance determinant of Tn4001 from *Staphylococcus aureus*: expression and nucleotide sequence analysis. *J. Gen. Microbiol.* **133:**3039–3052.

67. Sadaie, Y., K. C. Burtis, and R. H. Doi. 1980. Purification and characterization of a kanamycin nucleotidyltransferase from plasmid pUB110-carrying cells of *Bacillus subtilis*. *J. Bacteriol.* **141:**1178–1182.

68. Sakon, J., H. H. Liao, A. M. Kanikula, M. M. Benning, I. Rayment, and H. M. Holden. 1993. Molecular structure of kanamycin nucleotidyl transferase determined to 3 Å resolution. *Biochemistry* **32:**11977–11984.

69. Salauze, D., and J. Davies. 1991. Isolation and characterization of and aminoglycoside phosphotransferase from neomycin-producing *Micromonospora chalcea*: comparison with that of *Streptomyces fradiae* and other producers of 4,6-disubstituted 3-deoxystreptamine antibiotics. *J. Antibiot.* **44:**1432–1443.

70. Salauze, D., J.-A. Perez-Gonzalez, W. Piepersberg, and J. Davies. 1991. Characterization of aminoglycoside acetyltransferase-encoding genes of neomycin-producing *Micromonospora chalcea* and *Streptomyces fradiae*. *Gene* **101:**143–148.

71. Santanam, P., and F. H. Kayser. 1978. Purification and characterization of an aminoglycoside inactivating enzyme from *Staphylococcus epidermidis* FK109 that nucleotidylates the 4'- and 4"-hydroxyl groups of the aminoglycoside antibiotics. *J. Antibiot.* **31:**343–351.

72. Sarwar, M., and M. Akhtar. 1990. Cloning of aminoglycoside phosphotransferase (APH) gene from antibiotic-producing strain of *Bacillus circulans* into a high-expression vector, pKK223-3. Purification, properties and location of the enzyme. *Biochem. J.* **268:**671–677.

73. Sawaya, M. R., H. Pelletier, A. Kumar, S. H. Wilson, and J. Kraut. 1994. Crystal structure of rat DNA polymerase β: evidence for a common polymerase mechanism. *Science* **264:**1930–1935.

74. Schatz, A., E. Bugie, and S. A. Waksman. 1944. Streptomycin, a substance exhibiting antibiotic activity against Gram-positive and Gram-negative bacteria. *Proc. Soc. Exp. Biol. Med.* **55:**66–69.

75. Scheer-Abromowitz, J., T. J. Gryczan, and D. Dubnau. 1980. Origin and mode of replication of plasmids pE194 and pUB110. *Plasmid* **6:**67–77.

76. Shaw, K. J. 1997. Unpublished observations.

77. Shaw, K. J., H. Munayyer, P. N. Rather, R. S. Hare, and G. H. Miller. 1993. Nucleotide sequence analysis and DNA hybridization studies of the *ant(4')-IIa* gene from *Pseudomonas aeruginosa*. *Antimicrob. Agents Chemother.* **37:**708–714.

78. Shaw, K. J., P. N. Rather, R. S. Hare, and G. H. Miller. 1993. Molecular genetics of aminoglycoside resistance genes and familial relationships of the aminoglycoside-modifying enzymes. *Microbiol. Rev.* **57:**138–163.

79. Shaw, K. J., F. J. Sabatelli, L. Naples, P. Mann, R. S. Hare, and G. H. Miller. 1998. The application of molecular techniques for the study of aminoglycoside resistance. *In* N. Woodford and A. P. Johnson (ed.), *Molecular Bacteriology: Protocols and Clinical Applications*, vol. 15. Humana Press Inc., Totowa, N.J.

80. Shirafuji, H., M. Kida, I. Nogami, and M. Yoneda. 1980. Aminoglycoside-4'-nucleotidyltransferase from *Bacillus brevis*. *Agric. Biol. Chem.* **44:**279–286.

81. Siregar, J. J., K. Miroshnikov, and S. Mobashery. 1995. Purification, characterization, and investigation of the mechanism of aminoglycoside 3'-phosphotransferase Type Ia. *Biochemistry* **34:**12681–12688.

82. Tenover, F. C., D. Filpula, K. L. Phillips, and J. J. Plorde. 1988. Cloning and sequencing of a gene encoding an aminoglycoside 6'-N-acetyltransferase from an R factor of *Citrobacter diversus*. *J. Bacteriol.* **170:**471–473.

83. Thompson, C. J., and G. S. Gray. 1983. Nucleotide sequence of a streptomycete aminoglycoside phosphotransferase gene and its relationship to phosphotransferases encoded by resistance plasmids. *Proc. Natl. Acad. Sci. USA* **80:**5190–5194.

84. Thompson, P. R., D. W. Hughes, and G. D. Wright. 1996. Mechanism of aminoglycoside 3'-phosphotransferase type IIIa: His188 is not a phosphate-accepting residue. *Chem. Biol.* **3:**747–755.

85. Thompson, P. R., D. W. Hughes, and G. D. Wright. 1996. Regiospecificity of aminoglycoside phosphotransferase from Enterococci and Staphylococci (APH(3')-IIIa). *Biochemistry* **35:**8686–8695.

86. Trieu-Cuot, P., and P. Courvalin. 1983. Nucleotide sequence of the *Streptococcus faecalis* plasmid gene encoding the 3'5"-aminoglycoside phosphotransferase type III. *Gene* **23:**331–341.

87. Tsai, S. F., M. J. Zervos, D. B. Clewell, S. M. Donabedian, D. F. Sahm, and J. W. Chow. 1998. A new high-level gentamicin resistance gene, *aph(2")Id*, in *Enterococcus* spp. *Antimicrob. Agents Chemother.* **42:**1229–1232.

88. Ubukata, K., N. Yamashita, A. Gotoh, and M. Konno. 1984. Purification and characterization of aminoglycoside-modifying enzymes from *Staphylococcus aureus* and *Staph-*

ylococcus epidermidis. Antimicrob. Agents Chemother. **25:** 754–759.

89. **Van Pelt, J. E., R. Iyengar, and P. A. Frey.** 1986. Gentamicin nucleotidyltansferase. Stereochemical inversion at phosphorus in enzymatic 2'-deoxyadenylyl transfer to tobramycin. *J. Biol. Chem.* **261:**15995–15999.

90. **Waksman, S. A., and H. A. Lechevalier.** 1949. Neomycin, a new antibiotic active against streptomycin-resistant bacteria, including tuberculosis organisms. *Science* **109:** 305–307.

91. **Woodcock, J., D. Moazed, M. Cannon, J. Davies, and H. F. Noller.** 1991. Interaction of antibiotics with A- and P-site-specific bases in 16S ribosomal RNA. *EMBO J.* **10:** 3099–3103.

92. **Wright, G. D., and P. Ladak.** 1997. Overexpression and characterization of the chromosomal aminoglycoside 6'-N-acetyltransferase from *Enterococcus faecium. Antimicrob. Agents Chemother.* **41:**956–960.

93. **Wybenga-Groot, L., K. A. Draker, G. D. Wright, and A. Berghuis.** 1999. Crystal structure of an aminoglycoside 6'-N-acetyltransferase: defining the GCN5-related N-acetyltransferase superfamily fold. *Structure* **7:**497–507.

94. **Zhang, C.-C.** 1996. Bacterial signalling involving eukaryotic-type protein kinases. *Mol. Microbiol.* **20:**9–15.

Mechanisms of Resistance to Heavy Metals and Quaternary Amines

SIMON SILVER, RICHARD NOVICK, AND AMIT GUPTA

65

Resistances to toxic heavy metals in gram-positive bacteria (Table 1) have been studied for over 30 years, since Novick and Roth (42) reported multiple metal ion resistances on the "penicillinase plasmids" associated with early outbreaks of antibiotic-resistant *Staphylococcus aureus*. When restriction nuclease mapping became available, detailed physical maps of several such plasmids were developed (41). These included plasmid pI258, which contains, in addition to a determinant for β-lactamase, also those for mercury, arsenic, and cadmium resistance. These resistances have been studied at a physiological level and a molecular genetic level (e.g., references 61, 65, and 66). Studies of plasmid pI258 toxic heavy metal resistances are the most thorough for gram-positive bacteria, but closely related genetic determinants have been found for many other gram-positive bacteria. This chapter will cover four inorganic ion resistances (for As, Cd, Cu, and Hg) and also that toward organic quaternary ammonium biocide compounds (Qacs). Qac resistances were first found a quarter of a century ago and have been studied extensively by Skurray and coworkers (46). Quaternary ammonium resistance and most heavy metal resistances operate by efflux pumping of the toxic compound out of the cells. However, the basic structures and mechanisms of the pumps are quite different.

MERCURY RESISTANCE

Occurrence

Bacterial resistance to mercury compounds (Hg^r) is extremely widespread in industrially contaminated areas throughout the world and is often carried by transposons. Transposons carrying Hg^r have been found in permafrost bacteria that have been frozen for over 1 million years (40a). Pathogenic bacteria are likely to have acquired Hg^r from environmental organisms under selection owing to the use of inorganic Hg^{2+} in dental amalgams and of organomercurials in diuretics, as biocides (in particular, mer-bromin—sold in the United States as Mercurochrome—and thimerosal—sold as Merthiolate), in drugstore antiseptics, in storage solutions for suture catgut, and at rather high levels in hospital liquid hand soaps (50). By the 1970s in Japan, over 80% of hospital and environmental *Staphylococcus* isolates were resistant to mercury. Indeed, all the hospital-derived penicillinase plasmids (41, 42) contain operons conferring resistance to mercury and organomercurials, and 25% of some 800 early antibiotic-resistance plasmids from gram-negative bacteria also carried mercury-resistance determinants, presumably owing to selection by these various uses. When many of the clinical uses were eliminated during the unrelated mercury scares of the early 1970s, the frequency of mercury resistance in hospital isolates dropped to a few percent (50). Uses of mercury and organomercurials as biocides continue today (3), although information about commercial levels and usage is generally unavailable.

Comparative Structure and Function of Various *mer* Systems

The mechanism of bacterial mercury resistance has been the most thoroughly studied of the toxic inorganic ion resistances, starting with the demonstration that mercury-resistant gram-positive and -negative bacteria reduce mercuric ions to volatile mono-atomic elemental Hg^0. All functional bacterial mercury-resistance systems are organized as operons and have in common a transport system for importing Hg^{2+} into the cell's cytoplasm, a mercuric reductase, which acts on the imported Hg^{2+}, and a regulatory gene, *merR*, which controls transcription of the *mer* operon. Some systems have, in addition, an organomercurial lyase that releases Hg^{2+} from a variety of organomercurials. The importation of toxic Hg^{2+}, a required feature of all *mer* resistance systems, seems paradoxical; perhaps its function is to preclude the accumulation of toxic Hg^{2+} at the cell surface. The production of gaseous elemental mercury from Hg^{2+} as a resistance mechanism also seems strange but is likewise shared by all mercury-resistant bacteria tested, and the genes involved are all homologous. Within the staphylococci, the *mer* sequences are similar enough to cross-react by Southern hybridization (80), as are those of most bacilli (38). Between groups of gram-positive bacteria, however—*Bacillus*, *Staphylococcus*, and *Streptomyces* spp.—

Gram-Positive Pathogens, ed. by V. A. Fischetti et al.
© 2000 American Society for Microbiology, Washington, D.C.

TABLE 1 Heavy metal resistance systems in gram-positive bacteria

Heavy metal	Gene	Comments
AsO_4^{3-}, AsO_2^{-}, SbO^{+}	*ars*	Arsenate is enzymatically reduced to arsenite by ArsC. Arsenite and antimonite are "pumped" out by the chemiosmotic membrane pump ArsB.
Cd^{2+}	*cad*	Cd^{2+} (and Zn^{2+}) are pumped from bacteria by a P-type ATPase.
Cu^{2+}	*cop*	Chromosomally encoded Cu^{2+} resistance and homeostasis in *Enterococcus hirae* results from a pair of uptake and efflux P-type ATPases, which are governed by a novel transcriptional repressor and a separate activator (or chaperone) protein. Copper resistance encoded by a plasmid in *Lactococcus lactis* involves a two-component membrane sensor protein kinase plus a DNA-binding responder protein homologous to CopRS and PcoRS of gram-negative bacteria. The resistance mechanism has not been determined.
Hg^{2+}	*mer*	Hg^{2+} and organomercurials are enzymatically detoxified.
Pb^{2+}	*pbr*	Lead resistance appears to be due to accumulation of intracellular $Pb_3(PO_4)_2$ in *S. aureus*.

sequence similarities are too low to be detected by this means (23, 64, 79).

Figure 1 shows the basic structure of the mercury- and organomercurial-resistance systems for *Bacillus*, *Staphylococcus*, and *Streptomyces* spp. The resistance mechanism and genes are basically the same, but the order of the genes and the details are quite dissimilar. All three systems start with a regulatory gene, *merR*, and include *merA*, encoding Hg^{2+} reductase; *merT*, determining the transport function; and *merB*, encoding organomercurial lyase. The other open

reading frames (ORFs) shown are probably involved in the binding and transport of Hg^{2+}. *merA* and *merB* are adjacent in *Staphylococcus* spp. and *Streptomyces lividans* but are separated by a second transcriptional start site and a second regulatory gene in *Bacillus cereus* (21a, 23), and the *Streptomyces* operon is organized as two divergent transcription units. In *Bacillus mer* operons, there is a 1.1-kb gap between *merA* and the initially recognized *merB* gene (79). This region includes a second transcriptional start site, an ORF homologous to *merR* (25, 79) and a second *merB* gene (Fig. 1) (23, 79). The predicted amino acid sequence of the second MerR-like protein is rather different from the first (25), and the question of whether it responds to Hg^{2+} and activates *mer* transcription is open. However, the two putative organomercurial lyase gene products, MerB1 and MerB2, are likely to be functional.

MerR

The MerR proteins of *Bacillus* (25, 65) and *Staphylococcus* (65, 66) spp. are similar to the positively acting MerR protein of gram-negative bacteria that has been shown to twist and bend the operator DNA region in the presence of Hg^{2+} (45, 65), permitting the initiation of transcription. The MerR of *S. lividans* (10, 60), however, is a member of the class of metal-responding helix-turn-helix repressor proteins that includes ArsR, CadR, and SmtB (see below) and is the only one of more than 100 known MerRs that is a repressor rather than an activator. In the *mer* systems of gram-negative bacteria, *merR* is transcribed separately from the remaining *mer* genes, allowing tighter control of *mer* resistance genes than is possible with the gram-positive bacteria *Staphylococcus* and *Bacillus* spp., in which some transcription of the multigene *mer* operon must occur to provide MerR.

Transport

All three *mer* systems in Fig. 1 contain three genes (*merT*, ORF3/*merP*, and ORF4) whose products are thought to be involved in the importation of toxic Hg^{2+}. Mutation of genes in the transport system causes greatly diminished resistance, making the reductase gene effectively cryptic. This suggests that the primary target of Hg^{2+} lethality is outside the cytoplasmic compartment. There is tentative evidence that the products of the first genes are equivalent to the small MerT membrane transport protein of gram-negative bacteria, and the genes are therefore designated *merT*. MerP (ORF3) of *S. lividans* (10) is considered to be

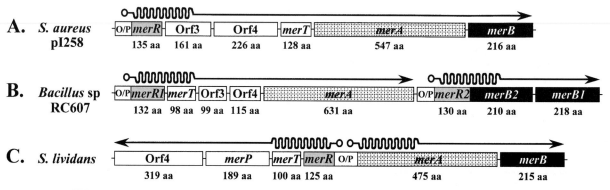

FIGURE 1 Comparisons of genes and products for mercury resistance in (A) *S. aureus* (plasmid pI258), (B) marine *Bacillus cereus* RC607 (chromosomal), and (C) *Streptomyces lividans* (linear plasmid).

equivalent to the periplasmic MerP of gram-negative bacteria, on the basis of N-terminal sequence homology to known signal peptides, even though it is twice the size. Computer analysis allowed recognition of a three-dimensional motif in the predicted product of *S. aureus* pI258 ORF3, related to the structures of thioredoxin and protein disulfide isomerases (17). Thus, the different ORF3s and ORF4s are thought to be involved in the importation of Hg^{2+}, although their roles have yet to be defined functionally.

Mercuric Reductase

The basic reaction mechanism for all mercuric reductases is NADPH-linked reduction of enzyme-bound flavin adenine dinucleotide (FAD), followed by reduction of a highly conserved dithiol cysteine pair in the active site, followed in turn by reduction of Hg^{2+} to Hg^0. Although the MerA proteins of gram-positive and gram-negative bacteria share overall only about 40% amino acid identities, if one considers only those amino acids that make contact with FAD or NADPH, over 80% are identical. Most mercuric reductases, including that of *S. aureus* plasmid pI258, start with an N-terminal 70- to 80-amino-acid domain that is considered to be involved in Hg^{2+} binding (62, 64) and is homologous to the periplasmic mercury-binding protein MerP. This region is missing in the *Streptomyces* protein (60), and there are two tandem copies in most bacilli tested (5, 38, 79).

The structure of the *Bacillus* mercuric reductase was solved by X-ray diffraction (57). As anticipated from the relatedness of the mercuric reductase sequence and biochemical mechanism to those of human glutathione reductase, the two structures are very similar. The enzyme is a homodimer, with each subunit containing a highly conserved active site with two critical cysteine residues. FAD is bound to each subunit, and there is an NADPH-binding site, necessary for electron transfer from NADPH to FAD to the substrate Hg^{2+}. The active site includes the redox-active disulfide region on one subunit and the substrate-binding site at the C terminus (including conserved vicinal Cys residues) of the other subunit. C-terminal tyrosine residues are found binding Hg^{2+} in the structure, and the tyrosines are involved in enzyme activity (51). The first 160 amino acids, containing the two hypothesized initial Hg^{2+}-recognition domains, have been proposed to serve as a "bucket brigade" intermediate carrier of the Hg^{2+} from the cell membrane to the C-terminal Hg^{2+}-binding site for electron transfer. This region is missing in the crystallographic solution for mercuric reductase (i.e., lacks a fixed position in the crystal), consistent with this proposed role.

Organomercurial Lyase

Organomercurial lyase cleaves the carbon-mercury bond in organomercurials by a proton-driven S_E2 reaction, releasing Hg^{2+}, which is then reduced to the volatile Hg^0 by mercuric reductase. The substrate specificity is rather broad. Some organisms (e.g., *B. cereus*; Fig. 1) possess two or more lyases that may have different specificities.

Genetic and Functional Diversity

Isolates of gram-positive bacteria from the methylmercury-polluted area of Minamata Bay, Japan (38), and from soil in mining regions in Russia (4–6) have provided interesting insights into the genetic diversity among Hg^r determinants throughout the world. For example, mercuric reductases with two N-terminal mercury-binding domains

were found in all tested Minamata Bay bacilli (38) and in most organomercurial- and mercuric-resistant bacilli from Russia (4, 5). However, mercuric reductases with a single N-terminal mercury-binding domain, similar to those of *S. aureus* and of many gram-negative bacteria (64), were found in Russian mercuric-resistant bacilli that lacked organomercurial resistance (4). Bogdanova and Mindlin (5) had earlier distinguished between these two classes of mercuric reductases in gram-positive soil isolates on the basis of immunological cross reactions and the ease of removal of the N-terminal sequence by trypsin.

Bogdanova et al. (6) additionally found mercuric reductase activity in a wide range of gram-positive soil bacteria (including *Micrococcus*, *Rhodococcus*, and *Arthrobacter* spp.) that are mercury sensitive and presumably lack genes for the Hg^{2+} transport pathway. These have a phenotype similar to that of the *merT* mutants mentioned above—that is, their inability to import Hg^{2+} would make the mercuric reductase effectively cryptic and account for their sensitivity.

Meissner and Falkinham (36) found a mercury-resistance plasmid, which also carries determinants of copper resistance and intracellular precipitation of copper sulfide (19), in an environmental isolate of *Mycobacterium* sp. This mycobacterial mercuric reductase was usual in that it used both NADH and NADPH, unlike the mercuric reductases of *Bacillus* and *Staphylococcus* spp., which function only with NADPH (49a). Subsequently, Steingrube et al. (69) identified mercury resistance and soluble mercury volatilization activities among a range of clinical *Mycobacterium* isolates. The frequency of mercury resistance was quite high in this collection, running up to 70% of 84 *Mycobacterium chelonae* isolates.

The clear conclusion is that mercurial-resistance systems using the same basic mechanism and its variants are found in all types of gram-positive bacteria, both clinical and environmental.

CADMIUM RESISTANCE

Cadmium resistance (Cd^r) is widespread among gram-positive pathogens and is usually associated with mobile genetic elements, which would suggest transfer from environmental bacteria. Cadmium compounds are used extensively in metallurgy and electronics and consequently are likely to be present in the environment, especially in industrially polluted areas. However, since cadmium is not (to our knowledge) used in hygiene or medicine, and since there is no known biological function for cadmium, which is toxic for all cells, the selective forces operating in the human and animal populations are not obvious.

There are several apparently different cadmium-resistance systems in a range of gram-positive pathogens and soil bacteria, including two well-differentiated Cd^r genotypes, known as *cadA* and *cadB*. All, however, seem to operate by cation efflux.

CadA Sequence and Structure

The CadA protein encoded by *S. aureus* plasmid pI258, the first discovered (42), was identified as an ATPase on the basis of its sequence (62, 63). Nearly identical *cadA* determinants were subsequently found on several large penicillinase plasmids (41), near the chromosomal determinant of toxic shock toxin, and within the chromosomal region responsible for methicillin resistance (80) in *S. aureus*

(SwissProt P37386; 89% amino acids identical to pI258 CadA) (12). Highly similar CadAs have been found on a plasmid-borne transposon in *Listeria monocytogenes* (30, 31) (SwissProt O60048; 65% amino acid identity), a plasmid of *Lactococcus lactis* (34) (GenBank GI 1699049; U78967; 62% amino acid identity), and the chromosome (or perhaps a linear plasmid) of the soil microbe *Bacillus firmus* (26) (SwissProt P30336; 80% identical amino acids). Those of staphylococci and bacilli represent the prototypes of a large subfamily of membrane heavy metal transporting enzymes (Fig. 2A), within the P-type cation ATPase family, the only type of transport ATPases that have a covalent phosphoprotein intermediate. Of these, only the CadA P-type ATPase of *Staphylococcus* sp. has been studied in detail.

The general structure of a heavy metal P-type ATPase is shown in Fig. 2A. There are eight membrane-spanning regions and three functional intracellular domains. The N-terminal domain frequently contains GMTCX$_2$C, a motif thought to coordinate the metal cation and therefore considered to function as a metal-binding element. The *S. aureus* pI258 CadA has one copy of this motif, which is homologous to the Hg^{2+}-binding motifs of MerP and the N terminus of mercuric reductase (62) (see above). The chromosomally encoded CadA of methicillin-resistant *S. aureus* has two such motifs or domains (12).

The second intracytoplasmic domain, between the fourth and fifth transmembrane segments, contains the

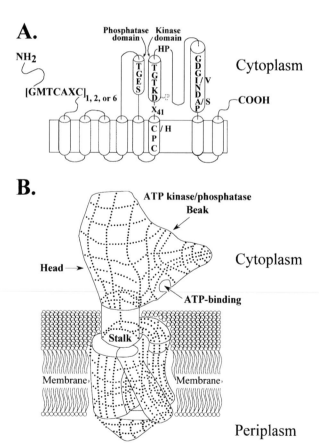

FIGURE 2 P-type ATPases for (A) efflux of heavy metals in gram-positive bacteria and for (B) efflux of Ca^{2+} in animal muscle. (Modified from references 64 and 84.)

phosphatase determinant, including a TGES tetrapeptide (64, 68), and the third intracellular domain represents the ATPase/kinase domain. The aspartic acid residue within the conserved DKTGT sequence on the ATPase/kinase region is phosphorylated and dephosphorylated during the catalytic cycle of cation transport. The kinase domain is large and highly conserved among P-type ATPases, with the conserved residues including some of those shown in Fig. 2A. In an 8-Å crystallographic structure of the animal muscle calcium ATPase (Fig. 2B), the phosphatase and kinase domains appear fused. Sequence homology analysis of the kinase domain of P-type ATPases has given more specific predictions for conserved regions (1).

The His-Pro dipeptide in the kinase domain is found in heavy metal P-type ATPases (Fig. 2A) but not in those for transport of potassium, calcium, or magnesium (68). The region C-terminal to the eighth transmembrane segment is intracellular and rather short in CadA but is longer and includes four additional transmembrane segments in calcium and magnesium P-type ATPases. The *Listeria*, *Lactobacillus*, and *Bacillus* CadA amino acid sequences, when aligned with that of *S. aureus* pI258, all have single short gaps at the junction of the N-terminal metal-binding domain and the ATPase domain.

Several other related heavy metal-type P-type ATPases are listed in GenBank as CadA-like. These are probably not cadmium-effluxing ATPases but paralogs with different divalent cations as substrates. They have considerably lower sequence similarity to the typical CadAs (40% or less overall amino acid identities) and are more closely related to P-type ATPases with substrate specificities for other cations, such as Zn^{2+}, Pb^{2+}, Ni^{2+}, and Cu^{2+} or Cu$^+$, that have recently been described.

CadA of *S. aureus* plasmid pI258 is one of the few bacterial P-type ATPases for which direct biochemical evidence of function is available. Cadmium enters *S. aureus* cells by a membrane potential-dependent system shared by Mn^{2+} and in resistant cells is then excreted by an energy-dependent process (76). Tsai et al. (74) demonstrated ATP-dependent uptake of radioactive Cd^{2+} and Zn^{2+} by inside-out membrane vesicles (equivalent to efflux by intact cells). Uptake activity was limited to vesicles from cells containing the *cadA* gene and was inducible by prior growth of the bacteria in the presence of cadmium salts. Tsai and Linet (73) incubated the same membrane vesicles with ^{32}P-labeled ATP and demonstrated the incorporation of radioactivity into a protein of the size expected for CadA. Radioactivity was incorporated only when the ATPase was activated by Cd^{2+}; ^{32}P in the protein was alkali labile, as expected for an aspartyl-phosphate ester, and was turned over on addition of nonradioactive ATP or ADP (73). In summary, the enzyme is a cation-proton antiporter in which an intracellular divalent cation is thought to be bound to the N-terminal metal-binding site, followed by phosphorylation of the conserved aspartate residue. This would change the conformation of the protein, resulting in the translocation of the bound metal ion to the extracellular space. In parallel, one or more protons move from the outside to the inside. Hydrolysis of the aspartyl phosphate by the intrinsic phosphatase activity would then restore the original conformation of the protein.

CadB

A second determinant of cadmium resistance was identified on the 30-kb *cadA*-containing plasmid pII147 and was accordingly designated *cadB* (41). After nearly identical Cd^r

determinants were found on several small, closely related rolling circle plasmids belonging to the pC194 superfamily (2, 15), it became clear that the pII147 *cadB* corresponds to one such small plasmid integrated intact into the pII147 genome.

There are currently four known *cadB*-like open reading frames, about 200 codons in length and sharing 84% overall sequence identity: that carried by pOX6 in *S. aureus* (16a), that described by Crupper et al. (15; GenBank U76550; PID g1916729), designated *cadD*, and that by Ben-Abdallah Chaouni et al. (2; GenBank U74623; PID g1658280) in other staphylococci, in addition to that of pII147. We propose renaming the *cadD* of Crupper et al. (15) as *cadB*, since it is clearly the same as the other *cadB*s.

Two additional sequences that are significantly (but less) similar to CadB were released by Chikramane and Dubin (12; GenBank L10909; PID g152979) and by Ivey et al. (26; PIR S28465; PID g280278). The first of these was reported to be involved in cadmium resistance in *S. aureus*, and the second to correspond to *qacF* for quaternary amine resistance (see below) in *B. firmus*. Both sequences are about 200 amino acids in length and 40% identical to CadB (CadD) of Crupper et al. (15). The *qacF* (possibly mislabeled and probably better called *cadB*) locus is on the large *Bacillus* plasmid carrying *cadA*. Further experimental efforts are needed to understand whether these are CadB orthologs or paralogs with different substrate specificity.

CadB is unrelated to CadA in sequence and is still of unknown mechanism. However, from the sequence it appears to be a small hydrophobic protein, perhaps passing across the membrane four or five times, consistent with a role as an efflux pump component (15).

Regulation of Cd^r

Cd^r determinants of both the *cadA* and *cadB* types are inducible at the level of transcription by subinhibitory cadmium salts, involving a classical repressor-mediated mechanism. Both pI258 and pOX6 Cd^r loci contain repressor genes, *cadC* (18) and *cadX* (16a), respectively, adjacent to the resistance genes. CadX is homologous to CadC, and both are members of the superfamily of metal-responding helix-turn-helix transcriptional regulators including ArsR (see below) and SmtB, the transcriptional regulator for cyanobacterial metallothionein (75), the first protein in this family whose structure has been determined by crystallography (14). CadC binds to the *cadA* operator/promoter region, blocking transcription in vitro, and is released by incubation with Cd, Bi, or Pb ions (18). On the basis of these results, we propose to rename CadC as CadR.

COPPER HOMEOSTASIS

Copper is widely used as an antibacterial/antifungal agent in agriculture and in water purification. For example, copper sprays are used on grapes, tomatoes, walnuts, and other commercial plants. North American, British, and Australian pigs are fed copper as a growth stimulant, which is thought to work by inhibiting intestinal microbes and facilitating the transition to solid foods on weaning. Given widespread environmental pollution with copper, owing to these practices and also to industrial and mining discharges, it is not surprising that environmental organisms have developed copper resistance as well as uptake mechanisms. At the same time, the clinical uses of copper are quite limited. Thus, the role of environmental copper in the de-velopment of resistance in pathogenic bacteria is unclear. However, copper is an essential trace nutrient, being required by many enzymes, and bacteria have evolved a copper homeostasis mechanism that involves uptake when the intracellular level is low and efflux when it is high and potentially toxic.

The best-understood copper transport systems, accounting both for resistance by efflux and for uptake, are those of the gram-positive pathogen *Enterococcus hirae* (for reviews see references 61, 64, and 68) (Fig. 3). Copper homeostasis is achieved by closely coupled uptake and efflux mechanisms, using P-type ATPases, such that uptake is activated by low internal copper and efflux by high internal copper. Two contiguous and cotranscribed genes, *copA* and *copB*, determine the uptake and efflux enzymes, respectively (43) (Fig. 3). The copper-transporting P-type ATPases of *E. hirae* have the basic properties of the P-type ATPases described above for cadmium. Thus, inside-out subcellular membrane vesicles isolated from *E. hirae* cells require ATP in order to accumulate ^{64}Cu$^+$. ^{110m}Ag$^+$ is also taken up, though the Cu efflux system does not confer resistance to silver (44, 67). The CopB protein, purified and reconstituted in phospholipid vesicles, formed an acyl-phosphate intermediate with radioactive ATP (81), as expected for P-type ATPases (68). The in vitro substrate for CopB is probably Cu$^+$ rather than Cu^{2+}, but whether copper is taken up initially as Cu^{2+} and subsequently reduced to Cu$^+$ or whether copper is reduced at the cell surface, before or concomitant with uptake, is not known. Predictably, *copA* mutants show increased resistance to copper and require higher levels for growth, whereas *copB* mutants are hypersensitive.

Copper-transporting P-type ATPases are members of the very widespread superfamily of metal cation-transporting enzymes. Both of the *E. hirae* enzymes appear to cross the membrane eight times (Fig. 2A and 3), which is standard for soft-metal ATPases, including the cadmium efflux enzyme described above, but is in contrast to 10 membrane-spanning segments found with the calcium (Fig. 2B) and magnesium ATPases. The copper-transporting enzymes of *E. hirae* can be compared with two human analogs of copper metabolism, in which mutations are responsible for the lethal hereditary diseases, Menkes syndrome and Wilson's disease (63, 68). The N-proximal metal-binding motif, GMTCX$_2$C (Fig. 2), is found in the CopA uptake ATPase but not in the CopB efflux enzyme. The Menkes and Wilson's ATPases each have six copies of this motif (63, 68), whereas CopA has a single copy. A 70-residue peptide corresponding to the fourth of the six metal cation-binding motifs of the Menkes copper ATPase was overproduced, purified, and its structure solved by nuclear magnetic resonance analysis (20) with bound Ag$^+$. This small protein domain has a tertiary structure consisting of four antiparallel β-sheet segments and two α-helical segments lying on top, with the GMTCX$_2$C motif protruding to the side like a thumb. This structure is also seen with MerP (above), suggesting that CopA will be similar. The Cu$^+$ efflux ATPase, CopB, lacks this cysteine-containing predicted metal-binding motif, but has three closely spaced copies of a methionine-histidine–rich metal-binding motif in a similar position near the N terminus, before the first predicted transmembrane segment (68).

The Cop system is regulated in response to both copper starvation (when the CopA uptake ATPase is needed) and copper excess (when the CopB efflux ATPase is needed)

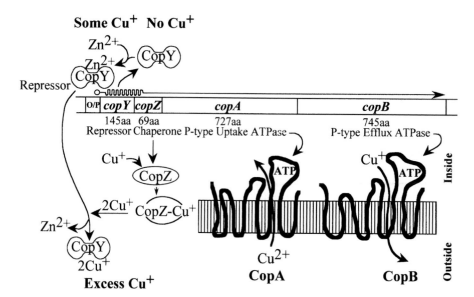

FIGURE 3 Transport and regulation of copper in *E. hirae*. The *copA* and *copB* genes determine copper uptake and efflux ATPases, respectively. The ATPases are shown passing across the membrane eight times (as in Fig. 2). Transcriptional control of the *cop* operon is governed by the upstream repressor (product of *copY*) at low Cu (inside) and activator (product of *copZ*) at high Cu (inside). Modified from reference 64.

(70). Regulation of *cop* operon transcription is governed by a unique pair of regulatory genes: *copY* (which determines a transcriptional repressor protein) and *copZ* (which determines an activator protein) (64, 70). CopZ is a small polypeptide with a $GMXCX_2C$ motif that is thought to account for copper binding. Both CopY and CopZ are metal-binding polypeptides. The CopY apo-repressor is inactive and fails to bind to the operator/promoter (Fig. 3) in vitro in the absence of Cu^+, as shown in DNaseI footprinting experiments (70). (Cu^{2+} is the cation stable under aerobic conditions in bacterial growth media, but this is thought to be reduced either concomitant with uptake or immediately after uptake, so that Cu^+ is the form of copper that regulates intracellularly and is substrate for the copper efflux ATPase.) At a moderate level of intracellular Cu^+, Zn^{2+} binds to a CopY dimer, converting it to a DNA-binding repressor that binds to the operator DNA site (70). At higher intracellular Cu^+ levels, the CopZ chaperone binds Cu^+, and CopZ-Cu^+ delivers two Cu^+ to CopY-Zn^{2+}, displacing the Zn^{2+} and forming a non-DNA-binding $2Cu^+$-CopY complex (12a). CopZ belongs to the newly identified family of Cu^+-delivering intracellular "chaperones" found in both bacteria and eukaryotes (12a). The CopZ-Cu^+-binding motif is homologous in sequence to earlier-recognized Cu^+-, Cd^{2+}-, and Hg^{2+}-binding motifs in proteins including periplasmic metal binding (MerP), intracellular enzymes (mercuric reductase), and membrane P-type ATPases (CadA and Menkes), as described above. The conclusion from sequence homology was borne out by the newly determined NMR structure of the CopZ protein, showing the close three-dimensional similarity to the metal-binding regions of the Menkes P-type ATPase protein and the Hg^{2+}-binding bacterial MerP protein (79a). The functioning of these two regulatory proteins explains the derepression of synthesis of both ATPases by very low external Cu^{2+}, then repression by intermediate levels, and finally activation of synthesis at higher levels of external

Cu^{2+} (44, 68) (Fig. 3). How this regulatory scheme would allow for net uptake or net efflux, as required, is not clear; presumably, there would be a second level of regulation, affecting either translation or activity of the CopA and CopB proteins.

RESISTANCE TO ARSENICALS AND ANTIMONY

Like copper compounds, arsenicals have been widely used as preservatives and antibacterials in agriculture, construction, and medicine. Arsenic compounds are also released into the environment during various industrial processes and are present in mine tailings. Chickens in North America are fed organoarsenicals as growth stimulants. "Chromated copper arsenate" is a widely used wood preservative. Domestic garden fungicides sometimes contain arsenic compounds. Remarkably, there are also unacceptably high levels of arsenic occurring naturally in drinking water from wells in lower Michigan and Wisconsin in the United States, as well as in extensive areas of India, Bangladesh, and Taiwan.

Whether or not arsenicals occurring in the environment or the clinic are responsible, bacterial resistance to arsenicals (As^r) is very widespread among pathogenic and environmental bacteria of all types, frequently in association with plasmids and other accessory genetic elements. First identified on *S. aureus* penicillinase plasmids (42), As^r determinants have more recently been found on the newly sequenced genomes of microbes as diverse as *B. subtilis*, *Escherichia coli*, and yeasts. Even the newly determined *Mycobacterium tuberculosis* genome sequence (13) has close homologs to functional arsenic resistance genes of other microbes, although these are yet to be tested experimentally. An interesting example is in *B. subtilis* 168, in which As^r is carried by the 40-kb SKIN element (for sigma K-gene insertion) (56, 71). This element is inserted in *sigK* and is removed by site-specific excision early in sporulation, reconstituting the gene. The SKIN element is not found in several other *Bacillus* strains, suggesting that it was acquired by horizontal gene transfer.

As with mercury resistance, fundamentally the same As^r genes are found on bacterial plasmids as on chromosomes (61, 64) (Fig. 4). However, the number and organization of the genes vary somewhat, resulting in considerable genotypic diversity. Four *ars* operons, three from gram-positive bacteria and for comparison the most thoroughly studied of the *E. coli* systems, are diagrammed in Fig. 4. These four, as well as all other known *ars* operons, have three genes in common: *arsR*, determining a transcriptional repressor that confers inducibility for the As^r system; *arsC*, encoding an enzyme, arsenate reductase, that converts arsenate to arsenite; and *arsB*, determining an inner-membrane chemiosmotic pump that effluxes arsenite and antimony. The prototypical *ars* operons, those of *Staphylococcus xylosus* plasmid p5X267 (52, 53) and its very close relative carried by *S. aureus* plasmid pI258 (28), contain just these three genes, which are therefore sufficient for inducible resistance to arsenate [AsO$_4^{3-}$; As(V)], arsenite [AsO$_2^-$; As(III)] (27, 29, 61), and antimony [Sb(III)]. In contrast, the *E. coli* plasmid R773 *ars* operon contains two additional genes, *arsA* and *arsD*, that are lacking in the known *ars* systems from gram-positive bacteria (Fig. 4). ArsD is a secondary repressor, having smaller effects than ArsR (64), and ArsA is a soluble ATPase that attaches to ArsB, converting it from a chemiosmotic (9) to an ATP-driven arsenite efflux pump (64). The ability of the ArsB protein to function alone as a chemiosmotic arsenite efflux transporter, or together with ArsA as an ATP-driven pump, is unique among known bacterial uptake and efflux systems. Most As^r systems, however, function as chemiosmotic pumps only (61, 64). The *Bacillus ars* operon contains a fourth (novel) gene, which is required for arsenic resistance (56), ORF2.

Arsenate Reductase

All bacterial As^r systems reduce arsenate to arsenite even though arsenate is less toxic than arsenite. In the absence of *arsC*, the *arsR-arsB* complex confers inducible resistance to arsenite and antimony(III) but not to arsenate (28). It seems counterintuitive to convert a less toxic compound to a more toxic one, but ArsC activity is closely coupled with efflux from the cells (29), so arsenite never accumulates intracellularly. Since arsenate is taken up as a phosphate analog, any arsenate efflux pump would probably have the undesirable side effect of depleting the cells of phosphate. It therefore seems likely that bacteria avoid this by reducing arsenate for export. This stratagem is evidently very old, evolutionarily, since the *Staphylococcus* and *E. coli* ArsC sequences do not show significant similarity, implying either divergent or convergent evolution, either of which could have happened only over an evolutionary time scale—which strongly implies that As^r evolved long before any human use or misuse. Furthermore, the two types of ArsC differ in the details of their respective functions: arsenate reductase of pI258 derives its reducing power from thioredoxin, while the arsenate reductase of plasmid R773 utilizes glutaredoxin. Conversely, glutaredoxin does not work with pI258 ArsC, and thioredoxin does not work with R773 ArsC. A basis for this difference may be the surprising finding that gram-positive bacteria lack the ability to synthesize glutathione, a capacity common to gram-negative bacteria, plants, and animals. The pI258 ArsC contains four redox-active cysteines. Presumably, two of these recycle between oxidized and reduced states, coupled to a redox cycle involving the cysteines of thioredoxin. In contrast, the R773 ArsC has only two cysteines,

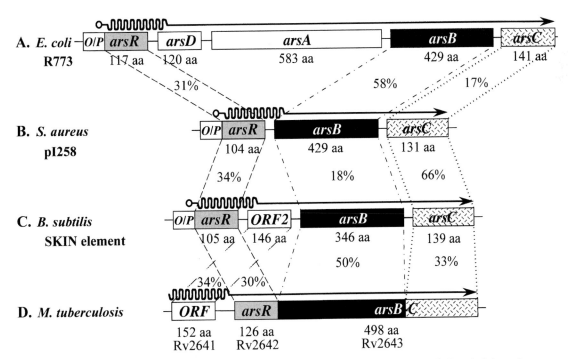

FIGURE 4 Comparison of genes and products for arsenic resistance in (A) *E. coli* (plasmid R773), (B) *S. aureus* (plasmid pI258), (C) *B. subtilis* 168 (SKIN insertion element in chromosome), and (D) *M. tuberculosis* (chromosome).

and probably only one of these is required for arsenate reduction, using glutaredoxin and possibly reduced glutathione (64). Nearly three dozen putative gene products from the genomic sequences of different microbes have been listed as arsenate reductases. The pI258 subfamily has members from gram-negative bacteria and even from *Archaea*. In the *B. subtilis* genome, there are two different candidate *arsC* genes, one with a product more closely related to that of pI258 and another to that of R773 (not shown). How these findings relate to the above-mentioned mechanistic features of ArsC remains to be determined.

In sum, the same *ars* operon conferring resistances to As(III), As(V), and Sb(III) appears to occur widely in gram-positive bacteria, but significant variations in the number and location of genes and in the degree of similarity in sequences (from very close to basically undetectable) have been found. Most of the *ars* systems have been identified by sequence homology only and require experimental testing of function. Among these is a probable As^r system from *M. tuberculosis* (13) (Fig. 4).

BACTERIAL RESISTANCE TO OTHER TOXIC METALS

In addition to resistances considered above, Ag$^+$, Co^{2+}, CrO$_4^{2-}$, Ni^{2+}, Pb^{2+}, TeO$_3^{2-}$, Tl$^+$, and Zn^{2+} resistances have been reported, more often for gram-negative than for gram-positive bacteria (61, 64). There is little information to date on the resistance genes or mechanisms. Those for Zn^{2+} and Pb^{2+}, incidentally, are probably not due to cross-resistance by *cadA*. An Ag$^+$ resistance determinant in *Enterobacteriaceae* has recently been cloned and sequenced (22), but none has yet been identified in gram-positive bacteria, although staphylococci and other gram-positive nosocomial pathogens have been exposed in clinical use to silver compounds. Silver nitrate has long been used as a topical cauterizing and antibacterial agent, and burn wards use silver sulfadiazine as an antibacterial. Although Ag$^+$ is a substrate for the Cu$^+$ efflux ATPase, CopB of *E. hirae*, Ag$^+$ resistance is unrelated to Cu resistance. Lead (Pb^{2+}) resistance in *Staphylococcus* sp. may involve intracellular precipitation as lead-phosphate granules (32). Chromate resistance associated with plasmids has been described for gram-negative bacteria, usually resulting from reduced cellular uptake, which probably represents efflux (61). Chromate reduction has also been described, for both gram-negative and -positive bacteria, sometimes involving cytoplasmic and sometimes membrane-bound chromate reductases (11, 61). Chromate reduction may increase resistance since reduced Cr(III) precipitates and is thus made biounavailable.

RESISTANCE TO QUATERNARY AMMONIUM COMPOUNDS

Qacs are potent synthetic bactericidal agents that are widely used as disinfectants in hospitals, industry, agriculture, and the home (3, 24). Eggs in the United States are washed in Qac-containing disinfectants; skin lotions, soaps and dishwashing detergents, and objects as diverse as children's toys and mattress pads often contain Qacs (33). The structures of two frequently used monocationic Qacs, benzalkonium chloride and cetrimide, are shown in Fig. 5.

Resistance to Qacs occurs widely in clinically important microbes, especially gram-positive bacteria (40b, 54, 55),

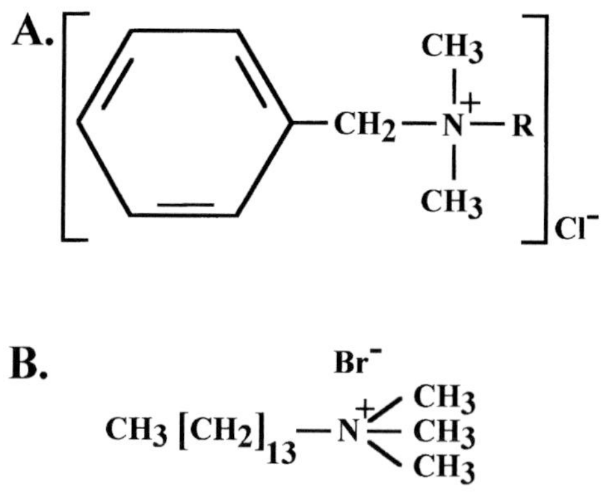

FIGURE 5 The structures of Qacs: (A) benzalkonium chloride (A) (R represents a mixture of alkyls ranging from C$_8$H$_{17}$ to C$_{18}$H$_{37}$) and (B) cetrimide.

including staphylococci, corynebacteria, *Enterococcus faecium*, *Streptococcus mutans* (54), *L. lactis* (78), and mycobacteria (35, 72). It is often found in bacteria isolated from foods and hygiene products (3). Qac resistance is frequently associated with plasmids, spreads by gene transfer, and severely compromises the utility of these agents as disinfectants. Qac resistance occurs primarily by energy-dependent efflux mechanisms of low specificity that also export other toxic substances, mostly cations. They are known as multidrug resistance (MDR) systems and occur in mammalian (cancer) cells as well as in bacteria.

We limit ourselves here to resistance to mono- and dicationic Qacs, recognizing that Qac resistance is representative of the much broader MDR phenomenon. As efflux systems of this type have been found with many bacterial species and relatively little is known of their mechanisms, it is useful to consider gram-negative and gram-positive organisms together. A number of different families of Qac resistance systems have been identified and are listed in Table 2. The nomenclature is unfortunately rather complex; some were originally identified on the basis of resistance to substances other than Qacs and therefore are not known as Qac resistance genes. In other cases, the same determinant has more than one name. For example, QacC and QacD are identical and are probably the same as Ebr (for ethidium bromide resistance). Russell (54) has reviewed the Qac resistance systems broadly, with special consideration of resistance levels for different compounds. Three general classes of Qac transporters have been identified and are diagrammed in Fig. 6.

QacA/B

The QacA/B is the prototype of a recently recognized family of large chemiosmotic drug/proton antiporters. These are over 500 amino acids in length, have 14 predicted transmembrane helices (Fig. 6C), and appear to function by a chemiosmotic electroneutral exchange of amphiphilic cationic drugs for protons across the membrane, driven by the transmembrane pH gradient (46, 48). About 30 ORFs putatively encoding chemiosmotic pumps with 14 membrane-spanning segments have been recognized (46) in sequences from bacteria and yeasts. The largest numbers are in gram-positive bacteria, including QacA/B of *S. aureus* and its homologs in mycobacteria, LfrA of *Mycobacterium smegmatis* (35, 72) and EfpA of *M. tuberculosis* and *Mycobacterium leprae* (16). The LfrA of *M. smegmatis* and

TABLE 2 Qac multidrug resistance efflux pumps in *S. aureus* and other bacteria[a]

Name (synonyms)	Size (aa)	Microbe	Accession no.[b]
Large Qac (no. of TMS = 14)			
QacA (QacB)	514	*Staphylococcus aureus*	P23215, AF053772
TcmA	538	*Streptomyces glaucescens*	M80674, P39886
EmrB	512	*Escherichia coli*	P27304, D90891
Mmr	466–475	*Streptomyces coelicolor, Bacillus subtilis*	P11545, Q00538
LfrA	504	*Mycobacterium smegmatis*	U40487
EfpA	530	*Mycobacterium tuberculosis*	Z95207
Smr (no. of TMS = 4)			
Smr, QacC, QacD, and Ebr	107	*Staphylococcus aureus*	P14319, M37888
QacH (Smr family but different)	107	*Staphylococcus* spp.	Z37964
EmrE	110	*Escherichia coli*	P23895
QacE	110	*Klebsiella aerogenes*	X72585, X68232
QacF (a CadB homolog)	197	*Bacillus firmus*	Z17326, S28465

[a] aa, amino acids; TMS, transmembrane segments; Smr, small multidrug resistance protein.
[b] GenBank database.

M. tuberculosis have been shown to confer resistance to Qac compounds (16, 72).

QacA, from *S. aureus*, confers resistance to monocationic disinfectants, to dicationic primary and secondary polyamines, and to structurally unrelated compounds such as acriflavine, chlorhexidine, diamidines, and ethidium bromide (37). QacB is very similar to QacA (seven differences out of 514 positions; 99% identities) but confers resistance only to monovalent Qacs and not to organic divalent cations such as chlorhexidine and diamidines (37). This difference has been localized to Asp-323, in the middle of the 10th membrane-spanning segment, which is replaced by Ala in QacB. Additionally, QacB has Glu-322 in place of Gly in QacA. These two dicarboxylic acid amino acids in the 10th transmembrane helix are considered to be involved in substrate binding (48); on the opposite face of helix 10 lie three proline residues, indicating restricted rotational flexibility. Certain mutations that alter single amino acids change the substrate range of QacA and QacB (48).

Smr

A second major family of Qac resistance determinants, also functioning as chemiosmotic drug/proton antiporters, is the Smr (small multidrug resistance) family. These consist of small proteins, 107 to 110 amino acids in length (Table 2), that form a four-α-helix membrane bundle (Fig. 6B) (1) and may assemble in the membrane as trimeric structures (Fig. 6A) (82, 83) so that a total of 12 transmembrane α-helices cooperate in forming a single transport complex. Closely related to Smr are QacC, QacD, and Ebr (46, 47, 49). Sasatsu et al. (55) have shown that two copies of *qacC* placed tandemly on a plasmid confer a higher resistance level (to ethidium bromide and to the quaternary amine benzethonium chloride) than a single copy. The Smr protein of *S. aureus* (21) and its homolog EmrE in *E. coli* (58, 82) were purified, reconstituted in proteoliposomes, and shown to function as chemiosmotic pumps. They are electrogenic proton/drug exchangers that function for Qac efflux by a process involving key glutamate residues (20, 47, 49). The α-helices of EmrE lie at an angle to the

membrane bilayer (1, 59), qualitatively similar to the more familiar, somewhat larger membrane protein, bacteriorhodopsin.

LmrP

A third major family, represented by the *L. lactis* LmrP protein, consists of proton-driven chemiosmotic drug efflux transporters with 12 predicted membrane-spanning segments, like *Bacillus* Bmr (see chapter 68, this volume).

LmrA

A fourth family is represented by LmrA, also from *L. lactis*, which is the first known bacterial MDR ATPase (Fig. 6A) (7, 8, 77, 78) and is homologous in sequence and function to the mammalian MDR. The 590-amino-acid LmrA polypeptide has an N-terminal domain that appears to cross the membrane six times plus a C-terminal intracellular ATPase domain (78). LmrA is homologous to both halves of the larger (1,280 amino acid) mammalian MDR ATPase (78), which probably arose by gene duplication and fusion. Indeed, the lactococcal LmrA complements human MDR activity when transfected into fibroblast cells (77). The bacterial LmrA protein, localized to the mammalian plasma membrane, confers resistance to a wide range of toxic compounds and is the first example of a prokaryotic membrane transport protein that can function in a mammalian cell.

Both LmrA and LmrP efflux amphiphilic cationic compounds with a broad range of structures. A novel model has been developed to explain the broad substrate specificity and the mechanism of both the LmrA ATPase and the LmrP chemiosmotic pump. Initially, the amphiphilic substrate is thought to partition to the outer leaflet of the cell membrane bilayer by lipophilicity mediated by the alkane chain or aromatic ring (Fig. 5) of the substrate. The Qacs or other compounds next "flip" by a slow process to the inner leaflet of the lipid bilayer (Fig. 6A). The substrate then rapidly diffuses laterally from the lipid layer to the LmrA or LmrP protein and is taken up by the protein, a process involving the cationic charge. Finally the Qac is effluxed to the exterior of the cell by the ATPase LmrA (Fig. 6A) or by LmrP. This model is referred to as a "hy-

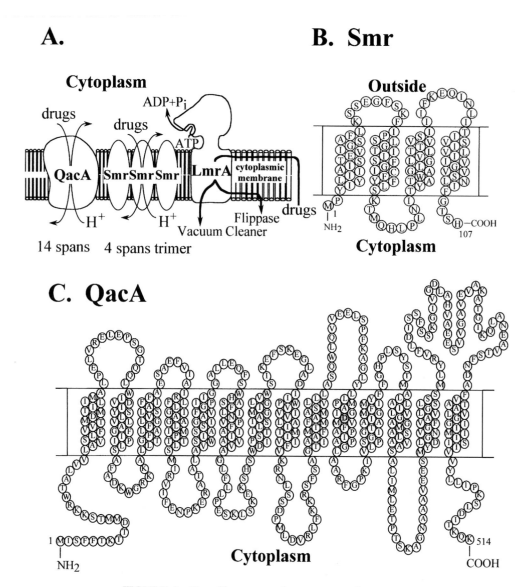

FIGURE 6 Qac efflux pumps of gram-positive bacteria.

drophobic vacuum cleaner" (Fig. 6A) if the toxic compound passes through the protein to the extracellular space or a "flippase" (Fig. 6A) if the toxic compound is transferred from the protein to the outer lipid leaflet, then to diffuse outward slowly (8, 77).

Finally, there is a large class of chemiosmotic proton/drug antiporters called the "major facilitator superfamily" that have 12 membrane-spanning, presumably α-helical regions. This superfamily includes many tetracycline-resistance efflux systems and the first bacterial MDR discovered, Bmr of *Bacillus* sp. (39, 40). These are considered elsewhere in this volume since they confer resistance to fluoroquinolones (see chapters 66 and 68).

In sum, mechanistic differences are associated with the differences in structures of the small and large Qac proteins. Paulsen et al. (46) consider the group of 14-transmembrane-segment chemiosmotic drug/proton exchangers a subgroup of the major facilitator group of transporters (most of which pass through the membrane 12 times). Thus, bacterial MDRs are not a family of sequence-homologous phylogenetically related transporters, but

rather the term is an umbrella for efflux transporter families that share wide substrate range but can differ in overall structure and in energy coupling mechanism. Whether there is a shared basic mechanism that allows a single protein transporter to recognize a range of substrates is not known.

In conclusion, it is clear that widely diverse bacteria have evolved a remarkable collection of mechanisms to protect themselves from toxic organic and inorganic compounds and that the sophistication of these mechanisms implies that they have evolved over a very long time—far longer than the existence of humans on the planet, let alone our use and misuse of these substances. Thus, these types of resistance are quite parallel to those of antibiotic resistance—commonly present as chromosomal genes in environmental organisms and carried by various types of mobile accessory genetic elements in pathogens of both gram-positive and -negative types. Resistance to these types of substances has probably been amplified by human activities leading to environmental pollution; but, unlike the situation with antibiotics, the selective forces that are presumably responsible for the transfer and prevalence of these resistances in pathogenic bacteria are often unclear.

The preparation of this chapter was supported by a grant from the U.S. Department of Energy. We thank colleagues involved in development of the ideas in recent years, including G. Ji, K. Nakamura, H. Nikaido, L. T. Phung, M. H. Saier, M. Sasatsu, and R. A. Skurray.

REFERENCES

1. **Aravind, L., M. Y. Galperin, and E. V. Koonin.** 1998. The catalytic domain of the P-type ATPase has the haloacid dehalogenase fold. *Trends Biochem. Sci.* **23:**127–129.

1a. **Arkin, I. T., W. P. Russ, M. Lebendiker, and S. Schuldinder.** 1996. Determining the secondary structure and orientation of EmrE, a multi-drug transporter, indicates a transmembrane four-helix bundle. *Biochemistry* **35:**7233–7238.

2. **Ben-Abdallah Chaouni, L., J. Etienne, T. Greenland, and F. Vandenesech.** 1996. Nucleic acid sequence and affiliation of pLUG10, a novel cadmium resistance plasmid from *Staphylococcus lugdunensis. Plasmid* **36:**1–8.

3. **Bessems, E., and P. M. J. Terpstra (ed.).** 1998. Hygiene and disinfection. *Int. Biodeter. Biodegrad.* **41:**167–312. (Special issue.)

4. **Bogdanova, E. S., I. A. Bass, L. S. Minakhin, M. A. Petrova, S. Z. Mindlin, A. A. Volodin, E. S. Kalyaeva, J. M. Tiedje, J. L. Hobman, N. L. Brown, and V. G. Nikiforov.** 1998. Horizontal spread of mer operons among Gram-positive bacteria in natural environments. *Microbiology* **144:**609–620.

5. **Bogdanova, E. S., and S. Z. Mindlin.** 1991. Occurrence of two structural types of mercury reductases among Gram-positive bacteria. *FEMS Microbiol. Lett.* **78:**277–280.

6. **Bogdanova, E. S., S. Z. Mindlin, E. Pakrov, M. Kocur, and D. A. Rouch.** 1992. Mercuric reductase in environmental Gram-positive bacteria sensitive to mercury. *FEMS Microbiol. Lett.* **97:**95–100.

7. **Bolhuis, H., H. W. van Veen, D. Molenaar, B. Poolman, and W. N. Konings.** 1996. Multidrug resistance in *Lactococcus lactis:* evidence for ATP-dependent drug extrusion from the inner leaflet of the cytoplasmic membrane. *EMBO J.* **15:**4239–4245.

8. **Bolhuis, H., H. W. van Veen, B. Poolman, A. J. Driessen, and W. N. Konings.** 1997. Mechanisms of multidrug transporters. *FEMS Microbiol. Rev.* **21:**55–84.

9. **Bröer, S., G. Ji, A. Bröer, and S. Silver.** 1993. Arsenic efflux governed by the arsenic resistance determinant of *Staphylococcus aureus* plasmid pI258. *J. Bacteriol.* **175:**3480–3485.

10. **Brünker, P., D. Rother, R. Sedlmeier, J. Klein, R. Mattes, and J. Altenbuchner.** 1996. Regulation of the operon responsible for broad-spectrum mercury resistance in *Streptomyces lividans* 1326. *Mol. Gen. Genet.* **251:**307–315.

11. **Campos, J., M. Matinez-Pacheco, and C. Cervantes.** 1995. Hexavalent-chromium reduction by a chromate-resistant *Bacillus* sp. strain. *Antonie van Leeuwenhoek* **68:**203–208.

12. **Chikramane, S. G., and D. T. Dubin.** 1993. Tn554: a *Staphylococcus aureus* chromosomal element encoding cadmium resistance determinants, and genes resembling the transposases genes of Tn554. National Center for Biotechnology Information, GenBank L10909.

12a. **Cobine, P., W. A. Wickramasinghe, M. D. Harrison, T. Weber, M. Solioz, and C. T. Dameron.** 1999. The *Enterococcus hirae* copper chaperone CopZ delivers copper(I) to the CopY repressor. *FEBS Lett.* **445:**27–30.

13. **Cole, S. T., R. Brosch, J. Parkhill, T. Garnier, C. Churcher, D. Harris, S. V. Gordon, K. Eiglmeier, S. Gas, C. E. Barry III, F. Tekaia, K. Badcock, D. Basham, D. Brown, T. Chillingworth, R. Connor, R. Davies, K. Devlin, T. Feltwell, S. Gentles, N. Hamlin, S. Holroyd, T. Hornsby, K. Jagels, A. Krogh, J. McLean, S. Moule, L. Murphy, K. Oliver, J. Osborne, M. A. Quail, M.-A. Rajandream, J. Rogers, S. Rutter, K. Seeger, J. Skelton, R. Squares, S. Squares, J. E. Sulston, K. Taylor, S. Whitehead, and B. G. Barrell.** 1998. Deciphering the biology of *Mycobacterium tuberculosis* from the complete genome sequence. *Nature* **393:**537–544.

14. **Cook, W. J., S. R. Kar, K. B. Taylor, and L. M. Hall.** 1998. Crystal structure of the cyanobacterial metallothionein repressor SmtB: a model for metalloregulatory proteins. *J. Mol. Biol.* **275:**337–346.

15. **Crupper, S. S., V. Worrell, G. C. Stewart, and J. J. Iandolo.** 1999. Cloning and expression of *cadD*, a new cadmium resistance gene of *Staphylococcus aureus. J. Bacteriol.* **181:**4071–4075.

16. **Doran, J. L., Y. Pang, K. E. Mdluli, A. J. Moran, T. C. Victor, R. W. Stokes, E. Mahenthiralingam, B. N. Kreiswirth, J. L. Butt, G. S. Baron, J. D. Treit, V. J. Kerr, P. D. Van Helden, M. C. Roberts, and F. E. Nano.** 1997. *Mycobacterium tuberculosis efpA* encodes an efflux protein of the QacA transporter family. *Clin. Diagn. Lab. Immunol.* **4:**23–32.

16a. **Dyke, K. G. H.** Personal communication.

17. **Ellis, L. B. M., P. Saurugger, and C. Woodward.** 1992. Identification of the three-dimensional thioredoxin motif: related structure in the ORF3 protein of the *Staphylococcus aureus mer* operon. *Biochemistry* **31:**4882–4891.

18. **Endo, G., and S. Silver.** 1995. CadC, the transcriptional regulatory protein of the cadmium resistance system of *Staphylococcus aureus* plasmid pI258. *J. Bacteriol.* **177:**4437–4441.

19. **Erardi, F. X., M. L. Failla, and J. O. Falkinham III.** 1984. Plasmid-encoded copper resistance and precipitation by *Mycobacterium scrofulaceum. Appl. Environ. Microbiol.* **53:**1951–1953.

20. **Gitschier, J., B. Moffat, D. Reilly, W. I. Wood, and W. J. Fairbrother.** 1998. Solution structure of the fourth metal-binding domain from the Menkes copper-transporting ATPase. *Nat. Struct. Biol.* **5:**47–54.

21. **Grinius, L. L., and E. B. Goldberg.** 1994. Bacterial multidrug resistance is due to a single membrane protein which functions as a drug pump. *J. Biol. Chem.* **269:**29998–30004.

21a. **Gupta, A.** 1999. RT-PCR: characterization of long, multigene operons and multiple transcript gene clusters in bacteria. *BioTechniques* **27,** in press.

22. **Gupta, A., L. Matsui, J. F. Lo, and S. Silver.** 1999. Molecular basis for resistance to silver cations in *Salmonella. Nat. Med.* **5:**183–188.

23. **Gupta, A., L. T. Phung, L. Chakravarty, and S. Silver.** Mercury resistance in *Bacillus cereus* RC607: transcrip-

tional organization and two new genes. Submitted for publication.

24. **Heir, E., G. Sundheim, and A. L. Holck.** 1995. Resistance to quaternary ammonium compounds in *Staphylococcus* spp. isolated from the food industry and nucleotide sequence of the resistance plasmid pST827. *J. Appl. Bacteriol.* **79:**149–156.

25. **Helmann, J. D., B. T. Ballard, and C. T. Walsh.** 1990. The MerR metalloregulatory protein binds mercuric ion as a tricoordinate, metal-bridged dimer. *Science* **247:** 946–948.

26. **Ivey, D. M., A. A. Guffanti, Z. Shen, N. Kudyan, and T. A. Krulwich.** 1992. The *cadC* gene product of alkaliphilic *Bacillus firmus* OF4 partially restores Na$^+$ resistance to an *Escherichia coli* strain lacking an Na$^+$/H$^+$ antiporter (NhaA). *J. Bacteriol.* **174:**4878–4884.

27. **Ji, G., E. A. Garber, L. G. Armes, C.-M. Chen, J. A. Fuchs, and S. Silver.** 1994. Arsenate reductase of *Staphylococcus aureus* plasmid pI258. *Biochemistry* **33:**7294–7299.

28. **Ji, G., and S. Silver.** 1992. Regulation and expression of the arsenic resistance operon from *Staphylococcus aureus* plasmid pI258. *J. Bacteriol.* **174:**3684–3694.

29. **Ji, G., and S. Silver.** 1992. Reduction of arsenate to arsenite by the ArsC protein of the arsenic resistance operon of *Staphylococcus aureus* plasmid pI258. *Proc. Natl. Acad. Sci. USA* **89:**9474–9478.

30. **Lebrun, M., A. Audurier, and P. Cossart.** 1994. Plasmid-borne cadmium resistance genes in *Listeria monocytogenes* are similar to *cadA* and *cadC* of *Staphylococcus aureus* and are induced by cadmium. *J. Bacteriol.* **176:**3040–3048.

31. **Lebrun, M., A. Audurier, and P. Cossart.** 1994. Plasmid-borne cadmium resistance genes in *Listeria monocytogenes* are present on Tn*5422*, a novel transposon closely related to Tn*917*. *J. Bacteriol.* **176:**3049–3061.

32. **Levinson, H. S., I. Mahler, P. Blackwelder, and T. Hood.** 1996. Lead resistance and sensitivity in *Staphylococcus aureus*. *FEMS Microbiol. Lett.* **145:**421–425.

33. **Levy, S. B.** 1998. The challenge of antibiotic resistance. *Sci. Am.* (March):6–53.

34. **Liu, C.-Q., N. Khunajakr, L. G. Chia, Y.-M. Deng, P. Charoenchai, and N. W. Dunn.** 1997. Genetic analysis of regions involved in replication and cadmium resistance of the plasmid pND302 from *Lactococcus lactis*. *Plasmid* **38:** 79–90.

35. **Liu, J., H. E. Takiff, and H. Nikaido.** 1996. Active efflux of fluoroquinolones in *Mycobacterium smegmatis* mediated by LfrA, a multidrug efflux pump. *J. Bacteriol.* **178:** 3791–3795.

36. **Meissner, P., and J. O. Falkinham III.** 1984. Plasmid-encoded mercuric reductase in *Mycobacterium scrofulaceum*. *J. Bacteriol.* **157:**669–672.

37. **Mitchell, B. A., M. H. Brown, and R. A. Skurray.** 1998. QacA multidrug efflux pump from *Staphylococcus aureus*: comparative analysis of resistance to diamidines, biguanidines, and guanylhydrazones. *Antimicrob. Agents Chemother.* **42:**475–477.

38. **Nakamura, K., and S. Silver.** 1994. Molecular analysis of mercury-resistant *Bacillus* isolates from sediment of Minamata Bay, Japan. *Appl. Environ. Microbiol.* **60:**4596–4599.

39. **Neyfakh, A. A.** 1997. Natural functions of bacterial multidrug transporters. *Trends Microbiol.* **5:**309–313.

40. **Neyfakh, A. A., V. E. Bidnenko, and L. B. Chen.** 1991. Efflux-mediated multidrug resistance in *Bacillus subtilis*: similarities and dissimilarities with mammalian system. *Proc. Natl. Acad. Sci. USA* **88:**4781–4785.

40a. **Nikiforov, V.** Personal communication.

40b. **Noguchi, N., M. Hase, M. Kitta, M. Sasatsu, K. Deguchi, and M. Kono.** 1999. Antiseptic susceptibility and distribution of antiseptic-resistance genes in methicillin-resistant *Staphylococcus aureus*. *FEMS Microbiol. Lett.* **172:**247–253.

41. **Novick, R. P., E. Murphy, T. J. Gryczan, E. Baron, and I. Edelman.** 1979. Penicillinase plasmids of *Staphylococcus aureus*: restriction-deletion maps. *Plasmid* **2:**109–129.

42. **Novick, R. P., and C. Roth.** 1968. Plasmid-linked resistance to inorganic salts in *Staphylococcus aureus*. *J. Bacteriol.* **95:**1335–1342.

43. **Odermatt, A., H. Suter, R. Krapf, and M. Solioz.** 1993. Primary structure of two P-type ATPases involved in copper homeostasis in *Enterococcus hirae*. *J. Biol. Chem.* **268:**12775–12779.

44. **Odermatt, A., H. Suter, R. Krapf, and M. Solioz.** 1994. Induction of the putative copper ATPases, CopA and CopB, of *Enterococcus hirae* by Ag$^+$ and Cu^{2+}, and Ag$^+$ extrusion by CopB. *Biochem. Biophys. Res. Commun.* **202:** 44–48.

45. **O'Halloran, T. V.** 1993. Transition metals in control of gene expression. *Science* **261:**715–725.

46. **Paulsen, I. T., M. H. Brown, and R. A. Skurray.** 1996. Proton-dependent multidrug efflux systems. *Microbiol. Rev.* **60:**575–608.

47. **Paulsen, I. T., M. H. Brown, S. J. Dunstan, and R. A. Skurray.** 1995. Molecular characterization of the staphylococcal multidrug resistance export protein QacC. *J. Bacteriol.* **177:**2827–2833.

48. **Paulsen, I. T., M. H. Brown, T. G. Littlejohn, B. A. Mitchell, and R. A. Skurray.** 1996. Multidrug resistance proteins QacA and QacB from *Staphylococcus aureus*: membrane topology and identification of residues involved in substrate specificity. *Proc. Natl. Acad. Sci. USA* **93:**3630–3635.

49. **Paulsen, I. T., R. A. Skurray, R. Tam, M. H. Saier, Jr., R. J. Turner, J. H. Weiner, E. B. Goldberg, and L. L. Grinius.** 1996. The SMR family: a novel family of multidrug efflux proteins involved with the efflux of lipophilic drugs. *Mol. Microbiol.* **19:**1167–1175.

49a. **Phung, L. T., and K. Nakamura.** Personal communication.

50. **Porter, F. D., C. Ong, S. Silver, and H. Nakahara.** 1982. Selection for mercurial resistance in hospital settings. *Antimicrob. Agents Chemother.* **22:**852–858.

51. **Rennex, D., R. T. Cummings, M. Pickett, C. T. Walsh, and M. Bradley.** 1993. Role of tyrosine residues in Hg(II) detoxification by mercuric reductase from *Bacillus* sp. RC607. *Biochemistry* **32:**7475–7478.

52. **Rosenstein, R., K. Nikoleit, and F. Götz.** 1994. Binding of ArsR, the repressor of the *Staphylococcus xylosus* (pSX267) arsenic resistance operon to a sequence with dyad symmetry within the *ars* promoter. *Mol. Gen. Genet.* **242:**566–572.

53. **Rosenstein, R., A. Peschel, B. Wieland, and F. Götz.** 1992. Expression and regulation of the antimonite, arsen-

ite, and arsenate resistance operon of *Staphylococcus xylosus* plasmid pSX267. *J. Bacteriol.* **174**:3676–3683.

54. **Russell, A. D.** 1997. Plasmids and bacterial resistance to biocides. *J. Appl. Microbiol.* **83**:155–165.

55. **Sasatsu, M., Y. Shibata, N. Noguchi, and M. Kono.** 1992. High-level resistance to ethidium bromide and antiseptics in *Staphylococcus aureus. FEMS Microbiol. Lett.* **93**:109–114.

56. **Sato, S., and Y. Kobayashi.** 1998. The *ars* operon in the *skin* element of *Bacillus subtilis* confers resistance to arsenate and arsenite. *J. Bacteriol.* **180**:1655–1661.

57. **Schiering, N., W. Kabsch, M. J. Moore, M. D. Distefano, C. T. Walsh, and E. F. Pai.** 1991. Structure of the detoxification catalyst mercuric ion reductase from *Bacillus* sp. strain RC607. *Nature* **352**:168–171.

58. **Schuldiner, S., M. Lebendiker, and H. Yerushalmi.** 1997. EmrE, the smallest ion-coupled transporter, provides a unique paradigm for structure-function studies. *J. Exp. Biol.* **200**:335–341.

59. **Schwaiger, M., M. Lebendiker, H. Yerushalmi, M. Coles, A. Groger, C. Schwarz, S. Schuldiner, and H. Kessler.** 1998. NMR investigation of the multidrug transporter EmrE, an integral membrane protein. *Eur. J. Biochem.* **254**:610–619.

60. **Sedlmeier, R., and J. Altenbuchner.** 1992. Cloning and DNA sequence analysis of the mercury resistance genes of *Streptomyces lividans. Mol. Gen. Genet.* **236**:76–85.

61. **Silver, S.** 1998. Genes for all metals—a bacterial view of the Periodic Table. *J. Ind. Microbiol. Biotechnol.* **20**:1–12.

62. **Silver, S., G. Nucifora, L. Chu, and T. K. Misra.** 1989. Bacterial resistance ATPases: primary pumps for exporting toxic cations and anions. *Trends Biochem. Sci.* **14**:76–80.

63. **Silver, S., G. Nucifora, and L. T. Phung.** 1993. Human Menkes X chromosome disease and the staphylococcal cadmium resistance ATPase: a remarkable similarity in protein sequences. *Mol. Microbiol.* **10**:7–12.

64. **Silver, S., and L. T. Phung.** 1996. Bacterial heavy metal resistance: new surprises. *Annu. Rev. Microbiol.* **50**:753–789.

65. **Silver, S., and M. Walderhaug.** 1992. Gene regulation of plasmid- and chromosome-determined inorganic ion transport in bacteria. *Microbiol. Rev.* **56**:195–228.

66. **Skinner, J. S., E. Ribot, and R. A. Laddaga.** 1991. Transcriptional analysis of the *Staphylococcus aureus* plasmid pI258 mercury resistance determinant. *J. Bacteriol.* **173**:5234–5238.

67. **Solioz, M., and A. Odermatt.** 1995. Copper and silver transport by CopB-ATPase in membrane vesicles of *Enterococcus hirae. J. Biol. Chem.* **270**:9217–9221.

68. **Solioz, M., and C. Vulpe.** 1996. CPx-type ATPases: a class of P-type ATPases that pump heavy metals. *Trends Biochem. Sci.* **21**:237–241.

69. **Steingrube, V. A., R. J. Wallace, Jr., L. C. Steele, and Y. Pang.** 1991. Mercuric reductase activity and evidence of broad-spectrum mercury resistance among clinical isolates of rapidly growing mycobacteria. *Antimicrob. Agents Chemother.* **35**:819–823.

70. **Strausak, D., and M. Solioz.** 1997. CopY is a copper-inducible repressor of the *Enterococcus hirae* copper ATPases. *J. Biol. Chem.* **272**:8932–8936.

71. **Takemaru, K., M. Mizuno, T. Sato, M. Takeuchi, and Y. Kobayashi.** 1995. Complete nucleotide sequence of a *skin* element excised by DNA rearrangement during sporulation in *Bacillus subtilis. Microbiology* **141**:323–327.

72. **Takiff, H., E. M. Cimino, M. C. Musso, T. Weisbrod, R. Martinez, M. B. Delgado, L. Salazar, B. R. Bloom, and W. R. Jacobs, Jr.** 1996. Efflux pump of the proton antiporter family confers low-level fluoroquinolone resistance in *Mycobacterium smegmatis. Proc. Natl. Acad. Sci. USA* **93**:362–366.

73. **Tsai, K.-J., and A. L. Linet.** 1993. Formation of a phosphorylated enzyme intermediate by the *cadA* Cd^{2+}-ATPase. *Arch. Biochem. Biophys.* **305**:267–270.

74. **Tsai, K.-J., K.-P. Yoon, and A. R. Lynn.** 1992. ATP-dependent cadmium transport by the *cadA* cadmium resistance determinant in everted membrane vesicles of *Bacillus subtilis. J. Bacteriol.* **174**:116–121.

75. **Turner, J. S., P. D. Glands, A. C. R. Samson, and N. J. Robinson.** 1996. Zn^{2+}-sensing by the cyanobacterial metallothionein repressor SmtB: different motifs mediate metal-induced protein-DNA dissociation. *Nucleic Acids Res.* **24**:3714–3721.

76. **Tynecka, Z., Z. Gos, and J. Zajac.** 1981. Energy-dependent efflux of cadmium coded by a plasmid resistance determinant in *Staphylococcus aureus. J. Bacteriol.* **147**:313–319.

77. **van Veen, H. W., R. Callaghan, L. Soceneantu, A. Sardini, W. N. Konings, and C. F. Higgins.** 1998. A bacterial antibiotic-resistance gene that complements the human multidrug-resistance P-glycoprotein gene. *Nature* **391**:291–295.

78. **van Veen, H. W., K. Venema, H. Bolhuis, I. Oussenko, J. Kok, B. Poolman, A. J. M. Driessen, and W. N. Konings.** 1996. Multidrug resistance mediated by a bacterial homolog of the human multidrug transporter MDR1. *Proc. Natl. Acad. Sci. USA* **93**:10668–10672.

79. **Wang, Y., M. Moore, H. S. Levinson, S. Silver, C. Walsh, and I. Mahler.** 1989. Nucleotide sequence of a chromosomal mercury resistance determinant from a *Bacillus* sp. with broad-spectrum mercury-resistance. *J. Bacteriol.* **171**:83–92.

79a. **Wimmer, R., T. Herrmann, M. Solioz, and K. Wüthrich.** 1999. NMR structure and metal interactions of the CopZ copper chaperone. *J. Biol. Chem.* **274**:22597–22603.

80. **Witte, W., L. Green, T. K. Misra, and S. Silver.** 1986. Resistance to mercury and cadmium in chromosomally-resistant *Staphylococcus aureus. Antimicrob. Agents Chemother.* **29**:663–669.

81. **Wyler-Duda, P., and M. Solioz.** 1996. Phosphoenzyme formation by purified, reconstituted copper ATPase of *Enterococcus hirae. FEBS Lett.* **399**:143–146.

82. **Yerushalmi, H., M. Lebendiker, and S. Schuldiner.** 1995. EmrE, an *Escherichia coli* 12-kDa multidrug transporter, exchanges toxic cations and H^+ and is soluble in organic solvents. *J. Biol. Chem.* **270**:6856–6863.

83. **Yerushalmi, H., M. Lebendiker, and S. Schuldiner.** 1996. Negative dominance studies demonstrate the oligomeric structure of EmrE, a multidrug antiporter from *Escherichia coli. J. Biol. Chem.* **271**:31044–31048.

84. **Zhang, P., C. Toyoshima, K. Yonekura, N. M. Green, and D. L. Stokes.** 1998. Structure of the calcium pump from sarcoplasmic reticulum at 8-Å resolution. *Nature* **23**:835–839.

Tetracycline Resistance in Gram-Positive Bacteria

LAURA M. MCMURRY AND STUART B. LEVY

66

Discovered in the late 1940s, tetracyclines are natural products of *Streptomyces* soil organisms. Forming a second wave of antibiotics, after penicillin, they and their semisynthetic derivatives came into rapid use because of their broad spectrum of activity and relatively low toxicity and cost. By 1971, tetracyclines represented 30% of all antibiotics consumed for human use (83). The emergence of tetracycline resistance in the mid-1950s, initially in gram-negative bacteria and then in gram-positive bacteria, resulted in the declining usefulness of tetracyclines (68). Today the tetracyclines are primarily used to treat Lyme disease, acne, rickettsia, chlamydia, and periodontal disease. They can sometimes be used against other infective strains whose susceptibility is known.

Because of the worldwide problem with resistance even to newer antibiotics, there has been recent interest, and some success, in developing new analogs of the tetracyclines that are effective against resistant strains, such as the glycylcyclines (8) (see below). Earlier reviews on tetracyclines and tetracycline-resistance mechanisms in general include references 10, 13, 25, 26, 50, 66, 83, 103, 113, and 120.

Note that in this chapter, we sometimes cite an article in which a primary reference is discussed, in place of the primary reference; in these cases, the word "see" is used before the reference number.

CHEMICAL PROPERTIES OF TETRACYCLINES

Tetracycline and most of its analogs have an aromatic D ring and three ionizable groups with pK_a values as shown in Fig. 1 (77). In an aqueous environment at pH 7.4, most of the molecules have approximately one negative and one positive charge on the A ring and an average partial negative charge on the oxygens near atoms 11 and 12. This is the form in which the drug is crystallized from water (122). When it is crystallized under conditions in which water is carefully excluded, the molecule has no charge at any location, but instead has intramolecular hydrogen bonding (122). This may be the form that crosses biological membranes. The uncharged form has been calculated from the

microscopic ionization constants for each group to be a surprising 7% of the total drug at pH 7.4 in water (91).

Groups at positions 5, 6, 7, 8, and 9 (Fig. 1) can be modified with some retention of activity, while the others cannot (10, 13, 83), except that removal of the group at position 4 does not affect activity against gram-positive organisms (10). The phenol-diketone region (positions 10, 11, and 12) is responsible for chelation of divalent cations (13). Activity against intact cells also requires the tricarbonylmethane group (positions 1, 2, and 3) of the A ring (13). Tetracycline chelates cations such as Fe^{3+}, Co^{2+}, Mn^{2+}, Mg^{2+}, and Ca^{2+} (in order of decreasing affinity [77]), and chelation plays an important role in tetracycline function. An illustration of a tetracycline-Mg^{2+} complex bound to a protein (the crystallized TetR repressor protein) can be seen in reference 47. In this case the net charge on the tetracycline moiety is -1.

ENTRY OF TETRACYCLINES INTO MICROBIAL CELLS

Gram-positive organisms are generally more susceptible to tetracyclines than are gram-negative ones, traditionally attributed to absence of an outer membrane in the former. There is an energy-dependent component of tetracycline accumulation by bacterial cells, above and beyond binding. This uptake derives from simple equilibration of the multiprotonated acid/base tetracycline with the transmembrane pH gradient portion of the proton motive force across the cytoplasmic membrane both in the gram-positive *Enterococcus faecalis* (85) and in the gram-negative *Escherichia coli* (see reference 91) without mediation by a transporter. In a cell in which the internal pH is 7.8 and the external pH is 6.1, it has been calculated that tetracycline should be concentrated twofold (91), although the energy-dependent concentration measured in gram-negative cells appears to be greater (78, 136; see also reference 131).

Gram-Positive Pathogens, ed. by V. A. Fischetti et al.

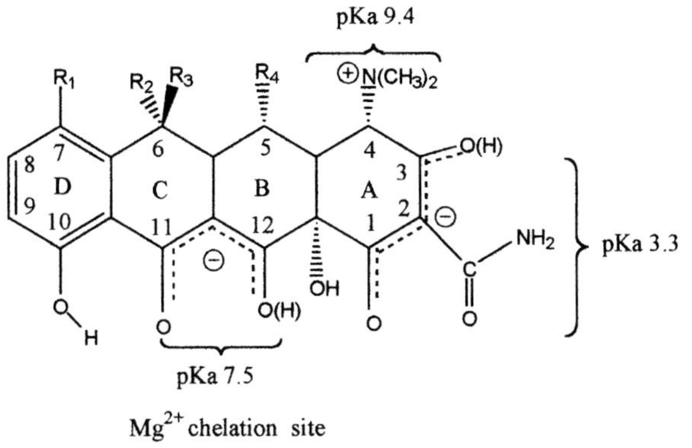

FIGURE 1 Structure of tetracycline and some of its clinically used analogs.

	R^1	R^2	R^3	R^4
chlortetracycline	Cl	CH_3	OH	H
doxycycline	H	CH_3	H	OH
minocycline	$NH(CH_3)_2$	H	H	H
oxytetracycline	H	CH_3	OH	OH
tetracycline	H	CH_3	OH	H

INHIBITION OF PROTEIN SYNTHESIS BY TETRACYCLINES

Tetracyclines are bacteriostatic antibiotics that act by inhibiting protein synthesis. These drugs inhibit the binding of aminoacyl-tRNA to the ribosome (124) in the acceptor (A) site (41); EF-Tu–dependent (enzymatic) binding is more susceptible to tetracycline than is nonenzymatic binding. These older studies have been summarized (25, 98). Most of the work involving protein synthesis has been done in E. coli, but the same picture was seen with the gram-positive Bacillus megaterium (28). The elongation process in protein synthesis has been reviewed (71, 139).

Tetracycline binds to ribosomes at a single high-affinity site of K_d of 1 to 20 μM (30, 40, 132) and at numerous sites with lower affinity (see reference 113). Thermal stability of the ribosomes increases upon binding (133). UV light-stimulated covalent photoincorporation of [^{3}H]tetracycline into ribosomes occurs mostly at protein S7 from the 30S subunit when careful correction for artifactual binding of tetracycline photolysis products is made (40). That the incorporation is physiologically meaningful seems likely since the labeling is both stereospecific and saturable, with a K_d similar to that of the high-affinity binding (40). More recent work shows that tetracycline is photolabeled not only to protein S7, but also to 16S rRNA, together resulting in loss of ribosomal function (93). Two of the three 16S rRNA residues to which tetracycline cross-links, G1300 and G1338, are located at the region of the rRNA protected by protein S7 from chemical attack, and the third (G693), as well as the residue A892 protected by tetracycline (see reference 113), are both at (or near) regions protected by tRNA (93). Whether S7, G963, G1300, and G1338 are close enough to form a single tetracycline-binding site is not known, but they are near the acceptor site. A concurrent enhancement of the chem-

ical reactivity of residues U1052 and C1054 (see reference 113) suggests a tetracycline-induced conformational change of the rRNA. However, two tetracycline analogs known to inhibit protein synthesis do not protect A892, while two others that do not inhibit protein synthesis do enhance U1052/C1054 (100), so these changes may not relate to inhibition of protein synthesis after all. Ribosomes reconstituted so as to lack the S7 or S14 proteins lose the tight binding of [^{3}H]tetracycline as assayed by filtration and sucrose density centrifugation, with a less dramatic decrease when proteins S3, S8, or S19 are absent (14). Except for S8, all of these 30S subunit proteins are linked topographically (11, 14) and may influence the avidity of the actual binding site for tetracycline. Cross-linking of tetracycline to loop V of the 23S rRNA (in the peptidyl transferase region, where the actual covalent peptide linkage event takes place) also occurs, and tetracycline also protects this loop against attack by a photoreactive tRNA derivative (93). However, loop V binding may not be relevant to the drug's inhibitory action since such cross-linking does not alter ribosomal activity (93).

Mutations involving ribosomal components associated with tetracycline resistance would be helpful in evaluating tetracycline's mode of action, but few have been found. In Bacillus subtilis a laboratory-created mutation at a locus called "tetA" caused tetracycline resistance, affected the electrophoretic migration of ribosomal protein S10, and gave resistant ribosomes (see reference 121). The locus mapped in the ribosomal gene cluster but has not, to our knowledge, been sequenced to verify that it is in the S10 gene. S10 is located near S7 and the other proteins mentioned above (11). In Propionibacterium acnes, clinical strains with 16S rRNA mutations have been described recently (105) (see "mutated rRNA" section below). The E. coli chromosome has seven copies of the gene for 16S RNA, which could interfere with selection of rRNA mutations.

In summary, tetracycline (likely as a tetracycline-divalent cation complex) interferes at the acceptor site of the prokaryotic ribosome, probably via high-affinity binding on or near protein S7 and certain residues of 16S rRNA. The structure of the (single?) binding site and how the tetracycline-ribosome interaction disrupts acceptor site function remain unknown. Tetracycline may sterically interfere with one or more points of attachment of the aminoacyl-tRNA.EF-Tu.GTP complex in the ribosomal acceptor site, or it may allosterically cause a deleterious conformational change of that site.

TETRACYCLINE-RESISTANCE MECHANISMS AND RESISTANCE DETERMINANTS: OVERVIEW

There are three different tetracycline-resistance mechanisms of clinical importance found among gram-positive organisms. They are active efflux, ribosomal protection, and mutated rRNA (103, 105, 113, 120). A relatively unimportant degradative mechanism has been identified so far only in gram-negative bacteria.

Multidrug transporters are able to efflux substrates that are chemically different (35, 96). Some from gram-negative bacteria include tetracycline in their repertoire, for example, AcrAB (E. coli) and MexAB/OprM (Pseudomonas sp.) (90). However, with the possible exception of an ABC efflux transporter (see below), no multidrug transporter from gram-positive bacteria has yet been reported to transport tetracycline.

We define a tetracycline-resistance determinant as a contiguous genetic unit encoding all the genes involved in the resistance. Mutations in genes encoding ribosomal proteins or rRNA are not counted as resistance determinants. Two determinants were historically classified as different if they showed no DNA hybridization with each other at high stringency. With DNA sequencing now routine, a >20% difference in structural protein sequence identity has typically defined a new determinant. Because some determinants have more than one gene, it was decided to make a distinction between the name of a gene and the class name assigned to the determinant (70). For example, the class P determinant is called Tet P; it has two genes, *tetA*(P) and *tetB*(P), encoding the proteins TetA(P) (efflux) and TetB(P) (ribosomal protection). The class M determinant is called Tet M; it has a single gene, *tet*(M), encoding a single protein, Tet(M). Since almost all single letters of the alphabet have now been used for the labeling of determinants, we have proposed a supplementary nomenclature for new determinants (69). Note that when searching databases, to retrieve intact a term containing parentheses, such as TetA(P), or a space, such as Tet M, it is necessary to type the term within quotation marks.

The emphasis in this chapter is on proteins found in gram-positive organisms. However, we have chosen to include selected information about certain related proteins from gram-negative organisms in cases where those proteins are better studied and offer insights.

TETRACYCLINE RESISTANCE BY ACTIVE EFFLUX

To date there are 19 known tetracycline-resistance determinants that encode active efflux of tetracycline as a resistance mechanism. Of these, eight are found in gram-positive organisms. The 19 determinants fall into six groups, based on the degree of identity of amino acid sequence of the effluxing proteins (Table 1). Between the groups there is little identity. Group 1 comprises classes A to E, G, H, probably I, J, Z, and an unnamed determinant from *Agrobacterium* sp., sharing about 45 to 78% identity (69, 113). The first and only member of group 1 to be found in a gram-positive organism is the class Z determinant, on a large plasmid (103, 126). Group 1 proteins have 12 predicted transmembrane α-helices, with a relatively long central nonconserved cytoplasmic loop connecting helices 6 and 7 (Fig. 2). Group 2 comprises classes K and L, 58 to 59% identical, predominating in gram-positive species but also found in *Veillonella* and *Haemophilus* spp. (103). Group 2 proteins have 14 predicted transmembrane helices (Fig. 2). Class L has two subclasses that are 81% identical, that from the *B. subtilis* chromosome and that from a variety of plasmids. Group 3 contains the chromosomally encoded OtrB and Tcr3 proteins (56% identical) with a topology like that of group 2 but with longer amino termini and longer loops 5-6 and 13-14 and, for OtrB, a longer C terminus (Fig. 2). Members of this group were discovered in the tetracycline-producing organisms *Streptomyces* spp. and do not follow the standard nomenclature for tetracycline-resistance determinants. Group 4 has only the TetA(P) protein, from *Clostridium perfringens*, with 12 putative transmembrane helices but no central loop (Fig. 2) (116). The class P determinant, on a conjugative plasmid, is unusual in having two overlapping genes (116). Group 5 has but a single member protein, Tet(V), from the *Mycobacterium smegmatis* genome, with 10 to 11 helices (31). Group 6 (unnamed determinant) from a *Corynebacterium striatum* plasmid, the only one expected to use ATP rather than a proton gradient as energy source, has two related putative ABC (ATP-binding cassette [see reference 35]) transporters, TetA and TetB, both of which are required for resistance and each of which has approximately five (or six?) putative transmembrane helices together with a carboxy-terminal ATP-binding cassette (125). Since this determinant also encodes oxacillin resistance, the proteins may actually constitute a multidrug exporter.

A low level of sequence identity is seen among all proton-coupled transporters, of which there are many (96). The first and second halves of group 1 Tet proteins, and the first six helices of group 2 Tet proteins, appear to have a common ancestor (109; see references 67 and 96). However, the different groups of tetracycline efflux proteins are more related to other transporters than to each other, and it is therefore likely that tetracycline efflux has evolved more than once (97, 115). The topology of these transporters can be predicted by computer-assisted hydropathy analysis, which locates stretches of hydrophobic residues sufficiently long to span a lipid bilayer as an α-helix. The experimentally determined topology for Tet efflux proteins has been consistent with such theoretical analyses, as shown for TetA(B) (57; see reference 113), TetA(C) (3), and Tet(K) (39, 48). In addition, Tcr3 is 25% identical to the lipophilic organic cation efflux protein QacA, which has a topology consistent with 14 transmembrane α-helices (see reference 96). The expected high α-helical content of TetA(B) has been seen in the purified protein by circular dichroism analysis (2).

With the presumed exception of the putative ABC transporter from *Corynebacterium* sp. mentioned above (125), all of the tetracycline efflux systems studied to date use a proton gradient across the cell membrane (74) as the energy source to drive tetracycline out of the cell against a concentration gradient. Calculations and experiments have shown that the efflux achieves an appropriately low cytoplasmic concentration of tetracycline (131). As a heuristic model, we can imagine that binding of tetracycline to a region of a Tet protein on the inside surface of the cell, when (and only when) coupled with binding of a proton to the protein on the opposite surface, results in a conformational change of the protein, allowing movement of tetracycline across the membrane. Then release of tetracycline to the outside and a proton to the inside occurs, and the protein returns to its pretransport conformation. Some questions of interest include: (i) Does the protein transport tetracycline in the unchelated or chelated form? (ii) What regions of the protein bind the two substrates (tetracycline and proton)? (iii) What is the conformational change(s)?

The properties of efflux systems are conveniently studied using cell-free everted (inside-out) membrane vesicles, in which an everted membrane proton gradient is generated by electron transport (from the addition of electron transport substrates such as lactate) or by the membrane F_0F_1 ATPase (with addition of ATP). That group 1 tetracycline-resistance determinants encode an energy-dependent ("active") efflux was initially demonstrated from the active uptake of [³H]tetracycline by everted vesicles from cells bearing the gram-negative determinants of classes A to D (80). The principles behind these better-studied

TABLE 1 Efflux tetracycline-resistance genes from gram-positive organisms[a]

Group	Gene(s)	aa[b]	Host of sequenced gene	GenBank nucleotide sequence accession no. (source)	% Identity of protein to that of pTHT15	Regulation	Reference	Locations of class members[c]
1	tet(Z)	384	Corynebacterium glutamicum	AF121000 (pAG1)	19 (related to tet(A-H) of gram-negative bacteria)	Inducible (repressor)	125	P**
2	tet(K)[d]	459	Staphylococcus aureus	M16217 (pNS1)	59	Inducible	92	C, P**
				S67449 (pT181)			43	
2	tet(L)	458	Bacillus stearothermophilus	M11036 (pTHT15)	100	Inducible[e]	51	P**
2	tet(L) (chromosomal)	458	Bacillus subtilis	X08034 (strain GSY908 chromosome; called "tetBS908"; identical in B. subtilis 168)	81	Inducible	110	C
3	otrB	563	Streptomyces rimosus	AF079900 (chromosomal DNA)	20	Unknown	79	C
3	tcr3 (tcrC)	512	Streptomyces lividans	D38215 (chromosomal DNA)	23	Unknown	29	C
4	tetA(P)	420	Clostridium perfringens	L20800 (pCW3)	20	Inducible	116	P**
5	tet(V)	419	Mycobacterium smegmatis mc^2	AF030344 (chromosomal DNA)	19	Unknown	31	C
6	tetAB	513 528	Corynebacterium striatum M82B	U21300 (pTP10)	(putative ABC transporter)	Unknown	125	P**

[a] A specific sequenced gene was chosen to represent each class. Information in the first eight columns applies to the gene. The last column indicates the locations of typical class members.
[b] Number of amino acid residues in protein.
[c] C, chromosome; P, plasmid; P**, conjugative plasmid.
[d] The two tet(K) sequences, one from 1986 and one from 1993, are identical; in intervening years, inaccurate sequences were reported (54, 84).
[e] Some plasmid-mediated Tet L determinants are constitutive (see text).

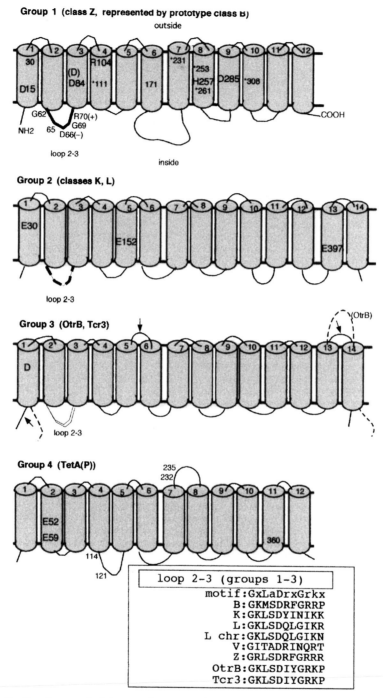

FIGURE 2 Predicted or verified (see text) transmembrane topology of tetracycline efflux proteins from groups 1 to 4 and (inset at bottom) consensus motif for loop 2-3 of proton-dependent transporters, followed by loop 2-3 sequences for Tet proteins of groups 1 to 3. The topologies of groups 5 and 6 are less certain (see text) and are not included. In group 1, the better-studied gram-negative TetA(B) substitutes for the gram-positive TetA(Z). Transmembrane "α-helices" (presumed) are gray and numbered. Charged residues predicted or known to be within transmembrane helices are shown with a large font and letter designation. In loop 2-3 of TetA(B), a "+" or "−" indicates the necessity of that charge at that site. Gly-62 and, to a lesser extent, Gly-69 probably cause a β-turn (see references 93 and 113). Other residues mentioned in the text are designated in small font without letters. All residues marked in TetA(B) except H257 are also found in TetA(Z); residue (D) in helix 3 is present in TetA(Z) but not in TetA(B). An asterisk denotes residues possibly involved in tetracycline binding in TetA(B). Arrows show regions in Tcr3 that are longer than similar regions in Tet(K) and Tet(L). A dotted line in group 3 similarly shows where OtrB has regions longer than those in Tcr3. In the inset, "L chr" refers to the protein from the chromosomal Tet L determinant.

efflux systems can aid in understanding the gram-positive systems.

A tetracycline-divalent cation complex is the species transported by those Tet efflux proteins examined to date. This point was suggested by the proportionality between the effectiveness of a given divalent cation in stimulating efflux and its affinity for tetracycline (149) and was proved by the cotransport of tetracycline and Co^{2+} for TetA(B) (149), Tet(K) (146), and chromosomal Tet(L) (46). A single divalent cation atom is transported per tetracycline molecule for TetA(B) (149) and for Tet(L) (24). Therefore +1 is the net charge of the actual substrate complex (tetracycline[−1] plus divalent cation [+2]). An energy-dependent efflux catalyzed by the TetA(P) protein can be seen in *E. coli* cells (116), but everted vesicles have yielded no activity. The K_m for TetA(B), Tet(K), and Tet(L) in vesicles is 10 to 50 μM, and the V_{max} is 1 to 30 nmol mg^{-1} min^{-1} (46, 80, 146, 149), which for TetA(B) corresponds to a k_{cat} of approximately 0.03 to 1 s^{-1}.

The number of protons used per substrate complex transported was found to be 1 for the gram-negative TetA(B) (142), meaning that the antiport exchange is electroneutral, that is, +1 (proton) comes into the cell while +1 (tetracycline-divalent cation) goes out. In the case of the chromosomally encoded Tet(L) from *B. subtilis*, however, transport of tetracycline is electrogenic, that is, it is energized by the electrical potential (it is also energized by the chemical potential) (24, 45, 46), indicating that more than one proton is exchanged per tetracycline-metal complex. This mechanism is relevant to the additional role of this protein in pH homeostasis under alkaline conditions (see the following section). Tet(K) is also a tetracycline-proton antiporter (49). Experiments using Tet(L) as a control show that Tet(K)-mediated tetracycline transport is electrogenic (45).

The protein residues involved in binding the tetracycline-cation complex are not known but may be hinted at by mutations that change the specificity or affinity of the protein. In the gram-negative TetA(B), mutations near the centers of transmembrane helix 4 (Gly-111), helix 8 (nonconserved Leu-253), and helix 10 (Leu-308) and toward the periplasmic end of helix 7 (Trp-231) enhance resistance to the normally inhibitory analog glycylcycline while decreasing resistance to tetracycline (134; see reference 113) (Fig. 2). Gln-54–Ala (helix 2) and Gln-261–Ala (helix 8) mutations raised the K_m for transport of tetracycline (see reference 113). The low-level photoaffinity labeling of TetA(B) protein with [³H]tetracycline disappears if Asp-285, in the center of helix 9, is mutated to Ala (59), showing that Asp-285 is involved, directly or indirectly, in substrate binding. A His-257–Asp mutation raises the K_m 4.5-fold (140), and an Asp-285–Asn/Ala-220–Asp double mutation raises it similarly (144). Therefore, residues that may be involved in substrate binding occur in transmembrane regions in both halves of the protein.

Tet proteins have several sequence motifs in common with other proton-dependent transporters (96). Particularly striking is the motif in the cytoplasmic loop connecting putative transmembrane helices 2 and 3 (loop 2-3) found in almost all such transporters (96, 97), including all of the proton-dependent tetracycline efflux proteins except TetA(P) (Fig. 2). Such a broadly conserved motif could be involved in transporter-proton interactions and their coupling to conformational changes, but Yamaguchi and coworkers have experimental evidence that this region may be a "gate" for tetracycline/metal entrance (143, 147). The

effect of mutations in this loop in TetA(B) has been reviewed (96, 113) (Fig. 2). Of five cysteine replacement mutations in loop 2-3 that allow activity, only reaction of Ser-65–Cys with sulfhydryl reagents such as *N*-ethylmaleimide (NEM) inactivates the protein, in proportion to the volume of the reagent molecule (see reference 113). Most interestingly, Ser-65–Cys becomes more accessible to the reagents in the presence of a tetracycline-divalent cation *even if* the protein also has an inactivating Arg-70–Ala mutation but *not* an inactivating Asp-66–Ala mutation (55). Therefore, Asp-66, but not Arg-70, may either bind tetracycline or cause a conformational change following binding elsewhere.

More recent studies show that conformational changes, including those accompanying binding of tetracycline, are transmitted between cytoplasmic loop 2-3 and other regions of the TetA(B) protein, both within and on the other side of the membrane (58, 118).

Cysteine scanning mutagenesis of TetA(B) has shown that positions reactive with NEM lie all along one side of transmembrane helix 2 (60). Since the NEM reaction reportedly requires the presence of water, the conclusion was that the surface of helix 2 might be part of an aqueous pathway for tetracycline/cation across the membrane (60). Similar experiments had shown little NEM reactivity within the putative membrane region of helices 3 and 9, confirming that they stretch across the membrane but are not pathways (see reference 60).

In the case of the gram-positive Tet(K), the motif in cytoplasmic loop 2-3 behaves differently from that of TetA(B): a negative charge at Asp-74, corresponding to the critically negative Asp-66 of TetA(B), is not essential (38). However, a GXXXDRKG sequence in cytoplasmic loop 10-11 of Tet(K) is a reduced version of the classical loop 2-3 motif. Asp-318 in the reduced motif is essential and cannot be replaced by Asn or Glu (38). It will be of interest to see to what extent this second loop behaves like the loop 2-3 motif, even though its location is so different.

Suppressor mutations that restore activity to mutant proteins have helped identify helices 7 and 9 as neighbors within the TetA(B) protein (see reference 113). Mutations in the first half of TetA(B) complement those in the second half on a second polypeptide, suggesting possible dimerization (see references 67 and 113). The two halves of group 1 Tet proteins do not, however, have independent functions, since a hybrid protein that has its first half from class B and its second half from class C, or vice versa, is inactive (107). An A/C hybrid, from more closely related classes, is active (107). Surprisingly, single mutations in either half of the inactive C/B hybrid can restore some tetracycline resistance (112). A mutation in the second half of TetA(C) can suppress a mutation in the first half, and vice versa (see reference 113). These results show that the two halves of the protein interact functionally.

Charged residues within hydrophobic transmembrane helices of transporters have proved to be irresistible targets for mutagenesis, since it is presumed that the energy cost in their maintenance within the membrane must imply a functional role. A negative charge at Asp-84 of the TetA(B) protein, in helix 3, is required for resistance, while changing Asp-285, in helix 9, to Asn or Glu eliminates all proton-coupled tetracycline transport (141); several changes can be made at Asp-15 without complete loss of activity (81). Arg-101, located just within transmembrane helix 4, is essential (56). On the other hand, His-257 can be mutated to other residues with retention of some activity (145) and is not even present in TetA(Z) (126). The gram-positive Tet(K) and Tet(L) proteins have three conserved transmembrane acidic (Glu) residues. In Tet(K) they are Glu-30 (helix 1), Glu-152 (helix 5), and Glu-397 (helix 13). The negative charge of Glu-30 and Glu-152 seems necessary in Tet(K), since replacement by Asp, but not Asn, maintains activity (37). Glu-397 cannot be replaced by Asp or Asn (37), so this residue may be critical. Random mutagenesis in gram-negative Tet(C) and selection for loss of resistance revealed the importance of some of the same residues mentioned above, as well as others (82).

The gram-positive class P efflux protein TetA(P) has two acidic residues within putative transmembrane helices. They are Glu-52 and Glu-59, both in helix 2 (Fig. 2). Both are critical since neither can be altered to Asp, Gln, or Lys without severe loss of activity (53). In the external loop 3-4, a negative charge at Glu-89 is needed for resistance (53). Site-directed substitutions at Glu-35 (external loop 1-2) and at Lys-214 and Arg-328 (charged residues within putative transmembrane helices) have little effect (53). Random mutagenesis identified mutations in two other loops that severely lower resistance without affecting the amount of transporter: Gly-114–Asp and Ala-121–Glu (cytoplasmic loop 4-5) and Glu-232–Lys and Asp-235–Asn (external loop 7-8) (4).

Both Tet(B) and Tet(K) can function if individually inactive fragments are present together in the same cell. This type of complementation is seen with the first and second halves of TetA(B) (108, 148). In the case of Tet(K), the second half of the protein together with the whole protein from which 62 consecutive amino acids are deleted near the carboxy terminus restore about 60% of the wild-type resistance (84). Therefore, certain domains of the proteins can be held together by strong noncovalent bonds.

Purified TetA(B) (117) or Tet(L) (24) reconstituted in liposomes (vesicles having a lipid bilayer, made in vitro from purified lipids) can perform tetracycline/proton antiport, showing that only the Tet proteins are necessary for this activity. In the reconstituted system, Tet(L) was reported to have a k_{cat} for tetracycline as great as 180 s^{-1} (24).

The mechanisms of the other gram-positive efflux determinants have been less well characterized. The gene from *Streptomyces rimosus*, identical or similar to *otrB* (79), results in a decreased accumulation of tetracycline in intact cells (94), although no further biochemical or genetic studies have been done. *otrB* (formerly known as *tet347* when its protein product was thought to have 347 residues) has recently been resequenced and postulated to encode a membrane protein of 563 residues and 14 transmembrane helices (79). The chromosomal *tet*(V) from *Mycobacterium* sp. causes active efflux from cells when cloned in multicopy (31).

TRANSPORT OF SODIUM, POTASSIUM, AND OTHER SUBSTRATES BY TETRACYCLINE EFFLUX PROTEINS

Some tetracycline efflux proteins from both gram-negative and gram-positive organisms can transport substrates other than tetracycline and protons. In gram-negative bacteria, a cellular defect for potassium uptake was twice unexpectedly discovered to be alleviated by TetA(C) but not by TetA(B) (33, 86). Strikingly, the first 97 amino acid resi-

dues (three helices) were sufficient for this activity (42, 86). Mutations in TetA(C) that eliminate tetracycline resistance still allow the potassium transport complementation (82).

In the gram-positive *B. subtilis*, screening for genetic loci whose disruption by Tn*917* made cells more sensitive to sodium or high pH pointed to the chromosomally located *tet*(L) gene (22). The effect is most obvious when the complete gene is deleted (23), which causes an enhanced susceptibility to tetracycline. Protection against high pH is completely restored by a cloned *tet*(L) gene and partially restored by a cloned *tet*(K) gene (23). The explanation for these results is that these Tet proteins can perform electrogenic Na$^+$/H$^+$ antiport and K$^+$/H$^+$ antiport (with radiolabeled Rb$^+$ representing K$^+$), so that more than one H$^+$ enters per exiting Na$^+$ or K$^+$, thereby maintaining the internal pH at about 7.7 even when the external pH is 8.5 (24, 45, 46). *tet*(L) appears to be the chromosomal gene most responsible for pH homeostasis mediated by Na$^+$(K$^+$)/H$^+$ antiport in *B. subtilis* (23). Purified Tet(L) reconstituted in liposomes catalyzes the Na$^+$(K$^+$)/H$^+$ exchange (24). The active sites on Tet(L) for transport of tetracycline and for monovalent cations by Tet(L) are different since K$^+$ and Na$^+$ do not interfere with tetracycline transport, and vice versa (46). Finally, K$^+$ can take the place of the proton in the Na$^+$(K$^+$)/H$^+$ antiport function of Tet(L) and even more so in that of Tet(K), resulting in the electrogenic uptake of more than one K$^+$ ion per single K$^+$ or Na$^+$ ion exported (45). That is, both Tet(K) and chromosomal Tet(L) catalyze the net uptake of potassium. This explains why mutations that reduce the activity of previously characterized "traditional" chromosomal K$^+$ uptake systems can be complemented by Tet(L) or Tet(K). Unlike TetA(C) (42, 86) and Tet(K) (44), Tet(L) does not appear to carry out potassium transport when a large carboxy-terminal region representing about 63% of the protein has been deleted (45).

The functioning of Tet proteins as monovalent cation antiporters may provide a selective advantage, e.g., with regard to pH homeostasis or potassium uptake, something to be considered in understanding the persistence of tetracycline resistance (45). Krulwich and coworkers also mention that these Tet proteins may eliminate most of the pH gradient component of the proton motive force, thereby lowering the pH gradient-dependent uptake of tetracycline in cells (21). Cells lacking chromosomal *tet*(L) grow slowly even at neutral pH, suggesting that Tet(L) may also transport an unknown metabolite (23). It will be interesting to see if any other Tet proteins have multiple functions as do Tet(C), Tet(K) and Tet(L), and to discover which regions of the proteins are involved in each function and how they are regulated.

TETRACYCLINE RESISTANCE BY RIBOSOMAL PROTECTION

Ribosomal protection is the most widespread of the tetracycline-resistance mechanisms. It is mediated by a cytoplasmic protein that reduces the susceptibility of ribosomes to tetracycline. Tet(M), Tet(O), and OtrA proteins have been shown to function as ribosomal protection proteins, while Tet(S), Tet(T), Tet(Q), TetB(P), Tet(W), and "Tet" are presumed to do so because of their related sequences (Table 2). "Tet" and OtrA, both from *Streptomyces* sp., are 65% identical (32). Four different (but unnamed) pre-

TABLE 2 Ribosomal protection tetracycline-resistance genes from gram-positive organisms[a]

Gene	aa[b]	Host of sequenced gene	GenBank nucleotide sequence accession no. (source)	% Identity of protein to Tet(M) protein from Tn916	Regulation	Reference	Locations of class members[c]
tet(M)	639	Enterococcus faecalis	M85225 (Tn916)	100	Inducible	123	C, P, T
tet(O)	639	Streptococcus mutans DL5	M20925 (chromosomal DNA)	76–77	Constitutive	63	C
	639	Campylobacter jejuni[d]	M18896 (pUA466) [M74450 upstream]			76 [135]	P**
tetB(P)	652	Clostridium perfringens	L20800 (pCW3)	38	Inducible	116	P**
tet(Q)	641 or 657	Bacteroides thetaiotaomicron	X58717 (conjugative transposon)	40	Constitutive	91a	C, P, T
tet(S)	641	Listeria monocytogenes	L09756 (pIP811)	79	Unknown	20	C, P, T
tet(T)	651	Streptococcus pyogenes A498	L42544[e] (chromosomal DNA)	44	Unknown	27	C
tet(W)	639	Butyrivibrio fibrisolvens[f]	AJ222769 (chromosomal DNA)	68	Inducible	6	C, T
"tet"	639	Streptomyces lividans 1326	M74049 (chromosomal DNA)	34	Unknown	32	C
otrA	663	Streptomyces rimosus	X53401 (chromosomal DNA)	33	Unknown	34	C

[a] A specific sequenced gene was chosen to represent each class. Information in the first seven columns applies to the gene. The last column indicates the locations of typical class members.
[b] Number of amino acid residues in protein.
[c] C, chromosome; P, plasmid; P**, conjugative plasmid; T, conjugative transposon.
[d] A tet(O) sequence taken from the gram-negative C. jejuni is included because of the availability of additional upstream sequences of interest.
[e] Not in GenBank, but in Genome Sequence DataBase (86a).
[f] Butyrivibrio sp. has recently been reclassified as a gram-positive-like organism (138).

sumed ribosomal protection classes from streptococci have also been cloned using degenerate PCR primers based on known ribosomal protection genes; they do not hybridize with each other or with class M, O, P, Q, S, or T (27).

Ribosomal protection was first described as one of two resistance mechanisms in *S. rimosus*, an organism that produces oxytetracycline. The efflux mechanism associated with the chromosomal *otrB* gene from this organism has been discussed above. The second chromosomal gene, now called *otrA*, encodes a ribosome-associated factor that causes resistance to tetracycline of in vitro protein synthesis and dissociates in high salt (94). More detailed studies have subsequently been done on Tet(M) and Tet(O). The 639-amino-acid Tet(M) protein is homologous to translational elongation factors EF-G and EF-Tu. The closest

identity is to EF-G, and much of the identity is with the GTP-binding domain at the N-terminal region (16, 75, 127). The other ribosomal protection proteins are also related to EF-G (127). The three-dimensional structure of EF-G is known and suggests a structure for ribosomal protection proteins (1).

Protein synthesis in crude extracts from cells bearing the Tet M determinant from Tn916 is resistant to tetracycline (15). The Tet(M) protein was overproduced, purified, and shown to confer resistance upon ribosomes from nonresistant cells (16). Like EF-G and OtrA, Tet(M) is released from resistant ribosomes by washing with high salt, and like EF-G, it has ribosome-dependent GTPase activity (16). However, Tet(M) is not an alternative EF-G since it cannot complement a temperature-sensitive EF-G in cells

(16), nor can it replace EF-G (or EF-Tu) during in vitro protein synthesis (18). The related Tet(O) protein is also a ribosome-dependent GTPase, and it binds both GDP and GTP (128). Site-directed mutations in Tet(O) at Asn-128, conserved in six different ribosomal protection proteins and known to correspond to important residue 142 in the GTP-binding site of EF-G, cause a decrease in Tet(O) activity, bolstering the notion that GTP binding is important to Tet(O) function (see reference 127).

Further studies showed that Tet(M) permits amino-acyl-tRNA to bind to the acceptor site of the ribosome in the presence of a normally inhibitory 50 μM tetracycline (18). That Tet(M) brings about the actual release of the drug, rather than just helping the ribosomes cope with bound tetracycline, was shown by the GTP-dependent loss of tetracycline from ribosomes within seconds of addition of Tet(M) (18). Tet(M) does not itself bind tetracycline (18). Similarly, Tet(O) reduces binding of tetracycline to ribosomes in the presence of GTP (and not GDP), and binding of radiolabeled GTP to Tet(O) is enhanced by ribosomes (132). A Tet(O)-GTPγ[^{35}S]-ribosome complex was identified by gel filtration (132), as was a [^{3}H]Tet(M)-GTP-ribosome complex (30). A nonhydrolyzable GTP analog catalyzes Tet(M) binding to ribosomes (30). Whether the energy from GTP hydrolysis serves to release tetracycline from the ribosome or to dissociate the ribosomal protection protein from the ribosome to allow its reuse has been investigated using the nonhydrolyzable GTP analog. In the case of Tet(M) the data were interpreted in favor of the first possibility (18) and for Tet(O) the second (132), although the difference may reflect the details of the experiments rather than the different proteins themselves.

The binding of Tet(M) to the ribosome is not affected by tetracycline but is inhibited by thiostrepton, which interferes with EF-G interaction with the ribosomal GTPase center. Moreover, EF-G and Tet(M) compete for binding to the ribosome, with Tet(M) having a greater affinity (30). The conclusion is that Tet(M) and EF-G have overlapping sites on the ribosome.

Chromosomal mutations at two loci, miaA and rpsL, reduce the effectiveness of ribosomal protection proteins (17, 129). miaA encodes an enzyme that catalyzes the first step in modification of A37 on those tRNAs that read codons beginning with U (9). A37 is just 3′ to the anticodon, and its undermodification leads to a decrease in the rate of elongation, an increase in reading errors at the first position of the codon, and a decrease in errors at the third position (9). Resistance due to Tet(M), but not to Tet(O), was reduced severalfold by mutations in miaA (17, 129). Mutations in rpsL, encoding protein S12 of the S30 ribosomal subunit, cause streptomycin resistance or dependence and reduce the ability of Tet(M) and Tet(O) to cause tetracycline resistance in cells, in one case as much as 32-fold (129). How the decrease in the rate of elongation and change in translational accuracy caused by mutations in miaA and rpsL might be related to their effects on the activity of ribosomal protection proteins is not known.

We can conclude that ribosomal protection proteins bind to ribosomes and somehow alter the ribosomal conformation to prevent tetracycline binding without harming protein synthesis. The hydrolysis of GTP to GDP may provide both energy for a conformational change transmitted to the ribosome and a mechanism for the departure of the protection protein from the ribosome to get out of the way of EF-G, which shares the Tet(M) binding site and whose action is required for the translocation step. The protection protein may cycle on and off the ribosome in synchrony with peptide bond formation. The effect of the miaA and rpsL mutations may involve the kinetics of sampling of aminoacyl-tRNAs by the decoding site. A specific model of the mechanism of action of the protection proteins has been proposed (132).

TETRACYCLINE RESISTANCE BY MUTATED rRNA

The first ribosomal mutation giving rise to clinical tetracycline resistance was described in 1998 in isolates of P. acnes (105). A change of G to C at a position cognate with nucleotide 1058 of E. coli 16S rRNA was seen in 15 of 21 resistant isolates (64-fold increase in mean MIC to tetracycline/doxycycline) and was not seen in any susceptible strains. When this mutation was re-created in rrnB, the E. coli gene for 16S rRNA, and cloned on a multicopy plasmid, E. coli bearing the plasmid was more resistant to tetracycline (and had a longer lag when grown without the drug, perhaps reflecting a slight loss of ribosome function). The mutation is in helix 34 of the 16S rRNA, very close to a residue that cross-links to the mRNA codon in the ribosomal acceptor site and near those residues 1052/1054 mentioned above ("inhibition of protein synthesis" section) that increase exposure upon tetracycline binding to the 30S subunit (105). Curiously, only the clinical strains isolated after 1988 had this mutation, while tetracycline-resistant mutants created in the laboratory did not have the mutation (105).

As noted earlier, a resistance mutation selected in the laboratory changed ribosomal protein S10 in B. subtilis, but this mutation has not been seen in clinical isolates.

TETRACYCLINE RESISTANCE BY DEGRADATION

Inactivation of tetracyclines is mediated by a normally cryptic, plasmid-borne tet(X) gene in the gram-negative anaerobe Bacteroides sp. The responsible protein appears to be an NADPH-dependent oxidoreductase that functions only aerobically, e.g., in E. coli (119), although the chemistry of the reaction is not known. Since this gene may have originated in gram-positive bacteria, as suggested from its close linkage to the gram-positive-related gene ermF (103, 119), future studies may reveal other such genes in gram-positive organisms.

TETRACYCLINE RESISTANCE BY UNKNOWN MECHANISM

tet(U), a plasmid-borne gene that gives low-level resistance to both tetracycline and minocycline in Enterococcus faecium and was tentatively categorized as related to ribosomal protection, encodes a predicted protein of only 102 amino acids with little sequence identity to any other tetracycline-resistance protein (102). This class U determinant may represent a new category of tetracycline-resistance mechanism, although the protein product is so unusual in size and sequence that recloning and resequencing might be worthwhile. The otrC tetracycline-resistance gene from Streptomyces sp. has not been sequenced, and its mechanism is unknown (103).

REGULATION OF EXPRESSION OF GROUP 1 EFFLUX (Tet Z)

The group 1 efflux determinants such as Tet Z, of which the prototype is Tet B carried by Tn*10*, each have two genes, *tetA* (transporter) and *tetR* (repressor), divergently transcribed from a common, complex regulatory region containing the promoters and two palindromic operators, O1 and O2 (47). In the absence of tetracycline, the repressor typically binds to both operators in such a way that transcription of *tetA* is inhibited more severely, while enough *tetR* expression occurs to maintain repression. In the presence of tetracycline at levels far below those inhibitory to protein synthesis, a repressor/tetracycline-divalent cation complex forms at a very high affinity (on the order of $K_d = 3 \times 10^{-4}$ μM), causing dissociation of the repressor from DNA with attendant transcription and translation of *tetA* and *tetR* (47). The three-dimensional structure of dimeric (gram-negative) TetR(D) protein complexed with tetracycline-Mg^{2+} (see reference 47) or in the absence of inducer (95) has been determined, permitting understanding of the switch mechanism by which the inducer works.

REGULATION OF EXPRESSION OF GROUP 2 EFFLUX (Tet K, L)

Resistance for Tet L determinants is typically inducible by tetracycline. For the plasmid-borne tetracycline-inducible Tet L determinant (of pTHT15), transcription initiates 120 nucleotides (nt) upstream of the Tet(L) translational start (51). Upstream from *tet*(L) in the transcript is the sequence for a putative leader peptide of 20 amino acids together with potential stem-loop mRNA structures (none a rho-independent transcriptional terminator), one that overlaps the leader peptide and another, downstream, that sequesters the ribosomal binding site (RBS) of Tet(L) (51, 114). Such a configuration suggested (51) that *tet*(L) might be regulated by translational attenuation, as are two other systems involving antibiotics acting on the ribosome, the *cat* chloramphenicol acetyltransferase and *erm* methylase genes (73). In the *cat* mechanism, the product of translation of the first six codons of the leader peptide inhibits the peptidyl transferase activity of the ribosome, causing ribosomal pausing. In the presence of the inducer chloramphenicol, the ribosome is then permanently stalled, resulting in unfolding of the downstream stem-loop that had masked the RBS of the Cat protein. A leader *cat* peptide truncated at codon 6 by introduction of a stop codon leads to constitutivity (73). The only evidence consistent with this model for plasmid-mediated *tet*(L), other than the characteristics of the upstream sequences, is that naturally occurring constitutive plasmid *tet*(L) genes have truncated leader peptides. A single-nucleotide difference between the sequence of pTHT15 and that of pLS1 (derived from pMV158) and pJH1 (bearing an identical determinant) results in leader termination after four residues (pLS1 [62], GenBank accession no. M29725; pJH1 [99], GenBank accession no. U17153). A 7-nt insertion for pAMα1 results in leader termination after five residues (52) (GenBank accession no. D26045).

The chromosomal *tet*(L) gene has two transcriptional start sites. The minor one, *b*, is the same as that used in the plasmid determinant, while the major start site, *a*, is 6 nt upstream (23). A mutation in chromosomal *tet*(L) resulting in a TG dinucleotide 14/15 nt upstream of *b* (this dinucleotide is present in the wild-type plasmid sequence)

converts *b* into the sole start site and enhances transcription and resistance about 10-fold (7). Induction of chromosomally mediated tetracycline resistance by tetracycline does *not* involve unmasking of a RBS and therefore does not occur by the type of translational attenuation described above (121). It appears rather to involve a putative mRNA stem-loop structure of −11.9 kcal mol^{-1} which overlaps the downstream half of the leader peptide sequence and whose stem ends just upstream of the Tet(L) RBS. Mutations altering this stem-loop structure reduce inducibility, while mutations in a smaller putative stem-loop that might have sequestered the RBS for Tet(L) have no effect (121). The *B. subtilis* mutation affecting ribosomal protein S10 prevents induction, so induction requires interaction of tetracycline with the ribosome (121). A stop codon placed by site-directed mutagenesis immediately following the first codon of the leader peptide permits induction, but removal of the first codon does not (121). The model presented is that tetracycline-promoted stalling of ribosomes during translation of the early codons of the leader peptide permits stabilization of the larger stem-loop, the role of which is to guide ribosomes from the leader to reinitiate translation at the RBS of Tet(L) (121). Since the RBS for the leader peptide is about 55-fold more efficient than that for the Tet(L) protein, induction results. A minor tetracycline induction of unknown mechanism also occurs at the transcriptional level (7, 23, 121).

Whether the plasmid-mediated Tet L determinant has the same mechanism of regulation as that determined experimentally for the chromosomal determinant is an open question. The stem-loop important for the induction of chromosomal Tet L (121) is similar to the stem-loop "B" postulated to form during induction of plasmid pTHT15 (51), although the complete set of upstream mRNA secondary structures predicted for the two kinds of determinants are not the same (GCG programs StemLoop and Mfold). The correlation of a truncated leader peptide with constitutivity for plasmid-mediated tetracycline resistance (as noted above) is consistent with both the translational attenuation and the reinitiation model.

As discussed earlier, the chromosomal *tet*(L) of *B. subtilis* is involved not only in tetracycline resistance but also in response of the cell to elevated pH and the presence of Na$^+$ and K$^+$. The Tet L determinant appears to be regulated by those factors in a manner different from that by tetracycline, which would make sense physiologically. Regulation may be transcriptional. When a sequence beginning 150 nt upstream from the transcriptional start site and extending to codon 381 of Tet(L) is fused in frame to LacZ, enzyme activity is regulated by all of the aforementioned factors, most dramatically by tetracycline (23). However, the promoterless *lacZ* gene at the 5′ end of a transposon Tn*917*, in turn inserted between the −35 and −10 elements of the *tet*(L) promoter, is not regulated by tetracycline but is regulated somewhat by the other factors. This is consistent with the tetracycline regulatory region's being in the leader peptide area and suggests that the control region(s) for the other factors, and possibly a second promoter, are 5′ to the interrupted promoter.

Because Tet(K) has many of the non-tetracycline-related functions of Tet(L), it might also be regulated by factors other than tetracycline, although there have been no studies on this point. Sequences upstream from *tet*(K) suggest a translational attenuation regulation of Tet(K) synthesis (54). The start site for *tet*(K) transcription on pT181 is 194 nt upstream from the protein initiation co-

plasmid in *Listeria* sp. and as part of an uncharacterized 40-kb unit on the chromosome in *Enterococcus* sp.; in the latter case, a resident plasmid caused conjugative transfer of the chromosome, thereby transferring Tet S (36). Tet W, located on the chromosome, is postulated to be transferred by conjugation in the rumen species *Butyrivibrio* sp. via a 40- to 50-kb transposon Tn*B1230* (6).

Conjugation of Tn*916* (bearing Tet M) is induced by tetracycline after excision and circularization of the transposon in the donor strain (19). Recent studies have shown that the mechanism involves induction of *tet*(M) by transcriptional attenuation, as described above, followed by readthrough into downstream genes, resulting, in turn, in up-regulation of a downstream promoter, P*orf7*. Since circularization of the excised transposon has placed the transfer genes downstream of this promoter, transfer is induced (19).

DEVELOPMENT OF DRUGS TO COMBAT TETRACYCLINE RESISTANCE

Because the tetracyclines have many desirable properties, efforts have been made to devise analogs not affected by the resistance mechanisms. Glycylcyclines are tetracyclines having an *N,N*-dimethylglycylamido group [–NH–CO–CH$_2$–N(CH$_3$)$_2$] at the 9 position and H at positions 5 and 6 (130). They inhibit bacterial cells that have either tetracycline efflux (group 1 or 2) or ribosomal protection systems and combat infections in mice (101, 130). They are not recognized as substrates by the TetA(B) efflux protein (see reference 113). They bind five times more strongly to ribosomes than does tetracycline and are correspondingly more inhibitory, which, however, only partially explains their effectiveness in the presence of ribosomal protection proteins (8). Glycylcyclines also work against tetracycline-resistant, rapidly growing mycobacteria (12). Interestingly, when a dimethylamino group replaces the *N,N*-dimethylglycylamido group at the 9 position, the analog becomes a substrate for the TetA(B) efflux protein (89).

Several analogs of tetracycline having a rather large hydrophobic group at the 6 position interfere specifically with the ability of the TetA(B) protein to efflux tetracycline (88, 89). One such analog can synergistically enhance the effectiveness of doxycycline against cells bearing efflux proteins from classes A, B, K, and L (89). In one case examined in more detail, the analog competes with tetracycline for binding to the TetA(B) protein, yet is transported poorly itself (as assayed via H$^+$ antiport) (87). Finally, one of these analogs was about eight times as effective as doxycycline against resistance mediated by Tet M (87).

Dactylocycline is a naturally occurring tetracycline analog glycosylated at position 6 that is active against tetracycline-resistant gram-positive organisms whose resistance mechanism was not described (137). Other screenings found that siderophores (Fe^{3+} chelators of biological origin) in the uncomplexed state inhibit both Tet K- and Tet B-mediated resistance (106), but the reason for this intriguing observation is not known.

SUMMARY

Tetracycline resistance is widespread across the bacterial kingdom. Tetracycline affects prokaryotes by binding to the ribosome and inhibiting protein synthesis. Three clinically important resistance mechanisms have emerged in gram-positive bacteria: active efflux, ribosomal protection, and ribosomal resistance. In the case of efflux genes *tetA*(Z), *tet*(K), *tet*(L), *tetA*(P), *otrB*, and *tcr3*, a membrane protein of 12 to 14 (putative) transmembrane helices uses the proton gradient across the cytoplasmic membrane to pump out a tetracycline-divalent cation complex. The putative ABC efflux transporter(s) encoded by genes *tetAB* from *Corynebacterium* sp. probably transports multiple drugs; each protein has an estimated five transmembrane helices and presumably uses ATP to energize efflux. Ribosomal protection proteins [genes *tet*(M), *tet*(O), and probably *tetB*(P), *tet*(Q), *tet*(S), *tet*(T), *tet*(W), *otrA*, and "*tet*"] are GTPases that bind to the ribosome, thereby preventing binding of tetracycline. Finally, a mutation in 16S rRNA, recently discovered in one organism, causes the ribosomes to be resistant to tetracycline.

Tetracycline resistance is often inducible by tetracycline. For class Z, studies of regulation in closely related determinants suggest that, when the repressor protein TetR(Z) binds tetracycline, it loses its hold on the operator regulating the efflux protein. For chromosomal class L, induction involves translational reinitiation. For classes K and plasmid class L, the induction mechanism is less clear but may involve a leader peptide and translational attenuation. For class M, induction probably occurs by transcriptional attenuation. In the *Streptomyces* spp. that synthesize tetracyclines, induction of tetracycline-resistance determinants may coordinate with drug production. The class K and L tetracycline efflux proteins also have an independent H$^+$(K$^+$)/K$^+$, Na$^+$ transport function and play a role in pH homeostasis. Regulation of these functions may differ from regulation of tetracycline transport.

Tetracycline-resistance determinants are often found on conjugative elements, both plasmid and chromosomal, allowing their ready transfer among widely different microorganisms. New tetracycline analogs effective against efflux and ribosomal protection resistance mechanisms are being developed.

Work in our laboratory is supported in part by grants from the National Institutes of Health. We thank all those who sent reprints or unpublished manuscripts, answered questions, or read and discussed this chapter, especially Trudi Bannam, David Bechhofer, Vickers Burdett, Terry Krulwich, Mark Nelson, Julian Rood, Andreas Tauch, Diane Taylor, Akihito Yamaguchi, and, most particularly, Teresa Barbosa.

REFERENCES

1. **Aevarsson, A., E. Brazhnikov, M. Garber, J. Zheltonosova, Y. Chirgadze, A. Al-Karadaghi, L. A. Svensson, and A. Liljas.** 1994. Three-dimensional structure of the ribosomal translocase: elongation factor G from *Thermus thermophilus*. *EMBO J.* **13:**3669–3677.

2. **Aldema, M. L., L. M. McMurry, A. R. Walmsley, and S. B. Levy.** 1996. Purification of the Tn*10*-specified tetracycline efflux antiporter TetA in a native state as a polyhistidine fusion protein. *Mol. Microbiol.* **19:**187–195.

3. **Allard, J. D., and K. P. Bertrand.** 1992. Membrane topology of the pBR322 tetracycline resistance protein: TetA-PhoA gene fusions and implications for the mechanism of TetA membrane insertion. *J. Biol. Chem.* **267:**17809–17819.

4. **Bannam, T. L., and J. I. Rood.** Identification of structural and functional domains of the tetracycline efflux protein

TetA(P) from *Clostridium perfringens*. *Microbiology*, in press.

5. **Barbosa, T. M.** Personal communication.

6. **Barbosa, T. M., K. P. Scott, and H. J. Flint.** 1999. Evidence for a recent intergeneric transfer of a new tetracycline resistance gene, *tet*(W), isolated from *Butyrivibrio fibrisolvens*, and the occurrence of *tet*(O), in ruminal bacteria. *Environ. Microbiol.* **1:**53–64.

7. **Bechhofer, D. H., and S. J. Stasinopoulos.** 1998. *tet*A(L) mutants of a tetracycline-sensitive strain of *Bacillus subtilis* with the polynucleotide phosphorylase gene deleted. *J. Bacteriol.* **180:**3470–3473.

8. **Bergeron, J., M. Ammirati, D. Danley, L. James, M. Norcia, J. Retsema, C. A. Strick, W.-G. Su, J. Sutcliffe, and L. Wondrack.** 1996. Glycylcyclines bind to the high-affinity tetracycline ribosomal binding site and evade Tet(M)- and Tet(O)-mediated ribosomal protection. *Antimicrob. Agents Chemother.* **40:**2226–2228.

9. **Bjork, G.** 1996. Stable RNA modification, p. 861–886. *In* F. C. Neidhardt, R. Curtiss III, J. L. Ingraham, E. C. C. Lin, K. B. Low, B. Magasanik, W. S. Reznikoff, M. Riley, M. Schaechter, and H. E. Umbarger (ed.), *Escherichia coli and Salmonella: Cellular and Molecular Biology*, 2nd ed. ASM Press, Washington, D.C.

10. **Blackwood, R. K., and A. R. English.** 1970. Structure-activity relationships in the tetracycline series, p. 237–266. *In* D. Perlman (ed.), *Advances in Applied Microbiology.* Academic Press, New York, N.Y.

11. **Brimacombe, R., B. Greuer, P. Mitchell, M. Osswald, J. Rinke-Appel, D. Schueler, and K. Stade.** 1990. Three-dimensional structure and function of *Escherichia coli* 16S and 23S rRNA as studied by cross-linking techniques, p. 93–106. *In* W. E. Hill (ed.), *The Ribosome: Structure, Function, and Evolution.* American Society for Microbiology, Washington, D.C.

12. **Brown, B. A., R. J. Wallace, Jr., and G. Onyi.** 1996. Activities of the glycylcyclines N,N-dimethylglycylamido-minocycline and N,N-dimethylglycylamido-6-demethyl-6-deoxytetracycline against *Nocardia* spp. and tetracycline-resistant isolates of rapidly growing mycobacteria. *Antimicrob. Agents Chemother.* **40:**874–878.

13. **Brown, J. R., and D. S. Ireland.** 1978. Structural requirements for tetracycline activity. *Adv. Pharmacol. Chemother.* **15:**161–202.

14. **Buck, M. A., and B. S. Cooperman.** 1990. Single protein omission reconstitution studies of tetracycline binding to the 30S subunit of *Escherichia coli* ribosomes. *Biochemistry* **29:**5374–5379.

15. **Burdett, V.** 1986. Streptococcal tetracycline resistance mediated at the level of protein biosynthesis. *J. Bacteriol.* **165:**564–569.

16. **Burdett, V.** 1991. Purification and characterization of Tet(M), a protein that renders ribosomes resistant to tetracycline. *J. Biol. Chem.* **266:**2872–2877.

17. **Burdett, V.** 1993. tRNA modification activity is necessary for Tet(M)-mediated tetracycline resistance. *J. Bacteriol.* **175:**7209–7215.

18. **Burdett, V.** 1996. Tet(M)-promoted release of tetracycline from ribosomes is GTP dependent. *J. Bacteriol.* **178:**3246–3251.

19. **Celli, J., and P. Trieu-Cuot.** 1998. Circularization of Tn*916* is required for expression of the transposon-encoded transfer functions: characterization of long tetracycline-inducible transcripts reading through the attachment site. *Mol. Microbiol.* **28:**103–117.

20. **Charpertier, E., G. Gerbaud, and P. Courvalin.** 1993. Characterization of a new class of tetracycline-resistance gene *tet*(S) in *Listeria monocytogenes* BM4210. *Gene* **131:**27–34.

21. **Cheng, J., K. Baldwin, A. A. Guffanti, and T. A. Krulwich.** 1996. Na⁺/H⁺ antiport activity conferred by *Bacillus subtilis* tetA(L), a 5′ truncation product of *tet*A(L), and related plasmid genes upon *Escherichia coli. Antimicrob. Agents Chemother.* **40:**852–857.

22. **Cheng, J., A. A. Guffanti, and T. A. Krulwich.** 1994. The chromosomal tetracycline resistance locus of *Bacillus subtilis* encodes a Na⁺/H⁺ antiporter that is physiologically important at elevated pH. *J. Biol. Chem.* **269:** 27365–27371.

23. **Cheng, J., A. A. Guffanti, W. Wang, T. A. Krulwich, and D. H. Bechhofer.** 1996. Chromosomal *tet*A(L) gene of *Bacillus subtilis*: regulation of expression and physiology of a *tet*A(L) deletion strain. *J. Bacteriol.* **178:**2853–2860.

24. **Cheng, J., D. B. Hicks, and T. A. Krulwich.** 1996. The purified *Bacillus subtilis* tetracycline efflux protein TetA(L) reconstitutes both tetracycline-cobalt/H+ and Na⁺(K+)/ H⁺ exchange. *Proc. Natl. Acad. Sci. USA* **93:**14446–14451.

25. **Chopra, I.** 1985. Mode of action of the tetracyclines and the nature of bacterial resistance to them, p. 317–392. *In* J. J. Hlavka and J. H. Boothe (ed.), *The Tetracyclines.* Springer-Verlag, Berlin, Germany.

26. **Chopra, I., P. M. Hawkey, and M. Hinton.** 1992. Tetracyclines, molecular and clinical aspects. *J. Antimicrob. Chemother.* **29:**245–277.

27. **Clermont, D., O. Chesneau, G. de Cespedes, and T. Horaud.** 1997. New tetracycline resistance determinants coding for ribosomal protection in streptococci and nucleotide sequence of *tet*(T) isolated from *Streptococcus pyogenes* A498. *Antimicrob. Agents Chemother.* **41:**112–116.

28. **Cundliffe, E., and K. McQuillen.** 1967. Bacterial protein synthesis: the effects of antibiotics. *J. Mol. Biol.* **30:**137–146.

29. **Dairi, T., K. Aisaka, R. Katsumata, and M. Hasegawa.** 1995. A self-defense gene homologous to tetracycline effluxing gene essential for antibiotic production in *Streptomyces aureofaciens. Biosci. Biotechnol. Biochem.* **59:** 1835–1841.

30. **Dantley, K. A., H. K. Dannelly, and V. Burdett.** 1998. Binding interaction between Tet(M) and the ribosome: requirements for binding. *J. Bacteriol.* **180:**4089–4092.

31. **de Rossi, E., M. C. J. Blokpoel, R. Cantoni, M. Branzoni, G. Riccardi, D. B. Young, K. A. L. de Smet, and O. Ciferri.** 1998. Molecular cloning and functional analysis of a novel tetracycline resistance determinant, *tet*(V), from *Mycobacterium smegmatis. Antimicrob. Agents Chemother.* **42:**1931–1937.

32. **Dittrich, W., and H. Schrempf.** 1992. The unstable tetracycline resistance gene of *Streptomyces lividans* 1326 encodes a putative protein with similarities to translational elongation factors and Tet(M) and Tet(O) proteins. *Antimicrob. Agents Chemother.* **36:**1119–1124.

33. **Dosch, D. C., F. F. Salvacion, and W. Epstein.** 1984. Tetracycline resistance element of pBR322 mediates potassium transport. *J. Bacteriol.* **160:**1188–1190.

34. **Doyle, D., K. J. McDowall, M. J. Butler, and I. S. Hunter.** 1991. Characterization of an oxytetracycline-resistance gene, *otrA*, of *Streptomyces rimosus*. *Mol. Microbiol.* **5:**2923–2933.

35. **Fath, M. J., and R. Kolter.** 1993. ABC transporters: bacterial exporters. *Microbiol. Rev.* **57:**995–1017.

36. **Francois, B., M. Charles, and P. Courvalin.** 1997. Conjugative transfer of *tet*(S) between strains of *Enterococcus faecalis* is associated with the exchange of large fragments of chromosomal DNA. *Microbiology* **143:**2145–2154.

37. **Fujihira, E., T. Kimura, Y. Shiina, and A. Yamaguchi.** 1996. Transmembrane glutamic acid residues play essential roles in the metal-tetracycline/H⁺ antiporter of *Staphylococcus aureus*. *FEBS Lett.* **391:**243–246.

38. **Fujihira, E., T. Kimura, and A. Yamaguchi.** 1997. Roles of acidic residues in the hydrophilic loop regions of metal-tetracycline/H⁺ antiporter Tet(K) of *Staphylococcus aureus*. *FEBS Lett.* **419:**211–214.

39. **Ginn, S. L., M. H. Brown, and R. A. Skurray.** 1997. Membrane topology of the metal-tetracycline/H⁺ antiporter TetA(K) from *Staphylococcus aureus*. *J. Bacteriol.* **179:**3786–3789.

40. **Goldman, R. A., T. Hasan, C. C. Hall, W. A. Strycharz, and B. S. Cooperman.** 1983. Photoincorporation of tetracycline into *Escherichia coli* ribosomes. Identification of the major proteins photolabeled by native tetracycline and tetracycline photoproducts and implications for the inhibitory action of tetracycline on protein synthesis. *Biochemistry* **22:**359–368.

41. **Gottesman, M. E.** 1967. Reaction of ribosome-bound peptidyl transfer ribonucleic acid with aminoacyl transfer ribonucleic acid or puromycin. *J. Biol. Chem.* **242:**5564–5571.

42. **Griffith, J. K., T. Kogoma, D. L. Corvo, W. L. Anderson, and A. L. Kazim.** 1988. An N-terminal domain of the tetracycline resistance protein increases susceptibility to aminoglycosides and complements potassium uptake defects in *Escherichia coli*. *J. Bacteriol.* **170:**598–604.

43. **Guay, G. G., S. A. Khan, and D. M. Rothstein.** 1993. The *tet*(K) gene of plasmid pT181 of *Staphylococcus aureus* encodes an efflux protein that contains 14 transmembrane helices. *Plasmid* **30:**163–166.

44. **Guay, G. G., M. Tuckman, P. McNicholas, and D. M. Rothstein.** 1993. The *tet*(K) gene from *Staphylococcus aureus* mediates the transport of potassium in *Escherichia coli*. *J. Bacteriol.* **175:**4927–4929.

45. **Guffanti, A. A., J. Cheng, and T. A. Krulwich.** 1998. Electrogenic antiport activities of the gram-positive Tet proteins include a Na⁺(K⁺)/K⁺ mode that mediates net K⁺ uptake. *J. Biol. Chem.* **273:**26447–26454.

46. **Guffanti, A. A., and T. A. Krulwich.** 1995. Tetracycline/H⁺ antiport and Na⁺/H⁺ antiport catalyzed by the *Bacillus subtilis* TetA(L) transporter expressed in *Escherichia coli*. *J. Bacteriol.* **177:**4557–4561.

47. **Hillen, W., and C. Berens.** 1994. Mechanism underlying expression of Tn*10* encoded tetracycline resistance. *Annu. Rev. Microbiol.* **48:**345–369.

48. **Hirata, T., E. Fujihara, T. Kimura-Someya, and A. Yamaguchi.** 1998. Membrane topology of the staphylococcal tetracycline efflux protein Tet(K) determined by antibacterial resistance gene fusion. *Biochemistry* **124:**1206–1211.

49. **Hirata, T., R. Wakatabe, J. Nielsen, Y. Someya, E. Fujihira, T. Kimura, and A. Yamaguchi.** 1997. A novel compound, 1,1-dimethyl-5-(1-hydroxypropyl)-4,6,7-trimethylindan, is an effective inhibitor of the *tet*(K) gene-encoded metal-tetracycline/H⁺ antiporter of *Staphylococcus aureus*. *FEBS Lett.* **412:**337–340.

50. **Hlavka, J. J., and J. H. Boothe (ed.).** 1985. *The Tetracyclines*. Springer-Verlag, Berlin, Germany.

51. **Hoshino, T., T. Ikeda, N. Tomizuka, and K. Furukawa.** 1985. Nucleotide sequence of the tetracycline resistance gene of pTHT15, a thermophilic *Bacillus* plasmid: comparison with staphylococcal Tc^R controls. *Gene* **37:**131–138.

52. **Ishiwa, H., and H. Shibahara.** 1985. New shuttle vectors for *Escherichia coli* and *Bacillus subtilis*. III. Nucleotide sequence analysis of tetracycline resistance gene of pAMα1 and ori-177. *Jpn. J. Genet.* **60:**485–498.

53. **Kennan, R. M., L. M. McMurry, S. B. Levy, and J. I. Rood.** 1997. Glutamate residues located within putative transmembrane helices are essential for TetA(P)-mediated tetracycline efflux. *J. Bacteriol.* **179:**7011–7015.

54. **Khan, S. A., and R. P. Novick.** 1983. Complete nucleotide sequence of pT181, a tetracycline-resistance plasmid from *Staphylococcus aureus*. *Plasmid* **10:**251–259.

55. **Kimura, T., Y. Inagaki, T. Sawai, and Y. Akihito.** 1995. Substrate-induced acceleration of N-ethylmaleimide reaction with the Cys-65 mutant of the transposon Tn*10*-encoded metal-tetracycline/H⁺ antiporter depends on the interaction of Asp-66 with the substrate. *FEBS Lett.* **362:**47–49.

56. **Kimura, T., M. Nakatani, T. Kawabe, and A. Yamaguchi.** 1998. Roles of conserved arginine residues in the metal-tetracycline/H⁺ antiporter of *Escherichia coli*. *Biochemistry* **37:**5475–5480.

57. **Kimura, T., M. Ohnuma, T. Sawai, and A. Yamaguchi.** 1997. Membrane topology of the transposon 10-encoded metal-tetracycline/H⁺ antiporter as studied by site-directed chemical labeling. *J. Biol. Chem.* **272:**580–585.

58. **Kimura, T., T. Sawai, and A. Yamaguchi.** 1997. Remote conformational effects of the gly-62–leu mutation of the Tn*10*-encoded metal-tetracycline/H⁺ antiporter of *Escherichia coli* and its second-site suppressor mutation. *Biochemistry* **36:**6941–6946.

59. **Kimura, T., and A. Yamaguchi.** 1996. Asp-285 of the metal-tetracycline/H⁺ antiporter of *Escherichia coli* is essential for substrate binding. *FEBS Lett.* **388:**50–52.

60. **Kimura-Someya, T., S. Iwaki, and A. Yamaguchi.** 1998. Site-directed chemical modification of cysteine-scanning mutants as to transmembrane segment II and its flanking regions of the Tn*10*-encoded metal-tetracycline/H⁺ antiporter reveals a transmembrane water-filled channel. *J. Biol. Chem.* **49:**32806–32811.

61. **Kornblum, J., and R. Novick.** Personal communication.

62. **Lacks, S. A., P. Lopez, B. Greenberg, and M. Espinosa.** 1986. Identification and analysis of genes for tetracycline resistance and replication functions in the broad-host-range plasmid pSL1. *J. Mol. Biol.* **192:**753–765.

63. **LeBlanc, D. J., L. N. Lee, B. M. Titman, C. J. Smith, and F. C. Tenover.** 1988. Nucleotide sequence analysis of tetracycline resistance gene tetO from *Streptococcus mutans* DL5. *J. Bacteriol.* **170:**3618–3626.

64. **Leng, Z., D. E. Riley, R. E. Berger, J. N. Krieger, and M. C. Roberts.** 1997. Distribution and mobility of the tetracycline resistance determinant tetQ. *J. Antimicrob. Chemother.* **40:**551–559.

65. Lepine, G., J.-M. Lacroix, C. B. Walker, and A. Progulske-Fox. 1993. Sequencing of a *tet*(Q) gene isolated from *Bacteroides fragilis* 1126. *Antimicrob. Agents Chemother.* **37**:2037–2041.

66. Levy, S. B. 1984. Resistance to the tetracyclines, p. 191–240. *In* L. E. Bryan (ed.), *Antimicrobial Drug Resistance*. Academic Press, Inc., New York, N.Y.

67. Levy, S. B. 1992. Active efflux mechanisms for antimicrobial resistance. *Antimicrob. Agents Chemother.* **36**: 695–703.

68. Levy, S. B. 1992. *The Antibiotic Paradox. How Miracle Drugs Are Destroying the Miracle*. Plenum Publishing, New York, N.Y.

69. Levy, S. B., L. M. McMurry, T. M. Barbosa, V. Burdett, P. Courvalin, W. Hillen, M. C. Roberts, J. I. Rood, and D. E. Taylor. 1999. Nomenclature for new tetracycline resistance determinants. *Antimicrob. Agents Chemother.* **43**:1523–1524.

70. Levy, S. B., L. M. McMurry, V. Burdett, P. Courvalin, W. Hillen, M. C. Roberts, and D. E. Taylor. 1989. Nomenclature for tetracycline resistance determinants. *Antimicrob. Agents Chemother.* **33**:1373–1374.

71. Lewin, B. 1997. Chapter 8, p. 179–212. *In Genes VI*. Oxford University Press, Oxford, U.K.

72. Lopez, P. J., I. Marchand, O. Yarchuk, and M. Dreyfus. 1998. Translation inhibitors stabilize *Escherichia coli* mRNAs independently of ribosome protection. *Proc. Natl. Acad. Sci. USA* **95**:6067–6072.

73. Lovett, P. S., and E. J. Rogers. 1996. Ribosome regulation by the nascent peptide. *Microbiol. Rev.* **60**:366–385.

74. Maloney, P. C., and T. H. Wilson. 1996. Ion-coupled transport and transporters, p. 1130–1148. *In* F. C. Neidhardt, R. Curtiss III, J. L. Ingraham, E. C. C. Lin, K. B. Low, B. Magasanik, W. S. Reznikoff, M. Riley, M. Schaechter, and H. E. Umbarger (ed.), *Escherichia coli and Salmonella: Cellular and Molecular Biology*, 2nd ed. ASM Press, Washington, D.C.

75. Manavathu, E. K., C. L. Fernandez, B. S. Cooperman, and D. E. Taylor. 1990. Molecular studies on the mechanism of tetracycline resistance mediated by Tet(O). *Antimicrob. Agents Chemother.* **34**:71–77.

76. Manavathu, E. K., K. Hiratsuka, and D. E. Taylor. 1988. Nucleotide sequence analysis and expression of a tetracycline-resistance gene from *Campylobacter jejuni*. *Gene* **62**:17–26.

77. Martin, R. B. 1985. Tetracyclines and daunorubicin, p. 19–52. *In* H. Sigel (ed.), *Metal Ions in Biological Systems*. Marcel Dekker, Inc., New York, N.Y.

78. McMurry, L. M., J. C. Cullinane, and S. B. Levy. 1982. Transport of the lipophilic analog minocycline differs from that of tetracycline in susceptible and resistant *Escherichia coli* strains. *Antimicrob. Agents Chemother.* **22**:791–799.

79. McMurry, L. M., and S. B. Levy. 1998. Revised sequence of OtrB (Tet347) tetracycline efflux protein from *Streptomyces rimosus*. *Antimicrob. Agents Chemother.* **42**:3050.

80. McMurry, L. M., R. E. Petrucci, Jr., and S. B. Levy. 1980. Active efflux of tetracycline encoded by four genetically different tetracycline resistance determinants in *Escherichia coli*. *Proc. Natl. Acad. Sci. USA* **77**:3974–3977.

81. McMurry, L. M., M. Stephan, and S. B. Levy. 1992. Decreased function of the class B tetracycline efflux protein Tet with mutations at aspartate 15, a putative intramembrane residue. *J. Bacteriol.* **174**:6294–6297.

82. McNicholas, P., I. Chopra, and D. M. Rothstein. 1992. Genetic analysis of the *tetA*(C) gene on plasmid pBR322. *J. Bacteriol.* **174**:7926–7933.

83. Mitscher, L. A. 1978. *The Chemistry of the Tetracycline Antibiotics*. Marcel Dekker, Inc., New York, N.Y.

84. Mojumdar, M., and S. A. Khan. 1988. Characterization of the tetracycline resistance gene of plasmid pT181 of *Staphylococcus aureus*. *J. Bacteriol.* **170**:5522–5528.

85. Munske, G. R., E. V. Lindley, and J. A. Magnuson. 1984. *Streptococcus faecalis* proton gradients and tetracycline transport. *J. Bacteriol.* **158**:49–54.

86. Nakamura, T., Y. Matsuba, A. Ishihara, T. Kitagawa, F. Suzuki, and T. Unemoto. 1995. N-terminal quarter part of tetracycline transporter from pACYC184 complements K^+ uptake activity in K^+ uptake-deficient mutants of *Escherichia coli* and *Vibrio alginolyticus*. *Biol. Pharm. Bull.* **18**:1189–1193.

86a. National Center for Genome Resources. Genome Sequence DataBase. http://www.ncgr.org/gsdb/

87. Nelson, M. L., and S. B. Levy. 1999. Reversal of tetracycline resistance mediated by different tetracycline resistance determinants by an inhibitor of the Tet(B) antiport protein. *Antimicrob. Agents Chemother.* **43**:1719–1724.

88. Nelson, M. L., B. H. Park, J. S. Andrews, V. A. Georgian, R. C. Thomas, and S. B. Levy. 1993. Inhibition of the tetracycline efflux antiport protein by 13-thio-substituted 5-hydroxy-6-deoxytetracyclines. *J. Medicinal Chem.* **36**:370–377.

89. Nelson, M. L., B. H. Park, and S. B. Levy. 1994. Molecular requirements for the inhibition of the tetracycline antiport protein and the effect of potent inhibitors on the growth of tetracycline-resistant bacteria. *J. Medicinal Chem.* **37**:1355–1361.

90. Nikaido, H. 1996. Multidrug efflux pumps of gram-negative bacteria. *J. Bacteriol.* **178**:5853–5859.

91. Nikaido, H., and D. G. Thanassi. 1993. Penetration of lipophilic agents with multiple protonation sites into bacterial cells: tetracyclines and fluoroquinolones as examples. *Antimicrob. Agents Chemother.* **37**:1393–1399.

91a. Nikolich, M. P., N. B. Shoemaker, and A. A. Salyers. 1992. A *Bacteroides* tetracycline resistance gene represents a new class of ribosome protection tetracycline resistance. *Antimicrob. Agents Chemother.* **36**:1005–1012.

92. Noguchi, N., T. Aoki, M. Sasatsu, M. Kono, K. Shishido, and T. Ando. 1986. Determination of the complete nucleotide sequence of pNS1, a staphylococcal tetracycline-resistance plasmid propagated in *Bacillus subtilis*. *FEMS Microbiol. Lett.* **37**:283–288.

93. Oehler, R., N. Polacek, G. Steiner, and A. Darta. 1997. Interaction of tetracycline with RNA: photoincorporation into ribosomal RNA of *Escherichia coli*. *Nucleic Acids Res.* **25**:1219–1224.

94. Ohnuki, T., T. Katch, T. Imanaka, and S. Aiba. 1985. Molecular cloning of tetracycline resistance genes from *Streptomyces rimosus* in *Streptomyces griseus* and characterization of the cloned genes. *J. Bacteriol.* **161**:1010–1016.

95. Orth, P., F. Cordes, D. Schnappinger, W. Hillen, W. Saenger, and W. Hinrichs. 1998. Conformational changes of the Tet repressor induced by tetracycline trapping. *J. Mol. Biol.* **279**:439–447.

96. **Paulsen, I. T., M. H. Brown, and R. A. Skurray.** 1996. Proton-dependent multidrug efflux systems. *Microbiol. Rev.* **60:**575–608.

97. **Paulsen, I. T., and R. A. Skurray.** 1993. Topology, structure and evolution of two families of proteins involved in antibiotic and antiseptic resistance in eukaryotes and prokaryotes—an analysis. *Gene* **124:**1–11.

98. **Peska, S.** 1971. Inhibitors of ribosome function. *Annu. Rev. Microbiol.* **25:**487–562.

99. **Platteeuw, C., F. Michiels, H. Joos, J. Seurinck, and W. M. de Vos.** 1995. Characterization and heterologous expression of the *tetL* gene and identification of *iso*-ISS1 elements from *Enterococcus faecalis* plasmid pJH1. *Gene* **160:**89–93.

100. **Rasmussen, B., H. F. Noller, G. Daubresse, B. Oliva, Z. Misulovin, D. M. Rothstein, G. A. Ellestad, Y. Gluzman, F. P. Tally, and I. Chopra.** 1991. Molecular basis of tetracycline action: identification of analogs whose primary target is not the bacterial ribosome. *Antimicrob. Agents Chemother.* **35:**2306–2311.

101. **Rasmussen, B. A., Y. Gluzman, and F. P. Tally.** 1994. Inhibition of protein synthesis occurring on tetracycline-resistant TetM-protected ribosomes by a novel class of tetracyclines, the glycylcyclines. *Antimicrob. Agents Chemother.* **38:**1658–1660.

102. **Ridenhour, M. B., H. M. Fletcher, J. E. Mortensen, and L. Daneo-Moore.** 1996. A novel tetracycline-resistant determinant, tet(U), is encoded on the plasmid pKQ10 in *Enterococcus faecium*. *Plasmid* **35:**71–80.

103. **Roberts, M. C.** 1996. Tetracycline resistance determinants: mechanisms of action, regulation of expression, genetic mobility, and distribution. *FEMS Microbiol. Rev.* **19:**1–24.

104. **Roberts, M. C.** 1997. Genetic mobility and distribution of tetracycline resistance determinants, p. 206–218. *In* D. J. Chadwick (ed.), *Antibiotic Resistance: Origins, Evolution, Selection, and Spread*. John Wiley and Sons, Chichester, U.K.

105. **Ross, J. I., E. A. Eady, J. H. Cove, and W. J. Cunliffe.** 1998. 16S rRNA mutation associated with tetracycline resistance in a gram-positive bacterium. *Antimicrob. Agents Chemother.* **42:**1702–1705.

106. **Rothstein, D. M., M. McGlynn, V. Bernan, J. McGahren, J. Zaccardi, N. Cekleniak, and K. P. Bertrand.** 1993. Detection of tetracyclines and efflux pump inhibitors. *Antimicrob. Agents Chemother.* **37:**1624–1629.

107. **Rubin, R. A., and S. B. Levy.** 1990. Interdomain hybrid tetracycline proteins confer tetracycline resistance only when they are derived from closely related members of the *tet* gene family. *J. Bacteriol.* **172:**2303–2312.

108. **Rubin, R. A., and S. B. Levy.** 1991. Tet protein domains interact productively to mediate tetracycline resistance when present on separate polypeptides. *J. Bacteriol.* **173:**4503–4509.

109. **Rubin, R. A., S. B. Levy, R. L. Heinrikson, and F. J. Kézdy.** 1990. Gene duplication in the evolution of the two complementing domains of Gram-negative bacterial tetracycline efflux proteins. *Gene* **87:**7–13.

110. **Sakaguchi, R., H. Amano, and K. Shishido.** 1988. Nucleotide sequence homology of the tetracycline-resistance determinant naturally maintained in *Bacillus subtilis* Marburg 168 chromosome and the tetracycline-resistance gene of *B. subtilis* plasmid pNS1981. *Biochim. Biophys. Acta* **950:**441–444.

111. **Salyers, A. A., N. B. Shoemaker, A. M. Stevens, and L.-Y. Li.** 1995. Conjugative transposons: an unusual and diverse set of integrated gene transfer elements. *Microbiol. Rev.* **59:**579–590.

112. **Saraceni-Richards, C. A., and S. B. Levy.** 1998. Single mutations in either domain of an inactive Tet(C/B) hybrid Tc efflux protein produce Tc resistance, abstr. V-87, p. 527. *Abstr. 98th Gen. Meet. Am. Soc. Microbiol. 1998*. American Society for Microbiology, Washington, D.C.

113. **Schnappinger, D., and W. Hillen.** 1996. Tetracyclines: antibiotic action, uptake, and resistance mechanisms. *Arch. Microbiol.* **165:**359–369.

114. **Schwarz, S., M. Cardoso, and H. C. Wegener.** 1992. Nucleotide sequence and phylogeny of the *tet*(L) tetracycline resistance determinant encoded by plasmid pSTET1 from *Staphylococcus hyicus*. *Antimicrob. Agents Chemother.* **36:**580–588.

115. **Sheridan, R. P., and I. Chopra.** 1991. Origin of tetracycline efflux proteins: conclusions from nucleotide sequence analysis. *Mol. Microbiol.* **5:**895–900.

116. **Sloan, J., L. M. McMurry, D. Lyras, S. B. Levy, and J. I. Rood.** 1994. The *Clostridium perfringens* Tet P determinant comprises two overlapping genes: *tetA*(P), which mediates active tetracycline efflux, and *tetB*(P), which is related to the ribosomal protection family of tetracycline-resistance determinants. *Mol. Microbiol.* **11:**403–415.

117. **Someya, Y., Y. Moriyama, M. Futai, T. Sawai, and A. Yamaguchi.** 1996. Reconstitution of the metal-tetracycline/H$^+$ antiporter of *Escherichia coli* in proteoliposomes including F$_o$F$_1$-ATPase. *FEBS Lett.* **374:**72–76.

118. **Someya, Y., and A. Yamaguchi.** 1997. Second-site suppressor mutations for the Arg70 substitution mutants of the Tn10-encoded metal-tetracycline/H$^+$ antiporter of *Escherichia coli*. *Biochim. Biophys. Acta* **1322:**230–236.

119. **Speer, B. S., L. Bedzyk, and A. A. Salyers.** 1991. Evidence that a novel tetracycline resistance gene found on two *Bacteroides* transposons encodes an NADP-requiring oxidoreductase. *J. Bacteriol.* **173:**176–183.

120. **Speer, B. S., N. B. Shoemaker, and A. A. Salyers.** 1992. Bacterial resistance to tetracycline: mechanisms, transfer, and clinical significance. *Clin. Microbiol. Rev.* **5:**387–399.

121. **Stasinopoulos, S. J., G. A. Farr, and D. H. Bechhofer.** 1998. *Bacillus subtilis tetA*(L) gene expression: evidence for regulation by translational reinitiation. *Mol. Microbiol.* **30:**923–932.

122. **Stezowski, J. J.** 1976. Chemical-structural properties of tetracycline derivatives. 1. Molecular structure and conformation of the free base derivatives. *J. Am. Chem. Soc.* **98:**6012–6018.

123. **Su, Y. A., H. Ping, and D. B. Clewell.** 1992. Characterization of the *tet*(M) determinant of Tn916: evidence for regulation by transcription attenuation. *Antimicrob. Agents Chemother.* **36:**769–778.

124. **Suarez, G., and D. Nathans.** 1965. Inhibition of aminoacyl-sRNA binding to ribosomes by tetracycline. *Biochem. Biophys. Res. Commun.* **18:**743–750.

125. **Tauch, A.** Personal communication.

126. **Tauch, A., A. Puehler, and J. Kalinowski.** DNA sequence and genetic organization of pAG1, a 19.8-kb R-plasmid of *Corynebacterium glutamicum* encoding a new class of tetracycline efflux and repressor proteins. In preparation.

127. **Taylor, D. E., and A. Chau.** 1996. Tetracycline resistance mediated by ribosomal protection. *Antimicrob. Agents Chemother.* **40:**1–5.

128. **Taylor, D. E., L. J. Jerome, J. Grewal, and N. Chang.** 1995. Tet(O), a protein that mediates ribosomal protection to tetracycline, binds, and hydrolyses GTP. *Can. J. Microbiol.* **41:**965–978.

129. **Taylor, D. E., C. A. Trieber, G. Trescher, and M. Bekkering.** 1998. Host mutations (*miaA* and *rpsL*) reduce tetracycline resistance mediated by Tet(O) and Tet(M). *Antimicrob. Agents Chemother.* **42:**59–64.

130. **Testa, R. T., P. J. Petersen, N. V. Jacobus, P.-E. Sum, V. J. Lee, and F. P. Tally.** 1993. In vitro and in vivo antibacterial activities of the glycylcyclines, a new class of semisynthetic tetracyclines. *Antimicrob. Agents Chemother.* **37:**2270–2277.

131. **Thanassi, D. G., G. S. Suh, and H. Nikaido.** 1995. Role of outer membrane barrier in efflux-mediated tetracycline resistance of *Escherichia coli*. *J. Bacteriol.* **177:**998–1007.

132. **Trieber, C. A., N. Burkhardt, K. H. Nierhaus, and D. E. Taylor.** 1998. Ribosomal protection from tetracycline mediated by Tet(O): Tet(O) interaction with ribosomes is GTP-dependent. *Biol. Chem.* **379:**847–855.

133. **Tritton, T. R.** 1977. Ribosome-tetracycline interactions. *Biochemistry* **16:**4133–4138.

134. **Tuckman, M., and S. Projan.** 1998. Characterization of TetA(B) glycylcycline-resistant mutants, abstr. C-97. *Program Abstr. 38th Intersci. Conf. Antimicrob. Agents Chemother.* American Society for Microbiology, Washington, D.C.

135. **Wang, Y., and D. E. Taylor.** 1991. A DNA sequence upstream of the *tet*(O) gene is required for full expression of tetracycline resistance. *Antimicrob. Agents Chemother.* **35:**2020–2025.

136. **Weckesser, J., and J. A. Magnuson.** 1979. Light-induced, carrier-mediated transport of tetracycline by *Rhodopseudomonas sphaeroides*. *J. Bacteriol.* **138:**678–683.

137. **Wells, S. J., J. O'Sullivan, C. Aklonis, H. A. Ax, A. A. Tymiak, D. R. Kirsch, W. H. Trejo, and P. Principe.** 1992. Dactylocyclines: novel tetracycline derivatives produced by a *Dactylosporangium* sp. *J. Antibiot.* **45:**1892–1898.

138. **Willems, A., M. Amat-Marco, and M. D. Collins.** 1996. Phylogenetic analysis of *Butyrivibrio* strains reveals three distinct groups of species within the *Clostridium* subphylum of the gram-positive bacteria. *Int. J. Syst. Bacteriol.* **46:**195–199.

139. **Wilson, K. S., and H. F. Noller.** 1998. Molecular movement inside the translational engine. *Cell* **92:**337–349.

140. **Yamaguchi, A., K. Adachi, T. Akasaka, N. Ono, and T. Sawai.** 1991. Metal-tetracycline/H+ antiporter of *Escherichia coli* encoded by a transposon Tn10: histidine 257 plays an essential role in H+ translocation. *J. Biol. Chem.* **266:**6045–6051.

141. **Yamaguchi, A., T. Akasaka, N. Ono, Y. Someya, M. Nakatani, and T. Sawai.** 1992. Metal-tetracycline/H+ antiporter of *Escherichia coli* encoded by transposon Tn10: roles of the aspartyl residues located in the putative transmembrane helices. *J. Biol. Chem.* **267:**7490–7498.

142. **Yamaguchi, A., Y. Iwasaki-Ohba, N. Ono, M. Kaneko-Ohdera, and T. Sawai.** 1991. Stoichiometry of metal-tetracycline/H+ antiport mediated by transposon Tn10-encoded tetracycline resistance protein in *Escherichia coli*. *FEBS Lett.* **282:**415–418.

143. **Yamaguchi, A., N. Ono, T. Akasaka, T. Noumi, and T. Sawai.** 1990. Metal-tetracycline/H+ antiporter of *Escherichia coli* encoded by a transposon, Tn10: the role of the conserved dipeptide, Ser65-Asp66, in tetracycline transport. *J. Biol. Chem.* **265:**15525–15530.

144. **Yamaguchi, A., R. O'yauchi, Y. Someya, T. Akasaka, and T. Sawai.** 1993. Second-site mutation of Ala-220 to Glu or Asp suppresses the mutation of Asp-285 to Asn in the transposon Tn10-encoded metal-tetracycline/H+ antiporter of *Escherichia coli*. *J. Biol. Chem.* **268:**26990–26995.

145. **Yamaguchi, A., T. Samejima, T. Kimura, and T. Sawai.** 1996. His257 is a uniquely important histidine residue for tetracycline/H+ antiport function but not mandatory for full activity of the transposon Tn10-encoded metal-tetracycline/H+ antiporter. *Biochemistry* **35:**4359–4364.

146. **Yamaguchi, A., Y. Shiina, E. Fujihira, T. Sawai, N. Noguchi, and M. Sasatsu.** 1995. The tetracycline efflux protein encoded by the *tet*(K) gene from *Staphylococcus aureus* is a metal-tetracycline/H+ antiporter. *FEBS Lett.* **365:**193–197.

147. **Yamaguchi, A., Y. Someya, and T. Sawai.** 1992. Metal-tetracycline/H+ antiporter of *Escherichia coli* encoded by transposon Tn10: the role of a conserved sequence motif, GXXXRXGRR, in a putative cytoplasmic loop between helices 2 and 3. *J. Biol. Chem.* **267:**19155–19162.

148. **Yamaguchi, A., Y. Someya, and T. Sawai.** 1993. The in vivo assembly and function of the N- and C-terminal halves of the Tn10-encoded TetA protein in *Escherichia coli*. *FEBS Lett.* **324:**131–135.

149. **Yamaguchi, A., T. Udagawa, and T. Sawai.** 1990. Transport of divalent cations with tetracycline as mediated by the transposon Tn10-encoded tetracycline resistance protein. *J. Biol. Chem.* **265:**4809–4813.

Chloramphenicol Resistance

IAIN A. MURRAY

67

CHLORAMPHENICOL STRUCTURE AND ANTIMICROBIAL ACTIVITY

Chloramphenicol was first introduced into clinical practice some 50 years ago, shortly after its discovery and isolation as a secreted secondary metabolite of *Streptomyces venezuelae*. In the absence of any resistance mechanism, it is a potent bacteriostatic agent that is active against the majority of gram-positive and gram-negative pathogens. Furthermore, as chloramphenicol equilibrates rapidly across biological membranes, including the blood-brain barrier, the antibiotic can be used to treat both extracellular and intracellular bacterial infection. The antibacterial function of chloramphenicol is a consequence of its strong inhibitory effect upon the peptidyl transferase activity of prokaryotic 50S ribosomes (4). Such inhibition does not extend to eukaryotic ribosomes, the activity of which appears insensitive to the presence of the drug.

The genetics and enzymology of the biosynthetic pathway in *S. venezuelae* that leads to the production of the chloramphenicol molecule have been elucidated over many years by Vining and coworkers (22). However, for the purposes of this chapter, it is sufficient that the reader be aware that the chemical structure of chloramphenicol is that of a derivatized 1,3-propanediol that carries *para*-nitrophenyl and dichloroacetamido substituents at C-1 and C-2, respectively (Fig. 1). The antibiotic is therefore a close structural homolog of the corynecins, a family of compounds produced by species of *Corynebacteria* wherein various nonhalogenated acyl groups replace the dichloroacetyl group present in chloramphenicol. When compared with many other antibiotics, the chloramphenicol molecule is both small (323 Da) and amphiphilic (being uncharged at physiological pH), properties that no doubt account for its ability to diffuse readily across membranes. The presence of the two chiral carbon atoms (C-1 and C-2) of chloramphenicol yields four potential diastereoisomeric forms. However, studies using such isomers produced via chemical synthesis demonstrate that only the D-*threo* compound, that which is synthesized by the producing organism, has antibiotic properties. A large number of additional synthetic analogs, in which one or more of the chemical substituents of the parent antibiotic are altered, have been used to dissect the detailed structure-activity relationships of chloramphenicol (21, 25).

CHLORAMPHENICOL RESISTANCE

Chloramphenicol resistance was identified in epidemic strains of gram-negative enteric bacteria within a few years of introduction of the antibiotic to medical practice and in gram-positive pathogens shortly thereafter. Such resistance, commonly but not invariably associated with the presence of enzymatic mechanisms of drug inactivation (see below), has subsequently been recorded in a diverse range of both benign and disease-associated gram-positive and gram-negative eubacteria and is now commonly observed in clinical isolates of many important human pathogens. This is still the case notwithstanding two decades during which administration of the antibiotic by physicians has been actively discouraged owing to toxicity problems associated with chloramphenicol therapy (25). Viewed from a late-20th-century perspective, it is all too obvious how the horizontal transfer of resistance genes within and between bacterial genera, coupled with the powerful selective pressure of widespread antibiotic use in medicine and agriculture, conspires to reduce the therapeutic utility of many antibacterial agents, including chloramphenicol. However, as such processes have been discussed extensively elsewhere (2, 24), in this chapter I focus on the cellular and biochemical mechanisms that gram-positive bacteria utilize to circumvent chloramphenicol antibiosis rather than on detailed analysis of the means whereby the genetic determinants of such resistance are disseminated.

Three generalized strategies whereby a bacterial cell might protect itself from the harmful effects of an antibiotic may be envisioned: (i) acquisition of mutations in the gene (or genes) encoding the target of antibiotic action that render it insensitive to the drug; (ii) reduced drug access to the target as a consequence of physical exclusion or active efflux of the antibiotic; and (iii) enzyme-catalyzed antibiotic inactivation. A fourth potential route to resistance, metabolic bypass of the site of antibiotic action, is

Gram-Positive Pathogens, ed. by V. A. Fischetti et al.
© 2000 American Society for Microbiology, Washington, D.C.

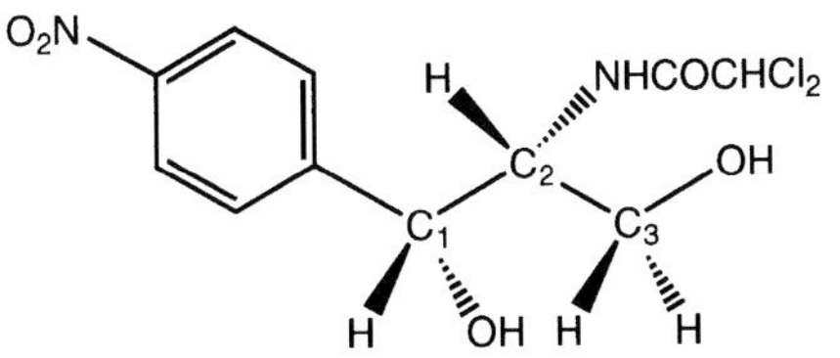

FIGURE 1 Structure of chloramphenicol. Chloramphenicol is a disubstituted propanediol containing two chiral centers at C-1 and C-2. Of four possible diastereoisomeric forms, only the naturally occurring D-*threo* compound (as depicted) is active as an antibiotic. The hydroxyl substituent at C-3 is the target of antibiotic inactivation by enzymic acetylation or phosphorylation (13, 14). Replacement of this hydroxyl group by a fluorine atom produces a potent antibiotic that is resistant to inactivation by these mechanisms (21).

irrelevant in the context of drugs that, like chloramphenicol, target components of essential and nonredundant pathways such as ribosomal protein synthesis. In the remainder of this chapter I address the extent to which each of these three mechanisms contributes to the overall scheme of chloramphenicol resistance in gram-positive bacteria.

Resistance as a Consequence of Target Modification

Resistance mediated by modification of the ribosomal target of antibiotic action appears to be of limited utility in protecting gram-positive organisms from inhibition by chloramphenicol. It is perhaps surprising, given the well-established precedent of streptomycin resistance (4), that no point mutations in genes encoding ribosomal proteins have been identified that abolish chloramphenicol binding. In *Bacillus subtilis*, a modest reduction in chloramphenicol sensitivity accompanies amino acid substitutions in ribosomal proteins L1, L12, and L15, but the degree of protection afforded is slight and likely to be insignificant were similar phenotypes to be generated in a pathogen. Why should this be the case? One possibility, favored by advocates of a primordial "RNA world" that predated protein-based cellular metabolism, is that the principal target of chloramphenicol action—by implication the catalytic component of the peptidyl transferase—is an RNA rather than a protein component of the ribosome. This is an attractive hypothesis given the broad spectrum of chloramphenicol action and the high degree of sequence conservation among prokaryotic rRNAs when compared with the relatively divergent ribosomal protein sequences. In addition, ribosome footprinting experiments demonstrate that the binding of chloramphenicol, and other antibiotics that inhibit peptidyl transferase activity, protects a specific region of 23S rRNA from chemical modification (12). However, although resistance to other antibiotics that inhibit ribosome function is commonly achieved via mutations within or posttranscriptional modifications of rRNA, notably among antibiotic-producing actinomycetes (1), this does not appear to be the case for chloramphenicol. In fact, ribosomal protein synthesis in *S. venezuelae*—perhaps the most likely source of a ribosomal resistance mechanism—remains sensitive to inhibition by chloramphenicol (10). The clear implication is that structural changes at the peptidyl transferase center that might abolish chloramphenicol binding are incompatible with satisfactory ribosome func-

tion, irrespective of whether the antibiotic acts at a site on rRNA, on ribosomal proteins, or on some combination of the two.

Resistance via Exclusion or Efflux

The observed MIC of an antibiotic often varies between bacterial genera or between different species within a genus irrespective of the presence or absence of a known drug-specific inactivation mechanism. Inhibitors that are characterized by an apolar or hydrophobic chemical structure such as rifampin or fusidic acid are, broadly speaking, more potent against gram-positive than against gram-negative bacteria. Also, *Pseudomonas aeruginosa* isolates are generally less sensitive to a broad range of antibiotics than are other gram-negative organisms. For many years it was generally assumed that such so-called "intrinsic resistance" reflects differences in the permeability of the cell wall or membranes of different bacteria and, in particular, that the gram-negative outer membrane provides a significant barrier to influx of certain antibiotics. However, with few exceptions (most notably species of *Mycobacterium*), the gram-positive bacterial cell envelope consists of a thick peptidoglycan layer that surrounds the cytoplasmic membrane and that does not provide an effective barrier to the diffusion of small molecules. Such bacteria are therefore unable to achieve resistance by denying an antibiotic access to the cell, particularly when, like chloramphenicol, that antibiotic can rapidly traverse the cell membrane. In fact, recent evidence suggests that antibiotic exclusion alone cannot account for the observed resistance even among gram-negative bacteria such as *Pseudomonas* sp. (16). Additional mechanisms are required, and it appears that one of the most common is active translocation of antibiotics out of the bacterial cell, facilitated by pump proteins localized in the cell membrane.

Facilitated efflux of chloramphenicol from cells has been recorded for a number of gram-positive genera and in all cases appears to be associated with the presence of one or more members of the so-called major facilitator superfamily (MFS) of membrane proteins. Such proteins, which have been identified in organisms ranging from bacteria to higher eukaryotes, couple import or export of a range of substrates to the transmembrane proton gradient (18). MFS proteins that are known to export chloramphenicol all belong to a subclass of these transporters which share a predicted tertiary structure containing 12 membrane-spanning polypeptide segments and which probably function via a drug/proton antiport mechanism (19). Those chloramphenicol efflux proteins that have been characterized biochemically fall into two general classes on the basis of substrate specificity. The first class comprises proteins that function as specific transporters of chloramphenicol and closely related compounds, while the second group are multidrug transporters that facilitate exclusion of both chloramphenicol and a structurally diverse range of antibiotics and other compounds.

To date, gram-positive chloramphenicol-specific transporters have been definitively identified only in the actinomycetes, the best-characterized being the chromosomal resistance determinant *cml* of *Streptomyces lividans* 1326. The DNA fragment carrying the *cml* gene is subject to amplification (via multiple tandem duplication events) when cells are exposed to chloramphenicol and confers high-level resistance (>200 μg/ml) when present in multiple copies (3). Homologs of *cml* are present in chloramphenicol-producing strains of *S. venezuelae* and are also associated with plasmid- and transposon-mediated chloramphenicol resistance in *Rhodococcus* and *Corynebacterium* spp., respectively. As species from each of these genera have been reported to synthesize chloramphenicol (or analogs thereof), it is tempting to speculate that these efflux proteins originally arose as a self-protection mechanism in the antibiotic-producing bacteria. The actual situation may be more complex as, at least in the case of the chloramphenicol producer *S. venezuelae* ISP5230, the *cml* homolog is known to be located adjacent to another gene (or genes) involved in chloramphenicol metabolism, and it is the latter rather than the former genes that appear to be the principal determinants of resistance (13). The data from genomic sequencing projects have revealed putative *cml* homologs in a variety of non-actinomycete gram-positive bacteria. The *B. subtilis* genome, for example, contains the predicted open reading frames *ydhL*, *ydeR*, and *ytbD*, which are described as being similar to *cml* and its homologs in *S. venezuelae* and *Rhodococcus fascians*, respectively (6a). However, the level of amino acid sequence identity between the *Bacillus* open reading frames and the *cml* proteins is only slightly greater than that revealed in comparisons with MFS proteins that act on non-chloramphenicol substrates. Further biochemical and genetic analyses are required to establish whether such gene products do indeed have a role in chloramphenicol efflux and resistance.

MFS multidrug efflux proteins have been identified in many gram-negative and gram-positive species, but only a few of these have been demonstrated to transport and confer resistance to chloramphenicol. The best-characterized proteins among gram-positive bacteria are those encoded by the *bmr* gene of *B. subtilis* and its homolog, *norA*, in *Staphylococcus aureus* (19). Like the *S. lividans cml* determinant, the chromosomal *bmr* gene is subject to amplification that results in increased drug resistance. Both *bmr* and *norA* confer resistance to a structurally diverse range of antibacterial agents that include antibiotics such as chloramphenicol, fluoroquinolones, and puromycin as well as ethidium bromide, acriflavine, rhodamine, and tetraphenylphosphonium compounds (17). Expression of *bmr* and *norA* is inducible by various transport substrates, transcriptional control in each case being mediated by regulatory proteins encoded at chromosomal loci adjacent to the transporter genes (19). *B. subtilis* also contains another inducible multidrug resistance gene, *blt*, closely related to *bmr* and *norA*, which confers resistance to a similar spectrum of drugs but also transports spermidine, a polyamine compound. Because *blt* is located in an operon involved in polyamine metabolism, it has been proposed that it functions primarily as a polyamine pump and that the other substrates, most of which are also cationic, are transported as a consequence of their similar ionization properties (15). It is not obvious why chloramphenicol, which is uncharged, should be a substrate for such transporters, particularly among staphylococci, in which high-level chloramphenicol resistance due to plasmid-encoded enzymic effectors (discussed below) is commonplace. Whatever the reason, it is almost certainly the case that similar inducible multidrug efflux mechanisms are active in other gram-positive pathogens. For example, active efflux of both chloramphenicol and norfloxacin—recently demonstrated in isolates of both *Enterococcus faecalis* and *Enterococcus faecium*—appears highly suggestive of the presence of a *norA* homolog in these species and may account for their intrinsic resistance to a range of different antibiotics (9).

Resistance as a Consequence of Antibiotic Inactivation

Enzyme-catalyzed alteration of drug structure is one of the most widespread and most effective means whereby bacteria circumvent the toxic effects of antibiotics. In principle, such resistance can be achieved either through antibiotic degradation or by modification of essential functional groups so that affinity for the cellular target of antibiotic action is lost. However, in the case of chloramphenicol, high-level resistance (>100 μg/ml) of both gram-positive and gram-negative bacteria is usually due to the latter mechanism. Such resistance is invariably associated with the presence of genes encoding variants of the bacterial enzyme chloramphenicol acetyltransferase (CAT). Unlike the β-lactamases, which are secreted and act extracellularly, the CATs are exclusively cytoplasmic proteins. CATs, as their name suggests, modify the antibiotic via the transfer of an acetyl moiety, and, in all known examples, the target of modification is the hydroxyl substituent at the C-3 position of chloramphenicol (Fig. 1) and the acetyl donor is acetyl coenzyme A (acetyl-CoA). The product 3-acetylchloramphenicol is unable to bind to ribosomes and is therefore devoid of antibiotic activity, as are two additional products, the 1-acetyl and 1,3-diacetyl adducts, which are generated in vivo by a combination of nonenzymic acetyl migration and a second cycle of CAT-mediated acetylation of the C-3–hydroxyl (14). All CAT variants are trimeric enzymes made up of identical 24- to 26-kDa subunits, and crystallographic analyses of substrate and product complexes reveal that chloramphenicol and CoA are bound at three equivalent active sites located at the subunit interfaces (Fig. 2) (7).

Details of the enzyme mechanism, which is common to all CAT variants, have been elucidated by Shaw and collaborators using a combination of kinetic, X-ray crystallographic, and site-directed mutational analyses (14, 20, and references therein). Acetyl-CoA and chloramphenicol approach the active sites from opposite surfaces of the CAT trimer, positioning the 3-hydroxyl group of the latter in close proximity to the imidazole side chain of His-195, an essential catalytic residue that is conserved in all of the 30-odd CAT variants identified to date. N$^{\varepsilon 2}$ of His-195, acting as a general base, abstracts the C-3–hydroxyl proton from chloramphenicol, thereby promoting nucleophilic attack of its oxygen at the adjacent carbonyl moiety of acetyl-CoA (Fig. 3). As a consequence, tetrahedral geometry is imposed upon the carbonyl group and partial (possibly formal) negative charge develops on the carbonyl oxygen, where it is stabilized by hydrogen bonding to the side chain hydroxyl of Ser-148, another absolutely conserved amino acid. This tetrahedral intermediate, which may approximate the transition state of the reaction, is further stabilized by additional conserved interactions (not shown in Fig. 3), which include indirect hydrogen bonds (i.e., via ordered water molecules) to the hydroxyl of Thr-174 and van der Waals' contacts between the methyl substituent of the intermediate and the side chains of hydrophobic amino acids. Collapse of the intermediate then yields the reaction products 3-acetylchloramphenicol and CoA. Although the reaction is freely reversible in vitro (when assayed in a coupled-enzyme system), the forward reaction is highly favored in vivo because the free energy of hydrolysis of the oxyester product (3-acetylchloramphenicol) greatly exceeds that of the corresponding thioester substrate (acetyl-CoA) (20).

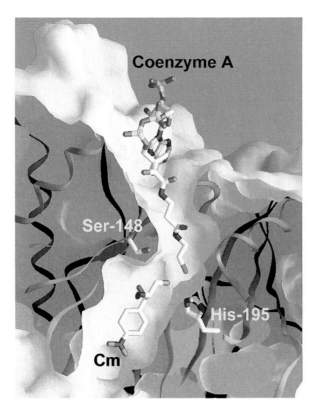

FIGURE 2 Structural basis of CAT-mediated chloramphenicol resistance. One of three identical active sites located at the subunit interfaces of the CAT homotrimer as revealed by X-ray crystallography of substrate (Cm, chloramphenicol) and product (Coenzyme A) complexes, respectively (7). Chloramphenicol and acetyl-CoA approach the catalytic center from opposite surfaces of the trimer via an extended tunnel in order to form a productive ternary complex, proximal to essential active-site residues that function as a catalytic base (His-195) and in transition-state stabilization (Ser-148)(14). Note that the dichloroacetyl substituent of chloramphenicol (see Fig. 1) has been omitted from the image to improve clarity. Image produced using the molecular graphics program GRASP (Nicholls, A., R. Bharadwaj, and B. Honig. 1993. *Biophys. J.* **64:**A166).

Inspection of GenBank and SwissProt databases reveals that 22 distinct CAT variants have been identified in nine different gram-positive bacterial genera, namely, *Bacillus*, *Clostridium*, *Enterococcus*, *Lactobacillus*, *Lactococcus*, *Listeria*, *Staphylococcus*, *Streptococcus*, and *Streptomyces*. However, as many of these sequences only show variation at a handful of amino acid positions, an alignment of eight sequences, including three variants each from *Clostridium* and *Staphylococcus*, is sufficient to represent the different classes of gram-positive CATs (Fig. 4).

As staphylococcal *cat* genes are commonly located on small (4 to 5 kb) multicopy plasmids and at least one clostridial variant, *catP*, is present on a transposon, it is not surprising that there is widespread evidence of horizontal transfer of these genes between bacterial genera. For example, homology searching of the above databases (unpublished results) reveals that *cat* sequences found on resistance plasmids isolated from *E. faecalis* and *Streptococcus agalactiae*, from *Lactococcus lactis* and *Listeria monocytogenes*,

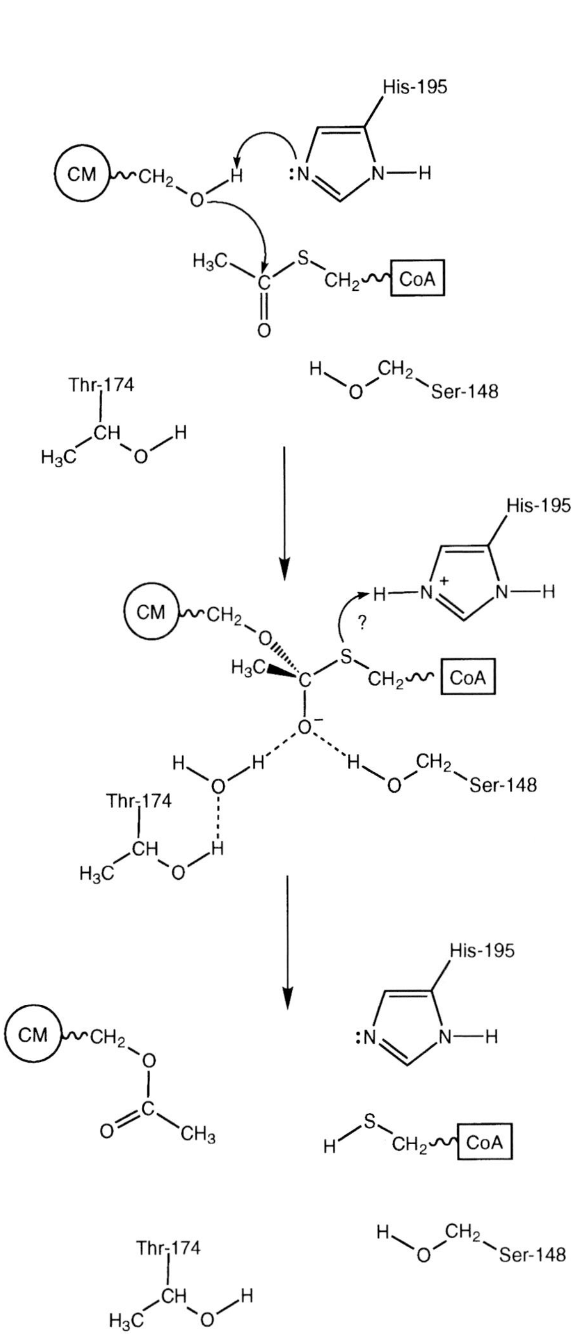

FIGURE 3 Mechanism of CAT-mediated chloramphenicol inactivation. In the ternary complex of CAT, chloramphenicol, and acetyl-CoA (upper panel), His-195 acts as a general base to abstract the proton from the 3-hydroxyl substituent of chloramphenicol, thereby promoting nucleophilic attack at the carbonyl of the acetyl substituent. This yields a tetrahedral oxyanion intermediate, which may approximate to the reaction transition state (central panel) and which is stabilized by hydrogen bond interactions involving Ser-148 and Thr-174. Collapse of the intermediate yields the product complex of CAT, CoA, and 3-acetylchloramphenicol (lower panel), the latter being unable to bind bacterial ribosomes and therefore devoid of antibacterial activity.

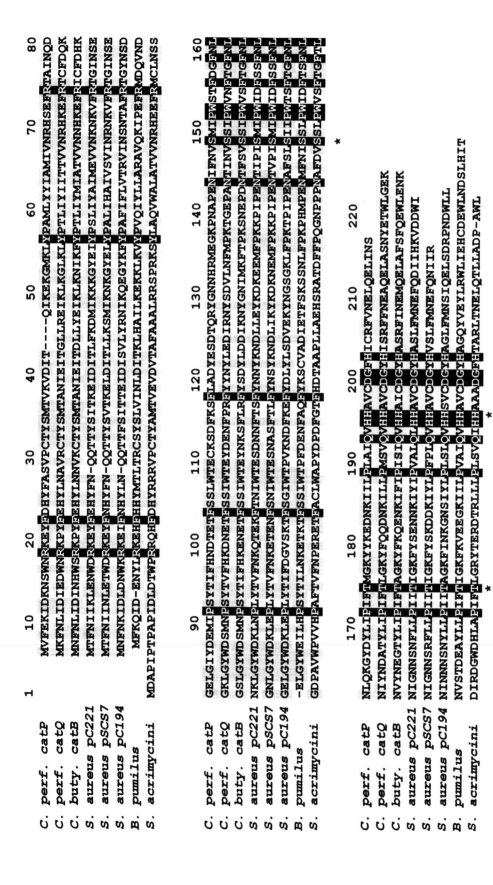

FIGURE 4 Alignment of amino acid sequences representing the eight different CAT protein subclasses present in gram-positive organisms. Three subclasses from clostridia and three from staphylococci are shown in addition to the CAT proteins of *Bacillus pumilus* and *Streptomyces acrimycini*. Amino acid residues that are absolutely conserved among all CAT isolates (both gram-positive and -negative) are depicted by vertical columns of white text on black background. Gray shading indicates positions where amino acid residues are conserved in four (or more) of the displayed gram-positive sequences. Critical active-site residues Ser-148, Thr-174, and His-195 are indicated by asterisks. The depicted sequences and their corresponding database accession codes are as follows: *C. perf. catP*, *Clostridium perfringens catP* gene (SwissProt cat2_clope); *C. perf. catQ*, *C. perfringens catQ* gene (SwissProt cat1_clope); *C. buty. catB*, *Clostridium butyricum catB* gene (SwissProt cat_clobu); *S. aureus pC221*, *S. aureus cat* gene from plasmid pC221 (SwissProt cat2_staau); *S. aureus pSCS7*, *S. aureus cat* gene from plasmid pSCS7 (SwissProt cat5_staau); *S. aureus pC194*, *S. aureus cat* gene from plasmid pC194 (SwissProt cat1_staau); *B. pumilus*, *B. pumilus cat-86* gene (SwissProt cat_bacpu); *S. acrimycini*, *S. acrimycini cat* gene (SwissProt cat_strac). Several additional staphylococcal *cat* genes have been identified, each of which encodes a protein that is almost identical to one of the three classes shown, and a resistance determinant isolated from *Clostridium difficile* is similarly related to *catP* (not shown). As discussed in the text, there is also clear evidence of horizontal gene transfer of *cat* genes between *Staphylococcus* and other gram-positive bacterial genera.

and from *Lactobacillus reuteri* are almost identical to the *cat* sequences of staphylococcal plasmids pC221, pSCS7, and pC194, respectively. Similarly, the aforementioned *catP*, associated with transposon Tn*4451* of *Clostridium perfringens*, is closely related to resistance determinants in *Streptococcus pyogenes* and in two gram-negative species, *Campylobacter coli* and *Vibrio anguillarum*. An intact and complete *catP* gene, though lacking the flanking sequences required for efficient transposition, was recently identified in 12 independent (and nonclonal) chloramphenicol-resistant clinical isolates of the gram-negative pathogen *Neisseria meningitidis* (5). Exactly how this sequence was transferred from gram-positive anaerobe to gram-negative aerobe remains to be determined, but the fact that it did so is testament to the remarkable promiscuity of *cat* genes. However, in contrast to the above, there is little evidence of horizontal transfer of the chromosomal *cat* genes of *Bacillus pumilus* and *Streptomyces acrimycini*, as *B. subtilis*, *Bacillus cereus*, *Bacillus anthracis*, and most actinomycetes generally lack CAT activity.

The expression of some gram-positive *cat* genes, notably the staphylococcal variants and the *cat-86* determinant of *B. pumilus*, is inducible by subinhibitory concentrations of chloramphenicol, by certain chloramphenicol analogs, and by some structurally unrelated antibiotics that also inhibit ribosome function. Such regulated expression of *cat* offers two potential advantages to the host cell: minimizing unnecessary protein synthesis and the avoidance of CAT-mediated acetyl-CoA hydrolysis, which invariably occurs in the absence of chloramphenicol. The common regulatory mechanism has been shown to be one of translational attenuation and has been investigated over many years by Lovett and coworkers (6). *cat-86* mRNA contains two consensus ribosome-binding site (RBS) sequences 5′ to the translational initiation codon. The most 5′ RBS directs production of a short leader peptide encoded upstream of the more 3′ RBS, at which *cat* translation is initiated. However, the latter RBS is sequestered in a stable mRNA hairpin, which precludes ribosome association. Induction of *cat* translation therefore requires destabilization of this secondary structure, and it appears that such destabilization is achieved by stalling of a ribosome, following translation of the first five amino acids of the leader peptide, at a position that overlaps the inverted repeat sequence. Such a mechanism implies that binding of inducers to the leader peptide-translating ribosome causes stalling but fails to account for the necessary positional specificity of the phenomenon. Recent evidence indicates that the specific position on the mRNA at which the ribosome becomes stalled is determined by the N-terminal pentapeptide (Met-Val-Lys-Thr-Asp) of the leader sequence that is itself an inhibitor of peptidyl transferase activity (8). In vivo it is likely that such inhibition only occurs in *cis* (i.e., is confined to those ribosomes translating *cat-86* leader peptide mRNA), as, in the context of antibiotic resistance, it would be self-defeating to regulate expression of *cat* by producing a pentapeptide that mimics the antibacterial activity of chloramphenicol.

Apart from O-acetylation, two further enzymatic mechanisms of chloramphenicol inactivation have been identified that may confer resistance in gram-positive bacteria. The first is hydrolysis of the *N*-dichloroacetyl substituent, and the second is phosphorylation of the C-3–hydroxyl group (Fig. 1). Both such activities have been reported in *S. venezuelae*, the chloramphenicol-producing organism. The existence of the hydrolase is suggested by the detec-

tion of derivatives of the expected hydrolytic reaction product (*para*-nitrophenylserinol) in *Streptomyces* culture (22). However, neither the hydrolase enzyme nor the gene that encodes it has been isolated, nor is there clear evidence linking such activity to chloramphenicol resistance. In the case of chloramphenicol phosphotransferase, both the enzymic activity—phosphoryl transfer from adenosine triphosphate to the C-3–hydroxyl of chloramphenicol—and the gene encoding it in *S. venezuelae* have been unambiguously assigned (13). Recombinant expression of this gene confers chloramphenicol resistance upon an *Escherichia coli* host, and it appears probable that the phosphotransferase activity is also the source of resistance in the chloramphenicol producer. The gene encoding chloramphenicol phosphotransferase in *S. venezuelae* ISP5230 is located adjacent to that which encodes a chloramphenicol efflux pump (*cml* homolog), implying that the two proteins act in a concerted fashion to provide protection from the antibiotic. One possibility, yet to be tested by experiment, is that chloramphenicol is exported from *S. venezuelae* in phosphorylated form and subsequently activated by extracellular phosphatases in a manner directly analogous to export of phosphorylated streptomycin by *Streptomyces griseus* (1).

The Past, Present, and Future of Chloramphenicol Resistance

What are the origins of the chloramphenicol resistance determinants described in the preceding discussion? Antibiotic-producing organisms are often cited as repositories of resistance determinants that may be acquired by other bacteria, but with chloramphenicol this does not appear to be the case: the genes encoding chloramphenicol-specific efflux pumps and chloramphenicol phosphotransferase have not been found in any gram-positive bacteria other than *Streptomyces* or closely related genera. Genes encoding CAT are found in *Streptomyces* spp. but not, to date, in species that synthesize the antibiotic. However, a candidate evolutionary antecedent of CAT exists in the form of the acetyltransferase (E2p) component of the pyruvate dehydrogenase multienzyme complex, a protein that catalyzes acetyl transfer from an enzyme-linked lipoyl prosthetic group to CoA. Although there is minimal conservation of primary amino acid sequence between CAT and E2p, the structures and catalytic mechanism of the two proteins are essentially identical (11). If CAT evolved from E2p (or at least shares a common ancestry), did it arise before or after chloramphenicol was introduced into clinical practice in 1948? Given that conservation of primary sequence between CAT and E2p is limited, and the fact that chloramphenicol-resistant clinical isolates of both gram-positive and gram-negative pathogens first appeared in the 1950s, it is probable that the resistance mechanism predated the discovery of the drug.

As mentioned previously, the toxic effects of chloramphenicol therapy have led, in many countries, to a marked decline or cessation in clinical use of the drug. In principle, this decline provides an opportunity to study the consequences of cessation of antibiotic use on the prevalence of bacterial resistance. If overprescription of antibiotics is a major driving force in the ongoing dissemination of resistance (23), can we now expect the prevalence of chloramphenicol resistance to fall? While it is to be hoped that this is the case, it seems likely that chloramphenicol-resistance genes will prove harder to get rid of than to acquire. The appearance, during the last decade, of a completely new class of bacterial enzymes that acetylate chloramphenicol but are unrelated, structurally and in part mechanistically, to CAT (14) suggests that chloramphenicol resistance is unlikely to disappear overnight. In addition, chloramphenicol usage, both by clinicians and as a component of over-the-counter topical medicines, remains common in some countries. Such usage as well as chloramphenicol use in animal husbandry, the full extent of which is difficult to estimate but likely to be large (24), may prove quite sufficient to maintain the current high frequency of resistance.

Finally, what are the prospects of overcoming chloramphenicol resistance by the route traditionally favored by pharmaceutical companies—producing a modified version of the drug that is not susceptible to the resistance mechanism? In fact, this compound already exists, at least in the case of CAT, which is arguably the only resistance mechanism of clinical significance. The 3-fluoro derivative of chloramphenicol (in which the C-3–hydroxyl group is replaced by fluorine; Fig. 1) is almost as potent an antibiotic as chloramphenicol, shares its ability to efficiently traverse biological membranes, and, lacking the target hydroxyl group, is entirely resistant to acetylation (21). The challenge for any would-be drug designer is, of course, to produce an analog of this compound that also lacks the toxicity associated with chloramphenicol. Ironically, were such a drug to be introduced into clinical practice, it is probable that chloramphenicol efflux proteins, long considered poor relations to CAT in the resistance stakes, would prove only too capable of stepping into the breach.

I am grateful to Jane Grasby, Peter Artymiuk, and Ewan Murray for their assistance with figure production and to Bill Shaw for helpful discussions regarding the early history of chloramphenicol and chloramphenicol resistance.

REFERENCES

1. **Cundliffe, E.** 1992. Self-protection mechanisms in antibiotic producers. *CIBA Found. Symp.* **171:**199–208.

2. **Davies, J.** 1994. Inactivation of antibiotics and the dissemination of resistance genes. *Science* **264:**375–382.

3. **Dittrich, W., M. Betzler, and H. Schrempf.** 1991. An amplifiable and deletable chloramphenicol-resistance determinant of *Streptomyces lividans* 326 encodes a putative transmembrane protein. *Mol. Microbiol.* **5:**2789–2797.

4. **Gale, E. F., E. Cundliffe, P. E. Reynolds, M. H. Richmond, and M. J. Waring.** 1981. *The Molecular Basis of Antibiotic Action*, 2nd ed. John Wiley & Sons, Inc., London, U.K.

5. **Galimand, M., G. Gerbaud, M. Guibourdenche, J.-Y. Riou, and P. Courvalin.** 1998. High level chloramphenicol resistance in *Neisseria meningitidis*. *N. Engl. J. Med.* **339:**868–874.

6. **Gu, Z. P., and P. S. Lovett.** 1992. Perturbing highly conserved spatial relationships in the regulatory domain that controls inducible *cat* translation. *Mol. Microbiol.* **6:** 2769–2776.

6a.**International *Bacillus subtilis* Sequencing Project.** http://bacillus.tokyo-center.genome.ad.jp

7. **Leslie, A. G. W., P. C. E. Moody, and W. V. Shaw.** 1988. Structure of chloramphenicol acetyltransferase at 1.75Å resolution. *Proc. Natl. Acad. Sci. USA* **85:**4133–4137.

8. **Lovett, P. S., and E. J. Rogers.** 1996. Ribosome regulation by the nascent peptide. *Microbiol. Rev.* **60:**366–385.

9. **Lynch, C., P. Courvalin, and H. Nikaido.** 1997. Active efflux of antimicrobial agents in wild-type strains of enterococci. *Antimicrob. Agents Chemother.* **41:**869–871.

10. **Malik, V. S.** 1979. Regulation of chorismate-derived antibiotic production. *Adv. Appl. Microbiol.* **25:**75–93.

11. **Mattevi, A., G. Oblomova, K. H. Kalk, A. A. Teplyakov, and W. G. J. Hol.** 1993. Crystallographic analysis of substrate binding and catalysis in dihydrolipoyl transacetylase (E2$_p$). *Biochemistry* **32:**3887–3901.

12. **Moazed, D., and H. F. Noller.** 1987. Chloramphenicol, erythromycin, carbomycin and vernamycin B protect overlapping sites in the peptidyl transferase region of 23S ribosomal RNA. *Biochimie* **69:**879–884.

13. **Mosher, R. H., D. J. Camp, K. Yang, M. P. Brown, W. V. Shaw, and L. C. Vining.** 1995. Inactivation of chloramphenicol by O-phosphorylation. *J. Biol. Chem.* **270:**27000–27006.

14. **Murray, I. A., and W. V. Shaw.** 1997. O-acetyltransferases for chloramphenicol and other natural products. *Antimicrob. Agents Chemother.* **41:**1–6.

15. **Neyfakh, A. A.** 1997. Natural functions of bacterial multidrug transporters. *Trends Microbiol.* **5:**309–313.

16. **Nikaido, H.** 1994. Prevention of drug access to bacterial targets: permeability barriers and active efflux. *Science* **264:**382–388.

17. **Nikaido, H.** 1998. Multiple antibiotic resistance and efflux. *Curr. Opin. Microbiol.* **1:**516–523.

18. **Pao, S. S., I. T. Paulsen, and M. H. Saier.** 1998. Major facilitator superfamily. *Microbiol. Mol. Biol. Rev.* **62:**1–34.

19. **Paulsen, I. T., M. H. Brown, and R. A. Skurray.** 1996. Proton-dependent multidrug efflux systems. *Microbiol. Rev.* **60:**575–608.

20. **Shaw, W. V., and A. G. W. Leslie.** 1991. Chloramphenicol acetyltransferase. *Annu. Rev. Biophys. Biophys. Chem.* **20:**363–386.

21. **Syripoulou, V. P., A. L. Harding, D. A. Goldmann, and A. L. Smith.** 1981. In vitro antibacterial activity of fluorinated analogs of chloramphenicol and thiamphenicol. *Antimicrob. Agents Chemother.* **19:**294–297.

22. **Vining, L. C., and C. Stuttard.** 1995. Chloramphenicol. *Biotechnology* **28:**505–530.

23. **Williams, R. J., and D. L. Heymann.** 1998. Containment of antibiotic resistance. *Science* **279:**1153–1154.

24. **Witte, W.** 1998. Medical consequences of antibiotic use in agriculture. *Science* **279:**996–997.

25. **Yunis, A. A.** 1988. Chloramphenicol: relation of structure to activity and toxicity. *Annu. Rev. Pharmacol. Toxicol.* **28:**83–100.

Mechanisms of Fluoroquinolone Resistance

DAVID C. HOOPER

68

Fluoroquinolones are antimicrobials that are widely used in clinical medicine. Initial members of this class were used largely for treatment of infections caused by gram-negative bacteria, but over time quinolones with increasing potency against gram-positive bacteria have been developed. With increasing use of quinolones for treatment of gram-positive bacterial infections, understanding of the mechanisms of quinolone resistance in gram-positive bacteria will be of increasing importance. In this chapter, I summarize current understanding of established mechanisms of resistance to this class in gram-positive bacteria. Much of what has been found in gram-positive bacteria has parallels in gram-negative bacteria, from which much of the original data were obtained. There are, however, important differences between gram-positive and gram-negative bacteria, both in target enzyme sensitivity and in the means by which efflux resistance mechanisms operate, that are of both clinical and fundamental importance.

ACTIVITY AND MECHANISM OF ACTION OF FLUOROQUINOLONES AGAINST GRAM-POSITIVE PATHOGENS

Structure-Activity Relationships

The earliest developed quinolones had their greatest potency against gram-negative pathogens, with less activity against gram-positive pathogens, but extensive studies of the structure-activity relationships of this class of compounds identified features that enhanced potency against gram-positive pathogens (12, 37). Some of these structural features have recently been incorporated into newly released congeners and additional compounds under development. Selected quinolone structures are shown in Fig. 1. All are based on a dual-ring structure with adjacent carbonyl and carboxyl groups at positions 3 and 4 of one ring, which are essential for activity. Additional substituents at other ring positions have been varied with effects on potency. The term fluoroquinolone derives from the presence of a fluorine at position 6, which is a feature of virtually all drugs of this class developed since the 1980s. Substit-

uents at positions 5 and 7, in particular, have been noted to have effects on activity against gram-positive bacteria. At position 5, addition of an amino group (sparfloxacin) or, to a lesser extent, a methyl group (grepafloxacin) improves activity against gram-positive bacteria relative to a hydrogen, which was commonly present in the initially developed fluoroquinolones (norfloxacin, enoxacin, ciprofloxacin, ofloxacin, and lomefloxacin). A piperazinyl ring was a common substituent at position 7, because it, in combination with a fluorine at position 6, substantially improved potency against gram-negative bacteria. Addition of one (grepafloxacin, ofloxacin, levofloxacin) or two (sparfloxacin) methyl groups to the piperazinyl ring enhanced activity against gram-positive bacteria, but an amino-substituted pyrrolidinyl ring at the 7 position of the quinolone rings (clinafloxacin) further enhanced activity against gram-positive bacteria relative to piperazinyl substituents. Derivatives of a pyrrolidinyl ring with an attached second ring (trovafloxacin, moxifloxacin, sitafloxacin) have also been exploited for enhancement of activity against gram-positive bacteria.

Type 2 Topoisomerases

Quinolones interact with both of the type 2 topoisomerases in eubacteria, DNA gyrase and topoisomerase IV, which are essential for bacterial DNA replication. Type 2 topoisomerases act by breaking both strands of duplex DNA, passing another DNA strand through the break, and resealing the initial broken strands (18, 62). This activity occurs at the expense of ATP, with ATP hydrolysis serving to reset the enzyme for another cycle of strand passage (4). DNA gyrase, which is composed of two GyrA and two GyrB subunits, encoded by the gyrA and gyrB genes, respectively, is the only enzyme that introduces negative superhelical twists into bacterial DNA. Negative superhelical twists are important for initiation of DNA replication, and they facilitate binding of initiation proteins. Positive superhelical twists accumulate ahead of the progressing bacterial replication fork and must be removed to avoid stalling of the fork. Both DNA gyrase and topoisomerase IV can remove positive superhelical twists, but only

Gram-Positive Pathogens, ed. by V. A. Fischetti et al.
© 2000 American Society for Microbiology, Washington, D.C.

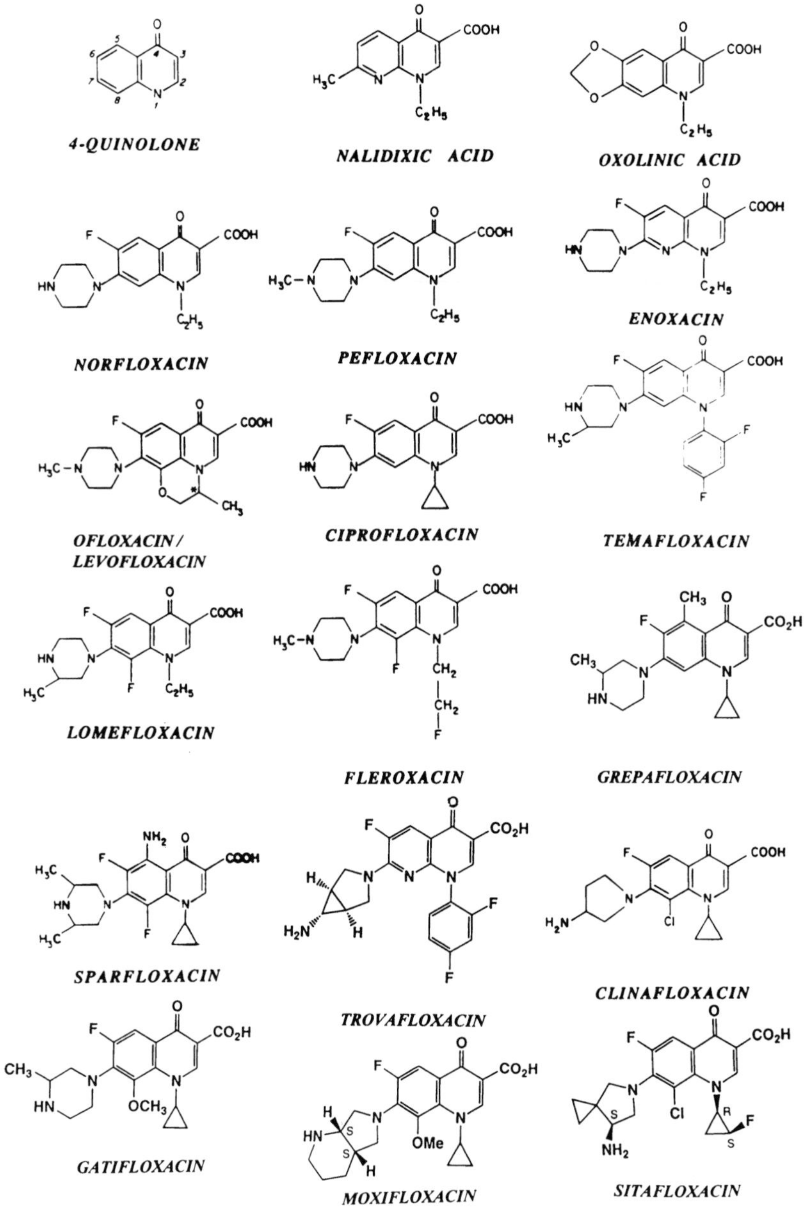

FIGURE 1 Structures of old and newer quinolones.

DNA gyrase can convert them directly to negative super-helical twists.

The two subunits of topoisomerase IV are homologous to those of DNA gyrase. ParC (or GrlA in *Staphylococcus aureus*) encoded by the *parC* gene is homologous to GyrA, and ParE (or GrlB in *S. aureus*) encoded by the *parE* gene is homologous to GyrB (14, 30, 31). Topoisomerase IV is the principal enzyme that removes the interlinking of daughter chromosomes at the completion of a round of DNA replication and allows their segregation into daughter cells (71). Although these physiologic roles for topoisomerase IV and DNA gyrase have been determined largely from studies of *Escherichia coli*, it is presumed that they will prove fundamentally similar in gram-positive bacteria. The catalytic properties of DNA gyrase and topoisomerase IV purified from *E. coli* and *S. aureus* appear to be similar for these two species, although the *S. aureus* en-

zymes function optimally in the presence of high concentrations of potassium glutamate (6), reflecting adaptation to the high intracellular concentrations of both of these ions in *S. aureus* (66).

Differences in Mechanism of Fluoroquinolone Action in Gram-Negative and Gram-Positive Bacteria

DNA gyrase was demonstrated to be a primary quinolone target in *E. coli*, based on genetic studies in which mutations in either the GyrA or GyrB subunit of this enzyme conferred increments in drug resistance (19, 22, 55). In contrast, for *S. aureus*, initial genetic studies showed that common first-step drug-resistance mutations were not in the subunits of DNA gyrase but were in a distinct genetic locus, *flqA* (61). *flqA* mutants were shown subsequently to have mutations in either the ParC (GrlA) (43) or ParE (GrlB) (17) subunits of topoisomerase IV. Whereas first-step mutants have mutations in topoisomerase IV, second-step mutants have mutations in DNA gyrase (13, 23). In addition, mutations in DNA gyrase subunits of *S. aureus*, when present in the absence of mutations in a subunit of topoisomerase IV, cause no change in susceptibility to most quinolones, except nalidixic acid (17). Thus, topoisomerase IV in *S. aureus* is a primary target of action of most quinolones.

These differences between *S. aureus* and *E. coli* appear to relate to the differences in the relative quinolone sensitivities of DNA gyrase and topoisomerase IV in the two species (6). For many quinolones, *E. coli* DNA gyrase is more sensitive than *E. coli* topoisomerase IV (24). In contrast, for at least some of this same group of quinolones, *S. aureus* topoisomerase IV is more sensitive than *S. aureus* DNA gyrase (6). Thus, the more sensitive of the two enzymes within a given species appears to determine the primary drug target indicated by genetic tests.

In two other gram-positive species studied, although data are less complete, a similar pattern is apparent for some quinolones. Ciprofloxacin-selected first-step mutants of *Streptococcus pneumoniae* have mutations in the ParC or ParE subunits of topoisomerase IV similar to those reported in *S. aureus* (47, 50). In addition, in *Enterococcus faecalis*, second-step but not first-step resistant mutants selected with ciprofloxacin have mutations in GyrA (32). Furthermore, in clinical resistant isolates, ParC mutations were found in low-level resistant isolates without a mutation in GyrA, and high-level resistant isolates had mutations in both ParC and GyrA, suggesting that first-step mutants may have mutations in topoisomerase IV (29). Thus, ciprofloxacin and other quinolones may have as their primary target topoisomerase IV rather than DNA gyrase in *S. pneumoniae* and *E. faecalis* as well as in *S. aureus*. This pattern, however, may vary with some quinolones in some species. Sparfloxacin-selected first-step mutants of *S. pneumoniae* have mutations in DNA gyrase, and the mutations in topoisomerase IV that affected ciprofloxacin activity have no effect on the activity of sparfloxacin (48). Thus, the results with sparfloxacin, as an exception to the pattern seen with ciprofloxacin and several other congeners, indicate that patterns of primary and secondary targets of quinolones will depend on drug structure.

RESISTANCE DUE TO ALTERED DRUG TARGET ENZYMES

Topoisomerase IV

Among antimicrobials in common clinical use for treatment of bacterial infections, quinolones and rifampin are distinctive for having predominant resistance mechanisms due to de novo selection of alterations in drug targets in contrast to acquisition of preexisting resistance determinants.

The largest body of information concerning quinolone resistance in gram-positive bacteria comes from studies of *S. aureus*. Mutations resulting in single amino acid changes in either the ParC or ParE subunits of topoisomerase IV cause four- to eightfold increments in resistance to ciprofloxacin and many other quinolones (13, 43). ParC mutations appear more common than ParE mutations among clinical resistant strains and have clustered in the amino

terminus, with position 80 being most commonly altered from serine (Ser) to either phenylalanine (Phe) or tyrosine (Tyr) (Table 1) (13, 15, 43, 51, 57, 58, 63, 69). Homologous mutations have been found in *parC* of strains of *E. coli* that are highly resistant and also have mutations in *gyrA*. Also commonly altered is position 84 from glutamic acid (Glu) to lysine (Lys) or leucine (Leu). Alterations in position 116 from alanine (Ala) to Glu or proline (Pro) are less common but also cause quinolone resistance similar to that due to mutations at positions 80 or 84. In addition, Ala-116–Glu/Pro mutations confer slight increases in susceptibility to novobiocin and coumermycin (16), coumarins that act as competitive inhibitors of the ATPase activity of topoisomerase IV and DNA gyrase. Mutations at position 116 are in proximity to the active-site Tyr-122. In *S. pneumoniae* similar *parC* mutations have been reported at positions 79 (Ser changed to either Tyr or Phe) and 83 (aspartic acid [Asp] changed to asparagine [Asn] or glycine [Gly]) in resistant mutants selected in the laboratory and clinical resistant isolates (Table 1) (20, 27, 38, 46–48, 59). Clinical resistant isolates of *E. faecalis* have also been recently shown to have similar mutations in ParC (Ser-80 changed to arginine [Arg] or isoleucine [Ile] and Glu-84–Ala). In one strain with low-level resistance, a mutation in ParC without a mutation in GyrA was found, and no GyrA mutants were found in the absence of a ParC mutation (29). Although genetic studies in *E. faecalis* have not yet proved the role of these ParC mutations in resistance, the analogy to the patterns found in *S. aureus* and *S. pneumoniae* in which genetic data are available strongly suggests that these ParC mutations contribute to resistance.

Mutations in ParE of *S. aureus* (17) and *S. pneumoniae* (50) have also been found in quinolone-resistant isolates. Changes at positions 470 (*S. aureus*, from Asn to Asp) and 435 (*S. pneumoniae*, from Asp to Asn) have been shown to cause quinolone resistance; other changes at positions 422 (Glu to Asp), 432 (Asp to Gly), and 451 (Pro to Ser) have been reported in resistant clinical isolates of *S. aureus* but not yet shown to contribute to the resistance phenotype (Table 1). No ParE mutations have yet been reported in *E. faecalis*. The Asn-470–Asp mutation of ParE, like the ParC Ala-116–Glu mutation, causes hypersusceptibility to coumarins (17). These mutations are in the same region also reported for resistance mutations in ParE of *E. coli* (7).

DNA Gyrase

DNA gyrase of *S. aureus* appears substantially less sensitive to several fluoroquinolones than *E. coli* DNA gyrase (6). Resistance mutations in GyrA in *S. aureus* and *S. pneumoniae* are generally not found in first-step mutants except those in *S. pneumoniae* selected with sparfloxacin, as noted above. Second-step mutations in either GyrA or GyrB, however, cause increased resistance in mutants with first-step mutations in ParC or ParE. Initial studies of resistant clinical isolates of *S. aureus* before the recognition of the role of topoisomerase IV in resistance commonly had GyrA and less often GyrB mutations (26, 54), but when ParC and ParE were also evaluated, GyrA or GyrB quinolone-resistance mutations have not been found in the absence of mutations in topoisomerase IV (14, 52). As noted above, *gyrA* single mutants constructed in the laboratory have no fluoroquinolone resistance but interestingly exhibit a fourfold increase in resistance to nalidixic acid, an early nonfluorinated quinolone with low potency against gram-positive bacteria (16), suggesting that nalidixic acid

TABLE 1 Mutations in the subunits of topoisomerase IV associated with quinolone resistance in gram-positive bacteria

Species	ParC			ParE		
	Amino acid position	Wild-type amino acid	Mutant amino acid	Amino acid position	Wild-type amino acid	Mutant amino acid
S. aureus	41	Val	Gly	422	Glu	Asp
	45	Ile	Met	432	Asp	Gly
	48	Ala	Thr	451	Pro	Ser
	80	Ser	Phe,[a] Tyr[a]	470	Asn	Asp[a]
	81	Ser	Pro			
	84	Glu	Lys,[a] Leu,[a] Val, Gly			
	116	Ala	Glu,[a] Pro[a]			
S. pneumoniae	79	Ser	Phe,[a] Tyr[a]	435	Asp	Asn[a]
	83	Asp	Asn,[a] Gly[a]			
	93	Lys	Glu[a]			
	95	Arg	Cys			
	137	Lys	Asn			
E. faecalis	80	Ser	Arg, Ile			
	84	Glu	Ala			

[a] Amino acids for which genetic data support a role for the mutation in causing resistance. Other mutant amino acids have been associated with resistance in clinical isolates.

has limited or no activity against *S. aureus* topoisomerase IV.

Mutations in GyrA have been clustered in a quinolone-resistance-determining region (QRDR) in the amino terminus that is homologous to similar regions in *E. coli* GyrA and ParC in which resistance mutations have been clustered. Similarly, resistance mutations in GyrB of *S. aureus* are in regions similar to those involved in resistance in *E. coli* GyrB and ParE (Table 2). Mutations in GyrA are found more commonly than mutations in GyrB and most often occur at positions 83 (*S. pneumoniae*) and 84 (*S. aureus*) with changes from Ser to Leu (*S. aureus*) or to Tyr or Phe (*S. pneumoniae*). Changes at positions 86 (*S. aureus*) and 87 (*S. pneumoniae*) from Glu to Lys are also commonly reported among GyrA mutants. All of the above-mentioned GyrA mutations and the less common Ser-85–Pro mutation of *S. aureus* have been shown to contribute to ciprofloxacin resistance in the presence of mutant topoisomerase IV. The Ser-83–Phe or Tyr mutation of *S. pneumoniae* also confers resistance to sparfloxacin in

TABLE 2 Mutations in the subunits of DNA gyrase associated with quinolone resistance in gram-positive bacteria

Species	GyrA			GyrB		
	Amino acid position	Wild-type amino acid	Mutant amino acid	Amino acid position	Wild-type amino acid	Mutant amino acid
S. aureus	84	Ser	Leu,[a] Ala, Val	437	Asp	Asn[a]
	85	Ser	Pro[b]			
	86	Glu	Lys,[a] Gly	458	Arg	Gln[a]
	106	Gly	Asp			
S. pneumoniae	83	Ser	Phe,[a] Tyr[a]	435	Asp	Asn[a]
	87	Glu	Lys,[a]			
E. faecalis	83	Ser	Arg, Ile, Asn			
	87	Glu	Lys, Gly			

[a] Amino acids for which genetic data support a role for the mutation in causing resistance. Other mutant amino acids have been associated with resistance in clinical isolates.

the absence of mutant topoisomerase IV (48), but the similar Ser-84–Leu GyrA mutation of *S. aureus* does not confer resistance to this or other fluoroquinolones studied in the absence of mutant topoisomerase IV (17). Other mutations, as listed in Table 2, have been associated with resistance in clinical isolates but have not yet been shown directly to contribute to resistance. Noteworthy is the similarity of the mutations in GyrA found in resistant clinical isolates of *E. faecalis* (Ser-83–Arg or Ile or Asn and Glu-87–Lys or Gly) to those found in *S. aureus* and *S. pneumoniae* (29, 32, 60). In the one study that evaluated both ParC and GyrA, no GyrA mutations were found in the absence of a ParC mutation (29). A GyrA subunit purified from another highly resistant but genetically uncharacterized clinical strain of *E. faecalis*, when combined with a wild-type GyrB subunit, was also shown to confer quinolone resistance on enzymatic activity (39). Thus, it is likely that resistance caused by GyrA mutation in *E. faecalis* is highly similar to that in *S. pneumoniae* and *S. aureus*.

GyrB mutations are found least often in laboratory and clinical resistant isolates and have been found at only two positions, 437 (*S. aureus*) or 435 (*S. pneumoniae*), with a change from Asp to Asn and at position 458 (*S. aureus*) with a change from Arg to Glu. No GyrB resistance mutations in *E. faecalis* have yet been reported. Data indicate that each of these mutations contributes to resistance in the presence of mutant topoisomerase IV.

Models for How Altered Drug Targets Cause Resistance

Direct data on quinolone interactions with topoisomerases come from studies of *E. coli* DNA gyrase, for which it has been shown that quinolones bind to a complex of DNA and DNA gyrase rather than DNA gyrase alone (53). The Ser-83–Trp mutation was also associated with reduced binding of norfloxacin to the gyrase-DNA complex (67). Thus, reductions in the affinity of drug for the enzyme-DNA target may mediate resistance for some classes of mutants, particularly those with mutations in and around position 83 (84 for *S. aureus* GyrA and 80 for *S. aureus* ParC) in the QRDR. The recently published X-ray crystallographic structure of a fragment of *E. coli* GyrA localizes the QRDR to a positively charged surface along which DNA has been modeled to bind (11). This region is adjacent to the two tyrosines (Tyr-122, one from each GyrA subunit) that are linked to DNA during strand breakage and might be considered a candidate for the site of quinolone binding. Thus, in one model, amino acid changes in the QRDR of GyrA (and by homology ParC) alter the structure of a quinolone-binding site near the interface of the enzyme and DNA, and resistance is effected by reduced drug affinity for the modified enzyme-DNA complex. As yet no topoisomerase crystal structures that include a quinolone molecule have been reported to enable assessment of this model.

Resistance mediated by mutations in ParE or GyrB might act by a different mechanism based on recent structural data. The crystal structure of a fragment of topoisomerase II of yeasts, which has homology with both GyrA (carboxy terminus) and GyrB (amino terminus), suggests that the region of quinolone-resistance mutations in ParE of *S. aureus* and *S. pneumoniae* is distant from the QRDR region of GyrA or ParC (5), at least in the one conformation of the enzyme that has been analyzed. The coumarin hypersusceptibility of ParE mutants suggests the possibility that impairment of catalytic function may contribute to quinolone resistance in these mutants (17). A similar phenotype was found in yeast mutants selected for amsacrine resistance that had mutations in the homologous domain of topoisomerase II and were shown to be catalytically impaired (64). Thus, in this model, mutations impair enzyme function as well as the formation of competent enzyme-DNA complexes that constitute the site of quinolone binding. Because ParC116 mutants also exhibit a coumarin hypersusceptibility phenotype and are in proximity of the Tyr-122 active site, they too might be postulated to have such a mechanism (16). No direct support for these hypotheses is yet available, and it remains possible that in other conformations of these enzymes the regions of GyrA (or ParC) and GyrB (or ParE) involved in quinolone resistance will be in proximity with each other and with bound DNA and so constitute the site of quinolone binding.

Interactions of Resistance Mutations Affecting Dual Targets

Stepwise incremental quinolone resistance often occurs by a series of mutations first in the primary enzyme target followed by mutations in the secondary enzyme target. Second and third mutations in primary and secondary targets may follow with further increments in resistance. The level of resistance conferred by an individual mutation is predicted to be determined by at least two factors, the extent to which the mutation alters the sensitivity of the mutant enzyme and the level of sensitivity of the other target enzyme. Thus, after mutation, whichever of the two topoisomerases (the mutant primary target topoisomerase and the wild-type secondary target topoisomerase) is the more sensitive determines the ceiling on the increment in resistance.

This sequence of resistance mutations implies that when there are differences in the sensitivity of the two target enzymes to a particular quinolone, a single spontaneous mutation may cause some increment in resistance and could be selected at a frequency typical of spontaneous mutations. In contrast, for a quinolone congener with equal activity against both topoisomerase IV and DNA gyrase, two mutations, one in each target, would need to be present simultaneously for the first step in resistance due to an altered target to occur, and thus resistance would be substantially less frequent.

RESISTANCE DUE TO ALTERED ACCESS OF DRUG TO TARGET ENZYMES

Quinolones must traverse the cell wall and cytoplasmic membrane of gram-positive bacteria to reach DNA gyrase and topoisomerase IV present in the cytoplasm. The cell wall is thought to provide little or no barrier to diffusion of small molecules such as quinolones, which have molecular weights around 300 to 400. Accumulation of quinolones by whole cells of *S. aureus* is nonsaturable under usual experimental conditions and likely occurs by simple diffusion across the cytoplasmic membrane. All active quinolones have a negatively charged carboxyl group at position 3, and most quinolones developed since 1985 have an additional positively charged group at position 7 (piperazinyl or pyrrolidinyl ring derivatives) and thus are zwitterionic. The proportions of positively charged, negatively charged, dually charged, and uncharged species of a given quinolone vary with pH, and it is presumed that it is the uncharged species that diffuses freely across the membrane and reaches equilibrium with the cytoplasm (44). Differences in the pH between the medium and the cytoplasm, thus, may affect partitioning of drug by altering the proportions of charged species that are "trapped" in the cytoplasmic compartment. These factors are likely responsible for the reductions in activity of zwitterionic quinolone congeners that occur below pH 7. Little is known about nonspecific binding of quinolones to bacterial cytoplasmic proteins or any compartmentalization of quinolones within the bacterial cell.

NorA of *S. aureus*

Acquired quinolone resistance that correlates with reduced quinolone accumulation in gram-positive bacteria has been identified in S. aureus and Bacillus subtilis and is due to active efflux of drug across the cytoplasmic membrane. In S. aureus, resistance of this type has been shown to be due to increased levels of expression of norA, a chromosomal gene that encodes a protein with 12 predicted membrane-spanning domains that is a member of the major facilitator superfamily of transporters (42, 70). Multidrug resistance (MDR) transporters of this type have broad substrate profiles, and NorA mediates pleiotropic resistance that includes ethidium bromide, rhodamine 6G, tetraphenylphosphonium (TPP), chloramphenicol, and hydrophilic quinolones such as norfloxacin, ciprofloxacin, and ofloxacin (41, 70). The activity of hydrophobic quinolones such as sparfloxacin and trovafloxacin is less affected by NorA. Cloned norA mediates a similar resistance phenotype in E. coli. In everted membrane vesicles prepared from E. coli cells containing cloned norA but not cells containing the vector plasmid alone, uptake of labeled norfloxacin (which represents drug efflux because of the reversed membrane orientation in everted vesicles) is saturable and dependent on the energy generated by the proton gradient across the membrane (42). Norfloxacin uptake in everted vesicles and the resistance phenotype associated with norA expression is inhibited by reserpine and verapamil (35, 42), which also inhibit other MDR transporters.

The level of expression of norA varies and may be regulated. Increased steady-state levels of norA mRNA and pleiotropic resistance are associated with single nucleotide changes upstream of norA in a putative promoter/operator region (28, 42). In addition, a mutant of S. aureus with norA expression induced by norfloxacin has been described (28). Resistance has also been associated with a single mutation in the norA structural gene (45), but little is known about structure-activity relationships of NorA. Also undefined are the normal functions of NorA in the cell. Although norA appears not to be an essential gene (25), based on the isolation of viable knockout mutants, it appears to be present commonly if not universally on the chromosomes of clinical isolates of S. aureus and coagulase-negative staphylococci (28). The extent to which mutations that increase norA expression or enhance its function contribute to resistance in clinical isolates of S. aureus is not yet certain. Blocking of NorA function with reserpine, however, increases quinolone susceptibility (42) and reduces the frequency of selection of resistant mutants with norfloxacin (35), suggesting a role for NorA in determining quinolone susceptibility even in wild-type staphylococci.

Bmr and Blt of *B. subtilis*

The Bmr protein of B. subtilis has 44% amino acid identity to NorA of S. aureus and also belongs to the major facilitator superfamily (40, 41). The expression of the chromosomal bmr gene results in a resistance phenotype similar to that due to expression of norA. Expression of bmr is induced by rhodamine 6G and TPP, two Bmr substrates, and bmrR, which is upstream of bmr, has homology to other transcriptional activator proteins (1). BmrR has been shown to bind the bmr promoter with increased affinity in the presence of these inducers, and the purified carboxy-terminal half of the Bmr protein has also been shown to bind rhodamine and TPP (34). A related transporter with an apparently identical substrate profile in B. subtilis, Blt, is not expressed in wild-type cells (2). Its expression is regulated by the product of the upstream gene, bltR, which is not induced by rhodamine. There is also an additional downstream gene, bltD, which is cotranscribed with blt and appears to encode a specific polyamine acetyltransferase (68). The role of this acetyltransferase within the cell is as yet unknown, but these findings suggest that bltD and blt may have related, specific physiologic functions in the cell.

Thus, fluoroquinolones, which are synthetic compounds, appear to be incidental substrates for NorA, Bmr, and Blt. Specific mutations or physiologic conditions that promote their expression will reduce drug activity and contribute to low-level resistance.

Other Efflux Transporters in Gram-Positive Bacteria

Other MDR-type transporters have been identified in gram-positive bacteria. These transporters include LmrP of Lactococcus lactis, Ptr of Streptomyces pristinaespiralis, and QacA/B of S. aureus, which belong to the major facilitator superfamily group, like NorA, Bmr, and Blt; Smr of S. aureus, which belongs to the small multiresistance family of transporters; and LmrA of L. lactis, which is the first identified bacterial MDR transporter of the ATP-binding cassette family. None of these transporters, however, appears to include quinolones in their broad substrate profiles (33). Recently, quinolone-resistant clinical and laboratory strains of S. pneumoniae have been shown to have reduced accumulation of quinolones that is reversible with reserpine, suggesting the involvement of an efflux system(s) in resistance in this organism as well (3, 8). In addition, most recently a norA homolog has been identified in S. pneumoniae, and its inactivation by recombination of a norA::cat insertion in the chromosome causes increased quinolone susceptibility (9). Although clinical isolates of quinolone-resistant S. pneumoniae are infrequent at present, it is noteworthy that as many as one-third of resistant isolates studied appear to have a reserpine-reversible resistance phenotype, suggesting a common role for drug efflux (10). Some highly quinolone-resistant clinical isolates of E. faecalis also appear to accumulate lesser amounts of norfloxacin than susceptible strains, suggesting a possible role for efflux in resistance in this organism as well. The genes responsible for this phenomenon and their role in resistance in E. faecalis, however, have not been defined (39).

OTHER POSSIBLE MECHANISMS OF RESISTANCE

No specific quinolone-degrading enzymes have been identified as a mechanism of resistance, but fungi that are capable of degrading quinolones through metabolic pathways have been reported (65). Recently, plasmid-mediated resistance to quinolones was identified for the first time in clinical strains of *Klebsiella pneumoniae* (36). The mechanism of this resistance is not yet known, but resistance was transferable on an MDR plasmid to other strains of *K. pneumoniae* and *E. coli*. No such plasmid-mediated quinolone resistance has been reported in natural isolates of gram-positive bacteria. One possible form of plasmid-mediated quinolone resistance in gram-positive bacteria, however, is due to MDR pumps. The QacA/B (49) and Smr (21) MDR pumps, which can cause resistance to other antimicrobials but not quinolones, are encoded on plasmids in *S. aureus*, and the *norA* gene cloned on a shuttle plasmid confers quinolone resistance in *S. aureus*, presumably owing to increased gene dosage (70). In addition, mutant resistant alleles of *grlA* and *grlB* of topoisomerase IV present on shuttle plasmids exhibit codominance over their chromosomal wild-type alleles and confer resistance when introduced into wild-type *S. aureus* (16, 17). Overexpression of *norA* and genes for topoisomerases from plasmids is known, however, to have toxic effects on the cell that may limit the fitness of resistant bacteria containing them (56). Thus, at present, quinolone resistance in gram-positive bacteria is attributable almost exclusively to chromosomal mutations.

REFERENCES

1. **Ahmed, M., C. M. Borsch, S. S. Taylor, N. Vazquez-Laslop, and A. A. Neyfakh.** 1994. A protein that activates expression of a multidrug efflux transporter upon binding the transporter substrates. *J. Biol. Chem.* **269:** 28506–28513.

2. **Ahmed, M., L. Lyass, P. N. Markham, S. S. Taylor, N. Vazquez-Laslop, and A. A. Neyfakh.** 1995. Two highly similar multidrug transporters of *Bacillus subtilis* whose expression is differentially regulated. *J. Bacteriol.* **177:** 3904–3910.

3. **Baranova, N. N., and A. A. Neyfakh.** 1997. Apparent involvement of a multidrug transporter in the fluoroquinolone resistance of *Streptococcus pneumoniae*. *Antimicrob. Agents Chemother.* **41:**1396–1398.

4. **Bates, A. D., M. H. O'Dea, and M. Gellert.** 1996. Energy coupling in *Escherichia coli* DNA gyrase: the relationship between nucleotide binding, strand passage, and DNA supercoiling. *Biochemistry* **35:**1408–1416.

5. **Berger, J. M., S. J. Gamblin, S. C. Harrison, and J. C. Wang.** 1996. Structure and mechanism of DNA topoisomerase II. *Nature* **379:**225–232.

6. **Blanche, F., B. Cameron, F. X. Bernard, L. Maton, B. Manse, L. Ferrero, N. Ratet, C. Lecoq, A. Goniot, D. Bisch, and J. Crouzet.** 1996. Differential behaviors of *Staphylococcus aureus* and *Escherichia coli* type II DNA topoisomerases. *Antimicrob. Agents Chemother.* **40:**2714–2720.

7. **Breines, D. M., S. Ouabdesselam, E. Y. Ng, J. Tankovic, S. Shah, C. J. Soussy, and D. C. Hooper.** 1997. Quinolone resistance locus *nfxD* of *Escherichia coli* is a mutant allele of *parE* gene encoding a subunit of topoisomerase IV. *Antimicrob. Agents Chemother.* **41:**175–179.

8. **Brenwald, N. P., M. J. Gill, and R. Wise.** 1997. The effect of reserpine, an inhibitor of multidrug efflux pumps, on the in-vitro susceptibilities of fluoroquinolone-resistant strains of *Streptococcus pneumoniae* to norfloxacin. *J. Antimicrob. Chemother.* **40:**458–460.

9. **Brenwald, N. P., M. J. Gill, and R. Wise.** 1998. Cloning of a novel efflux pump gene associated with fluoroquinolone resistance in *Streptococcus pneumoniae*, abstr. LB-4. *Program Abstr. 38th Intersci. Conf. Antimicrob. Agents Chemother.* American Society for Microbiology, Washington, D.C.

10. **Brenwald, N. P., M. J. Gill, and R. Wise.** 1998. Prevalence of a putative efflux mechanism among fluoroquinolone-resistant clinical isolates of *Streptococcus pneumoniae*. *Antimicrob. Agents Chemother.* **42:**2032–2035.

11. **Cabral, J. H., A. P. Jackson, C. V. Smith, N. Shikotra, A. Maxwell, and R. C. Liddington.** 1997. Crystal structure of the breakage-reunion domain of DNA gyrase. *Nature* **388:**903–906.

12. **Domagala, J. M.** 1994. Structure-activity and structure-side-effect relationships for the quinolone antibacterials. *J. Antimicrob. Chemother.* **33:**685–706.

13. **Ferrero, L., B. Cameron, and J. Crouzet.** 1995. Analysis of *gyrA* and *grlA* mutations in stepwise-selected ciprofloxacin-resistant mutants of *Staphylococcus aureus*. *Antimicrob. Agents Chemother.* **39:**1554–1558.

14. **Ferrero, L., B. Cameron, B. Manse, D. Lagneaux, J. Crouzet, A. Famechon, and F. Blanche.** 1994. Cloning and primary structure of *Staphylococcus aureus* DNA topoisomerase IV: a primary target of fluoroquinolones. *Mol. Microbiol.* **13:**641–653.

15. **Fitzgibbon, J. E., J. F. John, J. L. Delucia, and D. T. Dubin.** 1998. Topoisomerase mutations in trovafloxacin-resistant *Staphylococcus aureus*. *Antimicrob. Agents Chemother.* **42:**2122–2124.

16. **Fournier, B., and D. C. Hooper.** 1998. Effects of mutations in GrlA of topoisomerase IV from *Staphylococcus aureus* on quinolone and coumarin activity. *Antimicrob. Agents Chemother.* **42:**2109–2112.

17. **Fournier, B., and D. C. Hooper.** 1998. Mutations in topoisomerase IV and DNA gyrase of *Staphylococcus aureus*: novel pleiotropic effects on quinolone and coumarin activity. *Antimicrob. Agents Chemother.* **42:**121–128.

18. **Gellert, M.** 1981. DNA topoisomerases. *Annu. Rev. Biochem.* **50:**879–910.

19. **Gellert, M., K. Mizuuchi, M. H. O'Dea, T. Itoh, and J. I. Tomizawa.** 1977. Nalidixic acid resistance: a second genetic character involved in DNA gyrase activity. *Proc. Natl. Acad. Sci. USA* **74:**4772–4776.

20. **Gootz, T. D., R. Zaniewski, S. Haskell, B. Schmieder, J. Tankovic, D. Girard, P. Courvalin, and R. J. Polzer.** 1996. Activity of the new fluoroquinolone trovafloxacin (CP-99,219) against DNA gyrase and topoisomerase IV mutants of *Streptococcus pneumoniae* selected in vitro. *Antimicrob. Agents Chemother.* **40:**2691–2697.

21. **Grinius, L. L., and E. B. Goldberg.** 1994. Bacterial multidrug resistance is due to a single membrane protein which functions as a drug pump. *J. Biol. Chem.* **269:** 29998–30004.

22. **Hane, M. W., and T. H. Wood.** 1969. *Escherichia coli* K-12 mutants resistant to nalidixic acid: genetic mapping and dominance studies. *J. Bacteriol.* **99:**238–241.

23. **Hori, S., Y. Ohshita, Y. Utsui, and K. Hiramatsu.** 1993. Sequential acquisition of norfloxacin and ofloxacin resistance by methicillin-resistant and -susceptible *Staphylococcus aureus*. *Antimicrob. Agents Chemother.* **37:**2278–2284.

24. **Hoshino, K., A. Kitamura, I. Morrissey, K. Sato, J. Kato, and H. Ikeda.** 1994. Comparison of inhibition of

Escherichia coli topoisomerase IV by quinolones with DNA gyrase inhibition. *Antimicrob. Agents Chemother.* **38:** 2623–2627.

25. **Hsieh, P. C., S. A. Siegel, B. Rogers, D. Davis, and K. Lewis.** 1998. Bacteria lacking a multidrug pump: a sensitive tool for drug discovery. *Proc. Natl. Acad. Sci. USA* **95:**6602–6606.

26. **Ito, H., H. Yoshida, M. Bogaki-Shonai, T. Niga, H. Hattori, and S. Nakamura.** 1994. Quinolone resistance mutations in the DNA gyrase *gyrA* and *gyrB* genes of *Staphylococcus aureus. Antimicrob. Agents Chemother.* **38:** 2014–2023.

27. **Janoir, C., V. Zeller, M. D. Kitzis, N. J. Moreau, and L. Gutmann.** 1996. High-level fluoroquinolone resistance in *Streptococcus pneumoniae* requires mutations in *parC* and *gyrA. Antimicrob. Agents Chemother.* **40:**2760–2764.

28. **Kaatz, G. W., and S. M. Seo.** 1995. Inducible NorA-mediated multidrug resistance in *Staphylococcus aureus. Antimicrob. Agents Chemother.* **39:**2650–2655.

29. **Kanematsu, E., T. Deguchi, M. Yasuda, T. Kawamura, Y. Nishino, and Y. Kawada.** 1998. Alterations in the GyrA subunit of DNA gyrase and the ParC subunit of DNA topoisomerase IV associated with quinolone resistance in *Enterococcus faecalis. Antimicrob. Agents Chemother.* **42:**433–435.

30. **Kato, J., Y. Nishimura, R. Imamura, H. Niki, S. Hiraga, and H. Suzuki.** 1990. New topoisomerase essential for chromosome segregation in *E. coli. Cell* **63:**393–404.

31. **Kato, J., H. Suzuki, and H. Ikeda.** 1992. Purification and characterization of DNA topoisomerase IV in *Escherichia coli. J. Biol. Chem.* **267:**25676–25684.

32. **Korten, V., W. M. Huang, and B. E. Murray.** 1994. Analysis by PCR and direct DNA sequencing of *gyrA* mutations associated with fluoroquinolone resistance in *Enterococcus faecalis. Antimicrob. Agents Chemother.* **38:** 2091–2094.

33. **Lewis, K., D. C. Hooper, and M. Ouellette.** 1997. Multidrug resistance pumps provide broad defense. *ASM News* **63:**605–610.

34. **Markham, P. N., M. Ahmed, and A. A. Neyfakh.** 1996. The drug-binding activity of the multidrug-responding transcriptional regulator BmrR resides in its C-terminal domain. *J. Bacteriol.* **178:**1473–1475.

35. **Markham, P. N., and A. A. Neyfakh.** 1996. Inhibition of the multidrug transporter NorA prevents emergence of norfloxacin resistance in *Staphylococcus aureus. Antimicrob. Agents Chemother.* **40:**2673–2674.

36. **Martínez-Martínez, L., A. Pascual, and G. A. Jacoby.** 1998. Quinolone resistance from a transferable plasmid. *Lancet* **351:**797–799.

37. **Mitscher, L. A., P. Devasthale, and R. Zavod.** 1993. Structure-activity relationships, p. 3–51. In D. C. Hooper and J. S. Wolfson (ed.), *Quinolone Antimicrobial Agents.* American Society for Microbiology, Washington, D.C.

38. **Munoz, R., and A. G. de la Campa.** 1996. ParC subunit of DNA topoisomerase IV of *Streptococcus pneumoniae* is a primary target of fluoroquinolones and cooperates with DNA gyrase A subunit in forming resistance phenotype. *Antimicrob. Agents Chemother.* **40:**2252–2257.

39. **Nakanishi, N., S. Yoshida, H. Wakebe, M. Inoue, and S. Mitsuhashi.** 1991. Mechanisms of clinical resistance to fluoroquinolones in *Enterococcus faecalis. Antimicrob. Agents Chemother.* **35:**1053–1059.

40. **Neyfakh, A. A.** 1992. The multidrug efflux transporter of *Bacillus subtilis* is a structural and functional homolog of the *Staphylococcus* NorA protein. *Antimicrob. Agents Chemother.* **36:**484–485.

41. **Neyfakh, A. A., C. M. Borsch, and G. W. Kaatz.** 1993. Fluoroquinolone resistance protein NorA of *Staphylococcus aureus* is a multidrug efflux transporter. *Antimicrob. Agents Chemother.* **37:**128–129.

42. **Ng, E. Y., M. Trucksis, and D. C. Hooper.** 1994. Quinolone resistance mediated by *norA*: physiologic characterization and relationship to *flqB*, a quinolone resistance locus on the *Staphylococcus aureus* chromosome. *Antimicrob. Agents Chemother.* **38:**1345–1355.

43. **Ng, E. Y., M. Trucksis, and D. C. Hooper.** 1996. Quinolone resistance mutations in topoisomerase IV: relationship of the *flqA* locus and genetic evidence that topoisomerase IV is the primary target and DNA gyrase the secondary target of fluoroquinolones in *Staphylococcus aureus. Antimicrob. Agents Chemother.* **40:**1881–1888.

44. **Nikaido, H., and D. G. Thanassi.** 1993. Penetration of lipophilic agents with multiple protonation sites into bacterial cells: tetracyclines and fluoroquinolones as examples. *Antimicrob. Agents Chemother.* **37:**1393–1399.

45. **Ohshita, Y., K. Hiramatsu, and T. Yokota.** 1990. A point mutation in *norA* gene is responsible for quinolone resistance in *Staphylococcus aureus. Biochem. Biophys. Res. Commun.* **172:**1028–1034.

46. **Pan, X. S., J. Ambler, S. Mehtar, and L. M. Fisher.** 1996. Involvement of topoisomerase IV and DNA gyrase as ciprofloxacin targets in *Streptococcus pneumoniae. Antimicrob. Agents Chemother.* **40:**2321–2326.

47. **Pan, X. S., and L. M. Fisher.** 1996. Cloning and characterization of the *parC* and *parE* genes of *Streptococcus pneumoniae* encoding DNA topoisomerase IV: role in fluoroquinolone resistance. *J. Bacteriol.* **178:**4060–4069.

48. **Pan, X. S., and L. M. Fisher.** 1997. Targeting of DNA gyrase in *Streptococcus pneumoniae* by sparfloxacin: selective targeting of gyrase or topoisomerase IV by quinolones. *Antimicrob. Agents Chemother.* **41:**471–474.

49. **Paulsen, I. T., M. H. Brown, T. G. Littlejohn, B. A. Mitchell, and R. A. Skurray.** 1996. Multidrug resistance proteins QacA and QacB from *Staphylococcus aureus*: Membrane topology and identification of residues involved in substrate specificity. *Proc. Natl. Acad. Sci. USA* **93:**3630–3635.

50. **Perichon, B., J. Tankovic, and P. Courvalin.** 1997. Characterization of a mutation in the *parE* gene that confers fluoroquinolone resistance in *Streptococcus pneumoniae. Antimicrob. Agents Chemother.* **41:**1166–1167.

51. **Schmitz, F. J., B. Hofmann, B. Hansen, S. Scheuring, M. Lückefahr, M. Klootwijk, J. Verhoef, A. Fluit, H. P. Heinz, K. Köhrer, and M. E. Jones.** 1998. Relationship between ciprofloxacin, ofloxacin, levofloxacin, sparfloxacin and moxifloxacin (BAY 12-8039) MICs and mutations in *grlA, grlB, gyrA* and *gyrB* in 116 unrelated clinical isolates of *Staphylococcus aureus. J. Antimicrob. Chemother.* **41:**481–484.

52. **Schmitz, F. J., M. E. Jones, B. Hofmann, B. Hansen, S. Scheuring, M. F. A. Lückefahr, J. Verhoef, U. Hadding, H. P. Heinz, and K. Köhrer.** 1998. Characterization of *grlA, grlB, gyrA,* and *gyrB* mutations in 116 unrelated isolates of *Staphylococcus aureus* and effects of mutations on

ciprofloxacin MIC. *Antimicrob. Agents Chemother.* **42:** 1249–1252.

53. **Shen, L. L., W. E. Kohlbrenner, D. Weigl, and J. Baranowski.** 1989. Mechanism of quinolone inhibition of DNA gyrase. Appearance of unique norfloxacin binding sites in enzyme-DNA complexes. *J. Biol. Chem.* **264:** 2973–2978.

54. **Sreedharan, S., M. Oram, B. Jensen, L. R. Peterson, and L. M. Fisher.** 1990. DNA gyrase *gyrA* mutations in ciprofloxacin-resistant strains of *Staphylococcus aureus*: close similarity with quinolone resistance mutations in *Escherichia coli. J. Bacteriol.* **172:**7260–7262.

55. **Sugino, A., C. L. Peebles, K. N. Kreuzer, and N. R. Cozzarelli.** 1977. Mechanism of action of nalidixic acid: purification of *Escherichia coli nalA* gene product and its relationship to DNA gyrase and a novel nicking-closing enzyme. *Proc. Natl. Acad. Sci. USA* **74:**4767–4771.

56. **Sun, L., S. Sreedharan, K. Plummer, and L. M. Fisher.** 1996. NorA plasmid resistance to fluoroquinolones: role of copy number and *norA* frameshift mutations. *Antimicrob. Agents Chemother.* **40:**1665–1669.

57. **Takahata, M., M. Yonezawa, S. Kurose, N. Futakuchi, N. Matsubara, Y. Watanabe, and H. Narita.** 1996. Mutations in the *gyrA* and *grlA* genes of quinolone-resistant clinical isolates of methicillin-resistant *Staphylococcus aureus. J. Antimicrob. Chemother.* **38:**543–546.

58. **Takenouchi, T., C. Ishii, M. Sugawara, Y. Tokue, and S. Ohya.** 1995. Incidence of various *gyrA* mutants in 451 *Staphylococcus aureus* strains isolated in Japan and their susceptibilities to 10 fluoroquinolones. *Antimicrob. Agents Chemother.* **39:**1414–1418.

59. **Tankovic, J., B. Perichon, J. Duval, and P. Courvalin.** 1996. Contribution of mutations in *gyrA* and *parC* genes to fluoroquinolone resistance of mutants of *Streptococcus pneumoniae* obtained in vivo and in vitro. *Antimicrob. Agents Chemother.* **40:**2505–2510.

60. **Tankovic, J., F. Mahjoubi, P. Courvalin, J. Duval, and R. Leclercq.** 1996. Development of fluoroquinolone resistance in *Enterococcus faecalis* and role of mutations in the DNA gyrase *gyrA* gene. *Antimicrob. Agents Chemother.* **40:** 2558–2561.

61. **Trucksis, M., J. S. Wolfson, and D. C. Hooper.** 1991. A novel locus conferring fluoroquinolone resistance in *Staphylococcus aureus. J. Bacteriol.* **173:**5854–5860.

62. **Wang, J. C.** 1996. DNA topoisomerases. *Annu. Rev. Biochem.* **65:**635–692.

63. **Wang, T., M. Tanaka, and K. Sato.** 1998. Detection of *grlA* and *gyrA* mutations in 344 *Staphylococcus aureus* strains. *Antimicrob. Agents Chemother.* **42:**236–240.

64. **Wasserman, R. A., and J. C. Wang.** 1994. Mechanistic studies of amsacrine-resistant derivatives of DNA topoisomerase II. Implications in resistance to multiple antitumor drugs targeting the enzyme. *J. Biol. Chem.* **269:** 20943–20951.

65. **Wetzstein, H. G., N. Schmeer, and W. Karl.** 1997. Degradation of the fluoroquinolone enrofloxacin by the brown rot fungus *Gloeophyllum striatum*: identification of metabolites. *Appl. Environ. Microbiol.* **63:**4272–4281.

66. **Wilkinson, B. J.** 1997. Biology, p. 1–38. *In* K. B. Crossley and G. L. Archer (ed.), *The Staphylococci in Human Disease.* Churchill Livingstone, Ltd., New York, N.Y.

67. **Willmott, C. J., S. E. Critchlow, I. C. Eperon, and A. Maxwell.** 1994. The complex of DNA gyrase and quinolone drugs with DNA forms a barrier to transcription by RNA polymerase. *J. Mol. Biol.* **242:**351–363.

68. **Woolridge, D. P., N. Vazquez-Laslop, P. N. Markham, M. S. Chevalier, E. W. Gerner, and A. A. Neyfakh.** 1997. Efflux of the natural polyamine spermidine facilitated by the *Bacillus subtilis* multidrug transporter Blt. *J. Biol. Chem.* **272:**8864–8866.

69. **Yamagishi, J. I., T. Kojima, Y. Oyamada, K. Fujimoto, H. Hattori, S. Nakamura, and M. Inoue.** 1996. Alterations in the DNA topoisomerase IV *grlA* gene responsible for quinolone resistance in *Staphylococcus aureus. Antimicrob. Agents Chemother.* **40:**1157–1163.

70. **Yoshida, H., M. Bogaki, S. Nakamura, K. Ubukata, and M. Konno.** 1990. Nucleotide sequence and characterization of the *Staphylococcus aureus norA* gene, which confers resistance to quinolones. *J. Bacteriol.* **172:**6942–6949.

71. **Zechiedrich, E. L., and N. R. Cozzarelli.** 1995. Roles of topoisomerase IV and DNA gyrase in DNA unlinking during replication in *Escherichia coli. Genes Dev.* **9:**2859–2869.

Resistance to the Macrolide-Lincosamide-Streptogramin Antibiotics

BERNARD WEISBLUM

69

Biochemical mechanisms that mediate antibiotic resistance can be grouped under four main headings: (i) modification of the antibiotic, (ii) modification of the receptor, (iii) reduction of intracellular concentration of the antibiotic, and (iv) acquisition of a bypass enzyme that is not inhibited by the antibiotic. Genetic mechanisms that regulate antibiotic resistance can be classified as either negative- or positive-acting. The negative-acting mechanisms include (i) transcriptional repression, (ii) transcriptional attenuation, (iii) translational repression, and (iv) translational attenuation. Some forms of resistance are expressed constitutively, while others are inducible. The positive-acting mechanisms include (i) signal transduction via a histidine kinase-aspartyl phosphate cascade and (ii) the LysR family of helix-turn-helix DNA-binding proteins. Gene expression can also be regulated at the level of DNA replication through the selective amplification of genes that confer resistance or of enhanced replication of plasmids that carry them.

In this chapter, where possible, resistance is discussed in association with genes for which sequence information is available. A list of such genes is given in Table 1.

The macrolide-lincosamide-streptogramin B antibiotics constitute the MLS superfamily. As a group, they inhibit bacterial protein synthesis by their action on the 50S ribosomal subunit. Their site of action is close to the peptidyl transferase center of the ribosome, which is affected by each antibiotic in a distinguishably different manner. Erythromycin belongs to the group of antibiotics called macrolides, which in the strict sense are lactones with ring sizes of 12, 14, or 16 atoms. Erythromycin belongs to the 14-membered ring subgroup. The site and mode of erythromycin action on the 50S ribosome subunit overlap those of other macrolides, as well as those of the chemically distinct lincosamides and streptogramin B antibiotics. Thus, the study of resistance to erythromycin must also consider the manner and extent to which the lincosamide and streptogramin B families are involved. With respect to inhibitory action, the streptogramin A and B families interact synergistically. The structures of erythromycin and tylosin, which are 14- and 16-membered ring macrolides, respectively, together with those of lincomycin and streptogramins A and B are shown in Fig. 1. For an extensive compilation of the numerous macrolide antibiotics, including their classification, structure, and activity relationships, see Bryskier et al. (22). Streptogramins have been reviewed by Barriere et al. (11).

In the strict sense, MLS resistance denotes the original form of resistance to all macrolides, lincosamides, and streptogramin type B antibiotics, which is based on N^6 dimethylation of A2058 in 23S rRNA. Since the original description of MLS resistance proper, several forms of resistance to various MLS subset combinations have been reported. Moreover, the streptogramin A family, with a set of structures distinctly different from those of the streptogramin B family, also exerts its inhibitory action on the 50S subunit of the ribosome and synergizes with the inhibitory action of the streptogramin B family. The streptogramin A family will therefore be considered in this discussion because of its close connection with the streptogramin B family.

Forms of resistance in addition to the classic MLS type include mechanisms that involve only single classes or subclasses, such as (i) 14- but not 16-membered ring macrolides; (ii) 14- and 16-membered ring macrolides; (iii) lincosamides only; (iv) streptogramin B only; (v) streptogramin A only. Additional forms include (vi) resistance to two classes, namely, both 14-membered ring macrolides and streptogramin B antibiotics, and (vii) resistance to one or more of the MLS antibiotics by a transporter that has a broad spectrum of action extending to a wide range of structurally and functionally unrelated molecules. The range of different forms of resistance to MLS antibiotic subsets is based on diverse biochemical mechanisms other than posttranscriptional modification of their 23S rRNA target site. This chapter covers the diversity of resistance phenotypes, their biochemical basis, and mechanisms of genetic regulation that control their expression.

Gram-Positive Pathogens, ed. by V. A. Fischetti et al.
© 2000 American Society for Microbiology, Washington, D.C.

TABLE 1 Genes that mediate antibiotic resistance to macrolides, lincosamides, and streptogramins

Gene	Antibiotic specificity[a]	Enzymatic activity	Organism	Reference
SatA	Sa	Acetylase	*Enterococcus faecium*	99
vat	Sa	Acetylase	*Staphylococcus aureus*	7
vatB	Sa	Acetylase	*Staphylococcus aureus*	2
vatC	Sa	Acetylase	*Staphylococcus cohnii*	4
carA,B	MLSb	Combined	*Streptomyces thermotolerans*	39, 111
lmrA,B	MLSb	Combined	*Streptomyces lincolnensis*	139
myrA,B	MLSb	Combined	*Micromonospora giriseorubida*	55
srmA,b,C,D	MLSb	Combined	*Streptomyces ambofaciens*	96, 100, 111
tlrA,B,C,D	MLSb	Combined	*Streptomyces fradiae*	16, 137, 138
carA	M	Efflux ABC	*Streptomyces thermotolerans*	39
msrA	MSb	Efflux ABC	*Staphylococcus epidermidis*	102–105
oleB	M	Efflux ABC	*Streptomyces antibioticus*	8
srmB	M	Efflux ABC	*Streptomyces ambofaciens*	100
tlrC	M	Efflux ABC	*Streptomyces fradiae*	106
vga	Sb	Efflux ABC	*Staphylococcus aureus*	5
vgaB	Sb	Efflux ABC	*Staphylococcus aureus*	4
lmrA	L	Efflux pmf	*Streptomyces lincolnensis*	139
mefA	M	Efflux pmf	*Streptococcus pyogenes*	28
mefE	M	Efflux pmf	*Streptococcus pneumoniae*	120
mreA	M	Efflux pmf	*Streptococcus agalactiae*	27
acrAB	M++	Efflux pmf/complex	*Escherichia coli*	73
mdfA	M++	Efflux pmf/complex	*Escherichia coli*	37
mtrCDE	M++	Efflux pmf/complex	*Neisseria gonorrhoeae*	46
mgt	M	Glycosyltransferase	*Streptomyces lividans*	30, 129
ereA	M	Lactonase	*Escherichia coli*	94
ereA or ereB	M	Lactonase	*Staphylococcus aureus*	134
ereB	M	Lactonase	*Escherichia coli*	10
vgb	Sb	Lactonase	*Staphylococcus aureus*	6
vgbB	Sb	Lactonase	*Staphylococcus cohnii*	4
ermA	MLSb	Methyltransferase	*Staphylococcus aureus*	87
ermAM	MLSb	Methyltransferase	*Streptococcus sanguis*	52
ermAMR	MLSb	Methyltransferase	*Enterococcus faecalis*	91
ermB	MLSb	Methyltransferase	*Staphylococcus aureus*	36, 114
ermB-like	MLSb	Methyltransferase	*Streptococcus faecalis*	114
ermBC	MLSb	Methyltransferase	*Escherichia coli*	19
ermBP	MLSb	Methyltransferase	*Clostridium perfringens*	14
ermC	MLSb	Methyltransferase	*Staphylococcus aureus*	43, 53
ermCD	MLSb	Methyltransferase	*Corynebacterium diphtheriae*	51, 113
ermD	MLSb	Methyltransferase	*Bacillus licheniformis*	44
ermE	MLSb	Methyltransferase	*Streptomyces erythraeus*	15, 124, 125
ermF	MLSb	Methyltransferase	*Bacteroides fragilis*	98
ermFS	MLSb	Methyltransferase	*Bacteroides fragilis*	117
ermFU	MLSb	Methyltransferase	*Bacteroides fragilis*	47
ermG	MLSb	Methyltransferase	*Bacillus sphaericus*	86
ermGT	MLSb	Methyltransferase	*Lactobacillus reuteri*	121
ermIM	MLSb	Methyltransferase	*Bacillus subtilis*	85
ermJ	MLSb	Methyltransferase	*Bacillus anthracis*	63
ermK	MLSb	Methyltransferase	*Bacillus licheniformis*	68
ermM	MLSb	Methyltransferase	*Staphylococcus epidermidis*	71
ermP	MLSb	Methyltransferase	*Clostridium perfringens*	13
ermR	MLSb	Methyltransferase	*Arthrobacter luteus*	101
ermSF	MLSb	Methyltransferase	*Streptomyces fradiae*	60
ermTR	MLSb	Methyltransferase	*Streptococcus pyogenes*	112
ermZ	MLSb	Methyltransferase	*Clostridium difficile*	45
mdmA	MLSb	Methyltransferase	*Streptomyces mycarofaciens*	48
clr	MLSb	Methyltransferase, mono	*Streptomyces caelestis*	24
lmrB	MLSb	Methyltransferase, mono	*Streptomyces lincolnensis*	139
lrm	MLSb	Methyltransferase, mono	*Streptomyces lividans*	58
linA	L	Nucleotidyltransferase	*Staphylococcus haemolyticus*	20
linA'	L	Nucleotidyltransferase	*Staphylococcus aureus*	20
mphA	M	Phosphotransferase	*Escherichia coli*	92
mphB	M	Phosphotransferase	*Staphylococcus aureus*	65, 89
mphK	M	Phosphotransferase	*Escherichia coli*	64

[a] ABC, ATP-binding cassette; L, lincosamide; M, macrolide, M++, macrolides plus other, chemically unrelated antibiotics; Sa, streptogramin A; Sb, streptogramin B.

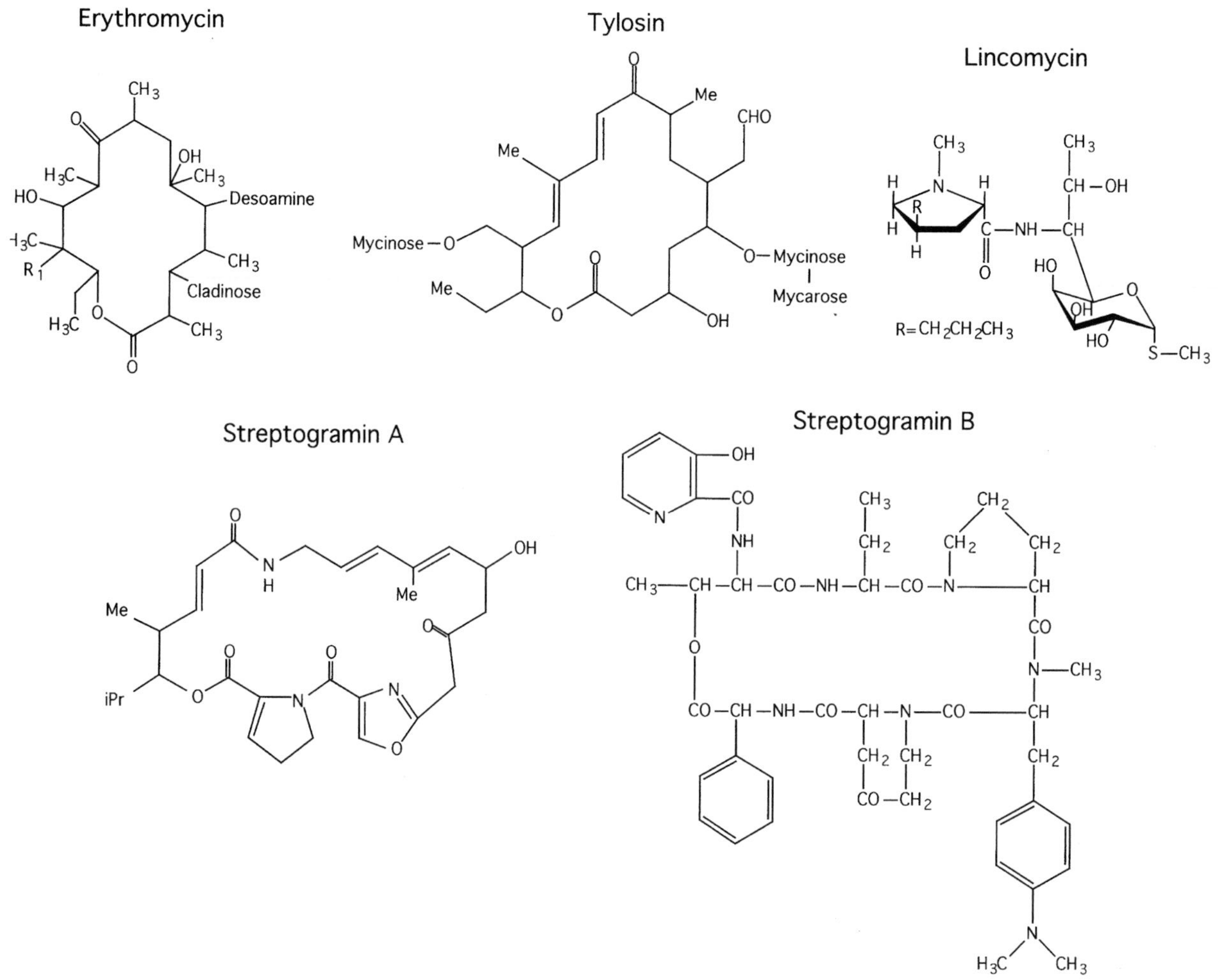

FIGURE 1 The MLS superfamily of antibiotics. Chemical structures of representative examples of the MLS antibiotics are shown: erythromycin, a 14-membered ring macrolide; tylosin, a 16-membered ring macrolide; lincomycin, a lincosamide; streptogramin A; and streptogramin B.

BIOCHEMICAL MECHANISMS OF RESISTANCE TO THE MACROLIDE-LINCOSAMIDE-STREPTOGRAMIN ANTIBIOTICS

Enzymatic Modification of the Antibiotic

Forms of antibiotic modification associated with acquired resistance include the following: (i) Macrolide ring esteratic cleavage (lactonase), mediated by *ereA* or *ereB*, which is limited to erythromycin and excludes 16-membered ring macrolides, has been reported in clinical isolates of *Escherichia coli* (9, 94). *ere*-like activity has also been reported in clinically resistant strains of *Staphylococcus aureus* (134). In contrast to the *E. coli* isolates, whose action was restricted to 14-membered ring macrolides, an *S. aureus* isolate was found capable of hydrolyzing both 14- and 16-membered ring macrolides. (ii) Macrolide glycosylation mediated by *mgt* has been observed in *Streptomyces lividans* (30, 129). (iii) Macrolide phosphorylation, mediated by the closely related *mphA* (92), *mphB* (65, 89), and *mphK* (64), has been observed in *E. coli*. *mphA* was found capable of phosphorylating only 14-membered ring macrolides, whereas *mphB* was found capable of phosphorylating both 14- and 16-membered ring macrolides. *mphA* differs from *mphK* at only five amino acid positions. (iv) Lincosamide-nucleotide adduct formation, mediated by *linA* and *linA'*, has been detected in clinical isolates of *Staphylococcus haemolyticus* BM4610 and *S. aureus* BM4611, respectively (20). Both LinA and LinA' modify lincomycin and clindamycin. Lincomycin is modified in the 3 position of the sugar moiety, whereas clindamycin is modified in the 4 position. Any of the four ribonucleoside triphosphates can serve as the 5'-monophosphoryl nucleotide donor to either of these sites. (v) Streptogramin A acetylation, mediated by *satA*, has been found in *Enterococcus faecium*, by *vat* (7) and *vatB* (2) in *S. aureus*, and by *vatC* in *Staphylococcus cohnii* (4). (vi) Streptogramin B ring esteratic cleavage (lactonase), mediated by *vgbB*, has been described (4) as a mechanism of streptogramin B inactivation by a clinical isolate of *S. cohnii*, which also carries *vatC*, making it resistant to both streptogramin A and B antibiotics.

Since glycosylation of biosynthetic precursors plays an important role in the biosynthesis of macrolides, it would be interesting to learn whether catalytic and recognition elements present in Mgt are related structurally to their counterparts in the enzymes that synthesize macrolides. Enzymatic mechanisms that modify macrolides appear to be limited to a relatively small number of clinical isolates, in comparison to the more commonly reported target site modification and efflux-based resistant isolates. The reasons that favor the emergence of target site modification and antibiotic efflux as the predominant phenotypes are not known. When these latter mechanisms occur in macrolide-producing strains, they are found in association, as in the tylosin resistance *tlrA,B,C,D* cluster in *Streptomyces fradiae* (16) or in the spiramycin resistance *srmA,B,C,D* cluster in *Streptomyces ambofaciens* (100). For reviews see references 131 and 132.

Modification of the Target Site: 23S rRNA

On the basis of its secondary structure, 23S rRNA has been subdivided into six domains (90). In *E. coli*, mutational sequence alterations in two of these domains, namely, II and V, have been implicated in resistance to erythromycin (34). In *S. aureus* and other pathogens, posttranscriptional modification involving dimethylation of A2058 located in domain V is seen. This methylation occurs posttranscriptionally at a site in the peptidyl transferase circle of 23S rRNA domain V, which corresponds to A2058, based on the *E. coli* numbering scheme (12, 70, 116). Modifications at A2058 can be due either to acquisition of a gene for adenine-N^6-methyltransferase, whose activity methylates A2058 posttranscriptionally, or by mutation of A2058 to any of the nucleotides. Both mechanisms result in ribosomes that bind MLS antibiotics with greatly reduced affinity. For reviews see references 131 and 132.

Posttranscriptional Methylation of Domain V (MLS Resistance)

The family of enzymes that catalyze posttranscriptional methylation of domain V are designated by the acronym Erm, which stands for erythromycin resistance methylase. As discussed in further detail below, ErmAM and ErmC contain an N-terminal domain that shares structural homology with *N*- and *O*-methyltransferases that act on other target molecules. The remaining one-third, which constitutes the C-terminal portion of ErmAM and ErmC, shows homology only with other Erm methylases and is therefore presumed to contain peptide sequences that recognize the substrate.

For a compilation of Erm isoforms, see Table 1. In its simplest form, MLS resistance is inducible by erythromycin (a 14-membered ring macrolide) but not by spiramycin (a 16-membered ring macrolide), lincomycin (a lincosamide), or pristinamycin I (a streptogramin B antibiotic). Dimethylation of A2058 leads to high-level resistance to all MLS antibiotics. There are exceptions to each of these attributes of inducible MLS resistance, as discussed below.

The Erm isoform, *lrm*, from *S. lividans*, monomethylates rather than dimethylates A2058, thereby conferring high-level resistance to lincomycin and low-level resistance to macrolides (58). The same holds for *lmrB* from *Streptomyces lincolnensis* (139) and *clr* from *Streptomyces caelestis* (24). In *S. lincolnensis*, which produces lincomycin, the newly synthesized lincomycin may act to autoinduce resistance to higher levels. The mechanism of regulation of the inducible form, which is discussed in detail below, is either translational attenuation, as in the case of *ermC* (43, 53), or transcriptional attenuation, as in the case of *ermK* (26, 68).

Mutations in Domain V

A second, related form of 23S rRNA alteration, one that produces intrinsic MLS resistance, is based on a mutational alteration of 23S rRNA at A2058, the same site at which posttranscriptional methylation confers traditional MLS resistance. This form of intrinsic resistance was first reported in clinical isolates of *Mycobacterium intracellulare* 23S rRNA (82) and has been attributed to G, C, or U at A2058. A noteworthy addition to the list of mutations that confer resistance is the changes at three successive positions, G2057A, A2058G, and A2059G, respectively, within the peptidyl transferase circle, which were identified in a collection of *Propionibacterium* spp. isolates by Ross et al. (104). For a more extensive listing of erythromycin-resistant mutants with alterations in 23S rRNA domain V, see reference 131. The susceptibility of inducible strains to noninducing MLS antibiotics in the clinical setting proved to be illusory because of the ease with which noninducing MLS antibiotics were able to select constitutively resistant mutants, leading to relapse (32, 130). For reviews, see references 72, 81, 131, and 132.

Posttranscriptional methylation of adenine was the first of the two types of 23S rRNA modifications to be discovered because of its occurrence in easily cultivated, rapidly growing isolates of *Staphylococcus* spp. and *Streptococcus* spp. Mutational alteration of A2058 can be expressed only in a background that contains a small number of 23S rRNA structural genes, e.g., *Mycobacterium* spp., *Helicobacter pylori*, *Mycoplasma* spp., *S. ambofaciens*, and *Propionibacterium* spp. A relatively low 23S rRNA gene copy number appears to favor the selection of such mutants in clinical use because of the inverse relation between dominance of such mutant genes and their copy number. For "low" rRNA copy numbers greater than one, e.g., *H. pylori*, which has two (127), or *S. ambofaciens*, which has four (96, 97), the emergence of resistance suggests that penetrance of as low as 25% might suffice to confer phenotypic resistance. An alternative explanation would be that the presence of a single resistance allele could serve as an entry-level modification that confers only a low level of resistance. As a result of internal recombination (gene conversion), replacement of sensitive alleles by resistant counterparts would increase the copy number of the resistant form.

Mutations in Domain II

Mutations in 23S rRNA have also been found between coordinates 1198 and 1247, located in domain II (31, 122). This sequence contains a 19-bp hairpin and loop located immediately upstream of an open reading frame that encodes a potential peptide MSLKV, "E-peptide" (122). The hairpin sequence contains a Shine-Dalgarno sequence, GGAGG. According to the model proposed, the short segment of domain II 23S rRNA functions as messenger for the E-peptide (123). The Shine-Dalgarno sequence is positioned upstream of the E-peptide open reading frame and would be expected, based on secondary structure predictions, to be sequestered and therefore inactive. Of the different kinds of domain II mutants that are theoretically possible, one kind would be expected to destabilize the secondary structure of domain II in a way that frees the Shine-Dalgarno sequence that is associated with the E-peptide open reading frame to initiate E-peptide synthesis more efficiently.

Translation of the E-peptide as a requirement for erythromycin resistance is corroborated by the observation that amber mutations in the E-peptide coding region eliminate erythromycin resistance, and suppression of these mutations restores it. The E-peptide also confers erythromycin resistance if it is expressed in *trans*, independent of its 23S rRNA context. In combinatorial modeling studies (123), it was determined that only a short peptide, three to six amino acid residues in length, was needed for E-peptide function. Attempts to model the E-peptide by combinatorial techniques with a five-amino-acid-peptide library indicated a preference for Ile or Leu at position 3 and for a hydrophobic amino acid at position 5.

Exogenously added, chemically synthesized E-peptide, at concentrations up to 1 mM, failed to confer erythromycin resistance. The high concentration (>1 mM) of E-peptide that would presumably be required to confer resistance, and the failure to detect E-peptide (limits not stated) in cell extracts, suggest that E-peptide may act optimally only while in the nascent state and associated with its birth ribosome. A question posed by this intriguing system is how can nascent E-peptide attached to tRNA ever confer resistance in vivo for the synthesis of anything other than E-peptide? Since ribosomes must synthesize E-peptide in order to become erythromycin resistant, how can they synthesize anything else in an erythromycin-resistant fashion if they can synthesize only one product at a time and that product is needed to confer resistance? A possible way in which E-peptide exerts its activity, therefore, is by an intramolecular reaction in which it migrates and attaches to an "out of the way" ribosomal site that does not preclude the synthesis of other proteins. The E-peptide data may represent an example of a cotranslational effect of a protein on its own synthesis while in the nascent state. The genetic data are compelling; however, the full appreciation of the activities of the E-peptide data will have to await confirmatory results from systems reconstituted in vitro.

Antibiotic Efflux Transporters

Transporters that mediate antibiotic efflux can be distinguished functionally according to whether they are driven by (i) proton-motive force (proton/antibiotic antiporter) or (ii) ATP hydrolysis (ATP-binding cassette [ABC] transporter). Transporters can also be distinguished according to whether they consist of (i) a single, integral membrane multidrug efflux protein, called the major facilitator with 4-, 12-, or 14-transmembrane domains, or (ii) multiple components.

The most complex transporters occur in gram-negative organisms. They are tripartite, multicomponent, and contain (i) a major facilitator, multidrug efflux protein, located in the inner cell membrane; (ii) a membrane fusion protein (MFP), located largely in the periplasmic space, as an extension of the major facilitator; and (iii) a porin, located in the outer cell membrane. The MFP is believed to project an extension into the cell membrane. For a review that covers specific examples of the different classes of transporters, see reference 73. Reviews that deal with the classes of multidrug efflux proteins and with the MFP group, separately, have been compiled (33, 95). It is important to distinguish between the tripartite multicomponent transporters, as described above, and a class of multicomponent ABC transporters in which the ATP-binding and transmembrane domains occur as distinct separate polypeptides. See reference 49 for a review of the ABC transporters.

Examples of macrolide efflux transporters include (i) M-type, (ii) L-type, (iii) S-type, (iv) MS-type, (v) actinomycete-type, and (vi) broad-spectrum, multi-drug-resistance (MDR)-type. This listing of transporters is based on the increasing diversity of antibiotics and other substances that are transported by members belonging to each class. Based

on the reviews cited above, there appears to be no connection between the diversity of a group of transported compounds and the structural complexity of the transporter. In published studies of M-, S-, and MS-type efflux transporters that are of clinical interest, specificity with respect to a wide range of MLS antibiotics is emphasized in order to identify alternative candidates for medical use. In contrast, studies leading to the discovery of actinomycete transporters (cited below) focus on the transport of the particular antibiotic produced by a *Streptomyces* strain under investigation, rather than on a wide range of unrelated antibiotics. Last, studies of efflux by MDR-type transporters (cited below) note erythromycin simply as an addition to a list of chemically dissimilar compounds whose efflux is mediated by the transporter.

M-Type Efflux

M-type efflux-based resistance was defined in the studies (118) in which it was reported that 75% (21 of 28) of erythromycin-resistant *Streptococcus pyogenes* strains and 85% (56 of 66) of erythromycin-resistant *Streptococcus pneumoniae* strains had the "M phenotype," which they characterized as resistance to 14- and 15-membered macrolides, and susceptibility to clindamycin and streptogramin B antibiotics. The respective genes, *mefA* from *S. pyogenes* (27) and *mefE* from *S. pneumoniae* (120), were cloned, functionally expressed, and sequenced. Comparison of their respective deduced amino acid sequences indicated that the two genes were 90% identical. Further analysis of their respective amino acid sequences showed the presence of 12 transmembrane domains. In the absence of the finding of an ABC sequence, we assume that MefA and MefE are driven by the proton-motive force. For the 19 MFPs for which sequence data were available, no member was identified in a gram-positive bacterium, archebacterium, or eukaryote (33). This would suggest that the MefA and MefE transporters belong to the simple category and that they facilitate efflux by means of a single polypeptide.

L-Type Efflux

Resistance to lincomycin/clindamycin mediated by *lmrA* in *S. lincolnensis* (33, 139) has been reported. LmrA has 12 membrane-spanning domains and resembles other proton/antibiotic antiporters. In contrast, the transporters found in macrolide-producing *Streptomyces* spp. belong to the ABC-based efflux ATPases (Table 1).

S-Type Efflux

Resistance to the streptogramin A family in *S. aureus* mediated by Vga and VgaB has been described (3, 5). Vga and VgaB share homology with other ATP-binding proteins and by inference can be classified as belonging to the ABC family of transporters. Efflux-based resistance to the streptogramin B antibiotics has apparently been described only as a part of MS-type efflux (described below).

MS-Type Efflux

A group of MS-type *Staphylococcus epidermidis* strains named MS has been described (102, 103, 105). These strains were reported to be resistant to 14-membered ring macrolides and streptogramin B but sensitive to 16-membered ring macrolides and lincosamides. A cloned 1.9-kb DNA sequence from one of their isolates contained the gene *msrA*, which conferred resistance, and its sequence revealed an open reading frame that encodes a 488-amino-acid protein whose regulation is mediated by translational attenuation (103), a mechanism of regulation similar to that which regulates *ermC*, as described below (Fig. 2).

Based on the similarity between the amino acid sequence of the 488-amino-acid MS transporter and the sequences of ABC transporters, Ross et al. (103) inferred that the MS transporter mediates the efflux of erythromycin and streptogramin B antibiotics with energy from the hydrolysis of ATP. The apparent absence of MFPs in gram-positive organisms would suggest that the MsrA transporters belong to the simple category of ABC transporters that facilitate efflux by a single polypeptide; however, hydrophobic transmembrane domains were apparently absent in MsrA, and it was postulated, therefore, that MsrA was only one member of an ABC transporter complex in which the transmembrane domains were located on a separate polypeptide (102, 103). Candidates for the auxiliary genes, *stpC* and *smpC*, encoded by the S. aureus genome, were tested by allele replacement-inactivation to determine whether their function was necessary for antibiotic efflux associated with MsrA. It was found that MsrA conferred MS resistance despite allele replacement of *stpC* and *smpC*. From these observations, it was concluded that S. aureus may have other auxiliary transmembrane proteins that can serve MsrA in the absence of *stpC* and *smpC*.

Phenotypically, resistance to 16-membered ring macrolides or clindamycin was not seen in MS strains, indicating that these antibiotics are not transported. Resistance to both erythromycin and streptogramin B was inducible by erythromycin, suggesting that erythromycin also induced resistance to itself, and that growth seen on solid medium in the presence of erythromycin was self-induced.

The MS-resistant strain of S. aureus that was reported by Jánosi et al. (57) was described as coresistant to erythromycin and streptogramin B but sensitive to clindamycin. Matsuoka et al. (77) showed that the N-terminal MS-resistance sequence of the MS-resistant S. aureus strain is identical to that of streptococcal MsrA to the extent of 31 amino acids, suggesting that the two transporters are closely related. Further studies (76, 88) showed that MS-type resistance previously described in S. epidermidis was also present in S. aureus, where it also conferred resistance to mycinamycin, a 16-membered ring macrolide. The MS transporter in S. aureus, therefore, appears also to transport both 14- and 16-membered ring macrolides and streptogramin B antibiotics. The strain studied by Matsuoka et al. (76) also carried an MLS determinant. Resistance based on the simultaneous presence of both antibiotic efflux and rRNA methylation resembles the multi-drug resistance seen in macrolide-producing actinomycetes (16, 100).

Actinomycete-Type Efflux

The genes *thrC* from S. fradiae (106), *carA* from Streptomyces thermotolerans (39), and *srmB* from S. ambofaciens (100) are all members of Streptomyces gene clusters that confer resistance to the 16-membered ring macrolide antibiotics they produce, namely, tylosin, carbomycin, and spiramycin, respectively. The three actinomycete genes were sequenced and their deduced amino acid sequences analyzed by Schoner et al. (111), who identified and aligned ATP-binding motifs resembling those found in ABC transporters. Last, Aparichio et al. (8) reported a gene, *oleB*, which specifies an efflux ATPase specific for 14-membered ring macrolides, in a Streptomyces antibioticus strain that produces oleandomycin. Whether *oleB* also

transports and confers resistance to 16-membered ring macrolides is not known. If we may extend the observed absence of MFPs to actinomycetes, we would expect that this group of transporters also consists of a single efflux protein.

It is noteworthy that ABC transporters are present both in Streptomyces spp. that synthesize antibiotics and in pathogenic staphylococci that provide the market for the antibiotics that the Streptomyces spp. produce (Table 1). These observations fuel speculative models that would attribute the origin of ABC transporters to antibiotic-producing Streptomyces spp.

Broad-Spectrum, MDR-Type Efflux

Edgar and Bibi (37) reported MdfA as "an *Escherichia coli* multi-drug resistance protein with an extraordinarily broad spectrum of drug recognition," which they found capable of transporting ethidium bromide, tetraphenylphosphonium, rhodamine, daunomycin, benzalkonium, rifampin, tetracycline, puromycin, chloramphenicol, erythromycin, certain aminoglycosides, and fluoroquinolones. On the basis of the observed loss of MdfA-mediated efflux in a mutant of E. coli, *unc*, which lacks F1-F0 proton ATPase activity, the authors inferred that MdfA belongs to the major facilitator subfamily that is driven by proton-motive force.

Jäger et al. (56) obtained a gene, named *cmr* (corynebacterial multidrug resistance), by shotgun cloning Corynebacterium glutamicum DNA and expressing it in E. coli. The gene *cmr* conferred resistance to a similar range of unrelated antibiotics that included erythromycin, tetracycline, puromycin, and bleomycin only in the heterologous host, E. coli, but not in the original host, C. glutamicum, even when expressed on a "multicopy plasmid." The basis for the selective expression of Cmr in a heterologous host has not been explained. Possibly, Cmr requires an MFP and porin, present in E. coli, for maximum expression, but that would not explain the role of an apparently inactive Cmr in C. glutamicum. A notable feature of the deduced structures of MdfA and of Cmr is the presence of 12 putative transmembrane domains, deduced by computer analysis, that resemble TetA and other proton/drug antiporter major facilitator proteins.

An additional example of the broad-spectrum type of multidrug transporters that can pump erythromycin is an AcrAB multicomponent homolog from Haemophilus influenzae (73, 109). In the case of the E. coli counterpart, AcrA is a proton/drug antiporter belonging to the major facilitator subfamily, and AcrB is the membrane facilitator protein. A porin that might interact with the AcrAB complex was not described. Another example is the MtrCDE multicomponent system from Neisseria gonorrhoeae (46), in which MtrD is the major facilitator component transporter, MtrC is the MFP component, and MtrE is the associated porin.

GENETIC MECHANISMS THAT REGULATE RESISTANCE

MLS Resistance

Erythromycin resistance is negatively regulated by attenuators. In studies on the *ermC* model system, carried by staphylococcal plasmids such as pE194, it has been shown

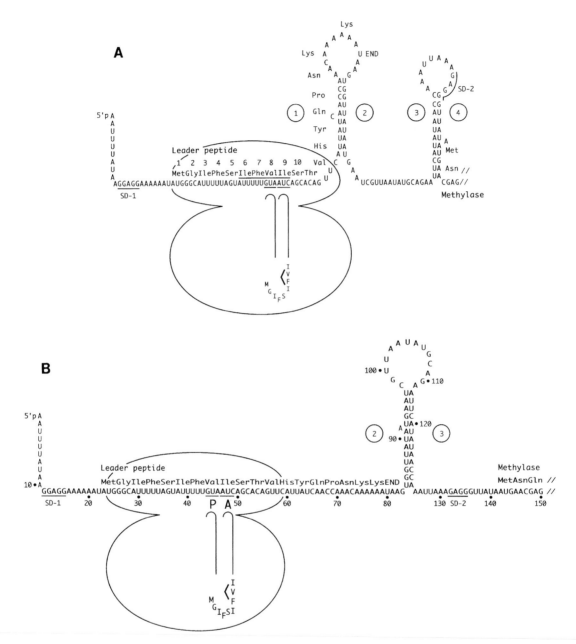

FIGURE 2 Proposed conformational transitions of the *ermC* leader sequence during induction. (A) Conformation of *ermC* mRNA leader sequence early during induction by erythromycin. The four major segments of the *ermC* attenuator are paired as 1:2 and 3:4, reflecting the temporal order of their synthesis. The ribosome is shown stalled during addition of Ile-9 to the growing leader peptide. See text for details. (B) Conformation of the *ermC* mRNA leader sequence in its fully induced state. As a consequence of stable complex formation between the erythromycin-ribosome complex and the *ermC* message, association between segment 1 and segment 2 is prevented. This favors association between segments 2 and 3, 2:3, which uncovers the ribosome binding site and the first two codons of the ErmC methylase encoded by segment 4. (C) Inactive conformation of the *ermC* mRNA leader sequence results from either removal of erythromycin or other inducer, or from maximal methylation of 23S rRNA and a maximum concentration of resistant ribosomes. The transition from the conformation shown in panel B to that shown here requires only that segment 4 associate with segment 1. A return to the ground state shown in panel A would require activation energy to dissociate 2:3, which would then enable a return to 1:2 and 2:3.

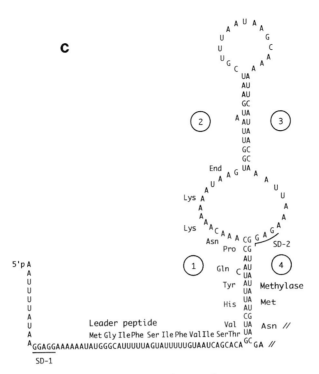

FIGURE 2 *Continued.*

that in the absence of erythromycin, the newly synthesized *ermC* mRNA folds into a conformation in which translation cannot be efficiently initiated (43, 53), as shown in Fig. 2. The 141-nucleotide *ermC* mRNA leader can assume at least three alternative conformations, two of which are shown. In its nascent form, the leader would be expected to assume the conformation shown in Fig. 2A. Induction increases the efficiency of ErmC translation owing to the conformational isomerization of the *ermC* message to the active form shown in Fig. 2B, in which SD-2, the translation start signal for methylase, and the first two codons of methylase are unmasked. The stalled ribosome is shown disrupting 1:2 at His-12. That the stalled ribosome perturbs intramolecular pairing of the *ermC* message at a site that corresponds to the His-12 codon is based on the experimental finding (54) that a mutation of C59A (H12N), leading to a mismatch at that position, resulted in constitutive expression of *ermC*. A third predicted conformation of the *ermC* message is one in which segment 2 pairs with segment 3 while segment 1 pairs with segment 4, as shown in Fig. 2C. This conformation would be associated with shutdown of the attenuator after removal of the inductive stimulus, or with low ribosome occupancy of segment 1 owing to a high degree of rRNA modification.

For a translational attenuator to function as proposed, erythromycin must stabilize the interaction of the ribosome with the message during synthesis of the leader peptide. The same type of interaction at the methylase structural gene would preclude the synthesis of methylase. Induction would only work if, under inducing conditions, the erythromycin-ribosome-mRNA-peptidyl-tRNA complex were to become destabilized, allowing a ribosome that is uncomplexed with erythromycin to synthesize methylase. This type of interaction is facilitated by sub-minimum inhibitory concentrations of erythromycin that support induction, i.e., between 1 and 10 ng of erythromycin per ml (133).

Indeed, erythromycin appears to inhibit the synthesis of proteins in either of two ways, depending on whether the protein-synthesizing ribosome-mRNA-peptidyl-tRNA complex dissociates or becomes stabilized. The first mode of erythromycin action is inferred from the work of Tai et al. (119), who observed the breakdown of polysomes in extracts from cells incubated with erythromycin, and from the work of Menninger and Otto (84), who observed the accumulation of peptidyl-tRNA complexes that were formed in the presence of erythromycin. In contrast, induction of *erm* under control of a translational attenuator requires stabilization of the polysome complex that synthesizes the leader peptide. Shivakumar et al. (115) reported a 10-fold increase in the functional half-life of *ermC* mRNA, while Sandler and Weisblum (110) reported a similar increase in the structural half-life of the *ermA* message. ErmC and ErmA are two structurally related homologous methyltransferases found in clinical isolates of *S. aureus*. They are both regulated by translational attenuation; however, their respective attenuators differ markedly in structure, but not function. The stabilized ribosome-mRNA complex and the footprint of the ribosome on the *ermC* message covering the leader peptide coding region were directly visualized by reverse transcriptase mapping of the *ermC* mRNA during induction (78). By varying the erythromycin concentration used for induction, the expected correlation between increased footprint intensity caused by the stalled ribosome and incremental unfolding of the attenuator stems was clearly seen.

In their review, Lovett and Rogers (74) discussed cotranslational regulation of protein synthesis, a process in which nascent protein functions as a *cis*-active regulatory element that modulates its own synthesis. We can consider nascent leader peptides to be cotranslational regulators that become optimally active when they are synthesized by ribosomes complexed with erythromycin. Two ways to study how the leader peptide sequence determines which antibiotics induce would be either to study natural variation, shown in Table 2 or to create variation by site-directed mutagenesis (79, 80). Both approaches have been tried with results leading to different conclusions. Of the 74 single-amino-acid mutants that were screened, some retained and some lost inducibility. The conservative change Val-8–Leu resulted in a loss of inducibility, whereas the conservative change Phe-7–Tyr retained inducibility. Of the mutations to Leu at each of the four positions IFVI, two retained inducibility (Ile-6–Leu and Ile-9–Leu) and two did not (Phe-7–Leu and Val-8–Leu). These observations point to a stringently determined specificity in which a single base change results in the loss of inducibility. In contrast, naturally occurring Erm leader peptides, listed in Table 2, show little sequence similarity, suggesting a much looser specificity.

Examples of natural variation of induction specificity that have been described include the following: (i) The lincosamide celesticetin (but not lincomycin) induces ermA (1), as well as ermC, and ermSV (62). (ii) ermSV from Streptomyces viridochromogenes is induced by both 14- and 16-membered ring macrolides (62). (iii) ermSF from S. fradiae is autoinduced by the 16-membered ring macrolide tylosin, which it produces during the late phase of its growth cycle (83). (iv) ermAMR from Enterococcus faecalis is induced by both 14- and 16-membered ring macrolides, with higher efficiency reported for 16-membered ring macrolides (91).

The mechanism of translational attenuation contrasts with that of transcriptional attenuation, in which the ribosomal pause is linked functionally to antitermination, another process in which the conformation of the message is linked to its expression. Transcriptional attenuation is well known to regulate amino acid biosynthesis (see reference 135 for a review) and may also regulate ermK. ermK mRNA synthesized in the absence of erythromycin consisted mainly of two size classes corresponding to rho-independent transcription termination sites located in the leader sequence of the ermK message (26, 68). Here, erythromycin would block translation of the leader peptide, allowing the message to refold in a conformation that would block formation of the transcriptional terminator stems.

The similarity between the region that includes the Shine-Dalgarno sequence of the ErmC methylase (AAA-GAGGG) and that of the site of methylation in 23S rRNA (GGAAAGA) has been noted (31a, 54) as a possible basis for ermC autoregulation by translational repression. According to such a model, ErmC above a critical cellular concentration binds to its Shine-Dalgarno sequence and inhibits its own translation.

MS Resistance

The DNA sequence of the MsrA open reading frame is preceded by a ca. 300-nucleotide leader sequence that encodes, sequentially, an upstream ribosome binding sequence, GGAGG; a putative eight-amino-acid leader peptide, MTASMRLK; and a noncoding region that contains four inverted complementary repeat sequences. Last, the MsrA open reading frame is preceded by its own ribosome binding sequence, AGGAG, that could be sequestered by secondary structure of the leader region and is presumed to become available if the upstream leader peptide sequence is occupied by a stalled erythromycin-ribosome complex. Since streptogramin B-type antibiotics cannot induce MS resistance but are transported by the

MS transporter, it should be possible to select constitutively resistant mutants, as in the case of the erythromycin-inducible, MLS-resistant strains. Ross et al. (102) deleted the proposed msrA attenuator region by recombinant methods and showed that the resultant construct expressed MsrA constitutively. This observation proves that the proposed attenuator region is a negative-acting element in the regulation of msrA. A proof of the detailed attenuator model that they postulated will require point mutants that destabilize the proposed set of stems and loops. By analogy to erm systems, such constitutively resistant mutants of msrA might also be expected to arise in clinical samples, where they would be expected to lead to relapse, for reasons similar to those that give rise to constitutively MLS-resistant strains.

NEW GENERATIONS OF MLS ANTIBIOTICS
Four Generations of Macrolides

The macrolides have evolved through four chemical generations since erythromycin became available for general clinical use in 1952. Erythromycin can be chemically modified to produce more useful antibiotics. The goals for improvement of its activity include increased acid stability, improved pharmacokinetics, a broader spectrum of action, and increased effectiveness against erythromycin-resistant strains. These improvements may be realized by discovering naturally occurring macrolides in which the desired attributes are already present or by chemical modification of erythromycin. Based on such improvements, four generations of macrolides are distinguishable.

First Generation: 14-Membered Ring Macrolides and the Emergence of Inducible MLS Resistance

Inducible MLS-resistant isolates of S. aureus began to appear shortly after the introduction of erythromycin into clinical practice. For details, see reference 131. Two other 14-membered ring macrolides, oleandomycin and megalomicin, were subsequently introduced into medical practice with marginal improvement in pharmacokinetics. Inducing activity comparable to that of erythromycin was found for both oleandomycin (42) and megalomicin (108). Because of a lack of significant clinical improvement by these two antibiotics, neither replaced erythromycin.

Second Generation: 16-Membered Ring Macrolides and the Emergence of Constitutive MLS Resistance

The first generation, the 14-membered ring macrolide erythromycin, induced resistance and was replaced by the second generation, i.e., 16-membered ring macrolides, which did not. The susceptibility of inducible strains to noninducing MLS antibiotics in the clinical setting proved to be illusory because of the ease with which noninducing MLS antibiotics were able to select constitutively resistant mutants, leading to relapse (32, 130). For reviews, see references 72, 81, 131, and 132.

Inducible erythromycin-resistant S. aureus strains apparently remained susceptible to 16-membered ring macrolides such as carbomycin, leucomycin, and spiramycin, suggesting that the larger macrolides would remain effective therapeutically, irrespective of the emergence of bacterial resistance to their 14-membered ring macrolide counterparts. Functionally, the 16-membered ring macrolides carbomycin, leucomycin, and spiramycin represent a

TABLE 2 *erm* leader peptides

erm	Amino acid sequence[a]	Reference(s)
ermC	MGIFS IFVI STVHYQPNKK	53
ermG1	MNKYS KRDA IN	86
ermG2	MGLYS IFVI ETVHYQPNEK	86
ermSF	MSMGI AARP PRAALLPPPSVPRSR	60
ermSV	MAANN AITN SGLGRGCAHSVRMRRGPGALTGPGSHTAR	61
ermCd	MLISG TAFL RLRTNRKAFPTP	51
ermJ/K	MTHSM RLRF PTLNQ	63, 68
ermAM	MLVFQ MRNV DKTSTILKQTKNSDYVDKYVRLIPTSD	52

[a]Each *erm* leader peptide sequence is aligned to emphasize the part of the sequence that corresponds numerically to IFVI in *ermC*. The alignment for the leader peptides with *ermC* is assumed but has not been experimentally verified. Only in *ermC* has it been shown that the 10th amino acid of the leader peptide can be replaced with a stop codon without loss of induction, whereas the same substitution at amino acid nine abolishes induction.

second generation of macrolides that were effective when first introduced into clinical practice because they could inhibit strains of *S. aureus* resistant to erythromycin and other 14-membered ring macrolides. Because the strains were clinically resistant to erythromycin, but susceptible to 16-membered ring macrolides, they were described as showing "dissociated resistance." It seemed logical that a common mechanism should serve to confer resistance to such closely related antibiotics as the 14- and 16-membered ring macrolides. In fact it does, but the original assumptions did not take into account that resistance was inducible and that different macrolides would induce with different efficiencies.

A second, unexpected property of the erythromycin-inducible (dissociated) strains was their apparent ability to mutate rapidly to high-level resistance to both 14- and 16-membered ring macrolides. These emergent resistant strains were described as displaying "nondissociated" or "generalized" resistance. On further examination, the nondissociated mutants were found to have become resistant to lincosamide and streptogramin B antibiotics as well (18, 25, 42). The nondissociated strains were simply constitutively resistant mutants of the inducible strain. Sequence studies of methyltransferase mRNA leader sequences from constitutively resistant mutants, from clinical and laboratory strains, suggested how these strains arose as a consequence of nucleotide sequence alterations. Such mutations act by destabilizing the secondary structure of the mRNA leader sequence, which, when stable and in the ground state, represses expression of methylase. Sequence analyses of constitutively resistant mutants explained the molecular basis for this clinically important phenomenon and also provided numerous critical tests of the attenuation mechanism. For reviews, see references 35, 72, 131, and 132.

Third Generation: Semisynthetic, Acid-Stable, Broader-Spectrum 14- and 15-Membered Ring Macrolides and the Emergence of Intrinsically Resistant *Mycobacterium* Spp. and *H. pylori*

A third generation of macrolides improved the acid stability, and therefore the pharmacokinetics of erythromycin, extending the clinical use of macrolides to *H. pylori* and *Mycobacterium tuberculosis*. Improved pharmacokinetics came at the price of selecting intrinsically resistant mutant strains with rRNA structural alterations. Expression of resistance in these strains was a surprise, explainable by low rRNA gene copy number, which made resistance dominant.

Third-generation, semisynthetic macrolides increased the acid stability of 14-membered ring macrolides and broadened the spectrum of macrolide use. Examples of these newer macrolides include clarithromycin, roxithromycin, and dirythromycin. The 15-membered ring macrolide azithromycin can also be included in this group. This group of macrolides offered enhanced chemical stability and therefore improved pharmacokinetics and a broader spectrum of action; however, they did not offer any improvement over erythromycin with respect to inhibiting cells resistant by any of the mechanisms cited above.

The third-generation macrolides incorporate alterations at the 9-keto and 6-OH positions that prevent the interaction between these two positions that occurs under acidic conditions. In the presence of acid, a hemiketal is formed between the C-6 and C-9 of the erythromycin ring system. The acidity of the stomach favors this reaction, which inactivates erythromycin, making it less effective as an oral agent. The chemical modifications at C-6, e.g., O-methylation (clarithromycin), or at C-9, e.g., substitution of 9-keto with 9-amino (dirithromycin), thus stabilize the antibiotic in the acidic environment of the gut, increasing its effectiveness when administered orally. In the case of azithromycin, improved pharmacokinetics owing to acid stability was combined with the introduction of a basic nitrogen atom into the macrolide ring (increasing its size to 15 atoms), leading to a possible improvement in the penetration of the gram-negative outer membrane (40, 126). An additional effect of these modifications is an extension of the effectiveness of these antibiotics to treat infections by organisms for which macrolide antibiotics were previously not considered to be agents of choice. These include *Mycobacterium avium-intracellulare* and *M. tuberculosis* as opportunists in the immunocompromised patient, and *H. pylori* as an etiologic agent in peptic ulcer.

The third-generation macrolides are indistinguishable from first-generation erythromycin with respect to their effectiveness in inducing resistance. However, the extended range of effectiveness introduced intrinsic MLS resistance (82, 127). This development was unexpected, because MLS resistance in clinical isolates had previously been shown to be due only to the posttranscriptional modification of A2058, specified by an acquired r-factor. The only previously known forms of intrinsic resistance, i.e., those due to the mutation A2058G, were seen in yeast mitochondrial ribosomes following selection with erythromycin or in *E. coli*, but only if the mutant rRNA gene was introduced on a multicopy vector.

Left to their own devices, staphylococci and streptococci treated with erythromycin failed to develop MLS resistance other than by the acquired route. The explanation for this became apparent when it was noted that *M. intracellulare* and *M. tuberculosis* have only a single copy of their 23S rRNA genes. As a result, mutations in the 23S rRNA genes in these organisms would be dominant. In contrast, in the presence of a higher number of 23S rRNA gene copies, a mutation in any given gene copy would be recessive and therefore not observed phenotypically.

Fourth (and Current) Generation: Ketolides, 14-Membered Ring Macrolides That Do Not Induce MLS Resistance

A fourth generation, the 14-membered ring ketolides, are the most recent development. Members of this generation are effective against inducibly resistant strains, and resistant strains have not yet been reported. Replacement of L-cladinose at position 3 in the macrolide ring with a keto group leads to a new class of semisynthetic, acid-stable 14-membered ring macrolides, 3-keto, 3-descladinosyl macrolides, or ketolides. Bonnefoy et al. (17) tested ketolides RU-64004 and RU-56006 in parallel with a set of earlier-generation macrolides (erythromycin, azithromycin, and clarithromycin). The two ketolides produced clear inhibition zones on solid medium that were truncated by the proximity of erythromycin, azithromycin, or clarithromycin. These observations are indicative of induction of resistance to RU-64004 by erythromycin, azithromycin, and clarithromycin. Ketolides do not induce MLS resistance but, like other noninducing MLS antibiotics, might select constitutively resistant mutants. Jones and Biedenbach (59), however, reported that the ketolide RU-66647 inhibits constitutively erythromycin-resistant strains of staphylococci, streptococci, and enterococci. It will be of interest to learn how ketolides interact with their cellular target, as well as how bacteria will evolve in response to members of this new group of antibiotics, which can inhibit both inducible and constitutively MLS-resistant isolates.

Two Generations of Lincosamides and Streptogramins

As in the case of the 16-membered ring macrolides, the lincosamides and streptogramin B antibiotics were considered candidates for use in treating erythromycin-resistant *S. aureus*. Semisynthetically modified lincomycin to yield clindamycin (2-chloro-2-deoxy lincomycin) improved the pharmacokinetics (21)) and broadened the spectrum of action of lincomycin but did not produce an antibiotic that would be effective in treating infections caused by the constitutive MLS-resistant strains.

Development of a second generation of streptogramins attempted to circumvent the relative insolubility of streptogramin types A and B in water, which limited their therapeutic use. Semisynthetic modifications were introduced into both groups of antibiotics and led to the synthesis of dalfopristin (streptogramin A) and quinupristin (streptogramin B), both of which are more soluble in water than their parent compounds, pristinamycin-II and -I, respectively. The use of these antibiotics in combination only partially overrides the resistance to quinupristin conferred by MLS-related ribosome methylation. Low and Nadler (75) reported that susceptible or inducibly resistant *S. aureus* cells are killed within 6 h of exposure to the combination, whereas MLS constitutively resistant cells remain viable at least 12 h when cultured under similar conditions in vitro. The molecular basis for these observations is not yet understood. We might expect that the use of dalfopristin plus quinupristin will become limited when MLS-resistant strains acquire genes that confer resistance to both dalfopristin and quinupristin. Allignet et al. (4) have reported an isolate of *S. cohnii* in which they found a plasmid carrying the resistance genes *vatC* (streptogramin A acetylase) and *vgbB* (streptogramin B lactonase). The emergence of resistance to the streptogramin A plus B combination should be a source of special concern since the combination dalfopristin plus quinupristin has been found to be useful in treating infections caused by vancomycin-resistant *E. faecium* (38).

INHIBITION OF Erm *N*-METHYL-TRANSFERASES, A NEW DIRECTION OF DRUG DESIGN

Inhibiting the catalytic activity of enzymes that confer macrolide resistance could hold the key to a class of drugs that sensitize resistant cells. Two approaches have been taken in the attempt to find such inhibitors: (i) high-throughput screening of chemical libraries in a search for inhibitors of methyltransferase activity and (ii) rational design of Erm methyltransferase inhibitors based on the X-ray crystallographic or nuclear magnetic resonance solution structure.

High-Throughput Screening for Inhibitors of ErmC *N*⁶-Methyltransferase

Clancy et al. (29) screened a chemical library of 160,000 compounds for their ability to inhibit *S. aureus* ErmC in vitro. In an attempt to distinguish between inhibitors that act at the catalytic center of the methyltransferase and those that act elsewhere, candidates that inhibited methyltransferase activity in the nanomolar to micromolar range were further tested for their ability to inhibit M.*Eco*RI [*E. coli* DNA-specific *N*⁶-(mono)-methyltransferase] and rat liver catechol-*O*-methyltransferase. Finally, candidates that selectively inhibited ErmC methyltransferase were tested, in combination with the 15-membered ring macrolide azithromycin, against *S. aureus*, *S. pyogenes*, *E. coli*, and *E. faecalis*. Of the candidates that were tested, seven had a 50% inhibitory concentration (IC_{50}) of <1 μM but were found to lack useful protective activity in an animal model system. Lead compounds, even if active only in vitro, might be useful as agents for selection of resistant methyltransferase mutants in order to learn how structural alterations in the enzyme affect its activity.

Rational Design of ErmAM *N*-Methyltransferase Inhibitors

The "structure first" approach requires the knowledge of an Erm structure together with lead compounds and the ability to identify sites at which the lead compounds can both bind to the enzyme and modulate its activity. Two physical structures are available for Erm proteins. (i) A solution structure of ErmAM based on nuclear magnetic resonance has been reported by Yu et al. (136). According to their model (Protein Database identification code 1YUB), ErmAM is organized as β-sheet and α-helical structural elements and was aligned with other methyltransferases as shown in Fig. 3. (ii) An X-ray crystal structure of ErmC has been reported by Bussiere et al. (23).

	1								180 181	246

(a) ErmAM   β1 α1 β2 α2 β3 α3 β4 α4 β5 α5 β6 α6 α7 β7 β8 β9 ‖ α8 α9 α10 α11

(b) VP39   α1 β2 α2 β3 α3 β4 α4 β5 α5 β6 ---- β7 — β9 ‖ α(8,9) α10 α11

(c) M.TaqI   α1 β2 α2 β3 -- β4 -- β5 α5 β6 α6 α7 β7 — β9 ‖ ------------

FIGURE 3 Comparative secondary structure alignment of methyltransferases. The ErmAM structure can be aligned with homologous counterparts of other methyltransferases. These include (i) ErmAM; (ii) VP-39, an mRNA cap-specific RNA adenosine 2′-O-methyltransferase (50); and (iii) M.TaqI, a DNA adenine N^6-monomethyltransferase (69). Structural elements β1 to β9 are common to methyltransferases, including catechol O-methyltransferase (not shown) and glycine N-methyltransferase (not shown). The structural alignment pattern implicates amino acid residues 1 to 180, organized as structural elements β1 to β9, as the location of the catalytic center of ErmAM, and the remaining four C-terminal helixes, α8 to α11, which span amino acid residues 181 to 246, as the part of ErmAM that interacts specifically with RNA. The structure of M.TaqI beyond β9 (dashed sequence) could not be aligned with C-terminal helixes α8 to α11.

The abrupt discontinuity in structural and functional homology for the three methyltransferases beyond β9 suggests that α8, α9, α10, and α11 are involved in RNA substrate recognition. Whether or not some part of the peptidyl transferase circle RNA sequence recognition specificity is also distributed within ErmAM structural units β1 to β9 is not known. A solution structure or crystal structure of ErmAM, together with its docked 23S rRNA substrate, would be helpful for identifying structural elements that play a role in the way in which ErmAM recognizes its substrate. The full-length (ca. 3,000-nucleotide) 23S rRNA may not be necessary for such an undertaking because a 27-mer (128a) and a 41-mer (67) ribo-oligonucleotide, shown in Fig. 4, based on parts of the peptidyl transferase circle sequence, are methylatable in vitro by the ErmE (34) and ErmSF (67) methylases, respectively. It will be of interest, using nuclear magnetic resonance and X-ray crystallography, to learn how helixes α8 to α11 interact with the 27- and 41-nucleotide minimalist substrates with the availability of milligram amounts of overexpressed recombinant Erm proteins, namely, ErmAM (136), ErmC (23), ErmE (128), and ErmSF (66).

CONCLUDING REMARKS

Erythromycin continues to preside over a macrolide family consisting of four living generations. With reference to the fourth generation, the flagship ketolide, 3-descladinosyl, 3-keto erythromycin, appears to be a "leaner and meaner" stripped-down version of erythromycin. Since calorically deprived and therefore leaner laboratory animals outlive their better-fed counterparts, will the same be true of ketolides? It is hard to believe that MLS-resistant pathogens that are ketolide susceptible will not eventually become ketolide resistant. Irrespective of how ingeniously nature or chemists design an antibiotic, there seems to be an efflux mechanism ready to bail it out. Lincomycin resistance by an efflux transporter, LmrA, has so far been described only in *S. lincolnensis* (139). The appearance of this mechanism in pathogens may only be a matter of time. Since individual resistance determinants for the synergistic streptogramin A plus B combination, dalfopristin plus quinupristin, have been found in clinically resistant strains, it may likewise only be a matter of time before strains carrying both determinants will be selected.

A new direction for antibiotic discovery is the combinatorial route. As applied to macrolides, polyketide synthase, the enzyme that is responsible for synthesis of the macrolide ring, has been rearranged at the genetic level to yield hybrid macrolide products, either by design (93, 107) or by chance (41). Since macrolides collectively are known to act on a wider range of biochemical processes than just 50S ribosome function, can we expect that the combinatorial approach will lead to the discovery of a bonanza of new antimetabolites?

I thank Joyce Sutcliffe and Richard Novick for critically reading the manuscript, and Carol Dizak and Rob Giannattasio for preparing the figures.

```
                              AAG 3'
(a)    CᴳCCCGCGACᴬGGAᶜ GGᴬ
       UᵤGGGCGCUG CCUᴀᵤCCᴳ
                              GA 5'


                  5' GGCᴬAGAᶜ GGᴬᴬᴬᴳ
(b)              3' CCG UCUᴀᵤCCᴄᴄᴬ
```

FIGURE 4 Minimalist 41-nucleotide RNA (67) and 27-nucleotide RNA (128a) substrates for in vitro methylation by Erm methyltransferase. The methylatable adenine residue is shown in bold. Unpaired nucleotides A, C (above), and AU (below), which appear as bulges, are the most critical sites in the minimalist sequence because changes at these locations are most effective in reducing methyl-accepting activity.

REFERENCES

1. **Allen, N. E.** 1977. Macrolide resistance in *Staphylococcus aureus*: inducers of macrolide resistance. *Antimicrob. Agents Chemother.* **11**:669–674.

2. **Allignet, J., and N. El Solh.** 1995. Diversity among the gram-positive acetyltransferases inactivating streptogramin A and structurally related compounds and characterization of a new staphylococcal determinant, *vatB. Antimicrob. Agents Chemother.* **39**:2027–2036.

3. **Allignet, J., and N. El Solh.** 1997. Characterization of a new staphylococcal gene, *vgaB*, encoding a putative ABC transporter conferring resistance to streptogramin A and related compounds. *Gene* **202**:133–138.

from *Bacillus anthracis* 590: cloning and expression of *ermJ*. *J. Gen. Microbiol.* **139:**601–607.

64. **Kim, S. K., M. C. Baek, S. S. Choi, B. K. Kim, and E. C. Choi.** 1996. Nucleotide sequence expression and transcriptional analysis of the *Escherichia coli mphK* gene encoding macrolide phosphotransferase. *Mol. Cells* **6:** 153–160.

65. **Kono, M., K. O'Hara, and T. Ebisu.** 1992. Purification and characterization of macrolide 2′-phosphotransferase type II from a strain of *Escherichia coli* highly resistant to macrolide antibiotics. *FEMS Microbiol. Lett.* **76:**89–94.

66. **Kovalic, D., R. B. Giannattasio, H. J. Jin, and B. Weisblum.** 1994. 23S rRNA domain V, a fragment that can be specifically methylated in vitro by the ErmSF (TlrA) methyltransferase. *J. Bacteriol.* **176:**6992–6998.

67. **Kovalic, D., R. B. Giannattasio, and B. Weisblum.** 1995. Methylation of minimalist 23S rRNA sequences in vitro by ErmSF (TlrA) N-methyltransferase. *Biochemistry* **34:** 15838–15844.

68. **Kwak, J. H., E. C. Choi, and B. Weisblum.** 1991. Transcriptional attenuation control of *ermK*, a macrolide-lincosamide-streptogramin B resistance determinant from *Bacillus licheniformis*. *J. Bacteriol.* **173:**4725–4735.

69. **Labahn, J., J. Granzin, G. Schluckebier, D. P. Robinson, W. E. Jack, I. Schildkraut, and W. Saenger.** 1994. Three-dimensional structure of the adenine-specific DNA methyltransferase M.*Taq* I in complex with the cofactor S-adenosylmethionine. *Proc. Natl. Acad. Sci. USA* **91:** 10957–10961.

70. **Lai, C. J., and B. Weisblum.** 1971. Altered methylation of ribosomal RNA in an erythromycin-resistant strain of *Staphylococcus aureus*. *Proc. Natl. Acad. Sci. USA* **68:**856–860.

71. **Lampson, B. C., and J. T. Parisi.** 1986. Nucleotide sequence of the constitutive macrolide-lincosamide-streptogramin B resistance plasmid pNE131 from *Staphylococcus epidermidis* and homologies with *Staphylococcus aureus* plasmids pE194 and pSN2. *J. Bacteriol.* **167:** 888–892.

72. **Leclercq, R., and P. Courvalin.** 1991. Bacterial resistance to macrolide, lincosamide, and streptogramin antibiotics by target modification. *Antimicrob. Agents Chemother.* **35:**1267–1272.

73. **Lewis, K.** 1994. Multidrug resistance pumps in bacteria: variations on a theme. *Trends Biochem. Sci.* **19:**119–123.

74. **Lovett, P. S., and E. J. Rogers.** 1996. Ribosome regulation by the nascent peptide. *Microbiol. Rev.* **60:**366–385.

75. **Low, D. E., and H. L. Nadler.** 1997. A review of *in-vitro* antibacterial activity of quinprustin/dalfopristin against methicillin-susceptible and -resistant *Staphylococcus aureus*. *J. Antimicrob. Chemother.* **39**(Suppl. A)**:**53–58.

76. **Matsuoka, M., K. Endou, H. Kobayashi, M. Inoue, and Y. Nakajima.** 1997. A dyadic plasmid that shows MLS and PMS resistance in *Staphylococcus aureus*. *FEMS Microbiol. Lett.* **148:**91–96.

77. **Matsuoka, M., K. Endou, S. Saitoh, M. Katoh, and Y. Nakajima.** 1995. A mechanism of resistance to partial macrolide and streptogramin B antibiotics in *Staphylococcus aureus* clinically isolated in Hungary. *Biol. Pharm. Bull.* **18:**1482–1486.

78. **Mayford, M., and B. Weisblum.** 1989. Conformational alterations in the *ermC* transcript *in vivo* during induction. *EMBO J.* **8:**4307–4314.

79. **Mayford, M., and B. Weisblum.** 1989. *ermC* leader peptide, amino acid sequence critical for induction by translational attenuation. *J. Mol. Biol.* **206:**69–79.

80. **Mayford, M., and B. Weisblum.** 1990. The *ermC* leader peptide: amino acid alterations leading to differential efficiency of induction by macrolide-lincosamide-streptogramin B antibiotics. *J. Bacteriol.* **172:**3772–3779.

81. **Mazzei, T., E. Mini, A. Novelli, and P. Periti.** 1993. Chemistry and mode of action of macrolides. *J. Antimicrob. Chemother.* **31**(Suppl. C)**:**1–9.

82. **Meier, A., P. Kirschner, B. Springer, V. A. Steingrub, B. A. Brown, R. J. Wallace, and E. C. Böttger.** 1994. Identification of mutations in the 23S ribosomal RNA gene of clarithromycin resistant *Mycobacterium intracellulare*. *Antimicrob. Agents Chemother.* **38:**381–384.

83. **Memili, E., and B. Weisblum.** 1997. Essential role of endogenously synthesized tylosin for induction of *ermSF* in *Streptomyces fradiae*. *Antimicrob. Agents Chemother.* **41:** 1203–1205.

84. **Menninger, J., and D. P. Otto.** 1982. Erythromycin, carbomycin, and spiramycin inhibit protein synthesis by stimulating the dissociation of peptidyl-tRNA from ribosomes. *Antimicrob. Agents Chemother.* **21:**811–818.

85. **Monod, M., C. Denoya, and D. Dubnau.** 1986. Sequence and properties of pIM13, a macrolide-lincosamide-streptogramin B resistance plasmid from *Bacillus subtilis*. *J. Bacteriol.* **167:**138–147.

86. **Monod, M., S. Mohan, and D. Dubnau.** 1987. Cloning and analysis of *ermG*, a new macrolide-lincosamide-streptogramin B resistance element from *Bacillus sphaericus*. *J. Bacteriol.* **169:**340–350.

87. **Murphy, E.** 1985. Nucleotide sequence of *ermA*, a macrolide-lincosamide-streptogramin B determinant in *Staphylococcus aureus*. *J. Bacteriol.* **162:**633–640.

88. **Nakajima, Y., L. Jánosi, K. Endou, M. Matsuoka, and H. Hashimoto.** 1992. Inducible resistance to a 16-membered macrolide, mycinamicin, in *Staphylococcus aureus* resistant to 14-membered macrolides and streptogramin B antibiotics. *J. Pharmacobio-Dynamics* **15:** 319–324.

89. **Noguchi, N., J. Katayama, and K. O'Hara.** 1996. Cloning and nucleotide sequence of the *mphB* gene for macrolide 2′-phosphotransferase II in *Escherichia coli*. *FEMS Microbiol. Lett.* **144:**197–202.

90. **Noller, H. F., J. Kop, V. Wheaton, J. Brosius, R. R. Gutell, A. M. Kopylov, F. Dohme, W. Herr, D. A. Stahl, R. Gupta, and C. R. Woese.** 1981. Secondary structure model for 23S ribosomal RNA. *Nucleic Acids Res.* **9:**6167–6189.

91. **Oh, T. G., A. R. Kwon, and E. C. Choi.** 1998. Induction of *ermAMR* from a clinical strain of *Enterococcus faecalis* by 16-membered-ring macrolide antibiotics. *J. Bacteriol.* **180:**5788–5791.

92. **Ohara, K., T. Kanda, T. K. Ohmiya, T. Ebisu, and M. Kono.** 1989. Purification and characterization of macrolide 2′-phosphotransferase from a strain of *E. coli* that is highly resistant to erythromycin. *Antimicrob. Agents Chemother.* **33:**1354–1357.

93. **Oliynyk, M., M. J. Brown, J. Cortes, J. Staunton, and P. F. Leadlay.** 1996. A hybrid modular polyketide synthase obtained by domain swapping. *Chem. Biol.* **3:**833–839.

94. **Ounissi, H., and P. Courvalin.** 1985. Nucleotide sequence of the gene *ereA* encoding the erythromycin esterase in *Escherichia coli*. *Gene* **35:**271–278.

95. **Paulsen, I. T., M. H. Brown, and R. A. Skurray.** 1996. Proton-dependent multidrug efflux systems. *Microbiol. Rev.* **60:**575–608.

96. **Pernodet, J. L., F. Boccard, M. T. Alegre, M. H. Blondelet-Rouault, and M. Guérineau.** 1988. Resistance to macrolides, lincosamides and streptogramin type B antibiotics due to a mutation in an rRNA operon of *Streptomyces ambofaciens. EMBO J.* **7:**277–282.

97. **Pernodet, J. L., F. Boccard, M. T. Alegre, J. Gagnat, and M. Guérineau.** 1989. Organization and nucleotide sequence analysis of a ribosomal RNA gene cluster from *Streptomyces ambofaciens. Gene* **79:**33–46.

98. **Rasmussen, J. L., D. A. Odelson, and F. L. Macrina.** 1986. Complete nucleotide sequence and transcription of ermF, a macrolide-lincosamide-streptogramin B resistance determinant from *Bacteroides fragilis. J. Bacteriol.* **168:**523–533.

99. **Rende-Fournier, R., R. Leclercq, M. Galimand, J. Duval, and P. Courvalin.** 1993. Identification of the *satA* gene encoding a streptogramin A acetyltransferase in *Enterococcus faecium* BM4145. *Antimicrob. Agents Chemother.* **37:**2119–2125.

100. **Richardson, M. A., S. Kuhstoss, P. Solenberg, N. A. Schaus, and R. N. Rao.** 1987. A new shuttle cosmid vector, pKC505, for streptomycetes: its use in the cloning of three different spiramycin-resistance genes from a *Streptomyces ambofaciens* library. *Gene* **61:**231–241.

101. **Roberts, A. N., G. S. Hudson, and S. Brenner.** 1985. An erythromycin-resistance gene from an erythromycin-producing strain of *Arthrobacter* sp. *Gene* **35:**259–270.

102. **Ross, J. I., E. A. Eady, J. H. Cove, and S. Baumberg.** 1996. Minimal functional system required for expression of erythromycin resistance by MSRA in *Staphylococcus aureus* RN4220. *Gene* **183:**143–148.

103. **Ross, J. I., E. A. Eady, J. H. Cove, W. J. Cunliffe, S. Baumberg, and J. C. Wootton.** 1990. Inducible erythromycin resistance in staphylococci is encoded by a member of the ATP-binding transport super-gene family. *Mol. Microbiol.* **4:**1207–1214.

104. **Ross, J. I., E. A. Eady, J. H. Cove, C. E. Jones, A. H. Ratyal, V. W. Miller, S. Vyakrnam, and W. J. Cunliffe.** 1997. Clinical resistance to erythromycin and clindamycin in cutaneous Propionibacteria isolated from patients is associated with mutations in 23S rRNA *Antimicrob. Agents Chemother.* **41:**1162–1165.

105. **Ross, J. I., A. M. Farrell, E. A. Eady, J. H. Cove, and W. J. Cunliffe.** 1989. Characterisation and molecular cloning of the novel macrolide-streptogramin B resistance determinant from *Staphylococcus epidermidis. J. Antimicrob. Chemother.* **24:**851–862.

106. **Rosteck, P. R., Jr., P. A. Reynolds, and C. L. Hershberger.** 1991. Homology between proteins controlling *Streptomyces fradiae* tylosin resistance and ATP-binding transport. *Gene* **102:**27–32.

107. **Ruan, X., A. Pereda, D. L. Stassi, D. Zeidner, R. G. Summers, M. Jackson, A. Shivakumar, S. Kakavas, M. J. Staver, S. Donadio, and L. Katz.** 1997. Acyltransferase domain substitutions in erythromycin polyketide synthase yield novel erythromycin derivatives. *J. Bacteriol.* **179:**6416–6425.

108. **Saito, T., and S. Mitsuhashi.** 1971. Antibacterial activity of megalomicin and its inducer activity for macrolide resistance in staphylococci. *J. Antibiot.* **24:**850–854.

109. **Sanchez, L., W. Pan, M. Vinas, and H. Nikaido.** 1997. The *acrAB* homolog of *Haemophilus influenzae* codes for a functional multidrug efflux pump. *J. Bacteriol.* **179:**6855–6857.

110. **Sandler, P., and B. Weisblum.** 1989. Erythromycin-induced ribosome stall in the ermA leader: a barricade to 5′-to-3′ nucleolytic cleavage of the ermA transcript. *J. Bacteriol.* **171:**6680–6688.

111. **Schoner, B., M. Geistlich, P. Rosteck, Jr., R. N. Rao, E. Seno, P. Reynolds, K. Cox, S. Burgett, and C. Hershberger.** 1992. Sequence similarity between macrolide-resistance determinants and ATP-binding proteins. *Gene* **115:**93–96.

112. **Seppala, H., M. Skurnik, H. Soini, M. C. Roberts, and P. Huovinen.** 1998. A novel erythromycin resistance methylase gene (*ermTR*) in *Streptococcus pyogenes. Antimicrob. Agents Chemother.* **42:**257–262.

113. **Serwold-Davis, T., and N. B. Groman.** 1986. Mapping and cloning of *Corynebacterium diphtheriae* plasmid pNG2 and characterization of its relatedness to plasmids from skin coryneforms. *Antimicrob. Agents Chemother.* **30:**69–72.

114. **Shaw, J. H., and D. B. Clewell.** 1985. Complete nucleotide sequence of macrolide-lincosamide-streptogramin B-resistance transposon Tn917 in *Streptococcus faecalis. J. Bacteriol.* **164:**782–796.

115. **Shivakumar, A. G., J. Hahn, G. Grandi, Y. Kozlov, and D. Dubnau.** 1980. Posttranscriptional regulation of an erythromycin resistance protein specified by plasmid pE194. *Proc. Natl. Acad. Sci. USA* **77:**3903–3907.

116. **Skinner, R., E. Cundliffe, and F. Schmidt.** 1983. Site of action of a ribosomal RNA methylase responsible for resistance to erythromycin and other antibiotics. *J. Biol. Chem.* **258:**12702–12706.

117. **Smith, C. J.** 1987. Nucleotide sequence analysis of Tn4551: use of *ermFS* operon fusions to detect promoter activity in *Bacteroides fragilis. J. Bacteriol.* **169:**4589–4596.

118. **Sutcliffe, J. A., A. Tait-Kamradt, and L. Wondrack.** 1996. *Streptococcus pneumoniae* and *Streptococcus pyogenes* resistant to macrolides but sensitive to clindamycin—a common resistance pattern mediated by an efflux system. *Antimicrob. Agents Chemother.* **40:**1817–1824.

119. **Tai, P. C., B. J. Wallace, and B. D. Davis.** 1974. Selective action of erythromycin on initiating ribosomes. *Biochemistry* **13:**4653–4659.

120. **Tait-Kamradt, A., J. Clancy, M. Cronan, F. Dib-Haji, L. Wondrack, W. Yuan, and J. Sutcliffe.** 1997. *mefE* is necessary for the erythromycin-resistant M phenotype in *Streptococcus pneumoniae. Antimicrob. Agents Chemother.* **41:**2251–2255.

121. **Tannock, G. W., J. B. Luchansky, L. A. Miller, and H. Connell.** 1993. Characterization of a plasmid (pGT633) from *Lactobacillus* that encodes resistance to MLS antibiotics. GenBank accession number M64090.

122. **Tenson, T., A. Deblasio, and A. Mankin.** 1996. A functional peptide encoded in the *Escherichia coli* 23S rRNA. *Proc. Natl. Acad. Sci. USA* **93:**5641–5646.

123. **Tenson, T., L. Q. Xiong, P. Kloss, and A. S. Mankin.** 1997. Erythromycin resistance peptides selected from random peptide libraries. *J. Biol. Chem.* **272:**17425–17430.

124. **Thompson, C. J., R. H. Skinner, J. Thompson, J. M. Ward, D. A. Hopwood, and E. Cundliffe**. 1982. Biochemical characterization of resistance determinants cloned from antibiotic-producing streptomycetes. *J. Bacteriol.* **151:**678–685.

125. **Uchiyama, H., and B. Weisblum**. 1985. N-methyl transferase of *Streptomyces erythraeus* that confers resistance to the macrolide-lincosamide-streptogramin B antibiotics: amino acid sequence and its homology to cognate R-factor enzymes from pathogenic bacilli and cocci. *Gene* **38:**103–110.

126. **Vaara, M**. 1993. Outer membrane permeability barrier to azithromycin, clarithromycin, and roxithromycin in gram-negative enteric bacteria. *Antimicrob. Agents Chemother.* **37:**354–356.

127. **Versalovic, J., D. Shortridge, K. Kibler, M. V. Griffy, J. Beyer, R. K. Flamm, S. K. Tanaka, D. Y. Graham, and M. F. Go**. 1996. Mutations in 23S rRNA are associated with clarithromycin resistance in *Helicobacter pylori. Antimicrob. Agents Chemother.* **40:**477–480.

128. **Vester, B., and S. Douthwaite**. 1994. Domain V of 23S rRNA contains all the structural elements necessary for recognition by the ErmE methyltransferase. *J. Bacteriol.* **176:**6999–7004.

128a. **Vester, B., A. K. Nielsen, L. H. Hansen, and S. Douthwaite**. 1998. ErmE methyltransferase recognition elements in RNA substrates. *J. Mol. Biol.* **282:**255–264.

129. **Vilches, C., C. Hernandez, C. Mendez, and J. A. Salas**. 1992. Role of glycosylation and deglycosylation in biosynthesis of and resistance to oleandomycin in the producer organism, *Streptomyces antibioticus. J. Bacteriol.* **174:**161–165.

130. **Watanakunakorn, C**. 1976. Clindamycin therapy of *Staphylococcus aureus* endocarditis. Clinical relapse and development of resistance to clindamycin, lincomycin, and erythromycin. *Am. J. Med.* **60:**419–425.

131. **Weisblum, B**. 1995. Erythromycin resistance by ribosome modification. *Antimicrob. Agents Chemother.* **39:** 577–585.

132. **Weisblum, B**. 1995. Insights into erythromycin action from studies of its activity as inducer of resistance. *Antimicrob. Agents Chemother.* **39:**797–805.

133. **Weisblum, B., C. Siddhikol, C. J. Lai, and V. Demohn**. 1971. Erythromycin-inducible resistance in *Staphylococcus aureus*: requirements for induction. *J. Bacteriol.* **106:**835–847.

134. **Wondrack, L., M. Massa, B. V. Yang, and J. Sutcliffe**. 1996. Clinical strain of *Staphylococcus aureus* inactivates and causes efflux of macrolides. *Antimicrob. Agents Chemother.* **40:**992–998.

135. **Yanofsky, C**. 1988. Transcription attenuation. *J. Biol. Chem.* **263:**609–612.

136. **Yu, L., A. M. Petros, A. Schnudel, P. Zhong, J. M. Severin, K. Walter, T. F. Holtzman, and S. W. Fesik**. 1997. Solution structure of an rRNA methyltransferase (ErmAM) that confers MLS antibiotic resistance. *Nat. Struct. Biol.* **4:**483–489.

137. **Zalacain, M., and E. Cundliffe**. 1990. Methylation of 23S ribosomal RNA due to *carB*, an antibiotic-resistance determinant from the carbomycin producer, *Streptomyces thermotolerans. Eur. J. Biochem.* **189:**67–72.

138. **Zalacain, M., and E. Cundliffe**. 1991. Cloning of *tlrD*, a fourth resistance gene, from the tylosin producer, *Streptomyces fradiae. Gene* **97:**137–142.

139. **Zhang, H. Z., H. Schmidt, and W. Piepersberg**. 1992. Molecular cloning and characterization of 2 lincomycin-resistance genes, *lmrA* and *lmrB*, from *Streptomyces lincolnensis* 78-11. *Mol. Microbiol.* **6:**2147–2157.

APPENDIX
Internet Addresses

Microbiology Resources

Microbiology (EINET)	http://www.einet.net/galaxy/Science/Biology/Microbiology.html
Microbiology Underground (Imperial College)	http://www.ch.ic.ac.uk/medbact/microbio.html

Microbiology Laboratories and Departments

Microbial Pathogenesis at NWFSC	http://research.nwfsc.noaa.gov/home-page.html
Department of Microbiology (Technical University of Denmark)	http://ftp.lm.dtu.dk/
Microbiology Department (University of Cape Town)	http://www.uct.ac.za/microbiology/
Department of Microbiology (University of Illinois)	http://www.life.uiuc.edu/micro/home.html
School of Dentistry (University of Pennsylvania)	http://biochem.dental.upenn.edu/

Microbiology Culture Collections and Databases

American Type Culture Collection (ATCC)	http://www.atcc.org
Culture and Germplasm Collections Listing	http://www.lm.dtu.dk/djour.html
Microbial Germplasm Database (MGD) at OSU	http://mgd.nacse.org/

Bacteriology

Actinomycetes-Streptomyces Resources (ASIRC) (University of Minnesota)	http://biosci.cbs.umn.edu/asirc/
Bacillus subtilis Database (Harvard University)	http://pbil.univ-lyon1.fr/nrsub/nrsub.html
E. coli Database Collection (ECDC) (Giessen University and CIBA-Geigy AG)	http://susi.bio.uni-giessen.de/ecdc.html
E. coli Genetic Stock Center (Gopher Server at Yale University)	gopher://cgsc.biology.yale.edu:70/1
Foodborne Pathogenic Microorganisms (FDA)	http://vm.cfsan.fda.gov/~mow/intro.html
The *Haemophilus influenzae* Rd Genome Database (HIDB) at The Institute for Genomic Research (TIGR)	http://www.tigr.org/tdb/mdb/hidb/hidb.html
The *Mycoplasma genitalium* Genome Database (MGDB) at TIGR	www.tigr.org/tdb/mdb/mgdb/mgdb.html

Staphylococcus aureus

Antimicrobial Techniques for Medical Nonwovens	http://www.microbeshield.com/
Combinatorial Chemistry	http://pubs.acs.org/
Cystic Fibrosis Foundation USA (CFUSA)	http://www.cfusa.com/
Harvard Health, newsletter: drug-resistant bacteria	http://www.harvard-magazine.com/
The Institute for Genome Research (site for the *S. aureus* genome)	http://www.tigr.org/
Staph Bacteria and Hospital-Based Infection (Lewin survey, 1997)	http://www.lewin.com/Press/staph611.htm
The Toxic Shock Information Service	http://www.toxicshock.com/.htm

Gram-Positive Pathogens, ed. by V. A. Fischetti et al.
© 2000 American Society for Microbiology, Washington, D.C.

Tuberculosis

MycDB: The Mycobacterium Database (Royal Institute of Technology, Stockholm)	http://www.biochem.kth.se/MycDB.html
Stanford Center for Tuberculosis Research	http://molepi.stanford.edu/
The TB/HIV Research Laboratory (Brown University)	http://www.brown.edu/Research/TB-HIV_Lab/index.html

Yeast

Candida albicans Database (Alces Server at University of Minnesota)	gopher://alces.med.umn.edu:70/11/candida
Saccharomyces Genome Database (SGD) (Stanford University)	gopher://genome-gopher.stanford.edu:70/11/Saccharomyces
Yeast Directory (Stanford University)	http://genome-www.stanford.edu/VL-yeast.htm

Parasitology

Malaria Database	http://www.wehi.edu.au/biology/malaria/who.html

Virology

Institute of Medical Virology (Bock Laboratories at University of Wisconsin-Madison)	http://www.bocklabs.wisc.edu/
Virology Servers List (Tulane University)	http://www.tulane.edu/~dmsander/garryfavweb.html
Virus Database Listing (Australian National University)	http://life.anu.edu.au/viruses/welcome.html

Biology and Informatics

Australian National University Bioinformatics	http://life.anu.edu.au/
Belgian EMBnet Node (BEN)	http://ben.vub.ac.be/
Biodiversity and Biological Collections WWW Server (Cornell University)	http://muse.bio.cornell.edu/
Bioinformatics (NWFCS/NOAA)	http://research.nwfsc.noaa.gov/bioinformatics.html
Biologists's Control Panel (Baylor University)	http://www.hgsc.bcm.tmc.edu/tools/
BioMolecular Research Tools (Iowa State University)	http://www.public.iastate.edu:80/~pedro/research_tools.html
Biosciences Global Gopher Server (Stanford University)	gopher://genome-gopher.stanford.edu/11/topic
The European Molecular Biology Laboratory (EMBL)	http://www.embl-heidelberg.de/
Harvard University Biological Laboratories	http://golgi.harvard.edu/genome.html
Medline (Submission Query Form)	http://ncbi.nlm.nih.gov/medline/query_form.html
Molecular Biology Databanks (New York University)	http://saturn.med.nyu.edu/srs/srsc
The Sanger Centre (Cambridge, U.K.)	http://www.sanger.ac.uk/
Scientific Toolkit for Molecular Analysis (Rockefeller University)	http://www/rucs/toolkit/toolkit.htm
Washington University at St. Louis (Genetics Dept.)	http://www.genetics.wustl.edu/
The World-Wide Web Virtual Library (WWW VL): Biosciences (Harvard University)	http://mcb.harvard.edu/Biolinks.htm

Computational Biology

BankIt: GenBank Submissions by WWW	http://www.ncbi.nlm.nih.gov/BankIt/index.html
Brookhaven National Laboratory (Biology Dept.)	http://genome1.bio.bnl.gov/
Computational Biology (University of Minnesota)	http://lenti.med.umn.edu/
Computational Molecular Biology at the NIH	http://molbio.info.nih.gov/molbio
EMBL Computational Services	http://www.embl-heidelberg.de/Services/index.html

The Entrez Browser (NCBI)	http://atlas.nlm.nih.gov:5700/Entrez/index.html
ExPASy Protein Sequence Server	http://expasy.hcuge.ch/
Genbank Database Query Form (NCBI)	http://ncbi.nlm.nih.gov/genbank/query_form.html
Genetic Analysis Software (University of Indiana)	gopher://ftp.bio.indiana.edu:70/11/IUBio-Software+Data/molbio
MBCR database links collection	http://condor.bcm.tmc.edu/databases.html
Molecular Biology Computational Resource (MBCR)	http://condor.bcm.tmc.edu/
The National Center for Biotechnology Information (NCBI)	http://www.ncbi.nlm.nih.gov/
NCBI: BLAST Notebook	http://www.ncbi.nlm.nih.gov/BLAST/index.html
Scientific Toolkit for Molecular Analysis (Rockefeller University)	http://www/rucs/toolkit/toolkit.html
Washington University Institute for Biomedical Computing	http://www.ibc.wustl.edu/

Sequence Analysis Programs

Cutter (Restriction Enzyme Mapping)	http://www.ccsi.com/firstmarket/cutter/
Genome analysis programs	http://pedant.mips.biochem.mpg.de
HMMER (Hidden Markov Models of DNA and Protein Consensus)	http://genome.wustl.edu/eddy/hmm.html
PSORT (Signal Sequence Prediction)	http://psort.nibb.ac.jp/
SCOP: Structural Classification of Proteins	http://scop.mrc-lmb.cam.ac.uk/scop/
Transmembrane Protein Segment Prediction (TMAP) (EMBL)	http://www.embl-heidelberg.de/tmap/tmap_info.html

Sequence Databases and Sequence Tools

Clostridium difficile genome project	http://sanger.ac.uk/Projects/C_difficile/
Codon Usage Statistics (Gopher Server)	gopher://ftp.bio.indiana.edu:70/11/IUBio-Software+Data/molbio/codon
HIV Sequence Database	http://hiv-web.lanl.gov
Human Genome Data Base (GDB)	http://gdbwww.gdb.org/
The Institute for Genomic Research (TIGR)	http://www.tigr.org/tigr_home/index.html
Listeria genome	http://www.pasteur.fr/units/gmp/Gmp_lm_stat.html
Molecular Biology Vector Sequence Database	http://vectordb.atcg.com/
The Nucleic Acid Database (NDB) (Rutgers University)	http://ndbserver.rutgers.edu/
Organelle Genomes (Gopher Server)	gopher://megasun.bch.umontreal.ca:70/11/Organelles
The Protein Data Bank (PDB) (Brookhaven National Laboratory)	http://pdb.pdb.bnl.gov/
Restriction Enzyme Database (REBASE)	http://rebase.neb.com/rebase/rebase.html
The rRNA Database	http://www-rrna.uia.ac.be/
Streptococcus pyogenes, Streptococcus mutans, and *Staphylococcus aureus* genomes	http://www.genome.ou.edu

U.S. Government Server Listings

Brookhaven National Laboratory (BNL)	http://www.tigr.org/tigr_home/index.html
The Center for Advanced Research in Biotechnology (CARB)	http://indigo15.carb.nist.gov/carb/carb.html
The Centers for Disease Control and Prevention (CDC)	http://www.cdc.gov/cdc.html
The Library of Congress	http://www.hpcc.gov/blue94/section.4.3.html
The National Aeronautics and Space Administration (NASA)	http://www.nasa.gov/
The National Center for Infectious Disease (NCID)	http://www.cdc.gov/ncidod/ncid.ht
The National Institute of Standards and Technology (NIST)	http://www.nist.gov/
The National Institutes of Health (NIH)	http://www.nih.gov/
The National Library of Medicine (NLM)	http://www.nlm.nih.gov/

The National Science Foundation (NSF)	http://www.nsf.gov/
The NSF FastLane	http://www.fastlane.nsf.gov/
The Northwest Fisheries Science Center (NWFSC)	http://research.nwfsc.noaa.gov/
The President (The White House)	http://www.whitehouse.gov/

Institutional Servers

The Jackson Laboratory	http://www.jax.org/
The Nobel Foundation	http://www.nobel.se/
The World Health Organization	http://www.who.ch/

Professional Associations

American Association for the Advancement of Science (AAAS)	http://www.aaas.org/
The American Association of Immunologists (AAI)	http://www-biology.ucsd.edu/others/aai
American Society for Microbiology (ASM)	http://www.asmusa.org/
FASEB (Information Services)	http://www.faseb.org/
The Society for General Microbiology	http://www.pharmweb.net/

Index